CERAMIC MATERIALS AND COMPONENTS FOR ENGINES

Proceedings of the
Third International Symposium

Las Vegas, Nevada USA
November 27–30, 1988

2 3 4 5—93 92 91 90

Printed in the United States of America.

International Advisory Committee

Executive Committee

The symposium was sponsored by The American Ceramic Society, Inc.

International Advisory Committee
Third International Symposium
Ceramic Materials and Components for Engines

Foreword

These proceedings of the 3rd International Symposium, Ceramic Materials and Components for Engines, are the result of serious work by many people. The authors are to be commended for their research and their willingness to present their results before their peers in an open technical meeting. The long term importance of a meeting such as the 3rd Symposium is measured in terms of the exchange of technical knowledge between the participants and the quality of the papers published as a result of the meeting. The former is difficult to measure, but the latter is easier to evaluate, if the papers are published together, as in this proceedings.

We decided early to require peer reviews of the manuscripts of the papers for the 3rd Symposium with the conviction that this would improve the quality of the papers written for this meeting. In 1986, I asked a number of colleagues in the United States to serve as reviewers for the papers for the 3rd Symposium, and they comprised the United States Symposium Committee. This committee included a total of 56 members, with a reasonably uniform representation between industrial and university researchers. I know that all worked hard, and sometimes had to do so under difficult time constraints. They have made an important contribution to the area of structural ceramics for engine applications. In many cases, a committee member reviewed six or more manuscripts and in some cases re-reviewed the same manuscript several times as the authors made the required changes.

The quality of the papers in these proceedings is largely due to the many hours of valuable time contributed by the members of the committee, whose names are listed below. It was a pleasure as meeting chairman to work with so many colleagues who contributed so much of their time to the success of the 3rd Symposium. Two other people contributed a great deal to the organization and quality of the 3rd Symposium. Mrs. Felicia Foust and Mrs. Donna Conger both did an outstanding job in coordinating the receipt of manuscripts, preparing the review documents, and maintaining an extensive database and communication system with the authors, reviewers, and staff of the American Ceramic Society. They both worked tirelessly in making the meeting a success. Their help was invaluable. Finally, the Co-Chairman, Dr. Matt Ferber, provided expert help and support plus great encouragement. His assistance is gratefully acknowledged.

Victor J. Tennery
Oak Ridge National Laboratory
February 21, 1989

United States Symposium Committee
Third International Symposium
Ceramic Materials and Components for Engines

Dr. Norman L. Hecht
University of Dayton Research Institute

Dr. Mark A. Janney
Metals and Ceramics Division
Oak Ridge National Laboratory

Dr. Curtis A. Johnson
Research & Development Center
General Electric Company

Dr. Nathan R. Katz
U.S. Army Materials Technology Laboratory

Dr. Richard L. Landingham
Lawrence Livermore National Laboratory

Dr. David C. Larsen
Corning Glass Works

Dr. Gordon Lewis
Department of Ceramic Engineering
Clemson University

Ms. Laura J. Lindberg
Allied-Signal Aerospace Company
Garrett Auxiliary Power Division

Mr. William D. Long
Wacker Chemicals (USA), Inc.

Mr. James W. MacBeth
Carborundum

Dr. J. J. Mecholsky, Jr.
Applied Research Laboratory
Pennsylvania State University

Dr. Gary L. Messing
Department of Materials Science & Engineering
Pennsylvania State University

Ms. Helen H. Moeller
Babcock & Wilcox

Mr. Arthur J. Moorhead
Metals and Ceramics Division
Oak Ridge National Laboratory

Dr. Dale Niesz
Center for Ceramics Research
Rutgers University

Dr. Arvid E. Pasto
GTE Laboratories, Inc.

Dr. Joseph L. Pentecost
School of Materials Engineering
Georgia Technical Institute of Technology

Dr. Michael Santella
Metals and Ceramics Division
Oak Ridge National Laboratory

Dr. Liselotte J. Schioler
Air Force Office of Scientific Research
Bolling Air Force Base

Mr. Robert B. Schulz
Office of Transportation Systems
U.S. Department of Energy

Dr. Jitendra P. Singh
Materials Science & Technology Division
Argonne National Laboratory

Dr. M. Srinivasan
Carborundum

Dr. Victor J. Tennery
Metals and Ceramics Division
Oak Ridge National Laboratory

Dr. John B. Wachtman, Jr.
Center for Ceramics Research
Rutgers University

Dr. James M. Wimmer
Materials Engineering
Garrett Auxiliary Power Division
Allied-Signal Aerospace Company

Opening Remarks
by
Victor J. Tennery, Chairman
3rd International Symposium
Ceramic Materials and
Components for Engines

Good Morning, and welcome to the opening of the 3rd International Symposium, Ceramic Materials & Components for Engines. I am Victor Tennery, the Program Chairman of the 3rd International Symposium, and I welcome you here today for the beginning of this exciting meeting on ceramic materials and their application in heat engines.

This meeting is held under the auspices of the American Ceramic Society, whose experienced staff have worked for many months in making the arrangements and planning the facilities which we will be using for the next three days. This meeting is, as you know, the third in a series which started in Hakone, Japan in 1983, where the 1st Symposium was held. This first meeting resulted in the publication of a proceedings containing 78 papers, and was considered to be such a success that a second meeting was planned for a later time. Our colleagues in the Federal Republic of Germany graciously offered to be the host for the second symposium, which was held in the spring of 1986 at Travemünde on the Baltic coast. The second symposium had a larger attendance than did the first meeting, and resulted in a proceedings containing 134 published papers. The meeting in Travemünde included a preliminary organizing meeting for the third symposium, and it was agreed that the third symposium would be hosted by the United States.

I agreed to serve as Program Chairman, and with the help of the Program Vice Chairman, Dr. M. K. Ferber of ORNL, the staff of the American Ceramic Society, and two very important Committees, the present meeting was organized, and I think will be a great success. The 3rd Symposium program and the papers which will be published in the form of a proceedings a few months from now are the result of significant work by two very important committees. These are the International Advisory Committee and the United States Symposium Committee.

The International Advisory Committee includes 15 members from eight countries and the Commission of European Communities. These members helped to publicize the meeting to their countrymen and provided great assistance in assembling the papers you will be hearing and seeing in poster form in the next three days. I express my appreciation to the members of the International Advisory Committee for their contributions to the content of this meeting.

Very early in the planning for the 3rd Symposium, we decided to submit the papers from the meeting to rigorous peer review, because we were convinced that this process would result in a stronger and higher quality program for you, the technical ceramic community. We also felt it very important to have multiple reviews for the papers in order to obtain a broad based evaluation of the manuscripts. The United States Symposium Committee was organized for the purpose of reviewing the manuscripts which the authors submitted. Members contributed many hours of their valuable time by reviewing as many as seven manuscripts each. In some cases, when re-reviews were required, these individuals reviewed ten or more manuscripts. We hope that you will find that our efforts have been worthwhile, and most importantly that you leave when this meeting is concluded with new knowledge and enthusiasm for the future potential for the application of advanced structural ceramics in advanced heat engines. I express my appreciation to the members of the United States Symposium Committee for their work in reviewing the manuscripts for the meeting.

The meeting is organized around essentially 12 technical topics, most of which we selected for the initial planning of the meeting. We rapidly realized that the number of abstracts submitted was so large that a three day meeting would require either a very large number of posters or several concurrent sessions. We decided to hold three concurrent sessions for most of the meeting in order to reduce the number of posters, and we did this based upon many requests which we received from authors. We have done everything possible to ensure that the meeting rooms are adequate and that the equipment required for the presentations function efficiently. The papers and posters you will see in the next three days come from 14 countries and, I think, represent an excellent picture of the state of research and development today for ceramic science and technology and the engineering applications for ceramic materials in advanced engines.

The session chairmen are organized to facilitate the operation of the technical sessions, and are ready and willing to give the speakers any assistance required. In many cases, members of the International Advisory Committee are serving with a United States colleague as co-chairmen for some of the sessions. In these cases, the United States member will assist the International Advisory Committee member in conducting the session.

I ask all speakers and session chairmen to please adhere as closely as possible to the published time schedule. Since generally three concurrent sessions are being conducted simultaneously, it is a very important courtesy to adhere to the schedule so that attendees can move between sessions and hear the papers they have selected.

Having read all of the manuscripts, I can assure you that in the next three days, you will all be hearing some very exciting papers, with new results being reported in the processing, properties and application of nitride, carbide, and oxide structural ceramics.

I wish you the best during the next few days. Let us now adjourn this introductory session, and begin the technical program.

Table of Contents

I. Processing-Powders

II. Processing-Composites

III. Processing-Rheological Aspects

IV. Processing-Nitride and Carbide Systems

V. Joining and Coatings

VI. Mechanical Properties

VII. Properties-Thermal and Mechanical Behavior

VIII. Properties-Creep and Fatigue Behavior

IX. Properties–Toughened Materials

X. Non-Destructive Evaluation

XI. Machining and Wear

XII. Engine Components

OPTIMIZATION OF A HIGH PURITY SILICON NITRIDE POWDER CONCERNING SINTERING AND MECHANICAL PROPERTIES

G. Franz, Bayer AG, PK-F, 4150 Krefeld-Uerdingen, FRG
B. Laubach, Bayer AG, PK-F, 4150 Krefeld-Uerdingen, FRG
U. Wickel, Bayer AG, PK-F, 4150 Krefeld-Uerdingen, FRG
G. Wötting, CFI GmbH&CoKG, 8633 Rödental, FRG
E. Gugel, CFI GmbH&CoKG, 8633 Rödental, FRG

ABSTRACT

Different powder properties of a high-purity silicon nitride powder were systematically varied to determine their influence on the sintering behaviour and on resulting mechanical properties at room and high temperatures. Differentiation of the overlapping effects of the powder variables was performed by use of statistical methods. According to the extent of the influence of the different powder characteristics evaluated, powder properties were specified in order to achieve a high sintering activity and optimized high temperature bending strength. These results were proven by tailoring such a silicon nitride powder.

INTRODUCTION

Sintering of silicon nitride has become a very attractive method for the production of dense, high-strength components. Progress which has been made during the last years was supported on the one hand by the improvement of Si_3N_4 starting powders, on the other hand by the development of sophisticated sintering techniques like sintering under increased nitrogen pressure or sinter-HIPping. The first point is essential for reaching high densities without simultaneous application of mechanical pressure such as during hot-pressing. The latter allows, in addition, the use of sintering temperatures higher than about 1800°C, providing the possibility to reduce the additive content necessary for a nearly complete densification.

Besides these general aspects, controversial statements are found in literature concerning critical Si_3N_4 powder characteristics which determine the sintering behaviour and the resulting mechanical properties (1 - 4). This seems to be due to several reasons. One important fact is that the processing conditions were often chosen arbitrarily and not adapted to specific powder requirements. This, however, is necessary especially if Si_3N_4 powders, produced by different methods, are to be compared. Simultaneously, the comparison of such different powders is superimposed by parameters which normally were not determined like crystallite size, agglomeration etc., which may become densification- and/or strength-controlling. This may lead to a random grading of powders, while appropriate processing and test conditions would have led to different assessments. A further point may be seen in the fact that often only the starting powder properties are taken into consideration, however, preparation of the starting composition (meaning homogenization with sintering additives) often changes powder characteristics markedly.

In order to improve Si_3N_4 starting powder properties, however, it is necessary to know the influence of the main powder characteristics on sintering and properties, excluding all other parameters discussed above by keeping them constant. This was realized by means of a special gas phase process which allows to vary powder properties in a wide range by slightly modifying the preparation conditions. Additionally, the processing conditions of the starting compositions as well as the additive content were kept constant, too. The amount of additives used is deliberately chosen rather low in order to see intrinsic differences in the behaviour of the powders, not suppressed by high amounts of liquid phase formed during sintering.

The variations in powder properties performed in this evaluation comprise the specific surface area, the oxygen content, the impurity content and the $\alpha/(\alpha+\beta)$-phase ratio. This is done by different post-treatment conditions of a single amorphous precursor material. Even with this procedure, however, superpositions of the effects of individual powder characteristics on the sintering behaviour and on resultant properties remain. Thus, by means of statistical methods, differentiations for the effects of individual characteristics are performed. This results in the definition of powder characteristics which extensively determine densification and resulting properties. According to this evaluation, a special Si_3N_4-powder grade is synthesized which exhibits all properties important to achieve a good sintering behaviour as well as optimized mechanical properties under the processing conditions used.

EXPERIMENTAL PROCEDURE

Table 1 summarizes characteristics of Si_3N_4 starting powders used in this evaluation. BET-values and oxygen contents were already derived from the starting compositions after the homogenization step with the sintering additives. The oxygen portion contributed by the additives is excluded by calculation. The ranking list of the powder characteristics (pointed out by the single underlined minimum values as well as by the double underlined maximum values in Tab. 1) before and after the homogenization step remains constant.

Table 1. Characteristics of Si_3N_4 starting compositions

Composition	BET m^2/g	O ② wt.%	C wt.%	Fe ppm	$\alpha/(\alpha+\beta)$ %
A	13.9	1.7	0.16	510	85
B	13.8	2.0	0.08	230	78
C	13.9	1.9	0.04	210	60
D	13.2	1.7	0.04	100	92
E	14.0	2.0	0.07	150	92
F	12.6 ①	1.7	0.05	360	92
G	17.3	2.4	0.06	120	92
H	15.6	1.9	0.05	95	91
I	13.4	2.2	0.05	510	92
J	14.6	2.8	0.06	330	92
K	18.5 ①	4.0	0.06	220	92

① Single underlined: Minimum value
Double underlined: Maximum value

② Oxygen contribution of the additives excluded

Each powder was mixed with 5 wt.% Y_2O_3 + 1 wt.% Al_2O_3 by wet-milling in 2-propanol, dried, sieved and pressed to plates of 75 x 60 x 6 mm^3 by uniaxial and cold-isostatic pressing. Sintering was carried out in a gas-pressure furnace, schedules applied comprised single-step cycles (between 1775°C/1 bar N_2 and 1950°C/12 bar N_2) and two-step cycles consisting of a pre-sintering period at low N_2-pressure and a high pressure step under 100 bar N_2. Plates were machined to samples dimensioned 3.5 x 4.5 x 45 mm^3 for the determination of bending strength at room- and high-temperatures (4-point-bending, spans 40/20 mm).

According to the reasons described, no principal relations between single powder properties, the sintering behaviour and the resultant properties could be indicated. Therefore, statistical evaluation methods were applied in order to take into consideration the overlapping effects of different powder variables. The three steps of this analysis are outlined in Fig. 1.

During the stepwise regression analysis, after each addition of a further variable to the fitting polynomial, the improvement of the fit to describe the experimental results is calculated in percent. The positive or negative nature of the variables' effect on the property considered is obtained from the linear correlation analysis. Finally, the contribution of the different variables (in percent) to describe the experimental results are used for further discussions as a representation of their extent of influence. The absolute sum of these data reflects the degree of description of the experimental results by the polynomial developed.

The authors are aware of the fact that the accuracy of this analysis is limited. Therefore, effects smaller than about 10 % should not be overestimated. It is thought, however, that the method applied is well suited to demonstrate general relations as well as to discriminate between overlapping influences of different powder characteristics.

RESULTS AND DISCUSSIONS

Sintering

To demonstrate the differences in sintering, Fig. 2 illustrates the best and the worst densification behaviour of the various starting compositions specified in Tab. 1.

It can be seen that evident differences result, ranging from incomplete densification even at the highest sintering temperature applied to nearly complete densification already at temperatures of about 1825°C (and 6 bar N_2).

Though the behaviour of certain compositions can be related to single powder properties, a more complete description of the densification behaviour affords a statistical analysis. Thus, fits with polynomials consisting of the variables BET, O, C, Fe and α-Si_3N_4 phase-content were performed. Results, demonstrating to which extent these powder characteristics determine the resulting sintering density are shown in Fig. 3.

First, it can be seen that the contribution of variables changes distinctly with increasing sintering temperature. At low temperatures, the O-content is of highest importance, followed by the BET-value. Presumably, the oxygen content determines the amount of liquid phase formed during early sintering stages, which controls the extent of rearrangement taking place. The liquid phase sintering mechanism proceeds with solution-reprecipitation, which is mainly controlled by the particle size, represented by the BET-value. Therefore, the influence of BET on the final density remains high up to high temperatures. High oxygen contents, in contrast, seem to be responsible for density reductions observed at high sintering temperatures. Fe- and α-content only show minor effects on the density. For the carbon content, however, a negative effect is found which probably results from reactions with the liquid phase at high temperatures, leading to volatile constituents like SiO and CO. By summing up the a b s o l u t e percentage values of the different variables, values of 94 % at 1775°C and 71 % at 1925°C result (see ΣΣ-graph in Fig. 3). Thus, at higher sintering temperatures, further factors not considered in the evaluation are important with respect to density. These might include factors like rising vapour pressures and steric hindrance of densification by grain growth. On the other hand, description of the experimental results by the polynomials evaluated is surprisingly good.

With the two-step sintering cycles, nearly complete densification was achieved with each composition. Due to small variations of the densities no distinct correlations with powder properties were found. This seems to result from the pressure-assisted densification being effective in this late stage of sintering, which depends to a minor extent on the starting powder properties. Nevertheless, powder properties and influences, as discussed above, remain of great importance. This is due to the fact that the gas-pressure sintering technique may only be applied successfully if during the low-pressure pre-sintering step closed porosity is reached.

Room-temperature bending strength

An identical evaluation was performed for the identification of powder properties which determine the room-temperature bending strength. In this case, however, a further factor supposed to influence strength, the sintering density, was added to the polynomial variables. The fitting procedure was performed for strength values obtained with samples sintered at 1875°C and by the two-step cycle. These samples showed a variation in data between 560 - 705 MPa and 600 - 810 MPa, respectively (see Tab. 2).

For both sintering conditions, best fits are obtained with the variables O, BET, α-phase content and density ρ after sintering. Other variables like C and Fe are of negligible influence. The contribution of the polynomial variables to strength control is represented in Fig. 4.

As can be seen, the magnitude of influence changes markedly with altered sintering conditions and the degrees of description of the experimental values by the polynomials evaluated are lower than for the densities. For samples sintered at 1875°C, the sintering density is found to exert the strongest influence on the r.t.-strength. Sintering density, however, depends markedly on the powder characteristics oxygen content and fineness, as shown before. Besides these, a further positive effect is found for the α-content which, in agreement with literature, is thought to favour the development of rod-like ß-grains responsible for high strength levels (5).

For the samples sintered by the two-step cycle, no more influence of the sintering density on strength was found, as the density values varied only slightly. The strongest effect results in this case for the oxygen content. It may be anticipated that this is due to the pressure-assisted densification mechanism being effective during the high-pressure step of the two-step cycle. High oxygen contents provide high amounts of liquid phase which favour the increase of the packing density, the interlocking of the grains as well as the healing of defects, improving room-temperature strength. Thus, the effect of oxygen primarily has to be seen in providing sufficient liquid carrier to allow the above mentioned mechanisms to take place.

High-temperature bending strength

Bending strength at 1100°C and 1300°C was determined only from two-step sintered samples. Characteristic values, comprising the results of all other samples, are given in Tab. 2, together with its corresponding r.t.-strength data.

Table 2. Characteristic bending-strength values of two-step sintered samples at room-temperature, 1100°C and 1300°C

Composition	Bending Strength σ ± s [MPa]		
	R. T. (5)*	1100°C (4)	1300°C (4)
D	601 ± 16	430 ± 13	290 ± 7
G	627 ± 19	483 ± 48	332 ± 28
I	730 ± 32	505 ± 31	367 ± 13
K	808 ± 17	497 ± 26	313 ± 7

*No. of samples tested

The common problem with relating the strength behaviour with single powder characteristics again is obvious. Thus, for these data the best fitting polynomials are calculated, too. As variables which still reveal non-negligible effects, however, only the oxygen- and the impurity content (which was used in this case) remain for both temperatures. Results are represented in Fig. 5.

As can be seen, the absolute percentage of describing the high-temperature strength values only reaches about 64 % for the

1100°C-values and 48 % for the 1300°C-values. This is thought to depend again on factors not considered in this evaluation, like the onset of subcritical crack growth due to the beginning of softening of the glassy phase (6). The temperature of softening ("transition region"), however, normally is not known and depends distinctly on type and amount of crystallization of the grain-boundary phase and the composition of the residual glass, respectively.

The relatively strong positive effect of the oxygen content on the strength at 1100°C, as can be seen in Fig. 5, seems to be associated with the promoting effect found for the room-temperature strength (see Fig. 4). With this respect it was discussed that rising oxygen contents favour densification and therefore have a positive effect. Fe and further impurities might also contribute by lowering the viscosity of the liquid phase. On the other hand, it has to be considered that high Fe-contents have deleterious effects on the oxidation resistance (7). To a minor extent, the same is true for the oxygen content. Thus, their concentration in Si_3N_4 starting powders should be minimized, though with this analysis positive effects result.

The correlation between powder properties and strength at 1300°C, however, results in contributions partially anticipated. The strongest, but in this case negative effect, is exerted by the impurity content, as is shown in the right part of Fig. 5. This may be related to a marked viscosity reduction of a glassy phase with increasing impurity contents. As already mentioned, all other starting powder characteristics are found to be of diminishing low influence.

CONCLUSIONS

It is shown that the sintering behaviour of Si_3N_4-powder-compacts as well as resulting mechanical properties can be described extensively by the starting powder properties. For the mechanical properties, the sintered density has to be added to the variables considered in order to achieve a higher degree of description. As anticipated, the sintering behaviour and resultant properties cannot be correlated with single powder characteristics but there are overlapping effects by different parameters. Therefore, the evaluation of densification- and strength-determining variables is done by calculating polynomials which best fit the experimental results.

With rising sintering temperature different parameters dominate the densification behaviour. During early stages of sintering, in this case the oxygen content is the rate-determining factor, while, at higher temperatures, the specific surface area, respectively the fineness of the powder is gaining increasing importance. For carbon, with rising sintering temperatures a rate-decreasing effect is found.

For the room-temperature strength, the most significant influence is found to be the sintered density, proceeded by factors which determine the microstructural development like the α-Si_3N_4 content. Depending on the sintering conditions applied, the powder fineness and/or the oxygen content are of great importance. For the high temperature strength, the oxygen content still exerts a strong influence mainly due to its effect on density. A strong negative effect, especially for strength at 1300°C, is found for the impurity content which may be related to the viscosity reduction of the glassy phase. This points out the necessity for a high-purity Si_3N_4-starting powder in order to achieve attractive high-temperature properties.

This evaluation also demonstrates that a single powder grade can hardly fulfill all demands concerning the sintering behaviour and the resultant properties. Thus, a compromise concerning the powder characteristics is necessary to suffice most requirements in accordance with the following more general guidelines (see also Tab. 1):

- the highest specific surface area as possible. This, however, is limited due to the unavoidable oxygen increase with rising fineness.

- a medium amount of oxygen in order to provide a high sintering rate but, simultaneously, to reduce temperature-sensitivity of the sintering density.

- a minimized carbon content in order to avoid hindrance of sintering.

- a minimized content of iron (and other impurities) to keep the viscosity of the remaining glassy phase high at increased temperatures.

- a high α-Si_3N_4 phase concentration to favour the development of ß-Si_3N_4-grains with high aspect ratios during sintering, exerting positive effects especially on bending strength at room-temperatures.

According to these objectives, a Si_3N_4-powder was synthesized by the gas-phase process, exhibiting the following properties:

Specific surface area	: 14 m^2/g	(16.0 m^2/g after processing with sintering additives)
Oxygen content	: 1.5 wt.%	(1.8 % after processing with sintering additives)
Carbon content	: 0.04 wt.%	
Iron content	: 100 ppm	
α-Si_3N_4 content	: 94 %	

Compacts, made of this Si_3N_4-powder, containing again 5 wt.% Y_2O_3 + 1 wt.% Al_2O_3, showed a high densification rate, leading already to a nearly complete densification by one-step sintering at 1825°C/6 bar N_2/1 h, comparable to powder H in Fig. 2. Room-temperature bending strength reached was 725 ± 20 MPa, strength at 1300°C still was 350 ± 15 MPa. Thus, the evaluation performed is thought to be an appropriate procedure to specify Si_3N_4-powder properties necessary to achieve tailored densification rates and ceramic properties.

REFERENCES

1 Wötting, G., Ziegler, G.: Powder Characteristics and Sintering Behaviour of Si_3N_4-Powders; Powder Met. Int. 18 (1986) 25 - 32

2 Wötting, G., Ziegler, G.: Sintering of Si_3N_4-Powder Compacts under high Nitrogen Pressure; in: P. Vincenzini (Ed.): "High Tech Ceramics", Proc. 6th CIMTEC (Milan, 1986); Elsevier, Amsterdam (1987), 983 - 992

3 Vandeneede, V., Leriche, A., Cambier, F., Pickup, H., Brook, R. J.: Sinterability of Silicon Nitride Powders and Characterization of Sintered Materials; in: S. Hampshire (Ed.): "Non-Oxide Technical and Engineering Ceramics", Elsevier Applied Science, London / New York (1986), 53 - 68

4 Itoh, N., Sasamoto, T., Sata, T., Komeya, K., Tsuge, A.: Impurities in Silicon Nitride Raw Materials and Correlation with Mechanical Strength of the Hot-Pressed Body; Yogyo-Kyokai-Shi 90 (1982) 209 - 213

5 Lange, F. F.: Fracture Toughness of Si_3N_4 as a Function of the Initial α-Phase Content; J. Am. Ceram. Soc. 62 (1979) 428 - 430

6 Lange F. F.: High-Temperature Strength Behaviour of Hot-Pressed Si_3N_4: Evidence for Subcritical Crack Growth; J. Am. Ceram. Soc. 57 (1984) 84 - 87

7 Singhal S. C.: Oxidation of Silicon Nitride and Related Materials; in: F. L. Riley (Ed.): "Nitrogen Ceramics", Nordhoff, Leyden (1977) 607 - 628

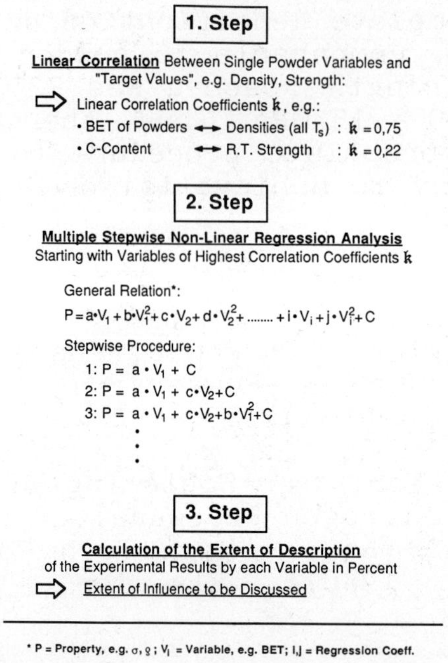

Fig. 1. Outline of the statistical evaluation applied

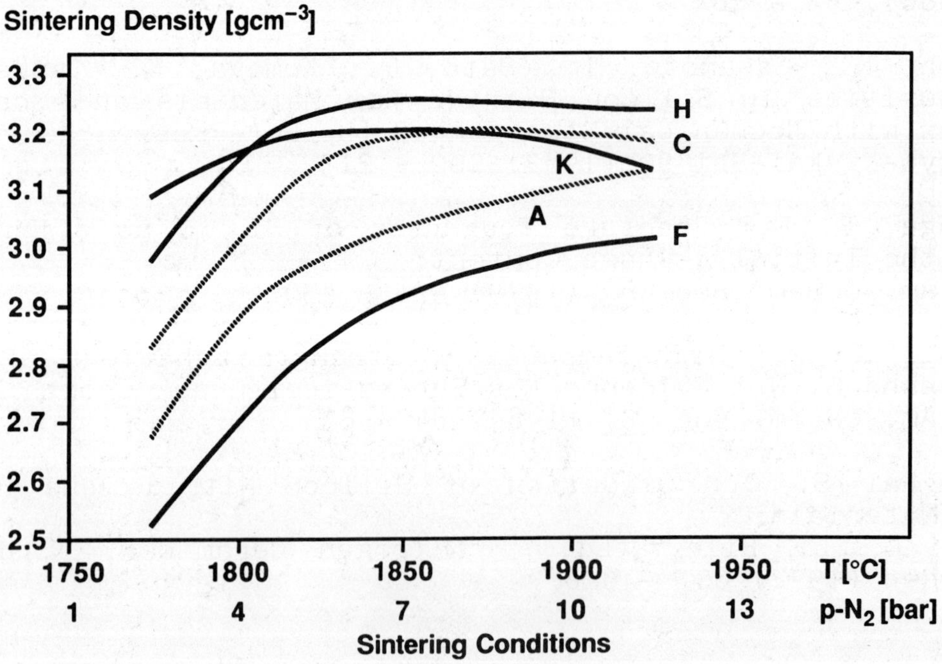

Fig. 2. Sintering densities as a function of sintering temperature; Time at temperature = 1 h

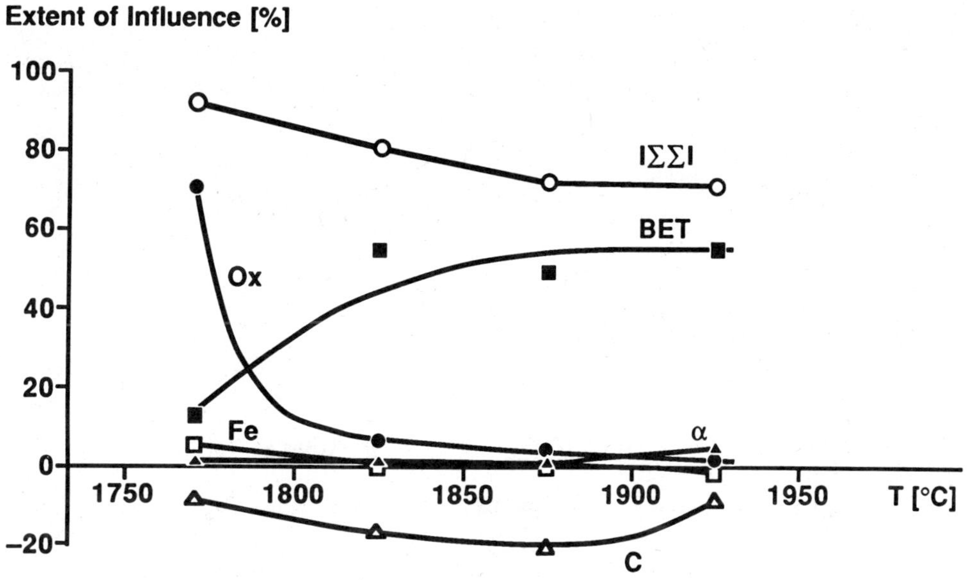

Fig. 3. Extent of (positive and negative) influence of different powder characteristics on the sintering density

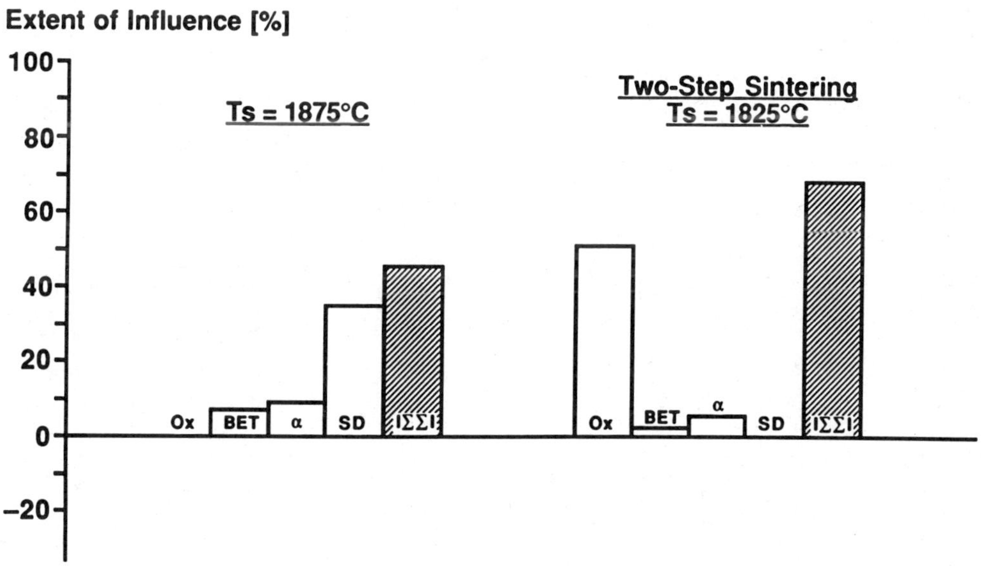

Fig. 4. Extent of (positive and negative) influence of powder characteristics and the sintering density on the bending strength at room-temperature

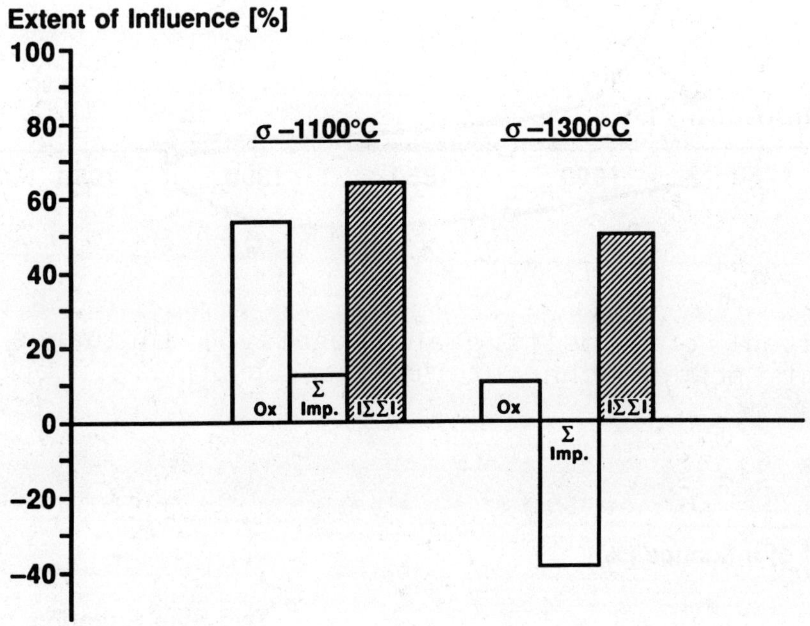

Fig. 5. Extent of (positive and negative) influence of powder characteristics on the bending strength at 1100°C and 1300°C

CONSOLIDATION BEHAVIOR OF HIGH PERFORMANCE CERAMIC SUSPENSIONS

Bruce E. Novich
Ceramics Process Systems Corporation
840 Memorial Drive
Cambridge, Massachusetts 02139

Dale H. Pyatt
Constructed Facilities Division
Department of Civil Engineering
Massachusetts Institute of Technology
Cambridge, Massachusetts 02139

ABSTRACT

Colloidal processing has been shown to produce low defect and uniform ceramic microstructures from submicron ceramic powders. These concepts were applied to colloidal pressing to determine critical design relationships for uniaxial consolidation of dense and uniform green bodies from colloidal suspensions.

Carefully controlled constant rate of strain consolidation experiments were carried out using Sumitomo AKP-30 alumina in water. The compression index decreased from 0.143 for a poorly dispersed alumina system to 0.077 for a well dispersed alumina suspension compression curve, indicating that the well dispersed system is stiffer in consolidation. The compression curves showed that as the degree of dispersion decreases, increased consolidation stresses are required to achieve a given particle packing density. The compression indices increased with increasing strain rate for well dispersed alumina suspensions. Permeability through the sample ranged from 3 E-8 cm/sec to 4 E-7 cm/sec, decreasing with decreasing void ratio during consolidation. Well dispersed samples gave lower permeabilities than did poorly dispersed samples over a given consolidation increment.

Coefficients of consolidation were non-constant over the experimental effective stress range, invalidating the general solution to the linear consolidation equation. An approximate incremental solution was applied which indicated rapid pressing cycles are possible by starting with a suspension having a high solids concentration. Application of this consolidation data to non-linear consolidation models is recommended for more exact prediction of consolidation time.

INTRODUCTION

Uniaxial and isostatic dry pressing are used to form over 95% of the electronic and structural components that are fabricated worldwide. Capacitors, ferrite magnets and cerdips are examples of the high volume dry pressed ceramic electronic components. Structural ceramic examples include sparkplug insulators, cutting tools and refractory brick. Potential applications exist in automotive products, for example, rollers and piston pins. Dry pressing is a preferred forming technique because it offers high volume production at low cost.

A key concern in the dry pressing of high performance ceramic components is the flow of powder into the compaction die, just prior to consolidation. Die loading can be particularly difficult if the powders are in the micron and sub-micron size range [1] as hydration and surface forces can cause ultrafine powder particles to agglomerate and exhibit poor free flow behavior. Inconsistent and non-uniform powder flow into the compaction die can result in particle packing variations leading to part warping and poor dimensional tolerance.

Most high performance ceramic powders for both uniaxial and isostatic pressing are intentionally agglomerated, usually by spray drying, in which powder particles are granulated with organic binders. The resultant granules, typically 100 microns in diameter, can flow readily into the compaction die in a uniform and reproducible manner. However, while die filling is greatly enhanced by powder granulation, remnant textures in the fired microstructures, originating from these consolidated agglomerates, can degrade the reliability and many high performance properties such as flexural strength [2]. Rhodes [3], Lange et al. [4] and Moffatt [5] are representative of the numerous investigators that have related processing textures in the green state to failure origins in the fired microstructure.

Colloidal processing, in which ceramic particle packing is controlled on the particle size scale and not the granule size scale, has been shown to give near perfect green microstructures. Particle coordination numbers can approach 12, to yield dense and uniform fired microstructures without excessive grain growth [6,7,8,9]. Two of the key attributes of colloidal processing as applied to pressing, are the high degree of particle dispersion leading to texture free microstructures and the high degree of uniformity and reproducibility associated with cavity filling.

The application of colloidal processing to the pressing of high performance ceramic components has been limited by the dilute nature of ceramic colloidal suspensions. Processing of colloidal suspensions has been typically limited to dilute suspensions , due to the historical inability to overcome interfacial and wetting forces of sub-micron ceramic particle surfaces that result in rheological limitations to suspension flow. Consolidation and dewatering of these suspensions can quickly become rate limited as the pore fluid is forced out of a consolidating finely porous compact, resulting in a low green density compact. If the sample is not fully consolidated the green density can be non-uniform. In general, more consolidation occurs at the drainage boundaries than internal to the sample, giving a non-uniform green microstructure. Low green density can also result in extended drying times, low green compact strength and large drying and firing stresses.

The critical step to overcoming the processing difficulties related to the consolidation of dilute suspension is the ability to make concentrated, flowable aqueous suspensions of narrow sized sub-micron ceramic particles in excess of 55 vol% of solids as shown by Cesarano and Aksay [10] and Novich and Occhionero [11]. Consolidation of a highly concentrated solids suspension will allow rapid consolidation kinetics, due to the reduced amount of pore fluid drainage necessary to achieve a high green part density. Smaller consolidation strains should translate to smaller stress gradients across the part resulting in more uniform microstructures.

The principal objective of this study was to identify the critical material- process relationships which govern the consolidation behavior of highly dispersed ceramic suspensions. Constant rate of strain consolidation (CRSC) experiments were carried out using high solids concentrated aqueous suspensions of sub-micron alumina as a function of initial suspension solids loading, degree of dispersion and uniaxial strain rate. Consolidation data was applied to consolidation theory to determine material process parameters. These processing parameters were applied to a pressing cycle design example.

CONSOLIDATION THEORY

The application of soil mechanics principals to the fabrication of high performance ceramics has been discussed by Janney 12, Onoda and Janney 13 and most recently Janney and Onoda 14. The reader is referred to Lambe and Whitman 15 for a comprehensive introduction to particulate mechanics. Janney and Onoda's work has princpally focussed on the application of critical state theory 16 to the forming of alumina/ water systems. Critical state soil mechanics is a comprehensive curve fitting model which can be used to describe the stress-strain behavior of high solids content (>35 vol%) and fully saturated particulate materials. The stress-strain paths can be modeled for both drained (pressing) and undrained (extrusion) systems in which these paths are analytically curve fit to yield empirical material constants which can be used to predict processing behavior. In the modeling of consolidation, Janney and Onoda showed how critical state soil mechanics could be applied to 1- dimensional consolidation (oedometer tests) and predict the volume change that would result after the application of a stress increment to the sample. However, critical state soil mechanics does not address the kinetics of consolidation, i.e. how much time will it take to achieve the volume change after stress application. The kinetics of consolidation will dictate process cycle time and part uniformity, which in turn control process economics and part quality.

Consolidation theory 17 is commonly used to describe the kinetics of volume change in a consolidating saturated particulate material. Terzaghi 18 first postulated that the effective stress, defined as:

$$\sigma' = \sigma_t - u \qquad (1)$$

effective stress = total stress - pore fluid pressure.

Terzaghi recognized that the effective stress, rather than the total stress controls the stress-strain behavior of saturated particulate systems, especially those related to volume change and shear strength. One very important application of the effective stress principal is consolidation, the time dependent decrease in volume change caused by permeability controlled drainage of the pore fluid. When a stress (load) is applied to a saturated particulate system, the load is carried by both the particle skeleton (effective stress) and the pore fluid (pore pressure). In response to this load, which causes a hydrostatic gradient across the sample, the pore fluid migrates through the skeleton toward the drainage boundary and out of the sample. As the pore fluid leaves the sample, the particle skeleton collapses effecting a volume change (consolidation). During consolidation, the pore pressure dissipates and the effective stress increases. The rate of pore fluid expulsion is described by the Kozeny- Carmen equation given as 12

$$k(cm/sec) = 980\ t_f^2\ e^3/(k_o\eta(3/r)^2(1+e)) \quad (2)$$

where k = coefficient of permeability
k_o = pore shape factor, approximately equal to 4.0
t_f = the tortuosity factor, approximately equal to 2.0
η = pore fluid viscosity, approximately equal to 1 cP
r = particle radius, which for AKP-30 is measured at 0.199 μm
e = void ratio, given as:

$$e = V_v / V_s = (1 / f_s) - 1 \quad (3)$$

where: V_v = volume of the voids
V_s = volume of the solids and
f_s = volume fraction of solids.

Equation 2 reduces to:

$$k\ (cm/sec) = 2.16\ E\ -7\ e^3 / (1 + e). \quad (4)$$

Terzaghi 17 gives the following governing equation for consolidation:

$$C_v \left(\frac{\partial^2 u}{\partial z^2} \right) = \frac{\partial u}{\partial t} - \frac{\partial \sigma_v'}{\partial t} \quad (5)$$

where z = the distance from the drainage boundary
σ_v' = the vertical effective stress
u = measured pore pressure
t = the hydrodynamic time lag
C_v = coefficient of consolidation.

Wissa et. al. 19 modified Terzaghi's equation for the 1-dimensional constant rate of strain condition:

$$C_v \left(\frac{\partial^2 \epsilon}{\partial z^2} \right) = \frac{\partial \epsilon}{\partial t} \quad (6)$$

where ϵ is the consolidation strain.

The compression index, C_c, relating consolidation strain to effective stress given by:

$$C_c = - (\Delta e) / (\Delta \log \sigma_v') \quad (7)$$

is measured as the slope from the log effective stress- void ratio curve, also known as the compression curve. The coefficient of consolidation can be determined directly from the compression index and permeability through the following relationship:

$$C_v = - (k\ \sigma_{va}'(1+e_o)/ (0.434\ \gamma_w\ C_c) \quad (8)$$

where γ_w = unit weight of the pore fluid and
e_o = the initial void ratio.
σ_{va}' = average vertical effective stress calculated as

$$\sigma_{va}' = (\sigma_v^3 - 2 u \sigma_v^2 + \sigma_v u^2)^{1/3} \quad (9)$$

The coefficient of permeability can be calculated from pore pressure measurements through the relation:

$$k = - (0.434 \varepsilon H^2 \gamma_w) / (2 \sigma_{va}' \log(1-u/\sigma_v) \quad (10)$$

where
ε = the strain rate
H = final sample height
u = measured pore pressure
σ_v= total vertical stress

Once C_v has been determined, the dimensionless time, T, defined as

$$T = C_v t / H_d^2 \quad (11)$$

can be calculated where t = consolidation time and H_d^2 = the drainage height, i.e. the longest distance water must travel to reach a drainage surface, and can then be applied to Spangler and Handy's 20 solution for the consolidation equation. This solution expresses the percent completion of consolidation for a given value of T. For example consolidation is 92% complete at T= 1 and 99% complete at T= 1.73. This approach is only valid over a constant C_v range and for small strains.

Consolidation experiments were carried out in which compression curves were measured, compression indices and permeabilities were calculated and coefficients of consolidation were determined for the alumina- water-Darvan C high performance ceramic system. Consolidation time values were calculated for 99% consolidation using the measured C_v values to provide a a cycle time estimate for the design of a wet pressing operation.

EXPERIMENTAL

Sumitomo AKP-30 alumina with a mean particle size[1] and distribution of 0.39 $\mu m \pm 0.11 \mu m$ and a nitrogen BET surface area[2] of 8.6 m^2/gm was used as the model ceramic for the consolidation study. Suspension for rheology and consolidation experiments were prepared by mixing the alumina powder with deionized water and Darvan C[3] dispersant in a mill containing alumina media. Suspensions were milled for 24 hours after which a sub-sample was used for rheological analysis and the remainder was used for consolidation testing. Rheological analysis was carried out using a rotoviscometer[4] at 20°C according to Novich and Occhionero 11.

Consolidation experiments were carried out in a constant rate of strain consolidometer (CRSC) as developed by Wissa et al. 19 and modified

---

[1] particle size and distribution measured by Nicomp photon correlation spectroscopy according to 23.

[2] surface area measured as a 100 point determination using nitrogen and a Quantachrome Autosorb-6.

[3] R.T. Vanderbilt Corporation

[4] Haake-Buchler CV-100/RV-100 rotoviscometer

by Pyatt 21 to accept fluid ceramic suspensions. Pyatt 21 discusses consolidometer calibration and proof testing. Consolidation experiments were carried out at strain rates of 1.2 E-3, 6.1 E-3 and 28.4 E-3 cm/min to determine strain rate effect on sample quality. Samples were uniaxially loaded with one directional drainage at the top of the sample. Pore pressure was measured at the bottom of the sample using pressure transducers. Discs, having the dimensions of 5 cm in diameter and 1.9 cm in height, were fabricated over an effective stress range of 0 to 100 kg/cm^2. These parts were sintered at 1480°C for 2 hours to full density. Green densities, measured after a 1000°C bisque fire and fired densities were measured by immersion. The fired discs were sectioned, polished and etched for scanning electron microscopy (SEM).

RESULTS AND DISCUSSION

Dispersion Quality

The degree of dispersion for the alumina suspensions was assessed by rotational viscometry according to Patton 22 in which the minimum viscosity condition represents the highest degree of dispersion. Dispersion experiments were carried out at 30 vol% solids over a shear rate range of 0 to 300 sec^{-1}. Figure 1 summarizes the flow behavior for the alumina- water- Darvan C system in which the minimum viscosity was observed at a Darvan C concentration of 0.75 wt% (total suspension basis). The flow behavior at 0.75 wt% dispersant was Newtonian with a minimum viscosity of 6 cP. Photon correlation spectroscopy 23 was used to determine that the viscosity increase for disperant concentrations less than 0.75 wt% was due to particle agglomeration. The increase in viscosity for dispersant concentrations greater than 0.75% was due to an increase in pore fluid viscosity contributed by the dispersant. Particle size analysis showed that the sample was well dispersed at Darvan C concentrations of 1 wt%, despite the increased viscosity.

Consolidation Tests

Carefully controlled constant rate of strain consolidation (CRSC) experiments were carried out to assess the effect of dispersion quality and strain rate on the consolidation behavior of these high performance ceramic suspensions. Figure 2 presents the consolidation curves for the alumina-water-Darvan C systems in which void ratio is plotted as a function of vertical effective stress for 0 , 0.25 and 0.75 wt% Darvan C samples at 1.2 E-3 cm/min vertical strain rate. Each curve shows the same trend of a steep curve at low stress followed by a flat curve at higher consolidation stress. The break in the curve between the two regions, occurring at $e = 0.9 \pm 0.1$, may represent a change in the compression mechanism from particle settling (a decrease in particle spacing to form a semi-solid) to particle packing (densification through improved particle packing under large stresses). This hypothesis is consistent with packing theory which predicts for cubic packing of perfect spheres, first contact would occur at e= 0.91. Since consolidation theory is applicable to small strains, only portions of the consolidation curve for $e < 0.9$ (vertical effective stress > 4 psi) were considered for consolidation theory analysis. The positions of the compression curves indicates that as the level of dispersion decreases, much higher consolidation stresses are required to achieve a given particle packing density. The values for C_c, the compression index, which were calculated for the 0%, 0.25% and 0.75 wt% Darvan C curves were 0.143, 0.124 and 0.077, respectively. This data agrees with Janney and Onoda 14 who report a C_c of 0.09 for their dispersed RC-HP alumina- water system. The trend of higher compression index with poorer dispersion indicates that dispersed systems are stiffer in consolidation than are flocculated systems. These values for compression index reflect that the alumina particle matrix is stiffer than typical clay, but are within the range of values (0.19 to 2.6) reported by Mitchell 24 for common clays.

Figure 3 shows the effect of strain rate on the consolidation behavior of the dispersed alumina- water- Darvan C system. Increased strain rates caused significant increases in pore pressure, which caused poor definition of the compression curve at low effective stresses. At higher effective stresses, greater than 50 psi, the curves are slightly different, showing a trend of increasing effective stress with increasing strain rate at constant void ratio. In terms of processing, this result indicates that when increasing strain rate, the effective stress that is applied to the ceramic compact must also be increased, to obtain a constant consolidation effect on microstructure. Compression indices for the highly dispersed system at 0.75% dispersant were 0.077, 0.098 and 0.128 for the 1.2 E-3, 6.1 E-3 and 28.4 E-3 cm/min strain rates, respectively. The compression indices increased with increasing strain rate. This result is consistent Smith and Wahls 25 and Leroueil et al. 26 CRSC tests on calcium montmorillonite and sensitive natural clays, respectively.

Sintering studies and microstructural analysis showed a significant effect of strain rate on part integrity and uniformity. Pressed discs which were formed at low strain rate, 1.2 E-3 cm/min, densified to 99% of theoretical density with uniform firing shrinkage with no part distortion and no part cracking. Figure 4 shows an SEM photomicrograph of a sintered disc fabricated at the lowest strain rate Discs pressed at the high strain rates, 6.1 E-3 and 28.4 E-3 cm/min, showed non- uniform densification across the sample thickness which resulted in cracking upon sintering. The drainage surface, the top surface, experienced less linear shrinkage than the bottom surface, causing the samples to distort and crack. This effect was most pronounced in the 28.4 E-3 cm/min strain rate. These results indicate that the strain kinetics were too fast for the drainage kinetics. Pore fluid could not escape from the particulate to relieve the pore pressure. This condition resulted in a sample with a consolidation gradient across the thickness, particularly in the $e > 0.9$ portions of the compression curves. To relieve this condition, a pressing cycle might consist of a low strain rate to $e = 0.9$ followed by a more rapid strain rate to $e = 0.67$ (60% of theoretical density). Also, by starting with a highly loaded suspensions, with a solids content close to the final green density, only a small amount of strain and a small amount of dewatering would be needed to reach a final consolidation state. Green density variations across the sample would be minimized and cycle times would be rapid relative to dilute suspension pressing.

The coefficient of permeability, k, was calculated using equation (10) and the results are plotted as a function of void ratio in Figure 5. The results show a decrease in permeability with decreasing void ratio, or increasing green density, consistent with Velazquez and Danforth's 27 work on the narrow sized colloidal silica. The data shows a trend of increasing permeability with decreasing dispersion quality at constant void ratio. These results are consistent with Aksay 7 and Aksay and Schilling 28 who showed that agglomerated green microstructures resulted in large open pore regions. Olsen 29,30 discusses that agglomeration will have the effect of increasing the effective particle size, by increasing the pore size between particle units. Consequently, larger flow channels will be present on agglomerated fabrics resulting in higher measured values of k than for the well dispersed system for the same void ratio, e. Equation (4) is also plotted on Figure 5 showing that the Kozeny-Carmen curve falls within the experimental data for the AKP-30 alumina-water-Darvan C system. The permeabilities ranged from 3.5 E-8 cm/sec at e = 0.65 (60% of theoretical) to 3 E-7 cm/sec at e = 1.75 (41% of theoretical).

The coefficient of consolidation was calculated using equation (8), the compression index and the coefficient of permeability. The results are plotted on Figures 6 and 7 in which the C_v varies over 4 orders of magnitude with void ratio. For systems where C_v is not constant, the consolidation equation can not be generally applied. Application of non-linear consolidation theory 31, 32, which consider changes of compressibility and permeability with void ratio, may be more appropriate. However, for small load increments, with C_v varying by only 50%, an average C_v can be used to give approximate results.

As an example, the compression curve in figure 2 for the strain rate of 1.2-3 cm/min and Darvan C content of 0.75 wt% was divided into 10 sections. A C_v was calculated for each section and applied to equation (11) to calculate the time to reach 99% consolidation, t_{99}, by assuming T = 1.78. The vertical strain during each consolidation increment, as calculated from the change in void ratio over the consolidation increment, was divided by t_{99} to give the maximum strain rate which would not be pore pressure dissipation limiting. The results of this calculation are listed in Table 1. The t_{99} values for each increment show that a 3 minute pressing time at a maximum strain rate of 1.5 E-2 cm/min would bring the alumina suspension from 59 volume % solids to 61 volume % solids, at which point the compact is solid.

Table 1. Consolidation Theory to Calculate Maximum Strain Rate

Δe	Δf_s	ΔH (cm)	C_v (cm^2/sec)	t_{99} (min)	ε_{max} (cm/min)
1.0-0.9	0.50-0.53	0.072	1.66 E-4	1,153	6.24 E-5
0.9-0.8	0.53-0.56	0.080	8.05 E-4	238	3.40 E-4
0.8-0.75	0.56-0.57	0.040	6.27 E-3	31	1.30 E-3
0.75-0.7	0.57-0.59	0.046	2.42 E-2	8	5.82 E-3
0.7-0.65	0.59-0.61	0.043	6.89 E-2	3	1.55 E-2

SUMMARY AND CONCLUSIONS

The consolidation behavior of high performance ceramic suspensions is strongly affected by the degree of dispersion and the consolidation strain rate. The compression index increased with decreasing degree of dispersion, indicating that higher consolidation stresses are needed to consolidate a flocculated suspension than a dispersed suspension for a given consolidation strain. Permeabilities increased with decreasing degree of dispersion, indicating that at constant void ratio, the flocculated alumina green microstructure has larger voids (flow channels) than the well dispersed sample. Consolidation coefficients, which were calculated from compression curve and permeability data, varied substantially with void ratio during consolidation invalidating the general application of the linear consolidation equation. Incremental application of the linear consolidation equation indicated that by beginning with suspensions with high solids loading, pressing cycles can be rapid and still produce a fully consolidated green microstructure.

ACKNOWLEDGEMENTS

The authors acknowledge Professor C.C. Ladd and Dr. J.T. Germaine of the Civil Engineering Department at MIT for their valuable input into this work and for program support from the MIT Ceramics Process Research Laboratory-Industry Consortium.

REFERENCES

1. DiMillia, R., Compaction Study of Spray Dried Alumina Agglomerates,PhD. Thesis, Alfred University, Alfred, N.Y., 1982.
2. Halloran, J.W., Role of Powder Agglomerates in Ceramic Processing, in Forming of Ceramics, Vol. 9, Mangels and Messing, eds., AcerS Pub.,Columbus, Ohio, 1984. pp. 67-75.
3. Rhodes, W.H., Agglomeration and Particle Size Effects of Sintering Yttria Stabilized Zirconia, J. Am. Cer. Soc., vol.64, 1981,pp.19-22.
4. Lange, F.F.,B.I.Davis and I.A.Aksay, Uniformity of Alumina-Zirconia Composites by Colloidal Filtration, J. Amer. Cer. soc., vol.66, no.10,1983, C190-C193.
5. Moffatt, W., Processing of High Strength Zirconia Ceramics, Ph.D. Thesis, Massachusetts Institute of Technology, Cambridge, Mass.,1986.
6. Lange, F.F., A.I. Aksay and B.I. Davis, Processing Related Fracture Origins: Part 3, Differential Sintering of Zirconia Agglomerates in Alumina/ Zirconia Composites, J. of Amer. Cer. Soc., vol.66, no. 6, 1983, pp. 407-408.
7. Aksay, I.A., Microstructure Control through Colloidal Consolidation, in Forming of Ceramics, Vol. 9, Mangels and Messing, ed., Amer. Cer. Soc., Columbus, 1984, pp. 94-104.
8. Barringer, E.A. and H.K. Bowen, Formation, Packing and Sintering of TiO_2 Powders, J. Am. Cer. Soc., vol.65, no. 12, 1982, pp. C199-201.
9. Barringer, E.A. na H.K. Bowen, Effects of Particle Packing on the Sintered Microstructure, Physics of Sintering, vol. 16, Amer. Cer. Soc., Columbus, 1985.
10. Cesarano, J. and I.A. Aksay, Processing of Highly Concentrated Aqueous Alumina Suspensions Stabilized with Polyelectrolytes, Poster #41-BP-86, 88th Annual Meeting of the American Ceramic Society, Chicago, Ill, 1986.
11. Novich, B.E. and M.A. Occhionero, Rheology of Concentrated Aqueous Suspensions of Narrow- Sized Alumina Powders, 3rd International Chemical Congress at the 195th National ACS Meeting, Toronto, 1988.
12. Janney,M.A., Plasticity of Ceramic Particulate Systems, Ph.D. Thesis, University of Florida, Gainesville, 1982.

13. Onoda, G.Y. and M.A. Janney, Applications of Soil Mechanics Concepts to Ceramic Particulate Processing, Adv. Powder Technology, ASM, Metals Park, Ohio, 1982.
14. Janney, M.A. and G.Y. Onoda, Particulate Mechanics of Highly Loaded Ceramic Systems, in Ceramic Powder Science, Advances in Ceramics Vol. 21, Messing and McCauley, eds., ACerS Publ., 1987, pp. 615-626.
15. Lambe, T.W. and R.V. Whitman, Soil Mechanics, Johm Wiley and Sons, N.Y., 1969.
16. Schofield, A.N. and C.P. Wroth, Critical State Soil Mechanics, McGraw Hill, London, 1968.
17. Terzaghi, K., Theoretical Soil Mechanics, John Wiley and Sons, N.Y., 1943.
18. Terzaghi, K., Erdbanmechanik, Franz Denticke, Vienna, 1925.
19. Wissa, A., J.T. Christian, E.H. Davis and S. Heiberg, Consolidation at Constant rate of Strain, J. Soil Mechanics and Foundations Division, ASCE, Vol. 97, No. SM10, 1972, pp. 1393-1413.
20. Spangler, M.G. and R.L. Handy, Soil Engineering, Harper and Row, N.Y., 1982.
21. Pyatt, D.H., Consolidation of Aluminum Oxide Particulate Systems, S.M. Thesis, Department of Civil Engineering, MIT, Cambridge, Mass., 1986.
22. Patton, T.C., Paint Flow and Pigment Dispersion, 2nd ed., John Wiley and Sons, N.Y., 1979.
23. Barringer, E.A., B.E. Novich and T.A. Ring, Colloidal Satbility of Ceramic Powders Using Photon Correlation Spectroscopy, J. of Coll. and Interface Science, Vol. 100, No. 5, 1984, pp. 400 -406.
24. Mitchell, J.K., Fundamentals of Soil Behavior, John Wiley and Sons, N.Y., 1976.
25. Smith, R.E. and H.E. Wahls, Consolidation Under Constant Rate of Strain, Journal of Soil Mechanics and Foundation Division, ASCE, Vol. 95, No. SM2, Proc. Paper 6452, 1969, pp. 519-539.
26. Leroueil, S., M. Kabbaj, F. Tavenas and R. Bouchard, Stress- Strain-Strain Rate Relation for the Compressibility of Sensitive Natural Clays, Geotechnique, Vol. 25, No. 2, 1985, pp. 159-180.
27. Velazquez, M. and S.C. Danforth, Casting of Monodisperse Colloidal Silica, Forming of Ceramics, Advances in Ceramics, Volume 9, Mangels and Messing, eds., American Ceramic Society, Columbus, Ohio, 1984, pp105-114.
28. Aksay, I.A. and C.H. Schilling, Mechanics of Colloidal Filtration, Forming of Ceramics, Advances in Ceramics, Volume 9, Mangels and Messing, eds., ACerS., Columbus, Ohio, 1984, pp. 85-93.
29. Olsen, H.W., Deviations from Darcy's Law in Saturated Clay, Soil Science of America Proceedings, 1965, pp. 135-140.
30. Olsen, H.W., Liquid Movement through Kaolinite under Hydraulic, Electric and Osmotic Gradients, AAPG Bulletin, Vol. 56, No. 10, pp 2022-2028.
31. Gibsen, R.E., G.L. England, M.J.L. Hussey, The Theory of One-Dimensional Consolidation of Saturated Clays, Geotechnique, Vol. 17, pp. 261-273.
32. Mesri, G. and Choi, K.K., Settlement Analysis of Embankments of Soft Clays, Journal of Geotechnical Engineering, ASCE, Vol. 111, No. 4, pp. 441-464.

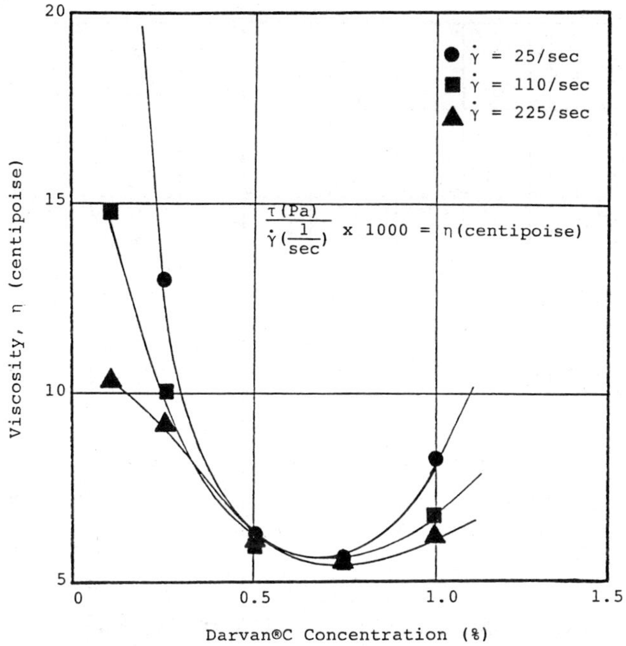

Figure 1. Viscosity versus Darvan C concentration at several shear rates for AKP-30 in water at a void ratio of 2.40.

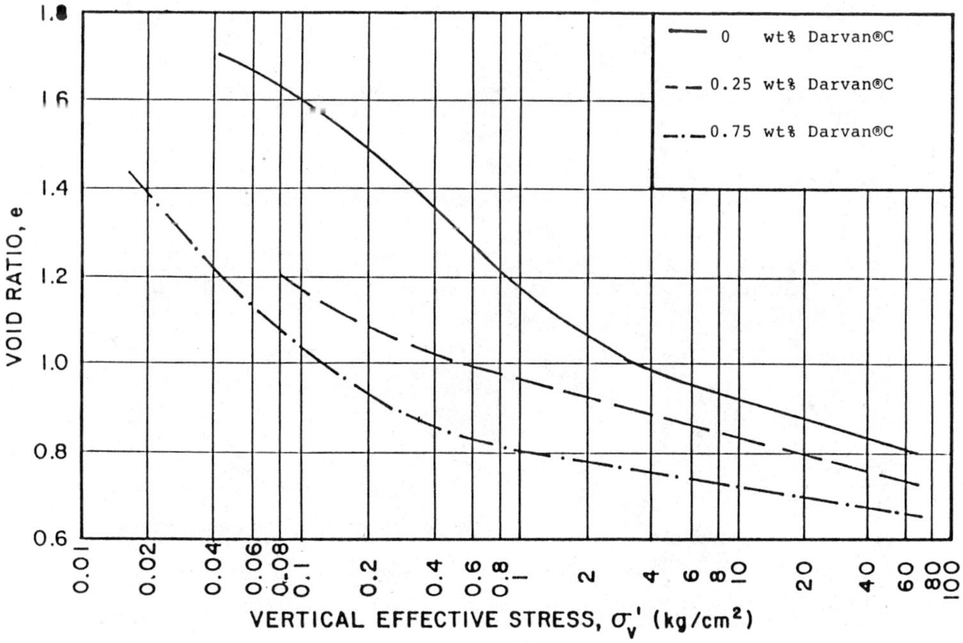

Figure 2. Compression curves from CRSC tests on AKP-30 dispersed with 0 wt%, 0.25 wt% and 0.75 wt% Darvan C and a strain rate of 1.2 E-3 cm/min.

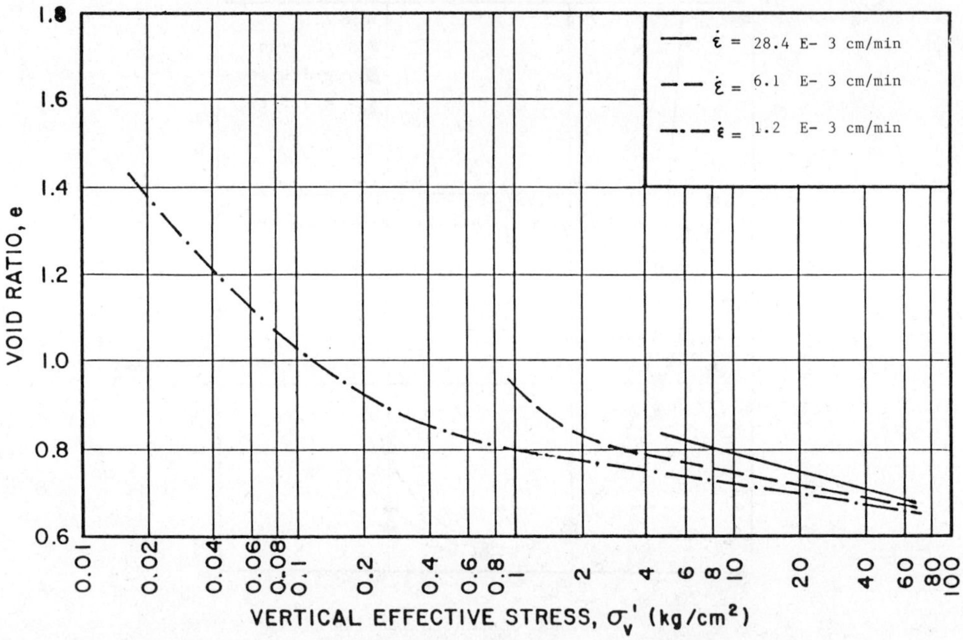

Figure 3. Compression curves from CRSC tests on AKP-30 dispersed with 0.75 wt% Darvan C at strain rates of 28.4 E-3 cm/min, 6.1 E-3 cm/min and 1.2 E-3 cm/min.

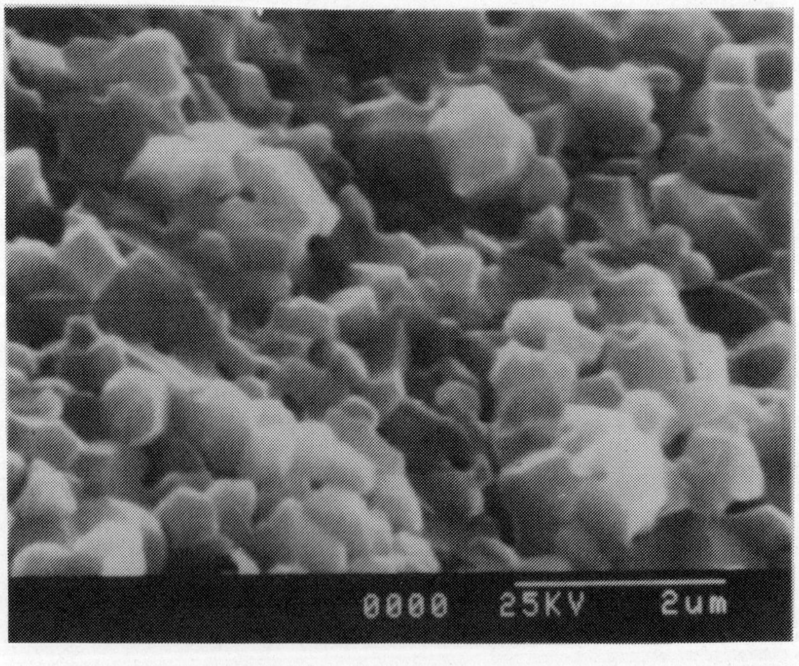

Figure 4. SEM photomicrograph of AKP-30 microstructure formulated with 0.75 wt% Darvan C and consolidated at 1.2 E-3 cm/min. Sintering temperature was 1325°C for 1 hour.

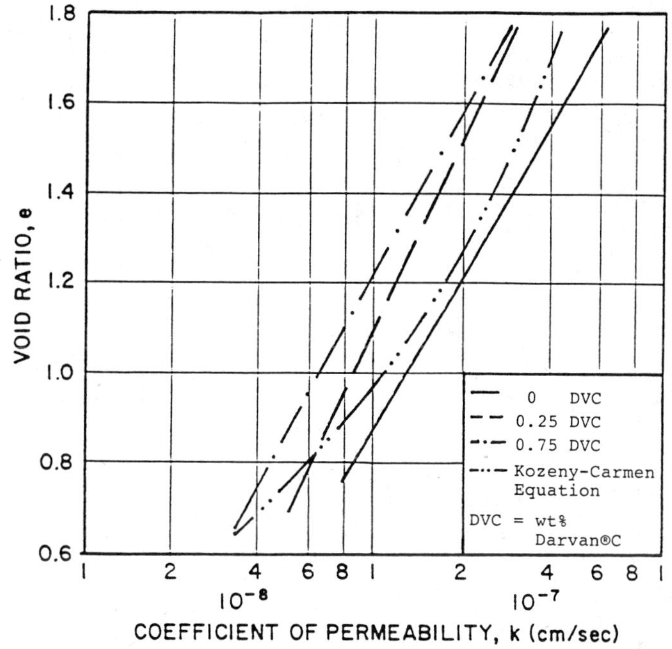

Figure 5. Coefficient of permeability versus void ratio from CRSC tests on AKP-30 dispersed with 0 wt%, 0.25 wt% and 0.75 wt% Darvan C and consolidated at a rate of 1.2 E-3 cm/min.

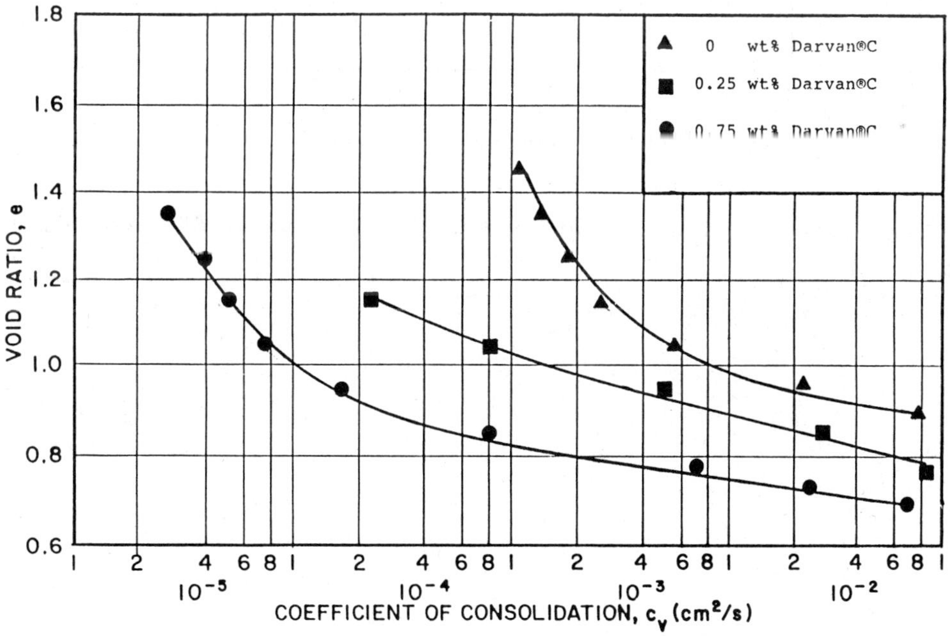

Figure 6. Coefficient of consolidation from CRSC tests on AKP-30 dispersed with 0 wt%, 0.25 wt% and 0.75 wt% Darvan C consolidated at a strain rate of 1.2 E-3 cm/min.

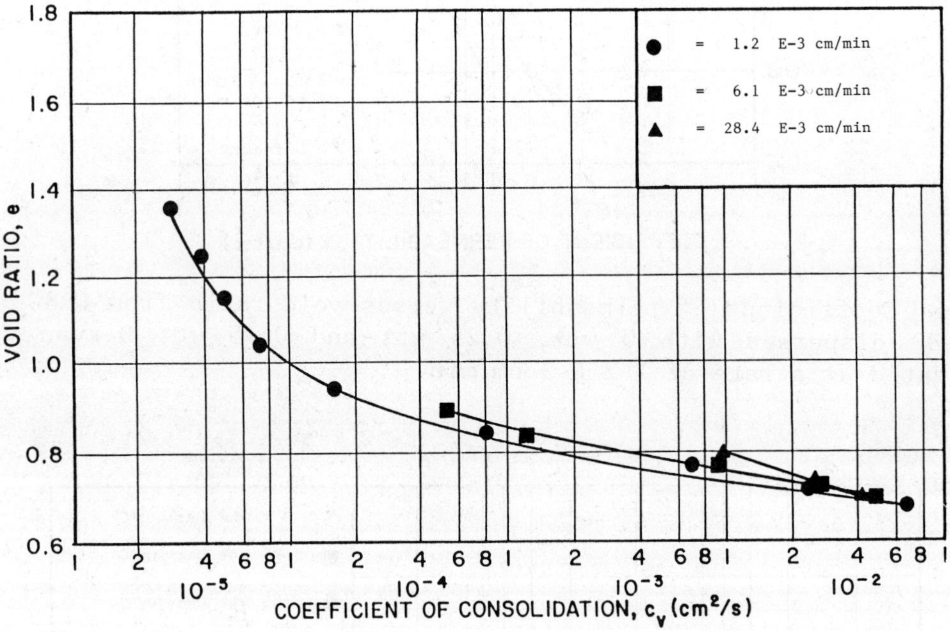

Figure 7. Coefficient of consolidation from CRSC tests on AKP-30 dispersed with 0.75 wt% Darvan C consolidated at 28.4 E-3 cm/min, 6.1 E-3 cm/min and 1.2 E-3 cm/min.

PROPERTIES OF CERAMICS MADE FROM CARBOTHERMALLY SYNTHESIZED SILICON NITRIDE

S. Natansohn

GTE Laboratories Incorporated
40 Sylvan Road
Waltham, MA 02254 U.S.A.

ABSTRACT

The synthesis of silicon nitride powders by the carbothermal reduction of silicon dioxide is described. Powders of various morphologies may be obtained by this technique by varying either the type of reactants used or the process conditions. The powder properties are useful in the fabrication of high-quality structural ceramics. The carbothermally synthesized silicon nitride powders are readily sinterable to densities approaching theoretical. The structural ceramics made from these powders have high mechanical strength at room and elevated temperatures as well as satisfactory oxidation resistance.

INTRODUCTION

The preparation of Si_3N_4 by the carbothermal reduction of SiO_2 in a N_2 stream was reported as early as 1895 and is thus one of the earliest methods of Si_3N_4 synthesis.[1] In recent years a number of investigators[2–10] have studied various aspects of this reaction. In fact, carbothermal reduction (CTR) is one of the four processes by which silicon nitride is produced commercially.[11] The emphasis of this report is on the assessment of the suitability of CTR-Si_3N_4 powders for the fabrication of structural ceramic components for applications such as advanced heat engine parts which are exposed to high thermal and mechanical stresses. To this end, the powders were evaluated primarily in ceramics containing 6 w/o Y_2O_3 as sintering aid, the PY6 formulation which is designed for high-temperature structural applications. Mechanical strength of the ceramics at room and high temperatures as well as oxidation resistance at elevated temperatures was determined and compared against ceramics made from commercial reference materials. The results show that ceramics of requisite properties can be made from CTR silicon nitride powders.

EXPERIMENTAL

The reactants used in this study were fumed silica and carbon black.* These materials have a surface area of about 200 m^2/g and a purity consistent with the requirements of structural ceramics. These components were mixed intimately as an aqueous slurry in a 2:1 carbon-to-silica weight ratio, which represents a five-fold carbon excess over the stoichiometric amount. In a typical preparation, the reactant mixture was placed in graphite boats and heated in a tube furnace at 1480°C for 4 hours in a stream of oxygen-gettered nitrogen flowing at a velocity of 2.5 cm/s; the $O_2 + H_2O$ content of the N_2 gas was less than 2 ppm. The reacted material was heated in air at 680°C for several hours to oxidize and volatilize the excess carbon residue.

The silicon nitride obtained in this manner was mixed in a 94:6 weight ratio with yttrium oxide to form a homogeneous powder. The powder was then either iso-pressed at 138 MPa and sintered at 1950°C under an overpressure of N_2 or hot-pressed at 34 MPa and 1725°C into dense billets from which bars were machined for the evaluation of the modulus of rupture (MOR) of the ceramics. The bars used for the MOR tests had a rectangular cross-section with overall dimensions of 0.127 x 0.254 x 2.54 cm. The four lengthwise edges were chamfered 0.025 cm at 45°. The bars were broken using four-point flexure with inner and outer spans of 1.02 cm and 2.28 cm, respectively, at a cross-head speed of 0.05 cm/min. The MOR of the ceramics was evaluated at room temperature, 1200°C and 1400°C.

Oxidation tests were conducted at 1000°C and 1350°C on polished bars such as those used for MOR testing. The samples were heated in ambient air atmosphere and their weight monitored at regular time intervals to an accuracy of $\pm 2 \times 10^{-6}$ g throughout an exposure period of at least 500 hours. They were placed on thin platinum wires to allow air access to the total surface area of the specimens.

RESULTS AND DISCUSSION

Reaction Mechanisms

The synthesis of silicon nitride by the carbothermal reduction of silica may be described by the overall reaction

$$3\,SiO_2 + 6\,C + 2\,N_{2(g)} \rightleftharpoons Si_3N_4 + 6\,CO_{(g)}\,. \quad (1)$$

In fact, the reaction proceeds via several intermediate steps, the key one of which is the formation of gaseous silicon monoxide by the reduction of SiO_2. A possible reaction route is

$$SiO_2 + C \rightleftharpoons SiO_{(g)} + CO_{(g)}, \quad (2)$$

but it was pointed out by Lee et al.[12] that the relatively fast kinetics and completeness of the CTR reaction are inconsistent with a solid-solid reaction. They postulated that the formation of SiO proceeds via the following two solid-gas reactions:

$$SiO_2 + CO_{(g)} \rightleftharpoons SiO_{(g)} + CO_{2(g)} \quad (3)$$

and

$$C + CO_{2(g)} \rightleftharpoons 2\,CO_{(g)} \quad (4)$$

---

* M-5 grade of fumed silica and Monarch 800 grade of carbon black, both manufactured by the Cabot Corporation.

The rate of both of these reactions is proportional to the surface area of the solid reactants, SiO_2 and C, respectively. The formation of Si_3N_4 occurs in the gas phase by the reaction of silicon monoxide and nitrogen:

$$3\ SiO_{(g)} + 2\ N_{2(g)} + 2\ C \rightleftharpoons Si_3N_4 + 3\ CO_{(g)}. \qquad (5)$$

Thermodynamic Considerations

A summary of the results obtained in the thermodynamic calculation (based on data in the JANAF tables[13] of the components present in the Si-O-C-N system at atmospheric pressure and in the temperature range pertinent to Si_3N_4 synthesis is given in Table 1. The results indicate that Si_2N_2O rather than Si_3N_4 is formed in a closed system, even in the presence of an excess of carbon and nitrogen gas. This result is consistent with the CTR reaction mechanisms discussed above; the buildup of the gaseous reaction products in a closed system would, according to Le Chatelier's principle, drive reactions (3,4, and 5) to the left, retard the formation of SiO and thus Si_3N_4.[5] Similarly, the formation of SiC, which is the stable species in this system at temperatures above 1500°C, is incomplete and occurs only at higher temperatures. This behavior can be explained by analogous considerations because the primary reaction for the formation of SiC in this system proceeds via the gaseous SiO species, namely

$$2\ C + SiO \rightleftharpoons SiC + CO. \qquad (6)$$

The inability to form Si_3N_4 under static reaction conditions has also been noted by other researchers.[14]

Silicon nitride is readily formed in an open system, one in which the gaseous reaction products can be removed from the reactant bed by a flowing gas. Complete conversion to Si_3N_4 occurs at 1400°C with an excess of carbon, 2.5 times over the stoichiometric amount. At 1500°C the SiO_2 is fully converted to SiC, which is the only solid product of the reaction. About half of the excess carbon reactant is in the form of CO, while the other half remains in the system as a solid. A variety of other carbonaceous species are formed but in concentrations which are orders of magnitude smaller than those indicated.

The addition of 10 m/o of H_2 to the N_2 stream does not affect the thermodynamic equilibrium yield of Si_3N_4 at 1400°C or that of SiC at 1500°C. A recent article[15] reported an increase in the rate of the carbothermal synthesis of Si_3N_4 in the presence of ammonia. This is equivalent to having hydrogen in the system because in the temperature range of the CTR reactions the ammonia is dissociated into its components, H_2 and N_2. Thus, in the specific example given in the last row of Table 1, the equilibrium mole fraction of NH_3 at 1400°C is 3.7 x 10^{-7}, while that of H_2 is 4.9 x 10^{-2}, some five orders of magnitude higher. The presence of H_2, however, is of great disadvantage because it leads to the formation of the toxic hydrogen cyanide in significant amounts. As shown by the data in Table 1, HCN is a very stable species in this system. It is formed from the excess solid carbon and, at the higher temperatures, in preference to CO and SiC as it becomes the prevalent carbon-containing molecule. Thus, safety considerations mitigate against the use of NH_3 or H_2 for the carbothermal synthesis of Si_3N_4 in any but the most strictly controlled laboratory environments. In this study, the results of a few exploratory experiments using an admixture of H_2 to the N_2 gas indicated no clear advantage in the reaction kinetics. Under the experimental conditions described above, full conversion (no unreacted SiO_2 found in the product) to Si_3N_4 is realized consistently within the indicated reaction time of no longer than 4 hours with yields in the range of 87% – 90%, based on the amount of Si in the reactant mixture. The lower than theoretical yield may be

Table 1. Summary of Thermodynamic Calculations*

MOLE RATIO OF REACTANTS				KEY PRODUCTS	MOLE FRACTION OF KEY PRODUCTS AT TEMPERATURE (°C)					
SiO_2	C	N_2	H_2		1400	1500	1600	1700	1800	1900
		Closed System								
1	2	0.67	—	C	0.214	0.188	0.0978	0.0	0.0	0.0
				SiC	0.0	0.0	0.0606	0.129	0.142	0.158
				SiO_2	0.0205	0.0	0.0	0.0	0.0	0.0
				Si_2N_2O	0.173	0.187	0.154	0.109	0.00619	0.0
1	5	14	—	C	0.184	0.184	0.107	0.107	0.107	0.108
				SiC	0.0	0.0	0.0519	0.0517	0.0514	0.0511
				Si_2N_2O	0.0263	0.0262	0.0	0.0	0.0	0.0
		Open System								
1	5	1400	—	C	2.14**	1.44**	1.44**	1.45**	1.45**	1.47**
				SiC	0.0**	0.704**	0.702**	0.697**	0.687**	0.657**
				Si_3N_4	0.236**	0.0**	0.0**	0.0**	0.0**	0.0**
				$CO_{(g)}$	1.42**	1.42**	1.41**	1.41**	1.41**	1.40**
1	5	1330	70	C	1.46**	0.277**	0.0**	0.0**	0.0**	0.0**
				SiC	0.0**	0.705**	0.696**	0.665**	0.592**	0.425**
				Si_3N_4	0.236**	0.0**	0.0**	0.0**	0.0**	0.0**
				$CO_{(g)}$	1.42**	1.42**	1.41**	1.38**	1.33**	1.26**
				$HCN_{(g)}$	0.673**	1.14**	1.43**	1.49**	1.62**	1.85**

*At 1 atmosphere

**The mole fraction values listed in the lower half of the table are all multiplied by a factor of 10^3; thus, the mole fraction of C in the open SiO_2-C-N_2 system at 1400°C is 0.00214.

ascribed to the removal of the gaseous SiO formed by reactions (2) and (3) from the reaction zone by the N_2 gas flow.

Powder Properties

Typical properties of a Si_3N_4 powder prepared by the CTR technique are listed in Table 2. This method of synthesis results in material with a high α-phase content,[2,5] and in this powder no β-phase was detected by x-ray diffraction techniques. The β-phase content depends on the reaction temperature and increases at higher synthesis temperatures; fully crystalline powders containing less than 5% of β-Si_3N_4 are obtained routinely.

Table 2. Properties of Si_3N_4 Powder Prepared by Carbothermal Reduction of SiO_2

Phase Composition:	
α-Si_3N_4:	100.0
β-Si_3N_4:	—
Surface Area (m^2/g):	3.9
Composition (w/o):	
Si	58.8
N	38.4
O	2.3
C	0.8
Impurities (ppm):	
Al	200
B	nd
Ca	nd
Cu	nd
Fe	90
Mg	100
Mn	<<50
Mo	nd
Ni	10
Ti	<50

nd — Not detected by emission spectroscopy

In addition to silicon and nitrogen, which are present in virtually a stoichiometric ratio, the minor components are oxygen and carbon. The former is present by design so as to form the appropriate composition with the Y_2O_3 sintering aid and facilitate the densification of the Si_3N_4 ceramic. The carbon present is a consequence of the synthetic route; the amount given here is the total carbon and, as will be shown later, is present as SiC. The free carbon, as determined by combustion at 900°C, is 0.02 w/o, about 2% of the overall total.

The concentration of the impurities listed in Table 2 is comparable to that found in commercial Si_3N_4 powders used for structural ceramic applications.[11] The impurity content is not process induced but is related to the purity of the starting reagents. Al and Mg, the two principal metallic contaminants, are not objectionable, particularly at these low concentrations, because they are frequently used as sintering aids in Si_3N_4 ceramics. Fe appears to be dispersed homogeneously in the powder as no evidence was found of the type of flaws associated with discrete Fe particles.[16]

Typical morphologies obtained in the CTR synthesis of Si_3N_4 powders are shown in Figure 1. A characteristic feature of this synthesis technique is that the powders are generally equiaxed in shape. The size of the crystallites depends on the reaction temperature, residence time, and reaction gas velocity; the specific surface area of the crystalline powders may be adjusted to values between 1 and 12 m^2/g. A powder consisting of equiaxed submicron particles synthesized at 1420°C is depicted in Figure 1(a), while that prepared under the conditions specified above is shown in Figure 1(b). It is also possible to produce well-crystallized powders with fully developed crystalline facets of obvious hexagonal symmetry [Figures 1(c) and 1(d)].

The reduction in nitrogen gas flow velocity during the reaction tends to result in larger, more uniform particles. A typical example, shown in Figure 2, depicts the particle size and morphology of a powder synthesized from the same components and at the same time/temperature conditions as that shown in Figure 1(b), except at half the gas velocity, i.e., 1.2 cm/s. The particle size distribution of this powder is shown in Figure 3. The uniformity of size and shape is ascribed to the fact that at slower gas velocities the nucleation of the individual Si_3N_4 crystallites occurs in a homogeneous manner and, under the less turbulent conditions, crystal growth occurs uniformly on the already formed nuclei in preference to additional nucleation.

Particle morphologies such as those shown in Figure 4 are also observed, but these are generally related to the type of carbon reactant used. Figure 4(a) depicts bridge-like structures consisting of small platelets of hexagonal cross-section in what appears to be a loose array. The structures in Figure 4(b) seem better crystallized, more cohesive, and solid; they show right-angle bends. Similar morphologies were observed by Szweda et al.[5] The x-ray diffraction patterns of all these powders are equivalent, showing α-Si_3N_4 as the dominant phase with an indication of β-phase presence as well, so that the various morphologies cannot be related to crystallographic differences. They may be caused by small differences in the metallic impurities of the various carbon reactants because the effect of such impurities on crystal morphologies is well established and used in whisker growth experiments.[17]

Properties of Ceramics

The summary of the results obtained in the evaluation of the flexural strength of sintered and hot-pressed ceramics made from CTR Si_3N_4 powders is given in Table 3. The density of both the sintered and hot-pressed billets exceeded 98% of the nominal value of 3.262 g/cm^3. The x-ray diffraction patterns of the ceramics indicate the presence of β-Si_3N_4 only, with no other crystalline phase detected.

The data on the experimental ceramics are compared with those obtained on reference materials made from GTE's SN-502 Si_3N_4 powder, which has been widely used and found acceptable in structural ceramic applications. The ceramic processing, sintering, and hot-pressing schedules as well as the test procedures were identical for both materials.

The room-temperature strength of the PY6 ceramics sintered from the two powders is comparable. However, at elevated temperatures the CTR material retains a much higher fraction of its initial strength. There is hardly any decrease at 1200°C and a loss of only 20% of the room-temperature strength at 1400°C. The ceramics hot-pressed from the CTR powder are stronger than the reference material throughout the test temperature range, from room temperature to 1400°C (Figure 5). The fraction of the retained initial strength is also higher.

Table 3. Mechanical Strength of PY6 Ceramics

	Sintered from		Hot-Pressed from	
	Reference Powder[18]	CTR Powder	Reference Powder	CTR Powder
Flexural Strength (MPa)				
Room Temperature:	620	600	790	970
1200°C:	480	590	710	810
1400°C:	380	480	455	645

Micrographic examination of the hot-pressed CTR ceramics (Figure 6) indicates that the material consists primarily of β-Si_3N_4, which is crystallized in the form of large, irregularly-shaped grains and large as well as small acicular grains with aspect ratios as large as 30:1 (designated as 1, 2, and 3 in Figure 6). No oxygen or carbon was detected inside the silicon nitride grains. An amorphous phase consisting of silicon, yttrium, and oxygen with a minor phase of silicon carbide was present in some of the intergranular spaces (designated as 4 in Figure 6). Silicon oxynitride was not detected either as a crystal or an amorphous phase. Small crystallites of silicon carbide (designated as 4 in Figure 6) of unique lamellar morphology were found within the crystalline matrix in contact with both silicon nitride and the amorphous intergranular phase. Examination of one such crystal [Figure 7(a)] by electron energy loss spectroscopy (EELS) identifies the presence of both silicon and carbon [Figure 7(b)]. The diffraction pattern [Figure 7(c)] identifies the major zone as {110} of β-SiC with evidence of twinning. There are also weaker diffraction spots which are typical of the planar defects that accompany polytypic transformations.

Various postulates have been made on the strengthening effect of SiC grains on Si_3N_4 ceramics,[19] but they generally involve the addition of a larger fraction of a SiC powder to the Si_3N_4 material prior to consolidation. In this case, it is difficult to ascribe the observed difference in the flexural strength of the hot-pressed PY6 ceramics to the presence of SiC in the CTR material because the amount, based on the total C concentration, is only 2.7 w/o. Tanaka et al.[19] noted an increase in the strength of sintered Si_3N_4/SiC composites with the addition of as little as 5 w/o of SiC, which they ascribed to the retardation of grain and pore growth. However, this observation was made on ceramics of relatively low density (~95% theoretical) and is not applicable to fully dense materials. No difference in grain size was detected in the materials under study so that strength differences are probably related to powder properties such as morphology, particle size, and particle size distribution. Watari et al.[20] reported room-temperature strength increases in hot-pressed Si_3N_4 doped with as little as 1000 ppm of carbon (~0.3 w/o SiC, if fully converted) and ascribe it to the strengthening of the grain boundaries by SiC. These data, too, were obtained on low density (<93%), low β-phase (46%) ceramics, and the improvement was not over a carbonless material but rather one containing a smaller amount of carbon (500 ppm). The presence of SiC in the grain boundary phase, detected in the materials under study, may account for the observed retention of strength at elevated temperatures because of retardation of grain boundary sliding under applied stress.

The data obtained in the comparative evaluation of the oxidation resistance of the sintered and hot-pressed PY6 ceramics are given in Table 4. The specific weight gain of the two materials after an exposure of 500 hours is comparable at both 1000°C and 1350°C. At the lower temperature the oxidation kinetics follow a linear dependence, while at 1350°C the data indicate a parabolic rate law. These observations are consistent with the elucidation of the oxidation mechanisms in such ceramics by Smith and Quackenbush.[21] The ceramic test pieces made from the CTR powder are indistinguishable from the reference material after the heat treatment, showing no unusual microstructural features nor indications of incipient failure. There are legitimate concerns about the oxidation stability of carbon-containing Si_3N_4 ceramics; catastrophic failures have been observed. However, in those materials carbon was present in its free form, which is prone to rapid oxidation with the evolution of gaseous components. In the ceramics made from CTR powders, the small amount of residual carbon exists as SiC, a material which has excellent oxidation resistance, and therefore the oxidation stability of the ceramic is not adversely affected.

Table 4. Oxidation Resistance of PY6 Ceramics

	Sintered from		Hot-Pressed from	
	Reference Powder	CTR Powder	Reference Powder	CTR Powder
500 hr at 1000°C:				
$\Delta W/A$ (10^{-3} kg/m^2):	0.66	0.71	0.38	0.38
K_L (10^{-10} kg/m^2 s):	4.1	3.6	1.6	1.4
500 hr at 1350°C:				
$\Delta W/A$ (10^{-3} kg/m^2):	1.3	1.1	3.4	2.4
K_p (10^{-12} kg^2/m^4 s):	1.0	1.6	7.0	3.2

CONCLUSIONS

The carbothermal reduction of silica is a convenient route to the synthesis of high-purity silicon nitride powders suitable for the fabrication of structural ceramic components. These powders are sinterable to parts with high densities. The flexural strength at room temperature and particularly at 1400°C is suitable for the requirements of advanced heat engine parts. The oxidation resistance of these ceramics is comparable to that of materials made from powders prepared by other routes and acceptable for this application.

ACKNOWLEDGMENT

The author thanks G. Czupryna and F. Scimeca for experimental assistance; C. Struck for thermodynamic calculations; M. Downey, F. Pink, and C. Beck for materials characterization; T. Emma for analytical electron microscopy; and last, but not least, J.T. Smith for continued support and encouragement.

REFERENCES

(1) D.R. Messier and W.J. Croft, "Silicon Nitride," in Preparation and Properties of Solid State Materials, Vol. 7, W.R. Wilcox, ed., Marcel Dekker, Inc., New York, NY. (1982), p. 132.

(2) K. Komeya and H. Inoue, "Synthesis of the α-form of Silicon Nitride from Silica," *J. Mater. Sci.* 10 (1975), p. 1243–1246.

(3) A. Hendry and K.H. Jack, "The Preparation of Silicon Nitride from Silica" in Special Ceramics, Vol. 6, P. Popper, ed., British Ceramic Research Association, Stoke-on-Trent, United Kingdom (1975), pp. 199–208.

(4) J.G. Lee and I.B. Cutler, "Reactions in the SiO_2-C-N System," in Nitrogen Ceramics, F.L. Riley, ed., Noordhoff, Leyden, Netherlands (1977), pp. 175–181.

(5) A. Szweda, A. Hendry, and K.H. Jack, "The Preparation of Silicon Nitride from Silica by Sol-Gel Processing," in Special Ceramics, Vol. 7, D. Taylor and P. Popper, eds., British Ceramic Society, Stoke-on-Trent, United Kingdom (1981), pp. 107–118.

(6) H. Inoue, K. Komeya, and A. Tsuge, "Synthesis of Silicon Nitride Powder from Silica Reduction," *Commun. Am. Ceram. Soc.* 65 (1982), C-205.

(7) M. Mori, H. Inoue, and T. Ochiai, "Preparation of Silicon Nitride Powder from Silica," in Progress in Nitrogen Ceramics, F.L. Riley, ed., Martinus Nijhoff, Boston, MA (1983), pp. 149–154.

(8) Shi-Chang Zhang and W.R. Cannon, "Preparation of Silicon Nitride from Silica," *J. Am. Ceram. Soc.* 67 (1984), pp. 691–695.

(9) V. Figusch and T. Licko, "Synthesis of Silicon Nitride Powder by Carbothermal Nitriding of Silica," in High Tech Ceramics, P. Vincenzini, ed., Materials Science Monographs 38, Elsevier, Amsterdam, Netherlands (1987), pp. 517–526.

(10) M. Mitomo and Y. Yoshioka, "Preparation of Si_3N_4 and AlN Powders from Alkoxide-Derived Oxides by Carbothermal Reduction and Nitridation," *Adv. Ceram. Mater.* 2 (1987), pp. 253–256.

(11) S. Natansohn, "Nondestructive Evaluation of Powders for Advanced Ceramics," in Non-destructive Testing of High-Performance Ceramics, A. Vary and J. Snyder, eds., American Ceramic Society, Westerville, OH (1987), pp. 73–88.

(12) J.G. Lee, P.D. Miller, and I.B. Cutler, "Carbothermal Reduction of Silica," in Proceedings of 8th International Symposium on the Reactivity of Solids, Chalmers, Sweden (1976), pp. 707–711.

(13) M.W. Chase, Jr., C.A. Davies, J.R. Downey, Jr., D.J. Frurip, R.A. McDonald, and A.N. Syverud, "JANAF Thermochemical Tables," 3rd ed., J. Physical and Chemical Reference Data, 14, Supplement No. 1 (1985).

(14) S.A. Badrak, T.S. Bartnitskaya, I.M. Baryshevskaya, T.Ya. Kosolapova, P.P. Pikuza, B.G. Trusov, and V.P. Turov, "Thermodynamic Calculations for the Assessment of Possible Ways of Formation of Silicon Nitride," Soviet Powder Metallurgy (Eng. Transl.) 20 (1982), pp. 643–647.

(15) B.G. Durham, M.J. Murtha, and G. Burnet, "Si_3N_4 by the Carbothermal Ammonolysis of Silica," *Adv. Ceram. Mater.* 3 (1988), pp. 45–48.

(16) A.E. Pasto, "Causes and Effects of Fe-Bearing Inclusions in Sintered Si_3N_4," *Commun. Am. Ceram. Soc.* 67 (1984), C178-C180.

(17) G.A. Bootsma, W.F. Knippenberg, and G. Verspui, "Growth of SiC Whiskers in the System SiO_2–C–H_2 Nucleated by Iron," *J. Crystal Growth* 11 (1971), pp. 297–309.

(18) C.L. Quackenbush and J.T. Smith, "GTE Sintered Silicon Nitride," *Am. Soc. Mechanical Engineers*, Paper No. 84-GT-228 (1984).

(19) H. Tanaka, P. Greil, and G. Petzow, "Sintering and Strength of Silicon Nitride–Silicon Carbide Composites," *Int. J. High Techn. Ceram. 1* (1985), pp. 107–118.

(20) K. Watari, K. Ishizaki, and K. Kawamoto, "Behavior of Carbon in Hot-Pressed Silicon Nitride Grain Boundaries and Its Influence on Mechanical Properties," *J. Ceram. Soc. Jpn, Int. Ed.* 96 (1988), pp. 128–134.

(21) J.T. Smith and C.L. Quackenbush, "Phase Effects in Si_3N_4 Containing Y_2O_3 or CeO_2:II. Oxidation," *Bull. Am. Ceram. Soc.* 59 (1980), pp. 533-537.

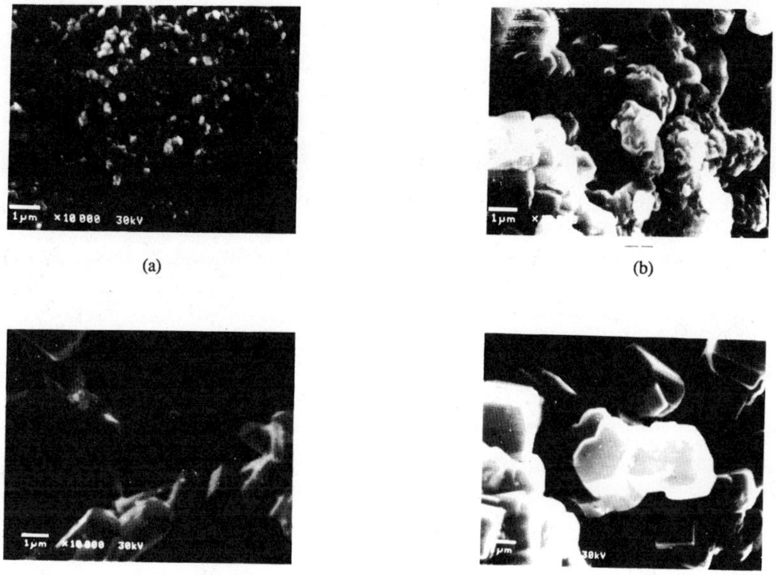

(a) (b)

(c) (d)

Figure 1. Morphologies of CTR Si_3N_4 powders.

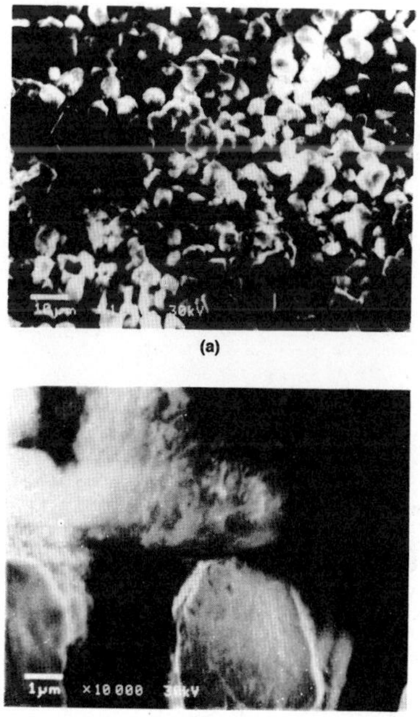

(a)

(b)

Figure 2. Particle size and morphology of CTR-Si_3N_4 synthesized at slow gas flow velocity; (a) 1000 X; (b) 10,000 X.

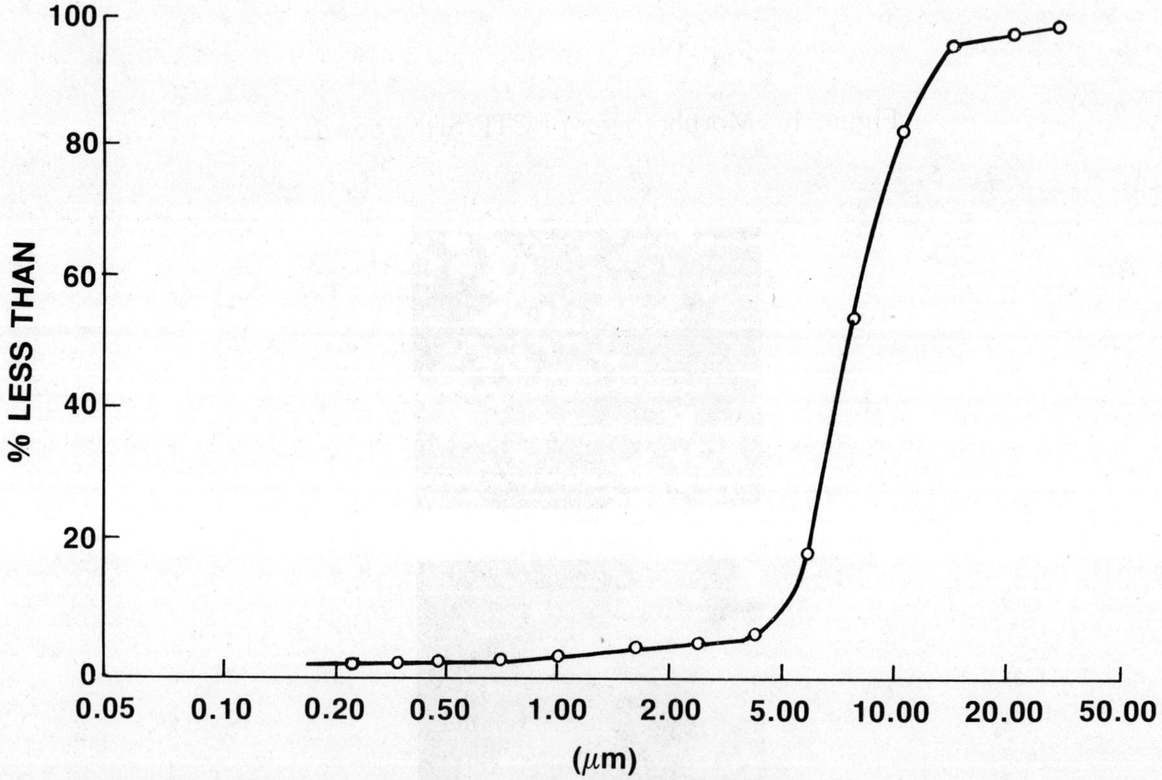

Figure 3. Cumulative particle size distribution of Si_3N_4 powder depicted in Figure 2.

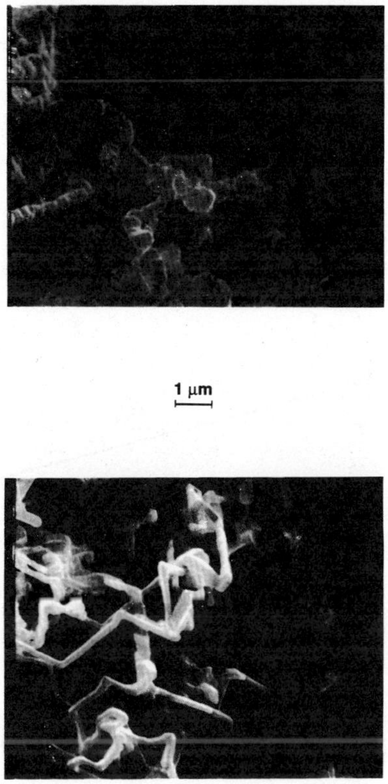

Figure 4. Bridge-like morphologies of certain CTR-Si_3N_4 powders.

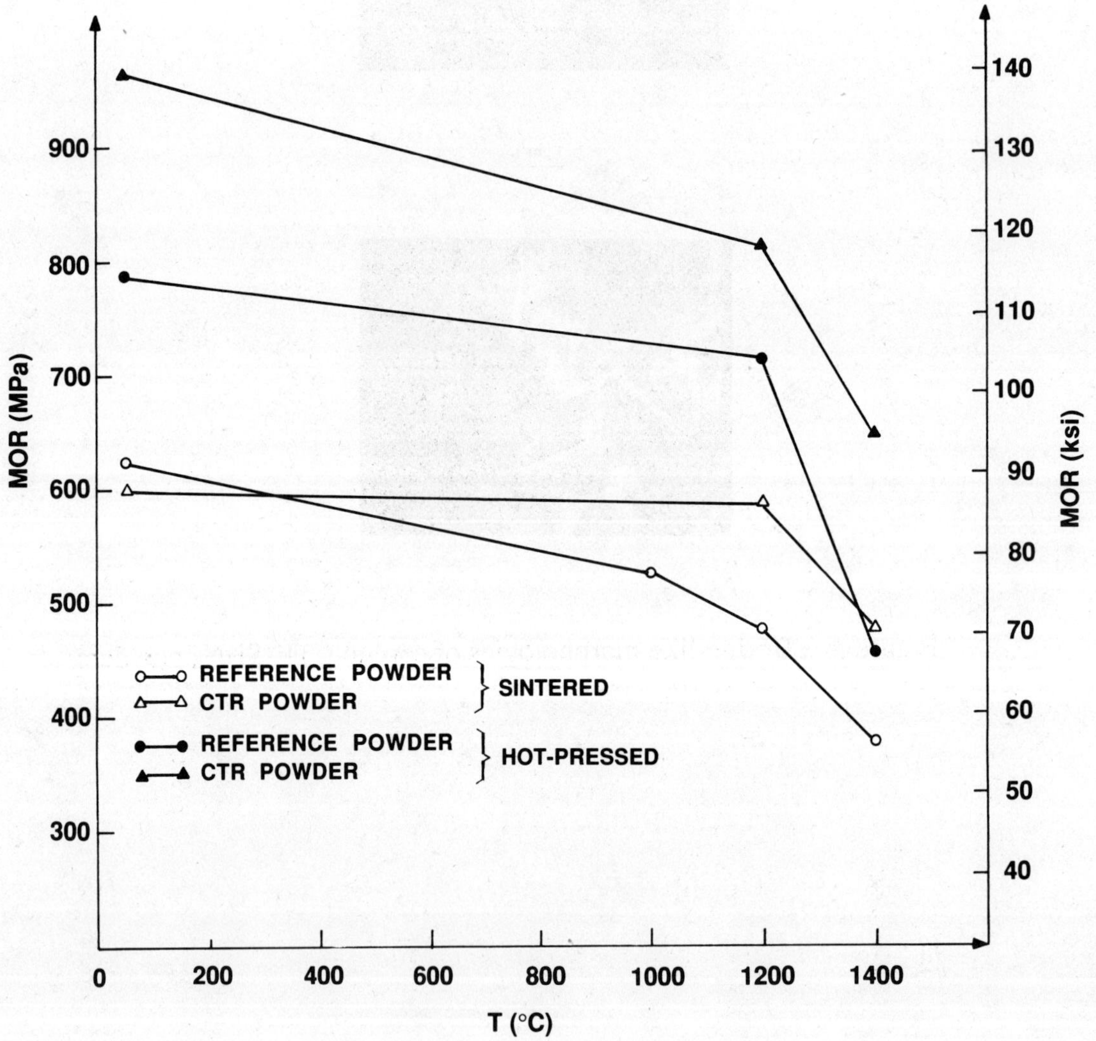

Figure 5. Temperature dependence of the flexural strength of the PY6 ceramics.

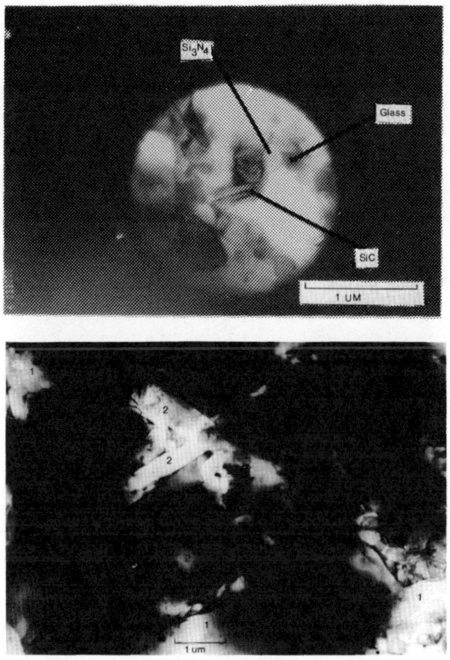

Figure 6. SiC crystallites in a Si_3N_4 matrix.

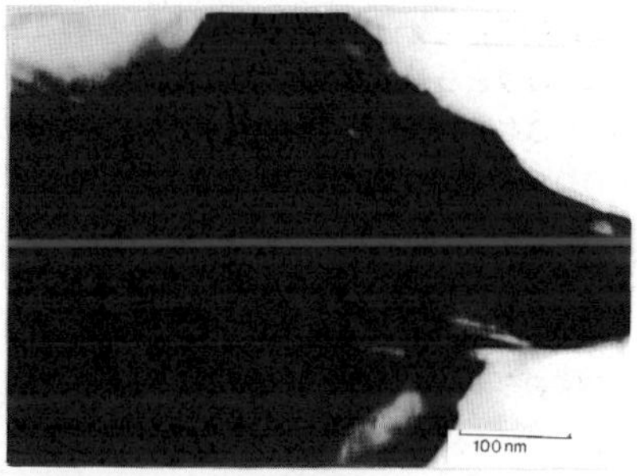

(a)

(b)

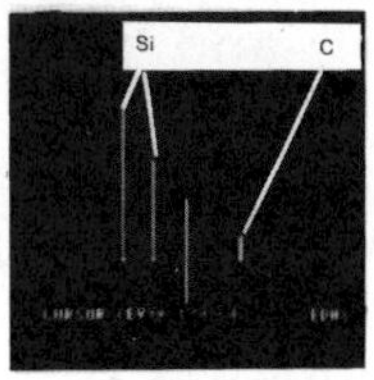

(c)

Figure 7. (a) SiC crystallite; (b) electron diffraction pattern of selected area of the crystallite; (c) electron energy loss spectrum.

CHARACTERIZATION OF SILICON NITRIDE POWDER FOR INJECTION MOLDING

Takahiko Honma, Hideyuki Masaki, and Shigetaka Wada
Toyota Central Research & Development Labs., Inc.
Nagakute, Aichi 480-11, Japan

ABSTRACT

In ceramic injection molding, ceramic powder characteristics, such as average particle size, particle size distribution, specific surface area, packing density and particle morphology, have a strong influence on injection moldability. Fourteen kinds of commercial silicon nitride powders were characterized. Five out of the fourteen kinds of powders were chosen and the apparent viscosities of their mixtures with a fixed organic binder system were measured. The apparent viscosity of the mixtures correlated reasonably well with the packing density and the specific surface area of the powders. Both of them are excellent parameters to estimate the injection moldability at high powder concentrations.

It was also shown that the powders suitable for injection molding have both a relative packing density more than 57 % and a specific surface area of about 10 $m^2 \cdot g^{-1}$.

INTRODUCTION

Using the injection molding forming technique, silicon nitride has recently been applied for engine components, such as turbocharger rotors[1] and prechamber combustion inserts[2]. The characteristics of a silicon nitride powder strongly influence not only the mechanical properties of a sintered body[3], but also the process conditions of injection molding. In particular, physical characteristics such as average particle size, particle size distribution, specific surface area, packing density and particle morphology have pronounced effects on the powder concentration and consequently the fluidity of mixtures of powders and organic binders.

Many studies[4-9] have been performed on the correlation between the powder characteristics and the viscosity of suspension. Farris[6] demonstrated the particle size distribution with the minimum viscosity by predicting the viscosity of multimodal suspensions from unimodal viscosity data. Mangels, et al.[7] studied the viscosity of concentrated suspensions of Si powder prepared by different millings in wax as a function of powder characteristics and experimentally confirmed Farris' prediction. Takahashi, et al.[8,9] investigated the effects of powder characteristics on the kneading and flow properties of highly filled Alumina-Thermoplastic Resin Systems and demonstrated that the relative compaction density was an excellent parameter to estimate the kneading and flow properties at high solid concentrations. They also demonstrated the validity of Farris' relation between powder concentration and viscosity.

There have been few reports relating powder characteristics and injection moldability for commercial silicon nitride powders. Therefore, with consideration of the above general concepts, the effect of the powder characteristics of commercial silicon nitride on kneading and fluidity was examined.

Silicon nitride powders are usually milled and/or provided with sintering aids in practical fabrications. Therefore, the characteristics of a powder mixture in practical fabrication will differ somewhat from those of a raw powder. The present work will offer useful data for the selection of a silicon nitride powder.

EXPERIMENTAL

Powder

Fourteen kinds of commercial silicon nitride powders were examined. Relative packing density (relative green density of powder compact), specific surface area, average particle size, particle size distribution and particle morphology of the as-received powders were characterized. Using an isostatic-pressed body at 0.3 GPa after die-pressing to $\phi 40 \times 5^t$ mm, the packing density was determined from its weight and dimensions. The relative packing density was calculated, assuming the theoretical density of silicon nitride as 3.2 $g \cdot cm^{-3}$. The specific surface area was measured using a single point, nitrogen adsorption BET technique. The average particle size and the particle size distribution were measured by Microtrac (Leeds & Northrup-Nikkiso, Submicron Microtrac SPA). The particle morphology was studied by scanning electron microscopy (SEM).

Five out of the fourteen kinds of powders were chosen, and the kneading torque and fluidity of the mixture of each as-received powder and a fixed organic binder system were measured.

Kneading

The mixing of each as-received powder and a fixed organic binder system was achieved using a Brabender type plastograph

(Toyoseiki, Lab-prastmill). Several powder concentrations in a range of 46 to 62 vol.% were examined. The proprietary organic binder system used was mainly composed of a thermoplastic polymer and paraffine waxes.

The mixture volume was 80 cm^3 against the space volume of mixer (100 cm^3). The kneading temperature was set to 160°C. The powder concentration at room temperature was used as an indication of the powder concentration. The powder concentration in the mixture during kneading and fluidity measurement was about 2 vol.% per 100°C less than that at room temperature. This is due to the fact that the thermal expansion coefficient of the organic binders is larger than that of the silicon nitride powder. The kneading speed was held constant at 30 rpm, and the kneading time was 90 minutes. The kneading torque in the steady state after 90 minute kneading was used for an index of the kneadability.

Fluidity

After kneading, the mixtures were cooled and pelletized for use as samples for fluidity measurement.

The apparent viscosity was measured with a capillary rheometer (Toyoseiki, Capillograph) using a 0.5 mm radius, 10 mm length die. The measurement temperature was 140°C. The apparent viscosity was calculated using Hagen-Poiseuille equation ($\eta_a = \frac{\pi R^4 P}{8LQ}$).

RESULTS

Powder Characteristics

The characteristics of fourteen kinds of commercial silicon nitride powders are given in Table 1. The relationship between the relative packing density and specific surface area of the powders is shown in Figure 1. The fourteen kinds of powders are roughly classified in four groups. No. 1 has a relatively small specific surface area, but a fairly large relative packing density. This behavior is assumed to be due to the large average particle size and the broad particle size distribution. Nos. 10, 11 and 12 are the smallest in relative packing density for corresponding specific surface area. This behavior is due to the sharp particle size distribution. Nos. 7 and 8 are the largest in both specific surface area and relative packing density. This apparently results from the small average particle size and inclusion of some content of extremely fine particles. The other eight kinds of powders were intermediate between the above mentioned grouping powders, and the specific surface area and the relative packing density are about 10 $m^2 \cdot g^{-1}$ and 55-65 %, respectively.

The average particle sizes of most of the powders were approximately 1 μm, and those of Nos. 1 and 14 were larger than those of the others.

Five kinds of powders, 9S (No.2), LC12 (No.8), H1 (No.9), E10 (No.10) and ESP (No.13), were chosen for the following experiments. 9S, H1 and ESP were chosen from the intermediate group (Figure 1). LC12 and E10 were chosen as the typical powders having a large specific surface area and a small relative packing density, respectively. Moreover, these five powders are commercially available on the scale of more than several ten kilograms per batch. 9 (No.1) was also typical, having a relatively small specific surface area, but a fairly large relative packing density. However, it was omitted in terms of poor sinterability.

SEM photographs of the five powders are shown in Figure 2. 9S, LC12 and H1 consisted of irregular particles whose sizes were distributed broadly. In particular, it was observed that LC12 contained some percentage of extremely fine particles. E10 consisted of spherical and considerably uniform fine particles. ESP had spherical particles like E10, but the particles were coarse, and the particle size distribution was broader than E10.

The particle size distributions of the five powders are shown in Figure 3. The distribution of E10 was sharp, while those of 9S, LC12 and H1 were broad. ESP had broader distribution than E10, while sharper distribution than the other powders. Among the five powders, LC12 included the largest content of fine particles below 0.3 μm. These results agreed with SEM photographs shown in Figure 2. Since the coarse particles measured by Microtrac were not observed in the SEM photographs, it was implied that all of the powders were agglomerated.

The powder characterization by particle size distribution seems to provide little quantitative information as it depends on the particle morphology, and is sensitive to agglomeration. However, both the relative packing density and the specific surface area are considered more effective for powder characterization, since they reflect the effects of average particle size and particle size distribution and are insensitive to agglomeration.

Kneading Torque

The relationships between the kneading torque and the powder concentration for the five powders are shown in Figure 4. 9S, H1 and ESP exhibited a low torque even at a powder concentration as high as 60 vol.%. However, both LC12 and E10 showed higher torque than the three powders even at powder concentrations as low as 56 and 52 vol.%, respectively.

Fluidity

Figure 5 shows the dependence of apparent viscosity on shear rate for 9S with various powder concentrations. The data for the other powders were omitted. In any powder, the apparent viscosity increased with the powder concentration. The slope of the apparent viscosity versus shear rate did not vary with the powder concentration.

With the exception of E10, Figure 6 shows the dependence of apparent viscosity on shear rate for the powders at a 52 vol.% powder concentration. E10 could be kneaded at a 52 vol.% powder concentration, but the mixture was too viscous to measure the apparent viscosity by the capillary rheometer. Therefore, the data for the 48 vol.% powder concentration was plotted in Figure 6 for E10. The slope of the apparent viscosity versus shear rate for E10, ESP, 9S and H1 were nearly the same. On the other hand, LC12 presented a steeper slope. The steeper slope is assumed to be due to rather high concentrations of extremely fine particles.

Figure 7 shows the relationship between the apparent viscosity and the powder concentration at a shear rate of about 1000 (s^{-1}), which is the middle value of the shear rate encountered in a mold in practical injection molding (between 100 and 10000 s^{-1}). The dependences of the apparent viscosity on the powder concentration were much the same except for E10. The apparent viscosities for 9S, H1, ESP and LC12 increased slowly up to a 60 vol.% powder concentration and were less than 0.5 KPa•s. The apparent viscosity of E10 was more than 1 KPa•s even at a 50 vol.% powder concentration.

DISCUSSION

The powder characteristics as related to injection molding were studied from the viewpoints of both kneading torque and fluidity in a powder concentration range of 46 to 62 vol.%.

High kneading torques were essential for E10 and LC12, as shown in Figure 4. The high kneading torque is related to the powder characteristics in Figure 1. The causes of this high kneading torque are the small relative packing density of E10 owing to the sharp particle size distribution (Figures 2 and 3), and the large specific surface area of LC12 because of the presence of extremely fine particles. On the other hand, 9S, H1 and ESP could be kneaded even at a powder concentration as high as 60 vol.% with a low torque.

The apparent viscosity of the mixture at a shear rate of 912 (s^{-1}), as encountered in practical injection molding, is related to the powder concentrations, as shown in Figure 7. 9S, H1, ESP and LC12 exhibited a low apparent viscosity, less than 0.5 KPa•s even at powder concentrations up to 60 vol.%. These powders have a relative packing density over 57 %, as shown in Figure 1. The large relative packing density seems to correspond to superior fluidity even at high powder concentrations in practical injection molding. However, the apparent viscosity of E10 increased abruptly at lower powder concentrations.

LC12 exhibited a steep dependence of apparent viscosity on shear rate, as shown in Figure 6. The apparent viscosity at low shear rates was higher than those of other powders. This result corresponded to the high kneading torque at low shear rates (Figure 4). The steep dependence of apparent viscosity on shear rate and the high apparent viscosity at low shear rates are assumed to be due to the presence of extremely fine particles which resulted in the large specific surface area. Also, the high apparent viscosity at low shear rates implies that use of LC12 would need a high torque for plasticization before injection in a common screw-type injection molding machine. Therefore, both a powerful kneader and a screwtype injection molding machine with high plasticizing capacity would be necessary for LC12.

Both the relative packing density and the specific surface area of the powder strongly affect the kneading torque and the apparent viscosity at high powder concentrations. The silicon nitride powders having both a relative packing density more than 57 % and a specific surface area of about 10 $m^2 \cdot g^{-1}$, such as 9S, H1 and ESP, shows easy kneadability and good fluidity.

CONCLUSION

1) Both the relative packing density and the specific surface area were excellent parameters to estimate the injection moldability of silicon nitride powders at high powder concentrations.
2) The powders suitable for injection molding, having both a relative packing density more than 57 % and a specific surface area of about 10 $m^2 \cdot g^{-1}$, were 9S, H1 and ESP.
3) It is suggested that LC12 needs both a powerful kneader and a screw-type injection molding machine with high plasticizing capacity because of a high apparent viscosity at low shear rates.
4) E10 will be suitable only when the powder concentration in the mixture is less than 48 vol.%.

ACKNOWLEDGMENT

We are grateful to Denki Kagaku Kogyo Co., Ltd. for measuring average particle size, particle size distribution and specific surface area of powders.

REFERENCES

[1]A. Watanabe, "Nissan, N-1 Type Ceramic Turbocharger", GTS

Jpn, 13 [52] 49-50 (1986).
[2]S. Kamiya, M. Murachi, H. Kawamoto, S. Kato, S. Kawakami and Y. Suzuki, "Silicon Nitride Swirl Lower-Chamber for High Power Turbocharged Diesel Engines", SAE Paper 850523 (1985).
[3]O. Abe, K. Saito, S. Kanzaki, H. Tabata, S. Kume and M. Machida, "Influence of Si_3N_4 Raw Powders Characteristics on Pressureless Sintering", R. Gover. Ind. Res. Ins., Nagoya, Jpn, 34 [11-12] 355-366 (1985).
[4]B. C. Mutsuddy, "Influence of Powder Characteristics on the Rheology of Ceramic Injection Molding Mixtures", Proc. Br. Cer. Soc., [33] 117-137 May (1983).
[5]M. J. Edirisinghe and J. R. G. Evans, "Compounding Ceramic Powders Prior to Injection Moulding", Br. Ceram. Proc., [38] 67-80 (1986).
[6]R. J. Farris, "Prediction of the Viscosity of Multimodal Suspensions from Unimodal Viscosity Data", Trans. Soc. Rheology, 12 [1] 281-301 (1968).
[7]J. A. Mangels and R. M. Williams, "Injection Molding Ceramics to High Green Densities", Cer. Bull., 62 [5] 601-606 (1983).
[8]M. Takahashi, S. Suzuki and H. Arai, "Mixing and Flow Properties of Alumina-Polypropylene Mixture System", J. Mater. Sci., Jpn, 35 [398] 23-28 (1986).
[9]M. Takahashi, S. Suzuki, E. Arai and H. Nitanda, "Effects of Powder Characteristics on Kneading and Flow Properties of Highly Alumina-Thermoplastic Resin Systems", J. Mater. Sci., Jpn, 36 [410] 29-35 (1987).

Table 1. Characteristics of Silicon Nitride Powders.

Silicon Nitride Powder				Relative Packing Density (%) at 0.3GPa	Specific Surface Area ($m^2 \cdot g^{-1}$)	Average Particle Size (μm)
Sample NO.	Maker	Grade	Production Method			
1	Denka	9	Nitridation of Silicon	63.4	3.8	11.40
2	〃	9S	〃	59.3	7.5	1.03
3	〃	9FW	〃	59.5	11.1	0.73
4	〃	GD	〃	59.9	7.2	1.23
5	Showa Denko	NU20	〃	55.3	9.4	1.61
6	〃	NU30	〃	62.4	12.2	1.62
7	Shin-Etsu Chemical	KSN10H	〃	62.4	26.0	0.77
8	Stark	LC12	〃	63.7	22.7	0.65
9	〃	H1	〃	63.1	10.9	0.90
10	Ube	E10	Diimide Decomposition	50.4	12.0	0.56
11	〃	E05	〃	53.9	5.1	1.10
12	〃	E02	〃	56.4	2.1	1.76
13	〃	ESP	〃	56.9	8.2	0.80
14	Toso	TS8	〃	56.9	12.4	3.52

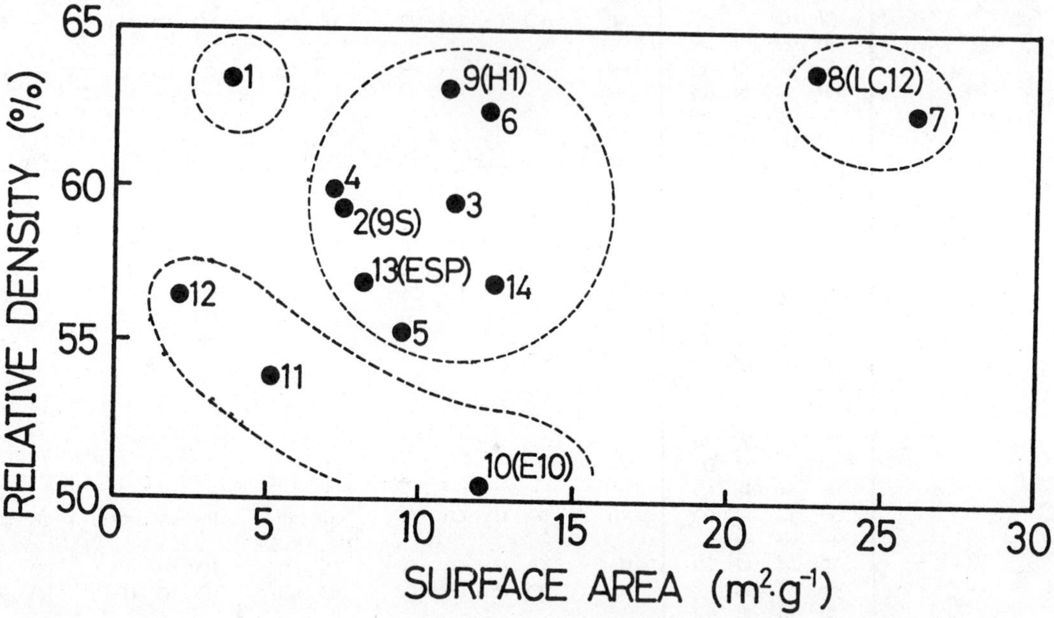

Figure 1. Relationship between relative packing density and specific surface area.

Figure 2. SEM photographs of the five powders (bar = 1 μm).

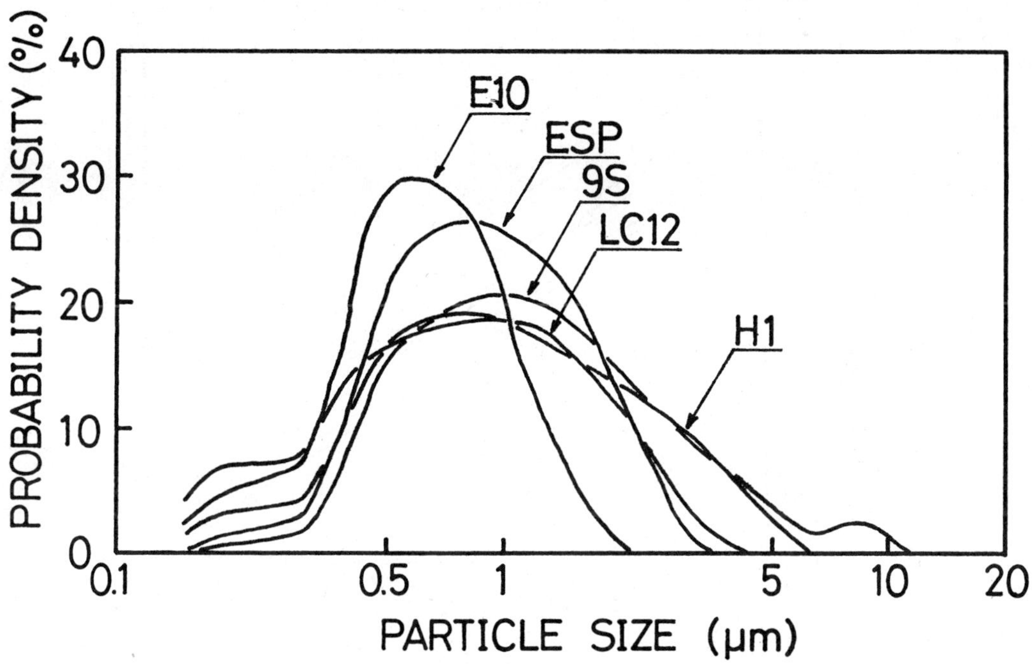

Figure 3. Particle size distributions of the five powders.

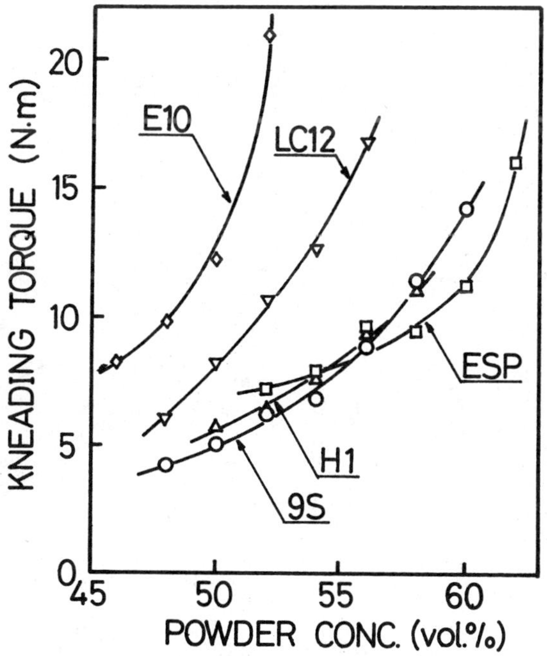

Figure 4. Relationships between kneading torque and powder concentration for the five powders.

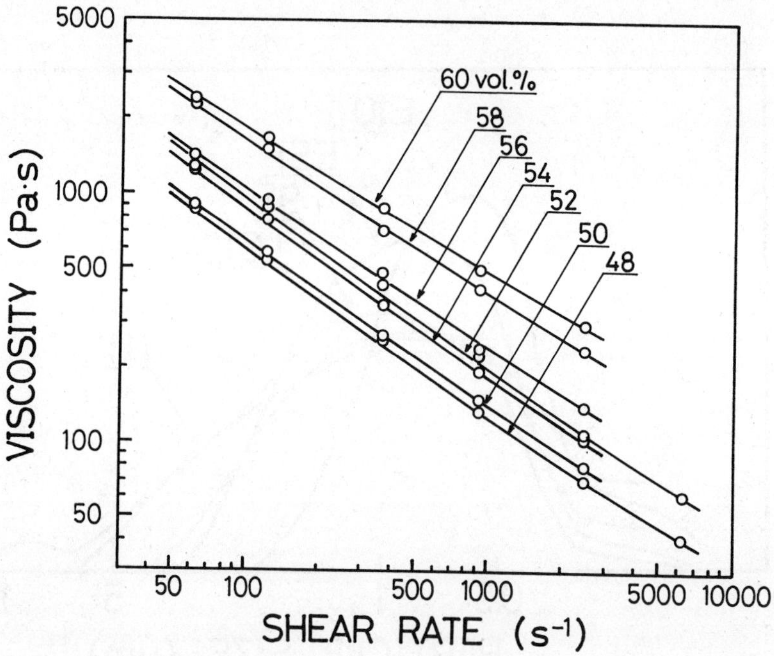

Figure 5. Dependence of apparent viscosity on shear rate for 9S with various powder concentrations.

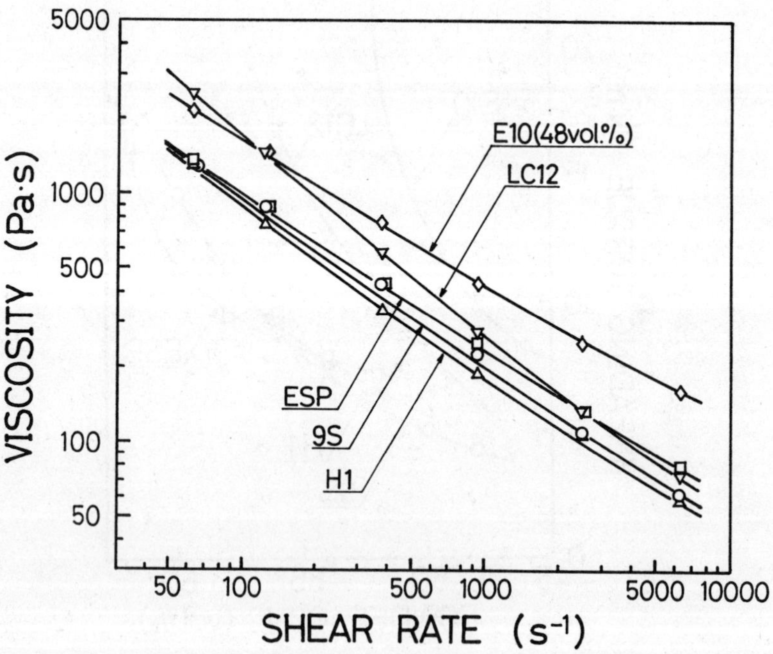

Figure 6. Dependence of apparent viscosity on shear rate for the five powders with 52 vol.% powder concentration (except E10).

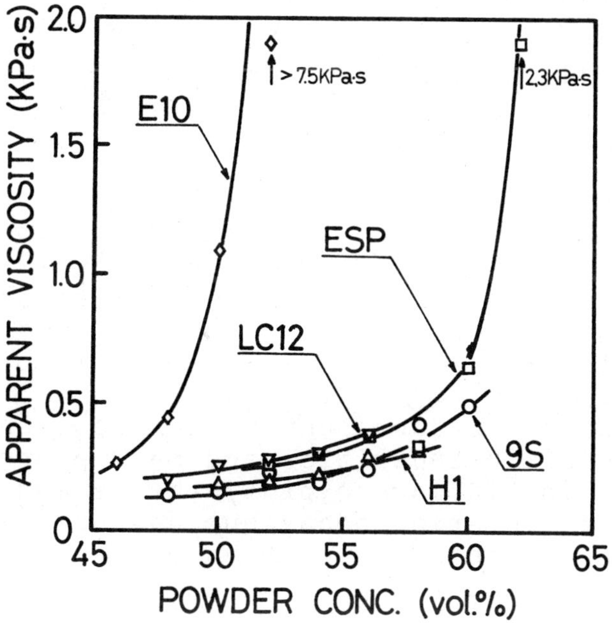

Figure 7. Relationship between apparent viscosity at the shear rate of 912 (s^{-1}) and powder concentration for the five powders.

CHARACTERISTICS OF PLASMA - DERIVED SILICON NITRIDE POWDERS

O. Schulz, D. Kandziora, H. Hausner

Technische Universität Berlin

ABSTRACT

Silicon nitride powders have been synthesized by the reaction between silane or silicon tetrachloride and ammonia in a rf - coupled argon plasma.

The powders which are predominantly amorphous show a low concentration of metallic impurities and have a grain size in the submicron range. The influence of synthesis and processing parameters on their sintering behaviour has been evaluated in a high-temperature dilatometer under an elevated pressure of nitrogen.

INTRODUCTION

The synthesis of sinterable silicon nitride powders can be achieved by the reaction between silicon metal and nitrogen (1), by a reaction in the liquid state between silicon tetrachloride and ammonia (2) or by gas-phase reactions (3). Due to the interest in high-purity powders, rf-plasma (4-8) and laser synthesis (9-11) have found increasing attention during the past years. Both methods yield high-purity powders of spherical morphology with high specific surface areas.

Only a few systematic studies have been done on the processing and sinterability of gas-phase derived silicon nitride powders although it is well known that the consolidation of these powders is difficult. Symons and Danforth (9) have stated that good sinterability is possible if the powders are unagglomerated and of spherical morphology with a crystallite size smaller than 25 nm. Schulz et. al. (8) have shown that plasma derived silicon nitride powders can be sintered to high densities after appropriate processing steps.

POWDER SYNTHESIS

In Fig.1 the reaction and deposition vessel of the plasma unit is shown schematically.

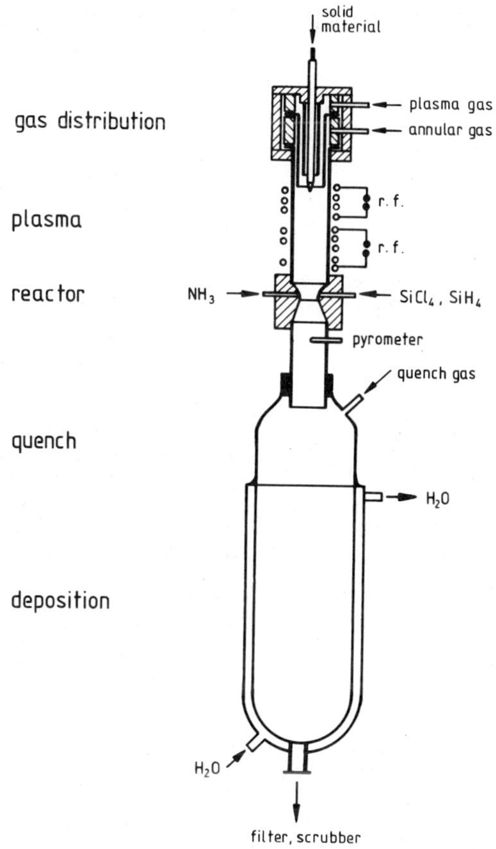

Fig.1 Plasma reactor (schematic)

The rf-coupled plasma extends into a reactor vessel, where the hot plasma-gas is used to effect the reaction between the gaseous reactants. The reaction products are cooled rapidly in the quench device. The deposition of the powder takes place in a deposition chamber and an attached filter unit. The argon plasma is generated with a 40 kVA high-frequency generator which operates in the frequency range between 4-6 MHz. The argon flow rate is approximately 1.5 m^3/h and the operating pressure 1.5 bar. The energy density can be varied between 10 and 50 MJ/m^3.. Silicon nitride powders have been synthesized starting from silane (diluted with nitrogen) or from silicon tetrachloride (diluted with argon) and ammonia according to the main reactions:

$3\ SiH_4 + 4NH_3 \rightarrow Si_3N_4 + 12H_2$ (SN-H Powder)

$3\ SiCl_4 + 4NH_3 \rightarrow Si_3N_4 + 12HCl$ (SN-Cl Powder)

Corresponding to the thermodynamics of the system $SiH_4 - NH_3$ and $SiCl_4 - NH_3$ different reaction conditions are neccessary for obtaining stoichiometric compounds (Table I)

Table I. Synthesis Conditions for Si_3N_4 - Powders

Silicon Compound	SiH_4	$SiCl_4$
Carrier Gas	N_2	Ar
Carrier Gas Conc. (vol.%)	98	80
NH_3 - Supply (m^3/h)	0.5	0.5
Gas Enthalpy (MJ/m^3)	12.5	25.0
Product Yield (g/h)	50	100

In the system $SiH_4 - NH_3$ a yield maximum has been found. If a certain energy input and the associated process temperature exceeds a certain limit an increasing amount of free metallic silicon is found in the powder formed by the decomposition of silane. In contrast, the yield in the system $SiCl_4 - NH_3$ increases with increasing temperature. This could be caused by a formation of substoichiometric reactive silicon chlorine compounds and a suppression of nitrogen and hydrogen containing intermediates (13). The upper yield limit is determined by the decomposition temperature of the silicon nitride. An incomplete reaction in the system $SiCl_4 - NH_3$ results in a high oxygen content in the powder caused by the reaction between non-reacted silicon tetrachloride and air-humidity during powder handling. In the system $SiCl_4 - NH_3$ ammonium chloride is formed through a side reaction between gaseous hydrogen chloride and ammonia. The ammonium chloride has to be removed from the powder by sublimation or by a washing step.

3. POWDER CHARACTERISTICS

The powder characteristics are summarized in Table II. The powders have a spherical morphology independent of the system investigated (Fig.2). Also shown in Table II are the characteristics of powder after the indicated heat treatments. The surface areas are high in comparison with commercial silicon nitride powders.

Table II. Characteristics of Plasma Powders

Reactants		SiH_4 +	NH_3	$SiCl_4$ +	NH_3
Thermal Treatment of Powder		-	1300°C 2h, N_2	400°C 0.5 bar	1450°C 2h, N_2
Particle Size	(nm)	260	317	245	305
Spec. Surface Area	(m^2/g)	65	32	49	34
Crystallinity		part	full	amorphous	part
ß - Si_3N_4	(%)	30	15	-	10
Nitrogen	(wt%)	35.2	37.2	36.1	34.8
Oxygen	(wt%)	2.1	4.0	2.7	5.0
Carbon	(wt%)	0.11	0.02	0.05	0.05
Met. Imp.	(ppm)	100^x	80	300	300

^{x}besides 8 wt% free silicon

The particles are agglomerated but the agglomerates can be dispersed easily. From the $SiCl_4$ - NH_3 reaction amorphous powders were abtained whereas the reaction in the SiH_4 - NH_3 system resulted in partially crystalline products with a significant amount of ß-Si_3N_4. With regard to the chemical composition the as-prepared powders had a nitrogen content of 90% of the theoretical value and an oxygen content of 2-3 wt.%. Metallic impurities (Fe, Al, Cr, Cu, Mg, Ca) are present in a quantity up to 300 ppm and are a matter of the purity of the reactants. The SN-H powders contain finely divided free silicon, the SN-Cl powders ammonium chloride and silicon-nitrogen compounds with a lower nitrogen content than stoichiometric Si_3N_4.

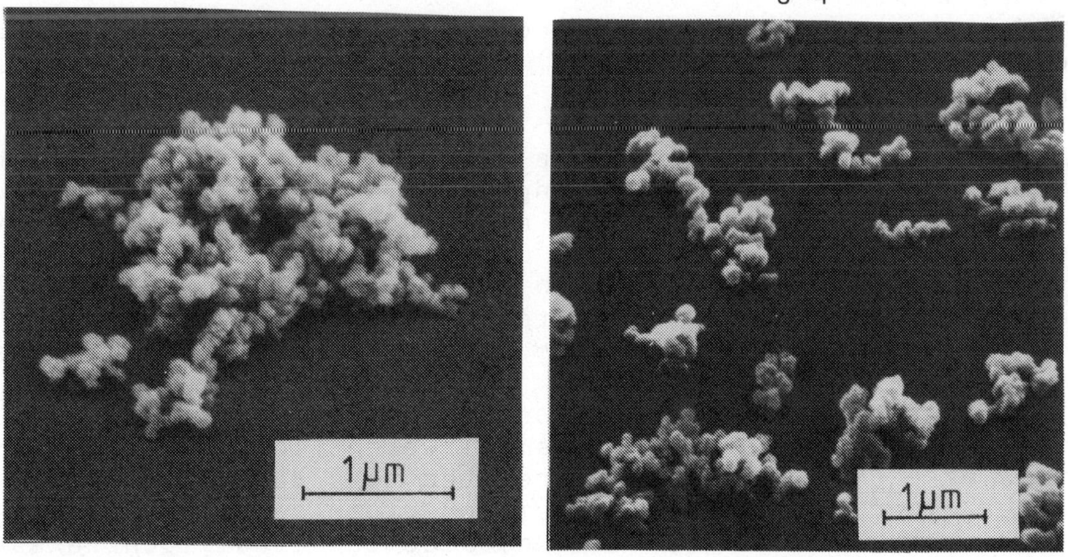

Fig.2 Morphology of Si_3N_4 - Powders
(left: powder from SiH_4-NH_3 right: powder from $SiCl_4$-NH_3)

The different powders were heat treated under nitrogen in order to transform the free silicon into silicon nitride or to remove the ammonium chloride and to crystallize at the same time the amorphous powders. (Fig.3)

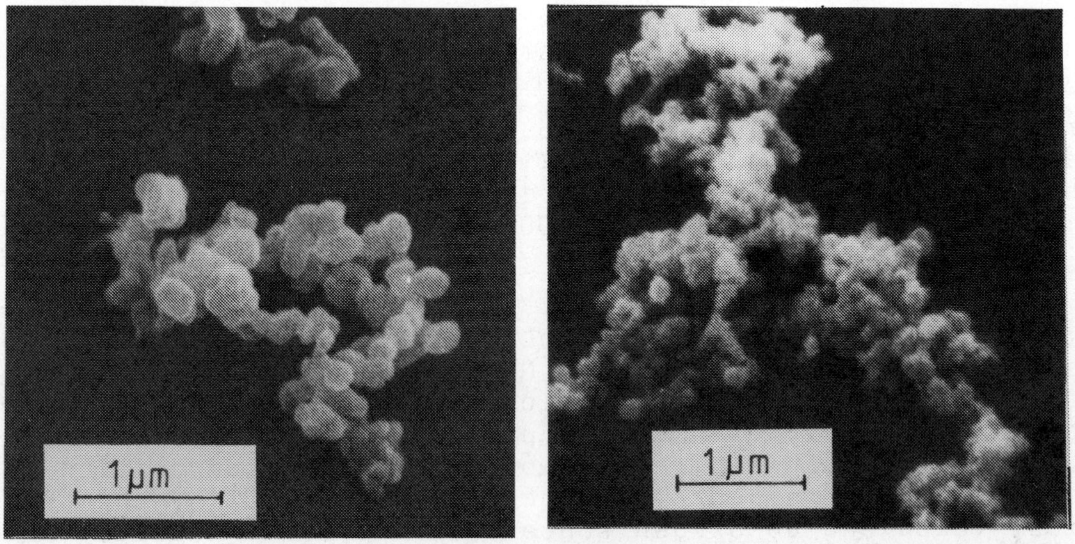

Fig. 3 Morphology of Si_3N_4 - Powders after thermal treatment under nitrogen (See Table II)
left: powder from SiH_4-NH_3 right: powder from $SiCl_4$-NH_3)

The SN-H powders were transformed into crystalline products already at temperatures of 1200°C, the SN-Cl powders only in excess of 1400°C (Fig.4)

The SN-Cl powders show in part a needle-like morphology which can be suppressed by high heating rates and short heating cycles. After the heat treatment of the amorphous SN-Cl powder the product contains more than 90% α- Si_3N_4, whereas the α-Si_3N_4 content in the heat treated SN-H powder is slightly lower. As expected the specific surface area of the powders is reduced after the heat treatment and the particles have grown in size.

In Fig.5 thermogravimetric curves of the powder types are shown. The SN-H powders show a small weight loss up to a temperature of 1200°C which can be related to a degassing of absorbed ammonia or the decomposition of imides. In the temperature range 1300 - 1500°C the nitridation of silicon takes place, associated with a change in colour of the powder from light brown to white.

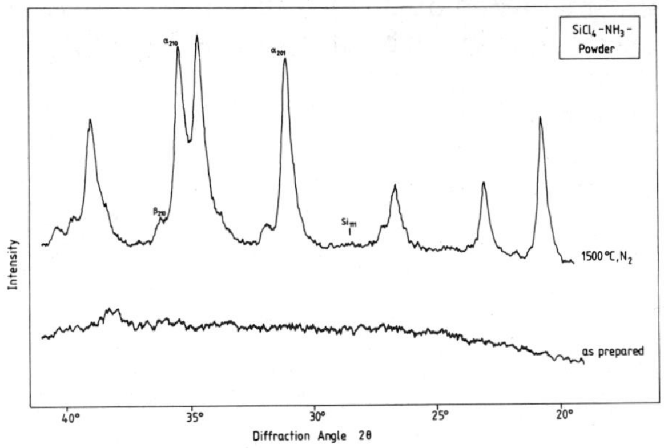

Fig.4 X-ray diffraction peaks before and after thermal treatment in nitrogen (SN-Cl powder)

The SN-Cl powder shows first weight loss in a region up to 350°C which can be attributed to the sublimation of ammonium chloride which has still remained in the powder. The higher weight loss in the temperature range 400-1200°C may be caused by the decomposition of imide compounds into the nitride and ammonia (13).

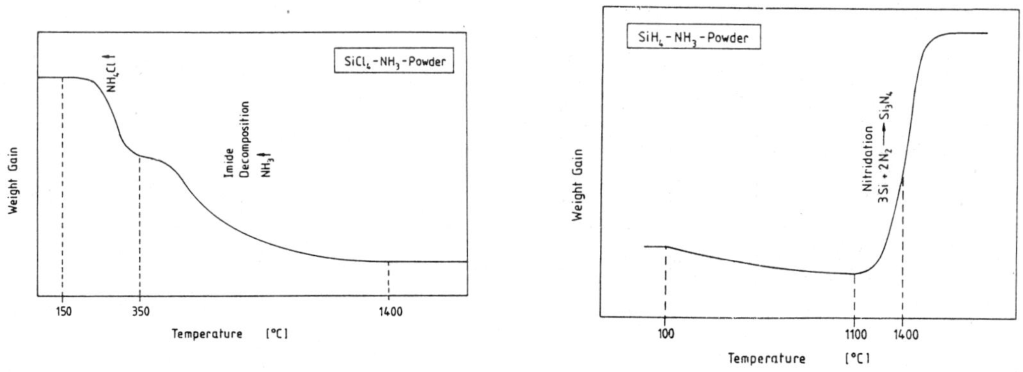

Fig.5 Thermogravimetric characteristics of the powders (qualitative)

The weight gain of the SN-H powder after the heat treatment in nitrogen was on the order of 5 wt%, the weight loss of the SN-Cl powder under the same conditions amounted to 1.5 wt%. Both powders show a higher oxygen content (Table II) after the heat treatment which can be avoided by using high purity nitrogen in this step. The flowability of the SN-H powders after the thermal treatment is improved in contrast to the SN-Cl powders with their needle-like morphology. It may be possible however, that the needle formation can be suppressed in the latter powder type by optimizing the conditions of the heat treatment.

SINTERING BEHAVIOUR

For the investigation of the sintering behaviour of the powders 1.5% MgO was added in the form of $Mg(NO_3)_2 \cdot 6 H_2O$ which was dissolved in isopropanol. The mixture was heat treated under nitrogen at 400°C for eight hours in order to decompose the nitrate to the oxide. The calcined powders were compacted with a presure of 250 MPa to cylindrical pellets of 10 mm diameter and 5.5 mm height.

The sintering behaviour was investigated in a high-temperature dilatometer under nitrogen pressure. A typical temperature - pressure cycle is shown in Fig. 6

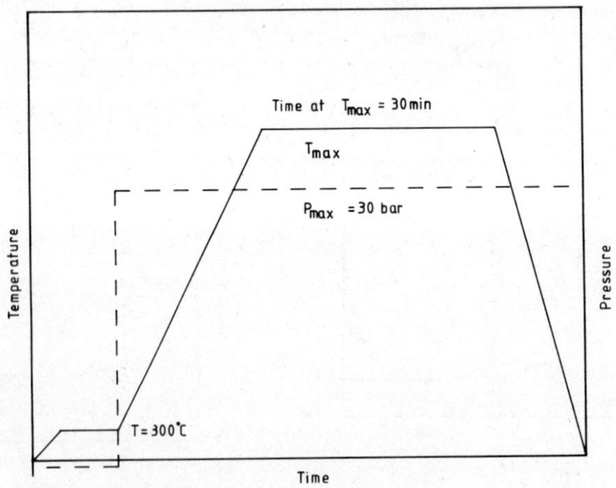

Fig.6 Temperature - pressure cycle for the sintering of the silicon nitride compacts

The heating rate for the SN-H powders was 44°C/min., for the SN-Cl powders 31°C/min.
The sintering kinetics of the different powders are shown in Fig.7 and 8

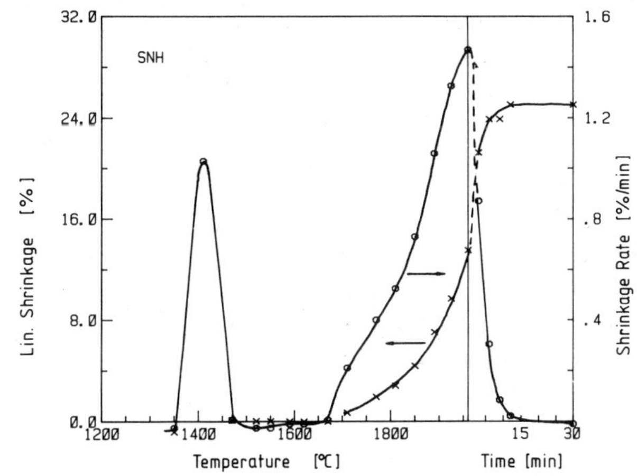

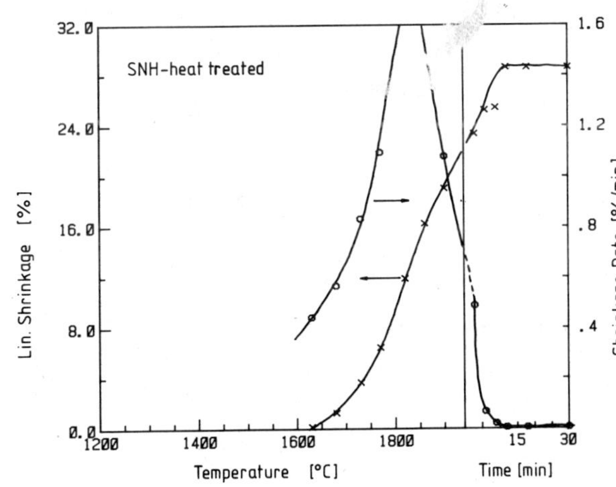

Fig.7 Sintering kinetics of SN-H powders (upper curve: as prepared, lower curve: after thermal treatment in nitrogen)

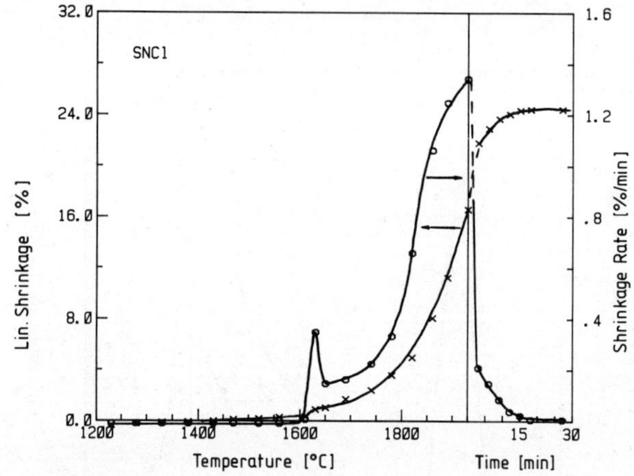

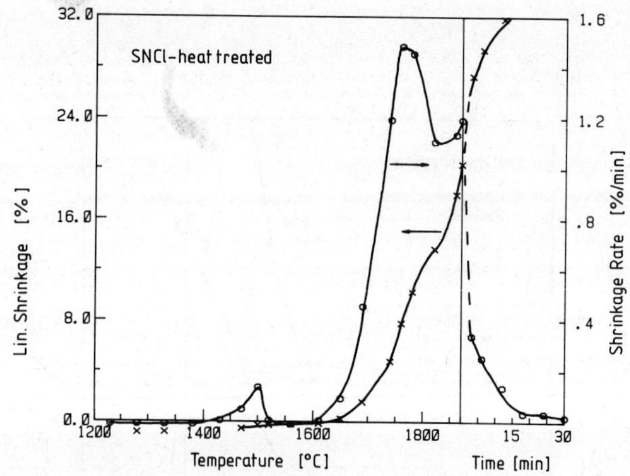

Fig.8 Sintering kinetics of SN-Cl powders (upper curve: as prepared; lower curve: after thermal treatment in nitrogen)

The maximum shrinkage rate of the as synthesized powders occur at a temperature of 1950 - 1960°C and is believed to be caused by a reaction of a magnesium oxide rich liquid (silicate) and the silicon nitride. In the heat treated powders this maximum is shifted to lower temperature and may be related to the oxygen content of the powder (Table III).

Table III. Oxygen Content of Powder and Temperature of Maximum Shrinkage Rate

Powder	Oxygen Content (wt%)	Temp.of Maximum Shrinkage Rate
SN-H (as prepared)	2.1	1960°C
SN-Cl (as prepared)	2.7	1950°C
SN-H (heat treated)	4.0	1820°C
SN-Cl (heat treated)	5.0	1760°C

The as synthesized SN-H powder which contains free silicon shows a maximum at 1400°C caused by the formation of liquid silicon. After the thermal treatment this peak is no longer present due to the transformation of the silicon into silicon nitride.
The SN-Cl powders show small maxima at 1630°C (as synthesized) or at 1500°C (heat treated). They may be caused by reaction between the magnesium oxide and a silica surface layer but at present it is not clear why similar peaks are absent in the SN-H powders.
In Table IV the green and sintering densities of the different compacted powders are shown. In general the green densities are low (below 40%th.d.), typical for gas-phase derived powders. Only the heat-treated SN-H powders can be sintered to a density of ~95% th.d. under the conditions already described. There seems to exist a close relationship between the crystallinity of the powder and its sinterability.

Table IV. Green and Sintering Density of Powder Compacts

Reactants	SiH_4 +	NH_3	$SiCl_4$ +	NH_3
Thermal Treatment of Powder	-	1300°C 2h,N_2	400°C 0.5bar	1450°C 2h,N_2
Green Density of Compacts (g/cm^3)	1.16	1.24	1.23	1.23
Sintered Density of Compacts (g/cm^3)	2.47	3.02	2.21 (cracked)	2.58

Whereas the SN-H powder is completely crystalline after the heat treatment the as prepared SN-Cl powder is fully amorphous and the two other powders are only partially crystallized. However several other reasons may be responsible for the difference in the sintering behaviour. In the case of the SN-H powder (as prepared), which contains free silicon the interfacial energy relationship in the silicon -silicon nitride system does not contribute to an enhancement in sintering. In the case of the SN-Cl powders the remaining concentration of volatiles (Fig. 5) will have a negative effect on sintering and the morphology of the SN-Cl powder after the heat treatment at 1450 °C will also impede densification.

In Fig. 9 the microstructure of the sintered SN-H powder is shown. It consists of a fine-grained silicon nitride with small sized pores. It can be expected that the sintered density can be increased by a higher concentration of sintering additives which is rather low with 1.5 wt% MgO in the powders investigated.

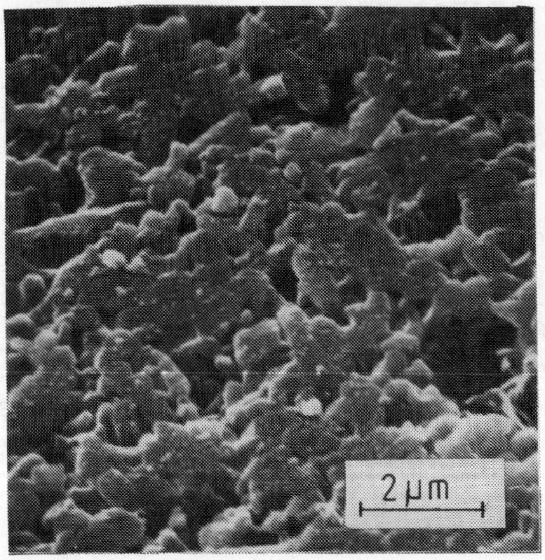

Fig.9 Microstructure of sintered silicon nitride (prepared from SN-H powder)

CONCLUSIONS

The characteristics of silicon nitride powders which have been prepared by rf-plasma synthesis via the SiH_4 - NH_3 or $SiCl_4$-NH_3

reaction were influenced significantly by the type of reactants, the operating parameters of the plasma and the post-synthesis processing steps. Under optimized conditions high-purity, sinteractive powders can be prepared.

ACKNOWLEDGEMENT

This work was supported by the Bundesminister für Forschung und Technologie (BMFT).

REFERENCES

(1) G. Schwier
Entwicklung ung Herstellung verbesserter Alpha-Silicium-nitridpulver für das Heißpressen und Sintern
in: Keramische Komponenten für Fahrzeug-Gasturbinen III, 55-69, Springer Verlag, Berlin (1984)

(2) K. Kasahi, K.Tsukuma, T. Tsukidate
"Method of Manufacturing High-Purity Silicon Nitride Powder"
US-Patent 4387079 (1983)

(3) A. Kato
"Vapor-Phase Synthesis of Ceramic Powders"
in: Advances in Ceramics, Vol. 21, 181-192, (1987)

(4) V.J. Jbberson, T. Hussain
"Synthesis of Ultra-Fine Silicon Nitrido in an R.F.-Plasma Reactor"
Proc. 7th ISPC, J. Timmermanns (ed.) Eindhoven, 692-696 (1985)

(5) J. Canteloup, A. Mocellin
"Synthesis of Ultrafine Nitrides and Oxynitrides in an RF-Plasma"
Special Ceramics 6, P.Popper (ed.) 209-222, (1975)

(6) G.J. Vogt, R.S. Vigil, L.R.Newkirk, M. Trkula
"Synthesis of Ultrafine Ceramic and Metallic Powders in a Thermal Argon RF-Plasma"
Proc. 7th ISPC, J.Timmermanns (ed.) Eindhoven, 668-673, (1985)

(7) Ch.W. Zhu, J.P. Yan
"Preparation of Ultrafine Si_3N_4 Powders in Radio-Frequency Plasma"
Proc. 7th ISPC, J. Timmermanns (ed.) Eindhoven, 657-661 (1985)

(8) O. Schulz, D. Kandziora, H. Hausner
"Plasma Synthesis and Sintering Behavior of Silicon Nitride Powders"
Ceramic Transactions Vol. 1 (1988)

(9) W. Symons, S.C. Danforth
"Synthesis and Characterization of Laser-Synthesized Silicon Nitride Powders"
Advances in Ceramics, Vol. 21, 249-256, (1987)

(10) G.W. Rice
"Laser Synthesis of Powders"
Advances in Ceramics, Vol.21, 229-236, (1987)

(11) W.R. Cannon, S.Ç. Danforth, J.H. Flint, J.S. Haggerty, R.A. Marra
"Sinterable Ceramic Powders from Laser-Driven Reactions"
J. Am. Cer. Soc. 65, 324-335, (!982)

(12) Y.Cang, R.M. Young, E. Pfender
"Thermochemistry of Thermal Plasma Chemical Reactions. Part. II. A Survey of Synthesis Routes for Silicon Nitride Production".
Plasma Chemistry and Plasma Processing 7, 299-316, (1987)

(13) R.R. Wills, R.A. Markle, Sh.P. Mukherjee
"Siloxanes, Silanes and Silazanes in the Preparation of Ceramics and Glasses"
Am. Ceram. Soc. Bull. 62, 904-911 and 915, (1983)

HOT ISOSTATIC PRESSING OF LASER SYNTHESIZED SILICON NITRIDE POWDER

W. Symons and S.C. Danforth, The Center for Ceramics Research,
Department of Ceramics, Rutgers University,
P.O. Box 909, Piscataway, N.J. 08855-0909

ABSTRACT

Laser synthesized silicon nitride powders of controlled characteristics were consolidated by hot isostatic pressing (HIP'ing). Experiments were conducted on an amorphous, stoichiometric powder which was 17 nm in size. The effect of atmospheric exposure on powder densification behavior was studied by processing samples either under ambient conditions (*exposed*) or under nitrogen atmospheres with less than 10 ppm combined oxygen and water (*unexposed*). Samples were HIP'ed from 1,600-2,200°C at 198 MPa of argon. HIP'ed specimens of *exposed* powders displayed high final densities, while *unexposed* samples did not.

INTRODUCTION

There has been considerable interest in developing ceramic materials for use in advanced heat engines. Possessing low density and excellent thermomechanical properties, ceramics provide a means for producing heat engines with efficiency ceilings far above what is presently possible with today's super alloys. One material, silicon nitride, has received considerable attention due to its high decomposition temperature (approximately 1880°C) as well as superb intrinsic creep properties.[1,2]

Exploitation of silicon nitride's intrinsic properties, however, is not presently possible. Current sintering practice relies on the addition of one or more oxide sintering additives (i.e. Y_2O_3, MgO, CeO, Al_2O_3, and BeO) to promote densification.[3,4,5] Densification is accomplished through a liquid phase sintering process by reaction of the additives with the native SiO_2 layer present on Si_3N_4 powders. Processing Si_3N_4 in this manner, however, results in the formation of a residual glassy grain boundary phase which softens at elevated temperatures, thus compromising the thermomechanical properties of the material.

Obviously, in order to take advantage of silicon nitride's intrinsic properties, it is highly desirable to densify Si_3N_4 without additives. The sintering difficulties associated with Si_3N_4 are attributed to various factors. Low bulk diffusion coefficients result from the strong directional bonding of this covalent material. Sublimation introduces additional problems of decomposition and variable stoichiometry as processing temperatures are increased. The net result is a high ratio of non-densifying sintering mechanisms (i.e. surface diffusion and evaporation-condensation) to those which promote densification (i.e. volume and grain boundary diffusion). Thus without additives, compacts of Si_3N_4 powder typically coarsen with little shrinkage.

Due to difficulties associated with sintering "pure" Si_3N_4, the material has been labeled "unsinterable". Si_3N_4, however, is not alone in this category. Silicon and SiC, also covalent in nature, exhibit similar sintering difficulties. It has been shown that Si and SiC can be densified, utilizing appropriate additives, without developing residual glassy grain boundary phases.[6,7,8] "Pure" silicon, 55 nm in size, has also been found to sinter, provided initial green densities are >42% ρ th.[9,10]

While "pure" Si_3N_4 has been labeled as "unsinterable", there is no apparent thermodynamic barrier to densification.[6] Utilizing Si as a model powder for sintering of covalent materials, it was determined that extensive sintering should develop in powder compacts when the ratio of diffusivity to particle radius is approximately 4×10^{-6} cm/s.[6,11] Assuming the ratio of diffusivity to particle radius for other covalent materials must equal or exceed silicon's value, Si_3N_4 powders less than 25 nm in size should sinter.

Densification may be further assisted through the application of pressure during the sintering cycle. It has been demonstrated that through pressure application, the energy available for densification can be increased by more than than 20 times.[12-15] Hot pressing (HP) and hot isostatic pressing (HIP'ing) are typical methods for applying pressure and temperature simultaneously. Miyamoto et al, recently demonstrated the densification of "undoped" commercial Si_3N_4 powders by HIP'ing.[16] Resultant bulk specimen properties (density, microhardness and flexure strength) were excellent. However, starting powders were processed in ambient atmospheric conditions prior to HIP'ing and likely contained significant oxygen levels.

Due to the higher free energy of formation possessed by SiO_2, Si_3N_4 will oxidize in atmospheres of oxygen and water vapor.[17,18] Recent work by Nilsen et al, has demonstrated that the hydoxilation kinetics of an amorphous laser derived Si_3N_4 powder are extremely fast when exposed to various relative humidity atmospheres at room temperature.[19] On exposure to oxygen atmospheres containing extremely low concentrations of water vapor, the kinetics were extremely slow. Water vapor attack on Si_3N_4 powders results in Si-OH layer formation. This layer may act as a sintering aid during HIP'ing. The fact that Miyamoto et al, reported the occasional identification of a secondary oxynitride phase in their HIP'ed samples seems to support this suggestion.[16]

In order to densify "pure" Si_3N_4 it seems imperative to: 1) synthesize a powder <25 nm in size, 2) pack powders uniformly into high green density compacts, 3) process the material to prevent the formation of an oxide layer, 4) assist densification through the application of pressure.

The objective of this work is to study the effect of oxygen contamination on the densification behavior of ultra-high purity laser-derived Si_3N_4 powders. In particular, stoichiometric, amorphous powders, approximately 17 nm in size, which were *exposed* and *unexposed* to the ambient environment were hot isostatically pressed.

EXPERIMENTAL PROCEDURE

A. Powder Synthesis

Si_3N_4 powders were produced via laser-driven gas phase reactions. Details of the process have been reported elsewhere, and therefore are only summarized here.[7,20-22] Si_3N_4 powders are produced by passing electronic grade SiH_4[1] and NH_3[2] orthogonally through a CO_2 laser[3] beam. Due to the strong absorption of the 10.6 μm CO_2 emission by SiH_4, reactants are rapidly heated, subsequently decompose and react to form Si_3N_4 powders. By careful manipulation of process variables (i.e. flow rates, reactor pressure, gas ratios and laser intensity) powders of controlled characteristics (size, stoichiometry and crystallinity) may be produced. A laminar flow of Ar[4] carries powders and reaction by-products from the reactor to a glass fiber collection filter.[5] The filter assembly is capable of being isolated from the reaction chamber and the ambient environment, permitting transport into a nitrogen atmosphere glove box.[6] In this way all post synthesis processing may be carried out in atmospheres containing a combined oxygen-water environment of <10 ppm. Powders processed in this manner are termed *unexposed.* The term *exposed* applies to powders which were exposed to the ambient environment after synthesis.

B. Powder Characterization

Synthesized powders were subjected to X-ray diffraction[7] and electron[8] diffraction analysis to determine crystallinity. Density was determined by He-pycnometry.[9] The surface area of powders was measured using the single point BET[10] gas adsorption method. Particle sizes were determined from BET-ESD calculation as well as

---

[1]Union Carbide Corp., Linde Division, Danbury, CT
[2]Matheson Gas Products, East Rutherford, N.J.
[3]Coherent Everlase 525-2, Sturbridge, MA
[4]Matheson Gas Products, East Rutherford, N.J.
[5]Balston Filter Products, Lexington, MA
[6]Vacuum Atmospheres, Model HE-43-2, Hawthorne, CA
[7]Siemens D500 Diffractometer, Germany
[8]Jem 100CXII, Joel Electron Microscope, East Brunswick, N.J.
[9]Model 1320, Micromeritics Instrument Corp., Norcross, GA
[10]Quantasorb, Quantachrome Corp., Syosset, N.Y.

TEM measurements. Stoichiometry was determined by XPS[11] analysis based on a 101.3 eV silicon 2p binding energy for stoichiometric Si_3N_4 powders.[23] Oxygen content was measured by neutron activation analysis[12] of exposed and unexposed powders. Wet chemical analysis[13] was utilized to determine the impurity content of the Si_3N_4 powders.

The activity of Si_3N_4 powders, *exposed* and *unexposed*, was determined by measuring the surface area of powders annealed in nitrogen as a function of temperature. Hand pressed pellets, with a green density of approximately 10% ρ th, were placed in a series of alumina crucibles and annealed in a nitriding furnace[14] using ultra high purity nitrogen. *Unexposed* samples were loaded in the crucibles within a nitrogen glove box and transferred to the furnace in a desiccator. Transfer times, from desiccator to furnace, were no longer than 15 seconds. Ramping rates of 25°C/minute were used and samples were held at temperature for one hour.

C. Hot Isostatic Pressing

Hot isostatic pressing was conducted using an argon[15] gas HIP.[16]. Cold isostatically pressed samples were encapsulated in Pyrex™ glass and then placed in graphite crucibles and loaded into the HIP. All cycles were conducted by first heating samples to 830°C, the softening point of Pyrex™, at a rate of 30°/min under 602 kPa (85 psi) pressure. After a 15 minute soak at 830°C, pressure and temperature were ramped to desired values.

1. Green Compact Formation. Pellets for use in HIP'ing studies were formed in a two stage pressing process. Synthesized powder was first hand pressed in a 19 mm diameter air hardened steel die. Pressures less than 350 kPa (50 psi) were utilized. On removal from the die, pellets were sealed under vacuum in rubber bags.[17] *Exposed* samples were sealed in two bags while *unexposed* samples were sealed in three. The entire process for *unexposed* powders was conducted in a nitrogen atmosphere glove box.

After bagging, samples were cold isostatically pressed[18] at 283 MPa (50 kpsi). *Unexposed* samples were removed from the glove box in their bags for this process. After CIP'ing *unexposed* samples were returned to the glove box. Subsequent to CIP'ing, samples were removed from the bags. The free surfaces of samples, *exposed* and *unexposed*, were ground on 600 grit SiC paper to remove any contamination received from the rubber bags.

2. Encapsulation. Pellets for HIP'ing were embedded in high purity boron nitride powder[19] inside Pyrex™ tubes of 21.5 mm ID × 25.4 mm OD. A Pyrex™ slug 19 mm in diameter was placed on top of the BN packing. The Pyrex™ tube was then connected to a high vacuum outgassing apparatus and outgassed at 5×10^{-6} torr and 600°C for 24 hours. The tube connection assembly of the outgassing station was capable of being isolated from the outgassing station as well as the environment. This permitted *unexposed* samples to be loaded within the glove box and transferred to the outgassing station without sample contamination. After outgassing, the Pyrex™ tube was fused to the Pyrex™ slug with an oxy-propane torch. Sealed cans were then placed into a 600°C oven and cooled slowly.

DISCUSSION OF RESULTS

A. Powder Synthesis and Characterization

The synthesis conditions utilized to produce Si_3N_4 powders by the laser process are listed in Table I. The

---

[11] Perkin-Elmer, Edison, N.J.
[12] IRT Corporation, San Diego, CA
[13] Luvak Inc., Boylston, MA
[14] Centorr Associates Inc., Suncook, NH
[15] Argon Zero Gas, Matheson Gas Products, East Rutherford, N.J.
[16] Autoclave Engineers, Erie, PA
[17] Trojan Latex Condoms, Carter-Wallace Inc., N.Y.
[18] Autoclave Engineers, Erie, PA
[19] HCM grade BN, Union Carbide Corp., Cleveland, OH

properties of powders produced under such conditions are presented in Table II. Figure 1 is a TEM micrograph of the as synthesized Si_3N_4. Table III lists the trace impurities of the powder as determined by wet chemical analysis. Particularly notable are the low iron, calcium, aluminum and carbon contents of these powders. Neutron activation analysis revealed that *exposed* powders possessed an oxygen content of 2.0 wt % with *unexposed* powders only possessing 0.5 wt %. FTIR work on more current powders indicates the oxygen levels of *unexposed* powders to be even lower. Results indicate that extremely "pure" Si_3N_4 powders of controlled characteristics may be produced by laser-driven gas phase reactions. Neutron activation analysis also indicates how dramatically the oxygen content of powders change as they are exposed to the ambient environment.

TABLE I: Si_3N_4 Synthesis Conditions

Gas Flow Rates:	
Silane:	72 cm^3/min
Ammonia:	320 cm^3/min
Argon: Annulus	600 cm^3/min
Chamber	1.5 L/min
Reactor Pressure:	720 torr
Flame Temperature:	1110°C
Laser Intensity:	5.6×10^3 Watts/cm^2

TABLE II: Si_3N_4 Powder Characteristics

Surface Area:	122 m^2/g
Particle Size:	
BET-ESD	17.0 nm
TEM	16.0 nm
He Pycnometic Density:	2.9 g/cm^3
Crystallinity:	Amorphous
Chemistry:	No Si Detected (X-ray, XPS)

TABLE III: Chemical Analysis of Si_3N_4 Powders

Element	Weight %
Cl	<0.010
S	0.007
Al	<0.001
Ta	<0.008
W	<0.008
C	0.072
Mg	<0.001
Y	<0.001
Ca	<0.001
Cu	0.001
Fe	0.001
Cr	<0.001
Ni	<0.001
Na	<0.001
B	0.003

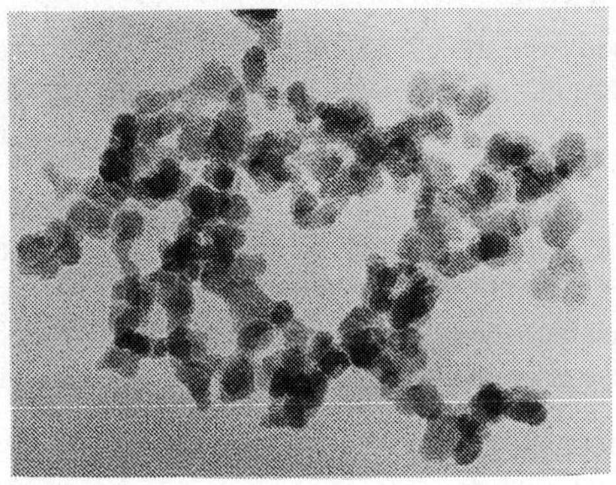

Figure 1: TEM Micrograph of as Synthesized Si_3N_4 Powder Bar=100 nm (▬▬▬▬)

While neutron activation analysis indicated how the oxygen content of synthesized powder changes on exposure to atmospheric moisture, annealing studies indicated how exposure affects powder activity. Figure 2 shows the activity of an *unexposed* and two *exposed* Si_3N_4 powders. The difference between the two *exposed* powders is that one was exposed to the atmosphere for only 48 hours prior to annealing (termed short exposure) while the other was *exposed* for a period of greater than two months (termed long exposure). Clearly the exposure of the powders to ambient surroundings dramatically affects their activity. *Unexposed* samples are extremely active, displaying a strong temperature dependence as well as a large activation energy for the mechanism responsible for the measured surface area reduction. However, when samples are exposed to the ambient environment, the temperature dependence for surface area reduction decreases. The decrease in temperature dependence is also a function of exposure time indicating that the chemistry and or structure of the *exposed* amorphous Si_3N_4 powder may be changing over long periods of time. X-ray diffraction and TEM analysis indicates conclusively that crystallization is responsible for the measured surface area reduction.

Particularly interesting results were found when powders were annealed ~20°C-50°C below the onset temperature for surface area reduction. In particular, powder compacts were observed to shrink during these soaks. Figure 3 shows the linear shrinkage exhibited by short term exposure powders as a function of time at temperature, for several annealing temperatures. Similar results were found for long term exposure powders as well as *unexposed* samples. The following list contains possible explanations for the observed behavior; 1) Weight Loss; Unlikely since the highest recorded weight loss of all samples tested was 5%, 2) Coarsening; This mechanism is capable of surface area reduction but cannot account for shrinkage, 3) Rearrangement; Due to the lack of mechanical stress to assist densification it is doubtful this would contribute to shrinkage, 4) Crystallization; While samples crystallized during annealing, conversion from 2.9 g/cm^3 to 3.18 g/cm^3 cannot account for the observed shrinkage, for 10%-15% shrinkage was observed in samples prior to crystallization, 5) Oxygen Pressure; Possibly attributable to *exposed* samples but not in the case of *unexposed* samples, 6) Sintering; It is possible that the small particle size and amorphous nature of the starting powder may permit sintering by a densifying mechanism. In particular, viscous sintering may be responsible for observed shrinkages prior to crystallization.

Presently it is unclear what particular mechanism is responsible for the observed shrinkage. The results are very encouraging if they do in fact represent the possibility of sintering Si_3N_4 (and perhaps other covalent ceramics) in an amorphous state.

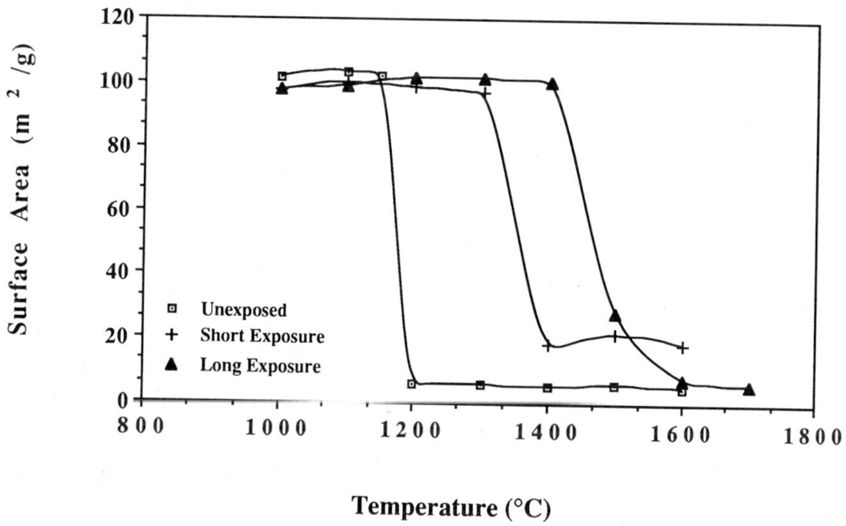

Figure 2: Surface Area vs Temperature for Si_3N_4 Powders

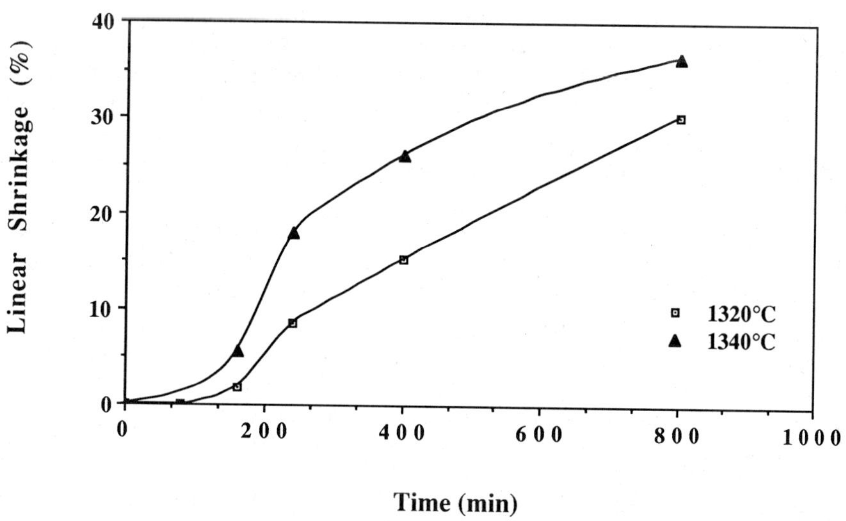

Figure 3: Linear Shrinkage vs Time for *Exposed* Si_3N_4 Powders

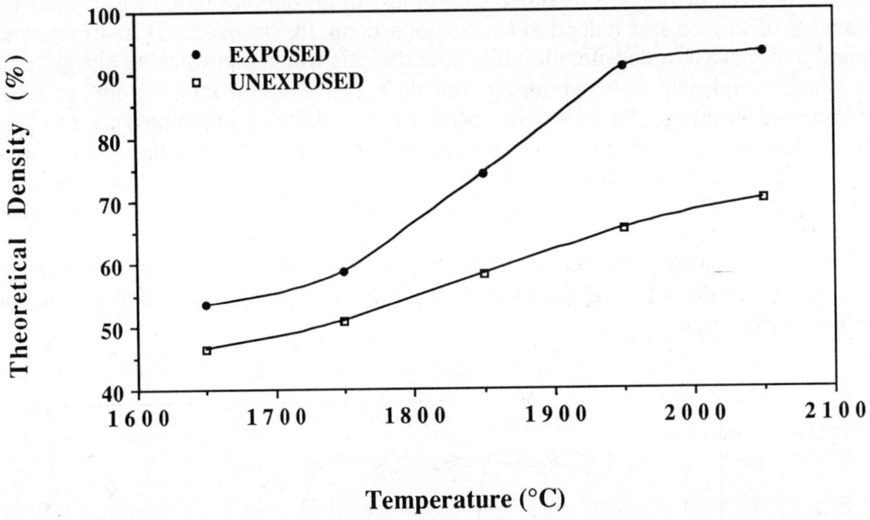

Figure 4: Density vs Temperature for *Exposed* and *Unexposed* Si_3N_4 HIP Samples (T=1950°C, t=60 min)

B. Hot Isostatic Pressing

HIP'ing final densities as a function of temperature at pressure for *exposed* and *unexposed* samples are illustrated in Figure 4. A pressing pressure of 198 MPa and hold time of 60 minutes was utilized. Density was determined by mercury displacement. Obviously the contamination of laser derived powders with oxygen from the ambient environment significantly alters the densification behavior of the powder. *Unexposed* samples attained a maximum density of 70% ρ th at 2050°C while *exposed* samples reached a significantly higher density of 94% ρ th. X-ray diffraction analysis of HIP'ed samples indicates both *exposed* and *unexposed* samples attain higher percentages of the β phase as the HIP'ing temperature increases. The kinetics of the phase transformation (α to β) between *exposed* and *unexposed* samples, however, appears to be different. Quantification of phase contents and conversion kinetics is currently being carried out.

Densification kinetics of *exposed* and *unexposed* samples for pressing times of 1-180 minutes at a temperature of 1950°C and pressing pressure of 198 MPa were investigated (Figure 5). *Exposed* samples are observed to be extremely active compared to *unexposed* samples. For both cases, the rate of densification falls off significantly after a time of approximately 20 minutes. X-ray diffraction analysis again indicates that the β content of samples increases as the time at temperature increases and that the kinetics of the phase conversion from α to β-Si_3N_4 differ for *exposed* and *unexposed* powders. *Unexposed* samples convert to 100% β-Si_3N_4 after 15 minutes while *exposed* samples became 100% β–Si_3N_4 only after an hour. Again, quantification of the phase ratios is currently being conducted. Figures 6 and 7 show the microstructural evolution of *exposed* and *unexposed* specimens respectively. After one minute at temperature, Figure 6A and 7A, individual particles are easily discernable for both *exposed* and *unexposed* samples. Particle sizes have increased to an average size of ~0.5 μm with some localized sintering observed in the *exposed* sample. After ten minutes at temperature, there is a large deviation between the microstructures of *exposed* and *unexposed* samples (Figures 6B and 7B). Individual particles are still quite discernable in the *unexposed* sample, while particle individuality is no longer apparent in the *exposed* sample. After one hour, Figures 6C and 7C, little change has occurred in the microstructure of the *unexposed* sample, while the *exposed* sample has entered final stage sintering. It is interesting to note that some localized sintering has occurred in the *unexposed* samples. Since the *unexposed* samples do not possess an Si-OH surface layer, while *exposed* samples do, this is indeed an encouraging result. Considering the starting density of samples (36% for *unexposed* and 42% for *exposed*) it is not surprising that the *unexposed* samples did not attain the high density of the *exposed.* This is especially true when one considers the extremely low bulk diffusion coefficients which operate in Si_3N_4. Obviously, the contamination of powders on exposure to the ambient atmosphere plays an important role in increasing the rate at which the dominate densification mechanisms operate. However, the need to create green compacts of increased density possessing unagglomerated microstructures is quite apparent. However, the current results indicate that laser derived Si_3N_4 powders can be densified. While exposure of synthesized powders to the ambient atmosphere enhances the

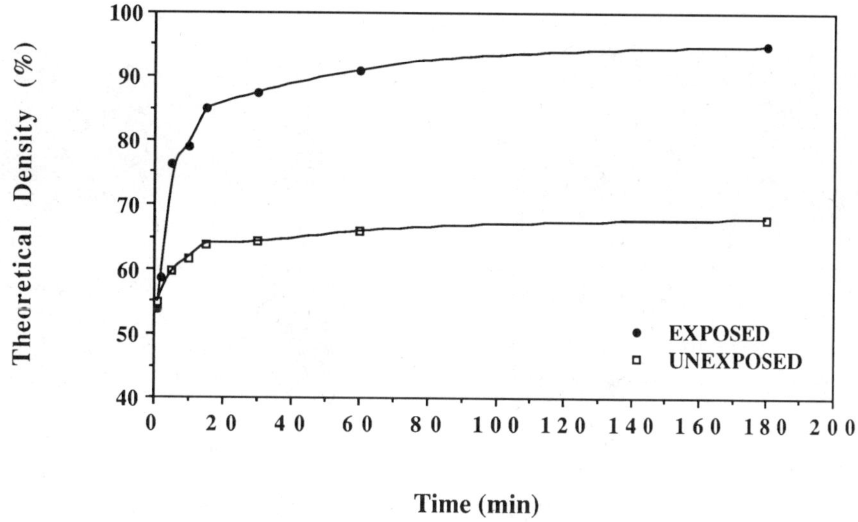

Figure 5: Densification Kinetics for *Exposed* and *Unexposed* Si_3N_4 HIP Samples (T=1950°C, 198 MPa)

Figure 6A: Time = 1 minute

Figure 6B: Time = 10 minutes

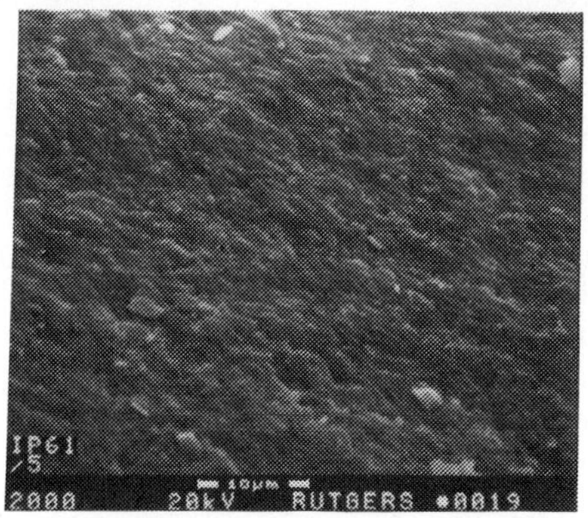

Figure 6C: Time = 60 minutes

Figure 6: SEM Fracture Surfaces of *Exposed* Si_3N_4 Samples HIP'ed for Various Times (T=1950°C, P=198 MPa)

Figure 7A: Time = 1 minute

Figure 7B: Time = 10 minutes

Figure 7C: Time = 60 minutes

Figure 7: SEM Fracture Surfaces of *Unexposed* Si_3N_4 Samples HIP'ed for Various Times (T=1950°C, P=198 MPa)

densification kinetics, *unexposed* powders are observed to display a limited amount of sintering. Through improved processing, (i.e. the casting of stabilized dispersions of *unexposed* as well as *exposed* powders), green bodies of improved microstructures may be utilized in HIP'ing to deliver superior HIP'ed densities as well as microsturctures.[24]

SUMMARY

Silicon nitride powders possessing unique characteristics may be synthesized via laser-driven gas phase reactions. Synthesized powders are extremely reactive with water in the ambient environment. During nitrogen annealing of synthesized powders, *exposed* as well as *unexposed* samples were observed to display linear shrinkages of up to 40%. Hot isostatic pressing of powders *exposed* to the ambient environment produced high final densities, ~94% ρ th, while *unexposed* powders were found to only attain a final density of ~70% ρ th for the same HIP'ing conditions. *Unexposed* samples, however, were observed to display some localized sintering suggesting that higher HIP densities may be attained if *unexposed* powders can be packed more efficiently and uniformly into higher green density samples.

ACKNOWLEDGEMENTS

The authors would like to gratefully acknowledge the financial support of the Rutgers Center for Ceramics Research. They would also like to acknowledge K. Nilsen for his assistance throughout this work.

REFERENCES

[1]. Pasto, A.E., "Causes and Effects of Fe-Bearing Inclusions in Sintered Si_3N_4," *J. Amer. Cer. Soc.*, 67,[9], 178-80, (1984).
[2]. Yeckley, R.M., "Stress Rupture Life of HIP Silicon Nitride," 83-C-88F, Talk presented at 12th Annual Conference on Composites and Advanced Ceramics, Cocoa Beach, FL, January, (1988).
[3]. Yamada, T., Shimada, M., and Koizumi, M., "Fabrication of Si_3N_4 by Hot Isostatic Pressing," *J. Amer. Cer. Soc.*, 60, [11], 1225-1228, (1981).
[4]. Terwillinger, G.R., and Lange, F.F., "Pressureless Sintering of Si_3N_4," J. Mater. Sci., 10, [7], 1169-1174, (1975).
[5]. Clarke, D.R., Zaluzec, N.J., and Carpenter, R.W., "The Intergranular Phase in Hot-Pressed Silicon Nitride: I, Elemental Composition," J. Amer. Cer. Soc., 64, [10], 601-607, (1981).
[6]. Greskovich, C. and Rosolowski, J.H., "Sintering of Covalent Solids," *j. Amer. Cer. Soc.*, 9, [7-8], 336-3343, (1976).
[7]. Danforth, S.C. and Haggerty, J.S., "Synthesis of Ceramic Powders by Laser Driven Reactions," Energy Laboratory Report, MIT-EL-81-003, July, (1980).
[8]. Haggerty, J.S., "Sinterable Powders From Laser Driven Reactions," Energy Laboratory Report, MIT-EL-82-002, September, (1981).
[9]. Moller, H.J. and Welsch, G., "Sintering of Ultrafine Silicon Powders," *j. Amer. Cer. Soc.*, 68, [6], 320-325, (1985).
[10]. Danforth, S.C., and Haggerty, J.S., "Mechanical Properties of Sintered and Nitrided Laser-Synthesized Silicon Powder," *J. Amer. Cer. Soc.*, 66, [4], C58-59, (1983).
[11]. Sawhill, H.T., "Crystallization of Ultrafine Amorphous Si_3N_4 During Sintering," Masters Thesis, MIT, Cambridge, MA, September (1981).
[12]. Hermansson, L., Nystrom, L., and Adlerborn, J., "Hot Isostatic Pressing of Silicon Carbide With No Sintering Agents," Ceramic Materials and Components for Heat Engines, Ed. by W. Bunk and H. Hausner, Deutsche Keramische Gesellschaft, 353-360, (1986).
[13]. Prochazka, S., and Rocco, W.A., "High Pressure Hot Pressing of Silicon Nitride Powders," *High Temperature, High Pressures,* 10, {1], 87-95, (1978).
[14]. Yamada, T., Shimada, M., Koizumi, M., "Densification of Si_3N_4 by High Pressure Hot Pressing," *Amer. Cer. Soc. Bull.*, 60, [12], 1281-1288, (1981).
[15]. Richardson, D.W., Modern Ceramic Engineering Properties, Processing and Use in Design, Ed. by G. Boothroyd and G. Dieter, Markel Dekker Inc., New York, 241-255, (1982).
[16]. Miyamoto, Y., Tanaka, K., Shimada, M., and Koizumi, M., "Survey of HIP Sintering Condition and Characterization of Dense Silicon Nitride Without Additives," Ceramic Materials and Components for Heat Engines, Ed. by W. Bunk and H. Hausner, Deutsche Keramische Gesellschaft, 271, (1986).
[17]. Peuckert, P., and Greil, P., "Oxygen Distribution in Silicon Nitride Powders," *J. Mater. Sci.*, 22, 3717-3720, (1987).
[18]. Singhal, S.C., "Thermodynamic Analysis of the High-Temperature Stability of Silicon Nitride and Silicon Carbide," *Cream. Int.*, 2, 123-130, (1976).
[19]. Nilsen, K.J., Rieman, R.E., and Danforth, S.C., "The Effects of Moisture on the Surface Chemistry and Dispersion Properties of Silicon Nitride in Imidazoline-Hexane Solutions," Ceramic Powder Science II, Ed. G.L. Messing, E.R. Fuller, H. Hausner, American Ceramic Society, Westerville OH, 469-476, (1988).
[20]. Cannon, W.R., Danforth, S.C., Flint, J.H., Haggerty, J.S., and Mara, R.A., "Sinterable Ceramic Powders from Laser-Driven Reactions: I, Process Description and Modeling," *J. Amer. Cer. Soc.* 65, [7], 324-330, (1982).
[21]. Cannon, W.R., Danforth, S.C., Flint, J.H., Haggerty, J.S., and Mara, R.A., "Sinterable Ceramic Powders from Laser-Driven Reactions: II, Powder Characterization and Process Variables," *J. Amer. Cer. Soc.* 65, [7], 330-335, (1982).
[22]. Symons, W., and Danforth, S.C., "Synthesis and Characterization of Laser-Synthesized Silicon Nitride Powders," Advances in Ceramics, Vol. 21, Ceramic Powder Science, Ed. by G.L. Messing, and K.S. Mazdiyasni, McCauley, J.W., and Haber, R.A., The American Ceramic Society Inc., Westerville OH, 247-256, (1984).
[23]. Vasquez, R.P., et al., "X-ray Photoelectron Spectroscopy Study of the Chemical Structure of Thermally Nitrided SiO_2," *Appl. Phys. Lett.*, 44, (10), 969, May, (1984).
[24]. Nilsen, K.J., Ph.D. Thesis, Rutgers University, Expected Completion September (1988).

CHARACTERIZATION OF THE COMPACTING BEHAVIOUR OF SINTERABLE α-SiC-POWDERS

R. Dietrich, LONZA AG, Basel, Switzerland
A. Kerber, G. Rünzi, LONZA-Werke, Waldshut- FRG

1. ABSTRACT

The mechanical strength and density of green bodies are influenced by the behaviour of powder particles under compressive forces. The elastic spring-back during unloading creates stresses and possibly cracks in the green body and thus a high risk of failure of sintered parts. The compaction behaviour of Acheson-α-SiC-powders produced by different methods, but nearly the same specific surface areas, is compared. A method to characterize the elastic spring-back of the compacts is presented. A correlation of these results to green density, pore size distribution and the likelihood of crack formation during green processing is discussed.

2. INTRODUCTION

The forming of green bodies can be a source of high rejection rates in ceramic processing. For this reason, it is worth knowing how the powder properties influence the mechanical strength of ceramic parts before firing. In order to eliminate the influence of binders, pressing agents and additives, all the experiments were conducted with pure silicon carbide powders. It is well known that while a binder increases the strength of a green body, the inherent weaknesses of a powder due to its morphology can not be completely overcome by optimizing the additives. On the contrary it is necessary to investigate the influence of particle size distribution and particle shape on the compacting behaviour, the green density and the pore size distribution which itself strongly affects the sintering behaviour.

3. EXPERIMENTAL PROCEDURE AND RESULTS

3.1 Powder Characteristics

The powders investigated were LONZA-powders and have been produced either by different milling procedures (powders 1,2 and 3) or blending of two fractions in the case of powder 4.
In Table I some granulometrical properties of the materials are presented. The geometric standard deviation

$$SD = \ln(x_{84}/x_{50}) = \ln(x_{50}/x_{16}) \qquad (1)$$

charactrizes the width and symmetry of a particle size distribution (PSD) and can be taken from a log-normal-distribution plot of the PSD, which was measured by a laser diffraction analyzer, type "Microtrac SPA" of "Leeds & Northrup".

Type	x_{50} (µm)	$SD = \ln \frac{x_{84}}{x_{50}}$	S_m (m^2/g)
1	0.47	0.54	15.5
2	0.46	0.59	13.7
3	0.76	0.87	18.5
4	1.0	bimodal	14.6

Table I Granulometrical Properties of α-SiC-powders

The specific surface areas S_m have been determinded by nitrogen adsorption measurements with a "Quantasorb Junior". Fig. 1 shows the cumulative mass distributions of the 4 different powders. Although the S_m-values are in a quite narrow range, the PSD's differ obviously. The difference between powder 1 and 2 and powder 4 can be explained by the higher amount of fines of powder 4. Powder 3 does not fit into this scheme, due to an incomplete deagglomeration of the very fine fraction of this powder.
Besides the granulometrical properties, the green density which can be achieved in dry compacting processes is an important powder characteristic. In Fig. 2 the green densities of the 4 different powders are shown as a function of the applied pressure in a die compaction process. The diameter of the tablets was 13 mm and the diameter to height ratio was 2 : 1. The green densities (GD) are given as a fraction of the theoretical density of silicon carbide, which is 3.21 g/cm^3.
All the materials show an increasing GD with elevated pressures, but the differing levels of densification are very obvious. All the subsequent results to be presented were obtained at a constant pressure of 400 MPa. In Fig. 3 the green densities are plotted versus the reciprocal value of the standard deviations of the PSD's. With increasing steepness of the distribution curves, the GD decreases very strongly. This effect may be explained by the increasing tendency of the fine particles to fill the interspaces between the big particles as the size distribution becomes wider.
The methods described above are routinely used to characterize ceramic powders. However, these methods do not allow one to predict the green strength and the tendency for cracks to form in the green body.

3.2 Principle of measuring the spring-back and the strength of green bodies.

The compacting and spring-back of the 4 different powders are characterized by measuring the force-displecement-curves during loading and unloading. A schematic of the experimental set-up is given in Fig. 4 The velocity of the upper punch was kept constant at 1 mm/min. for loading and unloading. The strength of the tablets was measured in diametrical compression using the same equipment without a die. The samples were loaded until the first crack could be identified on the shape of the force-displacement curve. Since the geometry of all tablets was the same, this breakage force was used as a comparative measure of tablet strength.

3.3 Force-Displacement-Curves

The compaction behavior of the different powders showed good reproducibility (SD< 4 %). As an example, two typical F-D-curves of the powders 1 and 3 are shown in Fig. 5. The applied forces correspond to a pressure of 400 MPa for both powders. The values of the displacements in this figure do not represent the whole compacting process. Since the initial stage of the process is characterized by large displacements and very low forces, the triggering limit of the force transducer was set at 500 N (F(s=0)=0.5 KN). For this reason conclusions on the green densities of the compacts from powders 1 and 3 cannot be drawn from this diagram. Although the shape of the curves is similar in the loading part, the spring-back behaviour during unloading is obviously different. Whereas powder 3 exhibits a redisplacement of only 0.13 mm, powder 1 shows a spring-back of 0.23 mm.
The curves for powders 2 and 4 lie between the curves presented in the figure and are not shown for clarity. The pictures of the tablets indicate that the magnitude of the elastic redisplacement has an influence on the strength of the compacts. The tablet with the lower spring-back does not show any cracks, whereas the tablet with the higher spring-back exhibits severe damage.
The mass related compacting energies are defined as follows:

$$E_{comp} = \frac{1}{m}\left[\int_{s=0}^{s=s_{max}} F_{load.}(s)\,ds - \int_{s=0}^{s=s_{max}} F_{unload.}(s)\,ds\right] \qquad (2)$$

with m = mass of the powder compact and
s = displacement

The differences in the compacting energies (Fig. 5) indicate that the movement of particles under compressive forces is hampered in the case of powder 1.

3.4 Microstructure

SEM-Photographs of the 4 materials are given in Fig. 6. The solid bar represents a distance of 5 micrometers. A comparison with the particle size distributions, measured by a laser diffraction analyzer, shows that the pictures are representative in terms of particles sizes. Powder 1 has some angular grains with a relatively high aspect ratio and a narrow PSD. Powder 2 shows nearly the same grain size distribution. However, angular grains cannot be detected and most of the particles have a roundish shape.
Powders 3 and 4 have a high amount of large grains, a fact which is not desirable if the strength of sintered parts and the tendency to excessive grain growth is taken into account. Whereas powder 3 has only rounded particles, in powder 4 some angular grains can be found.
Apparently, two factors are responsible for low spring-back:

1. rounded shape
2. broad particle size distribution

3.5 Pore Size Distribution and sintering activity of the compacts

It is well known that the width of the particle size distribution and the grain shape have an influence on the pore size distribution of the compacts. This can be correlated to the sintering behaviour of the material/1/. The pore size distributions of the green compacts have been measured by a mercury porosimeter at pressures up to 2.000 bars. As shown in Fig. 7, powder 1 has the narrowest pore size distribution which becomes gradually wider for the powders 2, 3 and 4. The first three powders can be sintered to densities $>$97 % of the theoretical density (TD), whereas powder 4 reaches only 92 % of TD. It can clearly be seen that the highest sintering activity, defined as the temperature at which maximum linear shrinkage rate occurs (T_{ms}), is correlated to a narrow pore size distribution. This finding is in accordance with results obtained by sintering monosized titania particles /2/ and /3/. The above mentioned sintering activities were measured by high temperature dilatometer tests.

3.6 Strength of green bodies

Due to the fact that uniaxially pressed tablets exhibit very inhomogeneous properties, the results were not evaluated according to the theory of a diametrical compression test which allows one to calculate the main stress values. Instead, the breakage forces of the tablets are presented in Fig. 8 as a function of the elastic spring-back.
A big difference in the breakage forces of green bodies made from powder 3 and the other powders can be seen. This fact can be explained by mircocracks in the green bodies 1, 2 and 4, which are generated during unloading due to the elastic spring-back of the compacts. Apparently the green strength of bodies 1, 2 and 4 is insufficient to sustain the stresses caused by the elastic recovery.

In body 3, two effects are responsible for the superior behavior:

1. Due to the wide grain size distribution and round shape of the particles, the number and area of contacts are larger and thus higher Van-der-Waals forces are acting resulting in a higher green strength.
2. Due to lower spring-back a lower inherent stress level is obtained after unloading.

4. SUMMARY

Table II gives an overview of the summerized results with respect to the spring-back behaviour, green density, strength of green bodies and the sintering behavior. An optimized ceramic powder should yield green bodies of high green density, high green strength and small pores with a narrow distribution resulting in a high sintering activity and high sintered density.
Table II shows that none of the powders meets all of these requirements completely. The powders which have a high green density tend to have excessive grain growth due to the relatively high amount of coarse grains and the reduced sintering activity, whereas the green bodies made from powders with a good sintering behaviour are possibly predamaged before firing.

Type		1	2	3	4
Spring-Back [% of Fin. Size]		3.38	2.70	2.35	2.51
Green Density [% TD] at 400 MPa		53.9	57.6	64.4	63.8
Breakage Force [N]		11	6	92	14
Sintering	T_{ms} [°C]	1960	1980	2090	2055
	Density [% TD]	> 97	>97	>97	> 92

Table II Summarized results

5. CONCLUSIONS

The results presented above clearly indicate that measuring the force-displacement-curves during the compacting process is a useful method to determine the compacting behaviour of ceramic powders. If this method is used in combination with the standard powder characterization techniques and sintering experiments, it is a powerful tool for the development of improved ceramic powders.

6. REFERENCES

/1/ J. Allemann, "Zusammenhang zwischen Porenstruktur und Sinterverhalten", Dipl.-Arbeit, Eidgenössische Technische Hochschule Zürich, 1988

/2/ H. Hausner, "Synthesis and Characteristics of Powders for Advanced Ceramics", Proc. of the 2nd Int. Symp. "Ceramic Materials and Components for Engines", 27-38, 1986

/3/ E.A. Barringer and H.K. Bowen, "Formation, Packing and Sintering of Monodisperse TiO2 Powders", J. Am. Ceram. Soc., 65, 12, C-199-201, 1982

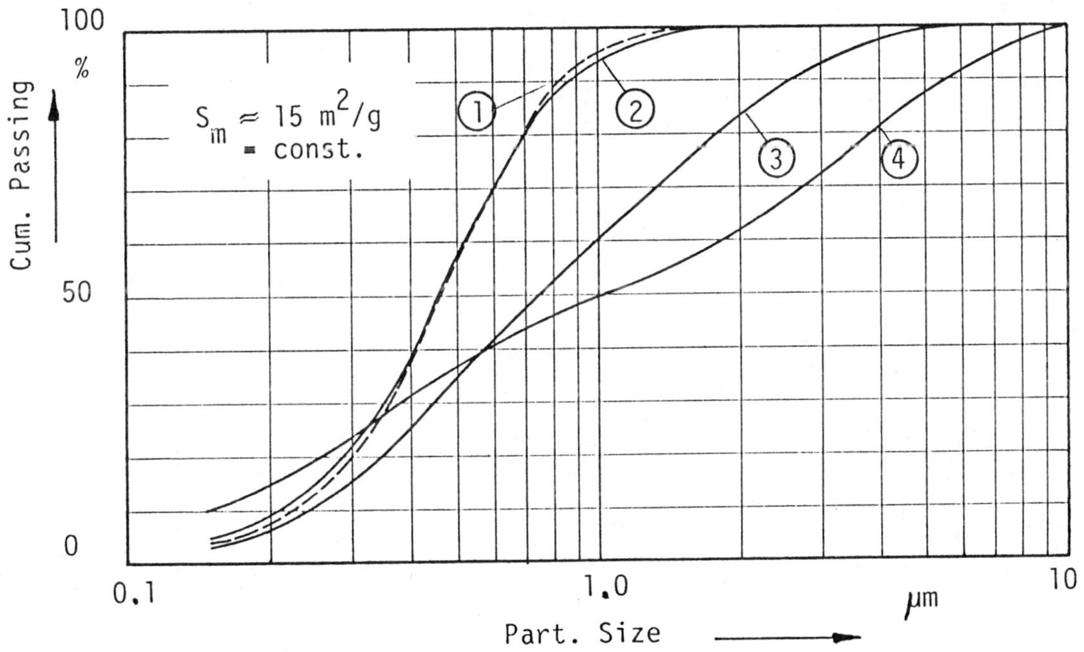

Fig. 1 **Particle size distributions of differently prepared α-SiC-powders, measured by Laser-Diffraction-Analyzer**

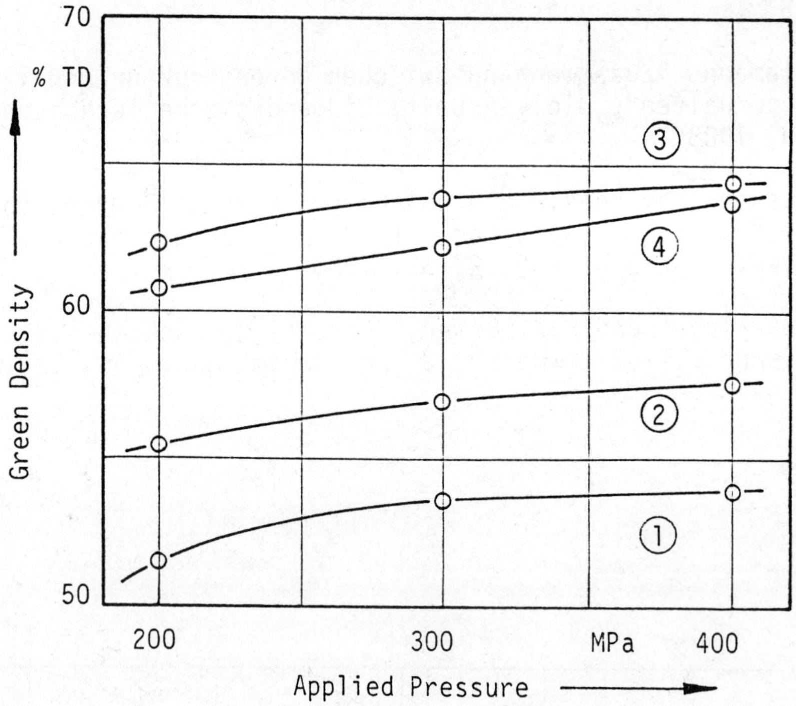

Fig. 2 **Green Densities and Applied Pressure**

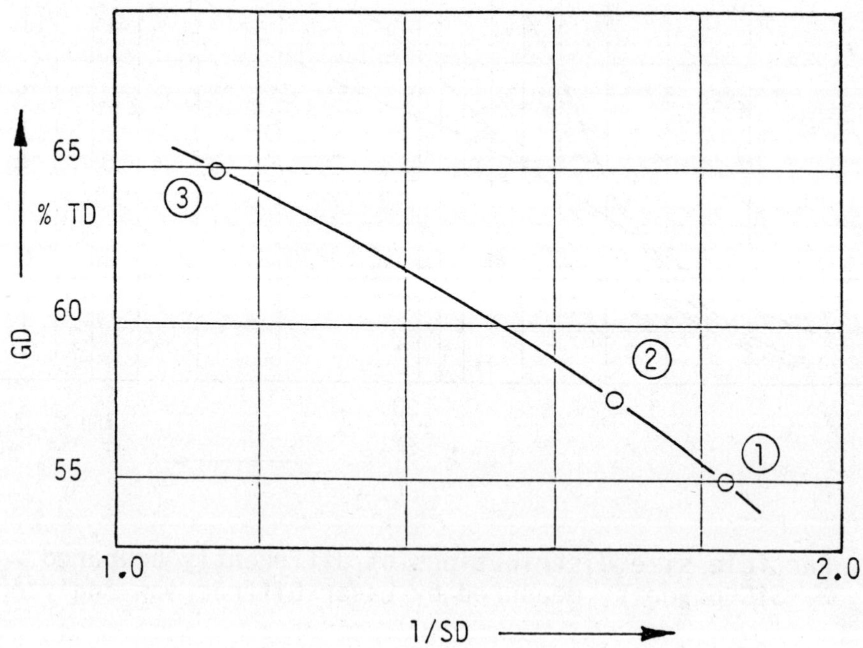

Fig. 3 **Green densities and steepness of particle size distribution**

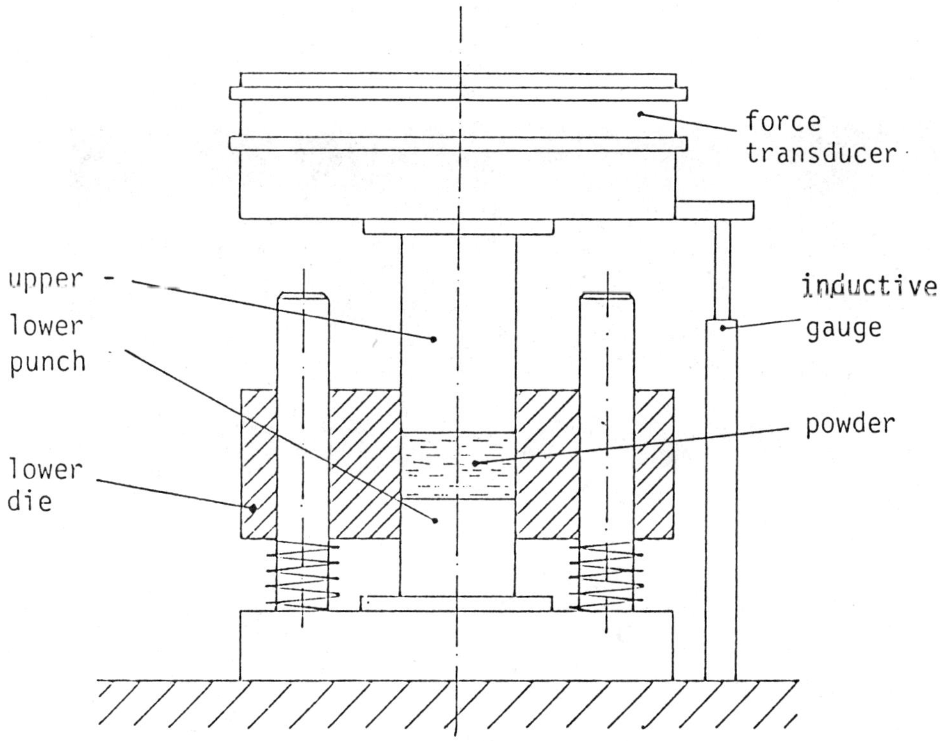

Fig. 4 Schematic of the experimental set-up

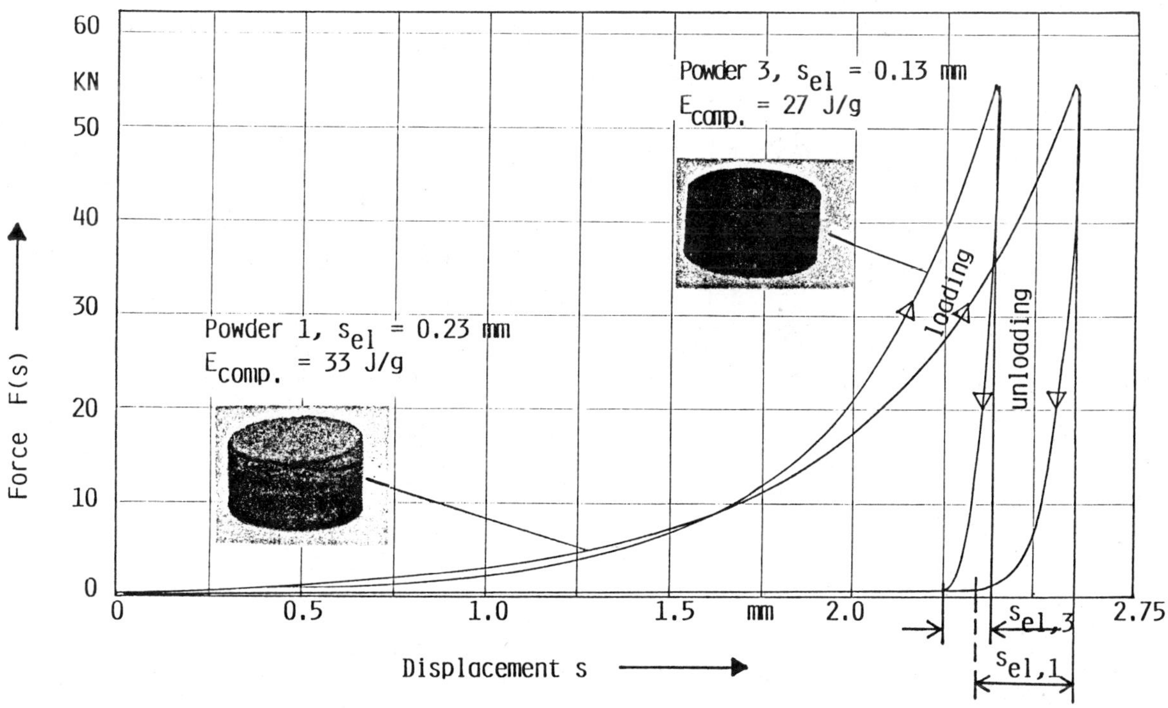

Fig.5 Force-displacement-curves of the compacting process

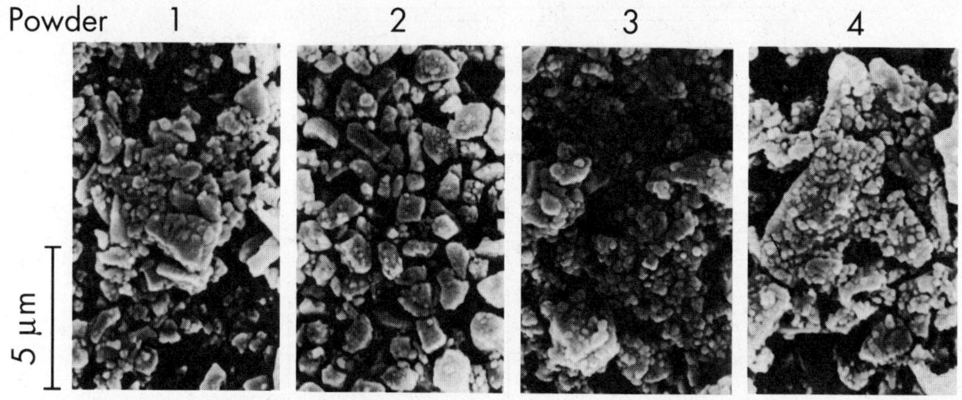

Fig. 6 SEM-photographs of the powders

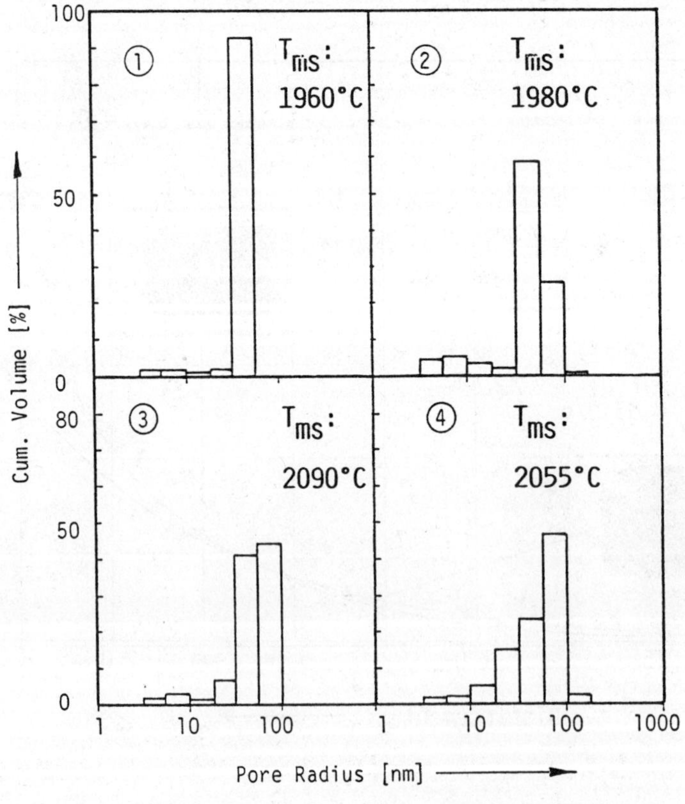

Fig. 7 Pore size distribution and sintering activity of SiC powder compacts

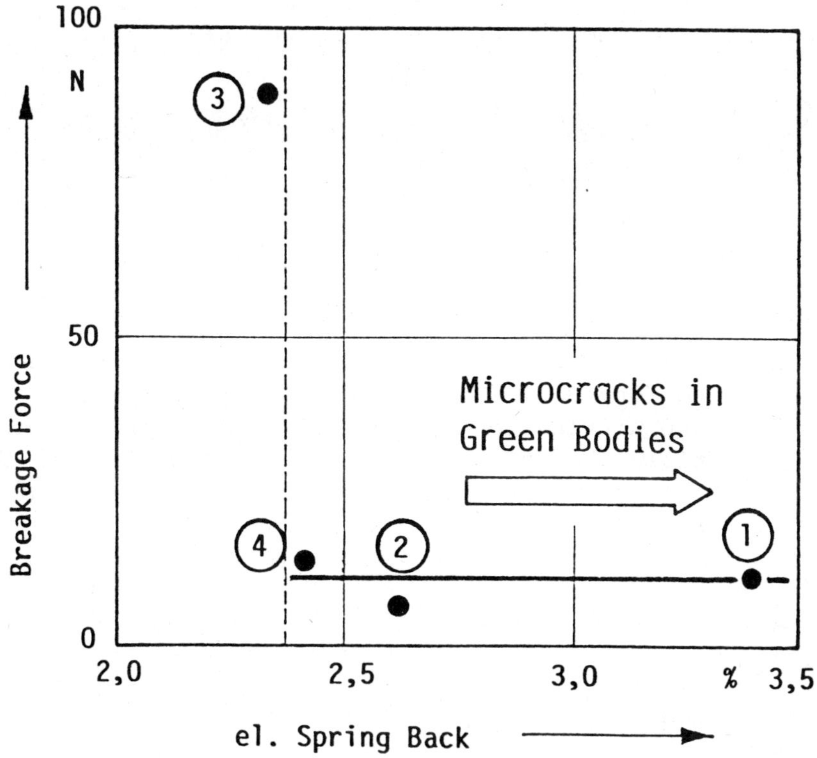

Fig. 8 Strength of Green Bodies and Compacting Behaviour

NUCLEATION AND GROWTH DURING CARBOTHERMAL REDUCTION AND NITRIDIZATION OF SILICA

W. Roger Cannon and Shi-Chang Zhang

ABSTRACT

In a previous paper it was observed that hexagonal prisms-shaped particles of uniform size formed with a linear growth rate during carbothermal reduction and nitridation of SiO_2 in excess carbon. A model is presented here to explain the uniform size of particles formed. The model suggests that a burst of nucleation occurs as the SiO content is initially high. This is followed by a growth period when the SiO content is lower.

INTRODUCTION

Carbothermal reduction of oxides is an important method of preparing non-oxide powders because of the low cost. A number of papers have described the synthesis of Si_3N_4 powders by carbothermal reduction and nitridation of high purity SiO_2 at about 1400^oC(1-7). In a previous paper the authors noted that the particles are very uniform in size. Fig. 1 is a micrograph showing both the hexagonal prism shaped Si_3N_4 particles and the precursor SiO_2 particles which are of a much smaller size and not distinguishable as individual particles. The carbon has been burned out.

Figure 1. Scanning electron micrograph of a mixture of Si_3N_4 and the precursor SiO_2 from reference 7.

LaMer [8] described a process, shown in Fig. 2, whereby monodisperse particles can be generated. In this process the concentration of reactants is increased to the point where nucleation can occur but nucleation uses up sufficient reactants that it no longer can continue and only growth continues. The process is often described as a burst of nucleation followed by growth.

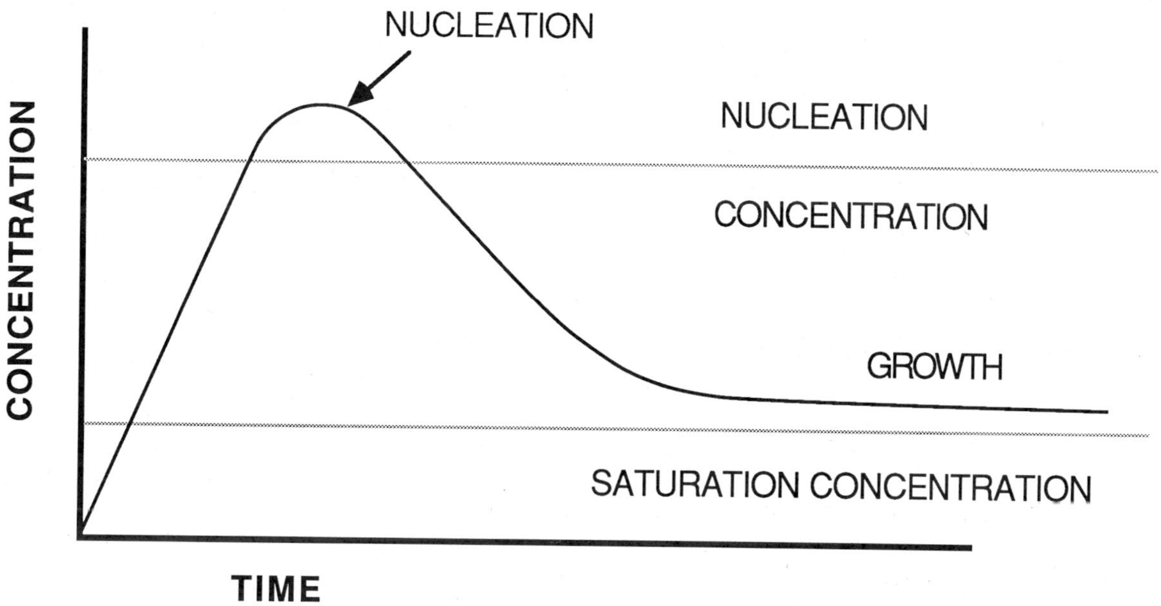

Figure 2. Schematic of the ideal nucleation-growth process for producing a narrow size distribution powder.

It was also shown in the previous paper that the rate of growth of the particles under the conditions of no nucleation and a constant rate of extension of the Si_3N_4 interface should follow the equation:

$$\frac{X}{2/7X + 1} = \frac{N}{m_o} \rho k_2 k_1{}^2 t^3 \qquad (1)$$

where X is the volume fraction of SiO_2 converted to Si_3N_4, N is the number of nuclei per unit volume, ρ is the theoretical density of Si_3N_4 (3.2 g/cm^3), k_2 is a geometric constant, k_1 is a growth rate constant, and t is time. Fig. 3 compares experimental results with the shape of the curve predicted from Equ. (1). The initial part of the curve is in agreement with this model. In the present paper thermodynamic calculations are used to show how this nucleation and growth pattern might occur in a silica-carbon powder bed.

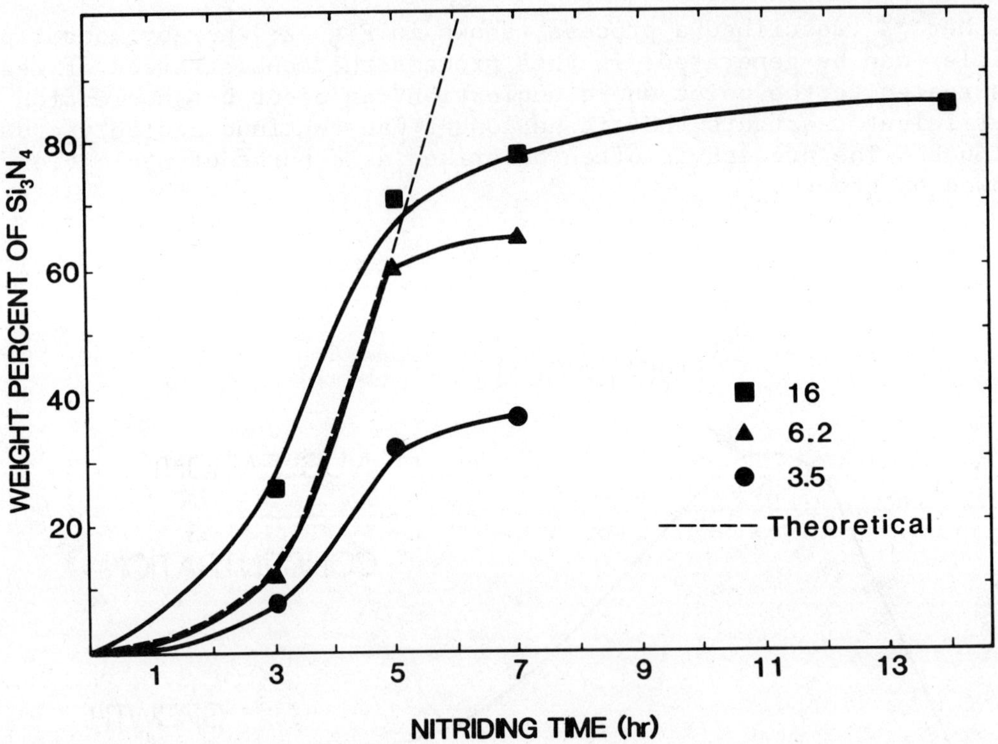

Figure 3. Weight percent Si_3N_4 vs time at 1400°C for samples with various C/SiO_2 ratios indicated from reference 7.

Nucleation and Growth Model

Between 1400°C and 1600°C α-Si_3N_4 forms from SiO_2 in an excess of carbon. There is strong evidence that the silicon precursor is SiO.(1) The temperature must be near 1400°C to maintain a high partial pressure of SiO. Sweda et al.(4) suggested because of the rapid reaction rate that SiO is formed by the reaction

$$SiO_2(g) + CO(g) \rightleftharpoons SiO(g) + CO_2(g) \qquad (2)$$

$$\log K_2\ (1400^oC) = -7.39$$

and

$$CO_2(g) + C(s) \rightleftharpoons 2CO(g) \qquad (3)$$

$$\log K_3\ (1400^oC) = 3.76$$

It was proposed(7) that nucleation of Si_3N_4 takes place on either the carbon or the SiO_2 surface. On the carbon surface the reaction might be

$$3SiO(g) + 3C(s) + 2N_2(g) \rightleftharpoons Si_3N_4 + 3CO(g) \qquad (4)$$

$$\log K_4\ (1400^oC) = 7.07$$

Two pieces of evidence suggest heterogeneous nucleation controls the rate of formation of Si_3N_4. Inoue et al[6] showed that by seeding the SiO_2/C mixture with submicron Si_3N_4 as a nucleating agent, the rate of formation of Si_3N_4 was greatly increased. Second, it was shown by Zhang and Cannon[7] that the particle size of the Si_3N_4 decreased as the specific surface area of the carbon and the silica increased, inferring that more nucleation sites were available, whereas Inoue et al[6] found that when they seeded the mixture that the rate of formation of Si_3N_4 did not increase with carbon specific surface area.

Although it has been suggested that reaction (4) controls the growth of the Si_3N_4 particles, alternatively it is suggested here and in the previous paper [7] that growth takes place by a gas phase reaction of molecules adsorbed onto the Si_3N_4 surface according to

$$3SiO(g) + 3CO(g) + 2N_2(g) \rightleftharpoons Si_3N_4 + 3CO_2(g) \qquad (5)$$

$$\log K_5 \ (1400^oC) = -4.21$$

According to reaction (5) the source of carbon is suggested to be CO rather C(s). Several features of Si_3N_4 growth support this point of view. Si_3N_4 particle growth rate and morphology was independent of the manner in which the carbon was distributed, ie. whether it coated the SiO_2 or whether it was distributed as discrete particles. In all cases the Si_3N_4 particles were well faceted hexagonal prisms of size much larger than the original carbon particles.(6)

The progress of nucleation and growth may be followed by considering the thermodynamics involved. In these calculations the JANAF Tables[9] are used, except for Si_3N_4 where the values of Hendry are used[10]. Initially SiO will be produced by reactions (2) until sufficient SiO is generated that nucleation according to reaction (4), for example, occurs. Growth will then proceed according to reaction (5). The overall reaction from the initial components is

$$3SiO_2 + 6C + 2N_2 \rightleftharpoons Si_3N_4 + 6CO \qquad (6)$$

$$\log K_6 \ (1400^oC) = -3.83$$

Reaction (6) indicates that 2 moles of gaseous N_2 must be supplied to the SiO_2 sample for 1 mole of Si_3N_4 and 6 moles of gaseous CO given off. The CO content in the powder bed will rise and the N_2 content fall as the reaction proceeds. The extent of the rise in CO content at the expense of N_2 depends on the diffusion rate of gases in and out of the powder bed. The ambient atmosphere is N_2. As the CO content rises the equilibrium P_{CO}/P_{CO2} will fall according to reaction (3) and as a result P_{SiO} will fall according to reaction (2). This is shown graphically in Fig. 4. However, a high P_{SiO} must be maintained for reactions (5) to continue to occur. The limiting CO/CO_2 ratio can be determined from the reaction

$$3SiO_2 + 6\ CO + 2\ N_2 \rightleftharpoons Si_3N_4 + 6\ CO_2 \qquad (7)$$

$$\log K_7 \ (1400^oC) = -26.39$$

For reaction (7) to continue to occur in the forward direction

$$P_{CO}/P_{CO2} \geq K_7^{1/6} = 2.5 \times 10^4$$

The equilibrium P_{SiO} is given by $P_{SiO} = K_2(P_{CO}/P_{CO2})$. At 1400°C when $P_{CO}/P_{CO2} = 2.5 \times 10^4$, $P_{SiO} = 1.0 \times 10^2$ Pa (1×10^{-3} atm). Thus at $P_{CO} = K_3(P_{CO}/P_{CO2})^{-1} = 0.22 \times 10^5$ Pa (0.22 atm) the P_{CO}/P_{CO2} ratio can no longer be maintained high enough for Si_3N_4 growth to continue (reaction (5)) and it will stop. It is likely that 0.22 atm. CO could be maintained in powder bed when CO is essentially the only gaseous product of the reaction. This is then the limiting point thermodynamically for growth to occur and may be called the CO saturation point. The saturation point is reached very early in the process.

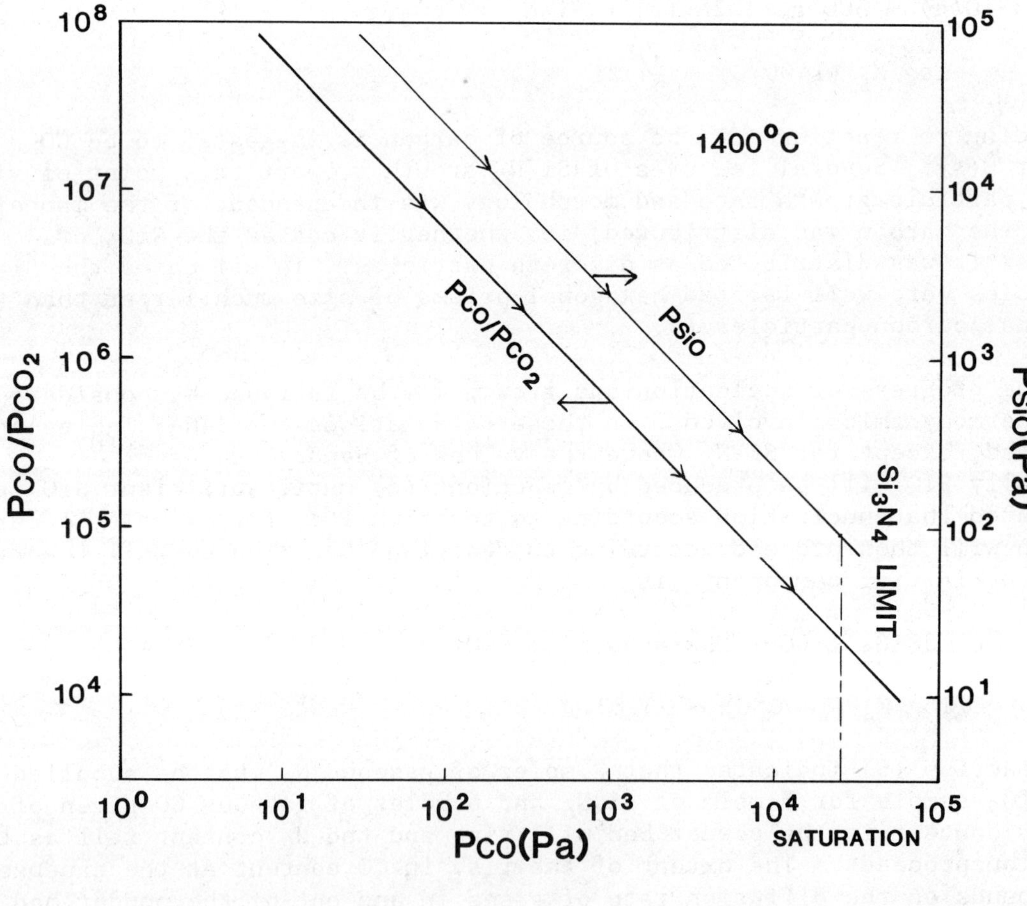

Figure 4. Equilibrium P_{CO}/P_{CO2} and P_{SiO} at 1400°C as P_{CO} increases.

The nucleation and growth curve shown in Fig. 5 was estimated by noting that 6 moles of CO are produced for every mole of Si_3N_4 and by making the following assumptions: (1) During the early part of the reaction all the CO and SiO produced stay in the powder bed and displace the nitrogen. (2) The powder bed is assumed to be 25% dense. (3) The oxygen content of the nitrogen is low $\ll$ 0.22 atm.

Initially CO is formed from oxygen in the atmosphere which may be as low as 10^{-10} Pa (10^{-5} atm.). Soon reactions (2) and (3) will occur which produces more CO and initially P_{CO} will equal P_{SiO}. The early part of the nucleation curve in Fig. 5 was calculated in the following way:

$$V_{SiO2(s)} = 1/3\ V_{gas} \tag{8}$$

where $V_{SiO2(s)}$ is the volume of SiO_2 in the powder bed and V_{gas} is the volume of gas filling the pores of the bed (the bed is assumed to be~25% dense). Assuming reaction (2) is controlling the build-up of CO and that the number of moles of SiO_2 used in the reaction equals the number of moles of CO in the atmosphere, then

$$X_{SiO2} = \frac{V_{SiO2}(\text{used})}{X_{SiO2}(\text{initial})} = \frac{1 \times 10^{-3} V_{CO}}{V_{SiO2}(\text{initial})} \tag{9}$$

and

$$P_{CO} = V_{CO}/V_{gas} \tag{10}$$

Combining (8), (9) and (10)

$$P_{CO} = P_{SiO} = 333\ X_{SiO2}$$

SiO will be produced by reactions (2) and (3) until sufficient SiO is generated that nucleation according to reaction (4), for example, occurs. At this point the CO content will rise and the SiO will lower until saturation, at which point the growth reaction will slow down. The fraction of Si_3N_4 formed during the growth process was calculated in a similar manner and shown in Fig. 5. The point at which nucleation ends and growth stops was chosen arbitrarily. Eventually a steady state condition should be achieved where the CO will diffuse out of the compact as N_2 diffuses in at a sufficient rate to keep the reaction going . Experiments show that the reactions are slow as expected from the Fig. 5 indicating that the P_{SiO} is maintained very low, near the critical value, and growth require many hours for completion. It is proposed that since nucleation requires some degree of super-saturation and that the growth takes place near the saturation so that very little nucleation occurs after this initial period. Calculations indicate (Fig. 5) that this is only after a few thousandth of a percent Si_3N_4 is formed. Thus a very uniform sized powder is produced.

These calculations illustrate how a kinetic process might be described using thermodynamics. No calculations have been made regarding the diffusion of CO out of the powder bed and N_2 into the powder bed but it was only assumed that the diffusion is slow and that a P_{CO} level of P_{CO} is at least 0.22×10^5 Pa (0.22 atm).

SUMMARY

Thermodynamic consideration along with the observation that gases diffuse slowly in and out of powder beds allow us to speculate on the mechanism of carbothermal reduction synthesis. In this instance the carbothermal reduction reactions were followed by considering the build-up of CO in a powder compact. A very short nucleation period followed by growth near saturation was thought to account for the uniform particle size.

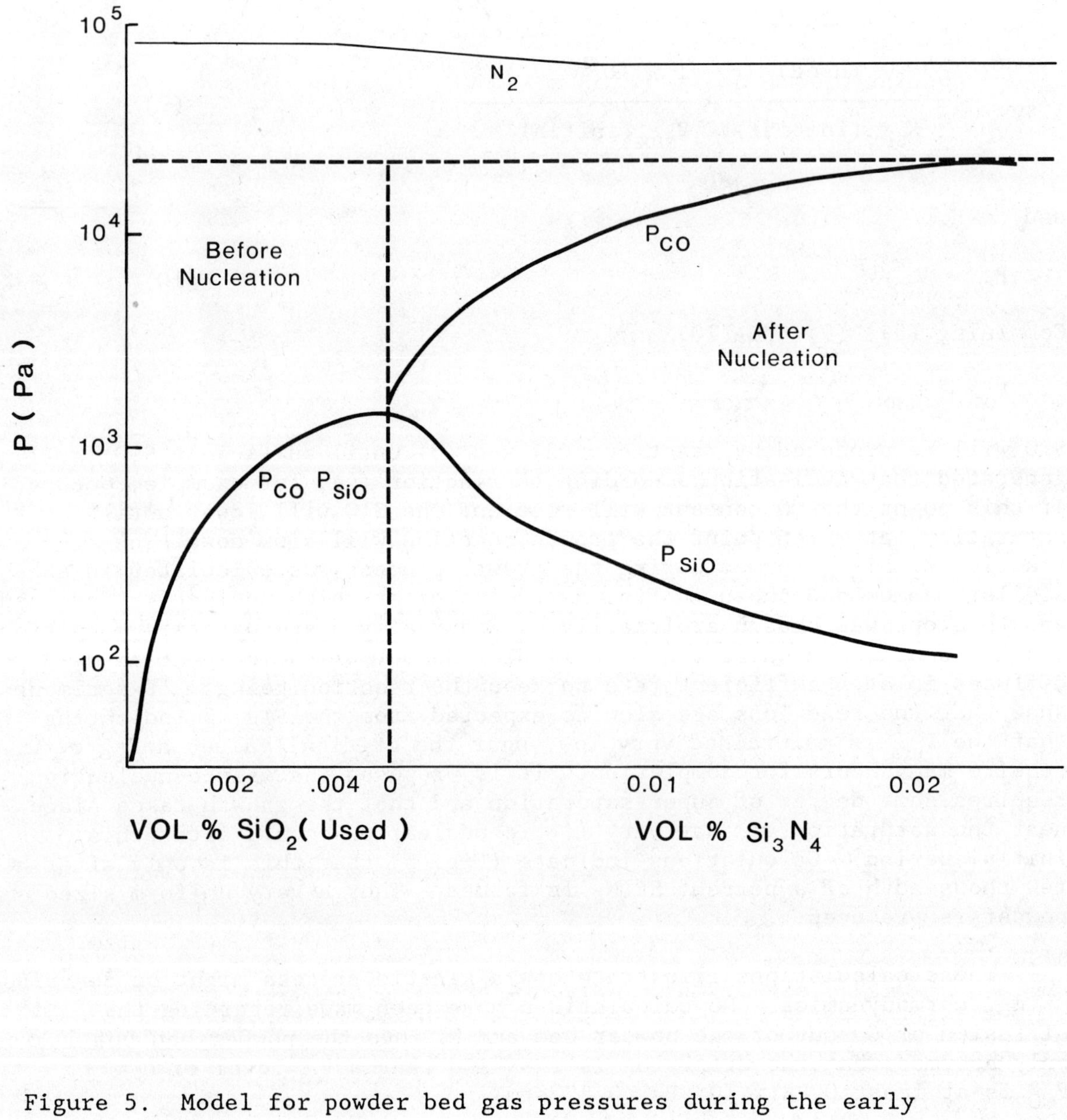

Figure 5. Model for powder bed gas pressures during the early growth of Si_3N_4.

REFERENCES

1. I. B. Cutler and W. J. Croft, "Silicon Nitride," Powd. Met. Int., 6, 92-94 and 143-46 (1974).

2. J. G. Lee and I. B. Cutler, in Nitrogen Ceramics, Edit by F. L. Riley, p. 175, Leyden: Noordhoft (1977).

3. A Hendry and K. H. Jack, "The Preparation of Silicon Nitride from Silica," p. 199 in special Ceramics, Vol. 6. Edited by P. Popper, British Ceramic Research Association, Stoke-on-Trant, England, 1975.

4. A Szweda, A. Hendry and K. H. Jack, "The Preparation of Silicon Nitride from Silica by Sol-Gel Processing," p. 107 in special Ceramics, Vol. 6. Edited by P. Popper, British Ceramic Research Association, Stoke-on-Trant, England, 1975.

5. K. Komeya and H. Inoue, J. Mater. Sci. Letters, "Synthesis of the α Form of Silicon Nitrid from Silica," 10, 1243-46 (1975).

6. H. Inoue, K. Komeya and A. Tsuge, "Synthesis of Silicon Nitride Powder from Silica Reduction," Comm. Am. Ceram. Soc., 65, C205 (1982).

7. S. C. Zhang and W. R. Cannon, "Preparation of Silicon Nitride from Silica," J. Amer. Ceram Soc., 67 [10] 691-695 (1984).

8. V. K. La Mer and R. H. Dinegar, "Theory, Production and Mechanisms of Formation of Monodispersed Hydrosols," J. Amer. Chem. Soc., 72 [11] 4847-56 (1950).

9. D. R. Stull and H. Prophet, JANAF Thermochemical Tables", 2nd Edition, National Standards Reference Data System, 1971.

10. A. Hendry, in Nitrogen Ceramics, Edit. by P. Popper, p. 183 Leydon Noordhoft (1977).

ZIRCON AND ZIRCON/ZIRCONIA STRUCTURAL CERAMIC MATERIALS FOR ENGINE APPLICATIONS

J. De Andrés and M.A. Rodríguez CERATEN S.A., Getafe (Madrid), Spain; J.S. Moya, I.C.V., Arganda del Rey (Madrid), Spain.

ABSTRACT

The effect of ZrO_2 content (0-20 vol %) on the sintering behavior, microstructure and mechanical properties (K_{1C}, σ_f) of zircon-zirconia composites has been investigated.

The mixtures were homogenized by attrition milling. Bars were isopressed at 200 MPa and heated between 1200°C - 1600°C.

The effect of ZrO_2 content on the sintering rate was evaluated by dynamic heating in a dilatometer.

When the ZrO_2 content is high, an increase in the sintering rate, final density and K_{1C} is obtained.

INTRODUCTION

It is obvious that applications for ceramic components in engines have increased in the last few years. This effort has been mainly done by obtaining high technology parts starting from expensive raw materials such as submicronic SiC, Si_3N_4, PSZ etc...

Zircon as a natural, abundant an unexpensive raw material with a high degree of purity is a potential candidate for engine components. It has already been stablished in literature (1,2,3) that zircon is a sinterable material with high chemical resistance and thermal stability up to 1600° C (4,5). However, its mechanical properties are not high enough, mainly toughness, as to be used for structural advanced ceramic parts.

Therefore, mechanical properties of zircon need to be enhanced. Among the different possibilities, reinforcement by the use of zirconia was chosen.

The aim of the present study is to analyse the effect of ZrO_2 on the sintering behavior and mechanical properties of zircon/zirconia composites.

EXPERIMENTAL PROCEDURE

The starting materials were high purity zircon powder* and monoclinic Zirconia powder**, with particle size distributions as shown in Fig. 1.
In order to establish the effect of ZrO_2 addition, three different batches were prepared :

1) Zircon (ZS)
2) Zircon + 10 vol % ZrO_2 (ZZ1)
3) Zircon + 20 vol % ZrO_2 (ZZ2)

After attrition milling the powders were dryed and isopressed at 200 MPa. A dilatometer furnace, at 10°C/min constant heating rate up to 1600°C, was used to study dynamic sintering of composites. The presurless isothermal treatments were performed in a computer controlled electrical furnace. Specimens were polished and thermally etched for microstructure analysis by SEM and EDX microanalysis. A three point bend method and microindentation tests (6) were made to determine bending strengh σ_f, and critical stress intensity factor K_{1C} , respectively. The Young's modulus was obtained by sonic resonance methods***.

---

* OPACIR S (QUIMINSA Spain). Impurities content (wt%) : Al_2O_3, 0.5; TiO_2, 0.2; Fe_2O_3 , 0.1.

** SCIS (M.E.L. England). Impurities content (wt%) : SiO_2, 0.2; TiO_2, 0.2; Fe_2O_3, 0.02.

*** GRINDO-SONIC MK_3 J.W. LEMMENS-ELEKTRONIKAN. V. LEUVEN. BELGIUM

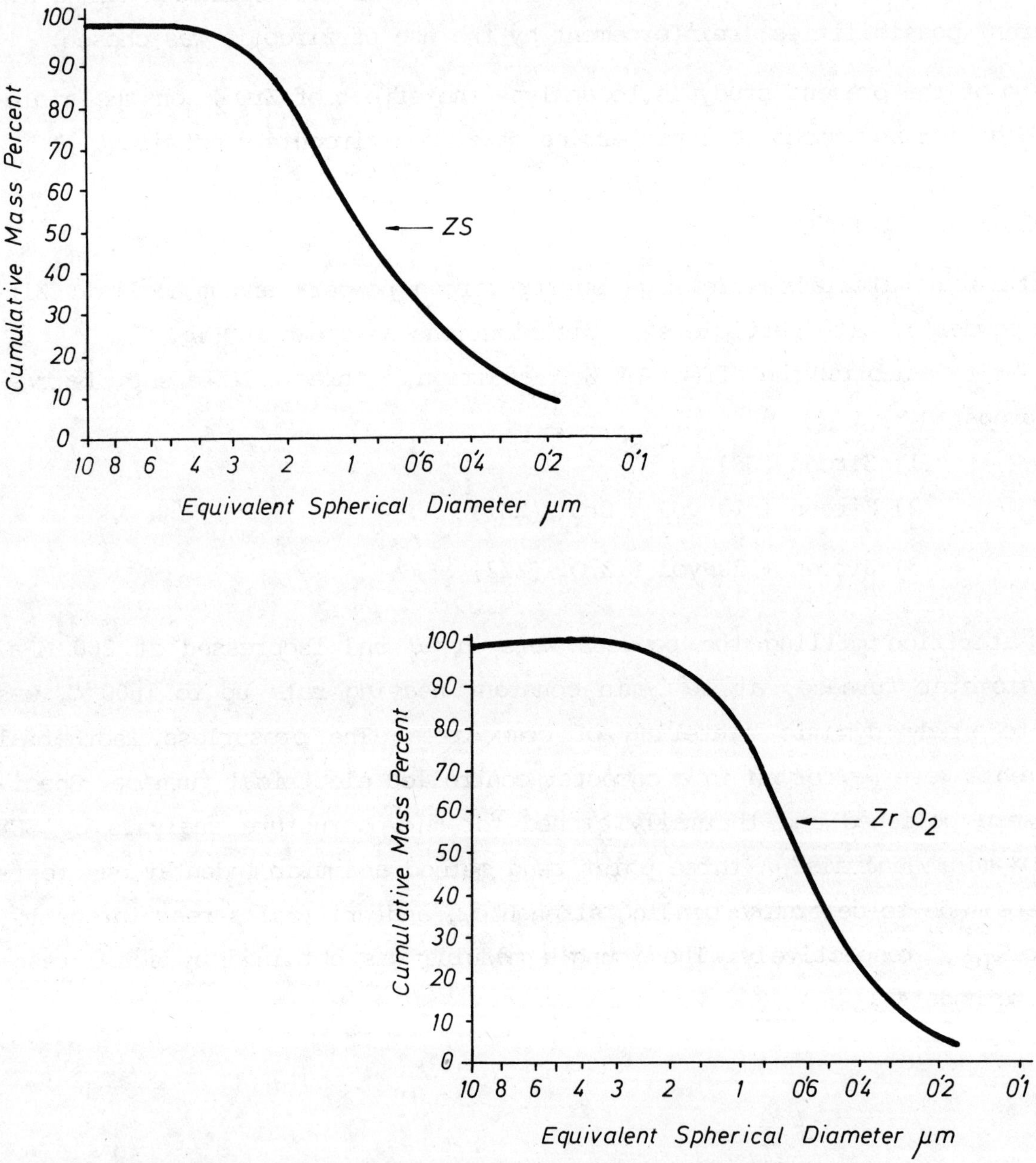

Fig. 1.- Particle size distribution of $ZrSiO_4$ and ZrO_2 raw materials

RESULTS AND DISCUSSION

As it can be observed from fig. 2 and 3, the sintering rate and densification for ZS and ZZ1 compositions show similar behavior. However, the ZZ2 composite shows a large increase in sintering rate at ~ 1300°C.

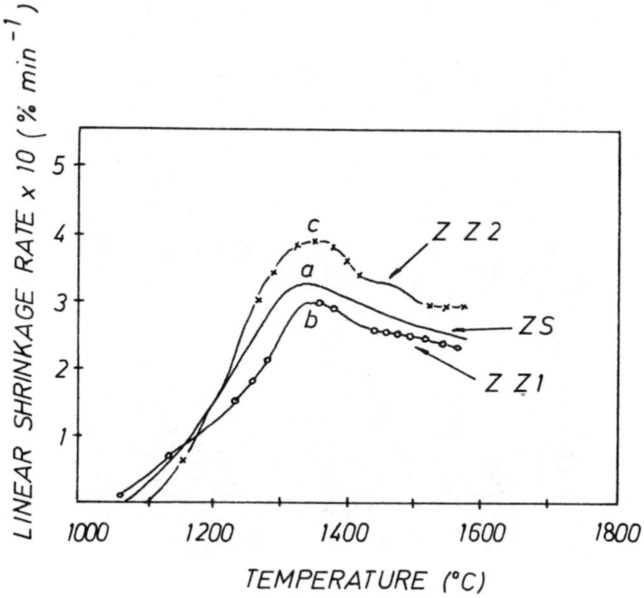

Fig. 2.- Linear shrinkage rate vs temperature for a) ZS, b) ZZ1 and c) ZZ2

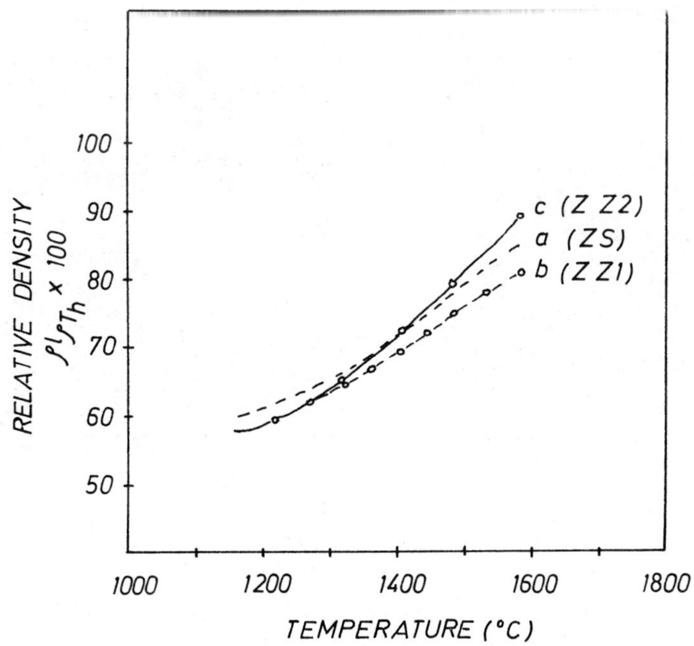

Fig. 3.- Relative density vs temperature for a) ZS, b) ZZ1 c) ZZ2

Table I and fig. 4 show that when the amount of ZrO_2 is ~ 20 vol % the sinte_ rability of the composite is enhanced and an increase of 50% is observed in K_{1C}

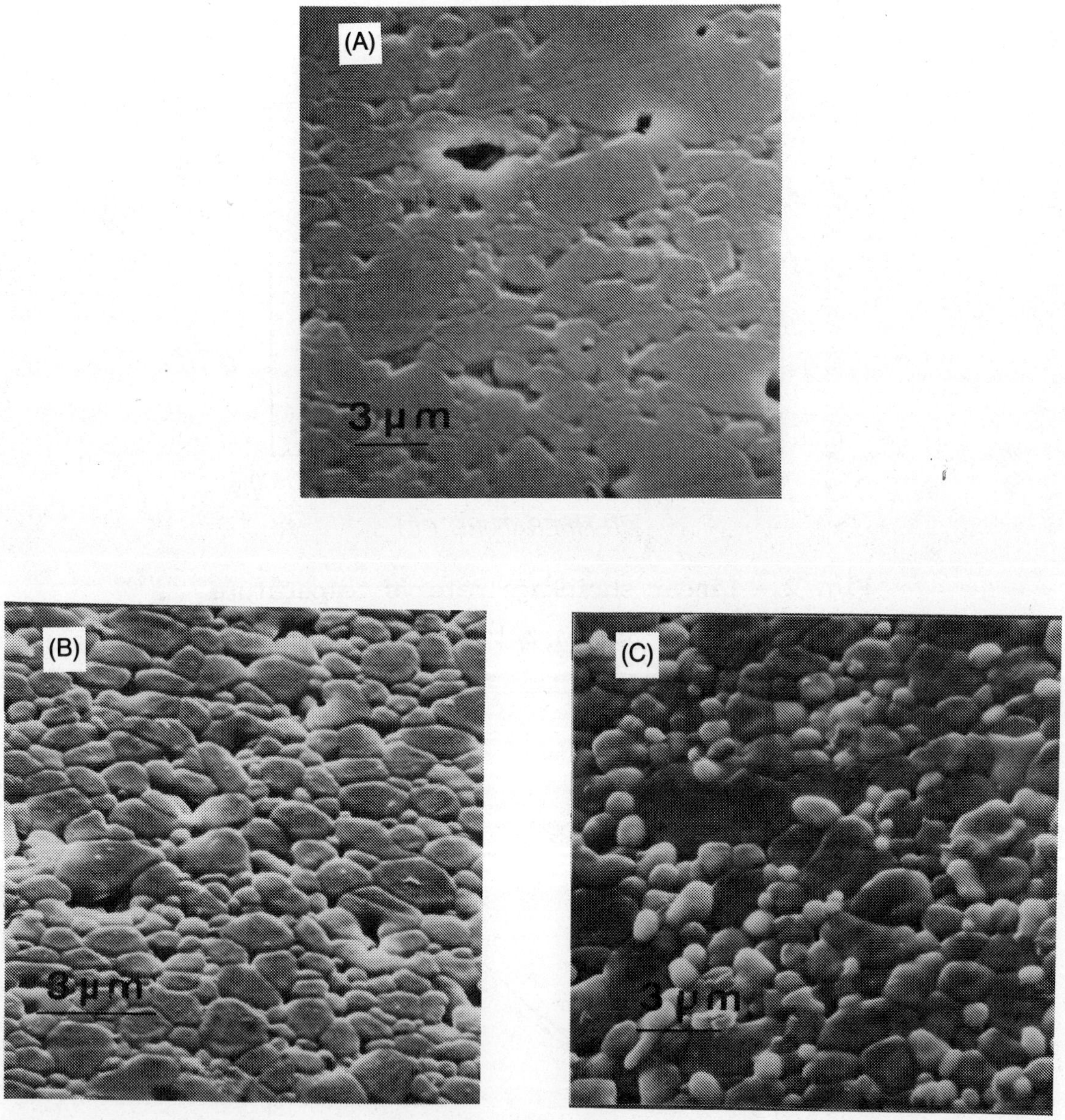

Fig. 4.- SEM photomicrographs of speciments (a) ZS; (b) ZZ1; (c) ZZ2, sintered to 1550ºC for 4 h.

Table I

Density and Mechanical properties corresponding to specimens sintered to 1550°C for 4 h.

Specimen	σ_f (MPa)	K_{1C} (MPam$^{-1/2}$)	$\rho/\rho th$ %	Young's Modulus(G.Pa)
ZS	288 ± 23	2.7 ± 0.2	96	243
ZZ1	230 ± 20	2.7 ± 0.2	96	217
ZZ2	260 ± 20	3.8 0.1	99	233

It can also be established from Fig. 5 that a dispersed phase of ZrO_2 inhibits zircon grain growth.

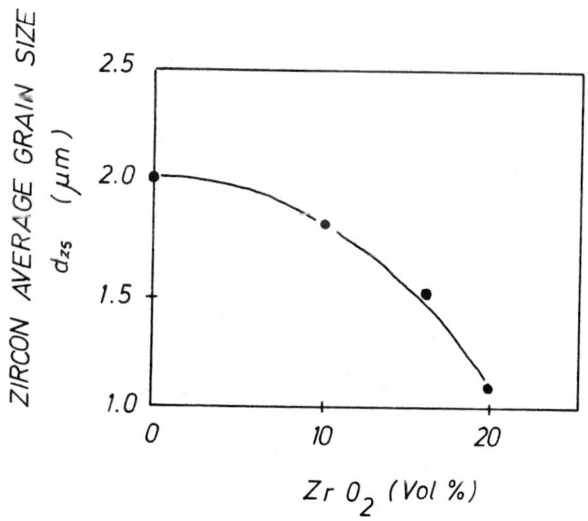

Fig. 5.- Zircon average grain size vs ZrO_2 content (vol %)

These results can be explained, in a similar way as previously reported by Moya and Osendi in the case of mullite/zirconia composites (7). The presence of a metastable solid solution of ZrO_2 at grain boudaries enhances grain boundary mass transport, leading to improved sinterability of the composite.

As there is a decrease in σ_f values (table I), and no significant fraction of t-ZrO_2 was detected by XRD analysis, the increase on K_{1C} is probably due to a microcracking effect. The complete understanding of this phenomenon is now being evaluated by TEM studies of grain boundaries in zircon/zirconia composites.

CONCLUSIONS

With the addition of amounts of zirconia (~ 20 vol %) nearly theoretically dense zircon bodies with increased K_{1C} values were obtained.

These results plus the material's thermal stability up to 1600°C, high chemical stability and low cost, suggest zirconia reinforced zircon is a potential ceramic material for engine applications.

REFERENCES

1.- R. Moreno, J.S. Moya, J. Requena.J. Mater. Sci. 5 (1986) 127-128.
2.- J. De Andrés, "Sinterización de Cuerpos densos de Circón. Efecto de las adiciones de TiO_2 y ZrO_2", M. Th. UNIVERSIDAD COMPLUTENSE DE MADRID (1986)
3.- J. De Andrés, S. De Aza and J.S. Moya, Ceram. Mat Comp. for Engines. Ed. W. Bunk, H. Hausner (1986) 299-306.
4.- P. Pena "Relaciones de fase en sistemas de óxidos refractarios de interés tecnológico. Sistema ZrO_2-Al_2O_3-SiO_2-TiO_2 (PH. Thesis) U. COMP. MADRID (1979).
5.- P. Pena and S. De Aza, J. Mater. Sci (1984) 135-142
6.- P. Miranzo and J.S. Moya, CERAM. INT. Vol 10 nº 4 (1984) 147-152.
7.- J.S. Moya and M.I. Osendi, J. Mater-Sci. 19 (1984) 2904-2914.

STUDY OF THE CONVERSION OF C+SiO_2 MIXTURES TO Si_3N_4 IN PRESSURIZED NITROGEN

Ekelund, M. and Forslund, B.
Department of Inorganic Chemistry, Arrhenius Laboratory,
University of Stockholm, S-106 91 STOCKHOLM, Sweden
Johansson, T*.
Kema Nord Industrikemi, S-84010 LJUNGAVERK, Sweden

ABSTRACT

The aim of the present study was to learn more about the reaction kinetics of carbothermal reduction of SiO_2 at elevated nitrogen pressures:

$$3\,SiO_2 + 6C + 2N_2(g) = Si_3N_4 + 6CO(g)$$

Pelletized intimate mixtures of C and SiO_2 were subjected to heat-treatment of fixed duration at different constant temperatures in the range 1400-1650 °C in a graphite furnace inside a pressure vessel. Different controlled nitrogen flows were used at different pressures in the range 0.1-6 MPa. The resulting phase assemblage and degree of conversion in each experiment were determined. The rate of formation of Si_3N_4 was found to be a function of the applied pressure, and also of the type of starting materials used. Thus, for the most reactive mixtures, made up of relatively fine carbon and coarse silica, the degree of conversion at ≈ 1500 °C increased with the pressure in the range 0.1 to 2 MPa. The pellets were found to be porous and uniformly reacted. In contrast, less reactive mixtures suffered a decrease of reactivity with increasing pressure, always turning into hard, dense bodies. That the samples could have a disposition for sintering to different degrees, competing with the nitride-forming reactions, is proposed as an explanation of the different trends observed.

INTRODUCTION

Synthesis of Si_3N_4 according to the reaction scheme

$$3SiO_2 + 6C + 2N_2(g) = Si_3N_4 + 6CO(g) \qquad (1)$$

was reported already in the 19th century[1]. It has been discussed frequently in the scientific literature during the last decades[2-5] as it might be an economically advantageous alternative to the presently dominating industrial process

$$3Si + 2N_2(g) = Si_3N_4 \qquad (2)$$

Recently it was recognized[6,7] that, by performing the carbothermal reduction and nitridation at elevated nitrogen pressures, high reaction temperatures (T > 1600 °C) could be used without initiating the unwanted formation of SiC, which defines an upper temperature limit between 1400 and 1500 °C to the process at atmospheric pressure[2,4]. The overall rate of reaction (1) will certainly increase with increasing temperature and, as it might seem, will also be positively affected by a high N/Si ratio, brought about by a high nitrogen pressure. However, it is not realistic to imagine that the true mechanism of reaction (1) is a solid-solid reaction between C- and SiO_2-particles in a powder mixture. Rather, as has been pointed out by previous investigators[3,5], it should be regarded as composed of elementary steps proceeding *via* the gas phase and including also species like SiO(g) and CO_2(g). This means that, in a powder bed or within pellets in a charge, gas phase mass transport must play an important role, the more so the higher the reaction temperature. Therefore, if gas phase diffusion controls the overall rate, it must be anticipated that raising the pressure beyond a certain limit will have a hampering effect on reaction (1). The present work was undertaken in order to study how the nitridation at high temperatures in mixtures of carbon and silica of different particle size are affected by (moderate) changes in nitrogen pressure.

EXPERIMENTAL

Method and Apparatus

To be able to compare the reactivities of different starting mixtures we chose a standard heat treatment procedure as follows:

The sample, typically weighing 0.3-0.5 g and charged in a carefully purged furnace inside a pressure vessel, was rapidly heated (≈ 20 deg/min) with simultaneous adjustment of the pressure in the range 0.1-6 MPa and the gas flow through the sample space (ϕ = 2-25 l/min). For each pre-selected annealing temperature the heating-up was regulated so as to avoid unnecessary overshoot. After a certain time (120 or 240 minutes) at the "plateau temperature" the furnace was turned off and the sample allowed to cool (≈ 30 deg/min down to 1000 °C, below that at a decreasing rate). The temperature of the "plateau" was controlled to within +/- 3 degrees, although occasional short deviations twice that large occurred. The temperature reading was judged to be accurate to within +/- 3 degrees. The reaction products were checked at room temperature by X-ray diffraction techniques using Guinier-Hägg focusing cameras, by elemental (N-, C-, O-) analyses** and by optical and electron microscopy (SEM). The carbon monoxide level in the gas leaving the sample space was continuously monitored during each run by means of an IR-detector*** at atmospheric pressure. The sample was weighed before and after the heat treatment, and the weight change served as a fast preliminary measure of the reactivity of a mixture.

The reactor consisted of a vertical, tubular graphite furnace (l=30, id=7 cm) coaxially mounted in a cylindrical pressure vessel (Fig. 1). The sample holder (Fig. 2) was a graphite crucible (h=35, id=35 mm) with a perforated bottom. It was fixed in a coaxial graphite susceptor tube, arranged in such a way that all gas leaving the autoclave was forced through the pelletized charge. This was supported by a perforated graphite foil, squeezed by a

Fig 1. Schematic drawing of the autoclave with graphite furnace. a) gas inlet, b) thermal insulation, c) graphite furnace, d) sample holder, e) pressure vessel, f) CO-detector, g) gas outlet

Fig 2. The sample holder. a) graphite furnace, b) sample holder with screw plug on top, c) sample, d) (W,Re)-thermocouple, e) susceptor with perforated lid, f)disc of porous graphite felt.

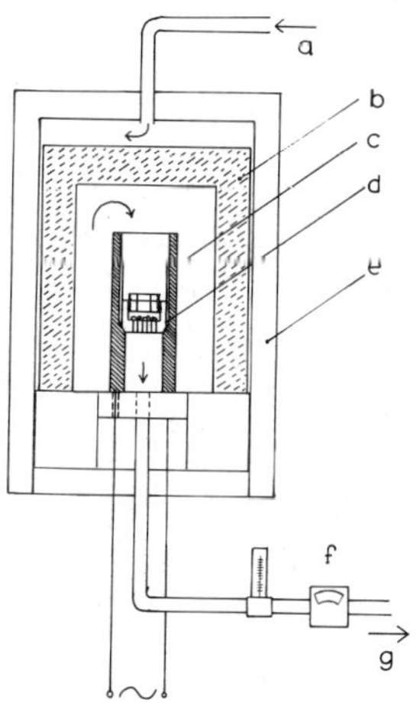

Fig.1.

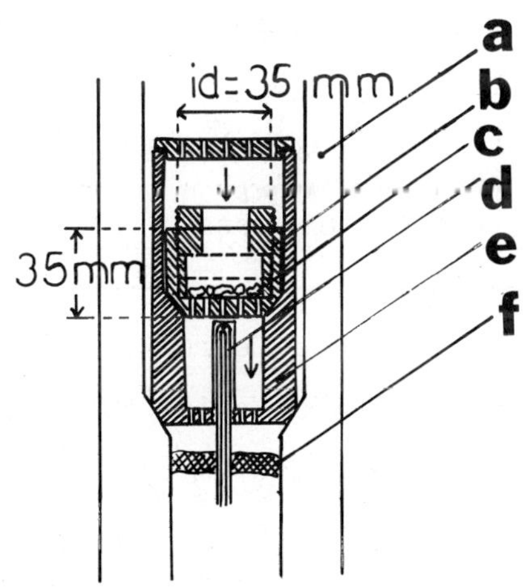

Fig.2.

graphite ring and a screw-plug on top. The gas leaving the sample space was forced to take the shortest way out of the autoclave *via* gas-tight graphite, ceramic and metal tubing, and so reached the CO detector after about 10 seconds' delay with the flows normally used (2-6 l/min). Each granular charge was spread on the supporting foil (A=7 cm^2) in a mono-layer. This was done to minimize the influence of reacting granules on their neighbours as in separate experiments the action of $CO_{(g)}$ has been found to appreciably hamper the nitridization reaction. For temperature measurement and control (W,Re)-thermocouples in protecting Al_2O_3 tubes were used, calibrated at different pressures by melting pieces of Au and Pd (contained in Al_2O_3 tubes) at the sample site.

Starting materials

Seven different starting mixtures were run and compared ; combinations of SiO_2-powders (>99.8 % SiO_2) and C-powders (C>92 %, H<6 %, S, O, N<1 %), each of three different mean particle diameters in the range 10nm-10μm. One mixture had been heat-treated under non-nitriding conditions ("pre-sintering") prior to charge. Table I is a list of the mixtures prepared, each with the mole ratio C/SiO_2=2.0 and designated by a letter A, B, etc., which will be referred to in the following. The granules were prepared from ball-milled (WC or SIALON balls) ethanol-slurries of the reactant powders in stoichiometric amounts. One type of pellet, of "macaroni" shape (d=1.5, l=3-8 mm), was prepared by extruding the wet, homogenized mixture through a perforated steel plate. The other type, formed approximately spherical granules (0.6< d <2 mm, sieved). In a few runs a crushed and sieved fraction (0.2< d <0.6 mm)of the "macaroni" granules was charged and in another, slices (l=0.5 mm) of these granules. The nitrogen gas, containing less than 10 ppm of total oxygen was used without further purification.

Table I. Starting materials

$C + SiO_2$ Mixtures	Designation
Medium C and Coarse SiO_2(quartz)[3]	A
Medium C and Medium SiO_2 (amorphous)	B
Fine C and Medium SiO_2	C
Fine C and Fine SiO_2 (amorphous)	D
Medium C and Fine SiO_2	E
Coarse C and Fine SiO_2[3]	F
Coarse C and Fine SiO_2[3](pre-sintered granules)	F'

3) 5 wt% α,β Si_3N_4 has been added

RESULTS

The reactants were charged in the form of pellets of a size and shape not much unlike a starting material for a large scale process. But the temperature range in this study had to be chosen lower than in a tentative industrial furnace, because the nitridation rate was found to be so rapid at T > 1600 °C and p=1-3 MPa that it was difficult to perform a comparative study of reactivities with reasonable precision. Tests with some mixtures at 1650 °C showed that the transformation to Si_3N_4 had virtually gone to completion in less than 2 hours. In this preliminary work we have therefore used the more convenient temperature range 1400-1500 °C.

In tables II, III and IV are listed a number of relevant parameter values for each run, together with the resulting relative weight change. The conversion number based on N-, C- and O- analyses as a rule tended to give a lower value than that calculated from the weight change, especially at lower pressures. The probable cause is a net evaporation from these samples of unnitridized SiO(g), swept away by the gas. From the X-ray powder photographs it could be stated that Si_3N_4, predominantly the α-modification was the only crystalline solid formed in all mixtures, save for traces of cristobalite found in the products of runs #1, #2 and #3.

Valuable information on reactions within a pellet can be gained through microscopic inspection of cross-sections after the run. The samples in this study turned out to show four types of cross-section appearances under the microscope:

I) a hard black pellet, showing a glassy lustre in reflected light and with a thin, grey, reacted surface layer (Fig. 3).

II) a hard black core and a porous, grey outer zone, with a sharp boundary in between (Fig. 4).

III) a rather porous matrix with a darker central and a light grey outer zone.

IV) a porous matrix of almost uniform grey colour (Fig. 5).

Common to all reacted granules is a more or less dense "fur", consisting of fibers and needles of α-Si_3N_4, covering their surfaces. As a rule, the graphite foil supporting the charge also became covered by a white felt of α-Si_3N_4, probably formed by the action of SiO(g), swept off the reaction mixture by the gas flow. The weight of this felt was, however, always found to be small compared with the total sample weight and the weight change.

Fig.3. Cross-section of a "macaroni" pellet, showing a glassy lustre in reflected light (Type I).
Pellet diameter: 1 mm

Fig 4. Cross-section of a "macaroni" pellet with an unreacted core and a grey, porous outer zone (Type II,III).
Pellet diameter: 1mm

Fig 5. Cross section of a homogeneously reacted, porous "macaroni" pellet (type IV). Pellet diameter: 1mm.

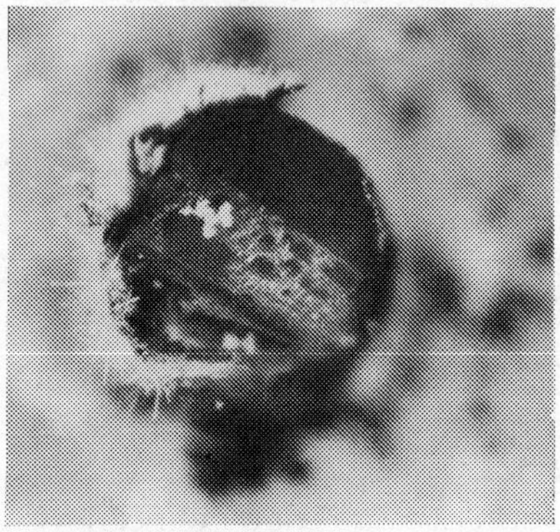

Fig.3.

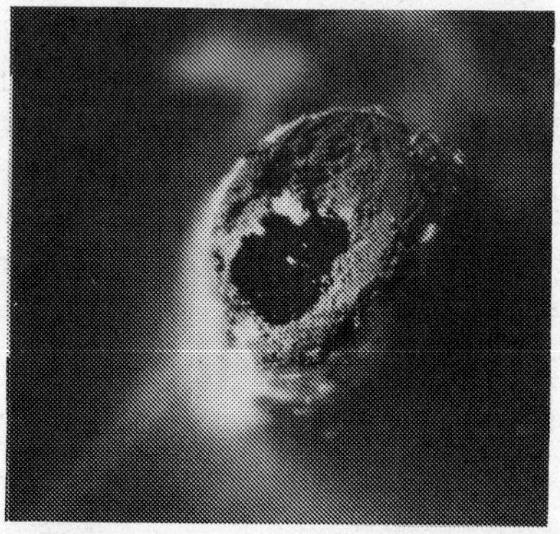

Fig.4.

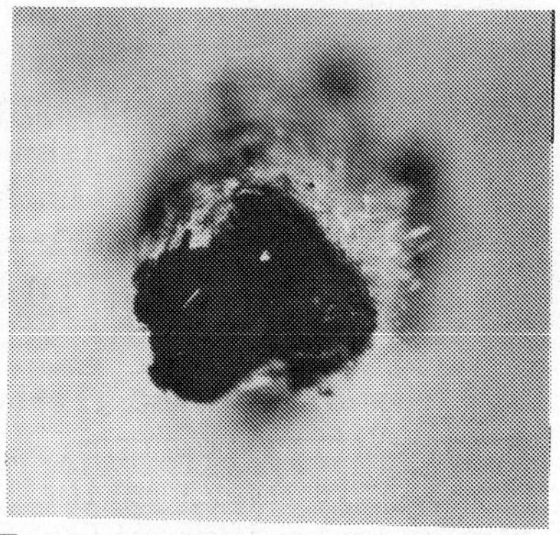

Fig.5.

A series of test runs with gas flows between 1 and 25 l/min had demonstrated that the degree of conversion is strongly dependent on the gas flow at low rates, < 2 l/min and p=3 MPa. The conversion is independent of ϕ at the maximum rate, where, evidently the effect of inter-pellet mass transport is negligible. A flow of 12 l/min at 3 MPa (or 6 l/min at 1.2 MPa, 2 l/min at 0.6 MPa) was found to be a good compromise; high enough to make the flow-dependence of the nitridation rate tolerable (*cf.* runs #10 and #11, Table II), yet low enough to leave some sensitivity in the continuous CO-measurements on the exit gas. For runs with $p(N_2)$ <0.6 MPa, gas mixtures of N_2 in Ar were used, keeping the total pressure = 0.6 MPa and ϕ = 2 l/min. For a given gas flow the degree of conversion also turned out to be a function of the charge weight, *i.e.* the coverage of the supporting graphite foil. This is exemplified in Table II by run #1, with a maximum charge (1 g)for a granular mono-layer, to be compared with run #2, with just half that much. For each mixture run in this study, a charge weight sufficiently small to make the reactivity independent of the sample size has been used.

Tables II and IV actually list subsets of comparative nitriding experiments. Some explore the effect of varying the annealing time at the plateau temperature (2 and 4 hours resp.), within others the pressure has been varied at a specific temperature (1410, 1470, 1490 °C). A comparison of conversion numbers for different mixtures at the same temperature and pressure, *e.g.* sets (#2, #4),(#5, #8, #9) at 2.8 MPa, 1470 °C and (#13, #16, #18, #19) at 0.6 MPa, 1490 °C reveals a remarkably wide spread in reactivity among these samples. The least reactive was found to be #19 , which is an "F"-type mixture, although heat-treated at ≈1200 °C for several hours in a separate run in order to promote sintering below the nitridization temperature.

The experiments listed in Table III have purposely been performed under conditions somewhat diverging from those of the standard procedure (*cf.* Table II). In #20 the plateau temperature was chosen a little below 1470 °C and furthermore approached more gently than usual (*cf.* #9).This was done to verify that the final degree of conversion and the sample properties are not very sensitive to the temperature program during the early (reactive) stage of a run. The test was motivated by the fact that a strictly reproducible heating procedure has not been maintained throughout this study. With a similar purpose the sample in #21 was annealed 1/2 hour at 1100 °C before it was brought to the plateau at 1470 °C. Evidently, (*cf.* #9) the uncertainty inherent in the annealing time of our experiments, due to the scatter in heating/cooling rates (≈ 20 deg/min), can be ignored to a first approximation. Runs #22 and #23 were performed, not in a pure nitrogen atmosphere but a CO/N_2 mixture with 1% and 3% CO, respectively. Evidently the degree of conversion is much diminished by 1% CO (*cf.* #13) at 1490 °C, and the nitride forming reaction(s) is completely quenched in 3% CO at 1450 °C. This is in accordance with the findings of Motoi and Sasaki[8], who observed that 5% CO mixed in the N_2 considerably reduced the nitridation rate at 1500 °C. Neither time nor temperature at the opening of the exit gas valve were strictly defined. They could vary a little from run to run, as could the time for adjustment of a steady gas flow through the sample holder. In #24 the valve was not opened until the furnace reached the plateau temperature, 1490 °C. It is seen that the conversion does not differ much from #14, when the valve was opened at 1100 °C. In spite of the extra supply of fresh nitrogen, the relative weight decrease was only slightly larger in the latter case and the main reason for this was probably that the plateau temperature in #24 turned out to be somewhat below 1490 °C. Nevertheless, it has been attempted throughout this study to keep all parameters as nearly equivalent as possible, especially within subset of runs, the results of which are to be compared and evaluated as a function of a particular parameter.

As expected a longer annealing results in a higher degree of conversion. With the mixture types studied, a substantial difference in weight change is observed between the 2 and 4 hour runs at 1410 °C ("C") and 1470 °C ("B"); and, as seen by microscopic inspection of the cross-sections (type II, Fig.4), the central, dark, unreacted zone shrinks as the reaction proceeds, while the outer, light grey, porous, reacted zone grows thicker. The picture is not that clear, when it comes to a comparison between Si_3N_4 formation rates at different pressures in the 0.1-6 MPa range. It can be seen from Tables II and IV, and from Figs 6 and 7 that the relative weight change decreases on a pressure increase for some mixtures, *e.g.* type C at 1410 ∘C between 0.6 and 2.8 MPa, while smaller granules of type C-mixtures show a positive pressure dependence of $-\Delta m/m_0$ at 1410 ∘C (Fig 7). This also holds for larger granules, type C, at 1490 ∘C, showing relatively high nitridation rates. Type B-samples have reactivities at 1470 ∘C between the extremes, decreasing with increase in pressure from 2.8 to 5.9 MPa. From the features of the curves in the diagram, Fig 6, it can be deduced that, at 1470/1490 ∘C, there is an optimal pressure for the nitridization rate around 2 MPa.

Fig 6. The relative weight change, as a measure of reactivity, for some of the mixtures, run 2 hours at 1470/1490 ∘C.
○ 1490 ∘C, Starting mixtures: type "C". (#15,#14,#13,#12)
▽ 1490 ∘C, Starting mixtures: type "E". (#17,#16)
▲ 1470 ∘C, Starting mixtures: type "B". (#6,#5,#7)
● 1470 ∘C, Starting mixtures: type "C". (#10,#9)

Fig 7. The relative weight change, as a measure of reactivity, for mixtures run 2 hours at 1410 ∘C.
□ Starting mixture: type "C" (#25,#26,#27)
■ Starting mixture: type "C", with grains 0.2-0.6 mm (#30,#29)

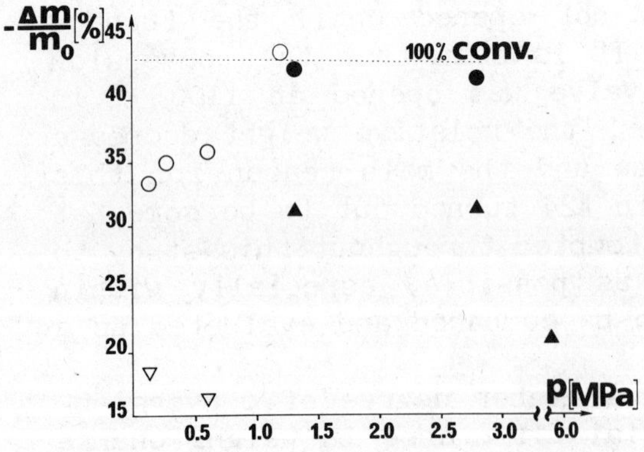

Fig.6.

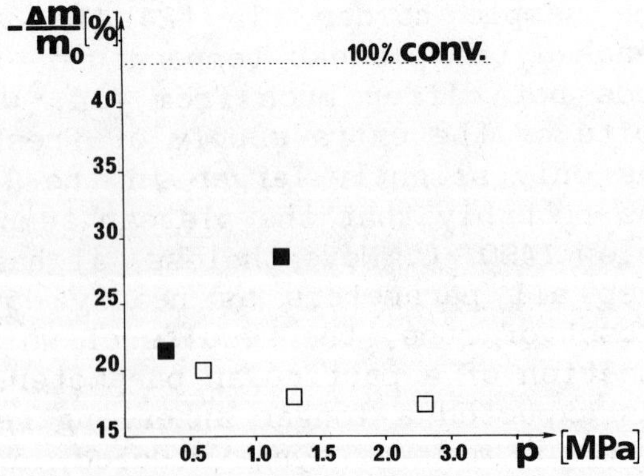

Fig.7.

Table II. Nitridation runs at 1470 and 1490°C.(RT = Residence time on the plateau, CS = Cross-section surface type)

Run #	Starting mixture type	Starting mixture weight g	Granule type	T °C	pN_2 MPa	ϕ $\frac{1}{min}$	RT h	Conversion $\frac{\Delta m}{m_o}$ (%)	Conversion $\frac{mSi_3N_4}{mtot}$ (%)	CS
1	A	1.001	"macaroni"	1470	2.8	12	4	28.7		IV
2	A	0.465	"macaroni"	1470	2.8	12	4	32.0	64.4	IV
3	A	0.778	slices (1≈.5 mm) of mac.	1470	2.8	12	4	32.9	65.9	IV
4	B	0.506	"macaroni"	1470	2.8	12	4	40.1	(*1)	II
5	B	0.509	"macaroni"	1470	2.8	12	2	31.4	55.3	II
6	B	0.521	"macaroni"	1470	1.3	6	2	31.2	53.1	II
7	B	0.567	"macaroni"	1470	5.9	25	2	21.1	32.1	II
8	D	0.394	"macaroni"	1470	2.8	12	2	30.1	51.3	III
9	C	0.487	"macaroni"	1470	2.8	12	2	41.9	(*2)	IV
10	C	0.471	"macaroni"	1470	1.3	5	2	42.6	(*3)	IV
11	C	0.389	"macaroni"	1470	1.2	20	2	43.3		IV
12	C	0.344	"macaroni"	1490	1.2	6	2	43.9	(*4)	IV
13	C	0.326	"macaroni"	1490	0.6	2	2	36.0		III
14	C	0.305	"macaroni"	1490 4)	.28	2	2	35.2	52.1	III
15	C	0.381	"macaroni"	1490 4)	.12	2	2	33.5	35.1	I
16	E	0.271	"macaroni"	1490	0.6	2	2	16.3	14.3	I
17	E	0.276	"macaroni"	1490 4)	.12	2	2	18.3	10.0	I
18	F	0.268	spheres 0.6<d<2 mm	1490	0.6	2	2	13.9	21.5	?
19	F'	0.284	pre-sinter "F"	1490	0.6	2	2	4.9	8.8	?

4) Ar/N_2 mixture.

(*1) 4.3% resid. C (*2) 2.1% resid. C (*3) 2.4% resid. C
(*4) 1.2% resid. C

Table III. Special runs at 1470 and 1490°C. (RT = Residence time on the plateau, CS = Cross-section surface type)

Run #	Starting mixture		Granule type	T	pN_2	ϕ	RT	Conversion		CS
	type	weight g		C	Mpa	l/min	h	$-\frac{\Delta m}{m_0}$ (%)	$\frac{mSi_3N_4}{mtot}$ (%)	
20	C	0.351	"macaroni"	1465	2.9	13	2	40.9		IV
21	C	0.293	"macaroni"	1100	2.9	22	2	43.0		IV
22	C	0.328	"macaroni"	1470 1490 (5	0.6	2	2	21.5	21.8	I
23	A	0.768	"macaroni"	1450 (6	2.9	2	4	2	0	I
24	C	0.326	"macaroni"	1485	.25	2	2	32.7	45.8	II

5) 1% CO in N_2

6) 3% CO in N_2

Table IV. Nitridation runs at 1410°C. (RT = Residence time on the plateau, CS = Cross-section surface type)

Run #	Starting mixture		Granule type	T	pN_2	ϕ	RT	Conversion		CS
	type	weight g		C	MPa	l/min	h	$-\frac{\Delta m}{m_0}$ (%)	$\frac{mSi_3N_4}{mtot}$ (%)	
25	C	0.310	"macaroni"	1410	0.6	3	2	20.1	24.5	II
26	C	0.482	"macaroni"	1410	1.3	6	2	17.9	24.0	II
27	C	0.425	"macaroni"	1410	2.8	13	2	17.5	24.3	II
28	C	0.304	"macaroni"	1410	2.9	12	4	30.5	52.1	II
29	C	0.317	grains	1410	1.2	6	2	28.8	43.8	?
30	C	0.370	0.2<d<0.6 mm(sieved)	1410	0.3	2	2	21.4	24.8	?

DISCUSSION

In this study of the Si_3N_4 formation rate in granulated $C+SiO_2$ mixtures under different nitrogen pressures, it has been found that, for a mixture with fine carbon but coarse silica grains ("C"), the reactivity increases with pressure from 0.1 to 3.0 MPa. The opposite was true for mixtures of coarse carbon and fine silica particles ("E"). Cross-section surfaces of product pellets made from type C raw materials typically appeared homogeneously reacted under the microscope (Fig 5), while pellets made from type E materials exhibited a light grey, completely (?) reacted outer zone, distinct from a hard, black unreacted core (Fig 4).

Fig 8. SEM picture of a type III pellet core.
Fig 9. SEM picture of a type I pellet core.

Fig.8.

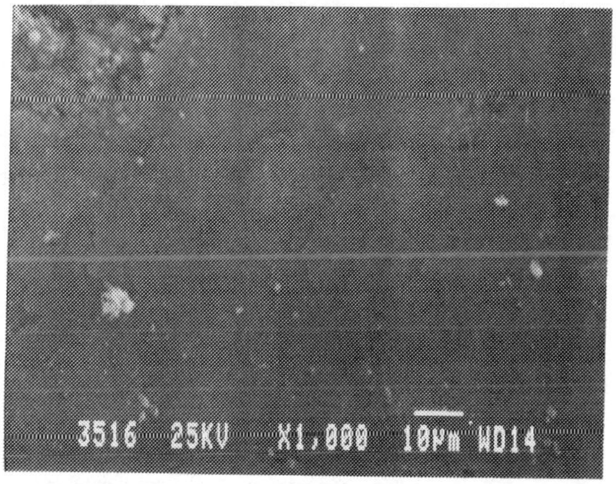

Fig.9.

SEM-pictures of the surfaces show that the core of a type III pellet (Fig 8) is quite porous, while an unreacted core of a type I pellet appears dense and homogeneous, also at high magnifications (Fig 9). The occurrence of a homogeneously reacted granule suggests that the overall process is largely controlled by chemical reactions, taking place simultaneously throughout the pellet, and not to any appreciable extent by mass transport of reactant or product species. Consequently, an increased nitrogen fugacity should enhance the nitridation rate. The cross-sections with two distinct zones, are indicative of a process controlled by intra-pellet diffusive transport of gaseous species. The chemical reactions proceed mainly at the zone boundary[9,10]. Gas phase diffusion coefficients are inversely proportional to the pressure, and so a decrease in overall reaction rate is to be expected when the pressure is increased. The crucial gas species might well be CO, leaving the granule through the pores, as the reaction rate has been found to be most sensitive to a change in p(CO) in the sample space (*cf.* #22, #23). A shortening of diffusion paths, brought about by reducing the pellet size, was enough to increase the relative importance of chemical reactions in samples #29 and #30, compared to #25, #26 and #27, and so to reestablish a positive $p(N_2)$-dependence of the nitridization rate (Fig 7).

It is not surprising that a mixture with small carbon grains ("C"), and thus large surface area and reducing capacity, is more reactive than one with coarser carbon particles ("B"). However, a small carbon grain size also seems to be a prerequisite for keeping a reacting mixture within the chemically controlled regime (IV,III) at 1470, 1490 °C ("C","D", in #9,#8 *cf.* "B", "E" in #6,#16). As an explanation of this fact, we prefer to think that a transition from chemical control to diffusion control occurs in the pellet due to pore closure, caused by sintering, mainly including SiO_2. Fine carbon particles in the starting mixture then might be more efficient in preventing sintering then coarse ones, perhaps for purely geometrical reasons (keeping the SiO_2 grains apart). That diffusion control may be dominant even in a type C-mixture at sufficiently low temperatures and nitridation rates supports the model of concurrent sintering phenomena, competing with Si_3N_4 formation. The drastically decreased reactivity in an F-mixture after "pre-sintering" (F',#19) provides further evidence. The different features characterizing product types I and II have probably arisen because nitridation of a type II granule has come nearer to completion, while in a type I granule sintering has proceeded further. This is why its black, unreacted core has assumed a dense, glassy appearance.

Run #15 at 0.12 MPa is noteworthy. In spite of a high reaction temperature, 1490 °C, the type C-mixture turned into a type I product. It is true that the degree of conversion is relatively low, $-\Delta m/m_0$ is only 33%, yet this is more than the 21% in #7, which yielded a type II product at 5.9 MPa. This indicates that sample properties in carbothermic nitridation of SiO_2 might be affected in more than one way by an increased nitrogen pressure. In an "A" mixture made up of quartz (#2), chemical control dominates($-\Delta m/m_0$ is independent of granule size, *cf.* #3) and persists in spite of low reactivity. This is probably due to crystalline SiO_2 being less inclined to sintering than an amorphous material.

The continuous CO-measurements showed that the carbothermic process accelerates at 1350 °C on heating. For all types of mixtures the reaction rate decreased drastically after just a short time (5-15 min) at the maximum temperature. This phenomenon has been observed by previous investigators[11,12]. Possible explanations are that initial close C-SiO_2 contacts within a virgin mixture cease to exist after a short time, or that the total SiO_2 surface diminishes as the reaction proceeds. The closing of pores, discussed above, may also contribute, as may the increasing resistance to diffusion when the reaction zone moves into the granule.

A continuation of the present work, mainly focusing on mixtures reacting under chemical control, is in progress at this laboratory. The objective is to obtain more information on the fundamental processes behind Si_3N_4 formation by carbothermic nitridation of SiO_2.

ACKNOWLEDGEMENTS

The authors wish to express their gratitude to Professor Lars Kihlborg,Department of Inorganic Chemistry, for helpful discussions and valuable comments on the manuscript. We also thank the following staff at the Arrhenius Laboratory: Hans-Erik Ekström and Bohdan Jelinski (construction and maintenance of the autoclave and furnace), Tomasz Niklewski and Gunvor Winlöf (electron microscopy) and Sven Westman (correction of the English of the manuscript). Advice and help, mainly with analytical problems, from Gunnar Hällgren and Bengt Gruvin at KemaNord is gratefully acknowledged. This work has received support from the Swedish Natural Science Research Council (NFR) and the Swedish Board for Technical Development (STU).

Footnotes:

*) member of the American Ceramic Society

**) performed by two independent commercial laboratories using fusion-gas analysis methods.

***) Leco Corp. Svenska AB

REFERENCES

1. H. Mehner, German pat. 88999, Sep 30, 1896
2. J.G. Lee amd I.B. Cutler, "Reaction in the SiO_2-C-N_2 system" p.175 in Nitrogen Ceramics. Edited by F.L. Riley, Noordhoff, Leyden, The Netherlands, 1977
3. A. Szweda, A. Hendry and K.H. Jack, "The Preparation of Silicon Nitride from Silica by Sol-Gel Processing", p.107 in Special Ceramics, Vol 7, Edited by P. Popper, Brittish Ceramic Research Association, Stoke-on-Trent, England, 1981
4. M. Mori, H. Inoue and T. Ochiai, "Preparation of Silicon Nitride Powder from Silica", in Progress in Nitrogen Ceramics, Edited by F.L. Riley, Martin Nijhoff, The Hague, The Netherlands, 1983.
5. Shi-Chang Zang and W.R. Cannon, "Preparation of Silicon Nitride

from Silica", J. Am. Ceram. Soc.,67 (10), 691-5 (1984)
6. T. Johansson, U.S. Pat. 4,530,825, July 23, 1985.
7. M. Ekelund B. Forslund, T. Niklewski, G. Eriksson G. Hällgren, T. Johansson and M. Hatcher "Establishment of the Si_3N_4/SiC Phase Boundary in the Si-C-O-N System, at Elevated Pressures", No.392,p.567 in Extended Abstract Catalogue 86-1, The Electrochemical Society's 169th Meeting, May 4-9, 1986, Boston, Ma, USA, 1986.
8. S. Motio and S. Sasaki, "Effects of N_2 and CO an the Syntetic Reaction of Silicon Nitride from Silica., Denki Kagaku 54, 170-2 (1986).
9. M. Ishida and C.Y. Wen, "Comparison of Kinetic and Diffusional Models for Solid -Gas Reactions", AIChE J. 14, 311-17 (1968).
10. J. Szekely and J.W. Evans, "A Structural Model for Gas-Solid Reactions with a Moving Boundary", Chem. Eng. Sci. 25, 1091-1107 (1970).
11. V. Figush and T. Licko, "Synthesis of Silicon Nitride Powder By Carbothermal Nitridation of Silica", p 517 in High Tech Ceramics, Edited by P. Vincenzini, Elsevier Publ. B.V., Amsterdam, The Netherlands, 1987.
12. I. Higgins and A. Hendry, "Production of β'-Sialon by Carbothermal Reduction of Kaolinite", Br. Ceram. Trans. J.,85, 161-6 (1986).

RELEVANCE OF EXCESS CARBON INCLUSIONS IN SINTERED SILICON CARBIDE

R. Hamminger
Hoechst AG
6230 Frankfurt/M. 80
Federal Republic of Germany

ABSTRACT

The role of C-additions in pressureless sintering of silicon carbide (SSiC) is commonly explained by the removal of SiO_2-layers on the starting powders. Practically, offering more C than necessary in the case of stoichiometric removal in order to result in a complete deoxidation process, inclusions of excess free C are a general feature of the microstructure of SSiC. This phenomenon was studied by high-resolution Auger electron spectroscopy on UHV-exposed fracture surfaces as well as by high-resolution transmission electron microscopy of (B, C)-doped SSiC-materials. The analytical results are discussed with respect to their relevance on materials evaluation and processing as well as on their impact on materials properties.

INTRODUCTION

The bonding of SiC is dominantly covalent with only 9 to 12% ionic character /1/. Thus, the resulting high bonding energies prevent diffusion and, consequently, sintering of pure SiC without applied very high pressure does not take place. In order to obtain high-density materials, pressureless sintering of SiC involves the use of special sintering additives. Most practicable are B and C /2, 3/, Al and C /4 - 6/ or even B, Al and C. Although many efforts have been undertaken to establish new types of additives, the addition of B or B-compounds together with C has remained most successful /7/. Be can also be used as an very effective sintering aid /8/ but plays no role in practice.

Although many proposals have been made in order to explain the means by which these additives activate sintering, no unambiguous theory about the effective sintering mechanism has so far been established. The earliest explanation about the effective sintering mechanisms of SSiC and the role of B- and C-additions was based on thermodynamic considerations /2, 9/. It has been assumed that B forms segregation layers at the grain boundaries which are responsible for a certain decrease of the grain boundary energy. Both, the SiO_2-coatings of the primary SiC-particles and the free Si (as a natural impurity element of SiC or being originated from the decomposition of SiC) are going to react with each other to form secondary SiC. During this procedure, the surfaces of the grains are free of any SiO_2-layers at least at an intermediate state. As a consequence, the energetic ratio of grain boundary to surface energy changes and the process of densification takes place.

A few years later it has been proposed /10/ that B reacts with C-additions during sintering by formation of B_4C, B_nC_m or even (B_nC_m + C)-compounds which may segregate as thin, possibly transient layers on the grains. The B-compounds are going to dissolve in SiC and consequently, by formation of gradients of concentration, are able to move along grain boundaries. Thus, the B-compounds may be completely dissolved during sintering and will disappear during heating up to maximum temperature. The final high density is then achieved by a volume diffusion process.

The explanation which has very recently been given /11/ is based on more general thermodynamic calculations. The sintering additives, principally, should not promote decomposition of SiC as well as they should not evaporate during sintering. Thus, it has been stated recently that many metal-type additives do not decompose SiC and can, at least theoretically, be used as densification aids in pressureless sintering of SiC.

However, the addition of C as graphite, C-black or pyrolysable resin was found to be necessary in any way of processing, irrespective of the type of doping elements used, i. e., B, Be and Al. Its role in densification is commonly explained by the deoxidation of the starting powders. In order to achieve a rather complete deoxidation process, practically there is more C-additon offered to the ceramic mass than necessary in case of stoichiometric removal of SiO_2. As a consequence, microstructural inclusions of excess free C are a significant but not very well-known characteristic feature of SSiC. This is due to the small spatial dimensions of the inclusions and the low atomic number of C which both cannot be analysed by routine methods. Though inclusions of excess free C have also been observed elsewhere /12, 13/ there was so far no detailed information on their relevance with respect to materials evaluation as well as on their impact on materials properties. However, it has been observed that non-stoichiometrical additions of excess C are influencing the microstructural development, mainly with respect to grain growth control /3/. In fact, microstructural inlcusions of excess free C may act as grain growth controllers due to the additional energy which is necessary to overcome the obstacles. This function of grain growth control of course is strongly dependent on the size, distribution and homogeneity of the inclusions. Consequently, SSiC-materials of high mechanical performance can only be obtained by homogeneous, fine-grained microstructures. Regarding the level of strength of SSiC which is about 450 MPa (4-point-bending), inclusions of this dimension observed are usually not under discussion with respect to the mechanical properties. In contrast, the inclusions of excess free C may have to be considered as detrimental in the case of their inhomogeneous distribution leading to the occurrence of agglomerate-type critical flaw dimensions. In fact, the principal goal of this contribution was to give a detailed description of the appearance and micromorphology of excess free C-inclusions in different (B, C)-doped SSiC-materials in order to comment their relevance on materials evaluation as well as on their impact on materials properties.

MATERIALS AND EXPERIMENTAL PROCEDURE

Specimens have been prepared by adding amorphous B* and C (in the form of furnace black or pyrolysable resin) to submicrometer alpha-SiC-powder**. After homogenisation by attrition milling bars with dimensions 10 by 20 by 50 mm have been cold isostatically pressed. Sintering has been performed in Ar. The procedure produced materials of high densities >3.1 g/cm^3 and homogeneous microstructures. In order to further support materials evaluation and processing by grain growth control during sintering, specimens have been prepared with different C-additive concentrations and were densified under identical sintering conditions.

The inclusions of excess free C were examined by means of imaging high-resolution scanning Auger electron spectroscopy*** on UHV-exposed fracture surfaces to avoid any detectable contamination. The optimum lateral resolution of the Auger equipment was less than 20 nm. The information depth of the Auger method, 0.5 to 3.0 nm typically, allowed monolayer-sensitive analysis without any preparation or contamination effects /14/. Reproducibility and detection sensitivity are in the range of 0.2 to 0.5 at.% in the case of light elements. The acceleration voltage of the primary electrons was typically 10 kV; the vacuum was about 8 x 10^{-10} Pa. Quantification of the Auger analysis was done using known elemental sensitivity factors.

To some extent the inclusions of excess free C were also examined by high-resolution scanning transmission electron microscopy**** on ion-thinned specimens.

* Grade I, H. C. Starck Berlin, Werk Goslar (FRG)
** A10, H. C. Starck Berlin, Werk Goslar (FRG)
*** Perkin-Elmer PHI 660
**** Philips 300 kV - CM30

RESULTS AND DISCUSSION

The microstructural inclusions which have first been observed by light microscopy occurring like pores were identified by X-ray diffraction to consist of excess free C in the form of graphite /14/. The graphite-type inclusions have been found to be very differently sized depending on the starting concentration of the C-addition as well as on the degree of homogeneity of its distribution in the green compact. The range of typical dimensions is from <0.1 up to about 5.0 µm in diameter at additive concentrations from 1.0 up to 8.0 vol%. The dependency of the additive concentration on the formation and size of the C-inclusions as well as its influence on grain growth is shown in Fig. 1, which is based on theoretical calculations (r_C = E/V; r_C = grain diameter, E = inclusion diameter, V = inclusion volume) /15/. As visible, in order to achieve a high quality, homogeneous product of an average grain size between 2.0 and 6.0 µm, the inclusions of excess free C should not exceed dimensions of more than 0.1 µm in diameter in order to ensure its grain growth controlling activity. This critical dimension of the C-inclusions is going to increase in the case of increasing contents of C-addition. For example, using 5 vol% C-addition the inclusions of 0.1 µm in diameter will lead to grain sizes of only about 2 µm, whereas the inclusions of 0.3 and 1.0 µm in diameter are going to result in grain sizes of 6 and 20 µm, respectively.

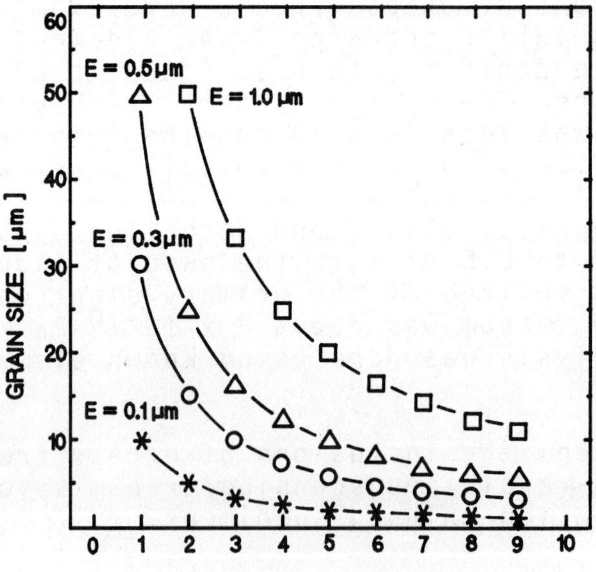

Fig. 1. Dependency of grain size (r_C) on inclusion diameter (E) and inclusion volume (V)

Practically, however, the microstructures of today's even high quality SSiC-materials exhibit inclusions of excess free C being larger than proposed theoretically (Fig. 1), as shown by fracture surface in Fig. 2. According to this observation it can be assumed that the inhibition of grain growth is practically not very effective. Again, this is a question of processing asking for an improved homogenisation and a better, i. e., very fine dispersion of the C-addition. So far, improvements in the grain growth control of SSiC have preferably to be seen in the sintering process and temperature profiling itself.

The dramatic influence of the concentration of the C-addition on the microstructural development and grain growth is shown by chemically etched surfaces in Fig. 3, the samples having been sintered under identical conditions. Exaggerated grain growth was observed when the total concentration of the C-addition was not sufficient to ensure complete deoxidation of SiO_2, the critical concentration being strongly dependent on the characteristics of the primary powders used.

Fig. 2. Fracture surface of SSiC

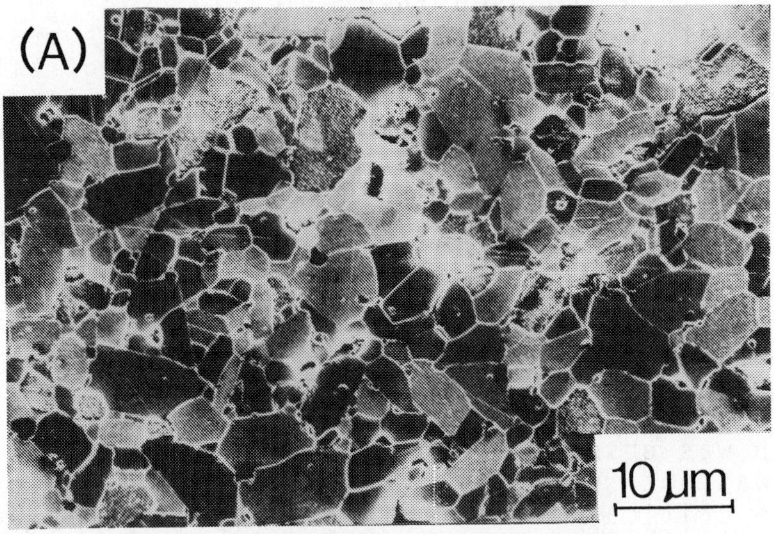

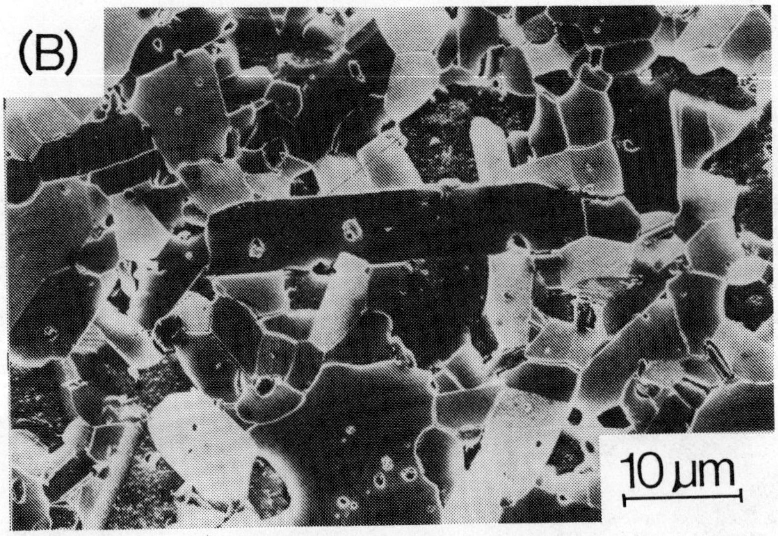

Fig. 3. Polished and etched surfaces of SSiC doped with (A) 3.5 wt% C and (B) 1.5 wt% C

Small-sized inclusions of excess free C as considered theoretically (Fig. 1) as effective grain growth inhibitors have been investigated by transmission electron microscopy. Especially inclusions of less than 0.1 µm in diameter have been observed preferably at triple point regions. The typical example of the micromorphology of an inclusion of <1.0 µm in diameter is shown in Fig. 4.

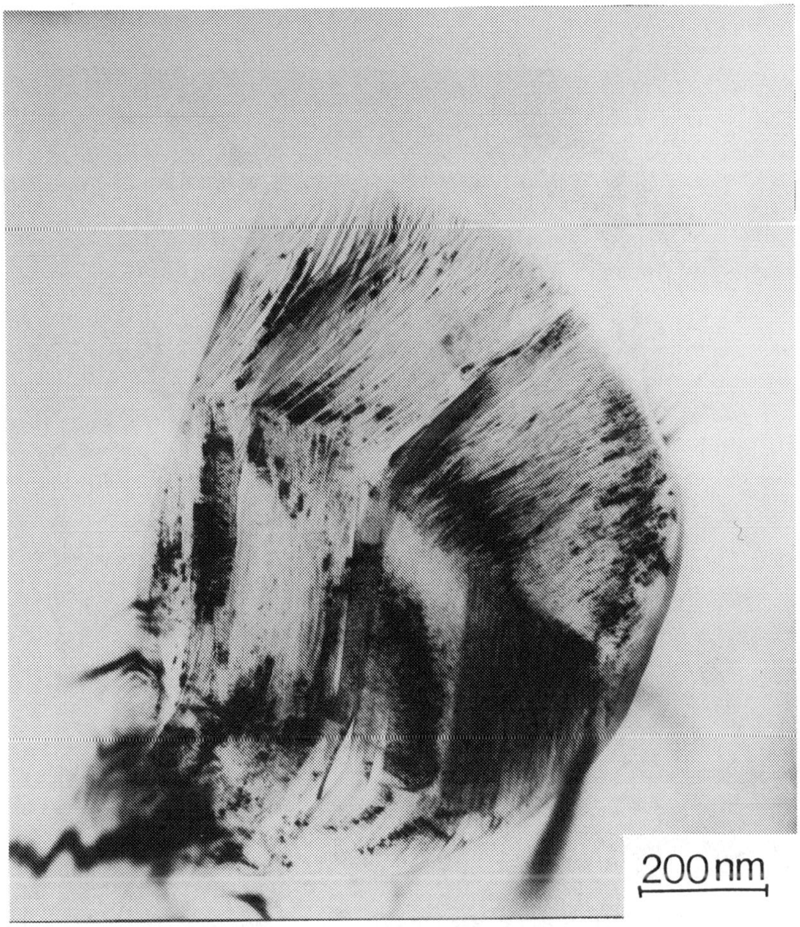

Fig. 4. Particle of excess free C (TEM)

Going into more detail by high-resolution observation, the micromorphology of small-sized inclusions of excess free C was found to exhibit an intricate, filamentary nature as shown in Fig. 5. As visible, selected area diffraction pattern of the inclusion revealed well defined rings from the graphite as well as some SiC-spots. In contrast, no evidence for the occurrence of amorphous C could be observed.

Details about the microchemistry, distribution and microheterogeneity of the inclusions of excess free C have been analysed by high-resolution scanning Auger electron spectroscopy. An example is given in Fig. 6 showing a typical fracture surface with differently sized inclusions and the appropriate distribution maps of C and Si. The coincidence of the

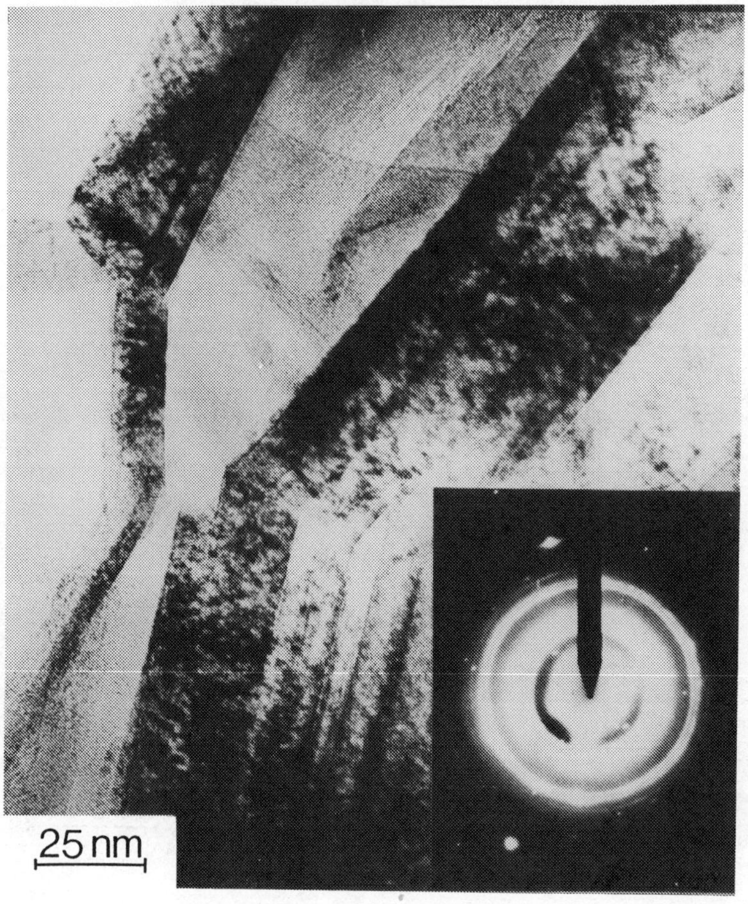

Fig. 5. High-resolution observation and selected area diffraction pattern of an inclusion of excess free C

dark particles in the micrograph and the maps (C-increases, Si-decreases) is conclusive. By point analyses, the only detectable elements are Si, C and B. The spectra of the marked spots in Fig. 6(A) are represented in Fig. 7(A), (B) and (C). Stoichiometric SiC was analysed at transgranular regions (Fig. 7(A)). Inside the inclusions, pure graphite-type C-peaks were observed (Fig. 7(B)). Some of the inclusions, in contrast, are graphite but also contain some B, the concentration being about 6 to 7 at.% (Fig. 7(C)). Precipitated B inside the graphite may be assumed to be present, as well as small portions of B_4C beside excess free C /13, 14/.

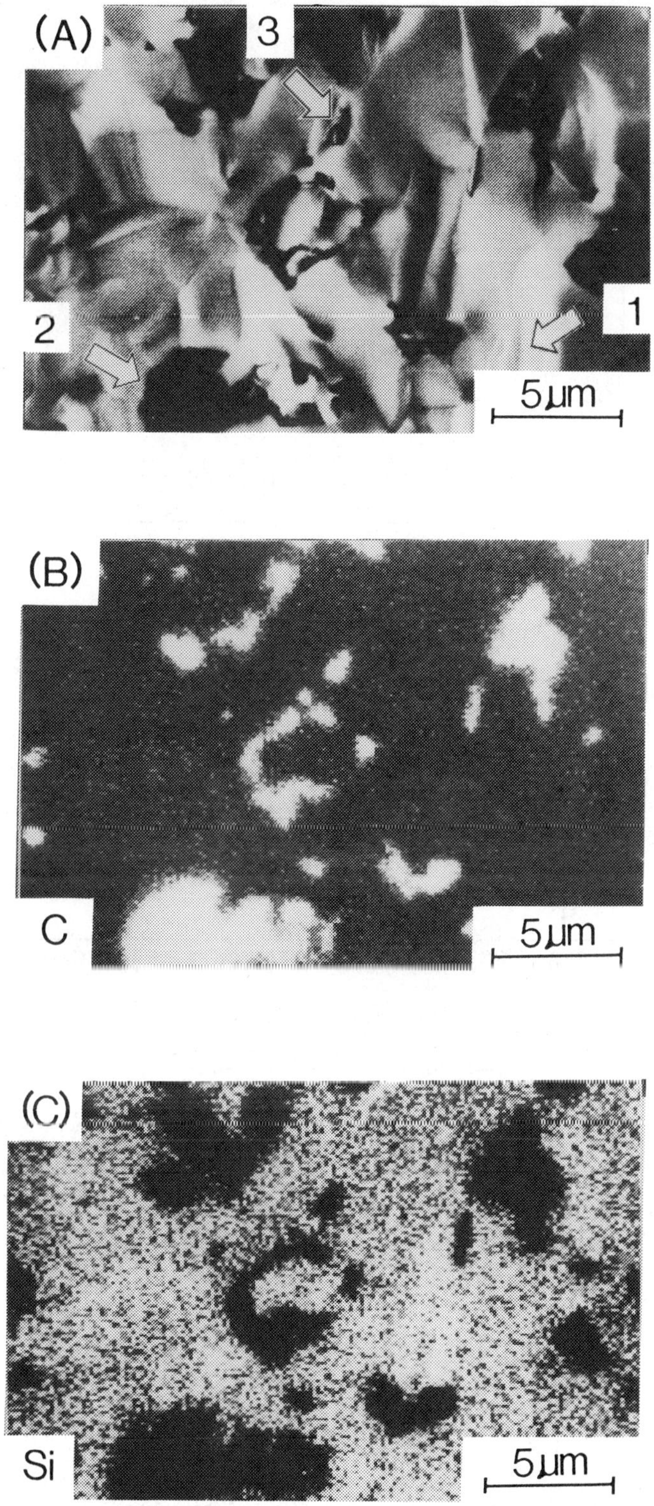

Fig. 6. Fracture surface of SSiC (A) and appropriate distribution maps for C (B) and Si (C) of (A)

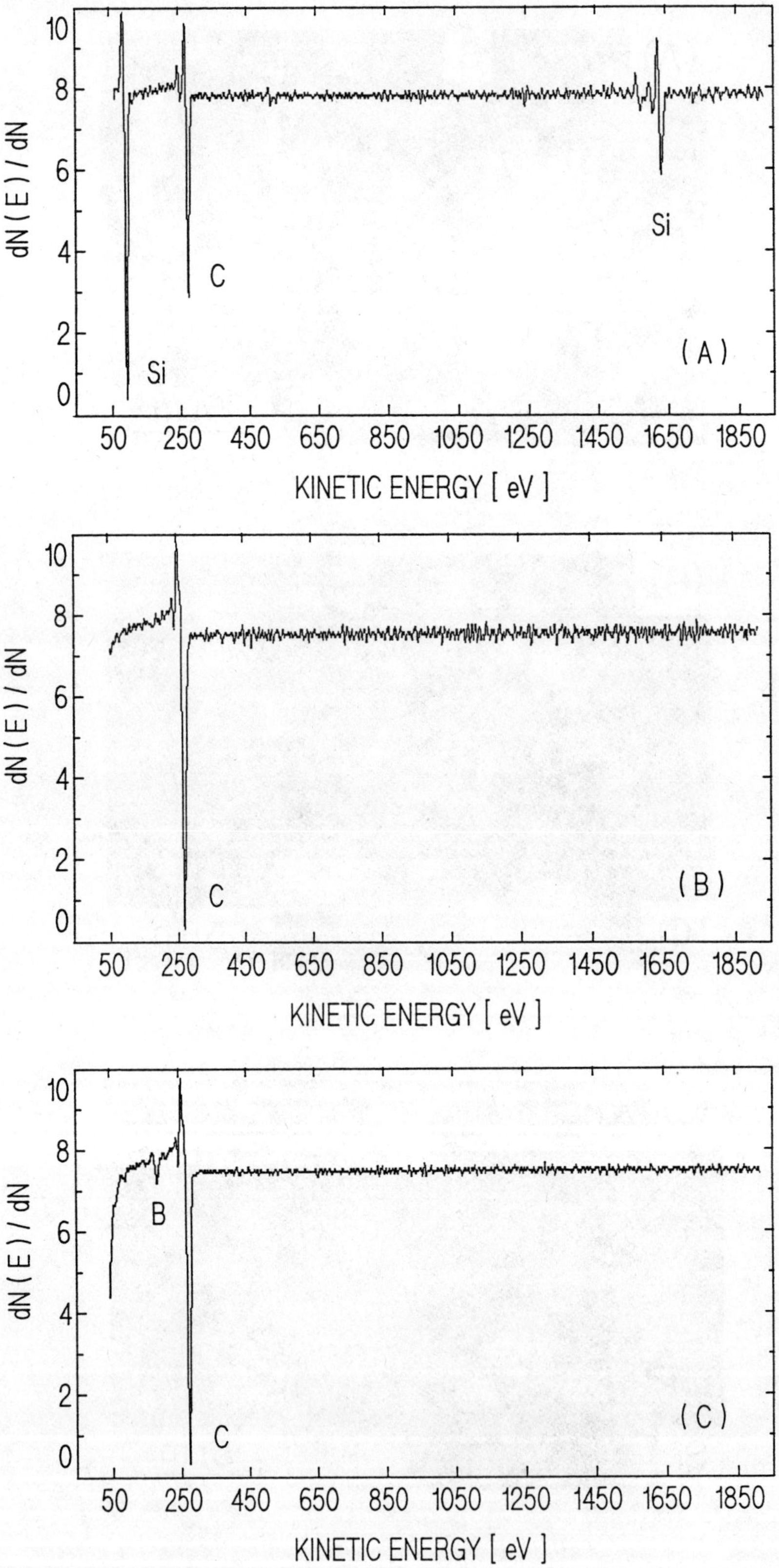

Fig. 7. High-resolution spectra of UHV-exposed fracture surface of SSiC (marked areas: (A) 1, (B) 2, (C) 3; Fig. 6 (A))

CONCLUSION

Though inclusions of excess free C have also been observed elsewhere, there was so far no detailed information on their relevance with respect to materials evaluation as well as on their impact on materials properties. Here, the analytical problems of the detection and analysis of partly very small inclusions consisting of the low Z element C even in combination with B have been overcome by the application of surface sensitive and high-resolution methods. The results of this methods have been obtained from final products of SSiC. It can be assumed, however, that equivalent agglomerates are also present in the powder compacts before sintering.

Regarding the level of strength of SSiC which is about 450 MPa (4-point-bending), inclusions of this dimension observed are usually not under discussion with respect to the mechanical properties. In contrast, the inclusions of excess free C may have to be considered as detrimental in the case of their inhomogeneous distribution leading to the occurrence of agglomerate-type inclusions of critical flaw dimensions. Beside the formation of agglomerates, it has to be taken into account that even high performance SSiC-materials are not free of single macro-inclusions of excess free C, which of course are going to act as critical flaw dimensions. The same is true, consequently, with respect to the occurrence of exaggerated grain growth. Finally, it can be assumed that inclusions of excess free C may influence the crack deflection behaviour as well as slow crack growth at high temperatures.

According to the observation of inclusions of excess free C which are much bigger than proposed in Fig. 1 it can be assumed that the inhibition of grain growth is practically not very effective. In any case it can be assumed that the efficiency of C-additions during sintering has to be improved by increasing the homogeneity of the materials. Generally, improved homogeneity means improved deoxidation of the primary powder particles, combined with a higher efficiency of the sintering additives and diminution of the number and size of inclusions of excess free C in the final product.

REFERENCES

/1/ A. R. Verma and P. Krishna, "Polymorphism and Polytypism in Crystals", John Wiley Inc., New York, London, Sydney, 92-135 (1966).

/2/ S. Prochazka and R. M. Scanlan, "Effect of Boron and Carbon on Sintering of SiC", J. Am. Ceram. Soc., **58**, 72 (1975).

/3/ S. Prochazka, "Sintering of Silicon Carbide", pp. 239-52 in Ceramics for High Performance Applications. Edited by J. J. Burke, A. E. Gorum and R. N. Katz. Brook Hill Publishing Co., Chestnut Hill, MA, USA, 1974.

/4/ W. Böcker, H. Landferman and H. Hausner, "Sintering of Alpha Silicon Carbide with Additions of Aluminium", Powder Met. Int., **11,** 83-85 (1979).
/5/ H. Hausner, "Pressureless sintering of Non-Oxide Ceramics", pp. 582-95 in Energy and Ceramics. Edited by P. Vincenzini, Elsevier, Amsterdam, 1980.
/6/ K. A. Schwetz and A. Lipp, "The Effect of Boron and Aluminium Sintering Additives on the Properties of Dense Sintered Alpha Silicon Carbide", Science of Ceramics, 149-58 (1980).
/7/ Y. Tajima and W. D. Kingery, "Solid Solubility of Aluminium and Boron in Silicon Carbide", J. Am. Ceram. Soc., **65,** C-27-C-29 (1982).
/8/ S. Prochazka, "Sintering of Silicon Carbide", pp. 421-31 in Mass Transport Phenomena in Ceramics. Edited by A. R. Cooper and A. H. Heuer. Plenum Press, New York, 1975.
/9/ C. Greskowich and J. H. Rosolowski, "Sintering of Covalent Solids", J. Am. Ceram. Soc., **59,** 336-43 (1976).
/10/ H. Suzuki and T. Hase, "Some Experimental Consideration on the Mechanism of Pressureless Sintering of Silicon Carbide", pp. 345-65 in Factors in Densification and Sintering of Oxide and Non-Oxide Ceramics. Edited by S. Somiya and S. Saito. Hakone, Japan, 1978.
/11/ K. Negita, "Effective Sintering Aids for Silicon Carbide Ceramic: Reactivities of Silicon Carbide with Various Additives", J. Am. Ceram. Soc., **69,** C-308-C-309 (1986).
/12/ R. F. Davis, J. E. Lane, C. H. Carter, J. Bentley, W. H. Wadlin, D. P. Griffis, R. W. Linton and K. L. More, "Microanalytical and Microstructural Analyses of B- and Al-Regions in Sintered Alpha-SiC", Scanning Electron Microscopy, **3,** 1161-67 (1984).
/13/ K. L. More, C. H. Carter, J. Bentley, W. H. Wadlin, L. Lavanier and R. F. Davis, "Occurrence and Distribution of Boron-Containing Phases in Sintered Alpha-Silicon Carbide", J. Am. Ceram. Soc., **69,** 695-98 (1986).
/14/ R. Hamminger, G. Grathwohl and F. Thümmler, "Examination of Carbon Inclusions in Sintered Silicon Carbide", Int. J. of High Technology Ceramics, **3,** 129-41 (1987).
/15/ W. D. Kingery, H. K. Bowen and D. R. Uhlmann, "Introduction to Ceramics", John Wiley and Sons, New York, Chichester, Brisbane, Toronto, Singapore, 457 (1985).

α'-SIALON CERAMICS CONTAINING MIXED MODIFYING CATIONS

M. Redington, K.P.J. O'Reilly and S. Hampshire.
Materials Research Centre, NIHE, Limerick, Ireland.

ABSTRACT

α'-sialons are solid solutions of α-silicon nitride with Al substituting for Si in the Si_3N_4 network in the structure and charge compensation occurring by "stuffing" other cations into interstitial sites in the structure. This paper reports on the formation of α'-sialon containing rare earth cations such as Nd, along with Ca or Li ions. The cations in the form of oxides initially act as densification aids but are then subsequently incorporated into the α-structure, thus reducing the amount of grain-boundary glass. The difficulties of densification of pure α'-sialons are outlined and the possibility of producing improved materials is discussed.

INTRODUCTION

Silicon nitride is difficult to sinter. At a temperature high enough for atomic rearrangement to occur, the material tends to decompose. Several methods, such as hot pressing (HPSN), reaction bonding (RBSN and SRBSN), hot-isostatic pressing (HIP) have been used to achieve good densification with low weight losses. However, these methods are not really suitable for conventional mass production and the production of complicated shapes. In this context, pressureless sintering is a more desirable production method.

Silicon nitride can be sintered via a solution-precipitation reaction. To enhance the formation of liquid, metal oxides are added which react with surface silica on the silicon nitride to give sufficient metal-silicon oxynitride liquid to enable a solution - precipitation to occur[1-3]. Metal oxides that are particularly useful are CaO, MgO and Y_2O_3. Unfortunately, these densifying aids usually form a second phase after sintering which can produce less than optimum mechanical properties, especially at high temperatures[1,2,4].

The use of alumina as an additive is a special case[5]. The AlO_4 tetrahedron is roughly the same size as the SiO_4 and this allows Si to be replaced by Al and N by O with valency compensation. Thus, there is a range of isomorphous substitituion where up to two-thirds of the Si can be replaced by Al and the β-Si_3N_4 crystal structure is retained. These ceramics are termed β'-sialons. If sialons are densified with metal oxides a large volume of liquid remains as a second phase, because the modifying cations cannot be accommodated into the β-structure[5,6]. Grain boundary modification, e.g. devitrification by post-sintering heat treatment, has resulted in ceramics with reasonably good mechanical properties[5,7].

An alternative route to pressureless sintering sialons is possible whereby the densifying aid, i.e. the metal oxide, is incorporated into the α-silicon nitride structure. This can occur when the metal oxide is Al_2O_3. Charge compensated substitution of Si and O by Al and N takes place and a modifying cation is accommodated into an expanded alpha-type Si_3N_4 crystal structure thereby forming α'-sialon ceramics[5,8]. The α structure differs from the β structure in that continuous channels parallel to the c axis in β are broken up into elongated holes in the α structure. These holes are cages in which another modifying or compensating cation (m) can be trapped. There are two such cages in the unit cell giving rise to a general structural formula of:

$$M_x\ Si_{12-(m+n)}Al_{(m+n)}\ O_n\ N_{16-n}$$

where $x < 2$, and M includes cations such as Li, Ca, Y and other rare earth ions.

Theoretically, the second phase can be eliminated by this accommodation; however, there are two main problems in sintering. Firstly, the region where α'-sialons are formed is towards the nitrogen rich end of the system and there is not much liquid available for densification. Secondly, as the reaction proceeds Al, O and the modifying cation, such as Y, are incorporated into the structure, and there is progressively less liquid to promote densification[9].

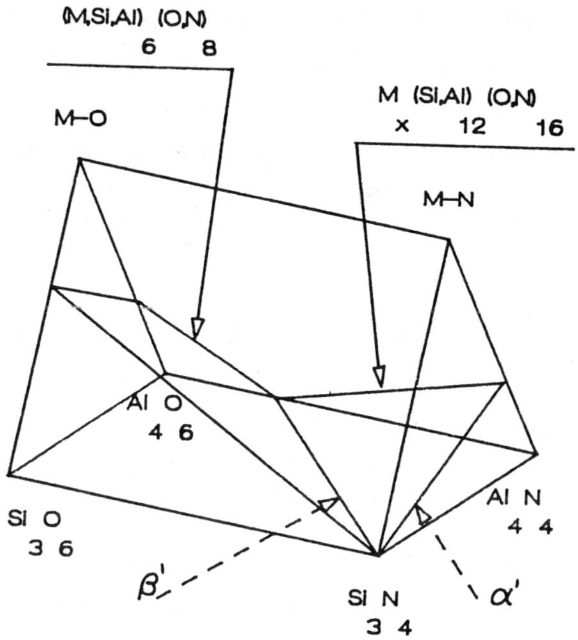

Fig. 1 The M-Si-Al-O-N Jaenecke prism showing the $(M,Si,Al)_6(O,N)_8$ and $M_x(Si,Al)_{12}(O,N)_{16}$ planes.

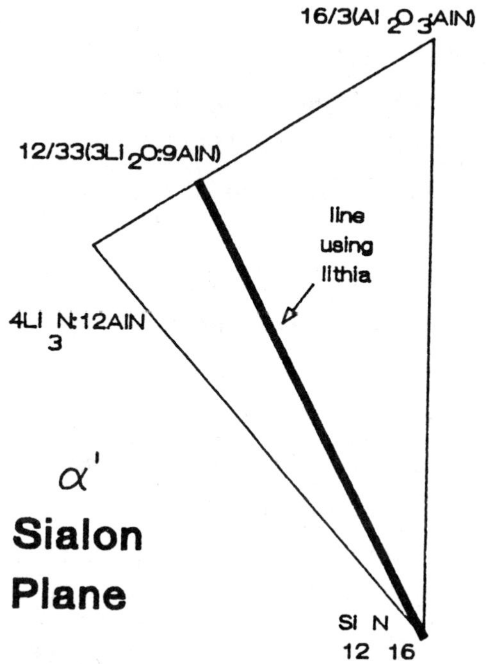

Fig. 2 The α-sialon plane in the Li-Si-Al-O-N prism.

This paper reports on the formation of α'-sialon ceramics using mixtures of two metal oxides. The metal oxides studied are calcia and neodymia, both known to be good α'-sialon formers, and lithia. Some characteristics of lithium based α'-sialons are initially discussed.

EXPERIMENTAL PROCEDURE

The compositional representation of sialon ceramics is adequately expressed using the Si-Al-O-N phase diagram based on equivalents. The addition of one modifying cation results in the use of a Jaenecke prism. See Fig. 1[10].

The compositions investigated were made up using mixes of Si_3N_4, AlN and metal oxides, but the carbonates of Li and Ca were used to form their oxides with compensation for CO_2 evolution. With the exception of the 'pure' lithium α'-sialon all compositions lie on the intersection line between the oxide-Si_3N_4-AlN plane and the Si_3N_4-4/3(Al_3O_3N)-MAl_3N_4 plane.

The silicon nitride (LC12 grade, particle size ~1μm) used in this investigation was supplied by H.C. Starck-Berlin (Goslar, W. Germany). The calcined neodymia used was 99.99% pure, supplied by Rare Earth Products (England). The lithium and calcium carbonate were supplied by BDH Chemicals. The particle size for the AlN was in the range 3-8μm.

The powders were mixed in iso-propanol, using a magnetic stirrer for thirty minutes, after which the alcohol was evaporated and dry mixing was continued for a further thirty minutes. Roughly 1g quantities of powder were uniaxially cold pressed to give pellets of 10mm dia. and 5mm height (approx.). The pellets were embedded in boron nitride in alumina crucibles and fired under 1 atm. N_2 in a lanthanum-chromite resistance furnace. Firing was carried out in the range 1450 - 1750°C for 30 minutes. The pellet and crucible were allowed to soak at 1200°C for thirty minutes prior to insertion into the hot zone of the furnace.

Densities were measured using a mercury displacement balance. The phases present in the fired pellets were determined using XRPD from diffraction profiles. Cobalt Kα radiation was used in conjunction with automatic peak detection. The lattice parameters of α′ were calculated using a minimum of ten strong peaks where these could be unequivocally identified.

The fracture surfaces were examined using scanning electron microscopy.

RESULTS & DISCUSSION

Li-sialons

Compositions on the α-sialon plane (Figs. 1 & 2) using Li_2O lie along the join $Si_{12}N_{16}$-$Li_8Al_{12}O_4N_{12}$. Compositions along this line, where the Li atomic content per unit cell is between 1 and 2, give mixtures of α′ and β′ on firing at 1750°C for 30 minutes. Prolonged heating at 1750°C for up to 4 hours causes any α′ to convert to β′. Weight losses > 6% are the norm and it is considered that Li loss causes a reduced α′ stability and a conversion to a more stable β′ form. This is confirmed by Hampshire et al[8] and Jama et al[11], where α′ contents rarely exceeded 30% using Li as the sole modifying cation. Jama et al.[11] have suggested that β′ and some eucryptite form on the

$Si_{12}N_{16}$-$Al_{16}O_{24}$-$Li_{48}O_{24}$ plane near the above compositions, although no eucryptite could be discerned in the present study.

High weight losses suggest that volatile Li_3N is lost before it can combine with Si_3N_4 to form $LiSi_2N_3$ as described below:

$$Si_3N_4 + 6Li_2O \longrightarrow 4Li_3N + 3SiO_2$$

$$Li_3N + 2Si_3N_4 \longrightarrow 3LiSi_2N_3$$

In the presence of Al, more Li is dissolved in a metal-silicon oxynitride liquid; however, on prolonged heating further Li loss occurs yielding β-sialon with a lower overall Li content.

In contrast, high yield Li-α′-sialons have been made with a starting composition having a higher Al content approximating:

$$Li_{1.48}Si_{8.00}Al_{4.84}O_{0.8}N_{15.47}$$

This composition does not lie on the α-sialon plane because the metal: non-metal ratio = 0.789 as opposed to 0.75 and it is closer to the AlN corner of the Jaenecke prism. Reduced weight losses, <1.5 wt%, coupled with an 85+% α′ yield indicate that the increased Al content has a beneficial effect on the retention of Li. Pure α′ compositions lying off the α-sialon plane at more oxygen-rich compositions have also been reported by Chatfield et al.[12].

Formation of α′ (α′ and β′) sialons using mixtures of modifying cations

Four equivalent compositions in the mixed cation α-sialon plane were prepared. Their starting compositions are given in Table 1, as formulae based on the α-$Si_{12}N_{16}$ structure. Composition 1 (Li) is used for comparison.

Table 1. α-sialon compositions investigated.

		M	Si	Al	O	N	Cation ratio	Title
0.	(Li)	1.48	8.0	4.84	0.8	15.47		Li("pure")
1.	(Li)	2.0	9.0	3.0	1.0	15.0		Li
2.	(LiCa)	1.0	9.5	2.5	1.0	15.0	Li:Ca=1:1	LiCa
3.	(LiNd)	1.0	9.0	3.0	1.0	15.0	Li:Nd=1:1	LiNd
4.	(CaNd)	0.75	9.124	2.875	1.0	15.0	Ca:Nd=1:1	CaNd

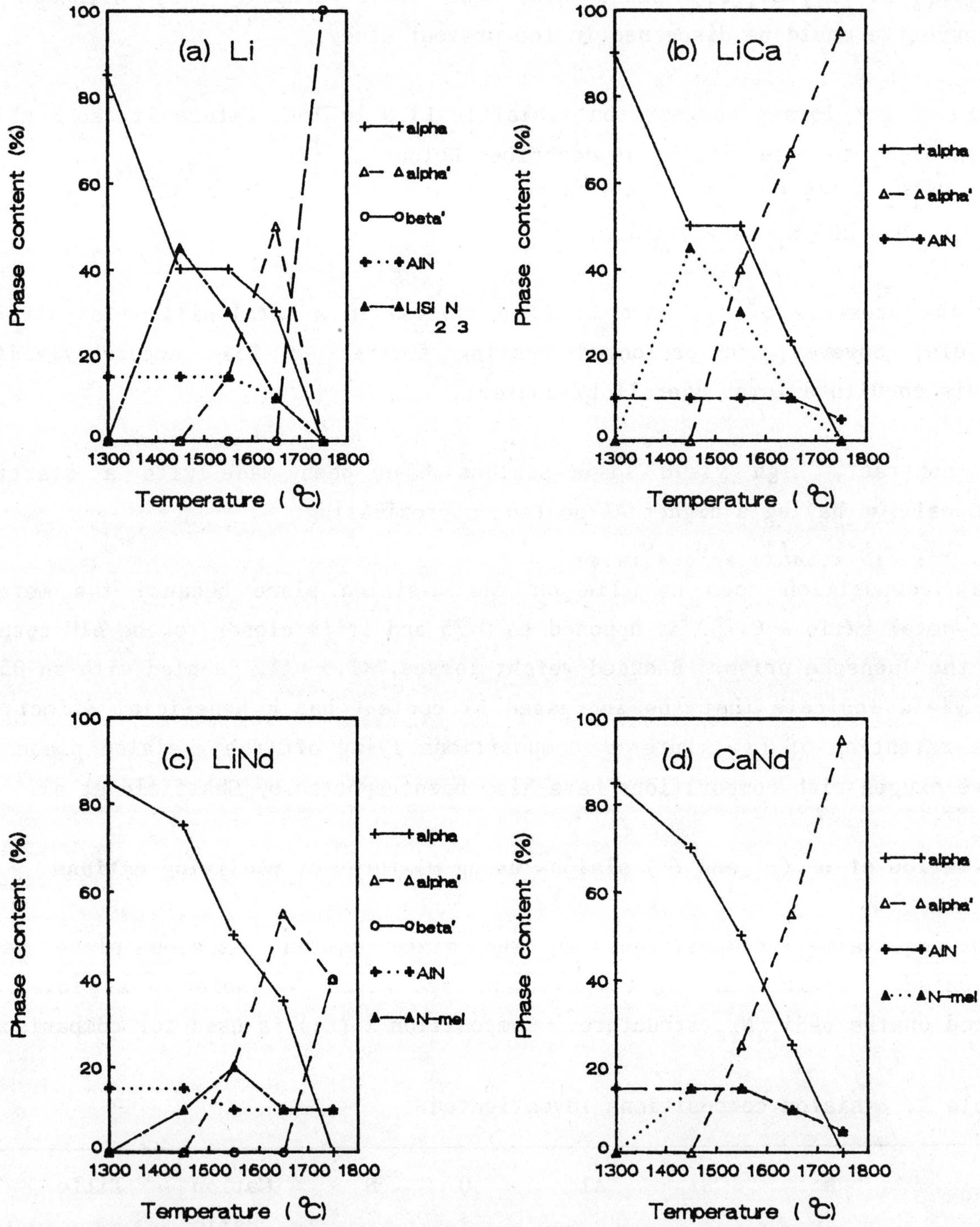

Fig. 3. Effect of firing temperature on phase content in α-sialon compositions investigated: (a) Li, (b) LiCa, (c) LiNd, (d) CaNd.

Fig. 3 shows the reaction sequences for these compositions in the range 1450 - 1750°C. Due to the slightly coarser nature of the AlN powder, some unreacted AlN was found to be present.

The two known good α′ forming cations when used in combination (Ca + Nd), give high yields of α′ at 1750°C. The major detected intermediate phase at 1450-1600°C was a Nd-N-melilite ($Nd_2Si_3O_3N_4$). Mixtures of both Li and Ca (LiCa) and Li and Nd (LiNd) act differently. The LiCa mixture gives a very good yield of α′ phase whereas LiNd resulted in a 50:50 ratio of α′ and β′. With such a concentration of Nd on its own, a higher yield of α′ would have been expected and, thus, the presence of Li changes the way in which α′ formation occurs. In the context of the suggested effect of Al on Li-α′ formation, it would seem that Ca in combination with Li offers superior performance over Nd+Li in the yield of α′. In complete contrast, the lithium only composition (Li) lying on the α-sialon plane, gives a maximum yield of α′(50%) at 1650°C and almost pure β′ at higher temperatures, suggesting that Li-based α-sialon compositions may be unstable at the higher temperatures.

On a practical level, weight losses for the mixtures of modifying cations were generally quite low compared to the lithium only composition (Fig. 4a). Also, for this latter composition (Li), bulk density is quite low compared with those for the mixed modifying cation compositions at all firing temperatures (Fig. 4b). This shows that lithium as a modifying cation on its own (present as Li_2O) is a poor sintering aid when compared to the cation mixtures, including those with Li.

The lattice parameters of all the α′-sialons were measured. The variations in lattice parameters with temperature are shown in Fig. 5. However, in the case of the Li only composition, lattice parameters could only be recorded for the Li-α′ formed at 1650°C. Even with such a small cation such as Li^+ (ionic radius 0.68 Å), there is a measurable expansion in the lattice parameters compared to α-Si_3N_4. This is reinforced by initial structure refinements on high yield Li α′-sialon (composition 0, off the α′-sialon plane) which indicate that the cell parameters for α′ formed at 1750°C are similar to the values obtained for Li-α′ found at 1650°C.

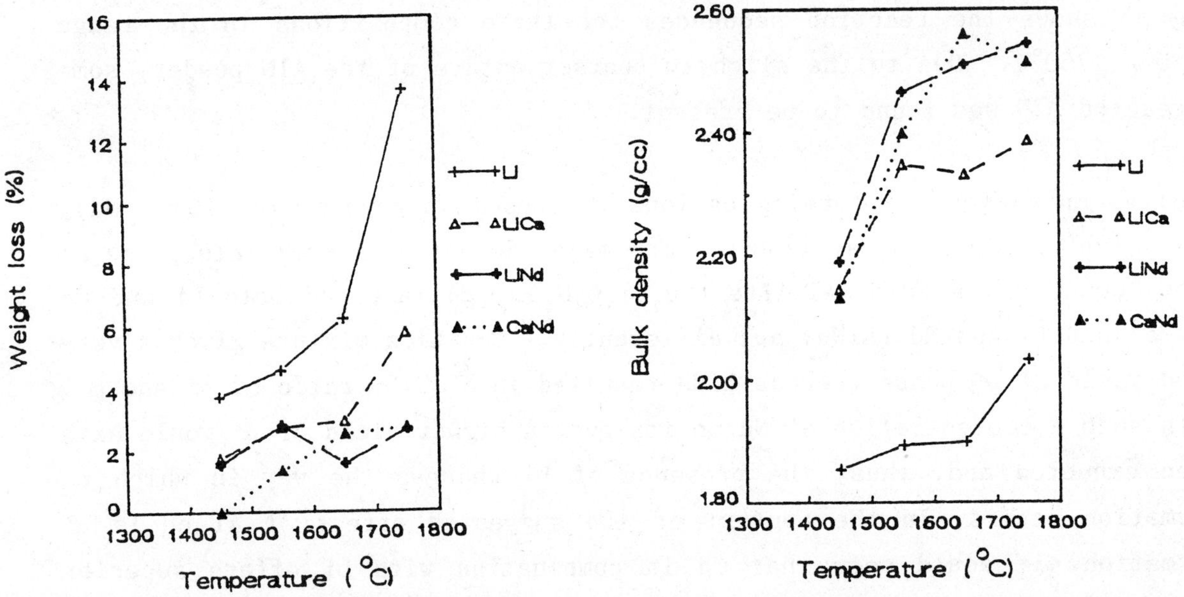

Fig. 4. Effect of firing temperature on (a) weight loss, (b) bulk density in α-sialon compositions with mixed modifying cations.

The change in the cell parameters for the mixtures show the same type of increase and then decrease as reported by Walls and Thompson[13] and Sun et al[14]. The increase in the c parameter for the LiNd composition seems quite large compared to the other mixed cation compositions. Walls and Thompson suggest that the initial cell dimensions are larger because the initial grain boundary liquid precipitates a cation-rich or oxygen rich α′. As the reaction proceeds, more Si_3N_4 dissolves and a more nitrogen rich α′ is formed.

It is not known whether structural rearrangement or atomic diffusion takes place as the reaction proceeds to give the final cell parameters. It is difficult to see how this could occur in the solid state. On the other hand, if nitrogen-rich α′ is deposited on more oxygen-rich crystals, diffractometer profiles should show considerable peak broadening for a range of differing α′-sialons formed during the course of the reaction. This is clearly not the case. Thus, the conclusion to be drawn is that silicon nitride dissolves more easily into the liquid at 1750°C than at 1650°C forming a homogeneous, more nitrogen-rich α′. Maximum cell parameters occur at higher temperature for Nd-α′ than for Ca α′. The cell

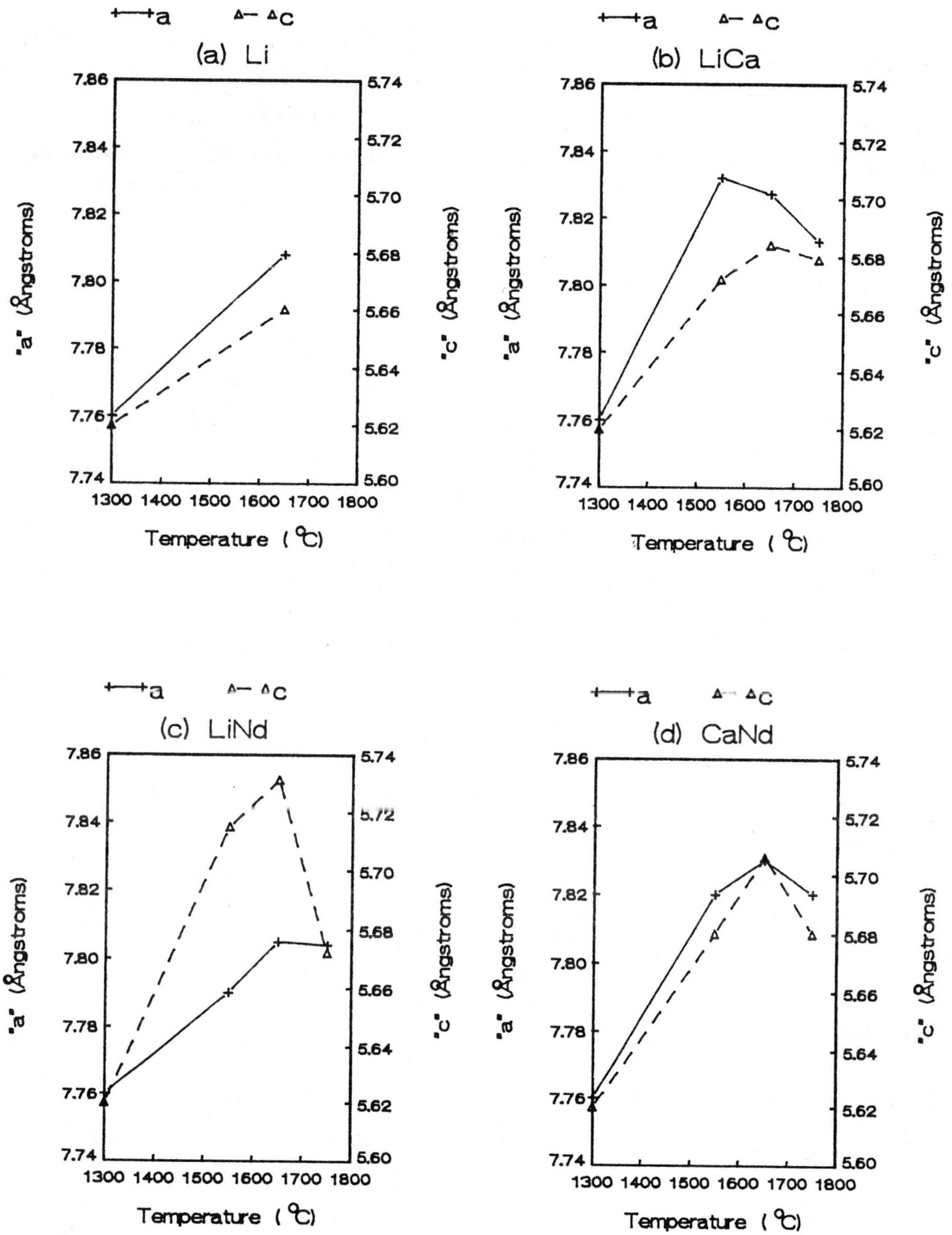

Fig. 5. Effect of firing temperature on lattice parameters of α-sialon compositions investigated: (a) Li, (b) LiCa, (c) LiNd, (d) CaNd.

maxima in the present study occur at 1550-1650°C, which is about 100°C greater than the temperature reported by Sun et al[13] for single cation α′ compositions. The temperature at which the cell parameter maxima occurs is related to the presence of the second cation although mixtures of modifying cations, added as metal oxides or carbonates, should, along with Al_2O_3, promote even lower temperature liquid formation.

Electron Microscopy

The fracture surfaces of two of the compositions were studied using scanning electron microscopy. The first of these was the LiCa sample. A typical micrograph is shown in Fig. 6a and this is similar to other α′-sialon microstructures observed by the authors. It has fine equiaxed grains in a tight structure. Extensive porosity of a fine nature is also evident, reflecting the low final value of bulk density recorded. The second sample was that of the Li composition(1). X-ray diffraction indicated almost pure β′ sialon. SEM of a fracture surface reveals the coarse structure (Fig.6b). A reasonable quantity of glass containing some bubbles is evident. Large pores, consistent with the coarseness of the microstructure and the high weight loss, are also observed. Qualitative examination indicates crystals with a high aspect ratio (>6). Fig. 6c is a view inside one of the large pores. Comparatively large plate-like crystals are visible, but it is difficult to characterise them fully. They are similar to 15R sialon[12] which is involved in the transformation of α′ and/or α to β′ during sintering.

CONCLUSIONS

Almost pure Li α′-sialons can be prepared from Al-rich compositions lying off the α′-sialon plane in the Li-Si-Al-O-N system. The use of Li_2O as a densifying aid does not appear to offer any benefits during the sintering of pure α′-sialon ceramics when used on its own since compositions on the α′sialon plane convert to β′ at higher temperatures. Li forms less stable α′-sialons in comparison to Ca. When used in combination with Li, Ca and Nd give improved yields of α′ and, in this context, the use of Ca as a mixed modifying cation offers superior performance over Nd.

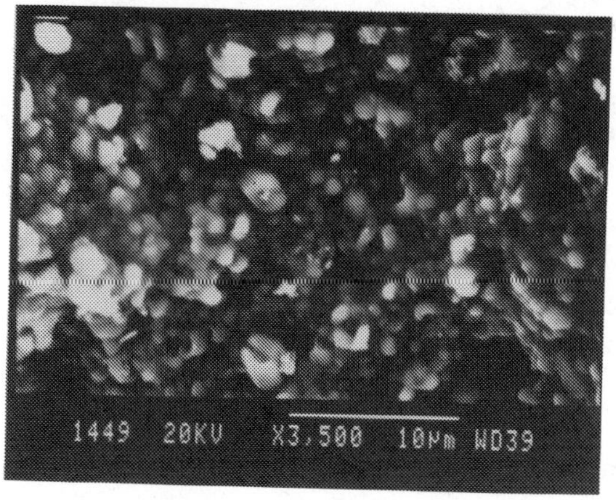

Fig. 6a SEM of (LiCa) composition showing equiaxed α-sialon grains.

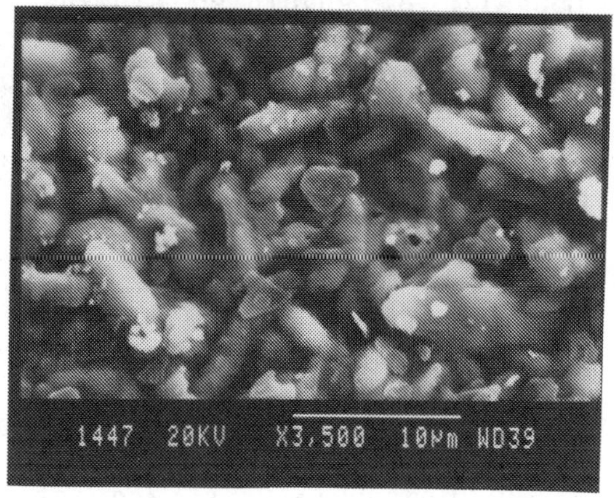

Fig. 6b SEM of Li composition (1) showing coarse β-sialon microstructure.

Fig. 6c SEM of Li composition (1) showing large pore with plate-like crystals evident.

ACKNOWLEDGEMENTS

The authors wish to acknowledge the assistance of Mr T Murdoch and Mr B Loughlin of Aughinish Alumina Ltd., Ireland, for the use of a Phillips X-ray diffractometer and also Dr M Pomeroy of NIHE, Limerick, for valuable discussion. This work forms part of a project funded by the National Board for Science and Technology (now EOLAS, the Irish Science and Technology Agency), Dublin Ireland.

REFERENCES

1. F.F. Lange, Am. Ceram. Soc. Bull., 62(12) (1983) 1369.

2. S. Hampshire, Metals Forum, 7(3) (1984) 162.

3. P. Drew and M.H. Lewis, J. Mater. Sci., 17 (1974) 262.

4. G. Gratwohl, Proc. 2nd Int. Conf. "Creep of Engineering Materials and Structures", eds. B. Wilshire and J. Owen, Pineridge Press (1984) p.565.

5. K.H. Jack, in "Non-oxide Technical and Engineering Ceramics", ed. S. Hampshire, Elsevier-Applied Science (1986) p.1.

6. M.H. Lewis, B.D. Powell, P. Drew, R.J. Lumby, B. North and A.J. Taylor, J. Mater. Sci., 12 (1977) 61.

7. M.H. Lewis, A.R. Bhatti, R.J. Lumby and B. North, J. Mater. Sci., 15 (1980) 103.

8. S. Hampshire, H.K. Park, D.P. Thompson and K.H. Jack,. Nature, 274 (1978) 880.

9. S. Slasor and D.P. Thompson, in "Non-oxide Technical and Engineering Ceramics", ed. S. Hampshire, Elsevier-Applied Science (1986) pp.223-230.

10. K.H. Jack, J. Mater Sci., 11(1976) 1135-1158.

11. S.A.B. Jama, D.P. Thompson and K.H. Jack, in "Special Ceramics 6", ed. P. Popper, B.C.R.A. (1975) p.299.

12. C. Chatfield, T. Ekstrom and M. Mikus, J. Mater Sci., 21(1986), 2297.

13. P.A. Walls and D.P. Thompson, in "Special Ceramics 8", ed. S.P. Howlett and D. Taylor, Brit. Ceram. Proc. No. 37 (1986), pp.35-50.

14. W.Y. Sun, P.A. Walls and D.P. Thompson, in "Non-oxide Technical and Engineering Ceramics", ed. S. Hampshire, Elsevier-Applied Science, (1986) pp.105-117.

PROCESSING IMPROVEMENTS IN A SINTERED Si_3N_4 MATERIAL

M.D. Pugh and R.A.L. Drew
McGill University, Montreal, Canada

ABSTRACT

The relationship between powder characteristics, processing parameters and mechanical properties of sintered Si_3N_4 has been investigated. The effect of surface finish of sintered bars on the bend strength was found to be small. Microscopy of fracture surfaces indicated that the majority of fracture initiation sites were related to microstructural inhomogeneities such as pores or low density regions rather than to machining flaws. In an attempt to reduce the size of these strength-controlling defects, various parameters in the processing route, including isostatic pressing pressure and burn-out of organics, were investigated and were found to affect the mechanical properties.

INTRODUCTION

The potential of structural ceramics in thermo-mechanical applications, such as heat engines, has been responsible for the development of a wide variety of materials, and among the most likely candidates are those compounds based on silicon nitride. These materials, which include the "sialon" ceramics, have high strength, excellent thermal shock resistance and good corrosion resistance. However, as with nearly all ceramics, silicon nitride is brittle and fails by fast fracture. Fracture initiation can occur from stress concentrations both at the surface of the sample and associated with internal flaws[1]. The mechanical properties of the material, such as strength and reliability, are strongly governed by the size, number and distribution of these flaws, which are most commonly present as pores, agglomerates, inclusions and surface cracks[2]. These strength-controlling defects are either introduced during fabrication of the solid body or by some post-fabrication surface treatment such as machining[3,4].

In order to increase the mechanical strength and reliability of ceramics, it is necessary to identify what type of defects are responsible for fracture and then reduce their size or, preferentially, remove them completely. This process entails several iterations due to the variety of defects encountered.

The present study identifies what type of defects appear to be predominant in causing failure in a sintered yttrium sialon material and then discusses the influence of various processing stages on improving the mechanical properties of the ceramic. Factors that have been seen to affect strength or reliability include isostatic pressing pressure, burn-out of organics and starting powder characteristics. In this way, optimization of the processing conditions and surface finish lead to enhanced mechanical performance.

EXPERIMENTAL PROCEDURE

A sialon composition, previously shown to have good mechanical properties[5], was selected from the yttrium sialon system. The powder mixture was comprised of Si_3N_4[(a)], Y_2O_3[(b)], Al_2O_3[(c)] and AlN[(d)], with a total additive content of 11.3 w/o. Homogeneous mixing of the powders was achieved by attrition milling in isopropyl alcohol. Earlier work[6] indicated that 30 minutes of milling with Si_3N_4 media in a plastic mill was sufficient for complete dispersion of the additives.

After milling, the powder slurry was microwave-dried to prevent additive segregation, followed by granulation through a 212 µm sieve. Modulus of rupture bars, with green dimensions 41 mm x 7.5 mm x 6 mm, were initially die-pressed at 17 MPa (~2500 psi) and subsequently isostatically pressed up to a maximum of 340 MPa (~50,000 psi). Where a burn-out treatment was included, samples were brought up to temperature in air and held for 1 hour.

The bars were packed in a boron nitride crucible; each bar being separated and surrounded by a Si_3N_4/BN powder bed to prevent decomposition. Sintering was carried out at 1800°C for 1 hour in a graphite resistance furnace with a slight nitrogen overpressure (0.1 MPa).

Surface preparation of the MOR bars involved longitudinal diamond grinding (240-grit) on both tensile and compressive faces, followed by bevelling of edges. Machining conditions were similar to those given in U.S. Military Standards[7]. Where indicated, some samples underwent subsequent grinding with 600-grit, followed by 6, 3 and 1 µm diamond lapping. Final bar dimensions were typically 36 mm x 7 mm x 3 mm.

The modulus of rupture was measured at room temperature using a 4-point flexure jig, with outer and inner spans of 30 mm and 15 mm, respectively, and a cross-head speed of 0.5 mm/minute.

After low power optical examination, fracture surfaces were ultrasonically cleaned, mounted and coated for scanning electron microscopy. Fracture initiation sites were more readily identified by tilting specimens at ~22°. The Weibull modulus was calculated as the slope of the line obtained when ln MOR was plotted against ln ln (1/1-f), where:

$$f = \frac{n\,(sample\ ranking)}{1 + N\,(total\ no.\ of\ samples\ tested)}$$

(a) LC12, H.C. Starck, W. Germany; (b) Y5603, Molycorp U.S.A; (c) Grade C, H.C. Starck, W. Germany; (d) A16SG, Alcoa, Canada.

RESULTS AND DISCUSSION

Machining Effects

The effect of surface treatment on flexure strength can be seen in Figure 1. There is

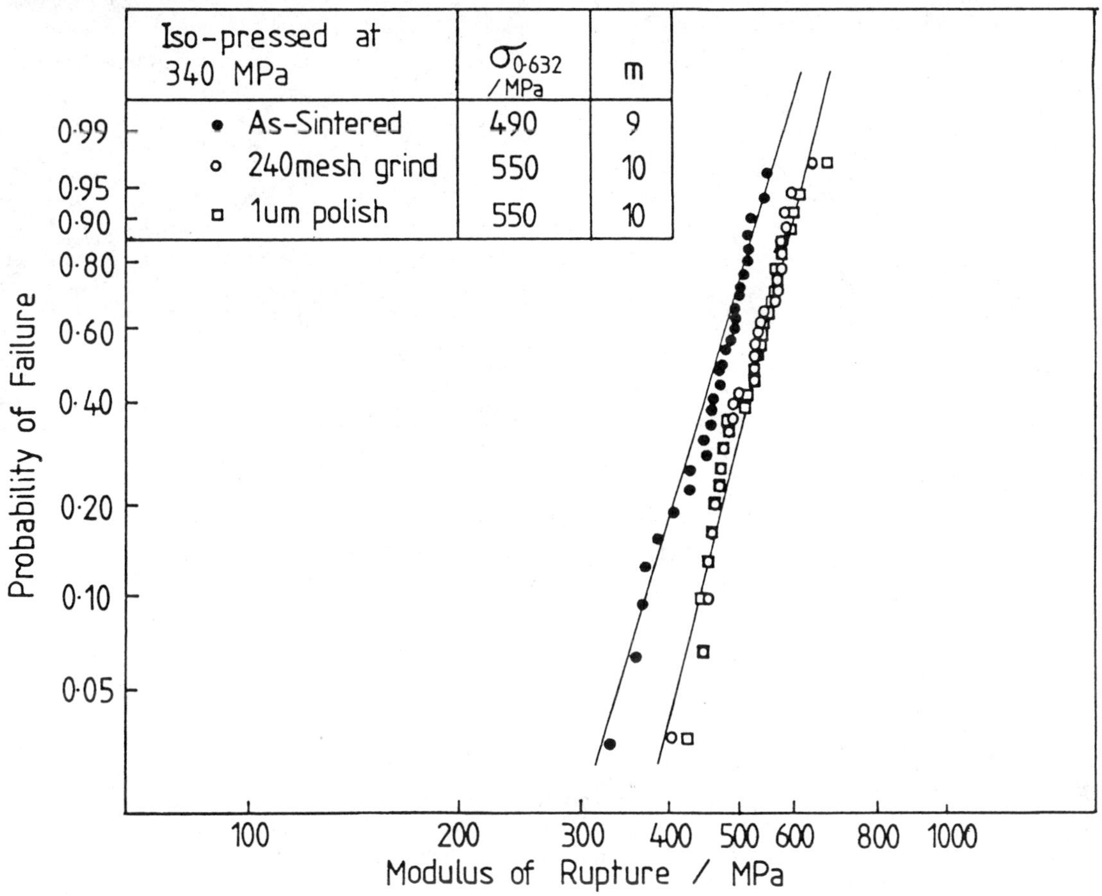

Figure 1: Effect of surface finish.

a slight strength improvement of ground or polished bars over those in the as-sintered state. However, this is thought to be due to the as-sintered bars not having flat and parallel load-bearing surfaces. The Weibull moduli (m) are ~10 for all treatments. There is negligible increase in either m or strength for bars that have been polished to 1 µm. Examination of the tensile surface of the machined (Figure 2a) and machined and polished bars (Figure 2b) shows that polishing has not removed or reduced the size of these relatively large (~ 30 µm) defects.

Fractography indicated that the majority of fracture initiation sites, especially those in low strength samples, were located at pores (Figure 3), agglomerates or low density regions, rather than at machining cracks. This indicated that any machining damage incurred was generally not as detrimental as internal flaws in the microstructure, unless the two defects combined.

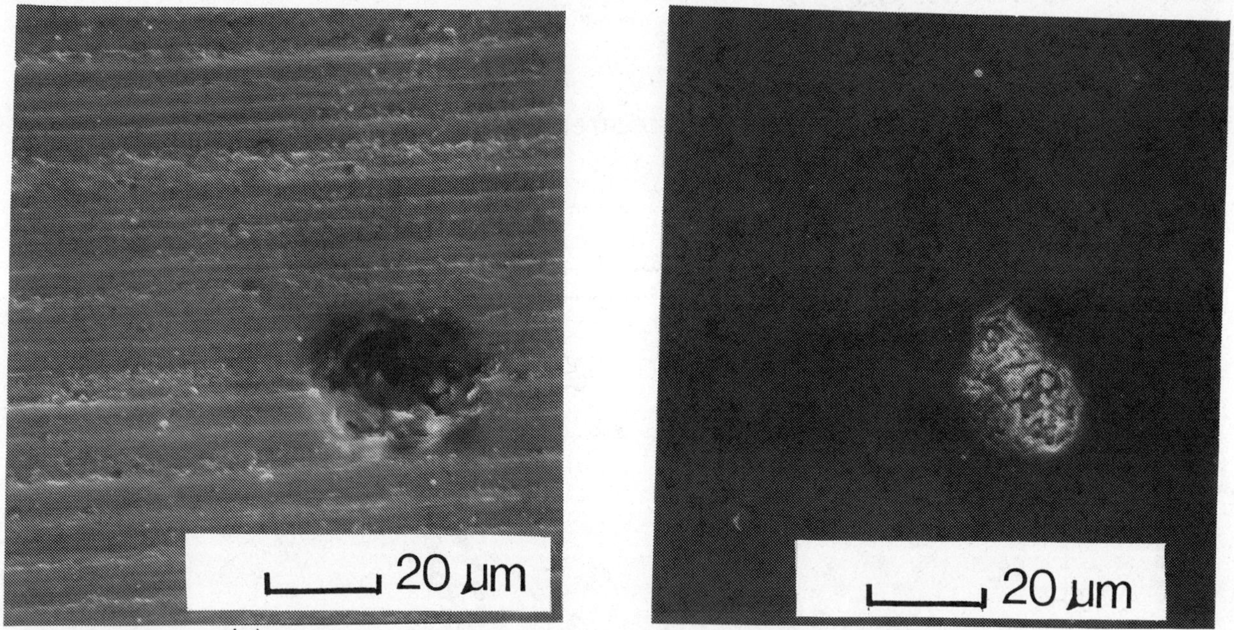

(a) (b)

Figure 2: Surface pores on (a) 240-grit ground, and (b) 1 micron polished sample

Isostatic Pressing Pressure

The effect of increasing isostatic pressing pressure on strength was investigated and the results are shown in Figure 4. The graph suggests that the pressing pressure has a significant effect on strength distribution, the highest pressing pressure (340 MPa) showing the least degree of bimodality. High isostatic pressing pressures have been known to be detrimental to the density and strength of alumina ceramics due to relaxation[8] but, as shown here and noted elsewhere[9], increasing pressing pressure up to 340 MPa is beneficial for silicon nitride materials. The increased pressure leads to higher green densities and this suggests that it reduces the size of pores or regions of low density, thus narrowing the strength distribution. However, fractography indicated that crack initiation at pores or voids was still common. On close examination, a variety of pore types could be identified. Cavities can be formed by differential sintering of agglomerates[4] and, as shown in Figure 5, regions of low density are also observed. The elongated β'-Si_3N_4 crystals growing across the pore are indicative of vapour-liquid-solid (VLS) reaction during sintering. The origin of this type of flaw is not obvious; the pore could either be formed by melt-out of an inclusion or by burn-out of some organic contaminant. Energy dispersive X-ray analysis did not indicate the presence of any metallic inclusions (Al,Si excepted) such as iron, which forms molten iron silicide at 1208°C and is often seen as tramp impurities in less pure Si_3N_4[10].

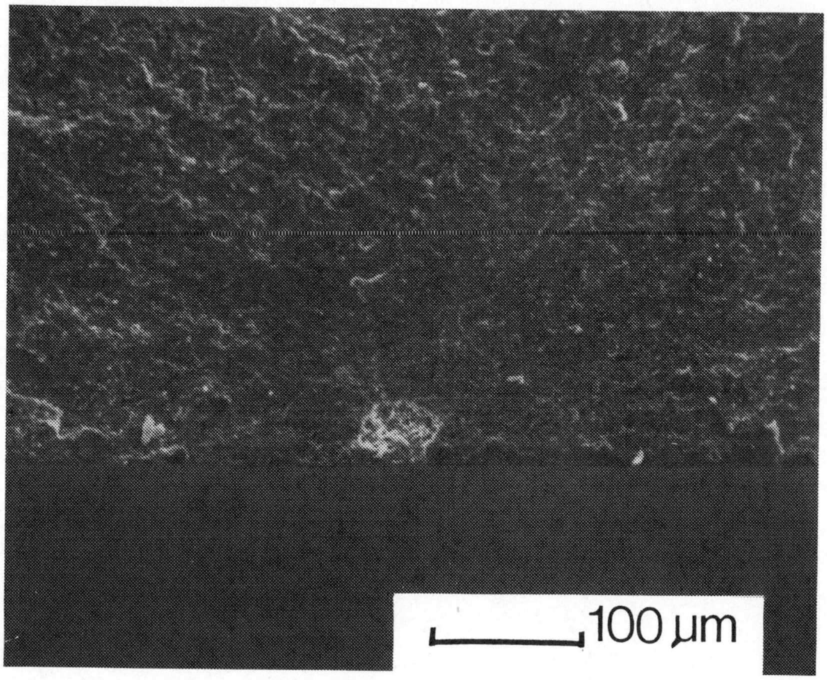

Figure 3: Fracture initiation site at pore.

Burn-out Treatment

The effect of subjecting green, die-pressed MOR bars to a burn-out treatment at either 500°C or 700°C for 1 hour in air, followed by isostatic pressing (at 340 MPa) is shown in Table I. Both burn-out treatments increase the green strength and m for the bars; the percentage increase being higher for burn-out at 700°C. By burning out at 700°C, the modulus has been doubled and the green strength increased by ~65%. Samples treated in a similar way were sintered, machined and tested in bending. There is a significant increase in characteristic sintered strength, from 647 MPa for non-treated bars, up to 706 MPa for those burnt out at 700°C.

It has been shown[11] that the green strength of ceramics is determined by interparticle bonding, flaw size and elasticity of the particle assembly. The glass transition temperature of yttrium silicate glasses is ~950°C and so it is assumed that no sintering or increase in interparticle bonding is taking place during the burn-out treatment at 500°C or 700°C. There was no significant change in green density (~2.07 Mgm^{-3}) between the different treatments, so no change in elasticity is expected. This would suggest that the green strength variation is due to reduction of flaw size or population. Fracture initiation sites were very difficult to identify in green bars, with no obvious defects such as those for sintered bars (see Figure 3). It appears that fracture initiates at locations of density variations in most cases. These density variations are removed during liquid phase sintering and are not a problem in the sintered material.

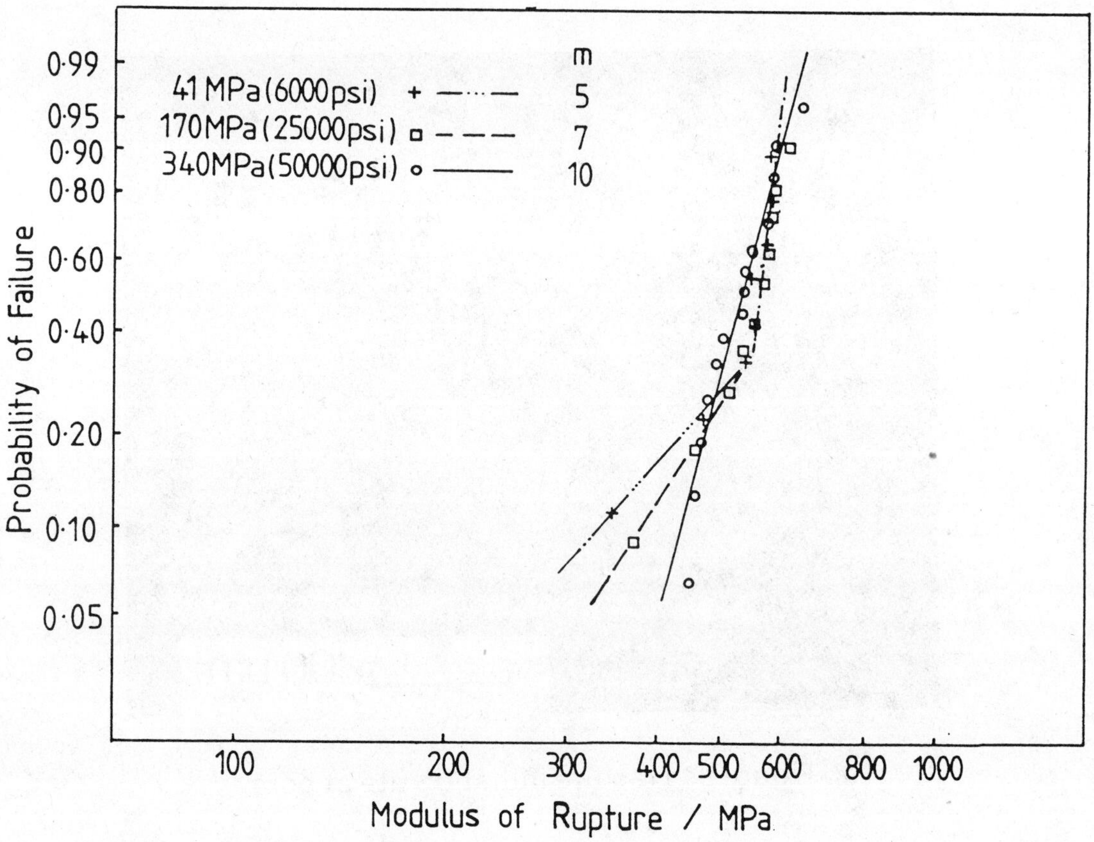

Figure 4: Effect of isostatic pressing pressure.

A possible explanation for the formation of discrete pores in the sintered body even after burn-out at 500°C is the presence of free carbon. By burning out organic contaminants, such as lint and plastic, and then isostatic pressing, such defects would be reduced in size and hence severity. However, if the burn-out period is too short or at too low a temperature for complete combustion, free carbon may be retained in the locality of the defect and, during sintering, will react with the liquid silicate phase to produce gaseous species such as SiO and CO.[12] This gas evolution within the sintering body may lead to poor sintering, bloating or pore formations. This would explain the observed strength behaviour; partial burn-out at 500°C giving a 7% increase in MOR and complete burn-out at 700°C increasing the strength by ~10%.

Si_3N_4 Starting Powder

In order to optimize the starting material, MOR measurements were made using four types of commercial Si_3N_4 powder. The powder characteristics and the strength of sintered bars (isopressed at 340 MPa) are shown in Table II and Figure 6, respectively. The UBE powder, although having the lowest specific surface area and relatively low green density, gives the highest sintered density (96.5% theoretical) and the highest strength. The narrow particle size distribution and small particle size (~0.1μm) of this powder increases its reactivity compared to the other powders

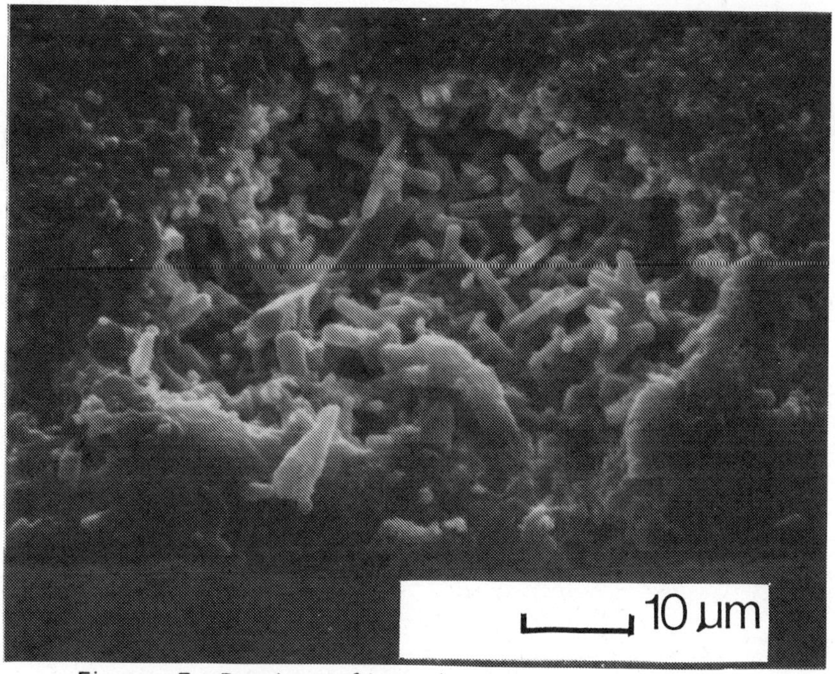

Figure 5: Region of low density causing fracture.

Table I: Strength Improvement after Burn-out

	Green		Sintered	
	Strength (MPa)	Weibull Modulus	Strength (MPa)	Weibull Modulus
No burn-out	3.7	4	647	8.0
500°C/1hr	4.7	6	690	8.0
700°C/1hr	6.1	8	706	13.0

* 10 samples tested in each case

and, hence, results in a higher fired density. The Toyo Soda powder also has a narrow size distribution but has a larger particle size than the UBE material. The LC12 Starck powder, made by direct nitridation rather than chemical synthesis (UBE & Toyo Soda), has a wide range of particle sizes, giving it a high green density and a good sintered density. However, it does contain some relatively large agglomerates which may be responsible for its slightly reduced strength. The Toshiba powder, although being a near-monosized powder, has a large particle size which results in very low green density and, hence, a low sintered density and strength.

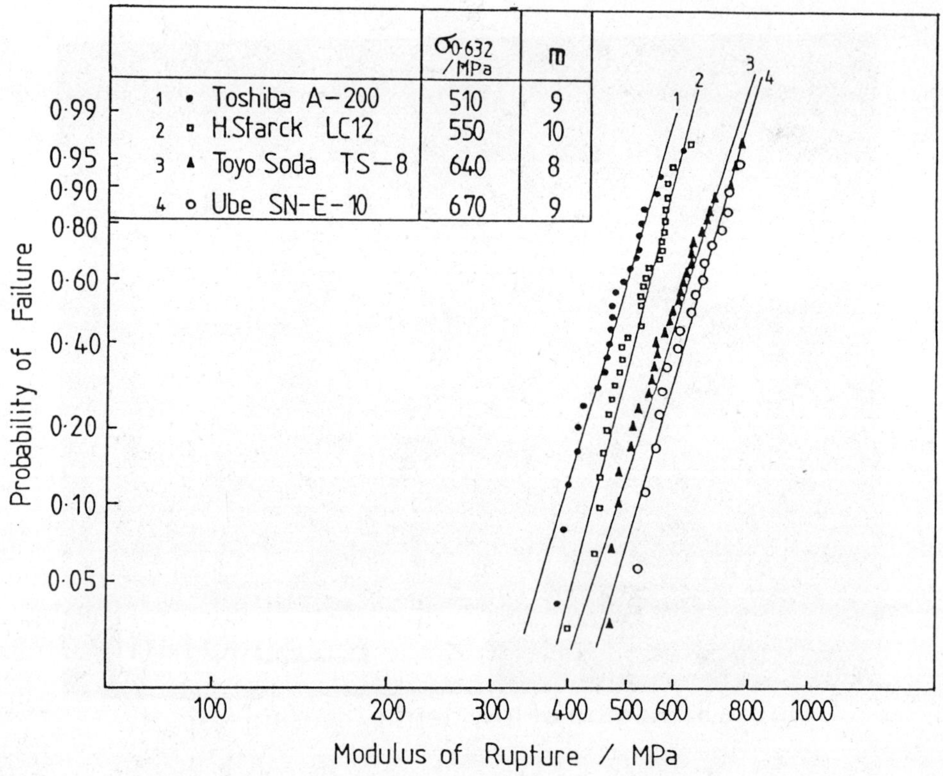

Figure 6: Effect of Si_3N_4 starting powder.

Table II: Silicon Nitride Powder Characteristics.

Powder and Grade	Manufacture Methods	S.S. Area (m^2/g)	Green Density (Mg/m^3)	% Theoretical Density	Mean Particle Size (μm)
H.C. Starck LC12	$Si + N_2$	23	2.01	96	0.3 (agg.)
Toshiba A-200	$SiO_2 + N_2 + C$	13.7	1.54	94	0.7
Toyo Soda TS-8	$Si(NH_2)_n$	10.4	1.85	95.5	0.3 (agg.)
UBE SN-E-10	$Si(NH_2)_n$	9.5	1.68	96.5	0.1

CONCLUSIONS

1. Surface grinding improves the strength of the ceramic due to more even load distribution in the 4-point bend apparatus and, because the strength-limiting defects are pores, polishing to 1μm does not significantly increase this value.
2. Increasing isopressing pressure narrows the strength distribution of the ceramic, increasing the Weibull modulus.
3. A burn-out treatment results in increased strength and Weibull modulus both in the green and sintered state, due to removal of organic inclusions prior to isostatic pressing.
4. The sintered strength is affected by the characteristics of the starting Si_3N_4 powder, with UBE powder showing the highest strength.

REFERENCES

1. J.P. Singh. "Effect of Flaws on the Fracture Behaviour of Structural Ceramics: A Review". Adv. Ceram. Mat., 3 (1) 1988, pp. 18-27.
2. A.E. Pasto, J.T Neil, C.L. Quackenbush. "Microstructural Effects Influencing Strength of Sintered Silicon Nitride." Ultrastructure Processing of Ceramics, Glasses and Composites. L.L. Hench and D.R. Ulrich (eds.), (Wiley) (1984), pp. 476-489.
3. K.R. McKinney, C.M. Herber. "Effect of Surface Finish on Structural Ceramic Failure." J. Amer. Ceram. Soc., 53 (9), 1970, pp. 513-516.
4. A.G. Evans. "Structural Reliability' A Processing-dependent Phenomenon." J. Amer. Ceram. Soc. 65 (3) (1982), pp. 127-137.
5. M.D. Pugh, L. Zarnon, R.A.L. Drew, "Controlled Processing of β'-Sialons Containing Y_2O_3 Additions". Proc. 1st Int. Conf. on Ceramic Powder Processing. G.L. Messing, E.R. Fuller, H. Hausner (eds.), Amer. Ceram. Soc. 1987, pp. 1051-1058.
6. L. Zarnon, R.A.L. Drew. "Dispersion Methods of Y_2O_3 in Si_3N_4 by Comminution". Submitted to J. Amer. Ceram. Soc., Oct. 1987.
7. Flexural Strength of High Performance Ceramics at Ambient Temperature, MIL-STD-1942 (MR) 21 Nov. 1983, p. 8.
8. T. H. Hare, H. Palmour. "Effect of Porcess Optimization Properties of Alumina Sintered under Rate Control". Ceramic Processing before Firing. G.Y. Onoda, L.L. Hench (eds.) (Wiley) (1978), pp. 307-320.
9. A.E. Pasto. "Current Research Issues in Silicon Nitride Structural Ceramics". Mat. Res. Bull., Oct/Nov. 1987, pp. 73-78.
10. H.R. Baumgartner, D.W. Richerson. "Inclusion Effects on the Strength of Hot-pressed Si_3N_4" in Fracture Mechanics of Ceramics I, R.E.C. Bradt, D.P.H. Hasselman, F.F. Lange (eds.) (Plenum), (1973), pp. 367-86.
11. K. Kendall, M. McNalford, J.D. Birchall. "The Strength of Green Bodies." Proc. Brit. Ceram. Soc. 37 (1986), pp. 255-265.
12. H. Knock, G.E. Gazza. "Effect of Carbon Impurity on the Thermal Degradation of a Si_3N_4-Y_2O_3 Ceramic". J. Amer. Ceram. Soc. 62 (12) (1979), p. 634-635.

MICROSTRUCTURE AND GROWTH MECHANISM OF SiC WHISKERS

Tian Jiemo*, Wang Lingling, Zhang Baoqing
Tong Xiaohua and Pan Guongguo

Fine Ceramic Lab. of Institute of
Nuclear Energy Technology,
Tsinghua University
P.O. Box 1021
Beijing, China

ABSTRACT

This paper describes the preparation and structure of β-SiC whiskers. The growth mechanism is discussed. The processing parameters are controlled to get cobweb or needle-like SiC whiskers. The crystal structures of β-SiC whiskers are investigated through TEM and X-ray diffraction.

Key words: SiC Whiskers, Whiskers.

INTRODUCTION

SiC whiskers have high purity and exhibit highly ordered single crystal structures. Therefore, their strengths are higher than SiC polycrystalline fibers. SiC whiskers have been produced, prepared and researched by many workers[1-5]. Knippenberg and co-worker[6] developed a process for producing alpha-SiC whiskers. Ryan[7] grew beta-SiC whiskers by the hydrogen reduction of CH_3 $SiCl_3$. Cutter[8] produced SiC whiskers with SiC powder from the SiO_2 and carbon present in rice hulls.Nutt[9] observed defects in silicon carbide whiskers made from rice hulls. Yamamoto[10] prepared SiC whiskers by mixing silica gel containing catalyst with furnace carbon black absorpting DBP (dibutyl phthalate). VLS process for producing SiC whiskers was also adapted by Shyne and Milewiski[11].
This paper describes the structure and growth mechanism of SiC whiskers, and the conditions of the growth of cobweb and needle-like whiskers are discussed.

EXPERIMENT

VLS process has been used for preparation of SiC whiskers. SiO_2 and carbon powder are used as starting materials. Particle sizes of SiO_2 powder range from 0.5 to 100μm. Their purity is 99.5% by weight. Catalyst (Nb_2O_5) has purity of 99.5% by weight. The average size of particles of Nb_2O_5 is 0.1μm. The amount of carbon black added is about 0.7 parts per part of SiO_2 by weight while the amount of Nb_2O_5 added is 0.01-0.1 part per part of SiO_2 by weight. The SiO_2, carbon black and Nb_2O_5 powders are wet mixed by milling in alcohol for 2-4 hours, this mixture is charged in a graphite reactor. At the same time, several graphite plates coated by Nb_2O_5 powder are placed in the reactor. The closed graphite reactor is then put into a furnace and heaed at 1600℃ for 3-5 hours under passing nitrogen (N_2) atmosphere of 0. 8-1. 0 atm. The pressure of nitrogen is used for controlling the supersaturated level of SiO and carbon in Nb_2O_5 melt. After the run is over, the furnace is cooled to room temperature. The resulting products include SiC whiskers and unreacted carbon powder. To remove the residual carbon, the products are put into a furnace and heated at 800℃ for 6 hours. Either cobweb or needle-like SiC whiskers are obtained with different post-treatment parameters.
X-ray sperctra of SiC whiskers were measured and their structure and morphology were examined with SEM and TEM.

EXPERIMENTAL RESULTS

X-ray diffraction spectra of SiC whiskers prepared show they are β-SiC whiskers(Fig.1).

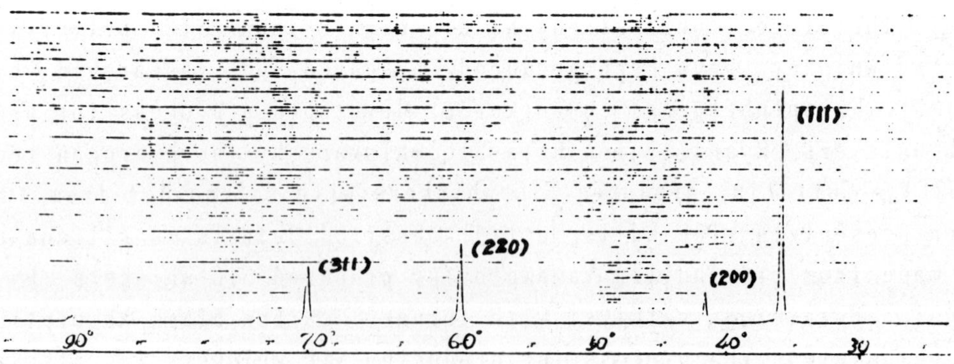

Fig.1 X-ray Diffraction Spectra of β-SiC Whisker
($Cuk_α$)

Cobweb β-SiC whiskers can be obtained when the particle sizes of SiO_2 powder are 10-100μm and the amount of Nb_2O_5 is less than 0. 01 part by weight of SiO_2. They have high yield, and their aspect ratios (length/diameter) range from 10-20 and their diameters range from 0.1-0.5μm (Fig.2).

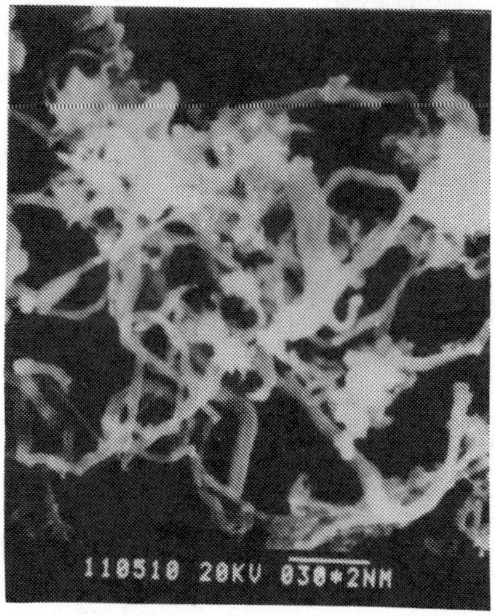

Fig.2 Cobweb SiC Whiskers

Needle-like SiC whiskers can be obtained by using SiO_2 powder of sizes less than 0.05μm and concentrations of Nb_2O_5 of approximately 0.1 part by weight of SiO_2. Their diameters are about 0.5μm while their lengths are in the range of 20-50μm (Fig.3). Upon heating for a longer time and at a higher temperature (>1600℃), longer β-SiC whiskers (Φ=0.1-0.5μm, L=100-200μm) can be obtained (Fig.4).

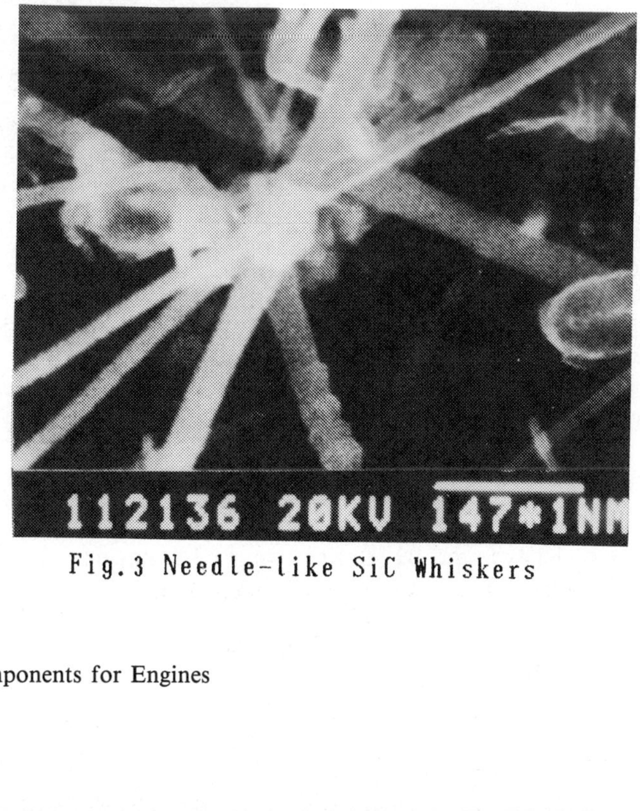

Fig.3 Needle-like SiC Whiskers

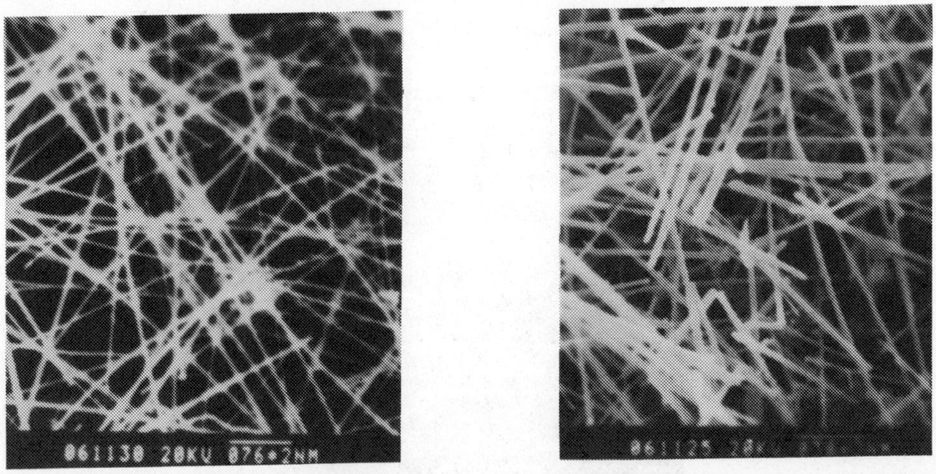

a b

Fig.4 Long β-SiC Crystal Whiskers

Several whiskers studied with TEM, exhibited very clear electron diffraction patterns (Fig.5), indicating that the whiskers contained little crystalline defects.

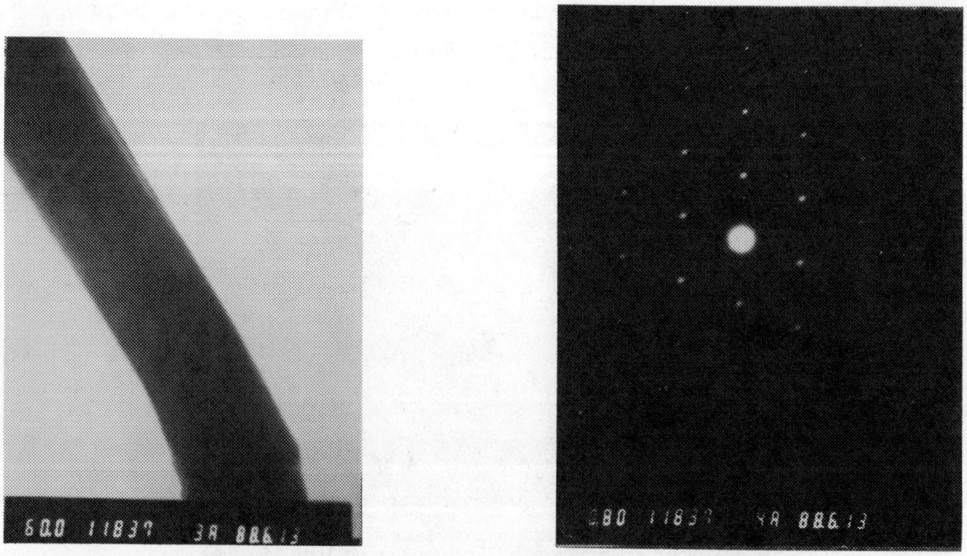

a b

Fig.5 Electron Diffraction Pattern of Needle-like SiC Whiskers(b, a)

In other cases β-SiC whisker imperfections were observed primarily in the form of dislocations (Fig.6, 7).

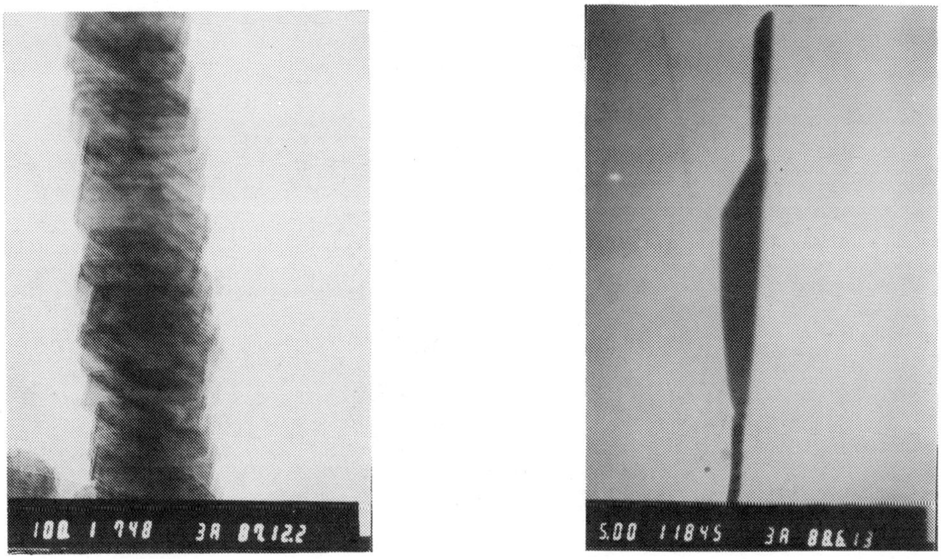

a b

Fig.6 The Stack of SiC Whisker Crystals

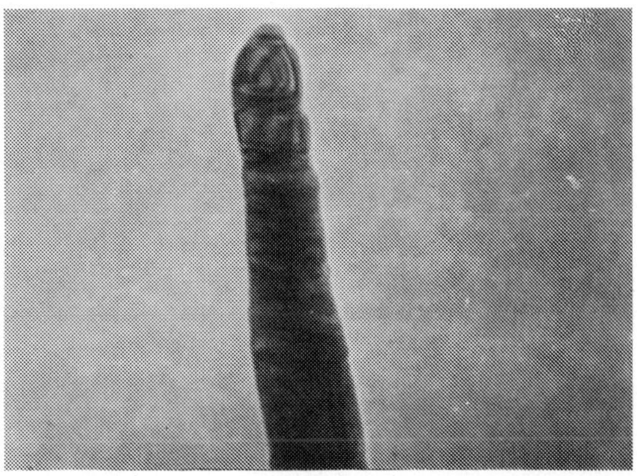

Fig.7 Screw Dislocations of Cobweb SiC Whisker Crystal

THE MECHANISM OF SIC WHISKER GROWTH

In order to get clear understanding of the mechanism of SiC whisker growth, their compositions were analyzed with X-ray electron energy spectra. Figures 8 and 9 show that Nb_2O_5 is located at the tip of needle-like β-SiC whisker.

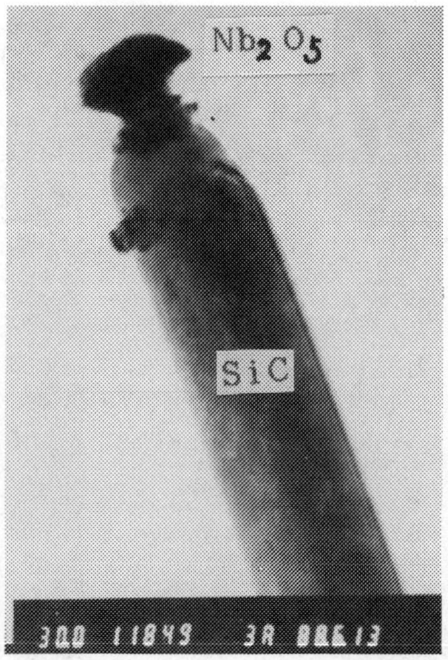

Fig.8 Needle-like β-SiC Whisker Crystal

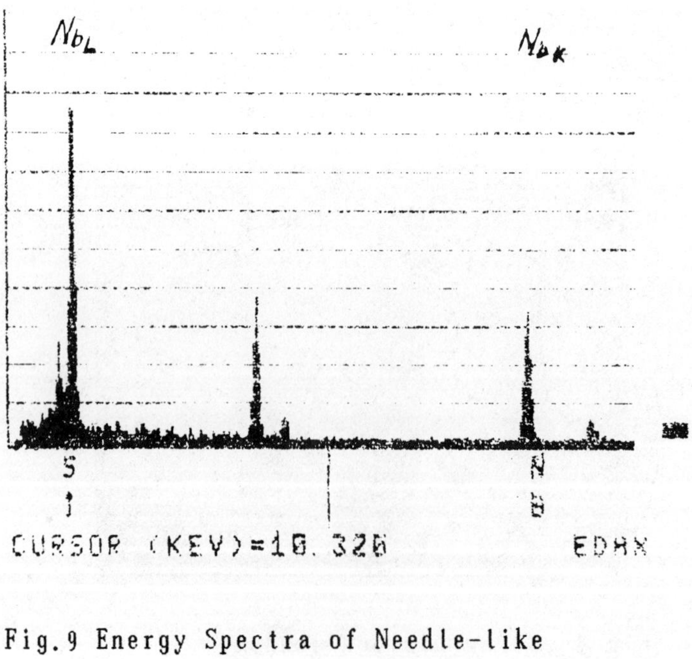

Fig.9 Energy Spectra of Needle-like β-SiC Whisker Crystal

In the process of SiC whisker growth the following reactions occur.

SiO_2 + C --SiO(g) + CO(g)

SiO(g) + 2C --SiC + CO(g)

These reactions illustrate the importance of SiO(g) in the growth process. According to Fig.8, we can image SiC whisker grows according to a liquid droplet model (Fig.10). If SiO_2 and carbon powder are mixed with catalyst powder, we can assume that carbon, SiO_2 and catalyst particles compose a mixed powder group. When temperatures reach the catalyst melting point, the catalyst becomes a liquid droplet and SiO_2 powders react with carbon to form SiO vapor. Finally SiC forms from the carbon and SiO(g) dissolved in the droplet. If SiC is supersaturated in the catalyst droplet, the SiC whisker grows.

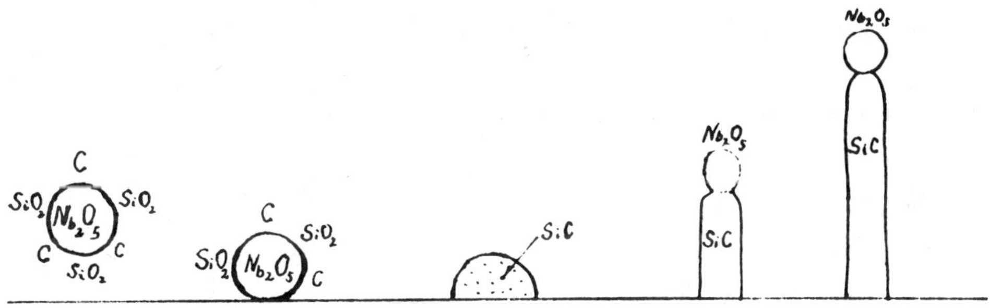

mixed powder group	Catalyst melts	SiC dissolved	SiC whisker grows out

Fig.10 Process of SiC Whisker Growth

Figures 2 and 7 indicate that a dislocation model is involved in the cobweb SiC whiskers. This type of growth is professed when the concentration of Nb_2O_5 is lowered. In this case, excess SiO(g) is formed which forms the dislocation growth. Additional details of this process are being studied. A number of factors can influence the SiC growth behavior. For example, if there is excess SiO_2, It will result in residual SiO_2 appearing in the products. Excess catalyst can lead to SiC whisker branching and kinking, while insufficient catalyst results in SiC powder mixed in SiC whiskers. Of course, it is important to control the starting materials and reaction temperature in this process.

If catalyst is coated on a graphite plate (or substrate), carbon atoms and SiO molecules are fed into catalyst droplet. When SiC becomes supersatured, SiC whiskers will grow. The more carbon vapor atoms, the more SiC whiskers grow.

SUMMERY

Cobweb or needle-like SiC whiskers were prepared. Reaction temperature and the quantity of SiO_2 and catalyst directly affected the SiC whisker growth and its quality. The structure and growth mechanism of β-SiC whiskers were discussed.

REFERENCES

1. J.V.Milewski, F.D.Gac et al, J.Materials Science 20(1985) 1160-1166
2. N.K.Sharma and W.S.Williams, J.American Ceramic Society. Vol.67(1984) No.11. 715-720
3. G.A.Bootsma, W.F.Knippenbery, J.Crystal Growth 11 (1971) 297
4. W.Dittmar and K.Neumann, Z.Elektrochem. 64 (1960) 297
5. G.W.Sears, Acta Met.1(1953) 45
6. W.F.Knippenberg and G.Verspui, "Silicon Carbide-1972" edited by R.C.Marshaill, J.W.Faust, Jr. and C.E.Ryan (University of South Carolina Press, 1973) P.108
7. C.E.Ryan, I.Berman et al, J.Crystal Growth 1(1967)225
8. I.B.Cutler, "Production of Silicon Carbide From Rice Hulls" US Pattent 3754076, Auges 1973
9. S.R.Nutt, J.American Ceramic Society Vol.67(1984) No.6 P.428-432
10. Akira Yamamoto, UK Patent Application GB 2116533A
11. J.J.Shyne and J.V.Milewski, "Method of Growing Silicon Carbide Whisker" US Patent 3622272, November 1971

THE EFFECT OF MOISTURE ON THE PROCESSING OF SILICON NITRIDE IN IMIDAZOLINE-HEXANE SOLUTIONS

K.J. Nilsen*, R. E. Riman, and S.C. Danforth Rutgers, The State University, Center for Ceramic Research, P. O. Box 909, Piscataway, NJ 08855-0909,
***Dow Chemical Co. Inc., Midland, Michigan 48674**

ABSTRACT

Ultrafine laser-synthesized Si_3N_4 powder was dispersed in hexane using Unamine O^{TM} as the dispersant. Anhydrous processing environments resulted in superior dispersion properties. The powder dispersion properties (rheological and visual) degraded if the powder was first *exposed* to ambient conditions. Surface studies (Fourier Transform Infrared Spectroscopy (FTIR), Karl Fisher Titration (KFT), and hydrolysis weight gain) indicate that the difference in the dispersion properties between powders processed in ambient and anhydrous conditions may be attributed to chemical modification of the dispersant molecule after adsorption onto the hydrated Si_3N_4 powder. The potential for achieving high green density compacts using dried dispersed powder has been demonstrated with uniaxial pressing.

INTRODUCTION

Silicon nitride bodies with optimum high temperature properties require high purity materials to avoid the formation of an undesirable glassy grain boundary phase. The covalent nature of Si_3N_4 demands a uniformly packed and highly dense powder compact made with ultra-fine particles to facilitate sintering.[1-3] Steric stabilization in a low dielectric solvent was chosen as the best route to create an uncontaminated, uniformly packed green body.[4,5] Laser synthesized powder was selected since its chemical and physical properties can be accurately controlled.[2,3] The characteristics of the laser synthesized Si_3N_4 powder used in this study are: 1) a BET surface area of 122 m^2/g, 2) a particle size of 17.0 nm (BET ESD) and 16.2 nm (TEM), 3) an amorphous structure (X-ray and electron diffraction), and 4) a stoichiometric ratio of Si/N (FTIR, ESCA, and wet chemical analysis). These characteristics have been interpreted to indicate that the Si_3N_4 powders are dense spheres.

The Si_3N_4-Unamine^{TM}-hexane slurry was selected as the best system for creating high purity, high density, uniformly packed green compacts for the following reasons: (1) the dispersions had a low relative viscosity, (2) possible oxygen contamination is minimized due to the low oxygen content in Unamine^{TM} dispersants and negligible oxygen contamination in hexane, (3) the Unamine^{TM} dispersant exhibits high surface activity essential for good dispersions, and (4) this surfactant is available commercially, and is commonly prepared by the reaction of oleic acid with 2-amino-ethanolamine.[6]

A review of silicon chemistry facilitates understanding and optimization of the dispersion behavior of the Si_3N_4-imidazoline-hexane system. The surface chemistry of Si_3N_4 powder prepared in various fashions and processed in aqueous suspensions or ambient air has been found to consist of an equilibrium between amino and silanol (Si-OH) groups.[7,8] Recently, Buska has shown the surface of an air exposed Si_3N_4 to consist primarily of Si-NH groups and silanol groups.[9] The species and reactivity of Si_3N_4 surface functional groups, not exposed to air or water, have not been widely studied.

Silica surface studies give an indication to the reactivity of silanol groups on Si_3N_4 surfaces. Room temperature silica surface studies indicate that condensation reactions between aliphatic alcohols and silanol groups on Si_3N_4 are highly unlikely. Instead, the adsorbate alcohol probably hydrogen bonds to silanol sites.[10] Based upon the literature of nonaqueous and aqueous solution chemistry of organosilylamines (viz. R_3-Si-NH_2) and silazanes (viz R_3-Si-NH-Si-R_3), amino groups would be expected to undergo hydrolysis and alcoholysis quite readily at room temperature. Many times, however, an acid initiator is essential for silazane alcoholysis. Silazanes have not been observed to undergo hydrolysis when the molecule was insoluble in neutral or basic aqueous solutions.[11] Other functional groups such as: C=C bonds, carbonyl groups (C=O), and amine groups (NH_x) have been shown to adsorb on to silica surface silanol groups.[12-14]

Exploring the chemistry of the functional groups present in the surfactant will help in understanding the interaction of Si_3N_4 with Unamine O™ surfactant. Unamine O™ is a 1-hydroxy-ethyl 2-heptadecenyl 2-imidazoline which consists of a five member heterocyclic ring with an alcohol at the one position and an aliphatic chain at the two position as depicted in the reaction below:[15]

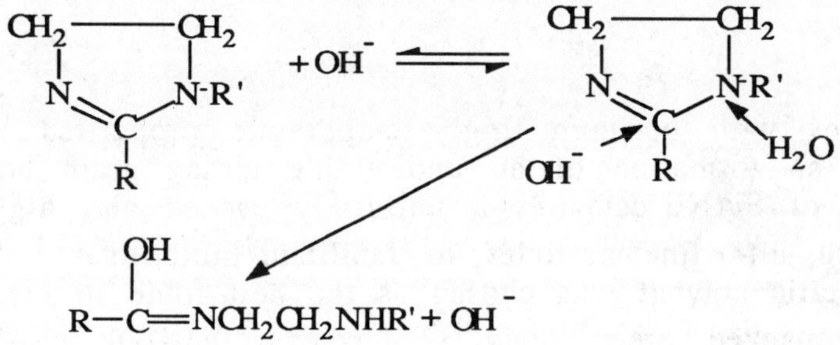

[1]

In acidic aqueous solutions at room temperature, the imidazoline has not been observed to undergo hydrolytic ring fissure.[6] However, the imidazoline ring is extremely reactive in the presence of water in non-aqueous and aqueous solutions that are neutral or basic. The imidazoline undergoes a base-catalyzed hydrolysis ring opening resulting in an amide reaction product (Reaction 1).[16]

This paper discusses the Si_3N_4 powder surface chemistry, the chemical interactions between the surfactant and the powder surface, the properties of the Si_3N_4-hexane-Unamine O™ dispersions, and the potential of this system to yield uniformly packed, high density green bodies. The techniques employed are: visual observations, rheological, gravimetric, titrimetric, spectroscopic, and compaction measurements. The results are then interpreted using chemical information from the literature sited above.

EXPERIMENTAL PROCEDURE

The laser synthesized Si_3N_4 powders were processed various ways to evaluate the effect of water on the powder's surface chemistry and dispersion properties. All powders referred to as *unexposed* were processed in a glove box* with a nitrogen atmosphere containing less than 10 ppm O_2 and H_2O. Powders referred to as *exposed* were introduced to air at room temperature ~20°C and ~50% RH for at least 72 hr. Powders referred to as *O_2 exposed* were subjected to a gas stream of 21% O_2 in nitrogen mixture at a flow rate of ~0.75 l/min.

ACS grade hexane** was used as received. Unamine O™*** was also used with no further purification. Solutions of imidazoline surfactant in hexane were prepared in air and dried over freshly activated molecular sieve (Type 4A).****

The quantity of chemisorbed and physisorbed water on Si_3N_4 powder was determined by a modified technique used for determining the hydroxyl content of amino alcohols.[17,18] The powder used for the study was an *exposed* powder that had been *heated and evacuated* at 200°C and 0.8 torr vacuum for 24 hours (to remove physisorbed H_2O), and subsequently stored in a glove box. The vessel used for the evacuation was faulty, and probably briefly exposed the powder to the atmosphere during transfer to the glove box. Physisorbed water on the powder was determined by the same method for measuring molecular water for insoluble powders (ASTM-203-63) except that 0.1 ml of Karl Fisher titrant was added past the conductometric endpoint of the pyridine solvent. This ensured that no hydrolysis of the powder would occur in solution.*****

Weight gain experiments were used to quantify the degree of hydroxylation and hydration of the Si_3N_4 powder in a humid atmosphere. The humidity of a flowing ultra-high purity nitrogen stream (0.75 l/min), was controlled by the adjustment of dew points as described by Wentzel.[19] The humidity was recorded with a hygrometer that was calibrated with a sling psychrometer. In a separate experiment, the oxidation of the Si_3N_4 powder was also measured by a weight gain experiment for an *unexposed* powder in an ultra high purity 21% O_2 in nitrogen gas stream at a flow rate of ~0.75 L/min.

Diffuse reflectance infrared spectroscopy (DRIFT)****** was used to characterize the powder surface chemistry and powder surface-dispersant interaction. All powder samples were prepared by mixing 5 wt% Si_3N_4 powder in dried KCl powder. A reference KCl background spectra was run for all cases, eliminating any effect the KCl would have on the sample spectrum. The *unexposed* samples were transferred to the FTIR and purged with ultra high purity nitrogen in the sample chamber to minimize air exposure. The *exposed* samples were run under dry air conditions. *Oxygen-exposed* powder was analyzed with the same procedure used for *unexposed* powder. The transmission spectrum of Unamine O™ was obtained using a smear technique on KBr disks under dry air conditions. All samples were scanned 100 times from 4000-400 cm^{-1} with a resolution of 2 cm^{-1}.

The rheological properties of the dispersion systems were assessed using relative viscosity measurements.******* The relative viscosity is defined as the viscosity of the suspension divided by the viscosity of the solvent-dispersant system.

---

* Vacuum Atmospheres, Hawthorne, CA.
** Fisher, Springfield, NJ
*** Lonza Corp. Fairlawn, NJ.
**** Union Carbide, Tarrytown, NY.
***** Fisher, Fisher Manual K. F. Springfield, NJ.
****** Perkin Elmer, Norwalk, Conn.
******* Haake Buchler, Saddlebrook, NJ.

The measurements of shear stress vs shear rate were performed on 1 vol% solids loaded dispersions which were mechanically agitated for 72 hr at ~20°C. The mechanical agitation was performed in a glove box for the *unexposed* powders. The measurements were taken over a shear rate range from 150 to 300 s^{-1}. Viscosity versus time data indicated that the brief exposure to air during the measurements had no effect on *unexposed* powder dispersion properties. Dispersions referred to as *water added* followed the procedure outlined above, except that 0.15 vol% water was added to *unexposed* slurries after 72 hours of agitation in a glove box. An additional 72 hours of agitation was performed on these *water added* suspensions before testing.

In order to evaluate the consolidation behavior of these dispersed powders, *unexposed* and *exposed* Si_3N_4-Unamine O™-hexane dispersions were first prepared (as above) and subsequently dried in a glove box. The dried mass was removed form the glove box and granulated in air using a mortar and pestle. The powders were then pressed in an evacuated (for 5 min) 1 cm diameter cylindrical die and pressed into ~1.5 mm wafers at pressures from 69-346 MPa (10-50 Ksi) in air. The bulk density of the green compacts was calculated based on measurements of sample dimensions and weight.

RESULTS AND DISCUSSION

The surface chemistry of the Si_3N_4 surfaces and their interactions with water and the Unamine O™ dispersant were examined to understand the dispersion and pressing properties of *exposed* and *unexposed* powder.

Surface Chemistry of *Unexposed* and *Exposed* Powders

The *unexposed* powder surface chemistry was characterized to understand the intrinsic, oxidative, and hydrolytic nature of the Si_3N_4 powder. Interpretation of the DRIFT spectra of the Si_3N_4 powder indicates that the surface of the *unexposed* powder is amino in nature. In Figure 1, Kubelka Munk absorption peaks at 1650 cm^{-1}, 1630 cm^{-1}, and 1550 cm^{-1} are assigned to an NH_2 bending, a free NH_3 bending and an NH bending mode respectively.[9,20,21] The absorption of the NH mode is dominant even though this mode has a weaker absorption coefficient[21] than the other two modes. This indicates that there is a predominance of silazane (Si_2-NH) species and a negligible concentration of silylamine (Si-NH_2) species on the *unexposed* Si_3N_4 powder surface.

Since triply bonded nitrogen surface species (Si_3N) are not directly detectable by infrared spectroscopy, the possibility exists that this specie is also present on the surface of the *unexposed* Si_3N_4. This is highly unlikely, however, due to the strained nature of this group and the reaction temperature (~1300°C) during powder synthesis.

Having determined that there is a large concentration of Si_2NH species, it was of great interest to learn whether these existed solely on the particle's surface or whether they might be bulk species as well. To answer this question, a calculation was made of the expected H content. Assumptions made in the calculation were: 1) a powder surface area of $120m^2/g$ and 2) the powders surface chemistry was Si_2NH. The 0.15 wt% H measured by combustion analysis indicated that there was a surface occupancy of 0.75 nm^2/Si_2NH site, a value which agrees well with the Karl Fisher and weight gain results. This result, combined with the TEM and powder density, serves as strong evidence that amino groups reside only on the powder surface, and not in the bulk of the powder.

Weight gain experiments were carried out in an attempt to characterize the reactivity of the Si_3N_4 powder's reactivity to the atmosphere. The results (Figure 2) indicate that the *unexposed* Si_3N_4 powder reacts very quickly with water vapor. The initial rapid weight gain regime (~1.5 wt%) can be attributed to the hydroxylation of silazane species following the reaction:

$$Si_2\text{-}NH + 2H_2O = 2Si\text{-}OH + NH_3 \qquad [2]$$

The rapid 1.5 wt% gain corresponds to 4 surface silazane species per square nanometer reacting to form silanol groups. Beyond 1.5 wt%, a slower weight gain regime results, which can be attributed to physically (or hydrogen bonded) adsorbed water. The amount of physically adsorbed water is seen to be proportional to the relative humidity.

The Si_3N_4 powder's surface coverage of silanol groups and water upon exposure to H_2O is also revealed by the Karl Fisher titration data. The *heated evacuated* powder yields a total water content of 3.73 wt%, while an *exposed* powder has a total water concentration of 5.72 wt%. The physisorbed water content of the *heated evacuated* powder is 1.73 wt%. Thus, the remaining 2.00 wt% corresponds to the concentration of Si-OH sites. This is considered to be in good agreement with the weight gain data described above.

Infrared spectra in the 1700 cm^{-1} to 1500 cm^{-1} range (Figure 3) reveals that the Si_3N_4 powder also exhibits limited reactivity with oxygen. The NH_2 and NH_3 absorption peaks present at 1650 cm^{-1} and 1630 cm^{-1} for the *unexposed* powder, are not observed for the *O_2 exposed* powder. However, the NH absorption at 1550 cm^{-1} remains in both spectra. This indicates that the small concentration of Si-NH_2 species present on the *unexposed* powder react to form Si-O groups after O_2 exposure. However, no characteristic Si-O stretching was detected at 1190 cm^{-1} for the O_2 *exposed* powder. This was expected since very few Si-NH_2 species exist on an *unexposed* powder surface.

Hydrolytic activity of the powder surfaces can be observed in more detail by comparisons of DRIFT spectra for the *exposed* and *unexposed* Si_3N_4 powder spectra as shown in Figures 1 and 4. The stretching regime of the *unexposed* powder shows the absorption due to N-H stretch at 3490 and 3395 cm^{-1}, whereas the *exposed* powder displays a freely vibrating Si-OH peak at 3760 cm^{-1} and a broad peak at 3350 cm^{-1} due to a combination of H_2O stretch, NH stretch, and associated Si-OH absorption (Figure 4).[8,10,20] The presence of molecular water for the *exposed* powder is confirmed by the appearance of a peak at 1635 cm^{-1} (Figure 1).[10] The peak at 1550 cm^{-1} indicates that silazane groups remain on the *exposed* Si_3N_4 powder surface. However, this peak was much less intense than the peak found in the *unexposed* powder spectrum indicating that hydrolysis (reaction 1) has taken place resulting in a lower concentration of silazane groups.

Imidazoline-Si_3N_4 (*Unexposed* and *Exposed*) Surface Chemistry

DRIFT spectra were obtained for both *exposed* and *unexposed* powders with adsorbed Unamine O™ (Figures 5 and 6 respectively). These spectra exhibit identical spectral features characteristic of the carbon backbone of the dispersant molecule.[5,20,21] However, the spectral features exhibited by the nitrogen bearing groups indicate that the dispersant interacts differently with *exposed* powder than it does with *unexposed* powder. The 1607 cm^{-1} spectral absorption in the *unexposed* case corresponds to the C=N bond[13,22] in the imidazoline ring structure. The presence of this absorption indicates that the ring structure of the dispersant has not changed upon adsorption onto the *unexposed* powder. In contrast, the *exposed* powder-adsorbate spectrum reveals no absorption at 1607 cm^{-1}, but instead shows two absorptions at 1645 cm^{-1} and 1545 cm^{-1} (corresponding to C=O stretch, and N-H/C-N coupling, respectively) indicative of a secondary amide.[20] These changes in the spectral features are attributed to the hydrolysis of the imidazoline ring. Ring fissure under neutral or base catalyzed conditions has been observed as cited earlier. Furthermore, DRIFT spectra of a reaction product derived from Unamine O™ - ammonia-ethanol solutions exhibit similar spectral features to the adsorbate found on the *exposed* powder.[23]

Based on the literature discussed earlier, the adsorption phenomenon observed in both the *exposed* and *unexposed* systems is believed to occur through hydrogen bonding. For the *unexposed* system, the hydrogen bonding is believed to take place between the acidic alcohol group of the Unamine O™ and the basic silazane surface groups of the powder. The *exposed* powder-adsorbate system is complicated by the hydrolysis reaction which produces various powder surface sites. In addition, ring fissure of the imidazoline results in various functionalities, such as secondary amides, amine and alcohol groups, which can interact with the surface groups found on the *exposed* powder surface.

Dispersion Properties

A series of visual observations on the powder dispersions combined with rheological data summarize the impact that water contamination has on the Si_3N_4 dispersion properties. *Unexposed* powder dispersions showed no visible sedimentation in a 0.5 hr time frame, while water contaminated (or *exposed*) systems exhibited rapid sedimentation (Table I). The *unexposed* system had a lower relative viscosity than the *exposed* system (Figure 7).

Possible reasons for the dispersion behavior to be adversely effected by water are: (1) incomplete coverage of the dispersant molecule due to competitive adsorption by water, (2) multiple attachment of the dispersant molecule, and (3) a change in the adsorbed layer thickness or a possible reduction in Hamaker shielding of the particles due to fissure of the imidazoline ring and/or modification of the surface groups on the powder. Interpretations of these results in terms of adsorption behavior and steric stabilization mechanisms will be the subject of a future publication.[24]

Processing

In an effort to determine the effectiveness of the hexane-Unamine O™ system, *exposed* and *unexposed* powders were dry pressed as described earlier. For the *unexposed* powder, extensive ammonia evolution due to hydrolysis (reaction 1) was detected by its pungent odor during the granulation process carried out in air. On the other hand, the *exposed* powder exhibited no noticeable odor during granulation. Compacts prepared from the *exposed*-adsorbate powder were easily removed from the die with no endcapping. In contrast, for *unexposed* powder samples, except those compacted at low pressure, extensive endcappping occurred. This behavior was believed to be due to pressure build-up of ammonia gas during the compaction process.

The results of these pressing studies are seen in Figure 8, where the bulk density is plotted against pressure for *unexposed*-adsorbate, and *exposed*-adsorbate powder. The results indicate that the *exposed* system has slightly higher densities than the *unexposed* system. Based on the rheological and visual data, this behavior was unexpected. Ammonia release during pressing and variations in the granulation process may have caused these discrepancies. To avoid the problem of ammonia evolution during processing, *unexposed* powder must be processed in inert atmospheres. The potential for forming uniformly packed green bodies from this system can, however, clearly be seen from the green density data presented. Without dispersion processing, *unexposed* and *exposed* powders cold isostatically pressed at a pressure of 347MPa (50 Ksi) only reach densities in the range of 35-40%.[25] On the other hand, with *unexposed* or *exposed* powders, use of a dispersion process followed by drying and die pressing yields green densities as high as ~50%. These processing procedures are far from optimized, and it is anticipated that even higher green densities may be realized by using more advanced powder consolidation methods.

CONCLUSIONS

FTIR spectra of *unexposed* powder show that the surface of the laser synthesized Si_3N_4 consists primarily of silazane groups (viz Si_2-NH). Weight gain experiments indicate that the *unexposed* surface reacts quickly with water. FTIR, KFT, and weight gain data indicate an equilibrium is established for an *exposed* powder, consisting of silanol, silazane, and adsorbed water species. FTIR spectra of *O_2 exposed* powder indicate that only the small constituent of Si-NH_2 species oxidize, whereas the predominant Si_2-NH surface species do not. The *unexposed* Si_3N_4 powder hexane-Unamine O^{TM} dispersions displayed significantly better rheological and flocculation behavior than the *exposed* powder system. The enhanced dispersion properties of the *unexposed* system (in comparison to the *exposed* system) are related to the chemical modification of the imidazoline molecule at the *exposed* powder surface. Processing results show that the dispersion system has great potential for yielding high density green compacts. A great deal of care must be taken in processing the *unexposed* system to avoid problems associated with hydroxylation of the powder surface.

ACKNOWLEDGEMENTS

The authors gratefully acknowledge the financial support of the Center of Ceramics Research. They also thank W. Symons and Dr. D. Denney for their invaluable discussions. Lastly, we thank A. Huerta for her laboratory assistance.

REFERENCES

1. H.T. Sawhill and J.S. Haggerty, "Crystallization of Ultrafine Amorphous Si_3N_4 During Sintering," Masters Thesis MIT, Cambridge, MA., Sept (1986).
2. W. Symons and S.C. Danforth, "Synthesis and Characterization of Laser-Synthesized Silicon Nitride Powders," in Ceramic Powder Science, Advances in Ceramics, Vol. 21, Ed., G. Messing, K. Mazdiyazni, J. McCauley, R. Haber, The Amer. Cer. Soc. Inc., Westerville, OH., 249-56, (1986).
3. W.R. Cannon, S.C. Danforth, J.H. Flint, J.S. Haggerty, and R.A. Marra, "Sinterable Ceramic Powders from Laser-Driven Reactions:1, Process Description and Modelling," *J. Am. Cer. Soc.* **65** [7] 324-30, (1982).
4. K.J. Nilsen, R. Riman and S.C. Danforth, "The Effect of Moisture on the Surface Chemistry and Dispersion Properties of Silicon Nitride in Imidazoline-Hexane Solutions," in Ceramic Powder Science II

Ceramic Transaction Vol.1, Ed., G. Messing, E. Fuller, and H. Hausner, The Amer. Cer. Soc. Inc., Westerville, OH., 469-76, (1988).
5. K.J. Nilsen, H. Wautier and S.C. Danforth,"Dispersion of Laser-Synthesized Si_3N_4 Powder," in Ceramic Powder Science, Advances in Ceramics, Vol 21, Ed. by G. Messing, K Mazdiyazni, J. McCauley, R. Haber, The Amer. Cer. Soc. Inc., Westerville, OH., 537-547, (1986).
6. K. Hofman, Imidazole and its Derivitives, Part 1, Interscience Publishers Inc., NY., (1953)
7. P.K. Whitman and D.L. Feke, "Colloidal Characterization and Modification of Silicon Nitride and SiC Dispersions," *Am. Cer. Soc. Bull.*, **65** [2] 89, (1986).
8. T.M. Shaw and B.A. Pethica, "Preparation and Sintering of Homogeneous Silicon Nitride Green Compacts," *J. Amer. Cer. Soc.*, **69** [2] 89, (1986).
9. Busca et al., "FTIR Study of The Surface Properties of Silicon Nitride," *Materials Chem. and Phys.*, **14**, 123, (1986).
10 M.L. Hair, Infrared Spectroscopy in Surface Chemistry, Marcell Dekker Inc., NY., (1967).
11. C. Earborn, Organosilicon Compounds, Academic Press Inc., NY., (1960).
12. K. Marshall and C.H. Rochester, "Infrared Study of the Adsorption of Oleic and Linolenic Acids onto the Surface of Silica immersed in Carbon Tetrachloride", *J. Chem. Soc., Faraday Trans., I,* [71], 1754, (1975).
13. A.R. Katritsky, Physical Methods in Heterocyclic Chemistry,Vol., IV, Academic Press, NY., 306, (1971).
14. A.K. Mills and J.A. Hockey,"Selective Adsorption of Methyl Esters on n-Fatty Acids at the Silica/Benzene and Silica/Carbon Tetrachloride Interface", *J. Chem. Soc., Faraday Trans. I.,* [71], 2398, (1975).
15. Product Information, Tertiary Amines, Lonza Corp., Fair Lawn, NJ.
16. B.G. Harnsberger and J.L. Riebsomer, "The Influence of Alkyl Substituents on the rates of Hydrolysis of 2- Imidazolines," *J. Heterocyclic Chem.*, **1** 188-192, (1964).
17. D.M. Smith and J. Mitchell Jr., Aquametry, Part III, John Wiley and Sons, NY., (1977).
18. D.M. Smith, J. Mitchell Jr. and W. Hawkins, "Analytical Procedures Employing Karl Fisher Reagent X. The Determination of Aliphatic Hydroxy Amines (Amino Alcohols)," *J. Amer. Chem. Soc.*, **66** 715-716, (1944).
19. A. Wexler, W. Wildhack, Ed., Humidity and Moisture Measurement and Control in Science and Industry, Rheinhold Pub. Corp., NY., 461-65, 1963.
20. L.J. Bellamy, The Infrared Spectra of Complex Molecules, John Wiley and Sons, NY., (1975).
21. K. Nakanashi and P.H. Soloman, Infrared Absorption Spectroscopy, 2nd Ed., Holden-Day Inc., San Fran., Ca., (1977).
22. M.J. Rosen and H.A. Goldsmith, Systematic Analysis of Surface Active Agents 2nd Ed., Wiley Interscience, NY., (1972).
23. M.J.D. Low and P.L.Lee, "Infrared Study of Adsorption *in situ* at the Liquid-Solid Interface", *J. Coll. Interface Sci.,* [45], 148, (1973).
24. K. J. Nilsen "The Effect of Moisture on the Surface Chemistry and Non-Aqueous Dispersion Properties of Laser Synthesized Silicon Nitride," PhD Thesis, Rutgers University, Piscataway, NJ, October, (1988).
25. W. Symons, "The Influence of Oxygen Content on the HIP'ing Behavior of Laser Synthesized Silicon Nitride Powders," PhD Thesis, Rutgers University, Piscataway NJ, January, (1989).

Table 1. Si_3N_4-Imidazoline-Hexan Dispersion Properties

Slurry Type	Viscosity (cP) 1 vol%	3 vol%	30 min Observation (3 vol% Solution)
Exposed	7.79	4.27	Settled into loose large flocs
Unexposed	2.40	2.20	Did not settle
Water added	4.90	2.45	Settled into loose large flocs

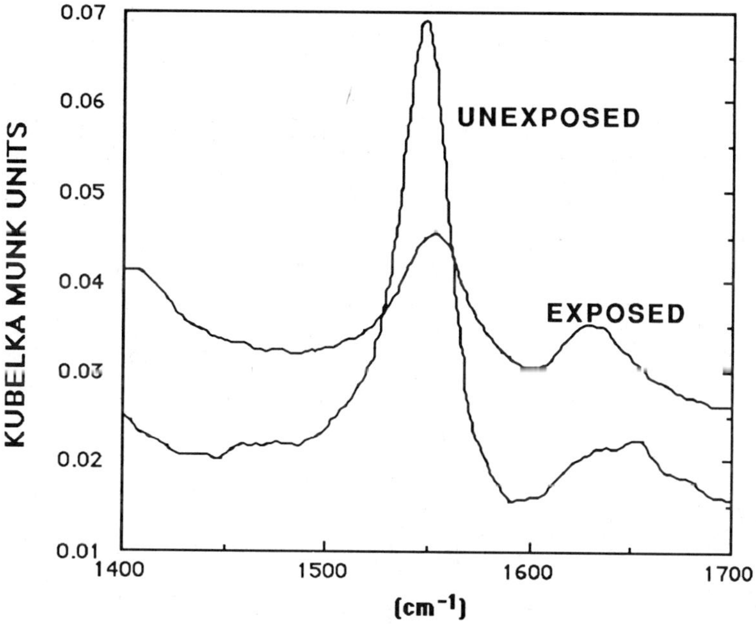

Figure 1. FTIR DRIFT Spectra of *Unexposed* and *Exposed* Si_3N_4 Powders from 1700 cm^{-1} to 1400 cm^{-1}.

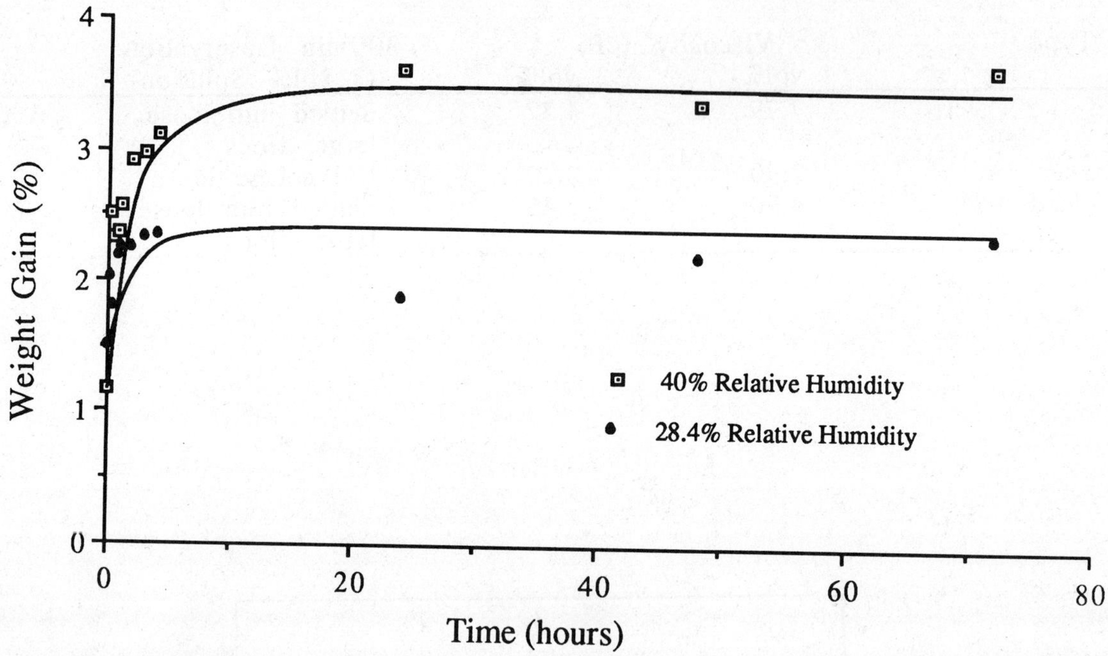

Figure 2. Weight Gain vs Time of Si_3N_4 Powders Exposed to Different Relative Humidities.

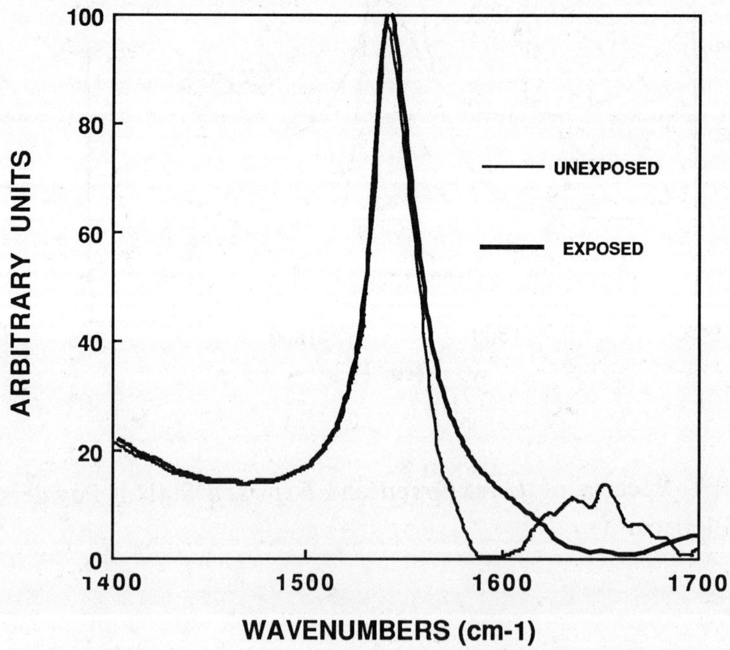

Figure 3. FTIR DRIFT Spectra of *Unexposed* and *O_2-exposed* Si_3N_4 Powders from 1700 cm^{-1} to 1400 cm^{-1}.

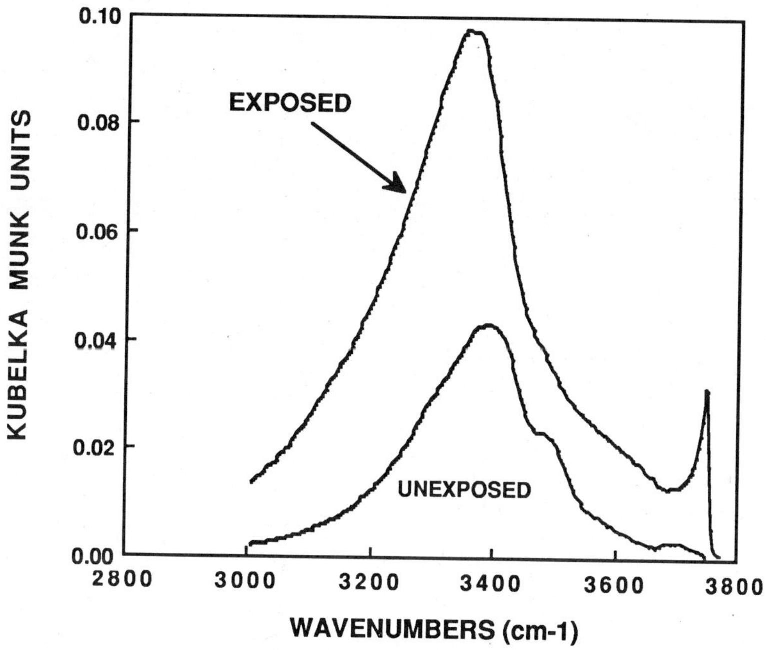

Figure 4. FTIR DRIFT Spectra of *Unexposed* and *Exposed* Si_3N_4 Powders from 3800 cm^{-1} to 3000 cm^{-1}.

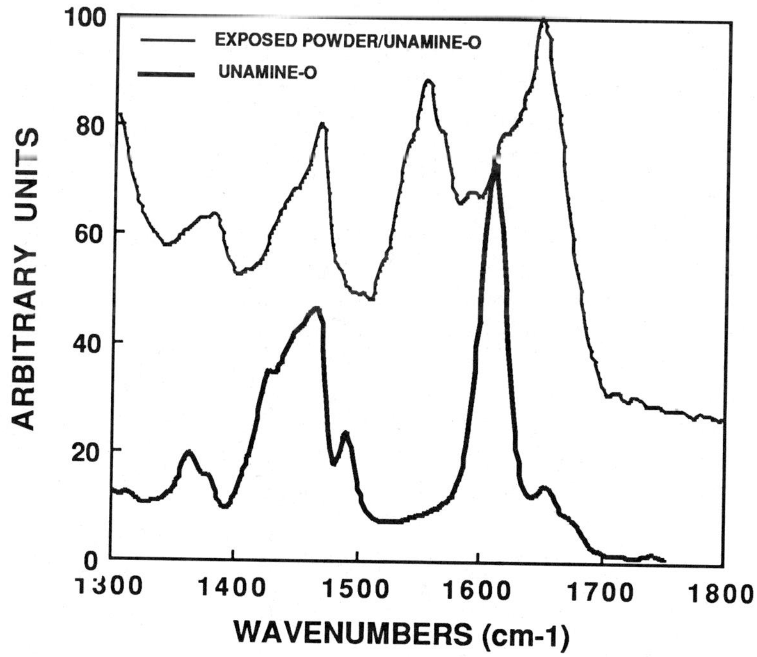

Figure 5. FTIR DRIFT Spectra of *Exposed* Si_3N_4 Powder with Adsorbed Unamine O^{TM}, and Pure Unamine O^{TM} from 1800 cm^{-1} to 1300 cm^{-1}.

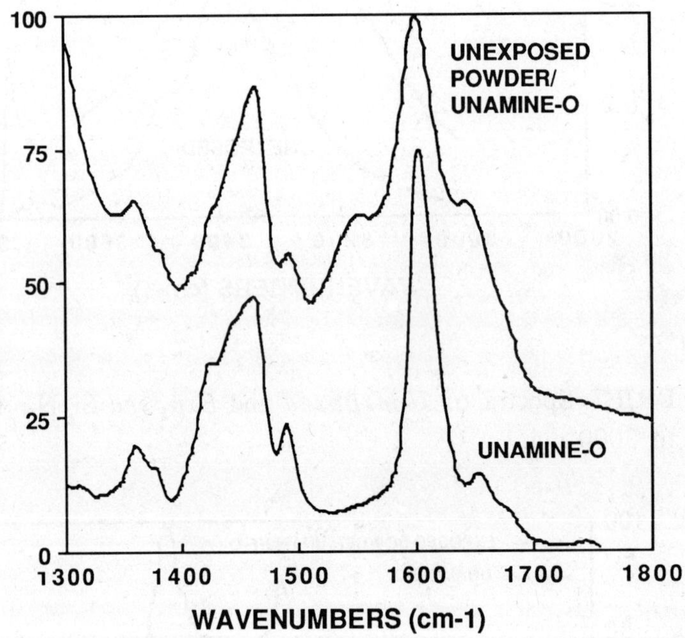

Figure 6. FTIR DRIFT Spectra of *Unexposed* Si_3N_4 Powder with Adsorbed Unamine O™, and Pure Unamine O™ from 1800 cm^{-1} to 1300 cm^{-1}.

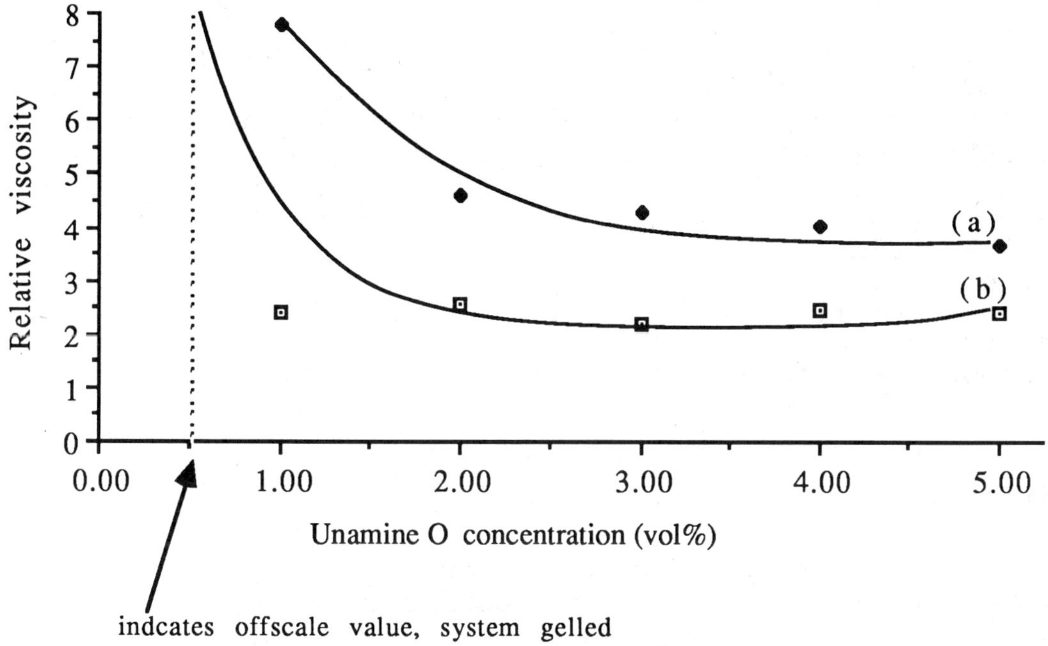

Figure 7. *Exposed* (a) and *Unexposed* (b) Si_3N_4 Slurry Viscosity vs Unamine O^{TM} Concentration.

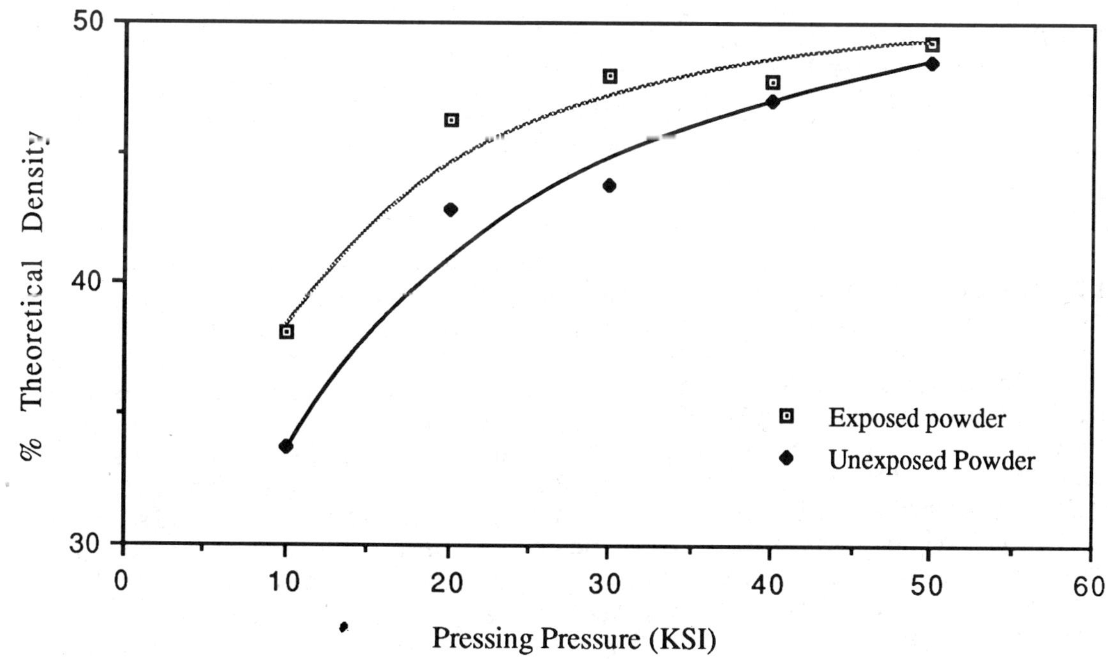

Figure 8. Packing Density of Si_3N_4 Powders that Were Dispersed, Dried and Die Pressed at Various Pressures.

INFLUENCE OF CERAMIC POWDER PROCESSING TECHNIQUES ON THE UNIFORMITY OF GREEN MICROSTRUCTURE

Vimal K. Pujari
Norton Company
Advanced Ceramics
Goddard Road
Northboro, MA 01532-1545

ABSTRACT

Green compacts were prepared by varying the solids loading in the suspension and employing alternative forming techniques, e.g. injection molding, slip casting (SC) and dry pressing (DP). Packing structure in the green compacts thus formed was characterized by mercury porosimetry (pore structure and green density), flexure test (green MOR) and SEM fractography (dispersion uniformity). An optimum solids loading in the suspensions was defined which yielded the smallest pore size, the highest density and flexure strength in the compact. Forming technique was also found to have significant influence upon the green compact characteristics. IM and SC processes (so called colloid processes) yielded much more uniform green microstructures than those produced by DP technique.

INTRODUCTION

In the development of high performance ceramics, the microstructural homogeneity of the densified component is of utmost importance. The uniformity of the sintered microstructure however is directly related to the green preform packing structure and density. Thus various factors that influence green microstructure in colloid forming techniques need careful examination. The three important factors which have been shown[1,2,3] to exert considerable influence may be listed as powder characteristics (e.g. surface area and size distribution), forming technique and powder loading. The effect of powder characteristics, especially, narrow vs. broad size distribution has been studied[1,2] extensively and was not explored here. Each approach, i.e. use of broad vs. narrow size distribution powder in the shape forming application, has its advantages and limitations. In the present study the other two factors i.e. solids loading in the suspension and the forming technique were explored in terms of their influence upon green microstructure uniformity. A relatively broad size distribution powder (unclassified), prepared by ball milling, was used throughout this study.

In the present study, the effect of solids loading on the green microstructure was studied using the injection molding process. Solids loading in a suspension influences[4] its rheological properties which in turn govern the casting/molding kinetics and resultant green microstructure. The green microstructure is characterized by its green density, pore size distribution, pore coordination number, etc. The size and number of these pores decrease during well controlled densification/sintering. Any inhomogeneity in the green microstructure may result in uncontrolled growth (rather than shrinkage) of some voids during sintering.

The importance of pore size distribution (in the green body) to the sintering process (kinetics) has been the subject of many investigations[5,6]. Lange and Davis[7] examined the change in the pore structure during sintering. The importance of pore size uniformity on the sintered microstructure has also been the subject of many investigators.[8,9] However, very limited effort has been made to correlate pore structure in the green compact (produced by rheologically different suspensions) to green and sintered strengths. The present study examines and establishes these correlations.

Green microstructure was also examined as a function of forming technique employed. Two colloid processes, injection molding and slip casting, were examined and compared with the base line dry pressing technique.

EXPERIMENTAL

The flow diagram correlating various experimental steps is schematically shown in Fig. 1. To study the effect of solids loading the experimental procedure consisted of a series of compounding tests (as employed in the injection molding process) with various powder/binder ratios. The powder used in this study was a commercially available silicon nitride powder (Kemanord Ind., Sweden) derived from the nitridation of silicon. The specific powder characteristics are shown in Table 1.

TABLE 1 Powder and Binder Characteristics

TYPE	AVG. SIZE* um	S.A.** (m^2/g)	PACKING DENSITY % (45 ksi)	BINDER
SILICON NITRIDE	0.8	12	57.5	THERMOPLASTIC BASED

(Kemanord Ind. Sweden)

* Sedigraph 5,000 ET, Micromeritics Ind. Corp.
** Digisorb 2600, Micromeritics Ind. Corp.

The compounding/mixing experiments were performed in a heated high shear mixer manufactured by Atlantic Research Corp., VA. All suspensions were prepared using a thermaplastic based binder system at 120°C with a mixing time of two hours. The details of the five IM compounds are shown in Table 2.

TABLE 2 Powder and Binder Characteristics

		MEASURED AS MOLDED DENSITY g/cm3			DEWAX DENSITY		
MIX NO.	POWDER LOADING v%	CALCULATED (THEORETICAL) A	MEASURED B	B/A X 100	Hg-POROSIMETRY g/cm^3	T.D.	MEDIAN PORE SIZE nm
1	57.7	2.26	2.23	98.67	2.02	61.39	60.3
2	59.7	2.31	2.29	99.13	2.034	61.82	61.8
3	61.7	2.36	2.33	98.73	2.057	62.52	54.4
4	64.2	2.42	2.37	97.93	2.034	61.82	62.6
5	66.3	2.47	2.42	97.97	1.902	57.81	86.5

The five IM suspensions were subsequently injection molded into test bars in order to evaluate their green strengths and resulting green microstructures. Green strength was measured using a four point bend fixture with specimen dimensions of 3mm x 4mm x 50mm. The green microstructure was examined by looking at the fracture surface of the debinderized specimens under SEM (for packing structure) and by mercury porosimetry (for pore size distribution and green density). Densities of as-molded bars were also measured by the Archimedes method and calculated by using the rule of mixtures. For the mercury porosimetry, dewaxed as well as bisque fired samples were employed.

The silicon nitride composition (with 4 w % Y_2O_3 as sintering aid) required HIP (Hot Isostatic Pressing) for densification. HIP, in general, masks the information required for a densification/sinterability study. Therefore, on a limited basis a sinterable TZP zirconia composition (4.5 w % Y_2O_3 stabilized) was also employed to study the effect of solids loading on the green and sintered microstructure. The zirconia powder used in this study was a melt derived powder manufactured at Norton Company. The experimental procedures for the zirconia powder were identical to those of silicon nitride. The mean particle size and the surface area of the powder was 0.6μm and 7 m^2/g, respectively. Modulus of rupture measurements were made both on as molded and sintered test bars.

Finally, in order to study the effect of forming techniques on the green microstructure, slip casting and dry pressing techniques were also examined using only the silicon nitride composition. The slip casting was performed using a well dispersed (pH 10) aqueous suspension with 70 w % solids loading. The slip was milled for an additional 12 hours and cast immediately thereafter. Besides surfactant no other binder was used in the slip preparation. Through the selection of proper surfactant, the solids loading (and, in turn, the viscosity) in the suspension was optimized at 70 w%.

For the dry pressing study, the silicon nitride powder was uniaxially pressed at 70 MPa in a lubricated die and subsequently cold isostatically pressed at 280 MPa. No binder was applied on the powder surface.

RESULTS AND DISCUSSIONS

Green Body Characterization

Different solids/powder loading in the IM suspension and alternate powder forming techniques resulted in green compacts with different green densities and pore structures. At first the effect of solids loading in the suspension will be described. Following that, various forming techniques will be examined.

Effect of Solids Loading

Mercury porosimetry results (median pore size) from IM and dewaxed specimens are shown in Table 2. Also shown in Table 2 are measured densities (both from Hg - porosimetry and Archimedes methods) of as molded and dewaxed test bars as a function of solids loading. The data in Tabel 2 represent average values measured from four to five samples.

The measured (B) and calculated (A) densities of as-molded bars were utilized to define the critical powder volume concentration (CPVC) following a procedure[10] developed by the author. Briefly, in this procedure the solids loading at which the ratio B/A reaches a maximum is described as the point of CPVC. At this point there is just enough binder to fill the interstitial pores of the powder compact. Later it will be shown how the pore structure of the green compact varies as we approach CPVC in the suspension.

The median pore size (from Table 2) is also plotted as a function of solids loading in Fig. 2. With increasing solids loading, the pore size reaches a minimum value of 54.4nm. With further increase in the solids content the pore size once again increases. The minimum value of the pore size is more than one order of magnitude smaller than the average size (0.8μm) of the starting powder. It is also interesting to note that the minimum in Fig. 2 corresponds to the CPVC (highest B/A ratio in Table 2) in the suspension. The green dewaxed density at this point also reaches its maximum value. The four point flexure strengths of as-molded green bars is also plotted in Fig. 2. The minimum in the pore size distribution corresponds to the maximum in the green strength. This suggests that the optimum green microstructure and moldability is achieved in the vicinity of CPVC in the suspension. The variation in the pore size with solids loading may be explained by examining the dispersion and rheological behavior of the suspension. Below CPVC, surplus binder within interstitial pores causes large interparticle spacing and hence larger median pore size after debinderizing. Furthermore, due to the lower viscosity and hence less intense shearing action all powder agglomerates are not totally broken down. Above CPVC, on the other hand, higher viscosity of the suspension causes higher shearing action during mixing which in turn results in smaller agglomerates. However, due to the insufficient binder content and the lack of interparticle lubrication, rearrangement of the particles to the closest packing configuration becomes more difficult which in turn results in larger median pore size.

The mercury porosimetry data of the dewaxed specimens are shown in Fig. 3 as frequency-pore size distribution plots for the three out of five IM compounds. The shift to smaller pore size distribution with increasing solids loading (up to CPVC), once again is obvious. However, above CPVC, (61.7 v/o) the pore structure once again coarsens and at 64.2 v% solids loading a large tail appears in the plot.

The packing structure in the green dewaxed body was further studied by examining the fracture surfaces under SEM. Fig. 4 shows the SEM micrographs of samples made from mix No. 1, 3 and 4 as shown in Table 1. Specimens 1 (57.7 v % solids) and 4 (64.2 v % solids) show jagged surface topography suggesting coarser pore structure and microstructural nonuniformity. Specimen 2 with 59.7 v % solids (not shown) shows improvement over specimen 1. As the solids loading is increased to 61.7 v% (specimen 3), a flat fracture surface morphology is observed with the smallest pores or agglomerates. This is confirmed previously by mercury porosimetry showing the smallest median pore size. Thus it may be surmised that the optimum pore size distribution (median size and uniformity) coincides with the condition when there is just enough binder to fill interstitial pores (i.e. CPVC).

The effect of optimum solids loading (in the suspension) was further confirmed using a (zirconia) sinterable composition as described in the previous Section. In Fig. 5, both the as-molded (green) and the sintered strengths are plotted as a function of solids loading. Once again, the maxima in both the green and sintered MOR corresponds to the optimum solids loading (CPVC) in the suspension.

It is important to mention here that besides influencing the green microstructure, solids loading in the suspension also impacts its molding and to a lesser extent dewaxing characteristics.

Effect Of Forming Technique

As described previously, the three forming techniques examined in this study were injection molding, slip casting and dry pressing. The results of the IM study were described in the previous section.

TABLE 3 Effect of Forming Technique on Green Microstructure

GREEN BODY CHARACTERISTICS	FORMING TECHNIQUE INJECTION MOLDING	SLIP CASTING	DRY PRESSING
Green Density (After binder removal if applicable) % T.D.	62.5	55.7	54.1
Total Pore Area m^2/g	24.820	20.096	22.622
Median Pore Size (nm)	54.4	85.1	92.8

Slip cast (SC) and dry pressed (DP) specimens were also analyzed by mercury porosimetry and SEM fractography (of the green fracture surface). The mercury porosimetry results from the specimens made by the three forming techniques are summarized in Table 3. SEM photomicrographs of the representative green samples from the three forming processes are shown in Fig. 6. For this comparison a specimen from mix no. 3 (Table 2) was used to represent the IM process. From Table 3 and Fig. 6, it may be concluded that the green microstructure obtained from the IM process is comparable (or better under optimal condition) to that of a slip cast specimen. For the SC process, the pore size (86.5nm) in general were found to be larger than those obtained from the IM process (54.4nm). Similarly, the agglomerate size in the SC specimen, as shown in Fig. 6, is four to five times larger than those observed in the IM samples. Larger pore size in the SC specimen is mainly due to the inherent flow channels of the process whereas the agglomerates are possibly due to the lack of shear action during consolidation. It is anticipated that the SC process can be improved further to reduce the size of the pores and the agglomerates approaching the level obtained from the IM technique.

Finally, the dry pressed green microstructure was compared with those of IM and SC specimen. Median pore size in the dry pressed specimen (92.8nm) was found to be larger than those of the other two processes. Also, the fracture surface of the DP specimen under SEM looks more jagged suggesting larger agglomerates/pores as compared to the IM and SC green microstructures. Larger agglomerates in the green body are expected to result in larger flaws after densification. The size of the hard agglomerates, on the other hand, in both the SC and DP specimens were found to be comparable. This is because both of these forming processes lack the shearing action of the IM process during consolidation.

From the aforementioned discussions, it is obvious that each forming process and the level of solids loading in the suspension yield different green densities and pore size distributions. The green density (as a per cent of theoretical density = 3.29 gm/cc) was therefore plotted as a function of the median pore size (Fig. 7). A linear relation between these two parameters is obtained irrespective of the consolidation technique or solids loading in the suspension. This is consistent with the published data in the literature.[11]

SUMMARY

1. Packing structure (and hence pore size) in the green compact is influenced by the concentration of powder in the suspension and also by the forming techniques employed.

2. Using the injection molding process, the effect of solids loading on the green compact pore structure was demonstrated. Suspensions with solids loading in the proximity of CPVC (critical powder volume concentration) and injection molded produced the most uniform green microstructure of the three techniques. The IM approach also showed the smallest median pore size of 54.4nm. This pore size is more than an order of magnitude smaller than the starting average powder size. Maximum values in the green and the sintered strengths were also found to coincide with CPVC in the suspension.

3. Three forming techniques, injection molding (IM), slip casting (SC) and dry pressing (DP), examined in this study produced green compacts with median pore sizes of 54.4nm, 86.5nm and 92.8nm, respectively. The overall shape of the frequency-pore size distribution plots for the three forming techniques, however, were similar.

4. Representative fracture surfaces of the green compacts made from the three forming processes were also examined by SEM. IM and SC green microstructures were found to be comparable. However, IM parts made with suspensions at CPVC showed smaller pores and agglomerates as compared to the SC

specimen. However, it is expected that with an improved dispersion in the slip, the SC green microstructure may be improved further.

As expected the dry pressed green microstructure looks more jagged suggesting larger agglomerates/pores as compared to IM and SC specimens.

5. The green density vs. median pore size relationship was found to be linear, irrespective of the suspension concentration or forming process used.

V. <u>REFERENCES</u>

1. I. A. Aksay, Advances in Ceramics, 9, 94-104, (1984).

2. E. A. Barringer and H. K. Bowen, "Formation, Packing and Sintering of Monodisperse TiO_2 Powders," J. Am. Cer. Soc. 65, C-199, (1982).

3. M. J. Edirisinghe, and J. R. G. Evans, "Properties of Ceramic Injection Molding Formulations," J. Mat. Sc. 22, (1987) 269-277.

4. R. J. Farris, Trans. Soc. Rheol. 12 (1958) 281.

5. T. Ikegami and Y. Moriyoshi, J.Am.Ceram.Soc., 67, 174-78 (1984).

6. M. Sararcan, C. H. Hsueh, and A. G. Evans, J.Am.Ceram.Soc., 66, 456-61 (1983).

7. F. F. Lange and B. I. Davis, Advances in Ceramics, 12, 679-713 (1986).

8. E. A. Barringer and H. K. Bowen, Ceram.Eng. and Sci. Proceedings, 5, 285-97 (1984).

9. M.A.C. Van de Graaf, K. Keizer, and A.J. Burggraat, Sci. Ceram., 10, 83-92 (1979).

10. V. K. Pujari, "Effect of Powder Characteristics on Compounding and Green Microstructure in the Injection Molding Process," Proceedings of the First International Conference on Ceramic Powder Processing, Nov. 87, Orlando, FL., Ceramic Transactions, Vol. 1, Ed. by G. Messing et al.

11. R. G. Frey and J. W. Halloran, J.Am.Cerm.Soc., 67, 199-203, 1984.

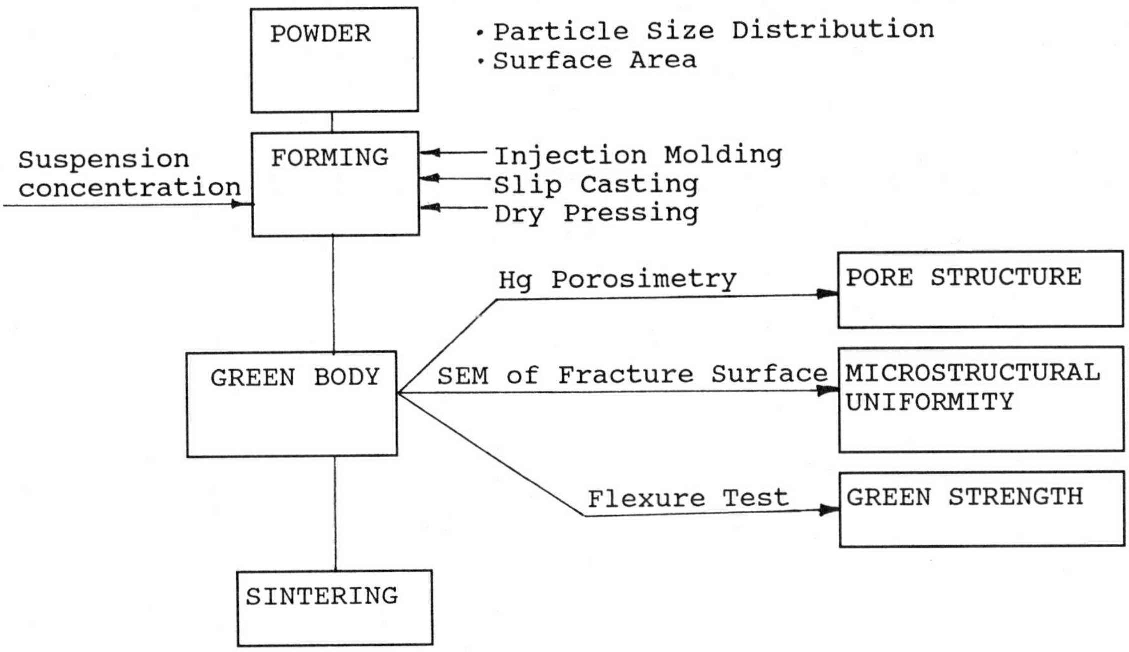

FIGURE 1 Schematic Flow Diagram of the Experimental Methodology

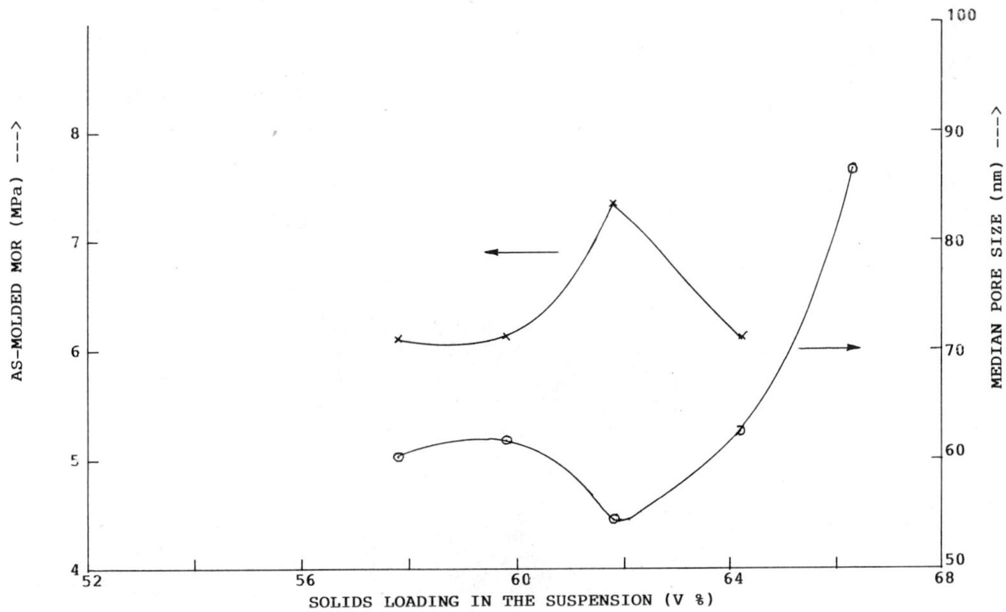

FIG. 2 GREEN MOR AND PORE SIZE AS A FUNCTION OF SOLIDS LOADING IN THE SUSPENSION

Figure 3.

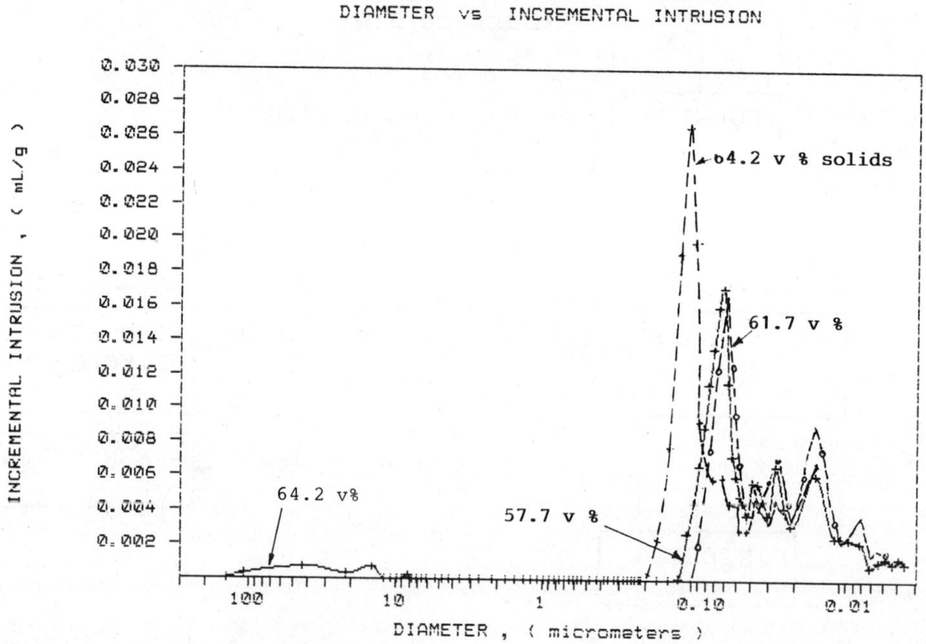

Fig. 4. Green Microstructure
of Injection Molded Dewaxed
Bars Made From Suspensions
With Different Powder Loadings

57.7 v % Powder

61.7 v % Powder

64.2 v % Powder

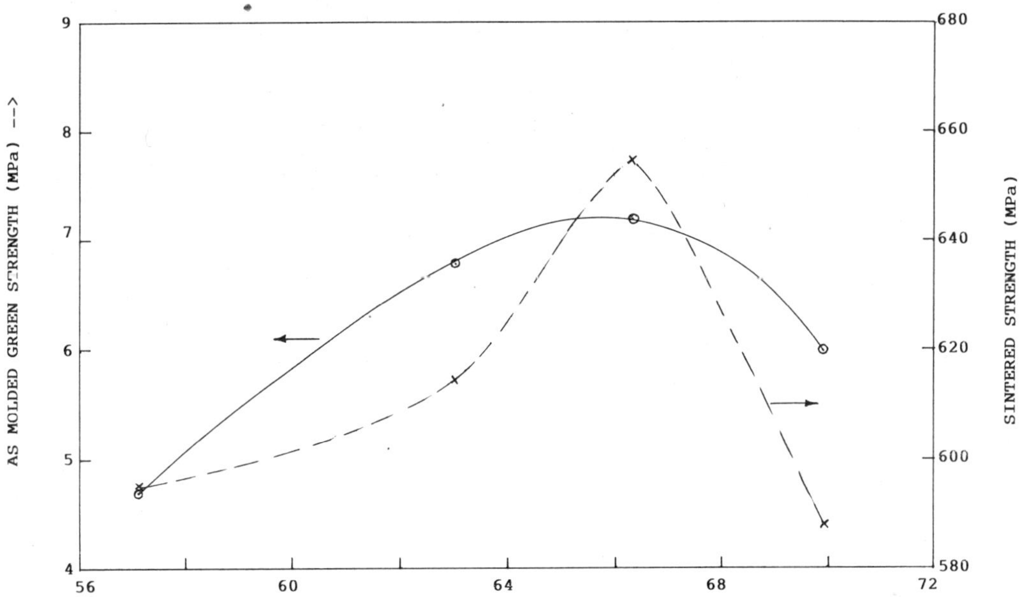

FIG. 5 Green and Sintered Strength as a Function of Solids Loading in the Suspension for zirconia composition

FIG. 6 Micrographs Showing Effect of Forming Techniques upon Green Microstructures Uniformity

a) Slip Cast

b) Dry Pressed

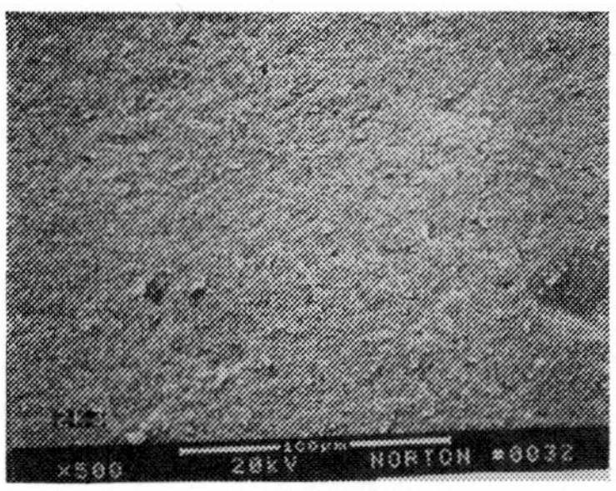

c) Injection Molded

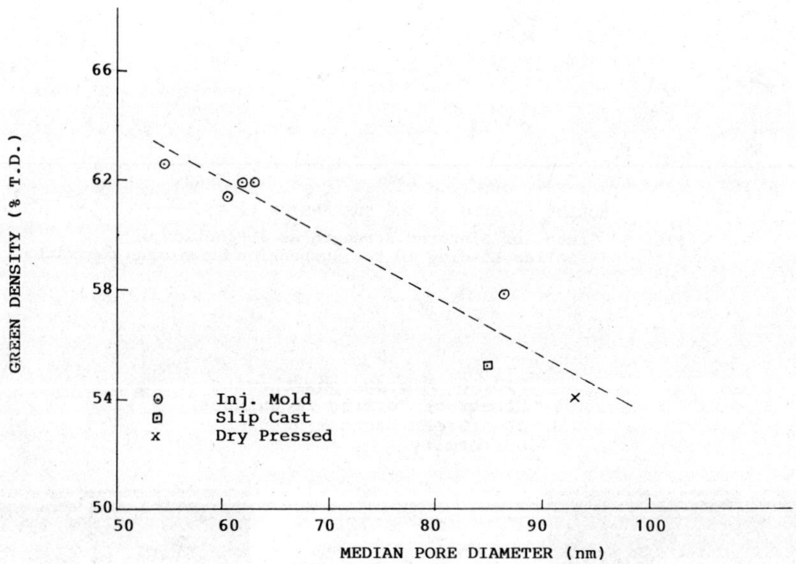

FIG. 7 Green Density vs. Pore Size

COMPACTION OF POWDERS. VII BINARY MIXTURE OF SILICA GEL AND MOLYBDENUM DISULFIDE.

I. Shapiro
Universal Chemical Systems, Inc.
Los Angeles, CA 90045-0124.

ABSTRACT

As part of the continuing study of the compaction phenomenon of powders, the ambient temperature compaction of a mixture of silica gel and molybdenum disulfide powders at several different ratios of ingredients was examined under low- and high-pressure conditions. The selection of these powders was based on their large differences in grain density values (porosity values) and in their ease of fragmentation. In the low pressure region a series of photomicrographs were taken of the powder mixture at various stages of the compaction process. In the high pressure region, measurements reveal the relationship of apparent density values for the mixture to those for the individual components. The high elastic compressibility of the compact points to the need for correcting compaction data for this elasticity effect. The use of porous materials in the compaction study opens new avenues of research and the data confirm previous findings that particles do not tend to slide past another and that there is no movement of particles lateral to the direction of applied force. The aspects of an equation that would more realistically describe the compaction phenomenon of powders are discussed.

INTRODUCTION

The principal objective of this general study of the compaction of powders is to formulate an equation which would relate or describe a characteristic of the powder as a function of pressure, and that such an equation would be applicable to a number of powders of different physical and mechanical properties. To this end, the consolidation phenomenon of a number of powders and powder systems is being studied. One such system described in this paper is the binary mixture of silica gel and molybdenum disulfide. The properties and the compaction behavior of the individual components have been described previously.[1,2] The large differences in grain density values and ease of fragmentation of the two materials complements the study of the binary system of thoria and molybdenum disulfide.[3] In this latter system, the grain density of thoria is much greater than that of the molybdenum disulfide whereas in the silica gel system the grain density of silica gel is much less than that of molybdenum disulfide. Thus the comparison of the two binary systems should be enlightening.

The present study was carried out by using similar experimental techniques of high- and low-pressure ranges with photomicroscopic viewing of the powder mass during the consolidation process. The results corroborate previous findings that particles do not slide past one another nor do the particles move in a lateral direction to the line of force of the applied load. Although both materials consolidate by fragmentation, even at relatively high pressures some of the original grains maintain their integrity.

The data for the compaction curves must be corrected for the elastic compressibility of the compact. For some powders this correction may be appreciable. All of these variables must be included in any equation developed to describe the compaction behavior of powders.

EXPERIMENTAL

Material

The details of the preparation of the starting materials and their various density values, i.e., true density, grain density, and bulk density, and their corresponding porosity values, have been described previously.[1,2] A summary of these material properties is shown in Table I.

Table I. Physical properties of powders of molybdenum disulfide and silica gel.

Property	MoS_2 [1]	SiO_2 [2]
True density, ρ_s, g/cm³	4.79	2.20
Grain density, ρ_{gr}, g/cm³	2.64	0.762
Grain porosity,%	44.9	65.4
Bulk density, ρ_B, g/cm³	1.56	0.500
Bulk porosity, %	67.4	77.3
Interparticle porosity, %	40.9	34.4
Mesh size designation	-100 +150	-100 +150
Log-geometric mean diameter, d_g, μm	146.0	138.0
Log-geometric standard deviation, σ_g	1.20	1.15

In both materials the grain porosity values are significant and greatly affect the compaction behavior. The two materials differ appreciably in their ease of fragmentation so any compaction curve must be able to account for such variabilities. In the case of molybdenum disulfide the grains collapse readily, i.e., at low pressures. The silica gel grains are more resistant but nevertheless disintegrate at moderate pressures. In the compaction process the apparent densities of the compact of the individual components exceed the grain densities of the material. The point at which the apparent density curve exceeds the grain density value for molybdenum disulfide occurs at 5 MPa; and for silica gel, at 100 MPa. Calculations of total void content as a function of pressure indicate that interparticle voids and intraparticle voids are both changing with pressure but the relative amounts at any pressure are still unknown. Known weights of the two powders (each sample taken from the same size group of -100 +150 mesh) were premixed gently to avoid accidental crushing of the individual grains, and then the entire powder mass was placed into the compaction apparatus. Tests were conducted on several different ratios of components.

Compaction Apparatus

For compaction of the powder mixture two different dies were used; however, the piston diameter in both was the same. In one case, an evacuable three-way split die of 1.25 cm diameter bore was used. The displacement opening of the cavity was measured by two precision depth gages mounted on the exterior of the die on opposite sides of the piston so that any error from canting of the piston was eliminated. Pressures up to 686.5 MPa ($7000\ kg/cm^2$) were applied incrementally to the die by means of a calibrated hydraulic press. The apparatus is calibrated for the compressibility of the piston under load -- correction for depth of cavity due to compressibility of the piston is critical.

In using the split die apparatus, known weights of the two powders were premixed and the mixture was then poured into the cavity. The apparatus was vibrated manually during the filling operation, then the piston was placed into position. The amount of powder placed into the split die was such that the initial height of the powder mass did not exceed 5 mm, which means the thickness-to-diameter ratio was kept below 0.4. A vacuum pump was connected to the outlet of the apparatus to minimize entrapment of air, especially at high loads. Pressure was applied in increments. After each application of incremental pressure, the pressure value tended to decrease slightly with time, with the greatest rate of decrease occurring initially. The pressure was brought back up to each predetermined value for each setting. The time for reaching steady state at each pressure level is relatively short, in the order of several minutes. The difference in the average of the two precision depth gages at two pressure settings indicates the change in thickness of the pellet after correction for compression of the piston had been applied.

At the end of each pressure experiment, the load on the piston was released, the die opened to remove any fine particles that may have been trapped between the piston wall and die, and the pellet was replaced in the die and compressibility measurements of the pellet were made over the same pressure range as the compaction experiment. (This procdure is to prevent false reading of the recompression experiment by cleaning all fine particles that tend to get between the piston and wall beyond the pellet-occupied cavity with its attendant possibility of seizing when the pressure load is first removed). As will be explained later, recompressibility of the pellet is an important factor in the compaction experiment.

For visual and photographic observation of the powder mix under pressure, a glass-lined die with the same size piston as the split die was used. The die body was composed of a glass sleeve encased in dental investment. An opening in the investment is made at the level of the pellet so the microscope tube and light can be inserted to the glass wall. The opening of the cavity between the movable and stationary pistons is measured by two precision gages arranged similarly to those for the split die. The glass sleeve must fit snugly with the piston; the failure rate of the glass sleeve is high (replaced for each experiment). A camera attachment (Zeiss Attachment Camera) to the microscope permits taking photomicrographs of the powder through the glass sleeve. The amount of powder is adjusted such that the initial separation of the pistons is approximately 1.5-2.0 mm, thus the entire length of the pellet would always be in full view of the microscope.

RESULTS

Low Pressure Compaction

In the low pressure compaction experiment the compaction of a mixture of molybdenum disulfide and silica gel was followed visually in the specially-designed die. The photomicrographic series is shown in Fig. 1. As pressure is applied, the height of the compact decreases. Since the two powders have contrasting colors, it is very easy to follow the course of the individual grains. The photographs clearly reveal that the grains or particles of either material do not slide past one another, nor is there lateral motion. Compaction takes place by fragmentation, but all particles maintain their relative positions as they move closer to each other in the direction of the applied force. As expected from the characteristics of the individual materials, the molybdenum disulfide grains disintegrate more easily than the silica gel grains.

The corresponding plot of apparent density vs applied pressure for this series of photomicrographs is shown in Fig. 2. For comparison the compaction curves of the individual components over the same pressure range are included. The weight ratio of the molybdenum disulfide to silica gel is 1.360. For describing the volume ratio of the two materials one must first ascertain what density values are to be used. For example, if one were to use the solid density values (ρ_s), the volume ratio of molybdenum disulfide to silica gel would be 0.625, which is equivalent to 38.5 vol % MoS_2 and 61.5 vol % SiO_2. The corresponding solid density of the mixture would calculate to be 3.20 g/cm^3. This condition would exist if all the void in the compact were ex-

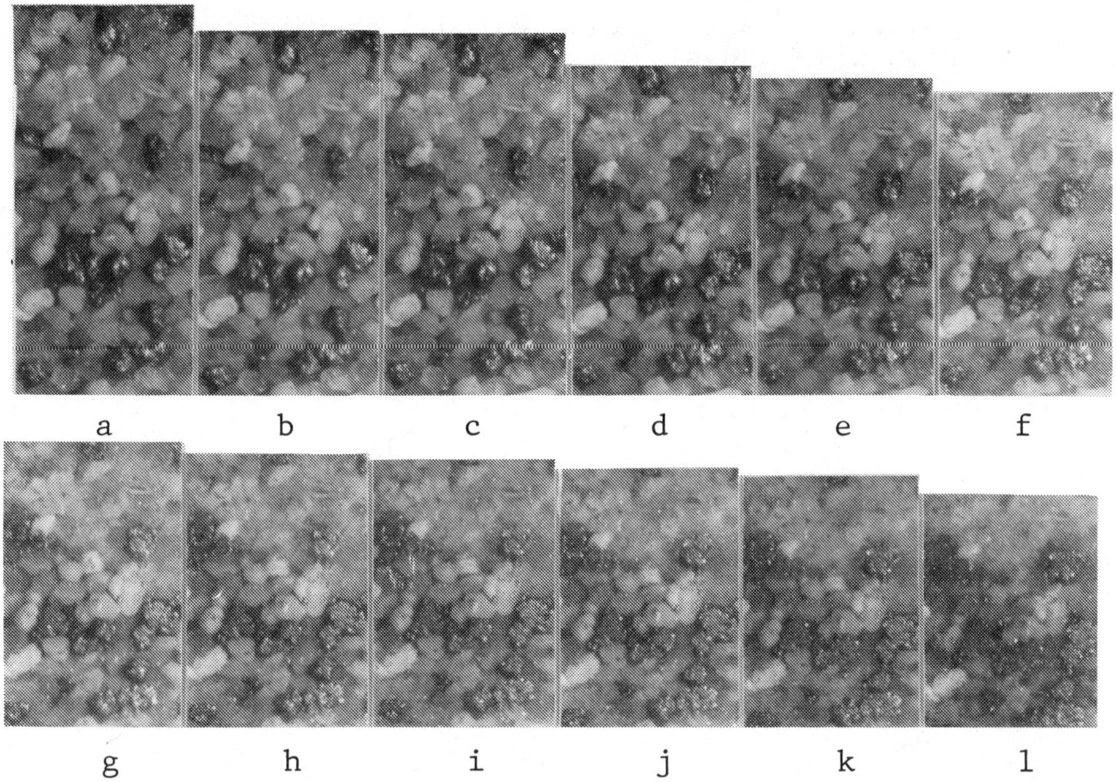

Fig. 1. Photomicrographic series showing compaction of mixture of molybdenum disulfide and silica gel as a function of pressure. Corresponding density vs pressure values are given in Fig. 2.

pelled. Under normal conditions at ambient temperature such a situation would not be realized. A more realistic approach to the volume ratio would be to use the grain density values. In this case the volume ratio would calculate to be 0.3925, which corresponds to 28.2 vol % MoS_2 and 71.8 vol % SiO_2. Now these values become significant in the initial filling of the cavity of the compaction apparatus. The bulk density of this mixture is measured to be 0.818 g/cm³ (zero load, Fig. 2). The interparticle porosity initially is 36.65 %, a value well within the expected range for this system. As pressure is applied, the grains began to fragment or disintegrate and both interparticle porosity and intraparticle porosity start to change. The exact ratio of this change at various pressures still needs to be measured.

High Pressure Compaction

Two high pressure experiments at different ratios of mixture of molybdenum disulfide and silica gel were conducted with the split-die apparatus at ambient temperature. As stated in the low pressure study, the porosity of the grains has a strong influence on the values of the volume of each component present. These various density values and corresponding volume content of the two materials are tabulated in Table II.

Table II. Apparent density/volume values for mixture of molybdenum disulfide and silica gel.

Property	Exp. No. 1	Exp. No. 2
Weight ratio MoS_2/SiO_2	2.654	1.3087
Apparent density, solid mix, g/cm^3	3.627	3.171
Volume content based on solid density values,		
vol % MoS_2	54.94	37.54
vol % SiO_2	45.06	62.46
Volume content based on grain density values,		
vol % MoS_2	43.37	27.42
vol % SiO_2	56.63	72.58
Bulk density of mix (zero load, Fig. 3), g/cm^3	1.054	0.850
Interparticle porosity of mix (initially), vol %	33.15	33.43

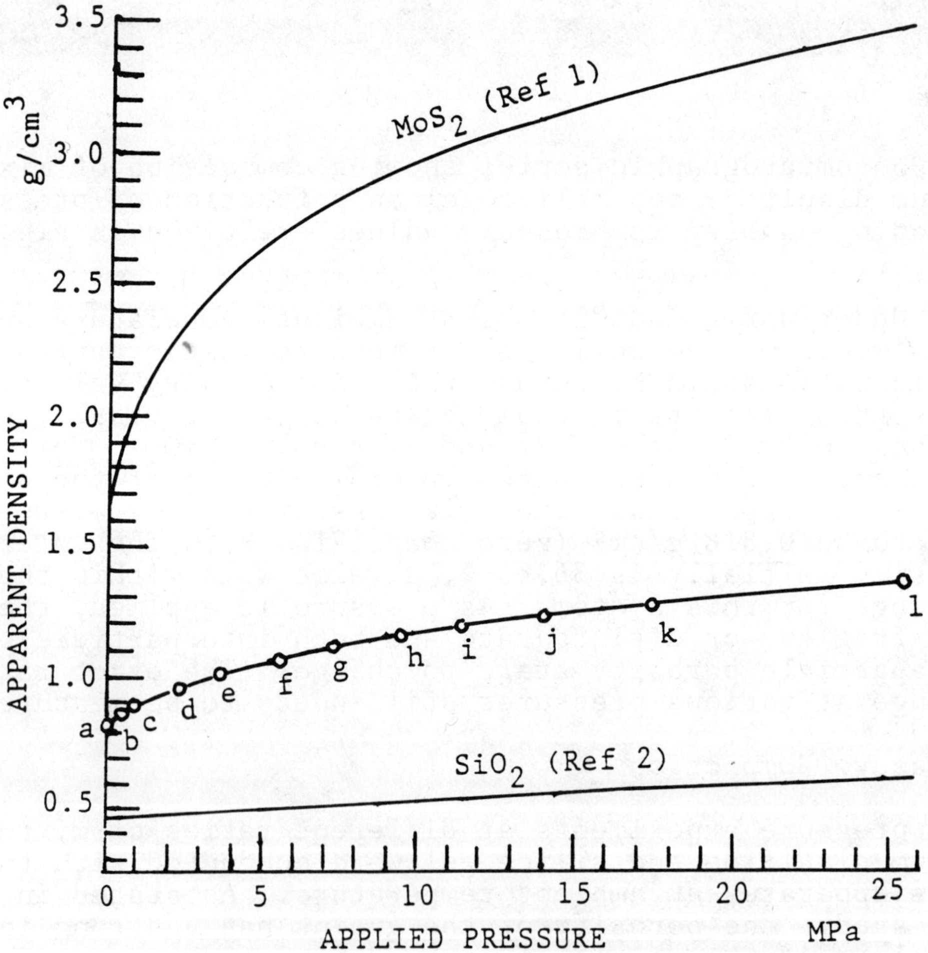

Fig. 2. Low pressure compaction of mixture of molybdenum disulfide and silica gel as compared to its components. Letters refer to plates in Fig. 1.

The experimental values of the observed apparent densities as a function of pressure for the two ratios of materials are shown in Fig. 3. The experimental points are designated by either a square (Exp. No. 1) or a circle (Exp. No. 2). At the conclusion of each high pressure compaction the pellet was removed from the die, then reinserted and recompressed. The recompression measurements for each experiment are designated by the symbol x. The raw compaction data have been corrected for the recompression of the pellet material as described in earlier papers of this series.[2,3] The corrected compaction curves are shown by dotted curves. The corrected compaction curves still represent the combined change in interparticle void and intraparticle void. Included in Fig. 3 are the values for the apparent density designated as "ρ_∞", which is an extrapolated value to an

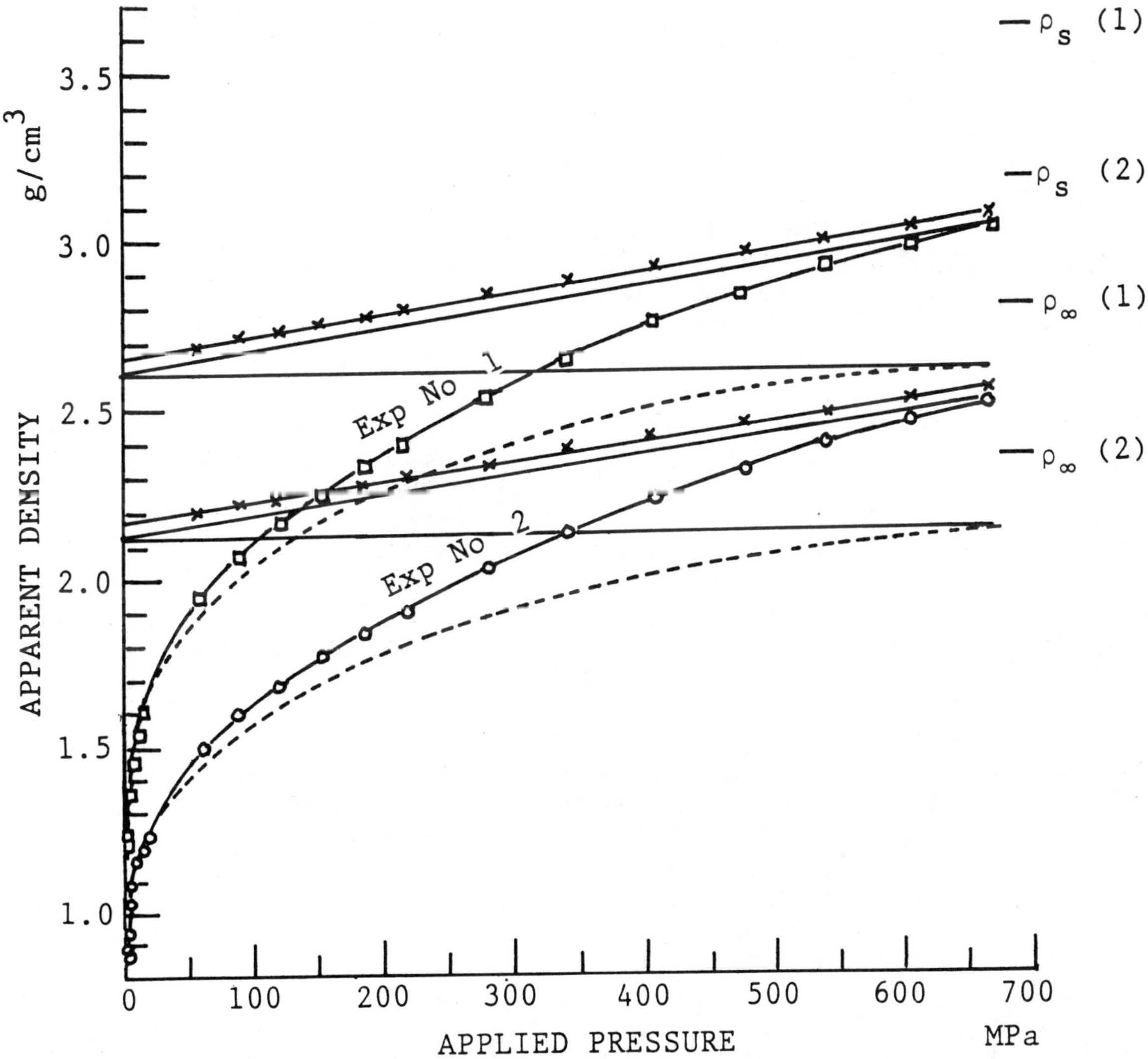

Fig, 3. High pressure compaction of two mixtures of molybdenum disulfide and silica gel showing correction for compressibility effect. Dotted curves represent corrected compaction curves. Values for concentration of components are given in Table II.

"infinitely" high pressure and represents a practical limit to which a pellet can be compacted at ambient temperature. This value is determined by a plot of the logarithm of the corrected relative reduction in volume as a function of the reciprocal of the pressure. The extrapolation of this curve to zero reciprocal pressure yields the infinite apparent density, ρ_∞. For Exp. No. 1 the value is calculated to be 2.80 g/cm³; for Exp. No. 2, 2.342 g/cm³. The calculated porosity at the highest pressure value obtained is 22.8% and 26.14%, respectively.

Pellets compacted to various pressure levels were also examined by means of the optical microscope. Photomicrographs of two views taken at the center of a pellet whose powder mix was compacted at 165 MPa are shown in Fig. 4. Plate a shows the silica gel dispersed in excess molybdenum disulfide; and Plate b, the molybdenum disulfide dispersed in excess silica gel. Plate a reveals several grains of silica gel which appear intact but the surrounding molybdenum disulfide grains are disintegrated. In Plate b both types of grains appear to be disintegrated.

DISCUSSION

The compaction behavior of the mixture of molybdenum disulfide and silica gel is similar to that of the individual components. From Fig. 1 we observed that during the consolidation process there is no visible slippage of particles (grains) past one another, and that there is no movement of particles lateral to the direction of applied force. Further, it can be shown from the curves in Fig. 3, that each component of the mixture behaves independently. At the high pressure levels, say, above 400 MPa, the molybdenum disulfide particles have been fully compacted to ρ_∞.[1] Since the weight of MoS_2 in the sample is known, one can calculate the volume occupied by MoS_2. From the apparent density at these pressure levels, the volume of silica gel can be ascertained. The calculated apparent density of the silica gel component is in reasonable agreement with the apparent density of pure silica gel.[2] Comparison of the apparent density values for silica gel in the two mixtures of Fig. 3 also is a good indication that the two components are acting independently. An analogous situation exists in the mixture of thoria and molybdenum disulfide.[4] Additionally, in some presently unpublished data for the binary mixture of tungsten and magnesium, the same type situation exists, namely, the two powder metals act independently. Thus, it can be concluded that in mixtures of components, each component acts independently. Hence any compaction equation that is described for a particular powder will also be applicable for a mixture of powders, except that there will be a separate term for each component. It should be noted that in all of these experiments the particle size of all materials is fairly uniform so that the initial packing of particles will be similar to each other. Another important factor in consolidation of powders is the need to separate compaction (irreversible process) from compressibility (reversible process). The importance of the compressibility effect is illustrated in Fig. 3. Obviously,

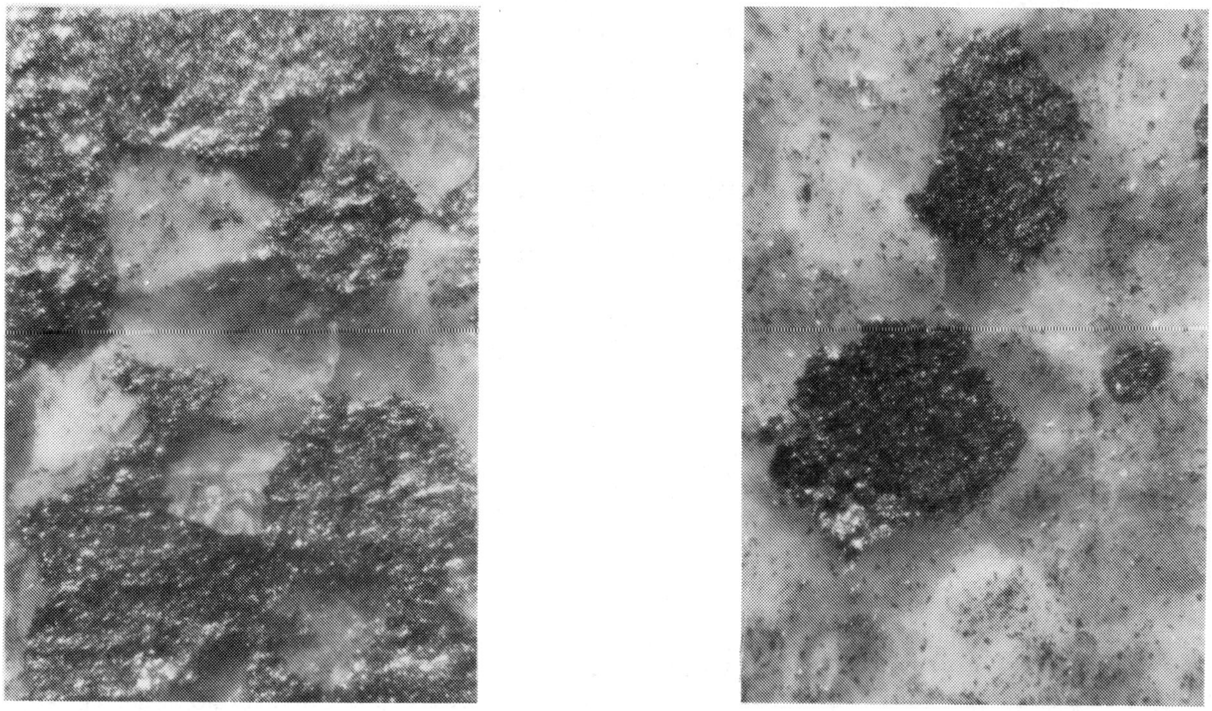

a b

Fig. 4. Photomicrographs of compaction at 165 MPa of mixture of molybdenum disulfide and silica gel. a: Silica gel dispersed in excess molybdenum disulfide. b: Molybdenum disulfide dispersed in excess silica gel.

ignoring the compressibility influence could lead to erroneous conclusions about compaction of powders.

In this study of the compaction of a mixture of powders, the premixing was such that isolated pockets rich in one or the other component were created. While this situation is not desirable, one can derive useful information such as illustrated in Fig. 4. With excess molybdenum disulfide some of the silica gel particles did not crush as in the case of the excess silica gel. This difference in behavior would be expected; crushing of silica gel particles in this situation requires silica gel to silica gel contact.

The use of the "infinite" apparent density, ρ_∞, obtained from extrapolation of the corrected compaction curve, has been described in an earlier paper.[4] Actually, the concept of an "infinite" apparent density value was first described by J. Heuberger[5] who was attempting to modify the "Shapiro-Konopicky Equation[6] in connection with pressing iron powders. The Shapiro-Konopicky Equation, first developed independently by Shapiro[7] and by Konopicky[8], works fairly well for easily deformed materials[6] This equation relates logarithm of porosity with pressure as shown in the following equation.

$$\ln P/P_o = -kp$$

where P = porosity at pressure p,

P_o = initial porosity,

and k = coefficient of powder compaction.

This equation is being used to develop a general material equation for compaction of powders, and represents the extreme case for plastically deformed material. The general equation will then be an expanded version of this equation and will contain terms relating to specific properties of the materials. Computer programs are being used to fit the variables in the equation to match the observed compaction data. Not only must the general equation fit the observed data, but also there must be a relationship of properties with the different materials.

CONCLUSION

Compaction of mixtures of powders can be treated as the sum of the compaction of the individual components. There is no visible slippage of particles past one another, nor is there movement of particles in a lateral direction to the applied force. Compaction curves must be corrected for the reversible compression effect. A general material equation for compaction of powders is being developed which utilizes the Shapiro-Konopicky Equation as a model. This latter equation represents a limiting case.

REFERENCES

1. I. Shapiro, "Compaction of Powders. III Molybdenum Disulfide," Materials Science and Engineering, 71, 333-340 (1985).
2. I. Shapiro, "Compaction of Powders. VI Silica Gel," in: Ceramic Transactions, Vol 1B, Ceramic Powder Science II, G. L. Messing, E. R. Fuller,Jr., H. Hausner, editors, The American Ceramic Society, Inc., Westerville, OH, 1988, 776-783.
3. I. Shapiro, "Compaction of Powders. IV Binary Mixture of Thoria and Molybdenum Disulfide," in: High Tech Ceramics, P. Vincenzini, editor, Elsvier Scientific Publishing Co., New York, 1987,38A, 601-608.
4. I. Shapiro, "Compaction of Powders. II Thoria," in: Deformation of Ceramics II, R. E. Tressler and R. C. Bradt, editors, Plenum, New York, 1984, 733-744.
5. J. Heuberger, "The Compressibility of Iron Powders," Trans. Chalmers Univ. Technol., Gothenburg, 98, 11-20 (1950).
6. W. D. Jones, "Fundamental Principles of Powder Metallurgy," Edward Arnold (Publishers), London, 1960, 284.
7. I. Shapiro and I. M. Kolthoff, "Compressibility of Silver Bromide Powders," J. Phys. and Colloid Chem., 51, 483-493 (1947).
8. K. Konopicky, "Parallelism of the regularities in ceramics and powder metallurgy," Radex-Rundschau, 141-148 (1948).

PART DISTORTION DURING BINDER REMOVAL

C.A. Sundback, M.A. Costantini, and W.H.Robbins
Ceramics Process Systems
Cambridge, MA 02139

ABSTRACT

Defects develop during binder burnout primarily because of part distortion. The formation of these defects can be directly detected by monitoring dimensional change during heating. In this study, thermal mechanical analysis was used to monitor dimensional change in a thermoplastic injection molded ceramic, heated at 0.5°C/min. The powder was a 0.6 μm alumina; the organic system contained polypropylene and two low temperature melting waxes.

Major part distortion occurred before and during periods of major weight loss. However, most defects were produced during rapid part distortion which accompanied rapid weight loss. Modifying the heating schedule to include a low temperature hold period minimized rapid part distortion and defect formation.

INTRODUCTION

Binder removal from filled ceramic green bodies can be a lengthy, low-yield process. The time required for binder burnout of thick-cross section, complex-shaped parts is often of the order of a week[1,2]. Yields can be low because of cracking, swelling, blistering, and/or delamination which occur as a result of part distortion[1,3].

These binder burnout defects could be predicted if the mass transport, heat transport, and chemical kinetic processes could be modelled as a function of temperature. However, because these processes are highly complex, modelling defect formation would be extremely difficult. Recently some understanding about these physical and chemical processes has been obtained from analyses of the organic solid residue and of the volatile products leaving the solid[4,5,6]. However, these analyses are tedious and expensive to conduct, and still the burnout mechanistic picture is not complete. Consequently, heating schedules are still typically developed using empirical analyses such as thermal gravimetric analysis (TGA).

Although not as versatile as a predictive model, a more appropriate analytical technique would be one that monitors part distortion, the cause of many binder burnout defects. Thermal mechanical analysis (TMA), commonly used to monitor linear dimensional changes in substances during heating, is easy to use, relatively inexpensive, and available in many thermal analysis laboratories.

In this study, the analytical techniques of TMA and TGA were used to monitor the dimensional changes and weight loss behaviors, respectively, of fine-grained, thermoplastic injection molded ceramic pieces during binder burnout. Major part distortion was found to occur before and during periods of weight loss. However, most binder burnout defects were produced during rapid part distortion which accompanied rapid weight loss. Modifying the heating schedule to include a low temperature hold period minimized rapid part distortion and consequently the extent of the binder burnout defects.

EXPERIMENTAL

Equipment

DuPont thermal analysis equipment was used in this study. The TGA was a DuPont 951 Thermogravimetric Analyzer. The experimental parameters used were: 50 cc/min flow rate of nitrogen purge gas and 1.0°C/min and 0.5°C/min heating rates. The lowest controllable heating rate was 0.5°C/min.

Thermal mechanical analyses were performed on a DuPont 943 Thermomechanical Analyzer; the major system components are shown in Figure 1. In this study, an expansion probe with a 1 gram load in the weight tray was used to monitor linear dimensional change. Vertical motion of the probe was detected by a linear variable differential transformer (LVDT). The expansion probe had a contact diameter of 2.54 mm. A small weight, 1 gm, was used to minimize probe penetration into the sample.

The TMA was purged with air to prevent condensation of volatiles onto sensitive instrument mechanisms. The gas, introduced into a steel tube on the back of the instrument, flowed down through the sample tube and out an opening close to the location of the sample.

Sample Specifications and Preparation

As detailed in Table 1, the green samples consisted of fine-grained, high-purity alumina ceramic with a thermoplastic binder system; the solids loading was 89 wt% solids (65 vol%). The thermoplastic binder was composed of an isostatic polypropylene and two low temperature melting waxes, wax1 and wax2. The ceramic and organic components had been mixed in a double-arm mixer and injection molded into a bar mold with cavity dimensions of 127 mm x 12.7 mm x 3.18 mm.

For this study, these molded bars were cut into squares, nominally 6.4 mm x 6.4 mm x 3.18 mm in size and 0.25 gm in weight. For each TMA test, one sample square was positioned in the sample holder so that the expansion probe contacted the approximate center of the sample and the probe monitored changes in the nominal 3.18 mm dimension. This sample surface in contact with the expansion probe was flat and smooth, since it was a molded surface.

TGA samples were obtained from cut squares, one TGA sample per square. A square was fractured and a chunk was broken off the fracture surface. This chunk, approximately 85 mg by weight, was used as a TGA sample.

Prior to thermal analysis testing, some green molded samples were heat-treated in a Thermolyne 6000 oven, which was purged with flowing air. The green samples were either placed in a crucible packed in 0.3 μm alumina powder or placed on a bare alumina substrate in this oven. The samples were heated at 1°C/min to 140°C and held for 16 hrs.

The thermal analysis curves presented in this paper are either curves which were representative or curves which were averaged over a number of experiments.

RESULTS AND DISCUSSION

Qualification of TMA

To be acceptable tools for binder burnout studies, analytical instruments must be capable of heating samples without introducing binder burnout defects and must also be sensitive enough to accurately monitor sample changes. Generally smaller sample sizes and faster heating rates are used in thermal analysis studies in comparison with actual binder burnout processes. Defect-free burnout can easily be achieved in TGA tests with 10 - 100 mg sample sizes. However the 3.18 mm thick TMA samples used in this study required slow heating rates in order to prevent sample cracks and delaminations.

Representative TMA traces for the studied thermoplastic ceramic system are shown in Figure 2 at heating rates of 1°C/min and 0.5°C/min. The shapes of the curves are similar at both heating rates. Gradual dimensional change occurred below 350°C. However, both samples experienced rapid dimensional change around 375°C. "Gradual" and "rapid" dimensional change denote moderate and very large slopes in the TMA curves.

Larger dimensional changes occurred in the more rapidly heated sample during the rapid weight loss periods at approximately 280°C and 370°C, as shown in Figure 3. The sample heated at 0.5°C/min appeared to be defect-

Table 1. Sample Formulation

Component	Wt%	Density (gm/cm^3)
Alumina*	89	3.98
Polypropylene**	7	0.90
Wax1	2	0.94
Wax2	2	0.92

* Alumina Specifications
A16-SG, Aluminum Company of America
0.6 μm average particle size
7 m^2/gm surface area

** Isostatic Polypropylene Specification
35 Melt Index

free; however, the sample heated at 1°C/min was cracked. Consequently, a heating rate of 0.5°C/min was used in this study.

The linear dimensional change monitored by TMA was in close agreement with the estimated shrinkage, based on density measurements. At a 0.5°C/min heating rate, the overall TMA linear dimensional change was approximately 0.71%; the linear shrinkage was estimated to be 0.57%, based on the density increase in the part from molding to binder burnout.

Explanation of TMA Diagram

Several noteworthy observations can be made about the TMA and derivative TGA curves shown in Figure 4, obtained at a 0.5°C/min heating rate. Note that significant TGA events often appeared to occur at higher temperatures than significant TMA events. Small differences in the temperature ranges of TGA and TMA events are not unexpected since different furnaces, flow paths, and sample sizes were used in the two types of experiments.

At temperatures less than 140°C, the sample experienced a gradual expansion of 0.9%, which was probably related to the thermal expansion of the organic components and trapped air. As listed in Table 2, pure wax1 and pure wax2 melted at 57°C and 65°C, respectively. Prior to melting, the linear expansions of wax1 and wax2 were 0.8% and 7.7%, respectively, based on TMA measurements. Below 65°C, the melting point of wax2, the ceramic/organic expansion was approximately 0.45% which is comparable to 0.53%, the ceramic/organic expansion estimate based on the concentrations and measured expansions of wax1 and wax2, listed in Tables 1 and 2.

Above 65°C, expansion continued until weight loss began at approximately 140°C. Assuming the waxes are fluid above 65°C, continued expansion was probably related both to the expansion of air compressed into the waxes during injection molding[1] and to the continued expansion of the organic components.

Sample shrinkage seemed to correlate with weight loss. The sample began to shrink as weight loss began, at approximately 140°C. Shrinkage continued up to approximately 350°C. Note that the TMA signal increased slightly beginning at 165°C, approximately the measured melting point of pure polypropylene (Table 2); this expansion may be related to the expansion of air compressed into the polypropylene during injection molding.

Table 2. Expansion and Melting Point Data of Binder Components

Component	Melting Temperature (°C)	Expansion (%)
Wax1	57	0.8
Wax2	65	7.7
Polypropylene	165	0.0

Between 140°C and 350°C, the weight loss rate was small, with a maximum rate of 0.015 wt%/min. However, at 370°C when the weight loss rate rapidly increased to 0.195 wt%/min, the sample rapidly expanded 0.4%. The sample then shrank as the rate of weight loss decreased. However, the measured dimension did not return to the pre-expansion value, suggesting that a small amount of deformation occurred during the rapid expansion.

Improvement in Burnout Schedule

Comparing the TMA and derivative TGA curves in Figure 4, major dimensional change occurred whether or not major weight loss was occurring. In the course of this study, cracks were observed in samples during periods of major dimensional change, again with or without accompanying major weight loss. However, the tendency for samples to crack was greater during periods of rapid dimensional change than during periods of gradual dimensional change.

Recognizing that rapid dimensional change generally occurred during rapid weight loss, the binder burnout schedule was modified to minimize the rate of dimensional change. Samples were held at 140°C for 16 hr on a smooth alumina surface or packed in 0.3 μm alumina packing powder prior to TMA analysis. The hold temperature of 140°C was chosen based on the reasoning that the melted waxes could be removed from the sample prior to the melting of the polypropylene, producing an open pore network for the escape of volatiles.

Samples which had been packed lost more weight during the heat treatment period than samples which were unpacked because of the capillary extraction force of the packing powder on the melted binder components[7]. The average weights lost from packed and unpacked samples were 1.4 wt% and 0.9 wt%, respectively.

All heat-treated samples had a core and rind structure. As expected, the organic level in the rind was lower than in the core; the organic levels in the rind and core were 5 wt% and 10 wt%, respectively, for a sample packed in powder during heat treatment.

Heat treatment of the samples at 140°C for 16 hrs significantly affected the dimensional response of the sample, as shown in Figure 5. Not surprisingly, heat-treated samples had significantly smaller dimensional increases below 140°C than the nonheat-treated sample.

However, between 150°C and 180°C, the behaviors of the heated treated samples were significantly different from the nonheat-treated sample. The heat-treated samples exhibited a significant expansion and contraction peak while the nonheat-treated sample showed only moderate dimensional change. The size of the dimensional change peak was larger for the unpacked sample which lost less weight during heat treatment than the packed sample. The cause is unclear for the change in the dimensional response of the heat-treated sample with respect to the nonheat-treated sample in this temperature range; the change could be related to the distribution of organics in the heat-treated structure or to a modification of the polypropylene structure during heat treatment.

Heat treatment significantly modified the dimensional response of the sample during the critical period when the rate of weight loss was high, approximately 350°C to 450°C. In comparison with the nonheat-treated sample, the unpacked heat-treated sample had significantly less dimensional change in this temperature range while the packed heat-treated sample showed no dimensional change in this temperature range.

The reduced dimensional response of the packed heat-treated sample is more dramatically illustrated by a direct comparison of the TMA curve and the derivative TGA curve of the core section, as shown in Figure 6. No dimensional change occurred after 300°C even when rapid weight loss was occurring. A highly probable explanation for the facile binder burnout process is that an open pore channel network developed during heat treatment through which volatiles readily escaped. Unhindered volatile escape reduced part distortion and, therefore, defect formation.

CONCLUSION

In summary, information about part distortion during binder burnout was obtained using the established, easy to use, relatively inexpensive, analytical technique of thermal mechanical analysis (TMA). For injection molded samples, major dimensional expansion and contraction were observed while heating, regardless of whether the sample was experiencing significant weight loss. Although cracks could be formed during gradual dimensional changes, defect formation was more commonly observed during rapid dimensional change. Heat treating the samples prior to TMA testing essentially eliminated the rapid dimensional change. Volatile escape appeared to be unhindered in the heat-treated sample during the rapid weight loss period, suggesting the presence of an open pore channel network after heat treatment.

REFERENCES

[1]C.L. Quackenbush, K. French, and J.T. Neil, "Fabrication of Sinterable Silicon Nitride by Injection Moulding," *Ceram. Eng. Sci. Proc.*, 3, 20-34 (1982).

[2]A. Johnsson, E. Carlstrom, L. Hermansson, and R. Carlsson, "Minimization of the Extraction Time for Injection Moulded Ceramics," *Proc. Brit. Ceram. Soc.*, 33, 139-47 (1983).

[3]"Injection Moulding Machine for New Ceramics"; Japan Steel Works (Hiroshima Plant), Trade literature, 1984.

[4]Y.N. Sun, G.W. Scheiffele, C.H. Do, M.D. Sacks, and J.W. Williams, "Thermal Degradation Mechanisms of Poly(Vinyl Butyral) Binder"; presented at the 89th Annual Meeting of the American Ceramics Society, Pittsburgh, 1987.

[5]Y.N. Sun, J.W. Williams, and M.D. Sacks, "Pyrolysis Behavior of Poly(Methyl Methacrylate)"; presented at the 89th Annual Meeting of the American Ceramics Society, Pittsburgh, 1987.

[6]S. Masia, P. Calvert, W. Rhine, and H.K. Bowen, "Effect of Oxides on Binder Burnout During Ceramics Processing"; submitted to *J. Mat. Sci.*, April 1988.

[7]I. Peltsman and M. Peltsman, "Low Pressure Moulding of Ceramic Materials," *Interceram*, 4, 56 (1984).

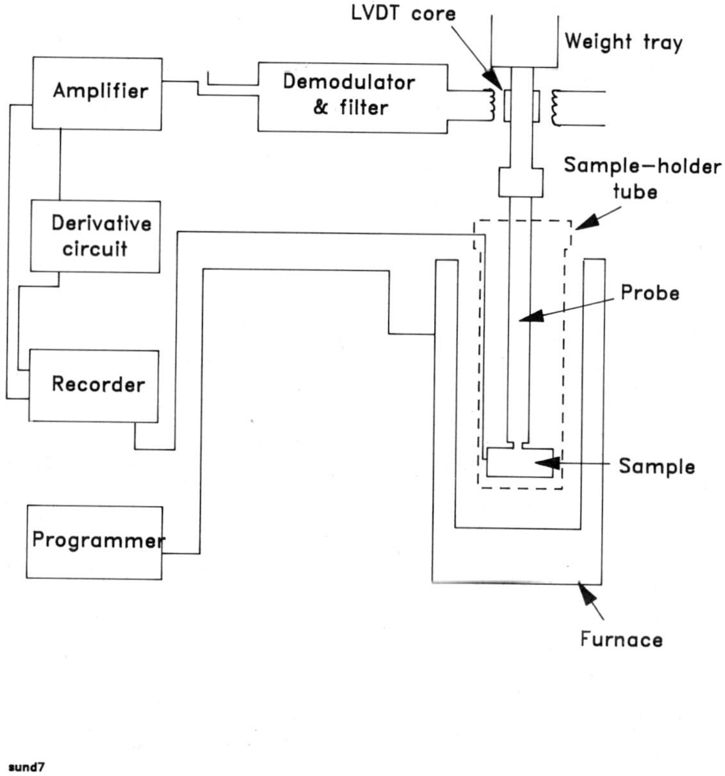

Figure 1. Schematic of DuPont 943 Thermomechanical Analyzer

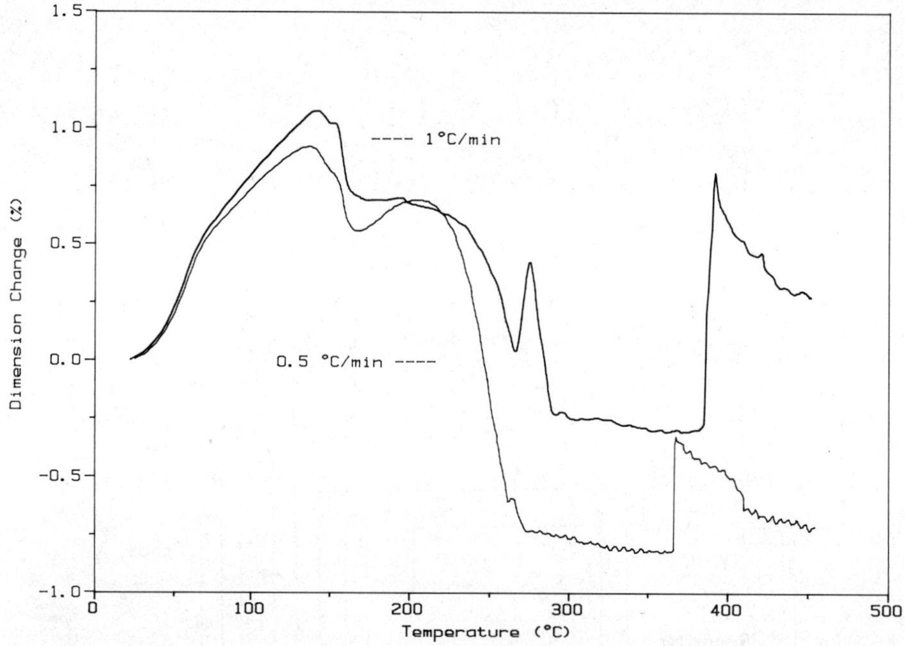

Figure 2. Dimensional Change Curves of Green Samples, Heated at 0.5°C/min and 1.0°C/min

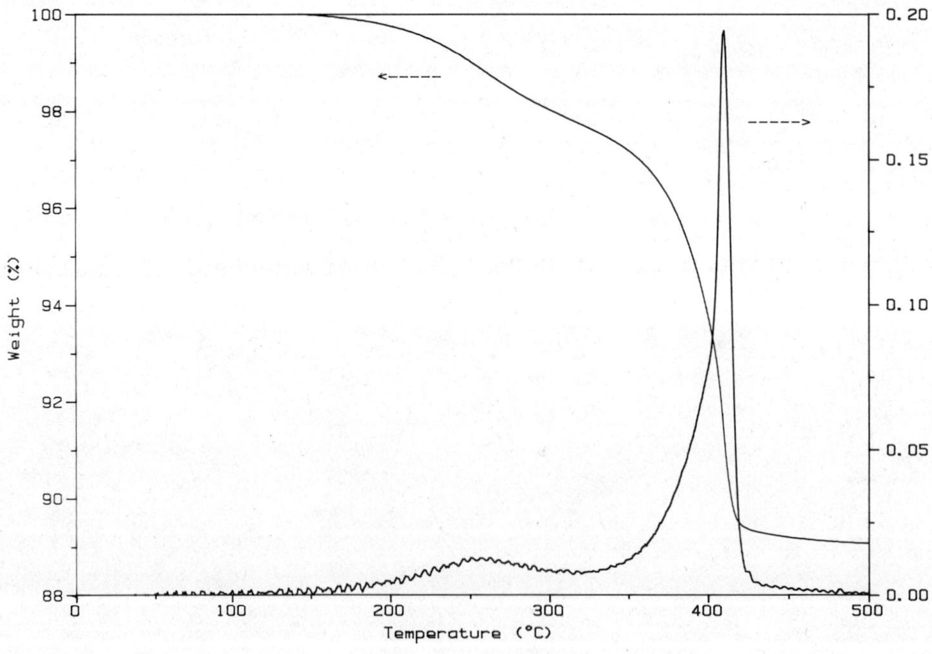

Figure 3. Weight Loss and Derivative Weight Loss Curves for Green Sample, Heated at 0.5°C/min

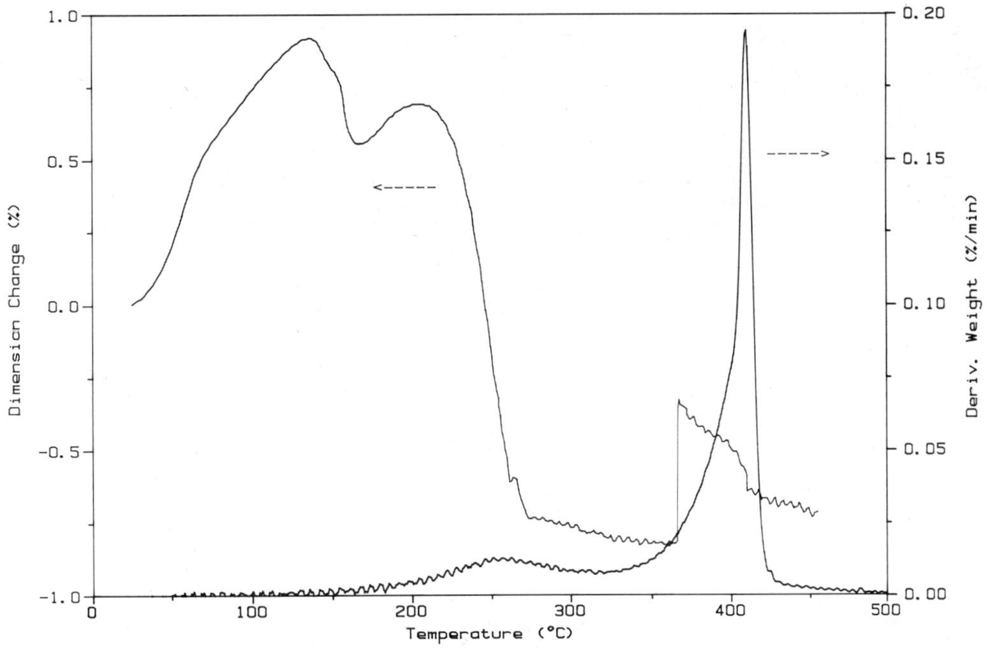

Figure 4. Dimensional Change and Derivative Weight Loss Curves for Green Sample, Heated at 0.5°C/min

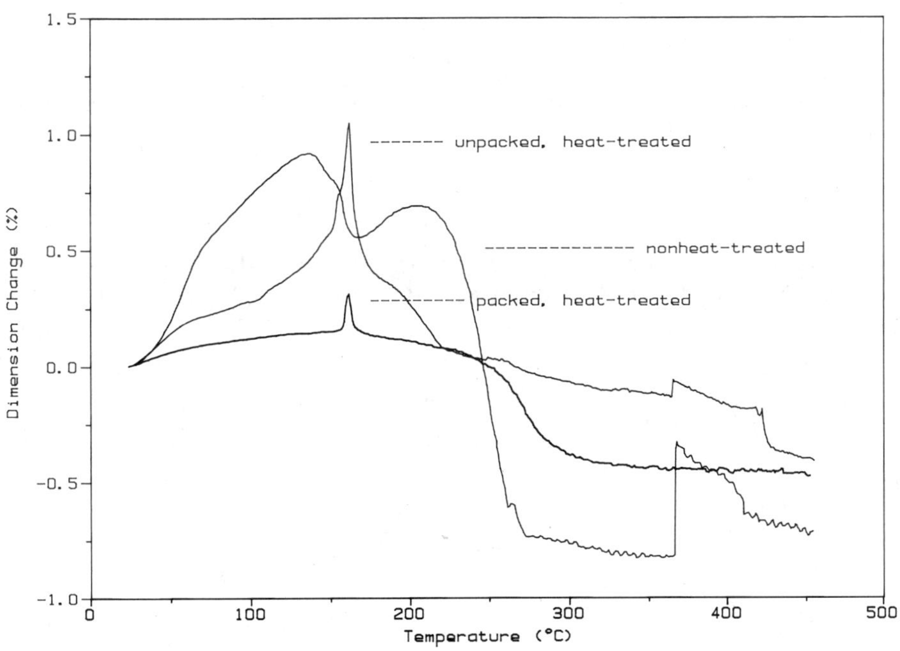

Figure 5. Dimensional Change Curves for the Following Samples: Green (Nonheat-Treated), Packed in Powder During Heat Treatment, Unpacked During Heat Treatment. TMA Tests at 0.5°C/min Heating Rate

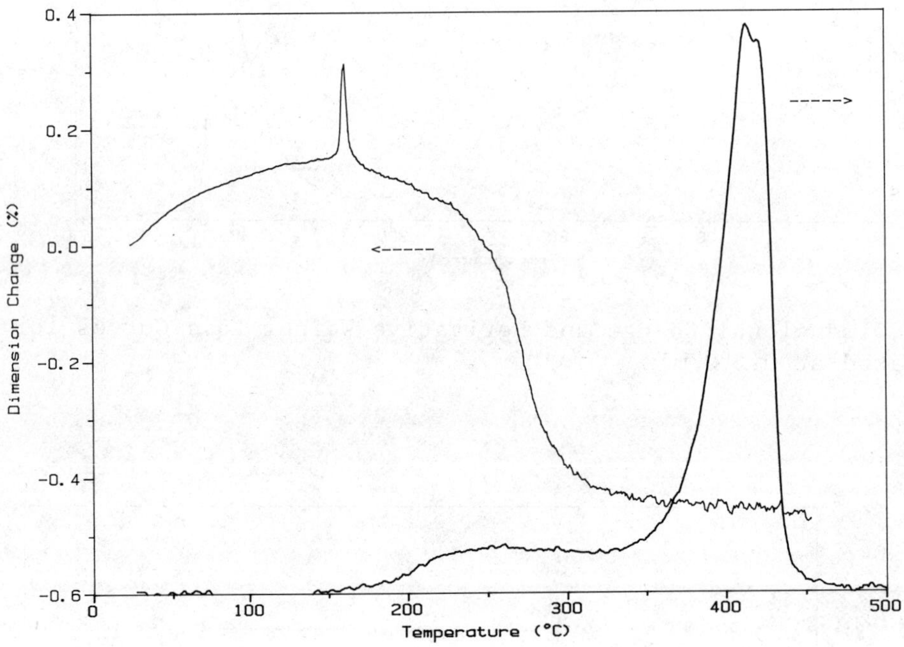

Figure 6. Dimensional Change and Derivative Weight Loss Curves for Samples Packed in Powder During Heat Treatment. TMA and TGA Tests at 0.5°C/min Heating Rate

STABILITY OF THERMAL BARRIER COATINGS

P. Vincenzini, CNR-Research Institute for Ceramics Technology, Faenza, Italy; G. Appiano, FIAT Research Centre, Orbassano, Turin, Italy; F. Brossa, CEC-Ispra JRC; S. Meriani, University of Trieste, Italy

ABSTRACT

Structural stability of ZrO_2-Y_2O_3 and ZrO_2-CeO_2 plasma sprayed thermal barrier coatings has been investigated by high temperature annealing and surface modification by electron irradiation. Thermal annealing resulted in phase transformations which are in substantial agreement with the stable/ metastable phase relationships in ZrO_2-Y_2O_3 and ZrO_2-CeO_2 systems. Electron surface treatment resulted in pore free recrystallized layers containing substantial cracks. A possible ferroelastic effect observed in the ZrO_2-Y_2O_3 coating after electron irradiation has been suggested to derive from domain switching driven by stresses originated by the extremely high cooling rate within the electron modified area.

INTRODUCTION

Zirconia Thermal Barrier Coatings (Z-TBCs) are main candidates for insulation of thermal engines. High temperature structural stability and stability against environmental attack of combustion gases containing corrosive metal impurities, especially Na, V, and P, are essential prerequisites for the above application. HIPing[1] and laser annealing[2,3] have been experienced in sealing the surface pores of Z-TBCs to prevent deep penetration of gaseous species or condensed corrosive liquids. Saturation of coating porosity with molten salts and hot corrosion (dissolution) by molten salt deposits have in fact been anticipated as potential failure modes of Z-TBCs.[4]

This study concentrates both on transformations resulting from high temperature annealing of ZrO_2-Y_2O_3 (Y-ZrO_2) and ZrO_2-CeO_2 (Ce-ZrO_2) TBCs and on their structural modifications driven by high energy electron irradiation.

EXPERIMENTAL

Thermal barrier coatings were produced by plasma spraying in air on valve steel substrate zirconia powders of composition ZrO_2-8wt%Y_2O_3* and ZrO_2-25wt%CeO_2-2.5wt%Y_2O_3.** An intermediate layer of plasma sprayed

*METCO 204 NS.

**METCO 205 NS.

Ni-17.5wt%Cr-2.5wt%Co-0.5Y bond coat was applied to improve adherence and durability. Compositions of the tested zirconia powder fall in the tetragonal (T) and monoclinic (M) + cubic (C) phase fields of room temperature equilibrium phase relationships of corresponding systems (Fig. 1).[5,6]

Annealing experiments were conducted in an electric heated oven in oxygen environment at 1100 to 1600°C and 10 to 100 h holding time. After annealing, the samples were left to slow cool down to room temperature into the furnace.

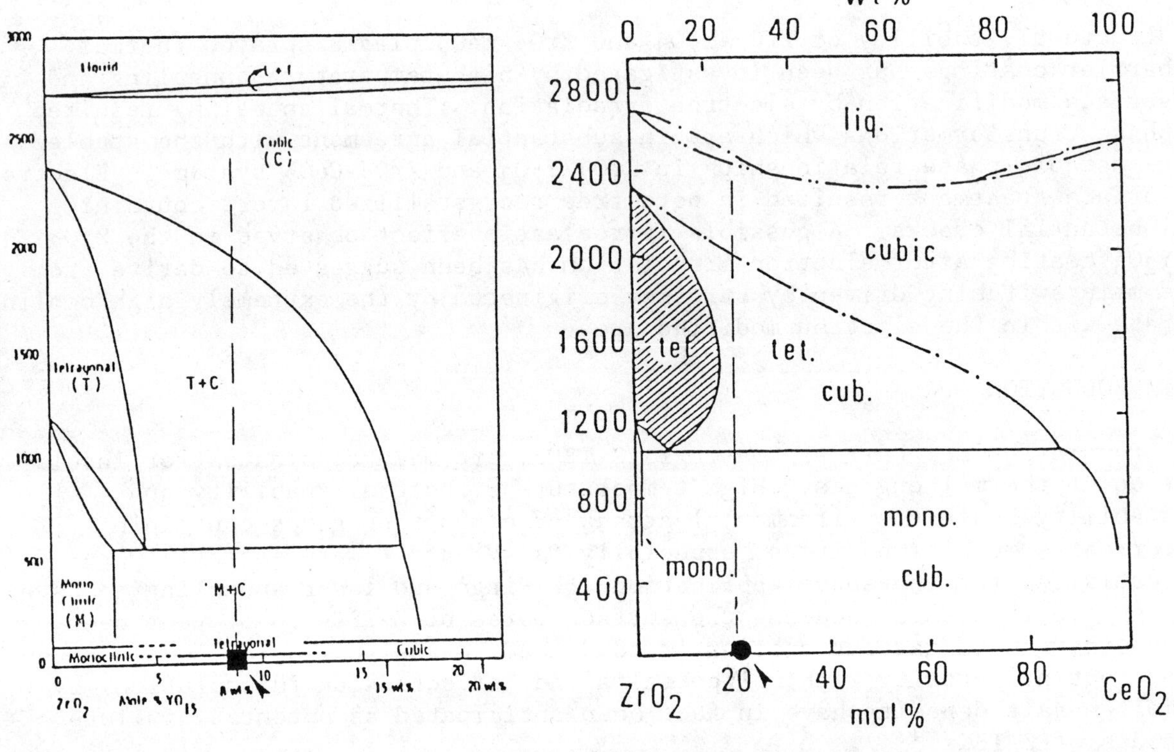

Figure 1. Equilibrium phase relationships on the ZrO_2-Y_2O_3 (Ref. 5) and ZrO_2-CeO_2 (Ref. 6) systems. Studied compositions are indicated.

The electron gun facility at the Ispra JRC (Fig. 2) has been used to irradiate Y-ZrO_2 and Ce-ZeO_2 TBCs deposited on metal substrated of 15 mm diam. The apparatus has an accelerating voltage of 10 to 30 KV and maximum power of 10 KW. The maximum acceleration voltage was used in our experiments. Electron gun irradiation parameters are summarized in Table 1.

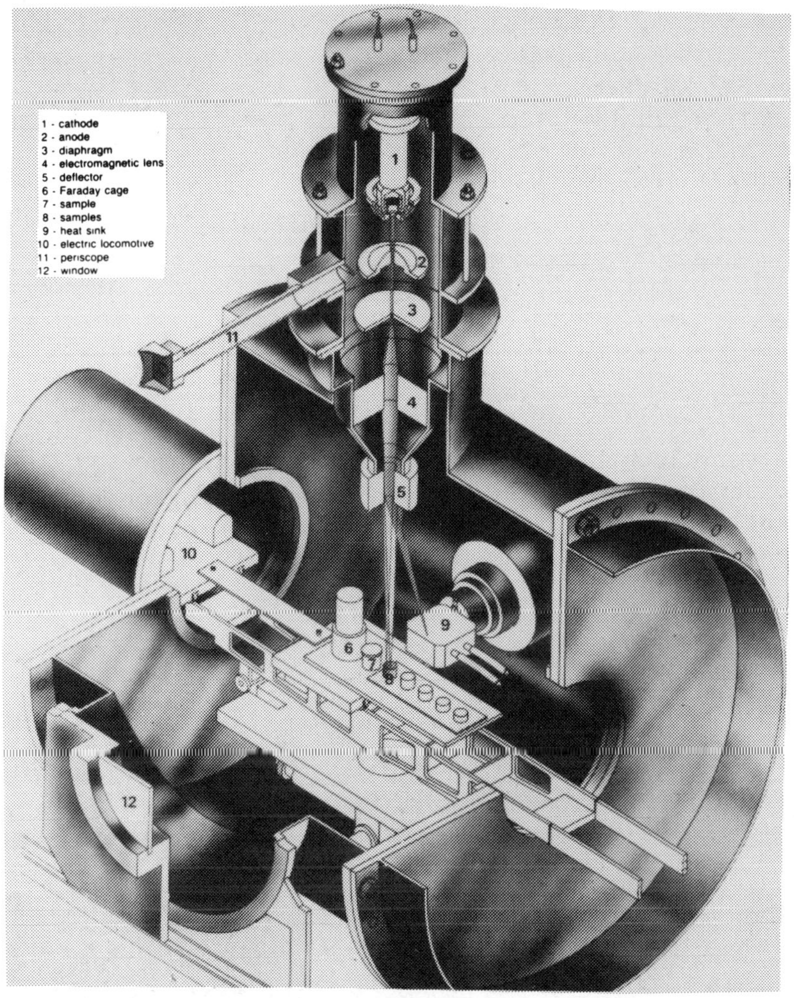

Figure 2. Electron gun configuration for irradiation tests.

Table 1. Electron gun irradiation parameters.

ELECTRON GUN IRRADIATION PARAMETERS	
VACUUM	10^{-4} TORR (DURING OPERATION)
IRRADIANCE	~12 KW/CM^2
IMPULSE DURATION	20.10^{-3} SEC
IRRADIATED AREA	0.8 CM^2
DEPTH OF PENETRATION	40-60 μM

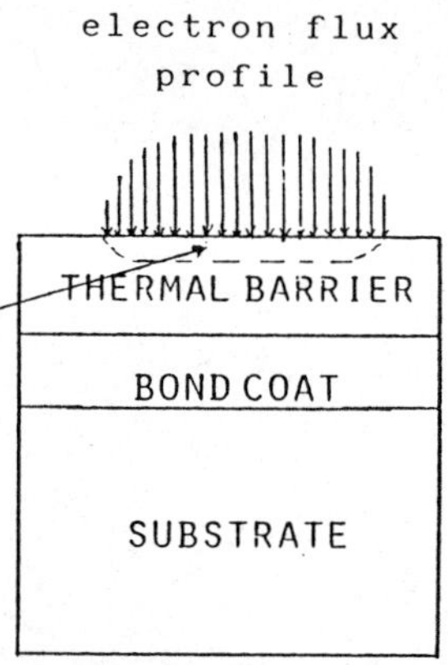

Because of the very low thermal conductivity of the TBC, electron impact resulted in temperatures far in excess of the melting point of zirconia s.s. up to a depth of about 40–60 μm. The brown appearance of the electron irradiated areas is shown in Fig. 3.

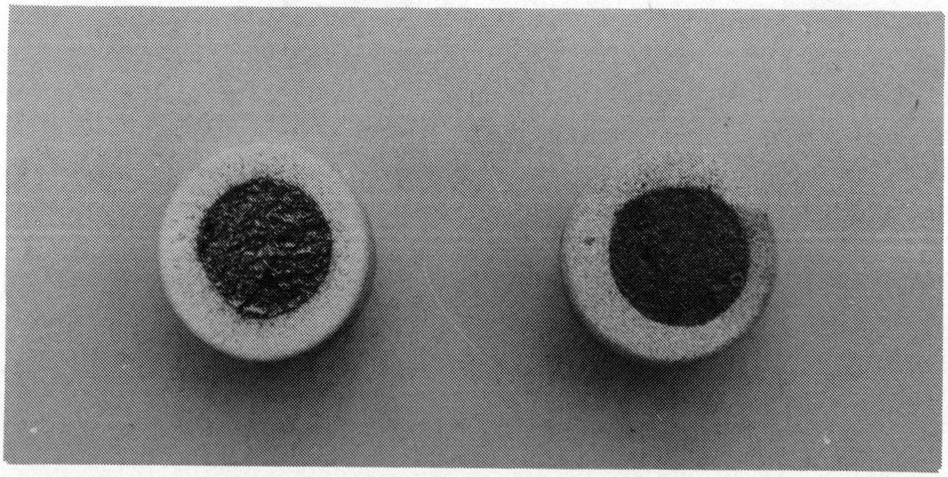

Figure 3. Electron irradiated samples. Dark zones indicate irradiated areas.

Annealed and irradiated samples were examined by XRD, SEM, and ESCA analyses.

SUMMARY OF RESULTS

Phase modifications occurred in the Y-ZrO_2 and Ce-ZrO_2 TBCs after 100 h annealing at different temperatures are summarized in Table 2.

Table 2. Summary of results of annealing experiments on Y-ZrO_2 and Ce-ZrO_2 thermal barrier coatings

ANNEALING	
ZrO_2-8.0 wt% Y_2O_3	
As sprayed	: nontransformable tetragonal T' (quenched from HT) minor amounts of M from T
1200°C 100h	: nontransformable tetragonal T' minor amounts of M from T
1400°C 100h	: high-yttria cubic ss C substantial M from low-yttria T
1600°C 100h	: distorted high-yttria cubic ss C minor amounts of M from low-yttria T
ZrO_2-25 wt% CeO_2-2.5 wt% Y_2O_3	
As sprayed	: distorted TZ°
1200°C, 1400°C, 1600°C } 100h	: TZ° with increased ordering at higher temperatures

The above transformation pattern is in substantial agreement with the well known stable/metastable phase relationships in the ZrO_2-Y_2O_3 system[7] and with the most recent findings on the metastable phase relationships of the ZrO_2-CeO_2 system studied under conditions of slow cooling rate.[8]

SEM micrographs of irradiated samples showing features of surface modification are shown in Figs. 4 and 5.

The most relevant effect deriving from electron irradiation is the melting of the irradiated area to a depth of some tens of microns (40–60 μm) depending on thermal conductivity of the material and local microstructure and geometry of the coating. During cooling, the molten material recrystallizes.

Two systems of cracks are generated during cooling because of tensile thermal stresses deriving from thermal expansion mismatch between the electron affected volume and the "cold" TBC interface. These develop either as cracks perpendicular to the sample surface within the electron-affected area and/or parallel to the sample surface at the boundary between the modified and unmodified regions.

Reduction of ZrO_2 and combined reduction of ZrO_2 and CeO_2 on electron irradiation is responsible for the brown color of the irradiated areas.

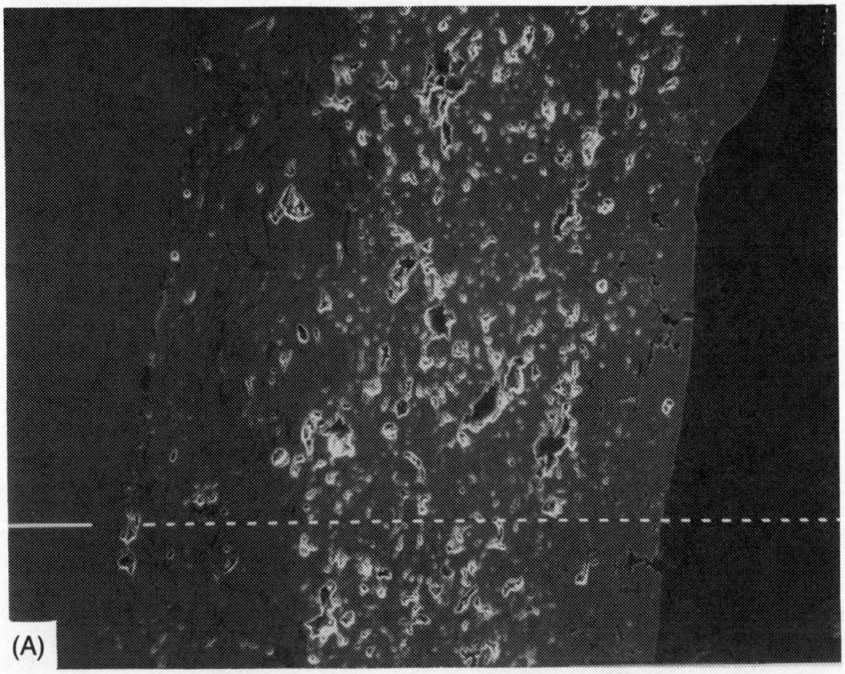

ZrO_2-Y_2O_3 (150 x)

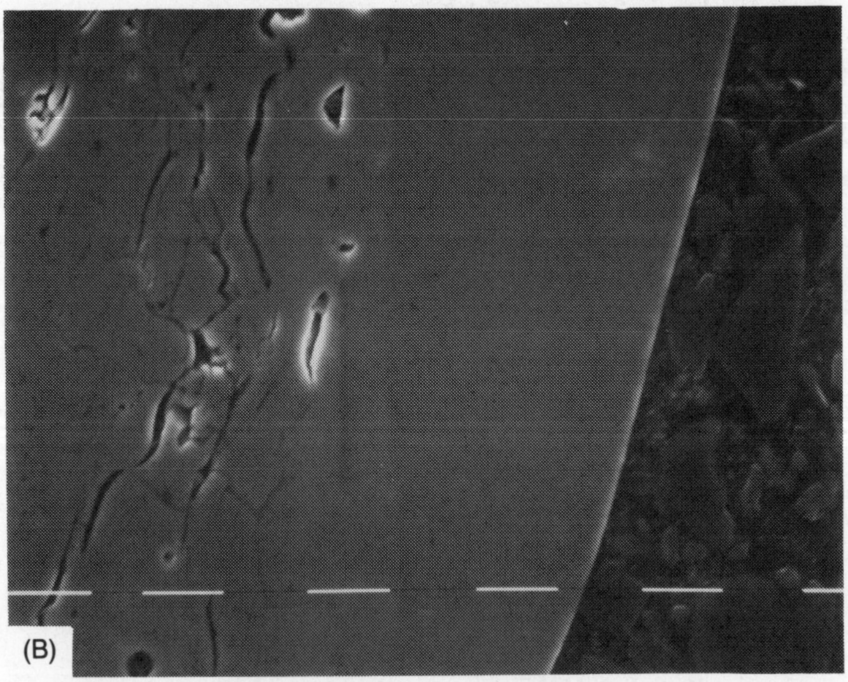

ZrO_2-Y_2O_3 (1250 x)

Figure 4. SEM micrograph of a polished cross section of irradiated Y-ZrO_2 sample. A) Enlarged view showing both the electron affected and the "virgin" areas and, B) A selected area showing almost complete sealing of surface pores and cracks generated at the electron affected/"virgin" interface.

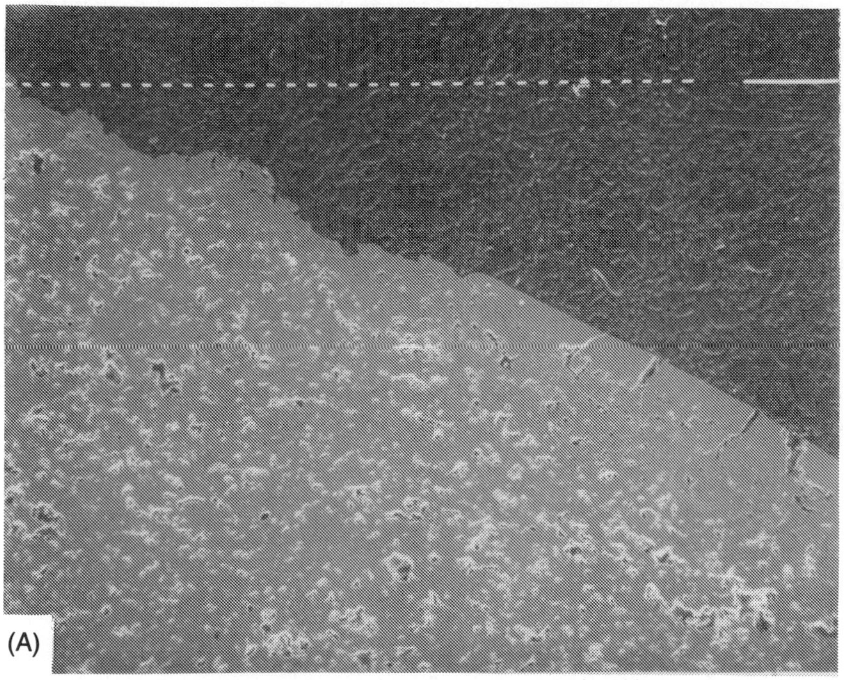

ZrO_2-CeO_2-Y_2O_3 (150 x)

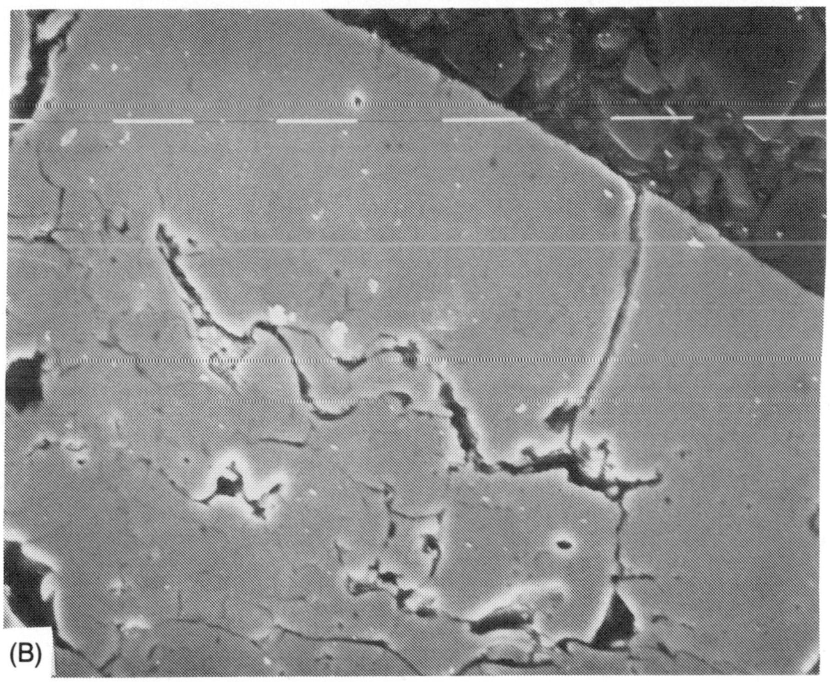

ZrO_2-CeO_2-Y_2O_3 (1250 x)

Figure 5. SEM micrograph of a polished cross section of irradiated Ce-ZrO_2 sample. A) Enlarged view showing both electron affected and "virgin" areas and, B) A selected area showing almost complete sealing of surface pores and two crack systems, perpendicular and parallel to the irradiated surface.

Although some evaporation from molten material is foreseen to occur under irradiation, no major selective evaporation seems to have occurred as shown by comparative EDAX analyses of electron affected and unaffected areas.

XRD analyses of the Ce-ZrO_2 TBC show that the cubic modification equilibrium phase at high temperature is frozen-in because of the extreme rate of quenching to room temperature. A preferential enrichment of Ce atoms possibly occurs on {200} lattice planes of cubic Ce-ZrO_2 as shown by XRD data.

Crystallization of a tetragonal nontransformable T′ phase with inverted intensities of the characteristic XRD doublets (002)(200/020), (220) (002/202) and (113)(131/311) compared to normal T′ has been observed in the Y-ZrO_2 TBC after irradiation during cooling (Fig. 6).

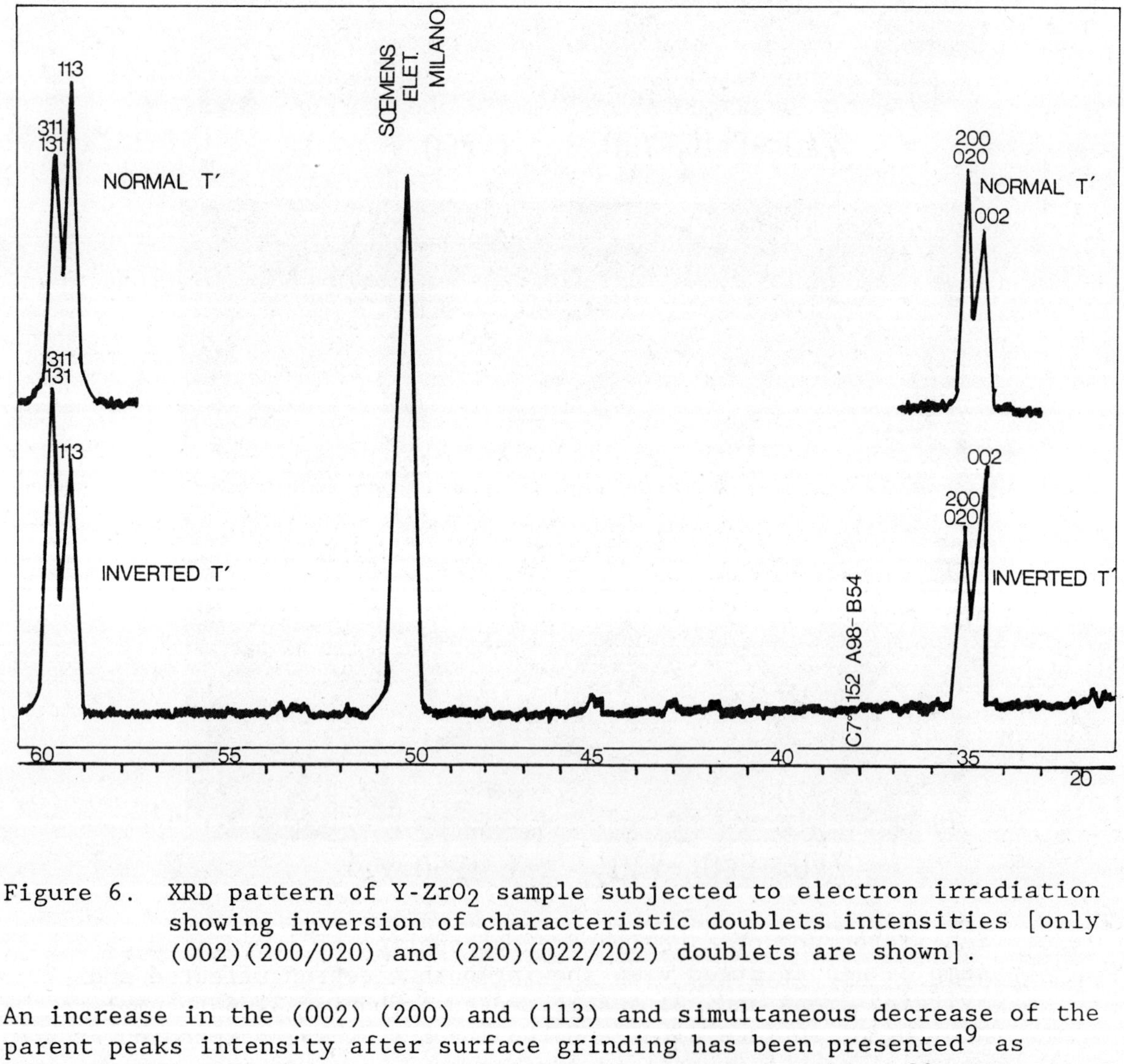

Figure 6. XRD pattern of Y-ZrO_2 sample subjected to electron irradiation showing inversion of characteristic doublets intensities [only (002)(200/020) and (220)(022/202) doublets are shown].

An increase in the (002) (200) and (113) and simultaneous decrease of the parent peaks intensity after surface grinding has been presented[9] as evidence for ferroelasticity in tetragonal zirconia. This effect may possibly account for the retention of high toughness observed in some Y-PSZ materials[10-13] above the monoclinic-tetragonal transformation temperature. In the present case, it is suggested that the substantial stresses which are generated within the electron-modified area because of the extremely high cooling rate can drive domain switching and alignment of c-axis of T' parallel to the direction of tensile stresses, the critical ("Curie") temperature for this effect possibly being the cubic-tetragonal transition temperature (about 2000°C for the present composition).

REFERENCES

1. H. Kuriabayashi, K. Saganuma, Y. Miyamoto, and M. Koizumi, "Effect of Hip Treatment on Plasma-Sprayed Ceramic Coatings Onto Stainless Steel," Am. Ceram. Soc. Bull., **65** [9] 1306-10 (1986).

2. G. S. Fischerman, C. H. Chen, J. M. Rigsbee, and S. D. Brown, Ceramic Engineering and Science Proceedings, **9**, 908-18 (1985).

3. J. R. Spann, R. W. Rice, W. S. Coblenz, and W. J. McDonough, "Laser Processing of Ceramics," in Emerging Process Methods in High-Technology Ceramics, R. F. Davis, H. Palmour III, and R. L. Porter Eds., Materials Science Research Vol. 17, Plenum Press, 473-503 (1984).

4. R. J. Bratton and S. K. Lau, "Zirconia Thermal Barrier Coatings," in Advances in Ceramics Vol. 3, Science and Technology of Zirconia, A. H. Heuer and L. W. Hobbs Eds., The American Ceramic Society, Inc., 479-83 (1981).

5. H. G. Scott, "Phase Relationships in the Zirconia-Yttria System," J. Mater. Sci. **10**, 1527-35 (1975).

6. E. Tani, M. Yoshimura, and S. Somiya, "Revised Phase Diagram of the System ZrO_2-CeO_2 Below 1400°C, J. Am. Ceram. Soc. **66** [7] 506-10 (1983).

7. R. A. Miller, J. L. Smialek, and R. G. Garlick, "Phase Stability in Plasma-Sprayed Partially Stabilized Zirconia-Yttria," Advances in Ceramics Vol. 3, Science and Technology of Zirconia, A. H. Heuer and L. W. Hobbs Eds., The American Ceramic Society, Inc., 118-130 (1983).

8. S. Meriani, "Metastable Tetragonal CeO_2-ZrO_2 Solid Solution," J. de Physique, Supplement au n.2, Tome 4é, pp. C1-485 (1986).

9. A. V. Virkar and R. L. K. Matsumoto, "Ferroelastic Domain Switching as a Toughening Mechanism in Tetragonal Zirconia," J. Am. Ceram. Soc. **69** [10] C-224-C-226 (1986).

10. R. P. Engel, D. Lewis, B. A. Bender, and R. W. Rice, "Physical, Microstructural and Thermodynamic Properties of ZrO_2 Single Crystals," in Advances in Ceramics, Science and Technology of Zirconia II, N. Claussen, M. Ruhle, and A. H. Heuer Eds., American Ceramic Society, Columbus, Ohio, 408-14 (1984).

11. D. Michel, L. Mazerolles, and M. Perez y Yorba, "Polydomain Crystals of Single-Phase Tetragonal ZrO_2: Structure, Microstructure and Toughness," ibid, 131-38.

12. R. P. Engel, D. Lewis, B. A. Bender, and R. W. Rice, "Temperature Dependence of Strength and Fracture Toughness of ZrO_2 Single Crystals," J. Am. Ceram. Soc. **65** [9] C-150-C-151 (1982).

13. D. Michel, L. Mazerollos, M. Perez y Jorba, "Fracture of Metastable Tetragonal Zirconia Single Crystals," J. Mater. Sci. **18**, 2618-28 (1983).

BEHAVIOUR OF THE FOG DROPS IN THE PROCESS OF SPRAY PYROLYSIS FOR PREPARING ULTRAFINE POWDERS

Dai Xiaming Tang Yuying Pan Guangguo Liu Saomin

Beijing Fine Ceramics Lab. Institute of Nuclear Energy Technology, Tsinghua University, China

ABSTRACT

The process of spray pyrolysis for preparing ultrafine powders has been studied to determine the behavior of fog drops which form. The probability of collisions among the fog particles has been calculated and the effects of this probability are discussed. This research has shown why void sphere particles are often observed during this process and conclusions have been reached as to how we avoid this phenomenon.

I. INTRODUCTION

As an important process for preparing ultrafine powders, the process of spray pyrolysis is being paid more and more attention[1-8]In our laboratory, this process is being researched. By this process some kinds of ceramic (such as ZrO_2, Y_2O_3, Al_2O_3) ultrafine powders have been prepared. The scheme of equipment is shown in Fig 1.

The movement and heat transport of the fog drops in the process of spray pyrolysis and their effects on the size and characters of the produced powders are very important for controlling the process and improving quality of the powders. In this paper, behavior of the fog drops in the process of spray pyrolysis is discussed.

II. THE MOVEMENT REGULAR PATTERNS OF A FOG DROP

1. The basical fluid mechanics parameters of a fog drop and the gas flow

The calculation will be given in detail in another paper, and only the results are given here.

i. Fog droplet size

In this work, the fog is produced by an ultrasonic sprayer. The diameters of the fog drops could be considered to conform to the canonical distribution and the average diameter could be calculated $\overline{D}=3\mu$. In order to simplify the calculation, according to the observation in the experimental practice, we could let Dmax=2D, Dmin=D/2

ii. Reynolds numbers

In our case, the fog drops are made of the dilute aqueous solution of a kind of zirconium salt and the carry gas is compressed air. The Reynolds number of a fog drop is Re=0.083 and the Fluid Reynolds number[9] for the carry gas is Ref=2600.

2. The equation of motion for a fog drop

Considering a fog drop which moves in a gas fluid horizontally, neglecting the influence of its gravity, one has[10]

$$dUp/dt=f(Re)/\tau_v\,(U-Up) \quad .... \quad (2\text{-}1)$$

where Up and U are velocities of the particle (fog drop) and carry gas respectively; f(Re) is an amendment factor, in the case of Re<1, f(Re)=24/Re; τ_v is the velocity relaxation time.

$$\tau_v = \frac{\rho_p D^2}{18\mu}$$

Where ρ_p is the density of the particle, μ is the viscosity index of the gas, D is the diameter of the particle. Putting the data that are measured and calculated in our condition into the equation, then

$$Up=U[1-\exp(-1.2\times10^3 t)] \quad ....(2\text{-}2)$$

which is the equation of motion for a fog drop in our condition. From (2-2), it is clear that after a very short time, the fog drop could obtain the same velocity as carry gas.

III. CALCULATION OF THE COLLISION-MERGENCE EFFECT OF THE FOG DROPS

In the process of spray pyrolysis, the concentration of the fog drop in the gas phase is very high. For example, in our condition it's about $n=7.8\times10^7$ cm^{-3}. So the collision and mergence effect of the fog drops should be considered. The causes for the collision and mergence among the fog drops could be those:

1. The difference of the velocity relaxation time

2. Brownian movement

3. The velocity difference of the fog drops in the direction of gravity.

It is assumed that in each case every two collided fog drops could be merged into one. The effects of the three causes could be calculated as follows:

1. The difference of the velocity relaxation time

During the velocity relaxation process caused by the abrupt change in the diameter of reactor or in the temperature field, the fog drops with different size will have different velocities, so they will collid with each other. It is approximately considered that after τ_v, these drops will all have the same velocity as the gas fluid.

Assumed D is the diameter of a fog drop. Then, during a velocity relaxation process, it could collid Z'il drops:

$$Z_{i1} = \eta n. (\pi/4)(D+\bar{D})^2(\alpha^2 - \alpha^2_{min}) \tau_v U \ldots (3\text{-}1)$$

where η is an efficiency coefficient of the collision, n is the number of the drops in a unit volume, $\alpha_{min} = D_{min}/\bar{D}$. In our system, during the fog enters the reactor before it has dried, the times of the relaxation processes should be less than 10. So the frequency of the collision mergence caused by the difference of the velocity relaxation time in our condition is

$$Z_{i1} < 3 \times 10^{-3}$$

2. The Brownian movement

We could treat the gas-fog mixture as a kind of gas which consists of two compositions one is the carry gas, other is the fog. Then it's easy to calculate the collision frequency of a "gas molecular"[11], the diameter of which is D, with other moleculars:

$$Z_{i2} = (\pi/4)\eta n(D+\bar{D})^2 \sqrt{\bar{v}^2} \Delta t \ldots\ldots (3\text{-}2)$$

where $\sqrt{\bar{v}^2}$ is the root mean square velocity, Δt is the flight duration of the drop before it has dried.

In our condition, $Z < 0.06$

3. The velocity difference in the direction of gravity

Using Stockes formula,[12] the terminal veloicty of the fog drop during its falling could be calculated and then the collision frequency by its fall could be calculated as follows:

$$Z_{13}=(\pi/4)\eta n(D+\overline{D})^2(Umax-Umin)\ldots(3-3)$$

where t has the same meaning as above, Umax, Umin are terminal velocities in the gravity direction for the maximum and minimum drops respectively and

$$Umax = K_1 r_{max}$$
$$Umin = k_1 r_{min}$$

here, $k \approx 1.19 \times 10^{6} cm^{-1} sec^{-1}$
in our case $Z < 3.4 \times 10^{-3}$
The total of above three terms, in our case, is

$$Zi = Z_{11} + Z_{12} + Z_{13} < 0.07$$

This shows that the influence of the collision among the fog drops is so slight that we needn't take it into account.

IV. THE SPLITTING BEHAVIOUR ANALYSIS OF THE FOG DROPS

The possible driving forces which causes the fog drops to split could come from two aspects, one is the circulation in each drop which is caused by flow dynamic reason; the other is the pressure of the bubbles which produced during heating the drops.

According to the theory and the experimental resuts[13] of the cloud physics, the first one could be neglected here, so we only discuss the splitting behavior caused by the processes of heat and mass transport.

From the calculation of the heat transport, it was found that the temperature relaxation time on the drop surface

$$\tau r \approx 7 \times 10^{-4} sec$$

and the temperature homogenization time in the drop is

$$\tau_i \approx 6.5\times10^{-3}\text{sec}$$

In our condition, the temperature of the fog drops can rise from room temperature to 100°C in a sec. Thus, it can be seen that in any moment, the temperature difference between gas and drops, between the surface and centre of the drops, all needn't be taken count of.

In the process of spray pyrolysis, fog drops are heated from outer surface. this means, according to the theory of liquid boiling[14], the process taking place here couldn't be a bubble boiling, but a relatively steady evaporation. Therefore, the H_2O moleculars in the surface of a drop runs into gas phase constantly, meanwhile the inner H_2O moleculars of the drop diffuse to the surface continuously. The ratio of the speeds of the two processes is very important for the particle size and the appearance of the produced powder.

The ratio of the speeds of the two mass transport processes could be attributed to the ratio of the speeds of the H_2O molecular diffusion in the gas and liquid phase. As

$$\alpha_g \approx 1\times10^4\alpha_l$$

the evaporation speed of H_2O molecular on the surface of the drop is much greater than the diffusion speed of molecular in the drops.

V. CONCLUSION

It is the major conclusion:

1. In a process of spray pyrolysis, especially in our condition, each fog drop could basically correspond to a produced ulerafine particle. The effect of collision-mergence among them could be neglected, they can't be split by the driving force which comes from fluid mechanics. This has been proved by the experimental results.

2. In general conditions, a bubble could be caused by the remaining water within the drop when its surface solidified at first. In this case, a hollow particle could be produced. This effect could be weakened even be avoided, if the process parameters are controlled properly (Fig. 2). Otherwise, the particle diameter becomes bigger and the shell of the hollow particle becomcs thinner, even brocken (Fig. 3).

Of course, the influence of the behaviour of the fog drops on the characters of the produced powders couldn't be clear without a discussion about the processes of chemical reaction and crystallization which would take place after the fog drops enter the chamber. We'll discuss these processes later.

REFERENCES

[1] Y.Hirata and A.Kato Chem. Sec. Jpn. 1979 No.10 PP. 1309-15

[2] M.Roy, R.R. Neurgaonkar Am. Cer. Soc. Bull 56[11]1023-24 (1977)

[3] 神崎修三，平尾喜代司，大津贺望，斋藤胜一，中川普兵衡，滨野健也
Yogyo-Kyokai Shi Vol.91 PP 81-86, 1983(Jpn)

[4] A.Kato Kagaku Sosetsu No 48 p26, 1985(Jpn)

[5] A.Kato, A.Takayama and Y.Morimisu "Preparation of silver particles by spray-pyrolysis technique" Chem. Soc. Jpn, No 12 pp 2342-44, 1985 (Jpn)

[6] 矢野丰彦，斋藤胜一，大西康司，川合和之
"Synthesis of ceramics fine particles by spray-pyrolysis technique", Ceramics Vol 7 PP. 88-94, 1986 (Jpn)

[7] T.Kumazawa, S.Kanzaki, J.Asanmi. O.Abe and H.Tabata "Sinterability of SiO - Al O powders prepared by spray pyrolysis", Yogyo-Kyokai Shi 94[5] 1986, pp. 485-490(Jpn)

[8] S.Inoue and A.Ono "Preparation of hydroxyapatite by spray-pyrolysis technique", Yogyo-Kyokai Shi, 95[8] 1987. PP 775-780 (Jpn)

[9] Howard. E. Hesketh.P.E 〈〈Fine particles in gaseous media〉〉 chap. 1 Ann. Anbor science 1979.

[10] G.Rudingor 〈〈Fundamentals of gas particles flow〉〉 Chap. 2, Elservier Scientific Publishing Company, 1980

[11] E.A.Strauff 〈〈Molecular physics〉〉

Vol. II Chap. 5, 1959(USSR)

[12] R.R.Rogers 《A short course in cloud physics》 Chap. 7, pergmon press, 1979

[13] R.R.Rogers 《A short course in cloud physics》 Chap. 7. pergmon, press, 1979

[14] E.A.Strauff 《Molecular physics》 Vol. III Chap. 2, 1959 (USSR)

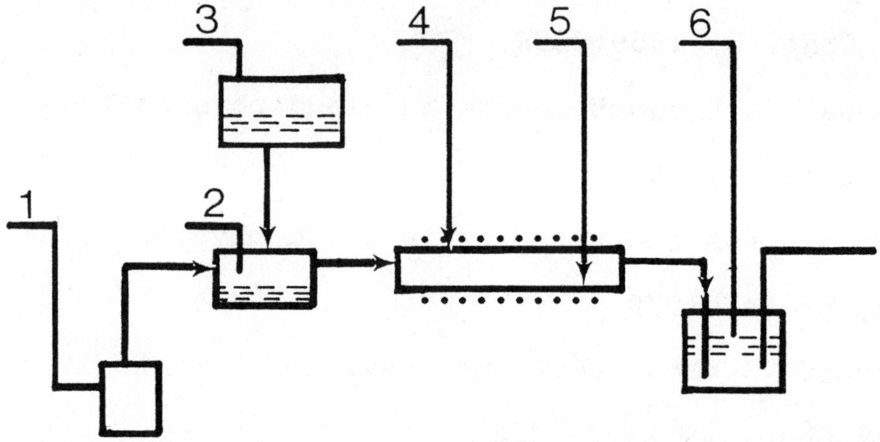

Fig. 1 The scheme of spray pyrolysis equipment

1. carry gas generator 4. electric furnace

2. sprayer 5. reactor

3. solution continer 6. powder collecter

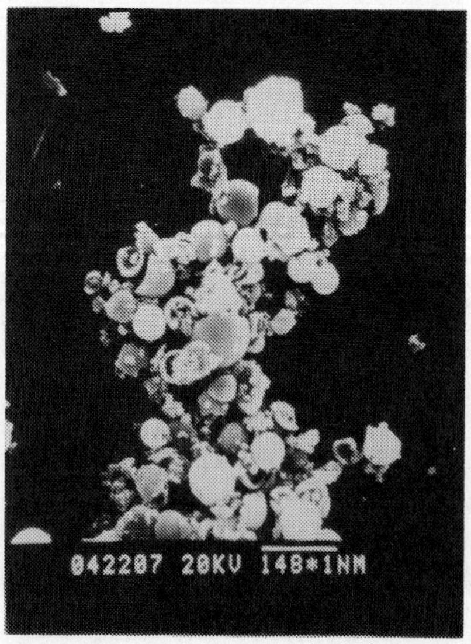

Fig. 2 The particles with good appearance

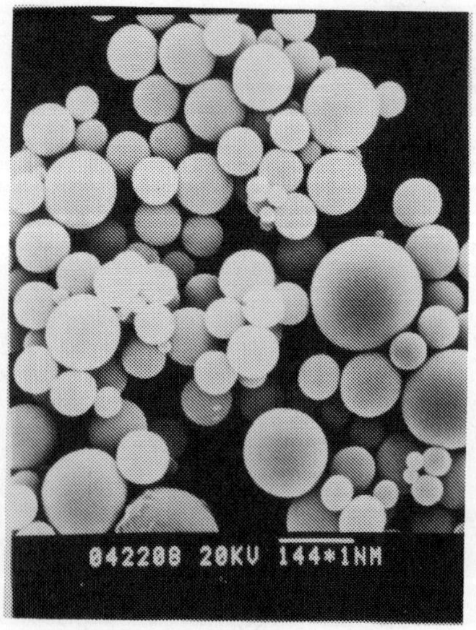

Fig. 3 The brocken particles

THE MECHANISM OF THE THERMAL EXTRACTION OF ORGANIC BINDERS FROM INJECTION MOULDED CERAMIC BODIES

Tao Zhang, Zhen Gang Chen, Zuo Zhao Jiang and Jian Guang Wu
Dept. of Materials Science and Eng.,
Tsinghua Univ.,
Beijing, China.

ABSTRACT

At various binder extraction stages, injection moulded bodies were cut into several parts and the losses of the organic vehicles were studied. The reactions among the organic vehicles and the dynamics of their thermal degradation of them were analysed. It turns out that with the change of temperature the distribution of binders is variable; using ceramic powders to cover the bodies, the distribution of organic binders in the bodies is changed.

At lower temperature the extraction process is controlled by diffusion and at higher temperature the process is controlled by thermal decomposition. The reactions among the organic vehicles can change their thermal decomposition rate.

1. INTRODUCTION

Ceramic parts in complex shapes are in increasing demand for many industrial productions as well as for high temperature applications in advanced heat engines. The needs for making complex ceramic compacts can be met by using injection moulding method satisfactorily. However this forming technique is restricted by the process of binder removal before sintering. The process of binder removal is time consuming and usually takes more than 7 days(1), and defects arise easily. Several approaches have been developed to reduce the extraction time and to minimize the flaws. Carlstrom(2) used RCE(rate control extraction) method, through the measurement of thermal balance to get isothermal weight loss in the same time. But, since the parts injection moulded are covered with ceramic powder, and the powder can absorb some organic binder diffusing from the parts, the weight change measured by thermal balance is usually not equal to the weight loss of the parts. B.C. Mutsuddy(3) studied the thermal degradation isothermally temperature. But the thermal degradation of organic binder is related to the thermal history, so the isothermal weight loss may not be useful for the practical process.

In other respects, since the mechanism of binder removal may be different in different stages, the maximum thermal weight loss allowed is different. So the weight loss is not the only factor which control the thermal extraction process.

The thermal extraction process is very complex. The organic binders in the parts degrade, diffuse to the surface, and react among themselves. Therefore studing the mechanisms and dynamics of the thermal extraction is helpful for improving the process.

2. EXPERIMENTAL PROCEDURE

The ceramic powders used were Al_2O_3 and Si_3N_4 powder, there specific surface was 2.63 m^2/g and 5.49 m^2/g respectively. The organic binders were ethylene-vinylacetate copolymer (EVA) 4--6 wt%, paraffin(PW) 4--6 wt%, stearic acid (St) 2--3 wt% and dibutyl phthalate (DBP) 2--3 wt%. 50 MPa injection pressure and a common open nozzle were used. The thermal extraction processes were conducted at atmosphere condition in a silicon carbide electric furnace.

The studies of the mechanism and dynamics during thermal extraction were done by two methods. First, using samples of different size, the sample was weighed continuouslly. If the thermal extraction process is controled by diffusion, the weight loss of smaller sample is larger than that of larger sample. If the thermal extraction process is controlled by reaction, the weight loss will not be affected by the size of the samples. This method is simple but not precise enough. Second, using segregating method, the samples for thermal extraction were 30*50*5mm and 40*50*8mm. At different temperature a sample was taken out and a part of 10*12*5mm and 10*12*8mm was cut from its center of it. This can be approximated as one dimension condition. These parts were divided into thin pieces with planer and the content of organic binders was measured in each piece with burning out method.

The heating rate during thermal extraction process was $10^{\circ}C/hr$. The rate of thermal degradation of the organic binders was not affected by heating rate when the heating rate is lower than $15^{\circ}C/hr$(4). Thus the experiment results can be used to understand the actual reaction.

3. RESULTS AND DISCUSSION

3.1 the thermal extraction process in the parts with different size

The relations between the weight loss of organic binder in the parts with different size and temperature are shown in Fig.1 and Fig.2. The weight loss curves are similar. It turns out that the thermal extraction process is mainly controlled by reaction. The thermal degradation of the organic binders are not affected by the kinds of ceramic powders. The dotted lines in Fig.1 and Fig.2 express the results calculated from addition of single organic binders. Comparing them with the measured results from the parts, it is shown that at the same temperature, the weight loss of the parts injection moulded sample is larger than that of the calculating results added from single organic binders, and the temperature of complete thermal extraction is much lower than that of EVA resin. The early experiment results on the reaction among the binders suggested that DBP can accelerate the thermal degradation of other organic binders.(4) In order to confirm this conclusion, a sample without DBP was measured. (Fig.3). At lower temperatures, since the pores are sealed by EVA resin, the oxidation effect and the diffusion of the degradation products in the sample are reduced. Therefore the adding result is larger than the result measured from the sample. but the finish temperature didn't change obviously. This result also shows that the thermal degradation of the mixes and the thermal extraction process is mainly controlled by reaction.

3.2 Analyzing the mechanism of thermal extraction with segregating method

In order to analyze the mechanism of thermal extraction more accurately, the distribution of organic binders in the samples during thermal extraction was measured by a segregating method.

The experimental results fore bare compacts are shown in Fig.4 and Fig.5.When the temperature is lower than 190°C, the content of the organic binder in the surface is less than that in the center of the sample. In this stage, DBP Stearic acid and paraffin degrade partly and the degradation of EVA resin is very small. The porosity in the sample is small and the resistance to diffusion is larger. Therefore the content of the organic binders in the center is larger than that in the surface and the process of thermal extraction is controled by diffusion.

When the temperature is higher than 210°C, the content of organic binders in the surface and center is smaller, and a maximum content is located at a distance about 1.2 mm to the surface. The main binders in the parts were EVA resin and paraffin. Two factors can increase the degradation rate, the oxidation effect and the accelerating effect of DBP. Obviously, The oxidation effect in the surface is larger than that in the center of the parts. The thermal decomposition temperature of DBP is lower. At the temperature lower than 210°C, since the diffusion resistance is larger the remainder decreases gradually from center to surface. In other words, the oxidation effect is stronger in the surface and the interaction of the organic binders is stronger in the center. Therefore, the content of residual organic binders is smaller in the surface and in the center of the parts, and the thermal extraction process is controlled by reaction. With the parts in different size, the same law has been obtained.

The results in the condition of using ceramic bed powder are shown in Fig.6. At the temperature lower than 190°C the results is similar to that in the condition without ceramic bed powder. The organic binder remained in the surface is less than that in the center of the parts. It indicates that the thermal extaction process is controled by diffusion. But when the temperature increases the content of the organic binders in the center is lower than that in the surface.

In order to analyze the effect of bed powder, the distribution of organic binders in the powder was measured (Fig.7). It shows that the thermal degradation product of the parts is absorbed by the powder partly. Especially when using r-Al_2O_3 powder, the absorbing amount is larger.

Therefore, when the parts are covered with ceramic bed powder since the powder absorbs some binder, the oxidation effect on the part is abated. In this stage, only the accelerating effect of DBP to the reaction is more important. So the content of the organic binder in the center is smaller than that in surface, and the thermal extraction process is controlled by reaction.

3.3 The dynamics study

The thermal degradation rate during thermal extraction can be analyzed by Coats-Redfern method[5].
The thermal degradation process can be experessedby Eq.(1).

$$\frac{dC}{dt} = K(1-C)^n \tag{1}$$

Where C=fraction of the binders decomposed at time t, n=order of reaction and k=rate contant given by the expression:

$$K = A \cdot \exp(-E/RT) \tag{2}$$

where A=frequency factor, and E=activation energy of the thermal degradation. For a linear heating rate of, say, a deg. hr^{-1}:

$$a = dT/dt \tag{3}$$

we can get:

$$\log\left(-\log\frac{(1-C)}{T^2}\right) = \log\frac{AR}{aE}\left(1-\frac{2RT}{E}\right) - \frac{E}{2.3\ RT} \qquad n=1$$

$$\log\left(\frac{1-(1-C)^{1-n}}{T^2(1-n)}\right) = \log\frac{AR}{aE}\left(1-\frac{2RT}{E}\right) - \frac{E}{2.3Rt} \qquad n\neq1 \tag{4}$$

For the experiment temperature range, the expression $\log\frac{AR}{aE}(1-\frac{2RT}{E})$ is sensibly constant, thus a plot of the left-hand of Eq.(4) against 1/T should result in a stright line of slop-E/2.3R for the correct value of n. With this method, the experiment results were regressed. The regression results are shown in Table 1. It turns out that the thermal extraction process can be expressed by one order reaction, and the process is mainly controlled by thermal degradation reaction.

Table 1. Activation Energy of Thermal Degradation

Name	Temperature (°C)	E (KJ/mol)	R
Si_3N_4 Part	230-400	35.00	0.980
Al_2O_3 Part	150-280	36.96	0.999
	280-400	36.91	0.999

Note: Regressed with n=1.

4. CONCLUSIONS

4.1 The segrating method is useful for analyzing the mechanism of the thermal extraction process. The process is controlled by diffusion when the temperature is lower than 190°C, and that is controlled by reaction when the temperature is higher than 210 C.

4.2 The oxidation effect and the interaction between DBP and other organic binders can accelerate the thermal degradation process. The Thermal dehradation rate of organic binders will decrease when the parts are covered with ceramic powder since it can change atmosphere.

4.3 The dynamics relation of thermal extraction can be expressed by one order reaction.

REFERENCES

1.B.C.Mutsuddy,Nonoxide Technical and Eng.Ceram., (1986) 397.
2.A. Johnsson, E.Carlstrom, L.Hermansson, and R.Carlsson, Advances in ceramrcs, (1984) 241.
3.B.C.Mutauddy,Nonoxide Techical and Eng. Ceram., (1986) 397.
4.Tao Zhang, ph.D Disseration, (1988).
5.A.W.Coats, J.P.Redfern, Nature, 201 (1964) 68.
6.Tao Zhang, Zuo Zhao Jiang, Jian Guang Wu and Zhen Gang Chen,J. of Silicate, to be pablished.

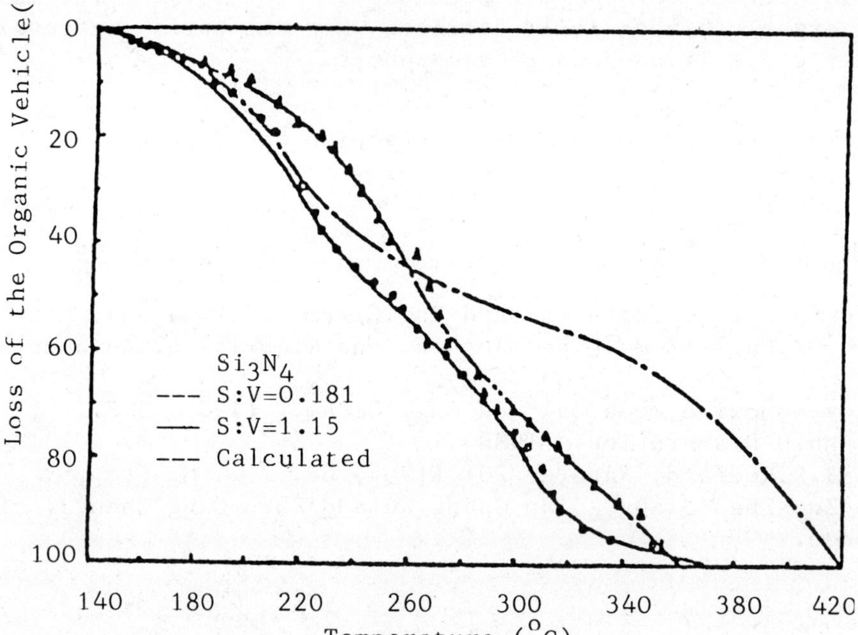

Fig.1 Thermal degradation rate of Si_3N_4

S:V=Surface area:Volume

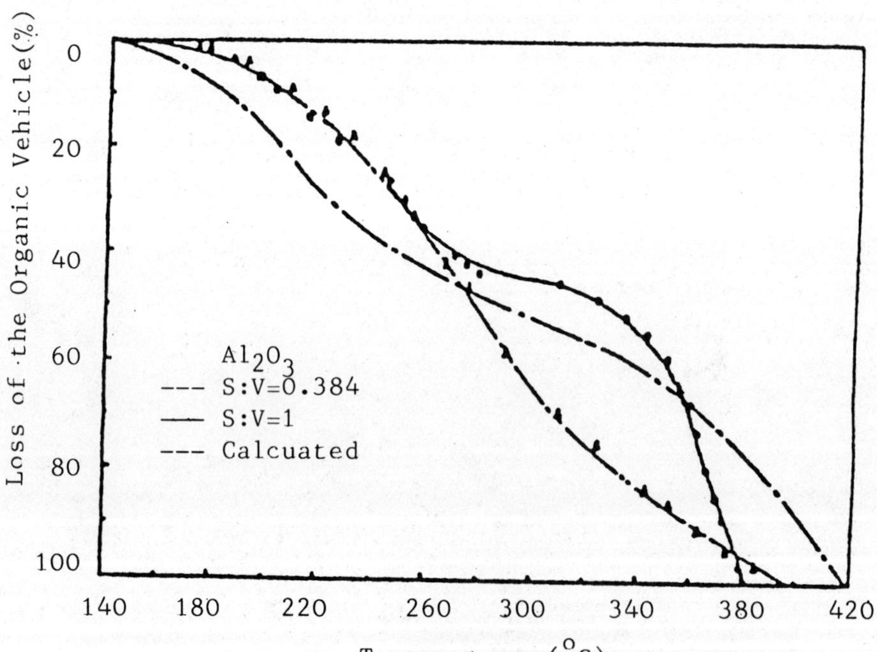

Fig.2 Tnermal degradation rate of Al_2O_3

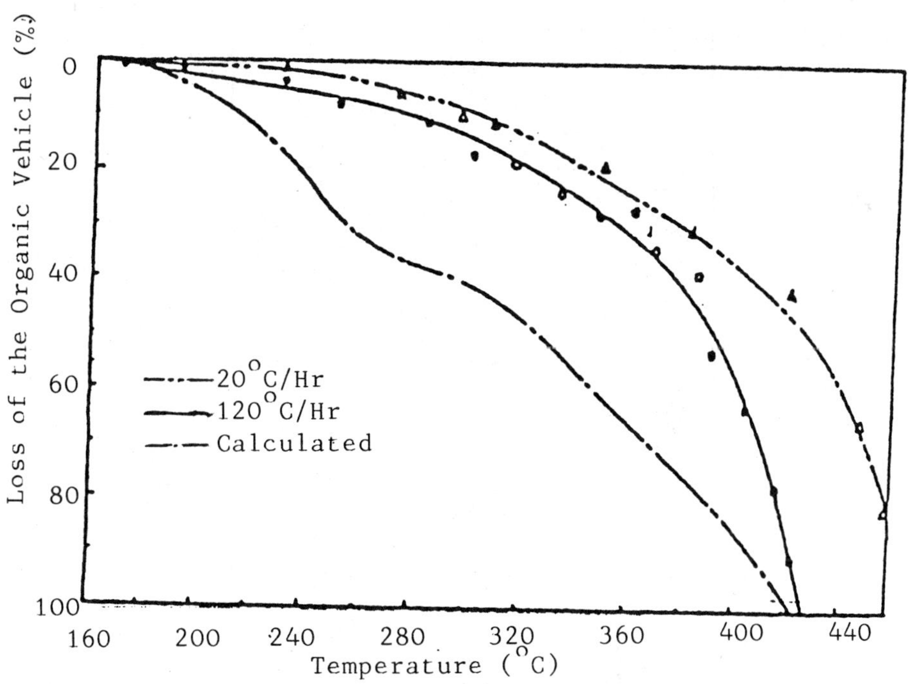

Fig.3 Thermal degradation of the part not containing DBP

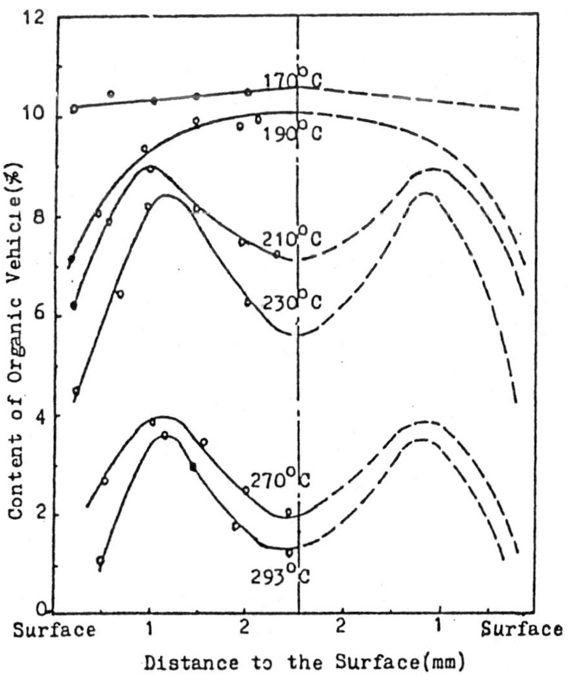

Fig.4 Distribution of organic binders in the parts

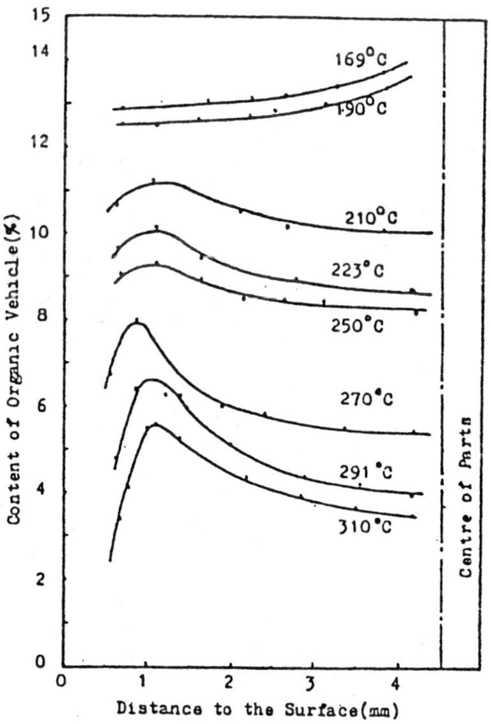

Fig.5 Distribution of organic binders in the parts

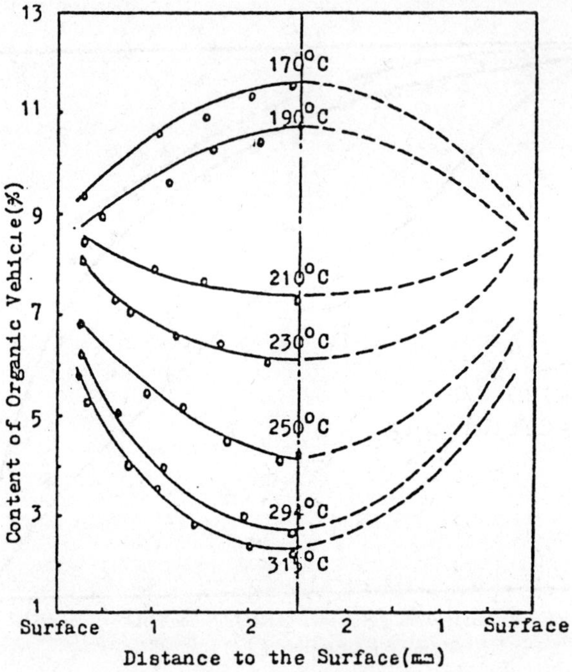

Fig.6 Distribution of organic binder in the parts with ceramic bed powder

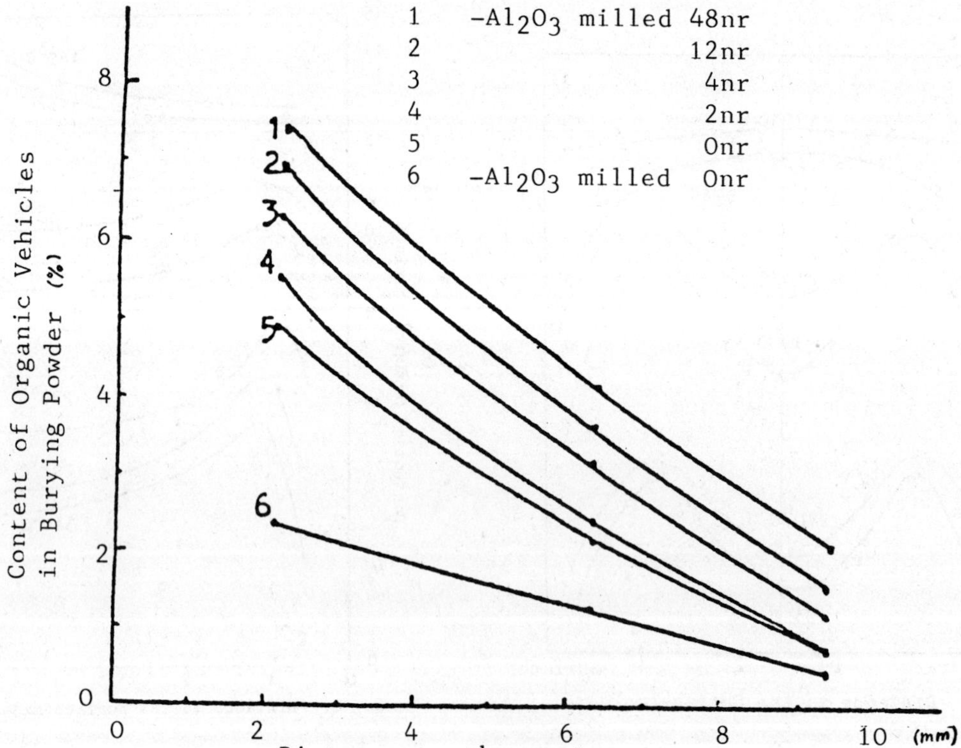

Fig.7 Distribution of organic binder in the bed powder

UNDOPED AND LOW-DOPED SILICON NITRIDE WITH WHISKER REINFORCEMENT MADE BY GLASS ENCAPSULATED HIP

H. T. Larker and J. E. Adlerborn;
ABB Cerama AB, Robertsfors, Sweden.

ABSTRACT

The use of encapsulation in combination with hot isostatic pressing (HIP) makes it possible to fully densify even high purity silicon nitride powders (total metal impurities <0.01 %). It is well known that oxidation rates and creep properties of pure silicon nitride are vastly superior to alloyed silicon nitride at very high temperatures. The unsufficient volume of liquid phase at consolidation temperature does however not allow the desireable formation of elongated ß-silicon nitride grains, common in high-doped or sialon type silicon nitride alloys. The fracture toughness of undoped silicon nitride will consequently be low, due to the low aspect ratio. Attempts at introducing whiskers of ß-silicon nitride or silicon carbide in undoped or low-doped silicon nitride have been made and physical properties and microstructure have been studied.

INTRODUCTION

Silicon nitride has been recognized as a main candidate for structural applications in advanced heat engines and many other devises. The intrinsic properties of silicon nitride such as high bond strength, high oxidation and corrosion resistance, also at high temperature and low thermal expansion are very important, e. g. for such engine components as turbine rotors.

A material processing technique that allows very pure silicon nitride to be prepared is chemical vapor deposition (CVD). Data reported in 1974-75 on such materials (1,2) show CVD-Si_3N_4 to exhibit excellent oxidation resistance and several orders of magnitude improved creep resistance over that of sintered silicon nitride materials. No creep deflection was observed in 3-pt bending after 198 h at 1560 °C at a maximum stress of 70 MPa. The CVD-technique has unfortunately not been found to be very suitable for the manufacture of complex shapes and large and thick components of silicon nitride. The superior properties reported on CVD-Si_3N_4 have however inspired the development of other techniques to manufacture parts of pure silicon nitride. Ultra high pressure methods using pressures of 3 - 5 GPa were shown successful in producing dense but small material samples in 1978 (3) and 1981 (4).

A processing technique which can combine flexibility in composition with flexibility in geometry and volume, is encapsulated hot isostatic pressing (HIP). It was already 1977 shown (5) that a commercially available silicon nitride powder, AME CP85, without sintering additives could be fully densified by glass encapsulated HIP. That powder quality, however, contained about 0.6 wt% metal impurities but purer powders resisted densification at the pressures and temperatures (300 MPa, 1750 °C) available at that time. Successful densification of additive-free silicon nitride powder with metal impurity content about 0.08 wt% was reported in 1986 (6) and with less than 0.01 wt% metal impurity content in 1987 (7, 8). In the latter two publications the required HIP parameters to obtain fully dense materials were 250 MPa gas pressure at 1950 °C for 2 h or 150 MPa at 2000 °C for 2 to 4 h, respectively. A very interesting finding in both these studies is that a higher flexural strength was measured at 1350 or 1400 °C, respectively, than at room temperature. The strength values at room temperature were, however, relatively low. Equiaxed grains were a characteristic feature of these materials and were in both studies considered to be one of the causes to the relatively low room temperature strength. The present authors are (as was also suggested in ref. 7) doubtful that the dissolution/precipition mechanism, which is a generally adopted sintering mechanism for silicon nitride (9), is a particularly important mechanism also in this case, because of the low content of liquid phase formers. It was in ref. 7 instead proposed that the sintering mechanism is a combination of a rearrangement of particles, plastic deformation and grain growth. Although a full conversion from α- to ß-Si_3N_4 was achieved no contribution to the fracture toughness of the material from the formation of elongated ß-grains appears to have been obtained.

One possibility to make a material containing elongated grains is to introduce fine ceramic whiskers which are carefully dispersed in the matrix material. With the commercial availability of good quality SiC whiskers several studies have already been published (10, 11, 12) usually utilizing uni-axial hot pressing and common amounts of sintering additives to the Si_3N_4 matrix. Recently a report on undoped silicon nitride with about 0.08 wt% metal impurities with 10 to 30 wt% SiC whiskers and using encapsulated HIP for densification was published (13).

In the present study ß-Si_3N_4 whiskers were introduced in an undoped silicon nitride matrix powder with less than 0.01 wt% metal impurities and in the same quality silicon nitride powder with 2.5 wt% Y_2O_3 and 0.2 wt% Fe_2O_3 as additives. In parallel, ß-SiC whiskers were introduced in the same type of matrix materials. The range 15 to 35 wt% of whisker additions was investigated.

EXPERIMENTALS

Silicon nitride powder of the quality SNE-10 produced by UBE Industries, Japan, was used as matrix material. It was selected because it is a commercially available material with comparatively low content of metal impurities, less than 0.01 wt%. Its characteristics are listed in table 1.

TABLE 1 Characteristics of raw material (suppliers data)

Property		UBE SNE 10	UBE-SN-WB	Tateho SCW-#1
		Powder	Whisker	Whisker
Surface area	(m^2/g)	13	--	--
Diameter	(µm)	--	0.1 - 1.5	0.05 - 1.5
Length	(µm)	--	10 - 20	20 - 200
Aspect ratio	(L/D)	--	20 - 50	20 - 200
Chrystal phase		97% α-Si_3N_4	100% ß-Si_3N_4	ß-SiC
Chemical analysis				
O	(%)	1.3	0.7	--
N	(%)	38.2	40	--
C	(%)	--	--	--
Al	(%)	<0.001	<0.003	0.32
Mg	(%)	--	--	0.13
Fe	(%)	0.005	<0.005	0.02
Ca	(%)	<0.001	<0.002	0.15
Y	(%)	--	0.72	--

In one series of experiments the matrix material was used without any additives and in another with 2.5 wt% Y_2O_3 and 0.2 wt% Fe_2O_3 as dopants. Silicon nitride ß-phase whiskers of quality SN-WB from the same supplier were used in one series of materials and Tateho SCW #1 silicon carbide whiskers from Tateho Corporation, Japan in another series of materials. Characteristics of both whisker materials are also listed in table 1.

The matrix powder was milled in a ball mill in methanol with silicon nitride milling media, the undoped material for 1 h and the doped for 24 h. The material was in both cases wet sieved through a 5 µm cloth and dried in a rotary evaporator followed by dry sieving through a 0.7 mm sieve. Parts of each of the two matrix materials were filled in polyvinyl bags and cold isopressed to blocks at 300 MPa pressure. Whisker reinforced material samples were prepared by mixing parts of the matrix materials prepared as described above with the necessary amounts of as-supplied silicon nitride or carbide whiskers. In all cases 25 wt% whisker compositions were prepared but with the doped matrix material also 15 and 35 wt% whisker compositions. The mixing was done in petroleum ether in a ball mill using silicon nitride milling media for 20 h. Ball milling has been shown to contribute to a very homogeneous microstructure of the composite (14). The mixtures were dried in a rotary evaporator and sieved through a 0.7 mm sieve. Blocks were as for the matrix materials without whiskers made by cold isostatic pressing at 300 MPa.

All the blocks were degassed in a vacuum oven at 750 °C, 0.1 Pa for 4 h. They were HIPed using the proprietary ABB Cerama glass encapsulation process, the blocks with undoped matrix at 1950 °C and 250 MPa pressure for 2 h and the blocks with doped matrix material at 1775 °C and 200 MPa for 1 h.

Test bars for bend tests were cut from the HIPed blocks. At room temperature were 3.0 by 3.0 mm bars tested in 3-point bend tests with 20 mm support. The bars were length-wise ground with ca 320 grit diamond wheel and the edges chamfered. The bend test bars were for high temperature tests similarly prepared but of 3.5 by 4.5 mm cross section. A four point bend test rig with lower/upper support distance 40/20 mm was used and the testing was made in air at 1350 °C.

Stress rupture properties were evaluated by the STSR (stepped temperature stress rupture) test (15) using the same 4-point test rig geometry and the same size of test bars as for the high temperature bend tests. Starting temperature was 1000 °C and the stress 150, 200 or 250 MPa. If the sample survived 24 h at 1000 °C the temperature was increased with 100 °C up to 1100 °C and tested for another 24 h and so on at 1200, 1300 and 1400 °C.

Fracture toughness was tested by the indentation technique on ground and diamond polished samples. Vickers indentation was made using a load of 10 kg and the Niihara equation (16) was used to calculate K_{IC}.

SEM (Jeol JS M-35) was used to study room temperature fracture surfaces. X-ray analysis was used to determine crystalline phases.

RESULTS AND DISCUSSION

The bulk densities of all samples but one reached above 99.5 % of the theoretical densities calculated from the compositions (considering the oxygen content of the starting powders). A sample containing 35 wt% silicon nitride whiskers had a slightly lower density, however still above 99.0 % of TD.

X-ray diffraction analysis showed ß-Si_3N_4 but no α-Si_3N_4 in the matrix material. ß-SiC but no other silicon carbide phase was found in any of the samples with SiC whiskers added. Traces of silicon oxynitride were found in all the samples, except the high purity silicon nitride material. Traces of $Y_2Si_2O_7$ of ß-type were present in the yttria doped material with 15 and 25 wt% ß-Si_3N_4 whiskers but of α-type in the sample with 35 wt% of these whiskers. The very low high temperature strength af the latter material may in part be explained by this observation.

The flexural strength of <u>undoped</u> Si_3N_4 was low at room temperature, but the addition of 25 wt% silicon nitride or silicon carbide whiskers gave a strength increase of 30 or almost 40 per cent respectively (Fig. 1). Figure 2 shows SEM micrographs of fracture surfaces of the silicon nitride and silicon carbide whisker containing materials. The structure of the material containing silicon nitride whiskers (Fig. 2a) has considerably coarser grains than the silicon carbide whisker containing material (Fig. 2b).

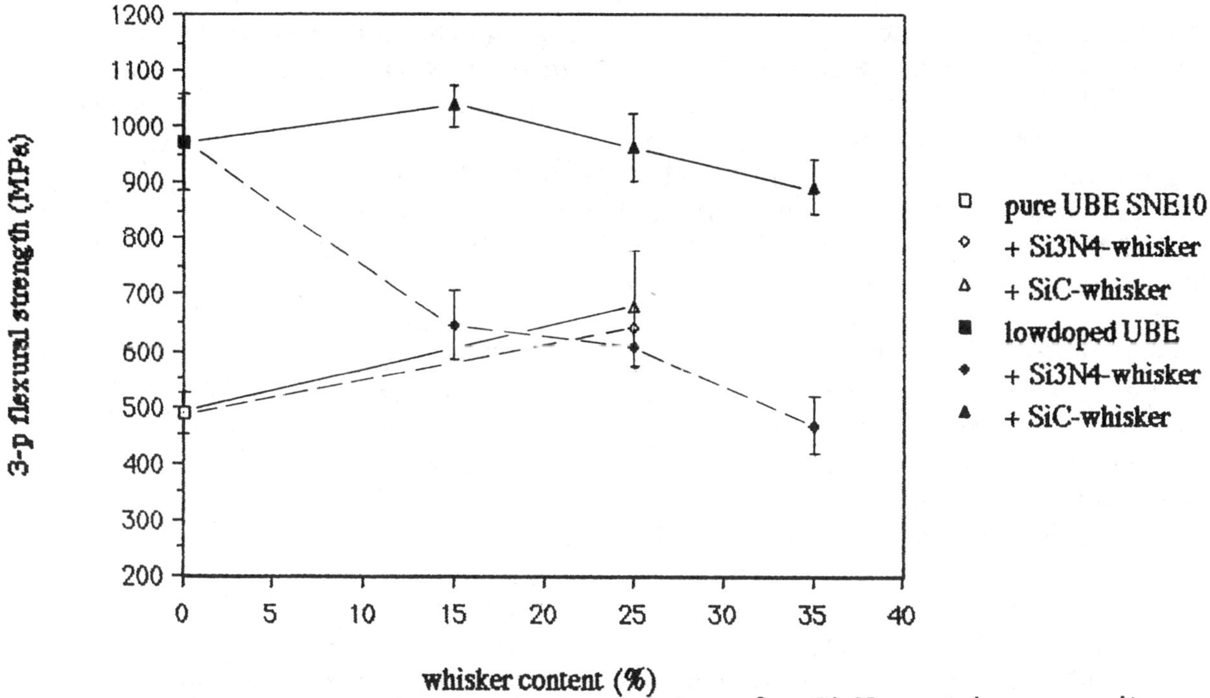

Fig. 1. Flexural strength at room temperature for Si_3N_4 matrix composites.

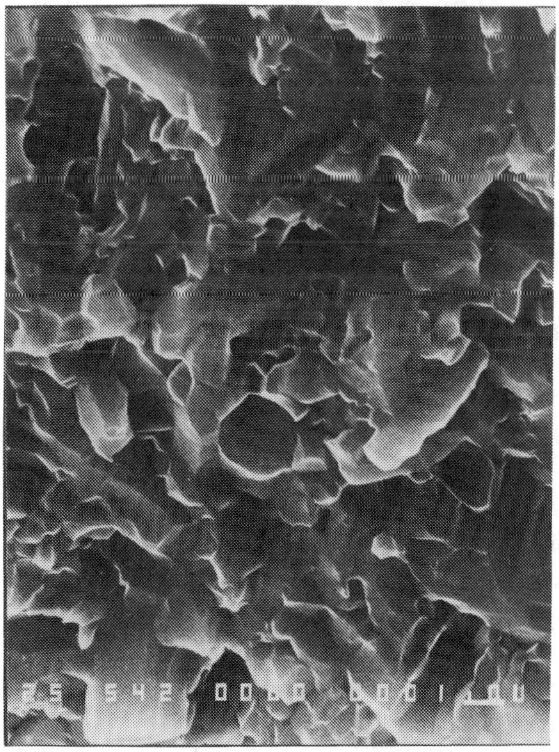

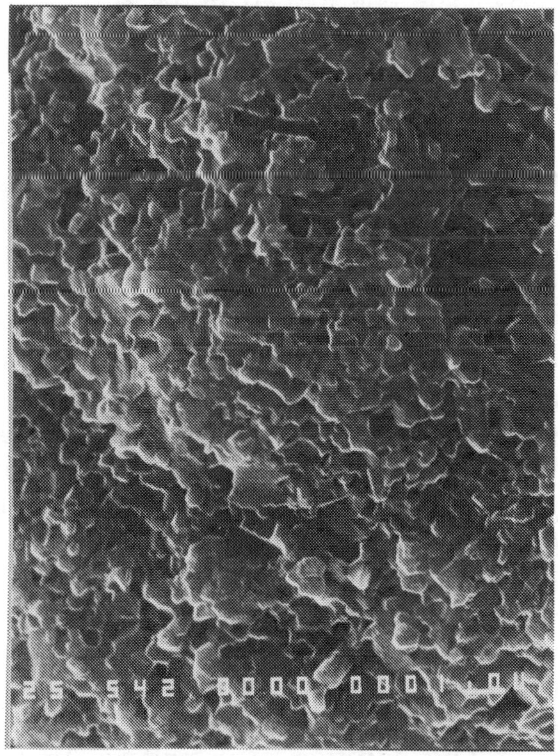

Fig. 2. SEM micrographs of fracture surfaces of 25 wt% ß-Si_3N_4 whiskers (Fig. 2 a), (left), and 25 wt% ß-SiC whiskers (Fig. 2 b), (right), in a matrix of undoped Si_3N_4. White bar 1 μm.

The grain size of the latter (fig. 2.b) is comparable to the grain size of the material without whiskers. The amount of Y_2O_3 present in the Si_3N_4 whiskers (cf table 1) is believed to have promoted the grain growth of the material containing these. The K_{IC} measurements (Fig. 3) show a higher increase for the Si_3N_4 whisker than for the SiC whisker containing material which may explain the small difference in strength.

At 1350 °C the undoped Si_3N_4 material without whiskers also in this investigation is stronger than at room temperature (Fig. 4) but the strength of both the whisker containing materials drops to about half their room temperature value. The STSR tests confirm the superiority at high temperature of the undoped Si_3N_4 material without whiskers (Fig. 5). It survived the entire test including 24 h at 1400 °C with an applied stress of 250 MPa, while there were premature failures of all the whisker containing materials.

The low-doped Si_3N_4 shows a high strength at room temperature (Fig. 1). It drops with 35 to 50 % in materials containing 15 to 35 wt% Si_3N_4 whiskers. As was the case with undoped matrix the fracture surface microstructure is relatively coarse (Fig. 6 a). The materials containing SiC whiskers show only slight changes in strength; an increase with 15 wt%, unchanged with 25 wt% and a slight decrease with 35 wt% ß-SiC whiskers. Very interesting is the decreased scatter band in strength observed. The m-value for the 15 wt% SiC whisker material is as high as m=34 (90% confidence interval m=23 to m=48). The microstructure (Fig 6 b) is as for undoped matrix with SiC whiskers relatively fine. There is no increase in K_{IC} observed except for the materials containing 35 wt% Si_3N_4 whiskers or SiC whiskers (Fig. 3).

The flexure strength at 1350 °C for the low-doped silicon nitride without and with 15 wt% ß-SiC whiskers is about 450 MPa (Fig. 4), a reasonably good value at that temperature. The higher SiC whisker contents give slightly lower strength values but the scatter band is small in all materials containing such whiskers. The materials with ß-Si_3N_4 whiskers exhibit much lower strength values, half to one third (for 35 wt% Si_3N_4 whisker content) of the 1350 °C strength values for this matrix material without any whisker additions or with SiC whiskers (Fig. 4). The STSR tests show the same tendency as the short term high temperature tests (Fig. 5). The material with 15 wt% SiC whiskers has properties comparable with the matrix material without whiskers and a higher content gives materials with only moderately reduced stress rupture properties. The silicon nitride whisker additions on the other hand seriously deteriorate also the long term high temperature properties (Fig.5) as was found for the short term properties (Fig. 4). Only the material with the lowest Si_3N_4 whisker content has any measureable stress rupture life above 1000 °C at 250 MPa applied stress level.

The addition of whiskers to the undoped Si_3N_4 matrix gives some improvement in both fracture toughness and room temperature strength for both whisker materials. In general, fiber reinforcement improves fracture

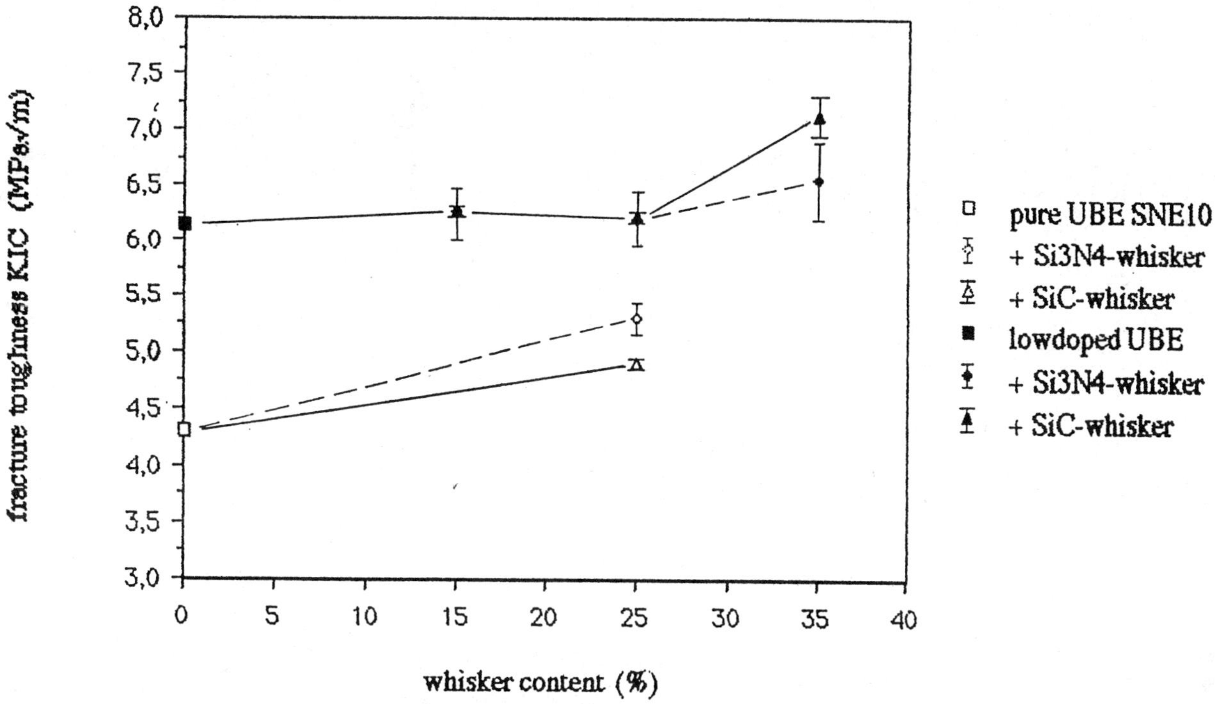

Fig. 3. Indentation fracture toughness at room temperature for Si_3N_4 matrix composites.

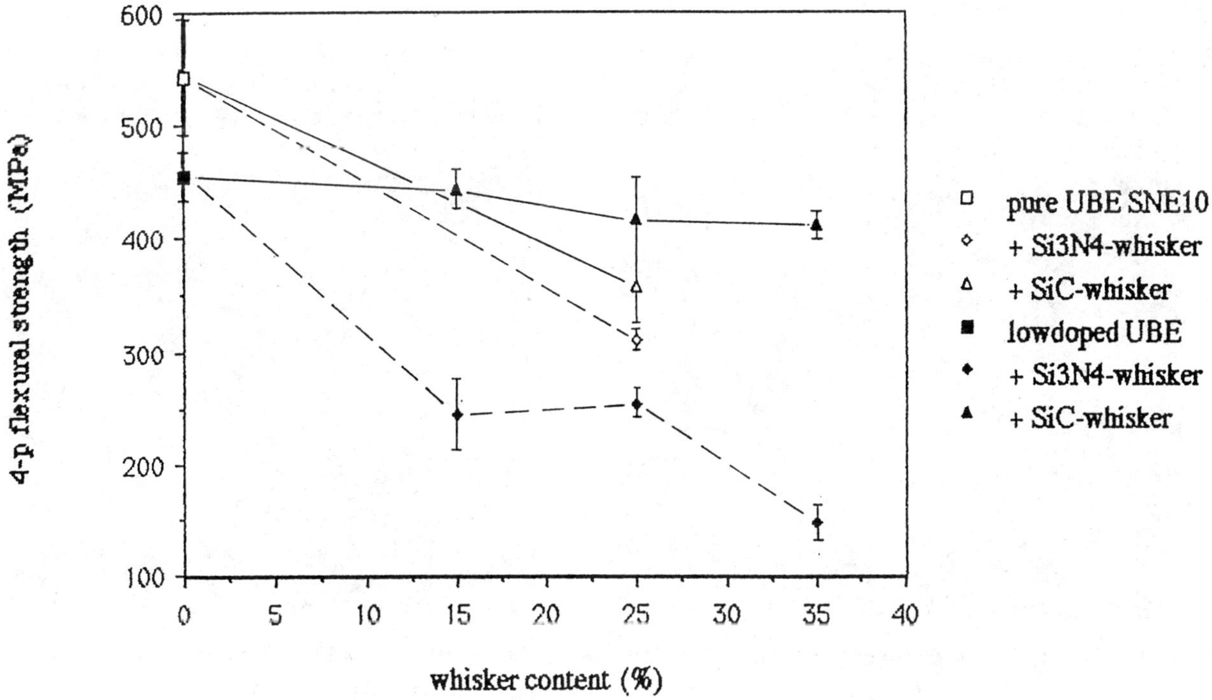

Fig 4. 4-pt flexural strength at 1350 °C for Si_3N_4 matrix composites.

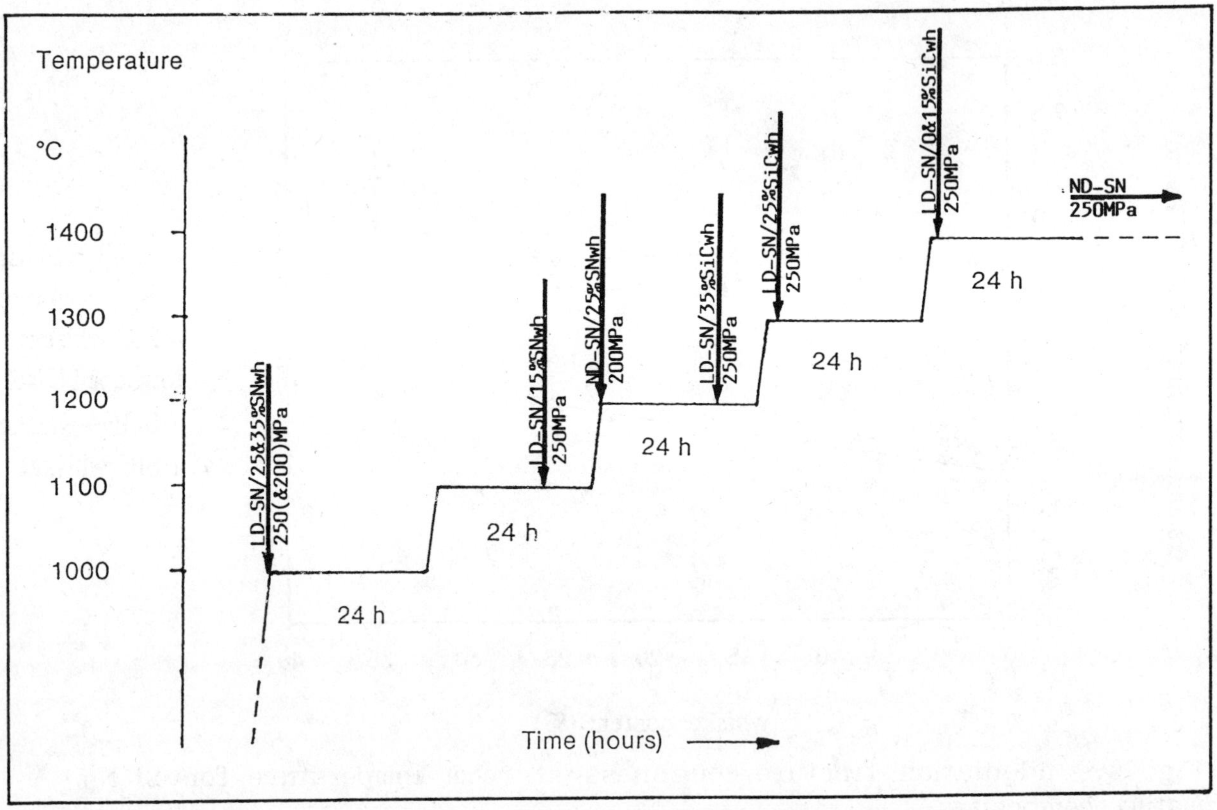

Fig. 5. Stepped temperature stress rupture (STSR)-data. ND-SN and LD-SN in the figure designate undoped and low-doped Si_3N_4 matrix respectively 25%SNwh or SiCwh means 25 wt% addition of ß-Si_3N_4 whiskers or ß-SiC whiskers and 250 MPa is the applied stress level (examples).

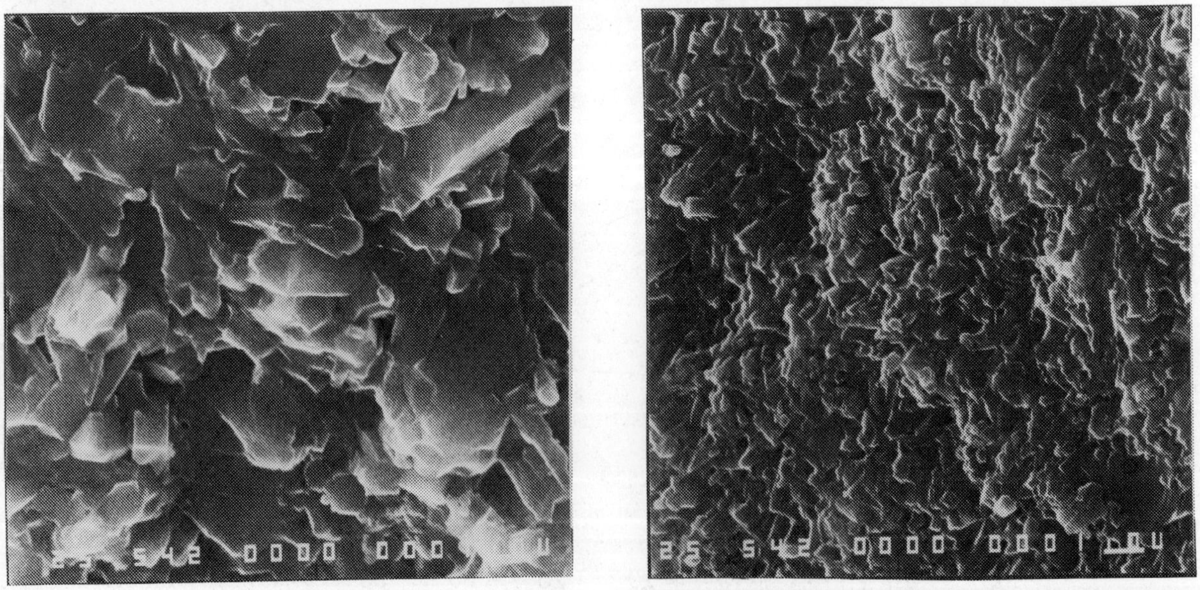

Fig. 6. SEM micrographs of fracture surfaces of 35 wt% ß-Si_3N_4 whiskers (Fig. 6 a) and 35 wt% ß-SiC whiskers (Fig. 6 b) in a matrix of silicon nitride with 2.5 wt% Y_2O_3 and 0.2 wt% Fe_2O_3 additions. White bar 1 µm.

toughness by mechanisms related to fiber sliding, crack deflection, crack bowing, and/or microcrack formation (17). The SEM observations give no support for mechanisms that require the material to plastically absorb energy, e.g. whisker pullout.

In reference 18 regarding Al_2O_3-SiC whisker composites, the influence of the whiskers on the linear elastic behaviour of the material is discussed. The stress peaks near the normally occurring defects in the matrix material could be reduced if a whisker is located there. Similar effects might be of importance also in the material combinations of this study, even if the thermal expansion mismatch, modulii of elasticity etc., are different. The directionality of properties in the ß-Si_3N_4 whiskers are probably of importance in this respect.

The observed increase in room temperature strength with addition of Tateho SCW #1 whiskers to undoped high purity silicon nitride does not appear to be in correspondence with the findings using the same grade whiskers in ref. 13, where a decrease was found. The explanation could be the more efficient mixing/milling procedure used in this case, which might also lower the aspect ratio of the whiskers.

The same mechanisms as for the undoped matrix are probably of importance for the low-doped matrix composites at room temperature. Regarding high temperature properties impurities in the systems are supposed to play a major role. The low-doped material obviously is not negatively effected at all with a 15 wt% SiC whisker addition but 25 or 35 wt% additions significantly deteriorate the stress rupture properties even if the short term, high temperature strength hardly is influenced. The undoped material on the other hand is strongly effected also regarding short term properties. It would be expected that this system has a much lower capacity to accommodate changes in composition, e.g. introduced as whisker impurities. The much higher processing temperature would enhance this behaviour.

The generally negative result obtained with the ß-Si_3N_4 whiskers should not lead to far-reaching conclusions. Pre-treatment of the whiskers, screening and different high temperature processing conditions, as well as improvements of the as-supplied whisker quality, could give completely different results.

CONCLUSIONS

A modest increase in K_{IC} with whisker content is obtained with both ß-Si_3N_4 whiskers or ß-SiC whiskers added to undoped or low-doped high purity silicon nitride. Both at room temperature and high temperature is the scatter band in strength narrowed with SiC whisker additions, for short term properties without impairing the strength. Only at room temperature for undoped Si_3N_4 matrix is the flexural strength increased with ß-Si_3N_4 whisker additions, however seriously reduced for low-doped matrix or at high temperature.

ACKNOWLEDGEMENTS

The partial financial support by the National Swedish Board for Technical Development is gratefully acknowledged. The SEM-work was performed by Lars Eklund at the Swedish Institute for Silicate Research.

REFERENCES

1. J.H. Rosolowski and C.D. Greskovich, "Ceramic Sintering", NTIS-report AD/A - 001 012. (Oct. 1974).

2. C.D. Greskovich, J.H. Rosolowski and S. Prochazka. "Ceramic Sintering".NTIS-report AD-A014-480. (July 1975).

3. S. Prochazka and C.A. Rocco. "High Pressure Hot Pressing of Silicon Nitride Powders". High Temperature - High Pressures 10(1):87-95 (1978).

4. T. Yamada, M. Shimada and M. Koizumi. "Densification of Si3N4 by High Pressure Hot Pressing". Am. Ceram. Soc. Bull. 60(12):1281-88 (1981).

5. H. Larker, J. Adlerborn and H. Bohman. "Fabricating of Dense Silicon Nitride Parts by Hot Isostatic Pressing". SAE Technical Paper No. 770335 (1977).

6. Y. Miyamoto, K. Tanaka, M. Shimada and M. Koizumi. "Survey for HIP Sintering Condition and Characterization of Dense Silicon Nitride without Additives". Proc. 2nd Int. Symp. on Ceram. Mater. and Components for Engines. Eds. W. Bunk and H. Hausner. Deutsche Keramische Gesellschaft, 271-78, (1986).

7. J. Adlerborn, M. Burström, L. Hermansson and H.T. Larker. "Development of High Temperature High Strength Silicon Nitride by Glass Encapsulated Hot Isostatic Pressing". Mater & Design, 8, (4), 229-32, (1987).

8. H. Okada, K. Homma, T. Fujikawa and T. Kanda. "Fabrication of Dense Si3N4 by Hot Isostatic Pressing". High Tech. Ceramics, ed. P. Vincenzini, Elsevier Sci. Publ., Amsterdam. 1023-32 (1987).

9. L.J. Bowen, "Densification Mechanisms and Development of Strength in Hot Pressed Silicon Nitride". Ph D thesis, University of Leeds, (1977).

10. P.D. Shalek, J.J. Petrovic, G.H. Hurley and F.D. Gac, "SiC Whisker/Si3N4 Matrix Composites", Am. Ceram. Soc. Bull., 65(2), 351-56 (1986).

11. S-T. Buljan, J.G. Baldoni and M.L. Huckabee, "Si3N4 - SiC Composites", Am. Ceram. Soc. Bull., 66(2), 347-52, (1987).

12. J.P. Singh, K.G. Goretta, D.S. Kupperman, J.L. Routbort and J.F. Rhodes. "Fracture Toughness and Strength of SiC-Whisker-Reinforced Si3N4 Composites", Adv. Ceram. Mater. 3(4) 357-60 (1988).

13. H. Takemura, Y. Miyamoto and M. Koizumi, "Fabrication of Dense Si3N4 - SiC Whisker Composite without Additives by HIP". Proc. Int. Conf. Hot Isostatic Pressing, ed. T. Garvare, Centek Publ. Luleå, Sweden, p. 329-36, (1988).

14. R. Lundberg, B. Nyberg, K. Williander, M. Persson and R. Carlsson, "Glass Encapsulated HIPing of SiC Whisker Reinforced Ceramic Composites", ibid, p. 323-27, (1988).

15. G.D. Quinn, "Characterization of Turbine Ceramics after Long-Term Environmental Exposure". AMMRC. Techn. Report 80-15, (April 1980).

16. K. Niihara, R. Morena, P.P.H. Hasselman, "Indentation Fracture Toughness of Brittle Materials for Palmqvist Cracks", Fracture Mechanics of Ceramics 5, ed. R.C. Bradt et al., Plenum Press, New York, p. 97-105 (1983).

17. R.W. Rice, "Mechanisms of Toughening in Ceramic Composites", Ceram. Eng. Sci. Proc. 2 [7-8] 661-701 (1981).

18. L. Björk and L. Hermansson, "Hot Isostatically Pressed Al2O3-SiC-whisker Composites", submitted for publication to Adv. Ceram. Mater. (1988).

MODIFICATION OF SiC WHISKER AND Si POWDER SURFACES USING ORGANOMETALLIC SURFACTANT

G. Krug and S. C. Danforth, Center for Composites Manufacturing Science and Engineering, Center for Ceramics Research, Rutgers University, Piscataway, NJ, 08855-0909

ABSTRACT

The surfaces of Si powder for reaction bonded silicon nitride (RBSN) and SiC whiskers (SiC_w) were modified using selected Si, Ti, Zr and Al containing organometallic surfactants. These organometallic surfactants were found to enhance dispersion and reduce the viscosity in a model injection molding formulation (mineral oil, Si powder, SiC_w and dispersant). Increasing alkyl R group lengths: CH_3, C_8H_{17}, and $C_{18}H_{37}$ for the Si based dispersants, were found to enhance dispersion and reduce the viscosity, thereby enabling use of higher solids loadings. The effect of metal type (Ti, Zr) was investigated, using organometallics with identical alkyl groups ($R=OC_{10}H_{19}$). It was concluded from results obtained through viscosity and sedimentation experiments that alkyl group chemistry determined the dispersant effectiveness while the metal type did not. These results suggest that organometallic surfactants can be used to enhance dispersion, reduce viscosity and potentially tailor the interfacial chemistry in whisker reinforced RBSN composites.

INTRODUCTION

Monolithic ceramics have sparked a great deal of interest as structural engineering components in heat engines and high temperature energy conversion systems due to their superior thermomechanical properties[1]. However, due to the intrinsic brittle nature of ceramics, component reliability has limited their use. The incorporation of reinforcement phases such as particulates, platelets, whiskers and continuous fibers are currently under investigation in an attempt to overcome these limitations. Specifically, whisker reinforced ceramics offer the potential of increased toughness and reduced the flaw sensitivity, thereby increasing component reliability [2-4].

Whisker reinforced ceramics are traditionally fabricated using one of the following methods: colloidal filtration, slip casting and hot pressing[5-7]. While each of these forming methods offers some inherent advantages, injection molding has been found to be a promising alternative[8]. In particular, injection molding enables fabrication of complex shapes in a rapid, automated manor. Additional advantages include close dimensional tolerances, and near net shape fabrication using RBSN.

In spite of the desirable properties whisker toughened ceramics can offer, they are currently limited by processing related flaws associated with poor dispersion techniques which results in whisker agglomeration or nesting.

Agglomerated suspensions are generally difficult to process [9,7] due poor flow behavior and high viscosity.

In the injection molding process, maintaining a low viscosity is particularly difficult as high solids loadings (Φ=55-75 vol %) are frequently used [10-12]. Under a given set of conditions; i.e. particle characteristics, binder formulation and temperature etc., the viscosity increases dramatically with increased solids loading [13,14]. With the incorporation of increasing whisker solids loading in the molding formulation, maintenance of low viscosity becomes even more difficult [15]. Therefore it is essential to use processing aids (dispersants) to achieve a low viscosity molding formulation with a homogeneous microstructure.

If fired/nitrided composite materials can be fabricated with homogeneous microstructures, then the next key issue is controlling the interfacial chemistry between the SiC_W and the RBSN matrix. Interfacial chemistry has been demonstrated to have a profound effect upon both the room and high temperature strength as well as the fracture toughness of ceramic composites [16-18]. As will be demonstrated in this paper, organometallic surfactants may provide a means of systematically addressing many of these key issues sited above, i.e. dispersion, low viscosity and controlled interfacial chemistry.

An organometallic surfactant is a monomer consisting of a metal atom coordinated by two different organic functional groups. A general formula for organometallic surfactants is:

$$R_nMX_{(4-n)}$$

Where: M is a metal, i.e. Si, Ti, Zr, and Al (where the 4 is replaced by a 3), X is a hydrolyzable alkoxy group which reacts with hydroxyl groups on the inorganic surface to form a covalent bond [19,20]. R is a nonhydrolyzable alkyl group. The chemistry of the alkyl group is important because it is effectively the new interface between the organic vehicle and the ceramic powder or whisker. The alkyl R group should have good solubility in the solvent, providing a means of producing a stericly stabilized, well dispersed, low viscosity injection molding formulation.

Historically [21] organometallics have been used as coupling agents to reduce viscosity, enhance dispersion, and as a mechanism to couple (bond) the filler and polymer in injection molded polymer composites. Although voluminous literature can be cited for interfacial chemistry required for filled polymers and metal matrix composites, little information is available for ceramic whisker reinforced non-oxide ceramic composites. Most of the literature in this area deals with fibers (and recently) whisker reinforced glasses, glass ceramics, and oxide matrix materials [22,23]. A thorough review of this work is beyond the scope of this paper. However, designing interfacial chemistry with the intent of optimizing the mechanical properties is important, in order to realize the full potential of ceramic-ceramic composites [24].

From the knowledge established from work on the pyrolysis of organometallic polymers as ceramic precursors [25], the potential exists to use organometallics surfactants as ceramic precursors to tailor the interfacial chemistry of injection molded SiC_W reinforced RBSN. This may be accomplished by pretreating the SiC_W and Si powder individually with chosen metal based organometallics. For example, a Ti, Zr or Al based organometallic may be reacted with the SiC_W and a Si based organometallic with the Si powder. In this way the potential exists to introduce select metal chemistries in a controlled and uniformly distributed fashion at the interface between the SiC_W and the RBSN matrix. However, prior to injection molding the dispersant effectiveness of the selected organometallics must be determined.

In order establish the dispersion behavior of the selected organometallic surfactants, a model injection molding system (mineral oil, Si powder, SiC_W ,and dispersant) was chosen. The dispersant effectiveness was evaluated from sedimentation and viscosity experiments. Results were correlated to increasing alkyl group lengths $R=CH_3$, C_8H_{17}, and $C_{18}H_{37}$ associated with the silicon based organometallics. Similarly, Ti and Zr based organometallics, containing identical alkyl groups, $R=OC_{10}H_{19}$, were used to determine the influence of metal type on dispersant effectiveness. An, Al based organometallic, with $R=OC_{18}H_{35}$ was also investigated. It was chosen because it has an alkyl R group nearly identical to the Si based organometallic.

EXPERIMENTAL

The Si powder* and SiC whiskers** were characterized by the following techniques: BET***, X-Ray Sedigraph***, He Pycnometry***, XPS****, and Diffuse Reflectance Infrared Spectroscopy*****(DRIFTS).

For the viscosity and sedimentation experiments 100 ml suspensions of SiC_W and Si powder in mineral oil were prepared. Dispersant concentrations are reported relative to the vol% solids. Samples were prepared using the following procedure: 1) the appropriate amount of mineral oil was weighed into a glass jar, 2) the dispersant (as received) was added using a microliter syringe, 3) the mineral oil and dispersant were mixed for 0.5 hr, and 4) the SiC_W or Si powder were added and mixed for 24 hrs in a low shear tumbling fashion.

After mixing, the Si and SiC_W suspensions were high shear mixed. The viscosity measurements were performed using a couette type viscometer****** . The following shear rate versus time schedule was used: 1) from 0 to 300 sec^{-1} in 5 min, 2) a 5 min hold at 300 sec^{-1}, 3) a ramp down from 300 to 0 sec^{-1} in 5 min. All viscosity values reported are those obtained at 300 sec^{-1} after the 5 min hold. The residual SiC_W suspension, not used for viscosity measurements, was poured into graduated cylinders for sedimentation measurements (after twenty days).

A) Si POWDER AND SiC_W CHARACTERISTICS

The Si powder used in these experiments had a average particle size of 8 μm as measured by X-ray sedigraph. The BET surface area was 1.55 m^2/g. Helium pycnometry gave a density of 2.34 g/cm^3. The SiC_W used had a diameter ranging from 0.2-1.0 μm and lengths ranging from 1.0 -100 μm. The measured density was 3.08 g/cm^3.

---

* Globe Metallurgical Inc., Selma, Al

** Grade SCW #1, Tateho ICD Group Inc., N.Y.,N.Y.

*** Micromeritics Norcross, GA

**** Perkin-Elmer, Physical Electronics Lab, Edison, N. J.

***** Perkin-Elmer Corp., Norwalk, CT.

****** Haake Buchler Saddle River, N.J.

The surface chemistry of the starting raw materials is critical to the use of organometallics as dispersants. The organometallic surfactants used in this study are listed in Table I. In order for these organometallic surfactants to form a covalent bond to the Si and SiC_W surfaces, a hydroxylated surface must be present. The hydroxyl groups are associated with the oxidized surface on the SiC_W and Si powder. XPS of the SiC_W and Si powder confirmed the presence of the oxide layer. Detection of the Si2p peak gave a strong measured intensity at binding energies of 102.7 and 105 eV for the Si-O bond on SiC_W and Si powder respectively. DRIFTS was used in an attempt to determine the nature of adsorbed species on the Si powder and SiC_W, however, the low specific surface area of the SiC_W and Si powder resulted in too low a signal-to-noise ratio to yield useful information.

As stated earlier, injection molding of whisker reinforced ceramic composites requires a high solids loading injection molding formulation which can flow readily into a complex shaped mold. This highly concentrated suspension of SiC_W, Si powder and molten polymer must have the proper rheological behavior, low viscosity, and be homogeneously dispersed.

TABLE I: ORGANOMETALLIC SURFACTANTS

ORGANOMETALLIC	FORMULA
SILANE*	
Methyltrimethoxysilane (MTMS)	$CH_3Si(OCH_3)_3$
Octyltriethoxysilane (OTES)	$C_8H_{17}Si(OC_2H_5)_3$
Octadecyltrimethoxysilane (ODTMS)	$C_{18}H_{37}Si(OCH_3)_3$
TITANATE**	
Neoalkoxy trineodecanoyl titanate (NTT)	$H_5C_2OTi(OOC_{10}H_{19})_3$
ZIRCONATE**	
Neoalkoxy trineodecanoyl zirconate (NTZ)	$H_5C_2OZr(OOC_{10}H_{19})_3$
ALUMINATE**	
Diisobutyl {oleyl} acetoacetyl aluminate (DOAA)	$(C_4H_9O)_2AlOCCH_3(CH)C(O)C_{18}H_{35}$

B) SiC_W DISPERSIONS

The effect on viscosity of using Si based organometallics with increasing alkyl R group length, for 1 vol% SiC_W in mineral oil suspension is presented in Figure 1. The expected trend was that increasing alkyl lengths ($R=CH_3$, C_8H_{17}, to $C_{18}H_{37}$) would result in a lower viscosity, due to increased steric stabilization associated with the longer alkyl groups. Although ODTMS ($R=C_{18}H_{37}$) resulted in the lowest measured viscosity, as expected, OTES ($R=C_8H_{17}$) resulted in higher viscosities than the MTMS ($R=CH_3$) treated suspensions. This behavior can be explained by the limited sensitivity of viscosity measurements at low solids loading. Higher solids loading viscosity measurements are more sensitive to increasing alkyl group lengths. This will be demonstrated later using higher

---

* Union Carbide Corp., N.Y., N.Y.

** Kenrich Petrochemicals Inc., Bayonne, N.J.

solids loading Si powder-mineral oil suspensions. The dispersion effectiveness at low solids loading can be more accurately determined through sedimentation experiments.

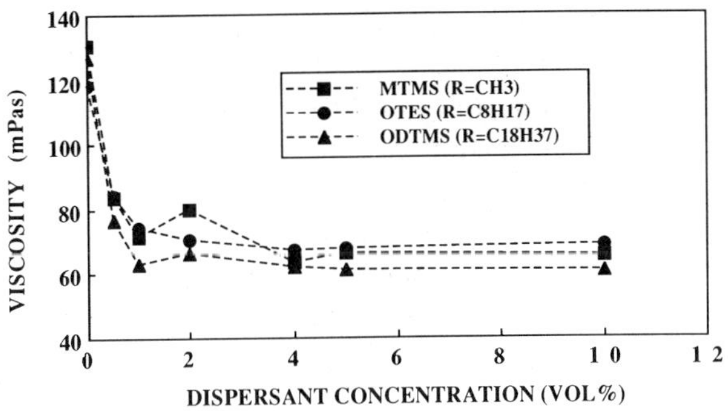

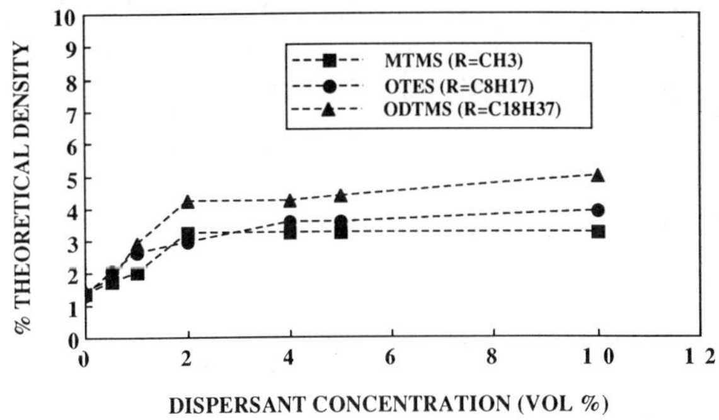

Figure 1. Influence of alkyl length on SiC_w suspension viscosity.

Figure 2. Influence of alkyl length on SiC_w sediment density.

Sedimentation experiments require low solids loading suspensions such that Stoke's settling takes place. For a well dispersed system, a higher sediment density will result compared to a system which contains agglomerated particles in suspension. In Figure 2, the % theoretical density is plotted for the same SiC_w suspensions used in Figure 1. The % theoretical density was approximately 1% for the 1 vol% SiC_w suspension with no dispersant. However at the 10 vol % (MTMS ($R=CH_3$), OTES ($R=C_8H_{17}$), and ODTMS ($R=C_{18}H_{37}$)) dispersant concentration level a 3, 4, and 5 fold increase in packing density resulted with increasing alkyl length respectively. From these results, it is clear that using the longest alkyl group ($R=C_{18}H_{37}$) results in a good dispersion yielding a high density SiC_w sediment. Because mineral oil is nonpolar, therefore steric stabilization is thought to be the active dispersion mechanism.

In the next set of experiments, the effect of metal type (Ti-NTT and Zr-NTZ) of the organometallic dispersants, with identical alkyl groups, $R=OC_{10}H_{19}$, was determined. Figures 3a and 3b show the results obtained from viscosity and sedimentation experiments performed on 1 vol % SiC_w suspensions in mineral oil. Using 5 vol % NTT and NTZ the density increased to 5.0 and 5.1 respectively

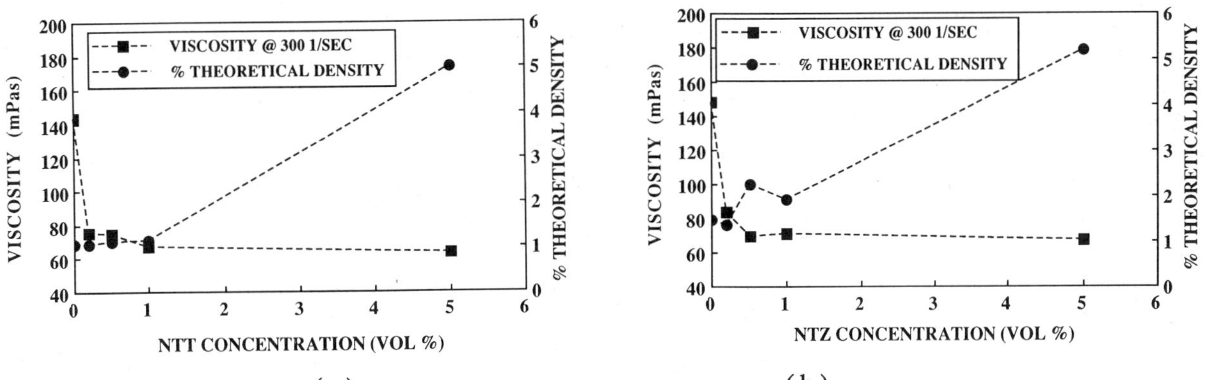

(a) (b)

Figure 3. Influence of metal type on SiC_w suspension viscosity and sediment density using: a)NTT,and b)NTZ organometallic dispersants.

over the untreated suspensions. For the untreated suspensions the viscosity was approximately 145 mPas. For suspensions treated with 5 vol% NTT and NTZ, the viscosity was 63 and 67 mPas respectively. Based on these results NTT and NTZ, the Ti and Zr containing organometallics with identical alkyl groups ($R=OC_{10}H_{19}$) are equally effective as dispersants for SiC_w in the model injection molding system.

The dispersion effectiveness of DOAA, an Al based organometallic with $R=0C_{18}H_{35}$, was also determined using the 1 vol % SiC_w-mineral oil system. This material was chosen because it has an alkyl R group nearly identical to the Si based ODTMS ($R=OC_{18}H_{37}$) organometallic. The results from the viscosity and sedimentation experiments are presented in Figure 4. Using 0 and 5 vol% DOAA, the viscosity decreased from 143 to 62 mPas respectively. At these same dispersant concentrations, the packing density increased four fold with the DOAA treated suspensions. Here again, it is the alkyl group chemistry which is important with respect to dispersion effectiveness, not metal type.

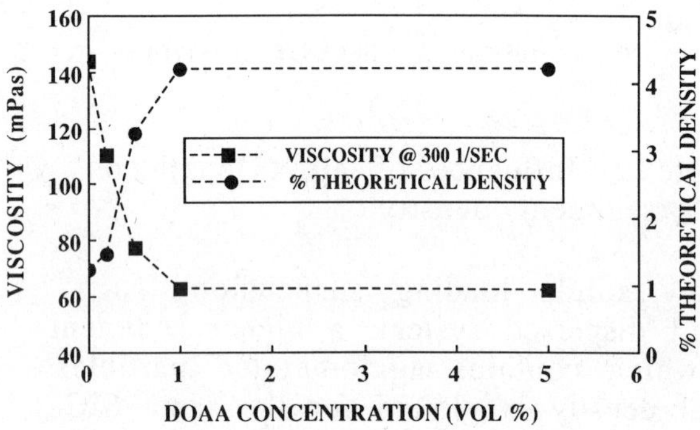

Figure 4. The influence of DOAA concentration on SiC_w viscosity and sediment density.

C) Si POWDER DISPERSION

Suspensions of 20 vol% Si powder in mineral oil were used to determine the effect of alkyl R group length upon dispersant effectiveness. The viscosity of 20 vol% Si powder suspensions containing up to 2 vol % dispersant were measured. These results given in Figure 5 are similar to those obtained for the 1 vol% SiC_w suspensions where the longest R group resulted in the lowest measured viscosity. At low solids loading it is difficult to determine dispersant effectiveness as a function of alkyl R group length. However, at increasing Si powder solids loading, the viscosity increases greatly, and the effect of increasing alkyl length was quite clear. These results are given in Figure 6. The viscosity of the 35 vol% Si suspension with no dispersant was 1700 mPas. With 1 vol% additions of MTMS ($R=CH_3$), OTES ($R=C_8H_{17}$), and ODTMS ($R=C_{18}H_{37}$), the measured viscosities were 987, 757, and 379 mPas respectively. As stated previously the injection molding of ceramics requires a high solids loading (55-75 vol%). As can be seen in Figure 6 the viscosity begins to increase greatly above the 35 vol% solids. Based on these results it is expected that at higher solids loading (55-75 vol%) larger decreases in viscosity will result using ODTMS ($R=C_{18}H_{37}$).

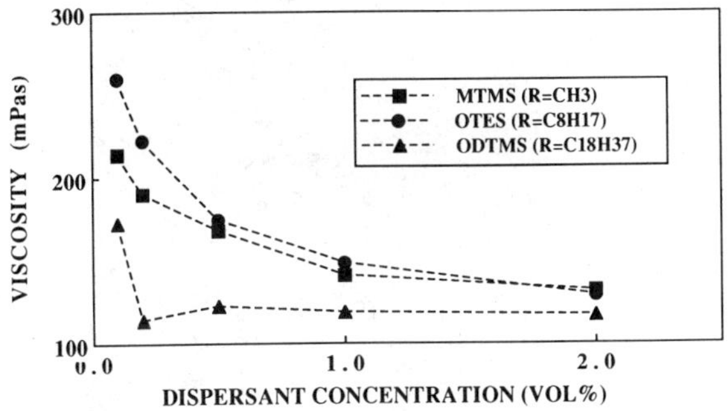

Figure 5. The influence of alkyl length and dispersant concentration on Si powder suspension viscosity.

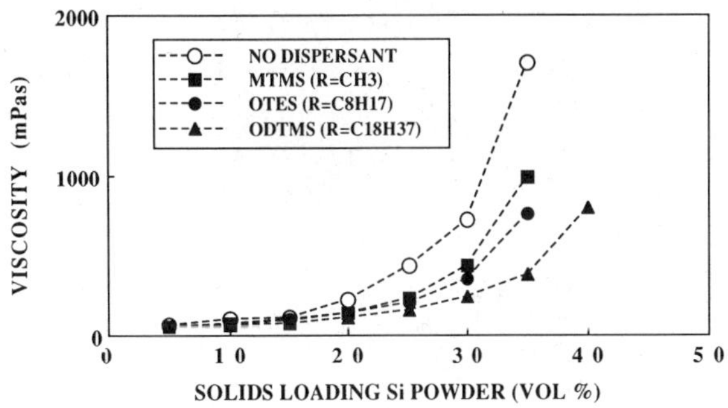

Figure 6. The influence of increasing alkyl lengths on suspension viscosity at higher Si powder solids loadings.

In order to determine the effect of metal type upon dispersant effectiveness, the same experiments were repeated using NTT an NTZ, both containing identical alkyl groups ($R=OC_{10}H_{19}$). First, the dispersant concentration necessary for optimum viscosity reduction was determined. Figure 7 shows the viscosity versus the concentration of NTT and NTZ for 20 vol% Si powder suspensions. The optimum dispersant concentration was 1 vol%, with viscosities of 105.3 and 105.4 mPas for NTT and NTZ respectively. Subsequently, Si suspensions (with Φ up to 40 vol%) were prepared with fixed 1% NTT and NTZ dispersant concentrations. The viscosity results are shown in Figure 8. NTT yields a slightly lower viscosity (496.2 mPas) than NTZ (546.5 mPas) at the 35 vol% Si powder solids loading level. As with the SiC_w systems discussed earlier, there is no distinct advantage of using Ti over Zr for viscosity reduction/dispersion purposes. In other words, both are equally effective because they contain identical alkyl R groups that result in a low viscosity, well dispersed system.

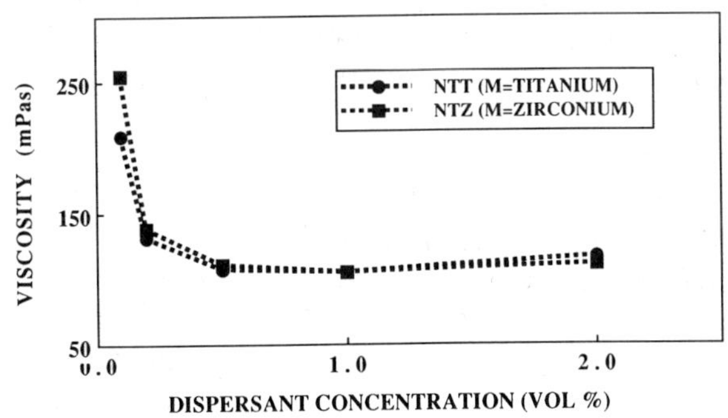

Figure 7 The influence of NTT and NTZ concentration upon Si suspension viscosity.

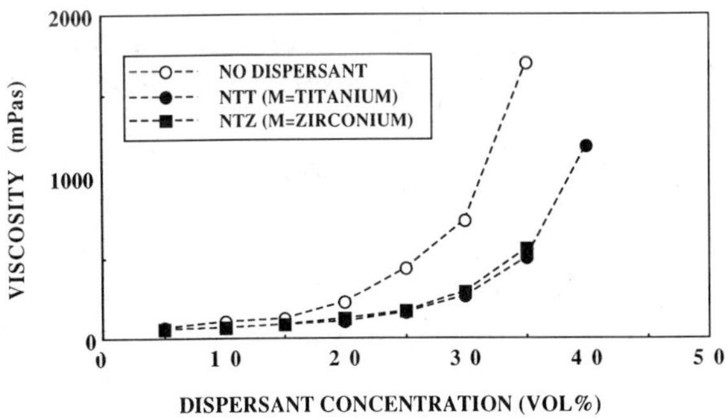

Figure 8. The Influence of metal type on Si suspension viscosity at higher solids loading.

A qualitative method of assessing a dispersants effectiveness is using optical microscopy on a untreated and 5 vol% NTZ treated Si powder suspension. Figures 9 a,b illustrate the effectiveness of NTZ to provide a good dispersion of Si powder in mineral oil, using this simple technique.

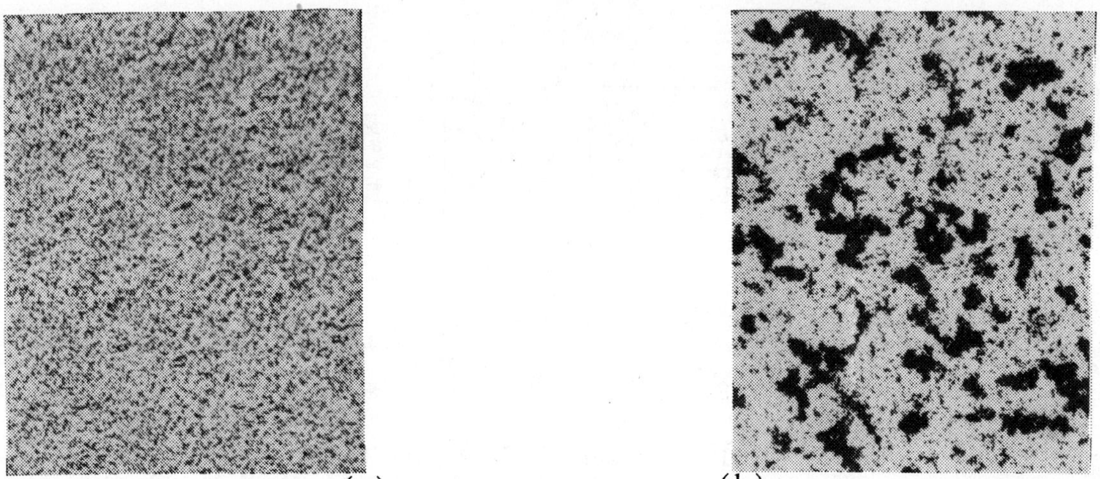

(a) (b)

Figure 9. 40X optical micrograph of Si powder in mineral oil with: a) 0 vol% NTZ, and b) 5 vol % NTZ.

SUMMARY

Injection molding SiC whisker reinforced RBSN requires a low viscosity formulation with excellent dispersion properties. The potential of various organometallics to fulfill these needs were investigated using a model injection molding system consisting of SiC_w, Si powder, mineral oil, and selected organometallic dispersants. Viscosity reduction and dispersion effectiveness increased with increasing alkyl R groups lengths, i.e. $CH_3 < C_8H_{17} < C_{18}H_{37}$, due to steric effects resulting from longer alkyl lengths. The effect of metal chemistry upon dispersion effectiveness was also investigated using NTT and NTZ (Ti, Zr based) organometallics having identical alkyl groups $R=OC_{10}H_{19}$. An aluminum containing organometallic having alkyl group $R=C_{10}H_{35}$, similar to the Si based ODTMS ($R=C_{18}H_{37}$) organometallic dispersant was also investigated. From these studies it can be concluded that the metal type had a negligible effect on viscosity reduction/dispersion enhancement of the systems investigated. By modifying the surfaces of the Si powder and SiC_w individually with selected organometallics, the potential exists to also optimize the mechanical properties by controlling the interfacial chemistry between the SiC_w and the RBSN matrix.

ACKNOWLEDGEMENTS

The authors wish to thank the National Science Foundation for supporting this work through the University of Delaware-Rutgers University Center for Composites Manufacturing Science and Engineering. The authors also wish to thank the following undergraduate students for their laboratory assistance: R. DeOliveira, S. Kwak and W. Fischer.

REFERENCES

1. A. C. Schaffhauser, E. L. Long, and V. J. Tennery, "Ceramic Technology for Advanced Heat Engines: Preliminary Program Plan," Oak Ridge National Lab Report ORNL/TM, Oak Ridge, TN, July, 1983.

2. B. R. Lawn and T. R. Wilshaw, "Fracture of Brittle Solids," Cambridge Univ. Press p 66, (1975)

3. G. C. Wei, and P.F. Becher, "Development of SiC Whisker Reinforced Ceramics," *Am. Ceram. Soc. Bull.*, **6** [2], 298-304, 1985.

4. M. Ruhle, B.J. Dalgleish and A. G. Evans "On the Toughening of Ceramics by Whiskers," *Scripta Metallurgica*, **21**, pp. 681-686,

5. F. Takao, W.R. Cannon and S.C. Danforth, "Non-Aqueous Colloidal Processing of Reaction Bonded Silicon Nitride-SiC Whisker Composites," *Advances in Ceramics*, **21** p 699, 1987.

6. J. R. Porter, F. F. Lange and A. H. Chokshi, "Processing and Creep Performance of Silicon Carbide Whisker Reinforced Alumina," Advanced Structural Ceramic, Materials Research Society, Symposia Proceedings vol., 78, pp. 289-294 1987

7. M. D. Sacks, H.w. Lee and and O. E. Rojas, "Suspension Processing of SiC Whisker-Reinforced Ceramic Composites," Presented at The First Conference on Ceramic Powder Processing Science, Orlando, Fla., Nov., 1-4, 1987.

8. M. J. Edirisinghe and J. R. G. Evans, "Review: Fabrication of Engineering Ceramics by Injection Molding. II. Techniques", *Int. j. High Technology* **2** [4] 249-278 (1986).

9. J. V. Milewski," Efficient Use of Whiskers in the Reinforcement of Ceramics," *Advanced Ceramic Materials*, **1**, [1], (1986).

10. J. A. Mangels, "Development of Injection Molded Reaction Bonded Si_3N_4," pp. 113-130 in Ceramics for High Performance Applications II edited by J. J. Burke, E. N. Lenoe, and R. N. Katz. Metals and Ceramics Information Center, Columbus Ohio,(1977).

11. P. A. Willermet, R. A. Pett and T. J. Whalen, Development and Processing of Injection Moldable Reaction Sintered SiC Compositions," *Am. Ceram. Soc. Bull.*,**57** [8] 744-747 (1978).

12. T. J. Whalen and C. F. Johnson, "Injection Molding of Ceramic," *Am. Ceram. Soc. Bull.*, **60** [2] (1981).

13. R. J. Ferris,"Prediction of the Viscosity of Multimodal suspensions From Unimodal Viscosity Data,". *Trans. Soc. Rheology*, **12** [1] 281 (1968).

14. J. A. Mangels and R. M. Williams," Injection Molding Ceramics to High Green Densities," *Am. Soc. Ceram. Bull.*, **62** [5] 601 (1983).

15. I. Tsao and S. C. Danforth, "Rheological Behavior of Injection Moldable Ceramic-Ceramic Composites," presented at The Third International Symposium, Ceramic Materials and Components for Engines, Las Vegas, Nevada, November 28-30, 1988.

16. B. A. Bender, D. Lewis III, W. S. Coblenz, R. W. Rice, "Electron Microscopy of Ceramic Fiber-Ceramic Matrix Composites-Comparison with Processing and Behavior," *Conf. Proc. Cer. Eng. and Sci.*, **5**, 513-529, (1984).

17. D. R. Clarke, "Grain Boundaries in Polycrystalline Ceramics," *Ann. Rev. Mater. Sci.*, **17**, pp. 57-74, (1987).

18. S. R. Nutt, and D. S. Phillips, "Interface Structure in Silicon Carbide-Silicon Nitride Composites," Presented at 1986 Annual TMS-AIME Meeting, New Orleans, LA., March, 1986.

19. E. P. Plueddeman, "Silane Compounds for Silylating Surfaces," in Silanes, Surfaces and Interfaces Edited by Donald E. Leyden, Gordon and Breach Science Pub., pp.1-25, 1986

20. D. E. Leyden and D. E. Williams, "Spectrochemical Characterization of Chemically Modified Surfaces," in Molecular Characterization of Composite Interfaces, Edited by Hatsuo Ishida and Ganesh Kumar, Plunum Press, New York, pp 377-386, 1985.

21. M. R. Rosen, "The Life History of a Silane Coupling Agent," *Journal of Coatings Technology*, **50**, [644], pp. 70-82, September, (1978)

22. J. J. Brennan and K. M. Prewo, "Silicon Carbide Fibre Reinforced Glass-Ceramic Matrix Composites Exhibiting High Strength and Toughness," *Journal of Materials Science* , **17**, pp. 2371-2383, (1982).

23. A. G. Evans, M. Rühle, B. J. Dalgleish, and M. D. Thouless," On Prevalent Whisker Toughening Mechanisms in Ceramics," in Advanced Structural Ceramics edited by P. F. Becher, M. V. Swain and S. Somiya, Materials Research Society, pp. 259-272, 1987.

24. P. F. Becher, C. Hsueh, P. Angelini, T. N. Tiegs, " Factors Influencing The Toughening Behavior of Whisker Reinforced Ceramics," in The Proceeding of Whisker-and Fiber-Toughened Ceramics Oak Ridge, Tennessee, pp.109-112, June 7-9 1988

25. 8. R. H. Baney, "Some Organometallic Routes to Ceramics," in Ultrastructure Processing Of Ceramics, Glasses, and Composites, edited by Larry L. Hench and Donald R. Ulrich, John Wiley and Sons Pub., pp.245-254, 1984.

GAS-PRESSURE COMBUSTION SINTERING OF TiB_2 BASED COMPOSITES

Seiji Ohmura and Yoshinari Miyamoto
The Institute Of Scientific and Industrial Research,
Osaka University, Ibaraki, Osaka 567, Japan
Mitsue Koizumi
Institute For Science and Technology, Ryukoku University,
Fukakusa, Fushimiku-ku, Kyoto 612, Japan

ABSTRACT

A new process of combustion sintering by gas pressure has been developed. In this way, the reactant encapsulated into a glass capsule is pressurized after softening the capsule by heating and then the combustion reaction is initiated by exothermic heat of the surrounding ignition agent. By this process, the ceramic composites of titanium borides have been synthesized and simultaneously densified above 97% density from the starting compositions of (100-x)(Ti+2B) + x Ti (x=10∿40wt%) at 40 to 100 MPa Ar pressures. The product phase included TiB2, Ti3B4 and TiB depending on the starting compositions, but no Ti metal remained. The grains were below 6 μm. The microhardness and fracture toughness are 20 GPa, and 4.1 MPa√m when fabricated from the reactant with x=10 under 100 MPa, and 19 GPa, 5.6 MPa√m under 40 MPa, respectively. This process is expected to be available for mass production of near-net shape components for a variety of ceramic based composites.

INTRODUCTION

The high-pressure combustion sintering (HPCS) is a new process to enable simultaneous synthesis and densification of ceramics by enhancing the self-propagating high temperature synthesis (SHS) under pressure [1]. The SHS is known as a highly exothermic reaction spontaneously propagating through the reactant and converting into a compound [2,3]. Many refractory metal borides, carbides, nitrides and silicides having high formation energies are easily formed by igniting the mixed powders of their constituent elements. The HPCS can be classified as reaction sintering, but has the following advantages not exhibited in conventional methods such as used for SiCSi, or Si3N4 :

1) Nearly full densification is possible.
2) Many material systems are applicable.
3) Reaction time is very short.
4) Long time firing is unnecessary.

Dense compacts of TiB_2 (95% density of theoretical)[1], SiC (92.6%)[4] and TiC (96.5%)[5] can be synthesized without sintering aids under a pressure of 3 GPa by the HPCS. However, the extremely high pressure, which was produced by means of a cubic anvil device, and therefore limited sample size was a problem for practical use of this process. In order to overcome these difficulties, in 1987 a more efficient process has been devised by employing gas pressure [6]. The isostatic gas pressure enables to form near-net shape components of larger scale.

This paper reports the process development of gas-pressure combustion sintering and experimental results on the simultaneous synthesis and densification of titanium boride ceramics. The boride ceramics are expected as advanced materials with eminent properties such as high melting point, high hardness, chemical inertness and high electrical conductivity for the applications to mechanical and structural uses under severe conditions.

EXPERIMENTAL

Development of Gas-Pressure Combustion sintering

The experiment of gas-pressure combustion sintering was conducted by using HIP equipment. Figure 1 illustrates the interior of the high pressure vessel where two electrodes made of Mo rod for ignition are installed through the bottom rid. The press-formed reactant covered with BN powder coating is encapsulated into a pyrex glass tube and vacuum sealed just like usual glass capsule method of HIP. The glass capsule is coated with BN powder again and embedded into the ignition agent which was charged in a graphite crucible. The BN coatings are to protect the glass capsule from the reactions with reactant and ignition agent. As a material for ignition agent, the mixed powder which produces a strong combustion reaction is used. Many kinds of ignition agents would be available and the product can be used as ceramic powders. Mixed powders of Ti and C are used at present, which can produce high temperature of 3210°K. The inside temperature of the glass capsule was measured when heated by this agent. A thermocouple was inserted into a pyrex glass tube charged with BN powder and the surrounding agent of Ti and C was ignited in air. The inside temperature was raised up to 1673°K within 20 seconds and high temperature above 1273°K was maintained for over 1 minute.

The combustion sintering is initiated as follows. First, the inside of the high pressure vessel is heated to about 973°K in order to make the glass capsule soften by using a HIP heater. The Ar gas pressure is applied to the appropriate level and then the ignition agent is burned by an ignition heater. It is expected that the use of isostatic gas pressure and ignition agent would lead the HPCS to a economic process to realize mass production of near-net shape components as illustrated in Fig.2.

Simultaneous Synthesis and Densification of Titanium Boride Ceramics

The powders of titanium metal with particle size below 40μm (99.5% purity) and amorphous boron below 2 μm size (95.0% purity) were used as raw materials. They were mixed in ethanol in various mixing ratios by adding excess titanium to the reactant of Ti + 2B composition as indicated with x in the following equation.

$$(100 - x)(Ti + 2B) + x\ Ti \quad (x=0 \sim 40\ wt\%)$$

The reactant was press-formed as a pellet 7 mm in diameter and 15 mm in length after drying at 413°K in vacuum. This reactant pellet was cold isostatically pressed at 100MPa, coated with BN powders, and degassed at 673°K for 2 hours in vacuum to eliminate water and volatile impurities which cause the porosity in the sintered body. The weight loss of the reactant increased by increasing the treatment temperature and saturated to 0.35 wt% at 623°K.

The glass encapsulated reactant was ignited according to the process described before. The solid state reaction of reactant slightly occurred by preheating to 973°K, but it little affected the combustion reaction. The Ar gas pressure was applied in the range from 10 MPa to 100 MPa. The ignition was carried out by passing a current of 100 amps for 2 or 3 seconds through a carbon ribbon heater which was laid on the top of the ignition agent.

The product phase was identified for a pulverized sample of the sintered body using a rotating anode X-ray diffractometer with CuKα radiation. The bulk density was measured by the Archimedes method. The relative density was calculated against the theoretical density of TiB_2 (4.53 g/cm3) since the densities of TiB_2, Ti_3B_4, TiB and Ti are almost the same value. The microstructure was observed by scanning electron microscopy (SEM) and transmission electron microscopy (TEM). The Vickers microhardness was measured under 19.6N load. The fracture toughness was evaluated by an indentation method with the same load [7].

RESULTS AND DISCUSSION

The formation energy of TiB_2 at 298°K is -293 kJ/mol which regulates the combustion temperature. The maximum temperature of product attainable for the reaction can be calculated under the assumption that the exothermic heat is used ultimately to raise the temperature of the final product in an ideal adiabatic system. Figure 3 shows a graphical estimation of the adiabatic combustion temperature (Tad) for TiB_2 by drawing the enthalpy curves of reactant and product. The exothermic heat can provide at least 40% of the heat of fusion, so that Tad coincides with the melting point of TiB_2 (3193°K).

When the reactant is ignited after preheating at 973°K, it is inferred that the fraction of product being fused will increase although the reaction temperature may not exceed the melting point of TiB_2 as shown in Fig. 3. Moreover, the adiabatic state of reaction seemed to be sustained for longer time due to heat supply from the ignition agent. The formation energy of TiC is 185kJ/mol and the adiabatic combustion temperature coincides with its melting point of 3210°K.

The actual temperature and the propagation velocity of the combustion reaction were measured for the pelleted powders of Ti + 2B by using a two color pyrometer and a high speed video, respectively. The samples with different packing densities from 50 to 70 % were ignited in an Ar gas atmosphere. They expanded about 50 % in volume but kept original shape after reaction. The maximum temperature was 3173°K ∿ 3273°K, which did not depend on packing density. The powdered product of TiB_2 was composed of uniform size particles around 5 μm and molten agglomerates were observed as seen in Fig. 4. These facts suggest that the reaction propagates almost in an adiabatic state. The propagation velocity increased with the packing density from 70 mm/s at 50% density to 150 mm/s at 70% density. The powdered product was composed of TiB_2 phase with a minor content of TiB as low as 1% by X-ray peak intensity.

Figure 5 shows the density of sintered bodies depending on the excess titanium when combustion sintering was conducted at 100 MPa. The densification was limited to 87% of theoretical density for the stoichiometric reactant with x=0, however, the products from the reactants with excess Ti over 5 wt% were densified above 95%. When 20 wt% excess Ti was added in the reactant, the density reached up to 99%.

X-ray diffraction patterns of the products are shown in Fig. 6. In the case of reactant with x=0, the product was composed of TiB_2 single phase. When excess Ti was added into reactant, the residual Ti metal was not detected, but additional phases of Ti_3B_4 and TiB appeared. The phase ratios changed with excess Ti, and Ti_3B_4 and TiB were the main phases at x=30 and x=40, respectively. The reaction heat would be decreased with increasing excess Ti because of lower formation energies of Ti_3B_4 and TiB than TiB_2 [8]. When the reactant with excess Ti was ignited by using a tungsten heat wire sealed into glass capsule in the previous work, Ti metal phase remained [9]. Consequently, the complete reaction of Ti and B is considered due to the heat supply from ignition agent. It is considered that the liquid phases of Ti_3B_4 and TiB with lower melting points than TiB_2 might enhance the densification. However, a further investigation is needed for understanding the densification process.

Figure 7 shows a TEM image of the microstructure of product with 98% density obtained from the reactant with x=10 under 100MPa. The secondary phase was not observed at the grain boundaries. It was found that a coherent relation of lattice image across the grain boundary spread in the sintered body. This coherent microstructure is noted as the characteristic feature which is essentially different from the microstructures formed by the conventional sintering of ceramic powders. Figure 8 shows the fractured surface of the compact fabricated by the same condition with the larger grains of a few μm size and smaller submicron ones dispersed. When the applied pressure was reduced to 40 MPa for the reactant with x=10, the grains became larger around 5 μm as seen in the fractured surface of Fig.9. The SEM photograph shows that the grains bonded tightly and the fracture occurred by both the transgranular and intergranular modes.

The reduction of applied pressure is desired for the development of this process to industrial use. Figure 10 shows the densities of the sintered bodies as a function of applied pressure. Two kinds of reactants with x=10 and 20 were supplied. The relative density of the product of x=10 was 97% at 40 MPa and that of x=20 was 98%. However, the densities were lower below 30 MPa although the reason is not clear yet. It may be possible to lower the applied pressure to less than 30 MPa for densification by further development of this process.

The Vickers microhardness and fracture toughness are plotted as a function of the reactant composition in Figs. 11 and 12, respectively. The compacts were fabricated at 100 MPa. The hardness decreased from 21 GPa at x=0 to 15 GPa at x=40 with increasing the contents of Ti_3B_4 and TiB phases with lower hardnesses than TiB_2. While the fracture toughness increased with x and reached near 5 MPa√m at x=40. The increase of fracture toughness may be due to the formation of composite structure with different phase particles.

The effect of applied pressure for the fabrication process on hardness and fracture toughness is plotted in Figs. 13 and 14. The hardness decreased with the decrease of applied pressure, but the fracture toughness increased. The influence of remaining pores is considered as an origin for these tendencies. It was also observed that the crack induced by the indentation propagated straightly through fine grains when fabricated at 100 MPa, but deflected frequently between larger grains when fabricated at 40 MPa.

SUMMARY

Dense titanium boride ceramics have been fabricated from the mixed reactants of (100-x)(Ti+2B)+xTi(x=10∿40wt%) by a newly developed process of gas-pressure combustion sintering. The densities reached over 97% of theoretical when fabricated under Ar gas pressures above 40 MPa. The products were composed of the mixed phases of TiB_2, Ti_3B_4 and TiB without leaving Ti metal phase and their contents depended on the starting compositions. The sintered body had a unique microstructure holding a coherent relationship between crystal lattices of each grains, which reflects that the densification occurred with formation of grains. The Vickers microhardness and fracture toughness were dependent upon the starting composition and applied pressure. The addition of other metal or ceramic powders to the reactant would produce new materials of ceramic based composites. The gas-pressure combustion sintering is expected to develop as an economic process available for the mass production of near-net shape components.

ACKNOWLEDGMENTS

The authors gratefully acknowledge Prof. K. Urabe in Ryukoku University who kindly performed TEM observation of sintered body. This work was supported by the grants from the joint research program between university and industry (Sumitomo Electric Industry Co., Ltd.) promoted by the Ministry of Education, Science and Culture, from the ISIR project of the new materials for efficient energy conversion, and from the research project in the Science and Technology Agency on the development of functionally gradient materials.

REFERENCES

1. Y. Miyamoto, M. Koizumi, and O. Yamada, "High Pressure Self-Combustion Sintering for Ceramics." Comm. A. Ceram. Soc., (1984) C-224-5
2. A. G. Merzhanov and I. P. Borovinskaya, "Self-Propagating High Temperature Synthesis of Refractory Inorganic Compounds," Dokl. Chem., (Engl. Trans.), 204 [2](1972) 0429-32
3. A. G. Merzhanov, "SHS-process : combustion theory and practice," archivum combustion, 1[1/2],(1981) 23-28
4. O. Yamada, Y. Miyamoto and M. Koizumi, "High Pressure Self-Combustion Sintering of SiC from Fine Silicon and Carbon Mixed Powders", Proc. 2nd. Int. Symp. on "Ceramic Materials and Components for Engines", Edited by W. Bunk and H. Hawsner, Deutsche Keram. Gessell., 1986, PP331-38
5. O. Yamada, Y. Miyamoto and M. Koizumi, "High-Pressure Combustion Sintering of Titanium Carbide", Comm. A. Ceram. Soc., 70 (1987) C-206-8
6. Y. Miyamoto and M. Koizumi, "Potentiality of High Pressure Combustion Sintering to Produce Ceramic Based Multi-Phase Components," Proc. Int. Symp. on "Sintering '87", Tokyo, 1987, in printing
7. A. G. Evans and E. A. Charles, "Fracture Toughness Determinations by Indentation", J. A. Ceram. Soc, 59 (1976) 371-71
8. W. F. Henshaw, A. Niiler and T. Leete, "Self-Propagating High Temperature Synthesis of Ceramics in Vacuum", Ceram. Eng. Sci. Proc., 4 [7-8](1983) 634-45
9. Y. Takano, S. Ohmura, Y. Miyamoto and M. Koizumi, "New Application of HIPing to Combustion Sintering of TiB2", Proc. 3rd Int. Conf. on Isostatic Pressing, MPR Publishing Services Ltd., London (1986) PP.21-1-11

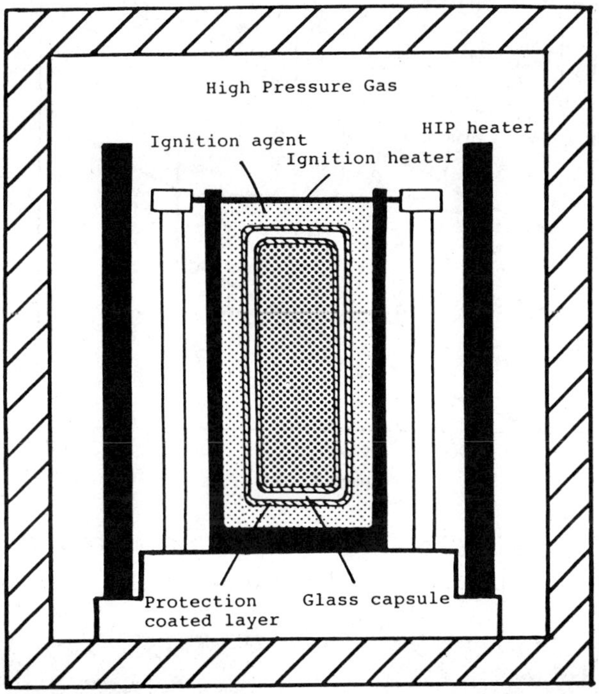

Figure 1. Schematic of apparatus used for gas-pressure combustion sintering.

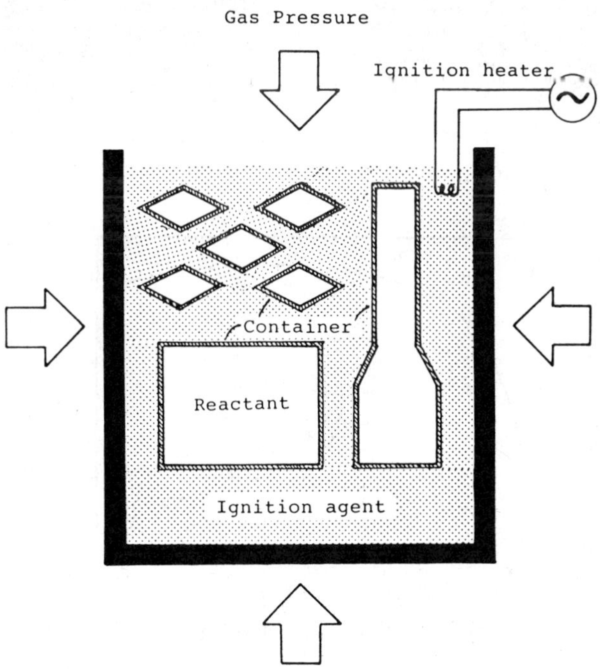

Figure 2. A model plant of gas-pressure combustion sintering for mass production of near-net shape components.

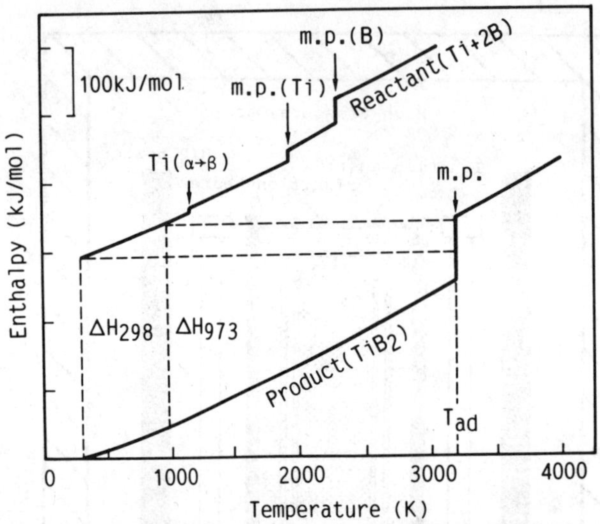

Figure 3. Enthalpy relation between the mixed reactant of Ti and 2B and the product of TiB_2.

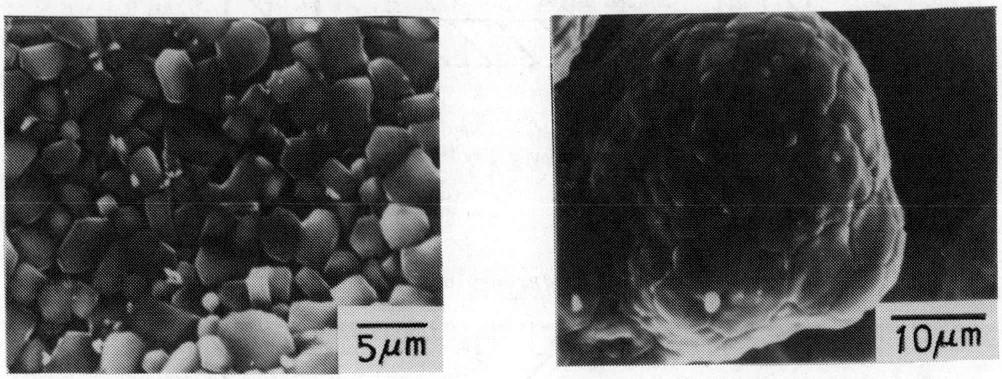

Figure 4. SEM photos of TiB_2 powders synthesized by the SHS in Ar Atmosphere.

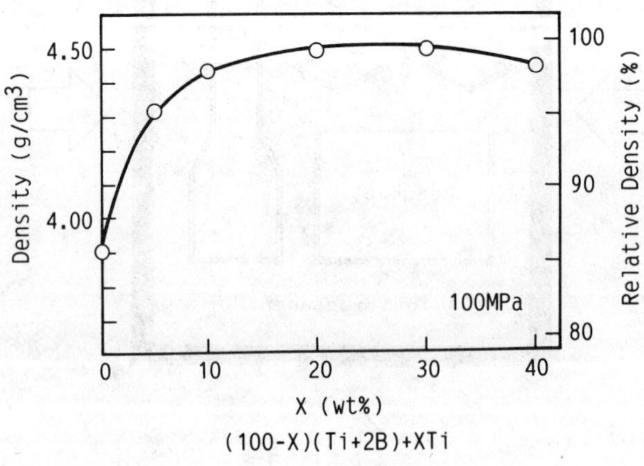

Figure 5. Density of titanium boride composites as a function of starting composition. The samples were fabricated at 100MPa.

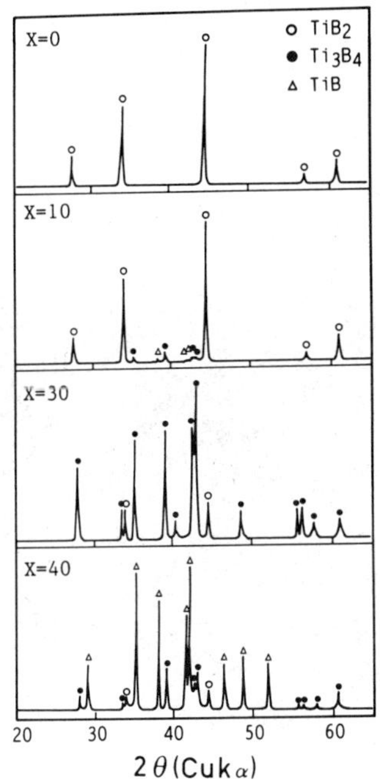

Figure 6. X-ray powder diffraction patterns of titanium boride composites.

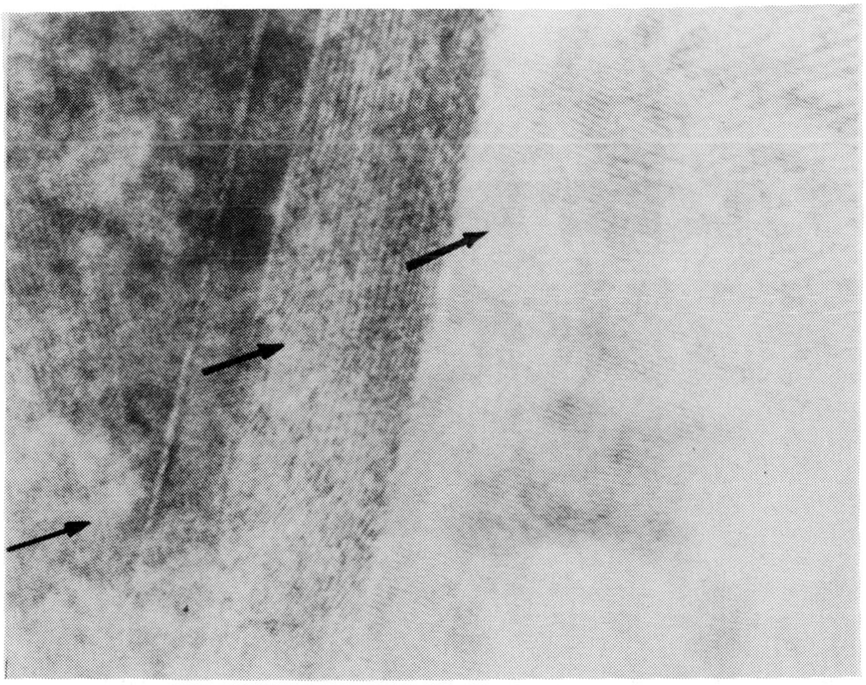

Figure 7. TEM image of microstructure of titanium boride composite fabricated from the reactant with x=o at 100MPa.

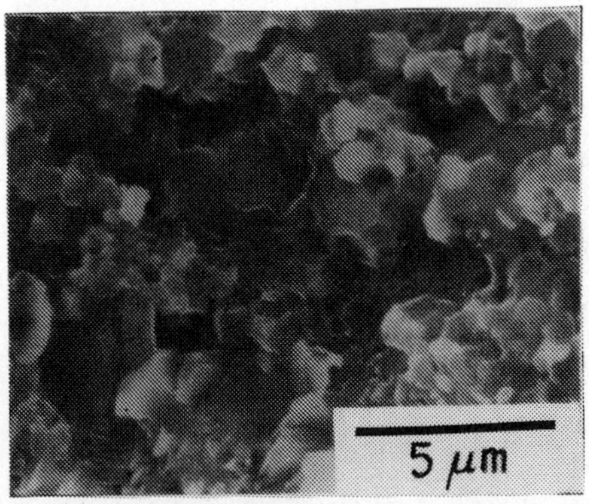

Figure 8. SEM photo of fractured surface of titanium boride composite from the reactant with x=10 at 100MPa.

Figure 9. SEM photo of fractured surface of titanium boride composite fabricated from the reactant with x=10 at 40 MPa.

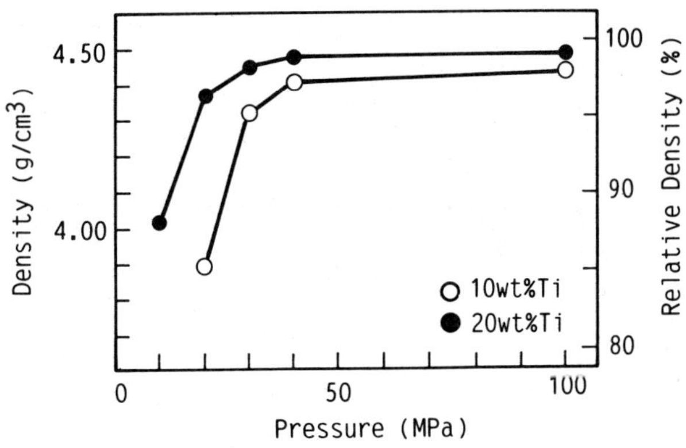

Figure 10. Effect of applying pressure on density for titanium boride composites fabricated from the reactants with x=10 at 40 MPa.

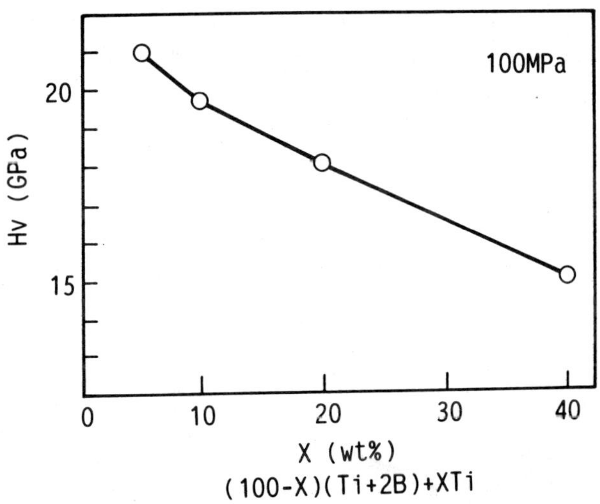

Figure 11. Vickers microhardness of titanium boride composites as a function of the starting compositions.

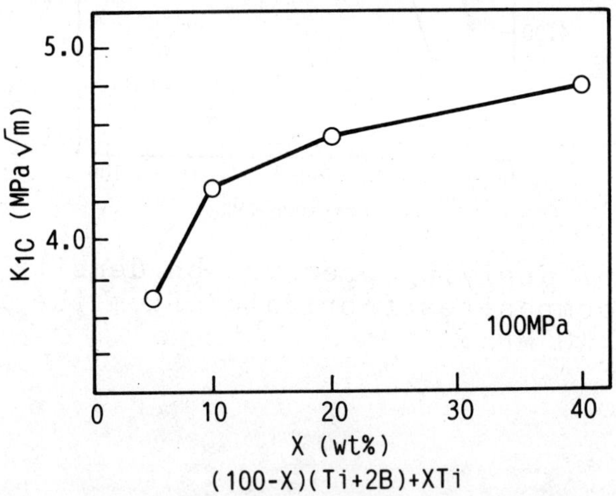

Figure 12. Fracture toughness of titanium boride composites as a function of the starting compositions. The samples were fabricated at 100MPa.

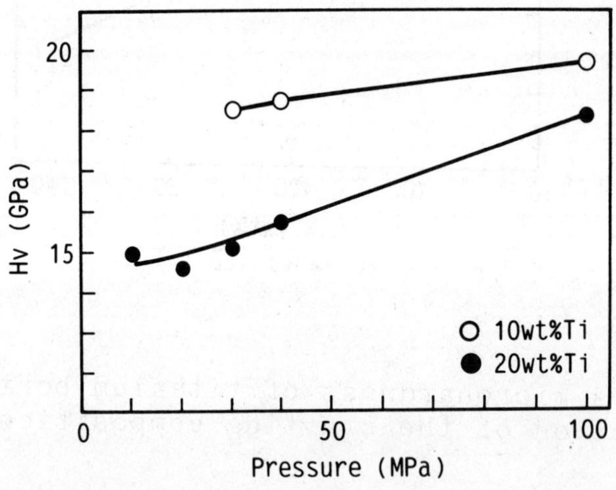

Figure 13. Effect of applying pressure on Vickers microhardness of titanium boride composites fabricated from the reactants with x=10 and 20.

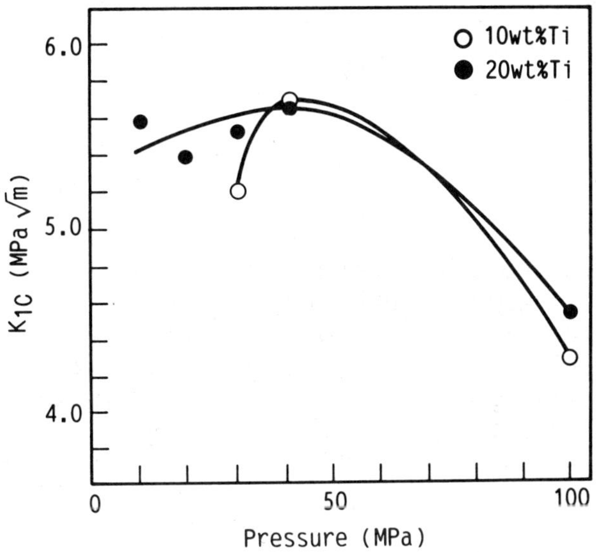

Figure 14. Effect of applying pressure on fracture toughness of titanium boride composites fabricated from the reactants with x=10 and 20 wt%.

PRELIMINARY STUDY OF SiC WHISKER/Si_3N_4 COMPOSITE

GU PEIZHI, ZHANG WEIRU, XIAO YIXIAN, and FAN QISHENG

Shandong Institute of Industrial Ceramics,
Zi-Bo, Shandong, China

ABSTRACT

Homogeneuous dispersion of SiC whiskers in silicon and silicon nitride matrixes by using wet process, ultrasonic and high speed homogenizing techniques has been achieved. Compared with the hot-pressed reaction bonded silicon nitride (HPRBSN) matrix, approximately 50% increase in fracture toughness and similar flexural strength for 10 vol% SiC whisker containing hot-pressed reaction bonded silicon nitride (SiC W/HPRBSN) composite have been obtained. Effects of additives on nitrided materials are discussed, mechanical properties and microstructure of SiC W/HPRBSN composites are evaluated.

INTRODUCTION

As an important structure ceramics, silicon nitride based materials have the advantages of high strength, excellent wear resistance and good thermal stability. However, its brittle nature and catastrophic failure in severe strees environment have limited the structural applications of those materials. Great interests in recent years have been focused on the development of SiC whisker or fiber reinforced Si_3N_4 composites to meet the needs of high fracture toughness, high reliability. Numbers of attempts and successes on that subject have been reported by authors in many contries (1,2,3,4).

Based on the consideration that the silicon based compact is easily to be machined into complex shapes, the silicon based composition was used as starting material to develop the Si_3N_4 matrix and SiC whisker composite in present study. The main purpose of the study is to search a possibility to develop the SiC whisker/Si_3N_4 composite with improved properties by nitriding the silicon based composition and then hot-pressing or, in the future, gas pressure sintering and HIPing the nitrided composition. In present study the test samples were fabricated by nitriding and hot-pressing the mixture of SiC whiskers and silicon based composition. For comparasion the Si_3N_4 powder was also used to prepare the SiC whisker containing hot-pressed silicon nitride (SiC W/HPSN) composite.

EXPERIMENTAL PROCEDURE

Preparation of whisker/powder mixture

The silicon powder used in this study has an average particle size of 5 μm determined by sedimentation method, BET surface area of 2.6 m^2/g, iron content of 1 wt%. The SiC whisker* chosen for the composite has an average diameter of 0.6-1 μm, aspect ratio of 20-40, crystalline of β 8F in major according to the commercial technical data. However the whiskers having larger diameter of 3 μm and length of 40 μm were observed with SEM as shown in Fig.1. Silicon nitride powder** with average particle size of 0.5 μm, BET surface area of 18 m^2/g, and α phase content more than 90 wt% was used. Sintering aids of $MgO-Al_2O_3$ (MA) spinel** and unstablized ZrO_2*** with particle size of 0.3 μm were used for both silicon and Si_3N_4 starting compositions.

Silicon powder of 75-90 wt% and sintering aids of 25-10wt% were ball milled in ethanol media for 8 hrs to prepare the matrix slurry. The proper contents of sintering aids MA spinel and ZrO_2 have been reported in previous study (5). Ethanol, ethyl-acetate, and ethanol with small amount of triethanolamine were tested as dispersant to select the one most suitable for the studied system.

The SiC whisker slurry was prepared by putting appropriate amount of whisker in a liquid dispersant and subsequently by ultrasonic treatment of 20 min and high speed homogenizing of 5 min. The two slurries were mixed and blended in a homogenizer for 10 min. With stirring, the mixture of slurries was heated to evaporate the solvent and then dried and sieved into 10 vol% and 20 vol% SiC whisker (calculation based on Si_3N_4 composition) and silicon powder mixtures.

The SiC whisker/Si_3N_4 powder mixture was prepared by the same procedure.

Nitriding and hot-pressing

The SiC whisker/silicon powder mixture was cold isopressed with a max pressure of 200 MPa. The disks of 50 mm diameter and 6 mm thickness were nitrided in a $MoSi_2$ resistance furnace in N_2/H_2 atmosphere for 50 hrs with the max temperature of 1400 °C which was determined according to the works by J.A.Mangels etc. (6).

The nitrided disks were then hot-pressed in BN coated graphite die in a graphite resistance furnace and N_2 atmosphere at the temperatures of 1650, 1680, 1700, 1730, 1750 and 1780 °C respectively with a holding time of 30 min. The pressure was applied gradually to reach the max of 25 MPa at the holding temperatures in all runs.

---

* Tateho Chemical, Japan.
** Synthesized by the authors.
*** Commercial available. China.

Material Property testing

The hot-pressed disks were diamond cut then grinded into the 3 X 4 X 35 mm bars for flexural strength testing and 2 X 4 X 35 mm bars for fracture toughness measurement. The densities of green bodies, nitrided and hot-pressed bodies were measured by mercurry immersing technique. Three-point flexural strength of specimen were measured at room temperature using an Electron Tensile Testing Machine with a span of 25 mm and a cross-head speed of 0.5 mm/min. Single edge notched beam technique was employed for fracture toughness measurement. A minimum of five bars was used for either of flexural strength and fracture toughness measurements for each corresponding hot-pressing temperature. The 3 X 4 X 35 mm testing bars were also used for thermal shock resistance testing. The nitrided samples were X-ray analysed to identify the phase composition. The morphology of fracture surface of selected samples were studied with SEM*.

RESULTS AND DISCUSSION

Mixing

Homogeneous dispersion of SiC whisker in ceramic matrix was achieved by using ethyl-acetate with small amount of nitrocellulose as dispersant and employing ultrasonic and high speed homogenizing treatments. The SiC whisker slurry obtained can be stable for 24 hrs without reagglomeration and rebundling of the whiskers, which was identified under optical microscope. Although good dispersing state of whiskers can also be obtained by using other dispersants such as ethanol and water, but the stability of the slurries was not satisfied. No aspect ratio change of SiC whiskers was found after 15 min of high speed homogenizing treatment. The wet process discribed by F.F.Lange and other authors (3,7,8) has also been successful in this study. Fig.2 shows a cross section of the green body of 10 vol% SiC whisker containing composite, indicating homogeneous distribution of whiskers in ceramic matrix.

Effect of oxide additive and SiC whisker on nitridation

It has been found that some oxide additives in silicon compact enhance the nitridation rate of silicon during nitriding especially at lower temperatures (9,10). The result is quite similar with the conclusion deduced for MA spinel and ZrO_2 used in this study. The increasing content of β phase in nitrided materials was obtained with the increasing of oxide additive content under the same processing parameters, that can be seen in Table I.

* Model TS-1, China.

Table I. Effect of oxide additive on nitridation of silicon compact*

Content of MA+ZrO_2 wt%**	$\frac{\beta}{\alpha + \beta}$ wt%***	Conversion of silicon to Si_3N_4 wt%
0	8	93
6	46	93
10	49	91
16	56	91

* Max temperature of nitridation 1400 °C.
** Calculation based on silicon nitride composition.
*** Mean value of five samples.

The green densities of SiC whisker/silicon compact and the conversion of silicon to Si_3N_4 in nitrided materials are listed in Table II. It seems from the result that the addition of SiC whisker dispersoid more or less inhibited the nitridation of silicon.

Table II. Effect of SiC whisker on nitridation of whisker-silicon compact*

Content of SiC whisker vol%	Green density % of T.D.**	Conversion of silicon to Si_3N_4 wt%
0	60	93
10 (ethyl-acetate)	57	84
10 (ethanol)	54	88
20 (ethyl-acetate)	52	86

* Max temperature of nitridation 1400 °C.
** Mean value of five samples.

The result is probably due to the lower green densities of whisker containing materials, oxide content, and oxygen impurity in N_2 gas used in this study, which causes the higher weight loss of the samples during long time of nitridation and results in lower conversion percent which was determined by the weight changes before and after the nitridation. Futher investigation is needed to detect the effect of SiC whiskers on nitriding behavior of silicon and to determine the optimum process parameters.

Densification

The 10 vol% SiC whisker/HPRBSN has an identical densification rate compared with the HPRBSN matrix during hot-pressing while those materials contain the same amount of sintering aids. But it is quite different for 20 vol% SiC containing composite. It is clear from the density data presented in Fig.3 that the densification (% of T.D.) for 10 vol% SiC whisker/HPRBSN composite is much higher than that for 20 vol% whisker/HPRBSN composite at every corresponding temperature while the hot-pressing parameters were similar for those composites. It also can be seen that the higher temperatures have not favoured the densification of above composites (result in lower density). It suggestes the 20 vol% SiC whisker containing composites may need longer hot-pressing time that was concluded by Sergej.T.Buljal etc., they found the higher the whisker content the lower the densification rate of the composite although the sintering aid contents were same.

Mechanical Properties

The three-point bend strength and fracture toughness values for HPSN, HPRBSN matrixes, SiC W/HPSN and SiC W/HPRBSN composites hot-pressed at 1730 °C are listed in Table III.

Table III Mechanical properties of SiC W/Si_3N_4 composites*

Materials**	Flexural strength MPa	Fracture toughness $MPam^{\frac{1}{2}}$
Si 10 (HPRBSN) matrix	660 ± 33	5.6 ± 0.3
Si 10 + 10 vol% SiC W (SiC W/HPRBSN)composite	620 ± 50	7.8 ± 0.3
Si 6 (HPRBSN) matrix	580 ± 21	3.4 ± 0.2
Si 6 + 10 vol% SiC W (SiC W/HPRBSN) composite	640 ± 57	5.1 ± 0.3
Si 10 + 20 vol% SiC W (SiC W/HPRBSN) composite	360 ± 74	4.0 ± 0.4
SN 6 (HPSN) matrix	800 ± 27	7.0 ± 0.2
SN 6 + 10 vol% SiC W (SiC W/HPSN) composite	850 ± 42	7.7 ± 0.3

* Materials were hot-pressed at 1730 °C.
** Si 10-silicon as atarting material, 10 wt% sintering aids based on silicon nitride composition.

Si 6-silicon as starting material, 6 wt% sintering aids based on silicon nitride composition.
SN 6- Si_3N_4 as starting material, 6 wt% sintering aids.

Composites containing 10 vol% SiC whiskers have identical or slightly increased strengths compared with the related matrixes. It is noticeable that the incorporation of 10 vol% SiC whisker into HPRBSN matrix has enhanced the fracture toughness by 40% to 50% for the composites containing different amount of sintering aids.

Compared with 10 vol% SiC whisker/HPSN composite, the 10 vol% SiC whisker/HPRBSN has lower strength and fracture toughness although the two materials had same amount of sintering additives, same whisker loading and similar densification levels as shown in Fig.4. The difference between the two composites is obviously due to the raw materials used in this study. The average particle size of silicon powder is 10 times of that of silicon nitride powder resulting in much larger β grains in SiC whisker/HPRBSN composite. Moreover the αSi_3N_4 content in nitrided compact was much lower than that in original starting Si_3N_4 powder, which was another reason to cause the lower strength and fracture toughness of SiC whisker/HPRBSN composites. It implicates that there is a potentiality to improve the mechanical properties of SiC whisker/HPRBSN composites by reducing the particle size of the starting silicon powder and optimizing the process parameters of nitridation.

The 20 vol% SiC whisker/HPRBSN composites show lower strength and fracture toughness because of their lower densities resulting from insufficient sintering at the hot-pressing temperatures used in this study. It may be possible to obtain the reinforced composites containing 20 vol% or more SiC whiskers by employing longer hot-pressing time and other proper sintering parameters.

The demonstration of thermal shock testing shows that the 10 vol% SiC whisker containing composites, both SiC whisker/HPRBSN and SiC whisker/HPSN, have improved thermal shock resistance, from Δ T 900 °C for the matrixes to Δ T 1300 °C for composites. No catastrophic rupture or spalling were observed when the samples were quenched into cold water from high temperature.

Microstructure

Fig.5 is a backscattered electron image of a fracture surface of as-nitrided 10 vol% SiC whisker/RBSN compact, showing a SiC whisker on the surface. It is clear from the image that the surface of SiC whisker has become coase and no whisker pull-out can be observed on the fracture surface. It seems the whiskers served as growth sites for Si_3N_4, and numbers of tiny silicon nitride crystallines deposited on the surface of SiC whiskers, which caused the morphology change of the whiskers. On the other hand, however, after hot pressing the pull-out of SiC whiskers can be observed on the fracture surface of SiC whisker/HPRBSN composite. Compared with the original whiskers, no much morphology change of the whiskers can be observed on the fracture surface of the composite as shown in Fig.6. It indicates that there still has had the toughening mechanism of whisker pull-out for SiC whisker/HPRBSN composite in spite of the growth of Si_3N_4 crystalline on the surface of SiC whiskers during nitridation.

The fractography of SiC whisker/HPRBSN material shows much rougher fracture surface than the HPRBSN matrix which indicates the crack deflection toughening mechanism has existed.

The interface debonding between SiC whiskers and the matrix can also be observed on the fracture surface of the composite as shown in Fig.7, that may be another factor contributes to toughening.

The evidence of anisotropy distribution of SiC whiskers in the whisker/HPRBSN composites has not been found since the cold isopressing was used, after nitridation the SiC whiskers had been randomly oriented in the matrix.

Fig.8 and Fig.9 are the fractography of SiC whisker/HPSN composite hot-pressed at 1730 °C, showing the crosslinking texture of fibrous SiC whiskers and β Si_3N_4 grains. It can be seen from the images that the surface of SiC whiskers remains smooth indicating no severe degradation occured during hot-pressing at 1730 °C.

CONCLUSIONS

Homogeneous dispersion of SiC whiskers in silicon and silicon nitride matrixes has been achieved by using wet mixing process, ultrasonic and high speed homogenizing techniques.

50% increase in fracture toughness for 10 vol% SiC whisker /HPRBSN composite has been obtained. The possible toughening mechanism factors are crack deflection, fiber pull-out and interface debonding.

Futher study should be focused on searching the proper sintering parameters suitable for 20 vol% or more SiC whisker containing composites.

ACKNOWLEDGMENTS

The authors thank Ms. Wang Xian and Ms. He Jingji for microstructure analysis, Mr. Du Dayaun for mechanical property measurements, and Mr. Ma Lieming for sample preparation.

REFERENCES

1) Peter D.Shalek etc."Hot-pressed SiC whisker/Si_3N_4 matrix composites", Am.Ceram.Soc.Bull.,65(2) 351-56 (1986).

2) Sergej T.Buljan etc."Si_3N_4-SiC composites", ibid.,66(2) 347-52 (1987).

3) Robert Lundberg etc."SiC monofilament-reinforced Si_3N_4 composites", ibid.,66(2)330-33 (1987).

4) D.K.Shetty etc."SiC monofilament-reinforced Si_3N_4 matrix composites",Ceramic Eng. & Sci. Proc., vol 6.No.7,8.1985.

5) Gu Peizhi etc."Sintering, properties and applications of Sialon-ZrO_2 and Sialon-BN fiber composites", 6th CMTEC proc.,1986.

6) J.A.Mangels etc."Effect of rate-controlled nitriding and nitriding atmospheres on the formation of reaction-bonded Si_3N_4", Am.Ceram.Soc.Bull.,60(6)612, 1981.

7) E.Carlstrom and F.F.Lange"Mixing of flocced suspensions", J.Am.Ceram.Soc.,67(8)c-169-170 (1984).

8) John R.Porter etc."Processing and creep performance of SiC whisker reinforced Al_2O_3", Am.Ceram.Soc.Bull., 66(2)343-47(1987).

9) H.R.Zhuang etc."Effect of Y_2O_3 and Al_2O_3 additives on the sintering behaviour and properties of SRBSN", New Inorganic Materials, published by Shanghai Institute of Ceramics, Shanghai, China, No.2,1984.

10) Fan Qisheng etc."Investigation of the nitriding and sintering of SRBSN with different oxide additives", to be published in "Industrial Ceramics", Shandong Institute of Industrial Ceramics, Shandong,China.

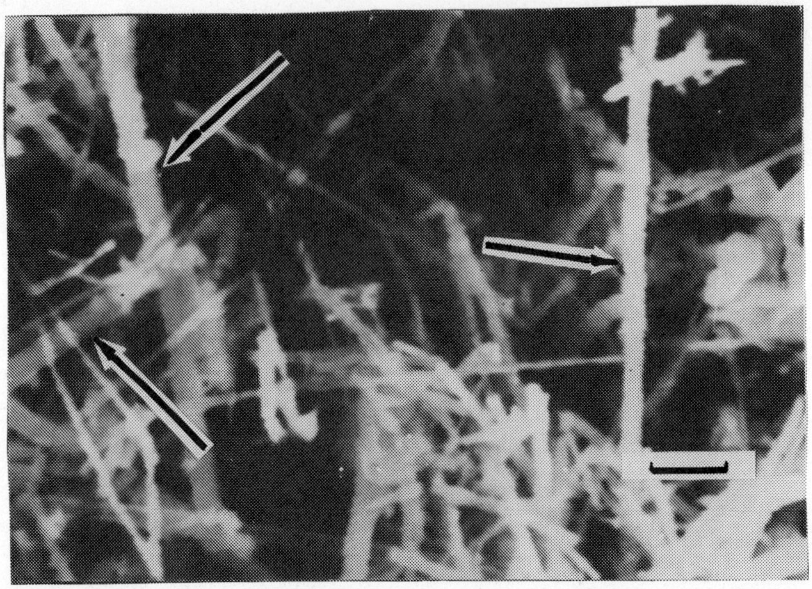

Fig.1. Observed SiC whiskers. Arrows point the whiskers with larger diameter. (bar = 5 μm)

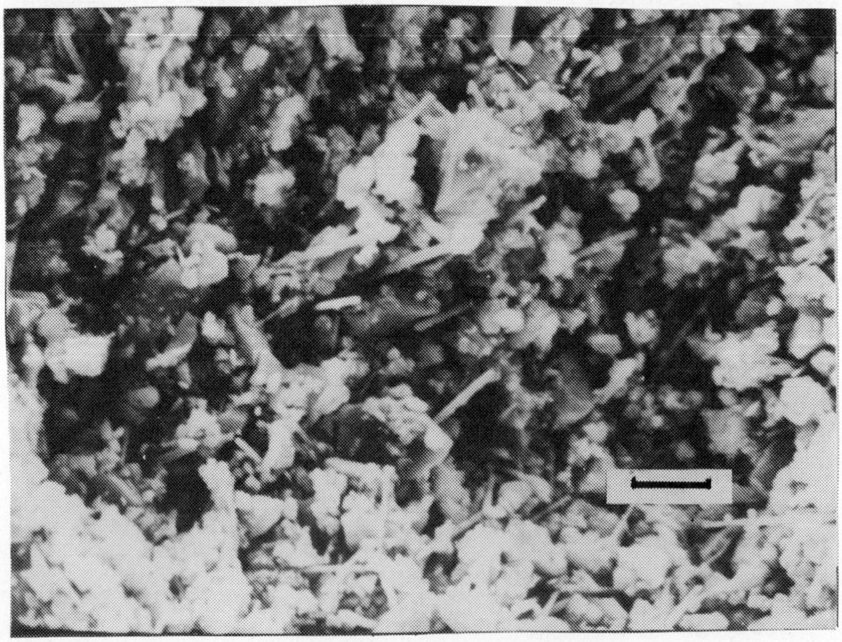

Fig.2. Back-scattered electron image of a cross section of the green body, showing homogeneous distribution of SiC whiskers. (bar = 10 μm)

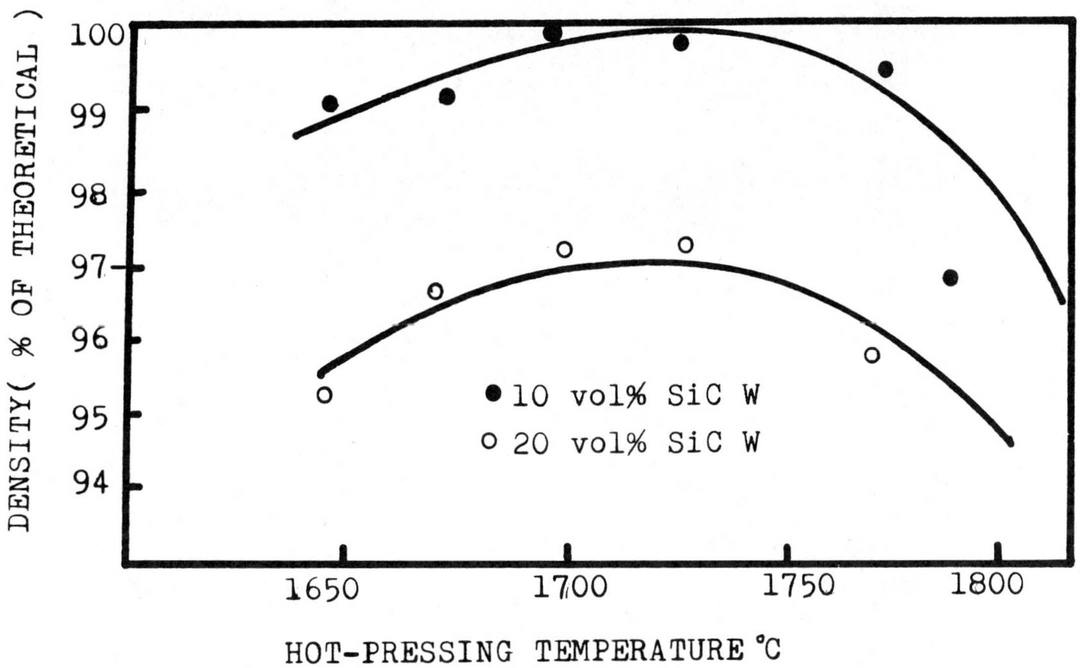

Fig.3 Densification of SiC whisker/HPRBSN composites.

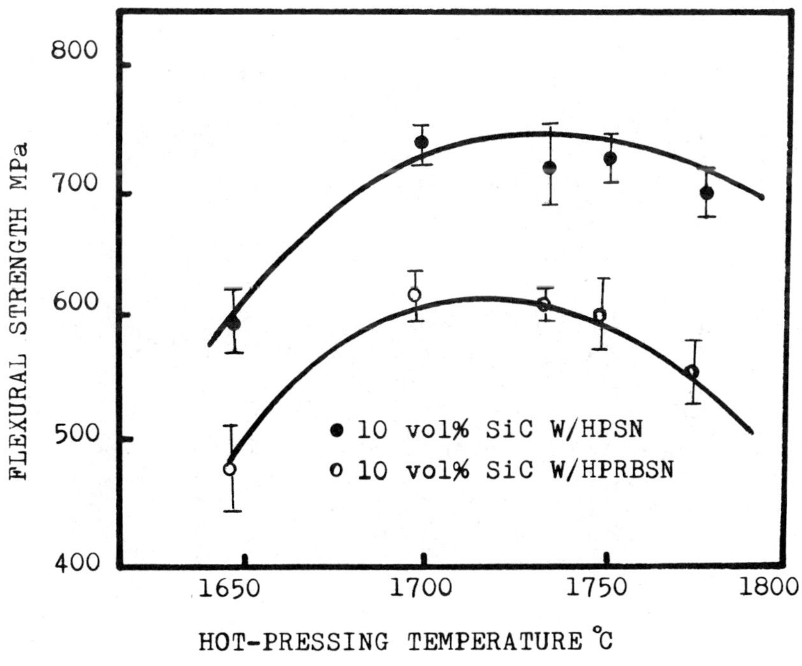

Fig.4 Flexural strengths of SiC whisker/HPRBSN and SiC whisker /HPSN composites. Sintering aid contents for both are 10 wt%.

Fig.5. Fracture surface of as-nitrided SiC whisker/RBSN compact. Arrow shows whisker surface.(bar = 2 μm)

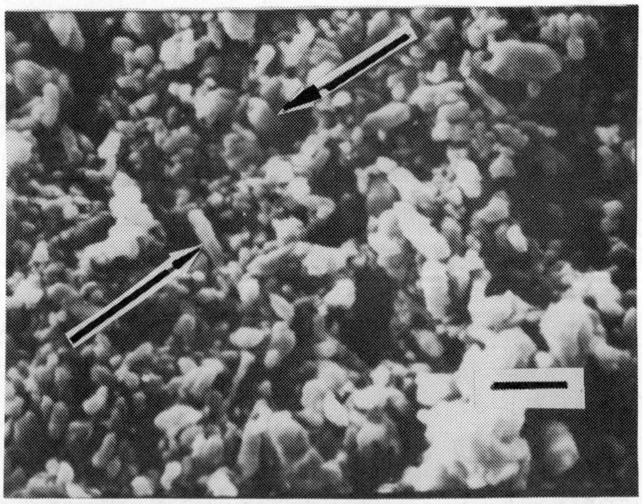

Fig.6. Fracture surface of SiC whisker/HPRBSN hot-pressed at 1730°C. Arrows indicate the whisker pull-out. (bar=3μm)

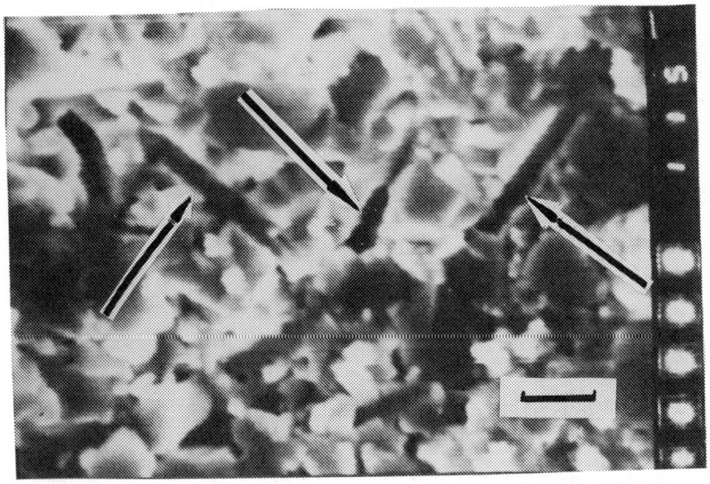

Fig.7. Fracture surface of SiC whisker/HPRBSN hot-pressed at 1730 °C. Arrows show the debonding between whisker and matrix. (bar = 1 μm)

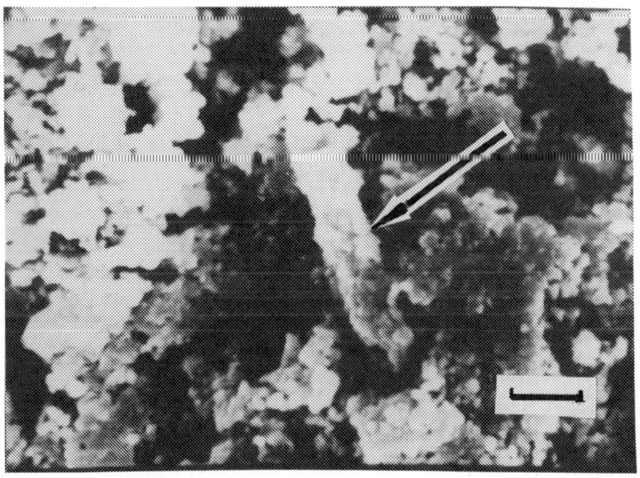

Fig.8. Fracture surface of SiC whisker/HPSN composite showing the texture of fibrous whiskers and β Si_3N_4 grains. (bar = 2 μm)

Fig.9. Smooth surface of SiC whiskers on fracture surface of the SiC whisker/HPSN hot-pressed at 1730°C. (bar = 1μm)

CERAMIC COMPOSITES PROCESSED BY TURBOMILLING

D. E. Wittmer and W. Trimble*
Dept. of Mech. Engr. and Energy Processes
Southern Illinois University
Carbondale, IL 62901

ABSTRACT

Agglomeration of fine ceramic powders and whiskers is a recognized problem area of ceramic processing. Traditional mixing techniques are time intensive and potentially damaging to the whiskers, often result in undesirable contamination, and are difficult to upscale. Techniques are required for improved processing of ceramics for heat engine applications which will provide dispersion/deagglomeration of composite materials without damage to the whiskers or causing significant contamination.

Turbomilling is a rapid, high-shear, wet grinding process originally developed to produce ultra-fine powders from relatively coarse particulate without the introduction of contamination. Recently it has been demonstrated that turbomilling can be used for improved dispersion, deagglomeration and homogenization of ceramic whisker-ceramic matrix composites. Evidence of the improved processing via turbomilling is observed from corresponding increases in strength and fracture toughness. Current results for ceramic whisker reinforced ceramic matrix composites will be presented.

INTRODUCTION

It is widely recognized in all areas of materials science that each individual processing parameter within every processing step can have a major effect on the final microstructure and resultant properties. This is especially true in the field of structural ceramics where the effect of flaw size and brittle fracture behavior make it essential to minimize flaw size and maximize failure strength in both monolithic and composite ceramics if they are to be utilized in the high-temperature applications for which they are intended. A major concern to those involved in the processing and production of both monolithic and composite ceramics is the deagglomeration of the as-received powders or whiskers, dispersion of the powders (or powders and whiskers), and finally homogenization of the resultant mixture.

The improvement in fracture toughness of monolithic alumina, mullite, and silicon nitride ceramics by the addition of SiC whiskers has been well documented.(1-14) In general, 30 to 40% increases in toughness often with degradation in strength (MOR by 4-pt loading), especially for high (>20 vol.%) whisker loadings, have been reported for these composites. Ideally, the materials used in processing these composites must first be deagglomerated and dispersed to produce a homogeneous mixture. This is best accomplished via wet processing, where the control of segregation during solvent removal is critical.

Conventional techniques, such as wet ball milling and the use of laboratory blenders, homogenizers and ultrasonicators, are usually employed to produce whisker/matrix slurries for slip casting or spray drying, or the slurries may be dried with agitation to thick suspensions and then dried, granulated, and screened, prior to pressure forming and consolidation. Extensive times are usually required for ball milling and can result in damage to the whiskers and incomplete dispersion, deagglomeration and homogenization of the whisker/matrix materials. The use of blenders, homogenizers and ultrasonicators, although typically used for processing of small laboratory sized batches, are not generally accepted as upscalable techniques for the processing of significant amounts of material and can create contamination in the form of the metal from the moving parts that are in contact with the whisker/matrix materials.

Turbomilling is a well documented grinding technique that is capable of producing ultra-fine powders from hard, abrasive materials.(15-21) The turbomilling equipment and technique have proven to be upscalable from laboratory sized units to large production sized turbomills which are currently being employed in the delamination and grinding of kaolins and the grinding of titania for the pigment industry.(16) Previous improvements in the turbomill construction(17) resulted in the development of an all polymer turbomill that does not induce contamination into the material being processed. More recent design modifications have resulted in an externally cooled, polymer lined construction with a range of capacities to allow the preparation of small laboratory sized to large pilot scale batches (0.5 liter to 6.0 liter mill volumes).(22)

In earlier work(18), the author reported the improved dispersion of boron and carbon as sintering aids in hot-pressed SiC, where the ultra-fine SiC powders were produced by turbomilling and the boron and carbon were added during the turbomilling process. Also, in the turbomilling of silicon nitride powder which contained a large fraction of whiskers[1], it was found that even after 8 h a significant fraction of whiskers remained. Based on this finding, a study was conducted to determine the feasibility of using the turbomill for improved deagglomeration, dispersion and homogenization of SiC whiskers in an aluminum oxide matrix.(22) Preliminary results showed improvement in both strength (by 4-pt bend test) and fracture toughness by a controlled surface flaw (CSF) techniques(23) on the order of 30% for a 30 vol.% Tateho[2] SiC whisker/RCHP-DBM alumina[3] composite processed by turbomilling.

This paper reports and discusses the preliminary results of utilizing turbomilling as a technique for improved processing of American Matrix[4] and Tateho SiC whiskers in alumina and silicon nitride[5] matrices.

PROCEDURE

A batch size of 500 grams of whisker and matrix materials was chosen for this study. Volume loadings of 20, 30, 35, 40 and 45 percent for both American Matrix and Tateho SiC whiskers in RCHP-DBM alumina and volume loadings of 15, 25, and 35 for both American Matrix and Tateho SiC whiskers in silicon nitride containing 2 wt.% RCHP-DBM alumina and 6 wt.% yttria[6] were processed by turbomilling. The SiC whiskers and matrix materials were batched and crudely blended in plastic bags. One liter of deionized, distilled water was added to the turbomill with the speed adjusted to 500 r.p.m., and 800 grams of PSZ[7] spherical media (2-3 mm in diameter) were added. The SiC whisker/matrix materials were then slowly added to the mill. The speed of the mill was then adjusted to 1200 r.p.m. and milling continued for 30 min. The mill was then discharged through a 60/200 mesh screen stack and the result slurry placed in a plastic container. For the SiC whisker/silicon nitride batches 100 ml of acetic acid were added during turbomilling to aid in flocculation

---

[1] *GTE SN-502 silicon nitride, GTE Products Corp., Towanda, PA*

[2] *SCW-1 SiC whiskers, Tateho Chemical Industries, Hyogo-ken, Japan.*

[3] *RCHP-DBM alumina, Reynolds Metal Corp., Chemical Division, Little Rock, AR.*

[4] *Grade 1 SiC whiskers, American Matrix, Knoxville.*

[5] *SN-E-10 grade silicon nitride, UBE Industries America, NY, NY.*

[6] *Yttria-99.99% 10 m^2/g, Molycorp, Inc., White Plains, NY*

[7] *PSZ spheres, Tohsoh, USA, Inc., Atlanta, GA*

of the resultant mixtures. The well flocculated slurries were allowed to settle for about 1 h and then dewatered to about 25 vol.% solids, prior to pressure casting a into cylindrical shapes (approximately 6.4 cm in diameter by 2 cm thick) using a low-pressure filter press. The pressure cast shapes were oven dried at 150°C and then hot-pressed[8] at 1750°C and 40 MPa for 30 min to 1 h, depending on the composition and consolidation behavior. The hot-pressed disks were then cut into bars 3 mm by 4 mm by 45 mm, the faces finished to 180 grit, and the corners chamfered at a 45° angle.[9] Test bars for 4-pt modulus of rupture testing were broken with the as-machined finish, while those for CSF toughness measurements were diamond polished on the tensile surface to a 6 micron surface finish. Loading for 4-pt bend testing and CSF was conducted on a fixture with an outside span of 40 mm and an inside span of 20 mm at a loading rate of 0.0254 cm/min. Indentations for the CSF and hardness measurements were made with a 5 kg load with a Vickers indenter. Prior to being measured optically at 1000X magnification, the indentation cracks were stained with an organic dye and then wiped clean with isopropanol. The hardness indentations and microcracks were then measured optically using polarized light with an interference objective and an adjustable digital filar. These measurements were confirmed using a similar optical arrangement and a standard 100 division filar.

RESULTS AND DISCUSSION

Due to problems with ORNL's hot-pressing equipment, only the results for 20 and 40 vol.% whisker loading of the RCHP-DBM and 15 and 35 vol.% whisker loading of the silicon nitride will be presented and discussed.

Turbomilling Process

Turbomilling as a means of dispersing the SiC whiskers in the respective matrices did not present a problem for any of the compositions studied. In previous research it was observed that the high shear created in the turbomill appeared sufficient to break up the strong whisker agglomerates, fracture the weak

---

[8] *Hot pressing done at ORNL, Oak Ridge, TN*

[9] *Machining done by BOMAS Machine Specialties, Somerville, MA*

whiskers with a minimum of damage to the stronger whiskers, and fragment some of the large hard particulate agglomerates.

In this research the concept was to create an environment in the turbomill where the whiskers and particulate would be dispersed by the high shear forces, but would be flocculated in the absence of the shear force. Upon discharge from the turbomill, the mixture would be a uniformly flocculated slurry that would not segregate upon settling; therefore homogeneous composite green forms could be fabricated by pressure filtration. Scanning electron microscopy and optical microscopy confirmed uniform flocculation of the whiskers and matrix materials in the green state (fig. 1). Well dispersed, randomly oriented whiskers, in the plane perpendicular to the hot-pressing direction, were observed by optical microscopy of the polished surfaces, at all whisker loadings. Typical optical micrographs are given in figs. 2 and 3.

SiC Whisker/Alumina Composites

For the SiC whisker/RCHP-DBM alumina composites it was not necessary to adjust the pH of the slurry to obtain well flocculated slurries. Stable suspension were produced that have remained in suspension for over 6 months with no apparent change in the floc size and no observable separation or sedimentation. The floc size was quite large (on the order of 0.5 to 1 mm) which settled rapidly to about 50 vol.% solids (covered by the balance of the supernatant water). It was important not to dewater these slurries to this high solids content prior to pressure casting, because the success of pressure casting in the pressure chamber used (no piston or pressure membrane) was dependent on maintaining a uniform pressure gradient through the casting until most of the mechanical water was removed. If the slurry was too thick it would cause "blow holes" through the center of the casting. A solids content of about 25 vol.% appeared to give satisfactory results, producing composites with bulk green densities of about 50% of the theoretical hot-pressed density.

SiC Whisker/Silicon Nitride Composites

For the SiC whisker/silicon nitride compositions it was necessary to reduce the pH of the mixture in the turbomill to about 4 to 5 by the addition of acetic acid, before significant flocculation was observed. These flocculated slurries also appear to have remained stable for over 6 months.

As with the alumina matrix composites, it was necessary to control the removal of the supernatant water to maintain a solids content of about 25 vol.% in the slurries before pressure casting. Since the floc size of these compositions appeared to be somewhat smaller than for the corresponding whisker loading of the alumina matrix composites, this made pressure casting somewhat more difficult. Initially a low pressure (about 5 psi) was applied to the chamber until the filter was completely blocked and then the pressure was slowly increase to about 25 psi until the casting was completed.

Hardness, MOR and Fracture Toughness[10]

SiC Whisker/Alumina Composites

Figure 4 shows the hardness values for the SiC whisker/RCHP-DBM alumina composites as a function of whisker loading. The hardness values for the American Matrix SiC whiskers are slightly higher than for the corresponding whisker loading for the Tateho SiC whiskers. This may be due to the differences in whisker diameters (0.2 to 0.5 microns for Tateho and 1-2 microns for American Matrix) and/or variations in the composition and volume of intergranular phases or the whisker/matrix interface.

MOR values were higher for the composites containing American Matrix SiC whisker compared to the composites containing Tateho SiC whiskers, as seen in fig. 5. At the 20 and 40 vol.% whisker loadings, average strengths of 540 MPa and 593 MPa, respectively, were measured for the American Matrix whiskers, compared to 525 MPa and 475 MPa, respectively for the Tateho whiskers. However, strength values for composites from both whisker sources are higher than for the alumina matrix alone (400 MPa), suggesting that load transfer is occurring and that the strength limiting flaws have been reduced in size. The increase in strength with increase in whisker loading for the American Matrix whiskers would indicate that possibly more load transfer is occurring and may be due to better packing of the larger diameter American Matrix whiskers compared to the smaller diameter Tateho whiskers.

Likewise, the fracture toughness values were considerably higher at both whisker loadings for composites made with the

---

[10]*These results are based on the average of 9 indentations for hardness, 3 MOR measurements, and 6 measurements of controlled flaw fracture toughness for each composition.*

American Matrix whiskers compared to the Tateho whiskers (fig. 6). Values of 6.1 and 7.3 MPa-$m^{1/2}$, respectively, were obtained for the composites containing American Matrix whiskers, compared with 5.4 and 5.1 MPa-$m^{1/2}$, for 20 and 40 vol.% whisker loadings, respectively, for composites containing Tateho whiskers, and 4.4 MPa-$m^{1/2}$ for the unreinforced alumina matrix hot-pressed under the same conditions.

Comparison of this data with reported values for the same whisker types and loading levels would suggest that, in general, higher strengths and fracture toughness values were obtained via the turbomilling and pressure casting process. Strength and fracture toughness data for a 20 vol.% Tateho whisker/RCHP-DBM composite prepared by high shear processing at ORNL has been reported as 380 MPa and 4.1 MPa-$m^{1/2}$, respectively. From the results of this research, it would appear that turbomilling (coupled with pressure filtration) produced composites with considerably higher strength and fracture toughness, compared to conventional high-shear processing, possibly due to better deagglomeration of the alumina and SiC whiskers, more uniform dispersion of the whiskers in the matrix, fracture of the weaker more faulted whiskers and lack of metallic contamination from the high-shear processing equipment.

SiC Whisker/Silicon Nitride Composites

As with the alumina matrix composites, the hardness values (fig. 7) for the SiC whisker/silicon nitride composites containing the American Matrix whiskers were slightly higher than for the Tateho whiskers. At the 35 vol.% whisker loading level, the Tateho whisker composite was only 95% dense, which would account for the low hardness value.

Strength values for the American Matrix whisker reinforced silicon nitride were significantly higher compared to the values for composites containing Tateho whiskers (about 20%) and the unreinforced matrix, as seen in fig. 8. Again the strength value for the 35 vol.% whisker loading for the Tateho whiskers was not representative of a fully dense composite; therefore, a direct comparison of properties for composites containing either whisker type at this loading level could not be made.

As shown in Fig. 9, at the 15 vol.% whisker loading level the American Matrix whisker composite had a fracture toughness value of 8.7 MPa-$m^{1/2}$, compared to 6.5 MPa-$m^{1/2}$ for the Tateho whisker composite and 5.4 MPa-$m^{1/2}$ for the silicon nitride matrix alone. At the 35 vol.% loading level, fracture toughness for the composite containing the American Matrix whiskers averaged 8.8 MPa-$m^{1/2}$. Therefore, it would appear that for the silicon nitride matrix there is little advantage in adding more than 15 vol.% whiskers for low temperature application; however, for higher temperature applications it would seem likely that strength and toughness retention would be greater for higher whisker loadings.

The results obtained for the strength of Tateho whiskers in silicon nitride are somewhat comparable to those reported in the literature by Buljan, Baldoni, and Huckabee(10) and Neil, et al.(11) for entirely different whisker and matrix sources.[11] Since both references reported that the tensile surface for the MOR bars was polished to a 1 micron finish, one would expect enhanced strength values and should not directly compare these values to those obtained for the as-machined surfaces tested in this study. Also, since there are likely differences in fracture toughness measurement techniques and processing conditions which may lead to more oriented microstructures, absolute values for toughness should not be directly compared, rather comparison to baseline measurements would give better indications of improvements in toughening. It is interesting that at 15 vol.% whisker loading composites from (10) would have a strength of about 750 MPa compared to 740 MPa for the 15 vol.% Tateho whisker composites from this study, while the reported fracture toughness for the 20 vol.% whisker composite by controlled surface flaw was only increased from a baseline value of 4.7 to 4.8 MPa-$m^{1/2}$ (2%) compared to 5.4 to 6.5 MPa-$m^{1/2}$ (17%) for the 15 vol.% Tateho whisker composites examined in this study.

The strength value of 865 MPa and fracture toughness value of 8.7 MPa-$m^{1/2}$ for the 15 vol.% American Matrix whisker silicon nitride composite represents a 31% increase and a 61% increase in toughness over the respective baseline values (660 MPa and and 5.4 MPa-$m^{1/2}$, respectively). These values represent a 14% increase in strength and a 61% increase in toughness compared with the respective values reported by (10) for a 20 vol.% whisker addition. The strength value of 860 MPa and fracture toughness value of 8.8 MPa-$m^{1/2}$ for the composite containing 35 vol.% American Matrix whiskers represents an increase in strength of 30% and toughness of 63% over the respective baseline values. These values represent a comparable strength and an

---

[11]*GTE SN-502 silicon nitride with 1.5 or 2.0 wt.% alumina and 6 wt.% yttria as the matrix material and SC-9 SiC whiskers Arco Chem. Co., Greer, SC (now Tateho America).*

increase in fracture toughness of 63% compared to the respective values reported by (10) for a 30 vol.% whisker addition, where the authors reported an increase of 27% over their baseline toughness.

These observed increases in toughness are not likely due to elongation of the beta-silicon nitride phase (which was not observed in etched polished sections) in the plane perpendicular to the hot-pressing direction, which has been observed by many researchers in monolithic silicon nitride following extended hot-pressing times. The elongated morphology of the beta phase is generally suppressed by the presence of the SiC whiskers, especially for the higher whisker loadings. In addition, the hot-pressing schedule for processing the composites evaluated in this study was insufficient to produce an elongated beta-silicon nitride, even at the low whisker loading level.

The results of this study indicate that increases in strength for low addition of SiC whiskers, comparable strengths for higher loadings of SiC whiskers and considerable increases in toughness, compared to values reported in the literature, can be obtained for SiC whisker reinforced silicon nitride matrix composites processed by turbomilling and pressure filtration.

CONCLUSIONS

Turbomilling has been demonstrated to produce deagglomerated, well dispersed, homogeneous SiC whisker reinforced alumina and silicon nitride composites. This improvement in processing over conventional milling techniques and non-scalable high-shear laboratory techniques was found to result in improved properties.

Strength and fracture toughness values for both SiC whisker reinforced alumina and silicon nitride matrix composites were found to be improved by the turbomilling and pressure filtration process used in this study. For both matrix materials, composites containing American Matrix SiC whiskers resulted in higher strength and toughness values, compared to the same whisker loading for composites containing Tateho SiC whiskers. A strength of 593 MPa and fracture toughness of 7.3 MPa-$m^{1/2}$ was measured for a 40 vol.% American Matrix SiC whisker/RCHP-DBM alumina composite, and a strength of 865 MPa and a fracture toughness of 8.8 MPa-$m^{1/2}$ was determined for a 35 vol.% American Matrix SiC/UBE silicon nitride composite, containing 2 wt.% Al_2O_3 and 6 wt.% yttria as sintering aids.

ACKNOWLEDGEMENTS

This research was sponsored by the U.S. Dept. of Energy, Assistant Secretary for Conservation and Renewable Energy, Office of Transportation Systems, as part of the Ceramic Technology for Advanced Heat Engines Project of the Advanced Materials Development Program, under contract DE-AC05-840R21400 with Martin Marietta Energy Systems, Inc. The authors would like to thank D.R. Johnson, Manager Ceramic Technology for Advanced Heat Engine Project, ORNL for his support and encouragement, and especially acknowledge T.N. Tiegs, ORNL for his assistance in specimen fabrication and testing, and many useful discussions. The authors also express their appreciation to H. Heystek and the Ceramic Research Group at the U.S. Bureau of Mines, Tuscaloosa Research Center for access to their research equipment. In addition, we are grateful to R. Nixdorf, American Matrix for supplying the Grade 1 SiC whiskers used in this study.

REFERENCES

1. T. N. Tiegs and P. F. Becher, "Sintered Al_2O_3-SiC-Whisker Composites," Am. Ceram. Soc. Bull., Vol. 66, No. 2, 1987, p.p. 339-342.
2. J. Homeny, W.L. Vaughn, and M.K. Ferber, "Processing and Mechanical Properties of SiC-Whisker-Al_2O_3-Matrix Composites," ibid, p.p. 333-338.
3. J. Porter, F.F. Lange, and A.H. Chokshi, "Processing and Creep Performance of SiC-Whisker-Reinforced Al_2O_3," ibid, p.p. 343-347.
4. R. Lundberg, L. Kahlman, R. Pompe, R. Carlsson, and R. Warren, "SiC-Whisker-Reinforced Si_3N_4 Composites", ibid, p.p. 330-333.
5. P. F. Becher and G. Wei, "Toughening Behavior in SiC-Whisker-Reinforced Alumina," J. Am. Ceram. Soc., Vol. 67, No. 12, 1984, C-267-269.
6. G. Wei and P. F. Becher, "Development of SiC-Whisker-Reinforced Ceramics," Am. Ceram. Soc. Bull., Vol. 64, No. 2, 1985 p.p. 298-304.
7. P. F. Becher, et al., "Toughening of Ceramics by Whisker Reinforcement"; Fracture Mechanics of Ceramics Vol. 7. Edited by R.C. Bradt, A.G. Evans, D.P. Hasselman and F.F. Lange. Plenum Press, New York, 1986, pp. 61-73.
8. Greenleaf Co. Trade Literature, 1986.
9. J. F. Rhodes, "Increased Toughness, Strength and Reliability Through Whisker Reinforced Ceramic Composites", Advanced Ceramics Proceedings: Processes for Toughening and Strengthening, Clemson University Continuing Engineering Education, Dec. 2-4, 1986.
10. S. J. Buljan, J. G. Baldoni, and M. L. Huckabee, "Si_3N_4-SiC Composites", Ceramic Bulletin, Vol. 66, No. 2, 1987, p.p. 347-352.
11. J. T. Neil, A. E. Pasto, and L. J. Bowen, "Improving the Reliability of Silicon Nitride", Advanced Ceramic Mat., Vol 3., No. 3, 1988, pp. 225-230.

12. P. D. Shalek, J. J. Petrovic, G. F. Hurley, and F. D. Gac, "Hot-Pressed SiC Whisker/Si_3N_4 Matrix Composites", Ceramic Bulletin, Vol. 65, No. 2, 1986, pp. 351-356.

13. J. P. Singh, et al., "Fracture Toughness and Strength of SiC-Whisker Reinforced Si_3N_4 Composites", Advanced Ceramic Mat., Vol. 3, No. 4, 1988, p.p. 357-360.

14. P. F. Becher and T. N. Tiegs, "Toughening Behavior Involving Multiple Mechanims: Whisker Reinforcement and Zirconia Toughening", J. Am. Ceram. Soc., Vol. 70, No. 9, 1987, p.p. 651-54.

15. I. L. Feld and B. H. Clemmons. Process for Wet Grinding Solids to Extreme Fineness. U.S. Pat. 3,075,710, Jan. 29, 1963. Assigned to U.S. Department of the Interior.

16. M. H. Stanczyk and I.L. Feld, "Comminution by the Attrition Grinding Process," Bu. Mines Bull. 670 (1980) 43 pp.

17. D. E. Wittmer, "Use of the Bureau of Mines Turbomill to Produce High-Purity Ultrafine Nonoxide Ceramic Powders", Bu. Mines RI 8854 (1984) 12pp.

18. D. E. Wittmer, "Ultrafine SiC Powder Produced by Turbomilling", Proceedings of SiC Symposium, Aug. 2-5, 1987, ACS Publications, in press.

19. J. L. Hoyer, "Effects of Turbomilling Parameters on the Comminution of Alpha-SiC", Bu. Mines, R.I. 9097 (1987) 9 pp.

20. J. L. Hoyer and A. V. Petty, "Fine Ceramic Powders Produced in the Bureau of Mines Turbomill", Ceram. Eng. and Sci. Proc., 6 [9-10] pp. 1342-1355 (1985).

21. S. V. Prochazka, "Grinding of Hard Substances to Submicron Grain Size", Advances in Ceramics, Vol. 21, Am. Ceram. Soc. Publ., 1987, Proc. of Ceram. Powder Sci. and Tech., Synthesis, Processing and Characterization Conf., Boston, MA, Aug. 3-6, 1986, pp. 311-320.

22. D. E. Wittmer, "Improved Dispersion Technique for Ceramic Whisker-Ceramic Matrix Composites", Ceram. Eng. Sci. Proc., 9 [7-8] pp. 735-40 (1988).

23. R. F. Cook and B. R. Lawn, "A Modified Indentation Toughness Technique", J. Am. Ceram. Soc., Vol. 66, No. 11,1983, C-200-201.

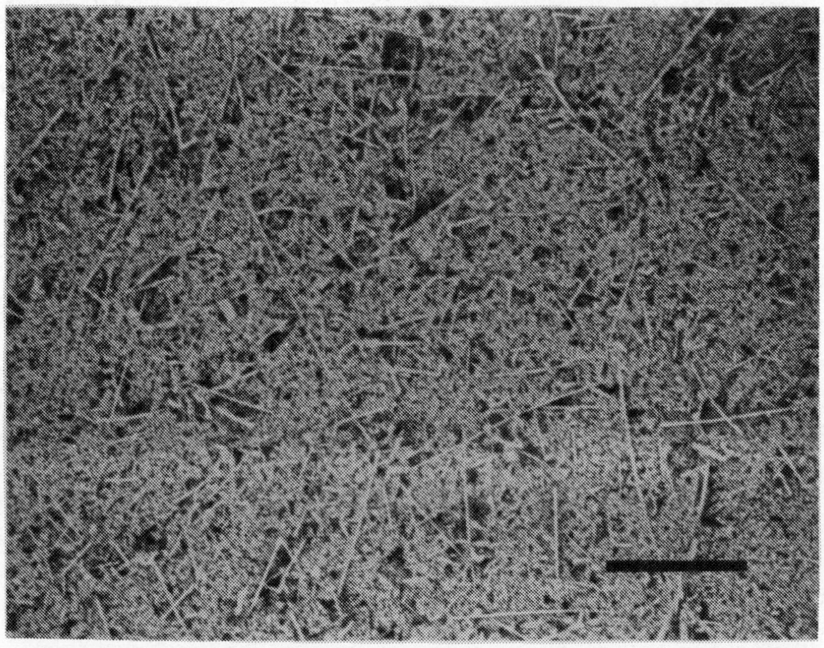

Figure 1. SEM of a 30 v/o Tateho SiCw/RCHP alumina composite after Turbomilling for 30 min. shows complete dispersion of the whiskers in the matrix powder. (bar = 20 um)

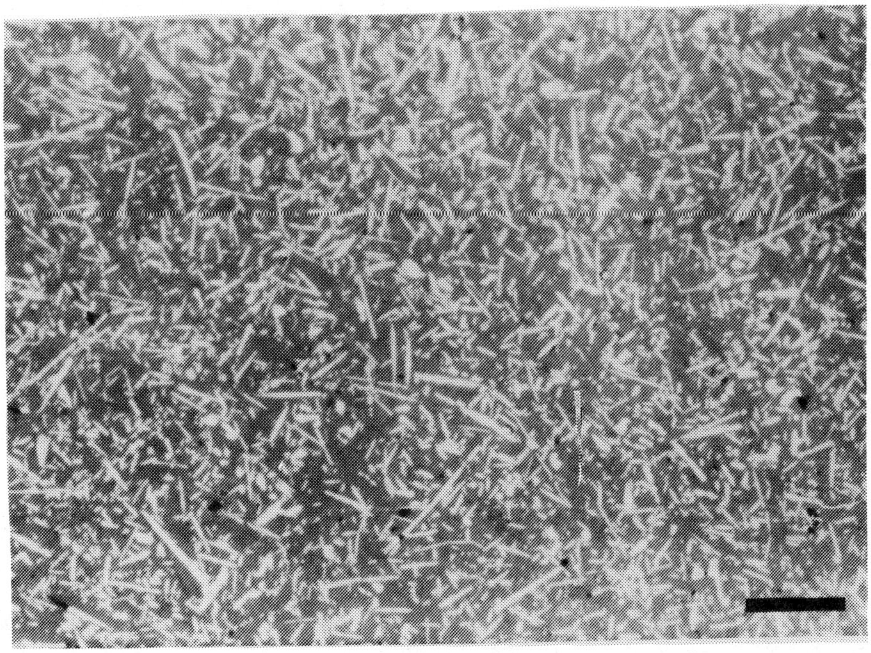

Figure 2. Optical micrograph of a polished section taken perpendicular to the hot-pressing direction for a 20 v/o American Matrix SiCw/RCHP alumina composite processed by turbomilling and pressure casting process, showing complete dispersion of the whiskers in the alumina matrix after hot-pressing. (bar = 50 um)

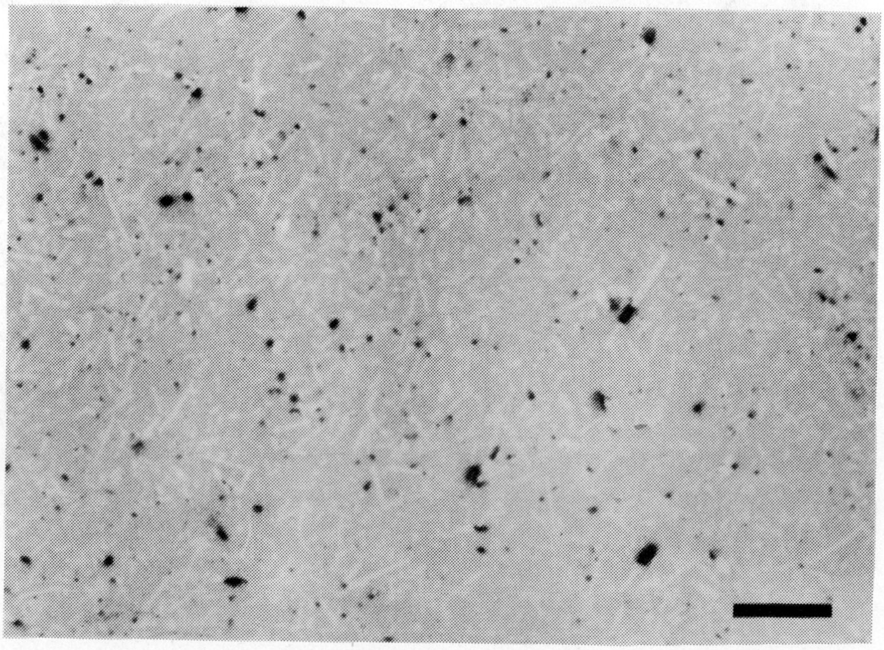

Figure 3. Optical micrograph of a polished section taken perpendicular to the hot-pressing direction for a 15 v/o American Matrix SiCw/UBE silicon nitride composite processed by turbomilling and pressure casting process, showing complete dispersion of the whiskers in the alumina matrix after hot-pressing. (bar = 50 um)

Figure 4. Vickers Microhardness Results for SiC Whisker/RCHP Alumina Composites

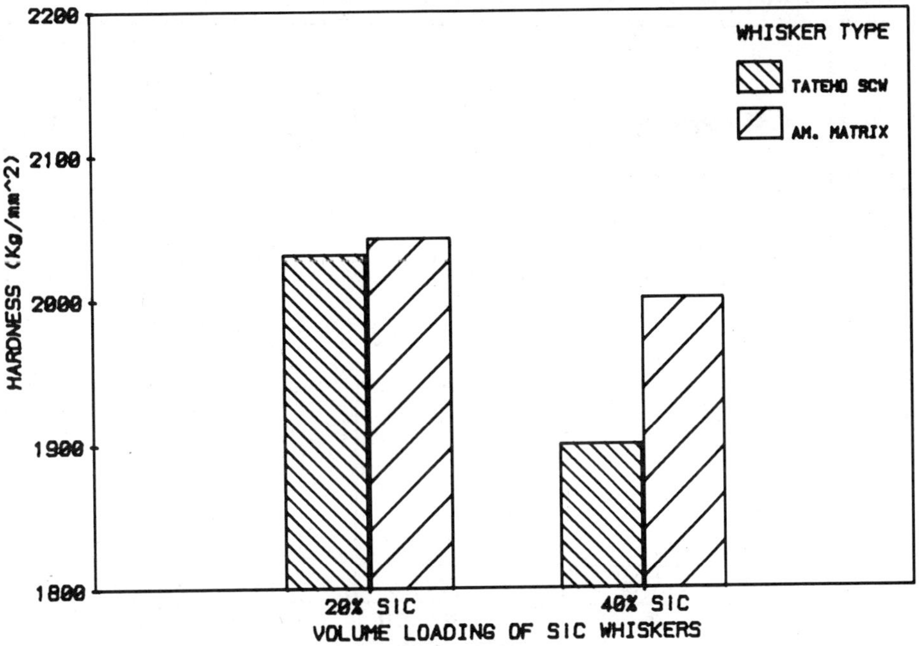

Figure 5. 4-Pt MOR Results for SiC Whisker/RCHP Alumina Composites

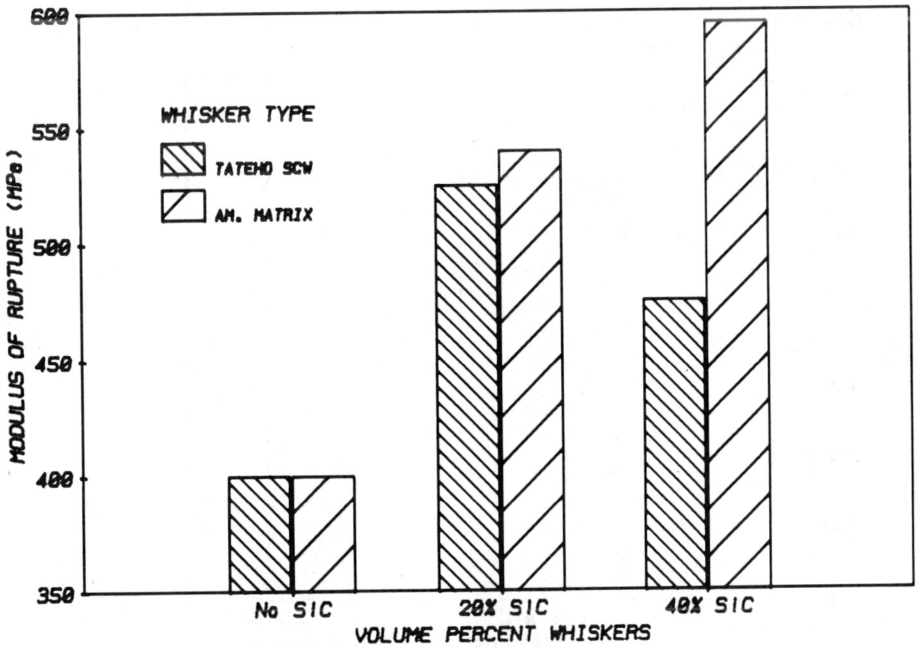

Figure 6. Fracture Toughness Results for SiC Whisker/RCHP Alumina Composites

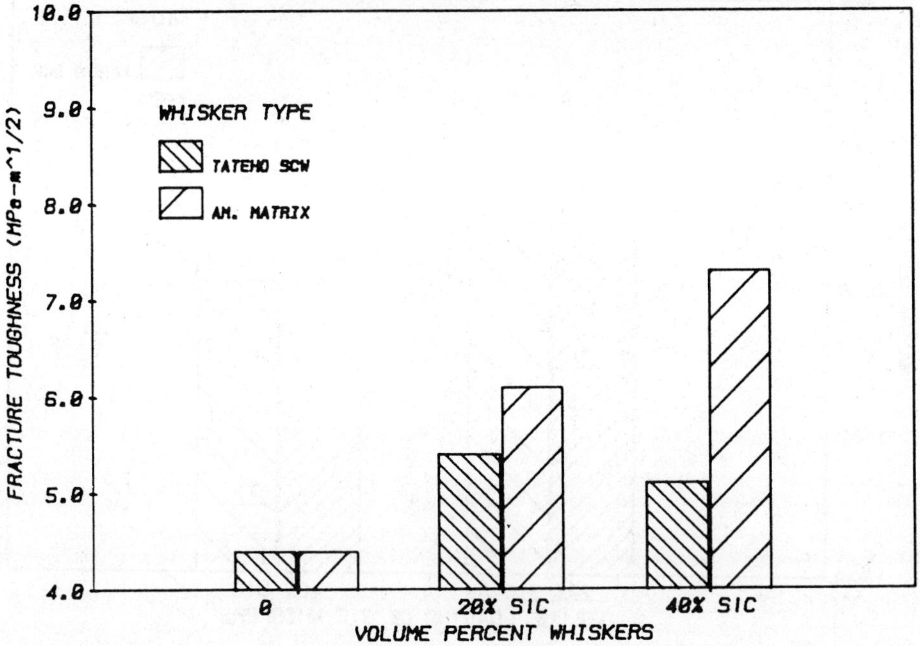

Figure 7. Microhardness Results for SiC Whisker/Silicon Nitride Composites

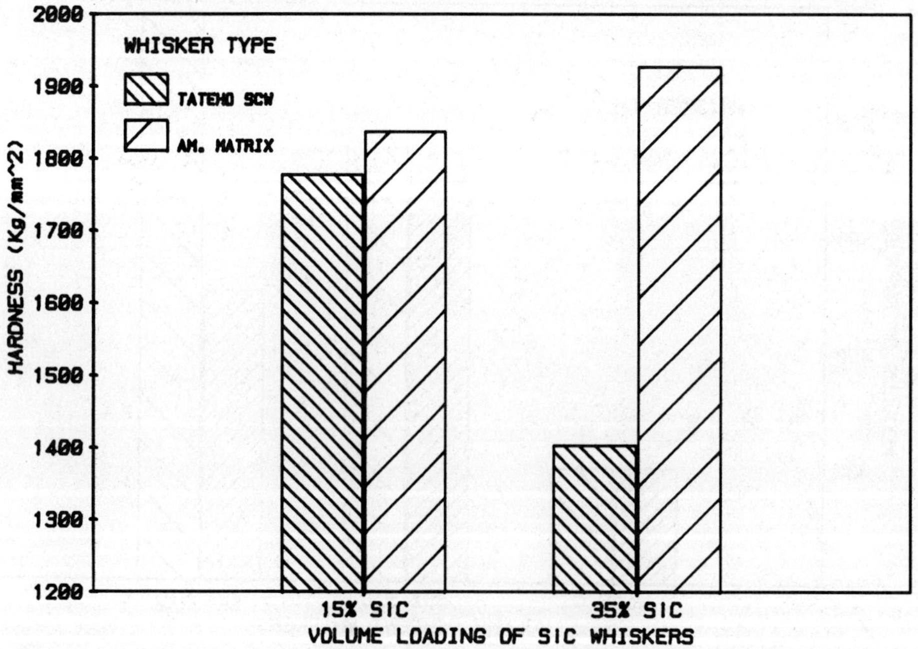

Figure 8. 4-Pt MOR Results for SiC Whisker/Silicon Nitride Composites

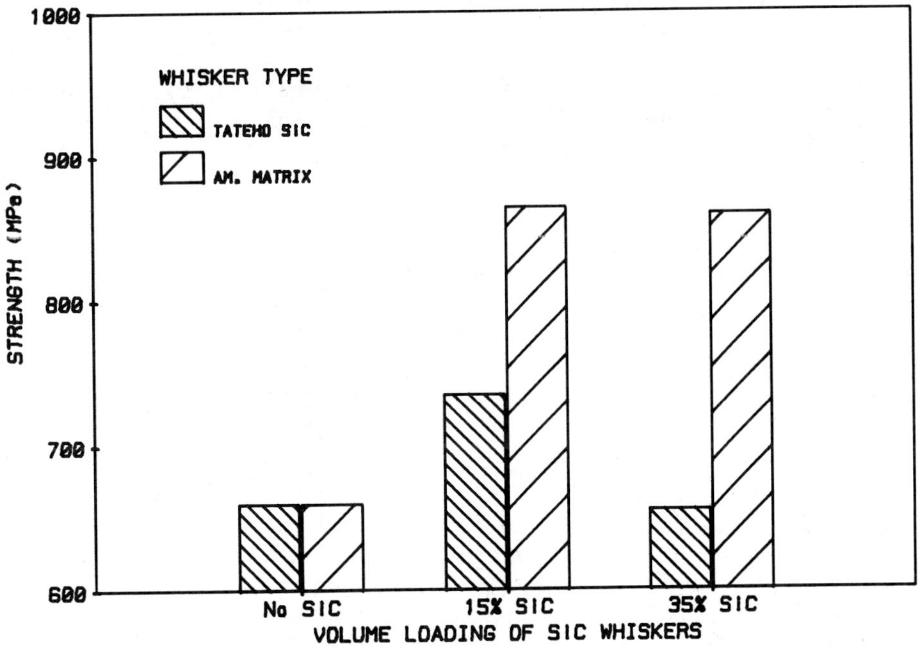

Figure 9. Fracture Toughness Results for SiC Whisker/Silicon Nitride Composites

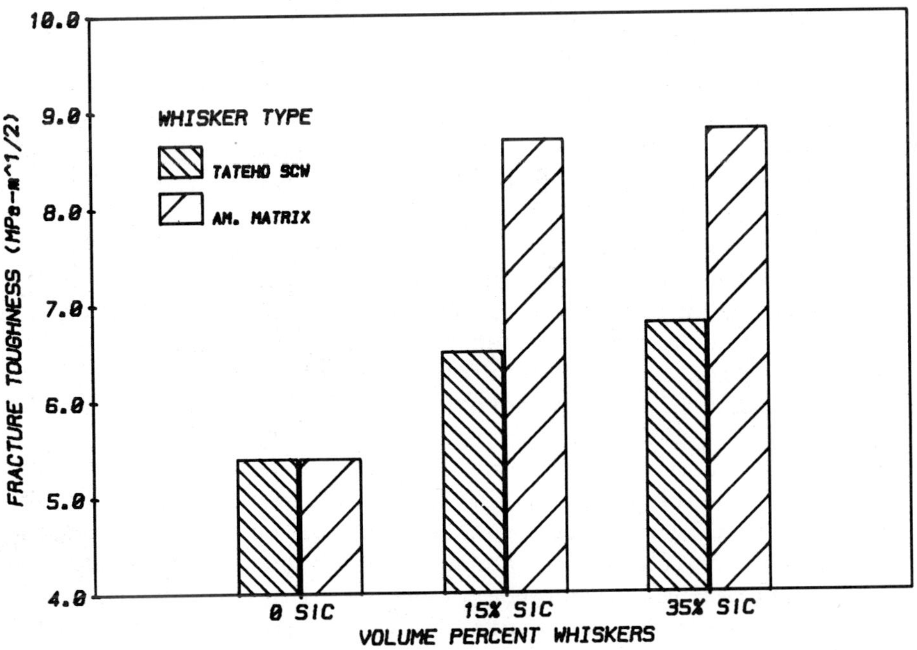

GRADED-CODEPOSITED ZrC-BN COATING FOR THE THERMAL PROTECTION OF CARBON-CARBON COMPOSITES

WITOLD KOWBEL
Materials Engineering, Department of Mechanical Engineering,
Auburn University, AL 36849

ABSTRACT

Carbon-carbon composites exhibit unique strength increase with increasing temperature. Current aerospace and military applications include rocket nozzles, various re-entry components and aircraft brake disks. Potential uses have been identified in future generation of military aircraft, advanced missile systems, and a variety of hypersonic aerospace vehicles. All of these applications take advantage of the high temperature capabilities of carbon and the benefits of fiber reinforcements. Reliable ultra-high temperature protection becomes a decisive factor prohibiting an extensive use of carbon-carbon composites in the above mentioned applications. Current protective coating systems are only effective up to the temperatures of 1500°C and are based on SiC. The ultimate application of SiC based coatings can not exceed 1800°C. This work represents a graded-composite coatings system able to provide thermal protection above 1500°C. The total coating is comprised of the diamond-like substrate infiltration, boron ion implantation of substrate, ZrC-C graded coating, and ZrC-BN composite coating. The graded and composite coatings were produced by the Chemical Vapor Deposition (CVD) technique. The microstructure, composition and mechanical properties were controlled by the two variables: input gas ratio and temperature. SOLGASMIX-PV program was used to find partial pressure for deposition which allow to obtain ZrC-C and ZrC-BN two phase fields.

I. INTRODUCTION

Carbon-carbon composites exhibit exceptional strength retention at temperatures over 2000°C [1]. They also exhibit low density and are nonstrategic materials. These properties make them extremely attractive for high temperature applications. Most of the high temperature applications involve extended periods of operations in oxidizing environment. Graphite and carbon materials are subject to oxidation degradation when utilized at high temperature in presence of oxygen. Effective utilization of these materials at high temperature requires protection from degradation including oxidation.

Ceramic coatings are currently used to protect carbon-carbon composites in h gh temperature applications. Ceramics compared with metals offer better capability to withstand higher temperatures and are more resistant to creep, oxidation and erosion. Ceramics coatings can be produced by plasma spraying, sputtering, electron beam evaporation, chemical vapor deposition (CVD), plasma assisted CVD, pack cementation, ion implantation and sol gel techniques. Plasma spraying is widely used to produce thermal barrier coatings. Plasma sprayed coatings are typically low density, high porosity stressed permeable structures. For thermal coatings the presence of porosity, microcracks and stress are beneficial since they increase the resistance to heat transport. However for oxidation and erosion protection plasma sprayed coatings are inherently non suitable because of the described problems. Additional problems with plasma sprayed coatings include poor adhesion between carbon-

carbon substrate and plasma sprayed coatings. Pack cementation coatings were abandoned as protection of carbon-carbon composites because of the reaction zone which extends considerably into the substrate and reduces load bearing capability across that zone. Physical vapor deposition techniques were explored in early work, however they were also abandoned because of low deposition rate. In addition they had a fibrous structure which is less stable on heating to high temperatures than is stress free material of equiaxed grain size [2]. The use of magnetic fields coupled with sputtering sources allows higher deposition which should remove one of the major disadvantage of sputtering, low rates. Chemical vapor deposition is currently used to produce a number of ceramic coatings for carbon-carbon composite high temperature protection.

In case of coatings for oxidation, wear and high temperature corrosion, or for prevention of reaction with the substrate, CVD is used because no other method can produce a dense impermeable coating of the desired coating material. The chemical vapor deposition process is complex, with many parameters regarding optimization in operation.

Oxidation protection of carbon-carbon composites depends on the regime of operating temperature. GA Technologies has developed an oxidation protective coating for the operating regime between room temperature and 1370°C, which showed oxidation resistance for hundreds of hours at these temperatures. This material designated VLG-25/IMCC-3 is composed of a carbon-carbon composite substrate with borate glass and an outer coating of CVD SiC. The estimated temperature limit for materials like VLG-25/IMCC-3 is about 1500°C, and is based upon the reduction of borate glass by carbon [3]. This is a result of CO pressure at higher temperatures greater than the total ambient pressure. Extending use of SiC based coatings can be accomplished using ceramic within the carbon material. Recent work at GA Technologies showed that SiO_2 can be stable in contact with silicon at temperatures up to 1800°C [4]. The 1800°C stability limit for SiO_2 with SiC represents the maximum temperature, below which SiC is useful as an outer coating, the same limits exist for Si_3N_4. Long term oxidation protection of carbon-carbon composites with temperature above 1800°C represent a real challenge. Outer coatings made of SiC and Si_3N_4 can not be used at temperatures above 1800°C because reaction products will destroy the SiO_2 layer necessary for oxidation resistance. HfB_2 and mixtures of HfB_2 and SiC were identified as highly oxidation resistant ceramic materials [10]. Coatings of these materials are currently being evaluated for short time oxidation resistance of carbon-carbon composites for temperatures above 1800°C. The use of coatings based on HfB_2 is beset by the rapid rate of oxidation. The use of these materials has to be only restricted to short time high temperature applications.

Another high temperature oxidation protection for C-C composites is the use of noble metal iridium, which was found to be a highly effective barrier to both oxygen and carbon diffusion at temperatures up to 2100°C. High cost and limited supply of iridium makes consideration of iridium impractical except for low volume applications. Several other ceramic coatings considered for high temperature protection include mixtures of HfC and HfO_2, Ta_2O_5 and HfB_2, TaB_2 and HfB_2 and SiC with ZrO_2 stabilized with CaO [5,6,7].

The key concern in coating performance is to match the thermal expansion coefficient for the coating with the substrate. In case of multilayered coatings this condition can be extended to include various coating layers. To assure good performance the difference of thermal expansion coefficients between adjacent layers and for the substrate should be within a factor $1x10^{-6}$/°C. The thermal stress resulting from the mismatch of the thermal expansion coefficients can be expressed as:

$$\sigma = \frac{E \alpha}{(1 - u)} \Delta T \qquad (1)$$

Where E, and μ
are Young's modulus, and Poisson's ratio and α thermal expansion coefficient difference between the substrate and the coating. Since carbon-carbon composite has a lower thermal expansion coefficient than the ceramics coating, the coating experiences tension or compression in thermal cycling. If we assume that at the deposition temperature the coating has a stress free state, then below the deposition temperature the coating is in tension and above the deposition temperature the coating is under compression. If we assume tensile strength of the coating is δ_f, then:

$$\delta_f = \frac{\alpha \Delta TE}{1 - \mu} \quad \text{and} \qquad (2)$$

$$\Delta T = \sigma_f \frac{(1 - \mu)}{\alpha E} . \qquad (3)$$

If we consider the microcracking to occur on the ZrC CVD coating as deposited at the temperature 1300°C we can used equation 3 to calculate temperature at which the onset of microcracking will start. The strength of CVD ZrC was reported by Kelly et al [8] to be $40 * 10^3$ psi. If the Young's modulus is $50 * 10^6$ psi, $\alpha = 6*10^{-6}$°C and Poisson's ratio for ZrC is 0.18 the corresponding $\Delta T = 100$. The thermal expansion mismatch between the ZrC coating and graphite substrate gives rise to the microcracking at high temperature, where carbon oxidation process is controlled by the diffusion kinetics. Graded coating lowers the difference in the thermal expansion coefficient between the adjacent layers leading to a much higher ΔT in the equation 3. The onset of microcracking can be shifted to the temperature regime where the oxidation is controlled by the surface reactions and the use of inhibitors provides sufficient oxidation protection. Application of the composite rule to calculate the thermal expansion coefficient of ZrC-C system shows that the values from $2.2 \times 10^{-6}C^{-1}$ to $7.5 \times 10^{-6}C^{-1}$ can be obtained depending on the phase fraction. Thus, in order to avoid thermal spallation, a graded coating would be a significant improvement. Opeka [11] showed that zirconium carbide can provide satisfactory oxidation protection for carbon-carbon composite for short time at surface temperatures approaching as high as 3200^{0}C. Zirconium carbide is also chemically compatible with carbon and has a low carbon diffusivity value at high temperature. Its oxidation resistance results from the formation of zirconium oxide. In order to solve the oxidation problems of carbon-carbon composites in the 1500°C to 1800°C temperature range, glass sealants are proposed. Particularly, boron oxide glass is considered due to its viscocity orders of magnitude lower than silica. Since vapor pressure of boron oxide at 1800°C is about 0.03 atm it may be possible to use borate glass sealants in conjunction with carbides in the temperature up to 1800°C. Application of codeposited ZrC-BN is intended to provide an extra sealing protection of the coating in case if some microcracking occurs. Boron oxide formed as an oxidation product of BN should act as the coating sealant.

II. Materials Used.

An industrially supplied graphitic carbon-carbon composite consisting of PAN based carbon fibers arranged in three-dimensional weave, supported with a coal for pitch matrix, was used. The composite has undergone several impregnation and graphitization steps to obtain a final density of $1.9/cm^3$.

III. Experimental.

Scanning Electron Microscopy (SEM) was performed on the coated and uncoated specimens in order to asses the morphology and analyze the chemistry, Jeol 840 SEM model equipped with ultra thin window Energy Dispersive X-Ray Spectroscopy (EDS) was used. Experiments were performed using 5kV voltage.

Transmission Electron Microscopy was performed on Jeol 1200 EX Transmission Electron Microscope equipped with Scanning Transmission Electron Microscope and ultra thin window EDS attachments.

The cross-section sample preparation method is similar to that described by Braceman and Sinclair [9]. Two or more slabs of the sample were put together with interface face-to-face using "crazy glue". A clamp was also used to improve the bonding strength between the samples. The glued sample was ground to a 3 by 30 mm cylinder and then cut using a low speed diamond saw. A 200 μm thick disc (3mm in diameter) was cut off with a piece of copper washer put on the top surface of the sample cylinder to avoid separation of interface during cutting. The slice was ground down using 600-grit SiC paper until approximately 100 μm thickness was obtained. The slice was then put on a steel cylinder with acetone soluble wax to perform dimpling. A small dimpling was very carefully performed at the final stage to obtain a small hole. A small hole is required to reduce ion milling time and to bias the sputtering rate difference between the substrate and the interface. Once a hole is obtained, the dimple is then polished to a mirror finish using the dimple polishing tool. Another washer was glued on top of a dimpled sample before taking the specimen off of the mounting cylinder to prevent breakage at the interface. If proper care was taken, 2-4 hour ion-milling at low angle (10 degree) should produce an electron transport area.

Selected area diffraction technique (SAD) was used to study the structural heterogeneity in the coatings and the interface. Also, microdiffraction (convergent beam) with beam size as small as 10nm was performed to determine the crystallite size and structural changes across the coating and the interface. The analysis of the convergent beam electron diffraction pattern was performed by using a computer code for pattern indexing as published by Noreyan [10].

Ion implantation with boron was performed using the Varion 200 Extrion model of ion implanter and the voltage of 100 kV. The boron depth profile and analysis was obtained with a Secondary Ion Mass Spectroscopy. Physics Electronics model operating at primary beam operating voltage 9.86 kV, raster size 100mm, and a secondary accelerating voltage of 4.5 kV was used.

Diamond-like coating was obtained using a magnetron sputtering system and a methane gas. Electron Spectroscopy for Chemical Analysis (XPS) was performed to analyze this coating. A Kratos 8000 XPS/Scanning Auger System was used with a Mg X-ray source operating in a vacuum of 10^{-9} Torr. Graded ZrC-C and codeposited ZrC-BN coating were obtained using a cold wall Chemical Vapor Deposition reactor. The ZrC-C system was obtained by using $ZrCl_4$, H_2, CH_4 and helium as a carrier gas. The flow rate of $ZrCl_4$ was controlled by the temperature of $ZrCl_4$ reservoir and the carrier gas flow rate. The flow rate of He, H_2 and $ZrCl_4$ was kept constant and the graded ZrC-C coating was obtained by varying the partial pressure of CH_4 between 3 x 10^{-1} atm and 10^{-3} atm. Salles [12] reported for similar deposition conditions and $P(CH_4)$ between 10^{-3} and 10^{-2} atm the ratio Zr/C to be between 1 and 0.65. Thus, the choice of methane partial pressures to obtain graded coating was based upon the thermodynamic calculations and the work of Salles. The deposition conditions for the ZrC-C graded coating are listed in Table 1. The codeposited ZrC-BN coating was deposited using $ZrCl_4$, BCl_3, H_2, CH_4, and NH_3.

The flow rate ratio of the $ZrCl_4$ vapor to the BCl_3 vapor was controlled by changing the ratio of their H_2-carrier gasses. The total gas flow rate of the H_2 carrier gas was kept constant. The deposition conditions for the codeposited ZrC-BN coatings are given in the Table 2.

The total coating system of the carbon-carbon composite is depicted in Figure 1. It comprises of the boron ion implanted carbon-carbon substrate, diamond like coatings graded ZrC-C coating and codeposited stoichiometric ZrC-BN coating. High temperature oxidation tests were performed by placing the coated specimens inside an alumina tube. Inductive coupling heating using a Lepel 50 kW RF generator allowed to heat the specimens up to 1800°C. Optical pyrometer was used to monitor the temperature.

IV. Thermodynamic Calculations.

The computer code, SOLGASMIX-PV [13] was used for the thermodynamic calculations. The CVD phase diagrams for Zr-C-Cl-H and Zr-C-Cl-H-B-N systems were estimated by these calculations. In case of Zr-C-Cl-H system the non-stoichiometric $Zr_x C_{1-x}$, was taken into account. The thermodynamic data for $Zr_x C_{1-x}$ were obtained by using the Gibbs-Duhem equation. All other thermodynamic data were taken from the Janaf tables [14]. All the species considered in this calculations are shown in Table 3. Similar calculations were performed for the second system. Table 4 allows all the species considered in obtaining ZrC-BN field of coexistence.

V. Results and Discussion.

1. Computer Calculations.

Figure 2 and 3 show the results of the calculated CVD phase diagrams. Figure 2 shows that the ZrC-C system contains : pure carbon, stoichiometric ZrC + C, nonstoichiometric ZrC and stoichiometric Zr + ZrC. At the 10^{-2} atm partial pressure of zirconium chloride a graded ZrC-C structure with a controlled ratio of $\frac{Zr}{C}$ in the coating can be obtained by varying the partial pressure of CH_4 between approximately 10^{-2} and 10^{-3} atm. Figure 3 shows that the codeposition of ZrC-BN can be accomplished by controlling the ratio $BCl_3/BCl_3 + ZrCl_4$. Thus the application of the SOLGASMIX-PV program is very useful in predicting the thermodynamically favorable conditions for the deposition and obtaining graded structures.

2. Uncoated and Silicon Nitride Coated Carbon-Carbon Composite

Figure 4 shows the SEM micrographs of the high temperature oxidized carbon-carbon composites. A severe degradation of the fiber-matrix interphase can be shown. This degradation can lead to a very significant loss of the interlaminar and tensile strength. Synergetic reinforcement of the matrix can be almost completely lost. Figure 5 shows a SEM micrograph of a low pressure CVD α -Si_3N_4 coating applied to the carbon-carbon composite. This type of a coating is currently used as a thermal protection for carbon-carbon composite. Figure 6 shows a SEM micrograph of the high temperature oxidation of Si_3N_4 coating. The protective SiO_2 layer which is very successful at the lower temperatures is disrupted and the coating is undergoing a rapid oxidation.

3. Boron Ion Implantation

The oxidation tests were performed on the ion implanted composite at 800°C and at 1800°C. The results at 800°C showed an significant effect in reducing the presence of the catalytic pits and thus inhibiting the oxidation. However, there was no significant effect detected after high temperature oxidation due to boron ion implantation. This is expected, because at low temperature the oxidation is controlled by the reaction while at high temperature the oxidation is controlled by the diffusion rate.

4. Diamond-Like Coating.

Figure 7 shows a diamond-like amorphous coating applied to the carbon-carbon composite. The diamond coated carbon-carbon composite was oxidized in the same way as Si_3N_4 coating. A significant difference can be observed, Figure 8. Surface of the diamond-like coating is much less degraded than the surface of Si_3N_4 coating. This is explained by the fact that a glass has extremely low porosity. Thus the oxygen accessibity is reduced and at high temperature the oxygen diffusion rate can be lowered.

5. Graded ZrC-C and Codeposited ZrC- BN Coatings.

As proposed graded ZrC-C and ZrC-BN coating was applied to the carbon-carbon composite, Figure 9. The effect of the deposition temperature is shown in the difference in the coating morphology. Analytical electron microscopy was applied to study the structural heterogeneity of the coating and confirmed what phases are present in the coating. Figure 10 shows a TEM micrograph showing small ZrC crystallites dispersed in the carbon matrix. Two electron diffraction techniques: selective area diffraction and the convergent beam electron diffraction techniques were used to analyze the structural heterogeneity. Figure 11 shows the corresponding electron diffraction patterns. The calculated d-spacings allowed to identify both graphite and ZrC. Computer code allowed a proper zone axis identification for the ZrC pattern. It was found to be (111). These techniques offer a tremendous advantage over previously reported use of the X-ray diffraction due to a much better lateral resolution.

There is very little work published in the open literature regarding application of graded coatings to carbon-carbon composite. Kelly et al [8] reported the use of graded CVD ZrC coatings for carbon substrates and showed that the graded coating was free of cracks and quite adherent. Hollabaugh [15] reported the use of a graded ZrC coating to coat particles. The only confirmation of the graded structure was obtained from a diffusive X-ray diffraction pattern, which suggested nonuniform stoichiometry in the coat. This work adds a new dimension to the graded coatings. First, computer calculations are used to predict the deposition parameters, Secondly, an elaborate crossectional analysis is provided to prove the existence of a graded coating. Crossectional analysis of the coated composite revealed that the inner layer contains only pyrolytic graphite. Intermediate layers contain a mixture of ZrC and graphite with different phase ratio along the coating crossection. The presence of only ZrC coating in the outer part is detected, Figure 12. Much larger grains as those shown in Figure 10 can be observed. Selected area diffraction pattern, Figure 13 shows only ZrC with no detected presence of graphite. The image analysis was used to quantify the phase ratio along the coating crossection, Figure 17. The variation in the ZrC/graphite ratio between 1 and 0 can be observed. Figure 17 provides also an additional evidence of the graded structure from the normalized Zr/C peak ratio as obtained from the ultra thin window EDS spectra. Thus the application of analytical electron microscopy confirmed that a graded ZrC-C coating was obtained under these deposition parameters. The most outer part of the coating should contain the mixture of ZrC-BN if the thermodynamic calculations were correct. Figure 14 shows a TEM micrograph of the codeposited ZrC-BN coating. Diffraction patterns were similar to those in Figure 11 and were identified as ZrC and BN. The application of conventional diffraction techniques to distinguish between hexagonal boron nitride and graphite is very difficult since BN $d_{oo2} = 3.33A^o$ and graphite $d_{oo2} = 3.36A^o$. In order to allow a very precise calculations of the lattice spacing and lattice constant a two beam convergent beam electron diffraction was used. This technique provided a perfect satisfaction of the Bragg's low condition, since the (hkl) spots lie on the Evald sphere. Application of this technique is shown in Figure 15, for the BN particles shown in Figure 14. This technique was also applied to the ZrC particles. The computer code mentioned above was used to perform very careful lattice parameters calculations. ZrC is a cubic system and the lattice parameter was found to be 4.697 A^o. The tabulated lattice parameter for ZrC is 4.696 A^o. Thus, the two beam convergent beam results showed that under the deposition conditions almost fully stoichiometric ZrC was obtained. For the BN system a = 2.505 A^o and C = 6.663 A^o were found which showed that almost stoichiometric BN was codeposited with ZrC.

6. Oxidation results.

The coating system was subject to the oxidation test as described above. SEM micrographs of the oxidized samples are shown in Figure 16. Almost full oxidation protection, combined with no cracking in the coating system as detected by the SEM can be observed. Compared to the oxidation of the silicon nitride coating, Figure 6, tested under similar conditions, it can be observed that the new coating system experiences much less thermal degradation. Lack of the microcracking proves that engineering importance of the graded coating concept. Since the carbon oxidation at high temperature is diffusion controlled, removal of the microcracking offers a tremendous advantage in the thermal protection of carbon-carbon composites. The oxidation protection of the coating is due to the formation of ZrO_2 and B_2O_3 as the oxidation products. Analytical Electron Microscopy was also used to analyze the oxidation products. The formation of zirconium oxide and boron oxide was detected in the outer part of the coating. Figure 18 shows the TEM micrograph of the outer part of the coating after the oxidation. Both zirconium oxide and boron oxide can be observed. The AEM of the underlying layers showed a growth of ZrC grains in the graphite matrix following the high temperature oxidation test. Rate of the oxidation of the coated carbon-carbon composite is controlled by the diffusion of oxygen through zirconium and boron oxides, and the evaporation rate of boron oxide. Since boron oxide has low viscocity it will provide a good sealing capability during a thermal cycling especially at the intermediate temperature range, where current sealing systems based on silica have failed.

VI. Conclusion.

This work showed an application of a complex graded-composite coating system as oxidation protection of C-C composite. This approach offers a fundamental advantage over currently used coatings: a/graded coating can be tailored to match the thermal expansion coefficient of the substrate and reduce or remove the thermal cracking: b/ Composite ZrC-BN coating as the codeposited CVD system offers a new dimension in the coating research and application ZrC-BN coating used in this work was used to successfully demonstrate that concept. This work shows an advantage of the use of analytical electron microscopy techniques to characterize the structural heterogeneity of the coating and to confirm the graded nature of the coating. Electron microscopy allowed also a distinction between stoichiometric and nonstoichiometric zirconium carbide.

VII. Acknowledgement.

This work was partially supported by a grant from NSF/EPSCOR.

References

1. Sheehan, J. E., paper presented at the 37th Pacific Coast Regional Meeting of the American Ceramic Society, San Fransico, 1984.
2. Webb, R.D., NASA Conference Publication, 2406, NASA, 1985, p. 149.
3. Gray, P.E., Sheehan, J.E., GA Technology Mc., Report, GA-A18446.
4. Shirorcky, G.H., Kaae, J.L., and Sheehan, J. E., paper presented at the 87th Annual Meeting of the American Ceramic Society, Ohio, 1985.
5. Opeka, M.M., Proceedings of the 11th Conference on Advanced Composite Materials, Cocoa Beach, Florida, 1987.
6. Rapp, R.A. Pierre, G.R., Vaught Corporation Report, R006CCR1200, 1981.
7. Sheehan, J.E., Kaae, J.L., and Shirocky, G.H., GA Technologies Report GA-A18786, 1987.
8. Kelly, E. Proceedings of the Third International Conference on Chemical Vapor Deposition, 1973, p.183.
9. Bravman, J.C. Journal of Electron Microscopy Techniques 1, (1984) 53.
10. Norayan, C., Journal of Electron Microscopy Techniques, 3 (1986) 151.
11. Opeka M.M., NASA CP - 2445, 1896, p.229.
12. Besman, T.M., ORNL/TM - 5775, 1977.
13. Salles, P., Proceedings of the Ninth International Conference on CVD, p. 1129, 1984.
14. Janaf Thermochemical Tables, 2d ed, Edited by D.R. Stull, Washington D.C. 1971.
15. Hollabaughm C.M., Journal of Nuclear Materials, 57 (1975) 325.

Table 1

Deposition conditions for the graded ZrC-C coatings

Deposition Temperature T dep/K

1400, 1600

Total gas pressure = 1 atm

$P_{(ZrCl_4)}$ = 2.5 x 10^{-2} atm

$P(CH_4)$ = 3 x 10^{-1} , 10^{-1}, 7 x 10^{-2}, 5 x 10^{-2}, 2.5 x 10^{-2}, 10^{-2}, 8.8 x 10^{-3}, 7.5 x 10^{-3}, 6.2 x 10^{-3}, 5 x 10^{-3}, 3.8 x 10^{-3}, 2.6 x 10^{-3}, 1.7 x 10^{-3}, 10^{-3} atm

H_2 Flow rate = 140 cm^3/min

He Flow rate = 820 cm^3/min

Table 2

Deposition conditions for the codeposited ZrC-BN coating

Deposition Temperature T dep/K

1500

Total gas pressure = 1 atm

$P(BCl_3/BCl_3 + ZrCl_4)$ = 0.4

Total flow rate = 600 cm^3/min

$\frac{NH_3 + CH_4}{H_2}$ = 2.2 x 10^{-2}

Table 3

Chemical Species in ZrC-C-CL-H System

(Gas Species)

Zr, Cl, Cl_2, ClZ_2, Cl_2Z_2, Cl_3Zr, CCl_4, H, H_2, HZr, CH, CH_2, CH_3, CH_4.

(Solid Species)

Zr_xC_{1x}, Zr, C

Table 4

Chemical Species in Zr-C-CL-H-B-N System

(Gas Species)

Zr, Cl, Cl_2, ClZr, Cl_2Zr, Cl_3Zr, CCl_4, H, H_2, HZr, CH, CH_2, CH_3, CH_4, NH, NH_2, NH_3, N, N_2, BN, BCl, BCl_2, BCl_3

(Solid Species)

ZrC, Zr_2C, B, BN

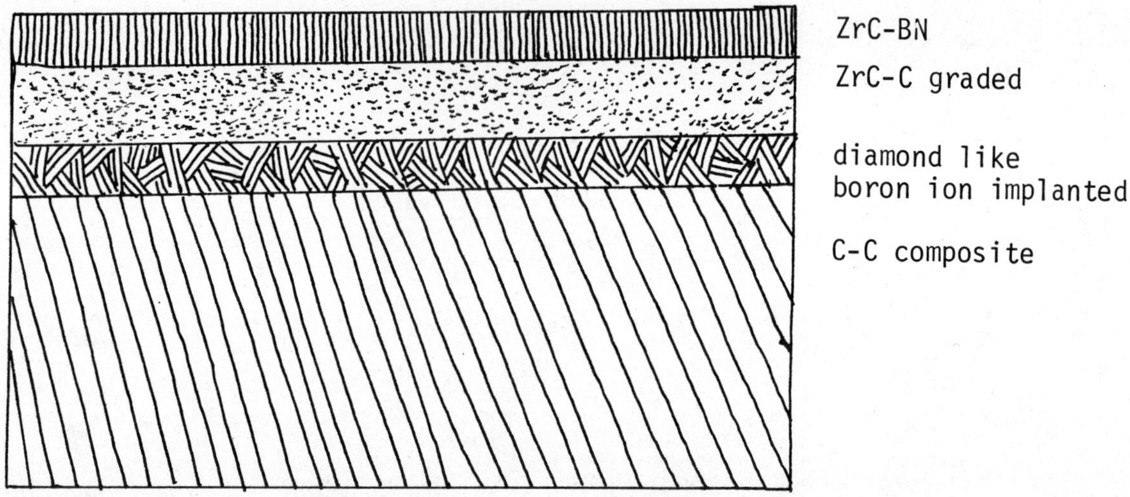

Figure 1. Schematic of the proposed protective system.

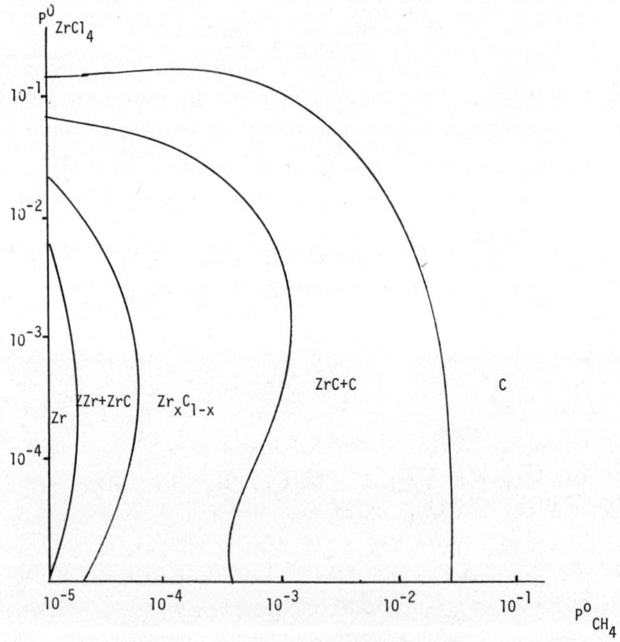

Figure 2. Calculated CVD phase diagram for the ZrC-C-Cl-H system, at the total pressure of 1 atm and the temperature of 1600°K.

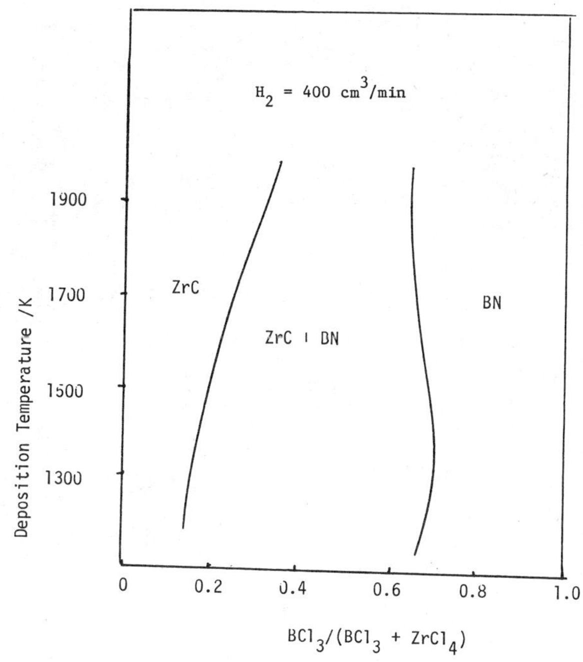

Figure 3. Calculated CVD phase diagram for the Zr-C-Cl-H-B-N system.

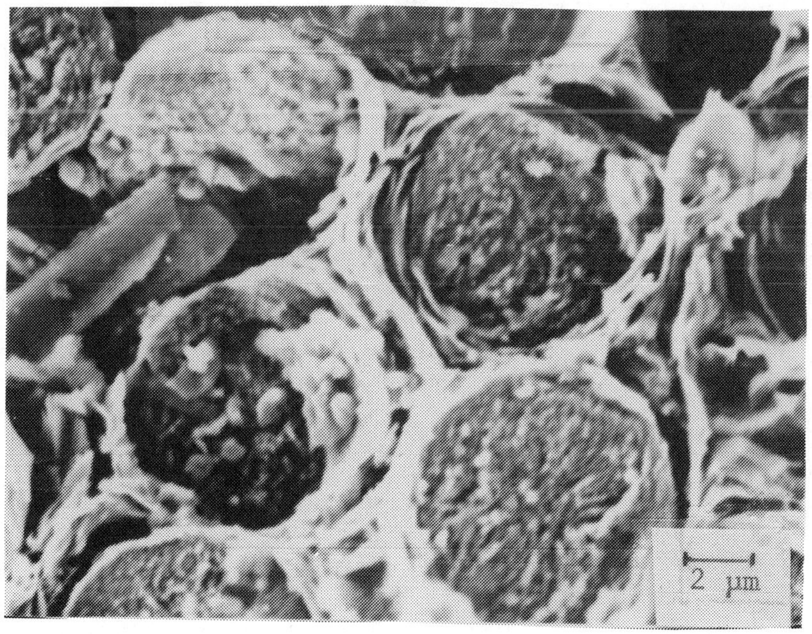

Figure 4. SEM micrographs of the transverse bundle of 3-D carbon-carbon composite after 20 min. air oxidation at 1800°C.

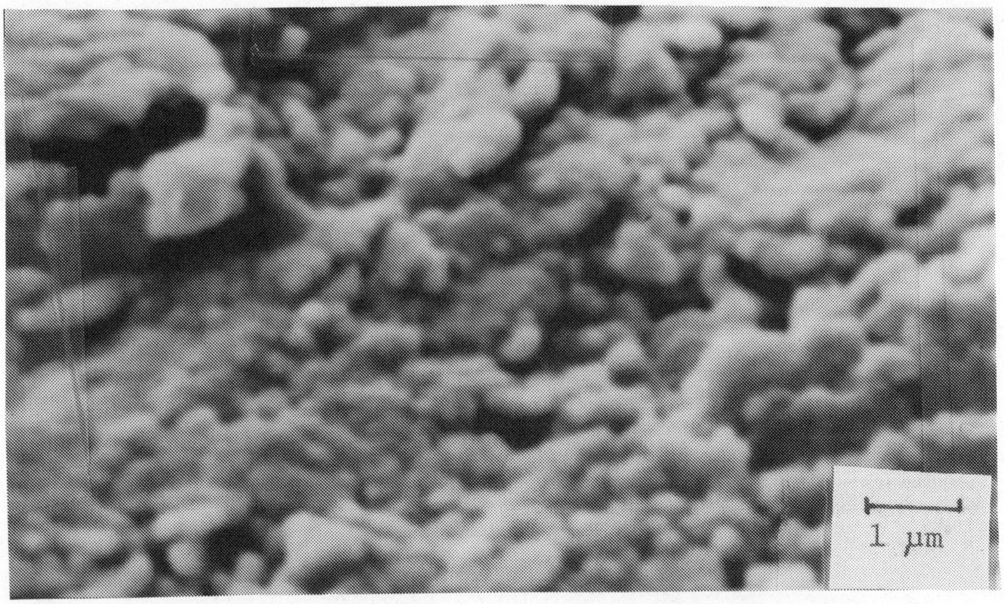

Figure 5. SEM micrograph of low pressure CVD Si_3N_4 coating.

Figure 6. SEM micrograph of Si_3N_4 coating after 20 minutes air oxidation at 1800°C.

Figure 7. SEM micrograph of diamond-like amorphous coating.

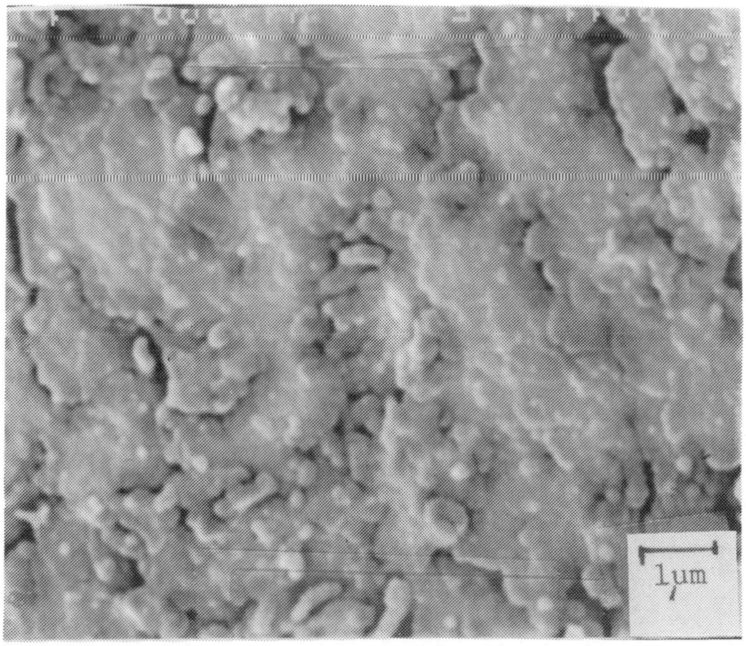

Figure 8. SEM micrograph of diamond like coating after 20 minutes oxidation treatment in air at 1800°C.

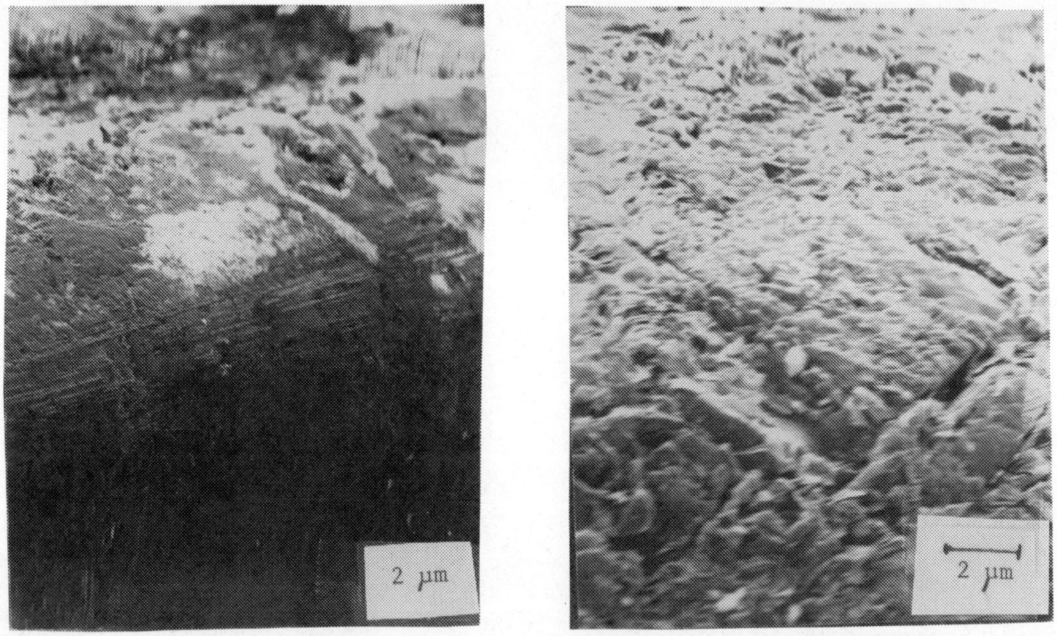

Figure 9. SEM micrograph of the graded C-C composite with the total protection system including: graded ZrC-C and Codeposited ZrC-BN.
a). ZrC-C deposited at 1400°K.
b). ZrC-C deposited at 1600°K.

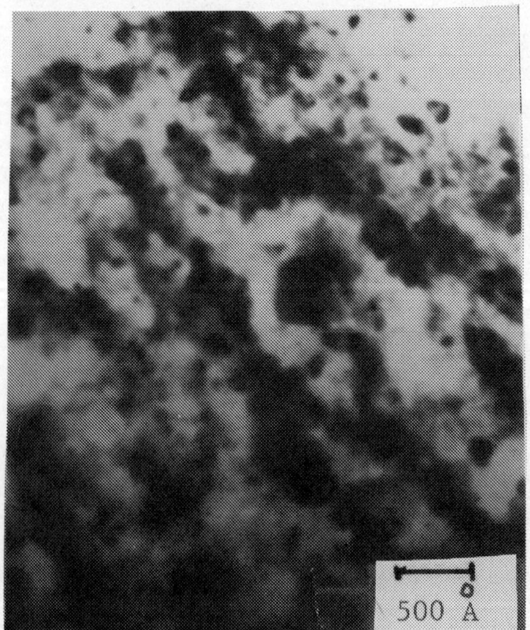

Figure 10. TEM micrograph of the graded ZrC-C coating where the dark particles represent disperse ZrC crystallites.

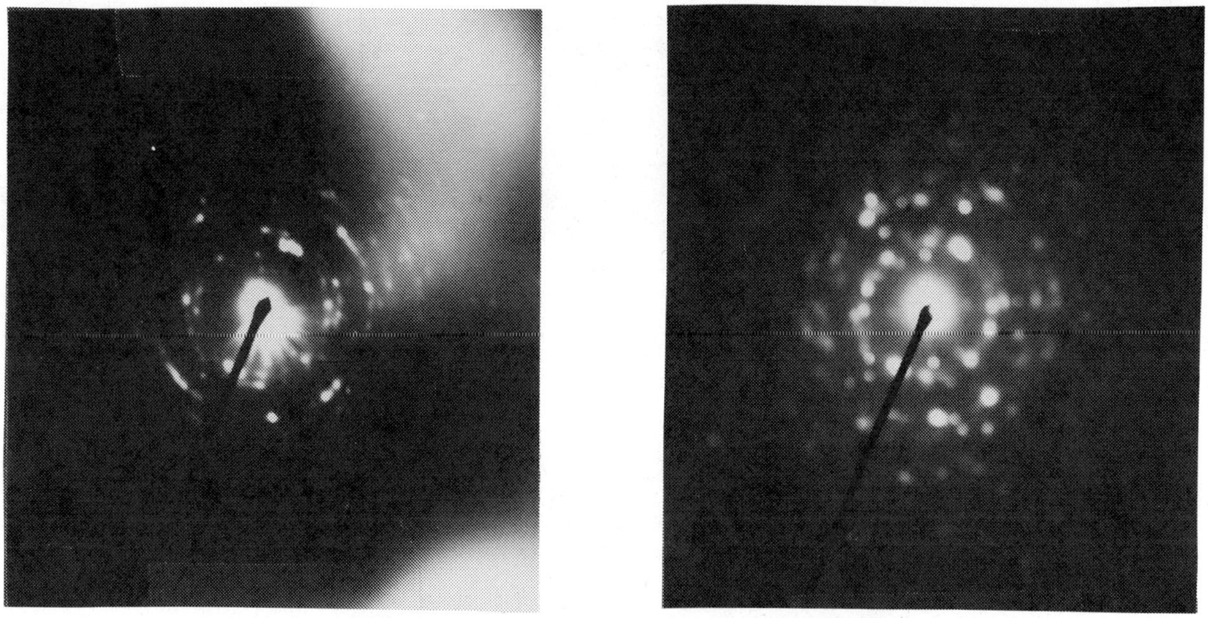

Figure 11. Electron diffraction patterns of the microstructure as shown in Figure 12 showing both graphite and ZrC crystallites.

a). Convergent beam electron diffraction pattern with (III) zone axis.

b). Selective area diffraction pattern.

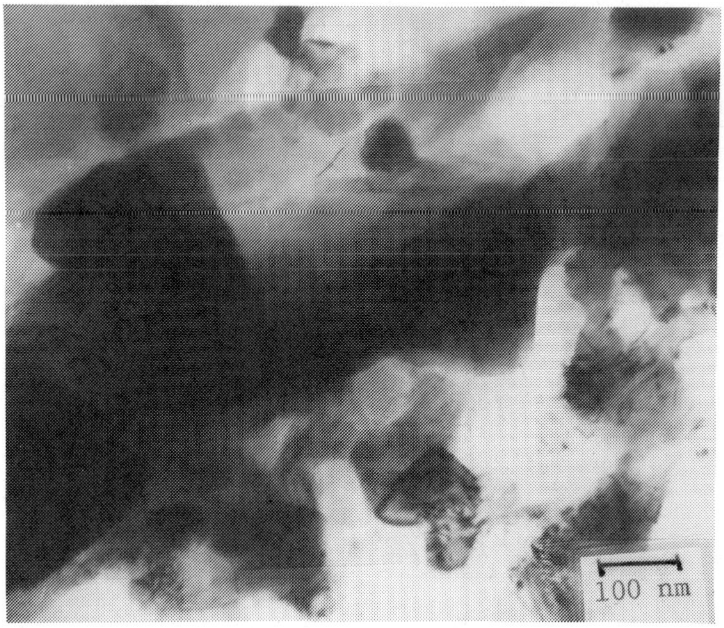

Figure 12. TEM micrograph of the ZrC structure within the coating.

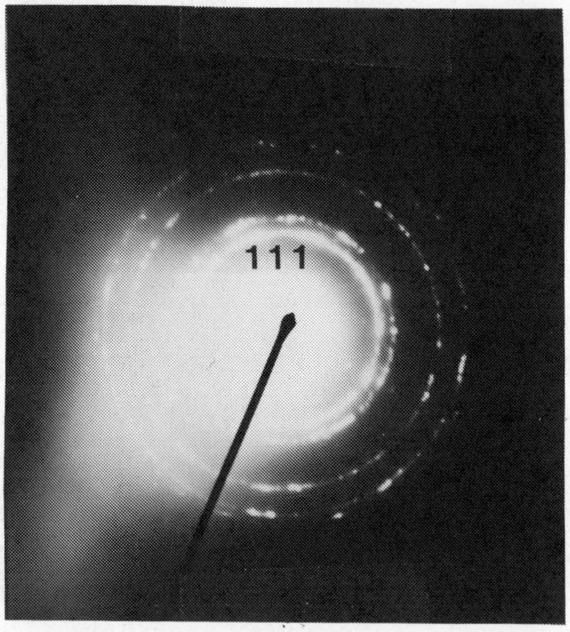

Figure 13. Selective area diffraction pattern of the ZrC structure on the coating, showing preferred orientation on (III) direction.

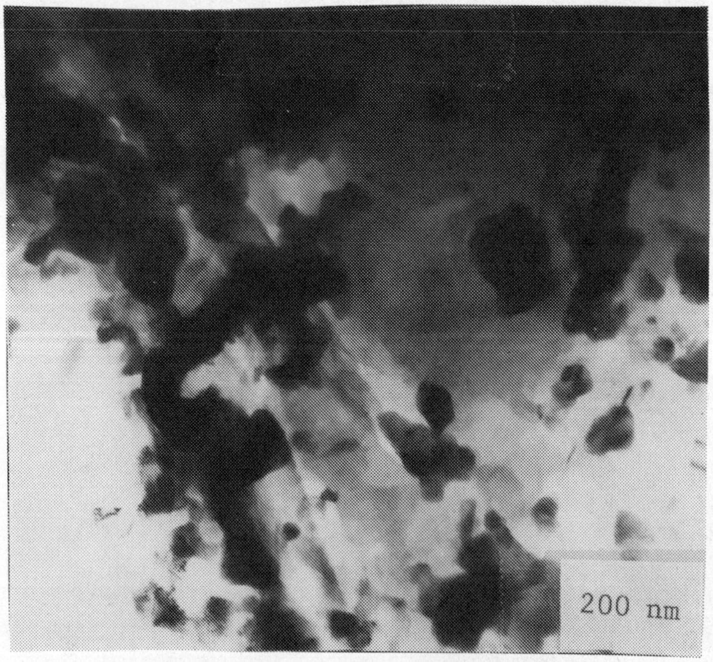

Figure 14. TEM micrograph showing BN dark particles immersed in the ZrC matrix.

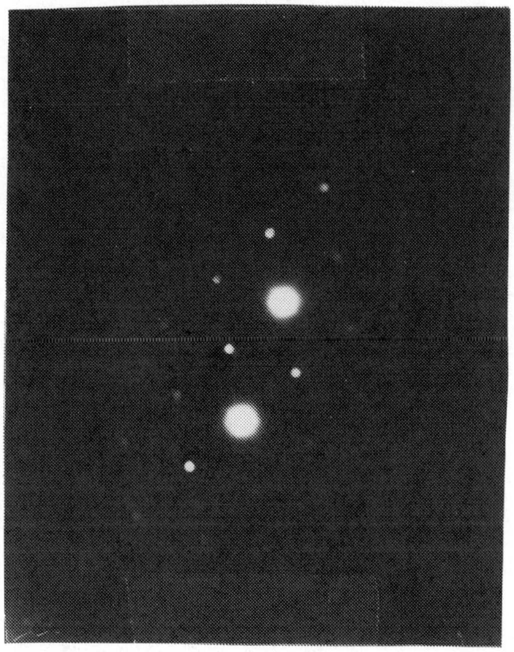

Figure 15. Double beam convergent beam electron diffraction pattern of the hexagonal BN structure as shown on Figure 16.

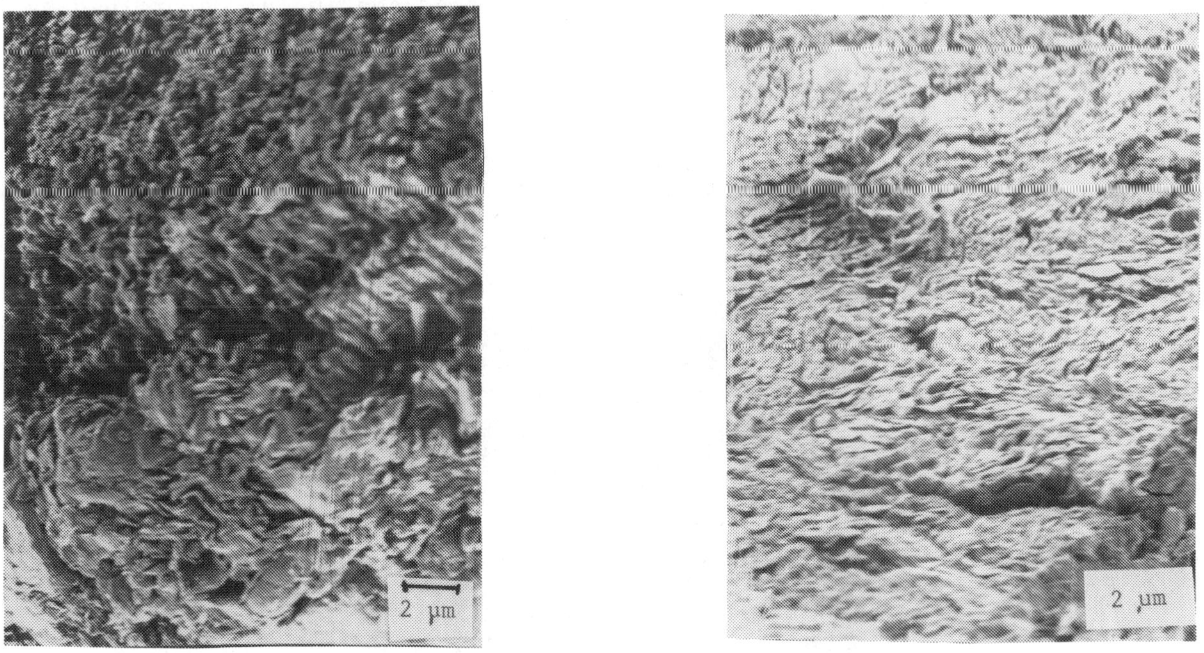

Figure 16. SEM micrographs at the coated carbon-carbon composite after an oxidation for 30 minutes at 1800°C, no cracking of the coatings can be observed.
a). ZrC-C, 1400°K.
b). ZrC-C, 1600°K.

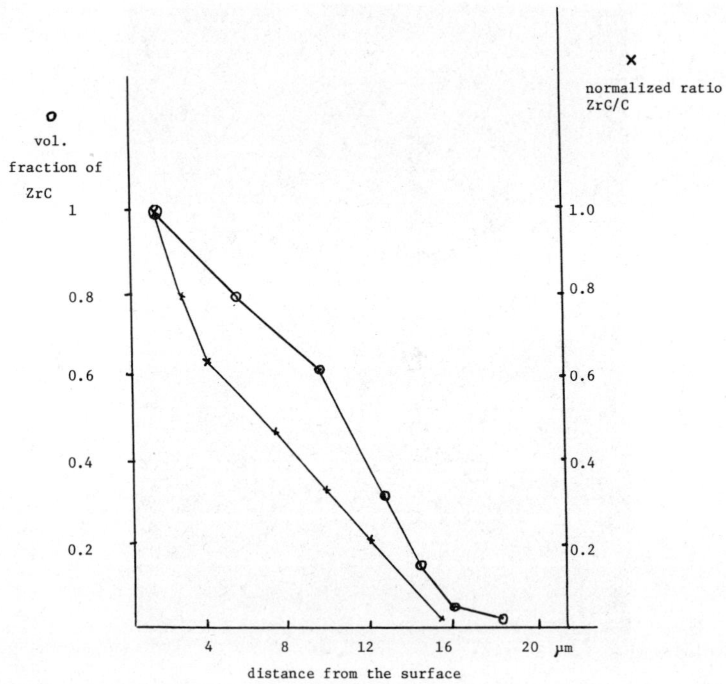

Figure 17. Phase fraction of ZrC and graphite in the graded coating as determined by the image analysis of TEM micrographs and normalized ratio of Zr/C in the graded coating as determined by the ultra thin window EDS.

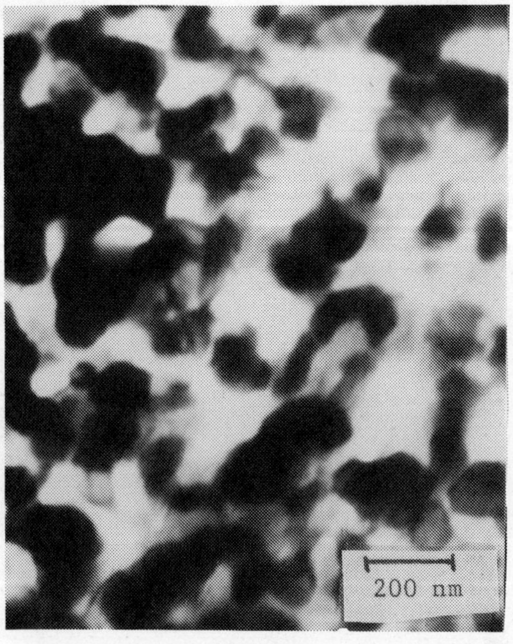

Figure 18. TEM micrograph of the outer part of the oxidized coating, showing boron oxide and zirconium oxide.

LIQUID PHASE SINTERED SILICON CARBIDE

Raymond A. Cutler and T. Barrett Jackson
Ceramatec, Inc.
2425 South 900 West
Salt Lake City, Utah 84119

ABSTRACT

Submicron silicon carbide was pressureless sintered to greater than 98% of theoretical density using alumina and yttria as liquid phase forming additives. The liquid phase allowed rapid sintering to occur at temperatures above approximately 1850°C. Short sintering times (less than 5 minutes) at 2000°C resulted in ceramics with hardness greater than 25 GPa, four point bend strength in excess of 400 MPa and indentation fracture toughness near 4 MPa·m$^{1/2}$ at room temperature. The lower sintering temperature and shorter sintering times, as compared to conventional pressureless sintered silicon carbide, limits grain growth (grain size was less than 5 μm). The sinterability of the SiC composites was shown to be dependent on the composition of the liquid phase. Advantages of liquid phase sintering for SiC-based ceramics and their potential applications in heat engines are discussed.

INTRODUCTION

Silicon carbide has been recognized as an important structural ceramic due to its excellent oxidation resistance, strength retention to high temperatures, high wear resistance, good thermal conductivity and relatively low coefficient of thermal expansion. Alliegro et al.[1] showed that SiC densified by hot pressing with 1.0 mol. % Al had excellent strength retention to 1400°C. Prochaska's sintering of SiC in the absence of applied pressure[2] opened the possibility of the mass production of relatively inexpensive SiC components. The use of SiC components in heat engines, however, has been hampered by the low toughness (2.5-3.0 MPa·m$^{1/2}$) and strength (300-400 MPa) of SiC. Efforts to improve the strength of SiC have been made by adding TiB_2 as a second phase[3], hot isostatically pressing (HIPing) using containment[4], or using liquid phase sintering[5]. In all three cases, the strength improvement is the direct result of limiting grain growth, thereby eliminating large grains which act as stress risers.

The advantage of liquid phase sintering lies in the ability to use particle rearrangement as a method of minimizing the critical flaw size. Lange[6] showed that the addition of 10 vol. % Al_2O_3 to SiC resulted in hot pressed SiC with an average grain size of 0.7 μm. Recent work at Ceramatec[5,7] has shown that rapid densification of SiC is possible using liquid phase sintering. The liquid phase resulted due to a reaction between Al_2O_3 and Al_4C_3 which produced a transient liquid (due to the Al_2O_3-Al_4O_4C eutectic) in route to forming Al_2OC[5,7]. Strengths in excess of 600 MPa were achieved by pressureless sintering, with additional strength improvement due to cladless HIPing[5]. However, the SiC-Al_2OC system was sensitive to processing (due to volatilization resulting from the carbothermal reduction of Al_2O_3) and required encapsulation during sintering[5,7]. The purpose of this paper is to report on an alternative liquid phase which allows pressureless sintering of SiC without encapsulation. The liquid phase sintered SiC has fine grain size, resulting in high hardness and moderate strength.

Omori et al.[8-11] have used a wide variety of rare earth oxides, usually in combination with alumina and/or boron compounds, to densify SiC. They maintain that these additives affect solid state sintering by lowering the surface energy of SiC, or increasing the diffusion rate by forming solid solutions with higher diffusion coefficients[11]. Negita[12] reports that other Japanese researchers are investigating the addition of yttria and rare earth oxides to enhance sintering of SiC. The Al_2O_3-Y_2O_3 phase diagram (see Figure 1) shows a liquid phase at temperatures as low as 1800°C[13]. Based on the premise of using an Al_2O_3-Y_2O_3 liquid phase to promote densification at temperatures and times similar to those used to sinter SiCAlON ceramics[5,7], the feasibility of sintering SiC without applied pressure or containment was evaluated along with its potential as a structural ceramic.

EXPERIMENTAL PROCEDURES

β-SiC* doped with 0.5 wt. % B and 4.0 wt. % C (conventional sintering aids) was prepared along with two other compositions using the same SiC source but liquid phases based on Al_2OC (SiC containing 10 wt. % Al_2OC[5]) and an equivalent volume fraction of Al_2O_3-Y_2O_3 (composition of Al_2O_3-20 mol. % Y_2O_3). All three compositions were processed identically by ball milling in hexane with WC-Co media, drying, screening, uniaxially pressing at 35 MPa followed by isostatic pressing at 200 MPa, and sintering without containment[5] in a graphite resistance heated furnace in 1 l/minute flowing Ar at temperatures between 1800 and 2100°C (isothermal hold time of 10minutes).

A second series of compositions were made with α-SiC** to investigate the effect of changing the alumina/yttria ratio in

---

*Betarundum (Ibigawa Electric Industry Co., Ltd., Ogaki, Japan).
**UF-15 (Lonza Inc., Fair Lawn, N.J.).

starting $SiC-Al_2O_3-Y_2O_3$ compositions. Eight compositions were made keeping the $Al_2O_3-Y_2O_3$ content constant at 10 vol. %. These compositions are shown in Table 1 below, along with theoretical densities (based on initial composition assuming that the density of garnet and perovskite yttrium aluminates can be predicted based on a simple rule of mixture) and phases expected after sintering based on the equilibrium phase diagram shown in Figure 1. All of the compositions were processed as described above. The sintering characteristics of the powders were initially investigated by heating all of the eight compositions at a rate of approximately 15°C/minute to 1000°C and increasing the rate to approximately 30°C/minute between 1000°C and the desired sintering temperature. The specimens were held at temperature for one minute and then furnace cooled to room temperature.

Density was measured by water displacement. Specimens (approximately 4 mm x 5 mm x 50 mm) for strength testing were ground with a 15 μm diamond wheel and broken in four point bending (40 mm support span and 20 mm loading span) at a constant cross head speed (0.5 mm/min.) using an universal testing machine*. Hardness was measured on polished specimens using a Vickers 136° diamond pyramid indenter attached to the universal testing machine by loading to 75 N (at 0.05 mm/min. and held 10 seconds before unloading). Indentation fracture toughness was measured using the calibration of Anstis et al.[14].

Table 1. Compositional Variation Within SiC-10 vol. % $Al_2O_3-Y_2O_3$

Yttria Content[a] (mole %)	Theoretical Density[c] (g/cc)	Expected Phases[b]
0.0	3.29	SiC, Al_2O_3
10.0	3.30	SiC, Al_2O_3, $3Y_2O_3 \cdot 5Al_2O_3$
20.0	3.32	SiC, $3Y_2O_3 \cdot 5Al_2O_3$, Al_2O_3
37.5	3.34	SiC, $3Y_2O_3 \cdot 5Al_2O_3$
50.0	3.35	SiC, $3Y_2O_3 \cdot 5Al_2O_3$, $2Y_2O_3 \cdot Al_2O_3$
66.7	3.37	SiC, $2Y_2O_3 \cdot Al_2O_3$
80.0	3.38	SiC, $2Y_2O_3 \cdot Al_2O_3$, Y_2O_3
90.0	3.39	SiC, Y_2O_3, $2Y_2O_3 \cdot Al_2O_3$

a. Yttria content relative to starting $Y_2O_3-Al_2O_3$ additive.
b. Assuming equilibrium conditions as shown in Figure 1.
c. Based on rule of mixtures of starting powders and neglecting media wear during milling.

*Model 1123 (Instron Corp., Canton, MA).

RESULTS AND DISCUSSION

Uncontained Liquid Phase Sintering

As shown in Figures 2 and 3, the liquid phase sintered materials sintered more readily than conventional SiC when heated to temperatures in excess of 1800°C with short (10 min.) isothermal holds. The liquid phase sintered materials showed rapid densification occurring between 1800 and 1900°C, while conventional SiC was densified more rapidly above 2000°C. The SiC-10 wt. % Al_2OC composition did not sinter to theoretical density due to the lack of containment[5]. The sintering of conventionally doped SiC is sensitive to the amount of carbon added[15]. A cursory investigation of different carbon levels (0.5, 1.0 and 2.0 wt. %) added to the boron-doped β-SiC did not increase its densification rate. One advantage of liquid phase sintered SiC is that materials which are not easily sintered using conventional sintering aids can be densified to near theoretical density, thereby eliminating the need for boron and carbon additions and allowing powders with high oxygen contents to be densified. The main drawback to the liquid phase additives identified to date, is that volatilization is at least an order of magnitude greater than with traditional sintering aids. Weight loss for the alumina/yttria-doped SiC specimens (see Figures 2 and 3) was between 4 and 7 wt. % between 1800 and 2100°C. While weight loss could be minimized by containment[7] during sintering, it was still possible to achieve identical densities with and without containment using alumina and yttria as sintering aids.

Effect of Liquid Phase Composition on Sinterability

Figures 4-7 show the effect of sintering temperature on shrinkage, density, open porosity, and weight loss, respectively. Figure 4 shows that alumina by itself is not as effective in promoting shrinkage as alumina-yttria compositions containing between 10 and 50 mole % Y_2O_3. Figures 5 and 6 show that increased shrinkage correlates excellently with improved density and decreased open porosity, as expected. These data not only show that the Al_2O_3-$5Al_2O_3 \cdot 3Y_2O_3$ (YAG) eutectic liquid allows liquid phase sintering, but they also suggest that yttria-rich liquids (liquids arising from $Al_2O_3 \cdot Y_2O_3$-$2Y_2O_3 \cdot Al_2O_3$ and $2Y_2O_3 \cdot Al_2O_3$-Y_2O_3 eutectics) do not promote sintering. This may be related to the wetting of SiC by these liquids or to the kinetics of forming the liquids. Weight loss, shown in Figure 7, was high, indicating that SiC is reactive with the liquid. Weight loss was nearly independent of composition, except at 2200°C where alumina-rich compositions showed higher weight loss than yttria-rich compositions. Weight loss is possible through reactions between SiC and Al_2O_3 such as

$$SiC + Al_2O_3 \longrightarrow SiO + CO + Al_2O \qquad (1)$$

which result in gaseous species. The reaction will proceed when

$$P_{SiO}P_{CO}P_{Al_2O} < \exp[-\Delta G^o{}_f/RT] \qquad (2)$$

where P is partial pressure, ΔG^of is the standard free energy of formation at temperature, T, and R is the gas constant. If one assumes that $P_{SiO} = P_{CO} = P_{Al_2O}$ then Reaction (1) will proceed when the gaseous reactants have a partial pressure less than 5.1×10^{-3} atm. at 2000K or 1.7×10^{-1} atm. at 2300K. These calculations indicate that Reaction (1) can occur in flowing Ar at high temperatures. There are other reactions which may occur, as well. An important point is that while weight loss in the Al_2O_3-Y_2O_3 system is not ideal, it is vastly improved over the SiC-Al_2OC system[5,7]. Work is ongoing on identifying liquid phases which form at lower temperatures to minimize the reaction between SiC and Al_2O_3.

In an effort to understand decomposition reactions, sintering parameters were investigated. The ramp time between 1000 and 2000°C was varied between 15 and 120 minutes. The weight loss increased with increasing ramp time for alumina-rich compositions. The reaction between alumina and silicon carbide is dominant and prereacting the alumina and yttria to form YAG (or Al_2O_3 plus YAG) prior to sintering reduces weight loss. Due to the preferential removal of Al_2O_3, as compared to Y_2O_3, density increases slightly with decreasing heating rate (i.e., increasing ramp time). Microscopy results show that weight loss occurs from the near surface region, as expected. Decomposition continues after densification has occurred. Decomposition increases with increasing isothermal hold time, as expected. Lower sintering temperatures permit longer hold times for the same amount of weight loss.

Characterization

Densities greater than 98% of theoretical could be obtained with either alpha or beta SiC having surface areas greater than approximately 5 m^2/g. The main advantage of the liquid phase is that rapid densification allows submicron grain size to be retained, as shown in Figure 8. The short sintering times also allow the retention of starting SiC polytypes and the elimination of large alpha SiC grains which typically weaken the material. The rapid sintering allows continuous sintering in a "pusher" type furnace, which is advantageous for economical production of high volume components.

Due to the low porosity and fine grain size of liquid phase sintered materials, high hardness (23-26 GPa depending on composition) resulted. These hardness values are comparable or superior to that measured on commercially available hot pressed or pressureless sintered SiC using the same technique. Liquid phase sintered silicon carbide containing 10 vol. % YAG was comparable or superior to conventional silicon carbide in solid particle erosion tests and pin-on-disk wear tests.

Indentation fracture toughness was approximately 4 $MPa \cdot m^{1/2}$. This is comparable to commercially available hot pressed SiC with Al_2O_3. There is a need to improve the fracture toughness of SiC and make it comparable with Si_3N_4. The room temperature strengths were in the range of 400 to 500 MPa. The fracture mode was primarily intergranular (see Figure 8), as opposed to conventional SiC which typically fractures transgranularly. Failure origins were voids, typically 25-50μm in diameter, which resulted due to decomposition reactions. These data indicate that liquid phase sintered SiC has room temperature mechanical properties comparable or improved over conventional silicon carbide.

A cursory look at elevated temperature strength was made by testing a SiC-10 vol. % YAG composition at 1200°C in air. The strength decreased from 443 MPa (std. dev. of 66 MPa) at room temperature to 336 MPa (std. dev. of 30 MPa) at 1200°C. Further elevated temperature testing (strength, creep, and oxidation resistance) is needed to establish the temperature range over which these liquid phase sintered silicon carbides can be used. It is readily apparent, however, that liquid phase sintered silicon carbide can be used in low temperature (<1000°C) applications such as wear parts. Applicable heat engine components include valves, valve seats, piston pins, cam followers, etc.

CONCLUSIONS

Simultaneous Al_2O_3 and Y_2O_3 additions to SiC, for compositions between approximately 10 and 50 mole % Y_2O_3 (balance Al_2O_3), were used to sinter SiC-10 vol. % Al_2O_3-Y_2O_3 ceramics to greater than 98% of theoretical density without containment. The liquid phase responsible for densification occurred in the same temperature range as the Al_2O_3-Al_4O_4C eutectic. The resulting microstructures were fine grained and exhibited high hardness, and fracture toughness similar to SiCAlON ceramics[5,7]. Further work is needed to limit decomposition, which appears to be primarily due to the reaction between alumina and silicon carbide. The primary advantage of the liquid phase is that rapid sintering limits grain growth and eliminates large alpha grains. Furthermore, the liquid phase sintering process is conducive to continuous throughput production of components for applications such as wear parts in heat engines.

ACKNOWLEDGEMENTS

This work was supported by DOE under the Small Business Innovation Research (SBIR) Program (contract DE-AC0384ER80191). Discussions with Professor Anil V. Virkar of the University of Utah, Professor D. Lynn Johnson of Northwestern University, and Andrew C. Hurford of Ceramatec are acknowledged. The technical assistance of Ted Valerio, Jim Reynolds and Kevin Rigtrup of Ceramatec is appreciated.

REFERENCES

1. R. A. Alliegro, L. B. Coffin and J. R. Tinklepaugh, "Pressure-Sintered Silicon Carbide," J. Am. Ceram. Soc., 39[11] 386-89 (1956).

2. S. Prochazka, "Sintering of Silicon Carbide," Proc. of the Conference on Ceramics for High Performance Applications, Hyannis, Mass., 1973, ed. by J. J. Burke, A. E. Gorum and R. M. Katz (Brook Hill Publ. Co., 1975).

3. C. H. McMurtry, W. D. G. Boecker, S. G. Seshadri, J. S. Zanghi and J. E. Garnier, "Microstructure and Material Properties of SiC-TiB_2 Particulate Composites," Am. Ceram. Soc. Bull., 66[2] 325-29 (1987).

4. S. Dutta, "Improved Processing of α-SiC," Adv. Ceram. Mater., 3[3] 257-62 (1988).

5. T. B. Jackson, A. C. Hurford, S. L. Bruner and R. A. Cutler, "SiC-Based Ceramics with Improved Strength," to appear in Silicon Carbide ed. by J. W. Cawley and C. Semler (Am. Ceram. Soc., Columbus, OH 1988).

6. F. F. Lange, "Hot-pressing Behavior of Silicon Carbide Powders with Additions of Aluminum Oxide," J. Mater. Sci., 10 314-20 (1975).

7. J. L. Huang, A. C. Hurford, R. A. Cutler, and A. V. Virkar, "Sintering Behavior and Properties of SiCAlON Ceramics," J. Mater. Sci., 21 1448-56(1986).

8. M. Omori and H. Takei, "Pressureless Sintering of SiC," J. Am. Ceram. Soc., 65[6] C-92 (1982).

9. M. Omori and H. Takei, "Composite Silicon Carbide Sintered Shapes and its Manufacturing," U. S. Patent 4,502,983 (March 5, 1985).

10. M. Omori and H. Takei, "Method for Preparing Sintered Shapes of Silicon Carbide," U. S. Patent 4,564,490 (Jan. 14, 1986).

11. M. Omori and K. Ohira, "Sintered Silicon Carbide Molding and Process for Production Thereof," U. S. Patent 4,569,921 (Feb. 11, 1986).

12. K. Negita, "Effective Sintering Aids for Silicon Carbide Ceramics: Reactivities of Silicon Carbide with Various Additives," J. Am. Ceram. Soc., 69[12] C-308-C-310 (1986).

13. T. Noguchi and M. Mizuno, Kogyo Kagaku Zasshi, 70[6] 839 (1967).

14. G. R. Anstis, P. Chantikul, B. R. Lawn and D. B. Marshall, "A Critical Evaluation of Indentation Techniques for Measuring Fracture Toughness: I, Direct Crack Measurements," J. Am. Ceram. Soc., 64[9] 533-38 (1981).

15. J. Hojo, K. Miyachi, Y. Okabe and A. Kato, "Effect of Chemical Composition on the Sinterability of Ultrafine SiC Powders," J. Am. Ceram. Soc., 66[6] C-121-24 (1984).

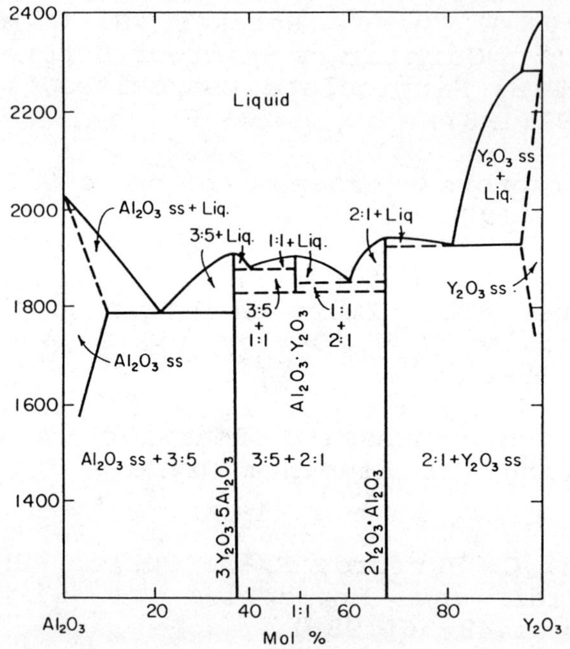

Figure 1. Equilibrium phase diagram between Al_2O_3 and Y_2O_3 after Noguchi et al.[13].

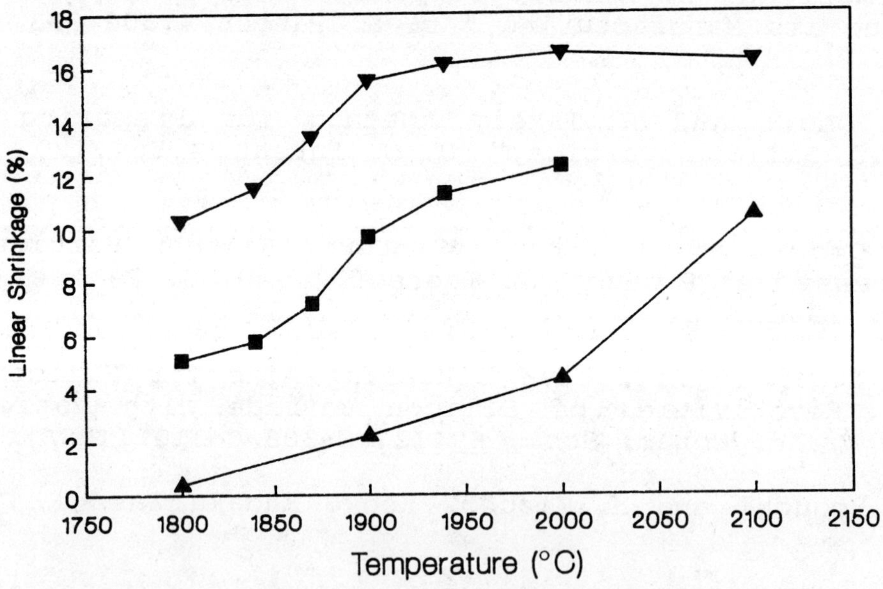

Figure 2. Comparison of linear shrinkage of SiC (triangles), SiC-10 wt. % Al_2OC (squares) and SiC with Al_2O_3 and Y_2O_3 as sintering aids (inverted triangles) pressureless sintered (10 min. isothermal hold) in Ar.

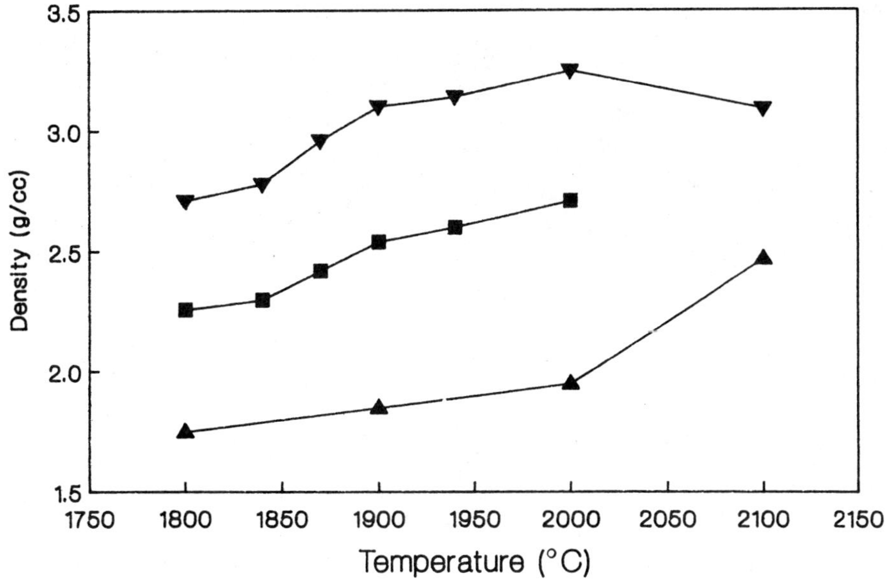

Figure 3. Effect of sintering temperature on density of the three compositions shown in Figure 2. Theoretical densities for SiC (triangles), SiC-10 wt. % Al_2OC (squares), and $SiC(Al_2O_3-Y_2O_3)$ (inverted triangles) are 3.21, 3.19, and 3.33 g/cc, respectively.

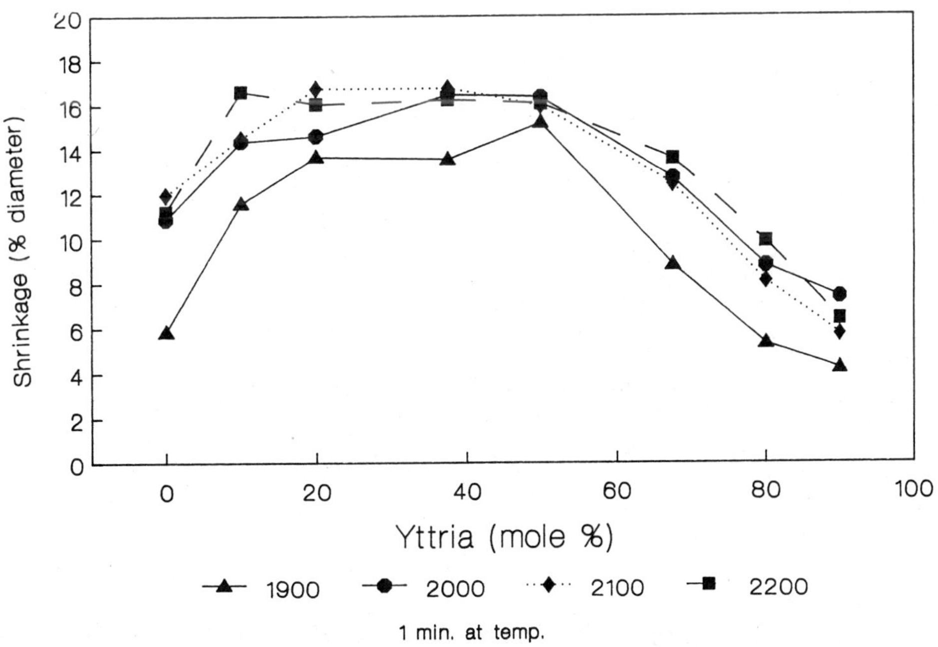

Figure 4. Shrinkage as a function of Y_2O_3 content in SiC-10 vol. % Al_2O_3-Y_2O_3 specimens sintered in Ar with an isothermal hold of one minute.

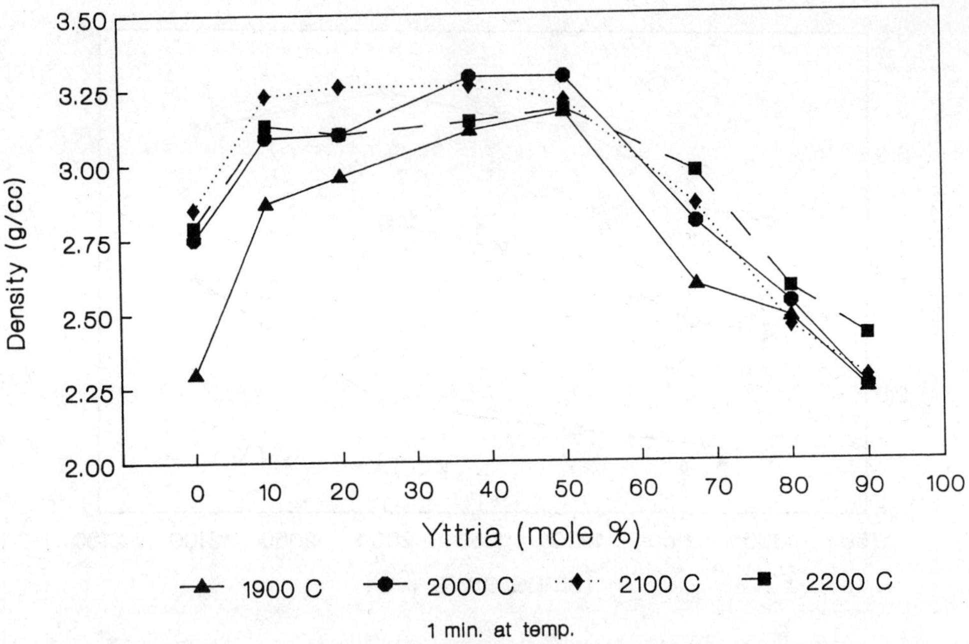

Figure 5. Density as a function of Y_2O_3 content in SiC-10 vol. % Al_2O_3-Y_2O_3 specimens sintered in Ar with an isothermal hold of one minute.

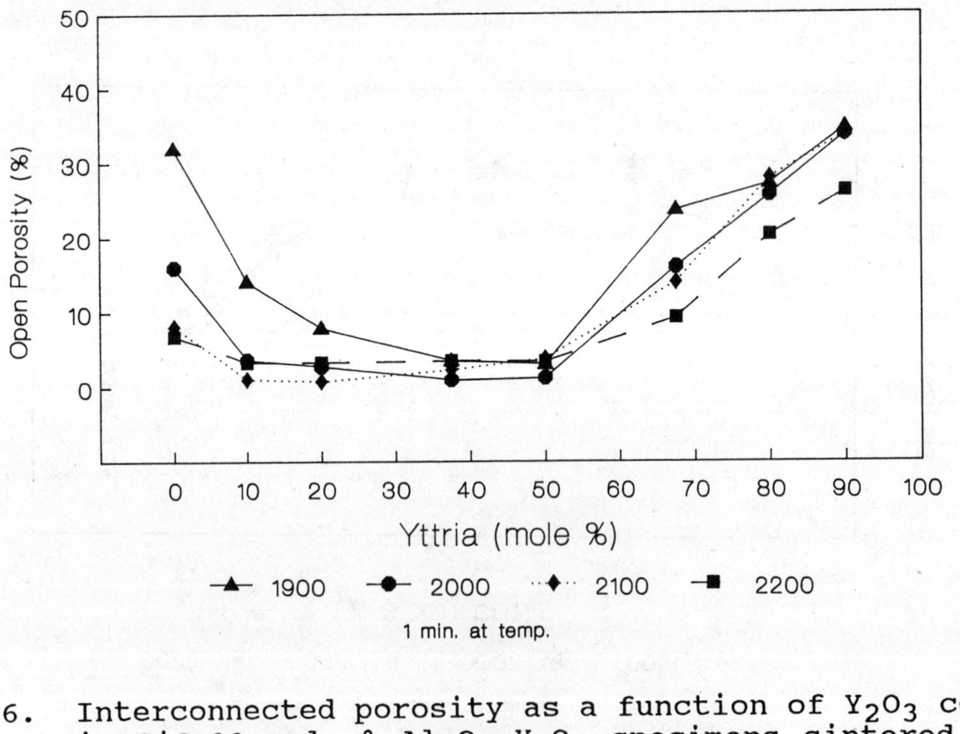

Figure 6. Interconnected porosity as a function of Y_2O_3 content in SiC-10 vol. % Al_2O_3-Y_2O_3 specimens sintered in Ar (one minute hold at temperature).

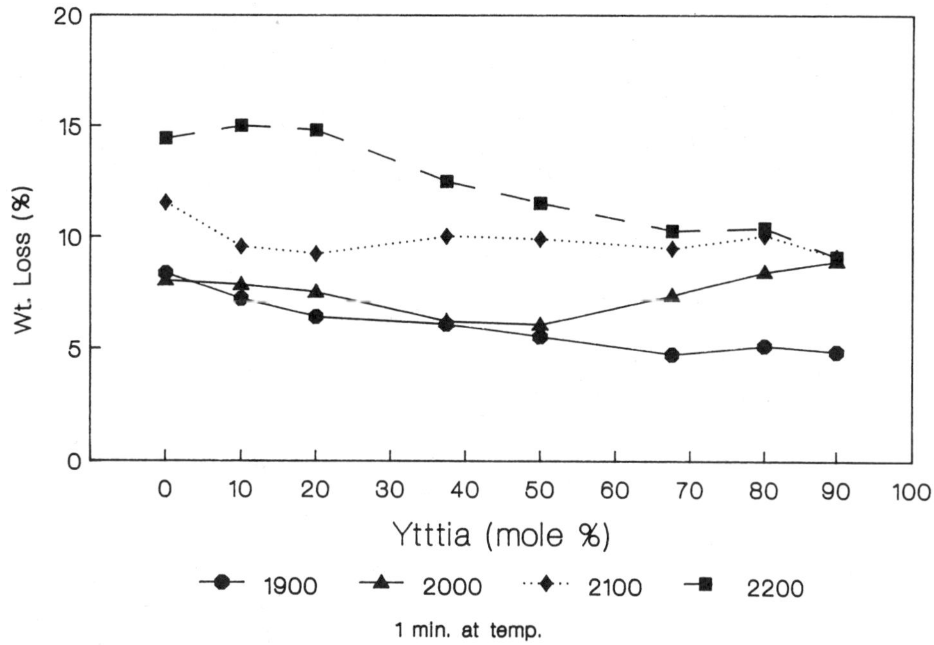

Figure 7. Weight loss as a function of Y_2O_3 content in SiC-10 vol. % Al_2O_3-Y_2O_3 specimens sintered in Ar with an isothermal hold of one minute.

Figure 8. Fracture surface of SiC-10 vol. % YAG sintered in Ar at 2000°C for 5 minutes. The SiC grain size is typically less than 1 μm.

REVIEW: COLLOIDAL PROCESSING OF SILICON NITRIDE CERAMICS

Peter Greil

Technische Universität Hamburg–Harburg, AB Technische Keramik, Harburger Schloßstr. 20, D–2100 Hamburg 90

Alwin Nagel*, Heinz Stadelmann and Günter Petzow

Max–Planck–Institut für Metallforschung, Institut für Werkstoffwissenschaften Pulvermetallurgisches Laboratorium, Heisenbergstr. 5, D–7000 Stuttgart 80

ABSTRACT

The influence of surface chemistry and solution chemistry on the stability versus flocculation of aqueous Si_3N_4 suspensions was evaluated. Surface hydrolysis and preferential adsorption of polyelectrolytic surfactants may stabilize multiphase Si_3N_4 suspensions with maximum zeta potentials of > −80 mV at pH 9/10. The negative surface charge at high pH is attributed to the pH–dependent dissociation of amphoteric silanol groups to anionic SiO(−) on the particle surface. The preferential adsorption of anionic surfactants with −COO(−) functional groups further increased the negative surface charge and hence slip stability. The critical flocculation concentration of trivalent Y(3+)–counterions from the sintering additives of 0.08 mmol/l at a pH 9 was not reached when the starting additive powders where prereacted to $Y_3Al_5O_{12}$. Slip casting of high concentrated aqeuous Si_3N_4 slips resulted in green compacts with high densities > 65 wt.% and uniform microstructures.

INTRODUCTION

Si_3N_4–based ceramics are of particular interest for high–temperature, high–stress applications in heat engines. Dense Si_3N_4–materials have been developed based on different sintering and hot–pressing or hot–isostatic pressing technologies, respectively. These materials exhibit high bending strength, good thermal shock resistance due to a low coefficient of thermal expansion and good resistance to oxidation compared to other high temperature structural materials. However, great differences of the material properties and component reliability still occur, which are primarily related to inadequate manufacturing technologies. The control of defects in the processing of ultrafine and pure Si_3N_4 powders to green compacts with uniform microstructure and high packing density is therefore considered as a key step in manufacturing of high strength and reliable Si_3N_4–ceramic components.

Ultrafine plasma and laser catalyzed Si_3N_4 powders with mean grain sizes of less than 100 nm and a very narrow grain size distribution were demonstrated to exhibit improved sintering properties that may be used for materials with better performance [1]. Surface related phenomena, such as agglomeration, formation of intergranular

* now with Feldmuehle AG, Plochingen, FRG

residual glassy phases, inhomogeneous distribution of sintering additives and surface adsorbed impurities generally gain in significance with decreasing grain size and increasing purity of the starting powders. Hence, advanced powder processing will depend more and more on colloidal chemistry methods, which allow better control of the surface related physical and chemical interactions by keeping the powder in a liquid suspension from the synthesis to the shaping process. Thus, contamination through adsorption of gases and organic molecules on the surfaces as well as uncontrolled agglomeration may be minimized. Wet chemical processing is necessary also for the processing of a new generation of multicomponent composite materials, where particulates, platelets, whiskers and fibers have to be uniformly dispersed and aligned, respectively, in a fine grained Si_3N_4 matrix. Particularly SiC–whisker and fiber reinforced Si_3N_4–matrix composites have shown superior fracture toughness and strength over monolithic materials, even though severe problems in the fabrication technology of dense products with complex shape still exist [2].

In the following, the colloidal processing of Si_3N_4 powders in aqueous suspensions will be considered. The attention will be focused on the surface and solution chemistry of Si_3N_4 in water and on the influence of the oxidic sintering additives on the slip stability.

POWDER PREPARATION AND SURFACE CHEMISTRY

A variety of novel chemical syntheses have been used to prepare Si_3N_4 powders, and there is increasing emphasis on using the preparative method to control the physical properties of the nitride. The sinterability of the powder and the strength of the sintered compact are dominated by particle size, chemical purity, phase composition, agglomeration behaviour and the type and amount of sintering additives. The methods used to prepare Si_3N_4 may be divided into conventional processes (direct nitridation of Si–powder and carbothermal reduction of SiO_2) and advanced reaction processes (sol–gel–conversion, liquid phase reactions of organometallic amides, gas phase reactions induced by plasma or laser and polymer pyrolysis) [3,4]. Powders manufactured by direct nitridation of Si at temperatures above 1300 ^{0}C usually have to be crushed and subsequently acid leached for purification. Powders from sol–gel conversion and liquid phase reaction at low temperatures may be used as grown. Gas phase reacted powders at $<$ 1200 ^{0}C are characterized by low crystallinity but extremely small grain size ($<$ 100 nm). Monosized, pure and loosely agglomerated Si_3N_4 powders were produced by laser–induced gas phase reactions [5]. These theoretically dense powders, which had particle diameters $<<$ 100 nm and oxygen contents of less than 0.1 wt.%, crystallized in N_2–atmosphere between 1300 – 1400 ^{0}C.

Adsorbed gaseous impurities such as CO_2 and H_2O as well as reaction products of surface oxidation and hydrolysis can be found on the particle surfaces, due to imperfect manufacturing. Impurities on the particle surface, however, are supposed to exert a dominant influence on agglomeration and dispersion behavior of fine Si_3N_4 powders in liquid suspensions. While the role of surface–adsorbed gases for processing of Si_3N_4 powders has not yet been systematically studied, sensitive oxygen analyses of the surface may provide information about the influence of oxygen contamination on the processing behaviour. Fig. 1 shows an X–ray photoemission spectrum of the 95–110 eV energy range for a Si_3N_4 powder produced by decomposition of silicon diimide (α–Si_3N_4 E–10, UBE Ind. Jp). The spectrum was deconvoluted into two peaks of equal half–width with an intensity ration of 6.2 to 1 and electron binding energies of 101.1 eV (Si 2p) and 102.8 eV (SiO_2/Si_2N_2O) [6]. Bulk chemical and surface sensitive XPS analyses of the oxygen distribution in various Si_3N_4–powders manufactured by different processes, reveal significant differences of the oxygen content in the bulk and on the particle surface. Table I summarizes the results of the oxygen distribution analyses obtained for various commercial Si_3N_4 powders. α–Si_3N_4 powders from the conventional manufacturing processes only show a minor part of the residual oxygen content of typically 0.9 – 2.5 wt.% to be concentrated on the particle surface, whereas an appreciable amount of approximately 0.8 –1.0 wt.% may be attributed to oxygen dissolved in the bulk. As stated earlier, nitrided powders undergo post–synthesis purification which includes chemical treatment to reduce the oxygen content. This may be the reason for the relatively lower surface concentration of O_2 in these powders. Si_3N_4 powders made by advanced reaction processes, especially the low temperature $Si(NH)_2$ decomposition powder, often show extremely low oxygen in the bulk (< 0.5 wt.%) while a higher oxygen concentration is found on the particle surface. Surface oxygen, however, may further be reduced by chemical treatment.

AQUEOUS SUSPENSIONS

From surface titration of Si_3N_4 in water suspensions it was concluded that the predominant titratable surface species on Si_3N_4 is silanol and that the adsorbed species is ammonia, most of which is physisorbed, but some of which is apparently chemisorbed [7] indicating the hydrolysis reaction

$$Si_3N_4 + 12\,H_2O \longrightarrow 3\,Si(OH)_4 + 4\,NH_3\,. \qquad (1)$$

Si_3N_4 powders are well known to exhibit an aging effect in water of pH 7 and to change the pH either to lower or higher values, which may be explained in terms of the surface composition and impurity content of the nitride powders. Hydrolysis of pure Si_3N_4 is expected to increase the pH because of the higher dissociation constant $pK_b = 4.75$ for NH_3 at rt

$$NH_3 + H_2O \longrightarrow NH_4^+ + OH^- \qquad (2)$$

as compared to $pK_{s1} = 9.51$ for the weak silicon acid

$$H_4SiO_4 + H_2O \longrightarrow H_3O^+ + (H_3SiO_4)^-\,. \qquad (3)$$

Acid/base reaction of Si_3N_4 containing a SiO_2–like surface layer, however, will be dominated by the hydrolysis of SiO_2 only

$$SiO_2 + 2\,H_2O \longrightarrow Si(OH)_4 \qquad (4)$$

and thus the dissociation of silicon acid results in a decrease of pH. The acid/base reaction of Si_3N_4 in water suspension may therefore be used to characterize the surface composition of the nitride powder.

Surface charging of the Si_3N_4 powders in water is likely to occur by a pH dependent dissociation of the weakly acidic silanol group on the particle surface according to

$$\left[-Si(OH)_2^+\right] \xleftarrow{H^+} \left[-Si(OH)\right] \xrightarrow{OH^-} \left[-SiO^-\right] + H_2O\,. \qquad (5)$$

The surface potential Ψ_0 of the Si_3N_4 powders dispersed in water may then be described by a modified Nernst–equation [8]

$$\Psi_0 = 2.3\frac{kT}{e}(pH_z - pH) - \frac{kT}{e}\left[\ln\frac{(SiO^-)}{(SiOH_2+)}\right], \tag{6}$$

with pH_z being the pH at the point of zero surface charge, where the concentration of $SiOH_2(+)$ and that of $SiO(-)$ equals zero. Thus, the effective surface potential will be controlled by the silanol site density on the nitride surface and their acid/base dissociation as a function of pH. Oxygen free Si_3N_4 exhibits an isoelectric point (iep) at a pH 6–7 which decreased to 4.2 after preoxidation. Simultaneously the amount of the zeta potential increased from –30 mV to –45 mV at a pH of 9 [9]. Fig. 2 shows the zeta potentials of pure and preoxidized Si_3N_4 (550 °C, 8 h) from this work and data given in [9,10]. Maximum zeta potentials are observed at pH values < 3 and > 8. With increasing zeta potentials, however, stability against rapid flocculation in aqueous solution will also increase due to electrostatic repulsion between the equally charged particles. Thus, Si_3N_4 may well be stabilized electrostatically against rapid flocculation in aqueous slips either at pH < 3 or > 8 having either a positive or negative surface charge, respectively.

A surface layer of silica–like composition may be generated on the Si_3N_4–particles by controlled preoxidation treatment, the thickness of which can be derived from the particle size and the oxygen pick up. Fig. 3 shows the relation of the zeta potential to the thickness of a hypothetical SiO_2–layer in a Si_3N_4–powder with a mean spherical grain size of 130 nm (LC–12, HC. Starck, FRG). The zeta potential was derived from electrophoretic mobility measurements (Zetasizer II, Malvern, GB) in the presence of an ammonia–polyacrylate with –COO(–) and $NR_4(+)$ functional groups (PC 33, Zschimmer + Schwarz, FRG) in a concentration of 1 ml/l. Up to a layer thickness of 1.4 – 2.0 nm, which is equivalent to 5 – 7 SiO_2 tetrahedra lengths, the zeta potential increases asymptotically to a maximum and remains constant thereafter indicating a total coverage of the underlying Si_3N_4–grain. Thus, the preoxidized Si_3N_4–powders can be treated similarly to an oxide powder, even though the iep at pH 4.2 is slightly higher than that of pure SiO_2 at pH 2.0 [11].

To demonstrate the basic significance of high surface potentials for the stabilization of submicron–sized Si_3N_4 powders against rapid flocculation in water suspension, the net interaction particle–particle potential was estimated by the combination of the repulsive potential due to electric double layer formation in water with the energy of attraction due to van der Waals forces [12]

$$V_{net} \approx 2\pi\epsilon\epsilon_0 a\Psi_0^2 e^{(-\kappa s)} - \frac{Aa}{12s}. \tag{7}$$

A is the Hamaker–constant, $\epsilon\epsilon_0$ denotes the relative dielectric constant of water and the absolute permittivity of free space, respectively. The variables a and s are the particle radius and surface separation and κ represents the reciprocal Debye– length, which is characteristic for the extension of the electric double–layer. The values used to establish the interaction potential given in Fig. 4 were $A = 2.26\times10^{-21}$ J, $\epsilon\epsilon_0 = 6.95\times10^{-10}$ C^2/Jm^2, $\kappa = 10^9$ 1/m and a = 300 nm, which may reasonably characterize the suspension of silica coated Si_3N_4 in water at 25 °C with an ionic strength of 10^{-5} mol/l of trivalent counterions. Although the estimate is based on a simple approach neglecting any steric interaction, particle size distribution and concentration effects, general tendencies for the flocculation behaviour of Si_3N_4 in aqueous suspensions may well be derived. Fig. 4 indicates a potential barrier preventing the Si_3N_4 particles from contacting one another only for a surface potential $\Psi_0 > \pm 30$ mV. At lower surface potentials, which are typical of oxygen free powders or significantly smaller grain sizes, additional steric repulsion forces have to be generated by specific adsorption of organic macromolecules in order to stabilize the Si_3N_4 powder against flocculation by a steric or combined electrosteric mechanism.

In general, Si_3N_4 suspensions can be stabilized electrostatically, but improvement of the suspensions to better meet the requirements necessary for defect–free processing is possible by incorporating polymeric addtitives. Generally, in aqueous media amphiphatic adsorption is the most frequent of adsorption processes encountered [13]. Amphiphatic molecules simply means that the molecule consists of two parts, each of which has an affinity for a different phase [12]. Amphiphatic macromolecules with R–COOH, R–CN, R–$CONH_2$, R–SO_3H or R–PO_4H functional groups may be used to provide an additional stabilization for water solvated Si_3N_4 powders with the hydrophobic hydrocarbon R–group being expelled from the water and the hydrophilic ionic group being attracted [13]. At high pH the functional group will be dissociated in water to an anionic type of agent. In aqueous media the adsorption of ionogenic surface active agents leads to an increase in the capacity of the electric double layer provided that the adsorption process does not destroy ionization. Anionic agents will tend to make the stabilizing potential more negative. The negatively charged carboxyl group –COO(–), for instance ,is believed to be adsorbed either on positively charged surface sites, via H–bridge bonding [14] or by a nucleophilic attack of the Lewis base –COO(–) on the strong Lewis acid Si(4+) ion. Thus, in the presence of the basic organic deflocculant PC 33, the formation of the surface silica layer resulted in a pronounced decrease in the magnitude of the zeta potential to a value of – 85 mV as compared to – 65 mV for the as received powder at pH 9 [9]. This indicates that the surface active deflocculant is preferentially adsorbed in the case of the preoxidized Si_3N_4 powder as compared to the as received powder. The nucleophilic attack mechanism seems to be more reasonable in explaining the preferential adsorption of the –COO(–) group at high pH where the silanol group is deprotonated because of the higher Lewis acid strength of Si in SiO_2 (ionic bonding) than in Si_3N_4 (covalent bonding); at lower pH the H–bridge bonding mechanism may be dominating.

The electrostatic stability of multiphase Si_3N_4 slips in water suspension may be restricted by the electrolytic effects of sintering additives necessary to promote a liquid–enhanced sintering process. Usually, oxides like Y_2O_3, Al_2O_3 or MgO are used which form positive counterions of high valency by the dissociation of their hydroxides in water

$$Me^{n+}(OH)_n \longrightarrow Me^{n+} + n\,(OH)^-, \qquad (8)$$

resulting in a reduction of the reciprocal Debye length κ [12]. For $Y(OH)_3$ and $Al(OH)_3$, as the hydrolysis products of Y_2O_3 and Al_2O_3 dissolution in water, the solubility products K_L

$$K_L = C_{Me^{n+}} \times C_{OH}^{\;3} \qquad (9)$$

are 10^{-23} and 1.9×10^{-33} $(mol/l)^4$, respectively. According to the Schulze–Hardy rule, the critical flocculation concentration (cfc) strongly depends on the surface potential Ψ_0 and the charge of the counterions z, as expressed by:

$$cfc = B\frac{\Psi^4}{A^2 z^2}, \qquad (10)$$

where A is the Hamaker–constant and B a geometrical constant. With Y_2O_3 as the additive, a Y(3+)–concentration of only 0.08 mmol/l was found to be sufficient to cause rapid flocculation of Si_3N_4 powders at a pH of 8 [9]. Similar cfc s of 0.01 – 0.1 mmol/l have been reported for other trivalent counterion concentrations [15]. Using the experimentally determined cfc for Y(3+) and the zeta potential dependence on pH given in Fig. 2, cfc s for trivalent counterions were estimated according to Equ. (10) as a function of pH in aqueous Si_3N_4–suspensions, Fig 5. The conditions for stability in the presence of dissociated cations may be found by comparing the pH–dependent cation concentration as defined by the solubility product with the critical flocculation concentration. Thus, flocculation will be prevented by using only such additives, which show a cation concentration less than the limit indicated in Fig. 5. For Si_3N_4 slip formation at high pH, Al(3+)–concentration is negligible whereas Y(3+)–cation concentration dominates the flocculation behavior. Either the pH has to be kept above approximately 8–9 or Y–containing compounds other than Y_2O_3 have to be used in order to decrease the Y(3+)–concentration. Thus, $Y_3Al_5O_{12}$ was prereacted from Y_2O_3 and Al_2O_3 powders at 1400 °C to reduce the concentration of dissociated Y(3+) because $Y_3Al_5O_{12}$ exhibits a much smaller solubility in water than Y_2O_3. It was successfully used in aqueous suspensions without causing Si_3N_4 flocculation at pH > 8 [9].

NON–AQUEOUS SUSPENSIONS

Problems related to multiphase oxide–nitride powder composition and surface oxygen pick up of the nitride may be drastically reduced by using non– aqueous suspension media. The non–aqueous processing of the nitride powders is characterized by the significance of the attractive van der Waals forces, which essentially determine the total potential energy of interaction between two suspended nitride particles in organic liquids [16]. It was concluded that the processing of Si_3N_4 without the use of additives is feasible for certain organic liquids when the attractive van der Waals force $V_A = -Aa/12s$ is less than a critical value, which depends on grain size, inhomogeneities and chemical additives. For extremely fine Si_3N_4 powders with a mean grain size of < 20 nm, good dispersion qualities were found in organic liquids for which $|V_A| < -5$ kT which may be further reduced to < –1 kT for larger particle sizes [16]. Some aromatic hydrocarbons such as benzene, toluene and xylene, C_6H_6, $C_6H_5CH_3$ and $C_6H_4(CH_3)_2$, and polar aliphatic acids such as propionic and butyric acid, C_2H_5COOH and C_3H_7COOH, may therefore be interesting for non–aqeous processing of Si_3N_4. Surfactants or polymers are required to provide sufficient steric hindrance to stabilize suspensions of Si_3N_4 in liquids for which $|V_A| > -10$ kT.

PRECIPITATION MIXING

The uniform distribution of sintering aids in the Si_3N_4 powder compacts is a crucial step in the fabrication of Si_3N_4 ceramics. Liquid–phase sintering experiments suggest that inhomogeneities introduced at this stage of processing remain in the material and may even become exaggerated during subsequent processing [17]. As inhomogeneities are likely to inhibit sintering and to degrade the mechanical properties and reliability of Si_3N_4 based materials, it is desirable to find powder processing routes that produce green compacts of exceptional homogeneity in order to produce materials with optimum properties.

Considerable advantages over dry and wet mechanical mixing routes are to be gained by using wet chemical methods, such as hydroxide precipitation from inorganic or organic salt solutions or from metalalkoxides [18]. At low pH precipitation of salts such as nitrates or oxalates by removing of the aqueous solvent may be used for homogeneous distribution of oxide additives, which are formed after thermal decomposition of the nitrate or oxalate, respectively, at temperatures < 600 °C. The finding that Si_3N_4 dispersions can best be formed at high pH values suggested that the alkalinity of such a dispersion could be used to form a metal hydroxide precipitate in situ by blending with an appropriate metal salt solution. Thus, hydroxides of Mg and Y were precipitated from nitrate solutions in $(C_2H_5)_4NOH$

$$Mg^{2+} + 2\,OH^- \longrightarrow Mg(OH)_2\downarrow, \qquad (11)$$

$$Y^{3+} + 3\,OH^- \longrightarrow Y(OH)_3\downarrow, \qquad (12)$$

where Y can also replaced by other metallic elements such as Al and rare earth elements [11]. Hydrolysis is limited to pH values above the critical value of approximately 8 as defined by the intercept of cfc and the concentration of dissociated metallic ions such as given for trivalent Y(3+) and Al(3+) anions in Fig. 5. Hydrolysis of metallalkoxides, single, double or polymetallic alkoxides, from alcoholic suspensions may be used for homogeneous distribution of sintering additives as well, for example

$$Al(OC_3H_7)_3 + H_2O \longrightarrow Al(OH)_3\downarrow + 3\,C_3H_7OH \qquad (13)$$

$$Mg\left[Al(OC_2H_5)_4\right]_2 + H_2O \longrightarrow Mg(OH)_2.Al(OH)_3\downarrow + 8\,C_2H_5OH \qquad (14)$$

The hydroxides are subsequently transformed to oxides by careful calcination of the dried powders at temperatures between 800 – 1200 °C. Coating of the Si_3N_4 particles with the sintering additives allows significant reduction of additive content because of homogeneous mixing on a molecular scale. Stabilization of coated Si_3N_4 particles in liquid suspensions may become easier, with the multicomponent powder mixture being then treatable as a single phase powder. Thus, heteroflocculation caused by particles with opposite surface charge in a multiphase powder mixture may be avoided. Future efforts will be directed on multifunctional surfactant polymers, which are able to stabilize the aqueous or non–aqueous suspension, supply the sintering additive elements and increase green strength as a binder phase. Thus, polymers based on advanced siloxanes/silazanes seem to be of high potential for improving the wet chemical processing of Si_3N_4 powders.

SLIP CASTING

Slip casting offers the advantage of being a near net shape consolidation process whereby green compacts can be prepared directly from wet mixtures without the necessity of a dry processing step that may lead to uncontrolled agglomeration of the powder. Whiskers and fibers can be aligned during casting or extruding which is of particular interest for composite materials with enhanced toughness. Slip casting of aqueous suspensions at pH > 9 is to be preferred because of a maximum zeta potential, decreasing concentration of positively charged counterions with increasing OH(–)–concentration according to Equ. (8) and adsorption of basic polyelectrolytic surfactants. Fig. 6 shows shear rate / shear stress curves of highly concentrated aqueous Si_3N_4 slips as a function of zeta potential. Viscosity of aqueous Si_3N_4 slips is significantly reduced with increasing zeta potential and no yield stress occurs in well deflocculated suspensions even at high solid concentrations > 40 vol.%. Final densities of > 65 % of the theoretical density could easily be attained in the green compacts. Pressure slip casting may further reduce the casting time, especially for larger components [19]. Drying of cast compacts with high packing density and small pore size may become a severe problem, because high capillary forces can cause cracking during drying. Thus, new polymeric surfactants may also help to reduce these forces by reduction of the surface tension of water. Rate controlled drying in a controlled atmosphere or drying by sublimation of the frozen liquid are being developed as alternative drying techniques, especially for larger components of complex shape.

SUMMARY

Powder processing of Si_3N_4 ceramics is characterized by the increasing use of new and extremely fine powders of high purity. Colloidal processing of well dispersed Si_3N_4 suspensions in aqueous or non–aqueous solutions is focused on the fabrication of green compacts with chemical homogeneity and high packing density. Stability against flocculation in aqueous suspensions primarily depends on a high zeta potential which may be increased at high pH by preferential adsorption of basic polyelectric surfactants on preoxidized nitride surfaces. Non–aqueous processing is focused on systems where only small van der Waals forces develop like in some aromatic hydrocarbons. Precipitation mixing may result in improved chemical homogeneity and minimized additions of sintering additives. Colloidal filtration and consolidation techniques have proven their advantage over pressing techniques for the formation of dense packed green compacts with better sintering properties. Future efforts will be directed on the development of multifunctional surfactants for aqueous and non–aqueous suspensions in order to facilitate the powder processing and to reduce potential sources for defect formation during processing.

Literature

[1] S. Futaki, Y. Shimizu, K. Shirasaki, Y. Moriyoshi, T. Sato, T. Sakai, J.Mat.Sci. 22 (1987) 4331.
[2] P.D. Shalek, J.J. Petrovic, G.F. Hurley, F.D. Gac, Bull.Am.Ceram.Soc. 65 (1986) 351.
[3] M. Hoch, K.M. Nair, Ceram.Bull. 58 (1979) 187., 191.
[4] D.L. Segal, Br.Ceram.Trans.J. 85 (1986) 184.
[5] S.C. Danforth, J.S. Haggerty, J.Am.Ceram.Soc. 66 (1983) C 58.
[6] M. Peuckert, P. Greil, J.Mat.Sci. 22 (1987) 3717.
[7] P.K. Whitman, D.L. Feke, Adv.Ceram.Mat. 1 (1986) 366.
[8] R. Hunter, Zeta Potential in Colloid Science, Academic Press, London (1981)
[9] A. Nagel, PhD thesis Univ. Stuttgart 1988
[10] M. Persson, L. Hermannsson, R. Carlsson, in Ceramic Powders, ed P. Vincencini, Elsevier Publ., Amsterdam, Neth. (1983) 735.
[11] T.M. Shaw, B.A. Pethica, J.Am.Ceram.Soc. 69 (1986) 88.

[12] P.C. Hiemenz, in Principles of Colloid and Surface Chemistry, ed. J.J. Lagowski, Marcel Dekker Inc.,New York (1977)
[13] G.D. Parfitt, Dispersion of Powders in Liquids, Elsevier Publ., Amsterdam, Neth. (1986)
[14] J. Ceserano III, I.A. Aksay, A. Bleier, J.Am.Ceram.Soc. 71 (1988) 250.
[15] E.F. Adams, in High–Temp. Oxides Part IV, Academic Press, New York (1971)
[16] A. Bleier, J.Am.Ceram.Soc. 66 (1983) C 79.
[17] D.R. Clarke, J.F. Flinthoff, in Proc. MRS Soc.Meet, Elsevier Publ., Boston, MA (1983) 29.
[18] H. Okamura, H.K. Bowen, Ceram.Int. 12 (1986) 161.
[19] F.F. Lange, K.T. Miller, Am.Ceram.Soc.Bull. 66 (1987) 1498.

Table I Oxygen distribution in various commercial Si_3N_4 powders as derived from bulk chemical and surface sensitive XPS analyses [5]

Powder preparation process	number of powders	total oxygen content (wt.%)	fraction of surface oxygen (wt.%)
$3Si + 2N_2 \longrightarrow Si_3N_4$	6	1.3 ± 0.2	30 ± 15
$3SiO_2 + 6C + 2N_2 \longrightarrow Si_3N_4 + 6CO$	2	2.5 ± 0.1	40 ± 20
$3Si(NH)_2 \longrightarrow Si_3N_4 + 2NH_3$	2	1.1 ± 0.3	70 ± 20

mean specific surface area of the powders 12 ± 3 m^2/g

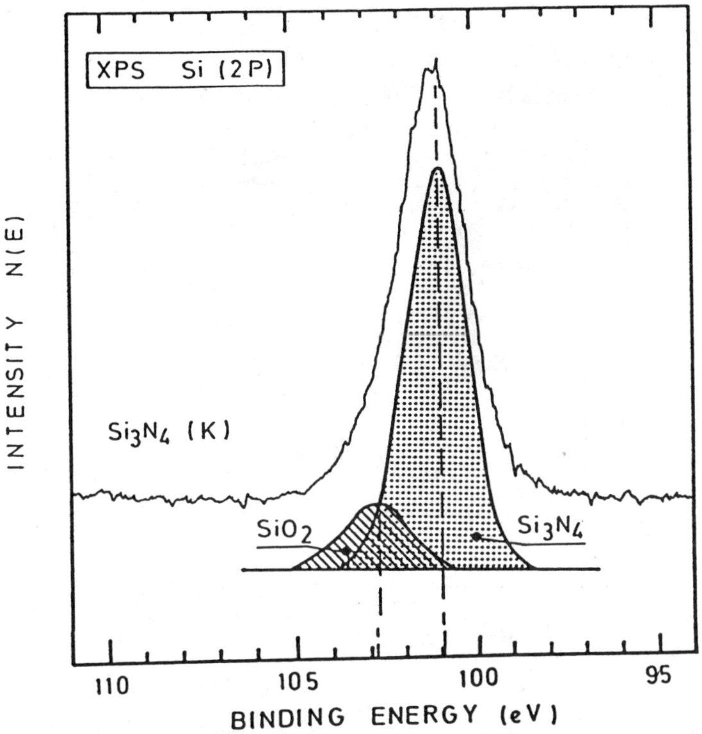

Fig.1 X–ray photoelectron spectrum of a α–Si_3N_4 powder (E–10, UBE) in the energy range 95 – 110 eV [6].

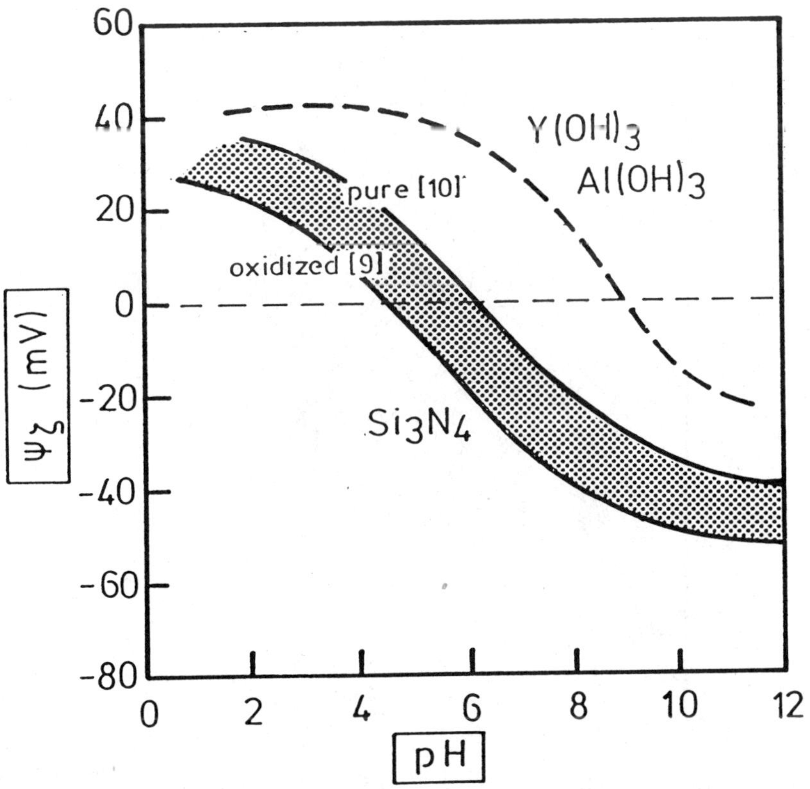

Fig.2 Zeta potential of pure and preoxidized (550 °C, 8h) Si_3N_4 powders [9,10] and the hydroxides of Y and Al in aqueous suspensions as a function of pH .

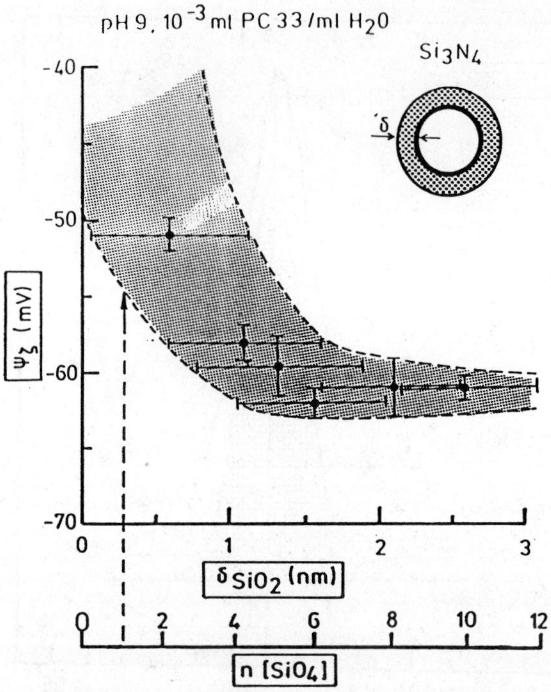

Fig.3 Zeta potential of Si_3N_4 powder in aqueous suspension at pH 9 as a function of surface oxide layer thickness (assuming SiO_2 composition) as derived from XPS measurements.

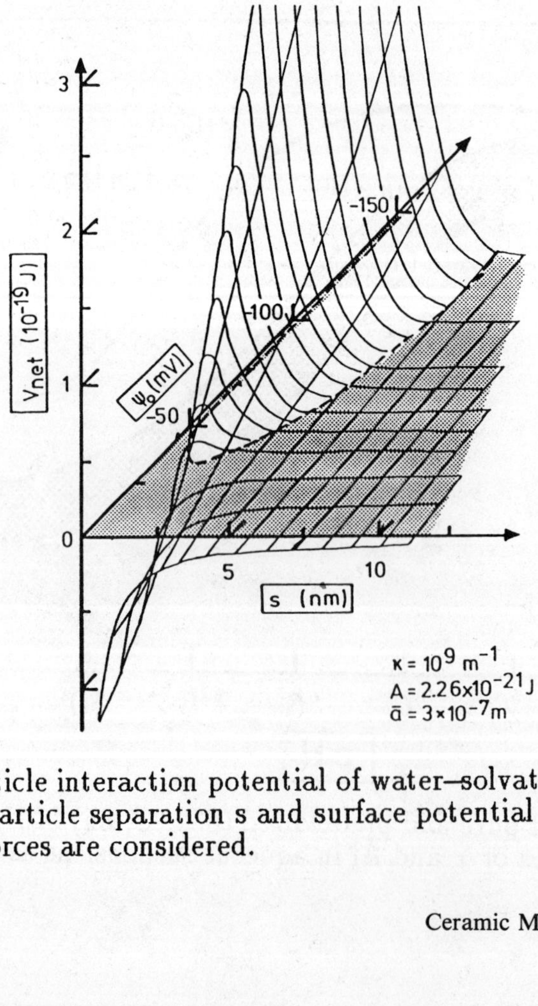

Fig.4 Particle–particle interaction potential of water–solvated Si_3N_4 particles as a function of particle separation s and surface potential Ψ_0. Only electrostatic forces are considered.

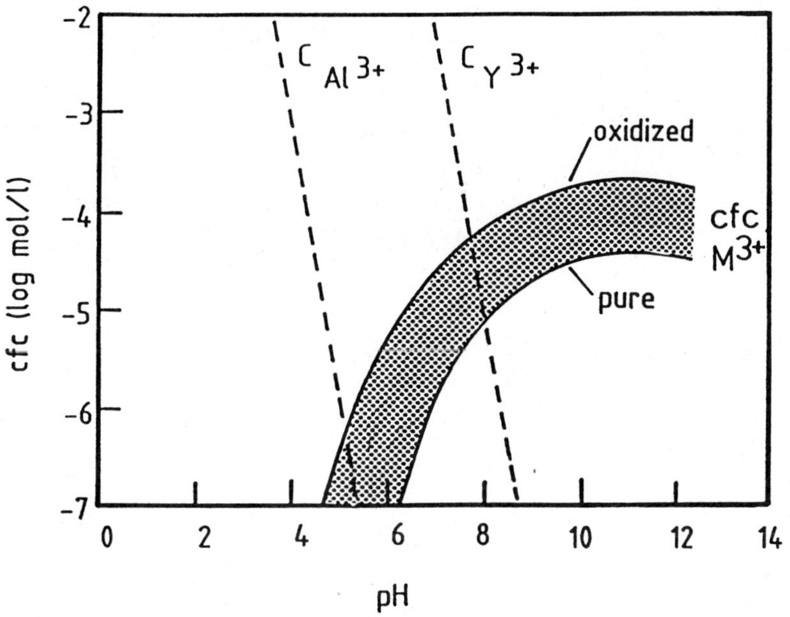

Fig.5 Estimation of criticial floccultation concentrations of cationic counter-ions in aqueous Si_3N_4 suspensions as a function of pH.

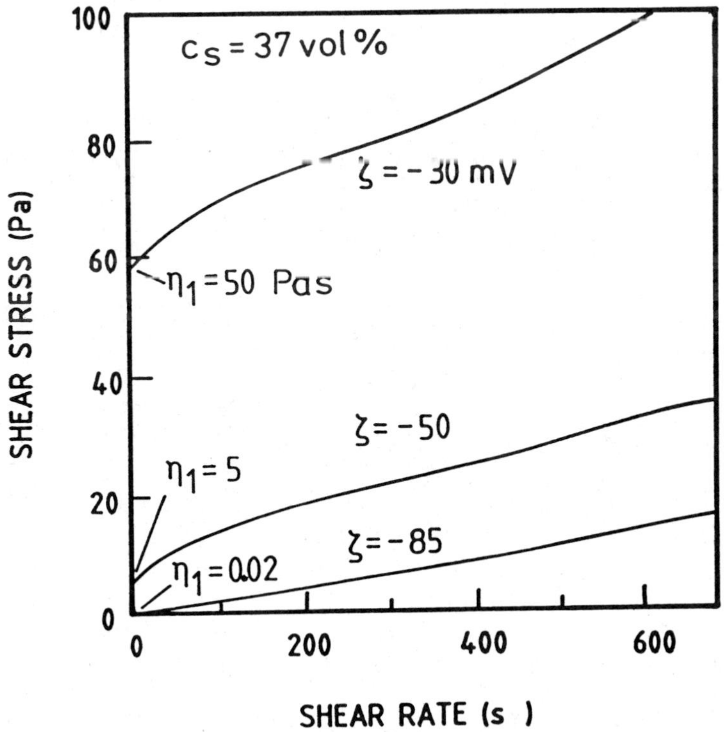

Fig.6 Flow curves of highly concentrated aqueous Si_3N_4 slips for three different zeta potential regimes.

PREPARATION OF CERAMIC COMPONENTS WITH COMPLEX SHAPE BY SLIP CASTING

FAN QISHENG, ZHAO HUEIQIN, ZHAO CHUANYI, AND GU PEIZHI

Shandong Institute of Industrial Ceramics,
Zi-Bo, Shandong, China.

ABSTRACT

The effect of pH, suspension additives, and solid loading on the viscosity of Si slurry with spinel MgO-Al_2O_3 (MA) and ZrO_2 as sintering aids has been investigated. Silicon powders with different particle size distribution were used for the study and the effect of particle size distribution was discussed. An aqueous slurry of Si powder with 10 wt% MA and 10 wt% ZrO_2 additives was used to slip cast components with complex shape such as engine rotors and valves. A slip cast body of an engine rotor with a homogeneous density distribution and 62% theoretical density was obtained.

INTRODUCTION

Due to its excellent properties such as high strength and oxidation resistance, silicon nitride based materials have been developed for applications in heat engines (1,2,3). Silicon nitride powder or silicon powder can be shaped by several methods: dry pressing, isostatic pressing, injection molding and slip casting. For complex shaped engine components, slip casting and injection molding are the best choices. Slip casting has advantages (4) compared to injection molding for making complex shape components because of higher densities, more homogeneous microstructure and lower cost. For example, slip casting combined with a fugitive wax process has been developed to make the sintered silicon nitride radial gas turbine rotor for the AGT 101 program (5,6).

Post sintered reaction bonded silicon nitride (PSRBSN) is one of approaches used to make silicon nitride based ceramics since this process has smaller shrinkage compared to sintered silicon nitride, thereby, reducing the formation of flaws and cracks during sintering.

The slip casting process is strongly influenced by the properties of the silicon powder and sintering additive suspension. Thus Ezis (6) and Sacks (7,8) have evaluated the effect of suspension preparation variables (i.e. solid loading, pH, ultrasonication) on both the silicon suspension properties and the resulting porosity of the cast bodies.

The objective of the present study is to investigate the effects of several parameters (i.e. pH, solid loading, type and amount of suspension additives and particle size distribution) on the properties of an aqueous suspension of Si with sintering additives of MA and ZrO_2 and to produce complex shaped components such as engine rotors and valves by slip casting.

EXPERIMENTAL

Material and characterization

The silicon powder "A" used in this study was dry ball-milling 1-3 mm Si powder for 20 hours in a steel mill with steel balls, using a charge ratio of 1:10 and a small amount of oleic acid as a dispersant. In order to obtain a gas free suspension, the resulting powder was aged in distilled water for two weeks. The resulting powder contained approximately 1.5 wt% ferric oxide. The silicon powder "B" was prepared by subjecting powder "A" to two hours additional vibration-milling. This powder was round contain about 2 wt% ferric oxide then as with powder "A", powder "B" aged in distilled water for two weeks to minimize out gassing.

Spinel $MgO-Al_2O_3$ synthesized from $MgCO_3$ and $Al(OH)_3$ at 1600 C in present laboratory and commercial ZrO_2 were employed as sintering additives.

The particle size distribution of silicon powder "A" and "B" were measured by a Centrifugal Particle Size Analyzer and are given in Fig.1 (curve A and B). Specific surface areas of spinel MA, ZrO_2 and powder "A" measured by SORPTY 1750 Rapid Surface Area Analyzer based on B.E.T. equation are 18 m^2/g, 23 m^2/g and 3 m^2/g respectively*.

Preparation and properties of aqueous silicon slurry

Slurry for slip casting was prepared by ball-milling. Mixtures of the silicon, MA, and ZrO_2 powders were mixed in a distilled water media for 1 hour. Ammonia and HCl solutions were used to adjust the pH value of the suspensions. The slurry was transferred to a stainless steel container and degassed with vacuum.

Viscosity measurement was performed at 20 C by a rotary viscometer at shear rate of 300 s^{-1}.

Mold and slip casting

The mold for slip casting was made from plaster and/or other porous materials with different porosity and pore size depending upon the positions in the mold. The slurry was pressed into the casting mold through a pipe by high pressure N_2 gas. The mold was contained in a sealed steel container to prevent leakge of the slurry during high pressure casting.

The densities of the dried cast bodies were determined with a Davidge densimeter by mercury immersion technique. The microstructure of cast bodies was investigated by SEM.

RESULTS AND DISCUSSION

Effect of pH value on viscosity and stability of slurry

Aqueous suspensions of non-plastic materials, such as Si tend to flocculate, so that controlling the stability and flowability of the suspensions by adjusting pH value of suspension and adding suspension additives is very important. The slurry mainly used in this study was produced from 80 wt% of silicon powder "A", 10 wt% of spinel MA powder and 10 wt% of ZrO_2 powder. The particle size and specific surface area of those materials have given above.

Because the silicon particles can be either positively or negatively charged, surface force on silicon particles and the Zeta potential will be changed when the pH of the slurry changes. This will result in changes of stability and flowability of the slurry. The relationships between viscosities and pH of the slurries studied are given in Fig.2. Curves A and B in Fig.2 are for slurries prepared from powders "A" and "B" respectively. For slurry A, viscosity decreases as the pH increases in pH range of 7.0 to 9.0, but it increases as pH increases in pH range of 9.0 to 10.0. This agrees with Sacks's result (7). Slurry A appears to be deflocculated best in pH range of 8.0 to 9.0. The reason for this is believed to be due to a high Zeta potential in this pH range (i.e. interparticle electrostatic repulsive forces between particles were large). Lower stability of the slurry with increased flocculation was observed to occur at both lower and higher pH.

Point S in Fig.2 shows Sacks's result which was obtained under conditions similar to those used in this study except the slurries in this study contain the sintering aids MA and ZrO_2 . It has been found (9) that the addition of MgO to either Si or silicon nitride suspensions strongly effects the rheological behavior of above suspensions. However the results of this study suggest that spinel MA does not significantly change the properties of the suspension.

Effect of particle size distribution on suspension viscosity and green density

As can be observed in Fig.1 powders "A" and "B" had different particle size distributions. The solid content of the slurries made from those powders differed markedly. Curve B in Fig.2 shows the viscosities of slurry B which has lower solid loading (39 vol%) compared with slurry A with higher solid loading (47 vol%). Due to its finer particle size if the solid content of powder "B" was increased to that used for

* The specific surface area of powder "B" was not measured.

powder "A", the viscosity of slurry B was too high to be measured. Curve B shows that within the pH range between 5.0 and 11.0 the viscosity of slurry B increases only slightly as the pH increases, indicating that the effects of Zeta potential and ionic strength on viscosity of slurry have became less important under the condition of lower solid content. This suggests that slurries with lower solid loading can be used for slip casting over a wide pH range which might lead to more ideal particle packing, higher cast density and a more free-flaw cast body. However, this does not appear to be the case since the density of the cast body from powder "B" with lower solid loading is lower (53% of T.D.) than that from powder "A" with higher solid loading (62% of T.D.). Differences in particle size distribution between powder "A" and powder "B" probably lead to the cast density variations. It has been noted that the maximum green density for slip casting article can be achieved when the particle size distribution follows the relationship (10):

$$P = 100(X/D)^n \qquad (1)$$

where P is the percent finer than any particle size X and D is the largest particle size in the distribution and n is a constant between 0.33 and 0.5. Ranges A' and B' in Fig.1 show the ideal particle size distribution ranges for obtaining the maximum green density which obey Eq.1 with the largest particle size contained in powder "A" and "B", to be 40 μm and 10 μm respectively. Comparing curve A with range A', it appears that powder "A" containing a higher percent of larger size particles and lower percent smaller particles than required by Eq.1. For powder "B", difference between curve B and range B' exists mainly in the middle particle size range. It is reasonable that the higher percent of middle size particles will decrease the density more seriously due to the lacking of smaller particles which can fill the interstices formed by the particles. Powder "A" and "B" were produced by ball-milling and vibration-milling respectively. This suggests that longer ball-milling of powder "A" may have modified the particle size distribution while vibration-milling tends to give a narrow particle size distribution which is not suitable for slip casting.

Suspension additive effect

Several types of inorganic or organic salts such as sodium silicate, sodium carbonate, sodium alginate, sodium tanninate, and sodium humate etc. can be added into Si slurry to achieve stable suspensions and thereby improving the flowability. Two organic salts were used in this study. Results are given in Fig.3. For suspensions with low concentrations of additive A, sodium alginate, (0.3-0.5 wt%), the viscosity decreases significantly as the additive content increases and for suspensions with higher concentrations of additive A (0.5-0.9 wt%), the viscosity increases as additive content increases. Additive B, sodium humate, behaves similar to those containing additive A, except the lowest viscosity occurs at 0.7 wt%. For suspensions with additivea A B, the lowest viscosity appears in 0.5 wt% and all values of viscosity are higher than those of suspensions containing additive A and additive B seperately, but the suspensions with additives A B are more stable. The cast body from this slurry is more easily removed and have higher strength.

Effect of solid loading and casting pressure on green density

The relationships between solid loading, casting pressure, and green density were investigated and results are shown in Fig.4 and Fig.5. Green density increases slightly as the solid loading in the slurry increases, with the highest green density (62% of T.D.) being obtained from 47 vol% solid loading. Green density also increases only slightly as applied casting pressure was increased from 0.2 to 0.8 MPa.

These results differ from other forming methods, such as dry pressing and injection molding, where green density is strongly effected by formimg pressure and solid content. It seems that the dispersed state of the suspension has provided suitable conditions for maximum packing and rearrangement of particles during casting, so that the application of pressure has little effect. However, the cast time was reduced and the cast body more easily removed from the mold when casting pressure was applied.

Fabrication of components by slip casting

Components with complex shape such as engine rotors and valves were prepared by slip casting with a Si slurry which contained spinel MA and ZrO_2. Such component is shown in Fig.6.

Mold

In the beginning, all parts of mold for slip casting were made from high strength plaster with a plaster:water ratio of 1:0.6. Because the thickness of the cast body is a variable depending upon the locations (the smallest thickness is 2 mm, the largest thickness is 60 mm of a rotor for example), it is quite difficult to obtain good components of a complex shape with a mold in which all parts were made from same material. If casting time is too long, the parts with smaller thickness will crack and if casting time is too short, the parts with larger thickness are still not solid. To solve this problem different porous materials were employed to make different parts of the mold and a thin film with smaller pore size was coated on the surface of some parts of the mold to control the casting rate. The mold structure for an engine rotor is shown in Fig.7.

Controlling of casting parameter

To avoid introducing bubbles into cast bodies, the rate of pouring slurry into the mold must be slow, and the slurry should be vaccum treated prior to casting. Fig.8 shows a typical void in a cast body caused by pouring slurry into mold too fast. After the mold was filled with slurry, the casting pressure was applied and increased in two steps. The period of time that the cast pressure is applied is important for the homogeniety of the cast body. In this study the green bodies of engine rotor with uniform density in the blades and the center of the rotor were obtained.

SUMMARY

As has been discussed, the addition of sintering aid MA does not significantly alter the properties of Si suspensions. Ball-milling produces a powder suitable for slip casting while vibration-milling tends to give a narrow particle size distribution which is less ideal. Casting pressure is less important to achieving high green density than suspension properties and solid loading. The slip-cast body of an engine rotor with 62% of T.D. and uniform density distribution has been obtained using plaster and/or other porous material mold. The results of this study show that slip casting is a candidate for fabrication of ceramic components with complex shape.

REFERENCES

1) D.J.CASSIDY and A.F.McLEAN "Advances and tends in processing complex ceramic components for heat engines", 6th CIMTEC, proceeding, P.2503, 1986.

2) D.W.RICHERSON "Processing, Microstructure, Properties Relationships for Automotive Structure Ceramics", Ceram.Eng. & Sci. Proc. Sept/Oct. 1986, P.1122.

3) J.P.TORRE, Y.BIGAY "Fabrication of Silicon Nitride Parts By Slip Casting", Ceram.Eng.& Sci.Proc. July/August 1986, P.893.

4) F.L.KENNARD "Ceramic Components Fabrication" Ceram.Eng.& Sci.Proc. Sept./Oct. 1986 P.1095.

5) A.EZIS, J.M.NICHOLSON "Method of Manufacturing a Slip Cast Article" U.S. Patent No: 4,067,943.

6) A.EZIS,J.T.NELL "Slip Casting of Silicon Based Ceramics" Am. Ceram. Soc. Bull. vol 58 P.883.

7) M.D.SACKS "Properties of Silicon Suspensions and Cast Bodies", Am. Ceram. Soc. Bull., vol 63 P.1510 1984.

8) M.D.Sacks etc."Properties of Silicon suspension and slip cast bodies", Ceram.Eng.& Sci. Proc., vol 6 P.1109 1985.

9) Fan Qisheng etc. "The effect of Adding MgO as a Sintering

Additive on Rheological Properties of Aqueous Silicon Suspensions", unpublished.

10) E.F.ADAMS "Slip-cast Ceramics", P.145-84 in High Temperature Oxide, Part lV, A.APLER editor, Academic Press, New York.

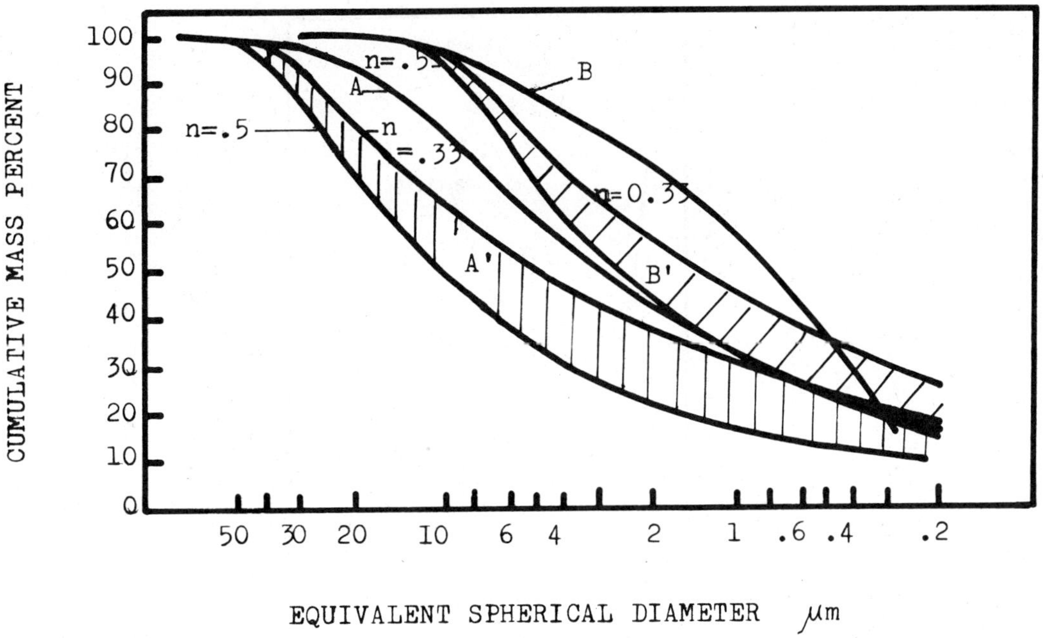

Fig.1. Particle size distribution of Powder "A" and "B".

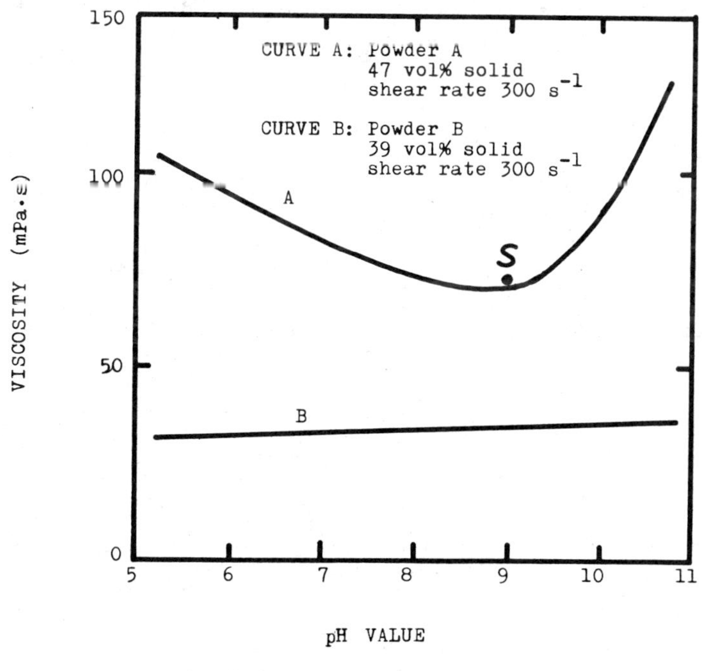

Fig.2. Plots of viscosity vs pH value for slurries A and B.

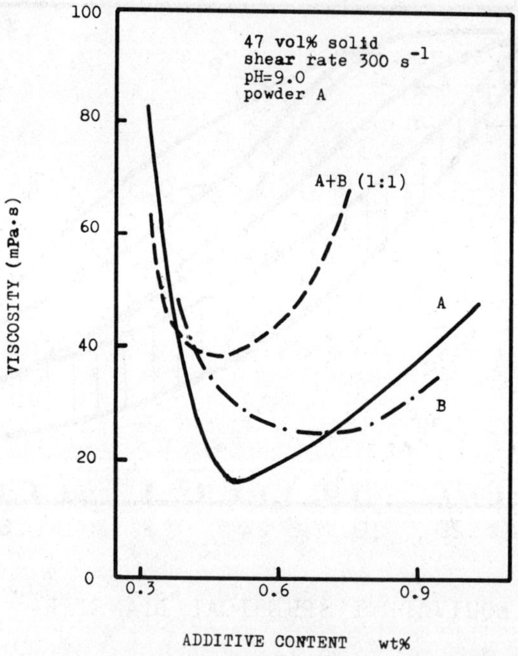

Fig.3. Plots of viscosity vs suspension additive content.

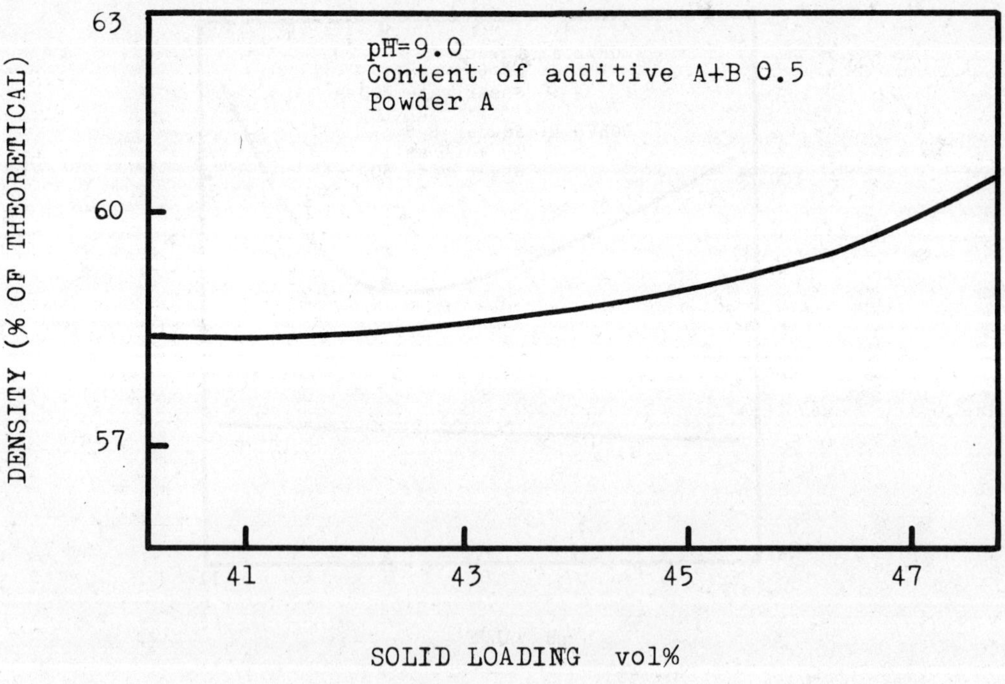

Fig.4. Plot of green density vs solid loading.

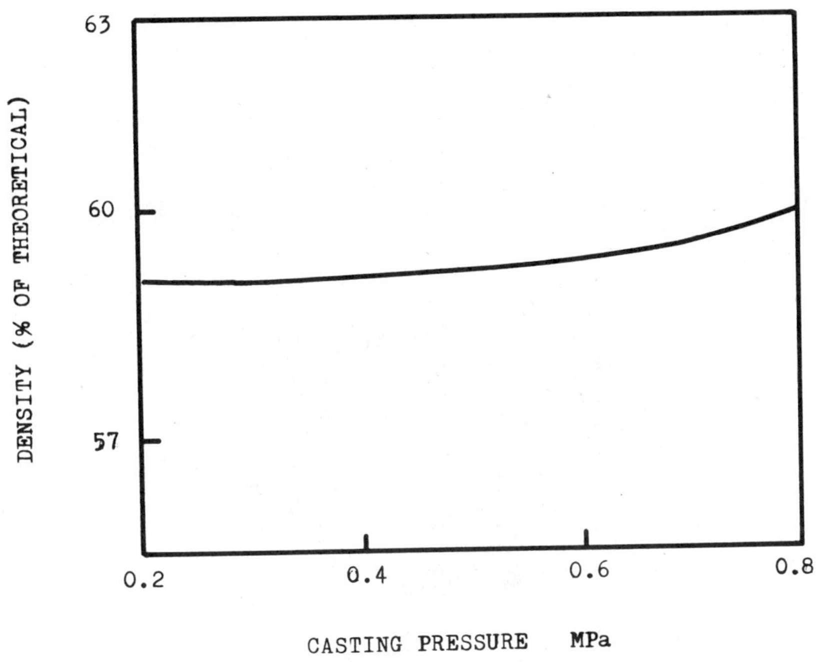

Fig.5. Plot of green density vs casting pressure.

Fig.6. Slip casted engine rotor.

Fig.7. Mold for engine rotor.

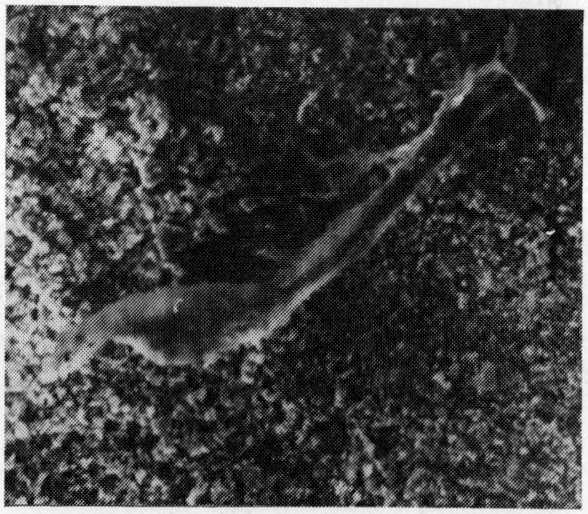

Fig.8. A void cast body.

OXIDATION MECHANISM OF ZrN AND ZrC POWDERS BY HIGH-TEMPERATURE HIGH-PRESSURE WATER

Masahiro YOSHIMURA, Makoto HAYAKAWA and Shigeyuki SŌMIYA
Research Laboratory of Engineering Materials, Tokyo Institute of Technology, 4259 Nagatsuta, Midori, Yokohama 227, Japan

ABSTRACT

The oxidation mechanism of ZrN and ZrC powders by high-temperature high-pressure water was investigated by using hydrothermal equipment. ZrC reacted with 100 MPa H_2O to yield monoclinic ZrO_2 and CH_4 above 240°C for 3h, while ZrN yielded monoclinic ZrO_2, NH_3 and H_2 above 190°C for 3h. The reactions were controlled by phase-boundary reactions. The produced ZrO_2 consisted of fine particles with about 10 nm and did not form a protective layer for ZrN and ZrC particles. They are different from the cases of Si_3N_4 and SiC powders where amorphous SiO_2 covers Si_3N_4 and SiC particles as a protective layer, in which the diffusion processes of oxidant and/or produced species control the oxidation.

INTRODUCTION

The oxidation/corrosion is one of the most serious failure in the practical use of carbides and nitrides as structural materials or cutting tools. We have been studying the effect of H_2O on the oxidation mechanism and kinetics of these materials. To emphasize H_2O, we have used high-temperature high-pressure water. Previously we reported the oxidation of Si_3N_4 and SiC by high-temperature high-pressure water(1-4). In the cases of Si_3N_4 and SiC, the reactions proceeded as follows;

$$Si_3N_4 + 6H_2O \longrightarrow 3SiO_2 + 4NH_3 \quad [1], \text{ and}$$

$$SiC + 2H_2O \longrightarrow SiO_2 + CH_4 \quad [2].$$

They are quite different from the oxidation in air or oxygen as follows;

$$Si_3N_4 + 3O_2 \longrightarrow 3SiO_2 + 2N_2 \quad [3], \text{ and}$$

$$SiC + 2O_2 \longrightarrow SiO_2 + CO_2 \quad [4].$$

The solid product is the amorphous SiO_2 which covers Si_3N_4 or SiC particles as a protective layer. Thus the rate-determined process is the diffusion process through the amorphous SiO_2. In contrast to this the cases of ZrN and ZrC are quite different as described below.

EXPERIMENTAL PROCEDURE

The starting materials were ZrN powder; Hermann C. Starck, -325 mesh, Zr:N= 1:0.91 (molar ratio), and ZrC powder; Hermann C. Starck, -325 mesh, average grain size= 3.2 μm (from the relative surface area), Zr:C= 1:0.98 (molar ratio). The transmission electron microscopy proved the grain size of ZrN powder to be about 1 μm and ZrC to be about 3 μm. Each sample weighed out about 100 mg was placed in an Au capsule of which top was narrowed by pressing to prevent the escape of solid materials but to allow the passing of fluids. The capsule was heated in a test-tube-type pressure vessel (volume 5.5 ml) under the pressure of 100 MPa of redistilled water at the temperatures of between 190° and 400°C for 1 to 16 h. After the treatment the sample in the pressure vessel was quenched by immersing into cold water, then the sample capsule was dried and weighed. Solid products were examined by X-ray diffraction and transmission electron microscopy. The oxidation rates were calculated from the weight gain or the integrated peak intensity of X-ray diffraction and compared with various model functions to determine the kinetics. Nessler's reagent was added to water remaining in the vesselto confirm whetherammonia was produced or not. Other gaseous products, collected above water through the drain valve, were analyzed by a gas chromatography.

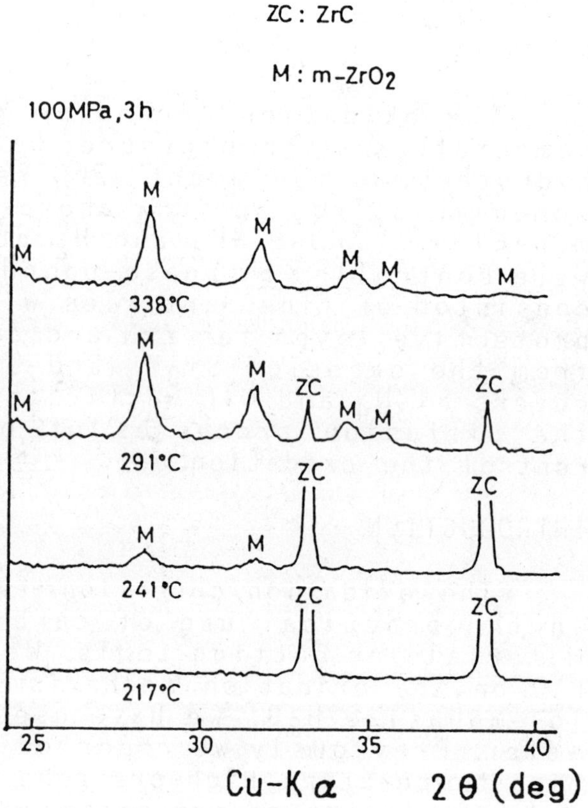

Fig. 1. X-ray powder diffraction pattern of the products by the hydrothermal oxidation of ZrC at various temperatures for 3 h under 100 MPa.

RESULTS AND DISCUSSION

ZrC reacted with H_2O to yield monoclinic ZrO_2 above 240°C for 3h, and above 340°C the reaction proceeded completely (Fig.1). The gaseous products were a large amout of CH_4 and small amout of H_2 (Fig.2), which must be due to the non-stoichiometry of zirconium carbide as the equation [5];

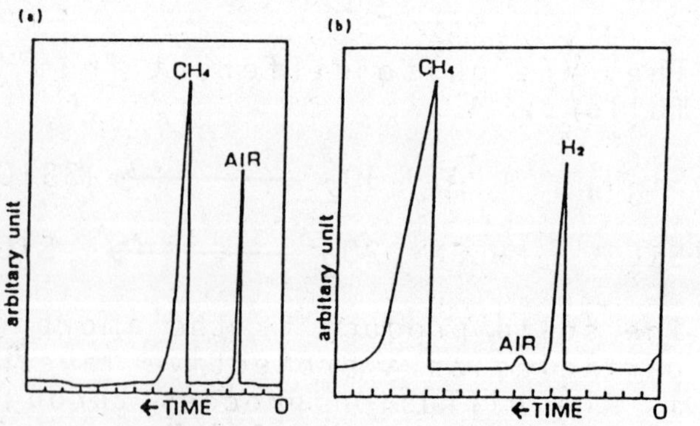

Fig. 2. Gas chromatography patterns of gaseous products by the hydrothermal oxidation of ZrC.
(a) carrier=He, column=activated carbon
(b) carrier=Ar, column=molecular sieve

$$ZrC_{1-X} + 2H_2O \longrightarrow ZrO_2 + (1-X)CH_4 + 2XH_2 \quad ---[5].$$

Among the following reactions[6-8], reaction[6] which has the largest negative free energy change appears to proceed in the conditions studied, where a large amount of excess H_2O remains in the system. These reactions, however, are competitive reactions, thus one must consider total free energy change for these reactions; to evaluate the thermochemical equilibrium.

$$ZrC + 2H_2O \longrightarrow ZrO_2 + CH_4 \quad ---[6],$$

$$ZrC + 3H_2O \longrightarrow ZrO_2 + CO + 3H_2 \quad ---[7],$$

$$ZrC + 4H_2O \longrightarrow ZrO_2 + CO_2 + 4H_2 \quad ---[8].$$

The equilibrium is accomplished when the free energy of the whole reaction system is to be minimum,

$$\Delta G = x\Delta G[6] + y\Delta G[7] + z\Delta G[8] + \Delta Gm \quad -------------[9],$$

where x, y and z are the ratios of which the reaction[6], [7] and [8] occurs respectively, ΔG is the free energies of the reaction system, $\Delta G[6]$, $\Delta G[7]$ and $\Delta G[8]$ are the free energies of the reaction[6], [7] and [8] respectively, and ΔGm is the free energy of the mixing, which is expressed by the following equation;

$$\Delta G = RT\ X(I)\ln X(I) \quad ------------[10],$$

where R is the gas constant, T is the temperature, and X(I) is the molar ratio of the produced gas(I) expressed by the following equation;

$$X(CH_4) = x/(x+4y+5z),\ X(H_2) = (3y+4z)/(x+4y+5z),$$
$$X(CO) = y/(x+4y+5z),\ \text{and}\ X(CO_2) = z/(x+4y+5z) \quad ---[11].$$

ΔG depend on the gas pressure;

$$\Delta G[6] = G(ZrO_2) + G(CH_4) - G(ZrC) - 2G(H_2O) + RT\ln\ (f_{CH_4}/f_{H_2O}{}^2) --[12],$$

$$\Delta G[7] = G(ZrO_2) + G(CO) + 3G(H_2) - G(ZrC) - 3G(H_2O) + RT\ln\ (f_{CO}f_{H_2}{}^3/f_{H_2O}{}^3) \quad --[13],$$

$$\Delta G[8] = G(ZrO_2) + G(CO_2) + 4G(H_2) - G(ZrC) - 4G(H_2O) + RT\ln\ (f_{CO_2}f_{H_2}{}^4/f_{H_2O}{}^4) \quad --[14].$$

where G(J) is the free energy of formation of the substance J under 1 atm, f is fugasity. In these condition we obtained the values of x, y and z which minimize ΔG to be $x=1.0$, $y=4.16\times10^{-10}$ and $z=2.5\times10^{-3}$ at 500K, respectively. Therefore we conclude the reaction [5] occurs in the conditions studied.

It should be noted that the oxidation by H_2O is quite different from the oxidation by oxygen(5,6) as the following equation;

$$ZrC + 2O_2 \longrightarrow ZrO_2 + CO_2 \quad ---[15].$$

ZrN reacted with H_2O to yield monoclinic ZrO_2 above 190°C for 3h, above 340°C the starting materials, ZrN, were disappeared (Fig.3). A very small peak appeared around $2\theta=30°$ between 190° and 400°C may indicate the formation of cubic phase, which contains nitrogen. The production of NH_3 was confirmed with Nessler's reagent. Other gaseous products in the vessel were proved to contain a large amount of H_2 but not to contain N_2 by gas chromatography. As a summary, the reaction was represented as follows;

$$2ZrN + 4H_2O \longrightarrow 2ZrO_2 + 2NH_3 + H_2 \quad ---[16].$$

This result is quite different from the oxidation in oxygen(7) or water vapor(8) above 600°C as equation [17] and [18].

$$2ZrN + 2O_2 \longrightarrow 2ZrO_2 + N_2 \quad ---[17], \text{ and}$$

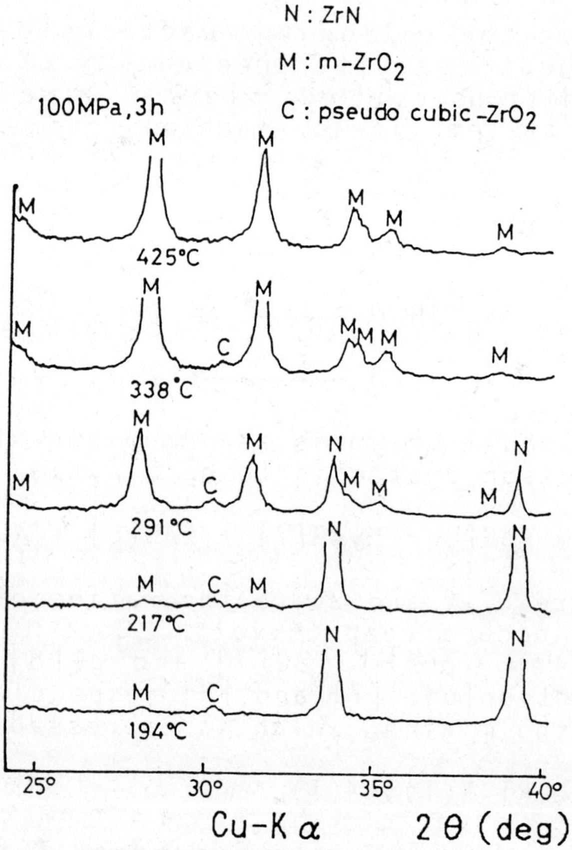

Fig. 3. X-ray powder diffraction patterns of the products of the hydrothermal oxidation of ZrN at various temperatures for 3 h under 100 MPa.

$$2ZrN + 4H_2O \longrightarrow 2ZrO_2 + N_2 + 4H_2 \quad ---[18].$$

In both cases of ZrN and ZrC, the oxidation rate calculated from the weight gain or integrated peak intensity of the X-ray diffraction was represented by a core-shrinking model (eq.[19]), not by a Jander-type model (not fitting in the early stage of the reaction) (eq.[20]);

$$[1-(1-\alpha)^{1/3}] - [1-(1-\alpha_o)^{1/3}] = kt \quad -----[19], \text{ and}$$

$$[1-(1-\alpha)^{1/3}]^2 - [1-(1-\alpha_o)^{1/3}]^2 = kt \quad -----[20].$$

where t is the time, α is the oxidation rate at t h, α_o is the oxidation rate at 0 h, and k is a constant. This indicates that the reactions were not controlled by diffusion processes but controlled by phase boundary reactions from the surface to the center of ZrC or ZrN grains. The oxidation rate was also represented by Avrami equation, when n=1 (eq.[21]);

$$-\ln(1-\alpha) + \ln(1-\alpha_o) = kt \quad -----[21].$$

This indicates that the nucleation of ZrO_2 controlled the

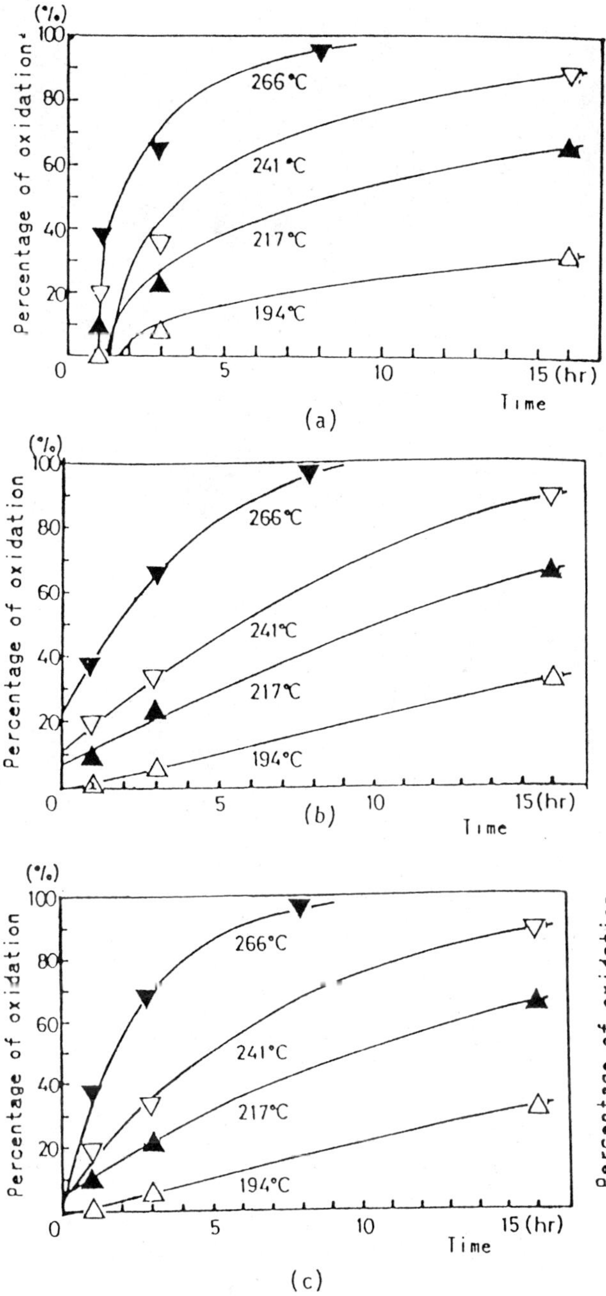

Fig.4 Isothermal oxidation of ZrN powders under 100MPa H_2O. Solid lines are drawn by;
(a) Jander-type equation,
(b) Core-shrinking model,
(c) Avrami's equation (n=1).

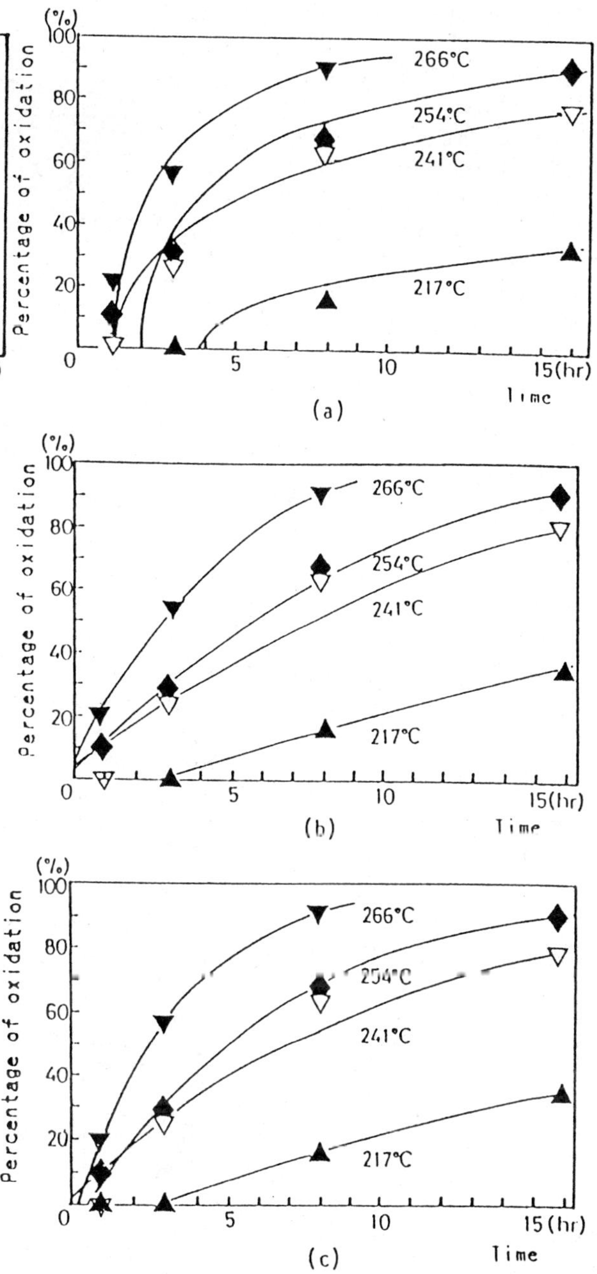

Fig.5 Isothermal oxidation of ZrC powders under 100MPa H_2O. Solid lines are drawn by;
(a) Jander-type equation,
(b) Core-shrinking model,
(c) Avrami's equation (n=1).

reaction rate and that the produced ZrO_2 nucleus little grew (Fig.4 and 5). Therefore the produced ZrO_2 seems to consist of very fine particles and does not form a protective layer for the oxidation of ZrN and ZrC powders.

Transmission electron micrographs revealed ultrafine ZrO_2 particles of about 10 nm as the solid products for the samples treated at 350°C for 3h under 100 MPa H_2O (Fig.6). It also supports the reaction-controlled mechanism rather than the diffusion-controlled one.

We propose here the reaction models for the oxidation of ZrN and ZrC powders in H_2O based on these results of model fittings and transmission electron micrographs (Fig.7). ZrN reacts with H_2O and then ultrafine ZrO_2 particles of about 10 nm nucleated on the surface of ZrN grains with about 1 μm or ZrC grains with about 3 μm, respectively. These ZrO_2 fine particles do not form a protective layer for ZrN or ZrC particles, but form clusters of particlesbecause of large volume expansions about 40% during the oxidation. The interface, where the reactions occur, of ZrN or ZrC particles goes inner to the center of the particles(9).

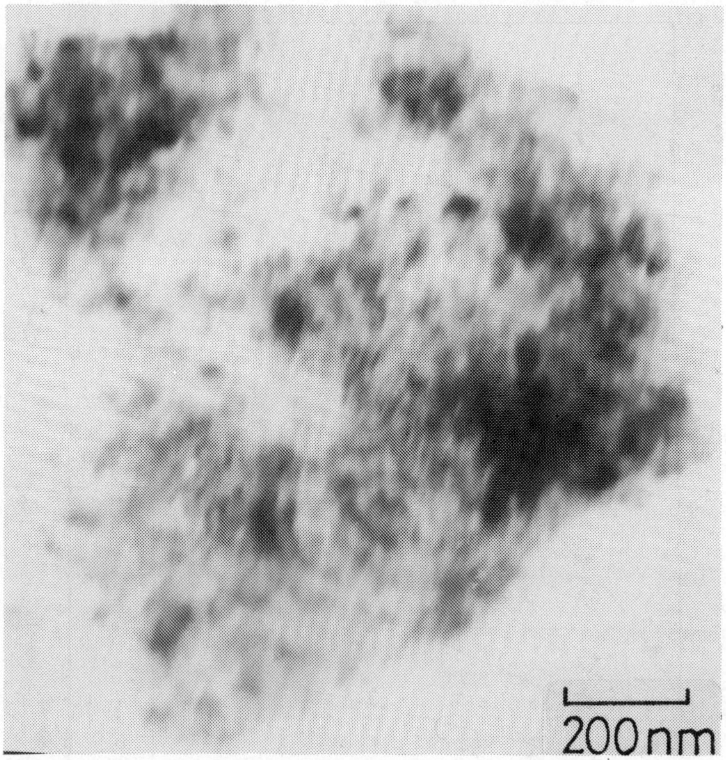

Fig.6 TEM photographs of the product after hydrothermal oxidation of ZrC powder at 350°C under 100MPa for 3h.

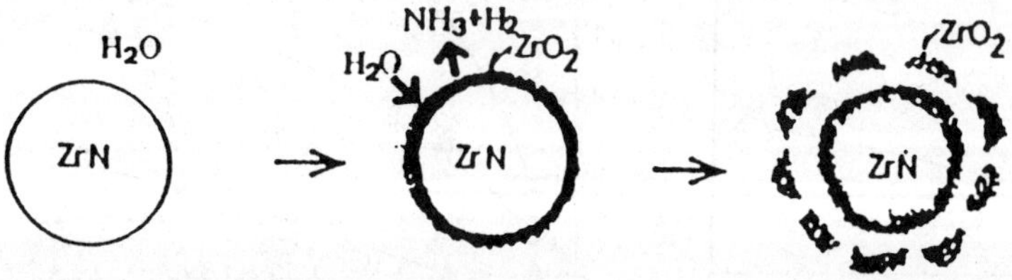

Fig.7 Reaction model of hydrothermal oxidation of ZrN powder.

As for the oxidation of ZrN and ZrC in oxygen, it has been reported that a parabolic rate law governs the initial stage of the oxidation of ZrC at an oxygen partial pressure in the range of 6×10^{-3} to 0.1 MPa(10,11), for the oxidation of sintered $ZrC_{0.9}$ between 1000° and 1500°C(12), or for the oxidation of ZrN thin films, with the activation energies Ea=193 to 222 kJ/mol for the oxidation of ZrC(10,11), and 238±8 kJ/mol for the oxidation of ZrN(14), respectively. These values are very close to the apparent activation energy of 180 to 218 kJ/mol reported of the oxidation of Zr metal(14-16).These results suggest that the oxidation of ZrN thin films or of ZrC in initial states is controlled by the diffusion of oxygen through the ZrO_2 layer. The oxidation of ZrN reported by Krusin-Elbraum(15) proceeded more slowly than the oxidation of Zr reported by Pemsler(16). This may have resulted from the decreased flux per unit time, because of the counter diffusion of produced nitrogen. Other reports suggested that the linear rate law governs the oxidation of ZrC(5-7,10-12) and ZrN(17), with varied apparent activation energies depending on reaction temperature; Ea=155 to 197 kJ/mol at 300° to 350°C(5), 76 to 92 kJ/mol at 350° to 450°C(6), 151 to 193 kJ/mol at 450° to 550°C(5,10), and 63 to 71 kJ/mol at 550° to 1000°C(6,12). The oxidation at 300° to 550°C is represented by the core-shrinking model or the Avrami's equation (5,10), which assumes that random nucleation and subsequent growth of ZrO_2 control the reaction. Between 550° and 1000°C Afonin et al.(12) suggested the rate-controlling step for the oxidation of ZrC is to be related to the chemisorption of oxygen on ZrC, because the activation energy for the oxidation of ZrC is close to the activation enthalpy of 54 kJ/mol(10) for the chemisorption of oxygen on ZrC. Based on these discussions, we can conclude that the activation energy in the oxidation of ZrC in oxygen is higher in the diffusion process than in the phase-boundary reaction process at any temperatures. Therefore the oxidation of ZrC should be governed by the parabolic rate law if it were accompanied by the formation of a dense ZrO_2 layer. However most of the reports suggests the linear rate law rather than the parabolic rate law indicates no dense

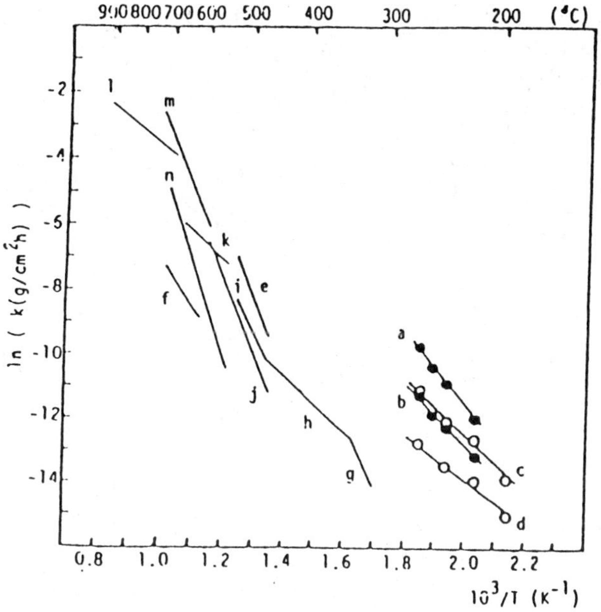

Fig. 8. Arrhenius plot for the surface contolled rate in the oxidatoin of ZrC and ZrN.

a : ZrC-H_2O ; E_a=99 kJ/mol (nucleation-controlled), Present work
b : ZrC-H_2O ; E_a=82 kJ/mol (core-shrinking model), Present work
c : ZrN-H_2O ; E_a=78 kJ/mol (nucleation-controlled), Present work
d : ZrN-H_2O ; E_a=63 kJ/mol (core-shrinking model), Present work
e : ZrC-O_2+H_2O ; E_a=193 kJ/mol, Bartlett, et al. (1963)
f : ZrN-H_2O (water vapor) ; E_a=126 kJ/mol, Caillet, et al. (1978)
g : ZrC-O_2 ; E_a=165 kJ/mol, Terekhova, et al. (1982)
h : ZrC-O_2 ; E_a=77 kJ/mol, Terekhova, et al. (1982)
i : ZrC-O_2 ; E_a=157 kJ/mol, Terekhova, et al. (1982)
j : ZrC-O_2 ; E_a=193 kJ/mol, Bartlett, et al. (1963)
k : ZrC-O_2 ; E_a=69.9 kJ/mol, Kuriakose, et al. (1964)
l : ZrC-O_2 ; E_a=63 kJ/mol, Afonin, et al. (1983)
m : ZrN-O_2 ; E_a=207 kJ/mol, Desmaison, et al. (1976)
n : ZrN-O_2 ; E_a=241 kJ/mol, Caillet, et al. (1977)

ZrO_2 layer is formed. The parabolic kinetics observed by Bartlett(10) during the initial stage of the oxidation may be attributable to the formation of the ZrO_2 thin film.

An Arrhenius plot of the rate constants gives apparent activation energies of 82 kJ/mol and 63 kJ/mol for the oxidation of ZrC and for ZrN by H_2O, respectively, on the basis of core-shrinking models, while on the basis of Avrami's equation (n=1) an Arrhenius plot of the rate constants gives apparent activation energies of 99 kJ/mol and 82 kJ/mol for the oxidation of ZrC and for ZrN, respectively (Fig.8). The oxidation of ZrN and ZrC by high-temperature high-pressure water, investigated in the present work, was initiated at much lower temperature than the oxidation in oxygen. One of the reasons are the different reaction mechanism that governs each reaction; that is, in high-temperature high-pressure water gaseous products are all hydrogen containing ones such as CH_4, NH_3 and H_2, while in oxygen gaseous products are CO_2 or N_2.

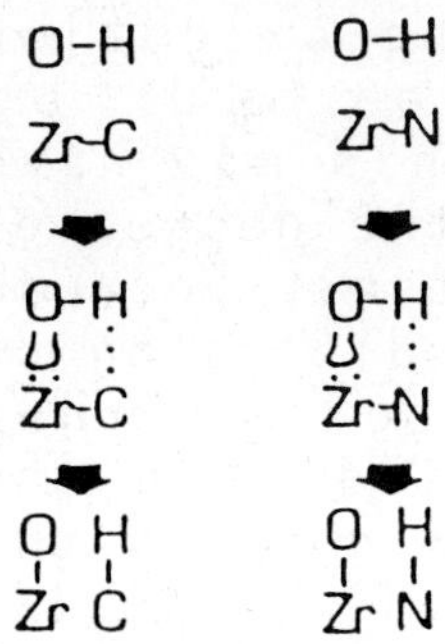

Fig. 9. Elementary reaction step in the hydrothermal oxidation of ZrC and ZrN.

The difference in the activation energies of about 20 kJ/mol for the oxidation of ZrC and ZrN by H_2O might be due to the difference in the reaction processes. Here we assume that Zr-O bonding and C-H bonding formed simultaneously (Fig.9). This corresponded to the assumption that the formation of ZrO_2 nucleus and the formation of CH_4 gas occurred simultaneously. The bonding energies of O-H, C-H and N-H bondings are 424 kJ/mol, 335 kJ/mol and 356 kJ/mol respectively (18). Therefore cutting O-H bonding and forming C-H bonding simultaneously need the energy of 89 kJ/mol, whereas cutting O-H bonding and forming N-H bonding simultaneously need the energy of 68 kJ/mol. Thus the formation of ZrO_2 nucleus from ZrC needs about 20 kJ/mol more than the formation of ZrO_2 nucleus from ZrN. Therefore the difference of about 20 kJ/mol in the activation energies for the oxidation of ZrC and ZrN by H_2O might be caused by the difference between the formation of C-H bonding and N-H bonding(19).

These mechanisms are completely different from those contemplated for the oxidation of SiC and Si_3N_4 powders by 100 MPa H_2O (1-4). In these cases the amorphous silica produced covered Si_3N_4 or SiC particles, and the overall reaction is controlled by the diffusion of the ions or molecular species through the amorphous silica layer.

CONCLUSION

ZrC and ZrN reacted by H_2O at lower temperatures above 200°C than by O_2 above 350°C. ZrC yielded monoclinic ZrO_2, CH_4 and H_2 by H_2O while by O_2 yielded monoclinic ZrO_2 and CO_2. ZrN yielded monoclinic ZrO_2, NH_3 and H_2, while by O_2 yielded monoclinic ZrO_2

and N_2.

The oxidation of ZrN and ZrC by H_2O are controlled by the phase boundary reactions, particularly the nucleation of ZrO_2, while the oxidations of Si_3N_4 and SiC by H_2O are controlled by the diffusion process in the amourphous silica layer.

the difference between the oxidation of ZrN and ZrC in H_2O and in oxygen may be due to the different reaction mechanisms. In H_2O the difference in the activation energies for the oxidation of ZrC and ZrN by H_2O might be caused by the difference between the formation of C-H bonding and N-H bonding.

REFERENCES

[1] M. Yoshimura, J. Kase and S. Sōmiya, J. Mater. Res.,1, 100-3 (1986).
[2] M. Yoshimura, J. Kase and S. Sōmiya, Yogyo Kyokai Shi (J. Ceram. Soc. Japan), 94, 129-34 (1986).
[3] M. Yoshimura, J. Kase and S. Sōmiya, "Chikka Keiso (Shin-Sozai Series)", Edited by S. Sōmiya, M. Yoshimura and M. Mitomo, Uchida Rokakuho, Tokyo, pp.121-34(1987).
[4] M. Yoshimura, J. Kase and S. Sōmiya, Proc. of the 2nd International Symposium, "Ceramics Materials and Components for Engine", Edited by W. Bunk and H. Hausner, pp.529-36 (1986).
[5] V. A. Terekhova, V. N. Kruglov and N. N. Bragin, Inorg. Mater.,18, 330-5 (1982).
[6] A. K. Kriakose and J. L. Margrave, J. Electrochem. Soc., 111, 827-31 (1964).
[7] M. Caillet, H. F. Ayedi and J. Besson, J. Less Common Met., 51, 305-22 (1977).
[8] M. Caillet, H. F. Ayedi and J. Besson, J. Less Common Met., 58, 37-46 (1978).
[9] M. Yoshimura, M. Hayakawa and S. Sōmiya, MRS Intl. Meeting on Advanced Materials, May 30 June 3, 1988, Tokyo, Japan, to be published.
[10] R. W. Bartlett, M. E. Wadsworth and I. B. Cutler, Trans. Met. Soc. AIME, 227, 467-73 (1963).
[11] L. C. Dufour, J. Simon and P. Barret, C. R. Acad. Sc. Paris, 265, 171-4 (1967).
[12] Yu. D. Afonin, V. N. Shalaginov and A. R. Beketov, J. Appl. Chem. USSR, 56, 2030-3 (1983).
[13] L. Krusin-Elbraum and M. Wittmer, Thin Solid Films, 107, 111-6 (1983).
[14] J. P. Pemsler, J. Electrochem. Soc., 105, 315-22 (1958).
[15] R. D. Misch, Acta Met., 5, 179 (1957).
[16] J. Belle and M. W. Mallet, J. Electrochem. Soc., 101, 339-42 (1954).
[17] J. Desmaison, M. Billy and W. Smeltzer, "Reactivity of Solids", Ed. by J. Wood, O. Lindqvist, C. Helgesson and N-G. Vannerberg, Plenum Press, New York and London (1977) pp.107-12.
[18] "Lange's Handbook of Chemistry", 13th ed., Ed. by J. A. Dean, McGraw-Hill Book Company, (1985) pp.3·128-31.
[19] M. Yoshimura, M. Hayakawa and S. Sōmiya, Seramikkusu Ronbunshi (J. Ceram. Soc. Japan), 96, 193-200 (1988).

THE INFLUENCE OF THE RHEOLOGY BEHAVIORS OF THE MIXES ON THE INJECTION PROCESS AND THE PROPERTIES OF CERAMIC COMPACTS

Tao Zhang, Zuozhao Jiang, Jianguang Wu, Zhengang Chen
Dept. of Mat. Scie, & Eng.,
Tsinghua Univ.,
Beijing, P.R.China.

ABSTRACT

The relation between the shear stress and viscosity of the ceramic injection moulding mixes, and the dynamic and static losses of pressure in the cavity during injection moulding were studied. The temperature distribution in the cavity was calculated. It turns out that the relation between shear stress τ and shear rate $\dot{\gamma}$ can be expressed by the equation $\tau = \eta_N \cdot \dot{\gamma}^n$. The change of pressure in the cavity is consistent with the change in temperature distribution. The density of the injection moulded bodies is not affected by the injection pressure under usual injection moulding condition.

1. INTRODUCTION

Injection moulding is receiving attention as a method of shaping engineering products from non-clay ceramic powder,(1) because it offers near net and complex shape production by automated process. In order to make the parts injection moulded with no defects and less center stress which will make binder removal and sintering process very diffcult, suitable processing parameters should be chosen. The following research work is helpful for the choice: 1) The rheology behavior of the ceramic injection moulding mixes. 2) The dynamic and static pressure loss and the temperature distribution in the cavity during injection moulding process. 3) The influence of injection pressure on the property of the green body injection moulded.

2. EXPERIMENTS

The ceramic powders used were Al_2O_3 and Si_3N_4 powders, their specific surface was 2.63 m^2/g and 5.49 m^2/g respectively. The organic binders were EVA resin, paraffin PW, stearic acid St and DBP.

The rheology behavior of the mixes was studied in capillary rheometry. The dynamic process of injection moulding was studied in injection machine. A spiral cavity whose cross section is a half circle (ϕ 14 mm) was used. Pressure sensors and thermocouples were mounted on different points in the cavity to measure the changes of pressure and temperature. The information of the changes of pressure and temperature with time during injection moulding were recorded by a computer.

3. RESULTS AND DISCUSSION

3.1 Rheology behavior of the mixes

The relations between shear rate and shear stress, relative viscosity η_N of six kinds of mixes with different composition(table 1) were shown in Fig. 1. The log shear stress against log shear rate graphs of all the formulations fitted a straight line relation with correlation coefficients R greater than 0.96. Therefore, the relation between stress τ and shear rate $\dot{\gamma}$ can be expressed by Eq. (1)

$$\tau = \eta_N \cdot \dot{\gamma}^n \qquad (1)$$

The slopes in Fig.1 (a) and (c) determine the values of n which are shown in table 1. Since n=0.23-0.30 (<1), the systems of the mixes are pseudoplastic fluid. Within the limit of experimental error the yield stress τ_s can be neglected.

3.2 The research on the dynamic process of injection moulding

The pressure in the cavity varies with time and position in the cavity. Typical experiment results recorded during injection moulding are shown in Fig. 2. The pressure P_1 in the place near the runner increases more quickly and is larger than the pressure P_2 in the place 420 mm away from the runner. It is due to that the melt of the mixes has higher viscosity and stronger resistance to the pressure transmitting. Therefore there is a larger difference of pressure in different point of a cavity. Analyzing the figures, as Fig.2, measured during injection moulding, dynamic and static pressure losses during injection moulding were made out. The dynamic pressure loss is caused by overcoming the resistance of melt's flow (Fig.3). With the increase of the melt temperature the viscosity of the melt decreases. Therefore the dynamic pressure loss decreases also. When the injection pressure is increased, the flowing velocity of melt increases. From rheology theory, obviously, the dynamic pressure loss will increase. But at higher temperature, the resistance to the melt flow decreases, the flowing velocity at initial stage does't change with the variation of maximum injection pressure, therefore the dynamic pressure loss is near constant(Fig.4).

Table 1 Regress Results of Rheology Testing

Name	Composition (wt%)					η_N	n	R
	Si_3N_4	EVA	PW	St	DBP			
SN1	85	4.5	4.5	6	0	58.6	0.23	0.980
SN2	85	4.5	7.5	3	0	58.7	0.26	0.996
SN3	85	7.5	4.5	3	0	130.0	0.25	0.989
SN4	85	4.5	4.5	3	3	70.0	0.23	0.990
SN5	85	5.25	5.25	3.75	0.75	9.24	0.30	0.960
SN6	85	6	6	3	0	73.7	0.23	0.981

Static pressure loss is the friction resistance when the melt stops flowing. It is the basis of analyzing the distribution of pressure and residual stress in the cavity. (Fig.5). With the increase of melt temperature, the friction resistance decreases. Therefore, the static pressure loss decreases also.But with the increase of injection pressure, the friction resistance increases, and the static pressure loss decreases.

The dynamic pressure loss is much smaller than static pressure loss, especially at higher temperature. For ceramic injection moulding, usually, the ratio of flowing lenth needed is smaller, therefore, the cavity can be filled in lower injection pressure. But with the increase of injection pressure, the static pressure loss increases quickly. If the injection moulding process is operated at higher injection pressure, the pressure distribution in the cavity will be less uniform, and the residual stress will increase. It is harmful to the property of the parts injection moulded.

3.3 The calculation of the temperature distribution in the cavity

the temperature distribution in half circle cavity was calculated with finite difference equation method. the boundary condition and the thermal diffusion coefficient was measured by experiment. The boundary condition is shown in Eq.(2), and the average thermal diffusion coefficient is $7*10^{-7}$ m^2/s. The calculating equation is:

$$\frac{T_{I-1,J}-2T_{I,J}+T_{I+1,J}}{(\Delta r)^2}+\frac{T_{I-1,J}-T_{I+1,J}}{2r_{I,J}\cdot\Delta r}+$$

$$\frac{T_{I-1,J}-T_{I+1,J}}{r_{I,J}^2\cdot(\Delta\theta)^2}=\frac{T'_{I,J}-T_{I,J}}{D\Delta t} \qquad (2)$$

$$T_s = 27 + 3.0t \quad t<6 \text{ s}$$

$$T_s = 47.6 - 0.43t \quad t>6 \text{ s}$$

where T_s = surface temp., r=radius, θ = angle, T=temp.,in the part.

A fraction of calculating results is shown in Fig.6. The results are consistent with the changes of the pressure measured in the cavity during injection moulding. These results can help us to adjust the parameters of injection moulding.

3.4 The relation between injection pressure and the density of the green parts

The relation between injection pressure and the density of the green parts is show in Fig.7. The densities of the green parts which were moulded under different injection pressure are similar. The extrusion expanding ratio of the mixes was measured in capillary rheometry. It shows that the extrusion expanding ratio is very small. Therefore the difference of first normal stress ($N_1=\tau_{11}-\tau_{22}$) is nearly equal to zero. In other words, the melt does't have plasticity and cann't be compressed. So the density of the part cann't be changed by injection pressure.

4. CONCLUSIONS

1) The relation between the shear stress and shear rate can be expressed by the equation $\tau = \eta_N \cdot \dot{\gamma}^n$ which characterizes the rheology behaviors of ceramic mixes. Due to n=0.23-0.30($<$ 1), the mixes belong to persudoplastic fluid.

2) The distribution of pressure in the cavity during injection moulding is not uniform. The change of pressure in the cavity is consistent with the change of temperature distribution in the cavity. Therefore, applying the experimental results of pressure and the temperature distribution calculated we can adjust the processing parameters to get uniform pressure distribution and smaller residual stress.

3) With the increase of injection pressure, the change of dynamic pressure loss is smaller, but static pressure loss varies greatly, and the difference of pressure on different point in the cavity increases. Therefore, using lower injection pressure is better unless the other requirements can not be satisfied.

REFERENCE

1. M.J.Edirisinghe, J.R.G.Evans, Journal of Materials Science 22 (1987) 269-277.

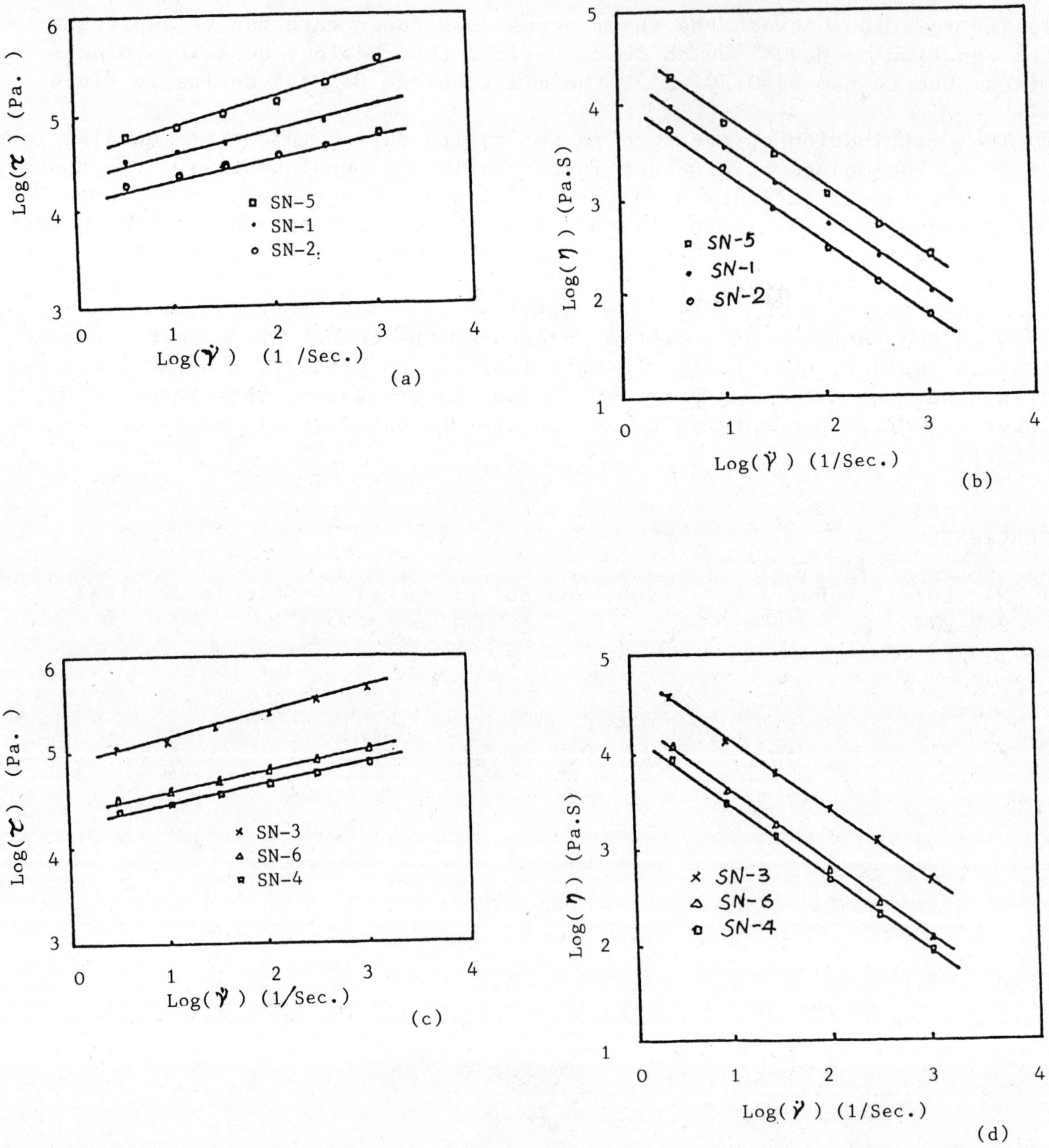

Fig.1 Rheology behavior of the mixes

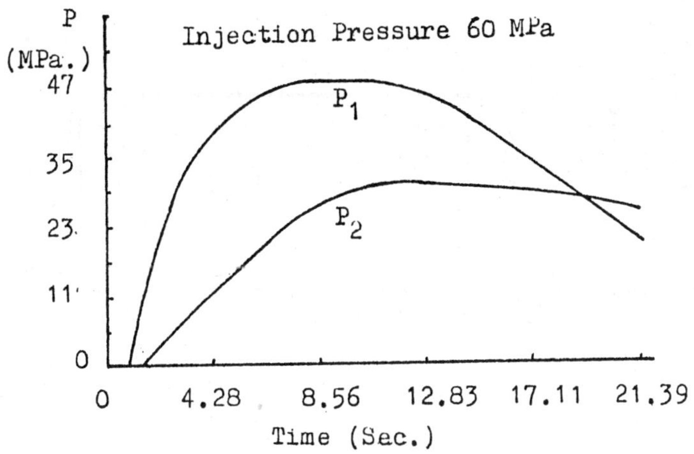

Fig.2 Changes of the pressure in the cavity with time

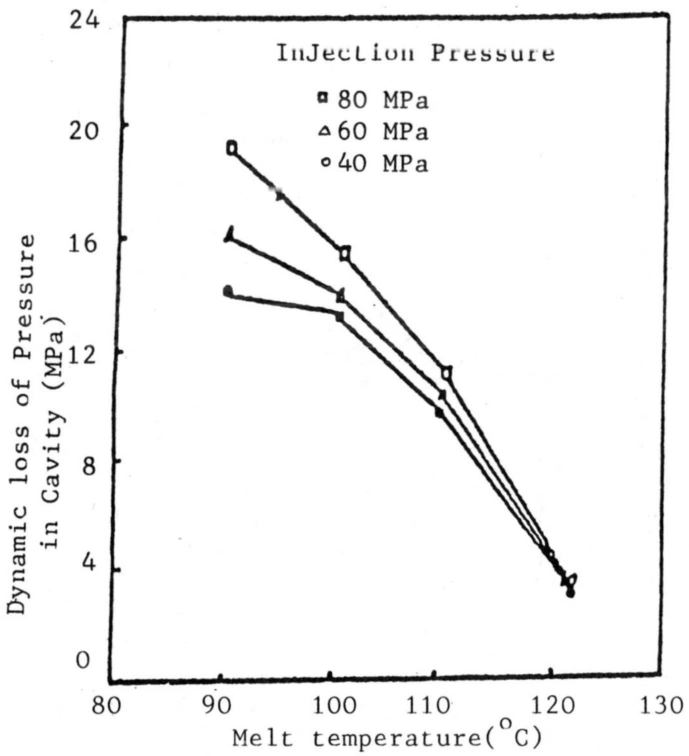

Fig.3 Dynamic pressure loss in the cavity

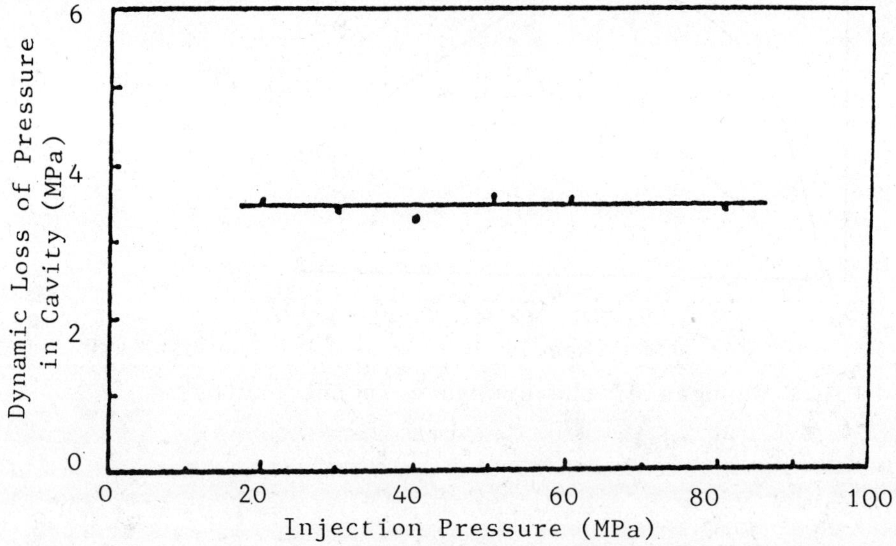

Fig. 4 Dynamic pressure loss in the cavity at T=120°C

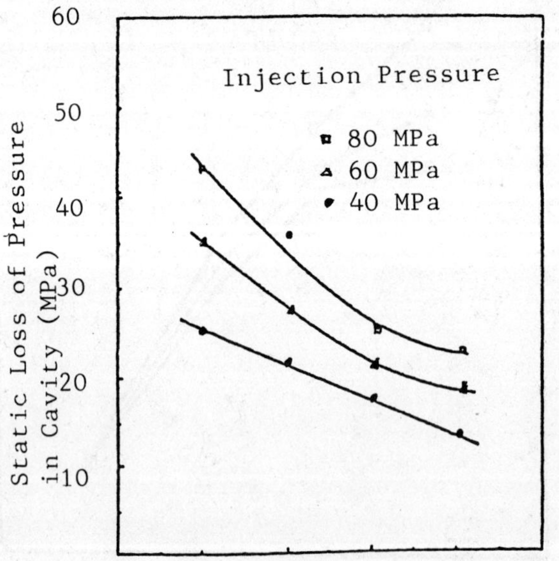

Fig.5 Static pressure loss in the cavity

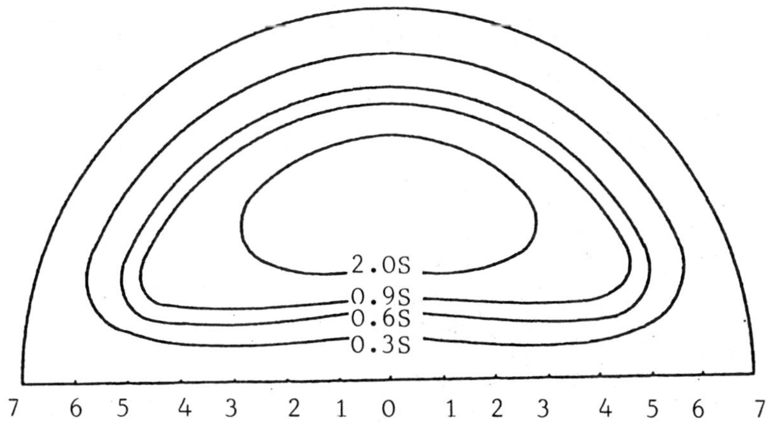

Temperature Distribution in Cavity (mm)
($T=100^oC$, $T_0=120^oC$)

(a)

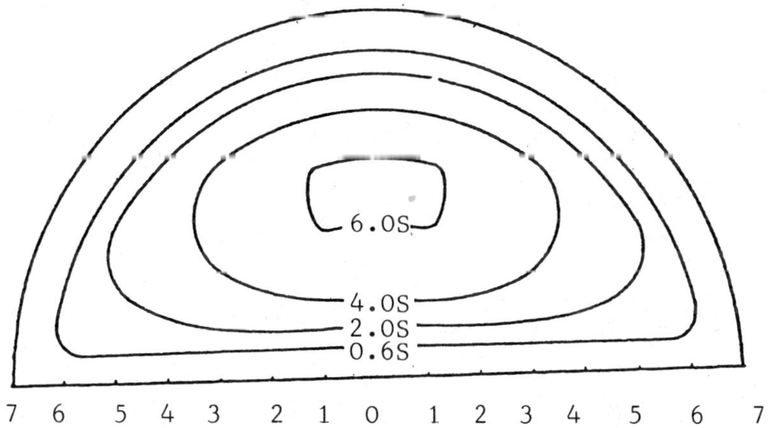

Temperature Distribution in Cavity (mm)
($T=70^oC$, $T=120^oC$)

(b)

Fig.6 The relation between temperature in the cavity and time

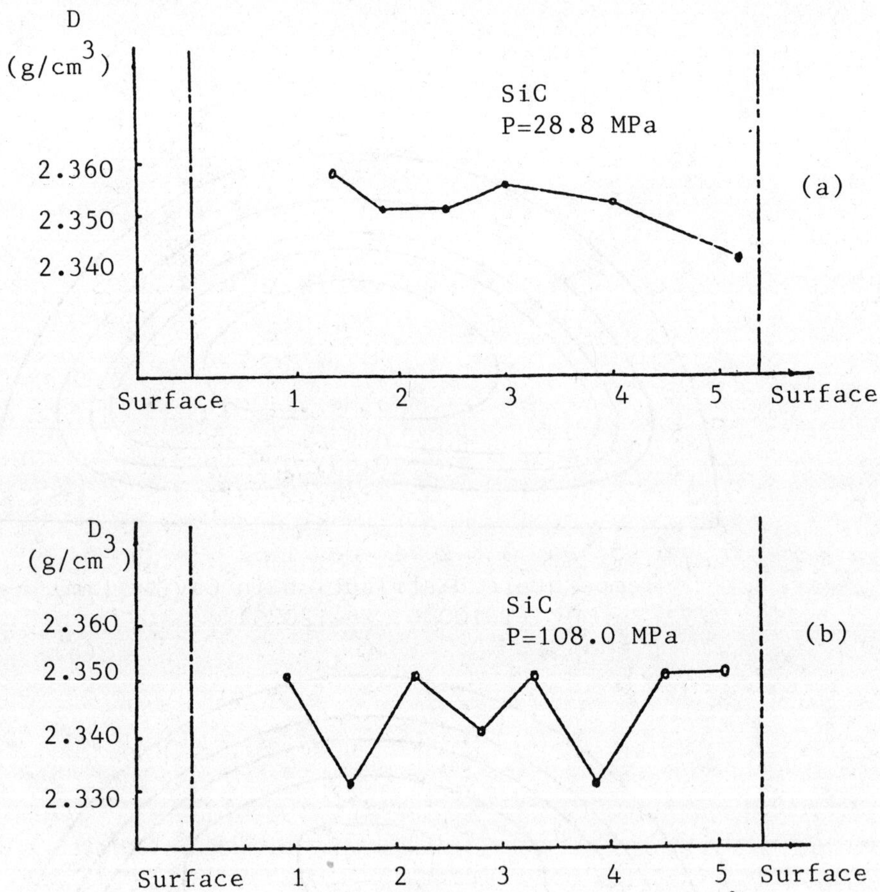

Fig.7 The distribution of density in the parts

RHEOLOGICAL BEHAVIOR OF INJECTION MOLDABLE CERAMIC-CERAMIC COMPOSITES

I. TSAO, D. DORAISWAMY, A. B. METZNER* AND S. C. DANFORTH

Department of Ceramic Engineering,
Center for Composites Manufacturing Science and Engineering,
Rutgers University, Piscataway, NJ 08855-0909

*Department of Chemical Engineering,
University of Delaware, Newark, DE 19716

ABSTRACT

The composite system of interest here is SiC whisker toughened reaction bonded Si_3N_4 formed via injection molding or extrusion. The rheology of Si powder and SiC whisker filled molten polyethylene was investigated by using dynamic and capillary viscometries. Results show that compounds with > 60 vol% solids exhibit a yield stress at low shear rate, and are shear thinning at high shear rate. Moreover, at 140°C in the capillary viscometry measurements, a nearly Newtonian behavior was observed for 60 vol% Si powder over the shear rate range from 200 sec^{-1} to 3500 sec^{-1}. Both wall slip and end effects were studied for the Si powder filled polyethylene system. Preliminary results indicate that end effects are negligible, yet wall slippage does exist. At low frequency (or shear rate), the slip effect becomes more pronounced as the parallel plate gap decreases.

INTRODUCTION

Whisker reinforced ceramic-ceramic composites have been shown to have great potential for high temperature structural applications. A great deal of work has been reported showing that superior mechanical properties (i.e. increased fracture stress, fracture toughness etc.) can be achieved by incorporating ceramic whiskers into a ceramic matrix.[1-6] In general, the common methods used for processing whisker reinforced composites are dry pressing and isostatic processing. In addition, other forming methods such as slip casting, extrusion or injection molding have recently received significant attention for ceramic-ceramic composites.[7-10] The difference between dry pressing and plastic forming techniques is that the former is easier but can only make simple shapes, while the latter can form quite complex shapes potentially with close tolerances. In addition, the liquid phases used for injection molding and extrusion, yield uniform dispersions and hence better microstructural uniformity than dry pressing.

The rheological behavior of filled systems for plastic forming processes is very complicated. The viscosity depends on filler size, shape, volume fraction and interaction between components.[11,12] In the past, much work on particle filled suspensions has been done. Several different models and equations were developed to explain the rheology of filled systems under very specific conditions, such as spherical powders and over a specific shear rate range etc. However, real systems, containing particles with irregular geometry are too complicated for these models. Although some successful theories and models are available for dilute systems, more effort is still necessary for concentrated systems. In recent years, as more interest has developed for polymer matrix composites, scientists have paid more attention to the rheology of fiber filled systems. However, very little work has been published on the rheology of powder-whisker-molten polymer mixtures, for injection molded ceramic-ceramic composites.[13-16] Although these systems show very complicated rheological behavior and modeling seems extremely difficult, fully understanding the rheological behavior during flow is a critical step towards improved processing of controlled microstructures and in turn improved reliability of the resultant mechanical properties.

In this work, the forming methods used are injection molding and extrusion. The rheological behavior of Si powder and SiC whisker filled molten polyethylene was characterized to determine the effects of total solids loading and powder to whisker volume ratio on the system's flow behavior. In addition, wall slip and end effects were also studied, because large errors due to end and slip effects will result in improper rheological properties interpretation and also influence whisker orientation during flow.

RAW MATERIAL

In this work, the polymer used is a linear low density polyethylene.* This polymer has a broad molecular weight distribution and density of 0.92 g/cm^3 (Table 1). The fillers used include Si powder** and SiC whiskers.*** The Si powder has a particle size distribution from 0.5 μm to 30 μm with an average diameter of 9 μm. The SiC whiskers have an aspect ratio up to 100 with average diameter of 0.5 μm. Table 1 also lists the physical properties of the solids.

Table 1. Physical Properties of Raw Materials

Polyethylene Wax

Grade	M_n	M_w	Viscosity (Pa-s)
A735	2600	8800	6 (140°C)

Si Powder

Diameter (μm)	Surface area (m^2/g)	Density (g/cm^3)
9.0 (average)	1.33	2.34

SiC Whisker

Grade	Diameter (μm)	Length (μm)	Density (g/cm^3)
SCW#105	0.2-0.8	1.0-100.0	3.15

* A735 polyethylene, Allied Signal Co., Morristown, NJ

** Globe Metal Co., Beverly ,OH

*** Tateho Co., Tokyo, Japan

EXPERIMENTAL PROCEDURE

All compounds investigated were mixed in a torque rheometer.**** The mixer has a 215 cm^3 capacity, with two counter rotating roller blades. The mixing conditions were fixed at 125°C and 100 rpm in order to achieve optimum mixing. The polyethylene was added into the preheated mixer. The Si powder and SiC whiskers were then sequentially poured into the mixing chamber after the polyethylene was completely molten. Mixing times were on the order of 2 to 4 hours depending on the total solids loading. The compounds were saved and processed into granules after the recorded torque reached a steady state level.

Two viscometers were used to characterize the rheological properties of the ceramic filled polymer systems. The complex viscosity***** was determined by the dynamic cone-plate and parallel plate methods.****** All measurements were made at 140°C. In the dynamic measurements, the material was pressed into a plate shape where the cone-plate has a gap of 49 µm with a cone angle of 5.7°. The parallel plate has the same radius as the cone-plate (1.25 cm). The parallel plate gap was varied to check for wall slip effects. In the capillary measurements, the extruder has four heating zones to control the temperature precisely (±2°C). A single screw delivers the granulated compounds into the capillary die. By measuring the pressure drop and melt mass flow rate, the apparent viscosity is determined. Dies with a diameter of 0.142 cm and three aspect ratios (5.0, 10.25 and 19.86) were used in order to study the end effect during flow.

RESULTS AND DISCUSSION

1. Dynamic Viscometry Measurements

In recent years, several investigators have been using dynamic viscometry to characterize concentrated system rheology.[18-21] One of the advantages of dynamic viscosity measurements is that the technique allows a description of the behavior of highly filled systems at close to the equilibrium state (without disrupting the particulate structure). In this paper, the results from dynamic cone-plate measurements will be discussed first. Figure 1 shows the results of measurements on 70 vol% Si powder in molten polyethylene. The complex viscosity of the molten polyethylene is strain independent (not shown in Figure 1), yet for the filled system it is strongly strain dependent. Because part of the shear energy is dissipated in the system, we can correlate the strain dependent behavior (non-linear viscoelasticity) to a close packed particulate network structure formed due to particles crowding each other. In addition, in Figure 1, the curve's slopes are -1 at low frequency, indicating a yield stress in the system. Shear thinning behavior appears at higher frequencies. This first stage analysis shows that this concentrated Si powder system is strain dependent below the yield point and then primarily shear rate dependent after the yield point.[22] Data for 60 vol% Si powder in molten polyethylene shows a behavior similar to that of Figure 1.

**** Haake Buchler Inc., Saddle Brook, NJ

***** $\eta^*=[(\eta')^2+(\eta'')^2]^{0.5}$, $\eta'=(G''/\omega)$, $\eta''=(G'/\omega)$, η^*=complex viscosity, G'=storage modulus, G''=loss modulus, ω=frequency

****** Rheometrics Inc., Piscataway, NJ

Different solids loadings of Si powder in polyethylene were characterized by dynamic measurement. Figure 2, shows 5 different curves representing solids loadings from 0 vol% to 75 vol%, measured at a fixed strain of 10%. From this figure, 0 and 30 vol% curves show Newtonian or slightly shear thinning behavior. There is no yield stress found, which means the polymer-filler interaction is dominant at low filler concentrations. High solids loading curves (60, 70 and 75 vol%) have a slope of -1 at low frequency, (i.e. show a yield stress) and then become shear thinning at higher frequency. A large difference in complex viscosity for different solids loadings (Figure 2), is found at low frequency; the difference decreasing as the frequency increases. The yield point of each curve shifts towards higher frequency as the solids loading increases. This results from the particles crowding each other and thus requiring higher energy for flow. Shear thinning behavior, after the yield point, indicates that less energy is needed to shear the system once the network structure begins to break down.

Figure 3 shows the relative complex viscosity as a function of the vol% of Si powder. The viscosity rises as the solids loading increases, increasing very rapidly near 75 vol% Si powder. The curve shifts to the right at higher frequency. These two curves deviate from the curve presented by Ferris[23] due to the irregular geometry of the particles being used here, also due to different shear rate ranges and different viscometry techniques used here.

A similar trend was also seen at a 70 vol% total solids loading, but, in this case the region shifts to the left with respect to the curve seen in Figure 4. Janney[9] also observed a similar phenomenon for Al_2O_3 powder-SiC whisker systems. The mechanism for the decreasing viscosity with increasing SiC whisker content at a fixed total solids concentration is not well understood at this time. However, in a review paper by Metzner,[11] a rigid rod model (comparable to polymeric liquid crystal fluids) used to predict fiber flow properties may give a reasonable explanation. According to this model, the rheology of the composite system strongly depends on the whisker orientation. At low whisker concentration, the viscosity increases as the whisker content increases and the orientation effect is less important. At a certain whisker concentration range, the whiskers tend to align along the shearing direction resulting in a lower viscosity. As more and more whiskers are incorporated into the system, the whiskers are no longer able to align because of entanglement, thus resulting in increasing viscosity with increased whisker loading.

The slip effect in these concentrated systems was also evaluated with the parallel plate method. In general, slippage is due to a thin wall layer of lower viscosity material formed at the boundary of the shearing wall. In Cohen and Metzner's paper,[24] more precisely, they define wall slip as follows. Slip refers to the abnormally low apparent viscosities which are exhibited by a fluid mixture in an inhomogeneous stress field or direct fluid-wall interactions, when compared with the viscosities measured in uniform stress fields in the absence of wall effects.

The first stage of this slip analysis was done on 60 vol% Si powder alone in polyethylene (the composite system will be evaluated in the near future). Figure 5-A shows the complex viscosities measured at a gap $< 200\ \mu m$ are lower than complex viscosities measured for a 200 μm gap. Figure 5-B shows that all the curves (i.e. viscosities) superimpose for gaps $\geq 200\ \mu m$. The low measured viscosities for gaps $< 200\ \mu m$ result from the slip effect along the wall. It can be concluded that the contribution of slip to the total flow at a given gap ($< 200\ \mu m$) decreases with increasing frequency (or shear rate) because of shear thinning behavior.[24] In addition, the degree of slippage increases with decreasing gap size.

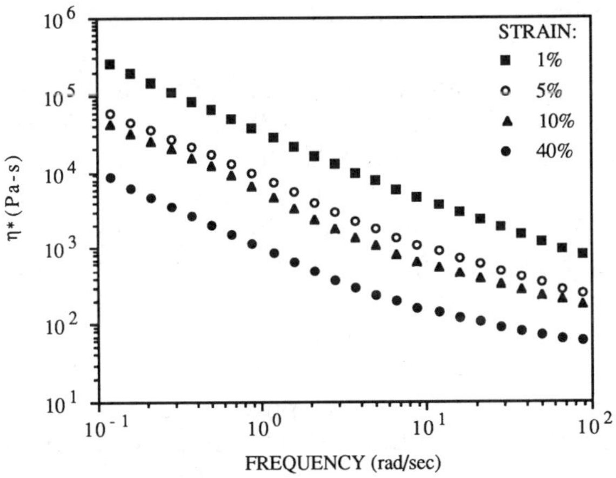

Fig.1 Complex viscosity (η*) vs. frequency for (polyethylene wax + 70 vol% Si powder) at different strains and 140°C.

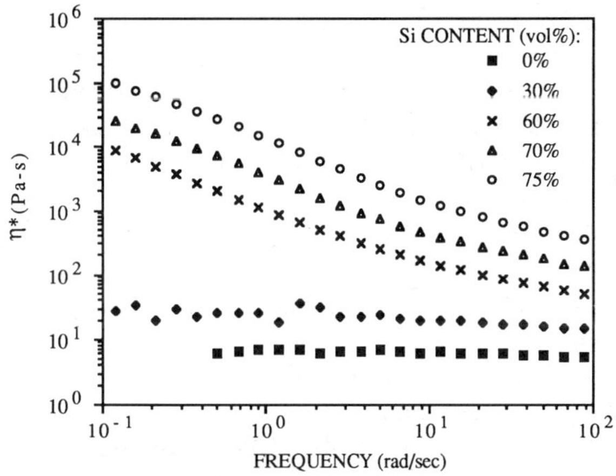

Fig.2 Complex viscosity (η*) vs. frequency for 5 different solids loading at 10% strain and 140°C.

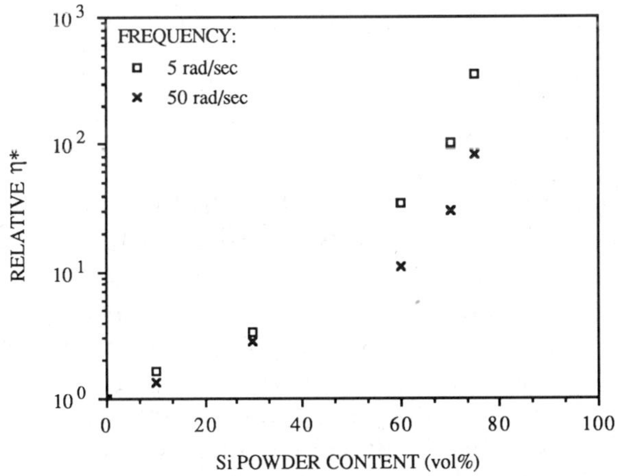

Fig.3 Relative complex viscosity as a function of Si powder content at 2 different frequencies and 140°C

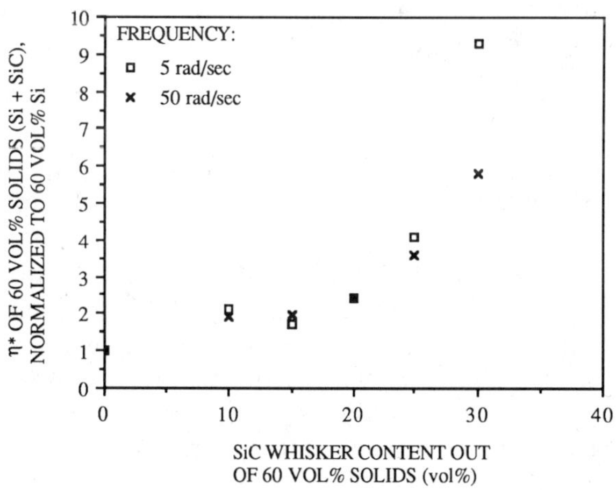

Fig.4 The complex viscosity of 60 vol% solids composite system (with Si and SiC_w) normalized to 60 vol% Si powder alone vs. SiC_w content (out of 60 vol% solids loading) at 10% strain and 140°C.

2. Capillary Viscometry Measurements

In the first half of this discussion, the dynamic data were presented and some interesting rheological phenomenon were discussed. The equivalent shear rate range used for the dynamic measurements was very low (<10 sec^{-1}). Thus capillary viscometry was also applied to investigate the rheology of the composite systems during flow at higher shear rates. The extruder type capillary viscometer used, which exposes the filled systems to shear rates from 200 sec^{-1} to 3500 sec^{-1}, approximates injection molding and extrusion.

Figure 6 shows the apparent viscosity vs. apparent shear rate for a fixed solids loading of 60 vol% Si powder at 140°C. The top curve (first run) represents the compounds directly from the mixer, and the second and third curves are the results of using recycled material (collected after the first and second measurements). The most interesting thing is that all three curves show Newtonian behavior (or very slight shear thinning) in the range from 200 sec^{-1} to 2500 sec^{-1}. Although the viscosity of the compounds decreases slightly with successive experiments, there is no clear evidence to show thixotropy, as the differences are quite small.

The end effect was studied for 60 vol% Si powder in polyethylene at 140°C. By definition, end effects result from non-steady laminar flow introduced by the end regions of the capillary tube. In addition, the pressure gradient will be influenced if the tube length is too short. The analysis was done according to the Bagley method.[25] In this technique, the true pressure drop was determined for various shear rates using a fixed tube diameter with various lengths. At a given shear rate, the equation is as follows:

$$\Delta P=4\sigma_w(L/D)+|P_e| \qquad (1)$$

where: ΔP is the measured pressure drop, σ_w is the shear stress at the wall, L/D is the aspect ratio of the capillary tube, and $|P_e|$ is the excess pressure drop due to end effects. In these experiments, the capillary tubes used had a fixed diameter of 0.142 cm with aspect ratios of 5.0, 10.25 and 19.86. By plotting ΔP vs. L/D, according to equation (1), the slopes and intercepts of the curves represent σ_w and P_e respectively. Figure 7 shows that the apparent viscosity vs. apparent shear rate curves are closely Newtonian for all three capillary diameters. Figure 8 shows that all curves intercept at an L/D of approximately 0. This result indicates that the excess pressure drop due to rearrangement of the velocity profile at the end regions is negligible for these concentrated systems because P_e is close to zero.

The flow behavior of the concentrated system containing SiC whiskers in capillary viscosity measurements was also characterized. As seen in Figure 9, the filled systems with SiC whiskers show strong shear thinning behavior compared to the powder only system. As the whisker content increases, the shear thinning behavior becomes more pronounced, and the viscosity difference increases as the shear rate decreases. The mechanism to explain this pronounced shear thinning behavior is due to whisker orientation. At low shear rate, the whiskers are randomly distributed in the matrix and tend to hinder flow. Then, the whiskers start to be aligned along the shearing direction as the shear rate increases. At a still higher shear rate, the whiskers become significantly oriented, resulting in a reduced viscosity. The evidence for whisker alignment effects can be seen in Figure 10. In Figure 10-A, where the fracture surface is parallel to the capillary axis, most whiskers appear to be lying in the plane, while for Figure 10-B, normal to the capillary axis, most whiskers are lying normal to the fracture plane.

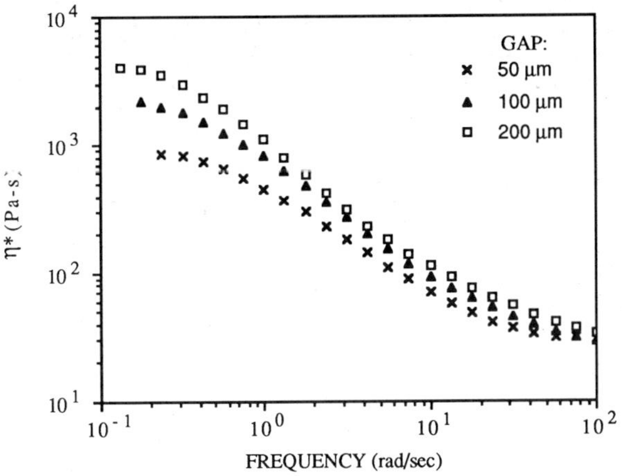

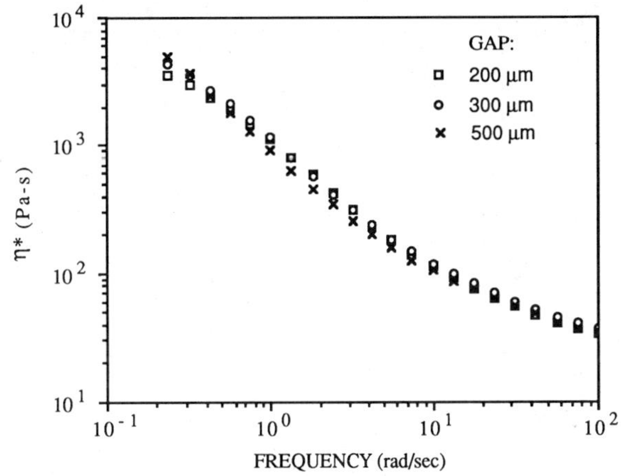

(A) (B)

Fig.5 Complex viscosity (η*) of 60 vol% Si powder system at 10% strain and 140°C: (A) parallel plate gap = 50, 100 and 200 μm, and (B) parallel plate gap = 200, 300 and 500 μm.

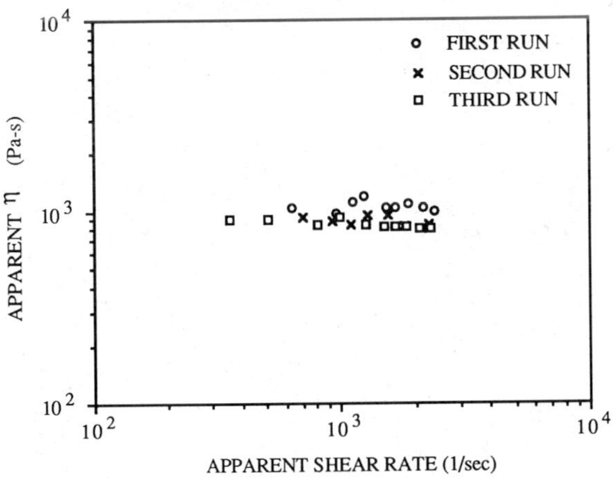

Fig.6 Apparent viscosity (η) vs. apparent shear rate for (polyethylene wax + 60 vol% Si powder) at 140°C.

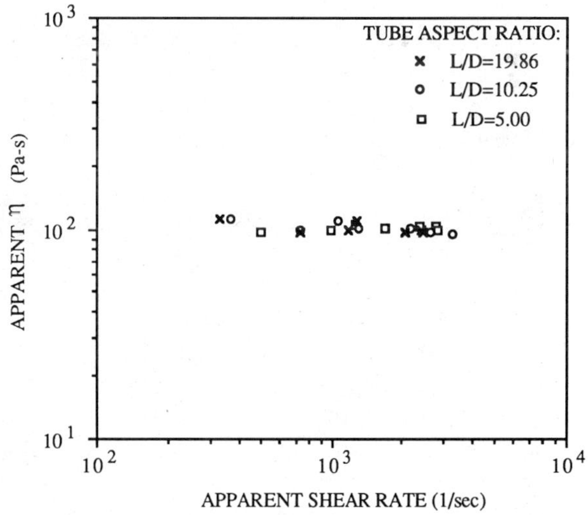

Fig.7 Apparent viscosity (η) vs. apparent shear rate for 60 vol% Si powder in polyethylene wax at 140°C and 3 different tube aspect ratios (tube diameter = 0.142 cm).

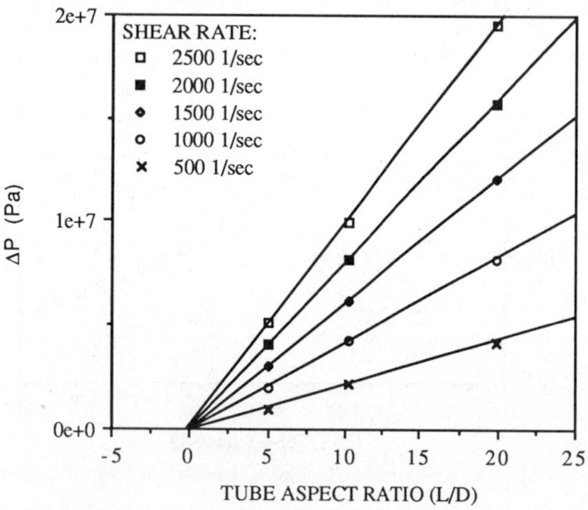

Fig.8 Measured pressure drop (ΔP) as a function of tube aspect ratio at different apparent shear rates for 60 vol% Si powder in polyethylene wax at 140°C (tube diameter = 0.142 cm).

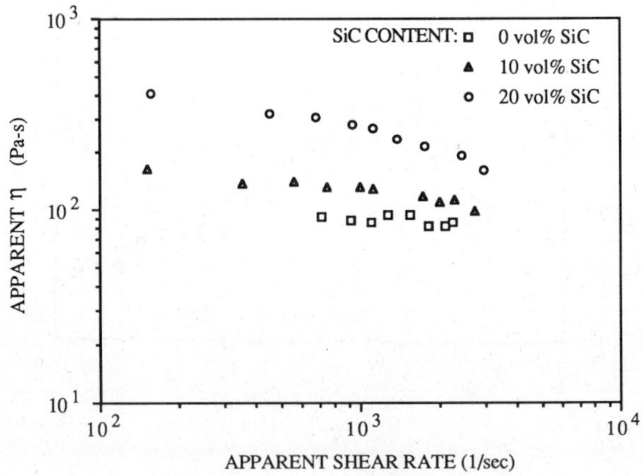

Fig.9 Apparent viscosity (η) vs. apparent shear rate for composite systems with a fixed 60 vol% solids loading in polyethylene wax at 140°C.

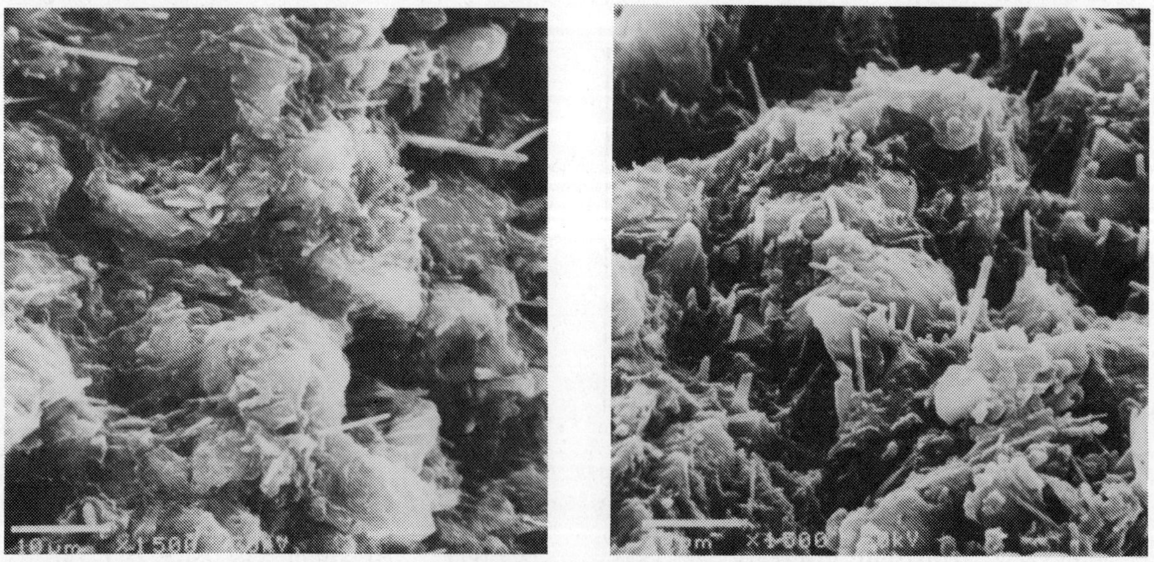

(A) (B)

Fig.10 Fracture surface of polyethylene wax + 60 vol% solids loading (90/10-Si/SiC_W) mixture: (A) fracture surface parallel to capillary axis,and (B) fracture surface normal to capillary axis.

SUMMARY

The composite system of interest here is SiC whisker toughened reaction bonded Si_3N_4 formed via injection molding or extrusion. The first part of this work shows that the powder and whisker filled system has a very complicated rheological behavior. At low shear rate (or frequency), the highly filled systems are shear history and strain dependent, resulting in a non-linear viscoelastic behavior. These concentrated systems, containing both powder and whiskers, show a yield stress and are strongly shear thinning over the entire shear rate range studied. For composite systems (at a fixed 60 vol% total solids loading) viscosity data show a local minimum (or plateau region) at around 10 to 20 vol% SiC whiskers at low shear rate (or frequency). The highly filled powder system, such as 60 vol% Si powder in polyethylene, shows a yield stress and shear thinning at low shear rate (or frequency), but eventually evolves to a Newtonian fluid region at high shear rate (injection molding range). The difference in rheological behavior between the powder filled systems with and without SiC whiskers (at high shear rate) is a result of whisker alignment. Preliminary data shows that the end effect is negligible, however wall slippage does exist.

ACKNOWLEDGEMENTS

The authors wish to thank the National Science Foundation for support of this work through the University of Delaware-Rutgers University Center for Composites Manufacturing Science and Engineering. The authors also wish to thank: Dr. A. Young for his invaluable assistance in this work, and T. Gryzmsky , R. Foss and G. Papatrefon for their laboratory assistance.

REFERENCES

1. P. F. Becher and G. C. Wei, "Toughening Behavior in SiC-Whisker-Reinforced Alumina," J. Am. Ceram. Soc., 67 [12] C267-C269 (1984).
2. G. C. Wei and P. F. Becher, "Development of SiC-Whisker-Reinforced Ceramics," Am. Ceram. Soc. Bull., 64 [2] 298-304 (1985).
3. F. D. Gac and J. J. Petrovic, J. V. Milewski et. al., "Performance of Commercial and Research Grade SiC Whiskers in a Borosilicate Glass Matrix," Ceram. Eng. Sci. Proc., 7 [7-8] 978-982 (1986).
4. J. V. Milewski "Short Fiber Reinforcements : Where the action is," Plastic Compounding, November, 1979.
5. J. Homeny, W. L. Vaughn and M. K. Ferber, "Processing and Mechanical Properties of Silicon Carbide Whisker/Alumina Matrix Composites," Am. Ceram. Soc. Bull., 66 [2] 333-338 (1987).
6. P. F. Becher, T. N. Tiegs, J. C. Ogle and W. H. Warcick, "Toughening of Ceramics by Whisker Reinforcement"; pp. 61-73 in Fracture Mechanics of Ceramics, vol. 7 Editted by R. C. Bradt, A. G. Evans, D. P. H. Hasselman, and F. F. Lange, Plenum Press, New York, 1986.
7. F. Takao, W. R. Cannon and S. C. Danforth, "Non-aqueous Colloidal Pressing of Reaction Bonded Silicon Nitride-SiC Whisker Composites"; Advances in Ceramics, 21, William J. Smothers ed., Am. Ceram. Soc., 699-708, (1987).

8. I. Tsao, G. Krug and S. C. Danforth, "Processing of Whisker Reinforced Ceramic-Ceramic Composites," Submitted to the 1987 Northeast Regional Meeting, "High Temperature Structural Composites: Synthesis, Characterization and Properties," Hoboken, NJ, May, 1987.
9. M. A. Janney, "Viscous and Viscoelastic Characterization of Whisker-Particulate Suspensions," Presented at 89th Annual American Ceramic Society Meeting, Pittsburgh, PA, April, 1987.
10. M. Sacks, H. W. Lee and O. E. Rojas "Suspension Processing of Alumina-SiC Whisker Composites," submitted to Advances in Ceramics.
11. A. B. Metzner, "Rheology of Suspension in Polymeric Liquids," J. Rheol., 29 [6] 739-776 (1985).
12. S. A. Khan and R. K. Prud'homme, "Melt Rheology of Filled Thermoplastics," submitted to Journal of Reviews and Chemical Engineering.
13. R. J. Crowson, M. J. Folkes and P. F. Bright, "Rheology of Short Fiber-Reinforced Thermoplastics and its Application to Injection Molding, 1. Fiber Motion and Viscosity Measurement," Polymer Engineering and Science, 20 [14] 925-933 (1980).
14. B. Chung and C. Cohen, "Glass Fiber-Filled Thermoplastics. I. Wall and Processing Effects on Rheological Properties," Polymer Engineering and Science, 25 [16] 1001-1007 (1985).
15. M. Sanou, B. Chung and C. Cohen, "Glass Fiber-Filled Thermoplastics. II. Cavity Filling and Fiber Orientation in Injection Molding," Polymer Engineering and Science, 25 [16] 1008-1016 (1985).
16. H. M. Laun, "Orientation Effects and Rheology of Short Glass-Fiber-Reinforced Thermoplastics," Colloid and Polymer Science, 257-269 (1984).
17. M. Takahashi, S. Suzuki and H. Avai, "Mixing and Flow Characteristics in Alumina-Thermoplastic Resin System for Ceramic Injection Molding," Presented at 89th American Ceramic Society Meeting, Pittsburgh, PA, April, 1987.
18. D. M. Bigg, "Complex Rheology of Highly Filled Thermoplastic Melts," Proc. Intl. Congress on Rheol., 429-437 (1984).
19. D. M. Bigg, "Rheological Analysis of Highly Loaded Polymeric Composites Filled with Non-Agglomerating Spherical Filler Particles," Polymer Engineering and Science, 22 [8] 512-518 (1982).
20. D. M. Bigg, "Rheological Behavior of Highly Filled Polymer Melts," Polymer Engineering and Science, 23 [4] 206-210 (1983).
21. J. H. Dow, M. D. Sacks and A. V. Shenoy, "Dispersion of Ceramic Particles in Polymer Melts," submitted to Advances in Ceramics.
22. D. Doraiswamy, I. Tsao, A. N. Beris, A. B. Metzner and S. C. Danforth, "The Rheology of Ceramic Suspensions," submitted to the Proceedings of the 10th World Congress on Rheology, Syndney, Australia, August, 1988.
23. R. J. Ferris, "Rheology of the Viscosity of a Multimodel Suspension from Unimodel Viscosity Data," Trans. Soc. Rheol., 12 [2] 281 (1968).
24. Y. Cohen and A. B. Metzner, "Apparent Slip Flow of Polymer Solution," J. Rheol., 29 [1] 67-102 (1985).
25. E. B. Bagley, "End Correction in the Flow of Polyethylene," J. Appl. Phys., 28 [5] 624-627 (1957).

DISPERSION AND STABILIZATION OF MULTICOMPONENT SILICON NITRIDE BASED AQUEOUS SUSPENSIONS

J.P. Pollinger
Garrett Ceramic Components Div.
Allied-Signal Aerospace Co.
Torrance, CA 90509

ABSTRACT

The characterization and development of Si_3N_4-Y_2O_3 aqueous suspensions for use in slip casting of complex shaped Si_3N_4 components using acoustophoresis is described. Using electro-static stabilization as the dispersing mechanism it was determined that pH 10.5 conditions resulted in suspensions with the best combinations of suspension and cast part properties, as determined by correlation of acoustophoresis results with viscosity, settling, and casting tests. Dilute suspension properties were correlated with high solids content suspension results.

INTRODUCTION

Forming of Si_3N_4 components for high strength-high temperature applications by slip casting using porous mold technology places many requirements on the aqueous suspension to be used. The suspension must result in a cast complex net-shaped part with homogeneous distributions of sintering aid and Si_3N_4 powder, uniform high green density, good dimensional control, no voids other than uniform part porosity, no agglomerates, and no dewatering or subsequent drying problems due to excessively small pore size or differential shrinkage. Assuming that powder composition and Si_3N_4/sintering aid particle sizes and distributions are constrained by the final desired microstructure and properties of the Si_3N_4 component, the aqueous suspension properties that may be adjusted to produce cast parts with desired properties are the degree of dispersion of the suspension and its solids content. Other factors influencing cast part properties, such as mold design and properties, pressure application, and other casting techniques such as centrifugal casting and electroacoustic dewatering, will not be discussed in this paper.

The dispersion of colloidal particles in an aqueous medium by electro-static and/or steric repulsion has been extensively discussed in the literature (1-5). High zeta potentials are generally observed to form well dispersed and stable suspensions at low solids contents. Particles with low zeta potentials exhibit floc formation, unstable suspensions, and rapid settling. For multicomponent colloid suspensions, powders with same-sign large magnitude zeta potentials exhibit well dispersed stable suspensions while powders with opposite sign zeta potentials exhibit hetero-coagulation, instability, and rapid settling(6). In cases where powders have very different densities or particle sizes, heterocoagulation may be desirable to prevent segregation due to differential settling rates(7-9).

Solids content also influences a number of suspension and cast part properties. Higher solids content suspensions may exhibit greater tendencies towards shear thinning and gelation because of decreased particle repulsive forces due to double layer overlap when particles approach separation distances on the order of double layer thicknesses(10,11). As a result, pourability and conformability of suspensions will be reduced. Segregation in multicomponent suspensions may be reduced with high solids contents because of higher viscosities and hindrance of settling due to greater particle concentration, which increases the amount of particle-particle collisions and reduces Brownian diffusion(12).

It was the objective of this study to determine the effects of degree of dispersion and solids content on aqueous suspensions and cast part properties of a typical high temperature-high strength multicomponent Si_3N_4 composition (95 wt% Si_3N_4 - 5 wt% Y_2O_3) by manipulating electro-static charges on particle surfaces, which can be achieved by control of pH(5), and varying solids content. The zeta potential of both Si_3N_4 and Y_2O_3 powders as a function of pH was determined and 5 vol% solids content suspensions of each powder, 95 wt% Si_3N_4/5 wt% Y_2O_3, and 50 wt% Si_3N_4/50 wt% Y_2O_3 suspensions were prepared and characterized by acoustophoresis, viscosity, and settling tests. It was felt that a 50/50 ratio of the two powders would provide more insight into the effects of combining the two powders than the 95 wt% Si_3N_4/5 wt% Y_2O_3 suspension, which might be dominated by Si_3N_4 properties. The effect of solids content on suspension properties and cast green density was examined by preparing Si_3N_4 suspensions ranging in solids content from 1 to 45.2 vol% and 37.7 vol% 95 wt% Si_3N_4/5 wt% Y_2O_3 suspensions at optimum dispersion conditions.

The recent development of commercial equipment to measure zeta potential using acoustophoretic techniques now allows the examination of concentrated colloidal suspensions. The acoustophoretic method used in these experiments utilizes the Ultrasonic Vibration Potential (UVP) effect, where an alternating pressure field is applied to a particle suspension. The density difference between the particles and liquid leads to a relative motion between the particles and liquid, which generates a periodic polarization of double layers and a resulting potential, which can be measured and is proportional to the electrophoretic mobility(13). The UVP signal is inversely proportional to the high frequency conductivity of the suspension. Since these experiments utilize low conductivity suspensions (no electrolytes were added besides acid and base additions for pH control, to reproduce slip preparation procedures) the large UVP signals can result in a large signal-to-noise ratio, which is desirable.

MATERIALS

Commercially available Si_3N_4 and Y_2O_3 powders were used in this investigation. Particle size distributions by laser light scattering, BET surface areas, and powder densities are listed in Table 1.

Table 1. Powder Characteristics

	Surface Area (m^2/g)	Particle Size Distribution (Size at Cum. Vol% Finer) 90%	50%	10%	Density (g/cc)
Denka SN-9S Silicon Nitride	7.2	4.4 um	0.94 um	0.38 um	3.2
Shin-Etsu SU Yttrium Oxide	9.5	1.3 um	0.56 um	0.27 um	5.0

EXPERIMENTAL PROCEDURE

Powders were prepared for zeta potential vs pH determination by preparing 5 vol% aqueous suspensions. The powder was placed in deionized water and mixed mechanically for 15 minutes. The suspension was then subjected to an ultrasonic horn[a] at 90 watts for 1 minute, then placed in an electrokinetic analyzer (ESA)[b], mechanically stirred and relative zeta potential [as calculated from ultrasonic vibration potential(UVP)] was monitored until it stabilized. The pH and zeta potential were then noted. The suspension was then titrated either to pH 2 or 10.5 using 2.0N HNO_3 or 2.0N NH_4OH, respectively, and zeta potential monitored continuously. Y_2O_3 was only titrated down to pH 7 because it dissolves under acidic conditions. Acid or base additions were made at 25 second intervals to allow for suspension equilibration. The isoelectric point was confirmed by monitoring the phase angle of the ESA probe, which switches 180° as the zeta potential crosses the isoelectric point. Samples were prepared for viscosity measurements and settling tests by preparing 5 vol% solids suspensions. Deionized water was adjusted to the desired pH (7, 8, 9, 10, or 10.5) and the powder added. After 15 minutes the samples were sonicated. The pH was maintained while the samples were aged until the zeta potential became stable. Viscosity was measured[c] using concentric cylinder sensors. Shear rates between 0 and 9 sec^{-1} were applied over a 3 minute ramp. Settling tests were performed by placing 15 ml of suspension into a sealed test tube. Final sediment heights and supernatent turbidity were noted after 48 hours.

Fifty wt% Si_3N_4/50 wt% Y_2O_3 mixtures of 5 vol% suspensions were prepared by premixing the dry powders, then adding them to deionized water preset at pH 7, 8, 9, 10, or 10.5. After the 50/50 samples had been prepared, 95 wt% Si_3N_4/5 wt% Y_2O_3 mixture suspensions were prepared at the same pH values for comparison and correlation with individual powders and 50/50 mixture results. The effect of solids content on cast density was examined by preparing Si_3N_4 suspensions at 1, 5, 10, 19, 37.7, and 45.2 vol% solids content. The pH of the deionized water was adjusted to 10.5 and the Si_3N_4 powder was added. The pH was maintained during powder addition. After all the powder was added, the suspensions were mechanically mixed for 15 minutes and then sonicated for 1 minute at 90 watts. The suspensions were further mixed for 1 hour and then

[a]Model W-385, Heat Systems Ultrasonics, Farmingdale, NY.
[b]Matec Instruments, Hopkinton, MA.

zeta potential monitored to verify stability. Viscosity was then measured and four discs at each set of conditions were cast using plaster base molds. The slip cast specimens were allowed to dry one day in air, then dried at 40°C for one day. Green density and percent porosity were measured using the Archimedes method. The effect of degree of dispersion on high solids content multi-component suspensions and cast part densities was examined by preparing suspensions of 37.7 vol% solids at pH 7, 8, 9, and 10.5 using Si_3N_4 and at pH 9 and 10.5 using 95 wt% Si_3N_4/5 wt% Y_2O_3. The suspensions were characterized by zeta potential, viscosity, and green cast part density.

RESULTS AND DISCUSSION

When placed in deionized water, the 5 vol% Si_3N_4 suspension produced a pH of 9.05 and a relative zeta potential of -14 mV. The calculated zeta potentials in these experiments must be considered relative values as the suspensions exhibit low aqueous conductivity (200-500 uS/cm^2). UVP signals are dependent on the complex high frequency conductivity in low conductivity solutions, but the ESA equipment is capable of only low frequency conductivity measurements. Thus the calculated zeta potentials are not absolute values but the relative change as suspension conditions are varied is directly proportional to the actual zeta potential for small changes in conductivity, which is the case for these experiments. No change in pH or zeta potential was observed during two hours of aging.

When placed in the ESA after mixing, the 5 vol% Y_2O_3 suspension exhibited a continual increase in zeta potential from 2 mV to 28 mV over 1.2 hours with a corresponding decrease in pH from an initial value of 7.28 to 7.00 over 1.2 hours and a decrease in UVP phase angle from -72 to 0°. At times greater than 1.2 hours, all three properties remained stable. A proposed explanation for the observed aging is that Y_2O_3 particle surfaces hydrolyze when placed in water.

The zeta potential of the individual Si_3N_4 and Y_2O_3 suspensions as a function of pH are shown in Figure 1. The Si_3N_4 exhibits an isoelectric point (IEP) at pH 4. Since the IEP of SiO_2 is pH 2, the Si_3N_4 surface is not simply SiO_2-like but must have some ionizable amine groups at the surface(9,14). Above pH 4 it exhibits a negative zeta potential that gradually increases in magnitude with pH. The Y_2O_3 powder exhibits an IEP at pH 9.25. At higher pH values it exhibits a negative zeta potential that increases rapidly in magnitude with pH. Below pH 9.25 it exhibits a positive zeta potential that increases rapidly with decreasing pH until pH 7, where dissolution begins. Based on Figure 1, it is expected that the Si_3N_4 powder should be well dispersed at pH values near 10.5 due to the large zeta potential (-23 mV). Y_2O_3 would be expected to be well dispersed at pH 7 (28 mV) or to a lesser extent at pH 10.5 (-19 mV) and poorly dispersed at pH 9.25 (0 mV). If the two powders were combined between pH values of 7 and 10.5, good dispersion due to mutual large repulsion forces would be expected at pH 10 to 10.5. Between pH 7 and 9.5, hetero-coagulation is expected due to oppositely charged zeta potentials and poor dispersion should result.

[c]Model RV/CV100, Haake-Buchler Instruments, Saddle Brook, NJ.

Viscosity, sediment height, and supernatent turbidity of each powder as a function of pH are also shown in Figure 1. The viscosity of the Si_3N_4 suspensions correlates well with relative zeta potential. As the zeta potential is increased, the viscosity decreases, indicating a more dispersed suspension. The viscosity of the Y_2O_3 suspensions also correlates well with zeta potential. The viscosity is very low at pH 7 (ξ= 28mV) and increases to a maximum at pH 9 (4mV), then drops off again at pH 10.5 (-19 mV). Sediment height also correlates well with zeta potential and viscosity for both powder suspensions. At high zeta potential and low viscosities, small sediment heights were observed, indicative of good dispersion while higher sediment heights were observed for lower zeta potential and higher viscosity conditions, indicative of floc formation and resultant poorer packing. Supernatent turbidity was rated on a scale from 1 to 5: 1 = clear, 2 = very slightly cloudy, 3 = slightly cloudy, 4 = cloudy, 5 = opaque. The ratings are subjective but observations indicate that differences between values were clearly discernable. The supernatent turbidity also correlates well with the other three parameters: zeta potential, viscosity, and sediment height. With high zeta potential, good dispersion is expected and is indicated by cloudy or opaque supernatent. Poor dispersion, indicated by low zeta potential, is indicated by clear supernatent, since flocs have formed and settled rapidly, trapping and dragging other particles down with them, with no mutual repulsion existing to keep particles suspended. The supernatent turbidity does track more closely with changes in zeta potential compared to the viscosity and sediment heights, which seem a little less sensitive to changes in zeta potential. Overall, for the two single component dilute (5 vol%) suspensions, all four measurements correlate well and indicate the relative degree of dispersion.

Properties of the 50 wt% Si_3N_4/50 wt% Y_2O_3 5 vol% suspensions as a function of pH are shown in Figure 1. For the 50/50 mixture suspension at pH 7, the relative zeta potential is the same (28 mV) as for pure Y_2O_3. Two mechanisms by which this result may occur are: 1.) the smaller and more numerous oppositely charged Y_2O_3 particles are attracted to the Si_3N_4 particles, heterocoagulation takes place, and the Y_2O_3 particles extensively cover the Si_3N_4 surface, masking the Si_3N_4 surface charge and resulting in a single component flocced system with all the particles exhibiting Y_2O_3 surface properties, or 2.) adsorption of soluble yttrium ions on the Si_3N_4 surfaces due to opposite charges resulting in well dispersed Y_2O_3 and Si_3N_4 particles with the same sign and magnitude zeta potential. The adsorption of soluble metal ions in colloid systems has been previously investigated(6,15-18) and yttrium has been specifically examined by Shaw and Pethica(9). But, since the ESA measures a single signal and this is a two component suspension, interpretation of signals is speculative without knowledge of the physical configuration of the powders in the suspension. Viscosity data indicates that the suspension seems well dispersed, consistent with the large relative zeta potential. Sediment height indicates hetero-coagulation has taken place though, since the composite height is greater than either of the component heights. Supernatent turbidity also indicates some coagulation is occurring. If soluble yttrium ions were coated on Si_3N_4 surfaces (soluble ion adsorption has been shown to occur in the timeframe of seconds(18)), a well dispersed suspension exhibiting Y_2O_3 and Si_3N_4 powders with the same charge would be expected and would result in low sediment heights and turbid supernatent, because ultrasonic treatment would be expected to break up flocs formed. Thus the most plausible explanation for the Y_2O_3-like zeta potential and floc formation is Y_2O_3 particle coagulation on Si_3N_4 particle surfaces.

At pH 8, the viscosity, sediment height, and supernatent turbidity all exhibit values close to pure Y_2O_3. Hetero-coagulation due to oppositely charged surfaces is expected and indicated by the high sediment height, low turbidity, and higher viscosity. At pH 9 and 10, all parameters indicate intermediate values between Si_3N_4 and Y_2O_3, except the supernatent turbidity at pH 9, which is clear and indicates that the Y_2O_3 flocs, then traps and drags the Si_3N_4 particles into settling. The intermediate values indicate that the small Y_2O_3 zeta potential does not induce large amounts of heterocoagulation, that the UVP signal is a combination of the two powders, and that one of the powder surfaces is no longer shielded by the other. At pH 10.5 both powders have large negative zeta potentials so mutual repulsion and good dispersion are expected. Zeta potential and sediment height exhibit values intermediate between the Si_3N_4 and Y_2O_3 values, while low viscosity and opaque supernatent indicate a high degree of dispersion. It is interesting to note that a pH 7 suspension may also be desirable if it ties up Y_2O_3 onto Si_3N_4 surfaces and lessens segregation that may occur due to density and particle size differences while exhibiting a viscosity and sediment height not much different from the well dispersed pH 10.5 suspension.

Dispersion properties of the 95 wt% Si_3N_4/5 wt% Y_2O_3 5 vol% suspension as a function of pH are also shown in Figure 1. Overall, parameters exhibit values very similar to pure Si_3N_4, which is expected due to the small Y_2O_3 addition and resultant small contribution to suspension properties. A few exceptions are noted though. At pH 7 zeta potential, viscosity, and supernatent turbidity exhibit the same values as pure Si_3N_4, but the sedimentation height is greater indicating that hetero-coagulation is taking place, expected due to the oppositely charged powder surfaces. At pH 8 the supernatent turbidity (clear) and greater sediment height again indicates that some heterocoagulation is occurring, but not enough to influence zeta potential and viscosity values. From pH 9 through 10.5 the zeta potential exhibits a slightly higher value then either of two components. The conditions for best dispersion of the 95/5 mix look to be at pH 10.5, due to highest same-sign zeta potential, lowest viscosity and sediment height, and most turbid supernatent.

The effect of solids content on pure Si_3N_4 suspensions is shown in Figure 2. Zeta potential determined by UVP decreases continuously with higher solids content above 10 vol% solids. The theory used to calculate zeta potential values from acoustophoretic measurements incorporates a linear dependence on solids content, but the dependence is nonlinear at solids contents above 10 vol% for many aqueous colloid suspensions due to the start of double layer overlap and hydrodynamic effects(11,19,20). Calculated zeta potentials at high solids contents are very useful, especially for comparison with same-solids content suspensions, but due to the nonlinearity must be carefully interpreted, compared with standards, but can not be directly compared with different solid content suspensions. Viscosity starts to increase appreciably above 19 vol% solids due to more frequent particle-particle interactions. The 37.7 and 45.2 vol% suspensions exhibit slight shear thinning behavior due to lower effective repulsion forces because of double layer overlap. All lower solids content suspensions exhibit newtonian rheology. The green density of cast compacts increase (and vol% porosity decreases) with increasing solids content as expected due to greater particle packing in suspension before casting. The point at which increasing shear thinning and/or dilatent rheologies start to hinder mold filling and cast part density have not yet been reached with a 45.2 vol% solids suspension. Besides producing higher green density parts, the higher solids content may be beneficial in reducing suspension settling and powder segregation because of hindered settling due to particle-particle contacts, and higher viscosities.

Figure 3 compares the effect of pH and Y_2O_3 additions to 37.7 vol% solids Si_3N_4 suspensions. Si_3N_4 suspensions prepared at pH 7 and 8 were too highly viscous and shear thinning to have properties measured or be pourable and conform to mold dimensions during casting, while suspensions at pH 9 and 10.5 had low enough viscosities and shear thinning tendencies to be measured and be pourable for casting experiments. The relative zeta potential for both the pure Si_3N_4 and 95 wt% Si_3N_4/5 wt% Y_2O_3 suspensions increases slightly from pH 9 to 10.5 (-14.4 to-14.5 mV and -13 to -13.7 mV, respectively) and decreases slightly with Y_2O_3 substitution for the same solids content. Although these changes are relatively small, they correlate with viscosity and cast green density measurements where it is observed that increasing the pH from 9 to 10.5 decreases viscosity and increases green density, which would be expected if degree of dispersion were increased. Substitution of 5 wt% Y_2O_3 lowers zeta potential and green density and increases viscosity, which indicates decreasing degree of dispersion, but the effect is very small at pH 10.5. This result is not predicted by results from the 5 vol% 95 wt% Si_3N_4/5 wt% Y_2O_3 suspension data but does correlate with results obtained in the 5 vol% 50 wt% Si_3N_4/50 wt% Y_2O_3 suspensions.

CONCLUSIONS

Dispersion conditions and solids contents that result in stable, castable aqueous suspensions and subsequent high green density slip cast 95 wt% Si_3N_4/ 5 wt% Y_2O_3 composition components have been achieved by: 1) producing well dispersed Si_3N_4-Y_2O_3 suspensions (generation of large like-charge electrostatic forces) by suspending the powders at pH 10.5 and 2) using high solids content suspensions to maximize suspension powder packing.

It was also observed that the acoustophoresis characterization technique, along with viscosity, settling, and casting tests, provides important information and insights on comparisons between dilute and concentrated multicomponent aqueous suspension and potential cast part properties, when correlated with each other.

REFERENCES

1. J.T.G. Overbeek, "Recent Developments in the Understanding of Colloid Stability," J. Colloid Interface Sci., **58** [2] 408-22 (1977).

2. J.T.G. Overbeek, "How Colloid Stability Affects the Behavior of Suspensions," in Emergent Process Methods for High Technology Ceramics, edited by R.F. Davis, H. Palmour III, and R.L. Porter, Plenum Press, NY, 25-44 (1982).

3. J Lyklema, "Interfacial Electrochemistry of Disperse Systems," ibid 1-24.

4. M.D. Sacks et. al., "Dispersion and Rheology in Ceramic Processing," in Ceramic Powder Science, edited by G.L. Messing, K.S. Mazdiyasni, J.W. McCauley, and R.A. Haber, Am. Ceram. Soc., Ohio, 495-516 (1987).

5. R.O. James, "Characterization of Colloids in Aqueous Systems," in Ceramic Powder Science, edited by G.L. Messing, K.S. Mazdiyasni, J.W. McCauley, and R.A. Haber, Am. Ceram. Soc., Ohio, 350-410 (1987).

6. T.W. Healy, G.R. Wiese, D.E. Yates, and B.V. Kavanaugh, "Heterocoagulation in Mixed Oxide Colloidal Dispersions," J. Colloid Interface Sci., **42** [3] 647-9 (1973).

7. E. Carlstrom and F.F. Lange, "Mixing of Flocced Suspensions," J. Am. Ceram. Soc., **67** [8] C169-70 (1984).

8. F.F. Lange, B.I. Davis, and E. Wright, "Processing-Related Fracture Origins: IV, Elimination of Voids Produced by Organic Inclusions," J. Am. Ceram. Soc., **69** [1] 66-69 (1986).

9. T.M. Shaw and B.A. Pethica, "Preparation and Sintering of Homogeneous Silicon Nitride Green Compacts," J. Am. Ceram. Soc., **69** [2] 88-93 (1986).

10. E.M. Deliso, A.S. Rao, and W.R. Cannon, "Electrokinetic Behavior of Al_2O_3 and ZrO_2 Powders in Dilute and Concentrated Aqueous Dispersions," in Ceramic Powder Science, edited by G.L. Messing, K.S. Mazdiyasni, J.W. McCauley, and R.A. Haber, Am. Ceram. Soc., Ohio, 525-536 (1987).

11. W. Albers and J.T.G. Overbeek, "Stability of Emulsions of Water in Oil: II. Charge as a Factor of Stabilization Against Flocculation," J. Colloid Sci., **14** [5] 510-18 (1959).

12. N. Ise, et. al., "Ordering of Charged Particles in Solution," Naturwissenschaften, **69**, S.544 (1982).

13. P. Debye, J. Chem. Phys., **1** 13 (1933).

14. D.L. Feke, "Stability and Rheology of Dispersions of Silicon Nitride and Silicon Carbide," NASA Contractor Report 179634, Lewis Res. Ctr., Cleveland, 1987.

15. R.O. James and T.W. Healy, "Adsorption of Hydrolyzable Metal Ions at the Oxide-Water Interface: I. Co(II) Adsorption on SiO_2 and TiO_2 as Model Systems," J. Colloid Interface Sci., **40** [1] 42-52 (1972).

16. R.O. James and T.W. Healy, "Adsorption of Hydrolyzable Metal Ions at the Oxide-Water Interface: II. Charge Reversal of SiO_2 and TiO_2 Colloids by Adsorbed Co(III), La(III), and Th(IV) as Model Systems," J. Colloid Interface Sci., **40** [1] 53-64 (1972).

17. G.R. Wiese and T.W. Healy, "Coagulation and Electrokinetic Behavior of TiO_2 and Al_2O_3 Colloidal Dispersions," J. Colloid Interface Sci., **51** [3] 427-33 (1975).

18. G.R. Wiese and T.W. Healy, "Adsorption of Al(III) at the TiO_2-H_2O Interface," J. Colloid Interface Sci., **51** [3] 434-42 (1975).

19. I.A. Aksay and C.H. Schilling, "Mechanics of Colloidal Filtration," ibid 85-93.

20. Matec Instruments ESA Literature, Hopkinton, Mass.

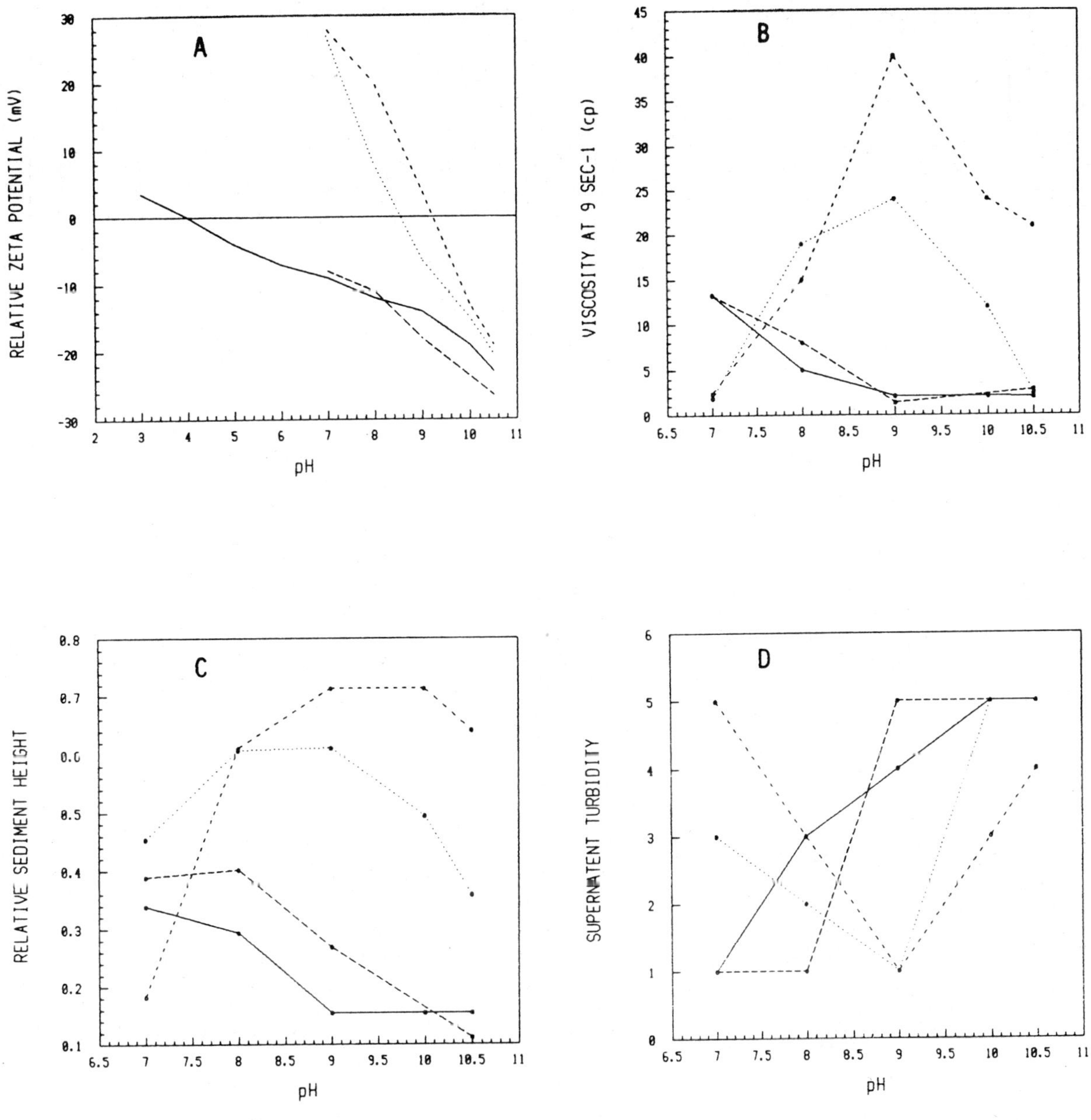

Figure 1. Comparison of properties of 5 vol% solids suspensions of Si_3N_4(—), Y_2O_3(- -), 95 wt% Si_3N_4/5 wt% Y_2O_3(--), and 50 wt% Si_3N_4/50 wt% Y_2O_3(·····) as a function of pH: (A) relative zeta potential, (B) viscosity, (C) sediment height, and (D) supernatent turbidity.

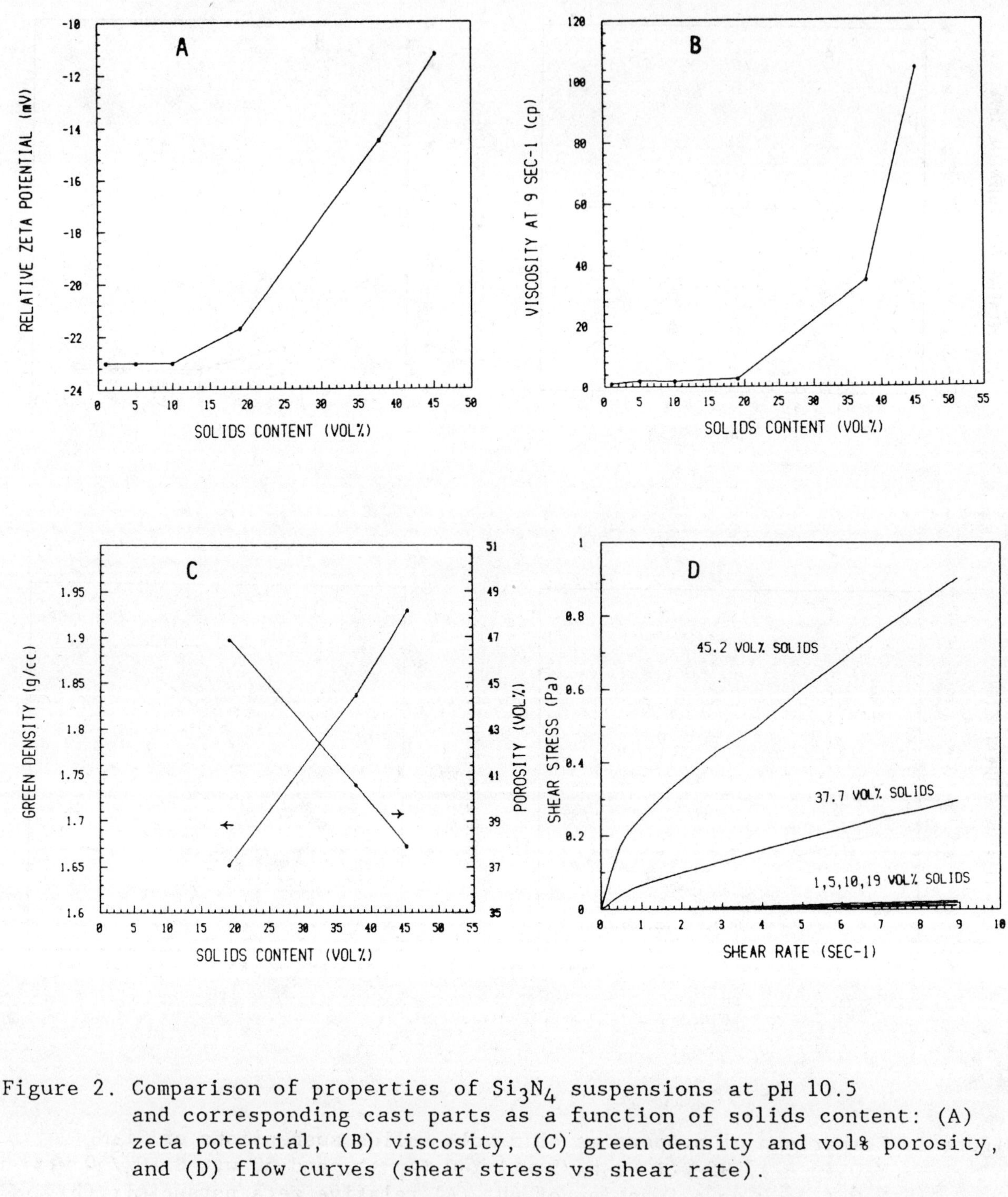

Figure 2. Comparison of properties of Si_3N_4 suspensions at pH 10.5 and corresponding cast parts as a function of solids content: (A) zeta potential, (B) viscosity, (C) green density and vol% porosity, and (D) flow curves (shear stress vs shear rate).

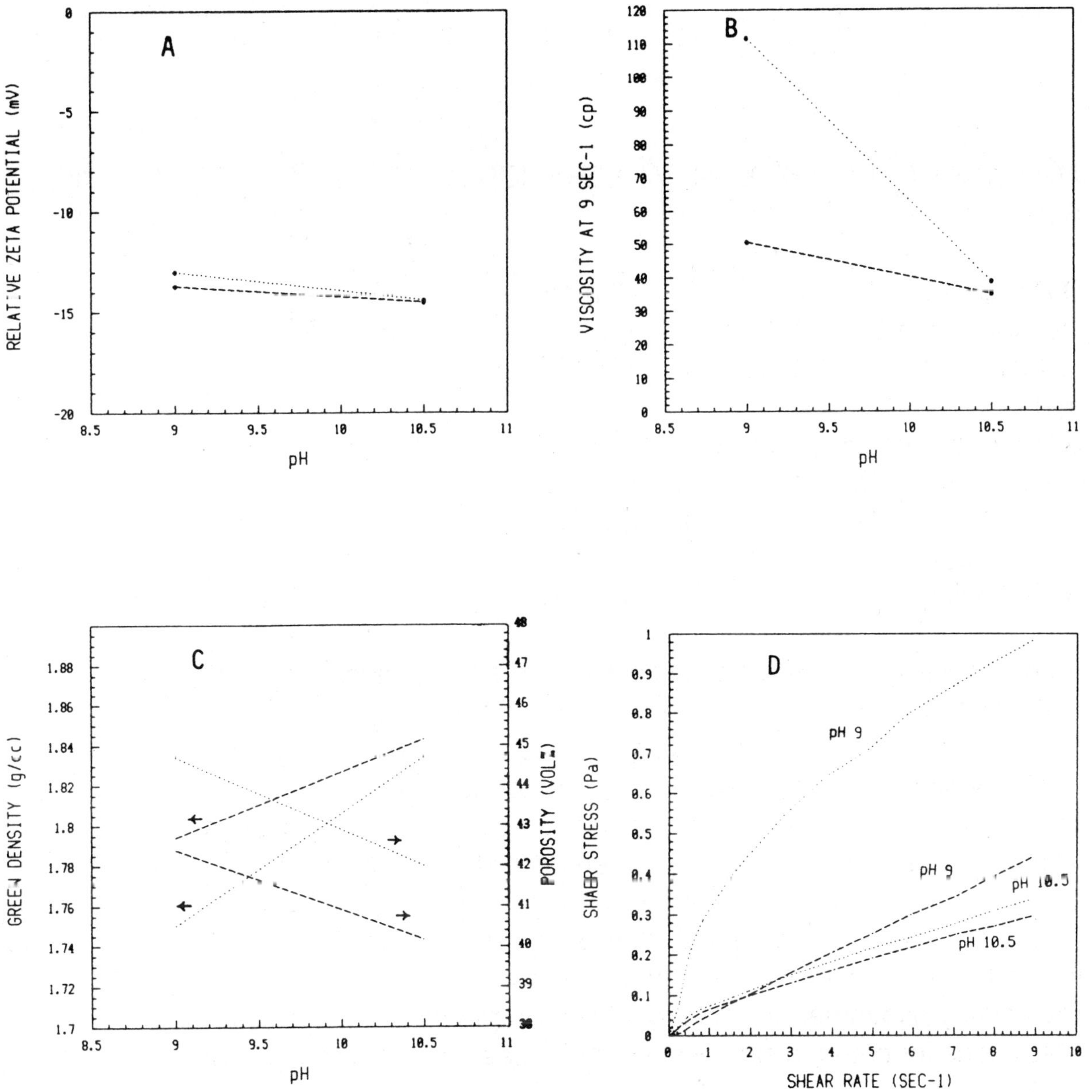

Figure 3. Comparison of properties of 37.7 vol% suspensions and corresponding cast parts of Si_3N_4(--) and 95 wt% Si_3N_4/5 wt% Y_2O_3(····) as a function of pH: (A) zeta potential, (B) viscosity, (C) green density and vol% porosity, and (D) flow curves (shear stress vs shear rate).

ANALYSIS OF THE RHEOLOGICAL BEHAVIOR OF CONCENTRATED CERAMIC SUSPENSIONS

D.Doraiswamy, I-L Tsao and S.C. Danforth
Department of Ceramic Engineering
Rutgers University, Piscataway, NJ 08854-0909
and
A.N. Beris and A.B. Metzner
Department of Chemical Engineering
University of Delaware, Newark, DE 19716

ABSTRACT

Highly concentrated ceramic suspensions in a low-molecular-weight polymer are used to manufacture whisker reinforced ceramic-ceramic composites by the injection molding technique. A detailed understanding of the rheology of these systems is crucial to the success of this operation. A simple non-linear model is presented which combines viscous, elastic and yielding phenomena in order to describe rheological data for concentrated suspensions of silicon particles in polyethylene. The model parameters are obtained solely using linear viscoelastic properties in oscillatory flow which are easily accessible experimentally. Preliminary experimental data are in good agreement with theoretical predictions for both the steady state as well as the dynamic behavior.

NOMENCLATURE

<u>Roman Letters</u>

D = viscous stress/elastic stress = $K\gamma_0{}^n\omega^n/(G\gamma_c)$

F(θ,n) = function defined in Eq. (7)

G = elastic modulus, Pa

G'_k, G''_k = k^{th} harmonic coefficients

G'_1 = storage modulus (abbreviated to G'), Pa

G''_2 = loss modulus (abbreviated to G"), Pa

K = consistency in power law model, $Pa.s^n$

n = power law index

S = dimensionless strain = γ_o/γ_c

Greek letters

$\dot{\underline{\underline{\gamma}}}$ = deformation rate tensor, 1/s

$\dot{\bar{\gamma}}$ = second invariant of deformation rate tensor, 1/s

$\hat{\gamma}$ = effective strain defined in Eq. (2)

γ_c = critical strain

γ_o= maximum imposed strain defined in Eq. (4)

η', η'' = dynamic viscosities defined in Eqs. (9) and (10), Pa.s

η^* = complex viscosity defined in Eq. (11), Pa.sτ = stress tensor, Pa

$\underline{\underline{\tau}}$ = stress tensor, Pa

τ = second invariant of stress tensor, Pa

τ_y = yield stress defined in Eq. (2), Pa

ω = frequency, 1/s

INTRODUCTION

The introduction of whiskers into a ceramic increases the fracture toughness of these systems and greatly reduces the sensitivity of these materials to microstructural defects. Consequently, whisker reinforced ceramic-ceramic composites have considerable potential for structural applications since they have a number of very appealing characteristics such as high strength, high toughness, light weight and, most importantly, high temperature resistance. Typical applications which exploit these properties are in the manufacture of turbine baldes and IC engine parts.

A proper description of the rheology of highly concentrated ceramic suspensions is essential in order to predict the process parameters of relevant to the injection molding of whisker reinforced ceramic-ceramic composites [1,2]. It is desirable to operate at as high a solids concentration as possible (>60 vol %) in order to minimize flaws that might occur due to particle rearrangement during the binder burnout or during the sintering or firing process. A detailed understanding of the rheology of these systems is essential in order to to relate pressure levels to injection rates and to determine the whisker orientation, residual stresses, and unwanted deformations that may occur in the molded part. In this paper a simple constitutive equation is presented (based on the concept of a limiting recoverable strain) in order to describe the non-linear strain dependent viscoelastic behavior of these materials. An analysis is then developed which enables the determination of the model parameters from conventional dynamic measurements. The resulting equations enable a correlation of the dynamic response with the shearing behavior in steady state flows which are usually encountered in injection molding.

A CONSTITUTIVE EQUATION FOR CONCENTRATED CERAMIC SUSPENSIONS

Highly concentrated ceramic suspensions, which are strongly influenced by particle-particle interactions, might be anticipated to exhibit a purely elastic response at low strains followed by a purely viscous response at high strains. These ideas are incorporated in the following simple equation:

$$\underline{\underline{\tau}} = \begin{cases} \left(\frac{\tau_y}{\dot{\gamma}} + K\dot{\gamma}^{n-1}\right)\underline{\underline{\dot{\gamma}}} & |\hat{\gamma}| = \gamma_c \\ G\underline{\underline{\hat{\gamma}}} & |\hat{\gamma}| < \gamma_c \end{cases} \tag{1}$$

The yield stress τ_y, in turn, is related to the critical strain γ_c at which the material first starts to exhibit viscous flow:

$$\tau_y = G\gamma_c \tag{2}$$

where G is the elastic modulus. The effective strain γ incorporates a recoverable strain in the model:

$$\frac{\partial \underline{\underline{\hat{\gamma}}}}{\partial t} = \begin{cases} 0 & |\hat{\gamma}| = \gamma_c \\ \underline{\underline{\dot{\gamma}}} & |\hat{\gamma}| < \gamma_c \end{cases} \tag{3}$$

where $\partial/\partial t$ is evaluated at a reference frame codeforming with the fluid [3]. The choice of reference frame does not affect the calculations for the simple flows considered in this work and will therefore not be further specified at this stage. Eq. (3) indicates that below the yield stress (or critical strain) the material behaves like a purely elastic solid and the entire strain is recoverable; above the yield stress, however, the maximum recoverable strain is limited to the critical strain.
Physically, this is equivalent to stating that the elastic deformation is recoverable while the viscous deformation is not, *i.e.* , the material is only capable of remembering the strain it undergoes as an elastic solid.
Eqs. (1) - (3) are simple, but significant, generalizations of the classical Herschel-Bulkley fluid [4] and are the proposed equations for concentrated

ceramic suspensions. A similar set of equations, though somewhat less general in form, has been proposed by van den Brule et al. [5] and Yoshimura and Prud'homme [6] in order to describe the rheology of concentrated dispersions and emulsions, respectively. A less formal approach has also been used to describe the rheology of unctuous materials by Elliot and Green [7].

The four parameters in the current model are the critical strain γ_C, the elastic modulus G, and the power law constants K and n. In the following section an analysis is presented in order to allow determination of these material properties from conventional dynamic rheological data.

DETERMINATION OF MODEL PARAMETERS FROM DYNAMIC MEASUREMENTS

In dynamic measurements a sinusoidal strain of the form

$$\gamma = \gamma_0 \sin(\omega t) \tag{4}$$

is imposed on the material and the resultant stress, which is also periodic, is then studied:

$$\tau = \tau_0 \sum_{k=1}^{\infty} [G_k' \sin(k\omega t) + G_k'' \cos(k\omega t)] \tag{5}$$

For linear viscoelastic materials the first harmonics are sufficient to describe the stress response, while for non-linear materials the higher higher harmonics can also become important. For linear materials the in-phase component of the resultant stress is characterized by the storage modulus G_1' which is a measure of the elastic response of the material while the out-of-phase component is characterized by the loss modulus G_1'' (abbreviated to G' and G", respectively, in the rest of this work).

For the kinematics outlined in Eq. (4), it is possible to obtain analytic expressions for G' and G" in terms of the dimensionless strain S, the dimensionless stress D, and the phase shift, θ [8,9]:

$$G' = \frac{2G}{\pi}\left[\frac{2-S}{S}\sin\theta + \frac{D}{S}\frac{(\sin\theta)^{n+1}}{(n+1)} + \frac{\theta}{2} + \frac{\sin 2\theta}{4}\right] \tag{6}$$

$$G'' = \frac{2G}{\pi}\left[\frac{2-S}{S}\cos\theta + \frac{D}{S}F(\theta,n) + \frac{3}{4} + \cos 2\theta\right] \tag{7}$$

The parameter θ is the phase shift introduced as a result of the elastic nature of the fluid. It depends on the imposed strain γ_0 and introduces a non-linear response in the fluid:

$$\theta = \cos^{-1}\left(1-\frac{2}{S}\right) \quad 0 \leq \theta \leq \pi \text{ for } S \geq 1 \tag{8}$$

The function $F(\theta,n)$ is defined by the following expression:

$$F(\theta,n) = \int_{\theta}^{\pi} \sin^{(n+1)}x \, dx \tag{9}$$

for the most general case, but reduces to an analytic expression in terms of an infinite converging series in terms of sine and cosine functions of and gamma functions of n for certain ranges of θ [8,9]. From Eqs. (6) and (7) it can be seen that the first harmonics G' and G" alone can be used to determine the material parameters of non-linear ceramic suspensions which can then be used to determine the higher harmonic coefficients. As will be shown later, the material parameters, once determined, can then be used to evaluate the shearing behavior of these materials in steady state flows. This feature is of considerable practical significance since, in general, dynamic data are far more easily obtained in the laboratory than steady state data in the range of shear rates commonly encountered in injection molding.

EXPERIMENTAL

Experimental dynamic as well as steady state data were obtained on a Rheometrics* RMS 800 mechanical spectrometer in a cone and plate geometry. The plate diameter was 25 mm and the cone angle was 5.7°. The experimental system investigated was a 70 vol % suspension of silicon particles** in polyethylene***. The average particle size of the silicon powder was 9 µm. The polyethylene exhibited Newtonian behavior with a viscosity of 6 Pa.s at a temperature of 140°C. Additional experimental details are provided elsewhere [10].

----

* Rheometrics RMS 800, Rheometrics, Inc., Piscataway, NJ

** Globe Metal Co., Beverly, OH

*** Allied Chemical Co., Morristown, NJ

RESULTS AND DISCUSSION

The dynamic viscosities η' and η'', and the complex viscosity η^* are often used in place of the moduli G' and G" and are defined as follows:

$$\eta' = \frac{G''}{\omega} \tag{10}$$

$$\eta'' = \frac{G'}{\omega} \tag{11}$$

$$\eta^* = \sqrt{\eta'^2 + \eta''^2} \tag{12}$$

Typical complex viscosity data are plotted in Fig. (1) in terms of an "effective shear rate" defined as the product of frequency, ω, and the maximum imposed strain, γ_0. It can be clearly seen that the curves for all the different strains (1% to 20%) are superimposed when the abscissa is defined in this manner. An initial slope of -1 is observed at low frequencies (corresponding to an elastic solid-like response) followed by a gradual transition to a viscous shear-thinning behavior.

The complex viscosity predicted by Eqs. (6) and (7) is plotted in Fig. (2) in terms of the same abscissa ($\gamma_0\omega$) for the following parameter values: G = 50,000 Pa, K = 400 Pa.s, n = 0.5 and γ_c = 0.005. The parameter values were determined by fitting the dynamic moduli (G' and G") using the 1 % strain curve and the other curves were generated using the same values. It can be seen clearly that the model predicts the proper superposition of the curves for the various strains in terms of the "effective shear rate".

Although the magnitude of the critical strain is very small, it can severely influence the flow patterns (and, consequently, the whisker orientation) in mold geometries of complex shape (like turbocharger molds) where abrupt changes in flow direction are possible. In some molding applications a random orientation of whiskers may be desirable while, in others, oriented whiskers might be preferred. The magnitude of the yield stress (or the critical strain) is related to the size of plug-flow region in the mold. Since the whiskers would tend to maintain a random orientation in the unyielded or plug-flow region, a good mold design could exploit this feature in order to obtain some control over the whisker orientation. The yield stress also determines the size of the largest molded part which would not slump under its own weight. It is therefore likely to be related to the green strength of the infired ceramic.

It is worth mentioning here that the individual magnitudes of the viscosity coefficients η' and η'' cannot be obtained from the curves shown here. For the data shown in Figures (1) and (2), the magnitudes of the dynamic coefficients are comparable over a wide range of frequency. This unusual behavior is not exhibited by the elastic Bingham model or the classical Herschel-Bulkley model. The modified Herschel-Bulkley model proposed here, however, predicts finite values for both the viscosity coefficients with a maximum error of ~ 25 %. This error might be attributed to an oversimplification of the model since the transition from an elastic to a viscous response may not occur at a point value of the strain, as assumed, but rather over a range of strains. It is planned to incorporate a smooth transition in behavior from an elastic to a viscous behavior in a further extension of the present work. Finally, it should also be noted that these preliminary data have not been corrected for slip effects which could introduce additional errors. Initial investigations, however, indicate that slip effects might not be very severe (causing an error in the apparent viscosity of less than ~15 %) [10].

Typical experimental steady state shearing data obtained in a cone and plate geometry are also plotted in Fig. (1) and the corresponding model predictions are depicted in Fig. (2). The parameter values used were the same as those used for the prediction of the dynamic behavior. It can be seen that the agreement between theory and experiment is very good. It thus appears that the model can predict the steady state behavior from dynamic measurements. It is intended to further check the validity of the model by obtaining high shear rate capillary data in the range of shear rates commonly encountered in injection molding. The close agreement between the steady state shearing data and the dynamic data is of considerable practical value since dynamic data are much easier to obtain than steady state high shear rate data.

CONCLUSIONS

A non-linear model has been presented for describing the rheological behavior of concentrated ceramic suspensions which combines elastic, viscous and yielding phenomena. A major feature of this new formulation is the incorporation of a recoverable strain. The recoverable strain is limited to a maximum value equal to the critical strain at which the transition from a purely elastic to a viscous shear thinning response occurs. Although the magnitude of the critical strain is usually very small, it might be expected to severely influence the flow patterns in molds of complex shape where abrupt changes in flow direction may be encountered. The model parameters are determined from conventional linear viscoelastic properties in dynamic measurements. Initial results for a 70 vol % suspension of silicon particles in polyethylene indicate good agreement between data and experiment for both the dynamic as well as the steady state shearing behavior. A very useful practical feature of the model is that it permits a prediction of the steady state behavior from dynamic measurements.

REFERENCES

[1] Tsao, I-L., Metzner, A.B. and Danforth, S.C., "Rheological Study of Highly Filled Ceramic-Ceramic Composites," Proceedings of the American Society for Composites, Second Technical Conference, September 23-25, Technomic, Lancaster, 1987, p. 316.

[2] Holman, R., Bigg, D.M., Mutsuddy, B.C., and Kurth, R.E., Advances in Ceramics, Vol. 9, 1984, The American Ceramic Society, 1984, p.259.

[3] Bird, R.B., Armstrong, R.C. and Hassager, O, Dynamics of Polymeric Liquids, Vol.1, 2nd Ed., John Wiley, New York (1987).

[4] Herschel, W.H. and Bulkley, R., Kolloid Z., 39, 291 (1926) as cited by Bird, R.B., Dai, G.C. and Yaruso, B.J., Rev. Chem. Engg., 1, 1 (1983).

[5] van der Brule, B.H.A.A., Dijksman, J.F. and Zander, P.J., Supplement to Rheologica Acta, Vol. 26, 186 (1988).

[6] Yoshimura, A.S. and Prud'homme, R.K., Rheologica Acta, 26, 428 (1987).

[7] Elliott, J.H. and Green, C.E., J. Textile Studies, 3, 194 (1972).

[8] Doraiswamy, D., Tsao, I-L., Beris, A.N., Danforth, S.C. and Metzner, A.B., Proceedings of the Xth World Congress on Rheology, Sydney, Monash University Printing Press, Vol. 1, 1988, p. 300.

[9] Doraiswamy, D., Tsao, I-L., Beris, A.N., Danforth, S.C. and Metzner, A.B., "A Rheological Model for Concentrated Ceramic Suspensions," in preparation.

[10] Tsao, I-L., Doraiswamy, D. and Danforth, S.C., "Rheological Behavior of Injection Moldable Ceramic-Ceramic Composites," paper to be presented at the 3rd International Symposium on Ceramic Materials and Components for Engines, Las Vegas, November, 1988.

ACKNOWLEDGEMENT

This work was supported by the National Science Foundation through the Center for Composites Manufacturing Science and Engineering.

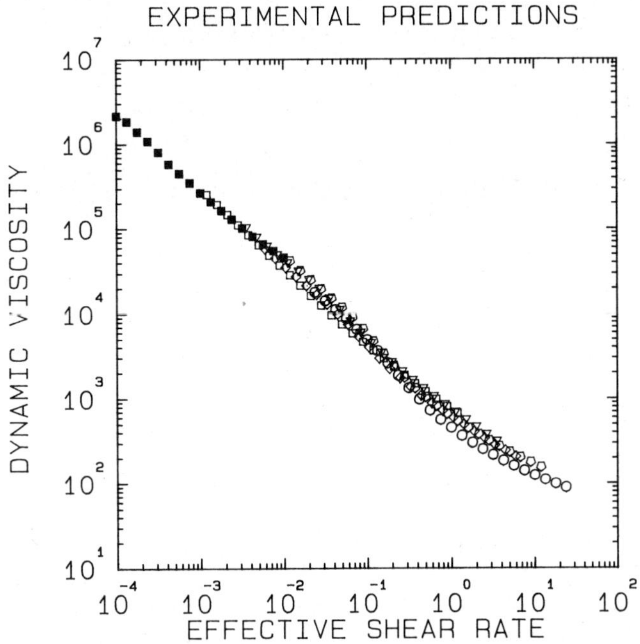

Figure 1: Experimental complex viscosity data and steady state shear viscosity data as a function of the "effective shear rate," $\gamma_0\omega$, and the steady shear rate, γ, respectively (symbols: O = 20% strain, ∇ = 10%, + = 5%, x = 3%, □ = 1 %, ■ = steady state).

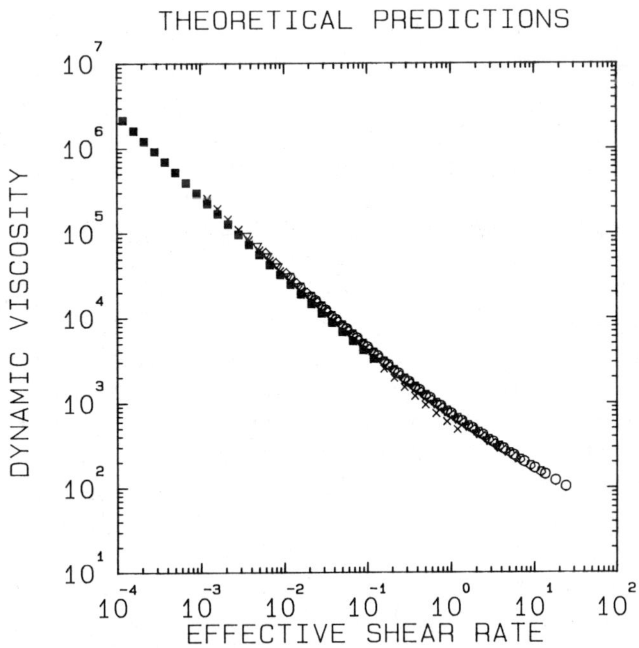

Figure 2: Theoretical predictions for complex and steady state shear viscosities as a function of the "effective shear rate," $\gamma_0\omega$, and the steady shear rate, γ, respectively (symbols are the same as in Fig. (1)).

SINTERING OF Si_3N_4-Y_2O_3-SiO_2 COMPOSITIONS WITH Mo_2C ADDITIONS

George E. Gazza
U.S.Army Materials Technology Laboratory,
Watertown, MA 02172

Barry Robinson
Geo-Centers Inc.
Newton Upper Falls, MA. 02164

ABSTRACT

Compositions of Si_3N_4-Y_2O_3-SiO_2 located on the Si_3N_4-$Y_2Si_2O_7$ join in the phase behavior diagram were sintered under high N_2 gas pressures at temperatures of 1850-1960C. Gas pressures up to 8 MPa were used in a two-step sintering method. Additions of Mo_2C (up to 1.0m/o) made to the base compositions appear to promote sintering of the samples. Grain size, morphology, and distribution of phases in the sintered materials were observed by SEM. Mechanical behavior, i.e., modulus of rupture, and stress-rupture, were determined at temperatures to 1250C as well as resistance to oxidation.

INTRODUCTION

The potential of silicon nitride for high temperature structural applications is well known. Of particular interest is the use of densification processes applicable to complex shapes. These processes include conventional sintering under 0.1 MPa of gas pressure, high gas pressure sintering (usually in the range of 1.0-10.0 MPa gas pressure), and hot isostatic pressing (100-300 MPa pressure). When relatively small amounts of additives are used for densification and/or when the particular additives used form high melting point/high viscosity liquids at the grain boundaries, high gas pressure processes are usually required for full densification.

In this study, sintering of compositions (shaded area of Figure 1) of Si_3N_4-Y_2O_3-SiO_2 near/on the Si_3N_4-$Y_2Si_2O_7$ join was investigated to determine the sintering parameters/conditions required for densification and resultant room temperature and high temperature properties. The densification process used was high gas pressure sintering with a two-step sintering cycle. A specific composition selected for study was 85.8m/o Si_3N-4.73m/o Y_2O_3-9.47m/o SiO_2. Previous studies (1-5) have shown that inadvertant impurities in silicon nitride, usually from powder processing or from starting materials, may produce defects in the densified product which limit strength of the material and affect high temperature mechanical behavior and oxidation resistance. Of particular concern was the observation that tungsten based inclusions provided consistent sources of cavitation in both MgO and Y_2O_3 doped Si_3N_4 (1). In these studies, the WC was introduced from the milling media and was present in amounts >3% with uncontrolled size and distribution. Other work (6) has reported no observable propensity for cavities to form at inclusions such as WSi_2 or WC. In a prior study (7), it was observed that the presence of WC impurity (up to approximately 2w/o) accelerated sintering and permitted full density of the specimens to be attained at lower temperatures than when WC was not present. The specimens appeared dense and very dark in color, similar in appearance to hot pressed silicon nitride. It was of interest to explore in the present study, the enhanced sintering effect using Mo_2C particles rather than WC as the use of lower density particles would improve their distribution in the silicon nitride matrix with the established powder mixing/milling procedures, and the reaction product, molybdenum silicide, has better high temperature properties than tungsten silicide. Also, other impurities usually present in WC milling media, e.g.,Fe, Co, could be reduced or avoided.

EXPERIMENTAL

The source of silicon nitride powder used for starting material was TOSOH Corp.(grade TS-7)* . Appropriate amounts of Y_2O_3** and SiO_2 (the oxygen content of each powder was used to estimate surface SiO_2 which was included in the calculation) were added to the Si_3N_4 powder to produce the composition 85.8m/o Si_3N_4-4.73m/o Y_2O_3-9.47m/o SiO_2. Preliminary studies (7) indicated that the presence of WC or Mo_2C particles in Si_3N_4-Y_2O_3-SiO_2 specimens accelerated densification of compacts and promoted the formation of beta phase Si_3N_4 at temperatures above 1650C. To further explore this observation and determine properties of the materials, Mo_2C particles were extracted from suspension after sedimentation of Mo_2C powder***, and were added to the Si_3N_4-4.73m/o Y_2O_3-9.47 SiO_2 composition in amounts up to 1.0m/o. The powder was ball-milled in ethanol using RBSN milling jars and hot isostatically pressed Si_3N_4 milling balls. It was dried and screened and uniaxially cold pressed into 5.08 cm diameter discs with a thickness between 0.63-0.94 cm. The discs were then cold isostatically pressed at 170 MPa to produce "green" densities of 1.62-1.68 g/cc. The specimens were placed in a RBSN crucible with a lid and embedded in Si_3N_4 cover powder. The exterior of the crucible was painted with a BN slurry. The crucible was placed in a graphite resistance heated, high gas pressure sintering furnace and subjected to a two-step sintering cycle (shown in Figure 2). Nitrogen gas pressures used were 1.0-1.5MPa in the first step, then increased to 7.5-8.0MPa in the second step. Temperature cycles varied from 1870/1880C to 1950/1960C for steps 1/2. Hold times were 60 minutes/30 minutes for steps 1/2 for each temperature cycle. A

* TOSOH Corporation, Tokyo, Japan
** Molycorp Inc., Louviers, Colo.
*** Alfa Products, Danvers, MA

60 minute hold time at 1200C followed the sintering cycle. Theaverage sintered density obtained for the composition cited using Toyo-Soda starting powder was 3.29±0.015 g/cc. The density values obtained for these sintered discs and the specimens machined from the discs were very consistent. The Toyo-Soda composition without Mo2C additions sintered to only 93-96% of the value obtained when Mo2C additions were used. Sintered discs were machined into specimens for physical, mechanical, chemical, and microstructural characterization.

Microscopy

Sintered specimens were microstructurally examined by SEM to determine grain size, morphology and distribution of minor phases. Grains were typically elongated with diameters of approximately 1-2um. Finer microstructures formed as the sintering temperature was reduced. Phase determinations found the Si3N4 to be essentially all beta-phase. A secondary phase dispersed throughout the microstructure was determined to be Y2Si2O7. This appeared to be present in gamma or beta polymorphic form. Preliminary examination of the grain boundary phase in sintered specimens examined at Oak Ridge National Laboratory by STEM (8) indicated that they were yttrium- and silicon rich, did not contain Mo, and were comprised of interlocking single crystal networks. The presence of fine, equiaxed particles containing Mo and Si by EDXA was also dispersed in the microstructure (Figures 3a & 3b). These particles were approximately 1-2um maximum diameter and found to be approximately 70w/o Mo and 30w/o Si. This represents a composition between Mo5Si3 and MoSi2. It has been reported (9) that MoSi2 is dimorphic and undergoes a eutectoidal transformation upon cooling below 1850C to yield the low temperature form, alpha MoSi2, plus Mo5Si3. Such compositions might also be formed by the reaction of Mo2C with SiO and/or Si3N4 where the Mo5Si3 particle may form a MoSi2 shell by further reaction with Si,SiO2, or SiO.

Room Temperature Modulus of Rupture (MOR)

Specimens were machined from densified discs for room temperature MOR tests. Specimens were machined to dimensions of 1.5mm x 2.0mm x30mm long. Four point loading was used with a 10mm distance between top load pins and 20mm between the bottom pins. Platen speed was 0.5mm/min. Strength values were highest when sintering temperatures between 1925-1960C were used and appeared to fall when sintering temperatures were lowered to 1850-1900C. Mean values ranged from 710 MPa ± 60 MPa and 550 MPa ± 80 MPa, respectively. The fracture surfaces of selected bars were examined to observe grain size and morphology and to determine fracture origins. As expected, the grain size increased somewhat at the higher sintering temperatures and larger prismatic grains developing in the microstructure were particularly evident.

Various flaw types were found for specimens which had

strengths on the low side of the distributions. A common flaw serving as a fracture origin was a large pore (Figure 4) usually 15-30um in diameter, which primarily is formed by localized, differential sintering of agglomerates present in the "green" compact. Another defect resulting from powder processing is a metallic inclusion. An EDX analysis indicates the presence of Fe, Ni, and Co in addition to Si. A defect believed to be introduced from the starting powder is an agglomerate with Cl and K present (Figure 5). Such impurities are not uncommon when Cl containing precursors are used to produce silicon nitride powder. Other flaws found at the fracture origins were machining flaws or atypically large grains. Some room temperature MOR tests were conducted with the survivors of stress-rupture tests to determine retained strength of the bars. Since the high temperature fixtures and the room temperature fixtures were of similar dimensions, loading of the bars for RT MOR was performed with the same bar orientation as for the stress-rupture tests. It was attempted to place the loading pins for RT MOR in the same location as the loading marks observed on the specimen surface from stress-rupture loading. Test conditions were the same as for MOR tests conducted on specimens machined from as-sintered discs. The stress-rupture survivors, however, had some curvature resulting from strain developed during the stress-rupture tests. Measured amounts of strain varied depending on several factors which included test temperature, time, and applied stress. Permanent strain values are listed in the discussion on stress-rupture. Another consideration in determining the retained MOR is the possible source of experimental error resulting from beam curvature. From curved beam analysis, it was shown (10) that if the beam curvature/specimen thickness ratio is large, >40, the error is determined to be small (<1%). In the present study, significant increases in strength were observed for stress-rupture survivors over those tested without stress-rupture exposure. Values ranged from 800-950 MPa. It is uncertain at this time whether strengthening resulted from microstructural changes, (e.g.,changes in flaw size/distribution, crystallization of boundary phases), produced by long term, high temperature exposure under stress or whether this effect is principally due to residual stress in the bars.

<u>Stress-Rupture Testing</u>

Specimens machined from sintered discs were tested to determine their behavior in stress-rupture in air at both intermediate (700-1000C) and high temperatures (1200-1300C). Specimen dimensions were similar to those used for MOR testing. A four point loading stress of 300 MPa was customarily used with occasional tests to 400 MPa. Stress-rupture testing was conducted using the temperature-time-stress parameters shown in Table 1. Most tests were conducted using cycles to 1250C and a stress of 300 MPa where total permanent strain values did not exceed 0.25%.

TABLE 1. Stress-Rupture Cycles (Temperature/Time/Stress) and Permanent Strain Measurements on Sintered Si_3N_4 Specimens.

Temperature(C)*	Time(hrs.)	Stress (MPa)	Perm.Strain(%)
700/900/1100/1250	24/24/24/150	300/300/300/300	0.10-0.15
900/1200/1250	100/150/250	300/300/300	0.10-0.20
900/1250	24/750	300/300	0.15-0.25
1200/1250/1300	150/150/100	300/300/300	0.25-0.35
900/1200/1200/1250	100/150/150/100	300/300/400/400	0.25-0.40
900/1250	150/150	300/400	0.20-0.35

* Cycles are identified as 1-6 from top to bottom of the list.

Twenty-seven specimens were stress-rupture tested with twenty-two surviving for the complete cycle time. Failures occurred in the first cycle at 700C, second cycle at 1200C and 1250C, and two specimens in the sixth cycle at 1250C. Failure origins were determined to be either large pores in the specimens, 20-30um diameter, small pore clusters, or agglomerate/impurity regions. Most specimens survived these stress-rupture conditions having a thin, coherent oxide layer on the surface with residual machining marks still evident. Permanent strain values were derived from measurements on the bars where values of distance between top loading points (L), bar thickness (t), and the rise (Υ) generated by the arc formed by the specimen curvature and the chord subtended by that arc, were used in the following formula:

$$\mathcal{E} = \frac{(4)(t)(\Upsilon)}{L^2} \quad (1)$$

The testing of stress-rupture survivors to determine retained strength was discussed in the Room Temperature Modulus of Rupture (MOR) section of this paper.

Oxidation

Specimens were oxidized in air at temperatures of 900C, 1200C, and 1250C and times up to 250 hours. Weight gains were determined by periodically weighing the specimens on a balance with a sensitivity of $2x10^{-5}$ grams. Values of the oxidation rate constant ($kg^2m^{-4}s^{-1}$) were calculated for each oxidation temperature. At 900C, values were in the range of $5x10^{-14}$ to $5x10^{-13}$ ($3x10^{-13}$ mean) while those at 1200C were found to be in the range of $7x10^{-13}$ to $2x10^{-12}$ ($1.1x10^{-12}$ mean) and 1250C inthe range of $1.6x10^{-12}$ to $3.7x10^{-12}$ ($2.7x10^{-12}$ mean). As was observed with the stress-rupture specimens, oxide formation on specimens was uniform and coherent with residual machining marks still visible.

DISCUSSION OF RESULTS

The addition of Mo_2C to Si_3N_4 compositions containing Y_2O_3 and SiO_2 was found to promote densification of the specimens at temperatures of 1850-1960C. The Mo_2C particle additions were kept fine (1-2um diameter) to avoid/minimize strength degradation as reported (3,4) for Fe-based inclusions. The resultant molybdenum silicide particles, formed from Mo_2C, were not observed to to be associated with fracture origins in MOR specimens. Best sintering temperatures appeared to be in the range of 1925-1960C. Although most specimens survived the established stress-rupture conditions, examination by TEM is necessary to determine whether cavitation occurred near molybdenum silicide particle sites. For long term applications, operating temperatures of 1200-1250C appear maximum for these materials if a small amount of creep, as indicated by the permanent strain values, is tolerable. Post heat treatments to crystallize grain boundary phases was not extensively explored but may improve creep resistance.

The mechanism involved in promoting densification, however, is not yet clear. Suspected formation of a lower viscosity liquid phase by molybdenum carbide reaction with the yttrium silicate phase does not appear likely. EDX analysis of areas of yttrium silicate phase near molybdenum silicide particles did not indicate the presence of Mo. The absence of such a reaction may be consistent with Lange's observation (5) that the presence of tungsten silicide (from WC) in hot pressed compositions prepared in the Si_3N_4-Y_2O_3-$Y_2Si_2O_7$ triangle (see Figure 1) did not produce surface pitting during oxidation. In MgO doped Si_3N_4, pitting is produced by enhanced oxygen diffusion through a low viscosity glassy phase formed by the reaction of magnesium silicate with WC impurity. Another mechanism being investigated involves the decomposition of Si_3N_4 by reaction of Mo_2C particles with Si_3N_4, SiO_2, and/or SiO. As indicated by Negita (11), the decomposition of Si_3N_4 by MoO_2 is favored by comparing standard free energies of reactions between Si_3N_4 and various metal oxides. Although Mo_2C by itself is not effective in promoting densification of Si_3N_4, it may act in a catalytic manner by preventing coarsening of the structure during sintering and promoting reaction between the Si_3N_4 grains and $Y_2Si_2O_7$ grain boundary phase.

Acknowledgment

The authors are grateful to G. Meyers for SEM and EDX analysis; A.Fuller for the stress-rupture data; and A.Zani for specimen preparation for microscopy.

Supported by the U.S.Department of Energy, Assistant Secretary for Conservation and Renewable Energy, Office of Transportation Systems, as part of the Ceramic Tecnology for Advanced Heat Engines Project of the Advanced Materials Development Program, under Contract DE-AC05840R21400 with Martin-Marietta Energy Systems, Inc.

REFERENCES

(1) Tighe,N.J., Wiederhorn, S.M., Chuang, T.J., and McDaniel, C.L., "Creep Cavitation and Crack Growth in Silicon Nitride", Proceeding of the International Conference on Plastic Deformation in Ceramics, Penn. State Univ., University Park, PA., 21-22 July 1983

(2) Wiederhorn, S.M., and Tighe, N.J., "Structural Reliability of Yttria-Doped Hot Pressed Silicon Nitride at Elevated Temperatures", Jour. American Ceramic Society, 66, [12], 884-889, 1983

(3) Pasto, A.E., "Causes and Effects of Fe-Bearing Inclusions in Sintered Si3N4", Communications of the Amer. Ceramic Society, Sept.1984, C-178-180

(4) Neil, J.T., Pasto, A.E., and Bowen, L.J., "Improving the Reliability of Silicon Nitride: A Case Study", Advanced Ceramic Materials, 3,[3], 225-30, 1988

(5) Lange, F.F., "Reaction of Iron With Si3N4 Materials to Produce Surface Pitting", Jour.American Ceramic Society, 61, [5-6], 270-71, 1978

(6) Clarke, D.R., "The Microstructure of Nitrogen Ceramics", Progress in Nitrogen Ceramics, F.L.Riley, Ed., NATO ASI Series, Martinus Nijhoff, Pub., 1983, 341-58

(7) Gazza, G.E., unpublished work

(8) Allard, L., personal communication

(9) Brewer, L. and Lamoreaux, R.H., II. Phase Diagrams", Atomic Energy Rev. Spec. Issue No.7 (Molybdenum:Physico-Chemical Properties of It Compounds and Alloys),p.320, , L. Brewer,Editor, International Atomic Energy Agency, Vienna, 1980

(10) Baratta, F.I., Matthews, W.T., and Quinn, G.D., "Errors Associated with Flexure Testing of Brittle Materials", MTL TR 87-35, July 1987

(11) Negita, K., "Effective Sintering Aids for Ceramics", Jour. of Materials Science Letters, 4, (1985), 755-758

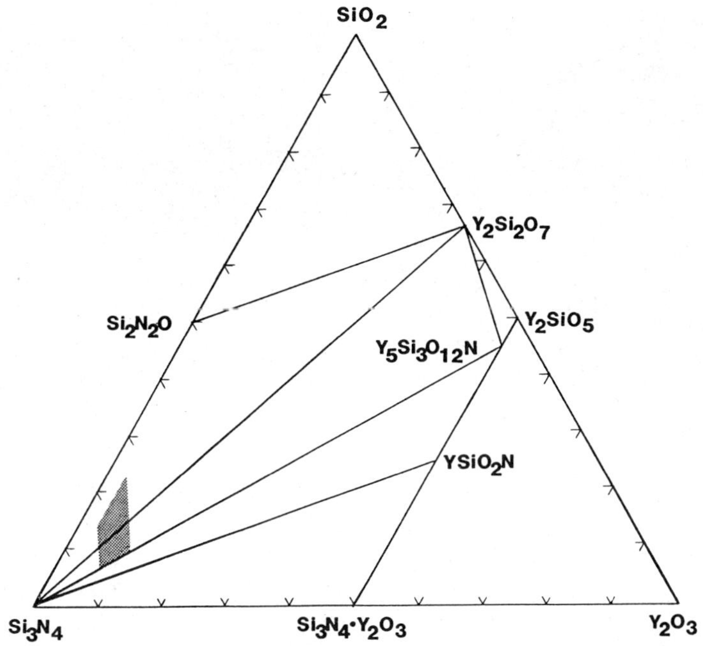

Figure 1. Phase diagram of Si3N4-Y2O3-SiO2 showing compositional area of interest (shaded area).

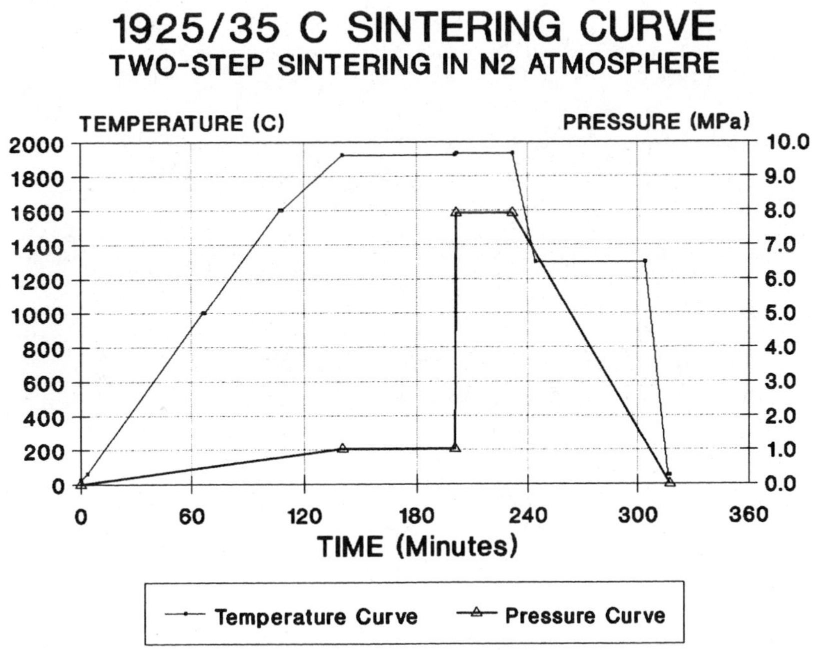

Figure 2. Typical two-step sintering cycle showing pressure and temperature parameters.

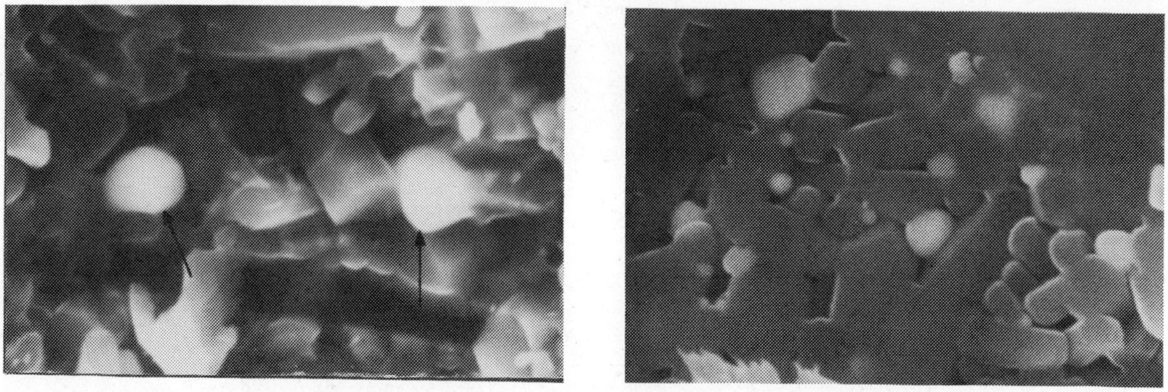

Figure 3. (a) Fractograph showing sites of molybdenum silicide particles (indicated by arrows) from reacted molybdenum carbide with silicon nitride (10,000X).
(b) Fractograph taken near fracture origin showing molybdenum silicide (white) located between silicon nitride grains (13,000X).

Figure 4. Large pore at fracture origin of bar tested at room temperature. Fracture strength was 511 MPa. (600X)

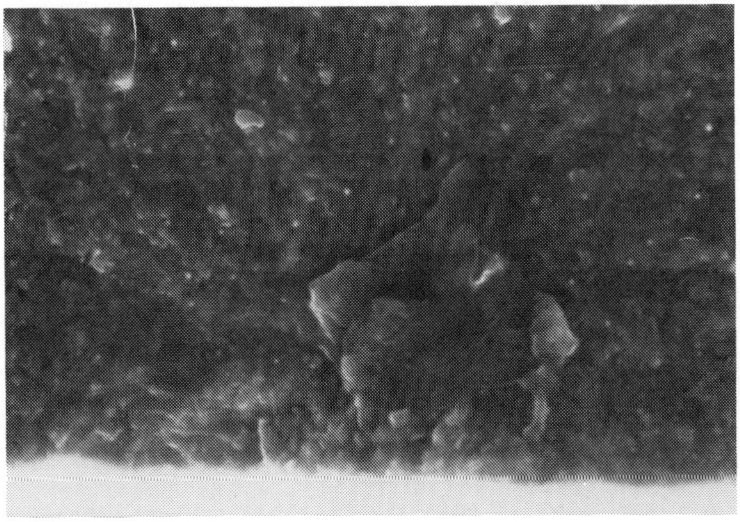

Figure 5. Agglomerate at fracture origin of sintered specimen. EDXA indicates presence of Cl and K. Fractrue strength was 506 MPa. (1400X)

MICROSTRUCTURE DEVELOPMENT OF SILICON NITRIDE CERAMICS

M. Peuckert, C. Boberski and P. Selgert
Hoechst AG, Materialforschung
D-6230 Frankfurt/M. 80, F.R.G.

ABSTRACT

Silicon nitride ceramics have been prepared by pressureless sintering at 1800 °C with Y_2O_3, La_2O_3, CeO_2, Pr_6O_{11}, Nd_2O_3, Sm_2O_3, Gd_2O_3, Tb_4O_7, Dy_2O_3, Ho_2O_3 and Er_2O_3 as sintering aids and in addition various amounts of Al_2O_3. The results suggest that Al_2O_3 acts as an inhibitor for glass phase crystallization, increases sintering density and lowers the aspect ratio of the β-Si_3N_4 grains. Crystallization of intergranular apatite phase increases room temperature strength, while a simple relation between nature of the glassy phase and flexural strength could not be observed.

INTRODUCTION

Silicon nitride can be sintered to dense ceramics of high strength at ambient and elevated temperatures by use of a vast variety of sintering additives, namely the oxides of aluminium, the alkaline earths, the rare earths, and some transition metals. Al_2O_3, MgO and Y_2O_3, and combinations of these are probably the most intensively studied additives. Despite numerous investigations the quest for the optimum Si_3N_4 sintering aid continues. This requires a detailed understanding of the influence of a single additive on densification and microstructure development and of the interactions in additive mixtures /1 - 3/. Therefore, we performed a systematic study on the pressureless sintering of Si_3N_4 using 11 rare earth oxides with or without additional Al_2O_3 as sintering aid.

Also the effect of the crystallization of amorphous intergranular phase on the strength of these SSN ceramics was investigated. Only very few studies have been made so far and they were restricted to Y_2O_3, CeO_2 and Al_2O_3 as sintering aids /4 - 8/. With these additives, it was generally observed that Y-silicates /4, 6/ or Y-garnet $Y_3Al_5O_{12}$ /7/ can be crystallized by heat treatment between 1100 and 1400 °C and that, thereby, high temperature strength is increased while strength at room temperature is lowered. No systematic study with other rare earths has been performed so far.

EXPERIMENTAL

Sample Preparation

Silicon nitride powder (average particle size d_{50} = 0.5 µm; BET = 21 $m^2 g^{-1}$; 95 % α-Si_3N_4) was intensively mixed with oxide sintering additives. Three series of samples were prepared: series I with rare earth oxides only, series II with rare earth oxide RE_2O_3 and Al_2O_3 in a molar ration of 2 : 1, and series III with rare earth oxide and Al_2O_3 in a molar ratio of 3 : 5. In all series the total amounts of additives was chosen such that the concentration of glass forming, sinteractive oxide was kept constant, i.e. the concentration of oxygen was always 0.2 mol per 100 g of SSN ceramic. This means in the case of Y_2O_3 additive combinations of 15.01 wt-% Y_2O_3, 10.01 wt-% Y_2O_3 + 2.28 wt-% Al_2O_3 (molar ratio 2 : 1), and 5.64 wt-% Y_2O_3 + 4.25 wt-% Al_2O_3 (molar ratio 3 : 5), respectively. Beside Y_2O_3 the following RE-oxides were studied: La_2O_3, CeO_2, Pr_6O_{11}, Nd_2O_3, Sm_2O_3, Gd_2O_3, Tb_4O_7, Dy_2O_3, Ho_2O_3, and Er_2O_3 (all oxides from Ventron, Karlsruhe, REO 99.99 %). Al_2O_3 used was of type A 16 from Alcoa (d_{50} = 0.5 µm). From the starting Si_3N_4 powder an additional 2.0 wt-% of SiO_2 was introduced. The premixed starting materials were wet milled in an attritor mill with isopropanol as medium; the resulting suspensions were dried and the dried powders isostatically pressed under 300 MPa to green bodies of approximate dimensions of 10 x 15 x 60 mm^3. These samples were then sintered in 0.1 MPa of N_2 with the following program: heat up in 1 h to 1800 °C, hold for 1 h at 1800 °C, then cool down in 3 h to room temperature. The sintered density was determined by the buoyancy method. Grain size and aspect ratio was measured on SEM pictures of overetched specimens as described in ref. /10/.

Post-sintering Annealing

Portions of the sintered samples were annealed in 0.1 MPa of N_2 for 24 h at 1350 °C in order to crystallize the amorphous intergranular glassy phase.

Mechanical Testing

From the sintered specimens test bars of 3.5 x 4.5 x 50 mm^3 were cut, polished (1 µm), and edges rounded and the 4-point bending strengths (20 mm/40 mm SiC fixture; 500 N s^{-1} loading) were determined. Data were averaged from at least 6 flexure bars per composition.

RESULTS AND DISCUSSION

As outlined in the experimental section, 3 series of rare earth doped Si_3N_4-ceramics have been prepared by pressureless sintering in N_2 at 1800 °C: the first without any, the second with little, and the third with a large additional amount of Al_2O_3. The results are summarized in Figs. 1 and 2. Almost all additive combinations yield satisfactory densification to more than 95 % of the theoretical density. An obvious correlation with the melting behaviour of the RE-Al-Si-O systems, nevertheless, is not found; Fig. 1 also shows the eutectic temperatures given in the literature /9/ of the ternary RE-Si-O and RE-Al-O systems. There is a general pronounced effect of the addition of alumina on densification, which can be understood by the decrease of the melting point of the quarternary compared with the ternary glasses. Two elements, Sm and Gd, seem to play a special role. Their behaviour is easily rationalized by the ceramic microstructure, Fig. 2. SSN-ceramics with Sm and Gd, without Al_2O_3, are characterized by very large aspect ratios /10/ of the β-Si_3N_4 grains. Fig. 3a - c shows SEM pictures of the corresponding microstructures. Our results, therefore, suggest that beyond a distinct aspect ratio of about 10, more pronounced in situ whisker formation can in fact occur, e.g. with Sm or Gd glassy phase. But, unfortunately such microstructures encounter self-inhibition of their densification. As Al_2O_3 is added the aspect ratio is lowered and density is increased, Figs. 1 - 3. Another effect of Al_2O_3 is shown in Fig. 2. With Y, La, Nd and Dy grain growth is accelerated while with Ce, Pr, Sm, Gd and Ho grain growth appears to be inhibited by Al_2O_3 addition.

For practical purposes the achieved bend strengths of the different SSN ceramics are, of course, of interest. The room temperature values are listed in Table 1. Considering the parameters density, aspect ratio and grain size, it is not surprising that a simple interpretation of mechanical strength versus chemical composition is impossible /1 - 3/. For the different elements the microstructure parameters are of changing importance. For example, in the case of Gd_2O_3 strength increases with Al_2O_3 content because of increasing density as discussed above. The strength decrease with Al_2O_3 content of the Dy samples can be rationalized by the decreasing aspect ratio, while the strength data of the Sm-SSN ceramics might be explained by a counteraction of aspect ratio versus density.

Table 1 also contains flexural strength values of SSN ceramics after 24 h annealing at 1350 °C. The data of (1 : 0) and (3 : 5) compositions are difficult to interpret since SSN without Al_2O_3 (1 : 0) contains partially crystallized intergranular phase even without annealing and in SSN with a large amount of Al_2O_3 (3 : 5) crystallization could not be detected by XRD /6/, see Fig. 4. Therefore, only data of (2 : 1) mixtures may be used to check the influence of the crystallization of the glassy intergranular phase. Except for Y, Ho and Er a general strength increase at room temperature due to glass phase crystallization is observed. Such an increase has been reported in the literature for Si_3N_4 with Y_2O_3 as intergranular phase /4/, but not for other compositions with Y_2O_3 and CeO_2 /7, 8/. The characteristic feature of the annealed ceramics shows up in the XRD analyses: in all samples apatite $RE_5(SiO_4)_3N$ is found except for RE = Y where $Y_2Si_2O_7$ and $Y_3Al_5O_{12}$ (garnet) and RE = Er where Er_2SiO_5 is detected. These findings become clearer when looking at the RE^{3+} ionic radii compatible with certain silicates and aluminates in the Si-Al-RE-N-O systems /9/. For most RE-oxides used in the present investigation the garnet phase does not exist. Though the reasons are not yet understood, it appears as if garnet and silicates (RE_2SiO_5, $RE_2Si_2O_7$) in the intergranular phase lead to a decrease while apatite yields an increase in strength.

CONCLUSIONS

Rare earth sintering additives in SSN ceramics influence densification, β-Si_3N_4 grain growth, and aspect ratio and thereby the strength of the ceramic in a complicated manner. Especially Sm_2O_3 and Gd_2O_3 yield large aspect ratios. Apatite $RE_5(SiO_4)_3N$ crystallizes from most RE intergranular glasses and enhances room temperature strength in contrast to garnet ($Y_3Al_5O_{12}$) or other silicate phases ($Y_2Si_2O_7$, Er_2SiO_5). Addition of Al_2O_3 hampers glass phase crystallization, enhances densification and lowers the aspect ratio.

ACHKNOWLEDGEMENT

The authors like to thank A. Bärreiter, G. Beck, U. Conrad, Dr. W. Herzog, Dr. E. Paulus, E. Weiler and F. Willems for their skilled assistance in the experimental work and Dr. F. Aldinger for many helpful discussions.

REFERENCES

/1/ W.S. Sanders and D.M. Mieskowski, "Strength and Microstructure of Sintered Si_3N_4 with Rare-Earth-Oxide Additions", Am. Ceram. Soc. Bull. 64, 304 - 309 (1985).

/2/ E. Tani, M. Nishijima, H. Ichinose, K. Kishi, and S. Umebayashi, "Gas Pressure Sintering of Si_3N_4 with Oxide Additions", Yogyo-Kyokai-Shi 94, 300 - 305 (1986).

/3/ E. Tani, S. Umebayashi, K. Kishi, K. Kobayashi, and M. Nishijima, "Gas-Pressure Sintering of Si_3N_4 with Concurrent Addition of Al_2O_3 and 5 wt-% Rare Earth Oxide: High Fracture Toughness Si_3N_4 with Fibre-Like Structure", J. Am. Ceram. Soc. Bull. 65, 1311 - 1315 (1986).

/4/ A. Tsuge, K. Nishida, and M. Komatsu, "Effect of Crystallizing the Grain-Boundary Glass Phase on the High-Temperature Strength of Hot-Pressed Si_3N_4 Containing Y_2O_3", J. Am. Ceram. Soc. 58, 323 - 326 (1975).

/5/ L.K.L. Falk and G.L. Dunlop, "Crystallization of the glassy phase in an Si_3N_4 material by post-sintering heat treatments", J. Mater. Sci. 22, 4369 - 4376 (1987).

/6/ J.T. Smith and C.L. Quackenbush, "Phase Effects in Si_3N_4 Containing Y_2O_3 or CeO_2: I, Strength", Am. Ceram. Soc. Bull. 59, 529 - 537 (1980).

/7/ J.C. Bressiani, "Study on the Glass phase Crystallization in Y_2O_3-Containing β-Sialon Ceramics", Thesis, Univ. Stuttgart (FRG), 1984.

/8/ J. Mukerji, P.K. Das, P. Greil, and G. Petzow, "Sintering of Si_3N_4 with Liquid in the System Ce_2O_3-AlN-SiO_2", Ceram. Int. 13, 215 - 221 (1987).

/9/ M.K. Reser (ed.), "Phase Diagrams for Ceramists", The Am. Ceram. Soc., Columbus, OH, Vol. I - III.

/10/ G. Wötting and G. Ziegler, "Characterization of Microstructural Variables of Sintered, Hot-Pressed and Hot-Isostatically Pressed Si_3N_4 and their Correlation with Mechanical Properties", Science of Ceramics 12, 361 - 370 (1983).

Table 1. Room Temperature Flexural Strength of SSN Ceramics with RE-Oxide and Al_2O_3 Additions before and after Annealing in N_2 at 1350 °C for 24 h.

RE-oxide Al_2O_3	0.2 mol-%		0.133 mol-% 0.067 mol-%		0.075 mol-% 0.125 mol-%	
		annealed		annealed		annealed
Y	413	588	612	584	546	563
La	-	-	-	-	518	528
Ce	-	-	323	439	374	443
Pr	495	475	491	494	-	-
Nd	-	-	386	410	-	298
Sm	448	480	538	554	317	274
Gd	276	310	377	392	443	419
Tb	-	-	507	555	-	-
Dy	563	486	547	561	491	500
Ho	-	-	654	574	542	629
Er	-	-	628	500	-	-

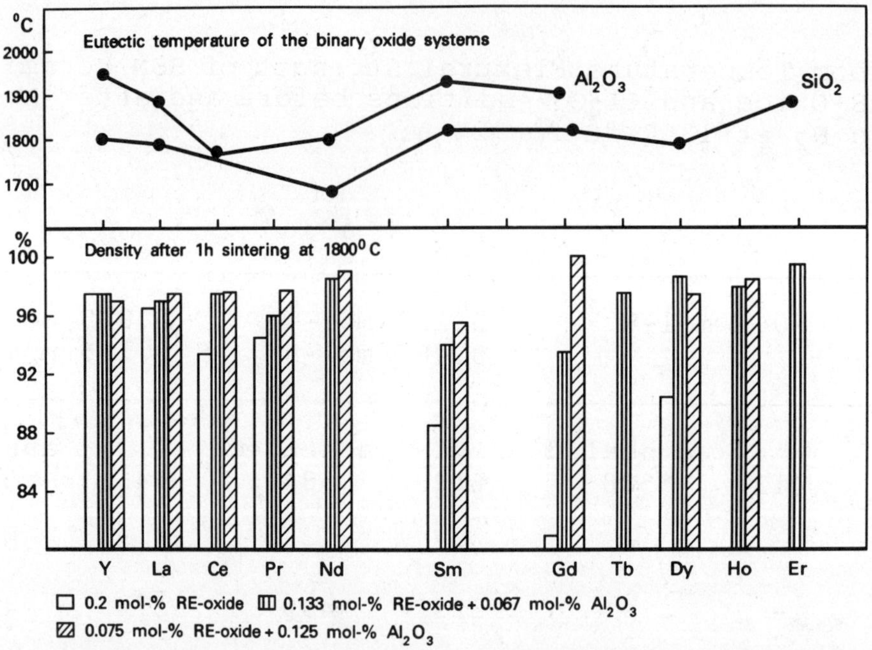

Fig. 1. Eutectic temperatures of the binary RE-Si and RE-Al oxide systems after /9/ (upper section), and relative theoretical density (lower section) after sintering for 1 h at 1800 °C in 0.1 MPa N_2; amount of additives is given in mol oxygen per 100 g SSN ceramic thus keeping oxygen concentation constant.

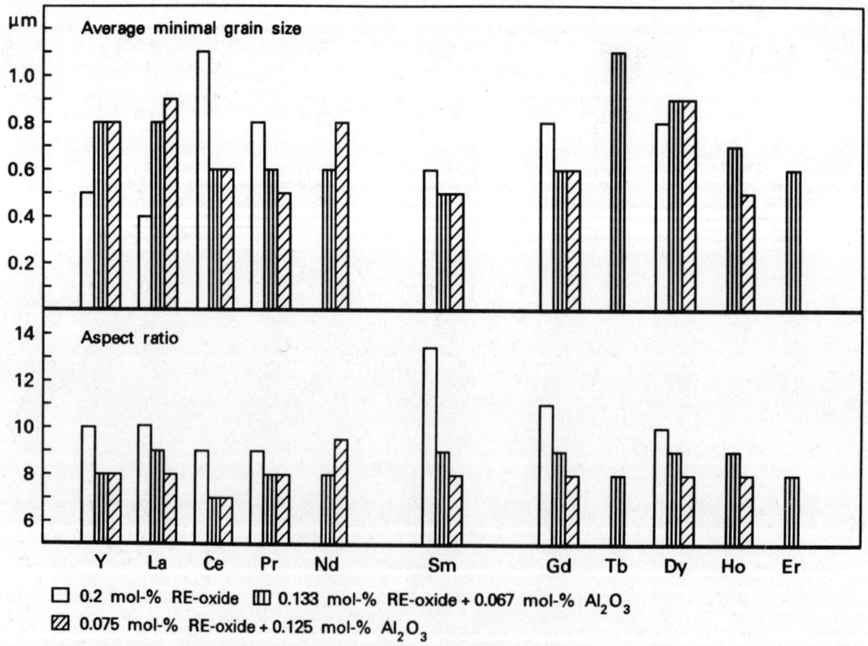

Fig. 2. Average minimal grain diameter and aspect ratio of β-Si_3N_4 in RE- and Al-doped SSN ceramic; for sample compositions see also Fig. 1.

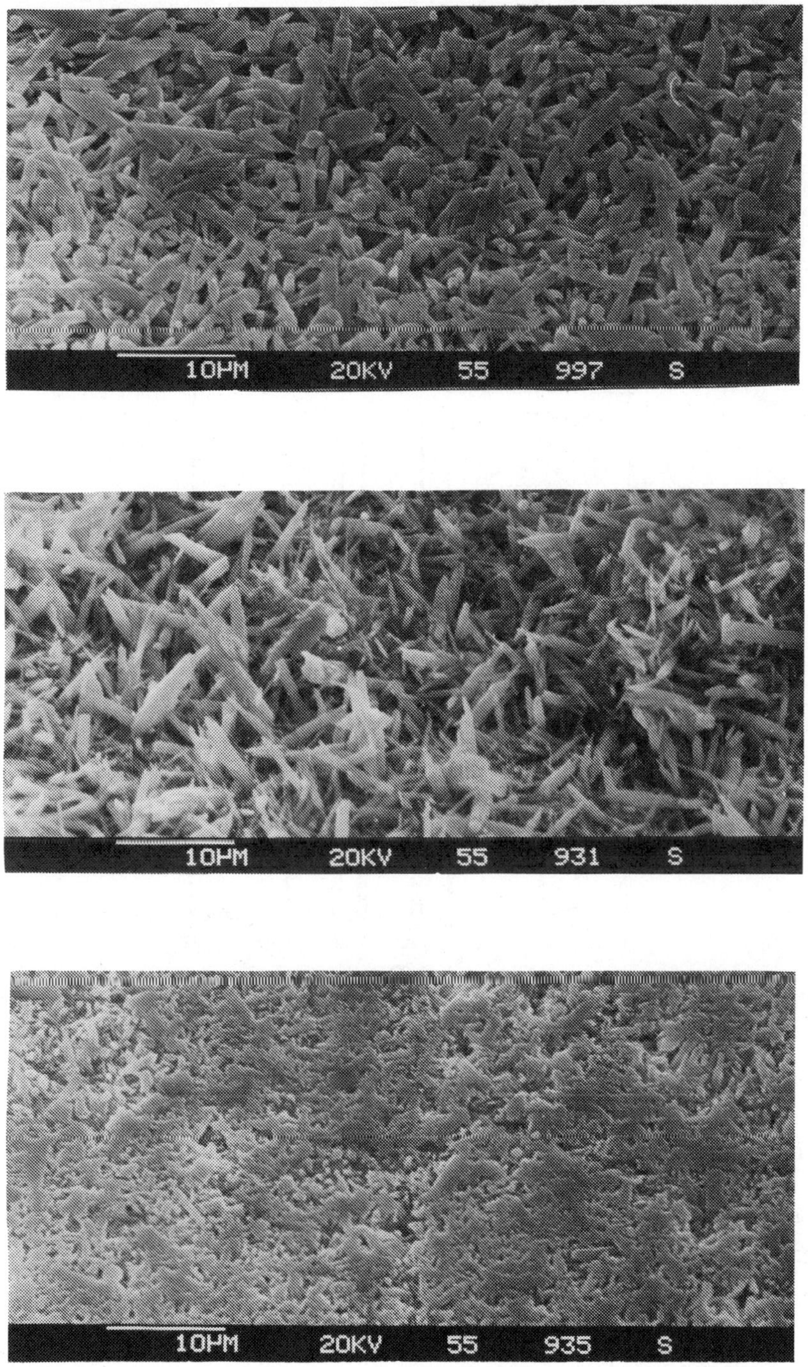

Fig. 3. SEM-micrographs of SSN ceramics with (a) 24.87 wt-% Dy_2O_3, (b) 24.17 wt-% Gd_2O_3, and (c) 16.07 wt-% Gd_2O_3 + 2.28 wt-% Al_2O_3 (polished, etched in molten NaOH).

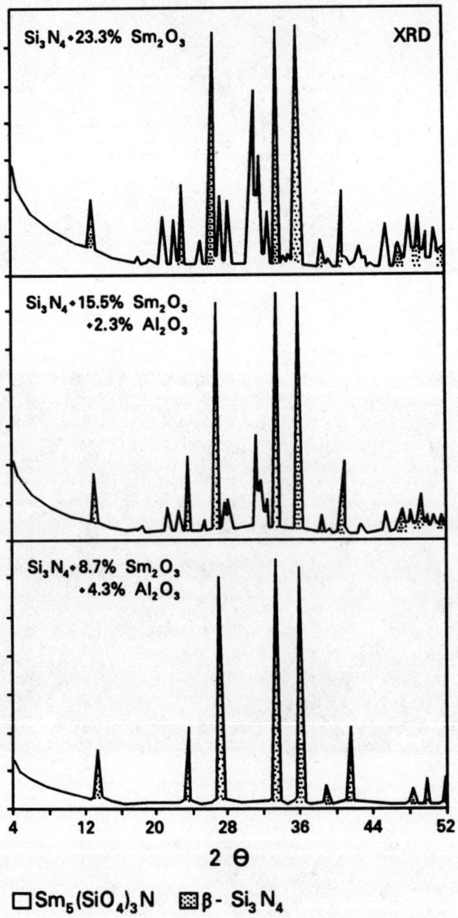

Fig. 4. X-ray diffractograms of SSN ceramics with different wt.-% of Sm_2O_3 and Al_2O_3 sintering aids after 24 h annealing in N_2 atmosphere at 1350 °C; β-Si_3N_4 and apatite $Sm_5(SiO_4)_3N$ are identified.

GAS PRESSURE SINTERING OF RBSN WITH YAG ADDITIVE*

W.L. Li, H.R. Zhuang, X.R. Fu and T.S. Yen
Shanghai Institute of Ceramics, Academia Sinica
Shanghai, China

ABSTRACT

Sintering behavior of a RBSN material with YAG additive under high gas pressures was studied at 1800-1930°C using a dilatometer. Densification processes were greatly affected by the sintering temperatures and the gas pressures applied. The rate-controlling step at 1800°C might be attributed to solution-reprecipitation, and at 1850°C and 1930°C the process might be diffusion. Specimens are densified at faster rates for the starting period under low gas pressures, while high gas pressures can enhance the sintering at final stage. Two-step gas pressure sintering is then needed for the fabrication of silicon nitride ceramics with good performance. The change of phase compositions and the microstructural features during sintering are also presented and discussed in relation to densification process.

INTRODUCTION

Gas pressure sintering has been developed as one of the most promising techniques for the fabrication of high performance silicon nitride ceramics by a number of investigators(1-3). By keeping the nitrogen pressure high during the sintering process, one can raise the sintering temperature without any significant decomposition of silicon nitride. Consequently, dense sintered bodies with minimum amounts of additives, which exhibits superior properties to those with high amounts of additives, can be obtained. Recently, gas pressure sintered silicon nitride ceramics from RBSN precursors have received a great deal of attention due to their low shrinkage and near-net-shape capability. Several engine components such as stators and radial rotors for gas turbines have been fabricated with this material(4-5). However, detailed studies on the sintering mechanism of RBSN under high gas pressures are not available. In this work, the gas pressure sintering behavior of a RBSN material with YAG additive was investigated at 1800-1930°C by using a specially designed dilatometer. The effect of sintering temperatures and gas pressures on densification processes was studied in detail. The phase composition changes and microstructure development were also studied.

* Project supported by the National Natural Science Foundation of China.

EXPERIMENTAL

The YAG powder was synthesized and ball milled in the laboratory and mixed with silicon powder (99% pure) according to the nominal composition (weight percent): 90% Si_3N_4, 10% YAG. After reaction bonding, the RBSN specimens with dimensions ϕ 6x60 mm were put in a graphite crucible and embedded in a powder mixture of Si_3N_4 and BN, then sintered at 1800-1930^oC and at 0.2-1.5 MPa nitrogen pressure in a gas pressure sintering furnace. Specimen length changes were measured by a differential transformer and automatically recorded. Phase compositions were determined by X-ray diffraction. Microstructural features were observed using a SEM equipped with an electron microprobe and a JEOL 200 CX transmission electron microscope.

RESULTS AND DISCUSSION

1. RBSN Characteristics

RBSN formed in a conventional nitridation cycle generally consists of 80-90% α-Si_3N_4 and less than 20% β-Si_3N_4. However, in the presence of YAG the phase composition of RBSN changed dramatically. X-ray diffraction analysis showed that the main crystalline phase after reaction bonding was β-Si_3N_4 solid solution (β'). The other minor phase was H, $Y_{10}(SiO_4)_6N_2$, where Al_2O_3 and Y_2O_3 needed for the formation of β' and H were probably supplied by the decomposition of YAG, Figs 1, 2 show the X-ray diffractogram and SEM picture of the RBSN with YAG additive. As can be seen from the SEM picture (Fig. 2), the YAG-doped RBSN is composed of very short columnar, near equiaxial grains. The average grain size is less than 1 μm, much smaller than that of starting silicon powder. The bulk density of the RBSN is 2.58 g/cm^3 and its porosity is 19.5%.

2. Effect of Sintering Temperature on Densification Processes

According to the kingery model[6] for liquid-phase sintering, the whole densification process can be divided into three stages: (1) particle rearrangement, (2) solution-diffusion-reprecipitation, (3) final pore closing and elimination. Solution-reprecipitation is considered as the main step in liquid phase sintering. Based on the experimental data on shrinkage measurements, one may get definite values of n as shown in the following equation indicating the rate-controlling process.

$$\frac{\Delta L}{L_o} = Kt^{1/n} \qquad (1)$$

where K is a constant determined by such parameters as temperature, liquid film thickness, coefficient of diffusion in the liquid phase, grain size and applied pressure etc, t is time. Fig. 3 shows the plots of $\Delta L/L_o$ vs t in log scale for RBSN at 1800-1930^oC under 0.2 MPa nitrogen atmosphere. Values of n are indicated in the figure. It is interesting to note that n values for specimens at 1800°C are close to 2, while the n values for specimens sintered at 1850°C and 1930°C are generally close to 3. The value of n as derived in the kingery model symbolizes the rate controlling step in the solution-diffusion-reprecipitation processes. However, at the same time, it is also a shape dependent factor. For spherical grains, kingery postulated that a value of 2 for n indicates solution-reprecipitation as the rate-controlling mechanism, and n=3 suggests that diffusion step is rate-controlling, while for lathshaped and prismatic particles, n values of 3 and 5 stand for the same rate-controlling steps. So a value of 3 for n, its explanation is not unambiguous. But judging from the near equiaxed grain shape of RBSN, it is more logical to suggest that diffusion is the rate-controlling mechanism at 1850°C and 1930°C during gas pressure sintering. From Fig. 3 it can be calculated, at 1800°C the second stage lasts 40 min, while at 1930°C it takes only 7 min for this stage. These facts further support the above mentioned suggestion, with raising the sintering temperature, the viscosity of liquid phase sharply decreases and the solution-reprecipitation process is accelerated, so the diffusion is rate-controlling.

3. Effect of Nitrogen Pressures on Densification Processes

The change in shrinkage of RBSN as a function of gas pressures applied at 1800°C was measured using the same apparatus. These plots are shown in two separate drawings (Figs. 4, 5) for clarity. The values of n and the duration of the second stage as calculated from the figures are listed in Table 1.

Table 1 n Values and Duration of the Second Stage at Different Pressures

Gas Pressures (MPa)	n Values	Starting and Final Time of the Second Stage		Duration (min)
		t_1 (min)	t_2 (min)	
0.2	2.2	8	40	32
0.5	1.9	7	50	43
1.0	2.4	15	66	51
1.5	2.1	13	50	37

It can be seen that the n values for RBSN at 1800°C under different nitrogen pressures are generally close to 2, this result indicates that under the pressure range of the present work, nitrogen pressures have no obvious effect on the sintering mechanism and solution-reprecipitation is the rate-controlling step. However, it is interesting to note that the proceeding of the different densification stages is greatly affected by the nitrogen pressures applied. With raising the applied pressure from 0.2 to 1.0 MPa, the starting time of the second stage (t_1) changes from 8 to 15 min, and the final (t_2) from 40 to 66 min. This means that the rearrangement and solution-reprecipitation processes have been prolonged as the gas pressure increases. It seems that the amount of dissolved nitrogen in the glass phase will be increased with the gas pressure, thus causing the glassy phase to become more viscous[7]. For liquid phase sintering, the densification rate during rearrangement and solution-reprecipitation is closely related to the viscosity of liquid phase. Apparently, the retardation of densification for the second stage is due to the increase in glass viscosity as the gas pressure raises to a higher level. It is interesting to note that when the nitrogen pressure raises from 1.0 to 1.5 MPa, the duration of the second stage reduces from 51 to 37 min. This result implies that the densification is enhanced at high nitrogen pressure. As is already known, the amount of nitrogen in the liquid phase is not unlimited. For the glassy phase in Y-Si-Al-O-N system, the limited amount of nitrogen was reported to be 15 at. %[8]. It is then reasonably assumed that the amount of nitrogen in liquid phase will reach a maximum with the gas pressure increased. As a result, the effect of gas pressures on the viscosity of glassy phase, and on the densification rate will be diminished. Meanwhile, the high nitrogen pressure may act as an additional driving force other than the capillary force of liquid phase between solid particles to promote the densification. As can be seen from Figs. 4 and 5, RBSN with YAG additive specimens are densified at faster rates for the starting period under low gas pressure, while high gas pressure can enhance the sintering at final stage. Obviously, the effect of nitrogen pressures on the sintering behavior at different stages is quite complex. The present work has discussed this problem from the relation between nitrogen pressure and liquid phase composition. According to our experimental results, two-step gas pressure sintering is considered more favorable for the fabrication of silicon nitride ceramics with high performance. The similar conclusion has been made by C. Greskovich taking other experimental procedures[3]. Moreover, high gas pressure can effectively depress the decomposition of silicon nitride during sintering. Fig. 6 shows the weight loss curve of RBSN heated at 1800°C for 3h. As can be seen, the weight loss sharply reduced with raising the nitrogen pressure. It has already been clear that the weight loss is mainly due to the decomposition of silicon nitride following two possible reactions:

$$Si_3N_4(s) \rightleftharpoons 3Si(l) + 2N_2(g) \qquad (2)$$

$$Si_3N_4(s) + SiO_2(l) \rightleftharpoons 4SiO(g) + 2N_2(g) + Si(l). \qquad (3)$$

Apparently, these reactions will have less tendency to the right when nitrogen pressure increases.

4. Phase Compositions and Microstructural Features

The phase compositions of RBSN sintered at 1800°C for 3h under different nitrogen pressures are listed in Table 2.

It can be seen that the main crystalline phase after sintering at 1800°C under different nitrogen pressures is β-Si_3N_4 solid solution, the other minor phases existed are B, K and M, these are probably crystallized from the liquid phase during cooling of the specimens. Among the minor phases, B exists in all cases, however, K phase crystallizes at lower nitrogen pressures, i.e.

Table 2 Phase Compositions at Different Nitrogen Pressures

Nitrogen Pressures (MPa)	Main Phase	Minor Phases
0.2	β'	B, K
0.5	β'	B, K
1.0	β'	B, M
1.5	β'	B, M

β'--β- Si_3N_4 solid solution

B-- Y_2SiAlO_5N

K--$Y_2Si_2O_4N_2$

M--$Y_2Si_3O_3N_4$

0.2 and 0.5 MPa, while M phase, which contains more nitrogen, crystallizes as the gas pressures of 1.0 and 1.5 MPa. These observations strongly suggest that with raising the nitrogen pressures in the furnace, the amount of nitrogen in the liquid phase also increases. Thus, the phase compositions of silicon nitride ceramics can be tailored by changing the nitrogen pressure applied and proper heat treatment afterwards. Fig. 7 shows the electron micrographs of RBSN sintered at 1800°C under different nitrogen pressures. After gas pressure sintering, the microstructure is composed of well-developed but fine textured elongated β' crystals intersected with each other. These figures also show that higher nitrogen pressure can depress the grain growth and make the matrix finer than the other specimen sintered at the same temperature.

CONCLUSIONS

Sintering behavior of a RBSN material with YAG additive under high nitrogen pressure has been investigated. From the results obtained the following conclusions can be summarized:

1. Densification processes are greatly affected by the sintering temperatures under 0.2 MPa nitrogen pressure. The rate-controlling step at 1800°C might be attributed to solution-reprecipitation, and at 1850°C and 1930°C the process might be diffusion controlled.
2. The effect of gas pressures on the densification processes is complex. Specimens are densified at faster rates for the starting period under low nitrogen pressures, and high nitrogen pressures can enhance the sintering at final stage. Two-step gas pressure sintering is more favorable for the fabrication of high performance silicon nitride ceramics.
3. High nitrogen pressures can depress the decomposition of silicon nitride at high temperatures and increase the density of sintered body. The grain boundary phase compositions of gas pressure sintered silicon nitride ceramics can be tailored to some extent by the nitrogen pressures applied.

REFERENCES

1. M. Mitomo, "Pressure Sintering of Si_3N_4", J. Mat. Sci., 11(6)1103-1107, (1976).
2. H.F. Priest, G.L. Priest and G.E. Gazza, "Sintering Si_3N_4 under High Nitrogen Pressure", J. Am. Ceram. Soc., 60(1-2)81, (1977).
3. C. Greskovich, "Preparation of High Density Si_3N_4 by a Gas-Pressure Sintering Process", J. Am. Ceram. Soc., 64(12)725-730, (1981).
4. J.A. Mangels and G.J. Tennenhouse, "Sintering Behavior and Microstructural Development of Yttrium-Doped Reaction-Bonded Silicon Nitride", Am. Ceram. Soc. Bull., 60(12)1306, (1981).
5. J.A. Mangels and G.J. Tennenhouse, "Densification of Reaction-Bonded Silicon Nitride", Am. Ceram. Soc. Bull., 59(12)1216-1218, (1980).
6. W.D. Kingery, "Densification During Sintering in the Presence of Liquid Phase: I", J. Appl. Phys., 30(3)301-306, (1959).
7. R.E. Loehman, "Preparation and Properties of Yttrium-Silicon-Aluminum Oxynitride Glasses", J. Am. Ceram. Soc., 62(9-10)491-494, (1979).
8. D.R. Messier and Alfreo Broz, "Microhardness and Elastic Moduli of Si-Y-Al-O-N Glasses", J. Am. Ceram. Soc., 65(8)C-123, (1982).

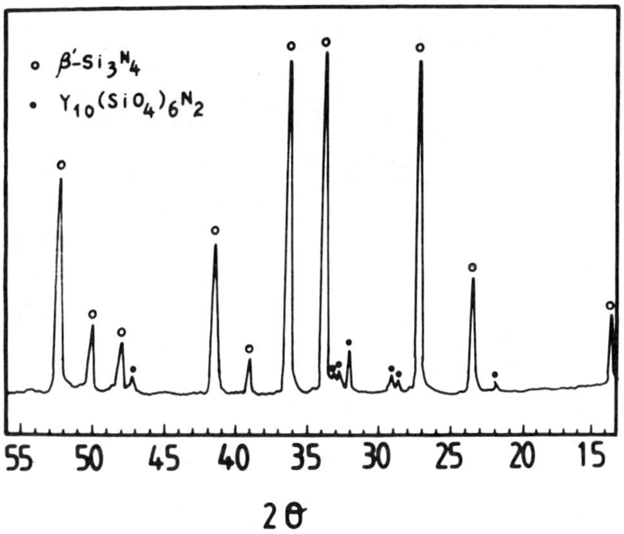

Fig. 1 X-ray Diffractogram for the RBSN with YAG Additive.

Fig. 2 SEM Picture for the RBSN with YAG Additive.

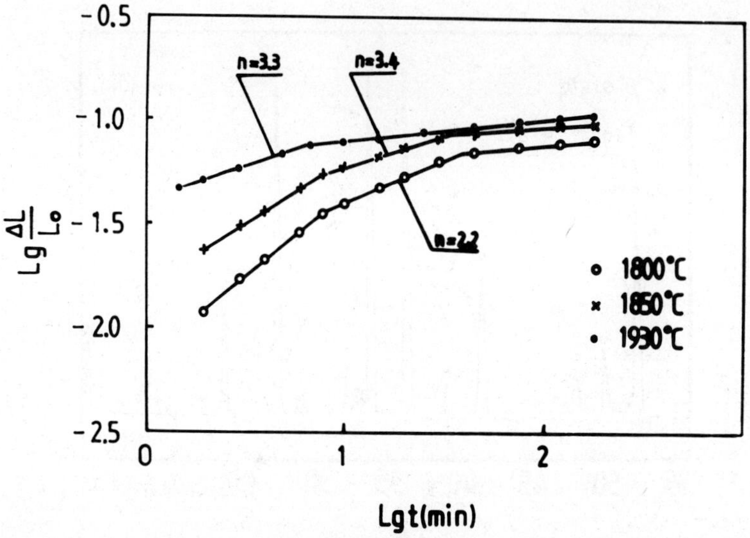

Fig. 3 Plots of LgΔL/L_o vs Lg t of RBSN at 0.2 MPa N_2 Pressure.

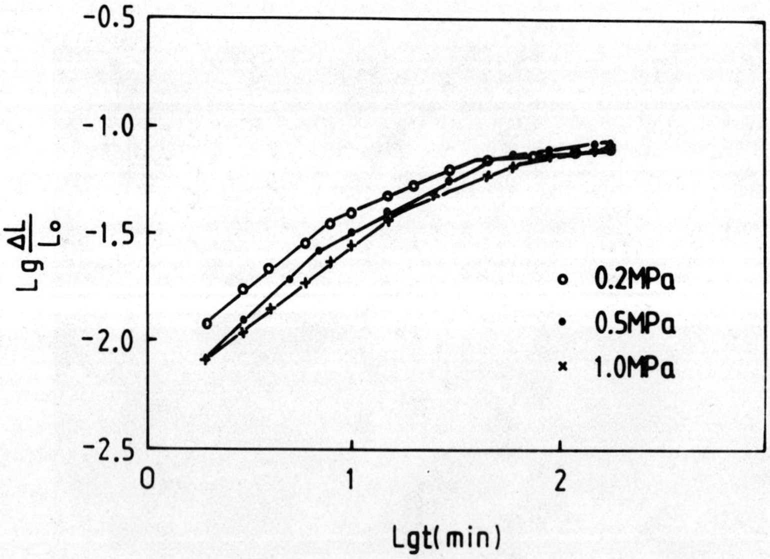

Fig. 4 Plots of LgΔL/L_o vs Lg t of RBSN at 1800°C and 0.2, 0.5 and 1.0 MPa Nitrogen Pressures.

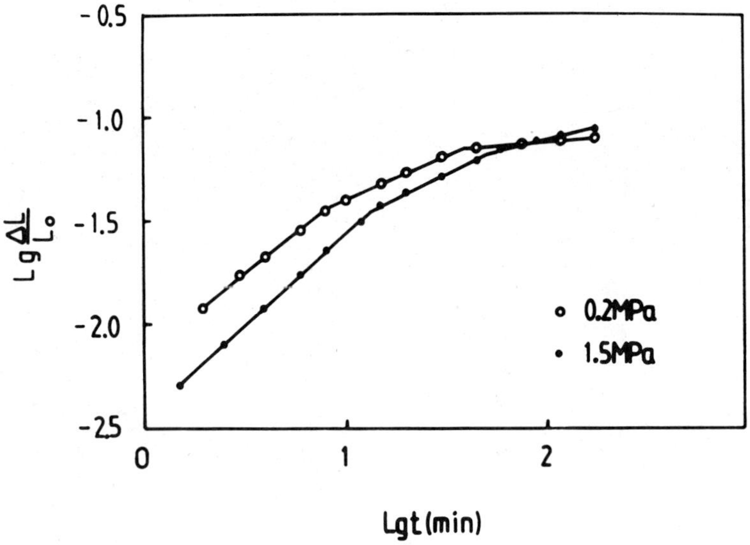

Fig. 5 Plots of Lg $\Delta L/L_o$ vs Lg t of RBSN at 1800°C and 0.2 and 1.5 MPa Nitrogen Pressures.

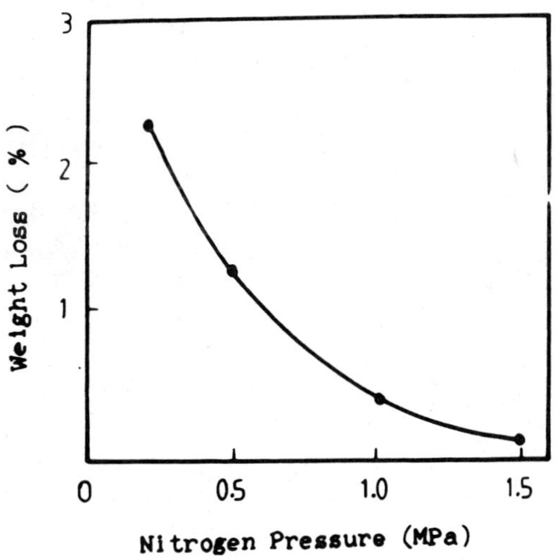

Fig. 6 Weight Loss as a Function of Nitrogen Pressures at 1800°C for 3 h.

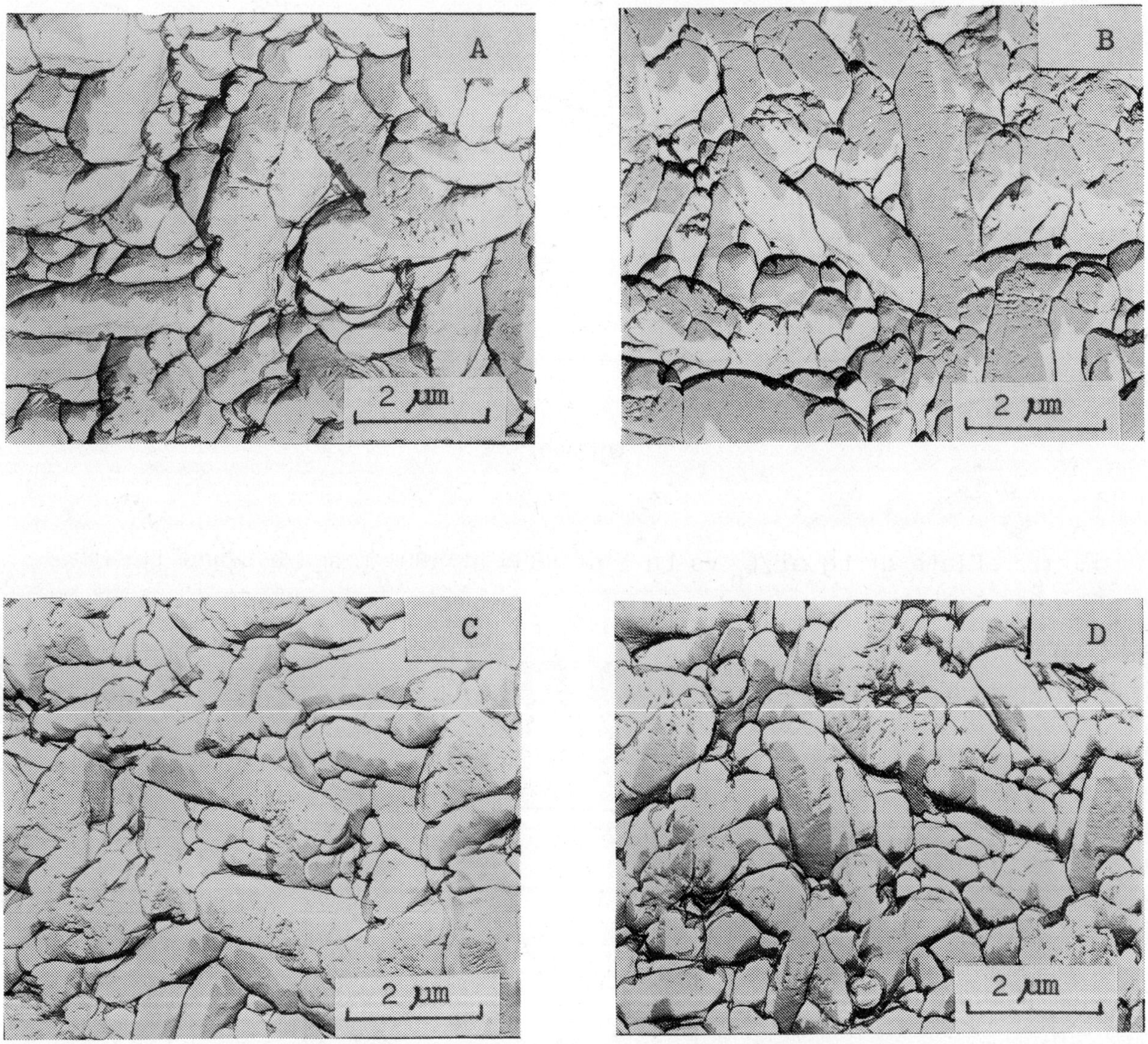

Fig. 7 Electron Micrographs of RBSN Gas Pressure Sintered at 1800°C for 3h.

A. 0.2 MPa B. 0.5 MPa C. 1.0 MPa D. 1.5 MPa

INFLUENCE OF CRYSTALLINE SECONDARY PHASES ON THE DENSIFICATION BEHAVIOR OF RBSN DURING POST-SINTERING UNDER INCREASED NITROGEN PRESSURE

H.J. Kleebe and G. Ziegler
DFVLR, German Aerospace Research Establishment
Materials Research Institute
D 5000 Köln 90, FRG

ABSTRACT

The influence of RBSN-characteristics, especially the crystalline secondary phases formed during the nitridation process, on the densification behavior of reaction-bonded silicon nitride and the resulting microstructure of SRBSN during post-sintering under increased nitrogen pressure is discussed. The type and amount of the secondary phases are controlled by the chemical characteristics of the Si-starting powder and the additive composition. Three different commercially available Si-powders which reveal particularly differences in the Fe- and Al-content were investigated, and a correlation between the microstructural characteristics of RBSN and of the post-sintered materials is presented. It is shown that besides micropore size distribution the crystalline secondary phases play an important part concern ing the densification behavior during post-sintering and the resulting microstructure of SRBSN.

INTRODUCTION

At present, the preparation of dense materials and near net-shaped silicon nitride components by the process of sintering or HIPing is state of the art[1-3,6]. Using reaction-bonded silicon nitride (RBSN) as starting material, the advantages of lower shrinkage rate during sintering and lower raw material costs, compared with Si_3N_4-powder compacts, have to be considered. In most cases the technique of pressureless sintering of RBSN as well as of Si_3N_4-powder compacts leads to an incomplete densification if low additive concentrations are used. With a higher amount of additives nearly complete densification of silicon nitride materials can be achieved, but a marked decrease of high-temperature mechanical properties of the pressureless sintered or post-sintered materials is ascertained[4-6]. Therefore, the technique of sintering under increased nitrogen pressure was developed. Gas-pressure sintering allows the use of higher sintering temperatures in comparison with pressureless sintering, as the incongruent decomposition of silicon nitride is suppressed[7-9]. Higher nitrogen pressure favors both the complete densification with lower additive concentration and the use of more refractory sintering aids.

Post-sintering of reaction-bonded silicon nitride (SRBSN) can be described as a two-step process with nitridation and sintering. The number of parameters which influence RBSN microstructure and properties increases, in comparison to gas-pressure sintering of Si_3N_4-powder compacts[10-13]. Thus, the way the densification behavior and the microstructural development of SRBSN are influenced by the characteristics of the RBSN-starting materials such as micropore structure and crystalline secondary phases which can be formed during the nitridation process is of great interest. Recent investigations show that the micropore structure of RBSN, which was systematically changed by the particle size distribution of the Si-powder, has a strong influence on the densification behavior and on the resulting mechanical properties of SRBSN. Fine micropores enhance the densification, particularly, if low additive concentrations are used[12,13]. In this paper the influence of the chemical characteristics of the Si-starting powders and of the type and amount of additives on the formation of the secondary phases during nitridation was studied. Different RBSN-materials which contain varying crystalline secondary phases were investigated to reveal the densification behavior and resulting microstructural development of SRBSN.

EXPERIMENTAL PROCEDURE

To prepare the three different RBSN-materials investigated, the addition of sintering aids to the Si-starting powder was performed by mixing the Si-powder plus additives in a roller mill and additionally in a high-speed ultratorax mixer. Three different additive compositions were used: 5 wt.% Y_2O_3, 5 wt.% Y_2O_3 + 1.1 wt.% Al_2O_3 and 10 wt.% Y_2O_3 + 2.2 wt.% Al_2O_3. The characteristics of the different Si-powders used are given in Table 1.

Table 1. Element concentration, specific surface area and average particle size of the Si-starting powders.

Si-Powder	Fe [wt.%]	Al [wt.%]	Ca [wt.%]	C [wt.%]	O [wt.%]	BET [m^2/g]	d_{50} [μm]
A	0.26	0.22	0.04	0.08	0.77	1.9	7.8
D	0.35	0.50	0.05	0.06	1.11	2.6	6.2
E	0.25	0.06	0.02	0.10	1.10	4.7	3.2

Si-powder compacts were prepared by uniaxial and cold isostatic pressing with a maximum pressure of 2500 bars. Nitridation of the Si-powder compacts was conducted under identical conditions: heating rate 6°C/h, maximum temperature 1420 °C, nitridation atmosphere 90 vol.% N_2 plus 10 vol.% H_2 with a gas pressure of 950 mbars, nitriding time 120 hrs. Micropore size distribution of the Si-powder compacts and of the RBSN-specimens was determined by mercury pressure porosimetry applying a maximum pressure of 2000 bars. The various crystalline secondary phases were analyzed using X-ray diffraction. The post-densification experiments under increased nitrogen pressure were performed in a gas-pressure sintering furnace designed for 100 bar nitrogen pressure and 2200 °C. For post-sintering of RBSN a cover powder and a two-step sintering cycle were used. During the first pre-sintering step the sintering conditions were 1875 °C, 90 min, 5 bars in order to achieve the stage of closed porosity in the pre-sintered specimens. In the second sintering step the nitrogen pressure was increased up to 100 bars, and sintering time and temperature were 90 min and 1920 °C, respectively. Microstructural development of the SRBSN-specimens was determined by scanning electron microscopy investigations.

RESULTS AND DISCUSSION

Secondary Phases

Systematic X-ray diffraction analysis showed that additive containing RBSN-materials form crystalline secondary phases during nitridation in all cases. The crystalline secondary phases determined represent a combination of H-, J- and K-phases. The estimated amounts of the crystalline secondary phases of the different RBSN-materials discussed in this paper are summarized in Table 2.

Table 2. Estimated amounts of crystalline secondary phases in various RBSN-materials using the starting powders A, D and E; additive content = 10 wt.% Y_2O_3 + 2.2 wt.% Al_2O_3;
xx = main content, x = small amount, (x) = traces.

	Crystalline Secondary Phases		
Si-Powder	**H-Phase** $Y_{20}Si_{12}O_{48}N_4$	**J-Phase** $Y_4Si_2O_7N_2$	**K-Phase** $YSiO_2N$
A	(X)	X	XX
D	XX	X	X
E	((X))	XX	X

For analyzing the crystalline structures in RBSN, the unequivocal differentiation of the J-phase $Y_4Si_2O_7N_2$ and the J'-phase $Y_{10}Al_2Si_3O_{18}N_4$ (as it is called in this text) is impossible because of α-Si_3N_4-interferences. Therefore, only the J-phase will be mentioned. Table 2 exhibits distinct differences in the main crystalline secondary phase content although all specimens were prepared under identical nitriding conditions and an identical additive concentration of 10 wt.% Y_2O_3 plus 2.2 wt.% Al_2O_3 was used. Material D with the highest impurity content shows the highest amount of H-phase, while in material E the secondary J-phase is predominant. In contrast with this, the highest amount of crystallized K-phase was analyzed in material A. This means that depending on the Si- starting powder characteristics, the only parameter varied during the RBSN-preparation, different main crystalline secondary phases are formed. The influence of differences in Fe, Al, and O contents in the Si-starting powders and in type and amount of additives on the formation of crystalline secondary phases may be summarized as follows:

- Increasing oxygen content of the Si-powder favors the formation of H-phase, as the Y/O-ratio is lowered.
- As the Y_2O_3-concentration is increased, there is a progressive change from H-phase to K-phase to J-phase.
- A high amount of metallic contamination leads to the crystallization of the H-phase.
- Addition of Al_2O_3 results in an increase of the secondary H-phase formation, too.

Consequently, the formation of crystalline secondary phases in RBSN during nitridation is markedly influenced both by the chemical characteristics of the Si-starting powder, and by the additive composition used.

As the formation of different crystalline secondary phases in RBSN occurs during the nitridation process, the question arises at what temperature these phases are formed. In order to determine this, two nitridation cycles were stopped at different lower temperatures. The first stop was at 1160 °C, the second stop at 1280 °C. In Figure 1 the corresponding X-ray diffraction patterns are shown.

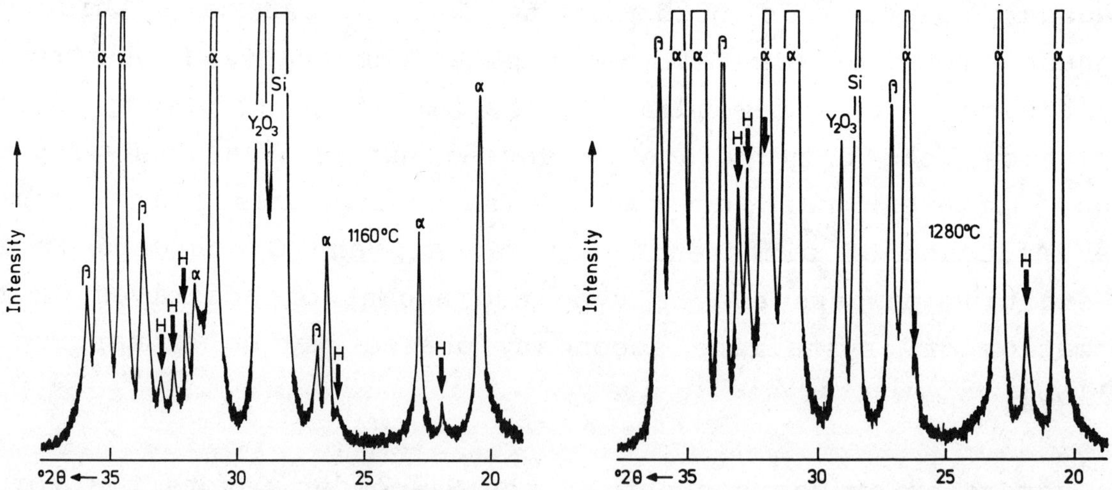

Figure 1. X-ray diffraction patterns for a stop at 1160 °C and at 1280 °C (material D).

In both X-ray diffraction patterns distinct patterns for the secondary H-phase are found, but with higher intensities in the second cycle (1280 °C). Moreover, Fig. 1 shows increasing intensities of α- and ß-Si_3N_4 and the secondary H-phase and a marked decrease of the Y_2O_3-concentration. This indicates a chemical reaction between yttria and silicon and/or silicon nitride during the nitridation process resulting in the crystalline secondary phases and the H-phase, respectively. Thus, the crystalline secondary phases in RBSN are formed in a very early stage of nitridation. Probably, during nitridation the formation of low melting glassy phases takes place, which recrystallize on cooling from nitridation temperature and form typical porous structures of the crystalline secondary phases found in RBSN. These processes are favored by a higher SiO_2-content and the low green density of the Si-powder compacts at the initial nitriding steps due to a higher N_2-permeability throughout the specimens (higher N_2-activity). From the experimental results the conclusions may be drawn that the chemical characteristics of the

Si-powder and the addition of sintering aids do influence the formation of low melting glassy phases and the crystallization of the secondary phases in RBSN. Since the thermal stability field of the secondary H-phase extends up to 1750 °C[14,15] the secondary phases in RBSN might be metastable phases from the thermodynamical point of view, but there is still only little information about the subsolidus field of these phases.

Densification

During sintering, the micropore structure of RBSN, contamination content of the Si-powder, and type and amount of additives influence the densification of RBSN. While a narrow pore size distribution with small average pore diameters favors densification, particularly in the temperature range up to about 1700 °C[13], a high amount of contamination elements, especially Fe, Al and O, vary both the composition and viscosity of the liquid phase during sintering. At the initial step of post-sintering these elements are able to support densification because a low melting glassy phase with a rather low viscosity is formed, but at elevated temperatures the problem of evaporation arises, which hinders densification[3]. Depending on type and amount of additives the first liquid phase is formed at the eutectic temperature and the liquid phase sintering process will start. All these parameters do influence the densification behavior of RBSN, but their effects are superimposed.

The appearance of crystalline secondary phases in RBSN with different thermal stability regions, in addition to the parameters mentioned above, should cause a variation in densification behavior during post-sintering. First, the yttria-content of the eutectic liquid is lowered, which results in a liquid phase with lower viscosity and on the other hand in a decrease of the liquid phase content because crystalline phases are formed. As the amount of additives has a more distinct influence on the densification than the additive composition, the densification process should be retarded. Second, depending on the thermal stability field of the secondary phases formed, only at rather high sintering temperatures are the crystalline phases dissolved and a homogeneous liquid present. This means that materials containing crystalline phases with a high thermal stability like the H-phase will show a rather low relative density after sintering. Besides these considerations it must be emphasized again that a superimposition of the different parameters occurs.

Figure 2 displays the densification traces of the RBSN-materials A, D, E as a function of sintering temperature and time. The results are obtained from a statistical evaluation method[10]. Significant differences concerning the achieved relative density after post-sintering of the three materials are obvious. Material E characterized by the lowest amount of metallic contamination, a rather small average micropore diameter (d_{50} = 0.07 μm) and concentrations in the J- and K-crystalline secondary phases shows the highest density after densification. Material A (d_{50} = 0.08 μm), with a higher Al-content, exhibits a similar sintering behavior as RBSN-material E. Compared to these materials, material D with rather high metallic contamination and H-phase content and a rather wide average micropore diameter (d_{50} = 0.20 μm) reveals at lower sintering temperatures the most rapid increase in density. However, with increasing sintering temperature and time the densification of material D becomes slower and the final density is low. This result cannot be referred to evaporation processes at higher temperatures since all weight loss data (see Fig. 2) are rather low and similar for each material. The results in Figure 3 are confirmed by pressureless and gas-pressure sintering experiments with Y_2O_3-doped materials for which evaporation processes can be excluded under the experimental conditions applied[16]. Also in these investigations material D, containing high H-phase concentration, shows the lowest final density.

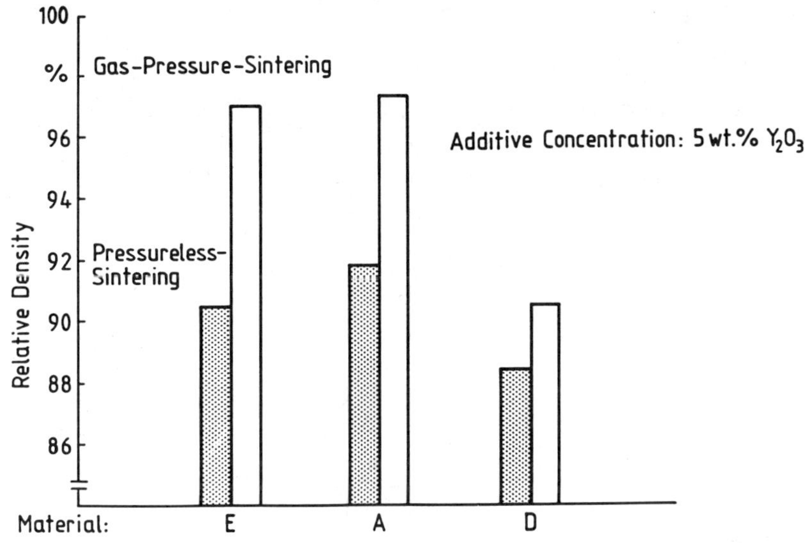

Figure 2. Relative density of RBSN-materials A, D, E depending on sintering temperature and time, additive concentration = 5 wt.% Y_2O_3 + 1.1 wt.% Al_2O_3; ΔG = weight loss (%).

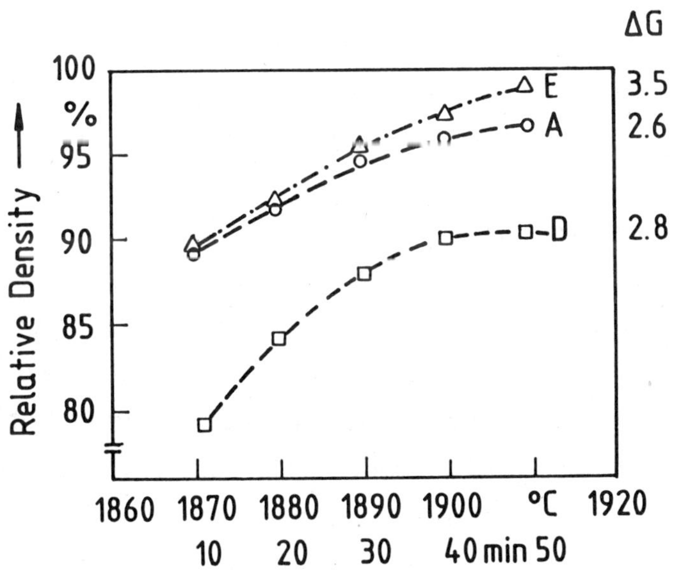

Figure 3. Relative density of RBSN-materials A, D, E after pressureless and gas-pressure sintering, additive concentration = 5 wt.% Y_2O_3.

Independent of a high crystalline secondary H-phase content in RBSN, the first eutectic liquid will form corresponding to additive composition and metallic contamination of the material in the beginning stage of sintering. This is indicated by the triangle in Figure 4, which displays an isothermal section through the phase diagram Si_3N_4-Y_2O_3-SiO_2 for 1630 °C.

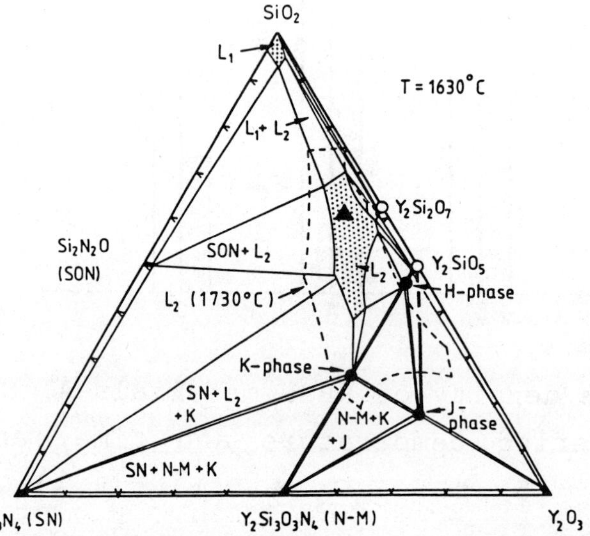

Figure 4. Isothermal section through the phase diagram Si_3N_4 - Y_2O_3 - SiO_2, broken line = liquid field at 1750 °C, triangle = eutectic liquid at 1550 °C[5].

Therefore, a material with a higher impurity concentration such as material D leads to a more rapid increase in density compared with material A and E in the early stage of sintering. Conversely, a high amount of crystalline secondary H-phase in RBSN with a thermal stability up to 1750 °C requires these crystalline phases to be reduced during post-sintering. For RBSN-materials with high crystalline H-phase content as main secondary phase it can be concluded that only at higher sintering temperatures and after a complete dissolution of the crystalline phases occurs, is a homogeneous liquid phase present throughout the sample to support densification via a liquid phase sintering process. Therefore, the crystalline secondary phases in RBSN - particularly the H-phase, but also the J- and K-phases - inhibit the densification during post-sintering.

Microstructure

The microstructure of SRBSN was also found to be influenced by the crystalline secondary phases in RBSN. Figure 5 displays a typical SRBSN microstructure after sintering under increased nitrogen pressure.

Figure 5. SEM-micrograph of polished and etched SRBSN-section (material D) with large elongated ß Si_3N_4 crystals, surrounded by a number of small crystals.

All SRBSN-materials investigated show in principle the same structural development illustrated in Fig. 5 with large elongated ß-Si_3N_4-crystals surrounded by a number of small crystals with a different aspect ratio. The various SRBSN-materials differ in the elongation of the large ß-crystals, the aspect ratio and fineness of the Si_3N_4-matrix crystals, depending on the Si-starting powders. One possible explanation for the formation of such a grain morphology during post-sintering is that the distribution of metallic contamination and the crystalline secondary phases are not uniform and (see Fig. 4) the initial eutectic liquid is present only in small local areas. In these small areas, solution, precipitation and grain growth will start, which leads on further sintering to the coarse grain growth in SRBSN mentioned above.

Thus, the crystalline secondary phases in RBSN favor both a limited densification and the grain growth of large elongated ß-Si_3N_4-crystals during post-sintering.

CONCLUSIONS

During nitridation of Si-powder compacts which contain sintering aids various crystalline secondary phases are formed. Using the X-ray diffraction method, a combination of H-, J- and K-phases was analyzed in all RBSN-materials. The concentration of the different phases is dependent on the chemical Si-powder characteristics and additive composition. The crystalline secondary phases are formed at an early stage of nitridation.

After post-sintering, SRBSN-materials with a rather coarse RBSN-micropore structure and additionally containing secondary phases with a high thermal stability field show the lowest density values. Crystalline secondary phases in RBSN inhibit densification during post-sintering as the crystalline phases have to be dissolved at elevated temperatures. Moreover, it can be concluded that crystalline secondary phases in RBSN favor the grain growth of large elongated ß-Si_3N_4-crystals during the post-densification process.

ACKNOWLEDGEMENT

Research grants from Deutsche Forschungsgemeinschaft (H.J.K.) are gratefully acknowledged.

REFERENCES

1. G. Wötting, G. Ziegler; Microstructural Development of Sintered, Hot-Pressed and Hot-Isostatically Pressed Silicon Nitride; Proc. of the 1st Int. Conf. "Ceramic Components for Engines", (1983) pp. 412-424, KTK Scientific Publishers, Tokyo.

2. A. Giachello, P. Popper; Sintering of Silicon Nitride at Atmospheric Pressure; in: Ceramics for High-Performance Applications III (1983) pp. 347-357, Plenum Press, New York.

3. G. Wötting, G. Ziegler; Influence of Powder Properties and Processing Conditions on Microstructure and Mechanical Properties of Sintered Silicon Nitride; Ceramurgia Int. 10 No. 1 (1984) 18-22.

4. F.F. Lange; Fabrication and Properties of Dense Polyphase Silicon Nitride; Am. Ceram. Soc. Bull. 62 No. 12 (1983) 1369-1374.

5. W. Braue, G. Wötting, G. Ziegler; Influence of Sintering Conditions and Mechanical Properties at Room and High Temperatures for Selected Y-Al-Si-O-N Materials; Science of Ceramics 13 (1986) C1-341 - C1-345.

6. G. Ziegler, J. Heinrich, G. Wötting; Relations between Processing, Microstructure and Properties of Dense and Reaction-Bonded Silicon Nitride; J. Mater. Sci. 22 (1987) 3041-3086.

7. C.D. Greskovich, J.H. Rosolowski; Sintering of Covalent Solids; J. Am. Ceram. Soc. 59 (1976) 336-343.

8. M. Mitomo; Pressure Sintering of Silicon Nitride; J. Mater. Sci. 11 (1976) 1103-1107.

9. H.F. Priest, G.L. Priest, G.E. Gazza; Sintering of Silicon Nitride under High Nitrogen Pressure; J. Am. Ceram. Soc. 60 (1977) 81.

10. H.-J. Kleebe, G. Wötting, G. Ziegler; Post-Sintering of Reaction-Bonded Silicon Nitride under High Nitrogen Pressure; in: Proc. VI. CIMTEC "World Congress on High Tech Ceramics" (1986), Milan, Italy.

11. H.-J. Kleebe, G. Ziegler; Nachsintern von reaktionsgebundenem Siliciumnitrid unter erhöhtem Stickstoff-Druck; Z. Werkstofftech. 18 (1987) 285-291.

12. H.-J. Kleebe, G. Wötting, G. Ziegler; Influence of RBSN-Characteristics on the Microstructural Development of Gas-Pressure Sintered Silicon Nitride; in: Proc. Science of Ceramics 14, Canterbury (1987), 407-412.

13. H.-J. Kleebe; Einflußfaktoren auf das Nachsinterverhalten von reaktionsgebundenem Siliciumnitrid; Dissertation, Universität Köln (1988).

14. R.R. Wills; Silicon Yttrium Oxynitrides; J. Am. Ceram. Soc. 57 (1974) 459.

15. R.R. Wills, S. Holmquist, J.M. Wimmer, J.A. Cunningham; Phase Relationships in the System Silicon Nitride-Yttria-Silica; J. Mater. Sci. 11 (1976) 1305-1309.

16. H.-J. Kleebe, G. Ziegler; Influence of the Starting Material on the Densification of RBSN during Gas-Pressure Sintering; submitted to PMI (1988).

CRYSTALLIZATION HEAT TREATMENTS OF SILICON NITRIDE CERAMICS AND GLASS-CERAMICS CONTAINING NEODYMIA

S. Hampshire, M. Leigh, V.J. Morrissey, M.J. Pomeroy, and B. Saruhan
Materials Research Centre, NIHE, Limerick, Ireland.

ABSTRACT

Pressureless sintering of silicon nitride ceramics can be easily effected with mixed magnesia/neodymia additions since a large volume of liquid is formed leading to good densification. Properties of the sintered product can be improved by post-sintering heat-treatment to crystallize the grain-boundary glassy phase. Crystallization of apatite phases occurs and the results are compared with crystallization in oxynitride glass-ceramics in which apatite phases are also formed.

INTRODUCTION

During the last two decades, there has been intense research activity on silicon nitride as a structural material for engine components in view of its high strength, maintained to elevated temperatures, high wear resistance and excellent resistance to thermal shock. These properties can only be realized in flaw-free, theoretically dense materials.

With regard to fabrication of silicon nitride ceramics, the only cost-effective route to the production of complex-shaped engine components is by pressureless sintering silicon nitride powder with oxide additives to form liquid phases [1,2] that allow more rapid diffusion of material and the attainment of near theoretical density.

Liquid phase sintering of silicon nitride involves three different stages[2]. Firstly, the oxide additives such as MgO or Y_2O_3 react with SiO_2, always present as a surface layer on the Si_3N_4 powder particles, and some of the nitride to form a M-Si-O-N liquid which allows PARTICLE REARRANGEMENT. Secondly[3], SOLUTION of the initial α-Si_3N_4 into the liquid takes place followed by DIFFUSION of material and subsequent

PRECIPITATION of β-Si_3N_4. Thirdly, if firing continues beyond the point of full transformation from α to β, then the liquid may still act to cause rounding and COARSENING of the grains. After subsequent cooling, the liquid either forms a glass as with MgO additive or crystallizes to form various oxynitrides as with Y_2O_3[4].

The physical processes are shown schematically in Fig 1. Rearrangement is induced by the liquid phase, after penetration into the interparticle space, by exerting a capillary pressure. Then, solution of material occurs at the α-Si_3N_4 contact points. Because of a concentration gradient, diffusion of Si and N occurs away from these points and, when conditions permit, precipitation of the more thermodynamically stable β phase occurs. With little or no β present in the original powder, a certain supersaturation must be reached before nucleation occurs followed by growth of the rod-like β grains. The resulting microstructure after cooling consists of interlocking β-Si_3N_4 grains with a residual secondary phase at the grain boundaries.

Densification of silicon nitride must always be considered in relation to the resulting microstructure and the most desirable, in terms of strength and fracture toughness, consists of interlocking β grains with high aspect ratios, very little or no glass phase, and crystalline grain boundary phases that are not oxidized at high temperatures. The factors which affect microstructural development in silicon nitride ceramics are crucial for subsequent engineering properties. The volume and viscosities of the sintering liquids can be varied by the use of mixed oxide additives such as MgO/Y_2O_3 and MgO/Nd_2O_3 combinations and by varying the total amount and the molar ratio of the additives. These can affect the densification, the α-β transformation, grain growth and morphology, and the type of secondary phase formed[5,6].

This paper reports on the promising results obtained with the $Si_3N_4/MgO/Nd_2O_3$ system and the improvements that can be achieved through post-sintering heat treatments to crystallize the grain boundary glassy phase. Comparisons are made with crystallization in oxynitride glass-ceramics containing neodymia.

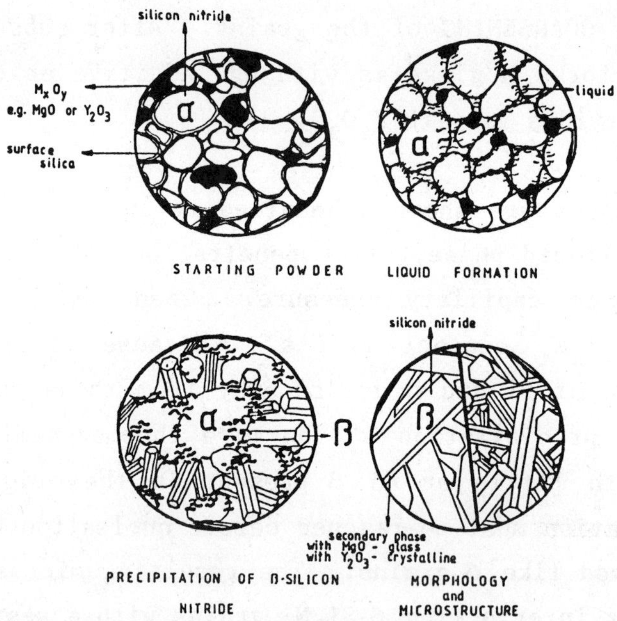

Fig. 1. Schematic diagram of microstructural development in silicon nitride sintered with oxide additives.

EXPERIMENTAL PROCEDURE

The silicon nitride powder (α/β = 95/5) was supplied by H.C. Starck-Berlin, Goslar, W.Germany (LC12 grade). Nd_2O_3 was 99.9% pure and supplied by Rare-Earth Products Ltd. MgO was a standard analytical grade powder.

Prior to cold-compaction of test bars (60mmx5mmx4mm) by die-pressing plus cold-isostatic pressing, the powders were blended in isopropanol until the alcohol was evaporated and dry mixing continued for a further thirty minutes in a WC mixer-mill.

For slip casting of silicon nitride bar samples from aqueous suspensions, magnesia must be added after casting, otherwise magnesium hydroxide is formed. A 0.3% by weight solution of an ammonium salt of polycarboxylic acid, Dispex A40 (Allied Colloids Ltd., Bradford, England) was prepared in distilled water. The neodymia was milled in a WC mixer-mill with the Dispex solution and then the required amount of silicon nitride was added to this mixture. The milled mixture was transferred from the container of the mixer into a beaker with an excess of 100ml distilled water. The mixture was left to stand for 48 hours after which the excess water was decanted. The pH of the slip was approximately 9. The slip was then cast into plaster-of-Paris moulds to make solid blocks (60mmx30mmx8mm). After a presintering at 1200°C in a nitrogen atmosphere (0.1 MPa) magnesium nitrate in the required proportion was impregnated under vacuum into the block samples. The samples were then heated very slowly (50°C/h) to 400°C to convert the magnesium nitrate to magnesia. After this, the prefired blocks were cut to produce the test bars (60mmx8mmx6mm).

Pressureless sintering was carried out in alumina crucibles in a nitrogen atmosphere (pressure 0.1 MPa) under a powder bed consisting of 50% BN and 50% Si_3N_4 + additive mixture having the same composition as the sample.

Gas-pressure (up to 210 MPa) sintering of test bars was carried out in a nitrogen atmosphere (furnace supplied by KCE of Rodental, W.Germany) under a BN powder bed containing silicon nitride sample holders.

Green and fired densities of the samples were measured with the use of a mercury displacement balance. The fired densities of the bars and porosities of all samples were measured by the use of a water absorption method.

The phases present after firing were analysed by means of X-ray diffraction with a Guinier-Hagg focussing camera and CuKα radiation.

Microstructural characterization was performed by the use of light and scanning electron microscopy (SEM) of etched samples. The grain size distribution and aspect ratio values were determined on etched SEM micrographs by means of a quantitative evaluation method after Wotting et.al[7].

RESULTS AND DISCUSSION

Previous work[5,6] has shown that, by use of mixed MgO/Nd_2O_3 additives to silicon nitride, there is a wide choice of combinations leading to favourable sintering and development of a beneficial microstructure. A typical fracture surface of a silicon nitride ceramic densified with

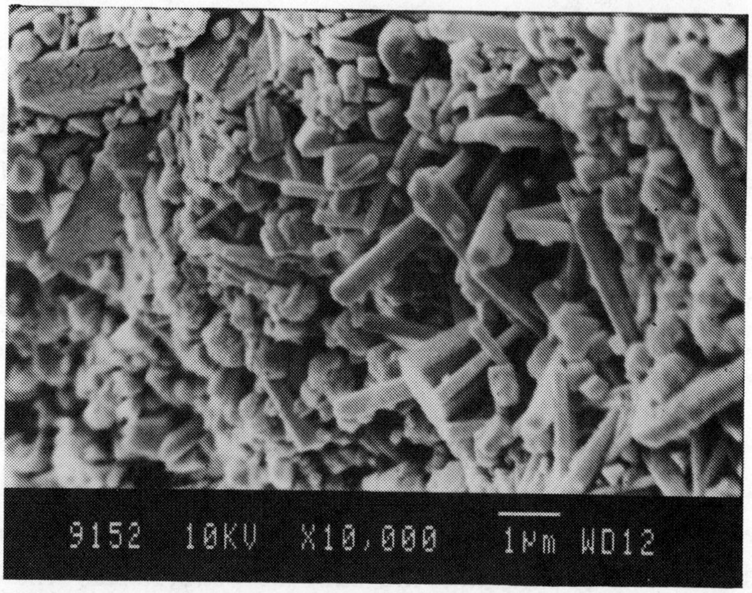

Fig. 2. Scanning electron micrograph of a fracture surface of silicon nitride densified with mixed MgO/Nd_2O_3 additives.

MgO/Nd_2O_3 additives is shown in Fig. 2.

The aspect ratios of β-silicon nitride grains after pressureless sintering at 1650°C for 2 hours with 10.5$^m/_o$ and 17$^m/_o$ of different $MgO:Nd_2O_3$ combinations are compared in Fig. 3. Full results for density, ρ_r, aspect ratio, $\bar{a}$, the extent of α-β transformation, α-β%, and the presence of secondary crystalline phases are shown in Table 1.

Table 1. Densification, α-β transformation, grain size and secondary phases for silicon nitride sintered with MgO/Nd_2O_3.

Additives: Molar Ratio $MgO : Nd_2O_3$	Additives: Total level ($^m/_o$)	ρ_R	$\bar{a}$	$\bar{d}$ (µm)	α-β %	Secondary Phases
1 : 1	10	92	3.3	0.7	80	Apatite
1 : 1	17	95	6.1	0.46	93	Apatite
4 : 1	10	94	–	–	80	Melilite (tr.)
4 : 1	17	91	5.4	0.41	98	Melilite (tr.)

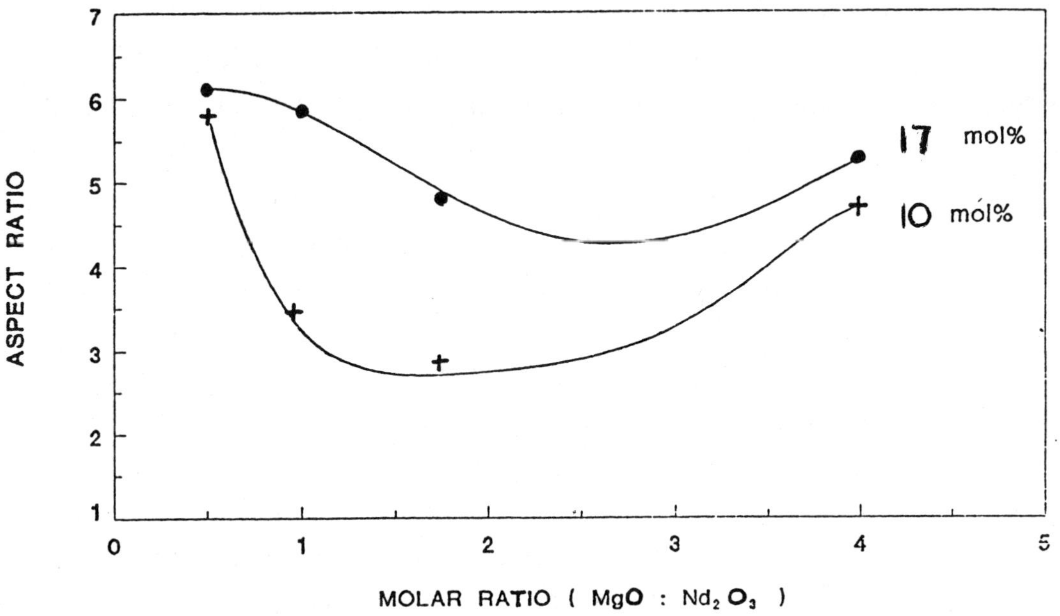

Fig. 3. Aspect ratios of β-silicon nitride grains after pressureless sintering at 1650°C for 2 hours with different MgO/Nd_2O_3 combinations.

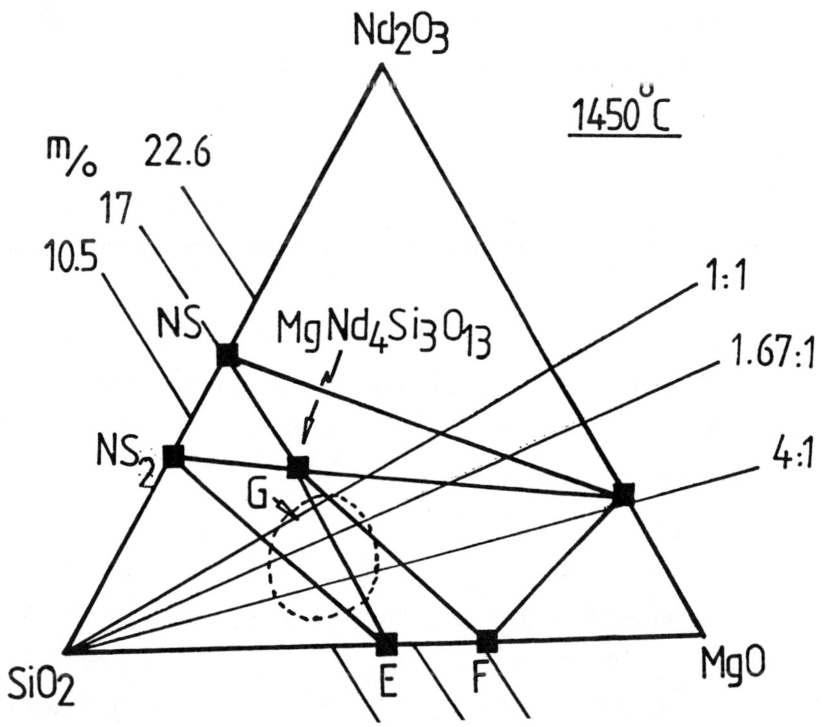

Fig. 4. The $MgO-Nd_2O_3-SiO_2$ ternary system.

With $10^{m}/_{o}$ of $MgO:Nd_2O_3$ in the ratio 1:1, the aspect ratio ($\bar{a}$) is 3.3 with a grain size ($\bar{d}$) of 0.7μm. With the same molar ratio (1:1) but a higher additive level ($17m/_{o}$), the grains are finer ($\bar{d}$= 0.46μm) and the aspect ratios are higher ($\bar{a}$=6.1). With the same total additive level ($17^{m}/_{o}$) but a higher molar ratio (4:1), grain size is similar ($\bar{d}$=0.41μm) but the aspect ratio is slightly lower at 5.4.

The secondary crystalline phases are shown in Table 1; the amounts formed are negligible suggesting that the original additives are present as part of a glass which is liquid at the sintering temperature. With the 4:1 molar ratio of $MgO:Nd_2O_3$, a trace of Nd-N-melilite ($Nd_2Si_3O_{12}N$) is detected while with a 1:1 molar ratio, Mg-Nd-N-apatite is observed. The apatite exists as a range of solid solution from $MgNd_4Si_3O_{13}$ to $Nd_5Si_3O_{12}N$. The equimolar ratio compositions should produce a grain boundary crystalline phase along a line close to the apatite composition as shown in Fig. 4 for the $MgO-Nd_2O_3-SiO_2$ system. However, the grain boundary compositions probably lie away from this oxide face within the five-component Jaenecke prism.

A silicon nitride composition containing 17m/o, of the equimolar ratio of $MgO:Nd_2O_3$ additives was chosen to observe the effects of processing conditions on the physical properties and microstructural characteristics of sintered silicon nitride materials. This composition, under pressureless sintering conditions at 1700°C, results in high relative density, full transformation to β-phase with high aspect ratio grains and a significant amount of secondary apatite phase at the grain boundaries.

A two-step gas-pressure sintering method was found[8] to improve the densification of the silicon nitride if the closed-pore stage is reached before the nitrogen pressure is increased.

High nitrogen pressures (210MPa) affect the grain morphology of β-silicon nitride. After two-step gas-pressure sintering at 1700°C, a coarser and low aspect ratio grain structure is observed. Room temperature strength of these silicon nitride samples is mainly controlled by aspect ratio and grain size of the β grains, except where large levels of porosity remain after sintering[8].

Above 1250°C, the presence of Nd-N-melilite as a grain-boundary phase causes a decrease in the high temperature strength of silicon nitride. These samples contain not only melilite but also residual glass. It is now well-documented that Y-N-melilite oxidixes in the range 900-1200°C with a high specific volume change which causes disruption of the ceramic. To avoid this problem in the present work, samples containing Nd-N-melilite were annealed during cooling in N_2 with 16 hour soaks at each temperature: 1375°C, 1275°C, 1175°C and 1075°C with a final cooling rate of 100°C h^{-1}. An apatite phase was the only grain-boundary crystalline phase in these materials observed by X-ray diffraction. Fig. 5 shows schematically the phase changes which occur, the melilite and glass reacting to produce apatite and β-silicon nitride.

Table 2 shows the compositions of 16 Mg-Nd-Si-oxynitride glasses (4 different Mg:Nd ratios x 4 different nitrogen contents) which were prepared by melting, in a BN-lined crucible under N_2, the appropriate combinations of oxides plus silicon nitride. For example, composition C with a Mg:Nd atomic ratio of 12:6.7 is close to the additive composition with $MgO:Nd_2O_3$ of 4:1. A qualitative assessment of viscosities reveals that viscosity increases with nitrogen content as found in previous investigations[9]. This is also reflected in an increase in microhardness with nitrogen content as shown in Fig. 6.

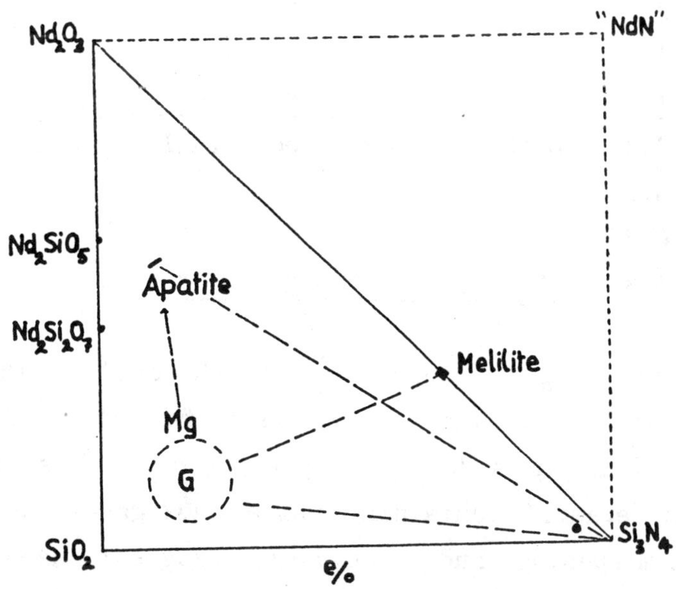

Fig. 5. Schematic diagram showing the phase changes during heat treatment of sintered silicon nitride material containing $MgO:Nd_2O_3$ additive.

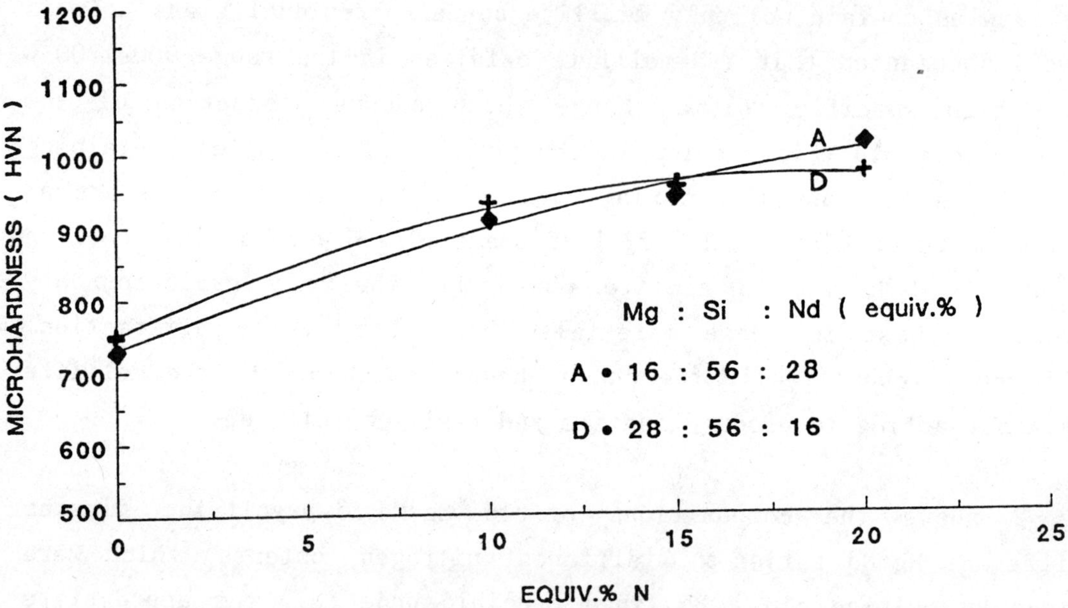

Fig. 6. Effect of nitrogen content on microhardness of Mg-Nd-Si-oxynitride glasses.

Table 2. Mg-Nd-Si-Oxynitride Glass compositions.

	Mg	Si	Nd	O	N		
A	16	56	28	80-100	0-20	equiv. %	
B	20	56	24	"	"	"	
C	24	56	20	"	"	"	
D	28	56	16	"	"	"	
A	8	14	9.3	40-50	0-6.7	atomic	ratios
B	10	14	8.0	"	"	"	"
C	12	14	6.7	"	"	"	"
D	14	14	5.3	"	"	"	"

Composition C containing 10 equiv.% N was subjected to the same heat-treatment schedule as described previously for the silicon nitride ceramics. Fig. 7 shows a SEM micrograph (polished section) of the resulting "glass-ceramic" containing white, hexagonal shaped crystals with a dendritic morphology and a few dark, hexagonal shaped crystals in a grey matrix. The white crystals are apatite.

Since the apatite exists as a solid solution over a range of composition:

$$Mg\,Nd_4Si_3O_{13} - Nd_{4.67}Si_3O_{13} - Nd_5Si_3O_{12}N$$

silicon nitrides containing these apatite phases at the grain boundaries are quite resistant to oxidation. The $Nd_5Si_3O_{12}N$ phase oxidizes[10] to give X_1-Nd_2SiO_5 and A-type $Nd_2Si_2O_7$ phases with a very small specific volume change (~3%).

In silicon nitrides with a Mg-Nd-Si-oxynitride glass as secondary phase, devitrification of the grain-boundary glass occurs during oxidation to give apatite and a reduction in the oxidation rate[11].

Thus, Nd-N-apatite is a beneficial secondary phase in silicon nitride ceramics and because of the extensive solid solution, it can be crystallized in a wide range of compositions with different $MgO:Nd_2O_3$ additive combinations.

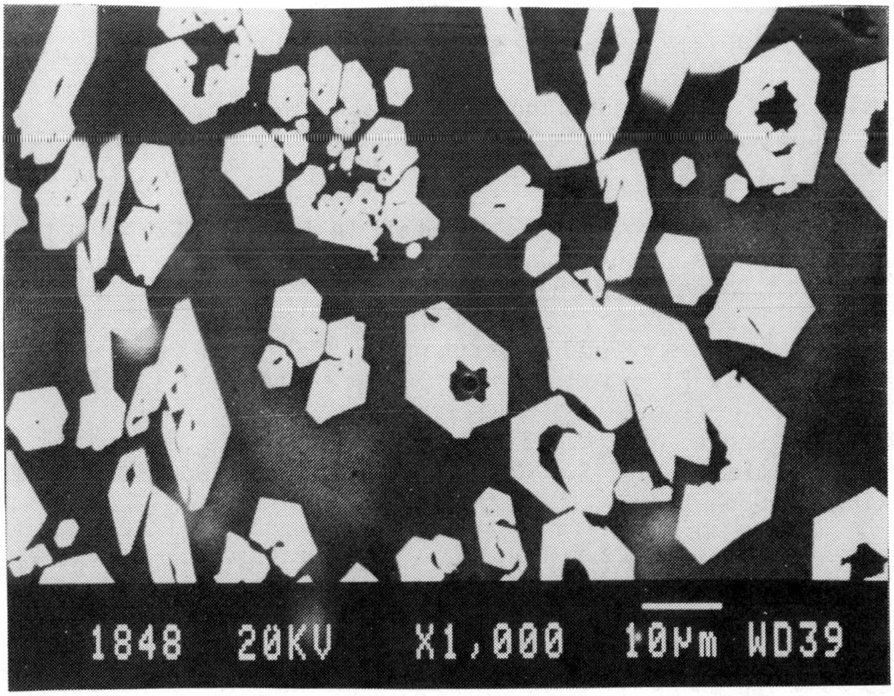

Fig. 7. Scanning electron micrograph of a polished surface of devitrified Mg-Nd-Si-oxynitride glass.

SUMMARY

Pressureless sintering of silicon nitride ceramics can be effected easily with a wide range of different mixed magnesia/neodymia additives combinations since a large volume of liquid is formed leading to good densification. By adopting a suitable post-sintering heat-treatment, the grain-boundary glass phase can be crystallized to give an apatite phase which exists as a range of solid solution.

Grain-boundary glass compositions have been prepared as bulk glasses. Microhardness increases with nitrogen content. Devitrification of these glasses results in the formation of apatite crystals. When melilite is formed in addition to the grain-boundary glass, heat treatment results in the transformation of this "unwanted" phase to apatite. The presence of apatite, rather than glass or other phases, results in lower oxidation rates in sintered silicon nitride ceramics.

ACKNOWLEDGEMENTS

This work forms part of a wider study of silicon nitride based materials that has been supported by the C.E.C., EOLAS (the Irish Science and Technology Agency) and the European Research Office of the U. S. Army.

REFERENCES

1. F.F. Lange, Am. Ceram. Soc. Bull.,62 (1983), 1369.

2. S. Hampshire, Metals Forum, 7 (1984) 162.

3. P. Drew and M.H. Lewis, J. Mater. Sci. 17 (1974) 261.

4. A.W.J.M. Rae, D.P. Thompson and K.H. Jack in Ceramics for High Performance Applications II, eds: J.J. Burke, E.N. Lenoe and R.N. Katz. Brook Hill Pub. Co., Chestnut Hill, Mass. (1978) p. 1039.

5. S. Hampshire, M.J. Pomeroy and B. Saruhan in Ceramic Components and Materials for Engines, Proc. 2nd. Int. Symp. eds. W. Bunk and H. Hausner, DKG, Bad Honnef. (1986) p.71.

6. S. Hampshire, M.J. Pomeroy and B.Saruhan in High Tech Ceramics Part, A, ed. P. Vincenzini, Elsevier Science Pub., Amsterdam, (1987) p.941.

7. G. Wotting, B. Kanka and G. Ziegler in Non-oxide Technical and Engineering Ceramics, ed. S. Hampshire, Elsevier-Applied Science Pub., Barking, (1986) p. 83.

8. B. Saruhan and S. Hampshire, Paper presented at the 5th Irish Durability and Fracture Conference, Cork (1987) to be published.

9. S. Hampshire, R.A.L. Drew and K.H. Jack, Phys. Chem. Glass., 26 (1985) 182.

10. J.K. Patel and D.P. Thompson, Brit. Ceram. Trans. J., 87 (1988) 70.

11. M.J. Pomeroy and S. Hampshire, Paper presented at the European Materials Research Soc. Symp., Strasbourg (1988) to be published.

KINETICS OF THE LOW-PRESSURE CHEMICAL VAPOR DEPOSITION OF SILICON CARBIDE*

Theodore M. Besmann and Michael L. Johnson
Metals and Ceramics Division
Oak Ridge National Laboratory
Oak Ridge, Tennessee 37831-6063

ABSTRACT

Measurements of the chemical vapor deposition (CVD) rates of <SiC> at low pressures have been made. Under the CVD conditions used here, these measurements showed a marked deviation from those previously made, indicating that different mechanisms are rate controlling. Thermodynamic calculations were performed which yielded likely important gas-phase species. These compared favorably with species determined by mass spectroscopy from the exit gases of the CVD system. Metallographic x-ray diffractometry and scanning electron microscopy techniques were used to characterize the deposited <SiC>.

INTRODUCTION

Silicon carbide has become a technologically important material because of its hardness and refractory nature and its optical and semiconductor properties. Films and coatings of <SiC> have acquired commercial interest as protective layers against oxidation, corrosion, and erosion. These are applied by a variety of methods, however, conventional chemical vapor deposition (CVD) or a variant is most typically used. Schlichting,[1] in his 1980 review of the CVD of <SiC>, for example, referenced 251 papers. In addition, composite structures utilizing chemical vapor deposited <SiC> as the matrix are currently being marketed. Significant development efforts, which point to a substantially greater number of potential applications, are also in progress.[2,3]

Chemical vapor deposition can be viewed as a series of sequential processes that include (1) introduction of reactant vapor by forced convection, (2) diffusion of reactants through a gas boundary layer to the substrate

*Research sponsored by the Office of Energy Utilization Research, Energy Conversion and Utilization Technologies (ECUT) Program, U.S. Department of Energy, under contract DE-AC05-84OR21400 with Martin Marietta Energy Systems, Inc.

**The following notation is used throughout this paper: () indicates gaseous species; {} indicates liquid species; <>indicates solid phase.

surface, (3) adsorption of reactive species onto the substrate, (4) surface reactions and diffusion that result in the inclusion of deposit elements into the coating, (5) desorption of product vapor species, (6) diffusion of product vapor species through the gas boundary layer to the bulk gas, and (7) removal of the product vapor species by forced convection.[4,5] The complex interactions of mass and heat transport, along with the chemical kinetics in CVD systems, have thus far prevented the development of fully predictive models for all but simpler, single-element CVD systems, such as the CVD of silicon.[6-8] However, various processes involved in CVD have been decoupled, and an understanding of some of these has provided valuable insight.

Surface processes are often important rate-limiting mechanisms in CVD. These are typically revealed under conditions that allow diffusive mass transport to be rapid or not applicable (molecular flow). Several reported studies, experimental and analytical, of the kinetics of these mechanisms in the CVD of <SiC> focused on determining activation energies and possible mechanistic interpretations.[9-14] The investigations vary in the CVD parameters used, which leads to some variance in the information obtained.

The object of this study was to characterize the surface kinetic processes in the CVD of <SiC> from (CH_3SiCl_3) (methyltrichlorosilane, or MTS) and hydrogen. An equilibrium analysis was performed to identify potentially important gas-phase species, and the results were compared with mass spectrometric analyses of the system exhaust gas. Measurements of deposition rate vs temperature and reactant concentrations under conditions that are likely to allow surface processes to be rate limiting (low pressure and temperature) provided kinetic information. The deposits were characterized by X-ray diffraction, metallography, and scanning electron microscopy (SEM).

EQUILIBRIUM CALCULATIONS

Chemical equilibrium thermodynamics, at first glance, appears to have little to do with chemical kinetics. The former is time independent and yields only the final equilibrium state, whereas the latter is time dependent and describes the rate of change. Equilibrium thermodynamic calculations are useful, however, for determining phases that are energetically allowed to form and the associated equilibrium partial pressures. Thus, they can also indicate which vapor species are present in high concentrations and are likely to be kinetically important.

A series of equilibrium calculations were performed over partial pressure-temperature space using an input concentration ratio of (H_2):(MTS) of 10:1 and a total pressure of 3.3 kPa. The F*A*C*T calculational system, with its resident library of thermodynamic values, was used.[15] Thermodynamic values for (MTS) vapor are absent from the F*A*C*T library, thus, they were introduced using the 298 K heat of formation determined by Bell et al.[16] of -549 kJ/mol and the heat capacity expression derived from Sin'ko et al.[17]

$$C_p = 96.939 + 75.868 \times 10^{-3}\,T - 1.369 \times 10^{6}\,T^{-2} - 21.539 \times 10^{-6}\,T^{2}\ (\mathrm{J\ mol^{-1}\ K^{-1}}),$$

where T is the temperature in Kelvin.

Equilibrium calculations for the CVD of <SiC> typically indicate carbon as a minor phase together with <SiC> as the major phase. However, in this study, carbon was specifically precluded from forming in the calculations, based on the observation that its formation is often kinetically inhibited for reactant Si:C ratios near unity at moderate temperatures (<1900 K).[9,18] With the use of this assumption, equilibrium compositions were computed that indicate that <β-SiC> is formed with the gas partial pressures shown in Fig. 1.

The calculational results displayed in Fig. 1 show that (H_2) and (HCl) are the major background species for the equilibrium system at all but the lowest temperatures. Below 1300 K, (CH_4) and ($SiCl_4$) are the dominant reactant species, with minor amounts of ($SiHCl_3$) and ($SiCl_3$) present. Above 1300 K, ($SiCl_2$) and (C_2H_2) are the more stable species, together with monatomic hydrogen and chlorine.

Under CVD conditions that cause a system to be rate-limited by surface adsorption, diffusion, or reaction, the gas phase may achieve an equilibrium composition in the absence of interactions with the deposited condensed phase(s).[4,13] Essentially, the gas phase is chemically isolated from the condensed phase(s) on the substrate because of slow, surface-related processes. Additional equilibrium calculations were thus performed assuming that no condensed phases were stable to explore potentially important partial pressure differences.(Fig. 2)

The results of calculations which assume that no condensed phases are stable differ substantially from those that allow <SiC> to form. Hydrogen still provides the base pressure, however, many more species have significant concentrations over the temperature range. Below 1200 K, (CH_4) is again important, but with ($SiHCl_3$) almost an order of magnitude larger than ($SiCl_4$). Above 1400 K, (HCl), ($SiCl_2$), and (C_2H_2) are again major species, however, the latter two exhibit partial pressures more than an order of magnitude larger than in the system in which <SiC> was allowed to form.

EXPERIMENTAL

Materials

The hydrogen and (MTS)* were of commercial grade, with the (MTS) having a reported purity of 98%. The substrates used were cylindrical POCO** graphite ~2.5 cm in diameter and 1 cm thick. The argon used for purging and maintaining the desired pressure was also of commercial grade.

Apparatus

Figure 3 is a schematic of the CVD system. Inlet gases are metered by mass flow controllers,*** and the (MTS) is controlled by a vapor source system.****

---

*Alfa Products, Danvers, MA.

**POCO EDM-3 Graphite.

***Model 1259B with controller model 247B, MKS Instruments, Burlington, MA.

****Tylan source 5, Tylan, Carson, CA.

The hydrogen and purge argon are passed through a titanium sponge-getter at 925 K to remove oxygen-containing species. The CVD exhaust gases pass through a soda-lime column and finally to an oil-displacement vacuum pump. Vacuum is controlled via a gas ballast system that uses a signal from a capacitance manometer* to control the flow of inert gas through a solenoid valve on the inlet side of the pump.

The reactor is a hot-wall system consisting of a fused silica tube (3.5 cm OD, 2 mm wall, 60 cm long) in a resistance furnace. The cylindrical substrate rests within the tube on a molybdenum foil holder with its front surface slightly off normal to the gas flow. The holder was designed to make only point contact with the substrate to minimize masking of the substrate surface. The holder also becomes quickly coated with <SiC> and was not observed to interact with the substrate or <SiC> coating. Experimental runs were typically 1 to 3 h in length.

Substrate temperature was measured with an optical pyrometer,** and corrections were applied for emissivity (determined through the use of a substrate with a black-body hole) and absorbance by the window and prism. The pyrometer was calibrated regularly and determined to be accurate at the melting point of gold (1338 K) to within 3 K. The pressure sensor was factory calibrated and zero adjusted before use. The mass flow controllers were also factory calibrated and checked before use with a wet-test meter. The (MTS) vapor source was checked for calibration using the difference in weight of {MTS} before and after a run. A quadrupole mass spectrometer*** was configured to sample the gas stream through a capillary tube that penetrates the system on the exhaust side of the reactor.

<u>Procedure</u>

The procedure used in deposition was designed to minimize unintended variation in the deposition parameters. Each substrate had its dimensions individually measured to account for any errors in machining that could lead to errors in surface area. The substrate was then loaded into the reactor, which was evacuated to <0.025 Pa to test its integrity. The titanium sponge-getter was then heated, and argon was allowed to flow through the reactor, which was maintained at 3.3 kPa. The reactor was then brought to temperature, and when the temperature had stabilized, deposition was initiated by replacing the argon with (MTS) carried by hydrogen. After a predetermined length of time the reactant flow was shut off, the power to the furnace was turned off, and argon was allowed to flow through the reactor during cooling.

Relative concentrations of gaseous species in the exhaust gas were determined by operating the mass spectrometer during a typical deposition run. Periodic data acquisition ensured that a steady state was being monitored.

Coating thicknesses were determined from the weight gain of the substrate, its surface area, and the density of <β-SiC> (3.21 g/cm^3). Blank runs at

---

*Type 122A (0-1000 torr) with controller model 250B, MKS Instruments, Burlington, MA.

**Model No. 95, Leeds and Northrup Co., Bergenfield, NJ.

***Model IQ 200, INFICON Leybold-Heraeus, East Syracuse, NY.

temperature under flowing argon indicated minimal weight change in the graphite substrates.

RESULTS

The influence of temperature and reactant concentration on deposition rate was investigated. Experiments depositing <SiC> from (MTS) and hydrogen were carried out under various conditions to isolate rate dependencies.(Table 1)

Table 1. Experimental CVD conditions and deposition rates [(MTS) flow rate equals 50 mL/min][a]

Run	Substrate temperature[b] (K)	Pressure (kPa)	(H_2) flow rate (mL/min)[a]	Run time (min)	Coating thickness (μm)	Coating rate (nm/min)
SiC-K-8	1297	3.3	500	60	15.8	264.0
SiC-K-9	1191	3.3	500	120	9.8	81.5
SiC-K-10	1245	3.3	500	60	12.3	206.0
SiC-K-11	1269	3.3	500	60	14.8	247.0
SiC-K-12	1167	3.3	500	180	11.1	61.6
SiC-K-13	1217	3.3	500	90	13.2	147.0
SiC-K-14	1217	3.3	250	80	11.4	143.0
SiC-K-15	1217	3.3	1000	80	11.9	149.0
SiC-K-16	1229	3.3	500	90	13.8	154.0
SiC-K-17	1205	3.3	500	120	15.8	130.0
GE-SiC-G-8	1341	101	500	60	21.2	353.0

[a]At 273 K, 101 kPa.
[b]Corrected.

Concentration dependence

At the pressure, temperature, flow rate, and range of concentrations investigated, the deposition rate of <SiC> is independent of (MTS) concentration.(Fig. 4) This zeroth-order dependence on reactant partial pressure implies that neither bulk gas flow, diffusion of reactants through a boundary layer to the substrate surface, nor adsorption of a reactant species is rate limiting. It is thus likely that under these conditions the substrate surface is saturated with reactant species, and that the controlling mechanism is a surface diffusion or reaction process. The possible mechanisms include diffusion of adsorbed atoms or molecules across the surface to an energetically favored position, such as a kink site, for inclusion into the coating. Also likely is a decomposition surface reaction that allows silicon or carbon to lose bonded hydrogen or chlorine.

Temperature Dependence

An Arrhenius plot of deposition rates determined at uniform conditions is shown in Fig. 5. It is possible to separate the data into two regions, with the lower temperature (<1245 K) values seen to have a larger slope. The lower temperature behavior is representative of a chemical kinetic rate-limiting mechanism, whereas the higher temperature rate-limiting process appears to have the smaller slope typical of gas-phase diffusion.[9,18-20] The lower temperature data were thus fit to a zeroth-order (due to lack of concentration dependence) Arrhenius relation:

$$r = A \exp(-E/RT),$$

where r is the deposition rate, A is the preexponential factor, E is the activation energy, and R is the Ideal Gas Law constant. The values computed from a linear least-squares fit of the data at <1245 K are E = 188 (±52) kJ/mol and ln (A) = 23.5 (±5.1) nm/min.

The magnitude of the effect of total pressure on the deposition rate behavior was assessed by comparison with the results of a run at 101 kPa, although at a somewhat higher temperature (Fig. 5). The result from the 101-kPa run can be seen to lie in the gas-phase diffusion limited region, agreeing well with the similar deposition rates at 3.3 kPa. Higher pressures also tend to shift the rate-limiting mechanism to gas-phase diffusion.[19]

A comparison of deposition rates based on thermodynamic equilibrium calculations and those experimentally determined can also be seen in Fig. 5. The upper curve of the figure represents a simple model of deposition which assumes that the entire heated area of the CVD reactor is uniformly coated under equilibrium conditions. As expected, the observed deposition rates are substantially less than the equilibrium values because of the constraining rate-limiting processes. In addition, the chemical kinetic controlled region is much more temperature sensitive.

Gas Composition

The concentrations of the species detected with the use of a quadrupole mass spectrometer are shown in Figs. 1 and 2 together with the equilibrium values and the input concentrations. The approximate partial pressures were determined from the mass spectrometric data by using the sum of the species' peak heights relative to those for hydrogen, the background gaseous species. A baseline mass spectrometric fragmentation pattern for (MTS) was obtained from the flowing CVD system with the furnace cold.

The gaseous species observed generally agree with those which the thermodynamic calculations indicate have significant (>1 Pa) partial pressures: (HCl), ($SiCl_4$), (CH_4), and ($SiCl_2$). The species ($SiHCl_2$) also appears; however, there are no available thermodynamic data for this species, and thus it was not considered in the calculations. The measured high partial pressure of (MTS) is likely due to the short residence time that prevented it from decomposing to its relatively low equilibrium partial pressure. Lack of detectable (C_2H_2), which would have a significant partial pressure at >1100 K, may also be due to the bulk gas not reaching the furnace temperature.

The observation that the rate-limiting mechanism is surface related indicates that the equilibrium partial pressures calculated, assuming no condensed phases form, are the most appropriate (Fig. 2). The measured high ($SiCl_2$) concentration supports this, because in Fig. 1, the calculated ($SiCl_2$) partial pressure is equal to or greater than that observed only at >1780 K, well above the substrate temperature of 1217 K. In the calculations for which no condensed phases were assumed to form (Fig. 2), the ($SiCl_2$) partial pressure also increases with temperature, however, under this assumption, the ($SiCl_2$) value exceeds the measured concentration at >870 K.

Phases/Microstructure

X-ray diffractometry identified the coatings as polycrystalline, cubic <SiC>. The typical preferred orientation of the (111) planes parallel to the substrate surface was also seen in the intensity shifts of the diffraction lines. With the exception of the graphite substrate, no phases other than <SiC> were observed by X-ray diffractometry.

Scanning electron micrographs of the surface of the <SiC> coatings display gently rounded, textured structures generally observed in relatively low-temperature deposits (Fig. 6). The fracture surface of the coating, also seen in Fig. 6, appears smooth, exhibiting some fine columnar features.

To delineate the coating grain structure, a polished cross-section was etched using a saturated water solution of NaOH and $K_3Fe(CN)_6$ mixed in equal volumes.[21] The etchant was heated to ~380 K, and the samples were etched for ~5 min. The microstructure revealed in Fig. 7 shows generally columnar, submicrometer grains. The grains also appear to radiate from some areas of the coating-substrate interface, perhaps because of surface nucleation phenomena. A small amount of light-colored banding is seen near the substrate, possibly the result of system upsets or some codeposition of elemental silicon.[21,22] Other than this banding, there is no evidence in the cross-section of a second phase. In addition, the density of the coating, as determined from the thickness measured from the cross-section and the weight gain of the substrate, approximates that of <β-SiC>, which is 40% greater than that of elemental carbon or silicon and thus precluding their presence in significant amounts.

DISCUSSION

A number of calculational equilibrium studies of the CVD of <SiC> from Si-C-Cl-H precursors have been reviewed by Schlichting,[1] including the most recent by Fischman and Petuskey,[12] and these are in general agreement with the results in Figs. 1 and 2. Gas-phase compositions can be inferred from the experimental study of Ivanova and Pletyushkin,[23] who analyzed condensed exhaust gases of the CVD of <SiC> from (MTS). They determined the major condensable product species to be {$SiHCl_3$}, {$SiCl_4$}, {MTS}, and {$(CH_3)_2SiCl_2$} from the CVD system operated at 1723 K. Their results thus confirm that (MTS) decomposes to independent silicon-bearing species in the CVD system, although at equilibrium the major silicon chloride at 1723 K is ($SiCl_2$) rather than ($SiCl_4$) (Figs. 1 and 2). Calculations indicate that ($SiCl_4$) and ($SiHCl_3$) are the dominant silicon species at lower temperatures, however, (Fig. 1 and 2), suggesting that the vapor equilibrated at a lower temperature (<1000 K) during cooling. This can also explain the presence of {$(CH_3)_2SiCl_2$}, which is unstable at elevated temperatures.

A number of studies of the kinetics of the CVD of <SiC> from chloride precursors have shown a variety of results (Table 2). With the exception of this study and that of Fedoseev et al.,[11] the CVD runs have all been performed at 101 kPa total pressure. In addition, where the relationship was investigated, it was determined that the rate-limiting mechanism was first-order with respect to (MTS) concentration. This is at variance with our observation and that of Stinton and Lackey[21] that the relationship is zeroth-order. The discrepancy may be explained by the relatively high flow rates in our system and the fluidized bed reactor of Stinton and Lackey, which help to saturate the substrate surface with reactant species, altering the rate-limiting process.

Table 2. Summary of <SiC> CVD studies

Si-C reactants	Reactant H_2/Si	Pressure (kPa)	Order[a]	T-range (K)	E (kJ/mol)	Reference
$SiCl_4 + CCl_4$	120	101		1373–1598	200	18
MTS	3.6	101	1st	1173–1273	120	14
MTS	26	101	1st	1473–1673	68	9
MTS	14	101	1st	1540–1773	209	10
MTS	2	20	1st	1480–1820	335	11
MTS	11	101		1523–1623	397–410	20
MTS	10	3.3	0th	1167–1245[b]	188(±52)	This study

[a]With regard to (MTS) concentration. No value indicates relationship not reported.
[b]Temperature range over which data were used to determine E.

There is also a wide range of reported activation energies (Table 2). Of the seven measured values, three are ~200 kJ/mol, two are ~400 kJ/mol, and there is one each at 68.2 and 120 kJ/mol. An assessment of the apparent CVD conditions gives no indication of a correlation of deposition parameters and activation energy, including whether hot-wall or cold-wall reactors were used. Hidden variables such as substrate surface character, gas-phase temperature, and system contaminants must therefore be the cause of the discrepancies.

There is a general consensus, based on thermodynamic preferences, that silicon and carbon are independently provided to the growing substrate surface to form <SiC>.[1,11–13,23] This consensus is supported by the observation, both in this study and in others, that preferential growth in the (111) plane suggests the formation of alternating layers of silicon and carbon atoms. In addition, the frequent observation of lamellar <SiC> CVD coatings may indicate the independent deposition of silicon and carbon, which yield alternating silicon- and carbon-rich layers.

It has been postulated that the rate-limiting chemical kinetic process involves the decomposition of the hydrocarbon species at the substrate surface based on the observation that the deposition of carbon is observed to be hindered.[9,18] Fischman and Petuskey[12] noted that the <SiC> and <SiC>-graphite CVD boundary is shifted toward the carbon-rich side from that predicted thermodynamically, suggesting that carbon deposition is suppressed. They propose that hydrocarbons are symmetric molecules and therefore have a lower sticking potential than do the asymmetric, polar silicon chlorides, thus giving adsorption of the carbon species a lower probability. This observation would not be applicable to the results of this study, however, because the deposition rate has been determined to be independent of reactant concentration. Thus, a relationship such as the Langmuir adsorption isotherm would not represent the data since it is concentration dependent. Fedoseev et al.[11] proposed that <SiC> deposition is limited by the heterogeneous reaction of the methyl radical. They base their conclusion on the similarity of the shape of the growth rate curve vs the partial pressure of hydrogen to that of the growth of graphite from methane. The same analogy is applied to the Arrhenius curves they report, which, therefore, also give an activation energy that is close to that for the deposition of graphite (~350 kJ/mol). The activation energies necessary for this mechanism, however, are almost a factor of 2 greater than that determined in our system.

CONCLUSIONS

Equilibrium calculations of the CVD of <SiC> from (MTS) yield partial pressures in reasonable agreement with the gaseous species concentrations observed with the use of mass spectroscopy. They indicate that the (MTS) reactant is relatively unstable and that deposition likely occurs via independent reactions of silicon chlorides (or silanes) and hydrocarbons.

The coatings deposited in this study had fine-grained, columnar structures and were cubic <β-SiC> with a preferred (111) orientation. The orientation is such that alternate layers of carbon and silicon atoms in the structure lie parallel to the substrate surface, further supporting the premise that silicon and carbon atoms are added independently.

Under the CVD conditions of this study, the deposition of <SiC> was determined to be independent of (MTS) or hydrogen concentration and has an activation energy of 188(±52) kJ/mol and a preexponential factor of ln (A) = 23.5(±5.1) nm/min. These observations lead to the conclusion that the likely rate-limiting process is surface diffusion or a surface reaction.

An assessment of the literature indicates that the rate-limiting process probably involves carbon species rather than those containing silicon. Significant further research, most likely involving in situ surface studies, will be required, however, to more closely identify rate-limiting mechanisms.

ACKNOWLEDGMENTS

The authors would like to thank E. George and P. Angelini for helpful reviews of the paper, J. Marsh for the metallographic work, H. Livesey for the graphics, C. A. Valentine for manuscript preparation, and R. H. Powell for editing the text.

REFERENCES

1. Schlichting, J., Powder Metall. Int. **12**(3,4), 141, 196 (1980).

2. Caputo, A. J., D. P. Stinton, R. A. Lowden, and T. M. Besmann, Ceram. Bull. **66**(2), 368 (1987).

3. Stinton, D. P., T. M. Besmann, and R. A. Lowden, Ceram. Bull. **67**(2), 350 (1988).

4. Spear, K. E., Pure Appl. Chem. **54**(7), 1297 (1982).

5. Spear, K. E., Proc. Ninth Intl. Conf. Chemical Vapor Deposition 1984, McD. Robinson et al., eds. (Pennington, NJ: The Electrochemical Society, 1984), pp. 81–97.

6. Coltrin, M. E., R. J. Kee, and J. A. Miller, J. Electrochem. Soc.: Solid State Science and Technology **131**(2), 425 (1984).

7. Coltrin, M. E., R. J. Kee, and J. A. Miller, Proc. Ninth Int. Conf. Chemical Vapor Deposition 1984, McD. Robinson et al., eds. (Pennington, NJ: The Electrochemical Society, Pennington, NJ, 1984) pp. 31–43.

8. Coltrin, M. E., R. J. Kee, and J. A. Miller, J. Electrochem. Soc.: Solid State Science and Technology **133**(6), 1206 (1986).

9. van Kemenade, A. W. C., and C. F. Stemfoort, J. Cryst. Growth **12**, 13 (1972).

10. Aggor L., and W. Fritz, Chem.-Ing.-Technik **43**, 472 (1971).

11. Fedoseev, D. V., V. P. Dorokhovich, A. V. Lavrent'ev, O. I. Zadoroshnyi, and I. G. Varshavskaya, Izvest. Akad. Nauk SSSR, Neorg. Mater. **12**(10), 1796 (1976).

12. Fischman G. S., and W. T. Petuskey, J. Am. Ceram. Soc. **68**(4), 185 (1985).

13. Fitzer V. E., and D. Hegen, Angew. Chem. **91, 316** (1979).

14. Brennfleck, K. E., Fitzer, G. Schoch, and M. Dietrich, Proc. Ninth Int. Conf. Chemical Vapor Deposition 1984, McD. Robinson et al., eds. (Pennington NJ: The Electrochemical Society 1984) 649–662.

15. Thompson, W. T., A. D. Pelton, and C. W. Bale, CALPHAD **7**(2), 113 (1983).

16. Bell, T. N., K. A. Perkins, and P. G. Perkins, J. Chem. Soc., Faraday Trans. I. **77**(198) 1779 (198).

17. Sin'ko, S. V., Yu. A. Pentin, G. S. Gol'din, and G. M. Kuramshina, Zh. Fiz. Khim. **59**, 2682 (1985).

18. Nickl, J. J., and C. von Braunmuhl, J. Less-Common Metals **25**, 303 (1971).

19. Besmann, T. M., and K. E. Spear, J. Electrochem. Soc.: Solid State Science and Technology, **124**(5), 790 (1977).

20. Ivanova, L. M., and A. A. Pletyushkin, Izvest. Akademii Nauk SSSR, Neorgan. Mater. **3**(10), 1817 (1967).

21. Stinton, D. P., and W. J. Lackey, Ceram. Bull. **57**(6), 568 (1978).

22. Gulden, T. D., J. Am. Ceram. Soc. **51**, 424 (1968).

23. Ivanova, L. M., and A. A. Pletyushkin, Izvest. Akademii Nauk SSSR, Neorgan. Mater. **4**(7), 1089 (1968).

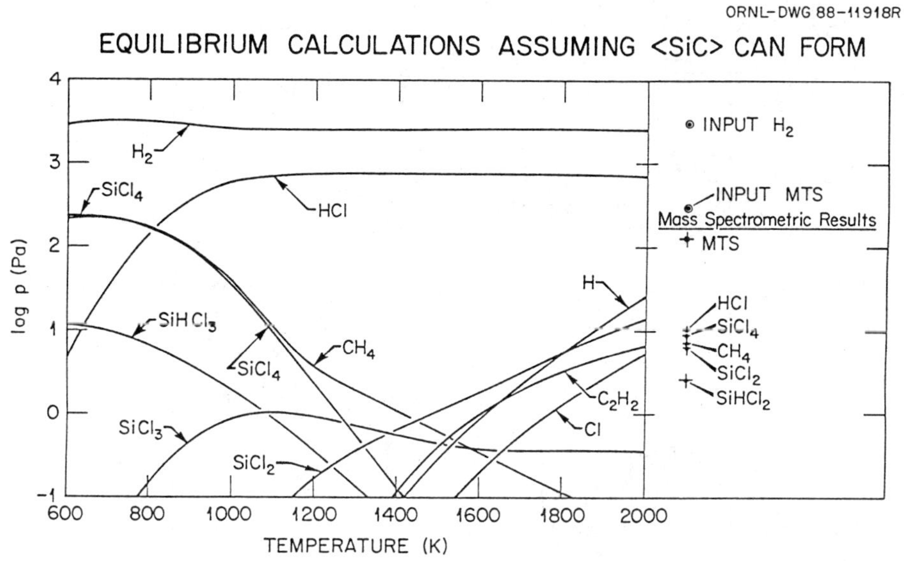

Figure 1. Partial pressure vs temperature plot for the CVD of <SiC>. The curves represent the equilibrium partial pressures, given the indicated input concentrations of (H_2) and (MTS), and assuming that only <SiC> is stable. Also shown are the concentrations derived from mass spectrometric measurements.

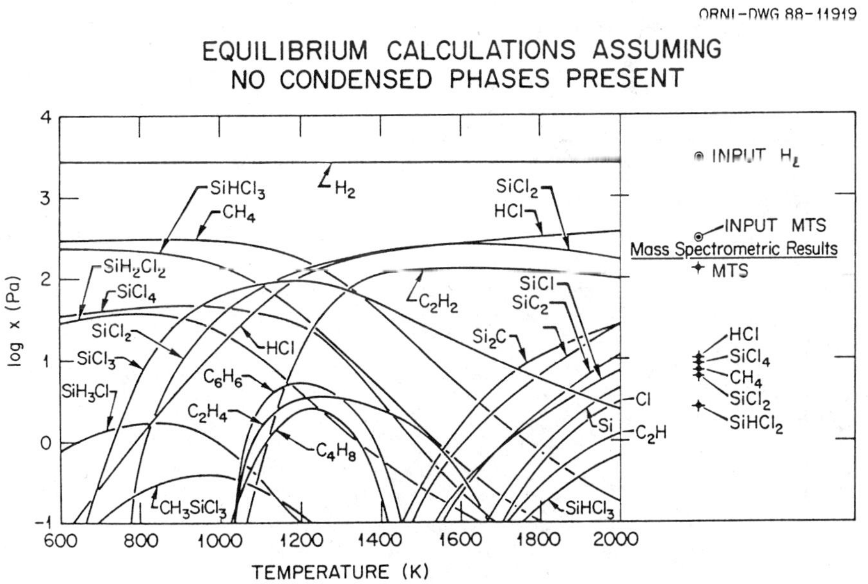

Figure 2. Partial pressure vs temperature plot for the CVD of <SiC>. The curves represent the equilibrium partial pressures, given the indicated input concentrations of (H_2) and (MTS), and assuming that no condensed phase are stable. Also shown are the concentrations derived from mass spectrometric measurements.

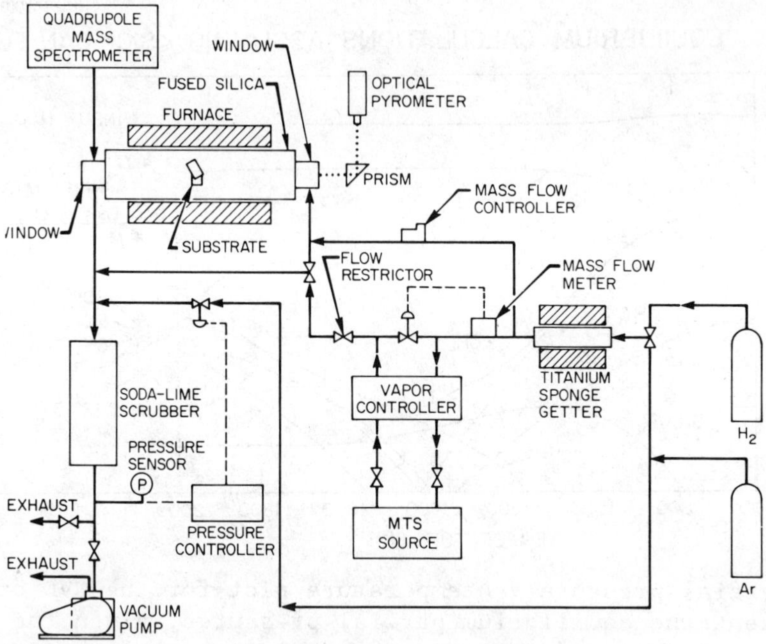

Figure 3. Schematic of the CVD system.

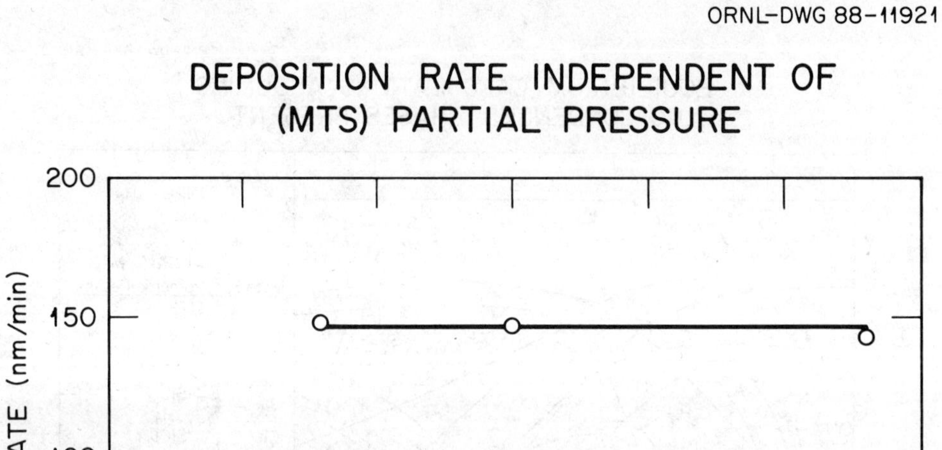

Figure 4. Input partial pressure of (MTS) vs <SiC> deposition rate at 1217 K, indicating a zeroth-order dependence (runs SiC-K-13, 14, and 15).

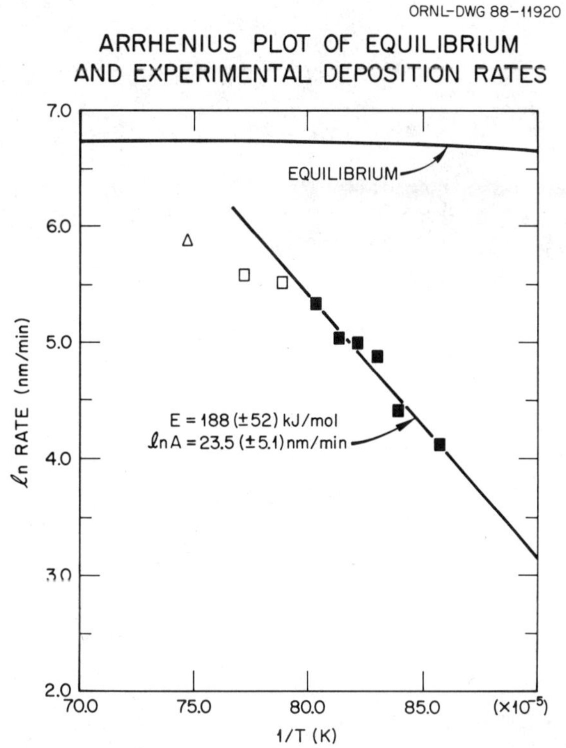

Figure 5. Arrhenius plot showing the experimentally determined <SiC> deposition rates and those predicted assuming thermodynamic equilibrium: solid squares indicate the data used to fit the straight line; open squares indicate higher temperature data not used in the fit; the open triangle indicates a higher-temperature, higher-pressure (101 kPa) deposition rate measurement.

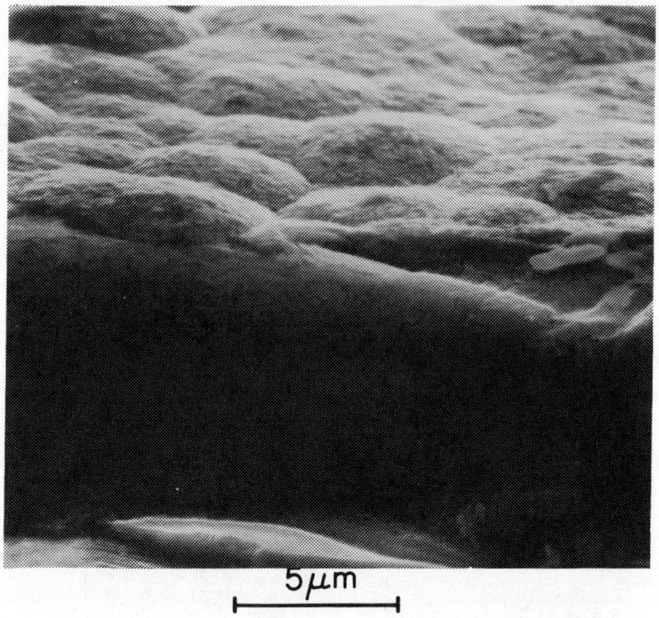

Figure 6. Scanning electron micrograph of the outer and fracture surfaces of a <SiC> coating on graphite (SIC-K-17).

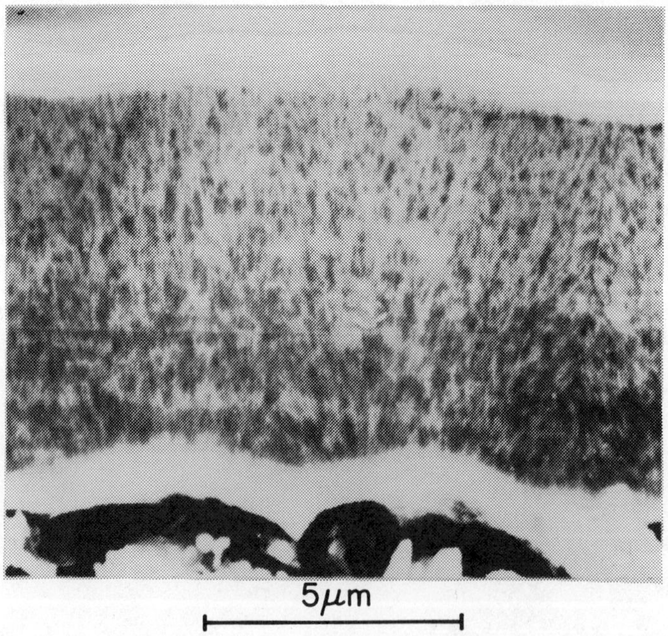

Figure 7. Scanning electron micrograph of a polished and etched cross-section of a <SiC> coating on graphite (SIC-K-15).

TIME DEPENDENT MECHANICAL BEHAVIOR OF SILICON NITRIDE CERAMICS ABOVE 1100°C

M. R. Foley and R. E. Tressler
The Pennsylvania State University
Department of Materials Science and Engineering
Ceramic Science Program
University Park, PA 16802

ABSTRACT

Slow crack growth and creep damage accumulation in silicon nitride ceramics are important processes which limit their usefulness in highly stressed components at elevated temperatures in heat engines. The threshold stress intensities for crack growth at elevated temperatures in air for a commercially available silicon nitride ceramic have been estimated from interrupted static fatigue tests. Tensile creep measurements for this material are being used to delineate the regimes of temperature and stress in which creep cavitation causes flaw nucleation and eventual failure.

INTRODUCTION

Silicon nitride-based ceramics are being studied for high temperature structural applications such as gas turbine engine components. By using these ceramics, the new high efficiency heat engines can be operated at temperatures higher than allowed by current super alloys. Although an excellent candidate for this application, silicon nitride-based ceramics are limited at elevated temperature due to their susceptibility to slow crack growth and creep damage.

Measurements of short-term strength, fracture toughness, slow crack growth and creep at high temperature have been reported for various silicon nitride ceramics[1-7]. In general, these properties are time, temperature and stress dependent and therefore, a detailed knowledge of these interrelationships are required to make long term reliability predictions. It has been shown[8-10] that there are two distinct regions of delayed failure for a typical silicon nitride ceramic at high temperatures. At low stress levels, rupture behavior is strain limited and failure occurs by creep rupture. At high stress levels the failure strains are very small and failure is governed by slow crack growth of preexisting flaws. For an MgO-doped HPSN a specific stress level has been reported by Grathwohl[8] dividing these two regions of behavior suggesting a threshold stress or stress intensity for crack growth. Below this threshold preexisting flaws apparently blunt, and either accommodated creep or cavitation creep control the time dependent strength.

In this paper our measurements of the threshold stress intensity for crack growth in an alumina yttria-doped silicon nitride using static loading tests are presented. In addition tensile creep experiments have been performed to characterize the behavior of this material below the threshold stress intensity for crack growth. Tensile tests were chosen to study the material behavior in the creep regime since the stresses are poorly defined in a creeping bend specimen[11].

EXPERIMENTAL PROCEDURE

1) Material and Specimen Design

The silicon nitride* examined in this investigation contained 2 wt% alumina and 6 wt% yttria as densification aids. The bulk density of this material was 3.24 - 3.26 g/cc[12]. The microstructure as shown in Figure 1 consisted of equiaxed grains of β-Si_3N_4 ranging in size from less than 0.25 up to 4 microns, elongated grains with aspect ratios ranging from 3:1 to 9:1, and an amorphous grain boundary phase.
Bend specimens were cut from rectangular blanks, approximately 20 x 30 x 5 mm in size using a diamond cut-off wheel. The major faces of the samples were surface-ground to a 600-grit finish. Tensile surfaces were hand-beveled using a 15 micron diamond disc to eliminate any edge-checking flaws that may have formed during grinding. The final test bar dimensions were approximately 4 x 2 x 30 mm.

Tensile creep specimens were cut from rectangular blanks, approximately 15 x 15 x 76.2 mm in size. The specimen design (Figure 2) was similar to the modified flat dogbone-shaped specimens developed by Wakai[13] and co-workers and also used by Carroll and Tressler[14]. The specimen design used in this study, however, eliminated the use of machined-in target arms for strain measurements using an optical extensometer. Target arms of the same composition as the specimen were attached to the specimen using platinum wire giving a gauge length of approximately 19 mm. The two holes in the specimen were used to pin the specimen to the loading system. It should be noted that the holes were double tapered and the holes were machined directly on the centerline of the specimen in order to avoid bending moments. The edges of the gauge length were polished using diamond paste down to 3 microns to minimize grinding flaws.

2) Loading Systems

a) Flexure

The four point bend test system consisted of alumina rams entering a furnace from the top and bottom and hot pressed magnesia-doped silicon nitride knife edges with upper and lower spans of 5.2 and 19.7 mm, respectively. A double plate and bearing assembly was used to allow alignment of the rams. The furnace was positioned between the lead screws of a computer controlled testing machine.

b) Tension

A schematic of the tensile loading system designed by Carroll and Tressler[14] is shown in Figure 3. The tensile specimen was attached to the two siliconized silicon carbide grips** using sintered alpha silicon carbide***pins. The pin and grip arrangement was located inside the furnace and resided at the test temperature. The two grips extended outside the furnace and were connected to the remainder of the load trains. The load was applied to the specimen by hanging weights from the loading platform.

* AY6 silicon nitride, GTE, Towanda, PA
** Crystar, Norton Company Worcester, MA
*** Hexalloy SA, Standard Oil Engineered Materials Co., Niagara Falls, NY

a)

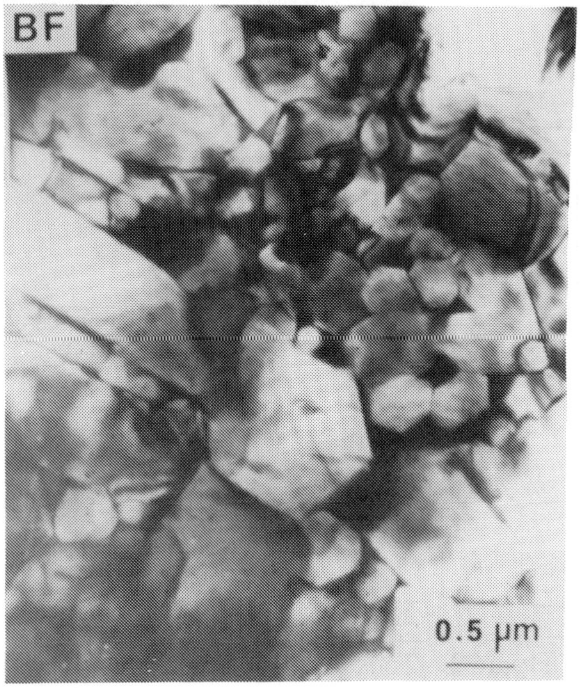

b)

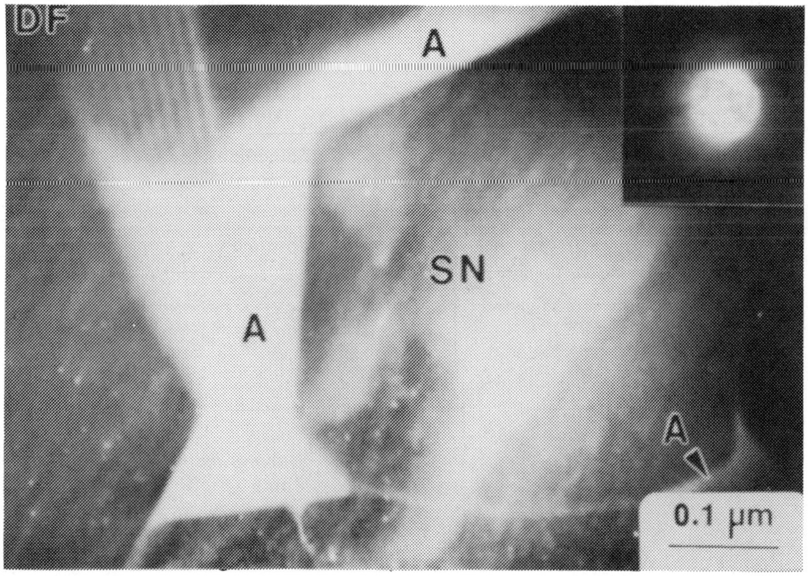

Figure 1. Transmission electron micrographs showing general mirostructural features by bright field imaging (a) and dark field imaging of amorphous grain boundary phase (b).

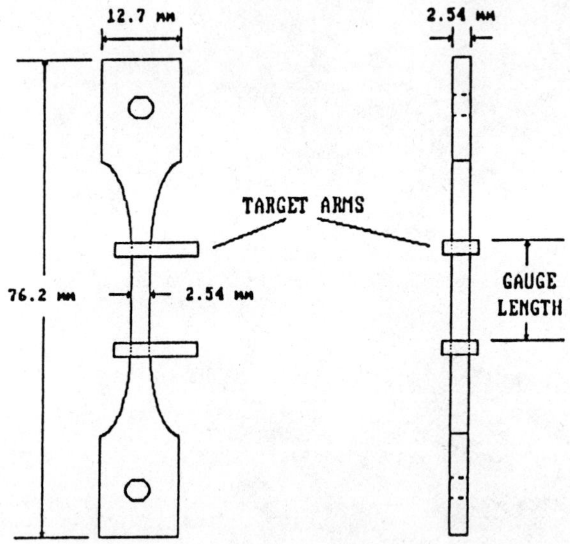

Figure 2. Tensile specimen design.

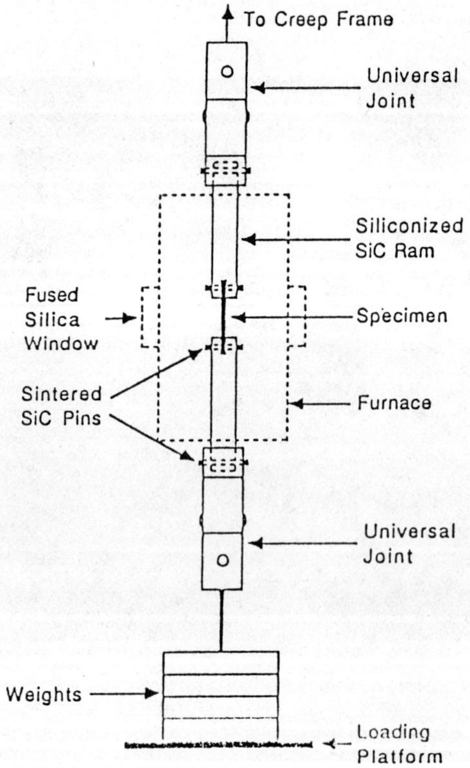

Figure 3. Loading system of creep apparatus.

The creep elongation of the specimen was measured using an optical extensometer*. The optical extensometer focussed on the two target arms on the tensile specimen through a fused silica window. The target arms were back-lighted using a halogen light source transmitted through a second fused silica window located in the back of the furnace. The optical extensometer detects the shadows of the two target arms using photo-cathode screens. As the target arms move apart during the creep test, the optical extensometer follows their movement, producing an output proportional to the displacement. The resolution of this measuring technique is approximately one micron at 1100-1200°C in air.

TEST PROCEDURE

a) Interrupted Static Fatigue Test

The interrupted static fatigue test used on the bend specimens in this study has been fully described by Minford[15,16] and co-workers. The test involves heating the sample to test temperature, rapidly loading to a predetermined static load level for a specified time. If the sample does not fail during the soak time, it is rapidly loaded to failure at the test temperature. this test directly measures the effect of an applied stress during static loading on the fracture stress of the material. The strength reflects any charges in flaw severity and gives one an estimate of the threshold stress intensity for crack growth in the material. Tests were performed on as-machined and oxidized samples at 1100 and 1200°C in air for ten hours. The stressing rate was 30-35 MPa/sec.

b) Creep Tests

Creep tests were performed at 1100 to 1200°C in air at stresses of 100, 150 and 200 MPa for up to 100 hours. Specimens were placed into the loading train under a prestress of 12 MPa before furnace heat up in order to seat the pins in the holes and minimize bending moments. The furnace was then heated to 1200°C in air for a minimum of 16 hours to oxidize the sample. The temperature of the furnace was then set to the test temperature (1100-1200°C) and held for another 8 hours to bring the entire loading system to thermal equilibrium before starting the creep test. Due to the sensitivity of the measuring technique to changes in ambient temperature, the entire loading train system was thermally isolated from the room. After oxidation and thermal equilibration of the sample and system, the sample was loaded to the desired stress level for the creep test.

RESULTS AND DISCUSSION

1) Threshold Stress Intensity for Crack Growth

Figures 4 and 5 summarize the interrupted static fatigue test results at 1100°C in air for the as-machined and oxidized (15 hrs @ 1200°C in air) silicon nitride samples, respectively. Plotted at an initial stress intensity of zero are samples fractured at 1100°C in air with no prior static loading. The symbols and error bars represent the mean fracture stress and 95% confidence interval, respectively. The threshold stress intensity, K_{Io}, is estimated as the lowest initial applied stress intensity at which at least one of the samples failed during static loading. The threshold stress intensity for crack growth for the as-machined and oxidized samples are 1.75 MPa·m$^{1/2}$ and 2.00 MPa·m$^{1/2}$, respectively. The range of K_{Io} values is due to the range of flaw sizes calculated from the initial strength distributions. Note also that, because there is no

---

* Model 200 x-2, Zimmer OHG, West Germany

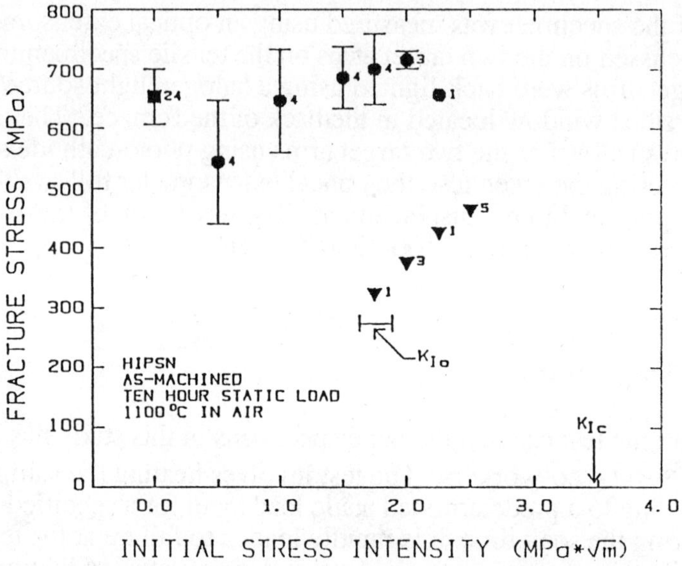

Figure 4. Effect of initial applied stress intensity on fracture stress of as-machined HIPSN at 1100°C in air, static load time: ten hours. (■ Tested after proof testing, no prior static loading; ● Tested after static loading; ▼ Failed during static loading; and 24 indicates the number of samples tested.)

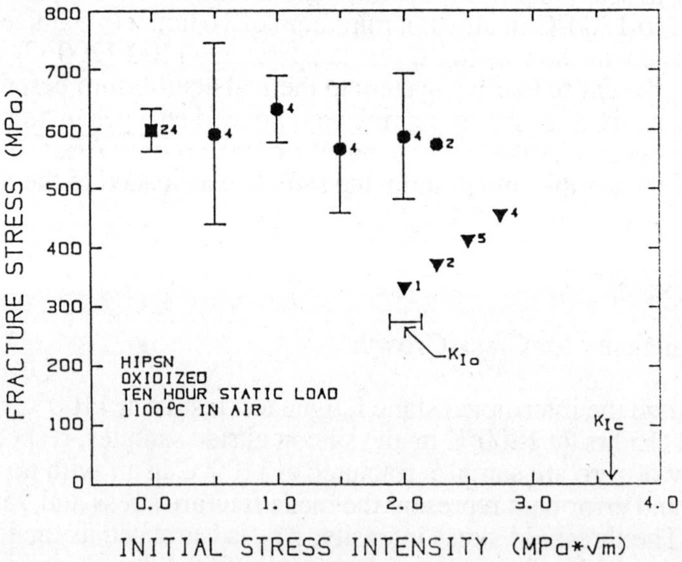

Figure 5. Effect of initial applied stress intensity on fracture stress of oxidized HIPSN at 1100°C in air, static load time: ten hours. (■ Tested after proof testing, no prior static loading; ● Tested after static loading; ▼ Failed during static loading; and 24 indicates the number of samples tested.)

significant change in mean fracture stress (indicated by the solid circle) with increased initial stress intensity up to K_{Io}, it is assumed that the original strength-limiting flaw population is unchanged.

The interrupted static fatigue test results for as-machined and oxidized samples at 1200°C in air are summarized in Figures 6 and 7. For the as-machined samples, a sharp decrease in mean fracture stress in observed with increasing initial applied stress intensity along with an apparent threshold stress intensity of 1.00 $MPa \cdot m^{1/2}$. The oxidized samples during static loading do not show any significant decrease in mean fracture stress with increasing initial applied stress intensity up to the K_{Io} of 1.50 $MPa \cdot m^{1/2}$.

The differences in K_{Io} values between the as-machined and oxidized materials during static load testing are apparently due to a microstructural change, and this effect is most apparent at 1200°C for this material. The decreasing mean fracture stress for the as-machined samples is probably due to the creation of a new flaw population which produced flaws that were more severe. This new flaw population could possibly have been created by creep cavitation at this temperature as identified by Tighe[7,17,18] and co-workers, Grathwohl[8] and Quinn[9,10]. Note that, because of the creation of new flaws, the sizes of which are not well-characterized, the actual initial applied stress intensity values used during testing were different from those calculated from the original flaw population. Therefore, it is more informative to record the effect of initial applied stress on the mean fracture stress for the as-machined specimens (Fig. 6).

The fact that the oxidized samples do not experience the same detrimental effects during static loading at 1200°C as did the as-machined samples presumably results from an oxidation-induced microstructural change. At this temperature it is well known[19-23] that additives and impurities such as Fe, Al and Y present in the grain boundary phase diffuse toward the surface. This rejection of impurities from the grain boundaries creates a more refractory grain boundary phase, and thus, produces a more creep resistant material.

2) Correlation between Flexural and Tensile Behavior

In order to delineate the regimes of temperatures and stress in which slow crack growth and creep cavitation are the dominant processes responsible for failure at high temperatures, tensile creep measurements were used. From the strengths of the bend specimens at these elevated temperatures, the strengths of tensile specimens can be predicted using Weibull statistics. In order to predict strengths on specimens with different volumes under stress, the effective volume must be calculated from the geometry of the specimen, the loading configuration and Weibull statistics. In the case of the rectangular cross section used in the bend tests (4 x 2 mm) and four-point bending test geometry (upper and lower spans of 5.2 and 19.7 mm, respectively), the effective volume (V_E) is given by the relationship for four-point bending[24]:

$$V_E = K\,V_B$$

where K, the load factor, is given by

$$K = \frac{\left(\frac{a}{b}\right)m + 1}{2(m+1)^2}$$

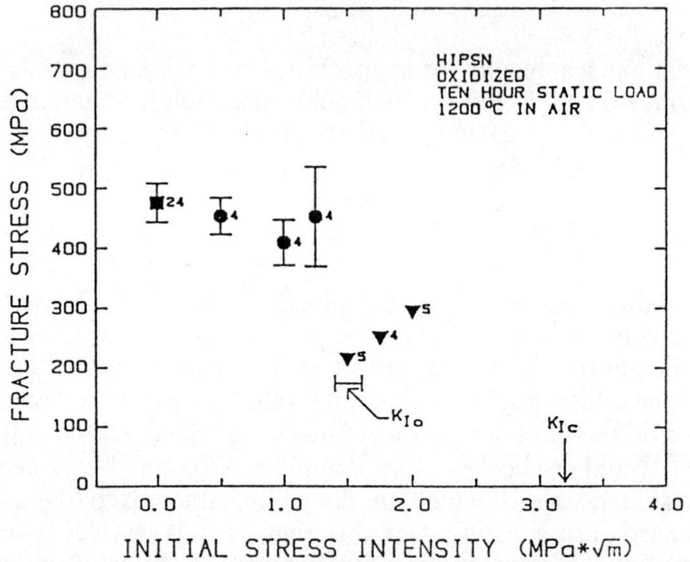

Figure 6. Effect of initial applied stress intensity on fracture stress of as-machined HIPSN at 1200°C in air, static load time: ten hours. (■ Tested after proof testing, no prior static loading; ● Tested after static loading; ▼ Failed during static loading; and 24 indicates the number of samples tested.)

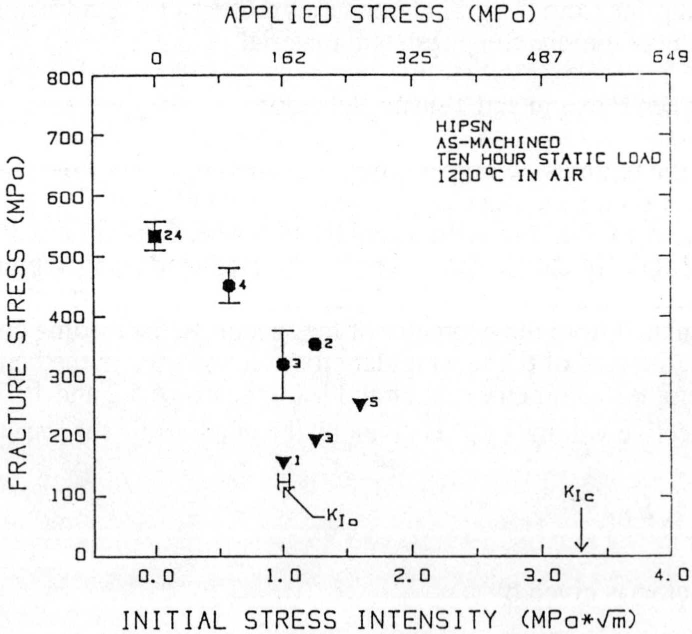

Figure 7. Effect of initial applied stress intensity on fracture stress of oxidized HIPSN at 1200°C in air, static load time: ten hours. (■ Tested after proof testing, no prior static loading; ● Tested after static loading; ▼ Failed during static loading; and 24 indicates the number of samples tested.)

where a is the upper span, b is the lower span, m is the Weibull modulus and V_B is the total volume under stress. Assuming that volume-distributed flaws dominate, the bend and tensile strengths are related by

$$\left(\frac{\sigma_T}{\sigma_B}\right)^m = \left(\frac{V_E}{V_T}\right)$$

where σ_T is the predicted tensile strength from four-point bend data, σ_B is the four-point bend strength and V_T is the total volume within the gauge length in a tensile specimen. The strengths of the tensile specimens can be estimated from the strength of the four-point bend specimens using the above relationship and the experimental Weibull parameters for the bend strengths. Because of the more stable behavior of the oxidized samples in bending, only oxidized specimens were used in tensile creep tests. The calculated applied stresses for fast fracture and for cracks to grow at the threshold stress intensity at 1100°C and 1200°C in air for oxidized tensile specimens are shown in Table 1. The Weibull moduli for the oxidized samples at 1100 and 1200°C were 7.0 and 5.0, respectively[25].

Table 1. Tensile strength and threshold stress predictions from flexural data on oxidized specimens

TEMP	FLEXURAL		TENSILE	
(°C)	$\bar{\sigma}_f$ (MPa)	σ_{th} (MPa)	$\bar{\sigma}_f$ (MPa)	σ_{th} (MPa)
1100	599	333	346	192
1200	476	224	238	112

3) Creep Results

Preliminary tensile creep tests were performed on oxidized specimens at temperatures from 1100 and 1200°C under stresses of 100, 150 and 200 MPa. Figure 8 summarizes some of the tensile creep curves at 1100°C in air for applied stresses of 100, 150 and 200 MPa. An initial primary creep regime (up to 10-15 hrs), followed by a secondary or steady-state creep regime was observed. The tests shown in Figure 8 were terminated prior to specimen failure except for the 200 MPa test which was part of a temperature-change test. Similar tensile creep curves have been observed for other silicon nitride-based ceramics[5,13,26,27] as well as for other ceramic materials[14,28]. No samples were tested to failure at 1100°C at these three stress levels and, hence, the presence or absence of tertiary creep cannot be ascertained.

The stress dependence of the steady state creep rate ($\dot{\varepsilon}_s$) at 1100°C was estimated by fitting to the Dorn equation:

$$\dot{\varepsilon}_s = A\,\sigma^n \exp(-Qc/kT)$$

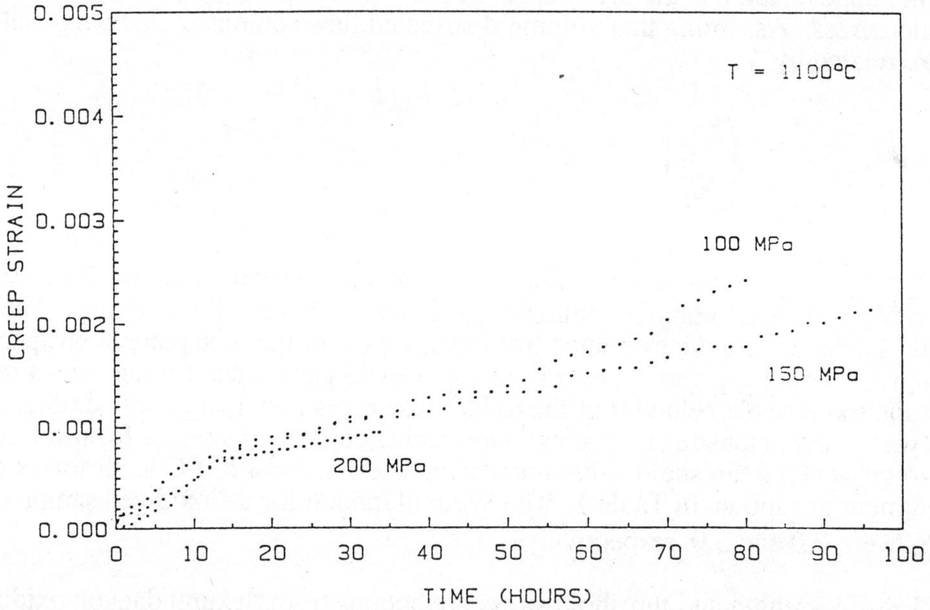

Figure 8. Creep strain versus time at 1100°C in air for specimens under stresses of 100, 150 and 200 MPa.

where A is a constant, n is the stress exponent for steady state creep, Qc is the apparent activation energy for creep and k is the Boltzman constant. Plotting the log $\dot{\varepsilon}_s$ versus log σ for different samples at constant temperature (Figure 9), no concrete dependence is obvious from this preliminary data. Other values of the steady state stress exponent for silicon nitride ceramics have been reported[5,8,29-31] and range from 1.5 to 2.4 at temperatures of 1200 to 1400°C in air. They all indicate that creep deformation is controlled by viscous flow of the grain boundary phase leading to creep cavitation, flaw nucleation and eventual failure. Grathwohl[8] has shown that above a distinct stress level it is slow crack growth that governs rupture behavior and below which rupture behavior is controlled by creep cavitation. It is obvious that more stress change tests and temperature change tests are needed along with microscopic analysis to characterize the creep behavior of this material.

Results of the bend data predict a value of applied stress of 192 MPa in tension at 1100°C to equal the threshold stress intensity for crack growth (Table 1). Thus, at stresses above 200 MPa one expects slow crack growth to dominate and very small strains to failure to be observed.

Two creep tests were conducted at 1150°C at 200 MPa. Figure 10 shows the creep curve for one of the tests. Failure had not occurred after 75 hours, but the test was terminated due to equipment malfunction. The second test led to specimen failure in approximately 4 to 5 minutes and no creep strain data was measured. These specimens seem to have bracketed the threshold stress intensity for crack growth at this temperature.
The predicted threshold stress for crack growth in tension is 112 MPa at 1200°C (Table 1). Tensile creep experiments at 1200°C resulted in failure after 3 minutes and 1 minute at stresses of 100 and 200 MPa, respectively. Further experiments at stresses near and below the predicted threshold stress are needed to confirm these approximate predictions.

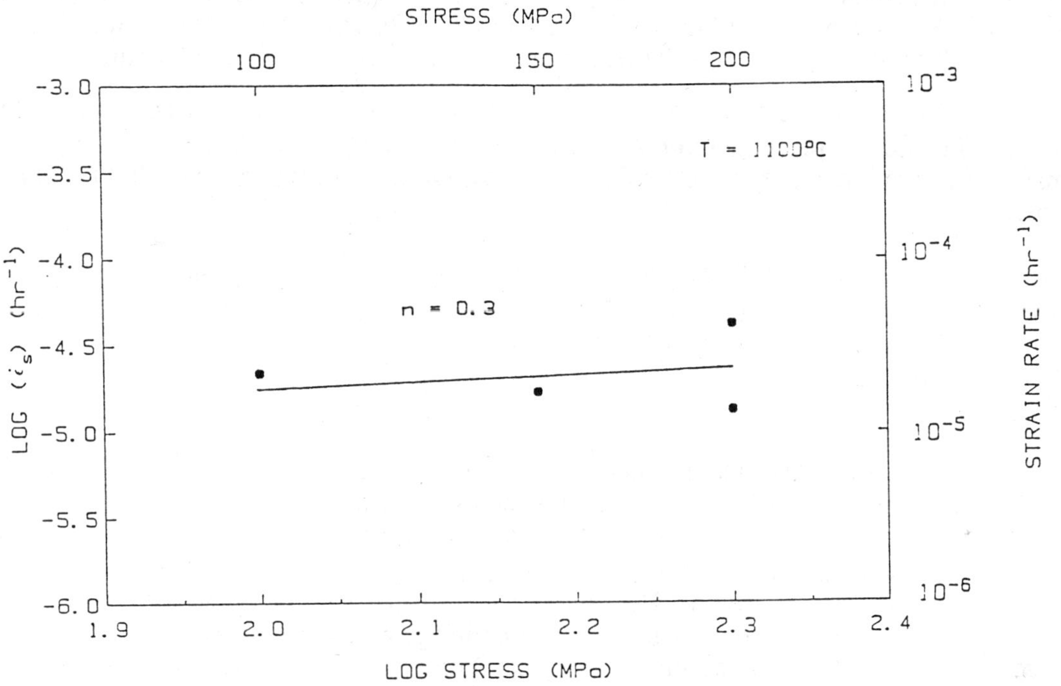

Figure 9. Log steady state creep rate versus log applied stress at 1100°C in air.

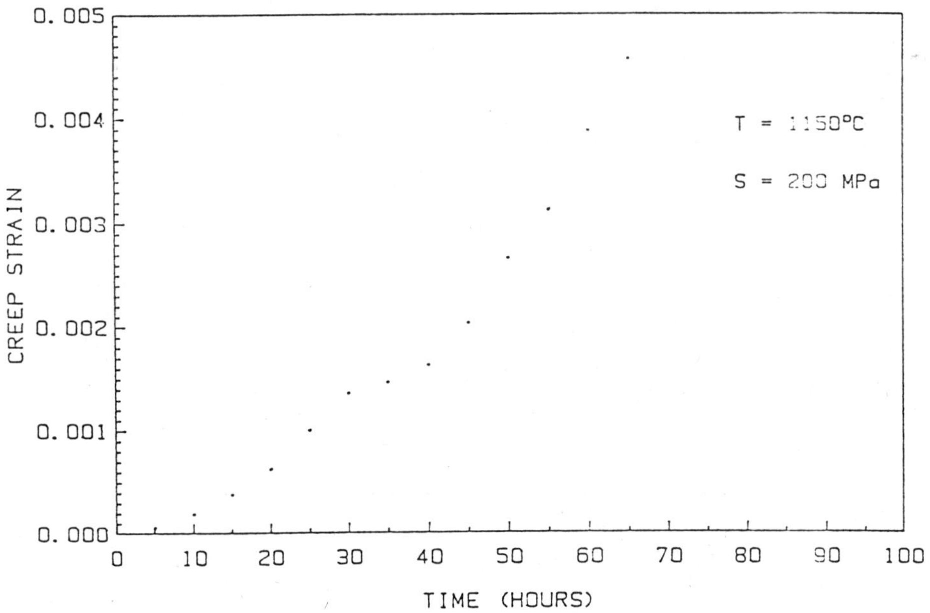

Figure 10. Creep strain versus time at 1150°C for a specimen under a stress of 200 MPa.

To estimate the activation energy for steady-state creep a temperature change test at constant stress (200 MPa) was performed from 1100 to 1175°C. The results are presented in Figure 11. Specimen failure occurred after a total test time of 92.05 hours and at a total strain of 0.444%.

From a plot of Ln $\dot{\varepsilon}_s$ versus 1/T, one can estimate the apparent activation energy for the Dorn equation. The data are presented in Figure 12. The data are too scattered and too few to estimate the activation energy with confidence. However, it is in the range of 300 to 500 kJ/mole.

Activation energies reported[5,8,29-31] for other silicon nitride ceramics for the temperature regime from 1100 to 1400°C range from 535 to 709 kJ/mole again suggesting that viscous flow of the grain boundary phase governs creep deformation.

SUMMARY

Interrupted static fatigue tests and tensile creep tests were performed in order to delineate the regimes of temperature and stress in which slow crack growth and creep cavitation are the dominate processes responsible for failure at high temperature. The threshold stress intensity for crack growth was estimated to be 1.75 and 2.00 MPa·m$^{1/2}$ at 1100°C in air for as-machined and oxidized samples, respectively. However, at 1200°C a strength degradation is observed for as-machined sample and is probably due to the creation of a new flaw population which produced flaws that were more severe by creep cavitation. The fact that the oxidized samples do not experience the same detrimental effect at 1200°C presumably results from an oxidation-induced microstructural change thus producing a more creep resistant material.

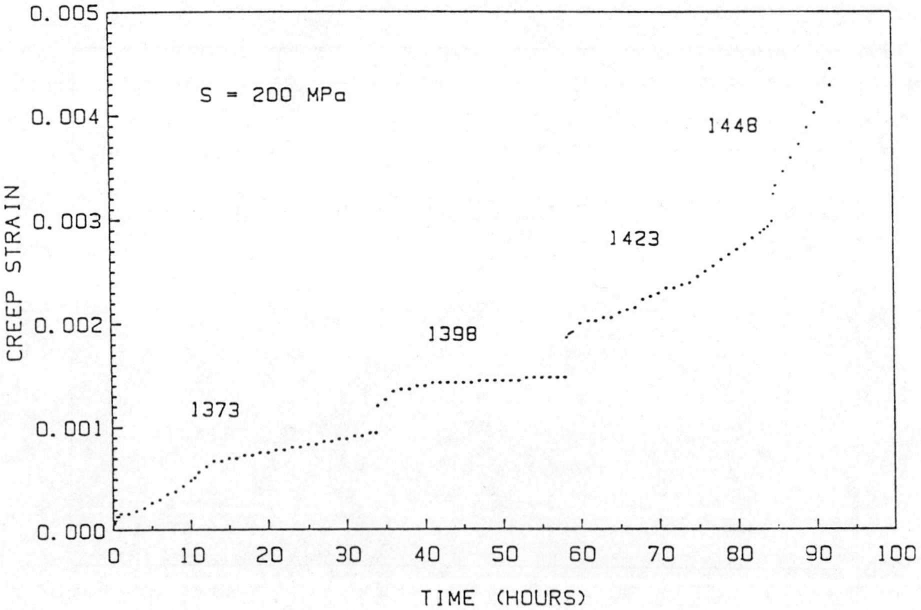

Figure 11. Creep strain versus time at 1100-1175°C for a specimen under an applied stress of 200 MPa.

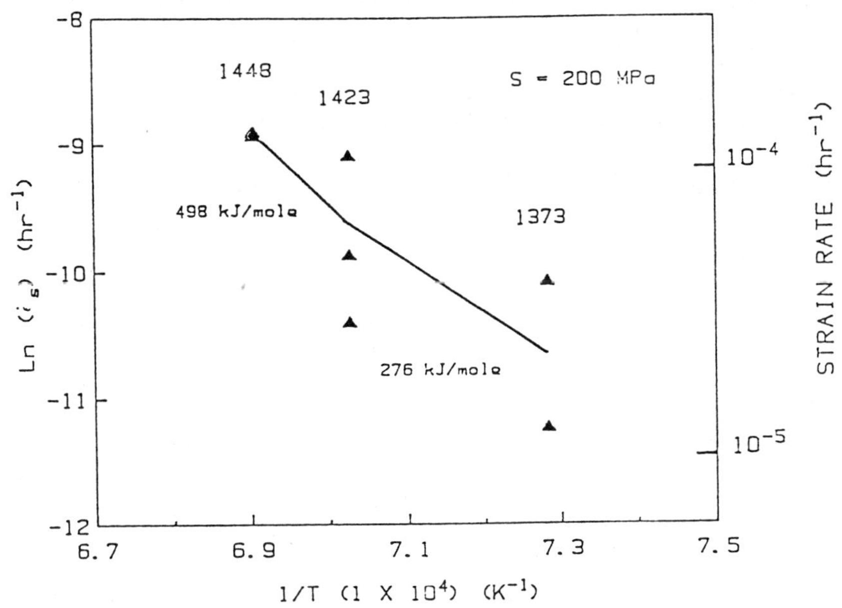

Figure 12. Ln steady-state creep rate versus temperature for an applied stress of 200 MPa.

The tensile creep behavior of the more stable oxidized silicon nitride was measured using a optical extensometer and a flat dogbone-shaped specimen. Tensile creep tests conducted at applied stresses near or below the predicted threshold stress at 1100°C exhibited typical primary and secondary creep behavior. No concrete dependence of stress on the steady-state creep rate was obvious from this preliminary data. Creep tests at 1150°C at 200 MPa show both creep behavior and slow crack growth rupture behavior suggesting that these specimens have bracketed the threshold stress for crack growth at this temperature. All creep tests performed at 1200°C led to very short term failure at applied stresses of 100-200 MPa. Thus, the threshold stress for crack growth is near or below 100 MPa as predicted by flexural data.

ACKNOWLEDGEMENT

This work was supported by the U.S. Army Research Office, Metallurgy and Materials Science Division.

REFERENCES

1. S. M. Wiederhorn, "Subcritical Crack Growth in Ceramics," pp. 613-46 in Fracture Mechanics of Ceramics, Vol. 2, Edited by R. D. Bradt, D. P. H. Hasselman and F. F. Lange, Plenum Press, NY, 1974.

2. G. D. Quinn and J. B. Quinn, "Slow Crack Growth in Hot-Pressed Silicon Nitride," pp. 603-636 in Fracture Mechanics of Ceramics, Vol. 6., Edited by R. C. Bradt, A. G. Evans, D. P. H. Hasselman and F. F. Lange, Plenum Press, New York, 1983.

3. T. M. Konushonis, "Crack Propagation in Silicon Nitride Ceramics," M. S. Thesis, The Pennsylvania State University, University Park, PA, May 1976.

4. T. E. Easler, "Environmental Effects in the Strength Distribution of Silicon Carbide and Silicon Nitride," M. S. Thesis, The Pennsylvania State University, University Park, PA, August 1980.

5. R. Kossowsky, D. G. Miller and D. S. Diaz, "Tensile and Creep Strengths of Hot-Pressed Si_3N_4," J. Mater. Sci., 10 983-997 (1975).

6. R. M. Arons and J. K. Tien, "Creep Strain Recovery in Hot-Pressed Silicon Nitride," J. Mater. Sci., 15 2046-2058 (1980).

7. N. J. Tighe, S. M. Wiederhorn, T. J. Chuang and C. L. McDaniel, "Creep Cavitation and Crack Growth in Silicon Nitride," pp. 587-604 in Deformation of Ceramic Materials II, edited by R. E. Tressler and R. C. Bradt, Plenum Press, New York, 1984.

8. G. Grathwohl, "Regimes of Creep and Slow Crack Growth in High Temperature Rupture of Hot-Pressed Silicon Nitride," pp. 573-86 in Ref. 7.

9. G. D. Quinn, "Static Fatigue in High Performance Ceramics," presented at ASTM Conference, San Francisco, CA (1982).

10. G. D. Quinn, "Fracture Mechanism Map for Hot Pressed Silicon Nitride," presented at Eighth Annual Conference of the Ceramic-Metal Systems Division.

11. D. F. Carroll, R. E. Tressler, Y. Tsai and C. Near, "High Temperature mechanical Properties of Siliconized Silicon Carbide Composites," in Tailoring Multiphase and Composite Ceramics, held at The Pennsylvania State University, University Park, PA, July 17-19, 1985.

12. M. R. Foley and R. E. Tressler, "Threshold Stress Intensity for Crack Growth at Elevated Temperatures in a Silicon Nitride Ceramic," to be published in Adv. Ceram. Mater., 2 [1] 1988.

13. F. Wakai, S. Sakaguchi, Y. Matsuno and H. Okunda, "Tensile Creep Test of Hot-Pressed Si_3N_4," presented at the International Symposium on Ceramic Components for Engines, Hakone, Japan, October 17-21, 1983.

14. D. F. Carroll and R. E. Tressler, "Effect of Creep Damage in the Tensile Creep Behavior of a Siliconized Silicon Carbide," to be published.

15. E. J. Minford and R. E. Tressler, "Determination of Threshold Stress Intensity for Crack Growth at High Temperature in Silicon Carbide Ceramics," J. Am. Ceram. Soc. 66 [5] 338-40 (1983).

16. E. J. Minford, D. M. Kupp and R. E. Tressler, "Static Fatigue Limit for Sintered Silicon Carbide at Elevated Temperatures," ibid., 11 769-73.

17. S. M. Wiederhorn, N. J. Tighe, T.-J. Chuang, C. L. McDaniel and E. R. Fuller, Jr., "Structural Reliability of Brittle Material at High Temperatures," NBS Technical Report for Project No. 5610464 (1983).

18. S. M. Wiederhorn and N. J. Tighe, "Structural Reliability of Yttria-Doped Hot-Presses Silicon Nitride at Elevated Temperatures," J. Am. Ceram. Soc. 66 [12] 884-89 (1983).

19. H. Du; private communication.

20. J. E. Siebels, "Oxidation and Strength of Silicon Nitride and Silicon Carbide," pp. 793-804 in Ceramics for High Performance Applications III Reliability, edited by E. M. Lenoe, R. N. Katz and J. J. Burke, Plenum Press, New York, 1983.

21. S. C. Singhal, "Oxidation of Silicon Nitride and Related Materials," pp. 607-26 in Nitrogen Ceramics, edited by F. L. Riley, Noordhoff International, 1977.

22. S. C. Singhal, "Oxidation and Corrosion-Erosion Behavior of Si_3N_4 and SiC," pp. 535-48 in Ceramics for High Performance Applications, edited by J. J. Burke, A. E. Gorum and R. N. Katz, Brook Hill, Chesnut Hill, MA, 1974.

23. D. R. Clarke and F. F. Lange, "Oxidation of Si_3N_4 Alloys: Relation to Phase Equilibria in the System Si_3N_4-SiO_2-MgO," J. Am. Ceram. Soc., 63 [9-10] 586-93 (1980).

24. C. A. Johnson, "Fracture Statistics in Design and Application," General Electric Technical Information Series Report No. 79CRO212 (1979).

25. M. R. Foley, "Flaw Behavior Near the Threshold Stress Intensity in a Hot Isostatically Pressed Silicon Nitride at Elevated Temperatures," M. S. Thesis, The Pennsylvania State University, University Park, PA May 1987.

26. F. F. Lange, E. S. Diaz and C. A. Andersson, "Tensile Creep Testing of Improved Si_3N_4, Am. Ceram. Soc. Bulletin 58 [9] 845-848 (1979).

27. H. Tanaka and Y. Inomata, "Primary Creep of Sintered Silicon Nitride," presented at the International Symposium on Ceramic Components for Engines, Hakone, Japan, 1983.

28. R. Morrell and K. H. G. Ashbee, "High Temperature Creep of Lithium Zinc Silicate Glass-Ceramics," J. Mater. Sci., 8 1253-1270 (1973)

29. G. Grathwohl and F. Thümmler, "Interaction Between Creep, Oxidation and Microporosity in Reaction-Bonded Silicon Nitride," Ceramurgia International 6 [2] 43-50 (1980).

30. S. V. Din and P. S. Nicholson, "Creep of Hot-Pressed Silicon Nitride," J. Mater. Sci., 10 1375-1380 (1975).

31. J. M. Birch and B. Wilshire, "The Compression Creep Behavior of Silicon Nitride Ceramics," J. Mater. Sci. 13 2627-2636 (1978).

TRANSIENT THERMOELASTIC FRACTURE ANALYSIS OF CERAMIC-TO-METAL BONDS: IMPLICATIONS FOR HEAT ENGINES

Klod Kokini
Purdue University
School of Mechanical Engineering
West Lafayette, IN 47907

ABSTRACT

The effect of a transient heating followed by cooling in a ceramic-to-metal bond interface was analyzed by considering a penny-shaped crack at the interface. The transient temperatures and deformations of the crack as well as strain energy release rates were calculated using the finite element method. The results were obtained for two ceramic thicknesses and it was shown that the thinner ceramic coating would be more resistant to failure. The effect of radial constraint on the structure was determined to be negligible. The implications of the results to engine conditions were discussed.

1. INTRODUCTION

Ceramic-to-metal bonds are developed for high temperature applications such as heat engines and gas turbines. Most of these are in the form of thermal barrier coatings which allow for larger engine temperatures and higher efficiencies. The advantages of such coatings have been discussed in reference 1.

A major obstacle to the successful use of such coatings has been the failure in the form of debonding and spalling from the substrate [2-5]. The failure region appears, in most cases to be near the interface between the ceramic and the metal. It is well known that thermal stresses are generated at such interfaces when dissimilar materials are bonded together [6]. Most of the studies in this area have considered steady-state excursions. More recently, thermal stresses in ceramic-to-metal bonds and glass-to-metal bonds have been analyzed by taking into account the transient nature of the heat transfer process [7-8].

However, for components of finite dimensions, the free edges of interfaces represent regions of stress singularities [9-10]. Furthermore, possible debonding defects or cracks that usually develop at the interface cannot be accounted for by using such analyses. In this case, it is preferable to use principles of fracture mechanics to determine the effects of a transient thermal load on the interface. Such analyses have been previously performed for steady-state thermal loads [11-13] and more recently by the present author, for the two-dimensional plane strain problems of an interface in a ceramic/metal bond subject to transient thermal load [14]. This analysis has shown that for slow cooling conditions, the largest strain energy release rate occurs at steady-state while for a center crack, the transient thermal load results in a maximum G which represents the worst condition. It was also shown that when the ceramic/metal structure is allowed to bend it is desirable to have a thicker ceramic layer at the center and a thinner one at the edge. The study of the case when the boundary is restrained showed that under these conditions the thick ceramic layer reached a large G value at steady-state. Also the effects of thermal and mechanical boundary conditions were studied [15]. A subsequent study considered the effects of material properties [16]. It was shown here that thermal expansion equality does not necessarily result in a smaller strain energy release rate for transient thermal loads.

For the case of engine components such as pistons and valves the thermal barrier coating deposited on the metal substrate has a circular configuration. This geometry causes the effect of a transient thermal load to be different than from a two-dimensional geometry. Consequently, in this paper, the effect of such a load is determined by using a fracture mechanics approach applied to a penny shaped interface crack.

2. DESCRIPTION OF PROBLEM

The geometrical configuration of the problem considered is shown in Figure 1. Two circular layers of ceramic and metal are bonded to each other. The configuration of a penny-shaped crack at the center of the interface is considered.

The ceramic is subjected to a transient thermal load through a heat transfer coefficient h. The lower surface of the metal and the sides are insulated. An important assumption is that the crack surfaces are insulated, i.e., no heat crosses the crack. This heat transfer process is shown in Figure 2.

The mechanical boundary conditions applied to the structure consist of constraining the lower surface of the metal from any deformations normal to it. Therefore, this surface is rigid and is not allowed to bend. In one case, the boundary on the outside radius is free to expand, while in a second case, the same boundary is prevented from deforming.

The problem is solved as a quasi steady-state thermoelastic problem. It is assumed that both materials are homogeneous, isotropic and linearly elastic. It is also assumed that no stresses exist in the materials at room temperature.

The problem is solved using the finite element method*. First the temperature distribution is obtained as a function of time. These are then used to calculate the displacements and stresses. The crack flank deformations (i.e. the opening and the closing of the crack shown in Figure 3) are used to determine a complex stress intensity factor which is then translated into a strain energy release rate [17]. The details of performing these calculations are described in previous publications and will not be repeated here [14,16].

---

* A program called ABAQUS, developed by Hibbitt, Karlsson and Sorensen, Inc., Providence, R.I. was used for the calculations.

3. RESULTS

The transient displacements and strain energy release rate for the cracked ceramic-to-metal bond shown in Figure 1 were obtained for only one crack length. It is known, however, that G becomes larger if the crack length gets larger. For the case considered, the crack length (crack radius a for the penny-shaped crack) was 1.5625 mm. The thickness of the metal layer was $L_m = 12.5$ mm. Two thicknesses of ceramic were considered. In one case, a "thin coating" of $L_c = 1.25$ mm was used, so that the thickness ratio was $L_c/L_m = 0.1$. In a second case, a "thick ceramic" layer of the $h_c = 12.5$ mm was considered and L_c/L_m was 1. The outside radius of the structure R was 50 mm.

The entire structure was taken from room temperature (20°C) to 600°C via a heat transfer coefficient of $h = 1500\ W/m^2K$ applied at the ceramic surface, until a steady-state uniform temperature was reached. Following this process, the ceramic was cooled at $h = 50\ W/m^2K$, which represents a much slower cooling process. This cycle attempts to approximate initial start up of an engine which occurs at a relatively large rate followed by total shut-off which results in cooling which could be faster than described here. The main difference between this analysis and engine conditions is that, at high temperatures, the distribution of the temperature does not become uniform, but rather results in some temperature gradient across the materials. The effects of this will be discussed below in relation to specific results which are calculated.

The two materials considered were zirconia and cast iron. The material properties for each are shown in Table 1 [11].

Table 1. Materials Properties of the Two Components

Zirconia	$k_c = 2.2$ W/mK
	$C_c = 460.6$ J/(kg·°C)
	$\rho_c = 5.9$ g/cc
	$E_c = 200$ G Pa
	$\alpha_c = 10.2 \times 10^{-6}$ /°C
	$\nu_c = 0.23$
Nodular cast iron	$k_m = 48.9$ W/mK
	$C_m = 418.4$ J/kg–°C
	$\rho_m = 7.3$ g/cc
	$E_m = 168$ G Pa
	$\alpha_m = 13.7 \times 10^{-6}$ /°C
	$\nu_m = 0.31$

3.1 Penny-Shaped Crack in an Unconstrained Cylindrical Ceramic/Metal Bond.

The case of a penny-shaped crack, at the center of the ceramic/metal interface with unconstrained edges is first considered. The resulting transient strain energy release rate G for the cases of the thin coating and thick ceramic are plotted in Figure 4. For the case of the thin coating, during heating, G increases, reaches a maximum value and then decreases to become zero at steady-state. During cooling, the same behavior is observed, except that the magnitude of the maximum G is smaller than during heating since the heat transfer coefficient in this case is smaller.

When the thickness of the ceramic layer is made equal to that of the metal, the behavior of G is seen to be similar to that of the thin coating. G steadily increases reaches a maximum and then decreases to a smaller value. As cooling is started a further increase occurs before G starts decreasing to become zero at room temperature. The important difference between the two thicknesses is that the magnitude of the maximum G for the thick ceramic layer is much larger than the one for the thin coating.

The reasons for this behavior can be analyzed by considering the deformations of the crack faces as shown in Figures 5 and 6. The two basic deformations of the crack are the sliding Δu and the opening Δv. For both thicknesses of coating the sliding and the opening increase, reach a maximum and then decrease during heating and cooling. However, for the thin coating, the largest deformation is the sliding, while for the thick ceramic, the largest deformation is the opening. The discontinuities in the behavior which are observed for the case of sliding in Figure 5 are due to the fact that a sign change occurs at these points which cannot be shown on the logarithmic plot.

The deformations of the crack are due to the temperature distribution at any instant of time. At the time when the thin coating experiences the largest G, the temperatures at selected points of the structure are shown in Figure 7. It can be noted that since the crack is an insulator, large temperature gradients result across it, thus causing the transient deformations. The gradients across the structure in the engine are expected to be somewhat larger [7]. For the case of the thin coating, the largest gradients occur in the direction parallel to the crack. This results in the large sliding displacement differential Δu shown in Figure 5. When the engine operates, the surface temperature fluctuates and this heat transfer process affects the temperature distribution inside the ceramic but without necessarily reaching the interface [18]. It is expected that this fluctuation will effect the Δv component and therefore cause some variation in the G value. This implies that the failure process for the thin coating is expected to be worse because of the turning on and off of the engine rather than the actual operation.

The temperature distribution at the time when G is maximum for the thick ceramic layer is shown in Figure 8. In this case, the largest temperature gradients occur in the direction perpendicular to the crack and interface. As a result, the maximum deformation of the crack is the opening as shown in Figure 6. Fluctuations of surface temperatures, in this case, are expected to contribute to the opening of the crack and therefore, result in a fluctuation of G that follows the fluctuation of the temperature.

The other expected difficulty with the thicker ceramic layer stems from the large temperature gradients near the top surface of the material as shown in Figure 8. The effect of these is to cause larger stresses near the surface which would result in surface cracks or subsurface cracks. On the other hand, it is clear that the thicker ceramic material does provide much better insulation while being prone to failure.

3.2 Penny-Shaped Crack in a Constrained Cylindrical Ceramic/Metal Bond

Most engine parts which are coated with ceramic are usually constrained to move in a cylindrical cavity where motion is allowed in the axial direction but could be prevented in the radial direction. In order to assess the importance of this constraint on the crack during a transient thermal load, the same conditions were applied to the structure with the thin coating except that the outside boundary was constrained from deforming in the radial direction.

The comparison between the transient strain energy release rate values for the constrained and unconstrained cases is shown in Figure 9. It can be noted that the transient maximum G's for the two cases are basically the same. This is due to the fact that during the transient process, it is the temperature gradients near the crack that mostly affect its behavior and the circumferential boundary is far enough that it does not have a significant effect on the maximum G. On the other hand, when the entire structure reaches equilibrium temperature at 600°C, the unconstrained ceramic/metal has a negligible G, while in the constrained case, a finite G of 0.09835 J/m^2 is obtained. While the deformations of the crack are not presented, this residual G is caused by a steady-state opening of the crack that results at this temperature.

4. CONCLUSIONS

The transient thermoelastic response of an interfacial penny shaped crack in a ceramic-to-metal bond was studied using the principles of fracture mechanics. It was shown that the deformations of the interface crack are made of an opening and a sliding component. These two deformations result in a transient maximum strain energy release rate which is larger than the steady state value. However, during the operation of an engine, the presence of a temperature gradient in the ceramic suggests that G would be in general large. The values of G obtained in this study are of an order comparable to published critical energy release rate values of ceramic/metal bonds [11]. These vary between 2.5 and 170 J/m^2 and in this case the largest G is 32 J/m^2 for the thick ceramic and 17 J/m^2 for the thin coating. However, it should be remembered that these values depend on the crack radius. In fact, when G_c is known for the ceramic/metal bond, this analysis can be used to calculate the critical crack length a_c under which catastrophic failure would not occur.

It is clear from these results that a thinner coating is preferable under the present loading and boundary conditions. However, even for the thin coating, the transient nature of the temperature distribution causes the shearing of the crack surfaces to change sign, thus causing several peaks during the turning on and off of the engine. Clearly, this behavior is detrimental to the reliability since it is expected to reduce the life of the ceramic-to-metal interface.

Finally, it was shown that the maximum G occurring during the transient thermal load is not affected by a radial constraint imposed on the outside boundary while the steady-state value becomes different.

It is believed that this analysis provides insight into the behavior of the interface between a ceramic/metal bond subjected to transient thermal loads which occur in an engine. More research needs to be performed however, to include residual stresses due to the manufacturing of the coating and to determine the effects of temperature fluctuations during operation of the engine as well as the effect of pressure loads on the ceramic. Such work is currently under progress.

ACKNOWLEDGEMENT

The support of the National Science Foundation under Grant No. MSM-8611455 is gratefully acknowledged.

REFERENCES

1. R. Kamo, W. Bryzik and P. Glance, "Adiabatic Engine Trends - Worldwide", SAE-SP 700, Technical Paper 870018, 1987.
2. G. McDonald and R. C. Hendricks, "Effect of Thermal Cycling on $ZrO_2 - Y_2O_3$ Thermal Barrier Coatings", Thin Solid Films, 73, 491-496, 1980.
3. R. A. Miller and C. R. Lowell, "Failure Mechanisms of Thermal Barrier Coatings", Thin Solid Films, 95, 265-273, 1982.
4. C. C. Berndt and R. A. Miller, "Failure Analysis of Plasma Sprayed Thermal Barrier Coatings", Thin Solid Films, 119, 173-184, 1984.
5. C. H. Leibert ande R. A. Miller, "Ceramic Thermal Barrier Coatings", I&EC Product Research and Development, 344-349, 1984.
6. B. Boley and J. H. Weiner, *Theory of Thermal Stresses,* Kreiger Publishing, Florida, 1985.
7. K. L. Hoag, D. P. Roehling, T. M. Yonushonis and B. E. Dulin, "Modeling of Thick Thermal Barrier Coatings", ASME Paper No. 88-ICE-20, 1988.
8. K. Kokini and R. W. Perkins, "Thermal Stresses in Annular Glass-to-Metal Seals Under Thermal Shock", AIAA Journal, 22 [10], 1472-1477, 1984.
9. K. Kokini and R. W. Perkins, "Free Edge Thermal Stress Singularities in Finite Concentric Cylinders", Journal of Computers and Structures, 19 [4], 531-534, 1984.
10. N. J. Pagano, "Free-Edge Stress Fields in Composite Laminates", Int. J. Solids and Structures, 14, 401-406, 1978.
11. P. O. Charreyron, N. J. Bylina and J. G. Hannoosh, "Ceramic-to-Metal Bonding from a Fracture Mechanics Perspective", in Fracture Mechanics of Ceramics, Blacksburg, VA, 1985.
12. C. J. Martin Moran, J. R. Barber and M. Comninou, "The Penny-Shaped Interface Crack with Heat Flow, Part I: Perfect Contact", J. Applied Mechanics, 50, 19-36, 1983.
13. J. R. Barber and M. Comninou, "The Penny-Shaped Interface Crack with Heat Flow, Part 2: Imperfect Contact", J. Applied Mechanics, 50, 770-776, 1983.
14. K. Kokini, "Interfacial Cracks in Ceramic-to-Metal Bonds Under Transient Thermal Loads", J. American Ceramic Society, 70, [12], 855-859, 1987.
15. K. Kokini, "Boundary Effects on the Interfacial Transient Thermal Fracture of Ceramic-to-Metal Bonds", to appear in *Advances in Ceramics,* 1988.
16. K. Kokini, "Material Compatibility in Interfacial Transient Thermal Fracture of Ceramic-to-Metal Bonds", to appear in J. Applied Mechanics, 1988.

17. R. E. Smelser, "Evaluation of Stress Intensity Factors of Bimaterial Bodies Using Numerical Crack Flank Displacement Data", Int. J. Fracture, 15, 135-143, 1979.
18. D. R. Tree, "Development of a Heat Flux Gauge for a Partially Insulated Internal Combustion Engine", M. S. Thesis, Purdue University, 1988.

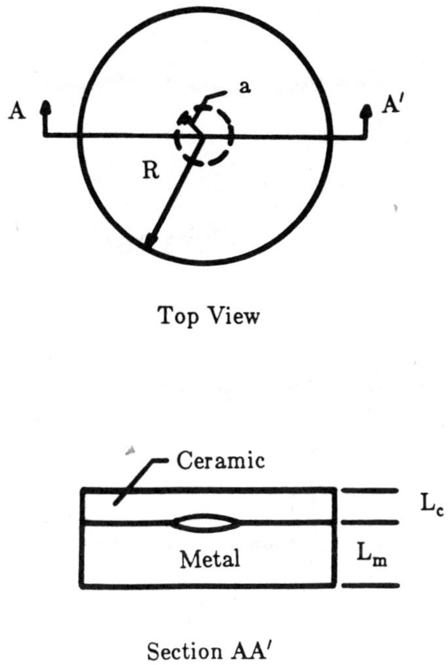

Figure 1. Schematic representation of penny-shaped crack in ceramic-to-metal bond.

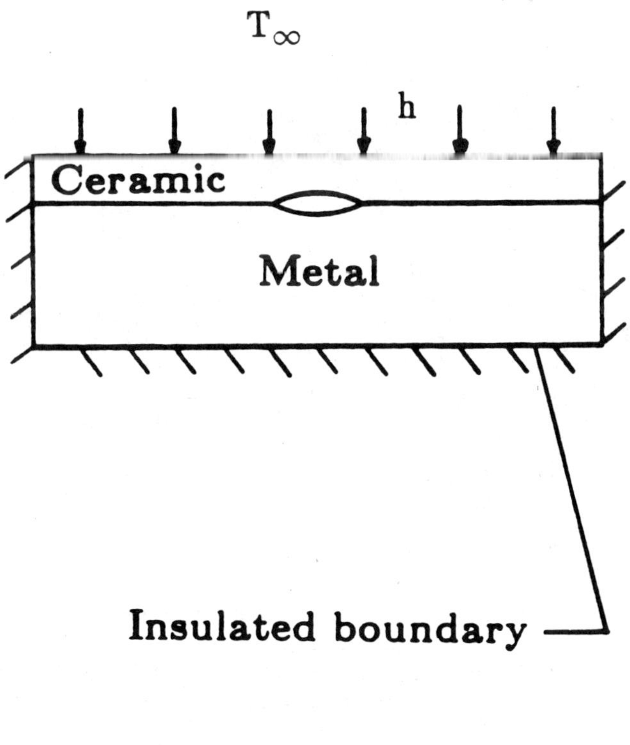

Figure 2. Schematic representation of heat transfer process.

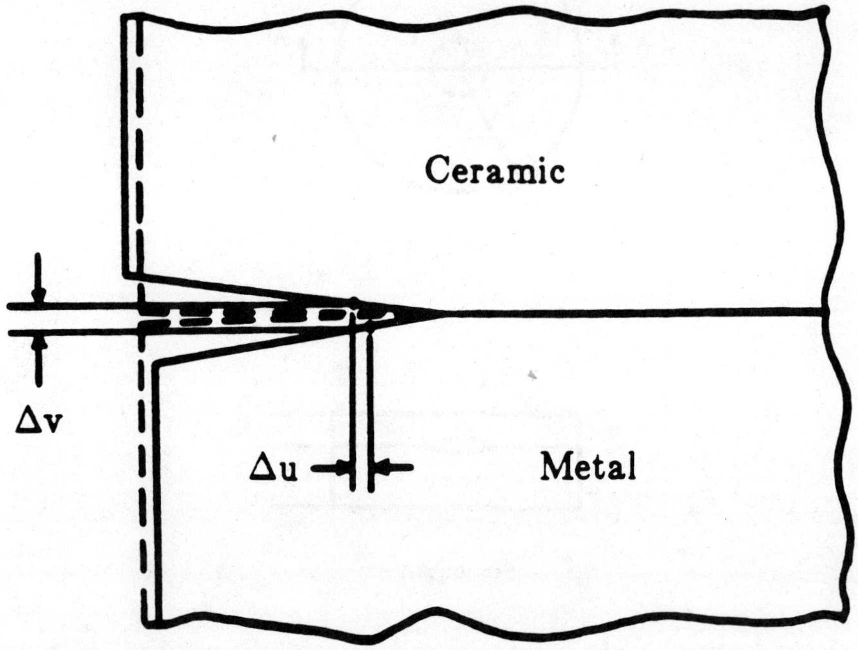

Figure 3. Deformations of the crack.

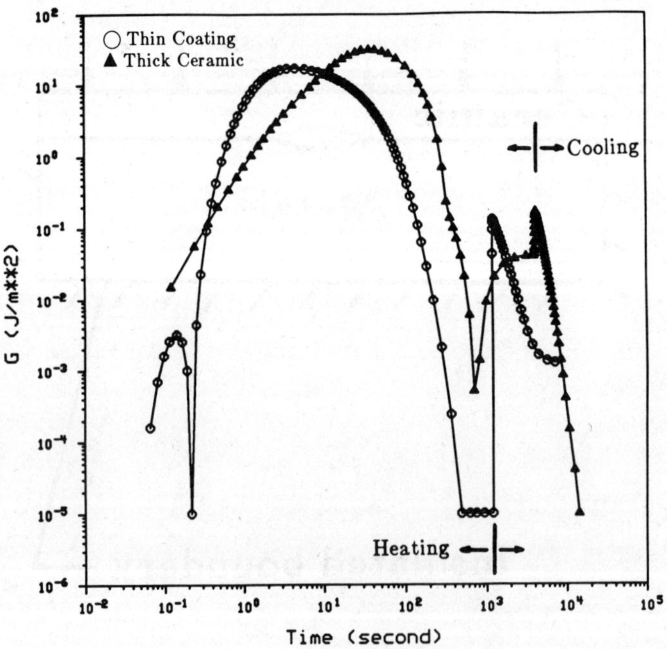

Figure 4. Strain energy release rate versus time for two ceramic thicknesses.

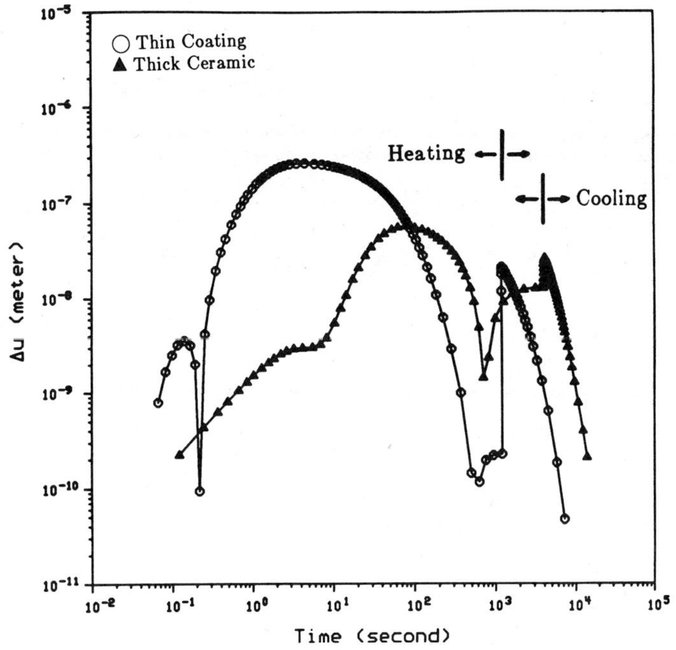

Figure 5. Sliding of the crack faces versus time at a distance from the crack tip of r/a = 0.05.

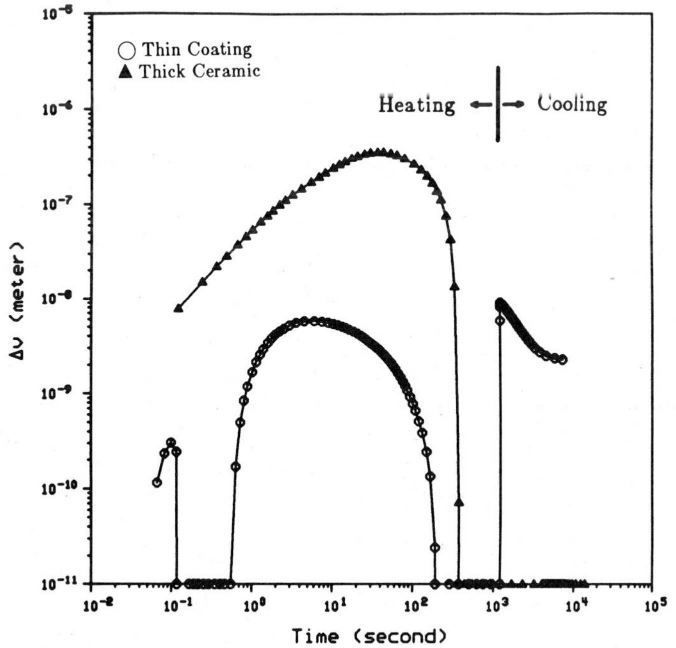

Figure 6. Opening of the crack faces versus time at a distance from the crack tip of r/a = 0.05.

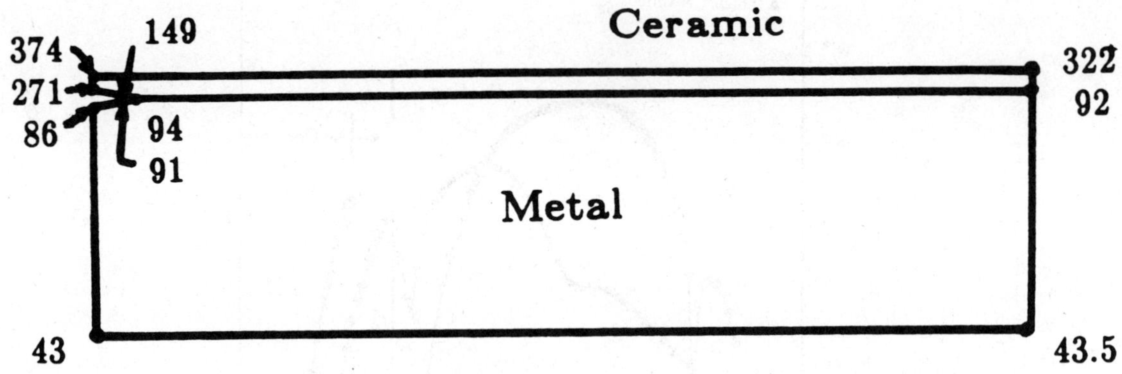

Figure 7. Selected temperatures for thin ceramic coating at t = 4.6 seconds when maximum G occurs (°C).

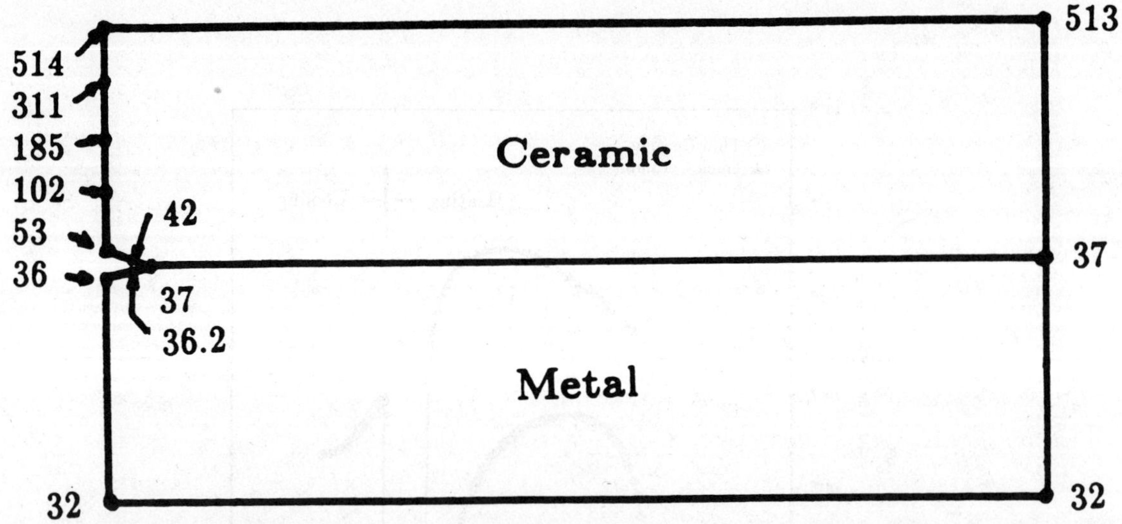

Figure 8. Selected temperatures for thick ceramic layer at t = 37.2 seconds when maximum G occurs (°C).

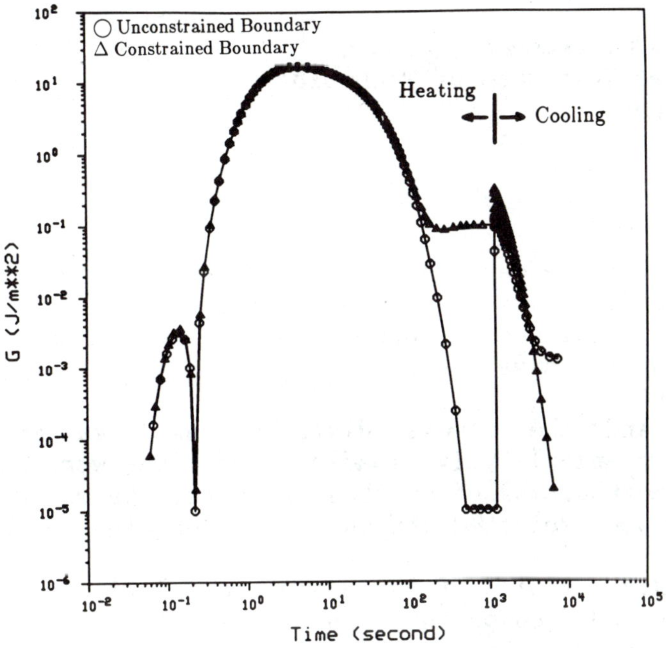

Figure 9. Comparison of transient G for the thin ceramic coating with and without radial constraint at the outside boundary.

EVALUATION OF A BRAZED JOINT FOR AN ENGINE VALVE

Y. Tzabari and D.G. Brandon
Technion - Israel Institute of Technology
Haifa 32000, Israel

ABSTRACT

In this contribution we describe the design process for a brazed joint in a pressureless-sintered silicon nitride engine valve. The valve, exposed in operation to the combustion process and the hot exhaust gases, was brazed to a stainless steel valve stem.

The temperature and the stress distributions resulting from simulated thermal and mechanical valve loading conditions were estimated using the finite element method. Based on these results, the materials and brazing parameters have been selected and successful bonding has been achieved.

The paper will give details of the design calculations, the non-destructive testing of the bonded component, the microstructural analysis of the joint and the results of proof-testing.

1. INTRODUCTION

The use of ceramics in heat diesel engines in growing [1,2]. The potential advantages of materials such as silicon nitride derive from their good strength at high temperatures and low thermal conductivity, in combination with their high resistance to corrosion, wear and thermal shock. Monolithic silicon nitride based ceramics have been used successfully for precombustion chambers, turbochargers and piston caps. However, monolithic sialon exhaust valves have been observed to fail at the lock groove [3]. The main factors responsible for this failure were the inertial, contact and bending stresses along the lock groove interface.

In order to minimise these concentrated stresses, the sialon valve, which is exposed to the hot exhaust gases, was brazed into a HNV-3 SAE valve stem. The design process included the component geometry, the selection of candidate materials and the brazing process parameters, and was related to the expected operating temperatures and stress distributions estimated from known engine performance using finite element methods.

2. DESIGN ANALYSIS

The exhaust valve is exposed to high temperatures and both thermal and mechanical loads. A heat transfer analysis based on the finite element method was applied to calculate the heat flux and the temperature field in the silicon nitride exhaust valve [4] as shown in Fig. 1. A stress analysis for the same valve was performed for the appropriate thermal and cylinder pressure loads. The inertial stresses, due to the rocker arm shift, can either be simulated [5] or estimated from the valve spring specification. The results of this stress analysis are shown in Fig. 2.

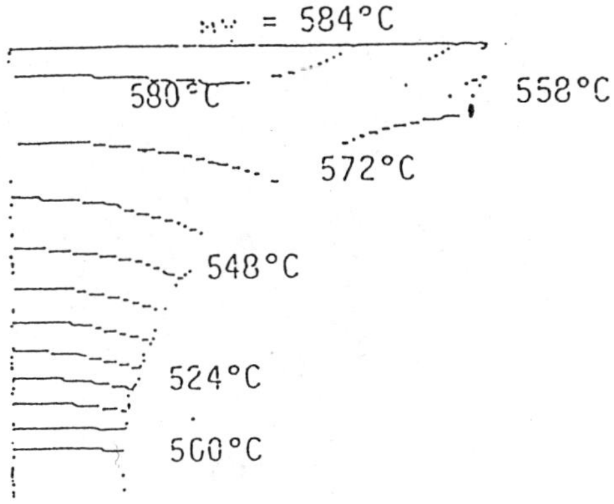

Fig. 1. Silicon nitride exhaust valve isotherms

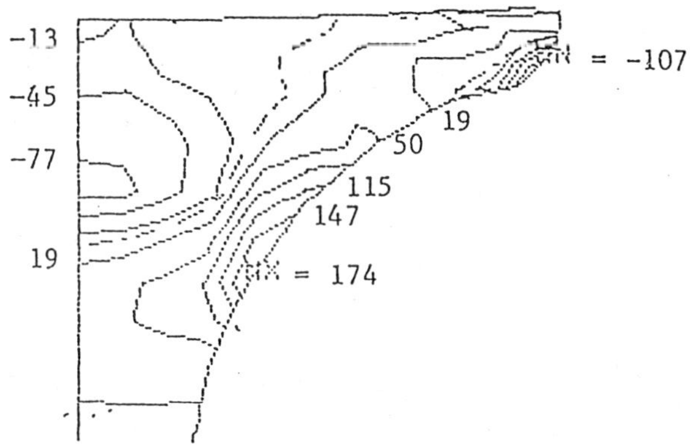

Fig. 2. Exhaust valve combined stress contours (MPa)

Based on these results, both the design concept and the geometrical dimensions of the brazed joint were evaluated (Fig. 3). A butt joint (Fig. 3a) will cause tensile stresses to be generated at the edge of the braze which will reinforce the operating stresses [6]. For the two other configurations (Fig. 3 b,c) the brazing process will itself generate shrinkage stresses, which can partially compensate the operating stresses.

Brazing the valve in the configuration of Fig. 3b provides an increased diameter in the brazing zone and reduces the stress concentration due to the geometry below that of the configuration of Fig. 3c. The selection of the braze material depends on the service temperature and the load in the brazing zone (l_2 in Fig. 3). It is recommended that l_1 be made as large as possible (high working temperature) or smaller to the end of the valve guide. This will help to minimize the bending force generated by inhomogeneity or asymmetry in the valve seat surface. The compromise solution finally selected was for the dimensions l_1 = 80mm and l_2 = 25mm. The final design for the silicon nitride exhaust valve is shown in Fig. 4. This design was for a Petter bench diesel engine, for which the other valve dimensions were derived from the original valve geometry.

In order to select the brazing material a preliminary thermal and stress analysis was performed to define the service temperatures and stresses. The results of this analysis indicated that the temperatures along the brazing zone were between 500 and 200°C, while the maximum principle stress did not exceed 400 MPa. This stress is actually decreased by the shrinkage stresses accompanying cool down from the brazing temperature without damage tolerances. A copper-silver braze [7] was selected for the preliminary study.

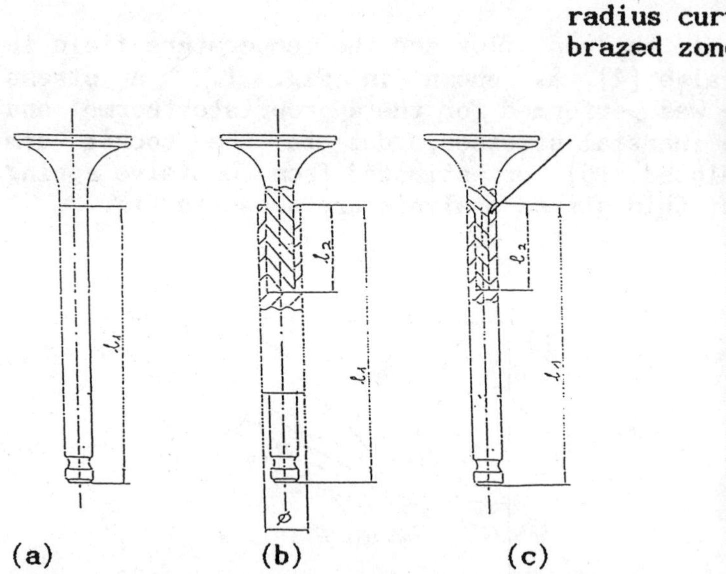

Fig. 3 Design concept and geometrical parameters

Fig. 4 Silicon nitride exhaust valve design

The thermal expansion coefficient, tensile elastic modulus and ultimate tensile strength of the braze material was measured (Fig. 5 and 6), and the final linear analysis was then made to predict the isotherms and stress contours, Fig. 7. The material properties used for this analysis are given in Table 1. The maximum principle working stress was calculated to be 470 MPa and occurred at the silicon nitride-braze interface.

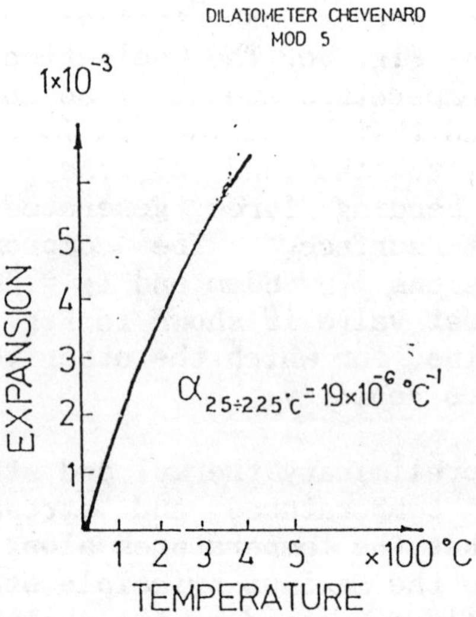

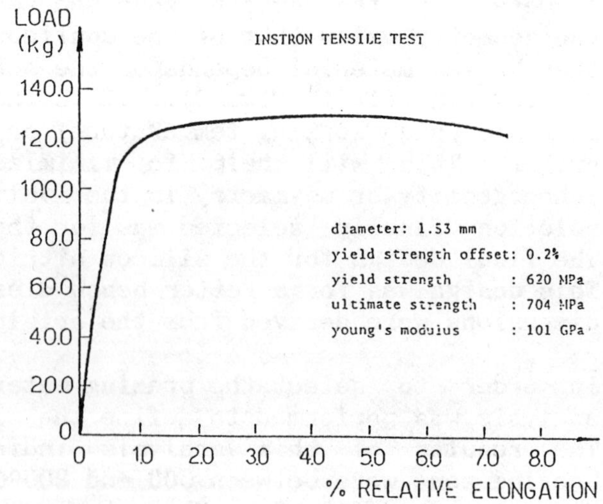

Fig. 5: The Ag-Cu braze alloy expansion as a function of the temperature °C.

Fig. 6. Load vs. relative elongation of Cu-Ag braze alloy at room temperature

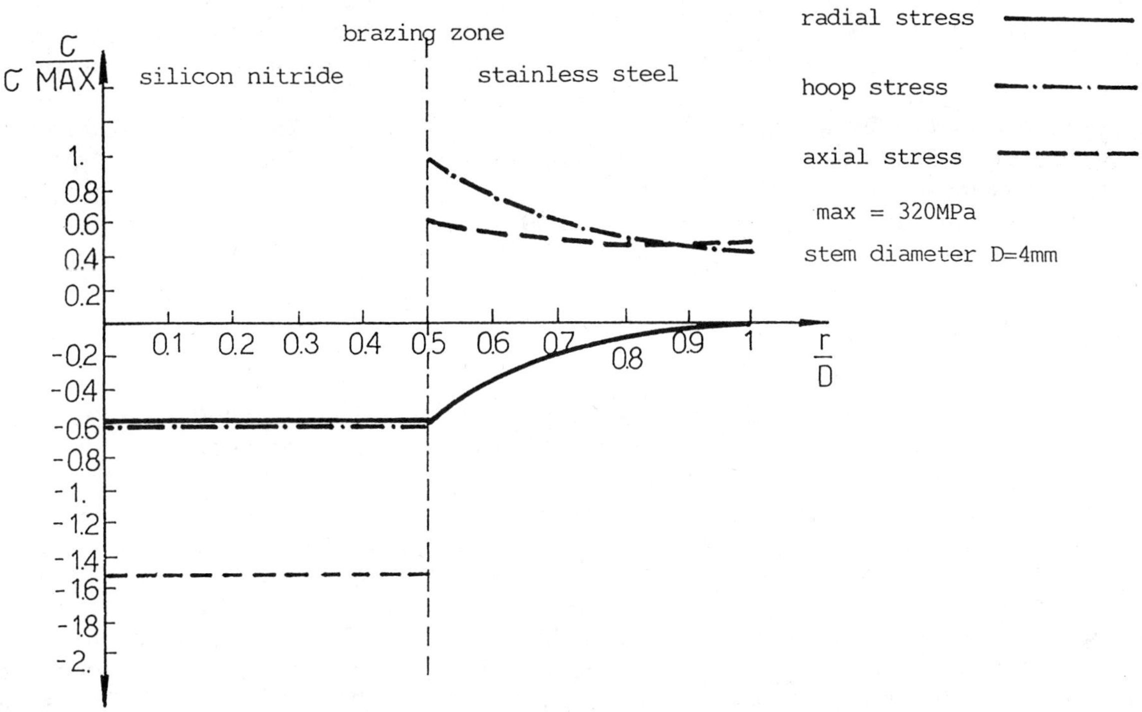

(a)

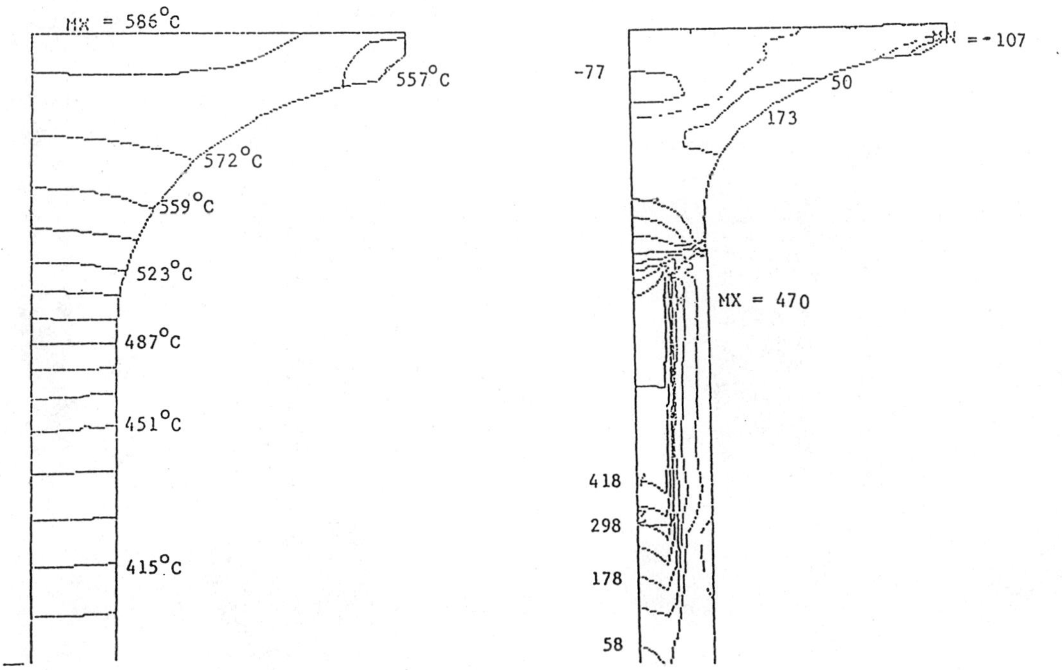

Fig. 7: Brazed exhaust valve analysis
(a) Shrinkage stresses
(b) Isotherms and working principle stress contours (MPa)

Table 1: Relevant materials properties.

	Silicon nitride	HNV-3 SAE valve stem	Cu-Ag Braze alloy
Density 10^3 kg.m^{-3}	3.2	7.9	9.4
Thermal conductivity w/m°C	21	25	390
Specific heat J/kg°C	620	500	380
Thermal expansion $°C^{-1} \times 10^{-6}$	3	13	19
Young's modulus GPa	290	210	101
Poisson's Ratio	0.23	0.3	0.35

3. BRAZING PROCESS

In earlier work [8] the wettability of silicon nitride by Ag-Cu alloys was examined. The best bond results were obtained at a braze thickness of 0.05 to 0.10mm. The valve was placed in an assembly jig to ensure that the dimensional tolerances were maintained (Fig. 8). The silicon nitride was fixed to the upper plate of the jig while the stainless steel stem, with a braze-containing top hole, was connected to the jig base. The ceramic valve was then lowered part way into the stem hole. The braze zone was induction heated to a temperature of 830°C, above the melting point of the braze, and the valve stem lowered the remaining distance into the locating hole, while the braze material flowed up to fill the space between the stem and the valve. The rate of insertion of the valve into the locating hole was controlled by preloading the top plate of the jig and no attempt was made to control either the heating rate or the temperature during brazing. The brazed valves, before and after final machining, are shown in Fig. 9.

Fig. 8: Assembly jig to ensure dimensional tolerances

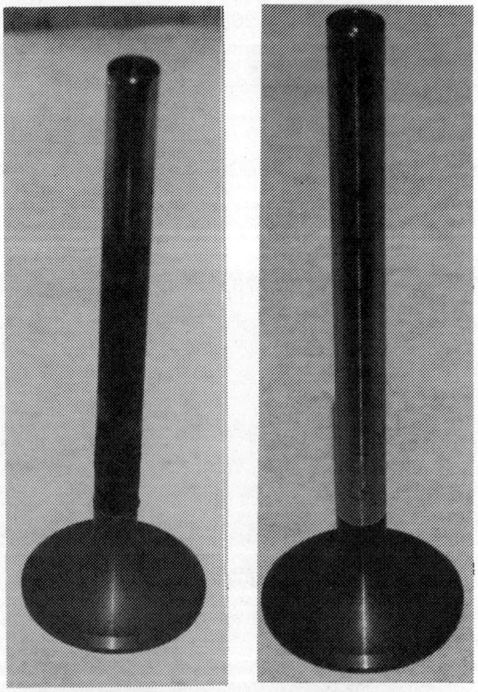

Fig. 9: The brazed valves before and after machining

4. NON-DESTRUCTIVE EVALUATION, MICROSTRUCTURAL ANALYSIS AND PROOF TESTING

X-ray radiography of the brazed valves indicated that the brazed layer was free of porosity, cracks and inclusions, but that the thickness was not uniform, indicating some eccentricity. The uniformity of the braze and freedom from defects were confirmed by a microstructural analysis of a cross section cut from one of the valves (Fig. 10). The brazed valves were also subjected to a thermal shock and thermal fatigue proof test, Fig. 11. The thermal shock rig used a propane gas burner which can function in either cyclic or continuous modes and cooling air controlled by a pneumatic system. The valves survived this test, for which the maximum temperatures were 780°C at the valve face and 400°C in the braze zone. During a mechanical proof test the silicon nitride valves failed at the radius of the brazed joint near the top of the steel stem, Fig. 12.

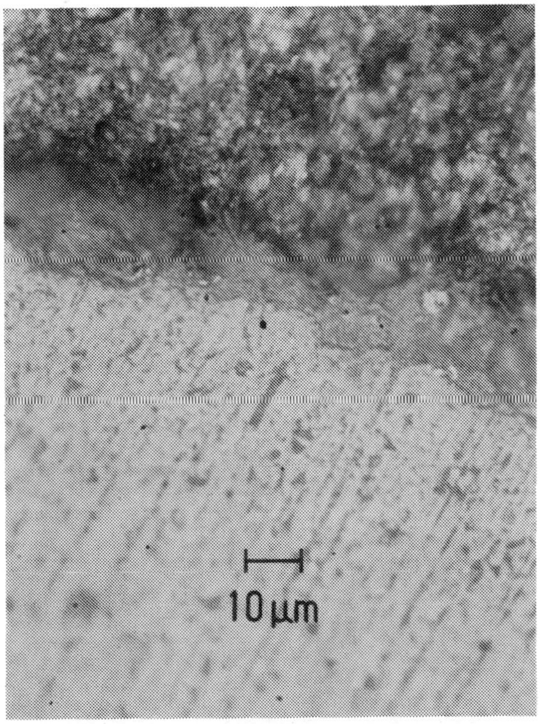

Fig. 10: The interface microstructure of a cross section cut from one of the valves.

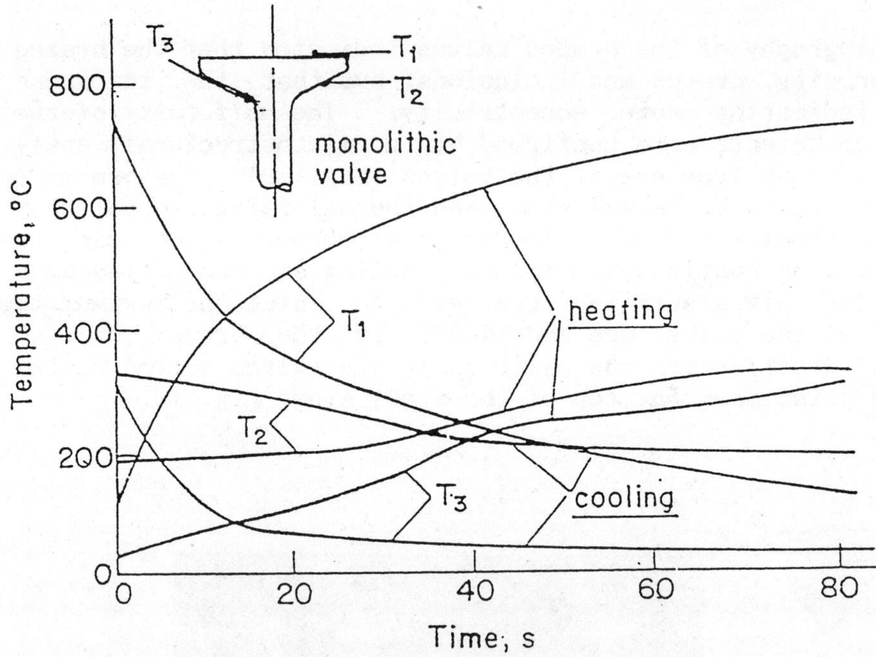

Fig. 11 Temperature variation with time during the thermal shock test for three selected points on the monolithic valve.

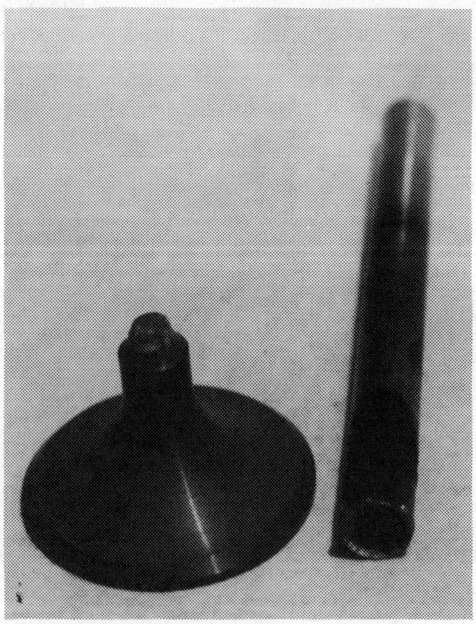

Fig. 12 Failure of the brazed valve under mechanical load.

5. SUMMARY

The operating conditions in service for a ceramic valve assembly made from sintered silicon nitride were calculated from the known engine operating conditions. High working stresses are generated at the silicon nitride stem due to the high working temperature. These stresses are reduced by the shrinkage stresses. Prototype valves manufactured according to the proposed design were found to fail at the brazed joint between the ceramic valve and the steel stem. The design of the joint could be improved by using a conical, tapered joint rather than the straight rod configuration used here.

REFERENCES

1. R. Kamo, W. Bryzik and P. Glance, "Adiabatic Engine Trends - Worldwide", SAE paper #870018 (1987).

2. O. Kamigaito, "Ceramic Components for Engines in Japan", SAE paper #870019 (1987).

3. M. Asnani and F.L. Kuonen, "Ceramic Valve and Seat Insert Performancein a Diesel Engine", SAE paper #850358 (1985).

4. Y. Tsabari, M. Gutman and A. Stotter, "Heat Transfer in a Ceramic Protected Valve", Heat and Mass Transfer in Gasoline and Diesel Engines Symposium", Dubrovnik, August 24-28, 1987.

5. T. Kosugi and T. Seino, "Valve Motion Simulation Method for High-speed Internal Combustion Engines", SAE paper #850179 (1985).

6. K. Sugahuma, T. Okamoto, M. Koizumi and M. Shimada, "Effect of Thickness on Direct Bonding of Silicon Nitride to Steel", J. Amer. Ceram. Soc., 68(123), C334-335 (1985).

7. B. Wielage, "High-Temperature Brazing of Ceramic Components", Ceramic Materials and Components for Engines, DKG 1986, p. 391.

8. A. Stotter, O.G. Brandon, Y. Tsabari, "Ceramic Technology for Adiabatic Diesel Engines", 3rd report, Technion, Haifa, march, 1987 (In Hebrew).

SOLID-STATE BONDING OF PARTIALLY STABILIZED ZIRCONIA TO METALS

C. Colin, D. Broussaud*,
Ecole Nationale Supérieure des Mines de Paris
Centre des Matériaux P.M. FOURT
B.P. 87, 91003 EVRY Cedex, FRANCE

F. Grivon, A. Wicker,
Laboratoire de Marcoussis
Centre de Recherche de la C.G.E.
Route de Nozay, 91460, MARCOUSSIS, FRANCE

ABSTRACT

Metal-to-ceramic junctions have been achieved by solid-state bonding of PSZ on cast iron and steel using an aluminium interlayer. This work aims at investigating the effects of bonding time and foil thickness on the RT tensile strength. Maximum temperature and pressure were fixed at 600°C and 25 MPa, respectively. Optimal bonding conditions are then outlined for each couple.

It was shown that short times could lead to fairly good bondings, strengths up to 60 MPa were obtained with fracture occuring in the intermetallic reaction layer. Evidence of diffusion at interfaces has been looked at using electron microscopy and microanalysis. The critical problem of residual stresses resulting from the thermal expansion mismatch has also been examined with regard to the extent of the bonding area. Finite element analysis revealed the prime role played by aluminium in the relaxation of stresses at the interface, provided elasto-plastic approach is considered instead of pure elastic.

INTRODUCTION

Whatever their strength, ceramics remain brittle and exhibit a low ability to deform, preventing their introduction as monolithic parts in metallic engineering structures as automotive engines. In order to improve their reliability, and to make up for their low workability, ceramics must be tightly combined with metals. Thus, ceramics will only be used for their specific characteristics and metals for their ductility.

Various techniques for joining ceramics to metals are available [1], namely, mechanical joining, brazing, fusion welding and solid state diffusion bonding. However, several important problems still remain unsolved affecting

* Now with Rhône Poulenc Research Center, Aubervilliers.

the strength of joints. The poor reactivity of ceramics towards metals limits the number of atomic bonds and the large thermal expansion mismatch responsible for stress concentration after cooling from the bonding temperature are two of the most critical aspects of joining ceramics to metals. Solid state diffusion bonding seems to be most promising because it is a simple process, and offers the possibility of relieving the stresses via the concept of buffer interlayers. Since these residual stresses are likely to cause fatal damage in the ceramic, a number of alternative solutions have been developed in this respect : from the use of soft metals as a single interlayer, primarily Aluminium [2-7] to the concept of laminated interlayers such as Aluminium/Invar [8], Titanium/Molybdenum [9], Zirconium/Niobium/ Zirconium [10], Aluminium/Titanium/Aluminium [11] and Nickel/Tungsten/Copper [12]. Hot-isostatic pressing (HIP) is also used and found to improve the bond strength with [13-16] or without [17] interlayers. Considering oxide ceramics only, these alternatives mainly concern alumina-to-steel bonding, so far very few studies have been carried out for the joining of zirconia to steel.

The major role of the interlayer on the residual stresses as calculated in the ceramic has been assessed for silicon nitride - and alumina - to steel junctions [12-13].Both the thickness or a combination of interlayers looks effective on relieving the residual stresses even when allowing for their elastic behaviour only.

In this paper, the diffusion bonding between a partially stabilized zirconia (PSZ) ceramic and two metallic alloys - steel and cast iron - using aluminium as single interlayer metal is presented. Aluminium displays the advantage of relieving the residual stresses resulting from the thermal expansion mismatch which is accommodated to some extent. Furthermore, aluminium has a high oxygen affinity and provides good bonds towards oxides. The purpose of the present study was to investigate the effects of bonding time and foil thickness on the tensile strength of the joints and to optimize the bonding conditions for each combination. The bond strength data are supplemented by information from microanalysis of polished cross-sections. Finite element analysis was also undertaken in an attempt to assess the effect of the extent of the bonding area on the residual stresses, allowing for both elastic and plastic behaviours of the materials to be joined.

EXPERIMENTAL PROCEDURE

Materials

The PSZ used in this work is a pressureless sintered material, ZN40 manufactured by Fedmühle, FR-Germany. It contains 3 Wt% MgO as a stabilizer and traces of HfO_2. The ceramic samples were supplied as-ground either in the form of pieces of size 12 x 12 x 6 mm^3 for tensile tests or in the form of discs of various thickness, 2, 4 and 8 mm and 30 mm diameter for deformation tests.

A standard XC 38 steel and a standard gray cast iron in use in automotive diesel engines were selected for the base-metals. These alloys were supplied as-machined in the form of rectangular pieces of size 12 x 35 x 6 mm^3 or in the form of discs 10 mm thick and 30 mm diameter.

Typical properties of the ceramic and the ferrous alloys are shown in Table I

TABLE I : Properties of materials at room temperature

properties / Materials	Density	Melting point	Coeff. of Thermal Expansion	Elastic modulus	Poisson ratio	Thermal conductivity	Yield Strength	Strength (*)
	$g\ cm^{-3}$	°C	$10^{-6}\ C^{-1}$	GPa		$W.m^{-1}\ K^{-1}$	MPa	MPa
ZrO_2 (PSZ)	5.74	2760	9.8	210	0.3	2.1	-	520
Al (1145 annealed)	2.70	660	25.5	69	0.34	230	34	90
Steel	7.82	-	14.5	205	0.33	54	965	1180
Cast iron	7.2	950-1230	13.0	120	0.3	50	300	400

* four-point bending strength for ceramic and tensile strength for metals.

From this table, it can be seen that ceramic and metal exhibit differing properties, especially elastic modulii and coefficients of thermal expansion. Therefore, the direct bonding of PSZ to metals would most likely induce fatal stresses.

Aluminium used as ductile metal interlayer is 99.5 Wt% pure and was rolled into 0.5, 0.2 and 0.1 mm thick sheets from which 12 x 6 mm^2 rectangular sections were punched for tensile assemblies. Deformation experiments were carried out using washers 30 mm outside diameter and varying inside diameters stamped from previously optimized sheet thickness. All materials were used as-received and were cleaned by ultrasonic in trichloroethane before being assembled.

Bonding Procedure

The parts to be bonded are stacked into a graphite die which was placed into a laboratory apparatus as indicated in Figure 1. The hot-pressing chamber consisted of a fused silica tube evacuated to about 10^{-2} Torr and the samples are heated up to around 600°C. Then the assemblies are subjected to a pressure of 25 MPa and held at bonding temperature from 15 to 60 min, the pressure is then released and the samples are allowed to cool under argon - 10 % H_2 atmosphere. Bonding temperature, pressure, atmosphere and cooling rate were kept constant in all experiments, these have been adjusted after studies carried out in-house and elsewhere [2, 6, 7]. In this work, only loading time, the thickness of the aluminium interlayer and bonding area were varied. The two types of joints achieved are shown in Figure 2.

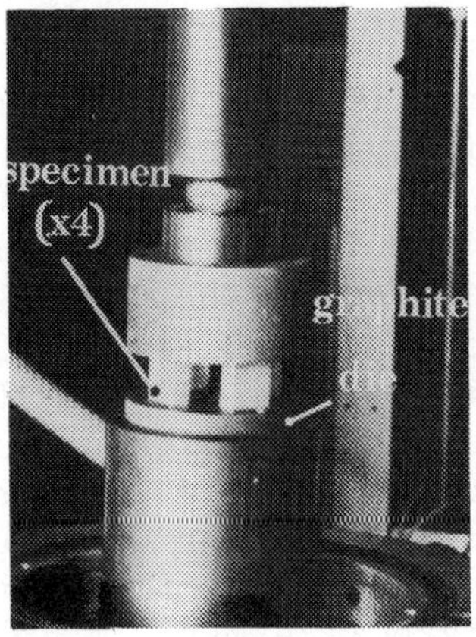

FIGURE 1 : Experimental bonding set-up.

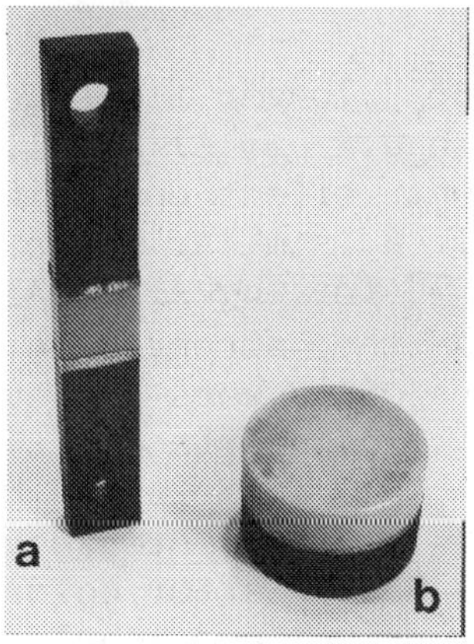

FIGURE 2 : ZrO_2/metal specimens after joining for :
(a) tensile tests
(b) deformation measurements.

Evaluation of Strength

The strength of the bonded specimens was assessed by tensile tests conducted at room temperature at a cross-head speed of 0.5 mm/min using a ZWICK machine . Care was taken for misalignment (cardan fixtures) to minimize the risk of sudden sample fracture. The ratio of the load of failure to the actual bonded area as measured after the test was used to characterize strengths. For optimal bonding conditions, the strength values are derived from triplicate tests.

RESULTS AND DISCUSSION

Strength

The effect of bonding time on the room temperature tensile strengths of the joints is illustrated in Table II. It was observed that all samples failed at the metal-aluminium interface rather than the ceramic-aluminium interface.

TABLE II : Experimental results of strength of PSZ/metal joints

initial interlayer thickness (mm)	Time (min)	Bonded area/ Available area (%)		Failure load/ Bonded area (MPa)	
		cast iron	Steel	Cast iron	Steel
0.1	15	75	94	34.4	23.6
	30	100	100	29.4	12.5
	60	100	-	26.1	2.4
0.2	15	94	100	26.0	34.6
	30	92	100	35.3	23.0
	60	100	100	39.2	13.9
0.5	15	100	100	63.9	46.9
	30	100	100	59.4	45.1
	60	100	100	62.2	44.4

The use of longer times generally leads to lower strengths except for cast iron with 0.2 mm Al foil. For 0.5 mm thick foils, increasing time has no significant effect on joint strengths, 15 min is already enough for maximum strength. A similar behaviour has already been outlined by Nicholas and Crispin [7] for the Al_2O_3/Al/Steel system. The other factor that has a marked effect on the strength is the initial thickness of the aluminium foils. The stronger joints were achieved using the thicker foils, they all fractured at the metallic interface. It was noticed that the joints made with steel using a 0.1 mm Al interlayer bonded for 60 min were not stable. Most of them fractured at the interface between PSZ and aluminium either when removing from the die or after holding in dried atmosphere prior to testing. Only one sample survived and could be evaluated for strength (2.4 MPa). This result indicates that a large residual stress has settled in the joint. So, these preliminary experiments have led to define optimal conditions i.e. holding time of 15 min for both base-metals assembled to zirconia using 0.5 mm thick aluminium foils. Such bonded samples could withstand tensile stresses up to 48.7 ± 4.2 for steel and 61.0 ± 1.9 for cast iron which is around 54 to 68% that of the ultimate strength of aluminium. Strength data are rather reproducible as pointed out with the runs using optimum bonding conditions which were triplicated. However, one must be cautious about other strength values which result from one single measurement. It can be speculated that bond strengths measured from thinner interlayers are less reliable due to the actual bonded area which is overestimated by the total surface. Finally, the precise evaluation of the strengths of the joints requires the knowledge of the stress field within a specimen when materials with different elastic modulii and Poisson's ratio are bonded. It has been reported that within a steel/ Al_2O_3/steel tensile specimen [15] similar to ours, the greater stress concentration calculated with single elastic model is within the ceramic close to the interface and was shown to be 16% higher than the applied stress for the joint under tension.

Micrographic Examination

Micrographs in Figure 3 show polished cross-section for the ZrO_2/Al/Steel (a,b) and ZrO_2/Al/cast iron (c,d) systems. No reaction layer, nor diffusion were observed between zirconia and aluminium by means of SEM and EPMA. Therefore, the interaction if any is on a submicron scale. The oxygen affinity of aluminium is sufficient for the bonding to be ascribed to a simple redox reaction.

Intermetallic compound layers were formed between aluminium and the ferrous alloys. There are two distinct layers in the reaction zone, one narrow (R1) on the aluminium side and the other thicker (R2) on the ferrous alloys side as evidenced micrographs 4b and d for steel and cast iron respectively. The average thickness of the overall reaction zone is around 100 μm at the aluminium/cast iron interface and around 200 μm at the aluminium/steel interface. Electron probe microanalysis has revealed that the composition of the main reaction zone (R2)is close to that of $FeAl_3$. It was also shown that Si and Mn to some extent have also diffused, being more pronounced with cast iron than with steel. Cracks have developed parallel and close to the reaction layer/steel interface. In the case of cast iron, graphite lamellas seem to prevent their progress in the reaction layer.

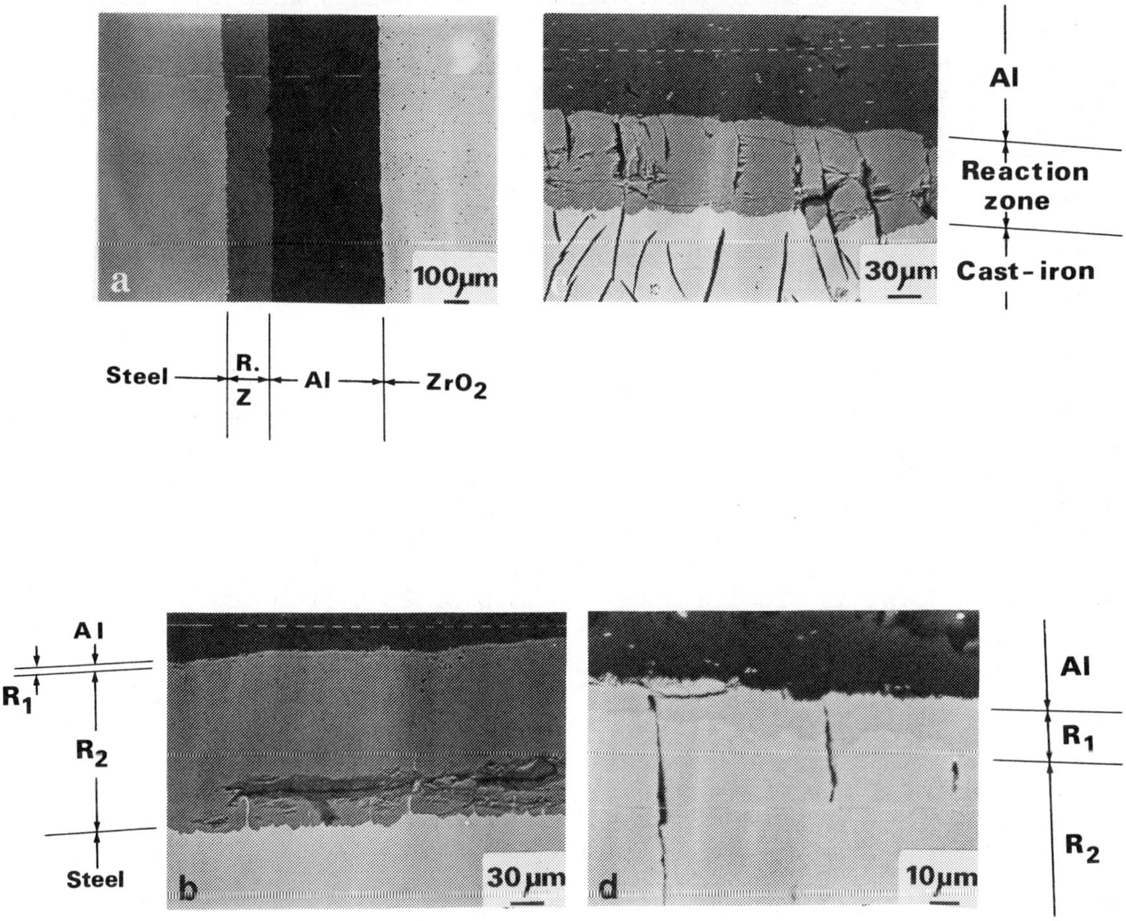

FIGURE 3 : Scanning electron micrographs of interfacial structures of ZrO_2/Al/base-metal joints (intact from tensile test)
(a, b) steel base (c, d) cast iron base

THERMAL STRESS ANALYSIS IN THE STEEL/ALUMINIUM/PSZ JOINT

As anticipated, a large stress concentration may be introduced in the joint during cooling down to room temperature from the bonding temperature as a result of thermal expansion mismatch between the dissimilar materials. In the present study it was aimed at assessing the distribution, nature and level of the stresses in the ceramic as effective bonding area varies, in order to minimize these stresses for practical applications. So as to validate this stress modelling, experiments have been carried out with ceramic discs bonded to a steel base as already shown in Figure 2. with the same geometry as in the model. The effective bonding area was controlled using a central bore 1 mm depth in the steel base. After bonding, the maximum deformation of the disc resulting from the residual stresses was measured via profilometry and averaged from four diameters. Experimental measurements were then compared to those obtained from the previously presented stress modelling.

The residual stresses generated in the junction were estimated by finite element method assuming that : (1) All materials behave elastically

(2) The base metallic and ductile metal interlayer elasto-plastically. The stress-strain behaviour was modelised through an isotropic temperature dependent hardening. Details of the constitutive equations and their implementing in a finite element code are given elsewhere [18].

In the former case, it is expected that the residual stresses will be overestimated, while the latter case should be more realistic. The model type was a cylindrical butt joint as illustrated by the finite element grid shown with boundary conditions in Figure 4. The physical and mechanical properties of each material versus temperature necessary for the calculation have been determined or compiled from available literature. The temperature difference is 580°C assuming a cooling-down from 600°C to 20°C. The calculations are carried out under the following assumptions :

1 - The joint is cooled down by step of 25°C from 600°C down to 50°C and by step of 15°C down to room temperature
2 - The initial stress of the joint is 0 MPa at 600°C
3 - The temperature of the joint is uniform at each calculation step.

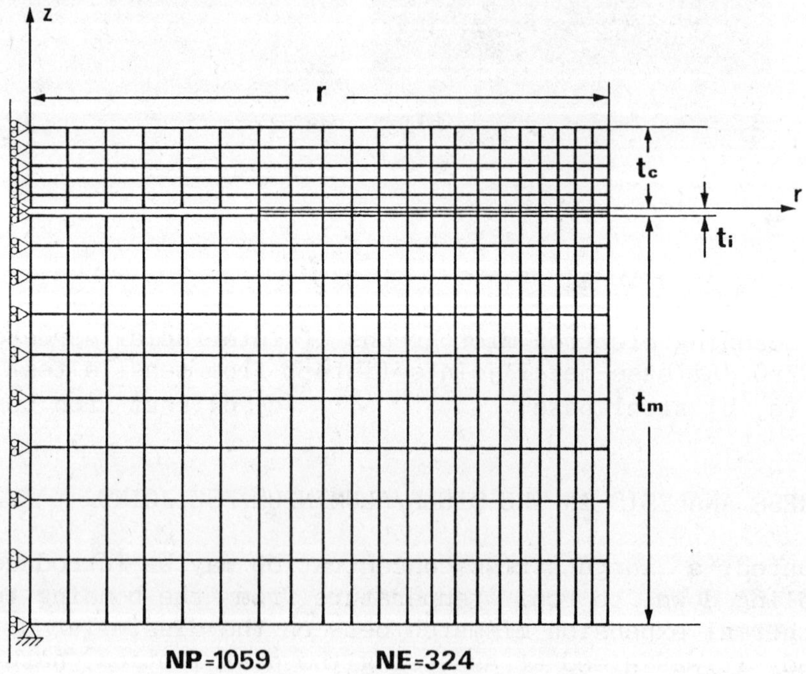

FIGURE 4 : Finite element grid with boundary conditions
r = 15 mm ; tc = 2 mm ; ti = 0.18 mm ; tm = 10 mm

Figure 5 shows a typical greater principal isostress map (hereafter called σ_p for a bonding area equal to 84%. Tensile stresses in zirconia have been mainly considered here, because it is the most important factor influencing the reliability of the joint. The maximum principal stress (σ_p max) appears on the external surface near the interface and this so remains for a bonding area ranging from 100% to about 55% as shown in Table III.

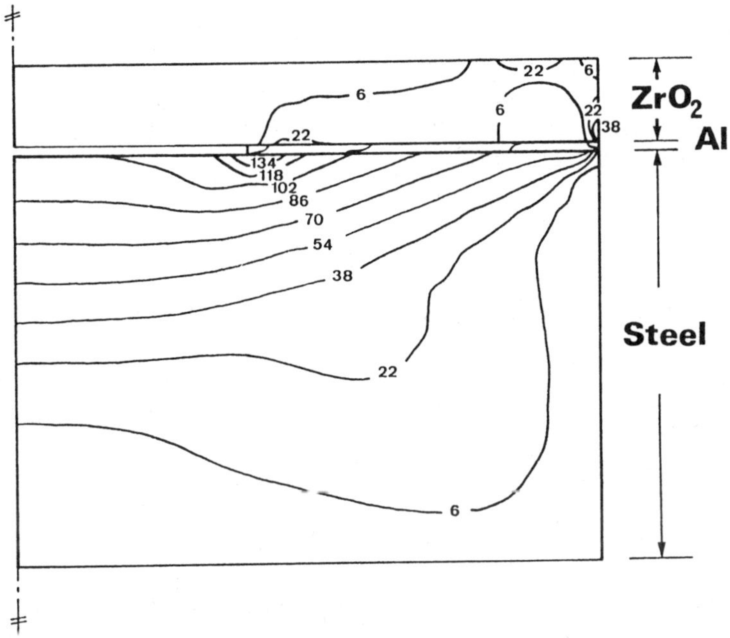

FIGURE 5 : Contour map of the greater principal stress $\sigma_{p\ max}$ = 40.2 MPa in ceramic

TABLE III : Maximum principal stress and location in ceramic

Bonded surface / Available surface (%)	100	84	64	51	36
Elastic $\sigma_{p\ max}$ (MPa)	151.4	171.6	252.2	317.2	509.0
location (r; z) mm	14.9 ; 0.6	14.9 ; 0.6	14.9 ; 0.3	10.8 ; 0.1	12.3 ; 0.1
Elasto-plastic $\sigma_{p\ max}$ (MPa)	32.7	40.2	50.2	-	45.8
location (r; z) mm	14.9 ; 0.3	14.9 ; 0.3	14.9 ; 0.3	-	12.8 ; 0.1

Below 55%, both elastic and plastic calculations indicate a change of the location of the maximum which still appears close to the interface but at the inside diameter of the interlayer. It is clearly shown that the plastic deformation of aluminium allows an important relieving of the stress field within the ceramic, it was calculated up to 16 to 30% according to cases. Furthermore, in the case of elastic calculation model, $\sigma_{p\ max}$ significantly decreases when increasing the bonding area, whereas using a plastic model it seems there exists an optimum ratio of the actual to available bonding area around 55% corresponding to the change of location of the maximum principal stress. In this latter case, the maximum stress decreases below 55 MPa i.e. about 10 times less than the fracture stress of zirconia (520 MPa) which is almost reached via elastic calculation.

As far as the deformation of the disc is concerned the diagrams of the Figure 6 illustrate both elastic and elasto-plastic calculations. The relieving of stresses by plastic deformation is translated by much lower levels of deformation. This correlated well with experimental measurements, the presented model can be considered as satisfying and probably improved allowing for the visco-plastic behaviour of the aluminium.

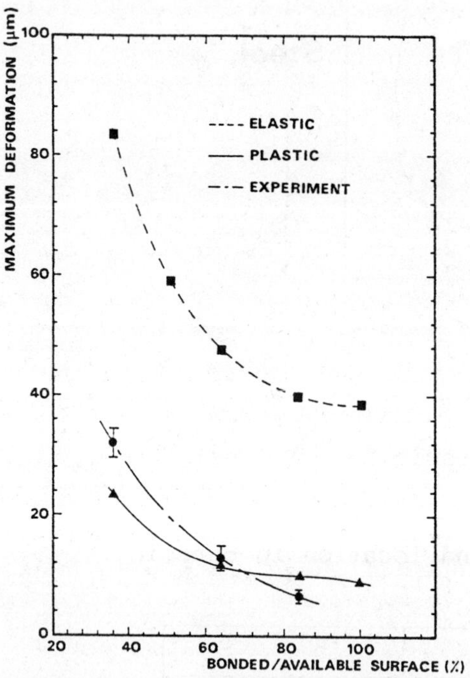

FIGURE 6 : Maximum deformation of the free surface of the disc ceramic as a function of the bonded area.

CONCLUSION

Solid state diffusion bonding was found to be a promising route for the achievement of good ceramic-to-metal junctions. The use of aluminium as interlayer metal between zirconia (PSZ) and either cast iron or steel resulted in maximum tensile strength of 60 MPa. Optimal conditions for bonding have been worked out, thick foils (0.5 mm) are more effective than thin foils due to a higher deformation ratio and the maximum strength is reached after relativity short times (15 min) at 600°C under 25 MPa.

Also in this work a thermo-plastic model has been applied to evaluate the level and the distribution of the residual stresses resulting from the expansion mismatch, allowing for the relaxation of stresses in the Al interlayer. This model has been validated via deformation measurements from butt joints. The plastic deformation of Al during cooling down lowers the level of stresses in the ceramic to acceptable levels for reliability. Thus this technique appears well adapted to components with simple shapes submitted to low tensile loads and operating at moderate temperatures. Works previously carried out from solid state bonded ceramics using Al interlayers have proved that such joints could be used up to 300°C and thermally cycled without damage [19]. Wear parts linked with the valve train in automotive engines are possible for application.

ACKNOWLEDGMENTS

This work has been supported by RENAULT-REGIENOV, thanks are due to V. Barrailler and G. Desplanches for their marked interest in our results. The contributions of G. Cailletaud for the numerical simulation and J.D. Bartout from Ecole des Mines for the microanalysis are also aknowledged.

REFERENCES

1 - M.G. NICHOLAS and D.A. MORTIMER, "Ceramic/metal joining for structural applications", Materials Science and Technology, Vol.1, N°9, 657-665,1985.
2 - K. SUGANUMA, T. OKAMOTO, M. KOIZUMI and M. SHIMADA, "Joining of silicon nitride to silicon nitride and to Invar alloy using an aluminium interlayer", Journal of Materials Science, Vol.22, N° 4, 1359-1364, 1987.
3 - K. SUGANUMA, T. OKAMOTO, M. KOIZUMI and M. SHIMADA, "Acoustic emission from ceramic/ metal joints on cooling", The American Ceramic Society Bulletin, Vol.65, N°7, 1060-1064, 1986.
4 - T. ISEKI and M.G. NICHOLAS, "The elevated temperature strengths of alumina-aluminium and magnesium-aluminium samples", Journal of Materials Science, Vol.14, N° 3, 687-692, 1979.
5 - J.T. KLOMP, "Bonding of metals to ceramics and glasses", The American Ceramic Society Bulletin, Vol.51, N°9, 683-688, 1972.
6 - M.G. NICHOLAS and R.M. CRISPIN, "The fabrication of steel-alumina joints by diffusion bonding", Proceedings of the British Ceramic Society, Engineering with ceramics, Edited by R.W. DAVIDGE, N°32, 33-40, March 1982.
7 - M.G. NICHOLA and R.M. CRISPIN, "Diffusion bonding stainless steel to alumina using aluminium interlayers", Journal of Materials Science, Vol.17, 3347-3360, 1982.
8 - K. SUGANUMA, T. OKAMOTO, M. KOIZUMI and M. SHIMADA, "Method for preventing thermal expansion mismatch effect in ceramic-metal joining", Journal of Materials Science Letters, Vol.4, N° 5, 648-650, 1985.
9 - F. HATAKEYAMA, K. SUGANUMA and T. OKAMOTO, "Solid-state bonding of alumina to austenitic stainless steel", Journal of Materials Science, Vol.21, N° 7, 2455-2461, 1986.

10 - G. ELSSNER, W. DIEM and J.S. WALLACE, "Microstructure and mechanical properties of metal-to-ceramic and ceramic-to-ceramic joints", Surfaces and interfaces in ceramic and ceramic-metal systems, ed. by Joseph Pask and Anthony Evans, Materials Science Research, Vol.14, Plenum press, New-York, 629-639, 1981.

11 - S. MOROZUMI, M. ENDO, M. KIKUCHI and K. HAMAJIMA, "Bonding mechanism between silicon carbide and thin foils of reactive metals", Journal of Materials Science, Vol.20, N° 11, 3976-3982, 1985.

12 - T. YAMADA, H. SEKIGUCHI, H. OKAMOTO, S. AZUMA, A. KITAMURA and K. FUKAYA, "Diffusion bonding SiC or Si_3N_4 to Nimonic 80A", High Temperature Technology, Vol.5, N°4, 193-200, November 1987.

13 - K. SUGANUMA, T. OKAMOTO, M. KOIZUMI and M. SHIMADA", Effect of interlayers in ceramic-metal joints with thermal expansion mismatches", Communications of the American Ceramic Society, Vol. 67, N° 12, C-256-C-257, 1984.

14 - K. SUGANUMA, T. OKAMOTO, Y. MIYAMOTO, M. SHIMADA and M. KOIZUMI, "Joining Si_3N_4 to type 405 steel with soft metal interlayers", Materials Science and Technology, Vol.2, N° 11, 1156-1161, 1986.

15 - K. SUGANUMA, T. OKAMOTO, M. KOIZUMI and M. SHIMADA, "Solid-state bonding of oxide ceramic to steel", Journal of Nuclear Materials 133/134, 773-777, 1985.

16 - K. SUGANUMA, T. OKAMOTO, M. KOIZUMI and M. SHIMADA, "Solid-state bonding of partially stabilized zirconia to steel with titanium interlayer", Journal of Materials Science Letters, Vol.5, N° 11,1099-1100, 1986.

17 - K. SUGANUMA, T. OKAMOTO, M. KOIZUMI and M. SHIMADA, "Effect of thickness on direct bonding of silicon nitride to steel", Communications of the American Ceramic Society, Vol.68, N°12, C-334-C-335, 1985.

18- H. BURLET and G. CAILLETAUD, "Numerical techniques for cyclic plasticity at variable temperature", Engineering Computations, Vol. 3, N° 2, 143-153, 1986.

19 - D. BROUSSAUD, A. WICKER and J.L. MAGNET, "Introduction of sintered α-SiC in diesel engines : example of a french experiment", Proceedings of the British Ceramic Society, Special Ceramics 7, Edited by D. TAYLOR and P.POPPER, N° 31, 19-25, June 1981.

DIFFUSION BONDING REACTIONS BETWEEN SILICON NITRIDE, SILICON OXYNITRIDE AND INCOLOY 909 BY HOT ISOSTATIC PRESSING.

Richard Larker, Bengt Loberg and Thomas Johansson
Engineering Ceramics Group
Department of Engineering Materials
Luleå University of Technology, Sweden

ABSTRACT

Solid state bonding by diffusion reactions is an attractive route to achieve joints that retain strength at high temperature. Intimate physical contact across the joint during bonding is important for the diffusion process.
Hot Isostatic Pressing (HIP) can enhance the interfacial area and thereby produce a stronger joint at a lower bonding temperature than other methods.
Silicon oxynitride, Si_2N_2O, is often formed on the surface of silicon nitride components during sintering and has several properties that might be beneficial as an interfacial layer against metals. Furthermore, if machining of the joint area of the Si_3N_4 component is to be avoided, the bonding reactions of Si_2N_2O to,e.g.,a superalloy is as important as for Si_3N_4 to the same alloy.
This work concerns the development of solid state bonding methods for the joining of silicon nitride to superalloys by HIP. It presents the first two steps in the development: The synthesis of pure dense Si_2N_2O for joining purposes and the first comparative study of the joining reactions Si_3N_4/superalloy vs. Si_2N_2O/superalloy.

1. INTRODUCTION

The introduction of engineering materials such as silicon nitride in demanding applications is often limited by the lack of adequate joining methods to other parts in the structure. This is especially true when advantage is taken of the superior high temperature properties of the ceramic. In those applications the large difference in thermal expansion characteristics makes it difficult to avoid thermal stresses that decrease the strength of the joint and might lead to complete failure of the component.
Silicon oxynitride, Si_2N_2O, has a tendency to form on the surface of a silicon nitride component during sintering. Several properties such as the coefficient of thermal expansion (CTE) and Young´s modulus make it an interesting material as an interfacial layer against metals. If machining of the joint area on the sintered component is to be avoided, the joining reactions for silicon oxynitride to the superalloy are as important as for the silicon nitride itself. This may prove to be a

cost-effective joining material for high-temperature applications such as engine components.
Earlier work on silicon oxynitride compounds by Washburn [1] predicts that some important properties such as the CTE and Young´s modulus differ less from those of a superalloy than the nitride and might therefore reduce thermal stresses.
The earlier work was, however, performed on materials containing large amounts of sintering additives and residual silicon nitride. More recent work by Huang et al [2] and Lewis et al [3] show a higher degree of conversion to silicon oxynitride, but this is still only possible with rather large amounts of additives. Sintering at a relatively low temperature to avoid excessive dissociation during pressureless sintering requires larger addition amounts.
Hot Isostatic Pressing of glass encapsulated powder bodies has the ability to prevent dissociation, thereby making it possible to sinter at a higher temperature using low amounts of additives [4]. Furthermore, the high driving force for densification is beneficial to increase the speed of the synthesis reaction:

$$Si_3N_4 + SiO_2 \Rightarrow 2\ Si_2N_2O$$

It was therefore considered to be both necessary for joining purposes and possible by HIP to develope a method to synthesize a dense material of almost pure silicon oxynitride.

2. SYNTHESIS OF SILICON OXYNITRIDE

Silicon oxynitride Si_2N_2O can be synthesized from a stoichiometric powder mixture of Si_3N_4 and SiO_2 during densification with Hot Isostatic Pressing using the following route:

- Mixing and ball milling of the powder mixture in water using milling media and jars of silicon nitride and polypropylene for 100 h.
- Drying of the powder using a thin-film evaporator and a heated vacuum chamber up to 250 °C.
- Forming of green bodies by Cold Isostatic Pressing (CIP) at 300 MPa of powder contained in PVC bags.
- Heat treatment of green bodies in a vacuum furnace up to 700 °C and 10^{-2} Pa to remove water and other adsorbed species from the powder surfaces.
- Densification using the proprietary ABB Cerama glass encapsulation and HIP technology. The process parameters were varied to improve the conversion into silicon oxynitride.

All samples were fully densified to the calculated theoretical density of each final composition.

2.1 Analysis

Studies of the synthesized compositions after HIP have been made by semi-quantitative X-ray diffraction analysis (XRD) using a Philips PW 1700 system equipped with PDF data base and pattern processing capability. An active dilution method (ADM) was used to improve the accuracy in the determination of the low residual amounts of hexagonal α - and β -phase compared with the major orthorombic Si_2N_2O, henceforth called γ -phase, in the samples. The synthesized dense sample of the highest Si_2N_2O content, obtained by HIPing at 1900 °C and 200 MPa for 4 h hold time, was consequently chosen for this measurement. The dense sample was pulverized and examined by XRD. Then 2 wt% of α -Si_3N_4 powder was added and the powders were well mixed before a second determination by XRD. A subsequent addition of α -Si_3N_4 was made up to a total addition level of 5 wt% followed by X-ray measurement. The resulting patterns were analyzed concerning the change of relative intensity of the α -phase peak compared to the γ-phase peaks.The following strong peaks were used:

α-Si_3N_4.:	D=2.547 Å;	hkl=<210>;	relative pattern intensity : 100%
β -Si_3N_4 :	D=2.492 Å;	hkl=<210>;	relative pattern intensity : 90%
γ -Si_2N_2O :	D=2.419 Å;	hkl=<002>;	relative pattern intensity : 80%
γ -Si_2N_2O :	D=2.383 Å;	hkl=<021>;	relative pattern intensity : 80%

First the α -phase peak height was plotted against the addition level to determine the initial α -phase content in the addition-free sample to be approx. 1 wt%. A straight comparison between the peak height of the α -phase peak and of the γ -phase peaks would, however, predict the α -phase to be approx. 2.5 wt%. Therefore the measured relative amount of α -phase determined by comparing these peaks should be divided by the factor 2.5 to avoid overestimation and give a more accurate determination of the low residual amount in the sample. The β -phase peak is assumed to behave in a similar way.

The yield of the Si_2N_2O HIP syntheses performed at different temperatures and hold times calculated by ADM is presented in figure 1.

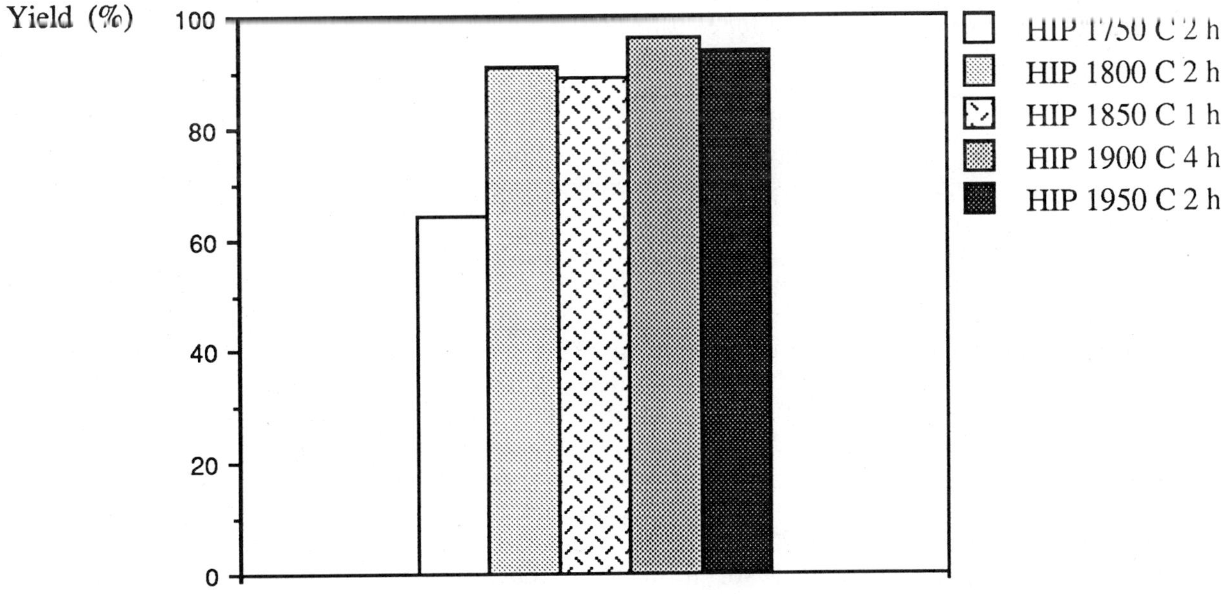

Figure 1. Silicon Oxynitride Yield achieved by HIP Syntheses.

2.2 Mechanical properties

Some mechanical properties of silicon oxynitride have been measured on the synthesized samples by bending tests and indentation techniques [5]. The results are:

- Young´s modulus: $E = 277$ GPa (measured by 4-point bending)
- Hardness: $H_v = 21$ GPa (Vickers diamond; 500 g; 10 s)
- Fracture toughness: $K_{IC} = 3.4\ \text{MPa} \cdot \text{m}^{1/2}$ (-"- ; 2000 g; 30 s)
- 3-P bend strength: $\sigma_f = 475$ MPa (20 mm span; 0.05 mm / min)
- Weibull modulus: $m = 5.7$ (22 specimens HIPed at 1850 °C)

Further evaluation is in progress [6].

3. JOINING WITH INCOLOY 909

To evaluate the possible benefit of using silicon oxynitride between the silicon nitride and the superalloy in the joint, the reactions between the ceramic materials and the alloy were examined. The silicon nitride alloy used is based on silicon nitride (UBE SN-E10) with 2.5 wt% yttria (Rhone-Poulenc) addition, designed for HIP fabrication of gas turbine components. The Incoloy 909 is chosen because of the good strength in the temperature region for a gas turbine wheel/axle joint intended (R.T. - 700 °C) combined with a low CTE compared with other high-temperature alloys, thereby reducing the thermo-mechanical stresses to a minimum. This alloy is therefore frequently used in such applications and thereby chosen in this investigation.The alloy composition is 42.0% Fe, 38.2% Ni, 13.0% Co, 4.7% Nb, 1.5% Ti and 0.4% Si.

The joining temperature should ideally not be much higher than the highest joint operation temperature to reduce thermo-mechanical stresses but high enough to give sufficient bonding strength. The high isostatic pressure during HIP enhances the contact area, thereby reducing the necessary temperature and time for the joint formation. A sufficient joining temperature might therefore be about 200 - 300 °C higher than the highest temperature during operation, or 900 - 1000 °C. The hold time at temperature and pressure was chosen as 2 h to give the diffusion processes enough time to give measureable reactions.

Ceramic samples of cylindrical shape of Si_3N_4 and Si_2N_2O were prepared by forming powders into rod shape by CIP followed by glass encapsulation and HIP to a final diameter of 7 mm. The rods were cut into pieces of 5 mm length and polished on the intended joint surface. The metal samples were cut out of a larger rod by Electro Discharge Machining (EDM) to pieces of the same dimensions as the ceramics and polished on the intended joint surfaces. To obtain more ceramic / metal interfaces from a certain amount of material, two pieces of ceramic material were mounted between three metal pieces to give a ceramic/metal rod with four ceramic/ metal interfaces in each capsule. The pieces were wrapped together in stainless steel foil and mounted in capsules made of mild steel tubes.

The benefit of using an intermediate layer of refractory metals was investigated by mounting 0.25 mm foil discs of titanium, zirconium and tungsten between three of the ceramic/metal pieces while leaving the fourth interface without an interlayer for the diffusion reactions dealt with in this paper. Four capsules, two with Si_3N_4 and two with Si_2N_2O ceramic pieces were evacuated and sealed. One of each type was HIPed at 900 °C and 200 MPa for 2 h hold time, the other two at 1000 °C and 200 MPa for 2 h. The capsules were cut in two halves in the axial direction by diamond saw to give a cross-section for microscopy and diffusion profile measurements.

3.1 Examination of joint cross-sections.

The cross-section of the joint was examined by light optical microscopy (LOM) and Scanning Electron Microscopy (SEM). The high residual thermo-mechanical stresses during cooling from joining temperature due to the differences in CTE caused cracking in the ceramics. This occured mainly in a convex manner typically some 200 µm from the interface due to the stress concentrations in this area.
The extension of the cracks was larger in the Si_2N_2O ceramics due to its lower fracture toughness partly caused by the limited synthesis experience.

The optical micrographs of the interfaces joined at 1000 °C is shown in figure 2.

Figure 2. Optical micrographs of ceramic/metal joints HIPed at 1000 °C.

3.2 Joint reaction analysis

The diffusion reactions in the interfaces between the ceramic and metal pieces during HIP were examined by SEM with Energy Dispersive X-ray Spectroscopy (EDS). The detectability of elements was limited to atomic weights higher than fluorine due to the Be window in the detector, i.e. the level of the important light elements nitrogen and oxygen in the ceramics could not be measured by this equipment.
The reaction zone was very thin in all four cases. The element distribution in the joint was examined by X-ray dot mapping at a microscope magnification of 3000 x. It was found that Nb and Ti were enriched at the interfaces close to Si_3N_4, especially at the higher joining temperature (1000 °C). Four dot maps of this interface with the distribution of Nb and Ti together with the major ceramic element Si and metal element Fe is shown in figure 3.

Nb

Ti

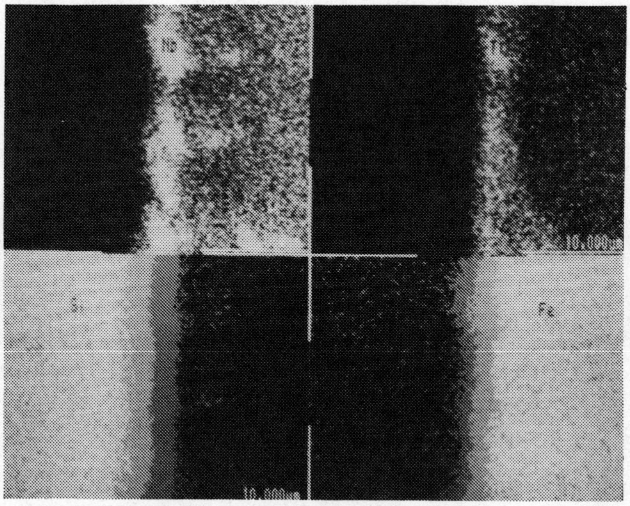

White bar = 10 µm

Si

Fe

Figure 3. Element Distribution in a Si_3N_4 / Incoloy 909 Joint HIPed at 1000 °C.

The diffusion profiles was determined by ZAF calculations of the element concentrations in 21 points situated at certain distances along a straight line across the interface far away from the capsule. The distance between the points close to the interface was chosen to be 0.5 µm in order to detect element gradients in the very thin reaction zones. The ZAF calculations compensate for the influence of difference in atomic weight (Z), absorbtion (A) and fluorescence (F) during EDS measurements, thereby allowing a quantitative evaluation of the results.

Plotting of the results is shown in figures 4.1 - 4.4 with the total measurements presented in the square in the upper left corners and the reactions close to the interface (< 5 µm) presented at a higher magnification in the center of the figures.

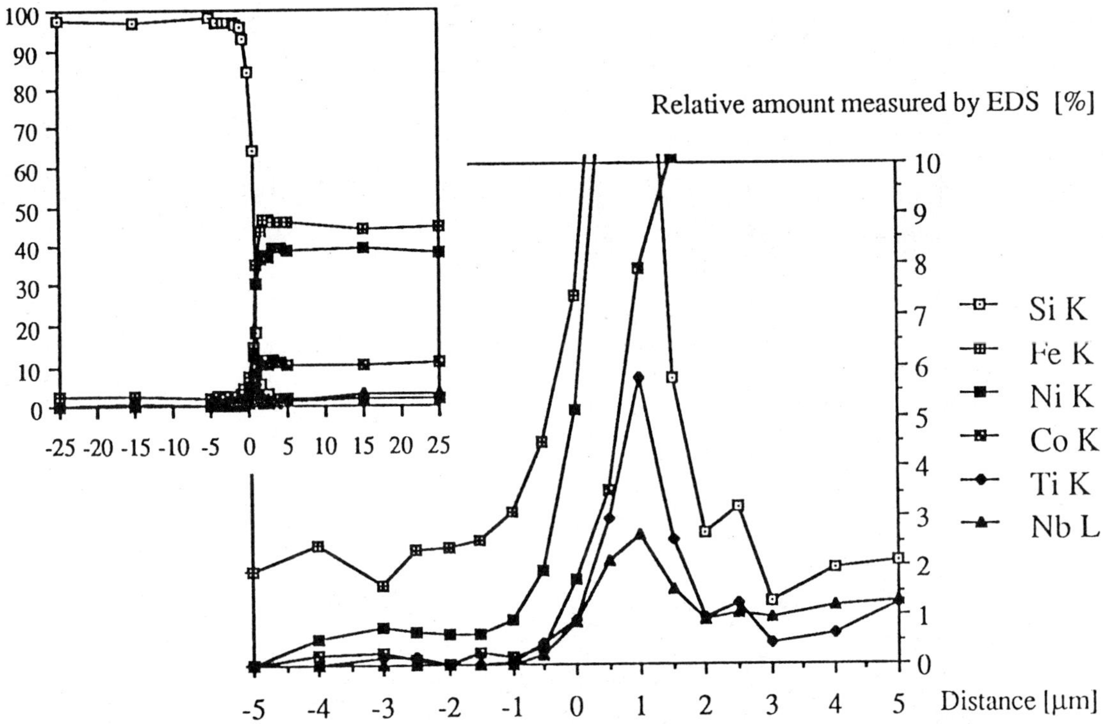

Fig 4.1 Si_3N_4 / Incoloy 909 HIP 1000 °C 200 MPa 2 h EDS-ZAF4

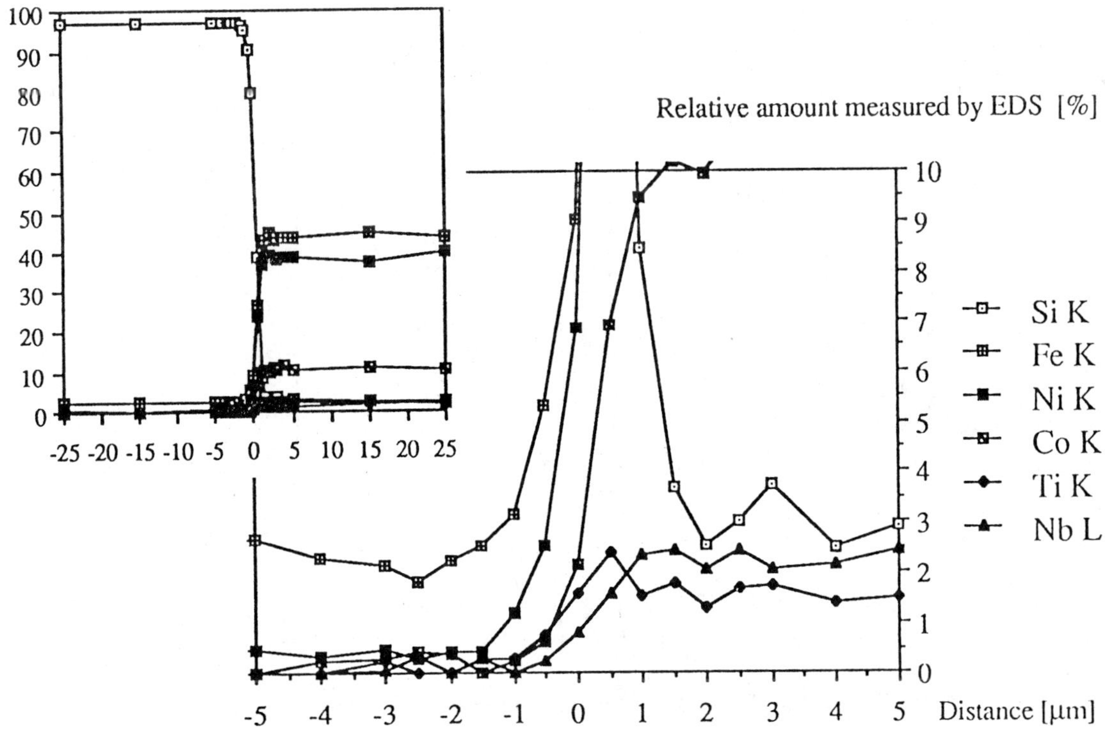

Fig 4.2 Si_2N_2O / Incoloy 909 HIP 1000 °C 200 MPa 2 h EDS-ZAF4

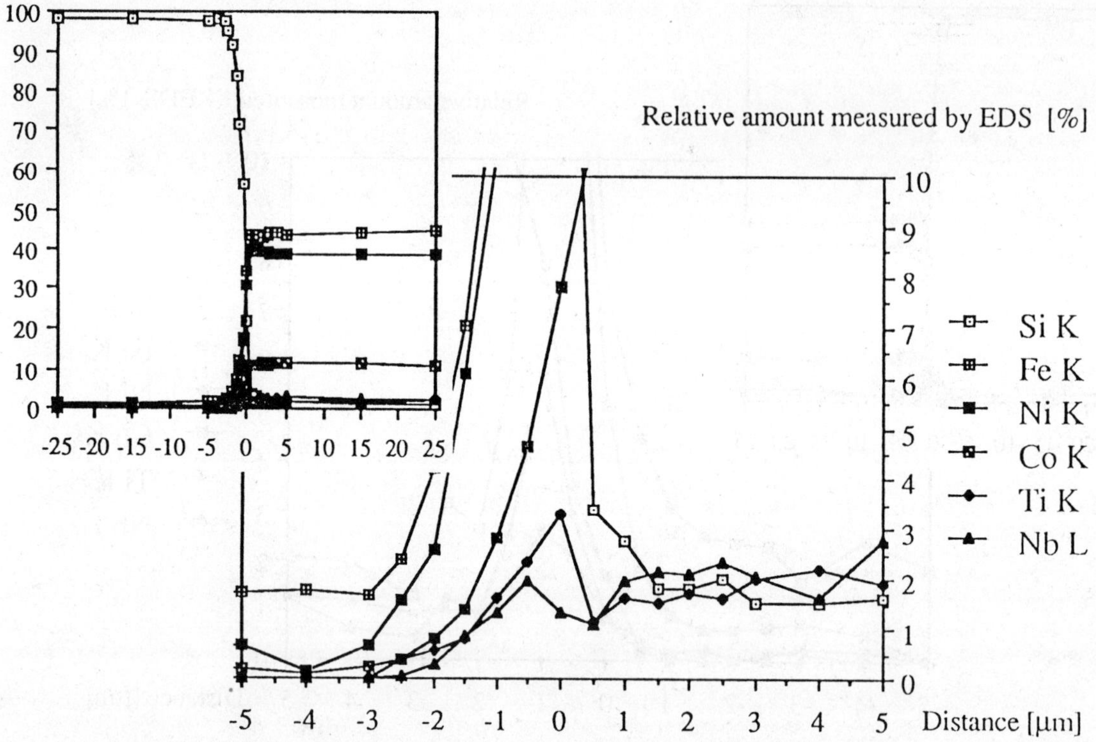

Fig 4.3 Si3N4 / Incoloy 909 HIP 900 °C 200 MPa 2 h EDS-ZAF4

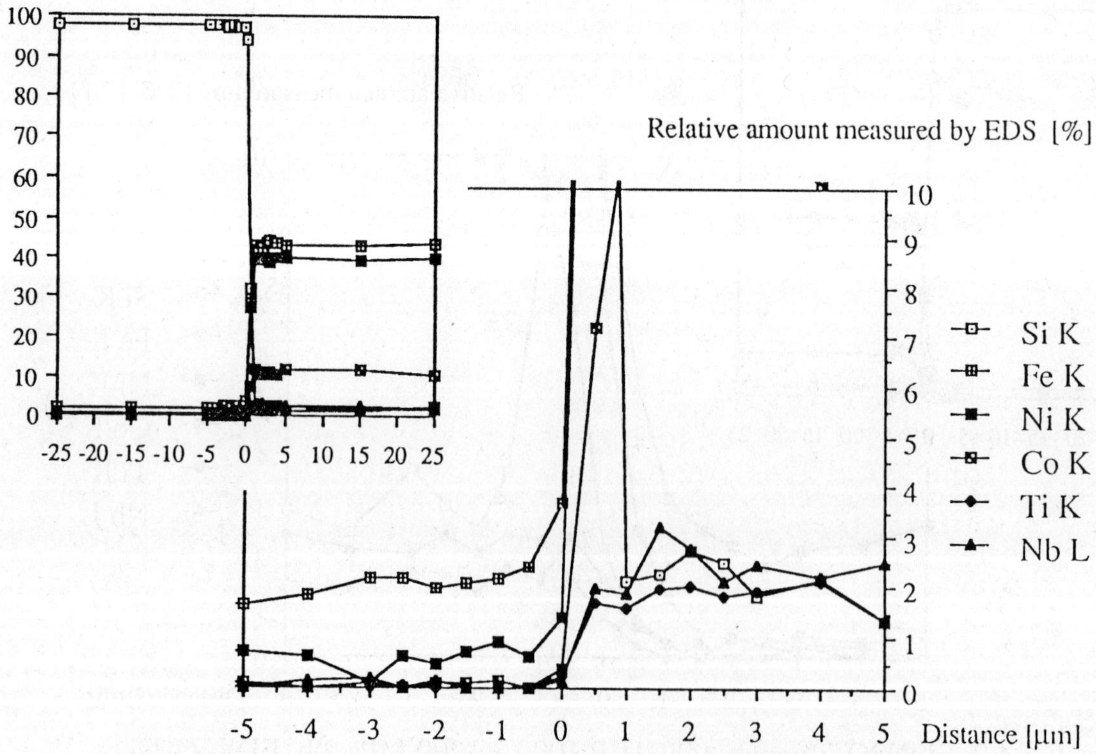

Fig 4.4 Si2N20 / Incoloy 909 HIP 900 °C 200 MPa 2 h EDS-ZAF4

5. DISCUSSION

The electron beam positioned at a certain point will exite characteristic X-rays from an interaction volume, which is typically 1 μm in diameter. The size is, however, dependant on the atomic weight of the material. The ceramic is composed of lighter elements (e.g. Si) than the metal (Ni, Fe, Co etc.) and the characteristic X-rays will therefore be exited from a larger volume in the ceramic.

The distance between the measuring points for the diffusion profiles close to the interface was 0.5 μm, which means that the interaction volumes overlapped. This has the effect that the real diffusion profiles are even steeper than the apparent ones in figures 4.1-4.4.

In the silicon nitride cases, the enhanced concentration of titanium and niobium in a thin layer (apparently about 2 μm in thickness) close to the initial interface is very pronounced, especially for the joint HIPed at 1000 °C. This behavior is not observed in silicon oxynitride cases.This difference could be explained by a higher stability of the silicon oxynitride in contact with the strong nitride formers Ti and Nb.

The formation of nitrides, silicides or oxides at the interface was not possible to analyze because of the limited detection capability of light elements explained in section 3.2. Thermodynamic calculations of this systems together with further EDS and Wavelength Dispersive X-ray Spectroscopy (WDS) analysis equipped with window-less detectors will be conducted.

According to our measurements, the Fe content in the ceramics had reached the same level of approx. 2% in all the samples, regardless of ceramic type or HIP temperature. This appeared to be highly improbable as the Fe content in the starting silicon nitride powder is < 0.005 %, and the levels in the fully densified ceramics are not significantly higher. The kinetics in this system does not allow Fe to diffuse into the ceramics to such an extent as the measurements implies. The high content of Fe in the ceramics was therefore assumed to be an artefact due to improper polishing. The probability of polishing contamination from the superalloy was however low in this case, as this should have given measureable levels of Ni and Co as well. Etching with HNO_3 followed by a second EDS measurement showed that the iron level in the ceramic diminished after this treatment, thus being a surface effect probably emanating from the soft mild steel capsule during polishing.
This result does not impair the reliability of the steep diffusional gradients determined in this study.

The evaluation of metal interlayers such as titanium, zirconium and tungsten between the ceramics and the superalloy is in progress. Preliminary results indicate that Ti and Zr have a higher tendency than W to react with both ceramics. The cracking in the ceramics caused by thermo-mechanical stresses is less pronounced in the joints with metal interlayers.

6. CONCLUSIONS

1/ Hot Isostatic Pressing is a useful method to obtain good conditions for solid state diffusion bonding between silicon nitride based ceramics and superalloys.

2/ Silicon oxynitride is more stable than silicon nitride against diffusion of reactive metals such as titanium and niobium from the superalloy.

3/ The properties of silicon oxynitride such as higher CTE, lower Young´s modulus and better chemical stability might be useful as an cost-effective intermediate joining layer for HIP joining of silicon nitride to superalloys for high-temperature applications such as engine components.

4/ Refractory metal interlayers seem to reduce thermo-mechanical cracking.

7. ACKNOWLEDGEMENTS

This work was financed by the National Swedish Board for Technical Development (STU). The valuable guidance during electron microscopy work and diffusion profile determinations by Ralph Harrysson is gratefully acknowledged.

8. REFERENCES

[1] Washburn, M.E.
Silicon Oxynitride Refractories
Ceramic Bulletin 46 (1967) p. 667-671.

[2] Huang, Z.K.; Greil, P.; Petzow, G.
Formation of Silicon Oxynitride from Si_3N_4 and SiO_2 in the Presence of Al_2O_3.
Ceramics International 10 (1984) n. 1 p. 14-17.

[3] Lewis, M.H.; Reed, C.J.; Butler, N.D.
Pressureless-sintered Ceramics Based on the Compound Si_2N_2O.
Materials Science and Engineering 71 (1985) p. 87-94.

[4] Larker, H.T.
Consistency is a Critical Problem for Ceramics - Can Hot Isostatic Pressing Give an Answer?
Interceram 36 (1987) n. 1 p. 45-47.

[5] Larker, H.R.; Hermansson, L.
Synthesis and Analysis of Silicon Oxynitride by Hot Isostatic Pressing.
Proceedings of the International Conference on Hot Isostatic Pressing -Theories and Applications; Luleå, Sweden June 15-16 1987
CENTEK Publishers 1988 pp 375-381.

[6] Larker, H.R.; Loberg, B.; Johansson, T.
Properties of HIPed Si_2N_2O (to be published)

STRENGTH AND MICROSTRUCTURE OF TITANIUM-VAPOR-COATED SILICON NITRIDE BRAZE JOINTS

M. L. Santella and L. C. Manley
Metals and Ceramics Division
Oak Ridge National Laboratory
Oak Ridge, Tennessee 37831-6096

ABSTRACT

A technique for brazing Si_3N_4 with metallic alloys was evaluated. The process involved vapor coating the ceramic with a 1.0-μm-thick layer of titanium prior to the brazing operation. The coating improved wetting of the Si_3N_4 surfaces to the extent that strong bonding between the solidified braze filler metal and the ceramic occurred. Braze joints of Si_3N_4 were made with Ag-Cu, Au-Ni, and Au-Ni-Pd alloys at temperatures of 790°C, 970°C, and 1130°C. The strengths of Si_3N_4-Si_3N_4 joints were determined by four-point flexure testing. The average strengths of joint specimens made with the Ag-Cu alloy were 326 ± 38 MPa at room temperature and 280 ± 37 MPa at 400°C. The average strength of joints made with the Au-Ni-Pd alloy was 264 ± 102 MPa at room temperature. Flexure strength as high as 474 MPa was obtained for a joint specimens made with the Au-Ni-Pd filler metal. The fracture paths in the joint specimens were almost entirely through the Si_3N_4 just beneath the bonded surface. Several joint specimens broke in the ceramic over 1 mm from the bond interface. Joint strength was not limited by adhesion of the titanium vapor coating, or braze filler metal strength. The shear strength of Si_3N_4-metal joints was low, and probably a result of high residual stresses due to the large thermal expansion mismatches between the ceramic and the metals.

INTRODUCTION

The development of ceramics as structural materials has served to focus attention on the importance of developing techniques for joining a particular ceramic to itself, other ceramics, and metals. Brazing, i.e. liquid phase bonding with metallic alloys, can produce ceramic-to-ceramic and ceramic-to-metal joints of high strength. Numerous studies have addressed the subject of brazing ceramics, and these works were recently reviewed by Nicholas and Mortimer[1], and Nicholas.[2] These authors noted that ceramic oxides, such as Al_2O_3, have been studied much more than any of the other ceramic materials, and that a considerable amount is known about the methods of obtaining strong bonding to oxide surfaces with brazing filler metals.

The emergence of silicon nitride as an important structural ceramic has created interest in developing joining techniques for this material, and the work on this subject has been reviewed by Johnson et al.[3] Various techniques have been used to join Si_3N_4 including solid-state bonding with and without intermediate bonding agents, liquid-phase bonding with glasses, and brazing with metal alloys. However, very little has been published on the brazing of Si_3N_4. Heap and Riley reported that Si_3N_4 can be brazed with silicon alloys.[4] Naka et al.[5] showed that amorphous $Cu_{50}Ti_{50}$ and $Ni_{24.5}Ti_{75.5}$ filler metals could be used to form strong braze joints of Si_3N_4. Also, Mizuhara and Mally[6] demonstrated that strong bonding to Si_3N_4 could be obtained with a Ag-Cu-Ti alloy.

The work presented here describes a method of brazing sintered Si_3N_4 to itself, to commercially pure titanium, and to the heat-resistant steel A286 with commercially available brazing filler metals. Wetting of the Si_3N_4 was accomplished by vapor coating the ceramic surfaces with titanium metal prior to brazing. Titanium was selected for a coating material because it was previously found to promote wetting and strong bonding for brazed joints of partially stabilized zirconia to itself and to metals.[7] The strengths of Si_3N_4-Si_3N_4 joints were determined by four-point flexure testing in air at room temperature and 400°C, and joint strengths up to 474 MPa were obtained. Several Si_3N_4-metal joints were shear tested at room temperature. The shear strengths of the Si_3N_4-metal joints were below 100 MPa.

MATERIALS AND EXPERIMENTAL DETAILS

The Si_3N_4 used for this work was produced by isostatic pressing and sintering (Kyocera SN-220). The density of this material was not determined, but is reported to be 3.2 g/cm³, with a porosity level below 0.1%. The sintering aids used for densification of the Si_3N_4 were not reported. The Si_3N_4 was brazed to itself and to two metals: A286, a precipitation-hardened heat-resistant steel; and, ASTM B265 Grade 1, commercially pure titanium. The nominal compositions of the braze filler metals used to form the joints are given in Table 1. The BAg-8a and BVAu-7 Gr 1 alloys were in the form of thin foil, 50 and 25 μm thick, respectively. The BAu-4 alloy was used as a 325-mesh powder.

All of the Si_3N_4 surfaces used for brazing were vapor coated with a 1.0-μm-thick layer of titanium by vacuum evaporation. The titanium vapor was produced in the vacuum chamber by melting a pure titanium source with an electron beam melting system operating, during steady state conditions, at 7 kV and 100 mA. Typically, the pressure inside the chamber at the time melting of the titanium began was 400 μPa and dropped to 50 μPa during evaporation. A shutter shielded the Si_3N_4 specimens from the titanium vapor until the pressure

Table 1. Braze Filler Metal Compositions

AWS-ASTM Designation	Composition, wt%						Brazing Temperature, °C
	Au	Ni	Pd	Ag	Cu	Li	
BAg-8a	--	--	--	71.7	28	0.3	790
BAu-4	82	18	--	--	--	--	970
BVAu-7 Gr 1	50	25	25	--	--	--	1130

inside the chamber stabilized. A standard quartz oscillation technique was used to monitor coating thickness during evaporation, and thickness was verified by reference weighing tabs. A typical evaporation time to produce a 1-μm-thick layer of titanium using our procedure was 30 min. The Si_3N_4 specimens were not preheated prior to coating. All specimens were cleaned ultrasonically with acetone followed by ethyl alcohol and then air dried before being placed in the evaporation chamber.

The arithmetic average roughness values of the various joint surfaces used for brazing were determined by profilometry to be $R_a = 4.05$ μm for the metallographic and shear test specimens, and $R_a = 0.54$ μm for the joint surfaces of the flexure test coupons. These surface roughness values were not affected by titanium vapor coating.

All brazing was done under a vacuum pressure of about 130 μPa at the temperatures shown in Table 1. The holding time at each brazing temperature was 5 to 10 min. Joint assemblies for metallographic examination and shear testing were made by securing together joint components and brazing filler metals with molybdenum wire. The assemblies made for flexure testing were held on edge in a graphite fixture with the brazed interface in the horizontal plane. No load was applied to any of the assemblies during brazing.

The shear strength of joints between Si_3N_4 and A286, and between Si_3N_4 and titanium were measured using a method described by Ritter and Burton[8] and subsequently modified by Moorhead and Keating.[9] For our experiments, 12.7 × 5 × 3.4 mm metal bars were brazed to 10 × 10 × 5 mm Si_3N_4 pads. The pads were held rigid and the metal bars were pushed off the pads at a crosshead speed of 8.5 μm/s using a fixture[8,9] fitted to an Instron machine.

The coupons used for making flexure test specimens were produced by a method similar to the one developed by Moorhead[10] except no attempt was made to control joint thickness and no load was applied to the joint assembly during brazing. Individual plates with dimensions of 25 × 14 × 3 mm were joined along the 25 × 3 mm edge. Afterward the brazed specimens were surface ground on both faces, and cut into flexure test bars with dimensions of 28 × 2.5 × 2 mm having a brazed joint in the center of each bar. Before testing the tensile faces on the bars were polished to a 1-μm finish and the edges of the bars were slightly beveled on a polishing wheel. Testing was done by four-point bending at a loading rate of 22.7 kg/s. The inner and outer spans of the testing fixture were 6.4 mm and 19.0 mm, respectively. The test bars broken in air at 400°C were equilibrated at the test temperature for 30 min prior to fracture.

The microstructures of cross-sectioned joints were examined optically and with a scanning electron microscope (SEM). Regions near the vapor-coated Si_3N_4 surfaces of the joints were analyzed using an x-ray energy-dispersive spectrometer (EDS). Fractured specimens were also examined in the SEM.

RESULTS

Wetting and Brazing of Si_3N_4

Although a detailed study of the wetting behavior of Si_3N_4 was not done, enough experiments were conducted to establish that the use of a titanium vapor coating on Si_3N_4 surfaces dramatically improved their wetting and brazing characteristics. Figure 1 is a photograph of a shear test specimen for a joint between Si_3N_4 and titanium showing that the Ag-Cu braze filler metal (BAg-8a) spread over the entire vapor-coated surface of the Si_3N_4, but not over the uncoated portions of the ceramic. Spreading of this Ag-Cu alloy was also obtained for joints between Si_3N_4 and the A286 steel. Fabrication of other types of specimens indicated that wetting of the titanium-vapor-coated Si_3N_4 by the Au-Ni (BAu-4) and Au-Ni-Pd (BVAu-7 Gr 1) filler metals was also excellent. For comparison, all three brazing filler metals were melted on uncoated specimens of Si_3N_4 under vacuum and in no case was wetting of the ceramic surface observed.

Microstructures of Si_3N_4-Si_3N_4 and Si_3N_4-metal Braze Joints

The overall appearance of the Si_3N_4-to-Si_3N_4 braze joints was similar regardless of the alloy and temperature used to form the bond. Typically the joints were 25- to 50-μm thick, contained very little porosity, and showed excellent wetting on the rather rough Si_3N_4 surfaces. Also, there was no indication of cracking in the Si_3N_4-to-Si_3N_4 joints after brazing at any of the three temperatures. The cross-sectional view of the microstructure of the joint brazed with the Au-Ni filler metal, shown in Fig. 2, typifies the appearance of joints made with the other filler metals as well.

The details of the interfaces between the Si_3N_4 and two of the filler metals are shown in Fig. 3(A) and 3(B). These cross-sectional views were formed with backscattered electrons in an SEM. Also, microanalysis of these regions was performed using EDS. The bond interface for the joint made with the Ag-Cu alloy is shown in Fig. 3(A), where the eutectic nature of this braze filler metal is clearly evident. The continuous band of material at the surface of the Si_3N_4 is the titanium vapor coating that was retained largely intact at its original thickness of 1 μm. The coating also showed no sign of debonding from the Si_3N_4 surface. The results of microanalysis by line scanning, shown in Fig. 3(A), and stationary electron probes indicated that there was no significant dissolution of the coating in the filler metal or chemical reaction of the coating with the Si_3N_4 substrate.

The bond interface region between the Au-Ni-Pd alloy and the titanium-vapor-coated Si_3N_4 is shown in Fig. 3(B). In this case the brazing alloy appeared to be single phased. The microstructure near the Si_3N_4 surface indicated the presence of a thin discontinuous reaction layer not clearly visible in Fig. 3(B). The trace of a microanalysis line scan is also shown in Fig. 3(B), and it indicated that while titanium was still localized near the Si_3N_4 surface the titanium coating apparently reacted considerably with the braze filler metal and no longer appeared as a distinct constituent of the microstructure. Elemental silicon was not detected in the braze filler metal although a thin reaction layer was apparent at the location of the original interface between the titanium and the Si_3N_4 surface. Despite these microstructural differences as compared to the joints made with the Ag-Cu and Au-Ni alloys, bonding of the Au-Ni-Pd filler metal to the Si_3N_4 appeared to be as good as that observed at the lower brazing temperatures.

The microstructure of the joint between Si_3N_4 and the A286 steel, shown in Fig. 4, was typical of the joints made with each of the metal components. The details of the microstructures near the metal interfaces varied with the particular alloy used. However, the microstructural features at the Si_3N_4

surfaces of the Si_3N_4-to-metal joints were identical to those found for the Si_3N_4-to-Si_3N_4 joints discussed above [Fig. 3(A)], and were not noticeably influenced by the type of metal alloy used to form the joint.

Flexure Tests of Si_3N_4-Si_3N_4 Braze Joints

The Weibull strength distribution of the Si_3N_4-Si_3N_4 joints brazed with the Au-Ni-Pd alloy and tested at room temperature is shown in Fig. 5. Twelve of these specimens were tested and their average strength was 264 ± 102 MPa with a Weibull modulus of 2.4. The spread in the strength values was 145 MPa to 474 MPa.

Specimens of the Si_3N_4-Si_3N_4 joints brazed with the Ag-Cu alloy were tested at room temperature and 400°C, and these data are shown in Fig. 6. At room temperature the average joint strength was 326 ± 38 MPa with a Weibull modulus of 7.7 and a spread in strength values of 258 MPa to 406 MPa. The average strength of the joint specimens tested at 400°C was 280 ± 37 MPa with a Weibull modulus of 6.4. The spread in these joint strength values was 237 MPa to 334 MPa.

Metallographic examination was performed on specimens from each testing condition. Representative fracture surfaces from a high strength and a low strength specimen from joints brazed with the Au-Ni-Pd alloy are shown in Fig. 7(A) and Fig. 7(B), respectively. Both surfaces are relatively rough in appearance and indicate that the fracture path traveled almost exclusively through the Si_3N_4 rather than through the filler metal or along one of the titanium-vapor-coating interfaces. In most specimens the fracture path was within about 50 μm of the interface between the titanium vapor coating and the Si_3N_4, although the fracture origin and path in several of the higher strength bars was 1 to 2 mm from the brazed surface. According to microanalysis done in the SEM, the light-colored features on the surfaces contain mostly titanium with indications of Au, Ni, Pd, and Si, and are probably remnants of the initial titanium vapor coatings. The higher strength bars generally had fewer of these titanium-rich regions visible on their fracture surfaces. Examination of mating fracture surfaces indicated that large titanium-rich features like the ones visible in Fig. 7(B) were associated with voids in the resolidified braze filler metal.

The appearance of the fracture surfaces of joint specimens brazed with the Ag-Cu alloy were the same regardless of whether they were broken at room temperature or 400°C. Overall these surfaces, shown in Fig 8(A), were relatively flat compared to those from the joints made with the Au-Ni-Pd alloy. A large fragment of the Ag-Cu braze filler metal is still attached to surface of the specimen shown in Fig. 8(A) indicating that the fracture path traveled through filler metal normal to the bonded surface. Microanalysis and the examination of cross-sectioned specimens, however, confirmed that the preferred fracture path was through the Si_3N_4 just beneath the coated surface. A higher magnification view of a fracture surface, Fig. 8(B), shows the appearance of the fractured Si_3N_4. Other features typical of the fracture surfaces of these specimens were titanium-rich regions and occasional voids in the braze filler metal.

Shear Tests of Si_3N_4-to-Metal Braze Joints

Two specimens each of Si_3N_4 brazed to titanium and A286 steel with the Ag-Cu braze filler metal were shear tested at room temperature. The configuration of the specimens before testing is shown in Fig. 1. The Si_3N_4 pads were held rigidly at the edge under the cantilevered part of the metal bar, and the load was applied at the rounded 'nose' of the metal bar in the direction of the cantilevered end. All of these specimens appeared to be free of cracks after brazing, although the existence of cracks in the untested specimens cannot be ruled out. The strengths measured under this shear loading arrangement were 97 and 100 MPa for the Si_3N_4-Ti joints, and 31 and 46 MPa for the Si_3N_4-A286 joints.

The fracture path of the Si_3N_4-Ti joints was mainly through the ceramic under the bond interface. Nugget-like fragments of Si_3N_4 remained attached to the titanium bars after testing, and the noses of the bars were slightly deformed by upsetting. One of the A286 joints fractured in a similar manner but the fracture path traveled much deeper into the Si_3N_4. The other A286 joint fractured near the interface between the steel and the solidified braze alloy, indicating poor adhesion of the Ag-Cu alloy to the A286. Neither type of joint gave any indication of poor adhesion of the titanium vapor coating to the Si_3N_4 surfaces.

DISCUSSION

This study showed that Si_3N_4 surfaces are not easily wet by common precious-metal-based brazing filler metals, and that vapor coating the Si_3N_4 surfaces with titanium prior to brazing improved their wetting characteristics and permitted braze filler metals to adhere strongly to the ceramic. The ability of vapor coatings, particularly of titanium, to improve the wetting characteristics of ceramics is well known,[11-13] and has been exploited for brazing partially stabilized zirconia to metals.[7] The vapor coating approach has not previously been applied to brazing Si_3N_4 and was selected for the present experiments because of the success of its application to joining zirconia. Using a metallic vapor coating rather than incorporating an active element like titanium directly into the brazing filler metal permitted readily available commercial braze filler metals to be used for joining.

Examination of the Si_3N_4 joints made with the Ag-Cu and Au-Ni filler metals indicated that the titanium vapor coating remained intact after the brazing operation and did not react significantly with the ceramic. For these two alloys it appeared that the improvement in wetting and brazing characteristics of the Si_3N_4 occurred because the titanium coating acted as a reaction barrier, isolating the ceramic surface from direct contact with the liquid metal. Consequently, the surface energy and chemical reaction conditions that determined the extent of wetting were those established between the liquid filler alloys and the titanium coating. Good bonding for these Si_3N_4 joints was obtained because the Ag-Cu and Au-Ni filler metals wet and bonded well to the titanium, and the titanium coating adhered well to the Si_3N_4.

For the Si_3N_4 joint brazed with the Au-Ni-Pd filler metal the titanium vapor coating reacted with the braze filler metal and the Si_3N_4, because of the higher brazing temperature and more favorable kinetic conditions for chemical reaction. The titanium, however, remained localized at the ceramic surface. In this case, wetting was accompanied by dissolution of the titanium coating in the braze filler metal and reaction of the titanium with the Si_3N_4. This observation implies that it may be possible to obtain the same result by adding titanium to the original BVAu-7 Gr 1 composition, making it a so-called 'active' brazing alloy.

Nickel and palladium form stable silicides, but because titanium was in direct contact with Si_3N_4 in these experiments titanium silicides and titanium nitride are the most likely chemical reaction products. The possible reactions for which thermodynamic data are available[14,15] are:

$$Si_3N_4 + 5Ti = Ti_5Si_3 + 2(N_2) \quad [1]$$

$$Si_3N_4 + 3Ti = 3TiSi + 2(N_2) \quad [2]$$

$$Si_3N_4 + 3/2Ti = 3/2TiSi_2 + 2(N_2) \quad [3]$$

$$Si_3N_4 + 4Ti = 4TiN + 3Si \quad [4]$$

and the free energies per mole of Si_3N_4 associated with reactions 1-4 at the temperatures used for brazing are given in Table 2. These data show that TiN, Ti_5Si_3, and TiSi can all form by reaction of titanium with Si_3N_4 at the three brazing temperatures used for this experiment. Consideration of the quantities ($\Delta G°_{Nitride} - \Delta G°_{Silicide}$) shows TiN is considerably more stable than the titanium silicides on a per-mole-of-titanium basis. Based on this analysis it is reasonable to assume that the thin discontinuous reaction layer observed at the Si_3N_4 surfaces of the joint made at 1130°C was TiN. This assumption is further supported by the work of Messier[13] who in a study of the use of titanium to enhance wetting of reaction-bonded Si_3N_4 by silicon concluded that the key to the effectiveness of titanium as a wetting aid was its tendency to form a TiN layer on Si_3N_4 surfaces. The general lack of porosity in the joints made at 1130°C despite significant chemical reaction also points toward TiN formation. Reactions of titanium with Si_3N_4 that result in titanium silicide formation also result in the formation of nitrogen gas which could be retained as pores in braze joints.

Table 2. Free Energies for Phases Formed by Reaction of Ti with Si_3N_4

Reaction Product	ΔG°(T), kcal		
	790°C	970°C	1130°C
Ti_5Si_3	-48.0	-61.6	-75.0
TiSi	-19.1	-35.9	-52.6
$TiSi_2$	45.9	32.3	18.7
TiN	-133.4	-130.7	-128.0

The data in Table 2 also indicate that TiN can form at the lower brazing temperatures if kinetics are favorable. No reaction layers were observed in the joints made at 790°C and 970°C but the presence of TiN cannot be precluded based on examination in the SEM. The reaction layer that formed between the titanium vapor coating and Si_3N_4 at 1130°C was just visible in the SEM at high magnification. If similar layers formed at lower temperatures they would be even thinner and consequently more difficult to image without the aid of more sensitive techniques.

The metallographic examination of the broken flexure test bars confirmed that adhesion of the titanium vapor coating to the Si_3N_4 surfaces after brazing was high, and suggested that joint strength was not limited by adhesion of the coating to the ceramic. The coupons used for producing the flexure bars were ground to their final dimensions before brazing, and no attempt was made to remove surface damage that may have resulted from machining. Surface cracks were never observed on metallographically prepared joint specimens, or on specimens examined in the SEM, but an exhaustive search for the presence of surface defects was not done. The proximity of the fractures to the interfaces between the Si_3N_4 and the titanium coating suggests that joint strength was controlled by near-surface microcracking in the Si_3N_4. Similar behavior has been reported by Mizuhara and Mally,[6] but in their experiments the machining damage observed was extensive and clearly visible.

The lines and Weibull moduli given in Fig. 5 and Fig. 6 were based on a least squares fit through the experimental data, and assumed a single population of flaws for each data set. The low Weibull modulus of 2.4 for the Au-Ni-Pd joints, Fig. 5, and the observation of large voids in some of the joint specimens suggests that the assumption of a single population of flaws may not be valid. The appearance of high-strength and low-strength joint specimens in areas where fracture occurred through the Si_3N_4 was similar, but as Fig. 7(B) shows porosity in the braze filler metal reduced the bonded area considerably in some specimens. This much unbonded area would undoubtedly reduce joint strength as well, and the occurrence and amount of unbonded regions did correlate well with lower strength. It is likely that flaws in the Si_3N_4 controlled joint strength only in the absence of significant porosity.

The source of the porosity in the joints made with the Au-Ni-Pd alloy is not known but it can result from a variety of sources including improper cleaning or handling of materials, nonuniform heating, or gas entrapment. Because the joints were made in vacuum a gas source which depended on chemical reactions during brazing would be required for gas entrapment to occur. As discussed previously, thermodynamics favors the formation of TiN rather than $[TiSi_x + y(N_2)]$ in the joints so it is unlikely that a gas source was available. Also, if porosity was produced directly by chemical reactions during brazing it would have occurred in all of the joints. The fact that porosity was not found in all the joints indicates that it probably was caused by factors related to cleaning, handling, or nonuniform heating.

The strength and metallographic data for the joints brazed with the Ag-Cu filler metal indicated that increasing the testing temperature to 400°C did not change the type of flaw responsible for failure. The higher Weibull moduli for these data, 7.7 at room temperature and 6.4 at 400°C, may be partly the result of the limited porosity in the joints made with the Ag-Cu alloy as compared to those made with the Au-Ni-Pd alloy. The intrinsic strength of the braze filler metal was not a factor in joint strength since fracture paths were through the Si_3N_4 at both room temperature and 400°C. The reduction of strength and Weibull modulus at 400°C was not accompanied by a change in fracture appearance, but may be related to residual stresses near the brazed surfaces that result from differences in thermal expansion coefficients between the braze filler metal and the Si_3N_4.[16]

Although the preferred fracture paths in joints made with both filler metals was through the Si_3N_4, the respective fracture appearances were different, as Fig. 7(A) and Fig. 8(A) show. This difference could be the result of the more extensive reaction of the titanium vapor coating with the Si_3N_4 that occurred at the higher brazing temperature used for the Au-Ni-Pd filler metal. In the joints made at 1130°C the titanium vapor coating reacted with both the Au-Ni-Pd filler metal and the Si_3N_4, and allowed the alloy to flow around irregularities on the ground ceramic surfaces and possibly to intrude into preexisting surface cracks and voids. For brazing at 790°C, however, the titanium coating remained intact. These differences in reaction details would result in different distributions of residual stresses near the brazed Si_3N_4 surfaces. Titanium has a thermal expansion coefficient (8.4 μm/m/°C) intermediate to those of the Si_3N_4 (3.2 μm/m/°C) and the brazing alloys (13-19 μm/m/°C), and could help to accommodate some of the residual stresses due to the difference in thermal expansion coefficients between the Ag-Cu alloy and the Si_3N_4 in the vicinity of the ceramic surface. The breakdown of the titanium coating at 1130°C means that the Si_3N_4 is more nearly in direct contact with the Au-Ni-Pd brazing alloy. This situation could result in higher residual stresses or a different distribution of residual stresses near the brazed Si_3N_4 surfaces, and could be the cause of the difference in fracture appearance as well as the tendency for more scatter in the strength data for the joints made with the Au-Ni-Pd alloy.

Shear testing of the Si_3N_4-Ti and Si_3N_4-A286 joints also confirmed that bonding of the Ag-Cu filler metal to the titanium-vapor-coated Si_3N_4 was very good, because in both cases fracture paths traveled through the Si_3N_4 far below the braze interfaces. The measured strength values cannot be taken as true shear strength values because of the residual stresses known to be in the joints[16], and because of the possibility that there were preexisting cracks in the Si_3N_4 joint components. The larger mismatch in thermal expansion coefficient between the A286 and Si_3N_4 ($\Delta\alpha$ = 16.5 μm/m/°C) would produce higher residual stresses in these joints compared to the Si_3N_4-Ti joints ($\Delta\alpha$ = 8.4 μm/m/°C) and was probably responsible for their lower measured strengths.

SUMMARY

Vapor coating Si_3N_4 surfaces with a 1-μm-thick layer of titanium was shown to be an effective way of promoting wetting and bonding during the brazing of this important structural ceramic at temperatures up to 1130°C. With brazing filler metals of Ag-Cu and Au-Ni (brazing temperatures of 790°C and 970°C, respectively) the titanium coating appeared to act as a reaction barrier on the Si_3N_4 surface, shielding the ceramic from direct contact with the liquid braze filler metal. At 1130°C the titanium appeared to dissolve in a Au-Ni-Pd braze filler metal and react with the Si_3N_4, but remained localized at the Si_3N_4 surface.

Flexure testing of Si_3N_4-Si_3N_4 joints established that the preferred fracture path was through the ceramic and indicated that joint strength was not limited by adhesion of the titanium coating to the Si_3N_4 surfaces or by the strength of the filler metal. Joint strengths as high as 474 MPa were obtained with a Au-Ni-Pd braze filler metal. For a Ag-Cu braze filler metal, joint strength as high as 334 MPa was obtained at 400°C. Porosity in joints made with the Au-Ni-Pd alloy may have limited their strength and contributed to the relatively large scatter in the strength data.

Brazing of Si_3N_4 to A286 steel, and to titanium, confirmed that adhesion of the titanium coating to the Si_3N_4 surfaces was excellent. The shear strengths of Si_3N_4-A286 and Si_3N_4-Ti joints were low, and undoubtedly the result of high residual stress levels in the joints due to the large thermal expansion coefficient mismatches between Si_3N_4 and the metals.

ACKNOWLEDGEMENTS

The authors thank D. Ray Johnson for his encouragement and interest; A. J. Moorhead and T. N. Tiegs for technical review of the manuscript; L. A. Zevenbergen for making the vapor coatings; and B. C. Leslie and J. R. Mayotte for metallographic assistance.

This work was sponsored by the U.S. Department of Energy, Assistant Secretary for Conservation and Renewable Energy, Office of Transportation Systems, as part of the Ceramic Technology for Advanced Heat Engines Project of the Advanced Materials Development Program, under contract DE-AC05-84OR21400 with Martin Marietta Energy Systems, Inc.

REFERENCES

1. M. G. Nicholas and D. A. Mortimer, "Ceramic/metal joining for structural applications," *Mater. Sci. Tech.* 1, 657-65 (1985).

2. M. G. Nicholas, "Active Metal Brazing," *Br. Ceram. Trans. J.* 85, 144-46 (1986).

3. S. M. Johnson, D. J. Rowcliffe, R. D. Brittain, and R. H. Lamoreaux, "Joining of Silicon Nitride," *Proceedings of the Second International Conference on Joining Ceramics, Glass, and Metals*, Bad Nauheim, West Germany, March 1985.

4. H. R. Heap and C. C. Riley, "Brazing Silicon Nitride," Brit. Pat. No. 1,310,997, 1973.

5. M. Naka, T. Tanaka, I. Okamoto, and Y. Arata, "Non-oxide ceramic joint made with amorphous $Cu_{50}Ti_{50}$ and $Ni_{24.5}Ti_{75.5}$ filler metals," *Trans. JWRI* 12, 337-40 (1983).

6. H. Mizuhara and K. Mally, "Ceramic-to-Metal Joining with Active Brazing Filler Metal," *Weld. J.* 64, 27-32 (1985).

7. M. L. Santella, J. P. Hammond, S. A. David, and W. A. Simpson, "Zirconia to Cast Iron Brazing for Uncooled Diesel Engines," pp. 235-41, *Proceedings of the Twenty-Third Automotive Technology Development Contractors'*

Coordination Meeting, Society of Automotive Engineers, Inc., Warrendale, PA, March 1986.

8. J. E. Ritter, Jr., and M. S. Burton, "Adherence and Wettability of Nickel, Nickel-Titanium Alloys, and Nickel-Chromium Alloys to Sapphire," *Trans. AIME* 239, 21-26 (1967).

9. A. J. Moorhead and H. Keating, "Direct Brazing of Ceramics for Advanced Heavy-Duty Diesels," *Weld. J.* 65, 17-31 (1986).

10. A. J. Moorhead, "Direct Brazing of Alumina Ceramics," *Adv. Ceram. Mater.* 2, 159-166 (1987).

11. S. Weiss and C. M. Adams, "The Promotion of Wetting and Brazing," *Weld. J.* 46, 49-s-57-s (1967).

12. E. F. Brush, Jr., and C. M. Adams, Jr., "Vapor Coated Surfaces for Brazing Ceramics," *Weld. J.* 68, 106-s-114-s (1968).

13. D. R. Messier, "Use of Ti to Enhance Wetting of Reaction-Bonded Si_3N_4 by Si," *Ceram. Eng. Sci. Proc.* 1, (1980).

14. I. Barin, O. Knacke, and O. Kubaschewski, *Thermochemical Properties of Inorganic Substances*, Springer-Verlag, Düsseldorf, 1977.

15. O. Kubaschewski and C. B. Alcock, *Metallurgical Thermochemistry*, 5th edition, G. V. Raynor, ed., Pergamon Press, New York, 1979.

16. M. L. Santella, "Brazing of Titanium-Vapor-Coated Silicon Nitride," *Adv. Ceram. Mater.* 3, 457-62 (1988).

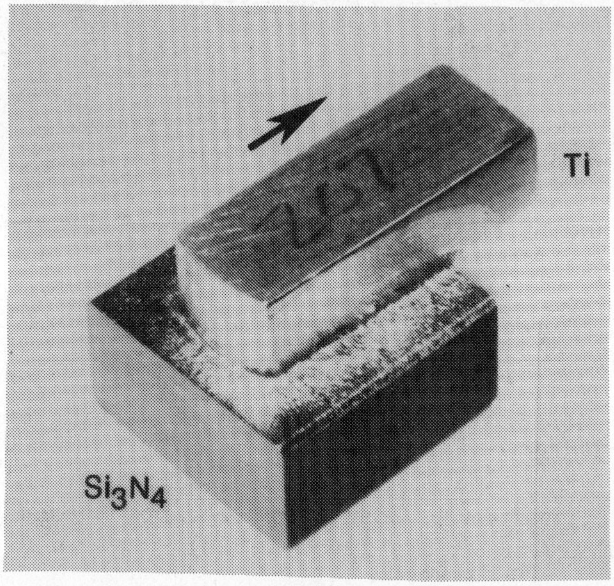

Fig. 1. Photograph of Si_3N_4-Ti joint showing spreading of the BAg-8a braze filler metal over the entire titanium-vapor-coated surface. Arrow indicates direction of loading for shear test.

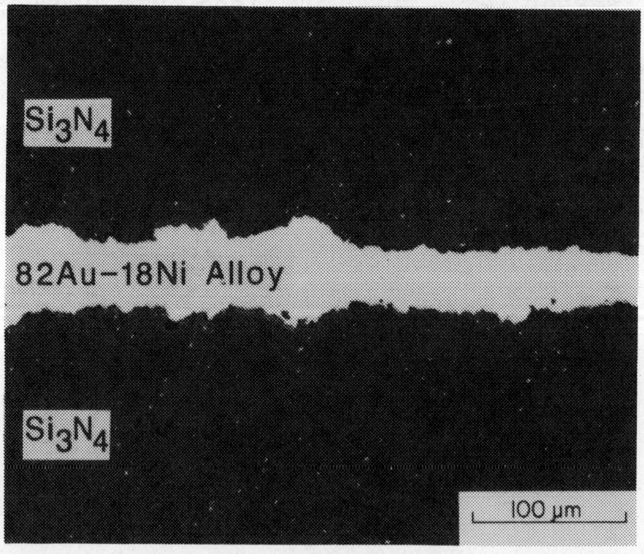

Fig. 2. Optical micrograph of Si_3N_4 brazed with a gold-nickel alloy.

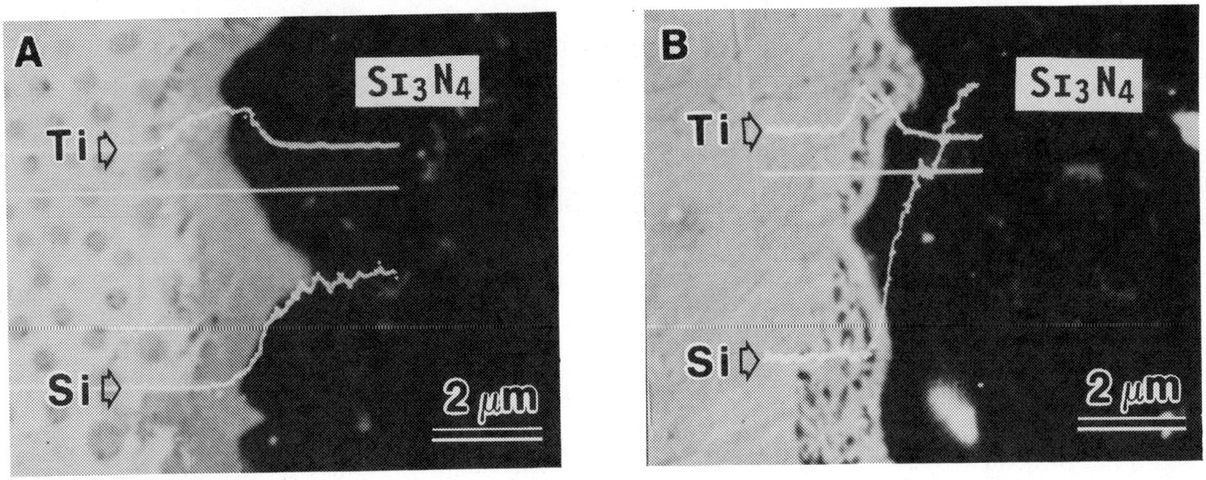

Fig. 3. Scanning electron micrographs of interface details for Si_3N_4 brazed with (A) silver-copper alloy at 790°C, and (B) gold-nickel-palladium alloy at 1130°C. The x-ray microanalysis line scans for titanium and silicon are indicated on each. Straight lines indicate locations of line scans.

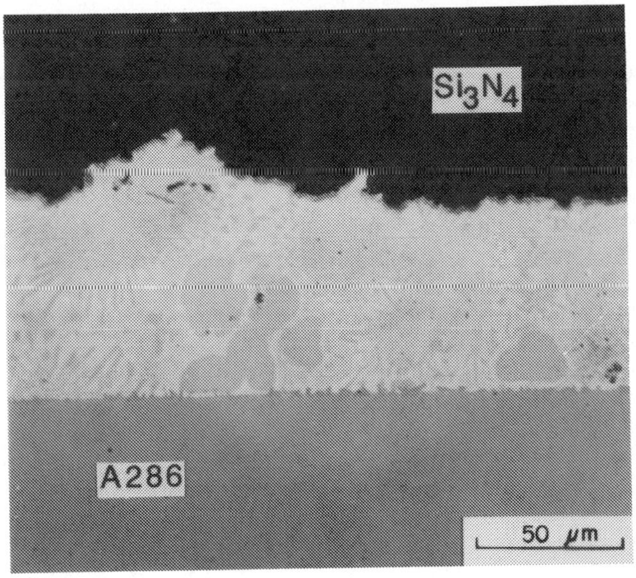

Fig. 4. Optical micrograph of Si_3N_4 brazed to A286 steel with silver-copper alloy.

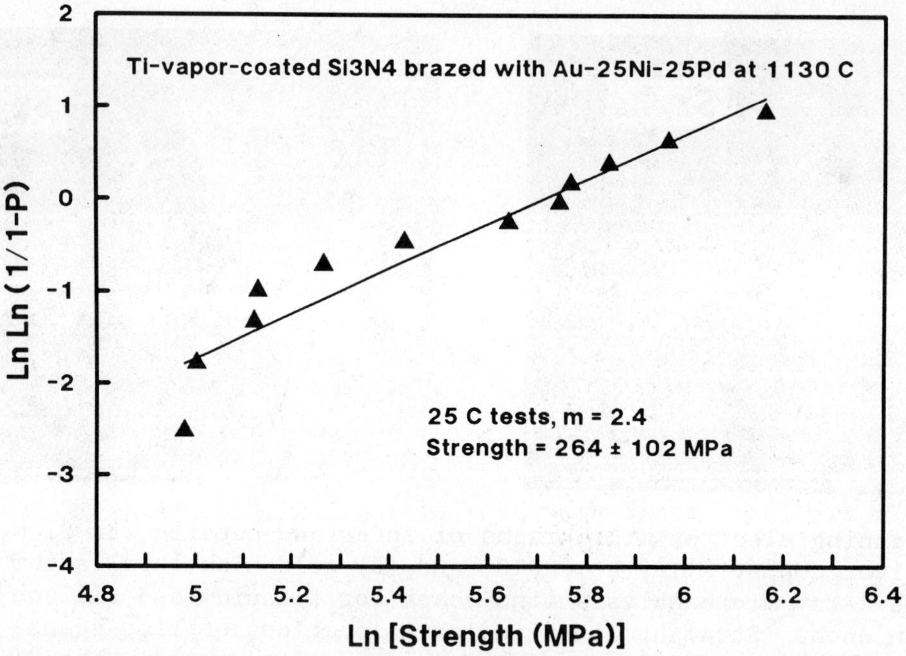

Fig. 5. Weibull plot for flexure strength specimens of Si_3N_4-Si_3N_4 joints made with Au-Ni-Pd braze filler metal at 1130°C.

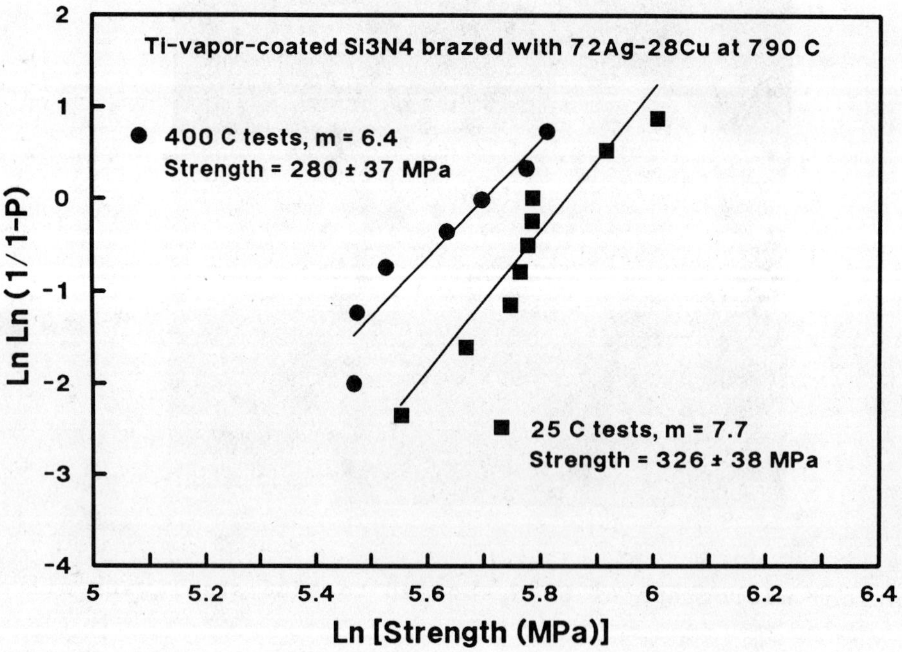

Fig. 6. Weibull plot for flexure strength specimens of Si_3N_4-Si_3N_4 joints made with Ag-Cu braze filler metal at 790°C.

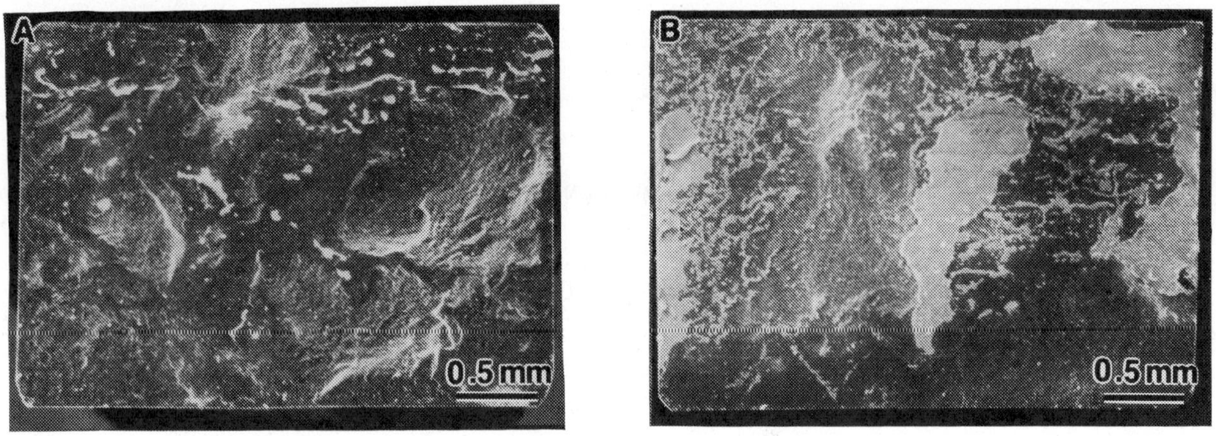

Fig. 7. Low magnification SEM views of broken flexure test bars from joints made with Au-Ni-Pd braze filler metal. (A) high strength specimen; (B) low strength specimen.

Fig. 8. SEM micrographs of broken flexure test bar from a joint made with Ag-Cu braze filler metal. (A) shows overall fracture appearance; (B) shows high magnification view of Si_3N_4 fracture surface.

NUMERICAL SIMULATION OF THE THERMO-MECHANICAL BEHAVIOUR OF THICK CERAMIC COATINGS.

E.Berger, N. Perrin, M. Boussuge, H. Burlet
Ecole des Mines de Paris
Centre des Matériaux P.M. FOURT
B.P. 87 - 91003 EVRY CEDEX, FRANCE

G. Desplanches
Régie Renault
Direction des Laboratoires Automobiles
92109 BOULOGNE BILLANCOURT, FRANCE

ABSTRACT

Light-alloy substrates coated by magnesium zirconate used as a thermal barrier have been investigated.

Bending tests carried out on coatings separated from the substrate allowed the strengths, the Young's moduli and the Poisson's ratios to be determined, for the magnesium zirconate itself as well as for the intermediate layers. These data were used in a finite element program, in order to predict the cracking of coated specimens submitted to a bending test.

Finally, from actual instrumented thermal shock tests, an equivalent simplified thermal shock model was defined. Thermo-elastic calculations performed using this approximation allowed the influence of producing or designing modifications on the thermal shock resistance to be evaluated.

INTRODUCTION

In order to have a reliable way of controlling the heat flow through the walls of the combustion chamber in diesel engines, and to reduce the pollution in exhaust gases,plasma spraying of ceramics on metallic parts is more and more seriously considered today. The successful use of this technique, however, requires the optimization of the processing as well as of the design of the coated components. With this aim, in a previous paper /1/, different types of mechanical tests and acoustic emission have been investigated, to characterize the mechanical properties of the coatings and of the interfaces.

This work explores a further direction : it concerns the numerical simulation of coated specimens submitted to mechanical and thermal stressing. In parallel, mechanical and thermal shock experiments were performed to gather the data needed by the calculation, and to validate the numerical predictions.

MATERIAL

The substrate studied consisted in an aluminium alloy containing 12 % of silicon, currently used for making pistons of automotive diesel engines. Its principal thermo-mechanical characteristics are summarized in Table 1. The magnesium zirconate (76 % ZrO_2, 24 % MgO) coating was plasma-sprayed by Renault with a self-acting torch equipped with a robot, which allows the variability of the handwork to be avoided. This isolating layer, containing a porosity varying between 5 % and 8 %, was bonded on the substrate using an intermediate layer. Its composition ranged from pure metal (95 % Ni, 5 % Al) to pure ceramic, through the thickness : this result is obtained by modifying the ratio metal/ceramic in the sprayed powder. The total thickness of the coating rised up to 2 mm. Figure 1 shows a typical microstructure of the whole coating.

TABLE 1 : Principal mechanical and thermal properties of the substrate and of the constituents of the thermal barrier.

materials	E_{US} (GPa)	E_{gauge} (GPa)	ν	σ_r (MPa)	α ($10^{-6}.K^{-1}$)	λ ($W.m^{-1}.K^{-1}$)	ρC_p ($10^6\ J.m^{-3}.K^{-1}$)
AS12UNG	76		0.3		21.2	150	2.6
NiAl	85	103	0.17	448	14.6	5.3	3.44
50/50	82	86	0.11	224	13.4	2.8	4
ZrO_2	65	28	0.024	54.5	10.7	0.8	5.3

FIGURE 1 : Typical microstructure of a coating.

BENDING TESTS ON COATINGS SEPARATED FROM THE SUBSTRATE

Parallelepipedic specimens of light alloy were plasma-sprayed with 3 different compositions of the powder : pure magnesium zirconate, pure nickel-aluminium, and a mixture of both (50 % ponderal each). The substrates were saw-cut close to the interface, and the coatings were machined by grinding to remove the entire substrate and to obtain specimens of 30 x 10 x 1 mm^3. All these operations were carried out with great care, in order to prevent any damage of the coating.

For each composition, some specimens were equipped with strain gauges of two different types : unidirectional along the length of the specimen and bidirectional 0-90°, the gauge lengths being respectively 4 mm and 2 mm. The instrumented specimens were submitted to 4-point bending (inner span 12 mm, outer span 28 mm) below their fracture strength, and the elongation of the gauge recorded as a function of the applied load. The test fixture used was "fully-articulated" : in particular, the loading points were free to rotate, to avoid the presence, on the side in tension, of compressive stresses due to friction.

This experimental apparatus allowed the Young's modulus and the Poisson's ratio of the 3 tested compositions to determine. Table 1 summarizes the obtained results. They can be compared with those obtained, on the same materials, by Gault et al. /2/ using the method of measurement of sonic waves velocity, also given in Table 1. A great difference between both values can be observed on the pure ceramic coating, the Young's modulus determined using the gauges remaining about 60 % lower.

This observation can be correlated with another one carried out on the load-strain diagrams recorded during bending tests. When cycling the specimens between two constant levels of load, the diagram exhibited a good reproducibility, except for the first loading. During this latter, the behaviour ofthe materials containing ceramic seemed unelastic, the Young's modulus decreasing up to reach the value obtained in the further linear cycles (which is mentioned in Table 1).

According to the litterature /3/, it can be thought that the first stressing of the specimens induced microcracks (in particular delaminations in planes perpendicular to the spraying axis), leading to a decrease of the apparent Young's modulus. On the other hand, the experiments performed using the ultrasonic method did not lead to the same microcracking because the very low strains applied to the material with this technique. Microcracking can hence be an explanation of the higher Young's moduli obtained with the ultrasonic technique, but this point however requires further consideration.

The values obtained for the Poisson ratio presented a good reproducibility but appeared very low compared to those usually determined on other materials : this phenomenon is probably in relation with the porosity and the strong anisotropy of this type of materials.

Finally, 6 specimens of each composition were broken in 4-point bending to measure their fracture strength : the results are given in Table 1. The value obtained for the ZM coating (54.5 MPa) can be compared with the strength measured along the spraying axis (12.7 MPa) and with this one evaluated on coatings bonded on the substrate (> 100 MPa) /1/. The anisotropy on the coating and the residual stresses present in the ceramic are certainly responsible of these differences.

NUMERICAL SIMULATION OF BENDING TESTS

The aim of the calculation was the simulation of 4-point bending tests previously performed on these materials /1/. During these tests, the coating, situated on the side in tension, exhibited cracks perpendicular to the principal tensile stress, successively initiated during the loading. The cracks propagated through the whole thickness of the ceramic coating (1.5 mm) and stopped close to the first particles of NiAl (at about 0.5 mm of the substrate).

Uncracked Specimen Simulation

Initially, the stress field was studied for an uncracked specimen and an elastic medium in plane strain state. The intermediate layer was supposed to be constituted of two different layers made of pure NiAl (0.1 mm thickness) and of a mixture in equal parts of NiAl and ZM (0.4 mm thickness). A single half of a specimen was calculated using a 2D-mesh, the elements being reduced in the coating itself : the mesh is represented in Figure 2. The applied load was choosen equal to the average load at which the first crack initiated during the experiments, corresponding to the actual strength of the coating bonded on the substrate.

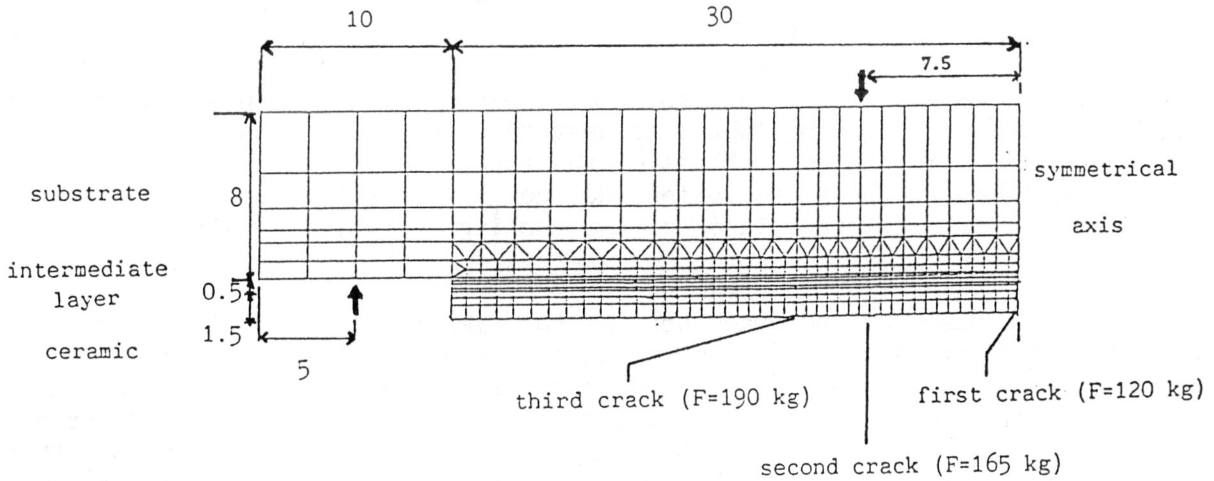

FIGURE 2 : Mesh used for the mechanical finite element calculation of a bending specimen (dimensions and locations of the cracks are mentioned).

This assumption should however be detailed, relating to the residual stresses due to plasma-spraying. As this mismatch stresses have not been yet determined in our case, it would be possible to estimate them from data collected in the litterature /4-7/. However, numerous authors have noted an important influence of some parameters on their values : composition /6/, behaviour /4-6/, roughness /8/ and cooling /7/ of the substrate, relative torch-substrate velocity /5-7/, sprayed particles characteristics /5/,... For this reason, we preferred, in this work, consider only the effects of the residual stresses on the strength of the coating. Hence, it is worth noting that all the calculated values of the stresses represent only the components due to the thermo-mechanical loadings and don't take into account the residual stresses.

The observation of the evolution of the principal tensile stress in sections along the thickness of the specimens allowed us to check that the stress field remained, despite the coating, very close to a pure bending one. The stress presented a maximum on the external side of the coating, its value being 90 MPa. The stress-steps corresponding to the change of material appeared to be always small compared to the associated stress. The difference between the higher calculated stress and the higher stress in the analytical solution for an homogeneous material (side in tension) was limited to 15 %.

Cracking Simulation

This part of our work consists in introducing cracks in the coating of the bending specimen, and to evaluate the influence of the cracking on the stress distribution in the coating.

Initially, a crack was supposed to be initiated in the symmetry plane of the mesh. From the point of view of the calculation, the crack was simulated by releasing the nodes in the crack plane, up to the first intermediate layer containing NiAl. This simple method of creating a crack in a mesh did not allow valid results to be obtained near the crack tip, because it neglects the plasticity (of NiAl as well as of the substrate) and the crack branching phenomena. However, these assumptions do not strongly influence the results obtained for the external face of the coating, situated far from the crack tip.

The 2nd (and also the 3rd crack because the symmetry of the calculation) was supposed to be created at the same value of the maximum tensile stress as the 1st crack, this stress being of course obtained for a higher load. Two values were deduced from the simulation : the position of the 2nd (and 3rd) crack, and the load corresponding to its initiation. The same calculation was then carried out for the 4th (and 5th) crack supposed created for the same maximum stress, and so on up to the 7th crack. The positions of the 5 first cracks and the values of the load at which they appear are summarized in Figure 2. Figure 3 represents the corresponding variations of the stress along the length of the specimen up to the 5th crack.

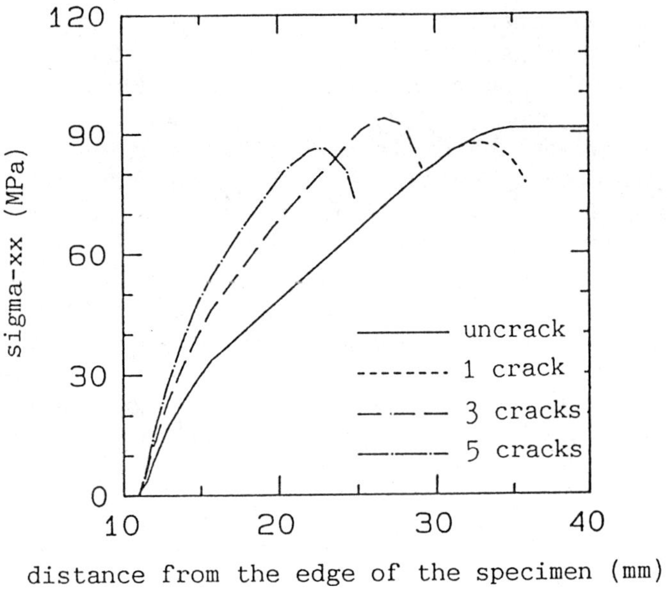

FIGURE 3 : Evolution of the outer fiber stress during the successive crack initiations.

The distances between the cracks for the 7 simulated cracks ranged between 4 and 8 mm, which can be compared with the experimental results obtained during 4-point bending tests conducted up to the fracture of the substrate /1/. In this case, the average distance between the cracks, measured for the 7 cracks nearest to the symmetry plane, is 4.5 mm. With respect to the scatter of the experimental data (standard deviation 1 mm), the calculated values appear in good agreement with the measured ones.

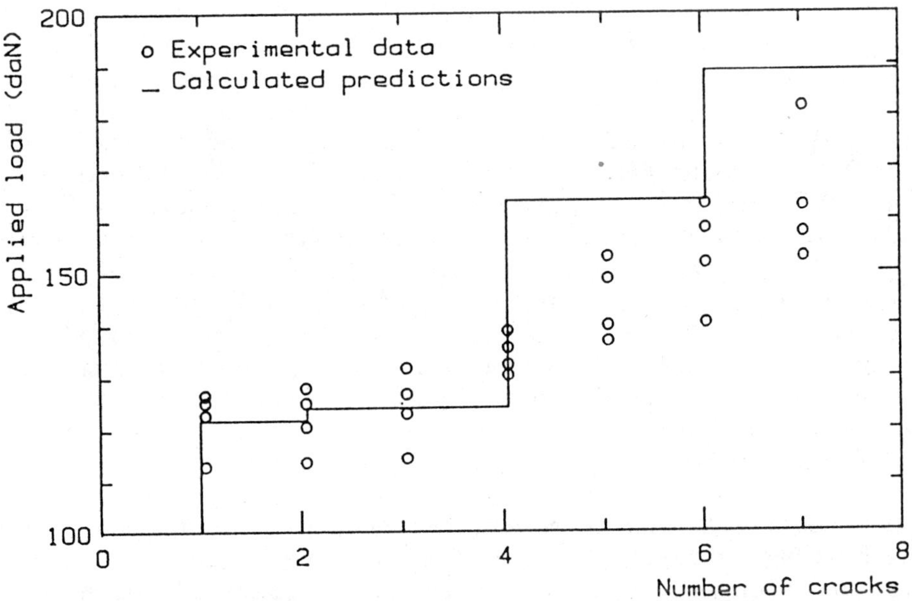

FIGURE 4 : Comparison of measured and predicted cracks initiations vs applied load.

Figure 4 compares the predicted loads associated to the creations of the first 7 cracks and the observed ones. From the 4th crack, the experimental loads are systematically lower than those obtained by calculation.

This effect can, however, be explained by the scatter in the strength of the ceramic coating, previously related by others authors /9/. For stresses lower than the average stress of the first crack creation, the probability of failure of the coating is not negligible, and, hence, cracking is apt to occur sooner during the loading, and, in addition, not exactly at the location of the maximum of the stress. The scatter observed for the experimental loads associated to crack initiations as well as for the cracks locations confirm this point. A statistical approach (based, for example, on the weakest link theory) would be necessary to take into account the discrepancy, in relation to the average values already produced by the calculation.

THERMAL SHOCKS

For evaluating the coating's resistance, Régie Renault used to perform thermal shock tests which are obtained by rapidly approaching an oxyacethylene torch to the coated end of cylindrical specimens.

The severity of the thermal shocks was evaluated using thermocouples instrumented specimens, for various torch-specimen distances and heating durations : during these tests, various types of crazing were observed. concerning the numerical simulation, different calculations were also performed on cylinders of 85 mm diameter and 10 mm length using axisymmetrical meshes. All these results are detailed in another paper /10/.

This study is only interested in a "reference thermal shock", which has been defined, from the experimental results, as the minimum one able to induce a cracking of the coating. Checking the invariability of the temperature and of the stress fields, the simulation was first simplified in order to reduce the calculation time : coarsening of the mesh (Figure 5) and of the time increments. Considering the limited difference of characteristics between the thin NiAl layer and the metal-ceramic composite one, only three materials were taken into account : metal, mixture in equal parts of NiAl and ZM and pure ZM. The cold end of the substrate was considered cooled by air at a constant temperature, whereas hot gases presenting a dome-shaped distribution of temperatures were supposed to heat the coating, all exchanges operating in accelerated convection conditions. Heat exchanges along the circumference were neglected. The thermal data used for the calculation are summarized in Table 1 /11/ and the boundary conditions in Figure 5.

The maximum tensile stress is always perpendicular to the spraying axis and, during the heating shock, the maximum of the stress in the ZM was found subsurfacic and along the symmetry axis of the specimen. Moreover, the shear stresses remain everywhere limited, allowing a cracking in tension of the coating to be supposed.

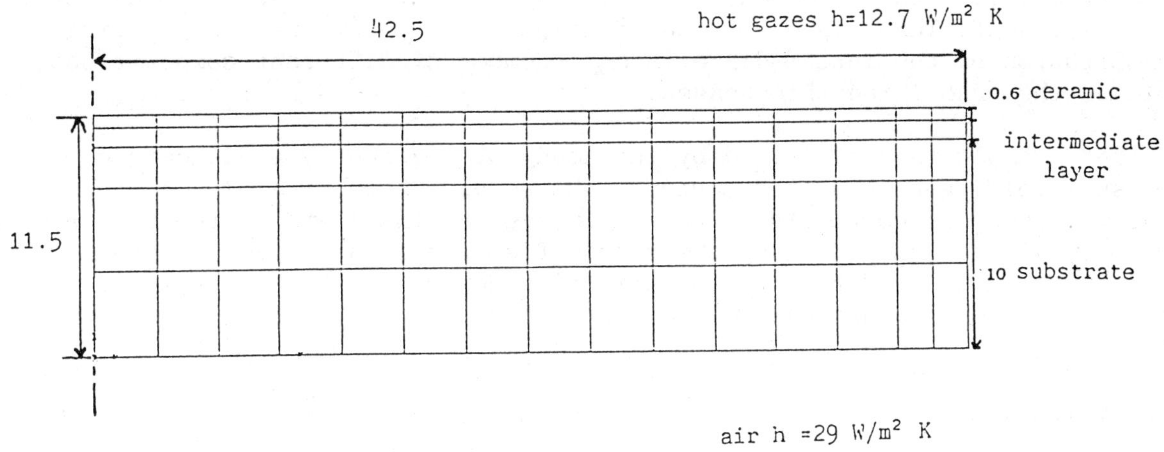

FIGURE 5 : Mesh used for the thermo-mechanical finite element calculation of the thermal shock (dimensions and thermal exchange conditions are mentioned).

Figure 6 represents the variation of the stress as a function of the depth, along the axis of the specimen. This example was obtained on a coating constituted of 0.8 mm mixture and 0.6 mm ZM. It is worth noting that both strengths of the mixture and of the ZM are reached approximately simultaneously in this case : the weakest layer is not systematically the ceramic one during a heating shock. The same thermal shock applied to a

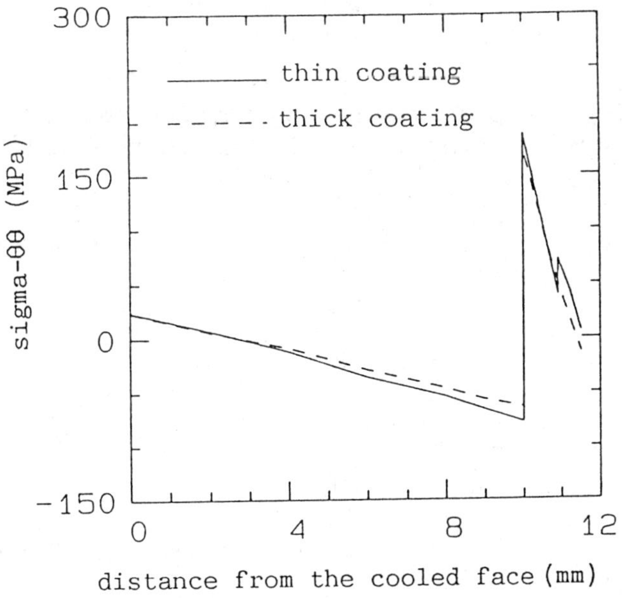

FIGURE 6 : Evolution of the hoop stress vs thickness along the symmetry axis of specimen.

thicker ZM coating (0.4 mm mixture and 1 mm ZM) is also represented in Figure 6 : it leads to an increase of the stress (15 %) in the ceramic coating itself. This type of calculation is useful for optimizing the constitution of the intermediate layer : number of different compositions, as well as the associated thicknesses.

Another example is constituted by the study of a geometry modification on the stress field. For example, various shapes of a bore in a piston crown were investigated. In particular, the influence of the slope of the bore and of the values of the radii on the stress field were estimated with respect to the geometrical conditions of a satisfying combustion : the obtained results are helpful for the designer of coated components.

CONCLUSION

The simulation of thermo-mechanical loading of coatings performed in this study gave results in good agreement with the experiments, although numerous approximations were used. A more accurate study would need a better knowledge of the materials : in particular, important problems remain concerning the anisotropy of the coating (from mechanical and thermal points of view) and the residual stresses. It would also be necessary to take into account the evolution of the calculation data during ageing, and to try to simulate the actual thermal and mechanical loading involved in an engine.

However, this work consisting in a first approach of the numerical simulation of ceramic coatings gives a proof of the interest of the calculation for designing the coating itself as well as the piston's crown.

ACKNOWLEDGEMENTS

This work has been supported by Régie Renault. The authors wish to thank Mr. Pouget for producing the plasma-sprayed samples and Mr. Gault of ENSCI for helpful discussions about the values of elasticity constants.

REFERENCES

/1/ COLIN C., BOUSSUGE M., VALENTIN D., DESPLANCHES G.
"Mechanical Testing of Plasma-sprayed Coatings of Ceramics", J. Mat. Sci., Vol. 23, 1988, pp. 2121-2128.

/2/ GAULT C., BOILEVIN S., DESPLANCHES G.
"Elastic Properties of Ceramic and Ceramic-metal Composite Thermal Barricr Coatings", Science of Ceramics 14, ed. D. TAYLOR, Canterbury, Sept. 7-9, 1987, pp. 389-394.

/3/ BERNDT C.C.
"Determination of Material Properties of Ceramic Coatings", Advances in Thermal Spraying, Montreal, Sept. 8-12, 1986, pp. 149-158.

/4/ MULLEN R.L., HENDRICKS R.C., Mc DONALD G.
"Finite Element Analysis of Residual Stress in Plasma-sprayed Ceramics", Cer. Eng. Sci. Proc., Vol. 6, 1985, pp. 871-879.

/5/ EATON H.E., NOVAK R.C.
"A Study of the Effects of Variations in Parameters on the Strength and Modulus of Plasma-sprayed Zirconia", Surf. Coat. Techn., Vol. 27, 1987, pp. 257-267.

/6/ HOBBS M.K., COOKE R.G., HARRIS B., REITER H.
"Surface Residual Stresses in Thermal Barrier Coatings", Brit. Cer. Soc. Proc., Vol. 39, 1987, pp. 119-132.

/7/ HOBBS M.K., REITER H.,
"Residual stresses in ZrO_2 - 8 % Y_2O_3 Plasma-sprayed Thermal Barrier Coatings, Surf. Coat. Techn., Vol. 34, 1988, pp. 33-42.

/8/ MULLEN R.L., HENDRICKS R.C., Mc. DONALD G.
"Interface Roughness Effect on Stresses in Ceramic Coatings", Cer. Eng. Sci. Proc., Vol. 8, 1987, pp. 559-571.

/9/ OSTOJIC P., BERNDT C.C.
"The Variability in Strength of Thermally Sprayed Coatings", Surf. Coat. Techn., Vol. 34, 1988, pp. 43-50.

/10/ PERRIN N., BURLET H., BOUSSUGE M., DESPLANCHES G.
"Testing and Modelling of Thermo-mechanical Resistance of Thermal Barriers", Plasma Spraying Conf., Paris, Nov. 14-15, 1988, proc. to be published.

/11/ DESPLANCHES G.,
"A Study of Plasma-sprayed Coatings for Use in Internal Combustion Alternative Engines", 1st Plasma Technology Symp., Lucern, May 18-20, 1988, Vol. 1, pp. 195-203.

CERAMIC COATINGS BY PLASMA FLAME SPRAY PROCESS ON DIESEL CYLINDER HEADS

Giovanni Aliprandi
Genoa University
Dept. Chemical Engineering
16145 Genoa   Italy

ABSTRACT

The purpose of this research was to verify the heat retention and fatigue resistance of diesel cylinder heads coated by plasma flame process with various thicknesses and qualities of ceramic materials.
After a series of tests on non-tempered secondary aluminium alloys,tests were carried out at a testing bench at 270°C for the wear resistance of tempered primary aluminium alloys.
First results were encouraging, though there is still a problem of the coating coming away after too few cycles.
A detailed report is made on results obtained, possible causes and conclusions reached after testing.

1) Introduction

The heads of automotive diesel and turbine engines made of aluminium are actually exposed to a stress which reaches the limits of endurance. The aim of this research was therefore to check on the possibilities of protecting these heads with a ceramic coating able both to resist repeated thermal shock as well as to operate as a thermal barrier.
Before this work was started, a specific method of evaluating the most important technological characteristics of ceramic coatings had been worked out in another laboratory (1) i.e.:
- tests on hot adhesive wear;
- tests for measuring the "thermal barrier" effect.

The results obtained from this first work , which are summarized further on, allowed trials to be carried out on the heads using those materials which had given the best performancces.

2) Results of the Previous Work (1)

Without going into details on the characterizing tests which were done in order to find the most suitable thickness for the coating and relative quality of ceramic materials as protection against wear and for thermal barrier, we shall report results obtained (plasma spray process was done with a Metco Gun type 7 MB; electrically driven at 80 Kw).
Table n.1 illustrates the materials under examination classified as:
- anti-stress coatings;
- coatings suitable for forming "thermal barriers".
The microstructural aspects of the various materials were studied by following the PWA (Pratt & Whitney Aircraft) specifications where provisions are made for controlling the presence of the various phases, porosity conditions, state of oxidation and mechanical type defects (cracks, flaking).Evaluation was made by comparing with micrographs used as references.
All the materials showed a distribution of the current phases, porosity and oxidation acceptable for comparison with PWA specifications. Generally speaking, there were no signs of flaking or cracking phenomenon in the interface of the metal base (with or without pre-coating) or in the coating itself, except for the W_4C coating which showed some signs of microcracking. As far as tests for hot wear were concerned, the damaging mechanism was attributed to detaching of particles in all cases, while there were no signs of deformation due to flaking.
Regarding quantities, the coating made of tungsten carbide offered a better resistance to wear than that of Cr_2O_3 and Al_2O_3, especially during the first part of the test. Furthermore, while the oxide coatings showed wear at a pratically constant rate ,in the case of tungsten carbide, this increased rapidly. This latter phenomenon is to be

Table 1. Coating Materials under Examination

Class of Utilisation	MATERIAL FOR COATING		THICKNESS		
	COATING	PRE-COATING	Pre-coatingmm	Coating mm	sign
ANTI-WEAR	Cr_2O_3	Ni + 4,5% Al	0,04	n.d.	1
	W_4C ; 12% Co	Ni + 4,5% Al	0,05	n.d.	2
	Al_2O_3 ; 13% TiO_2	Ni + 4,5% Al	0,07	n.d.	3
THERMAL BARRIER	Al_2O_3;(min.98,5%)	NO	-	0,20	4
		Ni+5,5% Al+5% Mo	0,15	0,35	5
	Al_2O_3; 25% kaolin	NO	-	0,15	6
	Al_2O_3; (min.94%)	Ni+5,5% Al+5% Mo	0,35	0,15	7
	Cr_2O_3	NO	-	0,20	8
	ZrO_2; 24% MgO	NO	-	0,15	9
	P.W.A. 253-1: ZrO_2; 24% MgO	80% Ni + 20% Cr 35% (80%Ni+20%Cr) 65% (ZrO_2+24% MgO)	0,18 0,09	0,18	10
	P.W.A. 253-3: ZrO_2; 24% MgO	95% Ni + 5% Al 35% (95%Ni+ 5%Al) 65% (ZrO_2 +24%Mg)	0,20 0,10	0,35	11

correlated with the presence of micro-cracks in the coating which facilitated the mechanism of wear by peeling. To sum up, the sample coated with Cr_2O_3 showed a performance which was half way between that of the W_4C and the Al_2O_3.

As far as measurements for the thermal barrier effect were concerned, tests were conducted (with a furnace temperature at 500°C and a minimum duration of 15 min.) on various coatings, including zirconia stabilized with magnesia. This material has already been used extensively for achieving thermal barriers, especially in combustion chambers of turbo reactors of aircraft.

The results confirmed that zirconia coating gave the best performance and it clearly showed the best thermal barrier effect according to the increase of the thickness of the coating. In other words, the zirconia coating with two layers of undercoating showed the same results as just one layer of magnesium zirconate with a thicker layer.

3) Experimental Part on Engine Heads

Based on the above mentioned experience, the most suitable coating was chosen for protecting engine heads made of aluminium.

A first head in primary aluminium alloy (AS9U1 i.e. with 9% silicon and 1% copper) was coated with 3 layers of ceramic material which had the following composition:

a) 80% Ni + 20% Cr	thickness: 0.1 mm	-layer underneath
b) 35% layer a)+ 65% layer c)	" : 0.15 "	-intermediate layer
c) 76% ZrO_2 + 24% MgO	" : 0.1 "	-outside layer

The head had been sanded beforehand with fine corundum; the coating was applied manually by Plasma Spray process with a Metco Gun, type 7MB, electrically driven at 80 Kw; distance of the support was about 10 cm.

The tests were carried out on the bridge between the valve seats according to the normal working conditions of thermal shock equipment. This equipment is essentially composed of:

- 1 metallic plate for anchoring the cylinder head, equipped with H_2O circulation which has the same functions as the engine block.
- 1 series of blowpipes for heating up the zone involved with a self recording of combustion gas flow and position.
- 1 series of nozzles for cooling the flow from the outside through spraying.
- A mechanical device equipped with a cylinder for moving the heating and cooling units.
- 1 network for feeding H_2O to the plate in order to cool the inside of the cylinder heads.

The equipment for measurements is made up of:

- Flowmeters for regulating the flow of combustion gas to each blowpipe.
- Flowmeters for regulating the flow of H_2O to the different cooling systems.
- Metallic plate incorporating a thermocouple to continuously record the temperature of the flame.
- Thermometer for measuring the temperature of the H_2O entering and coming out of the cylinder heads.
- Control panel.

26 seconds for heating : maximum temp. : 240°C
13 seconds for cooling : minimum temp. : 50°C.

After about two thousand cycles of thermal shock, the coating started flaking in the area impinged on by the flame; after 7200 cycles, there were no signs of cracking from one side to the other in the chamber (fig.1).

After this first result, the tests were repeated on three heads of secondary alloy (AS7Ȼ U3, i.e. 7% of silicon and 3% copper), with the aim of compensating for the increased cost of the alloy.

Each of the three heads was coated in a different way, viz.:

1) 0.1 mm thickness (Ni/Al);
2) 0.2 mm thickness (0.05 mm Ni/Al + 0.15 mm ZrO_2);
3) 0.3 mm thickness (0.05 mm Ni/Cr + 0.15 mm intermed. + 0.1 mm ZrO_2).

The thermal shock equipment was gauged at 270°C (30°C more) and each cycle lasted 36 seconds (23 secs. for heating and 13

secs. for cooling: i.e. with 3 secs. less for heating as compared to the previous test).
Before performing the tests on these heads, a comparative test was carried out on a head with identical characteristics, but without coatings: this cracked after 2000 cycles.
A third test was carried out on 4 heads in primary aluminium alloy (AS9U1) which is able to resist up to 7000 cycles without coating.
The four heads were coated in the same way, but the area around the four combustion chambers was coated in a different way. Viz.:

chamber 1 : 0.1mm Ni/Cr+0.1mm intermed.+0.1mm ZrO_2(tot.0.3)
chamber 2 : 0.1mm Ni/Al+ " " +0.1mm ZrO_2(tot.0.3)
chamber 3 : 0.1mm + " " +0.2mm ZrO_2(tot.0.4)
chamber 4 : 0.1mm + " " +0.3mm ZrO_2(tot.0.5)

Test conditions were exactly the same as for the previous tests.
The results were as follows: (figs. 2 3):

c.c. 1 : at 1370 cycles the coating showed signs of wear but, even at 5550 cycles there were no cracks.
c.c. 2 : at 1370 cycles, the surface was regular, at 3020 cycles there were signs of peeling; cracks appeared at 5550 cycles.
c.c. 3 : at 1370 cycles the surface was regular; at 3020 cycles, partial peeling: it appeared to be the chamber in the best condition; cracks appeared at 5550 cycles.
c.c. 4 : at 1370 cycles there was peeling; at 5550 cycles the peeling was worse but there were no cracks.

A series of measurements were then carried out to check if the coating had changed the microstructure of the support during the act of deposition.
Tests for Brinnell hardness near and far from the coating gave an identical value (110 HB).
Similar results were shown with tests for microhardness (Vickers HV 100 g) starting from the coated surface and gradually working away from it :128-130-124-126-121-131-124-128.

These measurements demonstrated that the projection by plasma on the aluminium support of a ceramic coating does not cause any variation in the microstructure of the metallic support.

4) Discussion on Results

The tests performed and results achieved are summarized in table n.2 and actually appear to be contradictory.
From an examination of these results we can derive the following data:

- the coating which showed by far the best results was the first one in chronological order, i.e. the one made of 3 layers, the first one in Ni/Cr and the third one of 0.1 mm of magnesium zirconate. The second test result was in Ni/Al as first layer, but with 0.3 mm of magnesium zirconate as the third layer.
- There was a noticeable difference in the behaviour of the two aluminium alloys. It is well known that the contemporary presence of copper and silicon is damaging as far as thermal shock is concerned: the stabilized secondary alloy AS7U3 is much less resistant than the hardened and temporary primary alloy AS9U1 (2000 cycles against about 7000 without coating).
  The first tests on the coatings seem to indicate that the duration is proportionally improved more in the secondary alloy than in the primary one (in fact it was possible to reach up to about 7000 cycles, the same as for non-coated primary alloy).
- A variation of 30°C in the test temperatures (from 240 to 270°C) creates greater wear.
- Peeling from wear of the coating will cause some particles of the ceramic material to pass into the ducts creating problems.

5) Conclusions

An analysis of test results lead to the following conclusions:

- the first tests on ceramic coating for heads of diesel

Table. 2. Summary of Tests and Results

Period	Type Alloy	Nr. test	TYPE OF COATING			
6/86	AS9U1 (primary)	1	Ni/Cr (0.1)+int.(0.15) + + ZrO2(0.1) Tot. 0.35 mm.	26+ 13	240/50	2000 cycles beginning peeling 7000 " non-passing crack
6/87	AS7U3 (secondary)	1	Ni/Al 0.1 mm	23+ 13	270/50	1500 cycles beginning peel 2400 " cracks(2chambers)
		1	Ni/Al(0.05)+ZrO2(0.1)=0.15	id.	"	1600 " beginning peel 3000 " cracks(1 chamber)
		1	Ni/Cr(0.05)+int.(0.15)+ ZrO2(0.1) Tot. 0.30 mm.	id.	"	Tests not reliable due to continuous halts.
12/87	AS9U1 (primary)	4	1°sam.0.1Ni/Cr+0.1 int.+ ZrO2(0.1) Tot. 0.3 mm.	23+ 13	270/50	1370 cycles:coating worn 5500 " wearing continues but no cracks
			2°sam.Ni/Al(0.1)+int.(0.1)+ ZrO2 (0.1) Tot. 0.3 mm.	id.	"	1370 cycles:regular;3020 beginning peeling; 5550 cycles passing crack
			3°sam.Ni/Al(0.1)+int(0.1)+ ZrO2 (0.2) Tot. 0.4 mm.	id.	"	1370 cycles:regular;3020 slight beg;5550 cyc.passing crack
			4°sam.Ni/Al(0.1)+int(0.1)+ ZrO2 (0.3) Tot. 0.5 mm.	id.	"	1370 cycles:beginning;5550 cyc.lot peeling

1

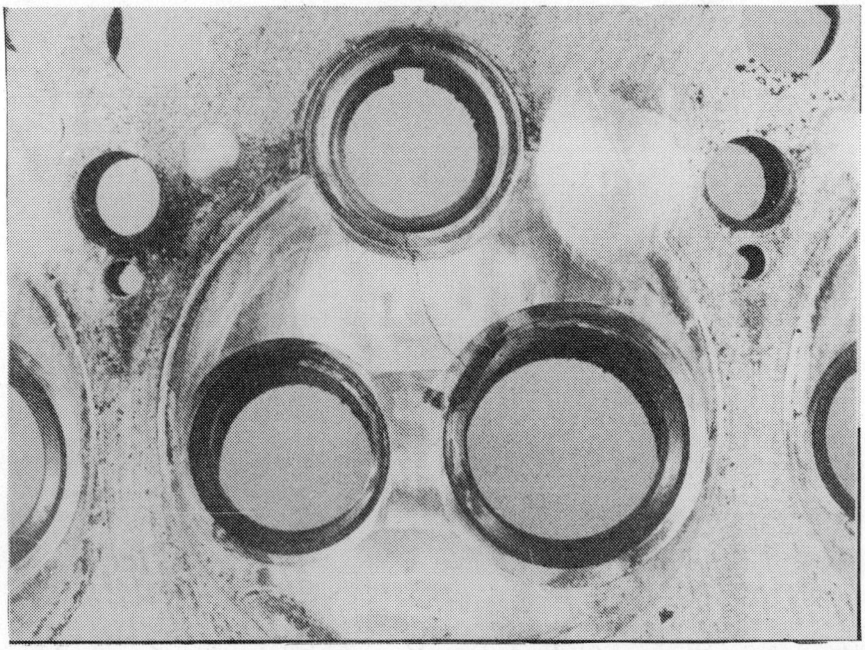

2

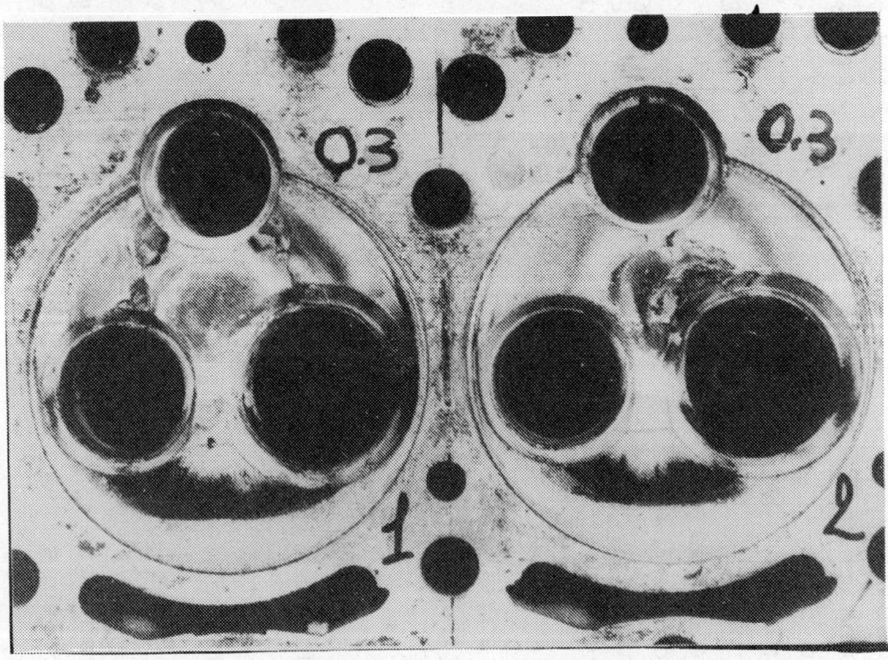

3

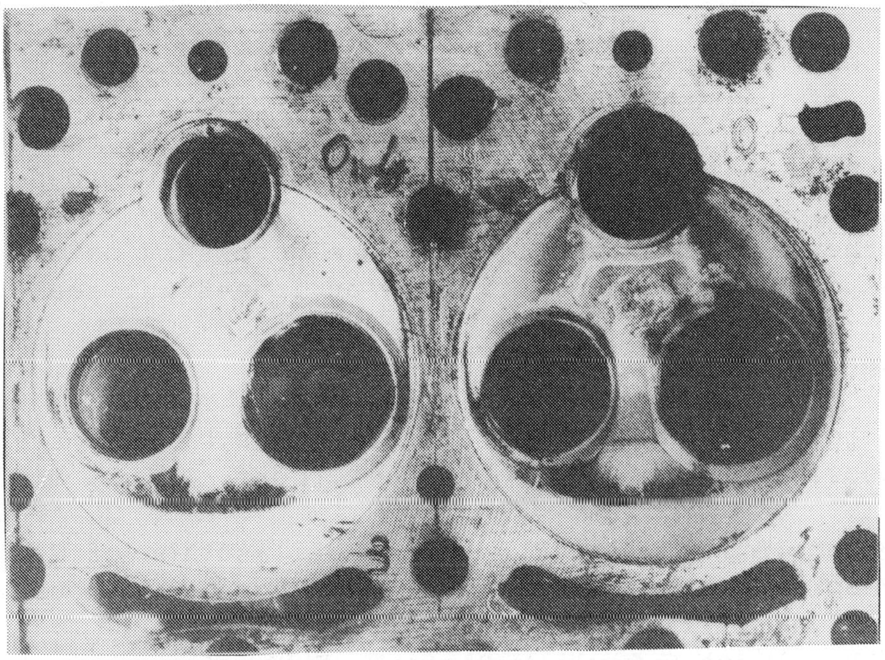

engines made of aluminium indicated that the ceramic material affords protection: it is still not clear if the life also depends on the type of aluminium alloy.

- Coating wear, after having been subjected to rapid thermal cycles, happens because of its own progressive peeling. In order to delay wear, it is necessary to concentrate on working conditions of the plasma coating, apart from the quality of the material, particularly regarding the different types of preparation of the support (sandblasting).
- The method of coating with three layers seems to be a valid one, however quality and thickness still have to be decided upon.

Many other experiments are therefore necessary before being able to offer a valid solution on an industrial scale.

REFERENCES

1. G.Aliprandi, G.Songini, "Ceramic Coatings by Plasma Process and Description of some Properties", Conf. A.I.M. Milan,Nov. 1985.
2. G.Aliprandi,A.Barbolini,P.P.Lauro,"Utilisation of Ceramic Plasma Spray Coatings in Aluminium Foundries". Industrial Ceramics, 7,4,215-216, 1987.
3. P.Boch,D.Fargeot.C.Gault,F.Platon"Int.Mag.High Temp/Refr.",18
4. R.McPherson, "Thin Solid Films", 83 (1981),297,310.
5. W.J.Gagnet, "Flame Spray Coating for Wear", Soc.of Autom.Eng., mid-year Meeting, Chicago, 19-23 May,1969
6. G.Wahl,LSingheiser,W.Thiele, "Thin solid films", 95,35-45 (1982)
7. G.W.Goward,"Materials and Coatings for Gas Turbine Hot Section Components" Proc.of Gas Turbine Mat.Conf.,Naval Ship Eng. Center, Washington,D.C.,1972.
8. A.R.Nicoll,"A Survey of Methods used for the Performance Evaluation of High Temp.Coatings",E.Lang ed.,Petten,Applied Science pubblishers.
9. I.A.Fisher,"Variables Influencing the Characteristics of Plasma Sprayed Coatings", Int.Met.Rev.,17,117-129,1972
10. R.A.Miller,"Oxidation-based Model for Thermal Barrier Coating Life",J.Am.Cer.Soc.,67,8,517-521, 1984.

INFLUENCE OF DIESEL ENGINE COMBUSTION ON THE RUPTURE STRENGTH OF PARTIALLY STABILIZED ZIRCONIA*

C. R. Brinkman, G. M. Begun, O. B. Cavin, B. E. Foster, R. L. Graves, W. K. Kahl, K. C. Liu, and W. A. Simpson

Oak Ridge National Laboratory
Oak Ridge, Tennessee 37831-6154

ABSTRACT

Results are reported for tests on partially stabilized zirconia (PSZ-TS and MS grade) bars exposed in the combustion chambers of two operating single-cylinder 0.825-L diesel engines. Specifics of test conditions and procedures are presented. Subsequent to exposure, the bars were subjected to four-point bending and the rupture strengths determined. The TS grade showed a decrease in average strength of 32%, while the strength of the MS grade decreased by about 9% in comparison to average behavior of unaged material. Results of X-ray diffraction analysis, Raman spectroscopy, isothermal aging studies, and ceramographic characterization are given to define reasons for material degradation.

INTRODUCTION

Engine exposure of commercially available and developmental ceramic materials in diesel engines was initiated at Oak Ridge National Laboratory (ORNL) in 1985.[1] The objective of this effort was to determine whether long-term exposure to the engine combustion environment would generate a change in mechanical properties. This was accomplished by exposing candidate structural ceramics to actual engine operating conditions and then examining them for changes by various nondestructive and destructive methods. Two types of partially stabilized zirconia (PSZ), Nilcra PSZ-TS and PSZ-MS, were selected for this study since PSZ has received considerable attention as a candidate material for various hot-section components in diesel engines.[2,3]

---

*Research sponsored by the U.S. Department of Energy, Assistant Secretary for Conservation and Renewable Energy, Office of Transportation Systems, and the Division of Materials Sciences, Office of Basic Energy Sciences, under contract DE-AC05-84OR21400 with Martin Marietta Energy Systems, Inc.

This is because PSZ material has relatively high strength, low thermal conductivity, good toughness and fair thermal shock resistance in comparison to many other ceramic materials.[4,5] However, there is some concern about the possibility of strength degradation mechanisms, such as aging and slow crack growth following prolonged periods of exposure to both temperature and stress at relatively low temperatures which can reduce both strength and toughness.[3,4,6] This material was chosen for study since exposure to the actual operating conditions found within the combustion chamber of our diesel engines was expected to produce both cyclic stresses and temperatures above 500°C but below about 900°C where substantial overaging is known to occur.

MATERIALS AND SPECIMEN PREPARATION

Two commercial grades of PSZ, MgO-PSZ-TS (thermal shock resistant) and MgO-PSZ-MS (maximum strength), were obtained from Nilcra Ceramics (USA) in the form of rectangular plates (101.6 x 101.6 x 6.35 mm). Rectangular modulus of rupture (MOR) specimen bars (50 x 4.5 x 3.5 mm) were subsequently cut from these plates. All grinding and polishing of the bars and chamfering of edges were performed in a longitudinal direction to a surface finish of about 0.25 μm. Both grades had an average grain size in the range of 30-50 μm. Specimens were measured and subjected to nondestructive examination, as will be discussed in a subsequent section, prior to testing.

Some uncertainty existed at the onset of this effort as to surface and internal temperatures attained in the bars during engine operation; for this reason some bars of each type were isothermally aged for 100 h in air at either 700, 750, or 800°C prior to rupture testing. This temperature range of 700 to 800°C was probably somewhat lower than peak surface temperatures (depending upon location), but was felt to be a realistic estimate of peak bulk temperatures. All subsequent rupture testing was performed at room temperature in air using a four-point bending fixture with an outer span of 40 mm and an inner span of 20 mm. Only specimens in the as-fabricated condition (non-aged) were subjected to exposure within the engines.

ENGINE TESTING RIG AND OPERATION

The earliest engine used for exposure testing was a single-cylinder, air-cooled, indirect-injection diesel (Deutz F1L511W) hereafter referred to as engine 1. This engine was a small, commercially available driver with 0.825-L displacement with a spherical prechamber. Access to this prechamber was made via the unused glow plug port (adjacent to the fuel injector) and via a machined port for pressure transducer instrumentation (Fig. 1). During this effort another otherwise identical two-cylinder version of the same engine was put into operation to double the exposure capabilities, and hereafter it is referred to as engine 2.

Exposure was conducted by "gripping" a pair of MOR bars side-by-side (separated by a thin shim) in a retainer. The retainer was then screwed into the instrumentation access port, as shown in Fig. 1.

Thus, two bars were exposed simultaneously to similar engine operating conditions. Size constraints of the prechamber permitted only 40 to 41 mm of the bar length to be cantilevered into the chamber. Orientation of the bars was such that direct impingement of burning fuel spray occurred.

Exposure times of 100 h were accumulated over several weeks of running several hours daily. Interruptions in exposure also occurred with random corrective maintenance on the engine/dynamometer tandem. The engines were run at a nominal constant speed of 1800 rpm and load of 41 to 43 $N\text{-}m_f$ torque on the two cylinders. Fuel consumption at these conditions was about 213 g/kWh, or a fuel energy input of about 21,480 W. Load was absorbed by a small eddy-current dynamometer.

TEMPERATURE MEASUREMENTS

The principal justification for performing the exposure test series centered around a belief that the thermal and gaseous environment of a combustion chamber cannot be accurately simulated in a laboratory oven. Clearly, one difficulty arises from the different time scale on which the specimens were being thermally cycled, i.e., many times per second from engine firing to daily start-up/shutdown operations. Also, the composition of combustion products might vary with time.

The surface and bulk material temperatures in the MOR bars were measured to allow assessment of any cause and effect relationship between temperature and strength degradation. Two individual methods were used to monitor bar temperatures during engine operation: passive maximum temperature (surface) measurements and internal thermocouples.

Templug Temperatures

The range of exposure temperatures was determined with Templugs.* Stainless steel mock-ups of two standard MOR bars were fabricated and drilled/tapped for surface mounting a Templug array. Templugs are small, special-alloy steel-threaded plugs that have well-defined thermal tempering curves. Hence, they can be used for peak temperature measurement. Installation of the 1.6-mm-diam plugs was accomplished over the length and periphery of each bar utilizing two different alloy grades to cover expected higher temperatures (550-852°C) at the exposed tip and lower temperatures (62-693°C) closer to the gripped end. Measurements made in this way indicated that temperatures exceeded the capacity of Templugs in several locations and were lower than the minimum discernable temperature elsewhere. Thus, it was concluded that temperature gradients existed along the bars during exposure and that surface temperatures exceeded 800°C.

---

*Available from Testing Engineers, Inc., P.O. Box 24075, 2811 Adeline St., Oakland, CA 94623.

Internal Thermocouple Temperatures

In an effort to measure internal bulk material temperatures, three PSZ-TS bars were center drilled (long axis) to depths of 15, 30, and 45 mm to accommodate internal thermocouples or optic pyrometer sensors. Fittings were fabricated for the retainer access port to permit the thermocouples to be inserted and sealed into the drilled bar cavities. Eccentricity of the holes with respect to the retainer's center axis was compensated for by a specially fabricated centering plug assembly. A 1-mm-diam, type "K" (chromel/alumel) thermocouple was used in all tests. For engine testing, the bars were allowed to reach a steady-state temperature over 1 h of running time at nominal load and speed. There was a significant temperature rise at start-up and a less severe cool-down at shutoff. The maximum temperature in the 45-mm-deep hole was 720°C after approximately 180 s of engine operation. The steady-state temperature was about 700°C after about 1 h. These same measurements were repeated on bars with the 30- and 15-mm drilled holes, and results are summarized in Table 1. The results show that a substantial temperature gradient existed along the longitudinal axis of the bars during exposure in the engine.

Table 1. Temperature measurements within three bars during engine operation

Hole depth (mm)	Engine			
	Maximum		Steady state	
	Temperature (°C)	Time (s)	Temperature (°C)	Time (h)
45	720	180	700	1
30	675	120	567	1
15	485	1,800	477	1

NONDESTRUCTIVE EVALUATION

Ultrasonic Examination

Four distinct ultrasonic techniques (all using pulse-echo ultrasonics) were applied to each bar. Three of the techniques monitored for flaw detection using high-frequency ultrasonics (50 to 75 MHz); whereas the fourth basically signatured the microstructure of each bar. Several of the bars contained demonstrable flaws, as detected by either volumetric or surface wave tests. However, the majority of the samples yielded no indications for any of the flaw tests. Although the size of the detected flaws was not determined, the conditions of the test were such that flaws having diameters of about 75 μm and larger could be detected.

Four of the PSZ-TS bars (engines 71, 81, 95, and 89) that were exposed to the combustion environment of the diesel engine were also subjected to ultrasonic evaluation following exposure. No additional indications were seen in the flaw tests, and the previously detected indications had not changed in size.

Radiographic Examination

The samples were radiographed in two orientations, with the radiation beam passing through the 3.5- and 4.5-mm-thickness. About ten irregularly shaped voids with a maximum dimension of 0.25 to 1 mm (0.010 to 0.040 in.) were detected in the samples (less than 10 total).

FLEXURAL STRENGTH

Results of flexural bar tests are given in Table 2 in the form of average strengths for each material condition. Also included are standard deviations, Weibull modulus, and "y" intercept values as calculated by both least squares and maximum likelihood methods of analysis. Individual strength values are plotted in the form of Weibull plots in Figs. 2 and 3 for engine exposed material. Figure 2 compares strength values of 12 PSZ-TS bars exposed in both engine 1 and 2 with results of 53 unexposed bars. Average strength values for the exposed material decreased by 32% in comparison to the unexposed material and the scatter in MOR values for the exposed material increased significantly, as can be seen in Fig. 2, and by modulus values in Table 2. An examination of the strengths of engine exposed TS bars indicated a tendency to cluster in that similar strength values were obtained for individual bars exposed as a pair. Figure 2 shows that bars exposed as pairs, e.g., 71 and 81 or 95 and 89, had strength values in the same half of the scatter band. This suggests that despite attempts to duplicate exposure conditions in terms of exposure time and engine operating conditions, differences occurred from one period to the next. Table 2 shows that isothermal aging decreased the average strength 7 to 10% for the TS grade material, but not to a level as low as that seen after engine exposure.

Figure 3 is a Weibull plot for the PSZ-MS data, which compares engine exposed strength values with results for as-fabricated specimens. Table 2 shows that engine exposure resulted in only a 9% decrease in strength and that there is not as much scatter in the data in comparison to that shown in Fig. 2 for engine-exposed TS grade material. Specimen numbers for individual bars are included in Fig. 3 to facilitate comparison of strength values. Test results appear to be inconclusive concerning the tendency for similar strength values for a given pair for this material. Note from Table 2 that prior thermal aging decreased the average strength of the MS grade material by 6 to 8% which was slightly less than the strength degradation in the engine-exposed material.

In an effort to isolate some of the surface degradation by chemical interaction from bulk specimen changes, for example, phase changes during engine exposure, two of the PSZ-TS bars had approximately 0.13-mm thickness of material removed by grinding and polishing from the width surfaces. This operation was done after engine exposure and prior to rupture testing and effectively restored the original surface finish. Average strength values for these two tests (Table 2) were 22% lower than that of the unexposed material, compared to a decrease of 32% for engine-exposed material. These results are only tentative due to the limited data available, but suggest bulk rather than extensive surface degradation.

Table 1 indicates that a substantial temperature gradient existed along the longitudinal axis of the specimen during engine exposure.

Table 2. Results of Tests Conducted on MOR Bars After Exposure to Several Conditions.[a]

No. of observ-ations	Material condition	Weibull modulus		y - intercept		Average strength (MPa)	Standard deviation (MPa)	Reduction in average strength[b] (%)
		Linear regression	Maximum likelihood	Linear regression	Maximum likelihood			
	PSZ-TS MOR bars							
53	as-fabricated	31.7	48.6	-202.5	-311.2	587.3	20.3	0
14	engine exposed for 100 h	3.5	4.2	-21.6	-25.4	401.8	120.5	32
23	preaged 100 h at 700°C	37.0	40.9	-233.9	-258.4	547.7	17.8	7
20	preaged 100 h at 750°C	15.1	14.7	-95.5	-93.2	539.1	42.0	8
20	preaged 100 h at 800°C	57.2	53.7	-359.2	-337.1	528.4	11.3	10
2	engine exposed for 100 h/ ground and polished	-	-	-	-	459.5	60.3	22
	PSZ-MS MOR bars							
40	as-fabricated	29.2	29.8	-191.1	-195.1	683.0	28.5	0
20	engine exposed for 100 h	19.8	22.0	-124.7	-141.9	625.0	38.7	9
20	preaged 100 h at 700°C	16.4	28.2	-106.8	-182.7	641.6	40.1	6
19	preaged 100 h at 750°C	19.3	24.7	-124.7	-159.6	629.4	36.5	8
20	preaged 100 h at 800°C	30.0	39.4	-194.1	-254.8	635.2	24.3	7

[a]Tests were conducted with a cross head speed of 8.47×10^{-3} mm/s or a strain rate of 1.1×10^{-4} s^{-1}.

[b]Reduction in strength in comparison to as-fabricated material.

Therefore, following the flexural tests, measurements were made on each MOR bar to determine the distance from the cold (gripped) end to the point of fracture. In the case of the TS grade, most fractures occurred about 12 to 15 mm from the cold end for specimens exposed in engine 2, and 33 to 35 mm from the cold end for specimens exposed in engine 1 (see Fig. 4); while in the case of the MS grade, fractures generally occurred in the central sections of the bar some 22 to 31 mm from the cold end. Failure locations of the aged and unexposed materials (both grades) tended to be at random locations between the two inner bearings of the loading fixture.

SURFACE EXAMINATION AND FRACTOGRAPHY

Following engine exposure, specimens were examined using optical and scanning electron microscopy (SEM) techniques. Figure 5 is an SEM photograph of the tensile surface of a PSZ-TS bar tested in the unexposed condition. Longitudinal grinding marks and pits where sintering voids intersected the surface are apparent. Figures 6 and 7 are representative photomicrographs of surfaces of specimens exposed in the engines. Specimens exposed in engine 1, i.e., PSZ-TS 71, 81, 95, and 89 were coated with an orange-to-black coating which varied in shade across an individual bar and also showed marked specimen-to-specimen variations. In some instances, e.g., specimen 71, this very adherent film had a sandpaper-like texture as shown in Fig. 6. The particulate matter shown in Fig. 6 appeared to be fused to the surface in that it could not be easily removed. Energy dispersive X-ray (EDX) analysis of this deposit showed it to contain increased concentration of iron, zinc, calcium, magnesium, and phosphorus relative to the fracture and ground surfaces of unexposed specimens. These contaminants may have been caused by either the fuel or engine oil.

Specimens exposed in engine 2 were generally found to be coated with a black nonadherent soot (carbon) or powder, which was readily removed by wiping. Typical surfaces near the fracture zone of specimens exposed in engine 2 are shown in Fig. 7. Figure 7 also shows representative sintering voids in the fracture surface of the specimen.

Optical and SEM fractography did not reveal any clearly defined internal fracture origins. Possibly surface pits and surface particulate matter (specimens exposed in engine 1) were the fracture initiation sites in most cases.

X-RAY DIFFRACTION ANALYSIS

Representative bars from each condition studied in this program were subjected to X-ray diffraction following four-point-bend rupture to measure the monoclinic phase content. Measurements were taken as near to the fracture surface as practical. Rupture strength of the individual bars is plotted as a function of monoclinic phase content for both the MS and TS grades in Fig. 8. Although considerable scatter in the data is apparent, there is a clear trend of decreasing strength with increasing monoclinic phase content. Note also that the greatest increase in this phase occurred in the TS material.

RAMAN SPECTROSCOPY ANALYSIS

Many of the bars analyzed by X-ray diffraction were also analyzed for apparent* monoclinic phase content by the techniques of Raman spectroscopy.[8,9] Measurements were taken normal to the fracture surfaces. Rupture strength is plotted as a function of apparent monoclinic content in Fig. 9. The results of Raman scattering studies confirm the results of X-ray diffraction and show a decrease in strength with increasing monoclinic phase content.

Measurements were also made normal to the major axis and in multiple locations along several of the specimens in the as-machined and engine-exposed condition prior to flexural loading. The surfaces were scanned and as-machined MS grade contained about 9%, as-machined TS grade 18 to 24%, and engine-exposed TS grade 15 to about 40% apparent monoclinic phase (Fig. 10). However, in the engine-exposed TS grade specimens most of the readings indicated an apparent monoclinic content of 15 to 26% with peak values of 31 to about 40% in the area of 13 to 17 mm from the cold end of the bars (Fig. 10). This is the area, as was previously indicated (Fig. 4) for bars exposed in engine 2, where failure occurred when the bars were subjected to four-point rupture loading. Therefore, it is suggested that the high-localized monoclinic phase content in this region was due to cyclic stresses produced in the cantilevered fixed bar or to some other engine unique stress riser condition that occurred during engine operation which subsequently induced failure in this area. Engine unique stress risers operating on the bars exposed in engines 1 and 2, seem to be a more probable cause since the postexposure failure sites were different as shown in Fig. 4.

SUMMARY AND CONCLUSIONS

Observations reported here demonstrate that PSZ-TS and PSZ-MS in the form of rectangular ceramic bars can be successfully exposed to the combustion chamber and environments of small operating diesel engines for periods of up to 100 h. The environment within the chamber was complex in terms of the chemical species present, it included water vapor which is known to degrade the strength of this material,[7] and may have contributed to some subtle flaw growth during engine operation. During post engine exposure MOR testing, bar fractures were noted to occur between 15 and 35 mm from the end of the bar gripped during engine exposure (Fig. 4), indicating possible differences in exposure conditions. From the Templug and results of thermocouple measurements given in Table 1, it was concluded that surface temperatures in bar rupture zones (engine 1) could have reached about 800°C which was high enough to degrade strength. Cyclic stresses were also probably high enough to produce strength degrading phase changes as indicated in Fig. 10.

---

*Percent monoclinic assuming sample to be 100% monoclinic plus tetragonal.

Two engines were used in this study. They were similar in most ways and were run to identical operating specifications, but there was some difference noted in the surface coatings of bars exposed in engine 1 in comparison to engine 2. This difference in coating produced in the case of the PSZ-TS probably also contributed to strength degradation and variability as well. However, it was believed that the primary degradation mechanism responsible for strength reduction was the localized increase in monoclinic phase content.

The principal conclusions concerning the effect of 100 h diesel engine combustion environment on partially stabilized zirconia ceramics are as follows:

1. Exposure of PSZ-TS caused a substantial decrease (32%) in mean flexural strength. This decrease was attributed primarily to an increase in monoclinic phase content as a result of exposure. Surface deposits and thermally induced or environmentally assisted microcrack growth may also have been important.

2. Exposure of PSZ-MS caused only a 9% decrease in mean flexural strength which was also attributed to an increase in monoclinic phase content due to exposure.

3. Based on retention of room temperature flexural strength, PSZ-MS is preferred to PSZ-TS for use in critical components of diesel engines.

Future efforts will be aimed at refinement of temperature measurements, a stress analysis of the exposed bar during engine operation, gas analysis of the combustion products and exposure of different materials and ceramic brazed joints to the engine environment.

ACKNOWLEDGEMENTS

The authors are grateful to E. L. Biddle for engine operation, R. H. Baldwin for conducting the MOR tests, and to M. K. Ferber for use of his software in statistical analysis of the data. The authors also wish to acknowledge D. R. Johnson and J. I. Federer for their many helpful suggestions in a technical review of the manuscript. Special thanks is also due to Billie Russell for her work in preparing the manuscript.

REFERENCES

1. R. L. Graves, W. K. Kahl, and E. L. Long, Jr., "Testing of Developmental Materials in Diesel Engine Combustion Chambers," The American Society of Mechanical Engineers, 345 E. 47 St., New York, NY, 10017, Paper No. 87-ICE-28.

2. R. Kamo and W. Bryzik, "Cummins/TACOM Advanced Adiabatic Engine," pp. 121-134, in *Proceedings of the Twenty-First Technology Development Contractors' Coordination Meeting P-138*, Society of Automotive Engineers, Warrendale, PA, March 1984.

3. M. K. Ferber, "Characterization of the Mechanisms Responsible for Time-Dependent Strength Variations in Partially Stabilized Zirconia," Ceramic Materials and Components for Engines (Second International Symposium), W. Bunk, H. Hausner, eds., pp. 583-592, Deutsche Keramische Gesellschaft e. V.: D-5340 Bad Honnel (1986).

4. D. C. Larsen, J. W. Adams, L. R. Johnson, A.P.S. Teotia, and L. G. Hill, Ceramic Materials for Advanced Heat Engines Technical and Economic Evaluation, Noyes Publications, Park Ridge, N. J., 1985, pp. 249-282.

5. W. J. Lackey, D. P. Stinton, G. A. Cerny, L. L. Fehrenbacher, and A. C. Schaffhauser, "Ceramic Coatings for Heat Engines Materials – Status and Future Needs," *Proceedings of International Symposium on Ceramic Components for Heat Engines*, Oct. 17-21, 1983, Hakone, Japan.

6. J. J. Swab, "Evaluation of Commercially Available Transformation Toughened Zirconias," Advanced Structural Ceramics, P. F. Becher, M. V. Swain, and Sk. Somiya, eds., Materials Research Society Symposia Proceedings, Vol. 78, pp. 137-146 (1987).

7. N. L. Hecht, "Environmental Effects in Toughened Ceramics," pp. 251-289, in *Ceramic Technology for Advanced Heat Engines Project Semiannual Progress Report for April through September 1986*, Oak Ridge National Laboratory, ORNL/TM-10308, March 1987.

8. V. G. Keramidas and W. B. White, "Raman Scattering of the Crystallization and Phase Transformations of ZrO_2," *J. Am. Ceram. Soc., in* 57, No.(22) 22-24 (1974).

9. D. R. Clarke and F. Adar, "Measurement of the Crystallographically Transformed Zone Produced by Fracture in Ceramics Containing Tetragonal Zirconia," *J. Am. Ceram. Soc.*, 65, 284-288 (1982).

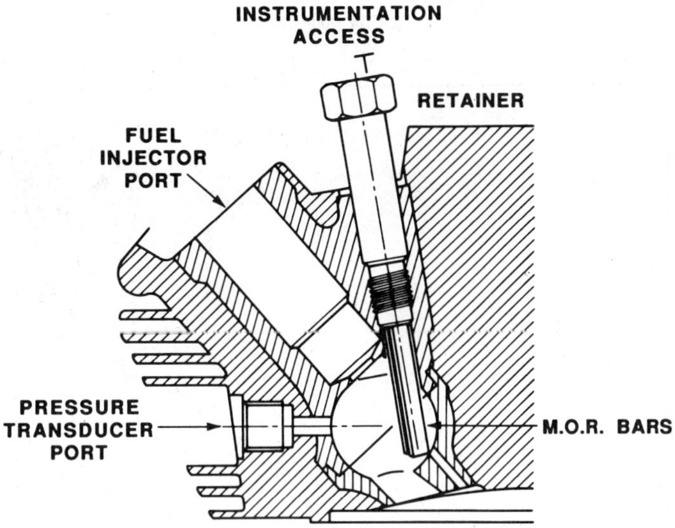

Fig. 1. Cross Sectional View of Deutz Cylinder Head Shows Bars' Position With Respect to Fuel Injector.

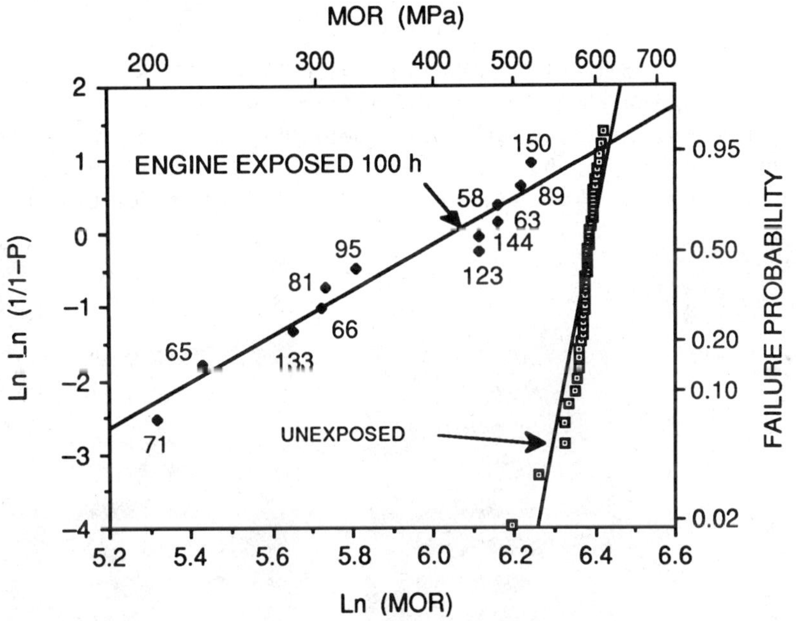

Fig. 2. Weibull Plot Comparing Modulus Of Rupture (MOR) Strength Data for Nilcra PSZ-TS Exposed to Diesel Engine Combustion Conditions and Unexposed Material. Numbers Indicate Bar Numbers and Specimens Exposed Together as Pairs Were as Follows: Engine 1, 71-81 and 95-89; Engine 2, 133-123, 144-150, 65-66, and 63-58. Specimens 151 and 152 Exposed in Engine 2 were Omitted for Clarity.

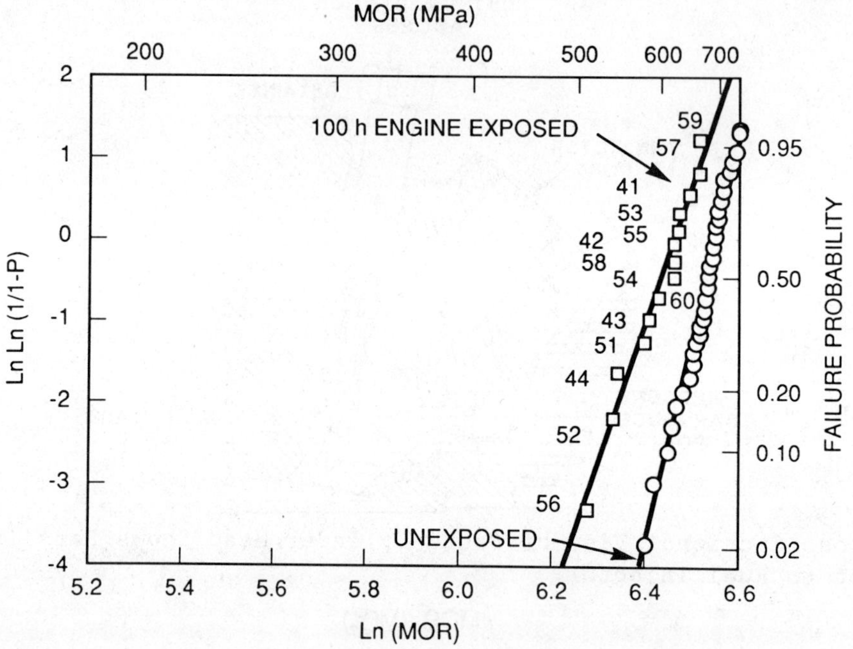

Fig. 3. Weibull Plot Comparing MOR Strength Data for Nilcra PSZ-MS Exposed to Diesel Engine Combustion Conditions with Unexposed Material. Numbers Indicating Bar Numbers and Specimens Exposed Together as Pairs Were as Follows: Engine 2, 51-52, 53-54, 55-56, 57-58, 59-60, 41-42, and 43-44. Specimens 45, 46, 47, 48, 49, and 50 Were Omitted for Clarity.

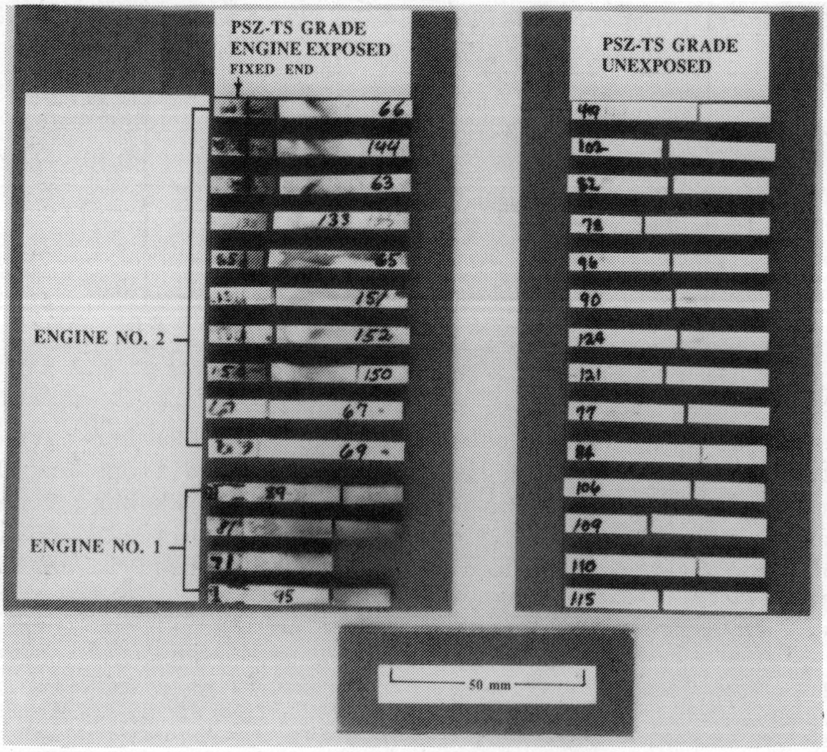

Fig. 4. Comparison of PSZ-TS MOR Bars Tested in the Exposed and Unexposed Condition.

Fig. 5. As-machined Surface of PSZ-TS Grade, Specimen 118, Flexural Tested in Unexposed Condition.

Fig. 6. As-exposed Surface of Specimen 71 of PSZ-TS Grade After 100 h in Diesel Engine 1's Combustion Chamber. Note Gritty Surface Condition.

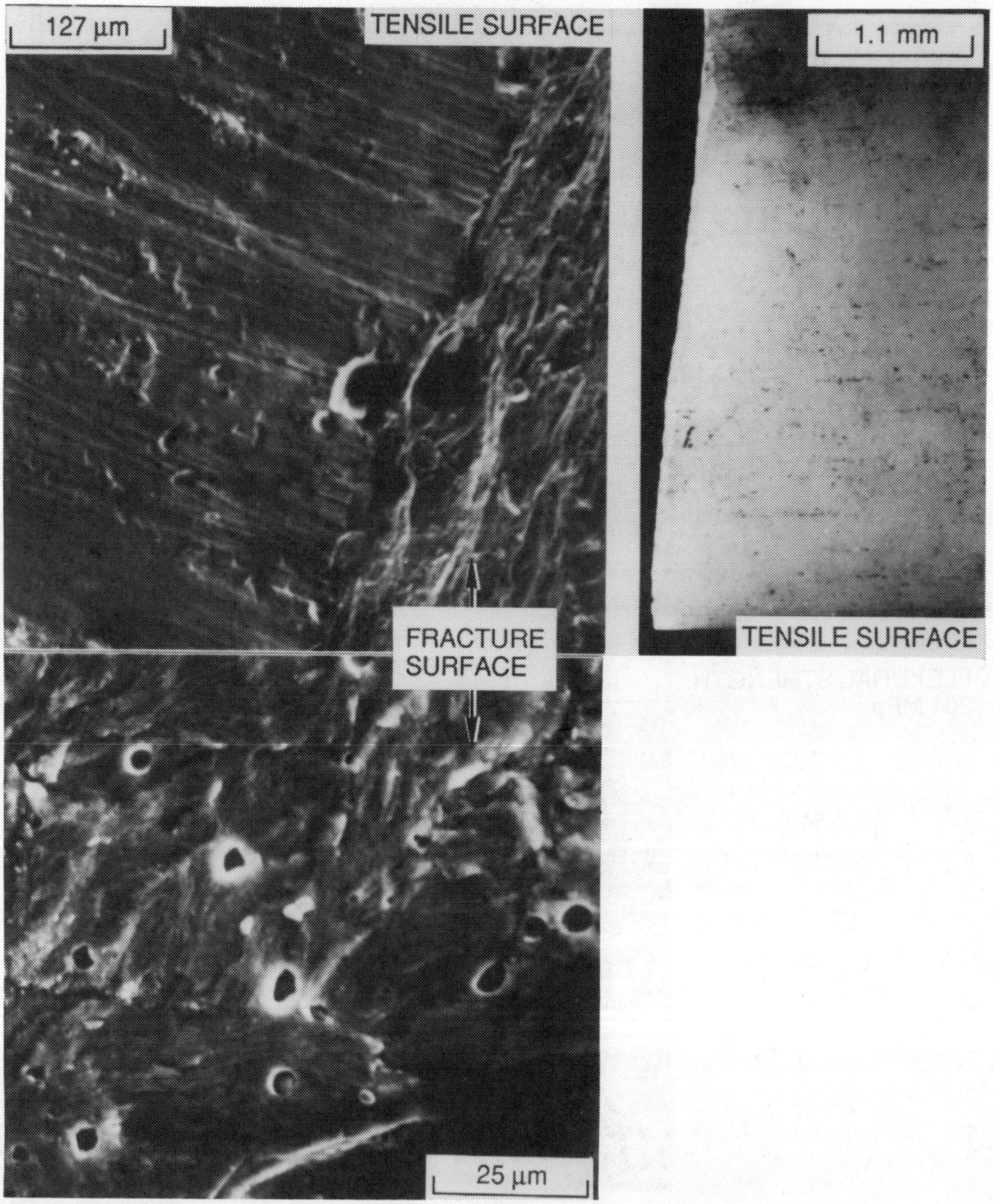

Fig. 7. Fracture and Tensile Surface Areas of PSZ-TS Specimen 65 After 100 h Engine 2 Exposure and Rupture Testing.

Flexural Strength: 228 MPa

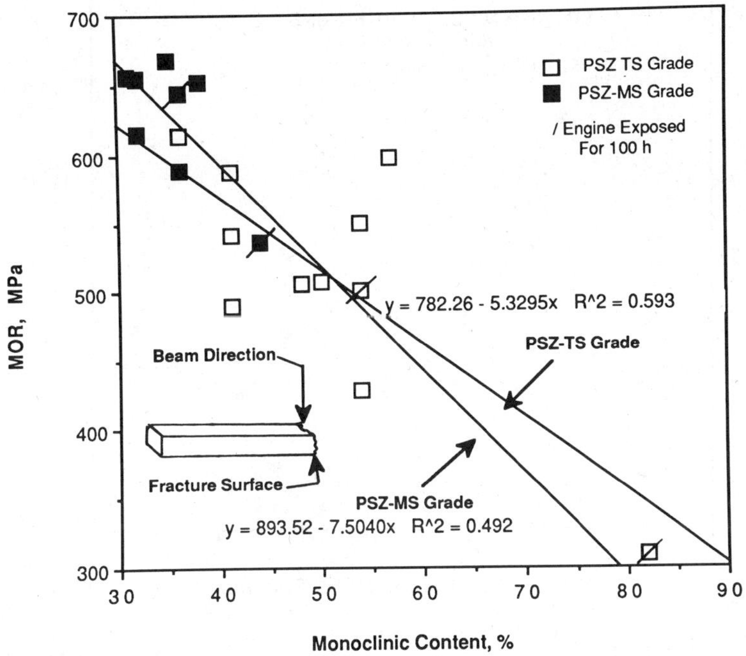

Fig. 8. MOR (Modulus of Rupture) for Individual Bars as a Function of Monoclinic Content From X-ray diffraction analysis.

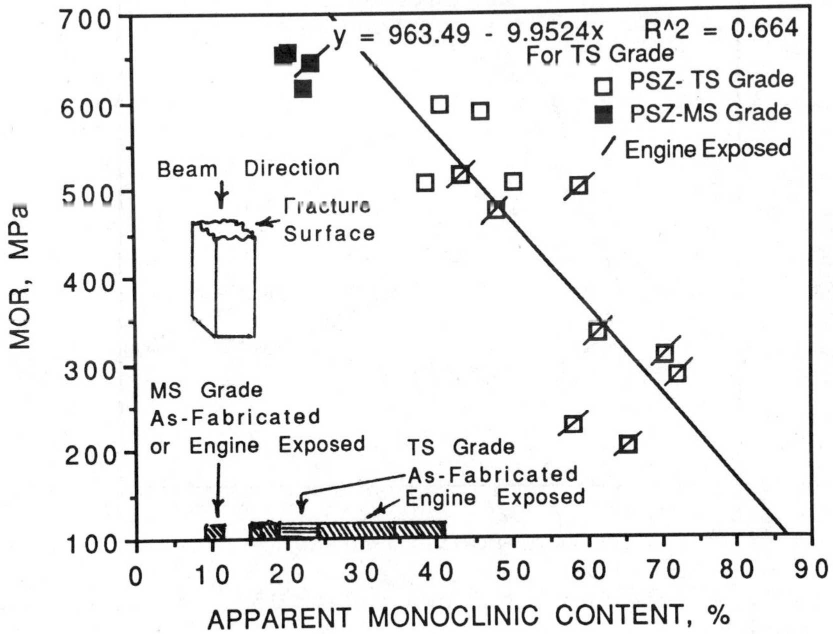

Fig. 9. MOR (Modulus of Rupture) for Individual Bars as a Function of Apparent Monoclinic Content as Determined at Fracture Surface by Raman Spectroscopy.

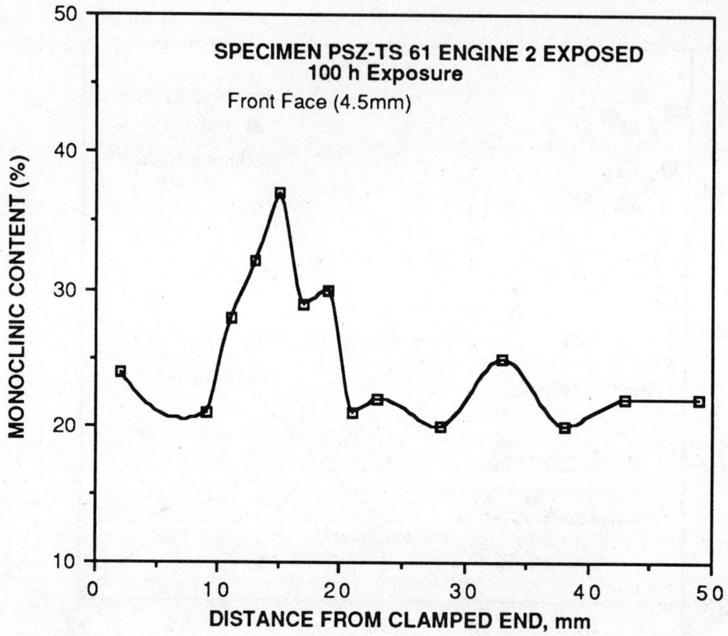

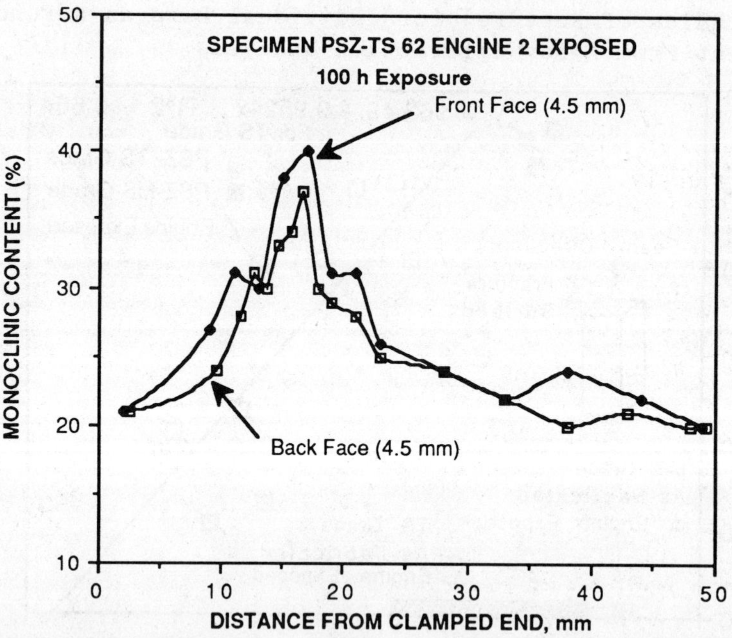

Fig. 10. Apparent Monoclinic Content Shows a Peak 15-17 mm From Clamped End of Engine-Exposed Bars.

Note: Monoclinic Content Assuming Only Monoclinic and Tetragonal Phases are Present.

CALCULATION OF THE WEIBULL STRENGTH PARAMETERS AND THE BATDORF FLAW-DENSITY CONSTANTS FOR VOLUME- AND SURFACE-FLAW-INDUCED FRACTURE IN CERAMICS

Shantaram S. Pai
W.L. Tanksley & Associates, Inc.
NASA Lewis Research Center
Cleveland, Ohio 44135

and

John P. Gyekenyesi
National Aeronautics and Space Administration
Lewis Research Center
Cleveland, Ohio 44135

ABSTRACT

This paper describes the calculation of shape and scale parameters of the two-parameter Weibull distribution through the least-squares analysis and maximum likelihood methods for volume- and surface-flaw-induced fracture in ceramics with complete and censored samples. Detailed procedures are given for evaluating 90-percent confidence intervals for maximum likelihood estimates of shape and scale parameters, the unbiased estimates of the shape parameters, and the Weibull mean values and corresponding standard deviations. Furthermore, this paper describes the necessary steps for detecting outliers and for calculating the Kolmogorov-Smirnov and the Anderson-Darling goodness-of-fit statistics and 90-percent confidence bands about the Weibull distribution. It also shows how to calculate the Batdorf flaw-density constants by using the Weibull distribution statistical parameters. The techniques described have been verified with several example problems from the open literature, and have been coded in the Structural Ceramics Analysis and Reliability Evaluation (SCARE) design program.

INTRODUCTION

The reliability and consistency of design strength data for ceramics is relatively poor when compared to the design data for materials like metals and plastics. The reason for this is that ceramics contain a large number of microcracks, with various sizes and orientations which, along with residual stresses from processing, cause a wide scatter in strength measurements. A number of commonly used uniaxially loaded test specimens can be used to measure ceramic fast-fracture strength over a wide range of temperatures. The Army Specification[1] describes a test method for determining the flexure strength, or modulus of rupture (MOR), of brittle beam test specimens at ambient temperatures. Because of the scatter in observed strength values, many samples (typically 30 or more MOR bars for a specified temperature) are required. However, an inherent weakness of flexure testing is that it emphasizes surface flaws relative to volume flaws. It also needs a greater order of magnitude of volume and area scaling between the specimen and the entire component geometry. To alleviate these problems, Oak Ridge National Laboratory[2] (ORNL) developed a hydraulic self-aligning grip system that can be used to perform tensile tests. With the tensile test specimen, the entire gage material volume (which is generally much larger than in the flexure specimen) is loaded to the same maximum stress.

Two main classes of fracture theory have been previously discussed in the literature. The first is derived from Griffith's flaw theory[3,4] which assumes the presence of cracks of specified sizes, shapes, and arbitrary orientations. There is always one crack with the least favorable orientation that causes failure. The second class consists of the weakest link theory (WLT) such as that of Weibull.[5] This theory is purely statistical in nature. Failure in ceramics is due to a single weakest flaw where the local stress reaches a critical value.

Weibull[6] assumed a unique cumulative strength distribution for the uniaxial fracture data obtained from simple test specimens and also proposed a method for calculating the failure probability in multidimensional stress states. His concept focuses on calculating the risk of rupture by averaging the normal tensile stress at any point in principal stress space. This approach is plausible, but rather arbitrary, and requires numerical modeling. Consequently, other models were proposed, and the most widely used model assumes that the principal stresses act independently (PIA).[7] However, the PIA hypothesis may lead to unsafe estimates of failure probability because it neglects the effects of the interaction of combined local principal stresses.[8]

Batdorf and Crose[9] subsequently proposed a new physically based statistical theory, wherein the material strength parameter is usually expressed in terms of a power function with appropriate flaw-density constants. Rufin et al.[10] further extended the Batdorf statistical theories to investigate effects of different flaw geometries and mixed-mode fracture criteria on overall material reliability. Gyekenyesi and Nemeth[11,12] developed a public domain computer program incorporating the preceding theories which could be coupled to a general purpose finite element code such as MSC/NASTRAN to predict the fast-fracture failure probability of ceramic components due to the presence of volume- and surface-type flaws.

The primary objective of this paper is to describe the estimation of the Weibull strength parameters and the Batdorf flaw-density constants for volume- and surface-flaw-induced fracture in ceramics. The evaluation of shape and scale parameters of the two-parameter Weibull distribution is explained initially, with the least-squares analysis.[13] Important design criteria should not be derived from small sample sizes because uncertainty is known to increase as the number of samples decreases. For small to moderately large sample sizes, distribution percentiles of the maximum likelihood estimates (MLE's) and unbiasing factors for the shape parameters have been computed and tabulated by Thoman et al.[14] Jeryan[15] developed a FORTRAN computer program to calculate the MLE's and other related statistical parameters for uncensored statistics. This paper summarizes the equations and procedures that are necessary for evaluating the MLE's of the shape and scale parameters, the Weibull mean values, and the corresponding standard deviations.[16] The Weibull log likelihood equations for censored statistics, developed by Nelson,[17] are used to calculate the MLE's for the censored data. This paper also evaluates 90-percent confidence intervals for MLE's of the shape and scale parameters and the unbiased estimates of the shape parameters for uncensored statistics. No rigorous studies are known to the authors that evaluate the confidence intervals and unbiasing factors for censored statistics. However, a Monte Carlo simulation is developed and tested for censored statistics, and the results are included in this paper.

An outlier is a data value that is far from the rest of the sample. Therefore, techniques are needed that detect outliers. Stefansky[18] describes a procedure for calculating critical values of the maximum normed residuals while testing for outliers. This technique is based on the normal distribution, and its application to Weibull data is of concern. However, it was used by Neal et al.[19] to calculate the critical values for different significance levels. Sample size-dependent polynomial approximating functions have been developed to compute the critical values at 1-, 5-, and 10-percent significance levels, and these functions are described in this paper.

Many goodness-of-fit tests based on empirical distribution functions (EDF) have been extensively discussed by D'Agostino and Stephens.[20] The Kolmogorov-Smirnov (K-S) test for goodness-of-fit is the most widely used in comparing an empirical distribution function with the distribution function of the hypothesized distribution for any finite sample sizes. Percentage points in K-S statistics for probabilities from 0.01 through 0.20 were tabulated by Miller[21] for different sample sizes. This work was further extended by Amstadter[22] in evaluating values for probabilities from 0.30 through 0.99. The Anderson-Darling[23] statistic A^2 has proven more sensitive to the discrepancies in the tail regions between the EDF and the cumulative distribution function. Lewis[24] discusses this for a fully specified distribution. This paper summarizes the steps to calculate these statistics by using polynomial approximating functions to represent available tabular values from the literature.

Confidence bands[25] on the Weibull line can be calculated and plotted[26] for different confidence levels by making use of the mean and the standard deviation of the sample. This paper also explains the calculation of 90-percent confidence bands about the Weibull distribution.

Also, several currently used theories that can be used for calculating the Batdorf flaw-density constants from the Weibull distribution statistical parameters are selectively included in this paper. The most widely used model among these theories assumes that the cracks are shear-insensitive. Different relationships are obtained, however, if shear-sensitivity of the flaws is included. Finally, all the statistical techniques summarized here were verified with several example problems from the open literature, including the problem of deep-groove ball bearings[14] and the recently obtained Elecktroschmelzwerk Kempten (ESK) hot isostatically pressed (HIPped) silicon-carbide[27] strength data.

PROGRAM CAPABILITY AND DESCRIPTION

Volume- and surface-flaw-based reliability analysis has been implemented in both the SCARE1 and SCARE2 versions of the postprocessor program.[11,12] Both versions of SCARE have broad capabilities: Temperature-dependent multiple-flaw populations, statistical material parameters, several crack configurations, and a number of well-known fracture criteria can be specified. Figure 1 shows the flowchart for the calculation of the shape and scale parameters of the two-parameter Weibull distribution, and many other related statistical quantities. For convenience, these latest additions to the SCARE program can also be used independently of the MSC/NASTRAN post-processing or without performing a reliability analysis.

INPUT INFORMATION

SCARE input requirements are grouped into three major categories: the Master Control Deck, the Specimen Deck, and the Structures Deck. For the calculation of the parameters, the data, which includes the fracture origins of the specimen, is entered in the Specimen Deck. Initially, for the statistical analysis of both complete and censored samples, each specimen fracture mode is observed by fractography or other means and identified as failed either by volume-flaw or by surface-flaw-induced fracture, V and S, respectively. If there is any uncertainty regarding a failure mode, it is input as an unknown failure mode U. The user can select any of the following three options of calculations for complete samples: Volume-flaw analysis only, surface-flaw analysis only, or both in one execution. For the last selection, two separate extreme fiber fracture stress data sets (which may be completely identical) are required for the analysis. The same user options are available for censored samples. When the third choice is selected, volume-flaw analysis will be done by treating surface-flaw and unknown-flaw failure-mode data as suspended items; surface-flaw analysis will be done by treating the volume-flaw and unknown-mode data as censored information.

OUTPUT INFORMATION

The output pertaining to the calculation of the Weibull strength parameters identifies the method of solution, the control index used for experimental data, the number of specimens in each batch, and the temperature of each test. In addition, the output echoes the input values of all extreme fiber fracture stresses with proper failure mode identification. Any data value that is far from the rest of the sample is detected as an outlier, and its corresponding significance level (1, 5, and 10 percent) is printed. Furthermore, the biased and the unbiased values of the shape parameter, the specimen characteristic strength, the upper and lower bound values at 90-percent confidence level for both the shape parameter and the specimen characteristic strength, the specimen Weibull mean value, and the corresponding standard deviation are printed for each specified temperature. For censored statistics, the values are generated first for volume-flaw analysis and, subsequently, for surface-flaw analysis.

The K-S test is done for each data point, and the corresponding K-S statistic factors D^+ and D^- are listed. Similarly, the K-S statistic for the overall population is printed along with the percentage significance level. The A^2 statistic, is also determined for the overall population, and its associated significance level is printed.

The next table of the output contains fracture stress data, the corresponding Weibull probability-of-failure values, the 90-percent upper- and lower-confidence band values about the Weibull line, and the median rank value for each data point. These statistical quantities are calculated by using either tabular values or approximating polynomial functions which are now incorporated into the SCARE program. Finally, the material parameters used in component reliability calculations are listed as a function of temperature. These include the biased Weibull modulus, the Batdorf crack-density coefficient, and the material Weibull scale parameter or unit volume and area characteristic strength, whichever is appropriate.

THEORY

The weakest link theory of material strength considers not only the size effect and loading system, but also the variation in fracture strength due to intrinsic defect distributions. It has been shown that experimental fracture strength data obtained from uniaxially loaded specimens is best approximated by the three-parameter Weibull distribution, defined by

$$P_f = 1 - \exp\left[-\int_V \left(\frac{\sigma - \sigma_u}{\sigma_o}\right)^m dV\right] \qquad (1)$$

where P_f is the probability of failure, σ is the applied tensile stress, σ_u is the threshold stress (location parameter usually taken as zero for ceramics), σ_o is the scale parameter with dimensions of stress x $(\text{volume})^{1/m}$, m is the shape parameter (or the Weibull modulus) which measures the degree of strength variability, and V is the stressed volume. A similar equation can be written for area-flaw-induced failure with area parameters replacing corresponding volume variables.

The Outliers

Before computing the estimates of the Weibull parameters, the available specimen fracture stress data must be carefully examined for outliers. Very often a data set may contain one or more values at the extremes which may not belong to the main trend of the overall population, and these are labeled as outliers. At the start of the outlier calculations, the sample mean $\bar{\sigma}_f$ and the sample standard deviation s are calculated. From these values, the normed residuals $\bar{r}_i$ for each sample are obtained as follows:

$$\bar{r}_i = \frac{\sigma_{fi} - \bar{\sigma}_f}{s}, \qquad i = 1,2, \ldots, N \qquad (2)$$

where N is the sample size. The normed residuals are normalized deviations of the data from the "center" of the data. The absolute maximum of the normed residual (MNR) statistic is compared to the critical value (CV) associated with the sample size. Critical values are calculated from the following equation:

$$CV = \frac{N - 1}{\sqrt{N}} \sqrt{\frac{t^2}{N - 2 + t^2}} \tag{3}$$

where t is the $(1 - \alpha/2N)$ quantile of the t-distribution with N-2 degrees of freedom, and α is the significance level. Separate polynomial approximating functions for sample sizes to 30 and above 30 have been developed for calculating CV at 1-, 5-, and 10-percent significance levels, and these functions are included in the SCARE program. If the MNR statistic is smaller than the three critical values, then no outliers are detected. However, if the MNR is larger than at least one of the three critical values, the corresponding data value with the MNR statistic is detected as an outlier with the appropriate significance level.

It is sometimes appropriate to omit the outliers and continue with the remaining analysis treating the reduced sample as a "new" sample or as a "censored" sample. But the rejection of outliers, strictly based on statistical analysis, can be misleading. Therefore, this technique of detecting the outliers should be used as a means of ascertaining the relative validity of the available data. Figure 2 shows the possible existence of a potential outlier at the high-strength range of the data set for Si_3N_4.[27] The two Weibull distributions shown in Fig. 2 are assumed.

Estimation of Statistical Material Parameters

The flexural test failure probability can be expressed in terms of the extreme fiber fracture stress σ_f, or MOR, using the Weibull form as

$$P_f = 1 - \exp\left[-C\sigma_f^m\right] = 1 - \exp\left[-C(MOR)^m\right] \tag{4}$$

where $C = (1/\sigma_\theta)^m$ and σ_θ is the volume/area specimen characteristic strength, or characteristic modulus of rupture MOR_o. In Eq. (1), σ_o is based on the unit volume, whereas in Eq. (4), σ_θ includes effects of the specimen volume. A sample may only contain data from specimens that failed either from a surface flaw or from a volume flaw. Such a sample is usually called a complete sample. In some cases, specimens in a sample may not fail by one failure mode only, but by several competing failure modes. The sample containing data from the mixture of failure modes and sometimes inclusive of unknown failure modes is referred to as a censored sample.

Many methods are available to estimate the statistical material parameters from experimental data. The success of the statistical approach depends on how well the probability function fits the actual material measurements, especially how accurate the value of m is. The two most popular models used to evaluate C and m are the least-squares analysis and the maximum likelihood methods. These statistical techniques are described in detail in the succeeding sections.

The Least-Squares Analysis

The survival probability of a ceramic specimen in flexure-strength testing is defined by Eq. (4). This equation can be linearized by taking the natural logarithm twice to obtain

$$\ell_n \ell_n\left(\frac{1}{P_s}\right) = \ell_n \ell_n\left(\frac{1}{1 - P_f}\right) = \ell_n C + m \ell_n \sigma_f \tag{5}$$

If the failure probability P_f is determined from conducting N tests and numbering the observed fracture stresses as $\sigma_1, \sigma_2, \ldots, \sigma_i, \ldots, \sigma_N$ in an ascending order, then

$$P_f(\sigma_i) = \frac{i}{N + 1} \tag{6}$$

where i is the median rank. A more sophisticated treatment using median rank regression analysis, which is used in the SCARE program, gives the following result:[26]

$$P_f(\sigma_i) = \frac{i - 0.3}{N + 0.4} \tag{7}$$

For the least-squares analysis, it is necessary to obtain the line of best fit with slope m and an intercept b which, as can be noted from Eq. (5), is equal to the natural log of C. By differentiating the sum of the squared residuals with respect to m and b separately, setting the results equal to zero, and then simultaneously solving for m and b, we obtain the desired results. With censored data, one cannot directly use the median rank regression analysis as given in Eq. (7) because of the presence of competing failure modes. Johnson[28] developed the rank increment technique, which is calculated from the following equation taken from Ref. 26:

$$\text{Rank increment} = \frac{(N + 1)\ (\text{previous adjusted rank})}{1 + (\text{number of items beyond present suspended item})} \tag{8}$$

For volume-flaw analysis, all MOR's designated as V's are considered as failure data; for surface-flaw analysis, the S's are considered as failure data. According to the procedure discussed in Ref. 26, the new adjusted rank values are obtained by adding the rank increment value to the previously adjusted rank. These adjusted rank values are then used to calculate the failure probability P_f by using median rank regression analysis (i.e., Eq. (7)).

The least-squares analysis is a particular case of the maximum likelihood method, if we assume that the error is normally distributed with zero mean and constant variance. Somerville and Bean[29] compared the least-squares analysis and the maximum likelihood methods for different sets of conditions. The study concluded that when the data fits the Weibull distribution, both techniques produce substantially the same results. The least-squares analysis is most useful in situations where the underlying probability distribution is not clearly established, and it is more robust for minor departures from the Weibull distribution. King[30] determined the accuracy of the estimates of the Weibull parameters by using the prediction interval method, and he showed that the sampling distribution of the regression coefficients follows a t-distribution. This is true only if the error is normally distributed with zero mean and constant variance. Therefore, the least-squares analysis method is usually not suited to calculate confidence intervals and unbiasing factors. For that reason, the maximum likelihood method has been extensively used in Weibull analysis and is discussed in detail next.

The Maximum Likelihood Method

The maximum liklihood method is often used because of certain intrinsic properties. The likelihood equation from which the MLE's are obtained has a unique solution. In addition, as the sample size increases, the solution converges to the true values of the parameters, and, therefore, the solution is an asymptotically normal and asymptotically efficient estimate of the parameters. Jeryan[15] compared the maximum likelihood method to the method of moments and the pseudo-least-squares method for random sample sizes with assigned values for the Weibull parameters. It was concluded that the MLE of the Weibull slope showed the least bias and variance. Another advantage of using the MLE for complete or censored samples is that no ranking functions or linear regression analysis are used in obtaining the results.

For the complete sample, the likelihood function is given by

$$L = \prod_{i=1}^{N} \left(\frac{m}{\sigma_\theta}\right)\left(\frac{\sigma_{fi}}{\sigma_\theta}\right)^{m-1} \exp\left[-\left(\frac{\sigma_{fi}}{\sigma_\theta}\right)^{m}\right] \tag{9}$$

By differentiating the logarithm of the likelihood function with respect to m and σ_θ, and then equating the resulting expressions to zero, we obtain the MLE's of m and σ_θ by solving the resulting simultaneous equations.

In the complete sample case, the MLE of the shape parameter m, denoted by $\hat{m}$, should satisfy the following equation:[26]

$$\frac{\sum_{i=1}^{N} (\sigma_{fi})^{\hat{m}} \ln(\sigma_{fi})}{\sum_{i=1}^{N} (\sigma_{fi})^{\hat{m}}} - \frac{1}{N}\sum_{i=1}^{N} \ln(\sigma_{fi}) - \frac{1}{\hat{m}} = 0 \tag{10}$$

The Newton-Raphson iterative technique is used to solve Eq. (10) and to calculate $\hat{m}$. Furthermore, the MLE of the Weibull parameter σ_θ, denoted by $\hat{\sigma}_\theta$, is evaluated by solving the following equation:

$$\hat{\sigma}_\theta = \left(\frac{\sum_{i=1}^{N} \sigma_{fi}^{\hat{m}}}{N}\right)^{1/\hat{m}} \tag{11}$$

However, if the sample is censored, the following equation is solved by iterative techniques to obtain the value of $\hat{m}$:[26]

$$\frac{\sum_{i=1}^{N} (\sigma_{fi})^{\hat{m}} \ell n(\sigma_{fi})}{\sum_{i=1}^{N} (\sigma_{fi})^{\hat{m}}} - \frac{1}{r} \sum_{i=1}^{r} \ell_n(\sigma_{fi}) - \frac{1}{\hat{m}} = 0 \tag{12}$$

In the preceding equation, r is the remaining number of specimens failed either by volume flaw or by surface flaw only. Finally, σ_θ is obtained from

$$\hat{\sigma}_\theta = \left(\frac{\sum_{i=1}^{N} \sigma_{fi}^{\hat{m}}}{r} \right)^{1/\hat{m}} \tag{13}$$

Incidentally, the MLE's of m and σ_θ for the volume-flaw analysis are denoted by $\hat{m}_V$ and $\hat{\sigma}_{\theta V}$; for the surface-flaw analysis by $\hat{m}_S$ and $\hat{\sigma}_{\theta S}$, respectively. Similarly the parameter C is denoted as $\hat{C}_V$ and $\hat{C}_S$. As shown in Fig. 1, the least-squares analysis is always done first, irrespective of the method of analysis chosen. However, if the maximum likelihood method is selected, the initial estimate of the shape parameter is obtained from the least-squares analysis. If no solution is found after 50 iterations, the maximum likelihood method is terminated, and the results from the least-squares analysis are used for further calculations in the program for both complete and censored samples.

Unbiasing Factors and Confidence Intervals

In the previously described maximum likelihood method, the MLE of the shape parameter is a biased estimate that depends on the number of specimens in the sample. Consequently, this parameter may not represent the true population. Thoman et al.[14] obtained the unbiasing factors for the MLE of m for complete samples by initially obtaining the MLE of m for complete samples. The unbiasing factors for the MLE of m in Ref. 14 depend on sample size. The unbiased estimate of m is obtained by multiplying the biased estimate of m with the unbiasing factors. The percentage points of the distributions of MLE's were evaluated and then used to calculate the confidence intervals for MLE's at several confidence levels for complete samples.[14] Reference 1 contains figures showing 90-percent confidence bounds for the Weibull modulus and the characteristic strength.

A method of obtaining definite confidence intervals for censored samples has not yet been developed because of the additional complexity of competing failure modes. However, the effects of competing failure modes were examined through a Monte Carlo simulation on a limited trial basis by the authors. The results indicated that the percentage points of the distribution asymptotically approached the complete sample case and depended on the number of failures for each failure mode.

It is general practice to obtain these bounds at 90-percent confidence level, and, therefore, 5 and 95 percentage points of distribution of the

MLE's of the parameters have been incorporated into the SCARE program, with data taken from Ref. 14. The following equations describe the manner in which the different limit values are calculated at 90-percent confidence level:

$$\hat{m}_{UPP} = \frac{\hat{m}_{BIAS}}{\hat{m}_{5\%}} \quad \text{and} \quad \hat{m}_{LOW} = \frac{\hat{m}_{BIAS}}{\hat{m}_{95\%}} \tag{14}$$

$$\hat{\sigma}_{\theta_{UPP}} = \frac{\hat{\sigma}_{\theta}}{\exp\left[-\frac{\hat{\sigma}_{\theta_{5\%}}}{\hat{m}_{BIAS}}\right]} \quad \text{and} \quad \hat{\sigma}_{\theta_{LOW}} = \frac{\hat{\sigma}_{\theta}}{\exp\left[\frac{\hat{\sigma}_{\theta_{95\%}}}{\hat{m}_{BIAS}}\right]} \tag{15}$$

where a minus sign has been used in Eq. 15 to account for the data format. Jeryan[15] used the unbiased value of the shape parameter MLE to calculate the upper and lower bounds, while Thoman et al.[14] used the biased value as done here.

The Weibull Mean and the Standard Deviation

After evaluating the MLE's of the Weibull parameters, one can obtain the Weibull mean μ and the standard deviation s by using the following two equations:[16]

$$\mu = \left(\hat{\sigma}_{\theta}\right)\Gamma\left(1 + \frac{1}{\hat{m}_{BIAS}}\right) \tag{16}$$

$$s = \left\{\left(\hat{\sigma}_{\theta}\right)^2\left[\Gamma\left(1 + \frac{2}{\hat{m}_{BIAS}}\right) - \Gamma\left(1 + \frac{1}{\hat{m}_{BIAS}}\right)\right]\right\}^{1/2} \tag{17}$$

where Γ denotes the gamma function. These equations are valid for both area- and volume-based flaw strength. Equations (16) and (17) indicate that as m increases, the mean of the distribution approaches the specimen characteristic strength, and the standard deviation approaches zero. The specimen characteristic strength σ_{θ} is related to the material scale parameter σ_o through the specimen effective volume/area relationship of

$$\sigma_{oV} = \sigma_{\theta V} V_e^{1/m_V} \text{ or } \sigma_{oS} = \sigma_{\theta S} A_e^{1/m_S} .$$

Goodness-of-Fit Tests

Graphical techniques can often be used to determine whether or not a set of observations belongs to a population distribution. However, subjective judgment is needed to test the goodness-of-fit of the data to the assumed distribution, and it is difficult to do this in many instances. Therefore, classical statistics tests are used in verifying the available data. In general, a statistic is a numerical value determined from a sample by using statistical computations. The difference between an empirical distribution function (EDF) and a given distribution function is called an EDF statistic. The supremum class statistic calculates the largest and smallest vertical differences between the two distribution functions. On the other hand, the quadratic class statistic measures the discrepancy between the two distributions through squared differences and the use of an appropriate weighting function.

The K-S statistic belongs to the supremum class and is very effective for small samples. To begin the K-S test, the sample is arranged in ascending order, and the EDF, $F_N(x)$, is obtained by using the expressions found in Ref. 20. The main intent of the K-S test is to calculate the well-known EDF statistic D, which is the maximum absolute vertical difference between the hypothesized cumulative distribution function $F(x)$ and the EDF function $F_N(x)$. As shown in Fig. 3, the statistic D^+ and D^- are the largest vertical differences when $F_N(x)$ is greater than $F(x)$ and largest vertical differences when $F_N(x)$ is smaller than $F(x)$, respectively. In the SCARE program, all three statistics are calculated by using the following expressions:[20]

$$\left.\begin{aligned} D^+ &= \left|\frac{i}{N} - F(x)_i\right| \\ D^- &= \left|F(x)_i - \frac{i-1}{N}\right| \qquad i = 1,2,\ldots,N \\ D &= \max(D^+ \text{ or } D^-) \end{aligned}\right\} \tag{18}$$

For ceramics design, the $F(x)_i$'s are equal to P_f's and are calculated by using Eq. (4) and the biased values of the MLE's of the Weibull parameters. In the censored samples, the cumulative survival probability is obtained by using Eq. (4) separately for volume and surface flaws, and then by multiplying the survival probabilities of the two concurrent flaw populations. Finally, the resultant predicted failure probability at each stress level is calculated by using the entire sample.

It is customary to express the EDF statistic D with an associated significance level α, which is the probability of obtaining a higher maximum absolute difference assuming that the data follows the theoretical distribution. This probability is a function of D and the sample size N. Miller[21] described a general formulation for evaluating α as follows:

$$\alpha = \frac{2}{\exp\left[2D^2 \dfrac{N}{\left(1 - \dfrac{0.5}{N}\right)^2}\right]} \tag{19}$$

For sample sizes of 30 or less, the values of α obtained by using Eq. (19) are equal to the values given in Ref. 22. However, for sample sizes greater than 30, a separate formulation is used, based on the discussion by Brunk,[31] and is given by

$$\alpha = 2 \sum_{j=1}^{\infty} (-1)^{j-1} \exp\left[-2j^2D^2\right]$$

or (20)

$$\alpha = 1 - \frac{\sqrt{2\pi}}{D} \sum_{j=1}^{\infty} \exp\left[-(2j-1)^2 \frac{\pi^2}{8D^2}\right]$$

In the SCARE program, the EDF statistics D^+ or D^- are calculated for every specimen fracture stress level. Then the EDF statistic D and the associated significance level are determined for the total sample size. A higher value of α indicates that the data fits the proposed distribution to a greater extent.

The A^2 statistic belongs to the quadratic class and measures the discrepancy between the EDF and the hypothesized cumulative distribution function by using the Cramér-von Mises formulation:[20]

$$Q = N \int_{-\infty}^{\infty} \{F_N(x) - F(x)\}^2 \psi(x)dF(x) \tag{21}$$

where $\psi(x)$ is a suitable weighting function applied to the squared difference. Anderson and Darling[23] introduced the A^2 statistic by defining $\psi(x) = [\{F(x)\}\{1 - F(x)\}]^{-1}$ in Eq. (21). This statistic is more suitable for identifying the discrepancies in the tail regions of the distribution. A new random variable Z, which is uniformly distributed between 0 and 1, is introduced so that $Z = F(X)$, when $F(x)$ is the true distribution of X. The Z_i's are then arranged in ascending order and are used to calculate the statistic A^2 which is given by[20]

$$A^2 = -N - \left(\frac{1}{N}\right) \sum_{i=1}^{N} (2i-1)\left[\ell n\{Z_{(i)}\} + \ell n\{1 - Z_{(N+1-i)}\}\right] \tag{22}$$

In this case, Z_i's are the predicted failure probabilities obtained from Eq. (4). It can be shown that for $N > 3$, the distribution of A^2 is sufficiently independent of the sample size.

Lewis[24] discussed the asymptotic distribution and gave the values for the A^2 statistic and the corresponding significance levels. We have developed several approximating polynomial functions to interpolate these values. If the A^2 statistic is too large, the EDF fits the proposed distribution poorly. The significance level for the sample is calculated by using the A^2 statistic and the corresponding polynomial function. Once again, the higher the value of α, the better the data fits the proposed distribution.

Confidence Bands on the Weibull Line

Kanofsky and Srinivasan[25] described a method for constructing a confidence band on a continuous distribution function which may have normal, exponential, and many other functional forms. They introduced and defined the random variable $\bar{L}$, (the statistic $\bar{L}$) as

$$\bar{L} = \max_x \left| F\left(x;\ \mu_p, \xi_p\right) - F(x;\ \bar{x}, s) \right| \quad (23)$$

where μ_p, ξ_p, $\bar{x}$, and s are, respectively, the population mean, the population standard deviation, the sample mean, and the sample standard deviation. The population parameters μ_p and ξ_p are unknown values. In Eq. (23), the statistic $\bar{L}$ is the maximum absolute difference between the two cumulative distribution functions. The distribution of the statistic $\bar{L}$ is obtained in two stages: (1) The closed-form expressions for $\bar{L}$ are derived, and (2) these expressions are used to derive the distribution of $\bar{L}$ by using Monte Carlo simulation. The random variable $\bar{L}$ is then used to construct different percentage confidence bands on $F(x;\ \mu_p, \xi_p)$ centered around $F(x;\ \bar{x}, s)$.

Abernethy et al.[26] extended the procedure to the Weibull distribution. Equation (24), with tabular data of K(N) from Ref. 26, is used in calculating the 90-percent confidence bands about the Weibull distribution for complete samples only. Thus,

$$\text{Confidence bands} = [F(x) - K(N),\ F(x) + K(N)] \quad (24)$$

where F(x) is the failure probability obtained by substituting the MLE's of the Weibull parameters in Eq. (4). The Kanofsky functions, denoted by K(N), depend on the sample size. We have developed two sample-size-dependent approximate polynomial functions to interpolate these values, and we have incorporated them into the SCARE program. The two terms in Eq. (24) give the lower and upper 90-percent confidence band values about the Weibull distribution, respectively. Figure 4 indicates that for HIPped SiC flexure bars, the fracture stresses lie between the confidence bands about the Weibull line. The value of the statistic $\bar{L}$ is 0.0829. It is also clear that the stress data fits a single Weibull distribution quite well.

Batdorf-Weibull Statistical Parameters for Reliability Analysis

In reliability calculations, the Weibull modulus m is directly used by all statistical fracture models. However, for the Weibull model the scale parameter σ_o is needed, while for the Batdorf approach the flaw-density coefficient k_B must be calculated. Both of these material constants are functions of m, C, and specimen geometry. Details of evaluating σ_o and k_B are summarized elsewhere,[11,12] but the results are included here to show how the two sets of constants are related. For volume-flaw analysis, using quadrant ($L_1 = 2L_2$), four-point MOR bar data with known geometry, σ_{oV} is calculated from

$$\sigma_{oV} = \left[\left(\frac{wh}{2}\right) \frac{(L_1 + m_V L_2)}{C_V (m_V + 1)^2}\right]^{1/m_V} \quad (25)$$

For shear-insensitive fracture criterion, k_{BV} is obtained from

$$k_{BV} = (2m_V + 1)\left[\frac{2C_V (m_V + 1)^2}{wh(L_1 + m_V L_2)}\right] \quad (26)$$

where w is the width of the rectangular beam, h is the height, and L_1 and L_2 are, respectively, the lengths between symmetrically placed outer loads and inner loads. By comparing Eqs. (25) and (26), we conclude that when using the normal stress failure criterion, k_{BV} and σ_{oV} are related by

$$k_{BV} = \left(2m_V + 1\right)\left(\frac{1}{\sigma_{oV}}\right)^{m_V} \tag{27}$$

However, for shear-sensitive fracture criteria when crack shape must also be specified, the format of Eq. (27) changes according to the criterion and the crack shape selected. In addition, if we note that $C_V = V_e \sigma_o^{-m_V}$, where V_e is the effective volume, Eq. (25) can be solved for V_e to obtain

$$V_e = \left(\frac{wh}{2}\right) \frac{(L_1 + m_V L_2)}{(m_V + 1)^2} = \frac{(m_V + 2)}{4(m_V + 1)^2} V \tag{28}$$

where $V = whL_1$, the total volume of the specimen. For uniaxial tensile loading, the effective volume is equal to the specimen gage volume V_g, which is at the same time the total specimen volume V of interest in that test.

Finally, the Batdorf surface-crack density coefficient k_{BS} is calculated from[12]

$$k_{BS} = \frac{2^{3/2-m_S}}{A_e} C_S m_S \left(m_S - 1\right) \int_{-1}^{1} (1 + \eta)^{m_S - 2} (1 - \eta)^{1/2} \, d\eta \tag{29}$$

where $\eta = [(2\sigma_f/\sigma_{cr}) - 1]$ and A_e is the effective area. Equation (29) can be integrated numerically by using Gaussian quadratures to evaluate k_{BS}. In pure tensile loading, the effective area in Eq. (29) is just equal to the specimen gage area A_g.

However, if we select shear-sensitive fracture criteria to convert MOR bar or tensile fracture data into Batdorf-model crack-function coefficients, Eqs. (27) and (29) will no longer apply. As an example, if we choose to model both volume and surface imperfections as Griffith cracks (usually conservative but not a good representation of flaws) and apply the total strain energy-release-rate criterion (self-similar crack extension) to predict impending fracture, the following equations are obtained:

$$k_{BV} = \left(m_V + 1\right)\left(\frac{1}{\sigma_{oV}}\right)^{m_V} \tag{30}$$

$$k_{BS} = \frac{m_S \sqrt{\pi}\, \Gamma\left(\frac{m_S}{2}\right)}{2\Gamma\left(\frac{m_S + 1}{2}\right)} \left(\frac{1}{\sigma_{oS}}\right)^{m_S} \tag{31}$$

In deriving Eqs. (30) and (31), collinear crack growth was assumed, and the results for typical values of m_V and m_S ($m_V = m_S = 10.0$) give considerably smaller Batdorf flaw-density coefficients than those obtained from using the previous shear-insensitive theory. Consequently, failure probabilities for a given structure will decrease when Eqs. (30) and (31) are used to characterize a material. Presently, a new effort is underway to derive closed-form solutions for the Bardorf coefficients by using out-of-plane crack extension criterion[32] with the penny-shaped crack (PSC) for volume flaws and the semicircular-edge crack for surface flaws. It is important to note that using the Batdorf theory requires no new material test procedures and data.

DISCUSSION AND EXAMPLES

Several benchmark problems were analyzed to calculate the required material strength parameters. The MLE's of the Weibull parameters for the results of endurance tests of nearly 5 000 deep-groove ball bearings[14] were compared and are included in table 1. For this problem, one data value (173.40×10^6 rev) is detected as an outlier at 1-percent significance level, the K-S statistic D is 0.1512 (Fig. 3) at 66.7-percent significance level, and the A^2 statistic is 0.3288 at 91.5-percent significance level. This indicates that the data fits the hypothesized Weibull distribution quite well, especially since only 23 specimens are available in this sample.

TABLE 1. - ENDURANCE TESTS OF THE DEEP-GROOVE BALL BEARINGS

[All estimates are biased estimates.]

Parameter estimates	Maximum likelihood estimates (MLE) of-	
	Weibull modulus, m	Characteristic strength, σ_θ, Revolution x 10^{-6}
Lieblein and Zelen[14]	2.230	80.00
Thoman et al.[14]	2.102	81.99
SCARE[11,12]	2.103	81.88

In addition, we have obtained the results of 80 SiC flexure bars tested at NASA Lewis and analyzed them by using our present enhancement to SCARE to calculate the MLE's of the Weibull parameters.[27] According to the results in table 2, the Weibull parameter values from SCARE match two other predictions very well and are in reasonable agreement with the other three, despite the use of different test operators, test equipment, and software to reduce the data.

TABLE 2. - WEIBULL PARAMETERS FOR FLEXURE STRENGTH DATA SET OF ESK HIPped SILICON CARBIDE BARS

[All estimates are biased estimates.]

Fracture data set	Maximum likelihood estimates (MLE) of-	
	Weibull modulus, m	Characteristic strength, σ_θ (MPa)
GTE Laboratories	7.16	541.17
Sohio	6.82	517.01
Allison	4.91	508.88
NASA Lewis	6.59	556.72
Garrett	6.60	554.01
SCARE	6.49	555.80

CONCLUSIONS

The general purpose, statistical, fast fracture probability code SCARE has been enhanced to include the details of calculating the shape and scale parameters of the two-parameter Weibull distribution by using the least-squares analysis and the maximum likelihood methods for volume- and surface-flaw-induced fracture in ceramics. If any sample, complete or censored, fits the Weibull distribution, then these two methods are the most powerful and appropriate techniques for obtaining the Weibull parameter estimates. In addition, for complete samples only, the maximum likelihood method enables the user to evaluate 90-percent confidence intervals for the MLE's of the Weibull parameters and the unbiased estimates of the shape parameter. Therefore, further research is needed in developing confidence intervals and the unbiasing factors for censored samples. The current addition to SCARE describes how to detect outliers and how to calculate Kolmogorov-Smirnov and Anderson-Darling goodness-of-fit test statistics and the 90-percent confidence bands about the Weibull distribution.

The fracture data of a material, especially in the limited-strength range used in laboratory tests, may fit many distributions other than the Weibull distribution. However, prior experience with ceramics suggests that the Weibull distribution is the best distribution to fit the failure data. Fractography is essential for proper identification of flaw origins. Independent Weibull equations are used for volume- and surface-flaw-based reliability analyses. In addition, the Weibull parameters are used to calculate the Batdorf flaw-density constants for shear-insensitive cracks using closed-form solutions. These constants are calculated for both volume- and surface-flaw-based reliability analysis. For shear-sensitive cracks, however, numerical procedures are used in the SCARE code, just as in multidimensional stress states.

REFERENCES

1. F.I. Baratta, W.T. Matthews, and G.D. Quinn, "Errors Associated with Flexure Testing of Brittle Materials," MTL-TR-87-35, U.S. Army Materials Technology Laboratory, Watertown, MA, July 1987. (Avail. NTIS, AD-A187470.)

2. K.C. Liu, and C.R. Brinkman, "Tensile Cyclic Fatigue of Structural Ceramics," in Proceedings of the 23rd Automotive Technology Development Contractors' Coordination Meeting, Society of Automotive Engineers, Warrendale, PA, 1986, pp. 279-284.

3. A.A. Griffith, "The Phenomena of Rupture and Flow in Solids," Philos. Trans. Roy. Soc. A, 221, 163-198 (1921).

4. A.A. Griffith, "The Theory of Rupture," in Proceedings of the 1st International Congress for Applied Mechanics, C.B. Biezeno and J.M. Burgers, eds., Springer, Berlin, 1924, pp. 55-63.

5. W. Weibull, "The Phenomenon of Rupture in Solids," Ingeniors Ventenskaps Akademien Handlinger, No. 153, 5-55 (1939).

6. W. Weibull, "A Statistical Theory of the Strength of Materials," Ingeniors Ventenskaps Akademien Handlinger, No. 151, 5-45 (1939).

7. A.M. Freudenthal, "Statistical Approach to Brittle Fracture," in Fracture, An Advanced Treatise, Vol. 2: Mathematical Fundamentals, H. Liebowitz, ed., Academic Press, New York, 1968, pp. 591-619.

8. S.B. Batdorf, "Some Approximate Treatments of Fracture Statistics for Polyaxial Tension," Int. J. Fract., 13 [1] 5-11 (1977).

9. S.B. Batdorf, and J.G. Crose, "A Statistical Theory for the Fracture of Brittle Structures Subjected to Nonuniform Polyaxial Stresses," J. Appl. Mech., 41 [2] 459-464 (1974).

10. A.C. Rufin, D.R. Samos, and R.J.H. Bollard, "Statistical Failure Prediction Models for Brittle Materials," AIAA J., 22 [1] 135-140 (1984).

11. J.P. Gyekenyesi, "SCARE: A Postprocessor Program to MSC/NASTRAN for the Reliability Analysis of Structural Ceramic Components," J. Eng. Gas Turbines Power, 108 [3] 540-546 (1986).

12. J.P. Gyekenyesi and N.N. Nemeth, "Surface Flaw Reliability Analysis of Ceramic Components with the SCARE Finite Element Postprocessor Program," J. Eng. Gas Turbines Power, 109 [3] 274-281 (1987).

13. A.S. Jayatilaka, Fracture of Engineering Brittle Materials, Applied Science Publishers Ltd., London, 1979, pp. 195-208.

14. D.R. Thoman, L.J. Bain, and C.E. Antle, "Inferences on the Parameters of the Weibull Distribution," Technometrics, 11 [3] 445-460 (1969).

15. R.A. Jeryan, "Use of Statistics in Ceramic Design and Evaluation," in Ceramics for High Performance Applications II, J.J. Burke, E.N. Lenoe, and R.N. Katz, eds., Brook Hill Publishing Company, Columbus, OH, 1978, pp. 35-51.

16. K.C. Kapur and L.R. Lamberson, Reliability in Engineering Design, John Wiley, New York, 1977, pp. 291-295.

17. W. Nelson, Applied Life Data Analysis, John Wiley & Sons, New York, 1982, pp. 333-395.

18. W. Stefansky, "Rejecting Outliers in Factorial Designs," Technometrics, 14 [2] 469-479 (1972).

19. D. Neal, M. Vangel, and F. Todt, "Statistical Analysis of Mechanical Properties," Engineered Materials Handbook, Vol. 1, Composites, ASM International, Metals Park, OH, 1987, pp. 302-307.

20. R.B. D'Agostino and M.A. Stephens, Goodness-of-Fit Techniques, Marcel Dekker, New York, 1986, pp. 97-193.

21. L.H. Miller, "Table of Percentage Points of Kolmogorov Statistics," J. Am. Stat. Assoc., 51 [273] 111-121 (1956).

22. B.L. Amstadter, Reliability Mathematics, Fundamentals, Practices, Procedures, McGraw-Hill, New York, 1971, pp. 76-79.

23. T.W. Anderson and D.A. Darling, "A Test of Goodness-of-Fit," J. Am. Stat. Assoc., 49, 765-769 (1954).

24. P.A.W. Lewis, "Distribution of the Anderson-Darling Statistic," Ann. Math. Stat., 32, 1118-1124 (1961).

25. P. Kanofsky and R. Srinivasan, "An Approach to the Construction of Parametric Confidence Bands on Cumulative Distribution Functions," Biometrika, 59 [3] 623-631 (1972).

26. R.B. Abernethy, J.E. Breneman, C.H. Medlin, and G.L. Reinman, "Weibull Analysis Handbook," PWA/GPD-FR-17579 Pratt and Whitney Aircraft, West Palm Beach, FL, Nov. 1983. (Avail NTIS, AD-A143100.)

27. V.J. Tennery, "IEA Annex 11 Management, Subtask 4 Results," Ceramic Technology for Advanced Heat Engines Project, Semiannual Progress Report, Oct. 1986-Mar. 1987, Martin Marietta Energy Systems Inc., Oak Ridge, TN.

28. L.G. Johnson, in The Statistical Treatment of Fatigue Experiments, Elsevier Publishing Co., New York, 1964, pp. 37-41.

29. P.N. Somerville and S.J. Bean, "A Comparison of Maximum Likelihood and Least Squares for the Estimation of a Cumulative Distribution," J. Stat. Comput. Simul., 14 [3-4] 229-239 (1982).

30. R.L. King, "The Determination of Design Allowable Properties for Advanced Composite Materials," GEC J. Research, 5 [2] 76-87 (1987).

31. H.D. Brunk, in An Introduction to Mathematical Statistics, Third Ed., Xerox College Publishing, Lexington, MA, 1975, pp. 445-447.

32. D.K. Shetty, "Mixed-Mode Fracture Criteria for Reliability Analysis and Design with Structural Ceramics," J. Eng. Gas Turbines Power, 109 [3] 282-289 (1987).

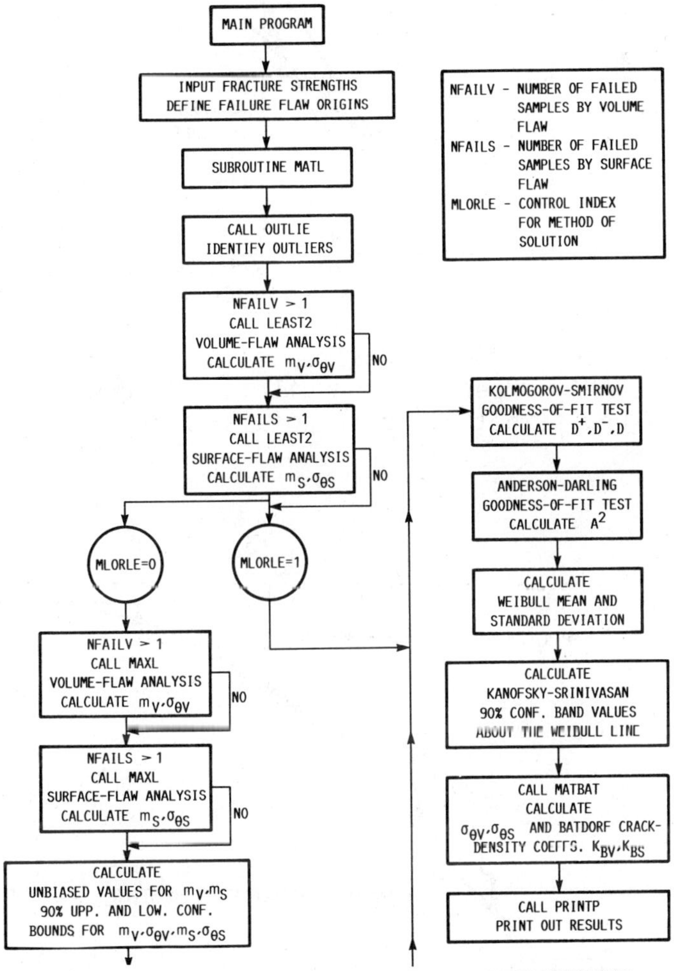

FIG. 1 FLOWCHART FOR MATERIAL STATISTICAL STRENGTH PARAMETERS CALCULATIONS.

Fig. 1 - Flowchart for material statistical strength parameters calculations.

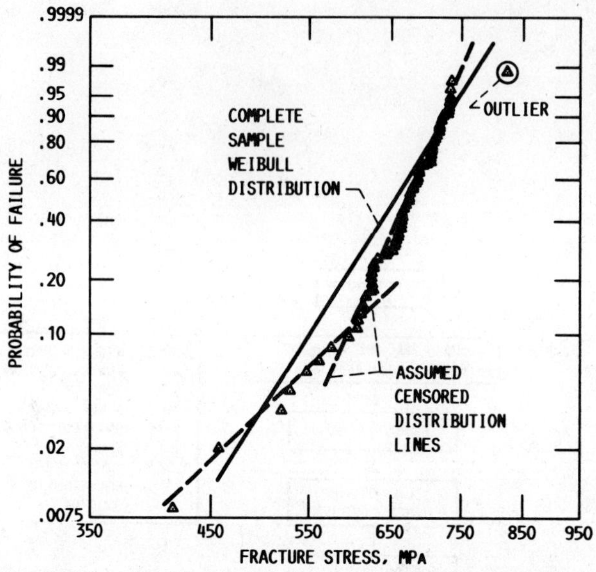

Fig. 2 - Complete sample, assumed censored sample Weibull distributions, and an outlier in ASEA CERAMA HIPped silicon nitride (Si_3N_4) (fracture stress data generated at NASA-Lewis; not all data points shown).

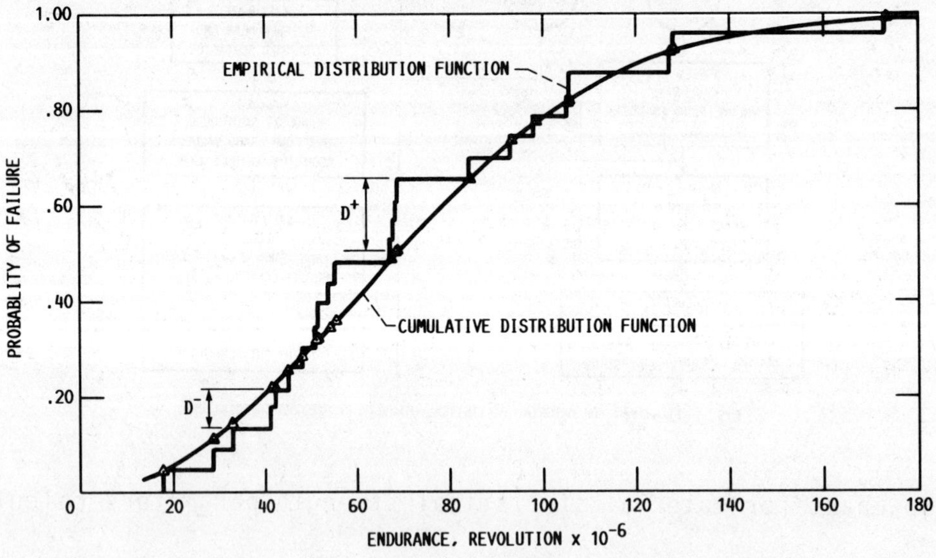

Fig. 3 - Empirical distribution function (EDF) of deep-groove ball bearings endurance tests.

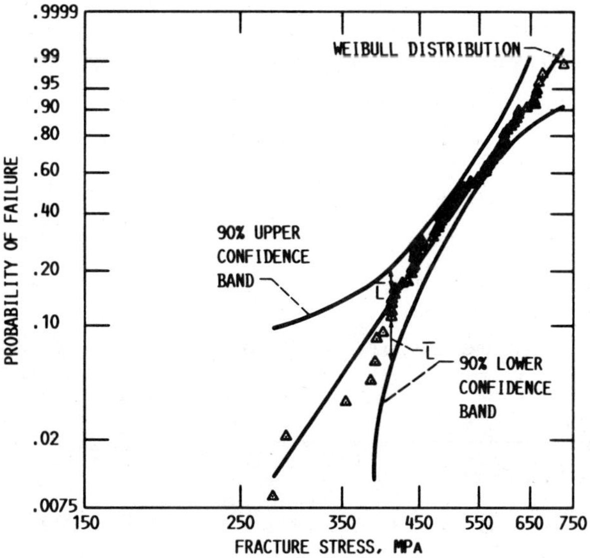

Fig. 4 - 90-Percent confidence bands about the Weibull line for ESK HIPped silicon carbide (SiC) (fracture stress data generated at NASA-Lewis; not all data points shown).

STUDY ON METHOD TO DETERMINE K_{IC} VALUE OF CERAMICS BY CHEVRON-NOTCHED 3-PT BENDING SPECIMEN

Zhenduo Guan   Genshi Sun
Material Seience & Engineering Department,
Tsinghua University
Beijing, PRC

ABSTRACT

By method of compliance calibration and regression analysis for different crack depths, the empirical expression of fracture toughness for brittle materials with definite specimen geometry of chevron-notched 3-pt bending loading configuration, was derived

$$K_{Ic} = \frac{P_c}{2W^{3/2}(1-\mu^2)^{1/2}\tan^{1/2}(\theta/2)} [17.96+20.71(a_o/w)+179.5(a_o/w)^2]$$

The effect of geometric dimension of specimen on K_{Ic} value is also discussed.

INTRODUCTION

The method of determining plane strain fracture toughness K_{Ic} of brittle materials by chevron-notched specimen has been reported in numerous investigations.[1-4] The specimen shape includes short rod, short bar cleavage specimen and 3-pt bending beam specimen (Fig.1) used in this study. The common shape characteristic relation between these specimens

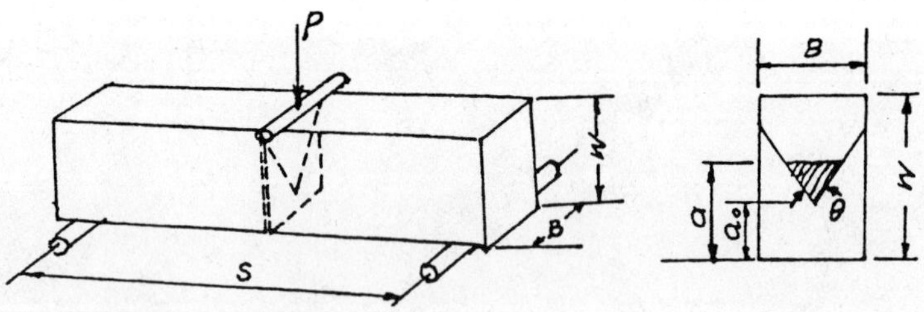

Fig.1 Chevron-notched 3-pt bending specimen

is that in the condition of fixed specimen dimension (length, width and thickness), with the measurable Chevron-notch angle θ and a_o/w (the relative length of Chevron-notch tip to base) as the main variables, the K_{Ic} expression can be determined by the method of compliance calibration.[3,4]

The advantages of this method are

(1) There exists a subcritical crack growth process before unstable fracture, when in the critical state, the behaviour of the crack growth is in accordance with that of the natural crack, this condition can not be satisfied by straight-through notched beam method (SENB).

(2) The critical crack length a_c of the Chevron-notched specimen described above is only related to the specimen geometry shape and not related to material. The K_{1c} expression doesn't include the value a_c. So this method is suitable to determine the K_{Ic} at high temperature.

(3) The specimen can be made conveniently. The slot is acquired by the diamond saw. The experimental procedure is similar to the bending strength test. The angle θ and value a_o are measured after fracture.

The specimen geometry size requirements are $\theta = 60^o \pm 1^o$, $a_o/w = 0.10 - 0.35$, $w = 1.5B$, $s = 4w$. The loading rate is restricted to 0.05 mm/min for sub-critical cracking consideration.

ENERGY METHOD FOR COMPLIANCE EQUATION

In the process of sub-critical crack growth, when the crack length a extends da, according to the Griffith energy principle, the required fracture surface energy is equal to released rate of elastic strain energy, i.e.

$$G_I = \frac{d(w-u)}{dA} \qquad (1)$$

where w is the work of loading, w=Py (y is the middle deflection), u is the elastic strain energy in the material, $u = 1/2(P.y)$, $dA = b \cdot da$, b is the crack width, $b = 2(a-a_o)\tan\theta/2$. Therefore, from E_q (1), we obtain

$$G_I = \frac{d(Py)}{2bda} = \frac{P^2}{2b}\frac{dC}{da} \qquad (2)$$

where C is the specimen compliance, C=y/p. In the critical state, E_q (2) becomes

$$G_{Ic} = \frac{P_c^2}{2b_c}\left[\frac{dC}{da}\right]_{a=a_c} \qquad (3)$$

where G_{Ic} is a material constant and equals the crack extension resistance, i.e. fracture surface energy.

If $G_I \geq G_{Ic}$, the crack will extend immediately. Because the crack front width b increases with the crack length a, if we combine E_q (2) and E_q (3), the crack extension condition will be obtained as follows

$$\frac{P^2}{2}\frac{dC}{da} \geq G_{Ic} \cdot b \qquad (4)$$

The crack front width b is a variable value during the cracking process. If

$$\frac{d}{da}\left(\frac{P^2}{2}\frac{dC}{da}\right) < \frac{d}{da}(G_{Ic} \cdot b) \qquad (5)$$

Then the extending crack will grow slowly and finally arrest. Only when the load increases and the left side of formulae (5) becomes larger than the right side (formulae (4) is also satisfied), the crack will extend again, which is the sub-critical crack growth process. In the critical state, formula (4) and (5) become equalities. Combining them we obtain

$$(a_c - a_o)\left[\frac{d^2C}{da^2}\right]_{a=a_c} = \left[\frac{dC}{da}\right]_{a=a_c} \tag{6}$$

This is the compliance condition equation of the Chevron-notched specimen in the critical state. The compliance measured needs to satisfy this equation.

COMPLIANCE CALIBRATION

Because the equations above have no intrinsic characteristic related to the material, and relate only to specimen dimensions and crack size, we use the Lc-4 hard aluminium alloy with the known elastic modulus to carry out the compliance measurement. A series of hard aluminium alloy specimens with different a_o was made. For every specimen, with the line cutting machine, we made artifical slot to get progressive depths a which imitates the crack extension depth. The slot width is 0.1mm. For different value a, measuring the specimen compliance in the condition of 3-pt bending, we acquire the experimental relationship between the compliance C and the crack length a for every a_o.Fig. 2 shows an example of our experiment. By curve regression

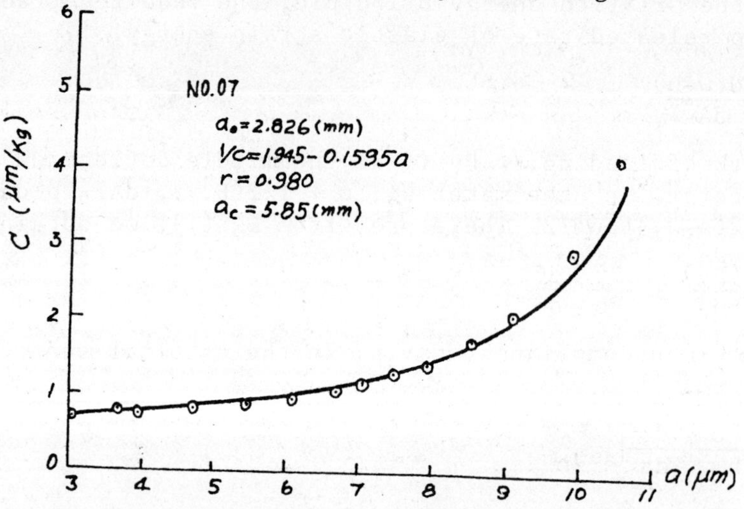

Fig. 2 Relationship between c and a of No.07 specimen

the constants A and B in the empirical equation

$$1/C = A + Ba \tag{7}$$

were determined as a function of a_o. Substituting E_q (1) for E_q (6), we have the relation between a_c and a_o

$$a_c = \frac{2}{3}a_o - \frac{A}{3B} \tag{8}$$

E_q (8) indicates that the critical crack length a_c depends only upon a_o and is independent of material.

Having obtained the compliance, we can get the G_{Ic} and K_{Ic} expressions by E_q (3). In dimensionless form we obtain

$$G_{Ic} = \frac{P_c^2}{2} \frac{1}{b_c E w} \left(\frac{dCEw}{da} \right)_{a=a_c} \qquad (9)$$

where E is the elastic modulus of material used in compliance, w is the specimen thickness. Putting E_q (9) into the plane strain fracture toughness formulae, we have

$$K_{Ic} = \left(\frac{E\,G_{Ic}}{1-\mu^2} \right)^{1/2} = \frac{P_c}{2w^{3/2}(1-\mu^2)^{1/2} \tan^{1/2}(\theta/2)} Y(a_o/w) \qquad (10)$$

where $$Y(a_o/w) = \left(\frac{Ew^3}{a_c-a_o} \right)^{1/2} \left(\frac{dC}{da} \right)^{1/2}_{a=a_c}, \text{ and } \mu = \text{Poisson's ratio.} \qquad (11)$$

Substituting E_qs(7) and (8) into E_q (11), we get

$$Y(a_o/w) = \left[\frac{27Ew^3}{4B(a_o+A/B)^3} \right]^{1/2} \qquad (12)$$

The corresponding data of $Y(a_o/w)$ and a_o/w were obtained and are shown in Fig. 3 by substituting values A and B for different a_o into E_q (12). By curve regression with respect to a_o, the empirical formulae of $Y(a_o/w)$ is obtained

$$Y(a_o/w) = 17.959+20.708(a_o/w)+179.543(a_o/w)^2 \qquad (13)$$

$$(r=0.998)$$

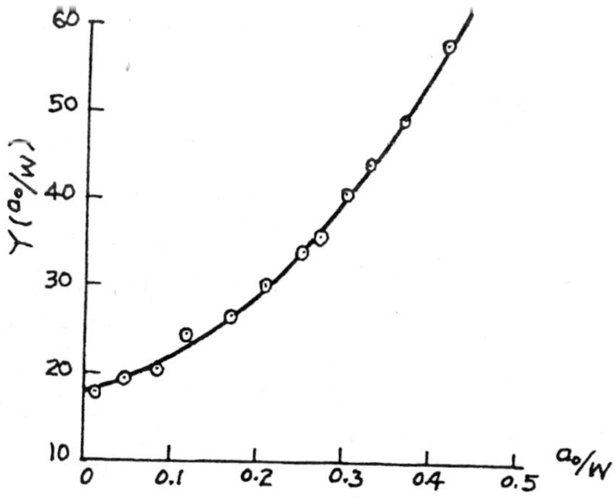

Fig.3 $Y(a_o/w)$ - (a_o/w) Curve

Using E_qs (10) and (13), if the Poisson's ratio is known, we can measure the critical load P_c by 3-pt bending test and then calculate the fracture toughness K_{Ic} with no necessity to determine the crack length a_c.

RESULTS AND DISCUSSION

Verified by experiments, this paper obtains the effects of the specimen geometry size on the K_{IC}, which is described as follows

(1) The range of value a_o/w used

Test indicates that when a_o/w varies at range of 0.05-0.41, it has no significant influence on K_{IC}. So we let a_o/w = 0.10-0.35.

(2) Angle θ

Test indicates that when $\theta = 60^o \pm 1.3^o$, the variation of K_{IC} is within the range of ±0.1%. Now we set $\theta = 60^o \pm 1^o$.

(3) The Chevron-notch slot width δ

H.Neuber[5] theoretically studied the stress concentration of a hyperbolic slot under the tensile stress and indicated that the degree of stress concentration has relations to b/ρ , where b is the specimen width at which slot is cut; b is corresponds to the critical crack width b_c at the Chevron-notch slot (described formerly). ρ is the slot curvature radius, which corresponds to half of the Chevron-notch slot width δ_c. In order to eliminate the error in the K_{IC} expression because of the different slot widths, it demonds

$$(b_c/\delta)_{\text{calibration specimen}} = (b_c/\delta)_{\text{testing specimen}}$$

b_c is a fixed value related only to the specimen shape and size, and for the calibration specimen (depth w=13.5 mm, δ=0.6 mm). When testing a specimen of depth w=6.75 mm, then δ must be $0.6X\frac{6,75}{13.5} = 0.30$ mm. By considering the ceramic specimen depth usually used is 6 mm, δ becomes 0.27 mm. In this paper, slot was cut by a diamond saw, slot width δ was measured as about o.28 mm.

(4) The K_{IC} values for the same HPSN were measured by the Chevron - notched, straight - through notched, and Knoop indented 3-pt bending beam method. The results are in Table 1.

From the table, it may be seen that similar K_{IC} values are obtained by Chevron-notched beam method and knoop indented beam method. For fine grain structural ceramics, such as HPSN, it has been reported that K_{IC} measured by SENB method has a tendency toward a larger value, which is also confirmed by our experimental results[6].

Table 1 Comparisons of K_{IC} results measured by three kinds of methods

	SENB method	Chevron-notched beam method	Knoop indented beam method
$K_{IC} \pm \sigma_{n-1}$	5.43±0.24	4.61±0.25	4.54 (Mpa $m^{1/2}$)
C_v	4.4%	5.4%	0.4%

SUMMARY AND CONCLUSIONS

1. Based on the good correlation factors obtained from the regression curves, when the chevron-notched beam method with compliance calibration is used to measure the K_{Ic} value of brittle materials, the experimental results are in good agreement with the theoretical deduction.

2. The plane strain fracture toughness K_{IC} expression obtained by Chevron-notched beam method with compliance calibration is written as

$$K_{IC} = \frac{P_c}{2w^{3/2}(1-\mu^2)^{1/2}\tan^{1/2}\theta/2}[17.959+20.708(a_0/w)+179.543(a_0/w)^2]$$

3. The requirements for the specimen geometry size are

$$B/w = 2/3, \quad w/s = 1/4, \quad \delta = 0.25\text{-}0.28 \text{ mm},$$

$$\theta = 60^\circ \pm 1^\circ, \quad a_0/w = 0.10\text{-}0.35$$

ACKNOWLEDGMENTS

The authors would like to express their appreciation to WenJie Si for preparation of this paper.

REFERANCE

(1) L.M.Barker " Theory for Determining K_{1c} for small, non-LEFM Specimens, supported by Experiments on Aluminum," Intern.J.Fracture, 15 (6) 515-536 (1979).

(2) D.Munz, R.T. Bubsey, and J.L. Shannon, Jr, "Performance of Chevron-notch Short Bar Specimen in Determining the Fracture Toughness of Silicon Nitride and Aluminum Oxide", J.Testing & Eval., 8(3) 103-107 (1980).

(3) Guan Zhenduo, "Study on K_{Ic} Measurement by Use of Short Chevorn-notch Bar Specimen", J. Chinese Silicate Soc., 12(4) 456 463 (1984).

(4) Du Sinmin, Guan Zhenduo," Compliance Calibration of Chevron-notched Cleave Specimen and Determination of K_{Ic} Values in Ceramic Materials", Ibid, 11(3) 314-324 (1983).

(5) H. Neuber, Theorie der ebenen Kerbwirkung, 51, Kerbspannungslehre,Zweite Auflage, Springer-Verlag, Berlin (1958).

(6) Guan Zhenduo, Du Sinmin," Comparison of Methods of Determining Fracture Toughness K_{Ic} of Ceramic Materials", J.Chinese Silicate Soc., 10(3) 262-271 (1982).

THE CHEVRON-NOTCHED BEND BAR TECHNIQUE FOR FRACTURE RESISTANCE MEASUREMENTS OF CERAMICS

Asish Ghosh, Michael G. Jenkins*, Kenneth W. White**, Albert S. Kobayashi and Richard C. Bradt
College of Engineering, University of Washington, Seattle, WA 98195. *ORNL, Oak Ridge, TN 37831. **Univ. of Houston, TX 77004.

ABSTRACT

Application of the chevron-notched bend bar specimen to the measurement of the fracture resistance of ceramics is described. The specimen and technique have the advantages that fracture toughness values, complete crack growth resistance, or R-curves and the work-of-fracture can all be readily obtained from a single test specimen. Furthermore, the method is easily applied to specimens at elevated temperatures, readily adapting to the non-contact LISG monitoring system. Measurements of a dense sintered SiC utilizing the technique are compared with results from other techniques and specimens.

INTRODUCTION

The increasing utilization of ceramics in design for structural applications has necessitated that materials engineers characterize the mechanical properties as thoroughly and as accurately as possible. The fracture resistance is of primary consideration[1]. Numerous experimental techniques exist to measure the fracture resistance of ceramics. Some of the methods are utilized more extensively than others for a variety of reasons, including the ease of sample preparation and also the simplicity of the measurment. However, the widespread acceptance, simplicity or the extensive use of a single technique, does not necessarily indicate the superiority of that method over others. The acceptance of any single technique must also be validated by its reproducibility and the agreement of its experimental values with those determined by other valid test methods.

Many of the experimental techniques which have been extensively used to measure the fracture resistance of ceramic materials have been recently reviewed by Sakai[2]. The techniques can be divided into microflaw and macroflaw methods. Two commonly utilized microflaw techniques are the controlled indentation-produced microflaw method(CFKI), where the flaw size is determined after breaking a strength specimen and the indentation microfracture technique which utilizes the crack lengths emanating from a high-load-level Vicker's micro-indentation. These microflaw methods utilize cracks or flaws which are of comparable dimensions to the actual strength controlling flaws of the material, and also often comparable in size to the microstructure. Macroflaw methods are the chevron-notched(CHV) and the straight-through-notched bend bar(SNBB) techniques (including the single edge precracked beam) the double torsion(DT) method, the double cantilever beam(DCB) technique and the compact tension sample. The macroflaw methods often have the advantage of ease of measurement of the crack lengths or in some instances may be constant K specimens that do not require actual crack length determinations.

In this paper the advantages of the chevron-notched bend bar technique and its ability to readily determine the fracture toughness, the crack growth resistance or R-curve and the total work-of-fracture of ceramic materials at elevated temperatures are addressed. The relative ease of using the chevron-notched bend bar method at elevated temperatures will also be discussed[3-6].

THE CHEVRON-NOTCHED BEND BAR

Figure 1 illustrates the chevron-notched bend bar specimen. The triangular nature of the chevron shaped section causes the macrocrack to readily initiate at the chevron apex yielding an "atomically-sharp" quasistatically extending macrocrack[7] at a low applied load. The feature of easy crack initiation at the chevron apex avoids the need to precrack the test samples, as is necessary for some other specimens including the single edge precracked beam specimen, double cantilever beam specimen, the double torsion specimen and the compact tension specimen. The apex of the chevron-notched test specimen gives rise to a high stress concentration such that crack initiation and propagation occurs at relatively low loads. Because of the low initiation load, the stored elastic strain energy in the specimen and the test machine is also low, which promotes the stable crack growth necessary for R-curve determination.

As the crack propagates the crack front continually increases in length. Since the edges of the crack front terminate at the edges of the sawed chevron section, the crack path is guided so that the area of the chevron is the area of the crack. This ensures that the crack propagation remains in the Mode I condition throughout. Numerical analyses have shown that the crack front is always in a plane strain condition [8,9]. Since in the chevron-notched beam specimen the crack front is constantly increasing, the

FIGURE 1 : THE CHEVRON NOTCH THREE POINT BEND SPECIMEN

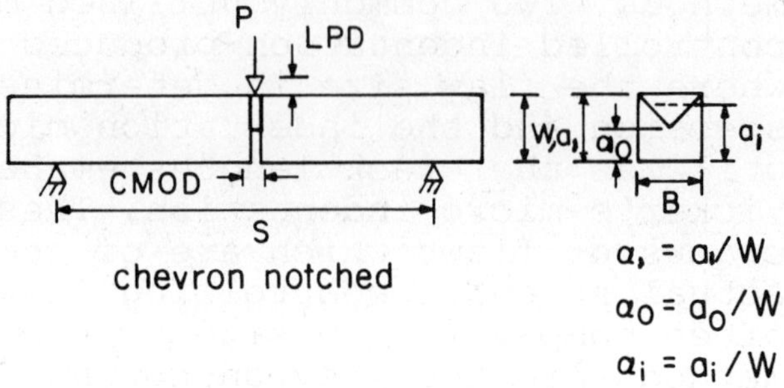

percentage of plane stress effect which might exist at the edges is negligible and a major portion of the crack propagation occurs in the plane strain condition. The testing load line for the chevron-notch technique is also in line with the crack propagation as shown in Figure 1 which avoids excessive bending of the specimen and assists with the easy alignment of the test specimen. These ensure the correct determination of the fracture resistance values.

As the crack progresses through the chevron, the crack front continually increases promoting stable crack growth. The area under the load displacement curve can be subdivided to assess incremental crack extension. Stress intensities at each incremental crack extension can be determined to yield the complete R-curve. Furthermore, the complete area under the load displacement curve can be used to determine the total work-of-fracture of the material, as descibed by Nakayama[10] and by Tattersal and Tappin[11].

As noted, the geometry of the chevron-notched bend bar specimen allows for the initiation and the stable growth of an "atomically sharp" macrocrack during the test. A schematic of a typical load displacement curve is shown in Figure 2. The specimen undergoes a linear load displacement curve until the crack propagation commences and then becomes non-linear just before it reaches P_{max}. This non-linearity is caused by the stable crack propagation. At P_{max}, a balance is achieved between the increasing crack area and the resistance to crack propagation by the material and the crack driving force from the external loading. At this point the geometric correction factor is at its minimum, Y_{min}, and therefore in combination with P_{max} gives a value of the critical stress intensity factor or the fracture toughness[12] for a flat R-curve material. Fracture toughness values can be calculated from the load maximum, P_{max}, and the minimum geometry correction factor, Y_{min}, according to the discussion by Pook[13] and by Munz et al[12].

$$K_{IC} = \frac{P_{max} \; Y_{min}}{B \; W^{1/2}} \qquad (1)$$

FIGURE 2 : TYPICAL LOAD VERSUS CMOD

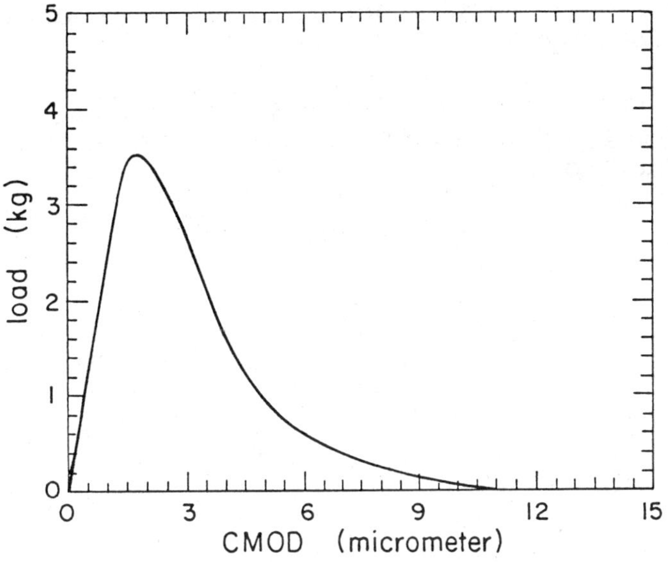

where K_{IC} is the fracture toughness, B is the specimen thickness and W is the specimen width. The relationship between the geometric correction factor, Y_{min}, and the stress intensity has been determined using a 3-D finite element analysis[5]. Once the maximum load for the load displacement curve is achieved, there occurs stable crack growth where the increasing resistance of the specimens to crack extension is just balanced by the crack driving force so that the crack propagates in a quasistatic, stable manner, instead of catastrophically as is characteristic of some other test techniques.

The load-displacement diagram along with a 3-D finite element analysis can be used to yield the crack growth resistance curve or the R-curve for ceramics. In the form of K_R, the R-curves can be plotted using the equations (2):

$$K_{Ri} = \frac{Y_i \, P_i}{B \, W^{1/2}} \qquad (2)$$

$$Y_i = \frac{1}{2} \frac{dc}{d\alpha} \frac{(\alpha_i - \alpha_0)}{(\alpha_1 - \alpha_0)} \qquad (3)$$

where P_i is the instantaneous load, B is the specimen width and W is the specimen thickness. The instantaneous compliance is given by c and the initial notch depth to width ratio is given by α_0. The instantaneous compliance is readily determined from the plot of dimensionless compliance [(E B CMOD)/P] versus the notch depth to notch width ratio (a/W). The slope of the plot at a particular (a/W) ratio gives ($dC/d\alpha$). The final notch depth to width ratio is given by α_1 and the instantaneous notch depth to width ratio where the geometrical parameter Y_i is calculated is given by α_i.

Crack growth resistance curves can also be determined on an energy basis using the area under the curve and the equations:

$$G_i = \frac{\Delta U_i}{\Delta A_i} \qquad (4)$$

$$A_i = (\alpha_i - \alpha_{i-1})\ W\ f(\ W,B\ ) \qquad (5)$$

where the G_i is the instantaneous strain energy release rate, ΔU_i is the instantaneous change in the strain energy as shown by the shaded area in Figure 3, W is the thickness of the specimen and f(W/B) is a function of the specimen width and thickness. This has been discussed at length by Jenkins et al[5]. Since the elastic modulus of the specimen material remains constant, changes in compliance that occur in the specimen during testing, can be accounted for in the calculation of the area under the load-displacement curve. The G_R-curves obtained are therefore an accurate estimate of the crack growth resistance of the material.

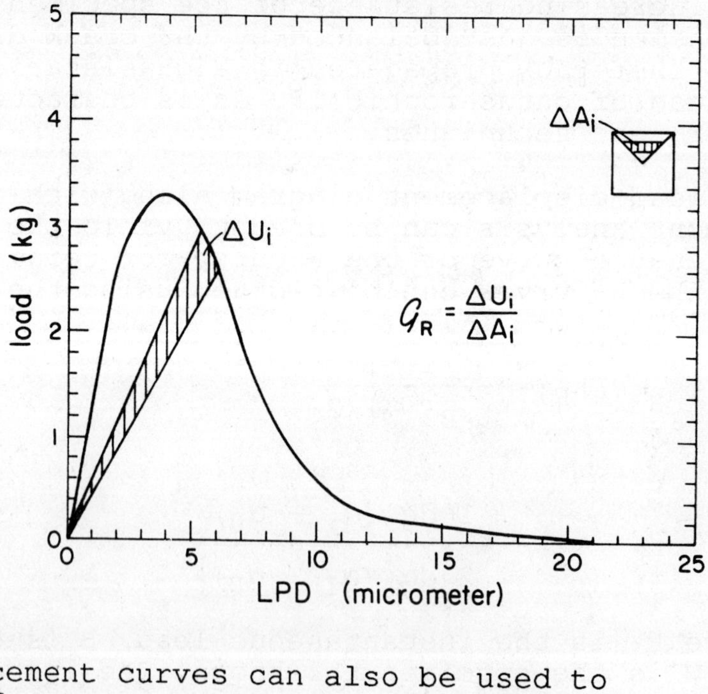

FIGURE 3 : SCHEMATIC OF THE G_R CURVE

The stable load displacement curves can also be used to determine the work-of-fracture as has been described by Nakayama et al[10] and Tattersal and Tappin[11]. The work-of-fracture is expressed by:

$$\gamma_{wof} = \frac{\int P du}{2A} \qquad (6)$$

where P is the load and $\int Pdu$ is the total energy as shown in Figure 4. A is the area of the crack face, also as shown in Figure 4. Apparent overall fracture toughnesses can also be estimated from the work-of-fracture values using the relationship given by the equation

$$K^2_{Ic} = \frac{2E\,\gamma_{wof}}{(1-\nu^2)} \qquad (7)$$

where γ is the work-of-fracture and ν is the Poisson ratio. For a flat R-curve material, the fracture toughness calculated by this equation should agree closely with the P_{max}-Y_{min} value. If, however, this value significantly exceeds the P_{max}-Y_{min} calculated fracture toughness, it is a strong indication of rising R-curve behaviour.

It has also been shown through finite element analysis that the fracture toughness calculated using the chevron notched technique is valid not only for brittle ceramic materials, but also for ductile materials[7]. This implies that many ceramic composites and ceramic materials exhibiting energy dissipative processes such as microcracking and phase transformations can also be evaluated using this technique. Microcracking ceramic materials may, however, exhibit a wide scatter in their fracture toughness values[4].

EXPERIMENTAL PROCEDURES

To provide an example for the validity of the chevron-notched bend bar technique, samples of a densely sintered alpha silicon carbide* were obtained from a commercial

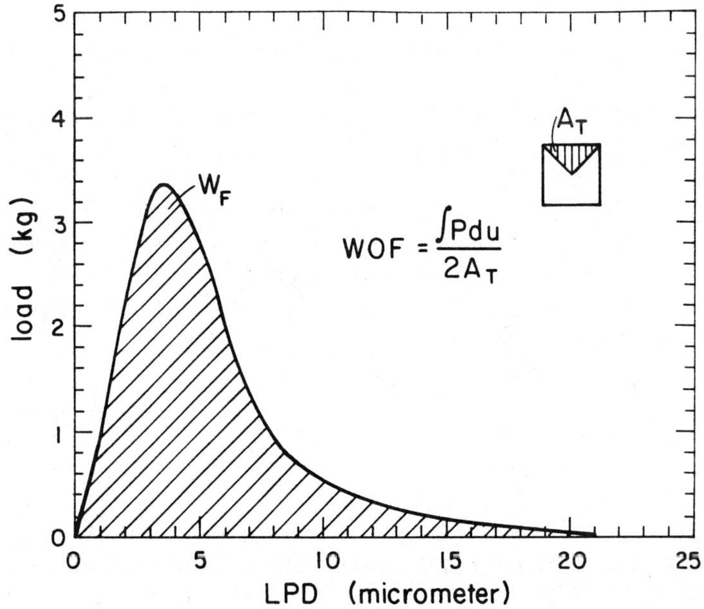

FIGURE 4 : SCHEMATIC OF WORK OF FRACTURE

*Hexoloy, Sohio Engineered Ceramics Company, Niagara Falls, NY

supplier in the form of bars of nominal dimensions 6mmX6mmX50mm. The Young's modulus of this silicon carbide is 430 GPa and the Poisson ratio is 0.17. The density of this sintered alpha silicon carbide was 3.16g/cc. Chevron notches were cut into these bar specimens at their midpoints using a diamond saw blade of thickness 0.010" on a commercial surface grinder. A schematic of the chevron-notched sample was previously shown in Figure 1. The notch depth to width ratio, α_0, was kept at 0.44 for all samples[15].

Fracture testing was completed by directly monitoring the specimen with the laser interferometric strain gage technique (LISG)[4]. Test fixtures were of the same dense silicon carbide material. A test machine cross head speed of 0.01 mm/min on a screw driven displacement controlled load frame was used to test the specimens with a three point bend fixture and a span of 40mm. The load-CMOD curve was analysed using a 3-D finite element program[5]. Elevated temperature testing was done from room temperature to 1400°C at intervals of 200°C using a furnace with SiC heating elements.

RESULTS AND DISCUSSION

A typical chevron-notched specimen load displacement curve for the SiC is shown in Figure 2. Crack growth was observed to be stable at all of the test temperatures. The values obtained for the fracture toughness of this silicon carbide using the chevron-notch beam technique are compared to fracture toughness values obtained by other investigations using different techniques in Table I. These clearly indicate that the fracture toughness values using other techniques, which also employ sharp cracks like the double cantilever beam technique and the double torsion technique, are essentially the same. Values of the fracture toughness at elevated temperatures using the chevron-notched technique also compare favorably with values obtained by the double torsion technique at elevated temperatures which indicates the usefulness of the chevron-notch beam technique for elevated temperature measurements. However, these results do not compare favorably with values obtained by the blunt crack techniques like the sawed notch straight-notch technique.

A typical crack growth resistance curve for this silicon carbide is shown in Figure 5[3]. This type of curve cannot be obtained using a specimen geometry or any technique which does not allow for extensive stable crack growth. Crack growth resistance curves, or R-curves are becoming increasingly important for characterizing ceramics which display an increasing resistance to crack extension with crack growth, especially when they do not exhibit a constant fracture toughness value[16]. Using the chevron-notch bend bar technique, crack growth resistance curves for materials

TABLE I. SUMMARY OF FRACTURE TOUGHNESS VALUES

REFERENCE	COMMENT	FRACTURE TOUGHNESS (MPa $m^{1/2}$) TEMPERATURE (^{o}C)							
		20	200	400	600	800	1000	1200	1400
Ghosh et al[6]	CHV	2.9	2.6	2.9	3.2	2.7	2.8	3.4	3.0
McHenry and Tressler[17]	DT	4.6				4.8 ($900^{o}C$)	5.0	5.4 ($1100^{o}C$)	
	CFKI	3.5				2.6 (900°)	2.6	2.6 ($1100^{o}C$)	
Faber and Evans[18]	DCB	2.5							
Srinivasan and Seshadri	CFKI	3.31							
[19]	SNBB	4.8			4.9	5.0	5.0	4.6	4.9
Coppola et al[20]	CFKI	2.7					2.7	3.4	
Munz et al[21]	CHV	2.96							
Henshall et al[22]	SNBB	3.5			3.3		5.5 ($1100^{o}C$)		5.8 ($1500^{o}C$)

which exhibit a non linear fracture resistance curve can be readily accomplished. The fracture resistance curve of this silicon carbide is shown in Figure 5. This SiC exhibits a flat crack growth resistance curve throughout. As reported elsewhere this fracture resistance remains constant with temperature through $1400^{o}C$[6]. Although the fracture toughness values of this SiC can be readily compared to values obtained from other specimen geometries, the crack growth resistance curves cannot be compared because they have not been reported using other techniques.

The work-of-fracture was calculated using Equation 6 and is summarized in Table II. The fracture toughnesses calculated from the work-of-fracture values are also given in Table II. The fracture toughness values which were obtained using this derivative method do not exactly agree with the values obtained for the silicon carbide using other techniques, with the exception of the room temperature results, consequently it is not a recommended procedure. However, the fact that these toughness values are generally about the same as the P_{max}-Y_{min} values substantiates the flat R-curve behavior of this material.

FIGURE 5 : ROOM TEMPERATURE G_R CURVE OF SiC

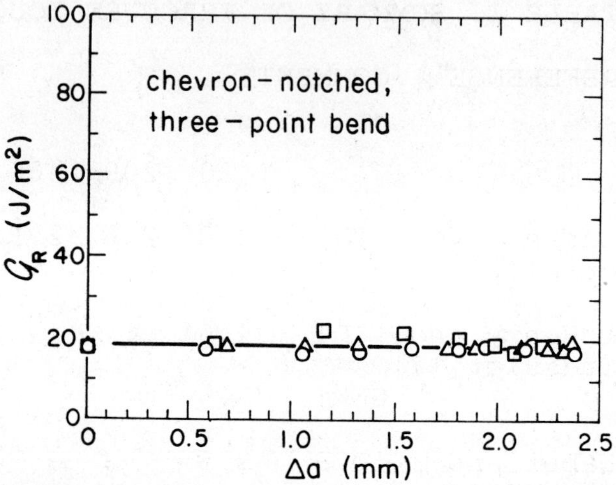

TABLE II. SUMMARY OF FRACTURE PARAMETER RESULTS

Temperature (^oC)	Average K_{Ic} (MPa $m^{1/2}$)	Average WOF (J/m^2)	K_{Ic} from WOF (Eq. 7) (MPa $m^{1/2}$)
20	2.91 ± 0.31	8.81 ± 0.64	2.82 ± 0.76
200	2.64 ± 0.33	6.39 ± 1.39	2.40 ± 1.12
400	2.88 ± 0.41	5.76 ± 0.80	2.26 ± 0.84
600	3.19 ± 0.49	7.98 ± 0.53	2.58 ± 0.68
800	2.71 ± 0.68	5.31 ± 1.03	2.10 ± 0.95
1000	2.76 ± 0.68	7.11 ± 2.22	2.42 ± 1.38
1200	3.37 ± 0.05	6.33 ± 1.69	2.26 ± 1.20
1400	3.03 ± 0.95	4.64 ± 0.54	1.93 ± 0.68
Overall Averages	2.94 ± 0.26	6.54 ± 1.38	2.35 ± 0.28

SUMMARY AND CONCLUSIONS

The chevron-notched bend bar technique is a reliable experimental method to obtain stable crack growth for the fracture resistance measurement of ceramics. The method yields results for a large "sharp" crack representative of an actual crack in the material and thus gives a realistic measure of the crack growth resistance of the material. Its advantage is that it can be used to measure the fracture toughness, the R-curve and the work-of-fracture from a single specimen. In addition to its use at room temperature, the technique can also be easily applied at elevated temperatures as it is readily monitored using the non-contact LISG methodology.

From the P_{max}-Y_{min} criterion the fracture toughness of brittle materials is readily measured using this technique. As above by the comparison of results for a dense SiC, this method yields comparable fracture toughness to other sharp crack techniques. For flat R-curve materials it is quite satisfactory.

The method also facilitates constructing the crack growth resistance curves for brittle ceramic materials. The crack growth resistance curve, or R-curve is an important fracture resistance parameter for those ceramic materials which exhibit non-linear load displacement curves. With the increasing importance of the use of structural ceramics which either microcrack or undergo phase transformations and also ceramic matrix composites, the chevron-notched bend bar technique is an excellent method to characterize the R-curve of the material.

The work-of-fracture is also readily calculated using the chevron-notch technique, although it is not always possible using other techniques. Although the work-of-fracture is usually considered to be a very useful parameter for many linear elastic ceramic materials, for non-linear fracture processes in many ceramic materials it is an important characterization parameter that is descriptive of the total energy absorption during crack propagation.

ACKNOWLEDGEMENTS

The research reported in this paper was supported by the Department of Energy, ORNL, Contract No. 86A-00209C. The authors are grateful to M. Srinivasan of Sohio Engineered Materials for assistance with providing the samples of SiC for which results are reported here.

REFERENCES

1)D. Broek, "Elementary Engineering Fracture Mechanics" 3rd Edition, Martinus Nijhoff Publishers, The Hague, (1982).

2)M. Sakai, "Evaluation Methods for Ceramics: Fracture Toughness of Polycrystalline Ceramics" Bull. Ceram. Soc. Jap., 23 [5] 480-488 (1988).

3)M.G. Jenkins, A. Ghosh, K.W. White, A.S. Kobayashi and R.C. Bradt, "The Fracture Resistance of a Sintered Silicon Carbide Using The Chevron-Notch Bend Specimens," Proceedings of the Symposium on Silicon Carbide held at Columbus, OH, Aug.1-3, (1987).

4)M.G. Jenkins, A.S. Kobayashi, M. Sakai, K.W. White and R.C. Bradt, "Fracture Toughness Testing of Ceramics Using a Laser Interferometric Strain Gage," Amer. Ceram. Soc. Bull., 66 [12] 1734-1738 (1987).

5) M.G. Jenkins, A.S. Kobayashi, K.W. White and R.C. Bradt, "A 3-D Finite Element Analysis of a Chevron-Notched Three Point Bend Fracture Specimen for Ceramic Materials," Int. J. Frac., 34 281-295 (1987).

6) A. Ghosh, M.G. Jenkins, K.W. White, A.S. Kobayashi and R.C. Bradt, "Elevated Temperature Fracture Resistance of a Silicon Carbide," accepted for publication by the J. Amer. Ceram. Soc.

7) L.M. Barker, "Theory For Determining K_{Ic} From Small, Non LEFM Specimens, Supported by Experiments on Aluminum" Int. J. Frac. 15 [6] 515-536 (1979).

8) J.C. Newman, Jr., "A Review of Chevron Notched Fracture Specimens" pg 5-31 "Chevron Notched Specimens, Testing and Stress Analysis" Ed. J.H. Underwood, S.W. Freiman and F.I. Baratta, ASTM STP 855, (1984).

9) I.S. Raju amd J.C. Newman, Jr., "Three Dimensional Finite-Element Analysis of Chevron Notched Specimens" pg 32-48 "Chevron Notched Specimens, Testing and Stress Analysis" Ed. J.H. Underwood, S.W. Freiman and F.I. Baratta, ASTM STP 855, (1984).

10) J. Nakayama, "Bending Method for Direct Measurement of Fracture Energy of Brittle Materials," Jap. J. of App. Phys., 3 [7] 422-423 (1964).

11) H.G. Tattersal, G. Tappin, "The Work of Fracture and its Measurement in Metals, Ceramics and Other Materials," J. Mater. Sci. 1 296-301 (1986).

12) D.G. Munz, J.L. Shannon, Jr. and R.T. Bubsey, "Fracture Toughness Calculation from Maximum Load in Four Point Bend Tests of Chevron Notch Specimens," Int. J. Frac., 16 R137-R141, (1980).

13) L.P. Pook, "An Approach to a Quality Control K_{Ic} Testpiece," Int. J. Frac., 8 103-108, (1972).

14) R.T. Bubsey, J.L. Shannon, Jr. and D. Munz, "Development of Plane Strain Fracture Toughess Test for Ceramics Using Chevron Notched Specimens," pg 753-771 "Ceramics For High-Performance Applications III: Reliability," Ed. E.M. Lenoe, R.N. Katz and J.C. Burke, Plenum Publishing Corporation, (1983).

15) D. Munz, R.T. Bubsey and J.E. Srawley, "Compliance and Stress Intensity Coefficients for Short Bar Specimens With Chevron Notches," Int. J. Frac., 16 4 359-374 (1980).

16)M. Sakai and R.C. Bradt, "The Crack Growth Resistance (R-Curve) of Non Phase Transforming Ceramics," Nippon Seramikkusu Kyokai Gakujutsu Ronbushi 96 [8] 801-809 (1988).

17)K.D. McHenry and R.E. Tressler, "Fracture Toughness and High Temperature Slow Crack Growth in SiC," J. Amer. Cer. Soc. 63, (3-4) 152-156, (1980).

18)K.T. Faber and A.G. Evans, "Intergranular Crack Deflection Toughening of Silicon Carbide," J. Amer. Cer. Soc., 66, (6) C94-C96, (1983).

19)M. Srinivasan and S.G. Seshadri, "Application of Single Edge Notched Beam and Indentation Techniques to Determine Fracture Toughness of Alpha Silicon Carbide," p 46-68 in Fracture Mechanics for Ceramics, Rocks and Concrete, ASTM STP 745, (1980).

20)J.A. Coppola, M. Srinivasan, K.T. Faber and R.H. Smoak, "High Temperature Properties of Sintered Silicon Carbide," The Carborundum Company, Niagara Falls, NY (1979).

21)D.G. Munz, J.L. Shannon Jr. and R.T. Bubsey, "Fracture Toughness Calculation from Maximum Load in Four Point Bend Tests of Chevron Notch Specimens," Int. J. Frac., 16, R137-R141, (1980).

22)J.L. Henshall, D.J. Rowcliffe and J.W. Edington, "Fracture Toughness of Single Crystal Silicon Carbide," J. Amer. Cer. Soc. 60, [7-8] 373-375 (1977).

OXIDATION OF SINTERED SILICON NITRIDE MATERIALS

L.Wang, C.He and J.G.Wu
Department of Materials Science and Engineering
Tsinghua University
Beijing, PRC

ABSTRACT

The oxidation and its effect on mechanical properties of sintered Y_2O_3-Al_2O_3 doped Si_3N_4 were investigated. The weight gain of the material oxidized at 1200°C for 1000h is 0.4mg/cm^2 which showed the material had excellent oxidation resistance. Oxidation at 1200°C in air greatly improved the high temperature strength of the material. The strength at 1300°C after oxidation at 1200°C for 1000h is 570MPa, which is 87% higher than that berore oxidation. The oxidation mechanism and the reasons for the improvement of high temperature strength are discussed.

INTRODUCTION

Si_3N_4 material is one of the candidate materials for engine applications, and its oxidation behavior has drawn world-wide attention. Many researches[1-5] have shown that the composition and its distribution and phases of Si_3N_4 ceramics change in an oxidizing atmosphere at high temperature. Therefore the mechanical properties of these ceramics are greatly influenced by oxidation.

Studies on oxidation of nitrogen ceramics have been conducted mainly on HPSN(Hot pressed silicon nitride), RBSN(Reaction bonded silicon nitride) and SSN(Sintered silicon nitride) with a single additive. Some consistent results have been obtained about rules of weight gain in oxidation and the diffusion mechanism. But differences exist among results about the influence of oxidation on mechanical properties. Some researchers believed that oxidation would improve the mechanical properties while some others thought that oxidation degraded the mechanical properties.

In this paper, the oxidation behavior of Y_2O_3-Al_2O_3 doped sintered Si_3N_4 and its influence on mechanical properties were studied.

EXPERIMENTAL

Starting Materials and Preparation

The powders used were α-Si_3N_4 (α>90%, BET area 6.4m^2/g, purity>95%) produced by Huaqing Ceramic Factory, Al_2O_3(99.9% purity, BET area 7.6 m^2/g) produced by Dahua Ceramic Factory, and Y_2O_3(99.9% purity, commercial). The three powders with the required compositions were mixed by ball milling.The mixed powders were shaped into compacts by biaxial and isostatic pressing (250MPa), and then embedded in a powder bed and sintered in a graphite furnace in a flowing N_2 atmosphere. The specimen size after machining and polishing is 3x4x36 mm. Table 1 shows the compositions, sintering and oxidation conditions.

Table 1 The Compositions and Conditions of Sintering and Oxidation

Material Batch	Composition(wt%) Si_3N_4	Y_2O_3	Al_2O_3	Sintering condition	Relative density	Oxidation condition
A	81.6	13.9	4.5	1750°C,2h	97%	1200°C,1000h in static air
B	80.0	15.0	5.0	1700°C,2h	97%	900°C, 300h in static air
C	same as A			1750°C,2h annealing at 1300°C in N_2, 4h	97%	1200°C, 1000h in static air

Oxidation

Oxidation was carried out in a furnace with SiC heating elements in static air. The specimens were removed from the furnace to weigh and measure the three-point bending strength (span 30 mm, cross-head speed 0.4 mm/min for R.T. test, loading speed 24.6kg/min for high temperature test,six specimens per condition were tested to generate the strength data).

RESULTS

Weight Gain

Fig.1 shows a graph of weight gain vs oxidation time. The curve can be fitted by the equation $y^n = kx$. For composition A, n = 1 when t is from 0 to 2h, n = 4 when t is from 2 to 200h, and n = 6 when t is from 200 to 1000h. The n value for B is nearly 1 from 0 to 300h. The weight gain of sample A oxidized at 1200°C for 1000h was 0.4 mg/cm^2, that of sample C oxidized at the same condition was 0.3 mg/cm^2. The weight gain of sample B oxidized at 900°C for 300h was 0.4 mg/cm^2.

Phases

The phases of the specimen before and after oxidation are summarized in Table 2. The surface of all three oxidized material batches consists of β'-Si_3N_4, G, $Y_2Si_2O_7$ and cristobalite. The phases at the interior of batch B and C before oxidation are the same as the ones after oxidation, while for batch A the interior phases after oxidation which include YAG and $Y_2Si_2O_7$ are different from those before oxidation.

Element Distribution

The cation contents of sample B are shown in Table 3. The contents of the unoxidized specimen are the same as its starting composition. After oxidation at 900°C for 300h the cation content at the interior of the specimen

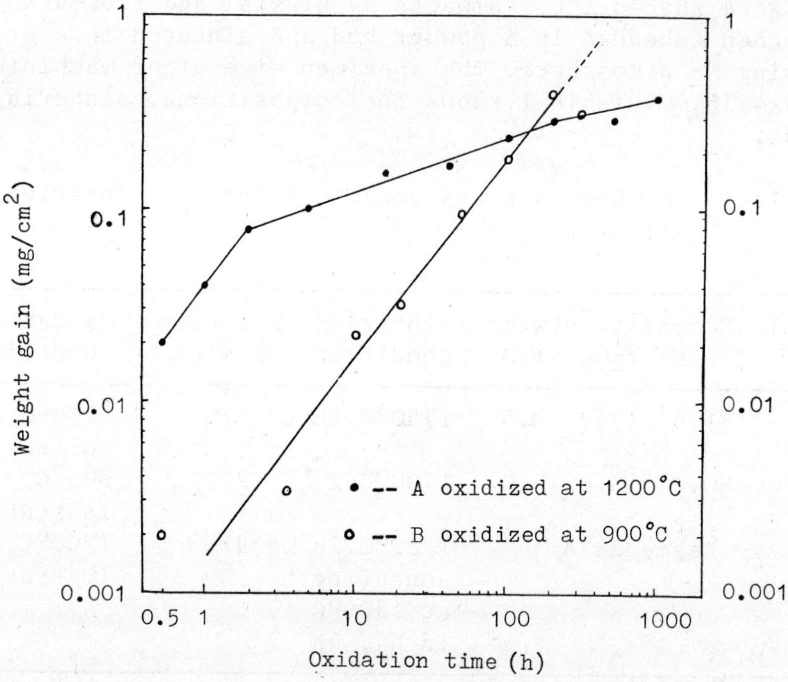

Figure 1. Logarithmic curves of weight gain vs oxidation time

Table 2. Summary of XRD analysis data

Material Batch	Oxidation Condition	Analyzing Zone	Phases*
A	unoxidized	surface	β', G
	1200^oC for 1000h	surface	β', G, $Y_2Si_2O_7$, cristobalite
		interior	β', G, YAG, $Y_2Si_2O_7$
B	unoxidized	surface	β', G
	900^oC for 300h	surface	β', G, $Y_2Si_2O_7$, cristobalite
		interior	β', G
C	unoxidized	surface	β', G, YAG, H, $Y_2Si_2O_7$
	1200^oC for 1000h	surface	β', G, $Y_2Si_2O_7$, cristobalite
		interior	β', G, YAG, H, $Y_2Si_2O_7$

* β'— Solid solution between β-Si_3N_4 and Al_2O_3
G — Glass YAG — $Y_3Al_2(AlO_4)$ H — $Y_5(SiO_4)_3N$

Table 3. Element contents (at%) of sample B

Element	Si	Y	Al	Ca	Fe	K
Before oxidation	80	13	7	--	--	--
After oxidation at 900°C for 300h, interior	83	12	5	--	--	--
After oxidation at 900°C for 300h, surface	77	17	2	2.3	tr.	1.7

is almost unchanged compared with that of the unoxidized specimen. But on the surface of the oxidized specimen the concentration of Y increases and that of Si decreases and impurity cations are found. It suggests that the outward diffusion of impurity cations occurs during oxidation. The content of impurity cations in sample B before oxidation may be beyond the sensitivity of EDAX.

Mechanical Properties

The influence of oxidation on the strength of the materials is shown in Fig. 2 to Fig.5.

For A composition a rapid decrease in the room temperature strength occured during the first 2h of oxidation at 1200°C (Fig.2) while that of B composition oxidized at 900°C showed the opposite tendency (Fig.4). With longer oxidation times the curves of both A and B become flat. The room temperature strength after oxidation is 97% of that before oxidation for material B (Fig.5) and 80% for A (Fig.3).

The high temperature strength of all three oxidized materials was higher than that of unoxidized ones. The strength at 1200°C of A and C increased 70% and 28% respectively by oxidizing at 1200°C for 1000h (Fig.2) while that of B increased only 15% by oxidizing at 900°C for 300h (Fig.4). The strength of A and C compositions rose rapidly during the first 5h of oxidation time at 1200°C and then remained unchanged with prolonged oxidation time (Fig.2).

The strength of the three materials remains essentially constant below 1000°C. At 1200°C, the strength of the oxidized samples, especially sample A, shows a much lower strength degradation than the unoxidized samples (Fig.3 and Fig.5). The oxidized batch A has a 570 MPa strength at 1300°C, which is nearly equal to its room temperature strength (613MPa). The above results suggest that oxidation at 1200°C for 1000h improve greatly the mechanical properties of the material at elevated temperature.

DISCUSSION

Oxidation Mechanism

Several chemical and physical reactions will occur during the oxidation of sintered Y_2O_3-Al_2O_3 doped Si_3N_4 at high temperature, among which the main one is:

$$Si_3N_4 + 3O_2 \longrightarrow 3SiO_2 + 2N_2 \quad (1)$$

The resultant SiO_2 crystallized partly to cristobalite and reacted partly with Y_2O_3 in the intergranular glass to form $Y_2Si_2O_7$:

$$Y_2O_3 + 2SiO_2 \longrightarrow Y_2Si_2O_7 \quad (2)$$

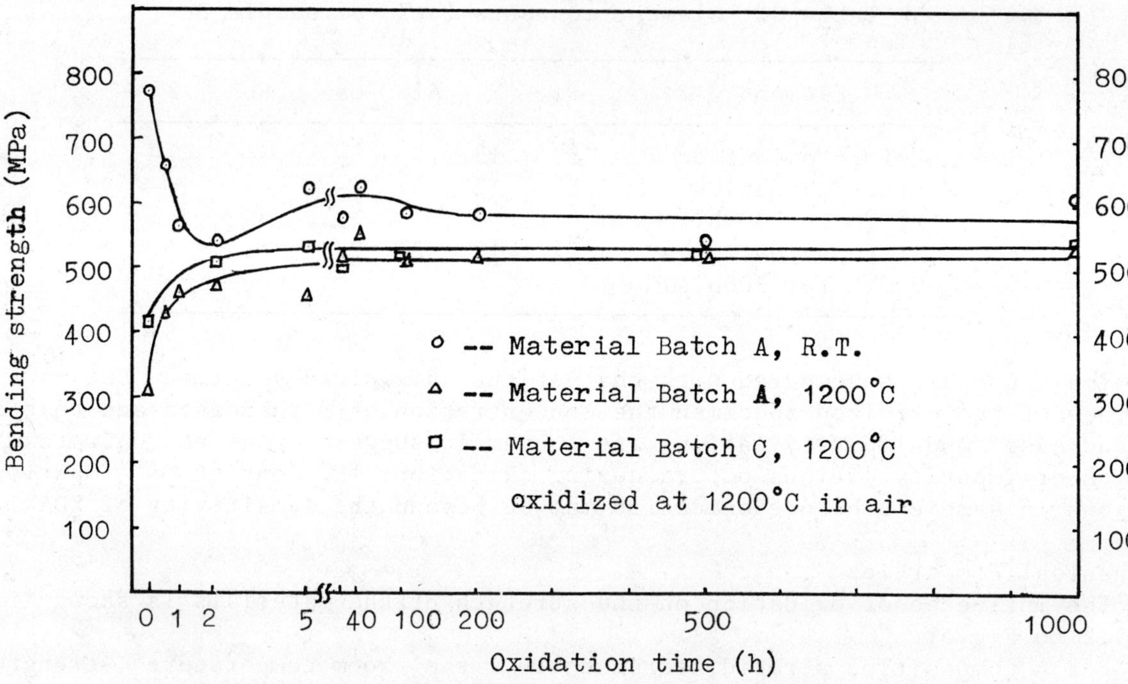

Figure 2. Effect of oxidation time on bending strength

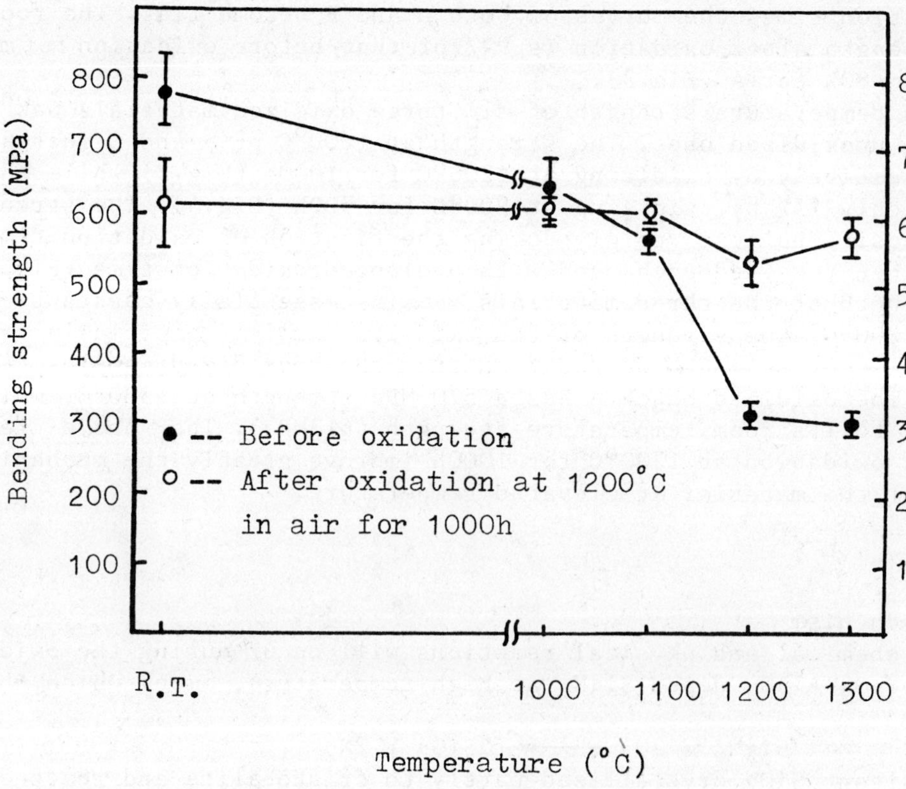

Figure 3. Bending strength of Material A vs temperature

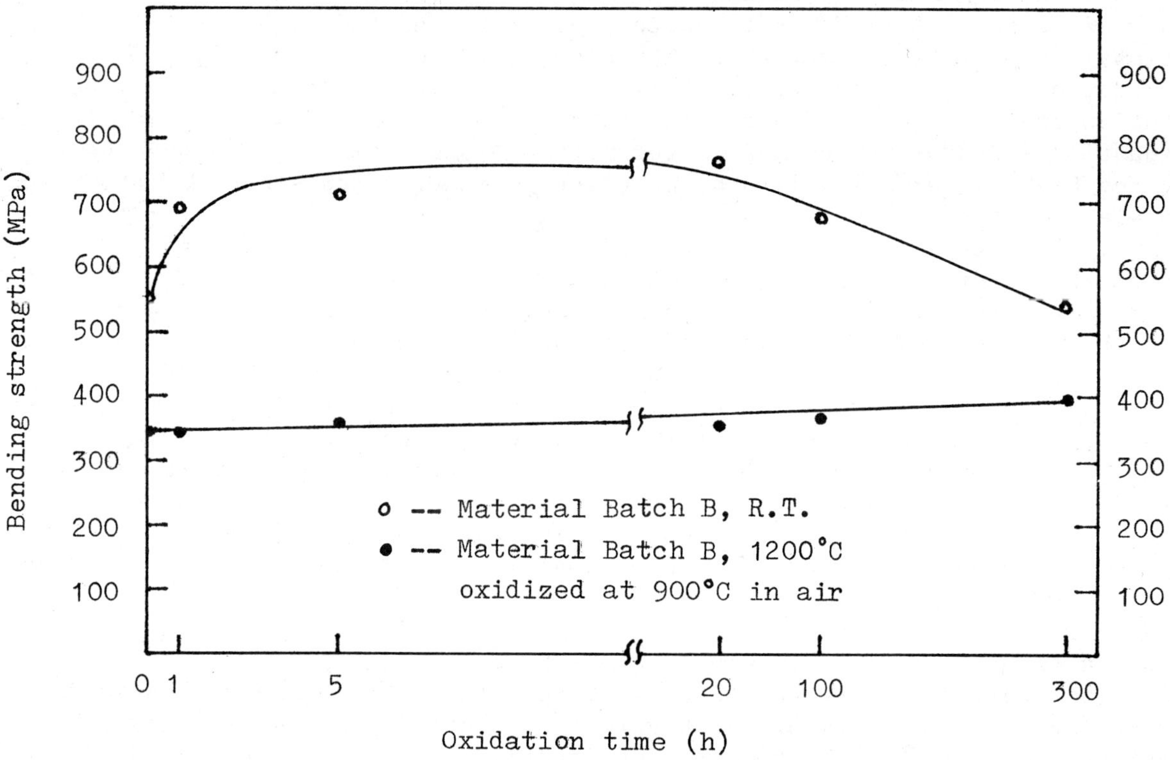

Figure 4. Effect of oxidation time on bending strength of Material B

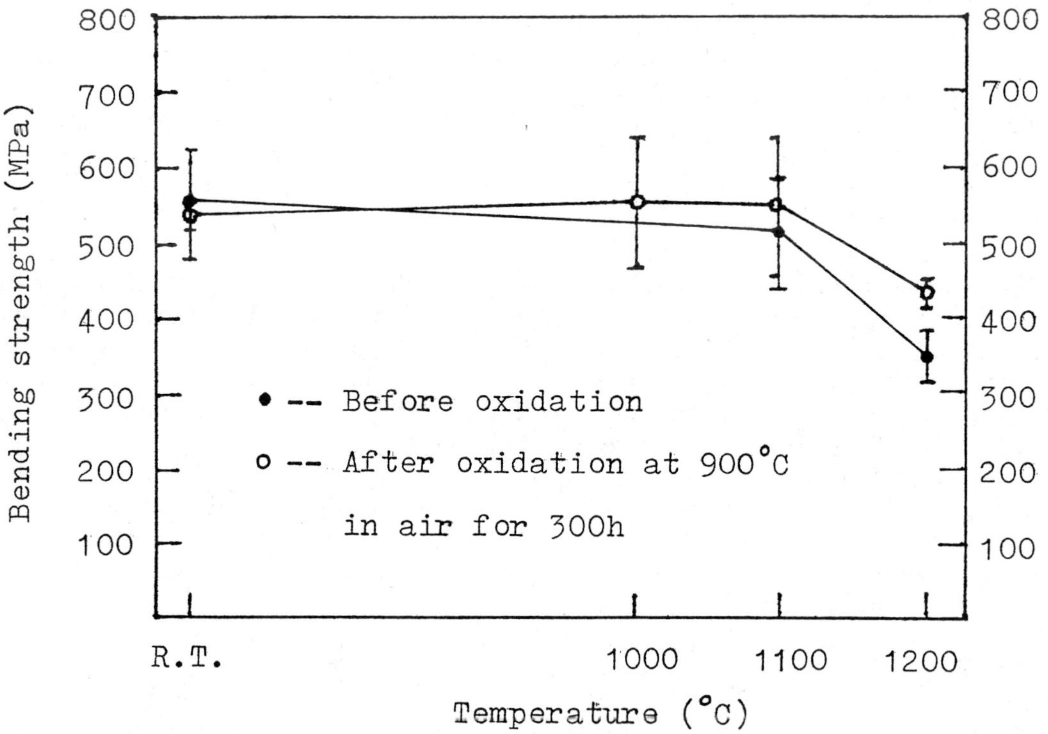

Figure 5. Bending strength of Material B vs temperature

At the beginning the oxidation reaction takes place rapidly becouse of direct contact of Si_3N_4 grains with O_2 in air, and the weight gain is proportional to oxidation time. When the oxide layer covers the surface fully, the oxidation diffusion of Si, Y, Al ions and impurity cations and the inward diffusion of O ions or molecules. The oxidation is slowed down and n value in equation $(W/A)^n = kt$ becomes larger with prolonged time. Therefore the sintered Si_3N_4 materials will show eccellent oxidation resistance if the oxide scale is not disturbed.

Table 3 shows that during oxidation Y ions and impurity cations diffuse from the interior to the surface oxide layer. This changes the intergranular composition and decreases the impurity content in the intergranular glass in the bulk material. Table 2 shows that the grain boundary glass in the interior of unannealed specimens crystallizes during oxidation at 1200°C but does not at 900°C . The phases on the interior of sample C have no apparent change during oxidation because the crystallization of the grain boundaries was almost completed during annealing before oxidation. Both oxidation of the surface and crystallization of the intergranular glass occurs during oxidation at the higher temperature while only oxidation occurs when the oxidation temperature is lower than the transformation temperature of intergranular glass.

Effect of Oxidation on High Temperature Strength

The present work showed the high temperature strength of sintered Si_3N_4 was improved by oxidation, which agrees with other reports[1]. But there exists some defferences about the explanation of this result. Three reasons are suggested as being responsible for the improvement in the high temperature strength through oxidation:

(a). Crystallization of the intergranular glass

The annealing treatment in N_2 at 1300°C given to sample C increased the strength at 1200°C over sample A from 309 MPa to 418MPa (Fig.2).The strength at 1200°C of sample A increased rapidly with oxidation time at 1200°C(Fig.2), while sample B (Fig.4) retained its strength at 1200°C during oxidation at 900°C. The explanation is that the intergranular glass crystallizes at 1200°C but does not at 900°C.

(b). Elimination of surface defects.

The strength of the sintered Si_3N_4 is highly sensitive to surface defects and microcracks. The elimination of defects and blunting of microcracks during oxidation give a positive effect on the high temperature strength. As shown in Fig.2, the strength of sample C which had been annealed before oxidation still increased during the first five hours of oxidation at 1200°C.

(c). "Purification" of the grain boundary.

The outward diffusion of impurity cations during oxidation "purifies"the grain boundaries in the bulk material and changes the composition of the intergranular glass. The high temperature behavior of the material is improved due to the increased viscosity of the intergranular glass as a result of the decrease in impurity cations.

CONCLUSION

1. The weight gain of sintered Si_3N_4 (doped with Y_2O_3-Al_2O_3) with oxidation time follow the expression $(W/A)^n = kt$. Oxidation of the Si_3N_4 at 1200°C consists of an active period, $\leqslant$5 hours and a passive period, $>$5 hours , the later suggesting a diffusion mechanism.

2. The Si_3N_4 material showed excellent oxidation resistance. The weight gain of the material oxidized at 1200°C for 1000h was 0.4 mg/cm^2.
3. Oxidation can improve the high temperature strength of the material. The strength at 1300°C after oxidation at 1200°C for 1000h was 570MPa, which is 87% higher than that before oxidation. Crystallization of intergranular glass, "purification" of the grain boundaries, and elimination of defects on the surface make an important contribution to the improvement of high temperature strength.

REFERENCE

[1]. F.F. Lange, B.I. Davis and M.G. Metcalf, Strengthening of Polyphase Si_3N_4 Materials through Oxidation, J.Mater. Sci., 18 (1983) 1497

[2]. Sunil Dutta and Bruno Buzek, Microstructure, Strength and Oxidation of a 10wt% Zyttrite - Si_3N_4 Ceramic, J. Am. Ceram. Soc., 67(2), (1984)89

[3]. Diane M. Mieskowski and William A. Sanders, Oxidation of Silicon Nitride Sintered with Rare - Earth Oxide - Additions, J. Am. Ceram. Soc., 68(7), (1985) C-160

[4]. R. K. Govila, Fracture of Flash Oxidized, Yttria - doped Sintered Reaction - Bonded Silicon Nitride, J. Mater. Sci., 22(1987) 1193

[5]. Johann E. Siebels, Oxidation and Strength of Silicon Nitride and Silicon Carbide, Special Ceramics 7, (1981) 793

MICROSTRUCTURES AND MECHANICAL PROPERTIES OF CeO_2-Y_2O_3/ZrO_2 CERAMICS

Tian Jiemo*, Zhan Baoqing, Wang Jialong
Wang Lingling, Tong Xiaohua and Guo Jiyou

Fine Ceramic Lab. of the Institute of Nuclear
Energy Technology, Tsinghua University.
Beijing Fine Ceramic Lab.
P.O.Box 1021, Beijing, China

ABSTRACT

Microstructures and mechanical properties of CeO_2-Y_2O_3/ZrO_2 ceramics were investigated using ultrafine CeO_2-Y_2O_3/ZrO_2 powder as the starting material. The structures of CeO_2-Y_2O_3/ZrO_2 containing 8 to 12mol%CeO_2 were composed of tetragonal phases. They exhibited an extremely high bending strength and fracture toughness, about 10^3 MPa and 15 $MPam^{1/2}$, respectively. The structural details of CeO_2-Y_2O_3/ZrO_2 were studied using TEM, SEM and X-ray diffraction analyses. The effects of the structure on mechanical properties are discussed.

INTRODUCTION

Y-TZP (Y_2O_3 stabilized tetragonal zirconia polycrystals), (Y_2O_3-Al_2O_3) -TZP and Ce-TZP have been studied by many workers[1-6]. (Y_2O_3-Al_2O_3)-TZP samples made using HIP processing showed higher bend strength and fracture toughness[7]. It is well known that there exists an extensive solid solution region with the tetragonal phase in CeO_2-ZrO_2 system. Ce-TZP phase is stable at near room temperature[8]. Therefore, Ce-TZP containing various amounts of CeO_2 content may be obtained. Tsukuma has reported that higher strength and fracture toughness of CeO_2-stabilized tetragonal ZrO_2 polycrystals was obtained[9] . Lange[10] has studied (CeO_2-Al_2O_3) -TZP ceramics and its transformation toughening.

This paper describes the microstructure of CeO_2-Y_2O_3/ZrO_2 ceramics produced from fine powder and their mechanical properties. The effects of the microstructure on the mechanical properties are discussed.

EXPERIMENTAL PROCEDURE

CeO_2-Y_2O_3/ZrO_2 powders were prepared by the coprecipitation technique using a mixed solution of $ZrOCl_2$, $Y(NO_3)_3$ and $Ce(NO_3)_4$ (Table 1). Then, the powders were calcined at 1000℃ for 2hr, wet-milled in ethyl alcohol for 24hr, dried, and then granulated through a 60 mesh screen. The samples were formed by cold isostatic pressing at 300MPa and sintered at 1450 to 1550℃ for 2 hr. Some samples were prepared by hot pressing at 1500℃ and 20MPa for 30 minutes.

The sample compositions were determined by EDAX(energy dispersive X-ray analysis). The bulk densities were measured by Archimedes method. The phase structure was analyzed by X-ray diffraction. The bend strength and fracture toughness samples were 3×4×40mm. The fracture toughness was measured by the chevron notched beam technique[9]. Chevron notches were 0.1-0.2mm. The X-ray diffraction was used to analyze the fraction of tetragonal-to-monoclinic transformation in the fractured samples.

RESULTS AND DISCUSSION

Composition of Phases

The phases present in the (CeO_2-Y_2O_3)/ZrO_2 samples were determined using X-ray diffraction. The results are given in Table 1 and Fig.1.

Table 1. Composition of the Phases of (CeO_2-Y_2O_3)/ZrO_2 *

Y_2O_3 (mol%)	CeO_2 (mol%)	TZP (%)	m-phase+ (%)
3	8	93	7
3	10	100**	0
3	12	100**	0

* : HP samples

** : In the samples, there are small amount of the precipitation phase of CeO_2

\+ : monoclinic phase structure.

From Fig.1, only a small amout of m-phase is found in HP Y-Ce-TZP containing 8mol% CeO_2 , but when the CeO_2 content exceeds 8mol% CeO_2, the excess CeO_2 precipitates in the t-phase (tetragnal phase) (Fig.1,b,c).

Sintered Y-Ce-TZP samples containing 8mol% CeO_2 are almost totally composed of the t-phase (Fig.2). But excess CeO_2(10mol%) only forms a small amout of CeO_2 precipitation phase in the Y-Ce-TZP (Fig 3). In the 2mol% Y_2O_3-12 mol% CeO_2/ZrO_2 samples sintered at 1450℃ or 1550℃ for 2hr, no m-phase was found (Fig. 4). They show that full t-phase materials are obtained with appropriate values of Y_2O_3 and CeO_2(Figs. 1-4).

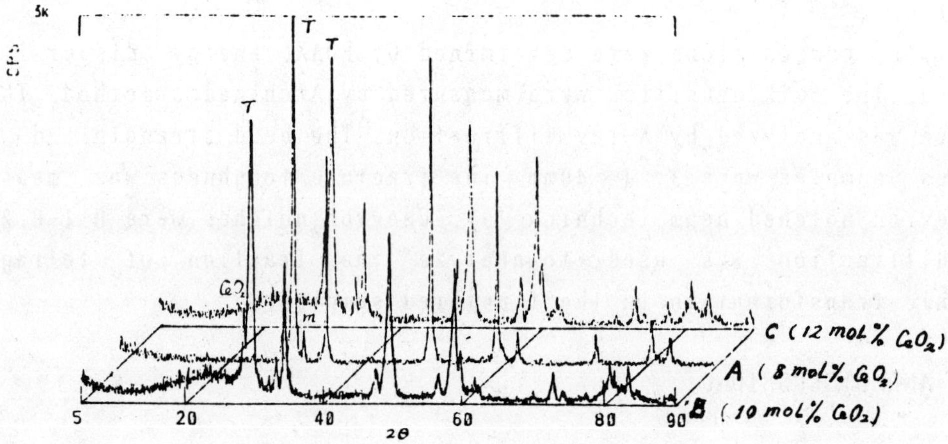

Fig.1. X-ray Diffraction Spectra of HP $(CeO_2-Y_2O_3)/ZrO_2$ (Cukα Radiation)

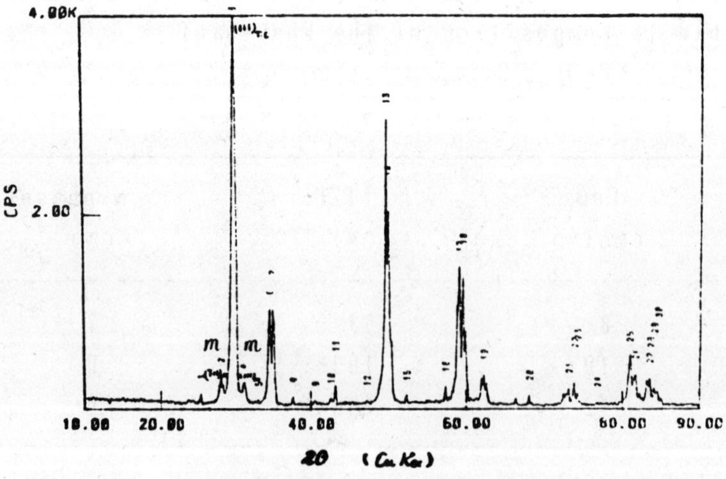

Fig.2. X-ray Diffraction Spectra of the Sintered (3mol% Y_2O_3-8mol%CeO_2)-ZrO_2 (Cukα)

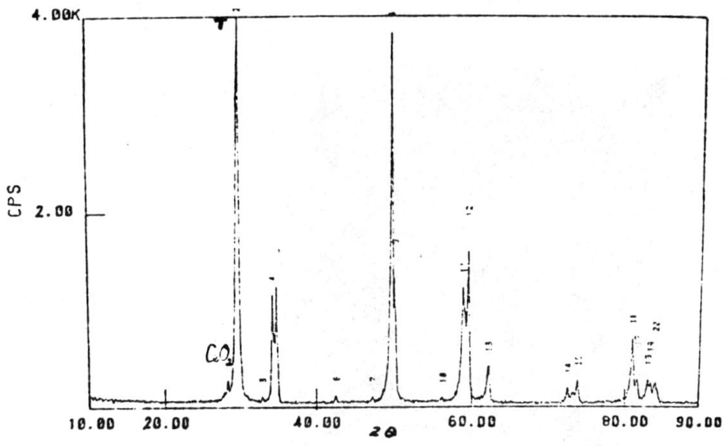

Fig.3. X-ray Diffraction Spectra of the Sintered ($3mol\%Y_2O_3$-$10mol\%$ CeO_2)-ZrO_2 (Cukα)

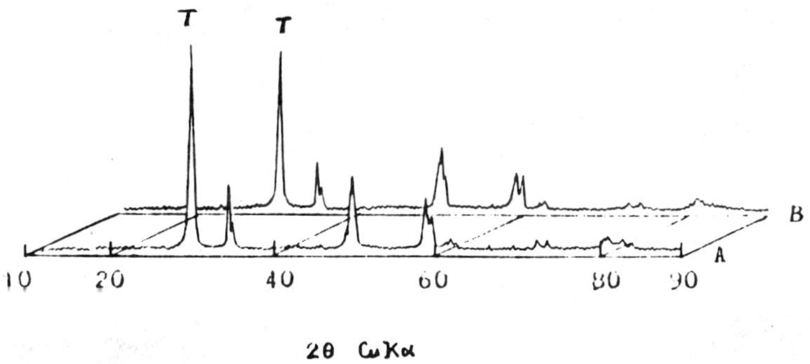

Fig.4. X-ray Diffraction Spectra of $2mol\%Y_2O_3$-$12mol\%CeO_2/ZrO_2$
A: sintered at 1450℃ for 2hr
B: sintered at 1550℃ for 2hr
(Cukα)

Microstructure and Morphology

The microstructure and morphology of Y-Ce-TZP were studied by TEM, SEM and EDAX methods. Figures 5 and 6 show the morphology and electron diffraction pattern of Y-Ce-TZP sample containing 3 mol% Y_2O_3 and 12 mol% CeO_2 respectively. Essentially no glass phase is found between grain boundaries in the SEM morphology (Fig.5). The electron diffraction pattern shows that it is a typical t-phase (Fig.6.b).

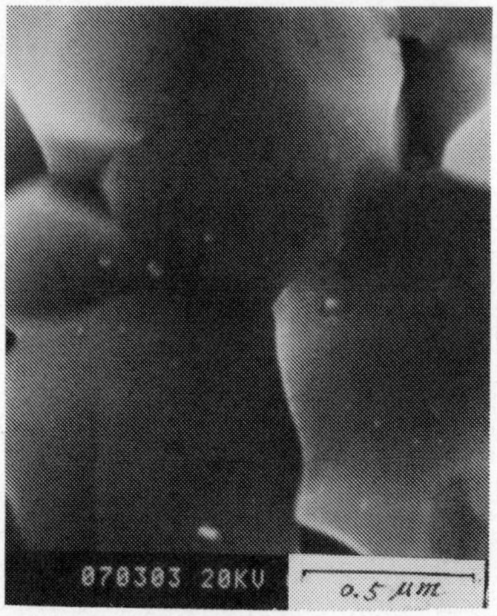

Fig.5. Morphology of
$3mol\%\ Y_2O_3+12mol\%\ CeO_2$-TZP

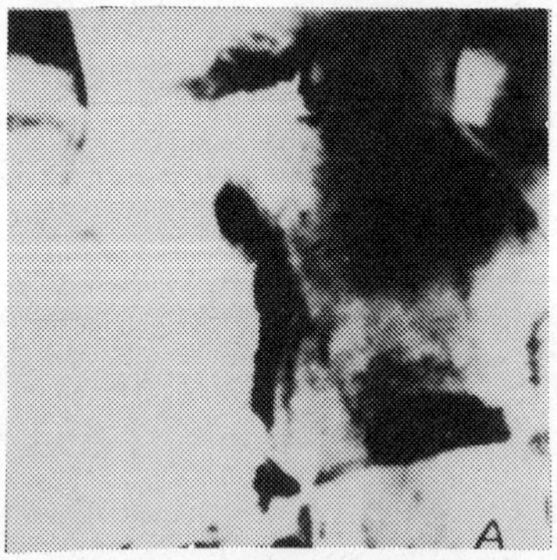

(a)

(b)

Fig.6. TEM Morphlogy (a) and Electron Diffraction Pattern (b)
of $3mol\%\ Y_2O_3-12mol\%\ CeO_2$-TZP

The grain sizes change with sintering temperature and the amount of CeO_2 in the sample. The average grain size of Y-Ce-TZP sintered at 1450℃ for 2hr is 1.5μm, and 2μm for the sample sintered at 1550℃ for 2 hr (Figs.7a,b).

a. sintered at 1450℃
for 2 hr

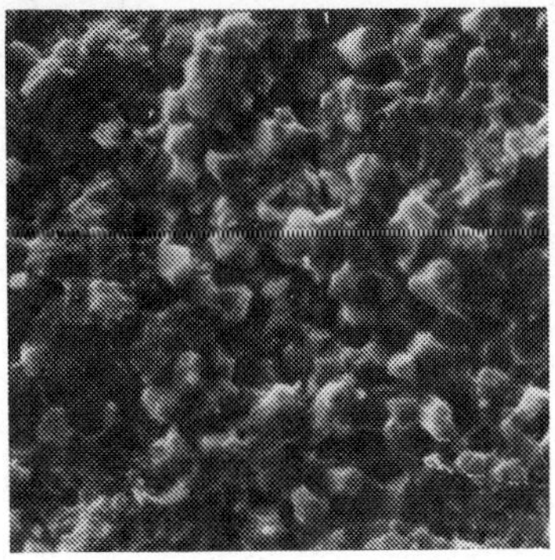

b. sintered at 1550℃
for 2 hr

Fig.7. Grain Sizes of Y-Ce-TZP Containing 12 mol% CeO_2 as a Function of Temperature (contained 3mol%Y_2O_3)

For the HP samples, the grain size depended on the CeO_2 content. The higher CeO_2 contents resulted in the smaller grain sizes. Grain sizes vary from 1μm to 0.4μm (Figs.8a,b,c).

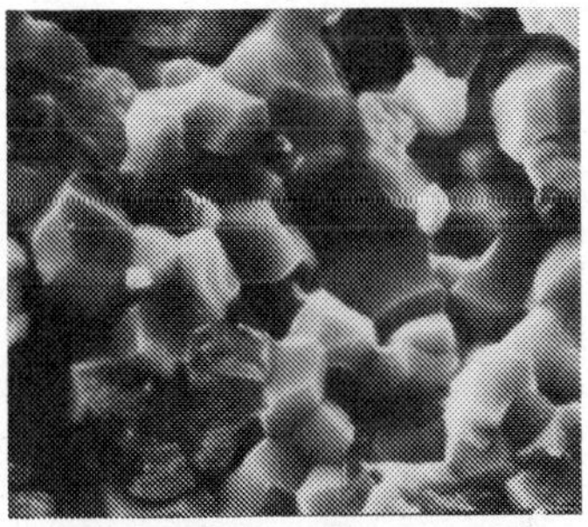

a. 3mol% Y_2O_3-8mol% CeO_2-TZP

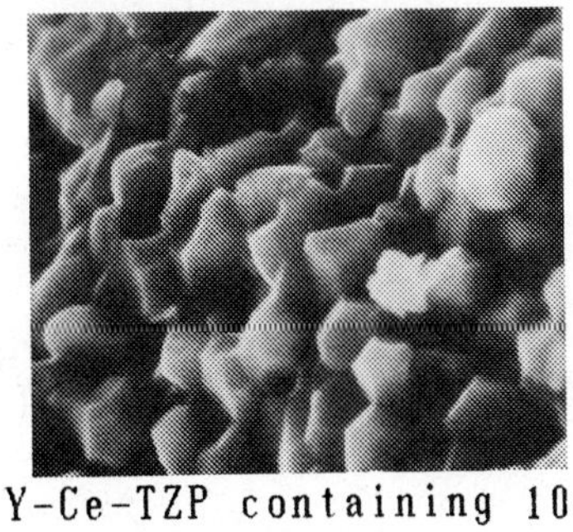

b. Y-Ce-TZP containing 10mol% CeO_2

c. Y-Ce-TZP containing 12 mol% CeO_2

Fig.8 Grain Sizes of Y-Ce-TZP Prepared by HP

The densities of Y-Ce-TZP sintered samples were found to vary from 6.10 to 6.20g/cm^3, while values of 6.20-6.25 g/cm^3 were obtained for HP samples.

Bending Strength and Fracture Toughness

The results of three-point bending stength and fracture toughness measurements are given in Fig.9 and Fig.10.

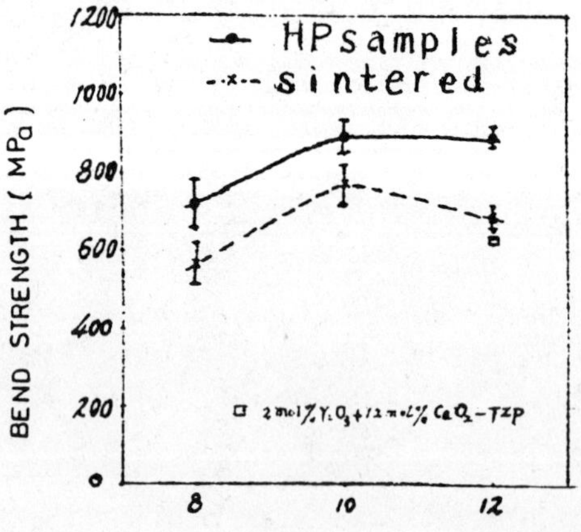

Fig.9. Bend Strength as a Function of CeO_2 Content (3mol% Y_2O_3)

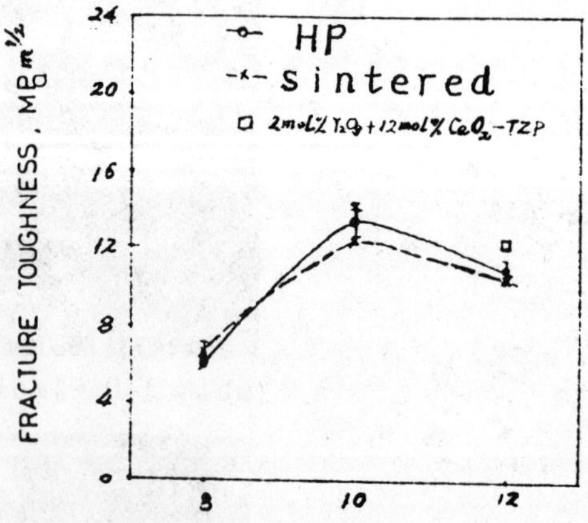

Fig.10. Kic as a Function of CeO_2 Content (3mol%Y_2O_3)

Table 2 Properties as a function of CeO_2 content (contained 3mol% Y_2O_3) in HP Y-Ce-TZP

Parameters	CeO_2(mol%)		
	8	10	12
Grain Size (μm)	1.2-1.5	0.8-1.0	0.3-0.5
Kic ($MPam^{1/2}$)	7	15	11
$\sigma_{bend\ strength}$(MPa)	700	940	935

The grain size effect on the strength and the toughness was seen. For grain sizes less than 1μm, high bend strength and medium Kic were obtained (Fig.8,9,10 and Table 2). For larger grain sizes the Kic and the strength were less. For samples with mixed large and small grain sizes, the bend strength and Kic values were low. Samples with uniform grain sizes of about 1μm and containing 3 mol% Y_2O_3-10 mol% CeO_2 in Y-Ce-TZP had the best bend strength and Kic values. Therefore, homogeneous grain sizes and control of absolute grain size by the processing parameters are very important for obtaining excellent mechanical properties.

Toughening Mechanism

The Y-Ce-TZP phase content was measured by X-ray diffraction on the fracture surfaces (Fig. 11, Table 3). When compared with unfractured samples the results indicate the m-phase content clearly increases. Therefore, stress-induced t-m transformation is a major toughening mechanism.

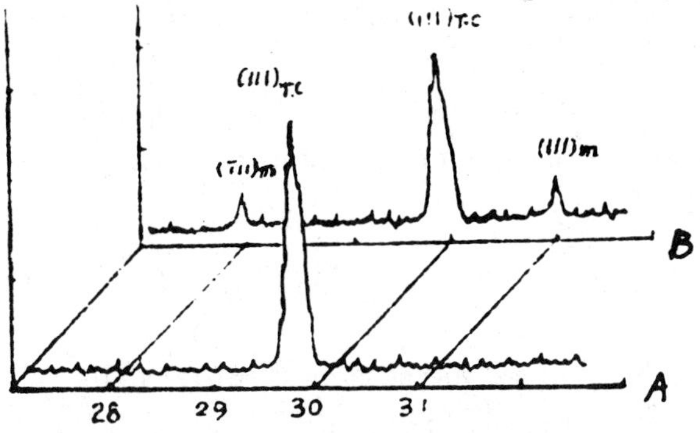

Fig.11 X-ray Diffraction Spectra of Before and After Fracture (A;B) (Cukα)

Table 3. m-phase Content in Y-Ce-TZP
Before and After Fracture Surface

3mol% Y_2O_3 + xmol% CeO_2-TZP	Before F.S m-phase (%)	After F.S m-phase (%)	(Kic $MPam^{1/2}$)
sintered samples			
12 (1450℃)	0	23	
12 (1550℃)	0	14	
HP samples			
8	7	16	(7)
10	0	12	(15)
12	0	10	(11)

Besides the t-m transformation-toughening, the addition of CeO_2 also strengthens the materials grain boundary because essentially no glassy phases were found at the grain bundaries.

SUMMARY

The following results were obtained:

1) Y-Ce-TZP containing about 8 to 12 mol% CeO_2 can form a stable t-phase easily.

2) The HP samples of 3 mol% Y_2O_3-10mol% CeO_2-ZrO_2 have higher mechanical properties, with bending strength and Kic reaching 940 MPa and 15 $MPam^{1/2}$, respectively.

3) The sintering temprature and CeO_2 content directly affects the grain sizes in Y-Ce-TZP.

4) In HP processed Y-Ce-TZP containing more than 8 mol% CeO_2, there were small amounts of CeO_2 precipitation

5) t-m transformation is the major toughening machanism.

REFRENCES

1. M.V.swain, L.R.F.Rose, J. American Ceramic Society, 51-18(1986)
2. M.Matsui, T. Soma, I. Oda, J.Am.Ceram. Soc.69[3] 198-202(1986)
3. Tsugio Satio, Shiro Ohtaki et al, J.Am.Soc. Lett. 5 (1986) 1140-1142
4. Koji Tsukuma, Masahiko shimada, Am. Ceram. Soc. Bull., 64(1985) 210-13
5. M.Ruhle, N. Claussen, Proc. 2nd Inter. conf, on the Sci. Tech of Zirconia, June 21-23 (1983)
6. T.W. Coyle, W.S Coblenz and B.A.Bander, Amer. Ceram. Soc. Bull 62(1983) 966
7. Koji Tsukoma and Kuniyoshi Ueda, J.Am.Ceram.Soc.68(1986) C56-58
8. F.F.Lange, J.Mater.Sci 17 (1982)255
9. Koji, Tsukuma, J.Materials Science 20(1985)1178-1184
10. F.F.Lange, J.Mater. Sci. 17(1982)225
11. D.Manz, R.T.Bubsey and J.L.Shannon, Jr.,J.Amer.Ceram.Soc. 63(1980)300

HIGH TEMPERATURE BENDING STRENGTH AND OXIDATION RESISTANCE OF β-SIALON PREPARED WITH MINIMAL GRAIN BOUNDARY PHASE

K. Yabuta, H. Nishio and H. Okamoto, Advanced Technology Research Center, NKK, Kawasaki, Japan; S. Umebayashi and K. Kishi, GIRIK, Tosu, Japan

ABSTRACT

The inherent inhomogeneity of conventional β-sialons prepared by powder mixing shows low strengths at high temperatures. To reduce this inhomogeneity, aluminum iso-propoxide solution was applied as a sintering additive instead of alumina and yttria powder. Consolidation, gas pressure sintering, and hot isostatic pressing provided a full dense β-sialon (z=0.5). This material exhibited 800 MPa 3-point bending strength from room temperature to 1300 ℃, and excellent oxidation resistance up to 1600 ℃.

INTRODUCTION

Ceramic gas turbines have been developed in which some components are used at temperatures up to 1350 ℃. Their performance must be achieved with oxidizing gas flow conditions in the presence of corrosive impurities from fuel and environment. Silicon nitride based ceramics have long been recognised as a candidate for such applications, because of their excellent strength and oxidation resistance properties. However, heterogeneity and porocity in microstructure of sintered materials are major problems. In general, starting materials, such as silicon nitride, alumina, aluminum nitride, and yttria powders are mixed. The poor dispersion of these powders often causes heterogeneous microstrutures. Pressureless or gas pressure sintering is a quite common process, but by this process often it is not possible to eliminate the porocity. To reduce this poor powder dispersion, sintering additives to form a liquid phase have been used. Yttria additions appear to be most successful in conjunction with alumina and some other materials. β-sialon thus prepared is composed of the main sialon phases, and crystalline or vitreous intergranular phases. The major problems of this material are fracture strength, creep rupture and oxidation resistance at temperatures above 1000 ℃.

Crystallization of glassy grain boundaries via post-sintering heat treatment improves high temperature properties in an oxidizing environment(1). However, this processing takes six days(2) or 1000 hours(3), which is unrealistic in fabricating most components.

Kishi and coworkers(4-8) applied a novel method in fabricating β-sialon; silicon nitride powder and aluminum alcoxide solution as starting materials were mixed, dried, calcined and hot pressed. The sintered material exhibited a markedly improved homogeneity in microstructure which is substantially composed of only crystalline phases consisting primarily of β- and o-sialon phase. This material resulted in a bending strength as high as 1200 MPa at 1200 ℃ (4-8). However, their hot pressing process had limited near-net shaping capability. The present work studies β-sialon fabrication or processing without liquid formation, and used hot isostatic pressing as the densification method; the primary concerns were high temperature bending strength and oxidation resistance.

EXPERIMENTAL PROCEDURE

Figure 1 shows the flow chart used to fabricate samples. Aluminum iso-propoxide was dissolved in a normal hexane and filtered to separate undissolved solid from the solution. α-silicon nitride powder, synthesized by the pyrolysis of

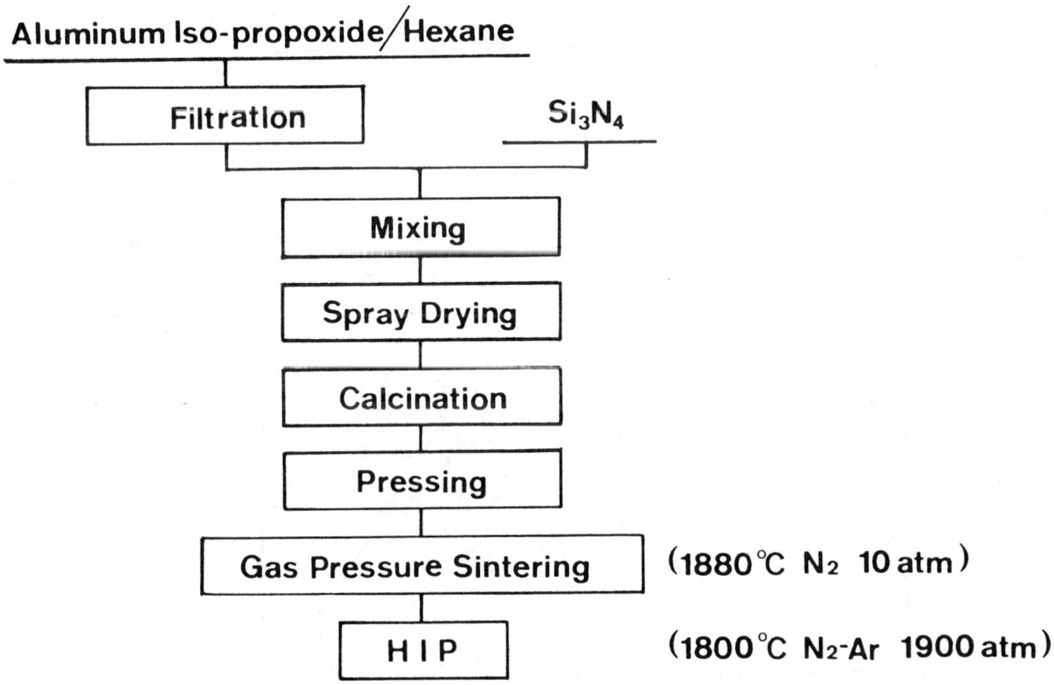

Fig. 1. Procedures for preparing β-sialon sintered materials.

silicon di-imide, was weighed to achieve a $Si_{5.5}Al_{0.5}O_{0.5}N_{7.5}$ composition, fed into this solution and ball-milled with silicon nitride balls. The resulting slurry was spray-dried to give granules, which were calcined at 900 ℃ for two hours in air, die pressed at 20MPa, and cold isostatically pressed at 300MPa. The compacts thus obtained were subjected to gas pressure sintering at 1880 ℃ and 1MPa in nitrogen for one hour and hot isostatic pressed at 1800 ℃ and 190MPa in mixed gas of 8% nitrogen and 92% argon for two hours. The sintered materials were then cut, ground and diamond-polished to give test bars with dimensions of 3×4×40 in mm.

Three point bending strengths were measured from room temperature to 1300 ℃. Weight gain was measured at both 1200 and 1350 ℃ in air. Changes in surface morphology with oxidation were obseved at 1200, 1350, 1500 and 1600 ℃.

RESULTS

<u>High temperature bending strength</u>

Figure 2 summarizes the bending strength results of β-sialon(NKK) mentioned above, and compares typical non-oxide

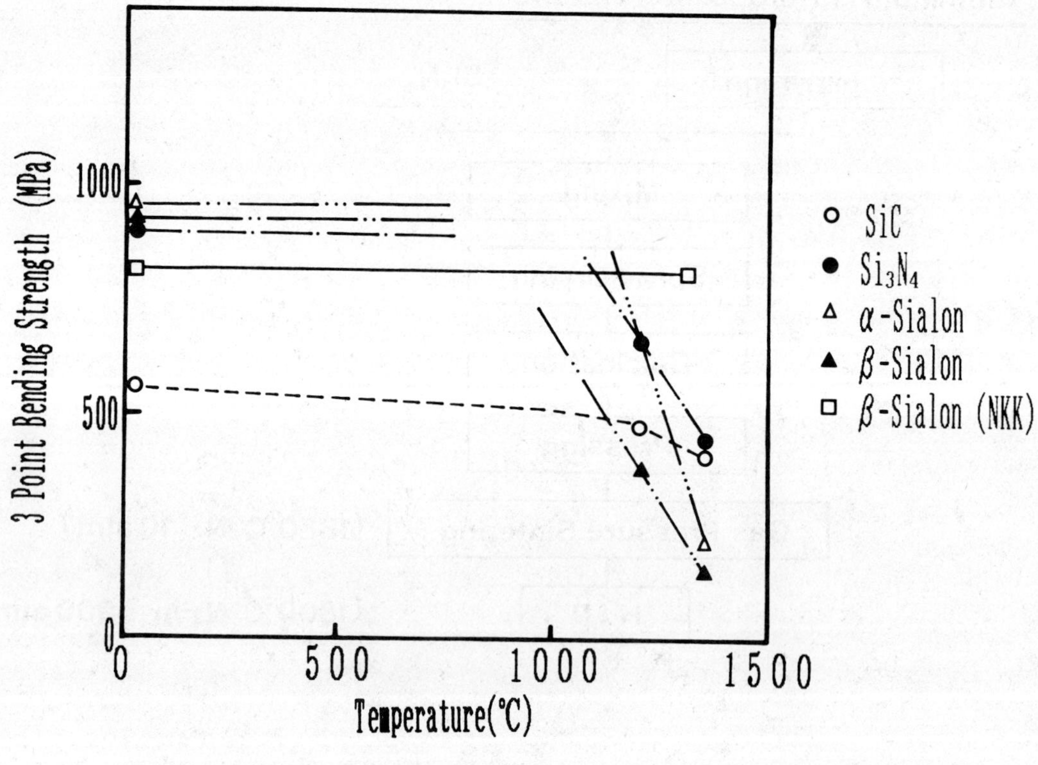

Fig. 2. Temperature dependence of 3-point bending strength of several materials.

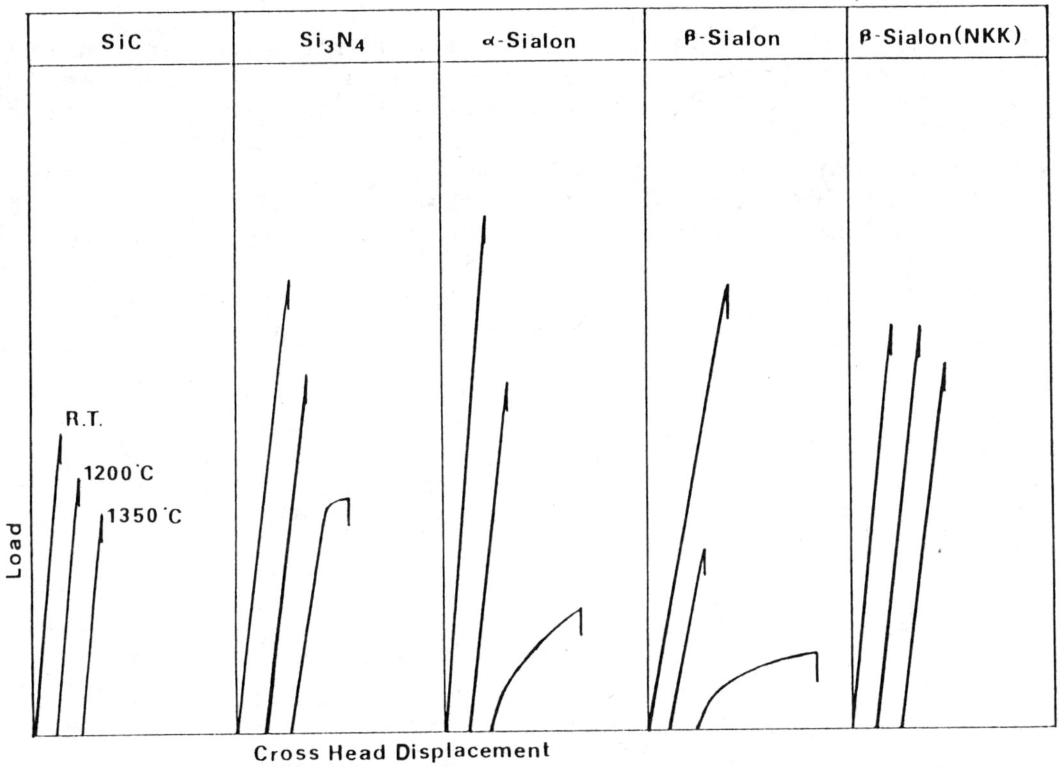

Fig. 3. Load-displacement diagrams of several materials at room temperature, 1200, and 1350 ℃.

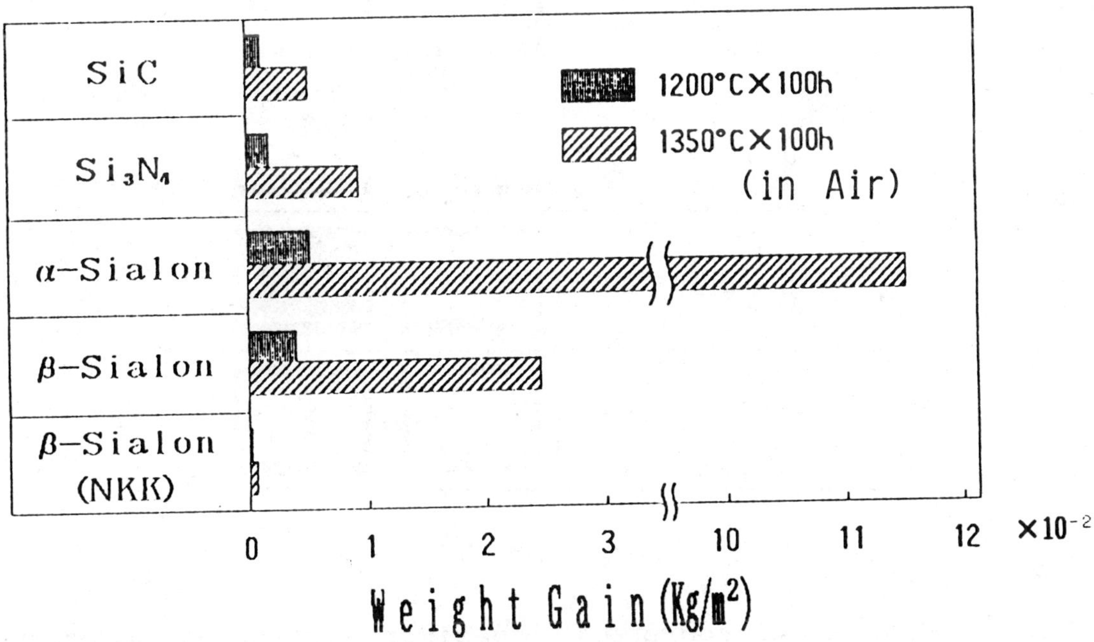

Fig. 4. Weight gain of several materials in air at 1200, and 1350 ℃.

ceramics available in Japan; silicon carbide, silicon nitride, α-sialon and conventional β-sialon. Note that β-sialon(NKK) exhibits 800MPa at room temperature and maintains this strength up to 1300 ℃. Although the bending strengths of silicon nitride, α-sialon and conventional β-sialon exhibit bending strengths superior to that of the present work at room temperature, they decrease in strength at temperatures between 950 ℃ and 1100 ℃, and have strengths less than 500MPa at 1300 ℃. Conversely, silicon carbide shows less reduction in strength above 1100 ℃, but the strength of 500MPa is much lower than that of β-sialon(NKK).

Figure 3 shows load-cross head displacement diagram of those materials at room temperature, 1200 and 1350 ℃. It is evident that β-sialon(NKK) exhibits no plastic deformation even at 1350 ℃. This behavior is very similar to that of silicon carbide. The other three materials exhibit plastic deformation prior to fracture at 1350 ℃.

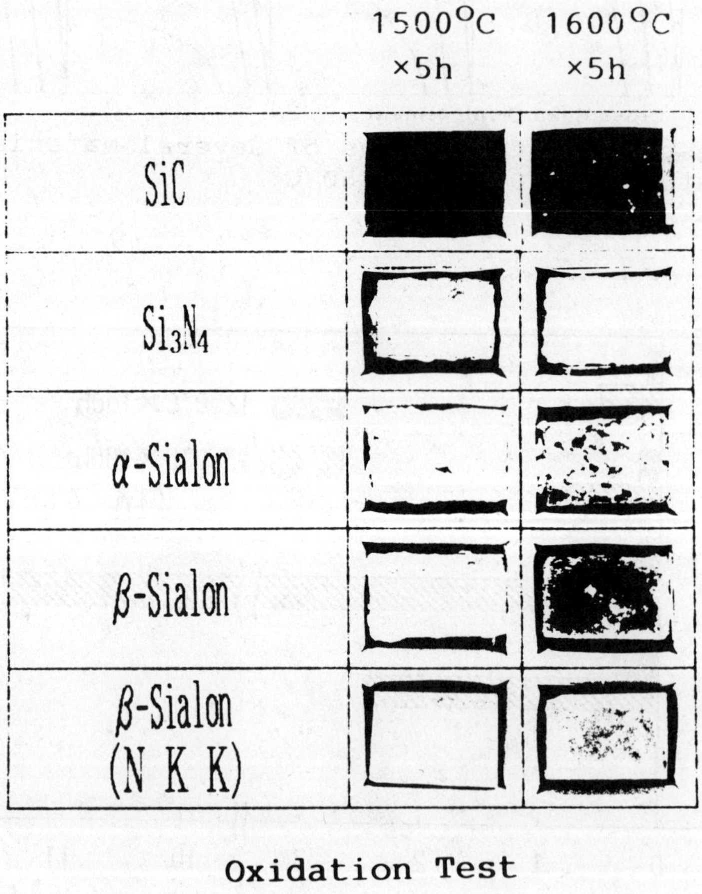

Fig. 5. Optical micrograph of specimen surface of several materials after oxidation in air at 1500, and 1600 ℃ for 5 hours.

Oxidation resistance

Figure 4 shows the weight gain of samples exposed at 1200 or 1350 ℃ in air for 100 hours. As shown in this figure, β-sialon(NKK) is not only superior to the conventional nitrogen ceramics in oxidation resistance, but also to silicon carbide which has been considered to be more oxidation resistant. The surface morphology of these samples was observed after heating to 1200 ℃ and 1350 ℃ for 100 hours. No changes were observed in β-sialon(NKK) and in silicon carbide at both temperatures. In silicon nitride, although no change was detected at 1200 ℃, a slight change in color to dark gray was observed at 1350 ℃. In α-sialon considerable swelling took place at 1350 ℃. In conventional β-sialon a change in color to gray was observed at 1200 ℃, and more apparent at 1350 ℃.

The results of oxidation experiments conducted at 1500 ℃ and 1600 ℃ are given in Figure 5 for the same materials. It can be seen that the effects described above are even more apparent; in silicon nitride, α- and conventional β-sialon large amounts of swelling and bubbles are observed at both temperatures. On the silicon carbide surface fine white spots

Fig. 6. Transmission electron micrograph of β-sialon(NKK).

are observed at 1500 ℃, and they became larger with the increase in temperature, as shown in the right top figure. Conversely, in β-sialon(NKK) no change is observed at 1500 ℃., and only slight change in color is observed at 1600 ℃.

β-sialon(NKK) demonstrated superior oxidation resistance up to 1600 ℃.

DISCUSSION

Figure 6 shows transmission electron micrograph of β-sialon(NKK). The grain boundary thickness is 2nm, and the size of triple junction is approximately of 5nm. In conventional β-sialon it has been reported that the size of triple junction is approximately 100nm(3), significantly larger than that of β-sialon(NKK). The minimized grain boundary and triple junction contribute to the high strength and the oxidation resistance at high temperatures. Essentially no grain boundary layer exists in contrast with the conventional β-sialon containing amorphous grain boundaries; thus no softening (Figure 3), and no oxgen diffusion path. In figure 7 a comparison of β-sialon(NKK) with previous β-sialons is given schematically.

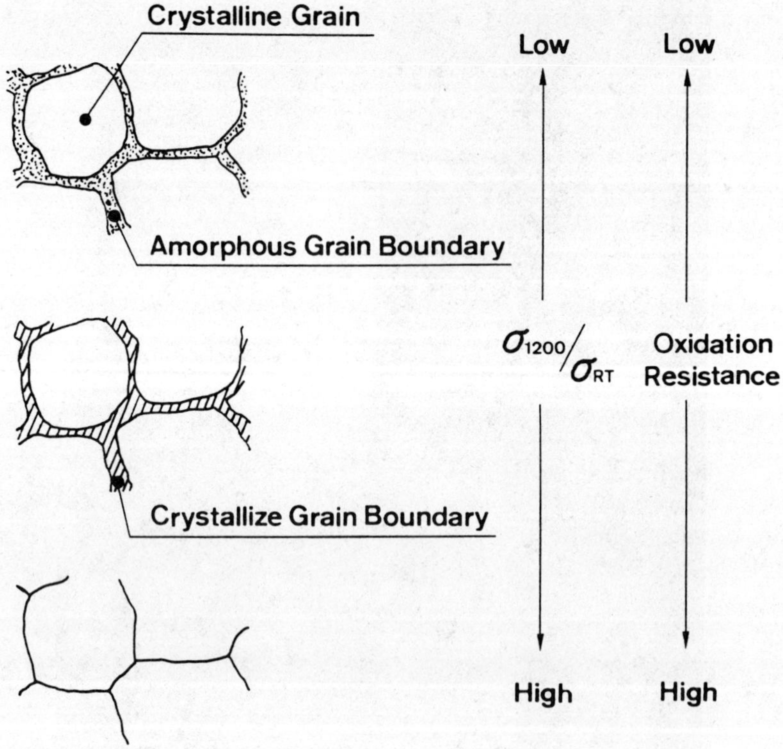

Fig. 7. Schematic of microstructure of β-sialon related to temperature dependence of bending strength.

Figure 8 shows fractgraphs of β-sialon(NKK) and silicon nitride; in β-sialon transgranular fracture is predominant contrary to the silicon nitride where the intergranular fracture is predominant and the pull-out of needle-like grain is observed. The fracture toughness of this β-sialon is only 3MPa $\sqrt{m}$, whereas that of silicon nitride is 6MPa $\sqrt{m}$. Such a low fracture toughness of β-sialon may restrict its application, and improvement of fracture toughness is the objective of our future work.

A typical application of this β-sialon is a control valve for blast furnace, where the valve is exposed to hot-air blast pressurized to 450KPa and flowing at a speed of more than

Fig. 8. Fractographs of β-sialon(NKK), and silicon nitride.

200m/s at temperatures of 1100–1300 ℃. The success in this application may encourage use of β-sialon(NKK) in gas turbine components.

CONCLUSIONS

β-sialon prepared by mixing silicon nitride powder and aluminum iso-propoxide solution, and hot-isostatically pressing pre-sintered material exhibits good bending strength up to 1300 ℃, and excellent oxidation resistance up to 1600 ℃.

This β-sialon contains little grain boundary phase, which contributes to this behavior.

Fracture toughness of this material is relatively low. However, this material has been successfully applied to a hot-air blast control valve used at 1100–1300 ℃. This success may encourage use of β-sialon(NKK) as engine components.

ACKNOWLEDGEMENTS

The authors wish to thank Mr. K. Yamada (NKK) for obtaining the transmission electron micrograph.

REFERENCES

1) B. S. B. Karunarante and M. H. Lewis, Journal of Material Science, 15(1980)1781.
2) D. P. Thompson, and W.-Y. Sun, P. A. Walls, Ceramic Materials and Components for Engines, Edited by W. Bunk and H. Hansner, Verlag Deutsche Keramische Gesellschaft, (1987)643.
3) M. H. Lewis, S. Mason, and A. Szweda, Non-oxide Technical Engineering Ceramics, Elsevier Applied Sci. Publrs. (1987) 175.
4) K. Kishi, S. Umebayashi, E. Tani, and K. Kobayashi, Journal of the Ceramic Society of Japan, 92(1984)231.
5) K. Kishi, S. Umebayashi, E. Tani, and K. Kobayashi, Journal of the Ceramic Society of Japan, 93(1985)629.
6) K. Kishi, S. Umebayashi, E. Tani, and K. Kobayashi, Journal of the Ceramic Society of Japan, 94(1986)179.
7) K. Kishi, S. Umebayashi, E. Tani, and K. Kobayashi, Journal of the Ceramic Society of Japan, 95(1987)630.
8) K. Kishi, S. Umebayashi, E. Tani, and K. Kobayashi, Journal of the Ceramic Society of Japan, 96(1988)698.

MECHANICAL AND THERMAL PROPERTIES AND THERMAL STABILITY OF REACTION-SINTERED Al_2TiO_5 PREPARED FROM HYDROTHERMAL POWDERS,

D.Peuckert , J.Hausselt and G.W.Kriechbaum, Degussa AG, Hanau, FRG; H.Schubert, Max Planck Institut für Metallforschung, Stuttgart, FRG

ABSTRACT

Al_2TiO_5 has a low thermal expansion and a low thermal conductivity. These properties make it a suitable material for thermal insulation and applications where thermal shock resistance is required. In order to stabilize the Al_2TiO_5 structure against decomposition into Al_2O_3 and TiO_2 additives such as SiO_2, B_2O_3, Fe_2O_3 etc. have to be added. Nevertheless, decomposition of stabilized Al_2TiO_5 cannot be avoided completely after long annealing at high temperatures. This may lead to a degradation of the mechanical and thermal properties. Al_2O_3/-TiO_2 mixtures doped with SiO_2 have been prepared by hydrothermal synthesis. These powders were reaction-sintered at different temperatures and times. Part of the samples were annealed at 900 to 1100°C up to 500 hours. The sintered samples as well as the additionally annealed samples were characterized with respect to micro-structure, phase composition and thermal properties. The influence of sintering and annealing conditions on the material properties, especially the phase stability of SiO_2 doped Al_2TiO_5 will be discussed.

INTRODUCTION

Its heat insulating capability together with its excellent thermal shock behaviour make aluminium titanate an interesting ceramic for engineering applications. These properties result from a microstructural peculiarity of forming microcracks between the grains during cooling from the sintering temperature. These microcracks, however, are responsible for the poor mechanical strength of the ceramic. Pure Al_2TiO_5 is stable in air above 1200-1300°C /1/. Below this temperature range decomposition into Al_2O_3 and TiO_2 occurs. Stabilizing agents, e.g., SiO_2, MgO, Fe_2O_3, Kaolinite, Mg_2SiO_4 and other oxides /1-4/, are known to retard the decomposition.

These oxides exhibit different influences on the formation, stability, microstructure, strength and expansion behaviour of Al_2TiO_5 /5/. With respect to the stabilization, SiO_2 seems to be one of the most effective additives.

Al_2TiO_5 ceramics can be prepared either by reaction-sintering of an Al_2O_3/TiO_2 powder mixture or by calcining the powder mixture at high temperatures, followed by milling and sintering. The latter route is the much more expensive and time consuming process. The properties of Al_2TiO_5 required for the use as an engineering ceramic are low thermal conductivity, low thermal expansion, thermal stability at temperatures up to 1100°C and mechanical stability.

In this paper the properties of Al_2TiO_5 reaction-sintered from a hydrothermally synthesized Al_2O_3/TiO_2 mixture will be described.

EXPERIMENTAL

Powder Preparation

The Al_2O_3/TiO_2 powder used for the experiments was prepared by coprecipitation and hydrothermal crystallisation followed by calcination. As starting materials for preparing the hydroxide gels, technical grade $TiCl_4$ and $Al_2(SO_4)_3 \cdot 16H_2O$ were used. The gels were prepared by coprecipitation with NaOH. The hydrothermal treatment was perfomed in a Hastelloyn lined stirring autoclave. The oxide suspensions were washed with distilled water. Powders were dried and calcined at 700 - 1000°C for 2 hours.

Sintering and Characterization

Sintering experiments were performed with isostatically pressed (500 bar) samples in an electrically heated furnace at various sintering temperatures between 1450°C and 1600°C and isothermal sintering times from 2 to 6 hours. Shrinkage behaviour as well as thermal expansion coefficients were measured with a dilatometer with Al_2O_3 as reference material.

The stability of the sintered Al_2TiO_5 was characterized by annealing the samples at temperatures of 900°C to 1100°C for up to 500 hours. Phase compositions were detected by x-ray diffraction with Cu-K radiation.

For metallographic preparation, samples had to be infiltrated with an expoxy resin. Microstructures were revealed by SEM.

RESULTS AND DISCUSSION

Powder Properties

The hydrothermally synthesized powder has a very high specific surface area of about 155 m^2/g and very small round particles of a few nanometers which are loosely packed in soft agglomerates with a mean size of about 10 µm as shown in Fig. 1. The powder consists of 55,5 % Al_2O_3, 41,5 % TiO_2 and 3 % SiO_2 (stabilizer) by weight. The phase composition is rutile, anatase and -alumina.

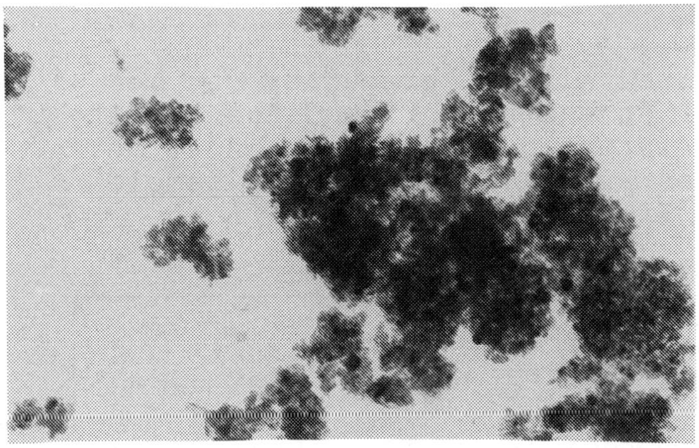

Fig 1:
TEM micrograph of the hydrothermal synthesized powder

0,1 µm

For sintering experiments these extremely fine powders were used. The green densities of the samples pressed from these powders are accordingly low (= 0,8 - 1 g/cm^3), and shrinkage during sintering is high.

Generally, Al_2TiO_5 parts were formed by slip casting or pressure slip casting. Calcining the powders at temperatures higher than 1000°C leads to a reduction of the BET surface area and to increased green densities of pressed samples of up to 1,6 g/cm^3. Calcining of the hydrothermal powders up to 1200°C does not affect the sintering activity.

Sintering

The densification of the hydrothermally prepared powder mixtures occurs by a reaction sintering process. In Fig. 2 the shrinkage characteristics are shown. The densification process is characterized by two maxima of the shrinkage rate. The first shrinkage period with a maximum of the shrinkage rate at 1080°C is attributed to sintering of the finely dispersed TiO_2 and is accompanied by about 9 % change in length.

In the temperature range between 1050°C and 1150°C the sintering rate is drastically reduced but increases again above 1150°C. At 1288°C the second sintering maximum occurs. Now the reaction to Al_2TiO_5 and the densification occur simultaneously. In contrast to the observations described in the literature /5/ no dilatation of the samples during the formation of Al_2TiO_5 has been found in our study. Though the formation of Al_2TiO_5 basically is accompanied by a volume increase, this effect is overcompensated by a high sintering rate at temperatures higher than 1150°C. A small amount of Al_2TiO_5 could be detected at 1250°C by X-ray measurements. The reaction was found to be completed at 1300°C after 4 hours of reaction time.

The hydrothermally synthesized Al_2TiO_5 precursor can be sintered to high densities. As shown in Fig. 3 the density increases with increasing sintering temperature and time. X-ray diffraction measurements indicated that all samples consist of single phase Al_2TiO_5 only.

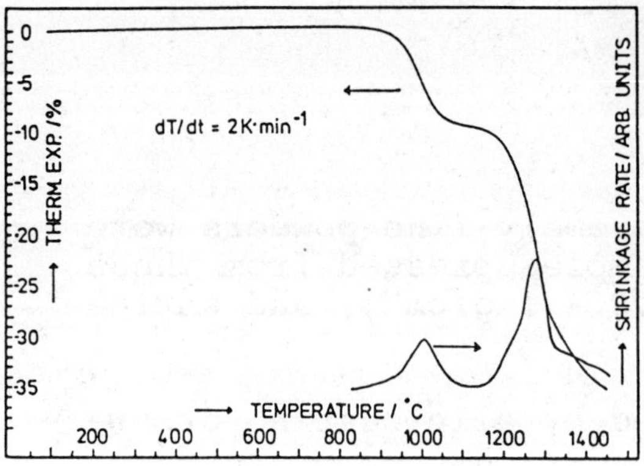

Fig. 2: Shrinkage behaviour of the Al_2O_3/TiO_2 mixture

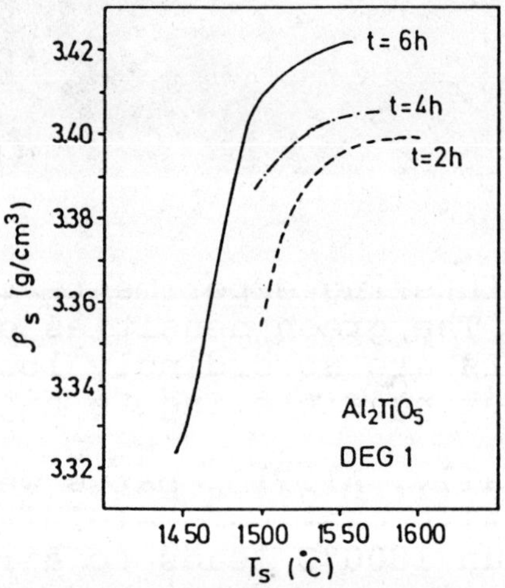

Fig. 3: Density of Al_2TiO_5 in dependence of sintering temperatures and time

Thermal stability

Al_2TiO_5 is thermodynamically stable at temperatures higher than 1200°C /6/. Unfortunately, grain growth of Al_2TiO_5 starts already at this temperature and limits the use as a high temperature material, e. g., in automotive engines. Between 700 and 1200 °C pure Al_2TiO_5 decomposes into TiO_2 and Al_2O_3. At 1100°C the maximum decomposition rate has been observed. By adding SiO_2 as stabilizer the decomposition rate is drastically reduced.

The long time stability of Al_2TiO_5 sintered for 2 hours at 1550°C has been tested up to 500 hours.

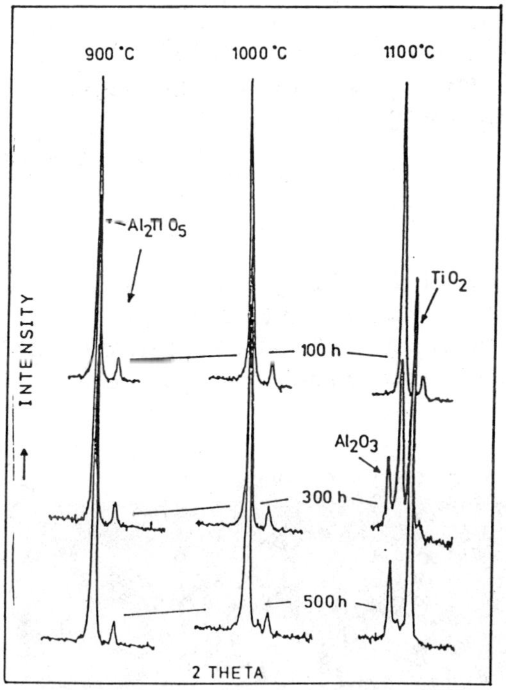

Fig. 4:
Phase composition of Al_2OiO_5 after different annealing conditions

At 900°C and 1000°C no decomposition could be detected after 500 hours of annealing in air, Fig. 4. At 1100°C the decomposition starts after 240 hours. The deckomposition rate of Al_2TiO_5 is not only dependent on the working temperature and the type of stabilizer, it was found that it also depends on the preparation of the samples, especially the sintering conditions. The samples sintered at different temperatures for different times, show different stability. The highest decomposition rate was found in the sample sintered at 1600°C followed by the sample sintered at 1450°C for 2 hours, whereas no decomposition has been observed within the samples sintered for 6 hours at 1450°C or 2 hours at 1500°C or 1550°C (Tab.1).

Tab.1: Phase compositions of Al_2TiO_5 after annealing at 1100°C for different times, measured by X-ray diffraction;

Sintering conditions	Phase composition As sintered	50 h	200 h	500 h
2 h 1450°C	ATO	ATO	ATO,TO AO (m)	AO, TO, ATO (w)
6 h 1450°C	ATO	ATO	ATO TO (w)	AO, TO, ATO (w)
2 h 1500°C	ATO	ATO	ATO	AO, TO, ATO (w)
2 h 1550°C	ATO	ATO	ATO	AO, TO, ATO (w)
2 h 1600°C	ATO	ATO TO (w)	AO, TO	AO, TO

Abbr.: ATO: Al_2TiO_5; AO: Al_2O_3; TO: TiO_2; w: weak; m: medium peak intensity; main phases are without index

Microstructure and Properties

The grain size distribution of sintered Al_2TiO_5 is very sensitive to the sintering conditions. With increasing sintering temperature and time the mean grain sizes increase and the grain size distribution becomes broader, Fig.5 and 6.

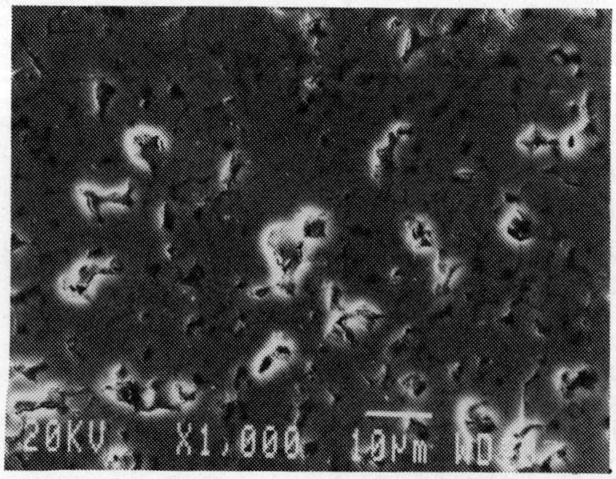

Fig. 5a

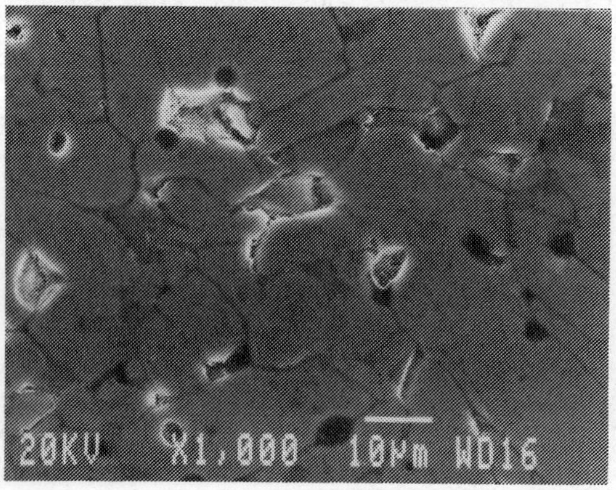

Fig. 5b

Fig. 5: Microstructure of Al_2TiO_5 sintered at 1450°C (a) and at 1600°C 2h (b)

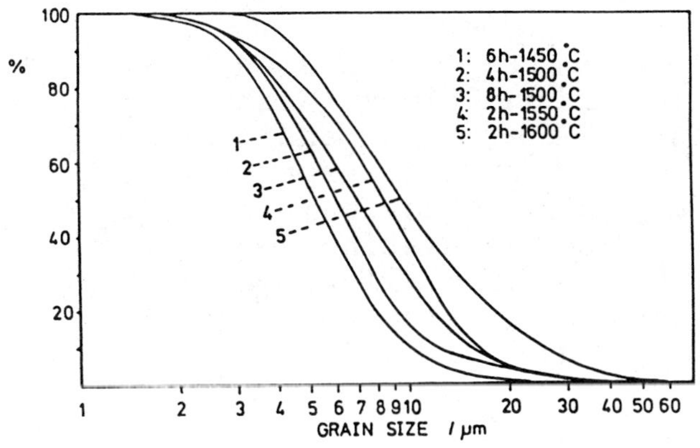

Fig. 6: Grain size distribution of Al_2TiO_5 sintered under different conditions

Fig. 6

Grain growth generally starts at quite low temperatures (1250°C) with the formation of the Al_2TiO_5 phase. The grain growth can be inhibited by additives, of which SiO_2 seems to be one of the most effective /7/.

In the micrographs, Fig. 5, the separation of the Al_2TiO_5 grains by microcracks can be observed. These microcracks are responsible for the exceptionally low bending strength of Al_2TiO_5. Samples sintered at 1550°C have a bending strength of only 20 MPa. Decreasing the grain size by sintering at lower temperatures seems to be a reasonable route to improve the bending strength of Al_2TiO_5 /8/.

Fig. 7 shows the microstructure of a sample sintered at 600°C and annealed for 200 hours at 1100°C.

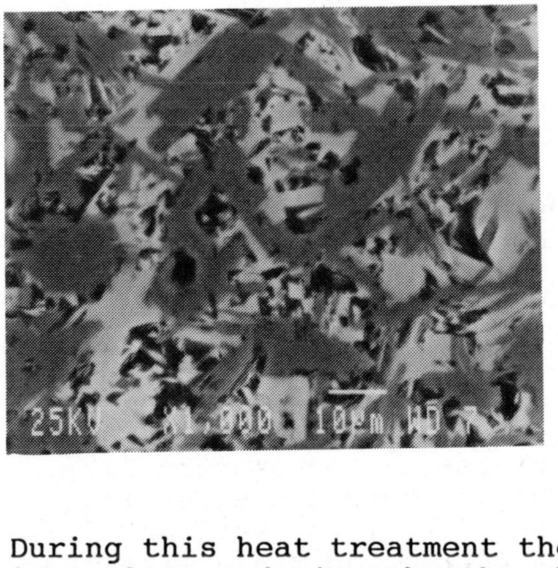

Fig. 7: Microstructure of Al_2TiO_5 completely decomposed into Al_2O_3 and TiO_2

During this heat treatment the sample completely decomposed into Al_2O_3 and TiO_2 whereby the microstructure changes, as compared to figure 5b. The different phases were detected using backscattered electrons.

The thermal expansion behaviour of Al_2TiO_5 depends on the microstructure, i. e., the sintering conditions. With increasing sintering temperature the thermal expansion decreases, Fig. 8. The coefficients of thermal expansion are listed in Table 2. The thermal expansion behaviour is characterized by a hysteresis between the heating and the cooling curve. On heating there is only a small temperature dependency of the expansion whereas on cooling the samples strongly contract first and then expand from 500 - 700°C to room temperature.

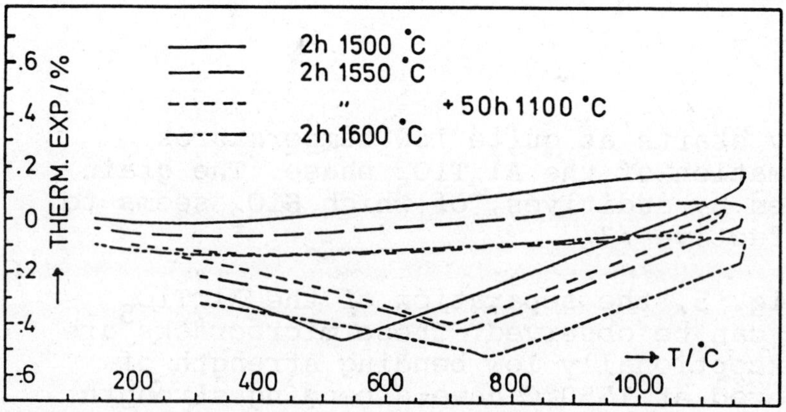

Fig. 8: Thermal expansion of Al_2TiO_5 sintered at different temperatures.

Tab. 2: Thermal expansion coefficients of sintered Al_2TiO_5

Sintering conditions	$\alpha \times 10^{-6}$/K RT-600°C	RT-1100°C
2 h 1500°C	1,04	1,87
2 h 1550°C	- 0,32	0,78
2 h 1600°C	- 1,84	- 0,40

After reheating the samples at 1100°C and annealing them 50 hours, the thermal expansion is decreased in the heating cycle (Fig. 8). In the case of decomposition during this heat treatment, an increase in the thermal expansion coefficient was observed.

The low thermal expansion of Al_2TiO_5 accompanied by an excellent thermal shock resistance results from the formation of microcracks upon cooling from the sintering temperature /12/. Microcracking is caused by the highly anisotropic expansion behaviour of Al_2TiO_5 in the different lattice directions /3,9, 10, 11/. This fact is enhanced with increasing grain size of Al_2TiO_5.

CONCLUSIONS

Hydrothermal synthesis is a suitable method for preparing very homogeneous and sinteractive powder mixtures which have not to be prereacted to Al_2TiO_5 before sintering. The sintered Al_2TiO_5 exhibits a good thermal stability at 1100°C and a low thermal expansion coefficient. With respect to the mechanical strength a compromise between a low thermal expansion and a good mechanical strength has to be accepted.

ACKNOWLEDEGEMENT

The investigations were financially supported by the German Ministry of Science and Technology (BMFT).

REFERENCES

(1) H.J. Pohlmann, K. Schricker and K.H. Schüller, "Untersuchungen an Werkstoffen im System Al_2O_3-TiO_2-SiO_2", Ber. Dt. Keram. Ges. 52 (1975) 179.

(2) D. Goldberg, "Systems Formed by Alumina with Some Oxides of Trivalent and Tetravalent Metals, Especially Titanium Oxide", Rev. Int. Hautes Temp. Refract. 5 (1968) 181.

(3) E. Gugel, P. Schuster, "Keramische Massen auf der Basis von Aluminiumtitanat", Tonind. Zeit. 98 (1974) 314.

(4) E. Kato, J. Takahashi, R. Kato and K. Hamano, "Fabrication of Al_2TiO_5 Ceramics, II. Stabilisation of Al_2TiO_5 with Cation Substitution", Rept. Res. Lab. Eng. Mat. 9 (1984)84.

(5) W. Lingenberg, Ph.D.Thesis, Technische Universität Clausthal (1985).

(6) A.M. Lejus, D. Goldberg and A. Revcolevschi, "New Compounds Formed Between Rutile TiO_2 and Oxides of Trivalent and Tetravalent Metals", C.R. Hebd. Seances Acad. Sci. Cer. C. 263 (1966) 1223.

(7) E. Kato, K. Daimon and K. Hamano, "Fabrication of Al_2TiO_5 Ceramics I", Rept. Res. Lab. Eng. Mat. 9 (1984) 75.

(8) Y. Ohya, K. Hamano and Z. Nakagawa,
"Fabrication and Mechanical Strength of Aluminium Titanate Ceramics Prepared from Synthesized Powders",
Yogyo Kyokai Shi 91 (1983) 290.

(9) B. Morosin and R. W. Lynch,
"Structure Studies on Al_2TiO_5 at Room Temperature and at 600°C",
Acta, Cryst., B28 (1972) 1040.

(10) W. R. Bussem, N. R. Thielke and T. Sarakauskas,
"Thermal Expansion Hysteresis of Aluminium Titanate",
Ceramic Age 60 (1952) 38.

(11) G. Bayer,
"Thermal Expansion Characteristics and Stability of Pseudobrookitetype compounds, M_3O_5",
Journ. Less. Common. Met. 24,2 (1971) 129.

(12) Ch. Hahn,
"Verbesserung von Aluminiumtitanat-Keramik",
Sprechsaal 118,12 (1985) 1157.

THERMAL SHOCK RESISTANCE CHARACTERIZATION OF CELLULAR ALUMINA CERAMICS

R.M. Orenstein and D.J. Green
Department of Materials Science & Engineering
The Pennsylvania State University
University Park, PA 16802

ABSTRACT

Cellular ceramic materials can find important uses in advanced heat engines. Strength, toughness and elastic moduli decrease in these materials as a function of density. Thermal shock resistance parameters can be modeled in order to predict environmental limitations and failure mechanisms due to thermal cycling. Specimens of a commercial lightweight alumina-mullite ceramic were quenched in water from temperatures ranging to 1000 C. Crack extension follows theoretical predictions if material properties of the solid phase are used in calculations. Theoretical predictions of macrocracking behavior in the bulk material are partially supported by experimental results, but quench medium infiltration into the bulk specimens limits attainable temperature differences. Thermal shock resistance on a macroscopic scale is higher than expected from calculations based on bulk properties. Thermal shock resistance of the solid phase can be tailored by altering the composition, microstructure and dimensions of the material.

NOMENCLATURE:

E	Young's modulus
E_o	Young's modulus of the uncracked material
G	shear modulus
h	surface heat transfer coefficient
k	thermal conductivity
K_{IC}	fracture toughness
R	thermal shock resistance parameter describing the critical temperature difference required for the initiation of damage
ΔT	maximum instantaneous temperature difference to which the material is subjected
α	linear coefficient of thermal expansion
B	Biot modulus
υ	Poisson's ratio
σ	imposed tensile or compressive stress
σ_f	stress level which results in material failure

INTRODUCTION

Lightweight ceramic materials possess properties which make them attractive for uses as diffusers, catalyst supports and particulate traps in advanced heat engines. In order to effectively engineer systems utilizing such materials, it becomes paramount to understand their behavior under thermal cycling. This paper will describe the results of thermal shock testing on a commercially available cellular alumina ceramic. Emphasis is placed on the ability to predict fracture behavior under imposed thermal stresses based upon measureable properties of the bulk material and of the solid phase.

Highly porous materials (i.e. those containing greater than 75% porosity) may be of several varieties. For engine applications, the network of the solid material (hereafter referred to as "struts") composing an open cell ceramic offers the advantages of high free flow volume with large effective surface area. Other lightweight ceramics are fibrous or closed cell in nature; however, only open cell ceramic materials will be considered in this study. In actuality, such a material must be considered as two separate phases; the solid network, which may or may not contain additional porosity, and the continuous void space. When considering purely mechanical stresses on a cellular system, the properties of the bulk material are of primary interest. These properties, however, are strongly influenced by the properties of the material composing the struts (1,2). Thermal stresses, by contrast, are caused by temperature gradients across a given portion of solid (e.g. the cross section of a strut). Resulting differences in thermal expansion or contraction across the solid give rise to stress profiles during heating and cooling. If the magnitude of the maximum stress allows the stress intensity factor of a crack to exceed K_{IC}, failure will occur. The value of this maximum stress for a large plate of a given material is commonly predicted by the formula

$$\sigma = E\alpha\Delta T/(1-\upsilon) \qquad (1)$$

For practical applications, ΔT is not an instantaneous temperature difference, but is influenced by the thermal conductivity, surface heat transfer coefficient and dimensions of the solid material. However, for the purpose of illucidating the failure mechanisms occurring in a cellular ceramic under thermal stresses, an approximately instantaneous temperature difference can be established.

The primary goal of this study is to attempt to model the behavior of a series of materials (of a variety of cell sizes and densities) when subjected to severe thermal shock. Eq. (1) is usually applied to materials which are at or near theoretical density, and which are assumed to have uniform porosity (if present) throughout the material. Commercially available cellular aluminas, along with the majority of other cellular compositions, possess complex microstructures. The solid phase contains a highly variable amount of porosity due to processing techniques employed. The attempt is made in this study to determine whether the properties controlling the initiation and growth of cracks caused by thermal stresses are those which can be measured from the bulk material, or those which must be measured from the isolated solid phase.

THEORETICAL MODELS OF THERMAL SHOCK BEHAVIOR

The behavior of a material under the influence of severe thermal stresses can be divided into two types of responses. The first response is the initiation of damage, as described by the thermal shock resistance parameter R:

$$R = \sigma_f(1-\upsilon)/E\alpha \qquad (2)$$

This is simply a rearrangement of Eq. (1), assuming failure occurs when $\sigma = \sigma_f$. We are concerned only with the finite temperature difference required to initiate fracture, and the material properties of interest are strength, Young's modulus, Poisson's ratio, and coefficient of thermal expansion. Of significance is the fact that the thermal stress does not reach a maximum instantaneously. The time dependence of σ is described by the Biot modulus, which is the product of the surface heat transfer coefficient and the dimension of the body divided by the thermal conductivity. However, suitable methods for measurement of such properties in these materials have not been developed; therefore, a method of rapidly quenching the specimens into water was employed. To perform calculations using Eq. (1), B was assumed to be $\approx \infty$.

The second response of concern is the stability of the crack propagation as a result of the thermal stresses. The extent of crack propagation has been described as a function of grain size (3), porosity (4,5), inhomogeneities (6,7), and initial damage (8). Two approaches have been taken to fundamentally describe crack arrest in thermally shocked ceramics. Hasselman (7) proposed that crack arrest occurs when the excess strain energy is solely converted into surface energy. Evans (9) performed studies which indicate that crack arrest occurs when the stress intensity factor of the crack falls below K_{IC}. The former approach leads to the conclusions that long cracks will propagate in a material having low fracture surface energy, high elastic modulus and high strength (i.e. small initial flaws and/or low initial flaw density) (5). The latter approach emphasizes the importance of fracture toughness as the property which determines the extent of crack propagation.

The experimental approach taken was to generate cracks in the materials by rapid quench from various temperatures. The onset of unstable crack growth in the bulk and strut materials is compared to that predicted by *R*, as calculated from measured properties. Reduction in strut strength following quenching is compared to visually observed crack morphologies.

EXPERIMENTAL PROCEDURE AND ANALYSIS METHODS

A 92% pure alumina material* was chosen for the experimental work in this study. Previous work (1) has shown that this ceramic possesses microstructural and mechanical properties which are consistent with the open cell models developed by Gibson and Ashby (10). Microstructural characterization of crushed strut material was accomplished by x-ray diffraction, flame emission spectroscopy and mercury porosimetry. Microstructural and macrostructural dimensions were determined by image analysis of scanning electron microscope (SEM) photographs. Eight combinations of cell size and relative density were studied, as outlined in Table 1. Specimens were made to shape by the manufacturer, and processed as a single batch to minimize density and microstructural variations.

Table 1. Characteristics of cellular alumina-mullite ceramics used.

Material Designation**	Mean Relative Density	Mean Cell Size (mm)	Mean Strut Outer Diameter (mm)	Mean Strut Inner Diameter (mm)
0711	0.11	4.0	0.80	0.26
0716	0.17	4.3	0.91	0.25
2008	0.09	3.0	0.44	0.18
2011	0.11	3.4	0.45	0.14
2016	0.18	3.3	0.68	0.17
2022	0.25	3.0	0.61	0.16
6511	0.10	1.1	0.16	0.05
6516	0.16	1.1	0.19	0.05

Room temperature elastic constants were measured on 5 samples of each type by the dynamic resonance method, using a modification of that described by Spinner and Tefft (11). Rather than physical coupling of the specimen to the driver and pickup, electromagnetic coupling was employed. Poisson's ratio was estimated from the Young's and shear moduli by the relationship

$$\upsilon = (E/2G)-1 \tag{3}$$

Since Poisson's ratio is believed to be invariant with density in these materials (10), a value of 0.21 (2) was used for calculations. Room temperature strength of 30 samples of each type of bulk material was determined by three point bending. Linear coefficients of thermal expansion between room temperature and 1000 C were determined on 3 samples of each type of bulk material by dilatometry. The average coefficient of thermal expansion over this temperature range was calculated from the mean slope of the expansion versus temperature curve.

A series of 5 specimens of each type were quenched from 50, 150, 300, 500, and 1000 C into icewater. They were held at these temperatures for 1 hour prior to quenching. The surfaces were

* Reticel™ Reticulated Ceramics, Hi Tech Ceramics, Inc., Alfred, NY

** First 2 numbers refer to pores per linear inch; second 2 numbers refer to % of theoretical density treated with Carfusin dye in order to reveal the extent of strut cracking when observed under the optical microscope. The specimens were also optically examined for macroscopic damage (i.e. cracks extending greater than one cell dimension). Final crack lengths and densities were estimated by image analysis of optical and scanning electron micrographs.

A series of specimens of the 0711 and 0716 materials were similarly heated and quenched, but from temperatures of 50, 100, 150, 200, 400, 600, and 800 C. The strength of 30 struts from the surface of each specimen, plus from control specimens, was measured by the method developed by Brezny, Green and Dam (12). The temperatures were chosen so as to provide a large number of data points in the temperature range of R calculated for the struts of these materials. To perform these calculations, the Young's modulus of the struts was estimated using the equation developed by MacKenzie (13):

$$E = E_0\{1 - 1.9\,(\text{porosity}) + 0.9\,(\text{porosity})^2\} \tag{4}$$

Due to the hollow nature of the struts, the porosity obtained by mercury porosimetry was corrected by the volume percentage of the hollow portion of each material. The Young's modulus of the uncracked material was theoretically determined using Hill's arithmetic average of the alumina and mullite fractions of the material. Young's modulus values of 380 and 220 GPa for pure alumina and mullite, respectively, were used. The coefficient of thermal expansion did not differ significantly between the various materials; therefore, the mean value of 6.20 x 10^{-6} C^{-1} obtained for the bulk materials was used for strut calculations. Likewise, the same value for Poisson's ratio (υ = 0.21) was used. Mean strength values from 150 fractured struts from each material were used to calculate R.

RESULTS AND DISCUSSION

Prediction Of Thermal Shock Behavior Based On Microstructure

Figures 1 and 2 show microstructures typical of these materials. The struts contain triangular holes running longitudinally, which are a result of the manufacturing process. The majority of cellular ceramic materials are processed by coating a polymer preform with a ceramic slurry. During the binder burnout process, the preform is also removed, leaving the struts hollow. Additional porosity and microcracking is present throughout the solid phase. Mercury porosimetry revealed a porosity of 23.1% by volume. Mean pore diameter was 0.1 μm. Spectrographic analysis revealed the phase volume fractions to be 92.4% alumina and 7.6% silica plus impurities. It is believed that the silica reacts with a portion of the alumina during processing to yield a mullite phase. This assumption is supported by the thermal expansion data, which is consistent with the linear coefficient of thermal expansion calculated by the Turner approximation (6) for such an alumina-mullite composition. Using values of 69.2% alumina and 30.8% mullite, E_0 was determined to be 320.6 GPa. Average grain size of all materials is 2 to 4 μm (1).

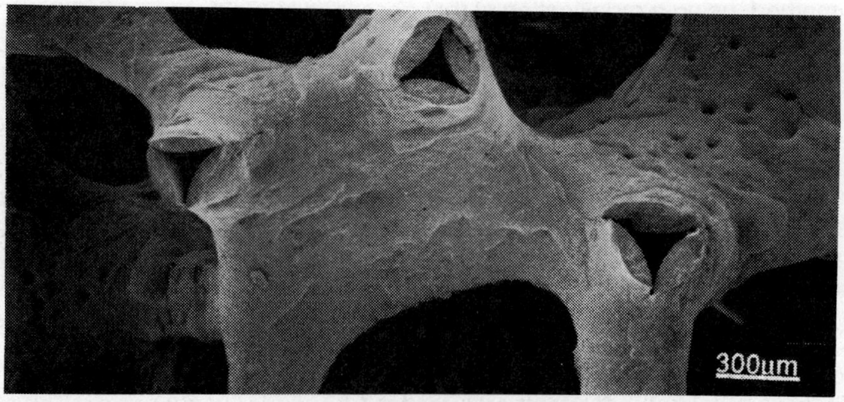

Fig. 1. SEM photograph of 2008 material. The triangular holes which result from preform burnout can be seen in the fractured struts. Additional large pores appear to the right and left sides of the photograph.

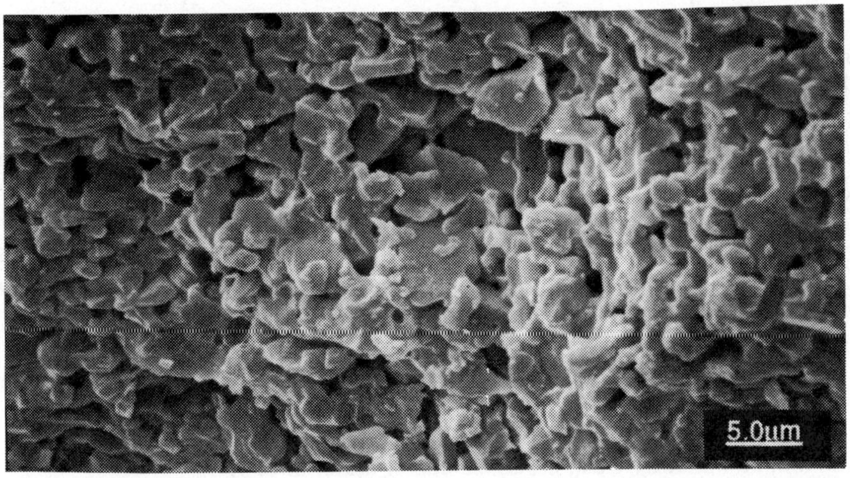

Fig. 2. SEM photograph of 2016 material. Numerous small pores are present in the sintered material, as described in the text.

Crack extension resulting from thermal shock is predicted to occur in an unstable manner within the struts. However, due to the high density of initial flaws and the steep thermal stress gradient, crack arrest is expected to occur within relatively short distances from the propagating flaw. The effect of stress fields from neighboring cracks is important in highly flawed materials (8,14); furthermore, crack arrest at open cells is likely in highly porous materials.

On a macroscopic scale, stable crack propagation as a result of thermal stresses is predicted on the basis of the very high density of initial flaws. The work of Brezny and Green (1) indicated that critical flaw sizes in the bulk materials are on the order of the size of a single cell; each cell can therefore be considered to be a flaw from which a crack can propagate. Hasselman's theoretical predictions (7,8) of increasing crack stability as flaw density increases is supported by numerous experimental results. Furthermore, infiltration of the quenching medium into the porous structure is likely to prevent the establishment of a significant thermal gradient on a macroscopic scale.

Calculated Thermal Shock Resistance Parameters

Table 2 lists property data and the calculated thermal shock resistance parameter *R* for the bulk and strut materials studied.

Table 2. Calculation of the thermal shock resistance parameter *R* for the bulk and strut materials.

Material	*E* (GPa)	σ_f (MPa)	*R* (C)
BULK MATERIALS:			
0711	2.9 ±0.2	1.8 ±0.3	78 (105-58)
0716	6.3 ±0.8	3.2 ±0.6	64 (93-44)
2008	1.3 ±0.4	0.8 ±0.2	83 (150-43)
2011	2.5 ±0.4	1.1 ±0.2	55 (84-37)
2016	5.3 ±0.4	1.9 ±0.3	46 (61-34)
2022	11.0 ±0.6	3.7 ±0.6	43 (56-32)
6511	1.0 ±0.1	0.3 ±0.1	37 (60-22)
6516	1.9 ±0.3	0.8 ±0.2	56 (84-45)
STRUT MATERIALS:			
0711	191.0	95.9 ±62.8	64 (112-21)
0716	192.0	117.6 ±99.8	78 (153-11)

The property data in Table 2 is presented as the mean value ± one standard deviation. The coefficient of thermal expansion value was [6.20 ±0.35] x 10^{-6} C^{-1}. R is presented as the value calculated from the mean property values; the values obtained by using the highest values of σ_f with the lowest values of E and α, and vice versa, are presented in parentheses. For the strut materials, the calculated values of E were taken to be constant, as was υ in all cases.

As can be seen, the calculated values of R are similar for the bulk and strut materials. Thus, an experimental measurement of R will be insufficient to distinguish between crack propagation through the bulk material and strut cracking. It was expected, however, that the observed crack morphology would give insight into the thermal stress failure mechanism.

Of great significance is a general calculation of B using typical values of h and k (6) for an alumina-silica refractory quenched into water. Dimensional analysis of the materials revealed that the Biot modulus can range from 0.5 to as low as 0.05 as strut cross sectional dimension varies. Consequently, even under severe quenching conditions, R for the solid phase can be expected to be higher than that calculated by Eq. (1) by at least a factor of two.

Results Of Thermal Shock Testing

The morphologies of the cracks which resulted from quenching indicates that fracture results from thermal gradients generated across the individual strut cross sections, and not the cross sections of the bulk specimens. Figure 3 shows a crack which propagated through the bulk of a specimen of 2022 material which was loaded in tension. It is evident that the crack propagated over a distance » one cell dimension. This is to be contrasted with Figures 4 and 5, which show typical crack morphologies following thermal shock.

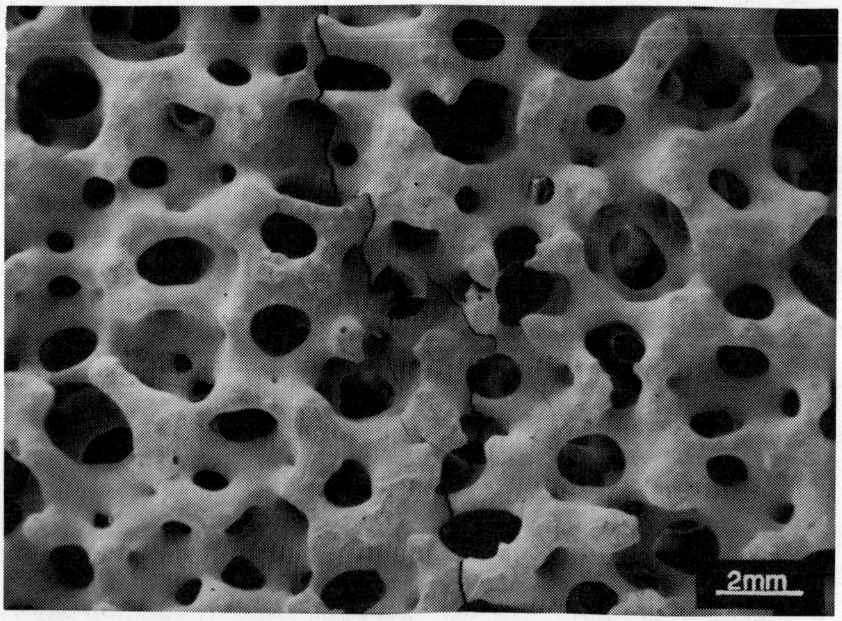

Fig. 3. Crack in a section of 2022 material loaded in tension to 3.9 MPa.

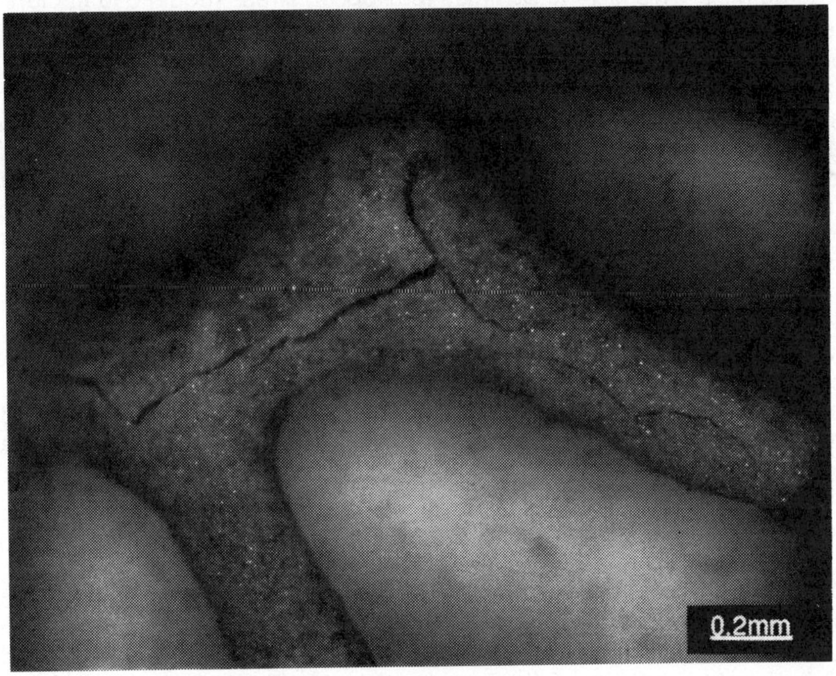

Fig. 4. Optical image showing crack pattern in a strut of 2011 material, quenched from 300 C into icewater.

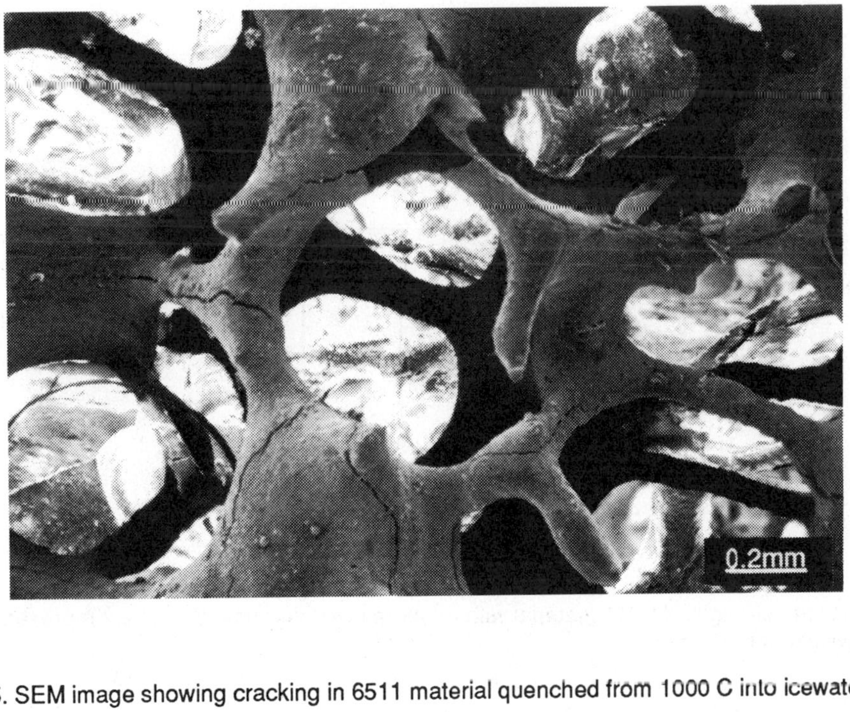

Fig. 5. SEM image showing cracking in 6511 material quenched from 1000 C into icewater.

As calculated from Eq. (1), the maximum thermal stresses on the bulk materials quenched from 1000 C range from 9 to 96 MPa. This is 14 to 30 times the applied stress required to fracture the bulk specimens. The assumption that $B \approx \infty$ is taken to be valid for the bulk specimens, as the specimen cross section (rather than the strut cross section) is used in the calculation.

The hypothesis that macrocrack propagation through the bulk material will not occur due to rapid cooling is thereby supported by the crack morphologies. Crack lengths > one cell dimension were only observed in the 6511 material quenched from 300 C and above. This is the lowest strength material of those tested, and it is likely that mechanical failure occurred due to boiling quenching medium. As can be seen from Figure 5, thermally-generated cracks in this material were typically shorter than a cell dimension, even from very large temperature differences. Crack lengths increased with temperature difference in all cases, with significant crack branching occurring from 300 C and above. It is unusual that the net-like pattern of interfering cracks usually observed in materials quenched from temperatures much above their critical temperature difference (14) was generally not observed, even at the 1000 C temperature difference. On the contrary, relatively isolated regions of cracking occurred in all materials. Cracks were observed in approximately 10% of the struts from 50 C, to essentially 100% of the struts from above 300 C. The Weibull moduli of the struts in these materials are low, on the order of 2-4 (1). Therefore, it can be predicted that only 10-20% of the struts will reach their critical temperature differences from 50 C, and over 90% of the struts will reach it from 150 C. When B is taken into account, these numbers may be reduced by over 50%; furthermore, the low Biot moduli account for the occurrence of intact struts in specimens quenched from the higher temperatures.

As can be seen in Figures 4 and 5, the majority of the cracks were oriented longitudinally to the strut axes, presumably because the strut surfaces experience mostly circumferential tension on rapid cooling. Cracks oriented transversely to the strut axes were only observed in the materials quenched from 1000 C. Even the majority of these, as is evident in Figure 5, appear to be extensions of longitudinal cracks in other struts and strut junctions.

The results of the strut strength measurements on the 0711 and 0716 materials are shown in Figures 6 and 7.

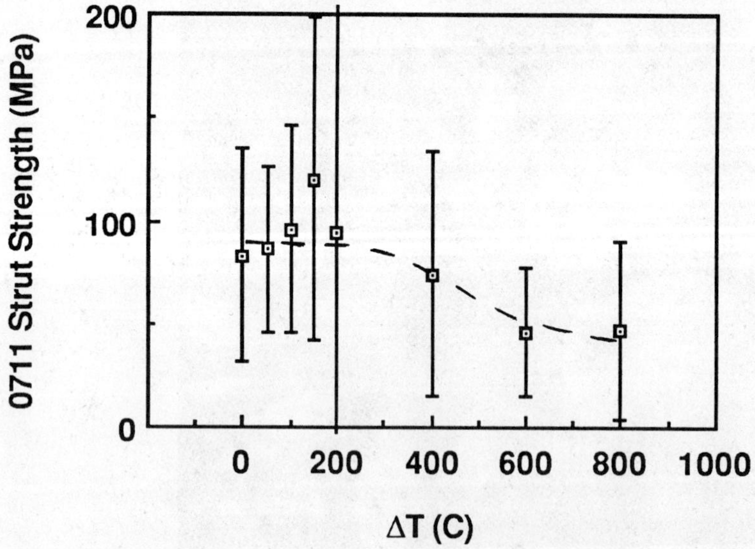

Fig. 6. Variation in strut strength of 0711 material with temperature difference. Error bars represent one standard deviation for 30 struts.

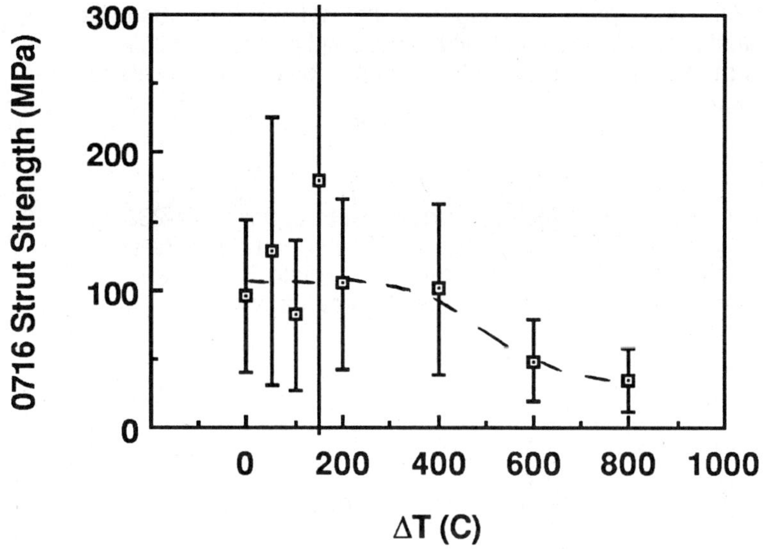

Fig. 7. Variation in strut strength of 0716 material with temperature difference. Error bars represent one standard deviation for 30 struts.

As can be seen, a large variability in strength of individual struts exists. The variability increases as a portion of the struts experience damage at the smaller temperature differences. The variability decreases as the majority of struts experience damage at temperature differences above 400 C. Statistical analysis indicates that a reduction in mean strength does not occur over temperature differences less than 400 C. This is in conflict with the observations of cracking at temperature differences as low as 50 C. However, calculated mean B for the 0711 and 0716 strut materials are 0.21 and 0.24, respectively. This indicates that mean R may be as high as 305 C for the 0711 material and 325 C for the 0716 material. In addition, it is hypothesized that the method of loading the struts perpendicularly to their axes (12) does not significantly increase the stress intensity factor of a crack which runs longitudinally along a strut. As temperature difference is increased, crack branching allows considerably higher stress intensity factors due to transversly running cracks.

When curves are faired through the data (dashed lines in Figs. 6 and 7), the reduction in mean strut strength resembles the response obtained by other workers (5,6,7,8,15) for highly flawed bodies which have been thermally shocked. Nevertheless, due to the statistical nature of flaws present in all ceramics, and to the considerable variation in flaw sizes present in these materials, measurements of strength following thermal shock are not considered by these investigators to be the most accurate method of characterizing damage. Further research is in progress to characterize the total flaw spectrum in the materials by dynamic resonance methods, both prior to and following quenching, in order to accurately determine the onset of damage and the degree of degradation following successive quenches. These experiments are expected to illucidate any relationships between thermal shock resistance parameters and relative density or cell size.

CONCLUSIONS

The commercial cellular ceramics evaluated have resistance to thermal shock on a macroscopic scale which is much higher than that predicted by theoretical modeling. It is proposed that infiltration of a water quenching medium is likely to prevent large thermal gradients from being established in open cell ceramic materials. Less severe quenches, such as oil or air, are expected to result in even smaller thermal gradients.

The onset of strut cracking is believed to be controlled by the nature of the solid phase composing the struts, particularly by the nature and amount of porosity contained within the struts. Therefore, the well known relationships between strength, Young's modulus, Poisson's ratio, fracture surface energy, and thermal conductivity which are used to predict the onset of thermal shock damage and the degree of damage resulting from thermal shock can be expected to predict the behavior of crack initiation and propagation within the solid phase of cellular ceramics. The fracture behavior of the struts will also dictate the reduction in strength of the bulk material as a result of thermal shock.

ACKNOWLEDGEMENTS

This study was supported by the Gas Research Institute, under contract no. 5084-238-1302. The authors acknowledge the students and staff of the Materials Science and Engineering Department who assisted in this study. Special thanks is extended to Rasto Brezny and Chuong Dam, who provided room temperature mechanical property measurements, and to Hi Tech Ceramics for supplying the samples evaluated.

REFERENCES:

1. R. Brezny and D.J. Green, "Fracture Behavior of Open Cell Ceramics," *J. Am. Ceram. Soc.*; submitted Jun. 1988.

2. H. Hagiwara and D.J. Green, "Elastic Behavior of Open Cell Alumina," *J. Am. Ceram. Soc.*, **70** [11] (1987) 811-815.

3. T.K. Gupta, "Strength Degradation and Crack Propagation in Thermally Shocked Al_2O_3," *J. Am. Ceram. Soc.*, **55** [5] (1972) 249-253.

4. R.D. Smith, H.J. Anderson and R.E. Moore, "Influence of Induced Porosity on the Thermal Shock Characteristics of Al_2O_3," *Am. Ceram. Soc. Bull.*, **55** [11] (1976) 979-982.

5. G. Ziegler, "Microstructural Aspects of Thermal Stress Resistance of High Strength Engineering Ceramics," *Z. Werkstofftech.*, **16** (1985) 12-18.

6. Introduction to Ceramics, 2nd Ed., pp. 822-830. Edited by W.D. Kingery, H.K. Bowen and D.R. Uhlmann. Wiley, New York,1976.

7. D.P.H. Hasselman, "Unified Theory of Thermal Shock Fracture Initiation and Crack Propagation in Brittle Ceramics," *J. Am. Ceram. Soc.*, **52** [11] (1969) 600-604.

8. D.P.H. Hasselman and J.P. Singh, "Analysis of Thermal Stress Resistance of Microcracked Brittle Ceramics," *Am. Ceram. Soc. Bull.*, **58** [9] (1979) 856-860.

9. A.G. Evans, "Thermal Fracture in Ceramic Materials," *Proc. Br. Ceram. Soc.*, **25** (1975) 217-237.

10. L.J. Gibson and M.F. Ashby, "The Mechanics of Three-Dimensional Cellular Materials," *Proc. R. Soc. London*, Ser. A, **382** (1982) 43-59.

11. S. Spinner and W.E. Tefft, "A Method for Determining Mechanical Resonance Frequencies and for Calculating Elastic Moduli from these Frequencies," *Am. Soc. Test. Mater. Proc.*, **61** (1961) 1221-1238.

12. R. Brezny, D.J. Green and C.Q. Dam, "Evaluation of Strut Strength in Open Cell Ceramics," *Adv. Ceram. Mater.*; submitted Feb. 1988.

13. J.K. MacKenzie, "Elastic Constants of a Solid Containing Spherical Holes," *Proc. Phys. Soc.* (London), B**63** (1950) 2-11.

14. H.A. Bahr, G. Fischer and H.J. Weiss, "Thermal Shock Crack Patterns Explained by Single and Multiple Crack Propagation," *J. Mater. Sci.*, **21** (1986) 2716-2720.

15. R.W. Davidge, Mechanical Behavior of Ceramics, pp. 118-131. Cambridge University Press, Cambridge, 1979.

THERMAL STABILITY OF Si_3N_4/SiC WHISKER COMPOSITES IN AIR

A.Bellosi, G.N.Babini
Reserch Institute for Ceramic Technology - Faenza, Italy

ABSTRACT

The oxidation behaviour of Si_3N_4/SiC whisker composites and of monolithic Si_3N_4 was investigated at temperature up to 1480°C. On the basis of parabolic kinetics and of the microstructural characteristics of the oxide scale, diffusion of Y through the grain boundary phase can be suggested as the rate governing step. At $T \leq 1100°C$, where Si_3N_4 is stable, weight gains result from the oxidation of SiC whiskers incorporated in the body, when open porosity is present.

The thermal stability of the SiC whiskers was also considered: oxidation starts at $\sim$400°C; diffusion of oxygen through the silica layer has been indicated as the slow step in the process.

INTRODUCTION

In order to decrease the brittle, flaw sensitive nature of silicon nitride, an increasing research activity on Si_3N_4-based composites has recently been performed. Among the reinforcement systems considerable attention has been focused on silicon carbide whiskers[(1-6)]. Studies are directed primarily towards the development and understanding of the mechanical properties of composites at both room and elevated temperature. The success of whiskers-reinforced Si_3N_4 used in high temperature applications will depend on the extent to which the matrix can prevent the oxidation of the whiskers[(7)]. Oxidation kinetics of monolithic Si_3N_4-based ceramics has been previously discussed in detail in terms of amount and type of additives and in relation to the viscosity of the grain boundary phase[(8-10)].

The purpose of this study is to evaluate the thermochemical stability of hot pressed Si_3N_4-SiC whisker composites under oxidizing conditions, in comparison with the behaviour of the Si_3N_4 baseline material under the same experimental procedure. Moreover, the oxidation resistance of the SiC whiskers is also considered.

EXPERIMENTAL PROCEDURES

Samples of the following compositions were uniaxially hot pressed at 30 MPa and 1750°C for 1 hour:

89 wt% Si_3N_4[1] + 3 wt% Al_2O_3[2] + 8 wt% Y_2O_3[3]	(AY)
80 vol% AY + 20 vol% SiC whiskers[4]	(AY-20W)
70 vol% AY + 30 vol% SiC whiskers	(AY-30W)

The values of some microstructural parameters and mechanical properties are summarized in Table 1(6).

Table 1. Characteristics of the hot pressed materials

Sample	Relative density %	Mean grain size (μm)	λ[5] $°C^{-1}$	HV_{500}[6] (GPa)	K_{IC}[7] ($MPam^{½}$)	E[8] (GPa)	σ[9] (MPa)
AY	99.8	0.9	3.18	20.7+ .9	4.8	340.0+.3	895+35
AY-20W	99.3	0.8	3.47	19.2+ .5	4.9	312.7+.4	735+34
AY-30W	95.0	0.7	3.62	13.0+1.2	5.3	322.6+.4	621+24

Oxidation tests

Short term ($\sim$30 hours) oxidation tests were carried out at 800-1480°C in flowing air on 10x10x2 mm polished and ultrasonically cleaned samples, diamond sawn from the hot pressed billets. Weight gain was continuosly recorded by using a T.G. apparatus capable of $2x10^{-5}$g resolution.

The oxidized surfaces were analyzed by X-ray diffraction scanning electron microscopy and microanalysis.

For a better understanding of the oxidation behaviour of the composites, the thermal stability of SiC whiskers was also evaluated, at temperatures up to 1400°C.

----

[1] Si_3N_4: LC12 Starck, West Germany; [2] Al_2O_3: S.G.A16 Alcoa, USA; [3] Y_2O_3: Rhone Poulenc, France; [4] SiC whiskers: Tokawhiskers,Tokai Carbon Co., Japan; [5] λ: thermal expansion coefficient (20-900°C); [6] HV: Vickers hardness, 500g load, 10 s. indentation time; [7] K_{IC}: evaluated by the equation proposed by Antis et al.(6); [8] E: Young's modulus determined by the resonance frequency method; [9] σ:modulus of rupture at room temperature measured by four points test on sample 30x3x3 mm, lower span 26 mm, inner span 13 mm, crosshead speed 0.5 mm/min.

RESULTS AND DISCUSSION

Oxidation behaviour of SiC whiskers

The reaction SiC whiskers $\longrightarrow SiO_2$ starts at about 400°C, is minimal in the range of temperatures from 400 to ~ 1000°C and very extensive above 1200°C.

The course of the reaction, expressed in terms of fractional reaction product (α) is reported in Fig.1.

A nearly complete reaction (~96%) was detected after 10 hours at 1400°C; X-ray diffraction analysis on the reaction product confirmed the presence of silica (mainly cristobalite).

The oxidation kinetics in isothermal conditions approximates parabolic behaviour:

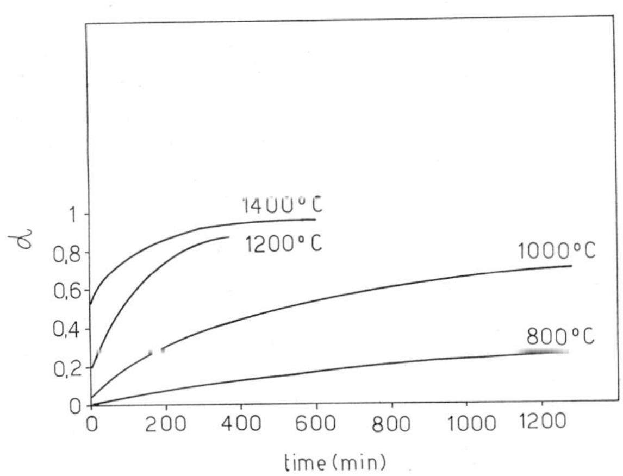

Fig.1.Fractional reactiom product (α) during oxidation of SiC whiskers in isothermal conditions.

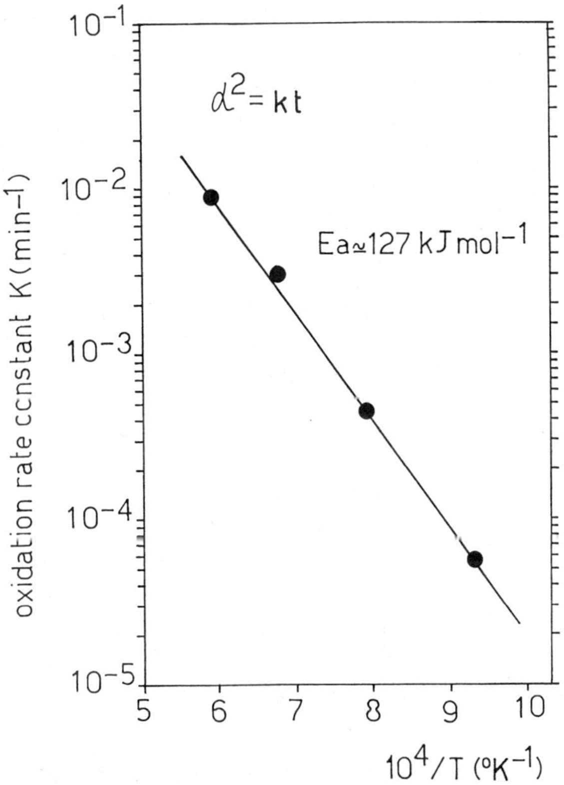

Fig.2.Arrhenius plot of the oxidation rate constants k for the oxidation of SiC whiskers

$$\alpha^2 = kt \qquad (1)$$

where α is the fractional reaction product, k(min^{-1}) is the oxidation rate costant and t is time.

The values of k, reported as a function of temperature in the Arrhenius plot of Fig.2,give an activation energy of 127 KJ/mole. Both kinetic and thermodynamic parameters account for a diffusion controlled process, where transport of oxygen through the pure silica layers parabolically grown during the oxidation is suggested as the rate-controlling step. In fact, the value of the activation energy accords with the data reported by Hotzfeld[11] and Norton[12] on the oxidation of pure SiC, for which permeation through the SiO_2 was indicated as the controlling mechanism of the phenomenology.

Oxidation behaviour of hot pressed materials

Experimental results: Fig.3 a,b show the weight gain against temperature during heating up (5°C/min) for the composites and the baseline material.

The weight gain starts at about the same temperature (400-500°C) as the oxidation reaction of SiC whiskers and, for

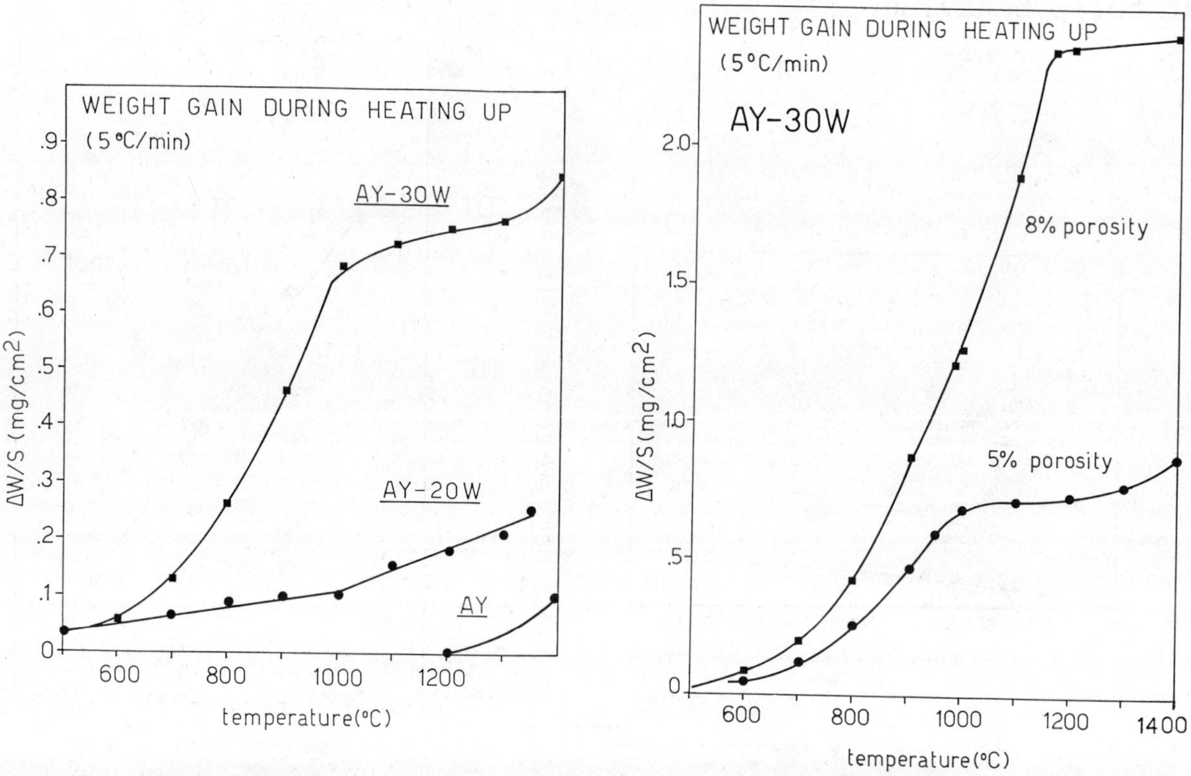

Fig.3. Weight gain during heating up of the Si_3N_4/SiC whiskers composites (a), and in evidence the effect of different porosity amounts (b).

composite AY-30W is strongly dependent on porosity and continues extensively up to ~1100°C; thereafter the oxidation is very slow, owing to the formation of a oxide layer, which originates from the oxidation of Si_3N_4 matrix and protects the SiC whiskers. The data suggest that the presence and amount of open porosity is important in controlling the oxidation characteristics of the composite. High densities need to be achieved to prevent oxidation of incorporated SiC whiskers.

Negligible weight gains were found for the monolithic Si_3N_4 in heating up regimes. In isothermal conditions, weight gains were not detected below ~ 1200°C for AY baseline silicon nitride, and

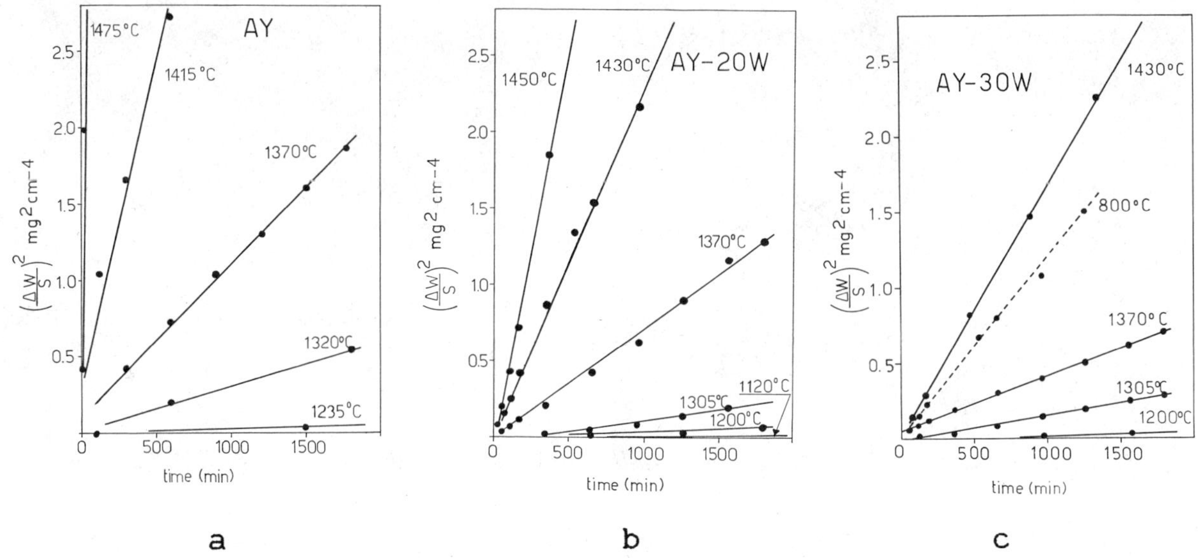

Fig.4. Parabolic weight gains for AY (a), AY-20W (b), AY-30W (c) samples under oxidizing conditions at various temperatures

below ~1100°C for AY-20W composite (Fig.4.a,b). On the contrary at temperature from ~800 to ~1000°C the thermal stability of the AY-30W composite is directly related to the oxidation of SiC whiskers which are present in the bulk and connected to the surface through open porosity channels.

The high weight gains (Fig.4 c) at 800°C of AY-30W material stand for an enhanced reaction of silicon carbide whiskers to SiO_2, that develops within the body of the sample, in fact no crystalline silica was detected on the surface. The morphology of the oxidized surface is the same as the starting material (Fig.5 a), but inside the pores the decomposition of SiC whiskers can be located (Fig.5 b), together with the formation of cracks and new open channels towards the bulk of the composite.

The isothermal weight gain against time curves ($1100 \leq T \leq 1480$°C) approximate the classical parabolic behaviour represented by the equation

$$(\Delta W/S)^2 = kt + b \quad kt \qquad (2)$$

where ΔW is the weight gain at time t, S is the geometrical surface exposed to the oxidation, $k = k_0 \exp(-Ea/RT)$ ($mg^2 cm^{-4} min^{-1}$) is the parabolic rate constant and b is a constant which account for the uncertainty of the exact definition of the beginning (t=0) of the isothermal parabolic process.

Plots of Eq.2 as a function of time are shown in Figs.4 a-c for AY, AY-20W and AY-30W samples, respectively. The analysis of the oxide scale by X-ray diffractometry shows the formation of cristobalite on samples annealed in air at T>~1100°C, and of

Fig.5. Surface of AY-30W composite oxidized 30 hours at 800°C.

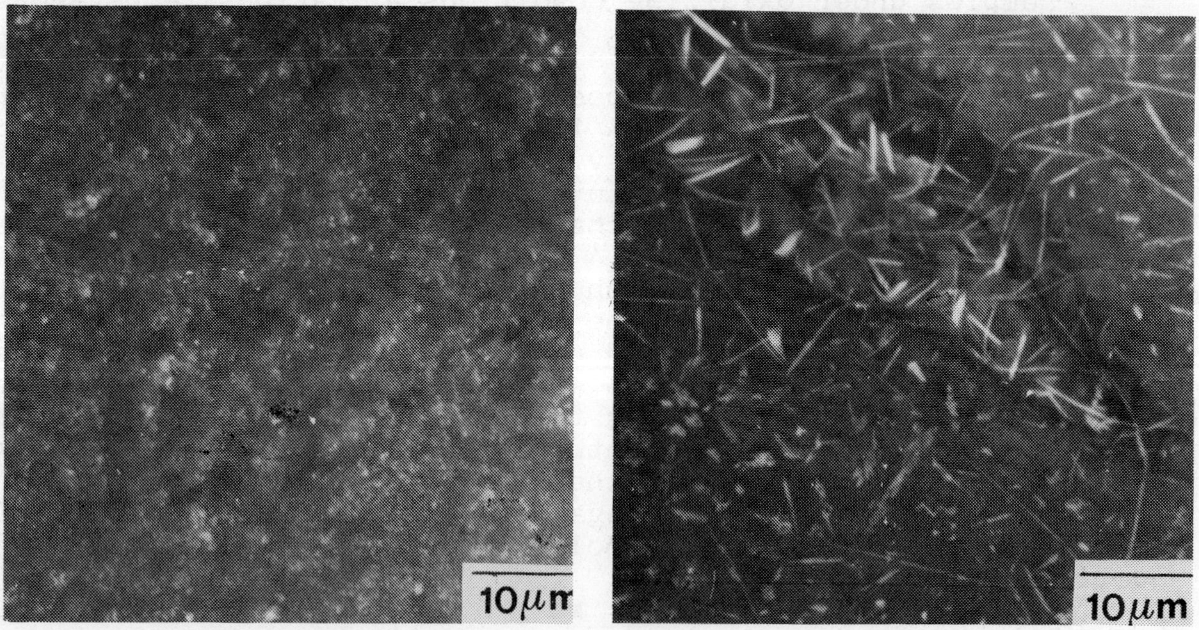

Fig.6. Surfaces of AY (a) and AY-20W (b) samples after oxidation for 30 hours at 1100°C : back scattered electron images showing the presence of Y rich phase.

$Y_2O_3.2SiO_2$. This yttrium silicate is detected after oxidation at $T>\sim$1100°C in AY-30W and $T>\sim$1200°C in AY and AY-20W samples.

The morphological and EDS analyses (Fig.6 a,b) after oxidation at 1100°C indicate a more evident diffusion of Y from the bulk to the surface of AY-30W, with the formation of a glassy layer where needle-like crystals ($Y_2O_3.2SiO_2$) segregate. This reaction product also fills the pores and prevents further oxidation of SiC whiskers. At higher temperatures the diffusion of Y increases and the oxidized surfaces (in Fig.7 a,b an example after oxidation of AY-30W at 1300°C) are covered by a scale with escape pores and bubbles and a lot of plate-like crystals (Y_2O_3) form in increasing amounts from AY to AY-20W and to AY-30W.

This feature is enhanced on the surface of samples oxidized at 1370°C: yttrium silicate crystals appear in a low concentration on AY oxide scale (Fig.8 a,b), which is however thicker and full of pores.

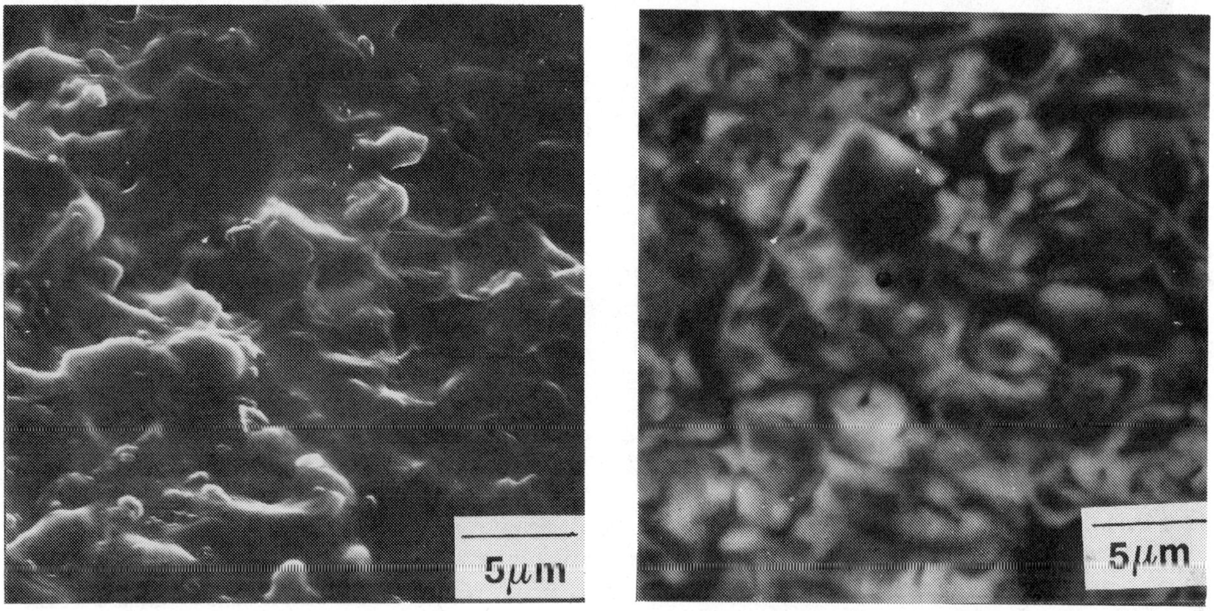

Fig.7. Surface of AY-20W composite oxidized 30 hours at 1300°C : a) morphology; b) back scattered electron image showing the high amount of plate-like $Y_2O_3.2SiO_2$ crystals.

Discussion of the results

The starting composition, amount and type of the secondary phases and the residual porosity are the main parameters of the materials which affect the oxidation behaviour. The compositional variation among the three tested materials is the decrease of the amount of Si_3N_4-phase in the composites.

Of the two sintering aids, Al_2O_3 and Y_2O_3, the main part of Al_2O_3 enters s.s. with silicon nitride; conversely, Y_2O_3 contributes to the formation of glassy or partially crystalline yttrium silicate phases, where impurities segregate.

At temperatures above 1100°C, where the contribution of the thermal instability of SiC whiskers to the overall oxidation process could be considered negligible, the same rate-governing mechanism can be suggested for the baseline Si_3N_4 and for the

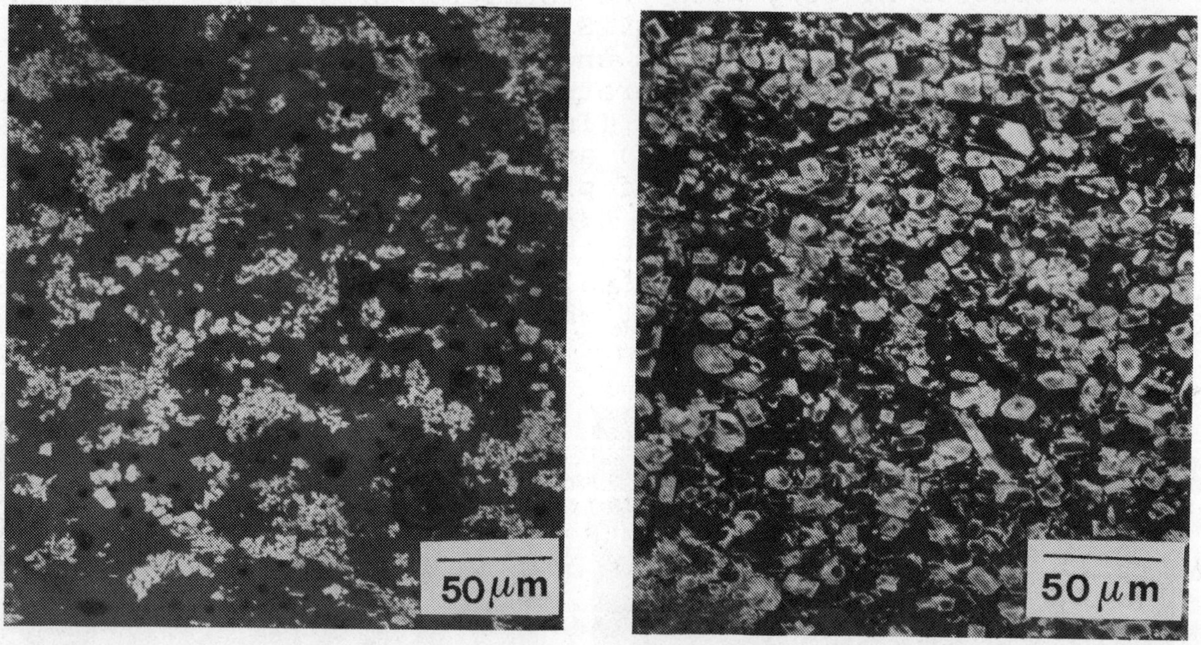

Fig.8. Back scattered electron image of the oxide product on AY (a) and AY-20W (b) materials after 30 hours at 1370°C.

Si_3N_4/SiC whiskers composites.

According to the experimental data and previous results(8,10), oxidation takes place through the classical mechanism where diffusion of additive and metal impurity cations to the nitride/oxide interface is the governing step.

Also the close link between the characteristics of the intergranular phase and the oxidation products can be associated with the diffusion of Y cations in an attempt to equilibrate the chemical composition of the oxide formed by the primary oxidation of Si_3N_4 grains and of SiC whiskers. Therefore, at temperatures in the 1100-1200°C range, on the surface of AY-20W and AY-30W materials,a SiO_2 content higher than that of AY is present as a consequence of the easy oxidation of SiC whiskers during heating up. This gives rise to an enhanced diffusion of yttrium from the bulk to the reaction interface in the composites: the oxidation rate constant k is the experimental parameter which accounts for it. Moreover the morphological and structural characteristics of the oxide scale formed after a thermal run of 30 hours at 1100°C (Fig.6 a,b) confirm that different amounts of materials have been transferred at the Si_3N_4/oxide reaction interface: nearly negligible diffusion in AY, slightly evident in AY-20W, clearly evident in AY-30W, where $Y_2O_3.2SiO_2$ crystals already form on the surface.

The values of $\Delta W/S$ for the composites, and consequently of k, determined from a mathematical interpolation of eq.(2), could be connected to the value of the considered surface (S): i) the geometrical area of the sample, or ii) the area of only the Si_3N_4-phase exposed to oxidation (i.e. substracting the area covered by the SiC phase).

In the first case the values of k, reported in the Arrhenius plot (Fig.9) result in apparent activation energies for oxidation ranging from 635 to 366 KJ/mole, this value decreasing with the increase of SiC content in the composites.

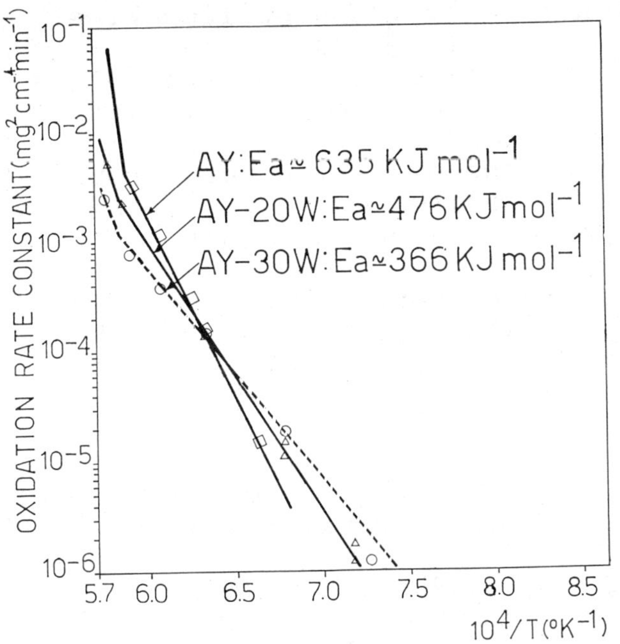

Fig. 9. Arrhenius plots of the oxidation rate constants relative to the Si_3N_4/SiC whiskers composite and to monolithic Si_3N_4.

In the second case, the values of the oxidation rate constants at T> 1300°C for the composites approximate the values of AY materials and, of course, the activation energy for all the materials approaches that (635 KJ/mole) of AY.

In any case, it can be stressed that at T>~1300°C the composites exhibit better oxidation resistance than the baseline Si_3N_4 material owing to a lower amount of oxidizable Si_3N_4. This give rise to smaller quantities of SiO_2 in the oxide scale which account for the observed morphology (Fig.7): on the oxidized surfaces of the composites, almost all the silica coming from the oxidation of silicon nitride enters the formation of $Y_2O_3 \cdot 2SiO_2$.

The drastic increase of k detected at very high temperatures (>~1430 °C), is associated to evident changes in the oxidation mechanism due to the decrease of the intergranular phase viscosity and consequently to an enhanced diffusion of Y and oxidation of Si_3N_4.

CONCLUSIONS

- Oxidation of SiC whiskers at 800≤T≤1400°C results in parabolic kinetics and the apparent activation energy for oxidation (~127 KJ/mole) suggests diffusion of oxygen through SiO_2 as controlling step.
- Oxidation tests up to 1480°C of Si_3N_4/whiskers composites and of the baseline Si_3N_4 material show that:
  a) At T<~1100°C, weight gain deriving only from the oxidation of SiC whiskers incorporated in the bulk, occurs when open

porosity is present.
The weight gain due to the whiskers on the surface is negligible. Therefore the thermal stability of the composites in this temperature range is determined by the amount of porosity and the volume % of SiC whiskers.

b) At T>~1100°C oxidation results in parabolic kinetics for all the systems.

c) Kinetics, morphology and composition of the oxide scale suggest diffusion of Y as the mechanism controlling the oxidation.

d) In isothermal condition, the oxidation rate from ~1200 to 1500°C is less for the composites owing to a lower content of oxidizable Si_3N_4-phase, and having SiC mostly oxidized during heating up.
Since more SiO_2 is already present due to oxidation of the whiskers at lower temperature, formation of Y-silicates closing off pores is more likely in the composites.

REFERENCES

1 - K.UENO, S.SODEOKA, "Fracture toughness of SiC Whiskers Reinforced Si_3N_4 Ceramics", Yogyo-Kyokai-Shi 94 (9) (1986) 981-985.

2 - P.D.SHALEK, J.J.PETROVIC, G.F.HURLEY, F.D.GAC, "Hot pressed SiC whiskers/Si_3N_4 Matrix composites", Am.Ceram.Soc.Bull. 65 (1986) 351-56.

3 - R.LUNDBERG, L.KAHLMAN, R.POMPE, R.CARLSSON, "SiC-whiskers Reinforced Si_3N_4 Composites", Am.Ceram.Soc.Bull. 66 (1987) 330-33.

4 - P.GREIL, G.PETZOW, H.TANAKA, "Sintering and Hipping of Silicon Nitride - Silicon Carbide Composites Materials", Ceramics Int. 13 (1987) 19-25.

5 - S.T.BULJAN, J.G.BALDONI, M.L.HUCKABEE, "Si_3N_4-SiC Composites" Am.Ceram.Bull. 66 (1987) 347-52.

6 - A.BELLOSI, G.DE PORTU, " Si_3N_4/SiC whiskers hot pressed composites" presented at the E-MRS Conference, Strasbourg, 31 May - 2 June 1988.

7 - T.J.CLARK, M.JAFFE, J.RABE, N.R.LACEY, " Thermal Stability Characterization of SiC ceramics fibers : I; Mechanical property and chemical structure effects", Ceram.Eng.Sci.Proc. 7 7-8 (1986) 901-913.

8 - S.C.SINGHAL, " Oxidation of silicon nitride and related materials", in NATO/ASI "Nitrogen Ceramics" edited by F.L.Riley (Noordhoff Publ. Comp., London (1976) pag.607.

9 - G.N.BABINI, A.BELLOSI, P.VINCENZINI, "A diffusion model for the oxidation of hot pressed Si_3N_4-Y_2O_3-SiO_2 materials", J.Mater.Sci. 19 (1984) 1029-1042.

10 - G.N.BABINI, A.BELLOSI, P.VINCENZINI, " Factors influencing structural evolution in the oxide of hot pressed Si_3N_4-Y_2O_3-SiO_2 compositions", J.Mater.Sci. 19 (1984) 3487-3497.

11 - K.HÖTZFELD, Acta Chem.Scand. 18 (1956) 1596.

12 - F.J.NORTON, Nature (London) 191 (1951) 859.

HIGH TEMPERATURE SILICON ALUMINUM CARBONITRIDES

R.J.Oscroft, P.Korgul and D.P.Thompson
Wolfson Laboratory
Department of Metallurgy and Engineering Materials
University of Newcastle upon Tyne U.K.

ABSTRACT

Phase relationships have been determined in the Al_4C_3-SiC-AlN system to explore the occurrence of new silicon aluminum carbonitrides and assess their suitability as high temperature engineering materials. The series of compounds of the type $Al_4C_3.nAlN$ (n=1-4) and $Al_4C_3.mSiC$ (m=1,2) which have been previously reported, have similar structures and some solid solution occurs between end-members with the same metal:non-metal ratio. In addition, the δ-phase of aluminum carbide which is a polymorphic modification of the normal 9R structure has an analogue along the Al_5C_3N-Al_4SiC_4 line in the form of a 12R modification. 4H silicon carbide which is stabilized by aluminum occurs widely in silicon carbide-rich compositions; so far it has not been possible to establish whether this can be represented by a specific range of compositions in the Al_4C_3-SiC-AlN system.

INTRODUCTION

Silicon carbide and silicon nitride are well established contenders for high temperature engineering applications. Silicon nitride is generally the preferred material up to temperatures of ≈1300°C because of its higher toughness (4-8 $MPam^{\frac{1}{2}}$). However the mechanical properties and the oxidation resistance of most nitrogen ceramics fall off rapidly above 1300°C either because of softening of grain-boundary glass or because Si_3N_4-metal oxide eutectics have been reached which cause local melting. Above this temperature, monolithic or SiC fiber-toughened SiC materials offer potential if good oxidation resistance can be preserved.

Both Al_4C_3 and AlN are refractory materials which have received limited attention as potential engineering ceramics for high temperature applications. In the search for the next generation of materials for use in the range 1500-2000°C, compositions in the Si-Al-N-C system are a possible choice. The present paper aims to review the compounds available in this system, clarify errors in the literature and assess the feasibility of preparing materials with good properties. The characterization of some new Si-Al-N-C phases is reported.

PREVIOUS WORK

No ternary compounds are known to exist in the SiC-Si_3N_4 and Si_3N_4-AlN systems. However along the SiC-AlN join, AlN has a 40% solubility in SiC and SiC a 20% solubility in AlN (see Figure 1)[1]. At very high temperatures the 2H SiC form transforms to a 3C modification which also contains AlN (up to 20%). AlN has a lower strength than SiC and rather poor resistance to oxidation at high temperatures; however it has excellent thermal stability. No other Si-Al-N-C phases are known to occur in the SiC-AlN-Si_3N_4 system but in the Al_4C_3-AlN-SiC system a range of compounds have been produced. Along the Al_4C_3-AlN join, Jeffrey and Wu[2,3] reported carbonitrides of the type Al_4C_3.nAlN where n=1-4 and along the Al_4C_3-SiC join similar compounds exist of formulae Al_4SiC_4 and $Al_4Si_2C_5$. Barczak[4] reported the existence of two forms of Al_4SiC_4 but his β modification is in fact a mixture of phases, mainly graphite, silicon carbide and the Al_3NCO phase prepared by Roebuck[5] also referred to as α'-Al_2OC by Yokokawa et al.[6] (see Table 1. Schneider et al.[7] studied the Al_4C_3-Be_2C-SiC system at 1860°C and found Al_4SiC_4 to be the only compound along the Al_4C_3-SiC join. The authors correctly indexed this phase on the basis of a hexagonal unit cell of dimensions _a_: 3.28 Å, _c_: 21.7 Å; they noted the similarity in X-ray patterns and unit cell dimensions between Al_5C_3N (also hexagonal with _a_: 3.281 Å, _c_: 21.67 Å [2,3]) and indeed it would be logical to assume that a solid solution occurred all the way between them. In a second paper, Schneider et al.[8] found that some solid solution occurred, but a complete solid solution was interrupted by a β phase the (hexagonal) unit cell dimensions of which were _a_: 3.164 Å and _c_: 5.084 Å.

Table 1. Reinterpretation of the X-ray Diffraction Pattern of β-Al_4SiC_4[4].

β-Al_4SiC_4 (after Barczak[4]) d (Å)	I	Al_3NCO (after Roebuck[5]) d (Å)	hkl	I	α'-Al_2OC (after Yokokawa et al.[6]) d (Å)
5.01	10	4.98	003	m	5.00
4.53	22	4.51	101	mw	4.51
4.00	13	4.01	012	w	3.99
3.38	30	Graphite 002			
2.75	100	2.73	110	vs	2.73
2.64	12	α-SiC			
2.55	48	2.53	015	w	
2.50	33	2.49	006	vs	2.49
2.42	22				
2.40	39	2.392	113	vs	2.40
2.37	24				
2.19	7				
1.866	19				
1.845	13	1.838	116	ms	1.840
1.588	39				1.660
1.578	22	1.574	300	s	1.576
1.421	20	1.417	1010/119	s	1.419
1.347	10				1.365
1.334	15	1.330	306	ms	1.332

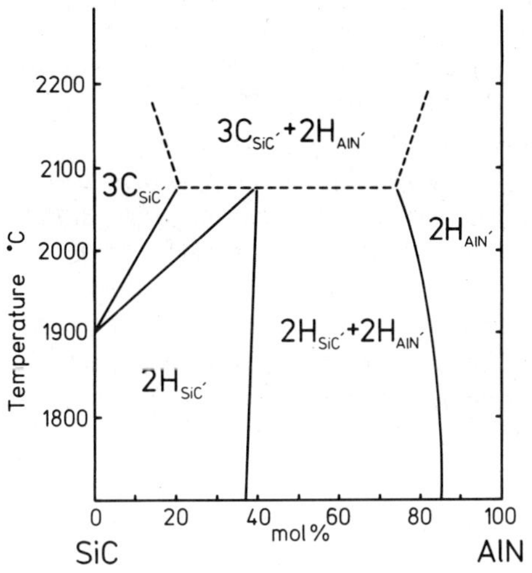

Figure 1. Behaviour diagram for the SiC-AlN system[1].

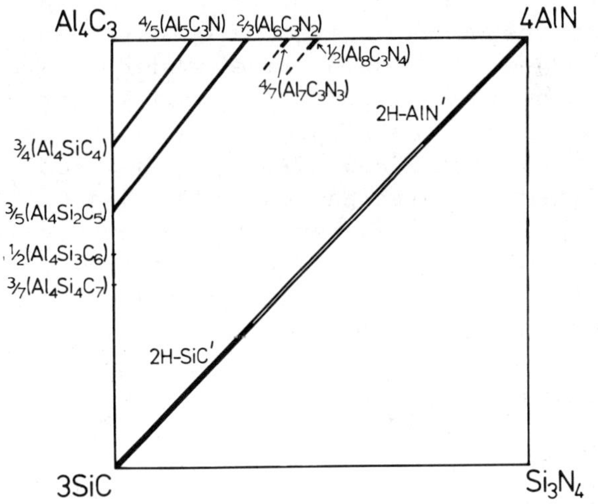

Figure 2. Phases in the Si-Al-N-C system.

These valuse are characteristic of a wurtzite type of phase and are virtually identical to those of Al_2OC (a: 3.168 Å, c: 5.092 Å). Since the authors comment on the presence of oxides in their products, this phase is not unexpected, but it is difficult to see how it can occur as a single phase at the composition suggested. A second phase was reported at the composition $AlN.Al_4C_3.SiC$ with (hexagonal) unit cell dimensions a: 3.25 Å, c: 40.18 Å. This corresponds to the half way point between $Al_6C_3N_2$ (hexagonal, a: 3.248 Å, c: 40.03 Å [2,3]) and the $Al_4Si_2C_5$ phase (hexagonal, a: 3.251 Å, c: 40.11 Å) reported by Inoue et al.[9]. The latter authors prepared single crystal specimens of both Al_4SiC_4 and $Al_4Si_2C_5$, confirmed the correctness of the unit cells deduced from X-ray powder work and showed how the structures were related to their compositions. The failure of Schneider et al.[8] to observe a complete range of solid solution between $Al_4Si_2C_5$ and $Al_6C_3N_2$ may be attributed to their lower temperature of preparation. A summary of the present knowledge of phase relationships in the Si-Al-N-C system is shown in Figure 2. Figure 2 does not indicate any region of stability for the 4H form of silicon carbide. It is well established[10] that this form is stabilized by aluminum but the precise amount of aluminum incorporation and even whether its composition can be represented on a diagram such as Figure 2 is at present unknown.

EXPERIMENTAL

Aluminum silicon carbides and carbonitrides were prepared either by direct reaction of carbides (and nitrides if appropriate) or by carbothermal reduction of mixed oxides, or by routes intermediate between these two extremes. Starting powders included Al_4C_3*, SiC^+, $AlN^\$$, $Si_3N_4^{\neq}$, Al_2O_3**, $SiO_2^£$, Si^{++} and $C^{\$\$}$ and in all compositions containing oxides, an extra 10 w/o of carbon was added. Powder mixes were ball milled in isopropanol with

* Aldrich, Laboratory Reagent Grade.
\+ Arendal, SIKA Grade I.
$ Starck-Berlin, Grade A.
≠ Starck-Berlin, Grade LC10
** Alcoa, Grade A17.
£ BDH, Precipitated.
++ Murex, High Purity Grade.
$$ Cabot Corporation, Monarch 1300.

the use of zirconia balls for 8 hours (pick-up typically less than 3%). Compacts were prepared by uniaxial and isostatic pressing and fired either embedded in boron nitride powder in a graphite crucible in a carbon resistance furnace in a nitrogen or argon atmosphere, or hot-pressed at 22 MPa in an inductively-heated graphite die. Heating rates in the two furnaces were in the range 30-50 ^{o}C/min. The products were analysed by X-ray powder diffraction with the use of a Hägg-Guinier XDC700 focusing camera and CuKα_1 radiation. High resolution electron microscopy was performed with JEOL 100C and 100CX instruments to observe diffraction patterns and lattice images of polytypoid phases. Microstructures were observed with the use of JEOL T20 and JSM35 scanning electron microscopes and cation contents were estimated with the use of X-ray energy dispersive spectroscopy (EDS). Densities were measured by displacement using a mercury balance.

RESULTS

Table 2 shows the results of firing compositions intended to give Al_4SiC_4 according to the equations:

$$2Al_2O_3 + SiO_2 + 12C \rightarrow Al_4SiC_4 + 8CO \quad \ldots (1)$$

$$Al_4C_3 + SiO_2 + 3C \rightarrow Al_4SiC_4 + 2CO \quad \ldots (2)$$

$$Al_4C_3 + SiC \rightarrow Al_4SiC_4 \quad \ldots (3)$$

Table 2. Preparation of Al_4SiC_4. (a) Furnace Conditions, (b) X-ray analysis.

(a) Furnace Conditions.

Run No.	Starting Materials	Furnace+	T(^{o}C)	t(min)	Atmos.	Weight loss (%)
1	$Al_2O_3/SiO_2/C$	HP	1900	60	Air	30
2	$Al_4C_3/SiO_2/C$	CF	2000	60	Ar	25
3	Al_4C_3/SiC	CF	2000	3	Ar	16
4	Al_4C_3/SiC*	CF	2000	3	Ar	13
5	Al_4C_3/SiC	CF	1900	3	Ar	11
6	$Al_4C_3/SiC$$	CF	2000	3	Ar	19
7	Al_4C_3/SiC	CF	2000	60	Ar	50

+ HP: Hot-pressed
  CF: Carbon Resistance Furnace

* 10% extra Al_4C_3 included in the starting mix
$ 2 w/o Y_2O_3 included in the starting mix.

(b) X-ray analysis

No.	Al_4SiC_4	$Al_4Si_2C_5$	$Al_4Si_3C_6$	$Al_4Si_4C_7$	δ-Al_4C_3	4H-SiC	12R	AlN	Al_2O_3	C
1					tr.	40	20	40		tr.
2			35	15		tr.	15	35	tr.	tr.
3	60	20				20				
4	100					tr.				
5	100					tr.				
6	15	35			10	20	20			
7	35	25		10		30				

Table 3. X-ray Diffraction Data for 12H, 21R and 12R Phases.

12H				21R				12R			
$Al_4Si_3C_6$				$Al_4Si_4C_7$				"Al_4SiC_4"			
a: 3.225 *c*: 31.775 Å				*a*: 3.120 *c*: 55.215 Å				*a*: 3.336 *c*: 31.29 Å			
hkl	d_{calc}(Å)	d_{obs}(Å)	I_{obs}	hkl	d_{calc}(Å)	d_{obs}(Å)	I_{obs}	hkl	d_{calc}(Å)	d_{obs}(Å)	I_{obs}
008	3.972	3.984	vvw	0015	3.682	3.688	vvw	003	10.44	10.43	m
100	2.800	2.788	s	0018	3.068	3.069	vvw	006	5.219	5.216	w
101	2.789			101	2.784	2.788	s	009	3.479	3.478	mw
102	2.758	2.758	m	012	2.773	2.776	ms	101	2.877	2.876	s
103	2.707	2.707	m	104	2.732	2.718	ms	012	2.841	2.841	ms
0012	2.647	2.648	vw	015	2.703	2.697	m	104	2.710	2.710	vw
104	2.641	2.630	vw	0021	2.630	2.630	mw	015	2.623	2.623	vvw
105	2.563	2.562	w	107	2.628			0012	2.610	2.611	mw
106	2.476	2.477	mw	018	2.584	2.582	w	107	2.427	2.427	ms
107	2.383	2.384	m	1010	2.476	2.477	mw	018	2.325	2.324	mw
108	2.289	2.290	m	0111	2.437	2.439	m	1010	2.123	2.123	w
0014	2.270	2.270	m	1013	2.330	2.331	m	0015	2.088	2.088	m
1010	2.101	2.101	vw	0024	2.301	2.301	m	1013	1.850	1.851	m
1011	2.011	2.010	vw	1016	2.169	2.169	w	0114	1.769	1.768	mw
1012	1.924	1.922	vvw	1019	2.012	2.010	vw	110	1.668	1.668	s
1013	1.842	1.841	m	1022	1.865	1.863	vw	113	1.647	1.647	vvw
110	1.617	1.616	s	0123	1.819	1.816	w	1016	1.620	1.621	ms
1017	1.555	1.556	mw	1028	1.610	1.616	m	0117	1.553	1.553	mw
1018	1.493	1.493	w	110	1.609	1.609	s	119	1.504	1.504	vw
1020	1.382	1.382	m	0129	1.572	1.573	w	0021	1.491	1.492	vw
1112	1.380			1031	1.501	1.500	vvw	021	1.443	1.443	vw
1115	1.285	1.284	w	0132	1.467	1.466	w	202	1.438	1.438	vw
				1121	1.372	1.372	w	205	1.408	1.406	w
				1124	1.319	1.319	m	027	1.375	1.374	vvw
								1115	1.303	1.303	m

Initial results were carried out in nitrogen gas but the products always contained aluminum nitride and/or aluminum nitride-rich phases showing that reaction was occurring between aluminum-containing starting materials and nitrogen possibly aided by excess carbon present. Subsequent experiments were therefore always carried out in argon. It is clear from Table 2 that the results of all the reactions are multiphase, but whereas runs based on Equations (1) and (2) gave no Al_4SiC_4, the route involving Equation (3) was much more successful. This is in good agreement with the results of Inoue et al.[9]. In Run 3, the presence of higher SiC-content phases is due to the oxidation of Al_4C_3 during powder processing and Run 4 which contained additional Al_4C_3 to compensate for this gave almost pure Al_4SiC_4. Run 5 shows that temperatures of either 1900 or 2000°C are satisfactory for this reaction and Run 7 shows that longer times are no advantage. The addition of yttria in Run 6, which was intended to produce an yttrium aluminosilicate liquid phase to homogenize the products and improve densification in fact had no effect.

All runs produce 4H SiC as either a major or minor impurity, and other aluminum silicon carbides were formed, namely $Al_4Si_2C_5$, $Al_4Si_3C_6$ and $Al_4Si_4C_7$. They occur because the oxygen content of the starting Al_4C_3 powder results in an SiC : Al_4C_3 molar ratio greater than unity. Their compositions are so similar that it is not surprising to get mixtures in the product. δ-aluminum carbide the alternative polymorph reported by Roebuck[5], occasionally occurs in the products presumably by transformation of unreacted 9R Al_4C_3. The phase labelled AlN' is a 1M : 1X composition close to the AlN corner of the AlN-SiC-Al_2CO plane. Further discussion of the phase labelled 12R "Al_4SiC_4" is given later; Table 3 gives X-ray diffraction data for the new 12H, 21R and 12R phases. Weight losses for all runs were large and in the region of 10-30%. This is mainly attributable to alumina on the surface of the aluminum carbide which dynamically oxidizes silicon carbide at high temperatures according to the equation:

$$SiC + Al_2O_3 \longrightarrow SiO + CO + Al_2O \qquad \ldots (4)$$

The fact that the overall composition is not displaced very far from the starting composition is because higher alumina contents reduce the overall aluminum carbide content as well as increasing the silicon carbide loss by Equation 4.

As a follow up to the runs described in the previous section, attempts were made to prepare pure $Al_4Si_2C_5$, $Al_4Si_3C_6$ and $Al_4Si_4C_7$. All runs were carried out in the graphite resistance furnace in argon. Table 4 clearly shows the difficulty of obtaining pure products. All runs were multiphase with large weight losses and similar phases were present as in Table 2. 6H-SiC remains in Run 13 because of incomplete reaction. Compositions with higher SiC : Al_4C_3 ratios (not shown in Table 4) were unsuccessful in producing Al-Si-C phases of higher m-value. However, Run 14 is included because this sample gave no polytypoid phases at all. Here weight losses had shifted the overall composition towards a final mix consisting of aluminium plus 4H SiC.

Table 6 shows the effects of small amounts of either metallic or non-metallic additives aimed at producing high-density products. The samples were either hot-pressed or pressureless sintered.

DISCUSSION

Since SiC, AlN and Al_4C_3 all form layered crystal structures, it is perhaps

Table 4. Preparation of $Al_4Si_2C_5$, $Al_4Si_3C_6$ and $Al_4Si_4C_7$.

No.	Starting Composition	T(°C)	t(min)	Wt. loss %	Al_4SiC_4	$Al_4Si_2C_5$	$Al_4Si_4C_7$	δ-Al_4C_3	4H-SiC	12R	6H-SiC	Al
8	$Al_4Si_2C_5$	2000	3	17	30	20		10	40			
9	$Al_4Si_2C_5$	1970	3	13	40	10		10	40			
10	$Al_4Si_2C_5$	1970	60	24	40		20		40			
11	$Al_4Si_3C_6$	2000	30	22	20	40			30	10		
12	$Al_4Si_3C_6$	2000	60	28	10	30			40	20		
13	$Al_4Si_4C_7$	2000	3	24	20	20			15	30	15	
14	$Al_4Si_3C_6$	2000	20	40				10	80			10

not surprising that all other compounds in the SiC-AlN-Al_4C_3 system are also layer structures exhibiting polytypoid behaviour. For convenience they are referred to here in terms of Ramsdell symbols, applied to the number of layers of non-metal atoms per c-repeat; thus, for example, compounds in the series Al_4SiC_4-Al_5C_3N are of type 8H, $Al_4Si_2C_5$-$Al_6C_3N_2$ are of type 15R and so on. This is consistent with previous nomenclature for polytypoid series[11]. Table 2 shows that more Al-Si-C phases occur along the Al_4C_3-SiC join than have been previously reported in the literature. Thus 12H ($Al_4Si_3C_6$) and 21R ($Al_4Si_4C_7$) modifications occur which match the compounds $Al_7C_3N_3$ and

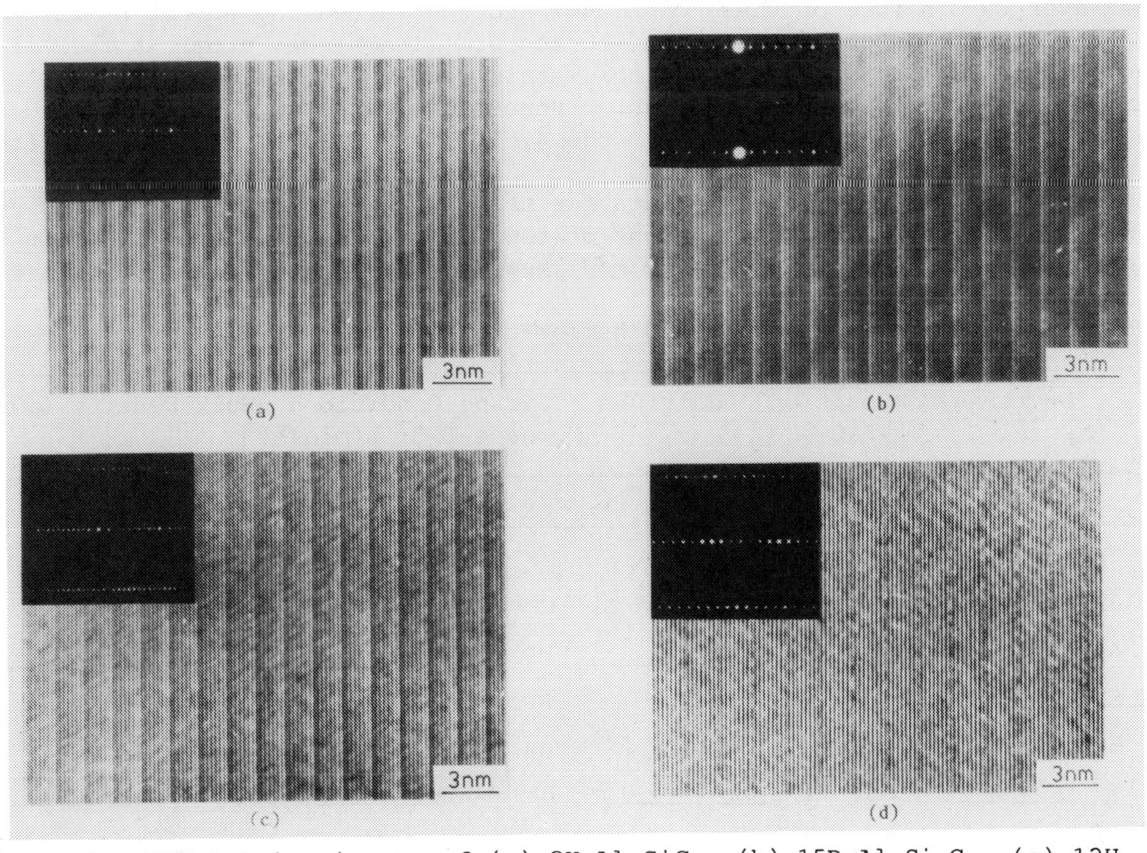

Figure 3. TEM lattice images of (a) 8H Al_4SiC_4, (b) 15R $Al_4Si_2C_5$, (c) 12H $Al_4Si_3C_6$ and (d) 21R $Al_4Si_4C_7$. In each case the electron beam is along [100].

Table 5. Unit Cell Dimensions of Aluminum Silicon Carbides and Related Phases.

Compound	Ramsdell Symbol	*a* (Å)	*c* (Å)	*c*/n (Å)	V/n (Å^3)
Al_4C_3	9R	3.330	24.89	2.766	26.6
Al_4SiC_4	8H	3.277	21.676	2.709	25.2
$Al_4Si_2C_5$	15R	3.251	40.108	2.674	24.5
$Al_4Si_3C_6$	12H	3.225	31.775	2.648	23.9
$Al_4Si_4C_7$	21R	3.210	55.215	2.629	23.4
SiC	4H	3.080	10.090	2.522	20.7
SiC	6H	3.078	15.102	2.517	20.7
δ-Al_4C_3	6H	3.408	15.923	2.654	26.7
"Al_4SiC_4"	12R	3.335	31.31	2.609	25.1
Mg_4ON_2	6H	3.549	15.991	2.665	29.1
Mg_4SiN_4	12R	3.442	31.326	2.611	26.8

and $Al_8C_3N_4$ which have the same M:X ratio and occur along the Al_4C_3-AlN join. Figure 3 shows TEM lattice images of all four Al-Si-C modifications, confirming the layer-type nature of these compounds and the increasing number of layers to the *C*-repeat as the SiC content increases. The higher atomic number of the metal atoms compared with the non-metals results in the domination of the TEM images by the higher Z elements. Hence the images show 5, 6, 7 and 8 layers per formula unit rather than the 4, 5, 6 and 7 layers as expected from non-metal derived Ramsdell symbols.

Table 3 lists X-ray diffraction data for the new phases (12H $Al_4Si_3C_6$ and 21R $Al_4Si_4C_7$). The cell dimensions of the Si-Al-N-C phases are listed in Table 5. In many specimens a 12R modification occurs which has a higher *a*

Table 6. Densification of Al-Si-C compositions.

No.	Starting Composition Al_4C_3	SiC	Additives	Firing Conditions Furnace	T ^{o}C	t(min)	Bulk Density (g/cm^3)
1	1	1	2 w/o Y_2O_3	CF	2000	3	2.01
2	1	2	1 w/o Fe, 1 w/o Si	CF	2000	60	1.71
3	1	2	5 w/o Fe, 5 w/o Si	CF	2000	60	1.65
4	1	4	-	HP	2000	60	3.09
5	1	9	-	HP	2000	60	3.16
6	1	1	1 w/o C	HP	2000	60	2.89
7	1	1	1 w/o Al + 1 w/o C	HP	2000	60	2.77

and lower c/n value than any of the phases mentioned above. The volume of the 12R cell (301.6 $Å^3$) is precisely 1.5 times that of either Al_4SiC_4 or Al_5C_3N (201.6 x 1.5 = 302.4 $Å^3$). The data are consistent with 12R being a 5M : 4X compound and with 8H and 15R being respectively hexagonal and cubic-stacked modifications of a common 4-layer block of structure (see Figure 4). This polymorphism of the 5M : 4X structures has analogies with the normal (9R) and δ (6H) forms of Al_4C_3 where the δ form has a larger a and a smaller c/n value than 9R but the unit cell volumes are very precisely in the ratio 6:9 i.e. 1:1.5. In all the tables, the composition of the 12R phase has been represented as "Al_4SiC_4" because no better compositional evidence is currently available and its structure is consistent with this formula. Table 5 also includes data for the 6H Mg_4ON_2 and 12R Mg_4SiN_4 phases reported by Roebuck & Thompson[11] which are isostructural with the 6H Al_4C_3 and 12R "Al_4SiC_4" phases. 4H was observed to be the major product at pure carbide compositions between $Al_4Si_4C_7$ and SiC. Careful analysis of high-4H samples with the use of EDS analysis indicated that the aluminum content was variable, a consistent feature of the 4H composition, and was in the range 1.2 - 2.3 a/o. Tajima & Kingery[12] measured the solubility of aluminum in 4H silicon carbide as 1 a/o and the c unit cell dimension of their most expanded cell was 10.090 Å. However, to some extent this is affected by trace amounts of other impurities in the silicon carbide and Tajima & Kingery's c_o parameter (corresponding to zero aluminium content), 10.082 Å, is higher than the value used by Mitomo et al.[13] (10.075 Å) and also higher than the value calculated from the best cell dimension of cubic SiC[14] (10.067 Å) on the assumption that the interlayer spacing in silicon carbide is the same in all polytypes. Taking 10.067 Å as a baseline for the 4H c-parameter and using Tajima & Kingery's slope of the

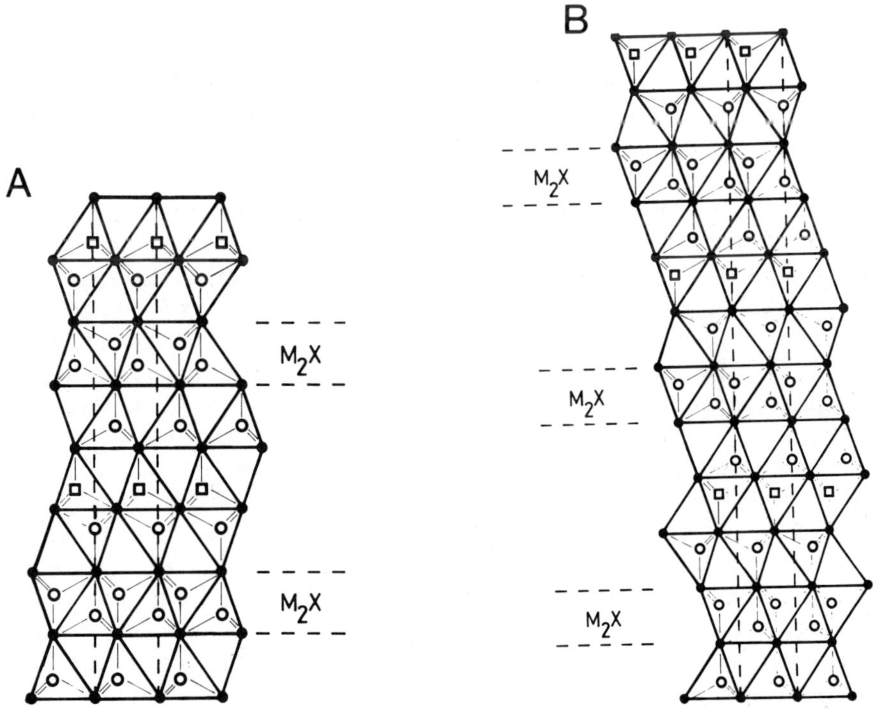

Figure 4. Schematic structure diagrams of (a) 8H Al_4SiC_4 (after Inoue et al.[9]) and (b) 12R "Al_4SiC_4" on the assumption that it is a cubic packed version of 8H. Al: open circles, Si: open squares, C: full circles.

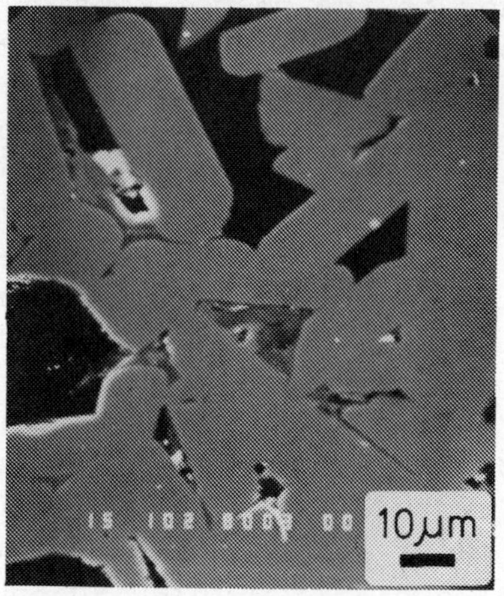

Figure 5. SEM micrograph of grains of 4H-SiC in Sample 14.

unit cell dimensions versus Al content graph, the 2.3 a/o aluminum limit calculated on this basis is in excellent agreement with the measured c value of 10.085 Å observed for our specimen.

Figure 5 shows an SEM micrograph of Sample 14 (see Table 4) which contains mainly 4H-SiC. This sample was prepared at the overall composition $Al_4Si_3C_6$ but on firing at 2000°C suffered a weight loss of 40%. The central region, surrounded by platelike crystals of 4H, is pure aluminum and the weight loss may be due to dynamic oxidation of aluminum carbide, removing the carbon as CO and leaving aluminum. This would leave the composition of the remainder of the specimen as Al plus SiC. Further work is in progress to see whether pure 4H-SiC can be produced at compositions along the Al_4C_3-SiC join and also to establish whether small amounts of nitrogen are beneficial for 4H formation.

Covalently bonded solids cannot be densified by pressureless sintering alone and the most successful route involves using additives which produce a high temperature liquid phase which allows densification by particle rearrangement followed by solution/reprecipitation. In the case of nitrogen ceramics, metal oxides (MgO, Y_2O_3 etc.) are successful additives because the resulting M-Si-O-N and M-Si-Al-O-N liquids can dissolve large amounts of nitrogen (up to 20 a/o) which readily facilitates solution/reprecipitation of crystalline nitride phases. In the case of silicon carbide, mixtures of boron and carbon have been successfully used to achieve densification; these contribute to the formation of a eutectic B-Si-C liquid which allows solution/precipitation of SiC[15]. Also, aluminum-based additions have successfully been used to achieve densification[10].

The various routes used for densification in the present study (see Table 6) have so far not yielded dense products but they give indications of how this might be achieved. Additions of yttria to an Al_4C_3-SiC starting composition with the aim of producing a grain-boundary Y-Si-Al-O-N glass show no indications at all of aiding densification, presumably because carbides have such a low solubility in oxynitride liquids. The same result is observed with Fe-Si additions even when added in quite large amounts. The remaining samples in Table 6 have been hot-pressed because this facilitates densification more readily than pressureless sintering techniques. Hot-pressing Al_4C_3/SiC mixtures without additives did give some densification. For example, the $Al_4Si_4C_7$ composition gave a density of 3.09 g/cm^3 and the $Al_4Si_9C_{12}$ composition gave a density of 3.16 g/cm^3. In fact additions of 1 w/o C or 1 w/o Al + 1 w/o C

to the Al_4SiC_4 composition gave densities of 2.89 and 2.77 g/cm^3 on hot-pressing which were no better than the additive-free compositions. Of course this might be due to oxidation of the carbon and volatilization of the aluminium and further experiments with larger quantities of additive are needed. However, the results show that the reaction between Al_4C_3 and SiC allows redistribution of material which under hot-pressing conditions facilitates densification. One possibility is that the eutectic composition in the SiC-Al_4C_3 system occurs close to Al_4SiC_4 at temperatures near 1900^oC and that the liquid phase so produced aids solution/reprecipitation.
The present results show that Al-Si-C materials can be densified and that, provided the oxygen content can be controlled, hot-pressing can be used to obtain fully dense products. However, further work is needed to explore how complete densification can be achieved by pressureless sintering techniques.

CONCLUSIONS

The present study has established the existence of several new Al-Si-N-C phases in addition to those already known in this system. Besides Al_4SiC_4 and $Al_4Si_2C_5$, two other Al-Si-C phases exist with compositions $Al_4Si_3C_6$ and $Al_4Si_4C_7$ and structures consistent with those of Al_4SiC_4 and $Al_4Si_2C_5$. In addition a second series of compounds occurs with only the 5M:4X member (12R) prepared so far; this phase is analogous to the δ-form of aluminum carbide. 4H SiC occurs widely as a secondary phase in many SiC-rich compositions in this system. It has so far been impossible to attribute a specific range of homogeneity for this phase and also to determine whether nitrogen can be accommodated as is the case with 2H SiC.
Densification studies indicate that at least 90% of theoretical density can be achieved by hot-pressing. Further studies are in progress to explore alternative densification additives in order to achieve fully dense products by pressureless sintering.
In all these materials, the most important parameter as regards high temperature performance is oxidation resistance; for the present compounds, optimum oxidation resistance is expected when a homogeneous surface layer of mullite is produced. Since each line of polytypoid phases (except the 8H series) has one composition where the Al:Si ratio is 3:1 as in mullite, (and even in the 8H series some modification of surface Al:Si ratio could be carried out to achieve the mullite value) it should be possible to precoat samples with a thin surface layer of mullite to achieve protection as regards oxidation. Clearly, mechanical properties can only be meaningfully assessed when samples have been prepared in a fully dense form; this will be reported in a later publication.

ACKNOWLEDGEMENTS

One of us (R.J.O.) gratefully acknowledges SERC funding in the form of a CASE award in collaboration with the BP Research Centre, Sunbury-on-Thames, UK.

REFERENCES

1. M.M.Patience, P.J.England, D.P.Thompson and K.H.Jack, "Ceramic alloys of Silicon Carbide with Aluminium Nitride and Nitrogen", in Proc. 1st Int. Symp. "Ceramic Components for Engine", Hakone 1983, Eds. S.Somiya, E.Kanai and K.Ando, publ. KTK Scientific, 473-479 (1984).

2. G.A.Jeffrey and V.Y.Wu, "The Structures of the Aluminum Carbonitrides,II" Acta Cryst., 20, 538-47 (1966).
3. G.A.Jeffrey and V.Y.Wu, "The Structures of the Aluminum Carbonitrides", Acta Cryst., 16, 559-66 (1963).
4. V.J.Barczak, "Optical and X-ray Powder Diffraction Data for Al_4SiC_4", J. Amer. Ceram. Soc., 44(6), 299 (1961).
5. P.H.A.Roebuck, "The Silicon-Aluminum-Oxygen-Nitrogen System", Ph.D. thesis University of Newcastle upon Tyne, 1978.
6. H.Yokokawa, M.Dokiya, M.Fujishige, T.Kameyama, S.Ujiie and K.Fukuda, "X-ray Powder Diffraction Data for Two Hexagonal Aluminum Monoxycarbide Phases", Amer. Ceram. Soc. Bull., 65(3), C40-1 (1982).
7. G.Schneider, L.J.Gauckler, G.Petzow and A.Zangvil, "Phase Equilibria in the System Al_4C_3-Be_2C-SiC", J.Amer. Ceram. Soc., 62(11-12), 574-6 (1979).
8. G.Schneider, L.J.Gauckler and G.Petzow, "Phase Equilibria in the Si,Al,Be/C,N System", Ceramurgia Int., 5, 101-4 (1979).
9. Z.Inoue, Y.Inomata, H.Tanaka and H.Kawabata, "X-ray Crystallographic Data on Aluminum Silicon Carbides α-Al_4SiC_4 and $Al_4Si_2C_5$", J. Mat. Sci., 15, 575-80 (1980).
10. K.A.Schwetz, F.Isemann and A.Lipp, "Injection-moulded Sintered Turbine Components of Aluminum-doped Alpha Silicon Carbide", Proc. 1st Int. Symp. "Ceramic Components for Engine", Hakone, 1983, Eds. S.Somiya, E.Kanai and K.Ando, publ. KTK Scientific, 583-94 (1984).
11. P.H.A.Roebuck and D.P.Thompson, "Polytypes in the Si-Al-O-N and Related Systems" in "High-temperature Chemistry of Inorganic and Ceramic Materials", Eds. F.P.Glasser and P.E.Potter, The Chemical Society, Special Publication No. 30, 222-8 (1977).
12. Y.Tajima and W.D.Kingery, "Solid Solubility of Aluminum and Boron in Silicon Carbide", Amer. Ceram. Soc. Bull., 65(2), C27-9 (1982).
13. M.Mitomo, Y.Inomata and H.Tanaka, "The Influence of Aluminum on the Stability of 4H- and 6H- Silicon Carbides at 2200°C", Mat. Res. Bull., 6, 759-64 (1971).
14. A.Taylor and D.S.Laidler, "The Formation and Crystal Structure of Silicon Carbide", Brit. J. Appl. Phys., 1, 174-81 (1950).
15. S. Prochazka, "The Sintering Process for Silicon Carbide: a Review", Tech. Rep. 81-CRD-314, General Electric Corp., Schenectady NY (1981).

ANALYTICAL MODELING OF ELEVATED TEMPERATURE MECHANICAL BEHAVIOR OF CERAMIC MATRIX COMPOSITES

Yuan Ruo Wang and Tsu-Wei Chou
Center for Composite Materials and
Department of Mechanical Engineering
University of Delaware, Newark, DE 19716

ABSTRACT

This paper studies the thermo-mechanical behavior of ceramic matrix composites. The thermal shock resistance of such composites is investigated through the analysis of three-dimensional transient thermal stresses and the application of composites failure criteria. The high temperature creep behavior of random short fiber reinforced composites is studied by means of advanced shear-lag approach and Eshelby's inclusion method.

1. INTRODUCTION

With the increasing application of fiber reinforced metal and ceramic matrix composites under severe environment, the thermo-mechanical behavior of such materials has received considerable attention. This is mostly because the temperature at which metal matrix composites (MMC) and ceramic matrix composites (CMC) could be utilized is much higher than that for polymer based composites.

In spite of the many favorable characteristics such as high-temperature strength, creep and corrosion resistance, ceramics display the unfavorable behavior of brittleness or notch-sensitivity. This can render such materials highly susceptible to catastrophic failure under conditions of thermal shock. The thermal shock resistance capability of monolithic ceramics has been studied since the 1950's [1-2]. However, there is a lack of understanding of the thermal shock resistance characteristics of CMC. Wang, Wang and Chou [3-4] first initiated the study of thermal transient effect by taking into account the thermal and elastic anisotropy of composites.

Analytical modeling of the creep of fiber reinforced composites has been performed since the 1960's [5-6].

Considerable effort has been made on the studies of the creep behavior of silicon carbide and aluminum oxide [7-8], because they are the most commonly used ceramic materials for gas turbines, nuclear containment materials and other high temperature engineering applications. The creep study of commercially available silicon carbide whisker reinforced aluminum oxide composites was first reported in Ref. [9]. In spite of several experimental studies, the modeling work on creep of CMC is still in its initial stage [10].

It is intended in this paper to gain a better understanding of the performance of CMC at elevated temperature, through the study of thermal shock resistance and analysis of creep behavior.

In order to evaluate the thermal shock resistance of high temperature composites, an accurate thermal stress analysis, a reliable failure study and a rational quantification of thermal shock resistance are required. The analytical approach needs to take into account fiber and matrix thermo-elastic and strength properties, fiber volume fraction and composites microcracking yield stress. A laminated composite is considered and the results are presented in "performance map" form developed by Chou and Yang [11].

A model of creep analysis has been first established for a unidirectional short fiber composites under off-axis tension. This model is then extended to consider biased and random fiber orientations. The advanced shear-lag approach and Eshelby's inclusion method form the basis of the analytical modeling work.

2. THERMAL SHOCK RESISTANCE

2.1 Thermal Shock Resistance of Isotropic Materials

For an isotropic solid sphere at the initial temperature $T=0$ and subjected to a thermal environment $T=T_o$ at $t=0^+$, the maximum compressive circumferential thermal stress of the sphere at the moment of application of the temperature T_o for high rate of heat convection between the surface of the sphere and environment is [12]:

$$\sigma_t = E\alpha T_o/(1-\upsilon) \quad (1)$$

where E=Young's modulus, υ=Poison's ratio and α=thermal expansion coefficient. A dimensionless stress parameter is introduced:

$$\sigma^* = S/\sigma_t \quad (2)$$

where S=tensile strength. Cheng [1] proposed an approximate relation between the σ^* and thermal conductivity K for low rate of heat convection between the surface of the sphere and environment:

$$\sigma^* = r_o h/3K \quad (3)$$

where r_o=radius of sphere and h=coefficient of heat convection.

Substituting Eq.(2) into (3), the thermal shock resistance parameter R is defined as

$$R = r_o h T_o / 3 = KS(1-\upsilon)/E\alpha \qquad (4)$$

Thus, the thermal shock resistance capability of a material is related to its strength as well as thermo-elastic properties.

2.2 Thermal Shock Resistance of Composite Materials

The solution of thermal stress field of composite materials induced by transient thermal boundary conditions can be expressed explicitly in only very limited cases due to the mathematical difficulties [3-4,13-14]. The failure mechanisms and failure criteria of composite materials are also very complicated [15]. Therefore, the relationship between thermal shock resistance capability and the structural and material characteristics for fiber composites can not be expressed explicitly in a simple form as Eq.(4) for isotropic materials. For the purpose of quantitative studies, the maximum allowable temperature change ΔT_{max}, without causing failure, has been taken as a measure of the thermal shock resistance capability of MMC and CMC, and expressed by means of "performance maps".

Two failure criteria, interlaminar failure criterion and matrix micro-cracking failure criterion have been adopted to predict the initial failure. Both are stress dominated criteria, and have simple mathematical forms ($\sigma_z<\sigma_{inf}$, $\tau<\tau_{inf}$ and $\sigma_x<\sigma_{mcy}$) and clear physical meanings (σ_{inf}, τ_{inf}=interlaminar tensile and shear failure strength, and σ_{mcy}=matrix micro-cracking yield stress). In the present analysis, the term "failure" is defined as the point at which the local stress state exceeds any of the failure strength state determined by the above failure criteria.

The analytical theory which is being applied to calculate the three dimensional stresses is based upon Refs.[3,14]. A symmetric cross-ply or angle-ply laminated plate subjected to sudden heating along its two long edges (Fig.1) is investigated. The three dimensional transient thermal stresses are obtained by means of perturbation method, and the magnitude of these stresses is proportional to the applied temperature. Consequently, the maximum allowable applied temperature ΔT_{max} is obtained from the stress analysis and failure criteria.

2.3 Numerical Examples

A four layer $(-\theta/\theta)_s$ laminated plate is taken as an example for numerical analysis. Two metal matrix and two glass-ceramic matrix composite systems are considered: presure-sintered SiC/Al, T300/Al, SiC/borosilicate glass(BG), and Nicalon/calcium aluminosilicate(CAS). The analysis shows that the initial failure of laminated glass-ceramic matrix composite is dominated by matrix micro-cracking, while the initial failure of laminated metal matrix composite is controlled by interlaminar failure.

The matrix micro-cracking yield strength is calculated based upon the relation $\sigma_{mcy}=E\varepsilon_{mcy}$. Here, ε_{mcy} is matrix micro-cracking yield strain and usually ε_{mcy} equals to 0.1-0.2% for glass-ceramic matrix. The interlaminar tensile and shear strengths are based on the composites delamination/splitting strength. The fiber and matrix thermo-mechanical properties are mainly adopted from Ref.[11], and the composites properties are calculated by means of rule-of-mixtures [16]. The initial stress state of the analytical model is assumed to be stress free. Residual stresses induced by fabrication due to the mismatch of thermal expansion of fiber and matrix are not considered.

The maximum allowable temperature changes, ΔT_{max} of the SiC/BG, SiC/Al, T300/Al and Nicalon/CAS composites for $(-45^o/45^o)_s$ symmetric laminates are shown in Fig.2.

3. CREEP ANALYSIS OF SHORT FIBER COMPOSITES

3.1 Basic Approach and Assumptions

For the analysis of creep behavior of random short fiber reinforced composites, an off-axis loaded unidirectional short fiber composite model is first considered. The applied load can be decomposed into three components (Fig.3). Assuming no interactions among σ_{11}, σ_{22} and σ_{12}, the problem is reduced to three independent sub-problems.

For the case of axial tension problem (σ_{11}), it is assumed that both fiber and matrix creep but with different rates. An advanced shear-lag approach is adopted for the analysis. For the transverse tension (σ_{22}) and shear (σ_{12}) problems, only the matrix is assumed to creep and the fiber remains linearly elastic. In other words, σ_{22} and σ_{12} can lead to creep deformation by matrix flow only. Eshelby's inclusion method is used to solve the stress field of the inhomogeneity problem.

The basic creep constitutive equation is assumed in the power law form.

$$\varepsilon' = A(\sigma/G)^n \exp(-Q/RT) \qquad (5)$$

where $\varepsilon' = d\varepsilon/dt$

σ(Mpa)	: applied stress
A(1/s), n	: constants
Q(KJ/mole)	: activation energy
R(KJ/mole.K)	: gas constant
T(K)	: absolute temperature
G(Mpa)	: shear modulus

The simplified forms of Eq.(5) for fiber and matrix,

respectively, are

$$\text{Fiber}: \varepsilon'_f = (\sigma_f/C_f)^{n_f}$$

$$\text{Matrix}: \varepsilon'_m = (\sigma_m/C_m)^{n_m} \tag{6}$$

where

$$C_i\,(\text{Mpa s}^{1/n_i}) = G_i A_i^{-1/n_i} \exp(Q_i/n_i RT) \qquad i = f,\ m$$

3.2 Creep Analysis

3.2.1 Axial Tension Problem

For applied load parallel to fibers, the advanced shear-lag approach is used to analyze the creep behavior of composites [10]. The creep rate can be obtained as

$$\varepsilon' = (\sigma/Cm)^{n_m} \exp\{a[x^{1+1/n_m} - (L/2)^{1+1/n_m}]\} \cos^{(2n_m)}\theta \tag{7}$$

where $a = \{[n_m^2/(1+n_m)]/2h\}[(1-\eta)(1-\beta)/2h]^{1/n_m}$, L=length of fiber, 2h=matrix width between two neighbouring fibers, θ=angle between fiber orientation and applied load and x is measured from the middle point of the fiber (Fig.3). $\beta(=\varepsilon'_f/\varepsilon'_m)$ is fiber/matrix creep ratio. β=0 indicates elastic fiber (no creep) and β=1 for fiber creeping at the same rate as the surrounding matrix. η is the fiber/matrix bonding-condition factor, i.e. η=0 for perfect bonding and η=1 for no bonding.

3.2.2 Transverse Tension and Shear Problem

For applied load perpendicular to fibers and the pure shear problem, the Eshelby's inclusion method is used to obtain the stress field in the matrix.

From Ref.[17] the stress field can be obtained as

$$\sigma_2 = (1 + \alpha_1)\sigma \sin^2\theta$$

$$\sigma_{12} = (1 + \alpha_2)\sigma \cos\theta\sin\theta \tag{8}$$

Thus, the creep rate for these two cases are

$$\varepsilon'_{m2} = (\sigma/C_m)^{n_m} (1+\alpha_1)^{n_m} \sin^{(2n_m)}\theta$$

$$\varepsilon'_{m3} = (\sigma/C_m)^{n_m} (1+\alpha_2)^{n_m} \cos^{(n_m)}\theta\sin^{(n_m)}\theta \tag{9}$$

where α_1 and α_2 are functions of material elastic properties and Eshelby's tensors.

3.2.3 Average Creep Rates

The average creep rates over the unit cell can be discussed according to the component stresses in Fig.3. For tensile stress parallel to fibers,

$$\varepsilon'_{ave1} = [\varepsilon'_m w + C_5 \int_0^{L/2} \varepsilon'_m(x)\, dx]/(w + L/2)$$

For stress perpendicular to fibers

$$\varepsilon'_{ave2} = \alpha_4(\sigma/C_m)^{n_m}(1+\alpha_1)^{n_m}\sin^{(2n_m)}\theta \tag{10}$$

For pure shear

$$\varepsilon'_{ave3} = \alpha_5(\sigma/C_m)^{n_m}(1+\alpha_2)^{n_m}\cos^{(n_m)}\theta\sin^{(n_m)}\theta$$

Here, α_4, α_5 and C_5 are constants determined from the unit cell geometric configuration.

Thus, the total composite creep rate in the loading direction is

$$\varepsilon' = \varepsilon'_{ave1}\cos^2\theta + 2\varepsilon'_{ave3}\sin\theta\cos\theta + \varepsilon'_{ave2}\sin^2\theta \tag{11}$$

3.2.4. Creep of Composites With Biased Short Fiber Orientation

The creep rate of composites with biased short fiber orientation along the loading direction can be approximated by

$$\varepsilon' = (2/\pi)\int_0^{\pi/2}\varepsilon'(\theta)W(\theta)\,d\theta \tag{12}$$

where $W(\theta)$ is the fiber orientation distribution function.

3.3 Numerical Examples

A SiC whisker reinforced Al_2O_3 matrix composites is taken as a numerical example. The material thermal and elastic properties are given in Table 1 [8,18].

The numerical results of Figs.4 show that for off-axis loaded unidirectional short fiber composites, the creep rate increases monotonicly as the fiber orientation angle θ increases for low fiber aspect ratio. For high fiber aspect ratio, the creep rate first decreases and then increases gradually with the fiber orientation angle. As the fiber aspect ratio L/d increases, the composites creep rate also increases. These results are obtained based upon the assumption that fiber creep rate does not surpass that of the surrounding matrix. Comparisons of the present analysis with experimental results are shown in Fig.5. Further improvement in the analytical model is under way. As the fiber aspect ratio L/d increases to infinite, it becomes the problem of unidirectional continuous fiber composites. It is under investigation.

CONCLUSIONS AND DISCUSSIONS

1. The thermal shock resistance of fiber composites can not be quantified in a simple explicit form as for isotropic materials. Therefore, the maximum allowable temperature change ΔT_{max} of fiber composites under thermal transient condition without causing failure is adopted as a measurement of the thermal shock resistance capability. The results for four composite material systems are given in Fig.2.
2. The creep analysis has provided explicit solution on the

stress-strain rate relation of random short fiber composites. The analytical technique is suitable for analyzing the high temperature creep behavior of fiber composites.

3. The SiC whisker reinforcement should be more creep-resistant than that of bulk SiC because of the single crystal structure of SiC whisker. All the experimental investigations have been performed for SiC bulk. There is no available technique and experimental data for creep study of SiC whisker.
4. The available data of activation energy Q from a variety of sources show significant scattering. The Q value has a great effect on the final results of creep rate of composites.

ACKNOWLEDGEMENT

One of the authors (T.W.C.) acknowledges the partial support of the Air Force Office of Scientific Research, under contract no. AFOSR-87-0383. Dr. Liselotte J. Schioler is the Program Manager.

REFERENCES

1. C.M.Cheng, " Resistance To Thermal Shock ", J. of the American Rocket Society, 1951, 21, 147-153.
2. W.D.Kingery, " Factors Affecting Thermal Stress Resistance Of Ceramic Materials ", J. of the American Ceramic Society, 1955, 38, 3-15.
3. Y.R.Wang & T.W.Chou, " 3-D Analysis of Transient Interlaminar Thermal Stress of Laminated Composites ", Symposium on Mechanics of Composite Materials 1988, June 20-22, University of California, Berkeley, California.
4. H.S.Wang & T.W.Chou, " Transient Thermal Behavior Of A Thermally And Elastically Orthotropic Medium ", AIAA Journal, 1986, 24, 664-672.
5. A.R.T.De Silva, "A Theoretical Analysis of Creep in Fiber Reinforced Composites", J. of the Mechanics and Physics of Solids, 1968, 16, 169-186.
6. H.Lilholt, "Creep of Fibrous Composite Materials", Composites Science and Technology, 1985, 22, 277-294.
7. R.Chang, " Creep of Al_2O_3 Single Crystals " J. of App. Physics, 1960, 31, 484-487.
8. C.H.Carter,Jr., R.F.Davis & J.Bentley, "Kinetics and Mechanism of High-Temperature Creep in Silicon Carbide: I, Reaction-Bonded", J. of the American Ceramic Society, 1984, 67, 409-417.
9. A.H.Chokshi & J.R.Porter, "Creep Deformation of an Alumina Matrix Composite Reinforced with Silicon Carbide Whiskers" , J. of the American Ceramic Society, 1985, 68, C144-C145.
10. J.Kim, " Thermo-Mechanical Behavior of High Temperature Fiber Composites " Ph.D dissertation, 1987, University of Delaware.
11. T.W.Chou & J.M.Yang, "Structure-Performance Maps of

Polymeric, Metal, and Ceramic Matrix Composites", Metallurgical Transactions A, 1986, 17A, 1547-1559.
12. S.P.Timoshenko & J.N.Goodier, Theory of Elasticity, McGraw-Hill Book Company, Inc., 1970
13. A.Y.Akoz & T.R.Tauchert, " Thermoelastic Analysis Of A Finite Orthotropic Slab ", J. of Mech. Eng. Sci., 1978, 20, 65-71.
14. Y.R.Wang & T.W.Chou, "3-D Analysis Of Transient Interlaminar Thermal Stress Of Cross-Ply Composites", ICCM-6, Vol.4, 1987, pp.383-393,
15. D.B.Marshall & A.G.Evans, " Failure Mechanisms In Ceramic-Fiber/Ceramic-Matrix Composites ", J. of the American Ceramic Society, 1985, 68, 225-231.
16. C.C.Chamis, "Simplified Composite Micromechanics Equations for Strength, Fracture Toughness and Environmental Effects", SAMPE Quarterly, July 1984, 41-55.
17. T.Mura, Micromechanics of Defects in Solids, Martinus Nijhoff Publishers, 1982.
18. A.H.Chokshi & J.R.Porter, "High Temperature Mechanical Properties of Single Phase Alumina", J. of Mat. Sci., 1986, 21, 705-710.

Table 1 Fiber and Matrix Creep Properties

	n	Q(KJ/Mol)	G(Gpa)	υ	A
SiC	5.7	711 ± 20	179	0.2	5.12×10^{29}
Al_2O_3	1.6–1.9	625 ± 70	154	0.23	6.09×10^{26}

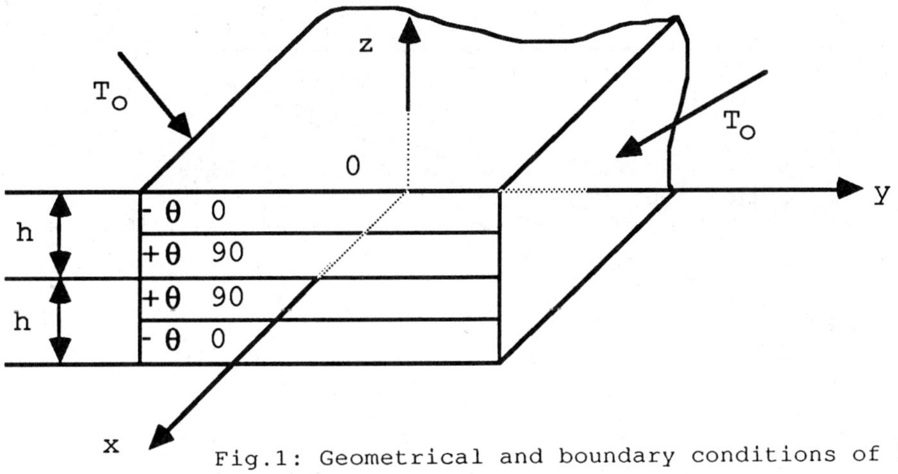

Fig.1: Geometrical and boundary conditions of laminate.

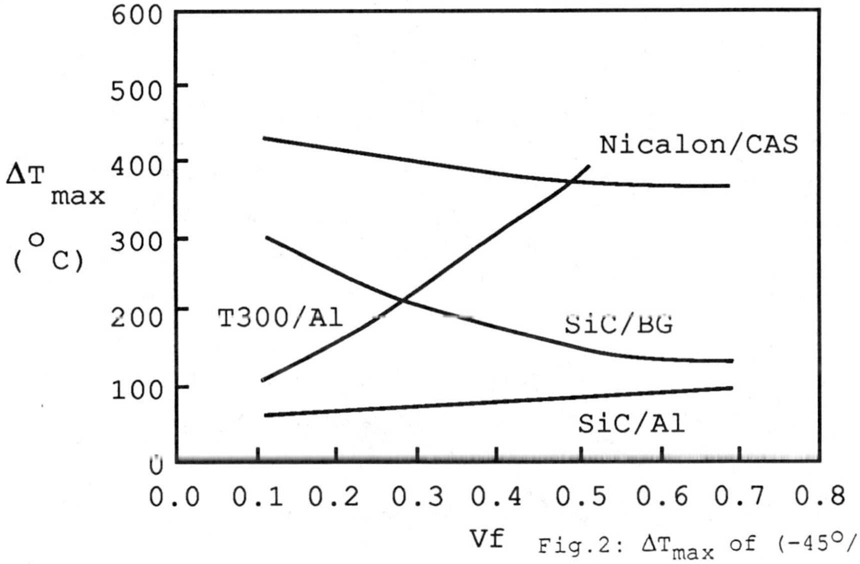

Fig.2: ΔT_{max} of $(-45^{\circ}/45^{\circ})_s$ laminates.

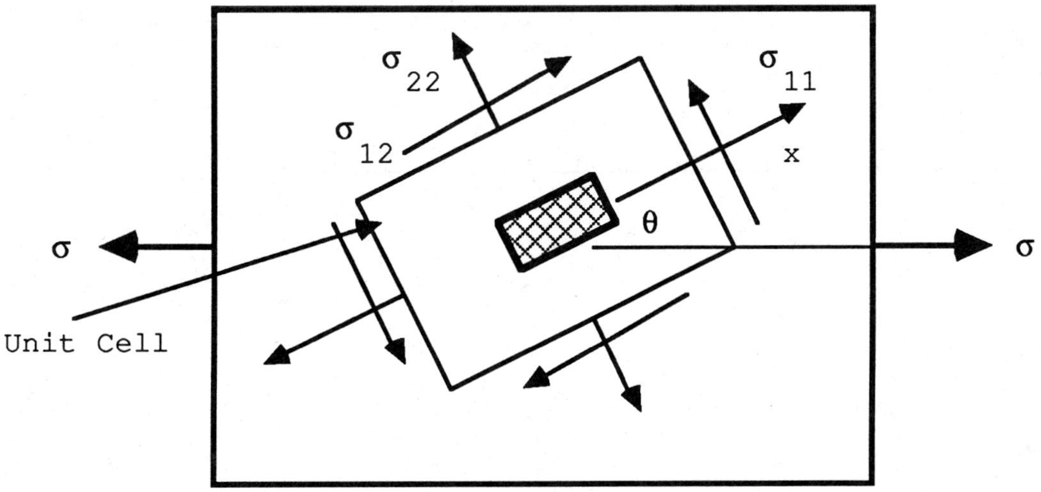

Fig.3: Model of creep analysis.

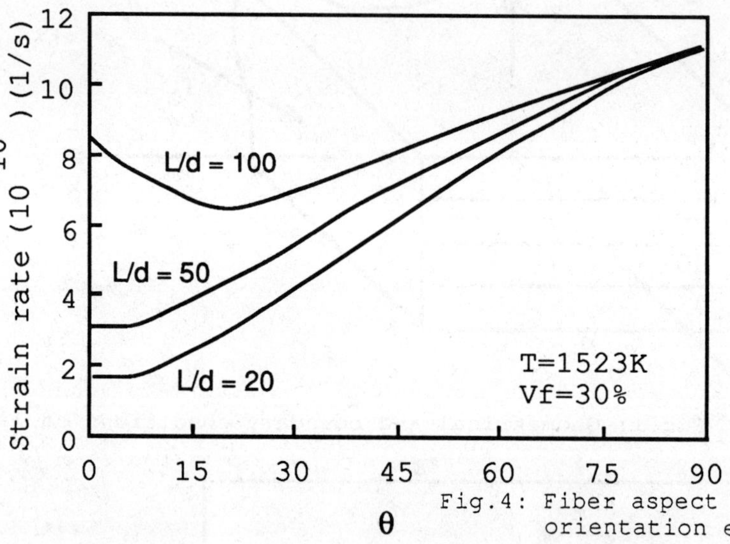

Fig.4: Fiber aspect ratio and fiber orientation effects on creep rate.

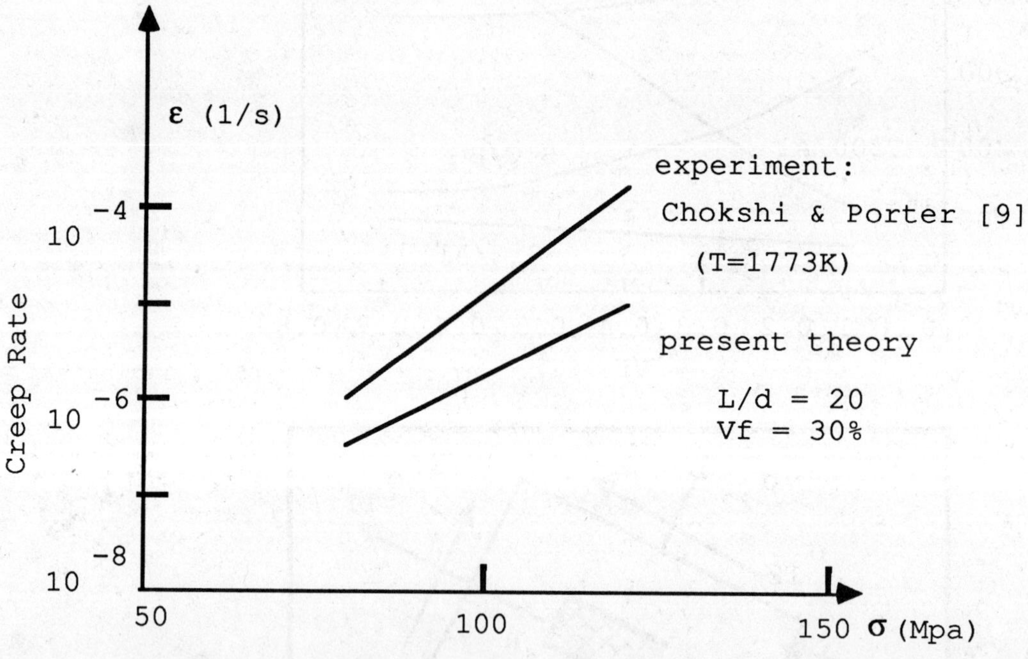

Fig.5: Comparison of analytical and experimental results.

HIGH-STRENGTH SILICON CARBIDES BY HOT ISOSTATIC PRESSING

Sunil Dutta
National Aeronautics and Space Administration
Lewis Research Center
Cleveland, Ohio 44135

ABSTRACT

Silicon carbide has strong potential for heat engine hardware and other high-temperature applications because of its low density, good strength, high oxidation resistance, and good high-temperature creep resistance. Hot isostatic pressing (HIP) was used for producing α and β silicon carbide (SiC) bodies with near-theoretical density, ultrafine grain size, and high strength at processing temperatures of 1900 to 2000 °C. The HIPed materials exhibited ultrafine grain size (0.3 to 5 μm for α-SiC and <1.00 μm for β-SiC). Furthermore, no phase transformation from β to α was observed in HIPed β-SiC. Both materials exhibited very high average flexural strength; as high as 650 MPa for α SiC and 673 MPa for β SiC. It was also shown that α-SiC bodies without any sintering aids (such as boron and carbon), when HIPed to high final density, can exhibit very high strength. Fracture toughness K_{IC} values were determined to be 3.6 to 4.0 MPa $m^{1/2}$ for HIPed α-SiC and 3.7 to 4.1 MPa $m^{1/2}$ for HIPed β-SiC. In the HIPed specimen, strength-controlling flaws were typically surface related. In spite of improvements in material properties such as strength and fracture toughness by elimination of the larger strength-limiting flaws and by grain size refinement, HIPing had no effect on Weibull modulus.

INTRODUCTION

Silicon carbide (SiC) has strong potential for heat engine hardware and other high-temperature applications. It combines low density, good strength, high oxidation resistance, and the capability to operate at temperatures in the range 1370 to 1650 °C (2500 to 3000 °F). Several reports indicate that silicon carbide can be sintered to high final density by pressureless sintering.[1-7] However, like other ceramics, sintered silicon carbide shows strength variability (i.e., low Weibull modulus owing to processing flaws). Densification by pressureless sintering also requires very high temperatures (~2150 to 2200 °C). In contrast, hot isostatic pressing offers many potential advantages such as use of lower sintering temperatures and lower amounts of sintering aids. This should yield improved microstructure, strength, and reliability.[8] Silicon nitride has been successfully hot isostatically pressed (HIPed) to high final density at about 1750 °C with glass encapsulation.[8-9] Silicon carbide requires higher temperatures (1900 to 2200 °C) for densification. Glass encapsulation softens at this high temperature and, because of its very low viscosity, tends to penetrate into the densifying specimens. Although work has been reported on glass and metal encapsulation for HIPing of silicon nitride preforms, very limited information exists about HIPing of silicon carbide with glass encapsulation.[10,11] Furthermore, information on the HIPing of SiC encapsulated with a refractory metal such as tantalum is virtually nonexistent. This paper reports the properties obtained by the hot isostatic pressing (HIPing) of α- and β-SiC compacts encapsulated in tantalum cans, including characterization of strength, microstructure, and toughness.

EXPERIMENTAL PROCEDURES

Two types of α-SiC and two types of β-SiC powders were used in this study. Chemical analyses of these powders are shown in Table 1. The α-SiC(I) powder* had no premixed sintering aids such as boron and carbon added; however, it did contain 1.74 wt % excess carbon in the as-received condition. On the other hand, the α-SiC(II)* powder had premixed sintering aids (0.6 wt % boron and 7.31 wt % carbon). Both β-SiC(I)** and β-SiC(II)*** powders had no premixed sintering aids (boron and carbon) but contained small amounts of free carbon (Table 1). Both α- and β-SiC compositions were cold formed by a slurry pressing technique.[12,13] The dried slurry-pressed compacts (4.67 cm in diameter by 0.6 cm thick) were wrapped with graphite foil, coated with boron nitride, and then placed in tantalum cans. The tantalum cans were outgassed for 6 to 8 hr at 1100 °C and then vacuum sealed. After a thorough leak check the cans were placed in a hot isostatic pressing furnace. An initial gas pressure of 14 MPa (2000 psi) was applied; then both temperature and pressure were increased simultaneously until the desired values were reached. The cans were HIPed at temperatures between 1850 and 2100 °C for 30 to 120 min under 138-MPa argon gas pressure. HIPed specimens were machined into test bars (2.54 by 0.64 by 0.32 cm), and the surfaces were ground with a 400-grit fine diamond wheel to a final surface finish of 8 rms. Density was measured by the water immersion method. Microstructural characterization was made by optical and electron microscopy. Flexural strength tests were conducted by four-point bending, with 0.95-cm inner and 1.87-cm outer spans. Testing was conducted at room temperature with a crosshead speed of 0.05 cm/min. Weibull modulus values were generated from the strength data, and the results were analyzed by using F- and t-tests. Fracture surfaces of selected test bars were examined by scanning electron microscopy to identify the strength-controlling critical flaws. Fracture toughness was determined by the Vicker's hardness indentation technique with a 2500-g load.

---

*H.C. Starck, West Berlin, FRG.
**Ibiden Co. Ltd., Ogaki, Japan.
***Superior Graphite Co. Ltd., Chicago, Illinois.

RESULTS AND DISCUSSION

Hot-Isostatic-Pressed α-SiC

The densification behaviors of HIPed α-SiC(I) and α-SiC(II) are shown in Fig. 1. The α-SiC(I) exhibited sluggish densification behavior, and this resulted in lower final densities at all temperatures from 1800 to 2100 °C.

The maximum density for this material (97 percent of theoretical) was achieved at 2100 °C (Fig. 1). By contrast, only 1900 °C was required to sinter α-SiC(II) powder (containing boron and carbon sintering aids) to a final density of 98 percent of theoretical. No further increase in final density was observed with further increases in temperature from 1900 to 2100 °C. The results suggest that, although it is difficult, α-SiC(I) powder can be HIPed to high final density without any sintering aids. The results are in good agreement with Homma et al.[10] and Larker et al.[11], who obtained high final density (96 percent of theoretical) in HIPed α-SiC without any sintering aids.

Microstructural characterization was made on polished and etched samples. Figure 2 shows photomicrographs of polished and etched α-SiC(I) HIPed at 1900 and 2000 °C for 30, 60, and 120 min. The microstructure essentially consisted of single-phase α-SiC matrix with residual porosities (determined from density measurements) varying from 14 percent for the 1900 °C, 30-min-HIPed specimens, to 5 percent for the 2000 °C, 120-min-HIPed specimens. The α-SiC grains were difficult to etch, and therefore the overall microstructure could not be resolved by optical microscopy. By contrast, clear microstructures with distinct grain boundaries were achieved in α-SiC(II) specimens (Fig. 3). Very little grain growth was observed in α-SiC materials HIPed between 1900 °C for 30 min and 2000 °C for 120 min (Figs. 2 and 3). The high-density α-SiC produced by HIPing exhibited an ultra-fine-grained microstructure with grain size varying from 0.2 to 5.0 μm. Grain morphology was equiaxed in both α-SiC materials.

The room-temperature flexural strengths σ_f of α-SiC(I) HIPed at 1900 and 2000 °C for various times are shown in Fig. 4. Average strengths ± 1 standard deviation (12 test bars) were 368±51 MPa, 444±51 MPa, and 488±40 MPa for specimens HIPed at 1900 °C for 30, 60, and 120 min, respectively. Final densities were 86, 88, and 91.5 percent of theoretical, respectively. Statistical analysis by the F-test (comparison of variances) indicated no significant differences among the variances. Student's t-test indicated that at the 95 percent confidence level there was no statistically significant difference in the room-temperature strength. At the higher HIPing temperature (2000 °C), average strengths were 551±98 MPa, 574±73 MPa, and 619±63 MPa for specimens having densities of 91.6, 92.7, and 95 percent of theoretical and HIPed for 30, 60, and 120 min, respectively. The F-test indicated no significant difference among the variances. Student's t-test at the 95 percent confidence level indicated no significant strength difference between specimens sintered for 30 min versus 60 min and for 60 min versus 120 min, but a significant difference in strength between specimens sintered for 30 min versus 120 min. Student's t-test also indicated that the room-temperature strengths of materials sintered at 2000 °C were statistically higher than 1900 °C strength at the 95 percent confidence level. The highest strength,

619 MPa, obtained in HIPed α-SiC(I) was higher than the strength (540 MPa) reported by Larker, et al.[11] for HIPed α-SiC material. The results suggest that α-SiC(I) containing no sintering aids (boron and carbon) can exhibit very high flexural strength when HIPed to high final density.

A density effect on strength or in other words a porosity effect on strength was observed in HIPed α-SiC(I) specimens. The strength of porous ceramics decreased nearly exponentially with increases in porosity. Various specific analytical relationships have been suggested for the effect of porosity on strength. Ryshkewitch[14] has shown that porosity and strength are generally empirically related by

$$\sigma = \sigma_0 \exp(-nP) \tag{1}$$

where

σ strength of porous specimen

σ_0 strength of fully dense specimen

n ranges from 4 to 7

P volume fraction pores

From this relation the volume fraction pores as a function of strength in HIPed α-SiC(I) was determined as shown in Fig. 5. Regression analysis yielded the following equation:

$$\sigma = 845 \exp(-5.64P) \tag{2}$$

Note that the constant in the exponent (n = 5.64) falls at the approximate midpoint between 4 and 7 and is in good agreement with Ryshkewitch's relation and that at zero porosity the strength would be 845 MPa.

The flexural strength data of α-SiC(II) (containing boron and carbon) are shown in Fig. 6. The average strengths (12 test bars) were 596±59, 614±61, and 625±65 MPa at 1900 °C and 650±58, 620±79, and 564±78 MPa at 2000 °C for sintering times of 30, 60, and 120 min, respectively. Statistical analyses such as the F-test and Student's t-test indicated no significant differences in strengths and variances. These strengths were statistically higher than the strengths found in α-SiC(I) specimens HIPed at 1900 and 2000 °C. However, at a final density level of 95 percent of theoretical for the α-SiC(I), the strength (619 MPa) of α-SiC(I) was statistically comparable with the strengths of α-SIC(II). The results suggest that α-SiC(I) (containing no sintering aids) can exhibit high flexural strength comparable to α-SiC(II) material when HIPed to the same high final density.

The data clearly indicated no systematic correlation between strength and Weibull modulus. Although a significant improvement in strength was achieved in hot isostatically pressed α-SiC over the sintered α-SiC strength of 428 MPa,[15] no improvement in Weibull modulus was obtained. For materials HIPed at 1900 to 2000 °C, the Weibull modulus was estimated to be 8 to 14 for α-SiC(I) and 8 to 13 for α-SiC(II). The Weibull modulus for pressureless sintered α-SiC ranges from 8 to 12.[15,16] Process-related critical flaws, such as large isolated pores and shrinkage cracks, typical of sintered materials[5] were not observed in HIPed materials. Typical fracture initiation sites were found to be surface related in HIPed α-SiC. These are shown in Fig. 7.

Hot-Isostatic-Pressed β-SiC

As-received β-SiC(I) and β-SiC(II) powders were used. The chemical analysis of these powders are shown in Table 1. Neither β-SiC(I) nor β-SiC(II) powders had premixed sintering aids. The β-SiC(I) powder was slurry pressed with 3 and 4 wt % carbon and 0.6 wt % boron; the β-SiC(II) powder was slurry pressed only with 3 wt % carbon and 0.6 wt % boron. On the basis of a preliminary HIPing study, a temperature of 1900 °C for 60 min was chosen for HIPing β-SiC compositions to near-theoretical density (97 percent). Typical microstructures of HIPed α-SiC(II) and β-SiC(I) are shown in Fig. 8. The submicrometer β-SiC grain size (<1.00 µm) was much finer than the typical microstructure of HIPed α-SiC (Fig. 3). Furthermore, the β-to-α phase transformation typical of high-temperature sintered β-SiC microstructure (Fig. 9) was not observed in the "low"-temperature-HIPed β-SiC (Fig. 8). The grain morphology was completely equiaxed and uniform.

The room-temperature flexural strengths of β-SiC(I) and β-SiC(II) HIPed at 1900 °C for 60 min are shown in Fig. 10. Average strengths (22 test bars) were 659±79 and 673±66 MPa for β-SiC(I) specimens with 3 and 4 wt % carbon, respectively. The strength of β-SiC(II) specimens with 3 wt % carbon was 645±100 MPa. Statistical analyses by the F-test (comparison of variances) and Student's t-test indicated no significant differences in variance or in room-temperature strength between β-SiC(I) and β-SiC(II) specimens. The results indicated that both commercially available β-SiC powders, when HIPed to high final density, yield material similar in strength to HIPed α-SiC. The strength data for both β-SiC(I) and β-SiC(II) were compared with those for HIPed β-SiC(II) specimens (Fig. 11). In general, no statistically significant difference in strength was found. The Weibull modulus was estimated to the 9.5 to 11 for β-SiC(I) and 7 for β-SiC(II) compositions, about the same as for α-SiC compositions. The strength-controlling flaws in HIPed β-SiC were generally surface related and similar to those observed in HIPed α-SiC. These are shown in Fig. 12. The highest-strength samples broke into numerous pieces, and specific failure origins could not be identified.

Fracture toughness K_{IC} was determined by direct crack measurements of Vicker's indentations in both α- and β-SiC HIPed to a final density of 97 percent of theoretical. Fracture toughness K_{IC} was estimated by using the relation derived by Evans,[17] which is as follows:

$$\frac{\log K_{1c}}{Ha^{1/2}} \; \frac{H^{2/5}}{E} = 16.32X^5 - 24.97X^4 + 11.23X^3 - 2.02X^2 - 0.34 - 1.59$$

where

K_{IC} fracture toughness

H   hardness

a   indentation radius

E   Young's modulus

X   log(c/a)

c   crack length

Fracture toughness values determined by this relationship gave results closest to existing K_{IC} data[16] determined by chevron notch and controlled-surface flaw techniques. The K_{IC} values ranged from 3.6 to 4.0 MPa $m^{1/2}$ for HIPed α-SiC and from 3.7 to 4.1 MPa $m^{1/2}$ for HIPed β-SiC. Statistical analyses such as the F-test indicated no significant difference in variances, and Student's t-test indicated no significant difference in K_{IC} between HIPed α- and β-SiC specimens. However, K_{IC} values for HIPed α- and β-SiC were higher than those observed in pressureless sintered α-SiC (3 to 3.6 MPa $m^{1/2}$). For HIPed α- and β-SiC and pressureless sintered α-SiC, the F-test indicated no significant differences among the variances, but Student's t-test indicated statistically significant differences in K_{IC} values. Although statistically significant differences were observed, the improvement was not of sufficient magnitude to be of practical significance. Higher K_{IC} values were related to smaller grain size in HIPed α- and β-SiC materials.[18,19]

CONCLUDING REMARKS

In this study both α- and β-SiC compositions, with and without sintering aids (boron and carbon), were HIPed to near-theoretical density (97 to 98 percent of theoretical). The HIPed materials had ultrafine grain size (0.3 to 5 μm for α-SiC and <1.00 μm for β-SiC compositions). Both materials exhibited high average flexural strength. Furthermore, α-SiC bodies without any sintering aids, when HIPed to high final density, exhibited very high strength. Also, higher fracture toughness was obtained in HIPed α- and β-SiC than in pressureless sintered α-SiC from an earlier study. The strength-controlling flaws, such as large voids and shrinkage cracks, typical of sintered SiC materials, were not observed in HIPed SiC. However, no improvement in Weibull modulus was observed over sintered SiC materials. Thus it is clear that although significant improvement has been made in strength via improved microstructure and minimization of processing flaws by HIPing, improvements in reliability and practically significant improvements in fracture toughness were not achieved. Thus other approaches such as compositing are needed to advance significantly beyond the current state of the art.

ACKNOWLEDGMENTS

The author thanks D.J. Kiser and M.R. Freedman for helpful technical discussions.

REFERENCES

1. S. Prochazka, "Sintering of Silicon Carbide," pp. 239-252 in Ceramics for High-Performance Applications, Edited by J.J. Burke, A.E. Gorum, and R.N. Katz, Brook Hill Publishing Co., Chestnut Hill, MA, 1974.

2. J.A. Coppola, H.A. Lawler, and C.H. McMurtry, "Silicon Carbide Powder and Sintered Materials From the Powder," U.S. Patent 4,123,286, 1978.

3. W. Bocker, H. Landfermann, and H. Hausner, "Sintering of Alpha Silicon Carbide With Additions of Aluminum," Powder Met. Int., **11** [2] 83-85 (1979).

4. M. Omori and H. Takei, "Pressureless Sintering of SiC," J. Am. Ceram. Soc., **65** [6], C92 (1982).

5. S. Dutta, "Sinterability, Strength, and Oxidation of Alpha Silicon Carbide Powders," J. Mater. Sci., **19** [4] 1307-1313 (1984).

6. S. Dutta, "Densification and Properties of α-Silicon Carbide," J. Am. Ceram. Soc., **68** [10] C269-C270 (1985).

7. K. Negita, "Effective Sintering Aids for Silicon Carbide Ceramics: Reactivities of Silicon Carbide With Various Additives," J. Am. Ceram. Soc., **69** [12] C308 (1986).

8. H.T. Larker, J. Adlerborn, and H. Bohman, "Fabrication of Dense Silicon Nitride Parts by Hot Isostatic Pressing," SAE Paper 770335, 1977.

9. H.T. Larker, "HIP Silicon Nitride," pp. 18-1 to 18-4 in Ceramics for Turbine Engine Applications, AGARD CP-276, AGARD, Neuilly-Sur-Seine, France, 1979.

10. K. Homma, F. Yamamoto, and H. Okada, "HIP Sintering of Silicon Carbide Without Additives," J. Ceram. Soc. Jpn. (Yogyo kyokai-Shi), **95** [2] 223-228 (1987).

11. H.T. Larker, L. Hermansson, and J. Adlerborn, "Hot Isostatic Pressing and Its Applicability to Silicon Carbide and Boron Carbide," Ind. Ceram., **8** [1] 17-20 (1988).

12. M.R. Freedman and M.L. Millard, "Improved Consolidation of Silicon Carbide," Ceram. Eng. Sci. Proc., **8** [7-8] 884-892 (1986).

13. J.B. Hurst and S. Dutta, "Simple Processing Method for High-Strength Silicon Carbide," J. Amer. Ceram. Soc., **70** [11] C303-C308 (1987).

14. E. Ryshkewitch, "Compression Strength of Porous Sintered Alumina and Zirconia," J. Am. Ceram. Soc., **36** [2] 65-68 (1953).

15. S. Dutta, "Improved Processing of α-SiC," Adv. Ceram. Mater., **3** [3] 257-262 (1988).

16. R.K. Govila, "High-Temperature Strength Characterization of Sintered Alpha Silicon Carbide," AMMRC TR-82-51, 1982. (Avail. NTIS, AD-A121437.)

17. A.G. Evans, "Fracture Toughness: The Role of Indentation Techniques," pp. 112-135 in Fracture Mechanics Applied to Brittle Materials, ASTM-STP-678, edited by S.W. Freiman, American Society for Testing Materials, Philadelphia, 1979.

18. B.D. Kruse and H. Hausner, "Fracture Toughness of Sintered Alpha Silicon Carbide," pp. 453-459 in Science of Ceramics, Vol. 11, edited by H. Hausner; Deutsche Keramische Gesellschaft, Germany, 1981.

19. C.M. Dannels and S. Dutta, "The Effect of Processing on Fracture Toughness of Silicon Carbide as Determined by Vicker's Indentations," J. Amer. Ceram. Soc., submitted for publication.

Table 1. - Analysis of As-received α-SiC and β-SiC Powders

Element	α-SiC(I)	α-SiC(II)	β-SiC(I)[a]	β-SiC(II)[b]
	Content, ppm			
Al	50	140	300	100
Cr	70	40	Trace	20
Fe	10	10	700	200
Ti	20	30	--------	-----
V	20	20	--------	-----
B	-----	[c]0.60	--------	-----
Free C	[c]1.74	[c]7.31	[c]0.65	[c]<2.0
Surface area (BET), m^2/g	31.47	22	19.8	10

[a]Ibiden Co. Ltd., Ogaki, Japan.
[b]Superior Graphite Co. Ltd., Chicago, Illinois.
[c]Unit is weight percent.

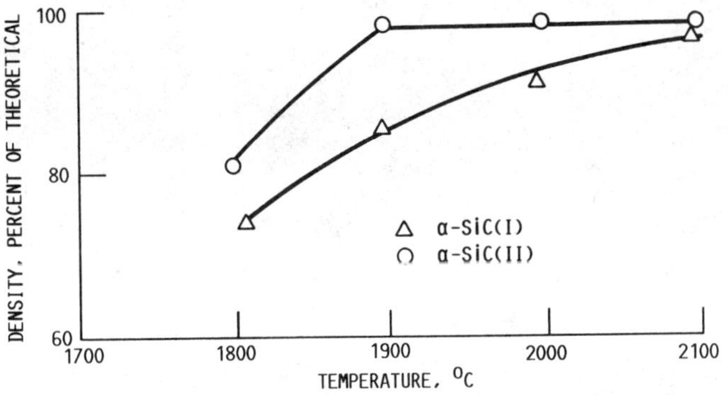

FIG. 1 RELATIVE DENSITY OF α-SiC POWDERS HIPED FOR 30 MIN AT DIFFERENT TEMPERATURES

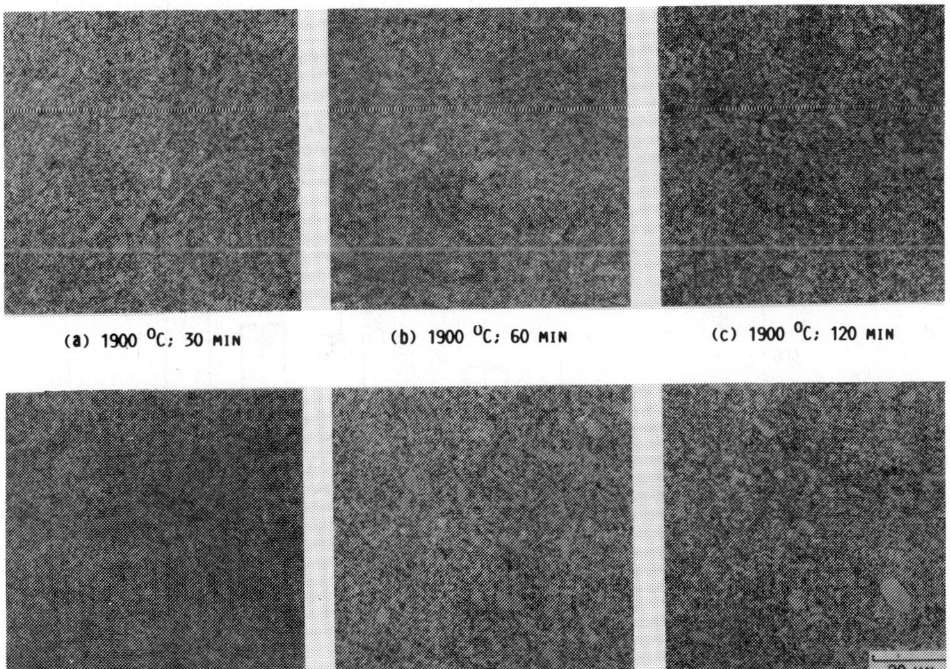

(a) 1900 °C; 30 MIN (b) 1900 °C; 60 MIN (c) 1900 °C; 120 MIN

(d) 2000 °C; 30 MIN (e) 2000 °C; 60 MIN (f) 2000 °C; 120 MIN

FIG. 2 MICROSTRUCTURES OF HOT-ISOSTATIC-PRESSED α-SiC(I)

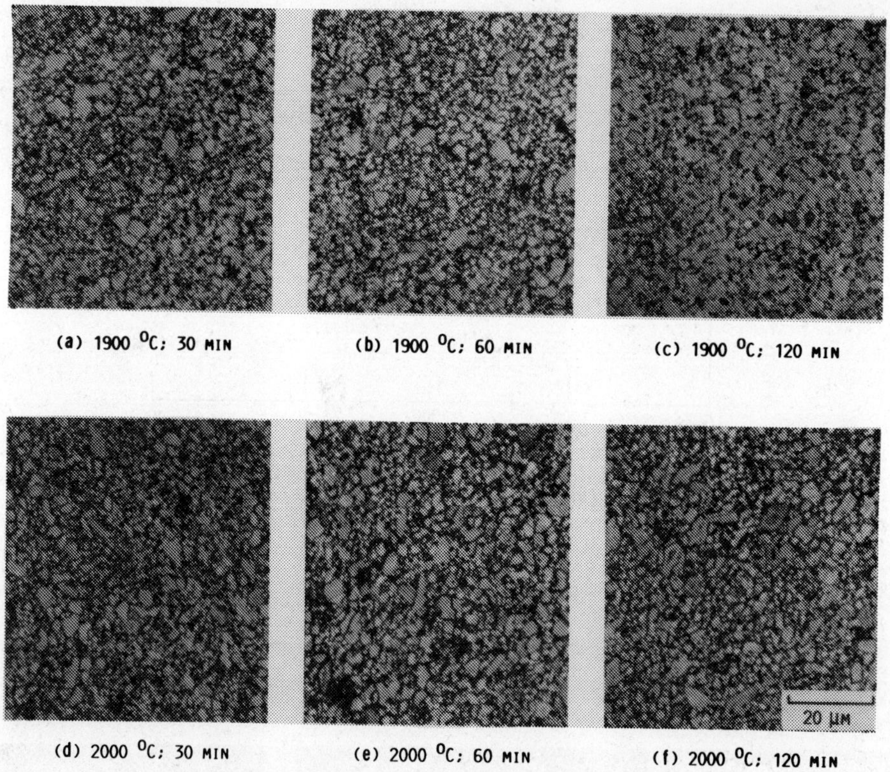

FIG. 3 MICROSTRUCTURES OF HOT-ISOSTATIC-PRESSED α-SiC(II)

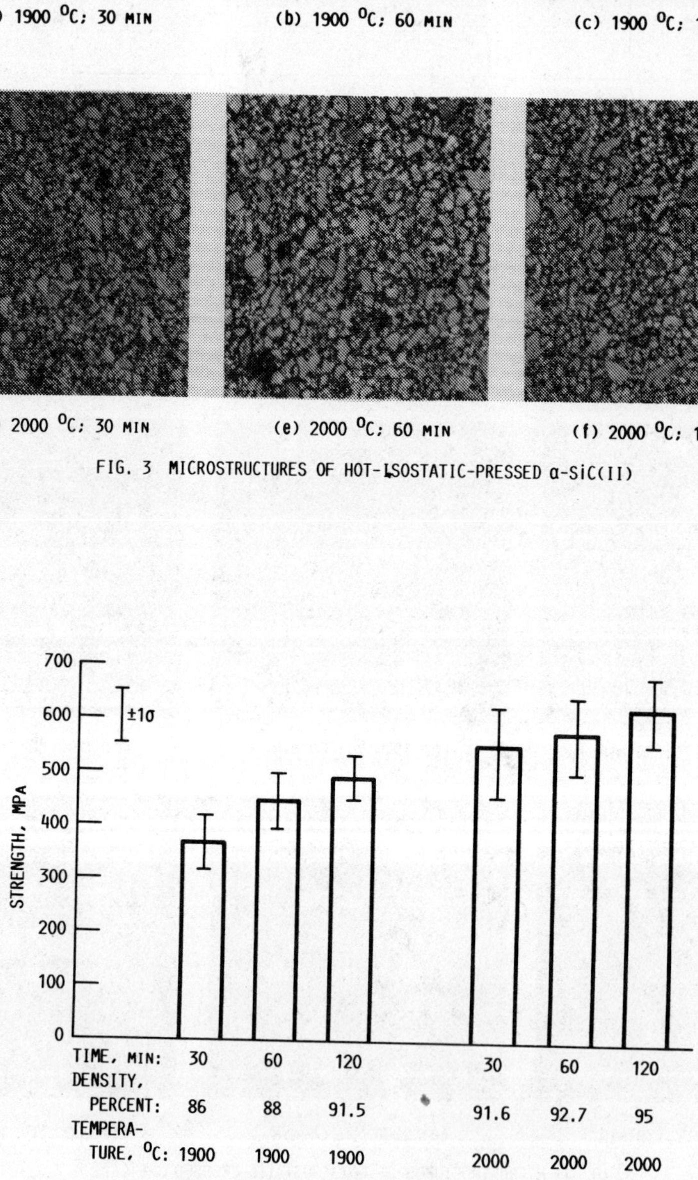

FIG. 4 ROOM-TEMPERATURE FLEXURAL STRENGTH OF α-SiC(I) HIPED AT DIFFERENT TEMPERATURES AND TIMES

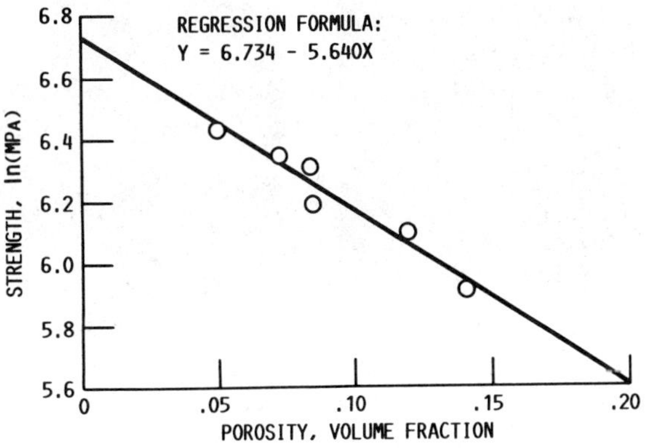

FIG. 5 ROOM-TEMPERATURE FLEXURAL STRENGTH OF HIPED α-SiC(I) AS FUNCTION OF POROSITY

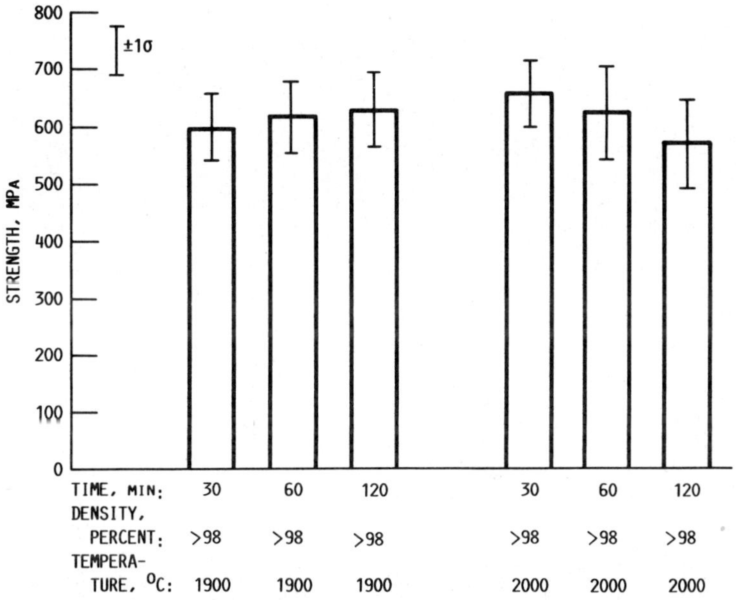

FIG. 6 ROOM-TEMPERATURE FLEXURAL STRENGTH OF α-SiC(II) HIPED AT DIFFERENT TEMPERATURES AND TIMES

(a) α-SiC(II); σ_f = 227 MPA; SINTERED AT 2150 °C FOR 30 MIN

(b) α-SiC(I); σ_f = 492 MPA; HIPED AT 2000 °C FOR 120 MIN

(c) α-SiC(II); σ_f = 480 MPA; HIPED AT 1900 °C FOR 30 MIN

FIG. 7 FRACTURE INITIATION SITES IN SINTERED AND HIPED α-SiC BARS

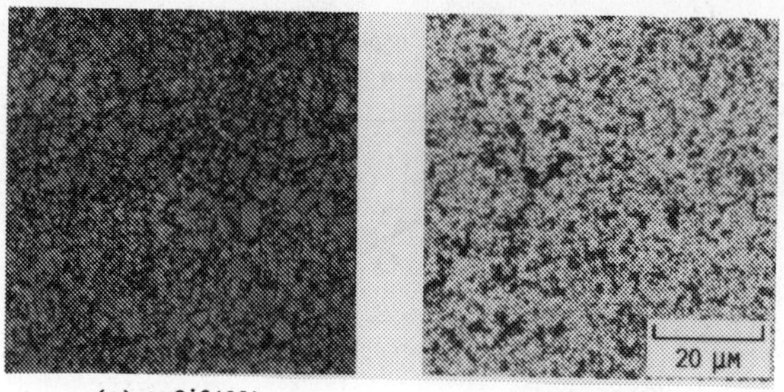

(a) α-SiC(II) (b) β-SiC(I)

FIG. 8 MICROSTRUCTURES OF α-SiC AND β-SiC HIPED AT 1900 °C FOR 60 MIN

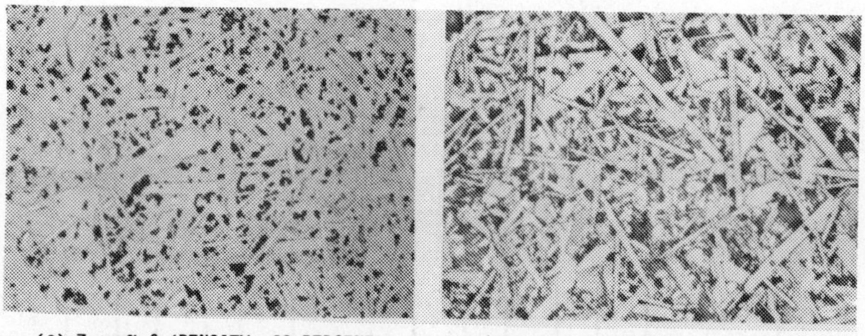

(a) 3 WT % C (DENSITY, 92 PERCENT) (b) 4 WT % C (DENSITY, 94 PERCENT)

FIG. 9 MICROSTRUCTURES OF β-SiC(I) SINTERED AT 2150 °C FOR 60 MIN

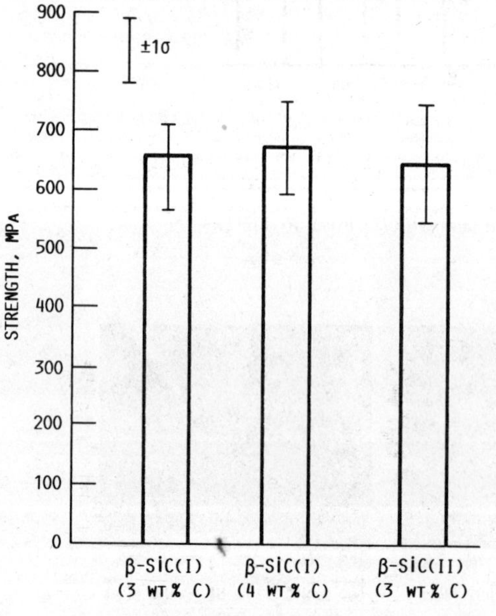

FIG. 10 ROOM-TEMPERATURE FLEXURAL STRENGTH OF β-SiC HIPED AT 1900 °C FOR 60 MIN

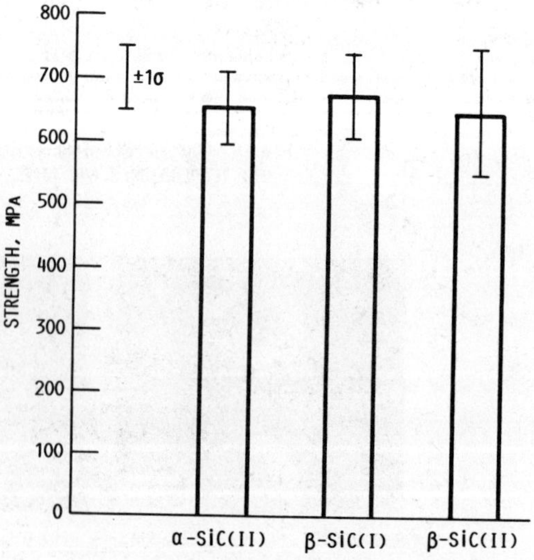

FIG. 11 ROOM-TEMPERATURE FLEXURAL STRENGTH OF α-SiC AND β-SiC HIPED AT 1900 °C FOR 60 MIN

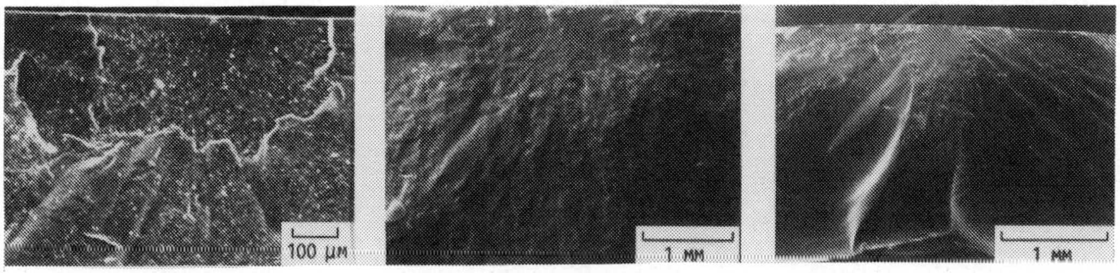

(a) β-SiC(I); σ_f = 700 MPa; 3 wt% C

(b) β-SiC(I); σ_f = 745 MPa; 4 wt% C

(c) β-SiC(II); σ_f = 506 MPa, 3 wt% C

FIG. 12 SURFACE FRACTURE INITIATION SITES IN β-SiC BARS HIPed AT 1900 °C FOR 60 min

INVESTIGATION OF TEST RELIABILITY IN K_{IC} DETERMINATION WITH KNOOP INDENTED 3-PT BENDING METHOD

Z. D. Guan and J. H. Gong
Material Science & Engineering Department, Tsinghua University, Beijing, PRC

ABSTRACT

By means of different loadings of Knoop indentations for each specimen, the fracture strength in 3-pt bending σ_f, is in good linear relationship with $(1/a)^{1/2}$, where a is the depth of the indentation. From the surface crack stress intensity factor:

$$K_{IC} = M_e \sigma_f (\pi a/Q)^{1/2} \quad ,$$

it can be shown that:

$$\sigma_f = K_{IC}\, Y^{-1} (1/a)^{1/2} \quad ,$$

where $Y = M_e(\pi/Q)^{1/2}$ with the given condition that Y is a constant.

Using linear regression method, the fracture toughness of two kinds of ceramics were obtained from the straight lines slope of the plot of σ_f V.S. $(1/a)^{1/2}$. Comparisons are made to results obtained using other methods.

INTRODUCTION

Fracture toughness is one of the major mechanical properties of structure ceramics. A method has been reported[1] in which a fracture toughness value, K_{IC}, can be derived from a 3-pt bending experiment using an artificial flaw as the source of fracture. The advantages of this method are that the shape and dimensions of the artificial flaw are similar to surface damage flaws. The length and depth of the crack are analogous to those in real structures. Therefore, the extension and fracture process of the crack under stress simulates the fracture behavior occuring in the real structure.

In recent years, numerous researchers have studied this problem[2,3]. Although the dimensions of the indentations and the shape of the specimens used differ from each other, the major factor affecting the accuracy and scatter of the data is the difficulty in measuring the dimensions of the indentations. Thus the obtained K_{IC} data is scattered, affected by artificial factors introducted by the inaccurate demension measurements. In this research, a series of indentation sizes were obtained by using different loads. The compressions were placed on the ceramic specimens which were subsequently fractured in 3-pt bending. A K_{IC} for the material was determined by evaluating the plot of bending strength σ_f V.S. indentation size. These K_{IC} values have proven to be comparable to those obtained using other methods.

For the indented 3-pt bending method, we used the formula for the critical stress field intensity factor of a surface crack[4]

$$K_a = M_e \sigma_a (\pi a/Q)^{1/2} \quad (1)$$

where σ_a is the bending stress of fractured section, a is the depth of the crack. M_e and Q are crack shape factors which can be obtained from the literature and are dependent upon the a/c ratio (c is half length of the crack) and the a/B ratio (B is the thickness of the specimen).

while at the same time, there exists another stress intensity factor due to residual stress[5]

$$K_r = X_r P_r / a^{3/2} \quad (2)$$

where X_r is a constant that depends upon geometric shape and material properties and has no relate to indented load. P_r is residual force.

The total stress intensity factor is:

$$K_I = K_a + K_r$$
$$= M_e \sigma_a (\pi a/Q)^{1/2} + X_r P_r / a^{3/2} \quad (3)$$

At critical condition

$$(K_I)_{crit.} = K_{IC} - M_e \sigma_f (\pi a/Q)^{1/2} + \frac{X_r P_r}{a^{3/2}}$$
$$= a^{1/2} [\sigma_f M_e (\pi/Q)^{1/2} + X_r P_r / a^2] \quad (4)$$

Let $Y = M_e (\pi/Q)^{1/2}$ then

$$K_{IC}\, a^{-(1/2)} / Y = \sigma_f + X_r P_r / a^2 Y \quad (5)$$

A series of experiments have been done for the relation between $a^{-(1/2)}$ and ultimate stress σ_f. The results show that they are in good linear relationship (see below). Therefore, from formula(5), we obtained K_{IC}/Y=constant and $X_r P_r / a^2 Y$= constant. If Y is constant for the given material, then P_r / a^2 is constant, which denotes the residual stress of the Knoop indentation.

Then K_{IC} can be calculated.

EXPERIMENTAL PROCEDURE

The materials used were hot-pressed silicon nitride HPSN and a Sialon. The specimen dimensions were 3x4x35mm. An indentation was made with a Knoop indentor at the middle point of the polished tensile surface of the specimen perpendicular to the length direction. Loads between 1500 and 4600g were used for each material. A 30 second duration was required to produce adequate (full) compressions. The specimens were then tested in 3-pt bending with a crosshead speed of 0.05mm/min and a bending span of 30 mm.

RESULTS AND DISCUSSION

In the condition that K_{IC} is a constant for a given material, there are two prerequests for using formula (1). First, M_e and Q must be independent of indentation size a, however this is restricted by a/c and a/B. M_e is affected by a/B which is small in magnitude and varies very little. Thus from the consulted literature data, M_e is almost a constant. Experiment results for parameter a/c are shown in table 1.

TABLE 1 Geometric parameters and test results of ceramic specimen

material	indented load (g)	a (mm)	2c (mm)	a/c	Q	B (mm)	M_e	Y	σ_f (Mpa)
HPSN	1500	.060	.142	.85	2.10	3.96	1.04	1.27	431
	2200	.068	.158	.86	2.12	3.96	1.04	1.27	399
	3000	.076	.179	.85	2.10	3.87	1.04	1.27	360
	3700	.084	.198	.85	2.10	3.87	1.04	1.27	360
	4600	.098	.223	.86	2.12	3.96	1.04	1.27	327
Sialon	2500	.055	.126	.87	2.16	2.89	1.03	1.24	310
	3000	.058	.132	.88	2.18	2.90	1.03	1.24	296
	4000	.062	.147	.84	2.08	2.96	1.04	1.23	273
	4500	.068	.160	.85	2.10	2.95	1.04	1.27	242

Table 1 indicates that M_e, Q and Y are constant at different levels of Knoop loading. The a/c ratio is also nearly a constant. Thus far, the first prerequest is satisfied. The second condition is the linear relationship between σ_f and $(1/a)^{1/2}$, so as to make sure that K_{IC} is a constant. From the bending strength data (shown in TAB.1), we obtained Fig.1 and Fig.2. The experimental results for both materials show that a good linear relationshop exists. Thus the second prerequest is also satisfied. Therefore, it appears feasible to use this method for K_{IC} determination. Rearranging formula (1) , we obtained

$$\sigma_f = K_{IC} Y^{-1} (1/a)^{1/2} \qquad (6)$$

In accordance with formula (6) and by the use of linear regression, we find

$$\sigma_f + 29.49 = 3.56\ (1/a)^{1/2},\ (r=0.996) \qquad (7)$$

for HPSN, and

$$\sigma_f + 364.87 = 5.02 (1/a)^{1/2},\ (r=0.998) \qquad (8)$$

for Sialon.

With an average value of Y, K_{IC} was determined from the slope of the plot of σ_f V.S. $(1/a)^{1/2}$.There are constant terms of 29.49 and 364.87 occuring on the left sides of formulas (7) and (8) respectively, making a distinction between this method and the one-point K_{IC} determination method (direct use of formula (1)). The slope and intercept of the straight line may be different for different materials, however the K_{IC} values only directly related to the slope.

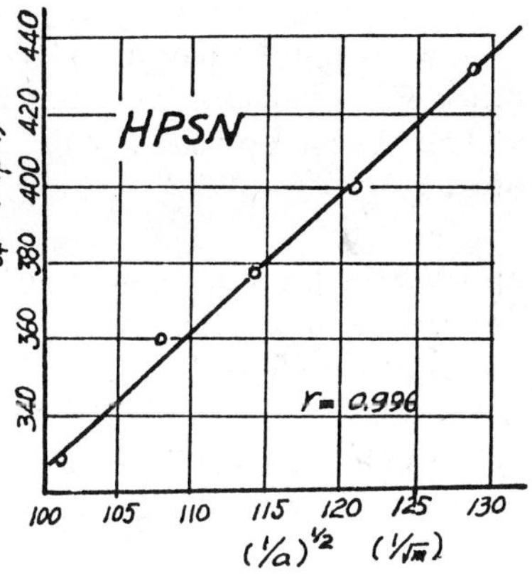

Fig.1 $\sigma_f-(1/a)^{1/2}$ relation of HPSN

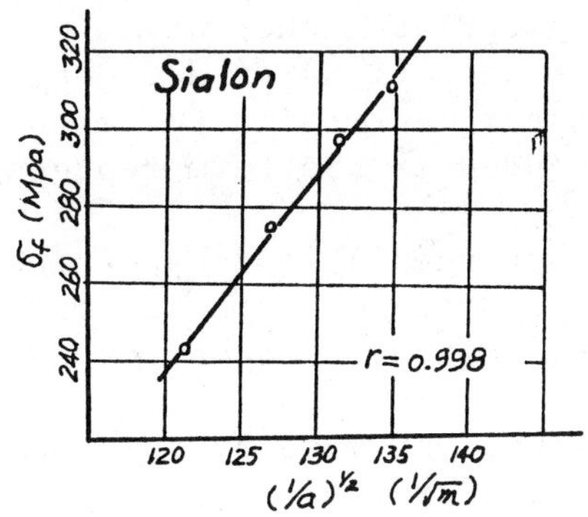

Fig.2 $\sigma_f-(1/a)^{1/2}$ relation of Sialon

A comparison of K_{IC} values acquired by this method and conventional ones respectively for three kinds of ceramics is shown in table 2. Other empirical data and experimental results from other methods are also given for comparison.

TABLE 2 K_{IC} values tested by different methods

material	$K_{IC} + \sigma_{n-1}$ (MPa)				
	regressive Knoop ind. method	direct use formula(1)	SENB method	chevron notched 3-pt beam method	literature data
$HPSN_1$	4.51±0.02	4.18±0.05			4.5-5.3
Sialon	6.32±0.10	2.73±0.13	6.14		
$HPSN_2$	4.54±0.02		5.43±0.24	4.61±0.25	

TAB. 2 indicates the comparison of the test data obtained with some conventional methods. The data show that the standard deviation for K_{IC} determined by this method is significantly smaller than for other test methods. The results of this method are more similar to chevron-notched 3-pt bending method, while the values obtained using the SENB method are relatively higher. The reason is due to the blunting effect at the notch-tip in the SENB test. The results of direct calculation using formula (1) from individual Knoop indented beam method are not correct in magnitude.

Emperical data given in the literature is on the same order as the results of this method.

SUMMARY AND CONCLUSION

1. It has been proven that for the two chosen materials, both of their a/c ratios are equal to 0.86, and their geometric shape factor $Y=M_e(\pi/Q)^{1/2}$

is a constant (1.26-1.27) for Knoop loads in the range of 1500-4600g.

2. For appropriate indentation dimensions, there is a good linear relationship between $(1/a)^{1/2}$ and 3-pt bending strength σ_f, making it possible to use linear regression to calculate the K_{IC} value from Formula (4), which is included in the term representing the slope.

3. Both the indentation shapes and dimensions approximate natural surface defects, therefore, the fracture toughness behavior occuring here approaches the extension behavior of natural cracks under stress.

4. K_{IC} values obtained by this method are independent with the residual stress produced by Knoop indentation. The constant terms of formulas(7) and (8) denote the effect of residual stress.

ACKNOWLEDGMENTS

The authors would like to express their appreciation to L.X.Zeng for preparation of this paper.

REFERENCE

[1]. Z.D.Guan, X.M.Du," Comparison of Methods of Determining Fracture Toughness K_{IC} of Ceramic Materials", J.Chinese Ceram. Soc., 10(3) 262-71 (1982)
[2]. J.J.Petrovic, R.Dirks, L.Jacobson, M.Mendiratta, "Effects of Residual Stress on Fracture from Controlled Surface Flaws", J.Am. Cer. Soc., 59(3-4) 177-8 (1976)
[3]. G.D.Quinn, J.B.Quinn, "Slow Crack Growth in Hot-pressed Silicon Nitride", Fracture Mechanics of Ceramics, V.6, 603-36 Edited by R.C.Bradt, D.P.H. Hasselmann, and F.F.Lange, Plenum, New York, (1982)
[4]. Chen Chi, et al., "Engineering Fracture Mechanics," Military Industrial Express, China, 133 (1977)
[5]. E.R.Fuller, B.R.Lawn, "Theory of Fatigue for Brittle Flaws Originating from Residual Stress Concentrations," J. Am.Cer.Soc.,66(5) 314-21(1983)

INFRARED SPECTROSCOPY FOR CHARACTERIZATION AND EVALUATION OF THE END PRODUCTS (C-S-H(III) AND C-S-H(II^2))

Francisa Hannawayya
Hannawayya's Laboratory for Chemical Research and Development,
Teknikhöjden, Drottning Kristinas väg 47, 114 86 Stockholm, Sweden.

ABSTRACT

Hannawayya originally gave the name C-S-H(III) to the end product of C_3S hydrate and C-S-H(II^2) to the end product of β-C_2S.

Absorption and reflection of infrared of C-S-H(III) and of C-S-H(II^2) with their different CaO/SiO_2 mole ratios were obtained in the 200-4000 cm^{-1} (wavenumber) regions. The absorption spectra of unhydrated C_3S is comparable to those of unhydrated β-C_2S.

The two main absorption spectra bands of SiO_4 are located in the 400-496 cm^{-1} and 800-1000 cm^{-1} regions. The absorption range of (OH) group of $Ca(OH)_2$ was found in the 3400-3600 cm^{-1} region, while the (OH) of water was observed around the 1650 cm^{-1} region. Generally SiO_2 showed two absorption spectra of strong bands at 460-490 cm^{-1} and of very strong band at the 950 cm^{-1} regions.

The absorption spectra of different CaO/SiO_2 ratios show the high lime Calcium-Silicate-Hydrate produces a stronger absorption in the region at 1400-1525 cm^{-1} than a comparable amount of lower lime Calcium-Silicate-Hydrate.

INTRODUCTION

Various tools, methods and techniques are employed for the anslysis, identification and characterization of cement and allied materials, e.g. microscopy and X-ray diffraction are the most commonly employed techniques for the study of raw materials of cement and cement and its hydration product. However, the application of the infrared spectroscopy to the investigation of cement and allied materials is rare and rather limited.

It is possible to use infrared spectroscopy as a tool in the broad quality control of Silicates (C_3S and β-C_2S) and, particulary, in assessing the formation of relative proportions of Calcium-Silicate-Hydrate (C-S-H) in silicate hydrates.

Infrared radiation is characterized either by its wavelength measured in microns, or by its frequency measured in wavenumbers (cm^{-1}). The range middle infrared (200-4000 cm^{-1}) has been used to a great extent for identification and characterization purposes.

The present investigation has been planned to study the characteristic variation of infrared spectra of the end products, C-S-H(III) and C-S-H(II2), as well as the unhydrated silicates (C_3S and β-C_2S) to identify the spectra of these end products and present new results, which could be of great benefit to the readers.

Finally, it is hoped that the publication will gave the way for wider adoption of this tool (infrared) by the industry, laboratories and test houses dealing with cement and allied materials.

History

About half the minerals known are Silicates. The Silicates are sometimes grouped as follows:

(1) Disilicates (RSi_2O_5) are salts of Disilicic Acid ($H_2Si_2O_5$);
(2) Polysilicates ($R_2Si_3O_8$) are salts of Polysilicic Acid ($H_4Si_3O_8$);
(3) Metasilicates ($RSiO_3$) are salts of Metasilicic Acid (H_2SiO_5);
(4) Orthosilicates (R_2SiO_4)are salts of Orthosilicic Acid (H_4SiO_4).

The selection absorption of infrared radiation by silicate minerals has been known since the early 1900's, when Coblentz /1-2/ obtained infrared absorption spectra of cleavage sheets or polished sections of many silicates. Coblentz examined the reflecting power of polished surfaces of minerals and found that the Silicates had reflection maxima in the region of 9-12 microns. He made the important observation that "The various Silicates have quite different reflection spectra".

In the 1930's Matssi and Krüger /3/ and Schaefer et al. /4/ published the infrared reflection spectra of some silicates minerals. Matssi and Broder /5/ examined plates cut from silicate crystals by infrared absorption. In the range 2-8 microns, Barnes et al. /6/ showed that the infrared absorption spectrum of a solid could be obtained by using a thin film of powder, provided that the powder particles had diameters smaller than the wavelength of the infrared radiation. Lecomte /7/, used a powder technique to obtain the infrared spectra of many inorganic compounds containing polyatomic anions such as PO_4^{3-}, SO_4^{2-}, and HCO_3^-. The powder technique has only recently been applied to the study of the silicates. Hunt et al /8/, obtained spectra of several silicates, used only particles having diameters less than 5 microns with the result that their silicate spectra show well defined absorption bands Launer /9/ has given a table summarizing the strong absorption bands of silicates with different Si-O motifs.

RESULTS AND DISCUSSIONS

The system Calcium-Silicate-Hydrate ($CaO-SiO_2-H_2O$, or C-S-H) is of important interest in pure and applied chemistry and also in geochemistry (the related chemical and geological properties of a substance).

The reaction products in system $CaO-SiO_2-H_2O$ vary widely subject to the CaO/SiO_2 mole ratio used.

Hannawayya /10/ originally gave the name C-S-H(III) to the Calcium-Silicate-Hydrate(III) or C-S-H(III) which is produced by bottle hydration of C_3S. The lower and upper limits of CaO/SiO_2 mole ratio of C-S-H(III) are 2.25 and 2.90, respectively. The composition of C-S-H(III) varies $CaO_{2.25-2.90}SH_{2.20-2.45}$. Hannawayya suggested that the upper limit of CaO/SiO_2 mole ratio should be 2.90 for C-S-H(III). This value never reached 3.0. The prior scientific evidence is equation No.1, reference /10/.

Bottle hydration is a method in which the anhydrous phase is mixed with an excess of water (usually more than 10 times its own weight) and is placed in a sealed polyethylene bottle mounted on a rotating wheel.

Hannawayya /10-11/ also originally gave the name C-S-H(I^1) and C-S-H(II^2) to the Calcium-Silicate-Hydrate which was produced by bottle hydration of β-C_2S hydrate. The lower and upper limits of CaO/SiO_2 mole ratio of C-S-H(II^2) are 1.56 and 1.88, respectively. The composition of C-S-H(II^2) varies $CaO_{1.56-1.88}SH_{0.92-1.66}$. Hannawayya suggested that the upper limit of the CaO/SiO_2 mole ratio should be 1.88 for C-S-H(II^2). This value never reached 2.0. The prior scientific evidence is equation No. 2, reference /11/.

The most important constituents in clinker-cement are the Silicates (C_3S and β-C_2S). These phases belong to the Orthosilicate group, C_3S being represented by the structural formula Ca_3SiO_5 or $Ca_3O(SiO_4)$. Therefore, infrared spectra of these Silicates are characterized by vibration mades of Ca-O motifs occurring in different frequency ranges. Accordingly, C_3S and β-C_2S will be considered individually.

1- INFRARED SPECTRA OF UNHYDRATED C_3S

The unhydrated C_3S compound is reported /12/ to be pseudohexagonal with an elementary cell of 9 molecules. Although the coordination number of calcium is 6, it is irregular in disposition with oxygen ions concentrated on one side of each Ca^{2+} leaving holes to accommodate another Ca^{2+} thereby causing higher lattice energy /12/. Lehmann and Dutz /13/ identified two strong bands at 943 and 893 cm^{-1} in C_3S spectra. A strong band at 934 cm^{-1} was observed in Alite corresponding to the 943 cm^{-1} band of C_3S. Though the infrared spectrum of C_3S has been reported by a number of workers, for the latest recorded infrared spectra of C_3S /12/, the main absorption bands are at 995 (strong), 903 (strong and broad), 846 (medium), 793 (shoulder), and 520 (strong) cm^{-1}. Tarte /14/ has studied the infrared spectra of orthosilicates and his conclusions are as follows: the SiO_4 is shown by a narrow band of weak or moderate intensity at 900-800 cm^{-1}. The region 1000-800 cm^{-1} broadly absorption.

The present author (Hannawayya believes that the fundamental bands of unhydrated C_3S is in terms of the fundamental modes of SiO_4 tetrahedron. When the SiO_4 group is part of a crystal lattice, the crystal plays a dominant role in determining which vibrations appears in the spectrum. The broad pattern of infrared absorption spectra of unhydrated C_3S is comparable to that of Dicalcium Silicate (β-C_2S) phase. However, (Fig. 1, C_3S-spectra) revealed the SiO_4 absorption band at 400-495 cm^{-1}. The transmission curve becomes very abrupt at 400 to 450 cm^{-1}. This is followed by a small rise in the reflection curve at 495 cm^{-1}, followed by strong intensity band from 510 to 530 cm^{-1}. After that a sudden very strong rise occurs in the reflection curve at 690 cm^{-1}. There are three strong maxima at 690, 705 and 720 cm^{-1}, respectively. Then there is another absorption of SiO_4 in the range of 1000 to 800 cm^{-1} where there is a distinct shift of a very strong intensity band from 810 to 830 cm^{-1} and the medium intensity at 880 to 950 cm^{-1} after a sudden rise again after 1000 cm^{-1}, there is deep, very strong maxima at 1320 to 1360 cm^{-1}, followed by a decrease to very strong sharp intensity in the region at 1470 Cm^{-1}, and medium at 1540 cm^{-1}, there is medium strong maxima at 1790 cm^{-1}, then the transmission shoulder-shape curve becomes weaker at increasing wavenumbers. In this regard the overall shape of the spectra range 1000 to 800 cm^{-1} is important. The absorption bands are listed in table 1.

2- INFRARED SPECTRA OF CALCIUM-SILICATE-HYDRATE(III) OR C-S-H(III)

The infrared absorption spectra of the end product C-S-H(III), particularly with the CaO/SiO_2 mole ratio varies from 2.75, 2.80 and 2.90, respectively, are shown in Fig. 1, together with the unhydrated C_3S.

The C-S-H(III) is described by Hannawayya /10/ as consisting of bundles of angular fibers intersecting at various angles. This product is nearly crystalline as supported by reference /10/.

Examination of the spectra of C-S-H(III) with the above varying ratios showed that all the patterns are sufficiently distinctive to be of value in identifying the hydrated C_3S, as C-S-H(III) is the end product of C_3S hydrate.

a- The Hydroxyl (OH) Groups Spectra

There are two kinds of hydroxyl (OH) groups in the spectra of Calcium-Silicate-Hydrate(III) or C-S-H(III), It has been found that (OH) groups absorb at different wavenumbers (cm^{-1}) depending on the degree of hydrogen bonding of the OH groups with each other or with other (O) atoms in the lattice. Where two or more (OH) groups are sufficiently close together, the hydrogen of the (OH) group may assume an oscillating position between two oxygen atoms with the formation of resonating bands between both oxygen atoms simultaneously, the so-called hydrogen band. Accordingly, the vibration which is characterized by a very strong broad region at 3400-3600 cm^{-1} belongs to the band of (OH) group of $Ca(OH)_2$. The position of this absorption band

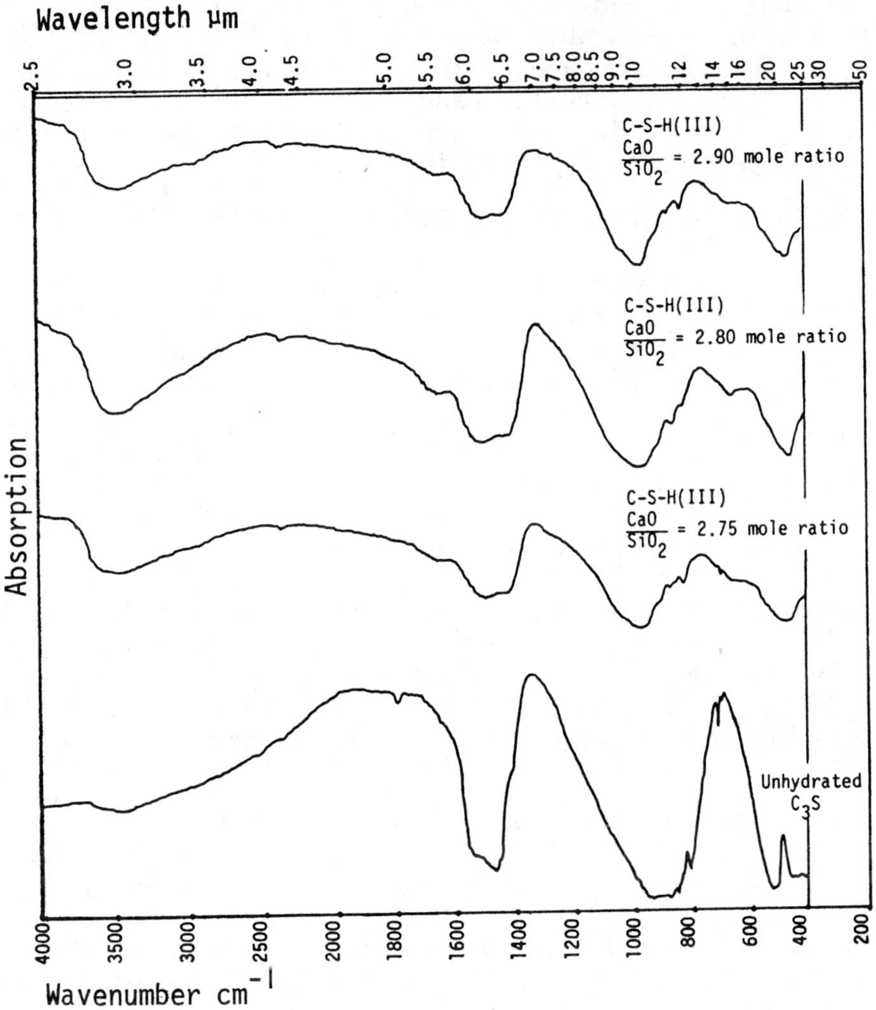

Fig. 1. Infrared absorption spectra of C-S-H(III), with its CaO/SiO_2 mole ratio varies from 2.75, 2.80 and 2.90, respectively, together with unhydrated C_3S.

is affected by the strength of bands formed by (OH) groups and is limited by their acidity. Therefore, the hydroxyl groups which are attached by a covalent bond and their absorption regions are revealed at 3400-3600 cm^{-1}. The other vibration of hydroxyl (OH) is a weak band observed around 1650 cm^{-1} in the spectrum, but there is no sharp band in this region. It is attributed to absorption of interlayer water in the 1650 cm^{-1} region.

b- The SiO_2 Spectra

The characteristic absorption bands of SiO_2 reveals no notable changes in all CaO/SiO_2 mole ratios of C-S-H(III), and generally SiO_2 reveals two main absorption regions, one of a strong band at 460-490 cm^{-1} and the other of very strong bands at 950-990 cm^{-1}.

All the CaO/SiO_2 mole ratios of C-S-H(III) spectrum include also the strong band at 1400-1525 cm^{-1}.

High lime Calcium-Silicate-Hydrate produces stronger absorption in the region 1400-1525 cm^{-1} than a comparable amount of lower lime Calcium-Silicate-Hydrate. Other region in the spectrum are definite bands with a strong shoulder at 750-780 cm^{-1} and strong absorption bands at 1310-1340 cm^{-1} (See Fig.1 of the spectra of C-S-H(III)). The characteristic absorption spectra showing the quality of their bands are presented in table 1.

3- INFRARED SPECTRA OF UNHYDRATED β-C_2S

Among the four known morphological forms of Dicalcium Silicate, the β-dicalcium Silicate (β-C_2S) is most relevant to Portland cement. The β-C_2S is a monoclinic modification of γ-C_2S with tetramolecular cell /15/. In this phase the oxygen Co-ordination of calcium is irregular, being six-fold for 4 Ca ions eight-fold for the remaining Ca ions /15/. The site symmetry cannot be used directly to determine the selection rules for the vibration of SiO_4 in β-C_2S as in γ-C_2S. The reason for recognition may be attributed to the known conversion of β-C_2S to γ-C_2S under pressure or impact as otherwise reported /16/. Other authors have made observations of the KBr spectra of β-C_2S, e.g. the spectrum of β-C_2S in Fig. 1, reference /17/ contains a group of vibrations of 503, 518 and 539 cm^{-1} and another group of vibrations at 844, 909 and 1010 cm^{-1}. D.M. Roy /18/ has also measured γ-C_2S and β-C_2S and obtained patterns between 700 to 1200 cm^{-1}. Launer /8/ has also measured the spectrum of β-C_2S at 620 to 5000 cm^{-1}. The spectra of Roy /18/ and Launer /8/ believe that there may possibly be more spectral detail to account for in well-crystallized material.

The present author Hannawayya has made an interesting study of infrared spectrum of unhydrated β-C_2S which (Fig. 2, β-C_2S-spectra) at the beginning a small rise in the reflection curve at 425 cm^{-1}, followed by a strong sharp absorption band at 520 cm^{-1}, then a very strong sudden rise in the reflection curve at 630 cm^{-1}. After that the curve drops gradually to three shatter nodes in the reflection curve at 655, 700 and 745 cm^{-1}, respectively. Then the absorption band in the range 840 to 995 cm^{-1} is broad with strong peaks at 840, 875 and 995 cm^{-1}, respectively. There is again a sudden strong rise in the reflection curve at 1370 cm^{-1}. In the slope there are three maxima at 1190, 1250 and 1290 cm^{-1}, respectively, followed by a very strong sharp band of intensity at range 1470 cm^{-1}. Again there is sudden rise in tne reflection curve at 1650 cm^{-1}, which weakly slows down. The absorption bands are listed in table 1.

4- INFRARED SPECTRA OF CALCIUM-SILICATE-HYDRATE(II^2) OR C-S-H(II^2)

C-S-H(II^2) described by Hannawayya /11/ as a lime-rich ratio of 1.55-1.88. When the solution was supersaturated with $Ca(OH)_2$ followed by precipitation of a crystalline product. the CaO/SiO_2 mole ratio was 1.88, which is described by the author (Hannawayya) as bundles of fibers of quite thin units in the form of cigars. Typical spectra of the C-S-H(II^2) together with unhydrated β-C_2S illustrated in Fig. 2, and the characteristic absorption spectra showing the quality of their bands was presented in table 1.

The two CaO/SiO_2 mole ratios. 1.56 and 1.88, of the product C-S-H(II^2) exhibit different infrared absorption between the 1100 to 850 cm^{-1} region, e.g. in the CaO/SiO_2 = 1.56 mole ratio yields a different pattern which gives rise to a few individual absorptions in the same range. Its results are a

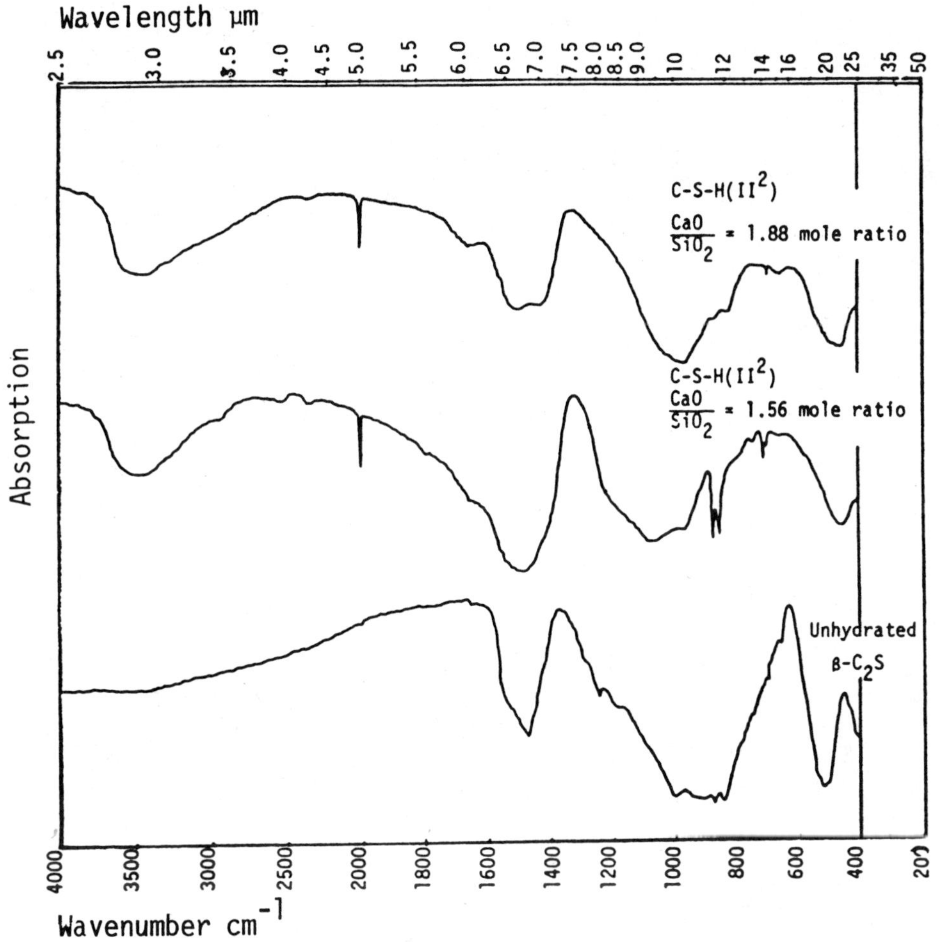

Fig. 2. Infrared absorption spectra of C-S-H(II^2), with its CaO/SiO_2 mole ratio varies from 1.56 and 1.88, respectively, together with unhydrated β-C_2S.

broad and diffuse absorption at 1100-1050 cm^{-1}. A strong pointed shallow hump is revealed at 890 cm^{-1} followed by a sudden drop with two additional sharp, intense absorption bands of very strong spectra at 875 and 855 cm^{-1}.

But in the CaO/SiO_2 = 1.88 mole ratio, the spectra in the 1100 to 850 cm^{-1} range provide individual absorption of a very strong band at 1000-950 cm^{-1}. After that a sudden medium shoulder is revealed at 865-825 cm^{-1}. In the above two CaO/SiO_2 ratios patterns, the SiO_2 reveals a medium strong band at 460 cm^{-1} and at 1000-950 cm^{-1} a very strong absorption band. In both C/S ratios is also a deep very strong hump at 1345-1320 cm^{-1}, and a strong band at 1500-1450 cm^{-1}. The medium strong hydroxyl (OH) groups in both C/S ratios spectra are located at 1655 cm^{-1} which is attributed to absorption of the interlayer water region, while the very strong broad region at 3540-3400 cm^{-1} belongs to the band of hydroxyl (OH) groups of $Ca(OH)_2$. See Fig. 2, table 1.

Table 1. Infrared (IR) spectra, data for the unhydrated Silicates (C_3S and β-C_2S) and the end products, C-S-H(III) and C-S-H(II^2), with their various CaO/SiO_2 mole ratio.

Wavenumber cm^{-1}, and the quality of their bands						
Unhydrated C_3S	C-S-H(III) C/S=2.75 mole ratio	C-S-H(III) C/S=2.80 mole ratio	C-S-H(III) C/S=2.90 mole ratio	Unhydrated β-C_3S	C-S-H(II^2) C/S=1.56 mole ratio	C-S-H(II^2) C/S=1.88 mole ratio
450 m 495 s, sh 510 s 530 s 690 vs,sh 700 m 705 s 710 vs 720 vs,sh 810 vs 820 m,sh 830 m 850 s 880 ms 950 ms 1120 vs, sh 1160 vs,sh 1470 vs 1540 m 1790 ms,sh	460 s 490 s 700 m 750 s,sh 780 s,sh 825 m 870 w 960 vs 990 vs 1310 vs, sh 1340 vs, sh 1420 vs 1500 vs 1655 mw,sh 3400 vs 3600 vs	450 s 660 m,sh 750 s,sh 780 s,sh 825 m 860 m 960 vs 990 vs 1310 vs, sh 1340 vs, sh 1400 vs 1525 vs 1650 m, sh 3400 vs 3600 vs	450 s 470 s 660 m, sh 750 s, sh 780 s, sh 825 s 870 m 950 vs 990 vs 1300 vs, sh 1340 vs, sh 1400 vs 1525 vs 1650 m, sh 3400 vs 3600 vs	425 ms, sh 520 s 630 vs, sh 655 m 700 m 745 m 840 s 875 s 995 s 1190 m 1250 s 1370 s, sh 1470 vs 1650 m, sh	460 ms 700 m 710 s 750 w 855 vs 875 vs 890 s, sh 965 s 1050 s 1090 s 1325 vs,sh 1470 vs 1500 vs 1655 ms 2000 s 3400 vs	460 ms 655 s } sh 700 s 825 m 865 m 950 vs } br 1000 vs 1320 vs } sh 1345 vs 1430 s } br 1500 s 1655 ms 2001 s 3400 vs

s = strong ms = medium strong w - weak br = broad
vs = very strong m = medium sh = shoulder

CONCLUSIONS

The present article contains an expansive study by infrared spectra for characterization and evaluation of the Silicates (C_3S and β-C_2S) and their end hydration products (C-S-H(III) and C-S-H(II^2)), respectively.

Absorption and reflection of infrared spectra of the end products C-S-H(III) and C-S-H(II^2) of varying CaO/SiO_2 revealed that the high lime Calcium-Silicate-Hydrate produces stronger absorption in the region at 1525-1400 cm^{-1} for C-S-H(III) than a comparable amount of lower lime Calcium-Silicate-Hydrate for C-S-H(II^2).

1- The Product C-S-H(III) of Varying C/S Ratios

The position of the hydroxyl (OH) absorption band is affected by the strength of hydrogen bands formed by (OH) groups and is limited by their acidity. Therefore, the (OH) group, which is attached by a covalent band and its absorption region, revealed at 3400-3600 cm^{-1}, belongs to the (OH) group of $Ca(OH)_2$. The other hydroxyl (OH) is a weak band observed around 1650 cm^{-1} region. It is attributed to absorption of interlayer water.

The SiO_2 spectra revealed two main absorption regions, a strong band at 460-490 cm^{-1} and the other with very strong bands at 950-990 cm^{-1}.

2- The Product C-S-H(II^2) of Varying C/S Ratio

The two C/S ratios of C-S-H(II^2) exhibit different infrared absorption between the 1100 to 850 cm^{-1} region, e.g. in the C/S = 1.56 mole ratio the pattern revealed only a few individual absorption band, but in the C/S = 1.88 the mole ratio provides individual absorption with a very strong band at 1000-950 cm^{-1}.

In the two ratios, the SiO_2 reveals a medium strong band at 460 cm^{-1} and at 1000-950 cm^{-1} a very strong absorption band.

The medium strong hydroxyl (OH) group in both C/S ratios spectra are located at 1655 cm^{-1} which depends on absorption of the interlayer water region, while the very strong band at 3540-3400 cm^{-1} region depends on the band of the hydroxyl (OH) groups of $Ca(OH)_2$.

3- Unhydrated C_3S and β-C_2S

The present author (Hannawayya) believes that the fundamental bands of unhydrated C_3S are in terms of the fundamental modes of SiO_4 tetrahedron. If the SiO_4 group is part of a crystal lattice, the crystal plays a dominant role in determining which vibration will be revealed in the spectrum. However, for C_3S, the SiO_4 absorption bands were revealed at 400-450 cm^{-1} and in the 800-1000 cm^{-1} region. But for β-C_2S, the SiO_4 revealed a strong sharp absorption at 520 cm^{-1} and another at 840-995 cm^{-1}.

Finally the above results of the present studies with thier interesting observations are shown in Figs 1 and 2 for the infrared spectra and the characteristic absorption spectra revealing the quality of their band presented in table 1.

ACKNOWLEDGEMENTS

The author (Hannawayya) is deeply grateful to Director General Sigvard Tomner, to Chief Engineer Paul Forsgren, to Chief of Adminisration Kjell Jegefors, to Head Manufacturing and Product-development Lenart Ståhl and to Department Director Carl-Otto Frykfors of the National Swedish Board for Technical Development (Styrelsen för Teknisk Utveckling, STU) for all help received in supporting grant for this research.

REFERENCES

/1/ Coblentz, William W., Investigation of Infrared Spectra, Parts III and IV, Washington, D.C.: Carnegie Institution of Washington (1906).

/2/ Coblentz, William W., Supplementary Investigation of Infrared Spectra, Parts V, VI and VII, Washington, D.C.: Carnegie Institution of Washington (1908).

/3/ Matossi, Frank, and Krüger, Hans, Das Ultrarote Reflexionsspektrum Von Silikaten: Zeit. F. Physik, 99, 1-23 (1936).

/4/ Schaefer, C., Matossi, F., and Wirta, K., Das Ultrarote Reflexionsspektrum Von Silikaten: Zeit. F. Physik., 89, 210-233 (1934).

/5/ Matossi, F., and Broder, O. Das Ultrarote Absorptionspektrum Einiger Silikaten: Zeit. F. Physik, 111, 1-17 (1938).

/6/ Barnes, R. Bowling, Bratain, R. Robert, and Firestone, R.S., Powder Filters for the Infrared: Phys. Rev., 47, 792 (1935).

/7/ Lecomte, Jean, Application Analytiques des Spectres Infrarouques de Poudres aux Comosés Minérauz et âPeau de Cristallisation: Analytical Chimica Acta, 2, 727-733 (1948).

/8/ Hunt C M. Infrared Absorption of some Compounds in the $CaO-SiO_2-H_2O$ System. Proceedings of the Fouth International Symposium on Chemistry of Cement. Washington, VI, pp. 297-305 (1960).

/9/ Launer, P.J., Amer. Miner. 37, 764 (1952).

/10/ Hannawayya F., "Study of the Structure and Crystallization Properties of Tricalcium Silicate (C_3S) and β-dicalcium Silicate (β-C_2S) Hydrate; and the New Calcium-Silicate-Hydrate (C-S-H(III), C-S-H(II^2), and C-S-H(I^1))". Printed in the 7th International Congress on the Chemistry of Cement, Paris Symposium Vol. II, p. 31-36 (1980).

/11/ Hannawayya F., "Study of the structure and Crystallization Properties of β-dicalcium Silicate (β-C_2S) Hydrate". Materials Science and Engineering Journal, 34, (1978), 183-197, Lausanne, Printed in the Nether-Land.

/12/ Ghosh, S.N. and Chatterjee, A.K. Journal of Materials Science 9, 1577-1584 (1974).

/13/ Lehmann H. and Dutz H., Infrared Spectroscopy Studies on the Hydration of Clinker Minerals and Cements. Proceedings of the Fourth International Symposium on the Chemistry of Cement, Washington, VI, 513-8 (1960).

/14/ Tare P., "Infrared Study of Orthosilicates and Orthogermanates-A New Method of Interpretation of the Spectra", Spectrochin Act. 18, 467-83, (1962).

/15/ Wyckoff R.W.G. "Crystal Structure", Vol. 4, Wiley, New York, (1968).

/16/ Yannaquis N. and Guinier A., Proceeding of the Fourth International Symposium on the Chemistry of Cement. Washington, Vol. 1, p. 21, (1960)

/17/ Della M. Roy, Journal American Ceramic Soc., 41, 293-299, (1958).

/18/ Hannawayya F., "Characterization and Properties of the New Calcium-Silicate-Hydrate (C-S-H(III) and C-S-H(II^2))". Paper presented at the Proceedings in the 10th International Conference on Cement Microscopy, San Antonio, Texas, U. S. A., (1988).

TIME-TEMPERATURE DEPENDENCE OF STRENGTH IN ALUMINUM NITRIDE

F. P. Skeele
Dow Chemical Company
Midland, MI 48674

M. J. Slavin and R. N. Katz
U.S. Army Materials Technology Laboratory
Watertown, MA 02172

ABSTRACT

Because of its unique combination of chemical, mechanical, and thermal properties, aluminum nitride (AlN) was investigated as a structural material for advanced heat engine applications. This paper will examine room temperature mechanical properties including hardness, fracture toughness, flexure strength and elastic constants of a hot pressed AlN. Time-temperature-stress behavior as elucidated by stepped-temperature, stress-rupture and isothermal stress rupture tests will be presented. A thorough fractographic examination of the strength limiting flaws and fracture characteristics of AlN will be discussed.

INTRODUCTION

Demanding high temperature applications have resulted in increased development and characterization of structural ceramics. Although not a new material, aluminum nitride (AlN) has not yet been exploited as a high temperature ceramic, probably due to the lack of substantial quantities of high quality material. However, with the development and wide spread interest in the thermal conductivity of AlN for electronic applications, the availability of excellent AlN powders has renewed interest in examining the high temperature mechanical behavior of this material for applications appropriate to its unique combination of chemical, mechanical, and thermal properties.

Room temperature values for elastic modulus, Poisson's ratio, bend strength, hardness, fracture toughness (K_{Ic}), thermal expansion and thermal conductivity have been reported by many investigators and commercial producers of AlN [1-10]. Some of these values are listed in Table 1.

Table 1. Summary of room temperature data reported on AlN.

$\rho^{\ddagger}$ g/cm^3	MOR MPa	Weibull Modulus	E GPa	μ*	Hardness GPa	K_{Ic} MNm$^{-3/2}$	K W/m•K	α x 10^{-6} K^{-1}	Process
3.26	370†		315	0.25	11.3-15.3	2.6			HP [1,2]
2.68	259								S [3]
					16.6				HP [4]
	265		345		12.0				HP [5]
3.27	235		350			3.5	60	5.7	? [6]
3.28	400		310				100	4.6	? [7]
3.3	450		281				135	4.5	S [8]
3.24	307±23	16.5					176	4.4	S [9,10]
3.25	340	21	330		11.8	3.0	170		S [10]

$^{\ddagger}\rho_{\text{theoretical}}$ = 3.26 g/cm^3 * μ = Poisson's ratio S = sintered ceramic HP = hot-pressed ceramic
† = strength measured in 3-pt bend, all other reported strengths did not report measurement technique.

High temperature properties have only been reported by a few researchers.[1,2] Thermal properties were studied by Boch et. al.[1] High temperature isothermal fast fracture strengths were measured in 3-point bending and showed a trend of decreasing strength up to 700°C and then remained roughly constant at higher temperatures (~250 MPa). It was also reported that a stress of 190 MPa, (75% of the strength) at 1000°C for 100 h exhibited only primary and steady state creep behavior in the material. Oxidation kinetics were reported and showed little occurred up to 1490°C (4 mg/cm^2) over a 24 h period.

Reported potential high temperature applications of aluminum nitride as coatings include a pump member applicable to a hot-chamber type aluminum die casting machine[11] and an intermediate coating on Si_3N_4 and SiC to help reduce contact-surface stress that can damage the surface and cause premature failure.[12] No monolithic applications of AlN for high temperature parts have been reported at the time of this writing.

The present paper describes the room temperature and high temperature mechanical properties of hot pressed pure AlN in detail and provides some insight into mechanisms involved in its mechanical behavior.

EXPERIMENTAL PROCEDURE

The aluminum nitride powder used in this study was produced via the carbothermal reduction of alumina:

$$Al_2O_3 + 3\,C + N_2 \xrightarrow[1600^\circ C]{} 2\,AlN + 3\,CO$$

The mean size of the powder was 2.3 with a standard deviation of 1.5 μm, measured with a Horiba CAPA 700 analyzer, with a crystallite size of ~ 0.5 μm, measured with a scanning electron microscope (SEM). The surface area of the powder was 3.0 m^2/g, measured by single point BET. A quantitive analysis of the metallic impurities measured by x-ray fluorescence is given in Table 2. The major impurities were carbon and oxygen at 0.06 and 1.0 weight percent, respectively.

Table 2. Impurity Content (ppm) of AlN powder.

Ca	Fe	Mg	Si	Other Metals
75	20	20	104	<10

This powder was hot pressed as-received in a N_2 atmosphere at 1800°C for 0.5 h at 21.3 MPa to form a tile 20.3 x 20.3 x 2.5 cm. The tile was machined into flexure specimens according to MIL-STD-1942(MR)[13] specimen configuration 'B'. Specimen size was 3.0 x 4.0 x 45 mm with a 45° chamfer along the four long edges.

Bulk density was measured on various flexure bars cut from the tile. Elastic moduli and Poison's ratio were also measure on the same flexure bars using a point contact pulse-echo ultrasonics technique. Grain size was measured according to the ASTM E 112-84 standard.

Room temperature flexural tests were performed to establish a reference strength and to characterize the strength limiting flaws. A universal testing machine (Model 1123, Instron Corp., Canton, Massachussetts) with a 5000 N load cell was used with four-point fixtures in accordance with MIL-STD-1942(MR)[13]. The spans were 40.0 mm and 20.0 mm and a crosshead speed of 0.5 mm/min was used. The relative humidity was 42% at 24°C. A sample size of forty specimens was evaluated at the U.S. Army Materials Technology Laboratory (hereafter designated as MTL) while a thirty specimen sample was evaluated at Dow Chemical Company.

Flexure data was analyzed with Weibull statistics.[14] The two-parameter Weibull modulus was determined by a least square fitted line through the strength data when plotted in a standard Weibull format. The rank percentage used for all the data presented is $p = (i - 0.5)/N$, where i is the i th data point and N is the total number of specimens. The Weibull modulus is the slope of the line.

Fracture toughness measurements were made on Chevron-notched specimens.[15,16] The geometry used was the short bar. Thin slots are machined longitudinally in the specimen, leaving a "V" shaped ligament in the intended crack plane. An opening load is applied to the mouth of the specimen, which causes a crack to initiate at the point of V. The constant widening of the crack front as is advances along the axis of the specimen causes the crack growth to be stable, thus, a real crack is created in the specimen just before the toughness measurement is taken. The fracture toughness measurement is made at the time of the load maximum when the crack is already in the central region of the specimen. Measurements were made using a Fractometer I (Model 4200, TerraTek Systems) with a 9.53 mm flatjack. The total test time used was about 30 seconds.

Hardness measurements where made with a Tukon® Microhardness Tester (Model 300, Acco Industries Inc.) using a Vickers indenter. Three flexure specimens were polished to a submicron finish and indented with a 500 gram load for a total of thirty measurements (i.e. ten measurements per specimen).

A stepped temperature stress rupture (STSR)[17] program was used to quickly screen the material while stressed to a range of test temperatures. This test was developed by MTL for simplified high temperature evaluation of materials still in their developmental stage. This test also isolates intermediate-temperature or load regimes where phase instabilities, oxidation, or other problems may degrade material performance. Four-point bend fixtures, machined from hot-pressed silicon carbide and having bearing spans of 40.0 and 20.0 mm, were used. A simple deadweight lever system applied the load into the furnace. The furnaces are described in detail in reference 18. The procedure, carried out in air, starts at 800°C after the furnace temperature is allowed to equilibrate for 5 min, after which the load is applied. After 24 h, if the specimen does not fail, the temperature is raised within 0.25 h to 900°C. This cycle continues for 1000°C, 1100°C , and 1200°C, but at 1200°C, the specimen cycle time is 72 h. In the event of failure the furnace is shut down. An arrow labeled with the applied stress denotes the time of failure on the STSR plot.

Based on the STSR results, the flexural isothermal stress rupture experiments were carried out at 800°C and 1200°C in air in several identical test furnaces. The furnace arrangement and loading procedure are identical to the step temperature stress rupture tests.

Specimens which did not fail during the STSR or conventional stress rupture tests were unloaded prior to cool-down. This was done to minimize frictional forces at the knife edges due to unequal specimen-fixture contractions. These specimens were then tested at room temperature to determine the retained strength. They were mounted in the room temperature fixture such that the the most highly stressed zone in the high temperature test was again loaded in a tensile manner.

Since the high temperature testing was done in air, some oxidation of the AlN was expected. Oxidation products were identified by x-ray diffraction of an oxidized flexure bar. To study the rate of oxidation more closely, one specimen was placed in a furnace and heated to 1200°C in air with no applied load. Weight measurements of the flexure bar were made periodically. The specimen was removed after 1000 h.

RESULTS AND DISCUSSION

The average bulk density measured on thirty flexure specimens was 3.23 ± 0.03 g/cm^3 (99% theoretical density). Young's modulus and Poisson's ratio, averaged from the same thirty specimens, was 323 GPa and 0.24, respectively. The average grain size was 2.25 ± 0.2 µm. The average Vicker's microhardness for thirty good indents (no spall or cracking) was 11.2 GPa with a standard deviation of 0.3 GPa. The average fracture toughness as measured on five Chevron-notched short bar specimens was 2.70 MNm$^{-3/2}$ with a standard deviation of 0.24 MNm$^{-3/2}$.

The two samples of flexure specimens yielded mean strengths of 288 MPa and 276 MPa with standard deviations of 26.5 and 31.1 MPa for sample sizes of 40 and 30, respectively. Figure 1 depicts the two parameter Weibull plots for both the 40 specimen and 30 specimen samples as well as the combined data set. The Weibull modulus for the 40 specimen sample was 13.2 with the characteristic strength of the specimen (σ_0) 300 MPa, while the Weibull modulus for the 30 specimen sample was 10.4 with the characteristic strength of the specimen 290 MPa. The confidence level (denoted as R^2) for both sets is 0.97. Combining both sets of data (Figure 1c) yields an average strength of 283 MPa, a standard deviation of 29.0 MPa, a Weibull modulus of 11.8 and a characteristic strength of 296 MPa. Here the confidence level is 0.99.

All specimens were subjected to a optical fractography up to a magnification of ~100 x. The flaws associated with the 40 specimen sample are listed in Table 3. Selected specimens were then photographed and prepared for examination in a SEM. Figure 2 shows the typical critical flaws found: graphite inclusions, agglomerates, and

Table 3. Resulting stress and corresponding flaw code for the room temperature flexure sample tested at MTL.

Stress(MPa)	Flaw	Stress(MPa)	Flaw	Stress(MPa)	Flaw	Stress(MPa)	Flaw
333	Inclusion	308	Inclusion	293	Inclusion	266	?
328	Inclusion	303	Agglomerate	293	Inclusion	263	Agglomerate
324	Pore	303	Agglomerate	290	Inclusion	261	Inclusion
323	?	303	Inclusion	289	Agglomerate	260	Pore
319	Pore	298	Agglomerate	285	Inclusion	256	?
318	Inclusion	297	Inclusion	281	Machine Damage	255	Inclusion
316	?	297	Agglomerate	276	Inclusion	248	?
315	Machine Damage	296	Pore	273	Inclusion	244	Inclusion
312	Inclusion	296	?	268	Pore	238	Inclusion
311	Inclusion	296	?	266	Machine Damage	233	Inclusion

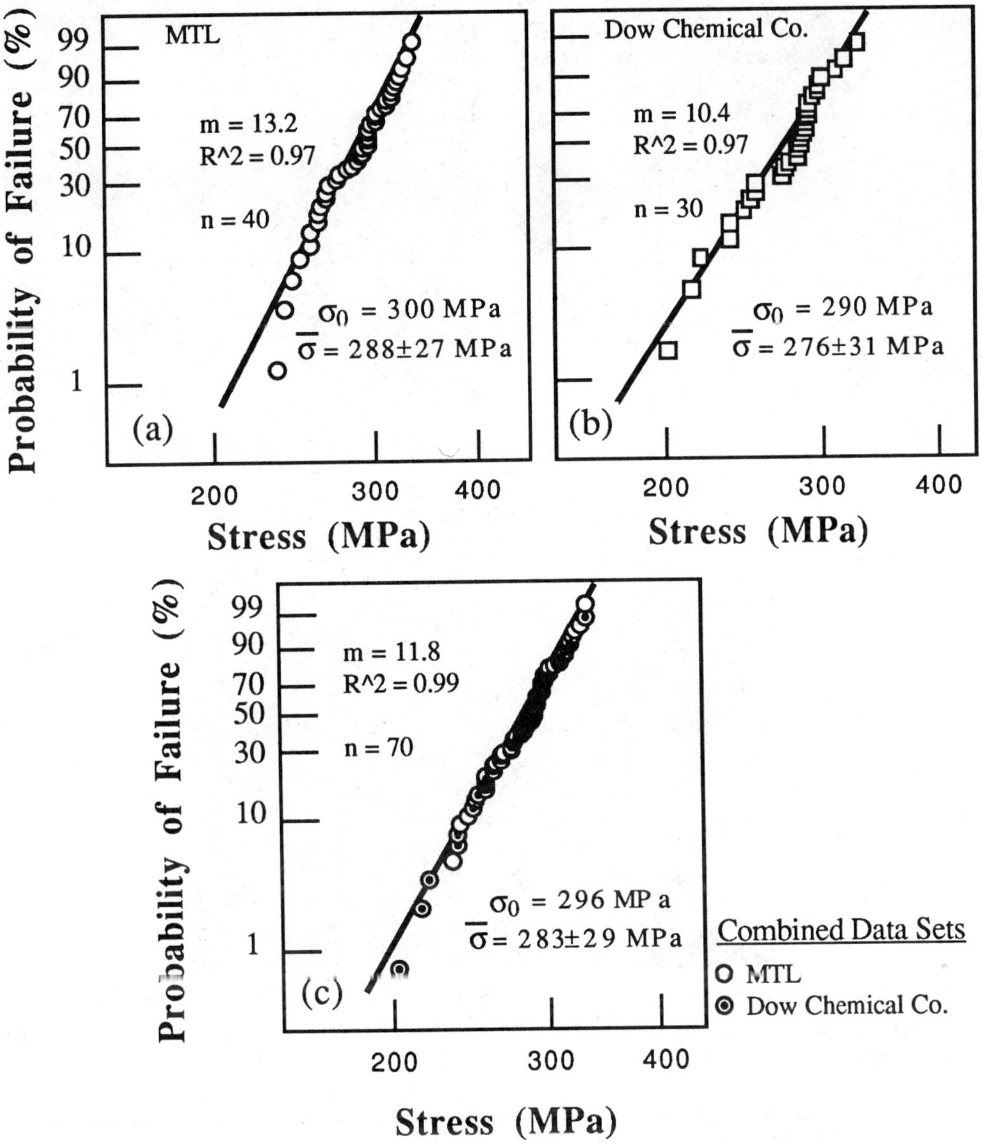

Figure 1. Weibull plots of AlN flexure strengths collected at (a) MTL and (b) Dow Chemical Company. Graph (c) shows the Weibull plot for both samples combined.

pores. These flaws are sintering defects that are volume distributed. Although machining damage was apparent on the specimens in the form of chatter and chamfer chips, only a few specimens failed from these flaws. Since the graphite inclusions made up over half of the observed critical flaws, and that they occurred in the volume of the material and not the surface, the powder was investigated as a source of the graphite inclusions. No graphite impurities were found in the powder after analysis by SEM and optical microscopy. Therefore, the graphite is believed to come from the hot press dies which may contain free graphite dust which is present during the powder loading stage.

Fifteen specimens were tested via the stepped temperature, stress rupture (STSR) procedure, Figure 3. Specimens failed on loading when subjected to stresses greater than 250 MPa. Eight specimens failed in a time dependent manner at 800°C and 1200°C at applied stresses between 175 and 275 MPa. Three specimens survived the entire temperature cycle with applied stress of 100 and 150 MPa (2). Permanent deformation of the survivors was negligible (<0.1% strain), however, extensive surface oxidation had occurred for specimens subjected to 1200°C. The retained room temperature strengths of the three survivors were 131, 102 and 173 MPa for the 100, 150 and 150 MPa applied stresses, respectively. The survivors had a mean weight gain of 0.028 g with dimensional increases averaging 1.2% and 1.8% in the base and height, respectively. Calculating from the

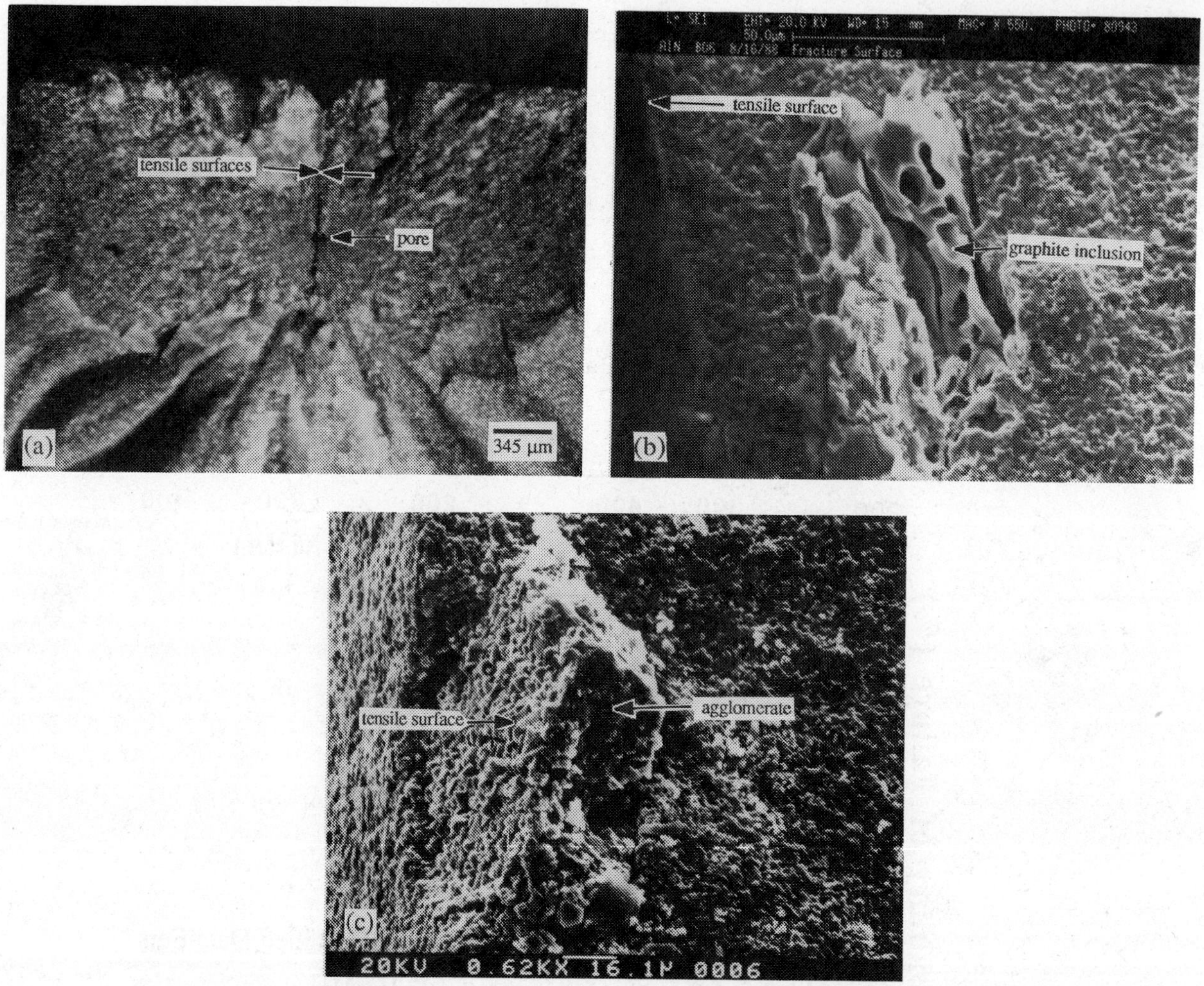

Figure 2. Room temperature fast fracture flaws, including (a) failure due to a surface pore (σ_{max} = 319 MPa), (b) failure due to a graphite inclusion (σ_{max} = 293 MPa) and (c) failure due to agglomeration (σ_{max} = 303 MPa).

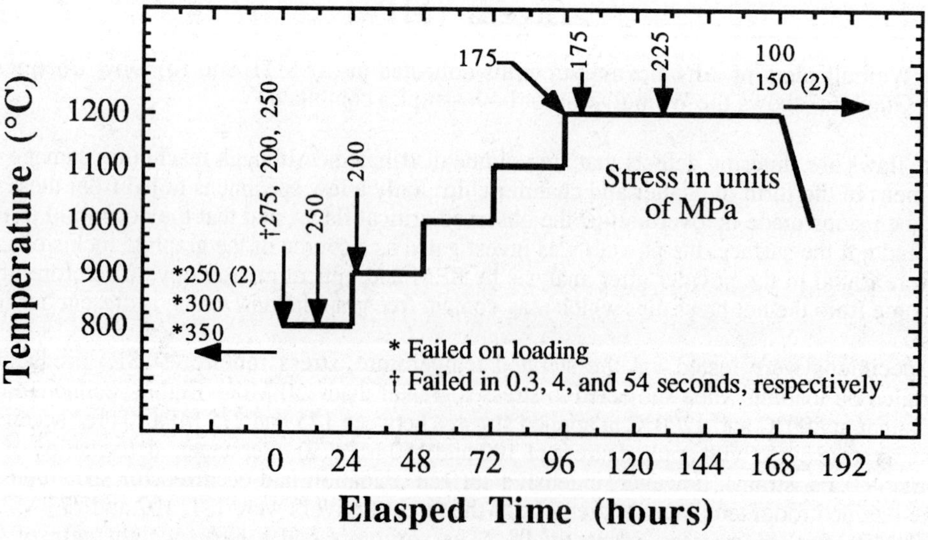

Figure 3. Step-temperature stress rupture results from 800° to 1200°C.

surface area of the flexure bar, this dimensional and weight increase would correspond to an oxide layer of approximately 15 µm, which was confirmed optically.

Based on the results of the STSR program which showed very little temperature dependence of strength between 800°C and 1200°C below the 200 MPa stress level, 800°C and 1200°C temperatures were picked for stress rupture experiments. Figure 4 shows the results of the 800°C and 1200°C isothermal stress rupture tests. Seventeen specimens were tested at 800°C and thirty-five specimens were evaluated at 1200°C. At 800°C six specimens failed on loading at applied stresses between 250 and 350 MPa. Four specimens survived 1000 h while sustaining stresses between 200 and 225 MPa.

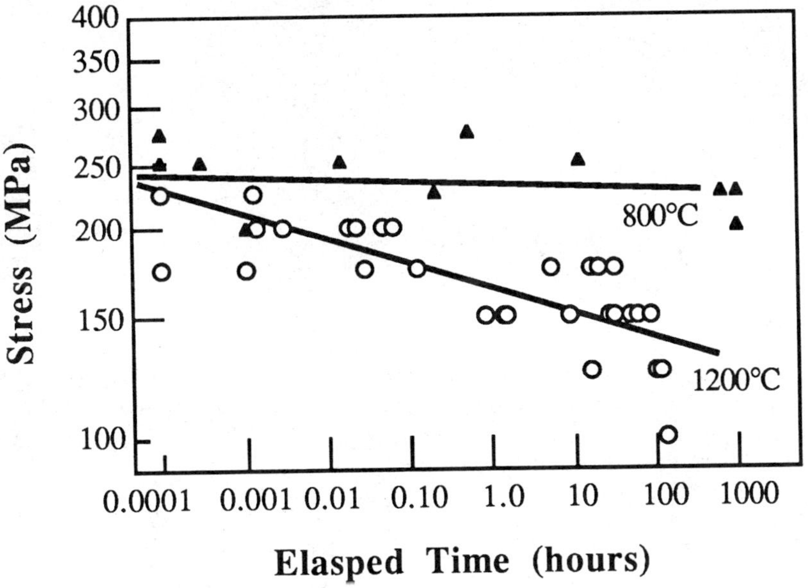

Figure 4. Isothermal stress rupture results. Specimens failing on loading are plotted at 0.0001 hours.

The retained room temperature strengths of the four survivors were 337, 290, 305 and 305 MPa for the 200, 225, 225 and 225 MPa applied stresses, respectively. The remaining seven specimens failed in a time dependent manner. There is considerable scatter in the times to failure but no indication of creep fracture or slow crack growth for these specimens. Some surface pitting was evident along with a glassy coating on the survivors though no specimen failed from these locations.

Four specimens failed on loading at 1200°C at stresses of 175 and 225 MPa. There were no survivors at this temperature. Thirty-one specimens failed in a time dependent manner though again there is no indication of creep fracture or slow crack growth. Severe oxidation of the longer time to failure specimens was common with oxide thicknesses to approximately 0.10 mm, measured optically. In these specimens some permanent deformation (~0.1% strain) occurred. The time dependent failure phenomenon may in fact be due to the gradual reduction in the cross section of the AlN specimen. The failure origins were identical to those at room temperature for all the high temperature specimens (refer to Figure 5) except those failing at 1200°C for moderate time under load (refer to Figure 6). In Figure 6, the specimen did not fail until the interface between the oxide layer and the AlN reached the agglomerate.

The oxidation was characterized by a weight gain and a volume increase from the original AlN specimen. The oxidation product was identified as α-Al_2O_3. On many fractured specimens, the oxide layer had flaked off. This is probably due to the specimen impacting the fixture on failure or because of thermal expansion differences (4.5 x 10^{-6} °$C^{-1}_{(20°C)}$ for AlN compared to 5.4 x 10^{-6} °$C^{-1}_{(20°C)}$ and 10 x 10^{-6} °$C^{-1}_{(1200°C)}$ for α-Al_2O_3). Experiments to evaluate the interface strength were beyond the scope of this study.

The rate of oxidation, Figure 7, shows a linear weight gain of 0.82 mg/hr which, for a 3 x 4 x 45 mm specimen, converts to ~ 3.1 mg/cm^2 for the initial 24 h period. Though somewhat lower than the oxidation rate reported at 1490°C[1] (4 mg/cm^2) the oxide layer does not prevent further oxidation of the AlN. Upon removal at approximately 300 h , the alumina layer had begun to flake off. Based on the oxidation rate measured during the experiment, the specimen (3 x 4 x 45 mm) would be completely converted to α-Al_2O_3 after approximately 500 h at 1200°C. When the specimen was removed from the furnace after 1000 h , the specimen had crept and failed

Figure 5. Specimen failed after 632 h at 800°C at a graphite inclusion. Note the pitted region extending from the surface into the now 2-D pore caused by inclusion pullout (σ = 225 MPa).

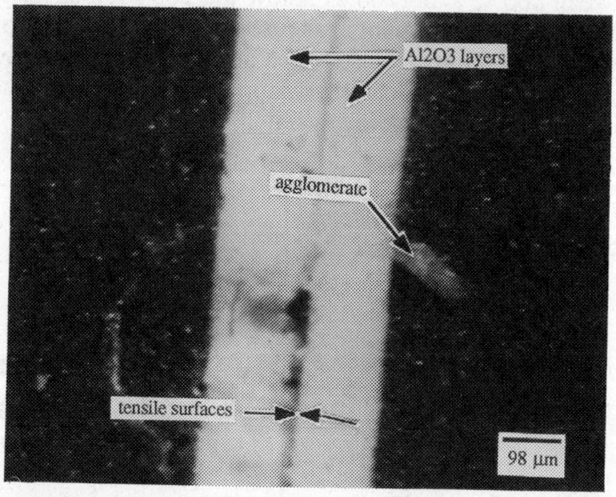

Figure 6. Specimen failed after 61 h at 1200°C at an agglomerate ~ 137 µm below the oxidized surface (σ = 175 MPa).

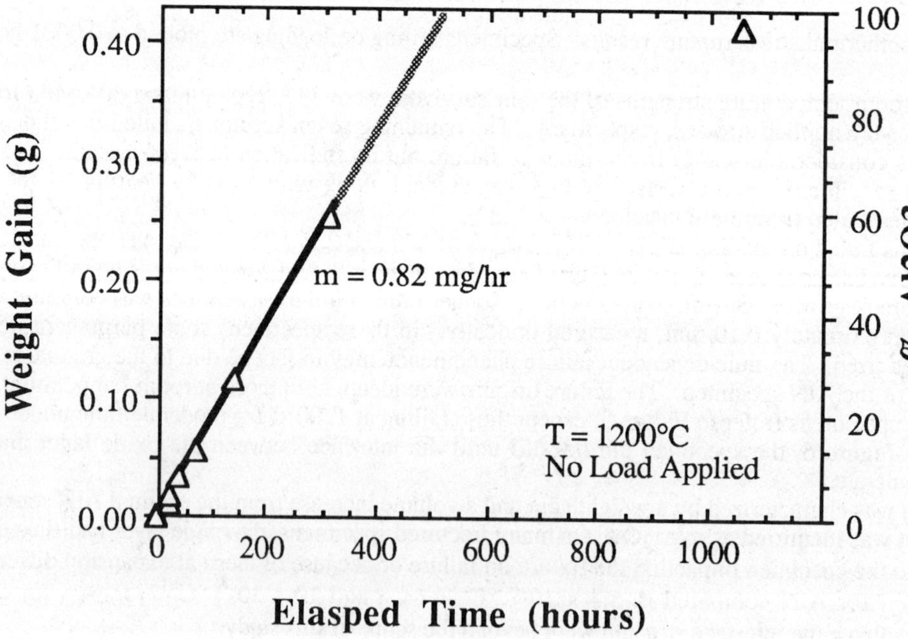

Figure 7. Oxidation rate of an AlN flexure bar to form α-Al_2O_3 at 1200°C.

under its own weight, Figure 8. The oxidize bar split into four sections indicating uniform oxidation from all sides with a poor bond between the oxide surfaces grown from adjacent faces.

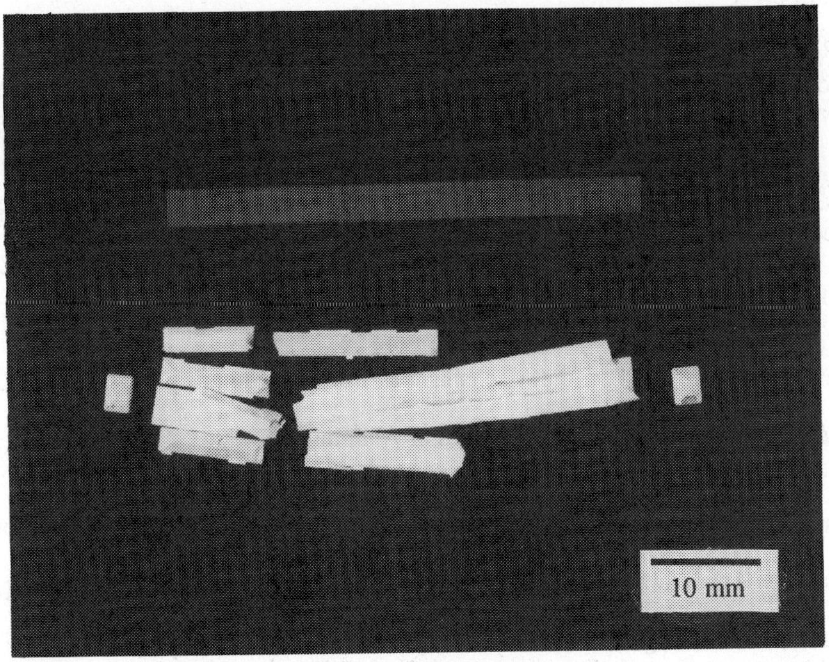

Figure 8. Virgin AlN flexure bar (top) oxidized 1000 h at 1200°C (bottom pieces).

SUMMARY AND CONCLUSIONS

A large aluminum nitride tile has been successfully hot pressed to near theoretical density without the use of additives to aid densification. The bulk mechanical properties, hardness, toughness, and strength, are very comparable to previously published data.

Table 4. Physical and Mechanical Properties of the AlN under investigation.

ρ g/cm^3	MOR MPa	Weibull m	E GPa	Poisson's ratio	Hardness GPa	K_{Ic} $MNm^{-3/2}$	K W/m•K
3.23±0.03	283±29	11.8	323±4	0.238±0.002	11.2±0.3	2.70±0.24	80 [19]

The high thermal conductivity is advantageous for applications requiring heat dissipation such as substrates and heat exchangers. The room temperature strength, with a Weibull modulus of 11.8, is comparable to reported values of for pure AlN. Failure in room temperature specimens originated at graphite inclusions (~60% of the time), agglomerates (~19% of the time) and pores (~15% of the time). Since these flaws were sintering defects, it is reasonable to expect improvements in strength with further refinement of powder processing procedures, which in turn should reduce the size and severity of the flaws. For more dynamic environments, an increase in the fracture toughness would be required. This would be best accomplished with the addition of a discontinuous reinforcement, either particulates or whiskers.

Little or no creep or slow crack growth was observed in specimens exposed to 800°C, however, a glassy phase was evident on the surface with some pitting. Limiting strengths were approximately the same as room temperature strengths. Evaluation of specimens exposed to 1200°C was complicated by oxide formation. High temperature applications (> 800°C) will be hindered by the extreme sensitivity to an oxidizing environment. The strength of AlN at temperatures between 800° and 1200°C should be investigated to determine the maximum temperature/stress regime before uncontrolled oxidation. Unlike glassy surface phases (oxidation) which often prevents further oxidation, the Al_2O_3 continues to grow at the expense of the AlN. Thermal conductivity would be little affected as long as the oxide layer remained thin. At higher temperatures, the ceramic would need to be coated to prevent the catastrophic oxidation of the part.

ACKNOWLEDGEMENT

This work was performed under a Cooperative Research and Development Agreement, written under the Federal Technology Transfer Act of 1986, between the U.S. Army Materials Technology Laboratory and the Dow Chemical Company.

REFERENCES

1. P. Boch, J. C. Glandus, J. Jarrige, J. P. Lecompte, and J. Mexmain, "Sintering, Oxidation, and Mechanical Properties of Hot Pressed Aluminum Nitride," *Ceramics International*, **8** [1] 34-40 (1982).

2. M. Billy and J. Mexmain, "Processing and Properties of Aluminum Nitride, a New Candidate for High Temperature Applications," *World Ceramics*, **2** 91-95 (1985).

3. K. Komeya, H. Inoue, and A. Tsuge, "Effect of Various Additives on Sintering Aluminum Nitride," *Yogyo-Kyokai*, **89** [6] 58-64 (1981).

4. M. Trontelj and D. Kolar, "On the Mechanism of Reactive Sintering of AlN," *Sp. Ceramics*, **6** 39-50 (1975)

5. K. M. Tayor and C. Lenie, "Some Properties of Aluminum Nitride," *J. Electrochem. Soc.*, **107** [4] 308-14, (1960).

6. Toshiba, TOSNITE™ Aluminum Nitride Product Literature (1987).

7. "Aluminum Nitride Substrate for Thermal Management in Microelectronics," Norton Technical Data, DS100, (1987).

8. N. Iwase, K. Anzai and K. Shinozaki, "Aluminium Nitride Substrates Having High Thermal Conductivity," *Toshiba Review*, No. 153, Autumn 1985.

9. N. Kuramoto, H. Taniguchi, and I. Aso, "Sintering Properties of High Purity AlN Ceramics," International Symposium on Ceramic Substrates and Packages, Denver, CO, USA, October 18-21, 1987.

10. N. Kuramoto, "Development of AlN Substrates," 15th New Ceramics Seminar: "From Seeds to Practical Applications," March 17-18, 1988 (Japan).

11. T. Ochial, M. Hirane, and H. Inoue, "Pump Member for Molten Metal," U.S. Patent 3,930,875 (1976).

12. V. K. Sarin, H. E. Rebenne, C. D'Angelo, and D. W. Oblas, "Development of Adherent Ceramic Coatings to Reduce Contact Stress Damage of Ceramics," ORNL/DOE Ceramic Technology Newsletter, **19**, [3-6] (1988).

13. U.S. Army MIL-STD-1942(MR), "Flexural Strength of High Performance Ceramics at Ambient Temperature."

14. W. Weibull, "A Statistical Distribution Function of Wide Applicability"," *J. Applied Mechanics*, [9] 293-297 (1951).

15. L.M. Barker, "Short Rod K_{Ic} Measurements of Al_2O_3," pp. 483-94 in Fracture Mechanics of Ceramics, Vol 3. Edited by R. Bradt, D. Hasselman, and F. Lange. Plenum, New York, 1978.

16. Chevron-Notched Specimens, Testing and Stress Analysis, Edited by J. H. Underwood, S. W. Freidmen, and F. I. Baratta, ASTM-STP 855, ASTM, Philadelphia, 1984.

17. G. D. Quinn and R. N. Katz, "Stepped Temperature Stress Rupture Testing of Silicon Based Ceramics," *Am. Ceram. Soc. Bull.*, **57**[11] 1057-58 (1978).

18. G. D. Quinn, "Guide to the Construction of a Simple 1500°C Test Furnace," Army Materials and Mechanics Research Center, AMMRC TR83-1, January 1983, NTIS ADA 125636.

19. Information obtained from Dow Chemical Company, 1988.

STATIC, CYCLIC AND DYNAMIC FATIGUE BEHAVIOR OF SILICON NITRIDE

Y. Tajima, K. Urashima, M. Watanabe and Y. Matsuo
Research and Development Department
NTK Technical Ceramics
NGK Spark Plug Co., Ltd.
2808 Iwasaki, Komaki, Aichi, 485 Japan

ABSTRACT

Static, cyclic and dynamic fatigue testing of a gas pressure sintered silicon nitride was carried out at several temperatures to study effects of temperature and loading conditions on the fatigue behavior. All the testing was conducted in 3-point bending for specimens without artificial flaws. The dynamic fatigue test results indicated that the material is highly fatigue resistant up to 900 °C but becomes susceptible to slow crack growth (SCG) at 1000 °C. Correlation among the results of the three kinds of tests were studied at room temperature and at 1000 °C. At room temperature, it is suggested that all the results can be correlated in terms of a single SCG parameter and that there is no significant enhancement of the SCG due to cycling. On the other hand, higher fatigue resistance for cyclic loading with higher frequency was observed at 1000°C, and it is considered to be due to a viscoelastic nature of the glassy grain boundary phase.

INTRODUCTION

Evaluation of fatigue behavior is essential to lifetime prediction of materials. There are several ways of characterizing the fatigue behavior; static, cyclic and dynamic fatigue tests with or without artificial flaws, and direct K_I-V measurements by fracture mechanics test techniques. Studies on the fatigue behavior of silicon nitride ceramics have recently become active, and there have been reports on the cyclic[1-5], static[2,4,6] and dynamic[6] fatigue behavior and K_I-V measurements[7,8]. Accelerated slow crack growth rate due to cycling, especially under the reversed loading conditions (tension-compression stress), has been observed at room temperature by several investigators[3-5,8], whereas Fett et al.[2] have reported that at 1200°C the lifetimes under cyclic load were higher than those predicted from the static fatigue lifetimes. Although the reported results agree to some degree, there seems to remain uncertainty. For example, some of the above studies[3,4,8] have employed artificial flaws, and there is no guarantee that the fatigue behavior is identical for the natural and artificial flaw specimens. Also, compositions and microstructure could influence the fatigue behavior at both room and high temperatures. Therefore, there is undoubtedly strong need to collect more data for various types of materials under well defined experimental conditions.

Gas pressure sintered silicon nitride (GPSSN) exhibits excellent mechanical properties and, thus, is a prime candidate material for the heat engine applications[9]. The distinguished features of the GPSSN include high strength and high reliability, which make the fatigue test easier without introducing artificial flaws. In the present study, the fatigue behavior of the GPSSN was evaluated under static, cyclic and dynamic loading conditions at room temperature and at higher temperatures. In the cyclic fatigue test, effects of a stress ratio (minimum stress/maximum stress) and a frequency were also evaluated.

EXPERIMENTAL PROCEDURE

The material studied was a GPSSN with the additions of alumina and yttria. It contains virtually no residual pores. Some of the physical properties at room temperature are listed in Table 1. Specimens for the fatigue tests were cut from blocks of the material and ground to rectangular bars (approximate dimensions 3 by 4 by 40 mm) with a 140-grit diamond grinding wheel.

All the fatigue tests were conducted in 3-point bending with a span of 30 mm. The static and cyclic fatigue tests were conducted at room temperature and at 1000°C in ambient air atmosphere. In the cyclic tests, the specimens were subjected to a sinusoidal cyclic stress with a stress ratio (R) of 0.1 or 0.5 and a frequency of 1 or 10 Hz. The loading conditions employed in this study are summarized in Table 2.

The dynamic fatigue tests were conducted at room temperature, 800°C, 900°C and 1000°C. Flexural strength (σ_f) was measured at two crosshead speeds (CHS's), 0.5 and 0.001 mm/min, with 10 breaks at each CHS.

Fracture surfaces of the selected specimens were observed by optical microscope and scanning electron microscope (SEM).

Table 1. Properties of GPSSN

Density (g/cm^3)	Young's Modulus (GPa)	Vickers Hardness (GPa)	Fracture Toughness by SEPB ($MPa \cdot m^{1/2}$)	Thermal Expansion Coefficient (/K)	Thermal Conductivity (W/m·K)
3.23	320	14.3	6.0	2.8×10^{-6} (r.t.-800°C)	27

Table 2. Loading Conditions of Static and Cyclic Tests

Temp.	Stress Ratio R	Frequency (Hz)
R.T.	1	-
	0.5	10
	0.1	10
	0.1	1
1000°C	1	-
	0.5	10
	0.5	1

RESULTS AND DISCUSSION

Dynamic Fatigue Test Results

The results of the strength measurements are summarized in Table 3, and Weibull plots of strength data are shown in Fig. 1. Weibull moduli are all greater than 15, indicating the material's high reliability. It is seen that the data at the two CHS's at each temperature line almost in parallel, and, thus, it is considered that the fatigue behavior can be reasonably evaluated under the present experimental conditions.

Starting with a power law relationship between slow crack velocity (V) and the stress intensity factor (K_I),

$$V = A \cdot K_I^N \qquad (1)$$

where A is a constant, one can obtain the following equation[10]:

$$\log \sigma_f = 1/(N+1) \log \dot{\sigma} + B \qquad (2)$$

where N is a fatigue parameter, $\dot{\sigma}$ is a stressing rate and B is a constant. If the stress-strain behavior of the material is purely elastic, the stressing rate is proportional to the CHS, and thus Eq. (2) can be rewritten as:

$$\log \sigma_f = 1/(N+1) \log(CHS) + C \qquad (3)$$

where C is a constant.

The N values calculated according to Eq. (3) are included in Table 3. These data suggest that the present material is highly fatigue resistant up to 900°C but becomes susceptible to slow crack growth (SCG) at 1000°C presumably due to softening of a glassy grain boundary phase. Therefore, further evaluation of fatigue behavior was carried out under the static and cyclic loading at room temperature and at 1000°C.

Table 3. Results of Dynamic Fatigue Test

Temp. (°C)	CHS (mm/min)	Mean Strength (MPa)	Weibull Modulus	N
R.T.	0.5	1090	15	94
	0.001	1020	16	
800	0.5	910	33	68
	0.001	830	15	
900	0.5	880	39	76
	0.001	810	23	
1000	0.5	830	17	15
	0.001	560	15	

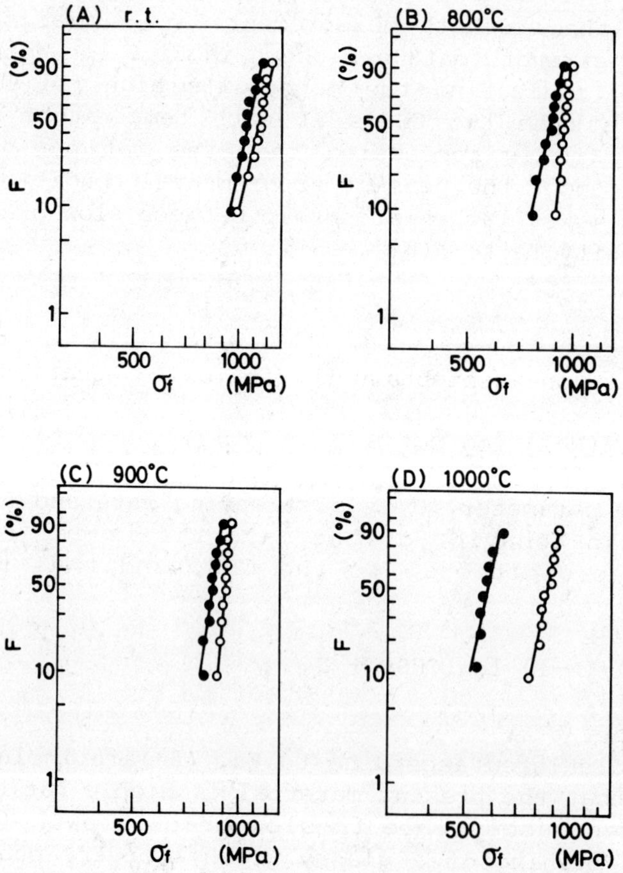

Fig. 1. Weibull plots of strength data at (A) room temperature, (B) 800°C, (C) 900°C and (D) 1000°C. Strength was measured at two CHS's, 0.5 (o) and 0.001 (●) mm/min.

Fatigue Behavior at Room Temperature

The static fatigue behavior is shown in Fig. 2, and the results of the cyclic fatigue test under different loading conditions are shown in Fig. 3-5. As shown in the figures, the present material exhibits high resistance to fatigue under any loading conditions in accordance with the dynamic fatigue results. The N values calculated by the least square method for the data of failed specimens in Figs. 2-4 are 54, 85 and 81, respectively.

Assuming that the crack growth is given only by K_I (Eq. (1)), it is possible to relate the static, cyclic and dynamic fatigue behavior[2,11]. Given the N value, the maximum stress for cyclic loading (σ_m) and the fracture stress for dynamic loading (σ_f), corresponding "effective" time to failure under the static loading with applied stress of σ_m and σ_f ($t_{c,eff}$ and $t_{d,eff}$) can be predicted by the following equations[2,11]:

$$t_{c,eff} = t_c/2\pi\int_0^{2\pi}[1/2(1+R+(1-R)\sin\phi)]^N d\phi \quad (4)$$

$$t_{d,eff} = t_d/(N+1) \quad (5)$$

where t_c and t_d are time to failure for cyclic and dynamic loading.

The effective time to failure was calculated using the N value of 54 and is plotted in Fig. 6 along with the static fatigue data. It is seen that there is no significant enhancement of the SCG rate due to cycling. Thus, it is considered that the mechanism for the fatigue is same for all the loading conditions and that the above prediction is valid under the present experimental conditions for the present material. Fracture surface observations revealed neither distinctive fracture origins nor SCG regions.

Several possible reasons may be considered for the discrepancy of the present results with literature data[3-5,8], i.e. no enhanced fatigue due to cycling in the present study. Materials are different in terms of compositions, porosity, microstructure, etc. Specimen configurations, preparation methods and testing conditions are also different. At present, it is not clear as to what is the most significant cause of the difference, and it should be clarified through further study in the future.

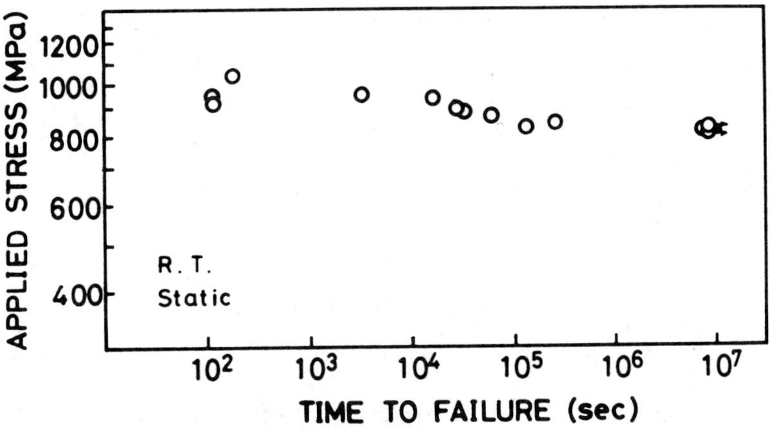

Fig. 2. Static fatigue at room temperature.

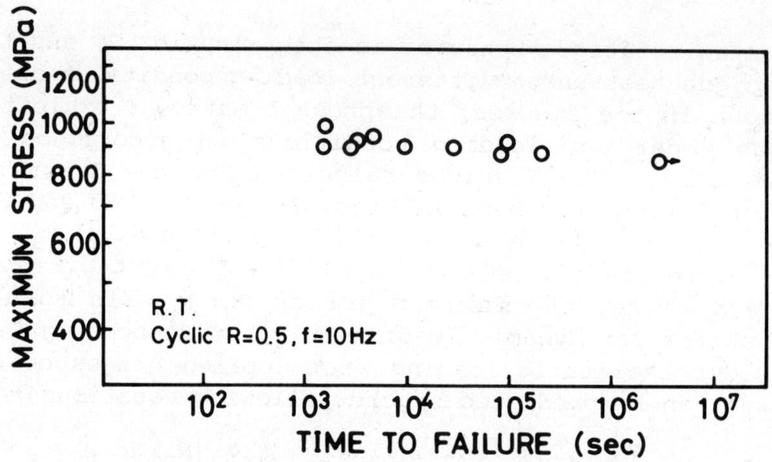

Fig. 3. Cyclic fatigue at room temperature. (R=0.5, f=10 Hz)

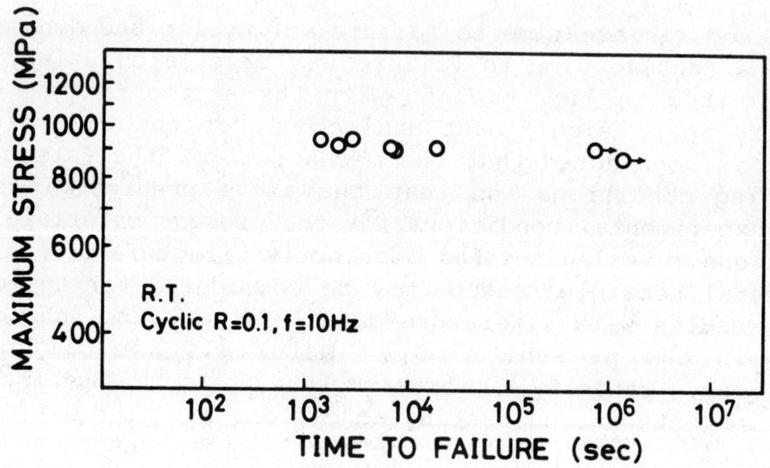

Fig. 4. Cyclic fatigue at room temperature. (R=0.1, f=10 Hz)

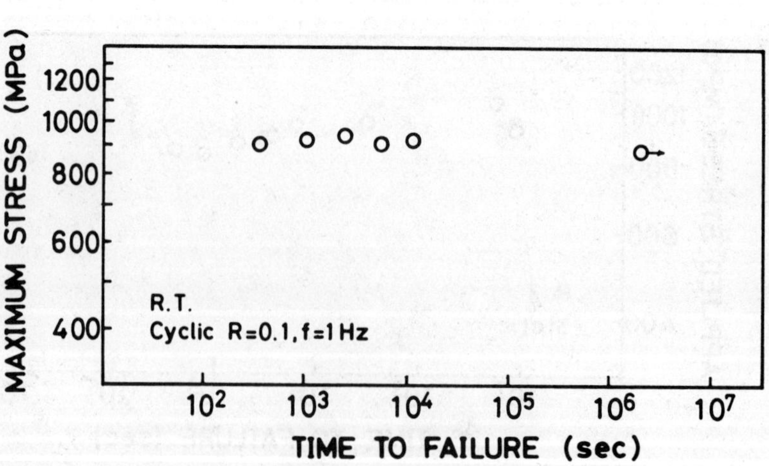

Fig. 5. Cyclic fatigue at room temperature. (R=0.1, f=1 Hz)

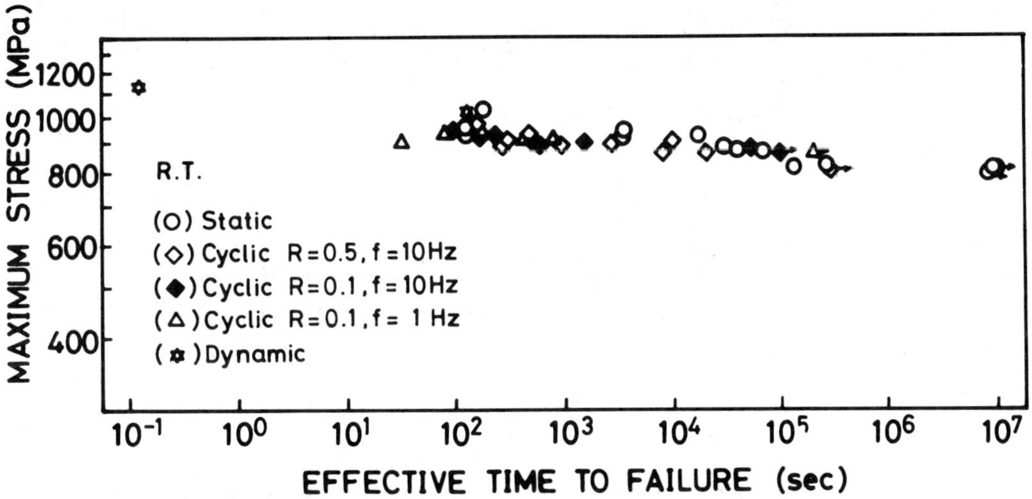

Fig. 6. Fatigue at room temperature. Effective time to failure for cyclic and dynamic tests were calculated according to Eqs. (4) and (5).

Fatigue Behavior at 1000°C

The results of the static fatigue test are shown in Fig. 7, and those of the cyclic test are shown in Figs. 8 and 9. It is seen from the figures that appreciable fatigue takes place at 1000°C. The N values calculated by the least square method for the data of failed specimens in Figs. 7-9 are 22, 39 and 30, respectively.

The effective time to failure was calculated in the same manner as in the previous section with the N value of 22, and all the data at 1000°C are presented in Fig. 10. Deviations from the static data are clearly seen for the cyclic data with f=10Hz, whereas predictions for the cyclic with f=1Hz and dynamic test data are in accordance with the static data. Fett et al.[2] have reported the same type of deviations for hot-pressed silicon nitride at 1200 °C. The loading conditions for their cyclic test were f=30Hz and R=0.2. They discussed the deviations with respect to different influences of the viscous glass phase on static and cyclic lifetime.

Frequency dependence indicated in this study could also be explained by the viscoelastic nature of the grain boundary phase. Fig. 11 shows the fracture surfaces of the differently loaded specimens which showed similar time to failure. Distinctive SCG regions are observed only for the static and cyclic with f=1Hz specimens (Fig. 11 (A),(B)). From these results it was concluded that, at high frequency, the grain boundary phase behaved more elastically and, thus, exhibited higher fatigue resistance.

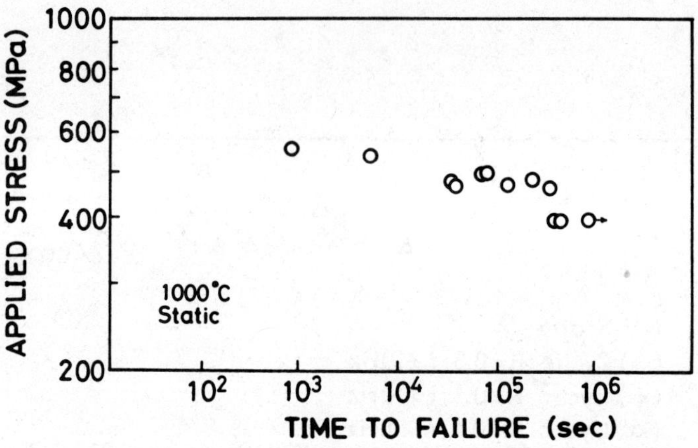

Fig. 7. Static fatigue at 1000°C.

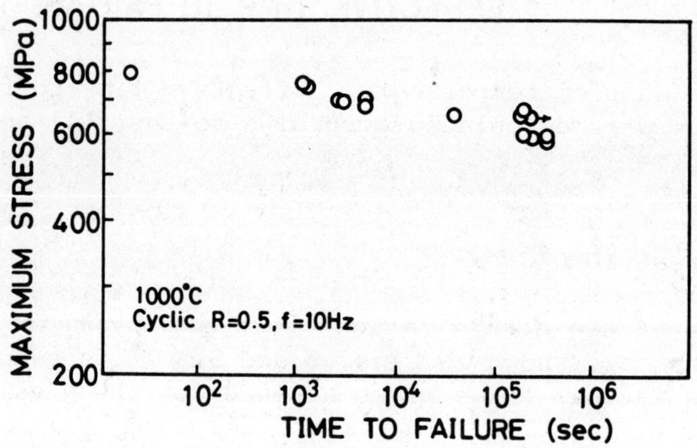

Fig. 8. Cyclic fatigue at 1000°C. (R=0.5, f=10 Hz)

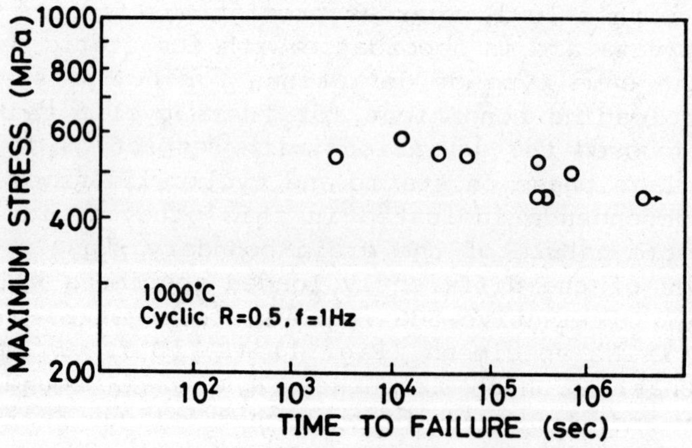

Fig. 9. Cyclic fatigue at 1000°C. (R=0.5, f=1 Hz)

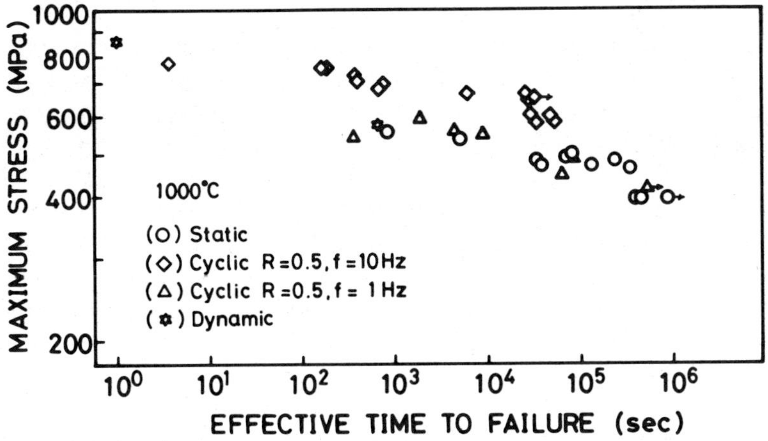

Fig. 10. Fatigue at 1000 °C. Effective time to failure for cyclic and dynamic tests were calculated according to Eqs. (4) and (5).

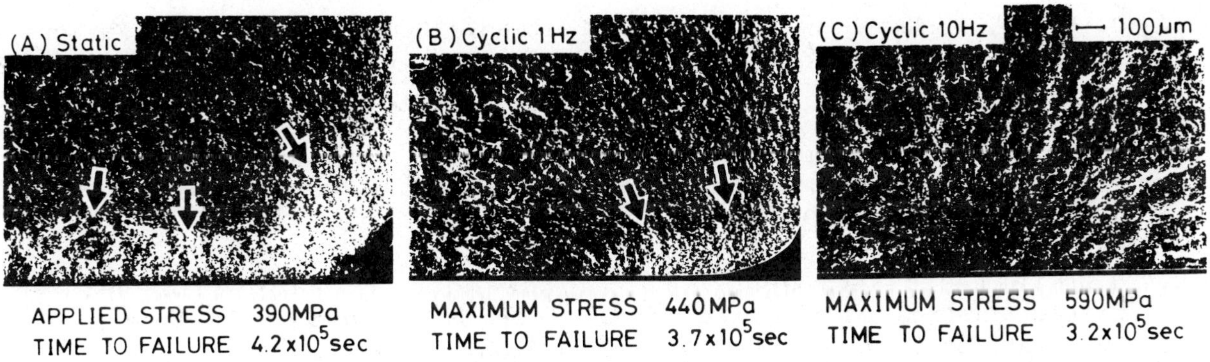

Fig. 11. Fracture surfaces of fatigue specimens tested at 1000 °C. (A) static and (B) cyclic with f=1 Hz specimens show distinctive SCG regions as indicated by arrows.

SUMMARY

Static, cyclic and dynamic fatigue behavior of the GPSSN was evaluated at several temperatures. The dynamic fatigue test results indicated that the material is highly fatigue resistant up to 900 °C but becomes susceptible to the slow crack growth at 1000 °C.

Correlation among the results of the three kinds of tests were studied at room temperature and at 1000 °C. At room temperature, it is suggested that all the results can be correlated in terms of a single SCG parameter and that there is no significant enhancement of the SCG due to cycling. On the other hand, the fatigue behavior becomes more complex at 1000 °C. Higher fatigue resistance for the cyclic loading with higher frequency is thought to be due to the viscoelastic nature of the glassy grain boundary phase.

REFERENCES

1. Y. Matsuo, Y. Hattori, Y. Katayama and I. Fukuura, "Cyclic Fatigue Behavior of Ceramics," pp.515-22 in Progress in Nitrogen Ceramics. Edited by F. L. Riley, Martinus Nijhoff Publishers, The Hague, The Netherlands, 1983.

2. T. Fett, G. Himsolt and D. Munz, "Cyclic Fatigue of Hot-Pressed Si_3N_4 at High Temperatures," Advanced Ceram. Mat., **1** [2] 179-184 (1986).

3. Y. Yamauchi, S. Sakai, M. Ito, T. Ohji, W. Kanematsu and S. Ito, "Fatigue Test for HP-Si_3N_4 Ceramics Indented with Knoop Pyramid," Yogyo-Kyokai-Shi, **94** [7] 631-35 (1986).

4. T. Kawakubo and K. Komeya, "Static and Cyclic Fatigue Behavior of a Sintered Silicon Nitride at Room Temperature," J. Am. Ceram. Soc., **70** [6] 400-405 (1987).

5. M. Masuda, T. Soma, M. Matsui and I. Oda, "Fatigue of Ceramics (Part 1)," Nippon Seramikksu Gakujutsu Ronbunshi, **96** [3] 277-83 (1988).

6. J. Chang, P. Khandelwal and P. W. Heitman, "Dynamic and Static Fatigue Behavior of Sintered Silicon Nitrides," Ceram. Eng. Sci. Proc., **8** [7-8] pp.766-777 (1987).

7. F. Wakai, H. Sakuramoto, S. Sakaguchi and Y. Matsuno, "Evaluation of Crack Propagation in Ceramics by Double-Torsion," J. Soc. Mat. Sci., Japan, **35** 898-903 (1986).

8. H. Kishimoto, A. Ueno and H. Kawamoto, "Crack Propagation Characteristics of Sintered Si_3N_4 under Static and Cylic Loads," J. Soc. Mat. Sci., Japan, **36** 1122-1127 (1987).

9. Y. Hattori, Y. Tajima, K. Yabuta, Y. Matsuo, M.Kawamura and T. Watanabe, "Gas Pressure Sintered Silicon Nitride Ceramics for Turbocharger Application," pp.165-172 in Ceramic Materials and Components for Engines. Edited by W. Bunk and H. Hausner, Deutshe Keramische Gesellschaft, 1986.

10. J. E. Ritter, J. P. Fahey and P. L. Duffy, "Slow-Stressing-Rate Testing of Ceramic Materials," Am. Ceram. Soc. Bull., **63** 1517-8 (1984).

11. R. W. Davidge, Mechanical Behaviour of Ceramics. Cambridge University Press, London, 1979; p.144.

ROLE OF OXIDATION IN CREEP AND HIGH TEMPERATURE FAILURE OF SILICON NITRIDE

O. Van Der Biest*, C. Weber+ and L. A. Garguet+.
* Department MTM, K.U. Leuven, Belgium
+ C.E.E. Joint Research Center, Petten, The Netherlands.

ABSTRACT

Creep and stress rupture behaviour of a HPSN (9 wt % Y_2O_3) was studied in air and nitrogen at 1400°C. This material has excellent creep and static fatigue resistance in comparison with other silicon nitrides. The intergranular phase consists of a crystalline silicon-yttrium-oxynitride with apatite structure. Cavitation at triple grain junctions is not observed after testing. Steady-state creep rates are only marginally influenced by the testing environment or by pre-oxidation, whereas lifetime is strongly influenced by these factors. At a stress of about 300 MPa, a static fatigue limit exists under inert conditions. Under oxidizing conditions, lifetime may be curtailed by formation of oxidation pits. Formation of these pits during pre-oxidation further reduces life. After lifetimes longer than 10^6 secs in air, another crack initiation mechanism appears which is related to to formation of a porous layer underneath the oxide scale. The implications of the observations with respect to in-service behaviour and design of improved materials is discussed.

INTRODUCTION

Previous work on high temperature behaviour of silicon nitride based ceramics has shown that oxidation may play an important role : e.g. Tighe and Wiederhorn (1) showed for a silicon nitride hot pressed with MgO that short term oxidation was beneficial for strength due to the healing of machining flaws, whereas after longer times high temperature strength was reduced due to pitting. Various authors (2,3) have demonstrated that creep resistance may be dramatically improved by a pre-oxidation treatment, by effectively removing creep enhancing impurities in intergranular regions.

The effect of oxidation on the overall failure process has not yet been studied in great detail. Environment may interact with the crack nucleation process as well as with crack growth. It has been suggested (4,5) that oxidation pits, in addition to their role as strength limiting defects, may be effective as crack nuclei during rupture at the relatively low stresses where components are expected to operate, but experimental evidence is not widespread. Subcritical crack growth (SCG) is often attributed to creep processes at the crack tip in silicon nitrides (6). Hence if SCG is lifetime determining, then clearly oxidation should have a beneficial influence on it. This has been shown for a monophase sialon by Lewis et al (2).

The purpose of the present work is to examine the role of oxidation in the failure processes of a silicon nitride hot pressed with Y_2O_3, which has excellent high temperature strength and creep resistance. Because of the potential importance of creep in crack growth as well as crack nucleation, the effect of oxidation on this process is also explicitly studied.

EXPERIMENTAL

Bar shaped specimens, with dimensions 3.5 x 4.5 x 50 mm^3 were cut from large hot pressed disks (diameter 180 mm, thickness 15 mm) so that the direction of the applied load on the bars was parallel with the hot pressing direction. The discs were hot pressed in consecutive runs from the same powder batch. It concerns an experimental batch, produced on request in 1984 by Feldmuehle A.G., Plochingen, Germany.
After hot pressing to near theoretical density, the composition of the discs was determined in weight percent as follows : Y_2O_3 : 9.0; MgO : 0.01; Fe_2O_3 : 1.0; ZrO_2 : 0.01; Al_2O_3 : 0.12; CaO : 0.05; balance Si_3N_4. The following phases were identified by X-ray diffraction : β-Si_3N_4, an apatite type phase $Y_{10}(SiO_4)_6N_2$ and $FeSi_2$.
Creep and stress rupture tests were carried out using a four point bending technique, with inner and outer span lengths of 20 mm and 40 mm respectively. Deformation of the sample with respect to the outer bearings is measured by a feeler rod connected to a high resolution transducer. The sample within a silicon carbide bending fixture, the extensometer and the pneumatic loading system with loadcell are all contained within a sealed alumina tube through which any gaseous environment can flow which is compatible with the construction materials. In the present experiments air and high purity nitrogen (99.99 % with less than 10 ppm O_2 and H_2O) are used. The pneumatic loading system with closed loop control electronics maintains a load which is constant within 2N.

RESULTS AND DISCUSSION

Microstructure before Testing

The structure consists of an interlocking mixture of elongated β-Si_3N_4 needles. A crystalline oxynitride phase with apatite structure is present as an intergranular phase. Examinations with TEM using high resolution lattice imaging techniques, failed to demonstrate the presence of amorphous phases at grain boundaries or at triple grain junctions.

Strength

The mean strength (10 samples) at room temperature is 643 MPa and is dominated mainly by machining flaws. This strength level is maintained to at least 1200°C. At 1400°C a mean strength of 500 MPa (9 samples) was determined.

Creep

Creep data obtained at 1400°C for a range of stresses are summarized in Figure 1. Data points represent stationary creep rates for individual samples determined in the conventional way (7) from the creep curves with a duration of at least 150 hrs. These curves are characterized by an initial transient response which may last up to 100 hrs at 1400°C.
Using the approach of Chuang (8), it has been shown (9) that the stress exponent in tension is approximately equal to the stress exponent in compression and is close to 1 as is evident also from Figure 1. From the variation of stationary creep rates with temperature over the range 1350°C to 1500°C an activation energy of about 650 kJ/mol has been determined.
The microstructure of crept samples is nearly indistinguishable from that of the as received material. General fine scale cavitation has not been observed. In view of the observed stress exponent and the small grain size, grain boundary transport is thought to be the main rate limiting creep mechanism (Coble creep).
Creep rates in nitrogen are also presented in Figure 1. Results show that the stationary creep rates are similar to those in air with a stress exponent of about 1.
A limited number of samples was pre-oxidized for 150 hrs at 1500°C and the oxide scale was removed before creep testing.The results (Figure 1) show that the stationary creep rates are not substantially changed by the pre-oxidation treatment.
On the basis of previous work on yttria densified silicon nitride (3), one would expect that creep rates are higher in inert environment, since any diffusion enhancing impurities would be retained in the intergranular region. Conversely, one would expect that creep rates are lower after pre-oxidation. The results presented here (Figure 1) do not confirm this prediction. A possible explanation for the behaviour of the present material may be that the yttrium-silicon apatite phase, which in view of our results is clearly very creep resistant, may have a higher solubility for some of the creep enhancing impurities such as CaO. There have been some suggestions for this in the literature (10). In fact this higher solubility of the apatite for impurities such as CaO, may also be the reason why amorphous phases which wet grain boundaries are not observed here.

Stress Rupture Tests

All tests were carried out at 1400°C, which is a target operating temperature for many silicon nitride components in heat engines. Lifetimes of samples in air and in nitrogen are presented in Figure 2. For the purpose of comparison with other materials, one may derive a value for the SCG parameter N of about 30 from the data in air by fitting a straight line to all data. This indicates that the present material has a higher static fatigue resistance than earlier hot pressed silicon nitrides (6).
In the procedure used to determine N, it is assumed that crack growth from a constant pre-existing flaw distribution determines lifetime. This is however a gross oversimplification in the present case. The data appear to indicate that at least two different mechanisms govern lifetime.

At high stresses (≈ 350 MPa) and correspondingly short lifetimes, testing environment does not have an influence. In this regime lifetime is indeed governed by growth of cracks from pre-existing flaws.

In contrast at lower stresses, samples tested in nitrogen are much longer lived than those in air. In particular at 300 MPa, most samples did not fail in nitrogen within the allowed time and so it appears that under inert conditions, a static fatigue limit exists, below which samples do not fail. Failure in nitrogen has been associated with those tests, where the ends of the samples had been attacked by an active oxidation mechanism. In these tests failure also occured towards the extremities of the sample and not in the center, which, by being shielded from the gasflow, remained free of corrosion. In air, the lifetimes of samples are controlled by the nucleation of cracks due to oxidation damage at or near the surface as will be detailed below.

A very wide spread in lifetimes is observed for samples which have been pre-oxidized at 1400°C for 500 hrs prior to loading (Figure 3).

Failure Mechanisms

Detailed metallography was used to distinguish some of the operating mechanisms. Optical microscopy on polished longitudinal cross sections of the samples showed small cracks emanating from the tensile edge of samples tested in air, whereas samples tested in nitrogen hardly showed any cracks at all.

As a means of quantifying the level of damage, the total crack length emanating from the tensile surface is plotted as function of time under load, for different "as machined" samples tested in nitrogen and air (Figure 4). Substantial damage starts to develop after about 10^6 secs under load in air. In contrast, only some isolated cracks, which are barely recognisable as such, are noted for samples tested in nitrogen.

Figure 5 compares the tensile face of specimens, tested in air and in nitrogen, whereby the superficial oxide layer on the former has been carefully polished away down to the oxide-matrix interface. Extensive cracking is only observed on the sample tested in air. Figure 6 shows a similarly prepared surface for a pre-oxidized sample which had seen a nominal (elastic) tensile stress of 275 MPa for 483 hrs. Some cracks emanate clearly from deep pits formed on the surface by oxidation. Apart from these, there are also other cracks present, which have a much higher density and for which a unique starting point cannot be identified. Comparison of samples with different loading times shows that the latter crack type is not observed for times less than 10^6 secs. On the other hand, cracks from oxidation pits may occur much earlier in life. The pits develop also during oxidation without load. Pre-oxidized samples which had failed from oxidation pits are marked in Figure 3 and it is clear that this can severely curtail their lifetime. Pre-oxidized samples which did not develop oxidation pits had lifetimes comparable with the "as machined" samples.

The origin of the oxidation pits, has not been unequivocally established. Both size and frequency of occurence of pits are variable. As already stated above, not all samples developed them. Most likely they are processing related, but the exact cause is not yet clear. For silicon nitrides densified with MgO, Lange [11] has established that iron rich inclusions can cause pitting, but for yttria densified silicon nitride such unique correlation could not be established. The iron content of the present material is rather high and iron rich particles, mainly as iron silicides, are dispersed throughout the microstructure. If these iron rich inclusions are the cause one would expect that the pits occur in each sample. The actual cause is still being investigated using a variety of approaches.

Some samples which had been pre-oxidized at 1500°C for 150 hrs and whose oxide scale had been removed, appeared to show a much reduced tendency for pit formation during testing at 1400°C. However, due to the relatively low number of samples involved and the statistical nature of the process, it could not be firmly established whether pre-oxidation at very high temperature constitutes a possible remedial treatment against pit formation.

The second crack nucleation mechanism which appears on a much finer scale after loading for about 10^6 secs on all samples tested in air, may be related to the formation of subscale porosity during oxidation. Figure 7 compares the surface region in cross section for samples tested in air and nitrogen. Elongated silicon nitride grains can still be recognized just underneath the actual oxide layer (Figure 7a), but the yttria rich intergranular phase has diffused into the oxide scale and a porous subscale layer has developed. In such layers, cracking on a fine scale can develop as observed in Figures 5 and 6.

GENERAL DISCUSSION

The stress range explored in the present work has been dictated by the requirement that failure times are not too long. These stresses are rather high with respect to the design stresses for components which are more likely to be around 100 MPa. One does not know then whether the failure mechanisms encountered are applicable to the lower stresses encountered in practice and whether an extrapolation to lower stresses and longer times of our data is justified. This is of course a general problem in testing of materials for long term high temperature service.

Among the many schemes which are possible to extrapolate stress rupture data to lower stresses and longer times, the Monkman-Grant relation is often found to be obeyed for many materials and for a wide range of stresses. This relationship states that the product of failure time (t_f) and steady state creep rate ($\dot{\epsilon}_s$) is constant. This product is plotted for our data in the lower stress range and it is evident that $t_f\dot{\epsilon}_s$ is not constant but a function of applied stress as suggested in Figure 8.

For further analysis of a suitable lifetime prediction scheme, particular at the lower stresses of practical interest, one needs information on crack growth rates and the factors controlling it. This information is presently being obtained and will be published shortly.

Assuming that the oxidation pitting is indeed related to an impurity in the material and can be eliminated by processing, one should consider the fine scale cracking due to subscale porosity as the more fundamental failure mode since it is linked with the basic microstructure of the material studied. It is presently being checked whether this mechanism is also operative at lower stresses, which are more relevant to high temperature components.

Our observations would also indicate that when designing the microstructure of silicon nitrides densified with yttria, it would be advantageous to have as low a volume fraction of intergranular phase i.e. as low an yttria content as is compatible with adequate densification rates. The distribution of the intergranular phase should also be as fine as possible.

In view of the requirement for near-net-shape forming, such material could best be processed by hot isostatic pressing.

CONCLUSIONS

The following can be concluded from the experimental work with respect to the influence of oxidation on creep and stress rupture behaviour of the silicon nitride hot pressed with 9 wt% yttria :
The oxygen content of the environment or pre-oxidation treatment influence only marginally the deformation rates.
For relatively high stresses, lifetime is governed by crack growth from pre-existing flaws. Environment does not play a significant role during the short life of the sample in this regime.
At lower stresses and longer lifetimes, a relatively high fatigue limit may exist under inert conditions.
Under oxidizing conditions, lifetime may be curtailed by formation of oxidation pits. Formation of these pits during pre-oxidation further reduces life.
After lifetimes longer than 10^6 secs, another crack initiation mechanism appears, which is linked to formation of a porous layer underneath the oxide scale.

REFERENCES

(1) N.J. Tighe and S.M. Wiederhorn, Fracture Mechanics of Ceramics, Vol.5, R.C. Bradt et. al. Editors, Plenum, N.Y., p.403,1983.
(2) M.H. Lewis, C.R. Heath, S.M. Winder, R.J. Lumby, Deformation of Ceramic Materials II, Plenum, N.Y., p 605, 1984.
(3) F.F. Lange, G.I. Davis, H.C. Graham, J. Amer. Ceram. Soc. Com., Vol 66, C-98, 1983.
(4) A.G. Evans, B.J. Dalgleish, Creep and Fracture of Engineering Materials and Structures, Proc. Third Int Conf., Swansea, Bath Univ. Press, p929, 1987.
(5) G.D. Quinn and L. Swank, J. of the Am. Ceram. Soc. Com., vol 66, C31, 1983.
(6) G.D. Quinn, J.B. Quinn, Fracture Mechanics of Ceramics, Vol 6, Plenum, N.Y., p603, 1983.
(7) G.W. Hollenberg, G.R. Terwilliger, R.S. Gordon, J. Amer. Ceram. Soc., Vol 54, p196, 1971.
(8) T.J. Chuang, J. Mater. Sc., Vol 21, p165, 1986.
(9) O. Van Der Biest, S. Valkiers, L. Garguet, P. Tambuyser, I. Baele, Brit. Ceram. Proc., Vol 39, p33, (1988).
(10) A.Rae, D. Thompson, N. Pipkin, K. Jack, Special Ceramics 6, British Ceramics Society, Stoke-on-Trent, p347, 1975.
(11) F.F. Lange, J. Amer. Ceram. Soc. Com., Vol 61, C270, 1978.

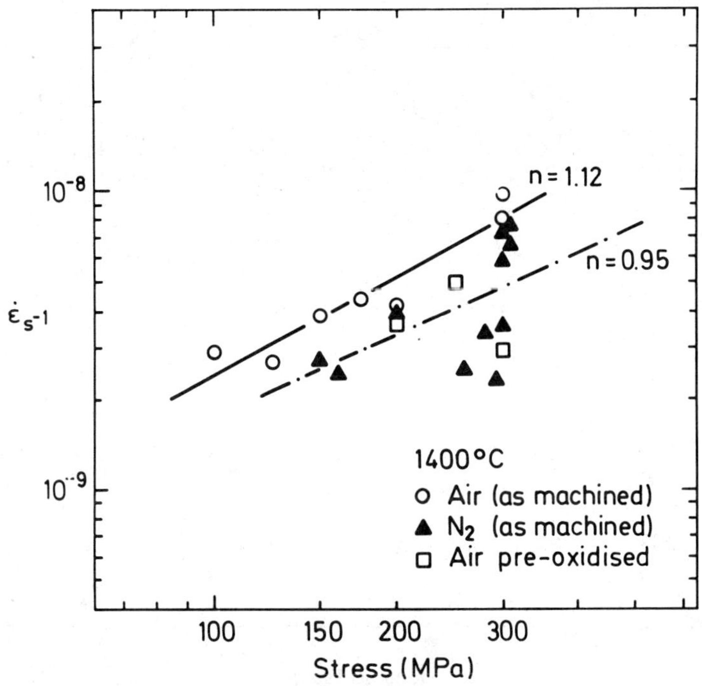

Figure 1 : Stationary creep rates at 1400 °C as function of stress in air, in nitrogen and after pre-oxidation for 150 hrs at 1500 °C. The solid line pertains to data in air. The interrupted line to data in nitrogen.

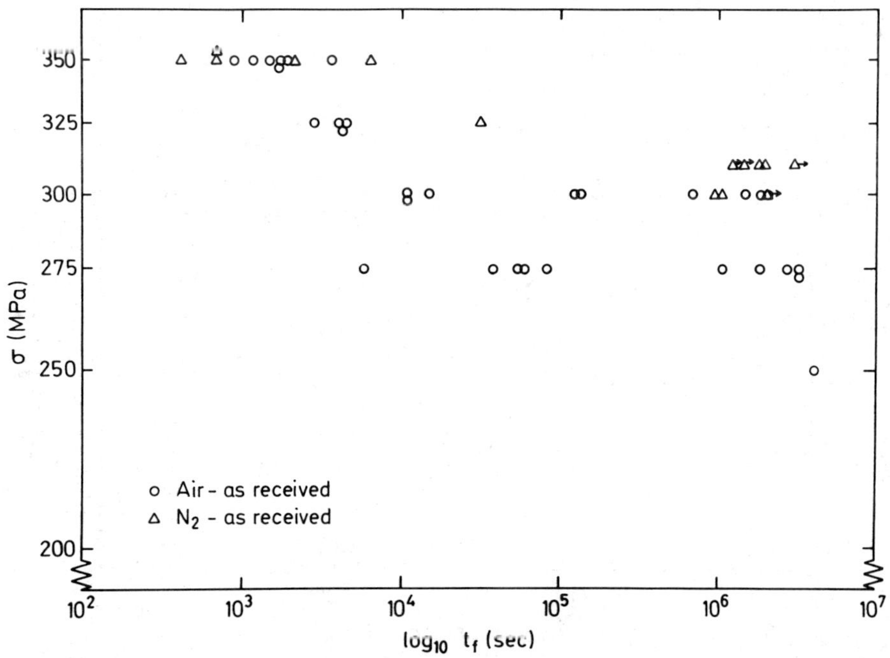

Figure 2 : Lifetimes at 1400 °C as function of stress in air and in nitrogen.

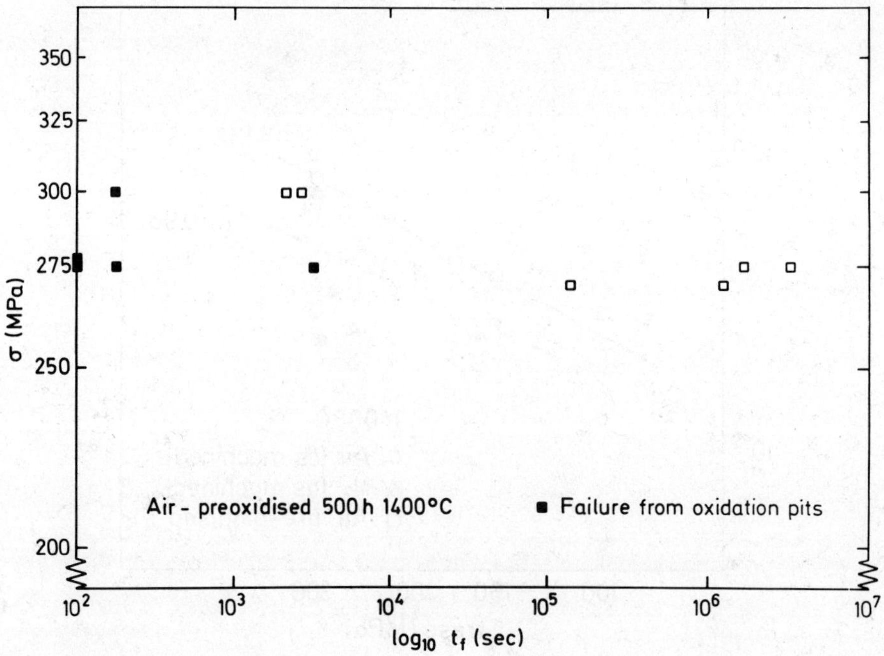

Figure 3 : Influence of pre-oxidation for 500 hrs at 1400 °C on lifetime at the same temperature.

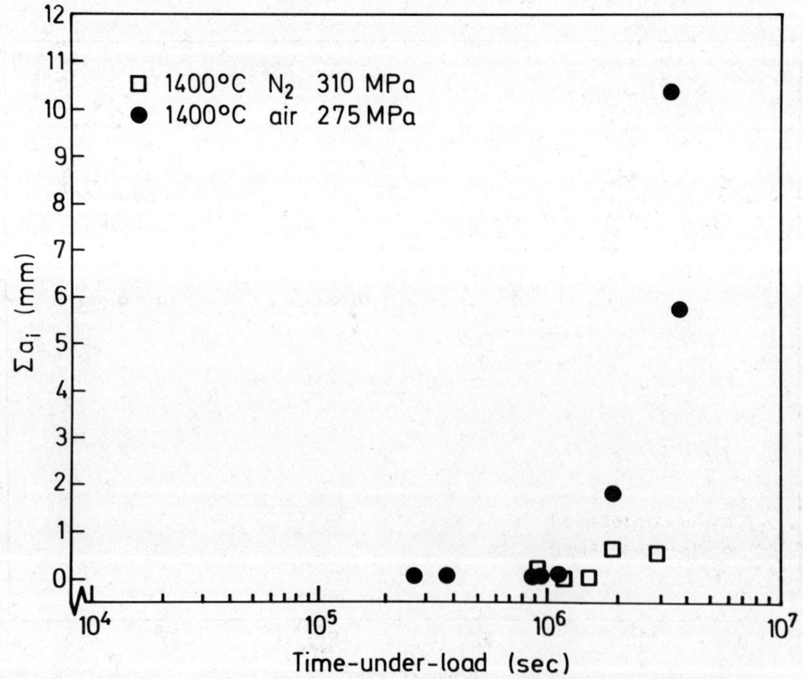

Figure 4 : Total crack length measured on longitudinal cros sections of "as machined" samples tested in air and nitrogen.

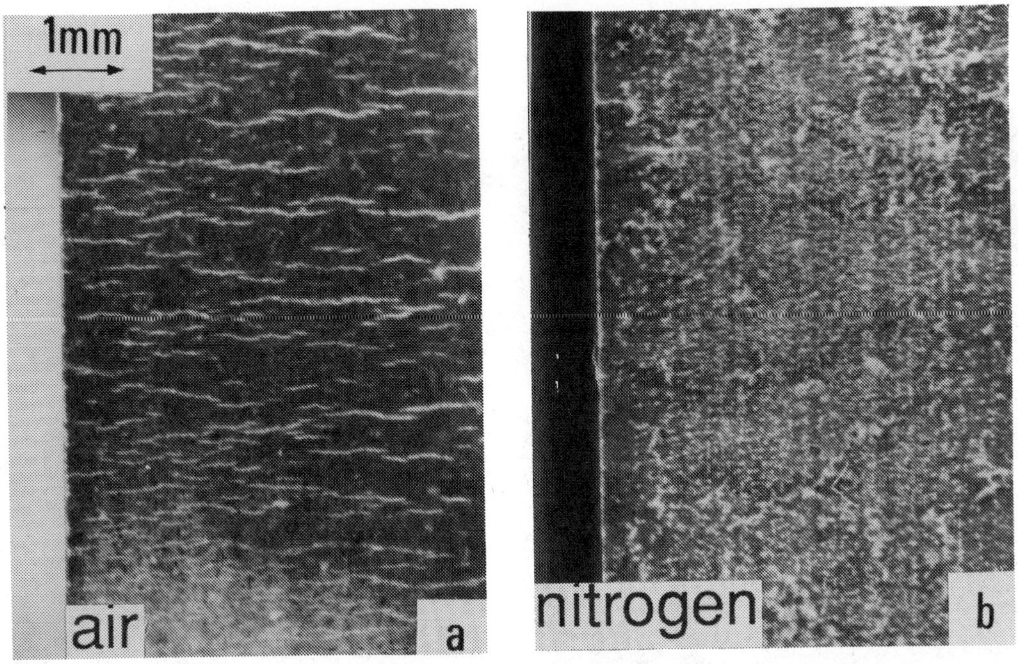

Figure 5 : Micographs of the tensile face of samples tested in air (a) and in nitrogen (b). Both were subjected to approximately the same stress for equal periods of time.

Figure 6 : Polished tensile face of sample just underneath the oxide scale (pre-oxidized 500 hrs at 1400 °C, loaded to failure after 483 hrs at 275 MPa). Some oxidation pits are marked by arrows.

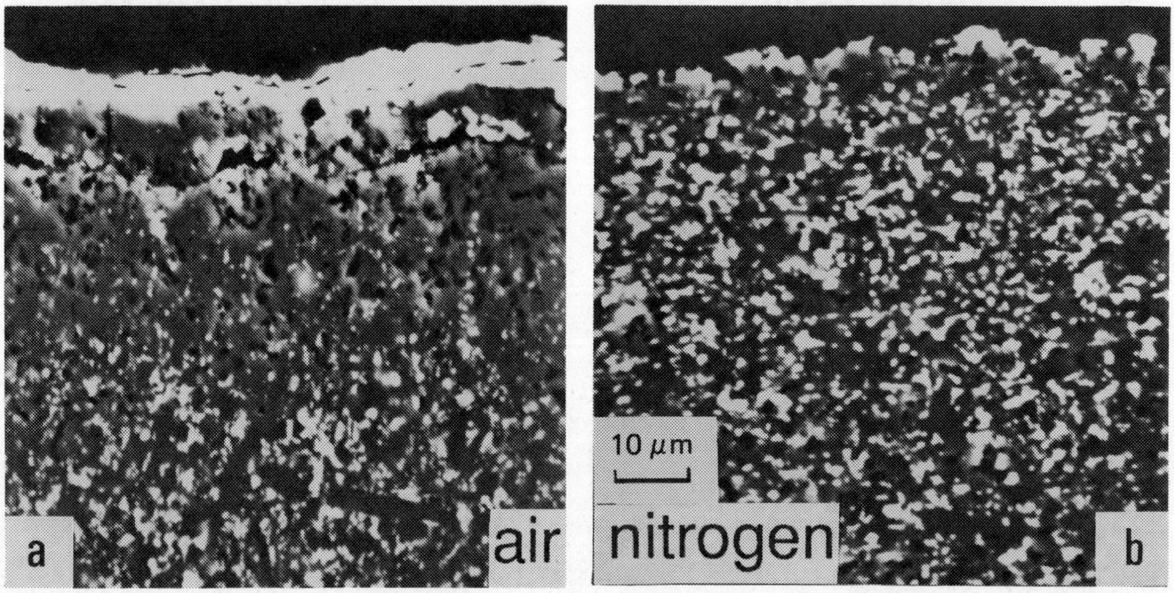

Figure 7 : SEM micrographs of surface scales in cross section (a) tested in air for 1.5 10^6 secs, (b) tested under nitrogen for 9 10^5 secs.

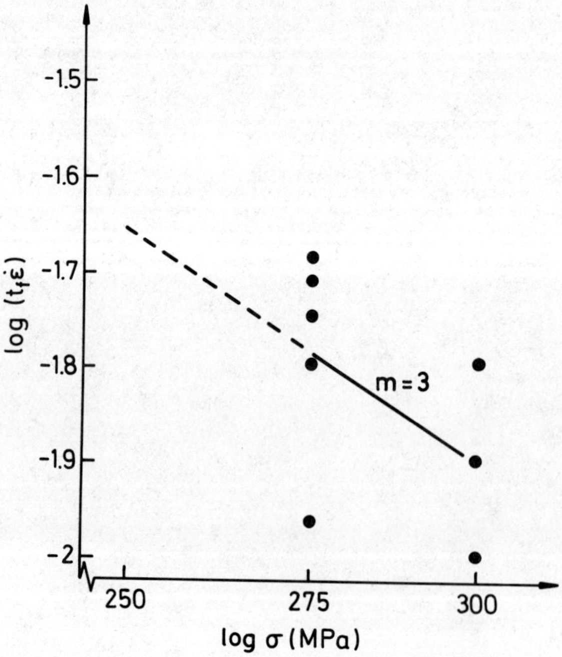

Figure 8 : Test of Monkman-Grant relation for "as machined" samples tested at 275 and 300 MPa in air.

GRAIN BOUNDARY TAILORING OF HIGH PERFORMANCE NITRIDE CERAMICS AND THEIR CREEP PROPERTY STUDIES

Y. R. Xu, T. S. Yen and X. R. Fu
Shanghai Institute of Ceramics, Academia Sinica
Shanghai, China

ABSTRACT

This investigation has attempted to design a grain boundary phase with yttria and lanthana additives which are highly refractory in the as-fired state and can be recrystallized to enhance the high temperature mechanical properties up to 1300-1400°C. High temperature fracture toughness and creep properties of these materials were studied, and the effect of alumina additions was evaluated.

INTRODUCTION

There are generally two main routes to develop silicon nitride ceramics for high temperature applications. The first is to design a highly refractory glassy grain boundary phase and the second is to recrystallize the grain boundary phase into crystalline phases that possess high melting points. Using Y_2O_3 plus La_2O_3 as complex additives may play a dual role in these two directions. Comparing with yttria plus alumina, as has already been used in some commercial products(1-4), the lowest eutectic temperature of the grain boundary phase can be raised by about 200°C when using Y_2O_3 plus La_2O_3(5-8). Moreover, a refractory $LaYO_3$ phase was developed at the grain boundaries by proper heat treatment which further enhances the high temperature property of the material. For comparison, the mechanical and creep properties of materials with additional alumina additives were also studied.

EXPERIMENTAL

Two types of raw materials were used for sample fabrication. One was the UBE powder from Japan which is reported to be a pyrolysis product from $Si(NH)_2$. The other was synthesized in the laboratory by nitridation of silicon powder and further ground by WC grinding media. Both powders have similar chemical compositions except the latter containing higher Fe_2O_3 and Al_2O_3 contents as impurities. The UBE powder is also much finer and more uniform although both are submicron in size (Table 1 and Fig. 1a and 1b).

The other additives used are 99.99% pure chemical reagents. Compositions of specimens were designated as shown in Table 2.
Specimens were prepared by hot pressing at pressures 25-30 MPa, and YL-1 can essentially attain full density at 1750°C for 60 min. or at 1800°C for 30 min. holding. The final conditions selected for fabricating YL-1 specimens were 1800°C and 45 min. holding, while for the rest with additional alumina were 1750°C and 60 min..

Table 1 Comparison of Silicon Nitride Powders Used

Powder	Chemical Composition (wt%)						α Content	Specific Surface
	Si_t	Si_f	N	O	Fe_2O_3	Al_2O_3	(%)	Area (m^2/g)
UBE	59.63	0.60	38.68	1.26	0.07	0.08	93	10.6
SN-II	59.58	0.50	38.57	1.31	0.85	0.61	95	2.4

Table 2 Compositions of Specimens

Specimen	Composition (mol%)				
	Si_3N_4	La_2O_3	Y_2O_3	SiO_2	Al_2O_3*
YL-1**	89.4	4.9	3.0	2.7	0
YLA-1	89.4	4.9	3.0	2.7	1.0
YLA-2	89.4	4.9	3.0	2.7	2.0
YLA-4	89.4	4.9	3.0	2.7	4.0
YLA-6	89.4	4.9	3.0	2.7	6.0

* Al_2O_3 was added in addition.

** YL-1 specimens were fabricated with both UBE & SN-II starting powders. The rest were with UBE powder only.

Mechanical strength and K_{IC} of these specimens at different temperatures were determined by normal procedures as described in detail in (9). The creep properties of these materials were intensively studied at temperature ranges from 1170°C to 1400°C and at stress ranges from 200 MPa to 300 MPa. Microstructural studies of specimens before and after creep tests were carefully pursued by transmission electron microscopy.

RESULTS AND DISCUSSION

The phase compositions of the as-processed specimens as studied by XRD are composed mainly of β-Si_3N_4 with minor or trace amounts of α-phase. All the oxides components turn over into a glassy phase situated at the grain boundaries. The microstructures of YL-1 and YLA-4 observed by TEM are shown in Fig. 2a and 2b. As shown by previous investigators(10-13), Si_3N_4 has a considerable solubility in the liquid phase at high temperatures. For nitride ceramics with yttria plus alumina additives, the ranges of solubility of N expressed as e/o (equivalent percent) varied from 15 e/o(12) to 33 e/o(13).

If we assume the nitrogen solubility in the glassy matrix phase as 20 e/o, then the volume fractions of the glassy phase in the specimens can be calculated as shown in Table 3, taking the density of the nitrogen containing glass to be 4.7 g/cm^3 (14).

Table 3 Glassy Phase in the Different Specimens
(Assuming N Content = 20 e/o)

Specimen	Si_3N_4 Solubility in the Glassy Phase (wt%)	Volume Percentage of the Glassy Phase
YL-1	6.8	15.7
YLA-2	7.3	16.8
YLA-4	7.9	18.0
YLA-6	8.4	19.1

The compositions of the glassy phase of different specimens were determined by EDAX attachment of TEM and the representative results of which are listed in Table 4. The positions of determinations are labelled on Fig. 3a and 3b. Since the accuracies of EDAX measurements are about $\pm$ 10%, the values obtained for different metallic elements in the glassy phase correlate reasonably well with what would be expected from the nominal compositions. Si content in the glassy phase is deleted because of the background influence of Si_3N_4.

Table 4 Compositions of Glassy Phase in YL-1 and YLA-4 as Determined by EDAX

Specimen	Position of Determination	Composition of Metallic Elements in Glassy Phase (at.%)		
		La	Y	Al
YL-1	1	63.5	36.5	
	2	66.8	33.2	
	3	63.3	36.7	
YLA-4	1	37.0	27.8	35.2
	2	36.8	28.1	35.2
	3	36.8	27.5	35.7
	4	36.9	27.0	36.1

The changes of mechanical strength of three kinds of specimens with temperature are shown in Fig. 4. Two of them were fabricated with Y_2O_3 plus La_2O_3 additives (YL-1), and the third had an additional 1 mol% Al_2O_3 (YLA-1). It is encouraging that the first two can retain their temperature strength up to 1300°C and their bending strength values at 1400^oC still maintain values from 680 MPa to over 800 MPa. The difference between these two is essentially caused by the difference of the raw materials used. The UBE powder apparently helps to evolve a finer microstructure with less defect in the specimens obtained. While the YLA-1 specimens, even with only 1 mol% additional Al_2O_3, show a decrease in strength from 1000^oC upwards. The effect of lowering the softening temperature by introducing alumina into the grain boundary glassy phase is remarkable and is probably the main cause for the drop of strength at higher temperatures. While the increase in volume percentage of glassy phase by introducing alumina is of secondary importance.

The fracture toughness of this type of nitride material is encouraging. Comparing with a material containing only yttria plus alumina additives, YLA-1 has higher K_{IC} values at high temperatures (Fig. 5), so do YL-1 specimens, whose room temperature $K_{IC} = 5.85 \pm 1.0$ MPa $m^{1/2}$ and at 1300^oC, $K_{IC} = 9.14 \pm 1.42$ MPa $m^{1/2}$. It is not intended to discuss in detail, but the features shown at the fracture surface tested at 1300^oC by SEM (Fig. 6) are interesting. Numerous small equiaxed crystals can be seen dispersed in the matrix and EPMA analysis indicated that they are devitrified crystals containing La and Y. Dispersion hardening and crack deflection[15] are the two possible mechanisms coming into play for the enhanced fracture toughness of this type of Y_2O_3 plus La_2O_3 containing nitride ceramics.

Better performance by using Y_2O_3 plus La_2O_3 as complex additives (without alumina) to obtain a more refractory grain boundary phase can be further evidenced by studying their creep properties. The creep rates at high temperatures increase significantly with increasing amount of additional Al_2O_3 in the additives (Fig. 7). YL-1 crept less than 0.1% (total strain) at 1220^oC, 200 MPa for 100 hrs. Even at 1360^oC under stress of 260 MPa, the total creep strain over 100 hrs amounts to only 0.6%. These materials are extremely creep resistant at high temperatures, which should be an important feature for their performance.

The creep mechanism of specimens with or without additional Al_2O_3 also changes as shown by the power factor value relating creep rate with stress (Fig. 8). Microstructural observations of the post-crept specimens by TEM were carefully undertaken. It may be discussed in brief that the mechanism of creep for YL-1 having power factor $n = 1.5$ is mixed[16,17], with diffusional viscous flow as the controlling factor. Grain boundary sliding and cavitation formation operate as concurrent mechanisms, especially at high temperature and high stress. Figures 9a and 9b show cavitation formation at a triple point and along grain boundaries. When YL-1 material is heat-treated, the stress power factor n drops to close to 1 as can be seen from Fig. 10. As observed under TEM, $LaYO_3$ recrystallizes readily from the grain boundary glassy phase either during heat-treatment or high temperature creep testing. Fig. 11 shows a typical recrystallized grain boundary in stripped form after creep. Electron diffraction gives two sets of patterns, one belonging to those of Si_3N_4 and the other from $LaYO_3$ (Fig. 12). It is very likely that with the crystallization of $LaYO_3$, the grain boundary glassy phase will decrease drastically in absolute amount (comparing with that shown in Table 3). This causes the dominant creep mechanism to change to diffusional creep. The change in slope (see Fig. 10) at high stresses ($\sigma >$ 350 MPa at 1360°C) indicates that there must be new elements introduced into the creep mechanism. Previous investigators had also observed such phenomenon under high stresses[18,19]. They suggested that cavitation became the controlling mechanism. When we investigated the specimens by high resolution electron microscopy, dislocations can be quite frequently observed (say for YL-1 specimen crept under 500 MPa at 1300°C) especially in someway associated with crackes (Fig. 13). So the dislocation mechanism should not be ruled out as a factor under such environments. This phenomenon warrants further investigation and discussion.

The mechanism of creep for specimens with additional alumina additives can be more clearly attributed to grain boundary separation and cavitation formation (Fig. 14). And the power factor is very much dependent on alumina content and temperature (see Fig. 8).

CONCLUSION

β-Si_3N_4 materials processed with yttria plus lanthana additives possess a highly refractory grain boundary phase at the glassy state that results in good mechanical strength retention up to 1300°C. The bending strength at 1400°C remains at a value of 680 MPa or 850 MPa depending on the starting nitride powders used. $LaYO_3$ crystalline phase can be readily recrystallized at the grain boundaries by heat treatment. The recrystallized grain boundary phase can further enhance high temperature creep properties and very likely other mechanical properties as well.

REFERENCES

1. A. Tsuge and K. Nishida, Am. Ceram. Soc. Bull., 57, 424(1978).
2. Q.T. Chen, F.Y. Wu and D.S. Yan, J. Chinese Silicate Soc., 10, 129(1982).
3. M.H. Lewis and R.J. Lumby, Powder Metall., 26, 73(1983).
4. A.E. Pasto, J.T. Neil and C.L. Quackenbush, in "Ultrastructure Processing of Ceramics, Glasses, and Composites," Eds. Henry and Ulrich (Wiley, 1984), pp. 476-489.
5. E.M. Levin, C.R. Robbins and H.F. McMürdie, "Phase diagrams for Ceramists", The Am. Ceram. Soc. Inc., 1969 Supplement, Fig. 2586.
6. E.M. Levin et al., ibid, Fig. 4419.
7. M. Mitomo, F. Izumi, S. Horiuchi and Y. Matsui, J. Mats. Sci., 17, 2359 (1982).
8. L.J. Gauckler, H. Hohnke and T.Y. Tien, J. Am. Ceram. Soc., 63, 35(1980).
9. Y.R. Xu, Ph.D. Thesis, Shanghai Institute of Ceramics, Academia Sinica, (1987).

10. K.H. Jack, "Nitrogen Ceramic", Ed. by F.L. Riley, (Noordhoff Intl. Publishing, Leyden, 1977), pp.643.

11. K.R. Shillito, R.R. Wills and R.B. Bennett, J. Am. Ceram. Soc., 61, 537 (1978).

12. R.E. Loehman, J. Non-Cryst. Solids, 42, 433(1980).

13. D.R. Messier and A. Broz, J. Am. Ceram. Soc., 65, c-123(1982).

14. Research Report, Inst. of Inorganic Materials Research, Japan, No.32, 45(1982).

15. A.G. Evans, Advances in Ceramics, 12, 193(1984).

16. R.M. Arons, J.K. Tien, J. Mater. Sci., 15, 2046(1980).

17. F.F. Lange, B.I. Davis and D.R. Clarke, J. Mater. Sci., 15, 601(1980).

18. R. Kossowsky, D.G. Miller and E.S. Diat, J. Mater. Sci., 10, 983(1975).

19. F.F. Lange, "Progress in Nitrogen Ceramics", Ed. F.L. Riley, (Martinus Nijhoff Publishers, Boston, 1983), pp.467.

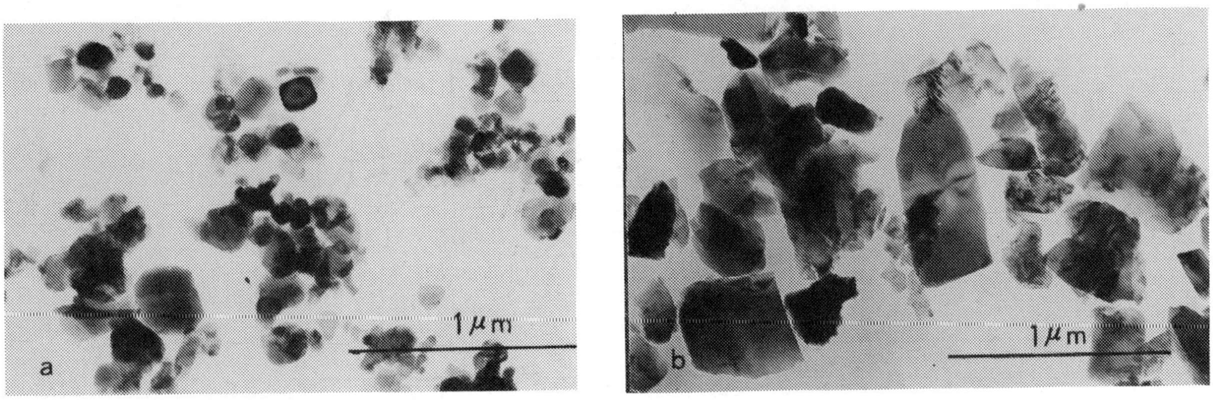

Figure 1 Morphology of Two Brands of Si_3N_4 Powders

1a UBE Powder, 1b SN-II Powder

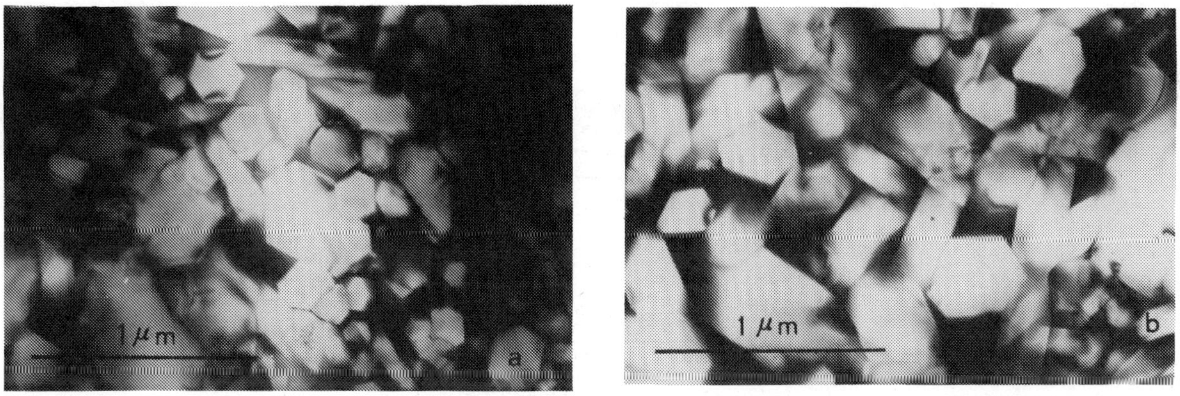

Figure 2 Microstructure of Silicon Nitride Specimens by TEM

2a YL-L Specimen, 2b YLA-4 Specimen

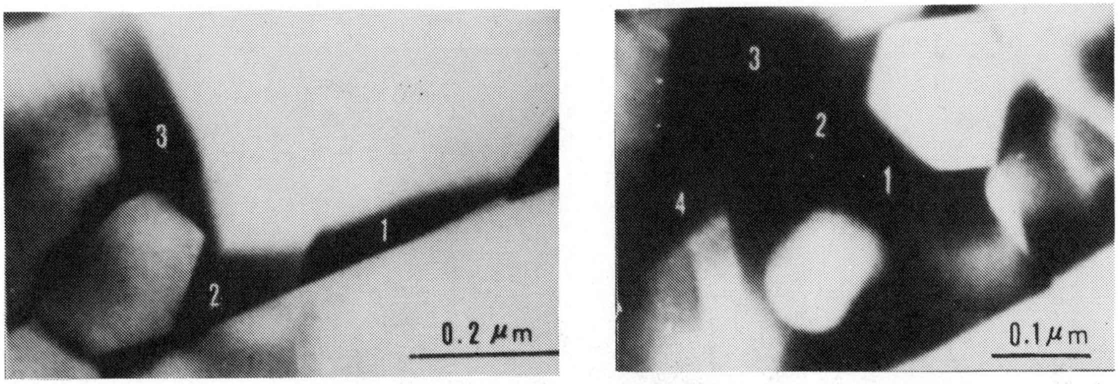

Figure 3 Bright Field Micrographs showing Positions where EDAX were Made

3a YL-1 specimen, 3b YLA-4 specimen

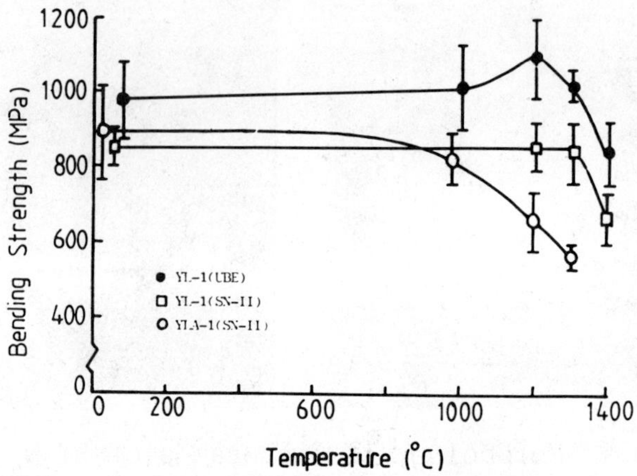

Figure 4 Transverse Strength vs Temperature for YL-1 and YLA-1

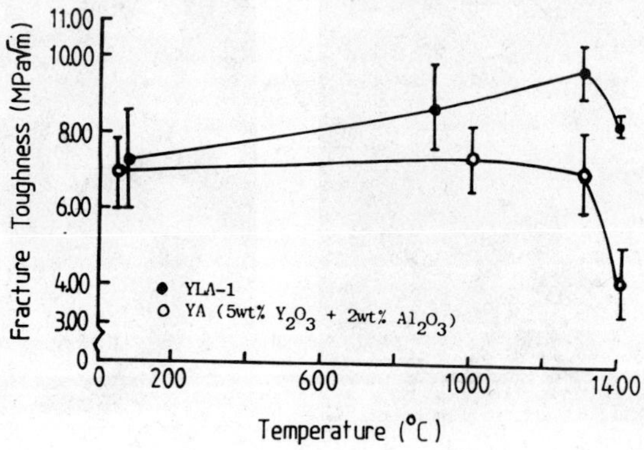

Figure 5 Fracture Toughness vs Temperature for YLA-1 and YA

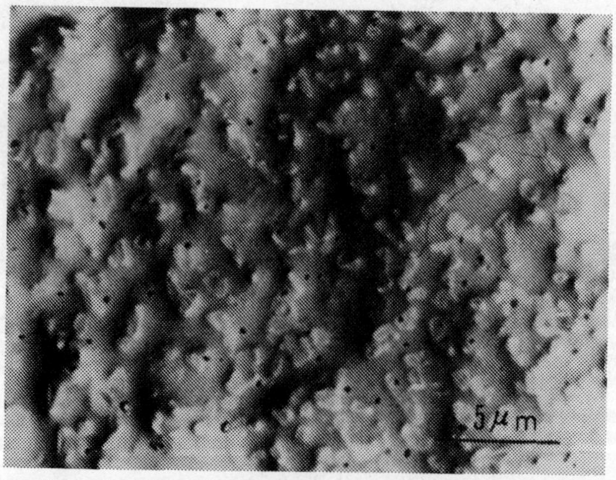

Figure 6 SEM Micrograph of the Fracture Surface of YLA-1 Tested at 1400°C

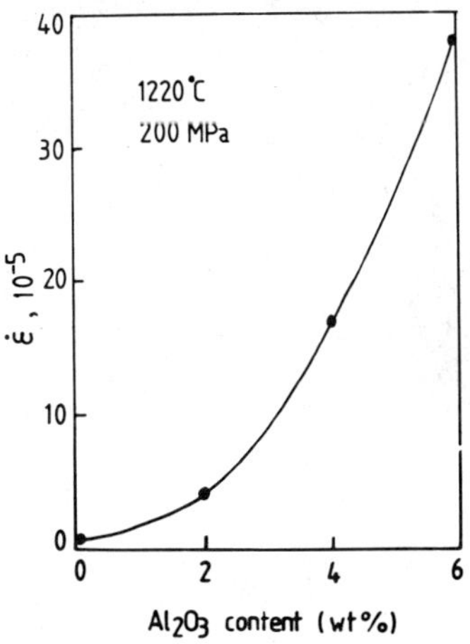

Creep rate vs. alumina content

Figure 7 Creep Rate vs Al_2O_3 Content in the Specimen

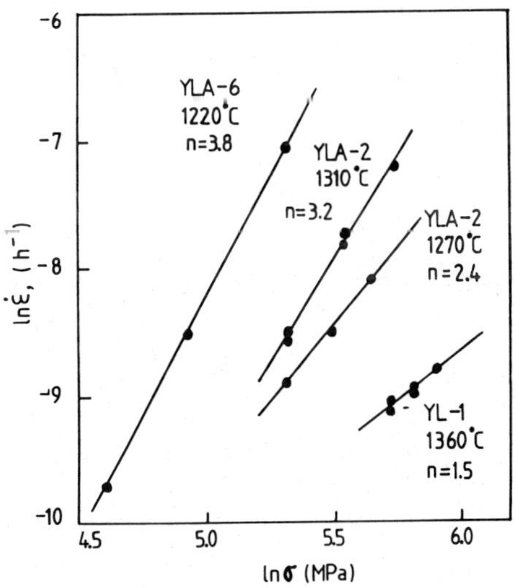

Creep rate as a function of applied stress

Figure 8 In $\dot{\varepsilon}$ vs In σ Relationship

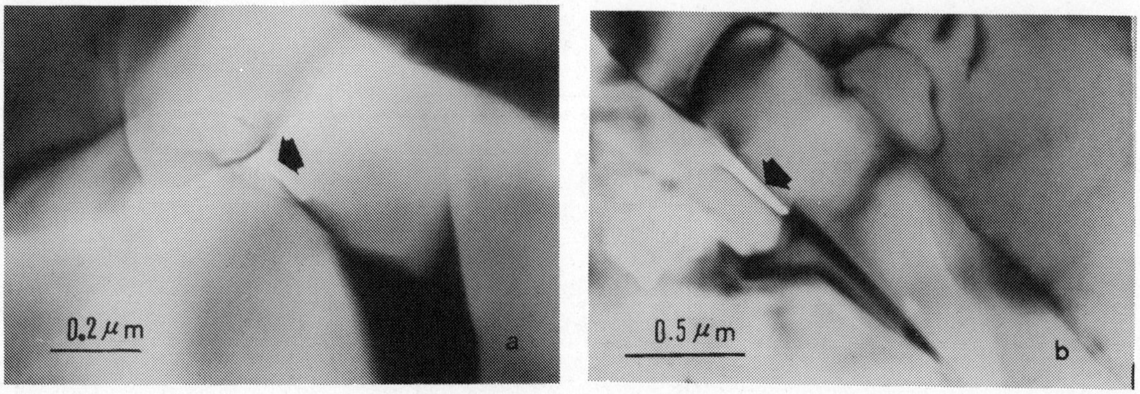

Figure 9(a,b) TEM Micrograph of YL-1 Crept at 1360°C for 38 hrs

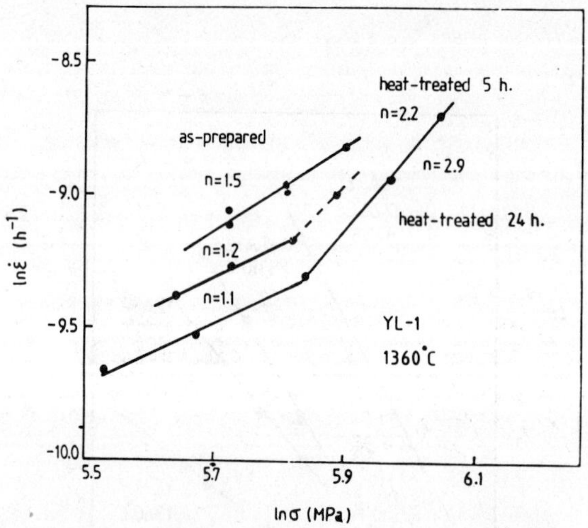

Figure 10 In $\dot{\varepsilon}$ vs In σ Relationship of Yl-1 after Heat Treatment

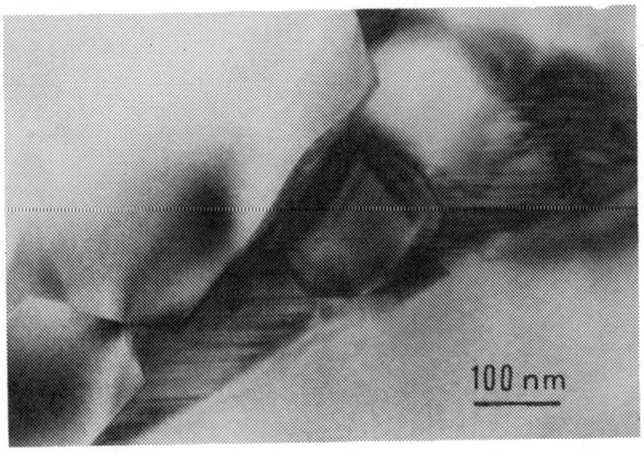

Figure 11 Recrystallized G.B. of YL-1 in Stripped Form after Creep

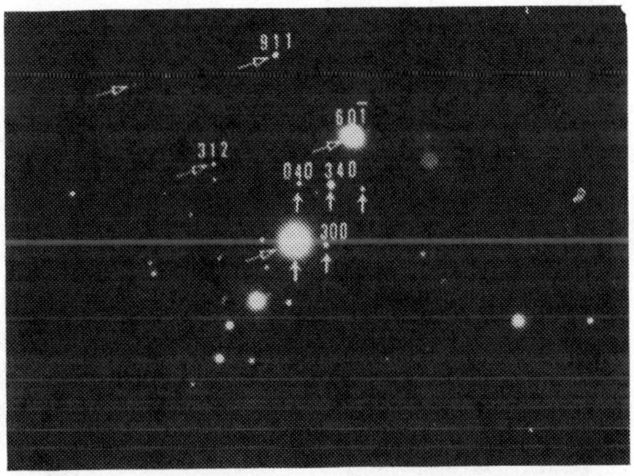

Figure 12 Electron Diffraction Patterns from the G.B. Region
(Hollow Arrows - β-Si_3N_4, Solid Arrows - $LaYO_3$)

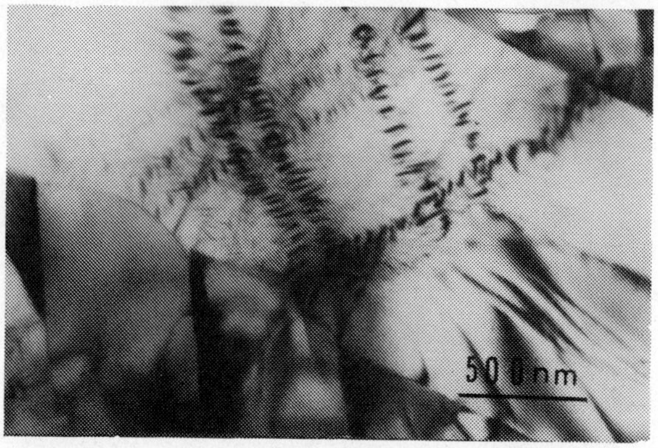

Figure 13 Complex Pattern of Dislocations Observed in YL-1 Specimen after Creep under 500 MPa at 1300°C

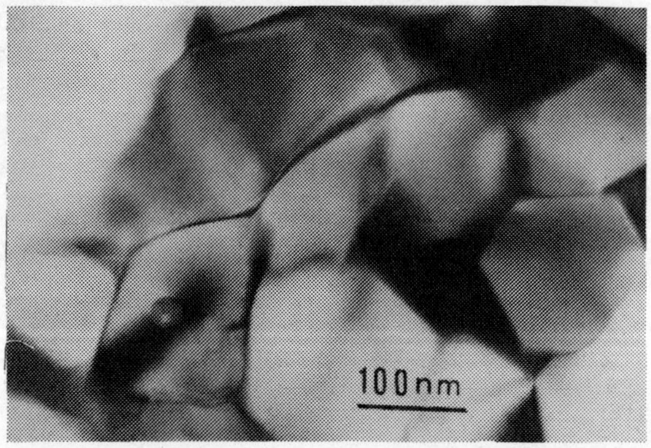

Figure 14 TEM Micrograph of YLA-6 Crept at 1220°C under 300 MPa

High Temperature Cavitation of HIP Silicon Nitride

Russell L. Yeckley and Kerry N. Siebein
Norton Company, Northboro, MA

ABSTRACT

Observations of cavitation in liquid phase sintered ceramics found that cavities nucleate and grow within the amorphous grain boundary. Failure is typically controlled by cavity growth through viscous flow mechanisms. Preliminary characterization of cavitation of hot isostatic pressed (HIP) silicon nitride with 4% yttria shows cavitation to occur primarily on two grain boundaries. The silicon nitride has a thin amorphous intergranular phase and yet analytical electron microscopy (AEM) indicates diffusional cavitation mechanisms are operative at 1370°C.

INTRODUCTION

Advances in encapsulation HIP technology have made possible commercialization of ceramics with complex shapes such as silicon nitride bearings. These improvements in the technology have renewed interest in the development of silicon nitride with low amounts of densification additive. The high temperature properties of the HIP densified silicon nitride are limited by the properties of the glassy intergranular phase. The glassy phase, in the form of a thin boundary film, softens at high temperatures and degrades the properties of the silicon nitride materials. Silicon nitride densified using magnesia additions is unacceptable for high temperature applications because the magnesia-silica glassy intergranular phase that forms softens at high temperatures [1]. The high temperature properties of silicon nitride were greatly improved by the addition of yttria as a densification aid due to the formation of a more refractory grain boundary phase [2]. A low volume fraction glassy phase also improves the high temperature properties, this is accomplished by using low amounts of additive and/or crystallizing the glassy phase [3,4].

The creep damage of silicon nitrides at elevated temperatures occurs in the glassy intergranular phase by cavity initiation, growth and coalescence to form cracks that eventually lead to failure [5-7]. Creep deformation occurs in silicon nitride materials by grain boundary sliding accommodated by diffusional creep and cavitational controlled mechanisms [8,9]. Diffusional creep involves the dissolution and redeposition of the solid phase through the liquid from regions of higher compression to those of lower, and is driven by a gradient in chemical potential [8,10]. Cavitational creep or viscous flow is the result of stresses in the grain boundaries that cause holes to nucleate in the glassy phase.

The cavities form in the grain boundaries and multigrain junctions and are completely contained within the glassy phase [5,6,11]. The cavities nucleate at inhomogenieties in the glassy phase such as impurities, vapor bubbles, dislocations and ledges [9]. The amount of cavitation increases as the viscosity of the glassy phase decreases and volume fraction of glassy phase increases [10].

The goal of this research is to develop a HIPed silicon nitride that will meet the requirements for turbine engines. Previous research on the creep deformation of silicon nitrides was performed on sintered and hot pressed materials containing various sintering aids [5,8,10,12]. This paper presents the preliminary results of an investigation into the deformation behavior of HIPed silicon nitride with 4% yttria.

EXPERIMENTAL PROCEDURE

The creep behavior of HIPed silicon nitride with 4% yttria, NTX154, was examined.

Flexure tests and stress rupture tests were performed to determine the creep behavior of the HIPed silicon nitrides. The flexure and stress rupture tests were conducted on 4 point quarter point fixtures with an outer span of 40 mm. The diamond machined flexure bar dimensions were 3mm by 4mm by 50mm. Flexure tests were performed at room temperature and 1370°C. The stress rupture test parameters were 1370°C and 300 MPa for 200 hours to simulate the life requirements for turbine engines.

Interrupted stress rupture tests were also performed the at 1370°C and 300 MPa; tests were suspended after 10, 50, 100 and 200 hours. The curvature of the tensile surface between the inner load points of the interrupted test bars was measured using a Tallysurf. The creep strain was calculated assuming that the curvature was an arc. The retained strength of the stress rupture bars was measured at 1370°C.

Bulk silicon nitride samples were prepared by diamond polishing for microstructural analysis. The samples were polished through 1 micron diamond paste and etched in molten KOH. Grain thicknesses were measured from SEM micrographs.

Specimen preparation for AEM was standard. Thin specimens were sectioned from the tensile surface of MOR bars with a diamond saw. Specimens were ground to 120 micron thickness and 3 mm disks were trepanned out using an ultrasonic disk cutter. The specimens were dimple ground to 10 micron thickness in the center and ion milled at 6keV and 12° tilt with argon ions. A thin layer of evaporated carbon was applied to the specimens to prevent charging.

A JEOL 200CX was used for the microstructural analysis and a Vacuum Generators HB5 STEM equipped with a Links windowless EDS detector was used for the quantitative microchemical analysis of the glassy intergranular phase. The measurements of the glassy phase composition were made in the multigrain junctions. The quantification of the EDS data was performed using the Cliff-Lorimer equation with adsorption corrections [13]. The k factors were determined experimentally using specimens of known composition. The EDS results are considered semi quantitative due to the error associated with the adsorption corrections and the k factor determination. Due to instrumental limitations, this technique was not capable of detecting low amounts of nitrogen. To confirm the light element analysis the oxygen concentrations were also calculated from stoichiometry using the intensity of the silicon and yttrium peaks.

EXPERIMENTAL RESULTS

a) Properties and Microstructure

The room temperature and 1370°C strength distribution of HIPed silicon nitride is shown in Figure 1. The strength distributions are similar; the strengths are controlled by surface damage. There was no evidence of slow crack growth on the fracture surfaces of the 1370°C test bars. Cracks were not observed on polished sections of the tensile surface and cross sections of the test bars.

The fine grain size of the HIPed silicon nitride is depicted in Figure 2. The grain thickness distribution is presented in Figure 3. The aspect ratio of the beta silicon nitride grains is approximately 3 to 4.

The as HIPed silicon nitride was examined using AEM. The material contains an amorphous intergranular phase in the form of a continuous grain boundary film. The typical width of the two grain boundaries is 30 nm. Figure 4 is a bright field image of the as HIPed material; the intergranular phase in this material is primarily amorphous. The composition of the glassy phase in the as HIPed material is presented in Table 1. Frequently the intergranular phase in the triple junctions and multigrain junctions was crystallized. Using selected area diffraction, the crystal structure of this intergranular phase was identified as triclinic alpha yttrium disilicate. The alpha yttrium disilicate is heavily twinned and the single crystals typically extend across several multigrain junctions. Figure 5 illustrates the morphology of an interconnected single crystal through the comparison of the bright field and dark field images.

b) Stress Rupture Behavior

The stress rupture behavior of silicon nitride with 4% yttria had two distinct behavior patterns shown in Figure 6. Some

Table 1
Semi-Quantitative Intergranular Phase Compositions
Determined by EDS

Material	Weight Percent		
	Si	Y	O
NTX154 (as HIPed)	15.1 ± 0.8	50.7 ± 5.1	34.3 ± 1.7
NTX154 (stress rupture)	16.5 ± 0.9	61.8 ± 6.2	21.7 ± 1.1

samples exhibited a short life, less than 10 hours at 300 MPa, while other samples survived 200 hours at 300 MPa at which point the stress rupture tests were suspended.

Interrupted stress rupture tests were run with strain measurements after the selected time interval followed by 1370°C fast fracture testing. The creep strain is plotted as a function of creep in Figure 7. The overall creep strain is very low and steady state conditions were not reached. The fast fracture test provided a relative comparison of the degree of accumulated damage. The retained strength was calculated assuming that the stress rupture bars were straight beams. The error associated with this assumption is less than 1.0 % for all samples. The results in Figure 8 indicate a slight decrease in the strengths of the stress rupture bars tested for 200 hours and that slight strengthening may occur up to fifty hours. Similar behavior has been reported with siliconized silicon carbide [14].

c) Cavitation

The 24 hour and 200 hour stress rupture samples were examined using AEM. Both specimens were cavitated. The cavity size and shape were similar in both specimens, but there were fewer cavities present in the sample tested for 24 hours. The cavity size distribution is presented in Table 2. Bright field micrographs of samples tested for 24 and 200 hours are presented in Figures 9 a and b respectively. The cavities are not contained within the thickness of the grain boundary channels. The cavitation is primarily along two grain junctions. The cavities appear to nucleate continuously on preferred grain boundaries or grain facets and are not homogeneously distributed. Cavity growth is very slow. A bright field image of the cavities in the specimen tested for 200 hours is presented in Figure 10; this micrograph illustrates shape and size of the cavities.

Large interconnected single crystals of alpha yttrium disilicate were present in some of the multigrain junctions after HIPing. Crystallization of alpha yttrium disilicate from the glassy

Table 2
Typical Cavity Size
Measurements made on the tensile surface of stress rupture bars tested at 1370 C, 300 MPa for 200 hours and 24 hours.

Time at 300 MPa (hrs)	Average Cavity length	Dimensions (nm) width
24	91 ± 33	35 ± 15
200	115 ± 87	49 ± 36

phase also occurred as a result of the stress rupture tests. The crystallites are twinned. In contrast to the crystallization occurring across several multigrain junctions in the HIP, crystallites form within isolated triple junctions during the stress rupture tests. A crystallite within a three grain junction of the sample tested for 200 hours is presented in Figure 11, the size of the crystallite is illustrated by comparison of the bright field and dark field images.

The composition of the glassy phase in a sample tested at 1370°C and 250 MPa for 200 hours is presented in Table 1. It is expected that the composition of this specimen is similar to the specimens tested at 1370°C and 300 MPa for 200 hours. The composition of the glassy phase in the stress rupture specimen contains less oxygen and more yttrium than that of the as HIPed material.

One sample had very low retained strength after 200 hours at 300 MPa. AEM analysis of this material revealed a high frequency of small inclusions in the grain boundaries, suspected to be silicon carbide. These inclusions appear to accelerate cavitation by raising the local stresses in the grain boundaries during grain boundary sliding. A cavitated grain boundary in the low strength sample is presented in Figure 12, the cavities are adjacent to the inclusion.

DISCUSSION

The goal of this work is to understand the creep behavior of HIPed silicon nitride, and to identify the microstructural mechanisms responsible for the short vs. long stress rupture life.

Detailed microstructural analysis was performed on the HIPed silicon nitride with 4% yttria. This material contains a low volume fraction of glassy intergranular phase due to low amount of additive, 4% yttria, and the crystallization of the glassy phase in the multigrain junctions during HIPing. The glassy phase in the as HIPed material forms a continuous boundary film. A twinned alpha yttrium disilicate structure crystallizes in the HIP, and the large crystals extend over many silicon nitride grains. The stress rupture specimens contain the same phases as the as HIPed material: beta silicon nitride, alpha yttrium disilicate, and a glassy intergranular phase. Additional alpha yttrium disilicate is crystallized during the stress rupture testing. The morphology of the newly crystallized alpha yttrium disilicate in the stress rupture samples is different from that in the as HIPed material. In the stress rupture samples, twinned single crystals of the alpha yttrium disilicate crystallize in isolated multi grain junctions.

The different morphologies of the alpha yttrium disilicate are due to the different crystallization conditions. Crystallization of the alpha yttrium disilicate during HIPing could begin above 1535°C as delta yttrium disilicate and transform to alpha on cooling. The yttrium disilicates have the following transformation temperatures [15]:

$$\alpha \xleftrightarrow{1225^\circ C} \beta \xleftrightarrow{1445^\circ C} \gamma \xleftrightarrow{1535^\circ C} \delta$$

The alpha yttrium disilicate crystallized during the 1370°C stress rupture tests transforms from the beta phase, which is the equilibrium phase at the test temperature. Twins form as the result of stress which produces highly localized strain in the lattice. The alpha yttrium disilicate crystallized in the HIP is more heavily twinned; this could be due to high HIPing pressure or the volume changes due to transformation of the delta to alpha phase. The composition of the glassy phase in the stress rupture bars is oxygen deficient compared to the glassy phase in the as HIPed material. The decrease in oxygen is attributed to the further crystallization of the alpha yttrium disilicate, which is oxygen rich, during the stress rupture tests.

The strength of the HIPed silicon nitride increased after short time under stress rupture as shown in Figure 8. Similar behavior was observed in siliconized silicon carbide [14]. The strength increase in the siliconized silicon carbide was attributed to blunting of machining flaws due to oxidation. The room temperature strength and 1370°C strength are controlled by surface flaws in the HIPed silicon nitride 4% yttria so blunting of machining flaws is expected to account for the strength increase.

The cavitation observations in the HIPed silicon nitride differ significantly from the cavitation behavior of liquid phase sintered (LPS) silicon nitride and LPS alumina [5,16,17]. Cavitation occurred primarily on two grain junctions in the HIPed silicon nitride as seen in Figure 3. The cavitation in hot pressed silicon nitride with 1% MgO was almost exclusively at multigrain junctions and triple junctions [5]. This work was performed at high temperatures where the intergranular phase has extremely low viscosity and cavitation occurs via viscous flowand solution reprecipitation mechanisms. But aluminas with 8 volume percent grain boundary phase tested in creep at 950°C to 1000°C formed crack like cavities [16,17]. Recent cavitation studies employing small angle neutron scattering (SANS) characterized cavitation in silicon carbide with a thin 20nm amorphous grain boundary [18]. The SANS results found that cavitation occurred within the grain boundary plane.

The cavitation behavior in LPS alumina and LPS silicon nitride support the cavitation models [5-7,11] based on the scenario: i) nucleation within the junction, ii) growth by viscous depletion to full facet sized cavities , iii) linking of cavities to form microcracks. The cavitation behavior of the HIPed silicon nitride does not fit the proposed models for LPS materials.

The cavities in as HIPed material presented in Figure 9 seem to have the equilibrium lens shape often observed in metals. The lens shaped cavities are typical of diffusion controlled cavitation. High temperature creep of polycrystalline alumina with no grain boundary phase results in cavities with equilibrium shaped, crack like and full facet morphologies [19]. There exists a transition from the equilibrium cavity shape to crack like cavities described by:

$$\frac{\sigma_1 b}{\gamma_s} = \frac{2.6}{f^*}\left(1 + \frac{1.2\Delta}{f^*}\right) \qquad (1)$$

where σ_1 is the local stress, 2b the cavity spacing, Δ is the ratio of surface diffusion to boundary diffusion, γ_s is the surface energy and f* the ratio of cavity length to cavity spacing [19]. The observed cavity morphologies and transition are consistent with models of diffusive cavity growth because when the surface diffusivity becomes too small to maintain uniform curvature the cavity shape becomes more crack like [19]. The crack like cavitation occurred in the alumina tested at 1750°C, HIPed silicon nitride may also exhibit crack like cavitation when tested at higher temperatures.

The dihedral angle α in Figure 13 is controlled by the ratio of γ_b, the boundary interfacial energy and γ_s, the grain energy according to the equation [20]:

$$\alpha = \cos\left(\gamma_b/2\gamma_s\right) \qquad (2)$$

The estimated angle in the HIPed silicon nitride is 37° which is much smaller than the 70° observed in metals. Even though the dihedral angle in the as HIPed silicon nitride is smaller than that observed in metals, the lenticular cavity shape with uniform curvature is typical of equilibrium cavities.

Lange has shown that in cavitational creep, grain separation is the rate controlling step in the growth of cavities [21]. The HIPed silicon nitride material did not exhibit any grain separation, the cavity growth is due to the diffusion of silicon nitride through the viscous grain boundary phase.

The cavity dimensions after 24 hours and 200 hours at 1370°C and 300 MPa are compared in Table 2. The larger deviation in cavity size and the increased cavity frequency in Figure 9 indicate that cavity nucleation is continuous. Continuous nucleation is typical for most proposed mechanisms.

In Figure 8 there is one failure at 45 ksi after 200 hours at 1370°C and 300 MPa. The lower strength of this bar was due to a high frequency of inclusions which increased cavitation. The increased cavitation due to the presence of inclusions has been reported in hot pressed silicon nitride [16]. The inclusions observed in this material were in the micron size range and resulted in a 40 percent strength reduction.

SUMMARY

The cavitation behavior of HIPed silicon nitride at 1370°C is distinctly different from previous investigations of other silicon nitride materials. The creep rupture behavior observed in HIPed silicon nitride is more typical of the creep rupture behavior observed in metals. The HIPed silicon nitride was densified with 4% yttria and contains a lower volume fraction glassy phase then silicon nitrides previously studied. In HIPed silicon nitride cavitation occurs primarily on two grain boundaries, is diffusion controlled and cavity nucleation appears to be a continuous process during deformation.

REFERENCES

1. K.H.Jack, Ceramics for High Temperature Applications, edited by J.J.Burke, A.E.Gourm, R.N.Katz, Brook Hill, Chestnut Hill, MA (1974) pp 265-86.
2. G.E.Gazza, J Amer Ceram Soc, 56 [6] 919730 p662.
3. A.Tsuge, K.Nishida, M.Komatsu, J Amer Ceram Soc, 58 [7-8] (1975) pp 323-326.
4. D.A.Bonnell, T.Y.Tien, M.Rhule, J Am Ceram Soc, 70 [7] (1987) pp 460-465.
5. J.E.Marion, A.G.Evans, M.D.Drory, D.R.Clark, Acta Metall, Vol 31, No. 10 (1983) pp 1445-1457.
6. R.L.Tsai, R.Raj, Acta Metall, Vol 3 (1982) pp 1043-1058.
7. A.G.Evans, A.Rana, Acta Metall, Vol 28 (1980) pp 129-141.
8. F.F.Lange, B.I.Davis, D.R.Clark, J Mat Sci, 15 (1980) pp 601-610.
9. R.Raj, J Eng Mat & Tech, april 1976, pp 132-139.
10. G.M. Pharr, M.F.Ashby, Acta Metall, Vol 31 (1983) pp 129-138.
11. M.D. Thouless, A.G.Evans, J Amer Ceram Soc, 67 [11] (1984) pp 721-731.
12. C.F.Chen, T.Y.Tien, Ceram Eng Sci Proc, 8 [7-8] (1987) pp 778-795.
13. G.Cliff, G.W.Lorimer, J of Microscopy, Vol 103, pt 2, March 1975, pp203-207.
14. E.R.Fuller,Jr., L.Chuck, T.-J.Chuang, A.L.Dragoo, R.F.Drause, N.J.Tighe, S.J.Schneider, NBS-5-12, 1983
15. K.Liddell, D.P.Thompson, Br Ceram Trans J, 85 (1986) pp 17-22.
16. S.M.Weiderhorn, B.J.Hockey and R.F.Krause,Jr. Ceramic Microstructures '88, ed. Pask and Evans, Plenum Pub. 1988.
17. K.Jakus, S.M.Weiderhorn, B.J.Hockey, J Amer Ceram Soc, 69 [10] (1986) pp 725-731.
18. H. Riedel, Fracture at High Temperatures, Springer-Verlag, Berlin, 1986, pp.67-84.
19. J. R. Porter, W. Blumenthal, A. G. Evans, Material Science Research, v. 14, Plenum Press, N.Y., 1981, pp.249-260.
20. J.Lankford, K.S.Chan, and R.A.Page, Fracture Mechanics of Ceramics Vol 7, ed. Bradt, Evans, Hassleman, and Lange, Plenum Press 1986
21. F.F.Lange, B.I.Davis, D.R.Clarke, J of Mat Sci, 15 (1980) pp 601-610.

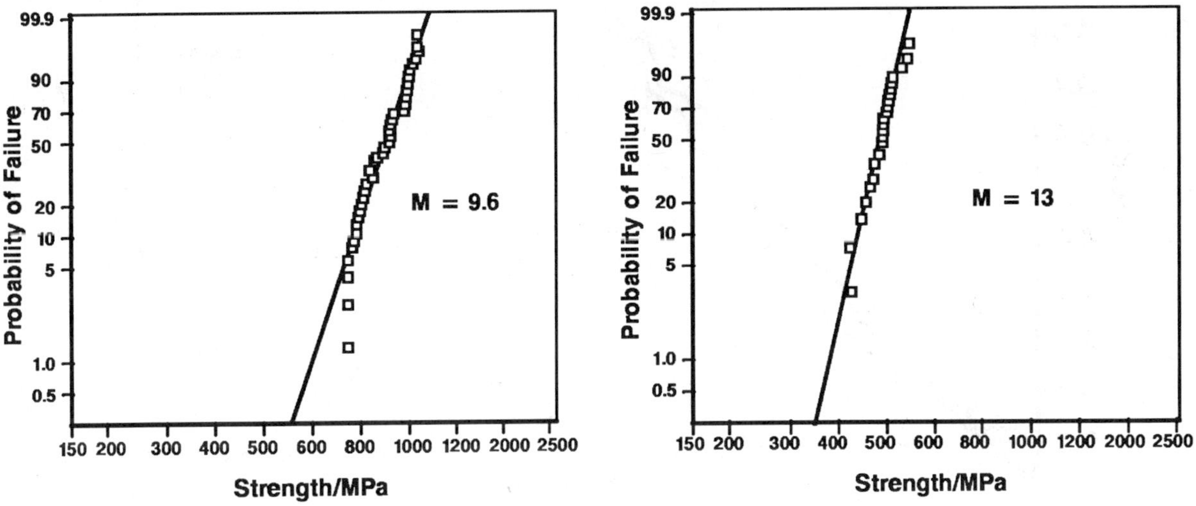

Figure 1. The room temperature (a) and 1370°C strength (b) distribution of HIPed silicon nitride.

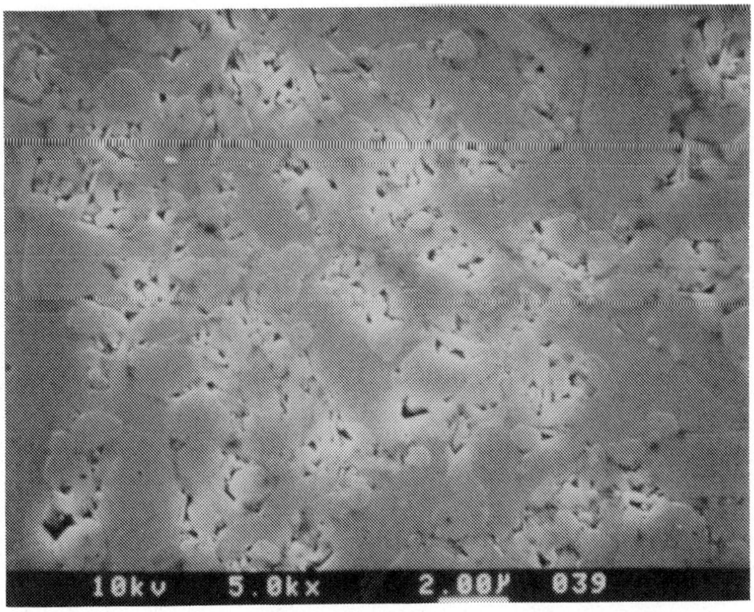

Figure 2. Secondary electron image of a HIPed silicon nitride specimen with 4% yttria. The specimen was chemically etched with molten KOH.

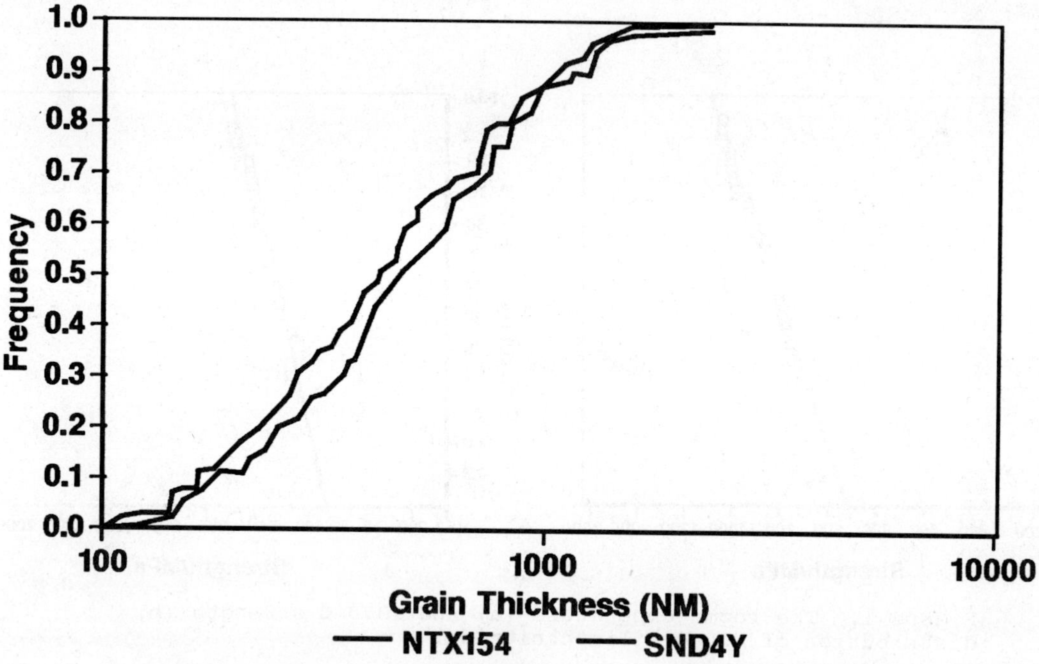

Figure 3. Grain thickness distributions for the HIPed silicon nitride with 4% yttria.

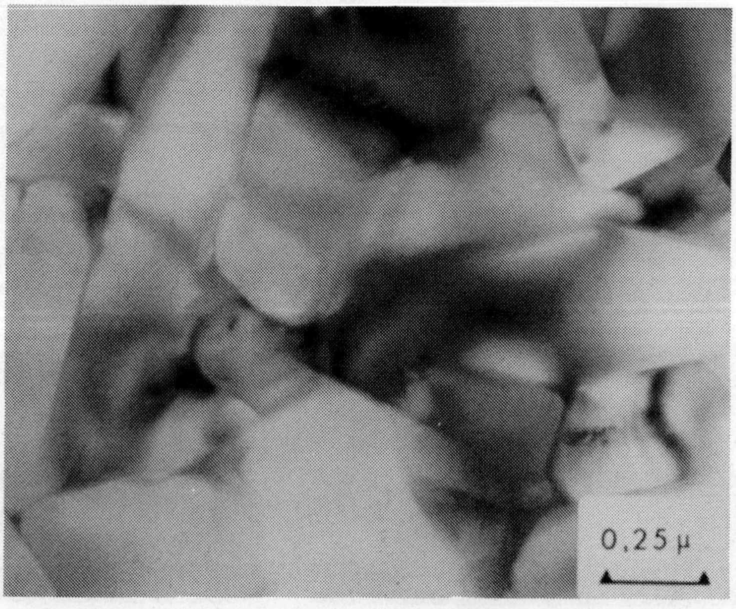

Figure 4. Bright field electron image of the as HIPed NTX154 silicon nitride. The intergranular phase is primarily amorphous, and it forms a continuous thin boundary film.

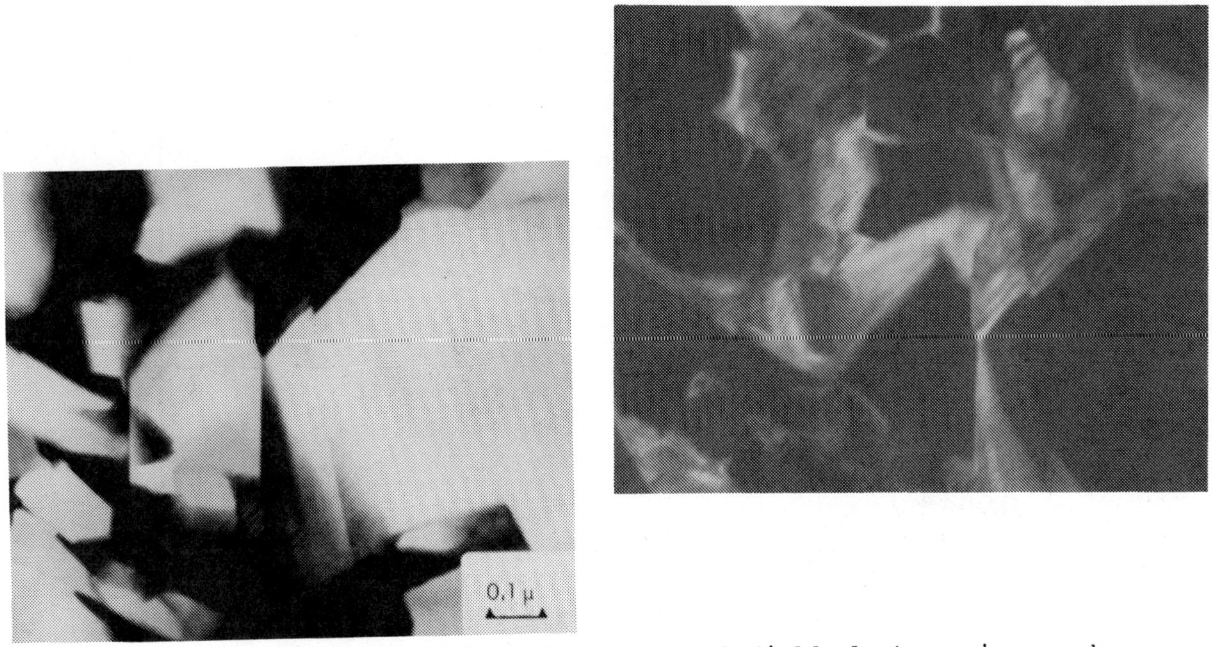

Figure 5. Bright field and dark field electron micrographs of the as HIPed silicon nitride illustrating the morphology of the crystalline intergranular phase. The glass crystallizes as large interconnected single crystals of alpha yttrium disilicate, extending across many silicon nitride grains.

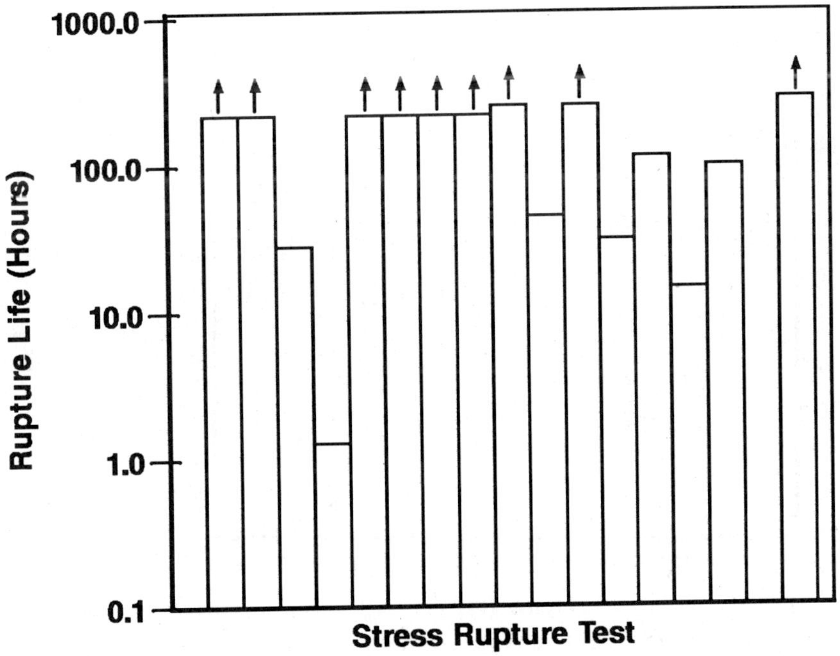

Figure 6. The stress rupture behavior of HIPed silicon nitride is variable. Some samples of the HIPed silicon nitride exhibit a short life, approximately 10 hours at 300 MPa, while other samples sustain 200 hours at 300 MPa.

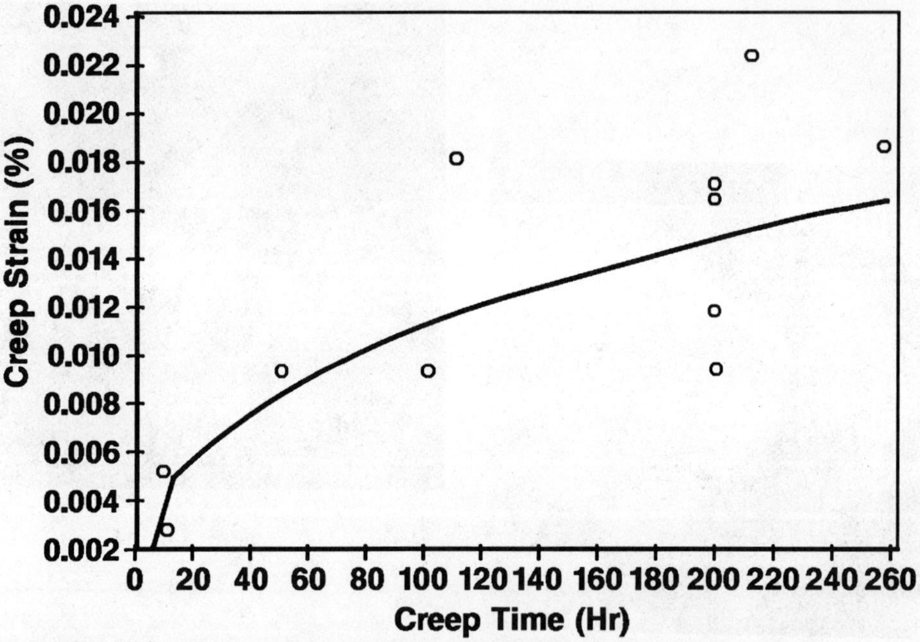

Figure 7. The creep strain is plotted as a function of creep for the interrupted stress rupture test bars. The overall creep strain is very low.

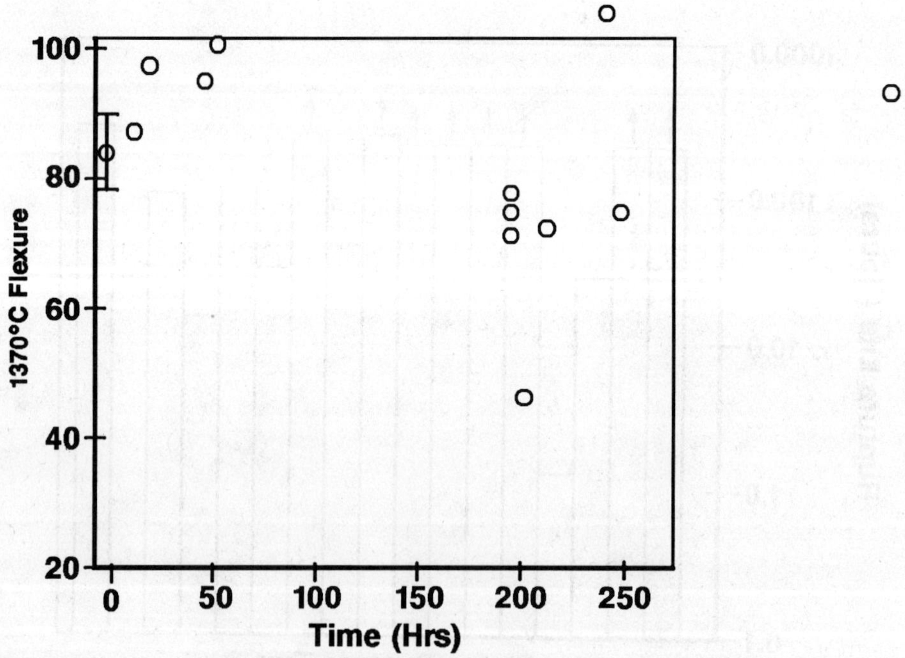

Figure 8. The 1370°C fast fracture test results show the retained strength of the HIPed silicon nitride after various stress rupture tests were interrupted.

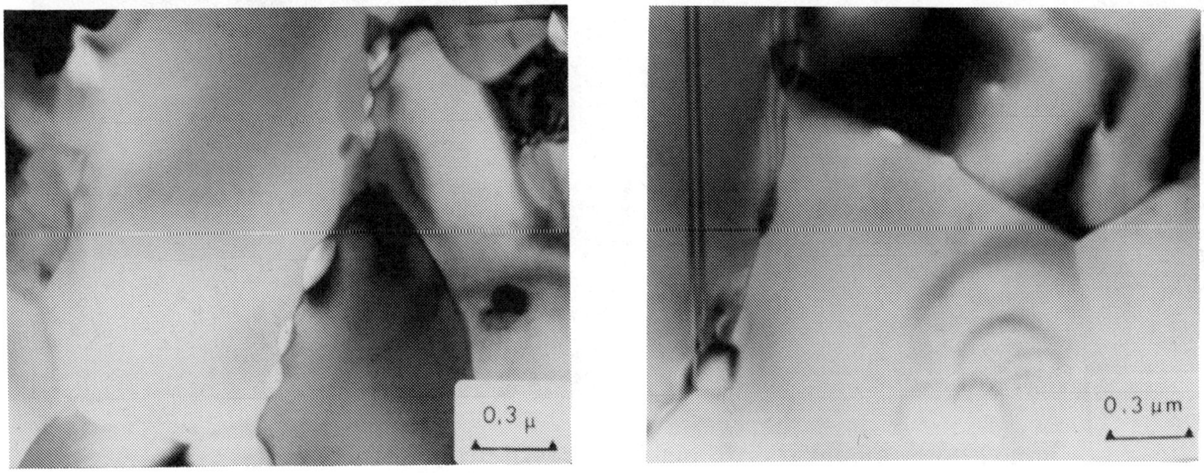

Figure 9. Bright field images of the microstructure of HIPed silicon nitride tested at 1370°C and 300 MPa for: a) 24 hours and b) 200 hours. The cavity size and shape is similar in both specimens, the sample tested for 24 hours contains fewer cavities than the 200 hour sample.

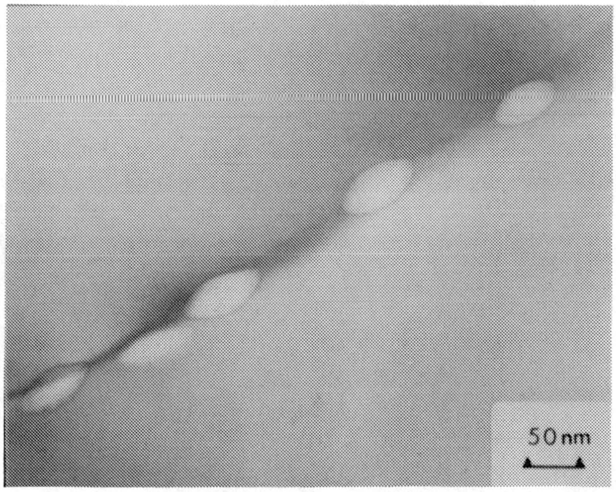

Figure 10. Bright field image of the cavities in the specimen tested at 1370°C and 300 MPa for 200 hours. The cavities are lens shaped, they are not contained within the grain boundary channel and are not homogeneously distributed.

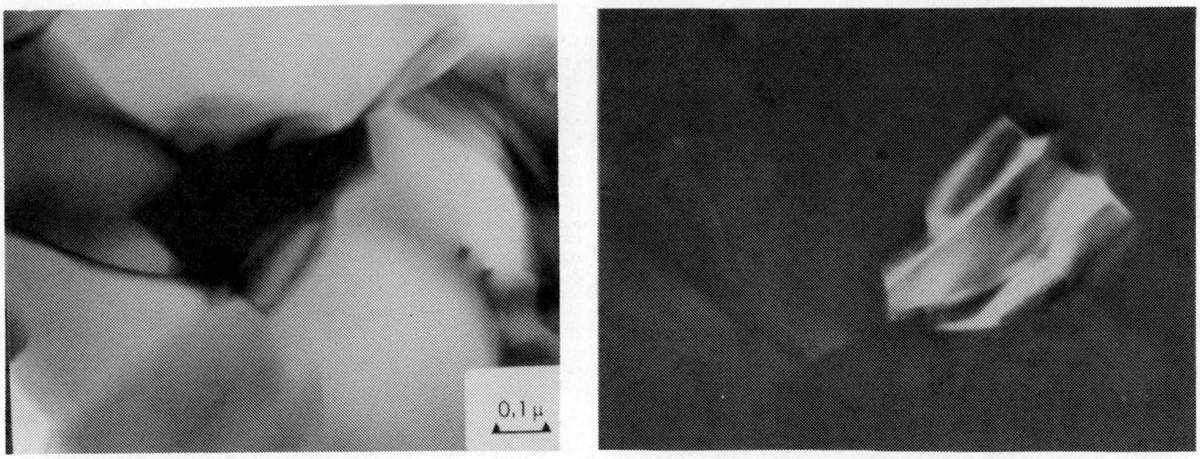

Figure 11. Bright field and dark field images of the crystalline intergranular phase in the 200 hour stress rupture sample. The morphology of a crystallite forming in an isolated triple junction is illustrated in the micrographs.

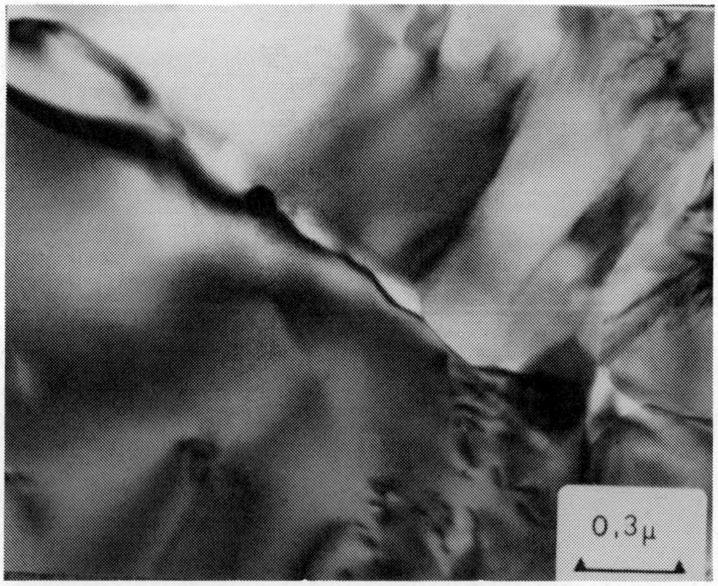

Figure 12. Bright field image of the sample that had low strength after being tested at 1370°C and 300 MPa for 200 hours. The low strength is attributed to the high volume fraction of inclusions in the grain boundaries.

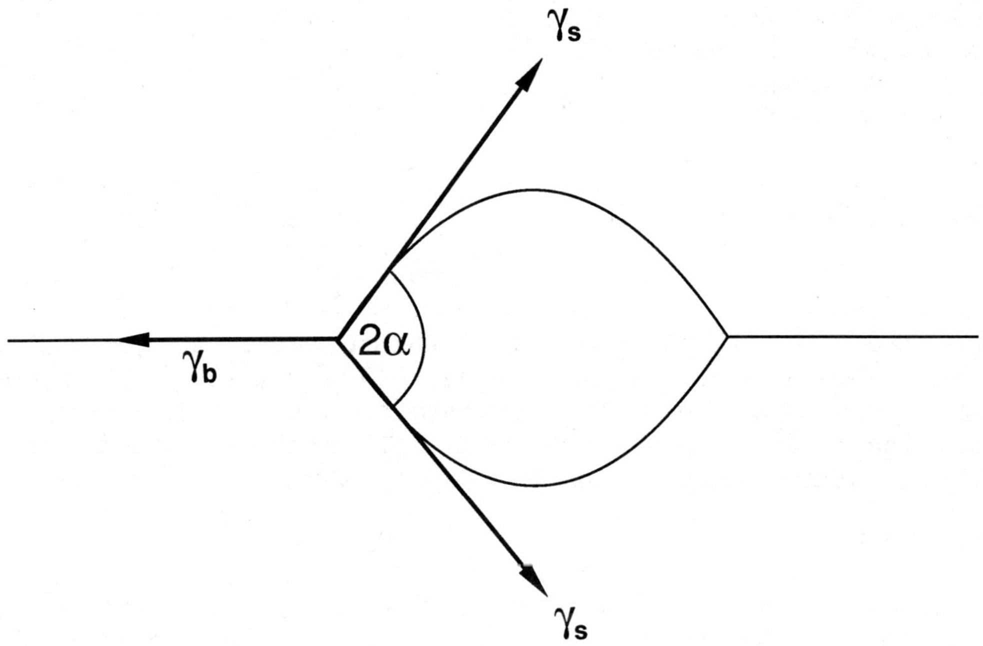

Figure 13. Sketch illustrating the relation between the interfacial energies and the cavity angle for a lens shaped cavity.

EFFECTS OF LOADING RATE AND TEMPERATURE ON DYNAMIC FRACTURE OF CERAMIC MATRIX COMPOSITES

K.H. Yang, A.S. Kobayashi, and A.F. Emery
Department of Mechanical Engineering
University of Washington
Seattle, Washington 98195

ABSTRACT

A hybrid experimental-numerical procedure was used to determine the dynamic fracture initiation toughnesses, K_{Id}, and the dynamic stress intensity factors, K_I^{dyn}, of alumina, TiB_2-particulate/SiC-matrix, and SiC_w/Al_2O_3-matrix composites at room and elevated temperatures under static and impact loadings. The K_{Id}'s at room and elevated temperatures were greater than the corresponding static fracture toughness, K_{IC}. The crack velocity versus K_I^{dyn} relations showed minor differences due to temperature changes but significant differences due to the change in loading rates.

1. INTRODUCTION

In many structural applications, ceramic components are subjected to severe loading rates such as particulate impact, which causes crack initiation, rapid crack propagation, and fracture. Hence, the use of ceramics as structural components in future heat engines requires the knowledge of the fracture resistance under impact loading at high temperature. Most impact studies of structural ceramics have been conducted for measurement of the dynamic fracture initiation toughness, K_{Id}, at room temperature. While K_{Id} was found to be greater than the corresponding static fracture toughness, K_{IC}, for alumina [1] and hot-pressed silicon nitride [2], K_{Id} was found to be equal to K_{IC} for silicon carbide [3]. Only hot-pressed silicon nitride has been impact tested at room temperature through 1400°C, where K_{Id} was found to be essentially constant up to 1200°C and dropped 20 percent at 1400°C [4]. No investigation on the dynamic fracture characterization, i.e., rapid crack propagation and crack arrest of structural ceramics, other than those of the authors [5-8], has been published to date.

The limited results on the room temperature dynamic fracture response of glass, reaction-bonded silicon nitride, and alumina [5-8] results showed that unlike polymers and metals, the rapidly running crack did not arrest when the driving forces, which in these studies were equated to the dynamic stress intensity factor (SIF), K_I^{dyn}, fell below a threshold value, i.e., the dynamic arrest stress intensity factor. However, no data describing the high temperature dynamic fracture response of ceramic matrix composite (CMC) materials has been published.

The purpose of this study is to report on the dynamic fracture response of one ceramic and two CMC materials at room temperature and at 1000°C/1200°C and to investigate the effects of dynamic loading rate and temperature on the dynamic fracture response.

2. METHOD OF APPROACH

Dynamic fracture testing of structural ceramics under the perceived operating conditions presents numerous experimental difficulties in monitoring simultaneously the time-varying applied load, which is imparted to the heated specimen, and the location of the propagating crack tip. A hybrid experimental-numerical procedure, which circumvents some of the above difficulties, was developed to study dynamic fracture response of brittle materials under impact loading and elevated temperature conditions.

2.1 Materials and Specimens

Material for the fracture specimen consists of commercially available alumina* and CMC materials, i.e., TiB_2-particulate/SiC-matrix** and SiC_w/Al_2O_3-matrix*** composites. The fracture specimens, which are single edge notched three-point bend specimens, were machined to the dimensions shown in Fig. 1. Three-point bend specimens were used because of their simplicity in geometry and relative ease in alignment at elevated temperature. The initial notch depth ratio was 0.5 for both specimens. The notches were machined by using a diamond-tipped saw blade with a nominal thickness of 0.25 mm, which produced a notch width of approximately 0.30 mm.

2.2 Experimental Procedure

The loading system consisted of a drop weight tower, which is mounted integrally with the furnace, with a maximum operating temperature of 1500°C as shown schematically in Fig. 2. The two port holes and one door, which are shielded with fused silicate glass panes to prevent the undesirable convection of air current in the furnace chamber, provided the passage of the laser beams. The mass of the steel impact body is 2.2 Kg. The small specimens were rapidly loaded by a drop weight with an impact velocity of 0.7 m/sec, which is greater than that of another published impact test of ceramics at high temperature [4]. The impact load was monitored by a load transducer at the top end of a push rod and outside of the furnace. A bottom load transducer, which was placed between the bottom end of the push rod and the fracture specimen, was used to validate the hybrid experimental-numerical procedure at room temperature. The two load transducer outputs were recorded on a transient wave form recorder.

A laser interferometric displacement gage (LIDG) [9] was used to measure the crack opening displacement (COD) near the prenotch tip. The optical principle of the LIDG is that interference fringes are generated by the reflection of an incident laser beam from two small reflecting microhardness indentations. Since the alumina and CMC materials used in this study lacked sufficient reflectivity to produce the optical interference fringes, highly reflective platinum tabs, typically 3.0 x 2.0 x 0.1 mm, were bonded near the crack tip of the specimen surface with Aremco Ceramabond**** adhesive shown in Fig. 3. Two indentations were located 0.4 mm apart along the machined crack and 0.5 mm behind the crack tip. The relative displacement of the two indentations, which is the COD, was measured by the interference fringe pattern motion. This motion was monitored by photomultiplier tubes and recorded on a transient wave form recorder. The motion of each whole-order fringe is equal to 0.91 μm of the crack opening displacement.

---

* AD-995, Coors Porcelain Company, Golden, Colorado 80401.
** Hexoloy-ST, Standard Oil Engineering Materials Col, Niagara Falls, New York 14302.
*** SA25, ARCO Chemical Company, Greer, South Carolina 29651.
**** Aremco Products, Inc., Ossining, New York 10562.

Crack extension was measured by a KRAK-GAGE* which was mounted on the other side of the specimen. Three specimens of each material were instrumented with a KRAK-GAGE and a LIDG and were impacted to failure at room temperature. This data was used to construct master curves, which related the measured COD and the measured instantaneous crack tip location during rapid crack propagation. These master curves can be used to locate the instantaneous crack length from the COD data at elevated temperature provided a proportional adjustment is made to accommodate for the changes in elastic moduli at elevated temperature. Two alumina specimens and three specimens of each CMC materials were impacted at 1000°C and 1200°C, respectively, to investigate the effect of temperature on the dynamic fracture of these materials.

Another two specimens of each material were statically loaded at room temperature to investigate the effect of loading rate on dynamic fracture of these materials. The maximum load at the onset of instability and the crack extension history were recorded using the KRAK-GAGE during rapid crack propagation.

2.3 Numerical Procedure

The crack extension history and the measured top impact load or the static maximum load were used to run the dynamic finite element code in its generation mode [10] to compute the dynamic fracture initiation toughness and the dynamic SIF during rapid crack propagation. K_{Id} was computed by using a calibrated COD procedure. K_I^{dyn} during rapid crack propagation was determined from the numerically computed dynamic energy release rate using Freund's relation [11].

For rapid crack propagation initiated under static loading condition, a similar hybrid procedure based on the experimentally determined crack extension history and the maximum load were used to drive the dynamic finite element code in its generation mode. The dynamic finite element analysis in turn yielded numerically K_{IC} and K_I^{dyn} during rapid crack propagation.

3. DYNAMIC FRACTURE INITIATION

Typical elevated temperature LIDG data for TiB_2-particulate/SiC-matrix composite is shown in Fig. 4. The onset of dynamic fracture initiation was easily recognized from the sudden change of the fringe motion. A simple average of the displacement of the two fringe patterns effectively removed any rigid body motion of the specimen. The near field crack opening displacement relation was used to calculate K_{Id} at the onset of fracture under dynamic loading condition. The crack tip loading rate of this impact test was of the order of 10^5 MPa$\sqrt{m}$/sec. Comparison of the results of this study with previous K_{IC} data [1,2-14] and K_{Id} [1] is made in Table 1.

The table shows that K_{Id} values are greater than K_{IC} at room temperature. While K_{IC} data of alumina used in this study is greater than the published K_{IC} values, K_{Id} value is close to the published K_{Id} value of alumina [1].

4. DYNAMIC CRACK PROPAGATION AND ARREST

The above procedure was used to determine K_I^{dyn} of the alumina and the CMCs during rapid crack propagation. The resultant crack velocity versus dynamic SIF relation of alumina is

* TTI Division, Hartrun Corporation, St. Autustine, Florida 32084.

shown in Fig. 5 where a insignificant difference is noted between the data for room temperature and 1000°C and insignificant difference is also noted between impact and static loadings at room temperature.

Figure 6 shows the resultant crack velocity versus dynamic SIF relation of TiB_2-particulate/SiC-matrix composite specimens where a small difference is noted between room temperature and 1200°C and significant difference is noted between the two loading conditions. The impacted three point bend specimen used in this study did not generate the lower dynamic stress intensity factor where crack arrest may have occurred. Further studies using different fracture mechanics specimens are necessary in order to resolve the existence or the lack of existence of a dynamic arrest stress intensity factor in CMC materials.

The same procedure was used to determine the dynamic SIF of SiC_w/Al_2O_3-matrix composite specimens. Figure 7 shows that the resultant crack velocity versus dynamic SIF relation was similar to the trend obtained in the TiB_2-particulate/SiC-matrix composite. While the trend of a decreasing crack velocity with decreasing dynamic SIF is observed under static loading, a propagating crack will continue to propagate where the dynamic SIF is substantially lower than the dynamic crack arrest stress intensity factor.

5. CONCLUSIONS

A hybrid experimental-numerical procedure was used to determine the dynamic fracture initiation toughness and dynamic SIF of alumina, TiB_2-particulate/SiC-matrix and SiC_w/Al_2O_3-matrix composites at room temperature and 1000°C/1200°C. The dynamic fracture initiation toughness values at room and elevated temperatures were greater than the corresponding static fracture toughness. The crack velocity versus dynamic SIF relation showed minor differences due to the change of temperatures but significant deferences due to the change of loading rates. The latter is a subject of future investigation.

ACKNOWLEDGEMENT

The results reported here were obtained under NASA contract NAGW-199.

REFERENCES

1. J. Duffy, S. Suresh, K. Cho, and E.R. Bopp, "A Method for Dynamic Fracture Initiation Testing of Ceramics," Division of Engineering, Brown University, May, 1987.

2. M.G. Mendiratta, J. Wimmer, and I. Bransky, "Dynamic K_{IC} and Dynamic Flexural Strength of HS-130 Si_3N_4," Journal of Materials Science, Vol. 12, pp. 212, 1977.

3. H. Abe, H.C. Chandan, and R.C. Bradt, "Low Blow Charpy Impact of Silicon Carbides," American Ceramic Society Bulletin, Vol. 57, No. 6, pp. 587-591, 1978.

4. S.T. Gonzy and D.L. Johnson, "Impact Fracture of Ceramics at High Temperature," Fracture Mechanics of Ceramics," Vol. 3, ed. by R.C. Bradt et al., pp. 495-506, 1978.

5. A.S. Kobayashi, A.F. Emery and B.M. Liaw, "Dynamic Fracture Toughness of Glass," Fracture Mechanics of Ceramics, Vol. 6, R.C. Bradt, A.G. Evans, D.P.H. Hasselman and F.F. Lange, eds. Plenum Press, New York, 47-61, 1983.

6. A.S. Kobayashi, A.F. Emery, and B.M. Liaw, "Dynamic Fracture Toughness of Reaction Bonded Silicon Nitride," Journal of American Ceramic Society, Vol. 66, No. 2, 151-155, Feb. 1983.

7. B.M. Liaw, A.S. Kobayashi, and A.F. Emery, "Effect of Loading Rate on Dynamic Fracture of Reaction Bonded Silicon Nitride," Fracture Mechanics Seventeenth Volume, ASTM STP 905, J.H. Underwood, F. Chait, C.W. Smith, D.P. Wilhelm, W.A. Andrews, and J.C. Newman, eds., American Society of Testing Materials, Philadelphia, 95-107, 1986.

8. A.S. Kobayashi and K.H. Yang, "A Hybrid Technique for High-Temperature Dynamic Fracture Analysis," to be published in the Proceedings of International Conference on Advanced Measurement Techniques, British Society for Strain Measurement.

9. W.N. Sharpe, Jr., "Interferometric Surface Strain Measurement," International Journal of Nondestructive Testing, Vol. 3, 59-70, 1971.

10. A.S. Kobayashi, "Dynamic Fracture Analysis by Dynamic Finite Element Method-Generation and Propagation Analysis," Nonlinear and Dynamic Fracture Mechanics, N. Perrone and S.N. Atluri, Eds., ASTM:AMD-35, 19-36, 1979.

11. B. Freund, "Crack Propagation in an Elastic Solid Subjected to General Loading—Non-Uniform Rate Extension," Journal of the Mechanics of Physics of Solids, Vol. 20, 141-152, 1972.

12. M.G. Jenkins, J.A. Salem, and S.G. Seshadri, "Fracture Resistance of TiB_2-Particle/SiC-Matrix Composite at Elevated Temperature," submitted to Journal of Composite Material, Nov. 1987.

13. C.H. McMurtry, W.D. Boecker, S.G. Seshadri, J.A. Zanghi, and J.E. Garnier, "Microstructural and Material Properties of SiC-TiB_2 Particulate Composites," American Ceramic Society Bulletin, Vol. 66, No. 2, pp. 325-329, 1987.

14. M.G. Jenkins, A.S. Kobayashi, K.W. White and R.C. Bradt, "Elevated Temperature Fracture Resistance of a SiC Whisker/Al_2O_3 Matrix Composite," to be published in Engineering Fracture Mechanics.

Table I.
Comparison of Fracture Toughnesses
of Ceramic and CMC Materials

Ref.	Fracture Toughness (Mpa $\sqrt{m}$)	Alumina		TiB_2 - particulate / SiC - matrix		SiC_w / Al_2O_3 - matrix	
		Temp.		Temp.		Temp.	
		Room	1000 °C	Room	1200 °C	Room	1200 °C
This Study	K_{Id}	5.7	5.1	8.3	7.6	7.9	6.9
	K_{IC}	4.3	-	6.8	-	6.6	-
[1]	K_{Id}	5.6	-	-	-	-	-
	K_{IC}	3.4	-	-	-	-	-
[12]	K_{IC}	-	-	4.6	4.6	-	-
[13]	K_{IC}	-	-	6.5-9.0	-	-	-
[14]	K_{IC}	-	-	-	-	6.2	5.0

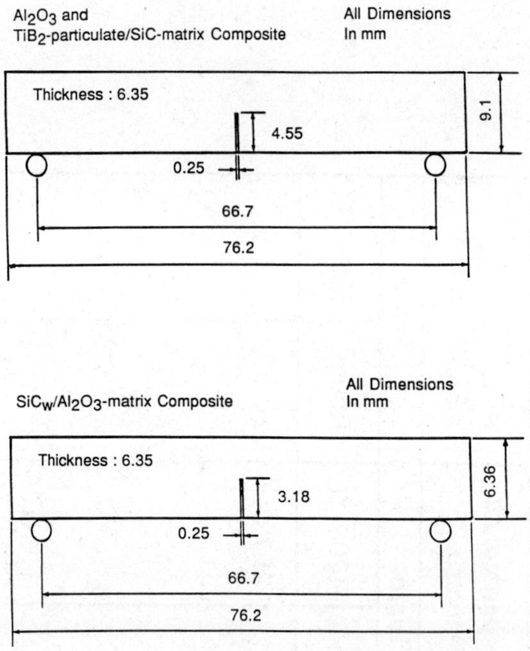

Figure 1   Three-Point Bend Specimen

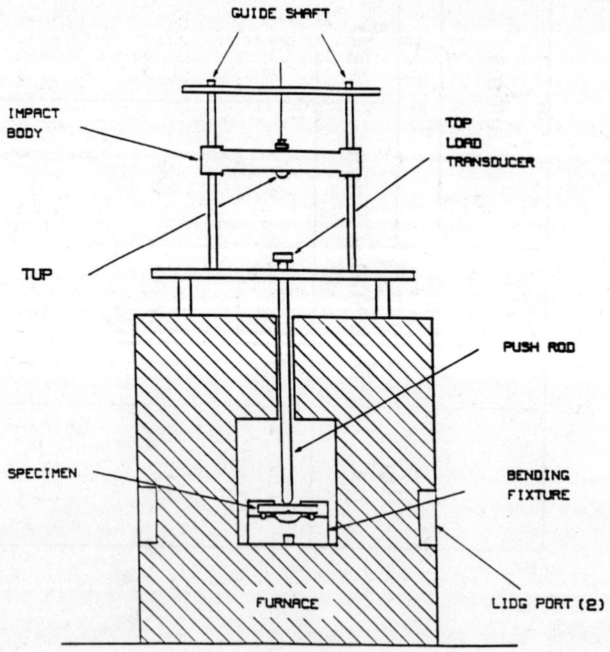

Figure 2   Schematic Diagram of Drop-weight Impact and Furnace with LIDG Ports

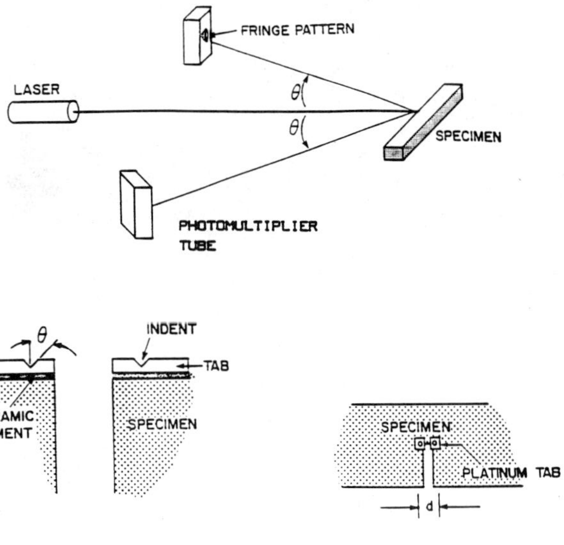

$$\Delta d = \frac{\Delta m_1 + \Delta m_2}{2} \frac{\lambda}{\sin\theta}$$

Figure 3 Schematic Diagram of Laser Interferometric Displacement Gage (LIDG)

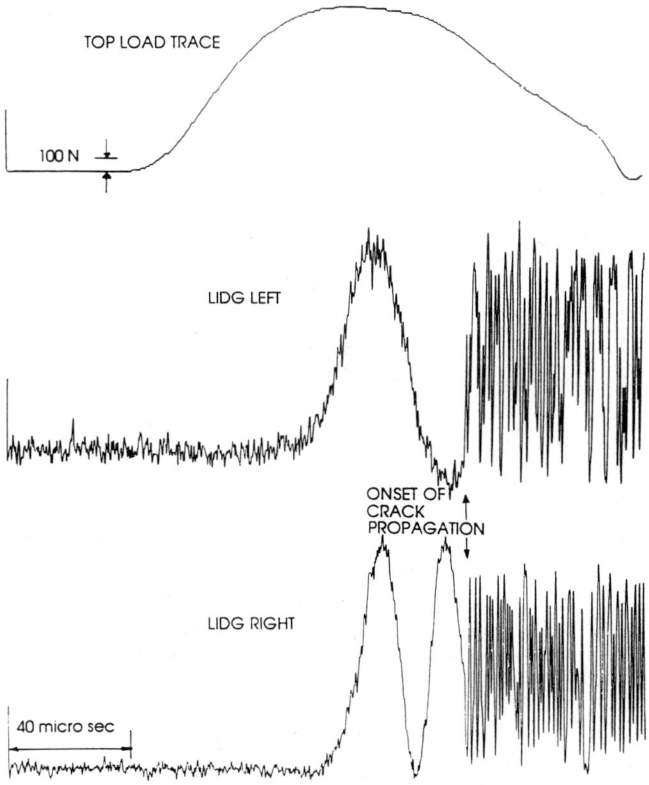

Figure 4 Time-Load Trace for Top Load Transducer and LIDG Data for Both PMTs at 1200°C, TiB_2-particulate/SiC-matrix Composite

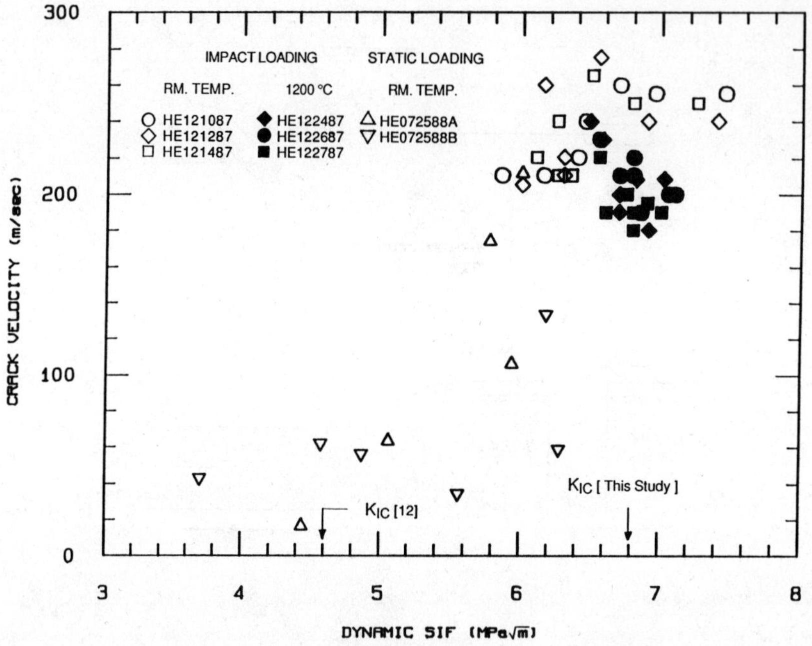

Figure 5 Dynamic SIF versus Crack Velocity Relation, Alumina

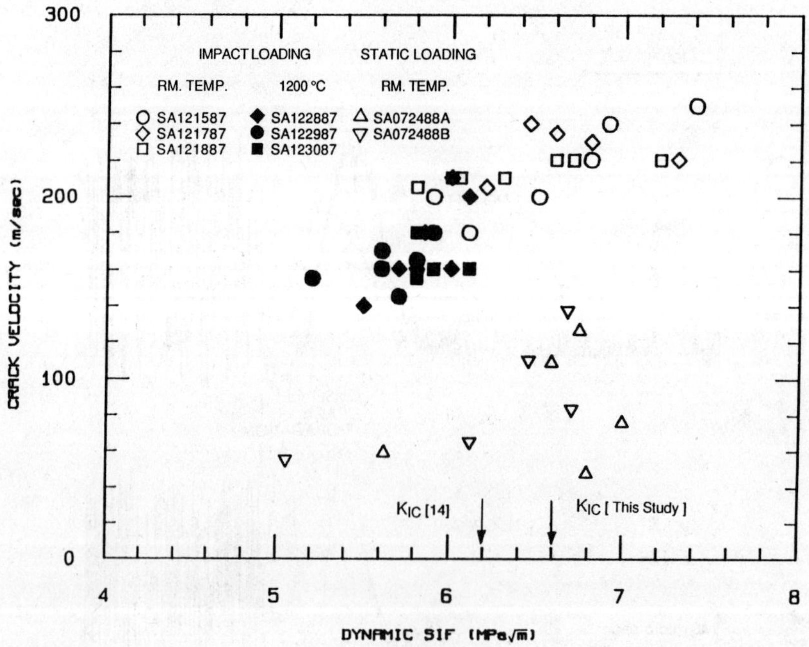

Figure 6 Dynamic SIF versus Crack Velocity Relation, TiB_2-particulate/SiC-matrix Composite

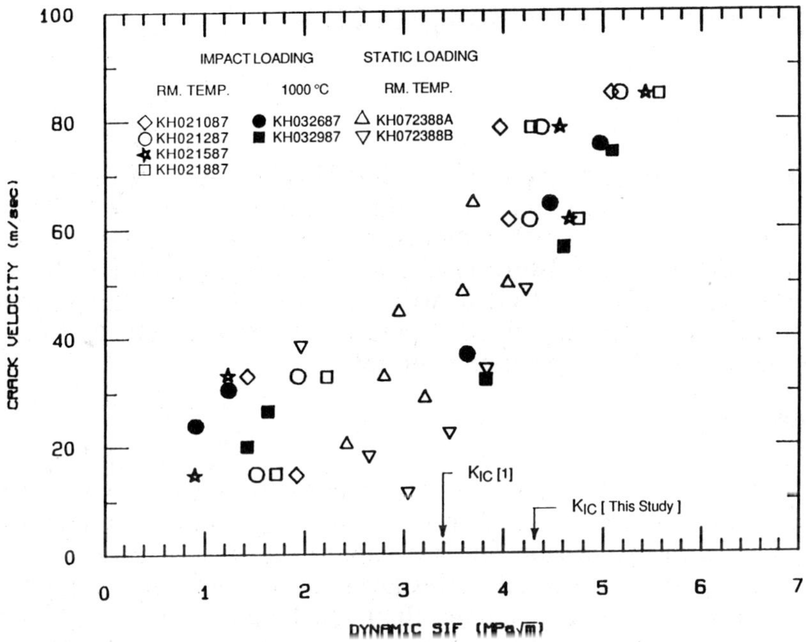

Figure 7 Dynamic SIF versus Crack Velocity Relation, SiC_w/Al_2O_3-matrix Composite

CREEP BEHAVIOR OF YTTRIA- AND ALUMINA-DOPED SILICON NITRIDE

A.E. Pasto, W.C. Van Schalkwyk, and F.M. Mahoney
GTE Laboratories Incorporated, Waltham, MA 02254

ABSTRACT

Four-point creep measurements were conducted in air on high-purity silicon nitride compositions containing yttria levels up to 6 w/o and alumina at levels up to 4 w/o. Specimens were subsequently examined ceramographically for microstructural effects. Tests were run for periods up to 1000 hours at temperatures from 1100 to 1300°C, under loads up to 276 MPa. Higher additive levels, in general, result in higher creep and crack growth rates, consistent with the increased amount and/or decreased viscosity of the secondary phase(s) present. A failure mechanism map is generated for the 6 w/o Y_2O_3 + 1.5 w/o Al_2O_3 composition material.

INTRODUCTION

Silicon nitride ceramics have been widely identified as candidate heat engine materials because of their physical, mechanical, and thermal properties. Components have been manufactured and tested for numerous high-temperature and stress applications in spark-ignited, diesel, and turbine engines. As these components operate in their design environments, they are subjected to conditions of varying temperature, stress, and atmosphere, all of which may have an effect on the component's life. The fast-fracture strength of a typical silicon nitride is high and relatively invariant to temperatures of about 1200°C. Thus fast-fracture-strength retention is not a concern. However, elevated temperatures bring about the possibility of other mechanisms which may become lifetime limiting, such as creep, slow crack growth, stress corrosion, and oxidation. The sudden changes of temperature due to transients such as startup or shutdown may also affect the materials adversely. Finally, independent of temperature, materials may fail in time due to generation of surface damage by wear or foreign object damage, impact, or erosion from contaminants in the lubricant.

Predicting the reliability of a high-temperature ceramic component over its intended lifetime is difficult, given the varied nature of its potential failure mechanisms. Only a relatively small body of work exists in this area, since the materials and their intended applications have been changing so rapidly. Techniques are now being developed to handle the complexity of the stress-temperature-failure mechanism modeling. An excellent example of the potential of this approach is the work of Soma, et al.[1] for silicon nitride turbochargers. Among the several mechanisms requiring detailed understanding are creep and slow crack growth.

This paper deals with creep behavior of silicon nitride which contains alumina and yttria, materials which are widely utilized structural ceramics.[2] Yttria is the primary densification aid, added to assist densification and promote the formation of the desired secondary phases to give the ceramic its strength and oxidation characteristics. Alumina is added to lower the sintering temperature.

MATERIALS

Silicon nitride compositions with Al_2O_3 content varying from 0 to 4 w/o and Y_2O_3 content varying from 4 to 6 w/o were tested. The compositions are labelled by a code showing the weight percent alumina followed by the weight percent yttria. A material containing 1.5 w/o Al_2O_3 and 6 w/o Y_2O_3 is, for instance, called A1.5Y6. Materials containing no alumina are labelled PY___. See Table 1 for a list of materials tested, their compositions, and fabrication methods. The basic silicon nitride source for all of the compositions except NC-132 was GTE SN-502 powder.

Table 1. Materials Description

Designation	Composition (w/o)				Densification Method
	Si_3N_4	Y_2O_3	Al_2O_3	MgO	
PY4	96	4			HIP
PY6	94	6			HIP
A1.5Y6	92.5	6	1.5		Hot-Pressed
A3Y6	91	6	3		Sintered
A4Y6	90	6	4		Sintered
NC-132*	99			1	Hot-Pressed

*Product of Norton Co.

Although the alumina-containing compositions exhibit excellent room-temperature strength, high-temperature (>1200°C) properties are believed to be marginal for certain applications (such as the AGT 100 engine rotor).[3] Previous research[4,5] has demonstrated that alumina addition to the Si_3N_4-Y_2O_3 system has profound effects on processing and mechanical properties; in particular, the high-temperature properties are strongly affected by increased alumina due to the lowering of the liquidus temperature and the increased secondary phase content.

To demonstrate the effects of Al_2O_3 addition on the properties of the Si_3N_4-Y_2O_3-Al_2O_3 system, test results on injection-molded and sintered silicon nitride are presented. Figure 1 shows the strength as a function of temperature for several of these compositions. It can be seen that the compositions containing higher amounts of Al_2O_3 have increased room-temperature strengths, but these strengths fall off rapidly at higher temperatures. At 1400°C, the PY6 composition retains 67%

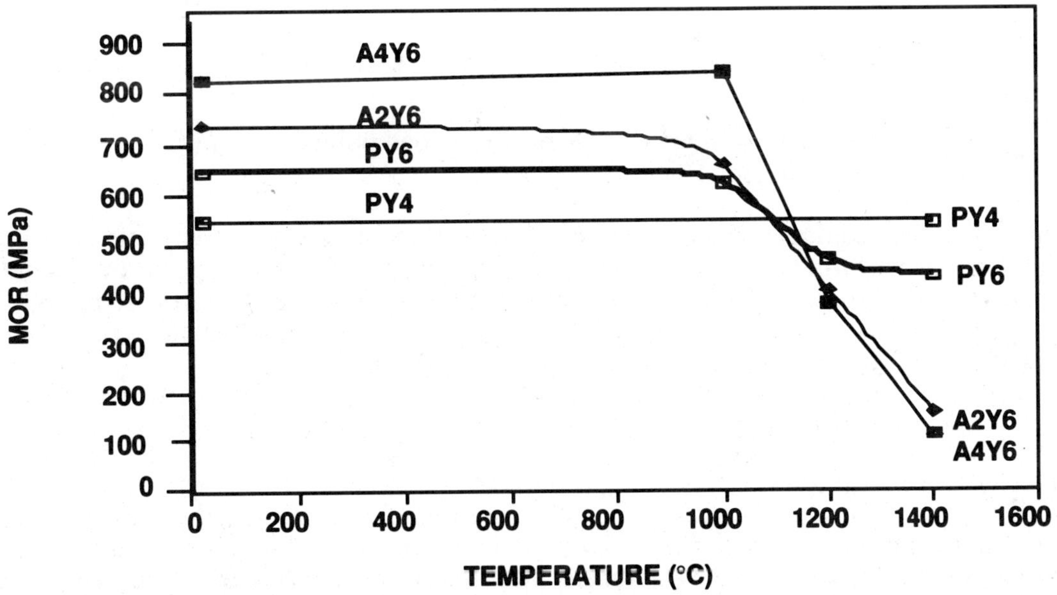

Figure 1. Fast fracture strength (modulus of rupture, MPa) for yttria- and alumina-containing silicon nitride.

of its room-temperature strength compared to less than 20% retained strength for the 2 and 4 w/o Al_2O_3 additions. Consequently, this composition has greater than three times the strength of either of the latter compositions at 1400°C. Concurrent with the loss in strength at high temperatures is an increase in deformation during testing for higher Al_2O_3 content compositions. The composition containing 4 w/o Al_2O_3 exhibits almost 450% deviation from linearity (a measure of the plastic deformation prior to fracture) compared to only 15% for the alumina-free PY6 composition. The significant decrease in deformation at 1400°C is indicative of a higher creep resistance.

These data indicate the benefit of using the PY6 composition where high-temperature (>1200°C) performance is desired. For even higher temperature capability, the PY4 composition has been developed. It is characterized by a flat strength-temperature behavior from room temperature to 1400°C (Figure 1).

The current investigation was initiated in order to more fully quantify the high-temperature mechanical behavior of these materials.

EXPERIMENTAL PROCEDURE

Test Specimen: The creep test specimen was a beam of nominal dimensions 1.27 mm x 2.54 mm x 25.4 mm. All surfaces were machined longitudinally to a 320 grit finish and chamfered on lengthwise edges.

Creep Test Apparatus: The specimens were tested in a four-point bend fixture made entirely of hot-pressed SiC. An Al_2O_3 probe connected to an LVDT measured beam deflection vs two reference probes at the inner load points. The tests were performed in air using a furnace based on a design by Quinn.[6] Temperature monitoring was performed by use of Pt/Pt-Rh thermocouples. Time, deflection of the beam, and temperature were recorded by a data acquisition system.

Calculation of Strain: Using deflection relative to the two inner load points, strain was calculated from the following relation,[7] assuming a >> d :

$$E_{max} = \frac{4hd}{a^2},$$

where

E_{max} = outer fiber strain
d = Deflection of the beam
a = Distance between the load points
h = Thickness of the beam

Accuracy of the Test Equipment: An analysis of the test apparatus' sensitivity and signal/noise discrimination was performed. Thermal variations were found to cause the greatest measurement uncertainty. The system detection limit was established to be a creep rate of $2 \times 10^{-10}\ s^{-1}$ in a 500 hour test and a signal to noise ratio > 3. Reproducibility experiments were performed on specimens cut from the same billet of material and stressed under identical conditions.

Creep Test Procedure: The specimens were heated in air to the planned test temperature under a constant 30 MPa stress caused by the load platform and loading ram. The test specimens were held at test temperature typically 1/2 to 1 hour prior to loading to the desired stress. At the conclusion of the creep test, the load was removed and the specimen cooled. After each test, deflection of the beam was measured with a digital linear gauge to confirm the final LVDT result. Physical measurement of the deformed beam was accomplished with a digital linear gauge and fixtures constructed at GTE Laboratories. Specimens were ceramographically prepared and evaluated using optical and scanning electron microscopic (SEM) techniques.

RESULTS AND DISCUSSION

Typical creep strain versus time curves show a rapid initial (primary) creep regime followed by a more constant (secondary) regime. The effect of stress is to increase the initial rates and total primary strains, as well as the secondary rate. Increasing temperature hastens the onset of secondary creep and increases the secondary rate. The effect of alumina content at constant yttria level is depicted on Figure 2 as a function of strain rate. Creep rate at 207 MPa increases with alumina content up to about 3 w/o, and then decreases.

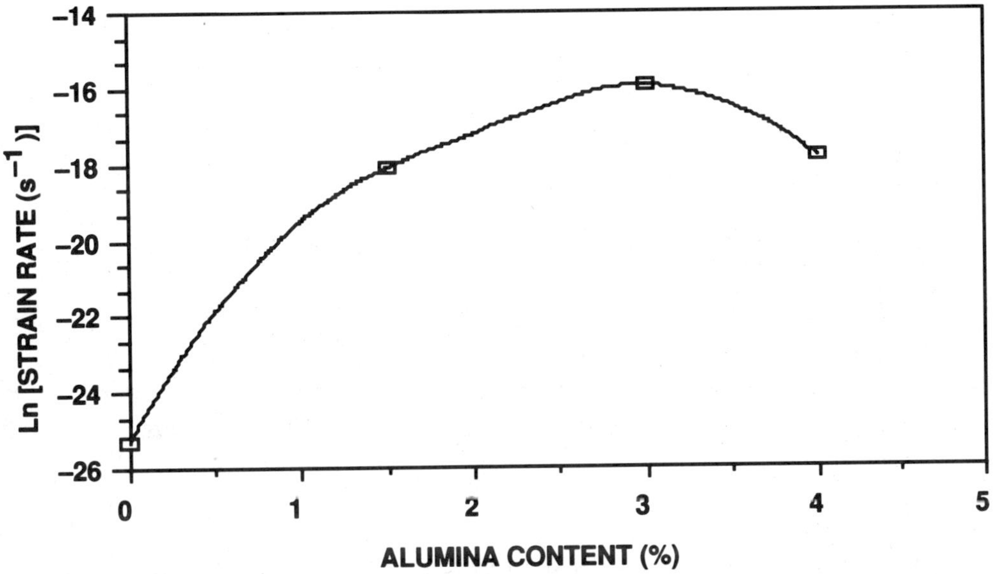

Figure 2. Effect of alumina content on creep strain rate of AY6 silicon nitride.

Reproducibility of the test data was excellent in most cases where the stress and temperature conditions were not extreme. The slope values calculated from five PY6 runs at 1300°C/207 MPa averaged $13.8 \pm 2.2 \times 10^{-10} s^{-1}$, whilc the two AY6 slopes measured at 1200°C/207 MPa agreed within 2%. However, as will be discussed later, at certain extreme conditions, deformation and premature failure occurred by subcritical crack growth rather than true creep. This behavior was common with A1.5Y6. For instance, one test (1200°C/207 MPa) was stopped by fracture after 97 hours (1.43% strain), whereas under similar stress-temperature conditions another was terminated by the operator after attaining a strain of 2.57% at 672 hours without failure. These results show that one must exercise caution when attempting to extrapolate creep rate data into new temperature or stress regimes.

The customary treatment[8] for creep data is to carry out creep testing until a steady-state rate is achieved, and then determine a constant creep rate. This latter datum is then analyzed for its temperature and stress dependency assuming that

$$\text{Creep Rate} = (\text{Const.})(\text{Stress})^n(\exp\text{-}E/RT) .$$

In order for this analysis to be applied, the creep strain rate must be shown to be constant with time (or strain). This was tested for the present study, at least for the non-extreme cases discussed, by plotting the instantaneous creep strain rate against time. Examples illustrating the approach to steady-state creep are shown for A1.5Y6 and PY6 in Figure 3. In neither case was a true steady state reached. However, the variation in strain rate with time was considered negligible, so that data analysis was performed using the above relationship. This behavior may be contrasted with that of Ernstberger, et al.,[9] who found orders of magnitude changes in strain rate within hundreds of hours.

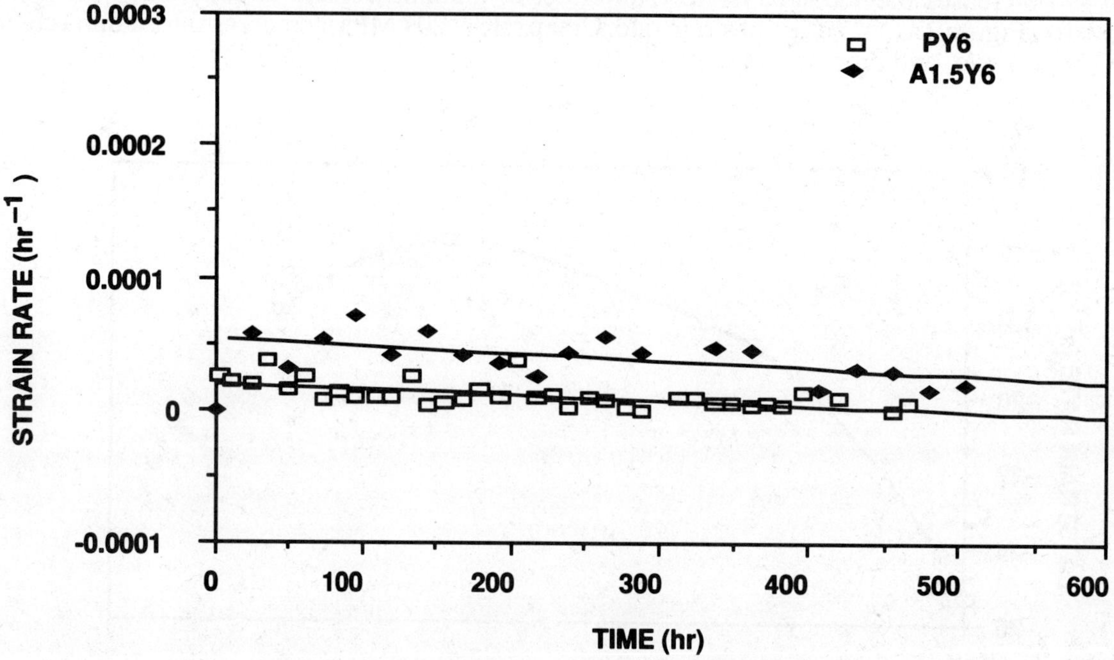

Figure 3. Strain rate-time plot for A1.5Y6 at 1200°C and PY6 at 1300°C, both at 207 MPa.

Strain was calculated from the raw displacement data and plotted against time. The slope of the strain-time curve was then calculated from a least squares straight line fit through the data over a large (tens or hundreds of hours) time increment, which was determined by a constant slope on the strain rate-time plot. Results of steady-state strain rate calculations are presented in Table 2 for most of the experiments to date.

The effect of temperature on creep rate is shown on Figure 4 for NC-132, A1.5Y6, A4Y6, and PY6. The plots can be fit by least squares regression to a straight line, the slopes of which are the activation energy for the processes controlling creep. The activation energies that result from this calculation are 443, 536, 469, and 632 kJ/mole for NC-132, A1.5Y6, A4Y6 and PY6 respectively. These values are within the range of activation energies noted in the literature for silicon nitride ceramics. For instance, Raj and Morgan[10] find a range of 546 to 630 kJ/mole. They show that these values imply that the mechanism controlling creep is solution-precipitation through the grain boundary phase.

A plot of Ln strain rate vs. Ln stress yields the stress exponent expected from the equation above. Figure 5 shows the data for A1.5Y6 and A4Y6 at 1200°C and for PY6 at 1300°C. The data for A1.5Y6 and PY6 are more complete, and it appears that the data are characterized by a non-constant stress exponent. This behavior is consistent with current literature on creep of silicon nitride,[9,11–15] which considers the influence on creep tests of several important factors. Some of these factors include the occurrence of simultaneous viscous grain boundary creep with cavitation creep; the fact that in bending, the neutral axis is shifting; and that the silicon nitride grains are being brought into contact on the compressive side as the glassy phase is being extruded out to the tensile side, thus shifting the load bearing region. This latter phenomenon has been illustrated for A2Y6 silicon nitride by the SEM work of Nolan.[16] The activation energies calculated, and the stress-dependence of the creep rate, are consistent with creep behavior dominated by grain boundary viscous processes, such as sliding and cavitation,[17] with simultaneous alteration of the intergranular phases by crystallization,[13] oxidation,[18] and rearrangement.[14] Microstructural examination by, e.g., TEM is required to assess the relative importance of these mechanisms for any particular material/temperature/stress conditions. This type of examination has not been performed on the subject materials.

Table 2. Strain Rates (x 10^{10} s–1) As a Function of Stress and Temperature for Several Silicon Nitride Compositions

Material/Test Temp. (°C)	Creep Rate at Stress Level (MPa)				
	69	138	207	241	276
PY4/1300			0.17		10
PY6/1200			0.10		
PY6/1250			1.3		
PY6/1275				5.7	
PY6/1300		8.9	14	22	37
NC-132/1150			5.7		
NC-132/1200		36	16		
NC-132/1250			240		
A1.5Y6/1100			1.4		
A1.5Y6/1150			5.3		
A1.5Y6/1200	23	45	140		
A3Y6/1200			1200		
A4Y6/1100			2.5		
A4Y6/1200		38	180		

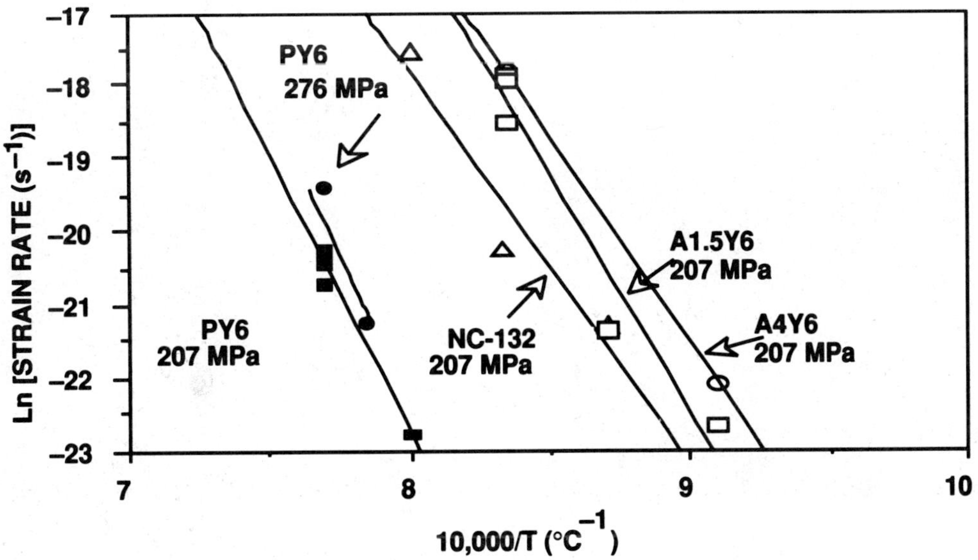

Figure 4. Effect of temperature on creep rate of silicon nitride.

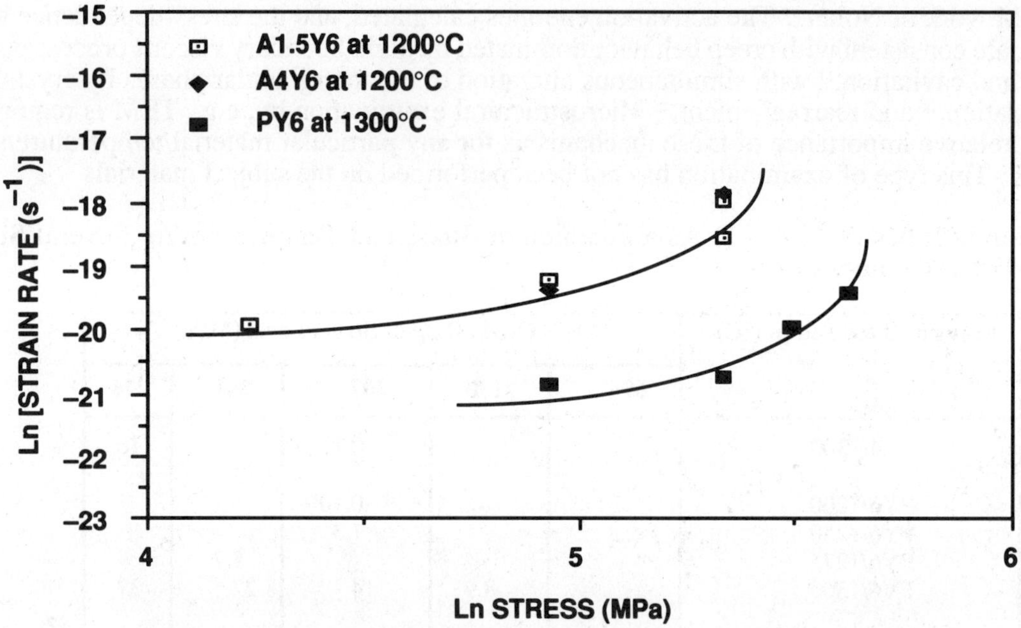

Figure 5. Effect of stress on strain rate of several silicon nitrides.

In certain extreme instances of stress and temperature, for instance at 1200°C and 207 MPa with A1.5Y6, the strain-time curves were linear from start to failure. Post-test examination of these specimens showed that subcritical crack growth had occurred instead of true creep. Typical fracture surfaces exhibited two distinct regions in the scanning electron microscope, as shown in Figure 6(a). The more reflective, rougher area is that generated during the crack growth events, characterized by extensive crack deflection and branching. The smoother area is generated during the fast fracture that indicated final failure. Figure 6(b) shows a second crack on the tensile surface of this bar.

Specimens undergoing true creep deformation showed no discernible deformation structures in the SEM. General features noted for those which underwent rapid creep included the presence of an oxide layer on the external surfaces, and the generation and linking of porosity into crack precursors. These features are illustrated in Figure 7.

Figure 6. Scanning electron micrographs of fracture surface of specimen exhibiting subcritical crack growth region. (A1.5Y6 tested at 1200°C, 207 MPa)

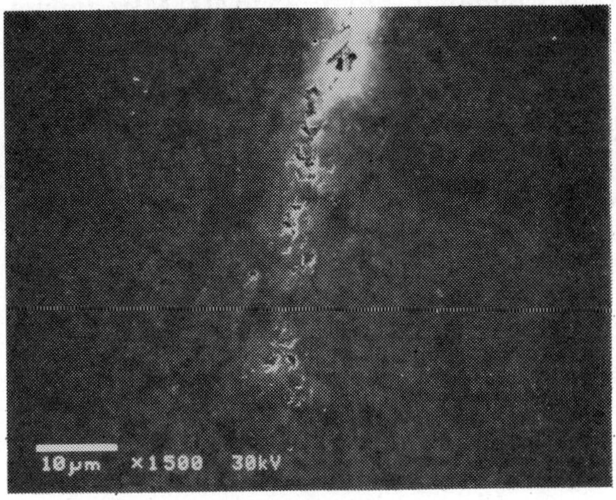

(a)

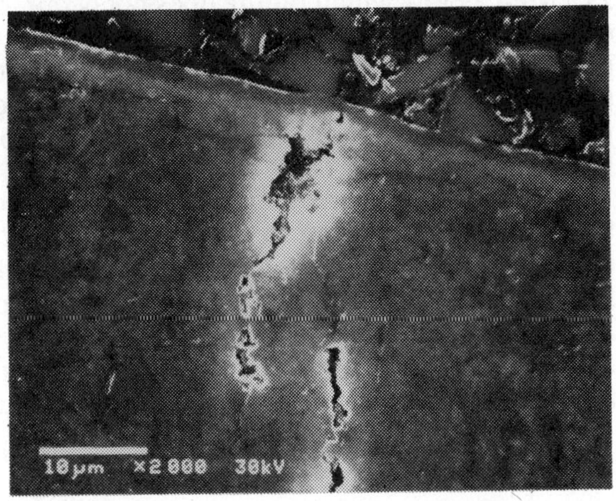

(b)

Figure 7. Scanning electron micrographs of polished surface normal to creep stress axis of A4Y6 specimen tested at 1200°C/207 MPa. (a) Exhibiting creep crack growth; (b) surface oxidation layer with void formation.

These results illustrate the existence of a critical stress above which subcritical crack growth behavior dominates fracture at any given temperature. For the A1.5Y6 composition, the transition occurs at or near 207 MPa at 1200°C. Below this stress level, creep failure will dominate. This interpretation is consistent with the results of other investigators who have studied this material.[19–20] Consideration of the results of the present study, along with the literature results at 1200°C[19] and the results at 1000°C,[19,20] allows the generation of a failure mechanism map[21] (Figure 8). This stress-temperature plot illustrates the predominant failure mechanism for A1.5Y6, not considering strain rate effects. Up to 900°C, this composition fails at about 650 MPa in fast fracture, with no measurable

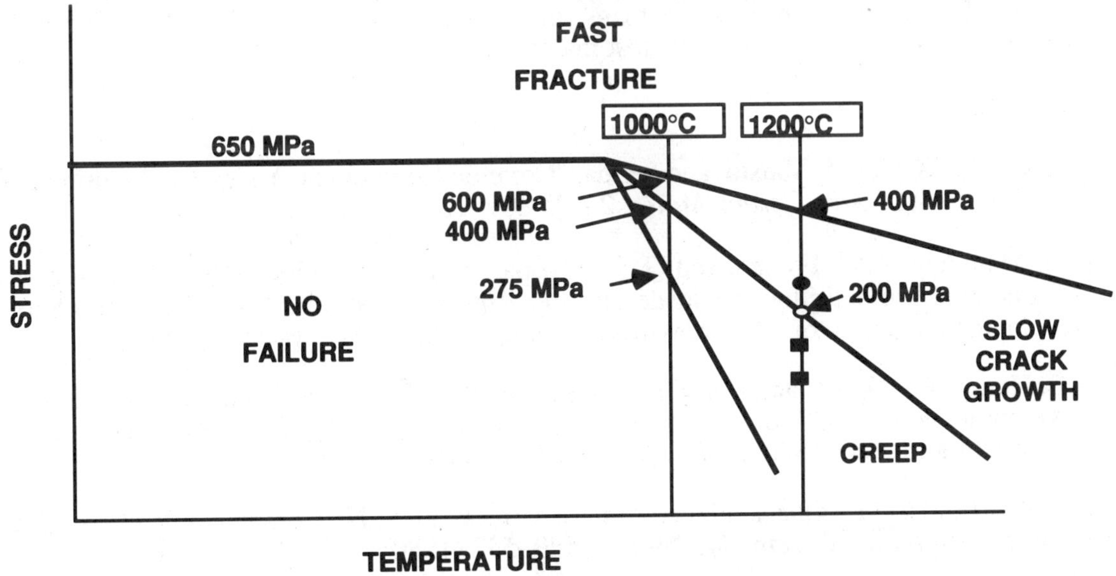

Figure 8. Schematic failure mechanism map for A1.5Y6 silicon nitride. Arrows point to data from references 19, 20. Data from current experiments shown as filled squares (creep dominant), as filled circles (subcritical crack growth dominant), and as open circles (transitionary behavior).

creep or subcritical crack growth behavior. At 1000°C, Khandelwal[20] reports a "threshold" level of 275 MPa, below which failure does not occur, due to crack blunting. From 275 to 400 MPa, failure occurs from creep damage accumulation, and from 400 to the fast fracture strength, subcritical crack growth predominates. At 1200°C, Quinn's data[19] show sub- critical crack growth predominating between 200 and 400 MPa, the fast fracture strength. Below 200 MPa, creep crack growth predominates. The current data, showing creep predominating at less than 207 MPa, confirms this finding.

Similar fracture maps can be constructed for other silicon nitride compositions, recognizing that the stress-temperature boundaries between failure mechanisms will shift in response to composition. The value of these maps is that, with an understanding of the stress-temperature-failure mechanism behavior, erroneous conclusions drawn from "accelerated testing" can be avoided. Accelerated tests, in which materials are tested at extreme temperatures or stresses to speed up data collection, must be confined to stress/temperature regimes where failure occurs via mechanisms relevant to the anticipated application environment.

SUMMARY AND CONCLUSIONS

Several different compositions have been developed which are capable of enduring the stress-temperature conditions of various heat engine applications. For applications at 1000°C or less, the A1.5Y6 composition can be expected to be utilized indefinitely at stresses up to 275 MPa, depending on the physical (e.g., tribological, atmospheric, etc.) environment. As temperatures exceed 1000°C, this composition becomes susceptible to time-dependent failure by creep crack growth at low stresses, and subcritical crack growth at higher stresses. Higher levels of alumina (up to 3 w/o) result in higher creep rates. For higher temperature application, the alumina-free materials are recommended. In these materials, lower yttria contents similarly result in reduced creep rates.

Finally, fast fracture, creep, and subcritical crack growth behavior can be rationalized on fracture maps, and used to predict failure modes for components under various operating stresses and temperatures.

ACKNOWLEDGMENTS

The authors would like to acknowledge the efforts of Tony Moschetti in supervising the construction and initial set-up of the creep test rigs, and personnel of the GTE-Prototype Engineering Center for the preparation of many of the test materials.

REFERENCES

1. T. Soma, Y. Ishida, M. Matsui, and I. Oda, "Ceramic Component Design for Assuring Long-Term Durability," *Adv. Ceram. Mater. 2* [4], 809–12 (1987).

2. F.M. Mahoney, K.W. French, and G. Bandyopadhyay,"Fabrication and Evaluation of AGT 100 Injection Molded Silicon Nitride Gasifier Rotors," Proc. 25th Automotive Technology Development Contractor's Coordination Meeting, SAE P-209, pp. 85–89 (1987).

3. H.E. Helms, P.J. Haley, and L.E. Groseclose,"AGT 100 Advanced Gas Turbine Technology Development Project," Proc. of the 24th Automotive Technology Development Contractor's Coordination Meeting, SAE P-197, pp. 137–146 (1986).

4. J.T. Smith and C.L. Quackenbush, "Phase Effects in Si_3N_4 Containing Y_2O_3 or CeO_2:I, Strength," *Bull. Am. Ceram. Soc. 59* [5] , 529–532 (1980).

5. C.L. Quackenbush and J.T. Smith, "Phase Effects in Si_3N_4 Containing Y_2O_3 or CeO_2: II, Oxidation," *Bull. Am. Ceram. Soc. 59* [5], 533–537 (1980).

6. G.D. Quinn, "Guide to the Construction of a Simple 1500°C Test Furnace," AMMRC TR 83-1 (1983).

7. G.W. Hollenberg, G.R. Terwilliger, and R.S. Gordon, "Calculation of Stresses and Strains in Four-Point Bending Creep Tests," *J. Am. Ceram. Soc. 54* [4], 196–199 (1971).

8. W.R. Cannon and T.G. Langdon, "REVIEW: Creep of Ceramics, Part 1 Mechanical Characteristics," *J. Mater. Sci. 18,* 1–50 (1983).

9. U. Enstberger, G. Grathwohl, and F. Thummler, "High-Temperature Durability and Limits of Sintered and Hot-Pressed Silicon Nitride Materials," *Int. J. High Technol. Ceram. 3,* 43–61 (1987).

10. R. Raj and P.E.D. Morgan, "Activation Energies for Densification," *J. Am. Ceram. Soc. 64* [10], C143–C145 (1981).

11. T. Fett, K. Keller, and D. Munz, "An Analysis of the Creep of Hot Pressed Silicon Nitride in Bending," *J. Mater. Sci. 23,* 467–474 (1988).

12. R.W. Evans, T. Murakami, and B. Wilshire, "Rate Controlling Processes During Creep of Silicon Nitride Ceramics," *Trans. J. Br. Ceram. Soc. 87* [2], 54–57 (1988).

13. D. Bethge, "High Temperature Creep and Slow Crack Growth Properties of HPSN as an Example of Ceramics Containing a Glassy Phase," *Int. J. High Technol. Ceram. 3*, 199–230 (1987).

14. R.M. Arons and J.K. Tien, "Creep and Strain Recovery in Hot-Pressed Silicon Nitride," *J. Mater. Sci. 15,* 2046–2058 (1980).

15. H. Cohrt, G. Grathwohl, and F. Thummler, "Non-Stationary Stress Distribution in a Ceramic Bending Beam During Constant Load Creep," *Res. Mechanica 10,* 55–71 (1984).

16. T. Nolan, Oak Ridge National Laboratory, Personal communication (1988).

17. F.F. Lange, "Evidence for Cavitation Crack Growth in Si_3N_4," *J. Am. Ceram. Soc. 62* (3–4), 222–223 (1979).

18. F.F. Lange, B.I. Davis, and D.R. Clarke, "Compressive Creep of Si_3N_4/MgO Alloys," *J. Mater. Sci. 15,* 601–618 (1980).

19. G.D. Quinn, "Static Fatigue of a Sintered Silicon Nitride," AMMRC TR 84-40 (1984).

20. P.K. Khandelwal, J. Chang, and P.W. Heitman, "Slow Crack Growth in Sintered Silicon Nitride," *Fracture Mechanics of Ceramics 8: Microstructure, Methods, Design, and Fatigue*, ed. by Bradt, Evans, Hasselman, and Lange, Plenum Press, NY, pp. 351–362 (1986).

21. G.D. Quinn, "Static Fatigue Resistance of Hot Pressed Silicon Nitride," *Fracture Mechanics of Ceramics 8: Microstructure, Methods, Design, and Fatigue*, ed. by Bradt, Evans, Hasselman, and Lange, Plenum Press, NY, pp. 319–332 (1986).

CREEP AND CRACK GROWTH RESISTANCE OF SILICON NITRIDE COMPOSITES

J.G. Baldoni and S.-T. Buljan

GTE Laboratories Incorporated
40 Sylvan Road
Waltham, MA 02254

ABSTRACT

Four-point creep tests were conducted at 1200°C to compare monolithic silicon nitride to silicon nitride composites containing silicon carbide dispersoids. The addition of large (8 μm) SiC particles was observed to increase the creep rate of the composite relative to the monolithic material. Silicon carbide whisker additions increased the resistance to creep, apparently by inhibiting grain boundary sliding. This effect was due to a reduction in the amount of sintering aids of the composite coupled with addition of acicular dispersoids to the microstructure. The susceptibility of whisker-containing composites to subcritical crack growth at 1200°C was also observed to be substantially reduced, demonstrating that these materials are flaw tolerant at elevated temperatures.

INTRODUCTION

Recent efforts toward increasing the strength and reliability of ceramics have centered on improving their fracture toughness through the tenets of composite design. Attempts to develop such improved materials have met with varying degrees of success.[1,2] It has been demonstrated that the fracture toughness and strength of silicon nitride, a material under active development for engine components and wear parts, can be increased considerably in the temperature range of 25°C to 1400°C.[3]

In many of the targeted applications, the time-dependent response of these advanced ceramics to the stress conditions imposed by the hostile operating environment to which they are exposed can dictate their success or failure. This study was undertaken to evaluate the resistance of silicon carbide whisker-toughened silicon nitride to creep and subcritical crack growth at elevated temperature.

Sponsored in part by the U.S. Department of Energy, Assistant Secretary for "Research Conservation and Renewable Energy, Office of Transportation Systems, as part of the Ceramic Technology for Advanced Heat Engines Project of the Advanced Materials Development Program, under Contract No. DE-AC05-84OR21400 with Martin Marietta Energy Systems, Inc."

EXPERIMENTAL PROCEDURE

Billets of monolithic Si_3N_4 and Si_3N_4 matrix composites containing SiC dispersoids were prepared by hot pressing. Characterization of the raw materials is given in Table 1.

The baseline Si_3N_4 material, AY6 (a composition of 92.5 w/o Si_3N_4 + 6 w/o Y_2O_3 + 1.5 w/o Al_2O_3), was milled, using silicon nitride media and screened prior to the hot pressing. For the particulate composite, SiC was added to the AY6 composition during milling. A settling column was used to beneficiate the SiC whiskers by removing large agglomerates and particulate matter which were found to be present in the as-received material. The SiC whiskers were blended into the milled AY6 powder and deagglomerated employing an ultrasonic probe. Such processing was found to yield good dispersion with minimal breakup of the whiskers. Hot pressing was carried out in BN-coated graphite dies in flowing nitrogen. The endpoint criterion for densification was a decrease in compaction rate to 2×10^{-4} cm/min of ram travel, which from experience has been found to yield high density (>99% of theoretical) Si_3N_4-based bodies.[3] It should be noted that this method of composite fabrication, that is, adding the SiC dispersoids to the prepared AY6 composition, results in a change in the amount of sintering aids (i.e., the Y_2O_3 and Al_2O_3 additives and the SiO_2 obtained from the surfaces of the Si_3N_4 and SiC powders) contained within the composites.

Mechanical property test samples were rectangular bars, 0.127 × 0.254 × 2.5 cm, machined from the billets with the tensile face perpendicular to the hot-pressing direction. The machine lay was parallel to the length of the bars and the edges were chamfered. The tensile face of each sample was wet-polished to a 1-µm diamond paste finish.

Table 1. Characterization of Raw Materials

BET Material	Surface Area (m^2/gm)	Particle Size (µm)	Impurity Analysis (w/o)				
			O_2	C	Mg	Ca	Fe
Si_3N_4*	4.0	0.5	1.50	ND	ND	ND	<0.05
SiCP**	0.3	8.0	0.31	0.60	<0.003	ND	0.07
SiCW***	3.3	0.5 dia.	1.78	0.36	0.2	0.2	<0.10

* (SN 502, (GTE)
** SiCP = SiC particle (Cerac)
*** SiCW = SiC whisker (Silar SC-9, Advanced Composites Materials Corporation)

Modulus of rupture and fracture toughness [controlled surface flaw technique[4]] tests were performed at room temperature in ambient air employing four-point loading, with inner and outer spans of 1.016 and 2.2286 cm respectively, at a crosshead speed of 0.05 cm/min.

Creep was evaluated via four-point loading in ambient air at 1200°C with imposed stresses ranging from 68.9 to 275.6 MPa (10 to 40 ksi) calculated from elastic beam theory. These conditions were chosen to simulate temperature/stress parameters which would be expected in heat engine applications,[5] the intended use of these materials. Descriptions of the test apparatus and procedures are detailed elsewhere in these proceedings.[6]

In order to investigate the subcritical crack growth resistance of these Si_3N_4-based ceramics, randomly selected test bars were precracked with a single 100 N Knoop indentation, which produces a surface flaw of approximately 125 µm in depth. The indented bars were then annealed at 1200°C in argon to relieve indentation stresses and subsequently creep tested.

RESULTS AND DISCUSSION

Mechanical property characterization, Table 2, showed that the addition of 30 v/o SiC particles (8 μm) to the AY6 monolith produced marginal increases in fracture toughness, while substantial toughening was obtained with the SiC whiskers. Due to the reduction in sintering aid content, the time required to achieve full density was extended by the addition of dispersoids. The whisker dispersoid further retards the rate of densification by inhibiting grain rearrangement during the sintering process, resulting in a fourfold increase in densification time to reach the same density level as the monolith.

Table 2. Characterization of Si_3N_4 Ceramic

Material	Densification Time (min)	Density Theoretical (% Theo)	K_{IC} (MPa·m$^{1/2}$)	MOR (MPa)
AY6	90	99.3	4.3 ± 0.1	778 ± 84
AY6 + 30 v/o SiCP	130	99.5	4.9 ± 0.2	885 ± 31
AY6 + 30 v/o SiCW	400	98.6	7.0 ± 0.2	894 ± 37

Variations in the densification parameters of the composites have been shown[3,7] to result in differences in the matrix grain size. It has also been demonstrated[3,8] that the fracture toughness of the SiC whisker composites is due to the combined effects of the particular matrix microstructure and the additional toughening imparted by the SiC whiskers.

Quantitative stereology of the densified bodies has revealed[7] that the grain size of the Si_3N_4 phase of these materials was in the range 0.3 to 0.5 μm, depending on the time at densification temperature, while the SiC whiskers had an average diameter of 0.5 μm and an aspect ratio of six.

Although there is a controversy[9,10] over the existence of true steady state creep for silicon nitride ceramics, examination of the data generated in this study indicated a constant creep rate after approximately 0.05% strain, Figure 1. Linear regression analysis of the creep strain-time data after 40 hr of testing yielded straight lines with correlation coefficients of greater than 0.95, with the majority being greater than 0.98. Therefore, the slopes of these lines were assumed to be steady state creep rates. Additionally, multiple tests conducted with a single material under identical conditions, Figure 2, showed a variation in the steady state creep rate of ±12%, indicating very good data reproducibility.

Figure 3(a) compares typical strain-time data at 1200°C and imposed stresses of 206.7 MPa (30 ksi) and 137.8 MPa (20 ksi) for the AY6 monolith and the AY6 + 30 v/o SiC whisker composite. The primary creep portions of the data were very similar. Once steady state creep was attained, the strain of the whisker-containing composite was lower. At 1200°C, the steady state creep rate of the whisker-containing composite was lower than its monolithic counterpart over the entire stress range investigated, Figure 3(b). The stress exponent from the relationship $\dot{\varepsilon} \propto \sigma^n$ of both materials had a value of approximately two over the range of conditions investigated. The value of the stress exponent is consistent with those obtained in other investigations of the creep behavior of Si_3N_4-based materials with different sintering aids using flexure,[11] tension,[12] or compression[13] modes of loading. Dislocation mobility in Si_3N_4 grains is unlikely below 1700°C.[12] The mechanism which has been ascribed to the creep of Si_3N_4 bodies containing an intergranular glass phase, the subject of this investigation, is grain boundary sliding controlled by the viscous flow of the glass phase. Due to geometric restraints, the motion of the intergranular phase is restricted by the Si_3N_4 grains, and a stress exponent greater than one is observed.[12] Since the reduced creep rate of the Si_3N_4-SiC whisker composite did not result in a change in stress exponent, it appears that the same mechanism may control the creep resistance of both materials. The whisker additions apparently retard grain boundary sliding and cavitation.

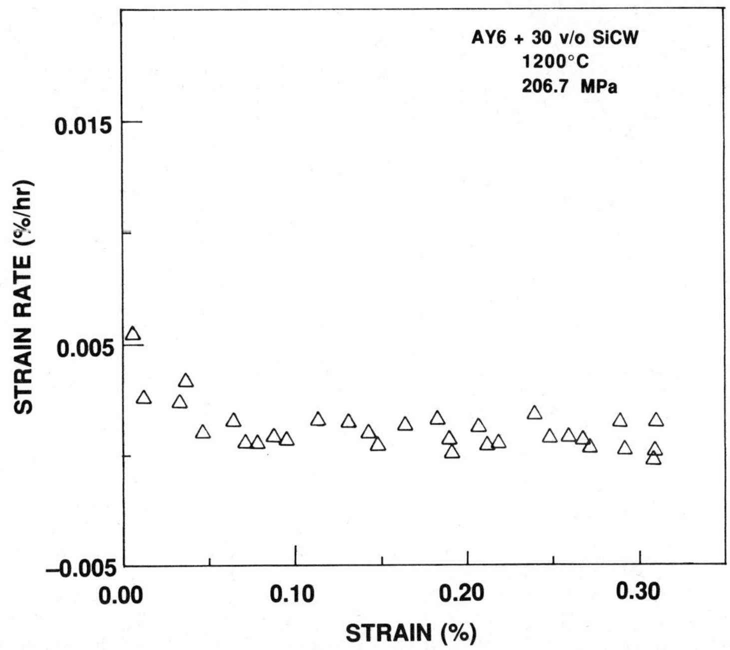

Figure 1. Effect of increasing strain on the strain rate of silicon nitride ceramics in four-point bending.

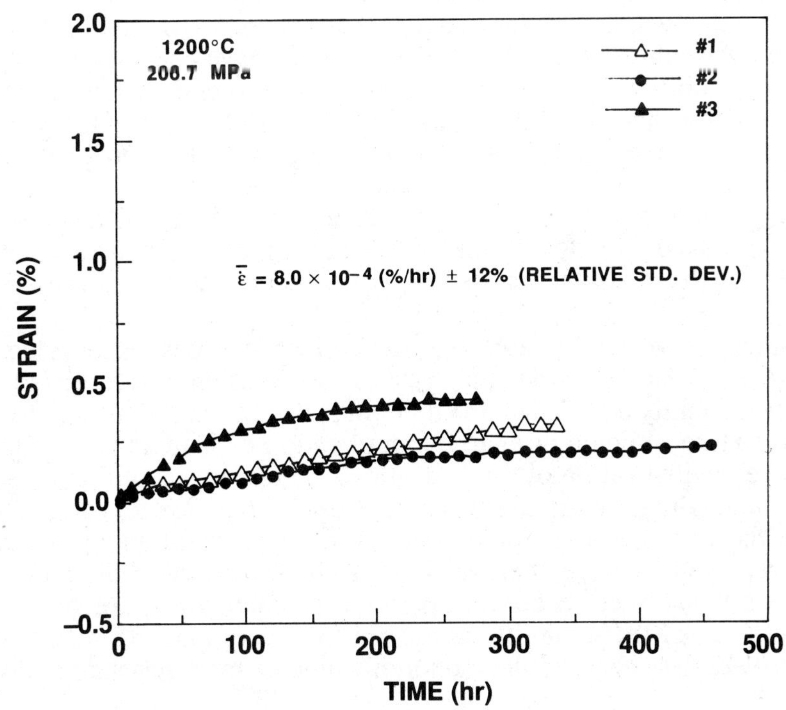

Figure 2. Reproducibility of creep data for an AY6-30 v/o SiC whisker composite.

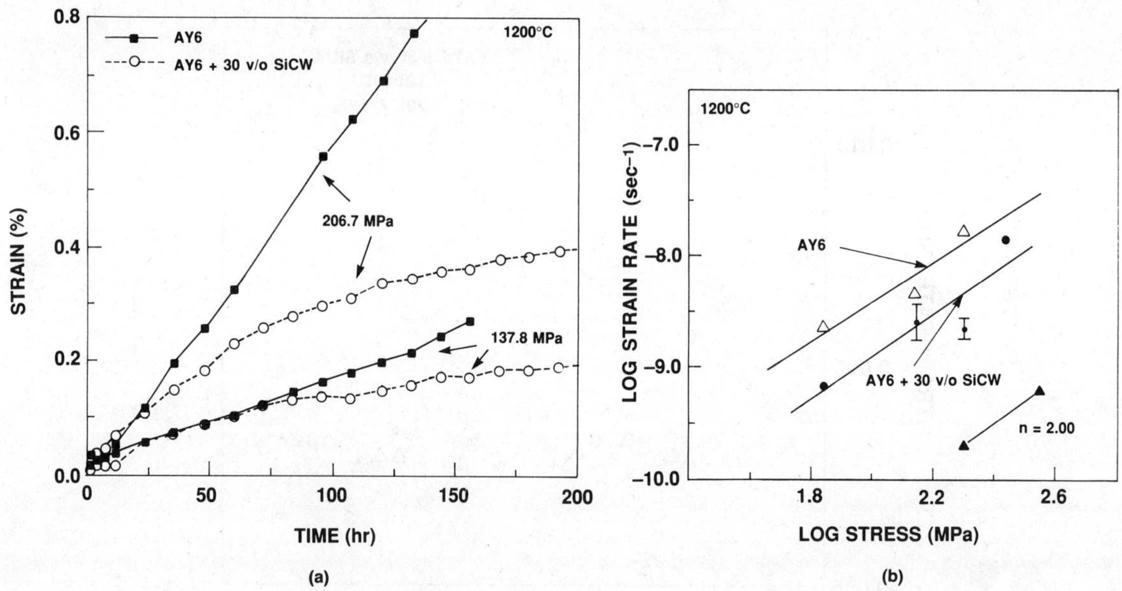

Figure 3. (a) Strain-time data and (b) strain rate data as a function of stress of monolithic AY6 and an AY6-30 v/o SiC whisker composite at 1200°C.

Data generated in the testing of the material containing 30 v/o SiC particles support this hypothesis, Figure 4. At 1200°C, the strain-time data of the particulate composite, Figure 4(a), showed a higher strain at a given time compared to monolithic AY6. Over the range of stress examined, the strain rate was consistently higher than that of the monolith, Figure 4(b), with a stress exponent of approximately two. Similar phenomena have been observed for MgO-fluxed (5%) Si_3N_4, where additions of 13 µm SiC particles were found to increase the creep rate at dispersoid concentrations above 15 v/o.[13] The two 30 v/o SiC composites contained the same reduced amount of Al_2O_3 and Y_2O_3, compared to monolithic AY6, although somewhat different total sintering aid contents, Table 3 (the difference owing to the SiO_2 contents of the SiC raw materials). It appears that replacing a substantial volume content (30%) of acicular β-Si_3N_4 grains with large equiaxed particles (8 µm SiC), while slightly increasing fracture toughness, detrimentally affects the material's resistance to time-dependent deformation at elevated temperature.

Other studies[14,15] of AY6 and AY6 + SiC whisker composites reported that at 1200°C in the stress range 50 to 350 MPa, the creep rate in compression of a composite containing 20 v/o SiC whisker was slightly higher than the monolith. The compression samples were annealed in N_2 at the temperature of creep prior to the application of load and subsequently tested in nitrogen. Extensive devitrification of the intergranular glass phase of the monolithic material occurred during annealing, while significantly less devitrification was observed in the composite. After the creep tests, the degree of crystallinity of the intergranular phase of the composite was reported to have increased. If grain boundary sliding controlled by the intergranular phase is assumed to be the operative mechanism of creep, these results, i.e., higher creep resistance of the monolith compared to the composites, follow since the more extensively devitrified glass phase of the monolith would be more resistant to deformation. The higher $SiO_2/Y_2O_3 + Al_2O_3$ ratio of the whisker-containing composite, Table 3, resulting in a more stable glass phase, accounts for this behavior. In the present study, x-ray diffraction of Si_3N_4-based samples crept for as long as 500 hr showed no indication of glass devitrification. It may be speculated that the intergranular glass phases of these materials are less prone to devitrification in an air environment than in nitrogen. Flexure testing in air at 1200°C, where glass devitrification is assumed to be minimal, clearly shows that the addition of 30 v/o SiC whiskers to AY6 increases resistance to creep.

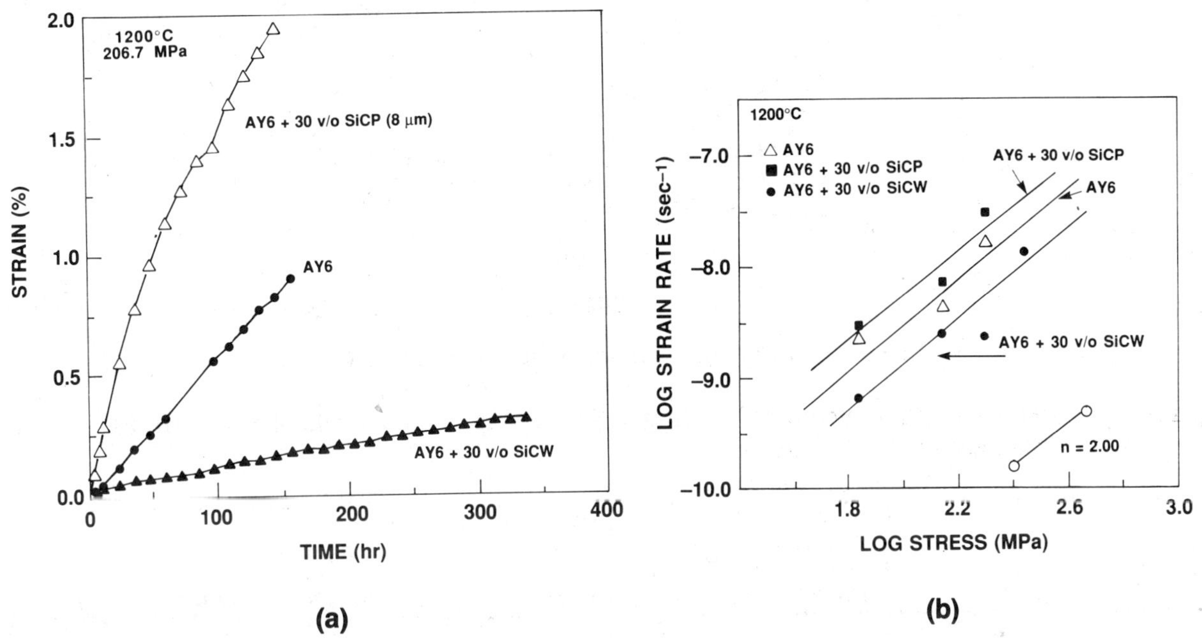

Figure 4. Effect of SiC dispersoid morphology on (a) strain and (b) strain rate.

In order to further evaluate the silicon nitride phase, a monolithic material with reduced amounts of sintering aids was prepared as an analog to the composite matrix. In this instance α-Si_3N_4 powder was milled with the prepared AY6 composition in a manner identical to the one used with the SiC particulate composite. This material required 110 min to densify. The density was 99% of theoretical with $K_{IC} = 3.9 \pm 0.1$ MPa·m$^{1/2}$ and MOR = 684 ± 25 MPa at room temperature. Limited testing at 1200°C showed that reducing the amount of sintering aids did significantly improve the resistance to creep at 206.7 MPa (30 ksi), Figure 5 and Table 3. The steady state strain rate of the AY6 + 30 v/o Si_3N_4 monolith was 60% less than that of the baseline AY6 material. These data also serve to more dramatically illustrate the effect of dispersoid shape on the creep rate of Si_3N_4-based composites. Although the resistance to creep of the matrix phase of the composites increased when the amount of sintering aids was reduced, this was detrimentally offset by replacing 30 v/o of the small acicular Si_3N_4 grains with equiaxed large (8 μm) SiC particles. Substituting larger (~2 μm equivalent diameter) acicular SiC whiskers for 30 v/o of the smaller (0.3 μm equivalent diameter) acicular β-Si_3N_4 grains amplifies the effect of reducing the amount of sintering aids and further increases the material's resistance to creep at 1200°C.

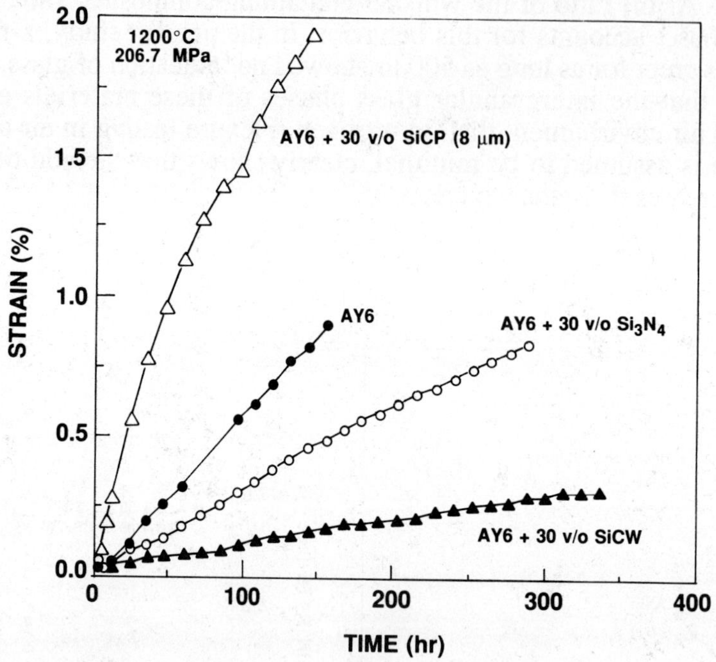

Figure 5. Strain-time data of silicon nitride ceramics.

Table 3. Creep Rates in Four-Point Bending and Sintering Aid Contents of Si_3N_4-Based Ceramics Shown in Figure 5

	Normalized Steady State Strain Rate (1200°C/206.7 MPa)	Sintering Aid Content (w/o of Composition)			
		Al_2O_3	Y_2O_3	SiO_2*	$SiO_2/Al_2O_3 + Y_2O_3$
AY6	1.0	1.50	6.00	2.60	0.35
AY6 + 30 v/o SiCP (8 μm)	2.0	1.05	4.23	2.00	0.38
AY6 + 30 v/o Si_3N_4	0.4	1.05	4.25	2.66	0.50
AY6 + 30 v/o SiCW	0.1	1.05	4.25	2.82	0.53

* Based on O_2 analysis of Si_3N_4 and SiC powders

It has been shown[16,17] that subcritical crack growth at 1200°C can contribute to the catastrophic failure of Si_3N_4 ceramics containing Y_2O_3 and Al_2O_3 sintering aids. The AY6 monolithic material which was precracked with a Knoop indentation and subjected to static loading at 1200°C failed within 20 min after the application of a 206.7 MPa stress. Examination of the fractured surface of the indented AY6 bend bar, Figure 6, revealed extensive slow crack growth prior to the onset of fast fracture. A precracked sample of the composite containing 30 v/o SiC whisker reinforcement responded to the imposed stress for over 400 hours in a manner analogous to that observed for standard, i.e., unindented, samples of this material. Similar tests conducted with indented and standard bend bars of composites from another study[3] with whisker additions of 10 v/o and 20 v/o exhibited the same behavior, Figure 7. At the three volume concentrations of SiC whisker additions examined, the indented samples had approximately the same steady state creep rates as standard bend bars, in spite of the presence of large surface flaws on the tensile faces, Table 4. The data also show that in bending at 1200°C with an applied stress of 206.7 MPa, the steady state creep rate of the AY6-based composites decreased with increasing content of whisker additions.

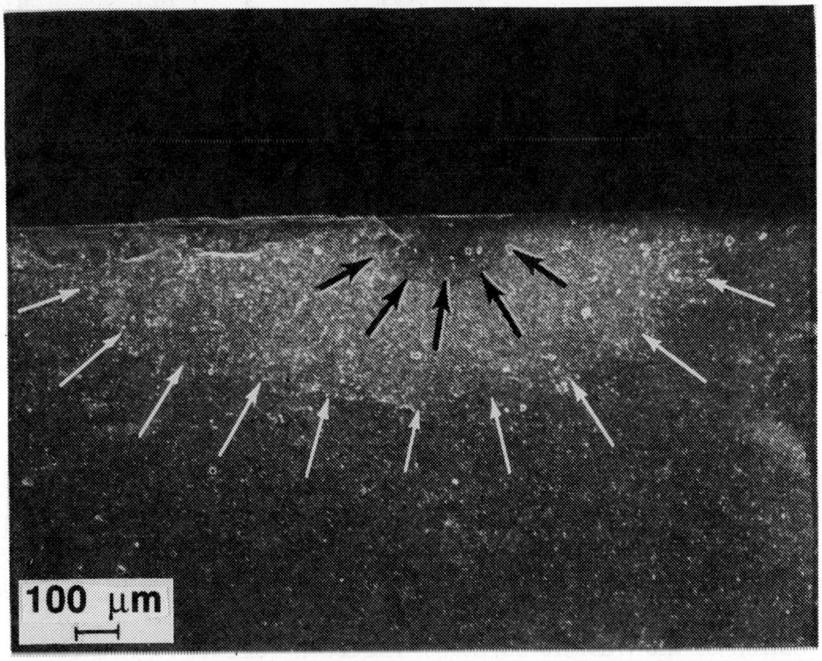

Figure 6. Subcritical crack growth in monolithic AY6 at 1200°C subjected to 206.7 MPa static stress. (Black arrows = original crack; white arrows = crack growth prior to fast fracture.

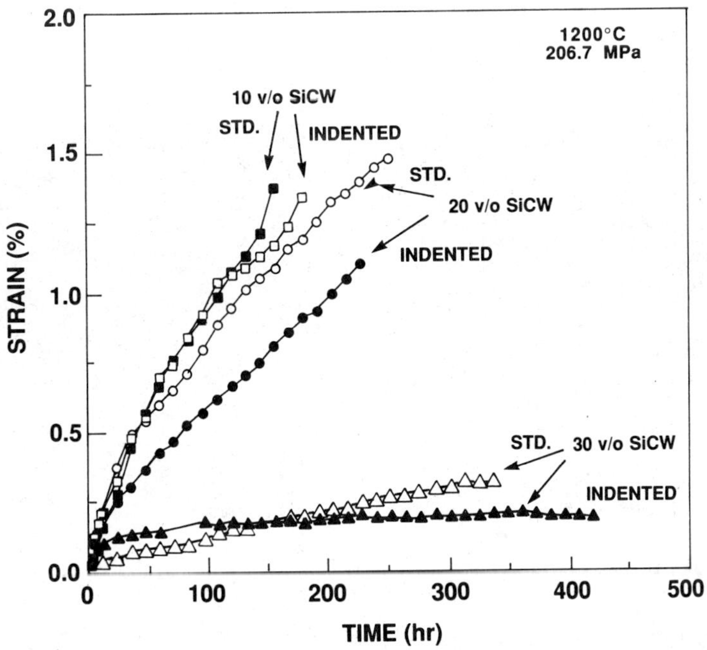

Figure 7. Comparison of precracked and standard bend bar subjected to creep at 1200°C.

Table 4. Steady State Creep Rates Obtained from Figure 7.

Material	Sample Type	Normalized Creep Rate
AY6 + 10 v/o SiCW	Standard Indented	10.4 10.1
AY6 + 20 v/o SiCW	Standard Indented	6.1 5.1
AY6 + 30 v/o SiCW	Standard Indented	1.0 1.0
AY6	Indented	Fractured in 20 min

Subsequent to these creep tests, the indented composite samples were broken in flexure at room temperature and the fracture surfaces examined. The surface oxide scale developed during the tests was clearly visible. Subcritical crack growth was not observed for any of the AY6-SiC whisker composites.

In order to elucidate the mechanism which imparts the increased subcritical crack growth resistance through the addition of the whisker dispersoids, indented bend bars of the AY6 monolith and the AY6 + 30 v/o SiC whisker composite were broken in four-point loading at 1200°C, employing a very low strain rate to generate fracture surfaces which approximate those which would have been observed in creep tests had the composite samples broken. The crosshead speed employed was 0.0005 cm/min, which is two orders of magnitude lower than that standardly used for MOR testing in this laboratory.

Examination of the fracture surfaces, Figure 8, reveals typical failure dominated by intergranular fracture for the monolithic AY6, while features for the AY6-SiC whisker materials give substantial evidence that the pullout of and bridging by the acicular dispersoids controls the low strain rate fracture of these composites.

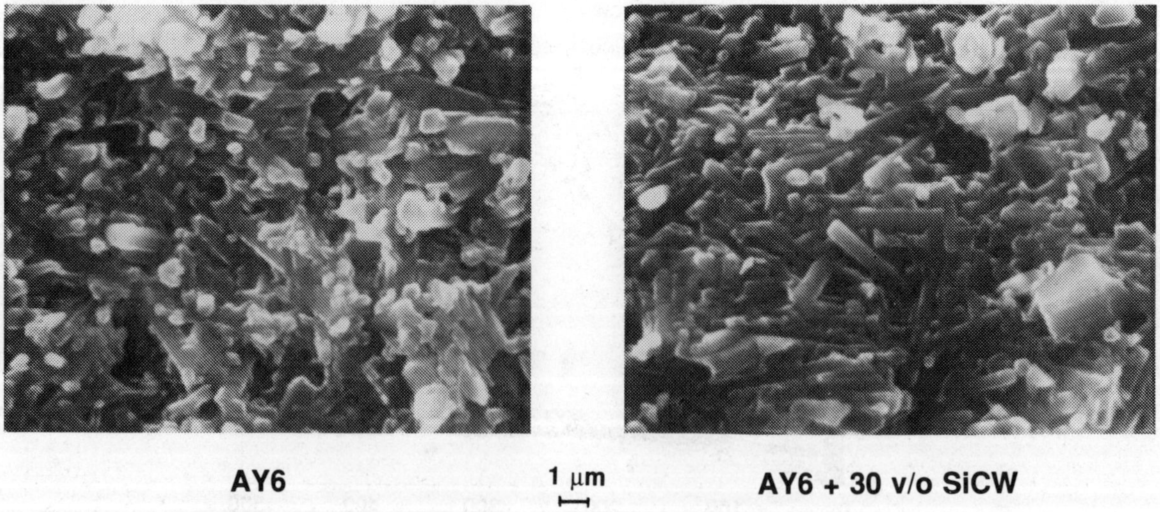

Figure 8. Fractographs of monolithic AY6 and an AY6-30 v/o SiC whisker composite broken in flexure at 1200°C at a crosshead speed of 0.0005 cm/min.

SUMMARY AND CONCLUSIONS

The addition of 30 v/o SiC whiskers to a (Si_3N_4 + 6 w/o Y_2O_3 + 1.5 w/o Al_2O_3) composition results in a composite ceramic body with substantially increased resistance to creep and subcritical crack growth at 1200°C in air in the stress range 68.9 to 275.6 MPa. The reduced content of sintering aid in the composites and the addition of acicular SiC dispersoids to the microstructure both contribute to the observed behavior. The whiskers appear to improve creep resistance by retarding grain boundary sliding. The evidence also suggests that whisker pullout/bridging is the operative mechanism which produces substantially increased resistance to slow crack growth, making the whisker-reinforced composite flaw tolerant at 1200°C. The substitution of equiaxed dispersoids for acicular matrix grains in silicon nitride ceramics containing intergranular glass phases can detrimentally affect time-dependent properties at elevated temperature.

ACKNOWLEDGMENT

The authors wish to acknowledge Mr. Michael Katsoulakos, who conducted the mechanical testing, and the GTE Laboratories Technical Publications Department who prepared the manuscript.

REFERENCES

1. G.C. Wei and P.F. Becher, "Development of SiC Whisker Reinforced Ceramics," *Am. Ceram. Soc. Bull. 64* [2], 298–304 (1985).
2. P.D. Sholek, et al., "Hot Pressed SiC Whisker/Si_3N_4 Matrix Composites," *ibid 65* [2], 351–356 (1986).
3. S.-T. Buljan, et al., "Si_3N_4-SiC Composites," *ibid 66* [2], 347–352 (1987).
4. J.J. Petrovic, et al., "Controlled Surface Flaws in Hot Pressed Si_3N_4," *J. Am. Ceram. Soc. 54* [3–4], 113–116 (1975).
5. D.W. Richerson, "Evolution in the U.S. of Ceramic Technology for Turbine Engines," *Am. Ceram. Soc. Bull. 64* [2], 282–286 (1985).
6. A.E. Pasto, et al., "Creep Behavior of Yttria and Alumina-Doped Silicon Nitrides," these proceedings.
7. S.-T. Buljan and G. Zilberstein, "Microstructure Development in Si_3N_4-Based Composites," *Mater. Res. Soc. Symp. Proc. 78,* 273–281 (1987).
8. S.-T. Buljan and J.G. Baldoni, "Silicon Nitride-Based Composites," in *Preparation and Properties of Silicon Nitride,* ed. D.A. Bonnell and T.Y. Tien (in print).
9. R.W. Evans, et al., "Rate Controlling Processes During Creep of Silicon Nitride Ceramics," *Trans. Br. Ceram. Soc. 87* [2], 54–58 (1988).
10. T. Felt, et al., "An Analysis of the Creep of Hot Pressed Silicon Nitride in Bending," *J. Mater. Sci. 23,* 467–474 (1988).
11. S.U. Dinand and P.S. Nicholson, "Creep of Hot Pressed Silicon Nitride," *J. Mater. Sci. 10,* 1375–1380 (1975).
12. R. Kossowsky, et al., "Tensile and Creep Strengths of Hot Pressed Si_3N_4," *J. Mater. Sci. 10,* 983–997 (1975).
13. J.M. Birch and B. Wilshire, "The Compressive Creep Behavior of Silicon Nitride Ceramics," *J. Mater. Sci. 13,* 2627–2636 (1978).
14. R.D. Nixon, et al., "Deformation Behavior of SiC Whisker Reinforced Si_3N_4," *Mater. Res. Soc. Symp. Proc. 78,* 295–302 (1987).
15. D.A. Koester, et al., "High Temperature Creep of SiC Whisker-Reinforced Ceramics," *Proc. Int. Conf. on Whisker and Fiber Toughened Ceramics, ASM* (in print).
16. G.D. Quinn, "Static Fatigue of a Sintered Silicon Nitride," AMMRC-TR-84-40 (1984).
17. P.K. Khandelwal, et al., "Slow Crack Growth in Sintered Silicon Nitride," in *Fracture Mechanics of Ceramics, Vol. 8,* ed. R.C. Bradt, et al., Plenum Press (1986).

HIGH TEMPERATURE CREEP CRACKING BEHAVIOUR WITH AN ADVANCED METHOD FOR DATA TREATMENT

Z. D. Guan W. J. Liu
Material Science & Engineering Department,
Tsinghua University
Beijing, PRC

ABSTRACT

High-temperature subcritical crack growth behaviour of HPSN has been studied by use of artificial knoop indentation defect. These were for actural using static and dynamic fatigue techniques. Better results have been obtained when modified formulae for static and dynamic techniques, in which the size of knoop defects was taken into account, were used. With an increase in temperature, the SCG parameters A and N increased and decreased respectively. Two SCG mechanisms, stress corrosion and cavitation, are discussed in detail.

INTRODUCTION

In recent years, high temperature structural ceramics have been developed and studied widely in many countries, so as to take advantage of its superior properties (such as refractoriness, high strength, resistance to oxidation and corrosion) to save energy. This demands a comprehensive understanding of high temperature deformation, oxidation and, especially fracture behaviour of ceramic materials. With an increase in temperature, the effect of sub-critical crack growth (SCG) in ceramics increases, it becomes one of the major process to determine the useful lifetime of ceramic materials.

In 1974, A. G. Evans studied the phenomenom of SCG of ceramics in room temperature and summarized the behaviour. The behaviour of subcritical crack growth can be described as[1]

$$da/dt = A\ K_1^N \tag{1}$$

Where da/dt and K_I are the crack growth rate and crack tip stress intensity factor respectively; A and N are SCG parameters, which depend on material composition, temperature and environment and so on. The experimental results of many kinds of ceramics verify the rule described above and it is found that the rule is also suitable at high temperature. At around 1300°C, A. G. Evans and S. M. Wiederhorn comfirmed the suitability of using Eq.(1) for hot-pressed silicon nitride (HPSN) at high temperature.[2]

Great interest has been given to SCG parameters and, studies are mainly concentrated on determining the SCG parameters. There are two methods for determining N and A, i.e. direct and indirect measurements. In direct measurement, the changes of crack length with time are measured and then the relationship between da/dt and K_I is determined by stress analysis. Indirect measurement is a method that the relationships between stress and duration time,or stress and stress rate are determined through experiments and then the results are treated by using the relation obtained from $E_q(1)$ to calculate the SCG parameters indirectly. Indirect measurement includes static fatigue (measuring the fracture lifetime t_f at different stress) and dynamic fatigue (measuring the fracture stress at different stress rate). The relations between them are derived from $E_q(1)$, and the simplified equation is written as[3]

$$t_f = A_1 \sigma^{-N} \quad (2)$$

$$\sigma_f = A_2 \dot{\sigma}^{1/(N+1)} \quad (3)$$

where both A_1 and A_2 include A and they depend on the original crack length, crack location and some other constant concerning stress regime. In practice, every specimen has a random distributed natural defects, so the corresponding values of A_1 and A_2 are not accordant between different specimens, and sometimes the disparity are significant. Therefore, mean value method obtained by many specimens is usually used to decrease the error as Eqs (2) and (3) are used. When a natural flaws is considered as the fracture source to determine SCG parameters, it requires a great deal of specimens and the value A can not be easily calculated from A_1 and A_2 because initial crack length a_o is unknown .

Some authors have used knoop indentation as a artificial defect instead of natural flaws to determine the parameter N, some good results have been obtained.[4] In their studies, the effect of the Knoop indentation size scatter to the results calculated was not taken into account when the same applied load and maintaining time for making indentation are used. The effects of indentation size to the values of A_1, A_2 and A are not considered also, and no method was proposed to calculate the parameter A.

This paper takes the Knoop indentation size into consideration and the data are treated by using the modified formulae derived from Eqs (2) and (3). The study of this kind of data treatment may provide an effective and convenient way to determine not only value N, but also value A.

THEORETICAL ANALYSIS

Two points must be considered when substituting natural flaws with an artificial one as the fracture source.

1. The shape of the artificial crack must be similar to natural one. This is why the knoop indentation is selected as artificial crack.

2. the stress intensity factor of an artificial crack can be determined. There have been some studies that show this condition can be satisfied only if the location of the knoop indentation is fixed.

In the condition of linear elastic, when the knoop indentation is located on the centre of tensile surface and perpendicular to the long axis of the three-point specimen, the crack tip stress intensity factor K_I can be expressed as[5]

$$K_I = M_e \sigma(\pi a/Q)^{1/2} \quad (4)$$

where σ and a are applied stress and indentation depth respectively, M_e and Q are indentation shape factors, they only relate to indentation depth and length ratio a/2c when a/B is less than 0.1 (B is the thickness of the specimens). The residual stress induced by indentation has analysed qualitatively. [6] [7] The influence of indentation-induced residual stress can be omitted before high temperature loading test.[8] Eq (4) can be simplified as

$$K_1 = Y \sigma a^{1/2} \quad (5)$$

Eq (5) is the relation usually used for K_I calculation. Y is geometry shape factor, $Y=M_e(\pi/Q)^{1/2}$. As long as the location and size of the knoop indentation meet the conditions described above and the ratio a/2c is constant, the geometry shape factor Y is a constant value.

With Eqs (5) and (1), the calculating formulae used in static and dynamic fatigue can be modified as follows.

1. Static fatigue expression including indentation size.

Substituting $K_I=Y\sigma(a)^{1/2}$ in Eq(1), we obtain

$$da/dt = A\,Y^N \sigma^N a^{N/2} \quad (6)$$

Assume the Y value doesn't change in the crack growth process (since the ratio a/2c doesn't change), after Eq(6) is integrated, (σ is constant in static fatigue) we obtain

$$[(2/(N-2)]\,(a_o^{-(N-2)/2} - a_c^{-(N-2)/2}) = A\,Y^N \sigma^N t_f \quad (7)$$

where a_o and a_c are initial crack and critical crack depth respectively, t_f is the fatigue lifetime. With consideration of $a_c \gg a_o$ and $a_o/2c_o = u$ (u is usually a constant for a given material), Eq(7) is simplified as

$$[(2c_o)^{1/2}\sigma] = A_1' \left(\frac{t_f}{2c_o}\right)^{-1/N} \quad (8)$$

where $2c_o$ = initial crack length $\quad A_1' = [2/(N-2)\,u^{(N-2)/2} A\,Y^N]^{1/N}$ (9)

A_1' is independent of the crack size and is a constant for a given material at a certain temperature. As long as the relation of $[(2c_o)^{1/2}\sigma]-(t_f/2c_o)$ is determined through experiment, the values of N and A_1' can be calculated more easily with Eq(8), and the value of A can be determined simultaneouly with Eq(9).

2. Dynamic fatigue expression including indentation size

Substituting $K_I=Y\sigma(a)^{1/2}$ and $d\sigma/dt=\dot{\sigma}$ in Eq(1), we obtain

$$a^{-N/2} da = \frac{AY^N}{\dot{\sigma}} \sigma^N d\sigma \quad (10)$$

Assume Y and u are constant, $a_c \gg a_o$ and $\sigma_f \gg \sigma_o$, we obtain Eq(11) by derivation similar to the static expression described above.

$$[(2c_o)^{1/2}\sigma_f] = A_2'[(2c_o)^{3/2}\dot{\sigma}]^{1/(N+1)} \quad (11)$$

where $$A_2' = [2(N+1)/(N-2)\ A\ Y^N\ u^{(N-2)/2}]^{1/(N+1)} \quad (12)$$

Similarly, A_2' is a constant at a certain temperature and independent of the crack size, the values of N and A can be calculated with Eq(11) and Eq(12) respectively.

Comparing Eqs(8) and (11) to Eqs (2) and (3), the modified expressions include the crack size effect. This makes the expressions more complicate, but as the A_1' and A_2' are constants and are independent with crack size, the data obtained and treatment are not troublesome. A more accurate results can be acquired with even one specimen in every constant stress or stress rate, therefore it is unnecessary to use the mean value method, in which several specimens are needed in order to delete the influence of the crack size scattering. So the method described in this paper makes it possible to use a few specimens to determine not only the value N and also the value A, which can not be easily calculated with the method used formerly.

EXPERIMENT DESCRIPTION

HPSN is used in this paper. The additives are 5 wt% Y_2O_3 and 2 wt% Al_2O_3.The sintering process is 1750°C/45min with the pressure of 250Kg/cm^2. The machining of the specimens includes cutting, grinding and polishing, and final polishing is done with W7 diamond paste on the tensile face and W28 diamond paste on the side faces. After polishing, the specimens are chamfered in the direction of angle 45° along the tensile edges so as to eliminate the defects produced during the machining. The final specimen size is about 4X3X36mm.

Knoop indentation, which is made with low-load hardness tester, is located on the middle of tensile face and perpendicular to the long axis of the specimen. Indented load varies from 0.5kg to 5.0kg and indentation time is 30 seconds. All the specimens are loaded in three-point bending with span 30mm. The Shimadzu Universal Testing Maching AG-2000A with high temperature accessories is used in all experiment. The load measurement accuracy and stress rate control accuracy of this machine are ±0.5% and ±0.1% respectively. All the data measurements are done in the air environment. In high temperature fatigue testing, the temperature varies from 1000°C to 1300°C. The specimens are maitained for 5 min at test temperature before testing so as to reach homogeneous temperature profile.

EXPERIMENTAL RESULTS

1. Knoop indentation morphology and mechanical characteristics

Knoop indentation morphology and mechanical characteristics are studied at room temperature. Before strength test, specimen indented in the way described above, the initial indentation size 2c and fracture stress σ_f of the specimen are measured and the relationship between them is built. The results are in Table 1. The curve of σ_f-$1/(2c)^{1/2}$ is drawn as fig.2, in a wide range

of 2c, the relationship of $\sigma_f - 1/(2c)^{1/2}$ is linear
Comparing this result with Eq(5), we know that $K_{IC}/Y(a/2c)^{1/2}\sigma_f$ is a constant, that is, Y is a constant with no dependence on indentation size as long as the ratio a/2c is constant.

Table 1. The relation between the length of Knoop indentations and the fracture stresses

σ_f(Mpa)	640	424	346	282
$2c_o$(μm)	62	125	171	245
$1/(2c_o)^{1/2}$	0.127	0.089	0.076	0.064

The indentation depth on the fracture profile is measured with optical microscope. For HPSN, the value of a/2c is about 0.42, which is similar to the results obtained by D. Quinn and J. B. Quinn[9]. Based on this results, Y=1.26 can be calculated with Eq(4). With the values of a/2c and Y, the SCG parameters of HPSN can be determined.

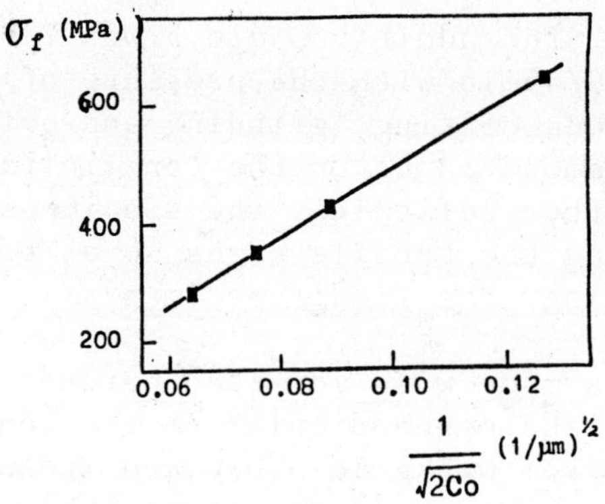

Fig.1 $\sigma_f - 1/(2c_o)^{1/2}$ curve of HPSN

2. Determining SCG parameters by static fatigue

The stress rupture test is done with the traditional technique, while the natural flawed specimens are considered. The results are shown in fig 2. The data are treated with Eq(2) to calculate the parameter N, at 1200°C and 1300°C, the value N is 22.5 and 18.5 respectively. In another experiment,the fatigue test is also done to use the knoop indentation as the fracture source. After removing of a few data which obtained when the fracture doesn't take place from the indentation, the results are treated with Eq(8) and drawn in Fig 3. At the temperature range of 1000°C and 1300°C, the value A is also calculated with Eq(9). Parameters N and A obtained by static fatigue technique are shown in Table 2.

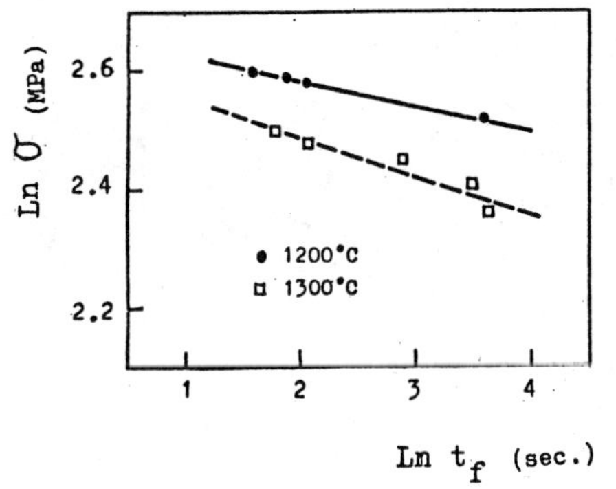

Fig.2 Static Fatigue Curve of Natural Flaw Specimen (HPSN)

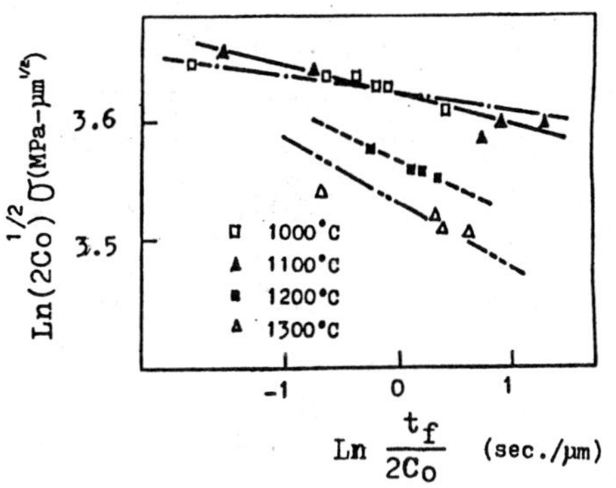

Fig.3 Static Fatigue Curve of Knoop Indentation Specimen (HPSN)

Table 2. Parameters N and A at Different Temperatures by Static Fatigue Technique

	1300°C		1200°C		1100°C		1000°C	
	N	A	N	A	N	A	N	A
Knoop Indentation	18.5 (r= 0.99)	10^{-16}	22.4 (0.93)	10^{-18}	39.1 (0.93)	10^{-29}	63.9 (0.85)	10^{-42}
Natural Flaw	15.5 (0.95)		24.5 (0.99)		——		——	

3. Determining SCG parameters by dynamic fatigue technique

At 1200°C and 1300°C, dynamic fatigue test is done to use the specimen with natural flaw or Knoop indentation. The results are treated following Eq(2) and Eq(11) respectively (fig.4 and fig.5), where every datum is the mean value of three tests for the natural flaw specimens, the data obtained when fracture doesn't happen from the indentation are removed.

The parameters N and A obtained by dynamic fatigue technique are tabulated in Table 3, where A is calculated with Eq(12).

Table 3. Parameters N and A at Different Temperature By Danymic Fatigue Technique.

	1300°C			1200°C		
	N	A	r	N	A	r
Knoop Indentation	14.1	10^{-13}	0.81	28.7	10^{-20}	0.83
Natural Flaw	13.4		0.76	19.6		0.80

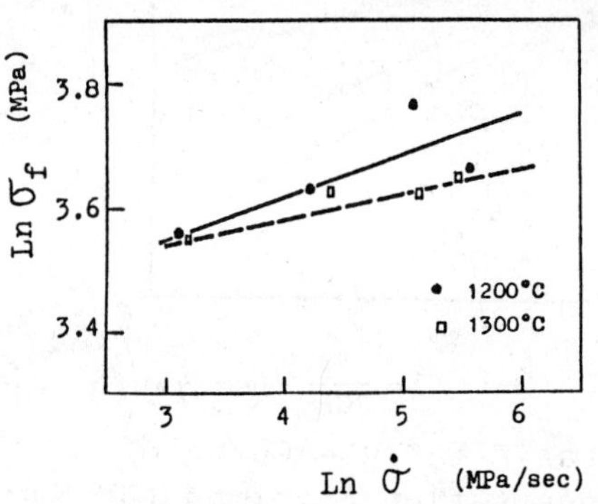

Fig.4 Dynamic Fatigue Curve with Natural Flaw (HPSN)

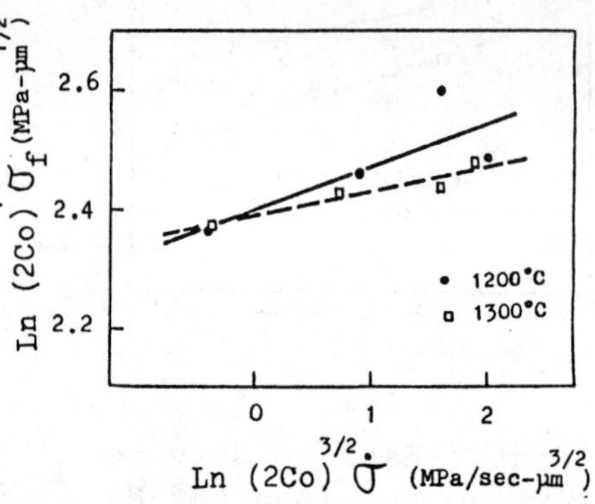

Fig.5 Dynamic Fatigue Curve of Knoop Indented Specimen (HPSN)

DISCUSSION

Before the knoop indentation replaces the natureal flaw as the fracture source, it is important to study the accordance of crack growth behaviour between them, i.e. whether the data obtained by the Knoop indentation specimen can be used for the specimen with the natural flaw. As mentioned in the theoretical analysis, it is difficult to determine the parameter A because of the unknown natural flaw size, so comparing N obtained by the two method is the only way to evaluate the effect of the two kinds of cracks. The authors[10] did the comparing test for the same material, it was found that there is no notable disparity between the values N obtained by the two kinds of crack specimens which treated by Eq(2). It was concluded tentatively that this two kinds of cracks have similar crack growth behaviour. The present study verifies this conclusion further.

Comparing Fig.4 and Fig.5, it can be seen that there is a very similar fluctuating tendency between the values measured by natural flaws and the ones measured by the artifical Knoop indentations. The values N obtained by static fatigue technique and dynamic fatigue technique are in Table 2 and Table 3 respectively, where includes the results of two kinds of cracks. No matter static or dynamic fatigue, there is no obviously different N between different types of cracks. Especially in the static fatigue techniques, it can be thought that the results obtained with the two kinds of cracks are similar with each other. The results above confirm the base view in this study, i.e. the SCG behavious of the Knoop indentation and the natural flaw are consistent and the known size Knoop indentation can replace the natural crack for SCG parameters measurement. Usually the linear regression correlation factor r can be improved after the Knoop indentations are used. As far as the results of static fatigue are concerned, the correlation factors are all greater than 0.93 at the temperature above 1100°C. This good results are favourable to practival application.

After using the Knoop indentation technique and taking the crack size into consideration, though the number of specimens in the test decreases, we obtain a better linear correlation factor because of eliminating the influence of random distribution both in size and location of the natural flaws. A better linear correlation factor indicates that the modified formulae and data treatment method are reliable.

In static and dynamic fatigue tests, the Knoop indentation size varies from 160μm to 270μm. In this range, a good result can be also acquired. It indicates that this method has a good suitability and can advance the study of component lifetime prediction when combined with the non-destructive inspection. In brief, this technique is a convenient and effective method.

The SCG phenomenon can be directly observed through the fractograph. At 1200°C and 1300°C, the SCG region size is large with length 3.0 mm and depth over 1.00 mm, indicating the important role of SCG in the fracture process. In Fig.6, the different morphology between SCG and brittle fracture region forms a clear line of demarcation. Generally speaking, SCG region is coarse and brittle region is smooth. There exists melted-like glass phase in SCG region which gives indirectly the evidence of the different crack growth rate of the two regions. The existing glass phase may be the result of SCG region's long time exposure to high temperature air and then surface oxidizing.

Fig.6 SEM Morphology of HPSN Fracture Surface at 1300°C

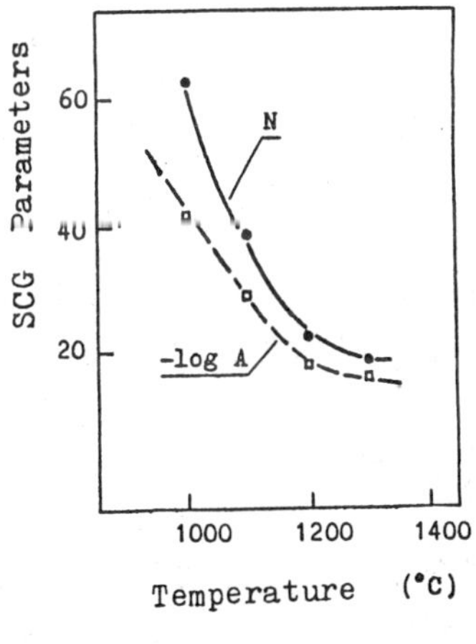

Fig.7 Temperature Dependent of SCG Parameters (HPSN)

Fig.7 shows the regulation of the SCG parameters changing with the temperature. With an increase in temperature, value N decreases and value A increases simultaneously. Especially when temperature is over 1200°C, the changes of N and A become smooth, indicating there are two kinds of SCG mechanisms playing the pricipal role in the different temperature range which devided by about 1200°C. There exists different activity energy for each SCG mechanism, showing the different sensitivity to temperature. In high-temperature mechanism,the activity energy is smaller,i.e. the temperature effect on SCG Parameters is less severe than that at lower temperature. Other authors have obtained the similar conclusion for HPSN [1]. In practical utilization , the transition temperature can be looked as the upper limit of applied temperature. When the temperature is over the limit, the creep and SCG behaviour of material deteriorate and the component will failure in a short period or mismatch with others.

There are two SCG mechanisms, one is the stress corrosion mechanism [11] which holds at all temperature range, the other is the creep fracture mechanism which acts only when the liquid phase exists at the boundary. [12] Considering viscosity of the boundary glass phase AL-Y-O begins to decrease at about 1200°C , we reckon that the stress corrossion mechanism acts before 1200°C and the creep fracture mechanism acts as the major role over 1200°C.

CONCLUSIONS

1. For given material,the ratio of length to depth for Knoop indentation is constant and no relate with indentation load. There exists a stable geometry shape factor in experimental range of this material, i.e. y=1.26.

2. Knoop indentation specimen has the same growth behaviour with the natural flaw one,a similar value N is obtained by this two kinds of cracks in fatigue testing.

3. After the Knoop indentation size is taken into account and the modified formula are used, the linear regression correlation factor is improved and the number of specimen used is decreased This technique, especially the static fatigue technique, has a large suitability range and is a convenient and effective method for determining SCG parameters N and A.

4. At around 1200°C, the changes of parameters with temperature become smooth for HPSN. The transition temperature may be related with the boundary glass phase softness, which is meaningful for both theoretical study and practical utilization.

ACKNOWLEDGENTS

The authors would like to express their appreciation to Wenjie Si for preparation of this paper.

REFERENCES

[1] A.G.Evans,"A Method for Evaluating the Time-dependent Failure Characteristics of Brittle Materials-and Its Application to Polycrystalline Alumina", J.Mater. Sci., V.7. 1137(1972)

[2] A.G.Evans, S.M. Wiederhorn,"Crack Propagation and Failure Prediction in Silicon Nitride at Elevated Temperatures", J.Mater. Sci., 9[2] 270-8(1974)

[3] R.K.Govila,"Material Parameters for Life Prediction in Ceramics", Ceramics for High-Performance Applications, ed. by J.J.Burke, A.E.Gorum, and R.N.katz, Brook Hill Pub. Co., Chestnut Hill, MA. 535-67(1983)

[4] J.E.Ritter, Jr., S.M.Wiederhorn,N.J.Tighe and E.R.Fuller, Jr.,"Application of Fracture Mechanics in Assuring Against Fatigue Failure of Ceramic Component," Ceramics for High-Performance Applications, ed. by J.J.Burke, A.E. Gorum, and R.N. Katz, Brook Hill Pub. Co., Chestnut Hill,MA.503-33(1983)

[5] Chen Chi, et al, Engineering Fracture Mechanics, Chinese military press. (1980)

[6] D.B.Marshall, B.R.Lawn,"Flaw Characteristics in Dynamic Fatigue: The Influence of Residual Contact Stresses", J.AcerS,63 (9-10).532-6(1980)

[7] E.R.Fuller, B.R.Lawn,"Theory of Fatigue for Brittle Flaws Originating from Residual Stress Concentrations", J.AcerS, 66(5) 314-21(1983)

[8] J.J.Petrovic, L.A.Jacobson, P.K.Talty, A.K.Vasudevan,"Controlled Surface Flaws in Hot-Pressed Si_3N_4", J.AcerS,58(3-4) 113-6(1975)

[9] G.D.Quinn, J.B.Quinn,"Slow Crack Growth in Hot-pressed Silicon Nitride," Fracture Mechanics of Ceramics, V.6, 603-36 Ed. by R.C.Bradt, D.P.H.Hasselmann, and F.F.Lange, Plenum Press, New York,(1982)

[10] Z.D.Guan, W.J.Liu,"Study on High Temperature Strength, K_{Ic} and Creep Fracture of Structural Ceramics", Mechanical Behaviour of Materials-V, Proceeding of the 5th intern. Conf., Vol.2.Ed.by M.G. Yan et al, Pergamon Press, 1277-82(1987)

[11] Z.D.Guan,Physical Properties of Ceramics, Ed.by Tsinghua press,1-59 (1981)

[12] A.G.Evans,"Creep Fracture in Ceramic Polycrystals",Creep and Fracture of Engineering Materials and Structures, Ed.by B.Wilshire et al.409(1981)

INVESTIGATION OF SELECTED Si_3N_4 AND SiC CERAMICS

N. L. Hecht,* D. E. McCullum, and G. A. Graves
University of Dayton
Dayton, Ohio 45469

ABSTRACT

The University of Dayton has been involved in a three-phase project to investigate the effects of environment on the mechanical behavior of commercially available ceramics being considered for heat engine applications. Phase I concerned with the evaluation of transformation toughened ceramics has been completed. Phases II and III are concerned with the evaluation of SiC and Si_3N_4 ceramics. Microstructure, chemistry, physical properties, and mechanical behavior at 25°C-1450°C are being investigated. A major focus of these evaluations is the flexural strength and dynamic fatigue of these materials from 1000°-1450°C.

INTRODUCTION

Since January 1985 the University of Dayton has been part of the Ceramic Technology for Advanced Heat Engines Program, through a subcontract with Martin Marietta at ORNL, to investigate the effect of environment on the mechanical behavior of commercially available ceramics being considered for use in heat engines.

In the first phase of this project the effects of environment on the mechanical behavior of commercially available ZrO_2 ceramics were investigated. In the second phase, two Si_3N_4 and one SiC were evaluated. Microstructure, chemistry, physical properties, and mechanical behavior at 25°, 1300°, and 1450°C were investigated. In the third phase, the tensile and fatigue properties of nine SiC and Si_3N_4 ceramics are being investigated at temperatures from 20-1400°C. Phases I and II have been completed and Phase III is in progress. The results of Phase II and the interim results of Phase III to date are described in this paper. The results of Phase I have been described elsewhere [1].

LITERATURE REVIEW

Since the late 1960's extensive effort has been underway to develop ceramic materials for use in heat engine applications. Major emphasis is on silicon nitride (Si_3N_4) and silicon carbide (SiC) ceramics. A significant body of literature has accumulated and it is apparent that a significant number of SiC and Si_3N_4 ceramics have been developed for high temperature structural applications. A large portion of the literature describes characterization studies and the property values obtained.

The literature was reviewed during the initial part of Phase II [2-26]. Property data available for the various SiC and Si_3N_4 ceramics were compiled according to fabrication category, a summary of which is presented in Table 1. Flexure strength, tensile strength, and elastic modulus property values are reported at both room temperature and 1400°C. The works of Larsen et al. [12-15], Quinn [23,24], Helms et al. [11], Richerson [26], and Govila [6-10] are particularly valuable in providing property data for a large number of these ceramics.

From the compiled data it can be seen that a considerable range of property values has been reported for the individual materials within any single process category. This wide range of values is probably due to the variations in chemical composition, starting powder size and morphology, and processing procedure used by the different material manufacturers. The large number of compositions and process variations are due, in part, to efforts to improve specific SiC and Si_3N_4 products.

At room temperature, the hot pressed Si_3N_4 and SiC materials have the highest strength. In general, the SiC materials retain a greater fraction of their room-temperature strength at elevated temperature, have somewhat better resistance to oxidation, and have greater creep resistance. The Si_3N_4 materials have lower coefficients of expansion and slightly higher fracture toughness.

Fracture origins at room temperature are reported to result from four sources: inclusions, very large grains, pores, machining flaws, or a combination of these. At elevated temperatures (above 1200°C) fracture origins usually occurred at oxidation pits, inclusions, machining flaws, or a combination of these. At temperatures above 1200°C time-dependent failure is reported for both Si_3N_4 and SiC [6-10]. However, the data in the literature are somewhat conflicting. Time-dependent failure appears to be more dominant for Si_3N_4 [14]. Subcritical crack growth, stress corrosion, and creep rupture are the primary mechanisms identified for time-dependent strength reduction.

EXPERIMENTAL PROCEDURE

For Phase II three materials (α-SiC, PY6, and XL144) were selected for evaluation. A brief description of these three materials is presented in Table 2. Rectangular test bars 51 x 6.4 x 3.2 mm were used for flexural strength measurements. The tensile surfaces of all test bars were ground, and the long edges of the tensile surface were chamfered.

Flexural strength measurements were made at 25°C, 1300°C, and 1450°C using an Instron Universal Testing Machine (Model 1123). The flexural strength measurements at 1300° and 1450°C were conducted in an ATS #3320 high-temperature furnace. Flexural strength was measured using a four-point bend test fixture and the specimens were loaded using machine crosshead speeds of 0.0064 and 0.00004 cm/sec. The data from the two different loading rates provided the basis for the dynamic fatigue analysis of the candidate materials. For the room temperature testing the bend fixture was made of steel and for the 1300° and 1450°C testing the bend fixture was made of SiC. The four-point bend fixture had an outer span of 3.81 cm and an inner span of 1.90 cm.

Fracture origin was determined by optical microscope (Nikon Epiphot Metallograph) and SEM (JOEL/SM - 840 with EG&G Ortec System 5000

Table 1. Summary of Reported Property Values for Si_3N_4 and SiC.

Material Type	Density (g/cc)	C.T.E.** 20°C-1000°C (10^{-6}/°C)	Young's Modulus GPa 20°C	Young's Modulus GPa 1400°C	Fracture Toughness (MPa $\sqrt{m}$)	Flexure Strength MPa 20°C	Flexure Strength MPa 1400°C	Tensile Strength MPa 20°C	Tensile Strength MPa 1400°C
Hot Pressed Si_3N_4 (HPSN)	3.18	3.3	300	175	4.8+	750	300*	375	150
(range)	3.07-3.37	3-3.9	250-325	175-250	2.8-6.6	450-1100	0-600	---	---
Sintered Si_3N_4 (SSN)	3.13	3.5	245	---	4.3	415	70	---	---
(range)	2.8-3.4	---	195-315	---	3.0-5.6	275-840	0-700	---	---
Reaction Sintered Si_3N_4 (RSSN)	2.4	2.9	175	155	3.6	200	250	170	140
(range)	2.0-2.8	2.5-3.1	100-220	120-200	---	50-300	0-400	70-210	---
Hot Pressed SiC (HPSC)	3.23	4.55	440	380	3.9	500	300	200	35-150
(range)	3.2-3.3	4.3-5.4	430-450	---	3.0-4.0	300-800	175-575	---	---
Sintered SiC (SSC)	3.1	4.5	395	372	4.0	375	380	---	---
(range)	3.0-3.2	4.4-4.8	375-420	300-400	2.5-6.5	275-535	240-450	---	---
Reaction Sintered SiC (RSSC)	3.0	4.33	360	275	---	310	190	77	---
(range)	2.9-3.1	4.3-4.4	350-375	200-320	---	175-450	70-450	---	---

*Above 1200°C dynamic fatigue shows slow crack growth.

+K_{Ic} increases at temperatures above 1000°C.

**C.T.E. = Coefficient of Thermal Expansion

Microanalysis System). The flexural strength values obtained were evaluated by Weibull analysis. Maximum likelihood estimates of the shape parameter, m (modulus), were determined.

The elastic modulus of three representative specimens of each candidate material was measured by a Grindo-Sonic (Model MR3ST) Transient Impulse/Elastic Modulus Apparatus. Hardness was measured on five representative samples from each candidate material using a Vickers microindent hardness tester. Thermal expansion was measured on four specimens from 25° to 1450°C using a Theta Industries Dilatronic II (Model 6024).

The microstructure and chemistry of the three candidate materials were also studied. Polished sections from representative specimens were viewed by optical microscope (Nikon Epiphot) and freshly fractured sections were viewed by scanning electron microscope (JOEL JSM-840). X-ray fluorescence (EDXA) spectra were obtained using the EG&G Ortec System 5000 attached to the JSM-840. Test specimens were also examined by x-ray diffraction (GE-XRD6). Fracture toughness was measured for the three materials by two techniques: the controlled flaw method (27) and the microindent method (28).

In Phase III nine ceramic materials (α-SiC, PY6, NT-154, SN252, SN-4, 147-3, ST, EC-152, and CVD) were selected for evaluation (see Table 2). To date five materials have been evaluated (SN-4, SN252, CVD, α-SiC, and NT-154). The characterization protocol used in Phase II was followed in the initial portion of Phase III with the exception of the flexural strength measurements. In Phase III flexural strength was measured according to the procedures described in MIL-STD-1942 (MR). The test specimens evaluated were 3 x 4 x 50 mm and the four-point flexure fixture had an outside span of 40 mm and an inner span of 20 mm. The dynamic fatigue behavior of these candidate materials from 1000°-1400°C was a major focus of this initial effort.

RESULTS

The density, elastic modulus, fracture toughness, and coefficient of thermal expansion measured for the candidate materials in Phase II and Phase III are presented in Table 3. The density measurements indicate that all of the materials are approximately 99% of the calculated theoretical density. The average density, elastic modulus, and coefficient of thermal expansion values measured were within the published range of values for these materials.

The fracture toughness values measured by the microindent technique tended to be higher than those measured by the controlled flaw method. However, both sets of fracture toughness values measured were lower than the values reported by the manufacturer.

The results of the flexural strength measurements made in Phase II for the three materials at 25°C, 1300°C, and 1450°C are presented in Table 4, as are the results of the Weibull analysis for this flexural strength data. The room-temperature flexural strength values obtained for α-SiC were in general agreement with published values. However, the strength values measured for the PY6 material were about 10% higher than published values, and the strength values for the first set of XL144 specimens were about 30% lower than manufacturer's published values. The second set of XL144 specimens had average flexural strengths more consistent with the values reported by the manufacturer.

Table 2. Candidate SiC and Si_3N_4 Ceramics Selected for Evaluation.

Sohio - Hexoloy α-SiC	Cold formed (injection molded or isostatic pressing) and pressureless sintered α-SiC with additions of B and Al
Sohio - Hexoloy ST	Cold formed and pressureless sintered α-SiC and TiB_2
GTE - PY6	Injection molded and hipped β Si_3N_4 with additions of Y_2O_3
Norton/TRW - NT154	Sintered and HIP'ed Si_3N_4
CVD - SiC	SiC prepared by chemical vapor deposition techniques
Howmet - SN-4	Cold formed (die pressing/isostatic pressing), sintered and HIP'ed Si_3N_4 with 10% oxide sintering aids (Al_2O_3, Y_2O_3)
Ceradyne - Ceralloy 147-3	Cold formed (isopressing/slip casting/injection molding) sintered Si_3N_4 with additions of Y_2O_3 and Al_2O_3
Kyocera - SN252	Cold formed (isopressed/slip cast/etc.) and sintered Si_3N_4 with additions of Yb_2O_3, Y_2O_3, and Al_2O_3 (note the sintering process is controlled to promote the growth of a whisker phase in the Si_3N_4 matrix)
NGK Spark Plug - EC-152	Gas pressure sintered Si_3N_4 with additions of Al_2O_3 and Y_2O_3

Table 3. Physical Property Measurements Obtained in Phases II and III.

	Materials					
Property Measurement	Sohio Hexoloy α-SiC	GTE PY6	Norton/TRW XL-144	Howmet SN-4	Kyocera SN-252	CVD Inc. SiC
Density (g/cc)	3.17	3.24	3.24	3.23	--	--
Young's Modulus (GPa)	427	293	314	204	--	470
Hardness (kg/mm^2)	2558	1458	1421	--	--	2600
Fracture Toughness CF/MI* (MPa·$\sqrt{m}$)	2.1/2.6	3.0/6.1	3.1/3.8	--	--	2.6
Coefficient of Thermal Expansion (10^{-6}/°C)	5	3.4	3.4	--	--	--
Project Phase Studied	II & III	II	II	III	III	III

*MI = microindent method

CF = controlled flaw method

+The fracture toughness values measured by the microindent method are found to be considerably higher (greater than 20%) than those measured by the controlled flaw method. It is reported that the lower values measured by the controlled flaw method are due to residual stresses introduced at the flaw. The very large difference in K_{Ic} values measured for the PY6 material by the two different methods could not be explained - Petrovic, J. J., et al., J. Am. Ceram. Soc., 59(3-4), pp. 177-179, 1976.

Table 4. Flexural Strength Measurements Made in Phase II.

Material	Temp. (°C)	Machine Crosshead Speed (cm/sec)	Strength (MPa)	Std. Dev. (MPa)	Weibull Modulus	Fracture Origin
αSiC						
	25	0.0064	331	41	9	Surface flaws.
	1300	0.0064	494	90	8	Surface flaws.
	1300	0.00004	500	77	8	Surface and edge flaws.
	1450	0.0064	500	52	10	Surface and edge flaws.
	1450	0.00004	455	59	9	Surface and edge flaws.
PY6						
	25	0.0064	641	69	11	Surface flaws.
	1450	0.0064	393	69	8	Surface, edge and subsurface flaws (mostly inclusions)+.
	1450	0.00004	241	36	7	Flaws and inclusions+ at or below the surface possible SCG.
XL144 (Batch 1*)						
	25	0.0064	538	97	6	Surface and edge flaws.
	1300	0.0064	604	44	13	Flaws and inclusions#.
	1300	0.00004	551	93	9	At or below the surface.
	1450	0.0064	453	53	12	Surface and edge flaws and subsurface inclusions#.
	1450	0.00004	213	12	6	At or below the surface SCG.
XL144 (Batch 2°)						
	25	0.0064	787	41	22	--
	1450	0.0064	537	76	7	--

*Specimens were machined from a hot pressed billet at the University of Dayton

°Specimens were prepared at Norton/TRW

#Inclusions - Fe, Cr

+Inclusions - Fe, Mo, Cr, Cu

The first set of XL144 specimens were machined and ground at the University. The second set of XL144 specimens were machined and ground by Norton/TRW. Both sets of specimens were machined to the same specifications. The average flexural strength of the second set was over 30% higher than the average room-temperature strength measured for the first set. The reason for the difference was not readily explained.

The flexural strengths measured for the three materials evaluated in Phase II at 1450°C were about 30% higher than the average strength values published for these materials. The two Si_3N_4 materials tested at 1450°C were observed to have undergone considerable oxidation when examined after testing. The α-SiC specimens did not appear to have undergone any observable oxidation. It is estimated that during flexural strength testing at 1450°C the test specimens are above 1000°C for about four hours. From the XRD analysis it was found that both cristobolite and $Y_2Si_2O_7$ formed on the surface of the Si_3N_4 test specimens.

Examination of the fractured test bars showed that at room temperature all of the α-SiC specimens failed due to tensile surface flaws. The PY6 and XL144 materials failed at both tensile surface and edge flaws. At 1450°C using the rapid loading rate it was observed that the α-SiC specimens failed at tensile surface and edge flaws and the PY6 material failed at tensile surface, edge, and subsurface flaws. The XL144 material tested at 1450°C failed primarily due to tensile surface flaws; however, some of the test specimens failed at edge flaws and at subsurface inclusions. It was further observed that failure initiation for the PY6 material at elevated temperature was also frequently at surface or subsurface inclusions. These inclusion sites were examined by EDXA and found to contain Fe,Cr, Cu, and Mo contaminates. The presence of Mo and Fe inclusions in sintered Si_3N_4 materials has been reported by Helms, et al. [11], Larsen [15], and Pasto [20]. Similar metal inclusions were also observed in the XL144 material.

For the α-SiC specimens tested at 1450°C and 1300°C using slow loading rates, fracture was initiated at flaws on the tensile surface or edges of the specimen (similar to the fast loading rate). However, in the XL144 and PY6 materials tested at 1450°C using the slow loading rate, fracture initiated at flaws or inclusions at or below the tensile surface and the failure process appeared to be initiated by slow crack growth. For the XL144 material tested at 1300°C using both fast and slow loading rates, fracture appeared to have been initiated at flaws and inclusions at or just below the tensile surface of the specimen. The flexural strength of the PY6 and XL144 ceramics at 1450°C is considerably lower for the slow loading rate, suggesting that these two materials are subject to time-delayed failure (fatigue).

In phase III the flexural strength of a second batch of Hexoloy α-SiC, Howmets SN-4, Kyocera SN252, CVD-SiC, and Norton/TRW-NT154 was evaluated. The results of these measurements are presented in Table 5. As shown in Table 5 the second batch of α-SiC had higher average room-temperature strength but lower average elevated-temperature strength. Flexural strength did not vary more than 50 MPa from 20°C-1400°C at both loading rates. As observed in Phase II, fracture origin for the Hexoloy material was due to surface and edge flaws. However, a majority of the flaws observed were found to be voids or pits at or just below the tensile surface. These pits or voids were not observed in the batch of specimens evaluated in Phase II.

Table 5. Flexural Strength Measurements Made in Phase III

Material	Machine Temperature °C	Flexural Crosshead Speed cm/sec	Strength MPa	Standard Deviation MPa	Weibull Modulus	Fracture Origin
α-SiC	25	0.004	402	52	9	Surface flaws
	1000	0.004	416	30	18	Surface flaws and pits
	1000	0.00004	412	37	12	Surface flaws, pits, and inclusions
	1200	0.004	435	53	9	Surface flaws pits
	1200	0.00004	441	6	10	Surface flaws pits
	1400	0.004	436	48	11	Surface and edge flaws and pits
	1400	0.00004	390	36	15	Surface and edge flaws and pits
SN-4	25	0.004	977	129	12	Surface flaws
	1000	0.004	744	91	--	Surface flaws
	1000	0.00004	514	47	--	Surface flaws possible SCG
	1200	0.004	402	40	--	Surface flaws
	1200	0.00004	312	15	--	Surface flaws possible SCG
	1400	0.004	190	--	--	Severe oxidation and blistering
SN-252	1000	0.004	526	31	20	At surface and edge due to large whiskers (rows 1000–1400, 0.004)
	1000	0.00004	466	26	20	
	1200	0.004	478	43	13	
	1200	0.00004	485	35	20	
	1400	0.004	459	29	19	
	1400	0.00004	498	29	21	Severe plastic deformation prior to fracture
CVD Sic	25	0.0064	554	120	--	--
	1400	0.0064	556	28	--	--
NT-154	25	0.004	907	79	15	Surface and edge flaws
	1400	0.004	610	30	23	Surface and edge flaws

The Howmet SN-4 material had very high room-temperature flexural strength and 1000°C fast fracture strength. However above 1000°C the strength decreased dramatically and at 1400°C the material was severely oxidized. The material also appeared to be susceptible to slow crack growth at 1000° and 1200°C.

Only a limited number of Kyocera SN-252 test specimens have been evaluated to date. However, the data obtained showed this material to be unique compared to the other Si_3N_4 materials evaluated. The SN-252 material contained significant quantities (5-6%) of Yb_2O_3 and very small quantities (less 0.5%) of Al_2O_3 and Y_2O_3. In addition the Si_3N_4 matrix contains a whisker-reinforcing phase dispersed throughout the matrix. This fibrous phase is developed during the sintering treatment by the use of a proprietary process. However, it was observed that most of the fracture origins were at large fibers/whiskers located at or just below the tensile surface. We have been advised that Kyocera is working to better control the whisker development process to prevent large whiskers from growing too large [29]. At the 1400°C slow loading rate, all of the specimens tested appeared to have undergone severe plastic deformation prior to fracture.

The preliminary testing of the CVD-SiC and NT-154 showed both materials to have impressive flexural strengths. The CVD-SiC had room-temperature and 1400°C flexural strength about 10 to 20% higher than the Hexoloy SiC evaluated in Phases II and III. The room-temperature flexural strength measured for the NT-154 is comparable to the highest values reported for hot pressed Si_3N_4. However, considerably more property data measurements are required for these materials.

CONCLUSIONS

Phase II of this project provided a limited evaluation for the three candidate SiC/Si_3N_4 materials. From the results obtained it was concluded that:

(1) In general, the property values measured were in agreement with published values for these materials.

(2) At 1450°C the Si_3N_4 ceramics investigated exhibited considerable susceptibility to slow crack growth (fatigue). The α-SiC ceramic did not show any susceptibility to slow crack growth at this temperature. At 1300°C slow crack growth was not observed for any of the materials evaluated. The measured flexural strength of the two Si_3N_4 ceramics at 1450°C was between 20 and 35% lower than the strengths measured at 25°C. The measured flexural strength of the α-SiC increased at 1450°C. At 1300°C the flexural strength of the XL144 Si_3N_4 ceramic decreased only 22% from its room-temperature value. Although the α-SiC had lower room-temperature strength than the Si_3N_4 ceramics, it maintained significantly higher strength values at 1450°C than the values measured for the Si_3N_4 ceramics. At 1300°C the α-SiC and the XL144 Si_3N_4 were more equivalent, and in fact, the XL144 had somewhat higher strength. The XL144 Si_3N_4 ceramic also appears to have greater reliability with its higher Weibull modulus values.

(3) The Hexoloy α-SiC material offers considerable promise for structural application at temperatures of 1450°C compared to the other two materials studied.

The results to date for Phase III indicate that the Howmet material does not appear to be effective for use at temperatures above 1000°C. For the other materials studied in Phase III considerably more data are required.

REFERENCES

1. Hecht, N. L., et al., "Environmental Effects on Toughened Zirconia Ceramics," Advances in Ceramics, Vol. 24, Science and Technology of Zirconia III, Am. Ceram. Soc., 1988.

2. Bradt, R. C. Pennsylvania State University, NASA-CR-165325, April 1984.

3. "Engineering Property Data on Selected Ceramics, Volume I, Nitrides," Metals and Ceramics Information Center, Battelle, MCIC-HB-07-Vol. I, March 1976.

4. "Engineering Property Data on Selected Ceramics, Volume II, Carbides," Metals and Ceramics Information Center, Battelle, MCIC-HB-07-Vol. II, August 1979.

5. Godfrey, D. J. "Silicon Nitride Ceramics for Engineering Applications," SAE paper 740238, 1974.

6. Govila, R. K. "Fracture and Intermediate Temperature Instability in Yttria-Doped Hot-Pressed Silicon Nitrides," Am. Ceram. Soc. Bull., 65(9), 1287-92, 1986.

7. Govila, R. K. "Strength characterization of yttria-doped sintered silicon nitride," J. of Mater. Sci., 20, 4345-4353, 1985.

8. Govila, R. K. "Phenomenology of fracture in sintered alpha silicon carbide," J. of Mater. Sci., 19, 2111-2120, 1984.

9. Govila, R. K. "High temperature uniaxial tensile stress rupture strength of sintered alpha SiC," J. of Mater. Sci., 18, 1967-1976,1983.

10. Govila, R. K. Uniaxial Tensile and Flexural Stress Rupture Strength of Hot-Pressed Si_3N_4," J. Am. Ceram. Soc., 65(1), 15-21, 1982.

11. Helms, H. E., P. W. Heitman, L. C. Lindgren, and S. R. Thrasher. Ceramic Applications in Turbine Engines, Noyes Publications, New Jersey, 1986.

12. Larsen, D. C. and J. W. Adams. "Property Screening and Evaluation of Ceramic Turbine Materials," ADA 142739, April 1984.

13. Larsen, R. P. and L. R. Johnson. Argonne, IL, ANL/CNSV-47, August 1984.

14. Larsen, D. C., et al., "Evidence of Strength Degradation by Subcritical Crack Growth in SiC and Si_3N_4," in Fracture Mechanics of Ceramics, Vol. 6, Evans, A. G., et al., Plenum Press: New York, 1984.

15. Larsen, D. C., et al., "Ceramic Materials for Advanced Heat Engines," Technical and Economic Evaluation, Noyes Publications: Park Ridge, New Jersey, 1985.

16. Lenoe, E. M., R. N. Katz and J. J. Burke, eds. Ceramics for High Performance Applications-III-Reliability, Plenum Press: New York, 1983.

17. Neil, J. T., K. W. French, C. L. Quackenbush and J. T. Smith. "Fabrication of Turbine Components and Properties of Sintered Silicon Nitride," ASME Paper #82-GT-252, 1982.

18. Pasto, A. E. in Ceramic Engineering and Science Proceedings, 5, 385-396, May/June 1984.

19. Pasto, A. E. J. Am. Ceram. Soc., 67(9), September 1984.

20. Pasto, A. E., J. T. Neil and C. L. Quackenbush. "Microstructural Effects Influencing Strength of Sintered Silicon Nitride," Proc. of International Conf. on Ultrastructure Processing of Ceramics, Glasses, and Composites, 1984.

21. Quackenbush, C. L. and J. T. Smith. "Sintering, Microstructure, and Properties of Si_3N_4- and SiC-Based Structural Ceramics," ASME Paper #81-GT-220, 1981.

22. Quackenbush, C. L., J. T. Smith, J. T. Neil, and K. W. French, "Sintering Properties, and Fabrication of Si_3N_4 + Y_2O_3 Based Ceramics," Progress in Nitrogen Ceramics, ed. by F. L. Riley, pp. 669-682, 1983.

23. Quinn, G. D., "Characterization of Turbine Ceramics After Long-Term Environmental Exposure," Report No. AMMRC-TR-80-15, Army Materials and Mechanics Research Center, April 1980.

24. Quinn, G. D. and J. B. Quinn. "Slow Crack Growth in Hot Pressed Silicon Nitride," in Fracture Mechanics of Ceramics, Vol. 6, Evans, A. G., et al., Plenum Press: New York, 1984.

25. "Reliability of Ceramics for Heat Engine Applications," Natl. Materials Advisory Board, NMAB-357, 1980.

26. Richerson, D. W. Modern Ceramic Engineering - Properties, Processing, and Use in Design, Marcel Dekker, Inc., New York and Basel.

27. Petrovic, J. J. and Mendiratta, M. G. "Fracture from Controlled Surface Flaws," Fracture Mechanics Applied to Brittle Materials ASTMSTP6785, W. Freiman, Ed., ASTM, pp. 83-102, 1979.

28. Evans, A. G., "Fracture Toughness: The Role of Indentation Techniques," Ibid, pp. 112-135.

29. Okajima, K., Kyocera America Inc., Personal Communication.

CORRELATION OF SUBCRITICAL CRACK GROWTH AND MICROSTRUCTURE IN HIP-SiC

W. Braue, J. Göring and G. Ziegler
DFVLR, German Aerospace Research Establishment
Materials Research Institute
D 5000 Köln 90, FRG

ABSTRACT

Two grades of hot isostatic pressed (HIP-)SiC - undoped and with 0.2 wt.% Al added as sintering aid - were tested with respect to subcritical crack growth. Mechanical testing established a distinct crack extension at T > 900 °C which was identified as intergranular fracture area at the fracture surface. Subcritical crack extension was found to be more pronounced in the Al-doped material. Microstructural characteristics were evaluated by means of analytical electron microscopy. In both HIP-SiC materials, nm-sized non-crystalline secondary phases were identified via convergent beam electron diffraction (CBED) at triple grain junctions which are regarded to promote subcritical crack growth. Small probe microanalysis of these amorphous phases revealed an aluminium silicate composition.

INTRODUCTION

HIPing of low impurity SiC starting powders is a well performed processing technique for economic production of net-shaped SiC structural ceramics with good mechanical and thermo-mechanical properties. HIP-SiC exhibits a uniformed, fine-grained microstructure which maintains relatively high strength at elevated temperatures. Moreover, reduction or even dispensing of sintering additives will generally improve materials properties such as creep or slow crack growth at high temperatures. Both the increase of thermal conductivity resulting in good thermal shock behavior as well as the significantly reduced scattering of strength data make HIP-SiC a strong competitor to sintered SiC materials[1].

Regarding engineering applications of HIP-SiC as structural component, subcritical crack growth at room and high temperatures defines a crucial materials property which describes the long-term failure behavior[2].

EXPERIMENTAL PROCEDURE

Materials and Microstructures

Two types of commercially available hot-isostatically pressed silicon carbide materials starting from genuine α-SiC powder were investigated: HIP-SiC with 0.2 wt.% Al, and such without any sintering additives. Both materials were fully densified.

Analytical electron microscopy was performed for a closer inspection of HIP-SiC microstructures. TEM foils were prepared by conventional ion beam thinning. In addition to a Philips EM 430 (300 kV), a Philips EM 400 TEM/STEM fitted with a field emission gun (100 kV) was used in particular for microanalytical experiments. During small probe microanalysis and CBED a cold stage was used to prevent specimen contamination.

Mean grain size of silicon carbide was between 1 and 2 microns with a few grains reaching up to 20 microns. Systematic TEM tilting experiments of α-SiC grains reveal 6H-SiC to be the dominant polytype in HIP-SiC (Figure 1). Typically α-SiC exhibits a high density of planar faults as indicated by intensive streaking in the corresponding diffraction patterns. Grain size distribution of the Al-doped material was found to be more inhomogeneous than in the undoped HIP-SiC which is due to discontinuous grain growth of the 6H-SiC promoted by aluminium addition[3].

A considerable amount of secondary phases predominantly located at triple grain junctions was observed in both HIP-SiC materials. The nature of low contrast triangular-shaped particles was confirmed as mostly non-crystalline by means of CBED. Their grain size may scale 0.5 microns in the Al-doped material. Amorphous phases similar in morphology - although to a lower amount and developing significantly smaller grain sizes - were identified in the undoped HIP-SiC. A more detailed characterization of the non-crystalline phases is given in section II (see Figure 4).

In addition, both HIP-SiC materials are characterized by graphite-rich areas (Figures 2 and 3). They consist of graphite-laths stacked parallel to a preferential orientation displaying varying degrees of structural disorder. Typically their grain sizes are considerably smaller than the corresponding average α-SiC grain dimension. Despite their intergranular character in HIP-SiC these graphite-rich areas are very similar to those reported from sintered (B+C)-doped α-SiC[4]. No graphite-rich porous triple grain junctions or any aluminium carbides which both are common features in Al-doped sintered SiC[5] were established in the HIP-SiC materials investigated.

Subcritical Crack Growth

Besides critical crack phenomena some non-oxide ceramic materials exhibit subcritical crack extension. Following previous investigations[6,7] subcritical crack growth is given by a power law,

$$v = A \cdot K_I^N$$

with the crack velocity given by v, the stress intensity factor by K_I and the crack extension parameters by A and N, respectively.

In this study subcritical crack extension was determined by dynamic fatigue experiments. Evans[8,9] has shown that fracture strength σ_f is related to the stress rate $\dot{\sigma}$. From a regression analysis of log σ_f versus log $\dot{\sigma}$ the crack velocity exponent N is given by the slope. The parameter A can be calculated by the intercept, the fracture toughness K_{Ic} and the inert bending strength σ_c. Flexural strength was determined by four-point bending tests (40/20 mm span) with different crosshead speeds between 5 mm/min and 0.5 microns/min. At least 10 specimens were used for evaluating the mean bending strength σ_f. The inert strength σ_c and the Weibull modulus m were measured at room temperature (30 specimens) with a crosshead speed of 5 mm/min. Fracture surfaces of almost all specimens tested were inspected via SEM.

RESULTS AND DISCUSSION

The results from the dynamic fatigue experiments of the HIP-SiC doped with 0.2 wt.% Al versus different temperatures are displayed in Figure 5 showing a consistently good correlation of log σ_f versus log $\dot{\sigma}$. Crack velocity parameters at room temperature, 900 °C, 1000 °C and 1200 °C are listed in Table 1. The inert bending strength was determined to be σ_c = 471 ±69 MPa with a Weibull modulus of m = 7.3. Room temperature fracture toughness was measured to be 4.5 $MN/m^{3/2}$ applying the hardness indentation method[10]. The Al-doped HIP-SiC exhibits only minor subcritical crack growth (N = 116) at room temperature. However, at temperatures exceeding 900 °C an intensive subcritical crack growth occurs. The value of N is almost constant over the temperature range between 1000 °C and 1200 °C (N = 30), but there is a distinct decrease in strength.

The HIP-SiC without sintering additives was tested at 1000 °C. In this case, calculation of the crack velocity parameters was not possible because of unsufficient linear correlation of log σ_f and log $\dot{\sigma}$ (Figure 6). Nevertheless some subcritical crack growth could be detected based on the evaluation of the intergranular fracture area. SEM fracture surface studies indicated a less pronounced subcritical crack growth than in the case of the

Al-doped HIP-SiC material. Moreover, these investigations emphasized that both materials actually follow the same fracture mechanism.

Fracture surfaces can be classified into three distinct areas (Figure 7). Area I covers the failure origin which in 85 % of the specimens investigated were accumulations of large grains and the rest, pores. There was no indication for a failure caused by machining. Around the failure origin a 100 % intergranular fracture mode was observed (area II) which was very pronounced in specimens tested at high temperatures. Area II is characterized by slow subcritical crack growth. With the crack becoming critical and extending with very high speed the intergranular fracture mode changes into a 100 % transgranular fracture surface (area III). With still increasing crack velocity there is a continuous change to a mixed mode fracture surface.

For both HIP-SiC materials it could be shown that the decrease in bending strength at higher temperatures and lower stress rate correlates well with an increase of the intergranular fracture area. In particular the decrease in bending strength of the Al-doped HIP-SiC between 1000 °C and 1200 °C could be well explained by a greater area of intergranular fracture at 1200 °C. For HIP-SiC without sintering additives intergranular fracture was only found for specimens tested at the lowest stress rate and was less extensive compared with the Al-doped material.

Characteristics of subcritical crack growth in both HIP-SiC materials as determined by dynamic fatigue experiments correlate well with both presence and location of non-crystalline phases in the microstructures. Small probe microanalysis from an approximately 50 nm-sized amorphous pocket in Al-doped HIP-SiC (Figure 4) reveals an enrichment of aluminium over silicon in the EDS spectrum and a significant oxygen K absorption edge in the ELS spectrum, respectively. Al was not found to be incorporated into the α-SiC matrix. No carbon was detected by means of ELS, thus inferring an aluminium silicate composition for the glassy phase. As there was no evidence for any aluminium in the graphite-rich areas, the glassy phase obviously constitutes the major sink for aluminium in HIP-SiC as long as only minor quantities are added to the batch or are intrinsic to the sintering powder employed.

Low refractory non-crystalline phases similar to those from HIP-SiC are known to trigger subcritical crack growth in various polyphase non-oxide structural ceramics. This effect is a characteristic feature in the Si_3N_4 system, but - although less pronounced - holds too for SiC materials. Severe strength degradation in various HIP-SiC materials is explained by low viscous non-crystalline phases[1] which will induce subcritical crack growth at elevated temperatures. Lange[11] postulated an aluminium silicate liquid in Al_2O_3-doped HIP-SiC. Thus in conclusion the amorphous secondary phases identified at triple grain junctions are regarded as the most critical microstructural feature for subcritical crack growth in HIP-SiC. As demonstrated in both the dynamic fatigue experiments and the microstructural characterization this mechanism holds for HIP-SiC irrespectively of a dopant added to the batch. It is reasonable to expect that even in the case of the undoped HIP-SiC the oxygen level of the genuine α-SiC starting powder - typically around 0.2 wt.% - if combined with impurities from the powder particle surfaces will balance to substantial amounts of silica-rich liquid phase mobilized in the high temperature regime during HIPing.

Compared to the impact of non-crystalline phases the influence of graphite-rich inclusions on subcritical crack growth in HIP-SiC is regarded of minor importance. In sintered SiC materials Shinozaki et al.[5] report the graphite-rich areas to remain very stable up to sintering temperature of 2000 °C. Moreover, there is evidence for even healing of microcracks within the graphite bundles during annealing[12].

CONCLUSIONS

Dynamic fatigue experiments have established substantial subcritical crack growth in HIP-SiC in excess of 900 °C irrespectively of a sintering additive added. The effect, however, was found to be more pronounced in the material doped with 0.2 wt.% Al which correlates well with the amount of non-crystalline phases identified at triple grain junctions in both HIP-SiC materials. Small probe microanalysis (EDS, ELS) revealed an aluminium-rich silicate composition which translates into a glassy phase with low viscosity and unsufficient refractoriness at elevated temperatures, thus promoting grain sliding effects and subcritical crack growth. Graphite-rich areas frequently observed in both HIP-SiC materials are regarded as less critical with respect to long-term failure characteristics of HIP-SiC at high temperatures.

ACKNOWLEDGEMENT

Small probe microanalysis on HIP-SiC was performed while W.B. was with the Center For Solid State Science, Arizona State University, AZ. Assistance and support from Prof. R.W. Carpenter were kindly appreciated. Helpful comments on the final draft were contributed by G. Quinn, MTL, Watertown, MA.

Research grants from Deutsche Forschungsgemeinschaft (W.B.) and Volkswagen Stiftung (J.G.) are gratefully acknowledged.

REFERENCES

1. K. Hunold, Sinterung und heißisostatische Nachverdichtung von Siliciumcarbid, Sinter-/HIP-Technologie, Gemeinschaftsausschuß für Pulvermetallurgie, 2-3 Dec. 1987, Hagen, FRG.

2. J. Göring and G. Ziegler, Subcritical Crack Growth for the Characterization of the Thermal Cycling Behavior of SiC and Si_3N_4-Materials, Science of Ceramics 14 (1987), 627-632.

3. W. Böcker, H. Landferman and H. Hausner, The Influence of Boron and Carbon Additions on the Microstructure of Sintered Alpha Silicon Carbide, Powd. Met. Int. 11 (1979), 83-85.

4. W. Braue, R.W. Carpenter and D.J. Smith, Analytical Electron Microscopy of Graphite-Rich Inclusions in Sintered α-SiC, submitted to J. Mater. Sci. (1988).

5. S. Shinozaki, R.M. Williams, B.N. Juterbock, W.T. Dunlon, J. Hangas and C.R. Peters, Microstructural Developments in Pressureless-Sintered ß-SiC Materials with Al, B, and C Additions, J. Am. Ceram. Soc. Bull. 64 (1985), 1385-1393.

6. J.W. Edington, D.J. Rowcliffe and J.L. Hemdshall, The Mechanical Properties of Silicon Nitride and Silicon Carbide, Part II Engineering Properties, Powder Metallurgy International Vol. 7 (1975), 136-147.

7. A.G. Evans and F.F. Lange, Crack Propagation and Fracture in Silicon Carbide, J. Mater. Sci. 10 (1975), 1659-1664.

8. A.G. Evans, Slow Crack Growth in Brittle Materials Under Dynamic Loading Conditions, Int. J. Fract. 10 (1974), 251-259.

9. A.G. Evans and H.J. Johnson, The Fracture Stress and its Dependence on Slow Crack Growth, J. Mater. Sci. 10 (1975), 214-259.

10. M. Majdic and G. Ziegler, Zusammenhang zwischen Rißlängen bei Vickers-Härteeindrücken und dem Bruchwiderstand keramischer Werkstoffe, Ber. Dt. Keram. Ges. 57 (1980),140-145.

11. F.F. Lange, Hot-Pressing Behavior of Silicon Carbide Powders with Additions of Aluminium Oxide, J. Mater. Sci. 10 (1975), 314-320.

12. N.J. Tighe, K.A. Hardman-Rhyne and Y.N. Lu, Characteristic Cavity and Microcrack Distributions in α-SiC, Si_3N_4, and ZrO_2, Proceedings of the 9th Annual Conference on Composites and Advanced Materials, Cocoa Beach, FL (1985), 835.

Materials	Temperature in °C	Crack Velocity Parameters N	lg B
HIP-SiC with Al	RT	116	3.47
	900	41	5.15
	1000	29	5.03
	1200	31	0.58
HIP-SiC without additives	1000	Low subcritical crack growth was observed, but could not be calculated	

Table 1. Subcritical crack extension parameters of HIP-SiC materials investigated with B given by $2/[AY^2(N-2)\ K_c^{N-2}]$ and Y as geometry factor

Figure 1. BF image demonstrating heavily faulted structure and discontinuous grain growth of α-SiC in Al-doped HIP-SiC. 6H-SiC polytype is revealed by further tilting into a $(1\ 1\ \bar{2}\ 0)$ orientation (see diffraction pattern inserted).

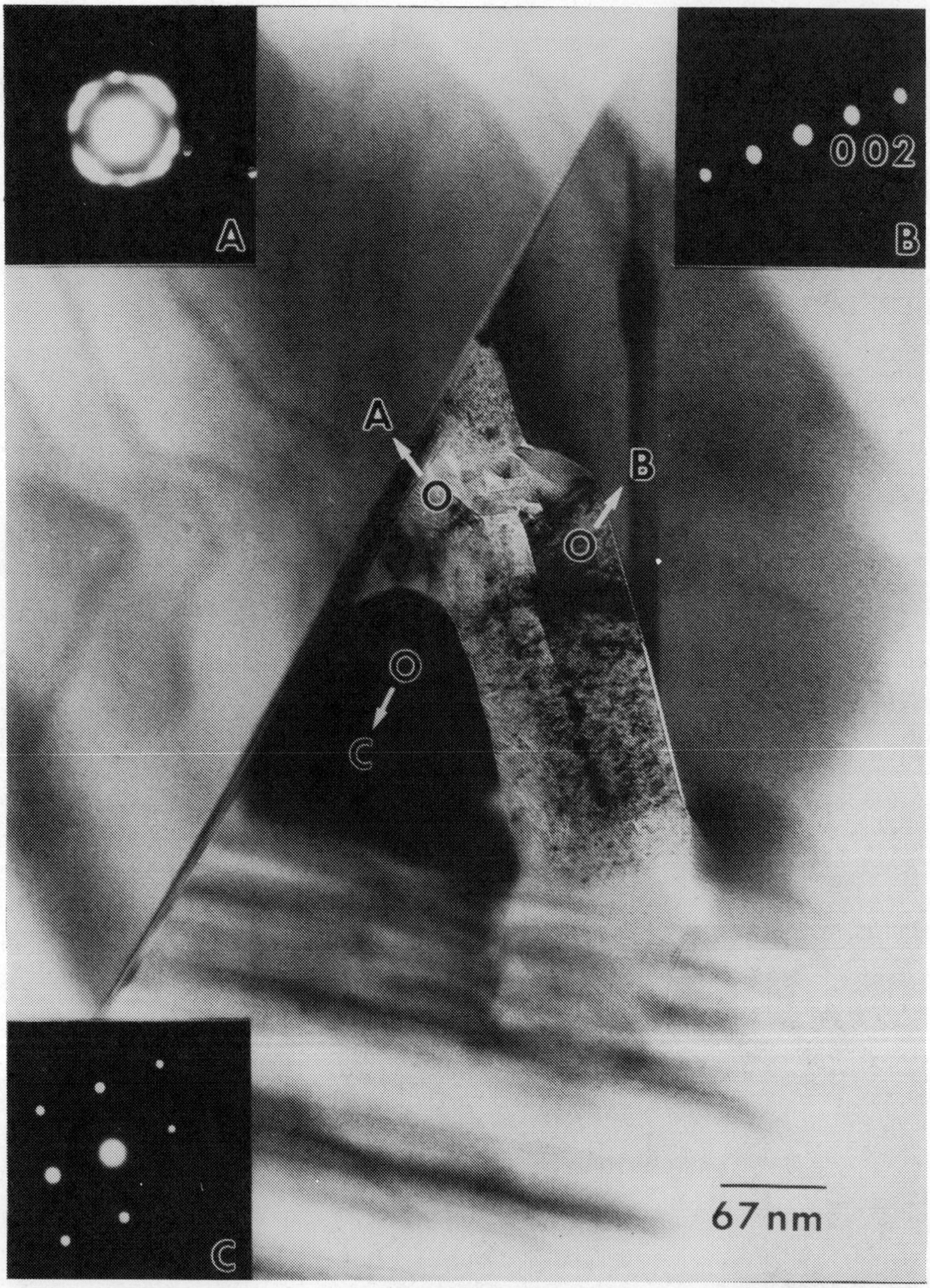

Figure 2. Graphite-rich area at triple grain junction in Al-doped HIP-SiC. CBED's taken from different positions refer to A) isotropic polycrystalline graphite, B) single graphite filament in preferential orientation and C) 6H-SiC close to ($\bar{3}$ 3 0 $\bar{1}$) zone axis.

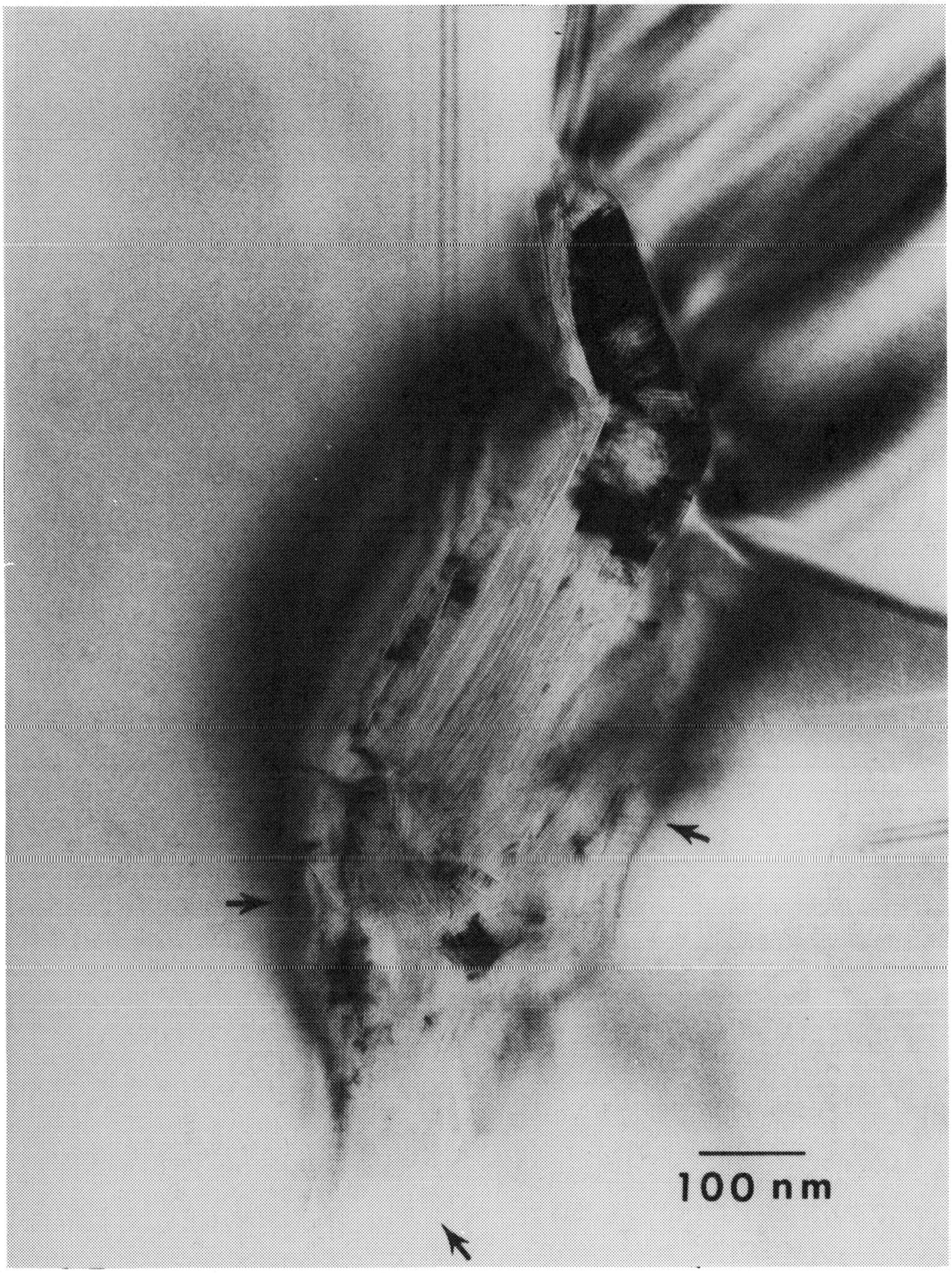

Figure 3. Graphite-rich area in undoped HIP-SiC showing deformed graphite extending into pocket of microcrystalline graphite filaments in different orientations.

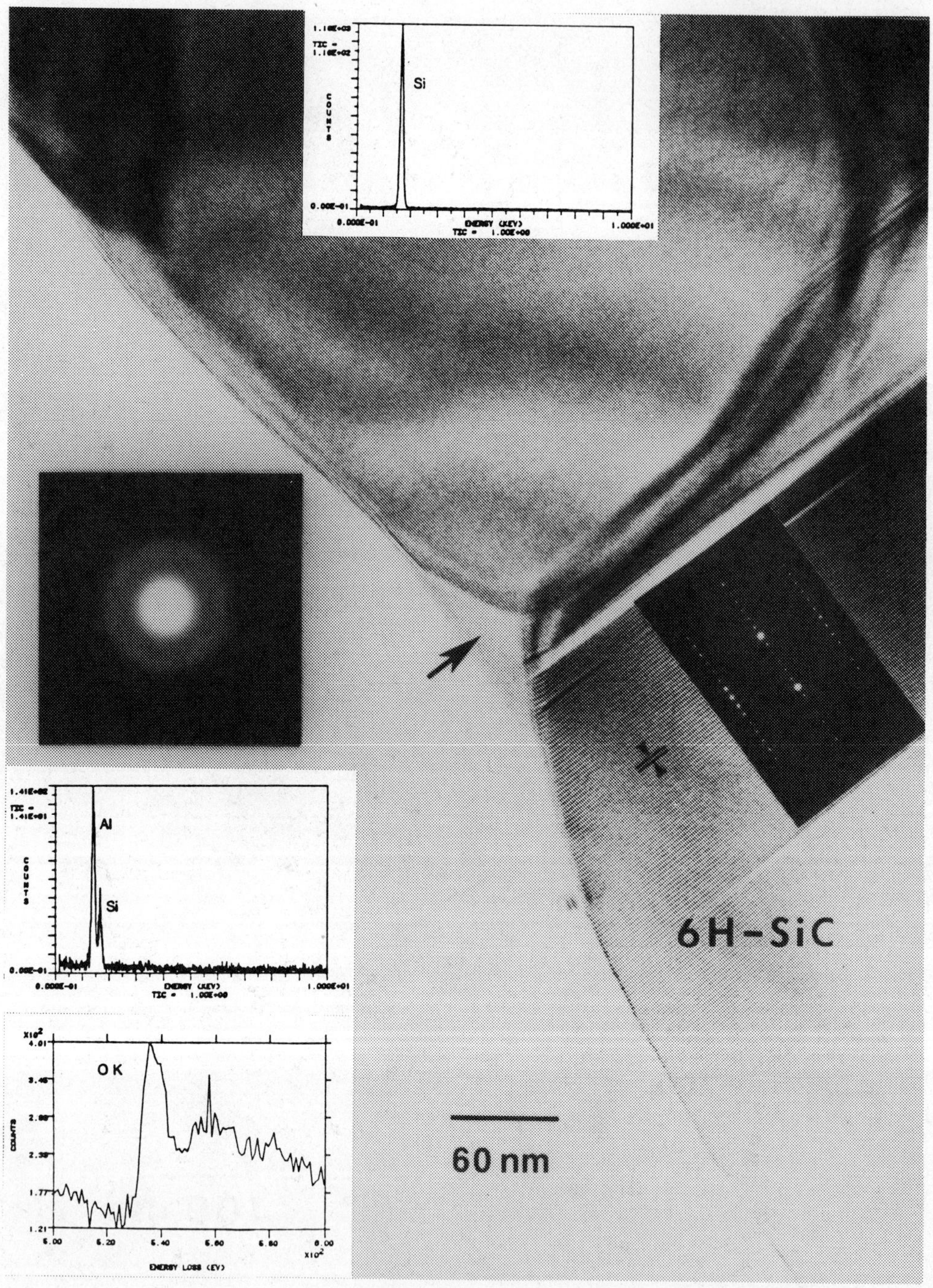

Figure 4. TEM BF image of non-crystalline secondary phase located at triple junction adjacent to 6H-SiC in Al-doped HIP-SiC. Note aluminium enrichment and oxygen-rich composition as revealed by EDS and ELS (focused TEM probe size approx. 10 nm).

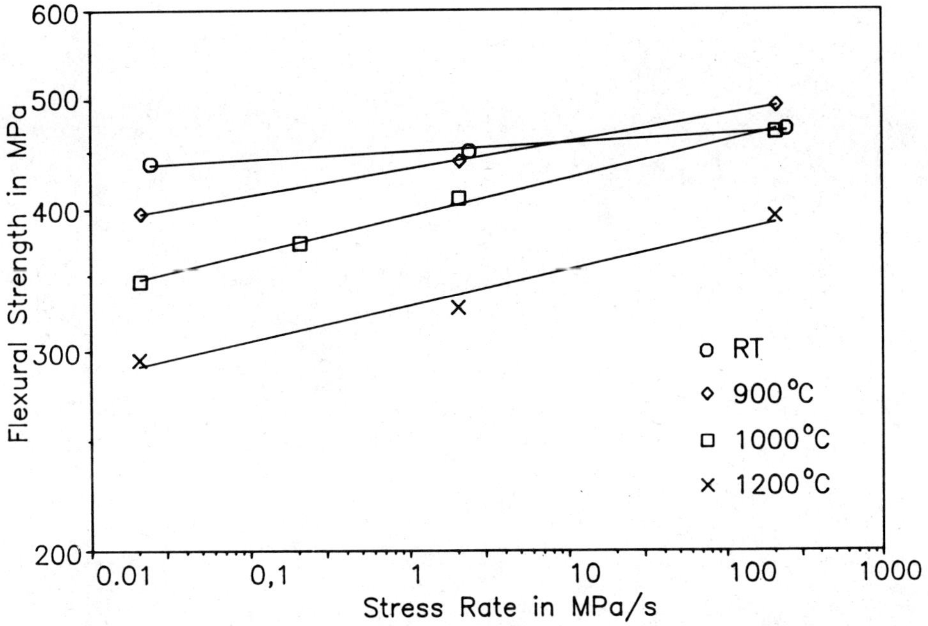

Figure 5. Dynamic fatigue tests of HIP-SiC doped with 0.2 wt.% Al performed at room and elevated temperatures.

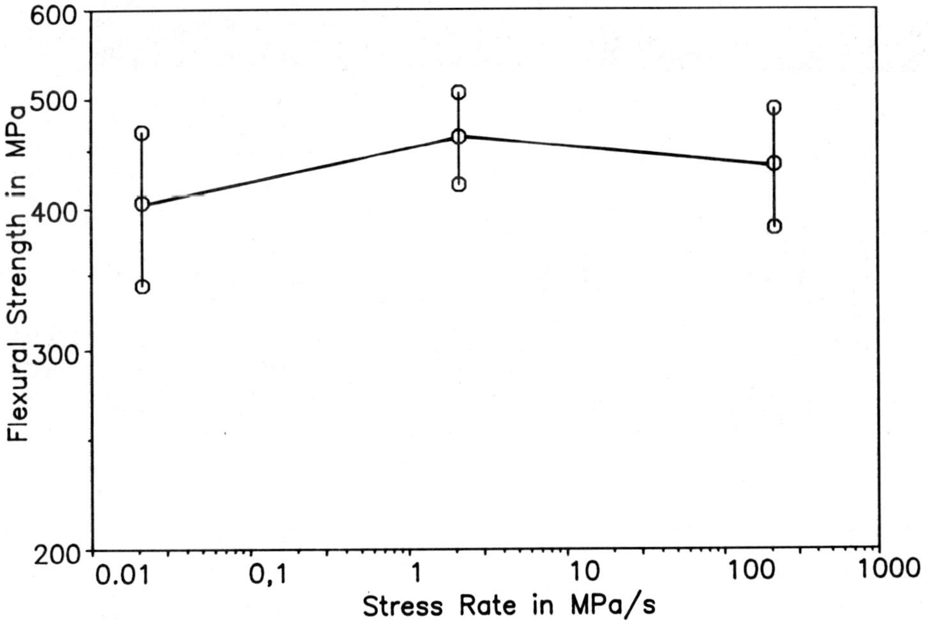

Figure 6. Dynamic fatigue tests of undoped HIP-SiC performed at 1000 °C.

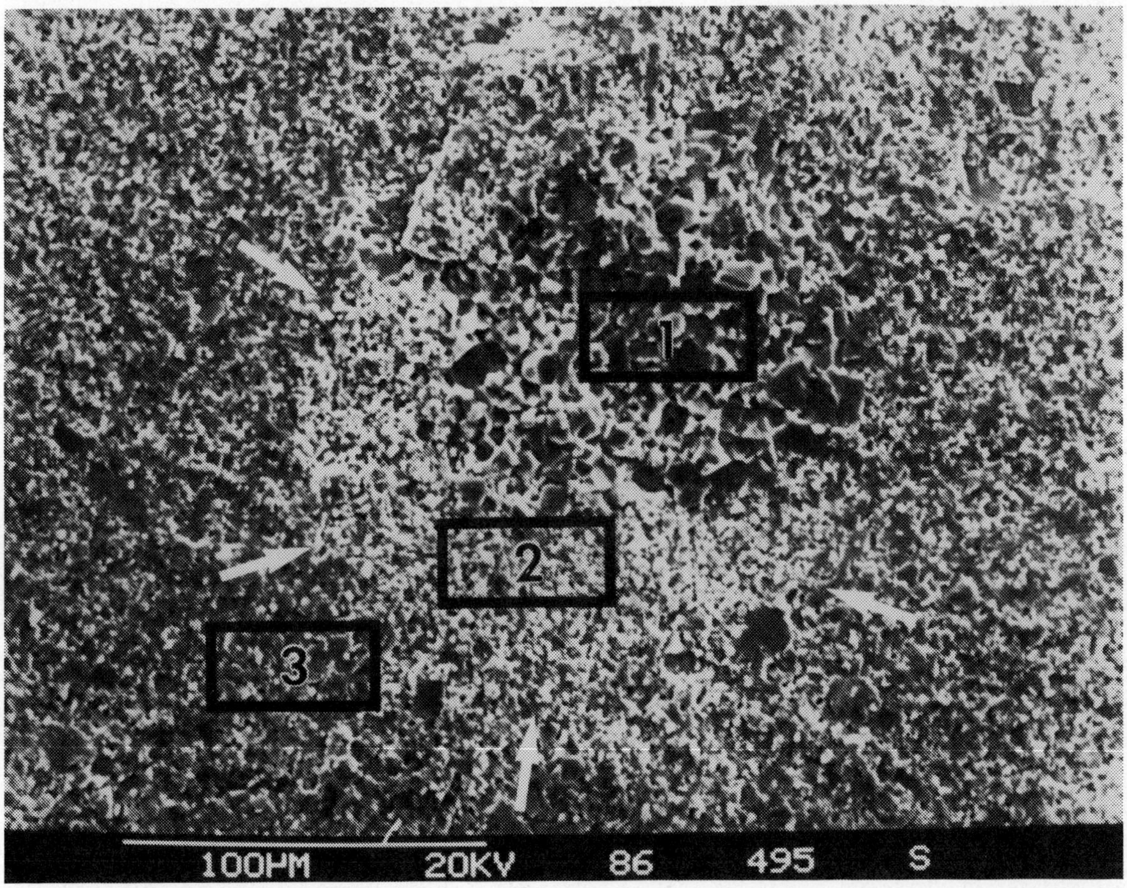

Figure 7. SEM fracture surface of Al-doped HIP-SiC after testing at 1200 °C and low stress rate. Areas indicated refer to failure origin and different mechanisms of crack propagation.
Area I: Exaggerated α-SiC grain growth (failure origin).
Area II: Intergranular fracture mode (subcritical crack growth), see arrows for extension of area II.
Area III: Transgranular fracture mode (critical crack growth).

STRENGTH DEGRADATION OF SILICON NITRIDE BASED CERAMICS BY HOT CORROSION IN ALKALI SULFATE AND CARBONATE MELTS

M. SHIMADA, T. SATO, Y. KOIKE AND T. ENDO
Department of Applied Chemistry
Faculty of Engineering
Tohoku University
Sendai 980, Japan

ABSTRACT

Six types of Si_3N_4 based ceramics, such as hot isostatically pressed Si_3N_4 without additives, hot pressed Si_3N_4, hot pressed sialons containing 0, 30, 60 and 100 % of α phase were corroded by K_2SO_4 and K_2CO_3 melts at 1150-1300°C and 925-1150°C, respectively. The surface chemical reaction-controlled shrinking core model adequately described the relationship between the weight loss of the specimen and time for the corrosion reactions in both K_2SO_4 and K_2CO_3 melts, and the apparent activation energies were 380-608 kJ/mol and 157-344 kJ/mol, respectively. The corrosion rate in K_2CO_3 melt decreased with increasing the contents of Al and Y elements in the specimens. The fracture strength of the specimens corroded by K_2SO_4 and K_2CO_3 melts degraded to 2/3 - 2/5 of the original values for up to 2 % of weight loss, and then was almost constant up to 30 % of weight loss. From the present experimental results, the prediction of time to rupture for the corroded specimens could be predicted.

INTRODUCTION

Since Si_3N_4 and SiC show excellent physical and mechanical properties, these ceramics have been studied for high-temperature structural applications, such as heat exchangers, hot gas turbines and magnetohydrodynamic (MHD) generators. However, these non-oxide ceramics are inherently unstable in an oxidizing atmosphere. Mayer and Riley[1], Erdoes and Altorfer[2], Tressler et al.[3], Bourne and Tressler[4], Ferber and Tennery[5], Becher[6] and Sato et al.[7] have investigated the corrosion behavior of Si_3N_4 ceramics in a variety of molten salt/gas environments, and reported that the corrosion bahavior was quite dependent on the chemical environments. Borne and Tressler[4] reported that exposures of hot-pressed Si_3N_4 and reaction-bonded Si_3N_4 to deep melts of NaCl or $NaCl+Na_2SO_4$ resulted in strength degradation of 30 to 60 %. The decrease in strength was due to a significant increase in the critical flaw size. Becher[6] reported that the degradation of the fracture strength of Si_3N_4 exposed to coal slags at high temperature depended on the chemical constitution of the slags. In general, it was considered that oxide additives such as MgO, Y_2O_3 and Al_2O_3 in the Si_3N_4 ceramics contributed to preferential attack via complex liquid silicate formation[8], but no systematic study has been reported for the effect of additives on the corrosion behavior. In the present study, a series of tests was conducted to investigate the effect of additives on the corrosion rate and strength degradation of Si_3N_4-based ceramics in K_2SO_4 and K_2CO_3 melts and the prediction of time to rupture for the corroded Si_3N_4 ceramics was examined.

EXPERIMENTAL

Hot isostatically pressed Si_3N_4 without additives, hot pressed Si_3N_4 and hot pressed sialons containing 0, 30, 60 and 100 % of α phase, denoted as HIP-Si_3N_4, HP-Si_3N_4, Sialon-0, Sialon-30, Sialon-60 and Sialon-100 were used as corrosion samples. The characteristics of these ceramics are summarized in Table 1. The amounts of α-Si_3N_4 phase was determined by Gazzara's method.[9] All samples were cut into rectangular coupons, 4 by 5 by 15 mm. In each experiment, a weighed sample and powdered reagent grade K_2SO_4 and K_2CO_3 were put into a test tube of highly purified alumina, 16 mm in diameter and 105 mm long, then

Table 1 Characteristics of the samples

Sample	Phase	Chemical compositions (at.%)					σ_{3b} (MPa)	K_{IC} (MPa·m$^{1/2}$)
		Si	N	Y	Al	O		
HIP-Si_3N_4	β	42.9	57.1	0	0	0	753	4.0
HP-Si_3N_4	β	41.2	54.8	0.86	0.76	2.44	870	5.3
Sialon-0	β	38.5	52.4	0.86	3.51	4.77	935	3.6
Sialon-30	$\alpha+\beta$	41.2	56.4	0.34	1.58	0.51	717	4.3
Sialon-60	$\alpha+\beta$	37.6	55.0	1.06	4.77	1.59	702	5.0
Sialon-100	α	34.2	53.5	1.74	7.94	2.65	552	4.9

α : α-Si_3N_4 phase, β: β-Si_3N_4 phase placed in an electric furnace regulated at the desired temperature. The amounts of K_2SO_4 and K_2CO_3 added were 5 times of the stoichiometry based on the equations (1) and (2).[7,10]

$$16Si_3N_4 + 27K_2SO_4 = 12(K_2O \cdot 4SiO_2) + 32N_2 + 15K_2S_{1.8} \quad (1)$$

$$24Si_3N_4 + 81K_2CO_3 = 18(K_2O \cdot 4SiO_2) + 48N_2 + 81CO + 126K \quad (2)$$

Nitrogen gas (>99.99% in purity) was injected into the tube at a rate of 20 ml/min throughout the reaction. After maintaining the desired temperature and time, the tube was removed from the electric furnace and cooled quickly to room temperature. The samples were washed in hot water in order to remove water soluble compounds such as K_2CO_3, K_2SO_4 and K_2O $4SiO_2$, dried, and weighed. The crystalline phase and microstructure of the samples were examined by X-ray diffraction analysis, infrared spectroscopy and scanning electron microscopy. Fracture toughness, K_{IC}, was determined by indentation fracture method. The fracture strength, σ_{3b}, of the sample was determined by 3-point bending test with a cross head speed of 0.5 mm/min and span length of 10 mm. Five specimens were fractured at each test condition.

RESULTS AND DISCUSSION

The time dependence of the degree of the wight loss for Sialon-60 in K_2CO_3 melts at 975-1150°C is shown in Figs. 1. A significant weight loss was observed above 975°C. In the heterogeneous reaction systems such as solid-liquid system, when the shrinking core model is applied and the surface chemical process is rate-limiting, the relation between the fraction of weight loss , α, and time, t, is expressed by Eq. (3).[11]

$$1-(1-\alpha)^{1/3} = kt/r_o \quad (3)$$

where r_o is the radius of the sample and k(m/sec) is the chemical reaction constant. Since the specimens used in the present study are rectangular coupons, r_o is defined as 3abc/2(ab+bc+ca), where a is the specimen width, b is the specimen thickness and c is the specimen length, respectively. As seen in Fig. 1, since the plots of $1-(1-\alpha)^{1/3}$ against t became straight lines, the surface chemical reaction controlled shrinking core model adequately described the relation between time and weight loss fraction of Sialon-60 in K_2CO_3 melts. Similar results were obtained for the corrosion of all specimens in both K_2CO_3 and K_2SO_4 melts. The chemical reaction constants, k, were determined by the slope of the straight lines. The

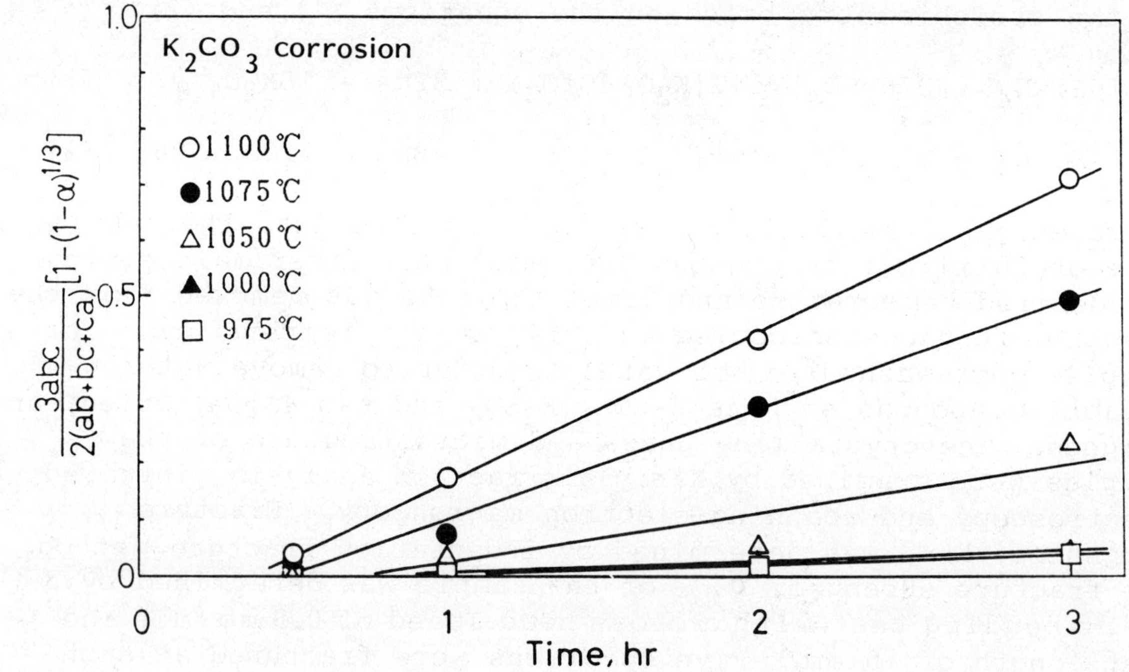

Fig. 1 Time dependence of weight loss for Sialon-60 specimens in K_2CO_3 melts exposed to nitrogen gas at various temperatures

apparent activation energies (ΔE) and frequency factors (A) in Arrhenius equation ($k=A\exp(-\Delta E/RT)$) for the corrosion rate of all specimens in both K_2CO_3 and K_2SO_4 melts were determined by the slope and intercept of Arrhenius plots of k and T. The results are summarized in Table 2. The apparent activation energies for the corrosion reaction of Si_3N_4 and sialons were 380-608 kJ/mol in K_2SO_4 melts and 157-344 kJ/mol in K_2CO_3.

Table 2 Frequency factors, A, and apparent activation energies, ΔE, for the oxidation of various Si_3N_4-based ceramics in K_2SO_4 and K_2CO_3 melts.

Sample	K_2SO_4 corrosion		K_2CO_3 corrosion	
	A (mm/hr)	ΔE (kJ/mol)	A (mm/hr)	ΔE (kJ/mol)
HIP-Si_3N_4	1.01×10^{15}	447	3.27×10^{6}	157
HP-Si_3N_4	7.66×10^{13}	424	3.75×10^{11}	320
Sialon-0	1.54×10^{18}	548	1.25×10^{9}	257
Sialon-30	2.46×10^{12}	380	8.00×10^{8}	250
Sialon-60	1.60×10^{16}	493	3.32×10^{12}	344
Sialon-100	4.21×10^{20}	608	2.85×10^{9}	275

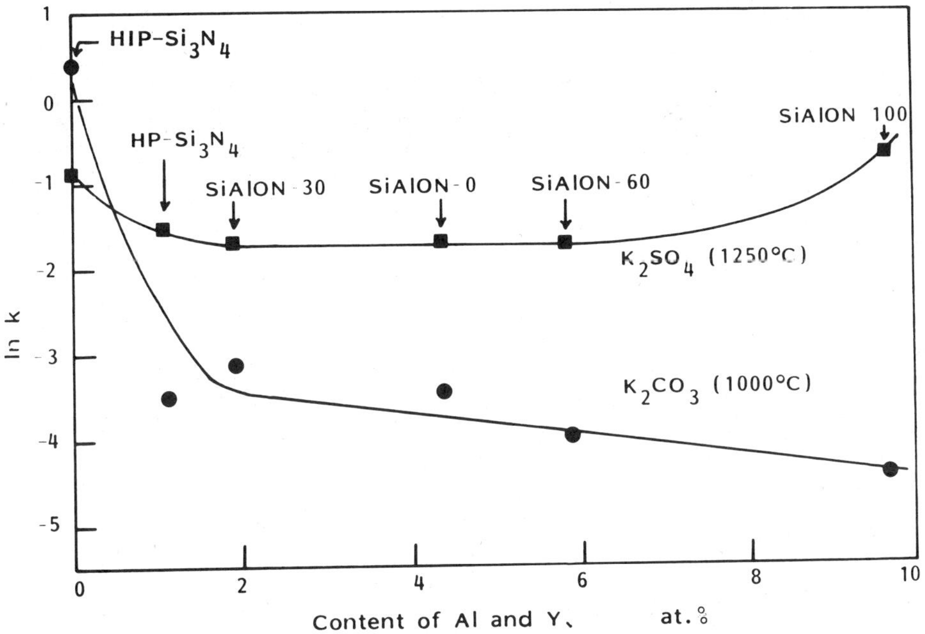

Fig. 2 Relationship between the corrosion rate constants and the total content of Al and Y for Si_3N_4-based ceramics in K_2SO_4 melt at 1250°C and K_2CO_3 melt at 1000°C

The corrosion rate seemed to depend on the contents of additives. The corrosion rate constant for each Si_3N_4 based ceramic in K_2SO_4 melts at 1250°C and K_2CO_3 melts at 1000°C are plotted against the total contents of aluminium and yttrium elements in Fig. 2. In both melts, the corrosion rate of HIP-Si_3N_4 without additives was significantly fast. The corrosion rate in K_2CO_3 melts decreased with increasing the contents of aluminium and yttrium ions, and the corrosion rate was in the order of HIP-Si_3N_4 >> HP-Si_3N_4 ≑ Sialon-30 ≑ Sialon-0 > Sialon-60 > Sialon-100. These results might be due to the protection of the corrosion reaction by Al_2O_3 and Y_2O_3. On the other hand, the corrosion rate in K_2SO_4 melts decreased with increasing the contents of aluminium and yttrium elements up to 1 at.%. It was almost constant up to 6 at.%, but increased again above 9.7 at.%. The corrosion rate in K_2SO_4 melts was in the order of HIP-Si_3N_4 ≑ Sialon-100 > Sialon-60 ≑ Sialon-0 ≑ Sialon-30 ≑ HP-Si_3N_4. By X-ray powder diffraction analysis on the surface of the specimens corroded in both K_2SO_4 and K_2CO_3 melts, no diffraction peak corresponding to corrosion product was detected. On the other hand, as shown in Fig.3, the adsorption peak at 1100 cm^{-1} corresponding to Si-O stretching was observed in the IR spectrum of the specimen corroded in K_2SO_4 melt, but

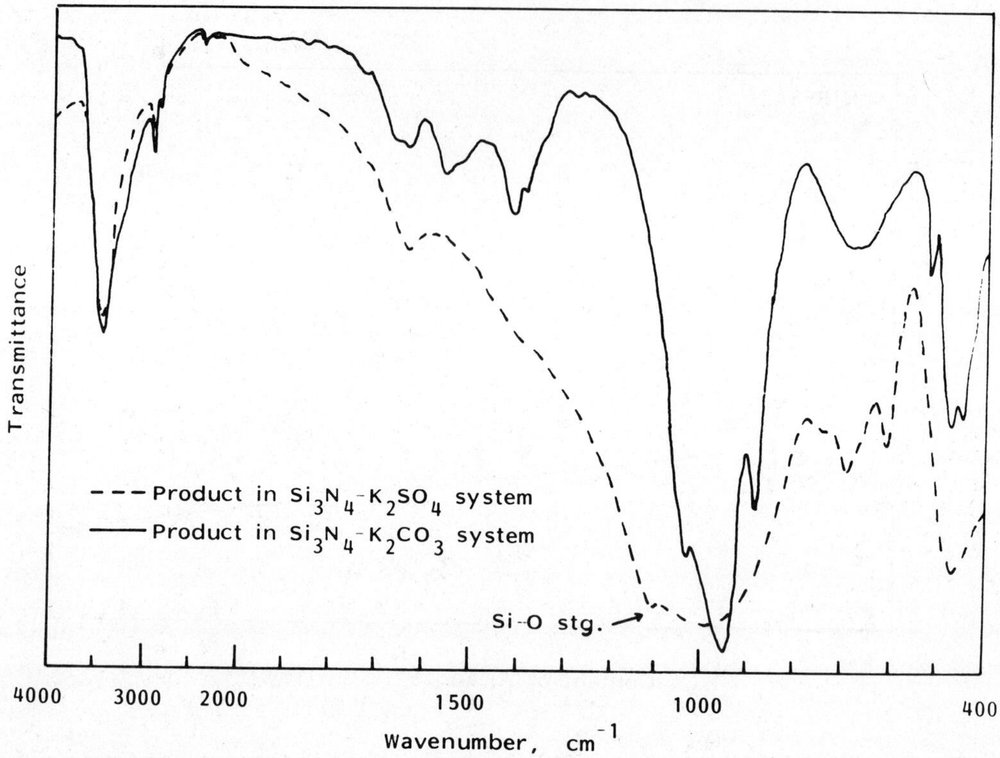

Fig. 3 Infrared spectra of HIP-Si_3N_4 corroded in K_2SO_4 and K_2CO_3 melts

was not observed in the sample corroded in K_2CO_3 melt. These results indicated that amorphous silica was formed on the surface of the specimen corroded by K_2SO_4 melt, and it played significant role on the corrosion reaction.

The scanning electron micrographs of the cross section of the Sialon-60 corroded in K_2SO_4 at 1150°C and K_2CO_3 at 925°C for 1 hr are shown in Fig. 4. Extensive roughening and the formation of large pits were observed on the surface of the corroded specimens. The corroded layer thickness of the specimens dipped into K_2SO_4 and K_2CO_3 melts were about 40 µm and 25 µm, respectively.

The fracture strength of the specimens corroded in K_2SO_4 and K_2CO_3 melts was determined by 3-point bending test using equation (4).

$$\sigma_{3b} = 3PL/2ab^2 \qquad (4)$$

where P is the applied load and L is the span length. The dimensions of samples of a and b the corroded specimens were measured without polishing. The fracture strengths determined

Fig. 4 Scanning electron micrographs of the cross sections of Sialon-60 corroded in K_2SO_4 melt at 1150^oC and K_2CO_3 melt at 925^oC for 1 hr

are shown in Fig. 5 as a function of weight loss of the specimens. The bending steregth of all specimens corroded in both melts degraded to 2/3 - 2/5 of the original values up to 2 % of weight loss, and then almost constant up to 30 % of weight loss. These results indicated that the new critical flaws about 2-6 times larger than original ones were formed at the initial stage of the corrosion reaction up to 2 % of weight loss, and then perhaps the flaws did not grow by the succesive corrosion reaction.

By assuming the formation of the semi-elliptical cracks on the surface, the flaw sizes, C, introduced by the corrosion in K_2SO_4 and K_2CO_3 melts were calculated by equation (5) using the values

$$C = K_{IC}^2/1.94^2\ \sigma_{3b}^2 \qquad (5)$$

of the fracture toughness, K_{IC}, before corrosion and the fracture strength, σ_{3b}, after corrosion shown in Table 1 and Fig. 5. The results are listed in Table 3.

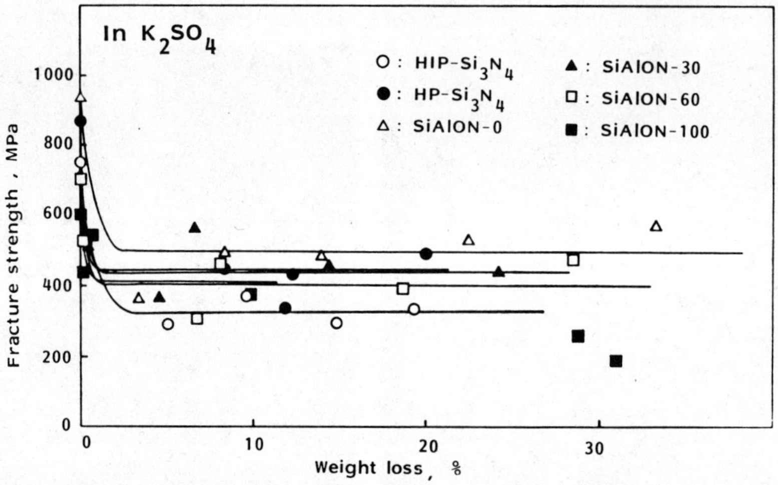

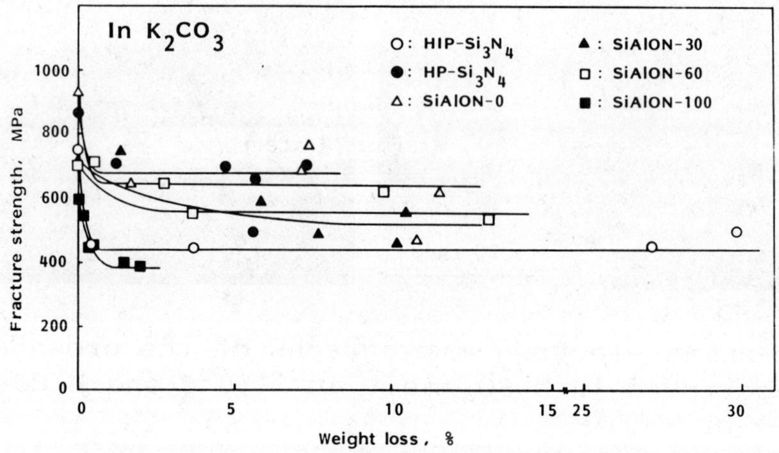

Fig. 5 Relationship between the bending strength and the weight loss for Si_3N_4 ceramics corroded in K_2SO_4 melt at 1100°C and K_2CO_3 melt at 925°C for various times

Table 3 Fracture strength and flaw length in various Si_3N_4-based ceramics corroded in K_2SO_4 melts at 1150°C and K_2CO_3 melts at 925°C

Sample	In K_2SO_4		In K_2CO_3	
	σ_{3b} (MPa)	C (μm)	σ_{3b} (MPa)	C (μm)
HP-Si_3N_4	450	40	680	20
HIP-Si_3N_4	320	36	450	16
Sialon-0	500	14	650	8
Sialon-30	440	25	520	18
Sialon-60	410	39	560	21
Sialon-100	420	36	380	44

The flaw sizes agreed well with the corrosion layer thickness observed by scanning electron micrographs (Fig. 4).

By assuming the isotropic shrinkage of the sample due to the corrosion reaction, equation (4) can be modified as equation (6).

$$P_t = 2ab^2(1-x)^3\sigma_{3bt}/3L = 2ab^2(1-\alpha)\sigma_{3bt}/3L$$
$$= P\sigma_{3bt}(1-\alpha)/\sigma_{3b} \qquad (6)$$

where P_t and σ_{3bt} are the fracture load and bending strength at time t and x is the degree of the shrinkage of the specimen. Since it is possible to estimate the flaw size, C, and the weight loss fraction, α, at time, t, for the specimen immersed in K_2SO_4 and K_2CO_3 melts by the present experimental results shown in Tables 2 and 3, the time to rupture could be predicted by equation (6) for each corroded specimen. The predicted time to rupture for the sample corroded at 1100-1200°C in K_2SO_4 melts and 1000-1100°C in K_2CO_3 melts under the load of 0.3P was calculated by using the initial sample size of 50 mm x 50 mm x 50 mm, where P corresponds to the fracture load of the uncorroded original sample. The results are listed in Table 4. It is notable that the resistance to the corrosion of Si_3N_4-based ceramics greatly increases by adding Al_2O_3 and Y_2O_3.

Table 4 The prediction of time to rupture for various Si_3N_4-based ceramics of 50 mm x 50 mm x 50 mm in K_2SO_4 melts at 1100, 1150 and 1200°C and K_2CO_3 melts at 1000, 1050 and 1100°C

Sample	In K_2SO_4			In K_2CO_3		
	1100°C	1150°C	1200°C	1000°C	1050°C	1100°C
HIP-Si_3N_4	270hr	69hr	19hr	4hr	2hr	1hr
HP-Si_3N_4	730	200	58	250	78	27
Sialon-0	2000	370	76	170	68	29
Sialon-30	620	190	64	140	59	26
Sialon-60	1800	390	94	270	80	25
Sialon-100	2100	330	57	410	150	61

CONCLUSION

1) The surface chemical reaction controlled shrinking core model could be applied to describe the relationship between the degree of the corrosion and the reaction time for the corrosion of

Si_3N_4-based ceramics in K_2SO_4 and K_2CO_3 melts.
2) The corrosion rate in K_2CO_3 melts decreased with increasing the contents of aluminium and yttrium ions in the specimen.
3) The corrosion resulted in the roughness on the surface of the specimens and degraded the fracture strength to 2/3-2/5 of the original values until up to 2 % of the weight loss, and then the fracture strength was almost constant up to 30 % of the weight loss.

ACKNOWLEGEMENT

This work has been supported in part by a grant-in-aid for Energy Research of the Ministry of Education.

REFERENCE

1) M. I. Mayer and F. L. Riley, "Sodium-Assisted Oxidation of Reaction-Bonded Silicon Nitride," J. Mater. Sci., 13, 1319-28 (1978).
2) E. Erdoes and H. Altorfer, Materials and Coatings to Resist High Temperature Corrosion. Edited by D. R. Holmes and A. Rahemel, Applied Science, London, 161-74 (1977).
3) R. E. Tressler, M. D. Meyer and T. Yonushonis, "Molten Salt Corrosion of SiC and Si_3N_4 Ceramics," J. Amer. Ceram. Soc., 59, 278-79 (1976).
4) W. C. Bourne and R. E. Tressler, "Molten Salt Degradation of Si_3N_4 Ceramics," Amer. Ceram. Soc. Bull., 59, 443-45 (1980).
5) M. K. Ferber and V. J. Tennery, "Behavior of Tubular Ceramic Heat Exchanger Materials in Acidic Coal Ash From Coal-Oil-Mixture Combustion," Am. Ceram. Soc. Bull., 62, 236-43 (1983).
6) P. F. Becher, "Strength Degradation in SiC and Si_3N_4 Ceramics by Exposure to Coal Slags at High Temperatures," J. Mater. Sci., 19, 2805-14 (1984).
7) T. Sato, Y. Kanno, T. Endo and M. Shimada, "Corrosion of Si_3N_4 in Molten Alkali Sulfate and Carbonate," Adv. Ceram. Mater., 2, 228-31 (1987).
8) J. L. Smialek and N. S. Jacobson, "Mechanism of Strength Degradation for Hot Corrosion of α-SiC," J. Amer. Ceram. Soc., 69, 741-52 (1986).
9) C. P. Gazzara and D. R. Meissler, Amer. Ceram. Soc. Bull., "Determination of Phase Content of Si_3N_4 by X-Ray Diffraction Analysis," 56, 777-80 (1977).
10) T. Sato, Y. Kanno and M. Shimada, "Corrosion of SiC, Si_3N_4 and AlN in Molten K_2SO_4-K_2CO_3 Salts," Int. J. High Technol. Ceram, 2, 279-90 (1986).
11) W.M. Mckewan, "Kinetics of Iron Ore Reduction," Trans. AIME, 212, 791-93 (1958).

EXPLORATORY HIGH-TEMPERATURE TENSILE AND CYCLIC FATIGUE CHARACTERIZATION OF COMMERCIAL MgO-PSZ*

K. C. Liu and C. R. Brinkman

Oak Ridge National Laboratory
P.O. Box 2008
Oak Ridge, Tennessee 37831-6155

ABSTRACT

Monotonic tensile fast fracture and tension-tension cyclic fatigue tests were performed on a commercial grade of MgO partially stabilized zirconia (MgO-PSZ) at room and elevated temperatures. Test results were analyzed and correlated as to strength, temperature, fatigue lifetime, and compared with (MOR) data when possible. Results showed that because of failure to incorporate phase transformation-induced plasticity, the MOR values were generally much higher than the true tensile strength in flexure. Fractographic examinations were made to study the fracture modes and failure origins of fracture initiation sites. The information was then used to calculate the critical fracture toughness based on the Griffith-Irwin theory. Comparisons of the calculated and available data showed reasonable agreement in some situations, but not in general. Limited test data obtained at temperatures above 800°C showed that tension-tension cyclic loading from low to high stress in increasing multiple steps raised the fatigue strength of the material above the monotonic uniaxial tensile strength.

INTRODUCTION

Partially stabilized zirconia (PSZ) sintered with a small addition of magnesia (MgO) has a number of attractive physical and mechanical properties that are ideally suited for diesel engine applications.[1] In addition to its high temperature strength, low thermal conductivity, good thermal shock resistance, and reasonably good wear resistance, this material exhibits good fracture toughness resulting from stress-induced phase transformation occurring in the vicinity of propagating cracks.

Successful utilization of this material as components in high temperature heat engines requires that key properties of the material be characterized so that the design criteria can be established. High temperature tensile characterization including fatigue-induced cracking and long-term stress rupture are especially important since engines operate under cyclic thermal conditions and the resultant critical stresses are in tension. However, most of the mechanical strength

---

*Research sponsored by the U.S. Department of Energy, Assistant Secretary for Conservation and Renewable Energy, Office of Transportation Systems as part of the Ceramic Technology for Advanced Heat Engines Project of the Advanced Materials Development Program, under contract DE-AC05-84OR21400 with Martin Marietta Energy Systems, Inc.

properties are usually evaluated in either a three- or four-point bending mode and the tensile strength of the material is calculated based on linear elastic beam theory. Criticism of the bend test method is that the maximum bending failure stress (modulus of rupture, MOR) may not reflect crack initiation from the largest flaw in the test volume, whereas the uniaxial failure strength is controlled by the largest defect in the test volume. Therefore, the latter is more representative of the true strength for practical design applications.

This paper illustrates that the correlation between the uniaxial tensile and MOR strength of MgO-PSZ is complex and the treatment may not be valid without incorporating nonlinear behavior resulting from the stress-induced transformation plasticity. For this reason, the design stress criteria inferred from the bend data are questionable for engineering design applications. Finally, it shows that the uniaxial tensile strength is governed by inherent flaws and can be correlated to flaw size through the use of the theory of fracture mechanics.

EXPERIMENTAL DETAILS

Material and Specimen

The material used in this experiment was a commercial grade of partially stabilized transformation toughened zirconia doped with 3.3-3.4 wt % MgO, heat treated for thermal shock resistance (TS grade).* The material was received in two product forms: short cylindrical rod blanks (16 mm in diameter and 145 mm in length) and square tiles (100 × 100 × 10 mm). Figure 1 shows typical microstructural features of the material with an average grain size of about 40-50 μm. It is a highly porous ceramic material containing uniformly distributed pores throughout the bulk of the material, ranging in size from 1-10 μm with an average pore size of about 3 μm. The buttonhead tensile specimen as shown in Fig. 2 was ground from the rod blanks. All circumferential grinding marks in the reduced gage section were removed by final longitudinal grinding along the specimen axis to a surface finish of about 16 rms. No polishing was given to the surface of the gage section of the specimen. The finished specimen had a 6.3-mm (0.25-in.) gage diameter and a 25.4-mm (1.0-in.) uniform gage length. Two rectangular MOR bar specimens (3.5 × 4.5 × 50 mm) were machined from the tiles and instrumented with strain gages on both sides of the flexure surfaces for four-point bend tests. X-ray diffraction analyses showed that a typical tensile specimen contained about 46% monoclinic phase and an MOR bar specimen about 35%.

Testing Equipment

All tensile tests were performed in air using a closed-loop electro-hydraulic controlled testing machine as shown in Fig. 3. The load train assembly was supported by a set of self-aligning coupling devices, also shown in Fig. 3, to eliminate detrimental bending stress resulting from misalignment in the load train assembly, hence, achieving a high degree of stress uniformity in the gage section of the tensile specimen. This condition is important to tensile testing because the tensile strength is sensitive to the inherent process flaw size and distribution in the bulk material. Details of the self-aligning devices are discussed elsewhere.[2]

*Nilcra Ceramics (U.S.A.), Inc., Elmhurst, IL.

Elevated temperatures were achieved during tensile testing of the ceramic specimen by induction heating, as shown in Fig. 3. Since the ceramic material was not electrically conductive, a clamshell-shaped silicon carbide susceptor was centered around and served to heat the reduced test section of the specimen. A more detailed description of the induction heating method can be found elsewhere.[3]

Tensile strain was measured directly from the uniform gage section of the specimen using a mechanical strain extensometer,[4] also shown in Fig. 3, capable of high temperature testing with accuracy within several microstrains. The outputs of load cell and the strain extensometer were recorded to establish the stress-strain relationship and engineering data such as moduli of elasticity.

All tensile tests were performed in load control using a triangular wave form as shown in Fig. 4(a). Cyclic loading was applied in tension with a minimum-to-maximum stress ratio, R, of about 0.1 and at a frequency of 0.5 Hz. A small number of tests were cycled in fatigue from low to high stress in multiple steps with increasing peak stress values, as shown in Fig. 4(b). The latter mode of testing is known as "coaxing" which frequently increases the fatigue endurance limit in metal by work hardening.

Two bend tests were performed at room temperature using a four-point bend jig with an inner and an outer span of 20 mm and 40 mm, respectively. Specimens were loaded at an apparent stress rate of about 50 MPa/s. The output signals from the load cell and each of the two strain gages bonded on the outer surfaces of the bend bar were recorded separately on XY-plotters to establish the relationships between the apparent flexural stress and measured strains.

TEST RESULTS

A total of eighteen tensile specimens were tested in either monotonic or fatigue loading at four temperatures: room temperature, 400, 800, and 1000°C. Test parameters and test results are summarized in Table 1 and data are plotted in semi-log plots of stress versus number of cycles to failure, as shown in Figs. 5 through 8. Closed symbols used in these figures indicate the lifetimes of the test specimens that ruptured at the indicated test conditions. Some specimens were tested from low stress to high stress loading in multiple steps as shown in wave form II of Fig. 4(b). Open symbols indicate the terminal points of each intermediate cyclic loading segment and that no rupture had occurred when the test was terminated at the indicated test condition. For the sake of clarity and simplicity, some open symbols representing intermediate loading were replaced by a broken zigzag line to indicate the loading history. A tensile fast fracture was considered to have occurred in a single cycle and the data points were plotted accordingly.

Because the number of test data is limited, the fatigue curve of the material at each test temperature is approximated by a shaded scatter band. Due to the limited specimens available, no room temperature fatigue tests were conducted. Instead, an approximate fatigue behavioral trend, as shown in Fig. 5, has been inferred from the 400°C fatigue curve which is illustrated in Fig. 6. It should be noted that the tensile datum of specimen 18 was excluded in the room temperature group, shaded area as shown in Fig. 5, due to the excessively large flaws that governed failure. A string of unusually large defects that measured about 1500 μm in length were obviously responsible for the low tensile strength failure.

Table 1. Results of tensile and cyclic fatigue tests of NILCRA's MgO-PSZ at elevated temperatures

Specimen number	Modulus of elasticity [GPa (10^6 psi)]	Tensile strength [MPa (ksi)]	Cyclic stress to failure MPa (ksi)	Number of cycles to failure	Intermediate cyclic stress [MPa (ksi)]	Number of cycles at intermediate cyclic loading	Total test time (h)
Test at Room Temperature							
15		326 (47.4)		1			0
16		448 (65.0)		1			0
22		338 (49.0)		1			0
18		272 (39.4)		1			0
Test at 400°C							
7		376 (54.5)		1			0
8		314 (45.5)		1			0
9			336 (48.8)	1,209			0.7
10	150 (21.7)		328 (47.5)	13			0
11	156 (22.7)		314 (45.5)	5			0
Test at 800°C							
2	167 (24.3)	292 (42.3)		1			0
3	156 (22.6)	269 (38.9)		1			0
4	148 (21.4)		279 (40.4)	3			0
5	156 (22.6)		254 (36.9)	2			0
6	160 (23.2)				241 (35.0)	45,126	
					260 (37.7)	124,946	
				90,837	279 (40.4)	88,970	
			295 (42.7)				194.4
1	159 (23.1)				235 (34.0)	77,071	
					256 (37.1)	48,709	
					271 (39.3)	132,555	
				30			
					287 (41.7)	88,970	
			300 (43.5)				193.0
Test at 1000°C							
12	148 (21.5)	244 (35.3)		1			0
13	145 (21.0)	253 (36.7)		1			0
14	147 (21.3)				212 (30.7)	49,120	
					228 (33.1)	82,315	
				40,979	242 (35.1)	129,036	
			270 (39.2)		254 (36.8)	43,725	191.8

The monotonic tensile data tended to show more scatter in strength at room temperature and at 400°C than at higher temperatures. However, it is not clear if this behavior is attributed to the transformation toughening activated by the tensile loading. The fatigue curves for 800 and 1000°C are not well defined due to the lack of data points in the intermediate-cycle range. Available data indicate that both fatigue curves are very flat, suggesting that fatigue-induced growth of microcracks in the bulk may be impeded by increasing crack resistance due to localized transformation-toughened microstructures.[5] However, the fatigue behavior beyond the test data remains uncertain and it may be modified by thermal aging as reported by Ferber et al.,[6] and/or environmental interaction.

Practically all the data points fall consistently within the scatter bands, except specimens 1, 6, and 14 that were step loaded and fell above the upper boundary of the scatter band, as shown in Figs. 7 and 8. Specific reasons for the high fatigue strengths in these specimens are currently being investigated. An obvious reason for the high fatigue performance was that the specimens were tested in coaxing mode, suggesting that some strengthening mechanism was involved. Short-term thermal aging (200-300 h as occurred during test) under moderate load, as reported by Ferber et al.,[6] may have contributed to the increase in fatigue resistance. However, overaging in excess of 1000 h was reported to be detrimental to the static fatigue lifetime.[6] Phase transformation toughening resulting from the tensile cyclic loading may have made little or no contribution to the increase in fatigue resistance in this case, because the toughening mechanism is relatively inactive at temperatures above about 700°C.[7]

Values of elastic moduli tabulated in Table 1 were determined from the stress-strain curves measured during the tests, using the mechanical strain extensometer described in the preceding section. For the purpose of comparisons with MOR data, specimen 22 was tested with two small resistance strain gages instrumented diametrically opposite to each other on the uniform gage section of the specimen. The outputs of the strain gages were plotted individually against the load cell output as shown in Fig. 9. Both curves are essentially identical, indicating that the specimen was loaded uniformly in tension. The stress-strain response was linear up to approximately 130 MPa with a modulus of elasticity of 201 GPa. A precise measurement of the proportional limit was difficult because of the gradual change in the slope of the curve as shown in Fig. 9. It was clear that the material deformed plastically and the permanent deformation was confirmed by post-test examinations. However, the extensive nonlinear deformation beyond the proportional limit observed at room temperature became less evident at failure as the temperature increased. For this reason, plastic strain ratcheting (accumulation of incremental plasticity) resulting from tension-tension cycling was deleterious to the fatigue performance at 400°C. In contrast, cyclic loading of this material at 1000°C exhibited only a perceptible amount of plastic strain accumulation by ratcheting at the beginning of cyclic loading. As the number of cycles increased, the ratcheting diminished due to hardening and the material behaved elastically, commonly called "shakedown" in metal fatigue studies. However, when the cyclic load was raised, the cycle of plastic strain ratcheting and shakedown reoccurred. Repeating this process in increasing steps, we demonstrated that it was possible to raise the fatigue strength of this material at temperatures above 800°C and to exceed the tensile fracture strength, as shown in Figs. 7 and 8.

Comparisons were made between the ORNL tensile data and the four-point-bend flexural strength data obtained from Nilcra data sheet[8] as shown in Fig. 10. Bend data for the same material obtained by other researchers[9] and results of two bend tests performed in our laboratory showed reasonably good agreement with Nilcra reported data. Pronounced differences between the tensile and flexural strengths observed at low temperatures were somewhat greater than general expectations, i.e., the difference calculated on the percentage basis with respect to the MOR values was not a constant. The two curves shown in Fig. 10 merged quite close to each other as temperature increased to 1000°C. The large difference in strength at room temperature may be due in part to the phase transformation plasticity resulting into a shift in neutral axis which is not accounted for in the calculation of bend strength using simple elastic theory. For this reason, a direct comparison between the tensile and MOR strength data of this material may not be meaningful, except for those data in the region where the material behaves as an elastic solid.

Measurements from an ORNL four-point bend test are shown in Fig. 11. The strains were measured from the strain gages bonded on the tensile and compressive surfaces and the apparent stress was calculated from the applied load, based on the standard bending formula for linear elasticity. The indication of uneven strain deformations measured on both surfaces of the MOR specimen strongly hinted that there was an internal stress redistribution through the depth of the MOR bar. This resulted from localized internal volume expansion of small constrained zirconia particles which transformed from the tetragonal structure to monoclinic in the bending stress field. Additionally, the nonlinear responses exhibited in Fig. 11 suggested that some plasticity had occurred in the surface region of the MOR bars. This implied that the internal stress distribution had changed from an idealized linear profile to a complex nonlinear one. Marshall[10] showed that the actual maximum bending stress in the MOR bar could be lower than the calculated apparent stress by as much as 40%. As the phase transformation became more difficult to take place at high temperatures, where the stress-strain response was practically linear to failure at 1000°C, the difference between the tensile and bend strength was shown to be marginal. Using a proportional correction factor varying from about 40% at room temperature to 0% at 1000°C, we estimated that the true failure stresses in the MOR bars might fall slightly above the uniaxial tensile strength within a scatter band of about 50 MPa. One would still expect the uniaxial strength to lie below the MOR data due to volume considerations.

DISCUSSION

The fracture surfaces of tested specimens were examined using optical and scanning electron microscopy (SEM) to determine the origins of fracture initiation sites and failure modes. Examinations showed that specimens tested at temperatures below 800°C often ruptured in a mixed mode of inter- and transgranular fracture. Extensive grain boundary decohesions were noticed also. In contrast, failure was transgranular in the specimens tested at 1000°C.

Careful examinations of all fractographs confirmed that in most specimens fracture was initiated by large voids in the bulk and subsurface regions. Fracture origins were usually found in the flat mirror region being surrounded by a zone of more rugged terrain featured with sharp crack branching and ledges. Topological features such as "river patterns" were used as a guide to pinpoint the exact location of the fracture origins. However, in some cases, the exact location and size of the fracture origin was obscured by multiple fractures and missing fragments. A large cluster of medium-sized porosity (about 10-20 μm) was also a potential candidate as a fracture origin. In this case, the general area of the fracture site could be recognized from the characteristic features of cave-like structures. However, the shape and extent of interconnecting defects were not discernable. A number of SEM micrographs of fracture initiation sites were selected and inserted in Figs. 5 through 8 to illustrate the general nature of the flaw defects that this material contained.

Performing precise measurements of fracture-initiating defects was often difficult because they varied irregularly in shape and character. There were hardly any two defects that looked alike. To facilitate fracture mechanics analysis, all the fracture-initiating defects were treated as plane elliptical cracks defined by "2a" as the major axis and "2c" as the minor axis. Results of the measurements are given in Table 2. The distance from the center of the fracture origin to the nearest specimen surface was measured as "b," as shown in Fig. 12.

Our experimental observations indicated that large voids or inclusions were constantly the major sources of tensile fracture failures. This suggests that the uniaxial tensile strength of engineering ceramics, which in a practical sense cannot be devoid of any process defects, is governed by the physical characteristics of the largest defects preexisting in the materials. An attempt is made here to illustrate that experimentally determined uniaxial tensile strength can be correlated reasonably well to the fracture toughness of the material, if the complex characteristic features of the source defects can be approximated by an analytical form.

According to the Griffith-Irwin theory of fracture mechanics, the critical fracture toughness, K_{Ic}, for a plane elliptical crack in an infinite solid subjected to a uniform uniaxial stress field, σ, at remote ends can be derived as follows:[11]

$$K_{Ic} = F_1 \, (\phi, \, c/a) \, K_{Io} \tag{1}$$

where

$$K_{Io} = 2\sigma \, (a/\pi)^{1/2} \tag{2}$$

where K_{Io} is the stress intensity factor for a plane circular crack of radius "a" in a solid under a uniform uniaxial tensile stress, F_1 is a variable boundary correction factor, and ϕ is the angle defined as shown in Fig. 12. The value of K_{Ic} in Eq. (1) is calculated using F_1 (max), the maximum value of F_1. The above expression remains valid if the crack is situated at the center of the circular cross section having a radius "r" that is much greater than "a." If the elliptical crack is located near the surface within the range of $c/b > 0.1$, K_{Ic} is given by

$$K_{Ic} = F_2 \, (\phi, \, c/a, \, c/b) \, K_{Io} \tag{3}$$

where F_2 is another correction factor which additionally accounts for the proximity to the specimen surface. For the sake of simplicity, the specimen surface is approximated as a straight edge in Eq. (3).

The values of K_{Ic} tabulated in column (h) were calculated using the ORNL tensile strength data listed in column (e) of Table 2 and were plotted in Fig. 13 also. In order to make meaningful comparisons with two sets of reported K_{Ic} data[8,12] which were known to have been obtained from bend tests, two sets of K_{Ic} values were calculated using the ratios of apparent bending strength (MOR) values determined from the curves given in Figs. 10 and 14(a) to the corresponding tensile strengths. The results were plotted in Figs. 13 (curve B) and 14 (curve D), respectively.

A comparison shows that the average K_{Ic} values (curve A) calculated from the tensile strengths are significantly lower than Nilcra K_{Ic} data (curve C) by about 55% at room temperature and by about 45% at 1000°C. The agreement between curve B, representing the average property of K_{Ic} calculated from MOR data, and curve C are respectable in view of the possible differences in monoclinic phase content and batch-to-batch variations in microstructure. The K_{Ic} data calculated from the tensile failure strength depicted by closed symbols in Fig. 13 are reasonably consistent, in particular, at high temperatures.

Comparative studies similar to the above were conducted also with data obtained by Tennery and his co-workers.[12] The MOR strengths used in K_{Ic} calculations are determined from the curve shown in Fig. 14(a). The K_{Ic} (curve D) values calculated from the MOR data were plotted in Fig. 14(b) for comparison with the K_{Ic} curve (curve E) determined from four-point bend tests with controlled flaws. A comparison shows that curve E is significantly higher than curve D in the lower temperature range below 600°C, but the two curves agree extremely well in the higher temperature range. Again, it is suspected that the differences in K_{Ic} at the low temperature might have been caused by batch-to-batch variations in monoclinic phase content.

Table 2. Summary of critical fracture toughness (K_{Ic}) calculation

(a) Specimen number	(b) Flaw size 2a x 2c (μm)	(c) Distance from edge b (mm)	(d) Ratio c/b	(e) Tensile failure strength (MPa)	(f) K_o (MPa.$\sqrt{m}$)	(g) F_1 (max) or F_2 (max)	(h) K_{Ic} (MPa.$\sqrt{m}$)
			Room Temperature				
15	100 x 100	0.55	0.09	326	2.60	1.00	2.60
16	60 x 60	0.05	0.6	448	2.77	1.03	2.85
22	155 x 230	0.14	0.82	338	3.35	1.07	3.58
18	1000 x 200	1.96	0.05	272	6.87	0.67	4.60
			Temperature = 400°C				
7	80 x 80	0.52	0.08	376	2.68	1.03	2.77
9	200 x 100	1.34	0.04	336	3.79	0.92	3.49
10	100 x 80	0.05	0.8	328	2.62	1.21	3.17
			Temperature = 800°C				
1	140 x 100	0.58	0.09	300	2.83	0.98	2.78
2	110 x 70	0.39	0.09	292	2.44	0.97	2.37
3	160 x 130	1.18	0.06	269	2.72	0.99	2.69
5	120 x 120	0.06	1.0	254	2.22	1.23	2.73
6	100 x 80	0.05	0.8	295	2.36	1.25	2.94
			Temperature = 1000°C				
12	120 x 120	0.86	0.07	244	2.13	1.0	2.13
13	130 x 80	0.06	0.67	253	2.30	1.3	2.99
14	90 x 90	0.05	0.9	270	2.04	1.23	2.52

CONCLUSIONS

Test results obtained from a series of tensile tests on MgO-PSZ at elevated temperature and two control MOR tests, along with fracture mechanics studies based on observed fracture origins, have led to the following conclusions:

(1) Monotonic tensile stress-strain curves show nonlinearity or apparent plasticity at temperatures below about 700°C; however, above this temperature the curve is linear indicating elastic behavior to near failure This nonlinear deformation was attributed to a stress-induced phase transformation and probably to some microcracking as suggested by Marshall.[10]

(2) Results of uniaxial tensile tests combined with additional evidences observed from two strain-gage-instrumented MOR tests suggest that the apparent tensile flexural stress calculated from linear elastic beam theory may overestimate the true maximum flexural stress by as much as 40% at room temperature due to apparent transformation-induced plasticity.

(3) For the same reason, the fracture toughness obtained from bend tests may also overpredict the intrinsic fracture toughness value of the material in excess of 25 to 40% depending on temperature. However, this conclusion is only tentative at this time due to the limited data available.

(4) Results of fractographic studies combined with fracture mechanics analyses imply that failure stresses in uniaxial tension are dictated most likely by the size, shape, and the location of the largest process flaw preexisting in the test volume.

(5) Unlike the fatigue behavior of metal alloys, the fatigue curves of this ceramic are very flat. This behavior suggests that fatigue-induced slow microcracks nucleated in the bulk may be impeded by increasing crack resistance due to localized transformation toughening.

(6) Results of cyclic step load fatigue testing (coaxing) at temperatures above 800°C indicate that the fatigue strength can be raised significantly higher than the uniaxial tensile failure strength.

In summary, the above observations clearly show value and necessity of tensile tests in the characterization of engineering ceramics.

ACKNOWLEDGEMENTS

The authors wish to thank Dr. M. K. Ferber for his valuable technical discussions, Drs. D. R. Johnson and M. K. Ferber for reviewing the manuscript, C. O. Stevens for performing the experiment, R. B. Crouse for performing the SEM work, and Billie Russell for preparing the final manuscript.

REFERENCES

1. R. Kamo and W. Bryzik, "Cummins/TACOM Advanced Adiabadic Engine," in Proceedings of the Twenty-First Automotive Technology Development Contractors' Coordination Meeting, P-138, pp. 121-34, Society of Automotive Engineers, Warrendale, PA, March 1984.

2. K. C. Liu and C. R. Brinkman, "Tensile Cyclic Fatigue of Structural Ceramics," in Proceedings of the Twenty-Third Automotive Technology Development Contractors' Coordination Meeting, P-165, pp. 279-84, Society of Automotive Engineers, Warrendale, PA, March 1986.

3. K. C. Liu, "Ceramic Specimen Heating by Induction Power," in Proceedings of the Conference on Mechanical Testing of Engineering Ceramics at High Temperatures, April 11-12, 1988, London, England; and in Int. J. of High Technology Ceram. (in press).

4. K. C. Liu, H. Pih, and D. W. Voorhes, "Uniaxial Tensile Strain Measurement for Ceramic Testing at Elevated Temperatures: Requirements, Problems, and Solutions," ibid (in press).

5. A. H. Heuer, "Transformation Toughening in ZrO_2 - Containing Ceramics," J. of Am. Ceram. Soc., Vol. 70, No. 10, pp. 689-98 (1987).

6. M. K. Ferber and T. Hine, "Fatigue Behavior of Partially Stabilized Zirconia Ceramics," in Proceedings of the Twenty-Fourth Automotive Technology Development Contractors' Coordination Meeting, P-197, pp. 201-08, Society of Automotive Engineers, Warrendale, PA, April 1987.

7. P. F. Becher and M. K. Ferber, "Mechanical Behavior of MgO-Partially Stabilized ZrO_2 Ceramics at Elevated Temperatures," J. of Mater. Science, Vol. 22, pp. 973-80 (1987).

8. Mechanical Properties Test Data Sheet Furnished by Nilcra Ceramics (U.S.A.), Inc.

9. N. L. Hecht, D. E. McCullum, D. W. Grant, J. D. Wolf, G. A. Graves, and S. Goodrich, "The Experimental Evaluation of Environmental Effects in Toughened Ceramics for Advanced Heat Engines," in Proceedings of the Twenty-Third Automotive Technology Development Contractors' Coordination Meeting P-165, pp. 299-310, Society of Automotive Engineers, Warrendale, PA, March 1986.

10. D. B. Marshall, "Strength Characteristics of Transformation-Toughened Zirconia," J. of Amer. Ceram. Soc., Vol. 69, No. 3, pp. 173-80 (1986).

11. D. P. Rooke and D. J. Cartwright, Compendium of Stress Intensity Factors, The Hillington Press, Uxbridge, Middx, 1976.

12. V. J. Tennery, M. K. Ferber, P. F. Becher, and S. B. Waters, "Characterization of the Fatigue and Slow Crack Growth Behavior of Partially Stabilized Zirconia Ceramics," in Proceedings of the Twenty-Second Automotive Technology Development Contractors' Coordination Meeting, P-155, pp. 409-18, Society of Automotive Engineers, Warrendale, PA, 1985.

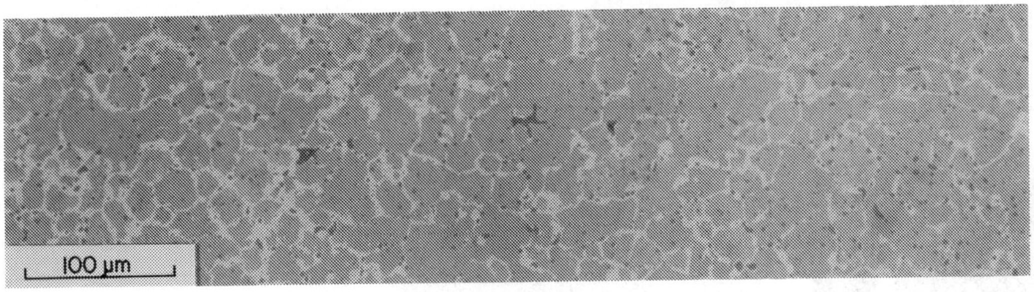

Fig. 1. Microstructure of MgO-PSZ used in this experiment. The grain boundaries are highlighted by eutectoid decomposition occurred during cooling. HF etched sample.

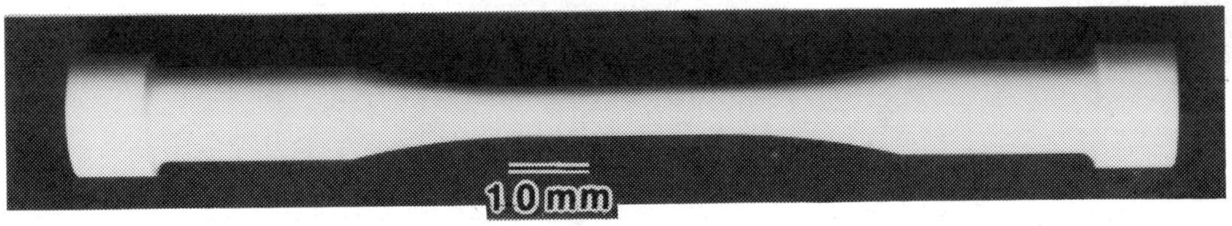

Fig. 2. Buttonhead uniform-gage tensile specimen made of MgO-PSZ.

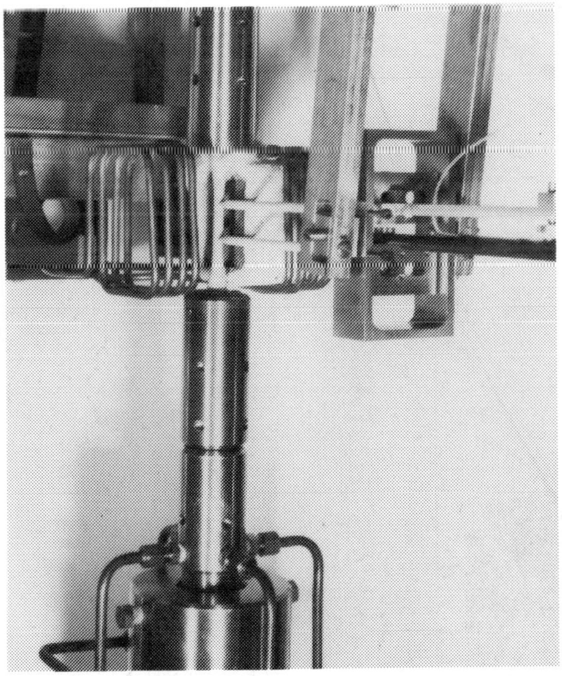

Fig. 3. Load-train column assembly and strain extensometer installed on universal testing machine for tensile testing of PSZ at high temperatures using induction heating.

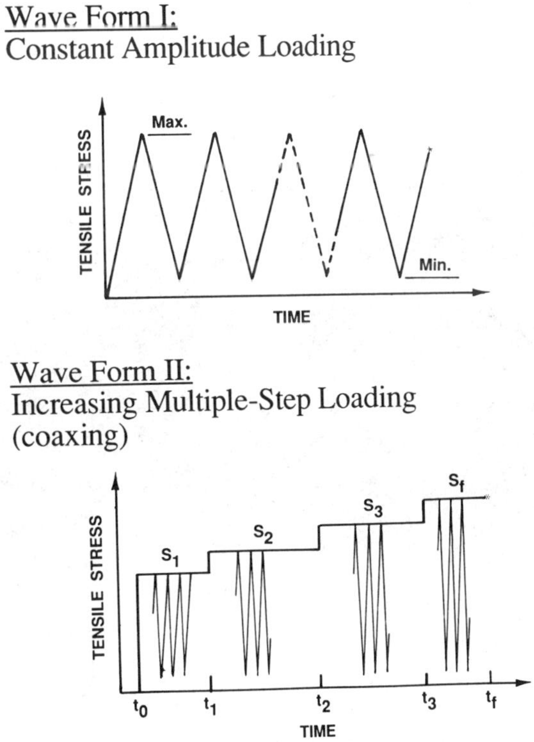

Fig. 4. Wave forms used in tension-tension cyclic fatigue testing.

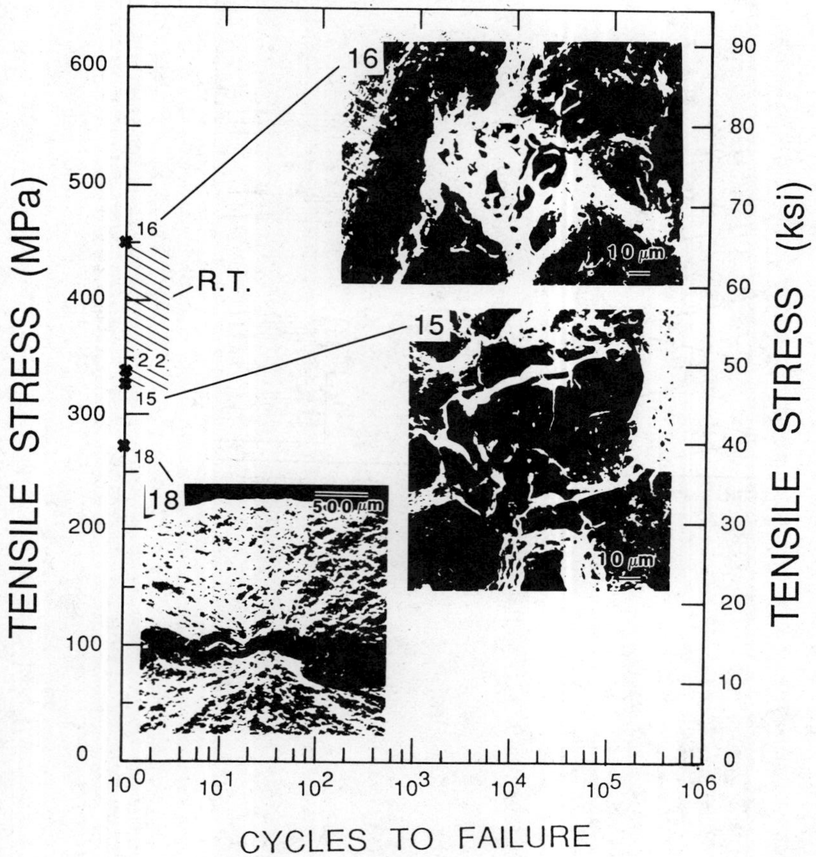

Fig. 5. Uniaxial tensile strength of MgO-PSZ at room temperature. Fracture origins are shown in the inserts. Unusually large process flaws embedded in specimen 18 acted as the fracture origin that substantially lowered its tensile strength.

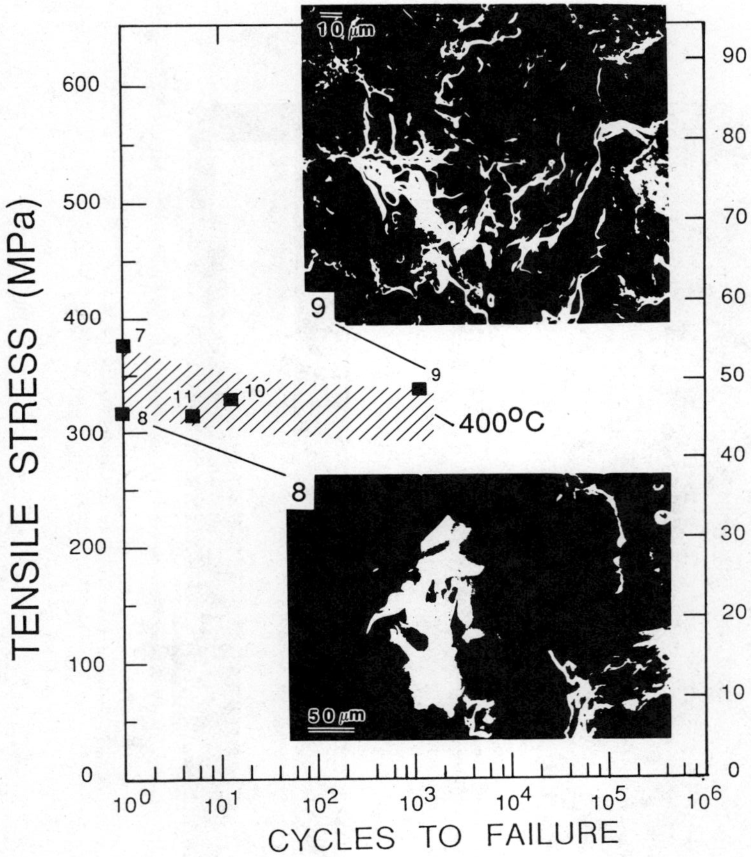

Fig. 6. Cyclic fatigue of MgO-PSZ tested at 400°C. Smooth grain surfaces visible inside the cavity were probed and shown to be rich in MgO.

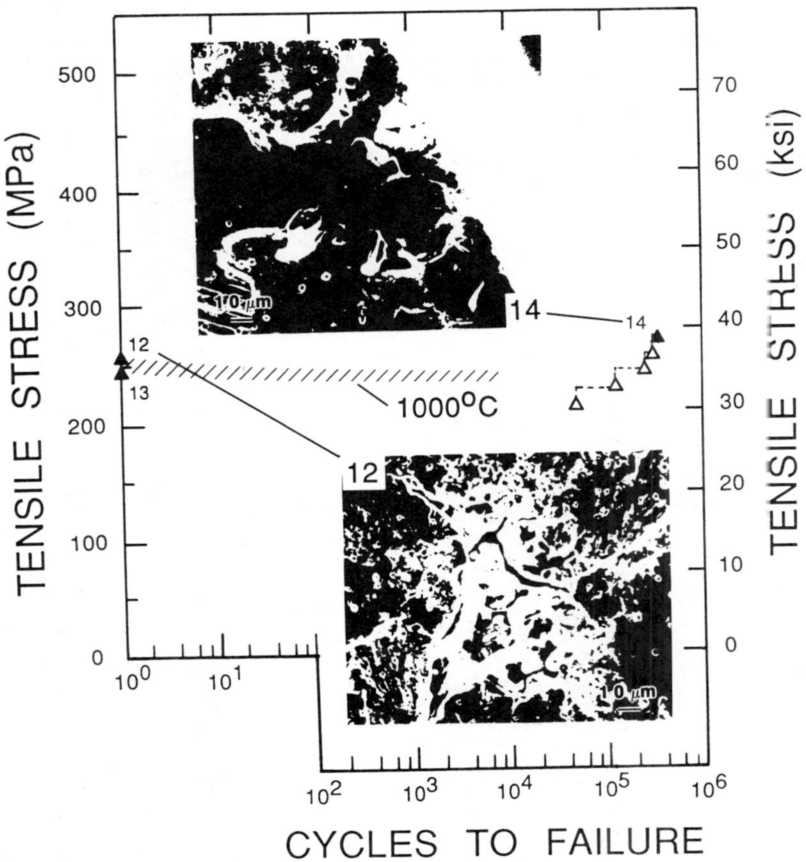

Fig. 7. Cyclic fatigue of MgO-PSZ at 800°C. The one initiated fracture in specimen 5 is located adjacent to the surface visible at the lower left corner of the SEM micrograph.

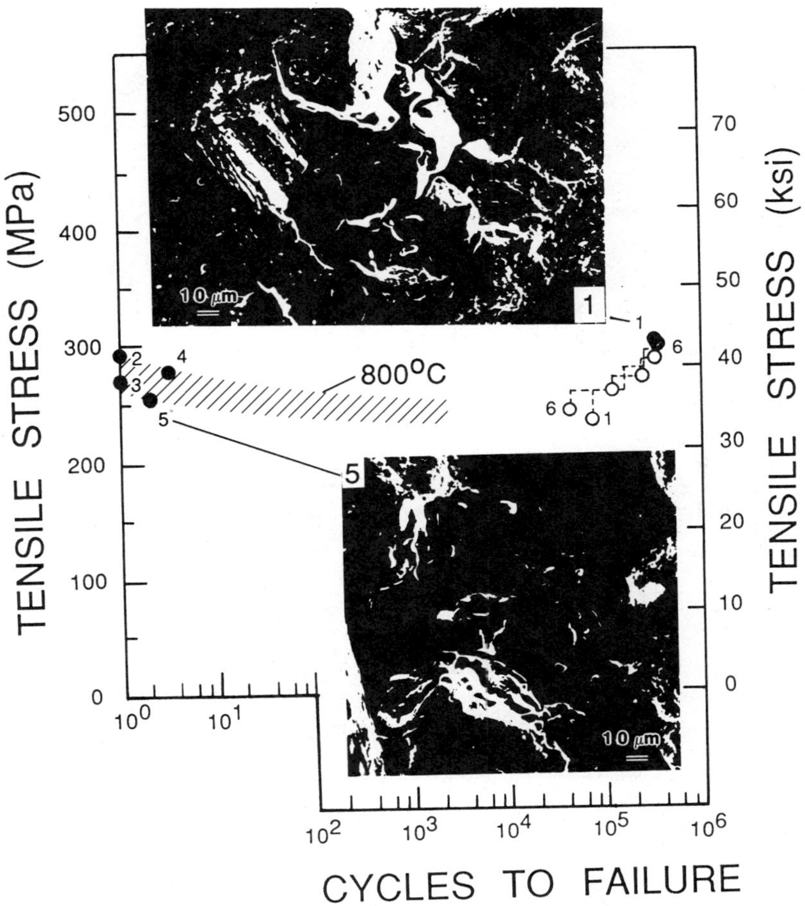

Fig. 8. Cyclic fatigue of MgO-PSZ at 1000°C. The shapes of the fracture origins of specimens 12 and 14 shown in the inserts are irregular and more complex compared with the others.

9.

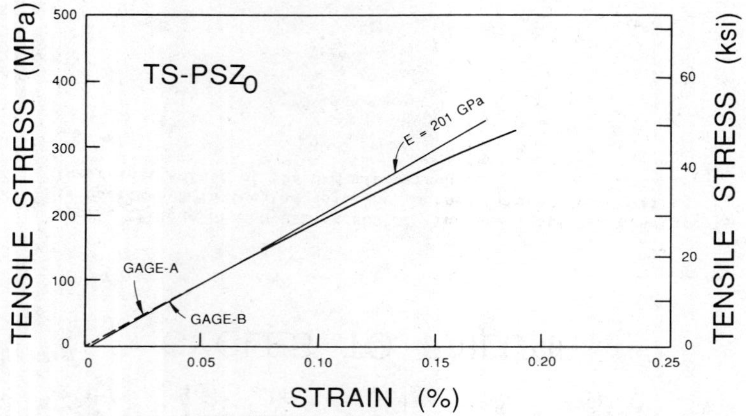

Fig. 9. Monotonic tensile stress-strain curves. Strains were measured from two strain gages cemented diametrically opposite to each other at the uniform gage section.

10.

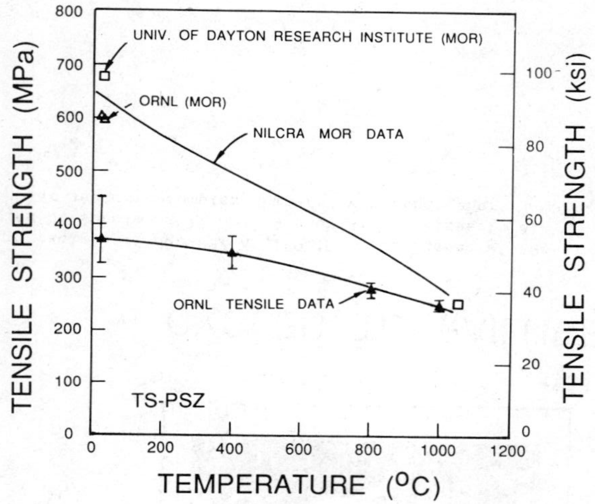

Fig. 10. Comparison between the ORNL tensile strength and the Nilcra four-point bend flexural strength of MgO-PSZ.

11.

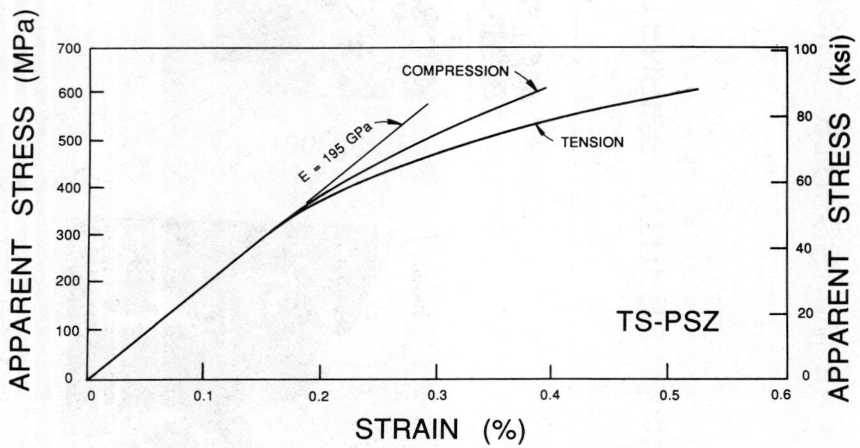

Fig. 11. Apparent flexural stress-strain curves for toughened MgO-PSZ in four-point bending. The strains were measured from the strain gages cemented on both sides of flexural surfaces.

12.

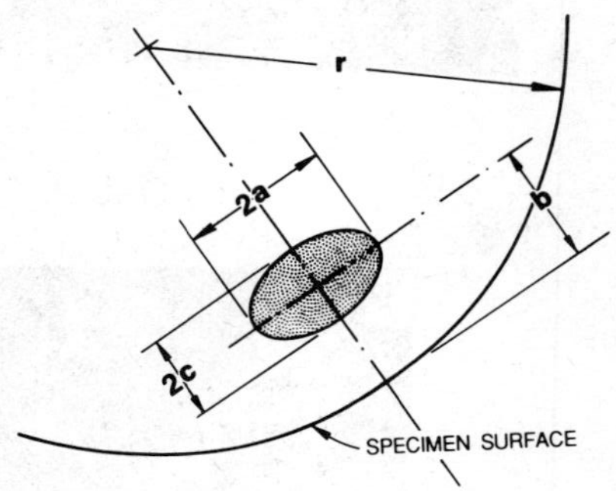

Fig. 12. Coordinate system and dimensions for elliptical fracture origin.

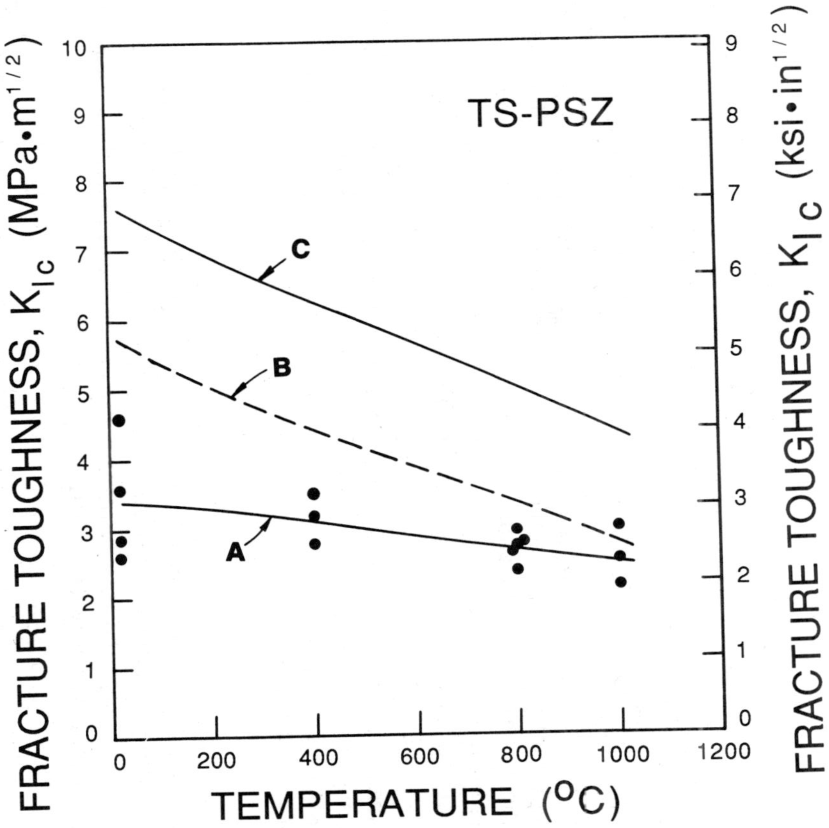

Fig. 13. Comparisons of bend-test determined and predicted K_{Ic} values: curve A – average K_{Ic} predicted from Eq. (1) using the ORNL tensile test data; curve B – average K_{Ic} predicted also from Eq. (1) but using Nilcra MOR data; curve C – Nilcra K_{Ic} data determined by bend tests.

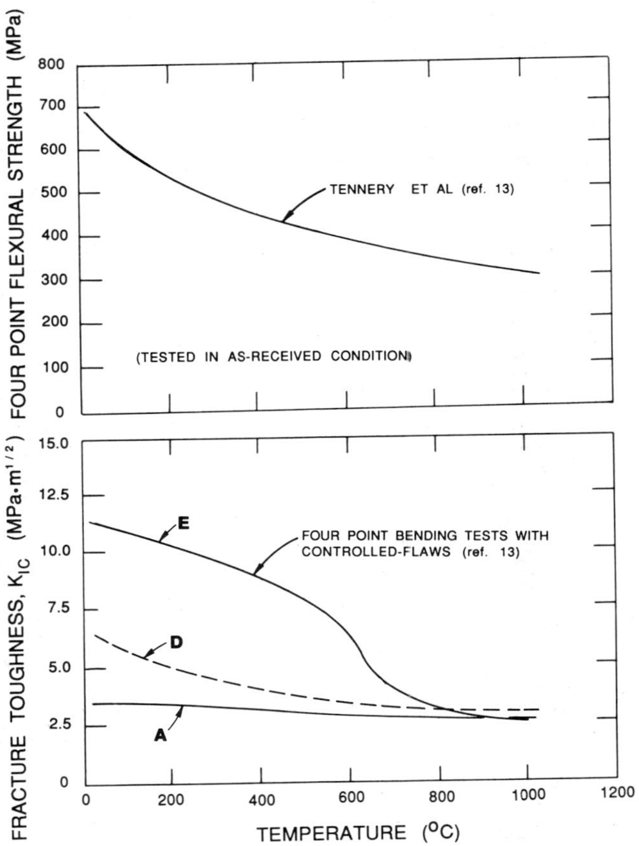

Fig. 14. (a) Four-point flexure strength of TS-PSZ ceramic versus temperature in air atmosphere, and (b) Curve D: fracture toughness predicted from Eq. (1) based on MOR values taken from (a); Curve E: fracture toughness obtained from bend tests with controlled flaws versus temperature.

SILICON NITRIDE-BASED CERAMICS FOR GAS TURBINE APPLICATIONS.

I.P.Tuersley, G.Leng-Ward, M.H.Lewis
Center for Advanced Materials Technology,
University of Warwick, Coventry CV4 7AL, England.

ABSTRACT.

A series of silicon nitride-based ceramics, under investigation for use in gas turbine applications above 1300°C, have been HIP'ped at 1725°C and 160 MPa. A range of tailored additive levels is being used in this programme in order to optimise microstructures capable of good creep and creep rupture performance without inhibiting fracture toughness. A preliminary survey of the ceramic microstructures and mechanical properties is presented.

1. INTRODUCTION.

1.1.

Applications within gas turbine engines present possibly the most arduous component conditions considered for engineering ceramics. In common with most heat engines, a gas turbine's efficiency may be directly related to its operating temperature, currently limited to approximately 1000°C by the use of nickel-based 'superalloys' for components such as turbine blades, discs and shrouds. The development of ceramic materials for use in these areas gives the potential for operating at temperatures of 1400°C and beyond, representing a significant gain in engine efficiency and specific fuel consumption. In addition, advantages such as low thermal expansion, (leading, for example, to a reduction in blade tip clearances) lower density (therefore lower induced stresses as well as a net weight saving) and non-strategic composition (availability and cost of constituents) have ensured that research into these materials has flourished in the last decade.

The advantages may be many and varied, but so are the associated problems. Not the least of these are the combinations of properties required by some of the most critical components. Whilst individual properties may often be met by some of the range of ceramic materials developed over the last twenty years, the combined requirements of such components still represent a formidable challenge to the ceramicist.

Of the monolithic ceramics that still offer a realistic solution to these problems, the Si-based materials (Si_3N_4, SiC) are leading contenders. Compared to other ceramics their high covalency crystals have intrinsic properties such as good oxidation resistance, high decomposition temperatures, low thermal expansion and resistance to plasticity, and they have been used widely as monolithics, or as both matrix and dispersed phases in the more recently developed composite ceramics.

It has been appreciated since the early development that the inherent strength and fracture toughness of ceramic materials may only be fully realised if the material is sintered to high or near-theoretical densities. In the case of SiC, with its high decomposition temperature, this is possible via solid state sintering with the addition of surface activating impurites such as boron and carbon. However, in the case of Si_3N_4, although solid state sintering is theoretically possible at 2100°C, decomposition at temperatures above 1800°C inhibits this process (1). Liquid sintering catalysts produce sufficient transport rates within the material in the 1700-1800°C region, but care has to be taken that secondary phase residues do not severely impair high temperature performance of the final material.

1.2. Principles of Composition selection.

Early studies of tailored compositions were based on the β'-Si-Al-O-N system. These early materials exhibited good strength and fracture toughness, but at high temperatures, creep resistance was poor due to the onset of cavitation within grain junction glass residues subjected to hydrostatic tension (2). This problem necessitated a tailoring of the chemistry of the minor intergranular phase such that it could be recrystallized to produce a stable diphasic microstructure. The situation is complicated by the need to create a low viscosity liquid phase at the sintering stage to allow anisotropic growth of the β' grains. Aspect ratios >5 have been shown to produce significant increments in K_{1C} due to enhanced crack deflection, crack-bridging and pull-out (3). This was achieved with later developments within the Y-Si-Al-O-N system, utilising the same substituted (β') major phase, with the minor phase being recrystallised as a yttrium aluminium garnet (*YAG*). The material exhibited a promising balance of creep properties and fracture toughness, but was limited to temperatures below about 1350°C due to oxidative degradation with the *YAG* and SiO_2 (formed as Si_3N_4 oxidizes) reacting to form a liquid in this temperature region; hence a reversion back to poor high temperature creep properties.

The choice of a sintering additive metallic oxide in the fabrication of β-Si_3N_4 ceramics is governed by a number of factors;

(i) Stability and viscosity of the M-Si-O-N oxynitride liquid sintering medium.

(ii) Softening temperature of any residual glass.

(iii) Ease of crystallization of the intergranular glass into an oxidation resistant phase.

Cations such as Ca^{+2}, Mg^{+2}, Nd^{+3}, La^{+3} and Y^{+3} are stable components of oxynitride liquids. Of these the lower valence metallic ions such as Mg^{+2} provide a lower viscosity sintering medium giving enhanced sinterability and densification kinetics. However, this is not important with HIP processing (hot-isostatic-pressing) as the high pressures and temperatures employed enable Si_3N_4 to be fully densified with relatively small liquid volumes based on Y_2O_3 and the rare earth oxide additives. The advantage of these M^{+3} metallic ions, in particular Y^{+3}, is that they form glasses with higher softening temperatures, important for the retention of high temperature mechanical properties. Small variations in the volume fraction of liquid phase at the HIP'ping stage have a significant effect upon the sintering kinetics and resulting microstructural morphology and are therefore also to be considered when tailoring the overall composition.

Finally, the ease of crystallization of the intergranular regions into a stable, oxidation-resistant phase is of crucial concern. Work by Lange et al (4) suggests that the intergranular oxynitride phases which oxidize with an appreciable volume change should be avoided. For example in the Si-Y-O-N system such phases can be avoided by fabricating compositions within the Si_3N_4-$Y_2Si_2O_7$-Si_2N_2O compatibility triangle (Refer to Figure 1). With reference to Figure 2, an enlargement of the Si_3N_4 corner of Figure 1, it can be seen that composition (A) made up of '96.5 wt% UBE Si_3N_4 +3.5 wt% Y_2O_3' lies just within this compatibility triangle. However, compositions with higher additive levels need extra SiO_2 additional to that present on the Si_3N_4 powder surface to bring the composition back beneath the $Y_2Si_2O_7$-Si_3N_4 tie line.

This programme, in conjunction with Roll-Royce PLC aims to optimise the microstructure of a Si_3N_4-based ceramic in relation to high temperature oxidation, creep and stress-rupture whilst retaining modest levels of strength and fracture toughness. Hot-isostatic-pressing (HIP) processing is used as means of achieving full density with three different volume fractions of sintering liquid and the chemistry of the latter is varied in a systematic way in relation to oxygen/nitrogen ratio and cation type with the object of achieving fully crystalline diphasic microstructures.

2. EXPERIMENTAL.

In the first HIP run seven billet compositions were fabricated, all with a constant 5wt% Y_2O_3 additive level. The oxygen/nitrogen ratio was varied by systematically replacing increasing proportions of the Si_3N_4 by 0, 2, 4 and 8 wt% SiO_2 respectively in compositions 1, 2 ,3 and 4, and, also by 5, 10 and 25wt% Si_2N_2O in compositions 5, 6 and 7.

In the second HIP run the compositions were tailored to give three different volume fractions of sintering liquid while maintaining a similar liquid composition. The Y_2O_3 additive levels tried were, 3.5 wt%, 5 wt% and 7.5 wt% in respectively compositions A, B, and C together with just enough additional SiO_2 in B and C to bring these compositions just below the Si_3N_4-$Y_2Si_2O_7$ tie line. Three equivalent Nd_2O_3 additive compositions , A′, B′ and C′ were also fabricated. Refer to Table 1 for a listing of all compositions and to Figure 2 for their positions on the behavioural $2Y_2O_3$-$4YN$-$3SiO_2$-Si_3N_4 phase diagram.

Table 1. Identification of ceramic compositions by weight % of starting powders, element, equivalent % of metal additive and equivalent % of nitrogen. (NB. eq%Y=3[Y]/(3[Y]+4[Si]), eq%N=3[N]/(3[N]+2[O])).

Id.	Weight %				Element					
	Y_2O_3 (Nd_2O_3)	Si_3N_4	SiO_2	Si_2N_2O	Y (Nd)	Si	O	N	eq%Y (Nd)	eq%N
1	5.0	95.0	-	-	.065	2.952	.202	3.866	1.62	96.65
2	5.0	93.0	2.0	-	.065	2.951	.297	3.802	1.62	95.04
3	5.0	91.0	4.0	-	.065	2.951	.395	3.737	1.63	93.42
4	5.0	87.0	8.0	-	.065	2.951	.592	3.606	1.65	90.14
5	5.0	90.0	-	5.0	.065	2.951	.269	3.821	1.62	95.51
6	5.0	85.0	-	10.0	.065	2.951	.338	3.775	1.63	94.37
7	5.0	70.0	-	25.0	.066	2.951	.547	3.636	1.64	90.89
A	3.5	96.5	-	-	.045	2.966	.172	3.886	1.12	97.14
A′	(5.1)	94.9	-	-	(.045)	"	"	"	(1.12)	"
B	5.0	94.0	1.0	-	.065	2.951	.249	3.834	1.62	95.85
B′	(7.3)	91.7	1.0	-	(.065)	"	"	"	(1.62)	"
C	7.5	90.0	2.5	-	.099	2.926	.374	3.750	2.48	93.76
C′	(10.8)	86.8	2.4	-	(.099)	"	"	"	(2.48)	"

The starting powder used was 'UBE SN E10' Si_3N_4 containing an analyzed oxygen content of 1.2 wt% (assumed to be present as SiO_2 on the powder surface). Impurity levels quoted by UBE for this powder are, Cl <100, Fe <100, Ca <50, and Al <50. It should be noted that the feasibility of representing compositions on such a greatly enlarged part of a phase diagram, Figure 2, depends on the accuracy of the analyzed oxygen content of the Si_3N_4 powder, their being no significant oxygen pick up during milling and on the levels of impurities such as Ca and Fe being low enough to ignore.

Appropriate mixtures of Si_3N_4 and Y_2O_3 or Nd_2O_3 were milled for sixteen hours in plastic cylinders using Si_3N_4 milling media and propanol as the medium. The milled mixtures were dried, sieved, and cold isotatically compacted into billets approximately 11 cm long and 3 cm diameter. The billets were HIP'ped at 1725° and 160 MPa for one hour at ASEA Cerama, Sweden.

The mechanical property testing to determine the modulus of rupture (MOR) and fracture toughness of the thirteen compositions was carried out at both Warwick University and Rolls Royce PLC,

Leavesden. Test bars were cut from the bulk of HIP'ped billets, polished to a 6 um finish and chamfered on the top surface edges. Room temperature MOR was performed in three-point bend and the results, averaged over six tests per composition. Fracture toughness (K_{1C}) values were determined via 4-point bend tests on indent initiated specimens (5). The high temperature MOR was performed at 1000°C, 1200°C and 1400°C in 3-point bend on a 'multiple shot' arrangement developed at Rolls Royce PLC, Leavesden (6).

3. RESULTS AND DISCUSSION.

3.1. Microstructure.

All thirteen ceramic compositions HIP'ped to close to full density (>99.3% theoretical density). X-ray powder diffraction traces of the HIP'ped billets show that complete transformation of the mainly α-Si_3N_4 starting powder into β-Si_3N_4 has taken place during sintering and densification. While ceramic 1 contains only Si_3N_4 as the major phase, as increasing amounts of Si_3N_4 were replaced by SiO_2 (ceramics 2, 3 and 4) or Si_2N_2O (ceramics 5, 6 and 7) in the starting compositions, correspondingly, an increasing proportion of β-Si_3N_4 was replaced by Si_2N_2O in the final densified ceramic. This is illustrated in the X-ray diffraction traces of billets 1, 2, 3 and 4 in Figure 3. This is consistent with the billet composition moving from close to the Si_3N_4-Y_2O_3 tie line progressively further into the Si_3N_4-Si_2N_2O-$Y_2Si_2O_7$ compatibility region.

Also evident in the X-ray diffraction traces in Figure 3 are small peaks from yttrium containing phases present in the crystalline intergranular regions. TEM microscopy shows that in all these HIP'ped ceramics the intergranular liquid present at 1725° has completely crystallized during cooling into the α-$Y_2Si_2O_7$ polymorph and/or variable amounts of an yttrium silicate phase termed 'y-yttrium silicate' and sometimes small amounts of the β-$Y_2Si_2O_7$ polymorph. The 'y-yttrium silicate' phase tended to form at the expense of the $Y_2Si_2O_7$ phase in the oxygen rich compositions '3'and '4'. The two most oxygen rich compositions (4 and 7) also showed evidence of some small silica rich globules within the intergranular regions. This suggests that the intergranular liquid composition does tend to become more silicon rich as well as Si_3N_4 being systematically replaced by Si_2N_2O as the composition is made more oxygen rich along the series 1-> 4 and 5-> 7.

The three variable Y_2O_3-additive compositions, A, B and C all HIP'ped to close to full density in the second HIP run. However, the intergranular liquid had crystallized fully during cooling into the 'y'-yttrium silicate phase rather than the α-$Y_2Si_2O_7$ phase found in ceramics 1 and 2 as illustrated in Figure 4. The three variable Nd_2O_3-additive compositions, A′, B′ and C′, also HIP'ped to close to full density with the intergranular regions fully crystallized into a neodymium silicate apatite phase. Before being cut into test bars billets of these six compositions were all annealed in a nitrogen atmosphere at 1280 C for 24 hours. This resulted in conversion of the 'y'-yttrium silicate intergranular phase into α-$Y_2Si_2O_7$ in A, B and C. For the Nd_2O_3 additive compositions the apatite phase was converted into the low temperature polymorph of $Nd_2Si_2O_7$.

The X-ray diffraction traces in Figure 4 illustrate these post-HIP annealing phase changes. Figure 4(a) and 4(b) show the Y_2O_3-additive ceramic B in respectively the 'as received' condition and after post-HIP annealing at 128O C. Similarly for the Nd_2O_3-additive ceramic B′. TEM micrographs in Figure 5 typically illustrate these intergranular microstructural changes for the Y_2O_3-additive ceramic B and also illustrate the very thin amorphous layer always left between an α-Si_3N_4 grain and the adjoining crystallized intergranular phase. Such thin amorphous layers are also present between 'y'-yttrium silicate, apatite, disilicate crystals and the β- Si_3N_4 in all the ceramics.

The types of silicate phase that crystallize in the intergranular regions of these Si_3N_4 ceramics depend on the cooling rate of the HIP cycle. There are four yttrium disilicate polymorphs with the temperature stability regions indicated as follows (from 7):

$$\alpha\text{-}Y_2Si_2O_7 \overset{1225°C}{\rightleftharpoons} \beta\text{-}Y_2Si_2O_7 \overset{1445°C}{\rightleftharpoons} \gamma\text{-}Y_2Si_2O_7 \overset{1535°C}{\rightleftharpoons} \delta\text{-}Y_2Si_2O_7$$

The 'y-yttrium silicate' phase is a phase distinct from the $Y_2Si_2O_7$ polymorphs, but with a very similar composition, stable at temperatures below ~1200°C and possibly stabilized by impurities (7). In the $Nd-Si-O-N$ system there are two neodymium disilicate polymorphs, L-Nd_2O_3 $\xrightleftharpoons{1480°C}$ H-Nd_2O_3, and a neodymium silicate apatite phase, $Nd_{9.333}(SiO_4)_6O_2$, which is more stable than its yttrium counterpart. The apatite phase composition may lie a small way between the oxide and nitrogen apatite, $Nd_3Si_5O_{13}N$. From the evidence of this work it appears that the cooling rates employed by the ASEA HIP unit favour the nucleation of the low temperature disilicate polymorphs, or, the 'y'-yttrium silicate or neodymium apatite phases if the disilicate phases do not have time to nucleate and grow. Small variations in the cooling rates experienced by specimens from one HIP run to the other may cause variations in intergranular crystallization, but post HIP annealing at above 1200°C should establish one particular thermally favoured disilicate intergranular phase.

In order to determine the impurity levels in the residual intergranular liquid at completion of a one hour HIP at 1725°C a small bar of the 5wt% Y_2O_3 additive ceramic (2) was taken back to 1650°C in a small experimental furnace for twenty minutes, then, rapidly cooled by removing from the furnace hot zone to the atmosphere in a few seconds. With this rapid cooling rate the intergranular oxynitride liquid did not have time to crystallize and the intergranular regions are completely glassy as illustrated in Figure 6 (a). The very small β-Si_3N_4 crystals distributed within the glass may have precipitated on cooling out of what was a high nitrogen content $Y-Si-O-N$ liquid present at 1725°C as the temperature dropped on cooling. The TEM EDX intergranular glass analysis shows Y, Si and O with no detectable level of impurites such as Ca and Fe. The light element detector is not sensitive enough to show any possible low level of nitrogen in this glass.

The microstructure of all the HIP'ped ceramics with compositions close to the Si_3N_4-$M_2Si_2O_7$ tie line consists of a very broadly bimodal distribution of large β-Si_3N_4 crystals typically in the dimension range of 1-5 microns dispersed in a matrix of smaller β-Si_3N_4 crystals mostly in the lower part of the range 0.1-1 microns. SEM micrographs in Figure 7 illustrate the variation in morphology with volume of sintering liquid for the Y_2O_3 additive ceramics A, B and C. The larger volume of liquid in the 7.5 wt% Y_2O_3 ceramic (C) has allowed the β-Si_3N_4 grains initially present and those first to precipitate more space to grow in the preferred elongated morphology. Typical aspect ratios of these grains are in the range of 5:1 to 8:1.With the lower liquid volume in A the scope for this elongated growth becomes restricted resulting in a less anisotropic morphology and slighly finer microstructure. The scale of microstructure and morphology of the equivalent Nd_2O_3 additive ceramics is very similar. The progressive replacement of a proportion of the Si_3N_4 by Si_2N_2O tends to result in a similar scale of microstructure but with a lessening in the anisotropic morphology.

4. PROPERTY EVALUATION.

The room temperature MOR results (three point bend) are plotted against the level of oxygen content in Figure 8. The fracture toughness results are similarly plotted against variation in oxygen content in Figure 9. The high temperature MOR results at 1000°C, 1200°C and 1400°C are plotted in Figure 10.

Figures 8 and 9 suggest that the effect of increasing oxygen content in the composition is to initially impair both strength and fracture toughness. There is also a marked difference in the effect of achieving this level of oxygen by means of SiO_2 or Si_2N_2O. Further addition of SiO_2 progressively restores the strength, whereas higher levels of Si_2N_2O reduce it. As ambient temperature properties will be governed to a greater degree by processing effects, this would indicate that the Si_2N_2O phase is so stable that it does not form a liquid early enough, or in sufficient volume to be as effective as SiO_2 as a sintering medium. High temperature strength, within the limits of experimental scatter, shows no appreciable difference between the compositions, but verify the advantages of the fully recrystallized yttrium-based matrix phase, as a strength of approximately 0.5 GPa is maintained at 1400°C.

The next stage of the programme investigates the effect of varying the level of M_2O_3 additive, and therefore both the volume fraction of sintering liquid and the amount of residual intergranular phase, using either Y_2O_3 or Nd_2O_3 as the additive. A similar series of tests were performed on identically prepared bars; both room- and high-temperature (MOR) (Figure 11), and room-temperature fracture toughness (plotted against Y or Nd levels in equivalent percent, Figure 12).

Figure 11 shows little difference from Figure 10, the corresponding results for the series varying in oxygen content, at temperatures up to 1200°C. This confirms that within this regime, fracture strength is determined to a greater degree by processing effects. There is however a greater variation in the range of results from the different compositions at room temperature. This may be explained by the changing morphology of the β-Si_3N_4 grains with increasing amounts of sintering liquid, demonstrated more fully by the fracture toughness results. At temperatures above 1200 °C, a clear division between those compositions sintered using Y_2O_3 and those containing Nd_2O_3 develops. At 1400 °C, there is a difference of between 0.15 and 0.25 GPa between corresponding M_2O_3 levels, the Nd_2O_3 samples showing the more marked degradation in strength down to a level of approximately 0.3 GPa. In both additive cases, the lower M_2O_3 compositions show a better retention of strength at this temperature, but it is worth noting that the lower level Nd_2O_3 additive does not perform as well as even the higher Y_2O_3 additive composition. This indicates quite clearly that at high temperatures, both the level and the chemistry of the intergranular phase becomes the strength-determining feature. Possible explanations for this involve differences in surface energies or interfacial cohesion of the various phases and are to be investigated further.

The fracture toughness results (Figure 12) show the expected rise in K_{IC} with increasing volume fraction of sintering liquid. This would occur as a result of increasing anisotropy of the β-Si_3N_4 grains enhancing crack-deflection, crack-bridging and pull-out as anticipated. The absence of any improvement in the fracture toughness of the *Nd* ceramics over their equivalent level *Y* ceramics indicates that the lower viscosity *Nd–Si–O–N* liquid does not give rise to any significant enhancement of β-grain elongation during the high temperature and pressure of the HIP process. This is born out by microstructural observation.

Steady-state creep properties of these compositions vary between 4.5 and 6.5 x $10^{-9}s^{-1}$ at 1300°C under 200 MPa of bending stress, with the Y_2O_3 sintered specimens exhibiting the lower creep rates. This suggests a greater transport rate within the Nd_2O_3-based intergranular amorphous phases.

Both the Y_2O_3- and Nd_2O_3-additive series of ceramics fabricated in the Si_3N_4-$M_2Si_2O_7$-SiO_2 combatibilty triangle show very good oxidation resistance after 100 hours at 1400°C in a laboratory furnace, although the *Y* ceramics do develop a thinner and more even protective oxide scale as illustrated by the micrographs in Figure 13. The exposed surfaces of the oxide scales of ceramic B and B′ are shown in (a) and (b) respectively, while cross sections through the oxide scales are shown in (c) and (d) respectively. The oxide scales consist of cristabolite (SiO_2) and white crystals of the respective $M_2Si_2O_7$ phase. The lower viscosity of the *Nd–Si–0* oxide layer allows blowholes to form at the surface caused by the evolution of nitrogen gas during oxidation. Both oxide surfaces show a crazed cracking pattern. This cracking occurs on cooling down from 1400°C, caused by the thermal expansion mismatch between the oxide layer and bulk nitride ceramic and/or by the β-α cristabolite transformation.

5. CONCLUSION.

The 5 wt% Y_2O_3 level/variable SiO_2 ceramics HIP'ped in this programme densified into a range of materials, from containing a wholly β Si_3N_4 matrix through to an approximately 2:1 Si_3N_4:Si_2N_2O matrix, depending on the oxygen level in the starting powder compact composition. The intergranular regions were crystallized into α-$Y_2Si_2O_7$ and /or 'y' yttrium silicate. The variable additive level composition ceramics formulated close to the Si_3N_4-$M_2Si_2O_7$ tie line and HIP'ped on a subsequent HIP run, densified into diphasic β-Si_3N_4 ceramics. After post HIP annealing at 1280°C the secondary intergranular was a low temperature $M_2Si_2O_7$ polymorph, $Y_2Si_2O_7$ or $Nd_2Si_2O_7$ depending on whether Y_2O_3 or Nd_2O_3 was used as the additive.

The intergranular regions were essentially fully crystallized except for a thin amorphous layer left at the β-Si_3N_4-intergranular phase interface. The type and distribution of phases found in the these HIP'ped materials is consistent with the original powder mixture taking into account, crucially, the oxygen level in the Si_3N_4 powder. This type of relationship, of course, is specific to the raw materials, processing and HIP conditions used, and needs estabilishing in order to maintain control over the composition of the residual intergranular regions.

The initial MOR and fracture toughness property measurements show no advantage in adding oxygen to the original compositions via Si_2N_2O additions rather than with SiO_2. Also there appeared no clear trend of improvements in replacing Si_3N_4 in the matrix with a proportion of Si_2N_2O.

Variation in M_2O_3 additive levels showed a progressive rise in fracture toughness with increasing volume fraction of sintering liquid. This however occurs at the expense of high temperature strength (above 1200°C) and steady state creep properties - more so if the sintering additive used is Nd_2O_3 rather than Y_2O_3. These results relate quite satisfactorily to the microstructures observed in the various compositions.

6. ACKNOWLEDGEMENT.

Financial support for research descibed in this paper has been gratefully received from Rolls-Royce PLC.

7. REFERENCES.

1. K.H.Jack, 'Progress in Nitrogen Ceramics', ed. F.D.Riley, NATO ASI SERIES E Applied Science (1983) p45.

2. B.S.B.Karunaratne and M.H.Lewis, J.Mater.Sc., 15 (1980) 449.

3. M.H.Lewis, G.Leng-Ward and C.Jasper, 'Sintering Additive Chemistry in Controlling Microstructure and Properties of Nitride Ceramics', Ceramic Transactions Vol.1, Ceramic Powder Science II (1988) 1019-1033.

4. F.F.Lange, S.C.Singhal and R.C.Kuznicki, 'Phase Relationships and Stability Studies in the Si_3N_4-SiO_2-Y_2O_3 Pseudoternary System', J.Am.Ceram.Soc., 60[5-6] (19770 249-252.

5. P.Chantikul, G.R.Anstis, B.R.Lawn and D.B.Marshall, J.Am.Ceram.Soc., 64 (1981) 539.

6. G.R.Essam, G.Syers, Proc.ICAS '88 Jerusalem, Israel.

7. K.Liddell and D.P.Thompson, 'X-ray Diffraction Data for Yttrium Silicates', B.Ceram.Trans.J., 85 91986) 17-22.

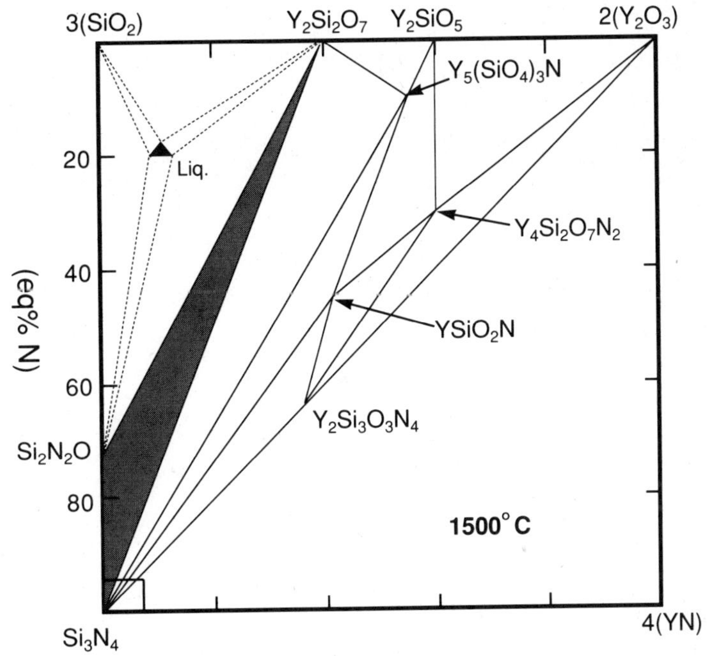

Figure 1. Y-Si-O-N system plotted in equivalent units as a Si_3N_4-$4YN$-$2Y_2O_3$-$3SiO_2$ phase diagram.

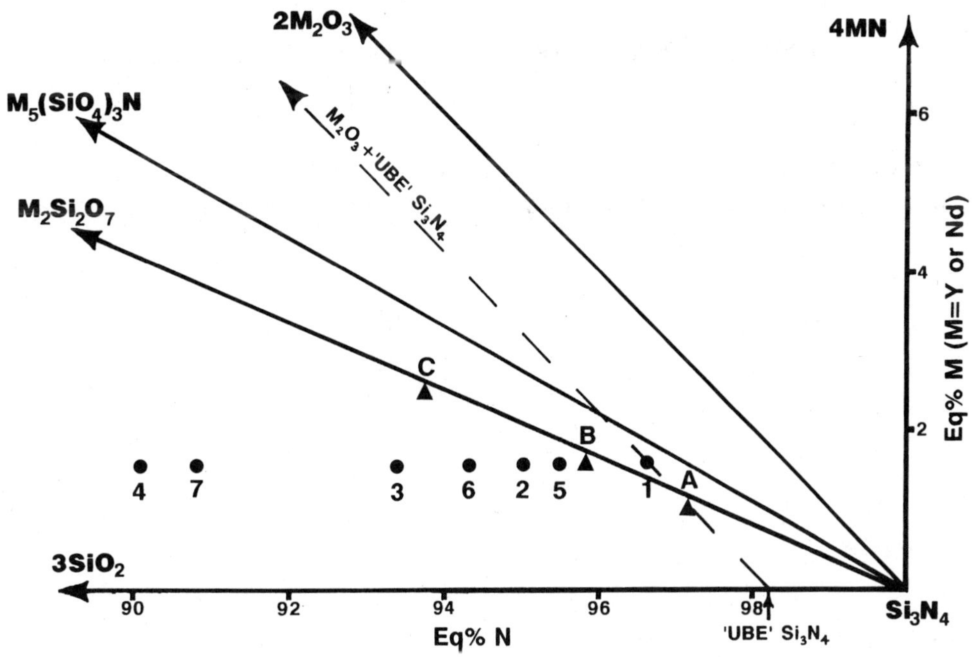

Figure 2. Enlargement of the Si_3N_4 corner of Figure 1.

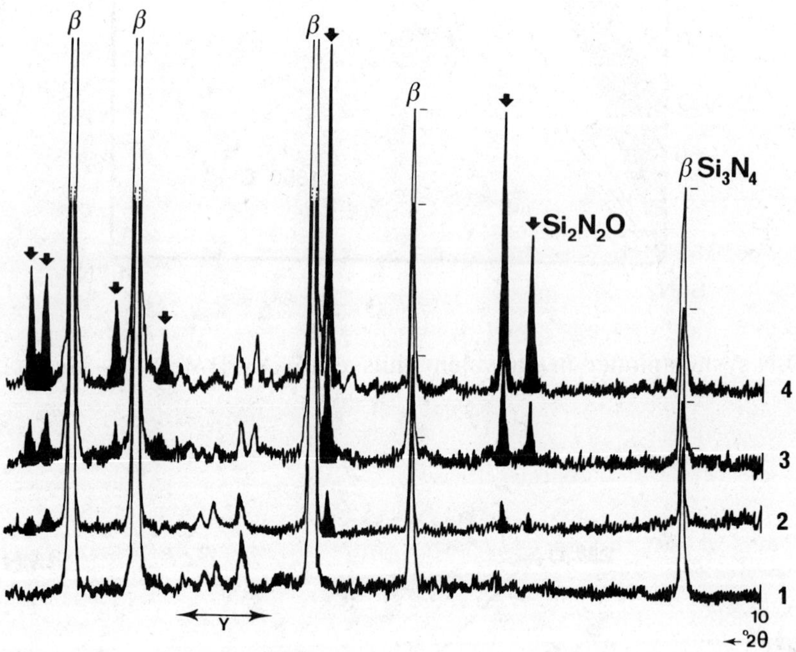

Figure 3. X-ray diffraction traces of HIP'ped ceramics 1, 2, 3 and 4, showing Si_3N_4 and Si_2N_2O peaks and the region of the main peaks from yttrium containing intergranular phases (<--> Y).

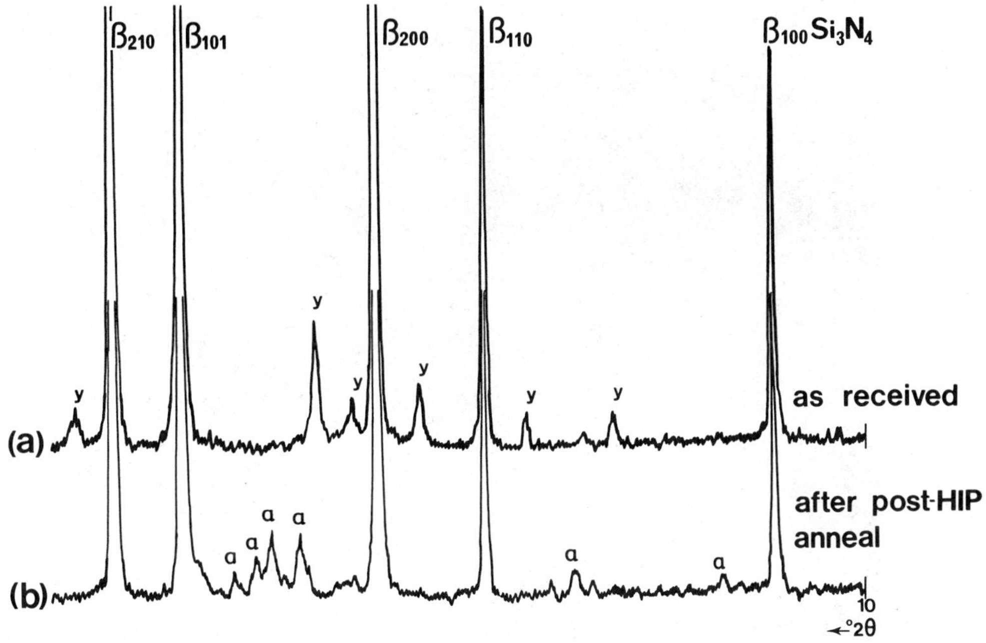

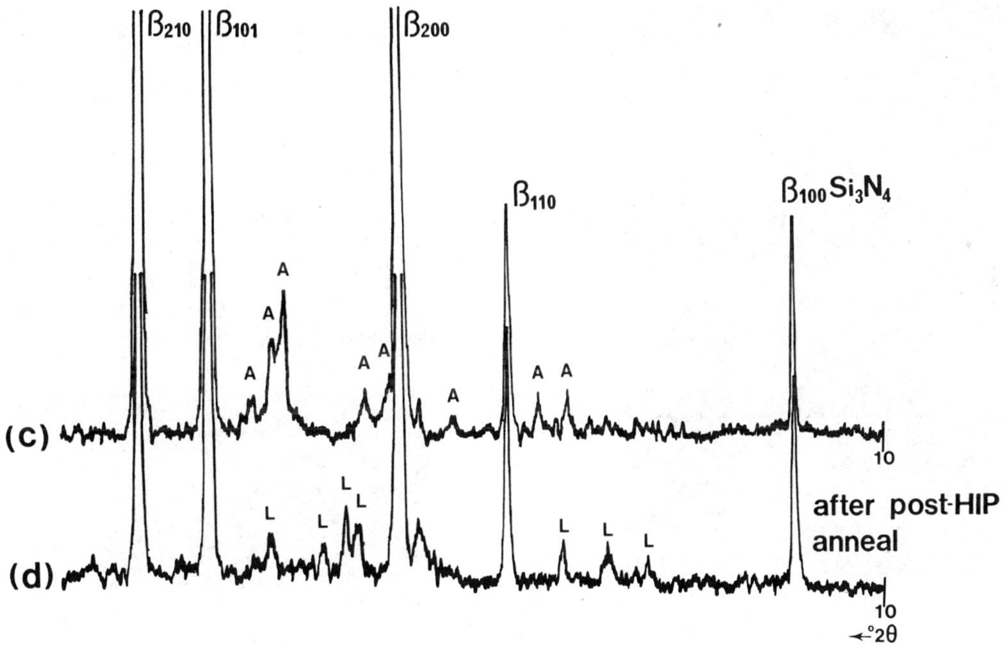

Figure 4. X-ray diffraction traces of HIP'ped ceramics B and B′, compositions close to the Si_3N_4-$M_2Si_2O_7$ tie line: Y_2O_3-additive material (a) in 'as received' condition and (b) 'after post-HIP anneal' at 1280°C (b); Nd_2O_3-additive material (c) in 'as received' condition and (d) 'after post-HIP anneal' at 1280°C.

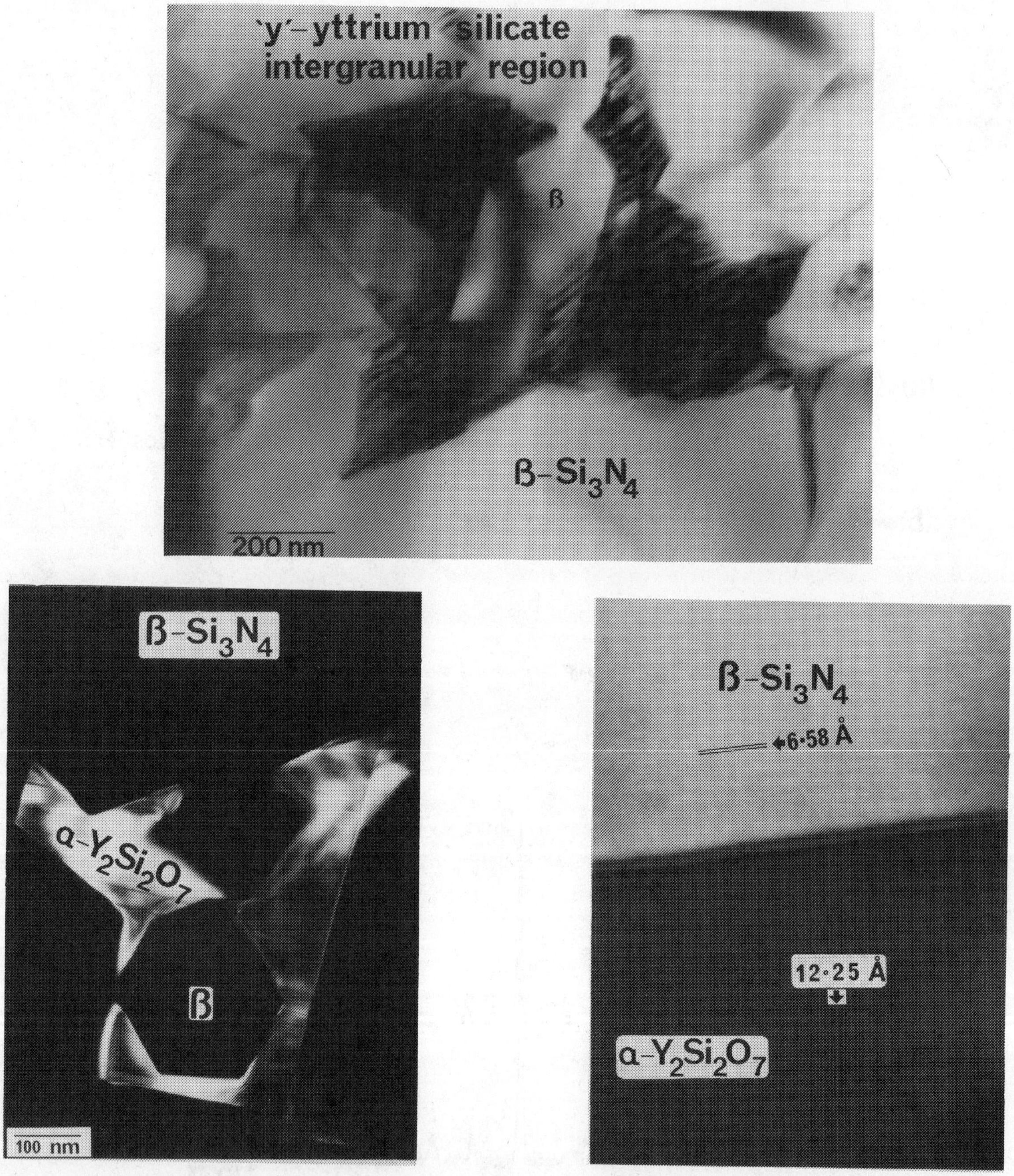

Figure 5. TEM micrographs of ceramic B showing in (a) the material as received with the intergranular region crystallized into the 'y'- yttrium silicate phase (dark mottled contrast). After annealing at 1280 °C, (b) shows an α-$Y_2Si_2O_7$ intergranular region in dark field contrast, and (c) a Si_3N_4-α-$Y_2Si_2O_7$ interface showing thin amorphous boundary layer.

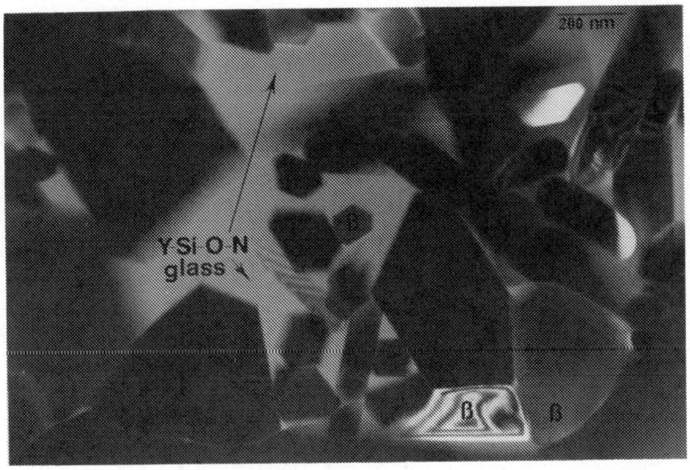

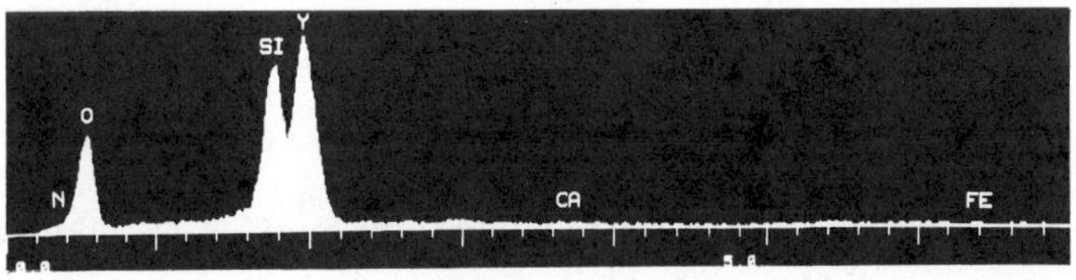

Figure 6. TEM micrograph of ceramic '3' after reannealing for 20 minutes at 165O°C followed by rapid quenching. A TEM EDX analysis of the residual intergranular glass is shown in (b).

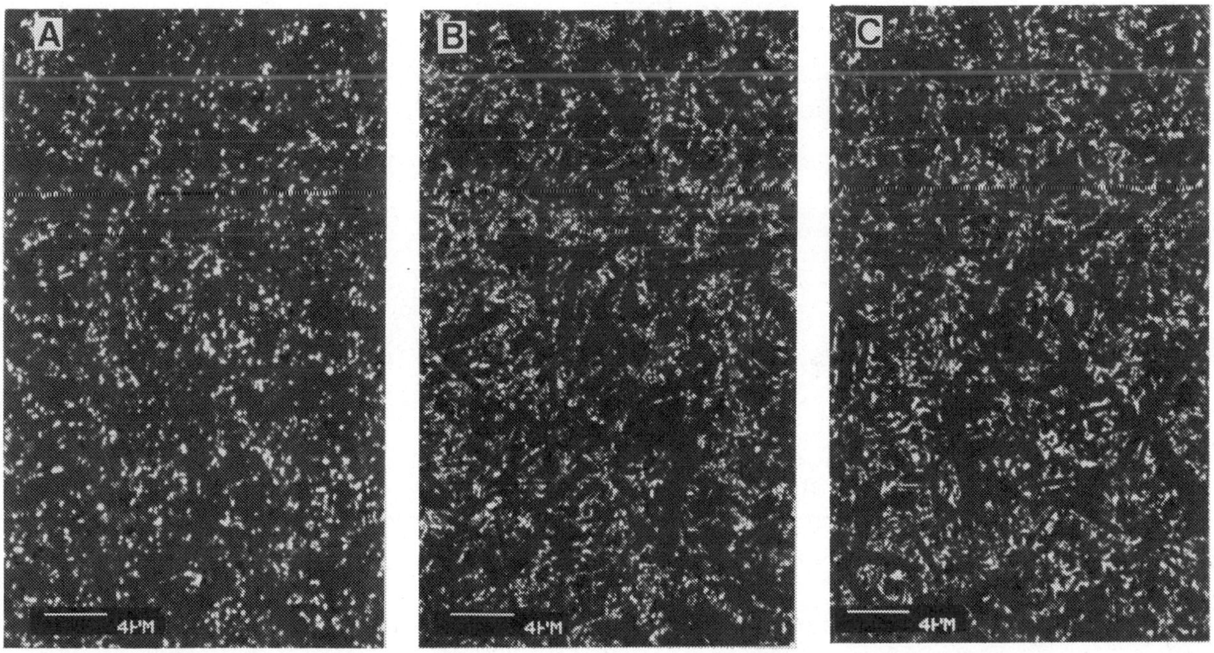

Figure 7. SEM micrographs of ceramics A, B and C (back scattered mode).

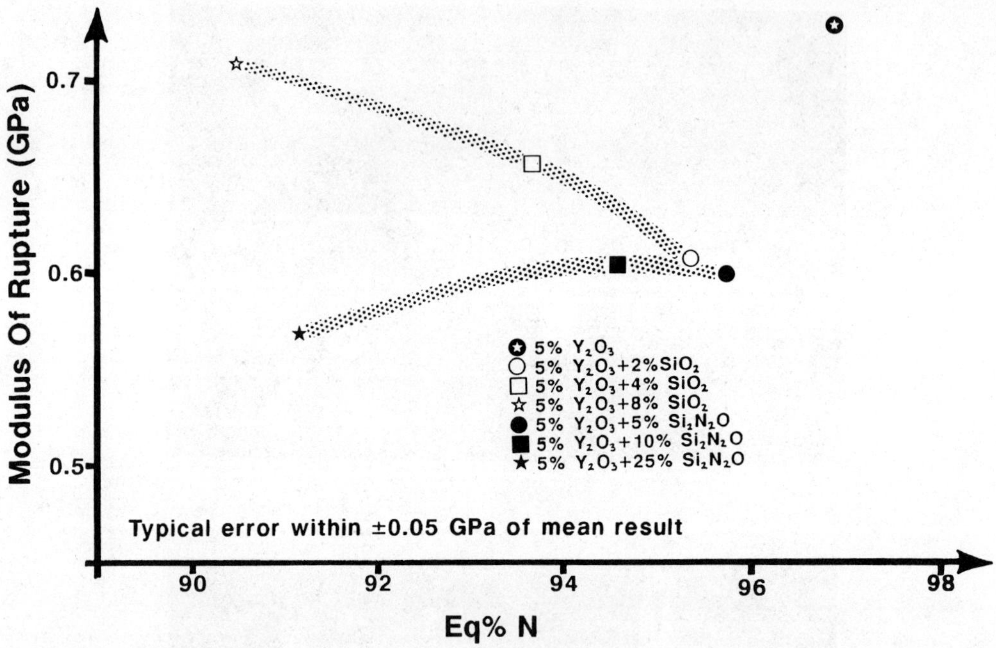

Figure 8. Room temperature MOR versus nitrogen content for 5wt% Y_2O_3-additive ceramics.

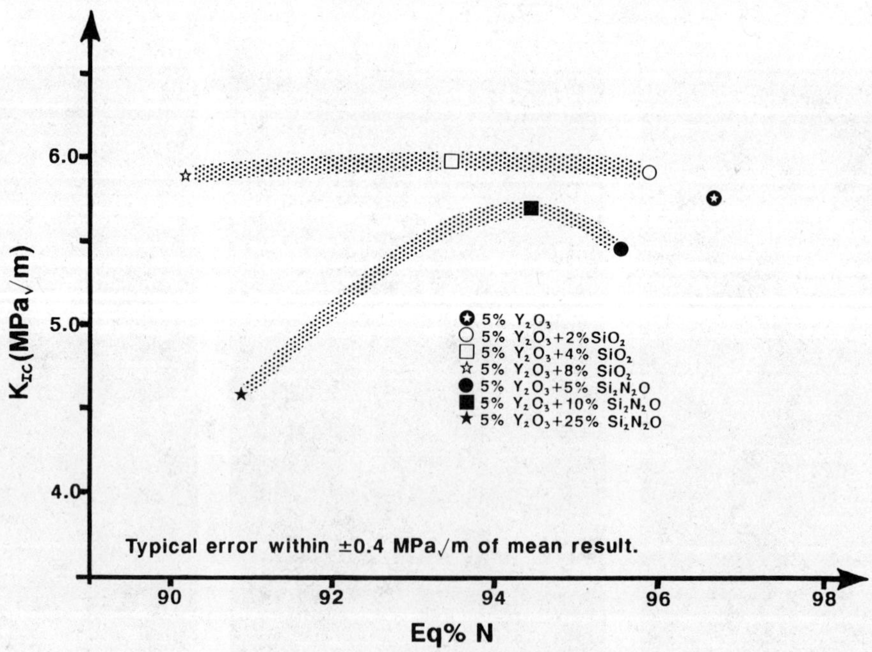

Figure 9. Fracture toughness plotted versus nitrogen content for 5wt% Y_2O_3-additive ceramics.

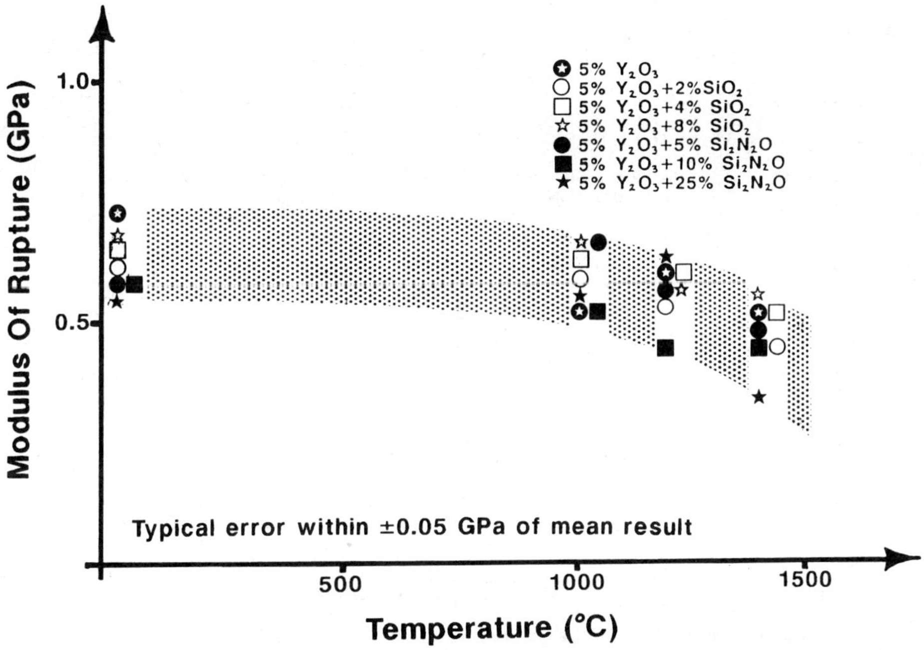

Figure 10. MOR versus temperature for 5wt% Y_2O_3-additive ceramics.

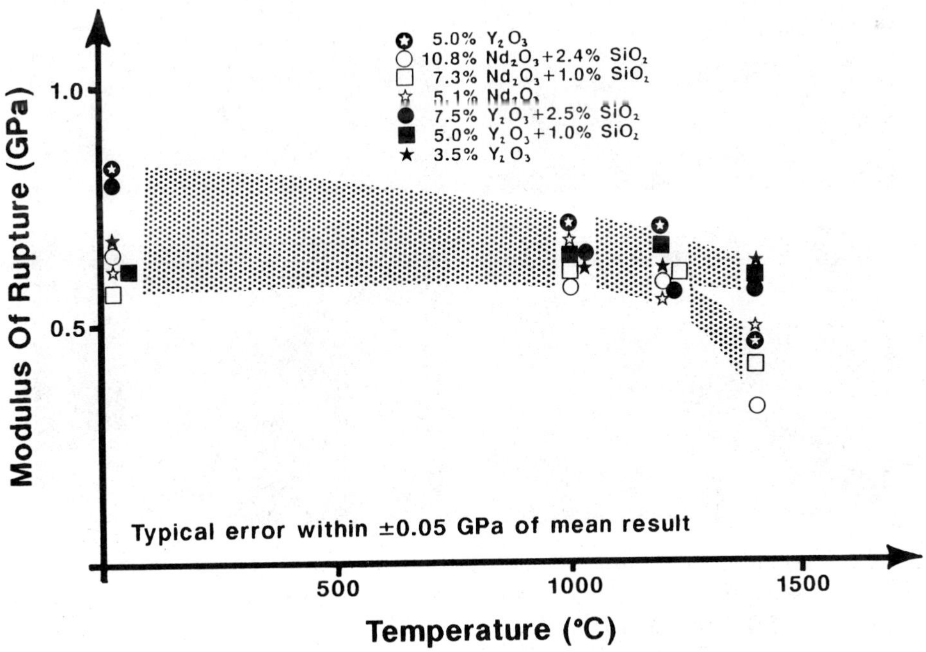

Figure 11. MOR versus temperature for Y_2O_3- and Nd_2O_3-additive ceramics A, B, C, A′, B′, and C.′

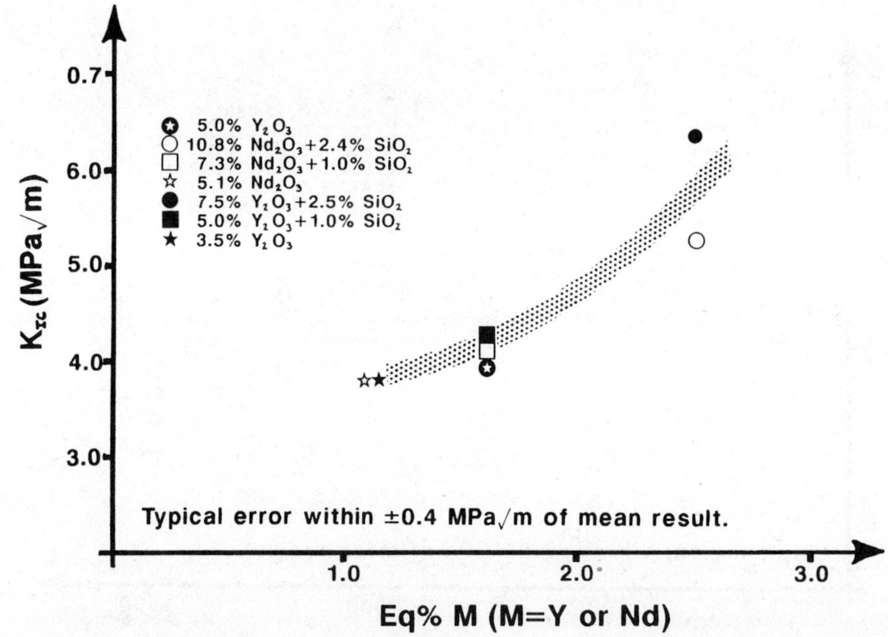

Figure 12. Fracture toughness versus metal additive levels for Y_2O_3- and Nd_2O_3-additive ceramics.

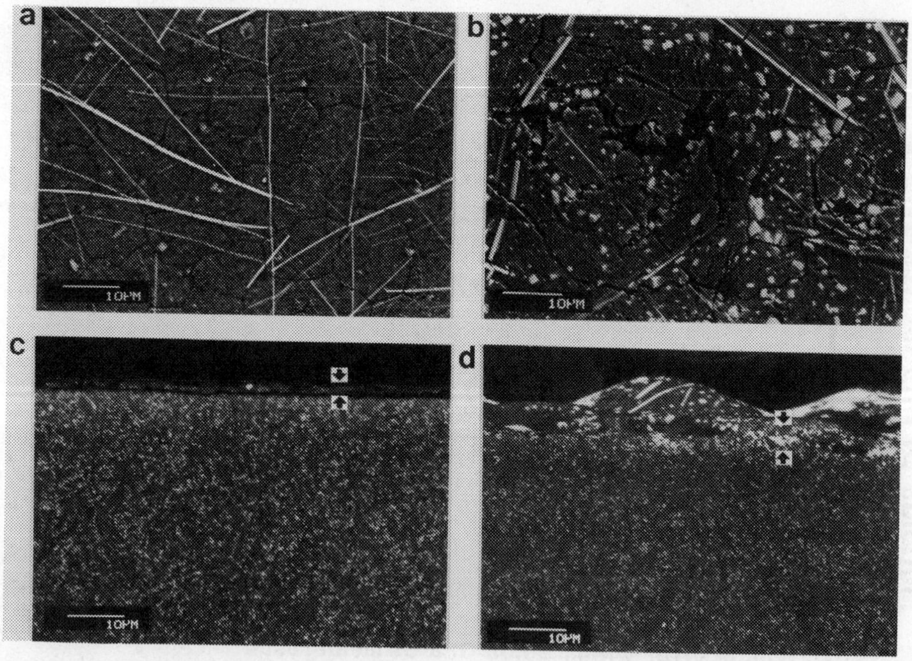

Figure 13. SEM micrographs of ceramics B and B′ after oxidation in air for 100 hours at 1400°C : (a) oxidized surface layer of B, (b) oxidized surface layer of B′, (c) cross section through oxidized surface layer of B, and (d) of B′.

BIAXIAL STATIC FATIGUE OF SILICON NITRIDE

G.Quinn and G. Wirth
DFVLR, Institut für Werkstoff-Forschung
D 5000 Köln 90, FRG

ABSTRACT

The equibiaxial static fatigue resistance of hot pressed silicon nitride was evaluated at high temperature. A ring-on-ring apparatus was used for stress rupture testing of disk specimens up to 1300°C in air. Equibiaxial lifetimes were substantially shorter than uniaxial lifetimes at similar stress levels. The strength degradation trends were identical however.

INTRODUCTION

The static fatigue failure of engineering ceramics is caused by a number of possible mechanisms including stress corrosion, slow crack growth, surface pit formation and creep fracture [1]. Hot pressed silicon nitride (HPSN) with magnesia sintering aid has been repeatedly analyzed due to its extremely consistent and reproducible behavior. For example, six different sets of flexural stress rupture data, from four laboratories, at 1200°C in air have given virtually identical results [2]. This is highly unusual and highlights the advantage of using this HSPN for crack growth and stress rupture studies.

Dozens of papers have been written on the static fatigue resistance of HSPN, but only recently has it been recognized that multiple mechanisms of time dependent failure can exist [3-6]. A meticulous study with flexural stress rupture experiments of up to two years duration has led to an extensive data base (Figure 1) culminating in the creation of a fracture mechanism map as shown in Figure 2 [3, 4].

The vast majority of static fatigue experiments for engineering ceramics have been in uniaxial stress states, indeed usually in four point flexure. Two studies have compared uniaxial and biaxial slow crack growth parameters for alumina, glass and a glass ceramic [7,8] and a vitreous bonded abrasive [9]. Testing was only at room temperature, and with short duration, dynamic fatigue experiments.

This study reports high temperature, long duration equi-biaxial stress rupture experiments for HSPN. HSPN was chose because of its high reproducibility and the possibility to examine more than one failure mechanism. The biaxial results are directly compared to comprehensive uniaxial results from references 3, 4 and 11. Biaxial testing was performed in both the slow crack growth and creep fracture regimes [Figure 2].

MATERIAL

The HSPN was NC 132 grade* and was made in 1977. This material has a magnesia sintering aid and has been previously characterized [10,11]. Disk specimens were cut from a single plate identified as billet L. This was made from the same powder lot and was pressed at the same time as billet A, C and P described previously [3,4].

EXPERIMENTAL PROCEDURE

Thirty disks were prepared with 45 mm diameter and 2.2 mm thickness. The disks were prepared by conventional slicing and diamond grinding, but with special care to minimize machining damage in the final steps. The disks then were intended to be lapped perfectly flat and to a depth of 25 microns from each side. The 25 microns is far greater than is necessary to produce a mirror finish, and was intended to eliminate typical machining damage inherent to the grinding process. This step was evidently <u>not</u> followed by the machine shop. Only a few microns were removed, producing a fine 1-2 rms microinch surface, which only managed to conceal the subsurface damage.

Seventeen four point flexure specimens (of size 2.1 x 2.8 x 45 mm) were also prepared and given identical preparation. The specimens were cut from the same billet as the disks. These precautions were taken to hopefully ensure the strength limiting flaws were identical in uniaxial and biaxial specimens.

The fixtures are shown schematically in Figure 3. Loading was via a ring-on-ring jig which produces an equibiaxial stress state. The rings were made from tubes of high purity recrystallized silicon carbide.** This material was chosen because of its high refractoriness and ready availability in tube form. Indeed, after all experiments, the appearance of the SiC had hardly changed and there was no chemical reaction evident at the load points. The top of each tube was crowned with a radius such that the support (outer) ring had a nominal diameter of 40 mm and the loading (inner) ring, 10 mm. A silicon nitride ball with a flat was used to apply the load centrally to the loading ring. The assemblage was aligned at room temperature and held in place by tack beads of a household acetate cement which subsequently burned off.

* Norton Company, Worcester, Mass., USA

** Crystar Grade, Norton Commpany

Deflection was measured from the midpoint relative to the outer ring. A silicon nitride ball contacted the specimen underside in order to ensure no contamination and a point location. The outer ring and middle point deflections were transmitted to a LVDT outside the furnace.

Stress rupture experiments were performed in a SiC resistance heating furnace equipped with a dead weight lever arm system.* The load was accurate to within 1 percent. Temperature was stabilized for 15-30 minutes prior to load application, which took 5-10 seconds+. Temperature was monitored by platinum thermocouple located 2 mm above the specimen. Temperature was held to within 3°C.

Six specimens were broken to determine a reference strength at room temperature. A screw driven universal testing machine** was used at a crosshead rate of 0.5 mm/min causing failures in 20-30 seconds.

All stresses were computed from the elastic stress formula:

$$\sigma = \frac{3P}{2\pi h^2} [(1-\nu)(a^2-r^2)/2b^2 + (1+\nu) \ln (\frac{a}{b})]$$

where b is the specimen radius, a the radius of the support ring, r the load ring radius, h the specimen thickness, P the load. Poisson's ratio ν was taken to be .26 at room temperature and .19 at 1200 and 1300°C.

The specimen and fixture sizes were chosen with specific criteria in mind. The loading rings were made of a high purity refractory, but weak silicon carbide. Load levels had to be kept low to minimize part breakage. It was desired for the sake of a valid comparison to uniaxial flexure, to have a similar Weibull volume or surface area, maintain the same stress gradient through the specimen thickness, and to maintain the same frictional constraint at the load points. The frictional constraint can lead to errors of the order of 5-20% in stress in both four point flexure and ring-on-ring testing. Both loading systems had fixed loading points, and it was desired to keep the magnitude of the friction effect the same. The friction error is directly proportional to the coefficient of friction and the specimen thickness. It was furthermore desired to keep the stress gradient in the ring-on-ring specimen comparable to the four point specimen. These considerations led to a requirement that the disk thickness (2.2 mm) be similar to the flexure specimen thicknesses of reference 11, which were 2.1 to 2.3 mm.

* Netzsch High Temperature Flexure Strength Tester Model 422

\+ The uniaxial specimens needed only 5 minutes to stabilize [ref. 11]

** Instron Model 1195 Canton, Mass., USA

The four point spans had been of the order of 40x20 mm, but such dimensions for the ring diameters in biaxial loading would have required very high loads. Thus an outer diameter of 40 mm was used but with an inner diameter of 10 mm. The area in the inner ring which has constant stress is thus 79 mm^2 which is comparable to the 56 and 71 mm^2 areas in the four point specimens from the earlier studies. Finally the error due to wedging stresses (due to excessive concentrated stresses under the loading ring) was minimized as a consequence of the system geometry. The absolute load to cause breakage was about ten times greater in the disk specimen, but the inner ring had six times more contact length, so the distributed load and thus the wedging stresses were comparable and rather low.

All of these preparations were successful as the fixtures were never damaged. The specimens broke in an ideal fashion, almost always within the middle gage area, and not under the load circle.

So in summary, the disk ring-on-ring test configuration is in essence a two-dimensional flexure test. As much as possible, testing conditions were chosen to permit a direct comparison to the one-dimensional, uniaxial flexure test.

RESULTS

The six disks broken at room temperature had an average strength of 501 MPa and a standard deviation of only 29. A Weibull two parameter analysis gives a modulus of about 25. Five of the fractures originated inside the gage area, the sixth outside. Figure 4a shows one of the former. Failure origins were readily observable based upon the crack branching patterns. Scanning electron microscopy unfortunately identified the strength limiting flaws to be primarily subsurface machining damage. Such damage appeared as 15-20 μm deep elongated semielliptical cracks which often overlapped. These occasionaly intersected at different angles creating a zig-zag defect at the failure origin. This was presumably due to a shift in orientation of the specimen during surface grinding.

The equibiaxial strength is substantially lower than the 738 MPa ($\pm$ 53 MPa) uniaxial strength which was measured on the 17 flexure bars. The Weibull two parameter modulus was 16 but could have been higher with the deletion of a few high strength points. Strength limiting flaws were either 15-25 μm deep semielliptical machining damage or WC inclusions.

The stress rupture results are shown in Figures 5 and 6. At both 1200 and 1300°C, biaxial lifetimes appear to be considerably shortened. The uniaxial stress rupture results are from Reference 2-4.

At 1200°C, the biaxial lifetime scatter is unfortunately much larger than the uniaxial results. Biaxial lifetimes are between 10 to 1000 times shorter. The short duration specimens (<0.2 hour) had small intergranular slow crack growth zones (100-500 micron size) changing to a transgranular fast fracture crack. Longer lived specimens had much larger crack growth zones. For example, the 250 MPa specimen at 0.4 hours had a 10 mm long SCG region.

The 1300^{0}C biaxial results are more clear since scatter in lifetime is dramatically reduced. Overall lifetime is shortened by a factor of about 20, but the static fatigue trend is comparable to the uniaxial results. The slope of a line through the biaxial results gives a SCG exponent of 8.4, whereas the uniaxial exponent is 9.9.*

The SCG zones at 1300^{0}C were much larger and more severe than at 1200^{0}C. Even the short duration specimens had large zones of the order of 10 mm length, which extended completely through the thickness. The specimens loaded at 200 MPa or less had SCG zones extending to the outer load ring and even to the specimen rim. It is clear that some of these fractures were due to creep fracture and not slow crack growth [Figure 2].

DISCUSSION

The reduction in lifetime for biaxial loading is not surprising. The slow growth cracks followed a very tortuous and winding path at both 1200 and 1300^{0}C, and numerous multiple crack patterns, even cracks growing at 90 relative to each other were observed. The cracks seemed to be active in all directions. The irregularity of the cracks suggests extensive microcrack interactions, or possibly the ability to circumvent pockets of resistance. Such crack freedom was noted by Pletka and Wiederhorn in their comparison of uniaxial and biaxial crack growth behavior [8].

The similarity in static fatigue strength degradation in biaxial and uniaxial modes of loading is reassuring since it would lead to simplified life prediction models. Service and Ritter [9] have also reported comparable static fatigue susceptibilities in uniaxial or biaxial stress states. Pletka and Wiederhorn [7,8] reported consistent susceptibilities in uniaxial and biaxial disk testing as well.

Although the biaxial lifetimes seem to be shorter, this is simply a consequence of the dramatically reduced fast fracture strength (501 MPa) of the biaxial specimens. Indeed, if the data in Figures 5 and 6 are normalized by the reference strengths at room temperature, then the static fatigue curves are comparable, and there is no biaxial lifetime shortening**.

A complicating factor is the extensive crack sizes in many of the biaxial specimens. These behaved in many respects like fracture mechanics specimens (such as double torsion). Double torsion experiments at 1200 and 1300^{0}C on the identical material were reported previously by this author [11]. A striking similarity in the scatter of results was reported: high

* Stress = constant x (time to failure) $^{-1/N}$; where N is the slow crack growth exponent. [ref. 2, 11]

** Indeed the biaxial curve would be shifted somewhat higher than the uniaxial curve varibility at 1200°C, good consistency at 1300°C. Variable zones crack growth resistance were reported as well [11].

The final complicating factor is related to the creep deformation. Deformation often exhibited primary, secondary and tertiary stages. The latter was undoubtedly due to a compliance effect as cracks became extremely large. Secondary creep deformation rates were never constant and always decreased, indicating no true steady state. Primary rates were as high as one hundred times faster than minimum rates. Overall deflections were of the order of 0.1 to 1 mm for most specimens. The analysis of the deformation to give strains and relaxed stresses has not yet been undertaken and may be diabolically complicated. A number of these specimens were loaded in conditions of creep fracture.

It was clear from the specimen fragments that creep and crack growth severely interfered with each other. That is, creep often blunted cracks and reduced stress intensities. Alternatively, crack growth created deformations that interfered with creep measurement. Indeed, itis a peculiar consequence of the ring-on-ring system, that as a crack grows in one direction the specimen will bend due to reduced compliance and thus will redistribute the ring loading to effectively reduce the stresses acting on the plane of the crack. Figure 7 shows the loading rings will not evenly distribute the load as the specimen deflects preferentially in one direction. Finally, at low stresses and high temperature, creep deformations are excessive and creep rupture becomes the dominant mechanism of fracture. As an extreme example, the specimen loaded to 150 MPa at 1300°C had numerous creep cracks and a central plug was almost punched out under the loading ring (Figure 8).

CONCLUSIONS

1. The biaxial stress rupture lifetime if HPSN appears to be drastically reduced compared to uniaxial lifetimes at high temperature. This is primarily due to a reduced fast fracture strength.

2. The biaxial fatigue trend of strength degradation is the same as the uniaxial trend.

3. Biaxial SCG cracks wend a tortuous path through the specimen, unlike biaxial fast fracture or uniaxial cracks.

4. The fast fracture strength limiting defects were the same in the biaxial and uniaxial specimens: machining damage and some WC inclusions.

5. The ring loaded disk specimen exhibited a wide range of behavior. Some fractures were ideal strength (i.e., small crack, no creep) type specimens, others were like fracture mechanics (large crack) specimens, and finally some as creep fracture specimens. This variable response, coupled with the inherent tendency for stresses to redistribute as the compliance changes due to large crack growth, indicates the ring loaded disk specimen has limited value. It should only be used for stress rupture experiments where crack sizes and deformations are small.

ACKNOWLEDGEMENT

The principl author wishes to thank Dr. R.N. Katz of US Army MTL who suggested doing the biaxial disk stess rupture testing almost ten years ago! The assistance of Mr. Bernd Kanka of DFVLR and Ms. Gail Meyers of MTL with scanning electron microscopy is appreciated.

REFERENCES

1. Quinn, G., "Review of Static Fatigue in Silicon Nitride and Silicon Carbide," Ceram. Eng. and Sci. Proc. 3 (1987) 77-98.

2. Quinn, G. and Swank, L., "Static Fatigue of Preoxidized Hot Pressed Silicon Nitride," Comm. Amer. Ceram. Soc. Feb. (1983) C31-2.

3. Quinn, G., "Static Fatigue Resistance of Hot Pressed Silicon Nitride," in Fracture Mechanics of Ceramics, vol.8, ed., R. Bradt, A. Evans, D. Hasselman and F. Lange, (Plenum Press 1986) 319-332.

4. Quinn, G., "Fracture Mechanism Maps for Silicon Nitride," in Proc. 2nd Int. Symp. of Ceramic Materials and Components for Engines, Lubeck, April 1986, DKG, 931-939.

5. Grathwohl, G., "Regimes of Creep and Slow Crack Growth in High Temperature Rupture of Hot Pressed Silicon Nitride," in Deformation of Ceramic Materials II, eds. R. Tressler, R. Bradt (Plenum Press 1984) 513-87.

6. Grathwohl, G., "Creep and Fracture of Hot Pressed Silicon Nitride with Natural and Artificial Flaws," in Creep and Fracture of Engineering Materials and Structures, ed. B. Wilshire, D. Owen (Pineridge Press, Swansea 1984) 565-77.

7. Pletka, B. and Wiederhorn, S., "A Comparison of Failure Predictions by Strength and Fracture Mechanics Techniques," J. Mat. Sci, 17 (1982) 1247-68.

8. Pletka, B. and Wiederhorn, S., "Subcritical Crack Growth in Glass Ceramics, in Fracture Mechanics of Ceramics," vol. 4, ed. R. Bradt, D. Hasselman and F. Lange (Plenum Press 1980) 745-759.

9. Service, T. and Ritter, J., "Dynamic Fatigue of a Vitreous Bonded Abrasive Tested in Four Point and Ring-on-Ring Flexure," Adv. Cer. Mat. 3 (1988) 49-51.

10. Quinn, G., "Characterization of Turbine Ceramics after Long Term Environmental Exposure," US. Army AMMRC Technical Report TR 80-15 (1980).

11. Quinn, G. and Quinn, J., "Slow Crack Growth in Hot Pressed Silicon Nitride, in Fracture Mechanics of Ceramics," vol. 6, ed. R. Bradt, A. Evans, D. Hasselman and F. Lange (Plenum Press 1983) 603-636.

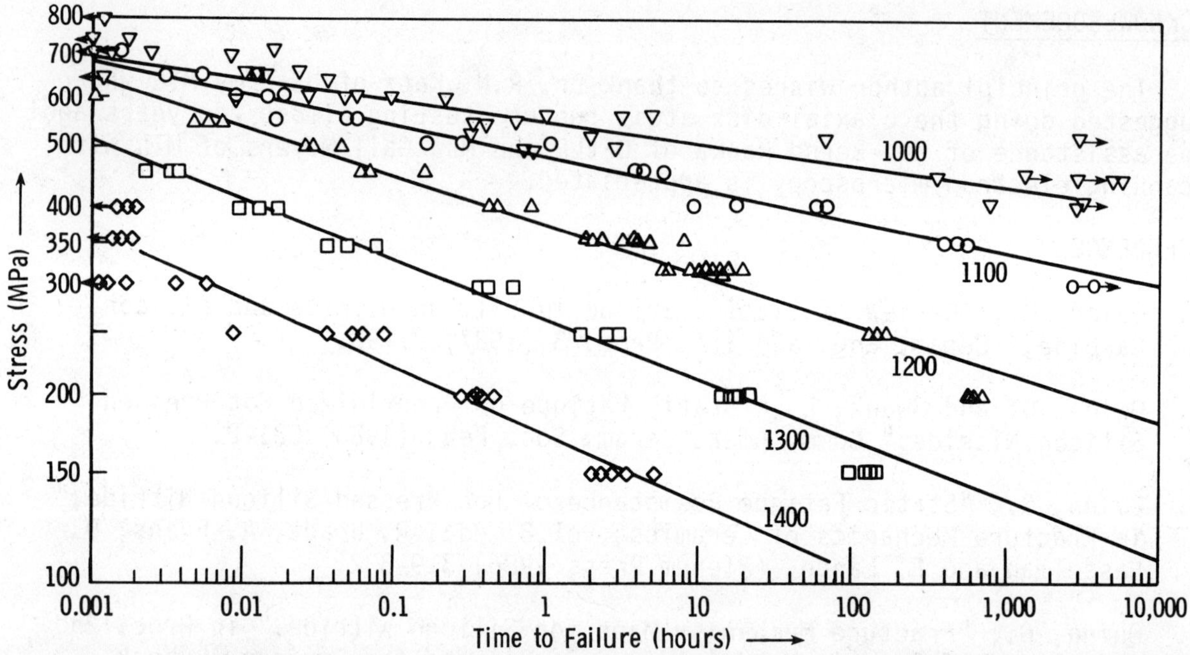

Figure 1. Flexural stress rupture of HPSN in air for as-machined specimens. References 3 and 4.

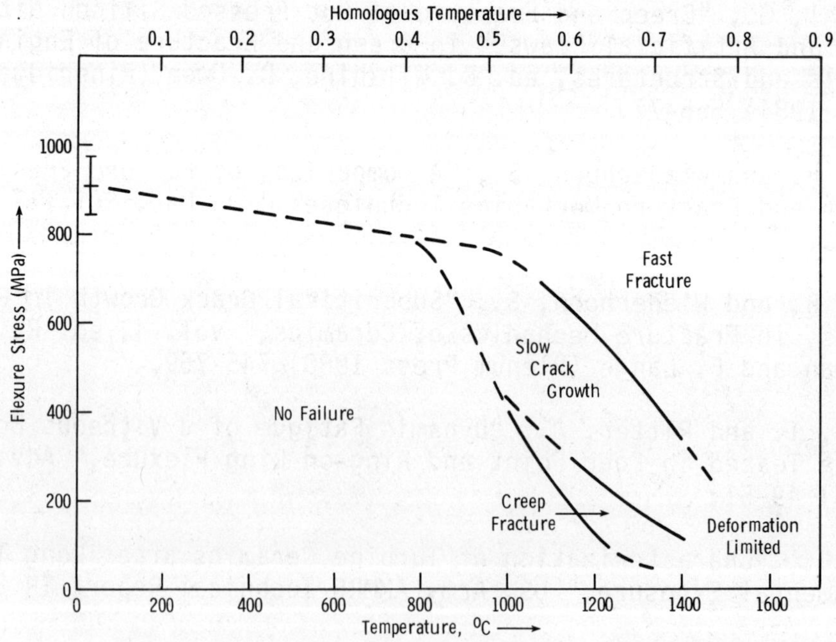

Figure 2. Fracture mechanism map for HPSN. References 3 and 4.

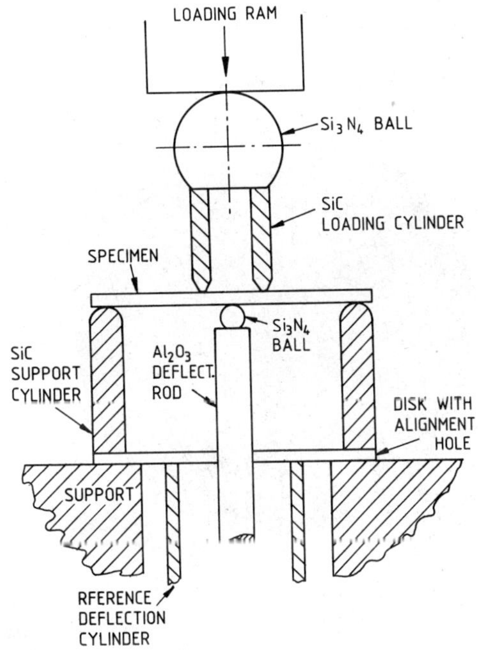

Figure 3. Schematic of the ring-on-ring equibiaxial fixtures.

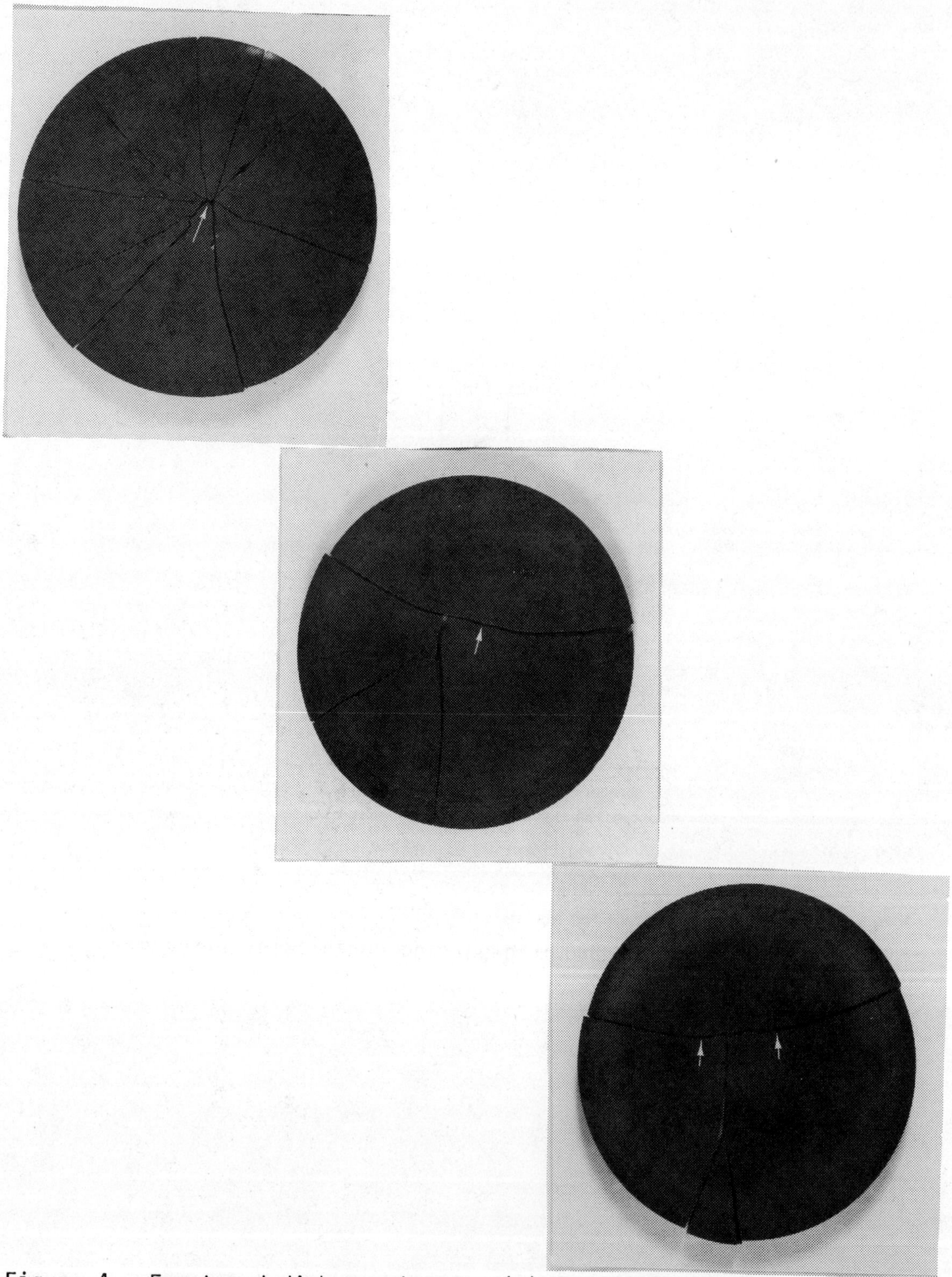

Figure 4. Fractured disk specimens. (a) is a room temperature fast fracture specimen. (b) is a specimen that failed at 0.2 hours at 1200°C, 300 MPa. A 1 mm diameter SCG zone was on the fracture surface (arrow). (c) was 2.5 hours at 1200°C, 300 MPa. The SCG crack was 8 mm long (arrows).

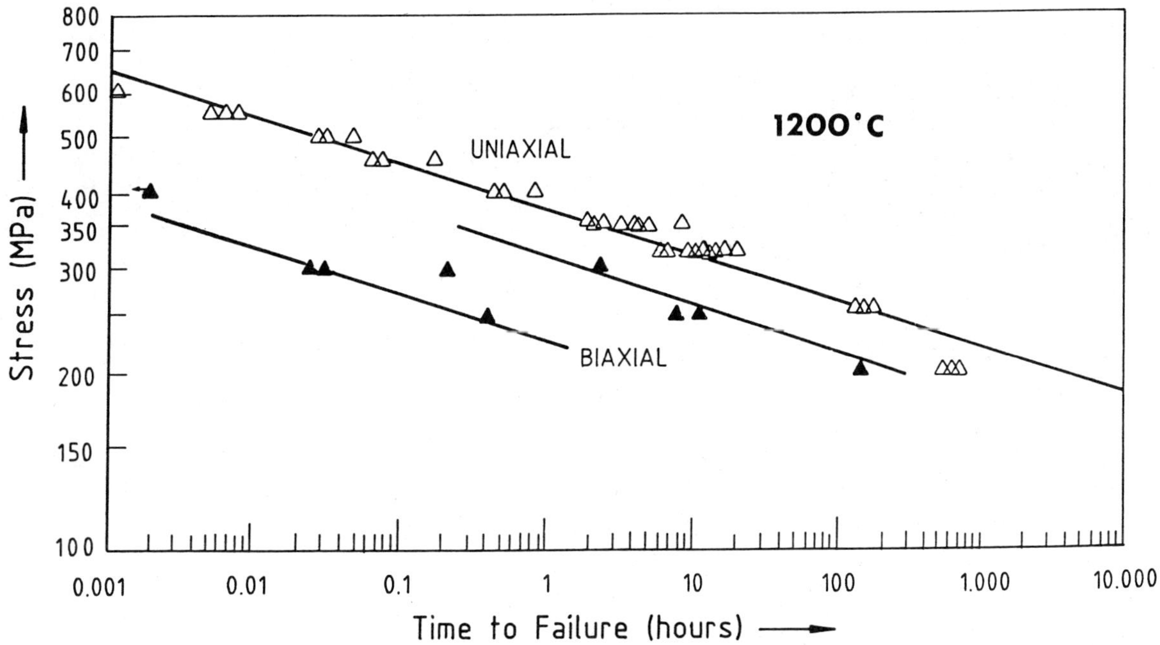

Figure 5. Biaxial stress rupture data (solid symbols) at 1200°C in air for HPSN. Uniaxial results (hollow points with fitted line) are given for comparison. The arrowed points are either failures on loading or an interrupted test.

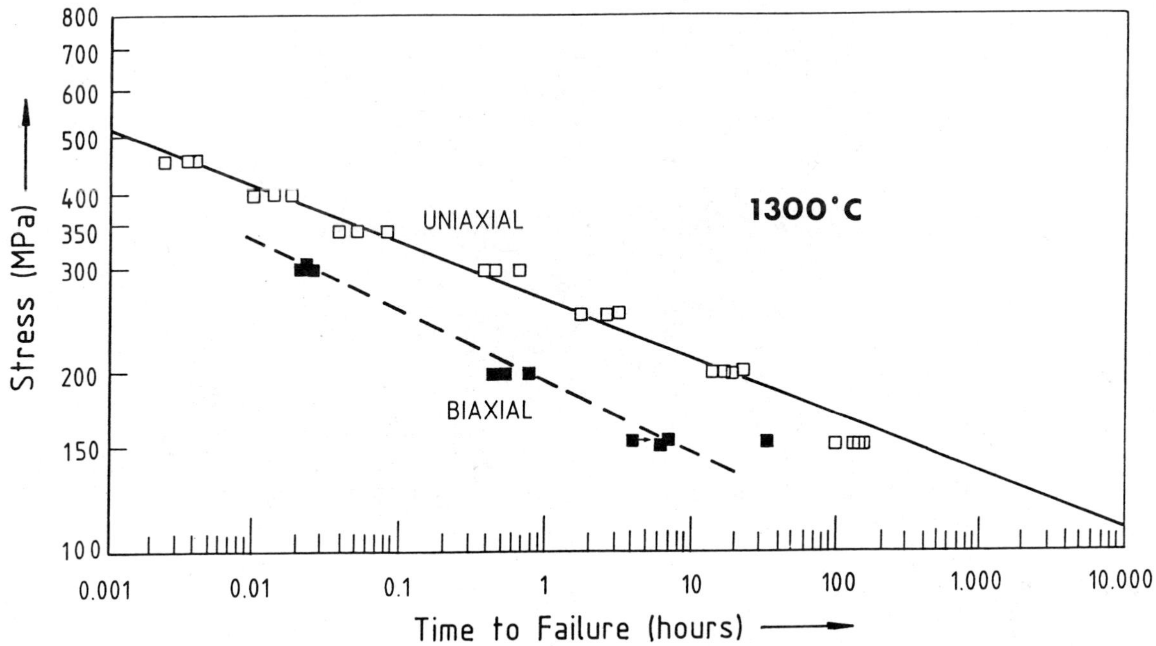

Figure 6. Biaxial stress rupture data at 1300°C in air (solid symbols). Uniaxial results (hollow symbols with fitted line) are for comparison. The arrowed points are either failures on loading or an interrupted test.

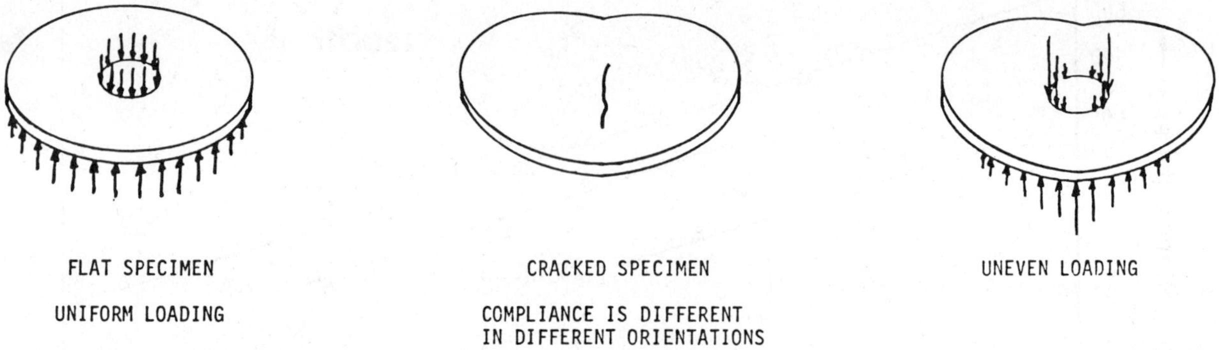

Figure 7. A cracked specimen has a nonuniform compliance which causes uneven loading.

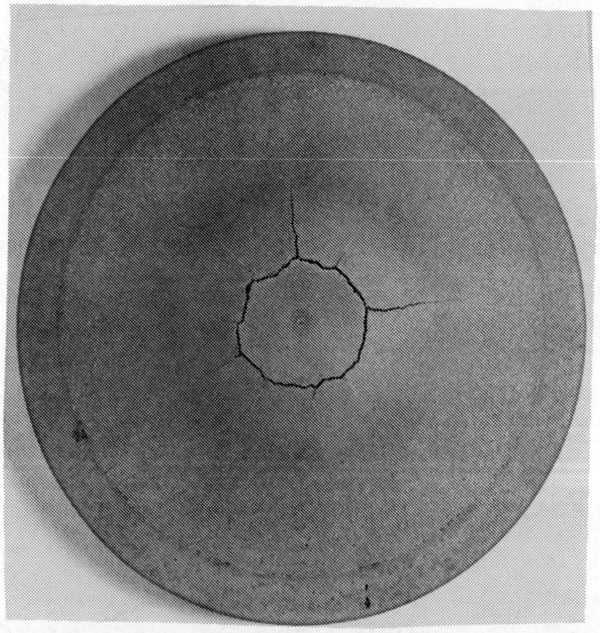

Figure 8. Specimen exhibiting excessive creep deformation and localized creep cracking at 32 hours at 1300°C with 150 MPa. The test was terminated just short of fracture when tertiary creep was well underway.

EFFECTS OF HOT CORROSION ON THE MECHANICAL PROPERTIES OF TETRAGONAL ZIRCONIA POLYCRYSTAL MATERIALS

Jeffrey J. Swab
U.S. Army Materials Technology Laboratory, Watertown, MA
and
Gary L. Leatherman
Worcester Polytechnic Institute, Worcester, MA

ABSTRACT

A variety of tetragonal zirconia polycrystal (TZP) materials were coated with various amounts of sodium sulfate solution and treated in air at 1000°C for 500 hrs. Room temperature flexure strength was determined before and after treatments on uncoated and coated specimens. Results show that Ce-TZP and alumina are resistant to corrosion by sodium sulfate, but Y-TZP is not. The Y-TZP had a strength loss of ~25% after 500 hrs. at 1000°C with a 10-20 mg/cm^2 coating. This loss is about 10% greater than that experienced by uncoated specimens when treated at 1000°C for 500 hrs. Examination of Y-TZP specimens with a heavier coating (50-75 mg/cm^2) showed no further loss of strength.

INTRODUCTION

The service environments projected for structural ceramics are quite hostile. In the gas turbine engine some of the structural ceramics will face hot corrosion: the attack by molten Na_2SO_4 (Tm=884°C) which condenses on engine parts when ingested NaCl reacts with sulfur impurities in the fuel. Marine propulsion gas turbine engines experience a similar lower temperature sulfate-induced corrosion. How long-term exposure to these corrosive environments effect the mechanical properties of ceramics is of critical importance to their successful application in engines as high temperature structural components.

Although at the present time monolithic tetragonal zirconia polycrystal ceramics (TZPs) are not being considered for structural applications such as gas turbine blades and other components in the high temperature regions (>1000°C) of gas turbine engines, this does not preclude the use of composites based on a TZP matrix from finding application in this environment. Moreover, the use of TZPs in non-structural applications in the high temperature region of gas turbines appears inevitable (e.g., thermal barrier coatings, etc.). Thus the effect of hot corrosion on the mechanical properties of TZPs may have important technological implications.

The actual mechanisms of hot corrosion in zirconia ceramics have been studied by several workers (1-4). However, there has been little work examining the interactions between hot corrosion and mechanical properties. Important properties that may be affected by hot corrosion are room temperature and elevated temperature strength, toughness and creep-rupture behavior. This paper presents initial results on the effect of hot corrosion on the room temperature and elevated temperature strength of two TZPs. An alumina ceramic has also been examined for comparison.

EXPERIMENTAL PROCEDURE

Three commercially available ceramics, an alumina+, a yttria-TZP# and a ceria-TZP$, were obtained and machined into bend bars 3 by 4 by 50mm long and carefully ground by a surface grinder such that the surface striations were parallel to the long axis. All four long edges were chamfered ~45° to a depth of ~0.15mm; the bars were machined according to Ref. 5. A minimum of 5 bars from each ceramic were used to determine the as-received room temperature strength. They were broken in 4-point flexure according to Army

+ AD-999, Coors Ceramic Co., Golden, CO

Z-191, NGK Insulators LTD, Nagoya, Japan

$ CZ203, Ceramatec, Inc., Salt Lake City, UT

MIL STD 1942, with inner and outer spans of 20 and 40 mm respectively, and a cross-head speed of 0.5 mm/min.

Three other groups of 5 bars from each ceramic were then subjected to one of the following treatments:

500 hrs. @ 1000°C with no coating
500 hrs. @ 1000°C with a 10-20 mg/cm^2 coating of Na_2SO_4
500 hrs. @ 1000°C with a 50-75 mg/cm^2 coating of Na_2SO_4

Coatings were applied my mixing anhydrous sodium sulfate (Na_2SO_4) with distilled water and applying the solution to one 4 by 50mm face of each bar. The coating was limited to a centered 30 mm span of the face. The bars were then heated using a hot plate, driving off the water and leaving behind a layer of Na_2SO_4. The weight of each bar was measured before and after application of the coating to ensure the amount was in the specified range.

Heat treatments were done in air, at laboratory ambient humidity (40-60%), in an unstressed condition on silicon carbide knife edges to assure uniform heating. The knife edges supported the bars well outside the area to be tested during 4-point flexure testing assuring no affect on the subsequent strength determinations. The coated surfaces were placed face-up during the heat treatments to minimize any loss of the coating.

Once the strength was determined the fracture surface of each bar was examined optically at low magnification and a scanning electron microscope (SEM) was employed for high magnification examination of selected fracture surfaces to determine cause of fracture.

Stepped-temperature stress-rupture (STSR) testing was done following the procedure outlined by Quinn and Katz (6). This type of test allows raid screening of a materials stress-rupture behavior over a wide range of temperatures and stresses using a small number of bend bars. The procedure involves loading a bend bar into a furnace containing a 4-point flexure fixture (inner and outer spans of 20 and 40mm respectively) and heating the furnace to 800°C in 2 hrs., in air, with no load applied to the bar. Upon reaching temperature a dead-weight load is applied. Should the bar survive

24 hrs. at this temperature, the furnace is then heated in (~10min) to 900°C and again the bar is allowed to soak for 24 hrs while under the same load. This cycle is repeated for 1000°, 1100° and 1200°C. If the bar fractures or excessive creep occurs, the power to the furnace is shut off by a microswitch. The time of fracture is then denoted on the STSR plot using an arrow, with the applied stress that caused fracture above the arrow. The symbols for the STSR plot are: (←) failure occurred upon loading at 800°C; (→) survived the full test cycle through 1200°C; and (↓) denotes time of failure between loading but before the full cycle is complete.

RESULTS AND DISCUSSION

Room Temperature Flexure Testing

The results of the room temperature flexure strength testing, Figure 1, show that all three materials exhibit very little strength degradation after exposure to molten sodium sulfate at 1000°C. The Ce-TZP is the most resistant followed by the alumina and the Y-TZP.

The slight strength reduction of the Ce- and Y-TZP after 500 hrs. at 1000°C with no coating can be attributed to the fact that these materials are transformation toughened ceramics and are susceptible to overaging when exposed to elevated temperatures for a prolonged time. The addition of a small amount of a rare earth oxide, such as CeO_2 or Y_2O_3, along with an appropriate processing technique, allows for the retention of metastable tetragonal zirconia at room temperature. However, when overaging occurs the tetragonal zirconia transforms to the stable monoclinic phase, resulting inn a reduction in strength (7-10). Alumina does not undergo this type of a transformation and thus shows no reduction in strength as a result of this exposure.

The addition of Na_2SO_4 to the long time/high temperature exposures does not cause any further strength reduction in the Ce-TZP and only minimal reduction in the alumina. However, the Y-TZP experiences a large reduction in strength. Hamilton and Nagelberg(1) have reported that molten vanadates "leach" Y_2O_3 from a ceramic causing degradation of the material. Others (2,3) have shown that SO_3 is highly reactive with Y_2O_3 and that the combined action of Na_2SO_4 and SO_3, under SO_3 partial pressures as low as 700 Pa at

700°C, caused degradation of the ceramic through Y_2O_3 depletion. This is a similar to the hypotheses proposed by Lange, et al.(11) to account for the catastrophic degradation of Y-TZP that can occur when it is treated between 150 and 400°C in the presence of water vapor. They have found that the Y_2O_3 reacts with the water vapor to form α-$Y(OH)_3$. The formation of this hydroxide depletes the Y_2O_3 from the tetragonal grains creating monoclinic nucelli which can grow spontaneously to transform the tetragonal grains to monoclinic. If the monoclinic grains become large enough microcracks will develop allowing further aging leading to degradation. Thus "surface corrosion" of Y-TZP may be in the form of Y_2O_3 depletion, resulting in the spontaneous tetragonal-to-monoclinic transformation of the zirconia and a reduction in strength beyond what can be attributed to overaging alone.

Ceria also undergoes a sulfation reaction (4), but the equilibrium SO_3 partial pressure at 850°C is 1200 Pa whereas for Y_2O_3 it is only $\sim$15 Pa(3). Thus the thermodynamics of the sulfation reaction is much more favorable in the case of Y_2O_3. Therefore, all things being equal, a CeO_2 stabilized zirconia should be superior to a Y_2O_3 stabilized zirconia in hot corrosion resistance.

The high purity alumina also exhibited good hot corrosion resistance. Gannon, et al.(12) found that the hot corrosion resistance of alumina refractories to molten K_2SO_4 increased with decreasing open porosity. The increased resistance was due to a reduction in the surface area available for corrosion. They also showed that control of the grain boundary chemistry played an important role in corrosion resistance. The effect of Na_2SO_4 on a glassy grain boundary phase should be similar to that of K_2SO_4 since alkali elements tend to behave similarly in silicate-based glasses. As a result, the minimal strength reduction in the alumina may be due to the corrosion of an, as of now, undetermined glassy grain boundary phase.

For all materials increasing the Na_2SO_4 coating does not cause further reduction in strength. For the Y-TZP this indicates that once a certain level of Y_2O_3 has been removed from the surface further corrosion via this mechanism does not occur in the time frame of these tests. In the alumina case the undetermined grain boundary phase is resistant to corrosion beyond a certain level.

Examination of the fracture surfaces of each materials showed that the type of flaw causing fracture did not change after exposure.

Stepped-Temperature Stress-Rupture Tests

Preliminary STSR tests on the Y-TZP material, Figure 2, show that the addition of a Na_2SO_4 coating has an adverse effect on the materials high temperature load carrying capabilities. This effect is pronounced at a 800°C under high loads (>250 MPa). Sodium sulfate melts at~884°C(13), thus during the 800°C step the Na_2SO_4 is probably able to deplete Y_2O_3 from the surface causing the previously mentioned transformation. At higher temperatures the coating would be a liquid and could fall off the tensile surface of the bar. (Due to the fixture design the tensile surface is face down during a STSR test.)

CONCLUSIONS

It was found that commercially available Ce-TZP has excellent resistance to strength degradation by molten sodium sulfate when compared to a Y-TZP. A possible mechanism of this strength degradation is the Y_2O_3 depletion of the Y-TZP surface which would allow the tetragonal-to-monoclinic transformation of the zirconia to occur spontaneously resulting in a reduction in strength above that seen due to overaging alone. High purity, dense alumina is also very resistant to corrosion. However, it does exhibit a slight strength reduction which may be attributed to corrosion of a undetermined glassy grain boundary phase.

Preliminary Stepped-Temperature Stress-Rupture results indicate that a Na_2SO_4 coating has an adverse effect on the high temperature load carrying capabilities of a Y-TZP. Further work is needed to determine the extent of this effect and the mechanisms by which fracture occurs at elevated temperatures.

REFERENCES

1, J.C. Hamilton and A.S. Nagelberg, "In Situ Raman Spectroscopic Study of Yttria-Stabilized Zirconia Attack by Molten Sodium Vanadate," J. Am. Ceram. Soc., 67, [10] 686-90 (1984).

2. R. Barkalow and F. Pettit, "Mechanisms of Hot Corrosion Attack of Ceramic Coating Materials," p704-14 in Proceedings of 1st Conference on Advanced Materials for Alternative Fuel Capable Directly Fired Heat Engines, CONF-790749, J.W. Fairbanks and J. Stinger, Editors, NTIS, Springfield, VA (1979).

3. R.L. Jones, D.B. Nordman, and S.T. Gadomski, "Sulfation of Y_2O_3 and HfO_2 in Relation to MCrAl Coatings," Metall. Trans., 16A, [2] 303-06 (1985).

4. R.L. Jones, S.R. Jones and C.E. Williams, "Sulfation of CeO_2 and ZrO_2 Relating to Hot Corrosion", J. Electrochem. Soc., 132 [6] 1498-1501 (1985).

5. "Flexure Strength of High-Performance Ceramics at Ambient Temperatures", Department of the Army MIL STD 1942, 21 Nov. 1983.

6. G.D. Quinn and R.N. Katz, "Stepped-Temperature Stress-Rupture Testing of Silicon-Based Ceramics", Am. Ceram. Soc. Bull., 57 [11] 1057-58 (1978).

7. J.J Swab, "Evaluation of Commercially Available Transformation Toughened Zirconias", p137-46 in Materials Reserach Society Symposia Proceedings, Vol. 78, Advanced Structural Ceramics. Edited by P.F. Becher, M.V. Swain and S. Somiya. Materials Research Society, Pittsburgh, PA (1987).

8. D.C. Larsen and J.W. Adams, "Long-Term Stability and Properties of Zirconia Ceramics for Heavy Duty Diesel Engine Components", prepared for NASA-Lewis Research Center, for US Department of Energy under Contract DEN 3-305, NASA CR-174943, (Sept 1985).

9. L.J. Schioler, "Effect of Time and Temperature on Transformation Toughened Zirconias", prepared for US Department of Engery under Interagency Agreement DE-AE-101-77, MTL TR 87-29 (June 1987).

10. N.L. Hecht, D.E. McCullum, D.W. Grant, J.D. Wolf, G.A. Graves and S. Goodrich, "The Experimental Evaluation of Environmental Effects in Toughened Ceramics for Advanced Heat Engines", Proceedings of the 23rd Automotive Technology Development Contractors' Coordination Meeting, p299-310, Society of Automotive Engineers, Warrendale, PA, March 1986.

11. F.F. Lange, G.L. Dunlop and B.I. Davis "Degradation During Aging of Transformation-Toughened ZrO_2-Y_2O_3 Materials at 250°C," J. Am. Ceram. Soc., 69 [3] 237-40 (1986).

12. R.E. Gannon, F.A. Hals and H.H. Reynolds, "Corrosion Studies in Materials for Auxiliary Equipment in MHD Power Plants", p212-24 in Corrosion Problems in Energy Conversion and Generation, Edited by C.J. Tedman, Jr.; Corrosion Division, The Electrochemical Society, Princeton, NJ (1974).

13. <u>Handbook of Chemistry and Physics</u>, 61st edition, pB-150 CRC Press, Inc., (1981)

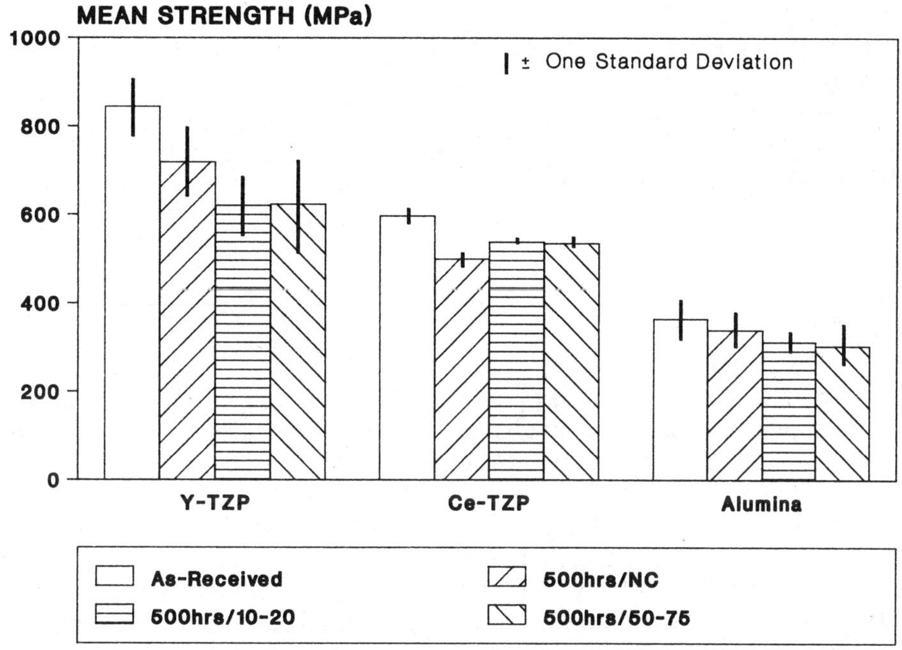

Figure 1, Strength After Exposure to Molten Sodium Sulfate

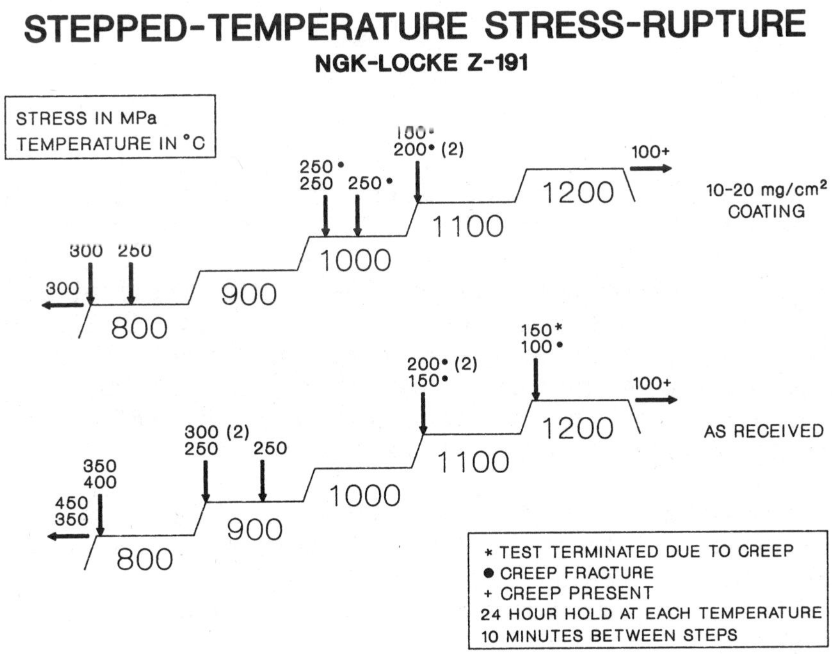

Figure 2, Preliminary Stepped-Temperture Stress-Rupture Results for a Y-TZP in the As-Received State and with a 10-20 mg/cm^2 Coating of Sodium Sulfate

CeO_2-ZrO_2 TOUGHENED Al_2O_3 CERAMICS OF HIGH STRENGTH AND TOUGHNESS

Giulio A. Rossi and Paul J. Pelletier
Norton Company
Advanced Ceramics Division
Goddard Road
Northborough, MA 01532

ABSTRACT

CeO_2-ZrO_2 toughened Al_2O_3 ceramics, with and without HfO_2 addition, were fabricated using chemically derived powders and different processing methods. Hot isostatic pressing was used in most cases after sintering to achieve virtually theoretical density. The ceramics have shown good strength (700-800 MPa) and toughness as high as 8.2 MPa.m$^{\frac{1}{2}}$ at room temperature. At high temperature (1000°C) about 70% of the strength and 55% to 75% of the toughness were retained. Although very little or no m-ZrO_2 was detected by XRD on ground and fracture surfaces, the decrease in toughness with temperature suggests the presence of transformation toughening at low temperatures. The strength was found to be limited by internal flaws and/or machining induced surface defects. Therefore, improved powder processing and shape forming methods are expected to result in even higher strengths. A fairly good homogeneity of distribution of the ZrO_2 phase dispersed in the Al_2O_3 matrix was shown by SEM.

1. INTRODUCTION

ZrO_2 toughened Al_2O_3 ceramics (ZTA) are a relatively new class of important composite materials which may find applications at low temperatures (i.e. wear components) and at high temperatures, for example in heat engines. The potential of these ceramics is due to the possibility of tailoring their properties to the application by changing, for example, ZrO_2 volume fraction, type and concentration of stabilizer (Y_2O_3, CeO_2, etc.), average grain size and size distribution. What makes these ceramics particularly attractive is the possibility of toughness retention over a wide temperature range, if multiple toughening mechanism contribute to the overall toughness.

Claussen(1) was the first to show that a dispersion of unstabilized ZrO_2 can enhance the toughness of hot pressed Al_2O_3. He reported a maximum value of 9.5 MPa.m$^{\frac{1}{2}}$, but also showed that the addition of ZrO_2 adversely affects the strength due to microcracking.

Becher(2) has measured peak strengths of 750 MPa for hot pressed samples of ZTA containing about 12 v/o unstabilized ZrO_2 and maximum fracture energy of 55 J/m^2 for 10 v/o ZrO_2.

Lange[3] has reported a maximum strength of 1150 MPa for a hot pressed ZTA material containing 30 v/o ZrO_2 stabilized with 2 m/o Y_2O_3 and a K_{IC} value of 7.43 $MPa.m^{\frac{1}{2}}$. In another paper[4] he showed that K_{IC} for pressureless sintered Ce-ZTA ceramics containing 30 v/o CeO_2-ZrO_2 is maximized for 12 m/o CeO_2 (K_{IC} = 8 $MPa.m^{\frac{1}{2}}$).

Tien et al.[5] have recently showed that ceramics in the system ZrO_2-HfO_2-Al_2O_3-Cr_2O_3 exhibit an average strength of about 490 MPa and average toughness at 7.4 $MPa.m^{\frac{1}{2}}$.

Boch et al.[6] have reported room temperature strength and toughness of 400 MPa and 6 $MPa.m^{\frac{1}{2}}$, respectively, for Ce-ZTA ceramics containing 15 v/o of 12 m/o CeO_2-ZrO_2. They have also shown that strength and toughness drop to 250 MPa and 3 $MPa.m^{\frac{1}{2}}$, respectively, at 900°C.

The present paper discusses the temperature dependence of the mechanical properties of Ce-ZTA ceramics, with and without HfO_2 addition, which was added to improve the toughness retention at high temperature.

2. EXPERIMENTAL

Two compositions were chosen for this study. The first composition consists of 20 w/o (12 m/o CeO_2-ZrO_2)-80 w/o Al_2O_3 and the second is the same as the first one, except that 20 m/o HfO_2-ZrO_2 replaces 100% ZrO_2.

The volume fractions of CeO_2-ZrO_2 and of CeO_2-ZrO_2-HfO_2 in the compositions described above are 0.14 and 0.13, respectively, calculated from the densities of the single oxides. Twenty m/o HfO_2 in HfO_2-ZrO_2 was considered enough to enhance the high temperature properties, without the risk of spontaneous microcracking of the fired ceramic on cooling caused by the t-m transformation.

Two types of CeO_2-ZrO_2 (12 m/o CeO_2) powders were used. The first one, designated TZ12CE, was purchased from a commercial source.[1] The second powder, designated GDZ12CE (GD stands for Gel Derived), was prepared by pouring a solution of Ce nitrate[2] and Zr oxynitrate[3] into excess concentrated NH_4OH under vigorous stirring. The gel formed was dried and calcined at 1000°C for 14 hrs. The calcined gel was then milled in a planetary mill with water and Y-TZP balls. The slurry was sedimented to remove the coarse fraction and oven dried.

1 Tosoh Corp., 1-7-7 Akasaka, Minato Ku, Tokyo, 107 Japan
2 E. M. Science, Cherry Hill, NJ 08034
3 Magnesium Electron, Inc., Flemington, NJ 08822

The CeO_2-ZrO_2-HfO_2 powder, designated GD(20HZ)12CE, was prepared with the same method. As a source of HfO_2, Hf nitrate[1] was used, which was dissolved in the Ce and Zr nitrates solution. the submicron Al_2O_3 powder used is commercially available.[2]

All powders were characterized for morphology (SEM), particle size distribution (Particle Size Analyzers[3,4]), single point BET specific surface area[5] and phase composition by XRD.[6]

The powders were not analyzed for impurities, but it is believed that contamination from mill and media is relatively low, since virtually no weight loss was found for mill and media after milling.

Four types of ceramics were made using the powders described above. Table I describes these materials, showing codes, type of powder, powder processing and firing conditions used.

Table I

Description of Materials and Production Method

Material	Powders Used	Powder Processing	Firing
Ce-ZTA-ET-S/H	TZ12CE SUM Al_2O_3	Mixing in ETOH then ETOH evap.	Sint. 1500°C/3 hrs. HIPed 1500°C/0.75 hr.
Ce-ZTA-FD-S	TZ12CE SUM Al_2O_3	Planetary mill then freeze-dry	Sint. 1500°C/3 hrs.
Ce-ZTA-GFD-S/H	GDZ12CE SUM Al_2O_3	Planetary mill then freeze-dry	Sint. 1500°C/3 hrs. HIPed 1500°C/0.75 hr.
Ce-ZHTA-GFD-S/H	GD(20HZ)12CE SUM Al_2O_3	Planetary mill then freeze-dry	Sint. 1500°C/3 hrs. HIPed 1500°C/0.75 hr.

1 Teledyne Wah-Chang, Albany, OR
2 Type AKP-30, Sumitomo Aluminum Smelting Co., Japan
3 Micromeritics Instrument Co., Model 5000ET, Norcross, GA 30093
4 Malvern Instruments, Inc., Model 4700C, Southborough, MA 01772
5 Micromeritics Inst. Co., Digisorb 2600
6 Philips Model XRG 3100

For material Ce-ZTA-ET-S/H, the TZ12CE and Al_2O_3 powders were ball milled with anhydrous ethanol[1] and ZrO_2 media in a polyethylene container and the slurry was subsequently dried using a vacuum rotary evaporator.[2] Alcohol was used to produce soft agglomerates and the rotary evaporator to minimize phase segregation.

In the case of material Ce-ZTA-FD-S, the TZ12CE and Al_2O_3 powders were mixed in distilled water using a planetary mill and Y-TZP balls for 2 hrs. and the slurry was immediately poured into liquid N_2 to prevent phase segregation. (Preliminary sedimentation experiments had shown that both powders are fairly stable against sedimentation at pH of about 6. Lower pH gave even better stability, but it was avoided to minimize possible leaching of CeO_2). The frozen mass was freeze-dried to produce a powder with soft agglomerates.

For materials Ce-ZTA-GFD-S/H and Ce-ZHTA-GFD-S/H the gel derived powders were mixed with Al_2O_3 using the planetary mill method described above, followed by freeze- drying.

All billets were made by die pressing at 69 MPa followed by isostatic pressing at 207 MPa.

Firing was done in air in a furnace with $MoSi_2$ heating elements to 1500°C for 3 hrs., with a heat-up rate of 100°C/hr. All billets, except those of material Ce-ZTA-FD-S, were subsequently HIPed in Ar at 1500°C and 207 MPa for 45 minutes in order to eliminate the residual closed porosity.

The densities of the green billets were calculated from weight and dimensions, whereas those of the fired billets were measured with the Archimedes method in water.

XRD patterns were obtained for as-fired and machined samples, as well as for the fracture surface of MOR bars broken at 25°C, 600°C and 1000°C.

The microstructure of the fired billets was obtained by SEM on polished and thermally etched samples.

The MOR was measured in 4 pt. bending on 3.2 x 3.2 x 30 mm chamfered bars, with spans of 25.4 and 12.7 mm and a cross-head speed of 0.5 mm/min. The strength was measured at 25°C, 600°C and 1000°C. For the high temperature tests, a hot pressed silicon carbide fixture[3] was used.

Fracture toughness was measured with the double torsion method at 25°C, 600°C and 1000°C.

1 J. T. Baker Chemical Co., Phillipsburg, NJ 08865
2 Model R110 Buchi Laboratory, 9223 Shawl, Switzerland
3 NORTON NC-203

3. RESULTS AND DISCUSSION

3.1 Powder Properties

Table II shows the physical properties measured for the powders used in this work.

Table II

Powder Properties

Powder	BET Area (m^2/g)	Phases (XRD)	d_{10} (nm)	d_{50} (nm)	d_{90} (nm)
TZ12CE	15.2	25% m 75% t	NA	NA	NA
GDZ12CE	NA	100% t	150	400	1700
GD(2OHZ)12CE	NA	8% m 92% t	<150	350	1300
SUM Al_2O_3	7.0	100% alpha	370	510	770

3.2 Densification Behavior

Table III presents the green and fired densities of the four different ceramics. The data of Table III show that the TZ12CE powder produces a billet with slightly better sinterability, in comparison with the GDZ12CE powder. The value of 4.38 for the third material, higher than the theoretical density, indicates that the actual composition is probably richer in ZrO_2 than the nominal one.

Table III

Densities (Mg/m^3) of the Billets Before and After Firing

Material	Green	Sintered	HIPed	Theor.
Ce-ZTA-ET-S/H	2.41 (56)	4.21 (98)	4.30 (100)	4.30
Ce-ZTA-FD-S	2.42 (56)	4.30 (100)	NA	4.30
Ce-ZTA-GFD-S/H	2.45 (57)	4.04 (94)	4.38 (102)	4.30
Ce-ZHTA-GFD-S/H	2.46	4.28	4.39	NA

Note: Numbers in parenthesis are percent of theoretical density.

3.3 Phase Composition and Microstructures

The microstructures of the four ceramics described in Table I were very similar. No large difference was observed between those of the ceramic made with the ethanol method and the others made by freeze-drying. Fig. 1 shows the microstructure of material Ce-ZTA-GFD-S/H. All materials contain clusters of CeO_2-ZrO_2 grains with maximum size of 5-6 ums. It is obvious that more processing is needed to improve the homogeneity of these ceramics.

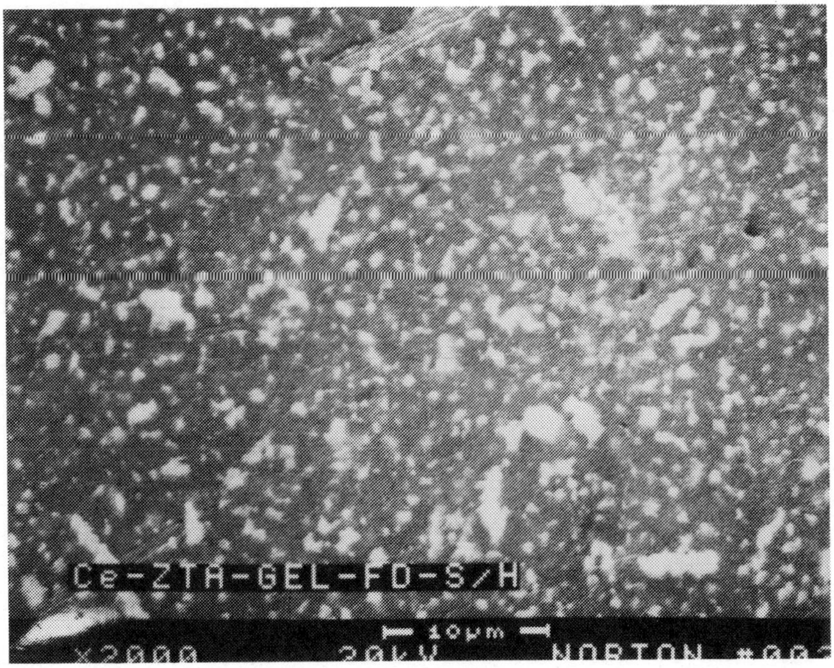

Fig. 1 Microstructure (SEM) of material Ce-ZTA-GFD-S/H polished and thermally etched at 1450°C/0.5 hr.

The XRD patterns of the sintered/HIPed billets have shown several interesting features. First, for the same billet (Ce-ZTA-ET-S/H), the as-HIPed orange surface showed much more m-ZrO_2 (70%) than the off-white interior (5%). It is believed that the orange color is the result of reduction in the HIP atmosphere, since the samples were uniformly off-white after high temperature annealing in air. Since the XRD pattern of the polished interior of the billet is not available, it is not possible to say whether the small amount of m-ZrO_2 in the interior of the billet was originally present or was formed as a consequence of the slicing stress. No matter what the answer is, however, this result shows that, unless the transformation is reversible, as suggested by Coyle et al.[7] for Ce-TZP ceramics and experimentally verified by Marshall and James[8] in Mg-PSZ, either the Ce-TZP grains are non-transformable under stress or the stress generated by the diamond blade is insufficient. The high m-ZrO_2 content of the reduced surface could be due to several reasons, i.e. CeO_2 depletion at the surface, preferential grain growth near the surface, or destabilization caused by reduction in the Ar atmosphere. It is known that the transformability of tetragonal CeO_2-ZrO_2 grains in Ce-TZP ceramics increases with increasing grain size and decreasing CeO_2 content.[9]

In order to understand whether or not the higher m-ZrO_2 content was limited to the surface layer, two XRD patterns were obtained from the two ground faces of a 50 x 25 x 1 mm sample (Ce-ZTA-GFD-S/H material) used in the fracture toughness measurement with the double torsion method. Of the two faces, one was orange and the other off-white. The former was obviously closer to the billet surface. From these patterns it appears that more m-ZrO_2 (15%) is present in the reduced face than in the other (3%). This result, similar to that found in the billet of the Ce-ZTA-ET-S/H material, shows that only a very small fraction of the grains are irreversibly transformed upon grinding in the oxidized material. The 3% m-ZrO_2 measured in this sample is lower than the 20% found by Boch et al[6] in a Ce-ZTA ceramic of the same composition. However, the microstructure of their material shows larger CeO_2-ZrO_2 grains, resulting from a higher sintering temperature (1600°C). The 15% m-ZrO_2 found in the reduced face, on the other hand, could have been partly originally present and partly could have been formed upon grinding. Since it is unlikely that preferential grain growth occurred near the surface at 1550°C in 45 minutes, which could explain the higher m-ZrO_2 content, it is possible that the reduced t-ZrO_2 (Ce_2O_3), is more easily transformable than the oxidized t-ZrO_2 (CeO_2), due to residual stresses. Further characterization is in progress to answer this question.

XRD patterns (short scan in the 25-35 2 theta degree range) were also obtained for fracture surfaces of the MOR bars (off-white color) of the Ce-ZTA-GFD-S/H and Ce-ZHTA-GFD-S/H ceramics, broken at 25°C, 600°C and 1000°C. No m-ZrO_2 was detected in these patterns. Since in Y-TZP's more m-ZrO_2 is found on fracture surfaces than on ground ones[10] this result was unexpected. Tsukuma and Shimada[11] have found 40% m-ZrO_2 on fracture surface of a Ce-TZP ceramic (12 m/o CeO_2) having a grain size of 2.5 um.

In the Ce-ZTA ceramics of this study, some m-ZrO_2 could be present on the fracture surfaces, but the surface roughness and the high volume fraction of Al_2O_3 make its detection difficult.

3.4 Mechanical Properties

Table IV shows the temperature dependence of strength for the four materials listed in Table I. These ceramics retain at 1000°C between 59 and 69% of the room temperature strength. A similar result was found by Tsukuma and Takahata (12) for a Y-TZP (2 mol.% Y_2O_3)/80 w/o Al_2O_3, which showed about 70% strength retention at 1000°C. The main reason for the good strength retention is believed to be a better toughness retention at high temperature, in comparison with the Y-TZP ceramics, as shown in Table V. For example, the pressureless sintered Y-TZP ceramics made in a parallel study have exhibited typical strengths of 1000 MPa at room temperature but only 350 MPa at 1000°C. After HIPing, values around 1500 MPa and 500 MPa were obtained, respectively. These results are in agreement with those reported by Tsukuma and Takahata (12), who showed that only 30% of the strength is retained at 800°C in a sintered Y-TZP (3 mol.% Y_2O_3). The relatively high strength values of the Ce-ZTA ceramics of this study seem to indicate that microcracking is absent in these materials. It is known that microcracking is likely to occur in a two-phase composite when the matrix contracts on cooling more than the dispersed phase and when the latter exceeds a critical size.[13] In the Ce-ZTA composites, however, the dispersed phase contracts more than the matrix.

Table IV

Temperature Dependence of Strength for Ce-ZTA Ceramics With and Without HfO_2 Addition

Material	25°C	#	600°C	#	1000°C	#
Ce-ZTA-ET-S/H	847 (12)	4	---	-	503 (22)	4
Ce-ZTA-FD-S	654 (13)	10	---	-	---	-
Ce-ZTA-GFD-S/H	741 (7)	7	489 (20)	3	508 (3)	4
Ce-ZHTA-GFD-S/H	703 (9)	6	505 (24)	5	462 (8)	5

Note: Strength values are in MPa. Numbers in parenthesis are % standard deviations. # = number of bars tested.

Fractographic analysis performed on the fracture surface of the MOR bars of the Ce-ZTA ceramics has the shown presence of several types of flaws, i.e. inclusions, voids and machining induced defects, particularly on the bar chamfers. Figure 2 illustrates one of the most commonly observed defects. In most cases alumina-rich, relatively large grains were associated with, or very close to, the fracture origin.

Fig. 2: Fracture surface (SEM) of an MOR bar of material Ce-ZTA-GFD-S/H showing an inclusion containing HfO_2 and Al_2O_3 near the surface (MOR = 695 MPa)

Table V shows the temperature dependence of fracture toughness for these ceramics. For the sake of comparison, K_{IC} values are also reported for a Y-TZP ceramic containing 2.5 m/o Y_2O_3.

Table V shows first that 55 to 75% of the room temperature toughness is retained at 1000°C, in contrast with the result found for Y-TZP ceramics without Al_2O_3, where only 25% is retained. Secondly, no significant improvement in room temperature toughness was obtained by alloying ZrO_2 with HfO_2, but a higher fraction was retained at 1000°C. Tien et al.(5) have shown that K_{IC} is maximized (7.4 $MPa.m^{\frac{1}{2}}$) when the $m-ZrO_2/t-ZrO_2$ ratio is around 2.5. They conclude that "multiple toughening mechanisms are active" in their system. Since in our materials $m-ZrO_2$ has not been detected on the fracture surface and only a small amount was found in the ground (oxidized) surface, in order to explain the relatively high toughness observed, in comparison to that of polycrystalline Al_2O_3 (about 4.5 $MPa.m^{\frac{1}{2}}$), it must be concluded that the t-m transformation could be reversible, as proposed by Coyle et al.(7), and/or that other toughening mechanisms, different from transformation toughening, operate in these materials, such as the ferroelastic domain switching.(15) The drop in toughness with increasing temperature suggests that transformation toughening exists at low temperature, as pointed out by Matsumoto et al(16), who state that in Ce-TZP ceramics "the contribution of transformation toughening to the overall toughness is negligible above 750°C".

Table V

Temperature Dependence of Fracture Toughness for Ce-ZTA and Y-TZP Ceramics With and Without HfO_2 Addition

Material	25°C	#	600°C	#	1000°C	#
Ce-ZTA-GFD-S/H	8.2 (9)	2	5.9 (5)	3	4.5 (4)	3
Ce-ZHTA-GFD-S/H	6.3 (7)	3	5.8 (6)	4	4.7 (3)	3
Y-TZP-S	8.8 (10)	5	---	--	2.2 (6)	5

Note: Toughness values in $MPa.m^{\frac{1}{2}}$. # = number of samples tested. Numbers in parenthesis are percent standard deviation.

4. CONCLUSIONS

The main conclusions which can be drawn from the present study are summarized below.

1) The Ce-ZTA ceramics, with or without HfO_2 additions, have exhibited good strength and toughness at room temperature and good retention of these properties at 1000°C, better than in Y-TZP ceramics.

2) The decrease in strength with increasing temperature appears to be caused by a similar drop in toughness, which suggests that transformation toughening is contributing to the overall toughness at low temperatures, even though no monoclinic ZrO_2 was observed in the fracture surface and very little in the (oxidized) machined surface. This suggests that the t-m transformation could be reversible at low temperatures.

3) More m-ZrO_2 was detected by XRD on reduced than on oxidized surfaces. A possible reason for this is easier transformability under stress of the reduced t-ZrO_2 (Ce_2O_3), but experimental verification is needed.

4) The addition of 20 m/o HfO_2 to ZrO_2 has resulted in better retention of toughness at high temperature, but the room temperature properties have been degraded. Further optimization is needed before firm conclusions can be drawn.

5) Fractographic analysis has shown the presence of several types of defects, whose elimination through improved processing should result in stronger ceramics, minimizing the need for hot isostatic pressing.

REFERENCES

1) N. Claussen, "Fracture Toughness of Al_2O_3 with an Unstabilized ZrO_2 Dispersed Phase", J. Am. Ceram. Soc., 59, 49-51, (1976).

2) P. F. Becher, "Transient Thermal Stress Behavior in ZrO_2-Toughened Al_2O_3", J. Am. Ceram. Soc., 64, 37-39, (1981).

3) F. F. Lange, "Al_2O_3/ZrO_2 Ceramic", U. S. Pat. 4,316,964 (2-23-1982).

4) F. F. Lange, "Transformation Toughening, Part 5, Effect of Temperature and Alloy on Fracture Toughness", J. Mater. Sci., 17, 255-262, (1982).

5) T. Y. Tien et al, "Toughened Ceramics in the System Al_2O_3: Cr_2O_3/ZrO_2:HfO_2", Int. J. High Tech. Ceram., 2, 207-219, (1986).

6) P. Boch, et al., "Renforcement Mecanique de Ceramiques Alumine-Zircone Avec Ajouts de CeO_2", L'Industrie Ceramique, N. 795, 451-58, (1985).

7) T. W. Coyle, et al., "Transformation Toughening in Large Grain-Size CeO_2 Doped ZrO_2 Polycrystals", J. Am. Ceram. Soc., 71, (2), C-88-C-92, (1988).

8) D. B. Marshall and M. R. James, "Reversible Stress-Induced Martensitic Transformation in ZrO_2", J. Am. Ceram. Soc., 69 [3], 215-17 (1986).

9) K. Tsukuma and M. Shimada, "Strength, Fracture Toughness and Vickers Hardness of CeO_2-Stabilized Tetragonal ZrO_2 Polycrystals (Ce-TZP)", J. Mater. Sci., 20, 1178-1184, (1985).

10) M. V. Swain, "Grain Size Dependence of Toughness and Transformability of 2 m/o Y-TZP Ceramics", J. Mater. Sci. Lett., 5, 1159-62, (1986).

11) K. Tsukuma and M. Shimada, "Strength, Fracture Toughness and Vickers Hardness of CeO_2-Stabilized Tetragonal ZrO_2 Polycrystals (Ce-TZP)", J. Mater. Sci., 20, 1178-1184, (1985).

12) K. Tsukuma and T. Takahata, "Mechanical Properties and Microstructure of TZP and TZP/Al_2O_3 Composites", Advanced Structural Ceramics, Vol 78, pp. 123-135, Ed. by P. F. Becher, M. V. Swain and S. Somiya, Pub. by Mat. Res. Soc., Pittsburgh, PA (1987).

13) F. F. Lange, "Fracture Mechanics of Ceramics", Vol. 2, R. C. Bradt, D.P.H. Hasselman and and F. F. Lange, Eds., Plenum Press, N.Y., 599, (1974).

14) R. Kamo and W. Bryzik, "CUMMINS/TACOM Advanced Adiabatic Engine", Ceram. Eng. Sci. Proc., 5, [5-6], 312-38 (1984).

15) A. V. Virkar and R. L. K. Matsumoto, "Ferroelastic Domain Switching as a Toughening Mechanism in Tetragonal Zirconia", J. Am. Ceram., Soc., 69, (10), C-224-C-226, (1986).

16) R. L. K. Matsumoto and R. J. Mayhew, "High Temperature Evaluation of New Ferroelastic Toughened Ceramic Materials", Final Report, Dept. of the Army, Contract No. #DAAL04-87-C-0058.

ACKNOWLEDGMENTS

The authors wish to thank Drs. E. Lilley, W. S. Coblenz, and J. B. Blum for their interest and for many stimulating discussions. This work was done as part of a research program sponsored by ORNL (DOE), Subcontract No. 86X22031C.

PREPARATION OF CE-TZP AND ITS MECHANICAL BEHAVIOR

S.G.Yang and K.Chen
Department of Inorganic Materials Science and Engineering
South China University of Technology
Wu Shan, Guangzhou
People's Republic of China

ABSTRACT

Transformation toughened ZrO_2 with 12mol% CeO_2(Ce-TZP) was investigated. Compared with Y-TZP, this material has coarser crystalline grains and correspondingly larger critical transition grain sizes for $t\text{-}ZrO_2 \rightarrow m\text{-}ZrO_2$ transformation, which had made it easier to prepare high performance Ce-TZP without the use of ultrafine starting powders. The addition of Al_2O_3 and TiO_2 was found to be very effective in lowering the sintering temperature and raising the mechanical properties at room temperature. The additions of 11-18wt% Al_2O_3 and 0.5-2wt% TiO_2 to 12mol%CeO_2-ZrO_2 gave rise to a remarkable increase in the fracture toughness, flexural strength, and microhardness. The effects of the additives are discussed. The relationship between the properties and microstructure was investigated with the aid of XRD, SEM, and EDAX methods. The maximum K_{IC}, σ_f and Hv were about 14MPa$\sqrt{m}$, 2000 MPa and 9GPa, respectively.

INTRODUCTION

The high performance ceramic Y-TZP hold one's considerable interest, but there are problems associated with it. These are: 1. an expensive manufacturing procedure 2. mechanical properties degradation after medium-temperature aging 3. severe creep at high temperature due to very fine grains of Y_2O_3-stabilized ZrO_2. To overcome these problems, it has been suggested that some other rare earth oxides such as CeO_2 should be used to replace or in combination with Y_2O_3 as the stabilizer(1,2). Actually the study of Ce-TZP had been started as early as 1984 and the good properties below room temperature were reported by T.K.Gupta(3). The transformation of Ce-TZP annealed in water was investigated in 1985 by T.Sato and M.Shimada, who used coprecipitated fine grain ceria-doped ZrO_2 powders as the starting materials and achieved Vicker's hardness and fracture toughness of 10GPa and 11-17MPa$\sqrt{m}$, respectively(4). In 1985, a Japanese patent claimed that good Ce-TZP had been achieved by using industrial grade ZrO_2 powders as the starting materials with Al_2O_3 and TiO_2

developed. The present study aimed at improving the mechanical behavior of Ce-TZP made from industrial raw materials through selecting CeO_2, Al_2O_3 and TiO_2 contents and investigating the relationship between properties and microstructure of the sintered bodies.

EXPERIMENTAL PROCEDURE

Industrial grade ZrO_2 powders and ultrafine ZrO_2 powders(lab-made) were used as the starting materials. The CeO_2(ultrafine powders) stabilizer had a purity of 99.95%. The additives used, Al_2O_3 contained 0.15wt% SiO_2, 0.5wt% MgO, 0.07wt% Fe_2O_3 and trace of CaO,and TiO_2 was chemically pure reagent. CeO_2 was wet mixed with ZrO_2 by alumina ball-milling. After drying, $ZrO_2(CeO_2)$ powders were calcined for 1 hours at 1250°C to form CeO_2-ZrO_2 solid-solution which would be helpful in the densification of the sintered body. Then Al_2O_3 and TiO_2 were added also by wet alumina ball-milling. After drying the resultant powders were isostatically pressed at 2000Kg/cm^2 to form square billets that were later cut into the slabs(6 by 6 by 50mm). The slab specimens then were sintered at 1400-1680°C for 1 to 10 hours in air. The bulk densities of the sintered bodies were measured by the Archimedes technique. The sintered slabs were ground(by grit carborundum) to the test size of 3-3.5 by 3-3.5 by 35-40mm and the surfaces that were to be indented were polished to mirror finish using W1 or W2 diamond pastes. The three-point-bend method was employed to measure the flexural strength with a loading speed of 0.4mm/min and a span of 26mm. The indentation technique (6) was used to estimate the fracture toughness(loads from 20 to 120Kg were used). The microhardness was also measured by Vicker's indentation at loads of 5 and 10Kg. The phases of the flat surfaces of the sintered samples were identified by X-ray diffraction analysis by scanning 2θ from 26 to 70°, and the phase contents were calculated by the formula supplied by R. C. Garvie(7). The fracture surfaces were observed by SEM, and the average grain sizes were determined from the pictures by the intercept method (8).

RESULTS AND DICUSSION

Characterization of as-sintered bodies

The samples which were prepared from industrial grade zirconia will be identified as CZ and CZA(with Al_2O_3 and TiO_2 doped) and that from ultrafine zirconia as FCZ. XRD analysis showed that tetragonal ZrO_2 crystal structure was present in all three materials as seen in Fig. 1. According to the CeO_2-ZrO_2 phase diagram(9), there would not be any cubic zirconia phase appearing below 12mol% CeO_2. When the amounts of ceria were lowered from 11mol%,the amount of monoclinic ZrO_2 increased rapidly with decreasing CeO_2 content in CZ samples, as shown in Fig. 2. Additions of Al_2O_3 and TiO_2 might have gone into solid-solution with $ZrO_2(CeO_2)$ grains or formed an amorphous phase, as there were not any other new diffraction peaks appearing in the CZA XRD patterns as is illustrated in Fig. 1-C.

The infulence of ceria content on Vicker's hardness,fracture toghness and flexural strength are shown in Fig. 3 and 4. It can be seen in Table 1 that 12mol% is the optimum content of CeO_2 for making high strength Ce-TZP, despite variations of Al_2O_3 and TiO_2 contents. Below 12mol% the metastable tetragonal zirconia could not be fully maintained and the transformation of t-ZrO_2 to m-ZrO_2 occurred during cooling causing microcracking and hence degrading the strength of the sintered body, whereas when stabilizer content was higher than 12mol%, the driving force for the stress-induced transformation was weakened and the transformation toughening effect lessened(10).

Obviously, the high alumina content was helpful in achieving high microhardness and toughness, as shown in Table 2 and Fig.5, which might be mainly due to the high hardness and elastic modulus of alumina. As for flexural strength, too much Al_2O_3 was not good (Fig. 6). This might be explained as that too high volume of alumina would considerably weaken the contribution of the CeO_2-stabilized ZrO_2,which possesses higher strength than Al_2O_3, to the whole sintered body. Considering both toughness and strength, the optimum amount of Al_2O_3 should be 11-12wt%.

Table 1 Flexural Strength of Ce-TZP(doped with Al_2O_3 and TiO_2, sintered at 1450°C for 5 hours)

Samples	Chemical Compositions	Strength(MPa)
CZA-8	10CZ18A1.5T*	535.89±111.31
CZA-9	12CZ8A1.5T	886.91±260.83
CZA-10	12CZ11A2T	2181.51±625.61
CZA-11	12CZ13A0.5T	1007.23±245.53
CZA-12	12CZ18A1T	848.15±74.92
CZA-13	16CZ8A2T	443.39±31.28
CZA-14	16CZ11A1.5T	496.62±129.61
CZA-15	16CZ18A0.5T	519.99±10.82

*10CZ18A1.5T means 10mol% CeO_2-ZrO_2 doped with 18wt% Al_2O_3 and 1.5wt% TiO_2.

Table 2 Effect of Al_2O_3 on The Microhardness of Ce-TZP

Samples	Sintering Condition	Hv(GPa)
12CZ	1560°C, 3h	6.67
16CZ	1560°C, 3h	7.80
12CZ13A	1450°C, 1h	8.64
16CZ18A	1450°C, 1h	10.28

Table 3 Mechanical Properties of FCZ, CZ and CZA

(A) FCZ (Contained 12mol% CeO_2 and 0.5wt% TiO_2)

Sintering Condition	Hv(GPa)	KIC(MPa$\sqrt{m}$)
1450°C, 1h	6.84	7.68
1450°C, 3h	7.00	7.96
1450°C, 5h	6.57	8.29
1500°C, 1h	6.85	7.35

(B) CZ (Contained 12mol% CeO_2 and 0.5wt% TiO_2)

Sintering Condition	Hv(GPa)	KIC(MPa$\sqrt{m}$)
1460°C, 3h	7.91	8.08
1500°C, 3h	7.00	7.96
1600°C, 3h	8.43	9.15
1680°C, 3h	8.32	10.36

(C) CZA (Contained 12mol% CeO_2,13wt% Al_2O_3 and 0.5wt% TiO_2)

Sintering Condition	Hv(GPa)	KIC(MPa$\sqrt{m}$)
1450°C, 1h	8.64	8.92
1450°C, 3h	8.21	8.73
1450°C, 5h	9.16	10.75
1500°C, 1h	9.06	11.42
1500°C, 2h	7.27	7.43

Microstructure of Ce-TZP

Fig. 7 and Fig. 8 show the microstructure of Ce-TZP made from ordinary and ultrafine zirconia powders. Fig. 9 shows that of Ce-TZP with Al_2O_3 and TiO_2 additions, in which the grains look a bit more rounded than without additives. This might be attributed to the action of a high temperature liquid phase produced from added Al_2O_3, TiO_2 and some SiO_2 impurity. The average grain size range from 2 to 5 μm under different fabricating conditions. About 4-5 μm is the optimum for the highest toughness and strength, as shown in Fig. 11 and Fig. 12. Compared with Y-TZP(11), this indicates that Ce-TZP has a coarser critical transition t-ZrO_2 grain size, which has made it possible to prepare high performance Ce-TZP ceramics without ultrafine starting materials, as shown in Table 3.
By applying the linear regression method to the experimental datas for the sintering temperature and optimum firing time(12), this study had calculated the apparent sintering activiation energy of CZA-10 sample as 107.03 Kcal/mol, which is about 40 Kcal/mol less than that of Y-TZP estimated by T. K. Gupta(12). The result is concurrent with the experimental phenomenon that the grain growth rate of Ce-TZP was much higher than that of Y-TZP(4) and hence CeO_2-stabilized zirconia has coarser crystalline grains.

Finally, by careful observation of the broken surfaces of the Ce-TZPs(Fig.8 and 9), it can be seen that there is a rather considerable proportion of transgranular fracture no matter whether the material was with Al_2O_3 doped or not, and even more, there are some fractured Ce-ZrO_2 grains that have the characteristic substructure viewed in detail in Fig.10. Whether such kind of substructure should be referred to layer dislocation or twins still remains uncertain due to lack of the sufficient evidence. It could be believed, however, that the Martensitic t-ZrO_2 →

m-ZrO2 transformation had occurred during fracturing. The transformation absorbed crack energy and thus toughened and strengthened the material remarkably.

Effect of Al2O3 and TiO2

Besides raising the mechanical properties, the addition of Al2O3 and TiO2 also apparently lowered the sintering temperature. It was found that the sintering of CZA materials could be carried on at as low as 1380oC in our work, which was impossible for CZ or FCZ systems. From Table 3 one can see that Ce-TZP with 13wt% Al2O3 has better hardness and toughness at relative lower firing temperature,compared to that without Al2O3. It should also be noted in Table 3-(C) that for CZA, KIC and Hv dropped from 11.42 MPa$\sqrt{m}$ and 9.06GPa to 7.43MPa$\sqrt{m}$ and 7.27GPa,respectively, as firing time was extended to 2 hours. Yet for CZ material, seen in Table 3-(B), 1600°C and 3 hours was still not long enough to reach these values for CZA. This implies that CeO2-ZrO2 has higher sinter activity after being doped with Al2O3 and TiO2. With the help of XRD and EDAX analyses, it could be deduced that some Al2O3 and TiO2 has gone into solid-solution with the ZrO2(CeO2) grains, which might result in Al^{3+} interstitials in the zirconia crystal lattice, as illustrated in following equation:

$$Al2O3 \xrightarrow{ZrO2} Al'_{Zr} + Al_i^{\cdot\cdot\cdot} + 3O_O$$

Such imperfections would have been helpful to activate the crystal lattices and hence enhance the rate of solid-state sintering. In Y-PSZ, it is a bit harder for Al2O3 to go into solid-solution with ZrO2(Y2O3) grains(13), since Y^{3+} ions had formed interstitials in the crystal lattice of ZrO2 and hence lessened the chances for Al2O3 to do the same. Whereas in Ce-TZP, Ce4+ ions did not take the positions of either vaccancies or interstitials. On the other hand, it can be seen that CeO2-ZrO2 grains look a bit more rounded after doped with Al2O3 and TiO2. This phenomenon suggested that there could have been some liquid-phase occurring during the sintering period,which would also be related to Al2O3 and TiO2 additives. So far,however,the actual mechanism of Al2O3 and TiO2 in TZP materials has not been clearly understood. The quaternary phase system CeO2-ZrO2-Al2O3-TiO2 is complicated,therefore further investigation of the CZA system is necessary.

CONCLUSION

High performance ceramic Ce-TZP has been obtained by selecting CeO2,Al2O3 and TiO2 contents. Since the critical transition grain size for t-ZrO2 to m-ZrO2 transformation is large, ultrafine powders as the starting materials are not necessary to achieve good mechanical properties, as compared to Y-TZP. The addition of Al2O3 and TiO2 are very helpful in lowering the firing temperature and raising the mechanical properties. The optimum amounts of additions have been determined to be 11-12wt% for Al2O3 and 1-2wt% for TiO2, but the doping mechanism has not been fully understood due to the limited knowledge about the CeO2-ZrO2-Al2O3-TiO2 quaternary system.

ACKNOWLEDGMENTS

The authors wish to thank professor S. T. Zhao and Y. Y. Wu for their help during the course of the experimental work.

REFERENCES

1. N. Claussen, "Microstructural Design of Zirconia-toughened Ceramics(ZTC)", Adv. Ceram. , Vol. 12, 1984, pp325-351
2. T. Sato,M. Shimada, "Control of the Tetragonal to Monoclinic Phase Transformation of Yttria-doped Zirconia Containing CaO,CeO2 and Al2O3 in Air and Water", CERAMIC MATERIALS & COMPONENTS FOR ENGINES, PROCEEDINGS OF THE SECOND INTERNATIONAL SYMPOSIUM, edited by W. Bunk et al, 1986, pp291-297
3. T. K. Gupta, C. A. Andersson, "Low Temperature Mechanical Properties of Ceria-alloyed Tetragonal Zirconia", Adv. Cryog. Eng., 30, 1984, pp367-373
4. T. Sato, M. Shimada, "Transformation of Ceria-doped Tetragonal Zirconia Polycrystals by Annealing in Water", Am. Ceram. Soc. Bull., Vol.64, No.10,1985, pp1382-1384
5. T. Koji et al, "Zirconia Ceramics",Jpn. Kokai Tokkeyo Koho, JP 60,246,261
6. K. Niihara et al, "Evaluation of KIC of Brittle Solids by the Indentation Method With Low Crack-to-indent Ratios",J. Mater. Sci. Lett., 1,1982,pp13-16
7. R. C. Garvie, P. S. Nicholson,"Structure and Thermomechanical Properties of PSZ in the CaO-ZrO2 System", J. Am. Ceram. Soc., Vol.55,No.3,1972,pp152-157
8. R. L. Fullman, "Measurement of Particle Size in Opaque Bodies" , J. Metal. , 3,1953,pp447-452
9. P. Duwez, F. Odell,"Phase Relationship in the System Zirconia-Ceria", J. Am. Cerm. Soc. ,Vol.33,No.9,1950,pp274-283
10. F.F. Lange,"Transformation Toughening(part 3)", J. Mater. Sci. ,17,1982,pp240-246
11. T. K. Gupta, F. F. Lange,"Effect of Stress-induced Phase Transformation on the Properties of Polycrystalline Zirconia Containing Metalstable Tetragonal Phase",J. Mater. Sci.,13,1978, pp1464-79
12. T. K. Gupta,"Sintering of Tetragonal Zirconia and Its Characteristics", Sci. Sinter. ,Vol.10,No.3,1978,pp205-216
13. T. Kosmac et al ,"Diffusion Processes and Solid-States Reactions in the Systems Al2O3-ZrO2(Stabilizing Oxide)(Y2O3, CaO,MgO)". Adv. Ceram. ,Vol.12,1984,pp546-551

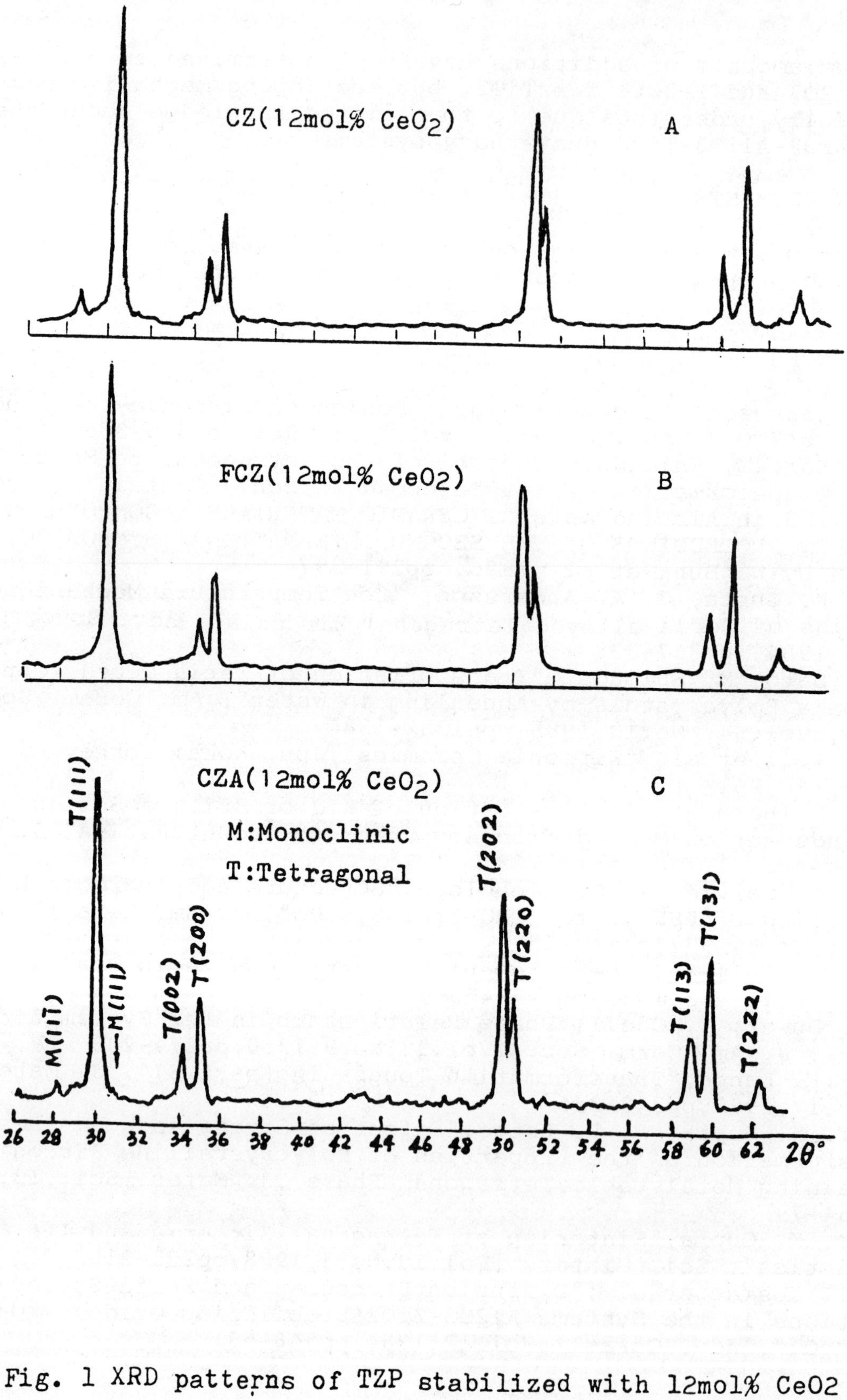

Fig. 1 XRD patterns of TZP stabilized with 12mol% CeO2

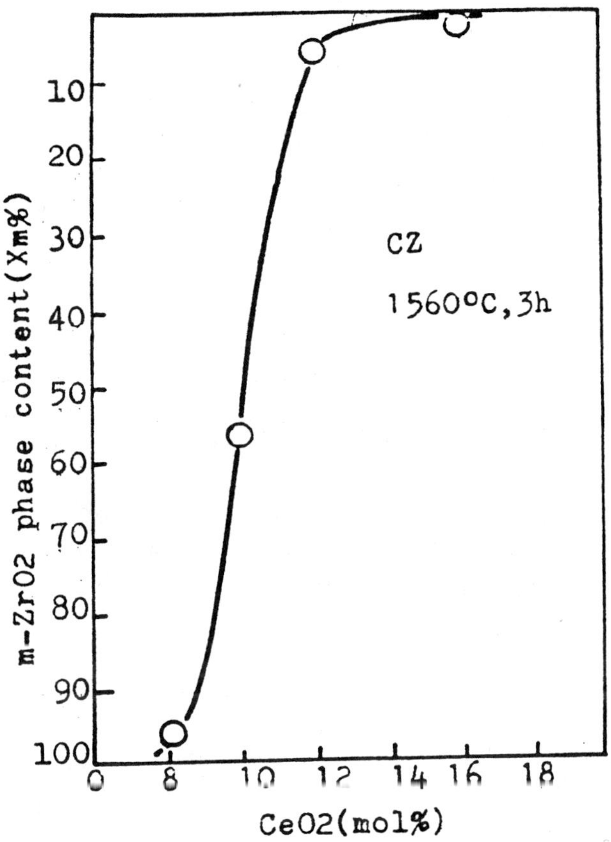

Fig. 2 Effect of ceria content on the m-ZrO_2 phase content in CZ

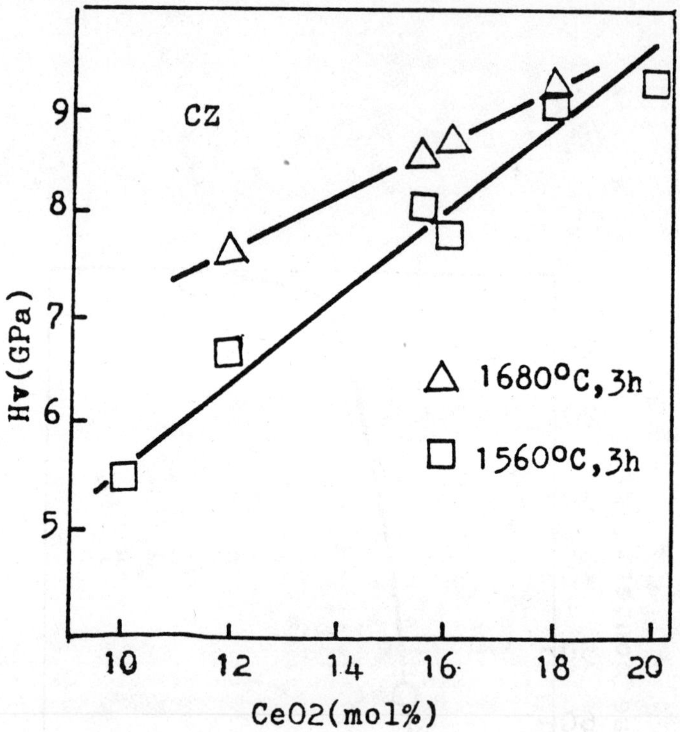

Fig. 3 Relationship between ceria content and microhardness

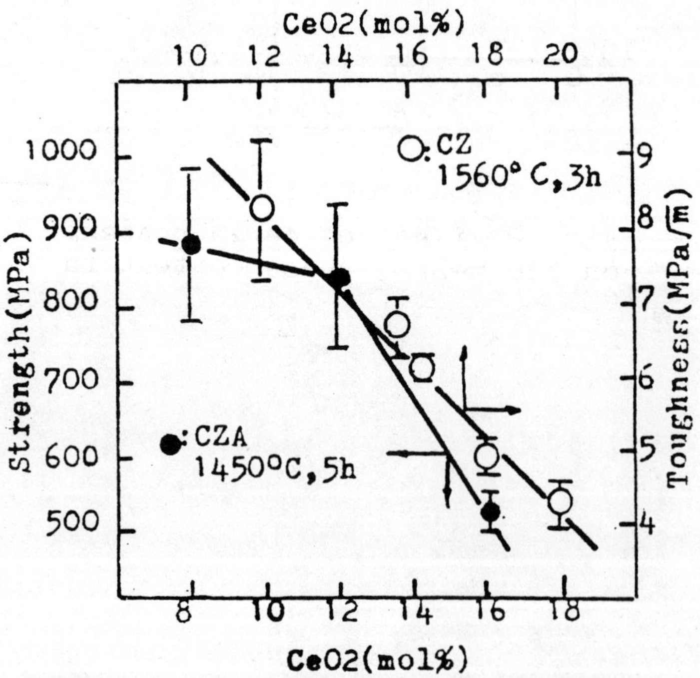

Fig. 4 Effect of ceria content on the strength and toughness

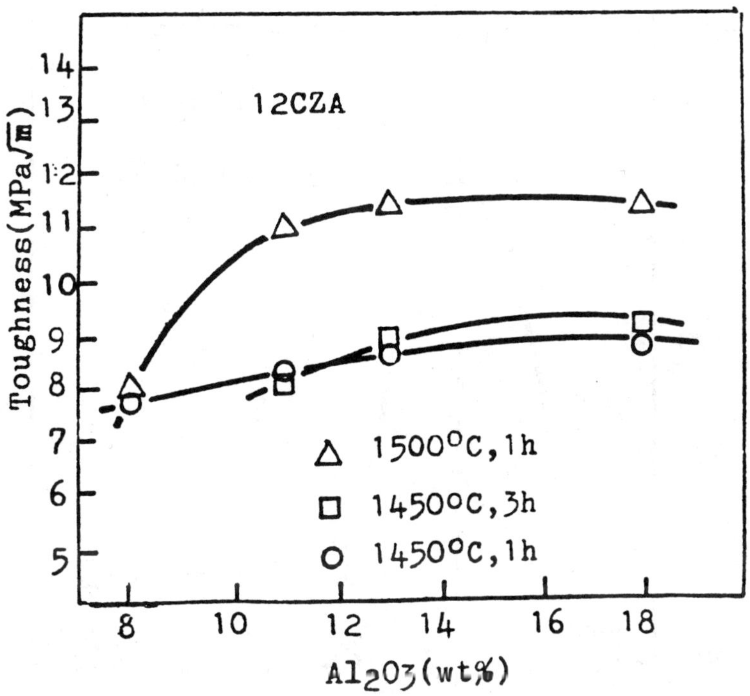

Fig. 5 Relationship between alumina content and fracture toughness

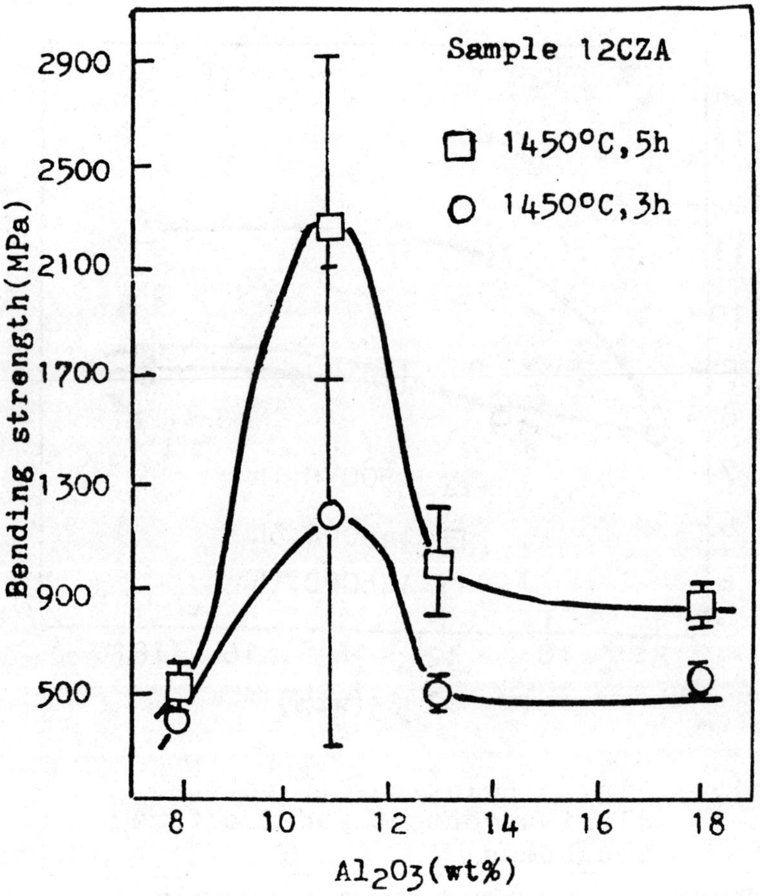

Fig. 6 Relationship between alumina content and flexural strength

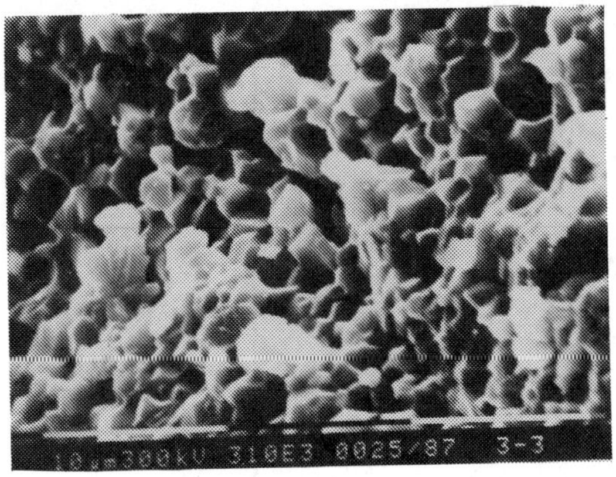

Fig. 7 SEM photograph of FCZ fracture surface(1460°C, 2h)

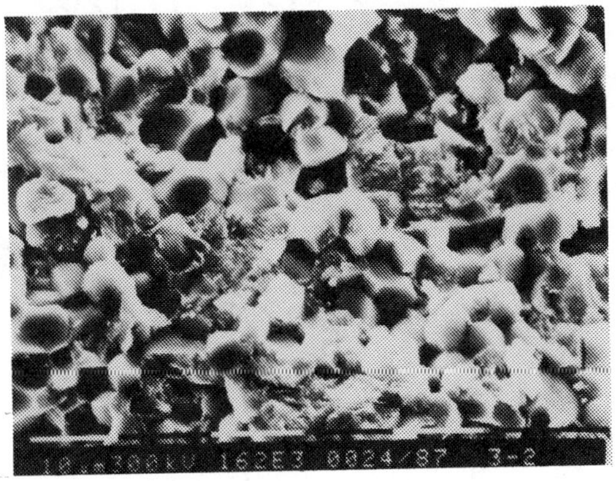

Fig. 8 SEM photograph of CZ fracture surface(1680°C,3h)

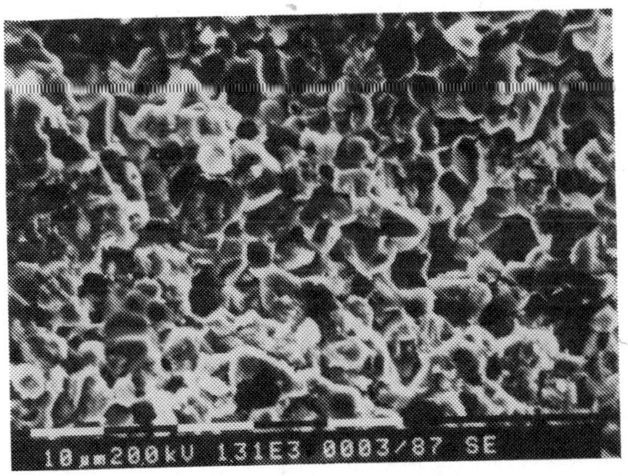

Fig. 9 SEM photograph of CZA fracture surface(1450°C, 5h)

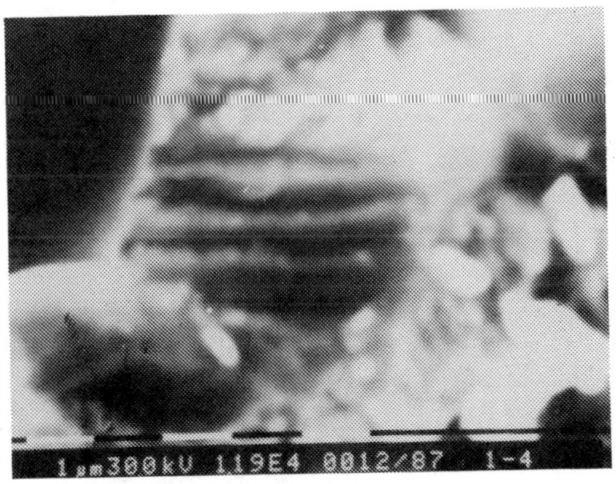

Fig. 10 Detail of the broken CZ grain (SEM,X11900)

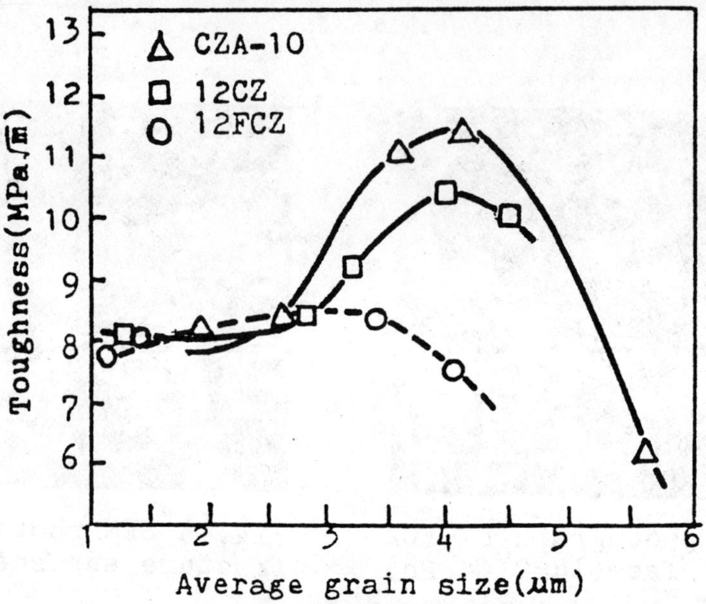

Fig. 11 Relationship between grain size and fracture toughness of FCZ, CZ and CZA

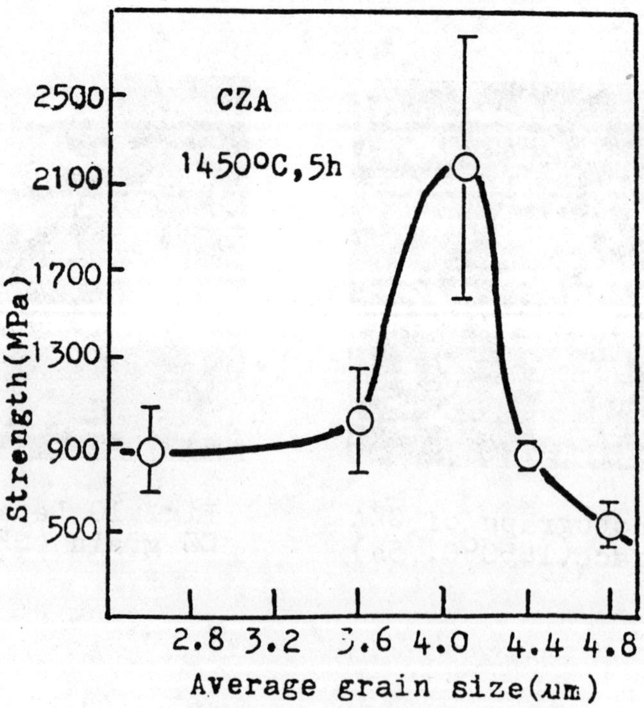

Fig. 12 Relationship between grain size and strength of CZA

STRENGTHENING OF OXIDE CERAMICS BY SiC AND Si_3N_4 DISPERSIONS

K. Niihara and A. Nakahira
The National Defense Academy
Physics Department
Yokosuka 239, Japan

ABSTRACT

The Al_2O_3-SiC, Al_2O_3-Si_3N_4 and MgO-SiC composites series have been investigated to improve mechanical and thermo-mechanical properties of Al_2O_3 and MgO ceramics. These composites were prepared by hot-pressing mixtures of fine powders at 1500 to 1900°C with 28 MPa applied pressure. The striking finding in these composites is that the fracture strength of Al_2O_3 and MgO ceramics was approximately 1.5 to 3 times increased by incorporating very fine SiC and Si_3N_4 particles and these strength improvement was also observed even at the temperature range from 1200 to 1400°C. Transmission electron microscopic observations revealed that the nano-size SiC and Si_3N_4 particulates were mainly dispersed within the Al_2O_3 and MgO matrix grains when the volume fraction of dispersions is less than 15vol%. The nanostructure-properties relationship will be discussed.

INTRODUCTION

Al_2O_3 and MgO ceramics are promising ceramics for many modern-day engineering applications because of their desirable properties including high refractoriness, good hardness and chemical stability. However, these oxide ceramics suffer from relatively low fracture toughness and strength, significant strength degradation at high temperatures and poor thermal-shock fracture resistance. Thus, many attempts have been made to resolve these problems of Al_2O_3 and MgO ceramics by incorporating the second phase dispersions.[1-10]

The ZrO_2 dispersion is observed to improve remarkably the fracture toughness and strength of Al_2O_3 due to its stress-induced transformation.[2,3] The significant increases in both fracture toughness and strength of Al_2O_3 and MgO are also achieved by incorporating the SiC whiskers.[4-11] The ZrO_2-reinforced composites show the strength degradation at relatively low temperatures, while the SiC-whisker reinforced composites are difficult to prepare the dense materials by pressureless sintering. To overcome these problems, therefore, we have been studying the Al_2O_3-SiC, Al_2O_3-Si_3N_4 and MgO-SiC nanocomposites, in which the SiC and Si_3N_4 particles are dispersed within the Al_2O_3 and MgO matrix grains.[12-15] In this paper, we describe the fabrication processing and nanometer-size structure (nanostructure) of Al_2O_3-SiC, Al_2O_3-Si_3N_4 and MgO-SiC composites and discuss the effects of nanostructures on their mechanical properties.

EXPERIMENTAL PROCEDURES

The γ-Al_2O_3 (average particle size of 0.4 μm from Asahi Chemical Co.), MgO (average particle size of 0.1 μm from Ube Industries Ltd.), two types of β-SiC (average particle sizes of 2 μm and below 0.3 μm from Ibiden Co.) and α-Si_3N_4 (average particle size of 0.5 μm from Starck Berlin) were selected as starting powders in this experiments. The compositions containing various volume fraction of SiC and Si_3N_4 particles were mixed by the conventional ball milling technique in the highly pure toluene with Al_2O_3 balls for 24h. The dried mixtures of Al_2O_3-SiC, Al_2O_3-Si_3N_4 and MgO-SiC were hot-pressed in N_2 atmosphere at 1500 to 1900°C with applied pressure of 28 MPa.

The densities of hot pressed composites were measured by an Archimedes method using toluene. The components of all composites were determined by X-ray diffraction analyses. The fracture surfaces and ion-beam thinned specimens were examined by optical microscopy, scanning electron microscopy (SEM), transmission electron microscopy (TEM) and also analytical transmission electron microscopy (ATEM).

Fracture toughness of these composites were evaluated by the indentation microfracture method with indenter loads of 0.49 to 18.6 N.[16] Hardness of the materials was measured using a high temperature Vickers hardness tester (Nikon, QM) up to 1400°C. Tests were carried out in vacuum at load of 4.9N and loading duration of 20s. The three point bending test was used to estimate the fracture strength under a crosshead speed of 0.5 mm/min specimen size: 3 x 4 x 45 mm, span: 30 mm). The tensile surfaces in the bending tests were perpendicular to the hot pressing axis and were polished to 1 μm diamond paste. The tensile edges were beveled.

RESULTS AND DISCUSSION

The densities of composites prepared in this works were strongly dependent on the volume fraction of particulate dispersions and hot-pressing temperature. For the Al_2O_3-SiC system, the near full densities were achieved at 1600°C up to 5.1 vol% SiC, 1700°C for 10.5 vol% and at 1800°C for 17.2 and 32.7 vol%. For the Al_2O_3-Si_3N_4 system, most materials were of near theoretical density at 1600°C. At higher temperatures, the densities appeared to decrease because of the formation of other phases such as β-SIALON. For the MgO-SiC composite, on the other hand, the densities of over 99% were observed for the hot-pressed materials at 1700 to 1900°C depending on the volume fraction of SiC particles.

X-ray diffraction analyses revealed that the Al_2O_3-SiC and MgO-SiC composites were composed of only α-Al_2O_3 and β-SiC and of only MgO and β-SiC, respectively and were free from other impurity phases. For the Al_2O_3 Si_3N_4 composite system, at the hot-pressing temperature of 1600°C and below, α-Al_2O_3 and β-$Si3N_4$ phases were observed with apparently 5 vol% of SIALON independent of initial Si_3N_4 content, however, above 1600°C the substantial increase of SIALON phase was observed.

As shown in Fig. 1, TEM observations indicated that most of the finer SiC particles, typically less than 0.2 μm, were predominantly dispersed within the Al_2O_3 grains and some larger particles at the grain boundaries Similar observations on the second phase dispersion were also observed for the Al_2O_3-Si_3N_4 and MgO-SiC composites. High resolution TEM observation revealed a notable fact that all the SiC and Si_3N_4 particles within the Al_2O_3 grains were very rounded but the SiC particles within the MgO grains were irregularly shaped. Figure 2 shows a typical high resolution TEM picture of the SiC particles within the Al_2O_3 grain. As is evident from Fig. 2, there was no obvious impurity phase at the interfaces. This observation was also true for the Al_2O_3-Si_3N_4 and MgO-SiC composites.

The hardness, strength and toughness tests of the Al_2O_3-SiC composites were made on the materials hot-pressed at 1600, 1700 and 1800°C for 5.1,

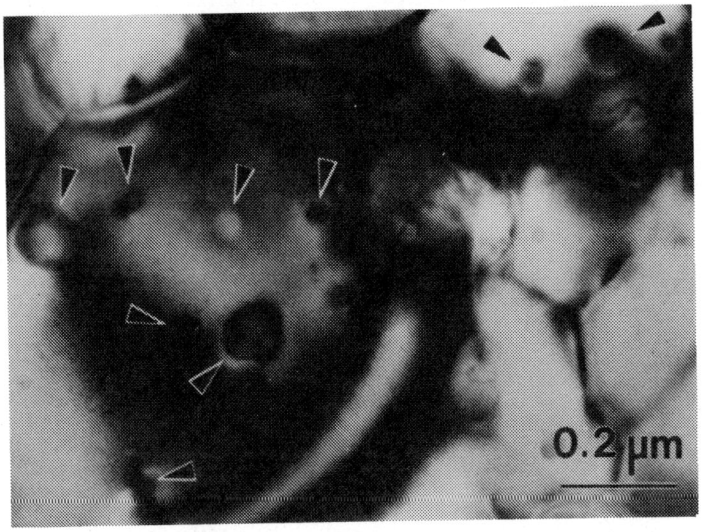

Fig. 1 Transmisssion electron micrograph of Al_2O_3-5vol% SiC nanocomposite. Arrows indicate the SiC particles within the Al_2O_3 grains and at the grain boundaries.

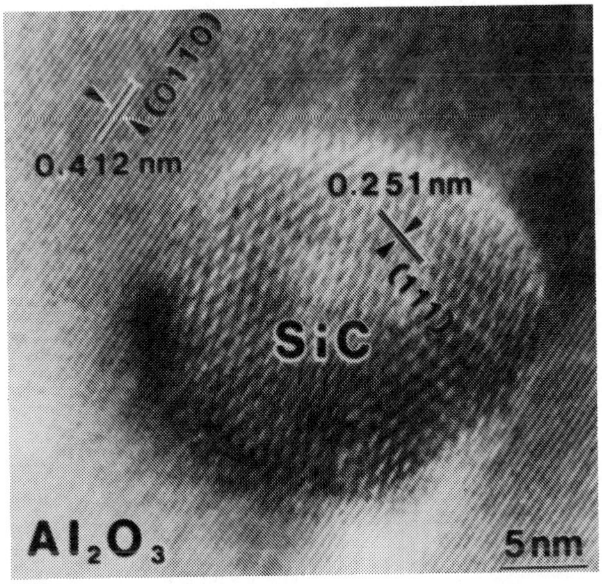

Fig. 2 High resolution transmission electron micrograph of the SiC particle dispersed within the Al_2O_3 grains for the Al_2O_3-5vol% SiC composite.

10.8 and 17.2 to 32.7 vol% SiC, respectively, which showed the near theoretical densities (over 99%) and were free from other impurity phases. For the Al_2O_3-Si_3N_4 and MgO-SiC composites, the mechanical tests were performed for the materials hot-pressed at 1600 and 1900°C, respectively.

The dependence of fracture strength and toughness of the Al_2O_3-SiC nanocomposite with volume fraction of SiC is shown in Fig. 3. The strength increases tremendously from 355 to 1017 MPa with only 5 vol% SiC, and remains slightly-decreased strength 870 MPa up to 32.7 vol% SiC. The fracture toughness shows a similar trend in which the K_{IC} values increases from 3.25 to 4.70 $MPam^{1/2}$ with 5 vol% SiC and keeps the constant value 4.35 $MPam^{1/2}$ up to 32.7 vol% SiC. For the Al_2O_3-Si_3N_4 nanocomposite, the fracture strength increased from 355 to 520 MPa with 16 vol% and was kept to be constant with increasing volume fraction of Si_3N_4. Similar trend was also obtained for the fracture toughness. The toughness was increased by the addition of Si_3N_4 from 3.3 to 4.3 $MPam^{1/2}$. The fracture strength of MgO was also improved from 330 to 540 MPa by incorporating 10 vol% SiC particles within the MgO grains, while the fracture toughness was increased from 1.7 to 2.8 $MPam^{1/2}$ with 10 vol% SiC.

The fracture surface observation by SEM revealed that the second phase dispersions within the matrix grains changed the fracture mode from the intergranular to transgranular fracture. It may be thought that the change of fracture mode is produced by the tensile hoop stress of over 1000 MPa around the inclusions within the matrix which is generated from the large thermal expansion mismatches. This high tensile stress will not produce the microcracks because the dispersed particle sizes is much smaller than

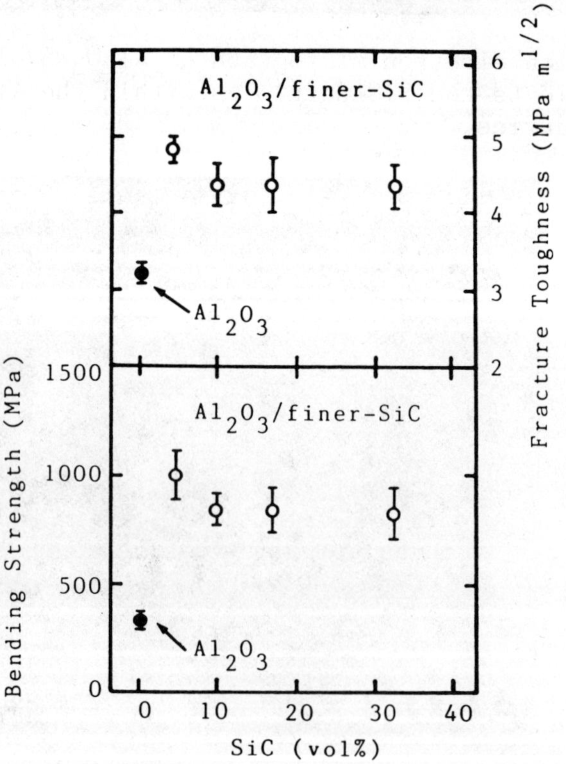

Fig. 3 Effects of SiC dispersion on the fracture toughness and strength for the Al_2O_3-SiC nanocomposites.

the critical particle sizes for microcracking.[5,17] Thus, these composites may be toughened primarily by the crack deflection, although the possibility of microcrack toughening can not be completely ruled out.

The critical flaw sizes estimated from standard fracture mechanics relationship[5] using the above-mentioned data are ∿22 and ∿9 μm for the monolithic Al_2O_3 and MgO, respectively, which are almost equal to those for the Al_2O_3-10vol% Si_3N_4 composite (∿19 μm) and for the MgO-10vol% SiC composite (∿10 μm). The strength increase at room temperature of these both composites, therefore, will be attributed to the improved toughness by the Si_3N_4 and SiC particles dispersed within the matrix grains. However, the notable strength improvement observed for the Al_2O_3-SiC composites (Fig. 3) can not be explained from only toughness increase. In the composite with 5 vol% SiC, the critical flaw size was decreased from ∿23 to ∿6 μm. This remarkable decrease in critical flaw size is probably due to the refinement of matrix grains and/or the compressive stress (over 1500 MPa) around the SiC particles which prohibits the grain boundary fracture of the matrix during cooling from the fabrication temperatures.[5,13]

Figures 4 to 6 indicate the temperature dependence of the fracture strength for the Al_2O_3-SiC, Al_2O_3-Si_3N_4 and MgO-SiC composites, respectively. The monolithic Al_2O_3 begins to exhibit significant reduction in strength above 1000°C whereas the Al_2O_3-5vol% SiC composite shows the remarkably improved strength up to 1000°C. As is evident in Fig. 5, much more notable improvement in high-temperature strength is observed for the Al_2O_3-Si_3N_4 composites. No strength reduction is observed up to 1300°C for the Al_2O_3-30vol% Si_3N_4 composite. Similar effect of the particle dispersion within the matrix grains on the high temperature strength was also observed for the MgO-SiC composites as indicated in Fig. 6. At the temperature range from 1000 to 1400°C, the MgO-30vol% SiC composite exhibits the considerably higher strength than that at room temperature.

It is well known that the grain boundary sliding and/or cavitation are responsible for the high-temperature strength degradation of the oxide

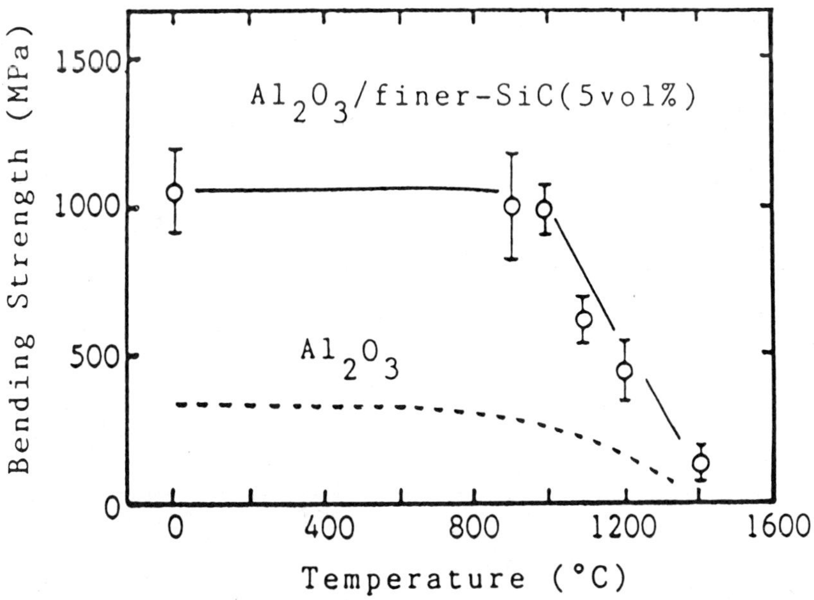

Fig. 4 Temperature dependence of the fracture strength for the monolithic Al_2O_3 and Al_2O_3-5vol% SiC nanocomposite.

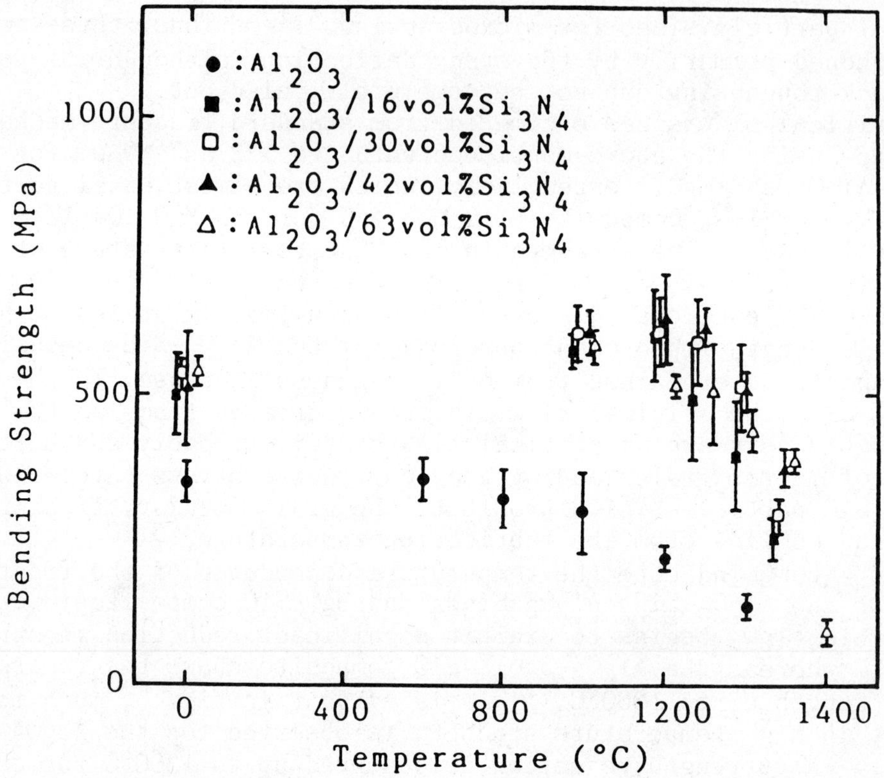

Fig. 5 Temperature dependence of the fracture strength for the Al_2O_3-based nanocomposites with various fractions of Si_3N_4 particles.

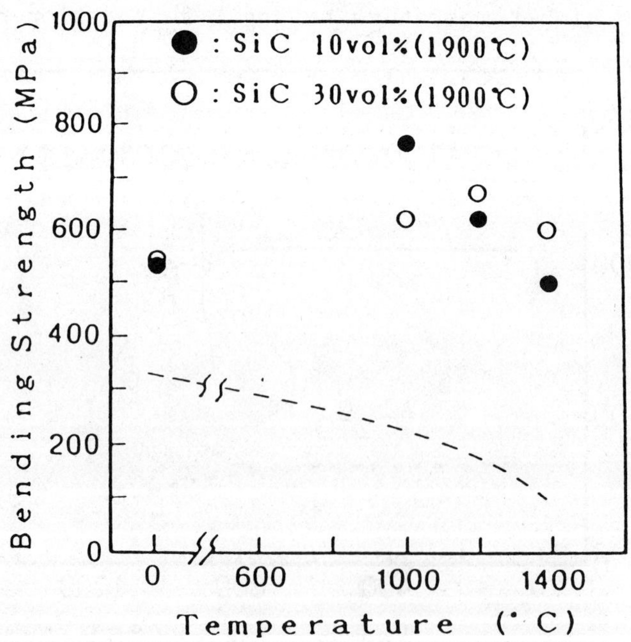

Fig. 6 Temperature dependence of the fracture strength for the monolithic MgO, MgO-10vol% SiC and MgO-30vol% SiC nanocomposites.

ceramics.[18] As above-mentioned, however, the dispersions within the matrix grains promote the transgranular fracture rather than intergranular. This phenomena were also observed at the temperatures above 1000°C. Thus, it may be concluded that the observed increase in strength at high temperatures is mainly due to the prohibitation of the grain boundary sliding or cavitation by the dispersions within the matrix grains. Another possible mechanism for the high-temperature strength increase is the creep prohibitation by the dislocation pinning with the fine dispersions within the matrix. This mechanism will be suggested by the fact that the hot hardness of the Al_2O_3-SiC and Al_2O_3-Si_3N_4 nanocomposites was significantly improved by the second phase dispersions, as is evident from Fig. 7.

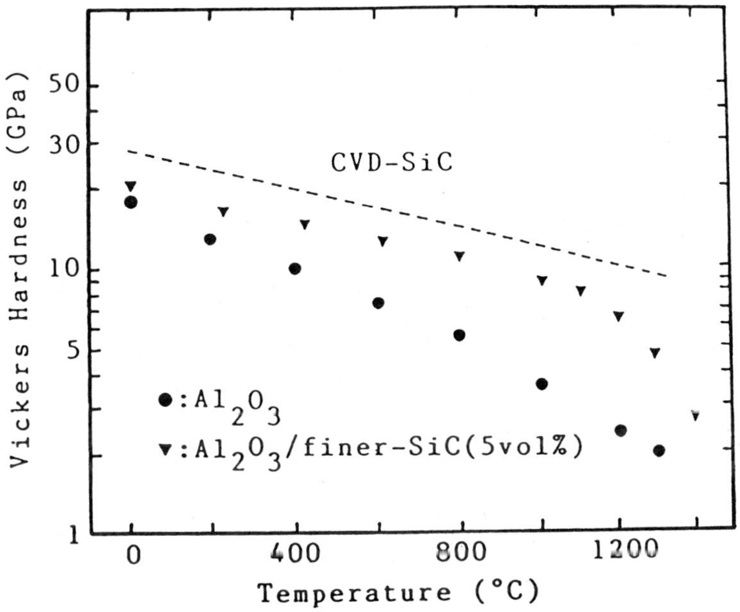

Fig. 7 Temperature dependence of the Vickers hardness for the monolithic Al_2O_3 and Al_2O_3-5vol% SiC nanocomposite.

Finally, we describe the thermal shock fracture resistance, ΔTc, of the Al_2O_3-based nanocomposites when quenched into water. The thermal shock fracture resistance, ΔTc, was approximately 200°C for the pure Al_2O_3, while it was about 420°C for the Al_2O_3-16vol% Si_3N_4 nanocomposite with the same specimen size as the pure Al_2O_3. This high ΔTc for the Al_2O_3-based nanocomposite is almost comparable with those for the SiC ceramics.[19]

CONCLUSIONS

The fracture toughness and strength of Al_2O_3 and MgO were significantly improved by incorporating the finer SiC and Si_3N_4 particles within the matrix grains. For the Al_2O_3-SiC nanocomposites, the maximum strength of over 1000 MPa was observed for the 5 vol% SiC. The second phase dispersions also increased drastically the high temperature strength of Al_2O_3 and MgO. For the Al_2O_3-30vol% Si_3N_4 composites, the strength degradation was not observed up to 1300°C. For the MgO-30vol% SiC composites, on the other hand, the strength degradation was not observed up to 1400°C.

ACKNOWLEDGMENTS

The present authors are grateful to Dr. M.V. Swain of CSIRO in Australia for greatly helpful discussions.

REFERENCES

1. R.P. Wahi and B. Ilschner, "Fracture Behaviour of Composites Based on Al_2O_3-TiC", J. Mater. Sci., 15[3] 875-885 (1980).
2. F.F. Lange, "Transformation Toughening", J. Mater. Sci., 17[1] 225-262 (1982).
3. N. Claussen, "Strength Strategies for ZrO_2-toughened Ceramics at High Temperatures", Mater. Sci. and Eng., 71[1] 23-28(1985).
4. F. F. Lange, "Strong, High-Temperature Ceramics", Ann. Rev. Mat. Sci., 4, 365-390(1974).
5. K. Niihara, A. Nakahira, T. Uchiyama and T. Hirai, "High-Temperature Mechanical Properties of Al_2O_3-SiC Composites", pp.103-116 in Fracture Mechanics of Ceramics, Vol.7, Edited by R.C. Bradt, A.G. Evans, D.P.H. Hasselman and F.F. Lange, Plenum, New York, 1986.
6. A. Nakahira, K. Niihara and T. Hirai, "Microstructure and Mechanical Properties of Al_2O_3-SiC Composites", Yogyo-Kyokai-Shi, 94[8] 767-772 (1987).
7. P.E. Becher and G.C. Wei, "Toughening Behavior in SiC-Whisker-Reinforced Alumina", J. Amer. Ceram. Soc., 67[12] C267-269(1984).
8. J. Homeney, W.L. Vaughn and M.K. Ferber, "Processing and Mechanical Properties of SiC-Whisker-Al_2O_3-Matrix Composites", Amer. Cera. Soc. Bull., 66[2] 333-338(1987).
9. T.N. Tiegs and P.E. Becher, "Sintered Al_2O_3-SiC-Whisker Composites", Amer. Ceram. Soc. Bull., 66[2] 339-342(1987).
10. T. Nishida, T. Shiono, H. Yamauchi and T. Nishikawa, "Fracture Energy of MgO-SiC Whisker Composite Ceramics", ZAIRYO, 36[1] 17-21(1987).
11. T. Uchiyama, S. Inoue and K. Niihara, "Multiple Toughening in Al_2O_3/SiC Whisker/ZrO_2 Composites", pp.193-200 in Silicon Carbide Ceramics, Edited by S. Somiya and K. Inomata, UCHIDA ROKAKUHO, Tokyo, 1988.
12. K. Niihara, "Ceramic Composites-Nanocomposites-", Electronic Ceramics, [9] 44-48(1988).
13. K. Niihara and A. Nakahira, "Development of Strong Al_2O_3-SiC Composites", Proc. of MRS Meeting on Advanced Materials, Tokyo, 1988, in press.
14. K. Niihara, A. Nakahira and T. Hirano, "Particulate Strengthened Ceramics, Nanocomposites", Proc. of US-Japan Seminar on the Processing of Advanced Ceramics, Seattle, 1988, in press.
15. K. Niihara, A. Nakahira and M.V. Swain, "Al_2O_3-Si_3N_4 Nanocomposite System", Int. J. High Technol. Ceram., to be contributed.
16. K. Niihara, A. Nakahira and T. Hirai, "The Effect of Stoichiometry on Mechanical Properties of Boron Carbide", J. Amer. Ceram. Soc., 67[1] C1-2(1984).
17. J. Selsing, "Internal Stresses in Ceramics", J. Amer. Ceram. Soc., 44[8] 419(1961).
18. A.G. Evans and W. Blumenthal, "High Temperature Failure in Ceramics", pp.423-448 in Fracture Mechanics of Ceramics, Vol. 6, Edited by R.C. Bradt, A.G. Evans, D.P.H. Hasselman and F.F. Lange, New York, 1983.
19. M. Midorikawa and T. Izeki, "Thermal Shock Testing of Dense SiC by Water Quenching", Yogyo-Kyokai-Shi, 94[1] 151-15(1986).

FRACTURE STRENGTH AND TOUGHNESS IN PARTICULATE REINFORCED GLASS COMPOSITES: EFFECT OF MICROSTRUCTURAL VARIATIONS

R. A. Haber, G. E. Hannon*, and J. B. Wachtman, Jr., Center for Ceramic Research, Rutgers University, Piscataway, NJ *Lanxide Corp., Newark, DE.

ABSTRACT

The mechanical behavior of dense particulate glass matrix composites was studied. A system containing a soda lime borosilicate frit and particulate aluminum oxide was chosen. Samples containing from 10 to 90 volume percent additions of alumina were fabricated by vacuum hot pressing. The mechanical properties of the composite materials were evaluated at ambient temperatures. The properties of interest were: fracture strength, fracture toughness, fracture energy, elastic modulus and crack extension force.

Experimental results show that an eight fold increase in strength over the parent glass is possible with 90 volume percent additions of alumina. The highest strength achieved for this composition was 529 MPa. Fracture toughnesses of the composite materials were determined using the strength method. Five indentation loads were used ranging from 5 to 100N. An exponential increase in fracture toughness values was observed as the alumina additions were increased up to the 70 volume percent additions. The composites containing 70 volume percent additions of alumina exhibited a fracture toughness of approximately 4.4 MPa*m^1/2. Fracture energy values were compared to those calculated from bowing and deflection models. Good correlation was observed when an additive fracture energy was considered resulting from bowing and deflection contributions.

INTRODUCTION

The majority of polycrystalline ceramics used in advanced applications contain a glassy phase. The amount of the glassy phase can vary from less than 0.4% to greater than 50%. The broad range of glass contents reflects the variety of applications for ceramic materials. The presence of a glassy phase into a ceramic (greater than 96% alumina) is most commonly an artifact of processing.[1] Moderate amounts of glass, 10 to 15 percent or greater, are found in bodies to which additives were used to aid in green body fabrication, reduce firing temperatures or contour selective properties.

Since alumina containing glass ceramics comprise a large segment of advanced ceramics and have been the subject of research for many years, it was chosen as the model system for this investigation. Extensive research has focused on modeling the fracture properties of alumina ceramics and alumina ceramics containing additions phases.[2-4] In general, most studies have simply modified the Griffith relation of fracture of brittle materials by fitting an empirical constant to observed data. Available models predict fracture properties for composites containing either high glass fractions (less than 40% alumina) or the high alumina fractions (greater than 80* alumina).[5-12] However, there does not exist a singular theory to explain the composite behavior over a large range of compositions, or microstructural variations. The current thinking has focused on modeling the fracture of composites in terms of understanding the toughness of a material rather than strength. THis is particularly advantageous in that fracture toughness models more easily be adapted to examine microstructural effects on the ease of crack propagation.

In nontransformation toughened systems, increases in fracture toughness in two phase ceramic systems have been associated with two distinct effects. The first is short range particle-crack interactions, which may include crack bowing and crack deflection.[13] The second microstructure-crack interactions takes place far behind the advancing crack front, a direct result of grain localized bridging or traction.[14] In the case of the later, the bridging model is used to explain materials which exhibit rising R-curve behavior during crack extension. In the case where materials do not exhibit R-curve behavior, the fracture models of crack front bowing and crack deflection models have shown good correlation in explaining the increase in fracture toughness of a brittle matrix as rigid inclusions are added.[15-16]

Deflection toughening occurs whenever interactions between the advancing crack front and an inclusion cause the crack to deviate from planarity by either tilt or twist processes. The consequence of such an interaction is the reduction in the stress-intensity factor at the crack tip. Deflection processes can occur in polycrystalline bodies where a grain boundary phase give rise to a residual strain at the interface of particles. The approach taken by Faber and Evans to evaluate the deflection process was to consider the local stress-intensity factor associated with the advancing crack.[16] Assuming that the crack front is subjected to mixed mode fracture processes, this relationship may be written as

$$E\,G = k_1^2\,(1-v^2) + k_2^2\,(1-v^2) - k_3^2\,(1-v) \qquad [1]$$

where v is Poisson's ratio, k_1, k_2, k_3 are local stress intensity factors and G is the average crack extension force across the crack front. Assuming as specific inclusion morphology this expression was further simplified. For inclusions having a spherical morphology, the net driving force for fracture due to the initial deflection component, G'_{sphere}, was found to be a function of the volume fraction of particulate and was expressed as

$$G'_{sphere} = (1 + 0.87\ V_f)\ G_c^m \qquad [2]$$

where G_c^m is the critical strain release energy rate for the matrix and V_f the volume fraction of inclusion. G' for inclusion morphologies having high aspect ratios was show to be greater than for spherical inclusions to twist and tilt deflection processes occurring. The implications of this model indicate that the relative increase in toughness of a matrix reinforced with low aspect ratio inclusions reaches a maximum at approximately 20 volume percent, after which further additions of a second phase have little affect.

Crack front bowing can be described as the interaction of a particle with a moving crack front, the intersection causing the crack front to be temporarily pinned.[15] It is only with the further addition of energy that the crack is able to move past the inclusions. Using an energy required to increase the length of the crack front, Lange described the crack extension force as the crack breaks away from its pinning site, G_c, as:

$$G_c = 2\ g_m\ (1 + (2\ c_m/3d)) \qquad [3]$$

where g_m is the fracture energy of the matrix, c_m is the matrix flaw size and d is the interparticle spacing. Further studies placed the restriction on the applicability of this model to pinning effectiveness, or inclusion's penetrability. Namely, that increases in a materials cracks resistance can only be obtained when second phase inclusions are not penetrated, or when fracture is intergranular. When the particles penetrability is exceeded the bowing mechanism becomes inapplicable.

While this relation does not account for residual stresses within a material arising from thermomechanical mismatches, it does, however, provide a relation between resistance to fracture and the spacing between inclusions. This can be further expanded to describe the effect of volume fraction and inclusion size by substituting the relations derived by Fullman[17] for mean free path for d in Equation[3] to obtain Equation[5].

$$d = (4r(1 - V_f))/(3 - V_f) \qquad [4]$$

$$G_c = 2g_m[1 + c_m(3 - V_f)/(6(1 - V_f))] \qquad [5]$$

where r is the radius of the spherical inclusion.

EXPERIMENTAL

In this study an alumina-glass system was selected where the volume fraction of alumina dispersed in the glass varied from 0 to 90% by volume in 10% increments. The alumina was a nominally pure alpha alumina with a narrow size distribution centered at 3 microns. The glass was chosen based on a match in coefficient of thermal expansion with alumina. This match was measured by diletometry. The composition was 14.1% SiO_2, 23.4% B_2O_3, 18.5% CaO, and 10.4% Na_2O, all values given in weight percent. The powders were comilled in water, filter pressed, dried and hot pressed. Hot pressing was necessary to minimize the reaction between alumina and glass at elevated temperatures. Hot pressing temperatures varied between 900 and 1100^oC, depending upon the volume fraction of alumina present . In all instances the observed porosity was not greater than 0 to 1%.

Samples for mechanical property evaluation bars were machined out of the hot pressed billets. All machined samples were polished to a 1 micron finish and subsequently annealed to relieve any residual stresses resulting from the machining process.

Young's modulus was measured using an ultrasonic technique. Utilizing the trough transmission method a 5 MHz pulse was passed through the polished, unsliced billet. The wave velocity is measured and using values for density and Poisson's ratio, the elastic modulus is determined.

Fracture strength was performed in four point bending. Fracture toughness determinations were made using the indented strength method developed by Anstis, et al.[18]. This procedure used bars of similar geometry to those used for transverse strength testing. To give the microstructural dependence of the fracture toughness, 5 indentation loads were used: 0.5, 1.0, 2.0 5.0 and 10.0Kg. A minimum of 5 bars per load per composition were tested. The stressing rate was approximately 100MPa/sec.

RESULTS

The entire range of hot pressed composites produced highly dense materials. The Young's moduli for the alumina/glass composites are given in Table 1. The hot pressed base glass was found to exhibit a modulus of 84 GPa. Upon alumina additions the composite modulus increased very slowly up to a value of 127 GPa at 50% additions. The Young's modulus increased exponentially upon further additions of alumina to a value of 258 GPa.

The fracture strength of the base glass was found to be 79 MPa. This value is consistent with other experimental values found in the literature.[19] Upon addition of alumina the fracture strength was found to increase rapidly, with the strength of the composite containing 20 percent alumina being twice that of the base glass, Figure 1. This trend was continued for all compositions up to the composite containing 90% alumina. The highest fracture strengths were found to be 524 and 529 MPa for the 80 and 90% alumina addition, respectively. This difference was not felt to be statistically significant.

The K_{IC} of the base glass was found to be 1.00 $MPam^{-1/2}$. The K_{IC} of the composite materials increased exponentially up to 70% alumina addition, 5.50 $MPam^{-1/2}$. At higher additions of alumina the K_{IC} leveled off to a value of 5.6 $MPam^{-1/2}$. Table 2 lists the K_{IC} for the various composites as a function of indentation load.

DISCUSSION OF RESULTS

In examining the applicability of the various toughness models discussed earlier, the indented strength data was examined. Lawn has shown that if the indentation strength parameter, $SP^{1/3}$, is plotted versus P, where S is the indented strength and P is the indentation load, a curve with a slope of zero constitutes a material whose strength is dependent upon flaw size.[18] This infers that the material does not exhibit R-curve behavior. Figure 2 illustrates this for the base glass and a 70% alumina composite. The remaining composites exhibited similar behavior. As a result of this it was concluded that the toughness of these materials was not controlled by the crack interface traction mechanism. This occlusion was not unexpected as studies by Cook et al.[20] have shown that for high alumina ceramics having a fine grains size, R-curve behavior was also not observed.

To examine whether deflection or bowing mechanisms were operative it was necessary to determine the fracture energy of the various materials. The fracture energy was calculated using the Griffith relation substituting values for fracture toughness, strength and Young's modulus. The fracture energy of the base glass was calculated to be 7930 $ergs/cm^2$.[15-21] Figure 3 shows the influence of incremental additions of alumina on fracture energy. The fracture energy of the glass as alumina on fracture increased to 70% addition where it reached a plateau for all subsequent additions. This plateau was believed to occur due to changes in the fracture behavior of the composites. Figures 4a and 4b illustrate representative fracture surfaces of the samples prior and after this plateau, namely a 60% alumina and 90% alumina reinforced composite. It was observed that at this plateau the occurrence of transgranular fracture began and increased with incremental additions of particulate.

If the crack deflection model is applied to the measured fracture energy data there was good correlation up to 50% particulate additions, ;e Figure 3. using

Equation 2, the crack deflection analysis for spherical inclusions, which most closely resembled the alumina particulate used, the maximum increase in toughness which could be expected was 1.87 times that of the matrix. Up to 50% alumina addition there was good agreement, however, after 50% alumina addition the fracture energy of the composites increased up by a factor of 5.4.

If the bowing model is applied to the measured data there is little correlation. However, it was observed that above 30% volume fraction the relative fracture energy of the composites stated that effects such as these could be expected and that the increase in fracture energy of a composite should be an additive relationship between deflection and bowing effects.[22] It can be clearly seen that if the contribution to fracture energy due to bowing and deflection are added, there is good correlation with measured data.

CONCLUSION

Even though it appears that particulate/glass composites offer simple model systems to study fracture behavior, their behavior is complex. During the study three mechanisms were examined to account for increases in composite fracture energy. These were: deflection, bowing and traction.

Due to the fine grain size of the particulate inclusions, evidence of traction was not observed. Direct correlation with bowing and deflection models showed little correlation with measured data. However, if the additive relation by Miyata, which considers both the contributions of deflection and bowing over the entire range of volume fraction, is considered good agreement exists with measured data. Deviation from this occurs at thigh volume fractions where fracture becomes predominantly transgranular.

REFERENCES

1. W.H. Gitzen, Alumina as a Ceramic Raw Material, Special Publication No. 4, The American Ceramic Society (1970).
2. J.R. Hellman, J. Matsko, S.W. Freiman and T.L. Baker, "Microstructure-Mechanical Property Relationships in 94% Alumina Ceramics," Tailoring Multiphase and Composite Ceramics, Plenum Press, New York, p367-69 (1986).
3. G.E. Hannon, "Alumina/Glass Composites: The Effect of Microstructure on Mechanical Properties", Ph.D. Thesis, Rutgers University, Piscataway, NJ 1988.
4. D.B. Binns and P. Popper, "Mechanical Properties of Some Commercial Alumina Ceramics," Proc. Brit. Cer. Soc., 6, 71-82 (1966).
5. A.G. Evans, "The Strength of Brittle Materials Containing Second Phase Dispersions," Philos.Mag., 26, 1327-44 (1972).
6. D.J. Green, P.S. Nicholson and J.D. Embury, "Fracture of a Brittle Composite," J. Mater. Sci., 13[5], 1657-61 (1979).

7. D. J. Green, "Fracture Toughness Predictions for Crack Bowing in Brittle Particulate Composites," J. Am. Cer. Soc., 55 [1] C-4 - C-5 (1983).
8. F. F. Lange, "Fracture Energy and Strength Behavior of a Sodium Borosilicate Glass-Composite," J. Am. Cer. Soc., 54 [12] 614-20 (1971).
9. N.M.H. Jinno, "Theoretical Approach to the Fracture of Two-Phase Glass-Crystal Composites," J.Mat. Sci., 7, 973-82 (1972).
10. M.P. Borom, A.M. Turkaol and R.H. Doremus," Strength and Microstructure of Lithium Disilicate Glass-Ceramics," J. Am. Cer. So.c, 58 [9-10] 385-91 (1975).
11. J.C. Swearengen, E.K. Beauchamp and R.J. Eagen, "Fracture Toughness of Reinforced Glasses," pp. 973-87 in Fracture Mechanics of Ceramics, Vol. 4 Edited by R. C. Brandt, D.P.H. Hasselman and F.F. Lange, Plenum Press, New York, 1978.
12. R.W. Davidge and T.J. Green, "Strength of Two-Phase Ceramic Glass Materials," J. Mat. Sci., 3 [6] 629 (1968).
13. K.T. Faber, "Toughening Mechanisms for Ceramics in Automotive Applications," Proceedings of the 12th Automotive Materials Conference, The American Ceramic Society, 409-39 (1984).
14. R. Steinbrech, R. Knehans, and W. Schaarwachter, "Increase of Crack Resistance During Slow Crack Growth in Alumina Bend Specimens," J. Mat. Sci. 18 265-70 (1985).
15. F. F. Lange, "The Interaction of a Crack Front with a Second Phase Dispersion," Philos. Mag., 22, 983-93 (1970).
16. K. T. Faber and A. G. Evans, "Crack Deflection Processes: I, Theory," Acta. Metal., 31 [4] 565 (1983).
17. R.L. Fullman, "Measurement of Particle Size in Opaque Bodies," Trans. AIME, 197 [3] 447-52 (1953).
18. G.R. Anstis, P. Chantikul, B.R. Lawn and D.B.Marshall, "A Critical Evaluation of Indentations Techniques for Measuring Fracture Toughness: II, Strength Method," J. Am. Cer. Soc., 64 (9) 539-543 (1981).
19. D.P.H. Hasselman and R.M. Fulrath, "Proposed Theory of Dispersion Strengthened Glass Matrices", J. Am. Cer. Soc., 49 [2] 68-72 (1966).
20. R.F. Cook, B.R. Lawn and C.J. Fairbanks, "Microstructure-Strength Properties in Ceramics: I, Effect of Crack Size on Toughness," J. Am. Cer. Soc., 68 (11) 604-615 (1985).
21. S.M. Wiederhorn, "Fracture Surface Energy of Soda-Lime Glass," from Materials Science Research, Volume 3, edited by Kriegel and Palmour, Plenum Press New York (1966).
22. N. Miyata, S. Ichikawa, H. Monji and H. Jinno, "Fracture Behavior of Brittle Matrix, Particulate Composites with Thermal Expansion Mismatch," from Fracture Mechanics of Ceramics, Vol. 7, Edited by R.C. Brandt, A.G.Evans, D.P.H. Hasselman and F.F. Lange, Plenum Press, New York (1986).

Table 1. Elastic Modulus of Alumina/Glass Composites Measured by Through-Transmission Ultrasonic Technique.

Volume Percent of Alumina	Young's Modulus (GPa)
0	84
10	88
20	94
30	114
40	122
50	127
60	124
70	200
80	222
90	258

Table 2. Fracture Toughness ($MPam^{1/2}$) of Alumina/Glass Composites as a Function of Indentation Load.

	Indentation Load (N)				
Volume Percent	5	10	20	50	100
0	1.19	1.11	1.22	1.18	1.04
10	1.42	1.24	1.34	1.42	1.48
20	1.48	1.46	1.47	1.48	1.46
30	1.76	1.83	1.67	2.09	1.88
40	2.36	2.35	2.41	1.76	1.89
50	1.97	2.85	2.30	2.56	2.44
60	2.92	3.51	3.57	2.90	2.67
70	4.15	4.75	4.57	3.50	2.97
80	3.03	4.14	4.18	4.69	4.14
90	4.44	5.33	4.72	4.29	3.71

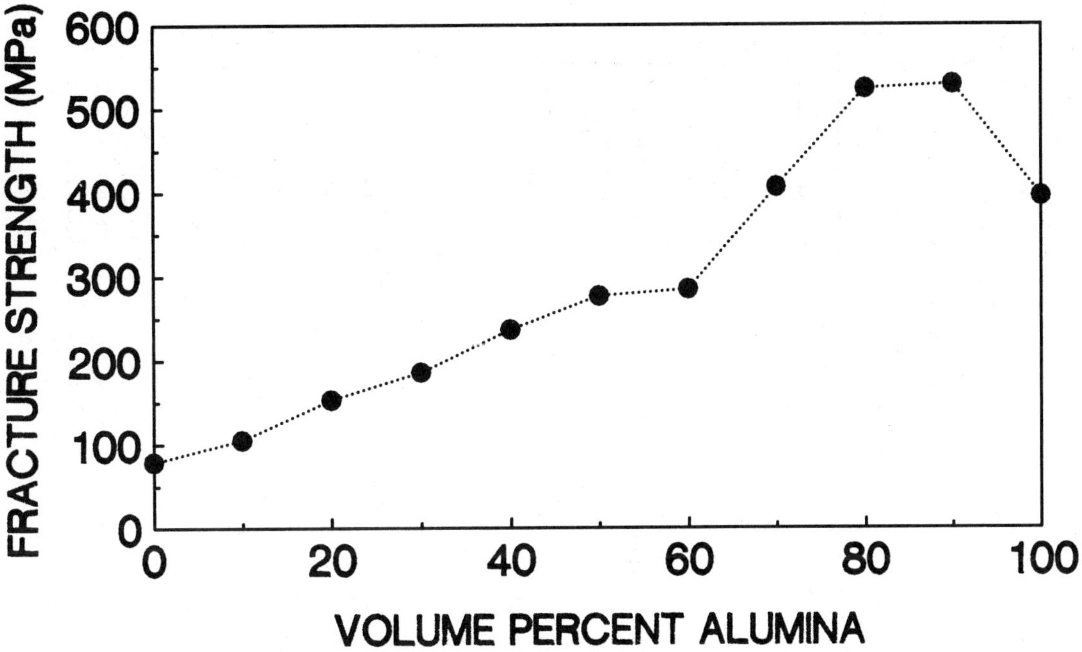

Figure 1: Fracture Strength of Alumina/Glass Composites as a Function of Alumina Content.

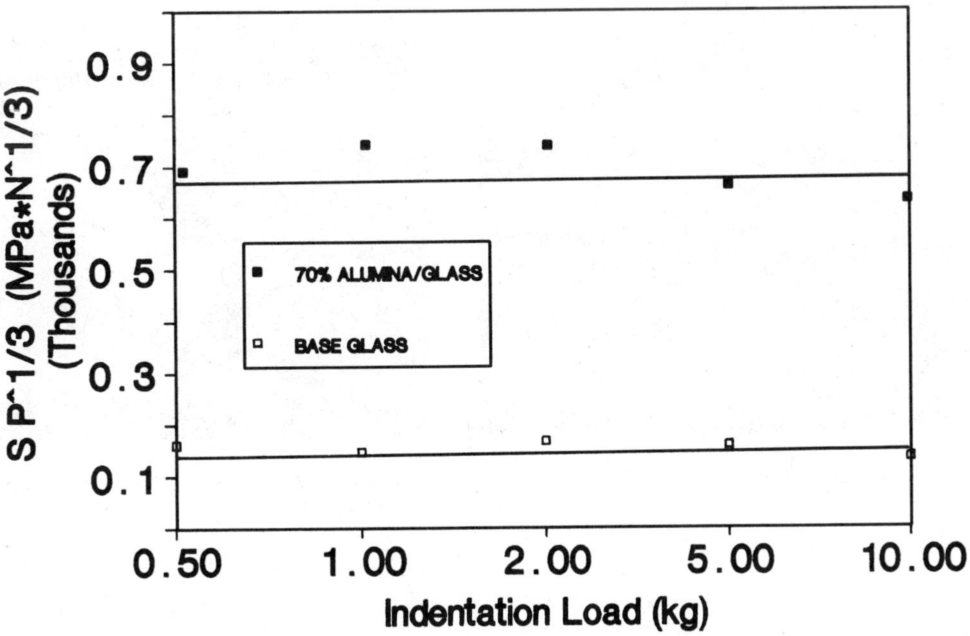

Figure 2: Indentation-Strength Parameter for Base Glass and 70% Alumina/Glass Composite.

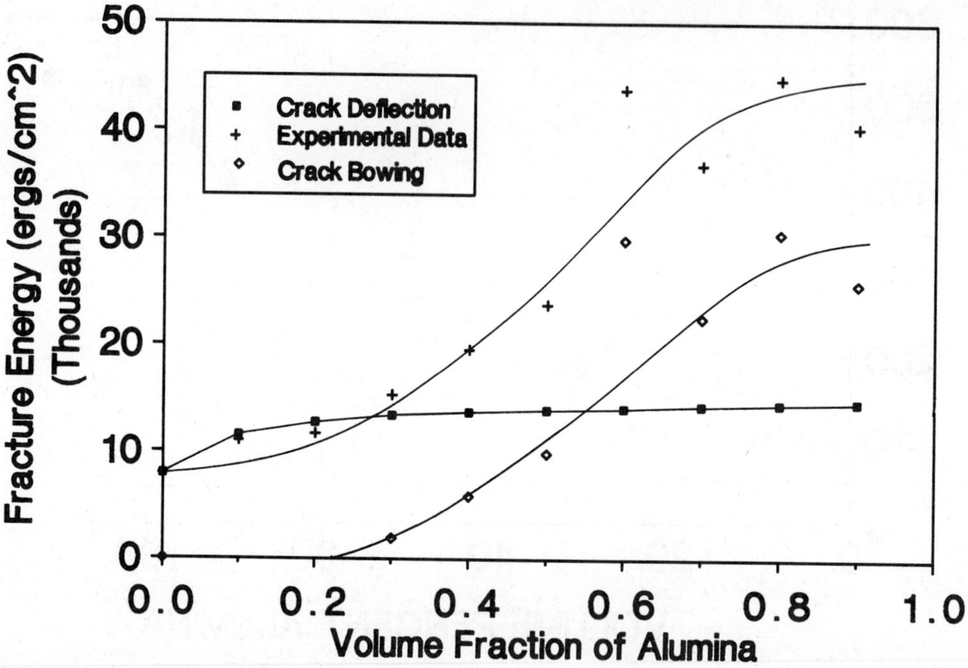

Figure 3: Observed Fracture Energy of Alumina/Glass Composites Compared to Contribution by Crack Deflection and Crack Front Bowing.

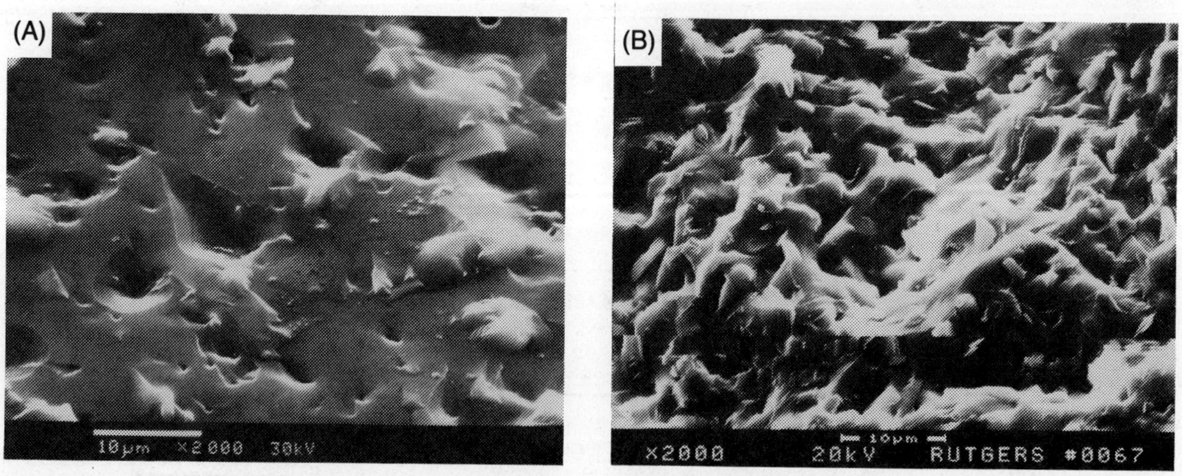

Figure 4: Fracture Surfaces of (a) 60% and (b) 90% Alumina/Glass Composites.

TAILORING OF PROPERTIES OF SiC WHISKER-OXIDE MATRIX COMPOSITES*

Terry N. Tiegs
Oak Ridge National Laboratory

ABSTRACT

Alumina and mullite reinforced with SiC whiskers have shown improved toughness and strength. The improvements are a result of crack-whisker interactions, such as crack bridging and deflection. The effectiveness of the interactions has been found to be functions of both the whisker physical characteristics and the surface chemistry. The important physical characteristics discussed are the diameter of the whiskers as well as the extraneous particulates associated with them. The whisker surface chemistry influences the bonding at the whisker-matrix interface. High toughness is observed when the bonding is limited and debonding can occur along the interface during crack propagation.

INTRODUCTION

Composites offer the possibility to improve the reliability of ceramics in high stress environments by increasing their fracture toughness. The higher reliability is particularly important for the demands required in advanced heat engine applications. The addition of SiC whiskers has been shown to increase the fracture toughness of alumina[1,2], mullite[3], silicon nitride[4] and zirconia[5]. Improvements in the strength of the composites has also been observed. The whiskers are acicular single crystals with

* Research sponsored by the U.S. Department of Energy, Assistant Secretary for Conservation and Renewable Energy, Office of Transportation Systems, as part of the Ceramic Technology for Advanced Heat Engines Project of the Advanced Materials Development Program, under contract DE-AC05-840R21400 with Martin Marietta Energy Systems, Inc.

nominal dimensions of 0.5 μm in diameter and 10-80 μm in length. They are currently fabricated commercially by numerous producers around the world.

Research into the mechanisms responsible for the observed toughening behavior in the composite materials shows that crack-whisker interaction resulting in crack bridging and deflection are the major toughening processes[6,7]. The result of these mechanisms is illustrated in Fig.1 which shows the fracture surface of a high toughness alumina-20 vol% SiC whisker composite. The whiskers are readily visible where debonding has taken place along the matrix-whisker interface. Whisker pullout and crack pinning are believed to also contribute to the overall composite toughness. Analytical models and available experimental evidence[6] indicates that the toughening effect will increase with increasing (1) whisker strength, (2) volume fraction of whiskers, (3) ratio of Young's modulus of the composite to that of the whisker, (4) whisker diameter and (5) ratio of the interfacial fracture energies of the matrix and the matrix-whisker interface. Unfortunately at the present time, the actual strength of the whiskers used in the current materials cannot be determined, so its effect on composite properties can only be estimated analytically. The effect of the whisker volume fraction on toughness and strength and how it can be varied has been reported numerous times before[1-3]. Results on the effect of the matrix Young's modulus have been reported previously[6] and since the alumina and mullite matrix systems are being examined, it is essentially fixed anyway. Consequently, the present discussion will address the effects of whisker diameter and interface properties and how they may be tailored.

Tailoring of whisker ceramic composites is also dependent on proper processing to maximize composite properties. Like monolithic ceramics, the whisker-reinforced composites are brittle materials showing low strain-to-failure and are sensitive to flaws in the materials. Thus the number and size of the flaws introduced either with the starting materials or during fabrication need to be minimized. The significant areas of fabrication determined to date that are important for reducing flaws will be discussed in the present paper.

EXPERIMENTAL PROCEDURE

Composite specimens were fabricated by blending alumina† or mullite†† powders with the SiC whiskers†††,* in hexane and dispersing the mixture with a high-shear homogenizer** . The mixtures were dried and hot-pressed in a graphite die at 1750-1850°C and 41 MPa (6 ksi) for 30 minutes. All of the composites fabricated contained 20 vol % SiC whiskers. Composites containing SiC particulates*** were fabricated in an identical manner.

Fracture toughness measurements were obtained using the multiple-indent flexure strength (MIFS) technique[8]. The technique employed flexure bars having nominal dimensions of 2.8 x 3.3 x >26 mm, with polished tensile surfaces that contained three Vicker's diamond indents (5 Kg load) to produce controlled flaws in the region of maximum tensile stress. The fracture strengths were determined by four point flexure on bars having the same dimensions as the MIFS specimens, but with 220 grit diamond ground tensile surfaces. Both the MIFS and flexural test specimens were oriented so that the tensile surface was perpendicular to the original uniaxial pressing direction.

Some whiskers were sedimented to remove extraneous particulates prior to incorporation into the matrix powders. The whiskers were dispersed at 2 vol % in deionized water with the high shear mixer and allowed to settle for 8-16 hours. Extraneous particulate was removed by simple decanting of the liquid above the whiskers. Typically, 10-25 % of the whisker material was removed in this manner.

RESULTS

As stated previously, theoretical solutions indicate that one of the important parameters affecting the fracture toughness is the whisker diameter. When the fracture toughness of alumina and mullite composites are examined as a function of whisker diameter, the relationships summarized in Fig. 2 are obtained. As shown, the improvement in the toughness of the alumina matrix composites is quite dramatic whereas the increase in the mullite matrix materials is

---

† Baikowski Corporation, Charlotte, North Carolina; Grade CR-1O
†† Baikowski Corporation, Charlotte, North Carolina; Grade CR-193
††† Advanced Composite Materials Corp., Greer, SC; Grade SC-9, Received 9/85
* Tateho Chemical Co., Japan; Grade SCW-1S, Received 1/86
** Brinkmann Instrument Co., Westbury, New York; Model PT 45/80.
*** H.C. Starck, Germany, Grade A-1

substantially lower. The contrast between the two matrices is mainly attributable to the large difference in the elastic moduli. This difference results in a much smaller bridging zone in the mullite composites as compared to the alumina materials (3-6 μm for alumina versus 1-2 μm for mullite at comparable whisker diameters[6,9]) and thus reduces the effectiveness of large diameter whiskers. Consequently, significant improvements in the fracture toughness of the mullite matrix system by only altering the whisker size will be difficult to achieve without substantial increases in whisker diameters. Other means, such as higher volume loadings and control of interface properties will have more effect at improving properties of the mullite composites.

Extrapolation of the curves in Fig. 2 for alumina composites to increase the fracture toughness by incorporating larger diameter whiskers is limited. This is because as the whisker size increases the alumina matrix is subject to microcracking due to the thermal expansion mismatch with the SiC. Other research[10] has shown that the microcracking is dependent on the particle size of the second phase inclusions and when microcracking occurs loss of strength is observed. To simulate the effect of various size whiskers, particulate SiC was sedimented into narrow size distributions and then incorporated into alumina-20 vol % SiC composites. The results on the flexural strength variation with particle size are summarized in Fig. 3. They indicate microcracking is not evident at particle sizes up to approximately 2.5 μm. Because of the limited data, it is not clear if microcracking is occurring in the composites with particles in the range of 2.5-4 μm. However, results on an alumina-20 vol % SiC composite made with whiskers having mean diameters of about 5 μm[§] exhibited very poor flexural strength as shown on Fig. 3 indicating microcracking was present. Although no microstructural evidence was obtained to confirm microcracking at the higher particulate diameters, it is evident that whisker diameters up to 2 μm should not result in strength degradation due to matrix microcracking in alumina composites. Since fracture toughnesses of >10 MPa$\sqrt{m}$ are predicted with whiskers larger than 1 μm, efforts to increase the diameters is currently of great interest. To realize that improved toughness, no degradation of other whisker properties, such as strength can occur. Because of negligible difference in thermal expansions between mullite and SiC, no microcracking problems are expected in those composites and much larger whiskers may be used. However, as previously shown increasing

§ Supplied by Los Alamos National Laboratory, designated 4,5 A VLS whiskers the whisker diameter in mullite composites does not result in dramatic toughness improvements.

For the crack bridging and deflection toughening mechanisms to operate in these composites, debonding along the whisker-matrix interface must occur during crack propagation. Other research[7,11], has shown that the whisker surface chemistry is important in determining the degree of bonding that occurs between the whisker and the matrix. In particular, it was determined that the presence of excess surface carbon on the whiskers results in a weak interfacial bond and high toughness alumina-SiC composites. The effect of excess surface carbon on the extent of crack-whisker interaction is illustrated in the contrast between Fig. 1 and Fig. 4. While extensive interaction is observed in Fig. 1, the relatively smooth fracture surfaces in Fig. 4 indicates little crack-whisker interaction and no inhibition to crack propagation. The fracture toughness measurements (8.3 MPa$\sqrt{m}$ versus 3.9 MPa$\sqrt{m}$ in Figs. 1 and 4, respectively) reflect these differences in behavior. Thus the surface chemistry of the whisker is a variable that must be known for proper control of the whisker-matrix interface to achieve debonding and obtain high toughness composites.

Recent results have shown that the whisker surfaces can also be modified by heat-treatments to change the surface chemistry. The heat-treatments under inert gas at temperatures >1000°C result in an increase of excess surface carbon[7]. X-ray photoelectron spectroscopy results shown in Fig. 5 illustrate the changes to the carbon on the whisker surfaces before and after heat-treatment. The low energy peaks at 282-283 eV are attributable to Si-C bonding, while the peaks at 284-285 eV are indicative of C-C bonds and excess surface carbon. The increase in surface carbon with the heat-treatment increased the fracture toughness of alumina-20 vol % SiC whisker composites from 3.9±0.2 MPa$\sqrt{m}$ to as high as 5.8±0.4 MPa$\sqrt{m}$. Examination of fracture surfaces from composites with surface-treated whiskers also showed increased crack-whisker interaction (Fig. 6).

While the toughness can be tailored by the appropriate selection of materials and processing techniques, the strength of the composites, like monolithic ceramics, is governed by the flaw sizes in the final materials. As the flaw size is decreased, improvements in the strength are observed. Extraneous material, such as SiC particulate and rice-hulls, have been reported to be associated with the currently available whiskers and result in strength degradation in alumina matrix composites[12].

A series of alumina-20 vol % SiC whisker composites were fabricated with whiskers that were treated in a variety of ways to reduce the flaw size present. The effect of the treatments on the flexural strength is given in Table 1. As shown, removal of material by simple sedimentation is an effective method to reduce the flaw size and improve the composite strength. Milling of the whiskers also reduced the average flaw size observed, but was not as effective as the sedimentation. Similar strength improvements by analogous techniques have been observed in silicon nitride-SiC whisker composites[13].

Improvement in strength is always important, however, we are also interested in whether the removal of the larger flaws introduced into the composites from the whiskers would improve the reliability of the ceramics. Examination of the Weibull modulus obtained on the composites with as-received and cleaned whiskers is shown in Fig. 7.

Table 1. Summary of results on effect of whisker treatments on the flexural strength of alumina-20 vol % SiC whisker composites.

Whisker Type	Flexural Strength (MPa)	Calculated Flaw Size[3] (µm)	Whisker Treatment
A[1]	541±63	140	As-received whiskers
A[1]	625±71	104	Sedimented once.
A[1]	741±77	59	Sedimented five times.
A[1]	660±97	83	Milled four hours.
B[2]	354±22	79	As-received whiskers.
B[2]	574±74	51	Sedimented five times.

[1]Advanced Composite Materials Corp., SC-9, Received 9/85
[2]Tateho Chemical Co., SCW-1S, Received 1/86
[3]Flaw size calculated by following: $c=(0.8 \times K_{Ic} / \sigma_f)^2$, where c is the flaw size, K_{Ic} is the fracture toughness and σ_f is the flexural strength.

These results show that the Weibull modulus changes very little with the various cleaning methods used. This implies that while the average flaw size was reduced, the distribution of flaw sizes remained nearly constant. In other words, while the vast majority of large flaws were removed some were still retained into the final composites. Improved techniques for eliminating the large flaws (>100 μm) associated with the whiskers need to be established and incorporated into composite fabrication if increases in the Weibull modulus of whisker-reinforced composites is to be achieved.

CONCLUSIONS

The fracture toughness and strength of ceramics can be improved by whisker-reinforcement. Alumina and mullite reinforced with SiC whiskers are currently under intense study because of improved fracture toughness and flexural strength as compared to the corresponding monolithic materials. While the results presented here are directed towards alumina and mullite, they have significance to other materials, such as silicon nitride, where oxide phases are present at the grain boundaries. This is because parameters, such as control of interface properties and minimizing flaw sizes, are pertinent to all whisker-reinforced ceramic composites.

The fracture toughness and strength improvement of SiC whisker-reinforced alumina and mullite is a result of crack-whisker interactions, such as crack bridging and deflection. The effectiveness of the interactions has been found to be functions of both the whisker physical characteristics and the whisker surface chemistry.

The fracture toughness was shown to increase with increasing whisker diameter. The effect was very dramatic for the alumina matrix composites and is limited only by microcracking of the alumina at large whisker diameters. Initial studies suggest that microcracking of alumina does not occur for whisker diameters up to 2 μm. The whisker surface chemistry influences the bonding at the whisker-matrix interface. High toughness is observed when the bonding is limited and debonding can occur along the interface during crack propagation. For this to occur, the presence of excess surface carbon was determined to be of particular importance. By changing the whisker surface chemistry, the bonding and thus the composite properties can be tailored. Heat-treatments were found to increase the amount of surface carbon and toughness improvements in alumina matrix composites were observed.

Removal of extraneous material from the SiC whiskers was found to be an effective method to reduce the flaw size and improve the flexural strength of alumina composites. However, these improvements in the strength did not result in significant increases of the Weibull modulus for the composites.

ACKNOWLEDGEMENTS

The author wish to acknowledge J. W. Geer for composite fabrication, H. Keating for preparing test specimens, and F. W. Christie for preparing the manuscript.

REFERENCES

1) P.F.Becher and G.C.Wei, "Toughening Behavior in SiC-Whisker-Reinforced Alumina," *J.Am.Ceram. Soc.*, 67[12]C-267-C-269 (1984)

2) G.C.Wei and P.F.Becher, "Development of SiC-Whisker-Reinforced Ceramics," *Am.Ceram. Soc. Bull.*,64[2]298-304 (1985)

3) T.N.Tiegs and P.F.Becher, "Whisker Reinforced Ceramic Composites"; pp.639-47 in *Tailoring Multiphase and Composite Ceramics*, Materials Science Research Series, Plenum Press, New York(1986)

4) P.D. Shalek, J.J. Petrovic, G.F. Hurley and F.D. Gac, "Hot-Pressed SiC Whisker/Si_3N_4 Matrix Composites," *Am. Ceram. Soc. Bull.*,65[2]351-56(1986)

5) N. Claussen, K.L. Weisskopf and M. Ruhle,"Tetragonal Zirconia Polycrystals Reinforced with SiC Whiskers," *J. Am. Ceram. Soc.*,69[3]288-92(1986)

6) P.F. Becher, C.H. Hsueh, P. Angelini and T.N. Tiegs,"Toughening Behavior in Whisker Reinforced Ceramic Matrix Composites," submitted to *J. Am. Ceram. Soc.*, 71[12]1050-61(1988)

7) P.F.Becher and T.N.Tiegs,"Toughening Behavior Involving Multiple Mechanisms: Whisker Reinforcement and Zirconia Toughening", *J.Am.Ceram.*,70[9]651-54(1987)

8) R. F. Cook and B. R. Lawn, "A Modified Indentation Toughness Technique," *Comm.Am.Ceram.Soc.*,11, C-200-1(1983)

9) P. Angelini, W. Mader and P.F. Becher, "Strain and Fracture in Whisker-Reinforced Ceramics," pp. 241-57 in *Advanced Structural Ceramics, MRS Symp. Proc., Vol. 78*, P.F. Becher, M.V. Swain and S. Somiya (Eds.) Materials Research Society, Pittsburgh, PA (1986)

10) F.F. Lange, "Criteria for Crack Extension and Arrest in Residual, Localized Stress Fields Associated with Second Phase Particles" pp. 599-609 in *Fracture Mechanics of Ceramics, Vol.2* , Plenum Press, New York (1974)

11) T.N. Tiegs, P.F. Becher and L.A. Harris, "Interfaces in Alumina-SiC Whisker Composites", p. 911-18 in *Ceramic Microstructures '86: Role of Interfaces* , Materials Science Research Series, Vol. 21, Plenum Press, NY (1987)

12) T.N.Tiegs and P.F.Becher, "Development of Alumina- and Mullite-SiC Whisker Composites: High temperature Properties,"pp.279-283 in *Proc.Auto.Tech.Contractors' Coord.Meeting* , 24rd, 1986, Vol.P-197, Society of Auto.Engineers, Inc., Warrendale, PA (1987)

13) S.T. Buljan, J.G. Baldoni, M.L. Huckabee,"Si_3N_4-SiC Composites", *Am. Ceram. Bull.*,66[2]347-52 (1987)

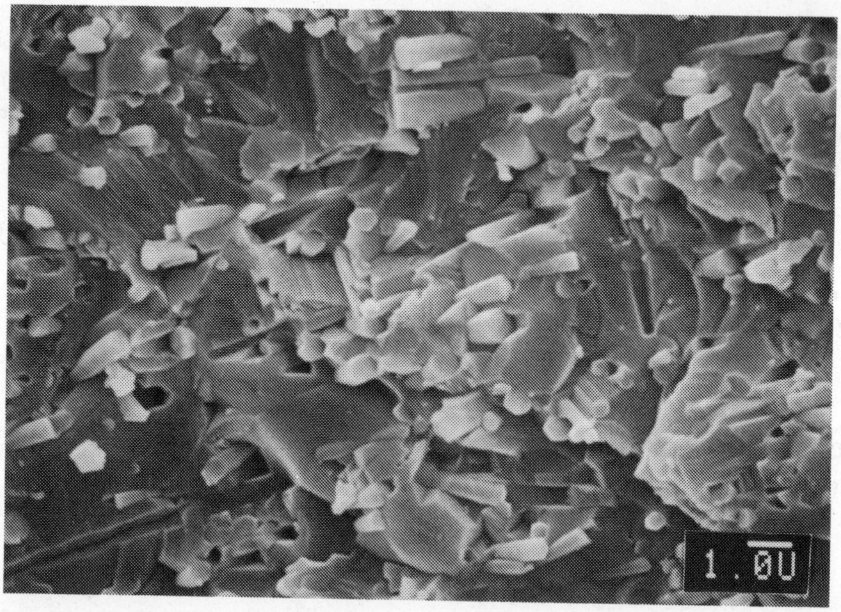

Fig. 1- Fracture surface of alumina-SiC whisker composite showing crack-whisker interaction. Composite made with Advanced Composites Materials Corporation SiC whiskers.

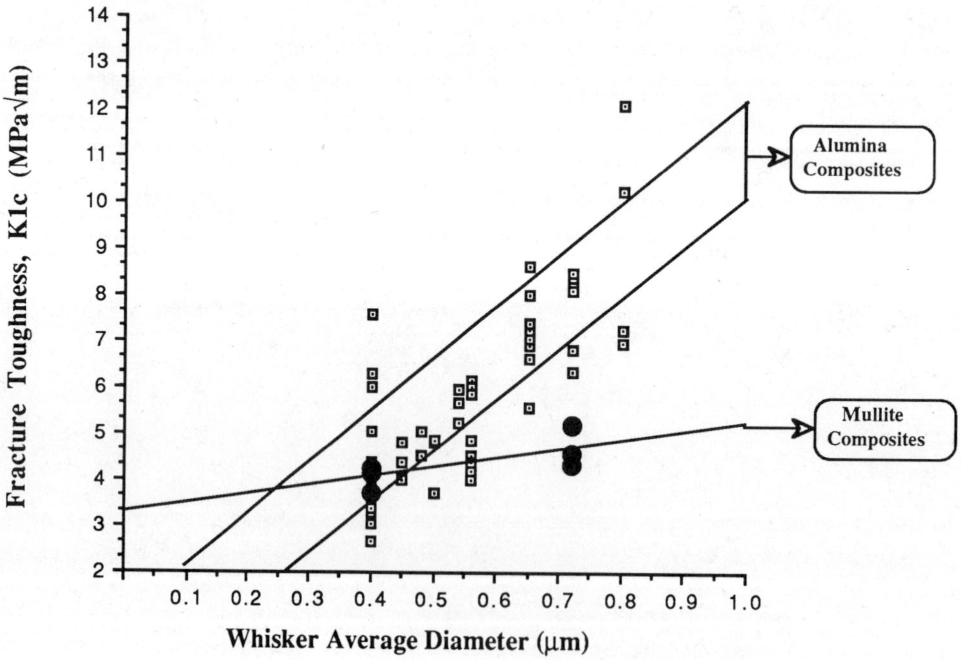

Fig. 2- Fracture toughness increases with whisker diameter in alumina and mullite matrix composites with 20 vol.% SiC whiskers. Composites made with Advanced Composites Materials Corporation SiC whiskers.

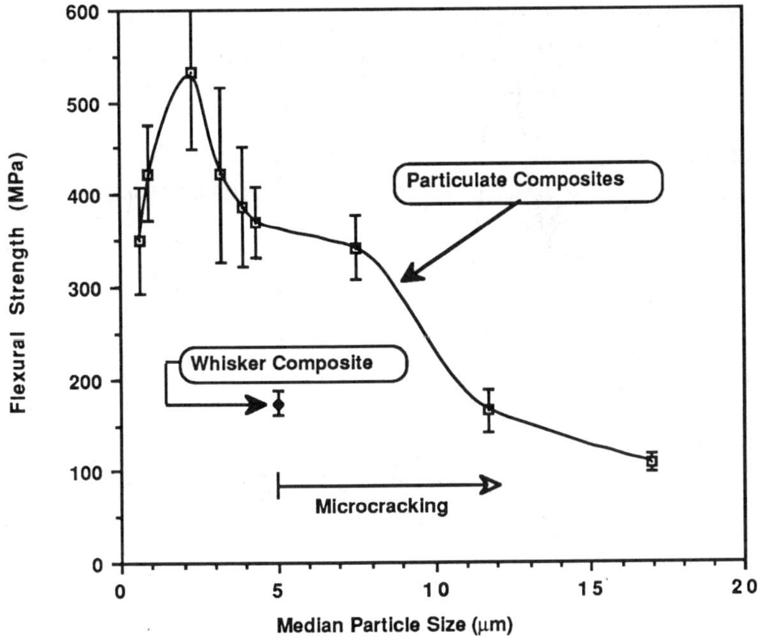

Fig. 3- Flexural strength of alumina-20 vol.% SiC Particulate composites with various mean particle sizes. The alumina-20 vol.% SiC whisker composite fabricated with whiskers supplied by Los Alamos National Laboratory.

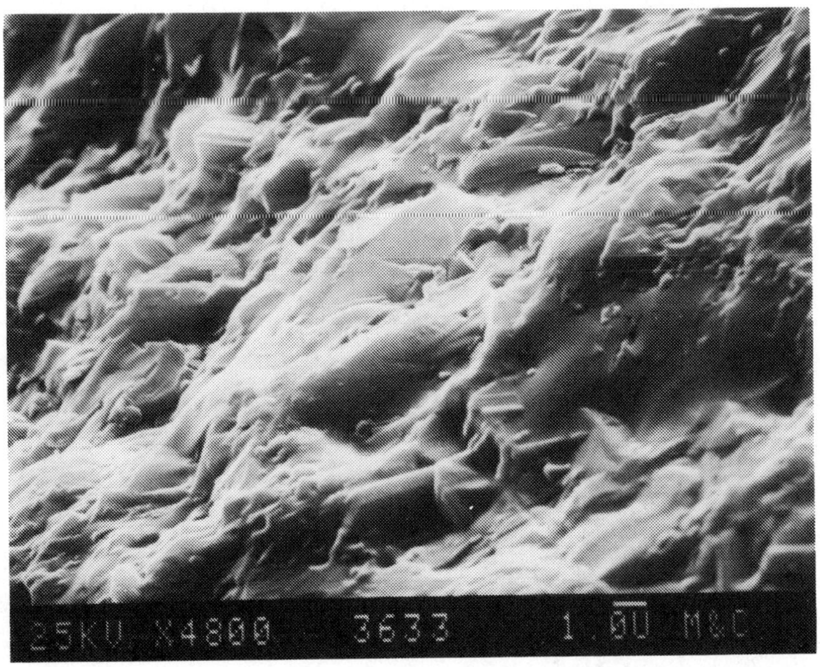

Fig. 4- Fracture surface of alumina-SiC whisker composite showing no crack-whisker interaction and no inhibition to crack propagation. Composite made with as-received Tateho Chemical Co. SiC whiskers.

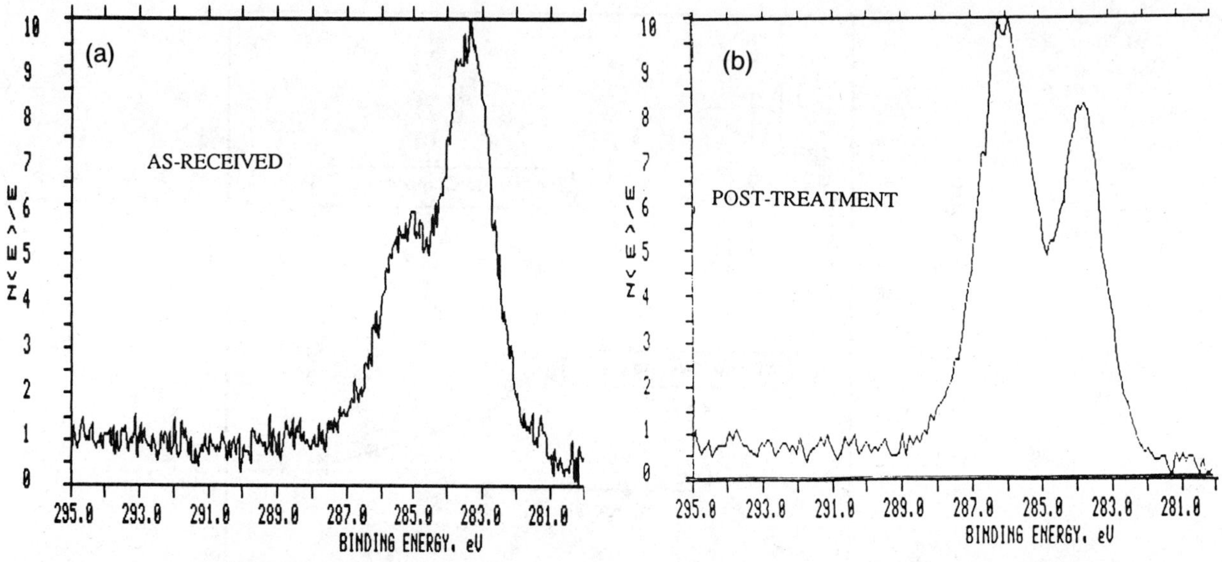

C 1s As-Received Surface **C 1s Heat-Treated Surface**

Fig. 5- X-ray photoelectron spectroscopy shows that heat-treatments of selected whiskers can increase the amount of surface carbon. (a) As-received whiskers, (b) Heat-treated whiskers.

Fig. 6- Fracture surface of alumina-SiC whisker composite fabricated with surface treated Tateho Chemical Co. SiC whiskers showing some crack-whisker interaction.

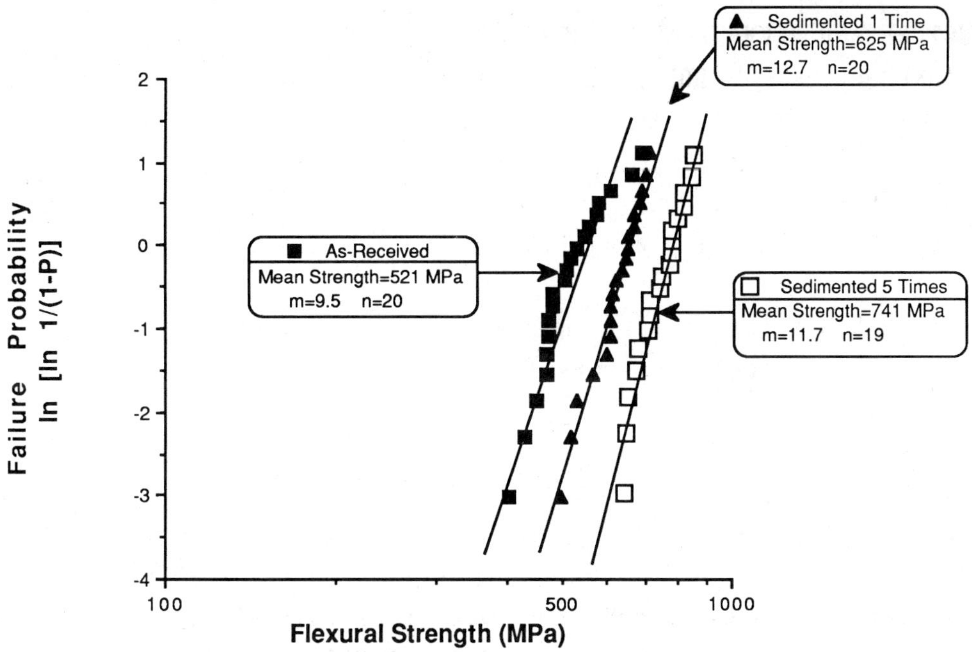

Fig. 7- Weibull plot of alumina-20 vol.% SiC whisker composites fabricated from as-received and whiskers sedimented to remove extraneous particulates. Composites fabricated using Advanced Composites Materials Corp. SiC whiskers.

MICROSTRUCTURE AND MECHANICAL PROPERTIES
OF SiC WHISKERS/CeO_2-Y_2O_3 DOPED ZrO_2 COMPOSITES

Tian Jiemo*, Zhan Baoqing,
Tong Xiaohua, Wang Lingling, Wang Jialong

Fine Ceramic Lab. of Institute of Nuclear
Energy Technology, Tsinghua University,
Beijing Fine Ceramic Lab.
P.O.Box 1021, Beijing, China

ABSTRACT

β-SiC whiskers/CeO_2-Y_2O_3 doped ZrO_2 composites with whisker contents of approximately 15 vol% were hot-pressed at 1500℃ to high densities. The addition of SiC whiskers was observed to increase fracture toughness of CeO_2-Y_2O_3/ZrO_2 ceramics. The ceramic micromorphologies are given by SEM photographs and EDAX analysis. They show β-SiC whiskers inserted into CeO_2-Y_2O_3/ZrO_2 matrix very strongly. The toughening mechanism of SiC whiskers in Y-Ce-TZP ceramics is discussed.

INTRODUCTION

Y-TZP(Tetragonal Zirconia Polycrystals) , Ce-TZP were studied by many workers[1-5]. Very small grain TZP has high-strength [6-8].Y-Ce-TZP has been discussed in our other work[9], which has shown that Y-Ce-TZP ceramics had higher bending strength and fracture toughness.
Fiber and whisker-reinforced SiC, Si_3N_4 and ZrO_2 matrix composites have been researched by Fitger, Lamic and Shalex[10-12]. Their experimental results show the additions of whiskers and fibers are an important way to increase ceramic toughness. The main purposes of the present work are to research the possibility of increasing the toughness of Y-Ce-TZP(CeO_2-Y_2O_3/ZrO_2) with SiC whiskers, and to discuss the relationship between microstructures and mechanical properties.

EXPERIMENTAL PROCEDURE

Cobweb and fiber SiC whiskers of 0.1-0.5 μm diameter, 4-10 μm in length were used for reinforcing Y-Ce-TZP. The whiskers were first dispersed by the ultrasonic vibration method in absolute alcohol.
The whiskers were mixed with CeO_2-Y_2O_3-ZrO_2 powder. The whisker-powder mixtures were dried in a drying box and were subsequently hot pressed at 1500℃ for 30 minutes in BN-washed graphite dies.
Rectangular bend bars (3×4×40mm) were cut from the hot pressed disks and the tensile surfaces were polished. Flexural strength and toughness were measured using these bars. The measurements of 3-point bending strength were performed using a 30mm span and a crosshead speed of 0.5mm/min. The fracture toughness was measured in a 3-point bending test using a 30mm span and a cross head speed of 0.05mm/min. The chevron notches were prepared with a diamond saw blade 0.2mm thick. Morphologies and structures were observed by TEM, SEM, EDAX and X-ray diffraction.

RESULTS AND DISCUSSION

X-ray Diffraction

SiC whiskers/Y-Ce-ZrO_2 composites were analyzed by X-ray diffraction. The results show Y_2O_3 and CeO_2 form a stable solution state with ZrO_2 which is called Y-Ce-TZP (Fig.1). In some hot-pressed (HP) samples, there was partial precipitation of CeO_2. The X-ray diffraction spectra of SiC appear in Y-Ce-TZP samples, however, the whiskers do not appear to affect the formation of Ce-Y-TZP(Fig.1).

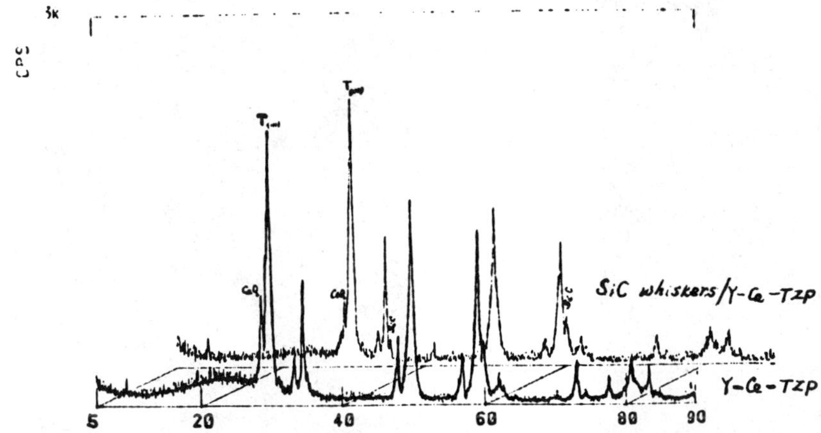

(Cukα)

Fig.1 X-ray Diffraction Spectra of Y-Ce-TZP with SiC Whiskers and Y-Ce-TZP

Structure and Morphology

The structure and morphology of SiC whisker are reported in another paper [13]. From fracture surface of Y-Ce-TZP, their morphologies were observed (Fig. 2-6) . Figures 2 and 3 show SiC whiskers appearing in the fracture surface. The SiC whisker fractured and the hole of SiC whisker pulling out are observed in figures 4 and 5.

Some whiskers have inserted into grain boundaries as shown as Fig.6a. SiC whisker was identified by EDAX spectra in Fig.6b. Fig.7 shows SiC whisker has inserted into Y-Ce-TZP grains.

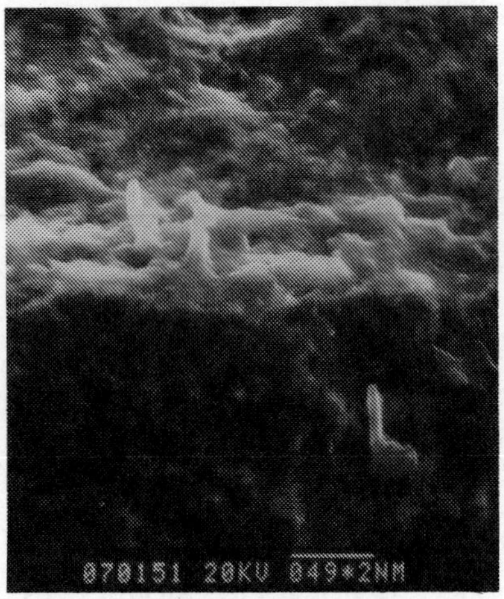

Fig.2 SiC Whisker in the Fracture Surface of Y-Ce-TZP

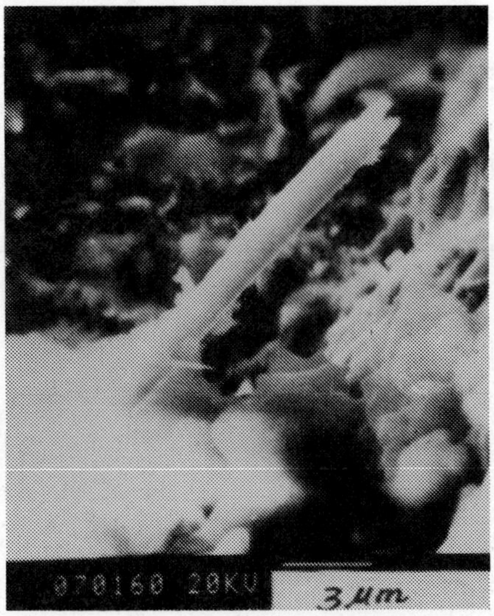

Fig.3 SiC Whisker in the Composite Matrix

Fig.4 SiC Whisker Fractured

Fig.5 The Hole of SiC Whisker Pulling Out

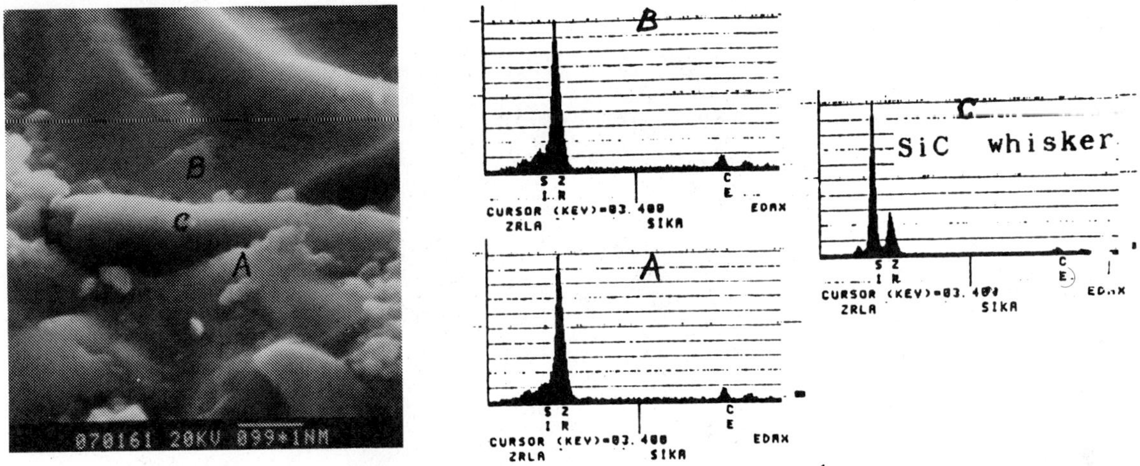

a b

Fig.6 SiC whisker has inserted into Y-Ce-TZP grain boundaries

a. SiC whisker and Y-Ce-TZP grains

b. Electron energy spectra of SiC whisker(C) and Y-Ce-TZP grains (A, B)

Fig.7 SiC Whisker and Y-Ce-TZP Grains

The interface reactions were found. Figures 8 and 9 indicate SiC whiskers react with the matrix. Fiber whiskers (Fig.8) and cobweb whiskers(Fig.9a, b) all strongly bonded together with matrix.

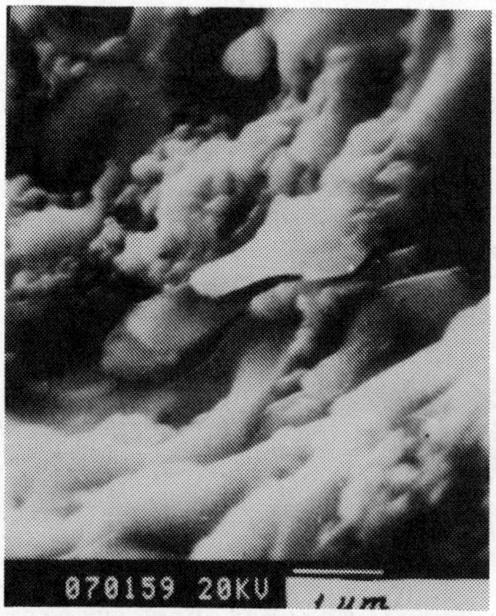

Fig.8 SiC Whisker and Matrix

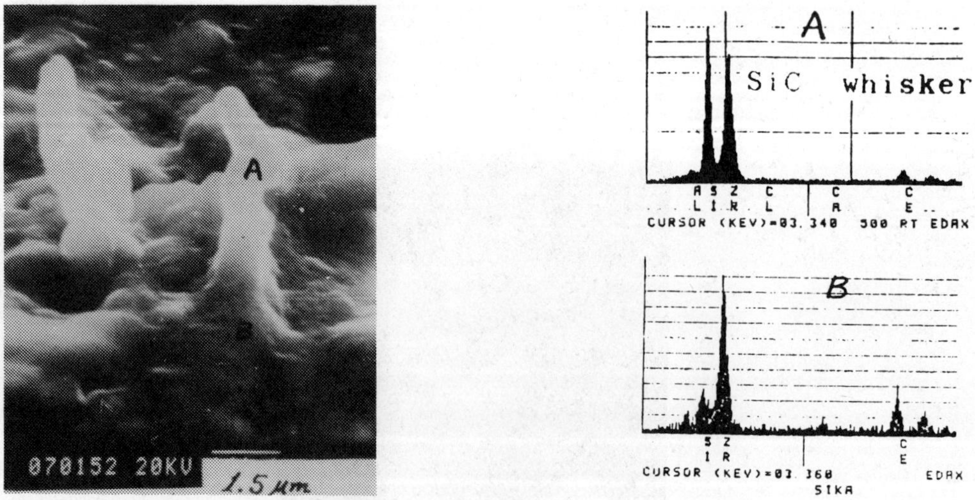

a b

Fig.9 SEM Morphology and Electron Energy Spectra of SiC Whisker and Matrix

Electron energy spectra of SiC whisker "A" and its root "B"(Fig. 9a) in Y-Ce-TZP grain are shown in Fig.9b. According to the appearance of Zr Peak in the spectrum of SiC whisker (Fig.9b), the interface reaction have occured between SiC and Y-Ce-TZP. At the same time, the SiC peak appears in Y-Ce-TZP grain (Fig9b,B). They show SiC whisker is bonding with Y-Ce-TZP grain.

Mechanical Properties

The bending strength and fracture toughness were measured using 3-point bending flexure at room temperature. The results of these test are given in Fig.10, and Fig.11.

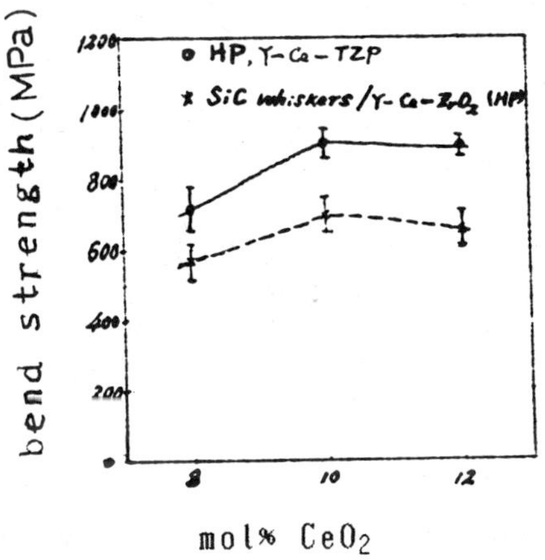

Fig.10 Bending Strength as a Function of CeO_2 Content (contained 3mol%Y_2O_3)

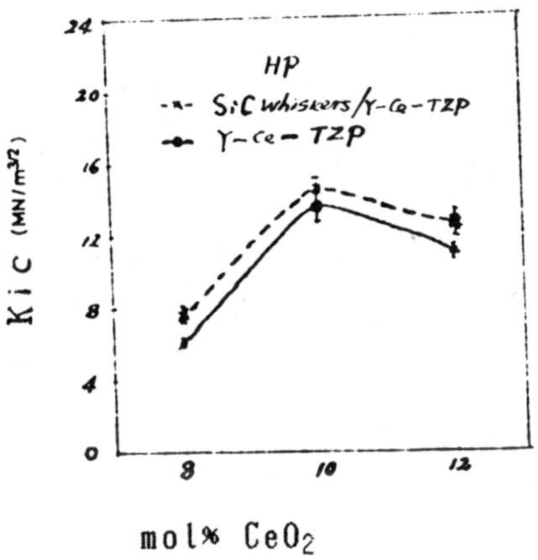

Fig.11 Fracture Toughness as a Function of CeO_2 Content (contained 3mol%Y_2O_3)

The experimental results shown in Fig.10 and 11 indicate that the addition of SiC whiskers to Y-Ce-TZP decreases the bending strength of Y-Ce-TZP for all compositions studied, but increases the fracture toughness. When the addition of 10mol% CeO_2 in SiC whiskers/Y-Ce-TZP, the maxima of bending strength and fracture toughness have been obtained. X-ray diffraction spectra of the fracture surface does not indicate the presence of m-phase ZrO_2.

That is to say there is no phase transformation during fracture process(Fig. 12) . Whisker debonding and pullout were rarely found. The fracture surface indicates there are intergranular fracture (Fig.4).

Because there are thermal expansion differences between the SiC whisker and the matrix, ($\alpha_{SiC}=(4-5)\times10^{-6}$/K, $\alpha_{matrix}=(8-10)\times10^{-6}$/K), there is a mismatch stress developed. This stress accounts for the reduction in bending

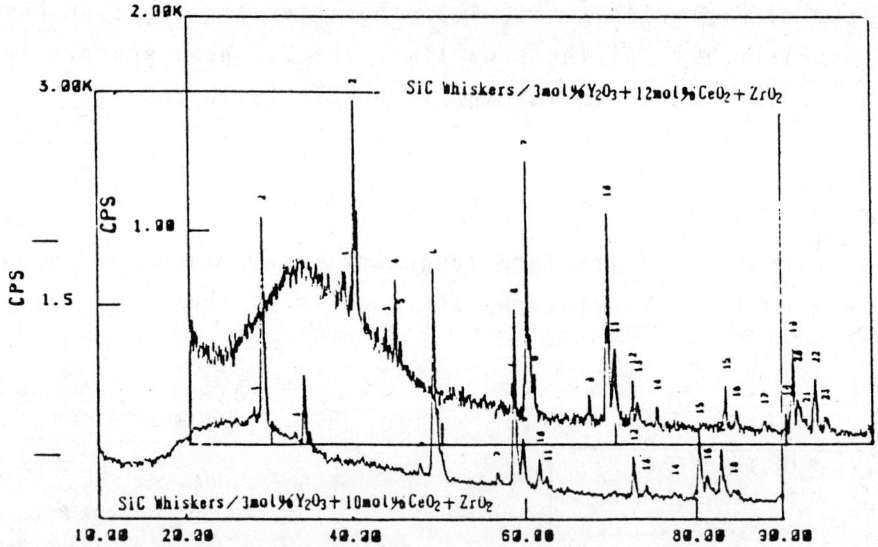

Fig.12 X-ray Diffraction Spectra of Fractured Surface of SiC whiskers/Y-Ce-TZP Composite

strength which is observed at room temperature.

The increase in fracture toughness is apparently due to the crack deflections around SiC whiskers[14-15] or the SiC whiskers interfering with the propagation of crack in the Y-Ce-TZP matrix.

SUMMARY

The experimental results show

1) The addition of SiC whiskers increased the room temperature fracture toughness of CeO_2-Y_2O_3/ZrO_2 .
2) SiC whiskers reacted with Y-Ce-TZP grains and formed the stronger boundaries.
3) SiC whiskers in Y-Ce-TZP matrix prevent t-phase from transformation during fracture process.

REFERENCES

1. M.V.Swain, L.R.F.Rose, J.American Ceramic Society, 511-18(1986)
2. M.Matsui, T.Soma, I Oda, J.Am.Ceram. Soc.69[3] 198-202(1986)
3. Tsugio Satio, Shiro Ohtaki et al, J.Am.Sci Lett.5(1986)1140-1142.

4. Koji Tsukuma, Masahiko Shimada, Am. Ceram. Soc.Bull.64(1985) 10-13
5. M. Ruhle, N.Claussen, Proc. 2nd Inter. conf, on the Sci.Tech of Zirconia June 21-23 (1983)
6. T.W.Coyle, W.S Coblenz and B.A.Bander, Amer, Ceram.Soc. Bull 62(1983)966
7. Koji Tsukoma and Kuniyoshi Ueda, J.A.Ceram. Soc.68(1985)C56-58.
8. F.F.Lange J.Mater. Sci. 17 (1982)255
9. Tian Jiemo, Zhan Baoqing, Wang Lingling, Tong Xiaohua et al, To be published in 3rd Inter.Symp.Ceramic Mater. & Components for Engines
10. Erich Fitger and Ranier Gadow.Am. Ceram. Soc. Bull., 65(1986)326-35
11. P.J.Lamic, G.A.Bernhart. et al, Am.Ceram.Soc. Bull.,65(1986)336-38
12. P.D.Shalex, J.J.Petrovic et al, Am. Ceram. Soc. Bull., 65(1986) 351-56
13. Tian Jiemo, Zhan Baoqiang, Wang Lingling, Tong Xiaohua et al, "Microstructure and Growth Mechanism of SiC Whiskers" to be published in 3rd Inter.Symp. Ceramic Mater. & Components for Engines
14. T.N.Tiegs, P.F.Becher, Am. Ceram.Soc. Bull.,66(1987) 339-42
15. D.Munz, R.T.Bubsey and J.L.Shannon, Jr., J.amer.Ceram. Soc 63(1980)300

CREEP BEHAVIOR OF SiC_w / Al_2O_3 COMPOSITES

David S. Liu, Azar P. Majidi
Center for Composites Manufacturing Science and Engineering
and Dept. of Mechanical Engineering
University of Delaware
Newark, Delaware 19716

ABSTRACT

Creep experiments were performed on a SiC-whisker reinforced alumina composite at temperatures of 1150°C and 1250°C in an air environment and at initial elastic stresses ranging from 130 MPa to 320 MPa. Reaction phases were found on the surfaces of all specimens after testing. Removal of the reaction layer revealed the presence of creep cracking on the tensile faces of several specimens. Whisker bridging and evidence of whisker pullout were found and are suggested to be factors responsible for the enhanced creep resistance of this material.

INTRODUCTION

Whisker reinforcement of polycrystalline ceramics has proved to be a popular method for overcoming the brittle nature of these materials for high temperature applications. Alumina reinforced with silicon carbide whiskers has been studied previously and found to have much improved properties over the monolithic matrix material in terms of fracture toughness,[1] strength,[2] and creep resistance.[3] While these improvements in properties are thought to be mainly due to such mechanisms as crack deflection and whisker pullout, less is known about the significance of these mechanisms at elevated temperatures and/or what other mechanisms become important. The objective of this paper is to present some experimental results and observations from a series of tests performed on a SiC-whisker reinforced alumina.

EXPERIMENTAL PROCEDURE

The material presently being studied is a commercially available composite containing 25% by weight silicon carbide whiskers in an aluminum oxide matrix.* Samples with the approximate dimensions of 2.5mm x 5mm x 50mm were cut and the edges beveled by the manufacturer from hot pressed billets. Hot pressing causes a preferential whisker orientation such that the whiskers are randomly oriented within planes perpendicular to the hot pressing axis.[4] Testing was performed such that these planes were oriented parallel to the direction of applied stress. Specimens were loaded in four point flexure at 1150°C and 1250°C in an air environment using sintered alpha-silicon carbide fixtures (with inner and outer loading spans of 20mm and 40mm respectively). Dead weight loading applied various initial elastic stresses ranging from 130 to 320 MPa. After heat up to the desired test temperature the furnace was allowed to soak for one hour in order to stabilize the temperature distribution. An optical extensometer # was used to focus in on the upper and lower halves of the test fixture thereby allowing the load point deflections to be continuously monitored throughout each test. Prior to testing the tensile surfaces were polished with 1/4 um diamond paste in order to remove any surface damage due to machining.

The side surfaces of some specimens were also polished and rows of Vickers indentations were placed along the depth of the beam in an attempt to measure the surface strains and location of the neutral axis after testing.[5] However, surface reactions (described subsequently) obscured the indentations making measurement impossible. Final outer fiber strains were then determined by measuring the radius of curvature of specimens which were cooled to room temperature under load and assuming that the neutral axis remained at the center of the beam during testing. For small deformations, the curvature will be approximately proportional to the load point displacement as detailed by Hollenberg et al.[6] Therefore, the ultimate outer fiber strain of ruptured specimens could also be calculated from load point displacement data gathered prior to actual failure.

RESULTS AND DISCUSSION

Mechanical Properties

Strain - time data are shown in fig 1. At 1150°C, all of the specimens underwent a period of primary transient creep where the strain rate decreased with increasing strain. The tests performed at 190 and 270 MPa applied stress settled into steady-state, secondary creep after approximately 60 and 20 hours respectively whereas the specimen tested under 320 MPa load fractured prematurely at a flaw in the specimen after 6.5 hours of testing. The specimen tested at 190 MPa did not fail after 231 hours of testing. At 1250°C, transient primary creep was again observed for all specimens along with secondary, steady-state creep behavior (fig. 2). The sample tested at 220 MPa underwent tertiary creep where the strain rate increased with increasing strain until fracture occurred after 24 hours of testing at a strain of 1.7%.

For comparison, the strain as a function of time and temperature is shown in fig.3 at an applied stress of 190 MPa. A strong temperature dependence is evident from the data which shows almost an order of magnitude of difference in steady state creep rate between the two test temperatures.

Surface Reactions

All of the test samples lost their mirror surface finish after testing. Microscopy revealed the occurrence of extensive surface reactions. At 1150°C, droplets of a glassy phase were observed to form on the tensile surface. Further reaction of the glassy phase and pore formation were found at the higher test temperature of 1250°C (fig. 4). The appearance of these surfaces are consistent with results obtained by Becher and Tiegs on a similar material system.[7] The glass formation is a result of the reaction:

$$SiC + \frac{3}{2}O_2 \rightarrow SiO_2 + CO$$

in an oxidizing environment. Also, at lower temperatures (<1000°C) mullite forms very slowly via the reaction:

$$SiO_2 + Al_2O_3 \rightarrow \text{mullite.}$$

At longer times and higher temperatures, however, the rate of mullite formation increases considerably, resulting in the altered appearance of the tensile surface exposed to 1250 °C.[7]

Creep cracking

Due to the reactions which occur on the surfaces of the specimens, mechanical polishing of the surfaces was required in order to remove the reaction layer and expose any creep damage which had occurred. Of the specimens which remained intact after testing, all exhibited considerable crack formation on the side and tensile surfaces (fig. 5). These cracks varied in length from 10 to 800 μm. However, the majority of the cracks found were less than 100 μm in length. Crack opening displacements rarely exceeded 2 or 3 μm.

The nucleation of cracks appears to occur preferentially in coarse grained regions of the composite. Chemical analysis of the matrix surrounding crack nucleation sites revealed relatively low whisker concentrations in most regions with the coarse grain structure. This is in agreement with a previous study on the origin of failure in tensile specimens of this system tested at room temperature.[8] In that study fracture initiation also occurred in regions with a coarse grain structure and low whisker density.

Evidence for whisker pullout, one of the primary toughening mechanisms in fiber reinforced ceramics at room temperature, was found along with bridging of cracks by whiskers (fig. 6). These mechanisms are believed to play a major role in the improved creep resistance of this composite as compared to the monolithic matrix material.[9] One interesting observation can be made at this point: while examples of crack bridging were not difficult to find in regions where cracking had occurred, neither were whiskers found bridging a single crack or protruding from crack faces over any appreciable distance along the longer cracks. Most examples tended to be isolated throughout the creep damage zone. This would seem to indicate that cracks tend to propagate either along grain boundary junctions with no whiskers lying across the boundary[10] or along regions deficient in whiskers. (In instances where several whiskers were observed bridging a single crack, the crack tended to be short and relatively blunt.) It is also possible that since cracks tend to nucleate in whisker deficient regions, crack coalescence is also occurring between these regions.

Whiskers seem to play a role in arresting the growth of cracks. Examples were found where cracks terminated in regions with a high whisker concentration (fig. 7). Creep cracks were also found lying side-by-side within a few microns of each other (fig. 8). If it is assumed that the growth of one of these cracks has been arrested by whiskers, these whiskers would then be supplying the necessary tractions for the second crack to open and propagate. Fig. 9 illustrates an example where whiskers appear to have actually precluded further growth of the primary crack and caused it to branch off into a different direction. These effects would improve the damage tolerance of whisker reinforced ceramics via larger strains to failure and improved fracture toughness.

CONCLUSIONS

1) The observed formation of the glassy reaction layer tends to reduce the overall creep resistance of the composite due to the inherently low viscosity of glass at these high temperatures. However, the formation of mullite also has the potential for increasing the overall creep resistance.[11] Whether these reaction phases serve as barriers to further oxidation of the interior or whether they serve to further degrade the creep properties of the composite is unclear at this point.

2) A large portion of the total creep strain can be attributed to the formation of creep cracks on the tensile surface. Nucleation of these cracks appears to occur in whisker deficient regions. These regions also tend to have coarser grain structures as compared to the surrounding matrix.

3) Crack arrest, whisker bridging and whisker pullout were found to be operative mechanisms which occur during creep crack growth and result in the improved creep resistance of the composite over the monolithic matrix material.

Acknowledgements

The authors thank the Center for Composites Manufacturing Science and Engineering at the University of Delaware for their funding and support for this research.

---

* Advanced Composites Corporation, Greer, S. C.
\# Zimmer, W. Germany

References

[1]Paul F. Becher, George C. Wei, "Toughening Behavior in SiC-Whisker-Reinforced Alumina," *J. Am. Ceram. Soc.*, **67** [12] C-267 - C-269 (1984).

[2]Joseph Homeny, Wallace L. Vaughn, Mattison K. Ferber, "Processing and Mechanical Properties of SiC-Whisker -Al_2O_3-Matrix Composites," *Am. Ceram. Soc. Bull.*, **66**,[2] 393-398 (1987).

[3]John R. Porter, Fred F. Lange, Atul H. Chokshi, "Processing and Creep Performance of SiC-Whisker-Reinforced Al_2O_3," *Am. Ceram. Soc. Bull.*, **66**,[2] 343-347 (1987).

[4]George C. Wei, Paul F. Becher, "Development of SiC-Whisker-Reinforced Ceramics," *Am. Ceram. Soc. Bull.*, **64** [2] 298-304 (1985).

[5]Ching-Fong Chen, Ph.D. Thesis, University of Michigan, 1987.

[6]G.W. Hollenberg, G.R. Terwilliger, R.S. Gordon, "Calculation of Stresses and Strains in Four-Point Bending Tests," *J. Am. Ceram. Soc.*, **54**, [4] 196-199 (1971).

[7]Paul F. Becher, Terry N. Tiegs, "Temperature Dependence of Strengthening by Whisker Reinforcement: SiC-Whisker-Reinforced Alumina in Air," *Adv. Ceram. Matls.*, **3** [2] 148-153 (1988).

[8]D.S. Liu, A.P. Majidi, "High Temperature Creep of SiC Whisker Reinforced Alumina Composites," to be published in 12th Annual Conference on Composites and Advanced Ceramic Materials, ed. J. Buckley, Am. Ceram. Soc., Cocoa Beach, FL (1988).

[9]Atul H. Chokshi, John R. Porter, "Creep Deformation of an Alumina Matrix Composite Reinforced with Silicon Carbide Whiskers," *J. Am. Ceram. Soc.*, **68** [6] C-144 - C-145 (1985).

[10]Karl Jakus, S.V. Nair, "Elevated Temperature Crack Growth in SiC Whisker Reinforced Alumina," to be published in 12th Annual Conference on Composites and Advanced Ceramic Materials, ed. J. Buckley, Am. Ceram. Soc., Cocoa Beach, FL (1988).

[11]W.D. Kingery, H.K. Bowen, D.R. Uhlmann, *Introduction to Ceramics*, 2d ed., John Wiley & Sons, New York, 1976, pp. 750-751.

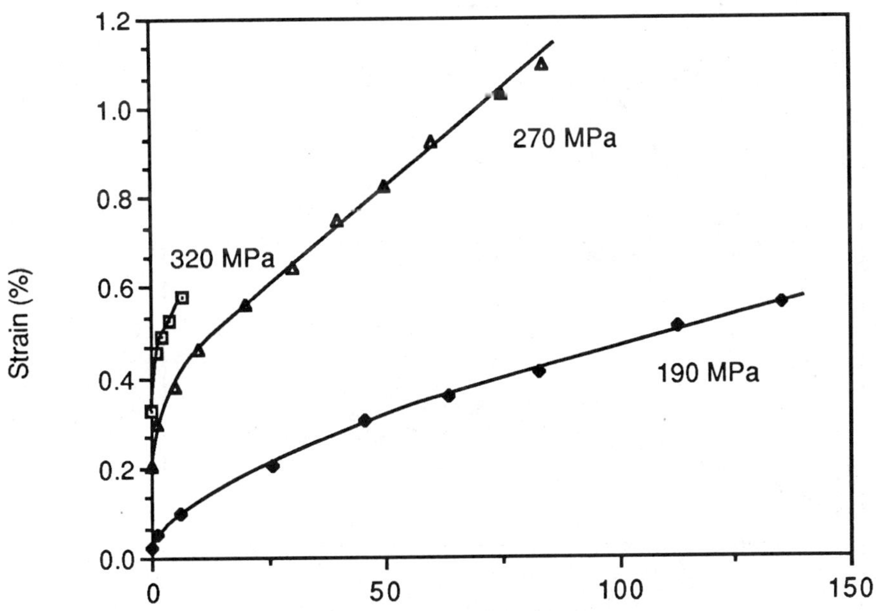

Fig. 1. Strain vs. time graph. T=1150 °C.

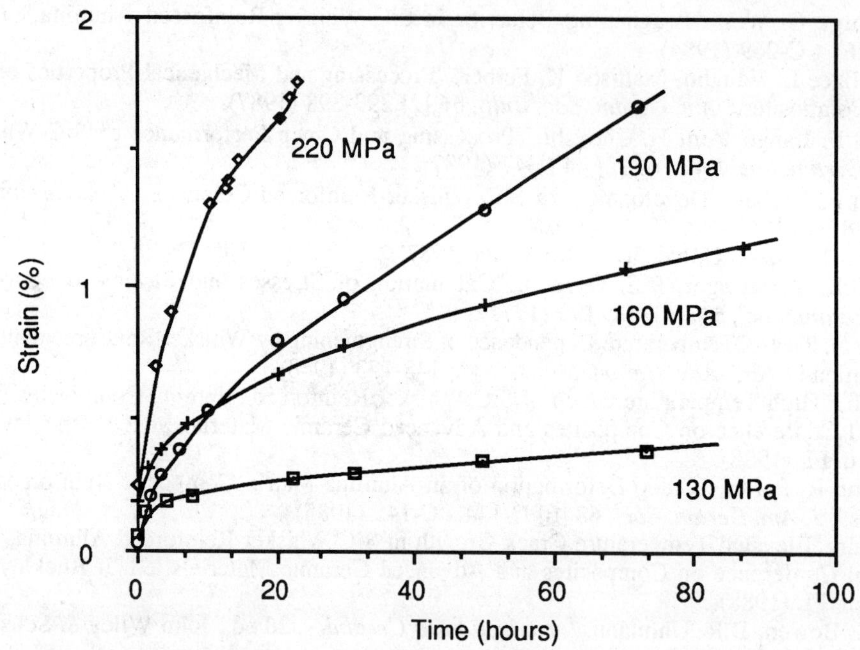

Fig. 2. Strain vs. time. T=1250 °C.

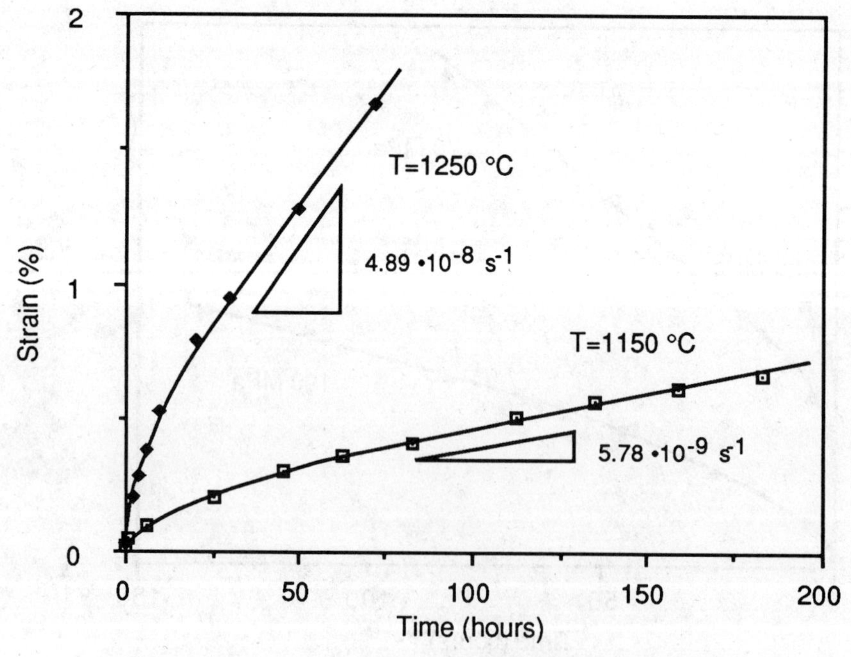

Fig. 3. Strain vs. time graph. Comparison between T=1150 °C and T=1250 °C at a stress of 190 MPa.

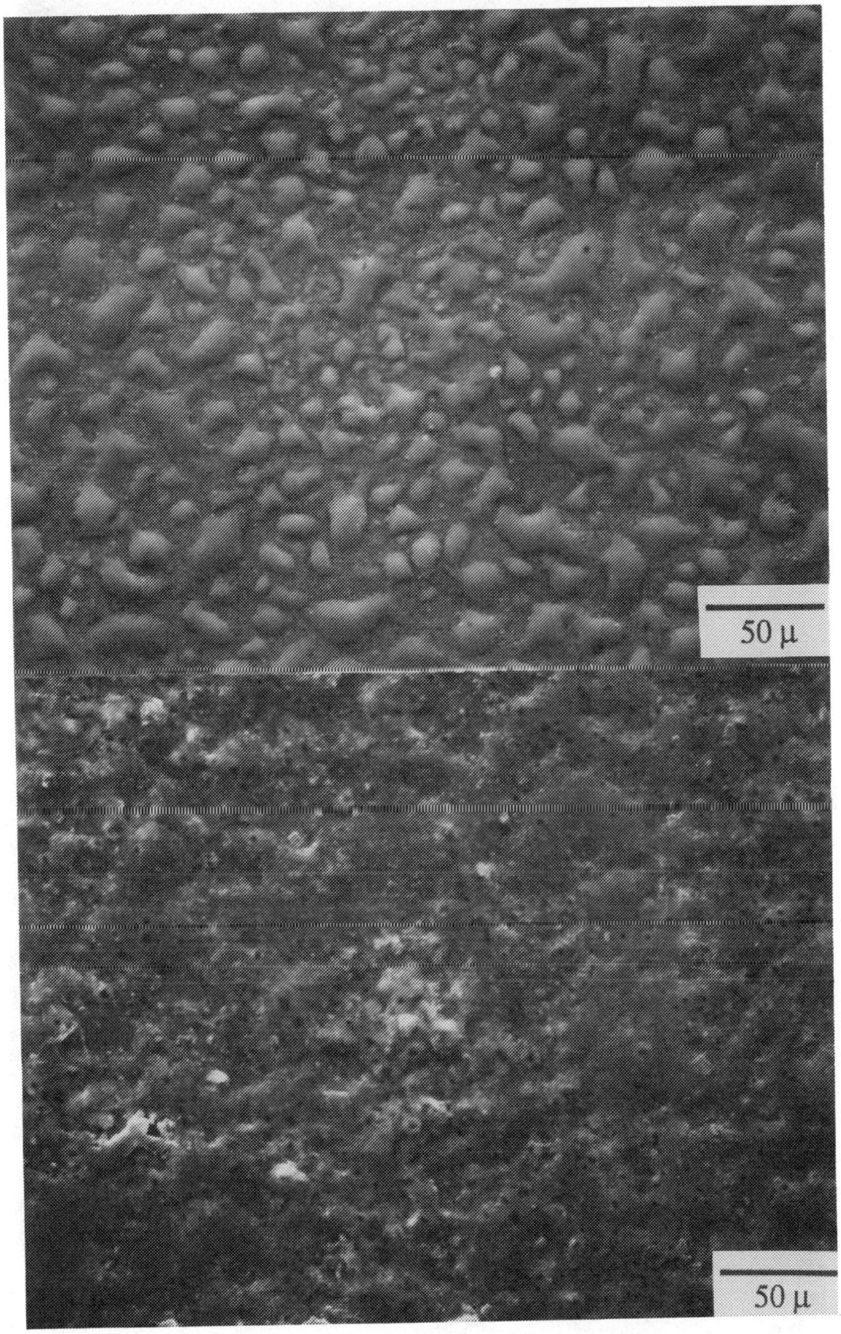

Figure 4. (a) Tensile face of crept bar showing glassy reaction phase. T=1150 °C, σ=270 MPa. (b) Tensile face, T=1250 °C, σ =160 MPa.

Figure 5. Polished tensile face of crept bar showing high density of creep cracks. T=1250 °C, σ=160 MPa.

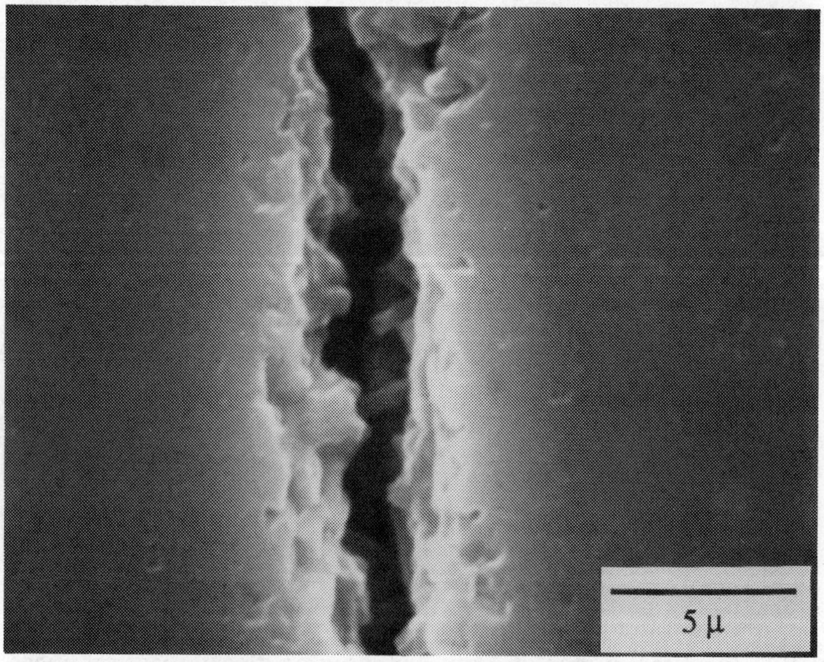

Figure 6. Blunt crack showing whisker bridging of the crack and protruding whiskers on the crack face. T=1250 °C, σ =160 MPa.

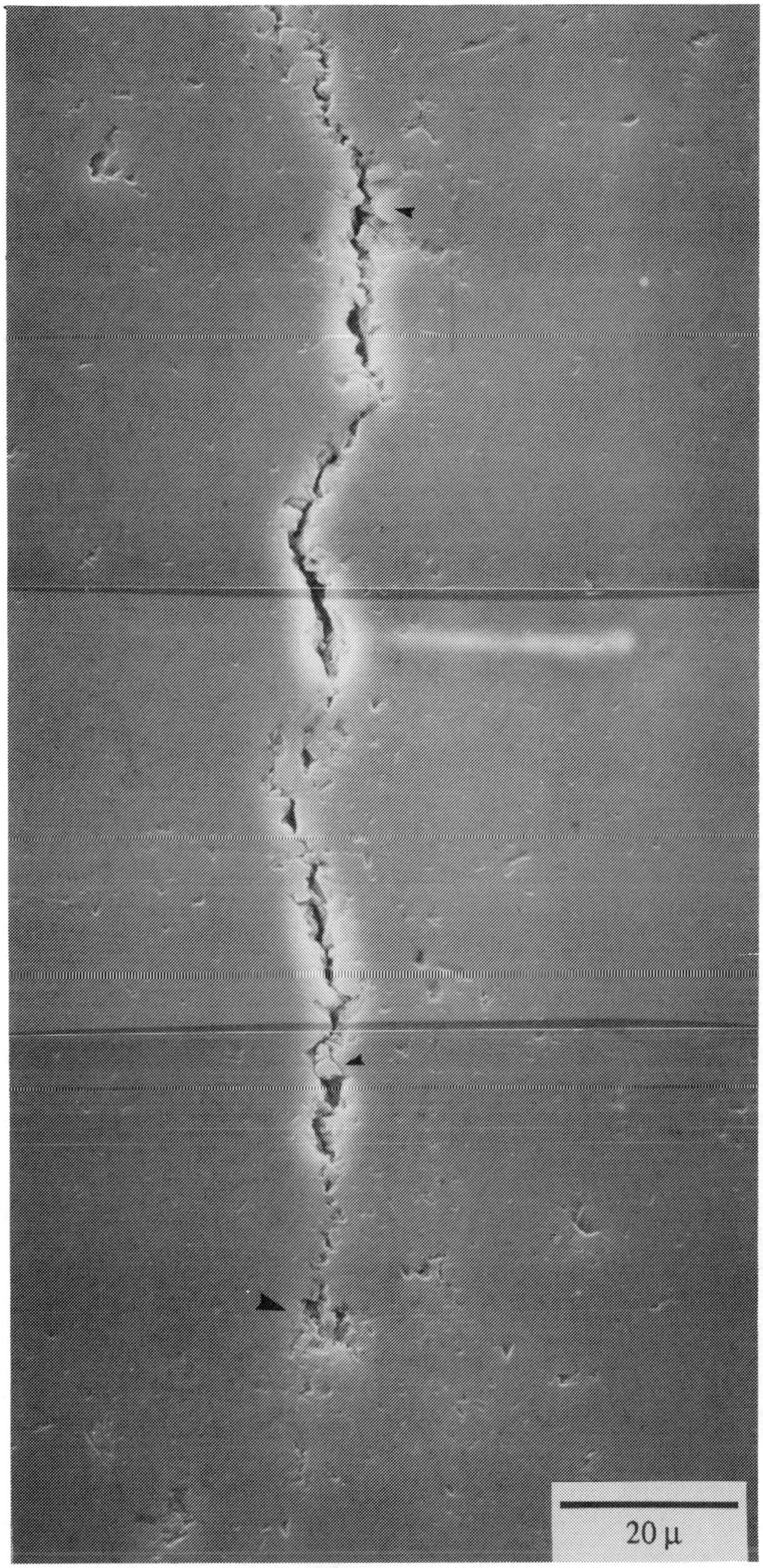

Figure 7. Large crack on tensile surface. Top arrows indicate large grains of alumina postulated to be crack initiation sites. Bottom arrow shows site of whisker concentration. T=1150 °C, σ=260 MPa.

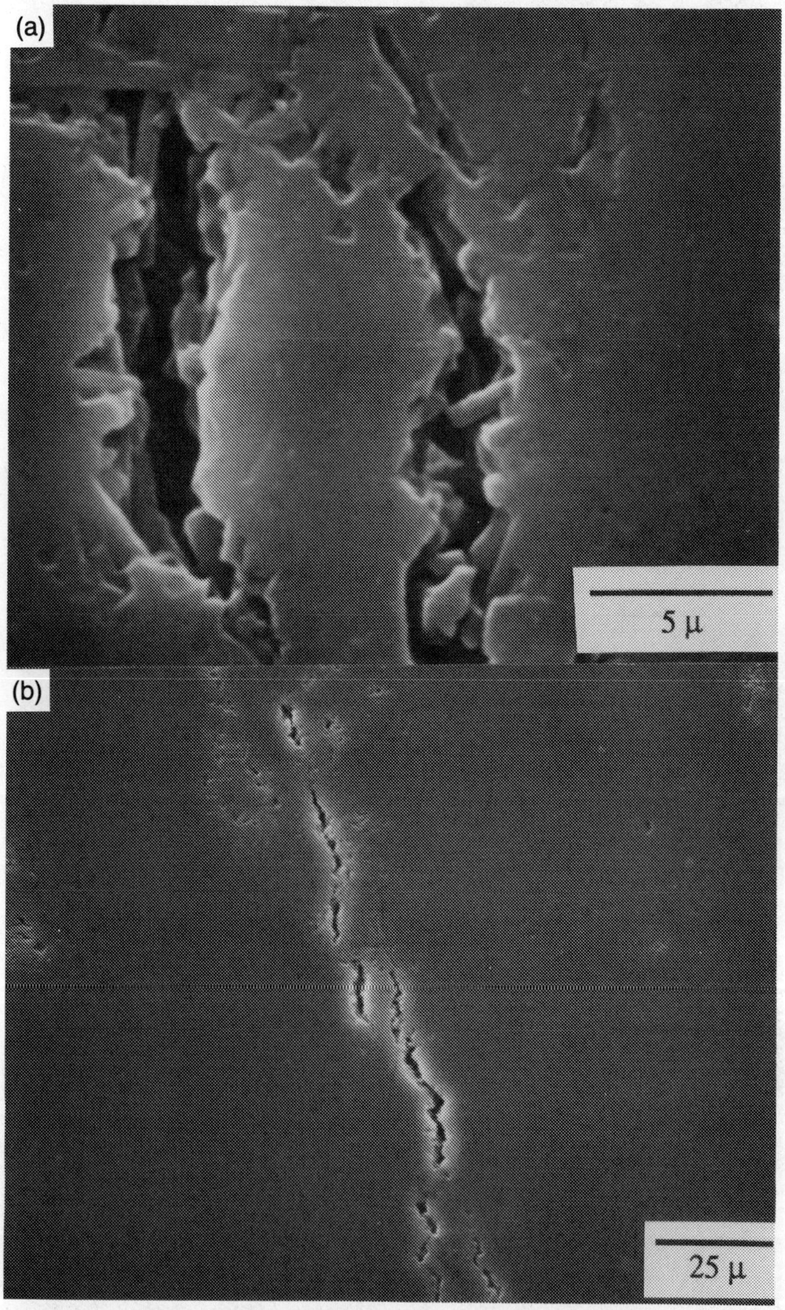

Figure 8. (a) Site of crack arrest and possible nucleation of a new crack adjacent to arrested crack. T=1250 °C, σ=160 MPa. (b) Lower magnification view.

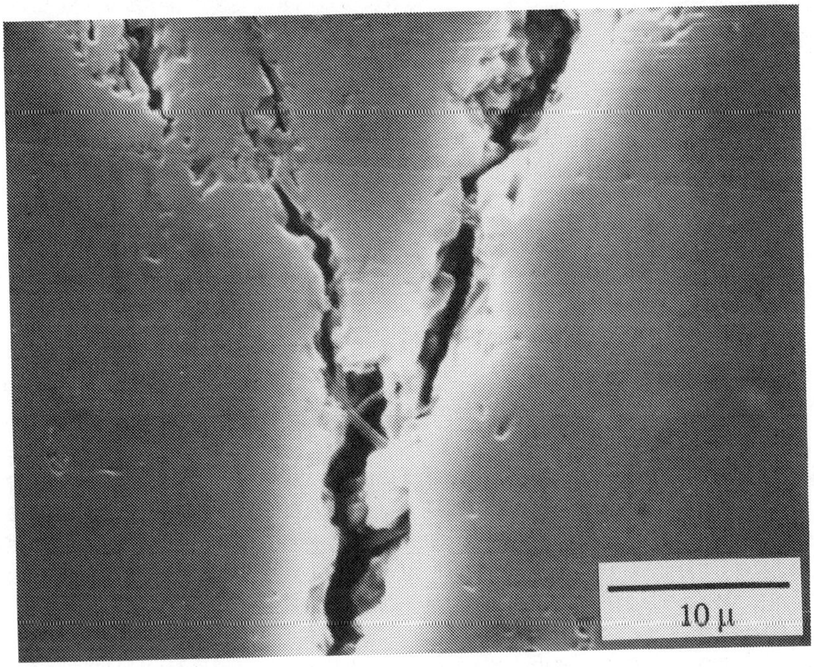

Figure 9. Whisker bridging and possible crack deflection. T=1250 °C, σ=160 MPa.

MECHANICAL PROPERTIES AND MICROSTRUCTURES OF Si_3N_4-BN COMPOSITE CERAMICS.

Toshihiko Funahashi*, Keiichiro Isomura,
Akira Harita, and Ryoji Uchimura
Technical Research Division, Kawasaki Steel Corp.
1, Kawasaki-cho, Chiba, Japan

ABSTRACT

A new type of composite ceramic of the Si_3N_4-BN system has been developed using a method which utilizes homogeneous mixing of ultra-fine constituent ceramic powders and gas pressure sintering. An essential feature of Si_3N_4-BN composite ceramics is that their mechanical properties can be tailored over a wide range by changing the BN contents. Another remarkable feature in practical use is that the composite ceramics are machinable while retaining relatively high strength, i.e., the bending strength of machinable composite ceramics with 10 and 20wt% BN is 350 and 200MPa, respectively. These values are more than two times higher than those of conventional machinable ceramics, such as Si_3N_4-BN composite ceramics made by a reaction sintering method or machinable mica ceramics. In addition, the thermal shock resistance is improved remarkably by increasing the BN content compared with Si_3N_4 ceramics.

INTRODUCTION

Silicon nitride (Si_3N_4) ceramics have excellent properties,i.e., high strength, relatively high toughness, and high wear resistance. The expansion of application of Si_3N_4 ceramics has been expected especially in the field of engineering ceramics. However, such properties as corrosion resistance to molten metal, thermal shock resistance and machinability are insufficient for their practical use as various machine parts. In order to improve these disadvantageous properties of monolithic Si_3N_4 ceramics, attempts have been made to disperse BN particles in the Si_3N_4 matrix as a second phase.

In the present study, a new type of composite ceramic of the Si_3N_4-BN system has been developed using slipcasting and gas pressure sintering instead of conventional production methods, such as hot-pressing 1)-3) or reaction sintering 4),5). The Si_3N_4-BN composite has the following advantages;
(1) Complete dispersion and mixing of raw materials have been achieved by the slipcasting method and hence the mechanical

properties of the Si_3N_4-BN composite have been improved compared with the conventional process.
(2) The composition of BN can be changed over a wide range, and the mechanical properties can be expected to be changed by BN/Si_3N_4 ratio.
(3) "Near net shape" forming can be achieved for machine parts with complicated shapes.
(4) Gas pressure sintering is economical compared with hot-pressing.

The Si_3N_4-BN composite ceramics developed in the present work have high bending strength, excellent thermal shock resistance, and machinability while retaining relatively high strength. This indicates that the composite ceramics can be applied to various machine parts in the future.

In this paper, the various mechanical properties of the Si_3N_4-BN system composite ceramics made by the slipcasting and gas pressure sintering method are described, and discussed in relation to their microstructures.

EXPERIMENTAL PROCEDURE

The experimental procedure is shown schematically in Fig.1. Two types of Si_3N_4 powders, designated as Si_3N_4(A) and Si_3N_4(B), sintering aid, and BN powder were used as starting materials. The properties of these Si_3N_4 powders are shown in Table 1. The Si_3N_4(A), produced by the imide-decomposition method, has higher purity and finer particle size than the Si_3N_4(B), produced by the direct nitriding method. The BN powder used as a starting material is also an ultra-fine (<1μm) powder. In this study sintering aid, which is ordinarily used for monolithic Si_3N_4 ceramic, was applied.

Various slurries consisting of these starting materials with different mixing ratio, dispersing agents, and water were prepared. The BN contents in slurries were adjusted to 0, 10, 20, 30, 40 wt% in the sum of Si_3N_4 powder, BN powder, and sintering aid. The green bodies (50mm by 50 mm by 10 mm) with various BN contents were slipcast in plaster molds. After drying and binder removal, sintering was carried out by heating for 2 hours under a nitrogen pressure of 0.9 MPa at 1800°C.

Test specimens (4mm by 3mm by 40mm) were obtained from the sintered body. Three point bending strength at room temperature and the bending strength after thermal shock by water-quenching treatment were measured. The thermal shock resistance was evaluated by measuring the abrupt decrease in the bending strength after water-quenching treatment from high temperature. The fracture surfaces after bending were observed by scanning electron microscope (SEM) to investigate the microstructures. In addition, general physical measurements, such as X-ray diffraction (XRD) analysis, porosity, Young's Modulus, thermal expansion coefficient etc. were carried out to investigate the relation between mechanical properties and microstructures.

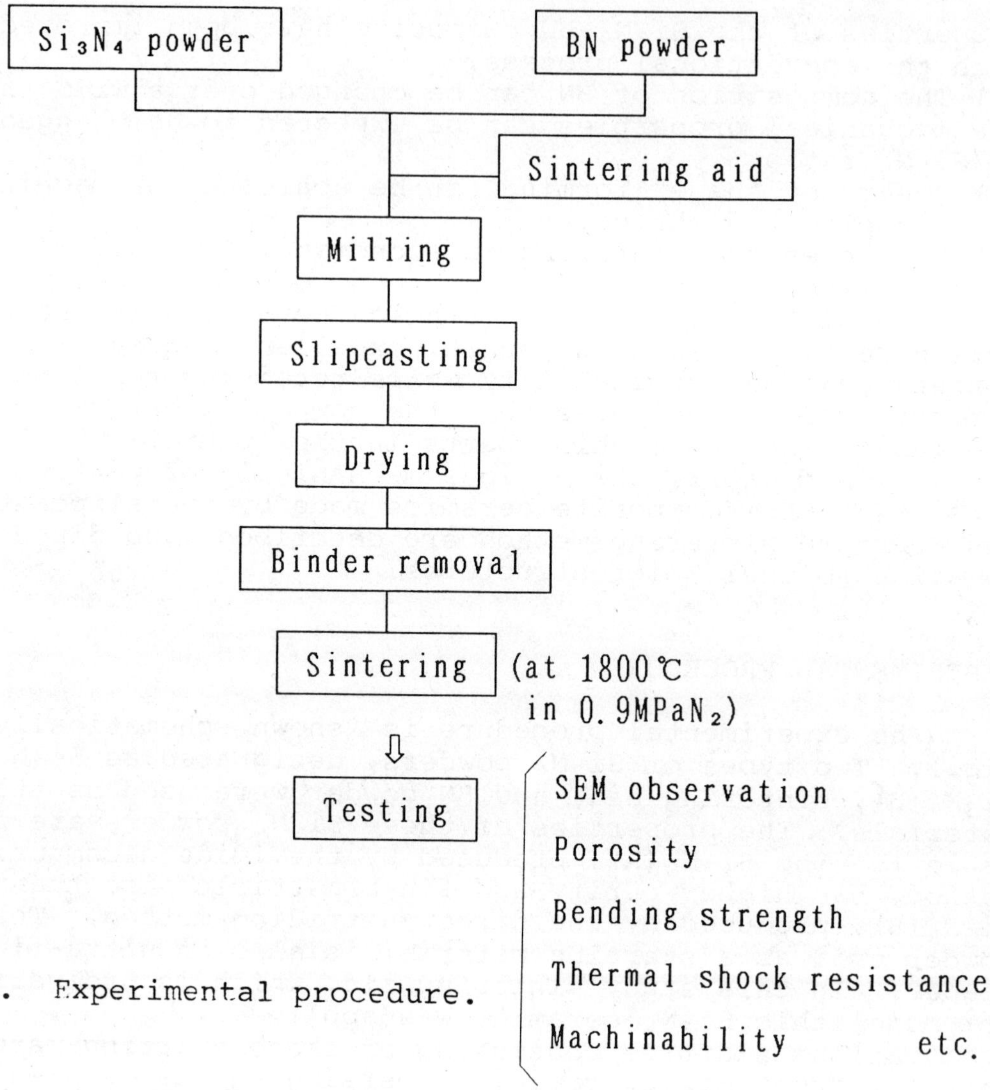

Fig. 1. Experimental procedure.

Table 1. Properties of Si_3N_4 powder.

		Si_3N_4 (A)	Si_3N_4 (B)
$(\alpha / \alpha+\beta) \times 100$ (%)		> 97	93
	Fe (ppm)	< 50	1800
	Aℓ (ppm)	< 10	1500
	Ca (ppm)	< 10	1800
	Mg (ppm)	< 10	< 100
Specific surface area (m^2/g)		10～11	7
Particle size (D_{50}: μm)		0.7	1.2

RESULTS AND DISCUSSION

Microstructure

The crystal phases identified by XRD in the present composite ceramics were all β-Si_3N_4 and hexagonal BN in the region of 10-40wt%BN, and no other crystal phase was detected. Fig.2 and Fig.3 show the microstructures observed by SEM on the fracture surface of the composite ceramics made by using Si_3N_4(A)powder and Si_3N_4(B)powder as starting material, respectively.

As shown in the matrix of the ceramics with 10wt% BN in Fig.2 and Fig.3, the microstructures are relatively dense textures and the growth of rod-like grains of β-Si_3N_4 is observed similar to the structure of Si_3N_4 ceramics 6),7). In the case of the ceramics with 30wt%BN, the microstructures show relatively porous and granular textures which consist of fine grains of β-Si_3N_4 and BN. As shown in Fig.2, the microstructure of the Si_3N_4(A)-20wt%BN containing fine Si_3N_4 powder is a rod-like grain texture similar to the ceramic with 10wt%BN. On the other hand, the ceramic of Si_3N_4(B)-20wt%BN containing coarse Si_3N_4 powder has a microstructure with a granular texture similar to the ceramic with 30wt%BN. The difference in the microstructures of the ceramics with 20wt%BN can be attributed to the sinterability of the Si_3N_4 powder, which mainly depends on the grain size.

Micropores are recognized in the SEM photographs of Fig.2 and Fig.3. Figure 4 shows the relation between porosity and BN content. Porosity was determined from the difference between volume density measured on the composite ceramics and

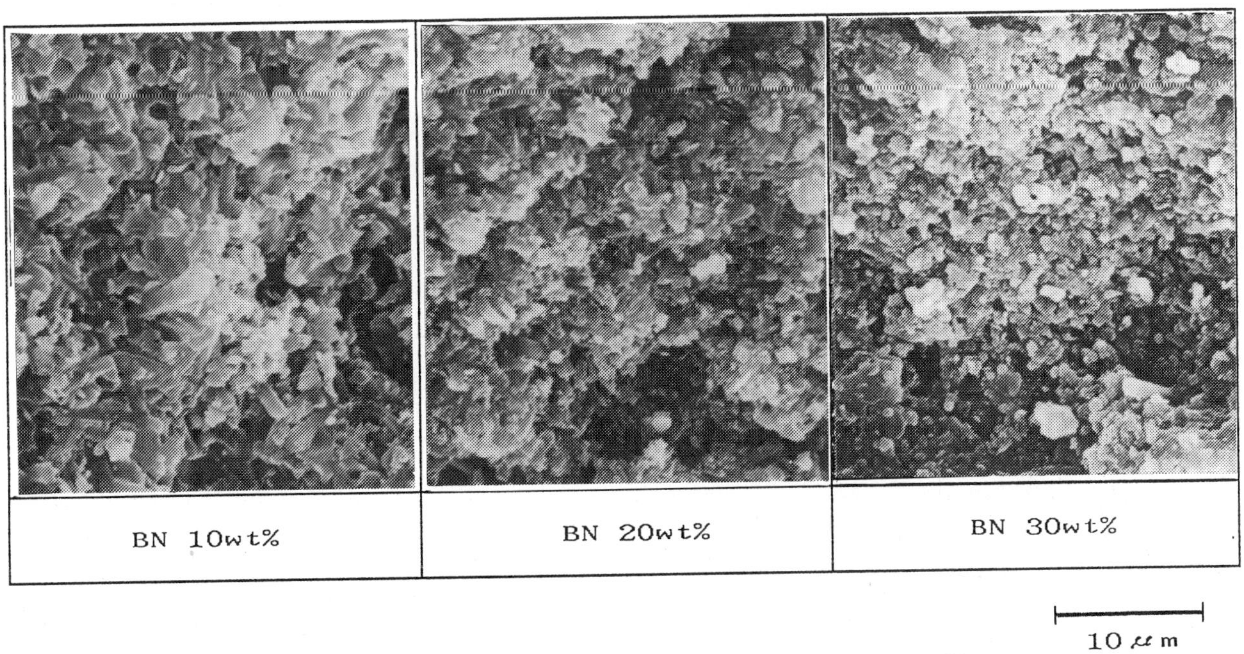

Fig. 2. SEM photographs [Si_3N_4(A)-BN].

Fig. 3. SEM photographs [Si_3N_4(B)-BN].

theoretical density calculated from composition of the composite ceramics. As shown in Fig.4, porosity increases with BN content, and this tendency is explained by the inhibition of densification by the BN component.

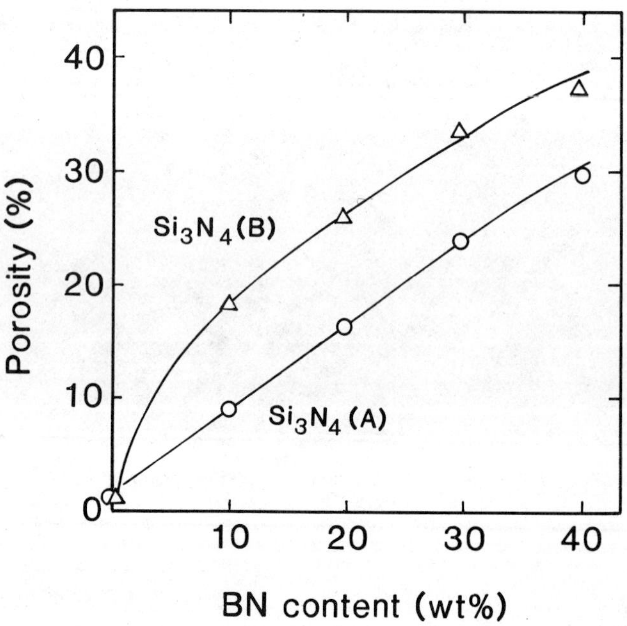

Fig. 4. Porosity of Si_3N_4-BN ceramics.

Bending Strength

Figure 5 shows the effect of BN content on the three point bending strength (σ_{b3}). Each point in Fig.5 is the average of twenty measurements. The decrease in bending strength with the increase in BN content can be explained by the increase in porosity as shown in Fig.4. The bending strengths of the Si_3N_4(A)-BN system are fairly higher than those of the Si_3N_4(B)-BN system. The difference in bending strength can also be mainly attributed to the difference of porosity.

One of the features of the present Si_3N_4-BN composite ceramics is high strength in spite of gas pressure sintering. The high bending strength of 530 and 320MPa is obtained in the case of Si_3N_4(A) 10 and 20wt%BN, respectively. Similarly, 350 and 200MPa are obtained in the Si_3N_4(B)-BN system. These values are higher than those of Si_3N_4-BN composite ceramics made by the reaction sintering method 4),5) by 2-4 times, and are almost equal to those of composite ceramics of the same system made by the hot-pressing method 2).

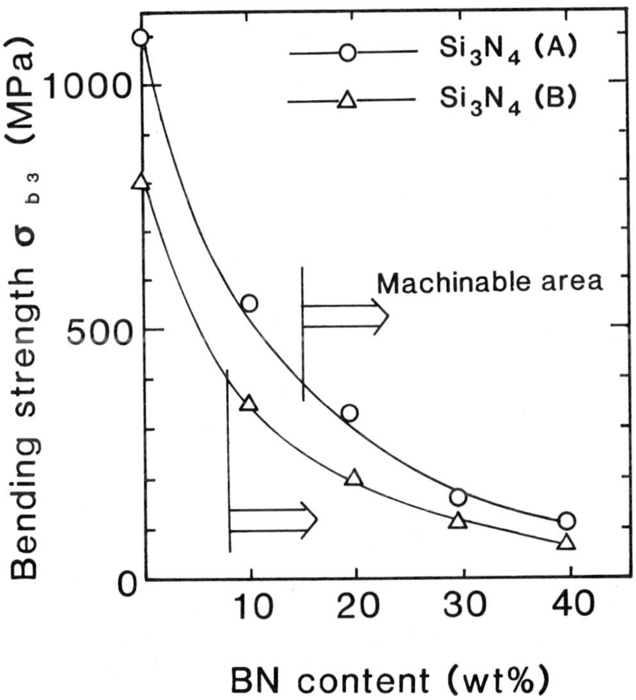

Fig. 5. Bending strength and machinability of Si_3N_4-BN ceramics.

Thermal Shock Resistance

Figure 6 shows the effect of BN content on thermal shock resistance(ΔT). ΔT is the temperature at which the bending strength decreases greatly after the thermal shock by water-quenching method. As shown in Fig.6, thermal shock resistance is improved as the BN content increases. ΔT is improved by 150-200°C and 300-350°C compared with dense monolithic Si_3N_4 ceramics by incorporating 10 and 30wt%BN, respectively.

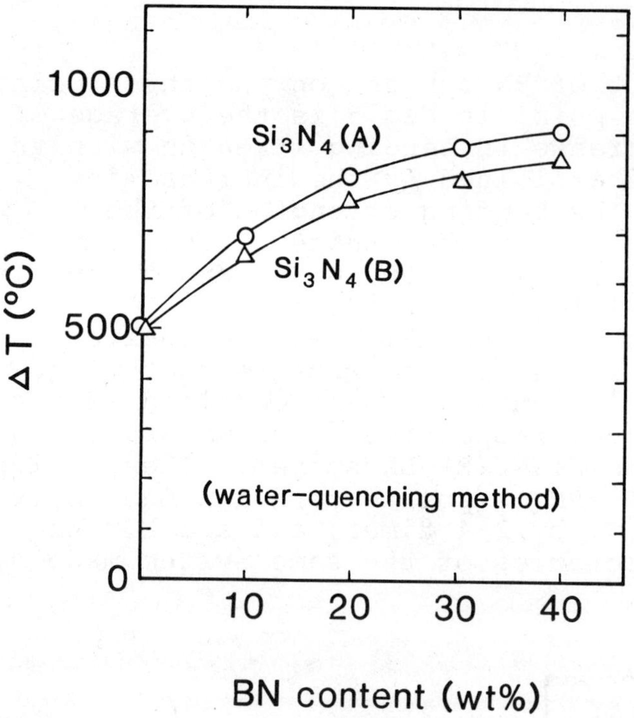

Fig. 6. Thermal shock resistence of Si_3N_4-BN ceramics.

In order to understand the thermal shock resistance behavior, the parameter (R) of thermal shock resistance expressed by Eq.(1) was applied. Table 2 shows Young's modulus, Poisson's ratio, and thermal expansion coefficient measured for composite ceramics of Si_3N_4(B) 10-40wt%BN.

$$R=S(1-\nu)/E\cdot\alpha \qquad (1)$$

S:tensile strength
ν:Poisson's ratio
E:Young's Modulus
α:thermal expansion coefficent

The calculated values of the parameter (R) are listed in Table 2, by using the bending strength instead of the tensile strength in Eq. (1). As shown in Table 2, the thermal shock resistance parameter (R) increases with the increase in BN content similarly to ΔT shown in Fig. 6. The improvement of the thermal shock resistance with the increase in BN content is mainly attributed to the reduction of thermal stress by the decrease in Young's modulus, Poisson's ratio, and thermal expansion coefficient.

It is obvious that the main factor which affects the thermal shock resistance is Young's modulus from comparison of Eq.(1) and Table 2. The decrease of Young's modulus with the increase in BN content is mainly due to the increase in porosity as shown in Fig.4.

Table 2. Various physical properties [Si_3N_4(B)-BN].

BN content (wt%)	Young's modulus (N/m^2)	Poisson's ratio	Thermal expansion coefficient(~1000℃) (1/℃)	Thermal shock resistance parameter (R) (℃)
10	1.50×10^{11}	0.21	3.4×10^{-6}	550
20	5.89×10^{10}	0.18	3.2×10^{-6}	850
30	3.41×10^{10}	0.13	3.1×10^{-6}	1120
40	1.71×10^{10}	0.13	2.7×10^{-6}	1280

Machinability

The composite ceramics in the present work have good machinability which is an important feature in practical use as engineering ceramics. The machinability of the ceramics was mainly tested by using a lathe with WC tools. As shown in Fig.5, the composite ceramics with more than 15wt%BN in the Si_3N_4(A)-BN system, and those with ≧ 10wt%BN in the Si_3N_4(B) BN system can be machined by a WC cutting tool. The appearance of machinability is considered to be due to the homogeneous dispersion of flaky BN powder. During the machining of this ceramic, microfractures introduced by the cutting tool propagate through dispersed flaky BN powder, then, the homogeneous dispersion and quantity of BN powder result in good machinability.

The machinability of this ceramic is such that even the ceramics with high strength over 300MPa can be machined by an ordinary WC cutting tool without using a diamond cutting tool. This is very advantageous for this ceramic to be applied to machine parts with complicated shapes.

SUMMARY

A new type of composite ceramics of the Si_3N_4-BN system has been developed by the slipcasting and gas pressure sintering method. The various mechanical properties of the composite ceramics were measured, and discussed in relation to the microstructures.

The following results were obtained;
(1) The microstructure of the composite ceramics with a low BN content(10wt%) had a relatively dense texture with rod-like grains of β-Si_3N_4 crystal. On the other hand, the microstructures of a high BN content (≧30wt%) ceramics had a granular texture. Porosity increased with increasing BN content.
(2) The bending strength decreased with the increase in BN content due to increase in porosity. High bending strengths of 530 MPa and 350MPa were obtained for the composite ceramics with 10wt%BN, composed of Si_3N_4(A) and (B) powder, respectively.
(3) The thermal shock resistance evaluated by the water-quenching method was improved with the increase in BN content. The thermal shock resistance temperature (ΔT) was improved by 150-200°C and 300-350°C compared with monolithic Si_3N_4 ceramics by incorporating 10 and 30wt%BN, respectively.
(4) The improvement of the thermal shock resistance with the increase in BN content was considered to be mainly attributed to the reduction of thermal stress by the decrease in Young's modulus, Poisson's ratio and thermal expansion coefficient of the composite ceramics.
(5) Composite ceramics with more than 15wt%BN are readily machinable and can be machined by a WC cutting tool. Good machinability was attained even for the ceramics having relatively high strength over 300MPa.

REFERENCES

1) K. Niihara, L. D. Bentsen, and D. P. H. Hasselman, "Anisotropy Effects in the Thermal Diffusivity of Si_3N_4-BN Composites", J.Am. Ceram. Soc. 64 C117-118 (1981)
2) K. S. Mazdiyasni and R. Ruh, "High/Low Modulus Si_3N_4-BN Composite for Improved Electrical and Thermal Shock Behavior", J. Am. Cerm. Soc. 64 7 415-419 (1981).
3) M. Iwasa and S. Kakiuchi, "Mechanical and Tribological Properties of Si_3N_4-BN Composite Ceramics", Yogyo-Kyokai-shi, 93 10 661-665 (1985)
4) A. Miyamoto, M. Ishikawa, M. Nishi, T. Ishizawa, and A. Shironita "Si_3N_4-BN Composite Ceramics and their Production Method." Jpn. Tokkyo Koho JP60-2 2,676 (85-22,676), 3 Jun 1985
5) K. Ichikawa and K. Katayama, "High Density Composite BN Ceramics and their Production Method" Jpn. Kokai Tokkyo Koho JP63- 40,771 (88-40,771), 22 Feb 1988
6) M. Mitomo and K. Mizuno, "Sintering Behavior of Si_3N_4 in 10 atm Nitrogen" Proceedings of Second International Symposium on Ceramic Materials and Components for Engines, Lübeck-Travemünde (FRG) 263-270 (1986)
7) M. Mitomo and K. Mizuno, "Sintering Behavior of Si_3N_4 with Y_2O_3 and Al_2O_3 Addition", Yogyo-Kyokai-shi, 94 1 106-111 (1986)

MECHANICAL CHARACTERIZATION OF SiC WHISKER-REINFORCED $MoSi_2$

D. H. Carter, W. S. Gibbs, and J. J. Petrovic

ABSTRACT

The mechanical characteristics of an intermetallic matrix with two different reinforcements were studied. The matrix material was $MoSi_2$, with either Los Alamos VLS SiC whiskers or Huber VS SiC whiskers. The purpose of the reinforcement was to provide toughening at ambient temperature and strengthening at elevated temperatures. The VLS whiskers greatly improved the yield strength of the matrix at 1200 °C, and also increased the room temperature fracture toughness of the matrix. The VS whiskers were added because they are much smaller in length and diameter, and therefore decreased the mean free path between whiskers, at the same volume fraction. The VS whiskers improved the toughness of the matrix at ambient temperature, and increased the yield strength of $MoSi_2$ at 1400 °C by 470%. The high strength of this new composite places this material in the realm of attractive engineering materials for high-temperature applications.

INTRODUCTION

Ceramic materials have some outstanding properties, such as high temperature strength, thermal shock and fatigue resistance, corrosion resistance, low density, and low thermal expansion which make them attractive materials for high temperature applications. For instance, if ceramics were used as engine components, the engine could be run at a higher temperature, and thus be much more efficient than it could be with metal components. This has sparked a great deal of interest in studying different ceramics for these types of applications, and has generated a wealth of knowledge of the properties of various ceramics.

Ceramics differ from metals in one very key aspect in that they are brittle, and do not show any yield upon loading at ambient temperatures. This lack of a stress-relieving characteristic, giving ceramics their brittle nature and low tolerance for flaws, has been a major drawback to using them in structural applications. One of the methods of dealing with this mechanical property, which can be referred to as its fracture toughness, has been the development of ceramic composites. Many authors[1,2,3] have summarized the mechanisms which may act in toughening ceramic composites. These include the transfer of load from the matrix to the fiber based on elastic modulus, microcracking or prestressing due to a difference in thermal expansion, crack deflection, and phase transformation toughening.

Ceramic composites still pose a challenge to the engineer, since the matrix is very brittle. There is a class of materials, however, which offers the advantages of a ceramic and also some of the beneficial mechanical characteristics of a metal. These materials are intermetallics, which at high temperature, have the excellent properties of a ceramic, but mechanically behave more like a metal, since they show yielding and stress-relieving characteristics. The focus of this work is on an intermetallic, molybdenum disilicide ($MoSi_2$), and composites based on this material.

Molybdenum disilicide has a melting temperature of 2010 °C, higher than the aluminides, and has excellent corrosion and oxidation resistance, almost as good as that of silicon carbide (SiC). Its oxidation resistance is due to the formation of a silica (SiO_2) layer, which acts as a protective film at high temperatures.[4] Mechanically, $MoSi_2$ behaves as a metal at high temperatures, since it undergoes a brittle-to-ductile transition at approximately 1000 °C. This has the advantage of giving $MoSi_2$ a stress-relieving characteristic, but since it undergoes creep and plastic deformation, its high-temperature strength is reduced. The other disadvantage with $MoSi_2$ is that it is brittle at lower temperatures.

Gac and Petrovic[5] showed that the addition of vapor-liquid-solid (VLS) SiC whiskers as a reinforcing medium may improve the fracture toughness at room temperature when the matrix is brittle, as well as increase the strength at elevated temperatures when the matrix is ductile. This work and others since then[6] have shown this to be correct. The improvements in mechanical properties of this material have been very encouraging. They have provided insight into the toughening and strengthening mechanisms, and also which of these mechanisms may provide the best results in this system. However, the strength and toughness values previously attained have not been attractive for most high-temperature structural applications.

In more recent studies, a new reinforcement for $MoSi_2$ has been examined. This new reinforcement is a vapor-solid (VS) whisker made by the Huber Corporation. It is a much smaller whisker, and it was thought that this may give improved strengthening due to the resultant shorter mean free path. The results from the mechanical testing of this composite will be analyzed to determine what mechanisms are involved and how to best utilize the advantageous properties of this intermetallic.

This paper will summarize results of both past and current studies in $MoSi_2$ composites, and discuss the goals of this research being performed at the Los Alamos National Laboratory. This summary is expanded upon by Carter,[7] where the details of the experiments can be found.

PROCEDURE

The procedure for fabricating both types of $MoSi_2$ composites was the same. The materials used were a pure $MoSi_2$ powder,[a] and 20 vol% of either Los Alamos VLS SiC whiskers[b] or Huber VS SiC whiskers.[c] The main difference between the two whiskers used is their size. In the final hot-pressed samples, the Los Alamos VLS whiskers were generally 100 to 200 microns long, 5 microns in diameter, and had an aspect ratio of about 20 to 30:1. The Huber VS whiskers, however, were generally 1 to 5 microns in length, had a diameter of about 0.1 microns, and had an aspect ratio of 10:1.

The powder and whiskers were weighed out and then blended dry for one minute in a small high-speed blender,[d] similar to a coffee grinder. The powder mixture was then placed in a grafoil-lined die for hot-pressing into disks. The disks measured 31.75 mm in diameter by 6.35 mm in thickness.

The samples were hot-pressed in argon at 1700 °C to 30.5 MPa, with a hold time at the peak temperature of about 5 minutes. The resulting densities were very good, typically between 97.5% and 98.4% of theoretical, which is 5.691 g/cm^3.

There was no apparent reaction between the SiC whisker and the matrix, as expected through thermodynamic calculations. The earliest study showed that the whisker was attacked, however this was attributed to iron and aluminum impurities present in the $MoSi_2$. This study also found substantial grain growth in the matrix, and attributed this to the impurities present. In more recent studies, a much purer powder has been used, and so the grain size of both the matrix and the composite were 5 to 10 microns, similar to the starting powder size.

Spectrochemical analyses indicated that the current $MoSi_2$ powder is significantly lower in Fe (300 wppm as compared to 1000 - 4000 previously), Al (100 wppm as compared to 1000 previously), Cr (50 wppm as compared to 250 previously), Ca (1 wppm as compared to 200 previously) and Co (<30 wppm as compared to 100 previously).

---

a. Stock #48108, 99.9% pure $MoSi_2$ powder from Alfa Products, Danvers, MA 01923, a division of Morton Thiokol, Inc.

b. Type 4-6A VLS beta-SiC Whiskers, Los Alamos National Laboratory, Los Alamos, NM 87545.

c. "XPW2 Silicon Carbide Whisker" from J. M. Huber Corporation, Borger, Texas, 79008-2831.

d. Ika Werk Model A109, Janke and Kunkle KG Staufen i Breisgan, West Germany.

Four-point bend strength and four-point bend Chevron-notched fracture toughness tests were performed on all three materials at room temperature, 1200 and 1400 °C. Further bend strength tests were performed on VS SiC whisker-reinforced $MoSi_2$ at 1000, 1100 and 1300 °C. Between 2 and 4 samples were tested at each condition. The bend samples measured 20 x 5.1 x 2.5 mm.

RESULTS AND DISCUSSION

The impurities present in $MoSi_2$ in an earlier work also caused the formation of a liquid phase at the grain boundaries during hot-pressing. This liquid phase, rich in silicon, aided the powder to completely surround the whiskers, whereas in the pure $MoSi_2$, the $MoSi_2$ powder did not flow evenly around every whisker. The porosity in each study was similar, however the earlier study exhibited a finer distribution of porosity, whereas the pure $MoSi_2$ caused more porosity surrounding individual whiskers. This difference caused the samples with the impure $MoSi_2$ to have a higher fracture toughness than those with the pure $MoSi_2$. The interface between the $MoSi_2$ and the whisker was weaker in the impure $MoSi_2$ samples, thus causing a higher fracture toughness.

Figures 1 and 2 show the $MoSi_2$ matrix and composite with VLS whiskers under polarized light. The size of the whisker is apparent. Also, it is evident that the whiskers break during hot-pressing. It can be seen that some whiskers are surrounded by pores. This occurs especially by two or more whiskers in very close proximity. This lowers the strength of the composite, since the matrix cannot transfer load to the whisker if there is a pore surrounding the whisker.

Figures 3 and 4 show the composite with Huber whiskers. The whisker clumps, in Figure 3, align in a plane perpendicular to the hot-pressing direction, as expected. There is a large amount of porosity within these agglomerations of whiskers. This is most likely the cause for the low strength of this composite at room temperature. The strength at high temperatures is not as sensitive to these pockets of flaws. It is evident that there is a large amount of SiC particulate along with the whiskers.

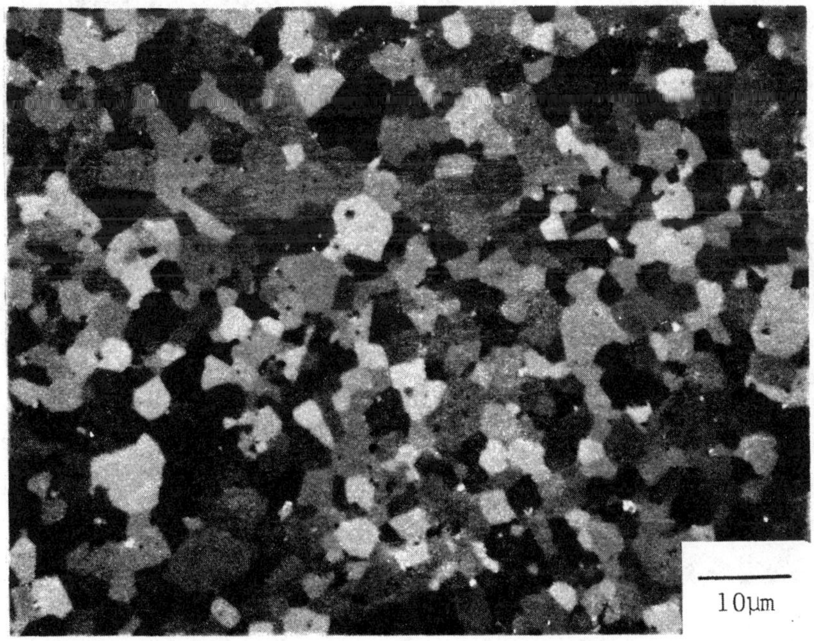

Figure 1: $MoSi_2$, Polarized Light

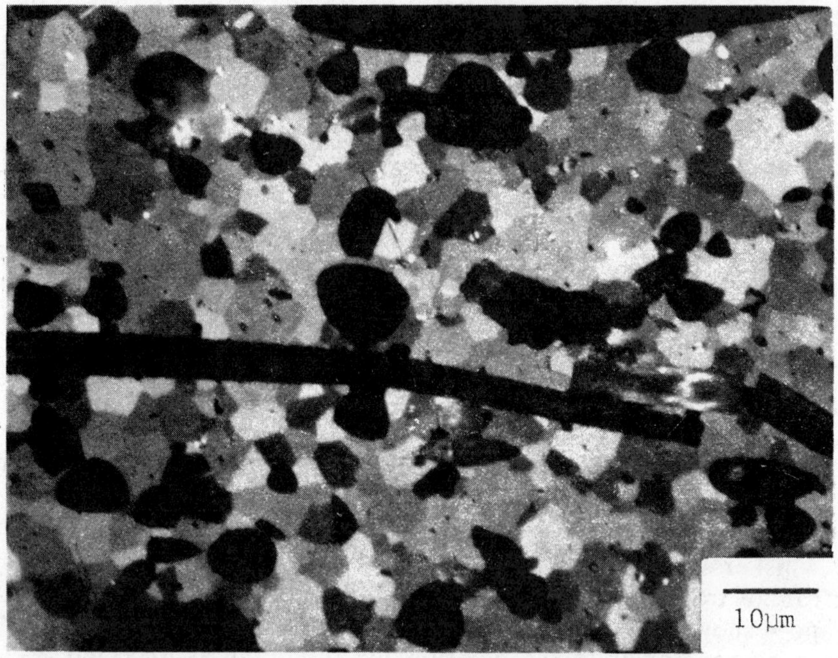

Figure 2: VLS SiC(w) - $MoSi_2$, Polarized Light

Figure 3: VS SiC(w) - $MoSi_2$

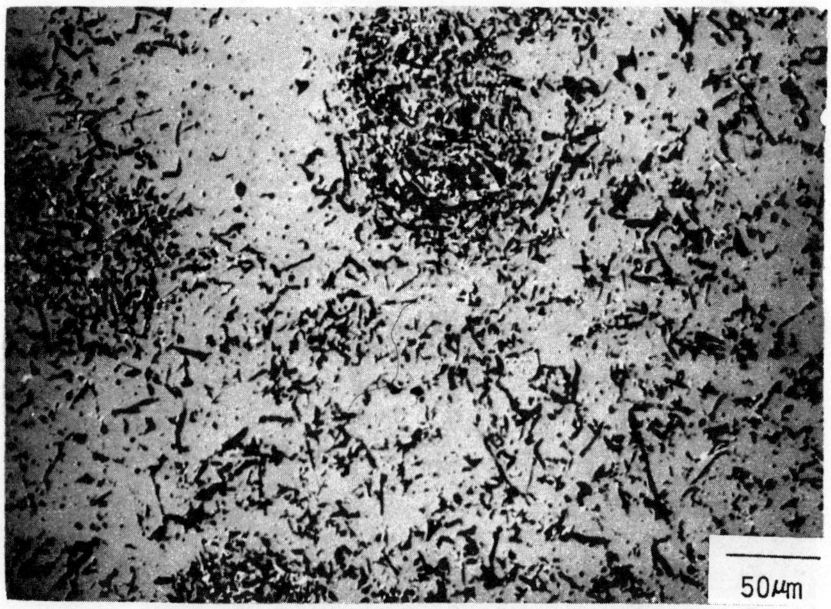

Figure 4: VS SiC(w) - $MoSi_2$, Polarized Light

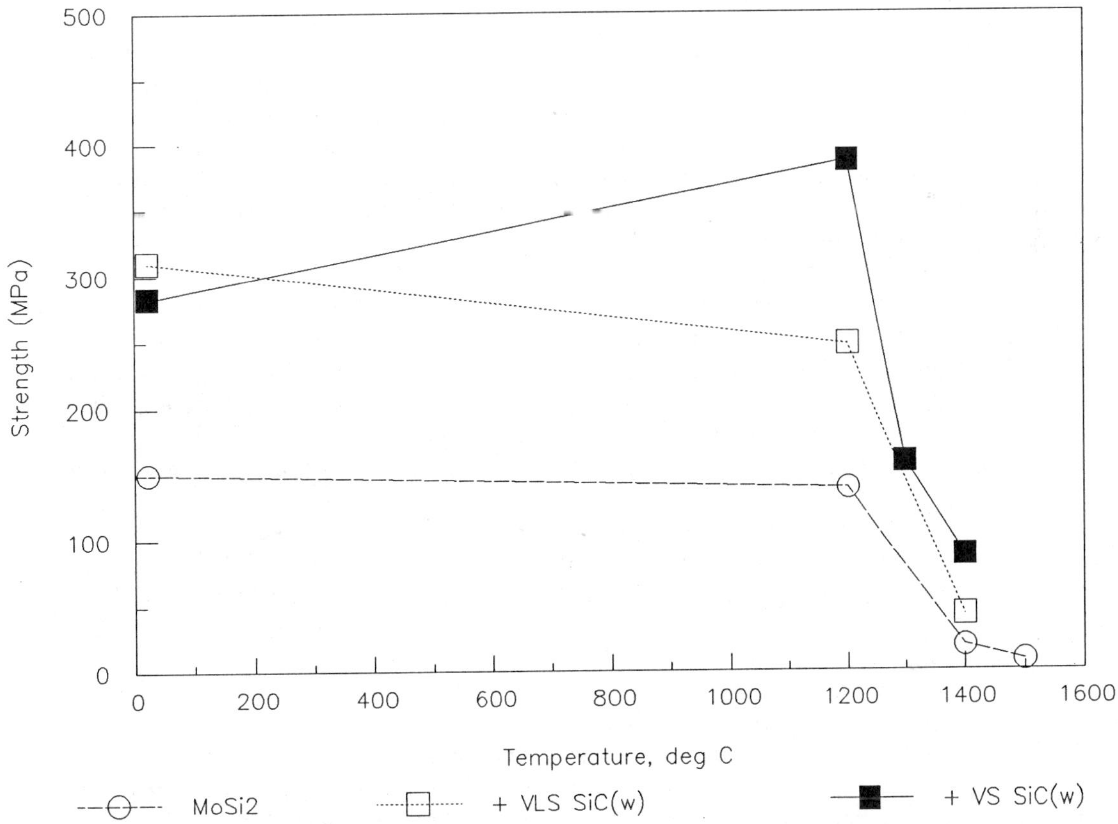

Figure 5: 0.2% Yield Strength as a Function of Temperature

The strength of the $MoSi_2$ at 1200 °C has been doubled by the addition of VLS whiskers, and tripled with the addition of VS whiskers. At 1400 °C the VS whiskers increase the strength of the matrix by 470%.

The new composite, using VS whiskers instead of VLS whiskers, improved the yield strength at 1200 °C, as shown on Figure 5. At elevated temperatures, the 0.2% offset yield strength for the materials are reported, since the samples underwent plastic deformation. Ultimate strength values above 1000 °C would not be an accurate measure of the strength of these materials because of this plastic deformation. At room temperature, the values on Figure 5 are actually the ultimate strengths, since the samples did not yield.

The strengthening provided by the VS whiskers is due to an entirely different mechanism than that which the VLS whiskers provide. Load transfer is a plausible model for high temperature strengthening by VLS whiskers, but not by VS whiskers. Recall that the VLS whiskers are at least twenty times longer than the VS whiskers. The important point is that the load transfer model requires a critical aspect ratio for optimum strengthening. This was calculated to be 12 for the purely elastic case, and 53 for the plastic case. This was found by applying a simple "rule of mixtures" to the elastic modulus, assuming the strain in the matrix and the strain in the fibers were equivalent.[8] The VS whiskers are much shorter, relative to the grain size of $MoSi_2$, and also have a smaller aspect ratio, and therefore are not long enough to provide adequate strengthening due to load transfer.

$MoSi_2$, like most ceramics, exhibits brittle behavior at room temperature due to the lack of independent active slip systems at room temperature. Neither the strength nor toughness of ceramics is controlled by dislocation motion, as with metals, because dislocations are either immobile, or are not mobile on enough slip systems to influence toughness.[3] Therefore, ceramics are made tougher by modifying the microstructure, and providing resistance to fracture via these microstructural modifications, and not by the movement of dislocations.

This reasoning cannot be applied to the behavior of $MoSi_2$ at high temperature. At elevated temperatures this material did not behave as a conventional brittle ceramic. It behaved more as a metal, in that it underwent plastic deformation, and exhibited creep and stress relaxation.

The bend test specimens bent to a very substantial degree at and above 1200 °C. However, the addition of SiC whisker reinforcement improved the materials load-bearing capability. Therefore, a different way of examining fracture in $MoSi_2$ must be used to fully understand its behavior, and how to control it. It is proposed that the mode of fracture in $MoSi_2$ at elevated temperatures is plastic deformation through dislocation motion. This could be due to either grain boundary sliding or dislocation plasticity.[9]

Very often in a two phase system such as this, one indication of grain boundary sliding, or diffusional creep, would be large void formations at boundaries between the two phases. Through SEM studies, there is no direct evidence of void formation at the grain boundaries.

Therefore, it is proposed that the deformation is due to classical dislocation plasticity, rather than grain boundary sliding, or diffusional creep. Transmission electron microscopy is presently being performed on these samples to locate regions of dislocation plasticity, and to provide more direct evidence to support this proposal.

The explanation for the increase in strength with the VS whisker composite, over the VLS whisker composite at elevated temperatures, follows readily from the model of dislocation plasticity. The mean free path is smaller for the VS whisker composite than for the VLS whisker composite, and thus strengthening due to dispersion strengthening is more effective. The VS whiskers are acting as "pinning sites" to control the dislocation motion.

Dieter[10] gives a summary of the strengthening mechanisms produced by a finely dispersed insoluble second phase in a matrix, which is applicable to this system. The degree of strengthening resulting from second-phase particles depends on the distribution of particles in the ductile matrix. For a constant volume fraction of reinforcement, a decrease in reinforcement size will decrease the average distance between the reinforcing particles, or the mean free path. This was one of the intentions of adding VS whiskers as opposed to VLS whiskers; to reduce the mean free path.

The VLS whisker composite, as would be expected from this discussion, showed a decrease in strength at 1200 °C. This is because the large whiskers are ineffective for controlling dislocations, since the mean free path is very large relative to the matrix grain size.

The strength at room temperature is very much dependent on the porosity of the sample, because there is no plastic deformation. The porosity for the VS samples was fairly high within the large groups of whiskers. These large whisker clumps acted as large flaws, and detracted from the composites room temperature strength. If these large clumps can be dispersed better, both the strength and fracture toughness will be improved, especially at room temperature. Note that it is not necessarily the clumps of whiskers themselves, but rather the porosity within these clusters which detracts from the room temperature strength. The elevated temperature strength is not as sensitive to these pores, because of the large degree of plastic deformation.

The low strength of the VLS composite at room temperature is also due to pores, but in this case the pores are surrounding individual whiskers. Figure 6 shows this porosity quite clearly. The matrix cannot possibly transfer load to the whisker if there are large gaps between the matrix and the whisker. This must detract from the material's strength.

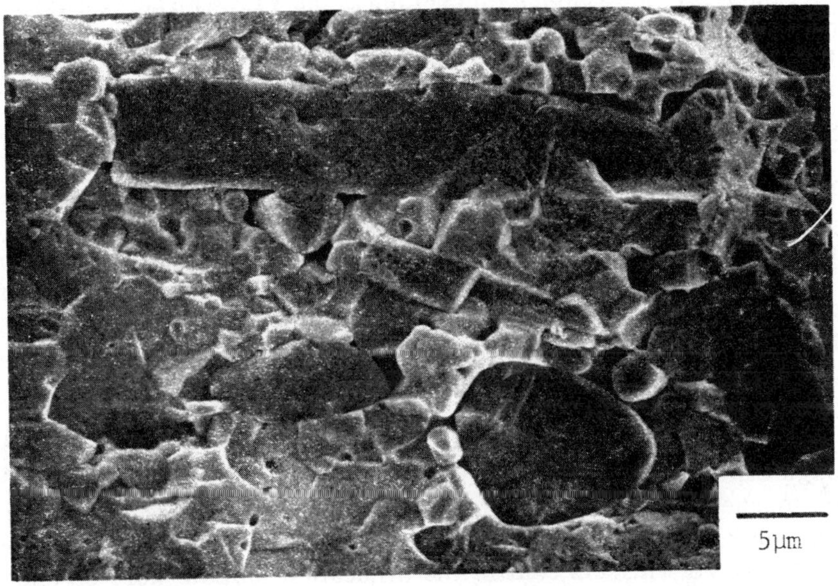

Figure 6: VLS SiC(w) - $MoSi_2$, Tested At Room Temperature, 2KX

The K_{IC} values are summarized in Table 1. Only the room temperature fracture toughness values can be considered K_{IC} values. At elevated temperatures, the fracture was not linear elastic. Nonlinear fracture mechanics must therefore be applied. The fracture toughness at elevated temperature is not reported as K_{IC}, but as work of fracture. The work of fracture values are discussed in detail by Carter,[7] and so need not be reported here.

Table 1: Room Temperature Fracture Toughness Data

	$MoSi_2$	$MoSi_2$ + VLS SiC(w)	$MoSi_2$ + VS SiC(w)
$MPa{\cdot}m^{\frac{1}{2}}$ ($ksi{\cdot}in^{\frac{1}{2}}$)	5.32 (4.84)	8.20 (7.45)	6.59 (5.99)

The fracture toughness results are also encouraging. It has been proposed that particle dispersion strengthening accounts for the dramatic increase in strength for the VS whisker composite at high temperature. Though this is the most important result from this study, the fracture toughness is also important, since it is directly related to the mechanisms of strengthening given above. The strengthening mechanisms discussed lead to an explanation of why the fracture toughness at room temperature was not improved by the VS whiskers as much as with the VLS whiskers.

At room temperature, the matrix is brittle, and so toughening must occur by one or more of the typical ceramic toughening mechanisms, such as crack deflection, crack bridging, or whisker pullout. For any of these mechanisms, the purpose is to interrupt the path of the propagating crack by lowering the stress field around the crack tip, and in front of the crack tip. It is desired to have long thin fibers in order to effectively stop cracks by the aforementioned mechanisms. The critical length to diameter ratio has been calculated by Gac et al.[8] and is close to that of the VLS whiskers.

The smaller size of the VS whisker is suited for controlling dislocation motion at high temperatures, but cannot provide as much toughening as the VLS whiskers do at room temperature, when there is no dislocation plasticity.

One of the mechanisms of toughening which has been observed in the VLS composite is crack deflection.[11] Other mechanisms are possible, such as crack bowing, branching, bridging, and microcracking. There is no evidence of whisker pullout, or debonding at room temperature. The microstructure of the room temperature test specimens show a brittle fracture, as expected.

When the whisker is surrounded by the matrix, the bond is normally very strong. This is shown explicitly in Figure 6. The crack propagated straight through the large horizontal whisker, leaving each half of the whisker bonded to the matrix. A strong bond is good for strengthening of many metal and polymer composites, but too strong of a bond is not good for the toughness of most ceramic composites since most reinforcements have a modulus on the same order as the matrix. Therefore the propagating crack would not be deflected, or the whisker would not show any debonding or pullout if the bond strength was too high.

At high temperature, the toughening mechanism provided by the VS whiskers is related to its strengthening mechanism. Since the fracture at high temperature is through plastic deformation rather than through brittle fracture, the control of dislocation motion will give a higher work of fracture than the toughening mechanisms designed for brittle fracture. Therefore, the work of fracture is much higher for the VS whisker composite than for the VLS whisker composite or the matrix because the VS whiskers control the dislocation motion more efficiently.

CONCLUDING REMARKS

The most important result from this study is the improvement in strength of $MoSi_2$ due to the VS whisker reinforcement. It has been found that the decreased mean free path with the VS whisker reinforcement as compared to the VLS whisker reinforcement served to increase the high temperature strength of $MoSi_2$ more than 450%. The reason for this is that at elevated temperatures the mode of failure of $MoSi_2$ is through dislocation plasticity, and the small VS whiskers acted in a dispersion strengthening mechanism, to inhibit this dislocation motion.

Since there is little or no dislocation motion in $MoSi_2$ at room temperature, the small whiskers had little effect on the mechanical properties at room temperature, specifically its fracture toughness. Though the VS whiskers did improve the fracture toughness slightly, the larger VLS whiskers are much better suited for the typical ceramic toughening mechanisms such as crack deflection.

In conclusion, the results from the mechanical tests performed in this work are extremely significant. It has been shown that $MoSi_2$ composites may indeed be useful materials for engineering applications at elevated temperatures. Substantial gains have been made in the understanding of the strengthening mechanisms important in $MoSi_2$ at high temperatures, and also the relevant toughening mechanisms at room temperature. These results are also very useful, in that they clearly dictate the logical path for future work on this system.

FUTURE WORK

First, it has been shown that a dramatic improvement in high temperature strength was obtained by utilizing a smaller whisker, and thus a shorter mean free path. A similar effect should be attained by the addition of fine SiC powder. This would enable a better control of the size of reinforcement. The clumps of whiskers introduced pores, which acted as flaws, and substantially decreased the room temperature strength. One solution would be to improve the method of processing to disperse these clumps of whiskers by using a wet-blending process, or one could simply add fine SiC powder instead.

Secondly, the VLS whiskers provided a better source of toughening at room temperature than the VS whiskers, because of their much larger size. Typical ceramic toughening mechanisms are dominant at room temperature, and so the composite must be designed with this in mind. The stress field at a crack tip must be minimized by alterations of the microstructure, rather than relying on inhibitors to dislocation motion, since dislocation plasticity does not exist at room temperature.

Therefore, the logical reinforcement for $MoSi_2$ would be a combination of VLS whiskers and fine SiC powder, which would take advantage of the relevant strengthening or toughening mechanisms at room temperature and at high temperatures.

There are other possible approaches which can be drawn from this work to improve the mechanical behavior of $MoSi_2$. First of all, the processing methods could be improved a great deal. Dry-blending is quick and easy, but does not give optimum properties. Also, one might attempt to increase the VLS whisker loading, which would probably necessitate a wet-blending technique. An increased whisker loading would help to decrease plastic deformation at high temperature, and also may help increase the fracture toughness at room temperature.

It may be more useful, however, to alter the composition of the composite. For instance, to further reduce the plastic deformation of $MoSi_2$ at high temperature, and thus improve its creep resistance, one might alloy the $MoSi_2$ with a more refractory material, such as tungsten disilicide.

REFERENCES

1. R. W. Rice, "Mechanisms of Toughening in Ceramic Composites," *Ceramic Engineering Science Proceedings.* **2** [7-8] 661-701 (1981).

2. D. K. Shetty, "Ceramic Matrix Composite," *Current Awareness Bulletin, Issue No. 118.* Metals and Ceramics Information Center, Battelle Columbus Laboratories, Columbus, Ohio, 1982.

3. S. M. Wiederhorn, "Brittle Fracture and Toughening Mechanisms in Ceramics," *Annual Review of Materials Science.* **14** 373-403 (1984).

4. J. Schlichting, "Molybdenum Disilicide As a Component of Modern High-Temperature Composites" (in German), *High Temperatures-High Pressures.* **10** [3] 241-269 (1978).

5. Frank D. Gac and John J. Petrovic, "Feasibility of a Composite of SiC Whiskers in an $MoSi_2$ Matrix," *Journal of the American Ceramic Society.* **68** [8] C200-C201 (1985).

6. W. S. Gibbs, J. J. Petrovic, and R. E. Honnell, "SiC Whisker-$MoSi_2$ Matrix Composites," *Ceramic Engineering and Science Proceedings.* **8** [7-8] 645-648 (1987).

7. David H. Carter, *SiC Whisker-Reinforced $MoSi_2$.* MIT Master's Thesis, 1988.

8. Frank D. Gac, John V. Milewski, John J. Petrovic, and Peter D. Shalek, "Silicon Carbide Whisker Reinforced Glass and Ceramics," *NASA Conference Publication 2406.* Proceedings of a joint NASA/DOD conference held at Cocoa Beach, Florida, January 23-25, 1985.

9. W. D. Kingery, H. K. Bowen, and D. R. Uhlmann, *Introduction to Ceramics 2nd Edition.* John Wiley & Sons, Inc., New York, 1976.

10. George E. Dieter, *Mechanical Metallurgy, 2d Ed.* McGraw-Hill Book Company, New York, 1976.

11. David H. Carter and George F. Hurley, "Crack Deflection as a Toughening Mechanism in SiC-Whisker-Reinforced $MoSi_2$," *Journal of the American Ceramic Society.* **70** [4] C79-C81 (1987).

FURTHER STUDIES OF THE LOW-TEMPERATURE OXIDATION PROBLEM IN YTTRIA-DENSIFIED SILICON NITRIDE CERAMICS

J.K.Patel and D.P.Thompson
Wolfson Laboratory,
Department of Metallurgy and Engineering Materials,
University of Newcastle upon Tyne, UK.

ABSTRACT

The catastrophic 1000°C oxidation behaviour of silicon nitride ceramics densified with >10 w/o yttria is now well known and several solutions to the problem have been suggested. However, the addition of silica (to avoid the formation of the melilite phase believed to be responsible for this behaviour) results in a material with worse mechanical properties especially at elevated temperatures. Alternatively a post-preparative flash-oxidation treatment is only a temporary solution and low p_{O_2} in-service conditions may result in the original problem recurring.
In the present paper it is shown that melilite is not the only cause of the catastrophic low-temperature oxidation behaviour and results are presented for additional silicon nitride materials which exhibit the effect. A more general explanation is afforded based on specific volume change such that if the total volume change is greater than 1 v/o, catastrophic oxidation will occur. Calculations for several sialon and silicon nitride ceramics show that yttrium densified materials are particularly bad because of the relatively low densities of yttrium silicates, whereas in equivalent rare earth systems, the specific volume changes are much less and the problem can in many cases be avoided.

INTRODUCTION

Catastrophic cracking of yttria-densified silicon nitride ceramics during oxidation at 1000°C was reported and examined by several authors some ten years ago[1,2]. More recently, additional studies have been carried out[3-10] which make the nature of the problem much clearer. The problem was believed to be due to the presence of crystalline grain-boundary phases in the final product (especially the nitrogen melilite phase, $Y_2O_3.Si_3N_4$) which started to oxidize at 1000°C and exhibited a very large volume expansion on conversion to yttrium silicate. This phenomenon is particularly important because it is the only disadvantage of what otherwise are excellent high-temperature materials. In Figure 1 for example (after Jack[2]), the lower specimen is of the same composition as the upper one (Si_3N_4 + 15 w/o Y_2O_3) but on oxidation at 1400°C

Figure 1. Silicon nitride hot-pressed with 15 w/o Y_2O_3 (a) after oxidation for 24 hours at 1000°C and (b) after oxidation for 24 hours at 1400°C.

gave very low weight gains and only a slight glaze appeared on the surface. Furthermore, the high eutectic temperatures at which liquid phase forms in these compositions combined with the ability of the yttrium silicon oxynitride phases to "mop up" residual metallic cations were additional benefits. The initial solution to the problem, i.e. modifying the composition by silica additions into a "safe" compatibility triangle which avoided the melilite phase[1] successfully overcomes the problem but at the same time the resulting materials do not exhibit quite the same high temperature properties because the overall composition is now nearer the eutectic in the system and liquid phase occurs at lower temperatures. In that sense, silica addition must be regarded as an unsatisfactory solution to the 1000°C oxidation problem.
The flash oxidation technique[11,12] is a more satisfactory solution to the problem. This consists of pre-oxidizing the specimen at a temperature above the catastrophic region so that the surface is covered with a silica-based protective glaze. On subsequent reoxidation at temperatures in the catastrophic range, the oxidized surface protects against oxygen ingress and catastrophic cracking is avoided. Again this method has disadvantages which dictate against it being a permanent solution to the problem. Firstly, if the material is used in an application where the surface is abraded by wear, it is likely that the pre-oxidized layer will be either removed or locally impaired. Subsequent exposure to air at 1000°C will once again activate catastrophic cracking. Alternatively, if the material is exposed to a reducing environment, the surface layer will be removed and catastrophic behaviour returns. This must also therefore be regarded as a solution wich is applicable only in selected circumstances.
The new input provided by the present work is in three main areas. Firstly it is shown that the problem is not merely due to melilite (or other Y-Si-O-N crystalline phase) because it can equally well occur in silicon nitride samples containing a glass as the only second phase. Secondly, the volume expansion occurring during oxidation has been put on a more quantitative basis than in previous investigations and this can be used to predict whether a particular glassy or crystalline second phase will result in catastrophic behaviour. Finally the merits of rare-earth Ln-Si-O-N phases are assessed as secondary grain-boundary phases for silicon nitride ceramics.

EXPERIMENTAL

Silicon nitride powder* was wet-milled in iso-propanol for 3 days with 10 w/o yttria$ and 2 w/o magnesia+ additions using zirconia media, and the resulting mix dried and compacted into pellets by die-pressing and cold isostatic pressing. Pressureless sintering was carried out in both carbon and tungsten element furnaces in nitrogen at temperatures in the range 1600-1750°C and fully dense products were obtained[13]. Similar specimens were also hot-pressed at 1700-1800°C and 30 MPa pressure either with or without the magnesia, and further compositions were hot-pressed with up to 14 w/o of La_2O_3≠ or Nd_2O_3≠ in addition to the 10 w/o yttria. Table 1 summarizes details of the specimens used. Samples were oxidized in a thermobalance at temperatures of up to 1500°C in air and weight gains measured as a function of temperature and time.

Table 1. Details of Specimens used in the Present Study.

Specimen No.	Composition (w/o) Si_3N_4	Y_2O_3	MgO	La_2O_3	Nd_2O_3	Furnace	T(°C)	Time(h)	Products by X-ray analysis β-Si_3N_4	Apatite	Glass
1	88	10	2			CR	1700	4	✓		✓
2	90	10				HP	1750	1	✓	✓	✓
3	88	10		2		HP	1750	1	✓	✓	✓
4	84	10		6		HP	1750	1	✓	✓	✓
5	80	10		10		HP	1750	1	✓	✓	✓
6	76	10		14		HP	1750	1	✓	✓	✓
7	88	10		2		HP	1750	1	✓	✓	✓
8	84	10		6		HP	1750	1	✓	✓	✓
9	80	10		10		HP	1750	1	✓	✓	✓
10	76	10		14		HP	1750	1	✓	✓	✓

CR: Carbon Resistance Furnace HP: Hot-pressed.

RESULTS

Figure 2 shows weight gains for silicon nitride densified with 10 w/o yttria and 2 w/o magnesia oxidized at various temperatures in the range 900-1400°C in air. Prior to oxidation, the specimens consisted of β-silicon nitride plus glass. Catastrophic behaviour (see Figure 3) occurs in the 1000-1100°C region and the improvement in oxidation due to passivation is clearly seen in the 1200-1300°C range. At higher temperatures ($>$1400°C) diffusion through the glassy surface layer causes significant weight gains. Similar specimens which had been completely devitrified by heat-treatment over a 6-day period at a range of temperatures between 1350 and 1150°C (each day the temperature

* Starck-Berlin, Grade LC10 $ Molycorp, 99.9%

\+ BDH, Analar Grade. ≠ Rare Earth Products, 99.9%.

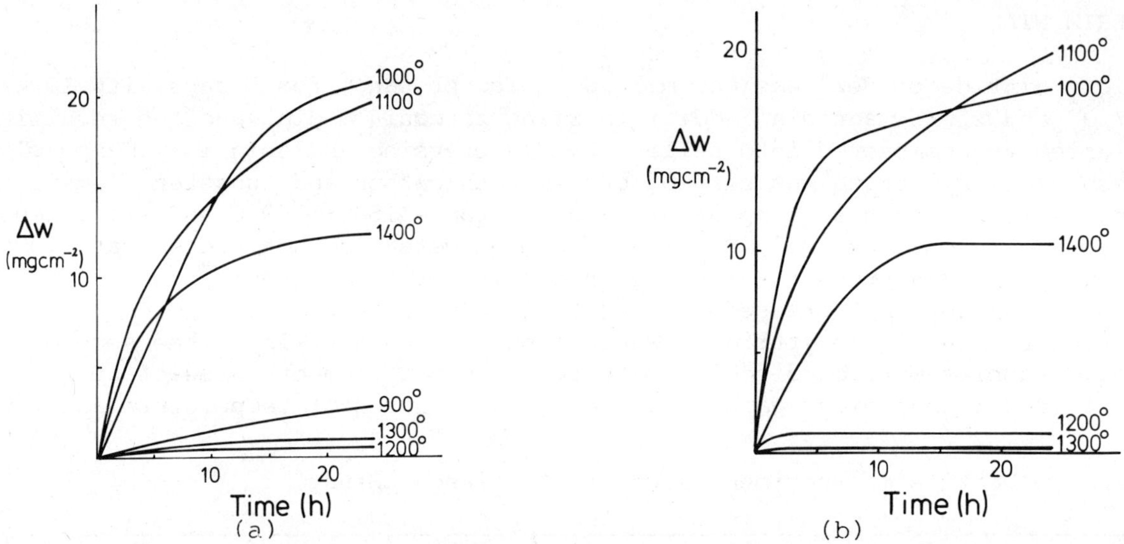

Figure 2. Weight gain versus time curves for Si_3N_4 materials densified with 10 w/o Y_2O_3 and 2 w/o MgO and oxidized for up to 24 hours in air at 900-1400^oC (a) as-sintered and (b) sintered and heat-treated to crystallize the glass into $MgYSi_2O_5N$.

was reduced in steps of 50^oC) were found to show similar behaviour (Figures 2,3). The devitrified specimen consisted of β-silicon nitride plus the nitrogen pyroxene phase, $MgYSi_2O_5N$[13]. A comparison of weight gains for glassy and devitrified materials after 24 hours is shown as a function of temperature in Figure 4.

Similar oxidation curves are shown in Figure 5 for silicon nitride hot-pressed with 10 w/o yttria and up to 14 w/o La_2O_3 and in Figure 6 for silicon nitride hot-pressed with 10 w/o yttria and up to 14 w/o Nd_2O_3. In these cases nitrogen apatite was the only crystalline phase in the products and formed because of the slow cooling rate in the hot press. The samples hot-pressed without rare earth oxide additions showed small weight gains when oxidized at 1000-

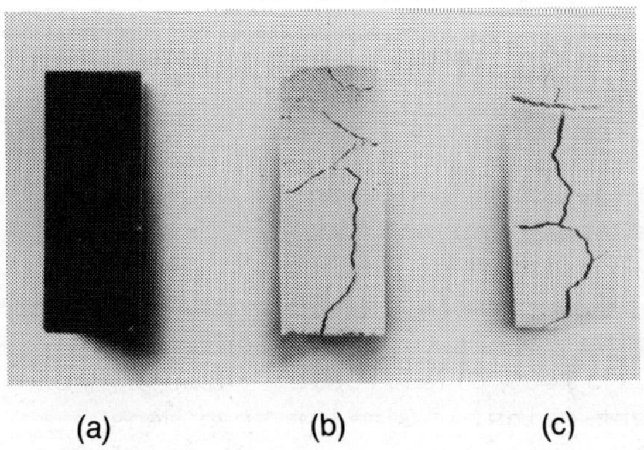

Figure 3. (a) Si_3N_4 materials sintered for 4 hours at 1700^oC with 10 w/o Y_2O_3 and 2 w/o MgO, (b) specimen as in (a) but after 1000^oC oxidation, and (c) as in (b) but heat-treated to crystallize the glass.

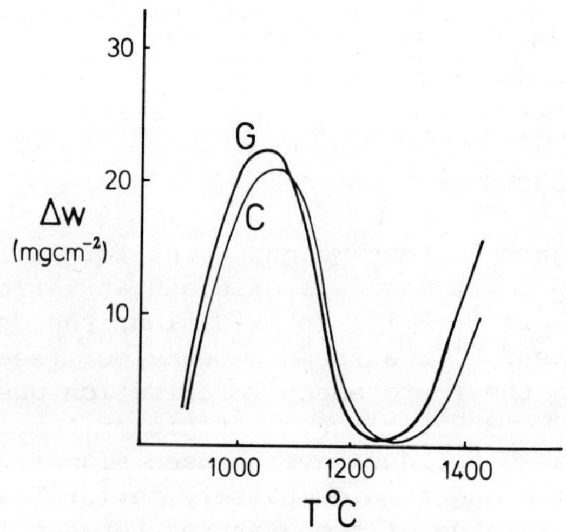

Figure 4. Weight gain versus temperature curves for the β-Si_3N_4 - glass (G) and β-Si_3N_4 - $MgYSi_2O_5N$ (C) materials shown in Figure 2.

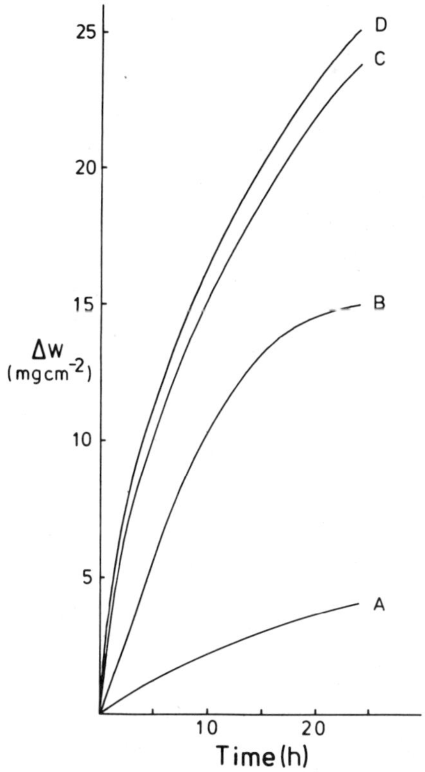

Figure 5. Weight gain versus time curves for Si_3N_4 - 10 w/o Y_2O_3 hot-pressed materials containing (A) 2 w/o La_2O_3, (B) 6 w/o La_2O_3, (C) 10 w/o La_2O_3 and (D) 14 w/o La_2O_3 after oxidation for up to 24 hours at various temperatures.

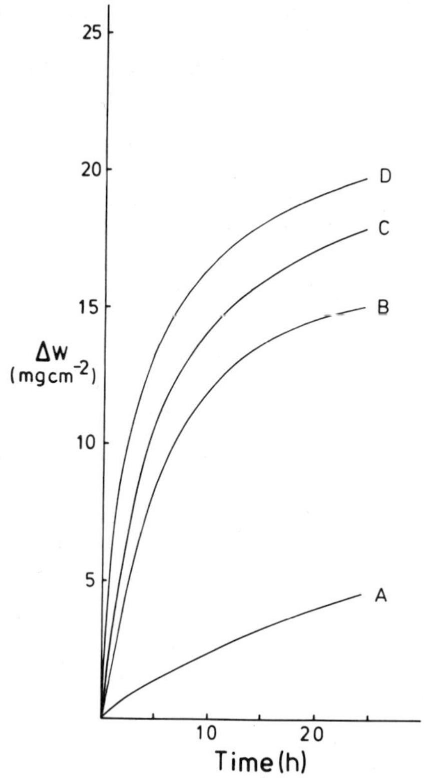

Figure 6. Weight gain versus time curves for Si_3N_4 - 10 w/o Y_2O_3 hot-pressed materials containing (A) 2 w/o Nd_2O_3, (B) 6 w/o Nd_2O_3, (C) 10 w/o Nd_2O_3 and (D) 14 w/o Nd_2O_3 after oxidation for up to 24 hours at various temperatures.

1100°C and no catastrophic cracking, whereas Samples (A)-(D) in both systems showed increasing levels of catastrophic cracking (see Figure 7). The problem is due not only to the nature of the grain boundary phase but also to the quantity present. The samples used for the 1000°C experiments shown in Figures 5 and 6 have good oxidation resistance at 1400°C emphasizing the desirability of avoiding the oxidation problem at lower temperatures.

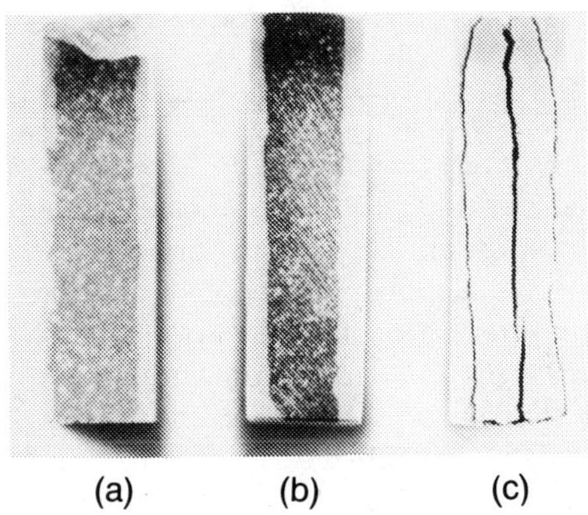

Figure 7. Samples of Si_3N_4 hot-pressed with (a) 10 w/o Y_2O_3, (b) 10 w/o Y_2O_3 - 2 w/o La_2O_3 and (c) 10 w/o Y_2O_3 - 14 w/o Nd_2O_3 after oxidation in air at 1000°C for 24 hours.

DISCUSSION

The equation for the oxidation of nitrogen melilite is:

$$Y_2Si_3O_3N_4 + 3\ O_2 \rightarrow Y_2Si_2O_7 + SiO_2 + 2\ N_2 \qquad \ldots (1)$$

In order to evaluate the specific volume changes it is important to specify which polymorphic forms of the oxidation products are used because $Y_2Si_2O_7$ occurs in at least 6 modifications[14] and SiO_2 also occurs in several forms. Data for the β form of $Y_2Si_2O_7$ and high (β) cristobalite have been used throughout the present discussion. In this case the volume expansion to achieve complete oxidation of the melilite phase (specific volume 141 $Å^3$) to β-$Y_2Si_2O_7$ and β-cristobalite is 33%. The value of 30% quoted by Lange[9] has probably been calculated using different crystalline forms of the products. However when the melilite is present as a second phase within a silicon nitride matrix, it is clearly important to use as a figure of merit the total volume expansion of the specimen (ΔV_{tot}) obtained by multiplying the figure obtained above by the volume proportion of melilite in the product. In the case of Si_3N_4 densified with 15 w/o Y_2O_3, the volume of melilite is $\approx$19%. The total volume expansion on oxidation of a Si_3N_4 - melilite composite containing 19 v/o melilite is therefore approximately 6%.
For the other three Y-Si-O-N phases, the equations for oxidation to yttrium silicates are:

$$\underset{\text{Apatite}}{Y_{10}(SiO_4)_6N_2} + 1.5\ O_2 \rightarrow Y_2Si_2O_7 + 4Y_2SiO_5 + N_2 \qquad \ldots (2)$$

$$\underset{\text{Wollastonite}}{YSiO_2N} + 0.75\ O_2 \rightarrow \tfrac{1}{2}Y_2Si_2O_7 + \tfrac{1}{2}N_2 \qquad \ldots (3)$$

$$\underset{\text{J-phase}}{Y_4Si_2O_7N_2} + 1.5\ O_2 \rightarrow 2Y_2SiO_5 + N_2 \qquad \ldots (4)$$

with corresponding specific volume changes of 11, 10 and 8% respectively. Even though these are much less than the value for melilite, it is consistent with the above calculations that catastrophic oxidation was reported for pure J-phase[15]. Again the discrepancies with the volume calculations of Lange[9] are due to different assumptions as regards the nature and crystalline form of the products.
Another important result is the volume expansion which occurs when silicon nitride is oxidized to silica. For the equation:

$$Si_3N_4 + 3\ O_2 \rightarrow 3SiO_2 + 2N_2 \qquad \ldots (5)$$

and assuming the β-form for silicon nitride and the β-cristobalite form for silica, the volume expansion is 85%. This may appear large but there must always be a significant expansion on going from a nitride to an oxide because 50% more oxygen atoms must be accommodated in the products. In the examples cited above, an oxynitride has been the starting material and the increase in total non-metal atoms is clearly less than when going from a pure nitride to a pure oxide. Of course Reaction (5) does not result in catastrophic behaviour because at temperatures where it occurs rapidly, the silica is starting to soften and plastic deformation prevents the build-up of stresses. Below this temperature, oxidation is so slow because it takes place on the surface and further oxidation must take place through the passivating surface layer. Insufficient oxidation takes place therefore to give rise to the catastrophic situation. This serves to emphasize that the calculations in (1)-(3) above do assume complete volume oxidation of the crystalline Y-Si-O-N phases whereas in silicon nitride and many other non-oxide materials, surface oxidation is the dominant process and passivation then controls the resulting performance.

The melilite oxidation calculation shows that a volume expansion of $\gtrsim$6% is more than sufficient to catastrophically oxidize silicon nitride. Again this is to be expected because a crude Young's Modulus calculation would indicate that silicon nitride materials with E=300 GPa would have to withstand stresses of $\gtrsim$6000 MPa to cope with 6% volume strains. However it is more important to define the upper level of strain tolerance in these materials.

If the criterion for catastrophic oxidation is that it occurs when a certain maximum total specific volume change has been exceeded, there is no reason why only melilite-containing materials should exhibit this behaviour. In fact it would be expected that any phase which showed a large volume change on oxidation or was present in large amounts would also illustrate the effect. This is the case in for example the Si_3N_4 + 10 w/o Y_2O_3 + 2 w/o MgO sample listed in Table 1 which consists of β-Si_3N_4 plus glass and does exhibit catastrophic oxidation as illustrated by the weight gain versus time curves shown in Figure 2 and the low magnification optical micrographs shown in Figure 3. The glass in these specimens is close to the composition $MgYSi_2O_5N$ and after heat-treatment at 1250°C for 24 hours can be substantially converted into the nitrogen pyroxene phase of this composition and structurally analogous to diopside, $CaMgSi_2O_6$. The oxidation behaviour for this sample is also catastrophic (see Figures 3 and 4) and in fact there is very little difference in behaviour between the glassy and crystalline materials.

The equation for the oxidation of crystalline $MgYSi_2O_5N$ is:

$$MgYSi_2O_5N + 0.75\ O_2 \longrightarrow \tfrac{1}{2}Mg_2SiO_4 + \tfrac{1}{2}Y_2Si_2O_7 + \tfrac{1}{2}SiO_2 + \tfrac{1}{2}N_2 \qquad \ldots (6)$$

and using the known specific volumes of the phases in this equation, the volume change on oxidation is $\approx$15%. Calculations show that the amount of $MgYSi_2O_5N$ in the product from a starting mix consisting of Si_3N_4 + 10 w/o Y_2O_3 + 2 w/o MgO is approximately 15 v/o. The total volume expansion of the Si_3N_4 - 15 v/o $MgYSi_2O_5N$ ceramic is therefore 2.3%. This is much smaller than the values calculated for Si_3N_4 ceramics containing 15 v/o of melilite. Similar calculations cannot be made accurately for the equivalent glassy material because an accurate value for the density of the glass is not available. However the overall volume change must be fairly similar because of the similarity in oxidation behaviour.

The remaining samples listed in Table 1 are silicon nitride materials densified with 10 w/o Y_2O_3 and up to 14 w/o of La_2O_3 or Nd_2O_3. They were hot-pressed to achieve full density because of the difficulty of obtaining fully dense samples by pressureless sintering. Figures 5 and 6 show the magnitude of the 1000°C effect for the two sets of materials and clearly it gets worse with increasing amounts of rare earth oxide additive. The most important feature of these specimens is that the additive after hot-pressing remains as an apatite phase in all specimens. Calculations of the total volume expansion of samples containing 14 w/o of rare earth oxide additive show that this is approximately 1.7% whereas the Si_3N_4 + 10 w/o Y_2O_3 specimen shows a total volume expansion of only 1%. From the micrographs shown in Figure 7, clearly the 14 w/o Nd_2O_3 specimen catastrophically cracks whereas the 0% rare earth oxide material is on the borderline of exhibiting this behaviour. This implies that a total volume expansion of 1% is closer to the maximum level of strain that can be tolerated and once again a rough Hooke's law calculation shows that this would generate a stress of $\approx$1000 MPa which is of the same order of magnitude as the strength of a fully dense silicon nitride ceramic.

One solution to the oxidation problem is to add small amounts of alumina. It has been known for many years that sialons do not exhibit 1000°C oxidation cracking but since yttrium densified sialons generally do not contain melilite as a secondary crystalline phase it was assumed that this was the explanation for the absence of 1000°C oxidation cracking. However in the present work it has been shown that the total volume change of the specimen is the more relevant criterion to use rather than the presence or absence of melilite and it is instructive to estimate volume changes for the oxidation of a sialon glass to compare with the data above. The limit of nitrogen solubility in Y-Si-Al-O-N glasses occurs at the approximate composition $Y_3Al_2Si_3O_9N_3$. Complete oxidation of this glass results in the formation of yttrium disilicate, mullite and YAG according to the equation:

$$Y_3Al_2Si_3O_9N_3 + 2.25\ O_2 \rightarrow 9/7Y_2Si_2O_7 + 3/14(3Al_2O_3.2SiO_2) + 1/7Y_3Al_5O_{12} + 1.5\ N_2 \quad \ldots (7)$$

Evaluating the volume change on the basis of a measured density of 3.72 g/cm^3 for the glass gives an overall value of $\approx$2%. Obviously the presence of $\approx$15% of this glass in a crystallized sialon ceramic will result in a total volume increase on oxidation which is far less than the 1% deduced above as the critical value for catastrophic behaviour. The difference in volume change for the oxidation of M-Si-O-N glass compared with M-Si-Al-O-N glass can be attributed to two factors. Firstly, the silica always present as an oxidation product of M-Si-O-N glass has now been diluted by a considerable proportion of mullite which is a higher density phase (3.5 g/cm^3 compared with 2.2 g/cm^3 for SiO_2). Secondly, the aluminum present in the oxides on the right hand side of the equation is in [6]-fold coordination as compared with [4]-fold coordination on the left hand side and this also results in a density increase on oxidation and hence less increase in specific volume.

In the work described above using mixed yttria/rare earth oxide additives for silicon nitride, it was noted that the main reason for the cracking was the quantity of the second phase rather than the specific volume increase for the apatite phase itself. In fact it was noticed that the latter actually decreased with increasing amount of rare earth additive and Table 2 shows the volume changes for oxidation of the four rare earth silicon oxynitride structure types. What is immediately obvious is that the volume changes are lower and increase with increasing rare earth atomic number. This is because the crystal structures of both Ln_2SiO_5 and $Ln_2Si_2O_7$ phases vary with atomic number and the structures adopted for the low-Z rare earths are more densely packed in comparison to the starting Ln-Si-O-N phases than is the case for the higher Z rare earth structures. Another advantage of rare earth oxide densified materials is that eutectic temperatures in Ln_2O_3-SiO_2 systems are lower than in the Y_2O_3-SiO_2 system and densification is therefore easier than with yttria. If fully dense samples can be obtained without the use of pressure, the data shown in Table 2 indicates that it should be much easier to avoid the 1000°C oxidation problem and further work is in progress to establish whether such materials can also retain good oxidation resistance at higher temperatures (1400-1500°C).

Table 2. Oxidation specific volume changes for melilite (M), wollastonite (W) apatite (A) and J-phase (J) oxynitrides in Ln-Si-O-N systems.

Element	Form of $M_2Si_2O_7$ (after[16])	Form of M_2SiO_5 (after[14])	M	W	A	J
La	A	X_1	19.3	2.5	3.7	1.6
Ce	A	X_1	23.3	-0.6	3.9	1.0
Nd	A	X_1	22.8		3.5	1.2
Sm	A	X_1	23.9		3.5	0.7
Gd	B	X_1	24.2		3.5	0.7
Dy	B	X_2	27.0			7.0
Ho	B	X_2	27.5			5.5
Er	B	X_2	26.0			6.0
Yb	C	X_2	34.7			6.5
Y	C	X_2	33.2	10.0	11.0	8.0

CONCLUSIONS

The catastrophic cracking exhibited during 1000^oC oxidation of silicon nitride samples densified with >10 w/o Y_2O_3 is not specifically a melilite effect as originally supposed but can be exhibited by any Si_3N_4 ceramic in which the total volume increase of the specimen during oxidation is >1%.
Compared with previous solutions of adding silica (which removes the problem but results in worse high-temperature performance) or giving a previous flash oxidation treatment (which is a temporary solution, satisfactory for non-reducing and non-abrasive applications) alternative solutions exist. Thus for example, addition of small amounts of alumina remove the problem but the resulting lower eutectics place an upper limit on high temperature performance. Reducing the total amount of additive effectively removes the problem but then densification may only be possible by hot-pressing or HIPing. A more promising solution appears to be the replacement of yttria by rare earth oxide additives because these have lower specific volume changes on oxidation, and also less additive (in mol % or v/o) is required for densification because of the lower eutectic temperatures in these systems. Both these factors should permit the low temperature oxidation problem to be reduced and perhaps eliminated. Further work is needed to explore whether such materials can indeed be satisfactorily fabricated by pressureless sintering techniques and whether they suffer from any other disadvantages which impair their performance as high temperature engineering materials.

REFERENCES

1. F.F.Lange, S.C.Singhal and R.C.Kuznicki, "Phase Relationships and Stability Studies in the Si_3N_4-SiO_2-Y_2O_3 Pseudoternary System", J. Amer. Ceram. Soc., 60(5-6), 249-252 (1977).

2. K.H.Jack in "The Role of Additives in the Densification of Nitrogen Ceramics", Final Technical Report for the European Research Office of the U.S.Army, Grant No. DAERO-76-G-067, November 1977.
3. R.K.Govila, "Strength Characterization of Yttria-doped Sintered Silicon Nitride", J. Mat. Sci., 20, 4345-53 (1985).
4. J.T.Smith and C.L.Quackenbush, "Phase effects in Si_3N_4 containing Y_2O_3 or CeO_2 I: Strength", Am. Ceram. Soc. Bull., 59(5), 529-32 (1980).
5. C.L.Quackenbush and J.T.Smith, "Phase Effects in Si_3N_4 containing Y_2O_3 or CeO_2 II: Oxidation", Am. Ceram. Soc. Bull., 59(5), 533-7 (1980).
6. F.F.Lange, "Fabrication and Properties of Dense Polyphase Silicon Nitride", Am. Ceram. Soc. Bull., 62(12), 1369-74 (1983).
7. R.R.Wills, S.Holmquist, J.M.Wimmer and J.A.Cunningham, "Phase Relationships in the System Si_3N_4-Y_2O_3-SiO_2", J. Mat. Sci., 11, 1305-9 (1976).
8. G.Q.Weaver and J.W.Lucek, "Optimization of Hot-pressed Si_3N_4-Y_2O_3 Materials" Am. Ceram. Soc. Bull., 57(12), 1131-6 (1978).
9. F.F.Lange, "Silicon Nitride Polyphase Systems: Fabrication, Microstructure and Properties", Int. Metals Review, 25(No. 247), 1-20 (1980).
10. H.Knoch and G.E.Gazza, "Effect of C impurity on the Thermal Degradation of an Silicon Nitride - Yttrium Oxide Ceramic", J. Am. Ceram. Soc., 62(11-12) 634-5 (1979).
11. R.K.Govila, J.A.Mangels and J.R.Baer, "Fracture of Yttria-doped, Sintered Reaction Bonded Silicon Nitride", J. Am. Ceram. Soc., 68(7), 413-8 (1985).
12. R.K.Govila, "Fracture of Flash-oxidized Yttria-doped Sintered Reaction Bonded Silicon Nitride", J. Mat. Sci. 22, 1193-8 (1987).
13. J.K.Patel and D.P.Thompson, "The Low-Temperature Oxidation Problem in Yttria-Densified Silicon Nitride Ceramics", Br. Ceram. Trans and J., 87(2), 70-3 (1988).
14. K.Liddell and D.P.Thompson, "X-ray Diffraction Data for Yttrium Silicates" Br. Ceram. Trans. and J., 85(1), 17-22 (1986).
15. M.Billy, "Kinetics of Gas-Solid Reactions", in "Nitrogen Ceramics", Proc. 1st NATO ASI, Canterbury 1975, Ed. F.L.Riley, publ. Noordhoff (Leyden), 635-42 (1976).
16. J.Felsche, "The Crystal Chemistry of the Rare Earth Silicates", Structure and Bonding, 13, 100-97 (1973).

TETRAGONAL ZTRCONIA POLYORYSTAL CERAMICS WITH HIGH TOUGHNESS AND STRENGTH

Tingkai Li, Zhijian Shen, Zishang Ding
Zhejiang University
Department of Materials Science and Engineering
Hangzhou, P.R. China

ABSTRACT

The effects of Y_2O_3 content on σ_f and K_{1c} of TZP have been studied. It is found that the σ_f and K_{1c} of TZP increased first and then decreased with the increasing Y_2O_3 content. There is an optimal addition of Y_2O_3 in which σ_f and K_{1c} reach a maximum. Additions of Al_2O_3 could increase k_{1c} and σ_f of TZP at the same time. This interesting phenomenon is explained by the following mechanism, the crack deflection and branch toughening caused by Al_2O_3 grains could enhance the transformstion toughening of ZrO_2. Controlling the content of AL_2O_3 and Y_2O_3 we obtained a TZP ceramic with K_{1c}=21.3 MPam$^{1/2}$ and σ_f=1578 MPa.

INTRODUCTION

Since the transformation toughening of zirconia was first noted by Garvie et al. [1] the mechanisms of transformation toughening have been extensively studied . The strengths of tetragonal zirconia polycrystal(TZP) materials have recently been reported as high as 2.5 GPa, although more typical maximum values vary between 1.0 to 1.5 GPa. The toughness of TZP has been shown to be more than 17 MPam$^{1/2}$[2], about five times that of general engineering ceramics. But for both TZP and PSZ ceramics it is found that the maximum toughness composition does not coincide with maximum strength of the materials[3] Heuer et al have studied the relationship between the toughness and strength of ZTC[4]. It was found that the strength increases initially with increasing toughness, and later the strength begins to decline, whereas the toughness continues to increase. Hence, the fabrication of zirconia toughended ceramics with high toughness and strength and the mechanism by which high toughness and strength are obtained are important subjects.

The present work is to investigate the effects of Y_2O_3 and AL_2O_3 content on mechanical properties of TZP materials, to prepare TZP materials with high toughness and strength and to understand the mechanism by which high toughness and strength are obtained.

EXPERIMENT

Powder and Specimen Preparation

Powders of α-Al_2O_3 and Y_2O_3 partially-stabillized zirconias for making of TZP ceramics were prepared on the basis of calcining high purity $Al_2(NH_4)_2(SO_4)_4 \cdot 24H_2O$ and coprecipitating mixed solutions of analytically pure $ZrOCl_2 \cdot 8H_2O$ and YCl_3. TZP samples were produced by hot-pressing (1600℃, 20MPa for 1h).

Measurements of transformation Volume and Width

The effective transformable volume fraction of ZrO_2 in TZP ceramics and the transformed volume fraction in fracture surfaces of the materials were measured by X-ray diffraction. The width of the transformation zone h was estimated by two wavelength method of X-ray diffraction of fracture surfaces (Cu K_α and Mo K_α). Calculations for these parameters are given as below,

a) the effective transformable volume fraction V_e

$$V_e = X_b - X_p$$

here, X_b and X_p are the volume fractions of ZrO_2(t) before and after a sample has been finely crushed and meshed.

b)the transformed volume fraction in fracture surfaces V_f

$$V_f = X_b - x_f$$

here, X_f is the volume fraction of ZrO_2(t) in fracture surfaces.

Measurements of Fracture Strength and Toughness

The fracture strength of TZP samples was measured in three point bending using a 20-mm span. The fracture toughness was evaluated by both micro-indentation and chevron notched beam technique. In the micro-indentation technique, a load of 100 N was applied to the specimen. In the chevron notched beam technique, chevron notches were machined to 0.21 to 0.24mm. Fracture toughness tests were performed by three point bending using a cross-head speed of 0.05mm/min.

RESULTS AND DISCUSSION

The Effects of Y_2O_3 Content on the Strength σ_f and Toughness K_{1c} of TZP

Figure 1 shows the variation of σ_f, K_{1c}, V_e and V_f of TZP with Y_2O_3 content. It is seen from this figure that as Y_2O_3 content increasses the fracture strength of the materials passes through a maximum and then decrease.
From microstructural analysis it is found that the grain size of ZrO_2 decreases with increasing of Y_2O_3 content when Y_2O_3 content is less than 4wt% ,and then increases with increassing Y_2O_3. Tho compositon of the TZP moves into the region of the cubic phase at 6wt% Y_2O_3, At 6wt% Y_2O_3 the microstructure shows larger grains present among smaller grains. Lange [5] considered these larger grains as cubic ZrO_2.When the Y_2O_3 content increases again the number and size of the larger grains increase (see Fig. 2 to 5). The correlation between grain size and Y_2O_3 content correlates to the strength of TZP. From TEM analysis it is found that when the grain size of the TZP ceramic is large, there are large grains of ZrO_2(m) surrounded by many small cracks at low Y_2O_3 contents (Fig. 6). Hence the strength of these ceramics are smaller. With increasing Y_2O_3content the volume of grains of ZrO_2(t) available for transformation toughening is increased. The grain size is fine and well distributed, and the strength σ_f is increased. When the Y_2O_3 content is increased further, large cubic ZrO_2 grains are present in the matrix,,and the relative content of ZrO_2(t) decreases. The volume fraction and width of the transformed

ZrO_2(t) in fracture surface of the materials also decreases (see Fig.1), and the strength σ_f decreased. It is also seen from figure 1 that as the Y_2O_3 content increases the toughness of TZP passes through a maximum whereas the volume fraction of transformed particales and the width of the transformation zone in the fracture surfaces decreases. These results are in agreement with following transformation toughening equation [6],

$$K=0.21E_cV_fe^Th^{1/2}/(1-\nu)+1.26v_fh^{1/2}\sigma_T(1-2\nu)/(1-\nu)$$

Because that ZrO_2 transformation toughening comes from two parts, One is dilational work done by the ZrO_2(t) to ZrO_2(m) transformation, another is the work done by the applied stress necessary to initiate the transformation. When Y_2O_3 content is smaller, the toughening only comes from transformation work. With the increasing of Y_2O_3 content both the transformation work and the work done by the applied stress necessary to initiate ZrO_2(t) transformation make contribution to toughening. When the Y_2O_3 content reaches a specific value, it is difficult ot induce ZrO_2(t) transform, so the toughness decreases.

The Effect of Al_2O_3 content on the Fracture Strength and Toughness of TZP

The effects of Al_2O_3 content on the fracture strength σ_f, toughness K_{Ic}, volume fraction and width of transformed ZrO_2(t) in the fracture surface of Y4-TZP are shown in Fig.7. It is seen from this figure that as the Al_2O_3 content increases the fracture strength and toughness of Y-TZP increase and pass through a maximum. Addtion of Al_2O_3 could increase K_{Ic} and σ_f of TZP. Analysis from toughening mechanics shows that the transformation toughening of ZrO_2 and crack deflection toughening and crack branching of Al_2O_3 could be additive. Because α-Al_2O_3 has a high elastic modulus and strength, addition of α-Al_2O_3 to the ceramic could make cracks deflect and branch when fracture of TZP materials take place. On the other hand, the amount of transformed ZrO_2 depends on crack length. hence, the crack deflection and branch toughening of Al_2O_3 could enhance the transformation toughening of ZrO_2. Analysis of fracture morphology by SEM shows the fracture surface of 4Y-95TZP (95Wt% 4Y-95 + 5Wt% Al_2O_3) is wavey, but the fracture surface of 4Y-TZP (4Wt% Y_2O_3 + 96% Wt% ZrO_2) is rather flat (as shown in Fig. 8 and 3). With increasing Al_2O_3 the relative volume of ZrO_2(t) decreases, and, on the other hand, Al_2O_3 grains could impede the transformation of ZrO_2(t) and make the volume fraction zone width of ZrO_2(t) smaller. As a result, the toughness of TZP decreases.

With additions of Al_2O_3 the strength of TZP increases slightly, also reported in references [3, 7]. From SEM microstructureal analysis (Fig. 8) the α-Al_2O_3 grains are columnar and well distributed in the ZrO_2 matrix. The size of Al_2O_3 grains (about 2μ) is larger than that of ZrO_2 grains (about 1μ). It is noted that the thermal expansion coefficient of Al_2O_3 which is about $8\times10^{-6}/℃$ [8] is smaller than that of ZrO_2(t) which is about $11\times10^{-6}/℃$ for a axis and $16\times10^{-6}/℃$ for c axis [9]. Hence, Al_2O_3 grains are under compression in the ZrO_2 matrix. It is not easy to form cracks between Al_2O_3 and ZrO_2 grains. By studying the boundary between Al_2O_3 and ZrO_2 using high resolution electron microscopy [10] it has been found that there is partial coherence structure in the boundary, so that the grain boundary possesses a high joint strength. The strength ot TZP does not decrease even though the grain size of Al_2O_3 is larger than that of ZrO_2. It may also be explained by the fact that the Al_2O_3 increases the elastic modulus of TZP since Al_2O_3 itself has high elastic modulus, and the crack deflection as well as branch toughening caused by Al_2O_3 grains enhances the transformation toughening of ZrO_2. Hence, the strength of TZP increases with the addition of Al_2O_3. However, when Al_2O_3 content is above a certain amount the grain size of Al_2O_3 will increase, the toughness of TZP will decrease, and the strength of TZP also decreases with increasing Al_2O_3 content. By controlling the content of Al_2O_3 and Y_2O_3 and applying interface treatment techniques, we obtained a TZP ceramic (4Y-95TZP) with K_{Ic}=21.3 $MPam^{1/2}$ and σ_f=1578MPa.

CONCLUSION

1. The σ_f and K_{Ic} of TZP increase first and then decrease with increasing of Y_2O_3 content, there is an optimal addition

of Y_2O_3 at which σ_f and K_{Ic} reach the highest value.
2. Addition of Al_2O_3 could increase K_{Ic} and σ_f of TZP at the same time. This interesting phenomenon is explained by the following mechanism, the crack deflection and branch toughening caused by Al_2O_3 grains could enchance the transformation toughening of ZrO_2.
3. By controlling the content of Al_2O_3 and Y_2O_3, we obtained a TZP ceramic with K_{Ic}=21.3MPam$^{1/2}$ and σ_f=1578 MPa.

REFERENCE

(1) R. C. Garvie, R.H.Hannick, and R. T. Pascoe, Ceramic Steel?,Nature 258(1975)703
(2)K. Tsukuma, and M. Shimada, Strength and Fracture Tougheness of Isostatically Hot-Pressed Composites of Al_2O_3 and Y_2O_3-Partially-Stabilized ZrO_2, J.Am. Ceram.Soc. 68(1)(1985)C-4
(3) M. V. Swain, Limitation of Maximum Strength of Zirconia-Toughened Ceramics by Transformation Toughening Increment J.Am. Ceram.Soc. 68(4)(1985) C-97
(4) A. H. Heuer, F. F. Lange, M. V. Swain, and A. G. Evans,Transformation Toughening, An Overview, J.Am. Ceram. Soc. 69(3)(1986) 1
(5) F. F. lange, Transformation-Toughened ZrO_2, Correlations Between Grain Size Contral and Composition in the System ZrO_2-Y_2O_3, J. Am. Ceram. Soc. 69(3)(1986)240
(6) Tingkai Li, Suxing Wu and Zishang Ding, Models of Transformation Toughening in ZrO_2-Alloys, International Conference Science of Ceramics 14 (1987) 183
(7) M. V. Swain, L. R. F. Rose, Strength Limitations of Transformation-Toughened Zirconia Alloys, J. Am. Ceram. Soc 69(7)(1986)511
(8) J.B. Wachtman, T.G. Scuderi, and G.K. Cleek Linear Thermal Expansion of Aluminum Oxide and Thorium Oxide from 100° to 1000° K, J.Am. Ceram. Soc. 45(7)(1962) 319
(9) R.N. Patil, and E.C. Subbarao, Axial Thermal Expansion of ZrO_2 and HfO_2 in the Range Room Temperature to 1400℃, J.Appl. Crystallogr. 2(1969)281
(10) Y. Wang, T. Li, and Z. Shen, Study of High Resolution Electron Microscopy (HREM) on the Boundary of Zirconia Toughened Ceramics (ZTC), International Conference on Composite Interfacess-Ⅱ (1988)

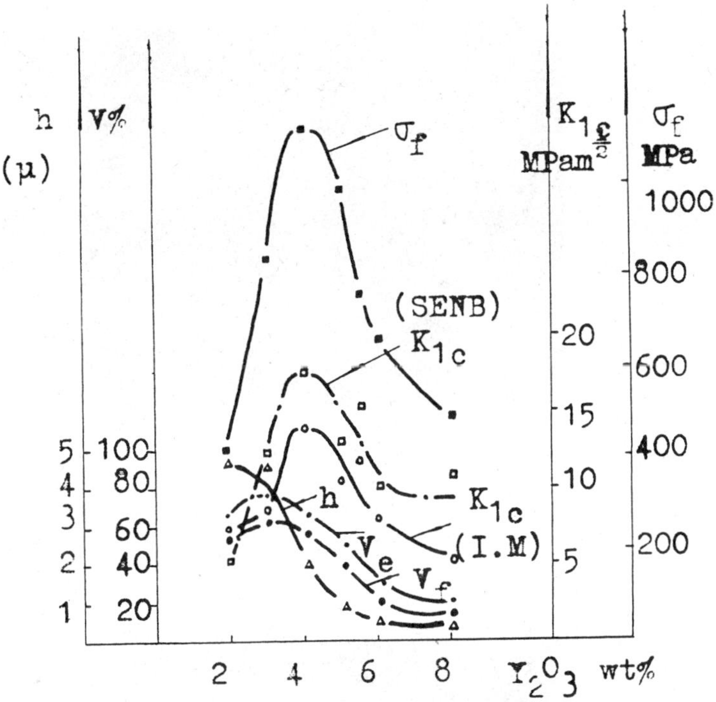

Fig. 1 Dependence of σ_f, K_{1c}, and V on Y_2O_3 content for TZP

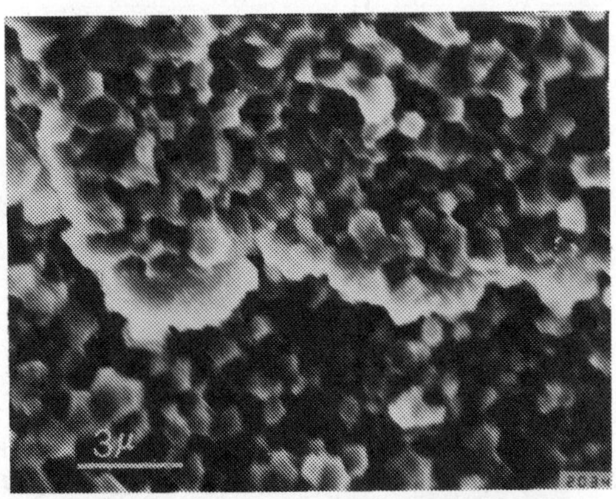

Fig. 2 SEM of Y2-TZP showing the fracture surface

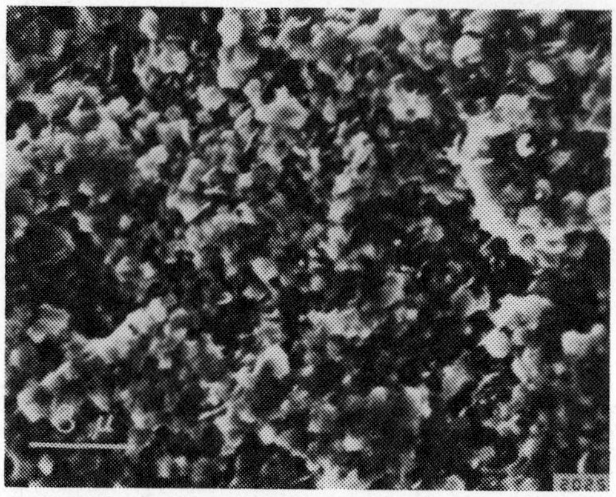

Fig.3 SEM of Y4-TZP showing the fracture surface

Fig.4 SEM of Y6-TZP showing the fracture surface

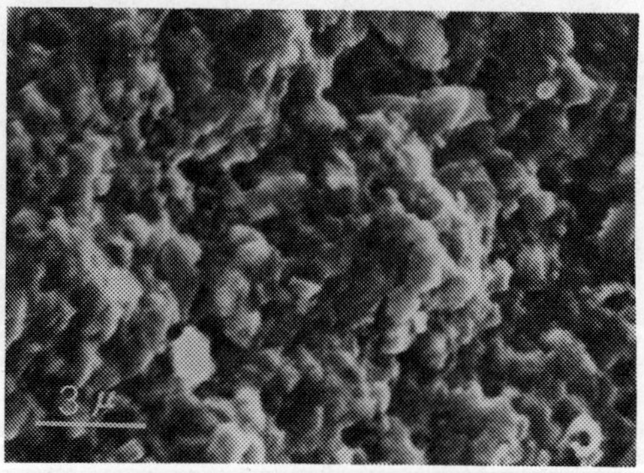

Fig.5 SEM of Y8-TZP showing the fracture surface

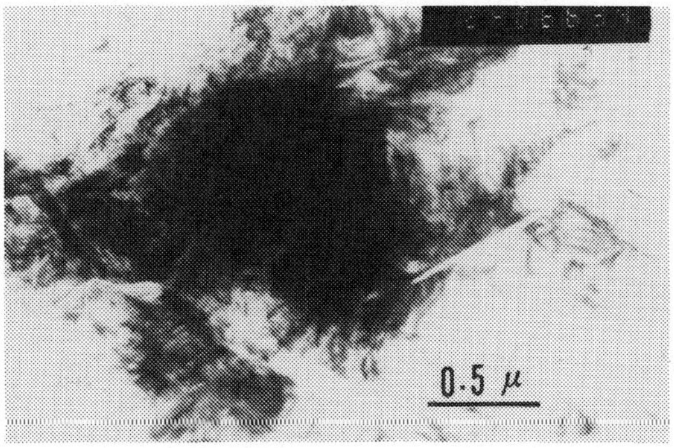

Fig.6 Microcrack in Y2-TZP

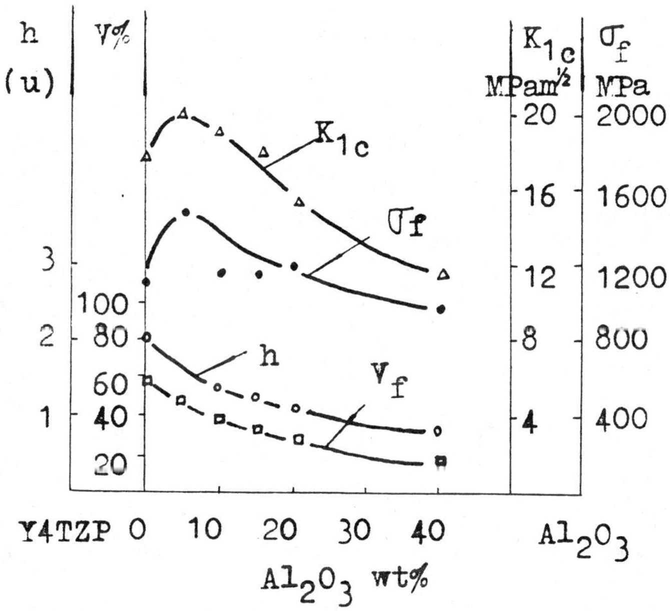

Fig.7 K_{1c}, σ_f, and V vs Al_2O_3 content in TZP

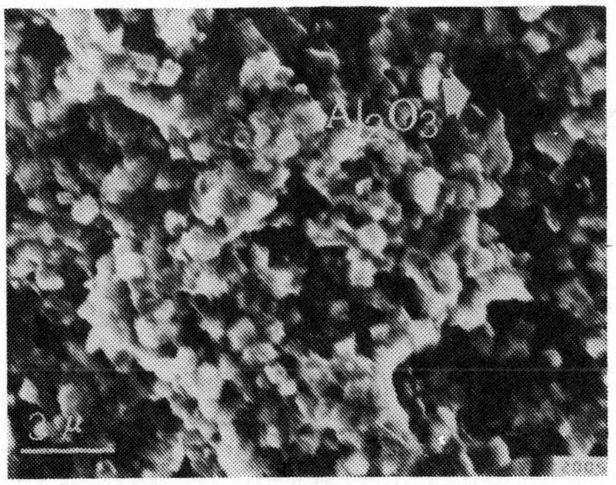

Fig.8 SEM of ZTA 4Y-95 showing the fracture surface

REVERSIBLE TRANSFORMATION MECHANISM IN Ce-TZP

Y. Huang, Z. Q. Zeng, L. L. Zheng and Z. Z. Jiang
Tsinghua Unversity
Department of Material Science and Engineering
Beijing, 10084, China

ABSTRACT

The phase transformation behavior of t-ZrO_2 to m-ZrO_2 for Ce-TZP under stressing or high radiation intensity of electron beam was investigated in detail by XRD and in-situ-TEM observation. XRD measurement showed that the transformation from t-ZrO_2 to m-ZrO_2 had occurred when tensile stress was applied on Ce-TZP containing 10 mol% CeO_2 samples and partial m-ZrO_2 was transformed back to t-ZrO_2 when the tensile stress was removed. Also, reversible transformation between t-ZrO_2 and m-ZrO_2 was demonstrated by in-situ-TEM observation under different radiation intensities of an electron beam. These results suggest that the ZrO_2 martensitic transformation mechanism in Ce-TZP is stress-assisted and deformation-induced. Laser-Raman-Microprobe measurement has shown that the transformation zone size h of Ce-TZP is much larger than that of other ZrO_2 transformation toughened ceramics and the distribution of m-ZrO_2 is not uniform in the transformation zone. The results suggest that the toughening mechanism of stress-induced transformation does not apply to Ce-TZP.

INTRODUCTION

Transformation toughened ZrO_2 ceramics are known for their high strength and high toughness which is attributed to the t-ZrO_2 to m-ZrO_2 transformation. Transformation from t-ZrO_2

to m-ZrO_2 is assumed to be irreversible and the stress-induced toughening mechanism is used to explain the enhancement of toughness for tetragonal polycrystalline ZrO_2 (TZP) ceramics. However, D. B. Marshall ect.[1] indicated that at room temperature stress-induced transformation is reversible in Mg-PSZ under certain conditions.

The aim of this paper is to demonstrate that the reversible transformation between t-ZrO_2 and m-ZrO_2 occurs under different stress and temperature conditions. XRD measurement, in-situ-TEM observations, and temperature-strain curve measurement were used in this investigation. The size and shape of the transformation zone h were determined by Laser-Raman-Microprobe.

EXPERIMENTAL PROCEDURE

In-situ-TEM observation

TEM observation of the martensitic, tetragonal to monoclinic transformation in zirconia, was used to provide the information on reversible transformation of zirconia under different radiation intensities of an electron beam. The material used in this study was a Ce-TZP containing 12 mol% CeO_2 with a grain size of 1.5 μm.

XRD measurements for loaded and unloaded specimens

The XRD was used to demonstrate the reversible transformation of zirconia under different stresses. The loading jig used is shown in figure 1. The material used in this study was a Ce-TZP containing 10 mol% CeO_2 with grain aize of 1.8 μm. The specimen surface was ground prior to examination.

Stress-strain and temperature-strain curve

The stress-strain and temperature-strain curve were obtained from four point flexure testing. Below 400°C strain gauges attached to the specimen surface were used to measure the stress-strain curve and the temperature-strain curve. The temperature-strain curve was obtained by heating the flexure specimens to 300°C.

Size and shape of transformation zone

Laser-Raman-Microprobe equipment was used to measure the size and shape of the transformation zone h. The distribution of monoclinic zirconia beneath the fracture surface was obtained by using an irradiating laser beam on the side surface of the fractured specimen (see Fig. 2). The diameter of the laser beam was 14 μm. The material used in this study was 10 mol% CeO_2 Ce-TZP.
The monoclinic phase concentration, Cm, derived from the Garvie and Nicholson relation[2], could be expressed directly in terms of the Raman intensities using the relation

$$Cm = \frac{Im^{181} + Im^{192}}{F \times (It^{148} + It^{264}) + Im^{181}\ Im^{192}}$$

where F is 0.97.

RESULTS AND DISCUSSION

In-situ-TEM observation

Figure 3 shows the TEM diffraction patterns obtained for the ZrO_2 ceramics under different radiation intensities of an electron beam. When the irradiating intensity of the electron beam was low, the tetragonal phase was predominate (see Fig. 3 (a) the diffraction pattern). As the irradiation intensity of the electron beam is increased, the tetragonal zirconia quickly transformed to the monoclinic phase. The diffraction pattern obtained for the monoclinic zirconia is shown in Fig. 3 (b). When the irradiation intensity of the electron beam was decreased to the former value, the monoclinic zirconia quickly transformed back to the tetragonal phase as shown in

Fig. 3 (c). It is clear that the electron beam provides the driving force for transformation of the ZrO_2 to transmitting grain of zirconia. Cycled change of the irradiating intensity lead to cycled transformation between the tetragonal and monoclinic phase in zirconia.

XRD measurement for loaded and unloaded specimens

In order to demonstrate the cycled transformation between the tetragonal and monoclinic phases, XRD measurement was used for loaded and unloaded specimens. The results of these measurements are presented in Table 1.

Table 1. XRD Results of 10 mol% CeO_2 Ce-TZP With Different Stress

loading	intensity of diffraction			M%	ΔM(t→m)	ΔM(m→t)	R%
	Im(11$\bar{1}$)	Im(111)	It(111)				
O	42	13	100	35.5	0	0	0
L	47	13	100	37.5	2.0	0	0
H	53	13	100	39.8	4.3	0	0
L	51	11	100	38.3	2.8	1.5	34.9
O	49	10	100	37.1	1.6	2.7	62.8

* O— no loading, L— low loading, H— high loading

$$R\% = \frac{\Delta M(m \rightarrow t)}{\Delta M(t \rightarrow m) + \Delta M(m \rightarrow t)}$$

$$M\% = \frac{Im(11\bar{1}) + Im(111)}{Im(11\bar{1}) + Im(111) + It(111)}$$

As the specimen was loaded in 3-point flexure and irradiated by x-ray with an area of 20 × 4 mm, the amount of phase change is small. But it is clear that the reversible transformation exists as stress was successively removed. These results suggest that the transformation from tetragonal to monoclinic is stress-assisted and deformation-induced martensitic transformation.

Stress-strain and temperature-strain curve

Cycled transformation between tetragonal and monoclinic occurs when the zirconia receives different energy such as stress--strain energy and temperature--heat energy. Fig. 4 shows both the stress--strain curve and the temperature--strain curve. When stress was successively applied on the zirconia, elastic deformation occurs initially and then transformation plasticity take place due to the driving force of transformation from $t\text{-}ZrO_2$ to $m\text{-}ZrO_2$ over transformation resistance. Within the range of transformation plasticity the amount of monoclinic zirconia phase increased due to increased stress and strain. When the stress was removed the strain was partially recovered (960 micro-strain left). Increased temperature provided the driving force for transformation from $m\text{-}ZrO_2$ to $t\text{-}ZrO_2$ increased and the residual strain gradually recovered. The amount of strain recovered reached 97% when the temperature was 280°C (see Fig. 4).

Relationship between transformation and temperature (T) and stress (σ)

The Ms of most TZP ceramics is less than room temperature, e.g., -25°C for 10 mol% CeO_2 Ce-TZP, -135°C for 11 mol% CeO_2 Ce-TZP [3]. If tetragonal zirconia transforms to monoclinic zirconia, a transformation driving force must be provided to overcome the transformation resistance force. The free energy change per unit volume may be written as:

$$\Delta G^{t\to m} = \Delta G^{t\to m}_{chem} + \Delta U^{t\to m}_{se} + \Delta U^{t\to m}_{s} + \Delta U^{t\to m}_{r}$$

Where ΔG_{chem} = the chemical-free energy change, is a function of temperature for constant composition

ΔU_{se} = the change of strain energy due to the applied stress and transformation of the zirconia

ΔU_{s} = the surface energy change

ΔU_{r} = the resistance of transformation due to friction between grains and other irreversible effects caused by transformation

Fig. 5 and Fig. 6 show the change of chemical free energy G_{chem} and elastic strain energy U_{se} with the change of temperature and stress.

From Fig. 5 and Fig. 6 we can see

$$G^{t}_{chem} = G^{m}_{chem} \qquad T=T_o$$

$$G^{t}_{chem} < G^{m}_{chem} \qquad T>T_o$$

$$G^{t}_{chem} > G^{m}_{chem} \qquad T<T_o$$

Increasing temperature above T_o , the chemical fraee energy change $\Delta G^{t\to m}_{chem} = G^{m}_{chem} - G^{t}_{chem} > 0$ and $\Delta G^{m\to t}_{chem} = G^{t}_{chem} - G^{m}_{chem} < 0$

For strain energy;

$$U^{t}_{se} = U^{m}_{se} \qquad \sigma = 0$$

$$U^{t}_{se} > U^{m}_{se} \qquad \sigma > 0 \text{ (tensile stress)}$$

$$U^{t}_{se} < U^{m}_{se} \qquad \sigma < 0 \text{ (pressure stress)}$$

$\Delta U^{t\to m}_{se} = U^{m}_{se} - U^{t}_{se} < 0$ and $\Delta U^{m\to t}_{se} = U^{t}_{se} - U^{m}_{se} > 0$, when tensile stress is applied.

In fact, the driving force for transformation from tetragonal to monoclinic is the resisting force of transformation from monoclinic to tetragonal.

If

$$\Delta G_{chem}^{t\to m} + \Delta U_{se}^{t\to m} + \Delta U_{s}^{t\to m} + \Delta U_{r}^{t\to m} \geqslant 0$$

$$\Delta G_{chem}^{m\to t} + \Delta U_{se}^{m\to t} + \Delta U_{s}^{m\to t} + \Delta U_{r}^{m\to t} < 0$$

additional energy is needed to overcome the transformation resistance $\Delta U_r^{t\to m}$ from tetragonal to monoclinic transformation.

As $\Delta G_{chem}^{m\to t} + \Delta U_{se}^{m\to t} + \Delta U_{s}^{m\to t} + \Delta U_{r}^{m\to t} < 0$

the reversible transformation from monoclinic to tetragonal spontaneously occurs. This material has the characteristics of superelasticity

If

$$\Delta G_{chem}^{t\to m} + \Delta U_{se}^{t\to m} + \Delta U_{s}^{t\to m} + \Delta U_{r}^{t\to m} \geqslant 0$$

$$\Delta G_{chem}^{m\to t} + \Delta U_{se}^{m\to t} + \Delta U_{s}^{m\to t} + \Delta U_{r}^{m\to t} \geqslant 0$$

all transformations between tetragonal and monoclinic zirconia would require additional energy to overcome the transformation resistance forces of both $\Delta U_r^{t\to m}$ and $\Delta U_r^{m\to t}$. This material has the characteristic of shape memory effect (see Fig. 4).
The results suggest that increasing stress is useful for transformation from tetragonal to monoclinic and increasing temperature is useful for transformation from monoclinic to tetragonal. Stress (strain energy) and temperature (heat energy) allow the transformation between tetragonal and monoclinic zirconia to take place. Therefore, $\Delta G^{t\to m}$ ($\Delta G^{m\to t}$) is a function of temperature and stress, $\Delta G^{t\rightleftharpoons m} = F(T,\sigma)$.

Toughness and transformation zone size

The zirconia transformation toughening mechanism is based on the linear elastic fracture model, as shown in Fig. 7 (4).

It is believed that the stress-induced transformation toughening mechanism is not reversible and the distribution of monoclinic zirconia phase in h is uniform. TEM observation results indicate uniform h in zirconia toughened ceramics. However, it was found by Laser-Raman-Microprobe that the distribution of monoclinic zirconia phase beneath the fracture surface was not uniform in the transformation zone h (see Fig. 8). As shown in Fig. 8, the zone size of Ce-TZP measured by Raman is much larger than that of other transformation toughening ceramics such as Y-PSZ is 5 μm, Mg-PSZ is 25 μm[5], **measured by XRD.** These results did not agree with the supposition that the distribution of monoclinic zirconia phase is uniform **in h.** So the stress-induced transformation toughening mechanism could not be used to explain the enhancement of the toughness for Ce-TZP. A new theory and toughening model needs to be found for Ce-TZP.

CONCLUSIONS

a) The $t \rightleftharpoons m$ reversible transformation occurs under different stress and temperature conditions for Ce-TZP.

b) The distribution of monoclinic zirconia phase in the transformation zone h is not uniform. The h for 10 mol% CeO_2 Ce-TZP is about 175 μm.

c) 10 mol% CeO_2 Ce-TZP has the characteristic of shape memory effect caused by zirconia transformation between tetragonal and monoclinic phase.

REFERENCES

1. D. B. Marshall and M. R. James, J. Am. Ceram. Soc., 69(3), 215-217, 1986.
2. R. C. Garvie and P. S. Nicholson, J. Am. Ceram. Soc., 55(6) 303-305, 1972.
3. Z. Q. Zeng and Y. Huang, "Determination of Ms for different Ce-TZP" unpublished work.
4. D. B. Marshall, A. G. Evans, and M. Drory, in "Fracture

Mechanics of Ceramics Vol. 6", 289-307.
5. M. V. Swain, Acta. Metall. Vol. 33, No. 11, 2083-2091,1985.

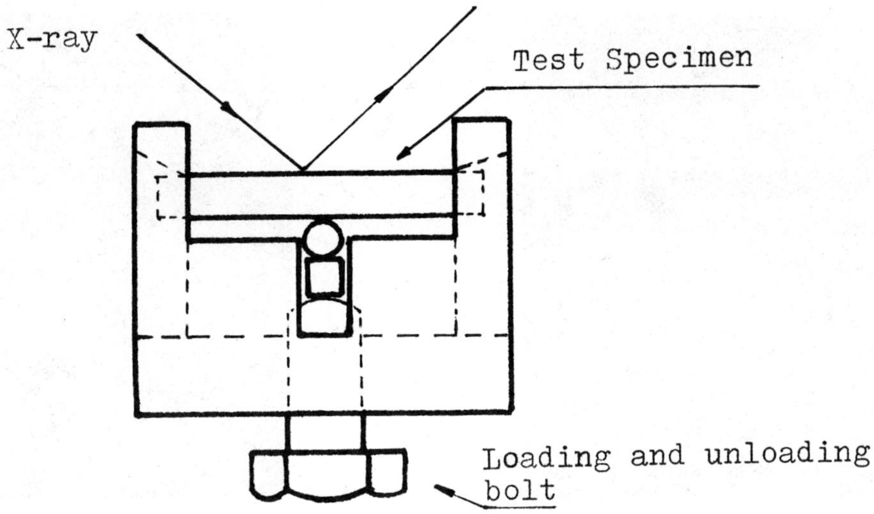

Fig. 1. Schematic diagram illustrating the loading and unloading

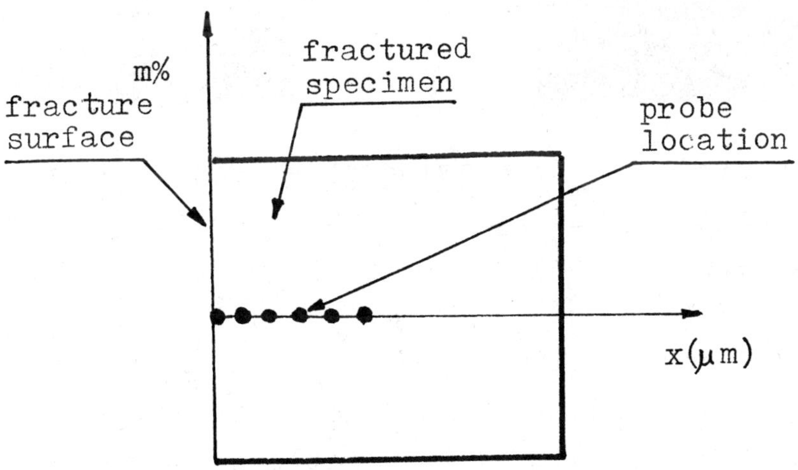

Fig. 2. Schematic diagram illustrating the manner in which the transformation zone's size and shape wereprobed using the Raman microprobe

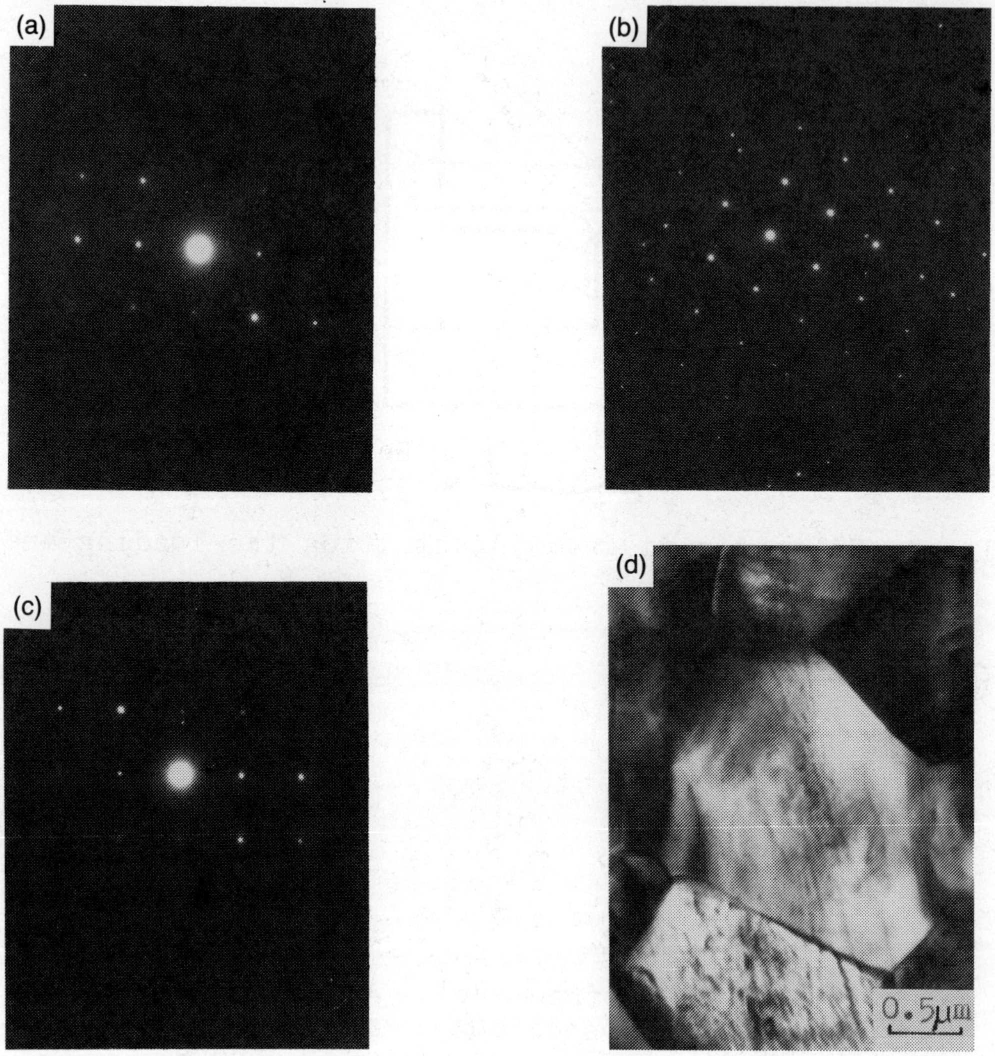

Fig. 3. In situ-TEM observation of electron diffraction for ZrO_2 grain under different radiation intensities of an electron beam

a) Zirconia diffraction pattern under low irradiation intensity (t-ZrO_2)
b) Zirconia diffraction pattern under high irradiation intensity (m-ZrO_2)
c) Zirconia diffraction pattern under low irradiation intensity (t-ZrO_2)
d) Diffraction grain

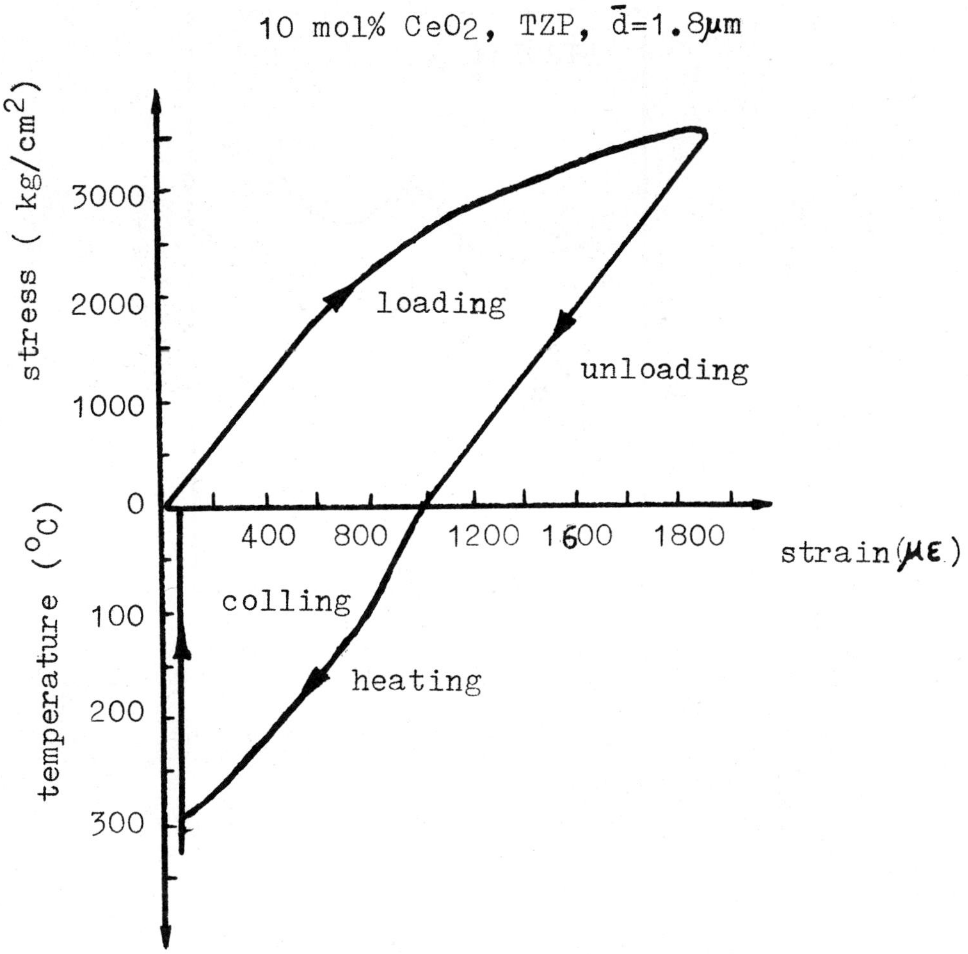

Fig. 4. Four-point flexure stress-strain curve and temperature-strain curve for 10 mol% CeO_2 Ce-TZP

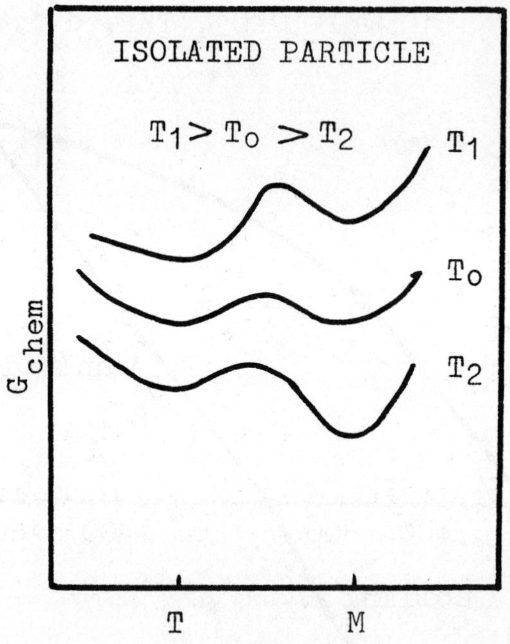

T— tetragonal, M— monoclinic

Fig. 5. Schematic diagram illustrating the change of G_{chem} with temperature

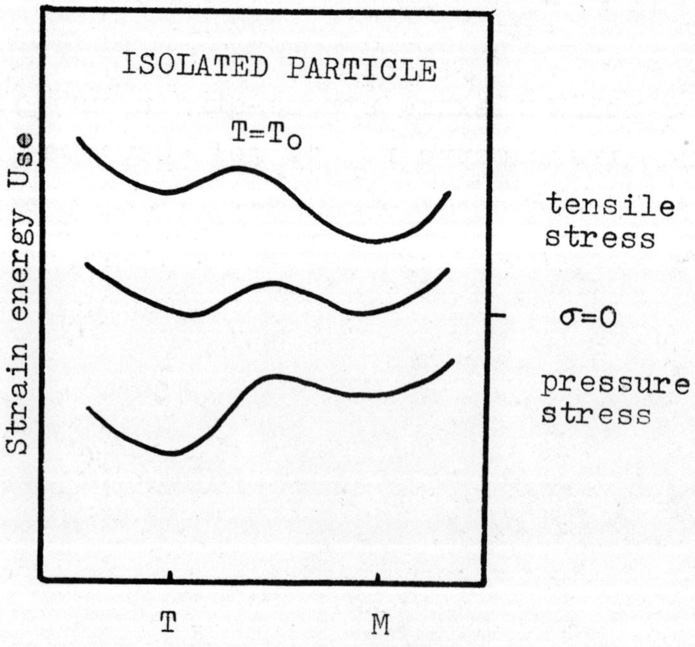

T—tetragonal, M—monoclinic

Fig. 6. Schematic diagram illustrating the change of G_{se} with applied stress

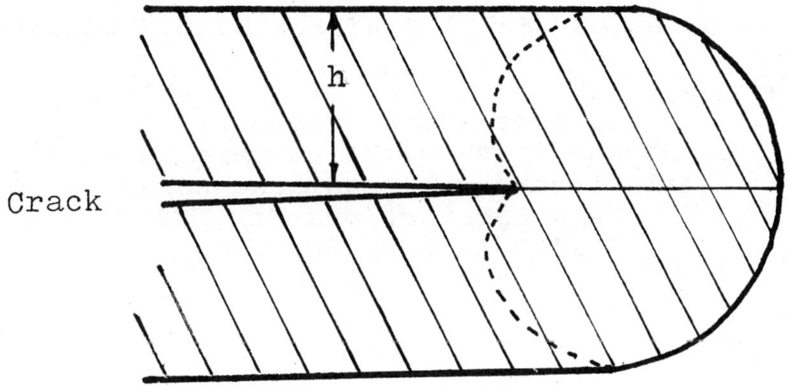

Fig. 7. Transformation zone defined by the dilatation contains transformed particles

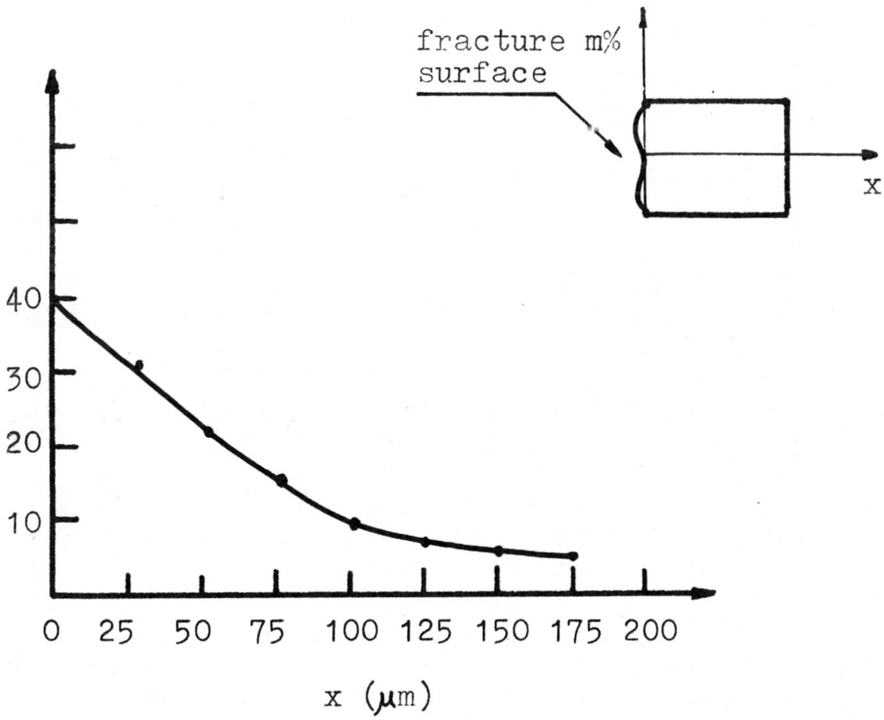

Fig. 8. Distribution of $m-ZrO_2$ on x direction of vertical fracture surface

TOUGHENING OF OXIDE CERAMICS BY RESIDUAL STRESS DOMAINS

E. Lutz and N. Claussen
Technische Universität Hamburg-Harburg
Arbeitsbereich Technische Keramik
Harburger Schloss-Str. 20
D 2100 Hamburg 90, FRG

ABSTRACT

A microstructure design concept is applied which enables high-strength ceramics to exhibit a pronounced R-curve behavior. This yields large work-of-fracture values and low strength degradation after thermal shock. The concept is based on relatively large (10-50μm) spherical zones dispersed in a strong matrix. The spherical zones expand relative to the matrix which introduces long-range tensile hoop stresses around the zones. Flaws propagating due to excessive stresses are deflected by branching toward the compressive zones and are arrested at or near the interface (stress boundary). R-curve mechanisms, similar to those operating in low-strength refractories, are acting (crack bridging , crack branching). The concept has previously been tested with Al_2O_3 and ZrO_2- toughened Al_2O_3 (1) and is here being tested with 2Y-PSZ, 3Y-TZP and Al_2O_3-alloyed TZP as matrix. Monoclinic ZrO_2 and monoclinic-ZrO_2-containing Al_2O_3 has been used as compressive zone material.

INTRODUCTION

The strength and toughness of various engineering oxide ceramics has been considerably enhanced in recent years applying different toughening concepts (1-6). However an enhanced K_{Ic} only leads to a minor increase in the critical temperature difference (ΔT_c) during thermal shock. This can be explained by the usually high thermal coefficient for expansion (α) and the temperature sensitivity of the toughness. Once ΔT_c is surpassed the retained strength (σ_r) is usually close to zero which is a direct consequence of the high strength, i. e. the large stored elastic energy (7,8).

On the other hand, it has long been known that low-strength refractories exhibit high ΔT_c or even no strength degradation at any thermal stress condition. Unfortunately, most low-expansion thermal-shock resistant oxides show low inherent strengths which usually cannot be sufficiently improved for any load-bearing engineering application. Thus the dilemma with most oxides seems to be engineering microstructures that exhibit either high strength or high thermal shock resistance but not both simultaneously.

An interesting correlation between thermal shock resistance and R-curve behavior was demonstrated by Swain (9,10) with differently treated Mg-PSZ ceramics. The relative strength loss σ_r/σ_f for ΔT just in excess of ΔT_c follows a simple proportionality (9,11,12):

$$\sigma_r/\sigma_f \sim \gamma_{WOF}/\gamma_{ic} \quad ,$$

where γ_{WOF} is the integral work-of-fracture and γ_{ic} the initiation fracture surface energy. Hence, the slope of the R-curve influences the strength degradation after a critical thermal shock. The range over which an R-curve is rising is controlled by the width of an energy-dissipative zone behind the crack tip (13,14). The slope is affected by the operating mechanisms which can be divided into short-range (stress-induced transformation, microcracking, and crack deflection) and in long-range effects (crack branching and bridging).

The present paper reports on the toughening of composite ceramics by long-range residual stress domains homogeneously distributed in a strong ceramic matrix (duplex composites) (1,15-17). Spherical polycrystalline inclusions of 10 - 50 µm in diameter were dispersed in a ceramic matrix material. The inclusions expand relative to the matrix during cooling. The strain mismatch and hence the stresses may result either from a phase transformation, thermal expansion difference, or chemical reaction which involve a positive volume change of the inclusions.

The concept was tested in the present work with tetragonal(t) ZrO_2 polycrystals (TZP) and Al_2O_3-alloyed TZP as matrix, and Al_2O_3 containing different amounts of monoclinic (m) ZrO_2 as inclusion material.

EXPERIMENTAL PROCEDURE

The powders for the matrix (component A) were either supplied by Toso, Japan, (2Y-TZP, 3Y-TZP, 3Y-TZP + 20, 40, 60 wt.-% Al_2O_3) in an agglomerated state or were prepared by spray-drying (Al_2O_3, ZTA) (18). Different mixtures of Al_2O_3 (CT 8000 SG, Alcoa) and 35, 50 and 80 vol.-% m-ZrO_2 (Dynazirkon F, Dynamit Nobel) and pure m-ZrO_2 were used as inclusion component B and were also prepared by spray drying. For further discussions they are labeled 35-ZTA, 50-ZTA, 80-ZTA and 100-Z, respectively.

Different compositions of A and B were dry tumble mixed for 12 hours. The agglomerate mixtures were cold isostatically pressed at 950 MPa. The green compacts were sintered at 1500° C for 2 h in air and HIPed at 200 MPa and at temperatures between 1400° and 1600° C for 10 to 120 min in Ar, Ar + 20 vol.-% O_2 and N_2.

The specimen dimensions were 3.5 x 3.5 x 45 mm. Strength and toughness were measured in 4-point-bending (span 24/8 mm) at a cross-head speed of 0.3 mm/min. Fracture toughness was measured by the strength-in-bending method (ISB) (19) with an indentor load of 300 N and, for comparison, by the direct crack length evaluation (ICL) (20) using loads between 100 to 300 N. Vickers hardness was measured with a load of 300 N. The elastic modulus was determined by the ultrasonic method. The m-phase fraction on polished and fractured surfaces was determined by the x-ray intensity ratio of m-(111)- and m-(111)-peaks to the t-(111)-peak (21). Thermal shock tests were carried out by quenching in 20°C water.

All R-curve experiments were performed in air with a three-point bending device (30 mm span) in a closed loop hydraulic testing machine. The sample dimensions were 9.65-10.72 x 2.65-3.15 x 35 mm. The displacement was directly measured and controlled by a Si_3N_4 pushrod placed at the lower surface of the specimen near the notch. The displacement rate of the system was held at 0.02 µm/N. The crack length was directly measured with a travelling microscope during the experiments. Crack length was also calculated from the extrapolated compliance curve and compared with the directly measured values. The initial crack was simulated by a 50 µm diamond saw cut with the depth chosen such that the crack-to-height ratio was between 0.52 and 0.66. Load-displacement curves were evaluated applying the energy concept (11).

RESULTS AND DISCUSSION

The t-m-transformation of ZrO_2 within the dispersed zones generates a stress field consisting of tensile hoop and radial compressive stresses around these zones which are under hydrostatic pressure (22-24). The extent of the stress field is controlled by the amount of the ZrO_2 particles transformed inside the B component on cooling from the sintering or HIPing temperature.

X-ray analyses on polished surfaces of several HIPed duplex composites showed that, in some cases, no or only a small fraction of the ZrO_2 particles inside B had transformed on cooling. As is well known, the martensite start temperature (M_s) of ZrO_2 particles depends on their size (25,26), the alloy content and the constraint pressure (27). In the present composites, M_s of the ZrO_2 particles inside B is probably be prevented by the constraint of B which increases with the amount of transformed ZrO_2 within the fully dense B zones. The same is true also for sintered composites when fully dense. Long-time HIPing (more than 10 min) at high temperatures (1550 - 1650°C) would result in the coarsening of ZrO_2 particles, hence would counteract the undesired stabilization.

Table 1. shows the m- and t-fractions in sintered and HIPed duplex composites with a 2Y-TZP matrix. In HIPed samples there is no difference in the m-ZrO_2-fraction between the duplex material and the matrix alone. The relatively larger increase of the m/(t+m)-ratio in the duplex material after fracture when compared to the respective data of the matrix is both due to the complete transformation in the B component and additional stress induced transformation of partially destabilized matrix areas adjacent to the B zones.

Fracture toughness, evaluated by measuring the indentation crack length , is strongly dependent on the indentor load and the local stress distribution. Thus the method is not suited to generate realistic K_{Ic} data (Table 1).

As evident from Table 1. and Fig. 1a and b the mechanical properties of the AB composites depend on the components A and B, the volume fraction and size of B, the content of m-ZrO_2 particles inside B, and the sintering and HIPing conditions. Strength, elastic modulus and hardness are reduced. All composites prepared so far behave in this characteristic way.

The fracture toughness exhibits a maximum and decreases when the tensile hoop stresses around B exceed the inherent strength of A, i.e. crack linking between the zones occurs. The stress domains activate different toughening mechanisms. A starting crack is deflected by twist and tilt out of its crack plane towards the inclusion. Video observation of the running crack during controlled fracture experiments revealed that the crack is arrested first at the interface between zone B and matrix A. A new crack starts behind the zone which further expands in completion of the transformation. Crack bridging takes place until the opening width compensates for the linear expansion of the zone. Depending on the local distribution of inclusions crack branching occurs between them. Zone-linking cracks develop parallel to the main crack also dissipating energy (Fig. 2. a and b).

These long-range toughening mechanisms lead to a marked load-displacement curve (Fig 3 a-d). As evident from Fig. 3 c and d the crack initiates and propagates already before the load maximum is reached. "Crack closure effects" cause a non-linear-elastic behavior in the first stage of loading and a slow and discontinuous load decrease with further crack propagation.

The difference between calculated and measured crack length and the high offset after unloading gives evidence for the deviation from linear elasticity (9). The more effective the stress domains are with respect to size, volume fraction and volume expansion of B zones, the steeper is the rise of the R-curve over a large crack length (Fig. 4).

These R-curves reflect a high ratio of γ_{WOF} to γ_{Ic}. The close relation between this ratio and the strength degradation after a critical thermal shock is shown in Fig. 5. and 6. The values obtained from the R-curves and the thermal shock experiments reveal the proportionality between σ_r/σ_f and γ_{WOF}/γ_{Ic}. The plots are lying on a parallel line shifted to lower ratios of γ_{WOF}/γ_{Ic} when compared with the results found by Swain (9). This is probably due to the energy evaluation method used.

CONCLUSIONS

The microstructure concept presented in this work, which leads to residual stress domains, results in a marked R-curve behavior combined with decreased strength degradation after a critical thermal shock. Since both thermal shock resistance and thermal insulation are improved with respect to pure Al_2O_3, duplex composites may be useful for hot engine components.

ACKNOWLEDGEMENT

Supported by the Stiftung Volkswagenwerk, FRG under contract No I/61 901 and 64 239. The authors thank Karl Kromp and Nicole Stange for kind assistance concerning the R-curve and thermal shock measurement.

REFERENCES

1. N. Claussen and G. Petzow. German Patent DE 32 33 019 (1982), US Patent 4, 506 024 (1983), Jap. Patent 64567/84.
2. A. G. Evans. *The New High Toughness Ceramics*, to be published, 1987.
3. A. G. Evans and R. M. McMeeking. *On the Toughening of Ceramics by Strong Reinforcements.* Acta Met. 34, 2435 (1986).
4. N. Claussen. *Strengthening Strategies for ZrO_2-toughened Ceramics at High Temperatures.* Mat. Sci. Eng., 71, 23 (1985).
5. I. Nettleship and R. Stevens. *Tetragonal Zirconia Polycrystals (TZP) - A Review.* Int. J. High Tech. Ceram., 3, 1 (1987).
6. W. R. Cannon. *Transformation Toughening Ceramics for Structural Applications*, to be published.
7. D. P. H. Hasselman. *Unified Theory of Thermal Stress Fracture Initiation and Crack Propagation in Brittle Ceramics.* J. Am. Ceram. Soc., 52, 600 (1969).
8. D. P. H. Hasselman. *Figures-of-merit for the Thermal Stress Resistance of High Temperature Brittle Ceramics: A Review.* Ceram. Bull. Int., 4, 147 (1978).
9. M. V. Swain. *R-Curve Behavior of Magnesia Partially Stabilized Zirconia and Its Significance To Thermal Shock,* in Fract. Mech. Ceram., 6, 355, Plen. Press, NY 1983.
10. M. V. Swain and R. H. J. Hannink. *R-Curve Behavior in Zirconia Ceramics*, in Advances in Ceram., 12, 225, The Amer. Ceram. Soc., Columbus, 1984.
11. D. R. Larson, J. A. Coppola, D. P. H. Hasselman, and R. C. Bradt. *Fracture Toughness and Spalling Behavior of High-Al_2O_3 Refractories.* J. Am. Ceram. Soc, 57, 417 (1974).
12. J. Homeny and R. C. Bradt. *Thermal Shock of Refractories.* Proc. of Conf. on "Thermal Stresses in Severe Environments", 343. Edt. D. P. H. Hasselman and R. A. Heller. Plenum Press, NY, 1980.
13. A. Bornhauser, K. Kromp, and R. F. Pabst. *R-Curve Evaluation with Ceramic Materials at Elevated Temperatures by an Energy Approach Using Direct Observation and Compliance Calculation of the Crack Length.* J. Mat. Sci., 20, 2586 (1985).

14. R. Steinbrech, R. Knehans, and W. Schaarwächter. *Increase of Crack Resistance during Slow Crack Growth in Al_2O_3 Bend Specimens.* J. Mat. Sci., 18, 265 (1983).
15. N. Claussen. *Umwandlungsverstärkte keramische Werkstoffe.* Z. Werkstofftechnik, 13, 138 (1982).
16. N. Claussen. *Microsrtructural Design of Zirconia-Toughened Ceramics.,* in Advances in Ceramics, 12, 325, The Am. Ceram. Soc., Columbus, 1984.
17. N. Claussen and E. Lutz. *Microstructural Design of Oxide Ceramic Composites for High-Thermal Shock Resistance.* Lecture on the International Symposium of Fine Ceramics, Arita, 1987.
18. D. Kumar, R. Ladisch, E. Lutz, and N. Claussen. *Pulveraufbereitung mit dem Sprühtrockner.* Cfi, 5, 141 (1988).
19. P. Chantikul, G. R. Anstis, B. R. Lawn, and D. B. Marshall. *A Critical Evaluation Of Indentation Techniques for Measuring Fracture Toughness: II, Strength Method.* J. Am. Ceram. Soc., 64, 539 (1981).
20. K. Niihara, R. Morena, and D. P. H. Hasselman. *Evaluation of K_{Ic} of Brittle Solids by the Indentation Method with Crack-to-Indent ratios.* J. Mat. Sci. Lett., 1, 13 (1982).
21. R. C. Garvie and P. S. Nicholson. *Phase Analysis in Zirconia Systems.* J. Am. Ceram. Soc., 55, 303 (1972).
22. J. Selsing. *Internal Stresses in Ceramics.* J. Am. Ceram. Soc., 44, 419 (1961).
23. R. W. Davidge and T. J. Green. *The Strength of Two-Phase Ceramic/Glass Materials.* J. Mat. Sci., 3, 629 (1968).
24. S. T. Lundin. *Studies on Tri-axial Whiteware Bodies.* Almquist and Wiksell, Stockholm, 1959.
25. A. H. Heuer, N. Claussen, W. M. Kriven, and M. Rühle. *Stability of Tetragonal ZrO_2 Particles in Ceramic Matrices.* J. Am. Ceram. Soc., 65, 642 (1982).
26. R. Garvie. *Stabilisation of the Tetragonal Structure in Zirconia Microcrystals.* J. Phys. Chem., 82, 218 (1978).
27. E. D. Whitney. *Electrical Resistivity and Diffusionless Phase Transformations of Zirconia at High Temperatures and Ultrahigh Pressures.* J. Electrochem. Soc., 112, 91 (1965).
28. Personal communication by M. Swain and M. Sakai

Table 1. Bending strength, fracture toughness (ICL and ISB with indentor load P), and m/(m+t)-ratio measured on polished and fracture surface of 2Y-TZP-based duplex ceramics sintered for 2 h at 1500° C and HIPed for 10 min at 1600° C. Superscripts a) and b) indicate zone size B of 16-32 µm and 45-65 µm, respectivly.

Material	σ_f [MPa]	K_{Ic}^{ICL} [MPam$^{1/2}$]		K_{Ic}^{ISB} [MPam$^{1/2}$]	m/(m+t)	
		P=100N	300N	P= 300 N	pol.	fract.
2Y-TZP						
sintered	785±92	9.7±0.6	7.7±0.4	5.87±0.12	0.03	0.45
HIPed	1007±148	12.5±0.4	9.9±0.6	6.87±0.37	0.01	0.51
2Y-TZP + 10 vol.-% 35-ZTA[a]						
sintered	462±35	15.4±2.5	11.4±1.5	9.25±0.21	0.03	0.48
HIPed	696±48	18.5±1.9	13.1±0.7	9.37±0.21	0.01	0.58
2Y-TZP + 10 vol-% 35-ZTA[b]						
sintered	348±12	no cracks	15.2±2.2	8.95±0.54	0.07	0.54
HIPed	627±56	no cracks	14.4±1.6	10.27±0.57	0.01	0.63
2Y-TZP + 20 vol.-% 35-ZTA[a]						
sintered	355±13	11.8±1.8	9.3±0.6	6.28±0.22	0.13	0.50
HIPed	592±19	no cracks	17.1±0.6	10.83±1.49	0.01	0.63

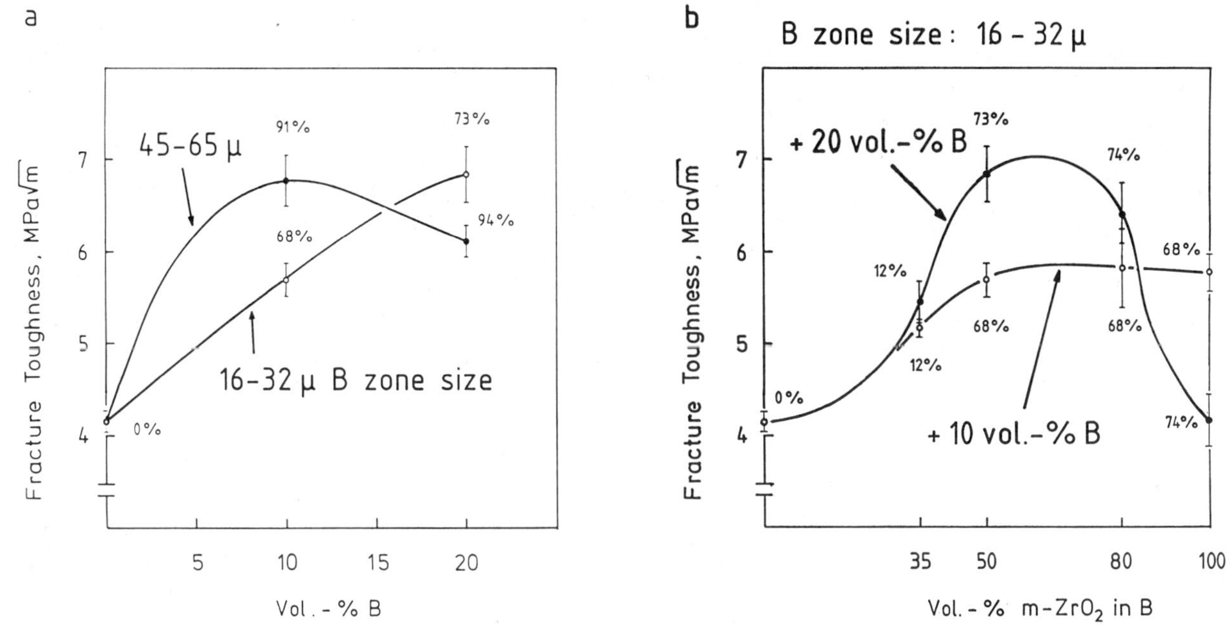

Fig. 1. Fracture toughness of duplex ceramics consisting of 3Y-TZP + 20 wt.-% Al_2O_3 matrix reinforced by 50-ZTA B zones sintered for 2 h at 1500° C. a) Dependence of the volume percentage of component B; b) Influence of the volume fraction of m-ZrO_2 inside B (16-32 µm). The percentage data refer to the m/(m+t)- ratios of ZrO_2 in the respective B zones.

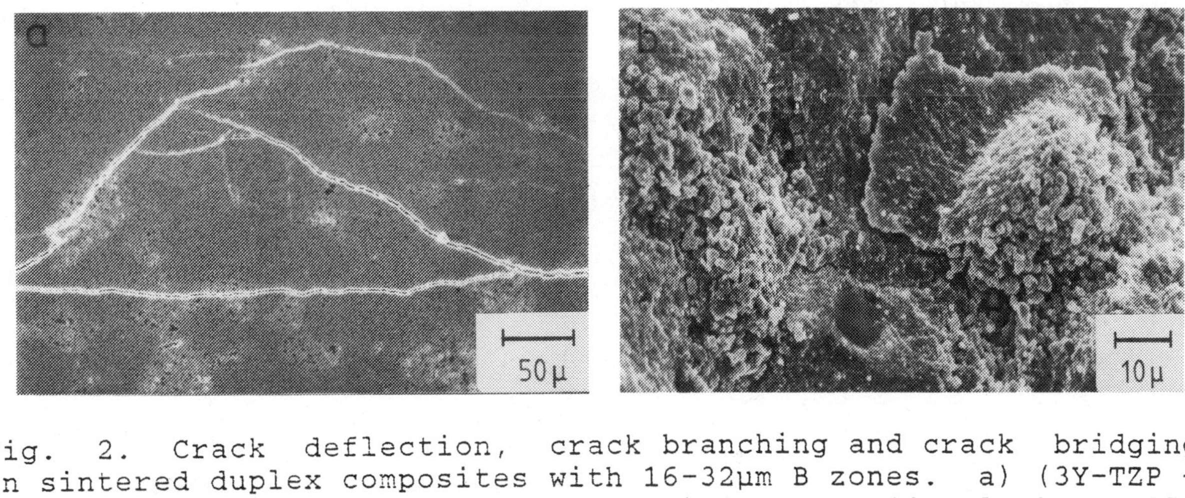

Fig. 2. Crack deflection, crack branching and crack bridging in sintered duplex composites with 16-32µm B zones. a) (3Y-TZP + 20 wt.-% Al_2O_3) + 20 vol.-% ZTA-80; b) 3Y-TZP + 10 vol.-% ZTA-35.

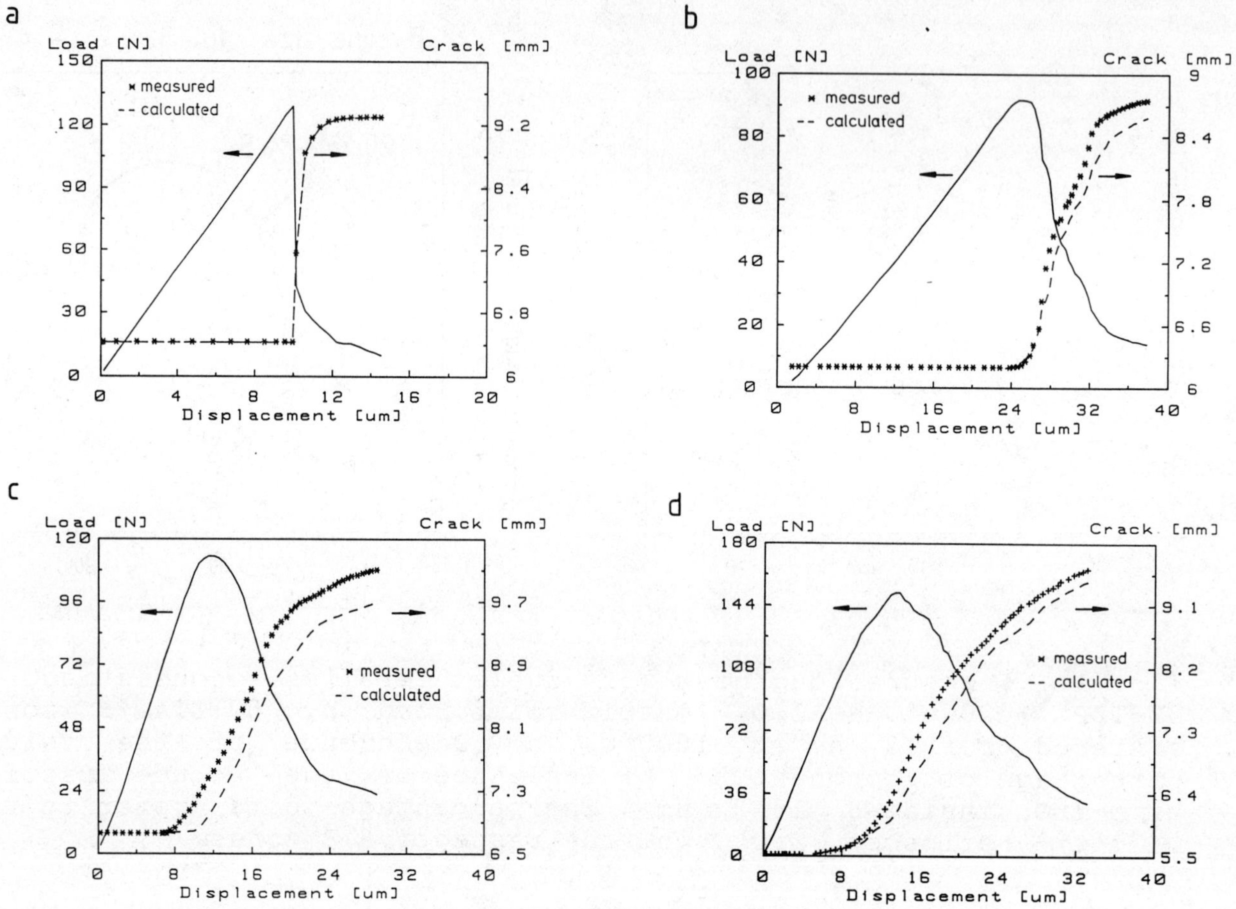

Fig. 3. Load-displacement curves of sintered 3Y-TZP+20 wt.-% Al_2O_3 reinforced with 20 vol.-% of 35-ZTA (a), 50-ZTA (b), 80-ZTA (c), and 100-Z (d). (B zone size is 16-32 μm).

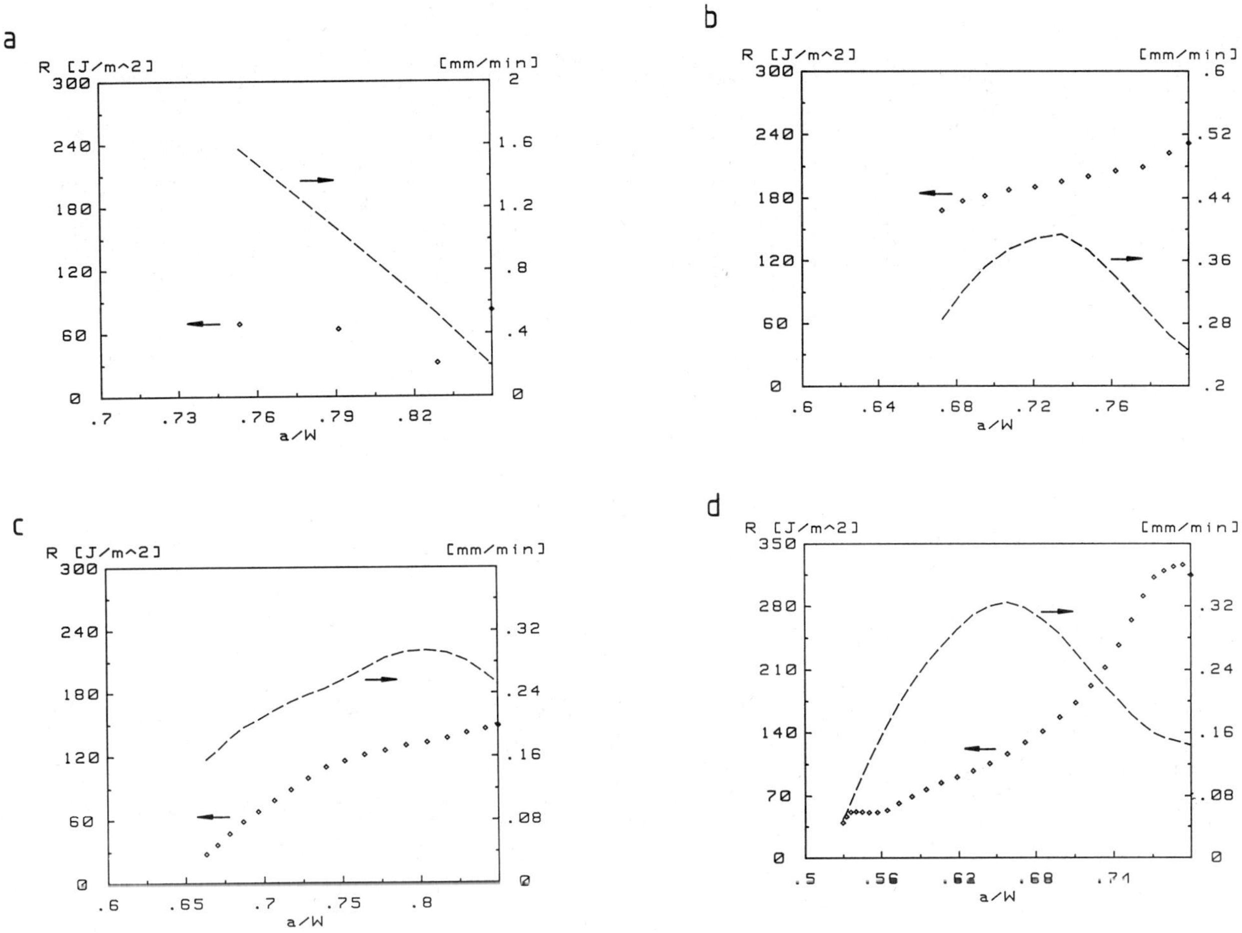

Fig. 4. R-curves resulting from the load-displacement curves (see Fig. 3.) calculated by the energy concept. The dashed lines show the crack velocity. "Concave-up" have also been found by other workers (28).

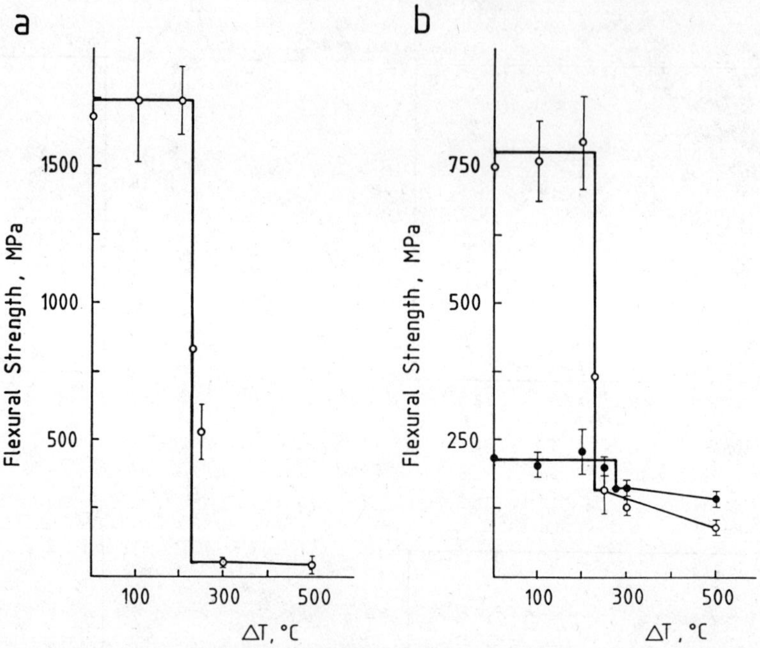

Fig. 5. Strength degradation after thermal shock of duplex composites HIPed at 1600 °C, 30 min in N_2: a) Matrix alone (20 wt.-% Al_2O_3 alloyed 3Y-TZP) and b) Matrix with 20 vol.-% B zones of size 16 - 32 µm consisting of 50-ZTA (open circles) and 80-ZTA (closed circles).

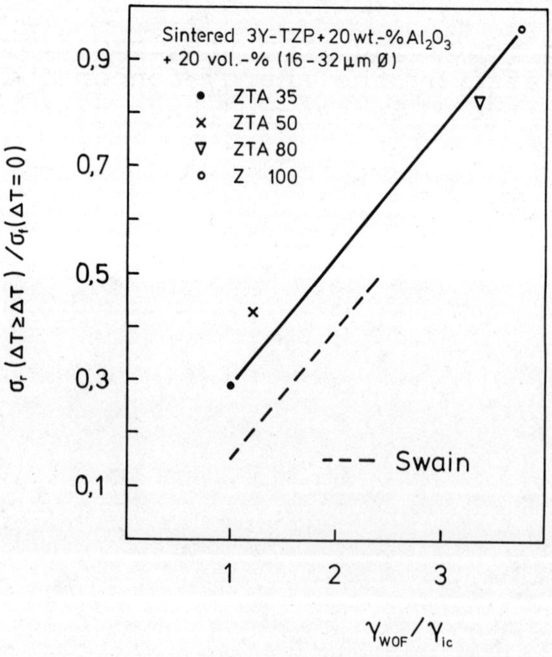

Fig 6. Relative strength degradation of several duplex composites versus γ_{WOF}/γ_{ic}.

Foreign Object Damage Resistance of Structural Ceramics

M. Masuda, H. Tsuruta, T. Soma, and M. Matsui
NGK Insulators, Ltd.,
Nagoya, Japan

Abstract

The equipment for the impact damage testing was successfully developed by using gun accelerators. The chipping fracture mode was investigated by using several kinds of Si_3N_4, SiC, Al_2O_3 and TZP. The critical velocity (which is the threshold impact velocity for fracture) was determined. The critical velocity of the chipping fracture mode was dependent on the K_{IC} and the strength of the target material. Chipping fracture mode also depend the hardness and the shape of the projectile. SiC shows the small resistance against FOD. The relationship between FOD resistance and mechanical properties of the ceramics and the mechanism of impact damage are discussed.

1. Introduction

Advanced ceramics are currently being evaluated for use as gas turbine components and turbocharger rotors. To successfully design the turbine rotors and the stators and utilize ceramics for those components, an understanding of the resistance against foreign object damage (FOD) is required. The FOD data on ceramic materials is minimal,[(1)~(5)] and the FOD mechanisms on those materials have not been characterized.

In the investigation of FOD phenomena, it is at first necessary to develop methods of accelerating projectiles up to 800m/s. Two types of accelerators are used to perform this task: a light gas gun accelerator and a propellant gun accelerator. In this study, two types of ceramic testing devices were developed to obtain the fundamental data on FOD resistance of ceramic materials.

The impact damage can occur under several fracture modes, as shown in Fig. 1. The chipping fracture mode was investigated using several kinds of ceramic materials. The mechanism of FOD under the chipping fracture mode is discussed from the experimental results.

2. Experimental Procedure

2.1 Testing Apparatus

Two types of particle projection devices were developed. One was a light gas gun which could accelerate the particles to any desired velocity in the range of 30 to 300 m/s. The other was a propellant gun which could propel the particles by burnt gas of powder to obtain the velocities in the range of 300 to 800 m/s. The specifications of the particle projection devices used in this investigation are summarized in Table 1.

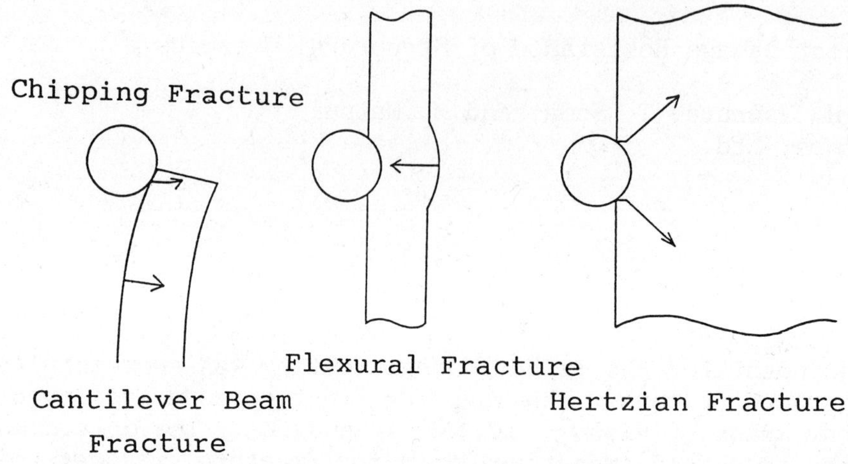

Fig. 1 Fracture mode of FOD.

Table 1. Particle Projection Device

	Low Speed Test Machine	High Speed Test Machine
Projection Method	Gas	Propellant
Available Velocity Range	30~300 m/s	300~800 m/s
Environment	RT in Air	RT in Air
Sabot Diameter	φ10mm	φ12mm
Sensor for Measuring Velocity	Photo Transistor	Coil

Table 2. Material Properties

	Toughness *1	Strength *2	Young's Modulus *3	Knoop Hardness *4	Pore Size *5 (μm)		
					mean(m)	standard deviation (σ)	m+2σ
	($MN/m^{3/2}$)	(MPa)	(GPa)	(GPa)			
SSN-A	6.8	530	240	12.0	2.69	1.86	6.41
SSN-B	5.9	880	265	14.1	1.71	1.12	3.96
SSN-C	6.4	1020	300	16.3	1.82	1.12	4.24
SSN-D	5.7	880	290	16.9	1.71	1.24	4.20
SSN-E	8.3	770	300	16.0	1.78	1.09	3.96
SSN-F	7.4	590	315	17.4	1.85	1.23	4.31
SSN-G	3.8	740	-	17.6	1.46	1.34	4.14
SSN-H	5.4	810	-	16.2	1.55	1.07	3.69
SiC-A	2.4	500	390	21.5	1.77	1.17	4.11
SiC-B	4.1	580	-	24.4	1.96	1.38	4.72
Al_2O_3	5.2	340	345	-	2.65	2.86	8.43
TZP	6.2	1030	205	11.4	1.53	1.13	4.19

*1: Single-Edge-Precracked-Beam method(6)
*2: 4-point bending test according to JIS R 1601
*3: Pulse echo overlap method according to JIS R 1602
*4: Load 300g
*5: Obtained by the image analyzer

2.2 Target Materials

Several kinds of sintered Si_3N_4 (SSN-A to H), sintered SiC (SiC-A), hot-pressed SiC (SiC-B), Al_2O_3, and TZP were selected for the target materials. The properties of these materials are summarized in Table 2.

Specimens used were machined plates with a width of 4 mm, a thickness of 1 mm and a length of 4 mm and the surface was finished by #800 grit size diamond wheel.

2.3 Projectiles

Steels, TZP, and brass spheres of 1 mm diameter were chosen to examine the effect of hardness of projectile on FOD. Furthermore, a cube of 4mg cast iron, which has the same weight as the steel ball was also used to examine the shape effect of projectile on FOD to simulate possible FOD in the gas turbine and the turbocharger, because the directional particles would strike the rotor in the gas turbine or the turbocharger.

2.4 Impact Tests

Impact was made on the edge of the ceramic plate to study the chipping fracture mode of FOD and the accuracy of the distance between center of the projectile and the edge of the target is less than 0.1mm. Figure 2 shows the schematic of chipping fracture test. The target was held as a cantilever beam of length 1mm. Specimens were observed to be chipped or not by the optical microscope after the impact testing.

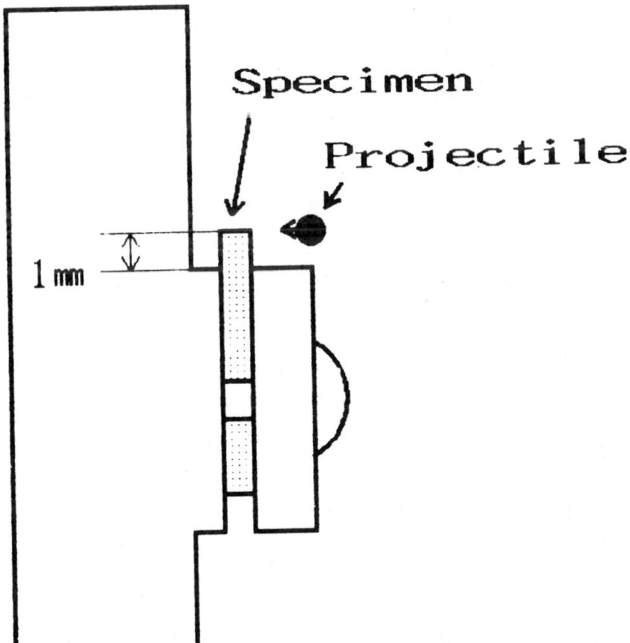

Fig. 2 Schematic of the test jig for chipping fracture test.

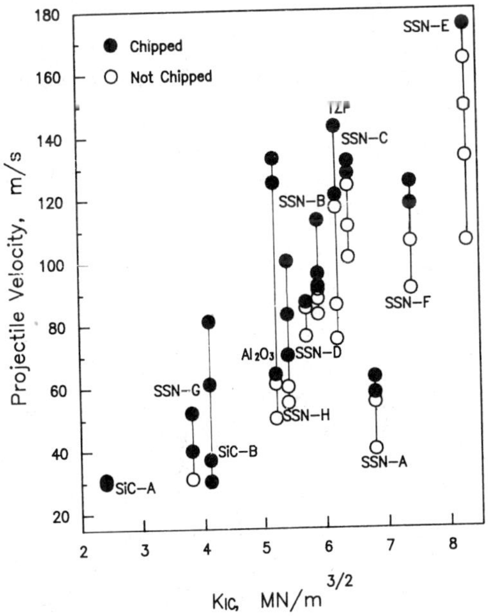

Fig. 3 The relationship between K_{IC} and impact velocity in chipping fracture test impacted by 1.0 mm diameter steel ball.

Fig. 4 The top view of the chipped target specimen impacted by 1.0 mm diameter steel ball. (a) SSN-A, (b) SSN-B, (c) SiC-A, (d) TZP. The arrow shows the impact direction.

3. Results

3.1 Chipping Fracture

The relationship between the fracture toughness of the target and the velocity of the steel projectile on chipping fracture is shown in Fig. 3. Chipping fracture occurred in both SiC targets at 30 m/s which was the lowest velocity in the low speed machine. However, except for SiC, the critical velocity that is the threshold velocity for chipping fracture was obtained. The typical fracture surfaces for chipping fractured specimens are shown in Fig. 4. Two types of fracture surfaces were observed. One is the crack conically propagating from the contact portion of the projectile to the reverse side as shown in Fig. 4 (b) and Fig. 4 (c). The other is the crack spherically propagating in the flank as shown in Fig. 4 (a) and Fig. 4 (d).

The relationship between the fracture toughness of the target and the critical velocity for impact with the steel ball is shown in Fig. 5. There appears to be a correlation between the fracture toughness and the critical velocity. The correlation coefficient (between fracture toughness and critical velocity) was 0.78. The relationship between the strength of the target and the critical velocity is shown in Fig. 6. There also appears to be a correlation between the strength and the critical velocity. The correlation coefficient (between strength and critical velocity) was 0.46. In general the critical velocity for chipping increased as the fracture toughness and the strength of the target material increased.

3.2 Effect of projectile

The effects of hardness of the projectile on the chipping fracture in two Si_3N_4 (SSN-A and SSN-B) are shown in Fig. 7. The critical velocity on the chipping fracture depended on hardness of projectile. In the case of SSN-B impacted with a hardened steel sphere (800Hv), the conical crack propagation was observed. However, SSN-B impacted with a TZP sphere (1000 Hv) resulted in the spherical crack as observed in SSN-A as shown in Fig. 4 (a).

Finally the effect of shape of the projectile on chipping fracture was examined using a 4 mg cube of cast iron projectile against SSN-B. A cube of cast iron was controlled to strike a half of one face of cube on the target. The result is shown in Fig. 7 by the square symbol. The critical velocity increased by 200m/s in the case of a cube of cast iron compared with a spherical projectile of similar hardness and same weight against SSN-B.

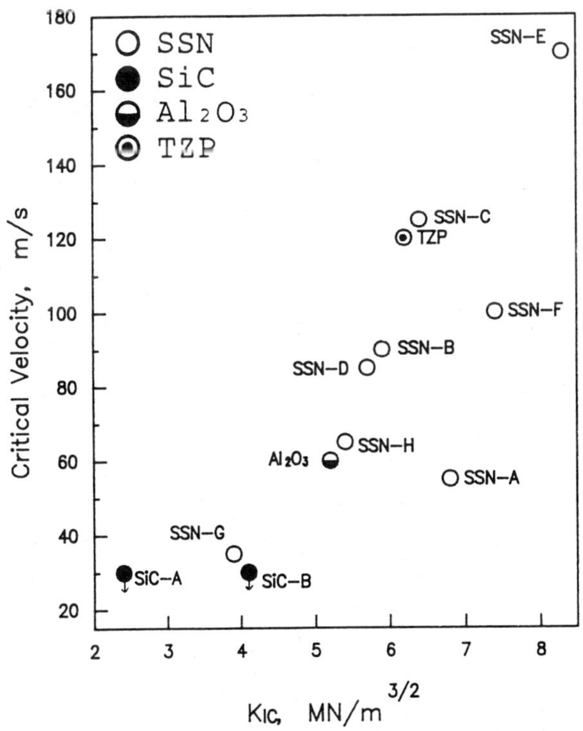

Fig. 5 The critical velocity in chipping fracture test as a function of K impacted by 1.0 mm diameter steel ball.

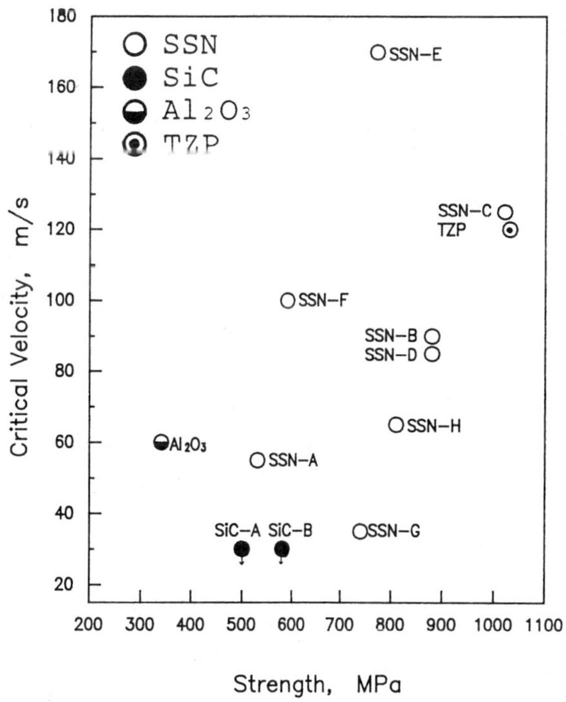

Fig. 6 The critical velocity in chipping fracture test as a function of strength impacted by 1.0 mm diameter steel ball.

4. Discussion

4.1 The effect of mechanical properties of target

The critical velocity strongly depends on the fracture toughness for impact in the chipping fracture mode. The defect which caused the chipping fracture is probably different from the fracture origin that determines the 4-point bending strength. Because the fracture origin of the 4-point bending is explained by the extreme value of pore size population, but impact tensile stress area would be too small to include such large pore in the impact tensile stress area. SSN-A, whose pore size is large, shows the lowest critical velocity among specimens which have similar fracture toughness value as shown in Fig.3. Therefore, it is suggested that the defect which induced the chipping fracture is related to the pore in the impact tensile stress area. Assuming the fracture occurs elastically, the maximum tensile surface radial stress(3) is given by

$$\sigma = \frac{(1-2\nu_1)}{2\pi} (5/3\pi\rho)^{1/5} (3/4k)^{-2/15} V^{2/5}, \qquad (1)$$

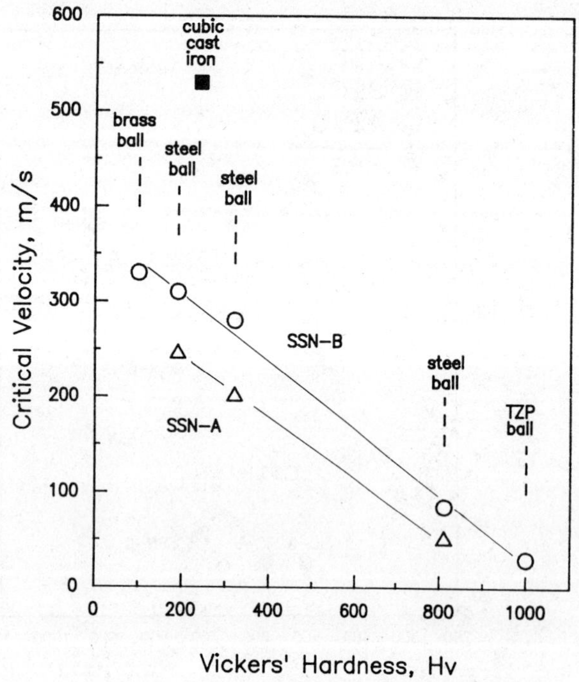

Fig. 7 Effect of hardness and shape of projectile impacted by 1.0 mm diameter steel, brass, and TZP ball and 4mg cube of cast iron.

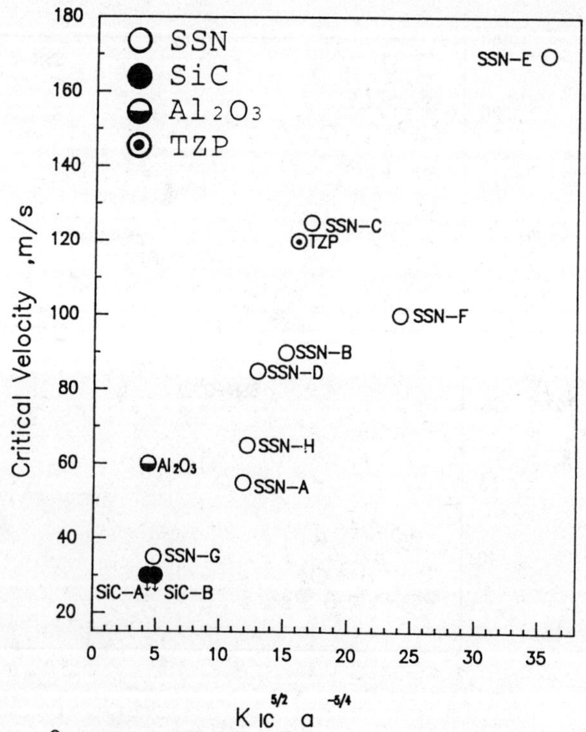

Fig. 8 Relationship between critical velocity and K a impacted by 1.0 mm diameter steel ball.

where

$$k = \frac{1-\nu_1}{E_1} + \frac{1-\nu_2}{E_2} ,$$

ν_1, ν_2 : Poisson's ratio of the specimen and the projectile,
E_1, E_2 : Young's modulus of the specimen and the projectile,
ρ : Projectile density,
V : Projectile velocity.

Assuming that the fracture occurs when σ reaches the critical value, σ_C, which is related to the fracture toughness, K_{IC}, as follows,

$$\sigma_C = \alpha K_{IC} / (\pi a)^{1/2} , \qquad (2)$$

where α is constant and a is the flaw size.

Combining Eq. (1) with Eq. (2) the relationship between projectile velocity and K_{IC} is given by

$$V_C \propto K_{IC}^{5/2} a^{-5/4} . \qquad (3)$$

Figure 8 shows the relationship between critical velocity and $K_{IC}^{5/2} a^{-5/4}$ using the effective pore size as the a value. The effective pore size is defined as the addition of the mean pore size and twice the standard deviation measured by using the image analyzer. The correlation coefficient between critical velocity and $K_{IC}^{5/2} a^{-5/4}$ is 0.89 and is better than between the critical velocity and the fracture toughness alone. Chipping fracture depends on both the fracture toughness and the pore size of the target material.

4.2 The effect of projectile

The critical velocity increased with decreasing hardness of the projectile in both tests (SSN-A and SSN-B).

Chipping fracture occurs when the radial stress at the edge of the contact area reaches a critical stress. The radial stress decreases with decreasing hardness and yield stress of the projectile. Therefore, if the projectile becomes softer, the higher speed is required to lead to the fracture of the specimen due to increasing the radial stress.

The critical velocity increased by 200m/s with using a cube of cast iron compared with a spherical projectile of similar hardness against SSN-B. When a cube of cast iron land a half of one face, the cube would immediately rotate. The increase in critical velocity might be due to the absorption of kinetic energy by rotation of projectile at impact.

5. Conclusion

The following conclusions were obtained about FOD for ceramics from this study.

1. Chipping fracture depends on the fracture toughness and the pore size of the target material.
2. Chipping fracture depends on the hardness and the shape of the projectile .
3. SiC shows small resistance against FOD.

References

1) A.G.Evans, Treat. on Mater. Sci. and Tec., 16, 1-68, 1979.
2) A.W.Ruff and S.W.Wiederhorn, ibid., 16, 69-126, 1979.
3) C.G.Knight, M.V.Swain and M.M.Chaudhri, J.Mat.Sci., 12, 1573-1586,1977.
4) S.Wada and N.Watanabe, J.Ceram. Soc. of Japan, 94 (11),1157-1162, 1986.
5) M.M.Chaudhri and S.W.Walley, Philos. Mag. A, 37(2), 153-165, 1978.
6) T.Nose and T.Fujii, J. Am. Ceram. Soc., 71 (5), 328-33, 1988.

DISLOCATION STRUCTURES WITHIN SPINEL CRYSTAL FRACTURE SURFACES

G.Y.Richardson, C.Cm.Wu
U.S. Naval Research Laboratory, Washington, D.C. 20375

R.W.Armstrong
Engineering Materials Group and Mechanical Engineering Department
University of Maryland, College Park, MD 20742

ABSTRACT

Nondestructive characterization of dislocation structures in fracture surfaces produced by three-point bend testing of spinel ($MgAl_2O_4$) single crystals has been accomplished by surface reflection x-ray diffraction topography (XRDT). Different dislocation structures were observed within the surface layers for {100}, {110} and {111} fracture surfaces. Scanning electron microscopy (SEM) was employed to characterize the external fracture surface features and for direct comparison with XRDT images of the fracture surface layer. A stereographic projection method illustrates the geometry of the fracture experiment as well as the x-ray diffraction geometry. The method appears to be useful for determining a priori the dislocation structures induced by the fracture process. Knowledge of the dislocation structures should prove valuable for fully understanding the meaning of fracture toughness stress intensity measurements.

X-RAY DIFFRACTION TOPOGRAPHY (XRDT)

Surface reflection x-ray diffraction topography (XRDT) is a powerful tool for the nondestructive characterization of the internal defect structure within bulk crystal surface layers, within a 50 microns depth, depending on the mass absorption coefficient, the incident x-ray intensity, and the diffraction geometry [1]. The surface reflection XRDT method, normally credited as the Berg-Barrett back-reflection technique, has been reviewed in several publications also describing the Lang transmission and Borrmann anomalous transmission methods [2-5]. The techniques have been shown to be capable of giving a complete characterization of a variety of crystal microstructures, including dislocations, stacking faults, ferromagnetic and ferroelectric domains, subgrain boundaries, slip bands, twin structures, segregated solutes, precipitates, and internal strain fields due to surface effects[2].

The XRDT methods have been applied in crystal growth [6,7] studies as well as to characterize the nature of semiconductor wafers and integrated circuits [8,9]. Residual deformation associated with indentation hardness testing [10] or cracking studies [11,12] have been described. Recently, the x-ray diffraction topography method has been applied to the nondestructive inspection of a nickel base alloy single crystal turbine blade [13]. Potential improvements in the technique allowing for rapid inspection of entire blades were suggested.

CRYSTAL SPECIMEN PREPARATION AND TEST EVALUATION

The spinel ($MgAl_2O_4$) single crystal used in this study was a transparent cylindrical commercial boule of approximately 3.5 cm diameter Czochralski-grown with an [001] growth direction. The Laue back reflection method was used to determine the orientation of the bulk crystal prior to cutting out fracture specimens. Individual specimens were diamond-sawed and ground to rectangular bars approximately 2 x 4 x 12 mm. Specimens were polished successively with 30, 15, 6 and 1 micron diamond pastes. The two edges along the specimen tensile surface were polished with a 6 µm diamond polishing wheel to prevent fracture from occurring at these sites. A Vickers indent at 300 grams load was centrally placed on the polished tensile surface so as to initiate fracture in a desired crystallographic direction. The bars, with a 12 mm span, were tested in three-point bending at a crosshead speed of 1 mm/min by the use of an Instron testing machine , Model 1122. Toughness determinations were made by the use of the indentation-fracture toughness technique for {100}, {110}, and {111} fracture planes [14]. Fracture surfaces were characterized by the use of SEM and XRDT employing the asymmetric crystal topography (ACT) technique [15] shown schematically in Figure 1.

In Figure 1, is shown a schematic diagram of x-rays emanating from a sealed tube and impinging on an asymmetrically cut crystal monochromator which is aligned so that the characteristic kα wavelength is Bragg-diffracted. The diffracted beam impinges, in turn , upon a crystal specimen at a glancing angle to the surface being investigated and is Bragg-diffracted from a predetermined crystallographic plane running more deeply into the specimen. An x-ray image intensifier is employed for real time viewing of the diffracted image. Minor adjustments to the crystal orientation are easily made to obtain a visual image of the entire surface prior to confirming the suitability of a chosen reflection by exposing a fast x-ray film (e.g. Kodak no-screen film) for typically 5 minutes. Finally a nuclear emulsion imaging plate is exposed over a time period ranging from 6 hours for {100} fracture surfaces to 24 hours for {111} surfaces, in accordance with the reflecting power in each case. Optimum resolution is achieved by: (1) selecting a diffracting plane in the specimen that allows a glancing incident angle for the x-ray beam, generally less than 10°, and gives a diffracted beam approximately perpendicular to the crystal surface layer being imaged; (2) by minimizing the vertical divergence of the incident x-ray beam [16]; (3) by minimizing the distance separating the imaging plate and the specimen while keeping the plate parallel to the incident beam without blocking the incident beam; and, (4) by the use of an imaging plate with a fine grain emulsion.

The x-ray source used here was a conventional copper x-ray tube (Phillips) operated at 30 KV and 20 ma. The source was used in a spot configuration (1 x 1 mm) for all of the topographs. Typically the distance from the crystal to the recording plate was 1-2 mm. The image was recorded by the use of Ilford nuclear plates, type L-4, having an emulsion thickness of 25 µm with a grain size of 0.25 µm. Bragg reflection from {511}-type planes were used to characterize the {100} fracture surfaces and {731 }-type reflections were used to characterize both the (110) and (111) surfaces.

A stereographic projection method of analysis [17] was employed to understand the x-ray and fracture experiments. Figure 2 is a standard ($\bar{1}$00) stereographic projection representing the x-ray diffraction conditions and, also, the cleavage fracturing process for this same (100) cleavage surface. The crack front is shown at the left side of the figure, spreading from the fracture origin which is the Vickers microindentation. Crack growth occurs in the [010] direction generating dislocations with Burgers vectors, $\vec{b}_1$ and $\vec{b}_2$, and line vector $\vec{\xi}$ as will be described. For our x-ray purpose here, two planes of diffraction are shown containing in each case the x-ray source, the diffracting plane normal and the diffracted image. The greater vertical divergence (non-parallel x-ray beams which are spread vertically) which would result from using an x-ray source in a line configuration is indicated on the x-ray stereograph by the larger circle of reflection. This configuration of the x-ray source would lead to somewhat poorer resolution of the dislocation lines compared to a spot source. The lower right corner of the figure shows the physical geometry for the ($\bar{5}\bar{1}$1) reflection. For optimum resolution, it is advantageous to have the diffracted x-rays exit normal to the surface being imaged, as described earlier, not at 45.6° as shown. However, the {511} plane was chosen for its strong diffracting power for spinel giving 45% relative intensity compared to the strongest reflecting plane of {311}-type. Because of the inclined beam direction, it becomes even more critical to have the recording plate as close to the specimen surface as possible, see also Figure 1.

SEM AND XRDT OBSERVATIONS

Values of the fracture mechanics stress intensity, K_{IC}, measured for the {001}, {011} , and {111} fracture planes were 1.2, 1.6, and 2.0 MPa(m)$^{1/2}$ respectively, in close agreement with reported values [18]. Figure 3 shows matched scanning electron micrographs and x-ray diffraction topographs of the spinel fracture surfaces. A different crack propagation direction was involved in each case, for example, [110] for the (001) plane in Figure 3-A. Fracture across the (001) plane appears to be the least complicated of the three surface types. It is the predominant cleavage plane in spinel. This fracture surface is relatively flat and featureless. The aligned dislocations (of enhanced x-ray intensity) with line vectors at 45° to the [110] are, therefore, positioned along the [010] direction. Additional "dots" of enhanced intensity, mostly observed in the lower half of the topograph, are probably the line intersection points with the fracture surface of individual dislocations whose line vectors, ξ, are along the [001] direction, which was the crystal growth direction.

Fracture along an average (110) plane is observed, in Figure 3-B, to be clearly nonplanar. The surface roughness indicates a somewhat tortuous fracture path and the dislocation line structure is clearly more disorganized. Long range deformation strains are probably indicated by the significant variations in enhanced x-ray intensity in several regions of the fracture surface, although surface shadowing effects also occur.

Fracturing along the average (111) plane, as shown in Figure 3-C, produced the roughest fracture surface in agreement with the highest value of K_{IC} measured. A somewhat more disorganized dislocation structure is indicated than shown for {011} fracturing. Note that the fracture surface in 3-C is approximately one half of that shown for the $(\bar{1}00)$ and $(\bar{1}\bar{1}0)$ specimens. Though giving a valid K_{IC} measurement, this smaller section helped to reduce the apparent surface roughness, thereby minimizing the effect of shadowing of the XRDT image.

FRACTURE STEREOGRAPH

Figure 4 shows a stereographic projection analysis developed in the present study [19] to describe the cleavage fracturing process, in this case, across the (100) plane for crack propagation in the [010] direction. Under tensile loading, shear stresses are generated in a modified way ahead of the crack tip including, for example, at 45° to the tensile axis. Dislocations with <110> Burgers vectors, specifically along the $[\bar{1}10]$ and [110] directions in the sense shown, are proposed to be emitted just in front of the crack tip to allow crack growth. For dislocation line vectors oriented along the ±[001] direction, as indicated in the Figure, (110) and $(\bar{1}10)$ slip systems would produce the simplest geometrical description of plastic flow for this case of [010] direction crack growth. Thus, in an XRDT image for this case, dislocation lines would possibly be observed to run along the ±[001] direction. A similar analysis of these same <110> Burgers vectors on {111} slip planes would give dislocation lines running along $[0\bar{1}1]$.

Figure 5 shows XRDT images giving evidence of ±[001] line directions for dislocations, in agreement with the fracture stereograph of Figure 4 leading to the expectation of {110} slip. Figure 5-A is a SEM micrograph of the $(\bar{1}00)$ fracture surface with the [010] crack propagation along the vertical axis. Figure 5-A_2 is a topograph of the same surface taken with a $[\bar{5}\bar{1}\bar{1}]$ reflection using a spot focus x-ray source. The diffraction geometry used in recording this image is described by the x-ray stereograph of Figure 2. Both Figure 2-A_2 and Figure 4 confirm that the dislocation line vectors are along ±[001].

DISCUSSION

Figure 6 shows a combination of x-ray diffraction and fracture stereographic information for the case of cleavage crack growth across the (001) plane in the $[\bar{1}10]$ crack growth direction. This is the situation for Figure 3-A_2 where the XRDT image is a $(\bar{1}\bar{1}5)$ plane reflection. In this case, slip in the [101] and $[10\bar{1}]$ directions on the $(10\bar{1})$ and (101) slip planes, respectively, produce dislocation line directions along the inclined [010] direction. Thus, Figure 6 and Figure 3-A_2 are in agreement. The operative slip systems deduced on a combined stress-state and XRDT imaging basis are of {110} <110> type for this spinel crystal.

Interestingly enough, {110} and {111} slip planes have been observed previously in spinel crystals depending on the crystal stoichiometry [20-23]. The {110}<1$\bar{1}$0> slip systems are reported to have the lowest critical resolved shear stress, less than the {111} and {100} systems, at temperatures below 800°C[24,25]. Glide on {111} is predominant for $MgO(Al_2O_3)_n$ stoichiometries of n=1 and n=3 but glide on {110} is also always present [20]. Glide occurs on {110} alone for n=2. The Burger s vector for both slip plane types is of <110> type . Thus, the possibility of {110} slip for any stoichiometry is in agreement with {110} slip being favored anyway for [010] direction crack growth. However, for [110] crack growth, as in Figure 6, {111} type slip is slightly favored over the observed {110} slip on a stress state basis[26]. This result gives some indication, perhaps, that the crystal stoichiometry is in fact in the direction of n being just greater than the theoretical value of n=1.0.

CONCLUSIONS

Different dislocation structures, revealed nondestructively in XRDT images of individual {001}, {011}, and {111} spinel crystal fracture surfaces, appear to relate to mode I fracture stress intensity measurements of 1.2, 1.6, and 2.0 MPa(m)$^{1/2}$ for the respective surfaces. The different patterns of dislocation lines observed for [010] or [110] crack propagation directions in {001} fracture surfaces agree with <1$\bar{1}$0>{110} slip predictions obtained from fracture stereographs.

ACKNOWLEDGEMENTS

The authors wish to thank Dr.S.B.Qadri for helpful advice and for use of x-ray equipment. Dr.M.S.Duesbery provided valuable discussions on dislocations in the spinel crystal structure.

REFERENCES

(1) B.K.Tanner, X-ray Diffraction Topography, Pergamon Press, New York, (1976).
(2) R.W.Armstrong and C.Cm.Wu, X-ray Diffraction Microscopy, in Tools and Techniques for Microstructural Analysis, J.L.McCall and W.M.Mueller (editors), Plenum Press, (1973)169-219.
(3) B.D.Cullity, Elements of X-ray Diffraction, 2nd ed. (1978), Addison-Wesley Publishing Co.,Inc..
(4) S.Weissmann, F.Balibar and J-F. Petroff (editors), Applications of X-ray Topographic Methods to Materials Science, Plenum Press, New York (1984).
(5) M.Fatemi, X-ray Diffraction Tools and Techniques for Solid State Devices, Review of Progress in Quantitative Nondestructive Evaluation, Vol. 5B, edited by D.O.Thompson and D.E.Chimenti, Plenum Pub. Co., (1986) 991-9.
(6) B.K.Tanner and D.K.Bowen (editors), Characterization of Crystal Growth Defects by X-ray Methods, Plenum Press, New York (1980).
(7) E.Kasper, P.Korczak, and H.Henkel, X-ray topographic analysis of dislocations in Czochralski-grown stoichiometric $MgAl_2O_4$ spinel single crystals, J.Mat.Sci., 9 , (1974)1696-1700.
(8) R.A.Forman, J.R.Hill, M.I.Bell, G.S.White, S.W.Freiman, and W.Ford, Strain Patterns in Gallium Arsenide Wafers: Origins and Effects, in Defect Recognition and Image Processing in III - V Compounds II, E.R.Weber (editor), Elsevier Science Publishers, B.V., Amsterdam, (1987) 63-71.
(9) S.B.Qadri, D.Ma, and M.Peckerar, Double Crystal X-ray Topographic Determination of Local Strain in MOS Device Structures, J.Applied Phys.Letters, 51, (1987)1827-29.
(10) R.W.Armstrong, X-ray Topography of Hardness Indentations, in Applications of X-ray Topographic Methods to Materials Science, S.Weissmann, F.Balibar and J-F. Petroff (editors), Plenum Press, New York, (1984) 295-300.
(11) G.Michot, A.George and G.Champier, Plastic Deformation at the Crack Tip in Silicon Crystals, in Applications of X-ray Topographic Methods to Materials Science, S.Weissmann, F.Balibar and J-F. Petroff (editors), Plenum Press, New York, (1984) 325-32.
(12) J.C.Bilello, Applications of Synchrotron X-ray Diffraction Topography to Fractography, in Applications of X-ray Topographic Methods to Materials Science, S.Weissmann, F.Balibar and J-F. Petroff (editors), Plenum Press, New York, (1984) 333-42.
(13) K.A.Green and R.E.Green, Jr., Application of X-ray Topography to Improved Nondestructive Inspection of Single Crystal Turbine Blades, Sixteenth Symposium on Nondestructive Evaluation, April (1987), Nondestructive Testing Information Analysis Center, Southwest Research Institute,13-21.
(14) P.Chantikul, G.R.Anstis, B.R.Lawn, and D.B.Marshall, A Critical Evaluation of Indentation Techniques for Measuring Fracture Toughness: II, Strength Method, J. Amer. Cer. Soc., 64,9 (1981) 539-43.
(15) W.J.Boettinger, H.E.Burdette, M.Kuriyama, and R.E.Green, Jr., Asymmetric crystal topographic camera, Rev. Sci. Instrum., 47, (1976) 906-11.
(16) W.T.Beard, T.S.Ananthanarayanan, and R.W.Armstrong, Line-Modified Asymmetric Crystal Topography (ACT) Technique, Sixteenth Symposium on Nondestructive Evaluation, April (1987), Nondestructive Testing Information Analysis Center, Southwest Research Institute, 23-32.

(17) R.W.Armstrong, X-ray Diffraction Topography Description via the Stereographic Projection, in Applications of X-ray Topographic Methods to Materials Science, S.Weissmann, F.Balibar and J-F. Petroff (editors), Plenum Press, New York, (1984) 33-43.
(18) R.L.Stewart and R.C.Bradt, Fracture of Single Crystal $MgAl_2O_4$, J. Mat. Sci., 15 , (1980) 67-72.
(19) G.Y.Richardson, Ph.D. thesis research, University of Maryland, in progress.
(20) N.Doukhan and B.Escaig, Dissociation of Dislocations in Spinels $MgO(Al_2O_3)_{1.8}$, , Journal de Phys.- Letters, 35, (1974) L-181.
(21) N.Doukhan, T.E.M. investigations of room temperature microplasticity in $MgAl_2O_4$ spinel, J. de Physique,40, (1979)L-603.
(22) J.Hornstra, Dislocations, Stacking Faults and Twins in the Spinel Structure, J. Phys. Chem. Solids,15, (1960) 311-23.
(23) T.E.Mitchell, L.Hwang, and A.H.Heuer, Deformation in spinel, J. Mat. Sci., 11, (1976) 264-72.
(24)S.H.Kirby and P.Veyssière, Plastic deformation of $MgO(Al_2O_3)_{1.1}$ spinel at $0.28T_M$: preliminary results, Phil.Mag. A, 41, (1980) 129-36.
(25) P.Veyssière and C.B.Carter, Dissociation of dislocations in $MgAl_2O_4$ spinel deformed at low temperatures, Phil.Mag.- Letters, 57, 4, (1988) 211-20.
(26) S.M.Ohr, An Electron Microscope Study of Crack Tip Deformation and its Impact on the Dislocation Theory of Fracture, J.Mat.Sci.Eng., 72, (1985) 1-35.

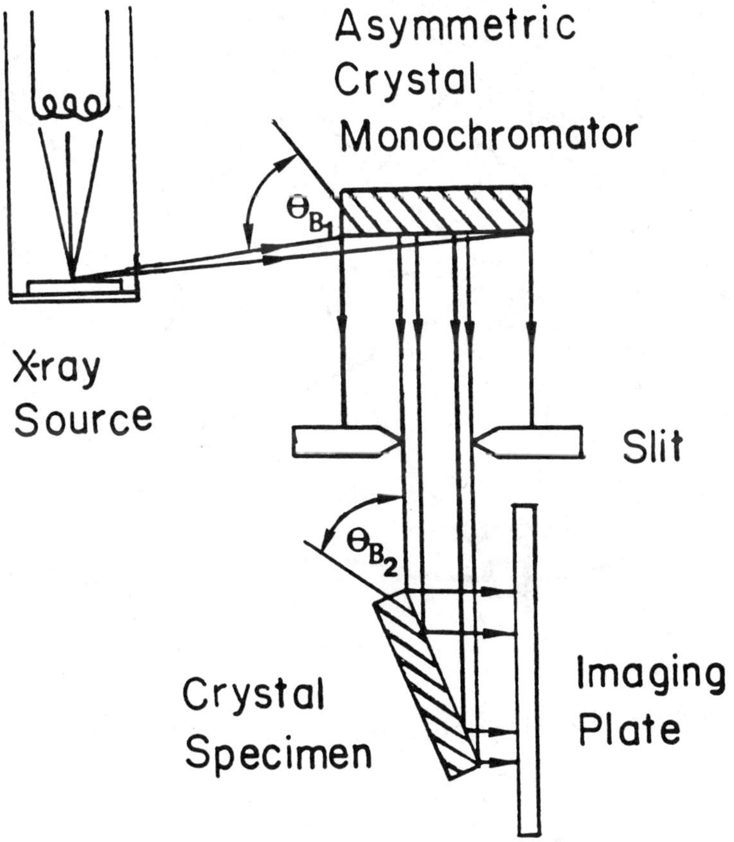

Figure 1 - Schematic diagram of the asymmetric double crystal topographic set-up.

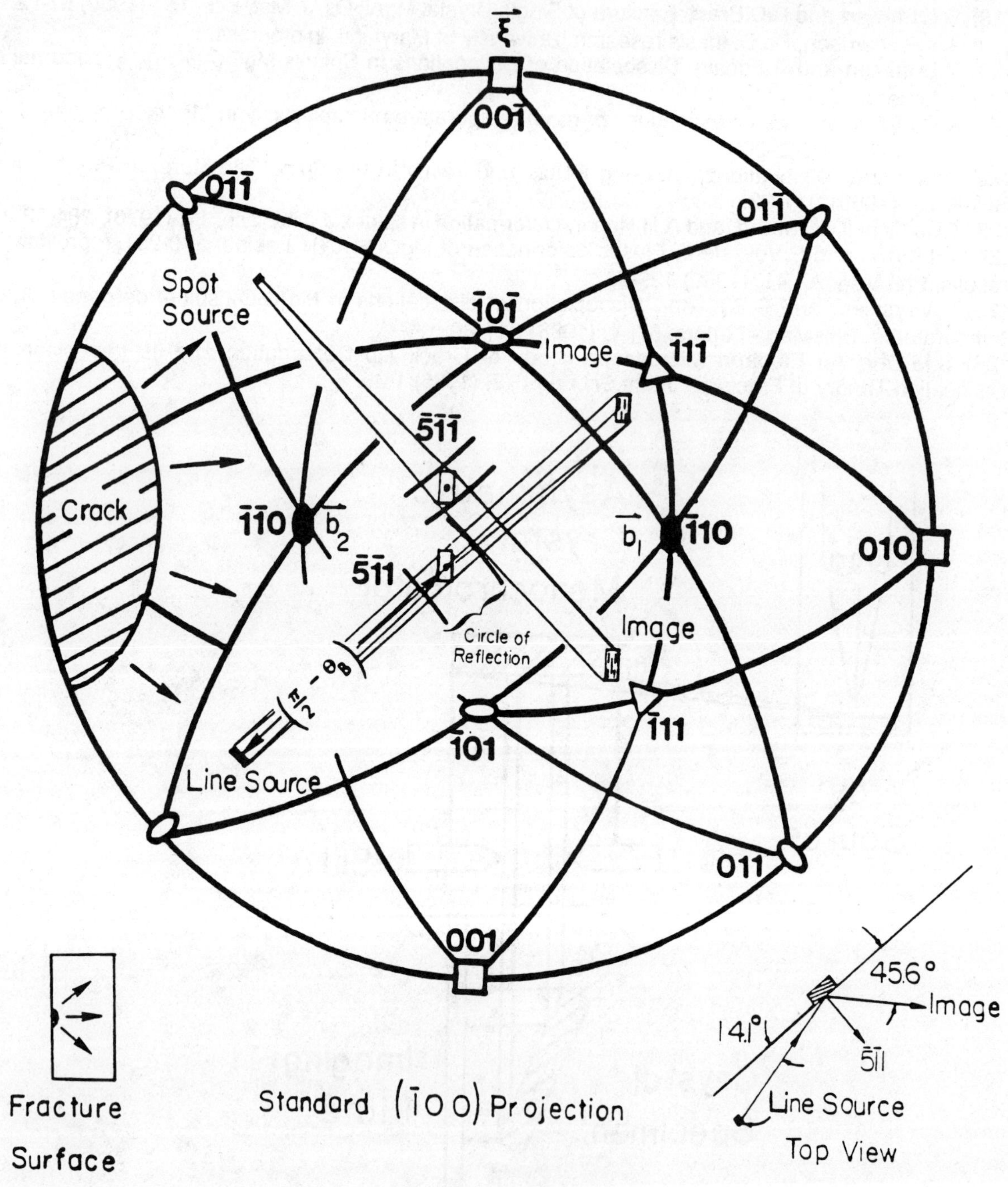

Figure 2 - X-ray stereograph illustrating the [$\bar{5}\bar{1}\bar{1}$] reflection used to obtain the topograph in figure 5 in which the source was arranged in the spot configuration. Also illustrated is a [$\bar{5}\bar{1}1$] reflection for a vertical configuration of the x-ray line source.

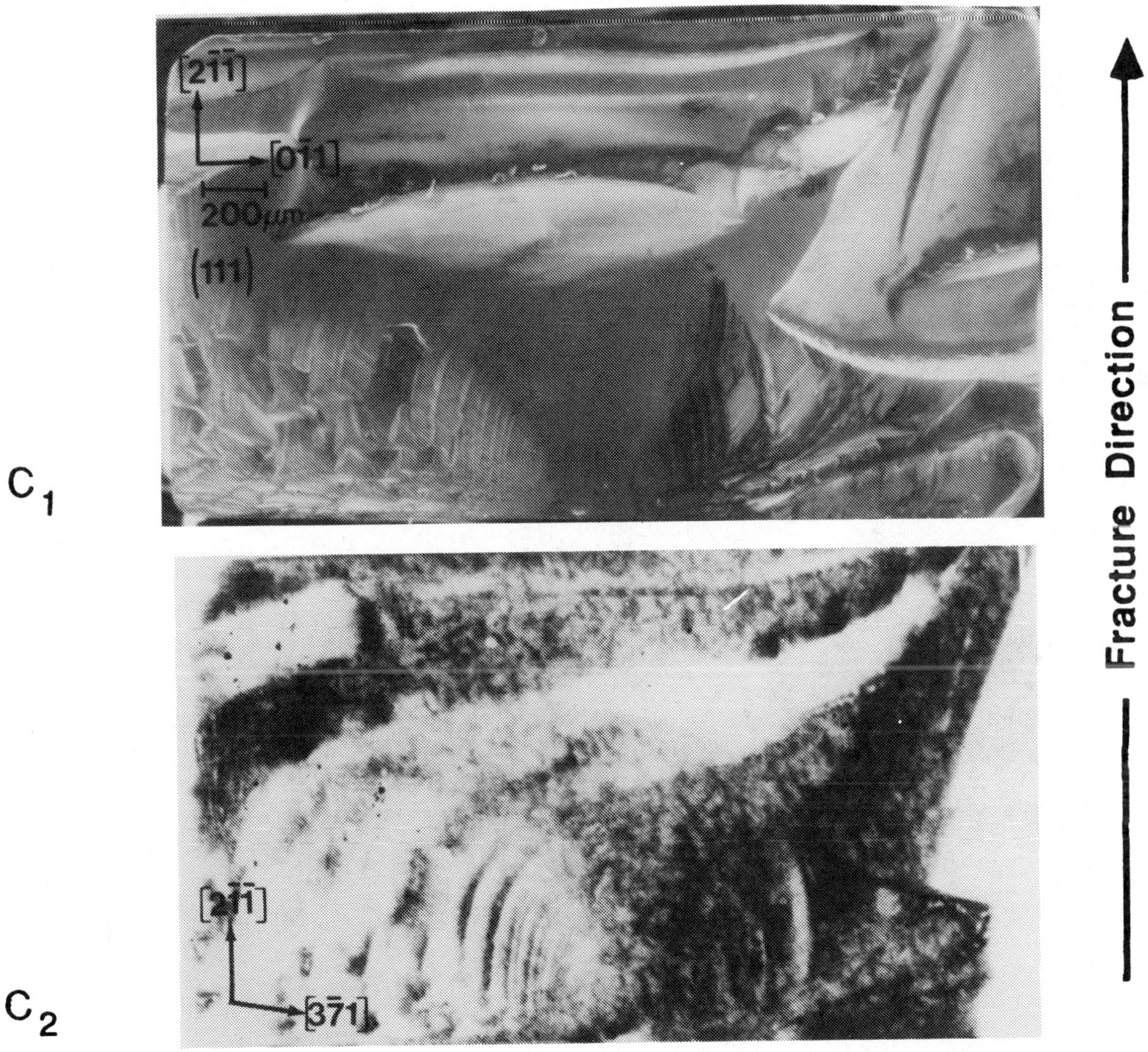

Figure 3 - C_1 Scanning electron micrograph of (111) fracture surface of spinel ($MgAl_2O_4$) with fracture direction indicated and - C_2 corresponding x-ray topograph taken with a [3$\bar{7}$1] diffraction vector.

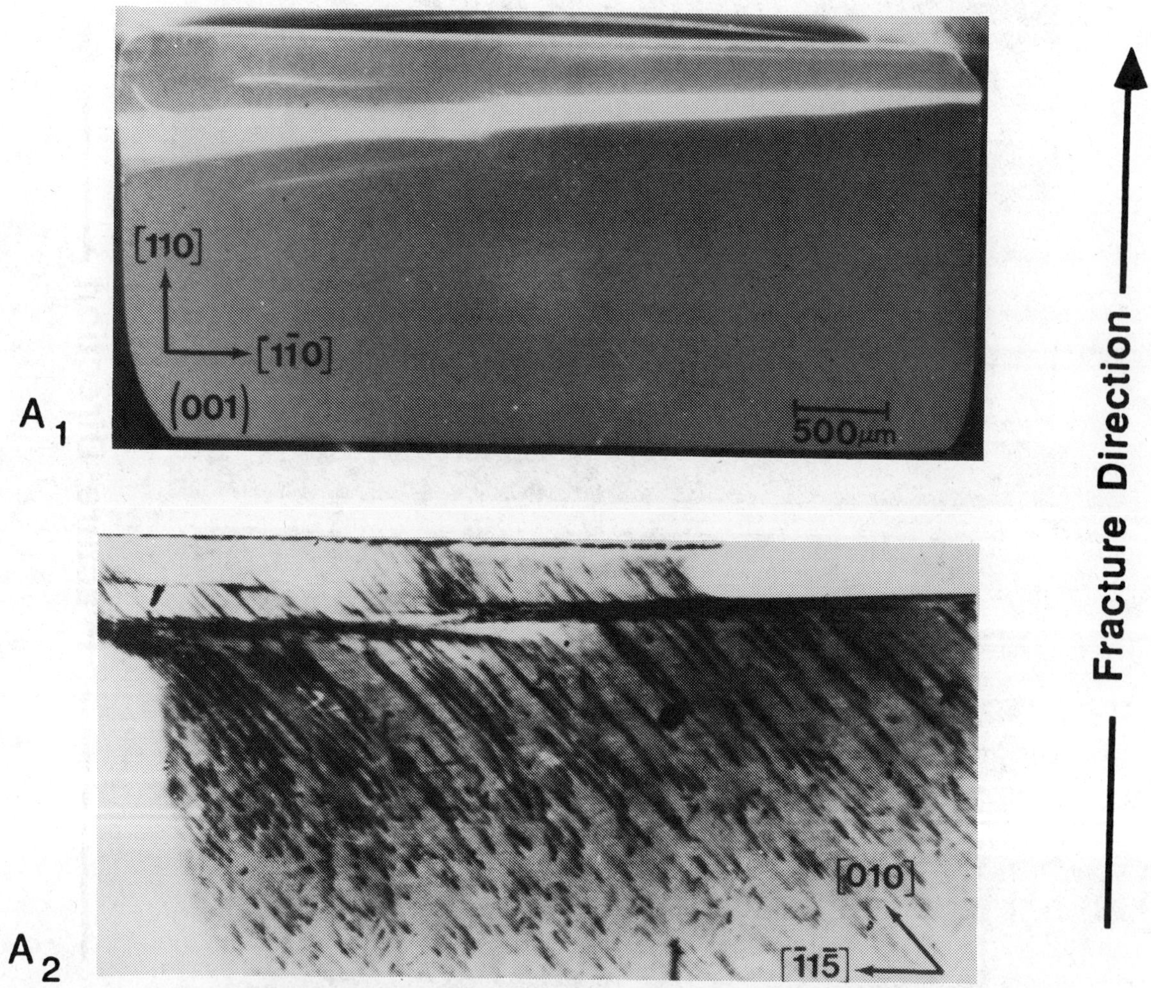

Figure 3 - A_1 Scanning electron micrograph of (001) fracture surface of spinel ($MgAl_2O_4$) with fracture direction indicated and -A_2 corresponding x-ray topograph taken with a [$\bar{1}1\bar{5}$] diffraction vector.

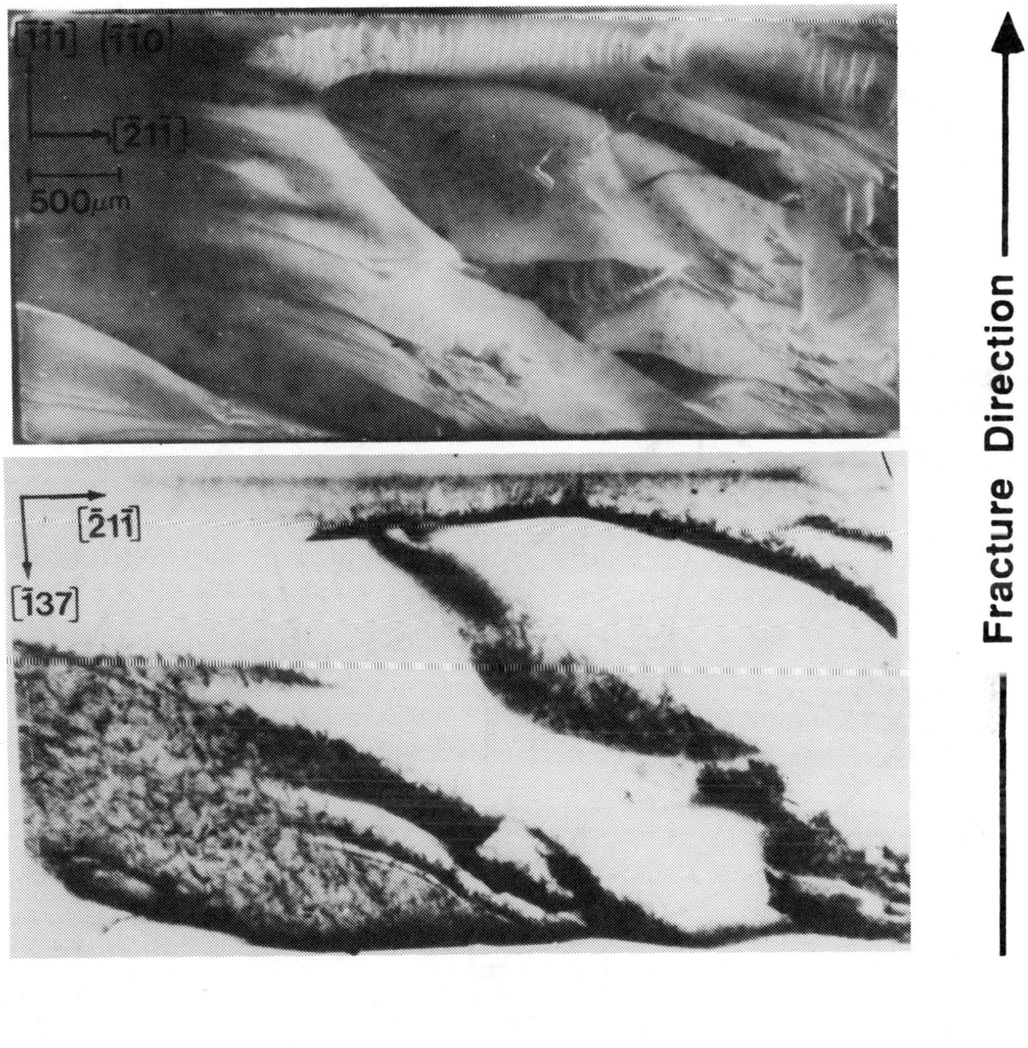

B 1

B 2

Figure 3 - B_1 Scanning electron micrograph of ($\bar{1}\bar{1}0$) fracture surface of spinel ($MgAl_2O_4$) with fracture direction indicated and - B_2 corresponding x-ray topograph taken with a [$\bar{1}37$] diffraction vector.

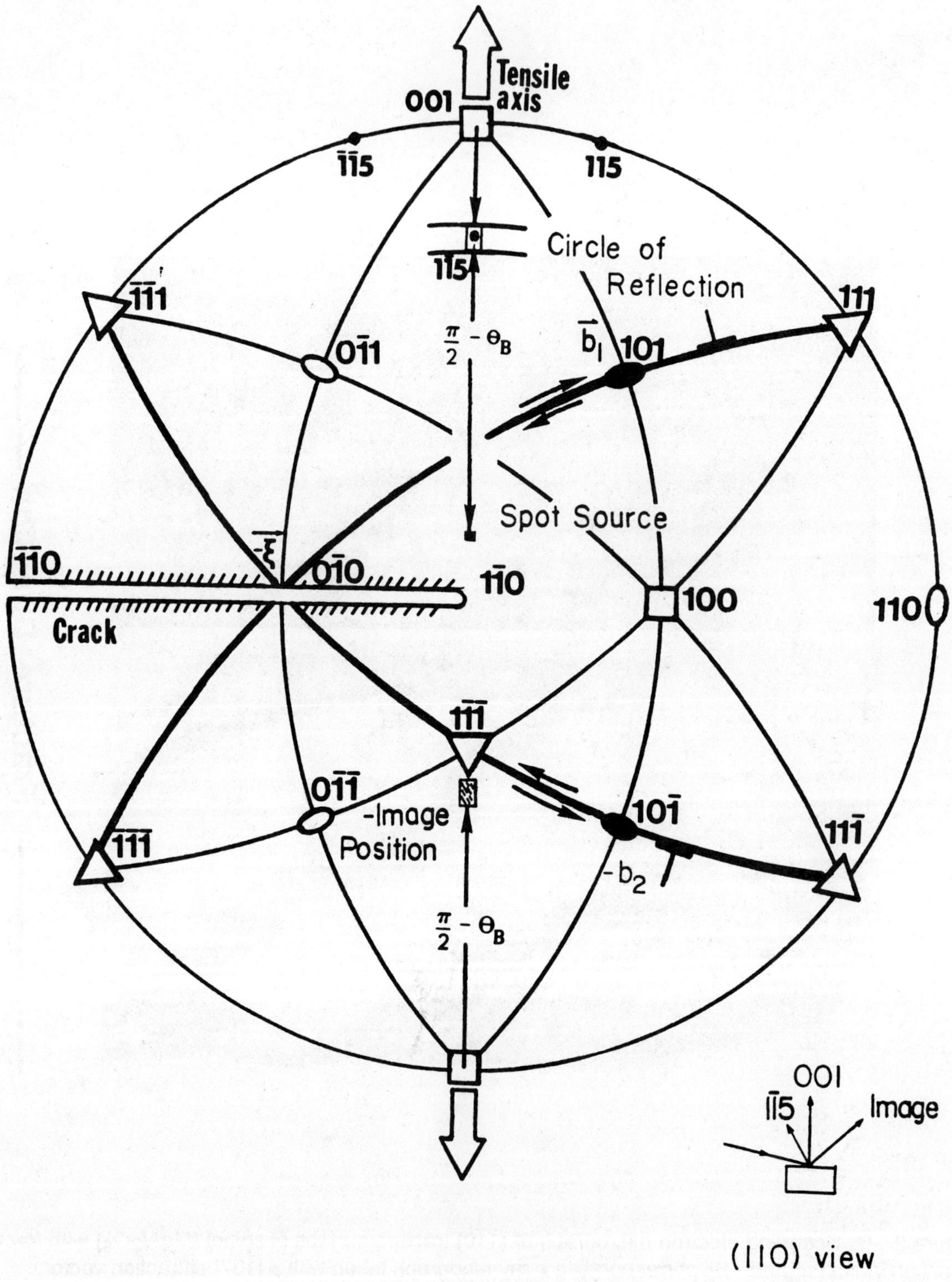

Figure 6 - Superposition of fracture and x-ray stereographs for a specimen with a (001) fracture plane and a [110]direction of fracture. Scanning electron micrograph and x-ray topograph of the fracture surface for this specimen is shown in figures $3A_1$ and $3A_2$ respectively. The crystal geometry for the x-ray topograph is also shown from a top (110) plane view in the corner drawing.

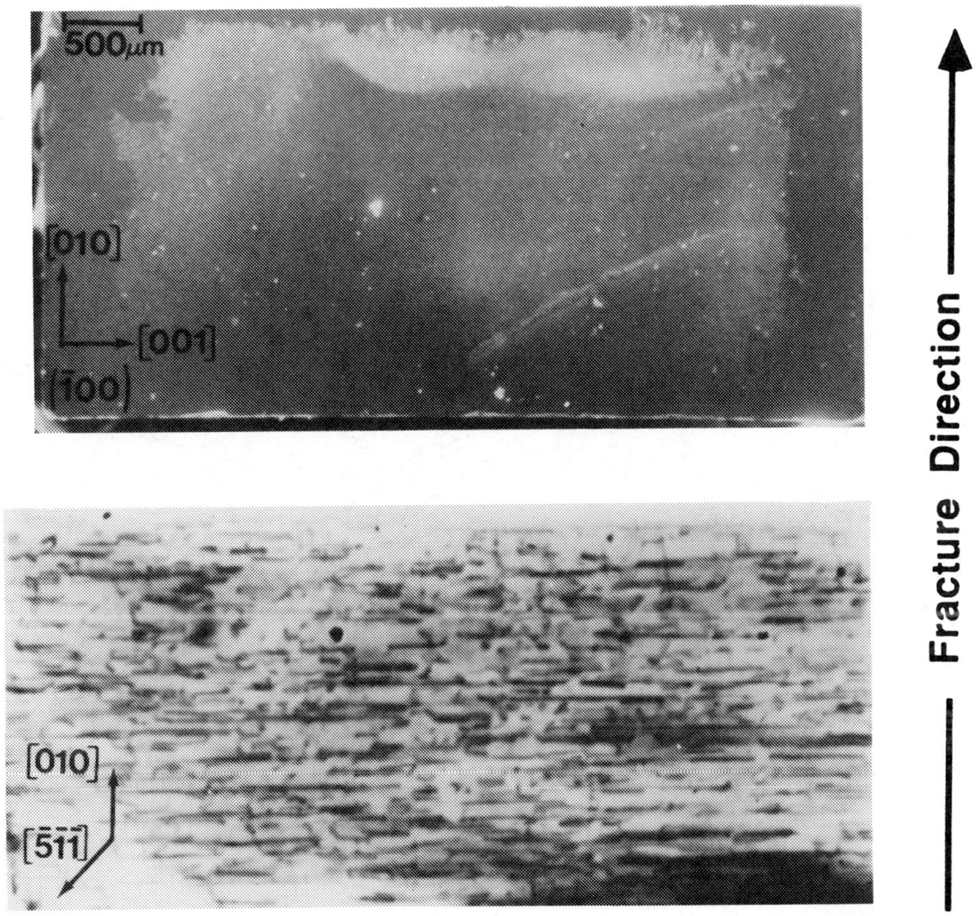

A_1

A_2

Figure 5 - A_1 Scanning electron micrograph of ($\bar{1}00$) fracture surface with a [010] fracture direction and - A_2 corresponding x-ray topograph taken with a [$\bar{5}\bar{1}\bar{1}$] diffraction vector.

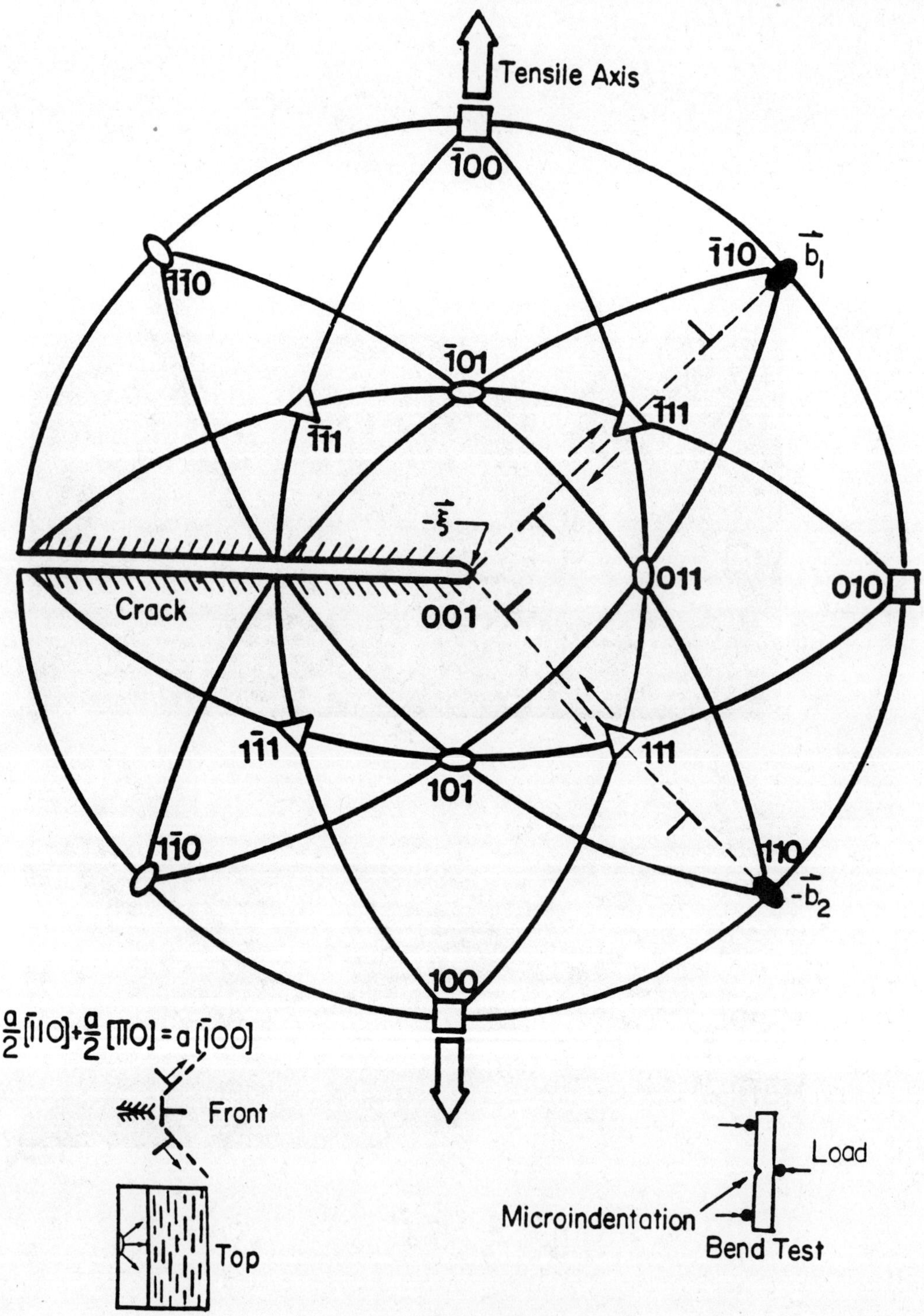

Figure 4 - Fracture stereograph incorporating the crystallographic geometry of the bend test illustrated in the schematic for (100) cleavage plane with a [010] direction of fracture. A model for the dislocation reaction occurring at the crack tip is also indicated.

NONDESTRUCTIVE EVALUATION OF CERAMICS BY X-RAY COMPUTED TOMOGRAPHY

H. Reiter, M. Maisl, and T. Scherer
Fraunhofer-Institut für zerstörungsfreie Prüfverfahren,
Universität, Gebäude 37, D-6600 Saarbrücken,
Federal Republic of Germany

ABSTRACT

Nondestructive evaluation methods are absolutely necessary to guarantee the reliability of ceramic materials and components for the optimization of the fabrication process as well as for controlling the production process in the line. Because of the small size of the strength controlling defects in ceramic materials a high detecting and resolving power of the NDE-methods is required. In scope of this the contribution discusses the microfocal X-ray and the X-ray computed tomography (CT) techniques. Results of investigations with a modified medical tomographic system and a high resolution tomographic unit (microfocal CT) with a resolution capability in the order of magnitude of 10 µm are presented. The paper discusses the components of a newly built microfocal CT-system and gives examples of its application on ceramic materials and components.

INTRODUCTION

High strength, low density, low electrical and thermal conductivity as well as high resistance against agressive mediums are excellent characteristics of ceramic materials. During the last few years an increasing interest in the application of ceramics as engineering materials has been observed especially because of the excellent combination of qualities - as opposed to metals - of specific weight, wear, and corrosion resistance at elevated temperatures. But fundamental disadvantages of these materials are their brittleness and therefore the lack of ability to release stresses due to mechanical or thermal loading by plastic deformation and furthermore the large scattering in the distribution of the strength values.

For the application of ceramics as engineering materials the available reliability is of decisive significance. The reliability can be improved by nondestructive testing (NDT) methods which are used for the optimization of the production process during a developing period of a new product and which are used later for the final control of the production line. In the literature /1/ the demand is emphasized for improved methods of NDT for the inspection and characterization of ceramics to enhance the reliability.

From a fracture mechanics point of view /2,3/, defects of a size of at least 100 µm lateral dimension can lead to failure depending on their kind and position. This number of 100 µm is one order of magnitude smaller than that valid for metallic materials. The detection and resolution of such small defects requires the application of advanced NDT methods.

For internal defects, especially in complex shaped parts, high resolution radiography with microfocal tubes is advantageous as opposed to other NDT-methods. Radiography yields a shadow picture of the object and consequently a superposition of the details in the object. All details of a real three dimensional object are projected to a 2-dimensional picture. Especially plane defects (crack like), orientations of defects and variations of microstructures in different directions can only be found by multi angle through radiation, finally leading to X-ray computed tomography. This technique yields the possibility to image only a single plane within an object and therefore gives information about the distribution and location of the defects within this object. X-ray computed tomography is a well known medical diagnostic technique. Commercial machines deliver cross section images with contrast resolutions better than 1% in a short time (in some minutes). But in general the obtainable geometrical resolution (mm) does not fit the requirements for the investigation of ceramics. Therefore a microfocal tomography system was developed.

This paper discusses the NDT of ceramics with the microfocal X-ray technique (microfocal radiography, MR), with a modified medical tomographic system (CT), and with a high resolution tomographic unit (micro-CT, MCT).

MICROFOCAL X-RAY TECHNIQUE

During the last few years the microfocal X-ray technique has been developed to a very effective method for the detection and resolution of small defects and inhomogeneities in low absorbing materials and components. In the X-ray tubes, a small focal spot size of about 10 µm could be realized by electromagnetic and electrostatic focusing techniques of the electron beam. Microfocal X-ray systems which can reach 200 kV and 1 mA at maximum were developed. With the increasing power of these units, their focal spot size increases as a matter of course. So we must expect at 100 watts a focal spot size of about 60 µm. But because of the low absorption of the ceramic materials it is possible to work with low photon energies and consequently with low powers. Most of our investigations of thin components are undertaken with 60 kV or 50 kV and 0.2 mA anode current. Therefore we have a small focal spot size and can use the projection technique which has two main advantages: the direct magnification and the reduction of secondary radiation. This makes small details visible and yields high contrast images with high signal-to-noise ratios.

The detectability of defects depends on their contrast, (difference of the absorption coefficients between defect and material) their size and the thickness of the object. Generally the contrast k in a radiograph is given by /4/:

$$k = \gamma \cdot (\log e) \cdot \Delta\mu \cdot \Delta x/(1+I_s/I_o)$$

γ = image medium constant (slope of the linear portion of the characteristic curve of the recording emulsion)
$\Delta\mu$ = difference in X-ray absorption for matrix and defect
Δx = difference of object thickness and defect size in beam direction
I_s/I_o = ratio of intensity of scattered to direct image producing radiation

In microfocal radiography the sample is separated from the recording medium and therefore I_s/I_o is significantly reduced, resulting in an increasing contrast. Because of the factor $\Delta\mu$ different defects of the same size lead to different contrast in the image. At 50 kV, a void in a SiC-part leads to a $\Delta\mu$ of about 2 cm^{-1} whereas a WC (tungsten carbide) inclusion leads to a $\Delta\mu$ of more than 100 cm^{-1} and accordingly to a contrast which is more than 50 times higher. Therefore a WC inclusion can be detected much more easily than a void of the same size. The linear absorption coefficient μ (and the difference $\Delta\mu$) decreases rapidly at higher X-ray energies resulting in a poor image contrast. For thin ceramic components it is therefore expedient and feasible to work with low energies which provide high image contrast.

Up to now the film seems to be the best recording medium concerning contrast and resolution. But film radiography is time consuming and expensive. The direct magnification which is possible with the microfocal X-ray systems helps to overcome the poor local resolution of some filmless recording mediums as for example of an image intensifier (0,3 mm inherent unsharpness). Because of the magnified details which are now imaged on the entrance window of the X-ray image intensifier its resolution capability must meet lower requirements than in the case of 1:1 magnification. In some cases filmless radiography with image intensifiers, solid state cameras and X-ray sensitive photodiodes will reach the film quality if an appropriate digital signal processing is added /5/. This includes averaging (enhancement of the signal-to-noise ratio), zero image correction, contrast enhancement, filtering, and false colour display.

In the following paragraphs, some examples are given for the investigation of ceramic components with our microfocal X-ray system. The microfocal X-ray tube is capable of working up to 200 kV and 1 mA anode current. The investigations are made at 50 kV and 0.2 mA in order to get a high image contrast and a small focal spot. At these low electrical power values, line structures in an X-ray lithography mask /6/ (line structures of 0,8μm gold on 3μm polyamid) down to 3μm could be resolved /7/. Measurements of the focal spot size with the wire method /8/ resulted in a value of 30 μm at 50W.

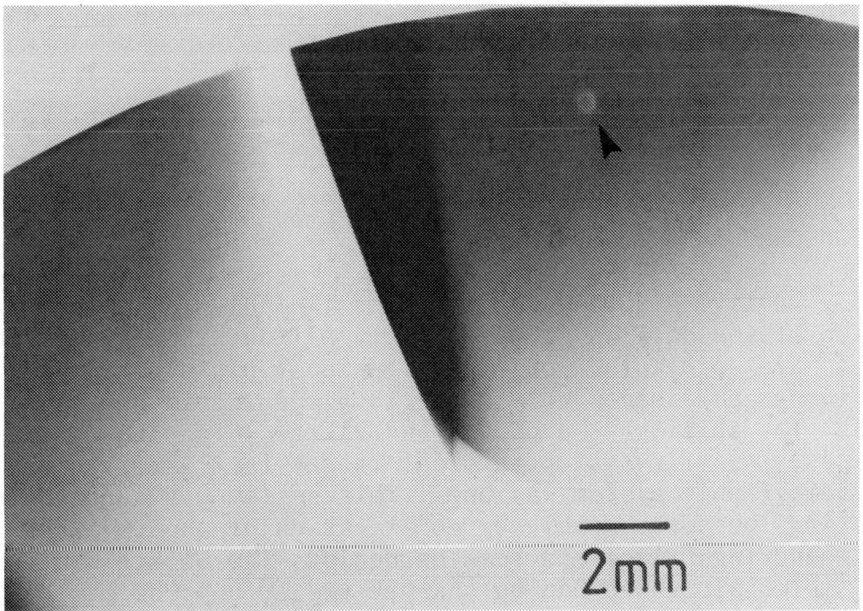

Fig. 1: Microfocal radiography of a HIPSiC rotor with an inclusion in one of the blades

Fig. 1 shows the result of an investigation of a HIPSiC turbocharger rotor. The direction of the penetrating radiation was parallel to the axis of the rotor. In a blade a large defect of about 500 µm could be detected. The defect looks like a ring of higher absorbing material. In the middle of this ring seems to be a lower absorbing region. According to the experiences with artificial defects in SiC and Si_3N_4 bending samples /9/ this behaviour is true of an iron inclusion which forms a reaction zone during the sintering process. During the investigation of more than 100 such turbocharger rotors many defects could be detected in each part. The size of the defects mostly lie between 50 and 500 µm. Most of the defects were of lower X-ray absorption like pores, also some cracks could be detected. Fig. 2 is an example of pores in blades of such rotors. The pore imaged in this figure has a diameter of about 60 µm.

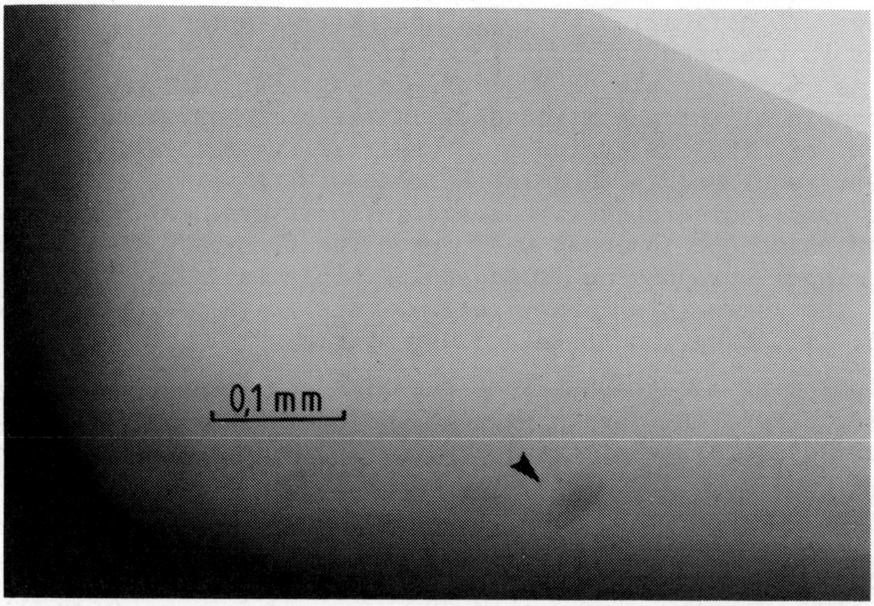

Fig. 2: Microfocal radiography of a HIPSiC rotor with a pore like defect in one of the blades

In Fig. 3 two defects are imaged in an axis of a rotor. This radiograph shows a crack with radial extension and a defect with lower X-ray absorption than the bulk material (void).

MEDICAL TOMOGRAPHIC SYSTEM

Radiography delivers a shadow picture of the object and consequently a superposition of the details in the object. All details of a three dimensional object are projected on a 2-dimensional picture. Opposed to that the computed tomography images only a single plane within an object. This technique has the ability to produce shadow free, high contrast images of object slices. In a first step in order to get familiar with the technique a medical tomographic system was used. This system was adapted for the material testing purposes mainly by modification of the collimation, adaption of the measuring time, and beam hardening. The object dimensions and the kind of materials which can be scanned are a function of the CT-system at hand and the interdependent contrast and spatial resolution /10/.

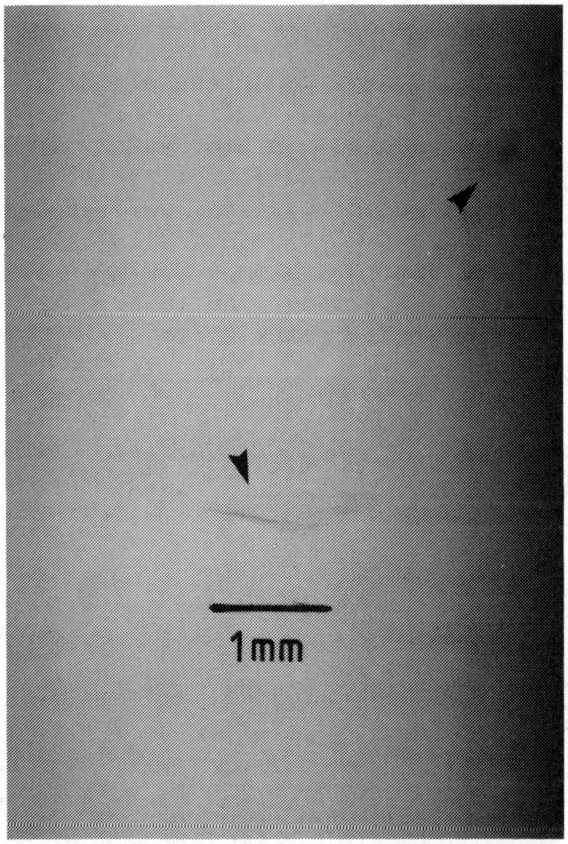

Fig. 3: Microfocal radiography of a HIPSiC rotor with two defects in the axis, a crack and a defect of lower X-ray absorption

This CT-system (Δ-Scan 50, Ohio Nuclear) is a second generation scanner (translation and rotation) with three detectors per slice, two simultaneous slices are possible. The focal spot size of the tube is 0.65 mm × 10 mm. The tube works at 125 kV and 30 mA. The images are reconstructed by 256 × 256 pixels. Theoretically an object can be reconstructed exactly from its projections, infinite number of projections and infinite small sampling provided. Practically the accuracy of an image reconstruction is limited by a number of parameters determining the spatial resolution of the system /11/: geometrical transfer function of the system (number of pixels, sample interval width w, geometry of detection (spot size, collimation)), number of angles, noise, dynamic range of the digitized data, reconstruction algorithm, detector real time response and machine vibration. In consideration of these parameters the medical CT-system achieves a resolution of 2 line pairs/mm.

For the detection of a defect, its contrast in the reconstructed image (cross section) must be significantly higher than the noise due to statistical fluctuations. The interdependency between contrast resolution and spatial resolution for a constant cross section thickness is given by $\Delta\mu/\mu \cdot w^2$ = const. In a given system a better spatial resolution will result in a poorer contrast resolution and vice versa. The relation allows an estimation about the material (absorption) which can be scanned dependent of contrast and spatial resolution. For the modified CT-system the range of materials covers polymers, composites, glasses, light metals and high strength ceramics.

The detectability of defects depends on the signal-to-noise ratio, the signal amplitude, on the line spread function of the system, contrast, size, shape and location of the defect. A SiSiC-disk of 80 mm diameter was used as a test object for the determination of the limit of detection of crack like defects. The test defects, slits, were produced by electrical discharge machining. The smallest slit detected in this rather thick test disk had the following dimensions: 0.125 mm × 0.9 mm × 2.5 mm, this corresponds to a volume of 0.28 mm^3. Fig. 4 shows the tomographic reconstruction of two test defects in this disk with a crack width of 190 µm and 195 µm.

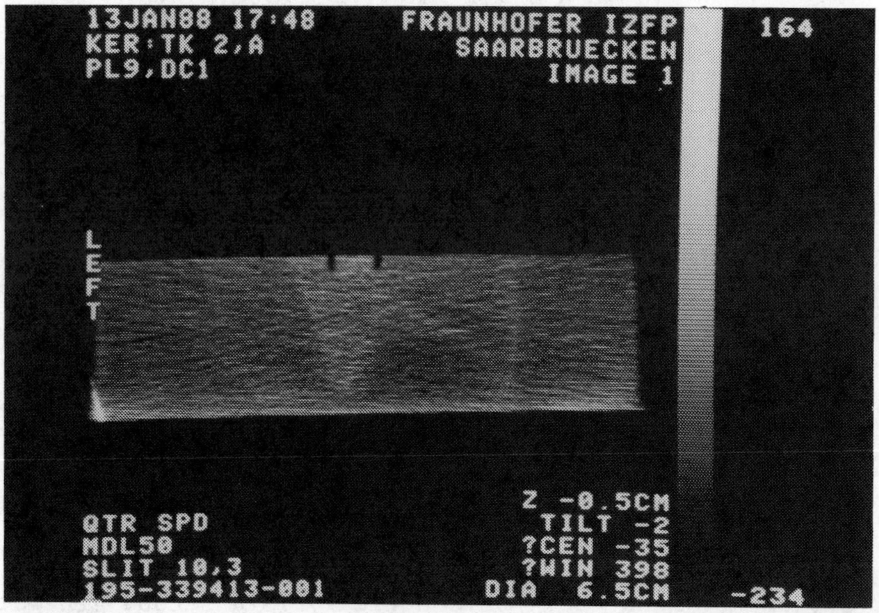

Fig. 4: Tomographic reconstruction of two test slits in a ceramic disk

Furthermore this picture shows a heterogeneous distribution of the density. Under the cracks the density is higher than in the edges. Much smaller crack sizes can be detected in thinner wall components. Fig. 5 shows a crack in an Al_2TiO_5 tube. The crack width is only 25 µm but it covers the wall thickness (4 mm) and the cross section thickness (5 mm) and therefore can be detected. In the figures 4 and 5 the cracks don't propagate exactly perpendicular to the cross section. Therefore they are imaged with a width up to three pixels.

The CT-system is a useful NDT tool for the investigation of materials and relatively large components. For ceramics, the material thickness can be 100 mm at 5% contrast resolution. The largest scan diameter is 450 mm. 4 minutes are needed for the registration of the data set and the reconstruction of the cross section. Consequently this method represents a rather quick imaging NDT-method.

HIGH RESOLUTION TOMOGRAPHIC UNIT

As indicated above the brittleness of the ceramic materials demand the detection of very small defects. These requirements cannot always be met by

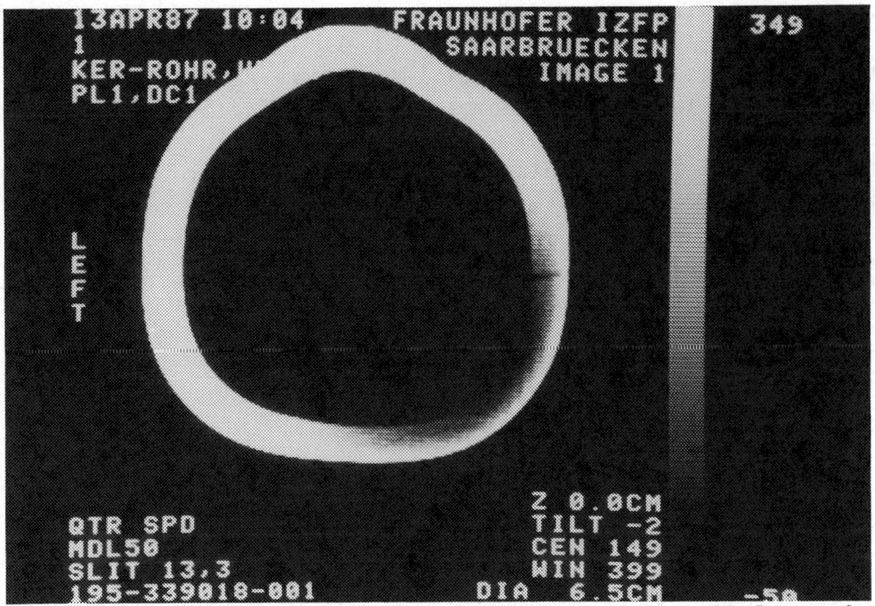

Fig. 5: Cross section of an Al_2TiO_5 tube, radial crack

the available commercial medical tomographic systems. In most cases their resolution is not sufficient and the thickness of the cross section must be reduced. The work with our medical system provides criteria for the conception of a high resolution tomographic system which will be discussed in the following paragraphs.

The set-up of our micro-CT-system is given in Fig. 6.

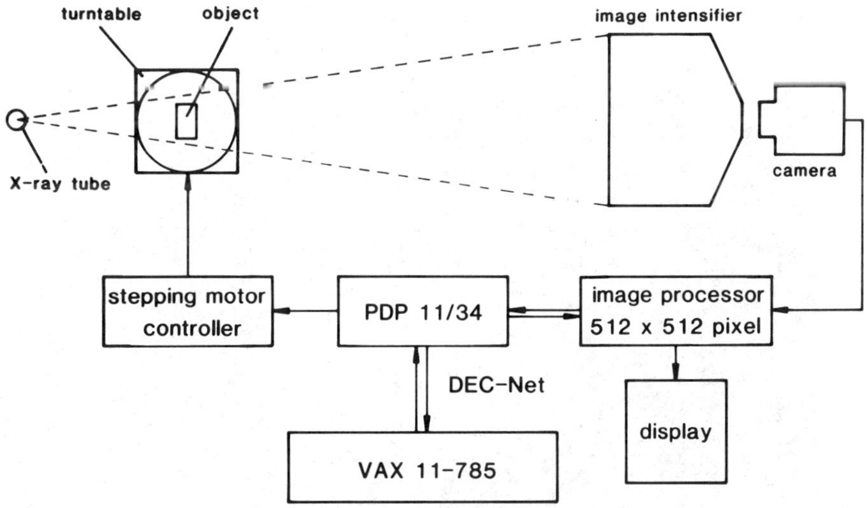

Fig. 6: Set-up of the micro-CT-system

A microfocal X-ray tube is used as X-ray source in order to decrease the geometrical unsharpness. The focal spot size of the X-ray tube changes between 10 and 100 µm depending on the voltage and anode current. The object is rotated in the fan beam of the tube. The stepping motor of the object desk is controlled by the computer. An image intensifier and a video camera are used as detector system. The details of the object are imaged onto the entrance window of the image intensifier with a direct magnification up to 30-fold. This direct magnification is possible due to the microfocal X-ray tube. The video signal of each projection is digitized with 8 bit. In order to get a better signal-to-noise ratio, up to 128 images can be added in real time. Only the central line of the 2-dimensional projections are used for reconstruction. After correction for tube and detector characteristics the data sets are reconstructed off line /12/. Here the filtered back projection algorithm is used.

At present the limits of our system are given by the following points: The objects must always be completely in the fan beam during the registration of the data set. The lines of the video signal are subdivided in 512 points resulting in a pixel matrix of 512 × 512 points. The smallest diameter of a reconstructed image is 10 mm leading to a pixel size of 20 µm. In this case the geometrical resolution can obtain values of 40 µm respectively 25 line pairs/mm.

Small pixel sizes and resolution capabilities require great care concerning the mechanical adjusting of the sample location and the rotation desk. The adjustment accuracy must be on the order of magnitude of µm. In case of the smallest pixel size (highest resolution) the power of the X-ray tube is limited to 40 watts at maximum, otherwise the focal spot size will increase inadmissibly.

The dynamic range of the detector system described is only about 20 dB. Therefore the difference between the unattenuated X-ray intensity and the intensity behind the absorbing object must not exceed the factor 100. This restricts the applicability of the system to low absorbing components.

Fig. 7 shows the result of the investigation of a test object in order to demonstrate the resolution capability of the system.

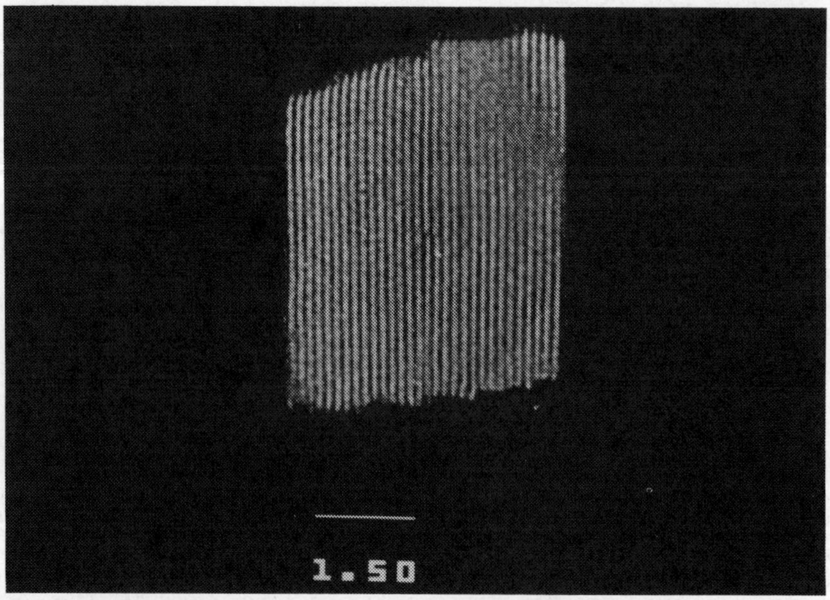

Fig. 7: Tomographic reconstruction of a test object with 50 µm thick aluminium foils

In this and in the following figures the scale is given in mm. The test object consists of aluminium foils and adhesive films each 50 µm thick and alternately glued together. Each of these laminates is clearly reconstructed. This is demonstrated once more in Fig. 8. Here the intensity distribution across the laminates is imaged. In this reconstruction the pixel size and cross section thickness was 21 µm.

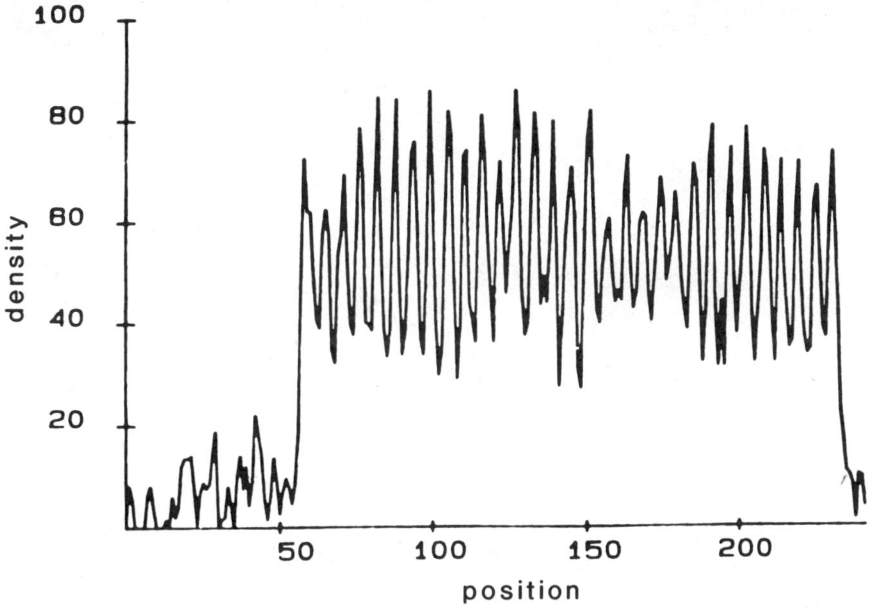

Fig. 8: Intensity distribution across the laminates of the test object

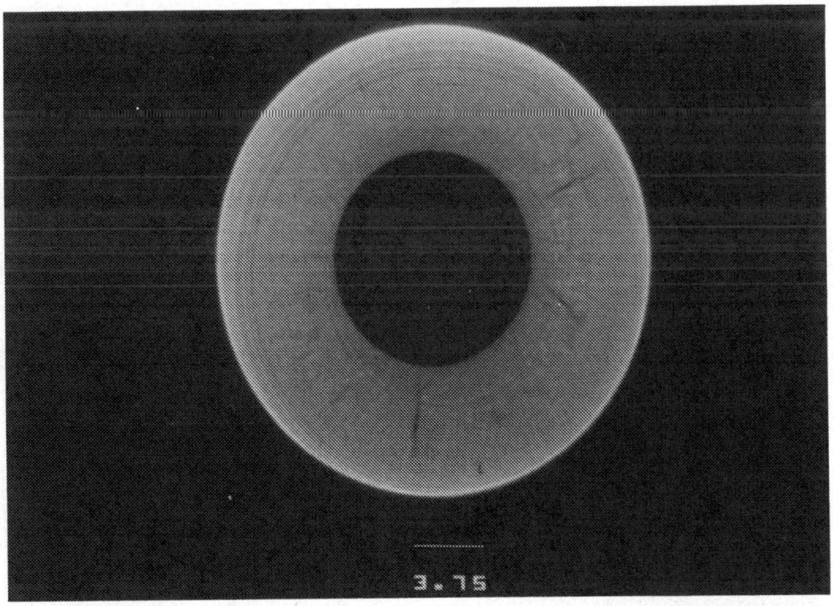

Fig. 9: Reconstructed cross section of a green ceramic part, radial cracks

Fig. 9 shows the reconstructed cross section of a green ceramic part of 30 mm diameter. The cracks are clearly visible, although the crack width (20 µm) is much smaller than the pixel size (75 µm). Another example for the investigation of a green ceramic component is given in Fig. 10.

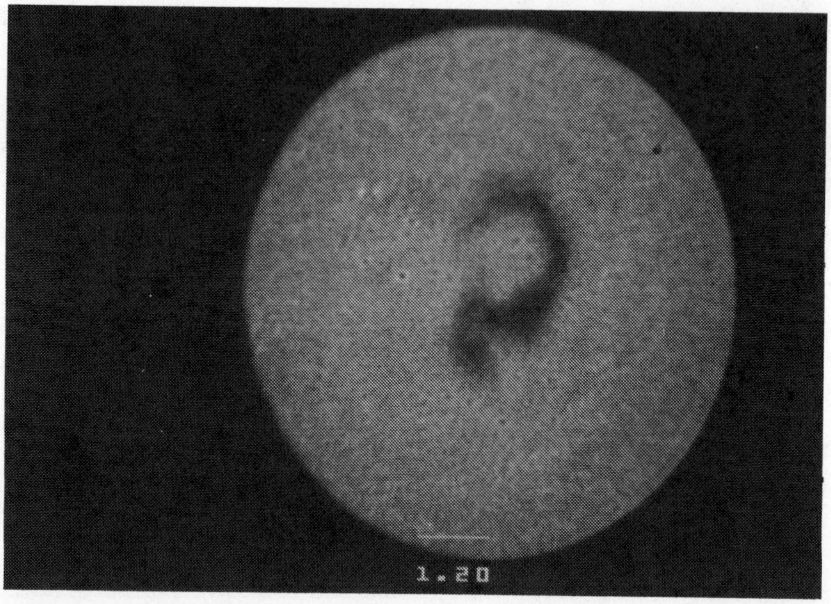

Fig. 10: Reconstructed cross section of a green ceramic part, crack and defects of 50 µm diameter

A large crack, voids, and inclusions of higher density on the order of magnitude of 50 µm diameter could be detected. The cross section almost lies parallel to the crack plane. In constrast to Figure 9 the material is very inhomogeneous. Pixel size and cross section thickness are 24 µm. A circumferencial crack could be visualized in the axis of a SiC turbocharger rotor, Fig. 11. Additionally there seems to be a higher density region in the middle of this axis whereas beyond the crack the density seems to be lower.

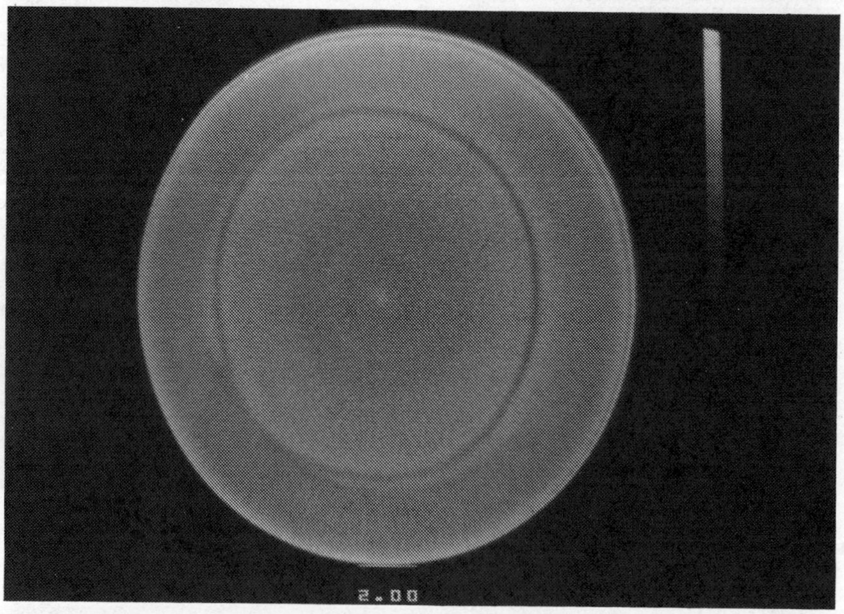

Fig. 11: Reconstructed cross section of an axis of a HIPSiC rotor, circumferential crack and density variations

CONCLUSIONS

The nondestructive testing of high-performance ceramics requires the application of advanced NDT-methods. X-ray methods are advantageous to other NDT-methods especially in complex shaped parts. High resolution radiography requires the application of microfocal X-ray tubes. The projection technique results in a direct magnification of the object details and in contrast enhancement by reduction of the secondary radiation. A further advantage of the small focal spot size is the high depth of focus of the radiographs. Any details inside an object will be imaged with practically the same sharpness.

Plane like defects (cracks) perpendicular to the radiation cannot be detected by radiography. Here only the multi-angle through radiation or computed tomography can represent a solution of the problem. Commercial and modified medical CT-systems are useful NDT tools for the relatively quick investigation of larger components, but they cannot meet the requirements to detect and resolve very small defects. For these purposes a high resolution tomographic unit as presented above is necessary. This system is capable of detecting and resolving in ceramic components details smaller than 50 µm. The efficiency of the system is not yet high enough. It can be increased by a detector system with higher dynamic range (for higher absorbing samples) and twice the number of elements resulting in a 1024 × 1024 pixel matrix (higher geometrical resolution). A parallel registration and reconstruction of the data must be realized to reduce the time for the imaging of a cross section. At present such a high resolution micro CT-system is being developed.

REFERENCES

/1/ R.W. McClung and D.R. Johnson
Needs Assessment for NDT and Characterization of Ceramics: Assessment of Inspection Technology for Green State and Sintered Ceramics.
Nondestructive Testing of High-Performance Ceramics, Conf. Proc., Eds.: A Vary and J. Snyder, ACSI, pp 33-51, 1987

/2/ A.G. Evans
Nondestructive Failure Prediction in Ceramics
In: Progress in Nitrogen Ceramics, Ed. R.F. Riley, Martinus Nijhoff Publ., Boston, 1983, 595-625

/3/ D. Munz, O. Rosenfelder, K. Goebbels, and H. Reiter
Assessment of Flaws in Ceramic Materials on the Basis of Nondestructive Evaluation
In: Fracture Mechanics of Ceramics 7, Eds.: R.C. Bradt, A.G. Evans, D.P.H. Hasselman, and F.F. Lange, Plenum Press, New York, 1986, 265-283

/4/ R.S. Sharpe and R.W. Parish
Engineering Application of Microfocal Radiography
In: Microfocal Radiography, Ed. R.V. Ely, Academic Press, London, 1980, pp. 43

/5/ K. Goebbels, H. Reiter, S. Hirsekorn, and W. Arnold
Nondestructive Testing of High Temperature and High Strength Ceramics Science of Ceramics 12, 1983, 483-494

/6/ J. Baumann
DGZfP-Jahrestagung Zerstörungsfreie Materialprüfung
28.-30.05.1984, Essen, P2-5

/7/ K. Goebbels and A. Reinhold
Advances in Microfocal X-Ray Equipment and State of the Art in High Resolution Defect Detection in Ceramics with X-rays
In: Ceramic Materials and Components for Engines, Eds.: W. Bunk and H. Hausner, DK6 Bad Honnef, 1986, pp. 857

/8/ H. Heidt and E. Nabel
Schnelle Methoden für die Überwachung von Mikrobrennflecken
DGZfP-Jahrestagung Lindau 1987, DGZfP Berichtsband 10, Teil 2, 675-682

/9/ O. Rosenfelder, D. Munz, H. Reiter, and K. Goebbels
Fracture Mechanical Failure Prediction of SSiC Based on Nondestructive and Fractographic Flaw Characterization
In: Ceramic Materials and Components for Engines, Eds.: W. Bunk and H. Hausner, DK6, Bad Honnef, 1986, pp. 1005

/10/ T. Vontz, K. Goebbels, M. Maisl, H. Reiter, and S. Hirsekorn
NDT for Small Defects by X-ray Tomography, Review of Progress in QNDE, Vol. 7A, Eds.: D.O. Thompson and D.E. Chimenti, Plenum Press, New York, 1988, pp. 389-397

/11/ T. Vontz
Röntgen- und Computer-Tomographie an technischen Werkstoffen und Bauteilen mit einem medizinischen System.
Diplomarbeit, Saarbrücken, 1985

/12/ M. Maisl und S. Hirsekorn
Rekonstruktion von Fehlstellen in Bauteilen mittels hochauflösender Röntgen-Computer-Tomographie
IzfP-Bericht 870201-TW, Saarbrücken, 1987

Detectability of Seeded Defects in HIP'ed Si_3N_4 Using Both Ultrasonics and Radiography

K.E. Amin, Advanced Ceramics, Norton Company, Northboro, MA
and
T. Derkacs and S. Matay, MMTC, Textron, Cleveland, OH

ABSTRACT

Both ultrasonic shear - wave C-scans, and microfocus radiography techniques were utilized to investigate the detectability limits for seeded inclusions, mainly Fe, WC, diamond, and voids in HIP'ed Si_3N_4. Those inclusions range in size from 50 to 300 µm. The probability of detection was determined at a 95% confidence level. The detectability of Fe inclusions was found to be better using ultrasonic techniques as compared to microfocus radiography film projection. The probability of detection of WC particles was found to be higher for film radiography than for ultrasonics at equivalent thickness sensitivity. Over the size range investigated there is an approximately linear relationship between the ultrasonic signal amplitude and the defect size for a given defect type.

Background

The mechanical reliability of structural ceramics is significantly affected by the presence of a variety of strength degrading flaws which are inadvertently introduced during processing. Reduction in the probability of failure of ceramic parts through improved processing and associated reduction in the flaw populations can be enhanced using non-destructive evaluation (NDE). To successfully and effectively implement NDE techniques two issues must be understood:

1)- the nature and size distribution of the structural flaws and

2)- the reliability of detection of such flaws by various NDE techniques.

At Norton Company, through our Si_3N_4 monoliths processing experience we have encountered the typical defects described in Table 1 with the size range indicated at various processing steps of the material. Those defects are mainly inclusions, voids, and cracks. Two of the most promising NDE techniques utilized for defect detection are radiography and ultrasonics, as discussed in References 1 to 8. Previous ultrasonic work on the detection of defects in ceramics has concentrated on demonstrating the sensitivity of the technique and evaluating its ability to characterize defects. Very little work has been done addressing reliability. Schuldies and Derkacs (Ref. 9), for example, inspected samples with a variety of defect sizes and types using both microfocus X-ray and ultrasonics. They did not calculate probability of detection, because they had no reliable way of either controlling the number of seeded defects or verifying results. What they concluded was that ultrasonic techniques were superior to radiography for smaller defects particularly those with lower x-ray absorption characteristics. Earlier work at Norton Company, NASA, GTE, and other organizations (Ref. 1, 4, 5, 6) show that the resolution of various radiographic techniques (real time digital imaging and microfocus projection film) is of the order of 50 to 300 μm depending on the type, orientation and location of the defects, as well as on the matrix material. Norton Company's previous work (Ref. 1) using both microfocus real time and film techniques shows that the sensitivity of detection of heavy metal inclusions is less than 2% of specimen thickness for Fe in green (molded) Si_3N_4 and 5% in HIP'ed material. The seeded particles, however, exhibited non spherical shape (roundness as high as 3) and a broader size distribution. In addition, simulating voids by using diamond particles did not yield conclusive data to practically assess the technique's sensitivity to detecting voids. NASA and others (Ref. 4 to 6) have published some data on the sensitivity of detection of voids seeded in very thin sample sections, <4 mm.

The present investigation was therefore initiated to determine the detectability limits for both ultrasonics and microfocus radiography techniques for more controlled narrow size seeded defects, larger sample size (higher number of seeded defects) and improved void simulation conditions. More importantly the intention was to clarify and deal with the problems that limit the ability to verify the actual defect population and recommend improvements in the techniques so that quantitative results and appropriate quality control information can be obtained using a combination of both ultrasonics and radiography techniques.

Method of Analysis:

Methods for analyzing the reliability of NDE inspection techniques are detailed in reference 7. In the routine inspection of components, there are four possible results:

o- A flaw is indicated where there is a flaw
o- No flaw is indicated where there is a flaw
o- A flaw is indicated where there is no flaw
o- No flaw is indicated where there is no flaw

The main objective of statistical evaluation of the reliability of non-destructive inspection is to estimate the probability of detection, P, with as few trial inspections as possible. This is done by defining a lower limit (lower bound), P_1, for the range of values that is expected to contain the true probability of detection. The degree of confidence, G, in the probability of detection is limited by the number of seeded flaws. The exact binomial distribution function is commonly used to find P_1 for a given total number of flaws/defects as follows:

$$1 - G = \sum_{X=S}^{N} \left[\frac{N!}{X!\ (N-X)!} \right] P_1^X \ (1-P_1)^{N-X} \qquad 1-$$

Where X is the total number of detected flaws and N the initial total number of flaws in the material. The probability of detection is also dependent on the sample size as illustrated in figure 1 where $P_1 = .99$ at $G = .95$ for detecting 500 defects out of 500 seeded defects as compared to $P_1 = .89$ for N = 25 and X = 25.

Radiographic detectability of defects is often expressed as a percent of thickness sensitivity as given by:

$$\text{Thickness sensitivity} = (D/T) * 100 \qquad 2-$$

Where T is the sample thickness in the direction of the x-ray beam, and D the defect size.

EXPERIMENTAL PROCEDURE:

Silicon nitride - 1% MgO tiles and discs were uniaxially pressed from a powder mixture of Si_3N_4 and MgO sintering aid. The tiles were seeded with either Fe, WC, or spherical shape particles simulating voids. The particle size distribution of those seeds was measured using Princeton Gamma Technics, PGT, digital image analysis system IV; see table 2. Two types of specimens were produced to simulate voids. One used a low density inclusion, diamond, and the other used an organic material, PMMA, that would burn out or vaporize during subsequent densification and thus leave a void. The particles were manually seeded on a tape in a square array each 5 mm apart near the center of the tiles in the quantity of either 25, 30 and or 300; see figure 2 as an example of such an array. The 5 mm spacing was selected to avoid signal overlapping and shadowing from individual particles during ultrasonic examination. In the case of seeded voids (PMMA) no tape was used. The particles were laid directly over the powder bed; additional powder was added and the sample cold pressed then hot isostatically pressed (HIP'ed). The samples with seeded PMMA were degassed before firing. Sample thickness was varied during the measurements using dummies of the same Si_3N_4 material.

A Ridge HOM X-161 X-ray microfocus unit was used for producing the microprojection film data. The operating parameters for the microfocus x-ray system were 50 to 90 KV with a beam current of .36 to .63 mA. The exposure times were varied from 5 to 30 minutes to achieve film densities in the range 1.80 to 2.20 using data from the characteristics exposure charts which we generated inhouse for

Agfa films. A phosphorus screen was placed in the path of the x-ray beam before it traveled through the sample which resulted in improved film contrast. The film exposures were conducted at magnifications of 2X to 11X. For image quality, exposures were made using the ASTM #5, 10, 15, 20, and 25 Aluminum penatrameters for each specific sample thickness. The exposed films were processed using an Agfa Structurex NDT 1 film processor unit. An 8 minute film processing cycle was used.

Ultrasonic data was obtained using an immersion shear wave technique. The instrument used is a commercial KB-6000 system with a computer controlled scanning system. The transducer is an Aerotech, alpha immersion transducer with a 6.35 mm diameter element and a focal length in water of 75.2 mm. The transducer has a peak output at 25 MHz. The focal spot size of the transducer was approximately .635 mm. The collected ultrasonic data included both amplitude and time at each position. The scan pattern was setup so that the data spacing is typically one half to one third of the beam diameter at focus. There is a deadband area at the top interface and at the edge toward which the transducer is angled to create the shear wave. The top deadband was inspected by turning the part over and repeating the inspection from the other side. For the larger parts where the defects were located near the center, the edge deadband did not extend into the defect area, so no further scans were needed. For the smaller samples, where the defects were distributed throughout the part, a second scan was made with the part rotated 180 degrees to inspect the area in the deadband of the first scan.

The ultrasonic data was analyzed using a computer program that allows data points to be clustered into groups representing individual defects. It also allows the location of the data in the scan field and statistical data about each cluster to be printed out. An example of the obtained C-scans for both Fe and WC seeded inclusions is provided in figure 3. The program provides better spatial resolution of data than is possible with a conventional C-scan and also allows separation of data into time layers. The analysis of the samples was carried out to provide information on the number of detected defects and their location. C-scans with different threshold ranges were plotted in different colors on the same paper. The C-scans were obtained only on samples which were either 3.00 or 6.35 mm thick since the method of adding dummy blank samples to a specimen to increase its thickness, as was done for x-rays, does not work for ultrasonics, because the interface between samples reflects the signal and acts as a defect.

A Fortran computer program was written to compute the lower bound probability of detection, P_1, from equation 1 above.

RESULTS AND DISCUSSION:

Detection of Inclusions:

Figs. 4 and 5 show the computed lower bound probability of detection, P_1, plotted versus the thickness sensitivity for both the WC and Fe seeded defects used in the present investigation period. Table 3 also summarizes the data obtained on all defects examined in this work. The defect sizes reported in this analysis are the originally seeded values. It should be noted that such values are higher than the actual defect size due to the effect of densification process on the bulk seeded defects; this will be discussed further when analyzing the data on voids detection. The probability of detection data was computed at 95% confidence level for measurements done using both microfocus projection and ultrasonic C-scans. The data in figure 4 indicate that the detection capability of microfocus film technique is comparable, if not slightly better for heavy metal (WC) inclusions than that of ultrasonic C-scans. The use of large WC markers around the boundaries of the seeded areas in most of the samples made it difficult to resolve the adjacent smaller seeded defects. Ultrasonics showed poorer spatial resolution than microfocus x-ray, but generally better sensitivity. However it is difficult to take advantage of this sensitivity to detect a small defect when it is spatially close to a large one. On the other hand the data generated on Fe seeded inclusions show clearly that ultrasonics offer better detection than microfocus film in the size range down to 47 µm; see Figure 5 and Table 3. The difficulty with Fe detection using microfocus radiography may stem from the fact that at the hot isostatic pressing temperatures the effective Fe particle size is reduced because of reaction with the Si_3N_4 matrix. This usually creates unsharp boundaries of different chemistry and density than the Fe particles; such boundaries usually have x-ray absorption characteristics similar to those of the matrix material. Ultrasonic velocity is still affected by the reaction zone and give rise to a signal. The iron rich reaction zones are often encountered in failure analysis of Si_3N_4 ceramics.

Detection of Simulated Voids:

Radiography data show a high percentage of detected diamond particles at thickness sensitivity of 4.0 with values dropping to less than 50% at thickness sensitivity around 2; see Figure 6. The probability of detection of diamond in the investigated size range was below .90. The ultrasonic C-scans apparently provided poorer probability of detection of diamond particles/voids and detected signals for defects outside the markers. This indicated the presence of a certain number of naturally occurring defects in those samples with a size detectable by the ultrasonics. Such results raised the question of whether all the defects detected within the markers region are the seeded defects or whether some are also natural defects. To provide a better understanding of the ultrasonic results, the amplitude data was also analyzed. The data from each scan was normalized in amplitude using the amplitude of a reference defect included in the scan; the reference defect is a 125 μm WC inclusion. The average measured amplitude for WC, Fe, and diamond inclusions is shown in Table 4. To provide a basis for comparison, the theoretical response from these inclusions was calculated. The approach was to use a conical approximation to the Bessel Function shaped beam profile and integrate the beam amplitude in the area of intersection of the beam and defect. The calibration constant of the system was calculated from the reference WC data and a reflection coefficient was calculated for each type of inclusion. The same calculation was also done for PMMA simulated voids data which is discussed below. Figure 7 is a plot of both the computed and measured data. It can be seen that for both WC and Fe the agreement is good, but for diamond it is not. The measured amplitudes are far below the expected values. This leads to two questions: why did we not detect diamond with C-scans, and what did we detect?. Since radiography data show indications of lower density particles at the seeded locations as shown in figure 8 for the case of the HIP'ed sample containing diamond seeds 300 μm in size; it is clear that such seeded defects are still there. Further metallurgical examination is underway to help provide some answers.

Both microfocus film radiography and C-scans were conducted on the samples seeded with both 50 and 100 μm PMMA particles. Radiography was able to detect around 20% of the 100 μm seeds and 0% of the 50 μm ones (see Table 3), while C-scans showed no defects. This defect size translates into 2 to 4% thickness sensitivity which according to previous investigations by GTE, NASA, and others (see e.g. references 5 and 6) should have been detected at a reasonably high probability. This leads to the conclusion that the HIP process drastically reduces the size range as well as closes up those voids. In addition, the defects were not spread in a uniform pattern as in the previous samples. They were somewhat randomly distributed throughout the volume; see the example of 255 μm PMMA seeds in a molded Si_3N_4 disc in Figure 9. Therefore we must ask the question, what is the probability that more than one defect is found in an area the size of the ultrasonic beam. If we look at the beam area as the source of a signal, the ultrasonic system simply determines the size of the reflection from each area sampled. In the case of this experiment, that area was an ellipse .635 mm by .94 mm. If a single defect gives a certain size signal, N defects will give a signal approximately N times as large. Therefore the ultrasonic indications must be looked at as a distribution of signals from one or more defects. The number of indications does not equal the number of defects unless the population of defects is small enough so that the probability of two or more being grouped within a beam diameter is very small. The probability of having more than one defect in any one single area is fairly high since the defect size is about one eleventh of the sample size. Using the sample seeded with 255 μm PMMA, an estimate of the expected distribution was made using the normal approximation to the binomial distribution and the numbers given above. The method was to calculate the probability of the number of defects being in each range. The observed distribution was estimated by assuming that the smallest signal represents a single defect, and calculating ranges up from there representing larger groups. The size of the ultrasonic beam was assumed to divide the specimen into 600 cells. At the circumference of the part, the edge effect makes one or two cells uninspectable. Therefore, the total number of cells inspected is in the range 450 to 520. The number of defects in the inspectable cells is expected to be in the range of 220 to 260. Therefore the probability of the defect being in the beam path is .53 to .62. Results are given in table 5. The question is what size are these voids. We assume they are smaller than 255 μm, but we would like to know by what fraction. There are two ways to look at the data. One is to compare the signal amplitudes to the computed theoretical curve, see Figure 7, and get a distribution of sizes for the 37 indications. The other is to assume that the smallest signal is from the void created by a single defect seed and that larger signals come from voids created by groups of 2 or more defects. If we make this assumption we can then calculate an amplitude range for each number of defects in a cluster and then count the number of occurrence of amplitudes in these ranges. This divides the 37 indications into groups representing different numbers of seeded defects. If we then multiply the number of occurrence times the number of seeded defects in the cluster, we get the total number of seeded defects detected. Results of the analysis are shown in Table 6. The main conclusion to be drawn from this analysis is that the voids resulting from HIP'ing of the 255 μm PMMA are around 20 μm. Our preliminary metallography examination did not detect these voids. The reason is that the samples were hot isostatically pressed while being encapsulated in glass. We had to remove the glass reaction zone by sand blasting before inspection, a process during which we lost surface pores. This leads to the second conclusion that the number of voids present in such samples is lower than the originally seeded particles (300). Clearly Figure 9 shows agglomerates of the PMMA seeds and a number larger than 2 PMMA particles per cell at the edge of the disc in the as prepared samples before firing, which may partially support the ultrasonic observation.

One of the main results of the present investigation was a better understanding of the difficulties involved in making probability of detection measurements on artificially seeded defects and the changes that might be needed to improve the quality of data. These problems fall into two general categories; specimen preparation and technique optimization. Another important result of this work is the ability to understand and establish a complimentary role for both techniques.

In the area of specimen preparation, success was achieved in preparing samples with known size defects at known locations. The interpretation of the ultrasonic signal was made sometimes difficult because of the shift in the location of the artificially seeded defects during processing. This also is affected by the purity of the matrix material. The second problem relates to the relative spacing of the defects particularly when there is a mix of large and small defects. This is evident from the contribution of the large markers in our samples to the ultrasonic signal which made it difficult to resolve the smaller seeded defects. In general reference samples with seeded defects which are suitable for radiography studies may not be optimum for ultrasonics and other NDE detectability work. The other issue relates to the ability to reliably measure the seeded defect size in the densified samples. Sample fractography does not always provide enough data; probably gradual proper sample sectioning near the location of such defects combined with x-ray microanalysis (EDS/WDS) and microstructural etching techniques will help provide needed data.

In some of our samples the difficulty of dispersing the fine particles, <40 µm, during sample preparation and subsequent seeding resulted in the presence of more than a single defect at the same location. Radiography usually reveals this; however more careful signal analysis is needed when using ultrasonics to minimize error.

In the area of technique optimization, the ultrasonic sensitivity in detecting defects is greatly limited by electronic noise due to both the hardware quality and the operating conditions when the instrument gain is turned to high levels. The authors have investigated the capability of other ultrasonic equipment that would be configured from commercially available components specifically made for inspection of ceramics. The conclusion of that study was that it is practical to reduce the electronic noise to a level where the system can achieve the same reliability on a defect one eighth of the size detected in this work. Such sensitivity might require a higher frequency, in the 30 to 50 MHz range and the use of a waveform digitizer and other instrumentation with larger bandwidth.

A different factor in instrumentation appears when examining thicker samples. The ultrasonic beam needs to be focused to concentrate the energy to see very small defects. The acoustic attenuation in dense Si_3N_4 is low, but velocity is very high, and therefore the focusing effect of the material is quiet high. The result is that the focal length for shear waves is a factor of about 4 less in the material than in water. Because transducers have a natural focal length, it becomes more difficult to obtain the needed focal length as thickness increases. If the defocused part of the beam is used the sensitivity drops drastically. The related problem is that the beam depth of field is relatively short. Therefore to perform a thorough high sensitivity scan on a thick ceramic specimen requires several scans at different focal depths with different focal length transducers. The problem is not that the sensitivity cannot be achieved, but that it is time consuming and expensive to do so.

Radiography hardware improvement can be achieved particularly when examining very small defects. If a smaller x-ray focal spot size is achieved, say around 2 µm, a five fold increase in magnification may be achieved over our existing setup without increasing the geometric unsharpness assuming other factors remain the same. Such a system is not yet commercially available.

CONCLUSIONS:

1- The probability of detection of Fe inclusions seeded in HIP'ed Si_3N_4 is higher for ultrasonic shear-wave techniques as compared to microfocus film radiography. This is valid particularly for defect thickness sensitivity of less than 2%.

2- In the case of seeded WC particles the reliability of radiography detection appears to be higher than that obtained using ultrasonics. It appears that any shift or a change in the seeded defect pattern geometry and/or location poses difficulty in getting highly reliable data interpretation.

3- Ultrasonics provide more sensitivity of detection of simulated voids than radiography, however signal analysis and data interpretation is tedious particularly when the samples contain a high density of defect population.

4- Radiography provides better spatial resolution than ultrasonics and provides an image of the defects from which the human mind can draw useful conclusions. With image processing it may even provide some quantitative information. However, it provides no depth resolution and is limited in sensitivity.

5- Ultrasonics provides poorer spatial resolution, which not only prevents separating the signals from two close defects, but also causes edge and near surface deadbands that require extra scans to eliminate. However, ultrasonics provides depth information, and theoretically has the potential to detect much smaller defects, and to characterize those defects through signal analysis.

6- Improvement in reference sample preparation and optimization of the measurement techniques as described in the text can significantly improve the reliability of inspection of ceramic materials.

ACKNOWLEDGMENT:

The authors would like to express appreciation to Mr. D. Patten for writing and debugging the computer program for the probability of detection statistics. We acknowledge the help of Dr. O. Kwon, Mr. G. Lambert, and Mr. J. Touhalisky for their help in sample preparation and experimental measurements.

REFERENCES:

1- Kamal E. Amin and Thomas P. Leo, "Radiographic Detectability Limits for Seeded Defects in Both Green and Hipped Silicon Nitride", Conference Proceedings on Nondestructive Testing of High Performance Ceramics, ASNT/ACS, August 25-27, 1987, Boston, MA, pp. 211-232.

2. Thomas Derkacs, "High Frequency Ultrasonic Inspection of Gas Turbine Ceramics", Proceedings of the Fourth Annual Conference on Composites and Advanced Materials". January 20-24, 1980, Cocoa Beach, FL, ISSN01296-6219, ACS, 1980, pp. 576-592.

3. Robert S. Gilmore, K.C. Tam, J.D. Young, and D.R. Howard, "Acoustic Microscopy from 10 to 100 MHz for Industrial Applications", Phil. Trans. Roy. Soc., London, A320, 1986, pp. 215-235.

4- Stanley J. Kilma and Alex Vary, "NDE of Structural Ceramics", presented at the International Gas Turbine Conference and Exhibit, Dusseldorf, Fed. Rep. of Germany, June 8-12, 1986, ASME 86-GT-279.

5. Daniel J. Cotter and William D. Koenigsburg, "Microfocus Radiography of High Performance Silicon Nitride Ceramics", Conference Proceedings on Nondestructive Testing of High Performance Ceramics< ASNT/ACS, August 25-27, 1987, Boston, MA, pp. 233-253.

6. Pramod K. Khandetwal, "Projection Microradiography of Ceramic Turbine Components, ibid.

7. P. E. Packman, S. J. Klima, R. L. Davies, J. Malpani, J. Mayzis, W. Walker, BGW Yee, and D. P. Johnson, "Reliability of Flaw Detection by Nondestructive Inspection", Metals Handbook, Vol 11, 8th ed., 1976, pp. 414-424, ASM, Metals Park, OH.

8. H. Berger and D. Kupperman, "Microradiography to Characterize Structural Ceramics", Materials Evaluation, Vol. 43, February, 1985, pp. 201-205.

9. J.J. Schuldies and T. Derkacs, "Ultrasonic NDE of Ceramic Components", ARPA/NAVSEA - Garrett/Ai Research Ceramic Gas Turbine Engine Demonstration Program Review, 21-2599(16), Maine Maritime Academy, Aug. 1-4, 1977.

Table 1– Typical Processing Defects / Microstructural Inhomogeneities in Ceramic Materials, (um)

Flaw Type	Powder Preparation	Suspension Preparation	Forming	Binder Removal	Consolidation	Machining
Agglomerates	1–50	1–10	1–5	--	--	---
Inclusions	10–100	10–100	10–100	--	up to 100	---
Voids	--	30–70	30–70	50–200	--	--
Density gradient	--	< 20%	< 20%	--	< 3%	--
Cracks	--	---	---	20–500	20–500	< 100
Sintering aid Distribution	< 30%	< 30%	< 30%	--	< 10%	--
Residual Stresses	--	---	< 80% of Strength	-- --	-- --	Subsurface < 60% strength
Elastic Anistropy			×	×	×	

Table 2– Particle Size Distribution of Seeded Defects, um

WC		Fe		PMMA		Diamond	
Ave	SD	Ave	SD	Ave	SD	Ave	SD
77	21	47	19	24	12	30	13
113	49	76	9	60	14	61	23
207	80	107	14	95	21	162	38
306	112	175	26	168	19	257	79
		336	51				

Table 3– Radiography and Ultrasonic Detection Data on Seeded Defects in Si3N4

Defect Type	Size, um	% Thickness Sensitivity	Ultrasonics % Detected	Ultrasonics PI	Radiography % Detected	Radiography PI
WC	77	1.21	50	.30	80	.62
	77	0.60	—	—	68	.49
	113	2.04	84	.72	84	.72
	113	0.44	—	—	48	.30
	113	0.90	—	—	92	.77
	207	3.70	100	.89	100	.89
	207	0.81	—	—	92	.77
	207	1.58	—	—	96	.82
	306	5.54	100	.89	100	.89
Fe	47	0.69	44	.27	0	0
	76	1.19	—	—	8	.01
	107	0.84	—	—	33	.21
	107	1.68	—	—	47	.39
	107	4.13	—	—	97	.90
	114	1.79	88	.72	60	.41
	200	3.14	100	.89	88	.72
	336	5.29	100	.89	100	.89
Diamond	61	.96	—	—	24	.11
	162	2.55	56	.38	—	—
	162	1.28	—	—	36	.20
	257	2.02	—	—	56	.38
	257	4.04	44	.27	92	.77
PMMA	95	3.16	—	—	20	.10–
	255	2.83	—	—	9	.06
	255	4.25	—	—	14	.11
	255	8.5	15–57		21	.20

Table 4– Average Amplitude of Ultrasonic Signal Obtained on Seeded Inclusions

Nominal Defect Size, um	Amplitude, mV Inclusion Type		
	WC	Fe	Diamond
300	4.21	2.38	.37
200	2.83	1.97	.41
100	1.83	1.40	---

Table 5- Expected and Observed Defect Cluster Distributions for Ultrasonic C-Scan Data of Hiped Si3N4 Containing 255 um PMMA Seeds

# of Defects in Cluster	Expected # of Clusters	Total # of Defects	Observed # of Clusters	Total # of defects
0	439	0	452-526	0
1	95	95	2	2
2	41	82	4	8
3	9	27	8	24
4	6	24	12	48
5	3	15	9	45
6	2	12	2	12
7	1	7	0	0
8	1	8	0	0
9	1	9	0	0
10	1	10	0	0
11	1	11	0	0
Total	**600**	**300**	**37**	**139**

Table 6– Distribution of Void Indications From Ultrasonic C–Scan of The Si3N4 Sample with 255 um PMMA Seeds

Amplitude Range mV	Number of Seeds Assumed in Void	Number of Indications in Range	Total # Seeds	Calculated Void Size um
0.00 – 1.89	1	2	4	20
1.90 – 3.18	2	4	8	40
3.19 – 4.44	3	8	24	60
4.45 – 5.72	4	12	48	80
5.73 – 6.99	5	9	45	100
7.00 – 8.26	6	2	12	120
Total		37	139	

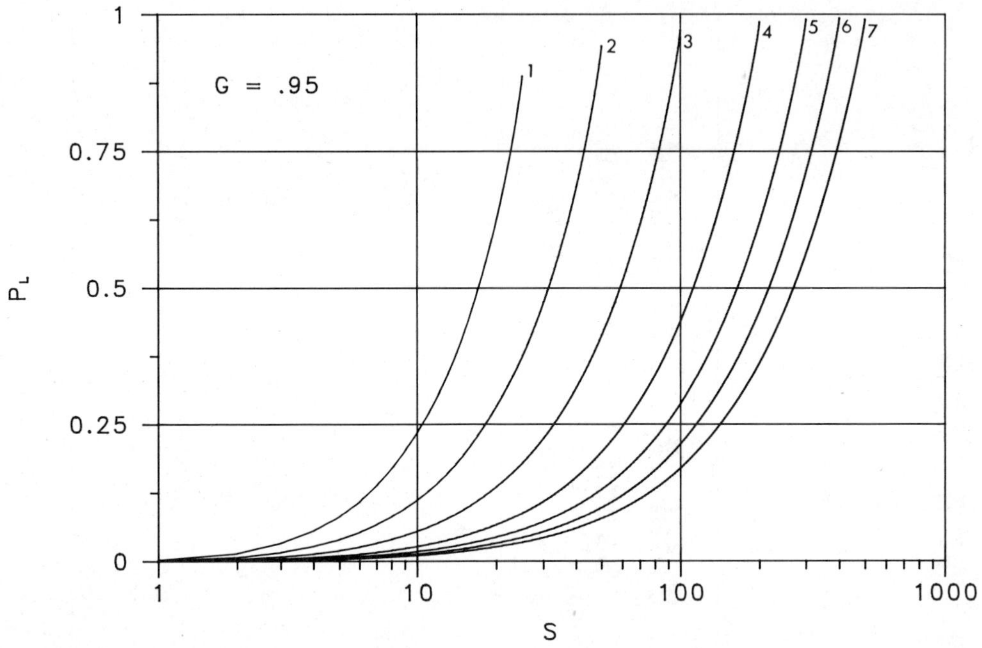

Figure 1- *Lower bound probability of detection Pl computed as a function of detected defects S and total number of seeded defects (sample size)`N .*

Curves : 1 2 3 4 5 6 7

N = 25 50 100 200 300 400 500

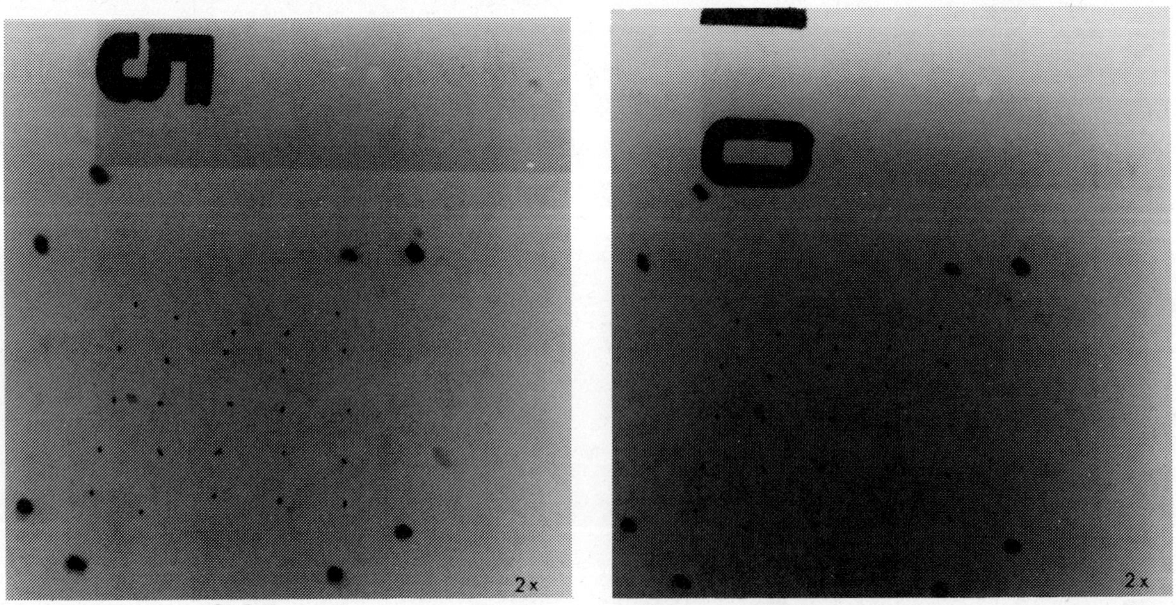

6.35 mm 12.70 mm

Figure 2- Microfocus radiographs of Si3N4 tile containing an array of 77 um size particles showing the effect of tile thickness on the detectability of such defects.

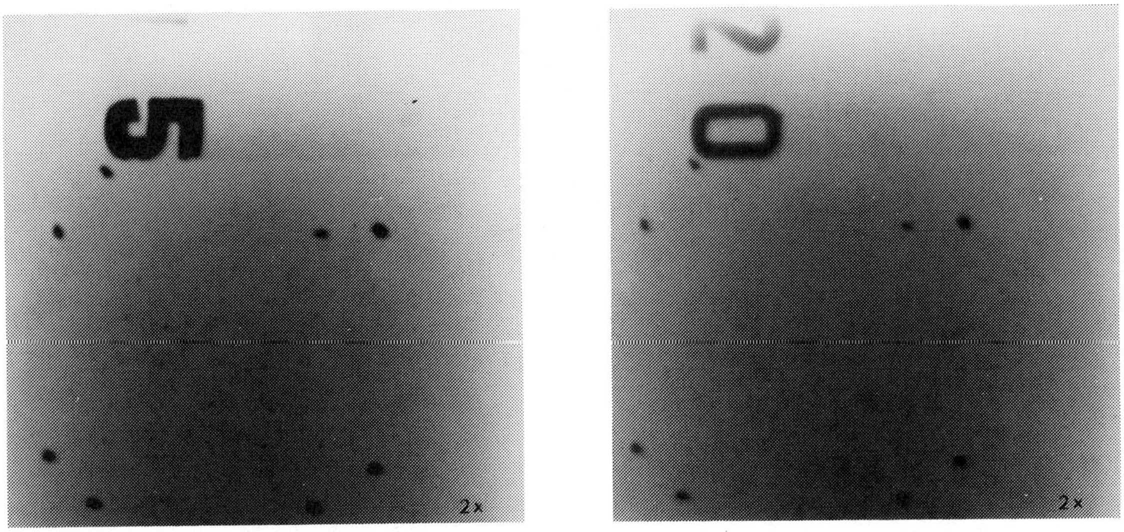

18.05 mm   25.4 mm

Figure 2- *Continued*

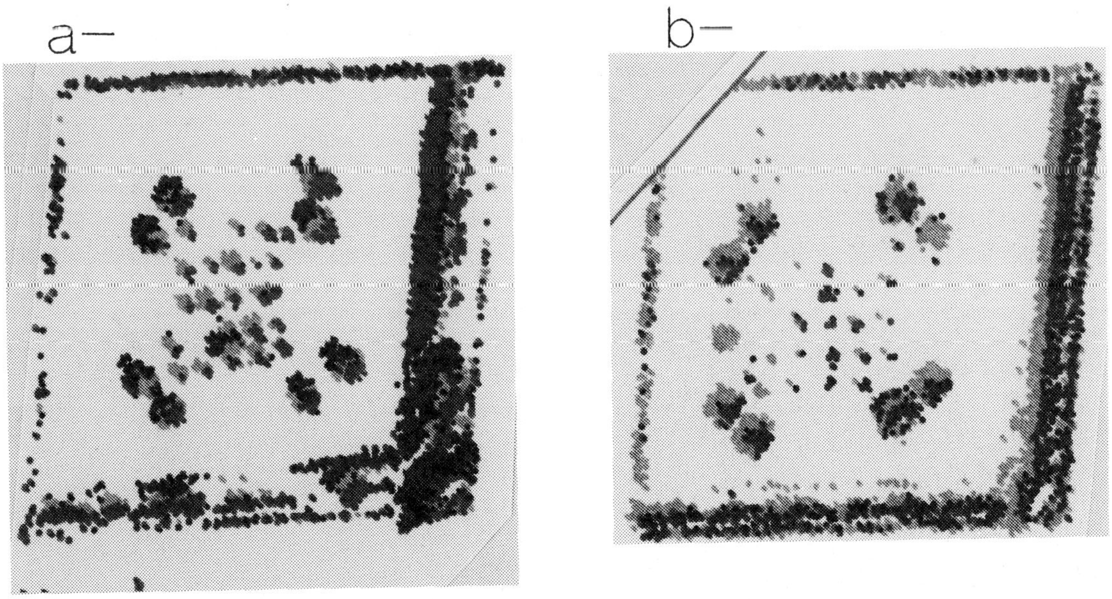

Figure 3- *An example of ultrasonic shear-wave C-scans of defects in Si3N4 tiles.*
a- WC seeded inclusions, size 113 um
b- Fe seeded inclusions, size 114 um

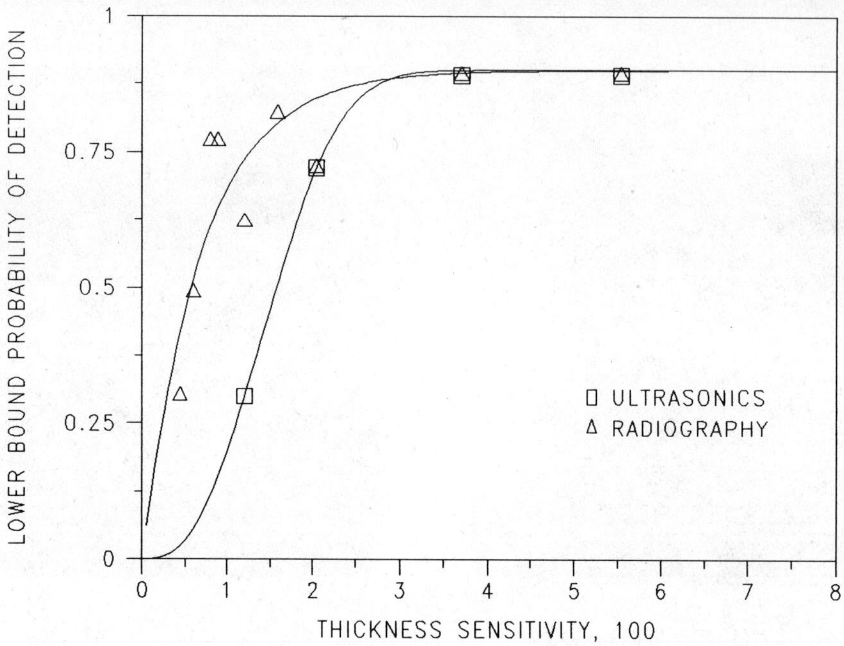

Figure 4- Lower Bound Probability of Detection of WC in HIP'ed Si3N4 As A Function of Thickness Sensitivity.

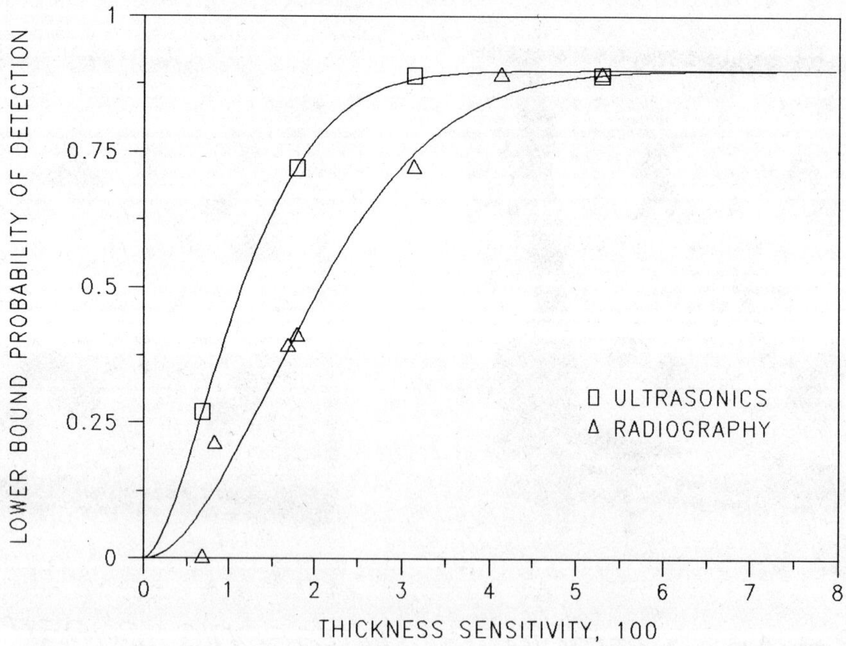

Figure 5- Lower Bound Probability of Detection of Fe in HIP'ed Si3N4 AS A Function of Thickness Sensitivity.

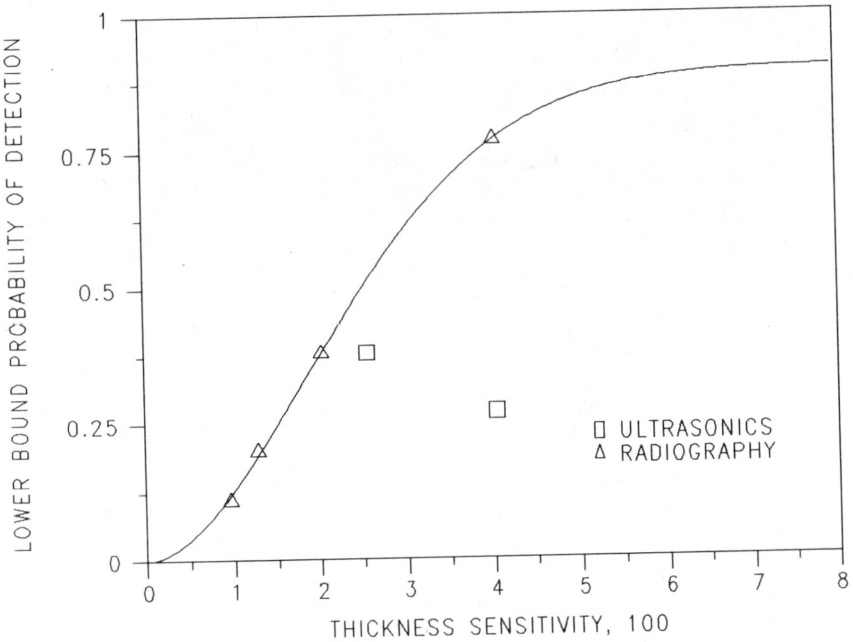

Figure 6– Lower Bound Probability of Detection of Diamond in HIP'ed Si3N4 As A Function of Thickness Sensitivity.

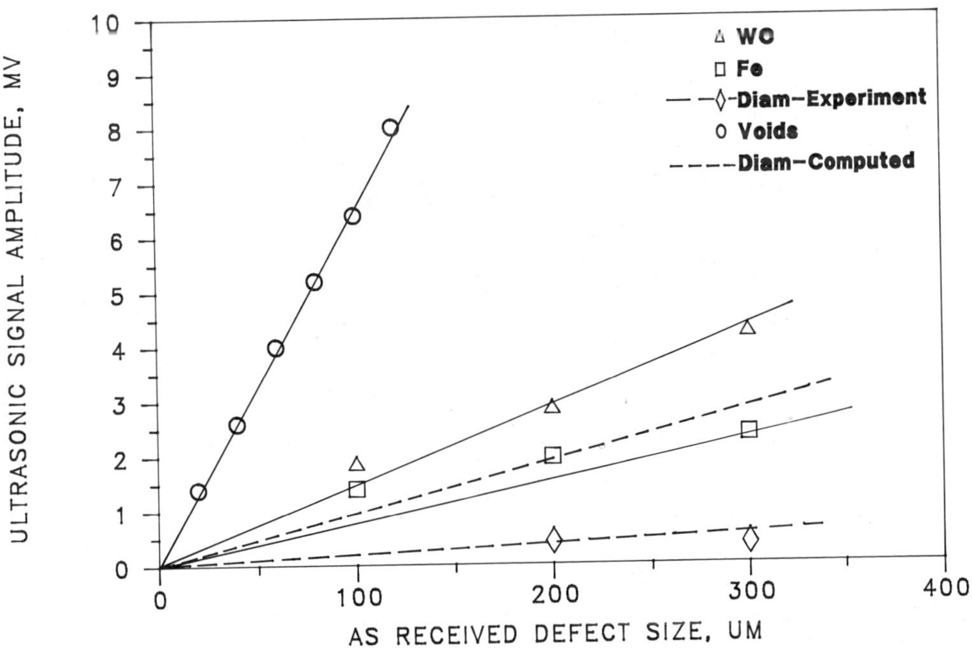

Figure 7– Measured Versus Computed Ultrasonic Signal As A Function of Defect Size

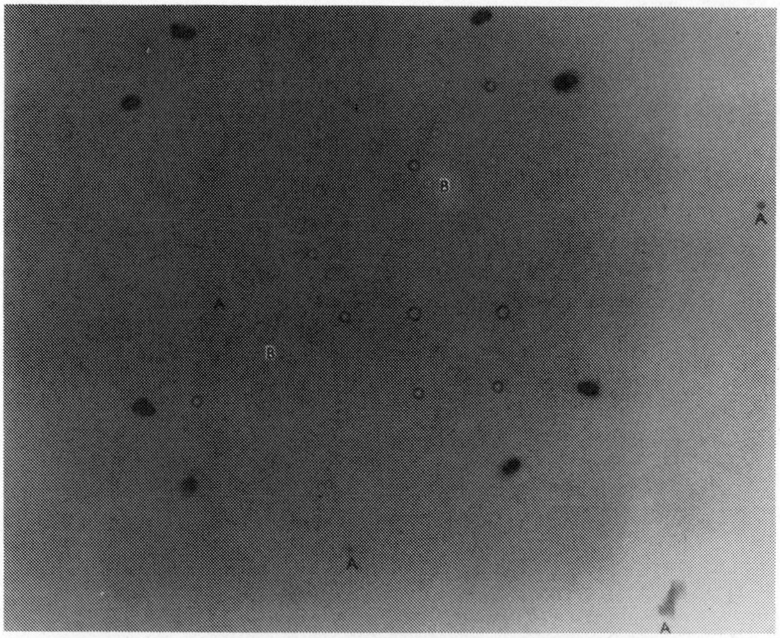

Figure 8- Microfocus Radiograph of Si3N4 Tile Containing Diamond Particles, 300 um in Size, Seeded in a Square Array. Those seeds are marked by circles to distinguish them from other non-seeded defects existing at locations such as A and B.

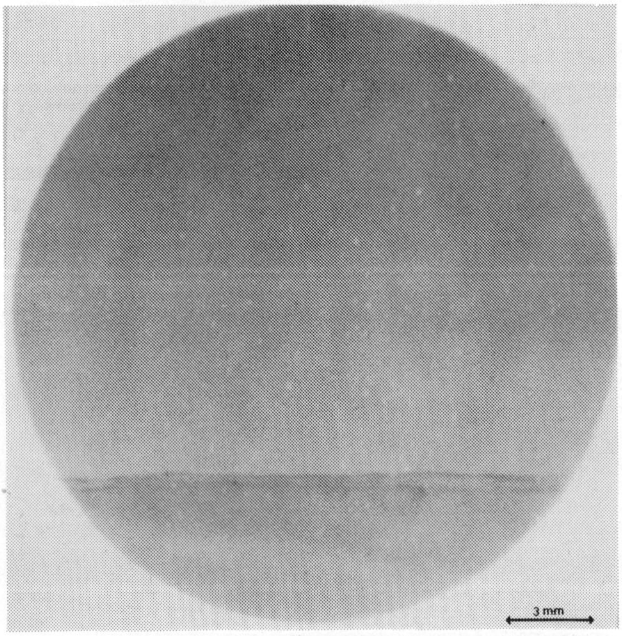

Figure 9- *Microfocus film radiograph of molded Si3N4 disc 2.60 mm thick containing 300 particles of PMMA with average size around 255 um. The particles are randomly seeded within the disc. Goemetric magnification is 8.*

X-RAY DENSITY MAPPING AND MATERIALS CHARACTERIZATION FOR A CERAMIC PRECOMBUSTION CHAMBER

A. Notea, Y. Tzabari, K. Rudich* & D.G. Brandon
Technion - Israel Institute of Technology
Haifa 32000, Israel

ABSTRACT

Current development of nondestructive techniques is primarily aimed at the extraction of quantitative information concerning characteristic features of the item examined. A feature of major importance is the material density, which may vary due to inhomogeneous compaction, sintering or composition. Knowledge of density variations is essential for better understanding of the production process as well as for quality control.

In this contribution, the results of film density obtained by x-ray radiography of ceramic precombustion chambers are described and compared with measured variations in hardness. Critical zones to be examined were identified by calculated temperature and stress distributions under service conditions.

1. INTRODUCTION

Mechanical features of ceramic products can be correlated with characteristic parameters detected by x-ray radiography. These parameters, resulting from the quality of the raw materials used and the process procedure, characterize the internal structure and inhomogeneity of the materials. The inhomogeneity flaws which are detected include cracks, internal voids, inclusions, and regions of different density or variable chemical composition.

The grey level image observed in the radiographic images represents the attenuation of the radiation in the product being examined. The attenuation is affected either by density or by the atomic number of the material. In the present contribution, the interpretation of the radiographic images taken from a ceramic precombustion chamber was attempted. The variations found in the attenuation suggest some variability in the manufacturing procedure. In order to relate the inhomogeneity to either density or atomic number changes, additional information was derived from hardness measurements. The severity of the observed inhomogeneity was related to the estimated temperature and loading conditions calculated from known service working conditions.

*Deceased Sept. 1988.

2. STRUCTURAL DESIGN

Over the past few years several studies of ceramic combustion engine components have been published, which include the development of a silicon nitride precombustion chamber for a diesel engine [1,2]. The primary considerations for the prechamber design are the geometrical definition, possible methods of assembly and the mechanical strength requirements.

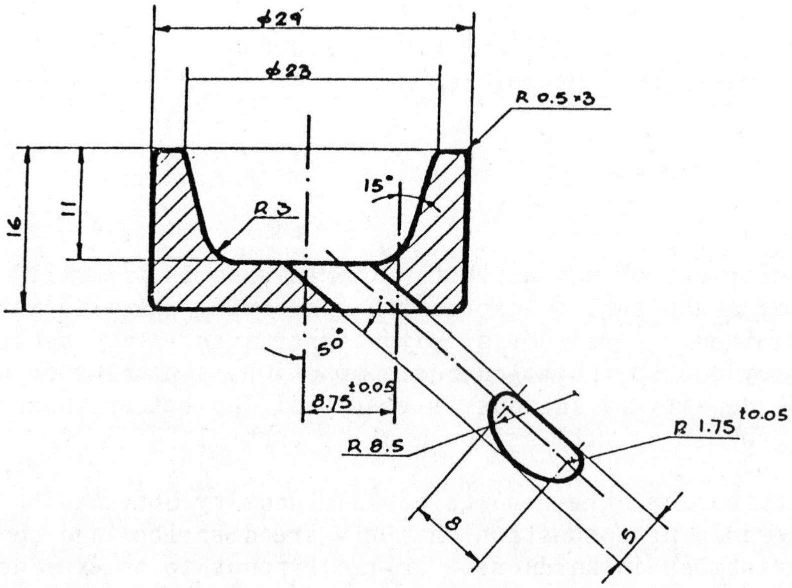

Fig. 1 Preliminary Precombustion Chamber Design. (Dimensions in mm)

A preliminary prechamber design is shown in Fig. 1, in which the shape has been kept as simple as possible, both in order to limit sources of stress concentration, and to simplify ceramic fabrication and subsequent machining.

In order to locate the critical area to be examined by radiography, the expected temperature and stress distributions were determined. The results indicated that the tensile hoop stress is built to maximum at the edge of the hole connecting the prechamber to the main chamber. Therefore, most of the radiographic examinations were directed at this area.

3. NONDESTRUCTIVE RADIOGRAPHIC RESULTS

The major drawback of ceramic materials is their unreliability associated with catastrophic failure due to brittleness, low shock resistance and nonuniform strength. Of all the operating components of an internal combustion engine, the precombustion chamber is subjected to the most severe thermal loading. In planning the procedures and methods of nondestructive testing and evaluation, it is therefore imperative to take into consideration all potential factors that might lead to failure such as:

1. Design induced failure due to the product geometry or its location in the assembly.
2. Production induced flaws and inhomogeneity due to inadequate process control, including inhomogeneities associated with the chemical distribution of sintering aids or the distribution of inclusions and impurities.
3. Flaws generated in service or actual failure of the product due to either the severity of the working conditions or as a result of production-induced inhomogeneities not detected at room temperature.

Due to the limited extent of the present investigation, only the detected variation in x-ray radiographic film density will be discussed.

3.1. The Digital Radiography Method

The concept of x-ray radiography is based on the interaction of radiation with the material of the examined component. The attenuation of x-rays penetrating the material is a function of several variables and only from the measurement of the transmitted radiation is it possible to differentiate between them. When no radiation reaches the film, it is completely transparent after development ("white"), while the dose of x-rays corresponding to the intensity of the incident, unabsorbed beam, can be adjusted to render the developed film completely opaque ("black"). Variations in either the material density, the atomic number of constituents or the thickness of the material, as well as the presence of voids or inclusions, then appear on the film as light gray levels of film density.

The optimal examination conditions [3] (energy, tube voltage, intensity and exposure time) are calculated for the material's elemental composition, density and thickness in order to obtain the film density which corresponds to the maximum gradient of the film response curve. Products of complex shape, namely parts having large variations in thickness or density of the constituents, require that the examined volume should be divided into regions and the irradiation conditions be optimized for each region. A set of radiographs obtained in this way reveals the different regions with approximately the same resolving power.

The radiographic image is described as a mathematical mapping of the examined component features into the film gray level variations. Once the mathematical relations are established, the interpretation involves the solution of the so-called 'inverse problem' [4]. For this purpose, the film density image is transformed to a digital image which renders to image processing techniques.

The radiographic image was digitized using an automatic microdensitometer (Photomation 1700, Optronics Inc.) controlled by a PDP 11/34 computer, with apertures of 25x25, 50x50, 100x100, 200x200 and 400x400 μm.

In this paper, the discussions are confined to the meaning of the variations found in the film density profiles. The noise in the studied profiles was suppressed by high frequency filtering.

3.2. Experimental Procedure

In the precombustion chamber, the circular area at the base of the chamber, identified as the critical area, has a thickness of 5 mm.

Based on the published manufacturer's data for the elemental composition of SYALON 101, the linear attenuation $\mu = \Sigma w_i \mu_i$ as a function of the beam energy was calculated, where w_i is the weight fraction of element i and μ_i is the mass attenuation coefficient in $cm^2 g^{-1}$. The optimal photon energy level of about 40 keV was obtained for the critical area and it was approximated by a 60 kV accelerating voltage. The exposure for film Structurix D4 (Agfa Gevaert), is based on responses measured for aluminium slabs and corrected for SYALON 101. In order to minimize geometric unsharpness, the focus to film distance was taken to be relatively large and the precombustion chamber was placed in close contact with the film packed in thin vacuum cassettes.

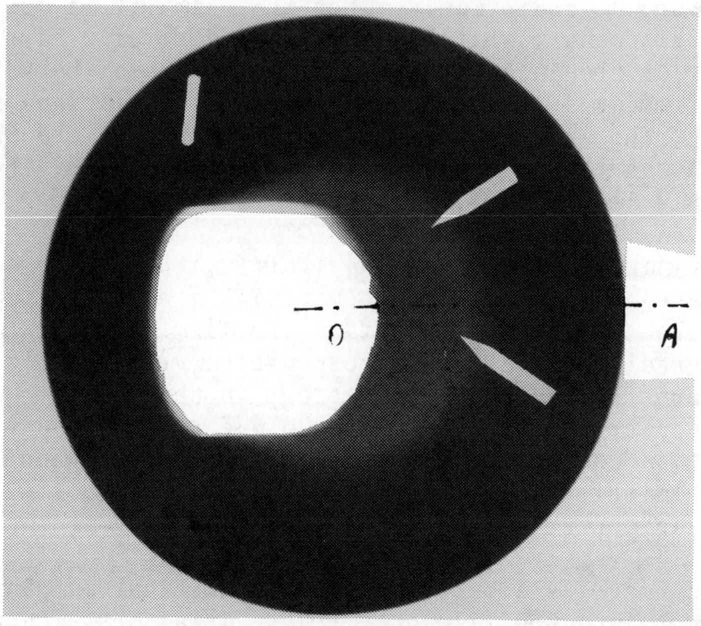

Fig. 2 X-ray radiograph of the critical area in the SYALON 101 precombustion chamber. The high attenuation ring is indicated by the arrows.

A set of six precombustion chambers were radiographed with a Radiofluor 360-072-700 unit. The x-ray beam was perpendicular to the prechamber's lower part, see Fig. 1, where the maximal stress is reached. In all cases a circular region of higher radiation attenuation was detected at the critical area, see Fig. 2. This ring indicates a failure in the chamber. The expected image, determined on the basis of the radiation chord length within the chamber material, shows a constant gray level at the base (where the thickness is 5 mm) and a gradual change occurs as the thickness increases.

The profile across the critical region (OA in Fig. 2) is shown in Fig. 3. The drop at the ring of high attenuation is clearly observed.

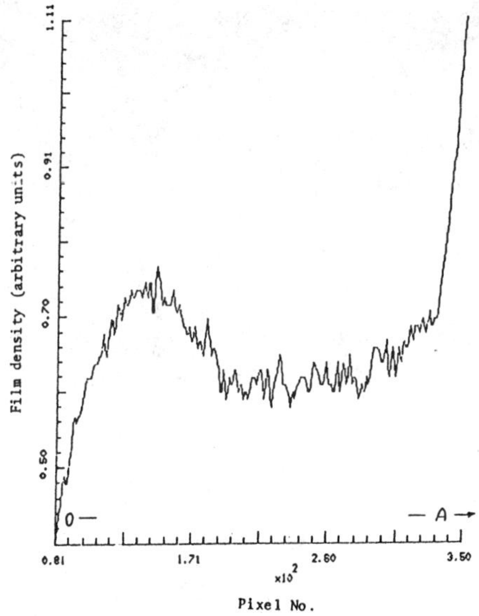

Fig. 3 Film density profile of the precombustion chamber along the OA line in Fig. 2

Additional measurements were performed using a microfocus real-time radiography system (Andrex BW155) with a geometric magnification of 5. The profile obtained across an image generated by summing up 256 frames is shown in Fig. 4. The agreement with the profile of Fig. 3 is obvious.

Fig. 4 Film density profile from microfocus real-time system.

At this stage, one of the precombustion chambers was sectioned, Fig. 5, and a 1 mm thick slice taken from the critical area was radiographed. The image showed inhomogeneity that could explain the integrated response shown by the profile (Figs. 2 and 3). It is interesting to note that the inhomogeneity could more or less be distinguished visually on the slice itself as areas of lighter shade.

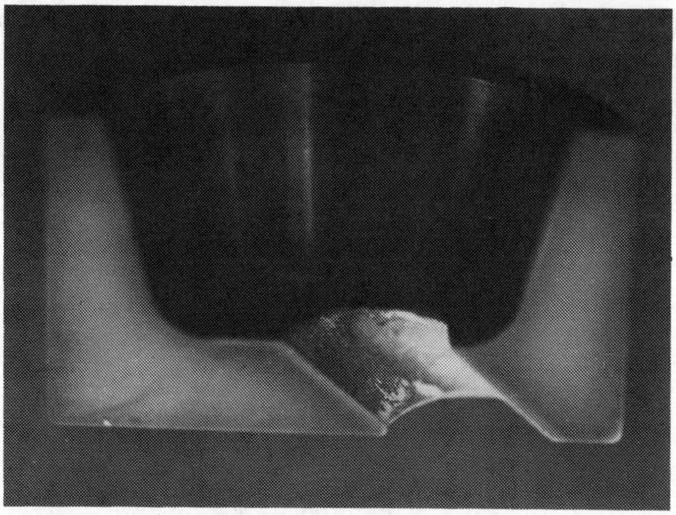

Fig. 5 Photograph of the sectioned ceramic precombustion chamber.

3.3. Interpretation of Results

The results obtained by the radiography show unexpected variations in film density at the critical area. The observed ring is interpretable as either an inhomogeneous density distribution or as variations in the heavy element distribution. The inhomogeneity in the density could result from incomplete sintering. While the heavy element distribution may be attributed to the migration during sintering resulting from a progressive elimination of porosity as sintering proceeds from one area to another, with the heavy element being segregated to the sintering front. Yttrium is the obvious candidate element in the present case, since it is the major constituent of the glassy phase responsible for liquid phase sintering in pressureless sintered silicon nitride.

The effects which have been observed are almost certainly a consequence of the sintering schedule, possibly associated with variations in green density due to pressure gradients in compaction. Such variations in starting density could be expected to affect the sintering kinetics, leading to preferential sintering in some areas followed by the formation of sintering gradients as described above. Of course, such effects would not necessarily lead to any variations in the volume fraction of the glassy phase. Since the oxynitride glass phase is known to have a high modulus and adequate strength, these phase inhomogeneities are unlikely to have much effect on the room temperature mechanical properties. They could, however, lead to in-service failure due to excessive creep.

In a future study, quantitative correlations between the material density, the mechanical properties and the radiographic analysis, will be established.

4. MICROHARDNESS MAPPING

Hardness mapping of the sectioned precombustion chamber was performed using a load of 1 Kgm on a Vickers hardness (Hv) microindenter. The hardness indentations were separated by at least 0.5 mm and were located at least 1 mm from the edge of the section. The approximate positions of the indentations are indicated in Fig. 6.

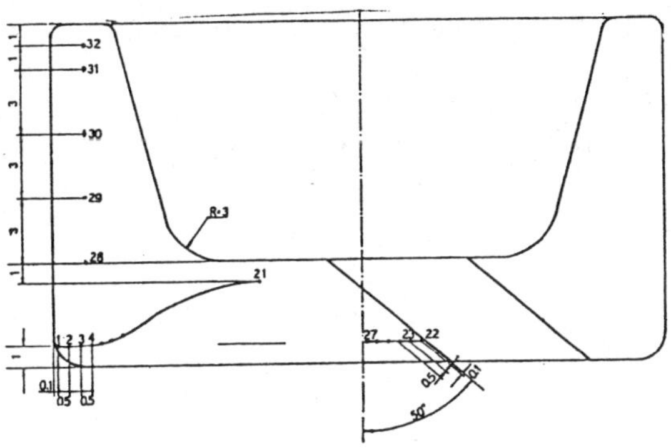

Fig. 6 The approximate positions (mm) of the indentations in the microhardness map of the sectioned precombustion chamber. (Each point is identified by Hv data sequence number).

The complete data map sequence is shown in Fig. 7 together with the 95% confidence limits assuming a Gaussian distribution. The average value for the hardness was 1529 $Kg.mm^{-2}$, with a coefficient of variance of better than 5%. From the data map, it can be seen that none of the regions fall outside the expected range of random scatter. In fact, the consistency of the data indicates that the hardness of the material is essentially constant, and unrelated to any variations in density. This result effectively rules out porosity as a possible cause of the density variations. However, since yttria-rich oxynitride glass may have both a hardness and an elastic modulus approximating those of β-Si_3N_4 (the primary phase), local variations in the glass content cannot be ruled out.

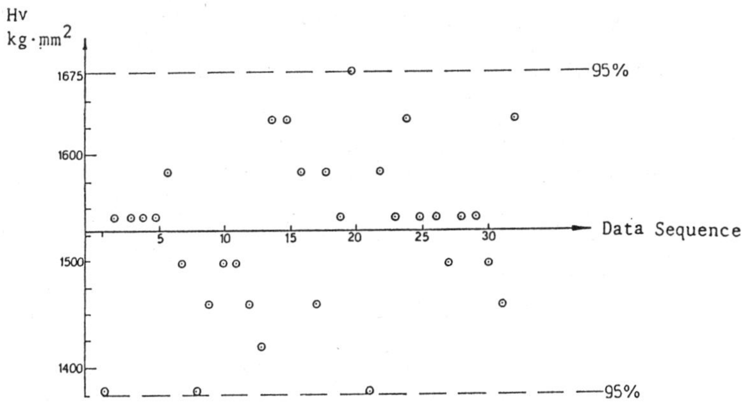

Fig. 7 Experimental scatter of the microhardness data sequence showing 95% confidence limits.

5. CONCLUSIONS

Information obtained by radiographic examination combined with additional experimental information may add considerably to the better understanding of the component manufacturing process. It is certainly important to identify critical zones where quality measurements will be more extensively performed. The results are important for production process development and no less valuable as a quality control tool at the production stage. This study is being continued in order to correlate the results with other test methods and includes a solution of the inverse problem involved in the quantitative interpretation of the radiographic results [4,5].

REFERENCES

1. H. Matsuka, H. Kawamura and S. Tonda, "Development of a Ceramic Pre-combustion Chamber", J.S.A.E. Review, March 1982.
2. S. Kamiya, M. Murachi, H. Kawamoto, S. Kato, S. Kawakami and Y. Suzuki, "Silicon Nitride Swirl Lower Chamber for High Power Turbocharged Diesel Engines", SAE paper #850523, March 1985.
3. A. Notea, "Evaluating Radiographic Systems Using the Resolving Power Function", NDT International, 16 (1983) 263-270.
4. A. Notea, "Phenomenological Inversion Approach for the Interpretation and Analysis of NDT Measurement Systems", NDT International, Dec. 1988.
5. A. Notea, D. Pal and M. Deutsch, "Density Distribution in Cylindrically Symmetric Objects from a Single Radiographic Image", in J.M. Farley and R.W. Nichols (ed.s) "4th European Conference on NDT", Vol. 3, p. 2093-2101, Pergamon Press, Oxford, U.K. (1988).

ACOUSTIC MICROSCOPY FOR NDE OF COMPLEX SHAPE COMPONENTS

Lawrence W. Kessler
Sonoscan, Inc.
530 E. Green Street
Bensenville, IL 60106
(312)766-7088

ABSTRACT:

Scanning Laser Acoustic Microscopy (SLAM) is an advanced form of ultrasonic imaging that produces high resolution images of samples in real-time over a frequency range from 10-200 MHz. Because of its speed, it can be used on-line or off-line for defect detection as well as for materials characterization. With proper fixturing, samples can be accommodated which are large in size, e.g. $1m^2$ or small, e.g. $1mm^2$. Complex shape samples such as cylinders and turbine blades have been accommodated successfully and are described here.

INTRODUCTION:

Ultrasonic inspection of complex shaped components has traditionally been achieved by means of contact transducers which are incrementally positioned over areas of interest and by liquid immersion ultrasonic scanning with positioning mechanisms for the transducer(s) and/or the sample. Such techniques, which are commonly known as C-Scan(1), are inherently limited by slow mechanical scanning and by uncertainties in knowing the precise acoustic beam trajectories within samples whose surfaces are not flat and parallel to the transducer. Acoustic microscopy, a relative newcomer to the field of ultrasonic testing, was originally thought to be restricted to tiny flat samples that have been specially prepared for examination as in metallurgy. This is not generally the case; in fact, with the Scanning Laser Acoustic Microscope (SLAM) in the frequency range of 10 to 100 MHz, where ultrasonic absorption losses in the coupling fluid can be circumvented by maintaining small working distances between the transducer and the sample, irregular shapes can be accommodated. (The absorption loss in water is approximately 0.22-22 db/cm over the frequency range indicated). In order to produce acceptable resolution within a sample, the surface roughness should be reasonably less than the subsurface flaw dimensions. In situations where it can be applied, the SLAM technique has the advantages of inspection speed, viz., 30 images per second and the ability to simultaneously image throughout the entire thickness of the sample through which the beam traverses. In addition, with SLAM there is no "inspection blind zone" beneath the surface of a sample which is a characteristic of C-Scan technology. Defects within this zone are very critical for brittle materials such as ceramics.

SLAM TECHNOLOGY:

The Scanning Laser Acoustic Microscope is a through transmission technique that insonifies the sample with a collimated plane wave of continuous ultrasound (rather than focused pulses). The energy is received, after selective and localized attenuation within the sample, by means of a focused scanning laser beam.[2-5] The output image is a true real-time shadowgraph image of features throughout the thickness of the part, somewhat analogous to real time x-ray.

As is characteristic of the shadowgraphic imaging mode, samples having flat and parallel surfaces do not alter the size or shape of a collimated acoustic beam unless there are regional changes in sonic velocity, which cause a lensing effect on the beam profile. For homogeneous materials, the images of isolated features and defects within the sample will have no size distortions except for diffraction effects. Curved surface samples, however, distort the once collimated wave in accordance with the surface shape and the laws of refraction at interfaces between materials with different refractive indices. Here, distortion also occurs to the images of internal defects. The acoustic beam distortion affects both focused C-Scan and shadowgraphic imaging techniques but there is a strong advantage to shadowgraphic imaging because of the relative ease of interpreting an image made with a collimated beam. A focused beam severely distorts as it crosses an irregular boundary and the abberations are not simple to predict or visualize.

In the SLAM technique, Fig. 1, a focused laser beam scans the surface of a plastic mirror (coverslip) which becomes perturbed by the particle motion of the bulk acoustic waves which arrive at the surface. The laser measures the amplitude of the acoustic wave on a point-by-point basis as it scans over the field-of-view; the output image is displayed on a CRT which is synchronized with the laser scan. The image brightness corresponds to the acoustic signal level at each point of the scan. The space between the coverslip and the sample surface is occupied by coupling fluid and this is usually kept to a minimum in order to maximize the signal level and minimize the loss of focus due to diffraction spreading. If the image becomes defocused significantly, the acoustic data are holographically reconstructed to restore focus to any selected depth within the sample.[6,7] With the holographic reconstruction procedure, only one set of acoustic amplitude and phase data are obtained, and from this, images at any depth are obtained.

EXAMPLE APPLICATIONS AND DISCUSSION:

Figure 2 is a photograph showing ceramic cutting tool inserts used by the metal-working industry. These particular samples are made of Al_2O_3 ceramic and they have flat and parallel faces. The thickness of each sample is approximately 3mm and, if the porosity is low, i.e. a few percent or less, the ceramic is quite transmissive to 100 MHz ultrasound. However, as the porosity increases, there is a monotonic increase in the acoustic attenuation, and therefore a darker image. Furthermore, a characteristic speckle pattern and texture may appear due to scattering of the ultrasound, which is not present at low porosity. Figure 3 shows 100 MHz acoustic transmission micrographs of two Al_2O_3 samples, one having higher porosity (3a) than the other (3b). Although the densities of these samples was not measured at the time, two other examples help to illlustrate the dependence of acoustic attenuation upon porosity [8]. In Fig. 4 a graph is shown of attenuation measured by SLAM, as a function of firing temperature for samples of dielectric ceramic material used to manufacture electrical capacitors. At the optimum firing temperature there is an attenuation minimum. Below the optimum, where porosity increases drastically, there is a strong attenuation increase. Above the optimum, grain growth also increases attenuation. but to a much smaller extent.

In any imaging system, the brightness and contrast controls of the CRT are adjusted by the operator to produce a good film exposure and a well-balanced view. Unfortunately, this process compensates out some of the quantitative aspects of the image needed to judge differences in acoustic transmission levels between the samples. To circumvent this problem, and to enhance the usefulness of the technique for materials characterization, the SLAM system uses two methods to quantify attenuation. First, the video signal is digitized prior to being displayed and the average pixel amplitude is measured over a user prescribed area of the sample. The average pixel level is then monitored as the conditions of the sample are changed. For example, a calibration sample may be used as a reference to make (relative) attenuation measurements. The second technique involves an electronic attenuation reference standard which is inserted into the electrical circuit path of the insonifying transducer. In this method, an average pixel level for a sample is established as above. Then when the sample condition is changed, the electrical attenuator is adjusted to restore the average pixel level to the previous reading. The advantage of this latter technique is that it is accurate over a wide dynamic range of signals, and large attenuation differences between samples.

Figure 5 illustrates the SLAM examination of an aircraft turbine blade which is made of a nickel based superalloy material. The surfaces are complex in shape and have not been prepared in any manner. The blade is constructed from two mating halves which are diffusion bonded together at the embossed areas shown in Fig. 6. The resulting image using an insonifying frequency of 30 MHz is shown in Fig. 7. In this image the bright horizontal stripes indicate the locations of bonds. The dark areas indicate either disbonds and internal air passages, which are used for cooling. The spatial nonuniformities within the bond lines, which is revealed as texture in the bright stripes, was typical of an acceptable bond as determined metallurgically. Unacceptable bonds were revealed as highly attenuating or completely absent acoustic transmission.

Figure 8 illustrates a typical setup for through transmission imaging of cylindrical tubes. Inner diameters from 5-100 mm have been examined in this way. A plane wave transducer is placed inside the tube and the laser beam scans the outside surface. The tube is translated and rotated to effect complete area inspection. Although SLAM reflection/refraction mode images have also been made by placing the transducer on the outside surface instead of inside, this will be discussed in a future paper. The effect of the cylindrical geometry is to cause a one dimensional line-focusing of the acoustic beam and a limited aperture of view compared to the plane wave situation. However, the angular field of coverage still incorporates about 15 degrees of useful view angle of the cylinder and this appears to be quite adequate for rapid screening. Figure 9 is a 30 MHz acoustic image of a metal tube which has a small notch along its axis. Note that the defect stands out with very high contrast against the bright background of transmission through the virgin material.

SUMMARY:

This brief paper illustrates a few applications of SLAM technology for characterizing materials and imaging samples of complex shape for real-time inspection. The principles are general and can be applied to samples of various composition including ceramics, metals, and composites.

REFERENCES:

1) Krautkramer, J., Ultrasonic Testing of Materials, Second Edition, Springer-Verlag, Berlin, 1977.

2) Kessler, L.W., Semmens, J.E., and Agramonte, F., "Scanning Laser Acoustic Microscopy: A New Tool for NDT", 11th World Conference of Nondestructive Testing, Las Vegas, NV, November, 1985, pp. 995-1002.

3) Kessler, L.W. and Yuhas, D.E., "Acoustic Microscopy - 1979", Proc. of IEEE, (invited paper), Vol. 67, No. 4, April, 1979, pp. 526-536.

4) Yuhas, D.E. and Kessler, L.W., "Defect Characterization by Means of the Scanning Laser Acoustic Microscope (SLAM)", Scanning Electron Microscopy, Vol. 1, 1980, pp. 385-391.

5) Kessler, L.W. and Oravecz, M.G., "Scanning Laser Acoustic Microscope (SLAM) Analysis of Advanced Materials for Internal Defects and Discontinuities", Proc. of Nondestructive Testing and Evaluation of Advanced Materials and Composites Conference, August, 1986, pp. 173-186.

6) Lin, Z.C., Lee, H., Wade, G., Oravecz, M.G., and Kessler, L.W., "Holographic Image Reconstruction in Scanning Laser Acoustic Microscopy", IEEE Transactions on Ultrasonics, Ferroelectrics and Frequency Control, Vol. UFFC-34, No. 3, May, 1987, pp. 293-300.

7) Yu, B.Y., Oravecz, M.G., and Kessler, L.W., "Multimedia Holographic Image Reconstruction in a Scanning Laser Acoustic Microscope", Presented at 16th International Symposium on Acoustical Imaging, Chicago, IL, June, 1987.

8) Kessler, L.W., Ewell, G.J., and Semmens, J.E., "High Frequency Ultrasonic Attenuation of Ceramic Capacitors As An Indicator of Quality", Proceedings of the 36th Annual Electronic Components Conference, Seattle, WA, May, 1986.

FIGURES

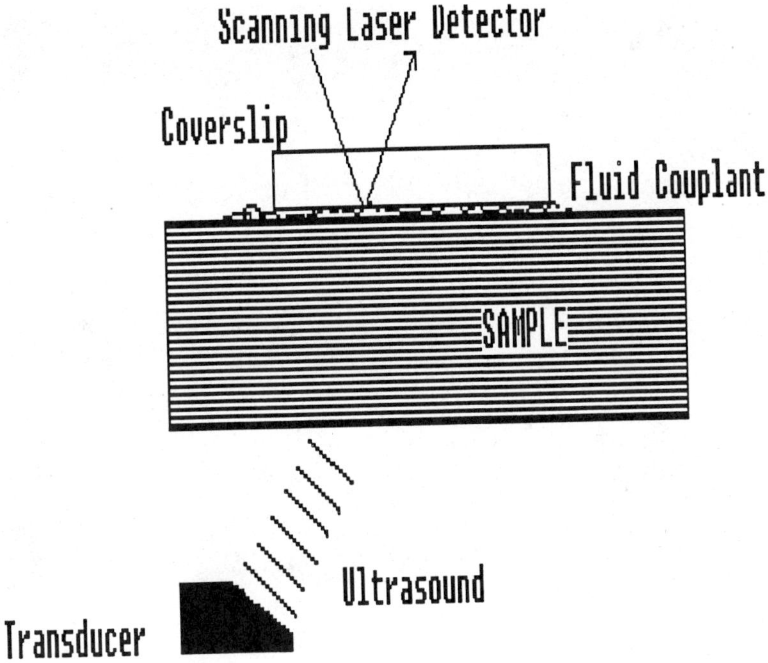

Figure 1 - Simplified block diagram of the Scanning Laser Acoustic Microscope

Figure 2 - Photograph of ceramic tool inserts

FIGURES

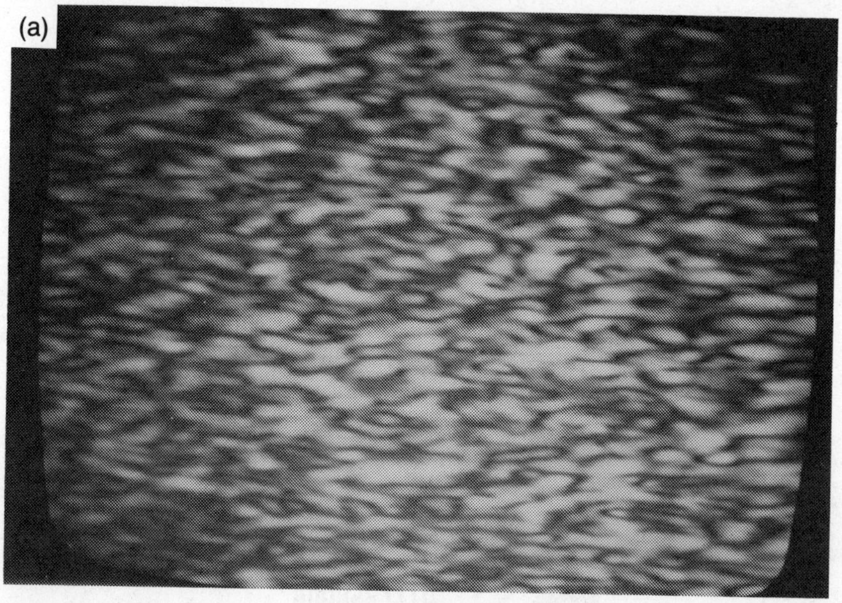

Figure 3 - SLAM micrographs of ceramic tool inserts, one of which (a) has higher porosity than the other (b). The porosity causes increased attenuation, therefore a darker image as well as a texture caused by speckle effects which arise from scattering. The horizontal field of view is 3 mm in these 100 MHz images.

FIGURES

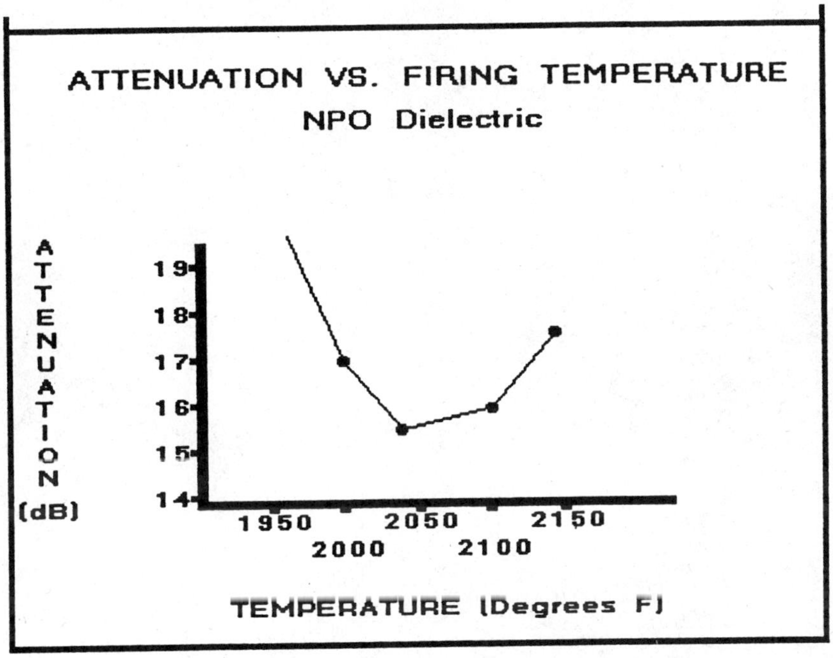

Figure 4 - Graph of ultrasonic attenuation at 100 MHz showing variations as a function of firing temperature for a ceramic dielectric material.

FIGURES

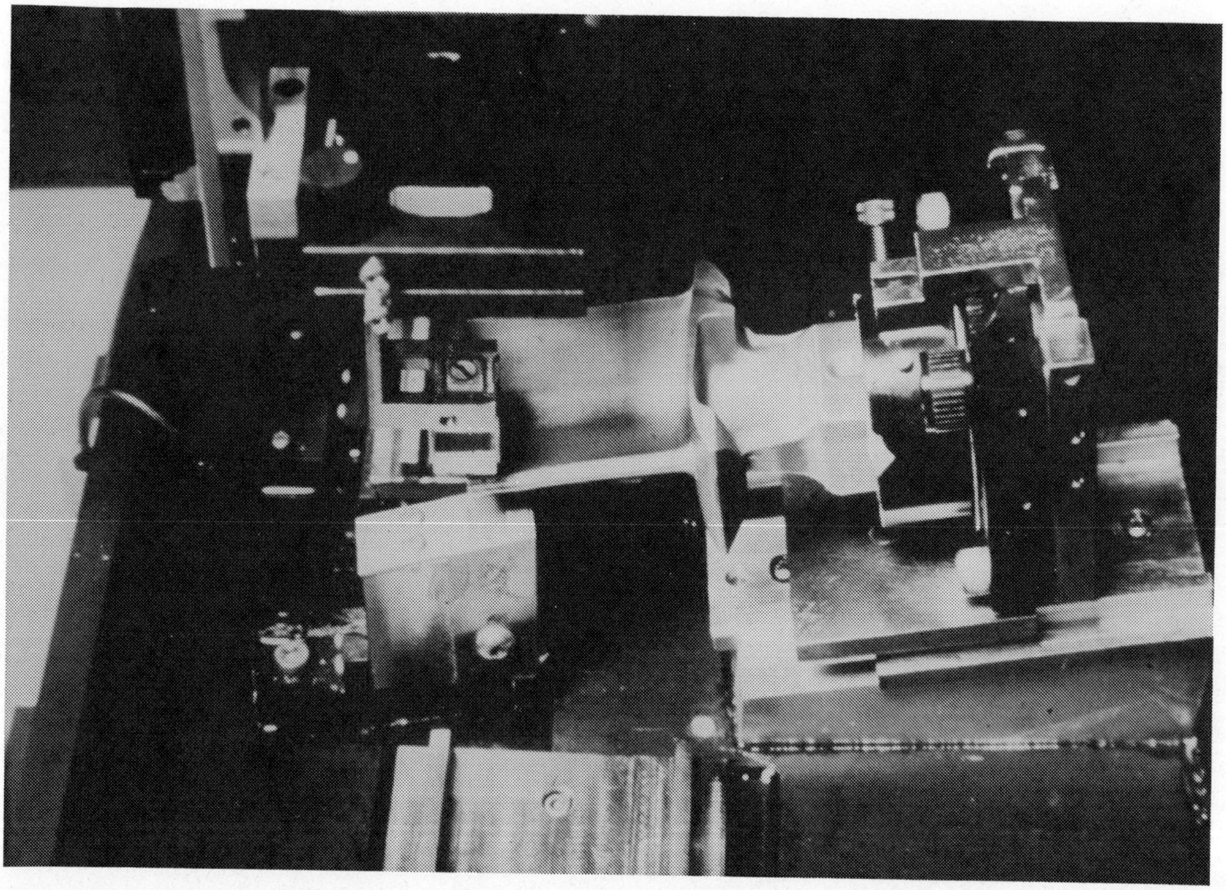

Figure 5 - Photograph illustrating the set up for SLAM examination of turbine blades. The blade is held in a fixture at a fixed angle and translated along a track whose position has been optimized for the shape of the part.

FIGURES

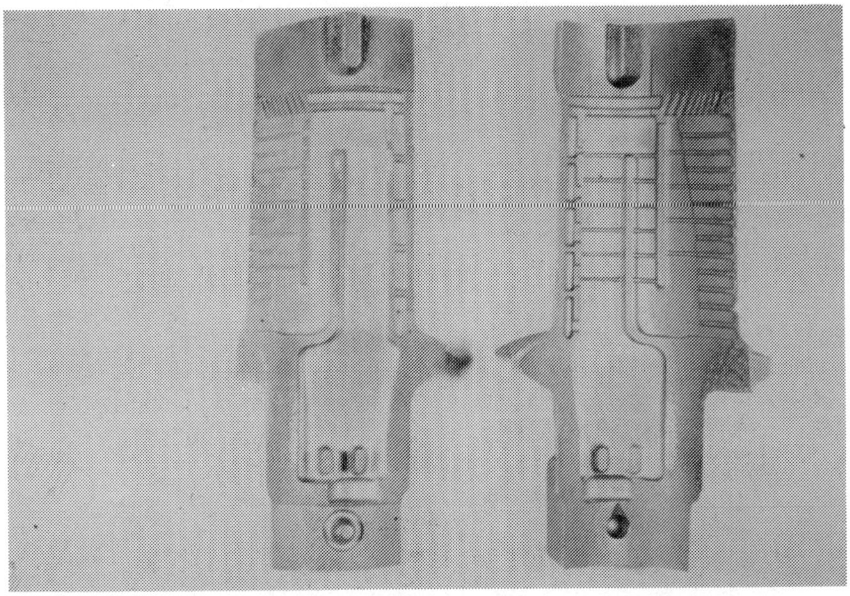

Figure 6 - Photograph of a two-part aircraft turbine blade prior to the assembly and diffusion bonding. Only the embossed areas of mating pieces should become bonded in order to have air passages within the blade for cooling purposes.

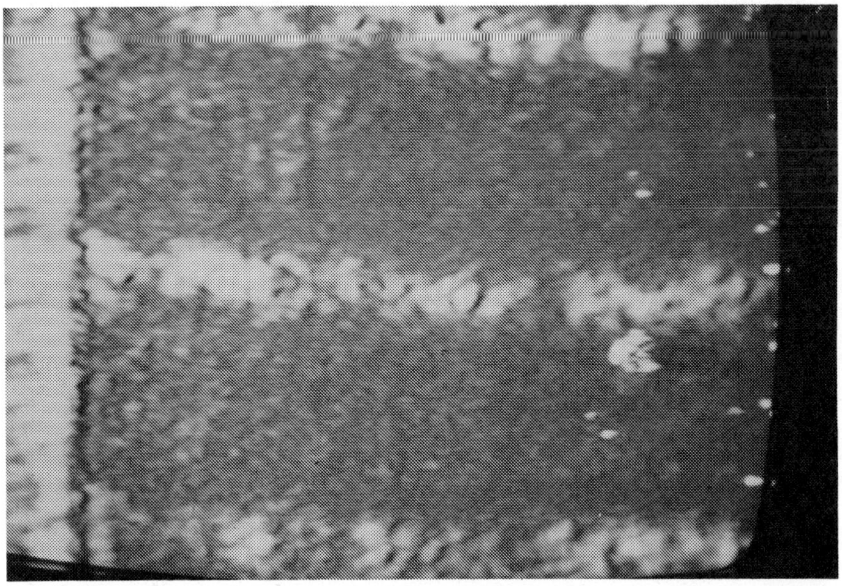

Figure 7 - SLAM acoustic image at 30 MHz showing horizontal areas of bond within the blade. The brightness of the image corresponds to the level of acoustic transmission across the bond interface. The horizontal field of view is 12 mm.

FIGURES

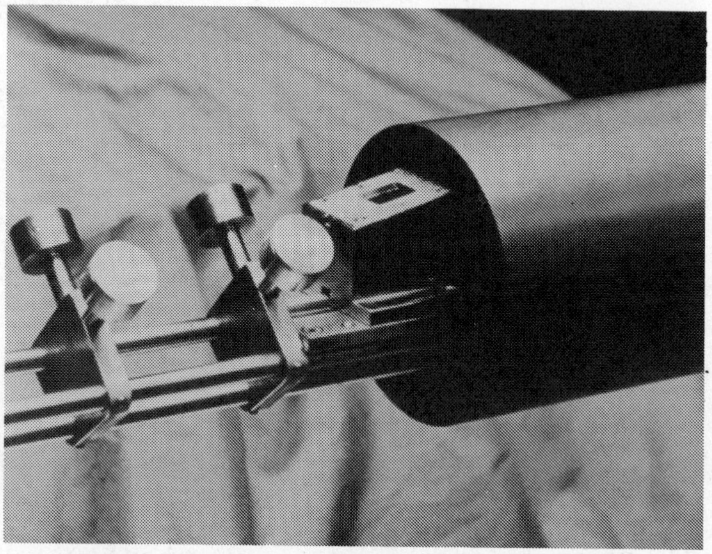

Figure 8 - Geometrical configuration for SLAM examination of cylindrical tubes.

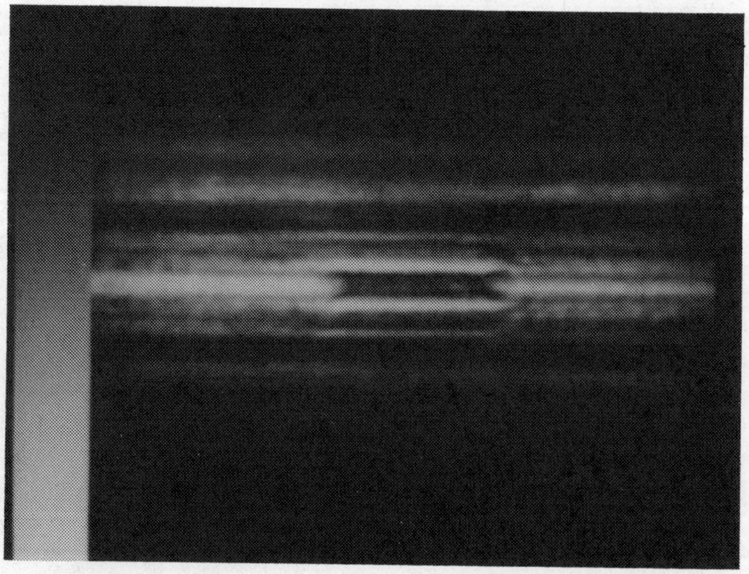

Figure 9 - SLAM acoustic image at 30 MHz showing a small defect within the wall of a tube. The restricted field of view in the vertical direction is due to the focusing of the sound beam by the sample curvature. The horizontal field of view is 14 mm.

INTERNAL STRESSES IN CERAMIC COMPOSITE CONSTITUENTS DETERMINED BY NEUTRON DIFFRACTION TECHNIQUE

J. P. Singh, D. S. Kupperman, and S. Majumdar
Materials and Components Technology Division
Argonne National Laboratory
Argonne, Illinois 60439 USA

ABSTRACT

The volumetric internal strains in the constituents of two ceramic composites, SiC whisker-reinforced Al_2O_3 matrix and monoclinic ZrO_2 dispersed in cubic ZrO_2 matrix, have been measured with the Argonne Intense Pulsed Neutron Source and Powder Diffractometer. The corresponding stresses were determined from strain measurements.

The internal strains in Al_2O_3/SiC have been observed to relax with increasing temperature, extrapolating to zero strain at ~1350°C. For ZrO_2 composites, 0.29% compressive strain was measured in monoclinic ZrO_2 particles for composites with 4 wt % monoclinic ZrO_2 content. This strain was observed to relax to 0.12% at 10 wt % monoclinic ZrO_2 content. This strain relaxation is probably due to the microcracking of the cubic ZrO_2 matrix as a result of large hydrostatic tensile stresses developed in the matrix at higher monoclinic ZrO_2 content.

INTRODUCTION

As a result of increasing demand for high-performance engines and structures, interest has increased in high-temperature materials with improved fracture toughness. Ceramic composites (both whisker reinforced and particulate) belong to a class of materials with potentially high fracture toughness. The high fracture toughness of the composites is believed to result from the interaction between crack and whisker/particulate inclusions. Several possible toughening mechanisms such as crack deflection,[1,2] whisker bridging,[3,4] whisker pullout,[5] crack pinning[6] and microcracking[7-9] have been proposed. The relative contribution of each of these mechanisms will depend upon the properties of the interface between the matrix and whisker/particulate inclusion as well as upon the state of stress at the interface. Therefore, understanding the state of stress at the interface is a necessary step for the development of improved ceramic composites. In this regard the neutron diffraction technique[10,11] can be a powerful tool for measuring bulk residual strains and stresses. In this paper, we present and discuss the results of an investigation in which residual strain and stress in the constituents of two ceramic composites, SiC whisker-reinforced Al_2O_3 matrix and monoclinic ZrO_2 dispersed in cubic ZrO_2 matrix, have been measured with the Argonne Intense Pulsed Neutron Source and Powder Diffractometer.

EXPERIMENTAL PROCEDURE

Specimens

The samples used in this study included Al_2O_3 matrix composites with 15 wt % SiC whisker-reinforcement and composites of a cubic ZrO_2 (with 8 mol % Y_2O_3) matrix with 5-19 wt % monoclinic ZrO_2 inclusions. The Al_2O_3-15 wt % SiC composites, supplied by ARCO,* were hot pressed into 30 x 25 x 5-mm billets. The SiC whiskers are ~10-30 μm long and ~0.75 μm in diameter. The whiskers have a high defect density that is due to stacking faults.[12] A typical micrograph of the fracture surface of the Al_2O_3-SiC composite (Fig. 1) clearly indicates fiber pullout. In Fig. 1, the pressing direction is from top to bottom.

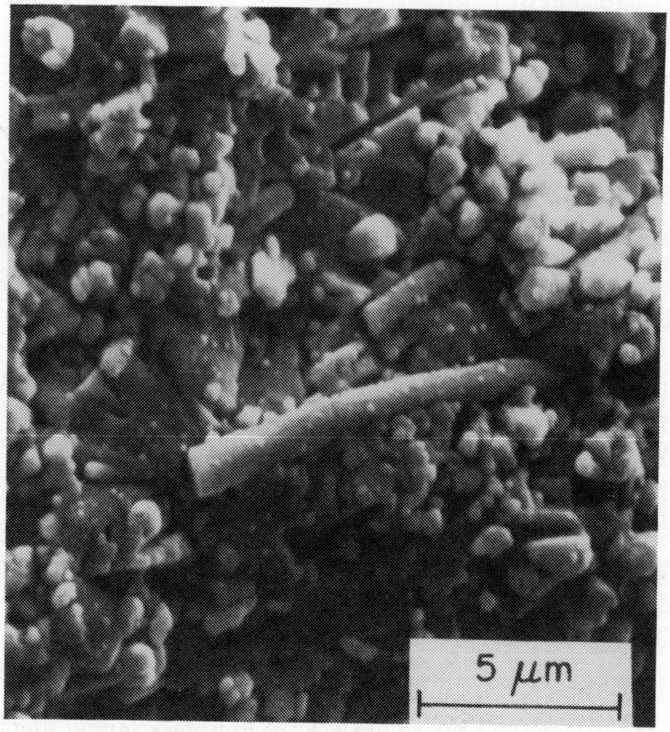

Fig. 1. Scanning Electron Micrograph of the Surface of an Al_2O_3-SiC Composite Specimen Showing SiC Whisker Pullout.

For the ZrO_2 composites, the powder mixtures of cubic ZrO_2 with monoclinic ZrO_2 were made by mixing cubic ZrO_2 powder with various amounts of monoclinic ZrO_2 (0-19 wt %) and an appropriate binder. The powder mixtures were wet-milled for ~16 h and were subsequently spray-dried. The rectangular bars (~5.1 x 0.6 x 0.3 cm) were pressed at ~103 MN/m^2 in a steel die using the spray-dried powder mixture. The bars were sintered at ~1450°C for 1-1/2 h in an air atmosphere and have a typical density of $\geq$95% theoretical. A typical micrograph of the fracture surface of cubic ZrO_2-4 wt % monoclinic ZrO_2 is shown in Fig. 2. Details of the experimental procedure have been published elsewhere.[13]

*ARCO Chemical Company, Greer, SC (courtesy of J. Rhodes).

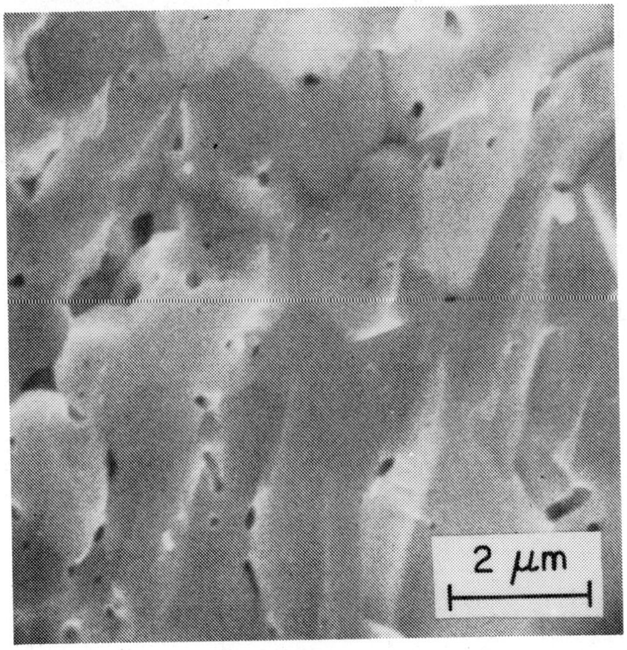

Fig. 2. Scanning Electron Micrograph of the Fracture Surface of Cubic ZrO_2-4% Monoclinic ZrO_2.

Neutron Diffraction Technique

Thermal neutrons with velocities of up to 1000 m/s are of interest for neutron diffraction. At these energies, the wavelengths are on the order of the lattice spacing, and the Bragg Law[14] of diffraction can be applied to neutrons as indicated below:

$$2d_{hkl} \sin \theta = \lambda_{hkl} ,$$

where d is the lattice spacing, 2θ is the angle between incident and diffracted neutron beams when a Bragg peak is detected, λ is the de Broglie wavelength of the neutron, and hkl are the Miller indices of the diffracting plane.

In the experiments described here, Bragg's law is used first to determine the lattice spacing d for a particular hkl reflection from Al_2O_3, SiC, cubic ZrO_2, and monoclinic ZrO_2 averaged over a volume of the strain-free powder. The stressed composites are examined next. The lattice strain associated with the hkl plane of a given phase in the composite is given by $e = (d - d_o)/d_o$, where d_o is the unstrained hkl spacing (for the powder) and d is the spacing for the composite. The main advantage of the pulsed source is that many diffraction peaks can be measured at the same time in many spatial directions.

The neutron pulses are generated at Argonne National Laboratory by accelerating protons to a very high energy (450 MeV) and directing them at a uranium target. The fast neutrons created by spallation are moderated and produce beams at various instruments such as the General Purpose Powder Diffractometer (GPPD), which is used for powder diffraction experiments. The pulse repetition rate is 30 Hz, with a peak thermal flux of about 4×10^{14} n/cm^2. The GPPD is about 20 m from the target. Neutrons are detected with banks of ^3He proportional counters (140 total) encircling the sample chamber on a 1.5-m scattered flight path at 20°, 30°, ±60°, ±90° and ±148° relative to the neutron forward direction. Data acquisition is accomplished with a PDP 11/34-Z8000 computer.

In view of the fact that d-spacings of (111) and (220) planes for cubic SiC structure are overlapping with those of (0006) and $(11\bar{2}0)$ planes for hexagonal structure, respectively, the crystallographic structure of the SiC whiskers is not clearly defined. Therefore, a cubic structure was assumed for the strain and stress analysis. This assumption should not result in large errors[10,11] in the analysis because the (111) and (220) planes of the cubic structure have the same lattice spacing as the (0006) and $(11\bar{2}0)$ planes of the hexagonal structure, and one of the $\langle 111 \rangle$ directions of the cubic structure and the $\langle 0006 \rangle$ direction of the hexagonal structure are both parallel to the whisker axis.

For SiC, the diffraction peaks of principal interest are the {111} and {220} peaks, because $\langle 111 \rangle$ and $\langle 2\bar{2}0 \rangle$ crystallographic directions are parallel and perpendicular, respectively, to the SiC whisker axes. In principle, the stresses in the whisker could be calculated from the residual strains in these directions. Because the {111} and {220} peaks of the cubic SiC were indistinguishable from the {0006} and $\{11\bar{2}0\}$ peaks of the hexagonal structure, the cubic structure was used in the analysis. Diffraction data for the cubic (311) planes were also obtained because these planes have no hexagonal analog. Subsequently, strain in $\langle 311 \rangle$ directions of the whiskers was obtained.

In the ZrO_2 composites, diffraction peaks were examined which are clearly identifiable in a mixed state (cubic and monoclinic). The most clearly defined, isolated Bragg peaks examined are the $\langle 200 \rangle$ in the cubic phase and the $\langle 112 \rangle$ and $\langle \bar{1}02 \rangle$ of the monoclinic phase.

RESIDUAL STRAIN AND STRESS MEASUREMENTS

During cooling of Al_2O_3-SiC and cubic ZrO_2-monoclinic ZrO_2 composites from fabrication temperatures, residual thermal strains and stresses develop at the interfaces of the matrix and second phases because of the differential lattice thermal expansion coefficients of the two phases. The lattice expansion coefficients of Al_2O_3 and SiC are 8.8×10^{-6}/°C and 4.6×10^{-6}/°C, whereas those of cubic and monoclinic ZrO_2 are 11.5×10^{-6}/°C and 7.15×10^{-6}/°C, respectively. Furthermore, for the ZrO_2 composites, a major cause of residual stresses is the tetragonal → monoclinic phase transformation of the monoclinic ZrO_2 inclusions and the associated volume change. As discussed before, the lattice strain in the constituents of the composites was calculated from the measured values of the unconstrained (powder) and constrained (composite) d-spacing for a given plane. Subsequently, model calculations[11] were made to evaluate residual stresses in the composites, and the calculated values were compared with the measured ones. In the calculations, the SiC whiskers were modeled as ellipsoids, and the monoclinic ZrO_2 particles were modeled as spheres.

Al_2O_3-SiC Composites

For Al_2O_3-SiC composites, the strain measurements were made in the temperature range of 25-1000°C. Figure 3 shows the volume-averaged residual strain in SiC and Al_2O_3 for various planes. The average strain in the Al_2O_3 matrix is smaller than that in whiskers. Although it is expected that large residual stresses and strains develop locally at the matrix-whisker interface, the volume-averaged strain in the Al_2O_3 matrix is small, particularly for the low volume fraction of whiskers. This is because the large interfacial stresses and strains in the matrix decay very rapidly with distance from the whisker. Currently, the neutron diffraction technique is not sensitive enough to detect the locally large residual strains at the Al_2O_3-SiC whisker interface. However, the resulting residual stresses at the interface can be calculated[11] on the basis of residual strains in the SiC whiskers. The residual strains in Fig. 3 can be extrapolated to zero values at ~1350°C, an observation that is consistent with the fact that the residual strains can be relaxed by creep mechanisms at high (~1350°C) temperatures.[15] Model calculations have been made to evaluate the strains in the individual directions of the <111>, <220>, and <311> families. From the strains in the individual directions, isotropic averages of the strains in <111>, <220>, and <311> directions have also been calculated. A comparison of the measured strains in the <111>, <220>, and <311> directions of the whiskers with the calculated strains is shown in Fig 4. The predicted isotropic average strains are within 30-40% of the measured values. In view of the simplified assumptions made for the model calculations, this agreement is reasonably good. With the assumption that no slippage or microcracking occurred at the Al_2O_3-SiC interface, the matrix stresses at the interface were calculated[11] from the computed values of whisker strain by applying Hooke's law. The stresses in the SiC whiskers and alumina matrix at the interface are shown as a function of temperature Fig. 5, where the axial and hoop directions correspond to [111] and [220] directions.

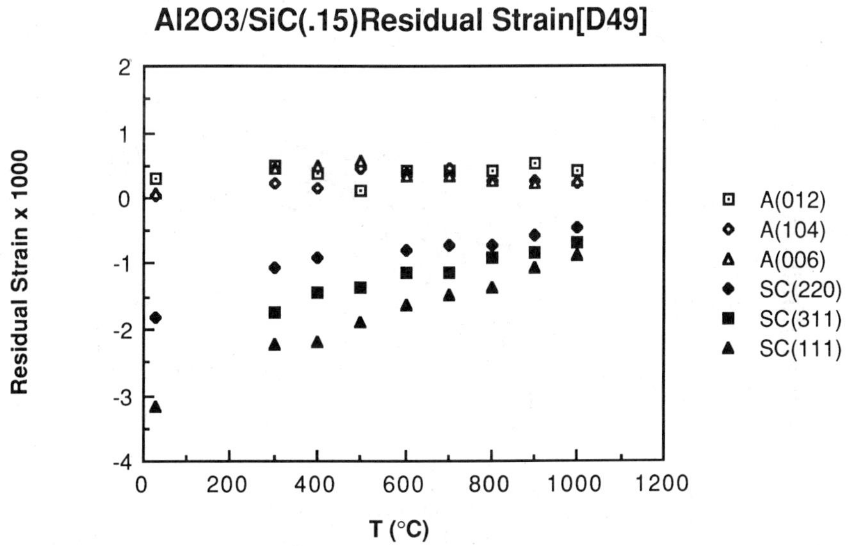

Fig. 3. Dependence of Residual Strain on Temperature for Various Planes of Al_2O_3/SiC Composites.

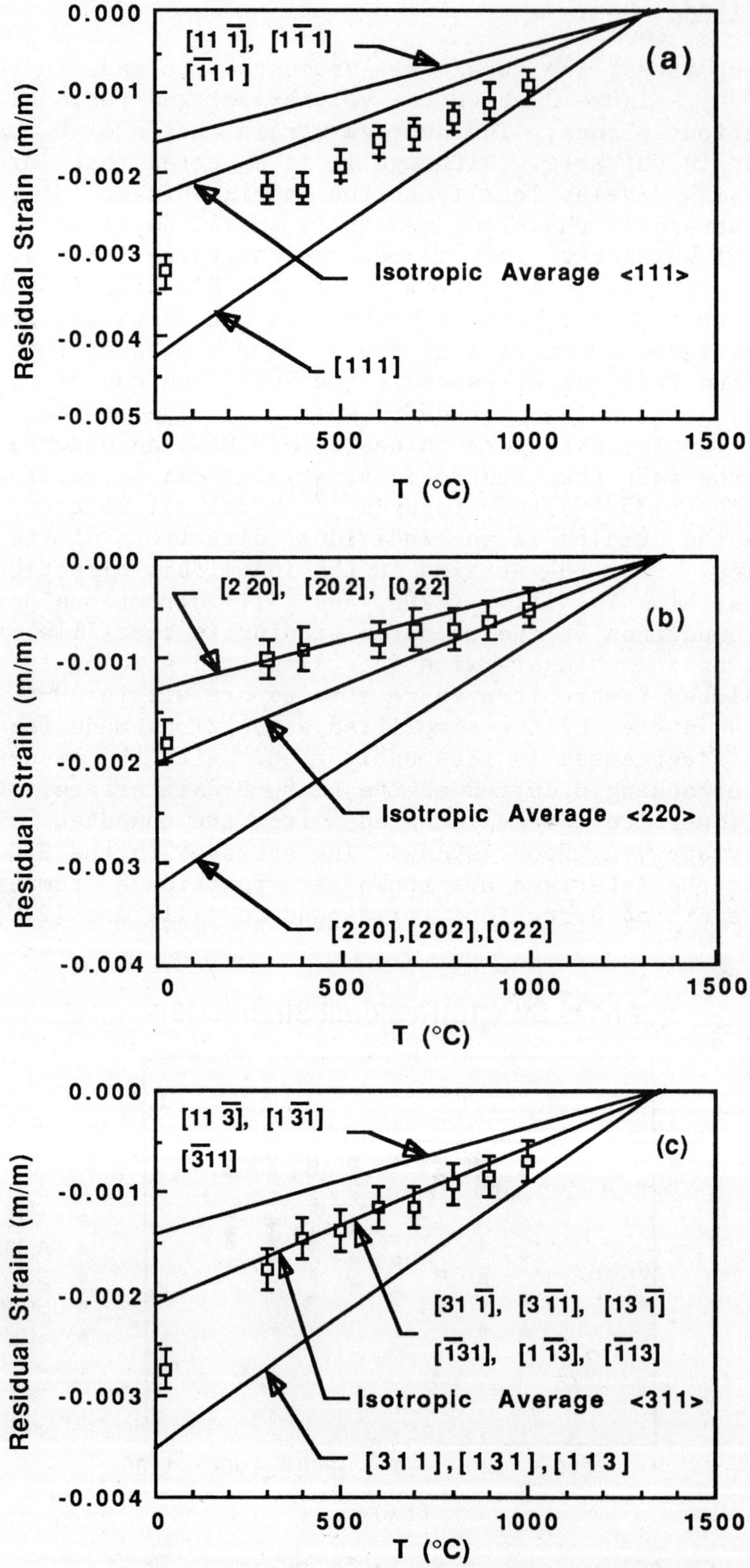

Fig. 4. Comparison of the Measured Strains (Symbols) and Predicted Strains in the Whisker (Solid Lines) in the (a) ⟨111⟩, (b) ⟨220⟩, and (c) ⟨311⟩ Directions.

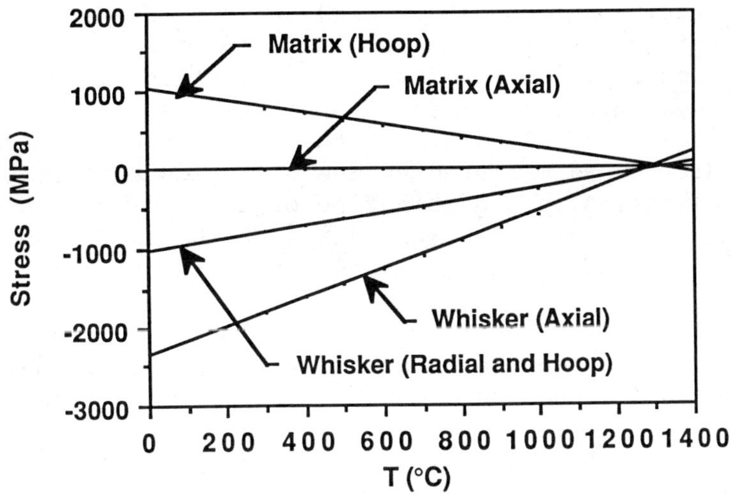

Fig. 5. Variations of Maximum Residual Stresses in the Whisker and the Matrix (at the Whisker-Matrix Interface) with Temperature.

Cubic ZrO_2-Monoclinic ZrO_2 Composites

Initially, strain measurements in the constituents of ZrO_2 composites were made for the composites containing 4 and 10 wt % monoclinic ZrO_2. The neutron diffraction data were obtained at room temperature (~25°C). Table 1 summarizes the measured strain values for different planes in the monoclinic and cubic ZrO_2 phases of the composites.

Table 1. d-Spacing and Residual Strain (±0.01%) Measured by the Neutron Diffraction Technique for Constituents of ZrO_2 Composites

Crystal Structure and Crystallographic Planes {hkl}	Powder d-Spacing (Å)	Sintered Composite with 4 wt % Monoclinic ZrO_2		Sintered Composite with 10 wt % Monoclinic ZrO_2	
		d-Spacing (Å)	% Strain	d-Spacing (Å)	% Strain
Cubic {200}	2.56802	2.56835	0.013	2.56885	0.032
Monoclinic {112}	2.02216	2.01632	-0.29	2.0197	-0.12
Monoclinic {$\bar{1}$02}	2.50015	-	-	2.49613	-0.16

Because the thermal expansion coefficient of monoclinic ZrO_2 is smaller than that of cubic ZrO_2 and a large volume expansion (4%) results from the tetragonal → monoclinic phase transformation at ~1070°C during the cooldown of the composite from fabrication temperature, large residual compressive strain in the monoclinic ZrO_2 phase and tensile strain in the cubic ZrO_2 phase are expected to develop. This is consistent with the data shown in Table 1 for the composite containing 4 and 10 wt % monoclinic ZrO_2 phase. Also, as expected, the residual stresses in the inclusion phase decreases and that in the matrix increases with increasing volume fraction of inclusion. However, an analysis[16] based on Eshelby model[17] predicts the residual strain in the inclusion to be a factor of 2-5 higher than the measured value. This difference in the predicted and measured values of residual stress is believed to be due to stress relaxation by a combination of creep, large- scale matrix microcracking, local microcracking of the matrix at the particle-matrix interfaces, and possibly the presence of a thin, soft layer at the particle-matrix interfaces.

Qualitatively, the dependence of measured residual strain agrees well with the measured [by the pulse echo technique[18]] values of the elastic modulus of the ZrO_2 composite as a function of monoclinic ZrO_2 content (Fig. 6). As is shown in the figure, the elastic modulus of the ZrO_2 composites remains almost constant for monoclinic ZrO_2 contents ≤ 4 wt %. This suggests the absence of any significant microcracking. For monoclinic ZrO_2 contents ≥ 6 wt %, the elastic modulus decreases by ~25%. This reduction in elastic modulus is probably due to large-scale microcracking of the cubic ZrO_2 matrix, because the elastic modulus is inversely proportional to the number and size of the microcracks.[19] The large-scale microcracking results from the high average hydrostatic tensile stress that is introduced in the bulk of the matrix when a large number of monoclinic particles are present.[16]

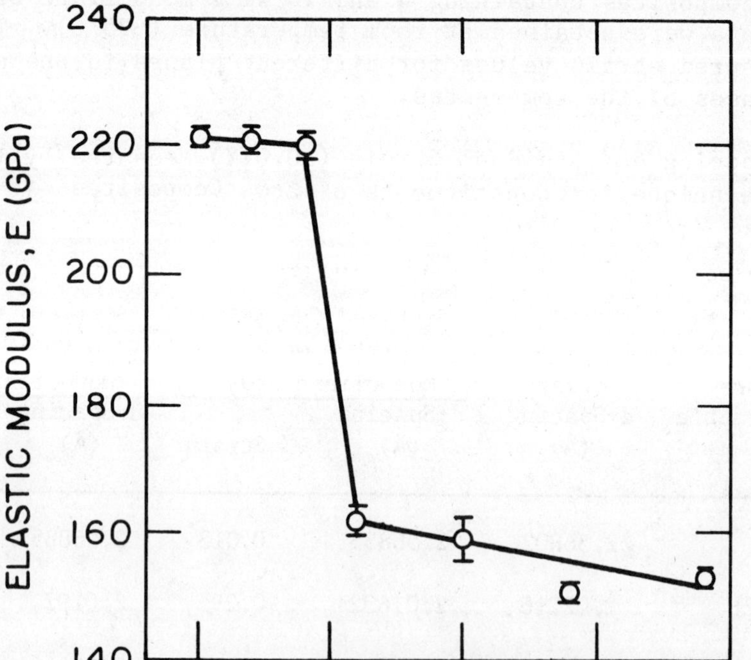

Fig. 6. Dependence of Elastic Modulus of ZrO_2 Composite on Monoclinic ZrO_2 Content.

In a model analysis by Majumdar,[16] the average hydrostatic stress in cubic ZrO_2 matrix (in the absence of any microcracking) was observed to depend on the monoclinic ZrO_2 content as shown in (Fig. 7). At ~6 vol % (6.4 wt %) monoclinic ZrO_2 content, the hydrostatic stress in the cubic ZrO_2 matrix is equal to ~200 MPa. This hydrostatic stress is equal to the flexural strength of the cubic ZrO_2 matrix (~200 MPa) and may cause fracture of the matrix. Therefore, large-scale microcracking will occur in the cubic ZrO_2 matrix at a monoclinic ZrO_2 content of $\geq$6 vol % (6.4 wt %).

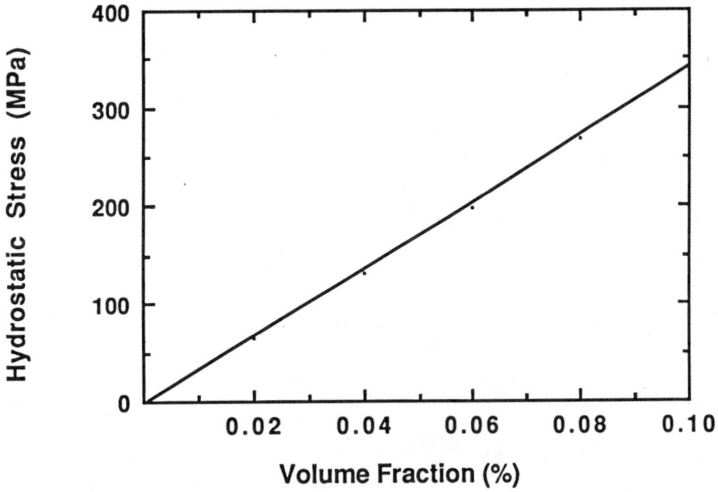

Fig. 7. Variation of Average Stresses in Cubic ZrO_2 Matrix with Volume Fraction of Monoclinic ZrO_2 Particles.

Although the present analyses are based on simplified assumptions, their results together with the measured elastic modulus clearly indicate that at high contents of monoclinic ZrO_2 (Fig. 7), the residual hydrostatic tensile stress can be of sufficiently large magnitude to cause large scale matrix cracking. This results in a partial relaxation of the residual strain. At low monoclinic ZrO_2 contents, large scale matrix cracking does not occur, and thus the residual strain is not substantially relieved.

CONCLUSIONS

These studies indicate that the application of neutron diffraction techniques to advanced composites is viable. Valuable information on the state of stress in the composite constituents can be obtained. This information can be used to optimize the design of ceramic composites for practical applications.

ACKNOWLEDGMENTS

This work has benefited from the use of the Intense Pulsed Neutron Source at Argonne National Laboratory and was supported by the U. S. Department of Energy under Contract W-31-109-Eng-38. The authors thank J. Faber for useful discussions and R. L. Hitterman and A. L. Bosak for their experimental assistance. Thanks are extended to R. B. Poeppel, Manager of the Ceramics Section, for his support throughout this work.

REFERENCES

1. K. T. Faber and A. G. Evans, "Crack Deflection Processes-I. Theory," Acta Metall. 31 (4) (1983) 565-76.
2. K. T. Faber and A. G. Evans, "Crack Deflection Process-II. Experiment," Acta Metall. 31 (4) (1983) 577-84.
3. A. G. Evans and R. M. McMecking, "On the Toughening of Ceramics by Strong Reinforcement," Acta Metall. 34 (12) (1986) 2435-41.
4. M. Ruhle, B. J. Dalgleish, and A. G. Evans, "On the Toughening of Ceramics by Whiskers," Scripta Metall. 21 (1987) 681-86.
5. I. N. Donald and P. W. McMillan, "Review: Ceramic-Matrix Composites," J. Mat. Sc. 11 (1976) 949-72.
6. F. F. Lange, "The Interaction of a Crack Front with a Second-Phase Dispersion," Phil. Mag. 22 (1970) 983-92.
7. A. G. Evans and K. T. Faber, "Crack-Growth Resistance of Microcracking Brittle Materials," J. Am. Ceram. Soc. 67 (4) (1984) 255-60.
8. T. W. Coyle, M. H. Guyot, and J. F. Jamet, "Mechanical Behavior of a Microcracked Ceramic Composite," Ceramic Eng. and Sc. Proc. 7 (7-8) (1986) 947-57.
9. R. W. Rice, "Mechanisms of Toughening in Ceramic Composites," Ceram. Eng. Sci. Proc. 2 (7-8) (1981) 661-701.
10. D. S. Kupperman, S. Majumdar, S. R. MacEwen, R. L. Hitterman, J. P. Singh, R. A. Roberts, and J. L. Routbort, "Nondestructive Characterization of Ceramic Composite Whiskers with Neutron Diffraction and Ultrasonic Techniques," Review of Progress in Quantitative Nondestructive Evaluation, 7B, Plenum Publishing Co., New York, (1988).
11. S. Majumdar, D. S. Kupperman, and J. P. Singh, "Determinations of Residual Thermal Stresses in a SiC/Al_2O_3 Composite Using Neutron Diffraction," J. Am. Ceram. Soc. (in press).
12. S. R. Nutt, "Defects in Silicon Carbide Whiskers," J. Am. Ceram. Soc. 67 (6) (1984) 428-31.
13. J. P. Singh, A. L. Bosak, C. C. McPheeters, and D. W. Dees, to be published.
14. A. J. Allen, M. T. Hutchings, C. G. Windsor, and C. Andreani, "Neutron Diffraction Methods for the Study of Residual Stress Fields," Adv. in Phys. 34 (4) (1985) 445-73.
15. "Alumina as a Ceramic Material," Ed. Walter H. Gitzen, The American Ceramic Society, Columbus, OH (1970).
16. S. Majumdar, unpublished work.
17. J. D. Eshelby, "The Determination of the Elastic Field of an Ellipsoidal Inclusion, and Related Problems," Proc. Roy. Soc. London Ser. A, 241 (1957) 376-96.
18. J. Krautkrämer and H. Krautkrämer, Ultrasonic Testing of Materials, Springer-Verlag, New York (1983).
19. R. L. Salganik, "Mechanics of Bodies with Many Cracks," Izv. Akad. Nauk SSSR, Mekh. Tverd, Tela, 8 [4] (1973) 149-58.

DENSITY AND POROSITY MEASUREMENTS OF GREEN AND SINTERED SILICON CARBIDE BY X-RAY TOMOGRAPHY

W. D. Friedman, P. K. Hunt
BP America, Cleveland, Ohio

R. W. Ohnsorg
Carborundum Co., Niagara Falls, New York

ABSTRACT

X-ray tomography is an easy and accurate method for nondestructive measurements of density and porosity distributions in green and sintered SiC making it a useful aid for process development and control. The detrimental effect of beam hardening on the density profile can be compensated for by embedding parts in water, sand, or potassium iodide solutions depending on whether they are green, porous, or sintered. Calibration for absolute density is done with samples of uniform density, or by comparing with changes in average density as is shown for sintered rotors before and after HIPing. Peak densities are also found to be within 1% of theoretical, even before HIPing. Green inspection is emphasized, e.g., determining the effectiveness of binder removal by impregnating the resultant porosity with xenon gas and expressing the change in X-ray absorption as an image of percent porosity.

INTRODUCTION

The current trend in modern manufacturing is not to place the emphasis on inspection of the final product, but to have a manufacturing process that, if it is controlled within observable limits, will produce the desired output with very few flawed parts.

Statistical techniques are used to monitor the process and determine if it is in control. However, this requires having relevant parameters that can be measured throughout the stages of processing, preferably at the earlier stages. The goal is to detect a deviation from the allowed statistical variation as soon as possible. Ceramic processing can only benefit from this approach when tools are developed to measure relevant properties in the green state. X-ray tomography has been suggested for imaging ceramics for many years [1-4]. Useful attributes are its high sensitivity to density variations and the quantitative nature of its images. Table 1 relates the critical properties that need to be examined at early processing stages to several variations of tomography that are suited for these inspections.

TABLE 1 Uses of X-Ray Computed Tomography for Ceramic Processing

PROCESS	CRITICAL PROPERTIES	INSPECTION TECHNIQUE
Forming Green Body	Binder Distribution	Dual Energy CT
-Injection Molding		
-Isopressing	Density Uniformity	CT
-Casting		
Binder-Removal	Distribution of Porosity	CT with Xenon Gas Impregnation
Sintering	Cracks	High Resolution CT
	Density Uniformity	CT
HIPing	Density Uniformity	CT

X-RAY TOMOGRAPHY

X-ray computed tomography (CT) has become a valuable medical diagnostic technique because it gives easy-to-interpret pictures of an internal cross section without the complications created by shadows of bones and other body parts that are not of interest. It is also very sensitive to small changes in density. These advantages are also important for industrial nondestructive testing. A feature not widely used by the medical community, but of great significance for ceramics, is the quantitative nature of the images. Each pixel in the display numerically represents the linear X-ray absorption at that point. With materials of uniform composition, this can be easily calibrated to yield a direct measure of the interior density. Spatial resolution can range from 0.25 x 0.25 x 2 mm to 1 x 1 x 10 mm in commercial medical machines with density accuracies of better than 1%.

X-ray techniques based on transmission respond to the linear attenuation coefficient μ where

$$\mu = \mu/\rho \times \rho \qquad (1)$$

The first term is the mass absorption coefficient and is a function of material composition and X-ray energy. It is not possible from a measurement at one X-ray energy or X-ray tube voltage to determine if μ has changed due to a variation in the material density (ρ), or whether the material composition changed thus modifying (μ/ρ).

The early application of CT to ceramics concentrated on sintered material where compositional variations are less likely. These problems required finding techniques for making accurate quantitative measurements of density with medical machines, which were being used on materials of higher absorption than they were designed for. Artifacts in the images made the data difficult to interpret. With the need to inspect green materials came the need to determine compositional variations. If measurements are made at several X-ray energies it becomes possible to separate out the effect on the transmission of μ/ρ and ρ. The most common technique for medical machines is to use two settings of the tube voltage and then express the transmission in terms of the thickness of two different materials, usually referred to as basis materials [5]. Pre-reconstruction methods perform the separation using the raw data and have the advantage of correcting for beam hardening at this stage. If the system does not provide access to the raw data it is still possible to do a post reconstruction separation using image data taken at two energies[6].

The raw data, which is the transmitted intensity at the high energy I_H and the lower energy I_L can be described by two simultaneous equations. The two unknowns which are the equivalent thicknesses of the two basis materials, d_1 and d_2, that will produce the measured transmission at the two energies. The equations 2a and 2b can be solved numerically by a table look-up process.

$$I_H (d_1, d_2) = \int_H I_O(E) \bullet \exp(-\mu_1(E) \bullet d_1 - \mu_2(E) \bullet d_2)\, dE \qquad (2a)$$

$$I_L (d_1, d_2) = \int_L I_O(E) \bullet \exp(-\mu_1(E) \bullet d_1 - \mu_2(E) \bullet d_2)\, dE \qquad (2b)$$

There will be one solution only if the equations are independent. This requires that μ_1 and μ_2 of the two basis materials change by different ratios as the energy changes. This is best achieved if photoelectric absorption is the dominant mechanism in one material at the lower energy, and Compton scattering dominates absorption in the other material when the X-ray energy is increased. Typical basis material pairs are water and calcium, or plastic and aluminum. The Siemens medical machine uses an upper energy of 125 kV. To get into the photoelectric regime requires a lower energy, 85 kV being chosen in the tradeoff between a larger photoelectric cross section and the poor signal/noise ratio found with reduced transmission at low energy. The problem is reduced in an industrial CT machine, since the part can be subjected to high X-ray dosage. Medical machines, however, do not permit operation at conditions harmful to a patient. Thus in spite of the many advantages of dual energy operation, there is still a need to operate medical machines in single energy mode to reduce image noise.

ARTIFACTS AND BEAM HARDENING

Images of industrial samples obtained on medical instruments frequently show artifacts due to a phenomenon known as beam hardening. Beam hardening occurs when polychromatic X-ray sources are used, as is the case for medical CT scanners. It is due to a preferential absorption of the lower energy X-rays as they pass through the sample. If a monoenergetic source were used, beam hardening could not occur. Beam hardening produces an apparent high density ring around the perimeter known as cupping, and also streaks along large path lengths. In addition, the measured attenuation coefficients are lower than the true values because the effective energy has shifted. These are problematic artifacts because their presence is not always obvious. Instrument manufacturers get around this problem by using compensating algorithms. But the algorithms have been optimized for the attenuation of X-rays by the human body and are not effective for materials having greater absorbing power. The problem is most severe for sintered materials; green ceramics are less affected. However since density uniformity is such a critical parameter for evaluating ceramics, techniques that alleviate some of the problem have been developed.

Early CT systems used water bags to surround the skull to prevent artifacts. This was extended in our case to immersing the samples in media having attenuating properties close to that of the sample. Cupping still occurs, but now it appears in the image at the perimeter of the immersion medium container instead of the perimeter of the sample. Two immersion media that have yielded good results are sand and aqueous potassium iodide solution. Sand has been most useful for green material while potassium iodide was required for sintered parts. The usual practice is to increase the concentration of the KI until its measured CT value is several hundred units below those for the part giving the background a dark appearance. The numerical values are also sufficiently separated on a histogram plot to prevent overlap between those for the part and the background. For large parts, which require large amounts of KI solution for complete immersion, the attenuation of the iodine leads to low transmitted flux and nosier images. Besides reducing the KI concentration, some improvement can be obtained by filling some of the path length with less absorbing material. A molded plug of paraffin was used successfully in the central region of a tubular duct. A more severe problem is the change in CT values with increasing KI concentration. This is shown in Fig. 1 where the observed CT value inside a sintered SiC part changes with concentration.

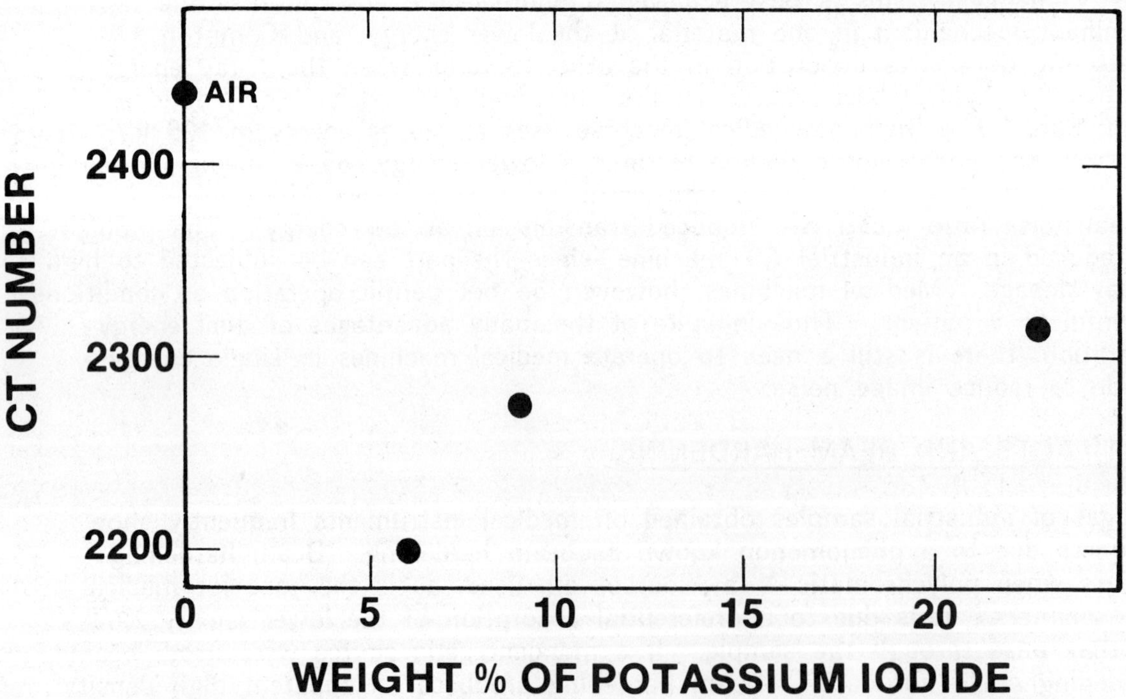

Figure 1 Change in CT Value of a Sintered SiC Part Immersed in KI Solutions

The initial drop in observed absorption is due to the X-ray beam spectrum becoming harder and thus less absorbing. There is a beam hardening correction built into the software which tries to raise the values in the central region. It is suspected that this is over compensating and causing the apparent increase starting at 6%. We find 9% KI gives good results for sintered parts and the result is less sensitive to changes in concentration. On dual energy machines, 3% was found to be sufficient. Calibration is complicated by the measured CT value of the immersed material changing with the density of the fluid. This also means the CT value will change as the part size varies. Thus the CT numbers must be calibrated for a given part immersed in the same fluid. Since the system is linear, calibration is not required to determine the amount of relative change in similar parts.

Another approach to compensating for cupping is to use a material closer to the absorption characteristics of the parts being examined for the calibration phantom, which is used to set the coefficients for the software beam hardening correction. This is most practical with your own machine, in our case a used Deltascan 100 by Technicare. The material must be known to be uniform, sometimes making it necessary to find a substitute material. Medical machines are always calibrated for people with water. We have used quartz to calibrate for rocks, and aluminum can be used for sintered SiC. Quartz scanned with a water calibration shows severe cupping in Figure 2a. If the beam hardening algorithm is calibrated for quartz, then it effectively does its job when quartz is rescanned in Figure 2b. CT numbers are the linear absorption coefficient normalized to zero for the calibration material and air corresponds to -1000. This is responsible for the change in the scale of the vertical axis.

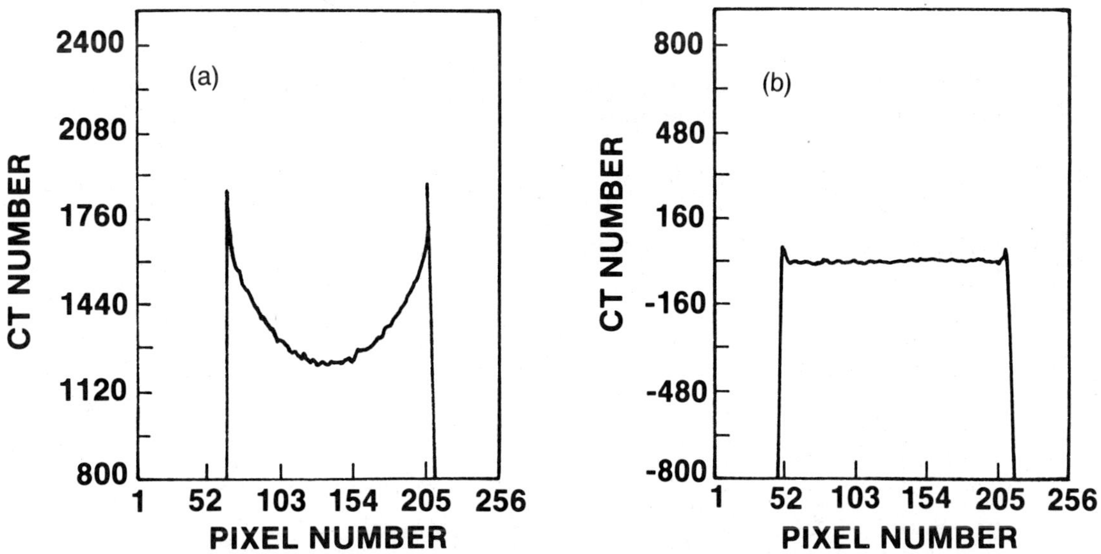

Figure 2 Line profile of CT numbers across 3" diameter fused quartz phantom, a) calibrated to water, b) calibrated to quartz.

Sometimes it is difficult to acquire uniform samples of the desired material. What is needed is the calibration constant α which converts CT numbers to actual density

$$\rho(g/cm^3) = \alpha(CT + 1000) \qquad (3)$$

It could be defined for nonuniform material if the CT value could be summed over all regions of the part

$$\bar{\rho} = \frac{M}{V} = \frac{\Sigma\rho_i\delta v}{V} = \alpha\delta v \frac{\Sigma(CT_i + 1000)}{V} \qquad (4)$$

This will not work when some regions are too thin to be read correctly, which was the case for a rotor with thin blades. This problem was solved by comparing CT data for the same part, before and after its density was increased by hot isostatic pressing (HIPing). It was noted that only the central region changed in density as depicted in Figure 3.

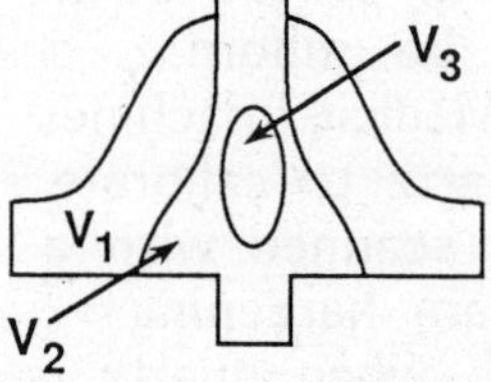

V_1 = Blade Volume - Fixed Density
V_2 = Outer Hub - Fixed Density
V_3 = Central Hub Volume - Changing Density

Figure 3 Schematic of Volumes Used in Model for Density Calibration

This part was modeled using a density ρ_1 for volumes 1 and 2 that does not change due to HIPing, and a variable density for V_3 whose change is responsible for the change in average density found by immersion measurements (Archimides principle). Thus the average density is found from the masses of each region

$$\bar{\rho} = \frac{\rho_1(V_1 + V_2) + \rho_3 V_3}{V_1 + V_2 + V_3} \qquad (5)$$

This can be expressed in terms of α from Eq (3), and the ratio of the variable density volume to the rest of the hub is defined in Eq (6)

$$k = V_3/V_2 \qquad (6)$$

to yield an equation with two unknowns α and k

$$\overline{CT_3} = \frac{1+k}{.6\ \alpha k}\ \bar{\rho} - f(CT_{max}, k) \qquad (7)$$

The average CT value for region 3 is measured before and after HIPing to solve for α. The result for 9 rotors was

$$\alpha = 9.321 \pm .04\ X10^{-4} \quad g/(cm^3\text{-CT numbers})$$

CT images of the hub region (Figure 4) show improved uniformity after HIPing, and with the calibration it was determined that the peak densities near the outer part of the hub were already close to theoretical density and thus would not be expected to change.

The images were taken under identical conditions: same part orientation and 9% KI solution. The KI solution also prevents artifacts caused by sharp edges particularly in parts with noncircular cross sections. The blades of the rotor are not visible since they are thin and the partial volume effect causes them to appear less dense. However, without KI, they produced artifacts that appear in the hub region.

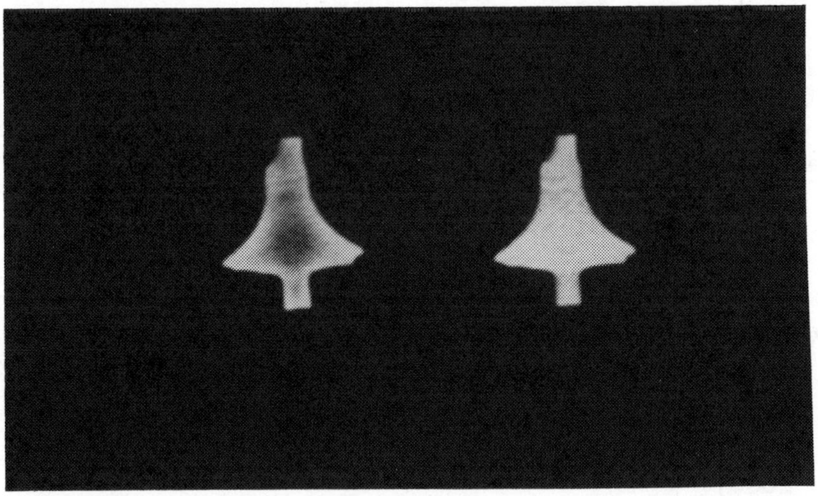

Figure 4 Tomography of SiC Rotor Hub Before (left) and After HIPing (right). The hub is 6 cm across and the darkest regions in the hub before HIPing represent densities that are 12% lower than the maximum density at the periphery

EFFECT OF DUAL ENERGY ON ARTIFACTS

One approach to correcting for beam hardening is to measure the transmission at two different accelerating voltages of the X-ray tube for many combinations of thicknesses of two basis materials. Since this calibration data is taken with the spectrum of the actual source, the details of the spectrum are not required, only that it be reproducible. Preprocessing the transmission data from the tomography system before doing reconstructions will allow the equivalent thickness of each basis material to be determined from the calibration data, thus it is the correct thickness and there are no errors due to beam hardening. Reconstructions are done using the log of the transmission, which is equivalent to the thickness for an image reconstructed for one basis material. Besides compensating for beam hardening, these images also have fewer streak artifacts emanating from edges in the object.

POROSITY MEASUREMENTS

Effective removal of the binder from green material is important for good sintering. Lower density regions detected in the center of the rotors suggested that the binder was not being fully removed from the core. This would be manifested as a region of lower porosity. Xenon impregnation has been suggested as a method for showing open porosity as the apparent density increases with xenon pressure in an X-ray tomography scan. If the porosity (p) is defined as,

$$p = \text{pore volume/total volume} \tag{8}$$

then the total X-ray absorption is a linear combination of the xenon in the pores and the material absorption.

$$\mu_T = p\,\mu_{xe} + (1-\rho)\,\mu_m \tag{9}$$

The test is done with the porous part in an aluminum pressure vessel and μ_{xe} is increased by raising the xenon pressure in the container. As long as the process is linear, the porosity can be found from the slope of the total absorption in the part plotted vs. the absorption in the pure xenon surrounding the part. Figure 5 shows that this slope is not constant due to beam hardening in the xenon. Data is more reliable for the higher xenon pressures and the slope or porosity is found to be lower for points near the center of the part where there could be residual binder.

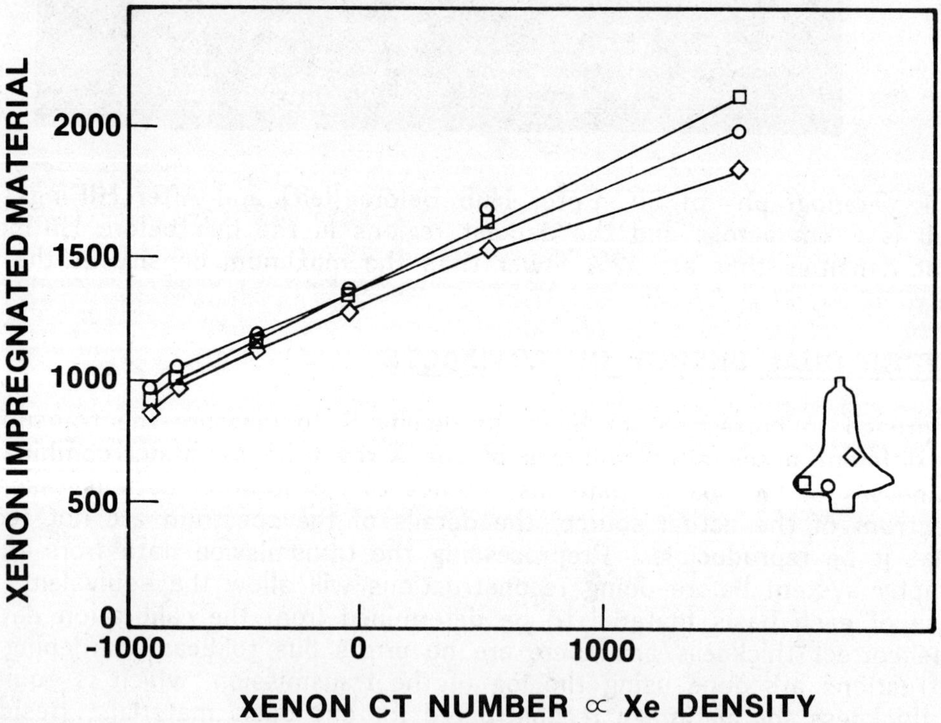

Figure 5 The variation of CT measured absorption in porous xenon impregnated material with increasing xenon pressure. The slope, and hence porosity, varies with where the CT values are measured in the hub.

An image computed by taking the pixel by pixel ratio of two CT images taken at the two highest xenon pressures will thus directly show the distribution of porosity in the part. The porosity is found to be 0.49 near the outside of the part. This value is close to that expected from the mix of powder and binders used for injection molding the part.

CONCLUSION

X-ray tomography is a powerful tool for characterizing ceramic materials. It offers opportunities to make unique measurements of density distribution and porosity distribution that are particularly important for green materials. However, caution must be exercised in interpreting results if there are possibilities of artifacts. Technique development and calibration are essential to insure high quality data. Tomography systems designed for industrial applications are just starting to become more available. They will allow the techniques developed for medical machines to be improved due to the greater flexibility of the industrial machines and the removal of the X-ray dose limitations.

ACKNOWLEDGEMENTS

This work was done in support of subcontracts with Allison Gas Turbine Division of General Motors Corporation (AGT100) and Garrett Auxiliary Power Division of Allied-Signal Corporation (AGT101), which are sponsored by the U. S. Department of Energy and managed by the NASA-Lewis Research Center.

REFERENCES

1. W. A. Ellingson and H. Berger, "Three-Dimensional Radiographic Imaging," in Research Techniques in Nondestructive Testing, Vol. 4, R. S. Sharpe ed., Academic Press, New York (1980), pp. 1-38.

2. W. A. Ellingson and M. W. Vannier, "Application of Tomographic Imaging to Structural Ceramics: Green-State Monolithics and Ceramic-Ceramic Composites," in Technical Digest of Topical Meeting on Industrial Applications of Computed Tomography and NMR Imaging, Optical Society of America, Aug. 13-14, 1984, Hecla Island, Manitoba, Canada, pp. MB6-1 to MB6-4.

3. J. W. Kress and L. A. Feldkamp, "X-ray Tomography Applied to NDE of Ceramics," Am. Soc. Mech. Eng., 83-GT-206 (1983).

4. P. Reimers and J. Goebbels, "New Possibilities of Nondestructive Evaluation by X-ray Computed Tomography," Mater. Eval., 41 pp. 732-737 (1983)

5. R. T. Alvarez and A. Macovski, "Energy Selective Reconstructions in X-ray Computed Tomography," Phys. Med. Biol., 21, 733 (1976).

6. H. J. Vinegar and S. L. Wellington, "Tomographic Imaging of Three-Phase Flow Experiments," Rev. Sci. Instrum., 58, pp. 96-107 (1987)

X-RAY MICROFOCUS AND DYE PENETRANT TECHNIQUES FOR CRACK DETECTION IN CERAMICS

H. Kölker, P. Henze, K.A. Schwetz, A. Lipp

ESK-GmbH, Kempten, W.-Germany

ABSTRACT

Capillary flaw inspection with liquid penetrant and x-ray microfocus radiology are applied and combined to detect microcracks and flaws in ceramic samples and turbocharger rotors produced by injection molding and sintering. Green and sintered bodies are inspected with similar detection limits. Cracks with a surface opening of less than 0.1 µm are easily detected. Crack fractography shows that they are completely filled with liquid penetrant. Inspection with x-rays is discussed with respect to the absorption coefficient and the sensitivity is presented as a function of anode voltage and material thickness. Image processing is applied to produce x-ray pictures on a TV-screen approaching the quality of pictures taken by photographic film methods. Examples of cracks in ceramic rotors are presented.

INTRODUCTION

For the development of new ceramic materials and components, appropriate testing procedures are required that can be applied during the various steps of component preparation. Thus specimens containing flaws can be sorted out at an early stage and time and costs can be saved. To fulfill their purpose such tests have to be non-destructive. Standard methods like ultrasonics, vibrational analysis and conventional x-ray applications can serve as first step routines, but they are not yet sufficient for ultimate requirements. Much better suited is the microfocus x-ray technique. It offers the advantage of being able to analyse both green and sintered bodies and gives the highest resolution of different defect types. It is well matched by liquid dye penetrants that give very clear information on surface cracks and flaws.
Since 1984 ESK has been used with a special microfocus tube, an image intensifier, a video camera and a monitoring system. In addition, the actual equipment includes a real time digital image processing unit for integration, different correction routines, and filtering. This paper reviews the recent improvements in our system.

SiC- and Si_3N_4 turbocharger rotors (1 - 6) are used to demonstrate these items. They are supplemented by standard bending specimens and AlN-slices (7) to discuss the resolution capability.

X-RAY MICROFOCUS ANALYSIS

Because of its integral nature x-ray radiography is helpful for a fast and effective detection of different defect types.
X-rays travelling through a sample will suffer a different absorption when they hit a pore or an inclusion with a different absorption coefficient than the rest of the material. A necessary prerequisite for good resolution is the use of a microfocus (8) as shown in Fig. 1 and already described by us in (9, 10).
The system is capable of high magnification which not only leads to a high geometrical resolution but also reduces x-ray scattering effects because of the large distance between the object and the image intensifier or film.
Thus, having established a situation where a sharp image can be formed on the screen even of very small objects without the danger of blurred shape (like in standard x-ray foci) one can proceed to discussing the contrast resolution. That is the detection limit based on the different absorption coefficient of an inclusion, crack, or pore with respect to the flawless material.

DETERMINATION OF THE LINEAR ABSORPTION COEFFICIENT

The intensity I of x-rays travelling through a material is attenuated according to the equation:

$$I = I_o \cdot e^{-\mu \cdot d} \qquad (1)$$

μ = linear absorption coefficient (cm^{-1})
d = thickness of the material (cm)

The linear absorption coefficient μ depends on the material and on the wave length of the radiation.
The mass absorption coefficient μ/ρ ($cm^2 \cdot g^{-1}$) can be taken from standard tables and transferred to the linear coefficient by multiplication with the density ρ ($g \cdot cm^{-3}$) of the material.
Unfortunately radiography uses white radiation, which means that at a given anode voltage a whole continuum of x-rays is produced, the so called "Bremsstrahlung". So at this stage it is completely unknown to which wavelength one should refer when looking at the absorption coefficients.
Fortunately however, the absorption strongly increases with increasing wavelength, so one can expect with sufficiently thick samples that the smallest wavelength will dominate the beam after it has left the material.

The smallest or cut-off wavelength $\lambda_{min.}$ is related to the anode voltage V (kilovolt) by the equation

$$\lambda_{min.} = 12.4 / V \text{ (Angstrom)} \quad (2)$$

So at a first approximation one can hope to get correct results when working with $\lambda_{min.}$. Fig. 2 proves that for SiC this is indeed so, at least in the voltage range 50-70 kV.
With this knowledge and Eq. 2, mass absorption coefficients from standard tables (11) as a function of wavelength were rewritten as a function of anode voltage (Table 1).

Table 1. Mass absorption coefficients ($cm^2 \cdot g^{-1}$) for various elements as a function of anode voltage

Anode voltage	Elements B	C	N	Al	Si	O
30	0.279	0.354	0.469	1.445	1.744	0.583
40	0.191	0.270	0.281	0.705	0.839	0.326
50	0.142	0.219	0.189	0.404	0.476	0.208
60	0.111	0.185	0.136	0.256	0.299	0.144
70	0.091	0.160	0.104	0.175	0.202	0.106
80	0.076	0.141	0.082	0.125	0.144	0.081
90	0.065	0.126	0.066	0.093	0.107	0.064
100	0.056	0.114	0.055	0.072	0.082	0.051

In a second step these values are transferred to the mass absorption coefficients of compounds - e.g. Si_3N_4 - by multiplying the absorption coefficients of the ingredient atoms with their mole fraction in the compound and adding them up. In a last step the result is multiplied with the density of the compound and thus transferred to linear absorption coefficients (Table 2).

Table 2. Linear absorption coefficient "μ" of various ceramics as a function of anode voltage. For practical purposes "μ" is given in (mm^{-1}).

Material	Anode voltage 30	40	50	60	70	80	90	100
B_4C	0.094	0.067	0.051	0.041	0.034	0.029	0.025	0.022
SiC	0.425	0.214	0.128	0.085	0.061	0.046	0.036	0.029
Si_3N_4	0.395	0.197	0.116	0.075	0.052	0.038	0.029	0.023
Si	0.406	0.195	0.111	0.070	0.047	0.034	0.025	0.019
AlN	0.356	0.179	0.106	0.069	0.048	0.035	0.027	0.021

In order to corroborate the assumption that the x-ray beam is dominated by the shortest wavelength of the initially white continuum, an experiment was performed with SiC plates of different wall thickness. For three anode voltages (50, 60 and 70 kV) the attenuation of the x-ray beam was calculated for different wall thicknesses using Table 2 and equation 1. The result is presented in Fig. 2 by the drawn lines and compared with experimental results (square dots) where the attenuation was measured using the digital image processing unit as a photometer.
The correspondence of theory and experiment is striking and corroborates the assumption above.

CONTRAST RESOLUTION

Our image processing unit can resolve intensity differences of one percent. If we assume a pore or a notch of depth Δd in an SiC-plate of thickness d, then the x-ray beam going through this pore is attenuated according to

$$I_1 = I_o \cdot e^{-\mu(d-\Delta d)} \qquad (3)$$

The neighbouring beam travelling through flawless material is attenuated according to

$$I_2 = I_o \cdot e^{-\mu \cdot d} \qquad (4)$$

The difference ΔI between these two intensities causes a certain contrast on the screen or film and is detected if

$$\Delta I/I_o = e^{-\mu(d-\Delta d)} - e^{-\mu \cdot d} > 1\ \% \qquad (5)$$

Based on prior experience we assumed that $\Delta d/d$ must be greater than 3 % to fulfill Eq. 5 for typical SiC samples. With this assumption $\Delta I/I_o$ was plotted as a function of "d" in Fig. 3 for different linear absorption coefficients expressed in anode voltages (kV) according to Table 2.

Fig. 3 shows that the contrast exceeds the 1 % mark indeed, but in addition it shows quite clearly that there are only certain ranges of wall thickness "d" where this relation holds.

To summarise the results:
a) a contrast resolution of 1 % corresponds to a 3 %-pore in SiC
b) to get optimal contrast a given wall thickness requires a certain voltage according to Fig. 3
c) the 3 %-pore relation is valid for a wide range of wall thickness. Thus, the detection limit of pores can be lowered well below the 100 µm-limit in thin samples and flaws should be detected by radiography that are responsible for catastrophic failure in typical engineering ceramics with a

bending strength in the range 300-700 MPa.

EXPERIMENTAL RADIOGRAPHIC RESULTS (IMAGE PROCESSING)

Fig. 4 shows the tungsten filament of a light bulb at a magnification of 30 x. The 10 µm filament is clearly resolved giving an indication of the geometrical resolution capability of this method. Metallic inclusions with a high absorption in ceramic materials are made visible equally well. The method can also be applied as an analytical screening for starting powders containing metallic impurities. As opposed to pores, inclusions with much higher absorption than the ceramic are only subject to the geometrical resolution. Their contrast against the background is always very high.

In order to corroborate the theoretical result that pores measuring 3 % of the wall thickness can be resolved ceramic samples were notched to different depths. Fig. 5 a shows an SiC standard bending specimen, 3,5 mm wide and 4,5 mm high. Three notches measuring 3.5 %, 2.2 %, and 1.3 % of the height (wall thickness) of the samples were cut across its surface. The microfocus picture at a magnification of 7 x with the beam travelling along the height of the specimen clearly resolves the 3.5 % and 2.2 % notch and still gives a hint at the 1.3 % notch. Fig. 5 b shows the corresponding intensity along the sample and across the notches clearly revealing that there is a signal at the 1.3 % notch above the noise level.

The same procedure as in Fig. 5 was carried out for a much thinner sample, an AlN-slice (7) 0.53 mm thick. Since the notches are very shallow (Fig. 6) it was not easy to get a very precise and uniform depth. So the figures given are somewhat averaged values. Again it is seen that a 4 % notch can clearly be resolved. In the corresponding intensity picture (Fig. 6 b) it is seen that all notches are well above the noise level.

Fig. 5 + 6 depict the capabilities and the limits of x-ray microfocus analysis. By comparison the image processing technique has considerable advantages over the film technique. Ceramic parts can be examined "on line" allowing manipulation during the analysis. A prominent example of this advantage is the search for cracks. A crack is very similar to a two-dimensional pore. Compared with the wall thickness of the ceramic component its third dimension is not worth while being mentioned so it usually cannot be detected in that direction. However, with online inspection the component can be rotated in front of the x-ray tube until the crack becomes visible. Furthermore, with image processing using shading correction and camera automation one can adjust the brightness of those parts that differ in thickness. With x-ray films there is always danger of overexposing thinner parts. Last but not least rotation yields the correct position of flaws inside the sample which is not possible with one film because one would only get the projection of

the flaw.
Fig. 7 and 8 show two examples of cracks in sintered SiC-turbocharger rotors. The cracks arose from residual stresses after injection molding and dewaxing. Flaws like these normally show up immediately after the dewaxing step so the component can be sorted out and the sintering step can be saved. The cracks can easily be distinguished from the traces of the rotor blades which are very prominent in Fig. 8.
Fig. 9 shows a sintered Si_3N_4-turbocharger rotor with a bore in the hub. Beside the higher intensity at the bore the image processing reveals other shaded areas of slightly increased intensity which is also corroborated by the intensity plot in Fig. 9 b. Obviously there are density variations in the hub.
Fig. 10 is the image of a SiC-nozzle before sintering. Beside the central bore it reveals some large inclusions (dark spots) and equally large pores (bright areas). The component, of course, was discarded.

DYE PENETRANT TEST

Capillary flaw inspection with liquid dye penetrants is a method for detecting those cracks and pores in ceramic materials that are open at the surface (12). The components are impregnated with a liquid dye and then washed with water. Thus, the surface is cleaned but the dye in the cracks remains untouched. Experience shows that after the cleaning procedure some of the dye is sucked out of the crack by capillary forces and spreads a few micrometers across the ceramic surface. By this the crack opening is optically enlarged. The dye will fluoresce when irradiated with ultraviolet light so even the smallest cracks with an opening of less than 0.1 μm could be made visible in our experiments. Fig. 11 shows two examples of sintered SiC-turbocharger rotors with cracks in the lower part of the hub. The method is very effective and reveals even those cracks that are not visible in the x-ray analysis. So it extends the applicability of nondestructive testing to the lower end of small flaws. Our experience shows that it can be applied equally well to moulded green bodies. For the purpose of good reproduction all photographs must be coloured because black and white films are overexposed by scattered ultraviolet radiation.

Fig. 12 shows the full potential of the method. Two SiC specimens for 4-point bending purposes were indented at the top with a Knoop diamond (Fig. 12 a) and the crack was then driven across the top surface and halfway down into the samples (Fig. 12 b) with a bridge method (13).
Investigation with a scanning electron microscope showed that the cracks were on the average only 0.1 μm wide but they can easily be seen with the liquid dye.

Cracks of this kind are created to measure the fracture toughness of materials (13).

It is interesting to see how the liquid penetrant can help to improve the accuracy of such measurements. Fig. 13 shows the fracture surface of two bending specimens after rupture in a 4-point bending arrangement. The right sample was impregnated after indentation with a Knoop diamond and the left sample after the Knoop-crack had been driven halfway through the samples with the aid of a bridge. The dye marks the cracks and the crack length which is important for evaluating the fracture toughness can be measured with better precision than without the dye.

SUMMARY

An overview of the non-destructive testing methods used at ESK to inspect SiC- and Si_3N_4-turbocharger rotors is presented. The paper focuses on x-ray microfocus analysis and liquid dye penetrants.

- A correlation between the linear absorption coefficient in ceramic materials and the anode voltage of the x-ray tube is established and the conditions for optimum contrast resolution are identified.
- Pores measuring more than 3 % of the wall thickness of ceramics parts can be identified by contrast. With metallic inclusions the limit is almost independent of the wall thickness and lies in the 10 μm range.
- On-line inspection and image processing prove to be a powerful analytical tool for finding flaws in engineering ceramics. The method is much more versatile than the film technique and the images are of comparable quality.
- X-ray analysis is supplemented by capillary flaw inspection with liquid dye penetrants. Surface flaws less than 0.1 μm wide can easily be detected. The method is extended to get more precision in fracture toughness measurements.

REFERENCES

(1) M. Langer, J.E. Siebels, H. Heinrich, K.A. Schwetz
SiC radial turbine rotor development - material, design, and fabrication influences
ASME 86-GT-151

(2) K.A. Schwetz, F. Isemann, A. Lipp
Injection moulded sintered turbine components of aluminium doped alpha silicon carbide
Proc. Int. Symp. Ceramic Components for Engines, 1983 Japan, 583-594

(3) K.A. Schwetz, W. Grellner, K. Hunold, A. Lipp, M. Langer
HIP-treated sintered SiC turbocharger rotors
Proc. 2nd Int. Symp.: Ceramic Materials and Components for Engines Lübeck-Travemünde, FRG (Eds. W. Bunk and H. Hausner) Verlag DKG (1986), 1051-1062

(4) M.F. Lasker
Ceramic turbine wheels for turbochargers
SAE Technical paper series 861130
(5) K. Katayama, M. Sasaki, T. Itoh
Development of ceramic turbine rotors
ASME 88-GT-282
(6) J. Greim, K. Hunold, K.A. Schwetz, A. Lipp
Injection moulded sintered turbocharger rotors; a comparison of SSiC with SSN and SRBSN
This meeting
(7) H. Kölker, W. Grellner
Aluminium nitride - a new substrate material
Ceramic forum international, 3/4-88, 75-79
(8) K. Goebbels, H. Reiter, S. Hirsekorn, W. Arnold
Non-destructive testing of high-temperature and high strength ceramics
Science of ceramics 12 (1984), 483-494
(9) W. Grellner, P. Henze, K.A. Schwetz, A. Lipp
Defect evaluation in SiC and Si_3N_4: Destructive and non-destructive tests
Proc. 2nd Int. Symp. "Ceramics Materials and Components for Engines"
in Lübeck-Travemünde, FRG, (Eds. W. Bunk and H. Hausner), Verlag DKG (1986), 815-823
(10) W. Grellner, K.A. Schwetz, A. Lipp
NDT of injection moulded green and sintered SiC turbine parts
ASME 86-GT-272
(11) International Tables for X-Ray Cristallography
Vol. III (1968), 162
The Kynoch Press, Birmingham, England
(12) Eindringverfahren, DIN 54 152 (German Industry Standard)
(13) R. Warren, B. Johannesson
Creation of stable cracks in hardmetals using bridge indentation, Powder Metallurgy 1984, Vol. 27, No. 1, 25-29

ACKNOWLEDGEMENT

Financial support by the German government (BMFT) is greatefully knowledged.

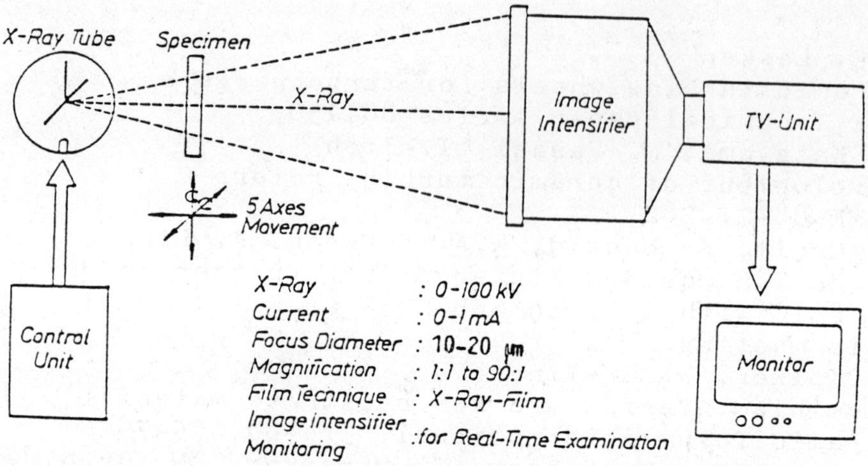

Fig. 1. Arrangement of ESK microfocus x-ray system

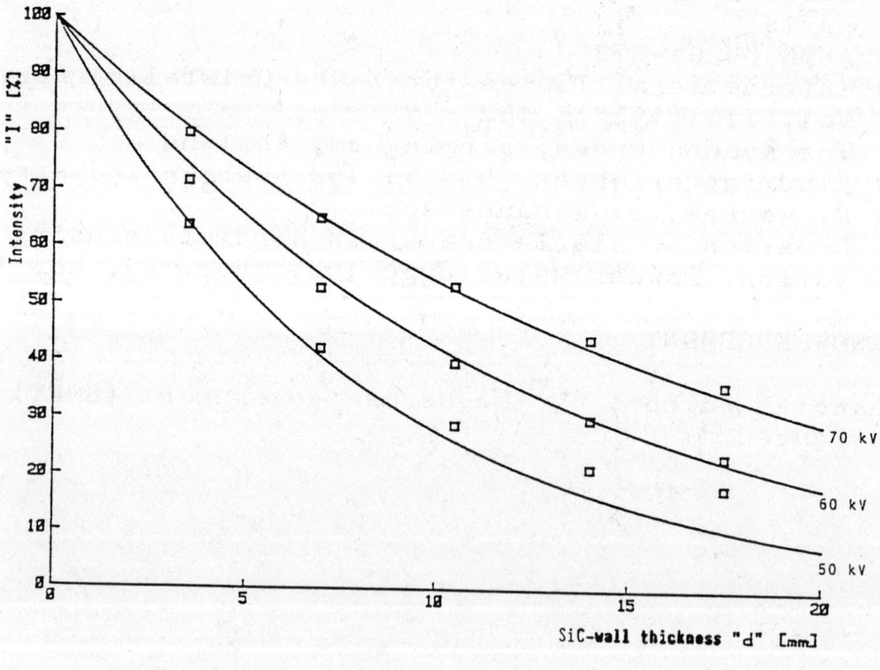

Fig. 2. Experimental verification (square dots) of the assumption (drawn line) that the x-ray beam leaving the sample is dominated by the cut-off wavelength of the initially white continuum.

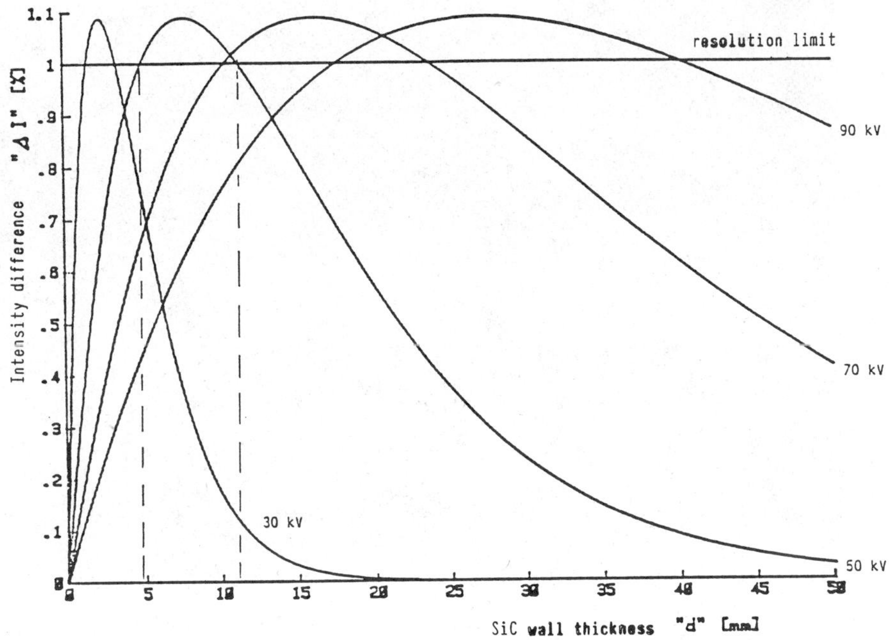

Fig. 3. Intensity difference ΔI between two neighbouring beams one of them going through a pore in the material measuring 3 % of the wall thickness (I_o = 100 %. see Eq. 5)

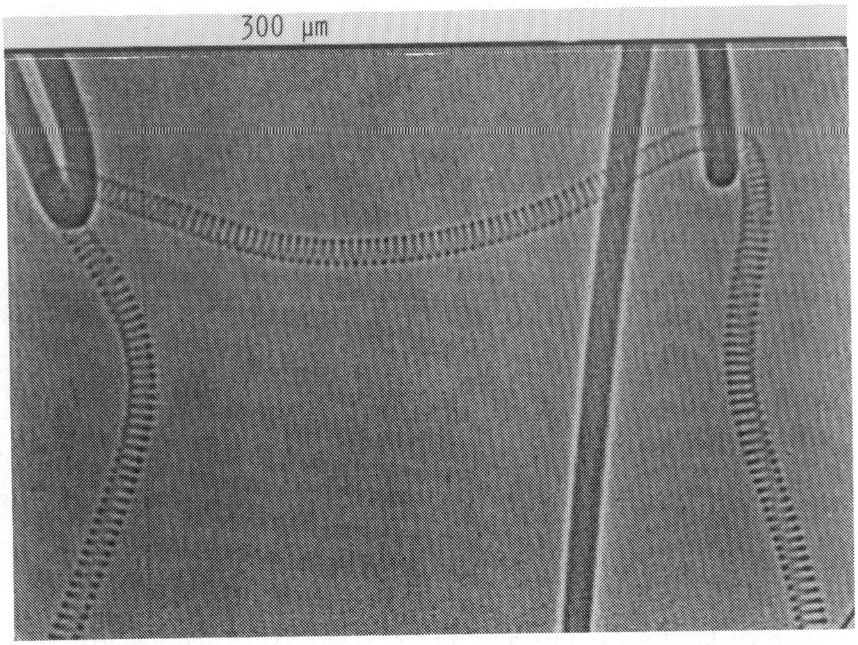

Fig. 4. X-ray microfocus image of a 10 µm tungsten filament (Magnification 30 x)

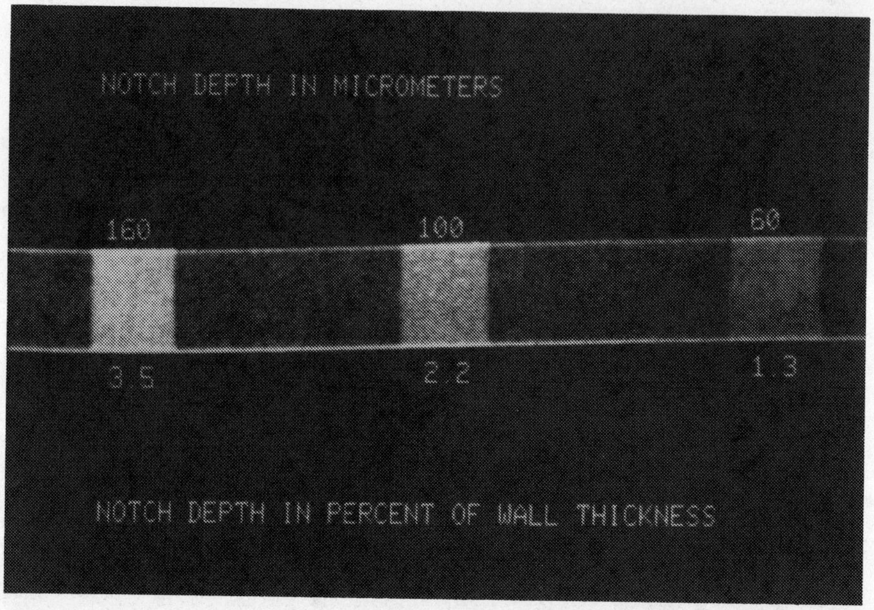

Fig. 5a. X-ray microfocus image of an SiC bending specimen with three notches across its surface.
Wall thickness = 4.5 mm. (Magnification 4.6 x)

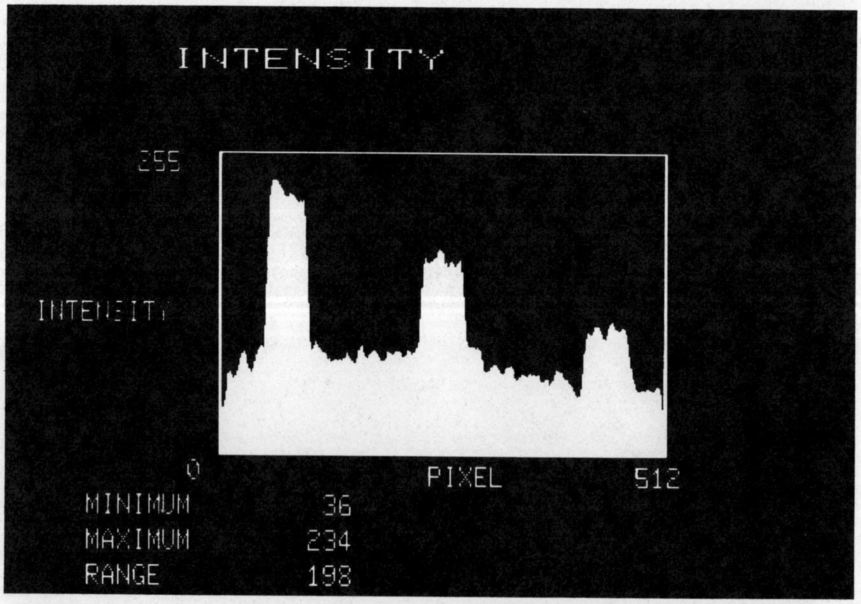

Fig. 5b. X-ray intensity across the SiC-sample of Fig. 5a.

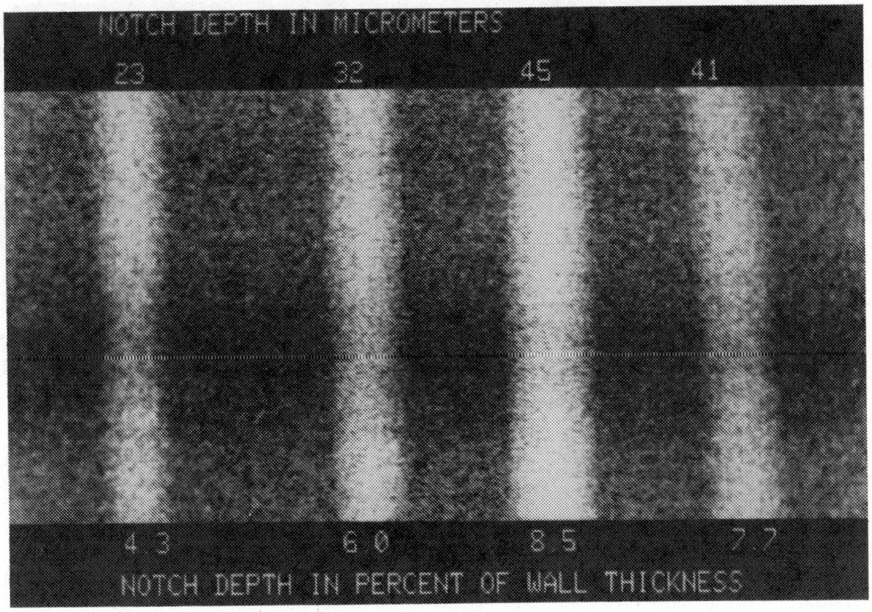

Fig. 6a. X-ray microfocus image of an AlN slice 0.53 mm thick with 4 notches across the surface. (Magnification 3 x)

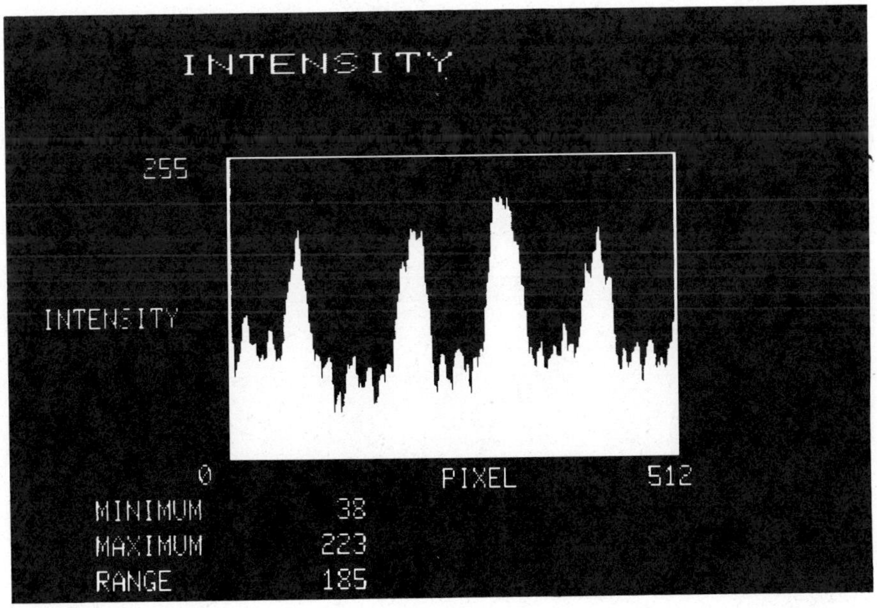

Fig. 6b. X-ray intensity across the AlN slice of Fig. 6a.

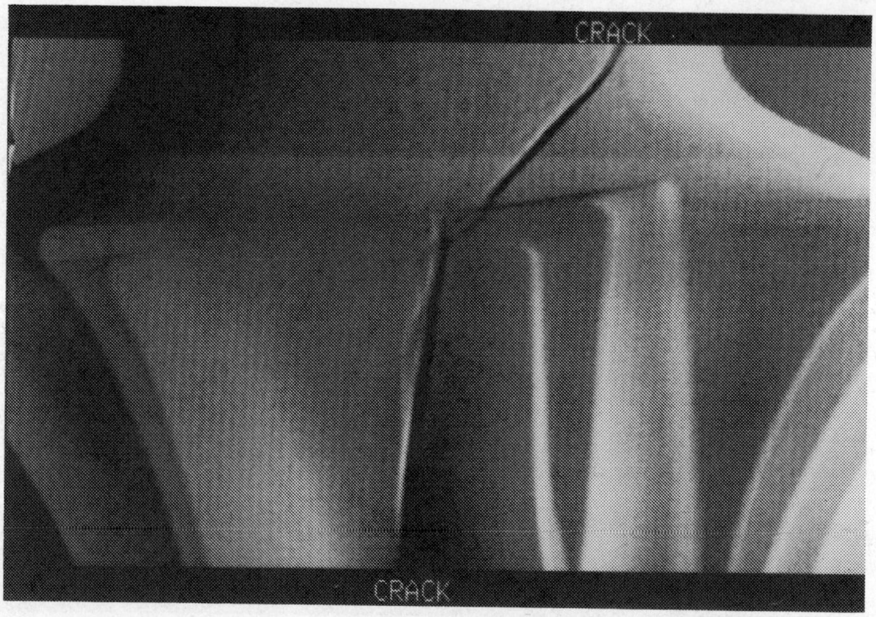

Fig. 7. X-ray microfocus image of cracks in the hub of a sintered SiC turbocharger rotor (Magnification 5 x)

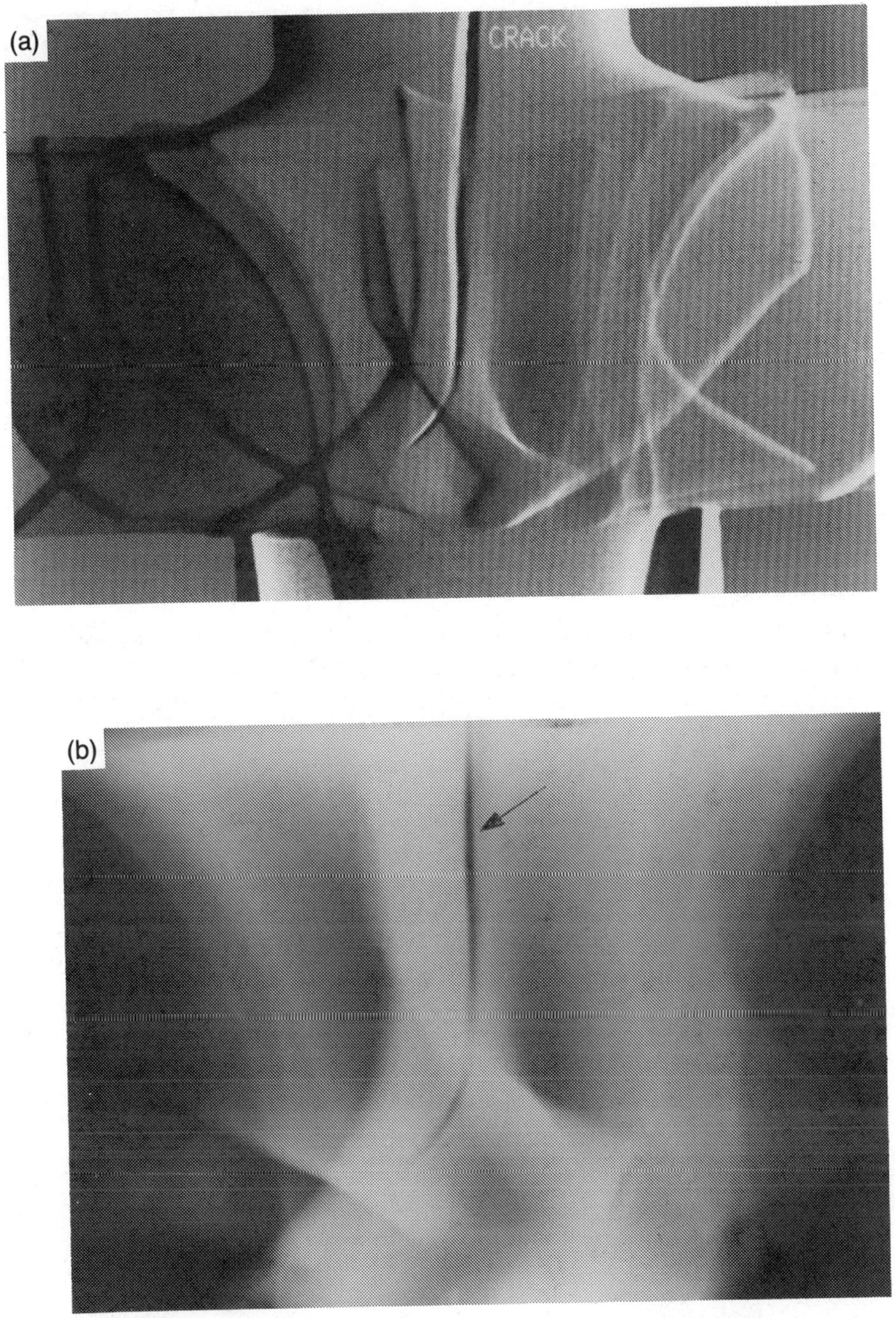

Fig. 8. X-ray microfocus image of a crack in the hub of a sintered SiC turbocharger rotor (Magnification 4 x)
a) Image processing
b) Film technique

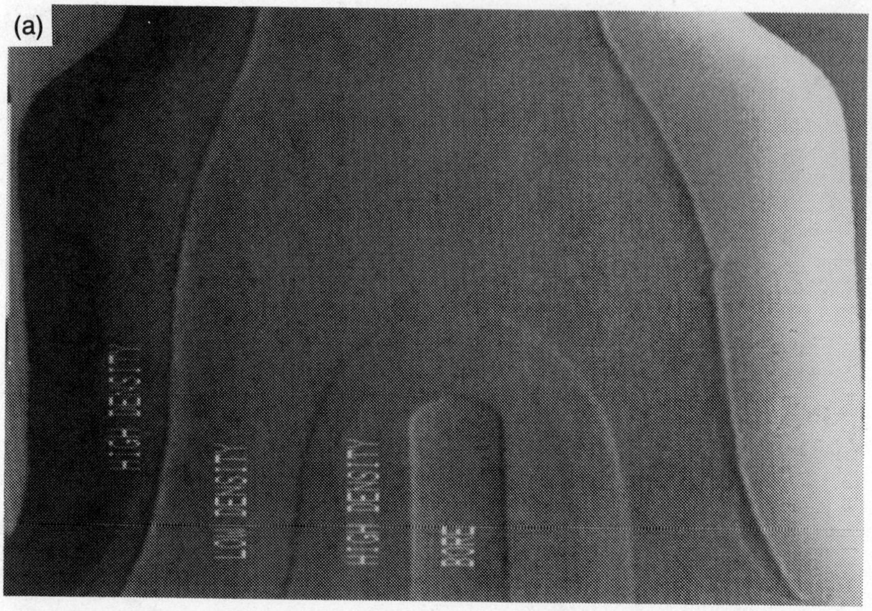

Fig. 9a. Density variations in the hub of a sintered Si_3N_4 turbocharger rotor (Magnification 5.5 x)

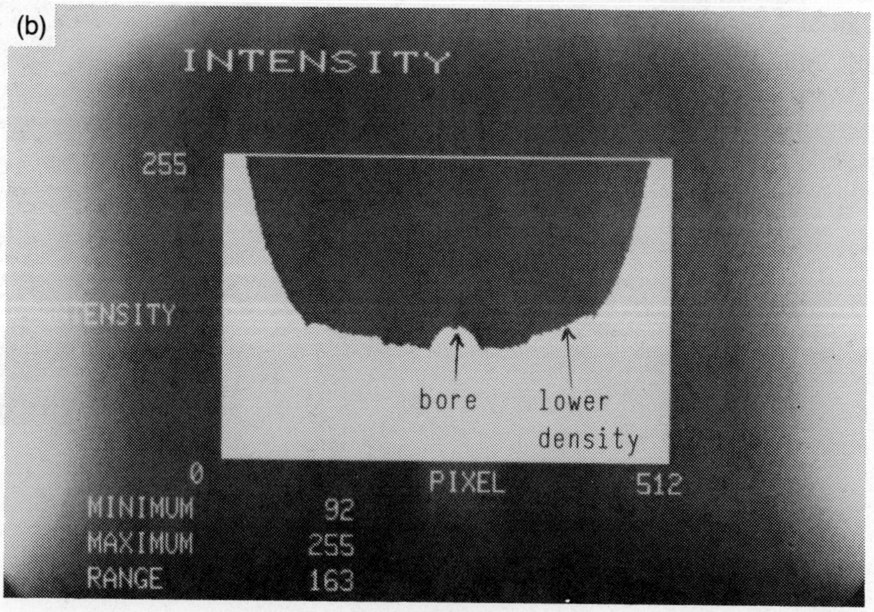

Fig. 9b. Intensity plot of Fig. 9a.

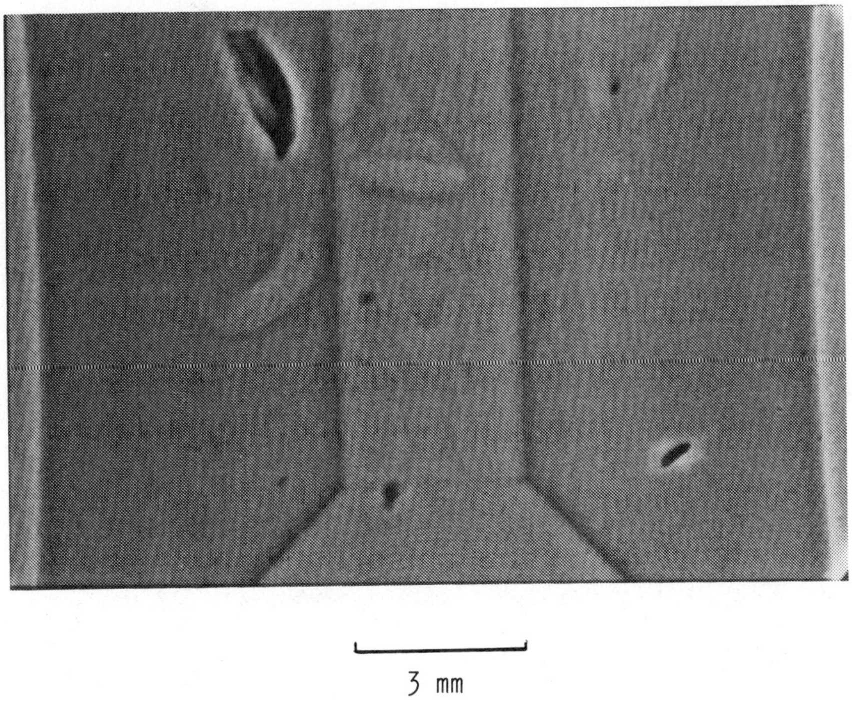

Fig. 10. X-ray microfocus image of a green SiC-nozzle.

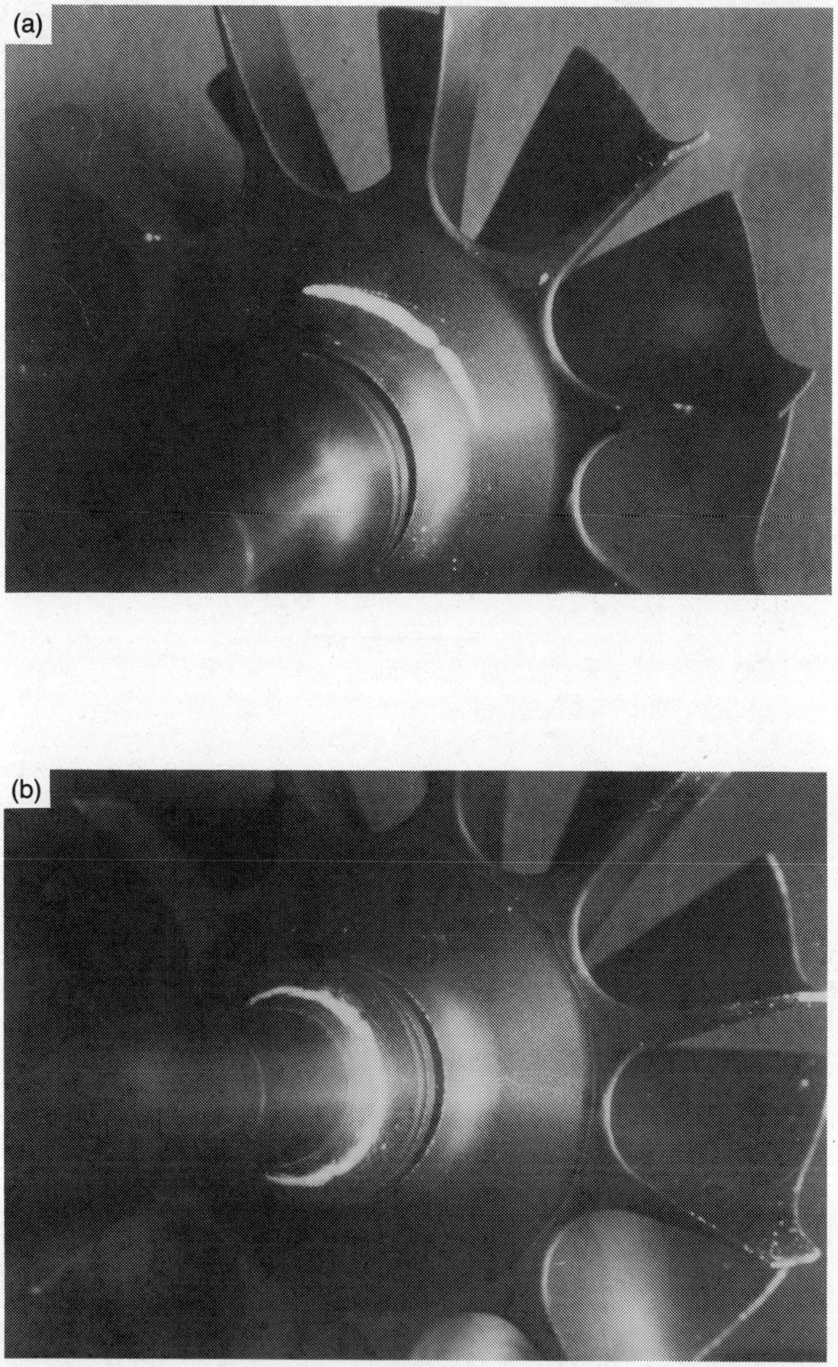

Fig. 11. Two SiC turbocharger rotors.
Cracks in the hub made visible with liquid dye penetrant.

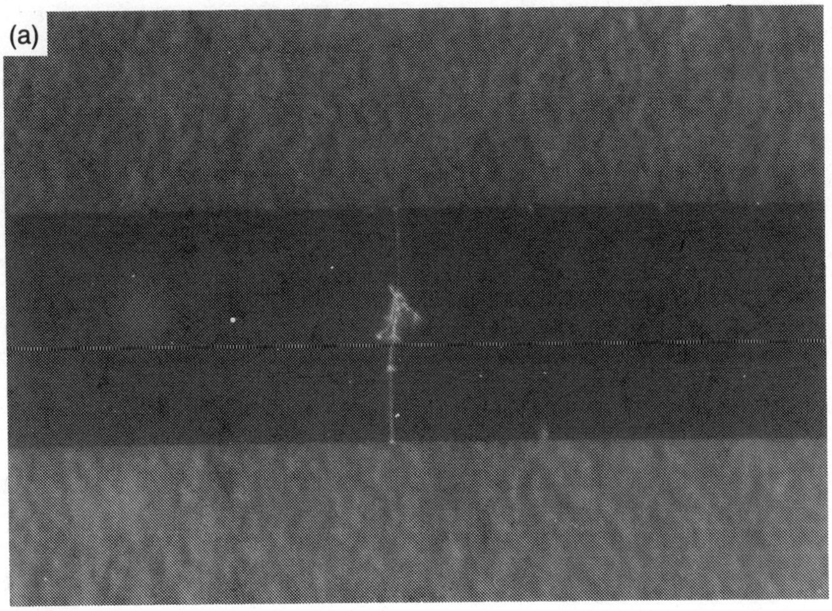

Fig. 12a. Top view of an SiC 4-point bending specimen. Knoop indentation and crack were made visible with liquid dye. (Magnification 7 x)

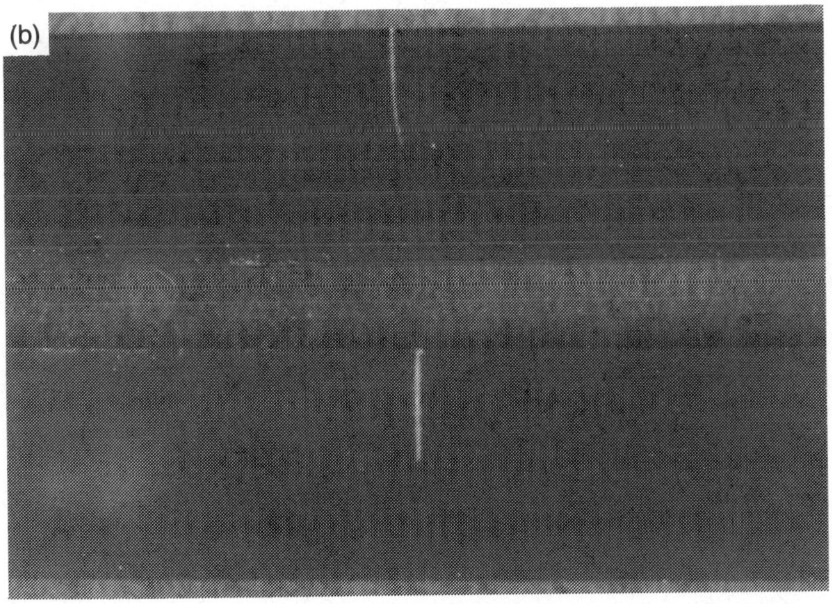

Fig. 12b. Side view of two SiC 4-point bending samples with cracks running half way through the samples filled with liquid dye. (Magnification 7 x)

Fig. 13. Fracture surfaces of two SiC 4-point bending specimens with different cracks before rupture test. The cracks had been impregnated with liquid dye.

NON-DESTRUCTIVE METHOD FOR EVALUATION OF PERMANENT MAGNETS

S. Patiño, E. Amano and R. Valenzuela
Instituto de Investigaciones en Materiales, UNAM
Apartado postal 70-360; 04510 México, D. f.

ABSTRACT

A method for the evaluation of ceramic permanent magnet is proposed. This method is based on the variations of the magnetic permeability as a function of temperature. The Curie point, an indication of the magnetized state of the sample, as well as a quantitative evaluation of its chemical homogeneity are obtained. We present some experimental results obtained on barium hexaferrite samples.

INTRODUCTION

Permanent magnet dc motors have experienced an impressive development in the last few years. This is due, in part, to the fact that permanent magnet motors are more efficient, less expensive and smaller than ac motors. dc motors are more efficient because ohmic losses are practically eliminated. Modern permanent magnets have allowed an important reduction in cost and size of dc motors. In particular, ferrite-based permanent magnets are less expensive and allow a decrease in circuit costs[1]. New permanent magnet based on rare earth amorphous alloys, while having a higher cost, lead to a significant reduction in size of the motor, due to their very high coercivity. These developments point toward a need for evaluation methods of permanent magnets.

In this paper, a method for the evaluation of ferrite permanent magnets is presented. This non-destructive method is based on the measurement of the magnetic permeability as a function of temperature up to the magnetic transition point, or Curie temperature. An analysis of the influence of the heating rate is also presented.

BASIS OF THE APPARATUS

In this experiment the sample, which is preferably in the form of a toroid, is used as the core of a transformer. A small, ac sinusoidal current is established in the primary coil. As a result of the electromagnetic induction, a voltage appears at the secondary coil. It can be shown[(2)] that, if the applied field generated by the current in the primary coil is low enough to remain within the reversible magnetization range, the amplitude of the secondary voltage is given by:

$$V_s = (\omega n_s n_p y/2\pi)(\ln D/d)\mu i \tag{1}$$

where ω = angular frequency
n_s = secondary coil turns
n_p = primary coil turns
y = sample thickness
D = sample outer diameter
d = sample inner diameter
μ = initial magnetic permeability
i = primary current

Thus, for a given sample, with a given number of turns in the primary and secondary coils, and a constant frequency and amplitude of the primary current, the expression (1) may be written:

$$V_s = K.\mu \tag{2}$$

with K = constant. In this form, variations in the magnetic permeability are detected as variations in the amplitude of the secondary voltage. If a rectifier is added to the secondary voltage, in conjunction with a thermocouple, a μ vs T curve can directly be obtained on a X-Y recorder. This apparatus is described in Reference 2, and is adapted for the measurement of soft magnetic ceramics, where initial permeability values are high enough to provide an easily detectable secondary voltage. For permanent magnets, where the initial permeability is very small ($\mu_R \sim 5$), the secondary signal is extremely low. In order to properly detect this signal, a band pass filter and a class A power amplifier is used in the secondary circuit. A saturation detector is proviede to prevent a saturation of the amplifier and hence a deviation from the linear response. The owrking frequency is fixed at 10 KHz, as this kind of magnetic material exhibits quasi-static properties up to several MHz.

RESULTS

Several samples of barium hexaferrite ($BaFe_{12}O_{19}$) were prepared from reagent grade $BaCO_3$ and Fe_2O_3 powders, by the normal ceramic methods.

Results of measurements on those samples were affected by several parameters. Among them, probably the most important is the heating rate. In order to verify the limits of heating rate which give reproducible results, we performed a series of measurements from room temperature, up to 480°C at heating rates ranging, from 1.1°C/min, up to 40.2°C/min with deviations from linearity on the heating rate which were typically smaller than 2%.

Initially, high-rate curves were used to determine the upper limit to obtain a reproducible permeability curve. However, a poorly defined Curie transition, which extends over ∿40°C resulted from the highest rate (40.2°C/min), Fig. 1. For example, at 18°C/min, the magnetic transition encompasses only ∿10°C, Fig. 2. At lower rates, 4.5°C/min, a peak appears just before the transition, Fig. 3. This is the Hopkinson peak, known from the permeability curves of soft ferrites. Finally, for the lowest heating rate, 1.1°C/min, Fig. 4, no significant difference with the 4.5°C/min curve was observed. Thus the upper heating rate to obtain a reproducible curve was fixed at 4.5°C/min.

The difference between curves on Fig. 1 and Fig. 4 can be interpreted on the basis of a thermal gradient within the sample. Due to a temperature distribution within the sample, while one fraction of it is below T_c, another one can reach it, and a third one is above T_c. The signal at the secondary coil is simply an average of all these contributions to the magnetic flux, and the fine effects (the Hopkinson peak, for example) are lost. The effect of the thermal gradient is equivalent to the effect of chemical inhomogeneity(3), as far as the response of the sample is given by a combination of "several" samples.

THE METHOD OF EVALUATION

The parameters that can be obtained by means of this method are the following:

1. Curie temperature.
2. Value of the magnetic permeability for heating and cooling runs.
3. Qualitative evaluation of Chemical homogeneity.

The Curie temperature value allows a determination of the sample composition. In the case of solid solutions, i.e., partial substitution of one of the cations, is is also possible to obtain a good evaluation of the amount of the substitution. For this, it is of course necessary that a calibration curve be available for the particular system.

The value of the initial magnetic permeability is indicative of the magnetized state of the sample. The permeability is proportional to the slope of the hysteresis loop(4) around the remanent state in a magnetized sample, and has a larger value for the unmagnetized case, where it is the true initial permeability. When the sample goes through the Curie temperature, magnetic ordering is broken and in the cooling process, a new magnetic domain arrangement is set when $T<T_c$. The value of permeability thus corresponds to the initial permeability value.

If the heating rate is lower than 4.5°C/min, the slope at T_c leads to an evaluation of the chemical homogeneity. As previously stated, an inhomogeneous sample behaves as a series of homogeneous samples with different compositions. The limits of inhomogeneity are indicated by the initial fall and the final part of the permeability curve. The perfectly homogeneous sample should give a vertical fall at T_c. This method is thus a very useful tool for the characterization of magnetic ceramics in general, and permanent magnets in particular.

ACKNOWLEDGEMENTS

The authors thank the technical collaboration of R. Reyes.

REFERENCES

1) R. J. Parker. "Recent Development on Permanent Magnet Application". pp 375-378. Proc. of the 3d. Int. Conf. on Ferrites. Ed. by H. Wantanabe. Center for Academic Publications, Japan (1981).

2) E. Cedillo, J. Ocampo, V. Rivera and R. Valenzuela. "An apparatus for the measurement of the initial magnetic permeability as a function of temper-

ature". J. Phys. E: Sci. Instr. 13 383-386 (1980).

3) A. Globus and R. Valenzuela. "Influence of Deviations from Stoichiometry on the Magnetic Properties of Zn-Rich Ni-Zn Ferrites". IEEE Trans. Magnetics MAG 11 No. 5 1300-1302 (1975).

4) M. A. Escobar, L. F. Magaña and R. Valenzuela. "Analytical Prediction of the Ferromagnetic Hysteresis Loop". J. Appl. Phys. 54 5935-5940 (1983)

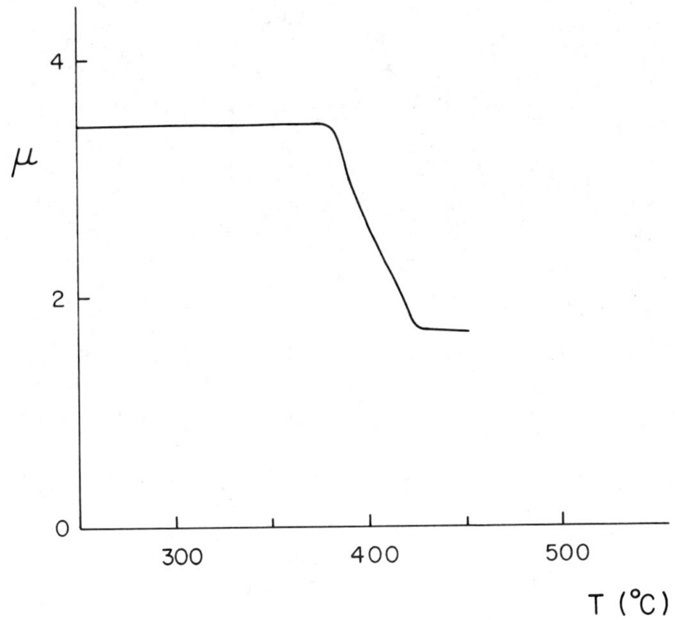

Fig. 1. The magnetic permeability as a function of temperature of a barium hexaferrite sample. The heating rate is 40.2°C/min.

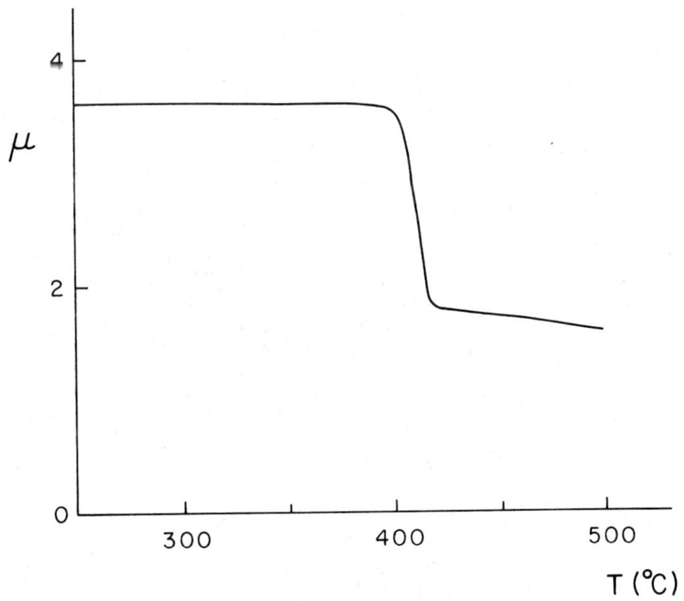

Fig. 2. The magnetic permeability as a function of temperature for a heating rate of 18°C/min.

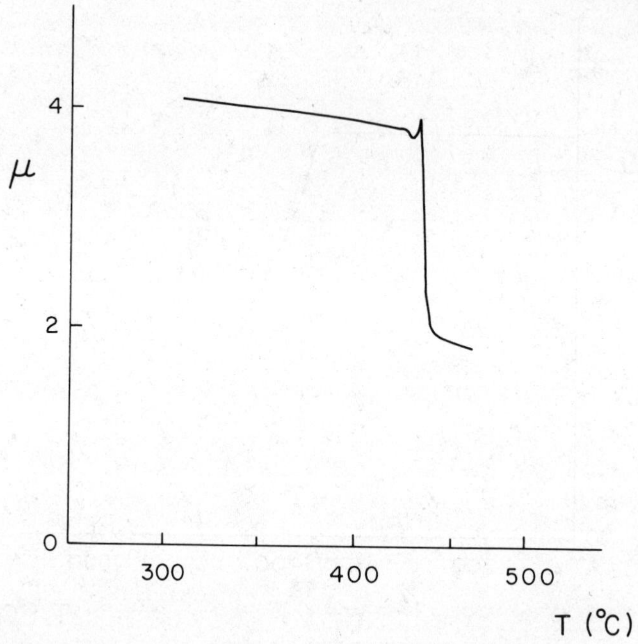

Fig. 3. Same curve then figs. 1 and 2, for a heating rate of 4.5°C/min.

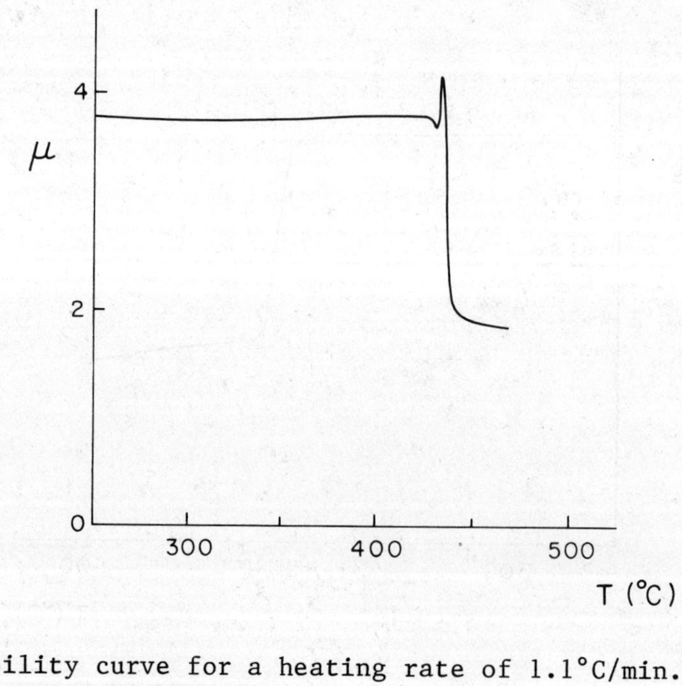

Fig. 4. Permeability curve for a heating rate of 1.1°C/min. This was the lowest rate investigated.

PROTON NUCLEAR MAGNETIC RESONANCE CHARACTERIZATION OF BINDER DISTRIBUTIONS IN Si_3N_4 CERAMICS*

P. S. Wong, W. A. Ellingson, R. E. Botto,** H. C. Yeh,† and J. P. Pollinger†

Materials and Components Technology Division
ARGONNE NATIONAL LABORATORY
Argonne, Illinois 60439

ABSTRACT

In the injection molding of Si_3N_4 ceramics, several problems associated with the interaction between organic binder and ceramic powder can cause defects in final parts. Many of these problems can be traced to poor distribution of the organic binder/plasticizer (B/P) used in the injection molding process. Nuclear magnetic resonance (NMR) is a chemically sensitive, noncontacting, nondestructive evaluation technique which is being studied for application to these problems. We have shown that the compositional variation of B/P concentration in Si_3N_4 ceramic samples can be examined with conventional ^1H NMR spectroscopy. The NMR spectra and proton relaxation times of multi-component polyethylene-based B/Ps that are used in injection-molded ceramics have been measured. Effects of temperature on the NMR parameters were also studied. Bulk characterization of the distribution of B/P has been shown to be observable with NMR imaging techniques.

INTRODUCTION

The efficient processing of high-performance, high-temperature structural ceramics planned for use in advanced heat engines requires a thorough understanding of the molecular interactions that occur in the early stages of the process. Indeed, processing science has received the highest priority in a recent congressional analysis of advanced ceramics technology.

---

*Work supported by the U.S. Department of Energy, Office of Fossil Energy, Advanced Research and Technology Development Materials Program, and the U.S. Department of Energy, Assistant Secretary for Conservation and Renewable Energy, Office of Transportation Systems, as part of the Ceramic Technology for Advanced Heat Engines Project of the Advanced Materials Development Program under Contract W-31-109-ENG-38.

**Chemistry Division, Argonne National Laboratory, Argonne, IL.

†Garrett Ceramic Components Division, Allied Signal Aerospace Company, Torrance, CA.

One ceramic processing method that holds the potential for low-cost, high-volume production is injection molding. In the injection molding process, the distribution of organic binders/plasticizers (B/Ps), the degree of dewaxing (i.e, completeness of removal of the organics), and the distribution of open (i.e., surface-connected) porosity left by the removal of the organics are important factors that affect local densification rates and hence mechanical properties. An uneven distribution of organic B/P has been identified as a potential cause of several types of injection molding defects.[1] Cost-effective nondestructive evaluation (NDE) methods to measure the distribution of the organic, the degree of dewaxing, and the distribution of the open porosity left after dewaxing would greatly aid the development of injection-molding technology, and of ceramics processing in general.

Nuclear magnetic resonance (NMR) is a chemically sensitive, noninvasive characterization technique which has been routinely used in chemistry, physics, and biology for over 35 years. In these areas, its main applications are in determining chemical composition, molecular or crystal structure, and dynamics of molecular motion.[2,3] The use of NMR to produce tomographic images of objects was reported in 1973.[4] Since that time, development of NMR imaging for medical diagnosis has proceeded at a furious pace. However, efforts to apply NMR imaging in the materials sciences have begun only recently.[5-7] NMR imaging clearly holds potential for aiding in the development of ceramics processing. It is a particularly promising technique for investigating the spatial distribution of organic B/Ps.

In the work reported here, we have measured the ^{1}H NMR spectra and proton NMR relaxation times of multi-component polyethylene-based B/Ps used in injection molding of ceramics. In addition, we have examined the effects of temperature on the NMR relaxation parameters and obtained preliminary results with NMR imaging techniques used to characterize the distribution of B/Ps. Injection-molded ceramics were emphasized for the following reasons, in addition to those noted above: (1) a reasonably high loading (~14-17 wt.%) of organic B/P is used in the process and thus there exists the potential for a reasonably large signal from the hydrogen protons in the organics, and (2) the very long times associated with B/P burnout provide clear economic justification for gaining detailed knowledge of the distribution of the organic binder.

Prior to the present project, limited work had been reported regarding the use of NMR imaging for ceramic applications. One problem with direct imaging of the organic B/P is that it is a "soft solid," i.e., wax-like. NMR measurements on solids generally require more specialized experimental techniques that place greater demands upon instrumentation than NMR measurements in fluids.[8,9] The absence of large-scale and isotropic molecular motion is the key element that complicates NMR measurement in solids, including soft solids such as organic B/P waxes.

The spin-lattice relaxation time, T_1, for the buildup of magnetic polarization of a nucleus directly affects the rate at which data can be accumulated from successive acquisitions of NMR signals. Thus, T_1 controls the total time that is necessary to acquire an image and, ultimately, the signal-to-noise (S/N) ratio that can be achieved. Because ^{1}H nuclei in soft solids such as B/Ps are expected to have short T_1's, spin-lattice relaxation is not expected to be a major problem. On the other hand, the spin-spin relaxation time, T_2, can significantly affect the spatial resolution and S/N ratio achieved in solids. T_2 values significantly greater than the echo time (TE) used in imaging experiments are desirable in order to avoid loss of signal intensity. Unfavorably short T_2's (requiring broad spectral bandwidths) lead to a low S/N ratio per unit of data acquisition time for imaging solids. Large spectral bandwidths make imaging-plane selection difficult and degrade the spatial resolution of frequency-encoded dimensions.

The shortest TEs that can be obtained with standard imaging equipment used for medical NMR imaging are in the range of 12-17 msec. Because the T_2's of polymeric B/Ps tend to be on the order of 1 msec or less, imaging of organic B/P material with conventional room temperature NMR imaging techniques would be difficult, if not impossible. This was confirmed by our earlier experiments.[10] However, with the use of a special NMR probe[6,10] we were able to obtain images of pure B/P at room temperature.

NMR SPECTROSCOPIC STUDIES

Because the polymeric organic B/Ps used in green-state ceramic compacts contain a high concentration of protons, they are potentially good candidates for NMR spectroscopic studies and NMR imaging. Prior to the imaging tests, we carried out a series of NMR spectroscopic studies to obtain information about the image quality that might be expected. Proton NMR spectra were recorded at 300 MHz (7.05 Tesla) with a Bruker Instruments Model AM-300 NMR spectrometer in the pulse-Fourier transform mode with quadrature detection. Samples of B/P were placed in 5-mm-dia NMR tubes for analysis. The 90° and 180° pulse widths (13 ± 0.2 μs and 26 ± 0.2 μs, respectively) were determined independently for each sample.

T_1 Measurement

The spin-lattice relaxation time, T_1, of a proprietary, multi-component polymeric B/P was measured by a standard inversion recovery pulse sequence, as shown in Fig. 1, where RD is the recovery delay time, τ is the variable delay time, and "180°" and "90°" are pulse angles for the inversion and sampling pulses, respectively. The measured T_1 value of B/P alone was ~320 msec. We subsequently used a pre-delay time of 1600 msec (5 x T_1) for the linewidth and T_2 measurements to ensure that the protons would relax to equilibrium magnetization before starting the next pulse sequence.

Room-Temperature Linewidth Experiments

We first conducted a series of room-temperature NMR spectral linewidth experiments (with a simple 90° pulse sequence). A Si_3N_4 injection-molding mix,[a] Garrett Ceramic Components Division mix GN10, containing a proprietary mixture of polyethylene-based materials was used. We used a set of 6

[a]Injection molding mix is the Si_3N_4 powder plus binder plus other proprietary materials, in a condition ready to injection mold.

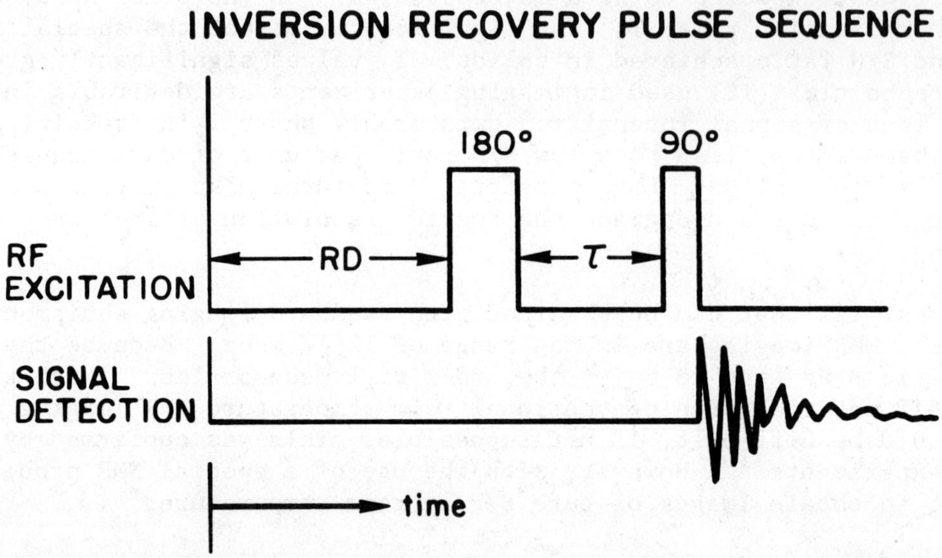

Fig. 1. Pulse Sequence in Inversion Recovery T_1 Measurements.

samples containing concentrations of 2, 5, 10, 13.5, 17, and 20 wt.% of B/P in ceramic powder. The relative NMR signal intensities from free induction decay (FID) were determined for the various B/P concentrations. The results are shown in Fig. 2. Any change in the chemical shift, ξ, of the resonance line is due to randomness and has no chemical significance. The chemical shift is calculated as follows:

$$\xi \text{ (in ppm)} = \frac{\text{observed shift (in Hz)} \times 10^6}{B_o \text{ (in Hz for specific nuclei)}} \qquad (1)$$

where

B_o = 300 MHz in a 7.1-T field for ^1H with gyromagnetic ratio of 42.6 MHz/T, and

ξ = 0 ppm for tetramethylsilane (TMS), which is used as a reference.

We also looked for T_2* differences among those samples. The relationship

$$\Delta_{1/2} \approx \frac{1}{\pi T_2^*} \qquad (2)$$

was used to calculate T_2*, where $\Delta_{1/2}$ is the full width at half maximum (FWHM) of the amplitude vs. frequency trace and T_2*[b] is the effective transverse

[b]If an aligned spin system is excited with 90± pulse, the net magnetization will be rotated into a plane which is transverse to the externally applied field. The transverse magnetization can be detected in this plane and will be seen to decay with a time constant given by T_2*.

relaxation time, which includes contributions from magnetic field inhomogeneities. The variations in FWHM and T_2* values with B/P concentration are shown in Fig. 3. From the data shown in Figs. 2 and 3, we note that three NMR parameters can be correlated with organic content in a Si_3N_4/B/P mix: the NMR signal strength from FID (Fig. 2) and the T_2* and spectral linewidth (Fig. 3).

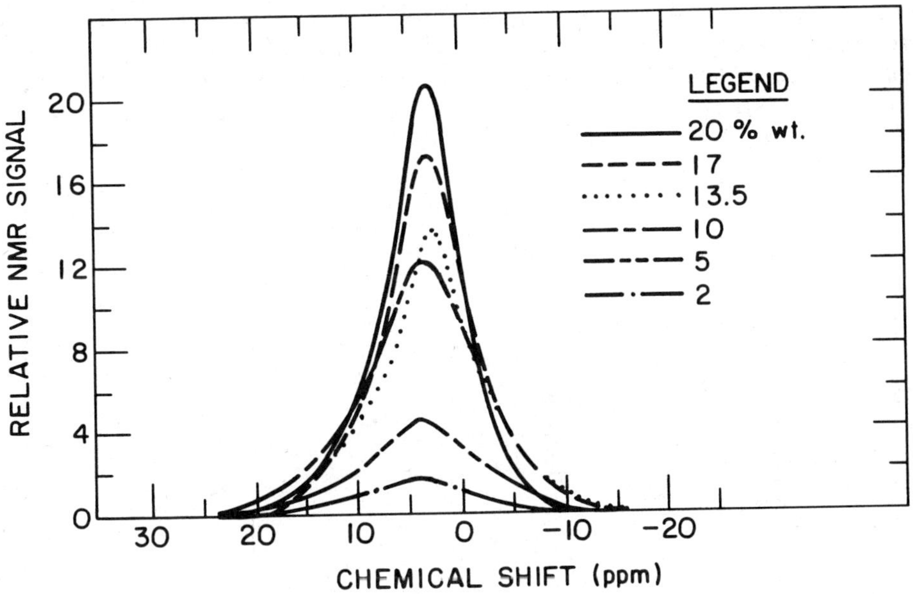

Fig. 2. Variation in NMR Signal Strength with Wt.% Organic Binder in Si_3N_4 Powder. Number of acquisitions = 140.

Fig. 3

Variation in FWHM and T_2* with Wt.% Organic Binder in Si_3N_4 Powder.

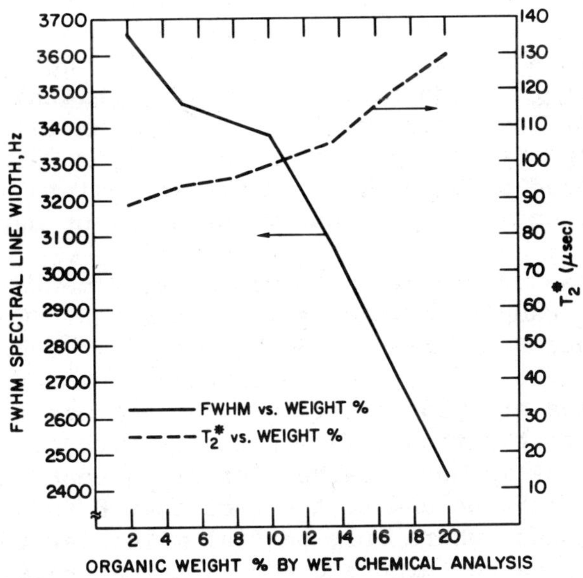

Variable-Temperature Linewidth Experiments

NMR spectral linewidths of abundant spins in solid materials are typically quite broad owing to dipole-dipole coupling interactions between nuclei.[8,11-13] Linewidths scale in magnitude according to the magnetogyric ratio of each spin of a coupling pair and inversely to the cube of the distance between discrete pairs.[3,14,15] An increase in molecular motion at elevated temperatures tends to average the dipolar coupling interaction between nuclei, causing a narrowing of the NMR linewidth. This behavior was demonstrated in experiments at 7.1 Tesla and 100°C. Two relatively sharp NMR resonances (FWHM of ~11 Hz) were obtained for the B/P alone, as shown in Fig. 4. The resonances seen at about 1.1 and 1.5 ppm respectively correspond to the methyl and methylene protons in B/P.

Fig. 4

Proton NMR Spectrum of B/P at 100°C.

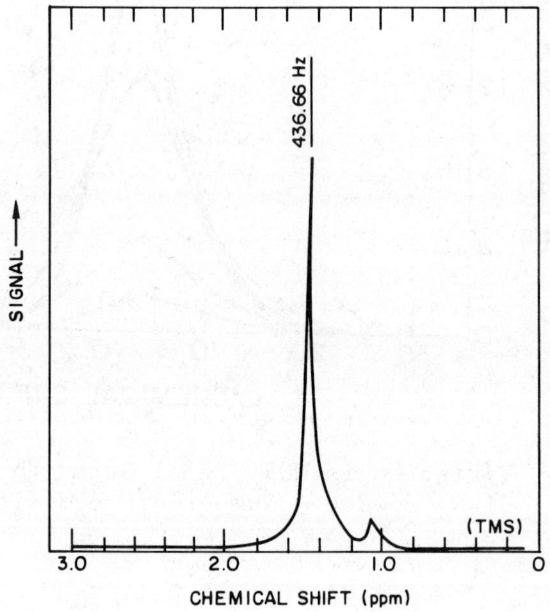

Proton NMR spectra recorded at 22°C, 40°C, 60°C, 80°C, and 100°C are shown in Fig. 5. The initial proton linewidths of ~1000 Hz in spectra recorded at the lower temperatures are substantially reduced above the binder melting temperature. In the spectrum recorded at 100°C, the resonance centered at about 1.5 ppm breaks up into several distinct lines representing magnetically inequivalent methylene groups.

We also carried out linewidth experiments at various temperatures on a Si_3N_4 sample containing 15.5 wt.% B/P. The relative NMR signal intensities of both B/P alone and Si_3N_4 + 15.5 wt.% B/P are plotted against temperature in Fig. 6. T_2* values estimated from the linewidth measurements for both samples are plotted against temperature in Fig. 7. It is interesting to note that although T_2* increases with temperature for B/P alone, a similar tendency is not observed for Si_3N_4 containing 15.5 wt.% B/P. This difference might be due to the fact that B/Ps are physisorbed to individual Si_3N_4 particles, where only partial molecular reorientation is brought about at the elevated temperatures. Alternatively, we may be seeing translational diffusion effects on the NMR relaxation times. These effects will be identified in a subsequent investigation by using a Carr-Purcell-Meiboom-Gill (CPMG) pulse method.

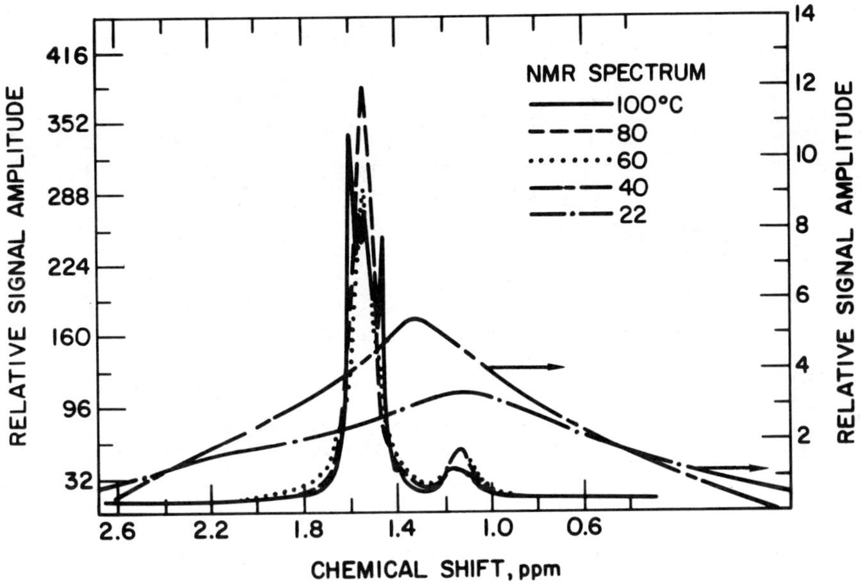

Fig. 5. Proton NMR Spectra of 100% B/P at Various Temperatures. Scale on the right is used for 22° and 40°C data.

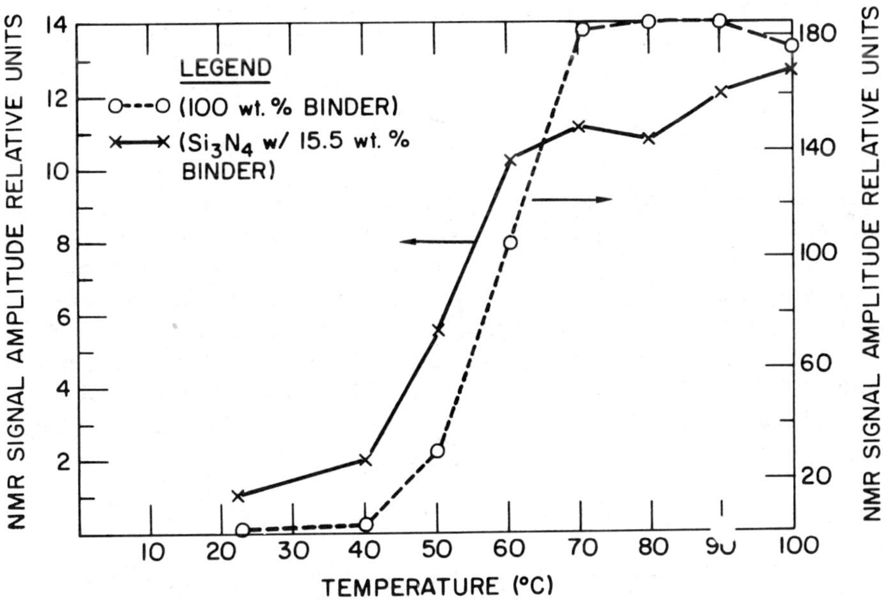

Fig. 6. Variation in Relative NMR Signal Intensities of 100% B/P and Si_3N_4 + 15.5 wt.% B/P with Temperature.

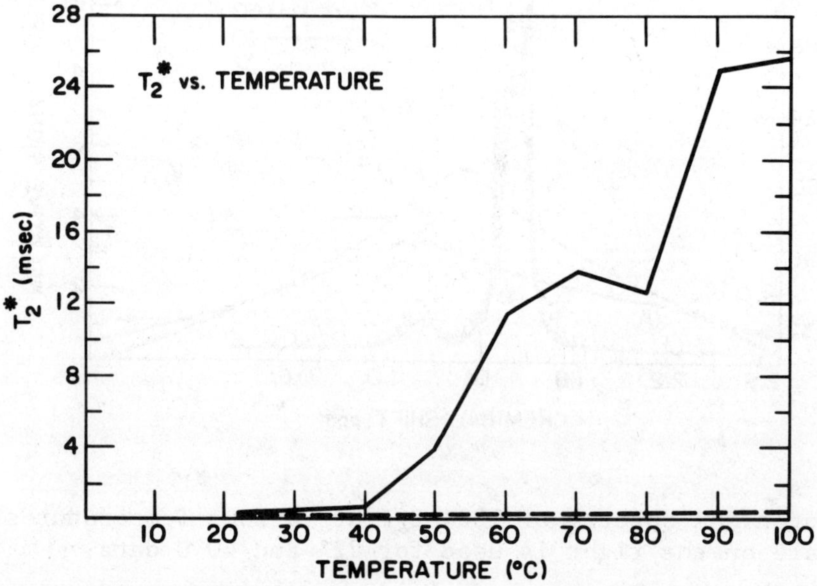

Fig. 7. Variation in T_2* of 100% B/P (Solid Line) and Si_3N_4 + 15.5 wt.% B/P (Dashed Line) with Temperature.

T_2 Measurements

In addition to the linewidth measurements, we carried out a series of T_2 measurements on the B/P alone. The Hahn spin-echo pulse sequence,[16] which is represented diagramatically in Fig. 8, was used to measure T_2. Inverting the transverse NMR magnetization through the application of a 180° pulse at some time TE/2 after the initial 90° pulse allows a direct measurement of T_2. The T_2 value obtained in this manner contains no contribution from magnetic field inhomogeneities; thus, $T_2 \geq T_2$*. The T_2 values are plotted against temperature in Fig. 9. The increase in T_2 with temperature indicates that the mobility of protons in the B/P increases with temperature.

DIRECT NMR IMAGING OF B/P DISTRIBUTION

In earlier work,[5,6,10] we showed that direct NMR imaging of the B/P distribution is possible if gradient pulses and RF pulse sequences are fast enough to compensate for extremely short T_2's. It was possible to circumvent the resolution-degrading effects of broad spectral lines and frequency-encoded projections by using special techniques that involved pulsed field gradients and phase encoding, which were not possible on a medical imager.

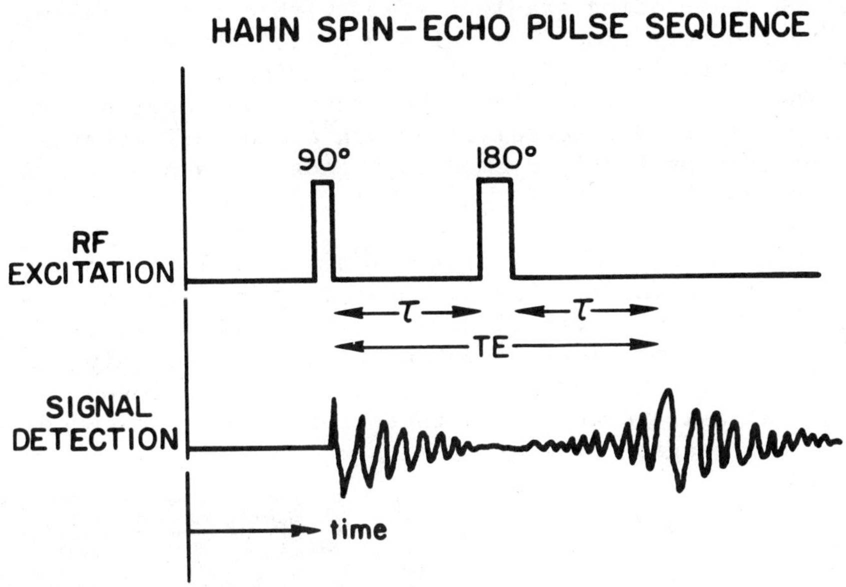

Fig. 8. Pulse Sequence in Hahn Spin-Echo T_2 Measurements. τ = variable delay time and TE = echo time.

Fig. 9

Variation in T_2 of 100% B/P with Temperature.

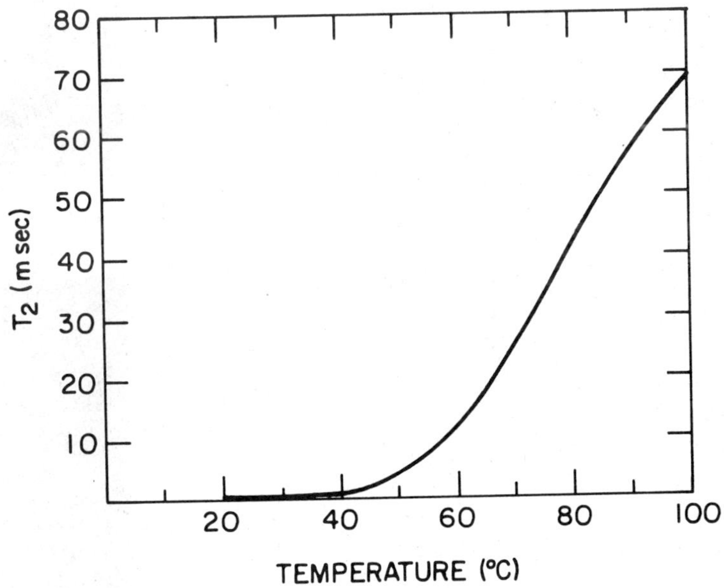

A green Si_3N_4 compact with 15.5 wt.% B/P, containing several drilled holes with diameters ranging from 1.1 to 4.8 mm, was used to test the imaging technique. Figure 10 shows a transaxial NMR image of this specimen. A two-dimensional spin-echo RF pulse sequence was used with no slice selection applied and a TE of 4.42 msec. The data were acquired by using 40 signal averages per phase-encoding gradient step, with a total imaging time of 29.3 min. The recovery time (TR) was 300 msec. The digital resolution in the image is 128 x by 128 y pixels, corresponding to 640 μm in either direction. The x resolution (along the frequency-encoding direction) will be degraded by the broad spectral linewidth of the B/P material. (The image does not look perfectly symmetric because of a nonlinearity in one field gradient coil.)

Figure 11 shows a transaxial NMR image of a green Al_2O_3 compact obtained with direct imaging methods. The 25-mm-diam cylindrical specimen had a length-to-diameter (L/D) ratio of 0.17 and the Al_2O_3 powder had a mean particle size of 4.5 μm. The concentration of B/P was only 2.5 wt.%. No slice selection (i.e., no tomographic imaging) was employed, and TE was 3.2 msec. Sixteen signal averages were taken for each phase-encoding gradient step, for a total imaging time of 17 h.

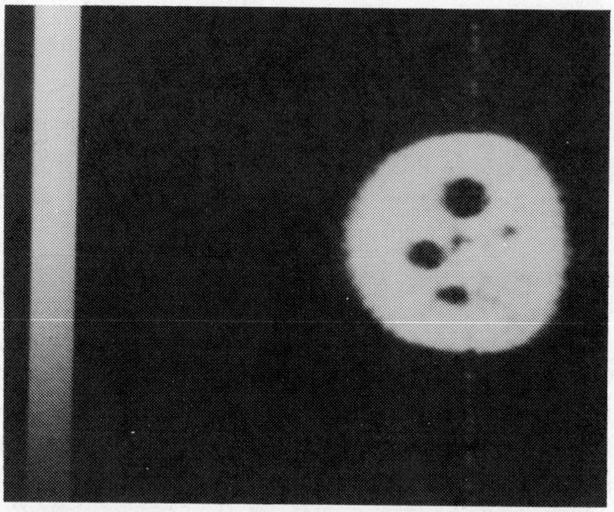

Fig. 10

Transaxial NMR Image of a Green Si_3N_4 Compact with 15.5 wt.% B/P, Taken with a Two-Dimensional Spin-Echo RF Pulse Sequence. TE = 4.42 ms; 40 signal averages = total imaging time of 29 min.

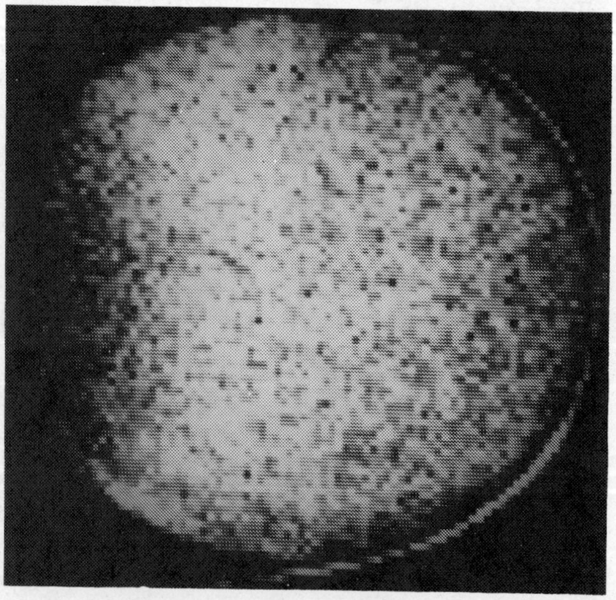

Fig. 11

Double Phase-Encoded Image of 25-mm-diam Al_2O_3 Sample with 2.5 wt.% B/P, Taken With Special Probe and TE = 3.2 msec. Image resolution is 128 x by 128 y pixels (310 and 340 μm, respectively). Total data acquisition time was about 17 h.

CONCLUSIONS

NMR characteristics of a selected multi-component polymeric binder typical of those used in injection molding of Si_3N_4 ceramics have been measured. A sharper NMR spectral linewidth is observed as the temperature increases, and this improves the spatial resolution in imaging experiments dramatically in the frequency-encoded dimension. Using the Hahn spin-echo pulse sequence, T_2 was shown to increase over the temperature range studied for the B/P alone, but not for a mixture of ceramic powder and B/P. This observation is tentatively attributed to solid-boundary effects in mixtures of ceramic powder and B/P. However, NMR signal intensity does increase with temperature; this increase would enhance the S/N ratio for imaging experiments.

The correlation between relative NMR signal intensity and B/P concentration in Si_3N_4 is much better for samples with high (i.e., >10 wt.%) B/P concentration. This finding indicates that large numbers of acquisitions are needed to improve the S/N ratio of samples with lower B/P content.

Direct tomographic imaging of organic B/Ps in ceramic materials can be accomplished by NMR providing that one uses appropriately designed RF coils (capable of transmitting short pulses to uniformly excite the broad spectral line of a solid material) and satisfactory gradient coils (capable of generating more intense and rapidly switched gradients).

ACKNOWLEDGMENTS

We are grateful to Arthur G. Kostka of the ANL Chemistry Division for performing the NMR linewidth and T_2 measurements. Helpful discussions with Arthur Thompson, also of the Chemistry Division, are acknowledged.

REFERENCES

1. H. C. Yeh, J. M. Wimmer, M. E. Huang, M. E. Rorabaugh, J. Schienle, and K. H. Styhr, Improved Silicon Nitride for Advanced Heat Engines, Ann. Tech. Report submitted to NASA by AiResearch Casting Co., Div. of Garrett Corp., NASA-CR-175006, October 1985.

2. E. D. Becker, High Resolution NMR, Academic Press, New York, 1980.

3. C. P. Slichter, Principles of Magnetic Resonance, 2nd Ed., Springer-Verlag, New York, 1978.

4. P. C. Lauterbur, "Image Formation by Induced Local Interactions: Examples Employing Nuclear Magnetic Resonance," Nature 242, 190 (1973).

5. W. A. Ellingson, J. L. Ackerman, J. D. Weyand, R. A. DiMilia, and L. Garrido, "Characterization of Porosity in Green-State and Partially Densified Al_2O_3 by Nuclear Magnetic Resonance," Ceram. Eng. Sci. Proc. 8(7-8), 503-512 (1987).

6. J. L. Ackerman, W. A. Ellingson, J. A. Koutcher, and B. R. Rosen, "Development of Nuclear Magnetic Resonance Imaging Techniques for Characterizing Green-State Ceramic Materials," in Proc. 2nd Intl. Symp. on the Nondestructive Characterization of Materials, Montreal, Canada, July 21-23, 1986, J. F. Boussiere, J-P. Monchalin, C. O. Ruud, and R. E. Green, Jr., eds., Plenum Press, New York, 1987, pp. 129-137.

7. L. B. Welch, S. T. Gronczy, M. T. Mitsche, L. J. Bauer, J. Dworkin, and A. Giambalvo, "Proton NMR Imaging of Green-State Ceramics," in Proc. Quantitative NDE Conference, La Jolla, CA, August 1-5, 1986, D. O. Thompson and D. E. Chimenti, eds., Plenum Press, New York, 1987, pp. 441-456.

8. M. Mehring, High Resolution NMR Spectroscopy in Solids, Springer-Verlag, Berlin, 1976.

9. D. E. Axelson, Solid State Nuclear Magnetic Resonance of Fossil Fuels, Multiscience Limited, Montreal, Canada, 1985.

10. W. A. Ellingson, J. L. Ackerman, S. Gronemeyer, and L. Garrido, "Nuclear Magnetic Resonance Imaging for Detecting Binder/Plasticizers in Green-State Structural Ceramics," in Proc. 16th Symp. on Nondestructive Evaluation, San Antonio, TX, April 21-23, 1987, pp. 231-240.

11. S. Emid and J. H. N. Creyghton, "High Resolution NMR Imaging in Solids," Physica 128B, 81-83 (1985).

12. S. Emid, "Ultra High Resolution Multiple Quantum Spectroscopy in Solids," Physica 128B, 79-80 (1985).

13. E. R. Andrew, "The Narrowing of NMR Spectra of Solids by High-Speed Specimen Rotation and the Resolution of Chemical Shift and Spin Multiplet Structures for Solids," Prog. Nucl. Magn. Reson. Spectrosc. 8, 1-39 (1971).

14. E. Fukushima and S. B. W. Roeder, Experimental Pulse NMR: A Nut and Bolts Approach, Addison-Wesley Publishing Company, London, 1981, pp. 2217-2295.

15. G. E. Maciel, "High Resolution Nuclear Magnetic Resonance of Solids," Science 226, 282-288 (1984).

16. E. L. Hahn, "Spin Echoes," Phys. Rev. 80, 580-594 (1950).

HIGH PRECISION MACHINING OF ADVANCED CERAMICS

Wilfried König, Matthias Popp
Fraunhofer-Institut für Produktionstechnologie IPT
Aachen, Federal Republic of Germany

ABSTRACT

Potential reduction of friction and wear through the use of advanced ceramic materials for engine components such as seals, bearings and valve adjusting plates is of great interest. Since there is no running-in effect with ceramic materials, however, an extremely high precision finishing process is required for such components. The complicated machining characteristics of ceramics pose additional problems.

The present paper discusses problems related to precision machining of advanced ceramics. A number of solutions with considerable potential in terms of reliable quality and efficiency are presented.

INTRODUCTION

High material property requirements have led to the use of ceramic materials for many applications in mechanical and process engineering.

The hardness and compressive strength of ceramic materials are superior to those of metals. Other advantageous performance characteristics are extremely high resistance to wear and corrosion and low specific gravity. One disadvantage of ceramics is their great brittleness; even the smallest defects in the component surface due to incorrect machining can lead to disastrous failure where load distribution under operating conditions is unfavourable.

Owing to the virtually complete absence of a run-in effect with ceramics, there is a need for extreme precision in the quality of finish of functional surfaces on components subjected to friction and sliding wear (Figure 1). For example, silicon infiltrated silicon carbide (SiSiC) bearings in pumps transporting corrosive and abrasive liquids require a roughness value of $R_a < 0.08$ µm and an out-of-flatness value of $f_e < 0.3$ µm. Similarly, zirconia (ZrO_2) or silicon nitride (Si_3N_4) rocker arm inserts and rams need a roughness value $R_a < 0.01$ µm.

In view of the advantages offered by ceramics for such applications, the question must be why they are not used more often, especially in larger-scale production.

Apart from the brittleness referred to above, the reason is to be sought in the extremely poor machinability of ceramics. This problem is compounded when more exacting component quality demands are imposed.

Only a limited range of possible machining processes can be employed for finishing. Precision machining under industrial conditions is limited to cutting with geometrically undefined cutting edges, e.g. grinding, honing, lapping and polishing.

Owing to the extreme hardness of ceramics, diamond generally has to be used as the cutting material; the combination of high hardness and strength, however, causes considerable wear to the diamond itself or to the tools. This entails higher machining costs and may lead to problems with accuracy-to-size-and-shape due to the high machining forces involved.

Finally, the low material removal rates which can be achieved result in long machining times and make it more difficult to attain an economically viable process.

The paper discusses the development of process parameters suitable for high precision machining of ceramics. Examples of plane and cylindrical machining are presented. The potential of each technology in terms of achievable surface finish and accuracy-to-shape is assessed.

POLISHING PLANE SURFACES

Flat polishing is a conventional machining process used to produce plane surfaces.
The workpieces to be machined are held by workpiece carriers between two polishing wheels rotating in opposite directions. Due to the planetary motion of the carriers and the loose guidance of the workpieces, this process provides advantageous machining without marking in a preferential direction (Figure 2).

Unlike grinding and honing, polishing is carried out with loose diamond grains held in suspension, which penetrate to the contact zone between the polishing wheel and the workpiece surface via boreholes in the upper polishing wheel.

Depending on the level of normal force applied and the material properties of the polishing wheel, diamond grains will be temporarily embedded in the polishing wheel during machining. This affects removal mechanisms and, particularly, machining quality.

If a hard polishing wheel is used, sliding and rolling motion of the diamond grains will predominate, and the normal force will press grains into the brittle workpiece material. Minute cracks are then induced in the surface of the workpiece, and material particles break away to produce material removal. Conversely, with a soft polishing wheel, diamond grains become embedded deeply in the wheel, removing material in a cutting process produced by the relative motion between the polishing wheel and workpiece surface.

The latter method is preferable in terms of optimum surface quality, since it enables roughness peaks left by the roughing process to be removed completely.
Flat machining of a silicon carbide workpiece shows the effects the use of different polishing wheel materials can have on achievable surface quality. Roughness as characterized by the average roughness height R_a is reduced from R_a = 48 nm to R_a = 7 nm when aluminium or polyurethane-impregnated polyester is substituted for cast iron (Figure 3).

The pliability of polyester is an additional factor allowing diamond grains to move away from the workpiece surface and preventing scratching by projecting grains.
The pores in the polishing wheel are large by comparison with the grain diameter and provide spaces large enough to hold free diamond grains. They are also advantageous in terms of chip removal and a constant supply of cooling lubricant and new diamond grains. They obviate the need for grooves of the kind used on non-porous polishing wheels.

Soft polishing wheels have disadvantages in terms of achievable accuracy-to-shape, however. The smooth material can easily be distorted under the influence of the normal force applied to the polishing wheel. Convex workpiece surfaces with more or less strongly rounded edges result.

When soft polishing wheels are used, there is a trade-off between surface quality and accuracy-to-shape. As shown in the example, the out-of-flatness of f_e = 0.5 μm by reference to the workpiece diagonal of 70 mm is increased to f_e = 1.2 μm.
The use of aluminium polishing tools offers a compromise between surface quality and accuracy-to-shape. Due to the low hardness of aluminium, the diamond grains can fix themselves deeply in the polishing wheel, which is a prerequisite for minute chip thickness. At the same time, the homogeneity of the material and its compressive strength, which is relatively high as compared to that of polyurethane impregnated polyester, avoid deformation of the polishing wheel with its disadvantages for workpiece accuracy-to-shape.

In general, the results document the impressive potential of this technology. The surface qualities achieved meet even the requirements for plane mirrors in material processing lasers.

In spite of the long machining time caused by the low effective velocity, this machining technology has considerable potential for rationalization. For example, a number of workpieces can be machined on both sides at the same time. This can be a particular advantage when plane parallelism is a factor, for example when machining bearing rings. The fact that this machining technology does not require any tool conditioning is an important process simplification.

FACE GRINDING WITH CUP WHEELS

Another technology used to machine plane surfaces is face grinding with diamond cup wheels (Figure 4).

The workpieces are clamped onto a rotating table and pass along the ring-shaped diamond layer of the cup grinding wheel. Depending on the arrangement of the workpieces on the table, infeed is either continuous or stepwise per table rotation.

An important parameter in precision grinding is the contact length, which is determined by the engagement conditions between tool and workpiece. It is greater for surface grinding with cup wheels than for peripheral grinding, where it is limited to few millimetres.

The resulting high overlap factor is an advantage, since it means that a larger number of cutting edges is used to machine the workpiece surface within the period of contact. This considerably improves the roughness values of the ground surfaces. This advantage can be further intensified by decreasing the rotational motion v_w of the workpiece or by increasing the peripheral speed of the wheel and widening the active grinding layer. Such grinding conditions result in a low chip thickness and, so far as is known at present, diminish the risk of surface damage from microcracks /1/.

Another advantage of face grinding can be seen in the twofold engagement of each workpiece as it crosses the ring-shaped grinding layer. The material removal rate is not, however, equally distributed between the outer and inner sides of the grinding layer. Most of the cutting is carried out by the outer layer, which engages first, whereas the inner layer is only active at the lowest chip thicknesses.

Apart from these advantages in quality, a unidirectional orientation of the grinding marks, making surface roughness dependent on machining direction, cannot occur, because of the final cross-shaped surface finish.

The highest surface roughness quality is achieved in spark-out. Machine deformation due to high machining forces relaxes during this period. Increasing spark-out time results in a diminishing cutting depth. Minute chip thickness results in a transition from chipping to plastic chip formation.

In the example shown, the process of spark-out results in an average roughness height of less than 0.035 μm for silicon carbide (Figure 5). It was possible to attain these values even with a relatively rough tool (D91 C50) in one machining step.

The close accuracy-to-shape achieved (f_e = 1.2 μm, referred to a measuring length of 55 mm) for the workpieces machined demonstrates the kinematic advantages of this technology. The larger the workpiece surface, the more effective the generating motion, in which the workpiece shape is produced by a straight line (projection of tool diameter). This substantially improves the productivity of this technique, as cup wheels can cover a large area, allowing several workpieces to be machined simultaneously.

An important criterion for the efficiency of a machining technology is the cost of tool conditioning. In this process, there is no need for profiling because of its generating kinematics. Wheels can, for example, be sharpened with a sharpening stone. It is, however, more effective to maintain the tool in a self-sharpening condition by matching the tool specification, process parameters and machining task. With cup grinding wheels, it is possible to encourage certain desired changes in the condition of the layer by selecting a special layer geometry and by applying different diamond layer specifications on the same tool /2/.

CYLINDRICAL GRINDING WITH A DIAMOND BELT

With cylindrical workpieces, external cylindrical plunge grinding is unfavourable in terms of process kinematics. Due to engagement conditions, there is only an extremely short contact length between workpiece surface and tool, leading to a small number of active cutting edges and a low overlap factor.

In contrast to face grinding, the average roughness height can be influenced less by choice of parameters but more by the grain size of the grinding wheel employed.

This necessitates a multi-step process strategy, featuring decreasing grain sizes, but entailing considerable effort for tool changes. Such a method is feasible but not efficient.

A multi-step belt-grinding strategy, however, proves to be far easier, since a rapid tool change can be made by inserting another type of grinding belt. Problems encountered with

conventional grinding wheel changes, such as concentricity errors or balancing-out difficulties, do not occur.
The use of non-endless diamond belts is positive for finishing in various respects (Figure 6) /3/:

For instance, inaccuracy problems due to the joints in endless belts can be excluded. As the tool is used only once, the belt is led over the contact trolley at a relatively low speed; the cutting speed required for this machining process is largely generated by workpiece rotation. As a result, the diamonds remain in the contact zone for a relatively long time, and are exposed to substantial wear.

In addition, the free space between the diamond grains will rapidly be clogged by particles removed from the workpiece surface. The continuous addition of new diamond grains, providing an optimum cutting tool at any time, is an advantage. There is no need for additional dressing and sharpening.

Oscillating belt motion can be used to improve machining quality. Higher cutting edge overlap can be achieved, with improved surface quality. At the same time, copying of the contact trolley contour on to the workpiece surface can be avoided.

The influence of oscillation on the final quality of the machined workpiece is clearly evident in the example (Figure 7). The average roughness height R_a = 0.28 μm can be reduced to R_a = 0.12 μm when a belt with a 9 μm diamond grain size is used. In general, roughness values R_a of 0.04 μm can be achieved by increasing machining times and particularly by using an extremely fine diamond grain of D 0.5 μm.

One disadvantage should be noted. With this type of machining principle, it is not possible to improve accuracy-to-shape, and a deterioration may even have to be accepted. Accuracy-to-shape therefore has to be generated in previous machining steps, for instance by grinding with grinding wheels.

SUMMARY

Many problems still accompany the production of ceramic components with surfaces of the highest precision. For this reason, there has been virtually no commercial exploitation of such processes in large-scale production.

Despite the limited range of available machining technologies and the complicated machinability of advanced ceramics, there are ways of improving both the efficiency of the machining process and the finished quality of the resulting workpiece.

By choosing the most suitable polishing tool material, for example aluminium, it is possible to produce extremely high quality in terms of surface roughness. Accuracy-to-shape drawbacks of the polishing process can virtually be eliminated.

Workpieces can be finished with cup grinding wheels if process parameters are adapted, enabling plane surfaces to be roughed and finished with a single tool. Face grinding therefore offers excellent potential for development as a future high efficiency finishing process.

Under certain circumstances, however, selection of different machining technology may be the appropriate alternative to improving the efficiency of a machining process. In peripheral machining of cylindrical workpieces, for example, grinding with diamond belts can help to make a multi-step machining process economically competitive through rapid, simple tool changes.

REFERENCES

1. Wemhöner, J.: Werkstoffgerechte Prozeßstellwerte - eine Voraussetzung für die wirtschaftliche Schleifbearbeitung von heißgepreßtem Siliziumnitrid. Industrieanzeiger 109(1987)81, 36-37.
2. Kleinevoss, R.: Spiegelglanz an keramischen Bauteilen. Industrieanzeiger 97(1987), 11-13.
3. Böckling, H-A.; Quambusch, A.: Neue Perspektiven zur Erzielung feinster Oberflächengüten unter Einsatz von Mikro Schleifmitteln. die maschine 12(1986), 33-34.

Fig. 1: Ceramic precision components

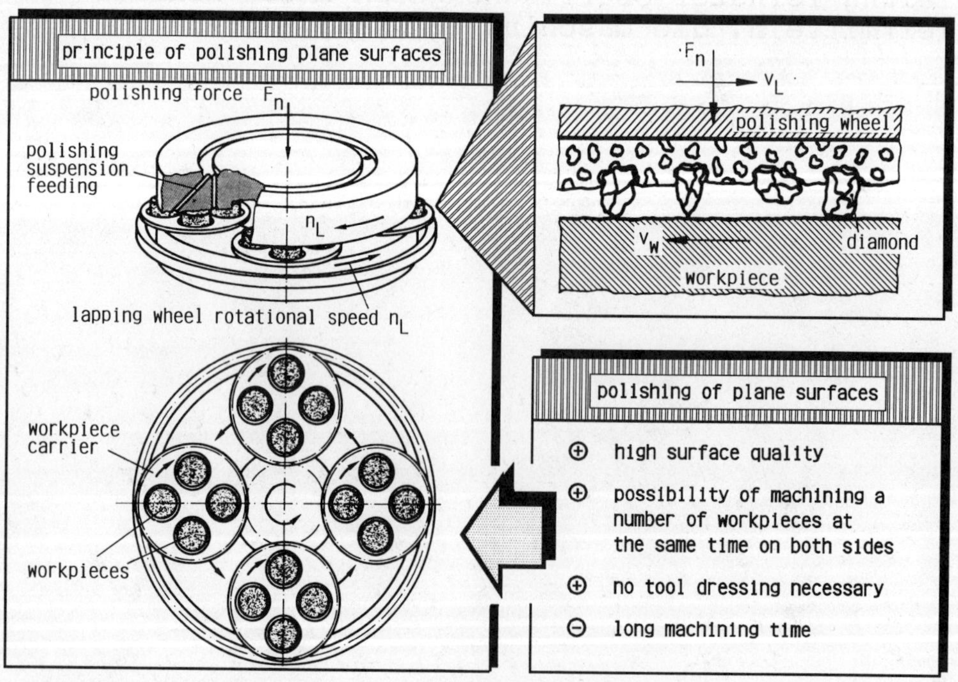

Fig. 2: Principle of polishing plane surfaces

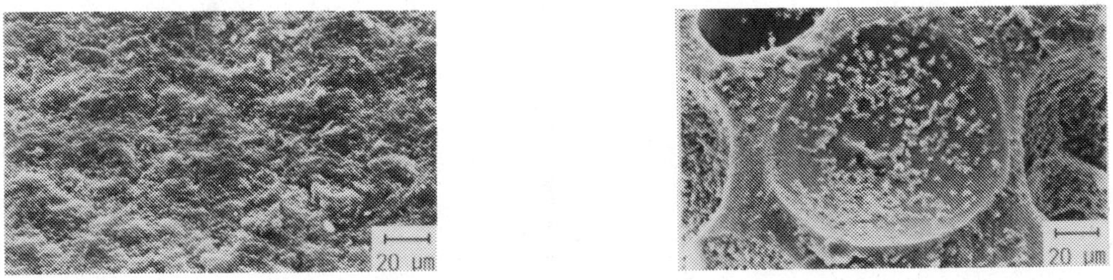

Fig. 3: Better surface finish but lost of accuracy by using soft polishing pads

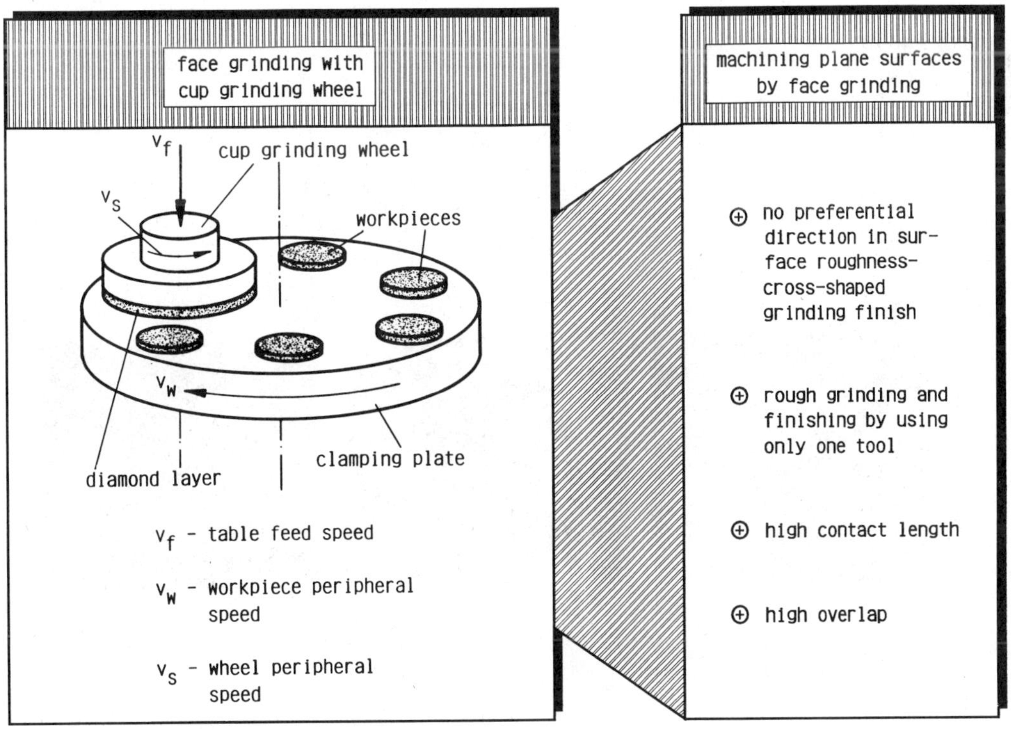

Fig. 4: Face grinding with cup grinding wheels

material	SiSiC
tool	D 91 C50 Bz
cutting speed	v_c = 32 m/s
infeed	a_p = 5 µm
dimension	50 x 50 x 5 mm
coolant	emulsion 3 %

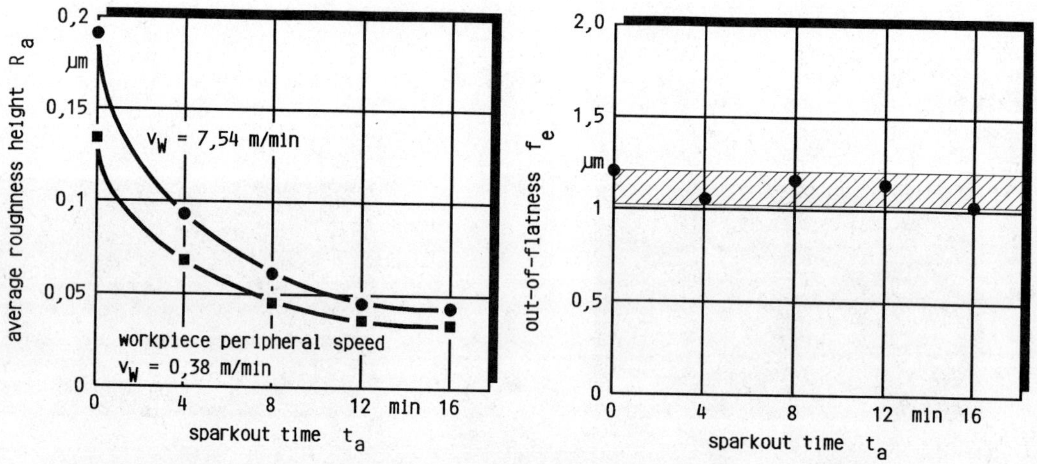

Fig. 5: Sparking out leads to highest surface quality

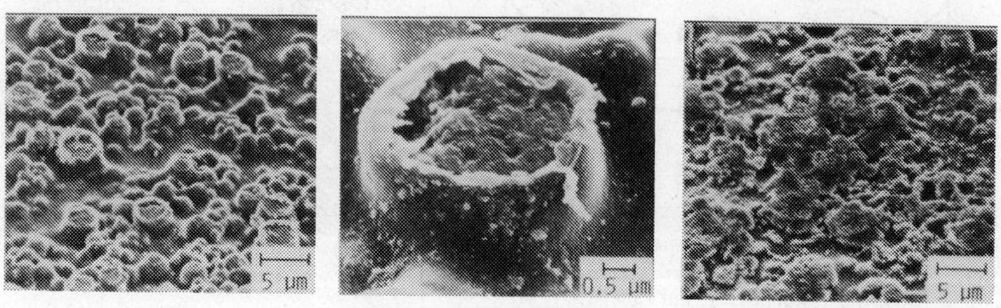

Fig. 6: Principle of belt grinding process

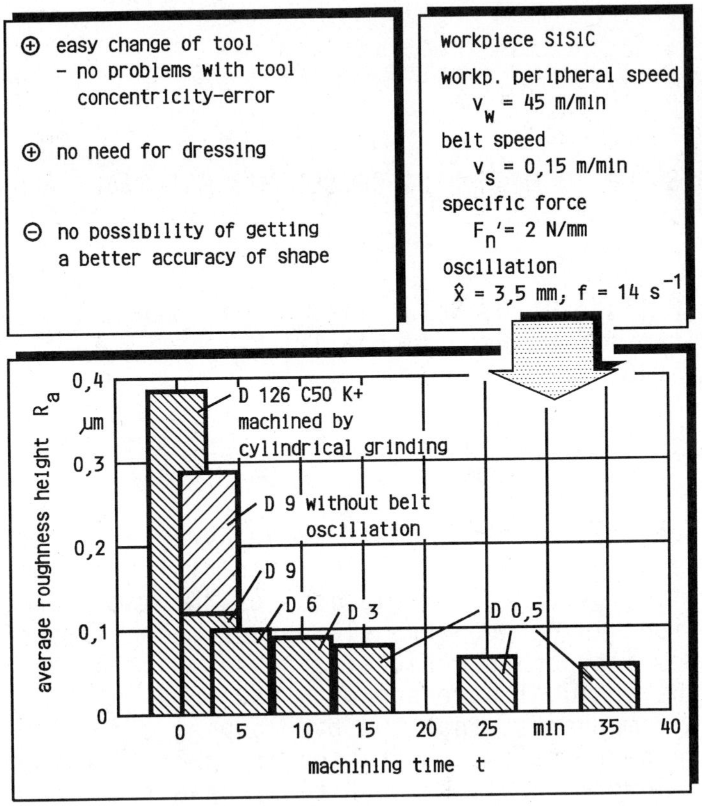

Fig. 7: Small grain size and high overlap lead to high surface quality in belt grinding

RESIDUAL STRESSES DUE TO MACHINING OF CERAMIC MATERIALS AND THEIR EFFECT ON STRENGTH

W. Pfeiffer, T. Hollstein, W. Berweiler, R. Prümmer,
Fraunhofer-Institut für Werkstoffmechanik,
D-7800 Freiburg, FRG

ABSTRACT

The investigations deal with non-destructive X-ray analysis of residual surface stresses due to maching of ceramic materials like alumina, zirconia and silicon carbide. Systematic variations of the machining conditions lead to a close relationship between the elementary mechanisms of the machining procedures and the introduced residual stresses. It can be derived from the correlation of the residual stresses with the bending strengths of the machined ceramic components, that two overlapping effects are of importance. Up to a certain extent, residual stresses are able to either improve or decrease strength. Severe machining conditions lead to a reduction of residual stresses and decreasing strength because of surface damages.

NOMENCLATURE

Grinding:

a_e = infeed
Q'_w = specific material removal rate
v_{ft} = feed rate in tangential direction
v_s = peripheral speed of grinding wheel

Lapping:

$F_{n''}$ = specific normal force
n_L = lapping disk speed
$\overline{d}_K$ = average grit diameter in µm.

Electro-discharge machining:

f_p = frequency of discharge
t_i = pulse length of discharge

General:

σ_o = characteristic bending strength
m = modulus of Weibull distribution
E = modulus of elasticity
ν = Poisson's ratio

INTRODUCTION

The role of residual stresses has been widely investigated for metals and is taken into account for the design of highly loaded components [1]. It has been found that the effect of residual stresses increases with increasing hardness of the material. Therefore, especially for ceramic materials, residual stress analysis should be obligatory to achieve sufficient reliability of ceramic components.

The aim of these investigations is to show the ability of the method of X-ray residual stress analysis to determine residual stresses in thin surface layers of ceramics due to various kinds of machining procedures. Combined with the results of statistically covered bending tests, the derived conclusions shall enable the manufacturer to perform high-quality surface engineering.

MATERIALS AND EXPERIMENTAL PROCEDURE

Sintered pure alumina, MgO-partially stabilized zirconia (PSZ) and sintered, silicon-infiltrated silicon carbide (SiSiC) samples with the dimensions 50 x 50 x 3 mm were investigated. The alumina and PSZ samples were lapped with B_4C abrasives using several different conditions. The SiSiC-samples were down-cut ground with a D151 C100K+ diamond grinding wheel (see [4]) or cut by wire electro-discharge machining at various energy rates. X-ray residual strain measurements were performed using a Siemens D 500 diffractometer, equipped with a position-sensitive detector (M. Braun, PSPC 50M), and controlled by a DEC 11/23-computer. X-ray elastic constants (XEC) were determined experimentally [2] for SiSiC. For alumina and PSZ, mechanically determined values of the elastic constants were used to calculate the stresses of the measured strains. The results of the X-ray stress determinations could be reproduced within $\pm$ 5 N/mm^2 accuracy. The complete measuring conditions are given in Table 1.

Table 1: Measurement data for X-ray stress analysis

Material	Lattice plane	Radiation	Penetration depth in µm	XEC($(\nu+1)/E$) in 10^{-6} mm^2/N
Al_2O_3	{2.2.0}	CrKα	11	3.1
	{4.0.10}	CuKα	33	3.1
ZrO_2	{6.2.0} cubic + {4.2.1} tetrag	CuKα	6	6.2
SiSiC	{2.0.10} 4Hα + {2.0.15} 6Hα	CuKα	25	2.7

More basic information about X-ray residual stress analysis of ceramic materials is compiled in [3].

It should be noted that the evaluated residual stresses are mean values of stresses integrated over the depth of penetration of the X-rays weighed by an exponential function due to absorption. Bending strengths of the machined samples were determined using the concentric-ring test with a loading ring diameter of 8 mm and a supporting ring diameter of 40 mm. According to the linear theory of plate bending, a constant maximum biaxial stress state prevails in the region within the loading ring. Due to the symmetry of this stress state, the "weakest" direction of the surface is always expected to lead to failure. Up to 30 samples with equally machined surfaces were broken at a loading rate of 2.5 kN/s. The statistical dispersion of the measured strength values was calculated by using the two-parameter Weibull distribution with the parameters characteristic strength σ_0 and Weibull modulus m.

RESULTS AND DISCUSSION

Lapping Of Alumina

Fig. 1 summarizes the results of X-ray residual stress determinations and bending tests of alumina in as-fired condition and after lapping with different normal forces and disk speeds. The two types of bars represent average residual stresses measured in surface layers with thicknesses of 11 and 33 µm, respectively.

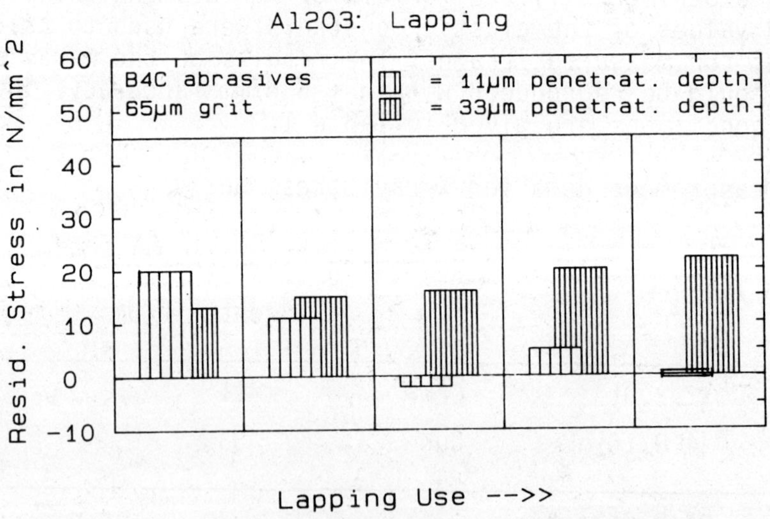

Fn"	[N/mm^2]	-	2	6	2	6
nL	[min^-1]	-	52	52	104	104
σ_0	[N/mm^2]	-	337	344	348	349
m		-	30	35	27	39

Figure 1: Lapping of alumina: Residual stresses due to different lapping forces ($F_{n''}$) and disk velocities (n_L). Lapping parameters and corresponding bending strength data below; measurement depth 11 µm and 33 µm.

At first, we realize that the as-fired surface of the alumina sample is not free of residual stresses. If we use the residual stress state of the as-fired condition as zero level, we find that lapping with a 65 µm grit produces small compressive residual stresses in a very thin surface layer compensated by tensile residual stresses in deeper regions. The amount of residual stresses increases with increasing normal forces (F_n'') and speed of the lapping disk (n_L). The corresponding characteristic bending strength data (σ_0) are marked below the bars. Although the differences of residual stresses and strength data are small, the compressive residual stresses in the thin surface layers seem to increase the strength of the lapped samples.

Lapping Of Partially Stabilized Zirconia (PSZ)

Lapping of PSZ was performed using B_4C abrasives with average grit diameters of 80 µm, 65 µm and 20 µm at constant normal forces (F_n'') of 4 N/mm^2 and at a disk speed (n_L) of 104 min^{-1}.

Fig. 2 shows that lapping of PSZ results in high compressive residual surface stresses. These compressive residual surface stresses are caused by the

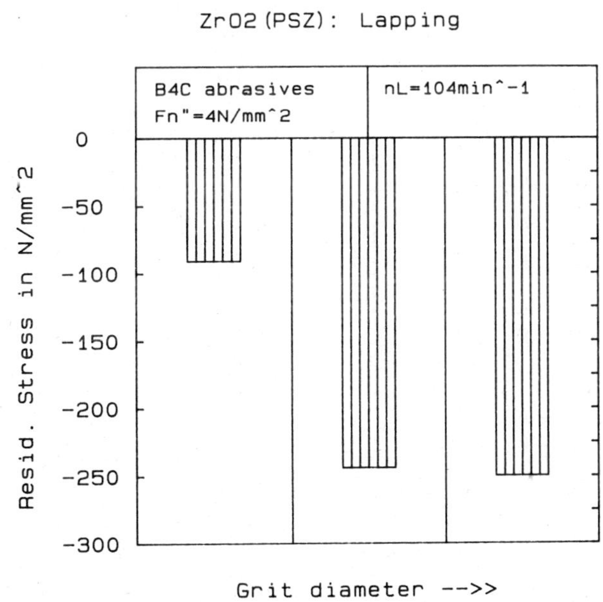

$\bar{d}_K$ [µm]	20	65	80
σ_0 [N/mm^2]	476	498	494
m	42	39	41

Figure 2: Lapping of zirconia (PSZ): Residual surface stresses due to various grit diameters ($\bar{d}_K$) of the B_4C abrasives. Corresponding lapping parameters and bending strength data below.

well-known effect of stress-induced phase transformation of tetragonal polytypes of ZrO_2 into monoclinic phases, combined with an increase of volume, due to lapping. The amount of the compressive residual stresses increases with harder loading conditions during lapping: Keeping the normal force F_n" constant at 4 N/mm^2, abrasives with higher grit diameters (80 µm) lead to lower areas of contact between grits and sample, and therefore to a higher local force during the lapping operation. For samples with a high level of compressive surface stresses, higher (increasing) characteristic bending strengths were measured. But the increase in strength is not as high as the amount of residual stresses would predict. Obviously, only a thin surface layer is tranformation-toughened by lapping.

Down-Cut Grinding Of Silicon-Infiltrated Siliconcarbide (SiSiC)

Grinding experiments on SiSiC were performed using various combinations of infeeds and feed rates in tangential direction. The results of part of these investigations are shown in Fig. 3. They were obtained with samples, which

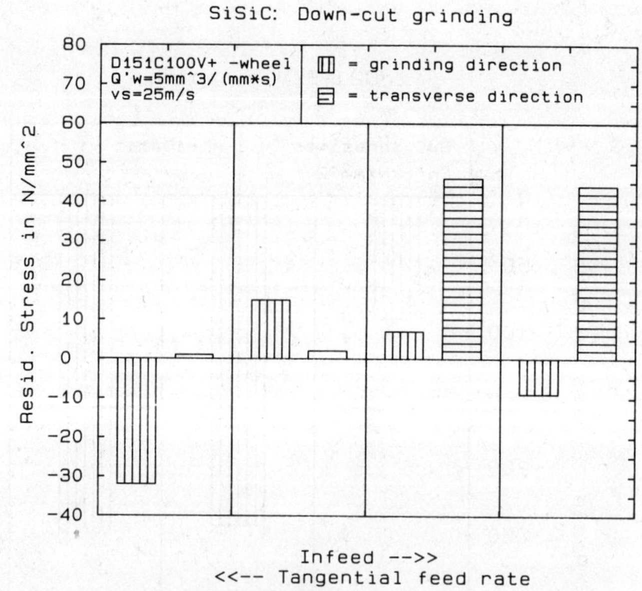

vft [mm/s]	167	50	20	10
ae [mm]	0.03	0.1	0.25	0.5
σ_0 [N/mm^2]	188	221	259	258
m	15.0	11.2	12.2	10.8

Figure 3: Grinding of SiSiC: Surface stress components parallel and transverse to grinding direction, due to different combinations of infeed (a_e) and feed rate in tangential direction (v_{ft}). Corresponding grinding parameters and characteristic bending strength data below.

were ground at a constant material removal rate of 5 mm^3/s per mm width of the grinding wheel. Residual stresses - representing the stress state of 25 µm thick surface layers - were determined in both the grinding direction and the transverse direction. The corresponding grinding parameters (v_{ft}, a_e) and bending strength data (σ_0) are summarized below the bar graph. The results show that grinding with large infeeds ($a_e \geq 0,25$ mm) and small tangential feed rates ($v_{ft} \leq 20$ mm/s) produces tensile residual stresses of about 50 N/mm^2 in the transverse direction (right half of Fig. 3) whereas after grinding with small infeeds and at high tangential feed rates no more stresses are found in transverse direction (left half of Fig. 3). Looking at the corresponding characteristic bending strength data (σ_0), we find a significant decrease of strength of those samples, where no more residual tensile stresses are found in transverse direction.

Grinding at small infeeds but at high tangential feed rates means that the chip thickness at the single grits of the grinding wheel is relatively high and the possibility of surface damage occurs [4]. Obviously, the grinding conditions used for those samples shown in the left half of Fig. 3 caused surface damage, creating surface cracks parallel to the grinding direction. As a consequence, the bending strength σ_0 decreases and the residual tensile stresses diminish in transverse direction. This conclusion becomes substantiated by Fig. 4, where residual stresses (transverse direction) and

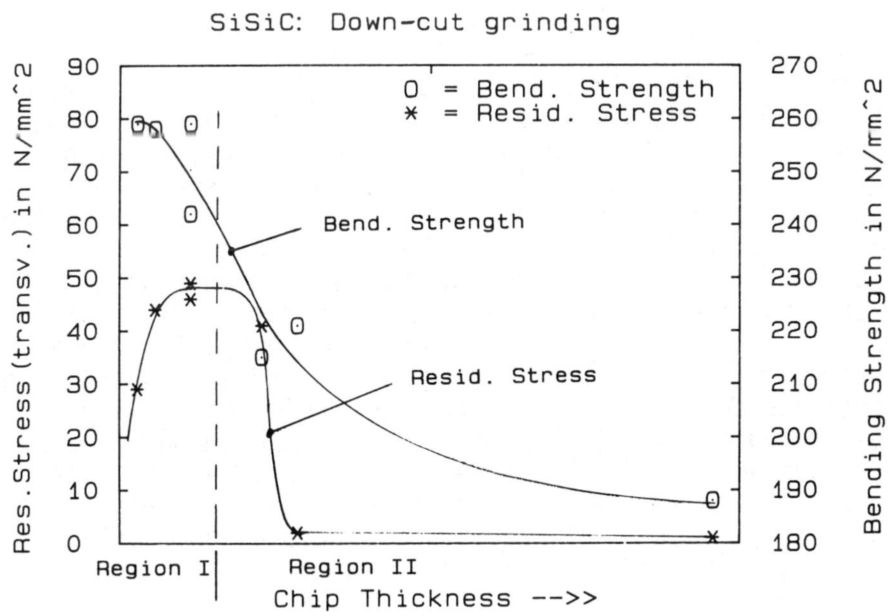

Figure 4: Grinding of SiSiC: Residual surface stress components (transverse direction) and characteristic bending strength data versus the chip thickness at the single grits of the grinding wheel.
Region I: Effect of residual stresses on strength,
Region II: Effect of surface damage on residual stresses and strength.

characteristic bending strength data of additional samples ground at different material removal rates are plotted versus a measure for the chip thickness at the single grits of the grinding wheel. Two different regions can be obtained: At low chip thicknesses (region I), we obtain increasing tensile residual surface stresses with increasing the chip thickness. These tensile residual stresses are likely to be produced by plastic deformation (compression) in transverse direction during the grinding process. In region I where SiSiC samples are ground at small chip thicknesses, an increase in residual tensile stresses is associated with a drop in bending strength.

Grinding at higher chip thicknesses at the single grits leads to surface damage [4]: First, the tensile residual stresses and bending strengths are diminished (region I/II). Second, when surface damage becomes more serious the tensile residual stresses are reduced, and the bending strengths of the machined material decrease due to damage (region II). Successful grinding of SiSiC should therefore be performed at low chip thicknesses which means low tangential feed rates and high peripheral speed of the grinding wheel in order to avoid surface damage and to reduce tensile residual stresses.

Electro-Discharge Machined SiSiC

Due to the high local energy rates at electro-discharge machining of SiSiC ceramics, large temperature gradients along the surface and through the depth are expected to produce residual stresses. Temperature gradients occur during rapid heating (discharge), followed by quenching as a consequence of the individual discharge pulses.

At low discharge frequencies (f_p) and pulse lengths (t_i), we find high tensile residual stresses in the direction transverse to the electro-discharge wire and low tensile residual stresses parallel to the wire (Fig. 5). At higher discharge frequencies and pulse lengths, the tensile residual stresses decrease and change to low compressive stresses. Two different effects seem to be important: First, temperature gradients within the surface plane are responsible for the different amounts of residual stresses in direction parallel and transverse to the electro-discharge wire. Parallel to the discharge wire, a more homogeneous temperature distribution leads to lower residual stresses. Second, increasing discharge frequencies and increasing pulse lengths, which mean shorter intervals between the discharge pulses, minimizes the time spans needed for rapid cooling and makes heat flow to deeper regions possible. The result is a more homogeneous heat distribution versus depth, and therefore, smaller residual stresses occur. Comparing the amount of tensile residual stresses in transverse direction with the bending strength data, we find a close correlation: Bending strength of electro-discharge machined SiSiC samples increases with decreasing tensile residual stresses.

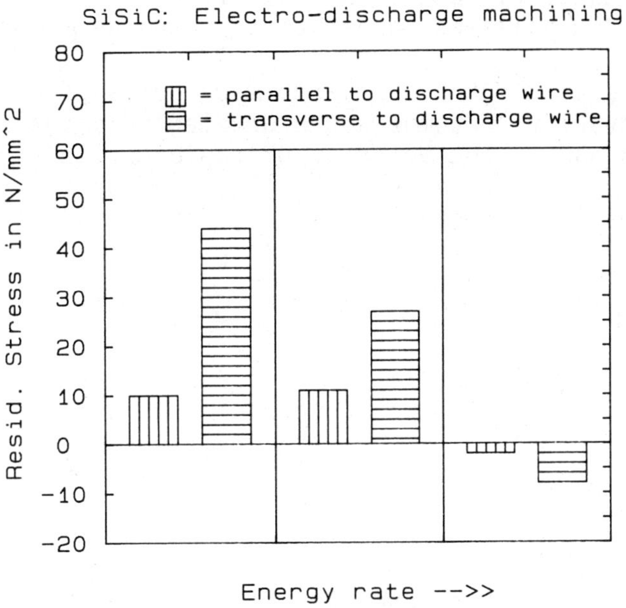

fp [kHz]	10	5	10
ti [µs]	0.6	2.0	2.0
σ_0 [N/mm^2]	174	180	198
m	5.3	9.6	8.4

Figure 5: Wire electro-discharge machining of SiSiC: Residual surface stress components parallel and transverse to the discharge wire. Corresponding machining parameters and characteristic bending strength data below.

SUMMARY

The present study deals with residual surface stresses determined by X-ray stress analysis of ceramic materials like alumina, zirconia (PSZ) and silicon-infiltrated silicon carbide due to various machining processes.

Lapping of alumina leads to small compressive residual stresses in a very thin surface layer compensated by small tensile residual stresses in deeper regions. The amount of residual stresses depends slightly on the conditions of lapping, and seems to influence the bending strength of machined samples.

Lapping of partially stabilized zirconia (PSZ) creates high compressive surface stresses due to the stress induced tetragonal to monoclinic phase transformation during machining. Abrasives with larger grit diameters lead to higher compressive residual stresses and to an increase of bending strength.

The investigations on grinding of silicon infiltrated silicon carbide (SiSiC) show that two main effects influence the residual stresses and the bending strengths. If the grinding parameters are chosen to guarantee that surface damage is excluded, the bending strength is reduced by increasing amounts of tensile residual stresses in the transverse direction. Non-optimized grinding conditions lead to surface damage which introduces flaws which both reduce tensile residual stresses and decreases strength.

Residual stresses of electro-discharge machined surfaces of SiSiC depend mainly on the severity of temperature gradients in the specimen during the discharge pulses and on the velocity of rapid cooling cycles between the discharge pulses. Low discharge frequencies and short pulses lead to tensile residual stresses which decrease the bending strength of samples.

ACKNOWLEDGEMENT

Machining of the ceramic materials was performed by J. Wemhöner, M. Popp, Fraunhofer-Institut für Produktionstechnologie, D-5100 Aachen, FRG. Part of these investigations has been sponsored by the Bundesministerium für Forschung und Technologie of FRG.

REFERENCES

[1] American National Standards Institute, Inc.: Surface Integrity; ANSI B211.1 - 1986.

[2] R. Prümmer, H.W. Pfeiffer-Vollmar; Determination of Surface Stresses of High-Temperature Ceramic Materials, Interceram 35 (1986) [4], 48-50.

[3] R. Prümmer, H.W. Pfeiffer-Vollmar; Determination of Surface Stresses of High-Temperature Ceramic Materials; Proc. of the British Ceramic Society, G.B., Stoke-on-Trent, 1984, 81-98.

[4] W. König, J. Wemhöner; Grinding SiSiC Ceramics: Achieving High Performance and Minimum Damage, see same proceedings.

IMPACT DAMAGE OF SILICON CARBIDE AND SILICON NITRIDE

Y.Hara and S.Hamada,
Engineering Research Center, Tokyo Electric Power Co. Inc.,
2-4-1, Nishi-tsutsujigaoka, Chofu-shi, Tokyo, 182 JAPAN
T.Minematsu, H.Takahashi and H.Abe,
Research Center, Asahi Glass Co. Ltd.,
1150, Hazawa-cho, Kanagawa-ku, Yokohama, 221 JAPAN

ABSTRACT

Ball contact damage was investigated for the case of static and impact loading for SiC and Si_3N_4. Impact tests using an air gun system were carried out in the range of 40 - 160 m/s impact speed. In the case of SiC specimens, main failures due to ball contact were Hertzian cracks or cone shape cracks. There are slight differences between the static and impact tests and the residual strength decreased gradually with increase in contact load. In the case of Si_3N_4 specimens with steel ball indent, there was no degradation of strength in either static or impact tests. On the other hand, radical degradation of strength occured by the impact with SiC and Si_3N_4 balls. Some deep median cracks were the cause of higher degradation of strength in contrast to SiC specimens. Differences in fracture between SiC and Si_3N_4 specimens seemed to be caused by the difference in strength or fracture toughness of the two materials.

INTRODUCTION

Tokyo Electric Power Co. Inc. has been developing a new type of gas turbine which is installed with ceramic parts for economy and diversified usage of fuels. Ceramic parts bring about the increase of intake gas temperature to 1300°C and the reduction of the amount of cooling air [1,2]. There are barriers to overcome in applying ceramics in this area. The FOD (Foreign Object Damage) problem is one of them. A recent study of ball contact damage of SiC has clarified the basic mechanism of FOD [3,4]; (1) Ball contact damages of SiC were Hertzian cracks or cone shape cracks, (2) The load of Hertzian crack initiation depended on the contact ball material; in the order of steel > SiC > Si_3N_4, (3) The residual strength vs. applied load curve coincided each other both for static and impact test conditions. This confirms the applicability of

Hertzian contact theory in the case of impact. In this study, Si_3N_4 was investigated under both static and impact contact conditions and compared with the results on SiC.

EXPERIMENTAL PROCEDURE

Test materials were in-house pressureless sintered SiC and Si_3N_4. The characteristics of the materials are listed in Table 1. Sintered blocks were ground to a bar specimen size of 7.5 x 10 x 85 mm. The method of surface finishing the specimen was #140 grinding. Test conditions are shown in Table 2. Three kinds of balls; steel, SiC, Si_3N_4 were used as indenters. The size of balls was 3 mm in diameter. Table 3 lists the mechanical properties of the balls. In static contact tests, the maximum load depended on the ball material; 10 kN with steel ball and 2 kN with SiC or Si_3N_4 balls, because of the difference of ball fracture load. The maximum impact speed was 160 m/s in impact tests. Three point bending tests were carried out to evaluate the MOR of specimens after contact tests. The number of specimens was 10 for each condition.

Figure 1 indicates the schematic test rig. An Instron type testing machine was used to apply the load in static contact tests. The cross head speed was 0.5 mm/min. An air-gun system was used as an impact test rig. An impact ball was set on a carrier made of juracon. When the electric valve opened, compressed air accelerated the carrier. The carrier stopped at the end of a barrel, which is 600 mm in length, the ball hit the specimen by inertia. The impact speed was changed by the pressure of compressed air (0.05 - 0.5 MPa) and was measured by three photo sensors.

Table 1 Characteristics of Material

Powder	β-SiC	α-Si_3N_4
Sintering aide	Al_2O_3	$Al_2O_3 + Y_2O_3$
Density	3.22 g/cc	3.23 g/cc
Young's Modulus	410GPa	304GPa
Poisson's Ratio	0.18	0.25
Vickers Hardness	25GPa	18GPa
Fracture Toughness (IF Method)	5.3 MPa·$\sqrt{m}$	7.2 MPa·$\sqrt{m}$
Average MOR	520MPa	990MPa

Table 2 Test Conditions

Material	Sintered SiC, Sintered Si_3N_4	
Contact	Static	Impact
Contact Load	1.5～10 kN	40～160 m/s
Ball Size	ϕ3.0 mm	
Ball Material	Steel, SiC, Si_3N_4	
Surface Finish	#140 Grind	
MOR	3 Point Bending Test (Span 60mm)	

Table 3 Characteristics of Ball Material

Material	Steel	SiC	Si_3N_4
Density (g/cc)	7.80	3.22	3.19
Young's Modulus (GPa)	210	420	290
Poisson's Ratio	0.30	0.16	0.25

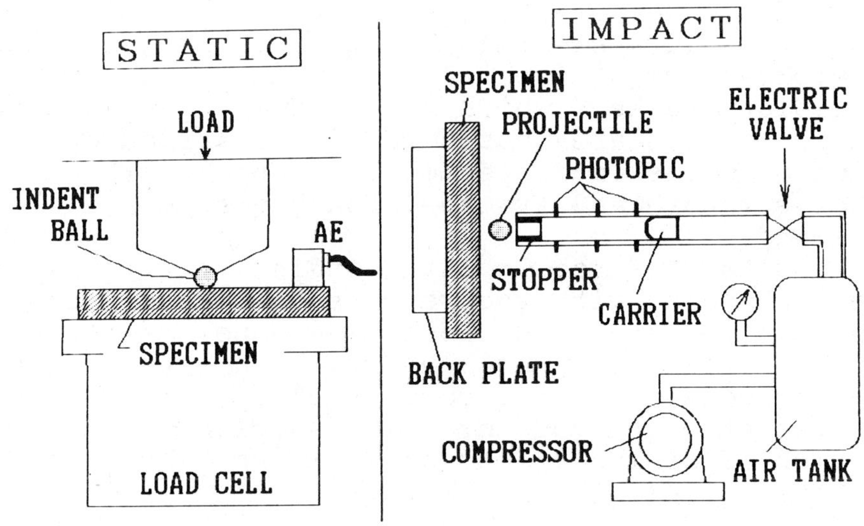

Fig.1 Schematic Test Rig

RESULTS

Relation between contact load and damage of SiC specimens

The relation between the residual strength and contact load for SiC specimens is plotted in Fig. 2 for both static and impact conditions. The contact load during impact was estimated with an assumption that Hertzian contact theory was applicable under impact conditions. The detail is shown in elsewhere [4].

Figure 2 indicates that the residual strength of SiC specimens decreased with the increase in load. Some of the specimens failed at the instant of impact. In the case of impact with steel balls, the results of static and impact tests apparently agreed and could be plotted on the same curve.

With regard to the effect of ball material, the degradation by Si_3N_4 ball impact was conspicuous compared with that by steel ball impact. Some specimens failed at the load of 6 kN when impacted with Si_3N_4 balls. In case of static loading test, the maximum load that is applied to SiC plates was limited to below 2 kN due to the failure of balls when Si_3N_4 or SiC balls were used. Therefore, the results of static and impact tests could not be compared.

Photograph 1 gives typical appearances of damaged surfaces. Concentric Hertzian cracks were formed and increased in number and size with the increase in load. When hit with steel balls, only Hertzian cracks were formed even at the maximum load of 20 kN (impact speed = 160 m/s) whereas median cracks which extended from the outer edge of Hertzian cracks were formed when hit with SiC balls (impact speed = 160 m/s) or Si_3N_4 balls (impact speed = 90 m/s).

High speed impact with Si_3N_4 balls formed hills with microcracks along the outermost rings of Hertzian cracks. The reason why Si_3N_4 balls caused the severest damage might be relevant to the order of the load at the initiation of Hertzian cracks; Si_3N_4 < SiC < steel [3]. Steel balls were conspicuously plastic deformed during impact.

The median cracks, which were formed by high speed impact with ceramic balls, are considered to be formed due to deeper penetration of balls into specimens, hence Si_3N_4 balls which have higher strength and fracture toughness might have caused severer damage.

The results of three point bending tests after contact damage were compared with the results of fractography. It was clarified that the strength loss was due to the innermost Hertzian cone cracks. Specimens with greater strength loss were accompanied with deeper cone cracks. In contrast to this results, the strength loss of Si_3N_4 specimens was caused due to the formation of median cracks as will be shown in the next section.

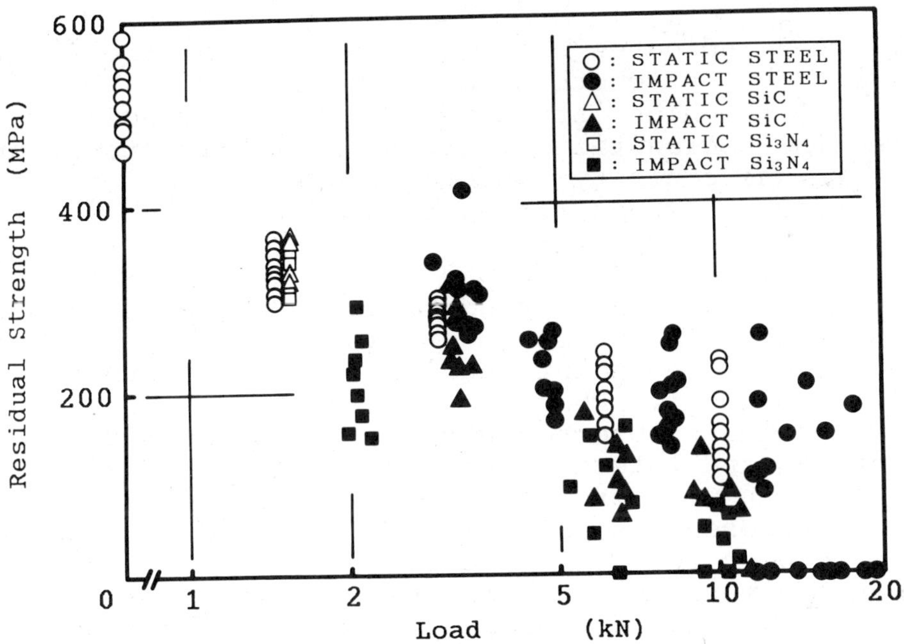

Fig.2 Relation between Residual Strength and Load for SiC

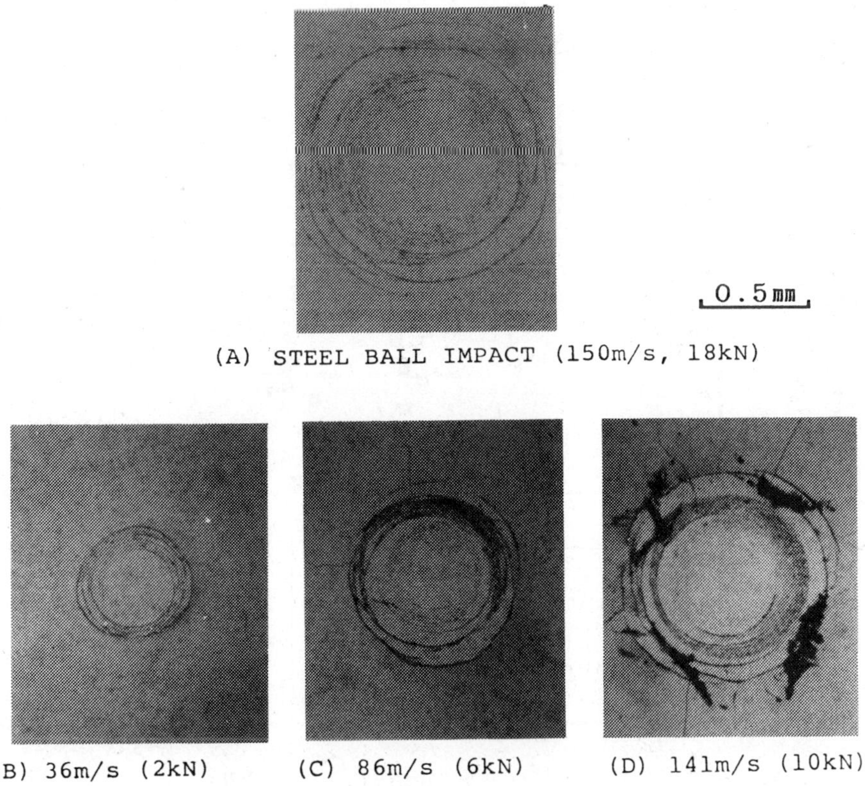

(A) STEEL BALL IMPACT (150m/s, 18kN)

(B) 36m/s (2kN) (C) 86m/s (6kN) (D) 141m/s (10kN)

Si_3N_4 BALL IMPACT

Photo.1 Typical Contact Failure for SiC

Relation between contact load and damage of Si_3N_4 specimens

Figure 3 indicates the relation between the contact load and the residual strength of Si_3N_4 specimens. When hit with steel balls, the high strength of Si_3N_4 specimens (1 GPa) prevented the degradation of strength both in static and impact tests. On the contrary, SiC or Si_3N_4 ball indent resulted in Hertzian cracks even at 1.5 kN, though reduction of strength did not appear. Obvious degradation of strength was observed at contact loads at 3 kN. Photograph 2 shows that high speed impact with Si_3N_4 balls formed hills with microcracks along the outermost rings of surface cracks. A cross section of a specimen after three point bending test showed that deep median cracks were present. Photograph 3 shows typical contact failure of Si_3N_4 specimens observed with a reflective microscope. Some median cracks and lateral cracks grew from Hertzian cracks or from hills with microcracks.

As mentioned above, the ball material affected the relation between residual strength and contact load in the case of Si_3N_4 specimens. It seems that the strong effect of ball material on ball contact damage of Si_3N_4 specimens was caused by following reasons: (1) High strength or fracture toughness of Si_3N_4 obstructs the growth of deep Hertzian cracks; even if it happened, it did not induce the degradation of strength. (2) Obstruction of growth of Hertzian cracks leads to the occurence of median cracks which have a higher stress concentration than that of cone shape cracks. This induces a greater reduction of strength.

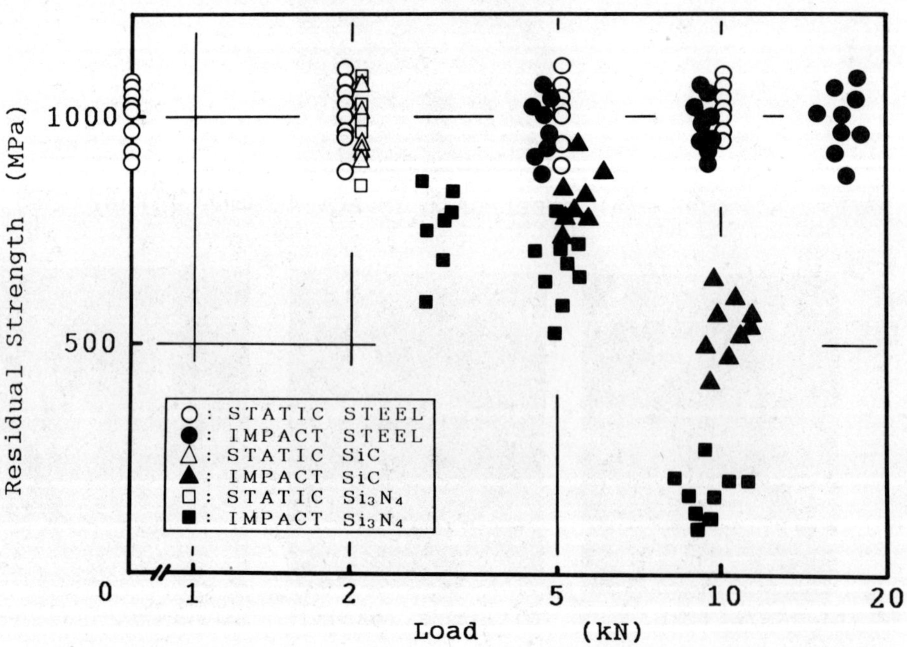

Fig.3 Relation between Residual Strength and Load for Si_3N_4

(A) Si_3N_4 BALL STATIC (2kN) (B) STEEL BALL STATIC (10kN)

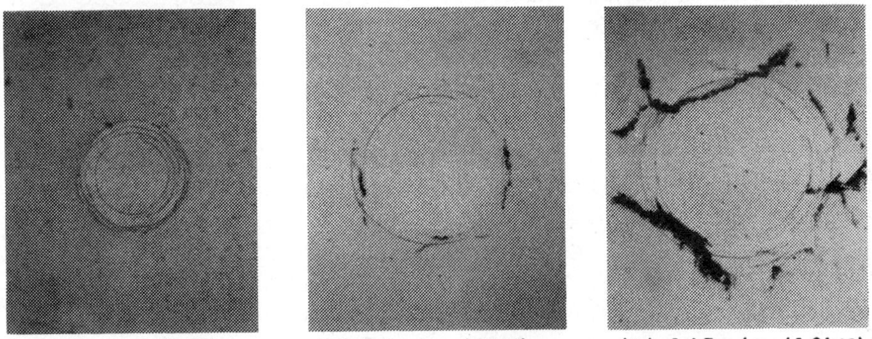

(C) 38m/s (3kN) (D) 95m/s (5kN) (E) 145m/s (10kN)

Si_3N_4 BALL IMPACT

Photo.2 Typical Contact Failure for Si_3N_4

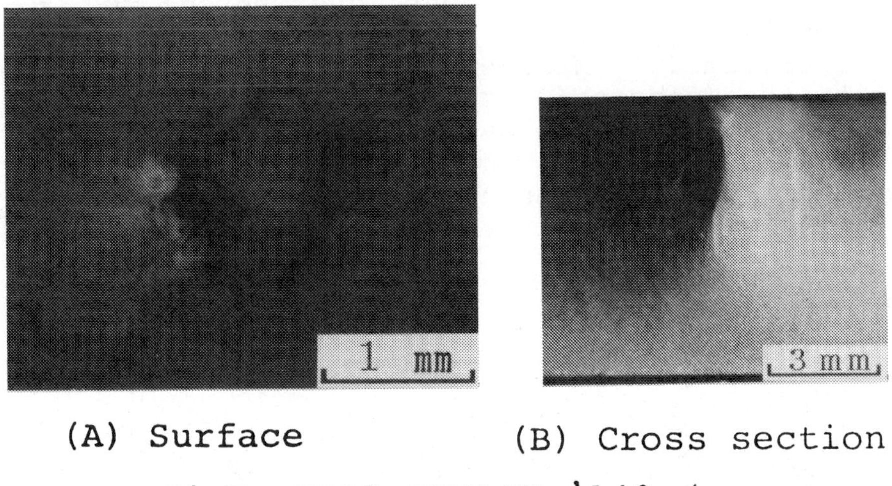

(A) Surface (B) Cross section

Si_3N_4 BALL IMPACT 140m/s

Photo.3 Typical Contact Failure for Si_3N_4 by Reflective Microscope

DISCUSSION

The numerical simulations of the static tests and the impact tests were carried out to clarify the effect of materials on the damage by the ball contact. The target material was SiC and the ball material was either SiC or Si_3N_4. The applied static load was 3 kN and the impact speed of the ball was 50 m/s.

The numbers of nodes and elements, including gap elements, were 2602 and 3393 respectively in the two dimentional axisymmetric finite-element model for the static stress analysis by MSC/NASTRAN. And the three dimensional finite-element model for the impact stress analysis by DYNA-3D, shown in Fig.4, had 1504 nodes and 696 elements. The time increment was taken as 5.8×10^{-4} μs and the computation was continued up to 3.6 μs, which was considered to be long enough to complete contacting.

Figure 5 and Fig.6 show the maximum principal stress distributions for the case of static and impact tests, respectively. Figure 6 shows the result at the onset of rebound (1.4 μs after the first contact) when the maximum tensile stress occures on the surface. Both figures show that although there are slight differences in the maximum principal stresses between SiC and Si_3N_4 ball contacts on quantities and distributions, they could not clarify the effect of ball materials on the damage by the ball contact.

In addition, Fig.6 shows that there is not only a high tensile stress field on the surface, which causes Hertzian cracks, but also a high tensile stress field around the contact area. It seems that the stress which occures around the contact area is the cause of the median cracks. In case of SiC specimens, a lower strength or fracture toughness of the material brought about the initiation of Hertzian cracks at low load, and then the increased load accelerated the growth of cone shape cracks. As for the Si_3N_4 specimens, the high strength or fracture toughness of the specimens obstructed the growth of Hertzian cracks, and median cracks were formed as the contact load was increased. Median cracks have a higher stress concentration than cone shape cracks and that induced a sudden drop of residual strength in case of Si_3N_4 material in contrast to SiC material. By taking the initiation and the extension of the crack into account, further calculations of the stress fields may clarify more clearly the differences in the fracture mechanism.

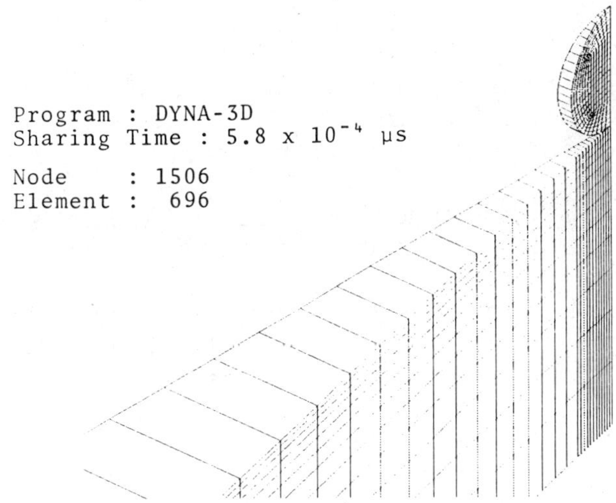

Fig.4 Finite Element Model for Ball Contact Stress Analysis

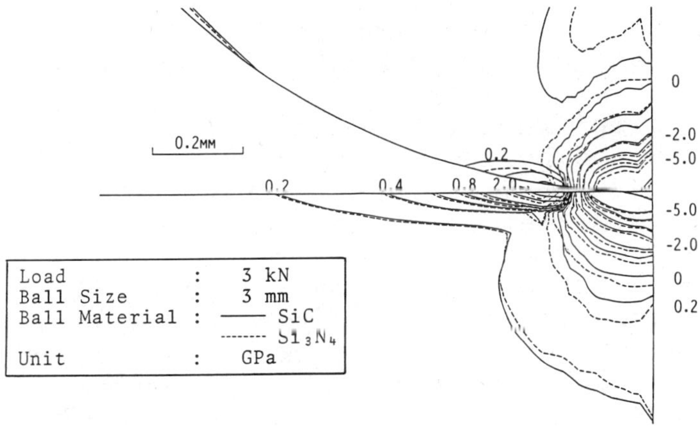

Fig.5 Maximum Principal Stress for SiC Specimen (Static)

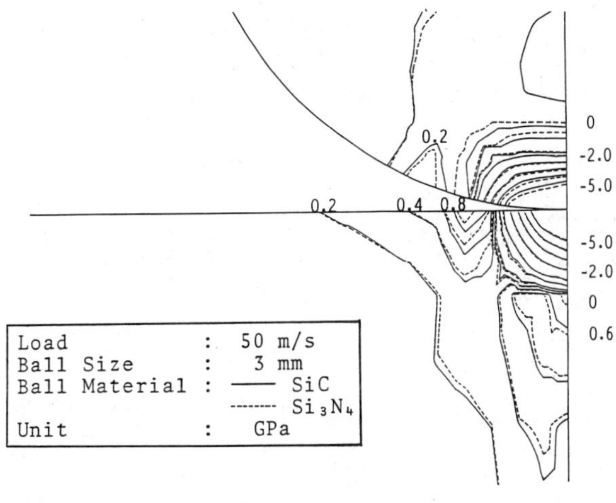

Fig.6 Maximum Principal Stress for SiC Specimen (Impact;t=1.4μs)

SUMMARY

In order to clarify the fracture mechanism of ball contact damage of SiC and Si_3N_4 materials, static and impact tests were carried out and following results were obtained.

(1) In case of SiC specimens, main failures with ball contact were Hertzian cracks or cone shape cracks. There are only slight differences between static and impact tests and the residual strength decreased gradually with increase in contact load. As to the effect of ball material, Si_3N_4 balls caused severer damage in the case of impact tests. It seems that the Si_3N_4 ball caused a deeper dent because it did not fail due to the high strength.

(2) In the case of Si_3N_4 specimens with steel ball indent, there was no degradation of strength in either static or impact tests. On the other hand, radical degradation of strength occured by the impact with SiC or Si_3N_4 balls. Some deep median cracks were the cause of higher degradation of strength in contrast to SiC specimens. Hertzian cracks which were induced in case of static contact tests with SiC or Si_3N_4 balls, did not affect the strength of specimens.

(3) Differences in fracture between SiC and Si_3N_4 specimens seemed to be caused by the difference in strength or fracture toughness of the two materials. Formation of deep Hertzian cracks in the SiC material reduced internal tensile stress which occured under the contact area. Hence, the applied load was spent in the extension of cone shape cracks. High strength or fracture toughness of Si_3N_4 material obstructed the growth of Hertzian cracks and median cracks were formed in the high tensile stress field under the contact area.

REFERENCES

(1) A. Ohkoshi, N. Watanabe, et al., Proceeding of Tokyo International Gas Turbine Congress, I-169 (1987).
(2) Y. Furuse, "Application of Ceramics to Electric Power Generating Gas Turbine", in abstract of International Symposium on Basic Technologies for Future Industries, March 22-25, 1988, 190 (1988).
(3) Y. Hara, T. Minematsu, et al., Proceeding of 1988 MRS International Meeting on Advanced Materials (in printing).
(4) T. Minematsu, H. Takahashi, et al., "Ball Impact Damage of Silicon Carbide", Asahi Garasu Kenkyu Hokoku (Repts. Res. Lab. Asahi Glass Co. Ltd.) 38 (1988).

HIGH STRENGTH CERAMIC COMPOSITES FOR ELECTRICAL DISCHARGE MACHINING

B. Calès, C. Martin and P. Vivier
Céramiques Techniques Desmarquest
78190 Trappes France

ABSTRACT

The recent improvements in mechanical properties of ceramics have led to the development of high strength and high toughness ceramic composites, the machining of which, using conventional techniques, is a rather complicated and expensive routine, especially when complex parts are manufactured. The electrical discharge machining is an attractive machining technique but requires electrically conductive ceramic materials. New conductive silicon nitride and alumina based composites with high mechanical properties were developed for this purpose by adding amounts of TiC and/or TiN particles in the ceramic matrix. The machinability and the physical properties of the composites are related to the content and the characteristics of the conductive TiC/TiN particles. The compositions can be adjusted to find the best compromise between high strength and electrical discharge machinability.

INTRODUCTION

The need for improving the mechanical reliability of advanced ceramics has been recently leading to the development of high strength and high toughness ceramic composites, such as particles reinforced silicon nitride or whiskers reinforced alumina, which are characterized by a 3-point bending strength of the order of 900 to 1000 MPa in a large temperature range and a toughness of 7 to 9 MPa$\sqrt{m}$ (1,2). However, these composites are extremely hard and their machining using conventional diamond tools is an inconvenient and expensive routine that highly contributes to the final cost of the ceramic composite parts. The use of electrical discharge machining (EDM) would be an attractive other routine but requires electrically conductive ceramics materials. The mechanisms of material removal by EDM have been recently analized by Petrofes and Gadalla [(3)]. In the last few years, several studies have been done to increase the electrical conductivity of ceramics. For instance, the electrical conductivity of SiC ceramics has been significantly increased by the addition of TiB_2 particles, leading to an EDM composite with material removal rates much more higher than for tungsten carbide [(4)]. It has also been shown that Si_3N_4 based composites with high electrical conductivity can be obtained by adding a conductive second phase such as carbide, nitride or boride particles [(5)] or silicon carbide fibers [(6)].

In the present study either TiN or TiC particles has been added to silicon nitride or zirconia-toughened alumina ceramics to enhance their electrical conductivity. In the case of silicon nitride ceramics, only TiN particles have been used because of the decomposition of TiC particles at high temperature during the sintering of silicon nitride under nitrogen atmosphere. Furthermore, for high TiC content, higher than 10 wt%, difficulties in sintering arise and a full densification of the material cannot be reached. Two grades of TiN powder was added to the silicon nitride matrix : the grade A with a mean grain size of 0.8 micron or the grade B with a mean grain size of 1.2 microns. In the case of alumina based composites, due to the sintering conditions under vacuum, the electrical conductivity was increased by addition of TiC particles having a mean grain size of 1.5 microns.

The composition of the two families of composites were adjusted in order to obtain the best compromise between a high strength and toughness and a high level of electrical conductivity.

EXPERIMENTAL PROCEDURE

Samples preparation

Silicon nitride based composites were obtained by mixing Si_3N_4 and TiN (grade A or B) powders with Y_2O_3 and Al_2O_3 used as sintering aids. The different powders were ball milled in liquid medium. Green parts, obtained by slip casting, were sintered under nitrogen atmosphere at about 1750°C [7] and densities higher than 97% of theoretical density are achieved for TiN content as high as 70 wt%.

The mixing step for the preparation of whisker-reinforced alumina is a rather complicated process. TiC particles were homogeneously mixed with alumina and zirconia powders and silicon carbide whiskers, using an adequate combination of ball milling, turbomixing and ultrasounds mixing (8). An original and patented routine has thus been developed for the preparation of conductive SiC-whiskers reinforced zirconia-toughened alumina with a very homogeneous microstructure. The green samples were hot pressed in graphite dies under vacuum at about 1600°C to achieve full density.

Samples characterization

The electrical conductivity of sintered ceramic composites were measured using either a complex impedance method for conductivities lower than $10^{-3}\ \Omega^{-1}\ cm^{-1}$ or a four-point d.c. method for higher conductivity values. The EDM tests were conducted on 5 mm thick samples on CHARMILLES TECHNOLOGIES machines.

The mechanical characterization consisted of three-point bend tests at room temperature.

ELECTRICAL CONDUCTIVITY AND MACHINABILITY OF CERAMIC COMPOSITES

Silicon nitride-based composites

The variation of the electrical conductivity of Si_3N_4-TiN composites is shown in figure 1 as a function of TiN content for the two grades A and B of TiN powder. Two domains clearly appear. For low TiN content lower than 37 wt% for grade A and lower than 45 wt% for grade B, the composites exhibit a high electrical resistivity and are insulating materials. For higher TiN contents, due to a percolation phenomenon, the conductivity rapidly increases up to the conductivity of pure titanium nitride. There is a marked effect of TiN grain size on the conductivity, especially for intermediate

TiN content (40 to 70 wt%), the conductivity being higher for the fine TiN powder (grade A). The probability to form a threefold network required for a percolation conduction is higher when smaller TiN grains are used.

Electrical discharge machining experiments were conducted on various composites with different TiN contents. For Si_3N_4-based composite having a conductivity lower than 2 x $10^2\ \Omega^{-1}cm^{-1}$ the EDM cannot be performed since no electric arcs are produced between the workpiece and the tool (wire or sink). For higher conductive composites electrical discharge machining is possible and a very good machinability is observed with brass wire for composites with a conductivity higher than 5 x $10^3\ \Omega^{-1}cm^{-1}$, using deionized water as dielectric. The material removal rate is two to three times higher than that for tungsten carbide and comparable to that for steel. The wire EDM characteristics are presented in Table 1. The machined surface roughness and the overburn thickness are directly related to the electrical

Table 1 : Wire EDM characteristics with a 0.25 mm brass wire

	Si_3N_4-TiN COMPOSITE	STEEL	TUNGSTEN CARBIDE
WIRE SPEED (mm/s)	80	60	80
CUTTING RATE (mm/mn)	3-5	4	1.5
ROUGHNESS (Ra μm) (COARSE CUTTING)	2.8-2.2	1.6	1.3
ROUGHNESS (Ra μ m) (FINISH CUTTING)	1.6-0.6	0.4	0.6

conductivity of Si_3N_4-based composite and are considerably reduced for composites with a conductivity higher than 2 x $10^3 \Omega^{-1} cm^{-1}$ (Fig. 2). Such a conductivity level requires a lower content of TiN particles using grade A instead of grade B. The surface roughness and the overburn thickness may also be reduced by combining coarse machining with finish machining (removal depth of a few microns) (Fig. 2).

Sinker EDM was performed with copper static die and oil as dielectric. The results confirmed the high machinability of the Si_3N_4-TiN composites (Table 2). The material removal rate is higher than for tungsten carbide and the die consumption is four times lower, thus leading to a better machining accuracy.

Alumina-based composites

Electrical discharge machining of SiC-whisker reinforced zirconia-toughened alumina requires a high content of an additional conductive phase, such as TiC particles. The electrical conductivity of SiC-whiskers composites containing only 30 wt% of whiskers is too low to perform EDM. Higher whisker contents give higher conductive composites but also lead to difficulties in sintering. A more appropriate solution for improving electrical conductivity consists in adding extra TiC particles, up to 30 wt% with respect to the zirconia-toughened alumina matrix (Fig. 3).

Wire EDM experiments show that machining is possible for composites having an electrical conductivity higher than 1 $\Omega^{-1} cm^{-1}$. This limit is about three orders of magnitude lower than for the machining of Si_3N_4-based composites. This is probably due to the difference in the characteristics of the matrices of the two families of composites. In the case of Si_3N_4-based composites, the Si_3N_4 matrix may be oxidized during EDM so that a higher electrical conductivity level is required for a good machining.

MECHANICAL PROPERTIES OF CONDUCTIVE COMPOSITES

Silicon nitride-based composites

Table 2 : Sinker EDM characteristics with a copper die and oil as dielectric

	Si_3N_4-TiN COMPOSITE	STEEL	TUNGSTEN CARBIDE
DIE POLARISATION (80V)	POSITIVE	POSITIVE	NEGATIVE
COARSE SINKER EDM			
INTENSITY (A)	10	10	–
DIE CONSUMPTION (%)	5	5	20
REMOVAL RATE (mm^3/mn)	60	70	30
FINISH SINKER EDM			
INTENSITY (A)	4	6	
DIE CONSUMPTION (%)	6	6	20
REMOVAL RATE (mm^3/mn)	18	22	12

As shown in figure 4, coarse TiN particles (grade B) can be added to the Si_3N_4 matrix up to 60 wt% (50 vol%) without modification of the room temperature fracture strength of the Si_3N_4 matrix. For higher TiN content (> 50 vol% grade B), strengthening, due to the well-known microstructure of Si_3N_4 with elongated ß grains, is no more effective and the fracture strength of the composite rapidly decreases.

When fine TiN particles (grade A) are used to increase the electrical conductivity of silicon nitride, a strengthening effect is observed for a TiN content of 50 wt% (Fig. 4). Such a strengthening may result from the thermal expansion mismatch between the Si_3N_4 matrix ($3 \times 10^{-6} K^{-1}$) and the TiN particles ($9 \times 10^{-6} K^{-1}$) that generates a compressive hoop stress around the TiN particles [(9)] and enhances a crack deflection mechanism. As shown on the SEM microstructures (Fig. 5), the fracture surface of Si_3N_4-TiN (grade A) composites is similar to that of pure silicon nitride with the same pull-out of elongated ß grains. Futhermore due to crack deflection mechanisms the fine TiN particles appear as protruding grains on the rough fracture surface. Only a few TiN grains appear to be cleaved by the crack.

For the composite containing TiN-grade B coarse particles, the SEM microstructure reveals a lot of cleaved TiN grains (Fig. 6) that did not contribute to a composite strengthening by a crack deflection mechanism. Therefore there is a critical size for the TiN particles to reinforce the composite.

Alumina-based composite

In the case of SiC-whisker reinforced alumina, the bending strength decreases when TiC particles are added. The thermal expansion mismatch between the alumina matrix and TiC particles is too small to induce a significant prestressing of either the matrix or the particles and leads to a composite strengthening. However, the resulting bending strength for high TiC content (30 vol%) is still high and is equal to 850 MPa (Fig.3).

CONCLUSION

Special grades of conductive ceramic composites, with either a silicon nitride or an alumina matrix have been developed for EDM. These composites exhibit high fracture strength, at least equal to that of the matrix, and high electrical discharge machinability. They are more easily machined than tungsten carbide and the tool consumption is quite lower. Machined surface with low roughness and high material removal rate have been obtained after optimization of the EDM conditions.

Therefore, these materials are expected to replace conventional silicon nitride or whisker reinforced alumina ceramics in a lot of applications where complex machined parts are required.

ACKNOWLEDGEMENTS

The authors thank ALLIAGES FRITTES METAFRAM and CHARMILLES TECHNOLOGIES for their help in the optimization of the EDM conditions.

REFERENCES

(1) M. RULHE, B.J. DALGLEISH and A.G. EVANS, Scripta Metall., 21, 681 (1987)

(2) G. ZIEGLER, J. HEINRICH and G. WOTTING, "Relationships between processing microstructure and properties of dense and reaction-bonded silicon nitride", J. Mat. Sci., 22, (9), 3041-3086 (1987)

(3) N.F. PETROFES and A.M. GADALLA, "Electrical discharge machining of advanced ceramics", Am. Ceram. Soc. Bull., 67 (6) 1048-1052 (1988)

(4) C.H. McMURTRY, W.D.G. BOECKER, S.G. SESHADRI, J.S. ZANGHI and J.E. GARNIER, "Microstructure and material properties of $SiC-TiB_2$ particulate composites", Am. Ceram. Soc. Bull., 66 (2) 325-329 (1987).

(5) E. KAMIJO, M. HONDA, M. HIGUCHI, H. TAKEUCHI and T. TANIMURA, "Electrical discharge machinable Si_3N_4 ceramics", Sumitomo Electrical Technical Review, 24, 183-190 (1985)

(6) N. TANARI, I. KONDO, K. UEENO and Y. TOIBANA, "Electrical discharge machining of Si_3N_4-SiC whisker composite ceramics", J. Ceram. Soc. Jap., 94 (12), 1231-1235 (1986)

(7) J.P. TORRE and Y. BIGAY, "Fabrication of silicon nitride parts by slip casting", Ceram. Engi. Sci. Proc. 7 (7-8), 893-899 (1986)

(8) B. CALES, P. MATHIEU and J.P. TORRE, "Preparation and characterization of whisker-reinforced zirconia toughened alumina", 813-818, in Science of Ceramics 14t edited by D. TAYLOR, The Institute of Ceramics, Stoke-on-Trent (UK) 1988

(9) R.W. RICE, "Mechanisms of toughening in ceramics matrix composites" Ceram. Engi. Sci. Proc., 2 (7-8) 661-681 (1981)

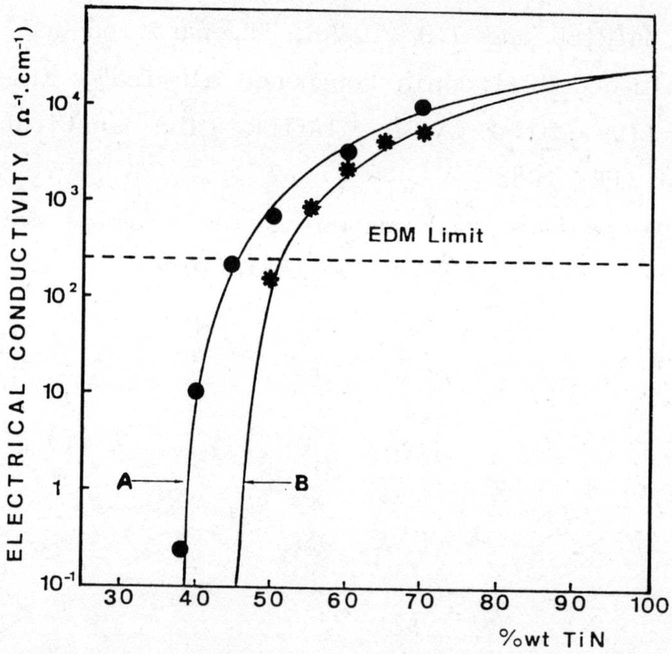

Figure 1 : Variation of the electrical conductivity of Si_3N_4-TiN composites versus the TiN content for TiN grade A and grade B.

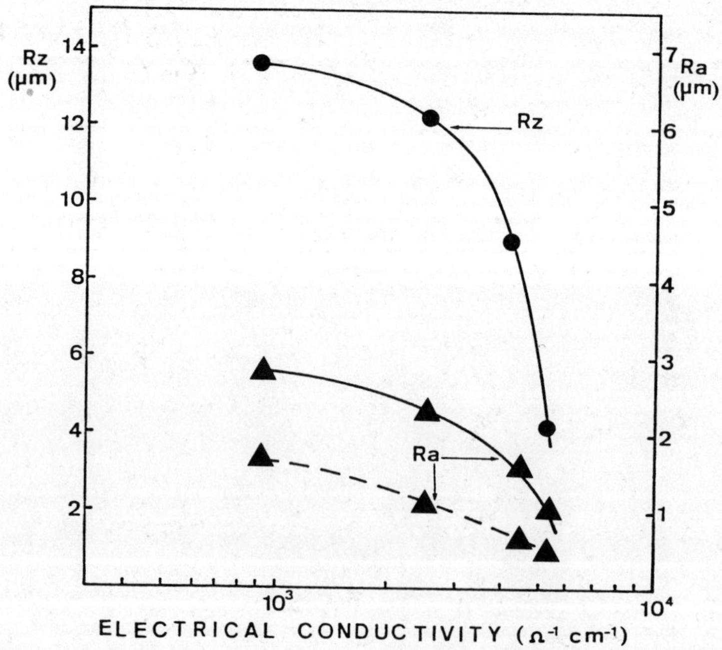

Figure 2 : Variation of the machined surface roughness versus the electrical conductivity of Si_3N_4-TiN composites. Straight lines correspond to coarse cutting and dotted line to finish cutting.

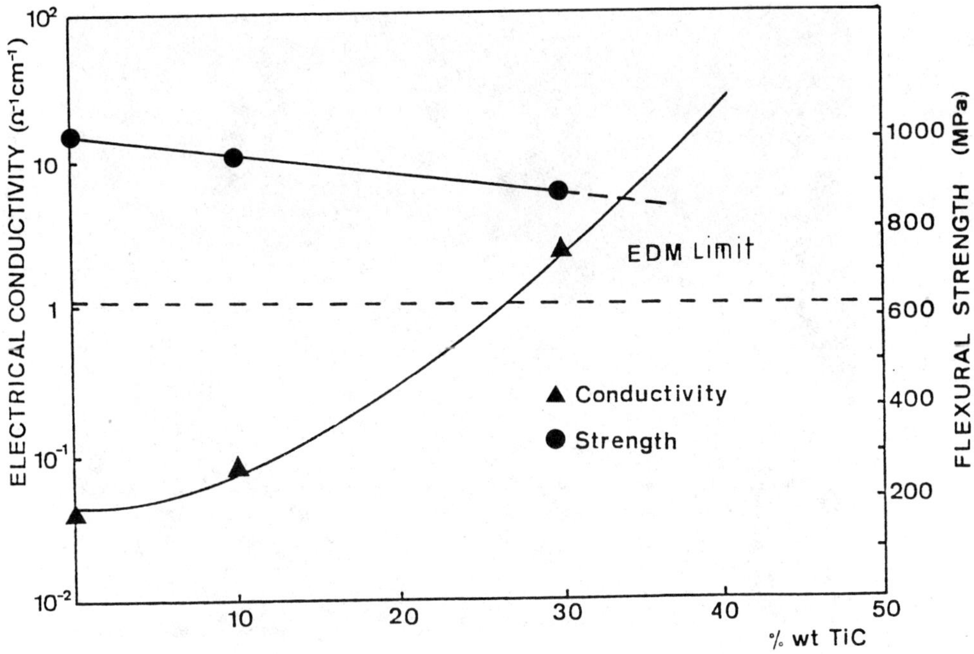

Figure 3 : Dependence of the fracture strength and the electrical conductivity of SiC whisker-reinforced alumina on the content of added TiC particles.

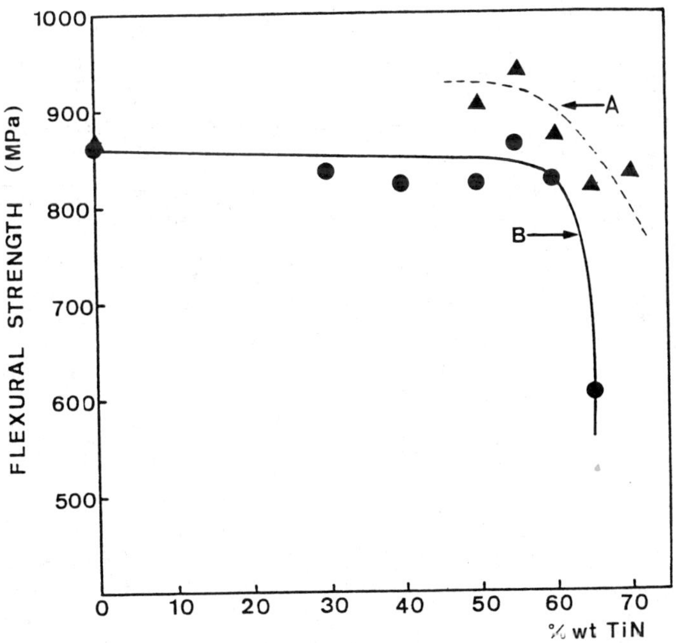

Figure 4 : Variation of the 3 pt. bending strength of Si_3N_4-TiN composites versus the TiN content (● : grade B - ▲ : grade A).

Figure 5 : SEM micrograph of Si_3N_4-TiN (grade A) composite.

Figure 6 : SEM micrograph of Si_3N_4-TiN (grade B) composite.

PREPARATION AND PROPERTIES OF Y-TZP ABRASION-RESISTANT GRINDING MEDIA

Yi-Yuan Wu & Shao-Tang Zhao
South China University of Technology
Structural Ceramics Res. Lab.
Dept. of Inorganic Materials of Sci. & Eng.
Wu Shan, Guangzhou, P. R. China 510641

ABSTRACT

In this study a wet-chemical method was used to titrate diluted ammonia at certain pH with the mixture of $ZrOCl_2 \cdot 8H_2O$ and YCl_3 solutions. High-purity ultrafine zirconia powder was prepared with this sol-gel process. The properties of the powder were identified with XRD, TEM and BET techniques. Grinding media were fabricated by isostatic pressing, followed by optimized-sintering to nearly 98% theoretical density. Some mechanical properties such as microhardness, fracture toughness, and bending strength were determined and microstructures examined. Comparison was made to indicate the superior abrasion resistance against commercial alumina grinding media(balls) and the validity of utilizing of zirconia grinding media was demonstrated.

PREFACE

The production of zirconia engineering ceramics necessiates high-purity and ultrafine raw materials. The traditional process used steel lined ball mills or vibration mills and steel grinding media, followed by acid washing to remove the ferrous contamination. High efficiency grinding machines have been developed in our university in the recent years, which can grind the powders to sub-micron sizes. So, it is very important to have a low wear, low contamination grinding media.

In the course of working upon development of ceramic materials and components for engines for abrasion applications, we used toughened zirconia material — the tetragonal polycrystal zirconia. The object of the work described in this paper was to develop a low abrasion grinding media as a tool to be used in the production of abrasion-resistant ceramic components and serves as a fabrication processing reference.

THE SELECTION OF FABRICATION PROCESSES

1. COMPOSITION

There is wide interest in research on toughened zirconia ceramics. It has been reported that the room temperature bending strength of Y-PSZ and Y-TZP may reach 1-2.4 GPa; fracture toughness K_{IC} 6-20 $MPa \cdot m^{1/2}$ [1-5]. The key process of preparing these kinds of materials is the selection of optimum stabilizing agents to sinter the products at certain temperatures and control the grain sizes beyond some critical values. With proper processing most of the transformable tetragonal zirconia will be left to room temperature to exert the utmost toughening function. Therefore, the particle size of raw materials are extremely critical. Among many methods, we choose the wet-chemical process to achieve high-purity and ultrafineness. The utilization of this kind of powder make possible to lower the sintering temperature and controllably uniform and fine-grained microstructure of sintered bodies.

According to the research work of S. U. Pliner et al [5], the maximum strength of the Y_2O_3-containing materials appeared in the interval of Y_2O_3 contents between 2.5-3.0 mol.%, as seen from Fig.1.On other hand,F.F.Lange[6] found in experiments when the Y_2O_3 contents were in the range of 2.5-7.5mol.% the fracture toughness K_{IC} decreased as the Y_2O_3 increased,as Fig.2. Since we use ultrafine raw materials such as 0.02-0.035 μm,from the view points of economy and processing, we choose 2.5 mol.% Y_2O_3 as partial stabilizer.

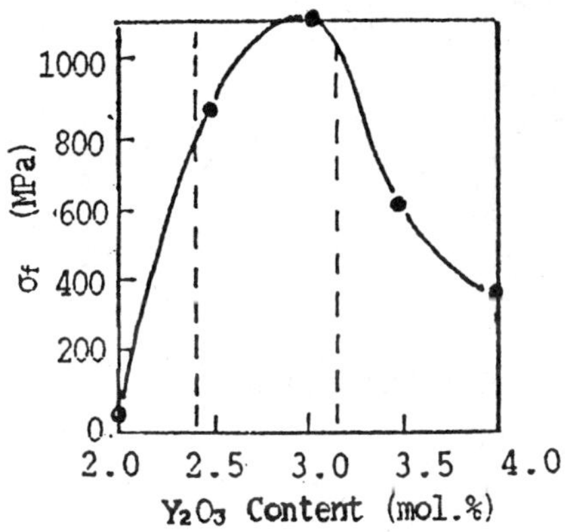

Fig.1 Strength vs Y_2O_3 (m/o)[5]

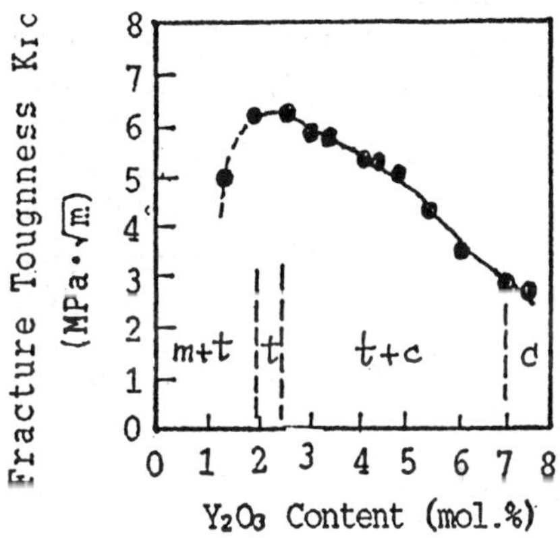

Fig.2 Fracture Toughness vs Y_2O_3 (m/o)[6]

2. PREPARATION OF ULTRAFINE POWDER

Sol-gel method was utilized to prepare ultrafine ZrO_2 powder,the flow chart is as Fig.3.

In order to have ultrafine particles in the reactants in aqueous reaction, $ZrOCl_2 \cdot 8H_2O$ was batched into 0.5 mol./l solution and mixed with adequate amount of YCl_3 solution to make a mixture of 2.5 mol.% Y_2O_3. The mixed solution was slowly and steadily dripped into the diluted ammonium hydrooxide solution and vigorously agitated. The pH of the mixture was kept at pH 10 during the reaction. The precipitate was washed with distilled water several times to remove Cl- group. Filtrated liquid was swung out with centrifugal machine. The precipatate was dried in a drying oven at constant controlled temperature, 80°C to 140°C, according to different cases. Extreme care had to be taken to keep drying evenly to avoid amergrating. Afterwards,vibration-grind the powder with alcohol for 3 or 4 hours, dried and screened to pass 60-mesh.

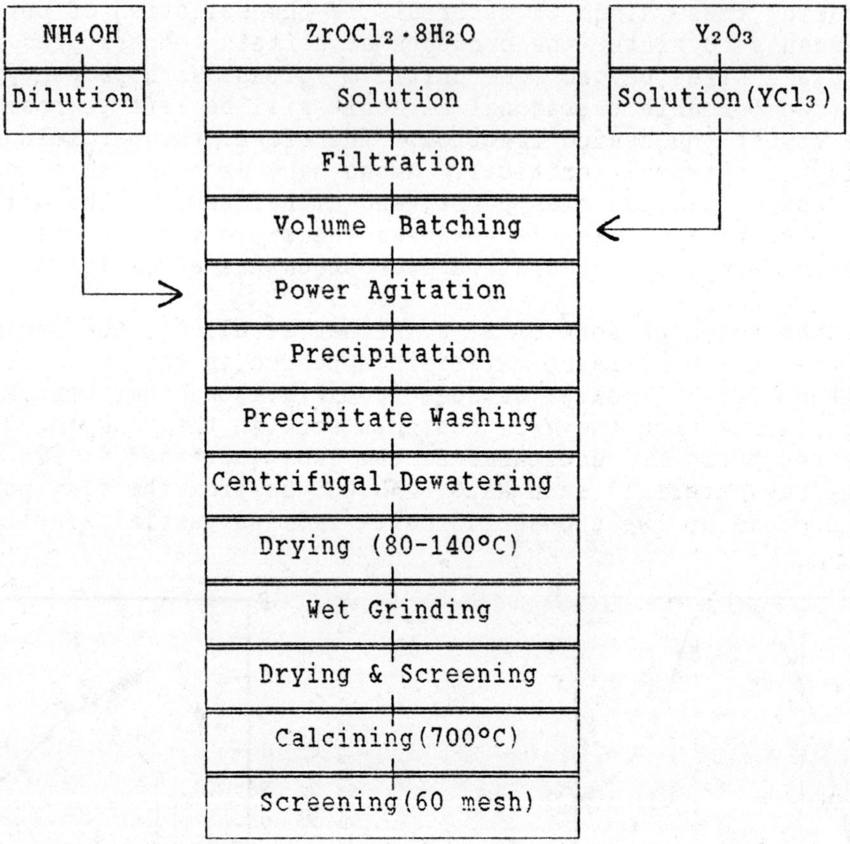

Fig.3 Flow-Chart of Y-TZP Processing in this research

3. FABRICATION METHOD

The fabrication of grinding media is cold isostatic pressing for uniform compacted in all directions in order to achieve higher densification and more uniform texture in the green body before sintering.
After forming was green body machining, electric furnace sintering.

EXPERIMENTAL TECHNIQUES

1. RAW MATERIALS CHARACTERISATION

(1)Particle Size and surface area
Particle size was determined under electronic microscope Model 100CX II and measurements on TEM photo were taken.
The calcining temperature has sifgnificant influences to the particle size of the powders. BET method was used to determine the specific area of the powders calcined at different temperatures.

(2)Thermal Analysis
Use DuPont 1090B Thermo Analyzer to examine the TGA and DTA characteristics of the powder. Heating Rate = 10°C/min.

2. DENSITIES AND SHRINKAGE

(1)Water Displacement method was to used in determining densities:

$$Do = Go/V = Go \times 100\% / (G2 - G1) \qquad (1)$$

(2)Shrinkages in the different stages of processing were calculated from the length measurement data.

3. MICROHARDNESS

The green body bars were cut into slices of 5-7 mm, two opposite surfaces were ground and sintered. The surfaces were polished with 360 and 500 mesh wet sandpaper successfully,and polished into mirror quality with #04, #02, W5 and W1 diamond polishing paste. Vickers' microhardness tester HV-120 was used to determine the indentation length without cracks and calculated the microhardness with the formula,

$$Hv = 1.8544\, P/(d \cdot d) \qquad (2)$$

where: P—Loading(kg); d—the average length of the two diagonals of the indentation, mm.

4. BENDING STRENGTH

The Bending Strength of specimens sintered was determined according the Chinese National Standard GB 6569-86,"Testing Method for Flexural Strength of High Performance Ceramics" [7]. Unusual data were discarded according to the deviation criterion in this standard. The testing parameters were: loading speed 0.4 mm/min.;span 30 mm;sample size: 36 x 4 x 3 mm.

5. FRACTURE TOUGHNESS K_{IC}

Direct Indentation Method was used to determine the K_{IC} values on an HV-120 Vickers' Microhardness Tester. The specimens previously described were taken to subject to load of 30 and 20 kg and the indentation lengths 2a and the lengths of cracks 2c were measured. We use the formula from K.Niihara [8] to calculate the values of K_{IC} $(0.25 < c/a \leq 2.5)$.

$$(K_{IC} \cdot \phi / H a^{1/2})(H/E \cdot \phi)^{2/5} = 0.035\, (l/a)^{1/2} \qquad (3)$$

where: ϕ— contraint factor, approximately equals to 3; H — Vickers' Hardness,Hv; E — Young's Modulus;c — the length of the cracks; a — the half and average length of the diagonal of the indentation. c = l + a.
As several investigators mentioned that this method was not a standard method,we only use it as a conditional comparative test within our parameters.

6. MICROSTRUCTURE OBSERVATION AND PHASE IDENTIFICATION

The bodies sintered at 1500 and 1550°C were observed under the SEM to measure the sintered grain sizes and texture. Plane samples were taken to subject phase identifation under The Model MAX IIA automatic X-Ray Deflectometer.

7.WEAR RESISTANCE OF GRINDING MEDIA

For the sake of examining the wear resistances of the grinding media, ZrO_2 grinding cylinders were placed in a 3 liter vibration mill to commercial alumina powder (by Bayer Method).
Some sintered corundum grinding balls were used to make comparison upon these two materials. Wear ammounts were determined for a period of grinding hours. The parameters of the grinding were: material/media/water = 1/3/2.5 (vol).

RESULTS AND DISCUSSION

1.DTA AND TGA DATA OF RAW MATERIAL

The DTA and TGA curve of the Y-TZP gel was shown in Fig.4. It is obvious that dehydration occurred gradually from about 100°C to 450°C. The peak of 454.5°C denoted crystallization.

2. SURFACE AREA OF THE POWDERS AND SHRINKAGE DURING SINTERING

The surface areas of the powders under different calcining temperatures were determined as Fig.5. Shrinkage curve is shown in Fig.6.

The surface areas of the powders decreased as the calcining temperature increased,and the slope of the curve decreased above 700°C. 700°C was chosen to reserve good sinterability to lower sintering temperature and balance the forming capabily and shrinkage of the green bodies.

3. MOROPHOLOGY OF THE POWDER

After calcination of 700°C,TEM photos (Fig.6, above) were taken and measured to get the average grain sizes. It was noticed that the average diameter was about 0.02 to 0.025 μm; quite uniform and well seperated. The rounded or nearly sphere shape make this powder easily achieves dense pack and uniformly sintered.

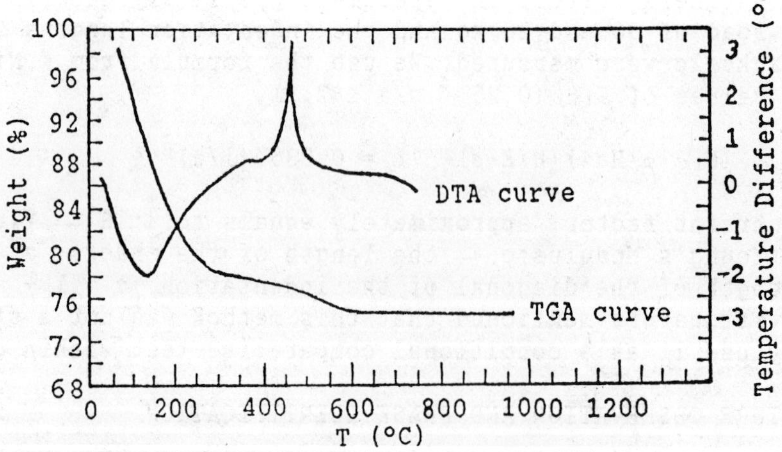

Fig.4 Thermal Analysis of Y-TZP Gel

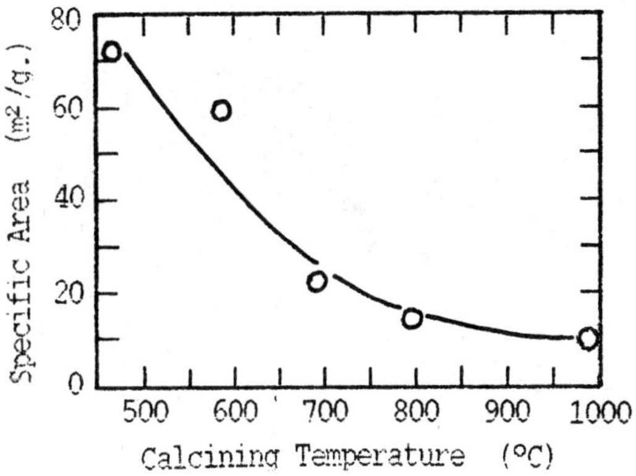

Fig.5 Spec.Area vs Calcining Temp

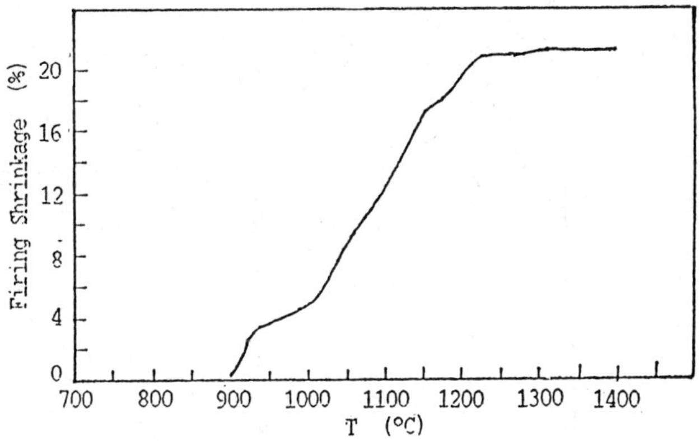

Fig.6 Shrinkage curve of Y-TZP body

4. PHASE EXAMINATION OF THE RAW MATERIALS AND SINTERED BODIES

The X-ray Deflection Patterns of the raw materials (700°C calcined) and sintered bodies (1500,1550°C) were shown in Fig.7,8 and 9. After sintering the main crystalline phase was tetragonal phase ZrO_2 when sintered to 1500°C, and monocline variety appeared in 1550°C sintering. The content of t-ZrO_2 sintered at 1500°C was nearly 100% and somewhat decreased at 1550°C.

5. MICRSTRUCTURE EXAMINATION OF THE SINTERED BODIES

The fracture saurface microstructures may be illustrated as Fig.10 to Fig.12, which represented sintered at 1500°C and 1550°C,respectively. In both cases polyhedral crystals predominated; grains well crystallized; textures were fine and uniform, the boundarires of the crystals were clear and sharp, there was a certain ammount of trans-crystalline fracture, which attributes good strength behaviour. The average grain size sintered at 1500°C is 0.25 μm, and 1550°C 0.74-1.0 μm. In order to retain more tetragonal zirconia phase to room temperature, and from the view point of increasing the abrasion resistance, it would be better to hold uniform and fine-grained bodies. So,we choose 1500-1530°C to sinter pieces of this composition,the thicker or bigger the pieces, the higher the sintering temperature.

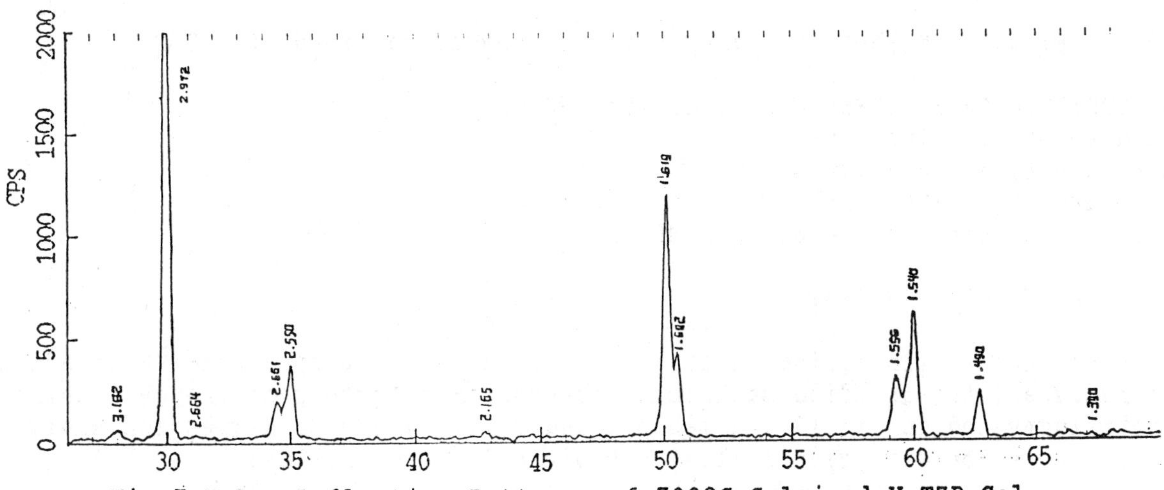

Fig.7 X-Ray Deflection Patterns of 700°C Calcined Y-TZP Gel

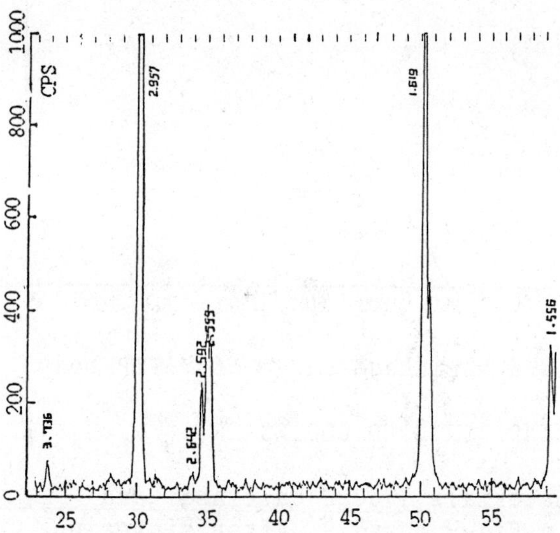

Fig.8 XRD Patterns of Y-TZP Grinding Media sintered 1500°C

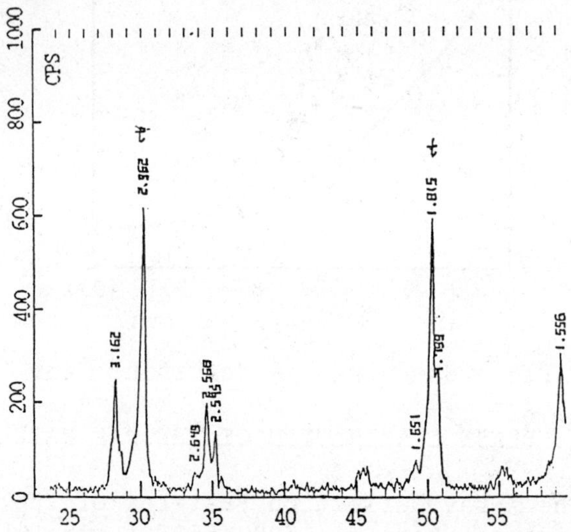

Fig.9 XRD Patterns of Y-TZP Grinding Grinding Media sintered 1550°C

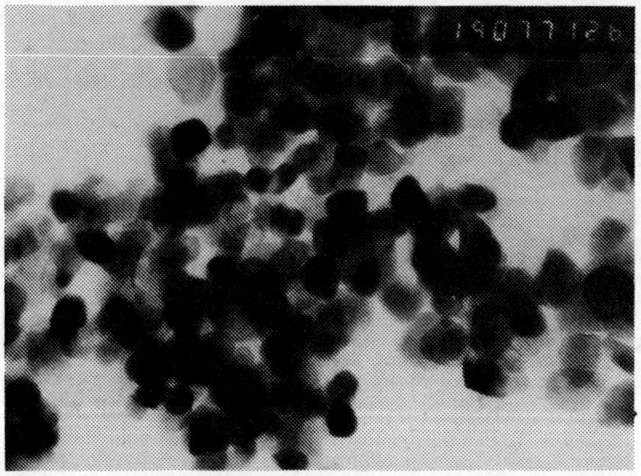

Fig.10 TEM Photo of Y-TZP ultrafinepowder calcined at 700°C

6. DENSITY,MICROHARDNESS AND FRACTURE TOUGHNESS

The determined values were:

bulk density = 5.81-5.94 g./cm^3;

microhardness = 1064-1197 MPa;

fracture toughness K_{IC} = 6.1-8.9 MPa·m^{½}

7. GRINDING ABRASION TEST

The technical alumina-grinding abrasion losses of the prepared Y-TZP grinding cylinders and the grinding balls manufactured in a factory in Shanghai,China were illustrated in Fig.13 and Table 1 and 2. Least Square Regression Analysis for the abrasion data gave following empirical relationships:

Fig.11 Fracture Surface SEM photo of Y-TZP Grinding Media sintered at 1500°C x 3hr

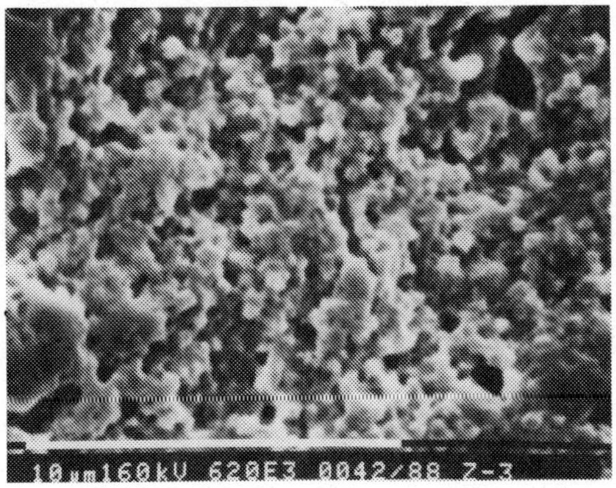

Fig.12 Fracture Surface SEM photo of Y-TZP Grinding Media sintered at 1550°C x 1hr

(1)The first stage of abrasion of alumina balls:

$$\Sigma\delta W = -1.4746 + 0.7644\ t$$

where: $\Sigma\delta W$—Weight loss during grinding, g.;t—grinding duration, hr. regression coefficient r=0.9870;regression deviation σ_r=1.6912.

(2)The second stage of abrasion of alumina balls:

$$\Sigma\delta W = 10.35 + 0.29\ t$$

r=0.9872; σ_r=1.0165.

(3)The all 8 points abrasion data for Y-TZP cylinders:

$$\Sigma\delta W = 0.1469 + 0.0346\ t$$

r=0.9816; σ_r=0.1710.

(4)The pilot run abrasion data for Y-TZP cylinders(not plotted):

$$\Sigma\delta W/W = -0.0545 + 0.005828\ t$$

where: W—initial weight of media, 2000 g. r=0.9895; σ_r=0.06244.

Table 1. Abrasion Test Data of Y-TZP and alumina grinding media (g.)

Gringing Duration(hr)	0	5	15	25	35	45	55	65
Aumina Balls	0	0.5	10.0	18.0	20.0	24.0	25.0	30.0
Y-TZP Cylinders	0	0.0	0.3	0.7	0.8	1.5	2.0	2.0

Table 2. Abrasion Test Data of Y-TZP Grinding Media in Pilot Run

Grinding Duration(hr)	0	14	62	133
Abrasion Loss (%)	0	0	0.25	0.75

The abrasion of Y-TZP grinding media is only about 1/9 of that of sintered alumina balls.

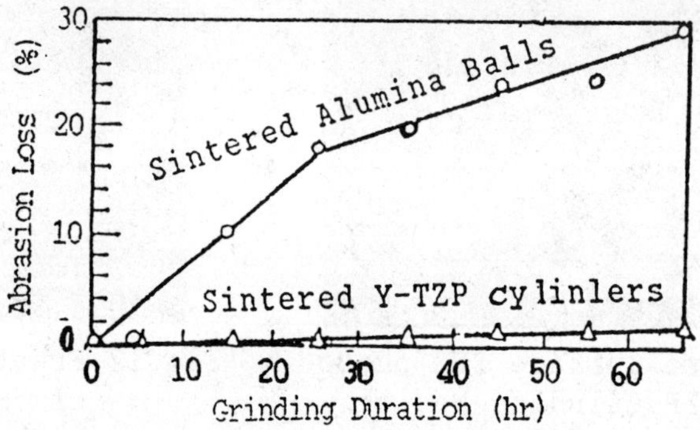

Fig.13 Abrasion ammount vs grinding duration of Y-TZP and alumina.

Table 3. 3-p bending strength data of Y-TZP Grinding Media

Sample Number	Range (MPa)	Average Value (MPa)	Standard Deviation (MPa)
17	1035-3948	2097	849

Table 4. Room Temp. Bending Strength of Y-TZP Ceramics from Various Sources

Composition (Y_2O_3 mol%)	Bending Strength (MPa)	Author or Company	Source	Remarks
2.5	2097	This Paper		d=5.81-5.94; $t\text{-}ZrO_2 \approx 100\%$
2.0	1300	Toyo Soda	[1]	Code 2Y: d=6.07; K_{Ic}=10
3.0	1200	-do-	[1]	Code 3Y: d=6.05; K_{Ic}=8
3.0	2400	-do-	[1]	Code Super: d=5.50; K_{Ic}=6
?	819	Matsumoto	[2]	d=5.49; K_{Ic}=8.0; $m\text{-}ZrO_2$=21%
?	556(230°C,400hr)	-do-	[2]	d=5.40; K_{Ic}=7.6; $m\text{-}ZrO_2$=41%
?	717(1000°C,24hr)	-do-	[2]	d=5.49; K_{Ic}=8.1; $m\text{-}ZrO_2$=12.5%
8.0	1200 500(at 800°C)	Kawanami	[3]	d=6.05; K_{Ic}=7-12
?	693 maximum	T.K.Gupta	[4]	$m\text{-}ZrO_2$=3%; $t\text{-}ZrO_2$=97%
3.0	1080	S.U.Pliner	[5]	S_{max}=0.4%; d_{av}=0.5-1 μm

8.STRENGTH

The three-point bending strength data of this Y-TZP grinding media are shown in Table 3. It is obvious that our results are quite excellent, but data are rather scattered, it seems that the achievement of high strength is extremely critical and needs carefull quality control in processing. The comparison of our data to Y-TZP from other sources is shown in Table 4.

CONCLUSIONS

1. The Y-TZP grinding media fabricated with ultrafine ZrO_2 powder from sol-gel method have excellent mechanical performance. It is abrasion resistant, low-contaminated and is well densified, so it is adquate to be used as components which necessiate high strength and abrasion-resistance at low temperatures. Moreover it is suitable to be ultilized as grinding media, especially for ZrO_2-containing ceramics and for some kinds of ceramic pigments and colors.

2. Wet-chemically prepared Y-TZP ultrafine ZrO_2 powder achieved ion-level mixture. It ensured uniformally distribution of stablizing agents, fine- and uniform-grained, highly reactive and sintered at much lower temperatures. fine-texture uniform structure grinding media could be made. The initial grain size, 0.02-0.035 μm, ensures sintered grain size <0.5 μm, and ensures large ammount of t-ZrO_2 retained to room temperature and raised fracture toughness of the material.

3. Isostatic pressing of the Y-TZP cylinders kept uniform pressure upon the pieces, gave same texture on the surface and inner the piece. The hardness, and strength of the bodies were superior to other conventional forming method.

4. The ultra-fineness of this powder gave low packing density of the powder and green density scattered, some troubles had been encountered in dimension controlling, but it could be solved by powder pre-pressing technique to stabilize the density.

REFERENCES

[1] H. Reh,"Die Hochleistungs Keramik -- II. Die Werkstoffe",Keramische Zeitschrift, [10] 600-5 (1986).
[2] R.L.K.Matsumoto,"Strength in Degraded Yttria-Doped Zirconia Polycrystals" J. Am. Ceram. Soc., 68 [8] C213 (1985); Adapted from Ceramic Source '87, ACerS(1985).
[3] T. S. Kawanami,"Toughened Zirconia Ceramics",Ceramics (in Japanese), 20 [6] 518-23 (1985).
[4] T. K. Gupta, Sintering of Tetra-Zirconia and Its Characteristics",Science of Sintering, 10 [3] 205-16 (1978).
[5] S. N. Pliner et al,"Tetragonal Zirconia High Strength Ceramics", Refractories (in Russian) [9] 19-20 (1986).
[6] F. F. Lange,"Transformation Toughening Pt.3: Experimental Observation in the ZrO_2-Y_2O_3 System".Rockwell International,Science Center
[7] Chinese National Standard GB 6569-86 "Testing Method for Flexural Strength of High Performance Ceramics" (1986).
[8] K. Niihara,"A Fracture Mechanics Analysis of Indentation Induced Palmqvist Crack in Ceramics", J. Mater. Sci. Lett., [2] 221-23 (1983).

MICROANALYTICAL CHARACTERIZATION OF WEAR DAMAGE IN AN ALUMINA-SILICON CARBIDE WHISKER COMPOSITE

C. S. Yust and L. F. Allard
Metals and Ceramics Division
Oak Ridge National Laboratory
P. O. Box 2008
Oak Ridge, TN 37831-6063

ABSTRACT

Wear damage in a silicon carbide whisker-reinforced alumina composite resulting from unlubricated, self-mated, pin-on-disk sliding in air at 800°C has been characterized using scanning Auger spectroscopy, transmission electron microscopy, and energy dispersive X-ray analysis. Using constant bearing pressure and sliding velocity, sliding for 2 km resulted in mild wear of the disk surface on which regions of silicon oxide were formed. At a sliding distance of 5 km, the wear rates measured still indicated mild wear, but pitting at the center of the disk wear track was observed. Subsurface damage of the alumina matrix beneath the disk wear track at 5 km sliding distance was characterized by transmission electron microscopy of cross sections. Wear debris observed on the wear surface was comprised of fine particulates of composition similar to that of the bulk material. The implications of the observations regarding wear mode and wear lifetime are discussed.

INTRODUCTION

The continuing trend toward increased operating temperatures for many types of industrial and transportation systems is sustaining the interest in ceramic materials for a wide range of applications. The potential for durability at elevated temperatures in severe environments makes ceramics attractive for high-temperature applications, while certain aspects of the mechanical behavior, particularly the tendency to brittle fracture, raises some concern.

Ceramics for use as heat engine components are useful not only because of their high-temperature strength, chemical stability, and high hardness (which has been taken to imply high wear resistance), but also because of their lower mass and/or lower density and the resultant improvements in the

*Research sponsored by the Office of Energy Utilization Research, Energy Conversion and Utilization Technologies (ECUT) Program, U.S. Department of Energy under contract DE-AC05-84OR21400 with Martin Marietta Energy Systems, Inc.

inertial response of rapidly rotating components. In many machine applications, ceramic components are used as dynamic rather than static members. Thus, they are exposed to varying stress states. Therefore, they must also be sufficiently resistant to wear under varying stresses to provide economically useful lifetimes. To date, studies of the wear behavior of ceramics suggest that acceptable wear resistance may be anticipated for some ceramics under controlled operating conditions [1,2].

The potential for fracture initiation and the corresponding transition from an acceptable mild wear mode to a catastrophic severe wear mode should be evaluated for expected specific load and temperature conditions. One of the mechanisms by which fracture may be induced is the accumulation of subsurface lattice strain in the region beneath the wear surface [3,4]. This paper reports the results of an initial investigation of subsurface damage effects in a silicon carbide whisker-reinforced alumina composite subjected to unlubricated sliding wear at contact stress levels comparable to those experienced by engine bearings and gears, although at much lower velocities. The subsurface deformation processes observed in this work can occur in many ceramic microstructures, especially those subject to high mechanical stresses at elevated temperatures. The performance of these materials over time cannot be adequately assessed without a background of knowledge of the nature of these cumulative subsurface damage effects.

MATERIALS AND EXPERIMENT

The specimens utilized in this study were prepared by hot pressing mixtures of an alumina powder (Baikowski CB-10) and silicon carbide whiskers (ARCO SC-9). The resultant bodies consisted of a fully-dense, fine-grained alumina matrix containing 10 vol % silicon carbide whiskers. The matrix grain size was 3 to 4 μm, and the whiskers were nominally 0.6 μm in diameter with a length-to-width ratio of about 30. Hot pressing produced a preferential alignment of the longitudinal axes of the whiskers toward a plane perpendicular to the pressing direction, although the whisker axis directions were random within that plane. The relation of whisker alignment to specimen orientation in these specimens was such that the plane of random whisker orientation was parallel to the wear surface of the disks. The spheres used as the pin member of the wear couple were prepared from individual hot-pressed cylinders of the whisker composite slightly larger in diameter than the finished ball diameter, 9.5 mm. The orientation of the whiskers within the spheres was not recorded. Further details on the processing and properties of this composite are available in the literature [5,6].

The elevated temperature wear tests were performed in a pin-on-disk test system incorporating a furnace arrangement capable of providing test temperatures to 1000°C [7]. The pin consisted of a sphere held in a rod, which itself was fixed in a yoke that was supported by a pair of strain-gauged cylinders serving as force transducers. The force transducers measured both normal and tangential force. Load was applied to the pin by dead weight loading. The test variables were applied stress, sliding velocity, sliding distance, and ambient air temperature. Only two of a series of experiments at 800°C are discussed in this paper, those worn under a contact area stress of 150 MPa and a sliding velocity of 0.5 m/s (83 cycles/s) for total sliding distances 2 km (15.9×10^3 cycles) and 5 km (67.7×10^3 cycles).

The geometrical arrangement of a sphere on flat configuration results in the formation of a small initial area of contact and a correspondingly large initial contact stress. In these experiments an initial load of 2.2 N was applied, resulting in an initial contact area diameter, calculated by the Hertz equation, of approximately 100 μm. The stress on this contact area is initially 280 MPa, but the area enlarges as sliding progresses, and the diameter and stress at the end of 1 km of sliding are typically 300 μm and 30 MPa. After the initial kilometer of sliding, sufficient load was applied to attain the desired contact stress level. Thereafter, the tests were interrupted at 1-km intervals to measure the contact diameter and adjust the load. The specimens discussed in this paper, therefore, experienced total sliding distances of 2 and 5 km, and sliding distances of 1 and 4 km under a stress of 150 MPa. These contact pressure levels correspond to those expected in common devices such as piston rings, gears, and bearings, although the velocities were necessarily much lower since these tests were performed without lubrication [8].

The disk specimen surfaces were polished to an arithmetic roughness, R_a, of 1-μm or less. The counterface specimens were spheres of 9.53 mm in diameter. The surfaces of the spheres were also finished to a 1-μm roughness. Polished surfaces were used to facilitate the detection of the onset of damage on the wear path.

Thin foil specimens for examination in the transmission electron microscope were prepared by cutting sections of the disk, which included portions of the wear track, and bonding the wear track surfaces of those sections with an epoxy to an unworn section of the same disk. Tungsten wires (0.05 mm in diameter) marked the wear track locations, as illustrated in Fig. 1. The bonded sections were inserted into an epoxy-filled thin wall metal tube. Thin sections of the bonded specimens were sliced perpendicular to both the plane of the wear track and the wear track itself. The cross-section slices were ground and polished to produce slices about 150 μm thick with a finished surface roughness of 1 μm. Disks 3 mm in diameter were core-drilled from the thinned slices so that the wear track region was slightly off-center. The disks were dimple-ground and ion-beam milled until a hole perforated the specimen and extended into the wear track region. The ion-milling process was monitored regularly, and the position of the hole edge was compared to an initial photograph of the track position, as determined by the tungsten wire marker. In the event the tungsten marker was ion-milled away prematurely, the wear track could still be accurately located with the photograph. Figure 2 shows the specimen after an initial hole was formed by ion-milling, and at the termination of the milling process. The position of the tungsten wire locator can be clearly seen in the cross section in Fig. 2 (arrow), while the track area is located in the inset by comparison even though the position of the wire is no longer evident.

RESULTS AND DISCUSSION

Wear coefficients for both experiments at 150-MPa contact stress were determined by measurement of the material loss at the ends of the pin tips. For specimen 135, worn for a total sliding distance of 2 km (67 min), the pin wear coefficient was 7.7×10^{-8} mm^3/N-m, and for specimen 197, worn for 5 km (167 min), the value was 3.8×10^{-8} mm^3/N-m. The disk specimens did not show evidence of material loss as determined by surface profilometry. A typical profilometer trace (Fig. 3), in fact, shows an accumulation of material at the edges and, to a small extent, at the center of the track, rather than a material loss. In this circumstance, a wear coefficient for the disk specimen was not calculated. The pin tip wear values and the disk wear surfaces are indicative of mild wear under the conditions applied. In addition, fracture was not observed on the disk wear surfaces at an optical magnification of 300X.

Scanning electron micrographs of both the surface of the wear track of specimen 135 and of the area adjoining the track indicate that a reaction product had formed on the SiC whiskers that intersected the specimen surface. Figure 4 shows the presence of such a reaction layer on the whisker surfaces in the region adjacent to the wear track. Auger analysis of these areas indicated that the reaction layer was composed of Al, Si, and O (Fig. 5a). Sputtering the surface to remove approximately 100 nm of material left a surface which was determined by the Auger analysis to be composed of silicon and oxygen (Fig. 5b). Finally, sputtering to a depth of 600 nm removed the reaction layer, exposing SiC, as shown in Fig 5c. Similar analysis of the debris layer on the wear track demonstrated that the debris layer was composed of the oxides of aluminum and silicon. Removal of this layer by sputtering exposed the original surface of the specimen which showed no visible evidence of wear.

Scanning electron microscopy (SEM) photomicrographs of specimen 197, which was worn at a contact stress of 150 MPa for 5 km, also experienced mild wear, based on both the calculated wear coefficient and on low magnification (300X) examination. At 3000X magnification (Fig. 6), the initiation of the removal of surface material by pitting was observed. Although some pits were found to lie across the width of the wear track, the largest pits and the greatest concentration of pits were present at the center of the track, where the Hertzian contact stress was the largest.

A transmission electron microscopy (TEM) micrograph of a typical region in the bulk of the specimen is shown in Fig. 7. The matrix grains and the silicon carbide whiskers are readily distinguished by both shape and substructure. The alumina grains are equiaxed and free of substructure such as dislocations, planar faults, or cracks. The SiC whiskers appear hexagonal in cross section or as rectangular sections nearly parallel to the surface. The internal whisker structure (planar defects normal to the whisker axis, inclusions, etc.) has been described in detail in the literature [9]. The as-formed composite microstructure does not contain evidence of any residual deformation effects remaining from the hot-pressing process. The material immediately beneath the wear track of specimen 135, worn for a total sliding distance of 2 km with the last kilometer at a contact stress of 150 MPa, also does not show any indication of lattice deformation. The total number of cycles of stress imposed on specimen 135 at the 150-MPa stress level was 1.59×10^4.

In contrast, the region beneath the wear track in specimen 197, worn for a total sliding distance of 5 km with the last 4 km at 150 MPa, contained very dense dislocation concentrations both in the near-surface grains beneath the wear track, and in grains at least 100 μm below the surface. The total number of cycles of stress imposed at this sliding distance was 6.77×10^4 cycles. The substructure of a grain about one grain diameter beneath the original wear surface is shown in Fig. 8. The nearly vertical linear features seen in the bright-field image, Fig. 8a, are dislocations lying on an active slip plane. The local strain effects accompanying the dislocation array are sufficiently disruptive to partially mask the dislocation images. A dark-field image (Fig. 8b) shows the dislocation lines more distinctly, although some of the local strain effects are still present. The diffraction pattern (Fig. 8c) is consistent with the <020> zone of the alpha-alumina structure (using three-index hexagonal notation for the rhombohedral structure). The diffraction pattern shows the dark-field imaging condition, and indicates that the normal to the plane of the photomicrograph is tilted about 2° away from the <020> zone axis, with the tilt approximately around the <210> direction, resulting in a nearly edge-on view of the basal planes. The observed dislocations are clearly lying on the basal plane (the minimum critical-resolved shear stress plane in alumina), and a series of planes on which slip has occurred are visible. Fractures appear to be present at the right side of the field of view in Fig. 8a.

The dislocation substructure shown in Fig. 9 was observed in a grain 100 μm beneath the surface (about 25 grain diameters below the surface). The dislocations at the center are emerging from a dense dislocation tangle at the left of the photo, some of which are shown in only weak contrast. The deformation is spreading between two whiskers, top and bottom, into an undeformed volume of material. Note also the fracture between alumina grains at the top of the field, apparently terminating at the silicon carbide whisker. While it is impossible to distinguish fractures formed during the wear process from those which might form in the thin foil as a result of the release of residual stress as the foil is thinned, the fact that cracks were not observed in the undeformed regions of the foil indicates that standard specimen preparation techniques do not introduce such artifacts.

The present observations show that dislocations can be formed in the subsurface regions beneath the wear track in an alumina-silicon carbide specimen worn at 800°C. The threshold number of cycles for dislocation introduction is greater than 15.9×10^{-3} cycles at a stress of 150 MPa, and a substantial dislocation concentration is present at 67.7×10^{-3} cycles of stress. The TEM photomicrographs also show evidence of residual lattice strain within individual grains, as indicated by contrast effects. The result of the accumulation of deformation and the accompanying lattice strain can be the formation of fractures as a stress relief mechanism. The formation of large strains as a consequence of lattice deformation and the failure of individual surface grains has been observed in the sliding wear of TiB_2 [3]. The formation of localized hot spots undoubtedly takes place during sliding, and the elevation of the lattice temperature may facilitate the introduction of dislocations. The existence of large dislocation concentrations suggests the initiation of damage due to mechanical strain, however, rather than the effects of thermal fatigue.

The SEM view of the wear track after 5 km of sliding, Fig. 6, shows the onset of pitting in the track, most severely at the center of the track. The magnitude of the contact stress is greatest at the track center, about 1.5 times the average value, and the lattice strain at track center would likewise be the greatest. Consequently, it is not surprising that pitting as a result of accumulated strain begins at the center of the wear track. An observation supporting the concept of surface grain failure is shown in

Fig. 10, in which the presence of fine debris is detected within an opening between grains in the near-surface region of the TEM specimen. The region containing debris most likely represents the intersection of the thinned specimen with the edge of a surface pit. The fine particulate material has the morphology and the composition previously observed for wear debris from a composite sample, as determined by energy dispersive X-ray analysis [9]. As pitting proceeds, the wear track surface will become more irregular, and larger debris particles forming at the interface will serve as third bodies promoting additional fractures and a full transition to severe, abrasion-assisted wear.

In a recent discussion of the mechanisms of wear in engineering materials, Ball stated that in many instances there is evidence of a two-stage wear process, the first stage being plastic strain accumulation in the lattice followed by second-stage microfracture [4]. In comparing the behavior of classes of materials, Ball considers that surface failure will occur in a hot-pressed alumina as the result of stress application to surface flaws prior to the accumulation of any lattice strain, and that the microstructure will provide little resistance to crack propagation. As a consequence, he classifies ceramics as materials with limited potential for wear resistance [4]. However, ceramics may be of greatest use in wear applications at temperatures in excess of 700°C, and at such temperatures a plastic response is possible in some ceramic microstructures, as has been demonstrated in the present results. In this case, the composite microstructure also serves to inhibit the propagation of surface initiated cracks, further enabling the accumulation of lattice strain in the near surface regions.

The sliding friction and wear response of most ceramics indicate that ceramics will require lubrication in practical systems. The use of lubricants does not eliminate the generation of stress fields within the contacting bodies, although it does modify their specific form and may decrease the magnitude of surface tensile stress. The observations reported in this work suggest that the subsurface response of ceramics may be a factor in determining component life. The evaluation of ceramic components for specific applications should consider the potential of the material to undergo deformation of the near- surface grains and the consequences of such a response at the anticipated service conditions.

CONCLUSIONS

The unlubricated sliding of an alumina-silicon carbide (10 v/o) whisker composite on itself in air at 800°C and an apparent contact stress of 150 MPa and sliding velocity of 0.5 m/s results in mild wear of both the pin and disk members of the wear couple. The pin wear rates are 7.7×10^{-8} $mm^3/N\text{-}m$ and 3.3×10^{-8} $mm^3/N\text{-}m$, respectively, at sliding distances of 2 and 5 km.

Surface profilometry and scanning electron microscopy confirm the mild wear of the disk wear tracks, although the initiation of pitting is observed at the center of the wear tracks of specimens worn at 0.5 m/s and an apparent contact area stress of 150 MPa.

Transmission electron microscopy of the region beneath the wear track shows the formation of dense dislocation tangles in the specimen worn at 150 MPa contact stress and 0.5 m/s sliding velocity after 6.77×10^4 stress cycles.

A specimen tested under the same conditions for 1.59×10^4 cycles did not show evidence of subsurface damage accumulation.

The accumulation of lattice strain in near-surface grains may result in strain relief by intragranular fracture, a concept which is supported by the observation of fine particulate debris in close proximity to grains which contained dense dislocation arrays.

The role of subsurface damage processes can be a significant factor in the service life of ceramic components, and will need to be evaluated for ceramics considered for tribological applications.

REFERENCES

1. K. F. Dufrane and W. A. Glaeser, "Wear of Ceramics in Advanced Heat Engine Applications," in Wear of Materials - 1987 ed. K. C. Ludema, American Society Mechanical Engineers, New York, 1987, pp. 285–91.

2. C. S. Yust, J. M. Leitnaker, and C. E. Devore, "Wear of an Alumina-Silicon Carbide Whisker Composite," Wear, **122**(2), 151–64 (1987).

3. C. S. Yust, "Low Speed Sliding Damage in TiB_2-Ni Composites," in Wear of Materials - 1983, ed. K. C. Ludema, American Society Mechical Engineers, New York, 1983, pp. 167–73.

4. A. Ball, "The Mechanisms of Wear and the Performance of Engineering Materials," J. S. Afr. Inst. Min. Metall. **86**(1), 1–13 (1986).

5. P. F. Becher and G. C. Wei, "Toughening Behavior in SiC-Whisker-Reinforced Alumina," Comm. Am. Ceram. Soc. **67**(12), 267–69 (1984).

6. G. C. Wei and P. F. Becher, "Development of SiC-Whisker-Reinforced Ceramics," Am. Ceram. Soc. Bull. **64**(2), 298–302 (1985).

7. C. S. Yust and L. F. Allard, "Wear Characteristics of an Alumina-Silicon Carbide Whisker Composite at Temperatures to 800°C in Air," Tribology Trans., in press.

8. M. N. Gardos, "On Ceramic Tribology," Lubr. Eng. **14**(5), 400–407 (1988).

9. S. R. Nutt, "Microstructure and Growth Model for Rice-Hull-Derived SiC Whiskers," J. Am. Ceram. Soc. **71**(3), 149–56 (1988).

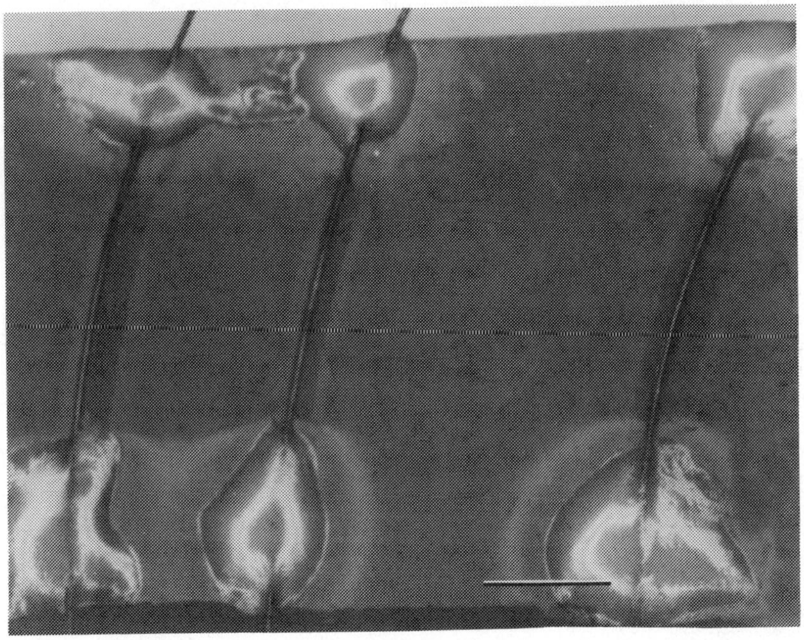

Figure 1. The position of the wear tracks on the disk surface was marked by 0.05-mm-diam tungsten wires secured with epoxy before the wear surface was bonded to the mating unworn composite section.

Figure 2. Transmission electron microscopy specimen 197 after the initial perforation; the circular hole (arrowed) in the residual bonding epoxy marks the wear path location. (Inset) Specimen after additional ion-milling to bring the thinned edge of the specimen beneath the wear path. Bar = 300 μm.

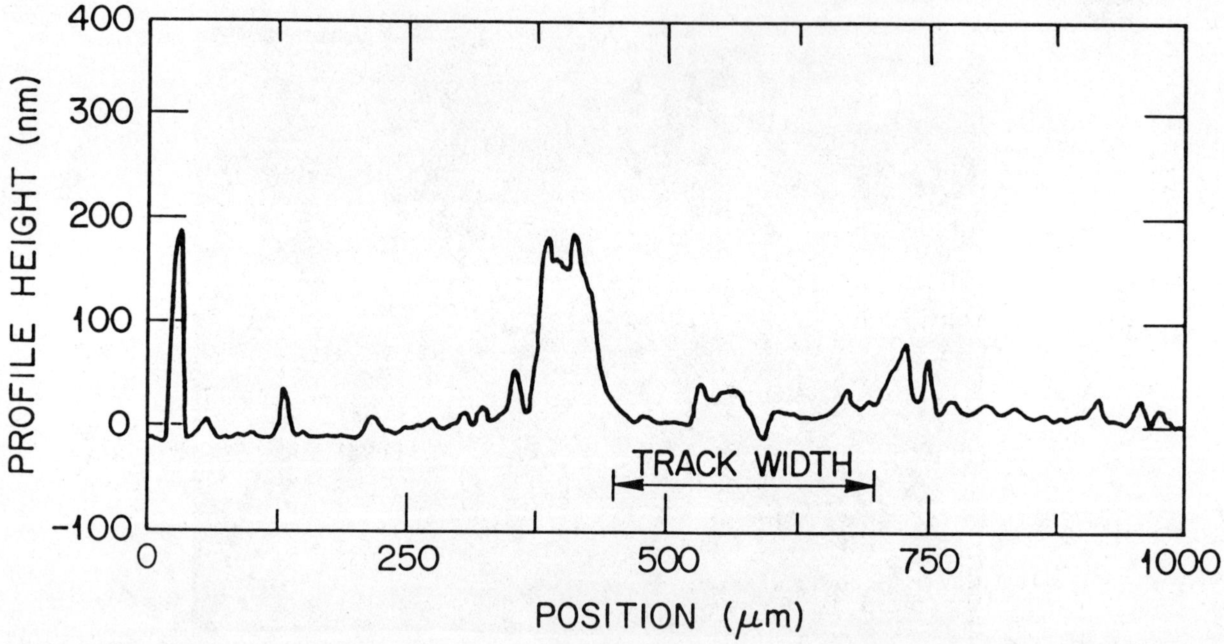

Figure 3. Typical profilometer trace of a wear track after test at 800°C. Debris is gathered at the edges of the track, and a small amount of debris is detected within the contact area on the track.

Figure 4. The formation of reaction product layers is observed on the silicon carbide whiskers which intersect the specimen surface adjacent to the wear track. Note the thermal etching of the alumina grain surfaces and boundaries. Bar = 1.5 μm.

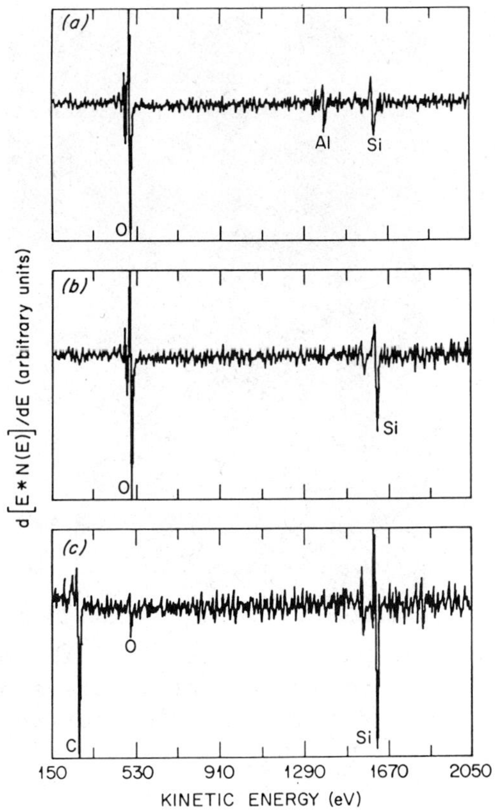

Figure 5. Auger spectra from the silicon carbide reaction product layer. (a) As formed, showing the presence of both aluminum and silicon oxides. (b) After removal of 100 nm by sputtering; aluminum is removed, and only silicon and oxygen are present. (c) At a depth of 600 nm, the oxide is removed and the silicon carbide whisker is detected.

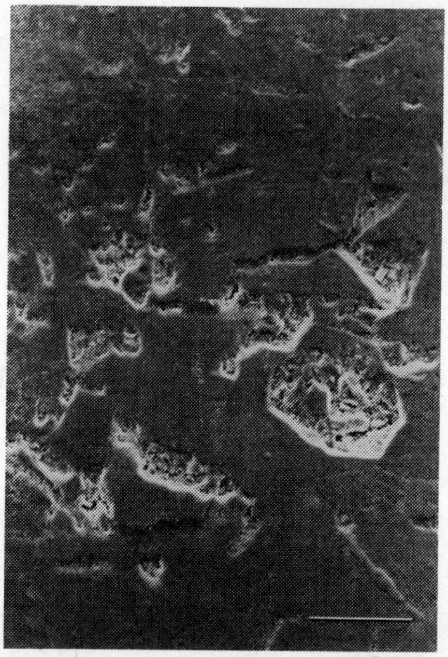

Figure 6. Scanning electron photomicrograph of the near-center region of the wear track of specimen 197 worn at a contact stress of 150 MPa for a 5-km sliding distance. The removal of material by surface pitting has been initiated. Bar = 5 μm.

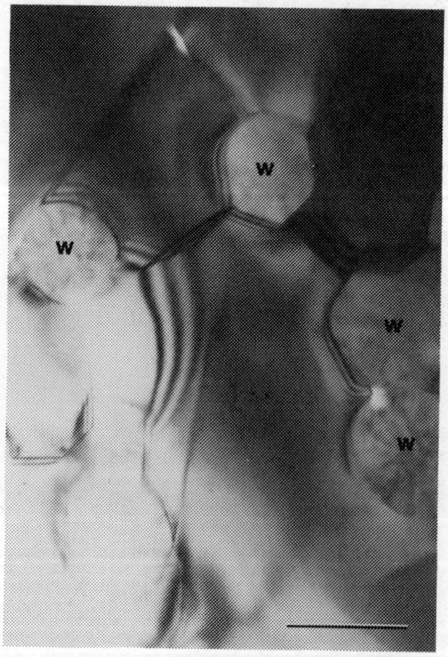

Figure 7. Transmission electron micrograph of a typical bulk region of the composite showing silicon carbide whiskers (w) and defect-free equiaxed alumina grains. Bar = 500 nm.

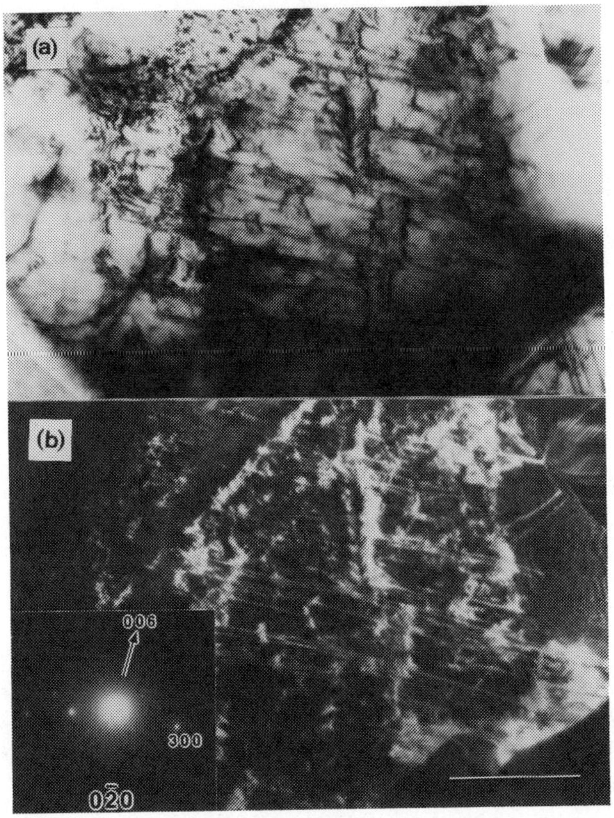

Figure 8. (a) Transmission electron microscopy photomicrographs of a near surface grain, showing dislocation content and associated strain effects. (b) Dark-field image of the dislocations within the grain. (Inset) Properly oriented diffraction pattern indicates the (300) imaging condition and that the linear dislocations lie on the basal plane. Bar = 500 nm.

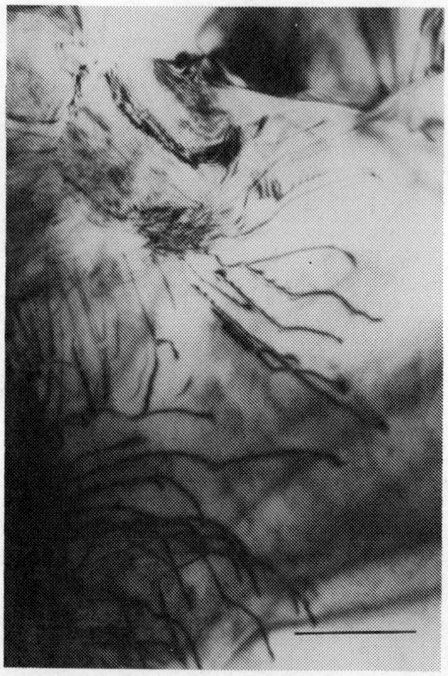

Figure 9. Transmission electron microscopy photomicrograph of dislocations in a grain approximately 100 μm below the wear surface. These dislocations are emanating from the edge of a dense dislocation tangle, left center. Bar = 500 nm.

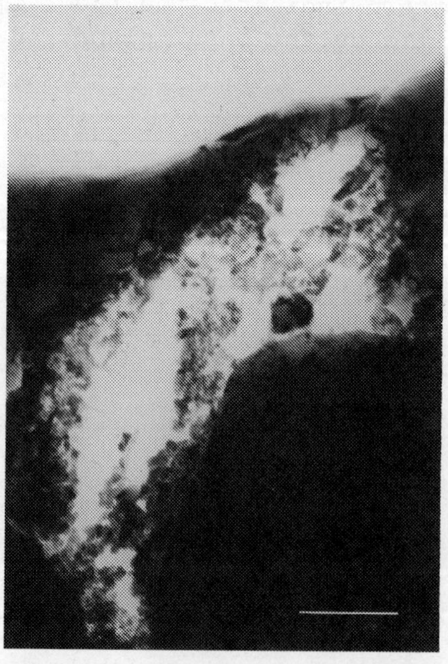

Figure 10. Transmission electron microscopy photomicrograph of fine particle debris accumulated in an opening at the wear surface. The opening most likely corresponds to the edge of a pit formed in the wear track. The top edge of the image is the wear surface; the small degree of rounding is an effect of the ion-milling process. Bar = 500 nm.

GRINDING OF SiSiC CERAMICS: ACHIEVING HIGH PERFORMANCE AND MINIMUM DAMAGE

Wilfried König, Jens Wemhöner
Fraunhofer-Institut für Produktionstechnologie
Aachen, FRG

ABSTRACT

SiSiC components almost invariably have to be finished by grinding. This entails problems both of reduced component strength (due to surface damage) and of the low material removal rate which can be achieved.

The paper presents methods for a SiSiC-adapted surface grinding strategy based on an analysis of process kinematics and a study of the basic mechanisms of material removal. Such methods ultimately lead to a considerable increase in both material removal rate and component strength.

NOMENCLATURE

a_e	mm	depth-of-cut
a_{ed}	mm	dressing depth-of-cut
a_{esh}	mm	sharpening depth-of-cut
C	kt/cm^3	diamond concentration
D	μm	diamond grain size
F_n	N	normal force
F'_n	N/mm	specific normal force
F_t	N	tangential force
G		grinding ratio
h_c	μm	depth-of-cut of the single cutting edge
l_k	mm	contact length
l_s	mm	grinding length
m		Weibull modulus
Q'_{sh}	$mm^3/(mm*s)$	specific sharpening stick removal rate
Q'_w	$mm^3/(mm*s)$	specific material removal rate
R_z	μm	average peak-to-valley height
V'_w	mm^3/mm	specific material removal
v_{fad}	mm/s	dressing feed rate
v_{ft}	mm/s	feed rate
v_{ftsh}	mm/s	sharpening feed rate
v_s	m/s	cutting speed

INTRODUCTION

Silicon infiltrated silicon carbide (SiSiC) components have to fulfill exacting requirements in terms of geometrical accuracy and low surface roughness in order to provide the anticipated machine and plant performance. As-infiltrated parts generally do not satisfy these requirements.

In most cases, grinding with diamond tools is indispensable as the final step in component production. Owing to the low ductility of SiSiC, grinding entails a particularly high risk of strength-reducing damage to the component surface. The grinding process is therefore virtually always carried out under very moderate conditions, resulting in long machining times. This leads to an increase in machining costs which substantially diminishes the competitiveness of SiSiC components as compared to steel or hard metal components.

The objective of this paper is to establish parameters for the surface grinding of SiSiC which will permit both a high material removal rate in the machining process and high strength of the ground components.

EXPERIMENTAL PROCEDURE

In the investigations described below, extruded SiSiC was machined with vitrified bonded diamond grinding wheels. The material supplied by *Sigri GmbH*, Meitingen, FRG, was in the form of 50 x 50 mm plates with a thickness of 5 mm. Grinding tests were carried out on a *Profimat 410* surface grinding machine (*Hauni-Blohm*). Diamond grinding wheels with a grain size of 151 μm, a concentration of 4.4 ct/cm^3 (C100) and a diameter of 400 mm supplied by *E. Winter & Sohn*, Norderstedt, FRG, were used in the tests. The diamond wheels were dressed with a diamond dresser and sharpened with a silicon carbide sharpening stick.

Prior to each test, 12 dressing strokes were made at a dressing depth-of-cut of a_{ed} = 2 μm per stroke and a dressing feed rate of v_{fad} = 2 mm/s. Sharpening parameters were a_{esh} = 10 mm and v_{ftsh} = 5 mm/s.

Emulsion at a concentration of 3 % was used as the cooling lubricant, with a flow rate of 160 l/m and a pressure of 8 bar. The cooling lubricant was filtered and re-circulated.

Strength was tested by the concentric-ring method /2,3/. Ten 50 x 50 x 3 mm specimens were fractured for each variation in the grinding process.

RESULTS

Figure 1 indicates the basic surface grinding parameters. The two most important parameters in the process of surface grinding are depth-of-cut a_e and feed rate v_{ft}. The product of these parameters is the specific material removal rate Q'_w, a measure of the workpiece volume removed per mm wheel width per second (Figure 2). In simplified terms Q'_w, may be regarded as the workpiece area crossed by the grinding wheel within the time unit indicated (hatched area in Figure 2).

Under industrial conditions, a specific material removal rate is almost always achieved by adjusting to a low depth-of-cut and a high feed rate. An identical value can, however, also be achieved by choosing a high depth-of-cut and a low feed rate. This variant is termed creep feed grinding. As far as the engagement process of the cutting edge is concerned, the essential difference between these two methods is the size of the contact area between the grinding wheel and workpiece.

In the case of creep feed grinding the contact length and thus the contact area are significantly larger. The volume of material to be removed is therefore distributed over a larger number of cutting edges, so that the depth-of-cut of a single cutting edge is lower. This has considerable effects on process behaviour and product quality.

Improving Grinding Ratio, Component Quality and Strength

Figure 3 illustrates the significant influence of different contact conditions on the cutting force during grinding. There are distinct rises in specific normal force not only with increasing material removal, as may be expected due to wear-induced dulling of the cutting edges, but also with increasing depth-of-cut. Since the grinding machines available for industrial machining of ceramic materials often possess only low rigidity, these high normal forces are one of the main reasons for the failure to exploit the creep feed grinding process.

The rise in cutting force with increased depth-of-cut can be explained directly by the engagement conditions of the tool. As the depth-of-cut increases, more cutting edges with smaller chip thicknesses are engaged simultaneously. Although this results in slightly smaller forces on the individual cutting edge, the overall normal force between the grinding wheel and the workpiece increases due to the greater number of cutting edges.

Engagement conditions also affect wheel wear. Wear was determined via the grinding ratio, which indicates how many mm^3 of workpiece volume can be removed per mm^3 volume of diamond layer. As the contribution of each cutting edge is smaller at a high depth-of-cut, wear will also be lower. This leads to lower grinding wheel wear, which is equivalent to a higher grinding ratio.

Since the effects of both process and tool behaviour are known, it is of interest to investigate the effect of engagement conditions on the component surface quality and strength of SiSiC. The left-hand section of Figure 4 shows surface roughness as a function of grinding length. Although there is a slight dispersion in the values, it is evident that a higher depth-of-cut and correspondingly slower feed rate lead to considerable improvement in surface quality. This is caused by the lower working engagement of the cutting edges, which results in lower surface roughness.

There are two complementary causes for the improvement in surface quality with grinding length. Following the dressing process there are very few active cutting edges on the outer surface of the grinding wheel. These edges, with their high depth-of-cut, score deep grooves in the workpiece surface during their first engagement, but dull or break out rapidly due to stress. Cutting edges further from the periphery are then engaged. Since there is a more than proportional rise in the number of cutting edges within the cutting area as the wheel radius decreases, the number of simultaneously engaged edges also rises sharply immediately after the start of the process. The result is a lower depth-of-cut of the individual cutting edges, leading to improved surface quality. There is a further reduction in surface roughness, and thus an improvement in surface quality, due to continued dulling of the cutting edges throughout the process.

As with roughness, creep feed grinding (Figure 4, right) has important advantages in terms of component strength. The characteristic strength of the specimens machined is increased by more than one third by grinding at a higher depth-of-cut. Strength was tested by the concentric-ring method /2,3/. The low working engagement of the individual diamonds in creep feed grinding evidently results in a material removal mechanism with advantages in terms of finished component strength.

Material Removal

Grinding tests were accompanied by single grit tests aimed at elucidating the basic mechanisms involved in grinding SiSiC with diamond cutting edges (Figure 5). The specific kinematic conditions at the individual cutting edge were identical for the grinding and single-grit grinding tests. For ease of handling, however, larger diamonds of the same type in a resin bond were used for the single grit tests.

When the chip thickness of the individual grain was low (h_c = 0.06 - 0.6 μm), i.e. under creep feed conditions, material was removed without brittle cracking. There were no changes in material behaviour even when the cutting edge crossed different parts of the structure (SiC-grains, free silicon). With greater chip thickness of the cutting edge (hc = 0.9 - 1.3 μm), i.e. within the range of low depth-of-cut and high feed rate for the real grinding process, large furrow-like cracks appeared in the workpiece surface. The cracks originated from the engaging diamond cutting edge but propagated beyond it in width and depth, so that larger workpiece particles broke away under the influence of the cutting force. It will be apparent that the resulting surface flaws may considerably reduce component strength.

Increasing Material Removal Rate

An initial summary of preferred parameters for SiSiC surface grinding indicates that parameters producing low depths-of-cut of the grinding wheel cutting edges tend to minimize surface damage.

The process chart in Figure 6 offers assistance in selecting material-adapted parameters. Assuming other parameters to be constant, a plus sign denotes a change in the same direction in linked parameters. Interpolating results to date in the process diagram, it is apparent that a decreased feed rate v_{ft} leads to a reduced depth-of-cut h_c of the individual cutting edge. However, the simultaneous rise in the depth-of-cut of the grinding wheel a_e due to the constant removal rate means that h_c is again increased.

Practical consideration indicates, however, that the reduction in the feed rate has a greater influence on the depth-of-cut of the single cutting edge h_c than does the increased depth-of-cut; creep feed grinding thus uses a larger contact area of the tool to remove the same amount of material, i.e. more cutting edges with lower depths-of-cut of the single edges (lower h_c) achieve the same material removal rate.

Apart from the option of choosing a tool with a different specification, the depth-of-cut h_c can be further reduced by increasing cutting speed. This exploits the fact that more cutting edges per time unit are engaged.

The material removal rate can now be raised by increasing the feed rate and cutting speed simultaneously while maintaining a constant ratio between them. The depth-of-cut of the single cutting edge is not altered. It may therefore be assumed that component strength remains constant in spite of the increased material removal rate. Figure 7 confirms this assumption. The characteristic strength remains approximately constant at about 265 MPa irrespective of the parameters. The slight difference in strength is within the range of measurement uncertainty.

In principle, workpiece surface roughness should also remain constant when the parameters are varied in this way. There is, however, an observable tendency for surface quality to improve as cutting speed increases. At the present stage of the study, this can be attributed only to dulling of the diamond cutting edges as a result of increased temperatures in the contact zone. This assumption is supported by the increase in tool wear with rising cutting speed. Tool wear was, for instance, 25 % higher at v_c = 50 m/s than at v_c = 25 m/s.

CONCLUSIONS

Figure 8 summarizes the individual results. The figure compares two processes: the normal industrial variant, i. e. a grinding process with low cutting speed and depth-of-cut and high feed rate, and a material-adapted process with parameters chosen for low depth-of-cut of single cutting edges.

The most important finding is certainly that the material removal rate can be doubled and the component strength simultaneously increased by more than 40 percent if the grinding parameters are adjusted to suit the material properties. The 35 % improvement in surface quality with unchanged tool life also favours the material-adapted process strategy.

A drawback of this form of grinding is, however, the eightfold increase in normal forces, imposing high demands on the rigidity of the machine tool employed. Nonetheless, the advantages of a machining technique geared to the specific features of SiSiC are so obvious that its use with appropriate grinding machines can be expected to lead to substantially increased efficiency in the manufacture of ceramic components.

ACKNOWLEDGEMENT

The authors are indebted to the "*Ministerium für Forschung und Technologie der Bundesrepublik Deutschland*" and to "*Sigri GmbH*", "*Hauni-Werke Körber & Co. KG*", "*Göhring Automation*" and "*E. Winter & Sohn*" for their generous support of investigations into the surface grinding of SiSiC.

The concentric-ring tests were carried out at the "*Fraunhofer-Institut für Werkstoffmechanik*" in Freiburg, FRG.

REFERENCES

/1/ Spur, G.; Stark, Ch.; Uhlmann, E.: Creep Feed Grinding of Non Oxide Ceramics with Diamond Grinding Wheels, ZwF 81(1986)6, S. 314-319 (in German)

/2/ DIN 52 292, Teil 1(Entwurf): Testing of Glass and Glass Ceramics, Determination of the Surface Strength, Concentric-Ring Test at Plate Shaped Specimens With Small Test Areas, Beuth Verlag, Berlin, 1983 (in German)

/3/ Soltesz, U.; Richter, H.; Kienzler, R.: The Concentric-Ring Test and Its Application for Determining the Surface Strength of Ceramics, Proceedings of the Satellite Symposium "Ceramics in Clinical Applications" of the World Congress on High Tech Ceramics, 6th CIMTEC, Milano, 1986

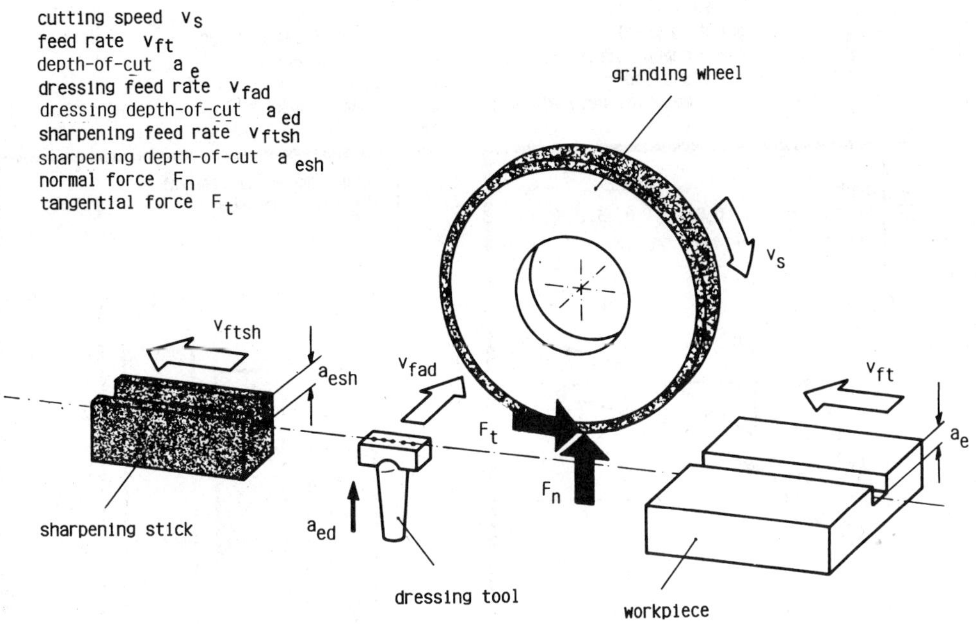

Figure 1: Basic Process Parameters of Surface Grinding With Diamond Grinding Wheels

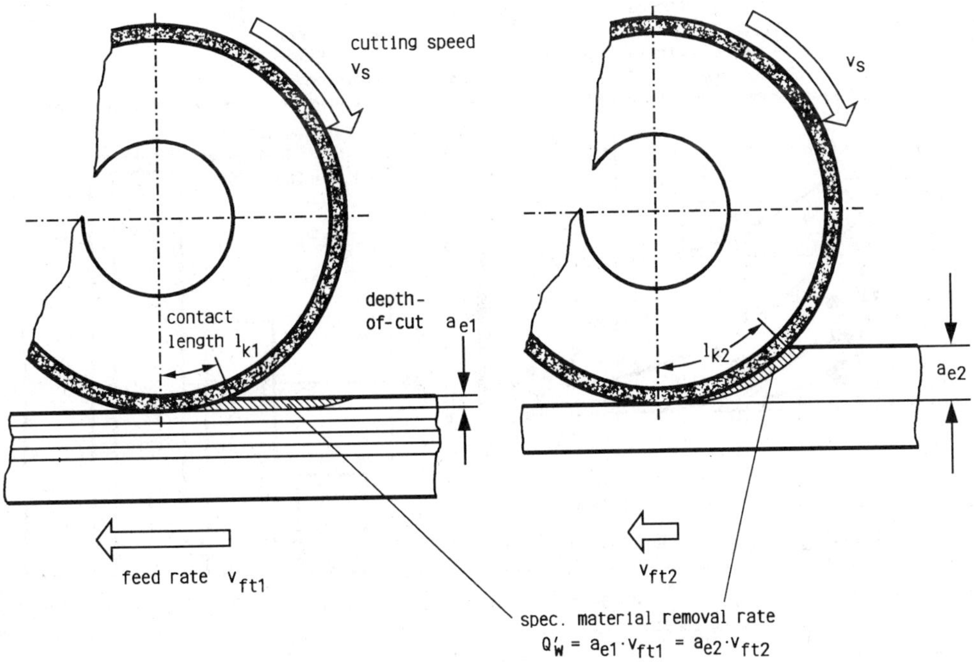

Figure 2: Contact Ratio in Surface Grinding

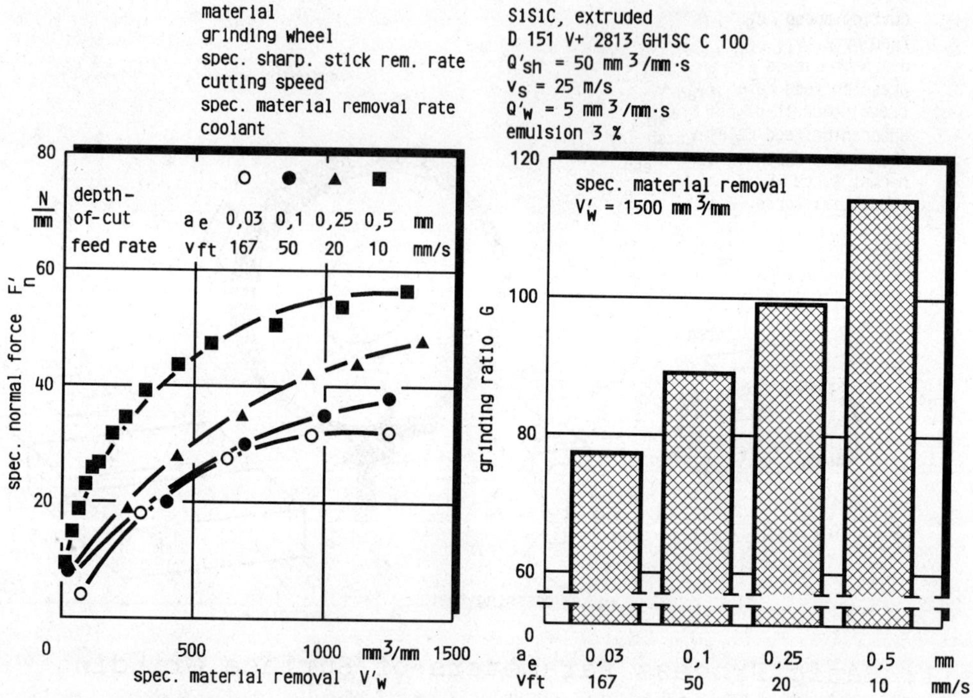

Figure 3: At Constant Material Removal Rate Normal Force and Grinding Ratio Rise With Depth-of-cut

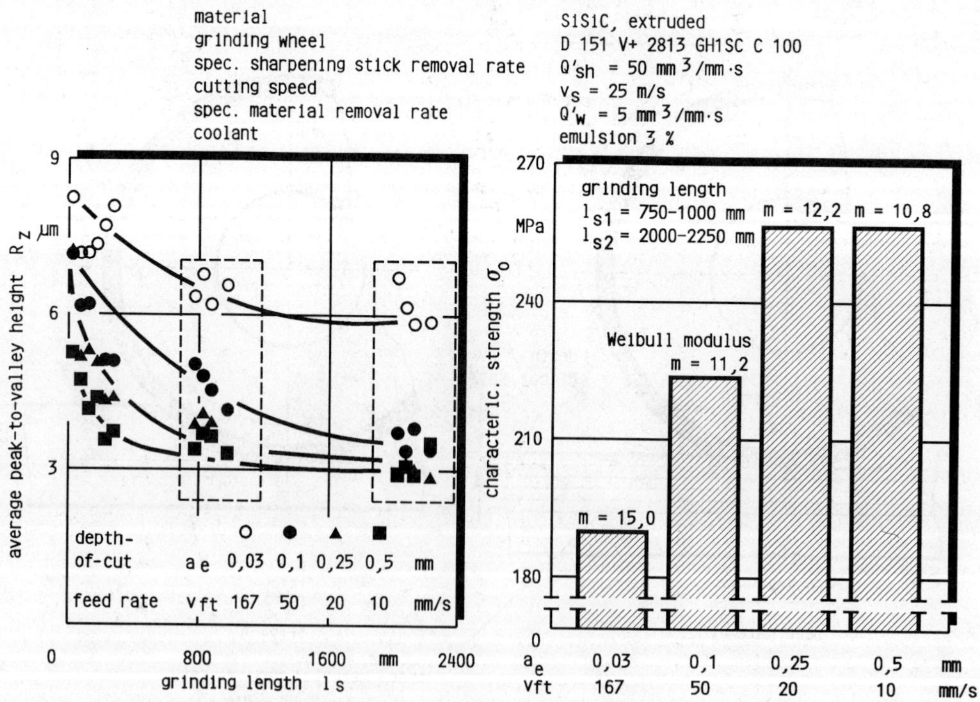

Figure 4: Surface Quality and Strength are Improved With Rising Depth-of-cut

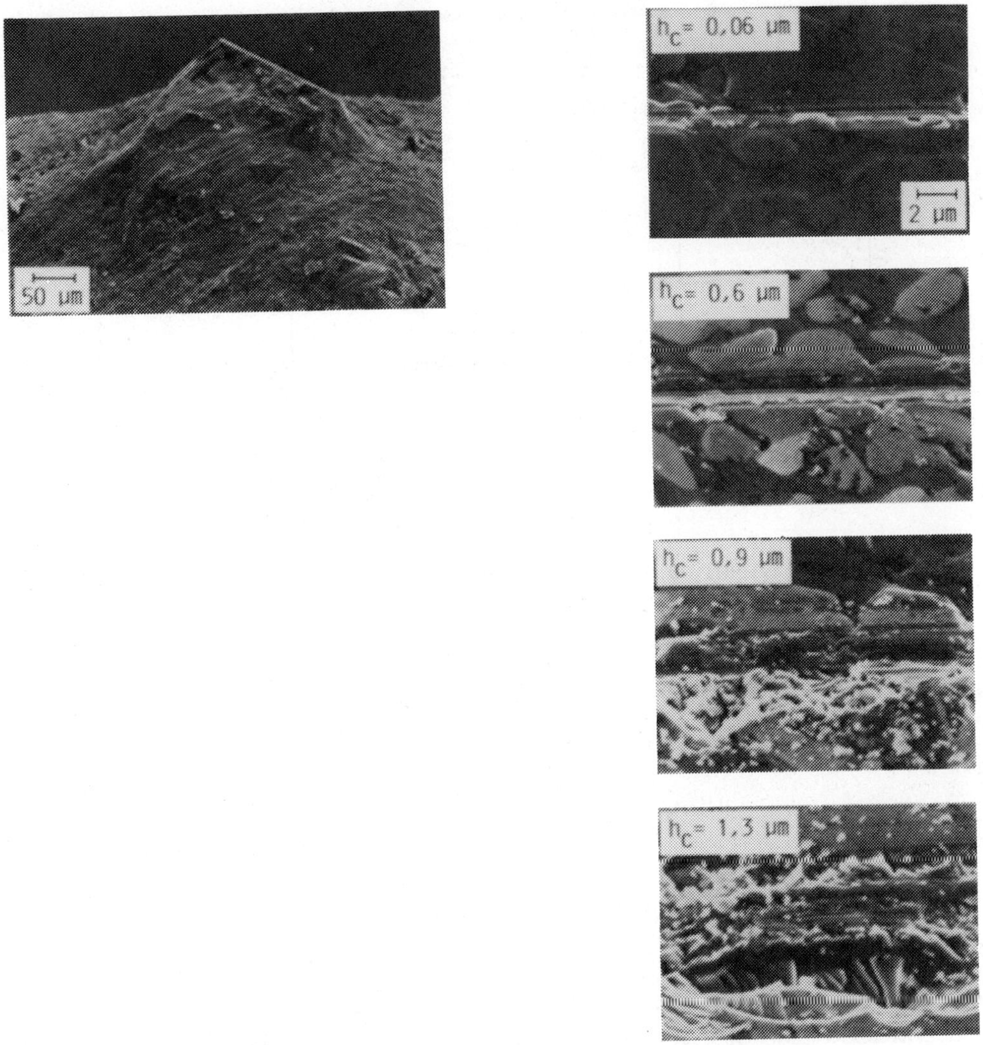

Figure 5: Single-grit Grinding of SiSiC

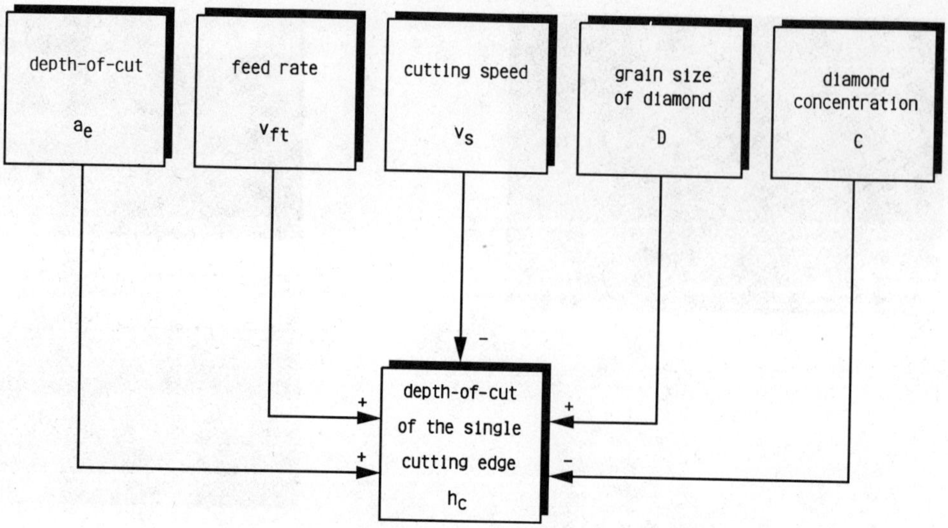

Figure 6: The Influence of Process Parameters on the Depth-of-cut of the Single Cutting Edge

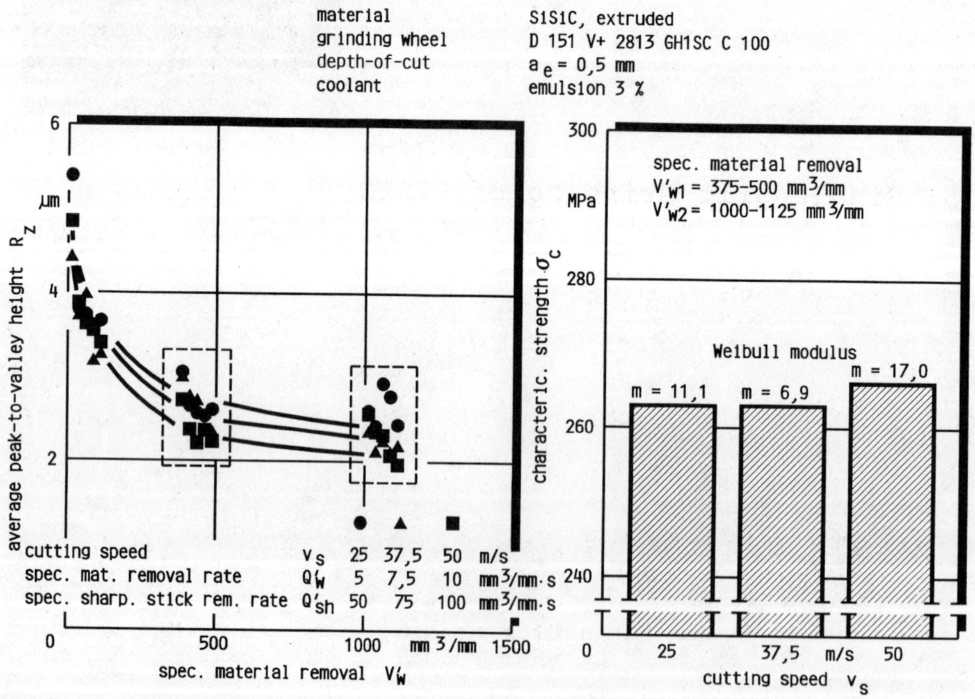

Figure 7: Surface Quality and Component Strength are not Diminished When Simultanously Increasing Material Removal Rate and Cutting Speed

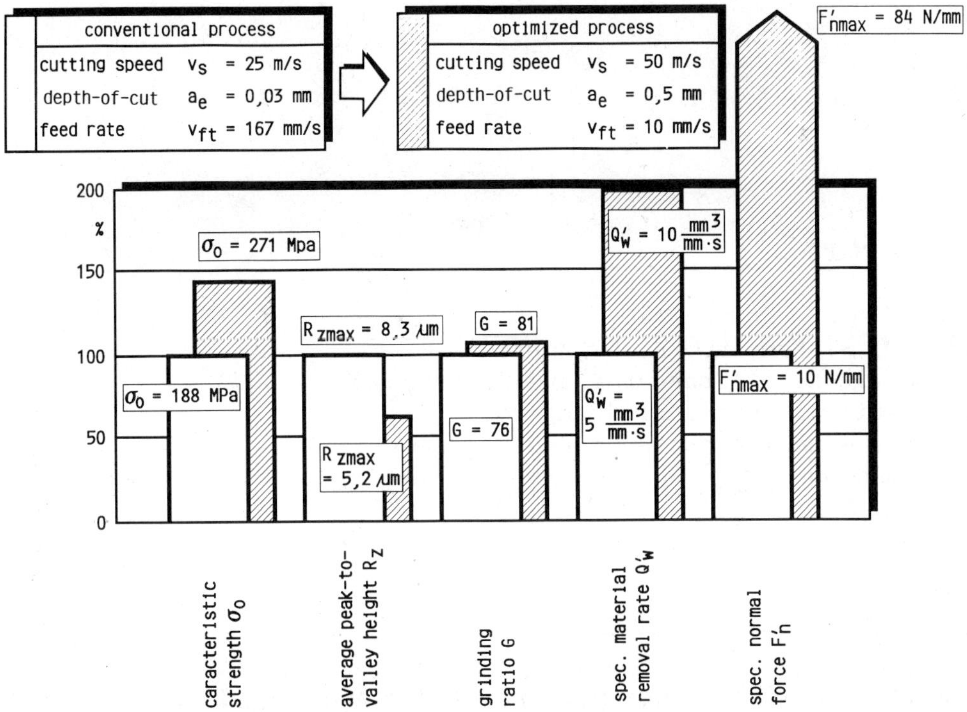

Figure 8: Comparison of Conventional and Material-adapted Grinding of SiSiC

CERAMIC WEAR MAPS

S. M. Hsu, D. S. Lim, and R. G. Munro
National Institute of Standards and Technology
(formerly National Bureau of Standards)
Institute for Materials Science and Engineering
Ceramics Division
Gaithersburg, MD 20899

ABSTRACT

New materials with unique properties are evolving from the considerable advances in materials science. These unique materials properties give rise to new technological opportunities for improved efficiency, higher power output, and longer lifetimes for applications such as heat engines. Introduction of new materials into technological development, however, is inhibited by a major barrier. Use of the new materials depends on the evaluation of the materials properties and the availability of appropriate design criteria.

Friction and wear characteristics of ceramics are crucial properties for their successful application to new engine designs. Tests for friction and wear performance usually are based on empirical field trials. These tests are very expensive. Laboratory measurements of tribological properties would be much less expensive. However, laboratory tests are specific to a particular wear tester and the specific operating conditions. Translation of laboratory results to industrial application and design is often difficult and fraught with uncertainty.

This paper describes a new systematic effort to measure and represent the wear characteristics of ceramics in a uniform and unified methodology. For any material, the methodology leads to a set of wear maps which collectively provide a comprehensive representation of the wear properties of the materials. The maps provide critical guidance to design engineers for materials selection.

INTRODUCTION

Wear is one of the key criteria for the selection of materials for heat engine applications. Unfortunately, most ceramics currently are optimized for their mechanical properties such as fracture toughness and hardness. These mechanical properties alone have proven to be unreliable guides for wear [1]. Generally, the wear of ceramics is a complex process involving physical, chemical, and system variables. Current understanding, for example, is inadequate to predict with reasonable accuracy, the wearlife of a ceramic material under any tribological conditions [2]. Design engineers faced with insufficient information must rely on their experience with prior generations of materials, i.e. metals. But, metals behave differently from ceramics. As a result, designers using ceramics must rely solely on empirical trials, conducted at great expense and difficulty.

This paper attempts to address this critical issue by designing a methodology that will provide a comprehensive representation of the wear characteristics of a material. A significant feature of this methodology is its departure from the historical approach of simulation testing. In the venerable historical approach, the actual mechanisms by which wear occurs under the applied conditions must be known before a successful simulation test can be developed. Then, the simulation test is used to develop a procedure for materials screening. This historical Mechanism-Simulation-Screening sequence limits the testing to only a narrow range of conditions that are specific to one end-use application of the material.

In contrast to the historical approach, it is proposed that a material be examined under a full spectrum of tribological conditions to establish the critical operating limits, regardless of the end-use application. The emphasis of this approach is firstly on determining what material characteristics limit or control the boundaries between critical and noncritical operating regions. This materials aspect of the wear map approach provides essential information both for materials screening and the materials design. Secondly, the method examines how the system and environmental variables may be used to enhance the useful operating regions. For example, could the effects of an increased operating temperature be offset by a change in the sliding speed?

The present paper is intended to be an introduction to the wear maps concept. In the process of this discussion, wear test results for alumina are used to illustrate the three dimensional wear maps concept.

The concept of "wear map" was first discussed by Tabor [2] who attributed the idea to Ken Johnson, inspired by the pioneering work of Ashby on "deformation maps" [3]. The original discussion centered on the need to predict wear. It was noted that wear occurs by various mechanisms: adhesion, abrasion, delamination, oxidation, fatigue, surface cell-formation, corrosion, erosion, and brittle fracture. Most research was found to use apparatus set up to create conditions under which a particular mechanism dominated. The results of those efforts could be interpreted to develop models which were mechanism and system specific. However, it was also noted that real mechanical systems usually involve several wear mechanisms which operate simultaneously and sometimes sequentially, and may involve mechanisms interacting with each other. It was suggested, therefore, that the predominant wear mechanisms should be categorized as a function of the operating parameters such as speed, load, and temperature.

Lim & Ashby subsequently published a paper on wear mechanism maps on steel under dry conditions using many literature data sources [4]. For the dry wear of steel, they identified four major wear mechanisms: seizures, melt-dominated wear, oxidation-dominated wear, and plasticity dominated wear including delamination. Based on these classification, they identified the localized "flash" temperature as a unifying concept and were able to calculate a "temperature map" and to derive the corresponding wear equations for the four dominant wear modes. The calculated results agreed with the observed data reasonably well. Since the publication of their paper in 1987, many researchers have begun to examine the concept of wear maps.

The present project began in 1986 in an effort to provide critical materials selection criteria and design guidelines for advanced gas-fired prime movers [5]. From the beginning, the emphasis of this project has been on the materials aspects of wear, and the rationale has been similar to Tabor's logic. Early in this work it became clear that a series of maps would be needed to fully represent the wear characteristics of ceramics.

For example, Fig. 1 is a two-dimensional graph of the wear of an alumina under two different loads, but at the same sliding speed. The figure clearly shows that this material undergoes a transition from a relatively low wear regime to a relatively high wear regime, and that where the transition occurs depends on the normal force acting at the tribocontact.

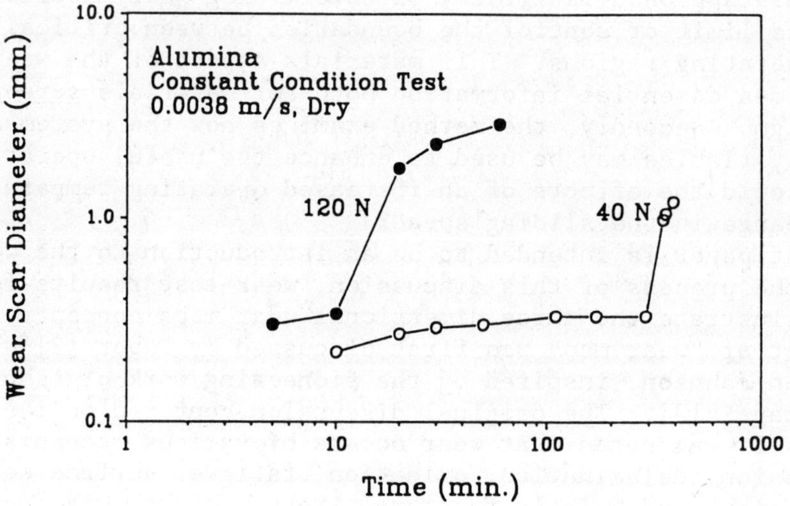

Fig. 1: Variation in the wear of alumina with normal force.

Fig. 2 is another two-dimensional graph which shows that the sliding speed of the wearing surfaces also has an effect on the wear transition. Sliding speed and load are both important design parameters, and either variable can induce a transition from low wear to high wear. If the purpose of the data is to provide design guidelines or material selection criteria, then it must be concluded that the effects of these variables should be examined simultaneously. In terms of the wear map concept, therefore, higher dimensional representations are preferred.

For a fixed material, the primary continuous variables in a wear test are sliding speed, normal force, temperature, and time. Thus, a five-dimensional representation would be preferred for a wear map. However, visual perception is limited to three dimensions and so, therefore, are practical wear maps. Hence, a general wear map in the present work consists of one axis for wear, and two axes chosen among the four continuous wear test variables. The specific choices and forms of the variables vary depending on the purpose of the map.

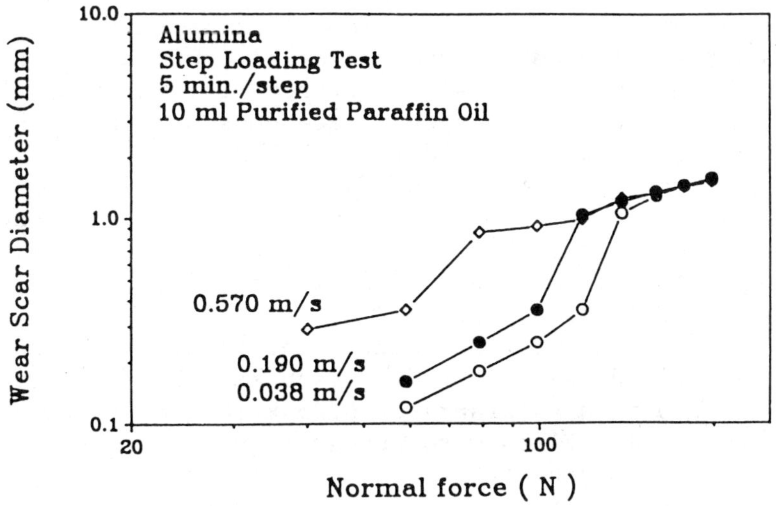

Fig. 2: Variation in the wear of alumina with sliding speed

EXPERIMENTS

Wear tests were conducted using a ball on three flats (BTF) wear test apparatus [6,7,8] that was developed as a variation on the widely used four ball wear tester (FBWT) [9,10,11]. In the conventional FBWT, the balls are held in a tetrahedral configuration. The base is formed by three balls that are held motionless during the experiment. The upper ball rotates about an axis normal to the base. This FBWT configuration provides three point contacts in an arrangement that is self-aligning. In the BTF, the stationary balls in the FBWT are replaced by three flats. The BTF design uses the same tetrahedral configuration to preserve the force vector resolution of the normal forces. Thus the BTF retains the advantages of self-alignment and reduces the number of balls required for each experiment. As a practical matter, flats are considerably easier to fabricate and to analyze than balls.

Two test procedures were used with the BTF apparatus for the present studies [7,8]. The first procedure was a constant condition test (CCT). In the CCT, the conditions that could be regulated by the experimenter, i.e. load, speed, ambient temperature, lubricant, and atmosphere, were held constant during the test. The CCT, therefore, was well suited to the measurement of time-dependent effects. The design of the CCT had the advantage of well defined test conditions. A disadvantage of this procedure was that a separate wear test had to be conducted for each set of conditions included in the study. The resulting large number of tests and the use of too many samples created concerns about cost and sample-to-sample variation.

The second procedure was the step loading test (SLT) which consists of a sequence of short CCT's in which: (a) The duration of each CCT step is fixed at a short time, on the order of 5-10 minutes; and (b) Each successive CCT has an applied load that is larger than the load of the previous CCT by a fixed increment, usually about 50 N. Thus, the SLT provided an indication of wear behavior over a range of conditions in a relatively short time. The SLT was more efficient in terms of time and cost than the CCT. Further, the potential sample-to-sample variation was less for the SLT than for the CCT since all steps in the SLT used the same material. One disadvantage of the SLT was that each constituent CCT had a different initial condition.

For the present study, sliding speeds were chosen in the range 0.0038 m/s to 0.57 m/s. Normal loads were varied from 19.8 N to 238 N. For the SLT, the normal load increment between steps was 49N. All tests were conducted at room temperature, approximately 25°C. For dry tests, no lubricant was used, but the tribocontact was purged with dry air at the rate of 0.75 l/min. The test duration ranged from 10 minutes to 400 minutes for the CCT, while the time per step for the SLT was 5 minutes.

Alumina (Al_2O_3) with the microstructural information and properties listed in Table I was used in this study. Each ball had a diameter of 1.27 x 10^{-2} m and was polished to a tolerance of 6.25 x 10^{-7} m with a surface roughness of 3.8 x 10^{-8} m. Flats were prepared by first cutting a disk from a rod with a 6.35 x 10^{-3} m diameter. Subsequently, the disks were polished to a mirror finish with a surface roughness of 4.0 x 10^{-8} m. The final thickness of each flat was 1.59 x 10^{-3} m.

Table I: Properties of the alumina material used in the present study

Property	Unit	Ball	Flat
Process method	--	Sintered	Sintered
Purity	%	99.5	99.8
Density	gm/cm^3	3.90	3.82
Grain Size	μm	3-15	2-10
Hardness (Vickers)	kg/mm^2	1670	1662
Elastic modulus	GPa	372	345
Poisson's ratio	--	0.22	0.22
Fracture toughness	MN/m^3/^{2}	4.5	4.5
Melting temperature	°C	2015	2015
Max. service temp.	°C	1750	1750
Specific heat (20°C)	J/gm°C	1.5	1.5
(800°C)		130	130
Thermal Conductivity	W/m°C		
(20°C)		30	30
(800°C)		8	8
Thermal Diffusivity	cm^2/s		
(20°C)		0.15	0.15
(800°C)		0.0012	0.0012

Previous work has shown that the procedure used to clean the specimens prior to a wear test can influence the test results [12]. There are two primary problems that may arise from the cleaning procedure. The first is that the solvents may fail to remove all the surface contaminants. This situation can occur for example, if the surface contaminants are not soluble in the chosen solvents. The second problem is that the solvent itself may produce deposits or contaminants while interacting with the surface. Analytical techniques must be used to establish the effectiveness for the prescription used in cleaning the test specimens. The cleaning procedure listed in Table II was found to give reliably prepared surfaces which were relatively free from undesirable contaminants.

Table II: The cleaning procedure used for the alumina samples prior to wear testing

1. Clean surface ultrasonically with pyridine.
2. Clean surface ultrasonically in a detergent solution.
3. Rinse specimen with deionized water while agitating by hand. Repeat several times.
4. Dry the specimen in a heated stream of air.
5. Bake specimen in oven for 12 hours at 550°C.
6. Immediately after baking, place the specimen in a desiccator until cooled to room temperature.
7. Specimen is ready for testing.

Friction was measured continuously during the test by recording the friction torque. A typical result for the coefficient of friction (COF) is shown in Fig. 3. Prior to the wear transition, the COF had a relatively steady value considering that the tribocontact was not lubricated. The mean value of the COF was about 0.27 with a variation on the order of 0.01. After the wear transition, the COF rose rapidly to a mean value of about 0.75 with the instantaneous value ranging from 0.55 to 0.85.

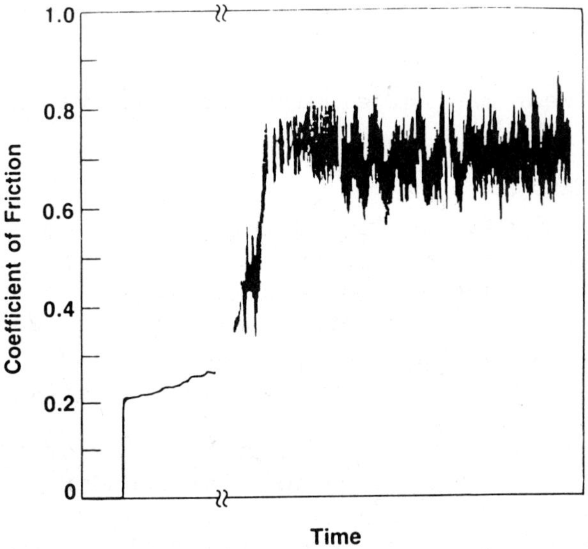

Fig. 3: Typical friction trace, before and after a wear transition in a dry wear test

Wear volumes were determined at periodic intervals during the wear test. The wear scars on the flat specimens in these BTF tests were closely circular in cross section. Hence, to determine the wear volumes, the diameters of the wear scars were measured using optical microscopy. Wear volumes were then calculated assuming that the worn region was a spherical segment that would conform to an unworn ball in the BTF.

A series of experiments was conducted to examine the wear of alumina sliding against the same alumina. A wear map constructed with sliding speed and normal force as the independent variables is shown in Fig. 4.

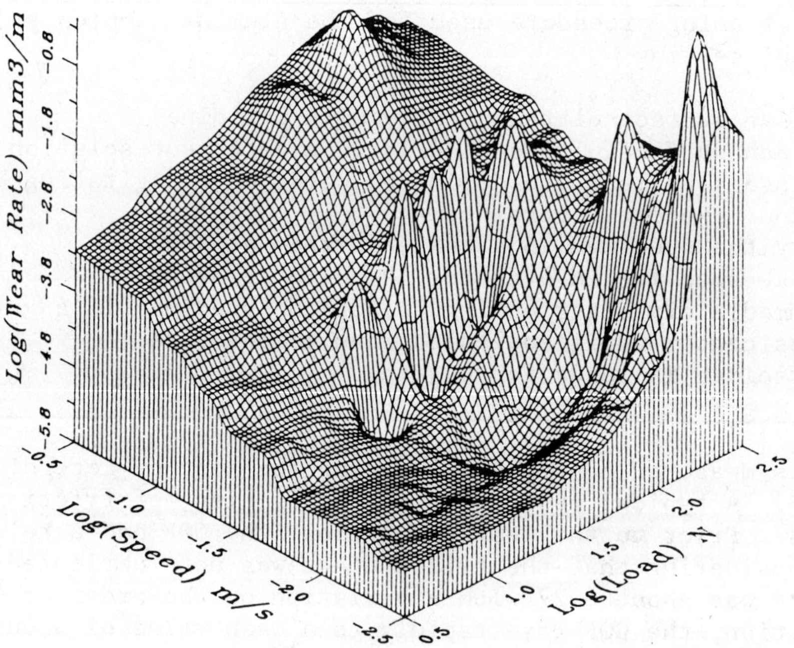

Fig. 4: A wear map for alumina under dry wear conditions

The wear map shows a low wear region strikingly bounded by a region of much higher wear. The transition between the two regions is quite steep. As a design guideline, it is important to recognize that the operating point of a wear component in an engine or any other machine is actually an operating range, with variations in both speed and load possible. The operating range must not allow excursions of the speed and load variables into any portion of the high wear regime.

Figure 5 shows another view of the wear behavior of this alumina, displaying wear as a function of time and load. Under repetitive wearing of the surface, there is again a steep transition to a high wear region, even when the initial wear is low. The endurance of this material under the test conditions depends strongly on the normal force.

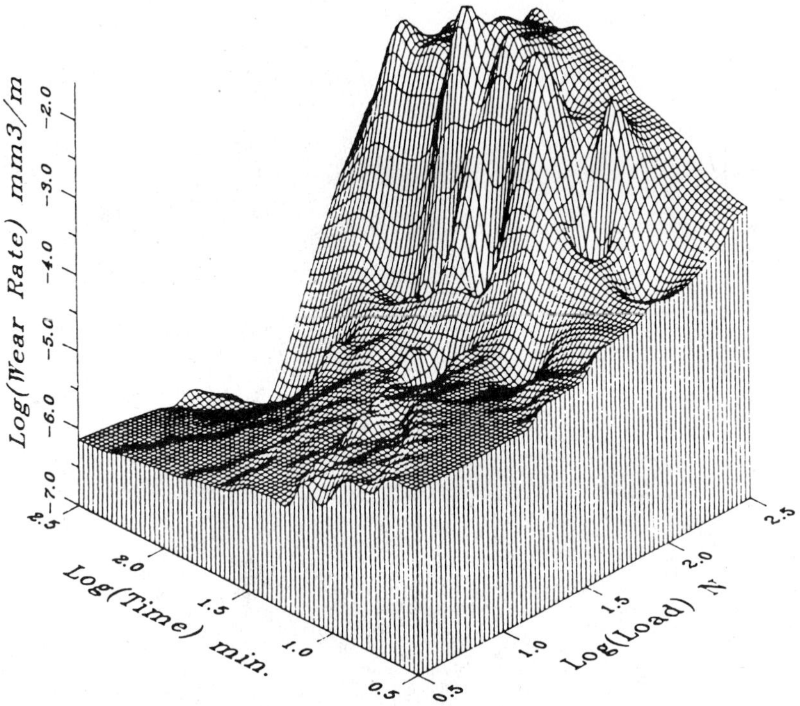

Fig. 5: Time- and load- dependent wear rate of alumina

CONCLUSIONS

Wear is a system's property. Friction and wear of a material depend on how they are measured (speed, load, contact geometry, motion), machine integrity (alignment, vibration), materials (defects, surface roughness, surface layers, property of the wear-couple), and environmental factors (lubricants, oxidation, gaseous atmosphere, surface reactions). Since wear is a transient destructive measurement, definition of the micro-contact conditions is difficult. Given these facts, is it possible to extract an "intrinsic" material property out of these extrinsic set of variables? The answer to this question given by the wear maps concept is: yes, within the context of a standardized procedure.

The first step is to divide wear into two parts: generic properties and system-specific properties. The generic properties assume that all tested materials are subjected to certain standard conditions and procedures. These procedures determine the amount of machine load (normal force and the shear stress) that the material can withstand and the speed limit at which the material usually fails by thermal stress. To measure the true performance property, dry sliding of the material on itself as well as the use of a nonreactive lubricant is needed. The use of a non-reactive fluid avoids excessive surface heating, which causes phase transformation and wear asymmetry [13], and provides a baseline for material performance under realistic conditions. For example, Fig. 6 shows the wear map of alumina when a purified paraffin oil is used as a nonreactive lubricant. Note especially how the effect of speed is reduced by the lubricant, while the

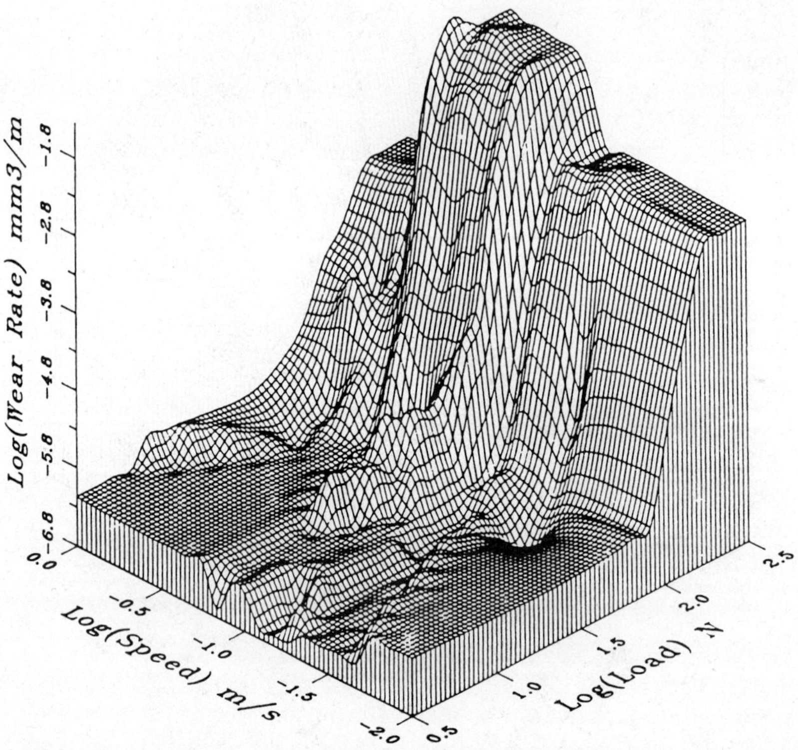

Fig. 6: Wear map for lubricated alumina

effect of the normal force is changed very little. To determine system-specific wear characteristics and to understand the true consequences of environmental factors, additional tests would be conducted to determine the deviation from this reference state of wear. The differences between the various sets of results determine the true performance characteristics of the material. Thus, the wear maps concept involves a series of maps constructed under three sets of conditions: dry wear, nonreactively lubricated, and application specific conditions.

The essence of the wear map concept is in its ability to reveal regions of distinct wear behavior and the dependencies of these regions on material variables and system parameters. Through a collection of graphical representations, the wear map concept, as defined here, is intended as a unified approach to several objectives. The most immediate use of a wear map is to identify the useful application ranges of a material. Upon combining results from several materials, the collected wear maps provide a guideline to materials selection. The same results viewed in the context of materials and processing parameters become a guide to new designs for these materials, optimizing parameters for wear resistance. The methodology of the wear maps enables different materials to be compared on the same basis. The collective information of a set of wear maps may uncover insights into the mechanisms of wear and, hence, wear control.

ACKNOWLEDGEMENT

The authors wish to acknowledge the generous support of the Gas Research Institute through a subcontract from its Center for Advanced Materials at The Pennsylvania State University. The helpful discussion of Mr. S. J. Dapkunas is gratefully acknowledged.

REFERENCES

1. S. S. Kim, K. Kato, K. Hokkirigawa, H. Abe, "Wear mechanism of ceramic materials in dry rolling friction", ASM J. of Tribology 108, 522-526, 1986.
2. D. Tabor, "Status and direction of Tribology as a science in the 80's: Understanding and prediction", Proceeding of Int. Conf. Tribology in the 80's, NASA CP 2300, 1, NASA, 1984.
3. H. J. Frost and M. F. Ashby, Deformation Mechanism Maps: The Plasticity and Creep of Metals and Ceramics (Pergamon Press, Oxford, 1982).
4. S. C. Lim and M. F. Ashby, "Wear mechanism maps", Acta. Metall. 35, 1-24 (1987).
5. R. G. Munro and S. M. Hsu, "Advanced Ceramics: A Critical Assessment of Wear and Lubrication", NBSIR 88-3722, (National Bureau of Standards 1988).
6. S. M. Hsu, R. G. Munro, and R. G. Gates, "Wear Test Methodology for Lubricated Ceramic Materials", Quarterly Reports of the ECUT Tribology Program, TRIB-ECUT-86-3 (Department of Energy, 1986).
7. S. M. Hsu and R. G. Munro, "Wear Mechanisms in Structural Ceramics", Quarterly Reports of the Center for Advanced Materials, pp. 96-117, June 1 - August 31, GRI Contract Number 5084-238-1302 (Gas Research Institute, 1987).
8. R. S. Gates, "Ceramic Tribology: Methodology and mechanisms of alumina wear", NBS-SP, (National Bureau of Standards, 1988).
9. S. M. Hsu, "Review of laboratory bench tests in assessing the performance of automotive crankcase oils", Lub. Engr. 37, 722-731 (1981).
10. R. S. Gates and S. M. Hsu, "Development of a four-ball wear test procedure to evaluate automotive lubricating oils", Lubr. Eng. 39, 561-569 (1982).
11. R. S. Gates and S. M. Hsu, "Development of an oxidation-wear-coupled test for the evaluation of lubricants", Lubr. Eng. 40, 27-33 (1984).
12. S. M. Hsu, R. G. Munro, R. S. Gates, and D. E. Deckman, "Wear Test Methodology for Lubricated Ceramics", Quarterly Reports of the ECUT Tribology Program, TRIB-ECUT-87-2 (Department of Energy, 1987).
13. P. A. Willermet, S. K. Kandah, "Wear Asymmetry - A comparison of the wear volumes of the rotating and stationary balls in the four-ball machine", ASLE Trans. 26, 173-178 (1983).

MARKETS FOR ADVANCED CERAMICS IN INTERNAL COMBUSTION ENGINES

J. M. Schoenung
IBIS Consulting
P.O. Box 4272
Irvine, California 92716

ABSTRACT

The potential market for advanced ceramic materials in internal combustion engines was investigated. The U.S. automotive engine market was characterized and was found to be very segmented, thereby offering the engine component supplier a variety of potential business strategies. The potential sales volumes and profit margins that the engine component supplier could derive from selling four ceramic components (valves, valve guides, cam rollers, and piston pins) were found to depend on the manufacturing costs, the acceptable selling prices, and the market volume for these parts. Under the most optimistic assumptions, the estimated size of this market is less than has been indicated by other projections. Although the market would not be large enough to support the ceramic industry, it could be a viable business for a few select companies.

INTRODUCTION

The worldwide market for advanced ceramic materials has been projected to grow to as much as 187 billion dollars (37 trillion yen) by the year 2000,[1] or only 17 billion dollars by the year 1995.[2] Yet others estimate that the market for advanced ceramics in the United States alone could be as much as 20 billion dollars by the year 2000,[3] or as little as 500 million dollars.[4] Regardless of size, the market for advanced ceramics is generally considered to consist of several potential and existing applications, ranging from electronics to heat engines.

Engine applications are one of the most publicized segments of the advanced ceramics market. Coverage of this subject has been extensive, ranging from scientific journals to public newspapers. For instance, entire issues of the American Ceramic Society Bulletin have been dedicated to the topic of ceramic engines.[5] Advanced Materials and Processing, Materials Engineering, Mechanical Engineering, Chemical Week, and Chemical & Engineering News have published papers in this area, as have Automotive Engineering and Automotive News.[2,3,6-10] But the increased awareness has spread beyond the scientific and engineering community. The business community has been confronted as well with articles in Fortune Magazine, the Christian Science Monitor, and the Wall Street Journal.[11-13] Even the general public has been notified of the potential application of ceramics to engines via local newspapers such as the Boston Globe and the Chicago Tribune.[14,15] The development of ceramics for engine applications has also been the focus of both international and domestic conferences.[16,17]

The significance of this market sector to the ceramic industry is best demonstrated, however, by the substantial amount of government funding that has been committed to the research and development of ceramic materials for engine applications. Past levels of government funding in this area, presented in Table 1, indicate that the United States, Germany, and Japan have invested more than 220, 50, and 10 million dollars, respectively, since 1971.

Not surprisingly, the attention given to the heat engine segment of the ceramics market has also appeared in the form of market projections specific to this application. A summary of some of the projections that have been published on the potential market for ceramics in engines are presented in Table 2.

Table 1. Government Funding for Ceramic Engine Programs ($ million U.S.)

Years	U.S.	Germany*	Japan**
1971-1973	10	na	na
1974-1977	22	13	na
1978-1980	54	12	na
1981-1983	96	13	10
1984-1986	38	14	na

* Currency conversion of 2.6 DM = 1 US$
** Currency conversion of 200 Yen = 1 US$
na: not available
Source: References 18 and 19

Table 2. Projections for the Potential Ceramic Engine Market

World Market			Source
15 billion dollars per year			[11]

U.S. Market ($ millions)	1990	2000	Source
	25-45	920-1300	[20]
	120		[21]
	100	1300	[10]
		1300	[14]
		204-988	[4]

Japanese Market (1990)	billions of yen	$ billions	Source
	280-420	1.4-2.1	[22]

In making these projections, only four of the authors explain how the projections were calculated. The world market projection shown in Table 2 is based on assumptions of the amount of ceramic material per engine, the price per pound of the material, and the number of engines produced. The first U.S. projection listed in this table assumes the percent penetration of ceramic engines into the engine market, as does the last U.S. projection listed. The Japanese market projection also assumes percent penetration of ceramics into either diesel engines or engine replacement parts.

For all of the above projections, the sources of the data used are not clearly referenced. In this paper, the data characterizing one segment of the engine market will be presented and these data will be used to project the potential market size for six representative ceramic engine components. In addition, estimates will be presented of the potential profitability resulting from the fabrication of these components using ceramics.

CHARACTERIZATION OF THE U.S. GASOLINE ENGINE MARKET

The segment of the engine market that was selected for this analysis is that of U.S.-produced gasoline passenger-car engines. This segment was selected because it represents the largest segment of the market in terms of number of engines produced. The following analysis focuses on the use of ceramics in this segment of the engine market, and does not encompass other potential applications such as diesel engines, gas turbine engines, or other than U.S.-produced gasoline engines.

The potential applications for structural ceramics in U.S.-produced gasoline car engines include wear components, valve train components, and reciprocating-mass parts. The introduction of these components would be primarily as part-for-part substitutes, as opposed to parts that require engine redesign; the market for these parts is characterized below.

In 1985, there were seven engine manufacturers in the U.S.. These are listed in Table 3 along with their respective engine production volumes since 1982.

Table 3. Production Levels of the U.S. Gasoline Engine Manufacturers

	MODEL YEAR			
COMPANY	1982	1983	1984	1985
General Motors	3288545	3668518	5224527	4729900
Ford Motor	1329420	1374725	1950656	2059533
Chrysler	711251	768230	1000170	1152922
American Motors	70898	168726	234159	166379
Honda of America	-	33508	117200	155102
Volkswagen of America	58229	61229	77448	67922
Nissan Motor	-	-	-	17368
	=======	=======	=======	=======
Total	5458343	6074936	8604160	8349126

Source: References 23-26

Each year these engine manufacturers produce several different engine models. The 29 different models and their respective levels of production for model year 1985 are listed in Table 4. Though over 8 million automotive engines were produced from these 29 models, more than half of the engine models were made in production runs of less than 200,000 units. The largest production run was just under 900,000 units. These data show that although the total market for automotive engines is large, the market for individual engines and engine components is segmented into more specialized markets.

The number of parts per engine type were calculated on the basis of the units produced and the number of cylinders per engine of the engine models made in model year 1985, and are presented in Figure 1. These data can be used to estimate the potential sales levels and profits that would result if a parts producer were to introduce ceramic components into various engines. Such estimates are presented in the next section.

POTENTIAL SUPPLIER SALES AND PROFIT LEVELS

The extent to which ceramic engine components will be able to penetrate the market characterized above will be determined by the potential for the following two criteria to be satisfied.[27] Both users and suppliers of these components must have an incentive for substituting ceramic components for conventional metallic components.

User incentives for changing to ceramic engine components include superior technical performance and cost reductions. Technically, ceramics exhibit better wear resistance, higher temperature strength, and lower density than metals. These properties can translate into enhanced engine performance. However, studies have shown that in the automotive industry, technical performance improvements are valuable to the user only if, ultimately, they result in a cost savings.[27,28]

The other requirement for successful substitution of ceramics into engines is that suppliers have an incentive to produce and sell these components. Essentially, this translates into the need for the supplier to be able to make a profit from selling these parts. Profits can only be realized if revenues are greater than costs. Sales revenues are determined by the selling price and the volume of parts sold. Manufacturing costs depend on the process, production volume, and several other variables.[29-32]

The volume of ceramic engine parts sold will depend upon the supplier's strategy. For instance, the supplier may choose to produce parts for engines that are made in small, large, or medium production runs. Alternatively, the supplier may choose to produce parts for more than one engine model so as to utilize economies of scale. Depending on which strategy is chosen, the supplier's potential revenues will vary.

To demonstrate the range of potential market sizes for ceramic engine components, two scenarios are presented. Within each scenario, six engine parts are considered: intake valves, exhaust valves, intake valve guides, exhaust valve guides, cam rollers, and piston pins.

Table 4. 1985 Model Year U.S. Gasoline Engine Production

COMPANY	ENGINE DISPLACE. (L)	ENGINE DISPLACE. (cu. in.)	CYLINDERS PER ENGINE	UNITS PRODUCED
General Motors				
	1.6	97	4	27943
	1.6	98	4	148379
	1.8	112	4	160644
	2.0	121	4	470162
	2.5	151	4	689530
	2.8	173	6	615328
	3.0	181	6	341635
	3.8	231	6	683578
	4.1	250	8	328532
	4.3	262	6	102844
	5.0	305	8	551485
	5.0	307	8	560338
	5.7	350	8	49502
Ford Motor				
	1.6	98	4	300907
	1.9	113	4	209200
	2.3	140	4	478225
	3.8	231	6	435519
	5.0	302	8	617490
	5.8	351	8	18192
Chrysler				
	1.6	98	4	15481
	2.2	134	4	894394
	2.6	159	4	74809
	5.2	318	8	168238
American Motors				
	1.4	85	4	83547
	1.7	105	4	66642
	4.2	258	6	16190
Honda of America				
	1.8	112	4	155102
Volkswagen of America				
	1.8	109	4	67922
Nissan Motor				
	1.6	97	4	17368

Source: Reference 26

The first market size scenario, shown in Table 5, assumes that the number of pieces sold per year are equal to the number of equivalent pieces consumed in the smallest production run in model year 1985. The second market scenario, shown in Table 6, assumes that the number of pieces sold per year are equal to the number of equivalent pieces consumed in the total production in model year 1985. The data in these tables are calculated assuming that the selling prices are equivalent to what the user would be willing to pay for the superior properties of the ceramic components, as identified by Schoenung,[27] thereby generating a high-low range for each market scenario. The potential market that could be generated from these ceramic engine parts, given the two extreme scenarios presented, range from less than one million dollars per year to more than 250 million dollars per year.

In order to estimate profits, it is also necessary to quantify the costs of producing ceramic engine components. Extensive work has been done elsewhere towards this end.[27,29,32] These studies have shown, among other things, that economies of scale exist in ceramic manufacturing processes. It can be assumed, therefore, that cost will also decrease with increasing cumulative production. As these costs decrease, the potential for suppliers to make a profit increases.

Table 5. Market from Smallest Production Run (M.Y. 1985)

COMPONENT	PRODUCTION (pieces/year)	LOW MARKET (millions $/year)	HIGH MARKET
Intake Valve	61,924	0.03	0.06
Exhaust Valve	61,924	0.03	0.06
Intake Valve Guide	61,924	0.02	0.04
Exhaust Valve Guide	61,924	0.02	0.04
Cam Roller	123,848	0.07	0.09
Piston Pin	61,924	0.04	0.05
		======	======
Total Market		0.21	0.34

Table 6. Market from Total Production (M.Y. 1985)

COMPONENT	PRODUCTION (pieces/year)	LOW MARKET (millions $/year)	HIGH MARKET
Intake Valve	46,961,800	23.48	46.96
Exhaust Valve	46,961,800	23.48	46.96
Intake Valve Guide	46,961,800	14.09	28.18
Exhaust Valve Guide	46,961,800	14.09	28.18
Cam Roller	93,923,600	56.35	70.44
Piston Pin	46,961,800	28.18	37.57
		========	========
Total Market		159.67	258.29

Figures 2 and 3 show two scenarios of cumulative profit, for the four engine components studied, as a function of cumulative production, starting from the first part manufactured.[1] The two scenarios presented in these figures correspond to different assumptions for target prices and powder prices, which are listed in Table 7. Both of the figures assume that the cost reductions with increasing cumulative production are derived from economies of scale.[27]

The first scenario for the cumulative profit curves, shown in Figure 2, corresponds to low target price and high powder price assumptions, i.e., the less optimistic but most realistic of the two scenarios. Under these assumptions, none of the ceramic components become profitable even with cumulative production as high as 100 million parts.

The second, and more optimistic, profit scenario, shown in Figure 3, corresponds to high target price and low powder price assumptions. Although it is reasonable to assume that powder prices will decrease with time and large volume demand, the assumption of maintaining sales at high target prices is optimistic in that it implies that suppliers of conventional engine components will not lower their prices with the intent of sustaining their market share. Acknowledging the implications of this optimistic scenario, the results indicate that all four of the engine components become cost competitive in this scenario, and that all four parts generate positive profits with cumulative production levels of less than 20 million pieces. At cumulative production levels of 100 million parts, the valves, valve guides, piston pins, and cam rollers represent cumulative profits of $55 million, $30 million, $29 million, and $28 million, respectively. The sum of these is $142 million, which can be compared to the sales volume projected in Table 6 for the total production of these parts in model year 1985, about $250 million.

Table 7. Target Piece and Powder Price Ranges for Cumulative Profit Scenarios

COMPONENT	MATERIAL	TARGET PRICE ($/kg)	POWDER PRICE ($/kg)
Valve	ZTA*	27.78/13.89	13.00/4.50
Valve Guide	ZTA	42.86/21.43	13.00/4.50
Cam Roller	ZTA	18.29/14.63	13.00/4.50
Piston Pin	Silicon Nitride	9.76/7.41	44.00/11.00

* Zirconia Toughened Alumina

[1] Cumulative profit at a given level of cumulative production is the summation of the profits (or losses) from the previous production plus the profit from the last part produced. A minimum is seen in cumulative production at the point where the piece profit equals zero. Cumulative profit becomes positive when the early losses have been compensated for by positive profit. The term profit used here implies revenues less direct costs.

CONCLUSIONS

The above discussion shows that, under optimistic assumptions, the supplier of ceramic engine parts could anticipate making a profit in this market. The magnitude of this profit, however, depends on the assumptions made with respect to the processing, the price of the raw material, and the user's acceptable component price. Even the most optimistic set of assumptions, (i.e., 100% penetration of ceramics into the market for cam rollers, valves, valve guides, and piston pins in U.S.-produced gasoline engines, sold at the highest acceptable price, and made from powders sold at the lowest feasible price), yields estimated profit and sales levels of only about $140 million and $250 million a year, respectively.

REFERENCES

1. Saito, S., "Present and Future World Market for High-Technology Ceramics," presented at World Congress on High Tech Ceramics, Milan, Italy, June 23-27, 1986.

2. Spalding, B.J., "Advanced Ceramics, The U.S. edge is in danger," Chemical Week, Feb. 6, 1985, pp. 38-42.

3. Sheppard, L.M., "Reliable Ceramics for Heat Engines," Advanced Materials and Processes, October 1986, pp. 54-66.

4. Johnson, L.R., Teotia, A.P.S., and Hill, L.G., A Structural Ceramic Research Program: A Preliminary Economic Analysis, ANL/CNSV-38, March 1983.

5. - Heat Engine Ceramics Issue, Am. Cer. Soc. Bull., 64 [2] Feb. 1985.

6. Sheppard, L.M., "Japan Sets the Pace in the Fine-Ceramics Race," Materials Engineering, Nov. 1983, pp. 19-23.

7. Harmon, R.A. and Beardsley, C.W., "Ceramics Heat Up," Mechanical Engineering, May 1984, pp. 22-35.

8. Sanders, H.J., "High-tech ceramics," Chemical & Engineering News, July 9, 1984, pp. 26-40.

9. - "Testing a small DI ceramic diesel engine," Automotive Engineering, Vol. 94, No. 8, August 1986, pp. 66-71.

10. - "Can U.S. crack the ceramic components market?" Automotive News, Sept. 29, 1986, p. D24.

11. Dizard, J.W., "The Amazing Ceramic Engine Draws Closer," Fortune, July 25, 1983.

12. Dole, C.E., "Ceramics make their break into the world of auto manufacturing," The Christian Science Monitor, Nov. 15, 1984, pp. 57-58.

13. Yoder, S.K., "Ceramic Engines Losing Support in Japan," The Wall Street Journal, Aug. 12, 1986, p. 33.

14. Patterson, G.A., "Technology reaches back to move ahead," The Boston Globe, Nov. 10, 1985, p. A1.

15. Van, J., "Experts Finding Myriad of Uses for 'Tough Clay'," Chicago Tribune, April 28, 1985, p. C1.

16. 13th Annual Materials Conference: "Processing of Automotive Ceramics," University of Michigan, Ann Arbor, MI. Nov. 6-7, 1985.

17. Second International Symposium and Exposition on Ceramic Materials and Components for Engines, Travemunde, West Germany, April 14-17, 1986.

18. Lenoe, E.M. and Meglen, J.L., "International Perspective on Ceramic Heat Engines," Am. Cer. Soc. Bull. 64 [2] Feb. 1985, 271-275.

19. "The Development of Structural Fine Ceramics," The Japan Industrial & Technological Bulletin, Special Issue No. 15, 1983. Japan External Trade Organization.

20. Charles River Associates, Technological and Economic Assessment of Advanced Ceramic Materials, Volume 1: Summary and Conclusions, NBS GCR 84-470-1, Aug. 1984.

21. Fisher, G., "Advanced Ceramics Accounting Begins," Am. Cer. Soc. Bull. 66 [1] 1987, 47-48.

22. Kimura, H., "The New Ceramics Industry in Japan: Present State and Future Prospects," LTCB Research, Monthly Economic Review No. 68, Sept./Oct. 1983 Special Issue, Long Term Credit Bank of Japan, Ltd.

23. Ward's 1983 Automotive Yearbook Forty-Fifth Edition. Ward's Communications Inc., Detroit, MI, 1983.

24. Ward's 1984 Automotive Yearbook Forty-Sixth Edition. Ward's Communications Inc., Detroit, MI, 1984.

25. Ward's 1985 Automotive Yearbook Forty-Seventh Edition. Ward's Communications Inc., Detroit, MI, 1985.

26. Ward's 1986 Automotive Yearbook Forty-Eighth Edition. Ward's Communications Inc., Detroit, MI, 1986.

27. Schoenung, J.M., "An Engineering and Economic Assessment of the Potential for Ceramics in Automotive Engines." M.I.T. Ph.D. Thesis. June, 1987.

28. J.M. Schoenung, "Ceramics in Automotive Engines: User Incentives for Substitution," 11th Annual Conference on Composites and Advanced Ceramic Materials, Cocoa Beach, FL, January 18-23, 1987.

29. J.M. Schoenung, E.P. Rothman, J.P. Clark and H.K. Bowen, "Cost Models for Ceramic Engine Parts," American Society of Metals Fall Meeting, Orlando, FL, October 1986.

30. J.M. Schoenung, "Silicon Nitride Powders: Investigating the Costs of Production," Fine Particles '88, Montreal, Canada, August 28-31, 1988.

31. J.E. Jacob, J.M. Schoenung, E.J. Lavernia, J.P. Clark, and N.J. Grant, "Cost Modeling RS Powders Produced by Inert Gas Atomization," Journal of Metals, October 1987.

32. Rothman, E.P., Clark, J.P., and Bowen, H.K., "Cost modeling of structural ceramics," Adv. Cer. Mater., Vol. 2, No. 1, 1987, pp. 34-38.

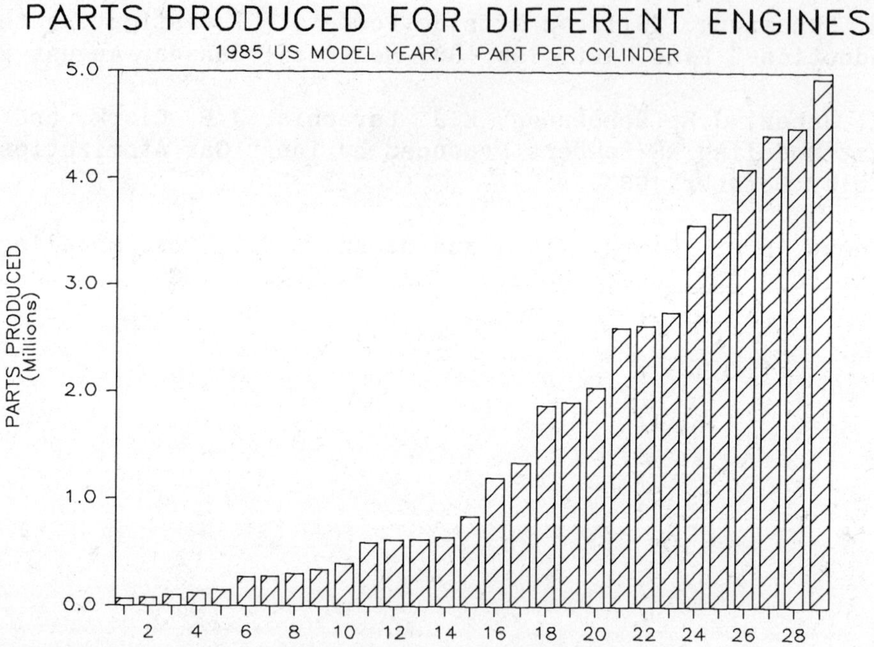

Figure 1. Engine Parts Production - One Per Cylinder
Number of parts produced for the 29 different engine types produced in the U.S. in model year 1985, assuming one of each part per engine cylinder.

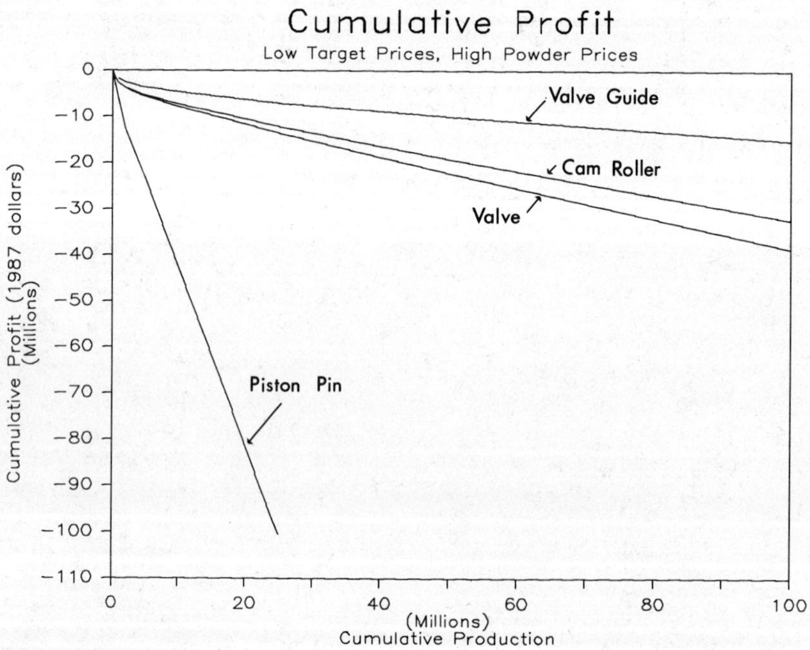

Figure 2. Cumulative Profit - Scenario A
Cumulative profit as a function of cumulative parts production for ceramic engine parts, assuming low target prices and high powder prices.

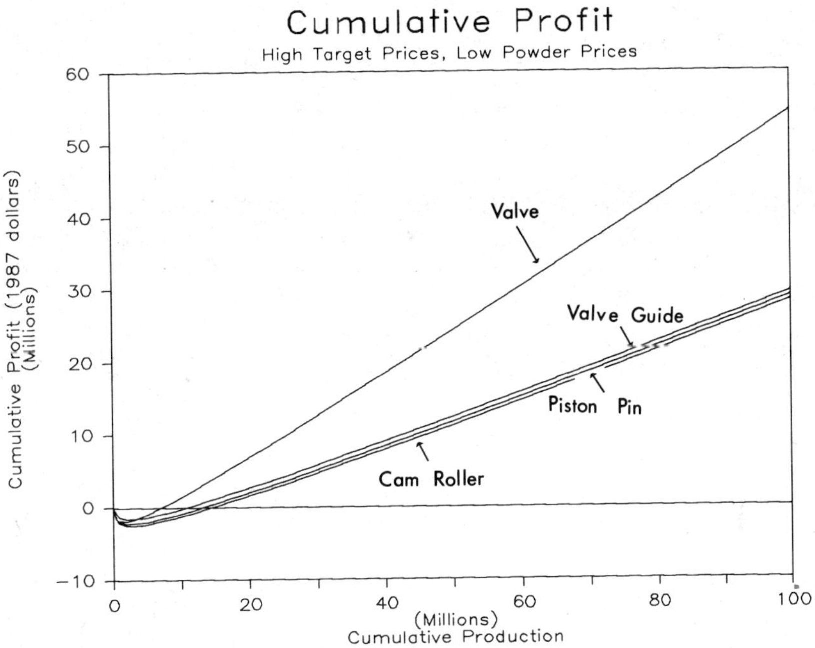

Figure 3. Cumulative Profit - Scenario B
Cumulative profit as a function of cumulative parts production for ceramic engine parts, assuming high target prices and low powder prices.

CRITICAL ISSUES FOR CERAMICS FOR GAS TURBINES

Dave Carruthers and Laura Lindburg
Allied-Signal Aerospace Company
Garrett Auxiliary Power Division
Phoenix, Arizona

ABSTRACT

Fast fracture and thermal shock properties of ceramics have proven sufficient to allow several successful feasibility demonstrations in gas turbines. However, the emphasis for future development must be on issues addressing long term durability and reliability. Currently, slow crack growth, creep, and impact sensitivity limit engine development progress. As these problems are overcome, the technology development issues will shift to cyclic fatigue, oxidation and corrosion behavior, and ultimately, to cost. The state of the art in gas turbine ceramic materials technology is reviewed and the goals of current and future development programs are projected.

BACKGROUND

The potential benefits of ceramic materials to gas turbine engines have been documented numerous times during the past decade. Early studies performed by Ford Motor Co. indicated that the use of ceramic components in an automotive gas turbine operating at 1370C (2500F) would double the horsepower and improve the fuel economy by 20 percent compared to a similar size all-metallic engine(1). Updates of these studies during the DOE sponsored Advanced Gas Turbine Programs by both Garrett(2) and Allison(3) projected a diesel fuel economy of 5.4 liter/100km (42.3 mpg) in a 1361 kg (3000 pound) vehicle over a federal combined urban and highway driving duty cycle. Ceramic benefits have also been projected for non-automotive gas turbine applications. During joint NASA/Army funded studies performed by Garrett, Allison, Williams International, Teledyne, and Avco Lycoming, the use of ceramics in auxiliary power units (APUs), commuter, helicopter, and missile engines was evaluated.(4). These studies concluded that the use of ceramics and/or ceramic composites could account for 23 to 65 percent of the projected fuel and cost savings achievable in the year 2000 using advanced technologies. In addition, Sunstrand Turbomach has projected improved turbine nozzle life during sand ingestion with the use of ceramic turbine stators. Under certain conditions, an advantage of up to 300 percent was observed compared to superalloys(5).

Additionally, it is projected that the use of lightweight, high temperature nonmetallic components will be required to meet Air Force engine goals for specific power. The successful application of ceramics to gas turbines, however, is an immense challenge to the industry. Significant accomplishments have been made toward the goal of ceramic engines yet sizable challenges in the areas of design, materials, fabrication, quality control and qualification remain to be resolved. This paper focuses on accomplishments to date and on the areas of technology requiring additional advancement.

DEVELOPMENT CHALLENGES

Before addressing the development challenges, it is beneficial to summarize the demands placed on ceramics by the gas turbine engine environment. These demands are quite severe and can vary considerably, depending on the application. The ceramic gas turbine applications considered include missile, automotive, military and commercial propulsion, and auxiliary power engines. Each of these applications has a different duty cycle, life requirement, operating temperature and stress envelope. The variety of engine lifetimes and power requirements is illustrated in Figure 1. Design lives can vary from 15 to 10,000 hours, with time at maximum power ranging from 1 to 40 percent. Most turbines operate the majority of the time at power levels below maximum power as illustrated. The duty

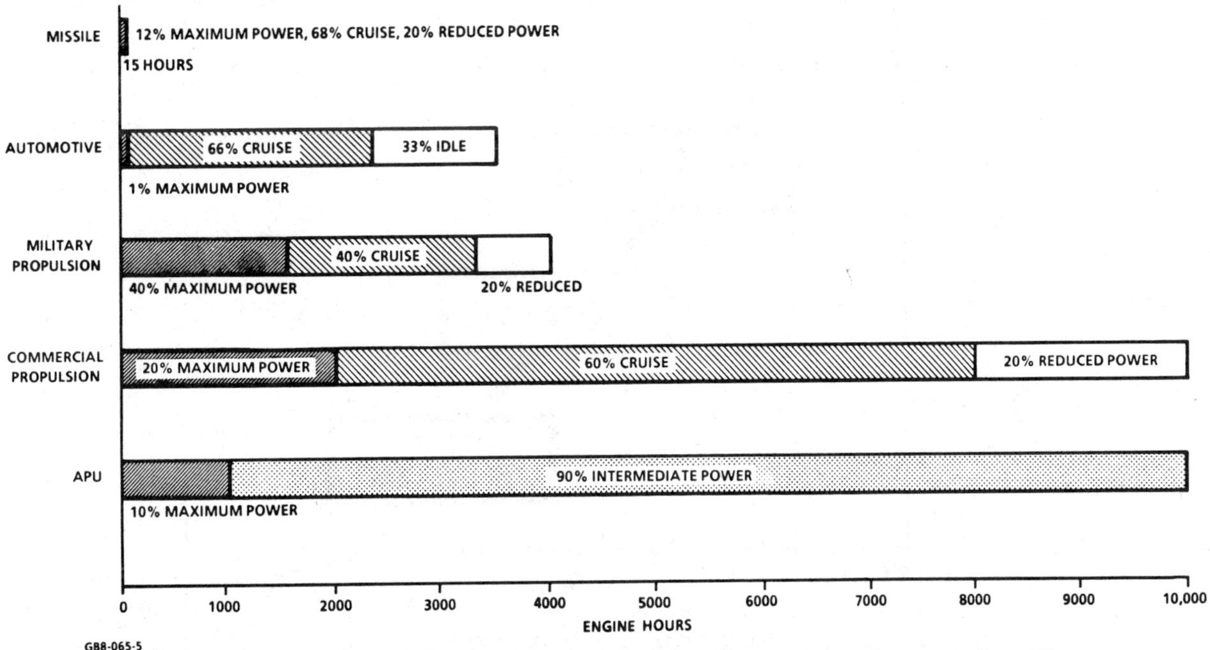

Figure 1. Duty Cycles and Lifetime Requirements Vary Considerably for Gas Turbine Engines.

cycle and engine life will dictate the strength, crack growth and creep rates, and environmental stability required of the selected material.

Although thermal transients also vary with engine and application, peak lightoff temperatures are generally reached in one to fifteen seconds. The severity of shutdown transients depends on the ram air or rolldown air temperature and differs significantly for regenerated and non-regenerated engines. Since thermal transients are often responsible for the peak stresses in ceramic components, they will typically dictate the degree of thermal shock resistance needed in the material and component design.

The introduction of advanced materials such as ceramics into gas turbines generally follows a sequence of development activities aimed at establishing feasibility, performance, reliability, and competitive cost. These activities are illustrated on a time line in Figure 2. As shown, these efforts typically overlap. First, the feasibility of both the design and the material to survive the severe duty cycle of the gas turbine must be established. As discussed previously, this duty cycle is application dependant, but in general, critical engine conditions include the thermal gradient induced stresses of lightoff, shutdown, and cyclic use, as well as sustained stresses and temperatures encountered during maximum and part power engine operation. Primary material properties affecting

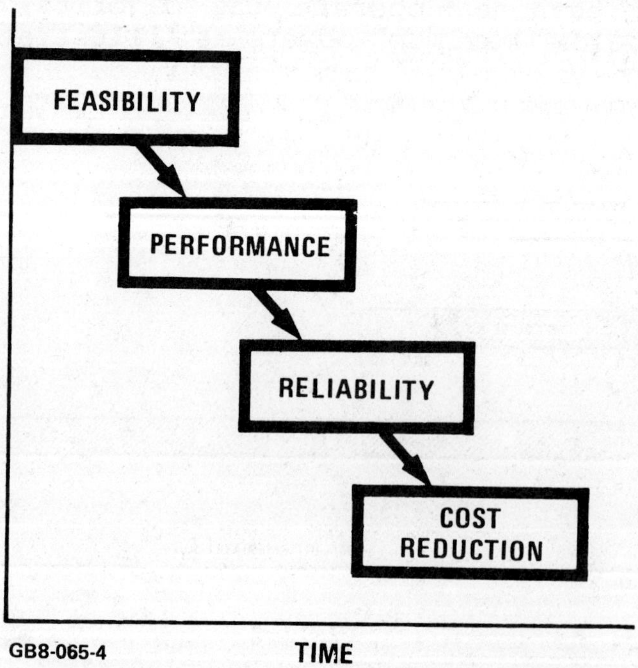

Figure 2. Introduction of Advanced Materials Into Gas Turbines Follows a Sequence of Development Efforts.

feasibility include strength, thermal expansion, thermal conductivity, elastic modulus, fatigue, oxidation, and corrosion. Feasibility also depends on how well the component design minimizes the thermal and mechanical stresses. Extensive analytical and experimental evaluation efforts are required to confirm all aspects of feasibility, including transient response, steady state, peak power, and extended life. Evaluations typically address these issues in a progressive manner.

Once feasibility is demonstrated, the issues of performance, reliability, and cost competitiveness must be demonstrated. Performance includes meeting power, fuel consumption, and/or specific power goals. Reliability includes demonstrating predictable engine durability under normal and adverse conditions. Cost competitiveness includes the initial cost as well as the cost of operation, which will vary with the nature of the application.

Although the development efforts of the past two decades have focused heavily on establishing the feasibility of using ceramics in gas turbines, recent work is also addressing the issues of performance, reliability, and cost. As experience in each of these four aspects is gained, so is a clearer understanding of the critical technical challenges which remain.

FEASIBILITY

Feasibility demonstration programs have highlighted engine lightoff as the initial design hurdle. The thermal shock during lightoff, the thermal growth mismatch between metallic and ceramic parts, and the potential for distortion in the metal parts each challenge the material as well as the ingenuity and analytical tools of the designer. Using thermal shock resistant components and forgiving attachments, significant achievements have been made through iterative design and material improvements. For example, in the AGT101 engine, normal design lightoffs to 1100C (2000F) in 5 seconds have been performed routinely.

However, engine operation is not limited to normal design conditions, and non-nominal lightoffs must be anticipated. Thus, the challenge to provide additional design margin still exists. Potential approaches to optimizing component designs for thermal shock resistance include reducing temperature gradients through the use of higher conductivity materials or thermal barrier coatings, the development of high temperature compliant interface materials for attachments, lower modulus materials and new attachment designs. The application of composite ceramic materials with increased resistance to catastrophic failure has significant potential to enhance reliability. Currently, the assessment of changes or improvements in thermal shock resistance requires expensive analysis or rig testing. Thus, an additional challenge is the development of a meaningful, low-cost method of characterizing ceramic materials for thermal shock resistance under conditions representative of gas turbine operation.

During the course of demonstration programs performed at Garrett, Turbomach, and Allison, impact resistance has been identified as a major technical barrier that must be overcome. Each company has experienced impact damage to ceramic parts during engine testing. Resolving the impact damage issue is being addressed in a four-pronged approach at Allied-Signal; thickening susceptible components, developing materials with improved toughness, development of high temperature particle separators and traps, and achieving a better definition of critical design and material parameters. The benefits of thicker radial turbine blades and particle traps have already been successfully demonstrated in automotive turbochargers. A high temperature particle separator is under evaluation in the DOE/NASA Advanced Turbine Technology Applications Program (ATTAP) at Garrett.

These feasibility demonstration programs have also highlighted cyclic operation as an important aspect of engine feasibility demonstration. Although Allison demonstrated numerous cycles of operation during the earlier Ceramic Applications for Turbine Engines program, and Garrett accumulated 155 successful engine cycles on the AGT101, uncertainty still exists about long term cyclic response of attachment designs, interfaces, and materials. Several recent material studies have highlighted the sensitivity of ceramic materials to cyclic fatigue[6,7,8]. Further understanding of this behavior and its impact on component design is needed.

Feasibility to achieve long engine life at maximum power conditions has also yet to be demonstrated. Under these conditions, life is a function of temperature, stress, atmosphere, and changes in the material defect population due to flaw nucleation and growth. Although engine life has been successfully demonstrated under a variety of conditions (Figure 3), these demonstrations have only confirmed feasibility for short life uses, such as missile applications. They have not yet confirmed readiness for automotive or typical aerospace use at operating temperatures and times required for performance payoffs. Limiting issues include insufficient creep and rupture properties of rotor materials, an incomplete understanding of the stress rupture and creep behavior of state-of-the-art materials, and an inability to accurately predict component life based on these properties. An example is the turbine rotor for the AGT101 engine. At the 1370C (2500F) turbine inlet temperature (TIT) maximum power condition, a critical condition occurs in the center of the rotor hub where the material temperature is near 1065C (1950F) and the biaxial peak stress is 210 MPa (30.5 ksi). Until recently, the rupture and creep properties of rotor materials have been insufficient for long life under

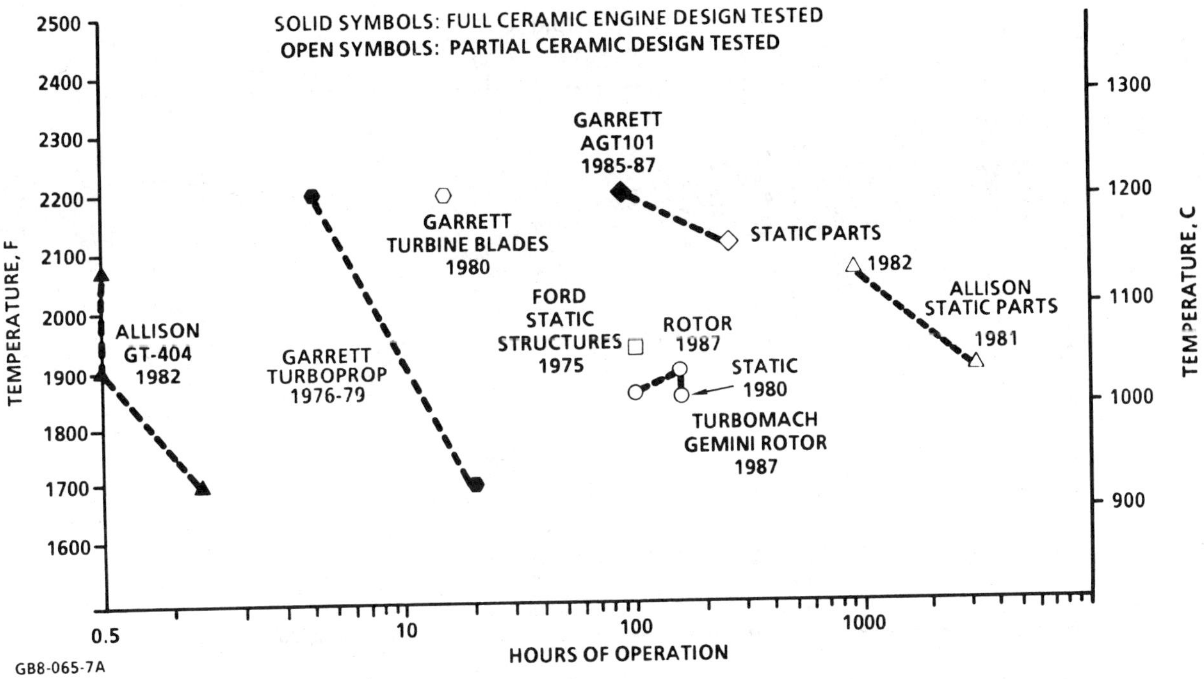

Figure 3. Engine Life With Ceramic Components Has Been Demonstrated Under a Variety of Conditions.

this condition. Additionally, the limited strength, rupture, and creep data available were obtained on small, unidirectionally stressed bend bars. The method of correlating such data to the large biaxially loaded rotor hub has not yet been verified.

While a number of unknowns still need to be addressed, the materials available for turbine use have improved dramatically over the past twelve years. Not only have the shape and size capabilities improved, but the fast fracture and stress rupture properties have also steadily improved. Figures 4, 5, and 6 show the progression of fast fracture, stress rupture, and cyclic durability properties for state-of-the-art silicon carbide and silicon nitride materials available in 1976, 1982, and 1988. The fast fracture flexure strength (Figure 4) of Standard Oil Hexoloy SA, sintered alpha silicon carbide has increased from 310 MPa (45 ksi) in 1976, to 415 MPa (60 ksi) in 1988. Sintered silicon carbide materials exhibit excellent strength retention at temperatures up to 1540C (2800F) and continue to retain most of their initial strength after extended cyclic exposure to combustor conditions[9] (Figure 6). Dense silicon nitride materials have higher strengths at low temperature than silicon carbide materials but decrease in strength with increasing temperature. In 1976, fully dense Si_3N_4 was only available in hot-pressed forms such as Norton NC-132. NC-132 had a room temperature flexure strength over 690 MPa (100 ksi) but the strength decreased rapidly above 1000C (1800F)[10]. Turbine components fabricated from NC-132 required grinding from billet material. Net-shape sinterable

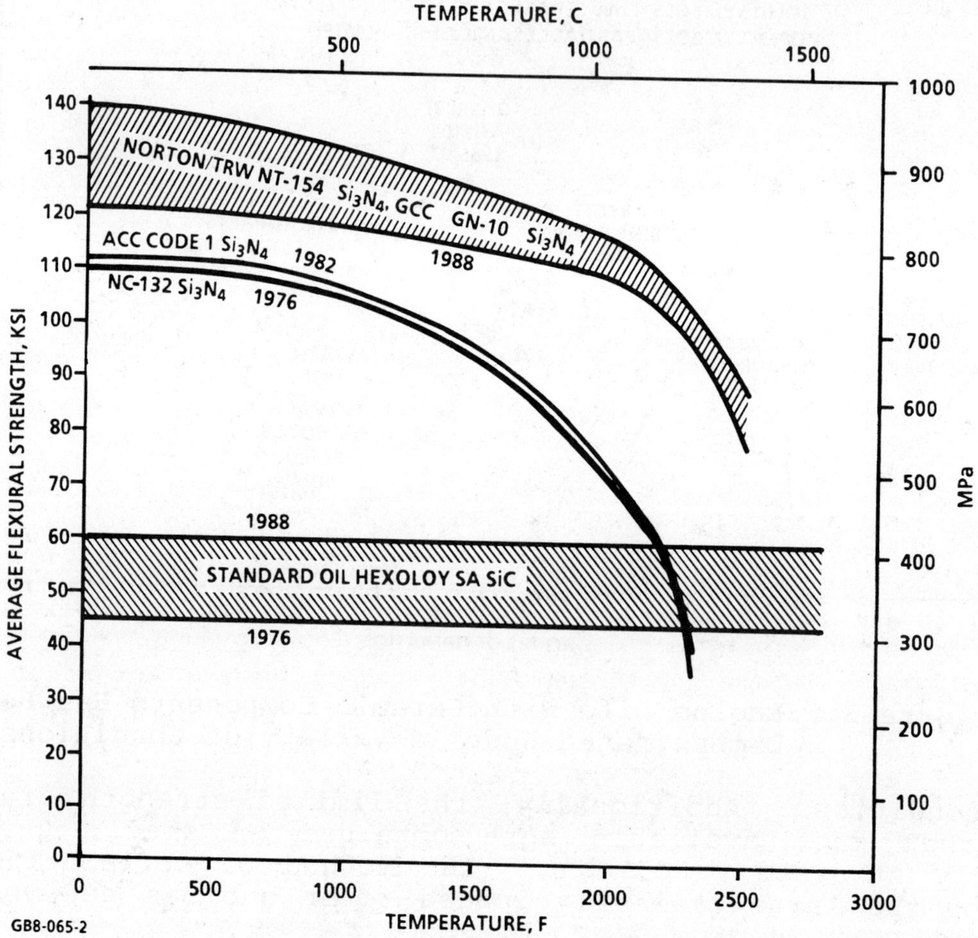

Figure 4. **The Strength Versus Temperature Properties of Advanced Ceramics Has Improved Significantly. (Allied-Signal Aerospace Data)**

silicon nitride components such as AiResearch Casting Company (ACC) Code 1 sintered silicon nitride became available in the late 1970's. These initial sinterable materials had the advantage of near-net-shape capability but exhibited similar strength loss at high temperature, Figure 4. Current state-of-the-art materials, such as Norton/TRW NT-154 and Garrett Ceramic Components (GCCD) GN-10 silicon nitrides, have the potential for net-shape capability, have room temperature strengths of 828-966 MPa (120-140 ksi), and retain their high strengths to higher temperature.

Figure 5 shows that the stress rupture life of silicon nitride materials has also improved considerably. ACC Code 1 achieved an improvement in stress rupture life over NC-132 and current state-of-the-art materials, e.g. GN-10 and NT-154, exhibit a large increase in both stress rupture and temperature capability. The early materials like NC-132 and ACC Code 1 did not have adequate stress rupture properties to satisfy the requirements of the AGT101 rotor at 1370C (2500F) TIT. The improved

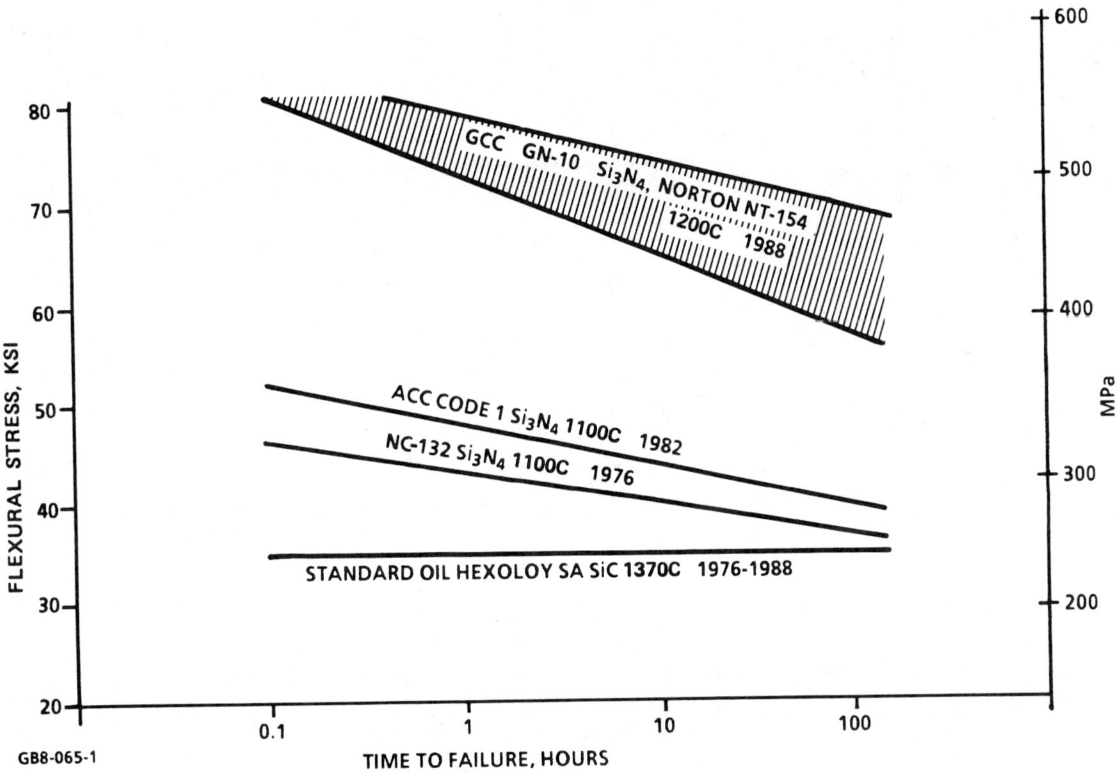

Figure 5. Stress Rupture Capabilities of Advanced Si_3N_4 Have Improved. (Allied-Signal Aerospace Data)

materials, NT-154 and GN-10, appear capable of withstanding the temperatures and stresses of the rotor at 1370C (2500F) TIT, at least for limited use.

The strength retention during long term cyclic exposure to combustion environments has also improved significantly for Si_3N_4(9,10), Figure 6. In the mid-1970's, only the reaction bonded form of Si_3N_4 was available in complex net-shapes. This low density Si_3N_4 maintained its strength after 1200C (2200F) cyclic exposure but its strength was insufficient for most turbine blade or rotor applications. Initial versions of sinterable fully dense Si_3N_4 showed significant reductions in strength retention due to pitting and thick, coarse oxide formation on exposed surfaces. The more refractory and stronger sintered silicon nitrides of the 1980's retain a greater portion of their initial strength. Sintered SiC material has consistently exhibited excellent strength retention after exposure to 1370C (2500F) and this durability has been maintained as higher strength SiC has become available(9,10).

PERFORMANCE

Analytically predicted improvements in performance have been demonstrated in both component and engine testing. Improved performance is sought through utilizing state-of-the-art aero-

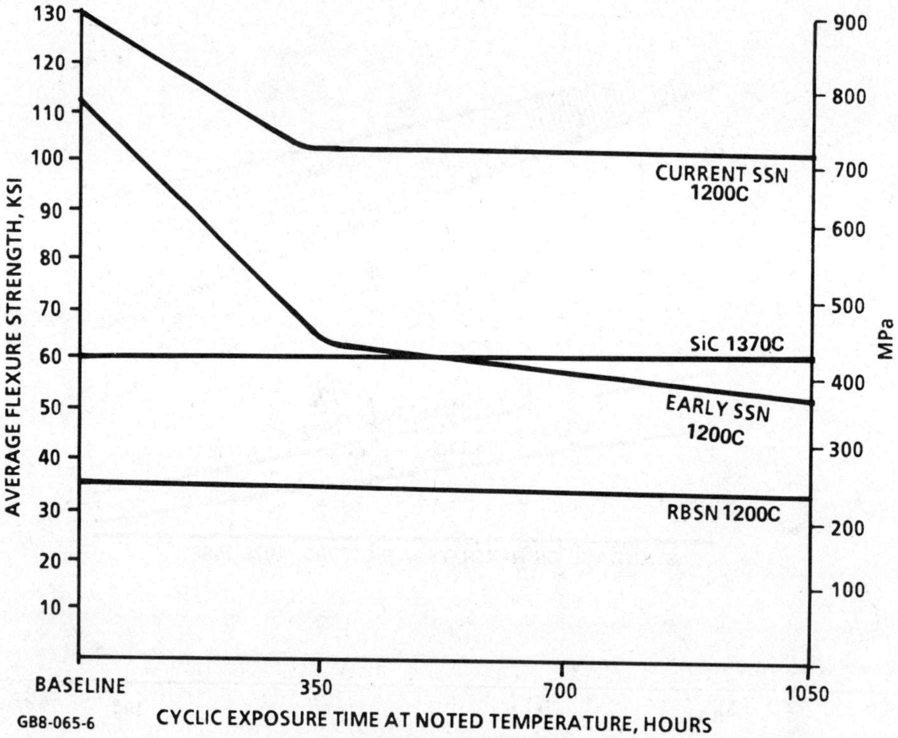

Figure 6. Advanced Si_3N_4 and SiC Retain Their High Strength After Cyclic Exposure. (Allied-Signal Aerospace Data)

dynamics, minimizing leakage, increasing the turbine inlet temperature, and reducing the cooling air requirements. Aerodynamic designs for turbine blades and vanes have typically been constrained in an effort to minimize stress from centrifugal loading and thermal shock. For axial turbine designs this has resulted in blades with minimal lean and twist, but otherwise turbine aerodynamic designs have not been constrained. Typical ceramic aerodynamic features achieved are illustrated in Figure 7. Within the structural constraints discussed, ceramic components have demonstrated satisfactory aerodynamic performance. The AGT101 ceramic radial turbine stage demonstrated an aerodynamic turbine efficiency in excess of 84 percent over a broad power range including idle, cruise, and maximum power. Ceramic designs have also demonstrated overall engine performance improvements. The Garrett TPE331-C turboprop engine demonstrated a 30-percent increase in shaft horsepower and a 7-percent reduction in specific fuel consumption (SFC) [11]. This was accomplished using uncooled ceramic parts operating at an average TIT of 1200C (2200F) with a rotor tip speed of 444 m/s (1457 fps). Improved performance is also a result of the increased tip speeds made possible by the high specific strength of ceramic materials. In 1980, the AGT101 rotor was designed for a maximum tip speed of 701 m/s (2300 fps), which exceeded the typical tip speed of a similar metallic rotor by 15 percent.

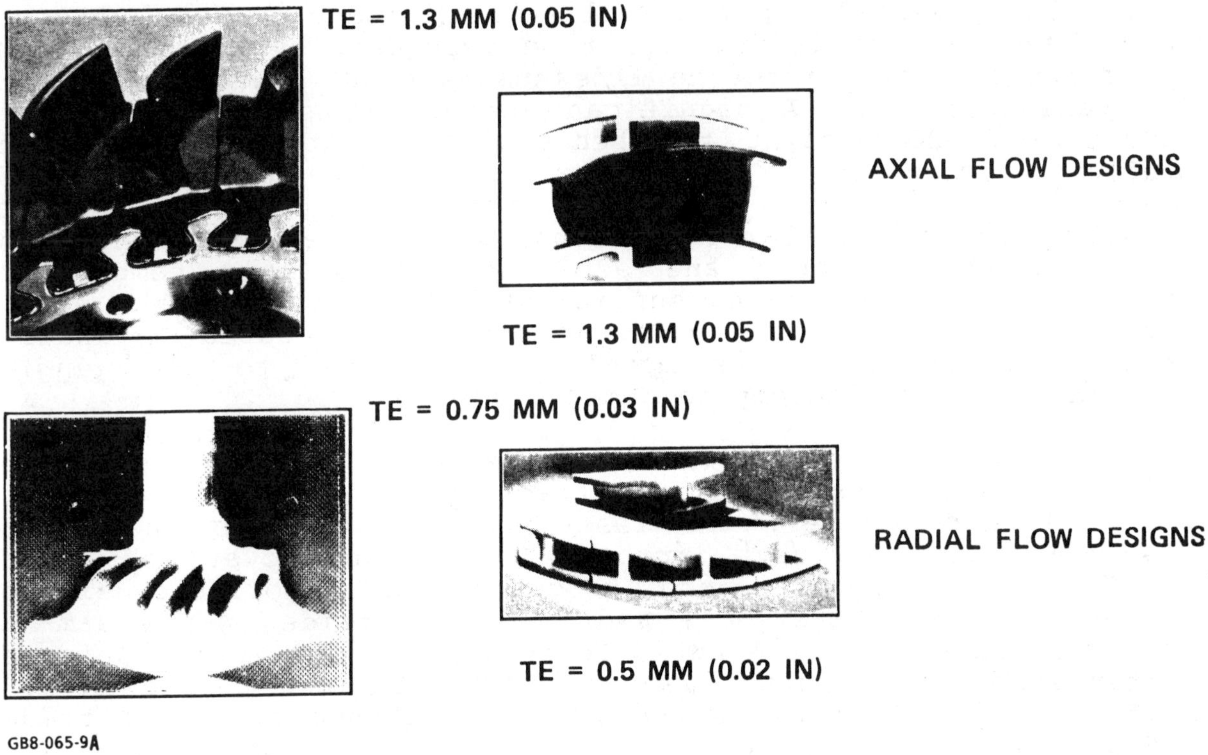

Figure 7. Desired Aerodynamic Features Have Been Achieved in Ceramic Components.

Currently, the overall engine performance of the AGT101 engine is limited by the temperature capability of the turbine rotor and the regenerator seals. Rotor material capabilities are steadily improving to meet the needs, as previously discussed. Needed are higher temperature metal seal substrates, monolithic ceramic seals, and improved low friction, durable, ceramic coatings. These are currently being developed under the ATTAP program at Garrett.

The effect of more impact resistant ceramic engine designs on engine performance has yet to be assessed. The use of thicker airfoil leading and trailing edges as well as particle traps will have some effect, although it is anticipated to be small. Ceramic turbocharger rotor designs have already incorporated thicker blade tips to improve impact resistance without significant loss in performance.

RELIABILITY

The issue of reliability is of increasing importance as feasibility and performance demonstrations continue to be successful. Reliability is an essential aspect of customer acceptance and of reduced user costs. Reliability is an issue of the design, the materials being used, and quality control[12]. Currently, analysis of component reliability is limited to the fast fracture mode under simple stress states and is based on a Weibull, weakest-link, probabalistic approach. Many attempts have been made to incorporate time dependent behavior and complex stress states into the life analysis but these have yet to be validated[13]. A significant effort is still required to understand and predict the behavior and interactions of the failure mechanisms in candidate gas turbine materials under realistic conditions. Models of fast fracture, and slow crack growth, as well as flaw initiation, growth, healing and coalescence under realistic and arbitrary stress and temperature states are needed. Also, sufficient data to support the statistical modeling of this behavior must be generated. Finally, the verification of the model's application to real engine hardware must ultimately be demonstrated.

In addition, materials with improved fast fracture strength, stress rupture properties, and fracture toughness are required. As mentioned earlier, the fast fracture strength and stress rupture properties have improved considerably over the years. In contrast, the fracture toughness of monolithic silicon carbide (2.5-3.5 $MPa \cdot m^{1/2}$) and silicon nitride (4-6 $MPa \cdot m^{1/2}$) has changed very little until recently with the introduction of titanium diboride particulate toughened silicon carbide, (3.5-4.5 $MPa \cdot m^{1/2}$), and microstructurally toughened silicon nitride, (6.5-10.0 $MPa \cdot m^{1/2}$), (Figure 8). Improved toughness has also been achieved by incorporating silicon carbide whiskers in silicon nitride and alumina, (5.5-9 $MPa \cdot m^{1/2}$), and silicon carbide fibers in silicon carbide or crystallized glass ceramics, (13-22 $MPa \cdot m^{1/2}$). Although ceramic composites have much higher toughness than the monolithics, issues such as net-and complex shape capability, rapid oxidation of the fibers,

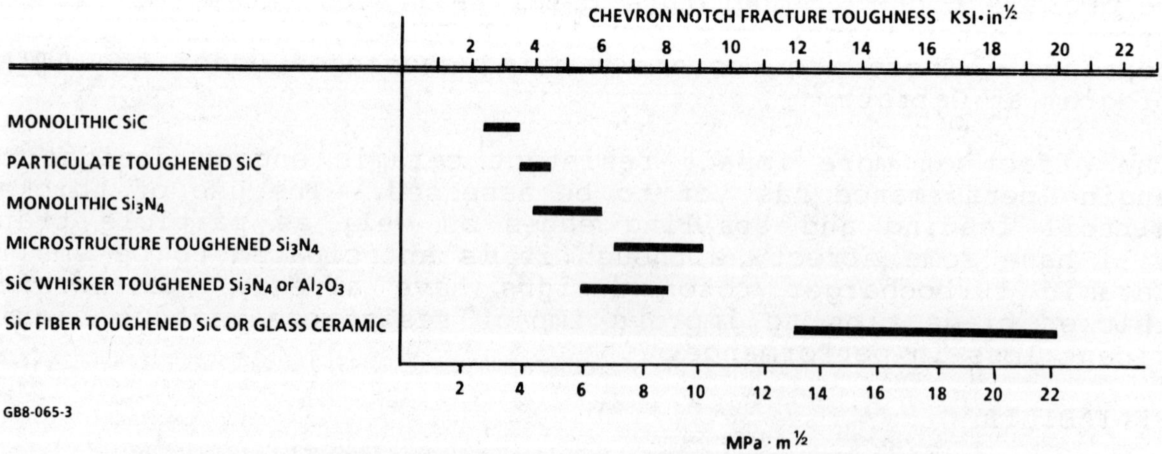

Figure 8. Improvements in Fracture Toughness Has Been Demonstrated With Ceramic Composites. (Allied-Signal Aerospace Data)

fiber/matrix interface stability at high temperatures, and lack of material property data limit the use of composites in gas turbine engines at this time.

Quality control aspects of reliability are directed towards assuring that the materials and properties of finished components are the same as those utilized during component design and life prediction analysis. Thus, quality issues must be addressed from design inception through component qualification. Quality control limitations, e.g., flaw detectability, must be considered during the design process. Parts must be designed such that critically stressed zones can be inspected or, at minimum, proof tested. Similarly, the use of in-process NDE and statistical process control are required to assure high yield along with product and material uniformity. The issue of incorporating NDE in design and fabrication has taken a lower priority in past programs due to the inability of NDE techniques to detect critical flaws in complex shaped parts and the lack of defined, critical and measurable parameters in the fabrication process. Thus, final component NDE methods such as radiography, ultrasonics, and dye penetrant have been widely used and proof testing has been heavily relied upon. This dependance on end item quality control is undesirable for large scale production use, although it continues to be used for automotive ceramic turbocharger manufacturing of over 7000 parts per month. From both reliability and cost reduction viewpoints, in-process NDE technologies must be incorporated into the design and manufacturing process.

The issue of reliability is the focus of major efforts currently funded by the DOE/ORNL Ceramic Technology (CT) Project and ATTAP programs. The DOE/ORNL CT Project is developing multidisciplinary efforts to enhance material development, measurement of material properties, process improvement and design methodology development. Specifically, the CT Project is supporting development of advanced heat engine composite materials at ORNL, Norton, GTE, and GCCD. Material behavior and test methodology development is being performed at the University of Dayton Research Institute, North Carolina A & T University, University of Illinois, University of Washington, NBS, AMTL, and at ORNL. Significant efforts addressing in-process NDE and process control are planned to be performed at Norton and at GTE, and two programs to develop and validate component design methods are scheduled to be performed at Garrett Auxiliary Power Division and at Allison Gas Turbine Division. The technologies from each of these programs will have significant influence on the DOE/NASA ATTAP effort. The ATTAP project will incorporate CT Project results and fund additional efforts in component design, data base generation, component manufacturing, and seal technology to demonstrate engine endurance and performance.

COST

To date, the issue of economics has been a minor aspect of the development of advanced ceramics for gas turbines. This will continue until ceramic engine feasibility, performance, and reliability are further demonstrated. However, developments in reliability will contribute to low cost production by enhancing product yield and design robustness.

Cost competitive ceramic gas turbine components require that parts be fabricated in net- or near net-shapes at production rates and yields to provide economies of scale. These will not be fully developed until the reliability justifies the commitment to production. However, prototype production experience will be essential in developing low cost and reliable manufacturing methods. Several Japanese ceramic manufacturers have initiated prototype turbocharger and bearing manufacturing efforts prior to the advent of low cost in an effort to gain valuable experience with ceramics.

Near net-shape capability is required due to the complex shapes typical of gas turbine parts, and due to the difficulty in machining these parts by grinding and/or ultrasonic machining techniques. Net-shape fabrication technologies have advanced significantly from the mid-1970's through the 1980's and should provide the means to achieve low cost. Figure 9 contrasts the labor intensive sequence used to machine turbine blades from a billet of hot pressed NC-132 in 1976 with the near net shape AGT101 rotor casting and final rotor product available with today's sintering technology. As previously mentioned, the process improvements developed during the CT and ATTAP programs will contribute to cost reductions.

SUMMARY AND CONCLUSIONS

Noteworthy accomplishments have been made toward the use of ceramics in gas turbines in the past two decades. Feasibility of surviving normal engine transient conditions and use temperatures over 1100C (2000F) have been confirmed repeatedly. Materials with impressive strength and elevated temperature durability are readily fabricated into the complex shapes required for gas turbines. Increased power at reduced SFC has been demonstrated, and no major barriers to the achievement of aerodynamic performance are envisioned. These accomplishments have established the technical base from which current and future efforts will embark. Emphasis can now shift from short life feasibility and performance demonstrations to long life engine reliability. To address reliability issues aggressive design methodology and material developments efforts have been initiated. Specific issues which need addressing include:

- Continued development of improved structural materials, addressing reduced data scatter, increased strength

1970's APPROACH TO TURBINE BLADES

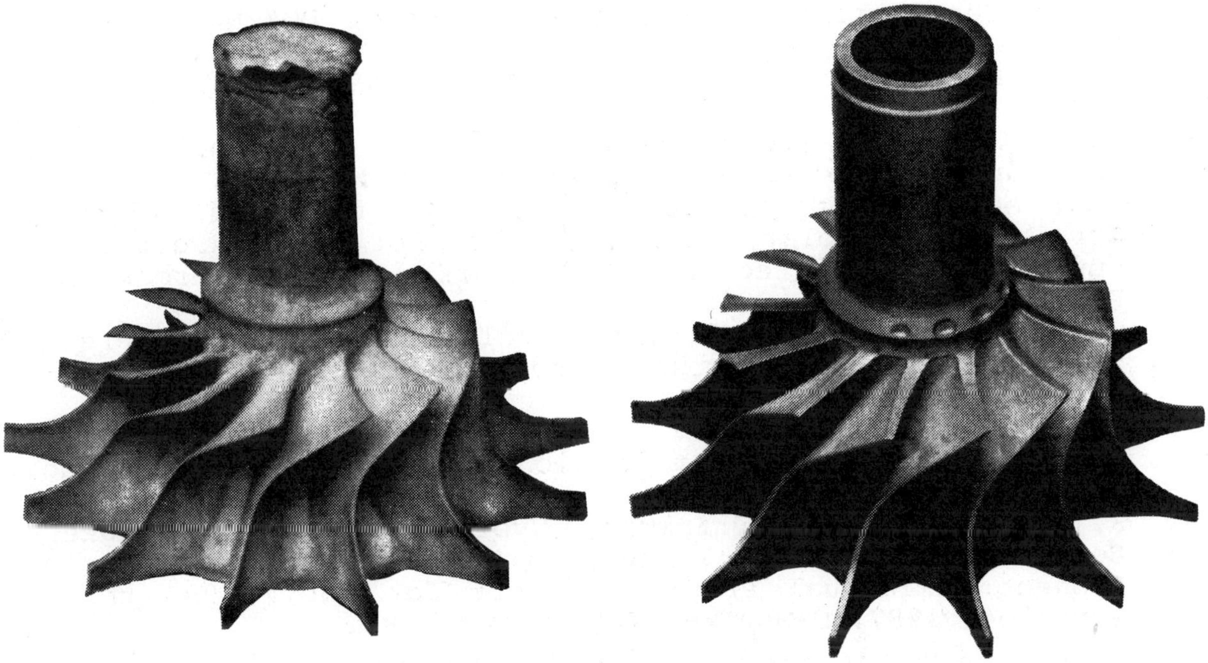

GB8-065-8

1980'S NEAR NET SHAPE FABRICATION

Figure 9. Fabrication Approaches Contribute to Component Cost Reduction.

retention at high temperature, reduced crack growth and creep, and improved toughness.

- o Development of low friction adherent coatings for high temperature regenerator seals.

- o Material data to support behavior modeling and life prediction analysis including fast fracture, crack growth, creep, oxidation, cyclic fatigue, and impact.

- o Improved and validated design methods capable of reliably predicting lives of complex components under multiaxial stress states under long term, cyclic, contact and impact conditions.

o Continued emphasis on producing consistent and uniform material properties in gas turbine components.

Successful results of these efforts will, in turn, provide the technology base from which cost competitive ceramic components and engine designs will emerge.

1. A.F. McLean, E.A. Fisher, Brittle Materials Design, High Temperature Gas Turbine, Final Report, AMMRC TR 81-14, March 1981.
2. Garrett Auxiliary Power Division, Advanced Gas Turbine (AGT) Development Project Final Report, DOE/NASA/0167-12, March 1988.
3. H. E. Helms, AGT 100 Project Summary, ASME 88-GT-223, June 1988.
4. M.R. Vanco et. al, An Overview of the Small Engine Component Technology Studies, AIAA-86-1542, 1986.
5. A.G. Metcalf, Application of Ceramics to Radial Flow Gas Turbines at Solar, in Ceramics For High Performance Applications, Edited by J.J. Burke et al. Brook Hill Publishing, 1974. p739.
6. T. Fett, G. Himsolt, D. Munz, Cyclic Fatigue of Hot-Pressed Si_3N_4 at High Temperatures, Adv. Cer Mat'ls, 1[2], 1986, p 179-184
7. S. Suresh, L. Han, J. Petrovic, Fracture of Si_3N_4-SiC Whisker Composites under Cyclic Loads, J. Am. Ceram. Soc., 71[3] 1988, C158-C-161.
8. K. C. Liu, C. R. Brinkman, Cyclic Fatigue of Toughened Ceramics, in Ceramic Technology for Advanced Heat Engines Project Semiannual Progress Report for April 1987 through September 1987, ORNL/TM-10705, March 1988.
9. L. J. Lindberg, Elevated Temperature Durability of Ceramic Materials, AIAA-88-3055, July 1988.
10. L. J. Lindberg, K. W. Benn, 3500-Hour Durability Testing of Commercial Ceramic Materials, Interim Report, DOE/NASA/0027-84/2, Dec 1984.
11. D. W. Richerson, K. M. Johansen, Ceramic Gas Turbine Engine Demonstration Program Final Report, Navy Contract N00024-76-C-5352, May 1982.
12. Reliability of Ceramics for Heat Engine Applications, NMAB publication NMAB-357, 1980.
13. Boulet, J.A.M., An Assessment of the State of the Art in Predicting the Failure of Ceramics, Final Report, ORNL/Sub/86-57598/1, 1988.

PROBABILISTIC DESIGN TECHNOLOGY FOR CERAMIC COMPONENTS

John S. Cuccio
Garrett Auxiliary Power Division
Phoenix, Arizona 85010

and

Joe H. Adams
Garrett Engine Division
Phoenix, Arizona 85010

and

Harry L. Kington
Garrett Auxiliary Power Division
Phoenix, Arizona 85010

ABSTRACT

This paper addresses probabilistic life prediction methods that are applicable to ceramic gas turbine component design. Current promising life prediction methods are discussed with emphasis on fast fracture and slow crack growth. Examples of turbine rotor life predictions made with these methods are provided. The inadequacy of these methods are identified along with areas for future research to support turbine engine design.

INTRODUCTION

Relative to the application of structural ceramics to the gas turbine, materials and process development have advanced more rapidly than life prediction capabilities. In order to design turbine components that effectively capitalize on the capabilities of advanced materials, ceramic life prediction research must focus on developing methods to predict failure modes that control component life. Due to the likelihood of interaction between failure controlling mechanisms, mechanistically based failure models with sound statistical treatment are required.

The function of life prediction methods is to predict ceramic component lives from specimen data. These predictions can be complicated by the differences between specimens and components, shown in Figure 1. Components can have complex stresses and temperatures, and be subjected to several failure mechanisms. Specimens usually have uniform stresses and temperatures, and are subjected to one or few failure mechanisms. Progress has been made in developing life prediction methods that can predict component lives from specimen data.

CHARACTERISTICS	SPECIMEN	COMPONENT
STRESS	UNIAXIAL, UNIFORM OR LINEAR GRADIENT	MULTIAXIAL, COMPLEX GRADIENT, CYCLIC
TEMPERATURE	CONSTANT, UNIFORM	CYCLIC, GRADIENT
FAILURE MECANISMS	TESTED INDIVIDUALLY	EXPOSED TO SEVERAL

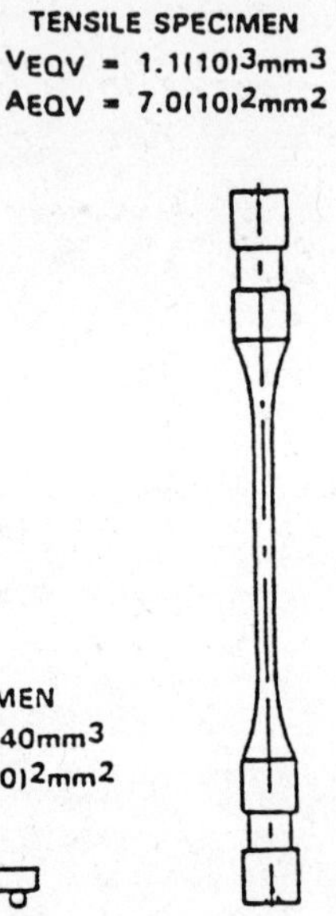

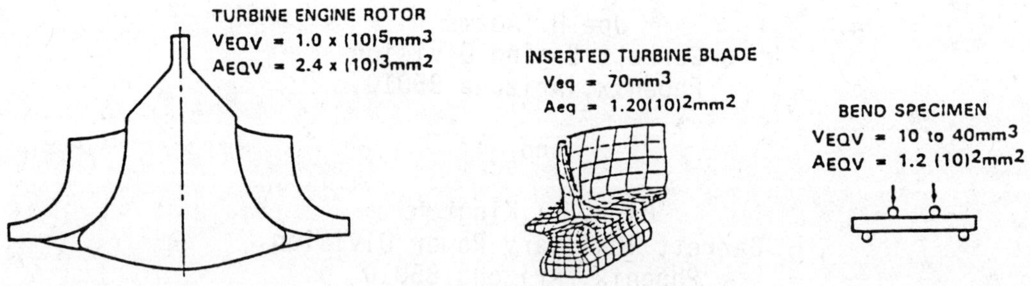

Figure 1. Differences between specimens and components complicate life prediction.

CURRENT LIFE PREDICTION METHODS

Design of reliable ceramic components requires methods that can accurately predict component failures. These methods must predict component lives from failure or damage modes including: 1) fast fracture, 2) slow crack growth, 3) creep, 4) oxidation/corrosion, 5) impact/erosion. Research has advanced life prediction methods to various levels of maturity.

The emphasis of this paper is on probabilistic methods which include fast fracture and slow crack growth. The current levels of technology as they apply to probabilistic component life prediction are provided in this section. A portion of the work reported in this paper was conducted under the Advance Turbine Technology Applications Project, Contract No. DEN3-335, sponsored by DOE and NASA.

Fast Fracture

Fast fracture is largely controlled by the distribution of material flaws in ceramics. Flaws common to ceramics include, pores, surface flaws, inclusions, large grains, and agglomerates. Fast fracture models have been proposed that consider the combined effects of flaw distributions and multiaxial stresses on failure probability.

Strength is affected by flaw size, morphology, physical properties, and interaction with the matrix. Flaws produce stress concentrations that reduce strength. These stress concentrations are a function of flaw size, flaw shape, and residual stresses in the matrix. Residual stresses occur during processing when the flaw and matrix, with different thermal expansion coefficients, cool. Identification of flaws and modeling their effect on strength forms the basis for ceramic strength prediction.

Methods for predicting the effects of flaw distributions on strength are usually based on fracture mechanics. Linear elastic fracture mechanics (LEFM) assumes that fracture occurs when the critical stress intensity (or fracture toughness), K_{IC}, is reached.

$$K_{IC} = Y \sigma \sqrt{a} \tag{1}$$

The applied stress at failure, σ, and flaw size, a, can be measured. Attempts (1,2,3,4) have been made to correlate flaw size to mode-I fracture toughness by determining the appropriate flaw shape parameter Y. This includes modeling flaws as, ellipses, spheres with a crack, and penny shaped flaws.

Evidence has shown that the LEFM model, "Eq. (1)", is not sufficient for predicting fracture of structural ceramics. Ceramics have been observed(5) to have small-flaw toughnesses on the order of 1/5 the toughness measured from large flaws. Observations of non-coplanar fracture also indicate that fracture is not uniquely controlled by K_{IC}. Multiaxial stresses and uniaxial stresses not normal to a flaw produce mode-II and mode-III shear deformation at a crack tip. Tests on preflawed Si_3N_4 (6) under mixed-mode conditions exhibit crack extension out of the initial flaw plane or non-coplanar fracture. Fracture in these tests occurred before K_{IC} was reached. Three theoretically based fracture criteria (7,8) with potential application to polycrystalline ceramics are:

- Strain Energy Release Rate
- Maximum Principal Stress
- Strain Energy Density

The application of these criteria has met with limited success. These models did not accurately predict fracture loads and crack direction identified from mixed-mode fracture testing(6), of Si_3N_4. Other (9, 10) attempts to extend these criteria to inherent flaw distributions in polycrystalline ceramics have also been inconclusive. Some test results(11,12) indicate that crack surface-roughness inhibits shear displacements between mating crack surfaces and affects the resistance of polycrystalline ceramics to fracture.

Considerable progress has been made in developing fast fracture methods. However, more study is required before fracture from inherent flaws exposed to component stresses and temperatures can be confidently predicted.

Slow Crack Growth

Research has shown that cracks can grow from initial flaws from time dependent processes. This time dependency, or slow crack growth, can cause fracture at loads significantly below fast fracture levels.

Insight into slow crack growth mechanisms has been obtained from analysis of high temperature fractures characteristics. In Si_3N_4 three characteristic fracture regimes, indicated in Figure 2, are observed(13) at temperatures below which creep dominates behavior.

REGION-1, Below a transition temperature, T_t, intergranular and transgranular fractures may be observed. Fast fracture methods should be applicable.

REGION-2, At temperatures approaching T_t, 1 um voids have been observed(13) near the fracture origin. It has been postulated that these cavities provide some stress redistribution around the flaw tip resulting in the increased toughness and strength that is often observed with increasing temperature. It has also been observed(14) that existing pores will blunt due to oxidation at these temperatures.

REGION-3, Above the transition temperatures strength and toughness decline rapidly, grain boundaries become strength limiting, and slow crack growth predominates. Fractures exhibit an intergranular slow crack growth region. It is proposed(13) that the lower viscosity of the grain boundary allows cavities to grow and coalesce into microcracks. Also, oxidation occurs more rapidly and produces a protective surface oxide that can prevent oxidation blunting of internal pores. Grain boundary sliding and slip in grains may also influence slow crack growth at elevated temperatures(15).

These findings indicate that several high temperature mechanisms have competing effects on ceramic strength.

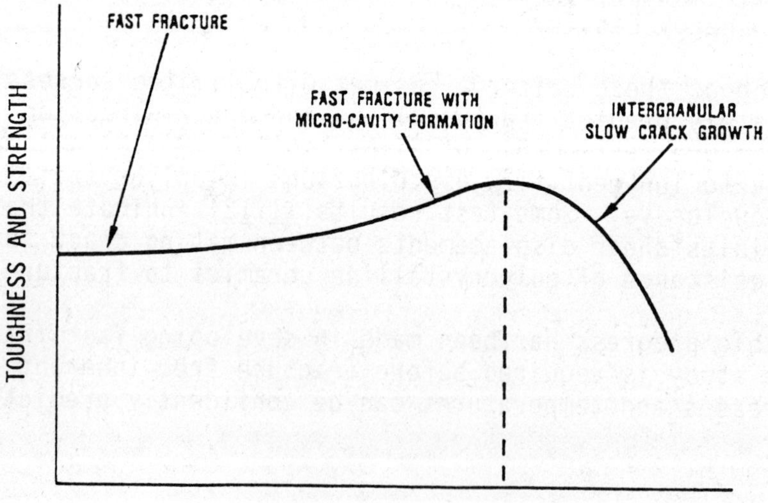

Figure 2. Strength and toughness are affected by changing fracture mechanisms with increasing temperature.

A power function is commonly used to model slow crack growth. This approach assumes that crack growth velocity is controlled by the crack tip stress intensity. The power function is:

$$V = A\, K_I^{\,n} \tag{2}$$

where V is the crack velocity, and A and n are material constants. This model was developed for glasses(16) but its application to polycrystalline ceramics has not been verified. Govila(17) ran double torsion, stress rate, and stress rupture tests and determined A and n independently for each data set. Agreement was not observed. The lack of agreement could be a result of the inapplicability of the model or the test methods.

Additional slow crack growth studies are needed to develop or verify a slow crack growth model that can be reliably predict ceramic component lives.

Statistical Methods

Probabilistic fracture models have been developed to predict fracture of ceramics. Most of this work has been directed toward predicting fracture from randomly distributed flaws exposed to multiaxial stresses. Limited work has been done to include the effects of slow crack growth.

Weibull statistics are commonly used to model ceramic strength distributions. The three parameter Weibull function is:

$$PS = \exp-\{[\ (\sigma - \sigma_u) \ /\ \sigma_o\ ]^{\ m}\} \tag{3}$$

Where, PS is the probability of survival, σ is the applied stress, σ_o is the characteristic strength, σ_u is the null strength, and m is the Weibull modulus. Further improvement in modeling strength data has been achieved by applying the Weibull to surface and volume strengths separately. The flaw types that dominate surface strength are often different than the flaws types that dominate volume distribution strength. Censoring(18) techniques are being used to determine surface and volume Weibull strength parameters from fracture data that has both surface and volume initiated failures. These parameters are being used (19,20) to predict both surface and volume component failure probabilities.

Methods have been developed to combine strength distributions and multiaxial fracture criteria. Batdorf(21) developed a model based on the assumptions that flaws can be represented as penny shaped cracks and that stress normal to a flaw controls failure. This model defines failure probability (PF) by:

$$PF = P_a\, P_b \tag{4}$$

where P_a is the probability that a flaw exists and P_b is the probability that it is oriented normal to a stress that will cause fracture. Batdorf later(22) extended his model to include mixed-mode fracture criteria. More recent work (10,20,23) has incorporated non-coplanar mixed-mode fracture criteria into Batdorf risk integration methods. While it is anticipated that these additions will improve fracture predictions, verification testing is needed.

Fast fracture and slow crack growth models have been combined to predict component lives. The slow crack growth power function "Eq.(2)" and the Weibull strength distribution model "Eq (3)" were combined by Govila(17) to give this risk of failure function:

$$\ln\,[1/(1-PF)] = \{[\sigma(B\,t\,\sigma^{2} + 1)^{-\frac{1}{n-2}} - \sigma_U]\;1/\sigma_C\}^{m} \qquad (5)$$

where $B = A\,Y^2\,(K_{IC})^{n-2}\,(n-2)/2$ and t = time. Before a combined failure function like this can be confidently used the underlying fast fracture and slow crack growth models must be verified.

Additional improvements in statistical theory and its application to ceramics are being pursued in other programs(24,25). The goals of these programs are to optimize test plans and data analysis techniques, develop methods to apply confidence bands to failure predictions, and integrate the effects of time dependent crack growth into fracture statistics.

Development of statistical methods for ceramic failure prediction has been largely theoretical. Experimental data is needed to evaluate the accuracy of these methods.

Component Failure Risk Analysis

Garrett Auxiliary Power Division developed an integrated ceramic component life prediction system. This system combines fast fracture, slow crack growth, and statistical methods with component stress and temperature analyses to predict failure probability as a function of time.

The failure risk integration process is shown in Figure 3. Finite element analyses are used to calculate component thermal and stress distributions. Risk of failure is determined for the volume and area of each element. The failure function "Eq. (5") is used in the risk integration. This function is separately applied to volume and surface failures. Parameters are determined from strength and stress rupture test data. The cumulative component survival probability (PS) is found with:

$$PS = \prod_{i=1}^{n} (1 - PF)\;; \qquad n = \text{number of elements} \qquad (6)$$

This function is evaluated for surface and volume failures.

This life prediction system accurately predicted failure rates of inserted turbine blades(25). Failure rates were calculated for NC132, hot pressed silicon nitride, blades exposed to a room temperature spin test. Spin testing of 205 blades to 54,200 RPM resulted in 6 failures. These failures were accurately predicted, as shown in Figure 4.

This life prediction system was latter applied to the AGT101 ceramic turbine rotor. The rotor was analyzed for operation at 100,000 RPM and 2500°F turbine inlet gas temperature. SN250, sintered silicon nitride, MOR and bend bar stress rupture test data(19) were used to determine fast fracture and slow crack growth function parameters. Predicted failure risks versus time are

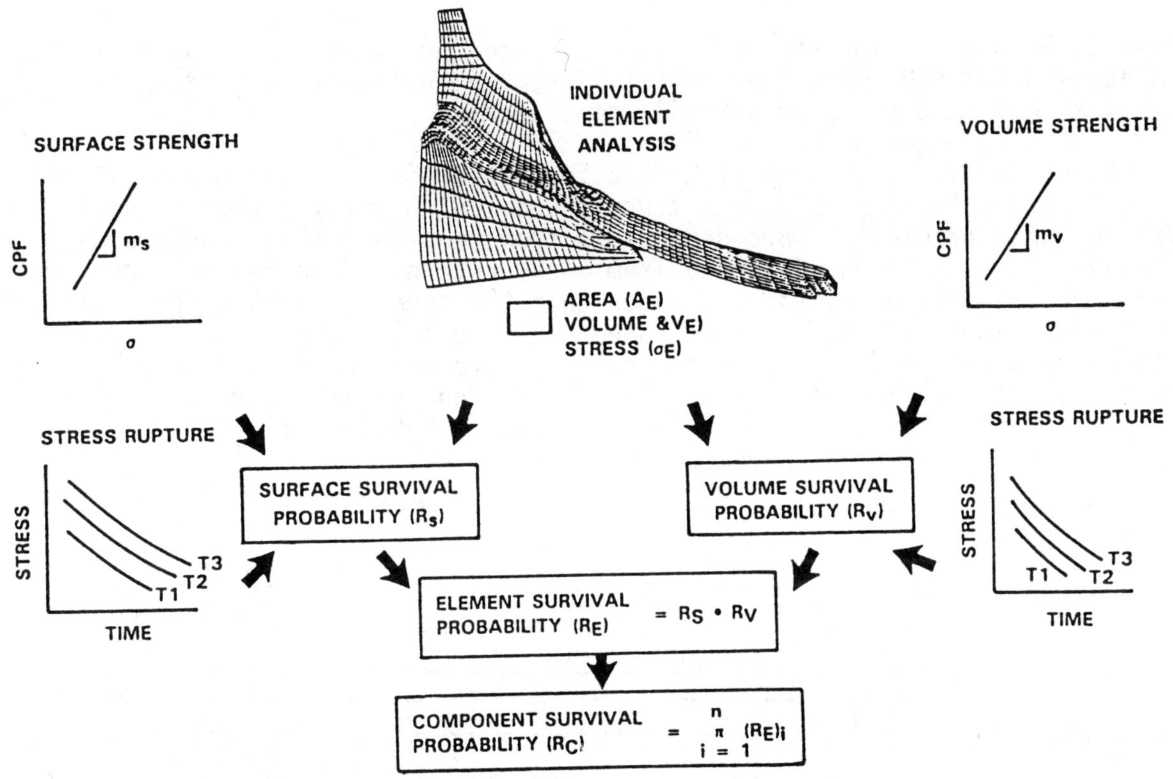

Figure 3. Schematic of failure risk integration process used for life prediction

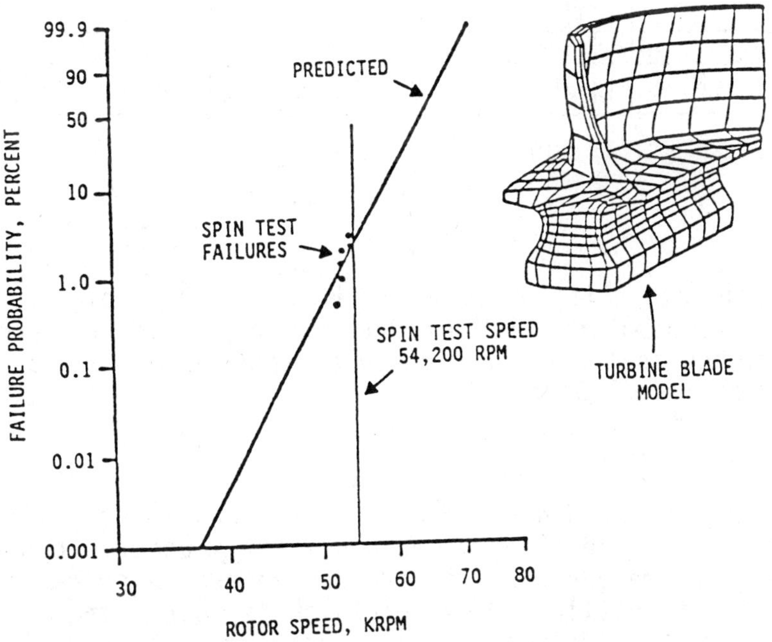

Figure 4. Turbine blade failures speeds were accurately predicted.

shown in Figure 5. The risk of volume failure, primarily from the rotor hub, dominates until 900 hours. At longer lives the surface risk exceeds the internal risk, due to the higher airfoil temperatures.

Rotor life was similarly predicted for operation at lower power, 70,000 RPM and 2200°F turbine inlet gas temperature. Slow crack growth parameters used for this prediction were determined from bend bar stress rupture data obtained with specimens sectioned from SN250 rotors. Specimen obtained from rotors had significantly shorter lives than the specimen used to generate data for the previous life calculation, shown in Figure 5. The lives calculated at this lower power condition with the shorter life properties are shown in Figure 6. A turbine rotor tested at these conditions in an engine failed after 12 hours. These limited results suggest that probabilistic life prediction methods are consistent with component capabilities.

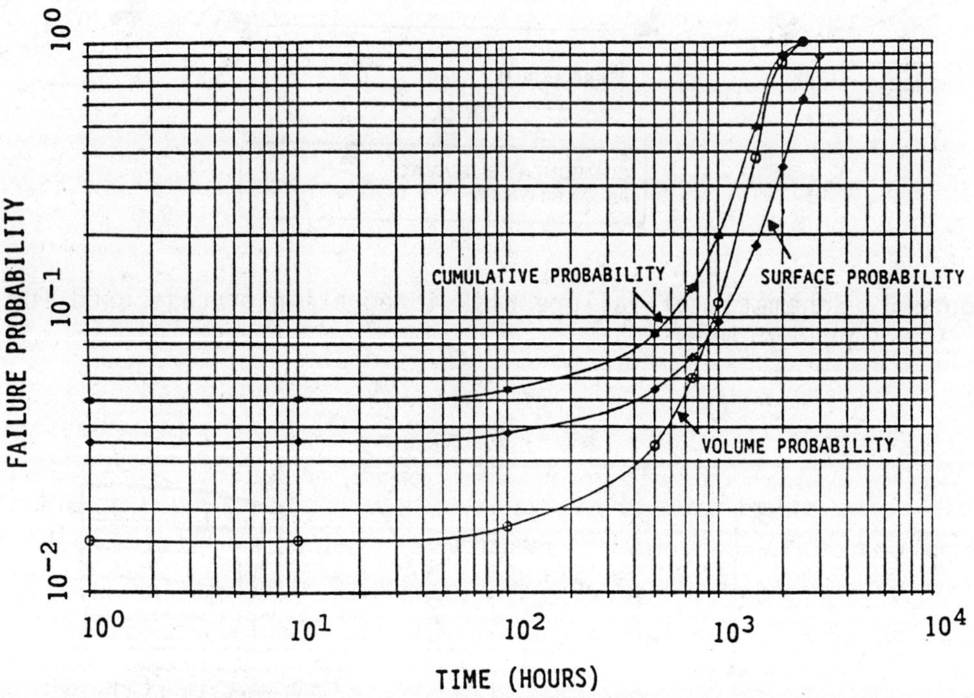

Figure 5. AGT101 rotor life was predicted for 100,000 rpm and 2500° F turbine inlet temperature. Slow crack growth properties were obtained from specimens that were fabricated as specimen. Points shown on the curves are calculated lives.

Additional development and verification of probabilistic life prediction methods are required before they can be confidently used to design ceramic components. Areas where advancements are needed are described in the following section. Garrett Auxiliary Power Division is addressing these needs in the NASA, ATTAP*; ORNL, life prediction program**; and internal research and development programs.

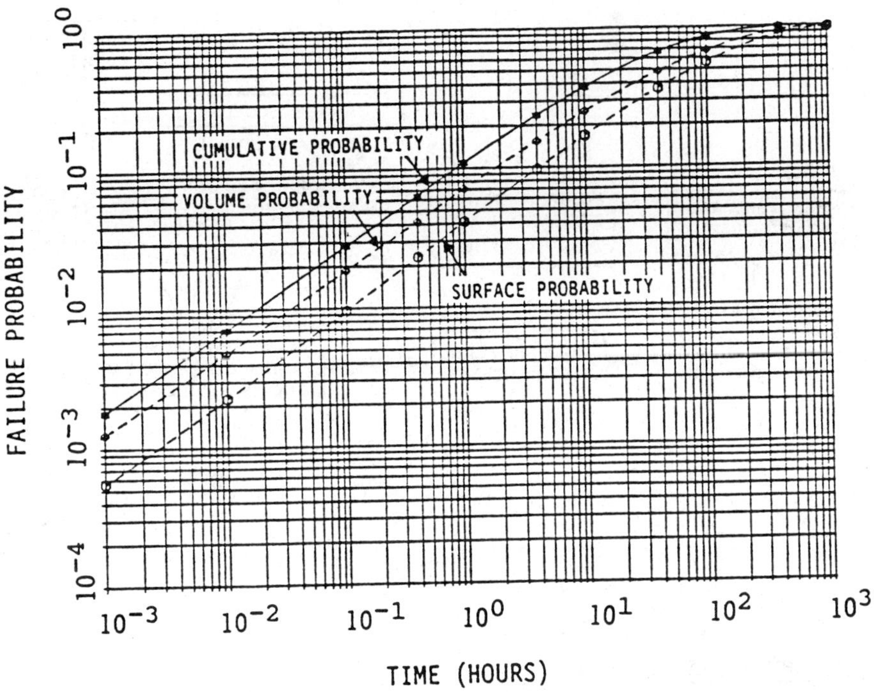

Figure 6. AGT101 rotor life was predicted for 70,000 rpm and 2200^0 F turbine inlet temperature. Slow crack growth properties were obtained from SN250 specimen sectioned from rotors. Points shown on the curves are calculated lives.

DIRECTION FOR FUTURE DEVELOPMENT

An approach for ceramic component life prediction and the methods needed to implement it are reviewed here. This approach for life prediction is an extension of the methods presented in the previous section.

A schematic of the approach being pursued for ceramic component life prediction is shown in Figure 7. This is a probabilistic approach for predicting component fast fracture and slow crack growth failures with provision for creep, oxidation/corrosion, and impact/erosion. Accurate modeling of flaw distributions and fracture resistance are central to this approach.

Advancements in life prediction methods are required to implement this life prediction approach. A summary of the methodology advancements required to implement each life prediction element is provided below.

---

*DOE/NASA, "Advanced Turbine Technology Applications Project (ATTAP)", contract number DEN3-335, started September 1987.

**DOE/ORNL, "Life Prediction Methodology for Ceramic Components of Advanced Heat Engines", expected to begin in 1989.

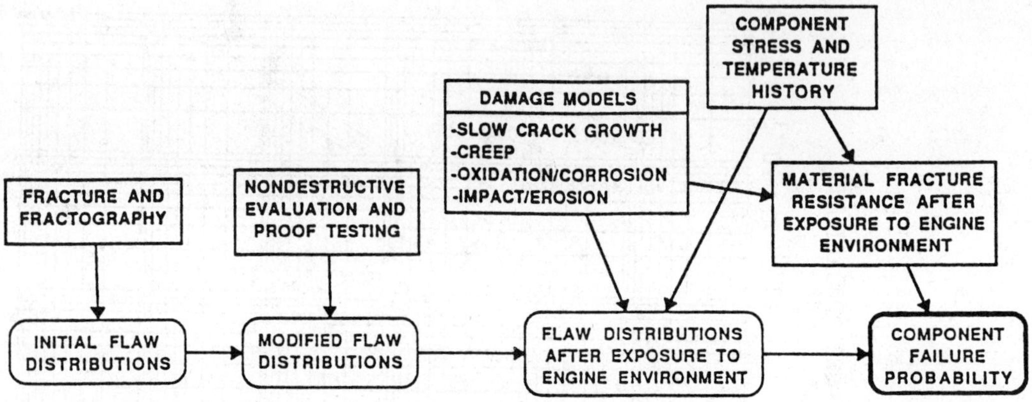

Figure 7. Life prediction methods are being developed to implement this approach for ceramic component life prediction.

Fracture and Fractography

This element provides accurate modeling of the flaw distributions that control strength and life. Although strength, not flaw, distributions may continue to be modeled, the flaws sampled during fracture testing must be representative of the flaws that control component strength and life. Advancements in fracture testing and fractography are needed in two areas: 1) improved definition of the type and number of specimens to use for fracture testing, and 2) definition of flaw characteristic that must be identified with fractography. More information about flaws is required to reduce the number of assumptions made about flaws when applying fracture models.

Fracture specimens must sample internal flaws to generated strength data that is useful for predicting the failure of high volume risk components like the AGT101 rotor. A minimum of 20% of the failures generated during fracture testing need to initiate internally before volume strength distributions can be determined. This conclusion is based on results from a Monti-Carlo simulations performed in conjunction with maximum-likelihood data anlaysis*. The percentage of internal failures produced is controlled by: 1) the ratio** of stressed-volume to stressed-area in a specimen, and 2) the distributions of flaws in the material being evaluated. The ratios of stressed-volume to stressed-area needed for 20% internal failures in SN250 and alpha silicon carbide were calculated to be .020 and .005 respectively. Ratios equal to or greater than these occur in either tensile or large bend specimens, as shown in Figure 8. More accurate definition of the specimen size requirements for generating volume strength distributions requires additional testing and analysis.

Fracture testing must also sample internal flaws from a sufficient volume of material. As shown in Figure 1 the AGT101 rotor has 10,000 time more stressed volume than a conventional bend specimen and 100 times greater stressed volume than a tensile specimen. This requires that 10,000 bend specimen or 100 tensile specimen be tested to sample the worst flaw expected in one rotor. Statistical methods being developed(26) to pool data from multiple

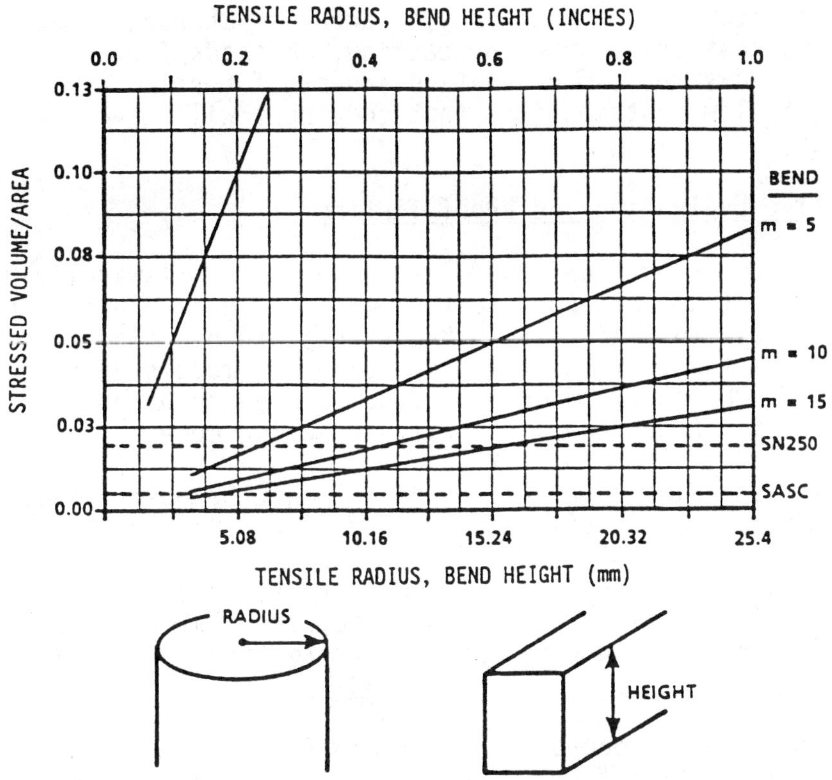

Figure 8. Analysis indicates that large bend specimens or tensile specimens can provide internal failures for SN250 and sintered alpha SiC, SASC. Dashed lines show stressed volume/area that will produce 20% internal failures, m is Weibull modulus.

specimen sizes may allow data from a few large specimen or component fracture tests to be combined with small specimen data to minimize the uncertainty produced by extrapolating small specimen data to large components.

More complete fractography is required to better characterize flaw distributions. The current application of fracture and crack growth models requires making assumptions about material flaws. Typically flaws are assumed to be flat cracks. Flaws such as agglomerates, inclusions, pores, machining damage, and large grains can affect strength(1) and life differently than an ideal flat crack. Additional work is required to identify the strength and life controlling characteristic of flaws that must be provided by fractography.

---

*Results from a Garrett Auxiliary Power Division internal research and development program

**This is the ratio of (V/A) were V = equivalent uniformly tensile stressed volume, and A = equivalent uniformly tensile stressed area

Fracture and fractography must provide accurate characterization of the flaw distributions that exist after processing. Methods for predicting the flaws screened by nondestructive evaluation and proof testing are needed to calculate the truncated flaw distributions that component will have when they are put into service.

Nondestructive Evaluation and Proof Testing

In this element of flaw based life prediction, flaws eliminated by nondestructive evaluation and proof testing would be determined. This requires developing models that accurately represent screened flaws and combine these models with the flaw, or strength, distribution models described in the previous section. NDE and proof testing studies need to be performed along with fracture testing to develop and verify screening models.

The flaw screening capabilities of NDE must be quantified. NDE cannot always screen a specific size and type of flaw. Models that represent the probability of detection (27) of flaw types, locations, and orientations with each NDE technique are needed. Development of probability of detection relationships requires performing NDE on geometries that are representative of ceramic components followed by destructive testing, or sectioning to identify flaws.

The flaw screening capabilities of proof testing must also be quantified. Methods for calculating flaw screening by proof testing must account for differences between proof testing and operating stresses and temperatures. Proof tests are often run at room temperature and do not precisely reproduce component operating stresses. Fast fracture models are needed to predict the flaws that are screened in each location in a component during proof testing. These models must also predict the orientation of the screened flaws. If the stresses produced during proof testing are in a different direction than the operating stresses, potential failure producing flaws will not be screened. If elevated temperature proof tests are used the potential detrimental effects of slow crack growth must also be considered.

Development of models that can predict the screening capabilities of NDE and proof testing will allow flaw screening capabilities to be taken advantage of during component design and allow components to be designed for better flaw screening.

Flaw Growth and Initiation Models

The purpose of this life prediction element is to predict changes in flaw populations over the service life of a component. Models for predicting flaw growth and initiation are needed. Only slow crack growth will be considered here, although flaw distributions can also be effected by creep, oxidation/ corrosion, and impact/ erosion.

Slow crack growth models must be applicable to cracks growing from inherent flaws exposed to engine conditions. These models must predict: 1) crack growth rates and directions from inherent flaws, 2) crack growth under a multiaxial stresses conditions, and 3) crack growth from surface and internal flaws. A slow crack growth model must be based on the mechanisms that control crack growth if it is to have these capabilities.

A model that can predict crack growth from inherent flaws is needed. The size of inherent flaws are typically .01 to .1 mm and can cause component failure when they reach 1 mm. Slow crack growth test, in which incremental crack size is measure, use precracked specimens with initial flaws that are 1 to 10 mm. Observations of cracks growing from inherent flaws are needed for slow crack growth model development.

A better understanding of slow crack growth mechanisms is also important. Understanding the mechanisms of slow crack growth such as, oxidation, grain boundary sliding, and micro-void coalescence, will allow more accurate crack growth prediction over a wider range of conditions. For example, if oxidation influences crack growth, surface flaws will grow at different rate than internal flaws. The effects of multiaxial stresses on crack growth will also be influenced by the controlling mechanisms. Statistical aspects of slow crack growth modeling should also be considered. Statistical methods are needed to account for the uncertainty in calculated model parameters, and may be needed for modeling the slow crack growth process in crack growth is an incremental phenomena not continuous. More emphasis must be put on microscopic evaluation of slow crack growth mechanisms.

Fracture Resistance Models

This element of the life prediction approach must predict the probability of fracture at any time in the life of a component. The probability of component failure is a function of the flaws that exist, state of stress and temperature, and ability of the material to resist fracture. Fracture testing of inherent flaws exposed to component representative stresses and temperatures is needed to verify or develop fracture models.

Fracture models, that have been developed largely from precracked specimen test results, must be extended to predicting fracture from inherent flaws. Models developed from precracked specimens may not be directly applicable to inherent material flaws such as, agglomerates, inclusions, pores, and large grains. These inherent flaws affect crack shear and opening displacements, and therefore crack tip stresses, differently than an ideal crack. Phenomena such as crack closure under shear loading will have less effect on stresses around an open pore than a tight crack with rough mating surfaces. Fracture testing of non-precracked specimens is needed to develop or verify fracture models that are applicable to components with inherent material flaws.

A fracture model must also be capable of predicting failure after a component has been exposed to other damage mechanisms. Slow crack growth, creep, and oxidation can change the severity of flaws through crack growth and crack tip blunting. Cracks can blunt by grain boundary sliding, cavity coalescence, or oxidation. A fracture model must be capable of accounting for these changes in flaw severity to predict component failures.

SUMMARY

Current probabilistic design technologies were reviewed, with emphasis on fast fracture and slow crack growth modeling. Examples of ceramic turbine blade and rotor strength and life predictions were provided. Although considerable progress has been made in developing life prediction methods, additional work is needed. Areas where additional work is need, with reference to an overall approach for ceramic life prediction, include:

- Defining specimen types and numbers required for design data generation
- Modeling the flaw screening capabilities of NDE and proof testing
- Developing models for predicting crack growth from inherent material flaws
- Verifying or developing fracture models to predict fracture from inherent flaws or flaws changed or generated by other damage mechanisms

Advancements in these areas are being pursued at Garrett Auxiliary Power Division.

REFERENCES

1. J. P. Singh, "Effects of Flaws on the Fracture Behavior of Structural Ceramics: A Review", Adv. Ceram. Mater. 3 [1], 1988, 18-27.

2. T. Hoshide, H. Furuya, Y. Nagase, and T. Yamada, "Fracture Mechanics Approach Evaluation of Strength in Sintered Silicon Nitride", Int. J. Fract. 26, 1984, 229-239.

3. A. G. Evans, D. R. Biswas, and R. M. Fulrath, "Some Effects of Cavities on the Fracture of Ceramics", J. Am. Ceram. Soc. 62 [1-2], 1979, 101-106.

4. A. G. Evans, "Microfracture From Thermal Expansion Anisotropy-I. Single Phase Systems", Acta Metall. 25, 1978, 1845-1853.

5. L. R. F. Rose, "The Crack-Size Dependence of Fracture Toughness in Ceramic Polycrystals", J. Am. Sci. Letr. 5, 1986, 455-456.

6. J. J. Petrovic, "Mixed-Mode Fracture of Hot-Pressed Si_3N_4", J. Am. Ceram. Soc. 68 [6], 1985, 347-355.

7. D. Broek, Elementary Engineering Fracture Mechanics, Martinus Nijhoff Publishers, The Hague, The Netherlands, 1982, 359-365.

8. Methods of Analysis and Solutions of Crack Problems, edited by G. C. Sih, Noordhoff International publishing, Leyden, The Netherlands, 1973, XXI-XLV.

9. D. K. Shetty, J.J. Petrovic, A. R. Rosenfield, and W. H. Duckworth, "Correlation of Uniaxial and Biaxial Fracture Strengths of Ceramics", Las Alamos Nat. Labs., DOE, W-7405-Eng-38, 1984.

10 A. C. Rufin, D. R. Samos, and R. J. Bollard, "Statistical Fracture Prediction Models for Brittle Materials", AIAA J. 22 [1], 1984, 135-140.

11. D. K. Shetty, A. R. Rosenfield, and W. H. Duckwoth, "Mixed-Mode Fracture of ceramics in Diametral Compression", J. Am. Ceram. Soc 69 [6], 1986, 437-443.

12. D. K. Shetty, A. R. Rosenfield, and W. H. Duckwoth, "Application of the Diametral-Compression (Brazilian Disk) Test", Engng. Fracture Mech. 26 [6], 1987, 825-840.

13. S.H. Knickerbocker, A. Zanvil, and S. D. Brown, "Displacement Rate and Temperature Effects in Fracture of a Hot-Pressed Silicon Nitride at 1100^0 to 1325^0C", J. Am. Ceram. Soc. 67 [5], 1984, 365-368.

14. "Creep Pheonomena in Reaction Bonded and Sintered Silicon Nitrides", NGK Technical Report, R-83-3, 1983.

15. N. J. Tighe, "The Structure of Slow Crack Interfaces in Silicon Nitride", J. Matr. Sci. 13, 1978, 1455-1463.

16. A. G. Evans, and S. M. Weiderhorn, "Proof Testing of Ceramic Materials-an Analytical Basis for Failure Prediction", Int. J. Fract. 10 [3], 1974, 379-392.

17. "Ceramic Life Prediciton Parameters", Ford Motor Co., DOE, AMMRC TR 80-18, 1980.

18. W. Nelson, Applied Life Data Analysis, John Wiley & Sons, New York, 1982, 356-386.

19. "Advanced Gas Turbine (AGT) Technology Development Project Final Report", Garrett Auxiliary Power Div., Doe/NASA, NASA CR-180891, 1988.

20. J. P. Gyekenyesi, "Scare: A Postprocessor Program to MSC/NASTRAN for Reliable Analysis of Structural Ceramic Components", ASME Paper 86-GT-34, J. of Engineering for Gas Turbines and Power,

21. S. B. Batdorf, and J. G. Crose, "A Statistical Theory for the Fracture of Brittle Structures Subjected to Nonuniform Polyaxial Stresses", J. Appl. Mech., June 1974, 459-464.

22. S.B. Batdorf, and H.L. Heinisch Jr., "Fracture Statistics of Brittle Materials With Surface Cracks", Engng. Fract. Mech. [10], 1987, 831-841.

23. J. P. Gyekenyesi, and N. N. Nemeth "Surface Flaw Reliability Analysis of Ceramic Components with the Scare Finite Element Postprocessor Program", ASME Paper 87-GT-69, Presented at the Gas Turbine Conference, Aneheim CA 1987.

24. W. P. Eatherly, "Advanced Statistics", ORNL/TM-9947, 1986.

25. C. A. Johnson, and W. T Tucker, "Advanced Statistical Concepts of Fracture in Brittle Materials", GE, ORNL/TM-9947, 1986.

26. "Ceramic Gas Turbine Engine Demonstration", Garrett Turbine Engine Company, ARPA/NAVAIR, N00024-76-5352, 1980.

27. A. P. Berens, and P. W. Hovey, "Flaw Detection Reliability Criteria", AFWAL-TR-84-4022, 1984.

THE ROLE OF NDE IN CERAMIC TURBINE ENGINE COMPONENT DEVELOPMENT

Janet Minter
Garrett Auxiliary Power Division
Allied-Signal Aerospace Company
Phoenix, Arizona

Parts of this work funded by DOE/NASA ATTAP Contract DEN3-335

ABSTRACT

A systematic approach to the use of nondestructive evaluation (NDE) throughout the ceramic engine component development cycle is necessary to achieve the high reliability required for turbine engine applications. Specifically, NDE techniques must be developed and applied to component design, development, fabrication, and test to ensure adequate materials characterization. This approach has the potential of reducing ceramic component cost by optimizing the initial design and eliminating unacceptable components early in the fabrication cycle. Although research NDE techniques have been developed that have the potential to solve the problems of ceramic component inspection, systematic application of these NDE techniques for inspection and process control has not yet occurred. The application of NDE techniques for both material characterization and flaw detection is discussed as well as an approach for applying NDE throughout the fabrication process to achieve highly reliable ceramic components.

INTRODUCTION

The combination of high modulus of elasticity, high strength at elevated temperature, and relatively low density give advanced ceramic materials the potential to increase turbine engine performance, reduce cooling requirements, and increase power performance.[1-3] Currently, the application of ceramic components in gas turbine engines is limited by inadequacies in material properties, fabrication processes, and design methods that can predict component life.[4] NDE can be an integral tool in the development of these technologies. The potential contribution to materials processing, component design, and component inspection make NDE a key factor in obtaining reliable components and, if implemented properly, reducing overall component costs.

Several approaches can be taken to increase the reliability of ceramic components. The most frequently used approach is to apply NDE and/or proof-testing to fully-processed components in order to eliminate those with unacceptably large flaws. Proof-testing has been heavily relied on in the past, however, proof testing is costly and does not always sufficiently duplicate service conditions. . . thus some flaws can be missed. Using final component NDE alone is limited because the minimum detectable flaw size for many techniques is not adequate for highly stressed ceramic components having complex geometries. The achievement of reliable, cost-effective components requires a ceramic design system within which NDE is integrated fully into component design, fabrication process control, and in-service inspection as shown in Figure 1. This change in philosophy from inspecting for quality to the use of NDE as a tool throughout the component design, fabrication, and test cycle is a major challenge.

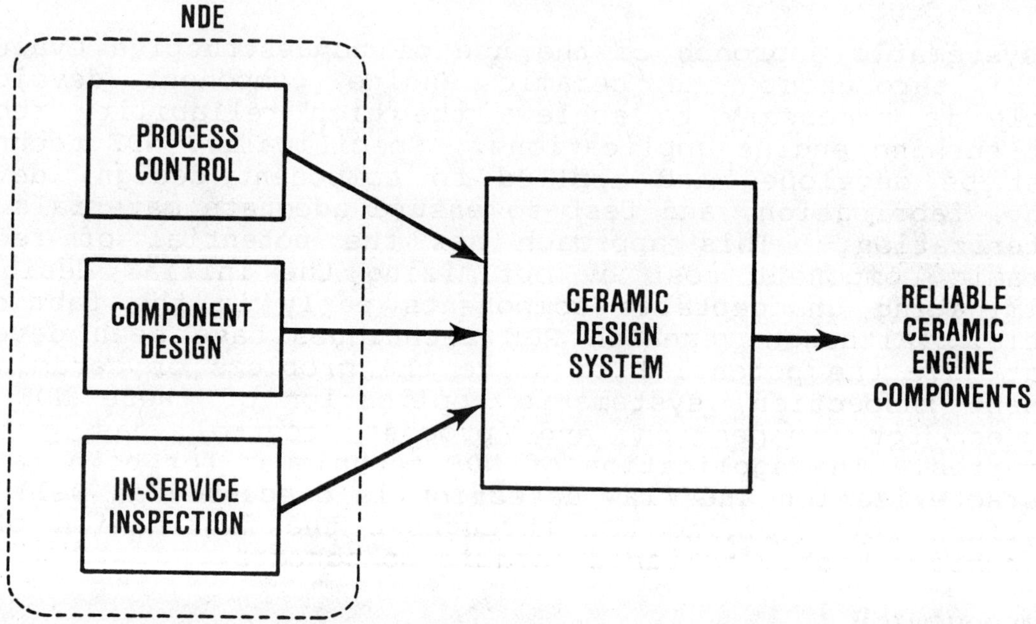

Figure 1. Ceramic NDE Challenge Requires Integrated Approach

NDE CHALLENGE

The strength of ceramic materials is controlled by the presence of surface and internal flaws such as voids, inclusions, large grains, density variations, and cracks which occur during both fabrication and machining. The size, type, and location of these flaws are the critical parameters that must be identified in order to predict component performance.[5,6] GAPD's* early ceramic gas turbine engine demonstration pro-

*Garrett Auxiliary Power Division, Allied-Signal Aerospace Company.

grams assessed the flaw detection capability of both conventional and developmental NDE techniques for fully dense ceramics through the evaluation of flat plate NDE standards seeded with defects.[7,8] As shown in Table 1, multiple techniques were studied under the DARPA* Ceramic Gas Turbine Engine Demonstration Program. These studies showed that intermediate and high frequency ultrasonic methods had the best resolution capabilities for internal defect detection of high and low density inclusions/voids. However, their application was limited due to the complex geometry of actual components. Surface inspection techniques capable of detecting small flaws were also limited to simple geometries. As a result, fluorescent penetrant and visual inspections were used on complex geometry components, rather than thermal wave or acoustic microscopy, methods which could detect much smaller flaws, but only on controlled surface finish test panels. Due to the lack in sophistication required to address complex geometries, limited NDE was used in ceramic development programs such as the NASA/DOE AGT101 program, which relied heavily on mechanical and thermal proof-testing to eliminate flawed components.

Table 1. Ceramic NDE Studies Show Advanced Methods Detect Small Flaws

INCLUSION SIZE (μm)	INCLUSION TYPE											
	WC		Fe		Bn		SiC		Si		C	
	RT	UT	RT	UT	RT	UT	RT	UT	RT	UT	RT	UT
500	X	X	X	X	0	X	0	X	0	X	X	X
250	X	X	X	X	0	X	0	X	0	X	X	X
125	X	X	X	X	0	X	0	X	0	X	0	X
25	X	X	X	X	0	X	0	X	0	X	0	X

X = DETECTED
0 = NOT DETECTED

During the last 8 to 10 years, the development of NDE as a science, as well as the vast increase in computing power and sophistication of instrumentation, have reduced many of the problems that initially limited the usefulness of NDE for component inspection. During that time, the scope of what NDE can potentially contribute to ceramic engine component devel-

*Defense Advanced Research Projects Agency

opment has considerably broadened. NDE can now be viewed as a materials characterization tool as well as a technique to find discrete flaws.[9,10] In addition, the quantitative capabilities of NDE have been improved.[11-13] For example, NDE can now quantify and characterize material anomalies in the 5 to 250 micron size range. Because NDE now has such wide potential applications, the GAPD ATTAP and Ceramic Component Life Prediction programs are both emphasizing the development and implementation of NDE techniques throughout the ceramic component design, fabrication, and test cycle.

Process Control NDE

To establish cost-effective, repeatable fabrication processes that can be scaled-up for high volume production, the approach to quality control, through final component NDE and proof-testing, must be altered to implement NDE early in the fabrication process. For some time, it has been recognized that the removal of flawed components early in the fabrication cycle can improve the cost-effectiveness of ceramic components. In addition, by identifying the source of the flaws, the process can be modified to reduce the size and quantity of flaws, thereby improving component reliability. For example, Figure 2 shows a ceramic rotor defect that was initially identified by ultrasonics and then characterized further by computed tomography. Since the defect was there prior to densification, early detection would have eliminated it from the fabrication and machining cycle.

CT OF Si_3N_4 ROTOR

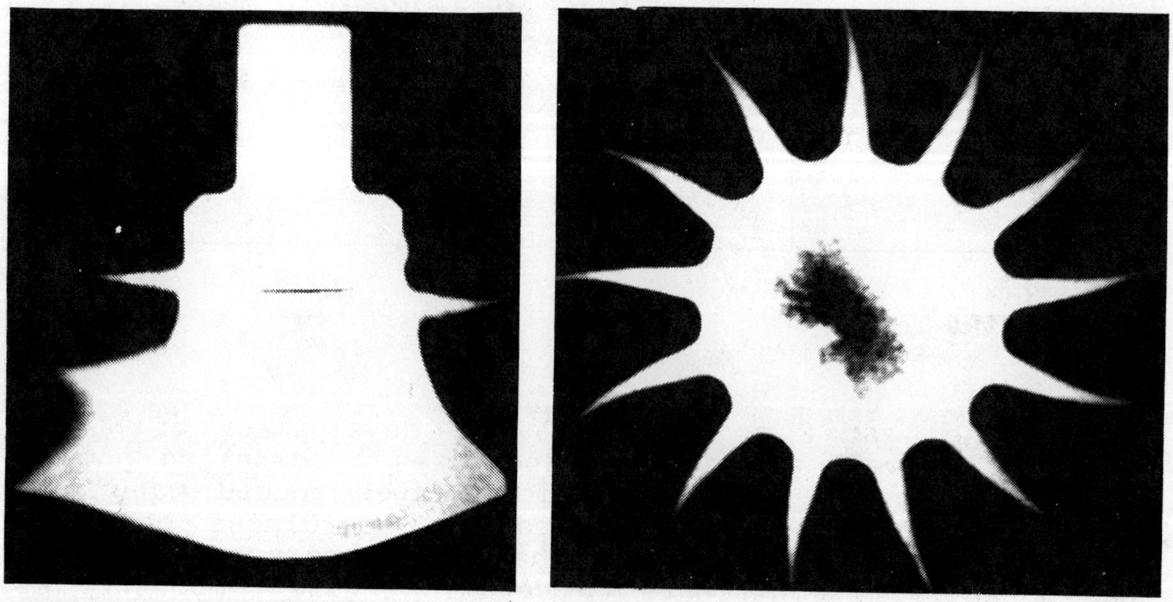

Figure 2. Detection of Flawed Components Eliminates Additional Costs of Further Processing

The broadened definition of NDE as a materials characterization tool can also favorably impact component yields and reliability. Johnson et al, have described the potential of applying NDE in this broader sense.[14] Material characteristics can be monitored during processing and subsequently correlated with desired properties. NDE can then be used as a quality control or process monitoring tool to attain components possessing the same NDE signature. For example, in injection molded components, poor binder distribution during mixing can lead to excessive porosity in the finished component. Nuclear magnetic resonance (NMR) techniques may be useful in detecting binder distribution variation in green ceramics.[15] In this case, the NDE will eliminate components with nonuniform binder distribution. Correlating green binder distribution data with the processing parameters can then be used to identify the cause of the nonuniformity.

Computed tomography and dual energy digital radiographic techniques show potential as process control tools also. These two techniques can be used for bulk internal inspection during the fabrication process to determine when discrete defects are introduced into the components, as well as small density variations, that may lead to premature component failure due to localized residual stresses caused by the density variation. As a green component NDE tool, dual energy digital radiographic techniques can be used to understand the difference between density variations caused by microporosity/voids and those caused by improper distribution of sintering aids, since the technique can differentiate between density and atomic number variations.

A key element in process control is a thorough understanding of how processing parameters control the quality of the finished component and how the parameters interact. If applied during initial fabrication process development, NDE can aid this goal by providing a tool to monitor component quality. This supplies the feedback required to optimize individual process parameters, as well as the entire process. Early application of NDE throughout the fabrication process will also provide the data required to select the most beneficial NDE techniques for the final process. That is, many potential NDE techniques can be applied and studied while fabrication process improvements are being implemented. The testing conducted to measure the processing improvement can also be correlated to NDE data to assess the applicability of the various NDE methods being applied. Thus, the feedback between both fabrication process and NDE expedites progress toward an optimum fabrication process. Producing components possessing a consistent NDE signature should result in a higher Weibull slope, a major goal of advanced processing. This higher Weibull slope can result in better reliability designs without an increase in average properties. The most preferred ceramic material is one with both increased average properties and a high Weibull slope as shown in Figure 3.

Component Design

As indicated in the previous section, close control of processing parameters by implementation of quantitative NDE techniques for both discrete flaw detection and material characterization can result in a uniform material and high Weibull slope. This, in turn, results in higher reliability designs. If uniform properties cannot be attained through process monitoring and control, then a somewhat different approach must be developed. In this case, the NDE signature of the material is correlated with destructive tests. This NDE signature can then be used as a predictive tool to establish component reliability. This approach has the added advantage that when the design methodology is developed, the NDE to support that methodology is also in place. During the component design phase, the geometry of critical areas can be optimized for NDE to help ensure the integrity of the final component. Correlating the NDE data with destructive data will also provide additional feedback for process improvements if the two processes are occurring simultaneously.

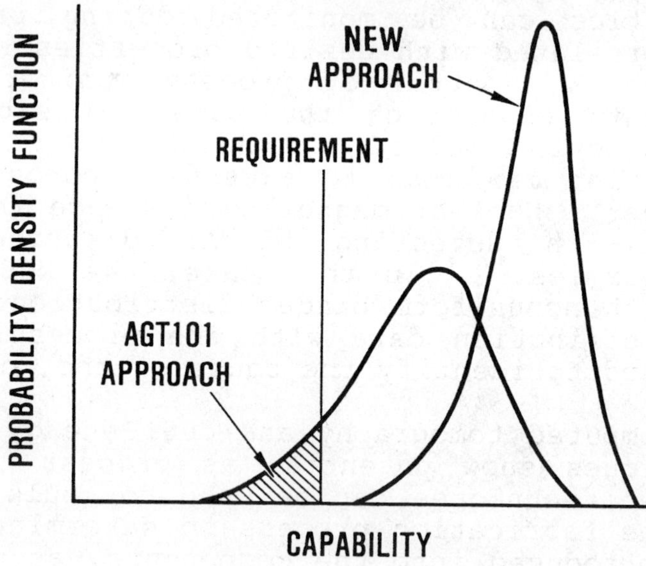

Figure 3. New Approach is Required to Achieve Reliable Components

To ensure component performance throughout the life of the engine, two approaches can be taken: use component designs and materials that have a large safety margin, or establish periodic inspections that remove components with flaws approaching the critical size from service. For the automotive industry, the first approach may be the most desirable since low maintenance is a major factor. For aerospace applications, the demand for high performance and the added criticality of operation mandates the second approach. Periodic inspections of engine components throughout the life of the engine, implementing similar techniques to those used during final component inspection, will assure the continued reliability of the ceramic components. The inspection interval is defined by the predicted life of the component as shown in Figure 4. From the component design analysis, the expected life of the component can be predicted based on the knowledge

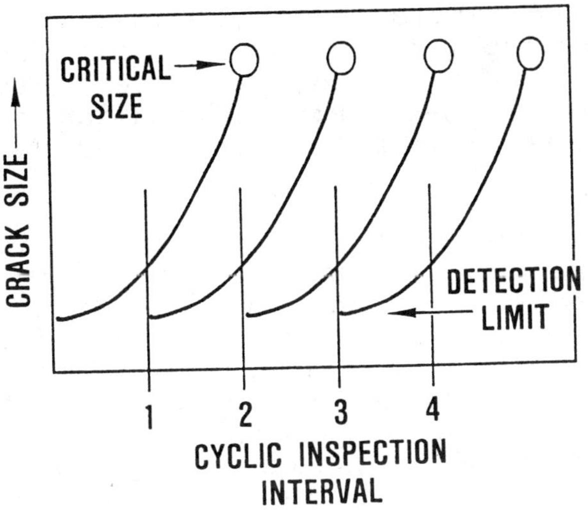

Figure 4. Damage Tolerant Design Directed to Exploiting Component Useful Life

gained through NDE that no material flaws exist above a certain detectable size and that those below that limit will not grow to failure during the inspection interval. Periodic component inspections are typically conducted throughout the life of the engine at inspection intervals of one half the time required for a flaw at the detectable limit to grow to critical size. This flaw or damage tolerant design approach fully exploits the useful life of the component without exposing the engine to undue risk of unpredictable failure. Implementation of this design approach requires that the geometry of critical areas be specifically designed for NDE.

Ideally, NDE can also be used to characterize the flaw distribution in the material being used to design the component. Currently, strength tests (usually flexural) are used to sample the flaw population and provide design data. This involves testing large quantities of specimens to ensure that the flaw population is adequately sampled. Once NDE detection limits and characterization capabilities are clearly defined, NDE methods can be used in addition to the destructive testing of test bars to characterize the material. Although some of the more sensitive NDE characterization and flaw detection techniques are limited to laboratory use on flat plates, these techniques may be useful in characterizing the flaw population in test bars prior to destructive testing, at least to the detection limit of the technique. This will allow more complete characterization of the flaw distribution rather than be limited to analysis of only the flaw causing failure.

Final Component and In-Service Inspection

The NDE methods developed to monitor the fabrication process and evaluate test specimens and components during development can aid in the selection of final component material, fabrication process, and inspection techniques which are consistent with the design requirements. Since both the NDE and fabrication process limits have been determined, selecting the material and fabrication process with the appropriate process controls and inspections ensures the selection of a material with predictable, reliable behavior. Final component NDE is predicated on the design, material, and fabrication requirements to assure high-quality and high-reliability components for the design. The final inspection results, when correlated with the in-process results, provide the necessary information to predict whether the component is capable of meeting final application requirements.

Another factor in determining whether and when to inspect is the overall benefit of the inspection versus the cost of the inspection. A thorough understanding of the fabrication and machining cost can help establish appropriate inspections for each component. Deming has developed a formula for determining the cost advantage of inspection at various stages of the fabrication/production cycle which compares the cost of inspection at a particular stage of fabrication to the value added by further processing.[16] The larger the ratio between the fabrication cost and inspection cost, the more cost-effective the inspection is when inserted prior to the value added operation. In the case of the turbine rotor defect discussed previously, the ratio of additional machining and proof testing versus inspection costs is 50 for development lot quantities. Clearly, even for small lots, the inspection cost is worth the investment. Based on inspection experience from metallic engines, this ratio can increase by an order of magnitude when the inspection is optimized for production. This is illustrated by the recent addition of computed tomographic inspection for internal wall thickness and core shift in cooled turbine blades. In this case, the time required to inspect a turbine blade during the development stage was 20 minutes. Through advancements in computed tomographic equipment and the development of specialized scanning techniques, the time required for this inspection has been reduced to 1 minute. With these improved inspection times the ratios of additional processing/versus cost of inspection is greater than 200 even when costs of the equipment and depreciation are included in the inspection cost. In both cases, the cost savings are conservative as they only include the costs of additional processing, which are easily determined. The cost to the producer of placing a nonconforming component in an engine are much greater, but far more difficut to quantify, and were not included in these examples.

The drawbacks of implementing the inspection after the component has been fabricated but prior to machining and final testing are:

- Value of fabrication is lost

- Component is not available for the engine build and production schedules are adversely affected.

Thus, from an engine manufacturer's viewpoint, earliest detection of nonconforming components is desirable. This approach will reduce the amount of inspection required on the final component and increase the confidence in the quality of the component. It will not eliminate the need for final component inspection as machining and qualification of the component can subsequently introduce defects in the component.

CONCLUSIONS

A synergistic approach to ceramic component design, development, and fabrication includes the implementation of NDE as a part of each step in the process. No one technique implemented at a single point in the process is likely to provide all the information required to produce reliable ceramic components. However, multiple NDE methods carefully selected and implemented throughout the design and fabrication process and correlated with one another will help achieve this goal. NDE as a materials characterization tool has been demonstrated, yet these techniques have not been applied systematically to the ceramic component fabrication process. Better materials characterization through NDE will help reduce final occurrence of nonconforming material, which will lead to the high Weibull modulus (Figure 2) ceramic materials require to meet the needs of advanced turbine engines.

REFERENCES

1 J.J. Burke, et al, "Ceramic for High Performance Application," Army Materials Technology Series, Brook Hill Publishing, (1974).

2 A.F. McLean, et al, "Brittle Materials Design, High Temperature Gas turbine," AMMRC TR 81-14 Final Report July 1, 1971-Aug. 31, 1979, Published 1981.

3 D.W. Richarson, "Evolution in the U.S. of Ceramic Technology for Turbine Engines," American Ceramic Society Bulletin (2), 1985).

4 "Ceramic Components for Turbine Engines," GTEC, AFWAL Final Report AFWAL-TR-83-8012, Volume III, Appendix B, (1982).

5 A.G. Evans, Fracture in Cermaic Materials: Toughening Mechanisms, Machining Damage, Shock, Part III, Chap 15, pp 364-384, Noyes Publications (1984).

6 "A Review of the Effects of Flaws on the Fracture Behavior of Structural Ceramics," Argonne National Laboratory, DOC, W-31-109-ENG-38, (1986).

7 D. Richerson and K. Johansen, "Ceramic Gas Turbine Engine Demonstration Program, Final Report," DARPA Contract Number N00024-76-C-5351, 1982.

8 Harold Helms, et al, Ceramic Applications in Turbine Engines, Noyes Publications, pp. 107-130, (1986).

9 G.Y. Baaklini, et al, "High Frequency Ultrasonic Characterization of Sintered SiC," Proceedings 11th Annual Conference on Composites and Advanced Ceramic Materials," January 1;8-23, 1987, Cocoa Beach, FL. Published by The American Ceramic Society, Inc. (1987).

10 W.A. Sanders, G. Y. Baaklini, "Correlation of Processing and Sintering Variables with the Strength and Radiography of Silicon Nitride," NASA Technical Memorandum 887251, (1986).

11 S.J. Klima, et al, "Nondestructive Evaluation of Structural Ceramics." NASA Technical Memorandum 88978, (1986).

12 S.J. Klima, et al, "Reliability of Void Detection in Structural Ceramics Using Scanning Laser Acoustic Microscopy." NASA Technical Memorandum 86945, (1985).

13 G.Y. Baaklini and D.J. Roth, "Radiographic Detectability Limits for Seeded Voids in Sintered Silicon Carbide and Silicon Nitride." NASA Technical Memorandum 87035, (1985).

14 D.R. Johnson, et al, "Needs Assessment for Nondestructive Teting and Materials Characterization for Improved Reliability in Structural Ceramics for Heat Engines," ORNL Technical Memorandum Number ORNL/TM10354, (1987).

15 L.B. Welsh, et al, "Proton NMR Imaging of Green State Ceramics," Review of Progress in Quantitative NDE, Vol. 6A, pp. 449-457, (1987).

16 E.P. Papadakis, "The Deming Inspection Criterion for Choosing Zero or 100 Percent Inspection," Journal of Quality Technology, Vol. 17, No. 3, pp. 121-127, July 1985.

CERAMIC MATERIALS AND COMPONENTS FOR ENGINES:
RESEARCH AND DEVELOPMENT IN THE CHINESE ACADEMY OF SCIENCES

T. S. Yen and X. R. Fu
Shanghai Institute of Ceramics, Academia Sinica
Shanghai, China

ABSTRACT

The Chinese Academy of Sciences has been following the research and development on high temperature structural ceramics for diesel engines for more than a decade. Materials of interest include silicon nitride, silicon carbide, transformation toughened zirconia ceramics and fiber reinforced ceramic composites. The principle areas of research are powder synthesis and characterization, materials and processing, machining and joining, thermal and mechanical properties study, nondestructive evaluation, and ceramic engine components fabrication and testing. The present paper is a short survey of what has been going on in this respect.

INTRODUCTION

At the end of the last decade, the Chinese Academy of Sciences, essentially the Shanghai Institute of Ceramics, decided to embark on a project in the research and development of advanced ceramic materials for diesel engines. The purposes are two-fold, one is to develop materials for engines, presumably for the 90's and possibly later, and the second is to conduct ceramic materials studies primarily for advanced structural applications. These materials must withstand the environment in the engines, however, other applications may be developed prior to or simultaneously with their use in engine applications. Therefore, the ceramic engine programme is really a program to promote the research and development of advanced ceramic materials with many-fold purposes - high temperature, high strength, increased toughness, anti-wear, anti-friction, and corrosion resistant. This is our philosophy for paying a great deal of attention to engine ceramic materials and components studies. The following is a short survey of what has been going on in this area.

SILICON NITRIDE CERAMICS

Among the advanced structural ceramics, silicon nitride seems to have attracted the most attention due to its overall superior properties leading to superior performance for heat engine applications. Research has been conducted along a number of directions, among them are:

1. to understand more fully the phase compatibility relationships of complex M, Si, Al / O, N systems thus providing the basis of tailoring nitride ceramics with desired major phases as well as the grain boundary phases[1].
2. detail studies on the reaction kinetics and thus the sequences between various components at different temperatures with the aim to provide further information for phase composition tailoring during material processing[2].

These two parts will not be included in this paper.

3. to improve the reliability of silicon nitride ceramics by ameliorating the strength degradation problem at high temperatures.
4. to develop near-net-shape fabrication technology for cost reduction in component manufacture.

These two approaches will be described in this paper. It should also be noted that several diesel engine components have been batch-produced in small quantities with satisfactory performance.

For near-net-shape fabrication, a post-sintered RBSN material has been developed, where preliminary machining can be performed at the intermediate stage and final machining can be reduced to a minimum amount. For the improvement in high temperature properties, research efforts have been focussed on grain boundary engineering by designing Y-Al-garnet as the grain boundary phase through heat-treatment of the post-sintered RBSN material. The phase compositions at different processing stages are shown in Table 1.

Table 1. Phase Compositions at Different Processing Stages

	After reaction bonding	After sintering at 1800°C, 3 h	After heat treatment at 1450°C, 5 h	After heat treatment at 1450°C, 25 h
Main phase	β'	β'	β'	β'
Minor phase	α	α'	α'	α'
	YAG	N'	YAG	YAG
Trace	AlN		K, N_{ss}	J_{ss}

β' -- β-Si_3N_4 solid solution α' -- α-Si_3N_4 solid solution α -- α-Si_3N_4

YAG -- $3Y_2O_3 \cdot 5Al_2O_3$ YAM -- $2Y_2O_3 \cdot Al_2O_3$ K -- $Y_2O_3 \cdot Si_2N_2O$

N -- $Y_3AlSi_2O_7N_2$ N' -- Y_2SiAlO_5N J -- $2Y_2O_3 \cdot Si_2N_2O$

N_{ss} -- solid solution between K and N phases

J_{ss} -- solid solution between J and YAM phases

The strength and fracture toughness of the final material obtained can be retained to 1400°C. Apparently, the recrystallization of YAG and the diminution in content of the glassy phase offer major contributions to the improvement of the mechanical properties[3].

Based on this material, the precombustion chamber for a diesel engine has been fabricated (Fig. 1). Rig tests were performed at the Shanghai Internal

Combustion Engine Research Institute. After that several hundred of these components were installed in truck and boat diesel engines for road tests. Information from car and boat users was collected from time to time. Lifetimes of the best ones exceeded 4,000 hrs with no trouble. Further work to improve the reproducibility of these components is the next target. If all goes well, precombustion cups are expected to be put into actual use by the middle of nineties.

Two possible routes can be followed to design sintered β-Si_3N_4 ceramics with high temperature performance. The first is to design a highly refractory glassy grain boundary phase and the second is to recrystallize the grain boundary phase into crystalline phases that possess high melting points. Using a combination of Y_2O_3 and La_2O_3 as additives may play a dual role in these two directions. By comparison with yttria and alumina, the combination of which has already been used in some commercial products, the lowest eutectic temperature of the grain boundary phase can be raised by about 200°C by using yttria plus lanthana. Moreover, the refractory $LaYO_3$ phase was identified at the grain boundaries by proper heat treatment in our experiments. The presence of this phase further enhances the high temperature properties of the material(4). It is encouraging that the silicon nitride ceramics with Y_2O_3 and La_2O_3 additives can retain their room temperature strength up to 1300°C and their strengths at 1400°C maintain values from 680 MPa to 850 MPa depending on the starting raw materials used. The merit of using Y_2O_3 and La_2O_3 to produce a much stronger grain boundary phase can be still better supported by studying the creep properties of these materials (Fig. 2). The creep tests were carried out on a 4-point bend specimen with dimentions 4 x 2 x 40 mm. Samples were creep tested between 1170°C and 1360°C at stresses between 200 and 360 MPa in air. The creep rates of specimens with rare-earth oxide additives only (YL-1) are much smaller than those with rare-earth oxides plus alumina (YLA-2, YLA-4 and YLA-6). The data for silicon nitride specimens with Y_2O_3 and La_2O_3 crept less than 0.1% at 1220°C over 100 hrs at a stress of 200 MPa. Even at 1360°C and 260 MPa for 100 hrs, the total deformation of specimens was only 0.6%(5). A more detailed discussion will be presented elsewhere(6).
A number of piston caps, the critical component for uncooled diesel engine, were fabricated with the above mentioned material design route. These piston crowns (Fig. 3) show good performance during rig tests in the Shanghai Research Institute of Ship Engines. Under full load and the highest rotation speed, they work for 250 hrs without any damage. Further work will be pursued to fabricate them by gas pressure sintering to further improve their reproducibility as well as to reduce the cost.

The third component fabricated and tested for diesel engines is the glow plug (Fig. 4). Silicon nitride ceramics are again selected with additives that form a glassy phase during hot pressing. This is the only processing method suitable for fabricating this type of component with a fine tungsten wire spiral embedded within the body. Certain details have to be worked out to insure the reproducibility of the geometric structure of the components. In-house rig testing conducted at the Internal Combustion Engine Research Institute gives a charge-discharge life of over 60,000 cycles. Moreover, the response time is superior to that of the conventional ones, as expected. An automatic hot-pressing setup will be designed for small scale production in the early nineties.

A valve guide made of silicon nitride ceramics was also fabricated (Fig. 5). Gas pressure sintering was chosen as the processing approach in order to insure the precise shape and size of the component. Rig testing in a diesel engine for 300 hrs showed that this part can work for a much longer time than the metal counterpart.

The fifth silicon nitride engine component under investigation is the turbocharger rotor. Reliability and reproducibility are even more important for this component. This means a lot of work on processing science and processing technology and the elimination of dangerous flaws through processing with the help of nondestructive evaluation. It will take additional time to understand and surmount these problems in a systematic way.

SILICON CARBIDE CERAMICS

Silicon carbide materials usually have slightly lower strengths than that of silicon nitride. However, the former can usually retain their strengths to temperatures higher than 1400^oC with much less pronounced subcritical crack growth. Silicon carbide ceramics also possess extremely good high temperature creep resistance combined with good oxidation resistance which make them promising candidates for long duration, high temperature applications. One weak point for SiC is its relatively lower fracture toughness. One of our research efforts has been to improve the fracture toughness by dispersion hardening with both TiC (15-30 wt%) and YAG (2-15 wt%) particulates. This has produced promising results. K_{IC} values $\geqslant$ 6 MPa $m^{1/2}$ and transverse strength of around 500 MPa to 600 MPa have been achieved. Using mild oxidation to smooth the surface defects, the room temperature strength can be upgraded from 500-600 MPa to 800 MPa[7]. The components under development consist of piston and cylinder liners for the uncooled engine design as well as the tappet part for the present day diesel engine (Fig. 8). A diesel engine fitted with a silicon carbide tappet has already run continuously for 300 hrs with no failure. Improvements in design for ceramic tappet as well as the choice of pairing materials to match with the ceramic parts are still areas of research.

TRANSFORMATION TOUGHENED ZIRCONIA CERAMICS

Transformation toughened zirconia ceramics offer the advantage of having lower thermal conductivity, a higher thermal expansion coefficient which matches with various metal parts, and higher toughness. We have made detailed studies in TZP materials, especially those with yttria. The toughening performance has a close relationship with the fraction of tetragonal phase that is readily transformable in the process zone during crack propagation. Again this is related to the yttria content and the optimum grain size distribution, which in turn are processing sensitive[8]. Since zirconia-toughened materials are only effective for intermediate temperature ranges, they may find limited application in engines with advantage. Three components were developed in the Shanghai Institute of Ceramics. They were a piston crown, an insulating headface plate, and a valve seat for the adiabatic diesel engine (Figs. 7,8,9). At this moment they are all still in the process of testing.

WHISKER OR FIBER REINFORCED CERAMIC COMPOSITES

Fiber or whisker reinforcement is an effective approach to improve the fracture toughness of ceramics. Researchers in the Shanghai Institute of Ceramics have been working on inorganic composite materials for more than a decade and some of them have found quite successful and unique applications, especially in our space technology[(9)]. However, for components in ceramic engines, only a few composite systems can be considered. At the moment, it seems that only SiC whiskers or fibers may withstand the severe environments prevailing in heat engines even when embedded in a ceramic matrix. Two materials projects are now ongoing, i.e. the SiC whisker/ZrO_2 and SiC whisker/Si_3N_4 composites. We plan to implement the latter system for turbocharger rotor fabrication and application by the middle of the next decade. Anyway, ceramic matrix composites are an area of research and development with diverse perspectives beyond those for engine applications.

NDE AND JOINING

These are also important for the development of engine ceramics. High voltage microfocus X-ray radiography, high frequency (at present 200 MHz) ultrasonic detection, scanning laser acoustic microscopy and acoustic emission for thermal fatigue and thermal shock damage are some of the techniques studied. With respect to joining, the brazing of silicon nitride ceramic turbochargers to steel shafts and the brazing of Y-TZP ceramic piston caps to aluminum alloy piston bodies are now under investigation.

In conclusion, we would like to point out that a decisive breakthrough for the application of ceramic components for engines has not been yet achieved on a worldwide scale. A long range R & D program is anticipated in order to obtain the necessary high reliability and reproducibility as well as the reduction in cost. This is especially true when the state-of-the-art reaches the stage of mass production. We feel that close cooperation and the joint efforts of peoples in the fields of materials development and processing engineering together with those in engine and component design will result in more effective progress in the future. It is even more encouraging to note that materials of this category will be of importance not only in engine applications but also in many other fields. In this sense, we look forward to cooperating with our conterparats from the other countries with common interests.

REFERENCES

1. W.Y. Sun and D.S. Yan,
   "Phase Equilibria Studies on Nitride Systems and Their Role for Material Compositional Optimiztion",
   Reviews of Solid State Sciene, 1(3-4), 493-504(1988).
2. W.Y. Sun, W.L. Li and D.S. Yan,
   "Control of Phase Distribution in O'- β' Sialon Composites", Scientia Sinica(Series A), 31(7), 888-896(1988).

3. H.R. Zhuang, W.L. Li, D.Q. Hua, S.L. Wen, Y.Y. Sun, X.R. Fu and T.S. Yen, "Microstructure and Properties of Post-Sintered RBSN with YAG as Grain-Boundary Phase", 368-376, "Proceedings of the First International Symposium on Ceramic Components for Engine", Eds. S. Somiya, E. Kanai and K. Ando, KTK Scientific Publishers, Tokyo, 1984.
4. Y.R. Xu, L.P. Huang, X.R. Fu and T.S. Yen, "Hot-Pressed Silicon Nitride Ceramics with Rare-Earth Oxides Additives", Scientia Sinica (Series A), 28(5), 556-560(1985).
5. Y.R. Xu, X.R. Fu and D.S. Yan, "Creep Behavior of Hot-Pressed Silicon Nitride Ceramics with Rare-Earth Oxides and Alumina Additives", Physica B 150, 276-282, North-Holland, Amsterdam, 1988.
6. Y.R. Xu, X.R. Fu and T.S. Yen, "Grain Boundary Tailoring of High Performance Nitride Ceramics and Their Creep Property Studies", to be published at the Third International Symposium. Ceramic Materials and Components for Engines, U.S.A. Nov. 27-30 (1988).
7. D.L. Jiang, J.H. Wang, Y.L. Li and L.T. Ma, "Studies on the Strengthening of Silicon Carbide Based Multiphase Ceramics. I. SiC-TiC System", to be published in the "Proceedings of E-MRS Conference, Spring Meeting", May 31-June 2, Strasbourg, France, 1988.
8. T.S. Yen, J.K. Guo, X.R. Fu and D.L. Jiang "High Temperature Structural Ceramics Development for Engines in the Chinese Academy of Sciences", 209-216, "Proceedings of the Second International Symposium. Ceramic Materials and Components for Engines", Eds. W. Bunk and H. Hausner, Verlag Deutsche Keramische Gesellschaft, Bad Honnef, 1986.
9. T.S. Yen "An Overview of Chinese Research on Processing, Microstructure and Properties of New Ceramic Materials", 1-35, "Proceedings of the First China-US Seminar. Microstructure and Properties of Ceramic Materials", Eds. T.S. Yen and J.A. Pask, Science Press, Beijing, China, 1984.

Fig. 1 Post-Sintered RBSN Precombustion Cups.

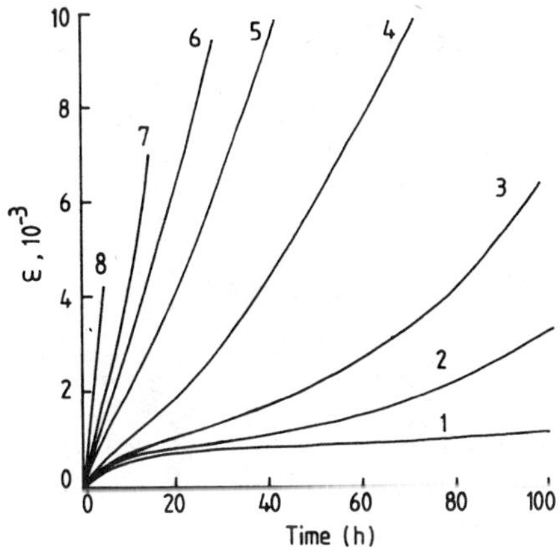

Fig. 2 Greep Strain-Time Curves of Silicon Nitride Ceramics at Constant Stresses and Temperatures. 1) YL-1 crept at 1220°C and 200 MPa; 2) YLA-4 crept at 1170°C and 300 MPa; 3) YL-1 crept at 1360°C and 260 MPa; 4) YLA-4 crept at 1220°C and 200 MPa; 5) YLA-4 crept at 1220°C and 300 MPa; 6) YLA-2 crept at 1310°C and 300 MPa; 7) YLA-6 crept at 1220°C and 300 MPa; 8) YLA-4 crept at 1260°C and 300 MPa.

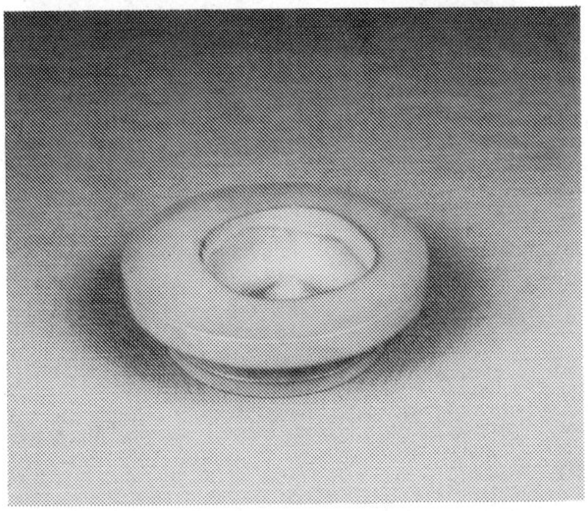

Fig. 3 Piston Crown Made of Silicon Nitride Ceramics Contains Rare-Earth Oxides.

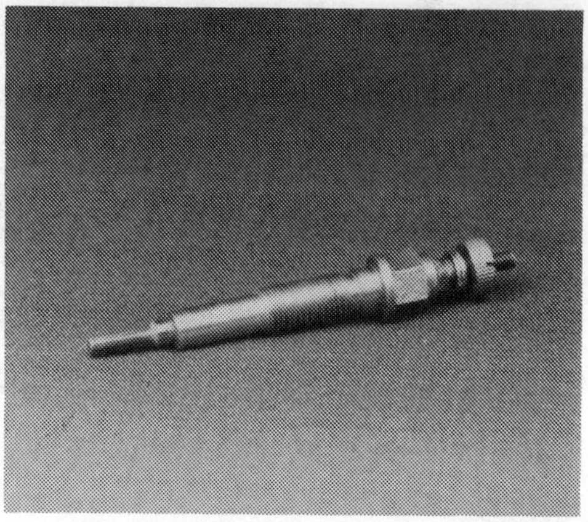

Fig. 4 Glow Plug Made from Hot-Pressed Silicon nitride Ceramics.

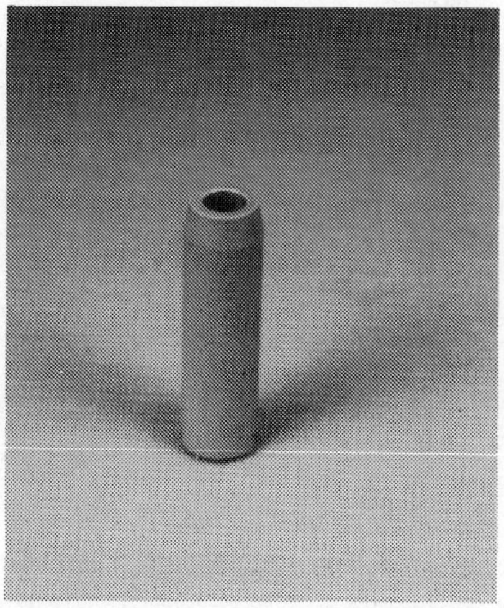

Fig. 5 A Silicon Nitride Valve Guide.

Fig. 6 A Silicon Carbide Tappet.

Fig. 7 Transformation Toughened Zirconia Piston Crown.

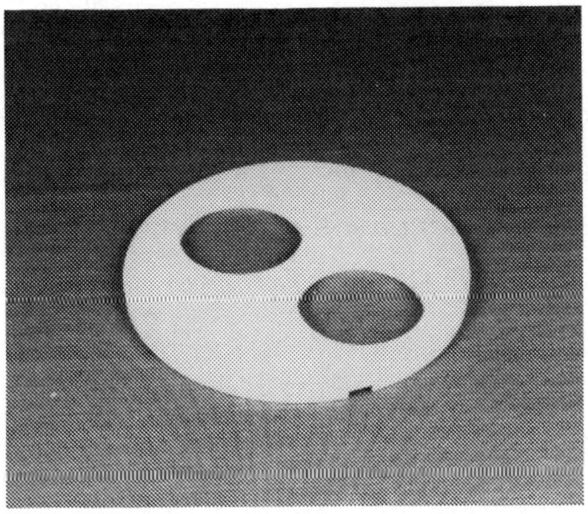

Fig. 8 An Insulating Cylinder Headface Plate Made from TZP Ceramics.

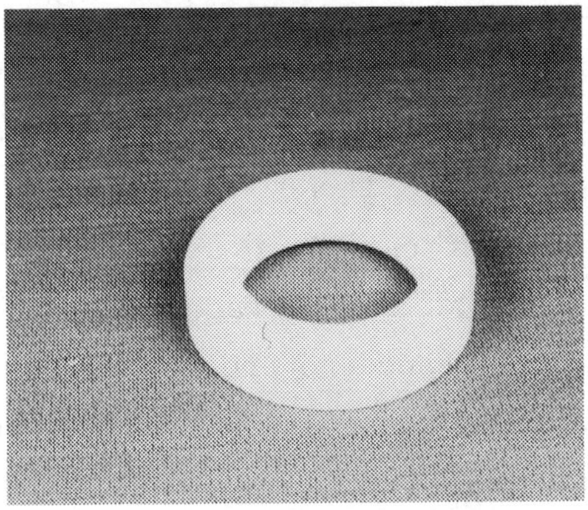

Fig. 9 Valve Seat Made of TZP Ceramics.

HEAT INSULATION COMPONENTS FOR GAS TURBINE ENGINE APPLICATIONS

B. Calès and P. Mathieu
Céramiques Techniques Desmarquest
78190 Trappes France

ABSTRACT

Advanced gas turbines require refractory insulating parts characterized by a low density, good thermomechanical properties and resistance to erosion by hot gases. For this purpose, a specific fiberous insulator has been developed using a blend of silica and alumina fibers. The bulk density may be adjusted between 80 and 400 kg/m^3. The 3-point bending strength of the fiberous material linearly increases with the bulk density. The thermal conductivity is very low. It increases with temperature and only reaches 0.18 W/mK at 1000°C for a density of 150 kg/m^3. These insulators may also be coated by a thin microporous layer which could be favourable to a good erosion resistance to hot gases.

INTRODUCTION

The development of automotive gas turbines has a requirement for surface-coated heat insulation components made of ceramic fiber material. The heat insulation system of a gas turbine is subjected to high thermal and mechanical stresses during the typical duty cycle [(1)]. Only fiberous insulators can be used under these conditions. Furthermore, the gas temperature at the combustion outlet is very high (1350°C) and the gas velocity can reach 180 m s^{-1} at the surface of the fibers. It is therefore necessary to improve the resistance to erosion by hot gases of the fiberous parts by an adequate coating. This coating will be subjected during service conditions to cyclic thermal shocks, so that a thermal shock resistant coating must be chosen.

Various fiberous insulators made of alumina, mullite or alumino-silicate fibers, with a bulk density ranging between 300 and 600 kg/m^3, has been studied by Heintze, Moergenthaler and Heckel [1]. They have characterized the 4-point bending strength and the thermal conductivity of these materials for various bulk densities. In the case of the pure alumina fiberous materials, the authors report a 4-point bending strength that progressively increases with the fibers density and a value of 1.5 MPa for a 300 kg/m^3 fiberous material is observed. The thermal conductivity of these fiberous insulators goes through a minimum value, which equal 0.18 W/mK at 1000°C, for a bulk density of about 250 kg/m^3. The thermal conductivity rapidly increases for higher or lower fibers density. Such a behavior is not in agreement with the results of Hayashi, Ohkawa and Mishikawa [2] which have specifically studied the room temperature thermal conductivity of alumino-silicate fiberous insulators and observed a continuous increase of thermal conductivity with bulk density.

The objective of this study was to develop a new fiberous insulator using a blend of silica and alumina fibers and an original manufacturing process for the bulk forming and the coating. The coating consists of a microporous layer impervious to liquid water. The experimental procedure is described below, as well as the mechanical, thermal and microstructural characteristics of the fiberous insulator.

EXPERIMENTAL PROCEDURE

The fiberous materials were prepared using chopped alumina and silica fibers, the size of which may be adjusted by a ball-milling operation. The size of the milled alumina and silica fibers is of the order of a few hundred microns. In order to precisely control the density of the fiberous parts between 80 and 400 kg/m^3, alumina and silica fibers were dispersed in a liquid medium with a basic pH, using an appropriate deflocculant. The liquid medium consisted of a water solution of a silica and boron oxide based inorganic refractory binder. The role of the refractory binder is to give high rigidity to the fiberous material over a large temperature range. The suspension of fibers was then filtered by an original pressure filtration process. The fiber cake is then dried using a microwave device and fired at high temperature. The firing temperature, typically between 1100°C and 1300°C, was carefully controlled. It must be high enough to get a local melting between alumina fibers, silica fibers and the inorganic binder just at the contact between fibers.

After firing, the fiberous part may be easily machined into complex shapes using conventional tools and machines.

A microporous layer, having the same composition as the fiberous material, may be deposited by spraying a mixture of milled fibers and powders onto the machined fiberous parts. The size of alumina and silica fibers in this mixture is precisely adjusted to prevent the porous layer from penetrating the fiberous part. The mean inter-fiber distance in the fiberous part has been measured, and it is about 30 microns for a fiberous part with a density of 300 kg/m^3 (Fig. 1). Therefore, short alumina and silica fibers larger than 30 microns must be used. Futhermore, in order to achieve a microporous and waterproof coating, the porosity of the surface layer is adjusted by adding fine alumina and silica particles. The pore diameter must be lower than the critical value Dc which is given by the following relation derived from Young and Laplace equation [3] :

$$Dc = \frac{4 \gamma \cos\Theta}{\Delta P}$$

where γ is the surface tension of the liquid, Θ is the liquid-solid contact angle and ΔP the pressure variation through the pore. With few simple hypothesis concerning the values of Θ and ΔP, the critical pore size for the microporous layer has been estimated to be of the order of a few microns. Such a porosity has been obtained by controlling the ratio of fine particles to short fibers in the suspension used for the coating.

The composition of the fiberous material, including the microporous layer, may be adjusted between 100% of alumina and 80 wt% silica - 20 wt% alumina. Due to their very low densities, 80 to 400 kg/m^3 (fiber volume fraction between 5 and 35%), the homogeneity of the fiberous parts has been controlled using X-ray tomography.

PHYSICAL CHARACTERISTICS OF THE FIBEROUS INSULATORS

Mechanical and thermal properties

The 3-point bending strength of the uncoated fiberous materials, with a ratio of silica/alumina of 4, has been measured at room temperature as a function of the density. The specimen size was 70 x 20 x 10 mm and the span dimension 50 mm. As shown in figure 2, it appears that the flexural strength of the fiberous material linearly increases with the fiber density and a value of 6 MPa was reached for a fiberous material with a density of 300 kg/m^3. The comparison with the results of Heinze and coworkers is difficult because no experimental detail is given (specimen size, span dimension).

In order to verify the isotropy of the material, different samples were also cut in the fiberous parts either parallel or normal to the filtration axis of the forming process. No difference was found between the two series of samples and the mechanical properties are not dependant on the forming process.

The thermal conductivity of the uncoated fiberous material with the same composition was measured by the transient hot wire method over a large temperature range for a fiber density of 150 kg/m^3. As shown in figure 3, the thermal conductivity increases with temperature. However the level of thermal conductivity remained very low and the thermal conductivity is only 0.15 W/mK at 800°C and 0.18 W/mK at 1000°C. These values are slightly lower than the thermal conductivity derived from the results of Heinze and coworkers [(1)] for a 150 kg/m^3 alumina fiberous insulator, i.e 0.21 W/mK at 800°C and 0.23 W/mK at 1000°C. At the minimum of the thermal conductivity v.s. bulk density variation a value of 0.18 W/mK has been measured by Heinze and coworkers for a 250 kg/m^3 alumina insulator(Fig. 4). As observed by Heinze et al., for low bulk density, the thermal conductivity at 1000°C decreases with the density of fiberous material. This behavior is different from that reported by Hayashi et al. [(2)] for the room temperature thermal conductivity of fiberous material. However, at room temperature the measured value of the thermal conductivity (λ = 0.057 W/mK) is in good agreement with the value $\lambda = 0.0599$ W/mK calculated with the experimental relation proposed by HAYASHI et al. for alumino-silicate insulators :

$$\lambda = a\lambda s + b\lambda f + \frac{(1-a-b)\ \lambda s\,\lambda f}{(1-y)\lambda s + y\lambda f}$$

with $a = 0.169\ V^{1.57}$

$b = (1-V)^{1.16}$

$y = 0.81 - 0.043\ \mathrm{Ln}V$

where λs and λf are the thermal conductivities of the fibers and the gas filled between the fibers respectively and V the volume fraction of fiber.

Due to their low thermal conductivity, these fiberous materials are very good thermal insulators. Their linear expansion coefficient has been measured using a precise differential dilatometer. No shrinkage was observed up to about 1200°C and the linear expansion coefficient between 20 and 1000°C is $7 \times 10^{-7} K^{-1}$. The dielectric properties of this fiberous insulator have also been analyzed at 1 MHz. The dielectric constant is $\varepsilon = 1.07$ with a low loss tangent $tg\,\delta = 4 \times 10^{-4}$.

Microstructural investigations

The microstructure of the fiberous insulators is shown in figure 5 for a bulk density of 150 kg/m^3. It may be noted that the microstructure of the material is made of a very homogeneous three dimensional network of randomly distributed short fibers having a uniform diameter. Due to the well controlled dispersion step of the manufacturing process, no fiber agglomerate was detected and no prefered orientation for the fibers was observed.

At higher magnification, the interfiber connections clearly appear (Fig. 6). There is a "soldering" of the different fibers which results from the local melting of the fiber in the refractory inorganic binder. This particular microstructure explains the high rigidity and flexural strength of these fiberous materials, even for a low bulk density of 80 kg/m^3. It can also be noted that the microstructure is a uniform close-meshed network (Fig. 1), the size of the "meshes" depending on the insulator bulk density. It is about 30 microns for a 300 kg/m^3 insulator.

The density of large fiberous parts, with a diameter of about 300 mm, has been controlled by X-ray tomography. The density mapping (Fig. 7) indicates a very uniform density, with a variation of fiber volume fraction between any two points lower than 1%.

The microstructure of the coated fiberous insulators is shown in figures 8 and 9. The microporous coating appears to be flat and with a constant thickness. With a 200 microns thick coating, the surface roughness of the fiberous part is quite low (Fig. 9), with a maximum roughness equal to Rmax = 120 microns. The external surface of the microporous layer is shown in figure 9. There is no continuous glass film at the surface but only large and separated area with melted inorganic binder. Such a microstructure would be favourable to a high thermal shock resistance and to a good erosion resistance to hot gases. These characterizations are presently under investigation using coated fiberous insulators.

CONCLUSION

An experimental procedure has been optimized to manufacture fiberous insulators with a very homogeneous density which can be adjusted between 80 and 400 kg/m^3. Large fiberous parts, as large as 300 mm in diameter and 300 mm high, can be produced by this technique.

The high rigidity and good flexural strength of these alumina plus silica fiberous insulators mainly result from the particular microstructure of these materials made of a very uniform and close-meshed three dimensional network. The thermal conductivity remains very low, even at high temperature. Therefore, these insulators used with a refractory coating are very promising for applications in gas turbine engine where insulators with low thermal conductivity and also a high erosion resistance to hot gases are required.

REFERENCES

(1) R. HEINZE, K. MOERGENTHALLER and J. KECKEL, "Development of surface-coated heat-insulation components made of ceramic fiber materials for automotive gas turbines" CEC Research Programme in the Material Sector : Substitution and Materials Technologies and Ceramics 1982-1985, Vol 1, December 1986.

(2) K. KAYASHI, H. OHKAWA and T. MISHIKAWA, "Production of thermal conductivity of ceramic fibrous insulators", J. Ceram. Soc. Jpn. Inter. Ed., 95, 842-847 (1987).

(3) A.W. ADAMSON, "Physical Chemistry of Surfaces", Interscicnce, New York (1967) .

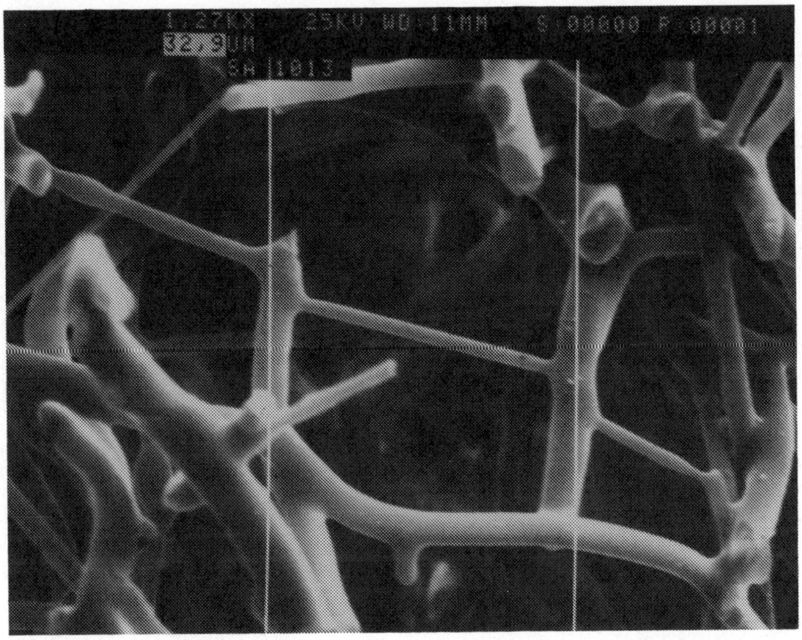

Figure 1 : Close-meshed network of a 300 kg/m^3 fiberous insulators.

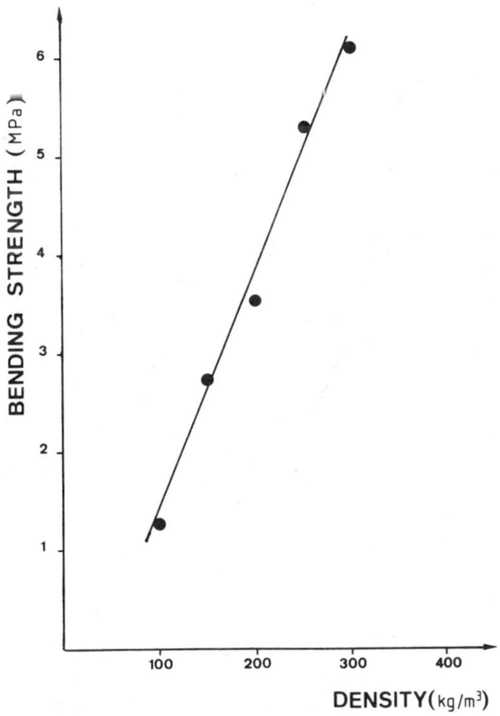

Figure 2 : Variation of the 3 point bending strength of fiberous insulators versus bulk density (SiO_2/Al_2O_3 = 4).

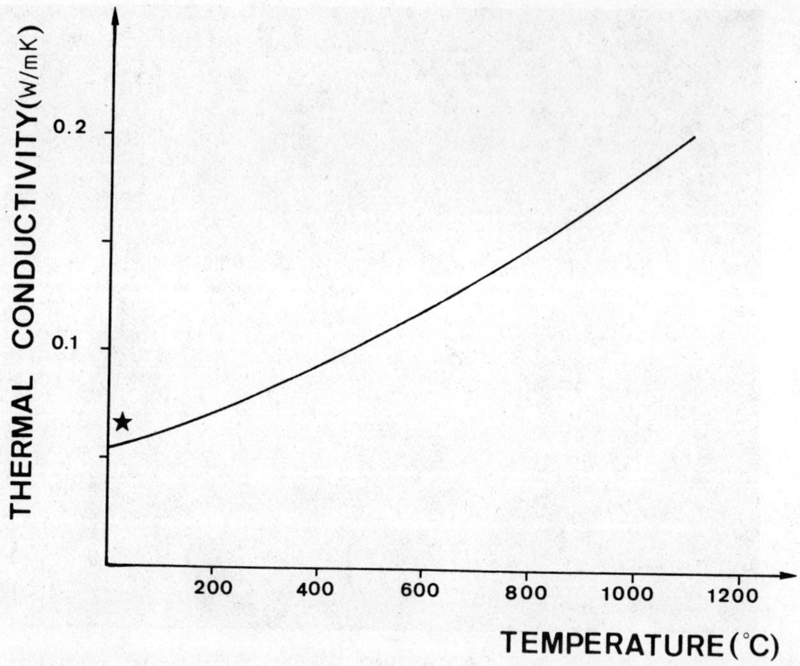

Figure 3 : Dependence on temperature of the thermal conductivity of a 150 kg/m^3 fiberous insulator (SiO_2/Al_2O_3 = 4). * : value calculated using the equation of K. Hayashi et al. (3).

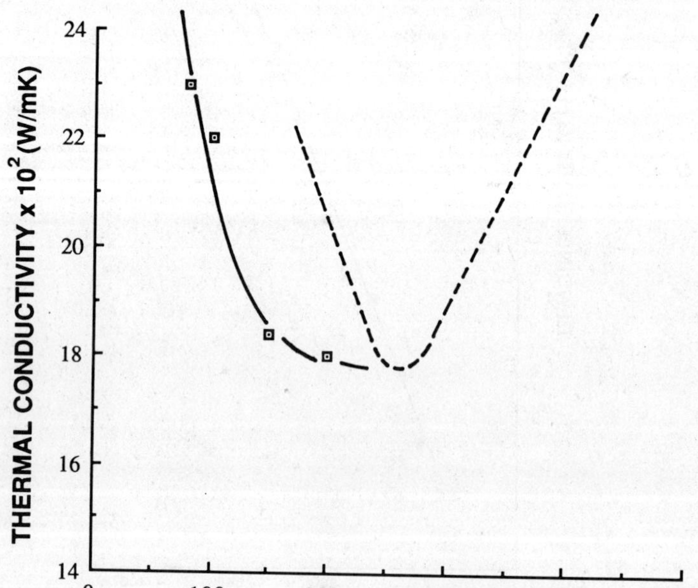

Figure 4 : Dependence of the thermal conductivity at 1000°C on the fiberous material density - dotted line : results of Heinze et al [1].

Figure 5 : SEM microstructure of a 150 kg/m^3 insulator.

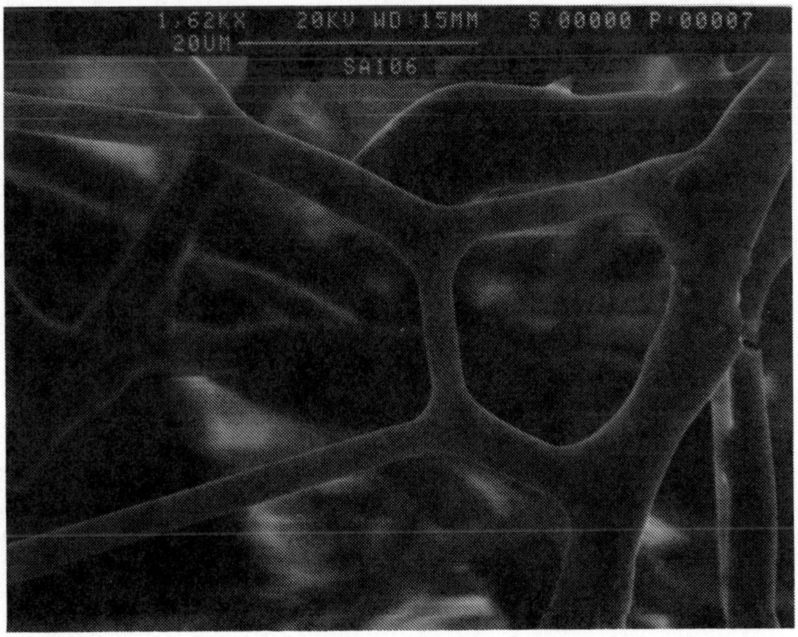

Figure 6 : Soldering contact between fibers.

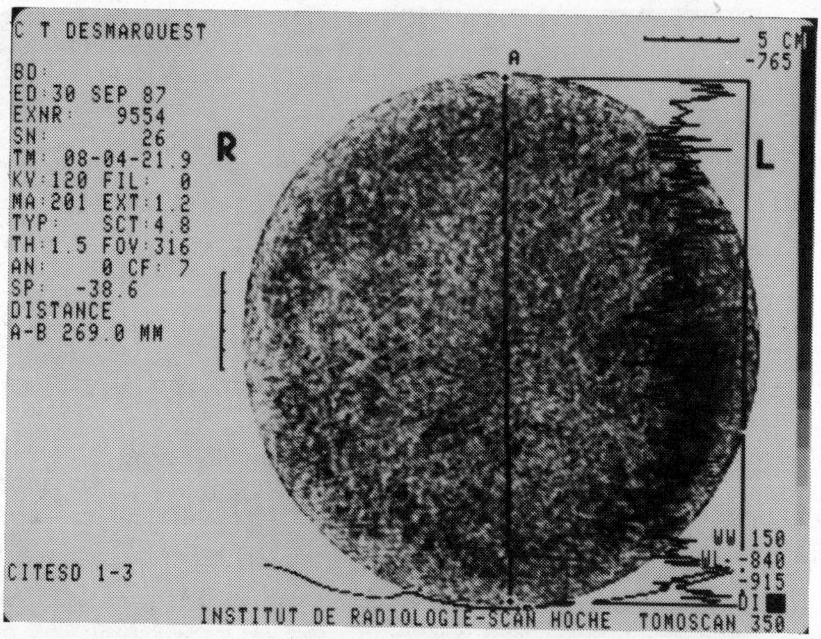

Figure 7 : X-ray tomography of a large fiberous part (300 mm in diameter).

Figure 8 : SEM microstructure of the microporous coating.

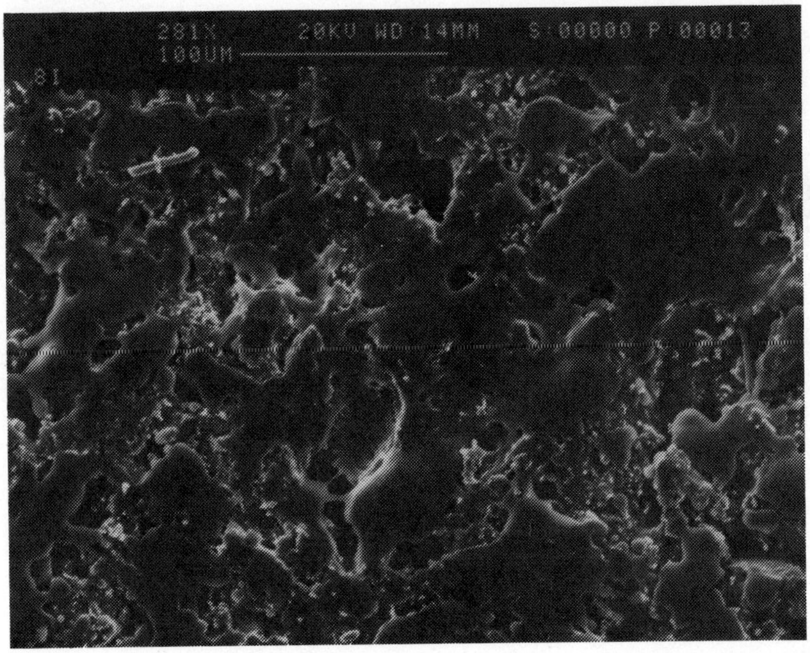

Figure 9 : External surface microstructure of the coating.

GAS PRESSURE SINTERED Si_3N_4 COMPONENTS IN SPARK IGNITION (SI) ENGINES

C. Suh
General Motors Corporation
General Motors Technical Center
30200 Mound Road
Warren, Michigan 48090

Y. Sato and T. Kojima
NGK Spark Plug Co., Ltd. (NTK)
14-18 Takatsuji-cho, Mizuho-ku
Nagoya, Japan

ABSTRACT

The development, design, and material processing of ceramic structural components, and its application for gasoline engines will be discussed. NTK developed a new gas pressure sintering technique which completely eliminates porosity due to thermal decomposition of silicon nitride, and enhanced quality of the material to meet the real world application. GM designed, built and demonstrated durability for several ceramic components including valves, piston pin, and light-weight monolithic piston. A vehicle installed with ceramic component engine was assembled and has traveled over 10,000 miles without component failure. Application of structural ceramics to spark ignition engines differs from conventional metallic component design techniques. Successful ceramic applications in heat engines need high quality materials, precision processes and computer design-evaluation techniques. The various ceramic components fabricated by NTK and tested by GM are shown in Fig. 1.

INTRODUCTION

Structural ceramics have attracted much attention recently, because of properties such as superior strength at high temperature, wear resistance, and low mass. The application of ceramic in heat engines is certainly a challenge for engine designers. In the past, the efforts were concentrated mainly in the area of the adiabatic diesel and the gas turbine with high temperature performance of turbine rotors. However, the progress of these activities has been disappointingly slow.

Even though considerable time and effort have been given to ceramic engine component developments, the actual applications are still in their infancy, and there are many areas yet to be explored. However, the progress has been steady, and it is certain that structural ceramics are now established. For example, ceramic glow plugs of diesel engines and ceramic turbocharger rotors have occupied 5% shares in the passenger car market in Japan. This trend stems from high material reliability due to the progress in improving sintering techniques, and further penetration is expected into more engine components.

Through experience, we have learned that for Spark Ignited (SI) engines the new high quality ceramics provide performance advantages over conventional metallic materials, if one uses quality material and proper design techniques. Simple replacement of a metal part with a ceramic part having the same shape almost always results in failure because the two have distinctly different properties.

DESIGN ANALYSIS

Regardless of the advantages, structural ceramics still have shortcomings mainly because of their brittle nature. They have linear behavior up to fracture, but little crack arrest capability; identical specimens under the same loading conditions may have considerably different strengths. This effect is attributed to the fact that the brittle materials contain a large number of microscopic flaws. The flaw distribution is statistical in nature; therefore, the shapes and sizes are all unique although the component has the same design. Normally, metallic component design is mainly determinative, while non-metallic component design is statistical.

For a ductile material, a simple stress analysis may provide enough information to evaluate the structural integrity; the area of high stress or strain of the structure is considered as a failure location of the component. However, this type of analysis is not sufficient for the brittle material design. Special rigid format of the MSC/NASTRAN computer program is written to combine thermal and elastic analysis at the same time.

A modified Structure Ceramic Analysis and Reliability Evaluation (S.C.A.R.E.) code has been used to evaluate the survival rate. The detail theory related to the code can be found in the references (4 and 5). Among other capabilities, an important part of design evaluation is to examine the failure modes carefully. Since the theory is based on "Weak Link Theory" proposed by Weibull and Batdorf (4), it is crucial to locate the weakest link within the whole structure. This may not be possible by using a regular stress or the probability display on the surface of structures, but is only possible by using the display with 3-D dissecting capability.

The upper part of Fig. 2 shows the octahedral stress distribution of a piston pin, and the lower part shows the failure probability distribution at the same loading condition. Normally the stress analysis will show the location where stress is low, high, and concentrated, and the place where the structure is likely to fail due to the limit of strength. If the evaluation is extended to statistical failure probability analysis, the location of the weakest area may not be the same with that of stress analysis, as seen already in Fig. 2. According to Batdorf's theory, the failure mode depends on the orientation of the angle of cracks, and the intensity of the stress components acting on the crack surfaces as functions of volume or surface area. Among other theories, the strain energy release rate criterion resulted, in most cases, in conservative assessment, and applied for the components evaluated in this study.

MATERIAL

The requirements of material properties for SI engines, gas turbines and Compression Ignition (CI) engines are unique. Low thermal conductivities and high thermal stiffness are not needed for the SI engine, but it is essential to have stable properties to 900°C, high mechanical strength, thermal shock resistance, corrosion resistance, and wear resistance. The lightness (low density) of the material is especially desired.

The properties of NTK's Gas-Pressure-Sintered (GPS) Si_3N_4 (EC-141) met the requirements mentioned above, and the reliability has already been proved, as for the past three years 150,000 turbocharger rotors were supplied and are in service without even minor defects.

The GPS technique elevates the limit of the thermal decomposition temperature of silicon nitride; the technique introduces nitrogen into the sintering furnace and pressurizes the chamber. This enables high-tempertature sintering, and thereby the sintered products are highly densified with a smaller amount of sintering aids, and produces high-reliability silicon nitride as shown by the properties in Fig. 3. Materials with lower fracture strength, fracture toughness and Weibull modulus than EC-141, barely sustained the engine start-up conditions, specifically GM's corporate durability testing conditions. The properties of EC-141 appear to be minimum threshold properties for SI engine application.

The HIP process could also be used to densify the Si_3N_4 and serves to eliminate pores formed during the sintering, but requires pre-sintering or encapsulation (see Fig. 4). The higher pressures used in HIP will result in higher cost and poorer mass productivity compared to GSP processing. The forming process is also important. In the case of valves, for example, slip casting, CIP or injection molding can be employed, and CIP was adopted with EC-141 material processing. The selection of the CIP method for this study was mainly due to the developmental nature of the project.

COMPONENTS TESTED

Valve Trains

The unique properties of ceramics, such as low mass and high mechanical strength, lift many barriers imposed by metallic components. The size of the valve and magnitude of the valve lift may be increased more than conventional valves if other constraints permit. The high valve lift will improve volumetric efficiency of the engine; simple substitution of ceramic components is not advantageous unless the design has been compromised. Ceramic valves are generally about 50% lighter in weight than metal valves, and the inertia loads are decreased accordingly. The reduction of valve mass yields reduction of spring rate and thus the valve train friction; and will improve fuel economy, noise level, and performance response. On the other hand, if the spring rate is kept the same, the limitation of engine speed may be increased with ceramic valves about 20% more, and will boost the power output. These features have already been examined in an engine design for racing vehicles. With 50% of valve mass and 30% of spring rate reduction, the fuel economy was improved about 3% when tested on the GM 2.0L engine. About 15% of valve train friction reduction results in 3% of total engine friction reduction and yields 1% of brake specific fuel consumption (BSFC) improvement.

Ceramic rocker-arm and tappets proved to be good for friction reduction, and Ogawa and others (6) also demonstrated the superiority of wear characteristics of rocker arm inserts.

The effect of reduced spring load has been reported by Updike and Nagle (3). Normally, valves move as prescribed by the cam profile, but if the engine moves at high speed, the valve inertia force sometimes exceeds the spring force, and starts showing abnormal behavior such as jump and bounce. Low mass ceramic valves suppress these abnormalities up to a much higher engine speed (see Fig 5A and 5B).

Piston Pin

The ceramic piston pin is about 50% less mass compared to the metal design, and it will reduce about 7% of the piston assembly mass if the piston design remains the same. This reduction also reduces 7% of the secondary shaking force by the piston assembly, and will improve the fuel economy about 1%.

Beside these factors, the strength of ceramic piston pins opens another possibility to optimize design of a low mass piston. If the pin bends under a certain load condition, the piston shows some strain which necessitates that the piston be heavy to lower the increased stresses. However, if the pin is stiff, the piston will be less strained.

Piston

The piston plays an important function as a reciprocating part and also has a very complicated configuration. The design and fabrication of a monolithic ceramic piston are certainly challenging. Normally a piston is made of an aluminum alloy; the most difficult part of the design mission is to overcome the weight barrier of ceramics against this lightweight material. We designed, fabricated and tested low mass ceramic pistons.

The ceramic piston survived under road load testing conditions. Knock and emission characteristics were also investigated. The main difference in the two pistons, aluminum and ceramic, lies mainly with the thermal expansion coefficients of the two materials; aluminum alloy has about nine times greater expansion coefficient compared to ceramic. This means that the ceramic piston may run at much higher temperature conditions without increasing the friction level, whereas it will be impractical for aluminum since the piston will expand more. For the same friction losses, one can run the engine at higher speed and power with a ceramic piston.

Various piston designs of diesel engines with ceramics have been surveyed by French (7). All the pistons discussed in the French report were ceramic coated metallic pistons.

OTHER PERFORMANCE EFFECTS

Effects on Knock

When a new material is used, particularly where the combustion flame is interfacing, it is of great concern whether any rough combustion may be observed. To have a smooth combustion, it is desirable to avoid knock. The thermal conductivity of silicon nitride, steel, and super alloy all change with temperature. The EC-141 ceramic has a higher thermal conductivity compared to the two other materials if the temperature is lower than 500°C. However the conductivities for the metals will be higher as temperature goes up beyond 700°C. Since the surface of valves and the piston crown area covered quite a large portion of combustion interfacing area, materials of different conductivities may have somewhat different temperature distributions.

First, the effects of lower thermal conductivity in these high temperature ranges on engine knock were investigated. The engine used was a 4-cylinder 1.8L displacement with OHC and water cooled. An experiment was conducted measuring the knock. The results at full load (WOT) are tabulated in Fig. 6; the level of knock was obtained by a frequency analyzer. There were no significant differences between the two cases of the metal and ceramic valves in advancing the ignition timing. As the ignition timing advances, however, the intensity of knock tends to become greater with ceramic valves than metal valves. The temperature measurements around the spark plug area for both cases showed little difference. From the above results, EC-141 ceramic valves gave the same knocking characteristics as metallic valves, but the intensity is slightly higher if there is any knock.

For the monolithic ceramic piston, the investigation is still under way, but the preliminary results indicated that the knocking phenomenon showed little difference between metallic and ceramic pistons. Other indications from the piston test were that the spark was required to be retarded and then advanced slightly when the engine is run from low to high speed, to maintain Maximum Brake Torque (MBT) conditions.

Effects on Emission

In engine design, one of the main goals is to limit exhaust emissions. Consequently, the effects of a new material on emission are always an important factor. A notable finding when the ceramic piston was tested, was the quantity of blowby; the magnitude was significant. Since the ceramic piston did not expand at operating thermal conditions as much as the aluminum piston, and we also used the same piston clearances for both piston designs, this created enough space to allow more blowby.

At this early developmental stage, we may contemplate that the quantity of emissions due to the ceramic components are not too different from a conventional material, at least at road load conditions. At high loading conditions, when the temperature rises significantly, this may change. These conditions are now under further investigation.

CONCLUSION

The present work demonstrated that if proper design, material and processing are used, monolithic ceramic components are liveable in the spark ignition engine. Differing from the efforts for low heat rejection engines or gas turbine engines, this study also pointed out another possibility of beneficial applications of ceramic in the area of the spark ignited engine.

The conclusions of this study were as follows:

1. To assure the designer's confidence, the value of the properties of Si_3N_4 should be as high as those of EC-141.
2. The valve system contributed a 3% fuel economy increase when ceramic valves are used.

3. Ceramic piston pin opens a new avenue for piston design since the material has higher stiffness, low mass, and low wear characteristics.
4. Monolithic ceramic piston design is challenging. The present design did not produce any friction increase nor scuffings on the bore surface. Since the thermal expansion of the ceramic is much less than metal, the combination of ceramic piston and cylinder liner should be ideal to minimize the bore distortion and to reduce the friction between the two.
5. Ceramic components have a tendency to increase the intensity of knock if there is any knock. The effects on emission due to the ceramics were also not a cause for apprehension.
6. The key of success for applications for ceramic heat engine components depends on a good design analysis.

REFERENCES

1. Y. Hattori, Y. Tajima, K. Yabuta, Y. Matuo, M. Kawamura, and T. Watanabe, "Gas Pressure Sintered Silicon Nitride Ceramics for Turbocharger Application", 2nd Intrnational Symposium Ceramic and Components for Engine, Lubeck-Travemunde, FGR, April, 1986.
2. K. Katayama, T. Watanabe, K. Matoba, and N. Katoh, "Development of Nissan High Response Ceramic Turbocharger Rotor", SAE Paper 861128.
3. S. Updike and P. Nagle, "Ceramic Valve Train Components", SAE Paper 880441
4. J. Gyekenyesi, "SCARE - A Post-Processor Program to MSC/NASTRAN for the Reliability Analysis of Structural Ceramic Components", NASA Technical Memorandum 87188.
5. J. Gyekenyesi, N. Nemeth, "Surface Flaw Reliability Analysis of Ceramic Components with SCARE Finite Element Postprocessor Program", NASA Technical Memorandum 88901.
6. Y. Ogawa and others, "Ceramic Rocker Arm Insert for Internal Combustion Engines", SAE Paper 860397.
7. C. French, "Ceramics in Reciprocating Internal Combustion Engines", SAE Paper 841135.

Fig. 1 A set of ceramic components

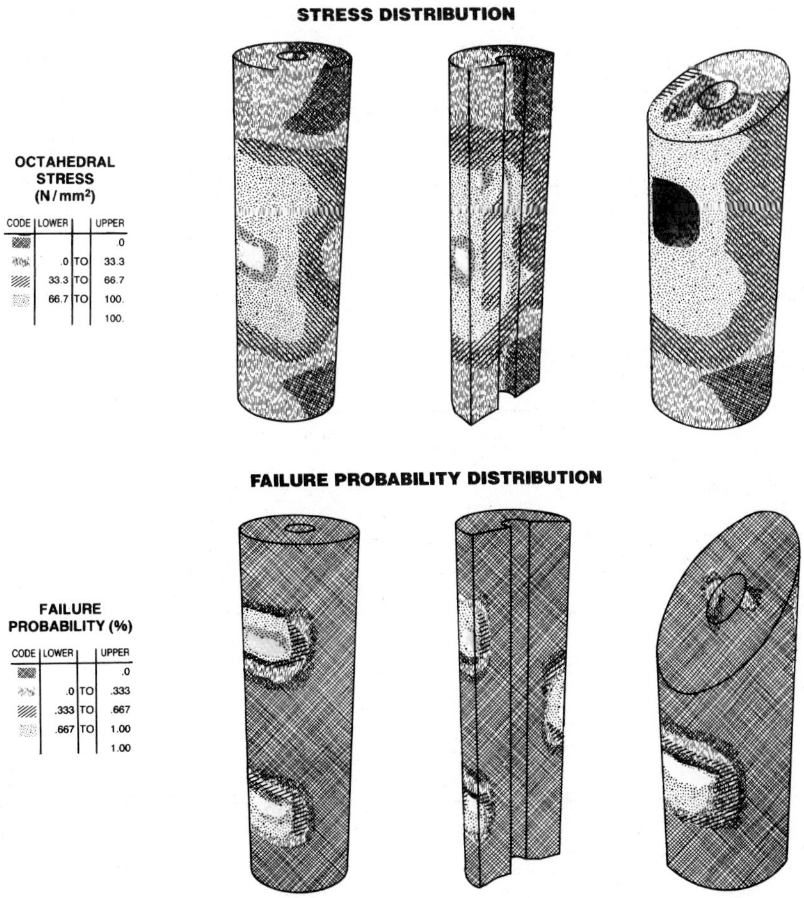

Fig. 2 Ceramic piston pin F.E.M. analysis

	MEASUREMENT CONDITION (R.T.= Room Temperature)	VALUE
Density (g/cm^3)	R. T.	3.22
Flexure strength (MPa)	R. T. 800°C 1000°C	900 850 740
Fracture Toughness ($MN/m^{3/2}$)	R. T.	7
Elastic modulus (GPa)	R. T.	310
Poisson's ratio	R. T.	0.24
Hardness (HR45N)	R. T.	87
Thermal expansion coefficient ($10^{-6}/°C$)	R. T.-800°C	2.8
Thermal conductivity (W/m·K)	R. T.	27
Oxid. Wt. Gain (mg/cm^2)	1000°C X 100 Hours	< 0.1
Weibull Modulus	R. T.	16

Fig. 3 Material properties of EC141 (GPSSN)

	GPS	HIP
Process	Mixing Forming Sintering	Mixing Forming Encapsulate / Pre-sintering Hipping
Pressure	~10 MPa	100~200 MPa
Temperature	1800~2100 °C	1600~2000 °C
Production Cost	moderate	high

Fig. 4 Comparison of GPS and HIP

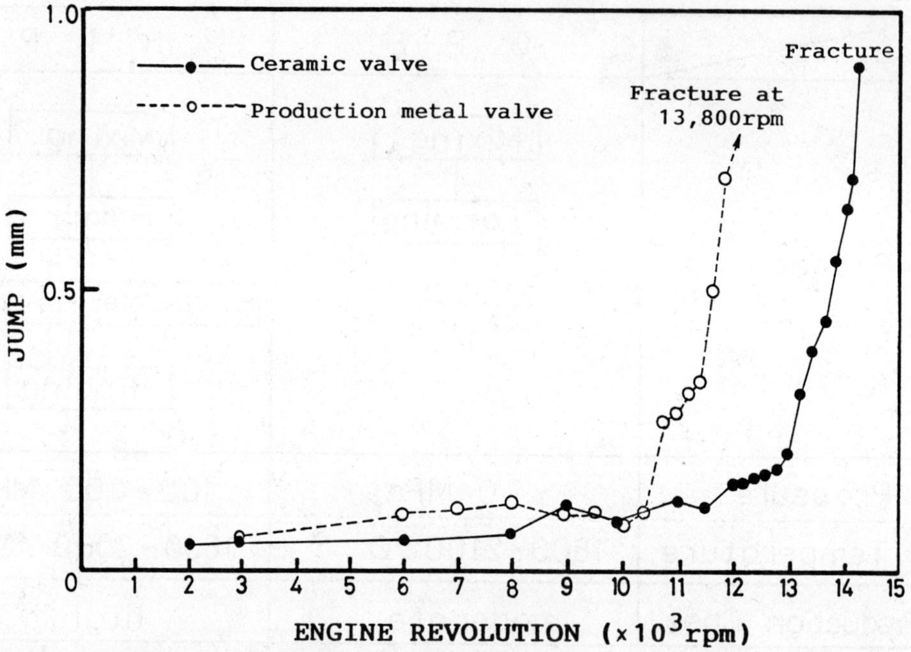

Fig. 5A Valve dynamic behavior about jump

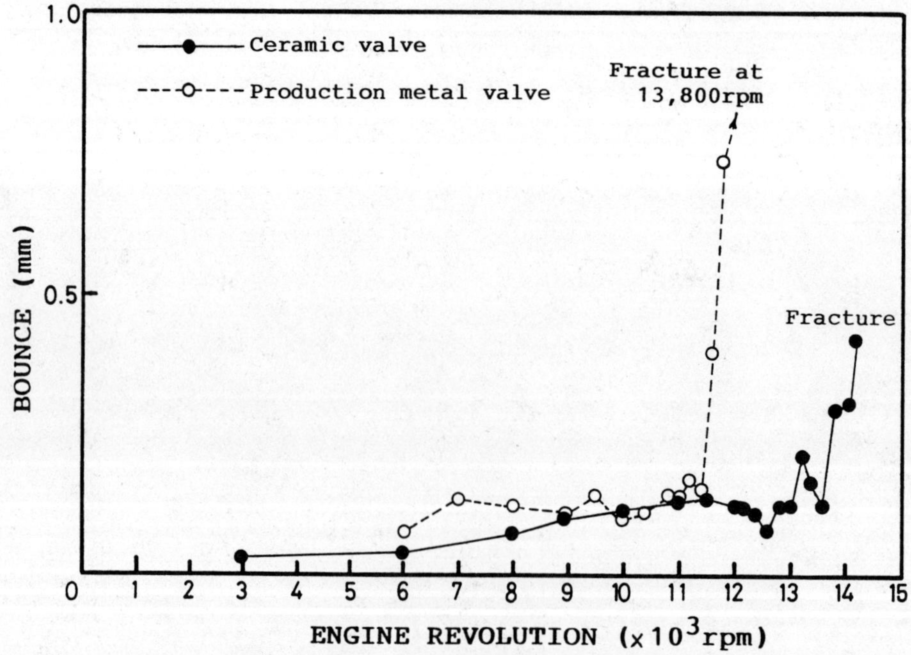

Fig. 5B Valve dynamic behavior about bounce

Load condition		Knock intensity	
Engine revolution (r.p.m.)	Ignition timing (°BTDC)	Ceramic	Metal
3000	X19	●	●
	X22	◆ ~ ▲	◆ ~ ▲
	X23.5	▲	▲
	X25	▲ ~ X	▲
4000	X23	●	●
	X24.5	◆	◆
	X26	▲	◆
5000	X22.5	●	●
	X24	◆	◆
	X25.5	◆ ~ ▲	▲
	X27	▲	▲
5600	X22.5	●	●
	X25	▲	▲
	X26.5	▲ ~ X	▲
	X28	▲ ~ X	▲ ~ X

● No knock ◆ Trace knock
▲ Light knock X Medium knock

Fig. 6 Results of knock test

RELIABILITY ANALYSIS OF A Si_3N_4 CERAMIC PISTON PIN FOR AUTOMOTIVE ENGINES

E. Amar, F. Gauthier, Renault, SCE 0853,
8-10 Avenue Emile Zola,
92109 Boulogne-Billancourt (France)

J. Lamon, Battelle-Geneva
7, Route de Drize
1227, Carouge/Geneva (Switzerland)

ABSTRACT

The data base required for the engineering reliability analysis of silicon nitride pins was determined. The reliability analysis was performed using the CERAM computer code, which uses the Weibull theory and a more fundamental model, the Multiaxial Elemental Strength Model.

The reference material was evaluated using 3-point and 4-point bending bars and biaxial flexure disks. Material reproducibility was then characterized on 4-point bending bars cut from as received piston pins. The influence of service conditions was finally determined on 4-point bending bars which were prepared from piston pin having run for 250 hours in an engine.

INTRODUCTION

Ceramics display characteristics which preclude the use of the conventional design methodology which is employed for metals. Ceramics exhibit brittle fracture and an unreliable mechanical resistance to fracture, as a result of the presence of a multitude of inherent minute flaws having a disordered distribution. Because of these flaws fracture strength is not an intrinsic criterion. But instead it exhibits an important variability which is affected by such parameters as the nature of the fracture inducing flaws and the stress state.

This paper shows an example of the use of an appropriate methodology for designing ceramic parts for automotive engines. This methodology is based upon the CERAM computer code for reliability analysis of ceramic components [1-4]. This paper reports the work at Renaud Co. in developing pistons pins of silicon nitride. The initial geometry was based on a stress analysis [5] and revised by engineering reliability analysis. Emphasis is on the investigation of the material and its properties which provided the data base for failure probability computations.

CERAMIC DESIGN METHODOLOGY

Appropriate methodology for designing ceramic components relies on the computation of mechanical reliability, based upon statistical approaches to brittle fracture. The calculated mechanical reliability permits the diagnosis of the current design and guides the decision to be made, if the component reliability is judged insufficient. Geometry must be changed first, until an acceptable reliability is reached. If the situation arises that geometry changes cannot afford reliability improvements, selection of a more appropriate material or improvement of the current service conditions will be guided by further reliability computations.

The well-known Weibull method which is routinely used for strength variability characterizations is statistical in essence. No direct use is made of the hypothesis that fracture is caused by the pre-existing microstructural flaws [6]. Recently a more fundamental probabilistic model was derived by incorporating fracture mechanics concepts into the statistical elemental strength approach [7, 8]. This Multiaxial Elemental Strength Model (MUEST) recognizes the flaws as physical entities and establishes the relationships between the basic mechanical response of flaws and the fracture probability of the bulk material. This model has been extensively described and validated in previous papers [2-4 and 6-9]. Both Muest and Weibull models are incorporated into a computer code CERAM developed by Battelle for designing ceramic parts and for reliability analysis [1]. This computer code is a post processor to finite element codes.

CERAM input requirements can be grouped into 2 categories. The first category consists of the stress distribution within the component. The second category defines the materials data which characterize the distribution of the fracture inducing flaws in the Muest model (flaw strength parameters) and the distribution of strength data in the Weibull statistics. These parameters are determined on small test specimens having well-defined geometries and stress states. They are derived from the comparison of the distribution of strength data with the distribution calculated using the Muest model or the Weibull method.

EXPERIMENTAL PROCEDURE

Material

The silicon Nitride selected for piston pin application was made by cold pressureless sintering. The composition is $Si_{6-z}\ Al_z\ O_z\ N_{8-z}$ with z equal to 0.5. The main properties are given in Table 1.

Test specimens

The test specimens were machined either out of reference billets (referred to as "billet" samples) or out of piston pin components (Table 2). Both billets and piston pins had been made by cold isostatic presssing and then sintered in exactly the same processing conditions.

Bars and disks were prepared from the reference billets. The bars were cut from the heart of billets in order to eliminate skin effects.

Table 1. Principal properties of the tested Si_3N_4.

Density	3.2 g/cm^2
Modulus of Rupture (4-point bending)	
RT	690 PMa
1000°C	690 MPa
1200°C	392 MPa
Fracture Toughness (SENB)	7.4 MPA$\sqrt{m}$

The test bars which were prepared from the as-received and engine tested piston pins had a different outer surface depending on their location in the piston pins: it was flat for the "heart" samples, whereas it was cylindrical for the "skin" samples, as it coincided with the circumference of piston pins.

For the determination of the data base relative to the reference material, and for the analysis of material reproducibility, the specimens were carefully polished, using an abrasive liquid (roughness Ra < 0.05 µm). For the determination of the surface characteristics of the component before and after engine testing, the tensile surface of the "skin" samples had the initial machining finish of piston pins (roughness Ra = 0.2 µm).

Mechanical testing

The specimens were tested on a tensile machine at room temperature and at a displacement rate of 0.2 mm/min. The tests were conducted as follows:

a) Tests on "billet" samples
The 3 point-bending tests used a steel fixture with a 60 mm span. The fracture always occurred in the vicinity of the center line of test bars.

The 4 point-bending tests were performed on the two parts about 35 mm long which were obtained after the 3 point-bending tests (upper span 12.5 mm lower span 24 mm).

The biaxial flexure of disks used the ring on ring technique with a loading ring of 2.5 mm in diameter and a supporting of 24 mm in diameter.

b) 4-point bending tests on samples from piston pins
The "heart" bars were tested on the same fixture as above. For the "skin" samples circumferential supports having the same 14 mm curvature radius as the outer surface were used. 3 randomly selected samples among the 7 available piston pins were engine tested.

Engine testing of piston pins

The dimensions of the ceramic piston pins are shown in Figure 2. The support span was 32 mm and the width of the connecting rod head was b =26 mm. A supercharged indirect injection diesel engine (2 litres - 4 cylinders) was used for the tests. This engine was selected because of the high load one reaches at maximum torque. The tests were performed at full loads and maximum power for 250 hours. The test durations were limited by piston ring failures.

The following loads were applied to these piston pins: a compressive load of 55,000 N at maximum torque at 3000 rpm and a traction load of 25,000 N at 4500 rpm at a mean effective pressure close to zero.

Table 2. Preparation of test specimens

	Sample origin				
	Billets			Piston pins samples	
	(70 x 28 x 28 mm)		(100 x 30 mm)	"heart"	"skin"
Samples	bars	bars	disks	bars	bars
Quantity	50	100	50		
Sample sizes	70 x 3 x 4	35 x 3 x 4	30 x 3	35 x 3 x 4	35 x 3 x 4
Tests	3 pt-bending	4 pt-bending	biaxial flexure	4 pt-bending	4 pt-bending
Surface finish	polished	polished	polished	polished	as ground

RESULTS AND DISCUSSION

The material data base required for the reliability analysis of components characterizes the statistical distribution of fracture strengths (Weibull method) or the distribution of the fracture inducing flaws (Muest model). In both cases it consists of two parameters including a shape parameter m and a scale factor (σ_{ow} in the Weibull analysis, and σ_{om} in the Muest model) [2]. These statistical parameters are derived from computations using the program CERAM. They are given by those values which allow the best fit of the computed failure probabilities with the experimental data to be attained. σ_{ow} and σ_{om} are obtained using the relevant Weibull or Muest models for failure probability computations.

For each fracture stress level, CERAM determines the failure probability contours and the total failure probability [1]. Figure 3 shows an example of CERAM output.

Tests on "billet"samples

The fracture strengths derived from the measured failure loads were plotted as a function of the cumulative failure probabilities. For the 3 point-bending specimens, the strengths were determined from the stress at the fracture origin. Linearity of the Log-Log plot of fracture distributions suggests that single flaw populations dictated failure in each set of strength data (Figure 4).

Fractographic observations of fracture origins indicated that failures were dictated by inclusions rich in silicon and by voids (Figure 5) which were predominantly located within the interior of specimens, except in the disks (Table 3). A single volume population dominated in the 3 point-bending specimens. In contrast two populations of surface and volume located flaws influenced the failure of the 4 point-bending and the biaxial flexure specimens (Table 3).

Table 4 shows that all the specimens exhibited comparable shape and scale parameters, indicating that the surface and the volume located flaws exerted similar influence upon failure. The 4 point and disk data were separated into surface and volume populations, and analysis was made on these 4 populations and then compared to the single 3 point loading and to the initial distributions. Both flaw populations were separated using the censored data method developed by Johnson [10]. This method provides an approximate initial estimate of the statistical parameters pertinent to both populations. The failure results for the 4 point-bending bars and the disks were reordered to conform with the 2 dominant populations. Comparison of the reranked populations with the single volume population observed on the 3 point-bending specimens reveals that the surface-dominated populations exhibit shape parameters comparable to the volume dominated populations ($m \sim 9$ see Table 4). These parameters are similar to those obtained when considering all the data. Moreover, the surface and volume strength data were randomly distributed. These results indicate that both populations exerted the same influence upon fracture.

Table 3. Fractions of fracture origins observed on the billet samples

	Surface %	Volume %	Inclusions %	Pores %
3 point-bending	3	97	47	53
4 point-bending	12	88	64	36
Disks	54	46		

In Table 4 it can be seen that the scale factors are dependent upon the model used. Moreover, the Weibull scale factors exhibit a significant dependence upon the testing conditions. In contrast, the Muest scale factors derived from the data determined on the 4 point-bending bars (uniaxial stress state) and from the disks (biaxial stress state) are very close. The discrepancy observed with the scale factors obtained with the 3 point-bending specimens may be attributed to shape parameter scatter. These results are in agreement with those which were reported in previous papers which showed that the Muest scale factors are stress state independent whereas the Weibull scale factors are significantly affected by the stress state [2, 3, 4, 6, 9].

The above data base was checked by using the statistical parameters determined from 3 point-bending conditions to predict the strength distributions of the 4 point-bending bars and the disks. Figure 6 shows that predictions and experimental strength data are in good agreement. Therefore these parameters were used for the subsequent reliability computations.

Analysis of reference piston pins

The strength distribution (Figure 7) obtained with the 24 "heart" samples is in excellent agreement with the distribution predicted from the materials data determined in 3 point-bending conditions on "billet" samples. This demonstrates the good reproducibility of this Si_3N_4 ceramic, which is an important prerequisite for achieving component reliability. Moreover, it is worth noting that the materials data pertinent to the "heart" sample strength distribution, were identical to those obtained for the 3 point-bending specimens (Table 5).

Fractographic examination showed that failure initiated either from the surface or the volume of the samples exclusively from inclusions rich in silicon (50% of failures) or iron (50% of failures). In contrast, only 2 failures from iron rich inclusions were detected in all the "billet" samples. However, based on the results shown by Figure 7, one may consider that this new population of iron rich inclusions has the same effect on failure as the silicon rich inclusions.

Table 4. Statistical parameters of the Si_3N_4 ceramic obtained on "billet" samples (volume analysis)

Testing conditions	Shape parameters (all data)	Scale parameters ($MPam^{3/m}$) Weibull σ_{ow}	Muest σ_{om}	Shape parameters (reranked data)
3 point-bending	9.8	97.0	84.1	volume : 9.8
4 point-bending	8.8	86.9	74.5	surface : 9.95 volume : 8.7
Disks	8.4	80.0	72.2	surface : 7.9 volume : 8.7

Conversely a significant strength decrease was observed on the "skin" samples when compared with the "heart" samples data (Figure 8). In the "skin" samples, 95% of the failures initiated from surface flaws induced by machining. Since the machining direction was parallel to the cross section of piston pins, the resulting flaws were perpendicular to the applied stresses, which significantly affected the strength of "skin" samples. The statistical parameters for the skin samples are much higher than for the billet samples (Table 5) as a result of a more homogeneous size distribution and decreased sizes for the machining flaws.

Analysis of engine tested piston pins

The strength distribution obtained with 32 "heart" samples from engine tested piston pins shows a slight strength increase when compared with the reference "heart" samples data (Figure 9) reflecting a change in the population of fracture inducing flaws. The heart of piston pins did not experience temperatures higher than 200°C on engine testing and since the flexural strength is constant up to 1000°C the strength differences should not be attributed to temperature.

The observation of failure origins shows that the failure inducing flaws consist exclusively of iron rich inclusions, which are predominantly located in the surface (67%) when compared with the reference "heart" samples (47%). The strengthening of "heart" samples may not be attributed to the presence of this third population of iron rich inclusions, since the analysis of the reference pins has shown in the previous section that iron rich inclusions and silicon rich inclusions had the same effect on failure. Moreover, there is no reason why this 3rd flaw population should become critical, since the piston pins for engine testing had been randomly selected. It seems therefore logical to attribute this strengthening effect to the increase of the fraction of surface controlled flaws. However, this point requires further investigation.

Table 5. Statistical parameters of the reference piston pins

	Shape parameter	Scale parameters Weibull	Muest
"Heart" samples (volume fracture)	9.78	97.0	$84.14\ MPa^{3/9.8}$
"Skin samples (surface fracture)	18.5	310	$298.3\ MPam^{2/18.5}$

In the "skin" samples from engine tested pins, failure occurred exclusively at the surface from machining defects. The strength distribution obtained with 64 specimens is identical to the distribution pertinent to the reference "skin" samples (Figure 10). However, surface roughness had been altered when compared with the reference specimens. Evidence of wear was shown by the pin parts which were in contact with the piston and with the connecting rod. Failure did not initiate preferentially within these surfaces. It seems that the initial machining defects remained predominant. Moreover, no metallic particles were trapped into the voids.

CONCLUSIONS

The reliability analysis approach used by Renault Co. for designing ceramic piston pins was presented. The work included material evaluation and determination of the data base required for reliability computations. Reliability computations were performed using the computer code CERAM which incorporates the Weibull and the Muest models. Both models predicted strengths which were in excellent agreement with experimental results. Moreover, according to logical expectation identical predictions were obtained with both models when the stress states were predominantly uniaxial.

This approach allowed important data on the performance of the considered silicon nitride to be obtained. It was shown that fracture was dictated by several populations including surface and volume located silicon rich inclusions and voids. These flaws exerted similar influence on strength.

These flaw populations were characterized by consistent statistical parameters when using different testing conditions. However, the Weibull scale factors exhibited a significant dependence upon the stress-state, whereas the Muest scale factors were not affected.

Silicon nitride exhibited a satisfactory reproducibility, since similar data were derived from specimens machined out of reference billets and specimens machined out of piston pins. Finally it appeared that engine testing for 250 hours did not affect the influence of the surface defects induced by machining, whereas a significant strengthening effect was observed. This strengthening effect was not satisfactorily elucidated. It may result from an increase in the fraction of surface located defects.

The next step of the design work will consist in calculating the reliability of the piston pins based upon the above data base, with the purpose of optimizing the design if the results demonstrate that reliability is not sufficient.

Acknowledgements

E. Amar thanks P. Bailloux (Renault) for his experimental help. J. Lamon gratefully acknowledges the assistance of D. Pherson (Battelle-Geneva) in numerical computations.

Geneva, 4 July, 1988
191-7111/JL/jd

REFERENCES

[1] J. Lamon: "Design of Ceramic Components: "CERAM", a Reliability Analysis Software Package".
Battelle Technical Report, April 1988.

[2] J. Lamon: "Ceramics Reliability: Statistical Analysis of Multiaxial Failure Using the Weibull Approach and the Multiaxial Elemental Strength Model". Gas Turbine and Aeroengine Congress, Amsterdam, The Netherlands - June 6-9-1988. ASME paper No. 88-GT-147.

[3] J. Lamon, D. Pherson: "Reliability analysis of C-Ring Specimens using the Multiaxial Elemental Strength Model".
90th Annual Meeting of the American Ceramic Society.
Cincinnati, OH, May 1-5-1988. Annual Meeting Abstracts. The Am. Ceram. Soc. Inc 176 (1988).

[4] J. Lamon, D. Pherson: "Reliability of brittle materials in Thermal Shock Heat Up". ibid [3].

[5] G. Simon, "Design of a Ceramic Piston Pin". Renault Technical Note No. 0812/ 86/185 - September 1986.

[6] J. Lamon: "Statistical Approaches to Failure for Ceramic Reliability Assessment".
J. Am. Ceram. Soc. 71 [2] 106-12 (1988).

[7] J. Lamon, A.G. Evans: "Statistical Analysis of Bending Strengths for Brittle Solids: A Multiaxial Fracture Problem".
J. Am. Ceram. Soc. 66 [3] 177-82 (1983).

[8] J. Lamon: "Statistical Analysis of Fracture of Silicon Nitride (RBSN) Using the Short Span Bending Technique".
Presented at the Gas Turbine Conference and Exhibit, Houston, Texas, March 18-21,1985. ASME paper No. 85-GT-151.

[9] J. Lamon: "Ceramics Reliability: The Short Span Bending Test for Evaluating Ceramic Multiaxial Failure".
Ceramic Materials and Components for Engines.
Proceedings of the 2nd International Symposium, Lübeck-Travemünde, FRG, April 14-17, 1986.
Edited by W. Bunk and H. Hausner, DKG.

[10] L.G. Johnson. "The Statistical Treatment of Fatigure Experiments". Elsevier Pub. Co., New York, 1964.

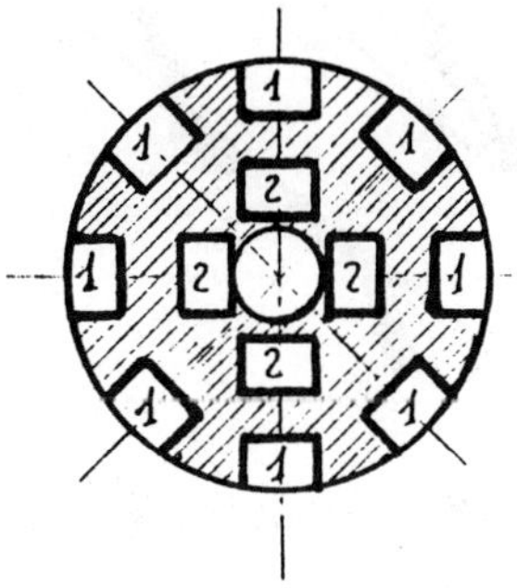

Figure 1. Sketch showing the location of the "skin" (1) and the "heart" samples (2) within the piston pins.

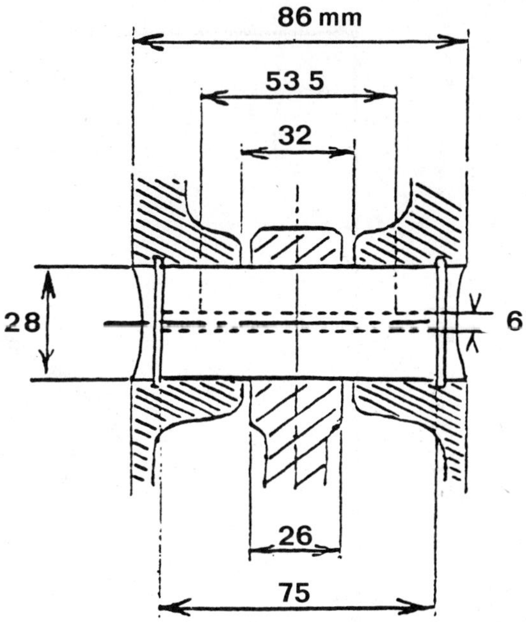

Figure 2. Sketch showing the piston pin loading on engine testing.

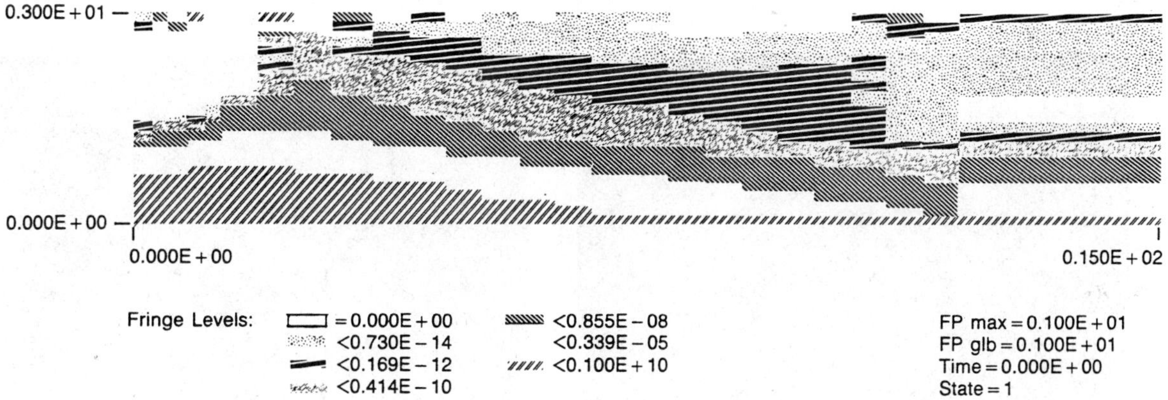

Figure 3. Example of probability contours computed for a disk using the CERAM program.

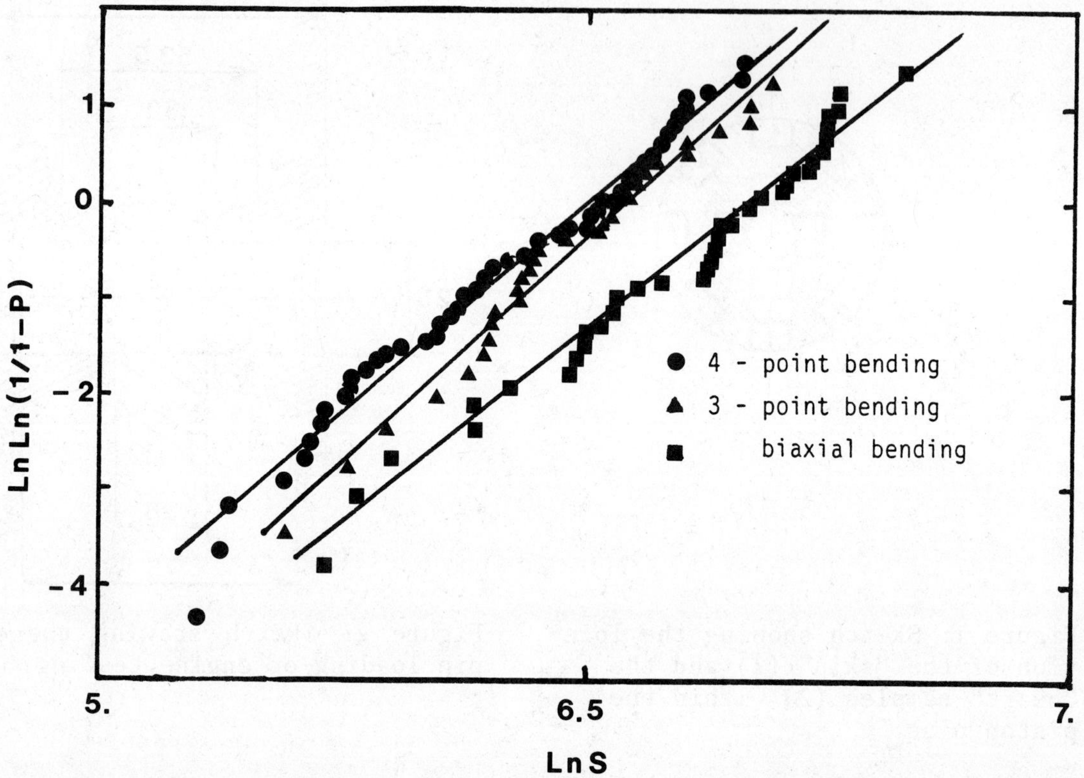

Figure 4. Strength distributions for the reference material.

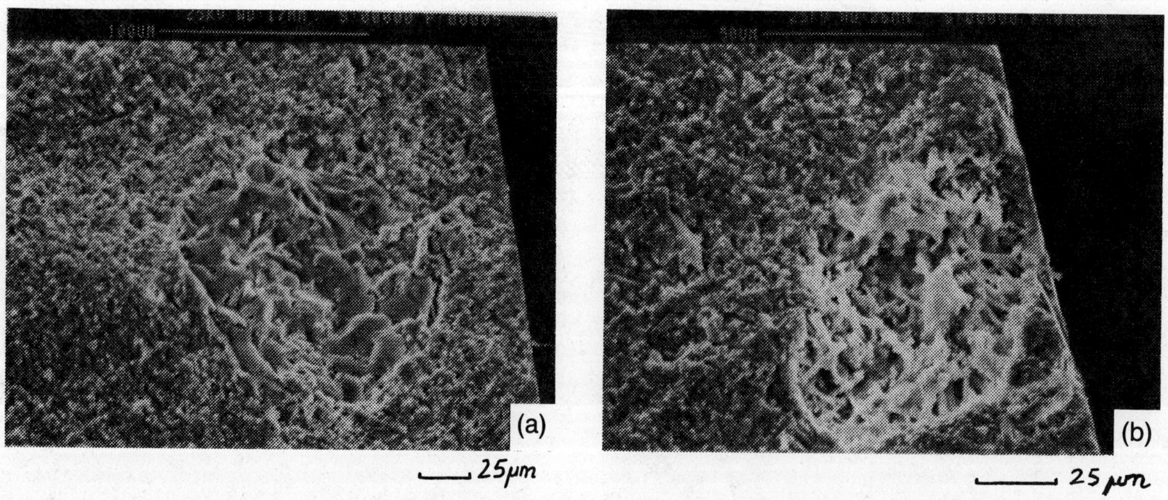

Figure 5. Scanning electron micrographs showing a large Si-rich inclusion (a) and a pore (b).

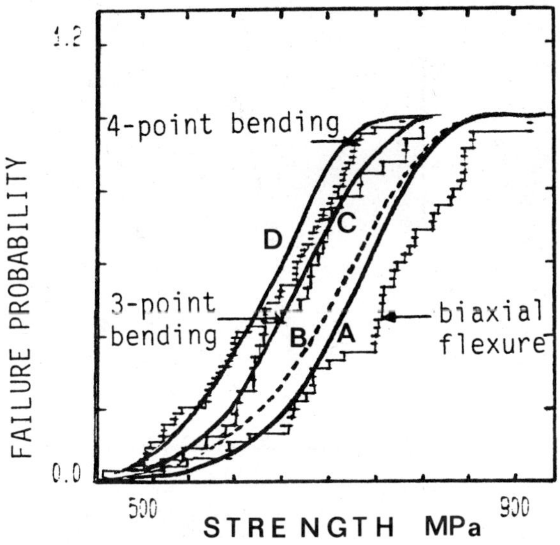

Figure 6. Strength distribution for the reference material. The lines (A) and (B) are the Weibull and Muest predictions of biaxial flexure strengths. Line (D) represents prediction of 4 point-bending data using both models. Line (C) is the base line.

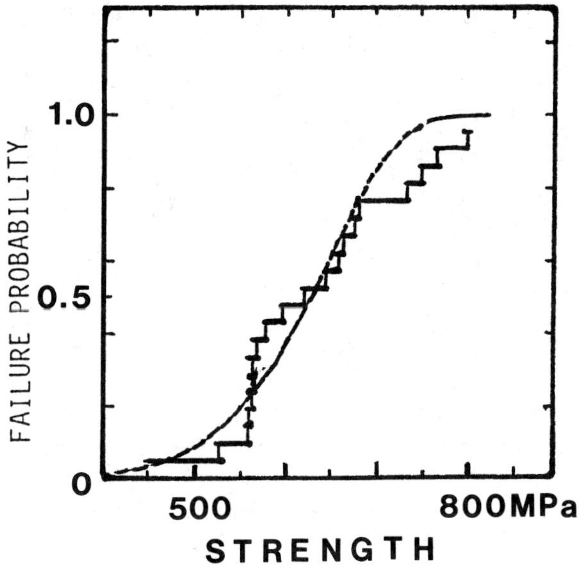

Figure 7. Failure strength distribution for the "heart" samples from reference piston pins. The solid line represents prediction using both models from 3 point-bending data.

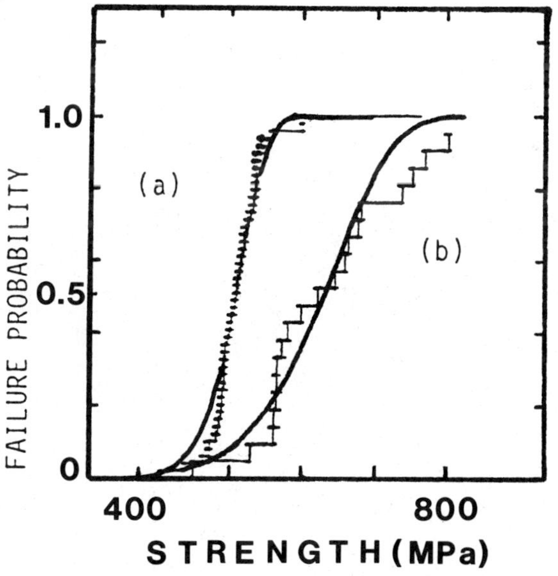

Figure 8. Strength distributions for the "skin" (a) and the "heart" (b) reference specimens. The solid lines represent predictions using both Weibull and Muest models based on 3 point-bending data for (b) and parameters given by Table 3 for (a).

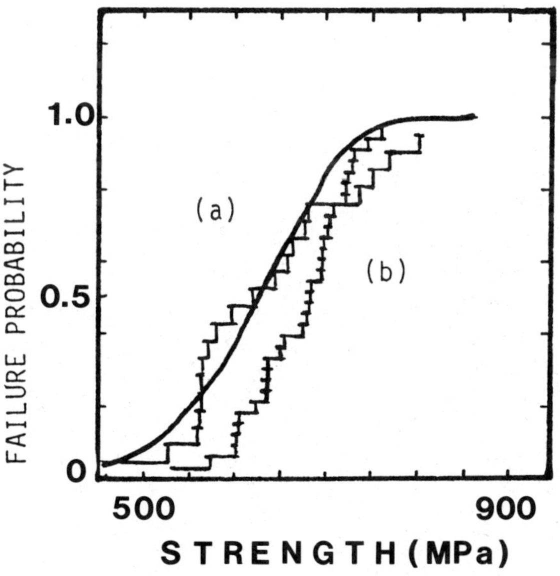

Figure 9. Strength distributions for for the "heart" reference (a) and "heart" engine tested (b) samples. The solid line represents prediction of the strengths for the engine tested samples using both models, based upon 3 point-bending data.

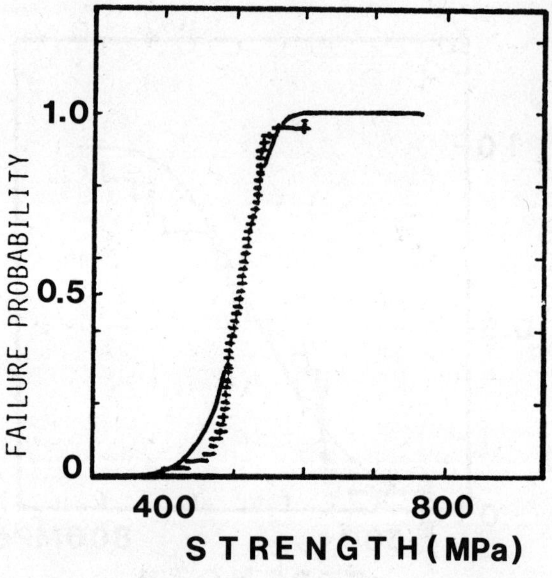

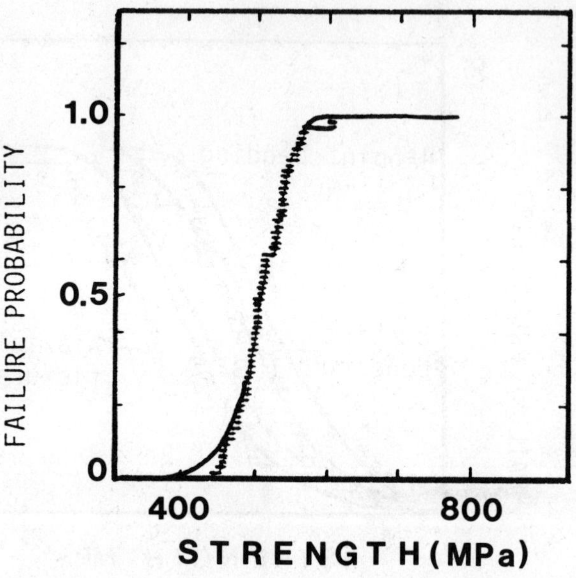

Figure 10. Strength distributions for the "skin" reference (a) and the "skin" engine tested (b) samples. The solid lines represent prediction using both Weibull and Muest models, based upon the parameters given by Table 3.

EMERGING CERAMIC COMPONENTS FOR AUTOMOTIVE GAS TURBINES

Harold E. Helms
Philip J. Haley
Allison Gas Turbine Division, General Motors Corporation
Indianapolis, IN 46227

INTRODUCTION

The automotive gas turbine (AGT) offers an alternative power plant to the automotive industry that has the potential for

- o improved fuel economy
- o alternate fuel capability
- o low emissions

while meeting life (reliability), performance, and cost requirements of the marketplace. The enabling technology to meet automotive requirements is that of emerging monolithic and/or composite ceramic components for hot flow-path parts in the AGT. This critical, high risk technology on AGT ceramic components is being evolved in joint Government/industry programs where prime AGT funding is being supplied by the U.S. Department of Energy and administered by the NASA Lewis Research Center, and new and improved ceramic materials work is administered by the Oak Ridge National Laboratories. In both the AGT programs and the ceramic materials programs, U.S. industry is providing significant cost share in joint programs with the Government and in separate activities that support and complement the joint programs.

Automotive gas turbines must offer the automobile industry significant improvements over currently produced power plants to justify their application in future vehicles. The key parameters that demand improvements are superior fuel economy with alternate fuel capability while meeting "clean air" (emission standards) and affordability demands of the marketplace. It has been conclusively established that metal AGT engines will not satisfy these criteria, especially the fuel economy requirement, because of limited turbine inlet temperature capability arising from chemical and strength limitations of existing superalloy metals. The only known current solution to this problem is to replace the hot gas flow-path components made of superalloy metals with components made of ceramics and ceramic composite materials. Achieving this technology breakthrough could permit turbine inlet temperature (TIT) to increase from 1038°C (1900°F) (current superalloy chemical and strength limit) to 1371°C (2500°F) (current target for ceramic components). This increase in TIT could improve

AGT fuel economy approximately 20% to 30% over that of current automotive powertrains. Benefits to worldwide transportation systems from such a new power plant are obvious and justify continued emphasis in many technical disciplines to achieve the potential of the AGT.

Since 1976, over 10,000 engine test hours and over 15,000 rig test hours have been achieved in developing ceramic components for AGT engines at the General Motors Corporation (see bibliography). These feasibility demonstrations using ceramic turbine section components prove the concept of the use of advanced ceramic materials in the AGT. The technological emergence of successful stationary components is easily identified. Also, the requirement of additional development for the most difficult rotating ceramic components (the turbine rotors) is easily identified. For this paper, it is expedient to classify ceramic components into four categories in the AGT engine development. These categories are as follows:

1. Simple stationary components
   - o Combustor body and rings
   - o Turbine vanes
   - o Flow-path shroud rings and couplings
   - o Couplings, seal rings
   - o Heat barrier rings
2. Complex stationary components
   - o Scrolls
3. Rotary heat exchangers (regenerators)
   - o Regenerator disks
   - o Bulkheads
4. Turbine rotors

The text of this paper discusses these four categories of AGT ceramic components.

CERAMIC COMPONENT EVOLUTION

The evolution of structural ceramic components for the AGT began with the requirements imposed for ceramic components to replace superalloy materials used in prior AGT engines. These requirements include strength at operating temperatures, dimensional control in cost effective production processes, and chemical and structural stability in the gas turbine environment. The development of iterative (between user and supplier) design and processing methodologies was identified as a key to successfully achieving usable ceramic components. Reduced to the reality of supplying ceramic components, these factors have defined the following actions to achieve an engineering basis for useful structural ceramic AGT components.

1. Application of new statistical design methodology.
2. Evolution of ceramic material standardized properties evaluation tests and the necessary design data base.
3. Development of new and improved cost effective ceramic component processing methods.
4. Development of new and improved ceramic powders/material chemistry and mixing for processing.
5. Development of new and improved inspection methods (NDE).

6. Identification of key coordination factors between user and supplier team members.
7. Education of management, designers, development teams, and the technical and political communities on ceramic component development.
8. Evolution of analytical tools for ceramic materials (reference: High Temperature Materials Laboratory at ORNL).
9. Proper selection of AGT component configurations to achieve successful ceramic components.
10. Demonstration in AGT engines of ceramic components proving the applicability of the evolving technologies with iteration to recognize and solve problems encountered.

This list defines a very large range of activities with development costs that will eventually total hundreds of millions of dollars. The structural ceramic product potential in future commercial and military business is many billions of dollars per year. Further, the technology being developed has applicability to all types of heat engines and to other applications outside power generation (e.g., tools, chemical processing, etc).

The U.S. Department of Energy, NASA, and the Department of Defense are Government agencies that helped to initiate advanced development programs on ceramic monolithic and ceramic composite components. The Office of Transportation Systems-Heat Engine Propulsion Division of the Office of Conservation and Renewable Energy, U.S. Department of Energy has provided continuous support for these projects since 1976 and must receive plaudits for their foresight, perseverance, and planning flexibility. U.S. commercial and military future products will reap large rewards from this investment. Progress achieved in this time period has been significant with applications of ceramic components beginning to satisfy the more simple marketplace demands in heat exchanger parts, valves, seals, cutting tools, bearings, turbocharger rotors, and improved armor plate in military vehicles. Enlarged use of these emerging materials is easy to foresee at this time.

AUTOMOTIVE GAS TURBINE DESIGN REQUIREMENTS

The automobile engine is a unique problem statement for a gas turbine because the average power usage is very small and transient power usage is very frequent when compared to either aircraft or industrial applications of gas turbine engines. The standard automotive driving cycle for the United States is 45% highway and 55% city driving. For the 160,934 kilometer (100,000 mile) life expectancy of an automobile (3,500 hr at an average speed of 46 kilometers per hour [28.6 miles per hour]), this corresponds to a time-averaged power requirement of 11 horsepower with no accessories running (14 hp with accessories) for an engine rated at 100 hp. This 11% average power compares to approximately 60% time-averaged hp for an aircraft engine or to 90% for an industrial gas turbine. Similarly, the automobile changes power setting an average of once every mile of use or 100,000 power changes in the automobile's lifetime. A typical aircraft will change power settings once every twenty minutes (10,500 cycles in 3,500 hr), and a typical industrial engine changes power settings less than once per hour (3,500 cycles in 3,500 hr).

Further, the automobile engine is required to accelerate from idle to high power in a very short time (less than 2 sec) to provide safe vehicle performance in traffic situations, while the aircraft engine can afford longer acceleration time (8 sec or more) and the industrial engine even longer (50 sec or more) since the rate of power increase is not as critical to either safety or required performance. These operating conditions are compared in Table I to show the "part power" and transient operating conditions of the AGT compared to turbine engines as commonly used in aircraft and industrial applications.

Table I. Comparison of Automotive, Aircraft, and Industrial Engine Operating Characteristics

	(1) Time-averaged rated hp usage (%)	(2) Approximate cycles of operation in 3500 hours	(3) Time for acceleration--idle to full power (sec)
Automotive	11	100,000	2 (or less)
Aircraft (Transport)	60+	10,500	8+
Industrial	90+	3,500	50+

Note: Exceptions to the average numbers shown in the table are recognized. These data represent broad spectrum numbers for generic turbine engine classifications.

Fuel efficiency in the AGT is extremely important. Studies that properly consider fuel efficiency combined with cost, reliability, and other performance issues always conclude that the AGT must include a heat exchanger (recovery of heat from exhaust gases) to meet or exceed current automotive spark ignition (SI) engine fuel economy. This heat recovery requirement dictates a very compact, efficient heat exchanger such as the ceramic rotary regenerator or a ceramic stationary recuperator. At the same time, the cycle pressure ratio usually optimizes at approximately 4.5-5.0 and the TIT optimizes at 1288°C to 1371°C (2350°F to 2500°F). The rotary ceramic regenerator has been chosen as the only high probability of success regeneration system to use in the AGT because no known successful ceramic stationary heat exchangers are within state-of-the-art technology today. In the AGT engine, the regenerator captures exhaust gas at 649°C to 982°C (1200°F to 1800°F), stores this heat in the heat transfer matrix, and transfers it as it rotates to the combustion system inlet. The exhaust gas from the engine is thus reduced in temperature to 200°F to 500°F over the operating power range, while the combustion system receives hot gas from the regenerator at 621°C to 954°C (1150°F to 1750°F) to which fuel is added to achieve the design TIT. Thus, the ceramic hot flow-path operates in a hostile thermal environment in the temperature range of 621°C to 1371°C (1150°F to 2500°F).

Table II summarizes these and other design requirements for the automotive gas turbine. Obviously, provision must be made for the containment of any high speed pieces in case of sudden structural failure, as is the case with current automotive powertrains. The overall costs of any alternate power plant must be competitive in terms of initial (sticker) vehicle price with current powertrain systems, thus implying production technologies for ceramic components that are high yield and cost effective. Maintaining powertrain life cycle costs at current levels additionally dictates stringent reliability/serviceability criteria for the AGT in general and the ceramic parts in particular.

Table II. Ceramic Automotive Gas Turbine Design Requirements

1. Design Life: 3500 hr; 100,000 thermal-speed change cycles; 33,600 low cycle fatigue (LCF) cycles (start/stop)

2. Temperature Range: 621°C to 1371°C (1150°F to 2500°F), idle to maximum power

3. Regenerator temperature range: 649°C to 982°C (1200°F to 1800°F), hot side and 93°C to 260°C (200°F to 500°F), cold side

4. Rotor containment at maximum burst speeds

5. Competitive initial cost and life cycle cost no greater than that of automotive powertrains currently on the market

6. Reliability and life comparable with powertrains currently on the market, including fewer than 0.25 major failures in 100,000 miles

*7. Emissions: must meet Federal Standards

8. Driveability suitable for safety and consumer considerations, including engine acceleration from idle to full power in less than two seconds

9. Noise and safety characteristics that meet Federal Standards

10. Ability to meet automotive maneuver and attitude requirements

*AGT engines can operate on various fuels, including diesel, jet, gasoline, alcohols, coal, etc. Emissions vary with fuel type.

The potential for meeting Federal emission standards with untreated exhaust using a variety of alternate fuels remains a key attraction of the automotive gas turbine. Using diesel fuel no. 2 as a design basis, the following target emission levels are achievable.

$$NO_x = 0.4,\ HC = 0.41,\ CO = 3.4 \text{ gm/mile}$$
$$\text{Particulates} = 0.2 \text{ gm/mile}$$

Other design requirements are that a gas turbine-powered vehicle meet all Federal noise and safety requirements as well as automotive maneuver and attitude requirements. These combined requirements (Table II) form a unique design envelope dramatically different from those of other gas turbine applications.

AGT COMPONENT CLASSIFICATION

Ceramic AGT engine components may be characterized in a relative sense by consideration of hot engine flow-path environmental parameters in conjunction with ceramic fabrication processes involved in their manufacture. Such environmental factors include applied stress, temperature exposure, and susceptibility to foreign object damage (FOD) from impact caused by entrained particles in the hot gas stream. Table III is constructed to classify ceramic AGT parts on a relative degree of application difficulty. Figure 1 shows the AGT 100 engine as a reference AGT to show the hot flow path utilizing ceramic components. Figure 2 shows a montage of AGT 100 ceramic components and the geometry of AGT 100 ceramic engine parts.

In Table III, components in categories I and III are considered at this time to be successful engineering developments. Component category II

Table III. Ceramic Component Application Parameters for AGT 100 Engine

	Max stress--ksi durability	Particle max impact velocity (fps) FOD	Max temperature--°C corrosion/erosion	Fabricability	Remarks
I. Simple stationary components					
Combustor rings, couplings, insulation rings, dome	5	50-400	1038 (1900°F)	Multiple choice	Current state of the art
Combustor body	15	300	1149 (2100°F)	Multiple choice	
Vanes	20	1500	1482 (2700°F)	Multiple choice	
II. Complex stationary components					
Scrolls	25	300	1149 (2100°F)	Slip casting-joining ceramic pieces-geometry complex	Requires joining development
III. Rotary heat exchangers					
Disk	0.2	150	1038 (1900°F)	Extrusion desired	Low coefficient of expansion and cost critical to success
Bulkhead	2	100	1038 (1900°F)	Slip cast OK	
IV. Turbine Rotors *	38	1670+	1204 (2200°F)	Thickness is crucial-flaws intolerable	Composites-toughness is critical

* Current monolithic rotors fail due to foreign objects generated from within engine (carbon from combustion process or from pieces of friable insulation). Combined centrifugal force field and thermal gradients from transient operation cause rotor to be an order of magnitude more difficult than any other component.

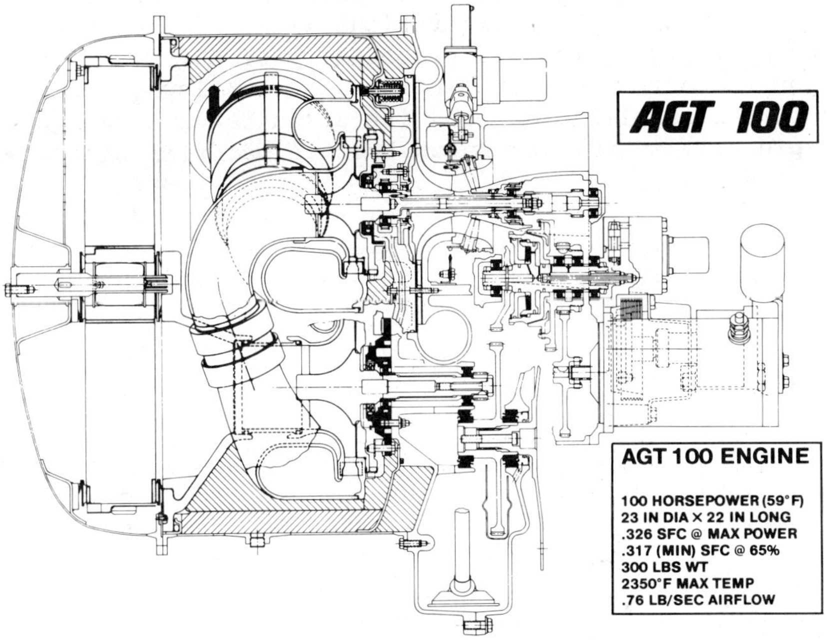

Figure 1. AGT 100 engine.

Figure 2. AGT 100 ceramic components.

(scrolls) is considered to require joining development, improved dimensional control, and improved flaw assessment to achieve structurally sound and stable components. Component category IV is considered to require improved impact resistance, improved microstructural uniformity, improved flaw assessment, improved dimensional control, and reduced process cycle times to achieve successful rotors. The production cost of all components is still unknown, but is considered to be resolvable with full scale development of processes such as those already achieved with spark plugs, oxygen sensors, turbocharger rotors, valves, and other ceramic components. Current overriding considerations with these structural ceramics is that design must provide a high probability of survival; processing must achieve required strength with minimum defects/flaws; testing must be realistic for the application; and iteration between design and application must properly adjust for known brittle material characteristics/properties.

In the following text each category of structural ceramic components is briefly discussed and key features of design or processing or test are identified.

Category I--Simple Stationary Components

Experience gained in over 10,000 hr of engine testing has shown that ceramic components of simple shape can perform extremely well. The shapes successfully used have evolved through probabilistic design methodology; iteration between design, processing, and rigs; and engine evaluations. Components that fit this category are as follows:

- Combustion liners, rings, and domes
- Turbine vanes
- Turbine vane support rings
- Turbine shroud rings, couplings, and seal rings
- Heat barrier rings

(See Figure 2 for AGT 100 components.)

In each of these components, special design features have been used to accommodate differential growth between the structural ceramic component and the mating component(s) (whether ceramic or metal). This allowance for differential growth, which recognizes differential rate of thermal response between flow-path parts because of thermal inertia and because of widely varying heat transfer to components during transient operating modes in the engine, has tended to reduce or eliminate induced bending and thermal stresses in the components. To illustrate this point the axial-flow turbine vane from Allison's CATE program (Ceramic Applications in Turbine Engine) is chosen as an example. Successful design has made the vane separate from inner and outer support flow-path rings (see Figures 3 and 4). The vane is immersed in the hot combustion effluent gas and the thin vane section responds to changing temperature very quickly with the full vane periphery surrounded by hot gas. At the same time inner and outer vane support rings are thicker than the vane (higher thermal inertia) and are exposed to changes in the hot gas stream flow path on only one side with basically smooth gas flow on the flow-path surface (thus, lower heat transfer input to the ring). The key design feature is that the vane is supported in pockets as opposed to normal metal vanes that are constrained by outer support walls. This arrangement results in the vane's capability to expand or contract with little or no induced stresses imposed by the vane support rings because of differential temperatures between the vanes and the vane support rings.

Similarly, the combustor body and the dome (Figure 5) at the fuel injection area are made as two pieces with diametral clearance to allow differential growth between the dome and the body. The joint between the body

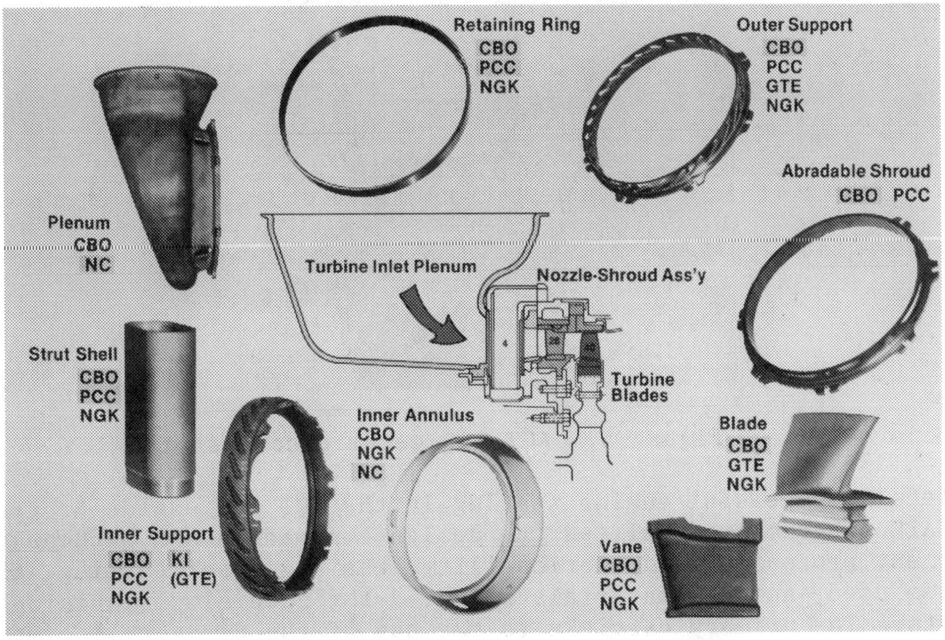

Figure 3. CATE ceramic components.

Figure 4. Assembled CATE ceramic vane set.

and the scroll is a spherical seat lined with a compliant layer material that accepts body or scroll tilting or distortion with no induced stress in the mating parts. The same design philosophy is used with the inner and outer vane support rings, which are attached to the metal support structure (radial keys accept differential growth). Also, the coupling rings, heat barrier rings, and seal rings are allowed to float rather than having severe constraint to differential growths. These design features minimize applied stresses and permit relatively thin and fabricable component shapes such as the vanes, the combustion body (a thin cylinder), the vane support rings (pockets for vanes--but relatively thin cylinders and the flow-path shroud rings [thin cylinders]), couplings (thin cylinders), and the piston ring seals.

Figure 5. AGT 100 ceramic combustor.

Completely successful engine testing in the AGT 100 engine program and in the CATE program has verified the design approaches and methodology, the component processing and fabricability, the rig proof testing, the NDE methods used, and the materials selected for such simple stationary components. The primary barriers to the use of the category I components in turbine engines are the educational and institutional barriers faced by any new material/design/process not proven in a production environment. Improved materials, processes, and designs as well as continued successful application experiences will further prove the established technology and enhance the eventual performance, reliability, and cost advantages of silicon carbide, silicon nitride, and zirconia simple stationary gas turbine components.

Category II--Complex Stationary Components, Scrolls

Figures 2 and 6 show the scrolls which deliver hot combustion gases from the combustor to the gasifier turbine or from the gasifier into the power turbine in the AGT 100 engine. These duct-like structures were designed to achieve uniform aerodynamic characteristics with minimal loss and minimal heat loss transition geometry. The fabrication of these shapes has been an extremely challenging technical task for ceramic suppliers. The necessity to make scrolls in separate pieces and then join these pieces into a single part has stimulated new technologies in the silicon carbide and silicon nitride materials. Several suppliers have faced this challenge of joining either three or five separate pieces together to form the required scroll geometry. While making the individual scroll parts has been successfully accomplished, the joining requirement has been a new and difficult task requiring either braze material development or a special second sintering operation. Only very limited success has been achieved to date in producing high strength continuous joints between mating parts of scrolls.

Three separate materials were chosen to make scrolls in the AGT 100 program. These were reaction bonded silicon carbide (The Norton Company); sintered alpha silicon carbide (Standard Oil Company), and silicon nitride (Kyocera International). Each company did produce complete scrolls. In

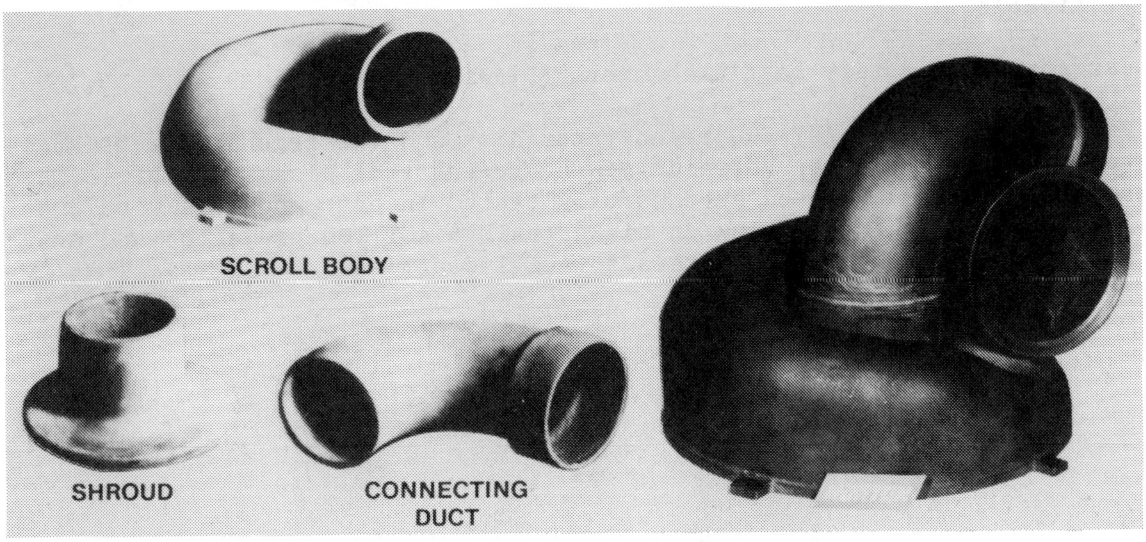

Figure 6. AGT 100 ceramic gasifier scroll assembly.

two cases, early scroll failures at the brazed joints were encountered in engine testing. In the third case, the Norton scroll produced survived more than 50 hr of engine testing before secondary damage from a rotor failure terminated the scroll evaluation test. The evident remaining challenge to producing future complex scroll geometries is that of developing substantially improved joining processes for separate monolithic ceramic parts into a single part that fulfills the required design function.

Problem solution to fabricating scrolls is believed to involve at least three development areas. These areas are:

1. Joining development--This involves either braze material/application evolution or other unique methods of achieving joints between separate ceramic parts.

2. Special tool development--This involves the holding or fixturing of individual parts to achieve proper dimensional control during the joining process.

3. Design/processing iteration--This involves the selection of separate part design to achieve optimum location of joints for producing the finished scroll for machining. This can affect assembly tooling, amount of joint area, and stress levels at joints.

A fourth consideration is the selection of starting materials as they may affect the ability to make joints. Fine grain powders may produce different joint wetting characteristics as compared to coarse grain starting powders when brazed joints are used.

Table III shows that for scrolls stress levels are moderate; air velocity is low relative to entrained particles for foreign object damage; average temperature is well within capabilities of silicon carbide and silicon nitride materials for 3500 hr life; and individual part fabricability is attainable.

Category III--Rotary Heat Exchangers (Disks and Bulkheads)

Figure 7 shows the AGT 100 regenerator disk and the regenerator bulkhead to which a seal on the hot side of the regenerator disk must mate. These two components have been successfully tested in past programs with over 10,000 hr of engine testing on disks (CATE & AGT 100 projects) and over 600 hr of testing on the bulkheads (AGT 100 project).

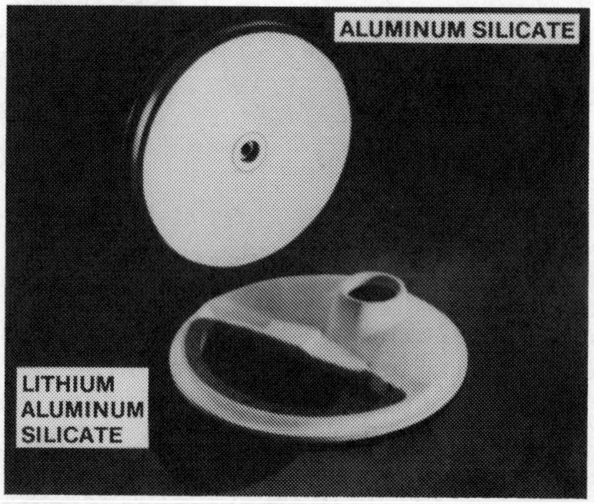

Figure 7. AGT 100 ceramic regenerator disk and bulkhead.

The regenerator disks successfully tested have been made by the Corning Glass Works from aluminum silicate (AS - aluminous keatite) material with a nearly zero coefficient of thermal expansion. (This is a remarkable, unique heat exchanger material). A unique spiral winding device produces the very compact heat exchanger surface (431 m^2/m^3 [1400 ft^2/ft^3]) with temperature capability to 1100°C (2012°F). Improvements in these disks can be made in the performance and reliability areas and it is believed that a major improvement can be made in the cost area. A single change, using an extrusion fabricating process instead of spiral winding, is believed to be capable of addressing all three of these areas.

Major progress in producing glass ceramic substrates for catalytic converters for SI engines by extrusion processing has been achieved in the past 5 to 10 years. A similar extrusion capability applied to regenerator disk processing would offer the following improvements:

- Uniform heat exchanger matrix geometry - performance--This feature offers lower pressure drop and higher, more uniform heat transfer in the matrix. These results yield nominally higher engine performance.

- Uniform heat exchanger matrix geometry - reliability--This characteristic eliminates nonbonded joints and nonuniform matrix wall thickness often found in the spirally wound regenerator disks. The few regenerator disk failures observed in the CATE and AGT 100 programs are believed to originate in thin walls produced in winding the disks or from nonbonded surfaces. This extruded component will thus improve reliability characteristics in disks.

- Order-of-magnitude faster processing--Spiral winding for regenerator disks involves variable tension equipment with green state glass ceramic material, which is a very slow, poor yield process method. Extrusion, by contrast, is a high production rate process that has the potential to reduce green state processing times by at least a factor of 10 and possibly by a factor of 30. This process development can show a major reduction in unit regenerator cost.

Thus, extruded regenerator disks have the potential to achieve improved performance, reliability, and cost targets for AGT engines. The current problem is to establish the processing capability to produce extruded regenerator disks.

The regenerator bulkhead shown in Figure 7 is a lithium-aluminum-silicate (LAS) material made by slip casting techniques. This part has proven itself in 600 hr of AGT 100 engine testing. While it is too early to project 3500 hr capability, no design, processing, or machining barriers are known at this time.

Category IV--Turbine Rotors

The AGT gasifier turbine rotor is an extremely challenging problem statement for ceramic materials. This fact is evidenced in the Table III conditions, which include the operating temperature environment (1288°C [2350°F] hot gas at blade tips to 649°C [1200°F] at metal shaft attachment--621°C [1150°F] temperature differential); the centrifugal force field induced by engine operation (43,000 rpm at idle to 85,600 rpm at full power); and transient engine operation (less than 2 sec from idle to full power--760°C [1400°F] TIT to 1288°C [2350°F] TIT). The combination of these conditions results in a component that is basically always operating in a tensile stress environment (ceramics are strong in compression and fail nearly always in tension); a component that sees extremely high mechanical and thermal stress gradients over short distances; a component that must be nearly free of critical defects; and a component that is very susceptible to FOD from particles entrained in the hot gas stream (because of high rotational turbine rotor tip speeds the impact energy from particles in the hot gas stream must either be absorbed or cause blade/rotor failure).

Although the turbine rotor tip speed is 503 m/sec (1650 ft/sec) (full power) in the AGT 100, Table III indicates that impact velocities at the blade tip can be 1670 ft/sec or higher, because of the possibility of a solid particle's trajectory due to "bouncing" causing its velocity to vectorially add to blade tip velocity to produce a relative velocity greater than the tip speed.

Such particle impacts add stress in the rotor as an additional tensile stress in a component already highly stressed in tension both centrifugally and thermally. The result of impact can be early failure in the rotor or the rotor blade (the rotor failure pulverizes the component and the blade failure causes loss of aerodynamic performance). Figure 8 shows an

Figure 8. AGT 100 ceramic gasifier turbine after chip ingestion.

example of a blade failure and Figure 9 shows an example of a rotor failure in the AGT 100 radial inflow turbine. AGT 100 engine testing suggests that the turbine rotors require the most continuing development because of their early failures in the engine from apparent FOD.

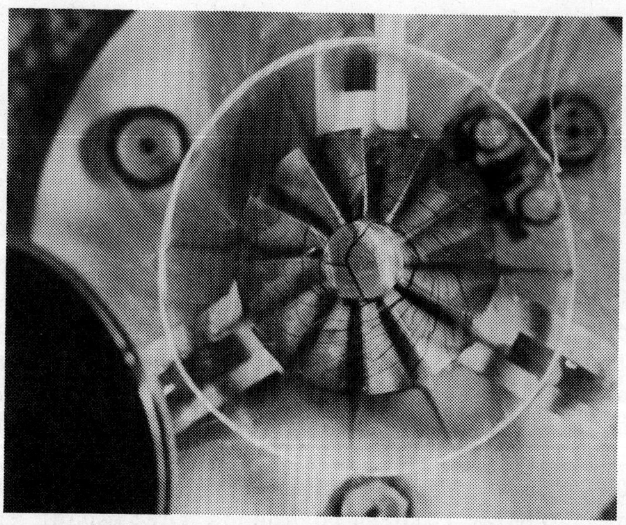

Figure 9. Ceramic gasifier rotor development.

The thickness of the AGT 100 rotor hub has made the rotor processing job very difficult and large variations in material microstructure have been observed between the rotor blade tips and the center portion of the hub. Also, many flaws have been observed in the center of the rotor hub, which suggests that improved processing must be developed before a reliable and reproducible radial inflow rotor can be achieved for the AGT. Rotor processing has shown large advances in the past five years from improvements in both ceramic materials' beginning powders and processing. Historically, powders have been suspended in polymers to make green component shapes for sintering. The "bake-out" of the polymer prior to sintering has been accomplished over very long cycle times to avoid excessive pore and flaw formations. New processes that either eliminate the polymer or use new or improved polymer chemistry are beginning to evolve.

Solving the rotor problems experienced in the AGT 100 engine demands carefully planned programs. Actions being taken and suggested include, but are not limited to, the following:

1. A lower tip speed rotor is being pursued in the DOE/NASA/GM ATTAP activity. This is the AGT5 rotor with a tip speed of 320 m/sec (1050 ft/sec) compared to the 503 m/sec (1650 ft/sec) AGT 100 rotor.

2. Design trade-offs on blade size (number of blades and blade thickness) and aerodynamics is being pursued.

3. Improved toughness ceramic materials for rotors are being evolved with the intent of achieving higher energy absorption before failure.

4. Improved combustion is being pursued to reduce the carbon formation in the off-design operating conditions in combustion systems. This hopefully will reduce or eliminate the prime source of foreign objects entering turbine rotors.

5. Improved engine insulation is being developed to eliminate a secondary source of foreign objects in the turbine section.

6. The turbine rotor stress field is being studied to optimize blade size and web size to maximize the margin between operating stress compared to failure stress at critically stressed areas. This includes the attachment area between the rotor and the metal shaft.

Current rotor problems are believed to relate to engine generated foreign objects, to nonoptimized blade designs and turbine wheel web and shaft attachment geometries, and to ceramic material toughness characteristics. Solutions for these problems demand continuing development in the integrated and iterative design/processing/materials/testing activity.

CERAMIC MATERIALS DEVELOPMENT

Major progress in materials used to produce ceramic components is being achieved in current industry, Government, and university activities. The concept of making heat engine and automotive gas turbine components from silicon carbide, silicon nitride, zirconia, glass ceramics, and other brittle materials was revolutionary to the ceramics industry and foreign to heat engine design as recently as 1975. Since that time the inevitable development of this concept has begun to be accepted and resources are being dedicated to evolving the capability to produce successful heat engine components made from ceramic materials. This has involved the creation of large groups of ceramic materials specialists who are starting to provide improved ingredients and processes to achieve specified material characteristics in components. This development also involves the design, acquisition, and use of a large amount of new and improved capital equipment. New and improved starting powders are becoming available and new and improved powder processing and component processing methods are evolving. New analytical tools for understanding the chemistry, microstructure, and materials properties are evolving as well as new and improved statistical design methodology. New and improved non-destructive inspection tools are beginning to evolve. Clever and innovative design geometries for heat engine components are beginning to evolve, while ceramic component test rigs and engines are being established that can evaluate ceramic components. This is an environment of high learning rates, new technology, and rapid decision making for those who integrate the emerging data base on ceramic materials to produce a product.

A simplified method for examining the rate of progress is to examine the strength at the temperature of ceramic materials (as measured by modulus of rupture) in 1980 and 1987. Figure 10 shows strengths versus temperature for silicon nitride and silicon carbide and the radical improvements that have been made in just seven years. It is believed that the potential exists to achieve similar progress in the next seven years and programs are in place to continue this creative and innovative breakthrough in ceramics technology. In the United States, the work being sponsored by DOE through the Advanced Turbine Technology Applications Project at NASA LeRC and the Oak Ridge National Laboratories (ORNL) Advanced Materials Development projects are examples of such outreach for new and improved ceramic monolithic and ceramic composite materials and components. Many other U.S. Government programs also add to this materials thrust. In Japan, similar programs sponsored by MITI (Ministry of International Trade and Industry) are being pursued for both materials and their application. In Germany and other European nations similar programs are in progress. This worldwide effort gives credibility to the expectation of continued materials improvement, the emergence of major improvements in ceramic component processing, and the emergence of widespread acceptance of heat engine ceramic components for useful, engineered products in the marketplace.

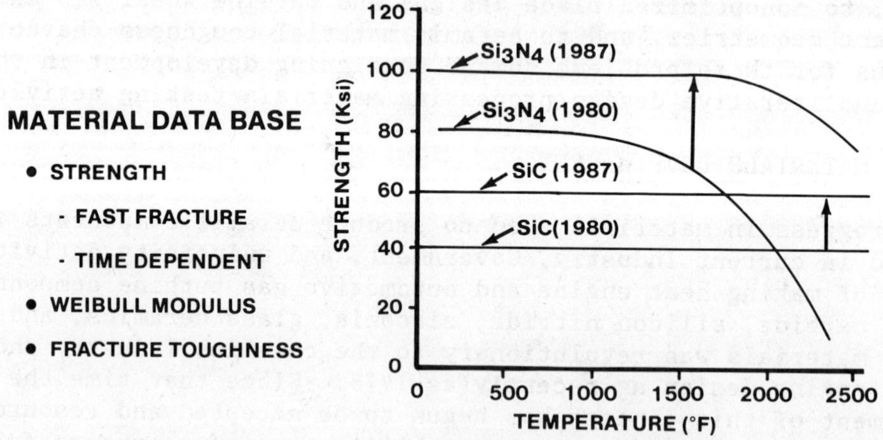

Figure 10. Ceramic materials development.

SUMMARY

1. Successful ceramic components for automotive gas turbines have been demonstrated for stationary simple shapes. These shapes include turbine vanes; turbine vane support rings; combustion liners, domes, and airflow control rings; and hot gas flow coupling rings, piston-type sealing rings, and thermal barrier rings.

2. Successful ceramic rotary heat exchangers and heat exchanger bulkheads for automotive gas turbines have been demonstrated. Extruded rotary regenerator disks can improve upon current state of the art with attendant potential to meet cost goals for the AGT.
3. Complex duct-like scrolls for hot gas flow have been demonstrated. Key among the characteristics requiring improvement is the joining of parts (ceramic to ceramic).
4. Turbine rotors for AGTs require a large amount of development to be successfully demonstrated. Areas requiring development are improved design of rotors; iteration between design and processing; improved ceramic material and processing capabilities (fewer defects, greater toughness, improved inspection, higher tensile strength); and improved engine design that reduces or eliminates foreign objects.
5. Ceramic materials development and AGT component application programs are now in place in the U.S., Japan, Germany, and other European countries which address solutions to remaining problems with rotors (vanes can be viewed as a part of rotor development), scrolls, and regenerators.
6. Progress in the ceramic components development area has been extremely significant and justifies major continuing development.
7. The spin-off technology now being applied to non-automotive gas turbine type heat engines and non-power plant applications is beginning to provide a base for future ceramic component development technology.

AGT's using ceramic components in the hot flow path are becoming a serious contender for future application in the automotive marketplace with the emergence of successful demonstrations of ceramic component technologies. Considerable development effort remains to demonstrate engine components, characterize materials, evolve improved processing techniques, and achieve promised performance with the AGT. However, the benefits from the use of ceramic components in terms of performance, reliability, and cost are so great that motivation to achieve successful AGT ceramics should produce early demonstrations of future power plants.

BIBLIOGRAPHY

Advanced Gas Turbine (AGT) Powertrain System for Automotive Applications, 1979-80, EDR 10327, DOE/NASA/0168-80/1, NASA CR-165178.

Advanced Gas Turbine (AGT) Powertrain System Development, 1980, EDR 10578, DOE/NASA/0168-81/2, NASA CR-165346.

Advanced Gas Turbine (AGT) Powertrain System Development, 1981a, EDR 10741, DOE/NASA/0168-81/3, NASA CR-165504.

Advanced Gas Turbine (AGT) Powertrain System Development, 1981b, EDR 10977, DOE/NASA/0168-82/4, NASA CR-167875.

Advanced Gas Turbine (AGT) Powertrain System Development, 1982a, EDR 11185, DOE/NASA/0168-5, NASA CR-168056.

Advanced Gas Turbine (AGT) Technology Project, 1982b, EDR 11443, DOE/NASA/0168-6, NASA CR-168325.

Advanced Gas Turbine (AGT) Technology Project, 1983a, EDR 11577, DOE/NASA/0168-7, NASA CR-174629.

Advanced Gas Turbine (AGT) Technology Project, 1983b, EDR 11682, DOE/NASA/0168-8, NASA CR-174798.

Advanced Gas Turbine (AGT) Technology Project, 1984, EDR 12070, DOE/NASA/0168-9, NASA CR-175018.

Advanced Gas Turbine (AGT) Technology Project, 1985, EDR 12344, DOE/NASA/0168-10, NASA CR-179484.

Advanced Gas Turbine (AGT) Technology Project, 1987, EDR 13295.

Proceedings of the Twentieth Automotive Technology Development Contractors' Coordination Meeting, Oct. 25-28, 1982, P-120, ISBN 0-89883-081-8.

Proceedings of the Twenty-Second Automotive Technology Development Contractors' Coordination Meeting, Oct. 29-Nov. 2, 1984, P-155, ISBN 0-89883-716-2.

Proceedings of the Twenty-Third Automotive Technology Development Contractors' Coordination Meeting, Oct. 21-24, 1985, P-165, ISBN 0-89883-734-0.

Proceedings of the Twenty-Fourth Automotive Technology Development Contractors' Coordination Meeting, Oct. 27-30, 1986, P-197, ISBN 0-89883-458-9.

Proceedings of the Twenty-Fifth Automotive Technology Development Contractors' Coordination Meeting, Oct. 26-29, 1987, P-209, ISBN 0-89883-469-4.

Advanced Gas Turbine (AGT) Technology Project, Final Report, 1988, EDR 13295, DOE/NASA/0168-11, NASA CR-182127.

Ceramic Applications in Turbine Engines (CATE), Final Report, 1984, EDR 11442, DOE/NASA/0017-6, NASA CR-174715.

INJECTION MOLDED SINTERED TURBOCHARGER ROTORS; A COMPARISON OF SSIC WITH SSN AND SRBSN

J. Greim, K. Hunold, K.A. Schwetz, A.Lipp
Elektroschmelzwerk Kempten GmbH
P.O. Box 1526
D - 8960 Kempten

ABSTRACT

This paper describes the state of development of silicon carbide (SSiC) and silicon nitride (SSN and SRBSN) for radial turbine rotors (K14 and K26) with integrated ceramic shaft and solid hub. The rotors differ only in the diameter of the blade rim (50 mm/K14; 65 mm/K26) and in the thickness and length of the shaft (98 mm/K14;117 mm/K26). The strength of the rotors was examined at room temperature (RT) and at 1000 °C.
The mean strength of the SiC rotors at RT was 404 MPa with a standard deviation s = 94 MPa, while the corresponding value at 1000 °C was 402 MPa with s = 74 MPa. The SSN rotors showed a significant higher strength at RT (673 MPa/ s = 56 MPa), which decreased to 407 MPa with an improved standard deviation of 29 MPa at 1000 °C. The SSN and SRBSN rotors survived tip speeds > 500 m/s at 1030 °C in a spin test machine. Since March 1988 ESK has been running a test car with a K26 SiC and a K26 SSN rotor.

INTRODUCTION

For almost one decade ESK has been involved in development and pilot plant production of ceramic rotors for turbocharged engines. Injection molding has evolved as the most economic way to manufacture monolithic SSiC as well as SSN and SRBSN turbocharger rotors. R & D was funded by the German Government (BMFT) and the Commission of the European Communities (EEC). There is a close R & D cooperation between the German turbocharger manufacturer, KKK on one side and the German automotive industry, Volkswagen and Daimler Benz on the other side.

The demands on a ceramic material for the turbowheel in charged Otto- and diesel engines are:

- high mechanical strength up to application temperatures superior to 1100 °C
- high oxidation and corrosion resistance
- thermal shock resistance and a
- low specific weight to reach a faster response of the turbocharger by reduction of the polar moment of inertia. FEM calculations and corresponding examinations showed, that the critical stress of a ceramic turbine wheel that occurs when starting from cold with subsequent acceleration

to full load, is so high, that nothing but the non-oxide ceramics silicon carbide and silicon nitride can stand it (1 - 3).

Not only the high-power Otto-engine, where exhaust gas temperatures of more than 1150 °C can occur, but also the diesel engine can profit from the application of non-oxide ceramic. Especially at frequent load cycles, typical for the running of passenger cars, the response of a metallic turbocharger is unsatisfactory up to now. If a non-oxide ceramic with only about 1/3 of the specific density of the metallic rotor is applied to the turbine rotor, the response time can be abridged distinctly. As less energy is set free in case of a ceramic rotor burst, the construction of the housing can have thinner walls. Thus an additional reduction of mass of the total exhaust gas turbocharger can be achieved (4).

Compared with the metallic rotors the ceramic rotors made of silicon carbide or silicone nitride stand out because of the reduction of mass and the better energy exploitation due to the possibility of increasing the exhaust gas temperature. Another advantage of a higher exhaust gas temperature, particularly in the diesel engine, is a reduction of the pollutant emission, i.e., especially smoke related particulates (5). The metallic super alloys used nowadays can be applied only at exhaust gas temperatures below 1100 °C. Aluminium-doped α-SiC, however, retains its high strength up to temperatures superior to 1400 °C. At temperatures below 1100 °C, however, Silicon nitride materials are distinguished by a higher strength and a better thermal shock resistance.

FABRICATION OF SSIC, SSN AND SRBSN ROTORS

For the pressureless sintering of SiC rotors (SSiC) a sinterable, aluminium (1 w/o) and carbon (1 w/o) doped α-SiC submicron powder is used (6). A α-Si_3N_4 submicron powder containing aluminium oxide and lanthanoxide or yttriumoxide (8 -10 w/o) as sintering aids is used for the gas pressure sintering under increased nitrogen pressure for SSN rotors.
The production of reaction bonded and subsequently post-densified SRBSN rotors is carried out with a Si-powder enriched with sintering additives and nitriding aids. The amount of sintering aids is less than required for SSN.
For each of these three starting powders an appropriate injection molding compound had to be developed. The organic binder content and the chemical composition of the plastification components must be compatible with the ceramic powder.

The injection molding mass represents a homogeneous dispersion of the ceramic powder in a system of binders of high and low molecular organic materials with a melting interval between 50 °C and 150 °C. The ceramic powder comprises about 80 w/o of the injection molding mixture. The system of binders consists of a mixture of thermoplastics, lubricants, resins, plasticizers, wetting and coupling agents and must be adapted to the injection molding as well as to the dewaxing process. The injection molding feed stock is compounded in a heated kneader at a temperature that corresponds to that of the binder component with the highest melting point to achieve the necessary homogeneity of the mass.

To judge the suitability of the injection molding compound, it is necessary to determine the length of the flow way in a spiral tool and the viscosity by means of a measuring extruder. To get reproducable results, injection molding is carried out in screw plasticating injection molding machines, controlled and adjusted by microprocessors. The optimum conditions can be determined by systematic variation of the injection parameters, temperature of mass and mold, injection velocity and value and time of secondary pressure. The injection cycle is controlled by measuring the inside pressure of the mold by means of a strap at the gate of the rotor (fig. 1).

Fig. 1: As Molded Turbine Rotor with the Strap for Measuring the Inside Pressure of the Mold

After injection molding the rotors are annealed in temperature controlled dewaxing furnaces. The removal of the organic binders takes place via a combination of various thermal reactions like fusing off, evaporation, sublimation, depolymerization and pyrolyzation and leads to a porous green blank, which shows, when suitable processed, neither shrinkage nor straining. The dewaxing process is the critical phase in the production of the rotors. When heating up at relatively low temperatures a critical stage must be passed, where open channels do not yet exist which would let escape the gaseous dewaxing products that have already built up. Thus the increased vapour pressure inside the injection molded part and the residual stresses produced by the enormous difference of wall thickness during injection molding may result in cracking.

NONDESTRUCTIVE INSPECTION BY X-RAY RADIOGRAPHY

To optimize the individual process steps during the development of complex ceramic components, in-process nondestructive inspection is necessary. X-ray microfocus radiography has been developed to a routine procedure (7 - 10) to detect cracks, inclusions and density variations during all phases of production (after injection molding, dewaxing, sintering and machining). On-line microfocus radiography, consisting of X-ray intensifier, camera and monitor, provides at 20fold magnification, a local resolution of about 8 µm, and a contrast resolution of about 3% of the wall thickness. Thus it is possible to detect inclusions and cracks inside the rotor with crack extensions below 50 µm within short time. By application of a 5-axes-manipulator, which allows examination of the object under slow rotation and recording of an existing crack in an optimum irradiation direction, even cracks smaller than 20 µm can be detected. It is possible to intensify the contrast via an image processing system, thus reaching nearly X-ray film quality, but saving the labour and expense of film processing. This image processing is executed off-line in about 30 sec. and can be recorded via thermo printer or video.

SINTERING OF SSIC, SSN AND SRBSN

The dewaxed SSiC rotors are pressureless sintered under vacuum or inert gas atmosphere in high temperature furnaces with graphite heaters at about 2100 °C. A sintered density of about 3.10 g/cm³ is reached, corresponding to about 97 % of the theoretical density (TD). During the sintering process a linear shrinkage of 16 - 17 % occurs, and tolerances of +/- 0.5 % can be kept. The sintering mechanism of SSiC is a pure solid state reaction without any evaporation or condensation processes involved and runs via grain boundary and lattice diffusion. The force to drive these diffusion processes is the high specific surface of the submicron-fine SiC powder. The reduction of the inner surface is equivalent to an increase of chemical links. Another increase in density of pressureless sintered SSiC to about 99 % TD can be achieved by hot isostatic pressing (HIP) at temperatures between 2000 and 2100 °C and a gas pressure of 2000 bar. Besides this increase in density, also strength, creep strength, oxidation resistance and wear resistance are improved by appropriate HIP parameters (11).

Densification of the dewaxed SSN rotors is achieved by liquid phase sintering, and runs via solution and precipitation in the glass phase. Each silicon nitride particle is covered by a thin (max. 10 nm) silicon oxide layer. Together with the oxide sintering aids a liquid phase is formed which enables a quick recrystallization of the initially equiaxed α-particles into a needle-like, mechanically strong and dense ß-silicon nitride. Upon cooling the liquid phase solidifies and strongly influences the material properties (12). Depending on the starting powder, the shrinkage is between 15 and 20 % and densities of 3.27 g/cm³ are obtained. For thermodynamic reasons it is appropriate to carry out the sintering process under increased nitrogen pressure to suppress the decomposition of silicon nitride at sintering temperatures between 1750 and 1950 °C.

At about 1900 °C silicon nitride would decompose completely under 1 bar nitrogen atmosphere.

The transformation from silicon to silicon nitride of the dewaxed SRBSN rotors takes place between 1300 and 1500 °C under nitrogen atmosphere. This process results without shrinkage in a porous silicon nitride blank of about 80 % TD, which is then completely densified, under similar sintering conditions as described for SSN. In this case the shrinkage is of the order of 7 - 8 %, which results in smaller tolerances than by SSN. Again, the ß-silicon nitride structure with a mechanically strong needle like structure is formed.

MACHINING OF SINTERED TURBOCHARGER ROTORS

After sintering the whole contour of the rotor is machined with plastic-bonded diamond disc in one step by clamping the rotor between two tips on a CNC plain grinding machine. The cylindrical tolerance of the two bearings on the shaft may not exceed 3 µm at a surface grade of R_Z = 1.6 µm. Fig. 2 shows the SSN K26 rotor in the different states of the production process.

Fig. 2: SSN Radial Turbine Rotor (d 65 mm)
From the Right to the Left: As Molded, as Sintered, as Machined

EXPERIMENTAL RESULTS

Component Strength

To evaluate the bending strength of the rotors specimens (2 x 4 x 34 mm) were cut from the shaft and the hub. These were lapped to a R_a-value of 0.3 µm. Subsequently the test bars were broken in a 4-point bending rig with an inner span of 15 mm and an outer span of 30 mm at a loading rate of 500 N/min at room temperature.
The strength of the SiC rotors was determined as 404 MPa with a standard deviation of 94 MPa and a Weibull modulus m = 4.4 (Fig. 3).

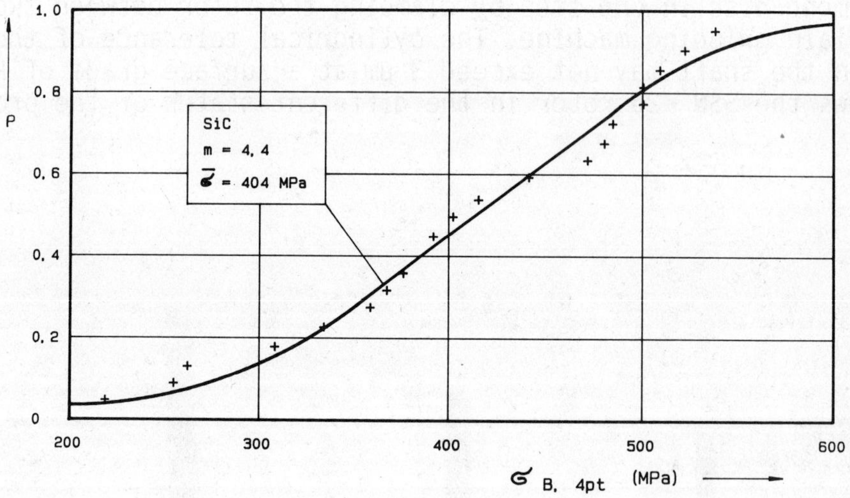

Fig. 3: Weibull Plot of Bending Strength (RT) for Test Bars Machined out of SiC Rotors

The SSN rotors showed a significant higher strength of 673 MPa with a standard deviation of 56 MPa and a Weibull modulus m = 11.0 (Fig. 4).

The high temperature strength at 1000 °C was measured at the same testing conditions described above. For SSN a decrease in strength to a mean value of 407 MPa was observed, while the standard deviation, however, improved to 29 MPa. The SiC samples remained at their strength level with a mean of 402 MPa and a standard deviation of 74 MPa.

For the SRBSN turbine material, only strength tests on injection molded test bars (3.5 x 4.5 x 45 mm) have been performed up to now.

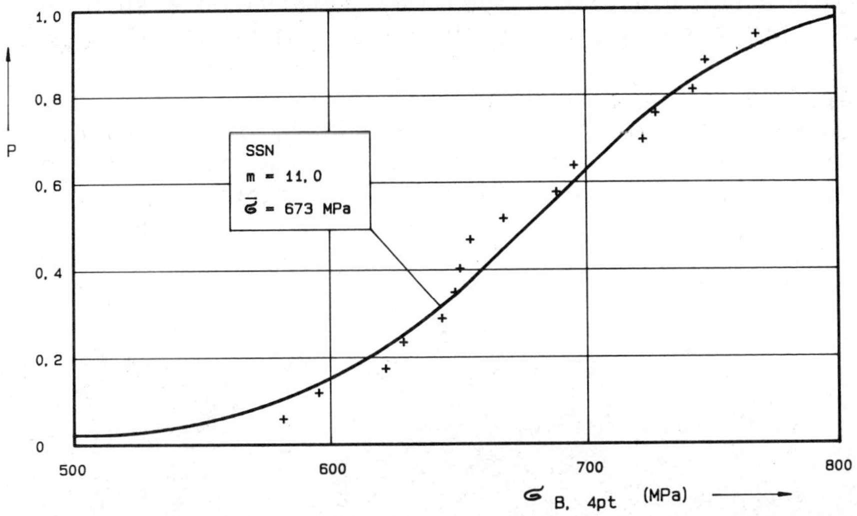

Fig. 4: Weibull Plot of Bending Strength (RT) for Test Bars Machined out of SSN Rotors

Table 1. Bending Strength of SRBSN Test Bars at Elevated Temperatures

Temperature	Bending Strength σ_B (MPa)	Standard Deviation S (MPa)
20 °C	698	39
900 °C	522	10
1100 °C	419	12
1200 °C	357	12
1300 °C	320	10

Test Condition	:	4-point bending test
inner span	:	20 mm
outer span	:	40 mm
stressing rate	:	500 N/min

Rotor Spin Test

For determination of the material strength under service conditions, the rotors have to undergo either a proof-test (at given velocity and temperature) or a bursttest (at given temperature) on a spin test machine (gif. 5). The spin test machine is driven by compressed air, and the hot gas is produced via fuel injection. Combustion temperatures up to 1300 °C and rotor tip speeds higher than 600 m/s can be achieved. The rotors are tested in conventional turbocharger housings. The metallic compressor impeller is fixed by means of a clamping device at the "cold" end of the shaft. Therefore the metal-ceramic-combination exerts only a pure compressive load on the ceramic shaft.

Fig. 5: Spin Test Machine for Burst- and Proof Tests

Up to now the following average and maximum burst speeds have been reached at 1030 °C:

Table 2. Average and Maximum Rotor Burst Speeds at 1030 °C

Rotor Type	Material	Average	Maximum
		Burst Speeds (m/s)	
K14	SSiC	422	458
(d 50 mm	SSN	449	518
	SRBSN	474	530
K26	SSiC	318	429
(d 65 mm)	SSN	431	518

These results demonstrate the superior potential of the silicon nitride materials compared to silicon carbide.

Automotive Tests

Since March 1988 ESK has been running a gasoline engined test car (Audi 200 turbo, Fig. 6) with a ceramic turbocharger rotor.

Fig. 6: ESK Test Car for Ceramic Turbocharger Rotors

In the first test cycle, the standard metallic turbocharger rotor was replaced by a K26 SSiC rotor. The test conditions were varied by a boost control up to a maximum boost pressure of 1 bar, which led to a maximum rotor speed of 96.000 rpm (326 m/s) and an exhaust gas temperature of 1030 °C.

After 10.000 on road kilometers the SSiC turbocharger rotor was replaced by a K26 SSN rotor in June 1988. During the following test cycles of the SSN rotor, the operating conditions of the metallic rotor were set up, such that the maximum rotor speed and the exhaust gas temperature increased to 110.000 rpm (374 m/s) and 1045 °C respectively. In the meantime the SSN rotor is running about 5000 kilometers without problems.

The driving tests confirmed the anticipated quicker response of the ceramic turbocharger compared with the heavy standard metallic rotor.

SUMMARY

During the development of monolithic silicon carbide and silicon nitride turbocharger rotors with integrated ceramic shaft ESK gained experience in cold and hot spin tests, evaluation of component strength and also in automotive tests with a gasoline engined car.
15.000 kilometers on-road tests with turbocharger rotors of both materials did not provide any problems and exhibited superior response characteristics compared to their metallic counterpart. At exhaust gas temperatures below 1050 °C, silicon nitride and SRBSN in particular feature a higher strength level and thus an increased reliability compared with silicon carbide. Nevertheless the automotive tests have demonstrated ample evidence that both silicon carbide and silicon nitride have met all requirements for a turbocharger material with the additional advantage of low specific weight.

ACKNOWLEDGEMENTS

Financial support by the Commission of the European Communities (EEC) and the German Government (BMFT) is gratefully acknowledged.

REFERENCES

1) M. Langer, J.E. Siebels, H. Heinrich, K.A. Schwetz: "SiC radial turbine rotor development - Material, design and fabrication influences" ASME.paper 86-GT-151, 31st. Intern. Gas Turbine Conf. and Exhibition, Düsseldorf, FRG, June 1986

2) K.A. Schwetz, W. Grellner, K. Hunold, A. Lipp, M. Langer: "HIP-treated sintered Silicon Carbide turbocharger rotors" Proc. 2nd intern. Symp.: Ceramic Mat. and Comp. for Engines, Lübeck-Travemünde, FRG (eds. W. Bunk and H.Hausner) Verlag DKG (1986), 1051 - 1062

3) P. Walzer, M. Langer: "Hochtemperaturkeramik für Verbrennungskraftmaschinen" MTZ 44 (1984), 6

4) R.P. Larsen: "Measuring the benefits and costs of ceramic turbochargers: The perspective from the driver's seat" SAE-paper 861129 (1986)

5) M.F. Lasker: "Ceramic turbine wheels for turbochargers" SAE-paper 861130 (1986)

6) K.A. Schwetz, F. Isemann and A. Lipp: "Injection molded sintered turbine components of aluminium doped alpha silicon carbide" in Ceramic Components for Engine (S.Somiya et al. Eds.), KTK Scientific Publ. Tokyo, Reidl, Dordrecht, Boston, Lancester (1984), 583 - 594

7) W. Grellner, P. Henze, K.A. Schwetz und A. Lipp: "Non-destructive Testing of SiC and Si_3N_4 Turbine Components" in: Proc. 3rd European Conf. on Nondestructive Testing, M.R. Caletti (Ed.), Vol. 5 (1984), 78 - 87

8) W. Grellner, P. Henze, K.A. Schwetz and A.Lipp: "Defect Evaluation in SiC and Si_3N_4: Destructive and Non-Destructive Tests" Proc. 2 nd Internat. Symp. Ceramic Materials and Components for Engine Lübeck-Travemünde, FRG (Eds. W. Bunk and H. Hausner) Verlag DKG (1986), 815 - 823

9) W. Grellner, K.A. Schwetz and A. Lipp: "NDT of Injection Molded Green and Sintered SiC Turbine Parts" ASME Paper No. 86-GT-272, 31st ASME Internat. Gas Turbine Conf. and Exh., Düsseldorf, June 1986

10) H. Kölker, P. Henze, K.A. Schwetz, A. Lipp: "X-Ray microfocus and dye penetrant tests of ceramic samples and turbocharger rotors" will be published in this conference proceedings.

11) K. Hunold "Sintering and hotisostatic postdensification of silicon carbide" Proc. Int. Conf. on HIP of materials, Antwerpen, April 1988

12) G. Ziegler, J. Heinrich, G. Wötting: "Review: Relationships between processing microstructure and properties of dense and reaction bonded silicon nitride" J. of Mat. Science 22 (1987) 3041 - 3086

DEVELOPMENT OF Si_3N_4 FOR GAS TURBINE APPLICATIONS

Antoni P. Taglialavore
Norton/TRW Ceramics
Northboro, MA 01532-1545

ABSTRACT

Activities associated with materials research are often concerned with composition and intrinsic properties as opposed to formability. To overcome this problem, Taguchi designed experiments were utilized to develop process parameters for a formable Si_3N_4 powder. Using these experiments, the transfer of a research-based material (prepared by isostatic pressing) to a component formable material (prepared by casting) was accomplished while preserving properties required for gas-turbine applications.

INTRODUCTION

Silicon nitride based ceramics are being researched and developed for high temperature structural applications such as gas turbine components. The use of these ceramics allows higher operating temperatures than current superalloys resulting in improved efficiency. Silicon nitride is an excellent candidate material for gas turbine use because of its strength-to-weight ratio, oxidation resistance, retained strength at elevated temperatures, and corrosion and thermal shock resistance.

Commercialization of silicon nitride for gas turbine applications requires highly reproducible processes by which researched material properties are preserved in a net-shape formed component. One objective of the Advanced Turbine Technology Applications Project (ATTAP),* is to develop and establish ceramic manufacturing processes capable of producing high quality and reliable gas turbine components. As a major subcontractor to Garrett Auxiliary Power Division,** Norton/TRW Ceramics has as its objective, the development and characterization of ceramic manufacturing processes capable of producing high quality and reliable Si_3N_4-based turbine rotors and stators for the AGT-101 gas turbine engine. Both components are being produced by net-shape forming processes. The rotor is being cast and the stator, injection molded. The material being employed is Norton/TRW's high temperature Si_3N_4, designated NT154. This material contains nominally 4 wt% Y_2O_3 and is prepared by hot isostatic pressing.

---

* United States Department of Energy Contract DEN3-335
** Allied-Signal Aerospace Co., Phoenix, AZ 85034

Following initial material research, it had properties as shown in Table 1 and Figure 1. These properties meet all the ATTAP material specifications and some of the ultimate project goals as shown in Table 2.

Table 1. NT154 Material Assessment and Qualification Properties

Physical Properties

Density(g/cc)	3.23 +/- 0.01
Elastic Modulus (GPa)	320
Shear Modulus (GPa)	126
Poisson's Ratio	0.273
Thermal Expansion Coefficient	$3.93 \times 10^{-6}/^{o}C$

Mechanical Properties

	Room Temperature	$1370^{o}C$
Modulus of Rupture (MPa)	920 +/- 153	531 +/- 46
Characteristic MOR (MPa)	985	552
Weibull Modulus	9.6 +/- 0.3	13.2 +/- 1.3
Fracture Toughness ($MPa \cdot m^{1/2}$)	3.6 +/- 0.3	N/A
Number of Samples	66	23

Thermal Properties

Temperature (^{o}C)	Specific Heat ($J/Kg \cdot {^{o}K}$)	Thermal Conductivity ($W/m \cdot {^{o}C}$)
25	661	37.6
500	1095	28.3
1000	1243	19.5
1400	1324	9.1

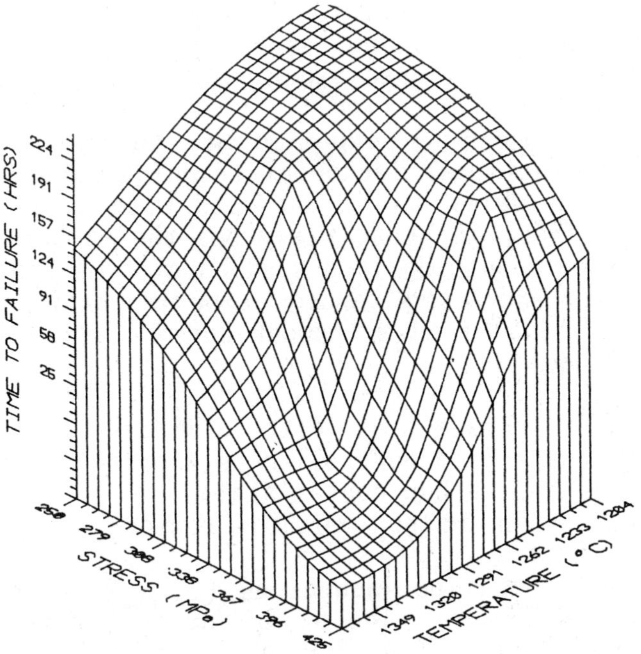

Figure 1. Stress Rupture Life Map of NT154

Table 2. ATTAP Si_3N_4 Material Specifications and Project Goals

Measured Property	Specification	Goal
Density (% of Theoretical)	99.5	99.5
Room Temp. Flexural Strength (MPa)	690	900
1370°C Flexural Strength (MPa)	517	690
Weibull Modulus (At All Temperatures)	8	18
Fracture Toughness ($MPa \cdot m^{1/2}$)	3.5	5
Flexural Stress Rupture Life (hrs)		
1260 °C - 350 MPa	150	500
1370 °C - 250 MPa	150	500

While the material property data of Table 1 and Figure 1 certainly make NT154 applicable for the ATTAP, the research-based material was quite difficult to net-shape form into components. With the ultimate goal of component commercialization, this situation was unacceptable. Therefore, the first issue to handle in the transfer of this material from the research to manufacturing stage was improving its formability while preserving mechanical properties. After this issue was resolved, reproducibility and further improvements in processing could be addressed. In this paper, experimentation on NT154 to make it net-shape formable while preserving mechanical properties is presented and discussed.

EXPERIMENTAL DESIGN

An efficient way of resolving issues of formability is through the use of designed experiments. Designed experiments are a general part of a systematic approach to experimentation.<1,2> In the experiment discussed in this paper, one of Taguchi's experimental designs was employed.<3,4,5> Taguchi designs are fractional factorial techniques which allow the effects of several process variables to be analyzed without performing every possible combination. Thus, it greatly reduces the number of trials needed to fully analyze a system resulting in a savings of time and money.

Formability was addressed through experimentation on powder processing. The broad objective of Si_3N_4 powder beneficiation, in addition to good compactibility, includes,

1. high specific surface area (4 to 25 m^2/g),
2. high alpha phase (and controlled beta phase) content,
3. low metallic and carbon contamination,
4. controlled oxygen and non-metallic content.

In this study, a modified Taguchi L4 ($2^3 \times 2^1$) experimental array was employed as shown in Table 3. The Si_3N_4 and Y_2O_3 blend represents various particle size fractions mixed together prior to milling. These two factors in concert with milling time allowed formability as well as the other objectives to be addressed concurrently. The outside factor was chosen to compare a net-shape forming process (casting) against a research process (isostatic pressing).

Table 3. Taguchi L4 (2^3 X 2^1) Powder Beneficiation Experiment Design

	GROUP 1 (INSIDE FACTORS)			GROUP 2 (OUTSIDE FACTOR)	
EXPERIMENT	A	B	C	D	
1	1	1	1	1	2
2	1	2	2	1	2
3	2	1	2	1	2
4	2	2	2	1	2

INSIDE FACTORS

A. Si_3N_4 Blend Set at Two Levels

B. Y_2O_3 Blend Set at Two Levels

C. Milling Time Set at Two Levels

OUTSIDE FACTOR

D. Forming Method
   1. CIP
   2. Cast

EXPERIMENTAL PROCEDURE

Powder blends were batched in five kilogram quantites in a hexagonal ball mill using reagent grade isopropyl alcohol and Si_3N_4 milling media. The slurry was passed through a 500 mesh sieve cloth before evaporating the alcohol. Processed powder was characterized for chemistry (Fe, C, O_2), surface area (BET) and particle size distribution (Sedigraph). The powder was formed by cold isostatic pressing (CIP) at 200 MPa and by casting using procedures developed for net-shape components. Samples were degassed and hot isostatically pressed using ASEA Cerama* proprietary encapsulation methods. Dense material was characterized for density, fracture toughness, flexural strength using four-point bend tests at room temperature and 1370 ^{o}C, and stress rupture life at 1370 ^{o}C and 300 MPa. Specimen geometry for all test bars was 3 X 4 X 50 mm, machined using MIL-STD-1942(MR). The inner/outer span of the test fixture was 20/40 mm, respectively, with a crosshead speed of 0.5 mm/min. Stress rupture testing was carried out in rigs from a design by Quinn.<6> Samples were heated to 1370 ^{o}C in air, stabilized, and then the load applied. Test were suspended at 200 hours if failure did not occur. Samples of broken room temperature test bars were polished and used to determine fracture toughness by indentation using the Anstis equation.<7>

---

* ASEA Cerama, Robertsfors, Sweden

RESULTS AND DISCUSSION

The results from each of the four experiments were analyzed in two ways. First, an Analysis of Variance (ANOVA) table was assembled for each set of response data. ANOVA ranks each of the response variables according to its percent contribution to variations in the processing factor under consideration. In this way, the importance of each processing factor may be evaluated as to its degree of influence on response variables. Second, the data was also analyzed by level averaging. This statistical technique predicts the results for all possible experimental combinations, whether or not they were actually performed in the fractional design. In this experiment, the data was simply level averaged for each of the two levels of each inside and outside factor. In short, ANOVA reveals the influence of a processing factor on the results, and level averaging allows selection between various levels of a particular processing factor.

ANOVA of the processed powder data is shown in Table 4. The Si_3N_4 blend influenced the particle size distribution most. Both surface area and oxygen content were influenced most by milling time. The Y_2O_3 blend did not significantly affect any of the response variables. In this way, future experiments need not consider Y_2O_3 blend as a processing factor. Conversely, error contributed most to both the carbon and iron contents. Either the proper choice of processing factors or experimental error itself disallowed determination of the influential processing factor. Future experimentation, therefore, will retain monitoring of these response variables.

Table 4. ANOVA of Processed Powder Data - Percent Contribution

Factor	Surface Area	Particle Size Distribution	Oxygen	Iron	Carbon
Si_3N_4 Blend	40	93	0	31	28
Y_2O_3 Blend	0	0	0	0	12
Milling Time	49	1	100	16	0
Error	11	6	0	53	60

ANOVA of dense material data is shown in Table 5. The Si_3N_4 blend contributed significantly to room temperature MOR. Once again, Y_2O_3 blend contributed little to the response variables. Milling time contributed significantly to stress rupture life in comparison to all other factors, except error which dominated. The outside factor, forming method, contributed significantly to density, fracture toughness, and 1370 ^oC MOR. Thus, the forming method plays a dominant role in material properties. These results accentuate the challenge in preserving research-based material properties (isostatically pressed) in a net-shape formed component (cast or injection molded).

Level averaging of the dense material data is shown in Table 6. The level averaging results allow selection of processing factor levels. For Si_3N_4 blend, level one is preferred due to superior 1370 ^oC MOR and stress rupture life. All other response variables were about equal. This blend

Table 5. ANOVA of Dense Material Data - Percent Contribution

Factor	Density	Fracture Toughness	Room Temperature MOR	1370°C MOR	Stress Rupture Life
Si_3N_4 Blend	2	0	34	19	1
Y_2O_3 Blend	0	0	11	6	1
Milling Time	0	4	2	0	32
Forming Method	82	79	6	63	10
Error	16	17	47	12	56

also had a broader particle size distribution and was quite formable by both casting and injection molding. For Y_2O_3 blend, level one is preferred because every response variable for this level is equal to or better than level two. Even though Y_2O_3 blend did not contribute significantly to the results as shown by ANOVA, the level averaging analysis allows selection of the better level. For milling time, level one is preferred based mainly on superior stress rupture life, with all other response variables being about equal. Milling time level one is lower than level two resulting in a coarser particle size distribution, lower surface area, and lower oxygen content. Finally, CIP forming is seen to be superior to casting by the level averaging analysis. The CIP material properties are equal to or better than those of the research-based material (Table 1 and Figure 1). Although cast material is inferior to CIP material, it still meets or exceeds the ATTAP component material specifications as shown in Table 3. Thus, component properties sufficient to meet the requirements of the application were preserved in a net-shape formable material.

CONCLUSIONS AND FUTURE DEVELOPMENT

1. Using Taguchi experiment design techniques, a net-shape formable NT154 powder was prepared which provides material properties adequate to meet ATTAP component material specifications.

2. Powder processing levels were set for pilot production. The pilot processed powder is easily formed by CIP, casting, and injection molding. Levels were set at,

   a. Si_3N_4 Blend Level One (broader particle size distribution),

   b. Y_2O_3 Blend Level One (better of the two by level averaging),

   c. Milling Time Level One (lower oxygen content and surface area).

3. Further investigations involving variations in powder processing and HIP conditions are being pursued. As an example, shown in Figure 2 is a three dimensional pictorial view of a Taguchi L9 (3^3 X 2^1) experimental array for powder processing which has been coupled to an L4 (2^3 X 2^1) array for HIP densification. This experiment is currently being performed at Norton/TRW Ceramics within the ATTAP. The interaction and optimization of both powder processing and densification are being pursued with the intent of reaching the Si_3N_4 component material goals of the ATTAP program (Table 3) in component form.

Table 6. Level Averaging of Dense Material Data

Factor	Density (g/cc)	Fracture Toughness (MPa x $m^{1/2}$)	Room Temperature MOR (MPa)	1370°C MOR (MPa)	Stress Rupture Life (Hrs)*
1. Si_3N_4 Blend					
- Level 1	3.22	3.7	940	563	134
- Level 2	3.22	3.7	1019	536	101
2. Y_2O_3 Blend					
- Level 1	3.22	3.7	1003	557	132
- Level 2	3.22	3.8	957	542	102
3. Milling Time					
- Level 1	3.22	3.8	992	550	168
- Level 2	3.22	3.6	968	549	66
4. Forming Method					
- CIP	3.23	4.1	1005	583	149
- Cast	3.22	3.5	958	534	86

* 1370 °C - 300 MPa

ACKNOWLEDGEMENT

The author gratefully achnowledges the financial support of the United States Department of Energy ATTAP subcontract from Garrett Auxiliary Power Division of Allied-Signal Aerospace Co., under prime contract to DOE (Contract No. DEN3-335). Special appreciation is given to Claude Brown, Jr. for his experimental design and data analysis, to Russell L. Yeckley for his scientific input, stress rupture testing, and the original research of NT154, and to Bryan J. McEntire for technical direction.

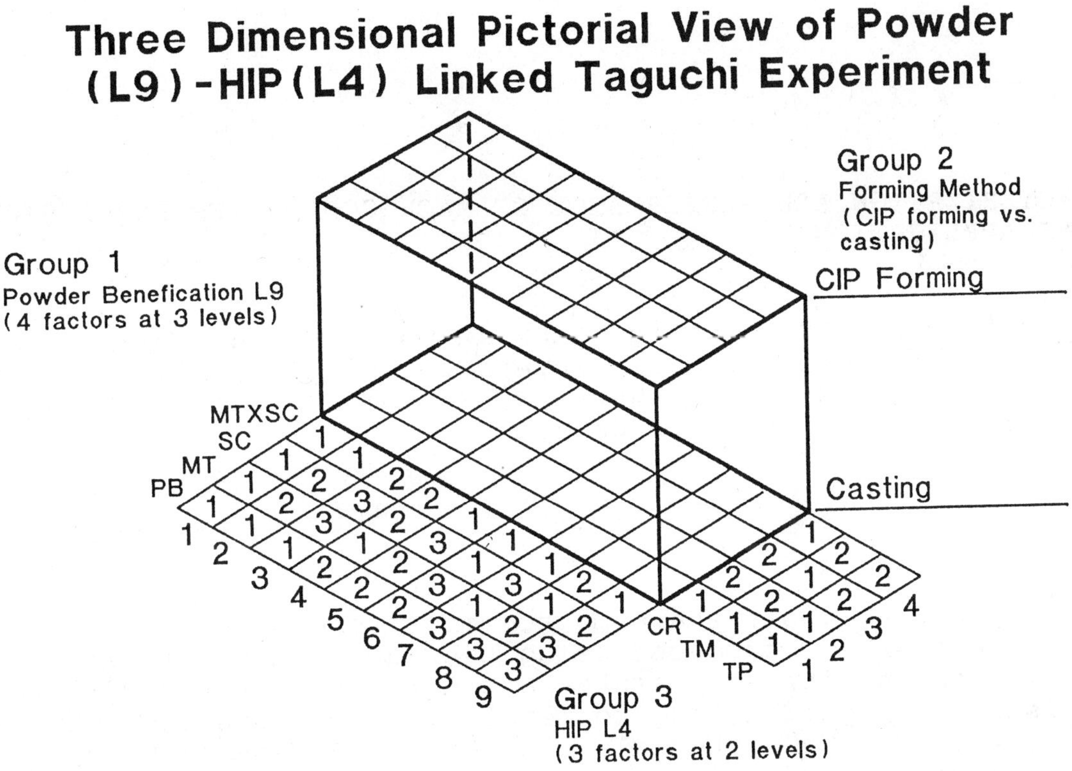

Figure 2. Three Dimensional Pictorial View of Powder L9 X HIP L4 Linked Taguchi Experiment

REFERENCES

1. D. C. Montgomery, Design and Analysis of Experiments, John Wiley & Sons, New York, NY, 1976.

2. O. L. Davies, The Design and Analysis of Industrial Experiments, Longman Group Limited, New York, NY, 1956.

3. G. Taguchi, System of Experimental Design Volumes I & II, Kraus International Publications, White Plains, NY, 1987.

4. P. J. Ross, "The Role of Taguchi Methods and Design of Experiments in QFD," Quality Progress, June 1988, pp 41-47.

5. B. Gunter, "A Perspective on the Taguchi Methods," Quality Progress, June 1987, pp 44-52.

6. G. D. Quinn, Guide to the Construction of a Simple 1500 ^{0}C Test Furnace - II, AMMRC TN 82-1, March 1982.

7. Anstis, Chantikol, Lawn and Marshall, "A Critical Evaluation of Indentation Techniques for Measuring Fracture Toughness: Indirect Crack Measurement," J. Am. Cer. Soc. (64) pp 533-538, Sept. 1981.

CHARACTERIZATION AND PERFORMANCE OF MACROCOMPOSITES FOR PISTON ENGINE APPLICATIONS

R. K. Dwivedi, J. R. Ramberg, Lanxide Corporation, Newark, Delaware; A. J. Gesing, Alcan International Ltd., Kingston, Ontario, Canada; and G. D. Webster, National Research Council of Canada, Ottawa, Ontario, Canada

ABSTRACT

Complex, near-net-shape macrocomposite components for piston engines were produced using Lanxide™ technologies. These macrocomposite components have ceramic matrix composite (CMC) surface layers bonded to metal matrix composite (MMC) substrates. The CMC materials consist of silicon carbide particles dispersed in an alumina matrix containing some residual metal. The MMC has an aluminum matrix with silicon carbide filler. Such a macrocomposite structure combines the attractive CMC features of high surface hardness, stiffness, and wear resistance with the higher toughness of the MMC substrate. The MMC renders the component resistant to catastrophic failure. To demonstrate near-net-shape component feasibility, macrocomposite camshafts for a Klockner-Humboldt-Deutz (KHD) F1L 511D single cylinder diesel engine were produced. Key properties of two CMC's used, composition C1 and composition C2, and the MMC core will be presented along with results of a short-term engine test directed toward evaluating the structural integrity of the camshaft. Additional engine application possibilities for macrocomposites such as piston pins are being explored.

INTRODUCTION

Within the last decade increasing attention has been focused on ceramics for applications in piston engines. Ceramics are particularly attractive materials for components subjected to wear and friction in the valve train because of their high wear resistance and low coefficient of friction compared to that of steel.[1-5] Despite the excellent wear and friction properties of ceramics, particularly those of silicon nitride, transformation toughened zirconia, and sialons, the introduction of ceramic components in valve train applications has been slow because of a number of their shortcomings,[5] which include:

- low fracture toughness compared to commercially used ferrous alloys
- high cost of production
- propensity to fail catastrophically
- low reliability

The macrocomposite concept was developed to overcome some of these shortcomings while utilizing ceramics' beneficial properties. Macrocomposite components with Lanxide™ ceramic matrix composite (CMC) surfaces bonded to Lanxide™ metal matrix composite (MMC) substrates offer the beneficial properties of CMCs, such as high wear resistance and low coefficient of friction, and yet are cost-effective, capable of being produced in very complex shapes, and capable of reducing the chances of catastrophic failure of the component.

This paper outlines the basic concept of macrocomposite components and demonstrates that Lanxide™ technologies can be used to produce such components in complex and near-net shapes. A prototype Klockner-Humboldt-Deutz (KHD) F1L 511D camshaft is used as a demonstration component. Preliminary property data on macrocomposite materials is briefly summarized. In addition, results of short-term engine testing of a macrocomposite camshaft are given.

MACROCOMPOSITE COMPONENTS

Macrocomposite components developed by Lanxide Corporation consist of a ceramic matrix composite (CMC) surface layer bonded to a metal matrix composite (MMC) substrate. A macrocomposite camshaft is schematically shown in Figure 1 to illustrate the concept. Both the CMC and the MMC can be tailored to yield a desired set of properties in the macrocomposite component. The macrocomposite concept is general and is applicable to a variety of component configurations and CMC/MMC combinations. However, in this paper the discussion will be limited to macrocomposite components consisting of CMC external surfaces, which are produced by the Lanxide™ directed oxidation method, bonded to internal MMC cores produced using particulate fillers. Figure 2 shows a cross section of a macrocomposite cylinder. In this and other illustrations, a coarse (24 grit) silicon carbide filler was used in the MMC; however, fine particulates have also been successfully used. Figure 3 shows some macrocomposite components produced using Lanxide™ technology. The piston pins are in the as-ground condition, while the camshaft is in the as-produced form.

The macrocomposite components offer many advantages including:

- light weight
- ease of processing compared to alternative ceramic components of the same degree of complexity
- low in-process dimensional changes
- cost-effective processing
- non-catastrophic failure
- high strength due to compressive residual stresses in the CMC layer provided by thermal expansion mismatch of the CMC and the MMC constituents

Prototype macrocomposite camshafts were produced to demonstrate the possible applicability of the macrocomposite concept for piston engine components.

PRODUCTION PROCESS

Lanxide™ ceramic composites are produced via a directed oxidation reaction of molten metals. In this process, the metal is oxidized to form a ceramic matrix reaction product containing some residual metal. This reaction product grows outward from the surface of the metal, and the metal is transported through the product to sustain the reaction at the growth front.[6,7] Filler materials placed in the path of the reaction can be incorporated in the matrix without being displaced. Filler materials preformed to shape are used to produce macrocomposites to near-net shape.

The production process for a prototype macrocomposite KHD F1L 511D camshaft is shown in Figure 4. A preform is slip cast using a silicon carbide slip, then dried and bisque fired to impart strength for subsequent handling. The preform is coated with an initiator on the inside and a barrier material on the outside. The initiator coating causes uniform directed oxidation, while the barrier material stops the directed oxidation process once the preform is fully infiltrated.

The coated preform is subjected to the proprietary macrocomposite formation process involving the directed oxidation of an aluminum alloy. The resulting macrocomposite camshaft is cooled, cleaned, inspected, ground, and re-inspected. Camshafts to be engine tested are shrink-fitted with a cast iron gear and inspected again.

Inspection of the camshafts at each stage generally consists of measuring their dimensions and scanning (by dye penetrant testing) for defects or cracks. Radiographic examination is carried out on as-produced camshafts to check for defects or cracks in the camshaft structure. After camshafts are ground and measured for dimensions, profiles of all three cams are checked and compared to ideal profiles (drawing specifications).

The thickness of the ceramic matrix composite surface can be varied widely; however, the CMC nominal thickness on the prototype camshaft was 5.4 mm.

To demonstrate scale-up capability and reproducibility of the process, more than 200 camshafts were produced. The reproducibility of surface details was excellent.

The dimensional changes during the camshaft production process were quite small compared to those of conventional ceramic processing methods. The overall dimensional changes (net expansion) during the process were less than 4% (3.7% in radial and 2.9% in axial dimensions), most of which occurred during bisque firing, before the macrocomposite formation step. The dimensional changes during the macrocomposite formation step were less than 1%. The process reproducibility was ±1.66% for diameter and ±1.36% for length (both expressed as percentage of average value of dimension). The surface finish, measured as average surface roughness, of the as-produced prototype camshafts was excellent (1.7 μm) and was comparable to that produced by investment casting of metals and slip casting of ceramics. Grinding with a 220 grit diamond wheel with resinoid bond reduced the average surface roughness to the range of 0.4 to 0.9 μm. Further reduction in average surface roughness could easily be achieved by grinding with a finer diamond wheel and lapping to provide average surface roughness down to 0.05 μm. The average surface roughness of the ground areas of the metal production model camshaft was 0.25 μm.

Figure 5 shows macrocomposite camshafts as-produced, as-ground, and fitted with a cast iron drive gear. The camshafts could be conveniently diamond ground without chipping. The macrocomposite camshafts were ground to the drawing specifications for a metal camshaft. Cam profiles were within 0.025 mm of the print profile.

PROPERTIES

Key properties of the ceramic matrix composite surface material are summarized in Table I. This table lists properties of compositions C1 and C2 ceramic matrix composites for comparison to cast iron, which is commonly used to produce camshafts. Unless otherwise noted, properties are at room temperature.

Composition C1 was the first Lanxide™ CMC surface layer composite developed and was used in most of the macrocomposite development program reported here, while composition C2, an improved material, is still under development for macrocomposite components. The primary difference between compositions C1 and C2 is in porosity--composition C1 has an undesirable level of up to 9.5 vol % porosity, whereas composition C2 has only about 1 vol % porosity. The C2 composite with less porosity has a significantly higher flexural strength and toughness. Figure 6 shows the plot of failure probability against fracture strength of composition C2 material. This plot was arrived at by analyzing the strength data of 30 samples tested in four-point bending using Weibull statistics. The Weibull modulus of the material was 11 while the characteristic strength was 159 MPa. All samples exhibited flexural strengths above 300 MPa while the maximum flexural strength measured was 481 MPa.

Figure 7 shows the microstructure of the composition C2 material, which consists of silicon carbide particles embedded in an alumina/aluminum matrix.

As Table I shows, most properties of compositions C1 and C2 compare favorably to those of cast iron: Lanxide™ composite materials show lower densities, higher Young's moduli, and higher scuffing resistances. The Lanxide™ composite materials also have lower toughnesses; however, since the composite material is bonded to metal matrix composites in a macrocomposite construction, the toughness difference may be less critical.

The properties of the metal matrix composite core material have not yet been fully characterized. To date particulates of various sizes have been used in MMC production; however, the process is flexible enough to allow the incorporation of fillers in other forms such as fibers, whiskers, and platelets. The properties of the MMC are critically dependent on the size, shape, and volume percent loading of the filler and on the composition and heat treatment of the matrix alloy. Depending on the particle size, these highly loaded metal matrix composites can have strengths as high as 400 MPa and fracture toughnesses of 12 to 16 MPa·m$^{1/2}$.

PERFORMANCE TESTING

The primary objective of the performance testing was to determine whether a macrocomposite camshaft could survive the environmental and loading conditions present in a piston engine.

The camshaft was tested at the Fuels and Lubricants Laboratory, National Research Council of Canada, Ottawa,

Ontario, Canada. The test engine had the following specifications:

Engine:	Klockner-Humboldt-Deutz
Model #:	F1L 511D
Combustion Cycle:	4 Stroke
Injector Type:	Direct Jerk Pump/Nozzle (Bosch)
Cooling Mode:	Air
Rated Power:	15 kW @ 2950 rpm
Rated Speed:	2950 rpm
Motoring DC Motor:	29.8 kW @ 1750 rpm

Oil temperature, injection pressure, load on two push rods and engine rpm were continuously monitored while the engine was running. The testing was carried out using Shell Rotella® T 15W 40 oil both under motored (driven by the DC motor) and fired (actual operation of the engine) conditions. The rpm of the engine was slowly increased in steps of 100 rpm and held at each level. Most of the testing was carried out at 1500 rpm (crankshaft rpm) and 85°C.

The macrocomposite camshaft was run against followers faced with wear inserts of Lanxide™ CMC composition F1, a material similar to composition C2 but with higher loading of silicon carbide (50-55%). The engine was run for approximately 100,000 camshaft cycles (200,000 crank shaft cycles) with most cycles accumulated at 1500 rpm (50% of the rated speed), 50% of the rated push rod load, and 30% of the rated engine torque.

The macrocomposite camshaft completed the 100,000 cycle test without showing any structural damage. Some wear damage was noted on the injector, intake and exhaust cams. None of the bearing surfaces showed damage. The wear noted was attributed to the undesirably high porosity of the composition C1 surface layer.

From this test it was concluded that the macrocomposite camshaft can be structurally sound for operation in engine loading conditions and environments. However, wear properties need to be further improved. The use of composition C2 and other improved low-porosity ceramic composites (e.g., composition F1) should significantly improve the wear performance.

CONCLUSIONS

The production and testing of prototype macrocomposite camshafts has demonstrated a number of features and potential advantages of the Lanxide™ macrocomposite concept, processing technology, and applicability to valve train components. Following is a discussion of significant conclusions.

1. The concept of macrocomposite components offers possibilities for combinations of the attractive features of CMCs (such as wear and erosion resistance, low coefficient of friction, high stiffness, high hardness, and high scuffing resistance) with features of MMCs (such as lower weight and ability to resist catastrophic failure). By varying materials and processing conditions, combinations of properties can be tailored to meet a variety of application requirements.

2. Macrocomposite components can be produced in large quantities in complex shapes with considerable ease.

3. The macrocomposite components produced in this study undergo <4% dimensional change during processing, of which <1% occurs during matrix growth. This is only a fraction of the dimensional change undergone by a component when processed by conventional ceramic processing techniques such as sintering. Furthermore, macrocomposite components can be produced with excellent surface finishes. The combination of low dimensional changes and good as-produced surface finish results in a unique near-net complicated shape making capability.

4. Properties of the ceramic matrix composite layer of macrocomposites can compare favorably with those of production valve train component materials such as cast iron. In addition, the metal matrix core can add to the component's strength (through compressive stresses in the bonding).

5. Macrocomposites can offer weight savings in valve train components. For example the macrocomposite camshaft is 40% of the weight of its steel or cast iron counterpart.

6. Preliminary tests indicate that macrocomposites can be structurally sound for operation in automobile engines, as short-term tests of the macrocomposite camshaft demonstrated.

ACKNOWLEDGMENTS

The authors express their appreciation for the assistance provided by David Friel, Robert Kantner, and George Altland of Lanxide Corporation in the production and characterization of macrocomposite camshafts; Brian Liko and Brian Mois of the Fuels and Lubricants Laboratory, National Research Council of Canada, Ottawa, Ontario, Canada for testing the macrocomposite camshaft; and Virgil Irick and Andrew Urquhart of Lanxide Corporation for reviewing this paper and making valuable suggestions. Thanks are also due to Jaime Sauerteig from Klockner-Humboldt-Deutz, Georgia, for providing part drawings and cam profiles.

REFERENCES

[1]A. F. McLean, Proc. 2nd Int. Symp. Ceram. Mat. and Components for Engines. W. Bunk and H. Hausner, eds., Verlag Deutsche Keramische Gesellschaft, Saarbrucken, Germany, 1986, 1023-34.

[2]H. Heinrich, M. Langer, and J. E. Siebels, "Experimental Results with Ceramic Components in Passenger Car Diesel-Engines," ibid., 1155-63.

[3]D. Fingerle, W. Gundel, and H. Olapinski, "Friction and Wear Reduction by Ceramic Components," ibid., 1191-1200.

[4]Y. Ogawa, M. Machida, N. Miyamura, K. Tashiro, and M. Sugano, "Ceramic Rocker Arm Inserts for Internal Combustion Engines," SAE Paper No. 860397, 1986.

[5]R. J. Lumby, P. Hodgson, N. E. Cother, and A. Szweda, "Syalon Ceramics for Advanced Engine Components," SAE Paper No. 850521, 1985.

[6]M. S. Newkirk, A. W. Urquhart, H. R. Zwicker, and E. Breval, "Formation of Lanxide™ Ceramic Materials," J. Mat. Res. **1** [1] 81-89 (1986).

[7]M. S. Newkirk, H. D. Lesher, D. R. White, C. R. Kennedy, A. W. Urquhart, and T. D. Claar, "Preparation of Lanxide™ Ceramic Matrix Composites: Matrix Formation by the Directed Oxidation of Molten Metals," Ceram. Eng. Sci. Proc., **8** [7-8] 879-85 (1987).

[8]W. J. Smothers, ed.; Ceramic Source '86®, Vol. 1. American Ceramic Society, OH, 1986, 351.

Table I: Properties of Lanxide™ Ceramic Matrix Composite Materials and Cast Iron

Property	Composition C1	Composition C2	Cast Iron
Density (g/mL)	3.1 ± 0.1	3.411 ± 0.007	7.09[a]
Phase Composition (vol %)	SiC = 34.2 Al_2O_3 = 46.1 Metal = 10.3 Porosity = 9.5	SiC = 35 Al_2O_3 = 54 Metal = 10 Porosity = 1	Tempered Martensite Metal Carbides[a]
Flexural Strength[b] (MPa)	287 ± 47 (25°C) 298 ± 47 (80°C)	396 ± 76	400-500[c] (Tensile)
Fracture Toughness K_{Ic}[d] ($MPa \cdot m^{1/2}$)	5.52 ± 0.58 (25°C) 8.58 ± 1.66 (80°C)	7.46 ± 1.11	37-45[e]
Young's Modulus (GPa)	231 ± 14	290 ± 7	83-211[e]
Poisson's Ratio	0.29	0.29 ± 0.05	0.17[e]
Coefficient of Thermal Expansion ($\times 10^{-6}$/°C)	7-8 (25-500°C)	7-8	8-19[e]
Thermal Conductivity (W/m·K)	70-90	70-90	46-52[e]
Coefficient of Dynamic Friction[f]	0.003-0.007[g]	0.003-0.007[g]	0.005-0.009[a]
Scuffing Load[f] (N) @ 1500 rpm	>1800[g]	>2220[g]	667[a]

[a] ASTM Grade M 4504 Malleable cast iron, surface hardened, properties measured at Lanxide
[b] Four-point bend strength using 3 x 6 x 50 mm bar samples
[c] ASTM Grade M 4504 Malleable cast iron, listed properties
[d] Chevron Notch four-point bend
[e] Reference 8
[f] Block-on-ring test, Exxon XD-3 Extra 15W 40 oil, lubricant, 75-85°C
[g] Against Lanxide™ composition F1, a material similar to composition C2 but with a higher proportion of silicon carbide

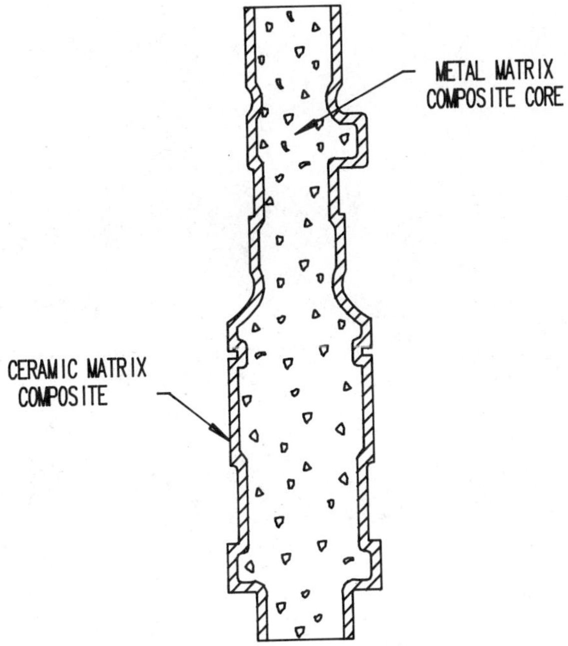

Figure 1: Schematic of a KHD F1L 511D prototype macrocomposite camshaft. The ceramic matrix composite layer is bonded to the metal matrix composite core.

Figure 2: Cross section of a macrocomposite cylinder showing the CMC surface layer and the MMC core.

Figure 3: Examples of macrocomposite components. The KHD F1L 511D camshaft shown is in the as-produced condition and piston pins are in the as-ground condition.

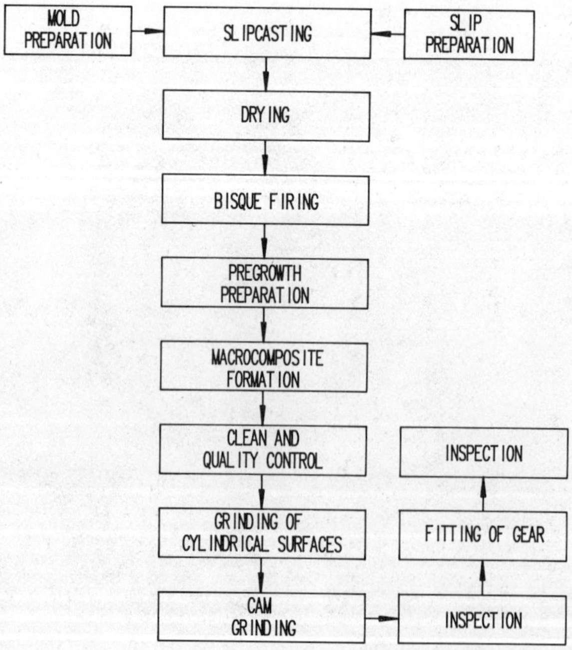

Figure 4: Flow diagram of the macrocomposite production process.

Figure 5: Prototype macrocomposite camshafts at three stages during processing. The part in the foreground is as-produced, the one in the center is as-ground, and the one in the back is fitted with a cast iron gear.

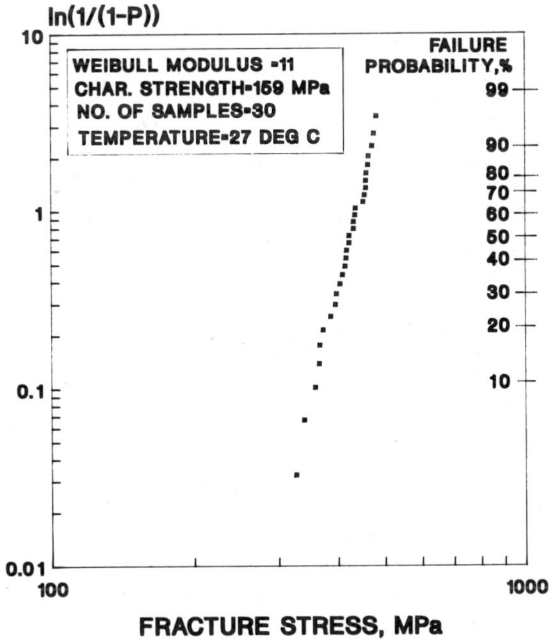

Figure 6: Plot of probability of failure versus fracture strength of composite composition C2 material.

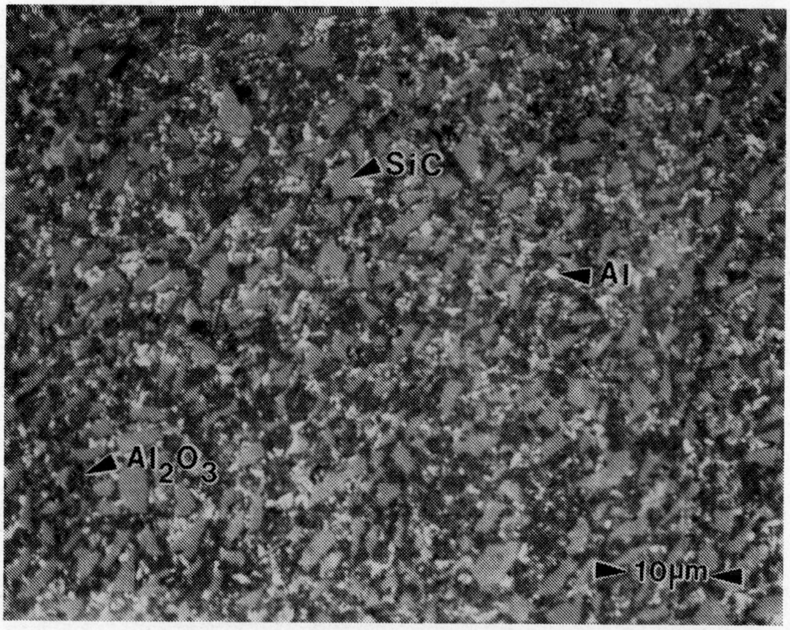

Figure 7: Microstructure of composition C2 composite. Silicon carbide particles are dispersed in an alumina/aluminum alloy matrix.

FABRICATION AND EVALUATION OF SILICON NITRIDE HEAT ENGINE COMPONENTS

G. Bandyopadhyay, F.M. Mahoney, D.J. Sordelet, K. French
GTE Laboratories Incorporated
Waltham, MA 02254

ABSTRACT

GTE Laboratories has successfully fabricated silicon nitride components for a variety of industrial and government heat engine programs. Evaluation of these parts and data from literature has revealed the difficulty in consistently attaining properties in large cross-section components equivalent to those demonstrated in small cross-section components. Recent efforts at GTE Laboratories have focused on improving large cross-section fabrication technology as well as the general reliability of heat engine ceramics. To achieve the necessary process and environmental controls, a Prototype Engineering Center (PEC) with clean-room powder processing operations has been established at GTE Labs. Significant improvements have been demonstrated in properties and performance of large cross-section ceramic components (e.g., turbocharger rotors) as a result of the improved process control, nondestructive evaluation, and the PEC facility.

1. INTRODUCTION

GTE Laboratories has been involved in fabricating silicon nitride heat engine components for nine years and has aided in demonstrating the feasibility of ceramic materials in high-stress, high-temperature applications through successful results in the CATE and AGT engine programs.[1,2] Despite this success, reliable fabrication of large cross-section (>12.7 mm) silicon nitride components such as radial and axial in-flow rotors remains a difficult challenge. A major concern is attaining properties in these components equivalent to those demonstrated in small cross-section parts such as CATE blades, AGT 100 vanes, and test bars.

Recent studies[3–5] have highlighted the differences in properties between small and large cross-section components fabricated for turbine applications. Larger cross-sections, in general, showed increased inhomogeneities which resulted in inferior and less reliable properties. Realizing these difficulties, GTE Labs' structural ceramics research has been focused on improving the large cross-section fabrication technology in addition to improving the reliability of heat engine ceramics.[6] A significant step towards this goal has been the establishment of a new Prototype Engineering Center (PEC) facility at GTE Labs with state-of-the-art ceramic powder processing and forming capabilities. The new facility incorporates numerous process and environmental improvements which have made a major impact on the fabrication capabilities and material properties of both small and large components. The improvements achieved in GTE Labs' component development programs as a result of the process improvements and controls are described in this paper.

In Section 2, results from the AGT 100 silicon nitride rotor fabrication program as performed by GTE Labs for Allison Gas Turbine Division (AGTD) of General Motors are described. These results represent properties of large cross-section, injection molded and hot-isostatically-pressed (HIPed) components obtained from a "laboratory process" (prior to the establishment of the PEC facility). The "laboratory process," as referred to in this paper, implies powder processing in limited quantities in conventional laboratory space. Although strict attention was given to protect the powder from foreign contamination, the desired level of cleanliness necessary for high-reliability structural ceramic components was difficult to achieve. Foreign contamination was sometimes detected in processed powder as well as in densified components. Batch-to-batch variations in the processed powder also limited the reproducibility of mechanical properties.

These limitations have been eliminated in the process now practiced in the PEC. The improvements incorporated in the "prototype process," as referred to in this paper, are summarized in Section 3. The impact of these improvements on properties of silicon nitride components have been significant; examples of property improvements in turbocharger rotors are described in Section 4.

2. AGT 100 SILICON NITRIDE ROTOR FABRICATION

A detailed description of GTE Labs' AGT 100 rotor fabrication effort has been presented previously.[4] Initial fabrication efforts started with GTE AY6 silicon nitride (Si_3N_4 + 6 w/o Y_2O_3 + 1.5 w/o Al_2O_3) and a prototype rotor configuration. Subsequently, the program shifted to GTE PY6 silicon nitride (Si_3N_4 + 6 w/o Y_2O_3) and engine configuration rotors. The change to PY6 was made to develop rotors with the improved high-temperature properties above 1200°C.

Components were fabricated by an injection molding and HIPing process.[4,7,8] A limited number of rotors were delivered to AGTD representing state-of-the-development components. The AY6 parts, delivered in 1984, generally contained visually observable surface blemishes such as fold lines and healed flaws. The PY6 rotors, delivered in 1987, represented GTE Labs' first attempt to form a large cross-section PY6 part. Although PY6 composition contained only 1.5 w/o less second-phase content than the AY6 composition, this small change caused surprisingly large differences in the overall fabrication process. The compounding, injection molding, binder removal, and densification cycles were all affected by this change, requiring optimization for the PY6 composition.

Table 1 summarizes the cold spin test data obtained by AGTD from the GTE Labs' AY6 and PY6 rotors. The developmental nature of these components and the limitations associated with the "laboratory process" caused significant variations in the rotor performance. Three out of five AY6 rotors achieved or surpassed the design speed (86,240 RPM), with an average burst speed of 106% for all five rotors. Five out of six PY6 rotors, on the other hand, achieved or surpassed the design speed. Two of the PY6 rotors were proof-tested to 102% and 104% of the design speed; the average burst speed of the remaining four rotors was 106%. Despite the variations in the rotor performance, the results successfully demonstrated the feasibility of fabricating complex large cross-section silicon nitride shapes by injection molding.

The spin test data, as shown in Table 1, indicated the need for process improvement and process control to improve the strength, reliability, and quality of the rotors. The room-temperature strength of test bars machined from these components also demonstrated significant scatter in strength and Weibull modulus (Table 2).* A key factor that contributed to the variations of the AY6 rotor material properties was the presence of undesirable alumina chips[8] due to the use of alumina milling media during powder processing. Silicon nitride media has been used in all subsequent powder processing. Additional process improvements and controls, such as the use of NDE and improved equipment to minimize the incorporation of

*All strength data reported in this paper were obtained in four-point bend tests in a SiC fixture with inner and outer loading points of 10.16 and 22.86 mm, respectively. The test bar size was 25.4 x 2.54 x 1.27 mm.

Table 1. Spin Test Data From GTE Silicon Nitride AGT 100 Rotors

Rotor Configuration	Si_3N_4 Composition	Fabrication Date	Density (% Theo.**)	Spin Speed* (RPM)	Test Type
Prototype Configuration	AY6	1984	98.8	110,800	Burst
			98.2	97,400	Burst
			98.8	92,600	Burst
			98.8	85,900	Burst
			98.8	68,500	Burst
Engine Configuration	PY6	1987	99.4	88,000	Proof
			99.4	90,000	Proof
			99.4	100,900	Burst
			99.7	90,700	Burst
			99.4	101,000	Burst
			99.6	74,000	Burst

* Design speed = 86,240 RPM
** Theoretical density = 3.262 g/cc

Table 2. Room-Temperature Strength of Test Bars Machined From AGT 100 Rotors

Rotor Composition	Rotor Number	Fabrication Date	Average Strength (MPa)	Weibull Modulus	Number of Samples
AY6	549	1984	787	7.5	27
	615	1984	842	22.5	9
PY6	A	1986	552	4.0	8
	B	1986	690	7.4	20
	C	1987	842	15.0	14

iron inclusions and the use of improved equipment and process optimization in the binder removal and HIPing procedures, have been implemented as they became available. The PY6 rotor properties demonstrate the improvement trend achieved as a result of the continuous focus of implementing process improvements. Figure 1 shows the strength data as a function of temperature for the three PY6 rotors, as well as the data from small cross-section parts (injection molded and HIPed test bars) fabricated in 1986. Note that the room-temperature strength of Rotor C fabricated in 1987 exceeded the strength of the small cross-section bars; however, the high-temperature strength was lower than the small cross-section test bars.

The difficulty of achieving comparable properties in large cross-section components is not unexpected from Weibull theory. The larger volume of material in a component such as the AGT 100 rotor increases the probability of occurrence of critical defects and material homogeneities within the part; the increased number and size of the defects lowers the average fracture strength. In addition, the larger volume of material broadens the size distribution of critical defects, which increases the variability of fracture strength. The increased variability decreases the Weibull modulus and reliability of the material.

To improve the material strength and Weibull modulus of a silicon nitride composition (e.g., AY6 or PY6), stringent control of the incoming powder and the powder processing steps are required to eliminate undesirable airborne contamination which would act as a flaw precursor. Other improvements in component quality must come from shape forming, densification, and quality-control procedures. These are the basis for the "prototype process" as developed in the PEC for fabrication of silicon nitride components.

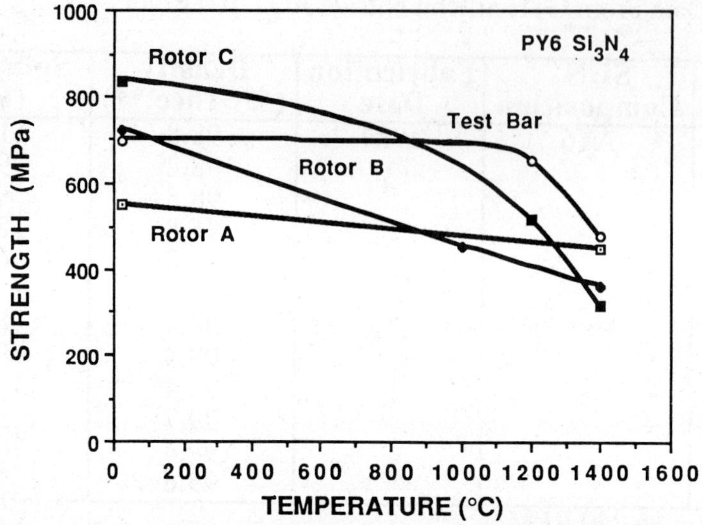

Figure 1. Material strength of test bars machined from GTE PY6 AGT 100 rotors. Data from injection molded and HIPed PY6 test bars are also shown.

3. DEVELOPMENT OF THE PROTOTYPE ENGINEERING CENTER

The PEC facility incorporates major changes related to new process improvements and the use of new equipment aimed at improving silicon nitride processing. One major improvement has been in the area of powder production and handling. A specially designed clean-room facility has been constructed solely for processing silicon nitride powders. Other improvements include the development of a proprietary binder removal process, increased sintering and HIPing capabilities, and stringent quality-control measures for process control and development.

Quality control includes extensive microfocus x-ray and visual inspection of components at various stages of processing. GTE Labs has developed the microfocus x-ray system for evaluation of a wide range of components and established flaw detection limits in silicon nitride ceramics.[9] The effectiveness of this inspection method has already been demonstrated for small cross-section parts such as test bars. The equipment and the technique are now being used extensively for large and complex shapes. The visual inspection of components generally includes examination at 20x magnification. Dye penetrant methods are used where appropriate. The primary surface defects are pits, pinholes, surface cracks, and fold lines from the injection molding process.

The effect of the prototype process on component properties has been quite significant. Examples of improvements achieved in a ceramic turbocharger rotor development program at GTE Labs are given in Section 4. The results are typical of those observed in several other shapes and sizes.

4. PROPERTIES IMPROVEMENT IN SILICON NITRIDE TURBOCHARGER ROTORS: A CASE STUDY

GTE Labs is collaborating with Volkswagen and Kuhnle, Kopp and Kausch (KKK) to develop ceramic turbocharger rotors for passenger cars. GTE Labs has supplied injection molded and HIPed AY6 silicon nitride rotors. The current rotors have an integral shaft design measuring approximately 150 mm in overall rotor length and 65 mm in tip-to-tip blade diameter. The cold-spin burst test results (as obtained by KKK) from rotors fabricated in two phases of the program show a significant improvement in performance (Table 3). The improvement is a result of the following factors:

1) Improved surface and internal quality of components due to major advances in NDE and surface inspection capabilities.
2) Improved material properties and reliability due to stringent process control and process innovation.
3) Improved machining and balancing.

The machining and balancing are strictly related to the specific shape and will not be discussed in this paper. The remaining two items are described here in greater detail.

Table 3. Cold Spin Burst Test Results From GTE AY6 Turbocharger Rotors

Program Phase	Time of Delivery	Average Rotor Density (% Theo. Density**)	Mean RT Burst Speed* (m/s)	Comments
I	11/86 – 2/87	>99%	384 ± 90	Met Phase I target. Revealed machining and mass balancing deficiency.
II	9/87 – 2/88	>99%	511 ± 70	Exceeded Phase II target by over 20%. Met all VW/KKK specifications.

* Average of 5 rotors
** Theoretical density = 3.262 g/cc

The Phase I rotors are representative of the laboratory process at the time of their fabrication (1986). Since that time, the surface and internal quality of complex-shaped components such as turbocharger rotors or turbine rotors have improved dramatically. The improvement in quality can be seen from results of nondestructive evaluation of large cross-section components of a proprietary shape (Table 4). Note that microfocus x-ray detectable cracks have been completely eliminated from the most recent set of components. Similar improvements have been observed in the turbocharger rotors (Table 5).

Table 4. Improvements in the Quality of Large Cross-Section Components

		Number of Microfocus and X-Ray Detectable Defects			
Component Area	Defect Types	June 1986	June 1987	July-August 1987	May 1988
Thick Section*	Cracks	20 ± 7	2 ± 3	1 ± 1	0
	Inclusions	28 ± 14	10 ± 6	5 ± 2	0
	Voids	6 ± 3	3 ± 2	1 ± 1	1 ± 1
Thin Section*	Cracks	12 ± 10	0	1 ± 1	0
	Inclusions	12 ± 6	1 ± 1	1 ± 1	1 ± 1
	Voids	8 ± 4	2 ± 2	4 ± 3	1 ± 1

Number of Visually Detectable Defects			
Component Area	Defect Types	June-August 1987	May 1988
Thick Section	Pits	3 ± 2	1 ± 1
	Cracks	3 ± 2	0
Thin Section	Chipped	4 ± 4	1 ± 1
	Broken	1 ± 1	0

*Detection limit in the thick section is ~150 μm and in the thin section ~50 μm.

Table 5. Comparison of Quality of Phase I and Phase II Turbocharger Rotors As Determined by Microfocus X-Ray Analysis

		Phase I	Phase II
Thick Section (Hub and Shaft)*	Cracks	4 ± 3	0
	Inclusions	2 ± 1	1 ± 1
	Voids	1 ± 1	1 ± 1
Thin Section (Blades)*	Cracks	1 ± 1	0
	Inclusions	3 ± 3	1 ± 1
	Voids	2 ± 2	1 ± 1

*Detection limit in the thick section ~ 125 μm. Detection limit in the thin section ~ 50 μm.

The improvement in the material properties and reliability of turbochargers manufactured by the prototype process is even more significant. Table 6 compares the fast fracture strength of test bars machined from companion Phase II turbocharger rotors fabricated by the laboratory process and the prototype process. The impact of the prototype process is reflected in the significant improvement in the material properties of Rotor #390-2. Note that similar properties have also been achieved in test bars fabricated by the prototype process.

Table 6. Material Strength of GTE AY6 Rotors/Test Bars

				MPa	
	Fabrication	Fabrication	RT	1000°C	1200°C
Component	Date	Process	$\bar{\sigma}$ ± SD	$\bar{\sigma}$ ± SD	$\bar{\sigma}$ ± SD
Rotor 191	1986	Laboratory	744 ± 83	690 ± 62	
Rotor 186-2	1987	Laboratory	814 ± 90	579 ± 48	338 ± 34
Rotor 195-2	1987	Laboratory	807 ± 76	510 ± 96	344 ± 41
Rotor 390-2	1988	Prototype	910 ± 34	703 ± 28	462 ± 28
AY6 MOR Bars	1/88	Prototype	924 ± 62	737 ± 48	427 ± 7
AY6 MOR Bars	4/88	Prototype	924 ± 62	710 ± 62	503 ± 14

Although the increase in strength in Rotor 390-2 is significant compared to those of the other rotors, more important is the marked decrease in the variability of the strength. Figure 2 shows the room-temperature strength distribution of bars machined from three rotors. The rotors fabricated by the laboratory process (#186-2 and #195-2) have Weibull moduli between 9 and 12, which is typical of large cross-section ceramic components. Rotor 390-2, however, displays an extremely high Weibull modulus of 34, which approaches the value of some metals. The high Weibull modulus gained in Rotor 390-2 was accomplished as a result of refinements and control gained in the fabrication processes in the PEC and reflects the ability to produce large cross-section components without increases in the size and distribution of critical defects.

The consistency of the PEC fabrication processes is also manifested in the uniformity of rotor material as shown in Figure 3. Material from the hub portions of Rotors 186-2 and 195-2 have lower strengths than the material from the rotor shaft areas, particularly at 1000°C. Rotor 390-2, however, displays uniform strength throughout the part.

The time-dependent strength of the turbocharger rotor material fabricated by the prototype process was measured at 1000°C. This data is critical because of the known susceptibility of silicon nitride ceramics to time-dependent strength degradation at elevated temperatures. Generally, under an applied load, the stress intensity at a crack tip causes the crack to grow at a rate described by

$$V = A(K_I/K_{IC})^N, \qquad (1)$$

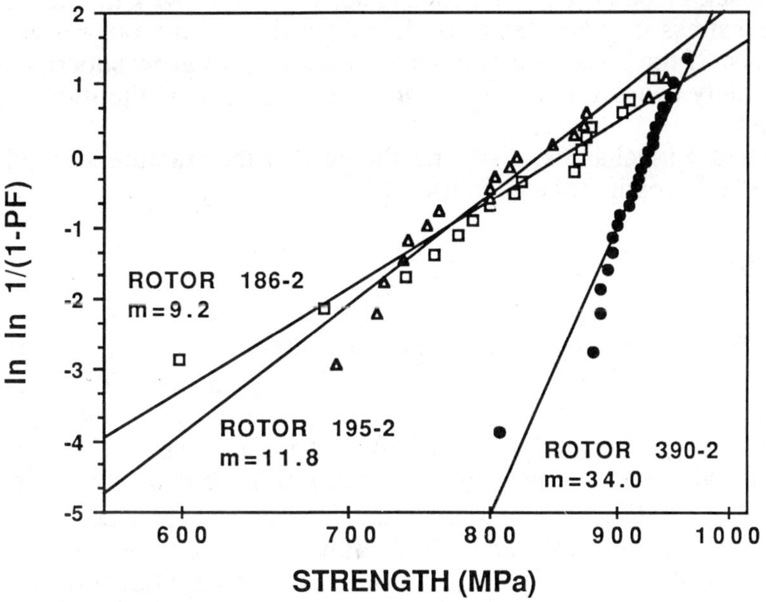

Figure 2. Comparison of material reliability in turbocharger rotors fabricated by the laboratory process (#186-2 and #195-2) and the prototype process (#390-2).

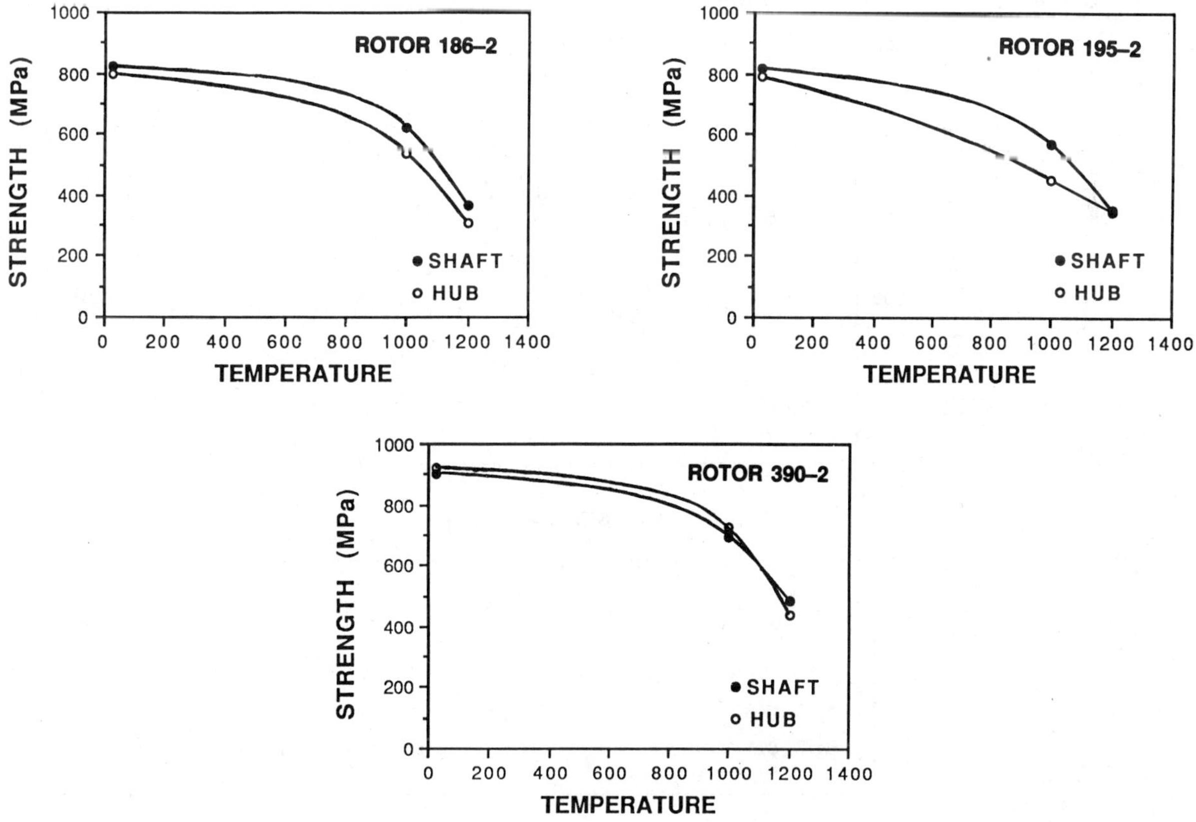

Figure 3. Material strength variability in hub and shaft areas of turbocharger rotors.

where V is the crack velocity; K_I is the stress intensity at the crack tip due to the applied load; K_{IC} is the critical stress intensity factor of the material (fracture toughness); and A and N are material constants. If the stress is maintained, the crack will grow subcritically to a size large enough to ultimately cause failure; the larger the crack grows, the more the strength of the material degrades.

From fracture mechanics, it can be shown that the fracture strength of a material is related to the applied stressing rate as follows:

$$\sigma_f = C\dot{\sigma}^{(1/(N+1))}, \tag{2}$$

where σ_f is the fracture strength; $\dot{\sigma}$ is the applied stressing rate; and C is a constant. The parameter N is the same constant as in Eq. (1) and is a measure of the material sensitivity to subcritical crack growth; low values of N indicate low resistance to subcritical crack growth. A linear regression of the slope of a log-log plot of fracture strength vs. stressing rate will therefore provide evidence of a material's resistance to time-dependent strength degradation. This data can also be used to predict the lifetime of components in specific stress conditions.[10]

Figure 4 shows the fracture strength as a function of stressing rate at 1000°C for several silicon nitride materials. The SN220M and AY6-1985 data are taken from tests conducted at AGTD.[11,12] The AY6-1988 data are from material machined from two Volkswagen turbocharger rotors fabricated by the prototype process. The significant increase in the N value shows the excellent resistance to subcritical crack growth (and therefore strength degradation) attained in recently fabricated components at GTE Labs. Although tests at GTE Labs were conducted in an argon environment, the high oxidation resistance of GTE silicon nitride and the relatively short duration of the tests (3 seconds to 1.5 hours) show that a significant improvement has indeed been attained in the crack growth resistance of the GTE silicon nitride.

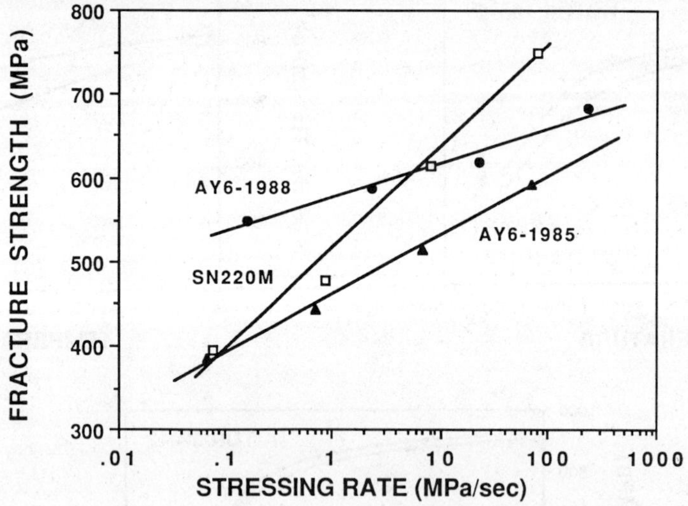

	TEST ATM.	N
AY6-1985	Air	15
AY6-1988	Argon	33
SN220M-1985	Air	9

Figure 4. Comparison of subcritical crack growth data from GTEL turbocharger material and data obtained from References 11 and 12.

5. SUMMARY

The feasibility of using GTE silicon nitride components in gas turbine applications has been demonstrated through CATE and AGT 100 programs. In an attempt to further improve the properties and performance of large cross-section components such as AGT 100 rotors and turbocharger rotors, GTE Labs has implemented major process controls through proprietary innovations, NDE, and a new Prototype Engineering Center facility equipped with environmental controls based on semiconductor "clean-room" principles. These changes have resulted in major improvements in properties and performance of ceramic components such as the turbocharger rotors.

ACKNOWLEDGMENT

The authors gratefully acknowledge Paul Fuce, Jean Gagne, Steve Quinn, Charles Landry, and Donna Thomas for experimental assistance at various stages of the experimental program. GTE Labs appreciates the cooperation of Volkswagen and Kuhnle, Kopp and Kausch with the turbocharger development program and thanks them for allowing publication of the rotor evaluation data.

REFERENCES

1. H.E. Helms et al., "Ceramic Applications in Turbine Engines," Noyes Publications, Park Ridge, NJ (1986).

2. H.E. Helms, "AGT 100 Project Summary," ASME Paper No. 88-GT-223, presented at the Gas Turbine and Aeroengine Congress, Amsterdam, June 6-9, 1988.

3. V.J. Tennery et al., "Application of Electron Microscopy for the Analysis of the Mechanical Behavior of Advanced Ceramics," ASME Paper No. 88-GT-72, presented at the Gas Turbine and Aeroengine Congress, Amsterdam, June 6-9, 1988.

4. F.M. Mahoney, K.W. French, and G. Bandyopadhyay, "Fabrication and Evaluation of AGT 100 Injection Molded Silicon Nitride Gasifier Rotors," *Proc. of the 25th Automotive Technology Development Contractors' Coordination Meeting*, published by the Society of Automotive Engineers, Inc., Warrendale, PA, April 1988, pp. 85-89.

5. G.L. Boyd and D.M. Kreiner, "AGT 100/ATTAP Ceramic Technology Development," *Proc. of the 25th Automotive Technology Development Contractors' Coordination Meeting*, published by the Society of Automotive Engineers, Inc., Warrendale, PA, April 1988, pp. 101-114.

6. J.T. Neil, A.E. Pasto and L.J. Bowen, "Improving the Reliability of Silicon Nitride: A Case Study," *Advanced Ceramic Materials 3 [3]*, 225-230, (1988).

7. G. Bandyopadhyay and K.W. French, "Fabrication of Near-Net Shape Silicon Nitride Parts for Engine Application," *J. Eng. for Gas Turbines and Power 108*, 536-539 (1986).

8. G. Bandyopadhyay and K.W. French, "Near-Net Shape Fabrication and Densification of Silicon Nitride," *Proc. of Workshop on Conservation and Substitution Technology for Critical Metals in Bearings and Related Components for Improved Bearing Performance*, Vanderbilt University, Nashville, TN, March 12-14, 1984.

9. D. Cotter and W. Koenigsberg, "Microfocus Radiography of High Performance Silicon Nitride Ceramics," *Proceedings of the Conference of Non-Destructive Testing of High Performance Ceramics*, Boston, MA, August 25-27, 1987, pp. 233-253.

10. S.M. Wiederhorn and J.E. Ritter, Jr., "Application of Fracture Mechanics Concepts to Structural Ceramics," *Fracture Mechanics Applied to Brittle Materials, ASTM STP 678, S.W. Freiman, Ed., Am. Society for Testing and Materials*, pp. 202-214,1979.

11. P.K. Khandelwal, J. Chang, and P.W. Heitman, "Slow Crack Growth in Sintered Silicon Nitride," *Fracture Mechanics of Ceramics 8*, R.C. Brandt, A.G. Evans, D.P.H. Hasselman, and F.F. Lange, ed., Plenum Press, New York, pp. 351-362, 1986.

12. J. Chang, P.K. Khandelwal, and P.W. Heitman, "Dynamic and Static Fatique Behavior of Sintered Silicon Nitrides," *Ceramic Eng. Sci. Proc. 8* [7-8], pp. 766-777 (1987).

A TWO-PIECE ARTICULATED CERAMIC PISTON FOR DIESEL ENGINES

K.Arakawa, M.Sugiyama, R.Shimazaki, S.Ito, and K.Ishisaka
Toyota Motor Corp.
Aichi, 471 Japan

ABSTRACT

In order to achieve high heat resistance and light weight, an articulated piston with a silicon nitride piston-crown and aluminum-alloy skirt has been designed for a DI diesel engine. From FEM analyses and test measurements, the stress levels of the main loads were determined. Component strength of the ceramic piston-crowns was ascertained from fracture tests and the capacity to withstand fast fracture assessed. Good agreement was obtained between predicted component strength based on weibull statistic and component strength test results. This indicates that the material in the components was statistically equivalent to that in the test specimens and that the analytical approach is correct. It also indicates that the piston-crowns were of high quality. A 300-hour engine test has been conducted without fracturing of the ceramic crowns.

INTRODUCTION

There have been continual demands to improve the power output of automotive engines. However, in operation, components of reciprocating engines suffer higher thermal and mechanical loadings in achieving higher power performance. Therefore, better heat resistant materials are required for reciprocating engines. For example, Al-alloy materials, which are conventionally used for pistons, become inadequate at elevated temperatures (Figure 1). There is therefore little choice except to use heavier metals such as cast iron. For this reason, research into light, heat resistant materials is of considerable importance.

Si_3N_4 is a most promising candidate material because it is both light in weight and highly heat resistant. The present study was conducted to investigate the possibility of applying this material to DI diesel engine pistons, which are moving parts subjected to high thermal load.

Some attempts to apply ceramic materials to pistons have previously been reported. Although assembling by bolts [1], cast-in [2] and shrinkage-fit [3] have been studied as methods of connecting a monolithic ceramic cap, bowl or insert, to a piston-head, problems of cracking [4,5] and relaxation [6] due to thermal expansion mismatch between metals and ceramics remain.

Accordingly, a piston combining a cast iron part and a ZrO_2 part, for which thermal expansion mismatch is very low, has been investigated [7,8], but it is hard to recognize any superiority in this structure because of weight gain and the relatively low strength properties of ZrO_2. Ceramic coating of the piston head poses a problem of peeling [9].

As all these problems are caused by thermal expansion mismatch between ceramics and metals, they are exacerbated under high thermal loads. The authors therefore concluded that an effective solution would be to adopt an articulated piston with an Si_3N_4 piston-crown and an Al-alloy skirt connected by a piston-pin. Figure 2 shows the design of the articulated piston with a ceramic piston-crown. The piston-crown is comprised of a combustion chamber and a pin-boss. Since there is no direct connection between metal and ceramic components and since the piston-pin is not subjected to high temperatures, there should be little problem of thermal expansion mismatch. Furthermore because the density of Si_3N_4 is almost as low as that of Al-alloy, the piston assembly is of light weight.

However, as the ceramic piston-crown directly bears the main piston loads, e.g. inertial force and combustion pressure, its durability has to be determined. This paper presents the results of investigations to assess the susceptibility of the ceramic piston-crown to fast fracture under engine operating conditions.

FEM STRESS ANALYSIS

A FEM analysis was performed in order to estimate the ceramic piston's capacity to bear the stresses due to inertial force and combustion pressure.

Engine loading conditions

A 4-cylinder, 1883cc experimental direct injection diesel engine with 83 mm bore and 87 mm stroke was used and turbo-charged to generate about 50 kW power. Thus, inertial force of reciprocation and combustion pressure are 6.8 kN and 15 MPa respectively, and these values were used in the analysis.

FEM analysis of stress due to inertial force

Figure 3 shows the FEM model of piston pin-boss and supporting structure. Z-axis acceleration applied to the model was 1.56×10^4 m/sec^2. Loads between the piston and piston-pin were propagated by spring elements.

The results are shown as a contour map of maximum principal stress in Figure 4. The peak value of 50 MPa occurs on the inner side of the pin-boss.

FEM analysis of stress due to combustion pressure

The FEM model of the piston and piston-crown assembly is shown in Figure 5. Combustion pressure was set up on the piston-head and gap elements were used to propagate loads. Figure 6 shows the contour map of maximum principal stress. The peak value of 120 MPa occurs on the under-surface of the combustion chamber section of the piston-crown.

Comparison with material strength

Room temperature strength properties of the Si_3N_4 material used are shown in Figure 7 and Table 1. Mean flexural strength (3-point bending) is 720 MPa, and Weibull modulus is 8.4. From these values, the mean tensile strength of a 6 mm diameter rod is calculated to be about 400 MPa.

It was tentatively assumed that the allowable stress was 1/4 of flexural strength, i.e. 180 MPa, or 1/2 of tensile strength of a 6 mm diameter rod, i.e. 200 MPa. Adopting these values as the lower limit for acceptability in non-destructive tests should ensure that the piston-crowns have a sufficient safety margin over the peak stress caused by inertial force or combustion pressure.

EVALUATION OF COMPONENT STRENGTH

In the case of ceramic materials, the strength of actual components (component strength) is usually much less than the material strength. Si_3N_4 piston-crowns were therefore made for testing so that the component strength could be measured and the results of the FEM stress analysis verified.

The strength of ceramic components generally depend upon the manufacturing process. To ensure that the piston-crown was as strong as possible, the following process was adopted. First, a cylindrical body was formed by cold-isostatic pressing. This was pre-sintered and rough-machined. After sintering, it was precision-machined to the final dimensions. Each test component was inspected for surface defects by means of fluorescent penetration and X-rayed for internal defects.

Verification of FEM stress analysis results

Because it is extremely difficult to measure the stress acting on a ceramic piston-crown in an actual engine, a test apparatus was devised to simulate the loads.

Figure 8-A shows the tension test apparatus designed to simulate the inertial force on the pin-boss. Stress was calculated at 4 points from measurements with strain gauges. Figure 9 compares this stress with FEM analysis results. The close agreement at each point indicates the validity of the analysis. The test apparatus for simulating combustion pressure by compression is shown in Figure 10. Stresses derived from strain gauge measurements coincided closely with FEM analysis (Figure 11), thereby confirming the validity of the analysis.

Pin-boss component strength

To assess the component strength of the Si_3N_4 piston-crowns, the pin-boss was secured on one side only (Figure 8-B) and tension applied until fracture occurred. Tests were conducted with pin-bosses of different dimensions (Table 2). Comparable FEM stress analysis were performed for each of the three sizes. Figure 12 is an example of the contour maps, which were very similar to each other and to Figure 4. In every case,the point of fracture origin coincided with the peak stress point, as may be seen from the examples shown in Figure 13. Accordingly, measured fracture loads were converted into fracture stresses using the FEM analysis result and the fracture stresses, which indicate component strength, were Weibull plotted (Figure 14).

For each size, the mean component strength value was about 400 MPa, which is eight times the value of the stress due to inertial force predicted by FEM analysis. This result suggests that the Si_3N_4 piston-crown is quite capable of enduring the inertial load in an actual engine.

Furthermore, in view of the high component strength and the fact that none of the fractures originated from a pore or inclusion, the manufacturing process adopted was thought to have produced piston-crowns of high quality.

Combustion chamber base component strength

Since it was impossible to fracture the base of the combustion chamber using the test hydraulic apparatus, component strength of this area was evaluated by 3-point flexural testing of bars cut out of a piston-crown. Figure 15 shows a Weibull plot of the results. The flexural strength of the test-pieces was found to be about the same as the flexural strength of Si_3N_4 material characterization test-bars with a minimum value of more than 500 MPa. Since the results were not dependent on the location from which the test-pieces were taken, it is reasonable to assume that the Weibull strength characteristics of the material in the combustion chamber base is equal to that in the pin-boss. Thus the mean component strength of the combustion chamber base is approximately four times the value of the predicted peak stress due to combustion pressure.

PREDICTION OF COMPONENT STRENGTH

Component strength of ceramic piston-crowns was predicted to provide a basis for comparison with the measured values.

Prediction method of component strength on the basis of Weibull statistics

Ceramic strength is generally analyzed by use of Weibull statistics [10]. Fracture probability is given by this theory as

$$F = 1 - \exp\left\{ - \int_V \left(\frac{\sigma - \sigma_u}{\sigma_o} \right)^m dV \right\} \qquad (1)$$

where m, σ_u, σ_o, and V are the Weibull modulus, the zero strength, the normalizing stress and the volume of component, respectively. From Eq. (1), it is possible to calculate fracture probability and to predict component strength.

E.q (1), can be expressed in terms of an effective volume V_e and the peak stress value σ_p

$$F = 1 - \exp\left\{ -\left(\frac{\sigma_p}{\sigma_o}\right)^m \times V_e \right\} \quad (2)$$

assuming $\sigma_u = 0$ and

$$V_e = \int_V (\sigma / \sigma_p)^m dV \quad (3)$$

As σ_p is also fracture stress, it follows that component strength can be predicted by calculating V_e. V_e can be calculated analytically in the case of flexural or tensile testing to evaluate material strength because of the simple shape of the specimen and the relative ease of calculating stress distribution [11].

However, it is difficult to apply these equations to predict component strength because V_e cannot be simply calculated due to the complexity of the component's shape and the difficulty of calculating stress distribution. It has been reported that V_e may only be calculated approximately on the basis of FEM stress analysis results [12].

The following method was therefore devised for calculating V_e directly from FEM stress analysis data in order to predict component strength.

(1)FEM stress analysis is carried out and the maximum principal stress $\sigma_{mxpr,i}$ of each element and the peak stress σ_p are obtained.

(2)Assuming that $\sigma_{mxpr,i}$ is uniform throughout each element, the effective volume of each element $V_{e,i}$ is given as

$$V_{e,i} = V_i \times (\sigma_{mxpr,i} / \sigma_p)^m \quad (4)$$

since V_i is the volume of each element.

(3)As the total effective volume V_e is the summation of $V_{e,i}$, V_e is given as

$$V_e = \sum V_{e,i} \quad (5)$$

Application to component strength (tension test of pin-boss)

Before applying this method to predict component strength, its accuracy was evaluated on a flexural test specimen. Weibull parameters were calculated (Table 1) from flexural strength data (Figure 7).

FEM stress analysis was conducted, and the predicted flexural strength obtained by this method is shown in Figure 16. Although increasing the size of the elements shifted the prediction slightly to higher and therefore less safe value, the predicted strength in both cases was close to the measured strength. The prediction method was thus deemed valid.

The component strength of the pin-boss was predicted from the stress analysis data (Figure 12). The prediction is compared with the measured values in Figure 17. Reading the mean fracture strength at the 50 % fracture probability point, the measured value is only 50 MPa lower than the predicted value. From this, it would appear that the strength limiting flow distribution in the piston-crowns is about the same as that for small test specimens and that the Weibull model and analytical approach employed in this work provides good results. It can also be stated that the quality of the Si_3N_4 piston-crowns is high, especially when the size of the pressing and the complex machining are considered.

ENGINE TEST

In view of the Si_3N_4 piston-crown's capacity to withstand inertial force and combustion pressure without suffering fast fracture as evidenced by the foregoing tests, an engine of the type used for the FEM analysis was equipped with these piston-crowns. Following a motoring test, a durability test at full loads and rated speed has been conducted. The engine test has been in progress for 300 hours to date without fracture of any of the ceramic piston-crowns (Figure 18).

CONCLUSIONS

1. A ceramic piston-crown with integral piston-pin boss has been designed for an articulated piston.

2. FEM analyses were made of the stresses due to inertial force and combustion pressure. The results agreed well with measured values obtained using apparatus set up to simulate these loads.

3. Average strength of the Si_3N_4 piston-crowns was found to be high as compared with the stresses calculated by FEM analysis, and was judged as being sufficient to resist fast fracture under actual operating conditions.

4. A method of predicting component strength curves directly from FEM stress analysis data on the basis of Weibull statistics was devised. The predictions agreed very well with experimental evaluations of the strength of components. This indicates that the statistical strength of the Si_3N_4 material in the manufactured piston-crown is approximately equivalent to that in the flexural test-bars. It also indicates that the quality of the Si_3N_4 piston-crown was estimated to be high.

5. The Si_3N_4 piston-crowns have completed a 300 hour engine bench test at full loads and rated speed without fracturing.

REFERENCES

(1)R.Kamo, M.Wood and W.Geary, "Ceramics for Adiabatic Diesel Engines", CIMTEC 4th, St.Vincent, Italy, June 1979.
(2)P.Walzer, H.Heinrich, and M.Langer, "Ceramic Components in Passenger-Car Diesel Engines", Mat. and Design, Vol.7, No.2, (1986), P.75-80.

(3)M.E.Woods and I.Oda, "PSZ Ceramics for Adiabatic Engine Components," SAE Paper, 820429, (1982).
(4)M.E.Woods, W.F.Mandler and T.L.Scofield, "Designing Ceramic Insulated Components for the Adiabatic Engine", Cer.Bull., Vol.64, No.2, (1985), P.287-293.
(5)R.Kamo and W.Bryzik, "Cummins/TACOM Advanced Adiabatic Engine", SAE Paper, 840428, (1984).
(6)R.Kamo and W.Bryzik, "Cummins-TARADCOM Adiabatic Turbocompound Engine Program", SAE Paper, 810070, (1981).
(7)D.L.Hartsock, "Partially Stabilized Zirconia Piston Bowl Reliability", ASME J. Eng. for GT and Power, Vol.109, Oct.1987, P.367-373.
(8)P.H.Havstad, I.J.Garwin and W.R.Wade, "A Ceramic Insert Uncooled Diesel Engine", SAE Paper, 860447, (1986).
(9)B.Schechter and R.L.Holtman, "An Investigation of Enhanced Thermal Barrier Coating Systems for Diesel Engine Components",22th Automotive Technology Development Contractors' Coordinating Meeting, SP-155, (1984), P.261-271.
(10)W.Weibull, "A Statistical Distribution Function of Wide Applicability", J.Appl.Mech., Vol.18, (1951), P.293-297.
(11)G.J.Desalvo, "Theory and Structural Design Applications of Weibull Statistics", WANL-TME-2688, May (1970).
(12)T.Takeuchi and N.Sasaki, "Evaluation of Ceramic Turbine Disk for Automotive Gasturbine Engines"(in Japanese), Proc. of The Annual Meeting of JSAE 802, 802B64, (1980), P.393-396.

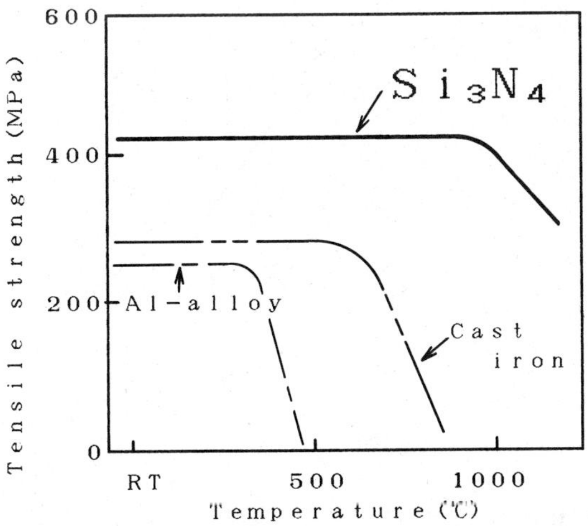

Fig. 1.
High-temperature strength of some materials

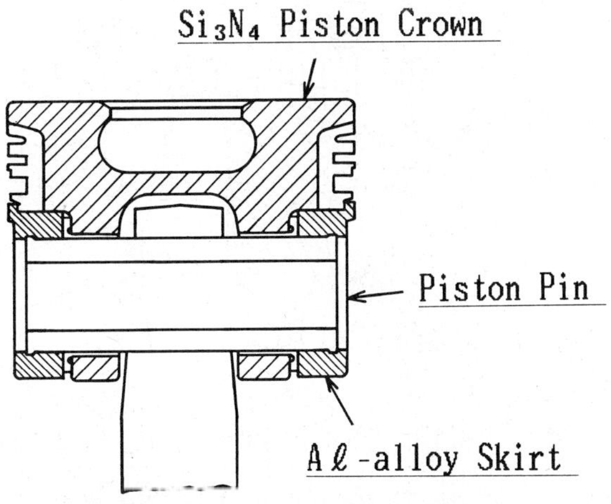

Fig. 2.
Schematic of an articulated ceramic piston

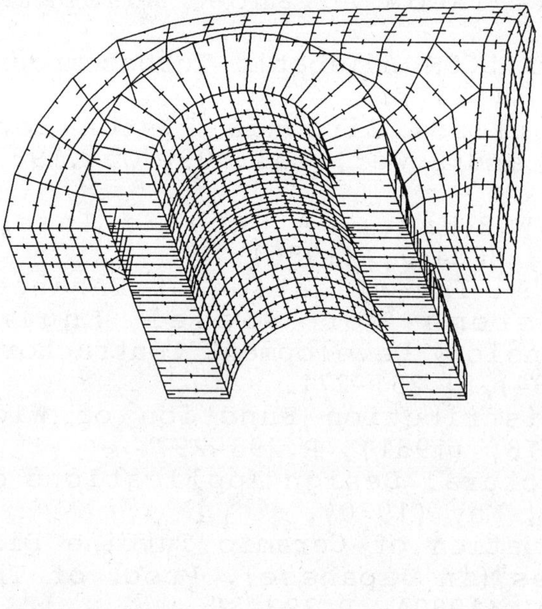

Fig. 3.
FEM model of piston pin-boss

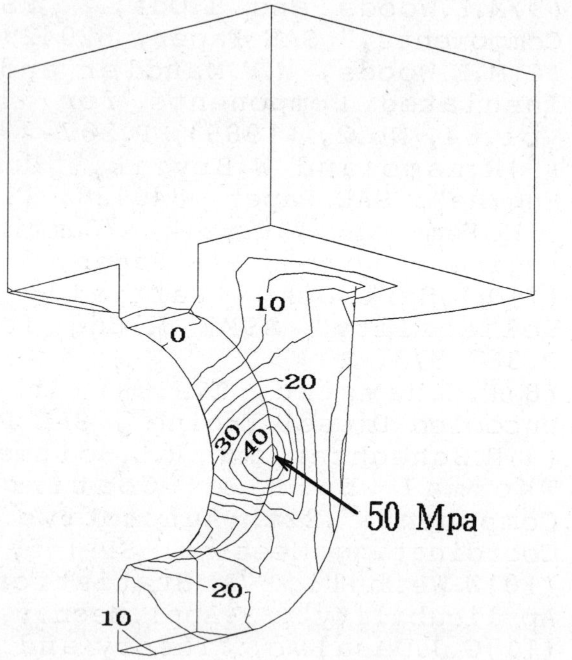

Fig. 4.
Maximum principal stresses in Si_3N_4 piston-crown caused by force of inertia

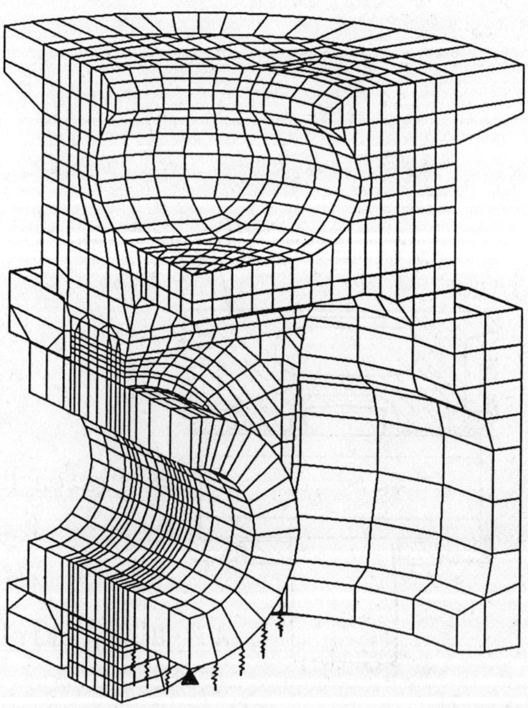

Fig. 5.
FEM model of piston

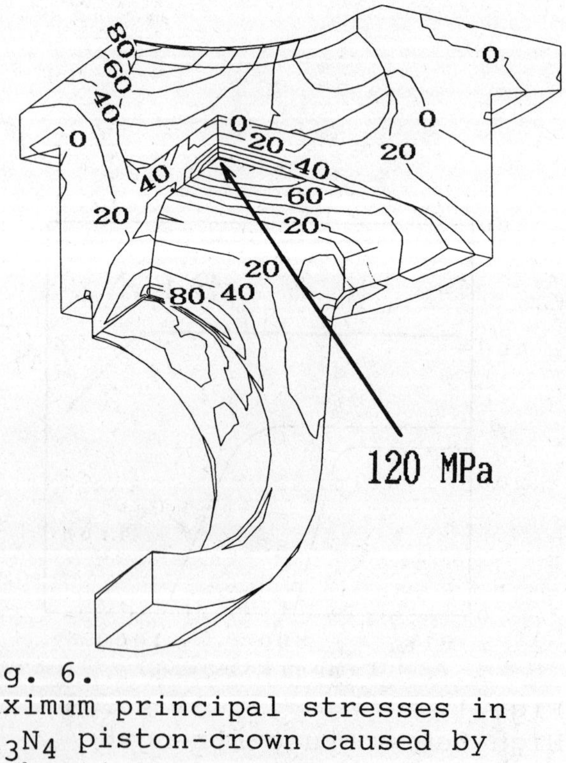

Fig. 6.
Maximum principal stresses in Si_3N_4 piston-crown caused by combustion pressure

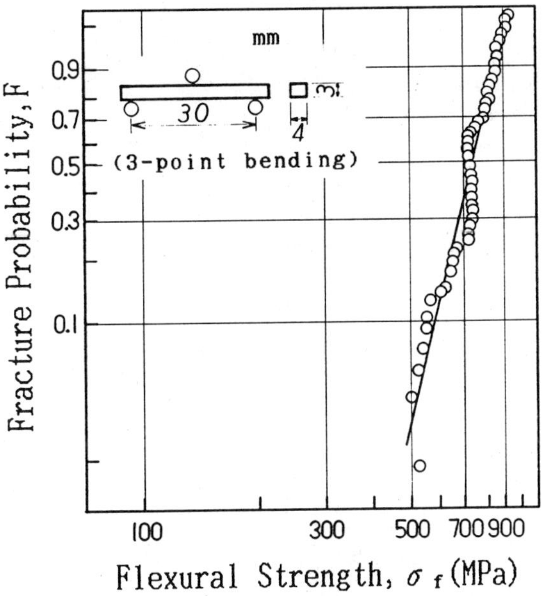

Fig. 7.
Si_3N_4 flexural strength

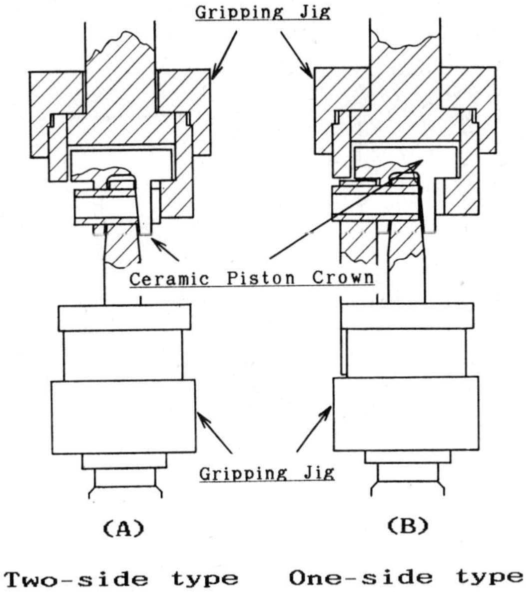

Fig. 8.
Schematic of pin-boss tension test apparatus

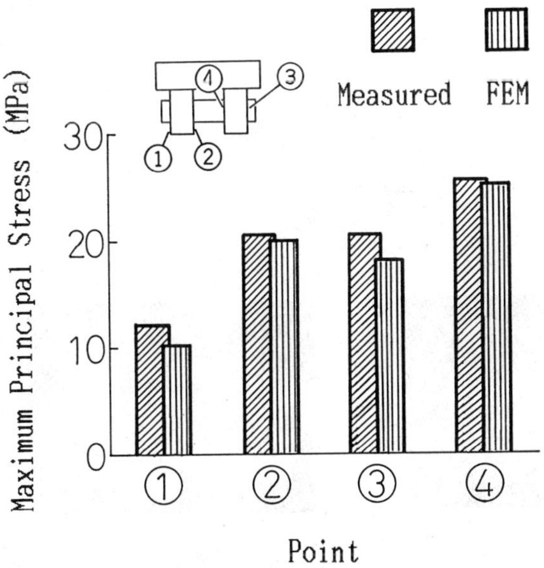

Fig. 9.
Comparison of measured stress with FEM

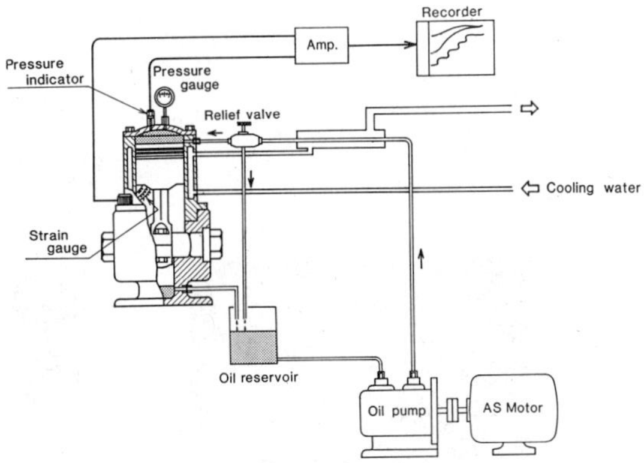

Fig. 10.
Schematic of compression test apparatus

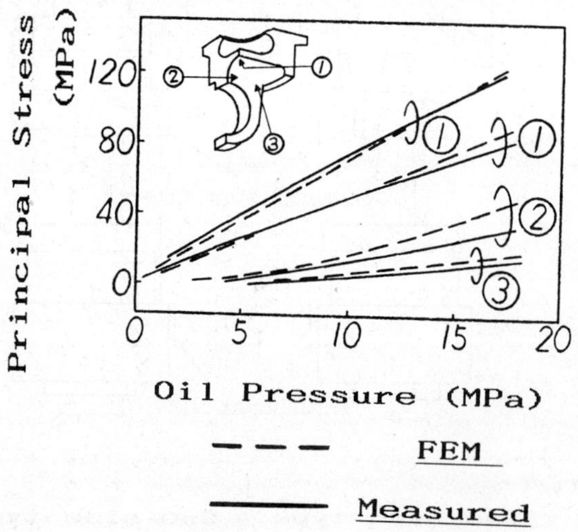

Fig. 11.
Comparison of measured stress with FEM

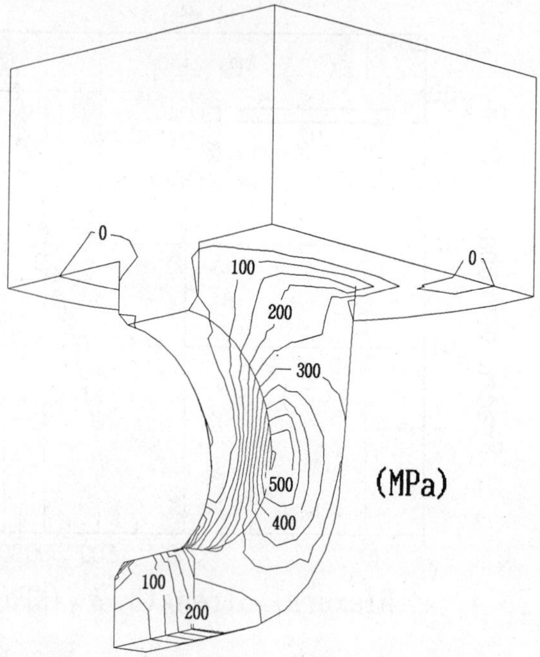

Fig. 12.
Maximum principal stresses on one-side tension test

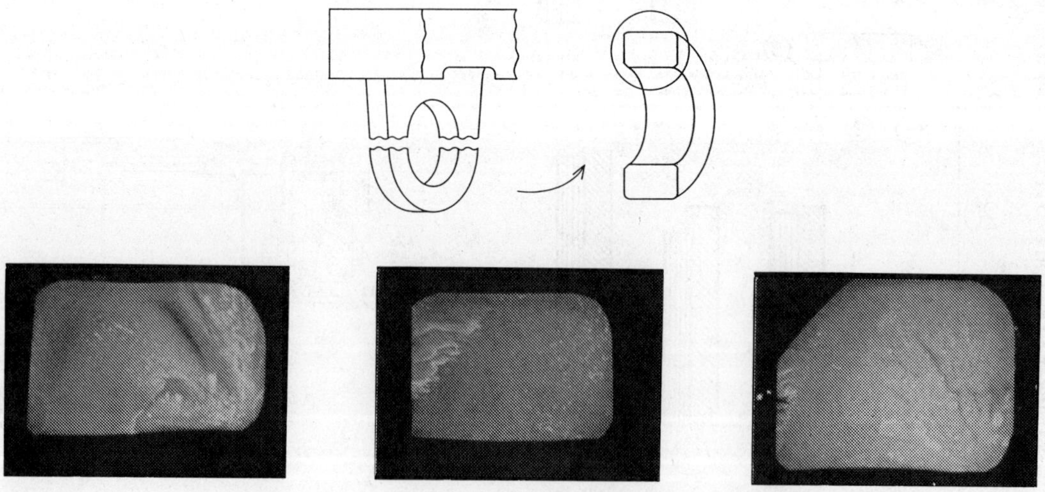

(A) No. 1 (σf =670MPa) (B) No. 2 (σf =310MPa) (C) No. 3 (σf =400MPa)

Fig. 13.
Observation of fractured surfaces

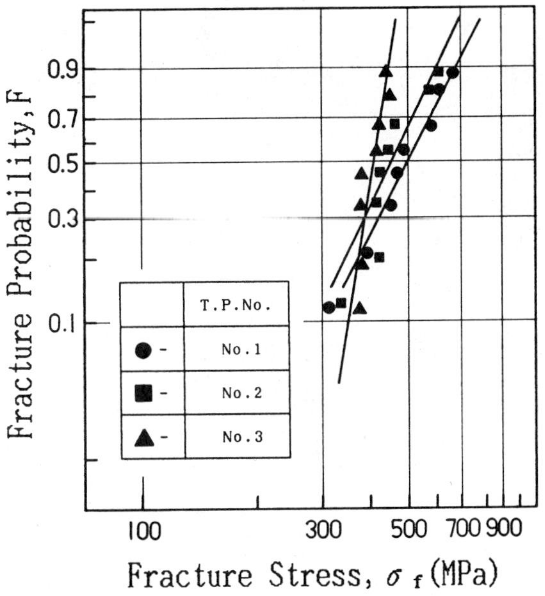

Fig. 14.
Component strength of pin-boss

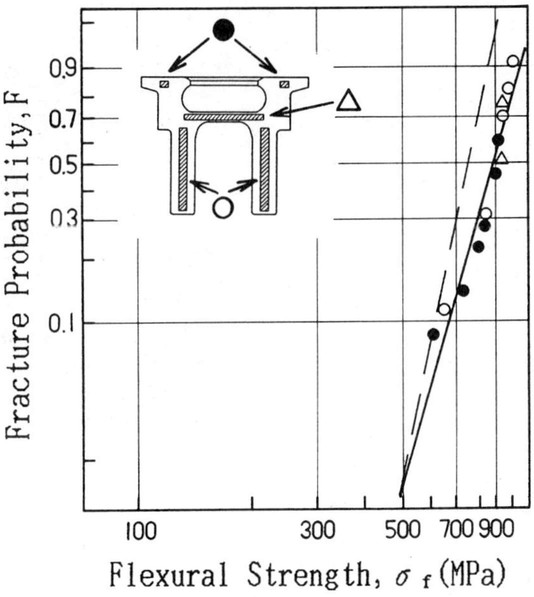

Fig. 15.
Strength of specimens cut from components

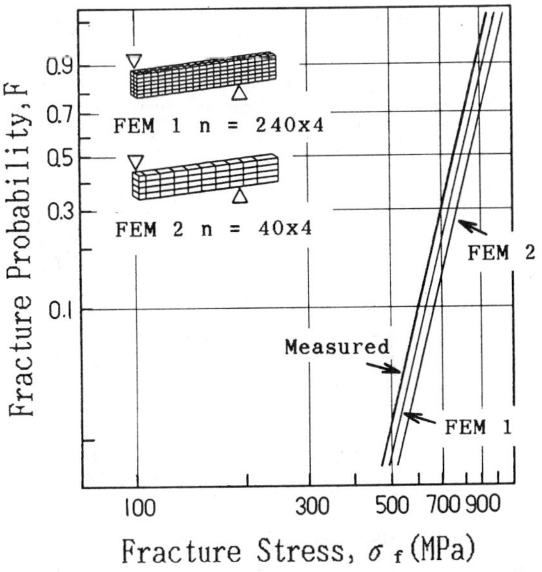

Fig. 16.
Application of prediction method to flexural testing

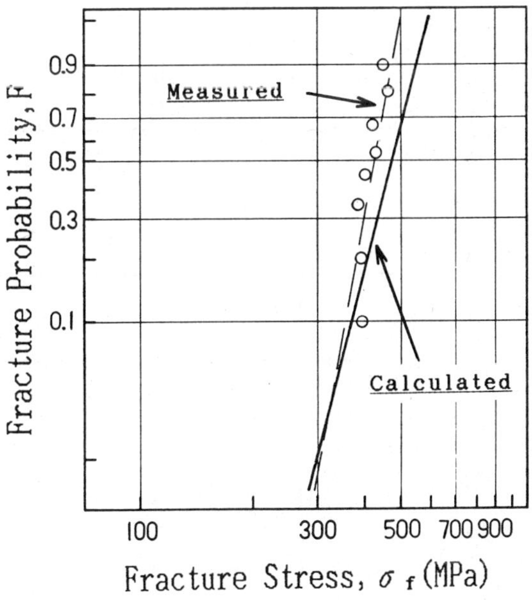

Fig.17.
Prediction of component strength

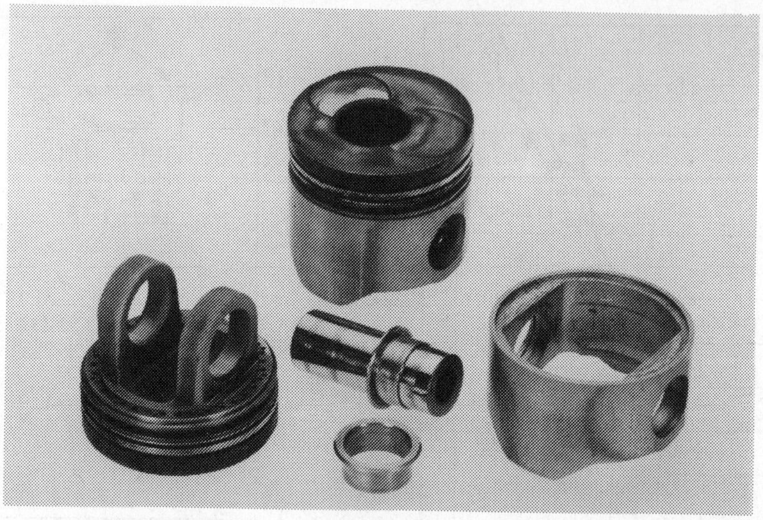

Fig. 18.
Photograph of ceramic piston after 300 hour engine test

Table 1.
Weibull parameters of the Si_3N_4 material

σ_f (MPa)	m	σ_o ($MPa \cdot mm^{3/m}$)	σ_u (MPa)
720	8.4	800	0

Table 2.
Dimensions of tested piston-crowns

	A (mm)	B (mm)
No. 1	12.5	R 20.5
No. 2	10.0	R 20.5
No. 3	12.5	R 23.0

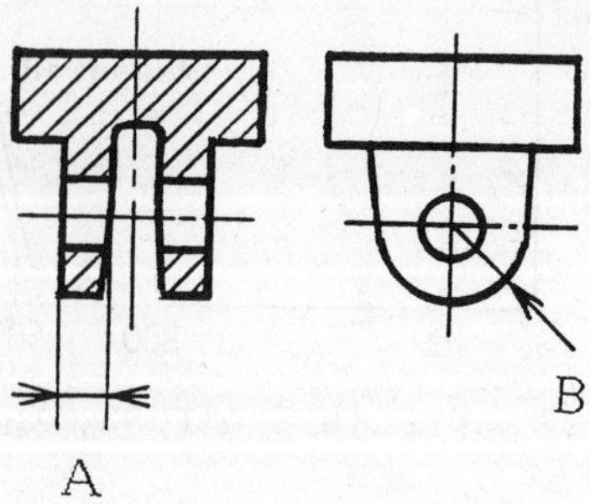

FABRICATION AND MECHANICAL RELIABILITY OF SiC/TiB_2 COMPOSITES

S. G. Seshadri, M. Srinivasan, J. W. MacBeth and M. O. Ten Eyck
The Carborundum Company
P. O. Box 832
Niagara Falls, NY 14302

ABSTRACT:

A review of the physical and mechanical properties of recently developed SiC/TiB_2 ceramic composite material is presented. This particulate composite exhibits enhanced mechanical and fracture properties over the monolithic α-SiC at temperatures in excess of 1000°C. The long term behavior such as subcritical crack growth and the retained strength after thermal exposure to air is also discussed. The conventional fabrication techniques such as isostatic pressing and injection molding and subsequent high temperature sintering are readily applied to the production of these composites for engineered products. In addition, the electro-discharge machinability provides exceptional flexibility in shapes required for automotive and armor applications. High temperature durability results with future applications in turbine engines are also discussed.

INTRODUCTION:

The general interest in ceramics in high temperature engines to enhance the thermodynamic efficiency has been well publicized over the past several years. Two major candidate materials for such applications have been identified as silicon carbide and silicon nitride.[1-3] There has been considerable progress in fabrication technology for complex shaped components of these materials suggesting their imminent use in heat engines and other high temperature structural applications.[4]

However, the use of these materials in actual applications is somewhat limited owing to their inherent brittleness. Many attempts to enhance the fracture resistance behavior of these materials while retaining the high temperature properties of the structural ceramics have resulted in some promising material systems.[5] In addition to the improved properties, the use of conventional and well-developed fabrication technology minimizing the occurrence of critical flaws in the stressed regions is expected to be a major factor in determining the potential of the new material systems for heat engine applications.

One new material system, namely, TiB_2 particulate reinforced sintered alpha silicon carbide, seems to fit these requirements very well. A particularly interesting feature is the sinterability of this composite without the application of external pressure such that complex shaped turbine components can be fabricated with ease, similar to pressureless sintered alpha SiC.[6-7] This paper will address the progress made in the development of this material with respect to the fabricability of complex components, as well as the enhanced properties at elevated temperatures.

SIC/TIB_2 COMPOSITES:

The SiC/TiB_2 ceramic is a two-phase particulate composite material containing discrete TiB_2 particles ranging from 1 to 5 μm in size dispersed in a finer grain alpha SiC matrix. The microstructure of this material obtained on a polished surface with no etching is given in Figure 1. The white TiB_2 phase is uniformly distributed throughout the matrix, with very little or no bonding present. Extensive transmission electron microscopy studies have not produced any evidence of the presence of bonding or additional phases in the TiB_2 /SiC interface.[8] This weak interfacial bonding also results in the pullout of the reinforcing TiB_2 particles due to the mechanical polishing procedure as seen in Figure 1.

The addition of TiB_2 to SiC matrix also causes a significant reduction in grain size of the matrix in the sintered microstructure compared to the sintered alpha SiC. It is thought that this decrease is related to the pinning of the grain boundaries by the TiB_2 particulates during densification and grain growth. The differential thermal expansion between SiC and TiB_2 causes weakening of the interface during cooling and when the particle size is large enough some limited separation is also seen. Consequently, this microstructure results in enhanced toughening due to crack deflection and modifications in fracture crack advance. Most of the properties of this composite are generally similar to SiC as seen in the Table 1.

Table 1. Typical properties of non-reinforced and TiB_2 reinforced sintered α-SiC.

Properties	Sintered α-SiC	SiC/TiB_2 Composite
Density, gms/cm^3	3.21	3.40
Young's Modulus, GPa (10^6 psi)	400 (58)	440 (64)
Flexural Strength, 25°C, MPa (ksi)	380 (55)	450 (65)
Fracture Toughness, 25°C, (SENB) MPa.m$^{0.5}$	4-5	6-8
Hardness (Vickers), GPa	28	29
Coefficient of Thermal Expansion, 10^{-6} /°C	4.02	4.60
Thermal Conductivity, W/m.°K	125	65

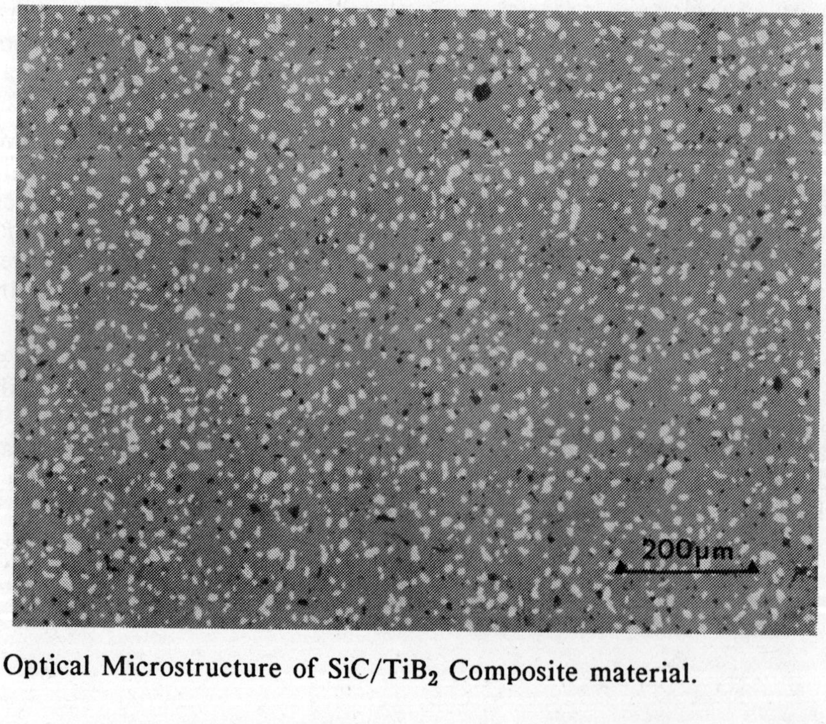

Figure 1. Optical Microstructure of SiC/TiB_2 Composite material.

FABRICATION OF COMPONENTS:

Considerable progress has been made in the fabrication technology of both large and complex shaped engine components by conventional ceramic forming techniques. In the past, the forming methods were focussed on making a limited quantity of reproducible parts with high strength and as close to near net-shape as possible to minimize final machining. Development of near net-shape fabrication methods is important for economical application in heat engines where close tolerances and good surface finish is essential for the required performance.

Cold isostatic pressing and injection molding (green forming) techniques have been demonstrated and extensively used for fabricating the α-SiC turbine components throughout the Advanced Gas Turbines (AGT) program.[4] The development work has continued now with the Advanced Turbine Technology Applications Project (ATTAP) extending these fabrication techniques to SiC/TiB_2 composites. While isostatic pressing followed by green machining has been primarily used to form SiC/TiB_2 components, the capability to fabricate components of this composite material by high volume methods such as injection molding is vital for economical application of this material.

The processing steps for the injection molding method for fabricating SiC/TiB_2 components is very similar to those used for sintered α-SiC. A thermoplastic compound roughly consisting of 50 volume % plastic and 50 volume % ceramic powders (α-SiC, TiB_2 and the sintering aids). This mixture is blended at elevated temperatures and molded on horizontal single screw machines. The binder removal and sintering steps are similar to those employed for sintered α-SiC. Figure 2 shows AGT-100 Power Turbine Vanes and the AGT-100 Gasifier Rotor illustrating both simple and complex shaped turbine engine components made by injection molding.

Another approach namely ram pressing, a well established process in the whiteware industry, is being developed for SiC/TiB_2 composites under the AGT-101 ATTAP project. This technique offers potential for high volume cost effective forming of geometries that can be pressed to shape such as the AGT-101 Backshroud shown in Figure 3. The present activities emphasize the development of mix preparation and flow behavior of the compound. The development of near-net-shape fabrication for this composite has utilized the conventional technologies which is essential in achieving close tolerances and relatively low cost for the heat engine components.

By varying the composition somewhat, the SiC/TiB_2 composite can be machined by electro-discharge machining (EDM). This composite has been successfully machined by both wire and sinker EDM techniques with cutting rates comparable to D-2 steel and aluminum.[9] Figure 4 shows rocker arm inserts as an example of potential automotive component shaped by the EDM method. The surface damage due to EDM has been shown to be much less than that due to conventional diamond machining techniques, resulting in higher Weibull modulus.[10] This unique capability offers additional flexibility in the final fabrication of components in the dense sintered state, without introduction of significant residual stresses.

MECHANICAL PROPERTIES:

The flexural strengths were measured up to 1400°C in static air using test specimens of dimensions 3.18 x 6.35 x 51 mm. under four point bending with inner and outer spans of 19 and 38 mm., respectively. These test specimens were cut from sintered plates and ground to size with edges chamfered prior to testing. A cross head rate of 8.5 mm/s was used. The fracture toughness was obtained using the single edge notched beam technique. These specimens were of the same dimensions as the flexural strength test specimens with a 2.5 mm. deep notch and were tested under four point bending with the same configuration.

The flexural strength of SiC/TiB_2 composite was found to be generally higher than that of the non-reinforced α-SiC material.[12] The flexural strength does not vary significantly with temperature up to 1400°C under fast fracture conditions (Figure 5).

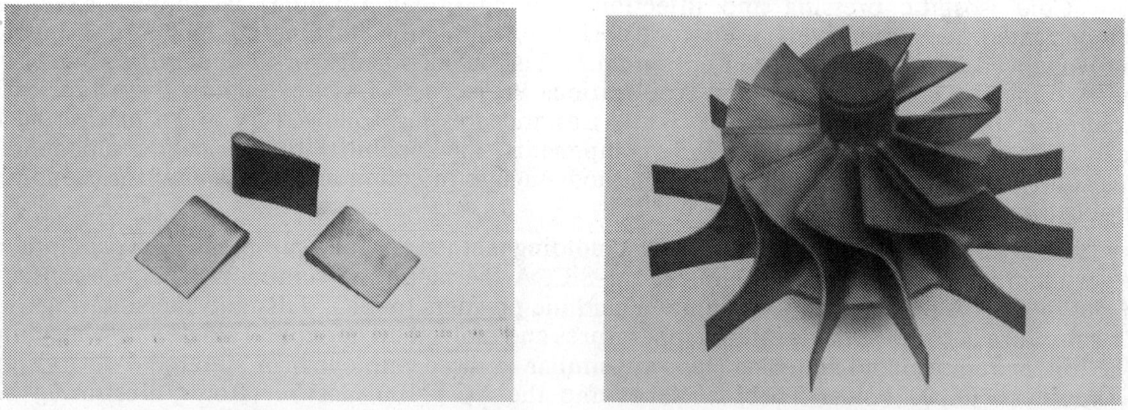

(a) (b)

Figure 2. (a) AGT-100 Power Turbine injection molded SiC/TiB_2 vanes.
(b) AGT-100 Gasifier Rotor - injection molded SiC/TiB_2.

Figure 3. AGT-101 SiC/TiB_2 Backshroud.

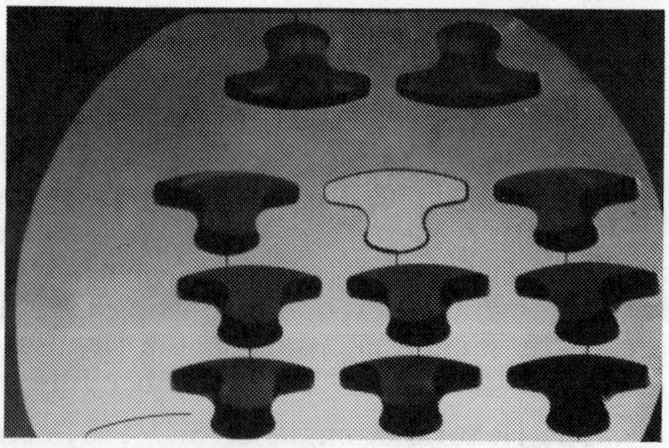

Figure 4. SiC/TiB_2 valve train rocker arm insert machined by EDM.

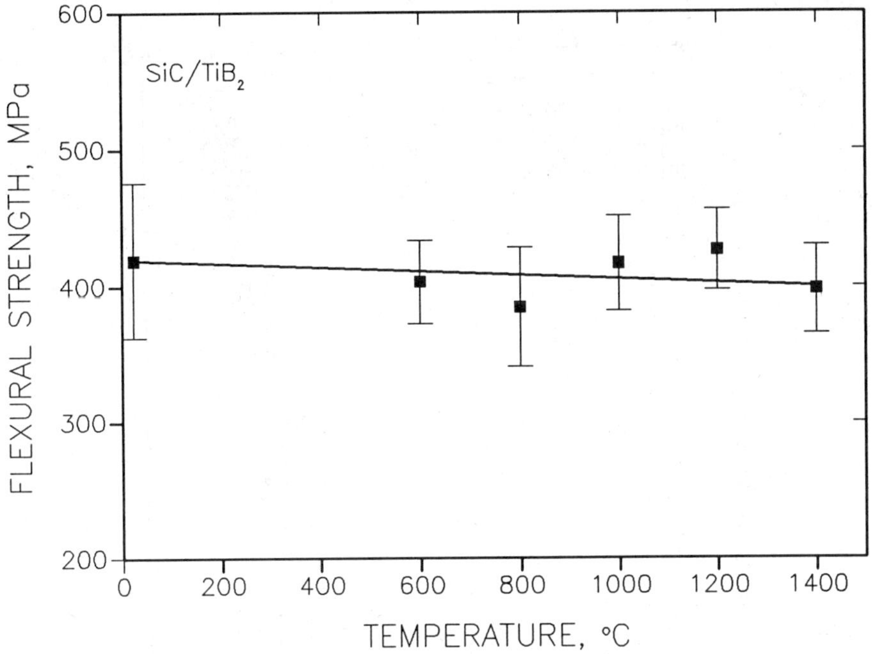

Figure 5. The variation of flexural strength of SiC/TiB_2 composite with temperature.

The strength distributions at selected temperatures are shown in the Figure 6. The Weibull modulus is greatly enhanced for this material resulting in much greater reliability than α-SiC. The Weibull moduli are generally in the range 12 to 18, especially at temperatures over 1000°C (Figure 7). This increase, combined with enhanced flexural strength, should result in greater tensile strength as well as better reliability for large components over the non-reinforced material.

The fracture origins in the SiC/TiB_2 material are similar to those found in the sintered alpha silicon carbide; agglomerates control failure at the lower end of the strength distribution. In general, the distribution of flaw types among machining, internal and surface flaws is similar to that found in sintered α-SiC. This observation is consistent with the processing methods employed which are almost identical to both these materials (Figure 8).

The effect of the addition of the TiB_2 particulates to the alpha SiC matrix is two-fold. In addition to the reduction in the grain size of the SiC matrix, the particles also provide enhanced fracture resistance for this material. The crack path through the microstructure clearly indicates transgranular mode in SiC matrix (similar to the non-reinforced material) and intergranular mode around the TiB_2 particles. The increased toughness has been identified earlier as a result of crack deflection and pinning mechanisms.[5] The fracture toughness values, measured by the single edge notched beam technique, is shown as a function of temperature in Figure 9. The small decrease in the fracture toughness at 600°C can be considered to be within the statistical variations encountered from batch to batch. The K_{Ic} from some batches have shown much higher values indicating the potential to improve this value significantly. However, the enhancement of fracture toughness due to TiB_2 reinforcement is retained at elevated temperatures up to 1400°C. This observation is consistent with the observations made by Jenkins et.al.[11] on this composite using the chevron notched beam technique to evaluate fracture toughness at elevated temperatures.

The variations in K_{Ic} are attributed to the TiB_2 particle dispersions resulting in non-uniform mean free path (inter-particle distance) between the reinforcing particulates and changes in the particle size distributions. The single edge notched beam technique is generally sensitive to the notch characteristics and may lead to some variation in toughness from batch to batch. Nevertheless, much of the variation results from localized differences in the microstructural

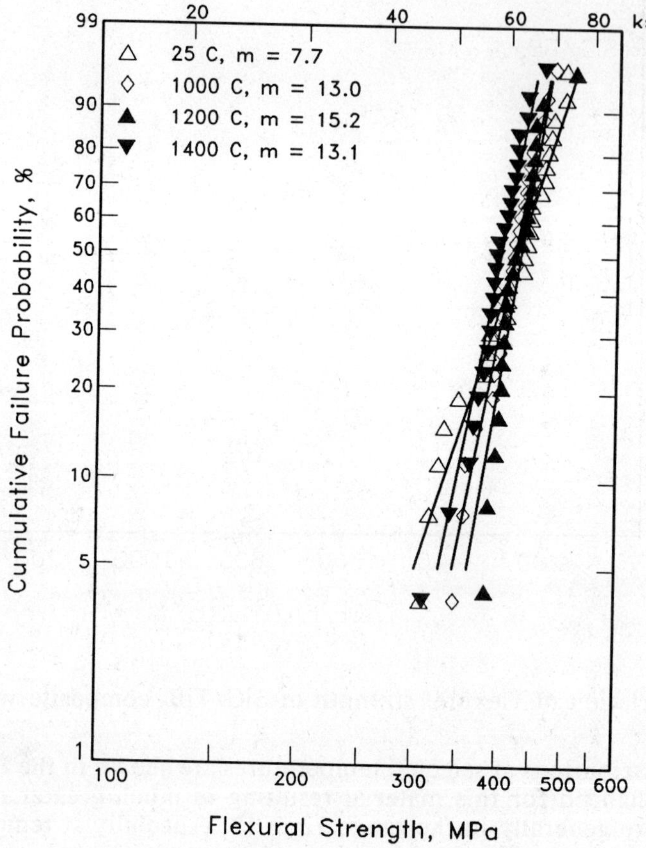

Figure 6. Flexural strength distributions in SiC/TiB_2 composite.

features, such as TiB_2 particle size and the distribution of the reinforcing particles. In addition to the crack deflection mechanisms, pinning was also observed especially around larger TiB_2 particles.

One critical need in engine applications is the long term durability as demonstrated by minimum or absence of slow crack growth of either pre-existing or newly-generated flaws during service. In order to obtain data on this aspect, dynamic fatigue tests were conducted in static air under flexure conditions.[13] The effect of stressing rate on the flexural strength distributions was found to be insignificant at 1000, 1200 and 1400°C. The results, analyzed by using trivariate regression procedures, yielded extremely high values (>200) for the slow crack growth exponent, n (Figure 10). The long term performance at elevated temperatures is also improved markedly over the non-reinforced material as the subcritical crack growth behavior is virtually non-existent up to 1400°C in air. The underlying mechanisms for the enhancement of resistance to slow crack growth is not known at this time as fractography work is underway.

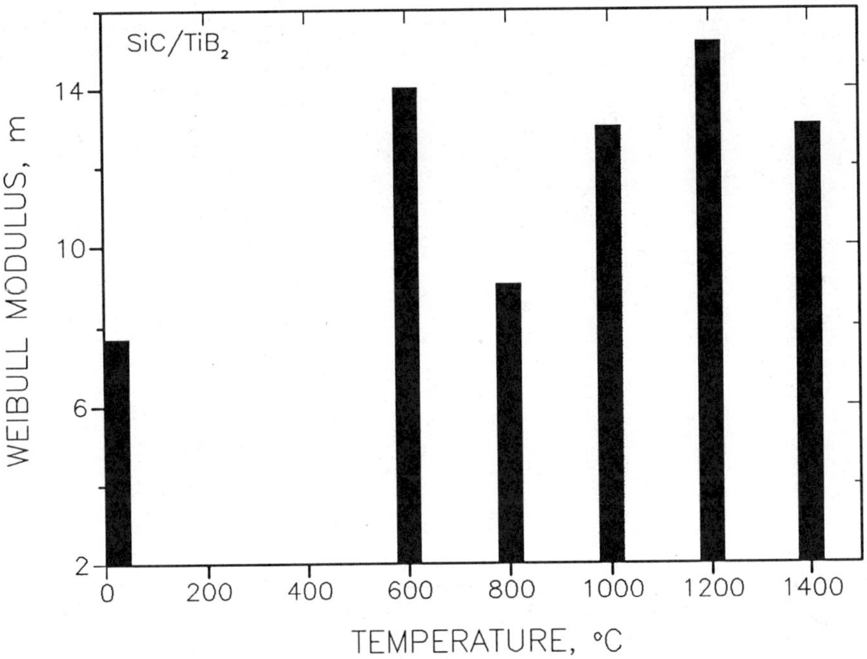

Figure 7. Variation of Weibull moduli of SiC/TiB_2 with temperature.

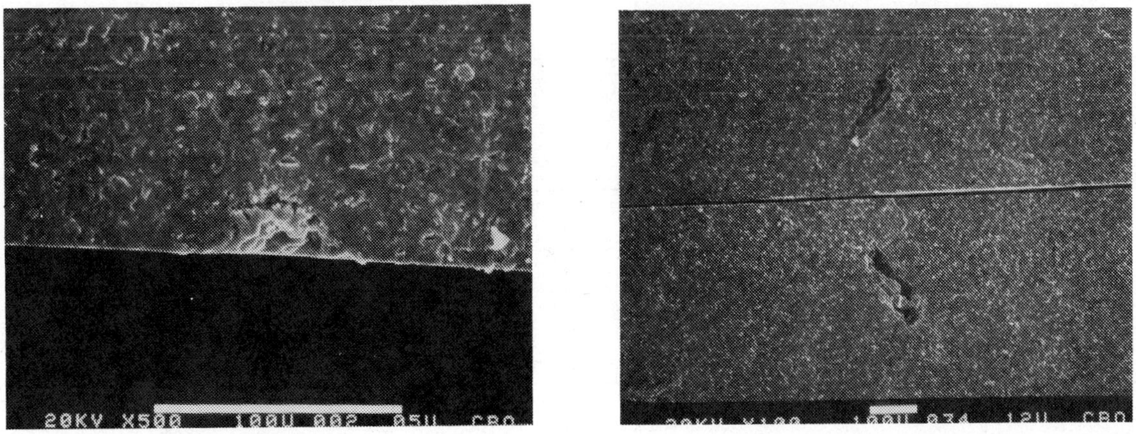

(a) (b)

Figure 8. (a) Poorly sintered agglomerate about 50 μm in diameter, located at the tensile surface.
(b) Elongated subsurface void, 150 μm long and 50 μm wide located 150 μm from the tensile surface.

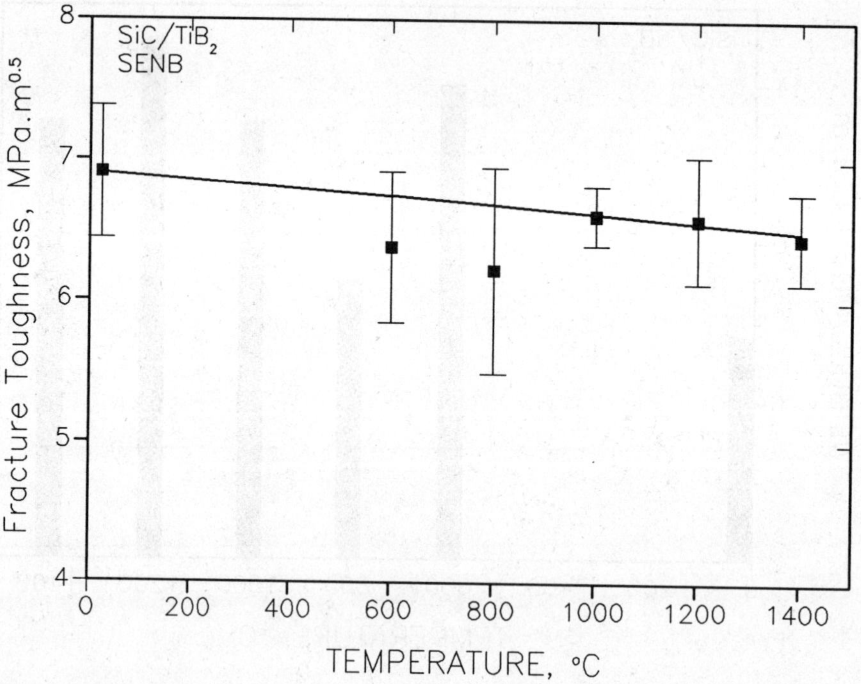

Figure 9. The SENB fracture toughness of SiC/TiB_2 composites.

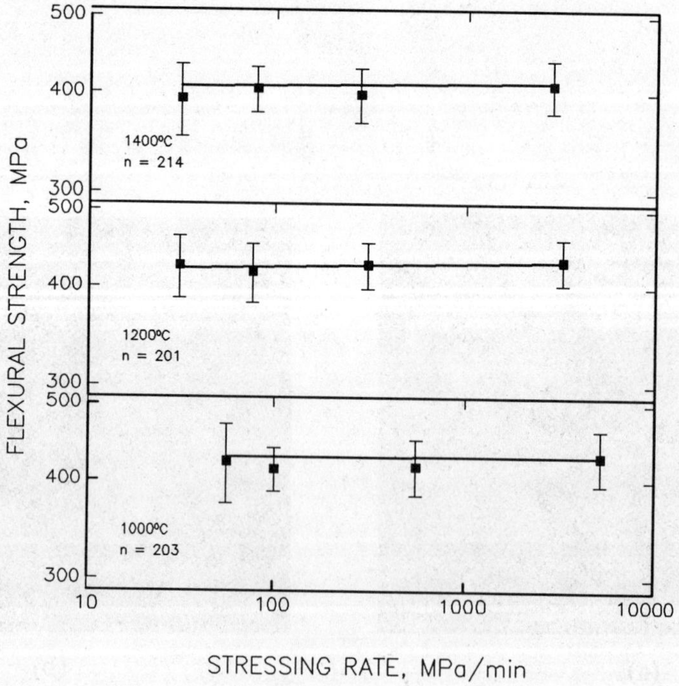

Figure 10. The dynamic fatigue results for SiC/TiB_2 composite indicating insignificant slow crack growth behavior.

However, owing to the increased tendency for TiB_2 to oxidize in air at higher temperatures, this material shows enhanced oxidation at temperatures greater than 1200°C. Within a short time a yellow surface scale is formed which has been identified to be TiO_2 (rutile) by X-ray diffraction techniques. The extent of oxidation is generally not significant up to 1200°C, but increases at higher temperatures. The static exposure to ambient air at these temperatures, however, results in the formation of a passive oxide layer which decreases the oxidation rate with time.[14] The oxidation rate falls steeply with time reaching very low values within a few hours.

The effect of static oxidation on the room temperature flexural strength of these materials was evaluated at 1000, 1200 and 1400°C (Figure 11)[13] The strength did not change with long term exposures at 1000 (1000 hours) and 1200°C (200 hours). The small increase in the retained strength, seen after short exposures below 1200°C, is within the statistical variations in the strength. However, significant reduction in strength was observed even after short term exposures at 1400°C. This observation is consistent with the oxidation behavior discussed earlier.

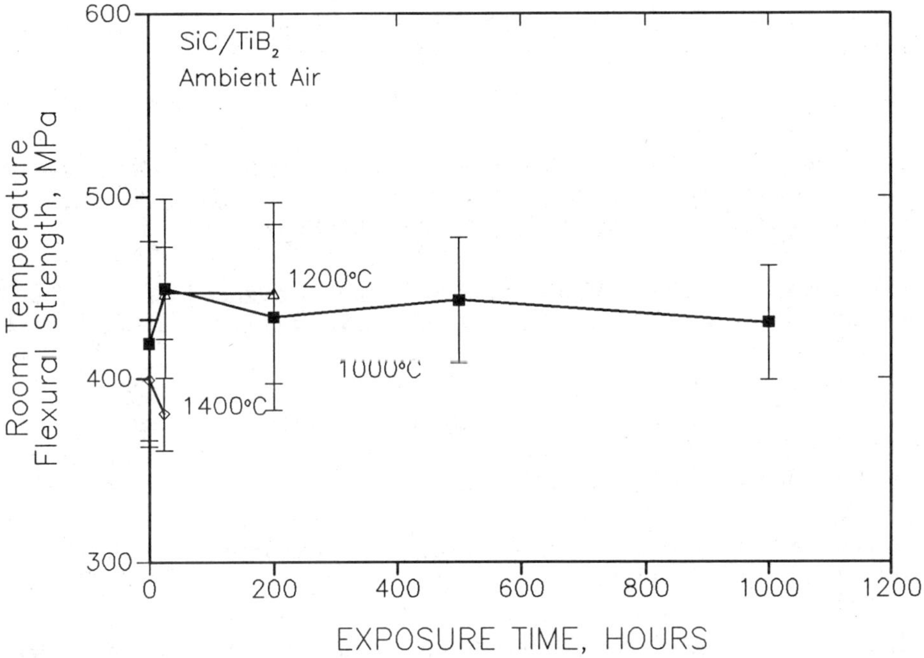

Figure 11. The effect of long term high temperature exposure to air on the retained strength at 25°C in SiC/TiB_2 composites.

CONCLUSIONS:

It has been shown that a ceramic particulate reinforced composite system consisting of TiB_2 particles in alpha SiC matrix can be fabricated into complex heat engine components by pressureless sintering. In addition, this system exhibits excellent properties with respect to both room temperature and elevated temperature strength, Weibull modulus, fracture propagation resistance, oxidation resistance and retention of strength after long term exposure to oxidizing environment. Also, dynamic fatigue experiments have indicated the absence of elevated temperature slow crack growth, thereby assuring higher lifetime in elevated temperature applications. More work needs to be done, however, in evaluating and understanding creep at elevated temperature, long term stress rupture/thermal cycling effects and fatigue properties.

The SiC/TiB_2 composite material is currently being evaluated for such engine components as wear inserts, roller followers and valve guides for piston engines and hot flow path components in turbine engines. In addition, other applications exist such as nozzles and bearing guides requiring wear resistance and armor applications with enhanced ballistic impact resistance. For successful application of this materials in heat engines, effective non-destructive evaluation procedures and accept/reject criteria need to be developed.

ACKNOWLEDGMENTS:

The authors acknowledge the efforts of many colleagues at the Technology and Structural Ceramic Divisions of the Carborundum Company who have contributed extensively to the development of this composite.

REFERENCES

1. Proc. 25th Automotive Technology Development Contractors' Coordination Meeting, P-209, Detroit, Mich., SAE-DOE, October 26-29, 1987.
2. H. E. Helms et al, "Ceramic Applications in Turbine Engines", Noyes Publications, Park Ridge, New Jersey, 1985.
3. G. L. Boyd and D. M. Kreiner, "AGT101/ATTAP Ceramic Technology Development", Proc. 25th Automotive Technology Development Contractors' Coordination Meeting, P-209, Detroit, Mich., SAE-DOE, October 26-29, 1987. p.101-114.
4. M. O. Ten Eyck, J. W. MacBeth, and T. B. Sweeting, "Advanced Gas Turbine Technology Development Project, Ceramic Component Developments", Topical Report, 1988. NASA/CR-180071.
5. W. D. G. Boecker, C. H. McMurtry, S. G. Seshadri, and M. Srinivasan, "Silicon Carbide and Sialon Ceramics for Structural Applications: Development Status and Potential", Proc. 2nd Intl. Symposium on Ceramic Materials and Components for Engines, Lübeck-Travemünde, 1986, p.689-696
6. C. H. McMurtry, W. D. G. Boecker, S. G. Seshadri, J. S. Zanghi, and J. E. Garnier, "Microstructure and Material Properties of SiC-TiB_2 Particulate Composites", Am. Ceram. Soc. Bull., **66**, [2] 325-29 (1987)
7. J. W. MacBeth, M. O. Ten Eyck, and R. W. Ohnsorg, "Non-oxide Ceramics for Advanced Heat Engine Applications", SAE International Congress and Exposition, Detroit, Mich., February 23-27, 1987. SAE Technical Paper No. 870468.
8. A. Zangvil, University of Illinois at Urbana, Private communication, 1986.
9. J. E. Garnier, unpublished work, Standard Oil Engineered Materials Company, 1986-87.
10. S. G. Seshadri, J. E. Garnier and K. Y. Chia, "Effect of Machining Techniques on the Mechanical Properties of SiC/TiB_2 Composites", 89th Annual Mtg., Am. Ceram. Soc., Pittsburgh, Pa. April 29, '87.
11. M. Jenkins, University of Washington, Private Communication, 1987.
12. S. G. Seshadri, C. H. McMurtry, J. E. Garnier, and W. D. G. Boecker, "High Temperature Properties of SiC/TiB_2 Particulate Composite Material", 39th Pac. Coast Regl. Mtg. of the Amer. Ceram. Soc., Seattle, Wash. Oct. 23, '86.
13. K. Y. Chia, S. G. Seshadri, and M. Srinivasan "high Temperature Performance and Reliability of SiC/TiB_2 Composites", 12th Annual Conference on Composites and Advanced Ceramics, Am. Ceram. Soc., Cocoa Beach, Fl. 1988.
14. S. G. Seshadri and M. Srinivasan, "Liquid Corrosion and High Temperature Oxidation Effects on Silicon Carbide/Titanium Diboride Composites", J. Am. Ceram. Soc., 71 [2] C-72 - C-74 (1988)

CERAMIC ENGINE VALVES

R. R. Wills
R. E. Southam
TRW Valve Division
Cleveland, OH 44110

ABSTRACT

Current trends in spark ignition engines are toward multivalve engines capable of operating at relatively high speeds to maximize performance and fuel economy at acceptable emissions levels. Structural ceramics, with their low density and good high temperature properties, appear to offer potential as an alternative valve material to the steels, superalloys, and titanium alloys currently in use. Progress to date on materials development, component design, component performance, and engine testing indicates that silicon nitride valves will operate satisfactorily.

INTRODUCTION

Spark ignition engines employ poppet valves to control both the flow of air into each cylinder and the removal of exhaust gases after each combustion cycle. The inlet valve head is usually wider than that of the exhaust valve and the materials used for each valve are different, principally reflecting the higher temperature properties required for the exhaust valve. Valve motion is usually controlled by a cam operated valve train system consisting of a cam, a tappet, a rocker arm, a pushrod, and a spring as shown in Figure 1. In this paper we briefly describe (1) some of the potential benefits of using ceramic valves, (2) materials selection criteria, and (3) component evaluation in dynamometer testing and in a passenger car.

BENEFITS OF CERAMIC VALVES

The replacement of a metal valve by a lighter weight ceramic valve reduces the mass of the valve train by the difference in weight of the two valves. A typical automotive stainless steel exhaust valve, such as 21-2N stainless steel, has a weight of 100 grams (0.22 lbs.) whereas the corresponding silicon nitride or silicon carbide valve is approximately sixty percent lower in weight. A valve train computer simulation program was used to examine the changes in forces resulting from the reduction in valve mass and to examine the engine speeds that caused valve "bounce." This condition

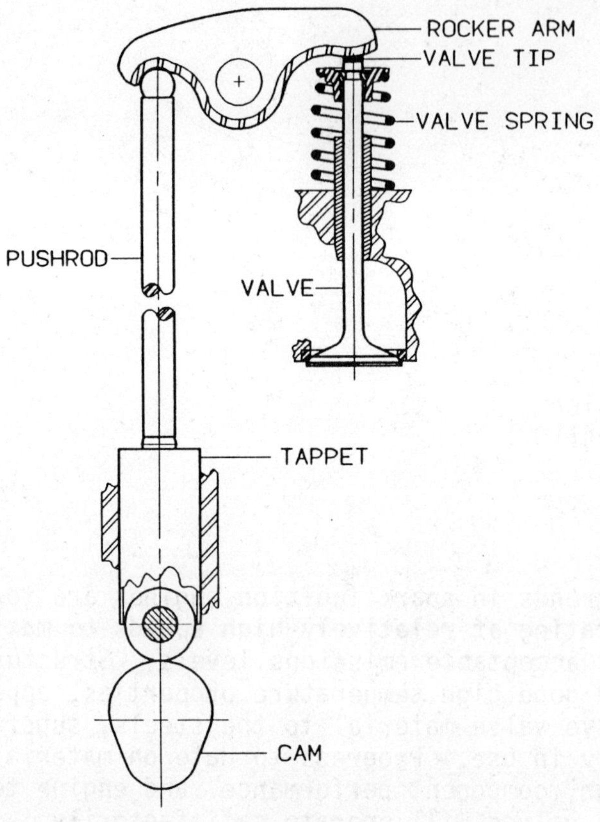

Figure 1. Elements of valve train in a pushrod engine.

occurs when the valve spring is no longer able to make the valve follow the cam.

Figures 2 and 3 show the velocities of the metal and ceramic valves at 4000 and 5100 rpm. At 4000 rpm the velocity profiles of both valves are similar, but at 5100 rpm the maximum velocity of the metal valve during the opening part of the cycle is higher than that of the ceramic. Furthermore, considerable oscillation occurs during seating at a 240 degree cam angle. This is the "bounce" condition. A similar condition is not achieved with the ceramic valve until 6000 rpm. A similar effect was also found by Allison [1] using a 1.9 liter engine to compare stainless steel and titanium valves. The valve "bounce" speed will, however, increase more with a structural ceramic valve than with titanium because of the density difference between these materials.

The lower weight of a ceramic valve means that valve seat loading will also be reduced. In addition, less loading is exerted on the cam. This may enable less costly seat insert, cam, and tappet materials to be used. A further possibility is to lower the spring loads since springs are generally chosen to ensure that valve motion is controlled by cam motion at maximum engine speed and not by the inertia of the components. Reducing the spring load lowers the friction in the valve train. This is typically 25 percent of the total engine friction for a passenger car [1].

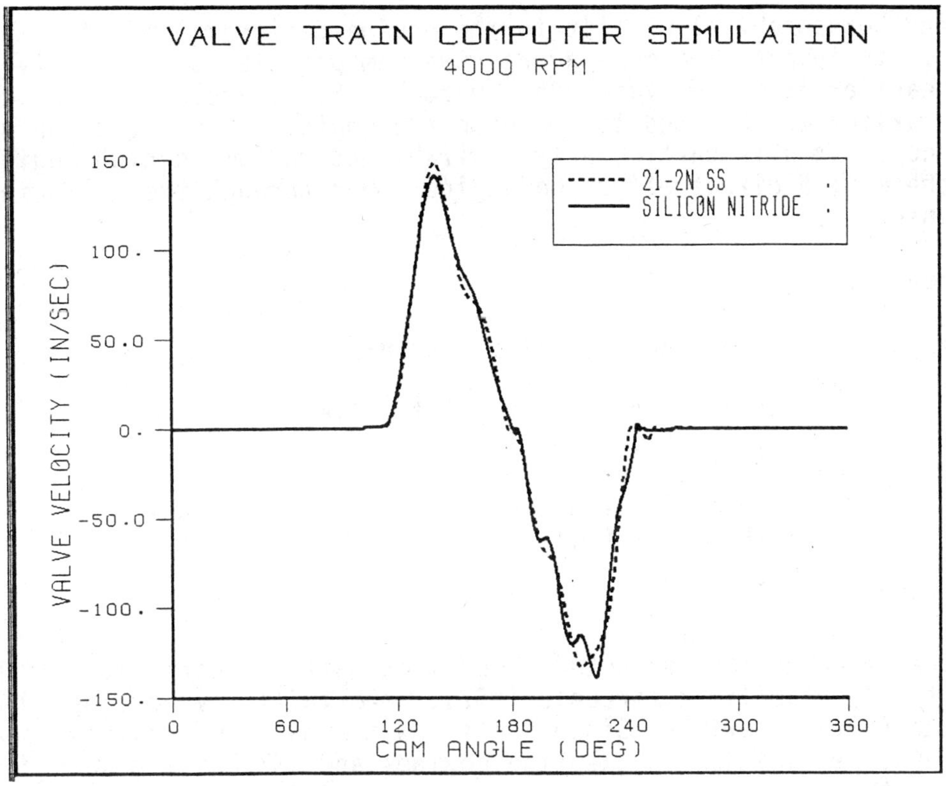

Figure 2. Velocities of 21-2N stainless steel and silicon nitride valves at 4000 rpm derived from computer simulation.

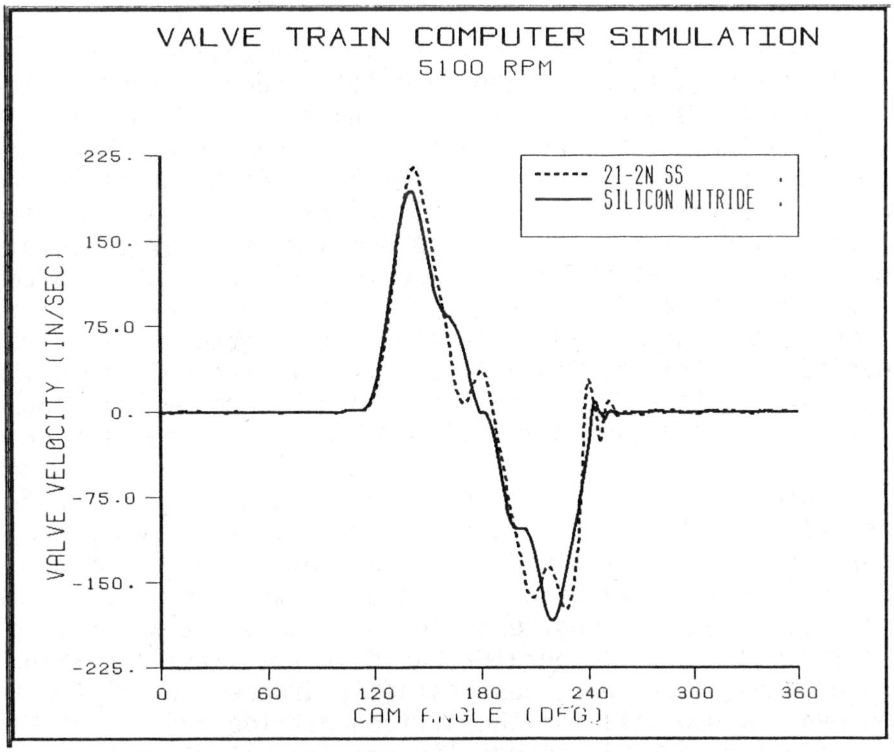

Figure 3. Velocities of 21-2N stainless steel and silicon nitride valves at 5100 rpm derived from computer simulation.

Another potential benefit to the engine designer using lighter weight valves is to incorporate more aggressive cam profiles so that valves can be opened earlier or closed later than normal. For example, opening the intake valves earlier can be used to increase the engine's air intake or volumetric efficiency. This is particularly desirable at medium to high engine speeds since there is a direct effect on engine power through the following relationship [2].

$$mep = J \rho_a e_v F Q_c \eta \quad \ldots\ldots\ldots\ldots\ldots\ldots\ldots\ldots\ldots (1)$$

where mep = the mean effective pressure
J = Joule's constant
ρ_a = density of air at inlet valve
e_v = volumetric efficiency
F = fuel air ratio
Q_c = heat of combustion per unit mass of fuel
η = thermal efficiency

MATERIALS SELECTION

The three primary structural ceramics, silicon carbide, silicon nitride, and partially stabilized zirconia (PSZ), were initially considered as candidate valve materials. Since PSZ's low thermal conductivity (see Table 1) might cause preignition in gasoline engines and PSZ has a higher density than the other two materials, it was selected only for potential low heat rejection diesel engines. However, preliminary testing [3] of PSZ valves in a diesel engine showed that PSZ lacked adequate thermal shock resistance.

The criteria for material choice was minimization of operating stresses. This was based upon both material properties and finite element analyses of silicon carbide and nitride exhaust valves operating under both steady state and transient state conditions. Steady state boundary conditions were established by correlating valve temperatures to those determined using temperature check valves. For a combustion chamber average gas temperature of 1832°F, silicon carbide is predicted to operate at lower temperatures than silicon nitride except at the valve face (see Figure 4). Heat conducted away from the valve face when it contacts the seat insert causes a temperature gradient across the valve head. The temperature gradient is greater for the silicon nitride because its thermal conductivity is lower than that of silicon carbide (see Table 2). The maximum thermal stresses are 130 MPa (20,000 psi) for silicon nitride and 90 MPa (13,000 psi) for the silicon carbide.

The second analysis was for the stresses generated under transient conditions. In this scenario, the cylinder head is suddenly cooled with cold water after running hot water under pseudo steady state conditions. These conditions were used to exaggerate the thermal shock conditions the valve would be exposed to during engine shutdown. The valve was initially assumed to be at the steady state temperature used in the first scenario. The cylinder head temperature was then instantly dropped to 100°F. Results showed 160 MPa (24,000 psi) for the silicon nitride and 117 MPa (17,000 psi) for the silicon carbide. These results are consistent with the thermal properties of these two ceramics.

Table 1. Typical Properties of Structural Ceramics

PROPERTY	UNITS	SILICON NITRIDE	SILICON CARBIDE	PARTIALLY STABILIZED ZIRCONIA
Density	g/cm^3	3.2	3.1	5.8
Flexural Strength (room temp.)	MPa	≈ 700	≈ 480	≈ 700
Young's Modulus	GPa	310	410	200
Fracture Toughness	MPa$^{0.5}$	5	3	6-15
Weibull Modulus		11	10	20
Thermal Expansion	10^{-6}°C	3.04	4.3	10
Thermal Conductivity	W/MK	31	83	2

*Room temperature properties

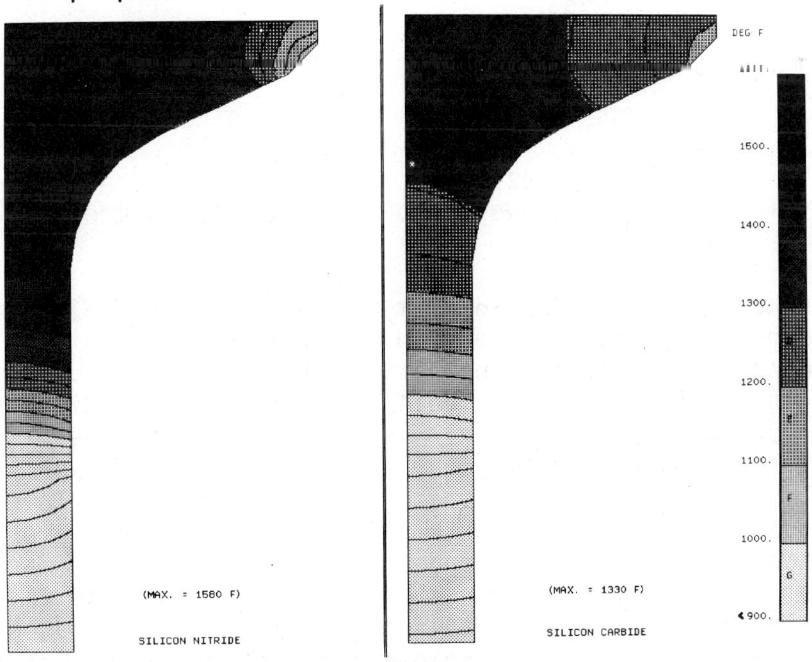

Figure 4. Temperature distributions across silicon nitride and silicon carbide valves operating under steady state conditions.

An engine valve is subjected to both mechanical and thermal loading and in the third scenario peak combustion pressure 8 MPa (1,160 psi) was applied to the valve operating at wide open throttle steady state temperatures and 0.127 mm (0.005 in) cylinder head tilt. Under these conditions, the silicon nitride valve operated at lower stresses than the silicon carbide (see Figure 5) principally because of its lower elastic modulus. This factor, together with the higher fracture toughness of silicon nitride, were the reasons for the selection of this material as the best ceramic valve material.

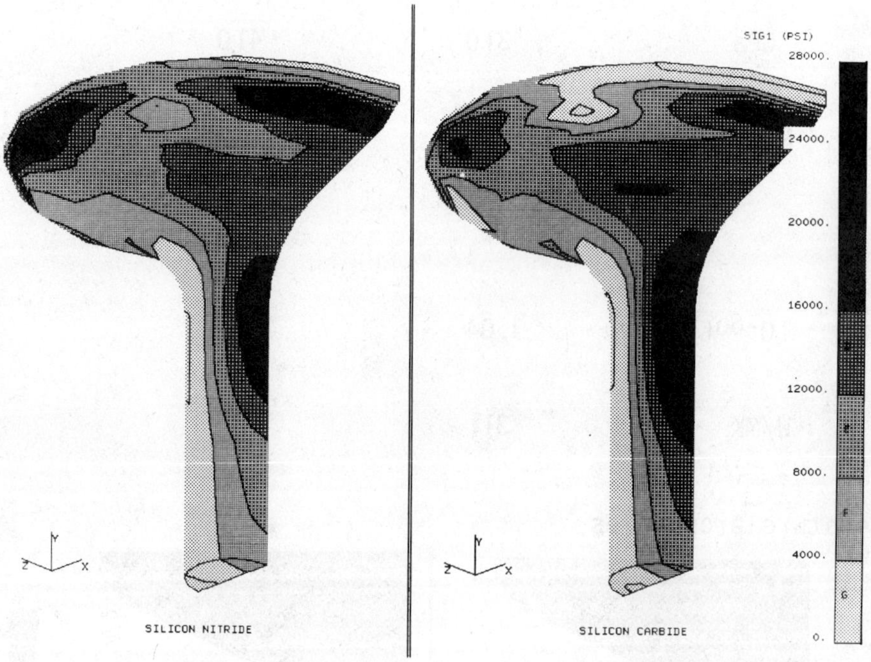

Figure 5. Stress distributions across silicon nitride and silicon carbide valves operating under combined thermal and mechanical loading.

Valves have been fabricated from a number of silicon nitride materials and, subsequently, tested in motorized rigs and firing engines. Not all of these silicon nitride materials performed satisfactorily. Failures tended to occur during engine cycling. Table 2 lists the properties of VMS-653, a sintered silicon nitride or sialon. This material has proven to be a satisfactory valve material. Fracture toughness data is expressed in two ways to reflect the diversity of measurement techniques employed today by numerous groups reporting data on this subject and the difference in local fracture toughness determined by indentation techniques and macroscopic fracture toughness measured by the single edge notched beam (SENB) technique. The dynamic fracture toughness (K_{ID}) was calculated from the characteristics of the flaw responsible for failure in the charpy impact test. The values for K_{ID} and K_{IC} by indentation are similar indicating that the fracture toughness of VMS-653 is not strain rate sensitive at room temperature.

Table 2. Properties of VMS-653*

Density (g/cm^3)	3.23
Thermal Expansion (C)	3.04 x 10^{-6}
Thermal Conductivity (watts/m k)	21
Young's Modulus (GPa)	300
K_{ID} (MPam $^{0.5}$)	4.73
K_{IC} (Indentation)	5
K_{IC} (SENB)	7.7
3 pt. Bend strength (MPa)	940
4 pt. Bend strength (MPa)	686
Weibull Modulus	11

*a sintered silicon nitride or sialon

COMPONENT TESTING

Dynamometer testing was performed using several gasoline and diesel engines; both light duty and heavy duty as shown in Table 3. The valves showed very little wear after testing and were in excellent condition. Deposits were easier to remove than on comparable metal valves. After testing in the passenger car gasoline engine, the valves were brushed, cleaned, and installed in a 1987 Oldsmobile Cutlass Ciera. All other components, except the locks, were production items. This vehicle, shown in Figure 6, has now accumulated over 20,000 miles without incident in a variety of driving conditions including cold start tests in which exhaust gas temperatures rise from below the freezing point to 450°C in 30 seconds. The replacement of metal valves by the silicon nitride valves has not affected the fuel consumption rate.

Table 3. Engine Conditions for Valve Durability Testing

Light Duty Diesel

1. Cyclic, 3600 rpm full load for 3 hrs. and 1500 rpm no load for one hour. Test period 100 hrs.

2. Cyclic, 3600 rpm full load for 10 mins., 3800 no load for 5 mins. Test period 1000 hrs.

Heavy Duty Diesel

1. Cyclic, 2100 rpm full load for 50 mins., 2375 rpm no load for 10 mins. Test period 1000 hrs.

Light Duty Gasoline

1. 4600 rpm full load constant speed for 100 hrs.

2. Cycling from 4100 to 4600 rpm at full load. Engine idled briefly every 14 mins. Test period 300 hrs.

Figure 6. Passenger car test vehicle for silicon nitride valves

VALVE TRAIN DYNAMICS

The trends expected from the valve train computer simulation program have been verified experimentally [4] using a 1987 General Motors pushrod engine. The red line speed for this engine was 5200 rpm. Tests conducted at this speed showed the standard metal valves to be oscillating on the seat inserts. The impact force on the seat insert was 5564 N (1251 lbs.). At the same engine speed, using silicon nitride valves and lighter spring loads (56 kg versus 76 kg), the impact force was reduced to 1098 N (245 lbs.). Despite using reduced spring loads, the valve "bounce" condition was not observed. Measurements of seat insert and valve guide wear, using both metal and silicon nitride valves also show that less wear occurs with the ceramic valves. For example, in an engine test operated under full load constant speed conditions average seat wear was reduced by a factor of 9 for the intake valves and 36 for the exhaust valves. This is consistent with the lower seating forces on the insert and may also be affected by reduced valve head flexing since silicon nitride has a higher elastic modulus than the ferrous alloys used as production intake and exhaust valves.

SUMMARY

Finite element analysis of silicon carbide, partially stabilized zirconia, and silicon nitride valves under steady state and transient state conditions showed that silicon nitride offered the best material properties. Good quality sintered silicon nitride valves passed extensive dynamometer testing and field trials in a 1987 passenger car. Potential advantages offered by these valves include reduced seat insert and valve guide wear, improved valve train dynamics, increased engine output, and reduced friction loss using lower spring loads.

ACKNOWLEDGMENTS

This work was sponsored by Norton/TRW Ceramics, a joint venture between Norton Company of Worcester, Massachusetts and TRW Inc. of Cleveland, Ohio.

REFERENCES

1. J. E. Allison, A. M. Sherman, and M. R. Bapna, "Titanium in Engine Valve Systems," J. Metals, March 1987, p. 15-18.

2. "The Internal Combustion Engine in Theory and in Practice," Volume 1, C. F. Taylor. Published by The Technology Press of the Massachusetts Institute of Technology and John Wiley and Sons, Inc., New York 1960.

3. M. Asnani and F. L. Kuonen, "Ceramic Valve and Seat Insert Performance in a Diesel Engine," Society of Automotive Engineers paper number 850358, 1986.

4. S. H. Updike and P. D. Nagle, "Ceramic Valve Train Components," Society of Automotive Engineers, paper number 880441, 1988.

QUALITY ASSURANCE OF HEAT ENGINE COMPONENTS THROUGH PROCESS OPTIMIZATION

Claude Brown, Jr., James W. Johnson, and George A. Fryburg
Norton/TRW Ceramics
Northboro, MA 01532

ABSTRACT

The development of a new business in ceramic heat engine components offers a unique opportunity to utilize off-line quality control techniques. The use of capability analysis for process reliability determination and experimental design for the improvement of this reliability are considered critical in pilot plant scale-up activities. In many aspects of the engine component prototype operation, designed experimentation has proven to be critical in the timely achievement of design goals and the assurance of high levels of reliability for engine components. The general approach to off-line quality control will be reviewed along with selected case studies.

INTRODUCTION

Successful establishment of ceramic engine component production requires excellence in materials, processing, design and delivery, all of which add up to excellence in product "quality." Assuring specific levels of quality at any stage of the production process requires a much more active role than does merely "controlling" the quality. In its full definition the term "Quality" describes both the characteristics of the final product (i.e, what you have) and the efficiency with which the product was created (i.e, how easily you got it). Thus, Quality Assurance implies not only that the application specific attributes be assured but that the method by which those attributes are created be optimized.

Off-line quality control, a term synonymous with quality assurance, is generally used to describe process optimization techniques whose goal is to make a process more efficient, controlled, and reliable. A process can be made more reliable if it is modified so that the product attributes created at that process are shielded from uncontrolled variations in the process parameters. Better process control can be achieved by fully defining the influence each parameter has on the attributes created at that process. Efficient processes result when the optimal settings are selected for the controlling parameters of a process. This allows one to achieve design requirements at suitable performance levels. Specific methods to achieve these goals are described along with two case studies which illustrate applications to material, process, and product development activities associated with the prototype production of ceramic engine components.

GENERAL DISCUSSION

Knowledge of the process technology is the key to controlling and thus assuring product quality. A process which is completely understood, fully qualified, and intelligently controlled at every critical step will impart a much higher degree of quality to the final product. This results in reduced cost, better design conformance, and increased reproducibility [1,2]. The advocated approach to quality assurance and off-line quality control can be described by these three different techniques:

1) Fractional Factorial Experimental Design
2) Response Surface Methodology
3) Process Capability Analysis

Experimental design is a general term which refers to the systematic approach to experimentation in one of many forms [3,4]. Fractional factorial experimental design is a specific family of techniques which allows the effects of several variables of a process to be analyzed without performing every possible experimental combination. Taguchi experimental design is one of these fractional factorial techniques which greatly reduces the number of trials required to fully analyze a system [5,6,7]. Data from such an experiment is usually analyzed using an ANOVA (Analysis of Variance) table in which each of the variables studied in the experiment is ranked according to its importance. Experimental variables are referred to as factors or parameters while observed results are termed responses or attributes. The factors studied by designed experiments are divided into two categories: control parameters and outside factors. Outside factors are process variables which will be defined as part of the process after an optimum setting is found. Controlling parameters are variables which control the result of a process once the relationships between them are defined.

Another important result from the analysis of a designed experiment is the level average information. This statistical technique allows predictions of the results for any possible experimental combinations whether or not it was actually performed in the fractional design. Matrices of this level average data provide a basis for modeling the response surfaces [8]. Response surface methodology is a technique where the relationship between two or more experimental variables and one or more experimental results is defined by three-dimensional surfaces. This response surface, sometimes referred to as a processing map, defines optimal regions of operation and/or increased stability.

Once the experimental design and optimization work are completed the process performance may be summarized using a process capability study [9]. This type of analysis will predict the long term performance of a process using normal statistics. This process capability can then be used to compare particular aspects of a process' performance to component specifications in order to judge the capability of the process.

The first case study, Sialon Sintering Behavior, illustrates how fractional factorial experimental design and response surface methodology are used to efficiently define and optimize a process in order to establish operational procedures. The second case study, Injection Molding of Wear Buttons, demonstrates how process capability analysis can be used to qualify an entire process for the achievement of specific component design goals.

CASE I: SIALON SINTERING BEHAVIOR

This case study illustrates how designed experimentation is employed as an accurate and efficient method to make process development decisions and establish operation specifications. The sintering behavior of a sialon composition is studied in order to understand the relationship between the processing parameters and important product attributes. Absolute control of these properties, component density, weight loss and shrinkage, has significant impact on the realized quality of the final component. This example illustrates the fundamental principles of engineered experimentation described in the previous section.

A detailed "processing map" is required for the efficient and reproducible operation of the sintering process. Such a processing map will allow the control of significant sintering parameters in order to predict and operate at optimum conditions while allowing process changes to be made with greater understanding and confidence. Another benefit is the opportunity to select operation parameters which not only achieve targeted results but which minimize the process' susceptibility to operational variation. The first step is to outline the processing parameters presumed to be significant. Next, a Taguchi style fractional factorial experiment is designed and includes a practical range of settings for these experimental parameters listed below.

1. Presence of Setter Grain
2. Crucible or Sagger Material
3. Powder Type/System
4. Sintering Temperature
5. Time at Sintering Temperature
6. Heating Rate to Sintering Temperature

Next we require information both to establish operational guidelines and to make decisions on the actual process design; thus, both controlling parameters and outside factors will be studied. The first three processing parameters, setter grain, crucible material and powder type, are considered outside factors. Our goal is to select optimum settings for these parameters and then define them as part of the process. The next three processing parameters, sintering temperature, sintering time and heating rate, are considered to be controlling factors since varying the settings of these parameters will be used to control the results of the sintering process.

Following completion of each of the experimental trials as dictated by the fractional factorial experimental design, standard ANOVA techniques are employed. This establishes the relative importance of each of the experimental parameters. Each parameter's influence in determining the variation of the three sintering responses of interest is calculated with the results of this analysis being summarized in three pie charts displayed in Figure 1.

Before constructing a processing map to relate the controlling parameters to the sintering attributes, the optimal settings for the outside factors are established. Considering the information displayed by the charts in Figure 1,

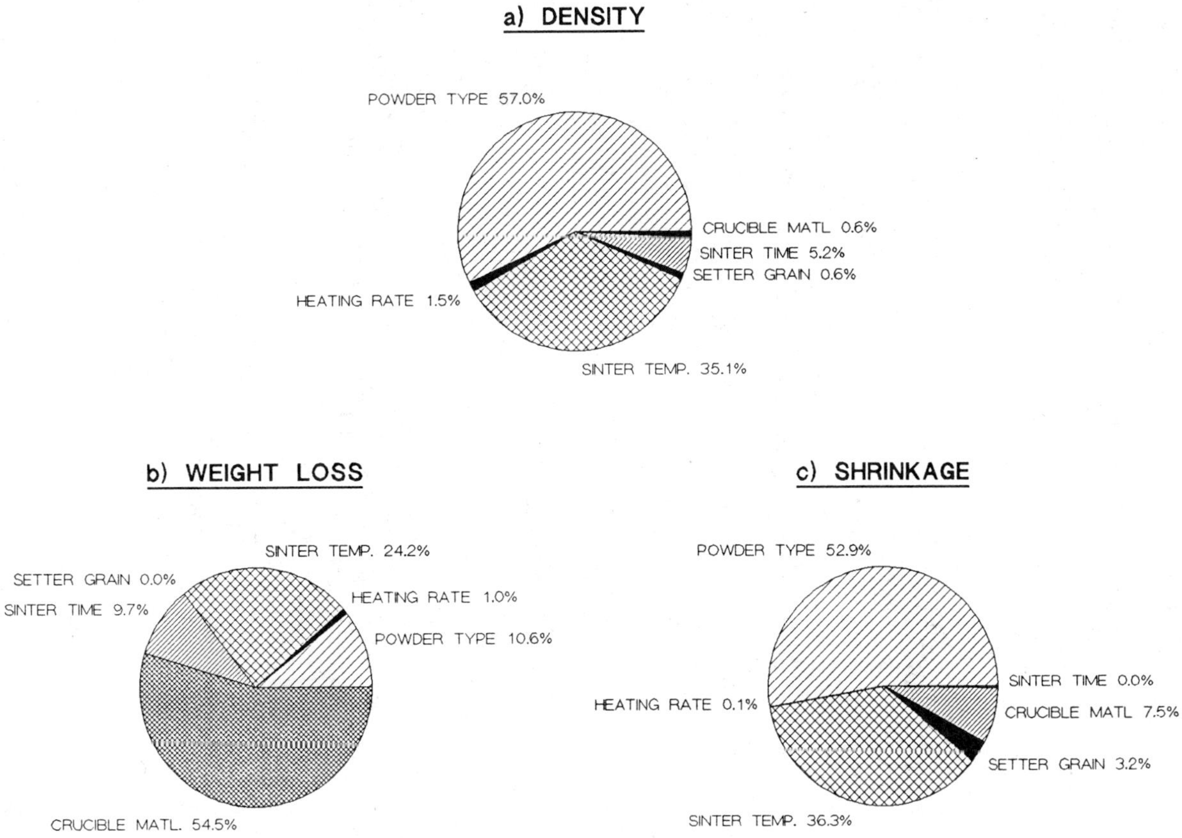

Figure 1: Sialon sintering characteristics [(a) sintered density, (b) weight loss, and (c) shrinkage] as influenced by various processing parameters. (ANOVA table data)

the presence of setter grain has no definitive effect on any of the sintering responses being monitored. However, the choice of crucible material has a significant impact on the weight loss experienced during sintering. Using level average data, the bar graph in Figure 2 illustrates the extent of the effect of crucible material selection on the weight loss of the sintered component. Since compositional control of the material is a prime factor in the performance of the component, a sialon crucible material is selected to help minimize the weight loss during sintering. In addition, the choice of powder type, particularly the binder system, has a large influence on both the density and shrinkage observed during sintering as quantified in figures 3 and 4. Consistent with the goals of 1) improving product quality by increasing final density and 2) minimizing dimensional shrinkage, the powder type denoted as "standard" is selected for this process.

With the selection of the outside factors, the statistical analysis of the same experimental data is re-computed in order to quantify the influence of the controlling factors themselves. This effort results in an updated ANOVA table and level average information which is used to generate new charts show-

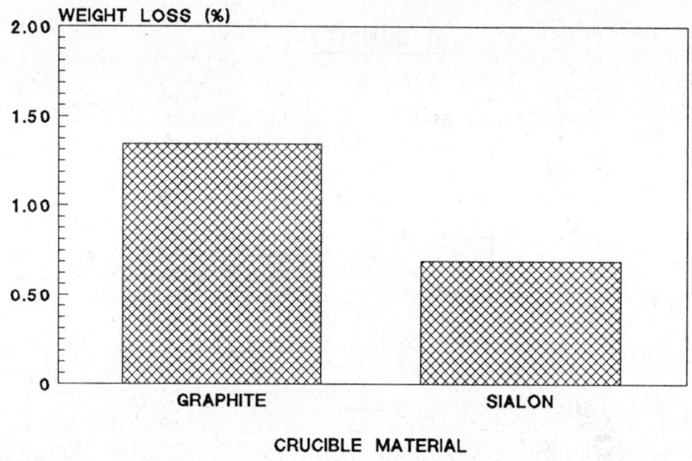

Figure 2: Sialon sintered weight loss as a function of crucible material. (Level average data)

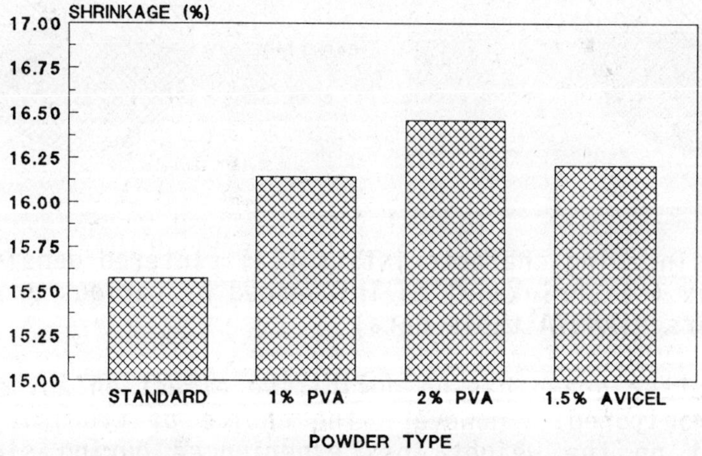

Figure 3: Sialon sintered shrinkage as a function of powder type. (Level average data)

ing the relative influence of the controlling parameters on the three sintering responses. The next step in analyzing the results of this designed experiment is to generate response surface plots using the level average data. These plots define the relationship between the significant sintering parameters and the responses of interest. Careful examination of such response surface models reveals the regions of optimal and/or stable process operation along with information upon which process change decisions and scale up plans can be based.

In the case of component shrinkage, sintering temperature is the only significant factor when judged against the error of the experiment and analytical

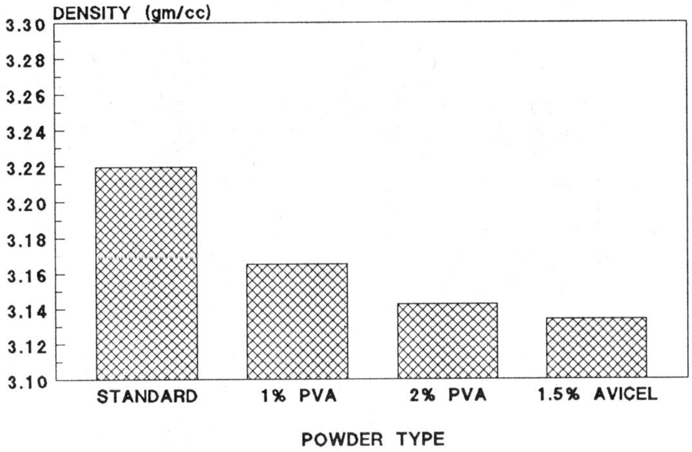

Figure 4: Sialon sintered density as a function of powder type. (Level average data)

technique. However, both sintering temperature and sintering time are considered significant influences on the weight loss experienced during sintering. While both weight loss and shrinkage are important sintering parameters to study, the goal is not to achieve a specific optimal value for these attributes since the process can be engineered to accept a previously established value. The goal of analyzing the response surfaces of weight loss and shrinkage is to locate the operational regions where the variations of these attributes caused by fluctuations in the controlling parameters are minimized (i.e., the goal is to define an operational region where the process is inert and highly reproducible).

Conversely, sintered density is an attribute of the sintering process where a region of maximum values is desirable since this can directly translate into increased reliability in component performance. After defining the settings of the outside factors, the modified ANOVA results indicate that both sintering time and sintering temperature are significant factors in determining final density. The controlling factor, heating rate, does not prove to be significant for any attribute of sintering over the rates studied here and therefore is set at a level judged appropriate by some other criteria. The response surface of sintered density as a function of both sintered time and temperature is illustrated by the surface and topographical plots in figures 5 and 6. The surface plot Figure 5 shows a region of maximum density at prolonged sintering times and higher sintering temperatures. This illustration, along with the contour plot in Figure 6, displays a very useful processing map for the sintering of this sialon composition. Based on knowledge of this response surface, many types of critical decisions can now be made with knowledge of the tradeoffs and with confidence in the predictions. For example, if a target density of 3.24 gm/cc is desired, any combination of temperatures and times along the 3.24 contour in Figure 6 are sufficient. If temperature was difficult to control, it is then better to operate along the 3.24 contour where the density is less sensitive to temperature fluctuations, i.e., at shorter times and higher temperatures. This allows for higher cosis-

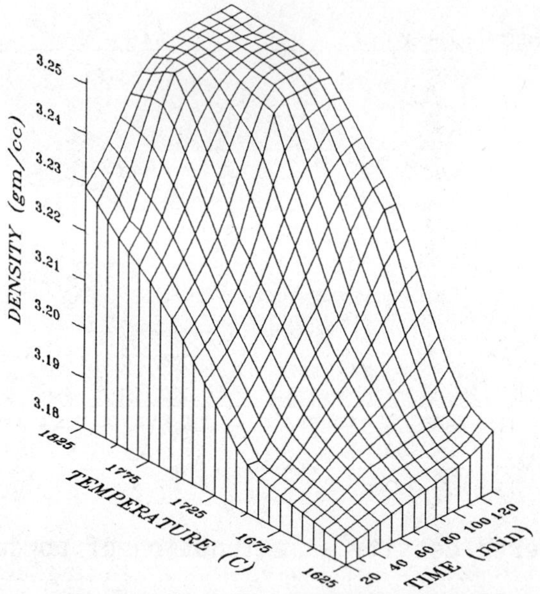

Figure 5: Surface plot illustrating sintered density as a function of both sintering temperature and time for standard powder. (Modified level average data)

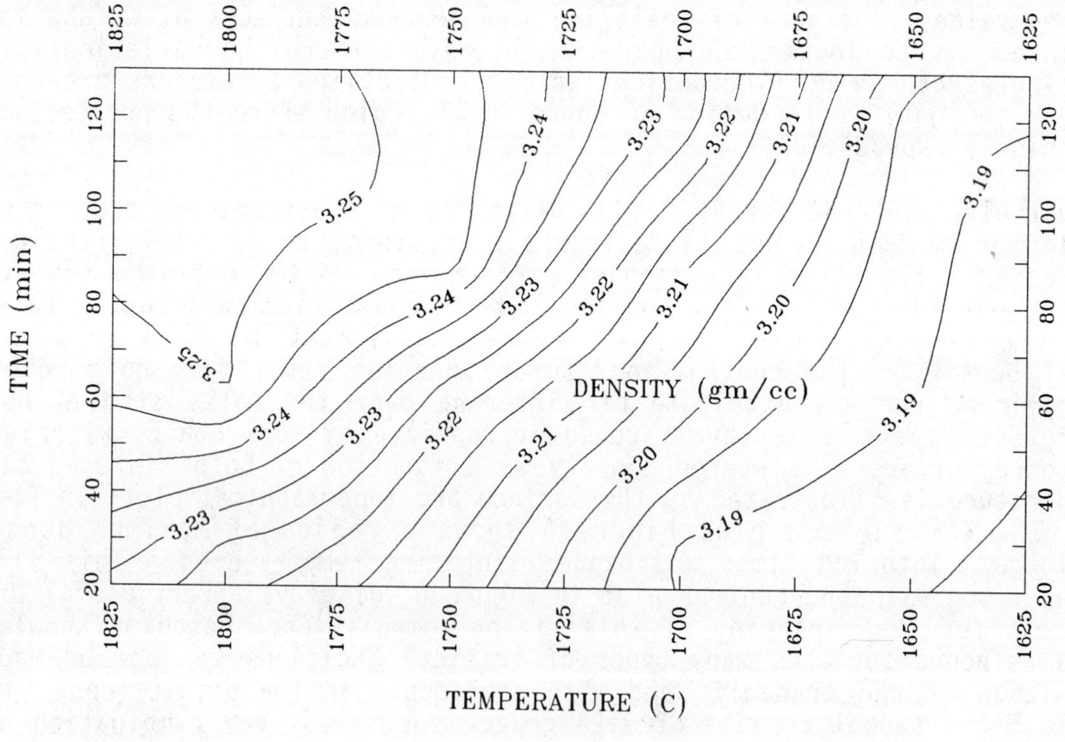

Figure 6: Contour plot illustrating sintered density as a function of both sintering temperature and time for standard powder. (Modified level average data)

tency and quality in the component. Knowledge of the response surface also allows efficiency based decisions to be made without compromising the component's performance which enhances the component's quality. An example of this is if a density of 3.23 gm/cc were the design target for a particular application, it may be achieved by sintering at 120 minutes at 1710 C or by sintering for only 60 minutes at 1765 C where the sintering time is reduced by half, shown in Figure 6.

CASE II: INJECTION MOLDING OF "WEAR BUTTONS"

The study of injection molded wear buttons presented here describes how another off-line quality control technique, process capability analysis, is used to refine and then qualify a process prior to its actual use in production. In this specific example, a prototype component is to be manufactured by the injection molding process in lots of approximately 1000 pieces. Although the injection molding process is well qualified and considered reproducible for several common components, the wear button, as shown in Figure 7, requires special consideration due to its small size, thin cross-sections and general configuration. In addition, the customer's dimensional requirements are firmly established, with very narrow tolerances.

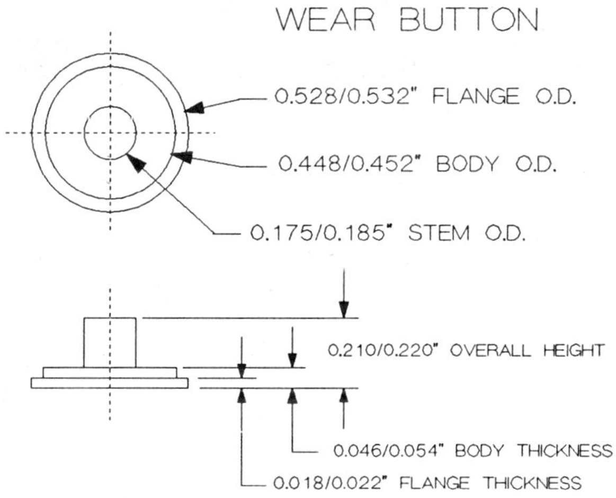

Figure 7: Design drawing for an automotive component used in a wear application, designated as "Wear Button."

Previous experimentation shows that the injection molding mix and operation procedures are stable and predictable and it is the mold's design and dimensional control which are the most critical processing parameters governing dimensional compliance of the injection molded components. A small scale evaluation run is conducted to establish long term process capability of the wear button injection molding process. Conventional process capability analysis, using normal statistics, confirms if the mold's dimensional and design features were adequate to assure compliance to the customer's design specifications for the component. Analysis of this 60 piece trial run indicates if any special quality control considerations or additional process engineering efforts are required prior to manufacturing of the order. The trial wear buttons are molded consecutively from a stable running injection molding machine and processed normally through to the densification stage. The parts are then examined in their final state for dimensional compliance and any unusual features.

For most of the dimensional attributes of the wear button, long term compliance to dimensional requirements is safely predicted using capability analysis (i.e., the predicted distributions were less than 75% of the total tolerance). Consequently, for these dimensional attributes sampling at final inspection is sufficient to confirm targeting. However, this is not the case for two dimensional attributes such as flange and body thickness.

Displayed in Figure 8a is the frequency distribution for the observed flange thickness of the 60 trial pieces. The capability analysis, using normal statistics, predicts a long-term spread of flange thicknesses more than twice the specified tolerance and that over 17% of the components would not meet the flange thickness requirement during the full production run. Based on previous experience in mold design of this size range, a spread in measurements this large is not expected; thus, the mold design itself is re-evaluated. The unusual result of the capability analysis flags an error in the mold design. The source of the large spread in flange thickness dimensions is identified as a mismatch of the mold's tooling insert which tends to tilt during use. The insert is remachined and the process capability study conducted to confirm correction of the problem. Figure 8b shows the frequency distribution of the flange thickness dimension following mold modification. The predicted distribution of measurements only consumes 60% of the tolerance.

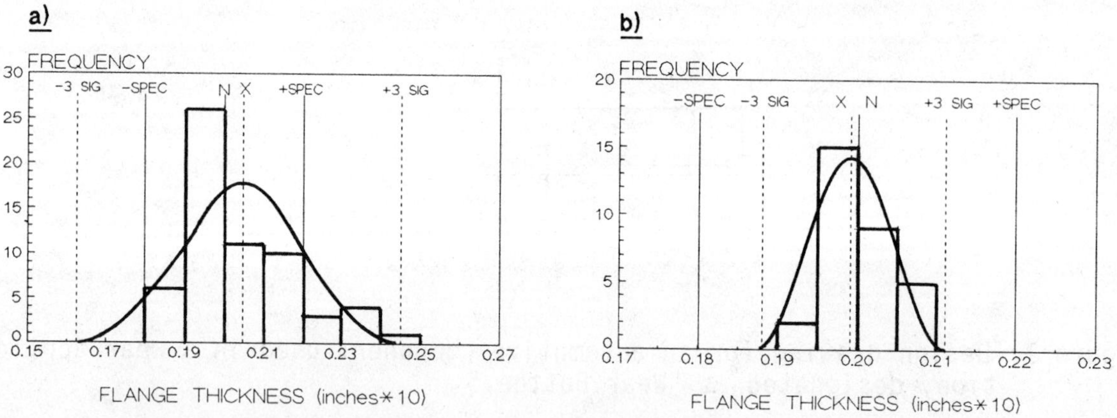

Figure 8: Frequency distribution for flange thickness measurements of "wear buttons" (a) during initial process capability study and (b) after process modification.

The other dimensional compliance problem uncovered by the capability analysis is the body thickness dimension. Figure 9a shows the frequency distribution of this dimension where the predicted long term range of measurements are 90% of the tolerance. This illustration also shows that the mean value of the predicted distribution is significantly different for the nominal value. In this case, the capability analysis indicates that the mold design is targeted

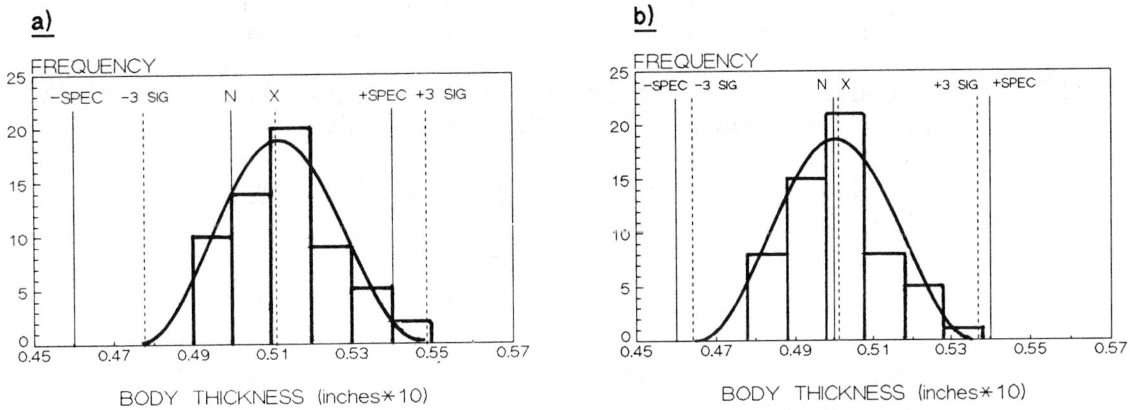

Figure 9: Frequency distribution for body thickness measurements of "wear buttons" (a) during initial process capability study and (b) after process retargeting.

0.00126 inches too thick for this particular geometrical feature and that the mold should be remachined prior to the production run. The mold remachining only adjusts the targeting of the body thickness dimension and not the spread, as shown in Figure 9b. Although within the upper and lower dimensional specification limits, the predicted spread of measurements for body thickness remains greater than 75% of the total tolerance. Current statistical process control techniques indicate that absolute inspection of this dimension is required at final inspection of the wear buttons.

CONCLUSION

Off-line quality control techniques greatly enhance the general effort to both improve and assure product quality of advanced ceramic components. Off-line quality control techniques have proven to be an indispensable set of engineering tools for the achievement of both material and component design goals. The implementation of these methods has allowed more efficient process development, manufacturing cost reduction, improvement of material properties, higher component quality and increased product reliability. No statistical technique is a substitute for a systematic approach to engineering or a functional understanding of the specific technology under consideration. However, when understood and used appropriately, these statistical tools can effectively compliment the required research efforts. This combination is proving essential in the goal of utilizing advanced ceramics for heat engines.

ACKNOWLEDGMENTS

The authors gratefully acknowledge assistance and experimental data provided by P. K. Caneen, L. D. Lynch, and B. J. McGeary, along with many other Norton/TRW personnel for their support of this approach to quality systems development for advanced ceramic components.

REFERENCES

[1] G. E. P. Box and S. Bisgaard, "The Scientific Content of Quality Improvement," Quality Progress, June 1987, pp 54-61.

[2] B. A. Jones, "Design Lab Experiments to Assure Product Quality," Research & Development, Dec. 1987, pp 54-58.

[3] D. C. Montgomery, Design and Analysis of Experiments, John Wiley & Sons, New York, NY, 1976.

[4] O. L. Davies, The Design and Analysis of Industrial Experiments, Longman Group Limited, New York, NY, 1956.

[5] G. Taguchi, System of Experimental Design Vol I & II, Kraus International Publications, White Plains, NY, 1987.

[6] P. J. Ross, "The Role of Taguchi Methods and Design of Experiments in QFD," Quality Progress, June 1988, pp 41-47.

[7] B. Gunter, "A Perspective on the Taguchi Methods," Quality Progress, June 1987, pp 44-52.

[8] J. A. Cornell, ASQC Basic Reference: How to Apply Response Surface Methodology, American Society for Quality Control, Milwaukee, WI, 1984.

[9] D. W. Thomas, Statistical Quality Control Handbook, Western Electric Co., 1956.

A THERMAL SHOCK FACILITY FOR TESTING DIESEL ENGINE CERAMIC MATERIALS

W.D. Alexander and F.J. Wallace University of Bath, England

ABSTRACT

As part of a continuing research programme into the use of ceramic thermal barriers in Diesel engines, at the University of Bath, a test facility has been designed and built to determine the thermal conductivity and the thermal shock behaviour of proposed ceramic materials. This facility was originally intended for the "in-house" testing of plasma sprayed, ceramic coatings so as to provide thermal conductivity data for finite element analysis of engine components. Subsequent development of the test rig extended its capability to provide additional data on the effects of long term thermal cycling of specimens. The thermal cycling feature of the apparatus provides a clean, controllable alternative to the methods currently used (such as flame heating and water quenching), particular attention has been focussed on the control of the heating source.The results included in this paper are of a preliminary nature, as the rig has only recently completed its commissioning phase.

INTRODUCTION

Initial research work at Bath University on low heat loss Diesel engines concentrated on the use of metallic heat shields with air gaps (1). These investigations provided a valuable precursor to the present effort which is aimed at the use of ceramics as the insulating medium. At the onset of the research programme, some eight years ago, there was, and to some extent still is, rather a dearth of accurate property data on the ceramic materials available. Since that time there have been marked advances and improvements in the quality of ceramics and in the property data available. This trend has resulted in increased interest in the use of ceramics for in-cylinder components.

Two distinct application routes may be identified when considering the use of ceramics for in-cylinder components. Generally this distinction is between those components manufactured in ceramic and bonded by some means to an existing metal body; and metal components which have ceramic deposited on them by some means, as a coating. Each of these routes has its own advantages and disadvantages, and as such, both monolithic insert and spray coating systems were included when drawing up the research programme.

Plasma spraying facilities within the university enabled coated test specimens to be produced, with control of spraying parameters and also powder compositions. It was therefore important to have a facility available which could rapidly test the thermal behaviour of the specimens and provide thermal property data for subsequent finite element analysis. The monolithic specimens, although characterised to a certain extent by the manufacturers, usually had little, or no data relating to the variation of thermal conductivity of the materials versus temperature, this further strengthened the case for an appropriate test rig.

Accurate material property data, both mechanical and thermal, and its variation with temperature, is essential for the finite element analysis of components, and as such has received due attention.The relationships which

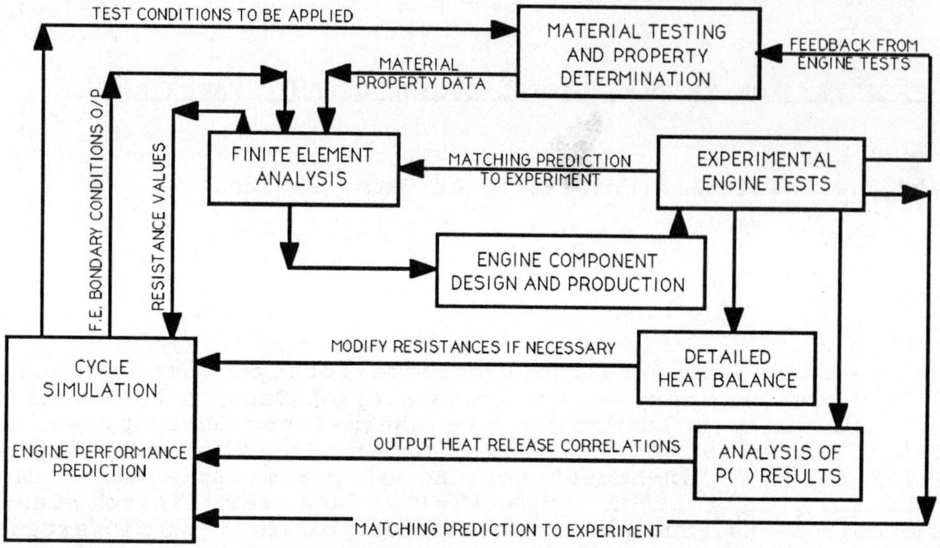

Figure 1 Project Activity Relationships

exist between the various project activities are shown in Figure 1. Several other property test facilities are in use but this paper will confine itself to a description of the thermal shock/fatigue rig and the associated thermal conductivity test facility.

THERMAL CONDUCTIVITY/THERMAL SHOCK TEST RIG

General

Figures 2a and 2b show the general layout of the test facility. The rig is basically comprised of two chambers, a heating section and a cooling section. For the case of thermal conductivity testing, only the heating section is used, whereas for the case of thermal shock/fatigue testing both chambers are used. A pneumatic ram, together with the pneumatic control box, provides movement of the specimen to and from the heating section. The heating section contains the quartz halogen lamps and their reflector assemblies, together with the necessary lamp power sensors. Cooling of the lamp reflector

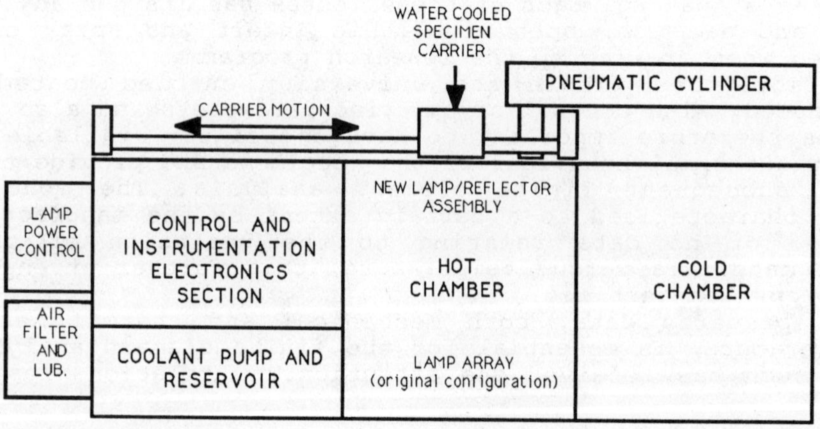

Figure 2a Schematic of Thermal Conductivity/Thermal Shock Rig

Figure 2b General Layout

assemblies and the specimen carriage is provided by a centrifugal water pump, reservoir and water to water heat exchanger, these items are housed at the bottom left of the rig. On the far left of the rig is the four channel phase angle controller which together with the power control circuitry is used to vary the lamp power. Also included is the automatic shutdown system which terminates operation of the rig should fault conditions arise or if the required number of cycles have been executed. Items common to both thermal conductivity and thermal shock will now be discussed followed by items specific to each of these, under appropriate headings.

Heating system. The method chosen for heating the specimen was that of quartz halogen lamps in specially designed reflectors, situated beneath the specimen.This choice is explained in the section on Thermal Conductivity. The original system was comprised of four separate lamps each with its own parabolic reflector, this allowed a very even distribution of flux over the specimen to be achieved. However, although this fitted the requirement for the

Figure 3 Lamp Reflector Assembly

Table 1 Surface Temperature Sensor Requirements

A	SMALL PHYSICAL SIZE	B	ABILITY TO WITHSTAND OPERATING CONDITIONS
C	FITTING OF DEVICE TO SURFACE SHOULD BE WITHOUT HEATING	D	FITTING OF DEVICE TO SURFACE SHOULD BE WITHOUT THE USE OF AN INTERMEDIATE LAYER
E	LINEARITY OF THE SENSORS OUTPUT IS DESIRABLE BUT NOT ESSENTIAL	F	SENSOR SIGNAL SHOULD BE AS LARGE AS POSSIBLE TO AVOID NOISE PROBLEMS
G	MATERIALS FOR PRODUCTION OF THE SENSORS SHOULD BE READILY AVAILABLE	H	ABILITY OF SENSOR TO WITHSTAND REASONABLE LEVELS OF SURFACE CRACK DAMAGE, WITHOUT CHANGING CALIBRATION

thermal conductivity tests, the maximum flux achievable for thermal shock testing was too low. This situation resulted in the design of a lamp reflector assembly specifically intended for thermal shock work, and was comprised of two, higher power lamps, held within a more compact single reflector unit. Figure 3 is a photograph of the lamp/reflector assembly. The mean flux at the specimen plane for the original lamp/reflector assembly was approximately $160KWm^{-2}$ with a maximum variation of ±5% , whereas the new lamp/reflector assembly achieved a mean flux value of approximately $450KWm^{-2}$ with a variation of ±10 %.

Control of lamp power. A commercially available phase angle power control unit was used in conjunction with the lamps to provide control of the heating flux at the specimen plane. The particular unit chosen has a 0 to 10 Volt analogue input control, which is fed from the power control box. Two main control modes are available from this box, namely :- specimen front face temperature control and a simulated convective type of control. The first of these has been primarily used for thermal conductivity test runs, to ensure rapid achievement of equilibrium conditions. The second control mode attempts to simulate the thermal conditions which result from a load step in the engine. A common requirement for both of these control modes was that the front face temperature of the specimen be known. A list of the features

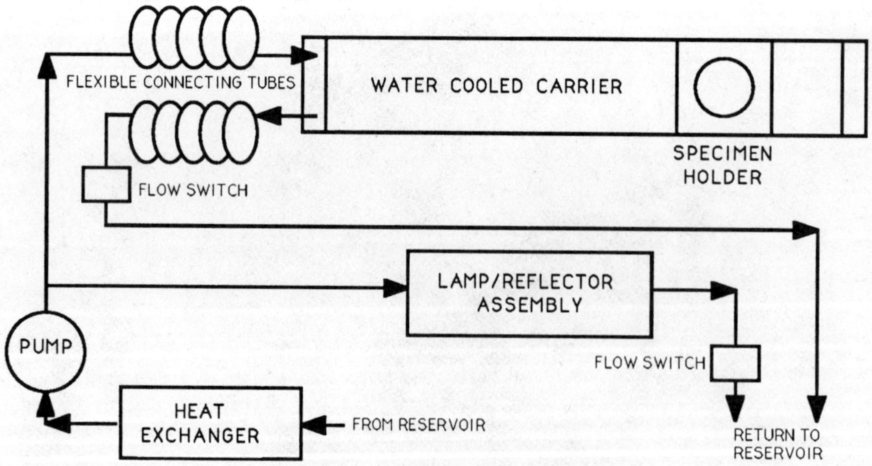

Figure 4 Main Cooling System for Rig

desirable in a front face temperature sensor are given in Table 1. Due consideration of the types of sensor available led to the choice of thin film thermocouples as the preferred form of sensor, and of the various methods of thin film deposition, vacuum sputtering appeared to give the best performance in the devices produced.

Cooling system. The rig cooling system, is shown in Figure 4. The primary, closed loop incorporates a coolant reservoir, centrifugal pump, and a water to water heat exchanger. The secondary open loop is comprised simply of a thermostatic valve which senses the primary reservoir temperature and controls the flow of mains water, from a header tank, through the heat exchanger to a drain. For thermal conductivity testing an additional cooling system is used to provide coolant for the calorimeter, this is discussed in more detail later.

Automatic shutdown system. This system was provided to allow unsupervised running of the rig, such that in the event of system failure, the rig would be safely shut down. When the rig is operating in the thermal shock/ fatigue mode this system allows a predetermined number of cycles to be executed after which the rig will automatically shut down. Trips are also included to shut the rig down in the event of operator error which might result in injury.

Items specific to thermal conductivity testing will now be discussed followed by those specific to the thermal shock mode of operation.

Thermal Conductivity

The determination of this material property is important not only from the point of view of analysing the heat loss from a component, but also the thermal stresses which will result in the material when exposed to engine operating conditions. The method chosen for the determination of thermal conductivity using this rig is a simple 1-D steady state type. A uniform heat source is applied to the front face of a simple disc specimen and a calorimeter is placed on the cold side to measure the throughput of heat. Front and back face temperatures of the specimen are monitored.

Specimen heating. The original design of heating source was that of a metal block with internally placed resistive heating elements. However, one of the problems associated with this type of heater is the non uniform contact which may occur through distortion of the heating block and specimen. This in turn results in non uniform heating of the specimen. The standard method of overcoming this, is to interpose a layer of compliant material between the heating block and the specimen. Unfortunately no materials were available which retained their compliant nature for the range of temperatures over which it was intended to test. It was necessary therefore to consider a non-contact form of heating the specimen. After due consideration of the non-contact methods of heating available, it was decided to opt for an incandescent lamp system, as outlined in the section on the Heating system.

Another important consideration regarding 1-D steady state thermal conductivity testing was that of edge losses from the specimen. Calculations by Somers and Cyphers (2) indicate that the the errors in thermal conductivity due to side losses from the specimen, are less than 1% for dimensions similar to those used in this apparatus. This has been borne out by F.E. analyses which have been carried out.

Specimen temperature and calorimetry. The measurement of the cold side temperature of the specimen is achieved using a small diameter sheathed thermocouple embedded in the base of the calorimeter (see Figure 5). Although not actually in contact with the back face of the specimen, the high thermal conductivity of the calorimeter block relative to the specimen ensures that the error produced by the placing of the thermocouple is small. Any distortion of the specimen may be taken up by a compliant layer of material between the specimen and the calorimeter block, as the temperature here is well defined and low. The measurement of the front face temperature is accomplished using a

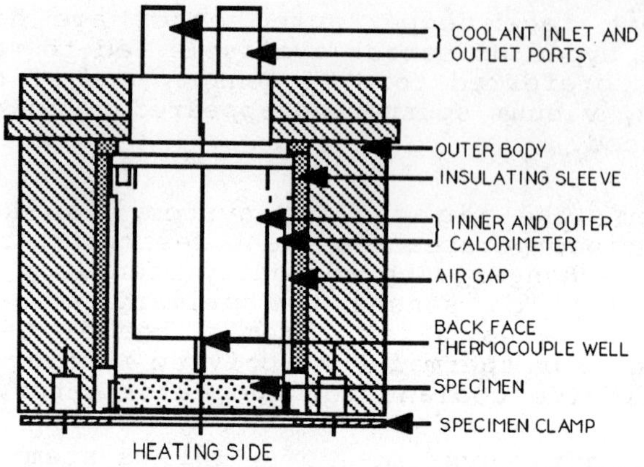

Figure 5 Thermal Conductivity Calorimeter

thin film thermocouple deposited on the specimen using sputtering. Initially nickel/gold thin films were used for this purpose, however it became necessary to use platinum/gold couples as surface temperatures were raised above about 450°C. The thin films used so far have been approximately 1-2µm thick and have thus imposed little error, in terms of temperature drop, on the specimen temperature monitored, while ensuring excellent transient response to temperature changes. The development of these thin film thermocouples is important not only for this test facility but for future monitoring of temperatures and heat fluxes within the engine.

The calorimeter has been designed to minimise errors in the measurement of heat from the specimen, and also to ensure an isothermal surface in contact with the specimen back face (see Figure 5). The coolant for the calorimeter is pure water, which is fed from a constant head reservoir. The inlet temperature of the coolant to the calorimeter is held constant using a water to air heat exchanger. Determination of mass flow of coolant through the calorimeter is monitored at intervals using a beaker, balance and electronic timer system. The temperature rise of the coolant on its passage through the calorimeter is monitored using sheathed thermocouples. These were chosen in preference to platinum resistance gauges on the basis of size considerations and immersion lengths. Matching of the thermocouples has ensured that an error of less than 0.1°C results in the measurement of differential temperature. An autozeroing circuit with a resolution of 1µV has been used for conditioning of the signal from the differential thermocouple pair.

Thermal Shock/Fatigue Testing

The low thermal conductivity of ceramic materials proposed for the insulation of low heat loss Diesel engines results in large temperature gradients and can lead to high thermal stresses, although to a certain extent this may be offset by the low thermal expansion of these materials over that of metals. However, it is insufficient to consider only the equilibrium temperature fields and thermal stresses. During load step transients which occur for typical Diesel engines, the thermal conditions imposed on the components change markedly and this may result in transient stress fields which are sufficient to exceed the material limits. Reference (3) gives details of analyses of an engine valve manufactured from two ceramic materials with markedly different properties, and highlights the need for both transient analysis and testing.

Traditional methods of testing for thermal shock were sufficient for ranking the materials in terms of thermal shock sensitivity but did not generally impose realistic levels of thermal shock as might be found in a typical engine. The purpose of this test facility was to enable a controlled thermal shock to be imposed on the ceramics proposed for use in engines. The use of non-contact radiant heating should allow separation of thermomechanical effects and those which arise from chemical/environmental effects. These latter effects can lead to confusion regarding the source of the failures which arise from testing.

Specimen motion. The specimen in this thermal shock rig is continuously cooled on the back face, and alternately heated and cooled on the front face to simulate typical thermal conditions which a component in an engine would be subjected to. As already outlined, the heating of the specimen is carried out with radiant sources, and then after a predetermined period, is moved to a cooling chamber within the rig. A rodless pneumatic actuator is used to provide motion of the specimen from the heating to the cooling zone, this device being particularily suitable where space is at a premium. Heating and cooling periods were designed to be independently variable from 10 seconds to 990 seconds using precision electronic timer circuits. A non-volatile counter gives the total number of cycles executed and incorporates a trip output when the desired number of cycles have been run.

Thermal shock heating control. The simulated convective type of control mode may be used during the heating of the specimens for thermal shock tests. This control mode requires that the front face temperature of the specimen be known as well as the heating flux at the specimen front face. The measurement of the specimen front face temperature has already been outlined. The heating flux at the front face of the specimen is derived from measurements of the light output from the lamps. A fibre optic cable is used to collect light from each of the lamps, this is then fed to a photodiode which senses the light output. Fibre optic coupling is used for two main reasons :- one is that it enables the electrically sensitive photodiodes to be situated remote from the electrically noisy environment of the lamps, and two, is that it enables the photodiodes to be placed in a temperature controlled enclosure, which is an essential requirement for stability of response. The photodiode/fibre optic response curve does not cover the entire range of wavelengths output by the lamps, but a check on this showed that the response is sufficiently linear for this purpose. The heat transfer coefficient and fluid temperature to be simulated are input to the power control box and internal circuitry calculates the lamp power required.Figure 6 schematically shows the circuit for the simulation of convective heat transfer.

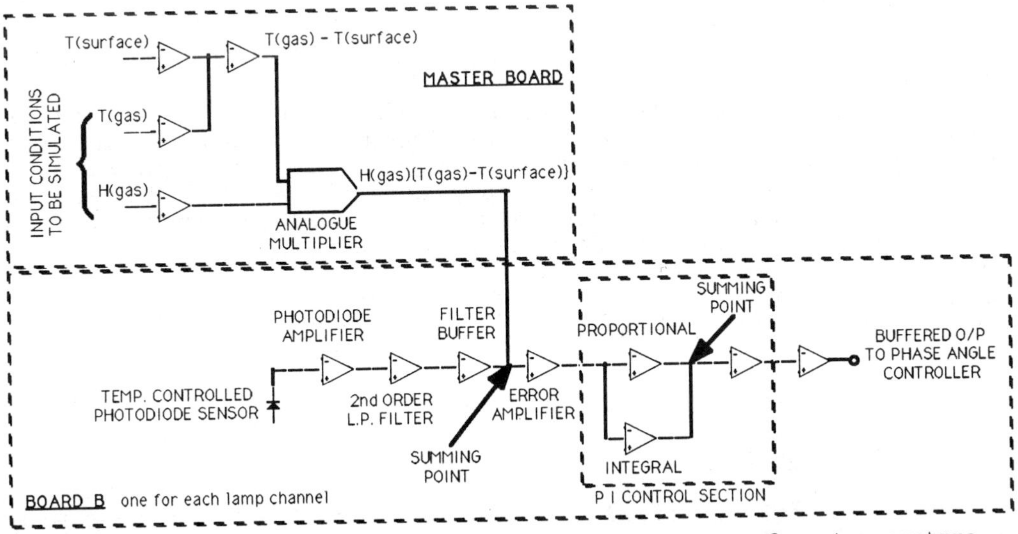

Figure 6 Convective Simulation Circuit Schematic

T(suface) - surface temperature
T(gas) - set cycle mean gas temperature
H(gas) - set cycle mean heat transfer coefficient

Down shock or cooling transients. Thus far only the heating transient has been considered, and as yet, further work has to be undertaken to examine the necessary hardware to simulate the cooling transient. This is a complex modelling problem involving both convective and radiative transfer to and from components within the combustion chamber together with potentially non linear heat transfer on the cooling side, this being dependent on the particular cooling transient being considered. The three main candidates for worst case cooling transient have been identified as :- a). full load to no load, engine motored; b). full load to no load, engine stopped; and c). full load to no load, engine idled. In cases a). and c). forced convective cooling is maintained at the back face of the specimen, however in b). although initially free convective transfer with the coolant may pertain, this may develop into boiling type transfer. In case a). although no fuel is combusting, compression work within the cylinder maintains the cycle mean gas temperature above 200°C. A drop in the cycle mean gas temperature from around 900°C to around 200°C accompanied by a drop in cycle mean heat transfer coefficients from around 500 $Wm^{-2}K^{-1}$ to 300 $Wm^{-2}K^{-1}$, presents certain difficulties in simulating. However the situation is not well represented by a water spray or immersion, and simulation of the cooling transient requires further research effort.

Acoustic emission. During the thermal shock testing of specimens an acoustic emission probe is attached to the back face via a waveguide. Large quantities of data are produced during a typical 1000 cycle run and data reduction has been necessary to ensure ease of interpretation. These results are to be presented in a future paper. Analysis of the acoustic emission results should yield valuable information relating to crack initiation and propogation, particularily for the coated spwcimens.

RESULTS

Finite Element Analyses

Simple axisymmetric, steady state, finite element analyses have been carried out to examine the effects of both uneven heating of the front face of the specimen, and uneven cooling of the back face, and the influence of these on the temperature field. Figures 7a through to 7c show the results.These theoretical results have been compared with equilibrium temperatures monitored on the test discs and show good agreement, the specimen centre point temperatures differing by not more than 5°C. However it should be remarked that the theoretical temperature fields are sensitive to the imposed boundary conditions, and more detailed measurements would have to be taken to be confident of such matching.

Thermal Shock Testing

As part of the commissioning trials of the rig, several 1000 cycle runs have been carried out. Thin film thermocouples have been deposited on all of the sample types used in these tests. The integrity of the platinum/gold thermocouples appears to be good on all samples so far tested. Materials tested to date include : LAS , cordierite and plasma sprayed zirconia. Figure 8 shows typical plots of specimen surface temperature versus time.

Thermal Conductivity Testing

This test facility is currently still being commissioned.

DISCUSSION

Finite Element Analyses (static)

The static thermal F.E. analyses carried out on the test specimens have proved valuable in the reduction of errors in the measurement of thermal conductivity, by identifying the sources of non 1-D flow and in the

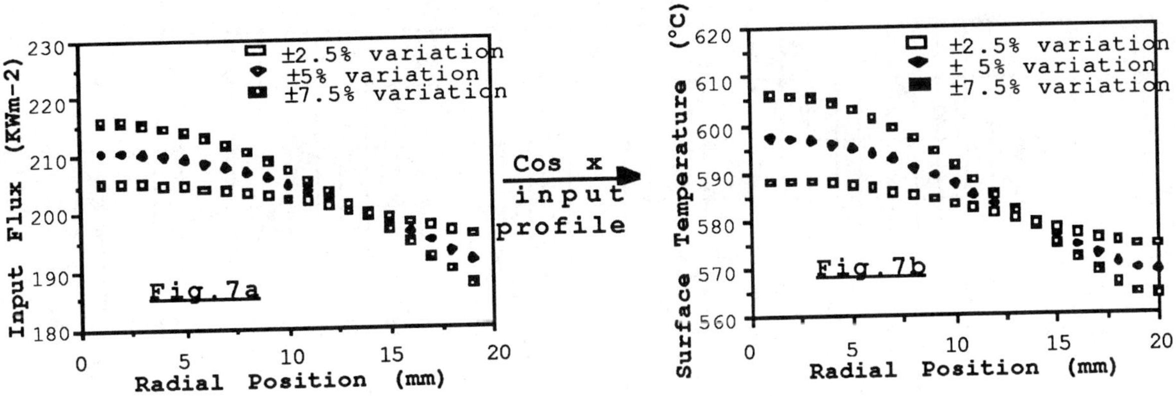

Figure 7a and 7b - Uneven Front Face Heating

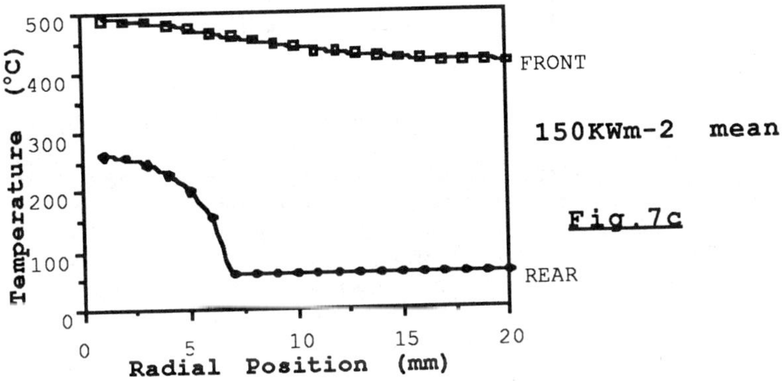

Figure 7c - Uneven Front Face Heating and Uneven Back
Back Face Cooling

positioning of the surface thermocouples. The thermal stress analyses have proved to be of value when considering the thermal distortion of the specimens and the effect that this has on the heat flow paths.

Experimental Temperature vs Time Data

Figure 8 shows typical front face temperatures obtained from specimens in the thermal shock rig.Glass ceramic 2 has lower thermal conductivity and slightly lower density than type 1, and this is reflected in the rate of rise of front face temperature, as would be expected. Matching of these, and similar results with transient F.E. models has shown the importance of correctly modelling the interface between the specimen and the cooled back plate on which it is mounted. This has a direct bearing upon the thermal modelling of ceramic components in general.
The performance of the thin film surface thermocouples, and the necessary connections to instrumentation will be important in the design of future probes for use in the test engines within the department.

SUMMARY

A test facility has been constructed which allows both thermal conductivity and thermal shock testing to be achieved. The thermal conductivity test facility should allow this property to be determined over a wide temperature range and provides a useful check on the results obtained using other methods.

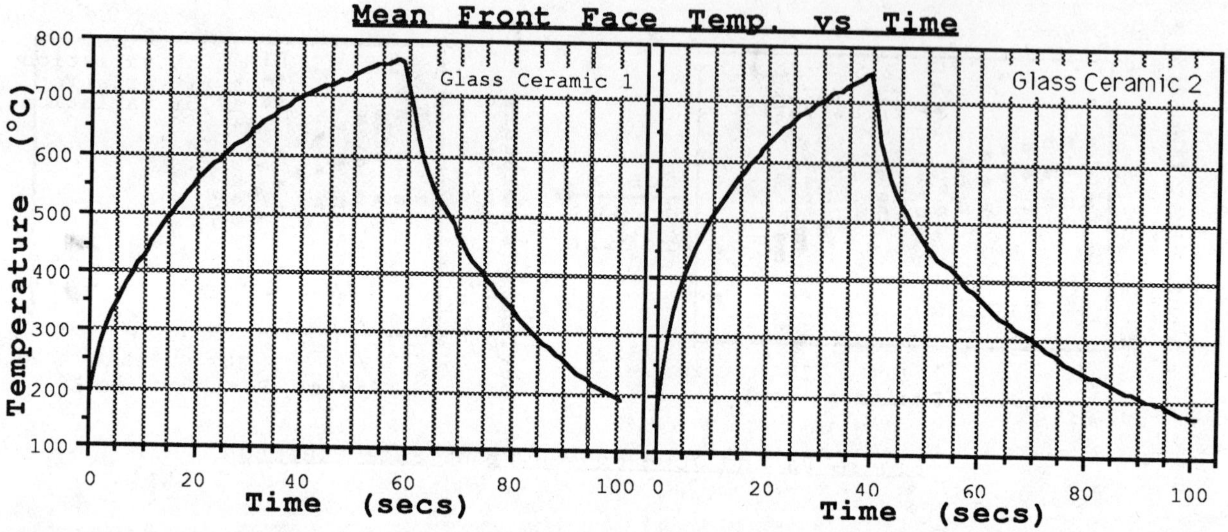

Figure 8 - Typical Specimen Front Face Temperature vs. Time Plots

The thermal shock facility of this rig should provide useful information regarding the behaviour of candidate materials, subjected to thermal transients similar to those actually found in the combustion chamber of a Diesel engine. The clean, controlled source of heating provided by the radiant source should allow detailed comparison of results obatined theoretically with those obtained experimentally. The results presented in this paper are of a preliminary nature, with the emphasis being on the design features which have been built in to the rig.

REFERENCES

1 Wallace, F.J., Kao, T.K., Alexander, W.D., Cole, A., Tarabad, M., "Thermal Barrier Pistons and Their Effects on the Performance of Compound Diesel Engine Cycles" S.A.E. Paper 830312, 1983

2. Somers, E.V., and Cyphers, J.A., "Analysis of Errors in Measuring Thermal Conductivity of Insulating Materials" Review of Scientific Instruments Vol.22, No.8 1951 p583-586

3 Kabat, D.M., Garwin, I.J., Hartsock, D.L., "Ceramic Valve Analysis, Reliability and Test Results" S.A.E. Paper 880670, 1988

PROPERTIES OF SILICON NITRIDE ROLLING BEARING FOR HIGH-SPEED AIR TURBINE

T.Mori*, K.Nishida**, K.Inoue***
H.Yasui****, H.Takebayashi****, K. Tanimoto****

ABSTRACT

Hybrid rolling element bearings (silicon nitride balls and steel races) were compared to conventional steel bearings for a selected air turbine.
It was determined that the steel bearing would seize in this application while the hybrid bearing would not seize.

INTRODUCTION

Hot isostatically pressed silicon nitride ceramics have excellent high temperature strength, low density and excellent wear resistance[1),2)].
The combination of these properties make silicon nitride an attractive material for rolling element bearing applications.
The low density of the silicon nitride material is particularly advantageous when applied to the rolling elements of high speed air turbine bearings.
In this case, the centrifugal forces[3)] generated by the rolling elements are a major contributor to the contact stresses generated in the bearing.
Improved high speed air turbines can be realized through the use of silicon nitride rolling elements.
The low density rolling elements greatly reduce sliding and centrifugal forces on the bearing race.
Because of a unique combination of material properties, dense silicon nitride ceramics have found application in rolling element bearings.
Silicon nitride rolling element bearings are now being used in application that conventional bearing steels can not be used because of temperature limitations or corrosion.
In the special case of high speed bearings the centrifugal force is defined by:

$$K = M(dp/2)\omega m^2$$

* Material and Components Group, Toshiba Co., 1-1 Shibaura 1-chome, Minato-ku, Tokyo, 105 JAPAN
** New Materials Engineering Laboratory, Toshiba Co., 8 Shinsugita-cho Isogo-ku, Yokohama, 235 JAPAN
*** Fine Ceramics Department, New Materials Division, Toshiba Co., 8 Shinsugita-cho Isogo-ku, Yokohama, 235 JAPAN
**** No. 1 Research Department, Koyo Seiko Co., Ltd., 24-1 Kokubuhiganjo-cho Kashihara, Osaka, 582 JAPAN

where K is the centrifugal force in Newtons, M is the mass of the rolling element in kilograms ωm is the angular velocity and dp is the pitch circle in meters.
The gyro moment of the ball is shown in the following equation[4]:

$$Md = I\omega w \omega m \sin^{\alpha}$$

where Md is the gyro moment in Newton-meters, I is the moment of inertia of the ball against the diameter in Newton-m-sec, ωw is a rotational angular velocity in radians per second, ωm is the revolutionary angular contact velocity (radians/sec) and is the contact angle in degrees.
From these equations it can be seen that decreasing the mass of the ball will decrease the centrifugal force and the gyro moment.

MANUFACTURING PROCESSES

In order to successfully use ceramic in a rolling element, the material must have an acceptable rolling contact fatigue life. Rolling element bearing materials must have:

(1) a high strength
(2) high hardness
(3) high toughness
(4) a high weibull modulus indicating few defects

In order to achieve a high strength and high weibull modulus, it is important to establish strict process engineering to control the microstructure of the ceramic.

Figure 1 shows the flow chart of the manufacturing process. Controlled amounts of yttria and alumina were added to the silicon nitride powder (shown in Figure 2) to aid in sintering.

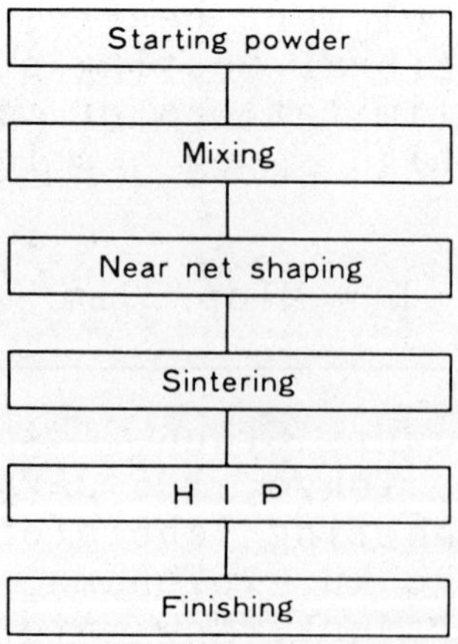

Fig. 1 Flow chart for manufacturing process of hipped Si_3N_4

Fig. 2 Scanning electron micrograph of the fine Si_3N_4 powder

After blending with a binder the powders were formed to near net shape by pressing in a mold.
Binders were removed in an inert gas furnace followed by sintering.
After sintering at 1973 to 2073 K to closed surface porosity, the silicon nitride was hot isostatically pressed for one hour at 2023 K using a 100 MPa nitrogen gas pressure.
A scanning electron micrograph of the finished polished silicon nitride ball is shown in Figure 3.
The material does not contain any residual porosity.
Figure 4[5)] shows the flexural strength of the hipped silicon nitride.

Table 1 compares the properties of the hipped silicon nitride and conventional high carbon chrome bearing steel (SUJ-2).

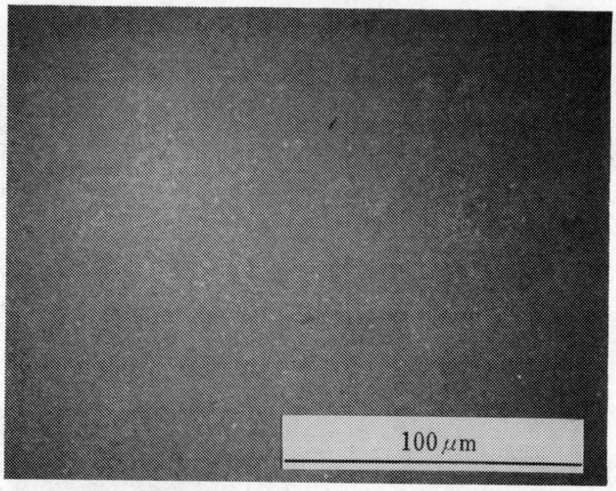

Fig. 3 Scanning electron micrograph of a polished hipped sintered Si_3N_4

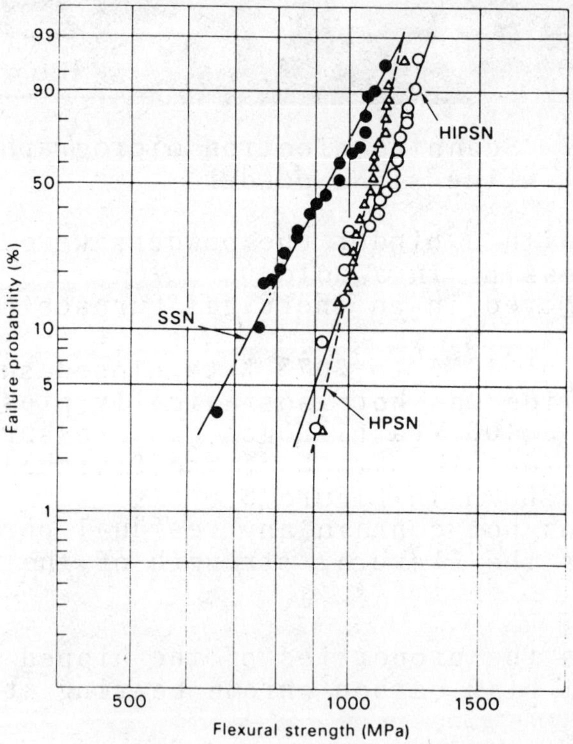

Fig. 4 Strength distributions at room temperature

Table 1 Properties of hipped si_3N_4 and high carbon chrome bearing steel (SUJ2)

Material properties	Hipped Si_3N_4	Bearing steel (SUJ2)
Heat resistance (K)	1073	453
Density (g/cm^3)	3.2	7.8
Young's modulus (GPa)	314	205
Poisson ratio —	0.26	0.3
Fracture toughness ($MPa\,m^{1/2}$)	7	—
Hardness Hv	1500	700
Flexural strength (3-Point) R.T. (MPa)	1078	—
Weibull modulus —	13	—
Coefficient of thermal expansion ($\times 10^{-6} K^{-1}$)	3.2	12.5

EXPERIMENTAL PROCEDURE

A schematic diagram of the high speed air turbine is shown in Figure 5.
The bearings are mounted on both sides of the axial air turbine.
Figure 6 shows a schematic of the air turbine control test system used in the evaluation of the bearings.
All bearings were lubricated by an oil mist.

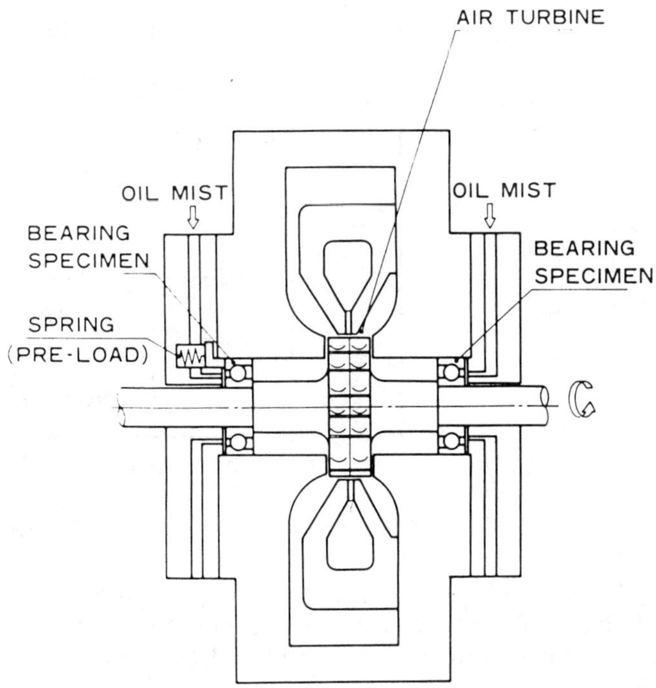

Fig. 5 Schematic diagram of testing equipment

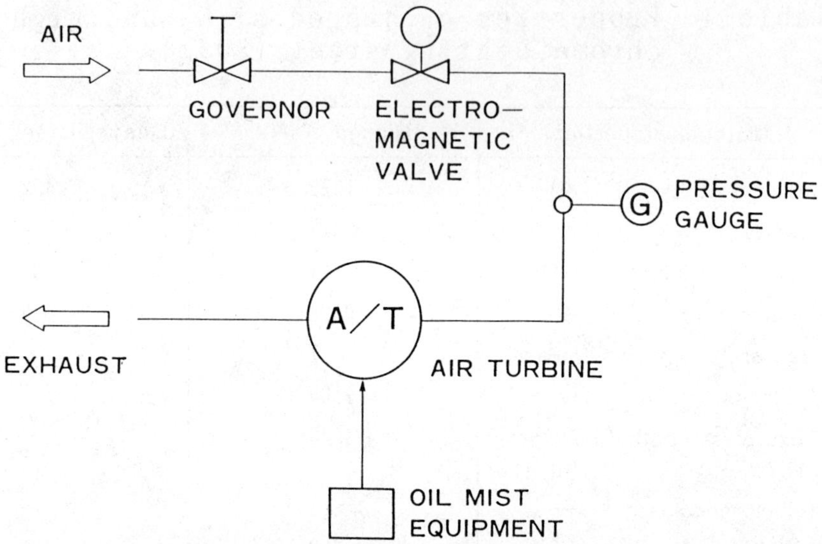

Fig. 6 Block diagram of experimental set-up

The bearing shown in Figure 7 were angular contact ball bearings with an inner diameter of 12 mm and a 20 degree contact angle.

Both hybrid bearings, steel rings and silicon nitride rolling elements, and all steel bearings were prepared for test.

(A)

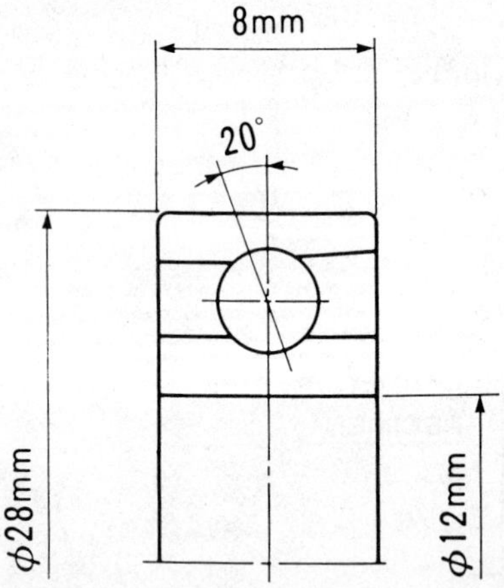

(B)

		Ceramic bearing	Steel bearing
Outer race & Inner race		Bearing steel (SUJ2)	Bearing steel (SUJ2)
Ball	Material	Hipped Si_3N_4	Bearing steel (SUJ2)
	Size	3/16"	3/16"
	Number of balls	13	13
Cage		Non-cage	Non-cage

Fig. 7 Bearing specimens

The bearings were tested according to the test conditions documented in Table 2.
The relationship between the supplied air pressure and the rotational speed of the bearing was determined during the tests.
Air pressure was increased at a controlled rate of 0.01 MPa every two minutes.
Acceleration and deceleration was controlled by opening and closing the electromagnetic valve on the test stand.

Using this method it was possible to investigate the acceleration response of the air turbine with the hybrid and metal bearings.

Table 2 Test conditions

Items		Condition
Revolution		0 ~ 120,000rpm
Axial load		147 N
Temperature		Normal temperature (R.T.)
Lubricating condition	Method	Oil mist (1×10^{-6} m³/s, 0.4MPa)
	Dynamic viscosity	30 m²/s (293K)

RESULTS AND DISCUSSION

Figure 8 depicts the relationship between the supplied air pressure and the turbine rotational speed.
As can be seen from this figure, the hybrid ceramic bearing operated at a higher speed than the conventional steel bearing at the same air pressure.
Acceleration response of the air turbine is shown in Figure 9. In this test, the steel bearing was found to be unstable in the 80,000 to 120,000 rpm speed range.
After the test, the steel bearings were found to have seized.
The hybrid bearings resulted in improved acceleration of the air turbine.
With the hybrid bearings the air turbine could accelerate to 120,000 rpm without seizure.
Smooth speed transitions were observed on the acceleration, Figure 9, and the deceleration, Figure 10.

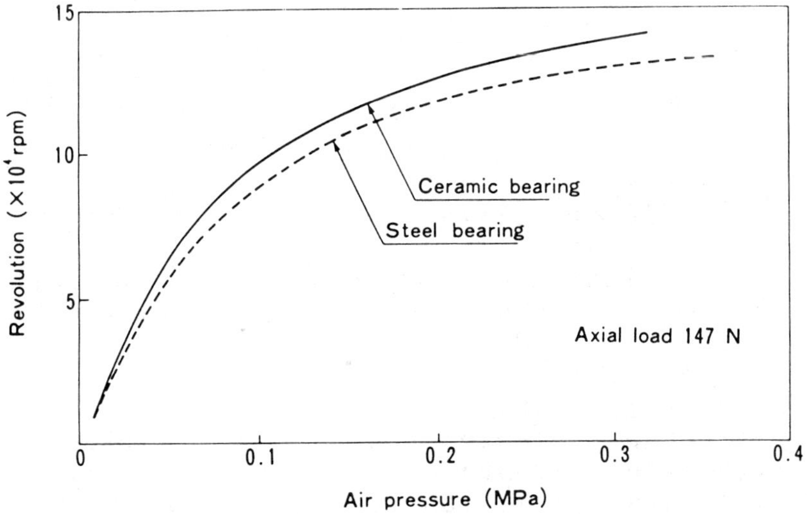

Fig. 8 Influence of supplied air pressure on the rotational speed of air turbine

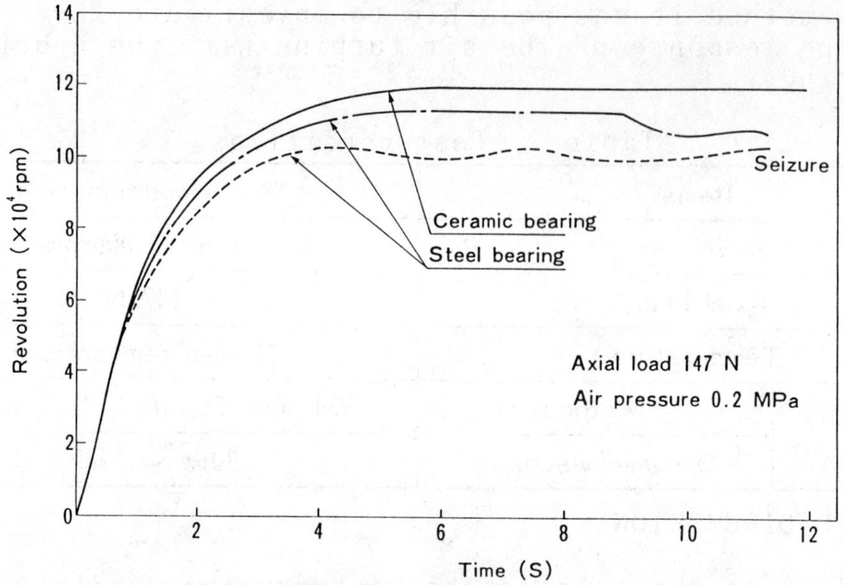

Fig. 9 Acceleration response

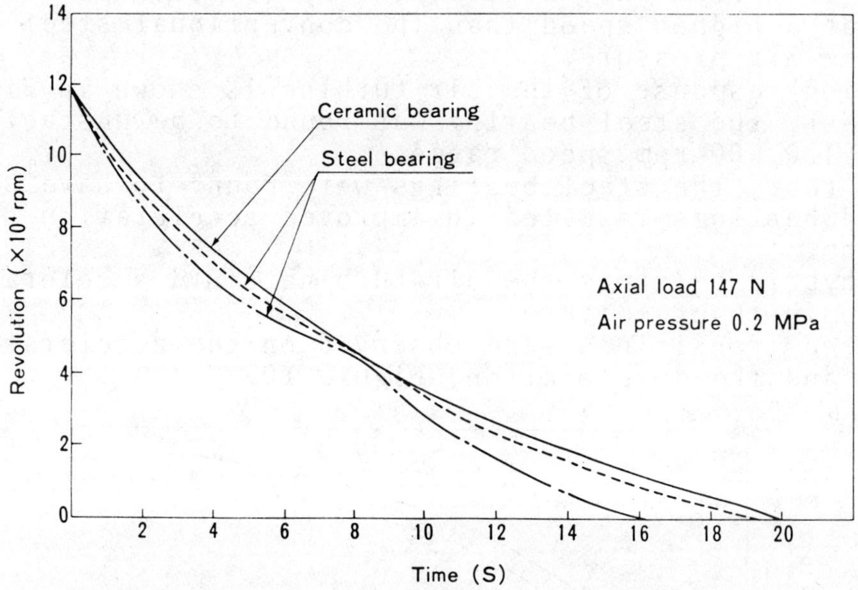

Fig. 10 Deceleration characteristics

These test results indicate that the centrifugal force, sliding of the ball because of gyro effects, and operation in marginally lubricated conditions are greatly improved through the use of silicon nitride rolling elements.
Additional acceleration and deceleration cycles of the hybrid and conventional bearings are shown in Figure 11.
The steel bearing quickly seized, within 1 to 2 cycles in marginally lubricated conditions.,
The hybrid bearing did not fail after 200 start stop cycles to the high speed.
Figure 12 shows the bearings after these tests.

The silicon nitride balls showed no evidence of wear, seizure or other phenomena after the 200 cycles.
The steel bearing balls had large marks indicating seizure after a few cycles.
These results indicate that the hybrid bearing concept with silicon nitride balls and steel rings is excellent for these types of bearing applications.

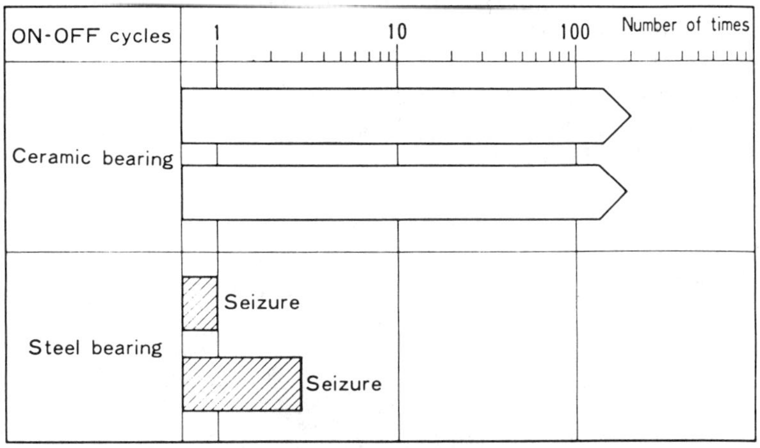

Fig. 11 Comparison of seizure resistance

Fig. 12 Some views of bearing specimens after testing

CONCLUSIONS

1) Hybrid bearings containing silicon nitride balls and steel races rotated at higher speed at lower air pressures than the conventional bearings.

2) Acceleration response of the air turbine was improved through the use of the hybrid bearings.

3) The air turbine containing the hybrid bearings had excellent speed stability.

4) Hybrid bearings appear to be more seizure resistant than steel bearings at high speed and under marginally lubricated conditions.

ACKNOWLEDGEMENTS

The authors wish to acknowledge Mr. J.C.Bentz for preparing the manuscript.

REFERENCES

1) R.J.Brook, F.L.Riley: MATERIALS & DESIGN, 5 (1984).

2) R.N.Katz, J.G.Hanoosh: Int. J. High Technology Ceramics, 1 (1985), 69-79.

3) Y.P.Chiu, H.Dalal: Army Mater. Technol. Conf. Vol. 2nd, (1974), 589-607.

4) A.Palmgren: Grundlagen der Wälzlagertechnik, 3 Auflage, S. 75, Franckh'sche Verlagshandlung, Stuttgart (1964).

5) K.Inoue, I.Ikeda: TOSHIBA REVIEW, 42 (1984), 61.

ZIRCONIA-TOUGHENED MULLITE AND ITS COMPOSITE REINFORCED BY SILICON CARBIDE WHISKERS*

J.Q. Tan, Q.M. Yuan, Z.F. Yang and Q.Liu
Tianjin University, Tianjin, China

ABSTRACT

Mullite is of current interest for possible application in the adiabatic engine as insulating material due to its low thermal conductivity, high thermal shock resistance and good chemical inertness. Of particular interest is the high strength and creep resistance at high temperature. The improvement in strength and K_{1c} value by introduction of zirconia transformation toughening makes this material even more promising. A number of ceramic head plates and cylinder upper liners have been manufactured from zirconia-toughened mullite (ZTM) and tested in a diesel engine by the authors in China. Preliminary results show that further improvement in mechanical behavior of ZTM is still necessary. Furthermore, the strengthening and toughening effects of incorporating silicon carbide whiskers are investigated.

I. INTRODUCTION

Mullite ceramics, especially when toughened with zirconia, possess characteristic properties, which allow them to be used as insulating components for adiabatic engines, namely: low values of density, Young's modulus, thermal conductivity and expansion; and hence, high thermal shock resistance[1]. Although their strength and toughness are lower than those of PSZ or TZP materials at room temperature, the properties are higher at elevated temperatures. However, applications of conventional

* Project supported by the National Science Funds.

mullite ceramics are limited because of inadequate strength and toughness. Efforts have been devoted to mullite/ZrO_2 composites containing increasing amount of ZrO_2(0 to 20 Vol%) for better densification, microstructure and applicable performance[2]. The introduction of ZrO_2 was proved to be favourable to the densification process of mullite materials and particularly to their mechanical behaviour at both low and high temperatures[3]. Different routes have been exploited for preparing mullite powders, but the preference of present authors lies on high-purity fused mullite for the practical and economic reasons, from which a number of ZTM ceramic components including head plate and cylinder upper liner, have been manufactured and tested in S135 type of diesel engine.

Attentions have also been given to whisker or fiber reinforcement of ZTM matrix, and the preliminary results are considered very promising for further improvement of strength and fracture toughness at various temperatures[4].

II. EXPERIMENTAL PROCEDURE

The experimental procedure is shown in Fig.1.

The starting materials used were high-purity fused mullite powder (average grain size<1 μm) and submicrometer ZrO_2 powder (0.1-0.2 μm). In order to obtain dense ceramic bodies fully sintered at temperature below 1600°C, the agglomerates in ZrO_2 and mullite powders must be eliminated as much as possible. Submicrometer mullite and ZrO_2 powders were dispersed individually in ethyl alcohol to form slurries. The preparation of mullite powders was carried out by milling in a ball mill with PSZ grinding media for a long time, and subsequent sedimentation under various pH conditions. In order to obtain the submicrometer powders, the slurry was first diluted to a concentration of 10 wt% solids, and then adjusted to acidic conditions of pH=2, then settled for 24 h. After that, the upper part of the suspension was siphoned and adjusted to pH=8.5-9, under which the suspension was unstable and the submicrometer mullite powders settled out. The powder produced was first washed with distilled water and then with ethyl alcohol. Finally, the powder was dispersed in ethyl alcohol giving mullite slurry.

The preparation of ZrO_2 slurry with ethyl alcohol was as follows: the $ZrOCl_2 \cdot 8H_2O$ was dissolved in distilled water under condition at pH=9. Additions of $NH_3 \cdot H_2O$ were made to bring the pH to 9 when the $Zr(OH)_4$ precipitated. The product was washed with water to eliminate the Cl^- and then with alcohol 3 times to obtain a satisfactory quality $Zr(OH)_4$ slurry.

The powder slurries of required particle size were mixed at the desired composition ratio in a ball mill for 4 h to reach high homogeneity, and then dehydrated at<100°C, calcined at 800°C for 1 h and milled by same route for 2 h with ethyl alcohol. After drying, the required mixture of ZrO_2 and mullite powders was prepared. The produced mixture of powders was pressed at 80 MPa, crushed for granules, and isostatically pressed at 350 to 400 MPa for shaping into green bodies of the cylinder upper liner, and then sintered at 1580°C for 4 h in air. The sintered ceramic bodies were machined by using diamond grinding instruments to achieve required shape and dimensions. The green bodies of head plate were sintered at 1550°C for 1 h by hot-pressing in Ar atmosphere and machined to fit the required shape and size.

Whisker reinforcing experiments were proceeded with ultrafine mullite powder prepared from the sol-gel method. Silicon carbide whiskers from Tokai Carbon Co. Ltd. Japan, was used as the starting material. The composition of mullite/25vol% ZrO_2/15vol%$SiC_{(w)}$ was mixed for 4 h in plastic pot with plastic milling media and ethyl alcohol. With the help of a small amount of surfactant, the whiskers were homogenously dispersed in a ZTM matrix. Dense ZTM matrix composites reinforced by SiC whiskers were obtained by hot-pressing at 1600°C for 1 h in a BN-coated graphite die with 28 MPa applied pressure. The resulting disks were cut into rectangular bars(26X5X2.5mm). Flexural strength and toughness of these specimens up to 800°C were measured. Microstructural information of these specimens were obtained by SEM.

III. RESULTS AND DISCUSSIONS

Table 1 shows the comparison of ZTM with Mg-PSZ and Y-TZP ceramics.

It can be seen from table 1 that the strength and toughness values measured at room and high temperatures for ZTM ceramics containing 15 vol%ZrO_2 could be increased by a factor of 2 in comparison with pure mullite ceramics. Hot-pressing the 15 vol% 2.5Y-TZP material improved the properties to a greater extent i.e. σ_{RT}=688 MPa, K_{1c}=8.61 MPa $\sqrt{m}$ which are close to that of Mg-PSZ, and even better high temperature strength $\sigma_{800°C}$=481 MPa (against $\sigma_{700°C}$=407 for Mg-PSZ). The operative strengthening and toughening mechanisms may be attributed to microcracking due to mismatch of thermal expansions of ZrO_2 and mullite and stress-induced transformation of tetragonal zirconia. Considering the application for adiabatic engine as insulating component, the choice of an optimum material was based on its excellent thermal insulating ability which requires a value of thermal conductivity as low as possible. It can be seen from table 1, that the thermal conductivities of ZTM and ZTM/15vol%Y-TZP are satisfactory and have no significant difference in comparison with TZP or PSZ materials.

Both Mg-PSZ and ZTM cylinder upper liners could withstand 106 h high loading operation without failure if compressive prestress was induced by proper "super-gain fitting", but the latter one was superior in the sense of low cost and simplicity of technology. Furthermore the stresses in the ceramic components during operation of the S135 engine were mainly thermal stresses. So the ZTM ceramics, which have much lower thermal stresses at the same temperature gradient, have an important advantage compared to Mg-PSZ. Therefore ZTM ceramics are promising potential candidate materials for application in ceramic engines.

Table 2 shows the properties of hot-pressed ZTM/$SiC_{(w)}$ ceramic composites. It can be seen from table 2, that the strength and toughness of the hot-pressed composites, containing 15 vol%$SiC_{(w)}$ and 25vol%ZrO_2, can be increased by 20% to 30% (σ_{RT}=510-530 MPa, $K_{1c(RT)}$=6-7 MPa $\sqrt{m}$, $\sigma_{800°C}$=400-420 MPa) in comparison with ZTM ceramics without SiC whiskers. The observation of fracture surfaces by SEM indicates that the energy for pull-out of the whiskers must be the main contribution to toughening (Fig 2).

IV. CONCLUSIONS

1. The strength and toughness at room and high temperatures for ZTM ceramics containing 15vol%ZrO_2 have been increased by a factor 2 in comparison with pure mullite ceramics, and reach values of σ_{RT}=350 MPa, $\sigma_{800°C}$=460 MPa, K_{1c}=3.4 MPa$\sqrt{m}$(RT), 4.5 MPa$\sqrt{m}$ (800°C).

2. The hot-pressed 15vol% 2.5Y-TZP material was further strengthened and toughened as follows: σ_{RT}=688 MPa, K_{1c}=8.6 MPa$\sqrt{m}$ (RT).

3. A ZTM cylinder upper liner withstood 106 h high loading operation in a diesel engine without failure when a compressive prestress was established in the component by "super-gain fitting".

4. The strength and toughness of the hot-pressed mullite composites, containing 15vol%$SiC_{(w)}$ and 25 vol%ZrO_2, have been increased by 20% to 30% (σ_{RT}=534 MPa, $\sigma_{800°C}$=403 MPa K_{1c}=6.0 MPa$\sqrt{m}$ (RT).

REFERENCES:

1. N.Claussen and J. Jahn, "Mechanical Properties of Sintered in Situ-Reacted Mullite-Zirconia Composites," J. Am. Ceram. Soc., 63[3-4] 228 (1980).

2. Moya and M.I. Osendi, "Effect of $ZrO_{2(ss)}$ in Mullite on the Sintering and Mechanical Properties of Mullite/ZrO_2 Composites." J.Mater. Sci. 2, 599-609 (1983).

3. Prochazka, J.S. Wallace and N.Claussen "Microstructure of Sintered Mullite-Zirconia Composites," J.Am Ceram. Soc., 66[8] c-125-c-127 (1983).

4. P.F. Becher and T.N Tiegs, "Toughening and Strengthening Behaviors in SiC Whisker- Reinforced Zirconia-Toughened Mullite" J. Am. Ceram Soc., 70[9]651-54 (1987).

1. Preparation of Components and Samples.

submicrometer mullite slurry in C_2H_5OH | $Zr(OH)_4$ Slurry in C_2H_5OH

↓

Mixing

↓

Drying

↓

Calcination

↓

Milling

↓

Drying

↓

Granulation

↓

Hot-pressing(1550°C, 1 h) (For head plate) | Isostatically pressing (for cylinder upper liner) → sintered (1580°C 4 h in air)

↓

Machining (to required shape and size)

2. Measurement

Strength: 3-point bending method (span=20 mm)

Toughness: Single edge notched bar test

Thermal conductivity: Method of longitudinal thermal flow in stable state

Fig.1. The experimental procedure

Table 1 Comparision of Properties of ZTM with Mg-PSZ and Y-TZP Ceramics

Material	σ_f (MPa)		K_{1c} (MPa$\sqrt{m}$)		λ 600°C
	R.T.	800°C	R.T.	800°C	Cal/sec.cm°C
ZTM (15 vol % ZrO_2)	350	460	3.4	4.5	0.00612
ZTM (M + 15 vol % ZrO_2) + 15 vol % 2.5 Y-TZP)	688	481	8.6	—	0.00874
Mg-PSZ (10 Mol % MgO)	680	407 (700°C)	8.9	5	0.00597
Y-TZP (2.5 Mol % Y_2O_3)	940	240	13.2	4.5	—
Pure M	180	—	2.4	—	0.00870

Table 2 Properties of Hot-Pressed ZTM/$SiC_{(w)}$ Ceramic Composites

Material	σ_f (MPa)		K_{1c} (MPa$\sqrt{m}$)	γ
	R.T.	800°C	R.T.	(J/m²)
ZTM	448	310	4.9	48
ZTM + 15 vol % $SiC_{(w)}$	534	403	6.0	78

(a) at room temperature for the $ZTM/SiC_{(w)}$ specimen

(b) at 800°C for the $ZTM/SiC_{(w)}$ specimen

Fig.2 SEM of Fracture Surface of Composites.

HIPRBSN PRECHAMBERS FOR THE DIESEL ENGINE

H.Gasthuber*, J.Heinrich**, J.Huber**, M.Steiner**, W. Bunk***

ABSTRACT

A silicon nitride prechamber has been developed for a Diesel engine in the 2 - 3 liter range. This chamber can be assembled into an existing cylinder head of a car without any alterations. The construction, made in 3 parts, takes into consideration the specific technological, material and practical needs. Temperature and stress distribution to be expected in the ceramic components were calculated on the base of a metallic standard chamber. The different processing steps, like injection molding, dewaxing, nitriding, encapsulation and hot isostatic pressing are described and the properties of the components are discussed. Material tests with HIPRBSN samples proved that the difference between the data observed and those calculated before was big enough to preserve the components. Non destructive and proof tests during virtually all steps of process were made to prove a high lifetime-expectation. After small design modifications some ceramic prechambers have been tested sucessfully in a fired engine under all working conditions for 300 hours.

1. INTRODUCTION

For Diesel engines the three different fuel injection methods direct injection, swirl chamber and prechamber are in use. In Mercedes-Benz car Diesel engines prechambers are exclusively installed since they guarantee good fuel economy, better driving comfort than all other systems and more advantageous conditions for particle emission limiting values than any alternative combustion processes. The prechamber is an important part of the combustion chamber, having approx. 20 percent of its total volume. Because of the high thermal load and expected opportunity of positive effects on the characteristic of the combustion process, the prechamber seems to be predestined for realization in ceramic [1 - 3].

The aim of this work therefore was to design a silicon nitride prechamber for a Diesel engine in the 2 - 3 liter range, to calculate temperature and stress distributions and to test these ceramic prechambers in a fired

* Daimler-Benz AG, Postfach 202, 7000 Stuttgart 60, FRG
** Hoechst CeramTec AG, Wilhelmstraße 14, 8672 Selb, FRG
*** DFVLR, Institut für Werkstoff-Forschung, Linder Höhe, 5000 Köln 90, FRG

all the holes are to be made in the injection molded or green body, and the ceramic hot pin is an integrated component.

3. THERMAL AND MECHANICAL CALCULATIONS

The ceramic prechamber was calculated with the finite element program "BETSY" [4] in respect of the likely stress temperature distribution.

Modelling for the temperature distribution made use of 4096 triangular ring elements and 2601 nodes, axially symmetrical (fig. 3). Part of the cylinder head is integrated in the model.

The calculation is based on the following assumptions:

- coolant temperature 90 °C
- combustion gas temperature 920 °C
- heat transfer coefficient 155 to 2200 W/m^2K
- engine speed 4600 rpm at full load.

The temperature distribution in the ceramic chamber thus calculated is presented in fig. 4. In contrast to the metallic chamber, the ceramic one is found to have a pronounced axial temperature gradient. This phenomenon is not only caused by the material used but also by differing design features. By a combination of material properties and design measures, it is possible to achieve practically any desired temperature profile at and within the prechamber.

The stress distribution was calculated by modelling the prechamber with 536 triangular intermediate nodal ring elements CTRIAX6 with 1215 nodes (fig. 5). In addition to the stationary temperature field as shown in fig. 4, stress in the form of a preload force of approximately 25 kN is acting on the chamber axially from above.

As the results of calculation indicate (fig. 6), the maximum stress for normal operating conditions is of the order of 230 N/mm^2 in the area of the prechamber seat surface so that it can be concluded that the component is adequately dimensioned in respect to material strength.

Even in the worst case scenario, the maximum reference stress would be 550 N/mm^2, which is unlikely to represent any problem in view of the high compressive strength of the material used. The deformation of the material under load of maximum thermal stress (full load at n = 4600 rpm) is presented in fig. 7 magnified by the factor 100.

engine under all working conditions. Another goal of this paper was the development of a processing technique for mass production of dense silicon nitride prechambers. For the production of components in narrow tolerances without post machining injection molding of silicon powder, dewaxing, nitriding, encapsulation and post densification by hot isostatically pressing was decided to be the most suitable procedure.

2. PRECHAMBER DESIGN

The prechamber is installed in the combustion chamber excentrically and at a slight angle, in a way that only its cap with the flame channels projects inward. At the top, the injection nozzle is screwed into the retaining ring holding the chamber in place. Its bottom edge lies on a tapered face of the prechamber (fig. 1).

Prechamber thermal stress is extremely high, notably in the area of the flame channels, as is corrosion. It is further increased by poor quality fuel. The material used limits the maximum temperature of the chamber, which is held at a tolerable level by fitting surfaces on the chamber neck and neighbouring cooling water passages in the cylinder head.

It did not seem particularly reasonable to reproduce a complete standard chamber in ceramic, therefore the following design concept was worked out (fig. 2):

- The chamber is divided into
  - ceramic prechamber bottom section
  - ceramic hot pin
  - metallic or ceramic top section.
- Putting compression stress on the ceramic parts by
  - supporting it at the combustion chamber end at the tapered surface above the neck
  - fixing it at the nozzle end by means of the metallic top section and a threaded sleeve which at the same time holds the injection nozzle in place
  - positive contact between ceramic bottom and metallic top sections by means of a spring element which absorbs longitudinal expansion in operation and assures a constant contact pressure.
- Foregoing subsequently machined fitting surfaces intended to enable the ceramic prechamber to be installed in the "as-fired" state, which, in the turn, offers the opportunity of manufacturing the part cost-effectively.
- In the experimental stage the parts are provided with flame channels and pin holes after the sintering process. Development, however, is aimed expressly at injection-molded finished parts. In other words,

4. PROCESSING

Injection molding was chosen as molding technique for the prechambers. This procedure allows the fabrication of components of complicated geometry to close tolerances, even in large quantities. Fig. 8 shows the production run for injection molded engine components made from hot isostatic densified reaction bonded silicon nitride (HIPRBSN) in a survey. A detailed description of the production run is given in [5] .

Starting materials for HIPRBSN engine components are silicon powders and sintering additives as required for the subsequent hot isostatic densification. In the milling process with which the production run begins, agglomerates of sintering additives are shattered and distributed homogeneously in the Si-starting powder. Moreover, the pressing properties of the Si powder are optimized by influencing grain size distribution and grain shape. Afterwards, the powder blend is mixed with organic additives like waxes, thermoplastics or resins. These organic additives have the same functional properties as binders, flow/slip/wetting and mold release agents. The blend of powder and plastification constituents is granulated according to a defined homogenization process and subseqently injection molded by conventional injection molding machines.

The organic components must be removed from the green compact before it can be transformed into silicon nitride by annealing in an atmosphere of nitrogen. This process is usually performed in oxidizing atmosphere at temperatures where the silicon is not oxidized. It is necessary to choose the organic components such that they can be driven out in the liquid or gaseous phase at different temperatures. If the binders, plasticizers, slip additives and mould separating agents are all baked out at the same temperature, then this would destroy the component. As far as possible, the loss in weight with raised temperature should be arranged to occur very evenly over a wide temperature range.

Reaction-bonded silicon nitride is manufactured by nitridation of the silicon powder compacts. The volume change of ca. 22 % leads to a reduction of the total porosity without a significant change in the total dimensions (ca. 0.1 %). This is the most attractive feature of this process, because the end product requires little or no final machining. The reaction of silicon to yield silicon nitride is usually performed in nitrogen at atmospheric pressure and at temperatures between 1000 °C and 1450 °C. A detailed description of the nitriding process is given in [6]. The reaction time for complete nitridation can be anything from a few hours to several days. The wall thickness of the components that are to be completely nitrided can only be up to ca. 3 cm, even if extremely long nitridation times are employed. This is because the extremely slow diffusion of nitrogen into thicker walls becomes the rate-determining step of the above reaction.

The components are densified in a high pressure autoclave using as pressure transmission medium gases like argon or nitrogen. For gas pressures up to 300 MPa the densification temperatures range between 1600 °C and 1900 °C. Holding times at end temperature and maximum pressure depend on the microstructural parameters of the starting material [7,8]. Usually they range between 15 minutes and 3 hours. For the densification of complicated shaped components the so-called "sinter-canning-technique" has been developed [9] . In this process a gastight layer is sintered onto the components matching it in composition and coefficient of thermal expansion (fig. 9). After the hot-isostatic densification these layers are removed by sand blasting.

5. MICROSTRUCTURE AND MATERIAL'S PROPERTIES

Starting with RBSN a fine grained microstructure is produced by HIP, the individual silicon nitride grains showing a relatively high aspect ratio (width/length) of approx. 5 - 6 (fig. 10).
As known by literature such "linked" microstructures lead to very high strength [8, 10, 11] . With almost theoretical density and a 100 % share of ß-silicon nitride the 4-point bending strength is 830 MPa (table 1). The strength values measured with samples worked out from prechambers are very high too, although these components have been tested as fired without any finishing. Furthermore, the c-ring samples show a very high oxidation resistance. After annealing at 1200 °C there has not been hardly any decrease in strength after 60 hours (table 1).

6. INSPECTIONS AND TESTS

All the delivered unfinished prechamber blanks were examined in respect of externally perceptible faults. X-ray pictures of the macrostructure were then taken to detect coarser faults prior to further processing.

Subjecting the prechamber to internal pressure is a realistic test which can be performed without great expense and which accurately simulates the stress distribution during operation. The stresses in the component remain moderate even if twice the ignition pressure is simulated; only major faults (shrink holes, pores, cracks) are likely to result in component damage. The maximum tensile stress occurring in the component corresponds to the tangential stress on the inside of the prechamber resulting from the respective ignition pressure, this being for the:

- o naturally aspirated engine approx. 40 N/mm^2 and for the
- o turbo charged engine approx. 50 N/mm^2.

The prechambers were tested at an internal pressure of up to 250 bar in a hydraulic pressure tester (fig. 11). None of the prechambers sustained any damage.

7. TEST RIG SETUPS

12 prechamber blanks manufactured from injection molded HIPRBSN were machined into parts ready for installation by grinding and ultrasonic treatment (fig. 12). Four of the prechambers underwent testing on the engine test rig in a 2-litre, 4-cylinder car diesel engine.

No problem was encountered in starting the engine after installing the ceramic prechambers. The engine was run for 7 hours at low load followed by a progressive increase to full load.

The striking features were

- o a power deficit of approx. 15 %
- o pronounced formation of smoke at full load.

To determine the reason for the loss of power, the injection timing was checked and then specifically altered by ± 5 cs° compared to the standard setting. This resulted, though, only in a further reduction in power output.

As the chambers did not reveal any signs of damage and the thermal characteristics of the ceramic material does not differ so greatly from those of the metal alloy used in the hot part of the standard chamber as to explain such a loss of power, the cause must have been geometrical variations.

For this reason, the original concept was modified slightly for the subsequent test setup; the prechamber was now divided functionally into four parts (fig. 13):

- o prechamber bottom section
- o prechamber top section
- o prechamber top section 2
- o hot pin

As was expected, no problem was encountered in installing three of these ceramic prechambers in the test rig engine. The fourth prechamber installed was a standard metal type. The test program (fig. 14) was the same as the first test run. Following a run-in period of 7 hours at low load, engine load was slowly increased.

This time, the 2-litre car diesel engine, equipped with the modified ceramic prechambers, produced its full power output of approx. 50 kW straight off. The engine started perfectly every time; rapid load cycles were achieved as easily as with the standard engine.

The test program was completed after 100 hours. Cylinder head and prechambers were disassembled for examination. The only damage ascertained was a fractured ceramic hot pin, which was neither attributable to the system, as such, nor did it cause any subsequent damage. Prechambers 2 and 3 were in perfect condition after running 100 hours.

The engine tests revealed the ceramic pin fixing to be a weak point, which was thereupon modified to achieve a material bond with the bottom section of the prechamber.

No problems were encountered with this modified ceramic prechamber. The endurance test program over 100 hours was successfully concluded.

8. OUTLOOK

Further tests are currently in progress to obtain information regarding the effects of the ceramic chamber on power output, fuel consumption and emission. By all appearances, though, no revolutionary changes are to be expected from mere material substitution.
But nevertheless, there are some perspectives for the realization of advanced design concepts, not having the limitations of the metallic materials used today.

9. REFERENCES

1. H. Matsuoka, H. Kawamura, S. Toeda: Development Of Ceramic Precombustion Chamber. JSAE Review, March 1984.

2. S. Sakurai, T. Matsuoka: Development Of Low Particulate Engine With Ceramic Swirl Chamber, SAE-Paper. No. 861407 (1986) Sept.

3. S. Nakatani, K. Yoshizu, T. Mitsunaga, A. Inoshita: Development Of All-Ceramic Swirl Chamber For Diesel Particulate Control. SEA-Paper, No. 861406 (1986) Sept.

4. FVV Forschungsvorhaben Nr. 209, 245, Integralgleichungsmethode, Abschluß-Berichtsheft 310, Band 1 und 2, Jhg. 1982.

5. M.Steiner, J.Huber: Development of a statistical proved injection molding method für RBSN, SRBSN and SSN, Final Report Nr. 03 ZC 1133, German Ministry for Research and Technology.

6. J.Heinrich: Influence of the silicon grain size on the microstructure and the mechanical properties of reaction bonded silicon nitride, Ber. Dt. Keram. Ges. 55 (1978), 238-241.

7. J.Heinrich, E.Backer, M. Böhmer: Hot isostatic pressing of Si_3N_4 powder compacts and reaction bonded Si_3N_4. J.Am.Ceram.Soc. 71 (1988) C-28-C-31.

8. G.Ziegler, J.Heinrich, G.Wötting: Review: Relationship between processing, microstructure and properties of dense and reaction-bonded silicon nitride. J.Mat.Sci. 22 (1987) 3041-3086.

9. J.Heinrich, M.Böhmer: Densification of porous ceramic comonents by hot isostatic pressing. German Patent 34 03 917.9, 1985.

10. F.F. Lange: Relation between strength, fracture energy and microstructure of hot pressed Si_3N_4. J.Am.Ceram.Soc. (1973), 518-522.

11. L.J. Bowen, T.G. Carruthers: Development of mechanical strength in hot pressed silicon nitride. J.Mat.Sci. 13 (1978), 684-687.

The authors thank Mr. Henn/DFVLR, Cologne for the encapsulation of the components and for performing the HIP experiments.

Tab. 1: Properties of HIP-RBSN determined with different test procedures.

test procedure / property	4-point bending, 30 mm outer span 10 mm inner span samples diamond ground	c-ring test, samples cut from precombustion chambers, surface as fired	c-ring test, samples cut from precombustion chambers, oxidized 1200°C/60 hrs.
fracture strength, MPa	830	660	650
Weibull modulus	10	10	9
Young's modulus, GPa	310	292	290
density, g/cm^3	3,26	3,26	3,26
% of theoretical density	99.7	99.7	99.7
ß-content, %	100	100	100

Fig.1: Installation of the metal prechamber into the car Diesel engine.

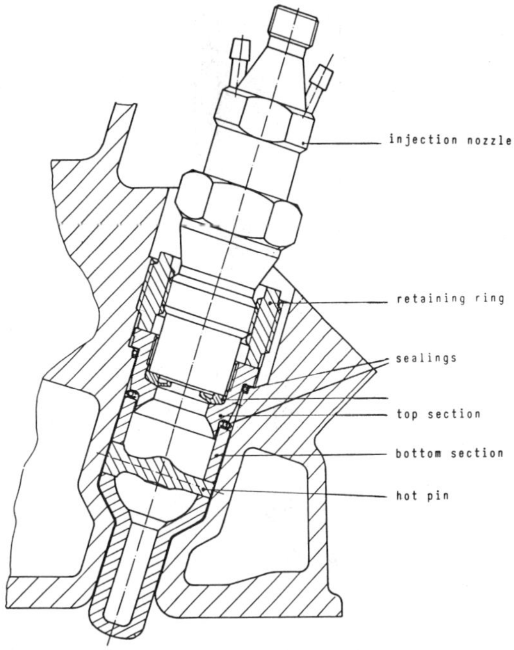

Fig.2: Arrangement of the ceramic prechamber in the cylinder head.

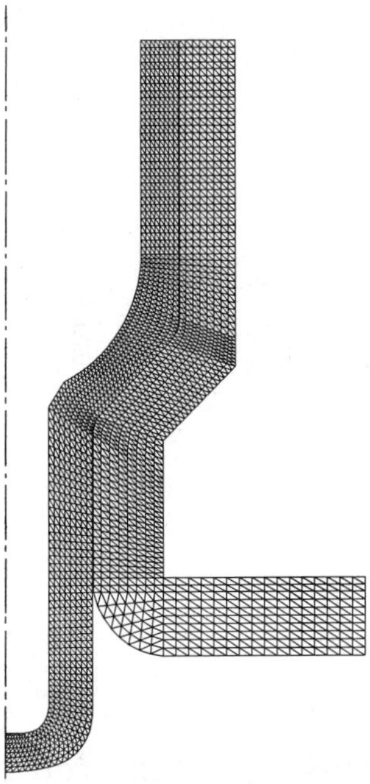

Fig.3: Node network for thermal calculations.

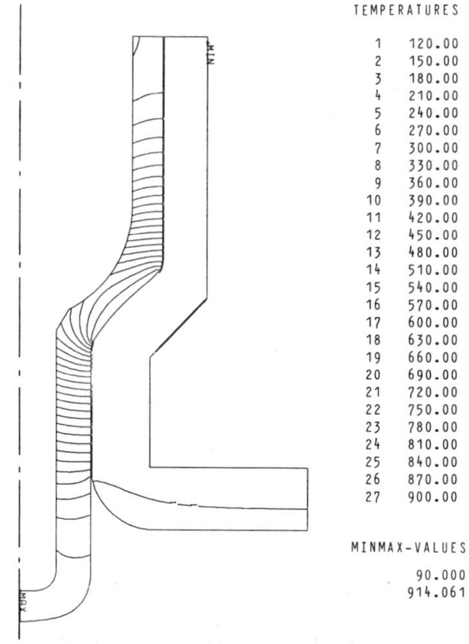

Fig.4: Temperature distribution in the ceramic prechamber.

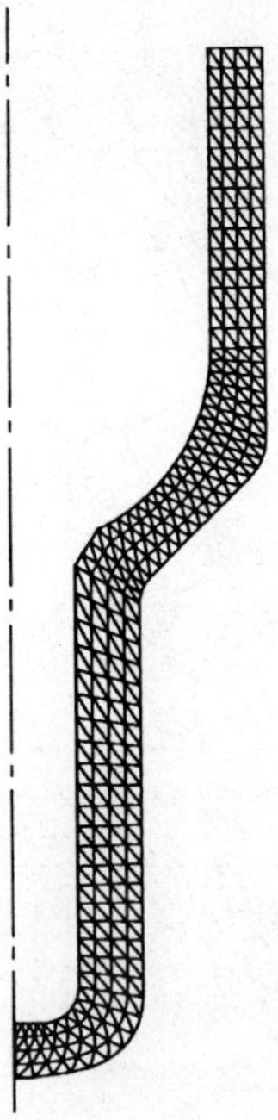

Fig.5: Node network for calculation of mechanical stress.

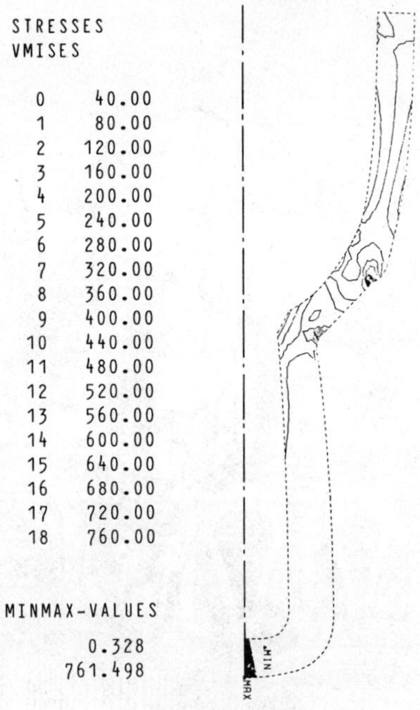

Fig.6: Equivalent stress distribution in the ceramic prechamber considering thermal and mechanical loads.

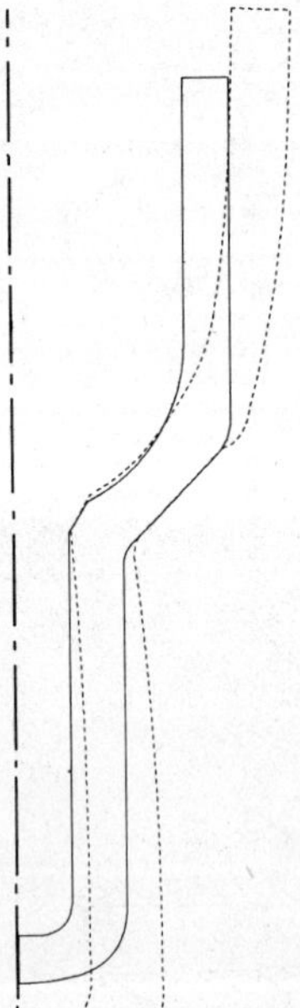

Fig.7: Calculated deformation of the ceramic prechamber caused by thermal and mechanical loads.

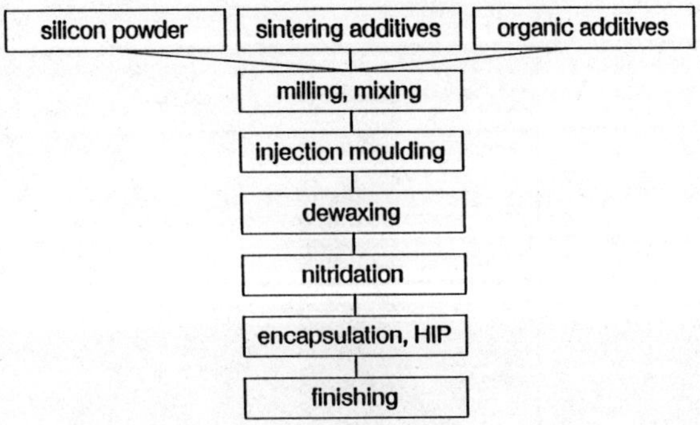

Fig.8: Processing steps for injection moulded HIP-RBSN engine components.

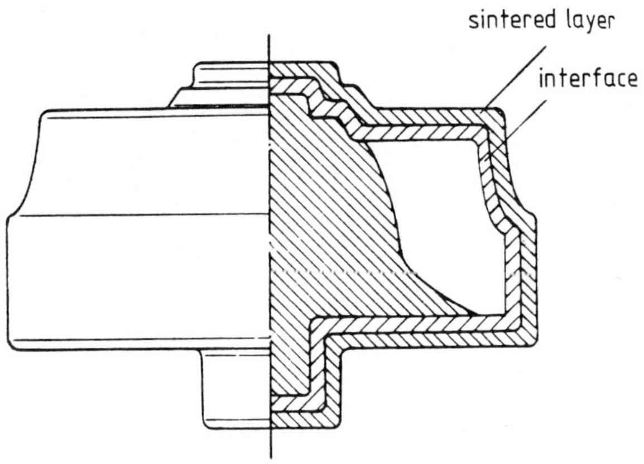

Fig.9: "Sinter-canning-technique" for the encapsulation of complicated shaped components.

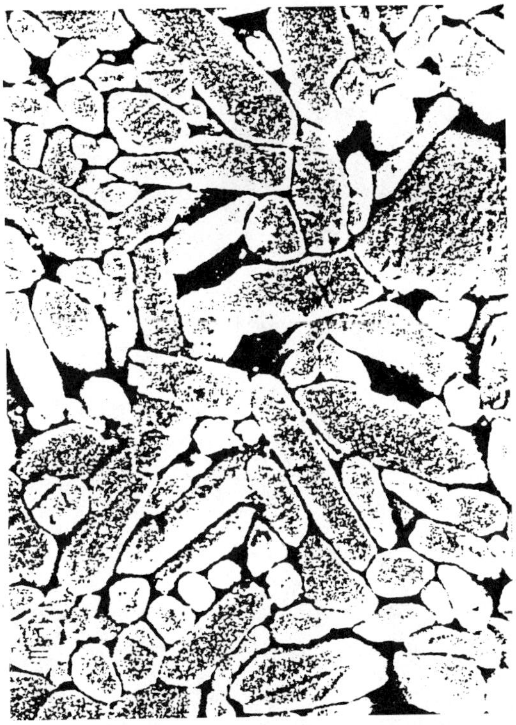

Fig.10: Microstructure of HIP-RBSN showing elongated ß-grains (etched microsection, REM).

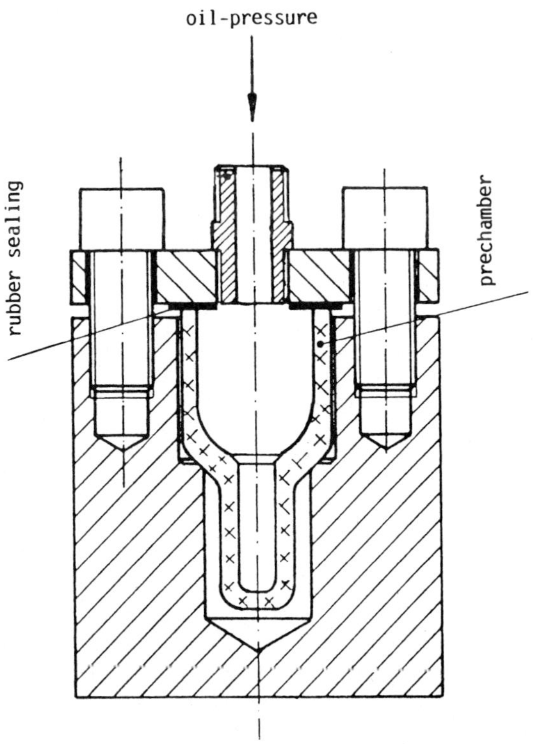

Fig.11: Hydraulic pressure tester

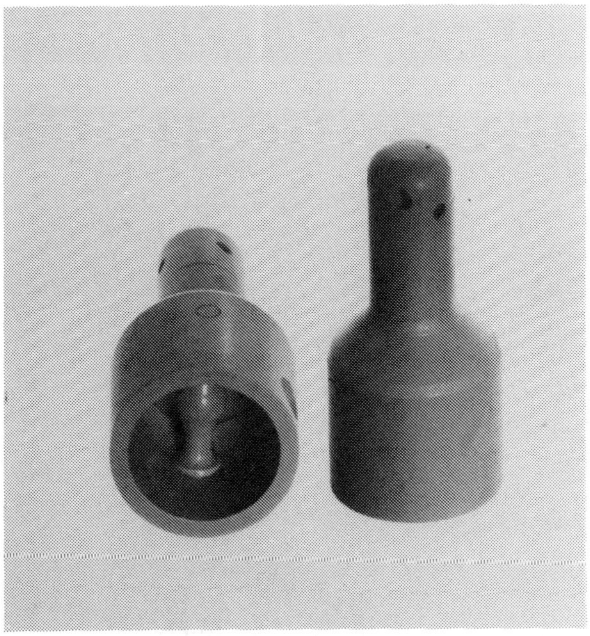

Fig.12: Ceramic prechamber with bottom pin

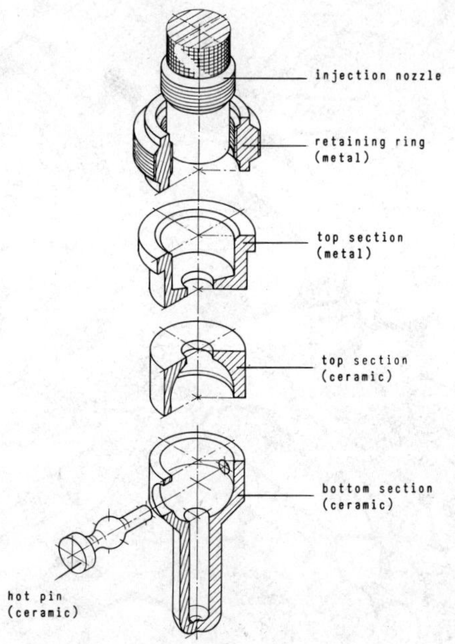

Fig.13:Exploded view of the succes-
fully tested prechamber
arrangement.

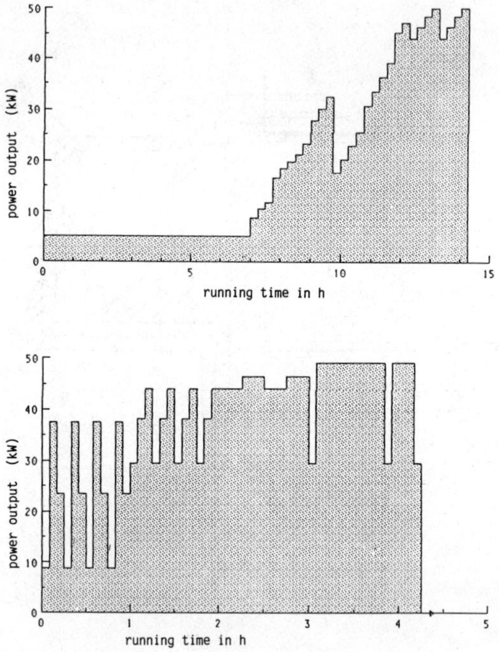

Fig.14:Test cycle for ceramic pre-
chamber in a 2 Liter car
Diesel engine.

SILICON NITRIDE CERAMIC WEAR RESISTANT MACHINE ELEMENTS

J.C. Bentz and T.M. Yonushonis
Cummins Engine Company
Columbus, IN U.S.A.

T. Aoba and Y. Fujimoto
Toshiba Corporation
Tokyo, Japan

ABSTRACT

Toshiba Corporation of Japan and Cummins Engine Company, Inc. in the United States have been developing potential near term applications of silicon nitride (Si_3N_4) ceramics for wear resistance in diesel engines. The primary activities have included a component selection process, development of low cost designs, materials, and manufacturing processes, and the evaluation of these components in bench and engine tests. These tests have confirmed that the development of the described injection molding process for the first of these components intended for production -- ceramic links and ball ends attached to the ends of steel links or shafts for valve and injector train linkages -- should lead to reliable components produced at low cost.

SILICON NITRIDE SELECTED

Properties of representative wear resistant ceramics are shown in Table 1. Alumina (Al_20_3) is very economical but it has a relatively low impact strength. Zirconia (ZrO_2) is not thermally stable to high temperatures or as wear resistant as other ceramic materials. Silicon carbide (SiC) has good wear resistance and corrosion resistance, but its impact strength is also relatively low. Silicon nitride has good wear resistance, corrosion resistance, and also has good impact strength. We have selected silicon nitride as the most promising candidate material for engine wear components.

STRENGTHENED SINTERED SILICON NITRIDE

Because silicon nitride is a covalent compound, it is difficult to achieve full densification without sintering aids. Magnesia (MgO) addition, as a sintering aid, was discovered in the UK in 1961 and a hot pressed material Si_3N_4-MgO was extensively evaluated as a high strength material. In 1967, Toshiba determined that the rare earth oxides such as yttria (Y_20_3) were more effective for strengthening. At present, rare earth oxide base additives are the most effective additives for high temperature properties. When silicon nitride powder is sintered with rare earth oxides such as yttria, the function and development goal of the sintering aid is based on the principle of developing an optimized microstructure for the sintered material. Elongated grain crystals have been shown to result in higher toughness. The aspect ratio of silicon nitride grains associated with the use of rare earth oxides seems to be higher than that of magnesia. At the same time the Y_2O_3-Al_2O_3 additions react with the silicon nitride resulting in a liquid phase, which promotes densification.

Table 1 Characteristics of typical engineering ceramics

Items	Unit	Si_3N_4	SiC	Al_2O_3	ZrO_2
Density	g/cc	3.2~3.3	3.1~3.2	3.90~3.99	5.9~6.1
Bend strength	MPa	700~1000 (R T) 450~850 (1200°C)	400~700 (R T) 400~700 (1400°C)	300~450 (R T) 150 (1000°C)	700~1200 (R T) 500 (800°C)
Thermal conductivity	W/m K	20~24	50~170	20	1.7~3.3
Thermal expansion coefficient	10^{-6}/K	3.2~3.4	4.3~4.8	8 ~ 9	9 ~11
Hardness	Hv	1400~1700	2400	1500	1300
Elastic modulus	GPa	300	400~480	320~370	150~250
Fracture toughness	MPa $m^{1/2}$	6 ~ 7	3 ~ 5	2 ~ 3	8 ~15
Advantage	—	•high strength •high thermal shock resistance •high wear & corrosion resistance	•high strength at high temperature •high corrosion resistance	• high corrosion resistance • low cost	• high fracture toughness

* Toshiba Corporation Technical Brochure

Both hot pressing and pressureless sintering are applicable for the densification. Both methods bring about high strength materials. To obtain a high strength material it is important to control the silicon nitride raw powder, the yttria and alumina composition, and sintering temperature profile. In 3-point bend strength testing, Toshiba has obtained approximately 1000 MPa for both hot pressed and pressureless sintered silicon nitride.

Of course, the strength of actual components is affected by their size and shape. It is necessary to note that different molding processes are required for different geometry and size components. This means that their functional strength will also vary unless precise process control has been defined for each component-molding process. For fabricating components which are of small size and/or complicated shape, injection molding has generally been selected because of easy powder-plastic mixture flow into the mold cavity from the gate. A photograph of two injection molded components, a fuel injector link and a ball end, is shown in Figure 1.

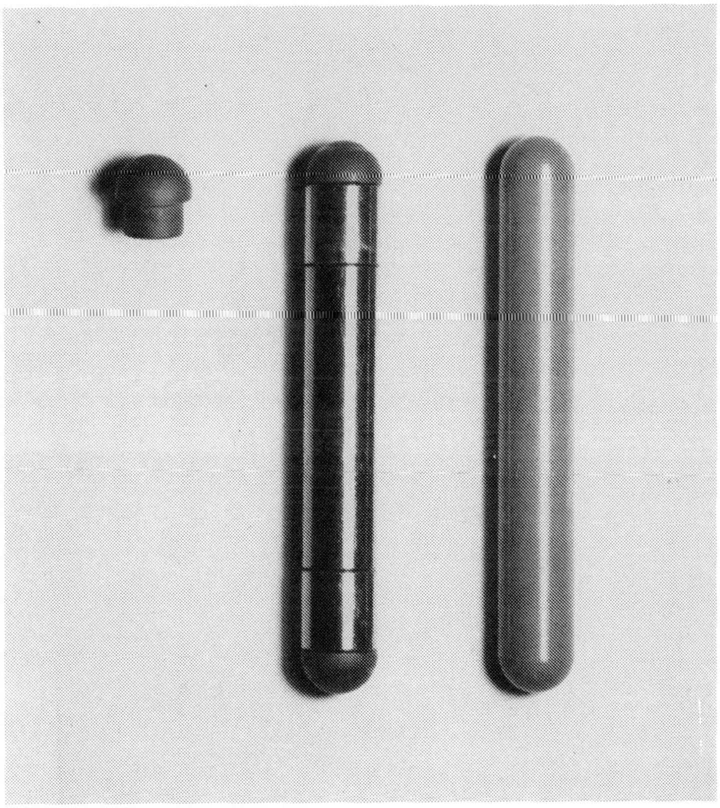

Figure 1 Injection molded ball end, a link assembly and a solid link.

FABRICATION PROCESS

The injection molding of silicon nitride wear components has been developed by Toshiba for several years. Because it is possible to control the composition and uniformity of the pre-mixed ingredients very precisely and also because this uniformity can be maintained by injecting the mixture into the mold rapidly at high pressure, this process is expected to produce net shape components. Limited finish grinding can be used to meet very tight tolerances. These processes are also suitable for automated mass production. The injection molding process, having been developed for the high volume production of plastics, can therefore be used to process ceramic components quickly and at low cost. The process is shown in Figure 2.

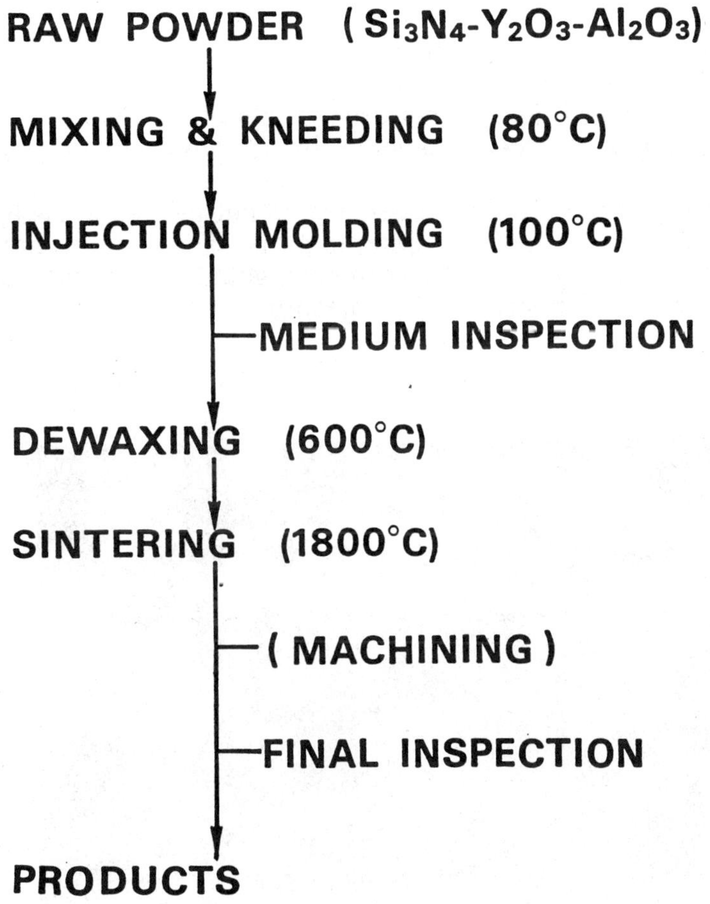

Figure 2 Injection molding process

This process involves mixing and kneeding a plastic resin, plasticizers, and other additives with the same volume of Si_3N_4-Y_2O_3-Al_2O_3 powder. The mixture is then formed into complex shapes by the use of improved plastic forming equipment. The molded parts are heated to remove the binder. After the dewaxing process, the parts are consolidated to over 98% relative density by heating to 1700 - 1800 Celcius in an atmosphere of nitrogen. In this process, the key components affecting success are selection of the binder, mold design, the pressure/temperature profile during the molding process, and the dewaxing process conditions. We have developed acceptable injection molding conditions by many experimental trials.

The initial injection rate of the 'ceramic powder - resin' molding compound must be very rapid to prevent premature cooling during injection. The injection pressure must also be high to fill up the mold homogeneously. However, the pressure must quickly be reduced after fill-up to avoid putting an unwanted high residual stress in the pre-form. The cavity pressure is then maintained at the lower pressure for the purpose of obtaining a uniform and stable shrinkage of the moldings upon sintering.

The principles of control of the injection molding machine are as follows:

1) Measurement of the injector pressure in the mold cavity.
2) Establish machine controls to develop the desired cavity pressure pattern.

Non-destructive flaw detection by fluorescent dye penetration and X-ray radiography is conducted on the injection molded and sintered parts. The sintered parts have a well developed microstructure which consists of the elongated grains and glassy second phases. Figure 3 shows a microstructure of injection molded silicon nitride which is very similar to that of die pressed.

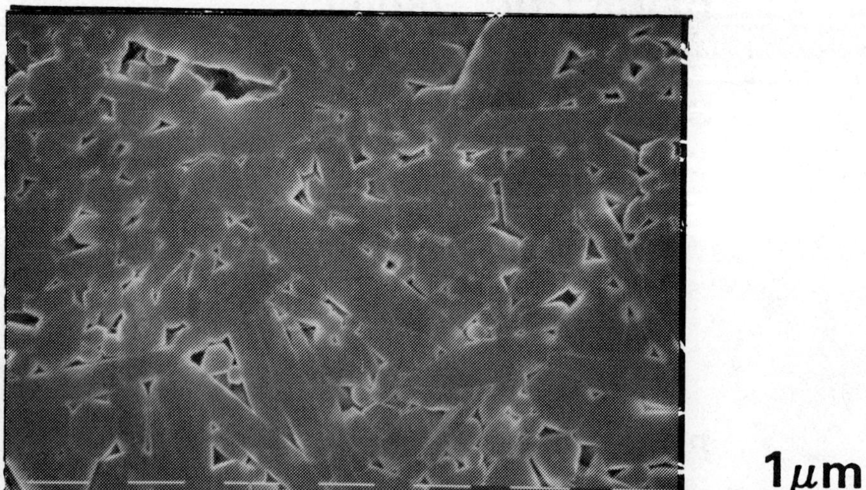

Figure 3 Microstructure of sintered Si_3N_4 (Injection molded)

Three-point bend strength and the weibull modulus for 3 x 4 x 40 mm specimen cut from an injection molded long bar specimen were 965 MPa and 9.4, respectively. Figure 4 shows the weibull plot of the strength values for injection molded silicon nitride.

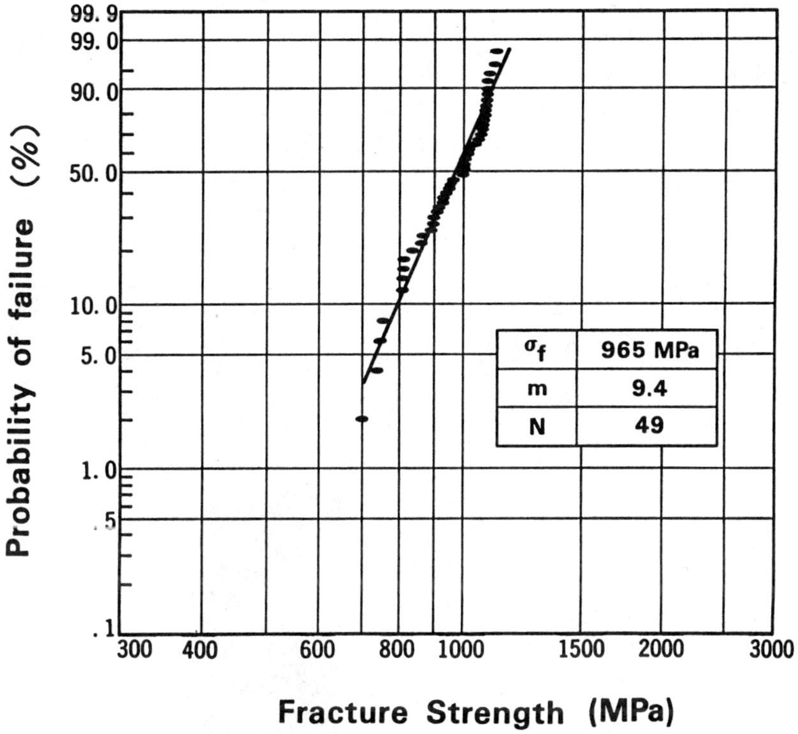

Figure 4 Weibull plot of the room temperature strength values of sintered Si_3N_4

From this data, tensile strength can be estimated to be approximately 544 MPa. This should provide adequate durability for the wear components in actual operation as this value far exceeds this maximum tensile stress value calculated by the finite element analysis model.

The dimensional accuracy for the sintered body is very high; so the fabricated wear parts may require no finish machining from the net sintered shape depending on the tolerance requirements of the component application. This sintering accuracy, enabling the part to be produced with little or no finish machining, will significantly lower finished component cost.

WEAR TEST RESULTS

The wear resistance of ceramics can be evaluated by many kinds of wear tests such as pin-on-disc, block-on-rings, and others. The Falex #1 test, ASTM Practice G77-83, a block-on-ring method was selected because of its convenience in evaluating the effect of oil condition, temperature, and pressure and other parameters important to the wear applications.

Figure 5 is typical of specific wear rate data for injection molded silicon nitride vs. steel compared to steel vs. steel. A 10% wear rate reduction has been attained by the silicon nitride components in fresh high quality engine oil; and a 90% wear rate reduction measured in used (degraded) engine oil. This is very significant because of the increase in the period of time before maintenance (adjustment) to the engine valve or injector train is required. A limited number of engine field tests of these components confirms that a significant improvement in the wear performance of the ceramic links can be realized.

FIGURE 5. FALEX 1 WEAR TEST RESULTS
Wear Coefficient (cubic mm/mm/N)

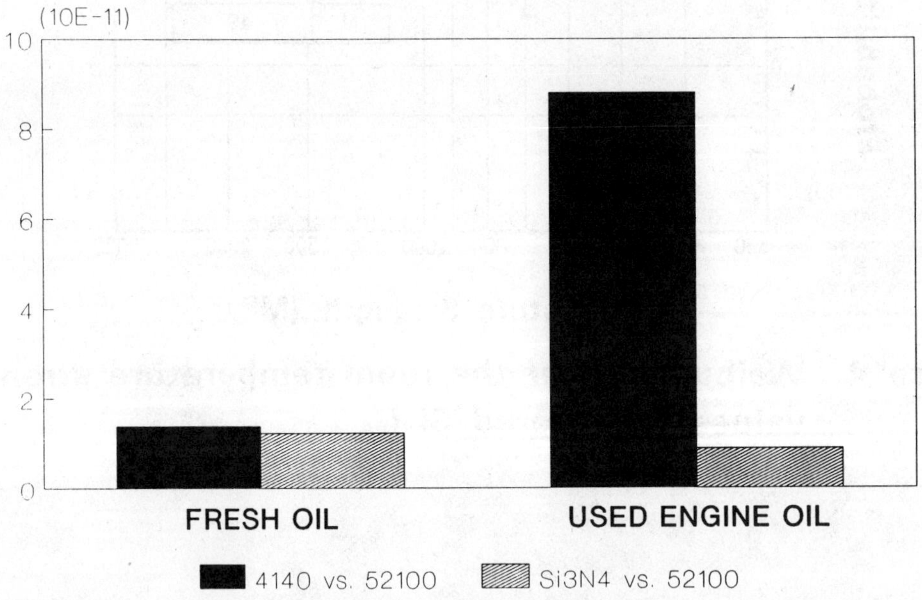

SUMMARY

Several components have been designed and fabricated from silicon nitride ceramic for use in heavy-duty diesel engines to reduce wear in highly loaded valve and injector train ball-socket locations. Component and mold designs have been developed and processing conditions optimized to permit the use of highly efficient injection molding technology to fabricate these components with either no machining or limited finish machining required to meet tight component tolerances. A variety of inspections, strength tests, rig and engine tests all confirm that it should be possible to produce these components at high volumes in a very cost effective manner.

SPHERICAL IMPACT DAMAGE AND STRENGTH DEGRADATION IN CERAMICS FOR AUTOMOBILE TURBOCHARGER ROTORS

Yoshio Akimune, Yasushi Katano, and Kazuo Matoba
Central Engineering Laboratories, Nissan Motor Co., Ltd.
1, Natsushima-cho, Yokosuka, 237 Japan

ABSTRACT

An investigation was conducted into spherical impact damage and strength degradation of silicon nitride and other candidate ceramics intended for automobile turbocharger rotor use. The surface damage morphology and strength degradation phenomena were different for each ceramic, resulting from differences in their mechanical property-related response behaviors. The impact damage morphology and strength degradation relationship is discussed on the basis of mechanical properties such as K_{IC}, Young's modulus, and hardness. The study determined that materials having higher Young's modulus and hardness relative to elastic response and higher K_{IC} relative to resistance to crack extension are preferred for turbine materials applications.

INTRODUCTION

The first commercial ceramic turbocharger rotor was utilized in automobiles in 1985, capitalizing on light weight and heat resistance properties.[1,2] Since the potential for catastrophic failure caused by impact damage was considered serious at that time, it was suggested that exhaust manifolds be regularly cleaned to eliminate the source of particles.[3] As a result, no impact damage-caused failures have been reported since the product's first appearance on the market. Turbocharger rotors are expected to be used at even higher temperatures to reduce fuel consumption, however, thereby producing another damage source such as oxide scale resulting from oxidation of the exhaust manifold. Furthermore, ceramics have also been developed for automobile gas turbine components functioning under high temperature operation(>1300°C). These developments clearly necessitate that in-depth studies of impact damage to ceramic materials be undertaken.

A considerable number of studies have been conducted to elucidate impact damage phenomena relative to turbine materials[4,5] as well as helicopter windshield glass,[6] and subsequent theoretical examinations into these phenomena based on Hertzian cone cracking and the indentation mechanism have also been attempted.[7-13] Moreover, a few studies have been performed to analyze the strength degradation-property relationship,[14-17] although the relationship remains to be clearly identified.

In the present study, impact damage testing is conducted on in-service silicon nitride and other candidate ceramics e.g. tetragonal zirconia polycrystals and silicon carbide. Surface damage morphology and strength degradation are then examined. Finally, the strength degradation phenomena and damage morphology were examined in an attempt to correlate these to fracture toughness and other properties.

EXPERIMENTAL PROCEDURES

Materials

Commercially available pressureless sintered silicon nitride (SSN)*, gas pressure sintered silicon nitride (GPSSN)*, and sintered silicon carbide (SSC)* were obtained. Hot pressed silicon nitride (HPSN) was produced by hot-pressing from commercial silicon nitride powder+(85 wt%), yttria(10 wt%), and alumina(5 wt%) powders. Tetragonal zirconia polycrystals(TZP) was also hot pressed from zirconia powder# containing 3mol% yttria under vacuum. As a spherical particle, TZP spheres (1.0mm in diameter)# were also obtained. The mechanical properties of these materials are listed in Table I.

The ceramic test specimens, whose dimensions are 8x50x3mm, were polished with two grades of diamond paste(6 µm and 3 µm) to eliminate machining damage to the surface and produce mirror-like flat and parallel surfaces.

TABLE I Mechanical Properties of Target Materials

Mechanical Properties	SSN	HPSN	GPSSN	TZP	SSC
DENSITY $\times 10^{-3}$ (Kg/m^3)	3.26	3.30	3.25	6.05	3.12
YOUNG'S MODULUS (GPa)	265	268	314	200	412
BEND STRENGTH (MPa)	530	600	850	1120	500
VICKER'S HARDNESS (GPa)	14.3	14.8	15.4	12.5	24.0
K_{IC} ($MPa\sqrt{m}$) (Indentation)[20]	3.5	4.7	6.0	7.5	3.0
POISSON'S RATIO	0.25	0.22	0.24	0.3	0.16

Impact Test

The experimental apparatus is schematically shown in Fig. 1. The TZP sphere was attached to the top of a sabot(ABS plastic) fitted with a small magnet. The sabot was then set in the pistol of the apparatus, and the He gas pressure (1.5--5.0 MPa) was allowed to rise to a selected level in the chamber. The diaphragm was then broken by the needle, permitting the gas to

---

FOOTNOTE
* NGK Spark Plug Co.,Ltd. Komaki, Japan :EC120(SSN),EC141(GPSSN),EC425(SSC)
\+ Herman Stark, H1, Goslar1, Federal Republic of Germany
\# Tosoh Co.,Ltd., ZYM, Tokyo, Japan

be released toward the pistol, driving the sabot to the end of the steel pipe. At the end of the pipe, the motion of the sabot was stopped, thereby ejecting the sphere toward the mirror surface of the target. Accordingly, the impact direction is parallel to the hot-press direction for HPSN and TZP.

The velocity of the sabot including the TZP sphere was analyzed by three co-centered electromagnetic detectors aligned coaxially with the pistol to measure the flight time. When the magnet in the sabot passed through the coils, direct current flow was recorded in the memory oscilloscope. The velocity was detected at each sabot firing based on time-of-flight principle. The tests were conducted at room temperature.

Post-impact Evaluation

After the impact test, the surface damage caused to the silicon nitrides was investigated by optical microscope and by scanning electron microscope (SEM), with depths and diameters of the crater measured by profilometer (Slone: DEKTAK). A four point bend test (upper span:10 mm, lower span:30 mm) with 0.5 mm/min cross-head speed was employed to measure the bend strength (cross section is 8x3mm). The impact site was set in the center of the outer span so that the maximum tensile stress was imposed on the damage area. The fractured surfaces were also examined by optical microscope and SEM.

RESULTS AND DISCUSSION

Surface Damage Morphologies

Surface damage morphologies are schematically shown in Fig. 2. Since the optical microscope was used for this observation, sharp and narrow microcracks were not visible although the surface damage transition is well described. Essentially, a large crater was produced at low impact velocity which then became larger, with both ring and radial cracks initiated at medium through higher velocities for SSN. On the other hand, in HPSN and GPSSN, craters were produced at low impact velocity, segmented ring cracks were initiated at medium velocity, with radial crack propagation at high velocity. For TZP, craters were produced at low impact velocity, while segmented ring cracks were initiated at medium through higher velocity. Clearly, damage morphology of SSC completely differed from that of others. Many co-centered ring cracks were produced on the surface from low to high impact velocities.

The theoretical contact diameters were calculated using Eqs.(1) and (2) following the Hertzian cone crack theory:(7-9)

$$2a=2R(5\pi V^2 \rho k/4)^{1/5}, \qquad (1)$$

$$k= (1-\nu p^2)/Ep + (1-\nu t^2)/Et, \qquad (2)$$

where, a is the radius of crater, ρ is the density, R is the radius of the sphere, and Ep, Et, νp, and νt are the Young's moduli and Poisson's ratio of the sphere and target, respectively. As noted earlier profilometry was used to measure the diameters and depths of the craters. From Figs. 3 and 4, SSN, HPSN and TZP show relatively large depths while GPSSN and SSC produced more shallow craters. SSN, HPSN, and TZP also reveal larger than theoretically expected diameters, while the diameter for GPSSN is quite similar to the theoretical calculation results. On the other hand, SSC shows a peak value due to ring cracks while is different from theory.

The onset velocities of these craters could not be characterized because the impact velocity could not be decreased below 80 m/s. Depth was found to be related to the diameter of the crater generated in each ceramic. Specifically, SSN and TZP exhibited an elastic/plastic response resulting in relatively large craters and segmented ring cracks at the circumference, with the radial cracks in SSN possibly being initiated by the target plastic deformation and/or elastic recovery following sphere rebounding.(5,14) On the other hand, SSC demonstrated an elastic response resulting in co-centered ring cracks, and HPSN and GPSSN presented an intermediate response between elastic/plastic and elastic as evidenced through small craters and ring cracks with radial cracks.

It is important to note in each material that the first contact was elastic which then changed to elastic/plastic during sphere penetration. Upon conclusion of sphere penetration or sphere rebounding, elastic recovery may have occured although the craters were not fully recovered, resulting in visible craters with ring cracks. Notably, radial cracks were also initiated when penetration exceeded a certain level. It is, therefore, concluded that the damage morphologies relate to material response behavior which is dependent on hardness and Young's modulus.(12-13) The radial cracks are thought to be dependent on plastic deformation and/or elastic recovery.(5,14) Typical damage morphologies are presented in Fig. 5(A)-(E).

Bend Strength Degradation

Fig.6 illustrates bend strength of the specimens after impact. Strength degradation was determined for SSN and SSC as a result of impact damage occurring near 100 m/s sphere velocity. For TZP and HPSN, strength degradation began around 140 m/s and 240 m/s, respectively. No degradation occured for GPSSN until close to the 280 m/s impact velocity. In each case, impact was thought to produce larger flaws and/or cracks than the inherent flaw sizes in the materials, leading to a degradation in strength.(15-17)

Fracture surface observation (Fig. 7(A)-(E)) reveals that a porous zone and median/radial crack were produced just under the impact crater for SSN. In addition, SSN indicates an elastic/plastic response resulting in a large crater and median/radial crack system. For low impact velocity, strength degradation was caused by the small crater produced on the surface. In this case the porous zone under the crater may act as the crack origin and directly relate to strength degradation. Either median or radial cracks may cause strength degradation when impact velocity becomes higher, although the origin of the fractures could not be identified on the fractured surfaces in this limited study.

The damage morphology of TZP is slightly different. A small crater produced in the TZP at low impact velocity did not originate a fracture. When the impact velocity exceeded 140 m/s, a large median crack was formed under the crater, which then resulted in strength degradation.

On the other hand, Hertzian cone cracks were initiated just under the impression, accompanied by ring cracks for HPSN, GPSSN, and SSC. Hertzian cone cracks result from a materials' elastic response, and are formed during impact loading for these materials. Although radial cracks were initiated for HPSN and GPSSN, their responses were elastic throughout and initiated cone cracks just under the impression when the impact velocity exceeded the critical value(Vc) of each. After a sphere rebounded or was crushed, radial cracks were formed near the surface resulting from target elastic recovery when the velocity was exceptionally high. These impact-initiated cracks may be related to strength degradation when the crack lengths exceed their inherent flaw sizes. For HPSN, a Hertzian cone crack was initiated in excess of 240 m/s, with subsequent strength degradation. The same phenomenon occurred at 280 m/s for GPSSN, while it arose under 100 m/s for SSC. Although the origin of the fractures resulting from the bend test were located either at the base plane of the cone and/or the radial crack near surface, the precise origin could not be identified from this limited research.

Impact Phenomena and Material Properties

Spherical particle impact produced two types of cracks in this limited number of materials. One was the Herztian cone crack and the other is the median/radial crack system. From the literature,[7-9,15-17] principal material properties found to be related to the Hertzian cone crack and median crack are Young's modulus (E), Vicker's hardness (Hv), fracture toughness (K_{IC}), Poisson's ratio (ν), and density (ρ), of both sphere and targets. Impact damage phenomena should be discussed in two processes, however, as crack initiation and crack propagation. In the remainder of this paper, E, Hv, and K_{IC} are explored to establish a correlation to the impact damage phenomena.

The large craters were produced on the surfaces as permanent impressions with hardness found to relate directly to the damage phenomena. E is dependent on the target response behavior. Fig. 8 is a crack initiation diagram involving E and Hv. When a Hertzian cone crack is formed, the material has an E and an Hv higher than 268 GPa and 14.8 GPa, respectively. When E and Hv are less than these values, median/radial cracks are initiated resulting from deep sphere penetration. Shockey et al.[5] indicated that the contact behavior transition in silicon nitride varies according to its room temperature and elevated temperature hardness. In the present experiment the contact behaviors of the materials are shown to differ depending on their hardnesses and young's moduli at room temperature. Elastic response may result when ratio of target hardness-sphere hardness(H_t/H_p) >1.2 and E_t/E_p > 1.3. Otherwise elastic/plastic response occurs in these experiments.

In 1924, Cardullo[18] suggested the relationship between hardness and Young's modulus in Eq. (3):

$$\text{Hardness} = (c)(\sigma_{ys}^{1-m})(E^m) \qquad (3)$$

where c and m are constants, and σ_{ys} is yield stress. The diameter and depth of the crater relate to E and σ_{ys}. The m-value is a factor related to the atomic binding force in single crystals[19], but it may also relate to the deformation behavior of polycrystals, with the m-value deciding the response behavior of the target. Therefore, E, Hv, and σ_{ys} of these materials dictate the target response and crack initiation morphologies.

Some researchers have suggested a relationship between material properties and critical velocity for strength degradation. In this regard, Wiederhorn and Lawn[17] have suggested Eq. (4) for Hertzian cone crack systems, and Evans[15] has reported Eq. (5) for median/radial crack systems:

$$Vc \propto (K_{IC}^{5/3})(E^{-7/6}) \quad \text{(elastic)} \qquad (4)$$

$$Vc \propto (K_{IC}^{2})(H^{-3/2}) \quad \text{(elastic/plastic)} \qquad (5)$$

Fig. 9, 10, and 11 show the logarithmic relationship of Vc-K_{IC}, Vc-E, and Vc-Hv. From Fig. 9, Vc increases with increasing K_{IC} values, although only two cases are presented: for Hertzian cone cracks and for median/radial crack systems. The slope for the Hertzian cone crack system closely matches the theoretical value,(17) although a difference from theoretical(15) is noticeable for the median/radial crack system. Figures 10 and 11 indicate a decreasing trend of Vc with increasing E and Hv values. The Hertzian cone crack and median/radian crack system relational tendencies agree with the literature data.(15,17) From the experimental results, we can offer suggestions concerning the relation between the material properties and Vc:

$$Vc \propto (K_{IC}^{1.54})(E^{-2.7}) \quad \text{(elastic)} \qquad (6)$$

$$Vc \propto (K_{IC}^{0.5})(H^{-2.5}) \quad \text{(elastic/plastic)} \qquad (7)$$

For turbine materials, Vc must be a higher value than the velocity at the turbine tip for regular turbocharger operation e.g. the turbine tip velocity is about 275 m/s at 94,000 rpm. Of the five materials used in this experiment, GPSSN produces the best resistance to impact damage and strength degradation. Turbine materials are required to be more resistant to impact damage. Thus, the preferred material is that which behaves elastically, resulting in Hertzian cone crack initiation, and has higher K_{IC} values for crack extension. Although these results may contradict each other because a higher E and Hv are necessary for elastic response even though they decrease Vc, materials which possess relatively high hardnesses, Young's moduli, and K_{IC} are preferred in future turbine materials.

SUMMARY AND CONCLUSION

An investigation was conducted into spherical impact damage and strength degradation occurring in silicon nitride and other candidate materials proposed for automobile turbocharger rotors. Surface damage morphologies and strength degradation behavior were found to be different for each ceramic. SSN and TZP demonstrated elastic/plastic response whereas HPSN, GPSSN, and SSC presented more elastic response.

Of the many properties, Young's modulus and Vicker's hardness were shown to relate to the response behavior and crack types of the materials. When E exceeded 268 GPa and Hv exceeded 14.8 GPa, for example, the target produced an elastic response in the form of a Hertzian cone crack, otherwise the target yielded an elastic/plastic response with median/radial cracks. Fracture toughness was determined to relate to Vc as a crack extension parameter, with the higher K_{IC} value increasing Vc whereas Hv and E values decreased Vc. Therefore a material having a higher E and Hv relate to Hertzian cone cracking, and K_{IC} relates to crack extension.

ACKNOWLEDGMENT

The Authors wish to express their sincere appreciation to T. Miyamura for his assistance with the impact test and pre- and post-impact evaluations.

REFERENCES

1 K. Katayama, I. Watanabe, K. Matoba, and N. Katoh, "Development of Nissan High Response Ceramic Turbocharger Rotor," SAE Technical Paper 861128, October(1986).

2 K. Matoba, K. Katayama, M. Kawamura, and N. Katoh, "The Development of Second Generation Ceramic Turbocharger Rotor-Further Improvements in Reliability, "SAE Technical Paper 880702,February(1988).

3 I. Matsuo and F. Nishiguchi, "The Development of Second Generation Ceramic Turbocharger," SAE Paper 880703 February(1988).

4 K. C. Dao, D. A. Shockey, L. Seaman, D. R. Curran, and D. J. Rowcliffe, " Particle Impact Damage in Silicon Nitride," Annual Reports, Part III, Office of Naval Research, Contract NO.N0014-76-057, May 1979.

5 D. A. Shockey, D. C. Erlich, and K. C. Dao, "Particle Impact Damage in Silicon Nitride at 1400 °C," J. Mater. Sci.,16 477-482(1981).

6 M. Abou el leil, F. Camaratta, and R. R. Digenova, "Impact Fracture of Thermally Tempered Glass Helicopter Windshields,"J. Am. Ceram. Soc., 68[1]c18-c21(1985).

7 H. R. Hertz, Hertz's Miscellaneous Papers, Chs. 5 and 6, McMillan, London, 1896.

8 F. C. Roesler, "Indentation Hardness of Glass as an Energy Scaling Law," Proc. Phys. Soc. London, Sec.B B69[1]55-60(1956).

9 B. R. Lawn and T. R. Wilshaw, " Review Indentation Fracture: Principles and Applications, " J. Mater. Sci.,10[6]1049-81(1975).

10 B. R. Lawn, A. G. Evans, and D. B. Marshall, " Elastic/Plastic Indentation Damage in Ceramics: The Median/Radial Crack System," J. Am. Ceram. Soc., 63[9-10], 574-81(1980).

11 B. R. Lawn and M. V. Swain, " Microstructure Beneath Point Indentations in Brittle Solids," J. Mater. Sci.,10[1],113-22(1975).

12 D. B. Marshall, " Geometric Effects in Elastic/Plastic Indentation, " J. Am. Ceram. Soc.,67[1], 57-60(1984).

13 S. S. Chiang, D. B. Marshall, A. G. Evans, "The Response of Solid to Elastic/Plastic Indentation I Stress and Residual Stress," J. Appl. Phys. 53[1], 298-311(1982).

14 A. G. Evans and T. R. Wilshaw, " Quasi-static Solid Particle Damage in Brittle Solids-I Observation, Analysis and Implication," Acta Metall., 24, 939-56(1976).

15 A. G. Evans, M. E. Gulden, and M. Rosenblatt, "Impact Damage in Brittle Materials in the Elastic/Plastic Response Regime," Proc. Roy. Soc.(London), Ser. A, 361, 343-65(1978).

16 B. R. Lawn, S. M. Wiederhorn, and H.H. Johnson, "Strength Degradation of Brittle Surfaces: Blunt Indenters," J. Am. Ceram. Soc., 58[9-10], 428-32(1975).

17 S. M. Wiederhorn and B. R. Lawn, "Strength Degradation of Glass Resulting from Impact with Spheres," J. Am. Ceram. Soc., 60[9-10]451-58(1977).

18 E. F. Cardullo, " The Hardness of Metals and Hardness Testing," Mechanical Engineering, N. Y.(10), 638-39(1924).

19 Y. Akimune and R. C. Bradt, "Microhardness Anisotropy in 15 mol% Yttria Stabilized Single Crystal Cubic Zirconia," to be published in Advances in Ceramics Vol 24.

20 B. R. Lawn and E. R. Fuller, "Equilibrium Penny-Like Cracks in Indentation Fracture," J. Mater. Sci.,10[12]2016-24(1975).

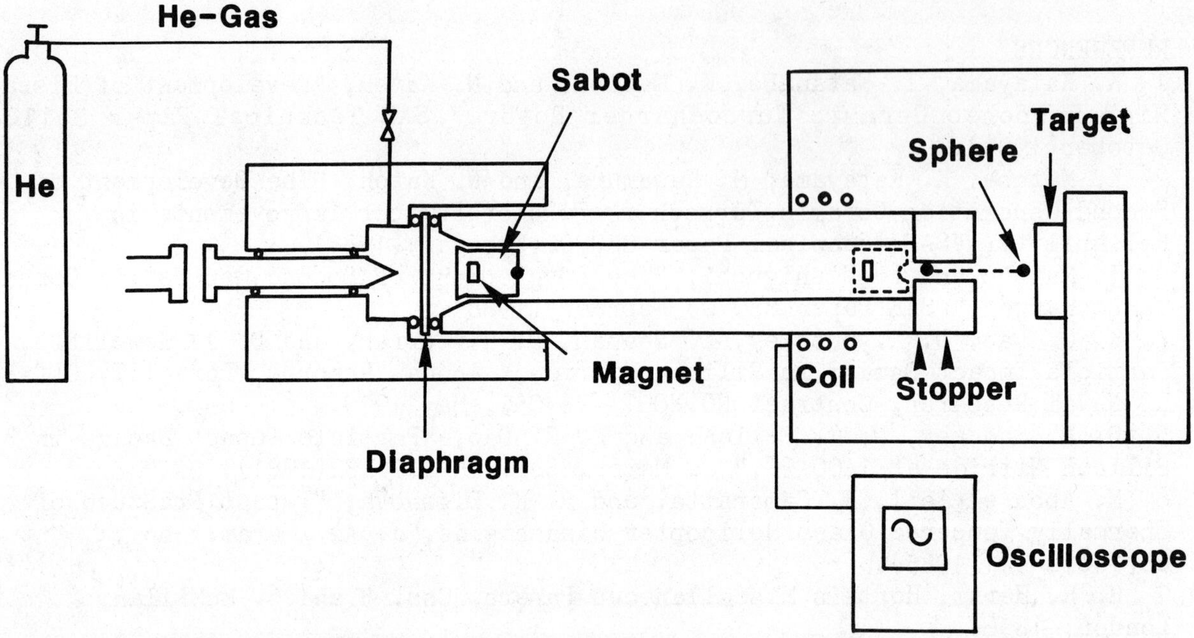

Fig. 1 Experimental Impact Test Apparatus

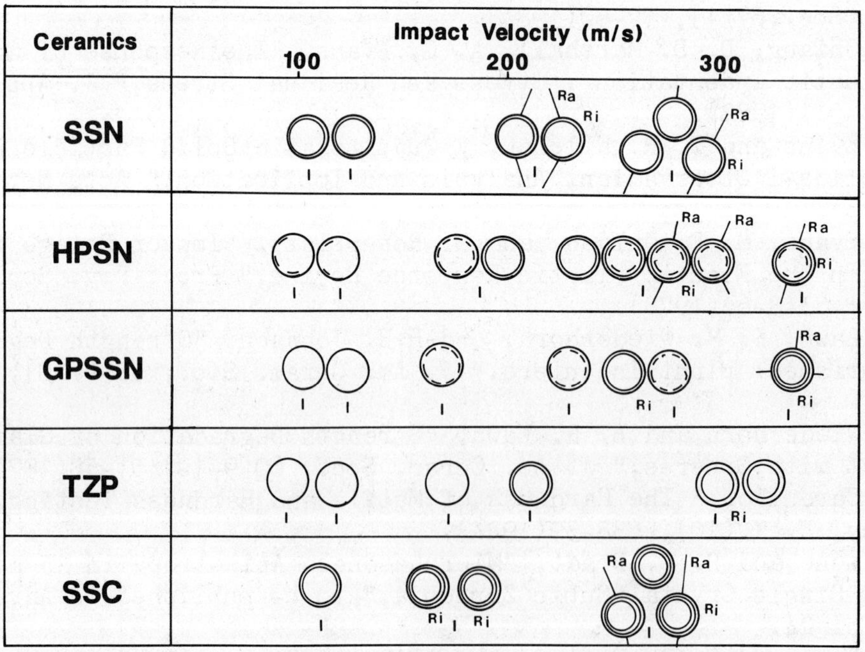

Fig. 2 Schematics of Surface Damage in Five Ceramics
(I: impression, Ra: Radial Crack, Ri: Ring Crack)

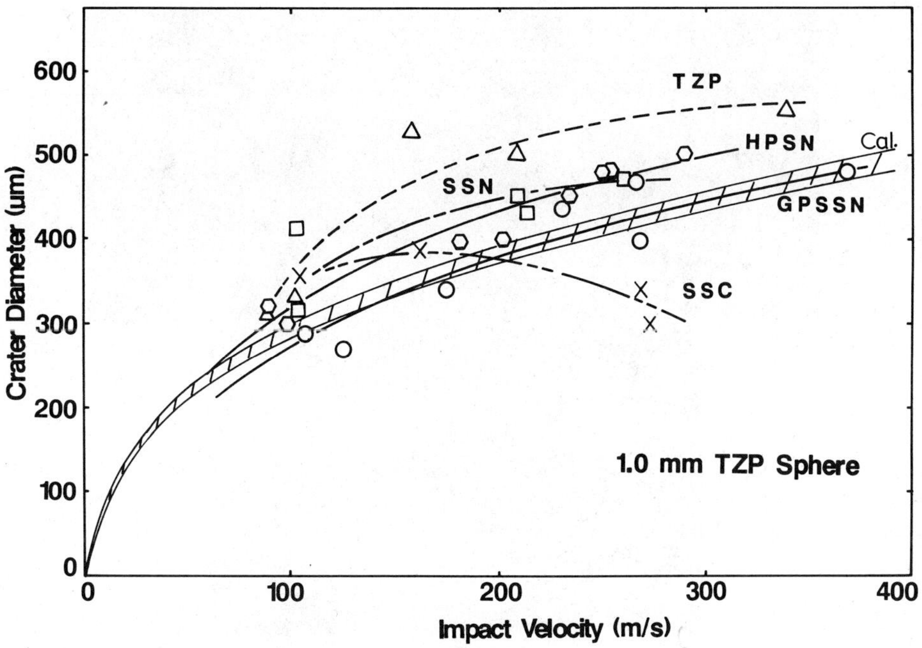

Fig. 3 Crater Diameters after Impact Test

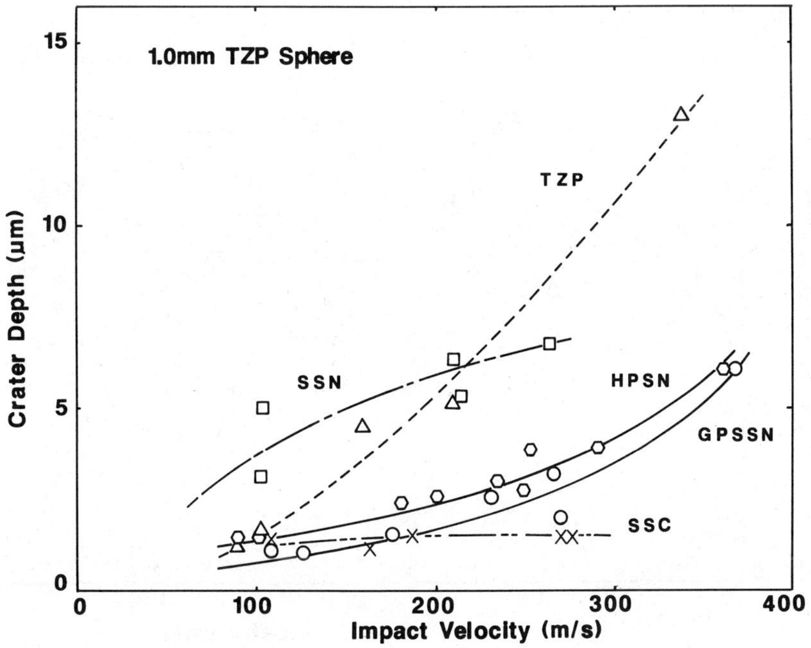

Fig. 4 Depth of the Crater after Impact Test

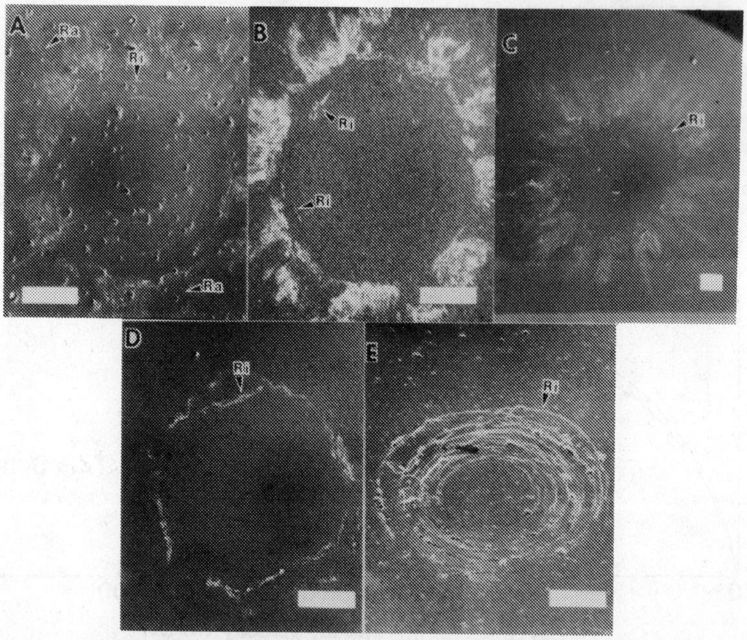

Fig. 5 SEM photographs of Typical Impacted Surfaces
(A)SSN at 285 m/s,(B)HPSN at 213 m/s,
(C)GPSSN at 280 m/S,(D)TZP at 214 m/s,
(E)SSC at 215 m/s (Bar 100 µm)

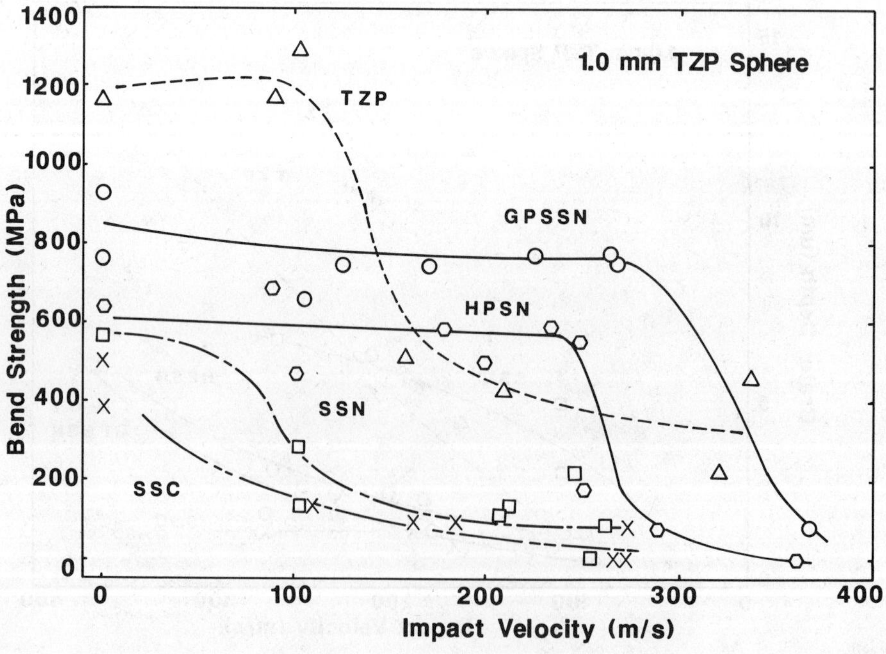

Fig. 6 Strength Degradation after Impact Test

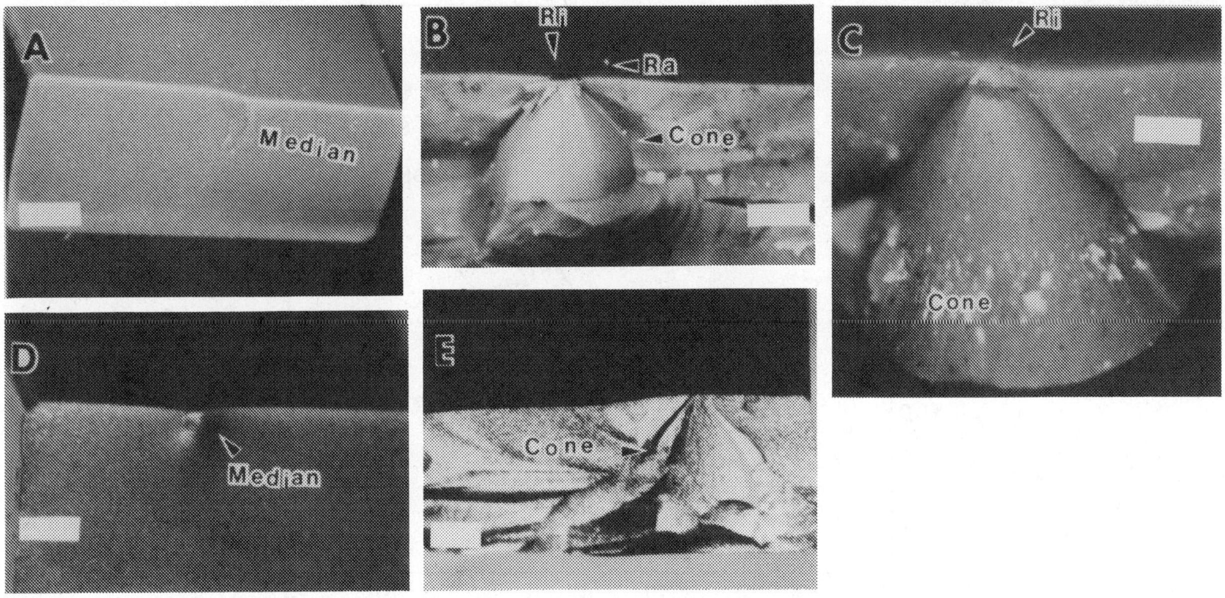

Fig. 7 Optical Micrographs of Typical Fractured Surfaces after Bend Test
(A)SSN at 103 m/s,(B) HPSN at 361 m/s,
(C)GPSSN at 368 m/s,(D)TZP at 208 m/s,
(E)SSC at 273 m/s (Bar 1 mm)

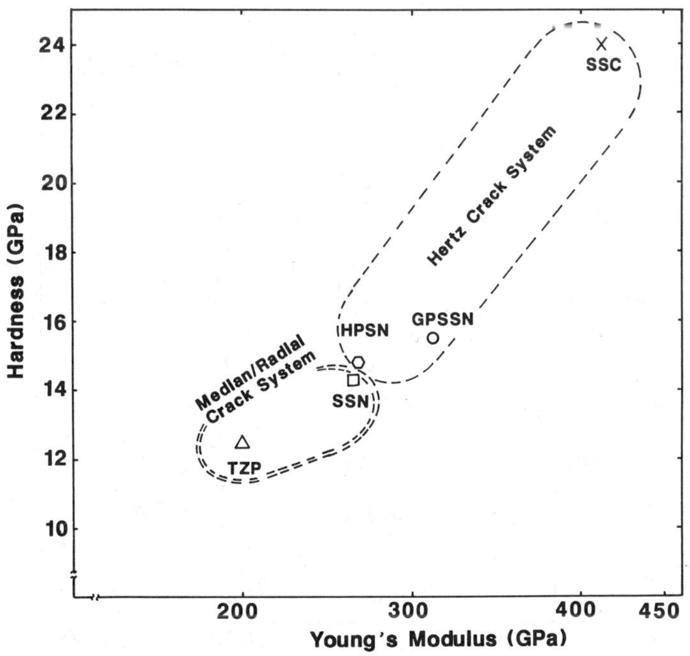

Fig. 8 Crack Initiation Diagram
-----Hertz Crack System
=====Median Radial Crack System

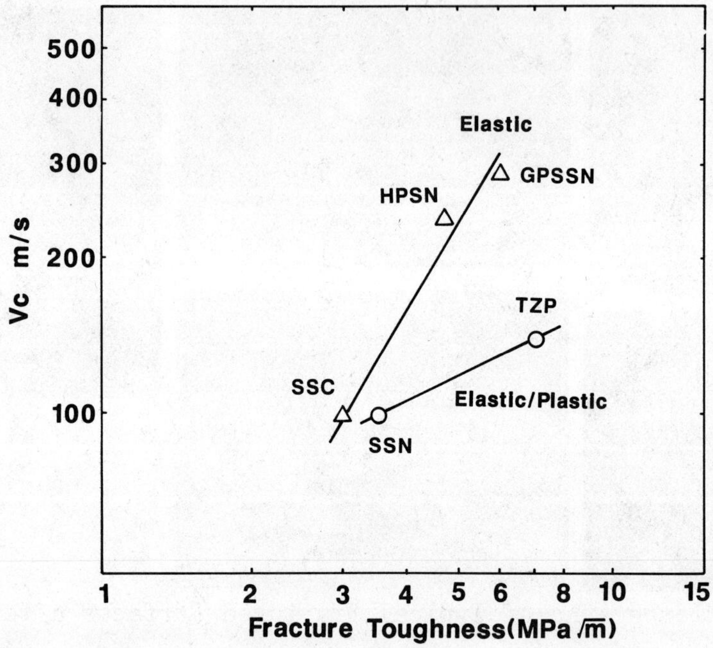

Fig. 9 Relationship between Vc and Fracture Toughness

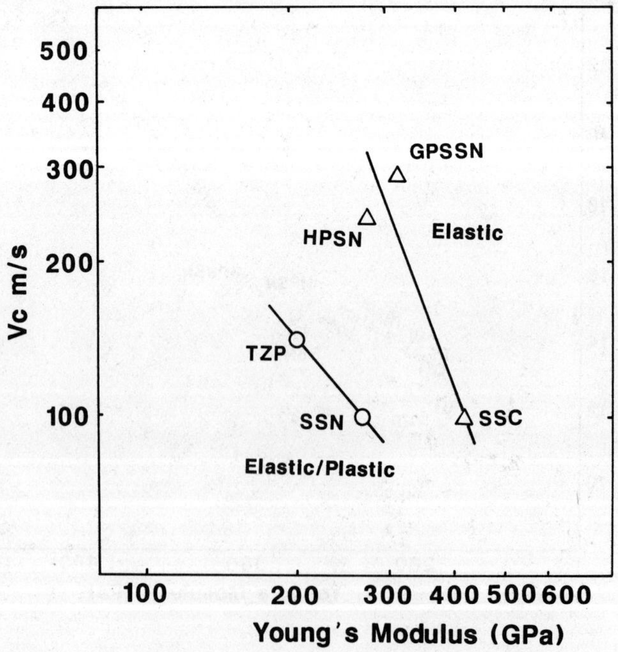

Fig. 10 Relationship between Vc and Young's Modulus

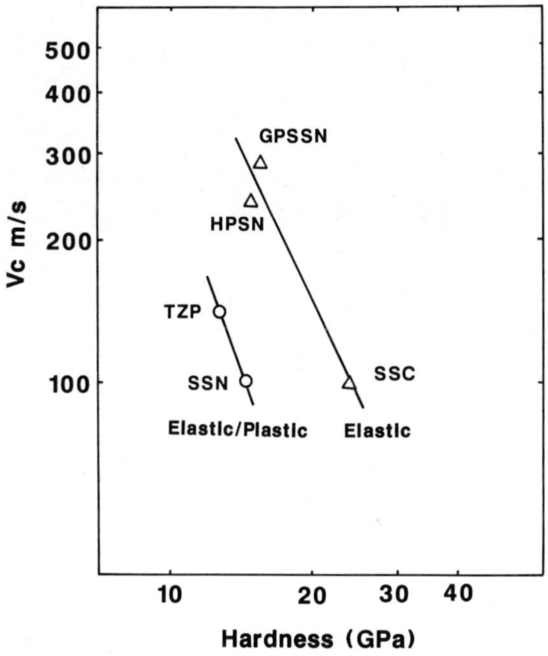

Fig. 11 Relationship between Vc and Hardness

NATIONAL PROGRAM IN SWEDEN FOR ENGINE CERAMICS

Lars Pejryd
Volvo Flygmotor
Aerospace Division
S-461 81 Trollhättan
SWEDEN

ABSTRACT

To stimulate industrial development of ceramics in Sweden, the Swedish Board for Technical Development (STU) has started a five year national program aiming for development of commercial products or prototypes in ceramic materials. Five different areas of interest have been identified and one of the areas is ceramics for engine applications.
To identify possible projects within each program area, pilot studies were undertaken. In this paper the result of the pilot study in the area engine ceramics and a brief overview of the national ceramic program is presented. Within the engine ceramic area four major projects with interest from both producers and users were identified. The projects span from ceramics for thermal insulation to fiber reinforced ceramics for gas turbine applications.

INTRODUCTION

Ceramic materials combine properties such as high mechanical strength and high corrosion resistance at high temperatures, high thermal chock resistance, low thermal conductivity and low density. This makes it potentially advantageous to use ceramic materials as alternatives to metallic materials in engines. This fact has been realised in many industrial countries, and in for example Japan, the United States and West Germany large programs on ceramics and application of ceramics in engines are in progress.
From the late seventies, the R & D work in Sweden on ceramics has increased substantially. This is true, both for industrially and governmentally financed R & D. However the volume of work is not comparable to that of the U.S., Japan and West Germany. The situation is most adequately described as a typically preindustrial development stage and much of the work has been directed towards basic research on material properties and process development. Successful appli-

cations have, however, also been demonstrated. For example, United Turbine, a company within the Volvo group, was in March 1982 the first company in the world to demonstrate a turbine engine with ceramic hot parts in a public road test. ABB Cerama has developed a technique for glass encapsulated HIP of ceramics and has sold licenses to use this technique to three U.S., one Japanese and one Swedish company.

As stated above, much of the work supported by the Swedish governement through the Swedish Board for Technical Development (STU) has been directed towards basic research. In an effort to stimulate industrial development and applied R & D of ceramics in Sweden, STU has started a five year national program aiming for development of commercial products or prototypes in ceramic materials. Five different areas of interest have been identified and one of these areas is ceramics for engine applications.

NATIONAL CERAMIC PROGRAM.

The potential of ceramic materials is appreciated by scientists and engineers working with materials development. However this appreciation is not so often shared by design engineers. Much of this is due to the lack of experience in working with these materials. The five year program has been started in part to overcome this problem and to stimulate industrial cooperation and development.

National R&D programs do often fall in one out of two categories. They are either mission type programs, like putting a man on the moon and SDI or programs mainly for educational purpose, for gaining and spreading knowledge. Very often one can see a correlation between the size of the country and the type of program most frequently carried out, large countries - mission type and small countries -educational type. In Sweden, a fairly small country, the national efforts in the ceramic field have till now been of the educational type. With this program aiming at products or prototypes a change towards mission type efforts can be seen.

In the program, with a total budget of 120 million swedish kronor (18,5 million US dollars), five different areas of interest have been identified. These areas are: ceramics for engine applications, ceramic armor, ceramic tools for forming of metals (excluding cutting tools), bioceramics and industrial ceramics (corrosion and wear resistant for industrial high temperature processes). In Table 1 a time table and cost plan for the five year program are shown.
Half of the costs for the program are to be financed by industry and half by the STU.

Table 1. Time and cost plan for the Swedish National Ceramic Program

Project	(Cost/year)/MSEK year 1	2	3	4	5	Total
Engine ceramics	6	7	5	5	7	30
Ceramic armor	5	11	10	2	2	30
Ceramic tools	5	7	7	7	4	30
Bioceramics	3	3	5	5	2	20
Industrial ceramics	2	3	3	2	0	10
Sum of costs.	21	31	30	21	17	120

ENGINE CERAMICS PROGRAM

To identify possible projects within each of the five different program areas, pilot studies were undertaken. The pilot study on engine ceramics showed that the greatest interest for joining a national program was from the engine producers, by companies within the Volvo group. Of the producers of ceramic materials great interest was shown by ABB Cerama and Sandvik Hard Materials, already working with engineering ceramics, but also by Ifö Electric, a more traditional ceramic industry. Universities and research institutes are of course also very interested in joining a national program on engine ceramics. Shown below are the four main project proposals where enough interest was generated from both materials producers and users and where possibilities for reaching the goal of finished prototypes or components within the five year period is believed reasonable. Final decisions on all projects have not yet been taken.

Project 1 — Fiber Reinforced Ceramic Application In Axial Gas Turbine (EU33)

Project 2 — Advanced Gas Turbine for Automobiles (AGATA)

Project 3 — Ceramic Combustor for Automobile Heater

Project 4 — Development of Ceramics for Thermal Insulation

Of the four projects proposed in this program, projects 1 and 2 are the Swedish part of European R & D cooperation projects within the EUREKA cooperation while projects 3 and 4 are totally Swedish projects.
Many of the intended research subjects within each of the four major

projects are of common interest to all projects. To ensure coordination and efficient use of the resources, a suggestion has been put forward to appoint a coordination group for the engine ceramic area. The proposed project organization is shown in figure 1.

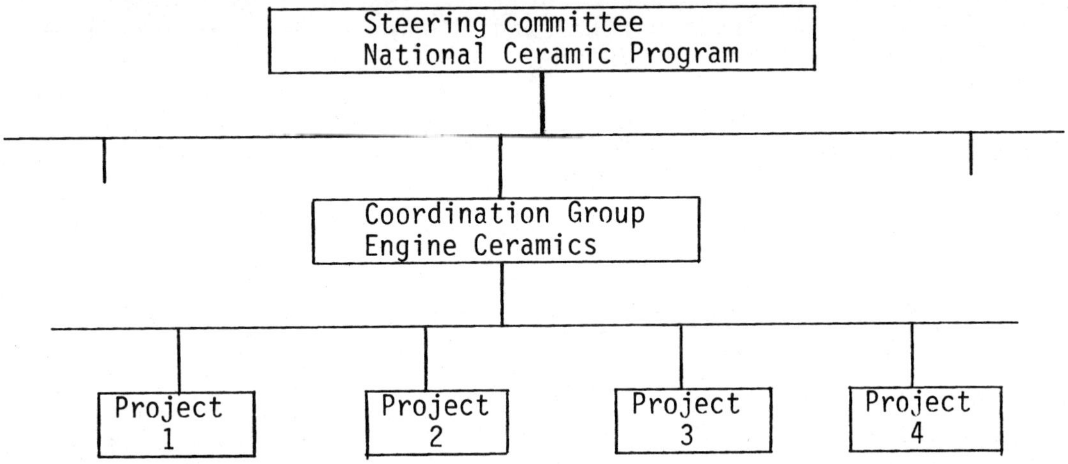

Figure 1. Project Organization of the Swedish National Program on Engine Ceramics.

ENGINE CERAMIC PROJECTS

Fiber Reinforced Ceramic Application in Axial Gas Turbine

This project is a joint project within the European EUREKA R & D cooperation between the following companies:

Volvo Flygmotor AB	Sweden
MAN Tech GmbH	West Germany
Alfa Romeo Avlo	Italy
SNECMA	France
SEP	France

The project aims at showing the possibility of applying fiber reinforced ceramics in gas turbines of 5-15 MW power rating. The five year project is budgeted at a total cost of 85 million French Francs (24.5 MUSD) and will involve development and testing of components such as stator vanes, turbine blades and combustion chambers. The material to be used is called CERASEP, a SIC/SIC composite made by chemical vapour infiltration (CVI), developed by SEP.

The Swedish part will involve the development of a combustion chamber and is planned to start from an existing combustion chamber of which the first part will be used more or less unchanged and to which a ceramic composite "tube" will be joined. The work will involve areas such as testing of mechanical and environmental properties of the materials, development of joining techniques, and design and combustion tests. The main part of the Swedish work will be performed by Volvo Flygmotor and additional work (within the national program) will be done at universities and research institutes. The financial split of the Swedish part of this project is Volvo Flygmotor 17 MSEK and STU 5MSEK.

Advanced Gas Turbine for Automobiles

This project is also a EUREKA project. The participants are shown below.

Automotive Industry	Gasturbine Industry	Ceramic Industry
Daimler Benz	Garrett	Hoechst (CER)
BMW		Bayer (CFI)
Volkswagen	Onera	CT Demarquest
Peugeot		Ceramiques et Composites
Volvo/United Turbine		Elektroschmelzwerk Kempten
		Grundzweig unt Hartman
		Plansee Metalwerk

Through these companies France, West Germany and Sweden are represented. The aim of the project is to develop ceramic components and systems for small gas turbines mainly for automobiles but also for other applications. The project is in its initial stage and in the preliminary work split the Swedish part, managed by United Tubine, is to be system calculations, and development of the axial turbine wheel and variable stator.
The total cost for this project (7 years) is about 300 MFF (50MUSD) The first phase (3.5 years) will cost about 170 MFF (29 MUSD). The Swedish share of the first phase will be about 18.5 MSEK (2.8 MUSD) of which 10.5 MSEK is for development of ceramic components. Half of this cost (5 MSEK) is expected to be financed by STU.

Work in Sweden on materials will be concentrated on silicon nitride based materials and will involve injection molding, slip casting and hot isostatic pressing as processing steps. Companies involved are United Turbine/Volvo, ABB Cerama and Sandvik Hard Materials. Universities and research institutes will also be involved.

Ceramic Combustor for Automobile Heaters

To increase the efficiency of heaters and to gain a wider experience in the use of ceramic materials, a project concerning the development of ceramic parts for heater products has been put forward. The project will involve the development and testing of relatively simple shaped tubular ceramic parts with thin walls. The materials used would have to be able to withstand rather high temperatures and thermal shock, while the mechanical loading is estimated to be rather low. The most likely materials to be used in the project would be SiC and Si_3N_4. The industries involved in such a project would be Volvo Flygmotor (producer of heaters) and ABB Cerama and Sandvik Hard Materials (material producers). The financial arrangement for this project is at present not defined. From the industrial side there are today no decisions to proceed with this project.

Development of Ceramics for Thermal Insulation

Apart from high strength materials, ceramics for thermal insulation are also of great interest to the autmotive industry. One of the most interesting materials for such purposes is aluminium titanate (Al_2TiO_5). This is largely due to its low coefficient of thermal expansion and there by high thermal shock resistance. The rather low strength value does, however, limit the use of Al_2TiO_5. However, preliminary work on sythesis of Al_2TiO_5 materials have shown a substantial increase in strength, and therefore a project aiming at the development of Al_2TiO_5 based materials, much stronger than those presently available, has been suggested.

The participants in this project are Volvo Technical Development, Ifö Electric and the Swedish Institute for Silicate Research. The cost for the project is estimated to about 4 MSEK for the first three year period with industry financing half of the cost.

CONCLUDING REMARKS

The four projects proposed above have been put forward to the steering committee of the national ceramic program. The committee has decided that all four fulfill the requirements to enter the national program. Work is now in progress to plan the projects in detail and to coordinate work of interest to several projects. Decisions on wheather all projects should proceed remain also to be made.

ACKNOWLEDGEMENTS

This work have been financed by the Swedish Board for Technical Development.

I thank Mr M Persson and Mr J Rehn for valuable cooperation during this work.

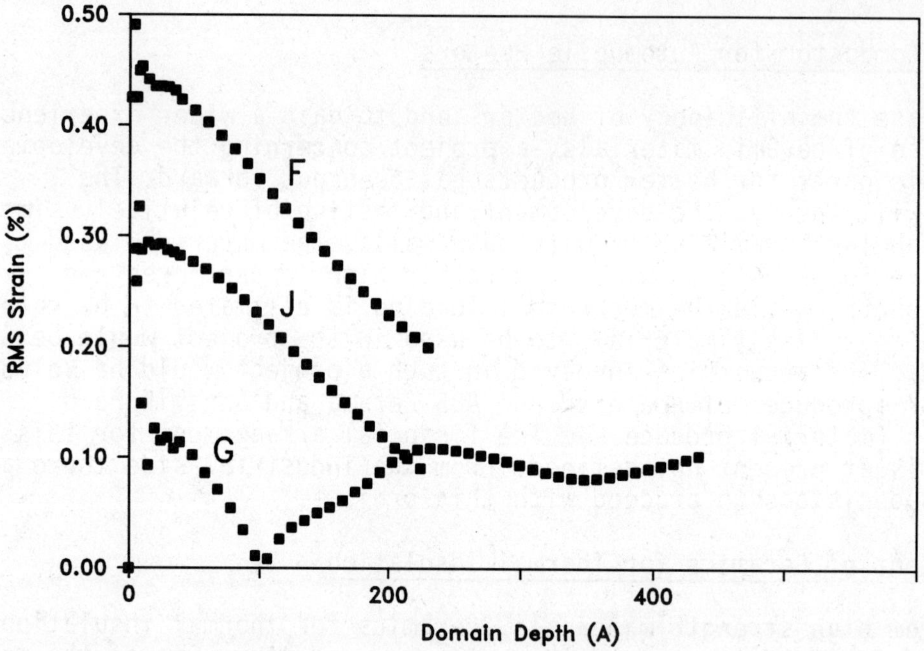

Figure 4
Nonuniform strain surrounding particles as determined by the Warren-Averbach technique.

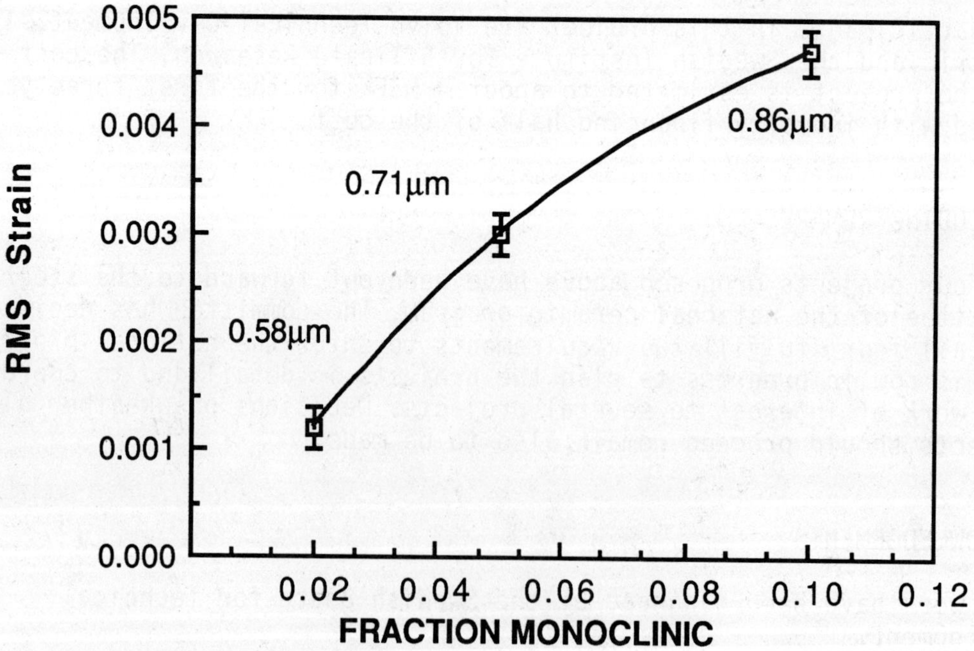

Figure 5
Maximum RMS strain vs fraction monoclinic.

ALUMINUM NITRIDE CERAMICS, A POTENTIAL MATERIAL FOR HEAT MANAGEMENT IN INTERNAL COMBUSTION ENGINES

Keith A. Blakely
J. Richard Schorr
Peter T. B. Shaffer

Advanced Refractory Technologies, Inc.
699 Hertel Avenue
Buffalo, NY 14207

ABSTRACT

Until recently aluminum nitride was not considered seriously as a practical engineering ceramic. Recent advances in the production of aluminum nitride powder, its fabrication, and properties have changed this. With the exception of hazardous beryllium oxide, combinations of properties that are routinely achieved with aluminum nitride are not achievable in any other ceramic.

Properties of aluminum nitride ceramics and cermets are presented and compared with those of other ceramics and engineering materials.

BACKGROUND

Near the turn of the century it was reported that molten aluminum would absorb nitrogen, react exothermically and form AlN (1). It was also known that AlN would react with cold water, hydrolyze and produce ammonia, thus forming the basis for the Serpek process for nitrogen fixation.

About 50 years later Taylor and Lenie (2) showed that AlN could be stabilized by heating to high temperature in nitrogen.

AlN continued to be of less than commercial interest for another 20 years. Syntheses, though straight forward in theory, were slow, energy intensive, and led to products of only intermediate purity. With recent increased availability of high quality powders, and associated decreases in costs, aluminum nitride is now being looked at with commercial applications in mind. Table 1 shows product specifications for presently available AlN powders as supplied by several producers.

Table 1. Specifications of Commercial AlN Powders

	Supplier				
Grade	ART A-100	Tokoyama Soda F	ESK	Starck A	Denka Pure
Aluminum	65.1	65.4	>63.5	---	---
Nitrogen	33.6	33.5	>30.8	>31.5	33.2
Carbon	<0.1	0.06	---	<0.1	<0.2
Oxygen	1.0	1.0	<3.0	<1.5	1.2
Metals	<0.03	<0.01	<0.8	<0.2	<0.1
Av. Size(um)	<4	<3	3-10	3-5	2-4
SA (m2/g)	2-5	3-4	---	1.5-3	2-4

SYNTHESIS

Three basic methods can be used to produce AlN. The elements can be reacted.

1) $2Al + N_2 \rightarrow 2AlN$

The oxide can be reduced by carbon additions in the presence of nitrogen or ammonia.

2) $Al_2O_3 + 3C + N_2 \rightarrow 2AlN + 3CO$

Various reactive aluminum compounds can be reacted.

3) $AlCl_3 + NH_3 \rightarrow AlN + 3HCl$

Reaction 1) offers the simplest method. Reactions must be carried out step-wise to coat the surface of the aluminum powder to prevent the exothermic reaction from overheating, melting the aluminum and forming large drops of molten aluminum that are slow to react completely. Metallic aluminum powders of only limited purity are available commercially. Long high temperature nitriding reactions lead to enhanced crystal growth, necessitating subsequent comminution with its associated impurity introduction.

The reaction is sufficiently exothermic that the material can be produced by SHS (self propagating high temperature synthesis) reactions. The resulting product, formed only at high nitrogen pressures, consists of a nearly dense sintered body (3).

Carbothermic reduction of alumina in the presence of nitrogen is the generally accepted approach. High purity aluminas are readily available. Careful low temperature oxidation removes the unreacted, excess carbon (4).

Reaction of aluminum trichloride (5), while ideal for laboratory production of the highest purity AlN has only limited commercial interest. The chloride hydrolyzes very readily, is corrosive, and yields over 2.5 times as much by product HCl as AlN product. In the laboratory, however, it is ideal since the aluminum trichloride can be purified by sublimation as a last step prior to reaction. The initial reaction product at lower temperature is a complex (AlCl3.NH3) which is converted to AlN by heating to higher temperatures.

FABRICATION

Recent literature is full of references to the sintering of AlN, effects of additives, impurities, etc. The two generally agreed on sintering aids are yttrium oxide (Y2O3) and calcium oxide (CaO). The former is favored in spite of its high cost when maximum thermal conductivity is required, the latter where applications are for refractory purposes.

It has been reported (6) that an apparent grain boundary cleaning mechanism may be responsible for the higher thermal conductivity of AlN sintered in the presence of yttrium oxide. Initially, the yttria sintering aid reacts with aluminum oxide on the surface of the aluminum nitride grains. These two oxides react, forming a YAG (yttrium aluminum garnet) phase which does not wet the AlN surface. As a consequence, the YAG retreats as a liquid phase to the interstices between the grains, leaving rain boundaries free of detectable impurity phases. These clean interfaces, together with a minimum of dissolved lattice impurities are directly responsible for the consistently high thermal conductivities of yttria sintered AlN.

Green bodies are routinely prepared by the usual fabricating conditions, pressing, extrusion, tape casting, and are sintered in nitrogen at temperatures ranging from about 1700° - 2000°C. The sintering conditions depend on the particle size, impurity levels, and sintering aid used. Recently, fully dense high conductivity AlN bodies of nearly 0.4 kgs have been sintered from injection molded parts molded at 75% green density.

PROPERTIES

One could cite properties for AlN, such as thermal conductivity, ranging over nearly an order of magnitude. The values selected are those readily achievable using carefully controlled, commercially viable conditions. Table 2 summarizes the properties of dense AlN fabricated parts.

Table 2. Properties of AlN

Density	3.26 g/cm3
Color	White
Hardness (Vickers)	1200 kg/mm2
Crystal Structure	Hexagonal, Wurtzite
	a=3.114, c=4.896
Electrical Resistivity	2 x 10 E12 ohm-cm
Dielect. Const. @ 1 MHz	8.9
Dielect. Const. @ 7 GHz	8.2
Dielect. Loss @ 1 MHz	0.001
Dielect. Loss @ 7 GHz	0.002
Dielect. Strength	10 kv/mm
Band Gap	6.2 EV
Thermal Expansion	
(RT to 200 C)	3.4 x 10 E-6/C
(RT to 400 C)	4.4
(RT to 600 C)	4.8
(RT to 1000 C)	5.6
Thermal Conductivity	
(RT)	185 w/m K
(100 C)	130
(theor.)	320
Use Temperature	
(air, oxidizing)	1100 C
(inert)	1800 C
Vapor Pressure (N2)	
725 C	5 x 10 E-12 mm
925	2 x 10 -8
1125	1 x 10 -5
1325	9 x 10 -4
2433	500 mm
2563	1 atmosphere
Young Modulus (E)	3.1 x 10 E4 kg/mm2
	(44 x 10 E6 psi)

Table 3. Properties of Typical Materials

	Density (g/cm3)	Thermal Conductivity (W/m°K)	Thermal Expansion (10 /°C)	Elastic Modulus (GPa)	Specific Modulus (GPa/g/cm3)
Aluminum	2.70	237	25	69	26
Aluminum Nitride	3.26	185	4.6	343	105
Aluminum Oxide	3.96	29	7.0	378	95
Beryllium Oxide	3.0	240	7.6	378	126
Boron Nitride	2.25	14	12.2	82	36
Copper	8.9	398	16.6	117	13
Magnesium Oxide	3.58	38	12.8	247	69
Molybdenum	10.20	140	9.0	316	31
Silicon Carbide	3.21	100	4.6	460	143
304 Stainless	8.0	17	17.3	192	24
Tungsten	19.35	187	8.0	405	21
Zirconium Oxide	5.89	2	6.5	247	42

In subsequent figures (Figures 1 - 6), properties of AlN are compared with other high conductivity ceramics and metals (Table 3), as well as the standard aluminum oxide and a common engineering metal, 304 stainless steel. Compared are density, thermal conductivity, elastic modulus, and thermal expansion. Specific modulus, defined as elastic modulus divided by density and specific thermal conductivity, defined as conductivity divided by density are also given.

It is interesting to note that AlN has one of the lowest densities, high thermal conductivity, high specific modulus, and the lowest thermal expansion. This combination of properties is not attainable with any other material listed, and offers very interesting potential for engineering both monolithic and composite AlN parts into heat engine applications.

PROPERTIES COMPARISON

Density (Fig.1): Weight is always a consideration in the efficiency of an engine, and at 3.26 g/cc, AlN is roughly 60% less than most steels and ferrous alloys. Of the ceramics being considered for engine applications, only silicon carbide and beryllia are lower. Silicon nitride is 6% greater, alumina is 22% greater, and zirconia is 80% greater.

Figure 1

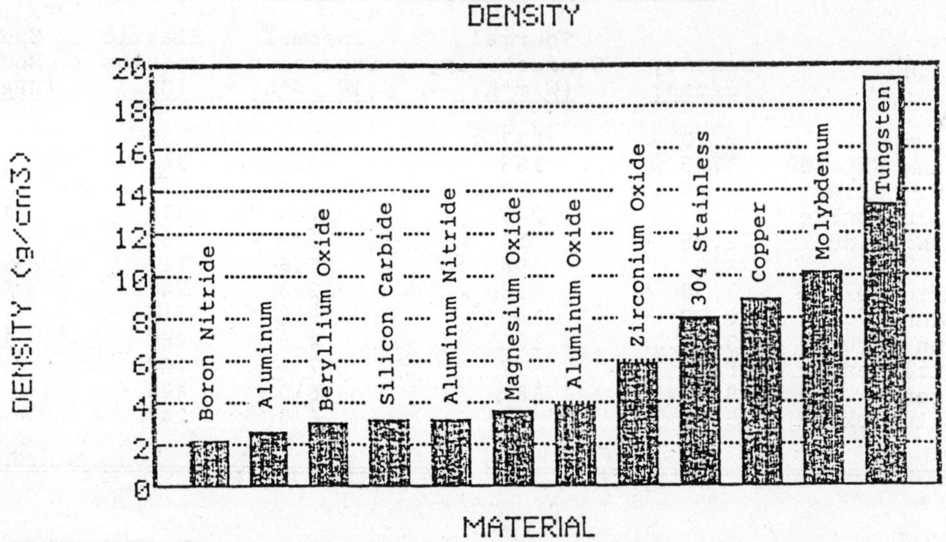

Thermal Conductivity (Fig. 2): While ceramics are generally considered to be notoriously poor conductors of heat (i.e. insulators), AlN is exceptional. At roughly 180 watts per meter-degree Kelvin, AlN is surpassed in thermal conductivity by only one other ceramic, beryllia (which is an expensive and highly toxic material), and copper and aluminum metal (both soft, low melting point materials). Thus, for areas where good mechanical properties at high temperatures are required, along with high thermal conductivity, AlN would appear to be the material of choice.

Figure 2

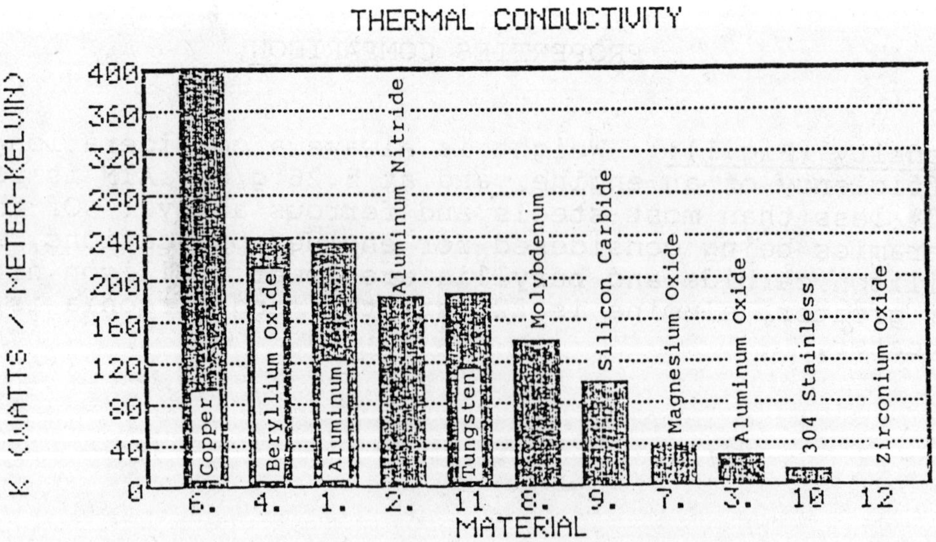

Youngs Modulus (Fig. 3): An advantage of introducing ceramic phases into metals, as is more fully discussed in the following section, is to increase the modulus of elasticity of the resulting composite. The increased stiffness provides numerous advantages in the design and operation of engine components and permits the incorporation of lighter-weight metals such as aluminum and magnesium, which have very low stiffness without such reinforcements.

Figure 3

YOUNGS MODULUS OF ELASTICITY

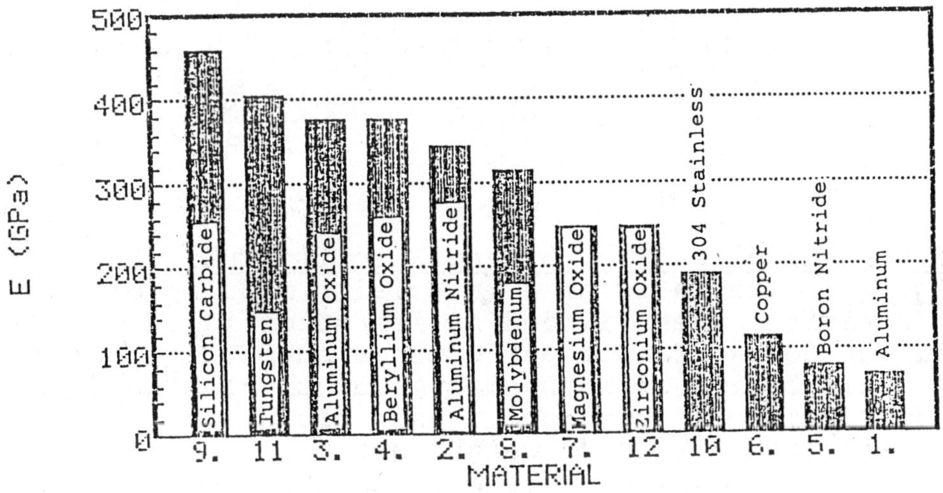

Thermal Expansion (Fig. 4): AlN and silicon carbide have the lowest thermal expansion coefficients of the materials listed in Table 3. The importance of this property is the fact that engine components do not typically operate at a steady state temperature. Thus, in determining tolerances on components and the interfacing between metal and ceramic parts, the expansion which occurs as temperatures rise (or subsequently fall) becomes an important consideration. The advantage of a low expansion material, such as AlN, is that very little dimensional change occurs over a broad temperature range, making it much easier to use.

Figure 4

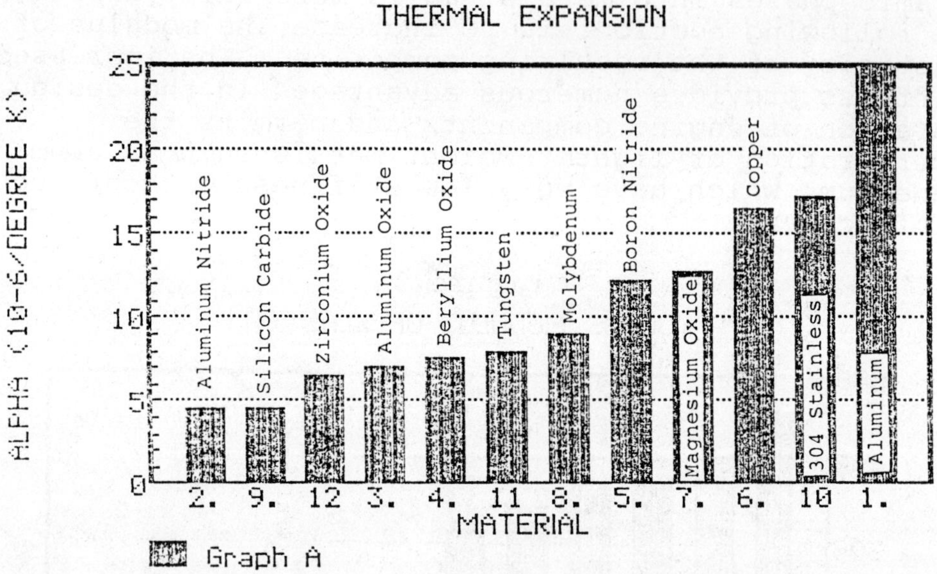

<u>Specific Modulus (Fig. 5) and Specific Thermal Conductivity (Fig. 6):</u> The ratio of a material's modulus or conductivity, when divided by its specific gravity, allows a direct comparison of performance on a unit weight basis. The desirability of both high modulus and high thermal conductivity have been previously discussed. Additionally, the benefits of reduced weight are well understood. Accordingly, high specific modulus and high specific thermal conductivity values are sought for engine applications. In both cases, AlN ranks third in the materials listed, and no other material is ranked higher in <u>both</u> categories.

Figure 5

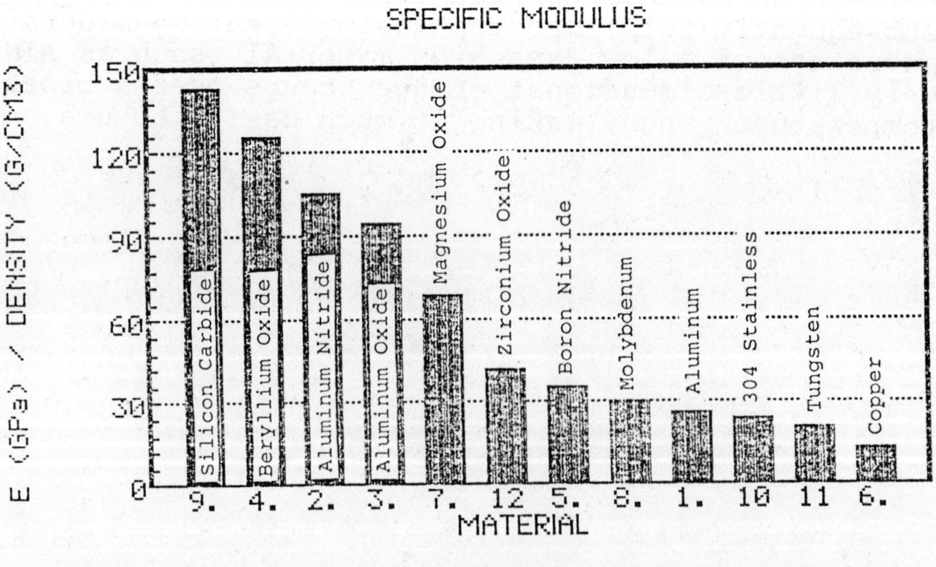

Figure 6

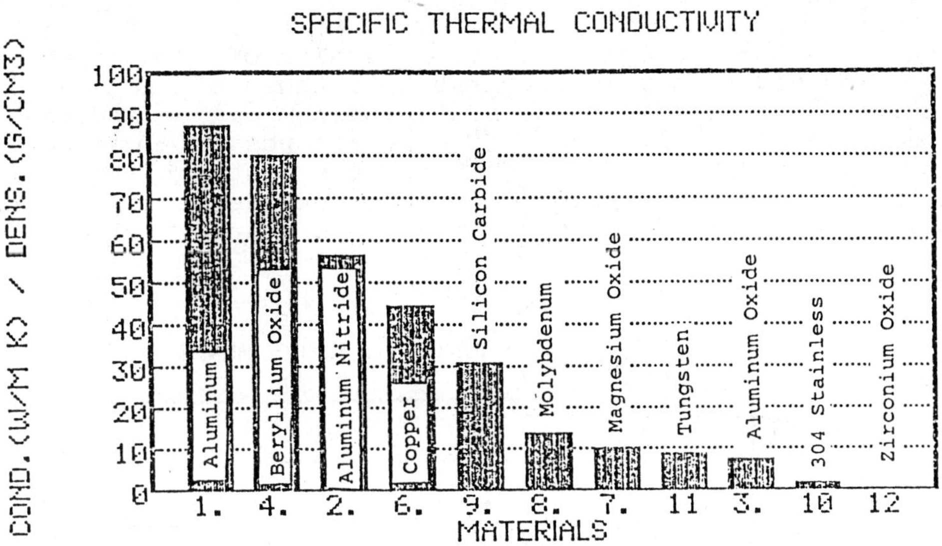

COMPOSITES

Silicon carbide-aluminum composites are presently being used to provide both increased stiffness and decreased thermal expansion than that of the matrix metal itself (9-11). Unlike SiC, AlN is stable in molten aluminum. Its density is similar to that of SiC, as is its thermal expansion. The somewhat lower elastic modulus may well be outweighed by the stability which will allow better and more prolonged mixing of castable composites and longer, more complete annealing to maximize matrix metal properties.

Recently, metal matrix composites (MMC) have been prepared using aluminum nitride in aluminum metal. Aluminum nitride has a rather unusual combination of properties. Not only is it stable in molten aluminum, but it is also wet extremely well by molten aluminum. Powder compacts of approximately 50% density have been successfully infiltrated with molten aluminum to produce fully dense bodies exhibiting excellent strength. Similarly, 75% dense unsintered preforms, formed by injection molding, have been successfully infiltrated with molten aluminum by squeeze casting.

Inasmuch as these procedures - extrusion, pressing, injection molding, infiltration, and squeeze casting - all can lend themselves to mass production and to near-net shape fabrication, these Al-AlN MMCs offer great promise for use in automotive applications. The high conductivity of both Al and AlN together with the high specific modulus of such composites opens an exciting new class of automotive cermets.

A variety of MMCs have been prepared using SiC particulates and whiskers in aluminum matrices (7, 8). From these, the effect on elastic modulus and thermal expansion of various fractions of the dispersed phases have been determined. By analogy, the properties of Al-AlN composites would be expected to follow those shown in Figures 7 and 8. Expansion of such composites can be reduced to levels comparable to those of the ceramic material. The elastic modulus of aluminum can be increased beyond that of steel. Both of these are possible without major sacrifices in density. In addition, the system Al-AlN will be stable, avoiding such problems as property degradation during annealing or during long term high temperature service.

Figure 7

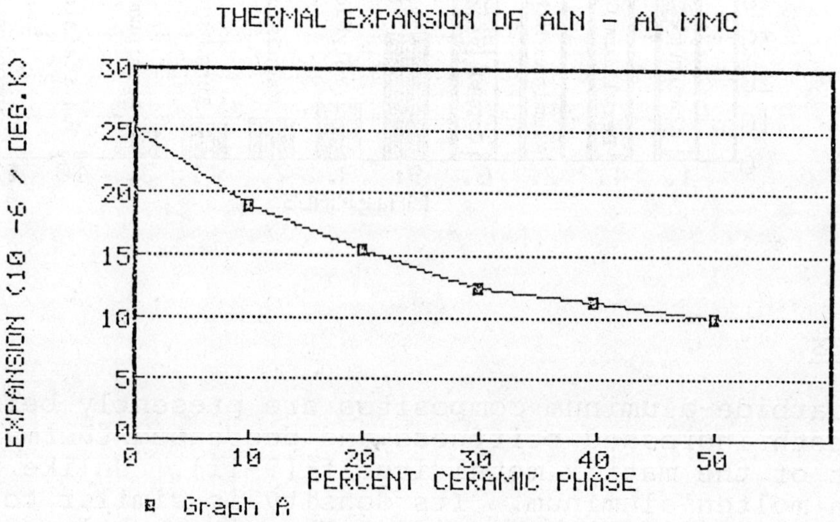

Figure 8

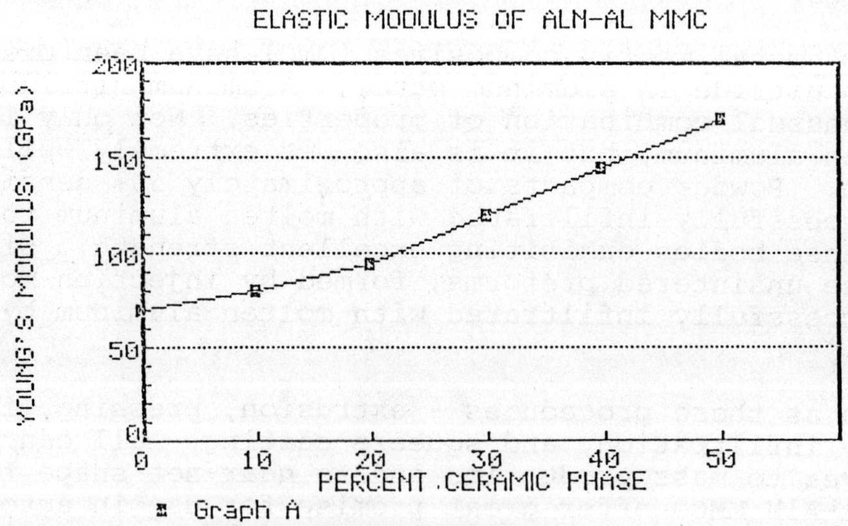

There have been cyrstallographic studies of the polymorphic (polytypic) behavior in aluminum nitride-aluminum oxynitride ceramic bodies (12, 13, 14). To the authors knowledge, however, these structural studies have not been correlated with measured properties. Aluminum nitride itself does not exhibit polytypism in the true form. Polytypes are a function of oxygen content.

To the best of the authors knowledge, aluminum nitride parts have not yet been applied to automative engines. Aluminum nitride is being evaluated as a substrate for ignition modules, and for spark plug insulators. These two applications are based on the high thermal conductivity and high electrical resistivity. Based on elastic modulus, thermal expansion, and thermal conductivity, aluminum nitride-aluminum MMC's are being evaluated for mechanical and thermal control. This MMC system combines excellent wetting characteristics and total chemical compatibility, along with high thermal conductivity and low density. In at least one instance, strengths in excess of 8×10^8 Pa (100,000 psi) have been reported.

CONCLUSIONS

Aluminum nitride, as a ceramic, exhibits numerous desirable properties for heat engine applications. Its unusual combination of high thermal conductivity, electrical resistivity, high modulus, and low density offer unique performance properties. Additionally, with the recent availability of comparatively inexpensive AlN in commercial quantities, a whole new class of high thermal conductivity ceramics and a family of stable Al-AlN metal matrix composites become available. Standard fabrication techniques - for both ceramics and metals - including injection molding, pressureless sintering, pressure infiltration and compocasting can be used to produce engine components of high performance and low cost which incorporate the desired properties of this unusual ceramic.

REFERENCES

(1) F. Fichter: (1907), Z. Anorg. Chem; 54, 322.

(2) K.M. Taylor and C. Lenie: (1960), J. Electrochem. Soc., 107 [4], 308-14.

3) J.B. Holt, D.D. Kingman, G.M. Bianchini: (1986), UCRL-93468, Proceedings of DARPA/Army Symposium, Daytona Beach, Florida, October 21-3, 1985.

4) N. Kuramoto, H. Taniguchi (Tokuyama Soda K.K.), 1986, U.S. Patent 4,618,592, October 21.

5) A.A. Pletushkin & NG Slavina: (1968), Inorg. Mater., 4 [6], 785-9.

6) P.T.B. Shaffer, S. Majorowski, J. Puszynski, S.K. Chiang, and D. W. Readey: "High Thermal Conductivity AlN", presented at the 90th Ann. Mtg, Amer. Ceram. Soc., Cincinatti, May 1-5, 1988.

7) J.E. Schoutens, Silicon Carbide Particulate and Whisker Reinforced Metal Matrix Composites Carbide and Tool J., 20 [4}, 11 (1988).

8) C. Zwehen, Metal Matrix Composites - An Overview, Carbide and Tool J., 20 [4], 7-10 (1988).

9) W.R. Mohn and J. Seman: "Instrument Grade Metal Matrix Composites for Imaging Infrared Guidance Systems", 3rd Annual ASM/ESD Adv. Composites Conference, Detroit, Sept. 15-17 (1987)

10) W.R. Mohn, Your MMC Product Will Stay Put: Res. and Dev., July, 1987.

11) W. R. Mohn, and D. Vokobratovich: Engineered Metal Matrix Composites for Precision Optical Systems, SPIE 31st Ann. Tech. Sym. an Optical and Optoelectronic Appl. Science, San Diego, CA, Aug. 16-21, 1987.

12) Corbin, N.D.: "Aluminum Oxynitride Spinel: A Review", U.S. Army Matl. Tech. Lab, Watertown, MTL-MS-87-3, July 1987.

13) McCauley, J.W., Corbin, N.D.: Phase Relations and Reaction Sintering of Transparent Cubic Aluminum Oxynitride Spinel. J. A. Cer. S. 62 (1-10), 476 (1979).

14) McCauley, J.W., Corbin, N.D.: High Temperature Reactions. Nitrogen Ceramics, Nijhoff Publishers (1983), p. 111.

RESIDUAL STRESSES IN THE Al_2O_3 MATRIX OF Al_2O_3-ZrO_2

G. Feng,T Tsakalakos, and W. Mayo
Department of Material Science

K. Wilfinger and W. R. Cannon
Center for Ceramics Research
Rutgers University, Piscataway, NJ 08855-0909

Residual stresses in the Al_2O_3 matrix of ground and unground surfaces of zirconia toughened alumina were measured with x-ray techniques. No net residual stresses were found on as-fired surfaces, but residual stresses were found on ground surfaces, increasing with severity of grinding and were proportional to the percent ZrO_2 undergoing the tetragonal monoclinic transformation during grinding. Principal stresses and directions with respect to the grinding direction were obtained. Also Warren-Averbach measurements were made on as-fired surfaces to determine the nonuniform stresses surrounding the particles. These stresses were also found to depend on the monoclinic content which in turn was controlled by the size of the particles.

INTRODUCTION

It is well known that surface grinding of transformation toughened ZrO_2 ceramics produces surface compressive stresses due to the tetragonal ZrO_2 transformation to the monoclinic phase. The transformation involves a volume increase of 3 to 5% which places the surface in compression. This effect was subsequently studied by several investigators. [1-4] It was shown that the ground Al_2O_3/ZrO_2 specimens can be twice as strong as annealed specimens. The strengthening is due to residual stresses opposing the applied tensile stresses. In addition to strengthening, compressive surface stresses are expected to improve other properties of transformation toughened materials, such as the resistance to contact damage.

The measurement of surface residual stresses has been performed by Green et al. [4,5] using an x-ray technique in which the Bragg angle is determined as a function the angle of rotation of the sample. A change in Bragg angle infers a strain in the lattice. Assuming that there are no stresses normal to the surface of the specimen, the magnitude of the compressive stress in the plane of the surface is determined. Green et al. [4,5], however, assumed that the stress is the same in all directions in the surface plane. In the present study such measurements are made in three different directions to determine any preferential direction of stresses with respect to the grinding direction. Also the extent of grinding induced transformation of ZrO_2 under varying grinding pressures is measured and correlated to the measured surface residual stresses.

Two sources of compressive surface residual stresses in these materials may arise: stresses due to plastic deformation during grinding and stresses related to the phase transformation near the surface as discussed above. One object of the current research is to determine which of these two dominates.

The stresses measurement was based on following equation

$$\sigma = KM \qquad (1)$$

where M is described below and K is the stress constant

$$K = - \frac{E \cot\theta_o}{2(1+\nu)} \qquad (2)$$

where θ_o is Bragg angle from the stress free sample. (In the current study the (11$\bar{1}$) reflection was utilized.) E is Young's modulus. ν is poisson's ratio. It was assumed that the material is elastically isotropic, and E = 344 GPa and ν = 0.25. K is essentially a machine constant and M is determined experimentally by the well known $\sin^2\psi$ technique, where ψ is the angular rotation angle with respect to the sample's surface as shown in Figure 1. In this method the Bragg peak shift is recorded as a function of sample tilt, ψ. When the peak shift is plotted versus $\sin^2\psi$ a straight line of slope M is obtained.

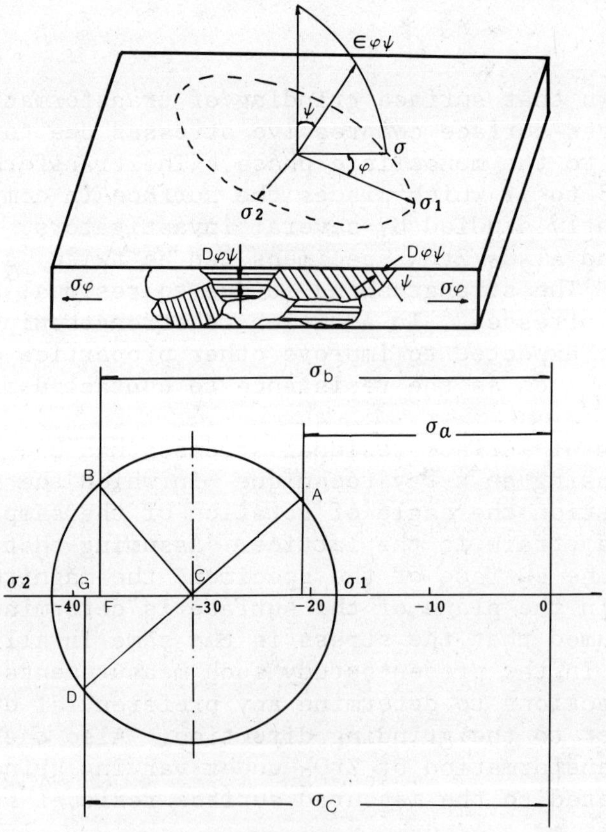

Figure 1

Geometric relationship between x-ray beam angle with the surface normal, ψ and angle φ with the principal stresses. Also shown is the Mohr's circle method of calculating principal stresses.

In addition to surface compressive stresses, local microstresses have also been determined [6] and are discussed in this paper. Residual compressive stresses in the surface of the material are balanced by tensile stresses in the interior. X-ray measurements of the surface will indicate a net strain in the surface by the shift in the Bragg angle. Local microstresses such as those arising from thermal expansion mismatch or the tetragonal to monoclinic transformation which are uniform throughout the surface will not result in a shift of the Bragg angle since the x-ray beam samples an equal portion of tensile and compressive stresses. For example, in a particulate composite in which the thermal expansion coefficient of the particle is greater than the coefficient of thermal expansion of the matrix, upon cooling from the sintering temperature the radial stresses are tensile but the circumferential stresses are compressive. The net result is that there is no shift in the Bragg angle but only a broadening of the peaks.

In this study we have applied the Warren-Averbach technique based on Fourier analysis of line broadening to determine the profile of nonuniform stresses within diffracting domains of the Al_2O_3 in a fired but unground surface. No attempt has been made to look at the internal stress state of the ZrO_2 particles.

EXPERIMENTAL PROCEDURE

Materials

Samples were prepared by sintering Al_2O_3 [a] -10% ZrO_2 [b] slip cast compacts 1600°C for one hour using a heating rate of 200°C/hr, then cooling to the room temperature at the same speed. Both the Al_2O_3 [a] and the ZrO_2 [b] were from a commercial source, but the ZrO_2 was carefully sized to eliminate any coarse tail.[7]

The sintered samples were found to contain 80% tetragonal ZrO_2 which was expected to transform to monoclinic during the later grinding process. The percentage of ZrO_2 in the monoclinic phase, which is referred to as amount of t-m transformation was determined from the areas under the (111) and (111) monoclinic peaks and (101) tetragonal peak [8].

The sample for grinding was mounted in a polymer disk which is 30 mm in diameter and 15mm thick. The disk can then undergo a heavy grinding with a grinding wheel without risk of breaking. Grinding conditions were adjusted from light grinding at which approximately 5μm of material was removed during each pass to heavy grinding at which 250μm of material was removed during each pass. At an intermediate level 20~30 μm was removed (medium grinding condition). A 17.8 cm diameter, 1.27 cm wide, 120 grit diamond grinding wheel was used in all studies. The surface speed was 33 m/sec.

Surface Stress Determination:

Surface residual stress measurements were made using the same x-ray diffraction technique as Green et al.[4,5]. CrKα (λ=0.2291nm) radiation was used which penetrated only to very shallow (<5 μm) depths.

Due to small experimental errors accurate values of M can only be determined by a multiple exposure method. In our studies four tilt angles between 0° and 45° were used and the slope was determined using the least squares fit.

As a result of grinding, x-ray peaks are extremely diffuse which makes it difficult to obtain peak positions accurately. Consequently to obtain the peak position accurately, the Gaussian curve-fitting method was used.[9] The observed x-ray intensities, Y, which were collected by computer-automated instrumentation were corrected for the LPA factor (Lorentz-polarization and absorption factor) given by [10],

$$\text{LPA factor} = \frac{(\cos 4\theta - 3)(1 - \tan\psi \cot\theta)}{1 - \cos 2\theta} \qquad (3)$$

Thus, the x-ray intensity, Z, corrected for the LPA factor is used in determining the peak position.

$$Z = \text{LPA factor} \cdot Y \qquad (4)$$

A comparison of the variation in peak position with both the parabola and the Gaussian curve fitting methods is shown in Table I, which shows that the peak position calculated by the Gaussian curve method had less variation than those calculated by the parabola method. In the first column, those data points with intensities greater than 20% of the maximum intensity were taken into account, whereas, in the second column, only the data points with intensities greater than 50% were considered.

Table I

Peak position calculations using parabolic and Gaussian methods based on the peak either from 20% of the maximum or 50% of the maximum.

	20% Max. Intensity		50% Max. Intensity	
	Parabola	Gaussian	Parabola	Gaussian
Peak Position $2\theta^o$	57.479	57.470	57.491	57.492
Standard Deviation, s	0.2652	0.2652	0.1379	0.1379

(3) Principal Stresses Determination:

Since the grinding induced residual stress may vary with direction, such a stress field can not be described without a determination of the stresses in all possible directions. The conventional way to handle this problem is to calculate principal stresses from which the residual stress in any direction can be calculated by using Mohr's circle shown in Figure 1 and the equation:

$$\sigma_\varphi = \frac{(\sigma_1+\sigma_2)}{2} + \frac{(\sigma_1-\sigma_2)}{2}\cos(2\varphi) \qquad (5)$$

where σ_1 and σ_2 are the two principal stresses (the principal stress normal to the surface is assumed to be 0) and φ is the angle with respect to σ_1.

To determine the principal stresses, we need to measure the stresses in three directions at 45° apart, which are termed σ_a, σ_b, σ_c.(note: σ_a is in the direction of grinding). The schematic diagram is shown in Figure 1. The principal stresses can then be determined by [9]

$$\frac{\sigma_1}{\sigma_2} = \frac{\sigma_a+\sigma_c}{2} \pm \left\{ \left(\frac{\sigma_c-\sigma_a}{2} \right) + \left(\frac{\sigma_a+\sigma_c-2\sigma_b}{2} \right)^2 \right\}^{1/2} \qquad (6)$$

Warren-Averbach Technique

Samples were prepared by carefully flattening the surface of the sample prior to sintering by grinding one side of the green sample on SiC paper.[6] In this way the surface on which x-ray measurements were made was flat after sintering without requiring grinding of the sintered sample. In order to perform the Warren-Averbach method it is necessary to obtain peak broadening information on at least two peaks of the same plane but different order. In this study the (113) and (226) planes were examined on as-sintered surfaces. Fourier transform coefficients are determined for the peaks which will allow stress profiles within domains and the domain sizes to be calculated. More details of the calculation are contained in reference 5.

RESULTS AND DISCUSSION

Surface Residual Stresses

Following sintering, the ZrO_2 is approximately 20% monoclinic and 80% tetragonal. After grinding, there was a dramatic change in the ZrO_2 x-ray peaks as shown in Figure 2. Table II gives the amount of transformation under these different grinding conditions.

Table II

Percent transformation (volume percent) vs grinding condition.

Grinding Conditions	t-ZrO_2,	M-ZrO_2,	T-->M Transformation
Before Grinding	80	20	
Light Grinding	71	29	9
Medium Grinding	56	44	24
Heavy Grinding	42	58	38

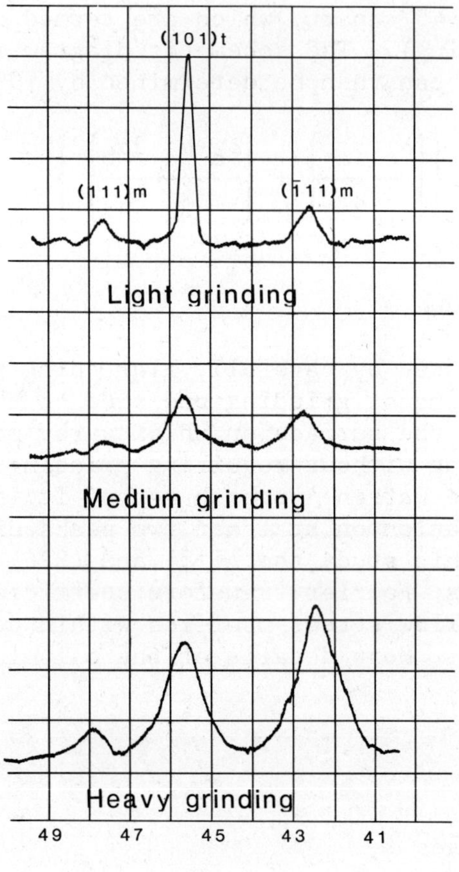

Figure 2

ZrO_2 X-ray diffraction pattern after grinding.

Clearly, as the grinding force increased the monoclinic phase content of the surface increased. Figure 3 shows that there is a linear relationship between the amount of transformation induced by grinding and the magnitude of the principal stresses. Thus the presence of transformed monoclinic particles is the dominant contribution to the surface residual stress. The σ_1 and σ_2 curves extrapolate to about 160 MPa at zero percent transformed. This is in approximate agreement with Lange et al.'s [11] results for residual stresses resulting from grinding of pure polycrystalline Al_2O_3. The latter stresses are at least in part due to plastic deformation in the ground surface and should be similar for Al_2O_3 and Al_2O_3-ZrO_2 but are of minor importance in unground surface of the Al_2O_3-ZrO_2.

Figure 3 also indicates an increasing anisotropy with increasing transformation to monoclinic during grinding. This may be an indication that certain crystallographic orientations transform preferably as a result of grinding stresses. Figure 2 also indicates a texturing of monoclinic particles as the ratio of peak heights of (111)m to (111)m increases with increased grinding pressure.

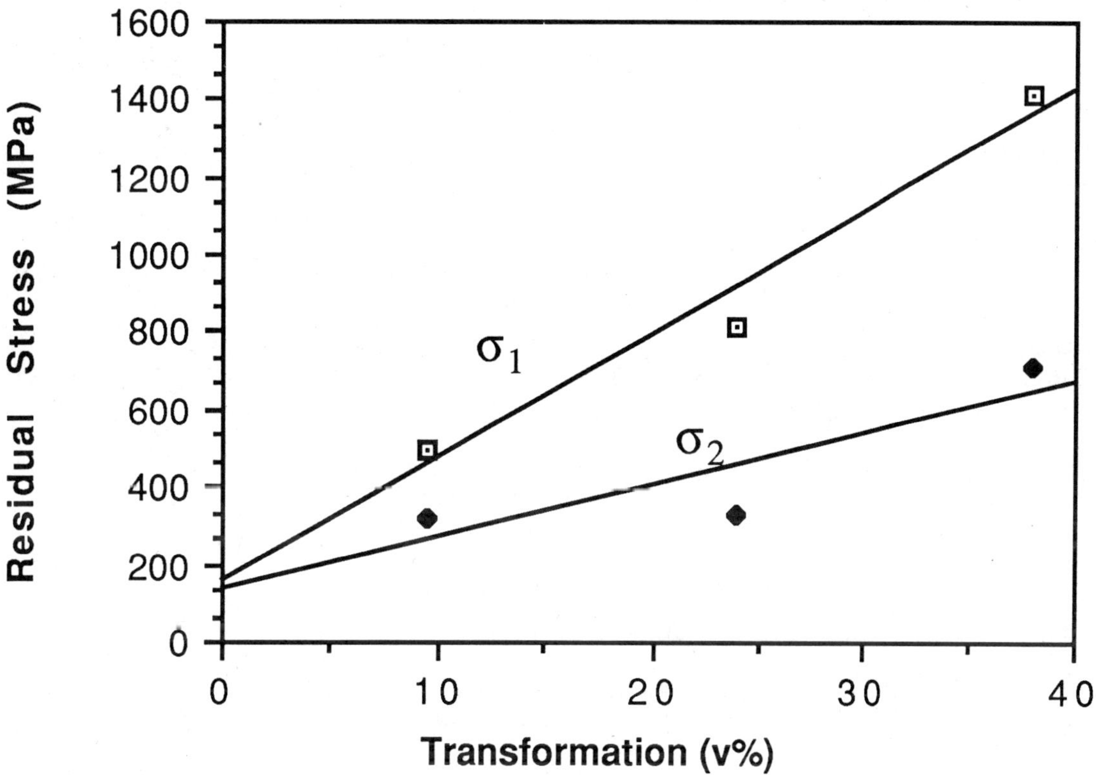

Figure 3

Magnitude of principal stresses after grinding.

Table III contains the raw data from Figure 3 and in addition the calculated surface stress using the following equation

$$\sigma_R = \frac{V_\nu E \epsilon}{(1-\nu)}[f_a - f_b]\left(\frac{t-t_g}{t}\right) \qquad (7)$$

where σ_R = the residual stress
V_ν = volume fraction of ZrO_2 in the composite
ϵ = the strain associated with the transformation
$f_a - f_b$ = the fraction of ZrO_2 transformed due to grinding

The agreement is quite good between the calculated and the average x-ray values, never the less, several simplifying assumption probably account for the disagreement the x-ray values. It is assumed in Equation (8) that the stress is isotropic in the surface plane which is incorrect as shown in Figure 6. It does not account for the decreasing monoclinic content and subsequently stress with increasing depth into the surface [4,5].

Table III

Stresses and principal stresses as a function of grinding conditions. These are compared with Equation (7).

Grinding Conditions	Vol% Transformed	Measured Stresses MPa σ_α	σ_b	σ_c	Principal Stresses MPa σ_1	σ_2	Calculated Stresses Equ. 7 MPa	Angle of σ to the grinding direction
light grinding	9	-341	-454	-477	-499	-318	-194	16.9^o
medium grinding	24	-522	-318	-613	-817	-318	-518	39.8^o
heavy grinding	38	-1339	-1249	-772	-1407	-704	-819	17.2^o

σ_a is in grinding direction. σ_b and σ_c are 45° apart with respect with σ_a.

Matrix Microstresses in the Unground Surface

Figure 4 shows the results of the Warren-Averbach analysis for measuring microstrains. The RMS strain is shown as a function of distance into the x-ray domains of the Al_2O_3 matrix in as-sintered Al_2O_3-ZrO_2.[6] The three curves represent three specimens prepared using controlled ZrO_2 size distributions. The measured maximum particle size of the ZrO_2 powder used in preparing the samples was 0.86 μm, 0.71 μm, and 0.58 μm for samples F, J and G, respectively. Microstructural examination indicated that the particle size changed little during sintering [7]. The distance into the domain is presumed to be the distance away from the ZrO_2 particle and the RMS strain profile is the average for the many domains sampled by the x-ray beam. Analytical solutions for stresses surrounding particles whose thermal expansion does not match the matrix all indicate that the level of stress at the surface of the particle, which is the maximum stress in the matrix, is not a function of the particle size. Since the particle size should not affect the maximum stress level, it is concluded that RMS strain level is dominated by the domains neighboring monoclinic particles.[6] Figure 7 indicates agreement with that point of view. It is expected that the stresses due to thermal expansion mismatch stresses are much smaller these stresses. It is not possible to determine the stresses associated with the strains without additional information but the magnitude of the strains indicates high stresses. These stresses may interact with cracks to deflect them or may result in microcracking.

Conclusions

In this study results are presented for residual surface stresses from grinding zirconia toughened alumina and for nonhomogeneous stresses (RMS strains) in an unground surface of zirconia toughened alumina. Measurements showed that there were no net residual surface stress in unground zirconia toughened alumina. Results have not been obtained yet for the nonuniform stresses in ground surfaces of zirconia toughened alumina but work is planned. In both of the above cases where stresses/strains were measured, the level of the stresses/strains was directly related to the extent of monoclinic transformation.

References

1. T. K. Gupta, "Strengthening by Surface Damage in Metastable Tetragonal Zirconia," J. Am. Ceram. Soc. 63,[1-2] 117 (1980).

2. N. Claussen and M. Ruhle, "Design of Transformation-Toughened Ceramics," Advances in Ceramics, vol. 3, ed. by A. H. Heuer and L. W. Hobbs, The amer. Ceram. Soc., Columbus OH, pp. 137-163.

3. J. G. Reed and A. Lejus, " Effect of Grinding and Polishing on Near-Surface Phase Transformations in Zirconia," Mater. Res. Bull.12 [10] 949-54 (1977)

4. David J. Green, F. F. Lange and Michael R. James, "Factors Influencing Residual Surface Stresses due to a Stress-Induced Phase Transformation," J. Am Ceram. Soc.,66 [9] 623-29 (1983).

5. Green, D. J., Lange, F. F., and James, M. R., "Residual Surface Stresses in Al_2O_3 ZrO_2 Composites," In "Science and Technology of Zirconia II," (N. Claussen, M. Ruhle and A. H. Heuer, eds.) Vol. 12, (1984) pp. 240-50 Am. Ceram. Soc., Columbus, OH.

6. W. R. Wilfinger, W. R. Cannon, and T. Tsakalakos, "RMS Matrix Strains in Transformation Toughened Alumina, to be published.

7. D. L. Porter and A. H. Heuer, "Microstructual Development in MgO-Partially Stabilized Zirconia (Mg-PSZ)," J. Am. Ceram. Soc. 62 [5-6] 298-305 (1979).

8. K. R. Wilfinger and W. R. Cannon, "Particle Interactions in Zirconia Toughed Alumina," submitted to the J. Amer. Ceram. Soc.

9. M. Kurita, "A New X-ray Method for Measuring Residual Stress and Diffraction Line Broadness and its Automation," NDT International, 20 [5] 277-84 (1987).

10. M. E. Hilley, "Residual Stress Measurement by X-Ray Diffraction - SAE J784A" Society of Automotive Eng., Warrendale, 1971. p 38.

11. F. F. Lange, M. R. James and D. J. Green, "Determination of Residual Surface Stresses Caused by Grinding Polycrystalline Al_2O_3," J. Amer. Ceram. Soc., 63, [2] C16-C17 (1983).

CERAMIC COMPOSITE VALVE SEAT INSERTS

D. J. Landini and H. D. Lesher, Lanxide Corporation, Newark, Delaware; and A. J. Gesing, Alcan International Ltd., Kingston, Ontario, Canada

ABSTRACT

Ceramic matrix composite exhaust valve seat inserts were produced to near-net shape using two Lanxide™ alumina matrix composite systems. One contained alumina filler particles while the other contained silicon carbide particles. The inserts were made in large numbers and were characterized for mechanical and thermal properties and evaluated in short-term engine tests. The near-net-shape fabrication process and related information on manufacturing consistency and reproducibility of properties within and among batches are described. The composite systems were found to have hot hardness values significantly greater than severe duty metallic valve seat inserts. The composite valve seat inserts were tested successfully for 100 hours of full load operation in a Klockner-Humboldt-Deutz (KHD) F1L 511D single cylinder diesel engine.

INTRODUCTION

The trend in today's automotive industry is toward improving performance and durability in both spark ignition and diesel engines. To accomplish these goals the development of advanced materials and improved engine designs is required. For example, in severe duty engines, especially diesel engines, nickel or cobalt based alloys have been used in valve seat inserts (VSIs) to extend the engine operating temperatures and service life above that of ferrous materials. Recently, ceramic monoliths and composites have shown considerable promise in reducing valve train wear in internal combustion engines, thus extending the service operating range.

The commercialization of ceramics for valve train components has been slow, however, because of the substantial challenge they present to the engine designer. Technical and economic issues must be addressed that will satisfy the concerns of the auto industry. The major technical issues revolve around the differences in properties between ceramics and metals. Perhaps the most striking of these is the significantly lower values of toughness of many ceramic monoliths compared to the alloys currently used in valve trains. The difference in toughness causes a fundamental difference in the failure mode of a ceramic and a metal. Ceramics fail, usually catastrophically, in a brittle manner from a single flaw subjected to a stress sufficiently high to initiate crack propagation. Metals, however, are typically much less susceptible to catastrophic failure since processes such as plastic deformation can locally redistribute the stresses. To encourage engine designers to use ceramics, it is important to demonstrate that adequate levels of strength and other properties can be achieved and reproduced consistently.

The major economic concern of the engine designer is the "cost to improved performance" ratio. Typically manufacturers will not pay a premium for a part unless significant benefits are recognized. It is therefore important for ceramic component manufacturers to minimize costly steps typical to many conventional forming and firing operations. The most costly of these steps is generally surface machining with diamond tooling to achieve the desired net shape of the final component and surface finish.[1] Near-net shape processing obviously has important advantages in reducing final component cost.

In consideration of these issues, goals for the demonstration valve seat insert program described here were: 1) to identify ceramic composite systems for use in the high impact, wear, and thermal environment provided by an exhaust valve seat and to develop reproducible processing techniques; 2) to compare properties of valve seat inserts produced by the Lanxide™ process with those of competitive materials and to demonstrate the reproducibility of these properties; 3) to conduct survival tests on the valve seat inserts in a diesel engine; and 4) to develop a cost model and determine preliminary cost estimates. The discussion to follow will focus primarily on goals two and three.

PRODUCTION PROCESS

Lanxide™ Directed Oxidation Process

The Lanxide™ directed oxidation process is schematically illustrated in Figure 1. The process involves the formation of a ceramic matrix composite reaction product by the oxidation of molten metal. The metal is transported through microscopic channels in the reaction product to sustain the reaction at the reaction product/vapor phase oxidant interface.[2] When beds or preforms of certain inert filler materials, such as fibers, whiskers, or particles, are placed in the path of the reaction, the directed oxidation process will proceed through the filler material without displacing it.[3] The result is a multiphase composite consisting of a ceramic matrix, a filler material, and some residual parent alloy.

The potential advantages of the Lanxide™ process are near-net shape forming capabilities, tailorability of the composite properties by modification of fillers and processing conditions, low processing costs, and scaleability to large and/or complex shapes.

Composite Compositions V1 and V2

The key design parameters used in selecting systems for exhaust valve seat inserts (VSI) are listed below:

- resistance to valve seat recession in engine environments
- compatible thermal expansion with the engine head to minimize VSI dropout or fatigue cracking of the head
- high elastic modulus and strength to prevent distortion during installation in the engine head or during engine operation
- adequate thermal conductivity to remove heat from the valve and seat areas
- reliable properties
- low cost
- process scaleability

The two systems chosen were an alumina-filled alumina matrix composite (composition V1) and a silicon carbide-filled alumina matrix composite (composition V2). Both systems offered potential advantages in valve seat applications in the areas of high temperature hardness, compressive strength, and corrosion resistance, yet they differed from each other in thermal conductivity, thermal expansion, and cost.

Test blanks of compositions V1 and V2 were produced in the simplified geometry of a ring nominally 41 mm outer diameter, 29 mm inner diameter, and 10 mm high. These shapes were used to generate reproducibility data on process parameters and selected properties. A large number of prototypes were also produced to the near-net shape of a Klockner-Humboldt-Deutz (KHD) F1L 511D single cylinder diesel VSI complete with beveled surfaces to demonstrate the amenability of the process to the more complex shape. Figure 2 shows a test blank, a near-net shape prototype VSI, and a diamond ground VSI.

The fabrication process used for the production of composite valve seat inserts is diagrammed as a flow chart in Figure 3. The process can briefly be described as follows. A ring preform of the alumina or silicon carbide particulates is produced by dry pressing. The ring is dried and bisque fired in air to remove organic binders and to provide sufficient strength for subsequent handling. The prefired ring is then coated with a growth initiator on one flat surface and a growth barrier on the remaining surfaces. The barrier coating serves to stop growth at the preform boundaries to define the ring geometry. The Lanxide™ directed oxidation process is then conducted by heating the preform in contact with molten aluminum alloy at 900°C in ambient air. Upon completion of the directed oxidation process, the composite ring is removed from the furnace and cooled, then grit blasted at low pressures to remove any residual barrier coatings or dross. Preliminary surface grinding is also done on the surface that was in contact with the molten alloy to remove remaining residual alloy. At this point the composite ring is ready for subsequent operations.

MICROSTRUCTURAL ANALYSES

Composition V1 and composition V2 ceramic composites consist of alpha alumina particles and silicon carbide particles, respectively, dispersed throughout a three-dimensionally interconnected alpha alumina matrix. Some residual metal alloy is also present. The amount of metal in these composites can be tailored and was selected in this instance to be relatively high (~15%). Representative microstructures from polished cross sections of each are shown in Figure 4. The phases present in the V1 composite are alumina (gray), and residual metal alloy (white). In this alumina filled composite it is difficult to distinguish between the alpha alumina filler particles and the alumina matrix due to the small grain size of the starting filler and the lack of visible grain boundaries in these optical micrographs. The phases present in the V2 composite consist of silicon carbide (light gray), alumina (darker gray), and residual metal (white). Optical image analysis of the V1 composite yielded average volume concentrations of 80.7 ± 2.7% alumina, 16.3 ± 2.3% residual metal, and 3.1 ± 2.0% porosity. For the V2 composite, average volume concentrations were typically 46.2 ± 1.3% silicon carbide, 36.8 ± 1.6% alumina, 15.6 ± 0.6% residual metal, and 1.4 ± 0.4% porosity.

PROCESS REPRODUCIBILITY

To evaluate the reproducibility of the valve seat insert process and subsequent VSI properties, a series of five

batches containing 60 VSI blanks per batch were produced. These blanks were examined for dimensional tolerances and selected physical and mechanical properties. This portion of the study was restricted to the V1 material. Results are presented in Tables I and II.

The composite VSI dimensions were very consistent from batch to batch and within each batch (e.g., the outer diameter varied by less than 0.5%). There were only small in-process dimensional changes in going from the pressed preform size to the final composite size; therefore, near-net shape can be achieved more easily than by traditional processing methods, which typically involve high shrinkages.

The physical and mechanical property data in Table II also illustrate the consistency of the composition V1 composite. The strength, hardness, and density do not vary significantly from batch to batch.

DETAILED PROPERTY DATA

Composition V1 and composition V2 composites were prepared in the form of plates (~50 x 50 x 10 mm) to provide samples for more detailed characterization of physical and mechanical properties. These properties are listed in Table III along with those of competitive metal VSI materials. The table reveals that these first-generation Lanxide™ composites offer an interesting combination of properties that make them attractive for VSIs and other engine components.

The V1 and V2 composites offer several potential advantages over metal VSI materials. Perhaps the most important of these is hot hardness. A plot of hot hardness for the Lanxide™ composite VSI materials and several metallic VSIs is shown in Figure 5. As apparent from the figure, the values of hot hardness for the Lanxide™ composites are significantly higher than those of either Stellite-6 or Eatonite, two alloys designed for severe service environments with values of hot hardness established as acceptable for use in most current engines.[6] Seat recession of a VSI is highly dependent on the hardness of the material, especially at operating temperatures; therefore, composition V1 and V2 composites should offer substantial improvements in VSI service life. Another advantage is that the thermal conductivities of the V1 and V2 composites are comparable to or greater than those of the metals. This should prevent localized overheating of the valve seat area and increase the removal of heat from the valve.

The coefficients of thermal expansion (CTE) of the Lanxide™ composites approaches that of many of the metal alloys used in an engine. In comparison with other ceramics, the result should be less stress generated on the composite component or on metal in contact with the composite due to less thermal expansion mismatch. In the case of a VSI the amount of interference fit used to insert and hold it in place is substantially reduced; therefore, lower stresses are generated on the VSI and on the engine head. This reduces the chance of cracking of the VSI or yielding of the engine head between the exhaust and intake valves.

The measured flexural strengths of the composition V1 and V2 composites is lower than that of some other candidate engineering ceramics such as silicon carbide or silicon nitride.[7] However, preliminary in-house stress analyses indicate that the VSI does not experience significant tensile stresses during operation but instead undergoes predominantly compressive stresses. Flexural strength therefore may not be extremely critical to VSI performance.

The strength distribution listed in Table III was determined using Weibull statistics, which are based on the weakest link theory of failure.[8] The Weibull modulus, m, is indicative of the variability of the flaw size or flaw distribution. A higher value of m typically implies a more uniform distribution of flaws or a highly homogeneous population of flaws. Plate glass typically has an m value of 1 to 2, while some steels have m values as high as 50. The measured Weibull modulus of 20 is quite high for ceramic or ceramic composite materials and is indicative of a very uniform flaw size, evenly distributed within this composite. The implication of the data is that the production process used for these composites is very reproducible. This can be important from a design and cost standpoint.

In summary, the combination of properties engineered into these Lanxide™ composite compositions--high hot hardness, thermal conductivity, and expansion coefficient plus adequate strength, together with excellent reproducibility and dimensional accuracy--appear to be very attractive for the VSI application.

ENGINE SURVIVAL TESTING

Engine testing was used to evaluate the survivability of compositions V1 and V2 valve seat inserts under actual operating conditions. The air-cooled engine selected for the test was a KHD F1L 511D single cylinder diesel engine equipped with an Eaton Model 66 dynamometer. This engine offered conveniences of having the head and cylinder wall separate from the engine block and of having a greater thermal expansion mismatch between the aluminum head and the composite VSI as opposed to that which would exist between the composite VSI and a cast iron head; thus, a more rigorous trial for the composite VSI was provided.

Prior to testing, the engine was outfitted with thermocouples to monitor head temperatures at numerous locations and exhaust gas temperatures near the exhaust valve outlet. Near-net-shape VSIs of V1 and V2 composites were diamond machined and installed into KHD heads by shrink fitting using standard installation procedures as outlined in the KHD F1L 511D service manual. The valve mating surface of the VSI was then diamond polished (30 μm) to provide a close fit with the valve. The heads were remounted and testing was initiated. Two test modes were used: a steady state mode and a cyclic mode. Steady state testing consisted of running the engine at a 60% rated load for six 16-hour intervals (~100 h). Cyclic testing consisted of an eight hour period during which the engine was cycled between full throttle, full load conditions to full throttle, no load conditions in a stepwise sequence. During cyclic testing two cold starts to full load conditions were also run. Engine operating conditions measured for V1 and V2 VSIs and for the standard KHD VSIs (ferrous alloy) are shown in Table IV. Maximum engine head temperatures were typically in the range of 150 to 170°C for the Lanxide™ composites, and slightly lower for the KHD standard VSI. Exhaust gas temperatures were in the 350 to 450°C range for the steady state testing but ranged as high as 585°C in the cyclic test.

After engine testing, the valve seat inserts, valves, and heads were examined visually and by optical microscopy for signs of wear, cracking, or other damage. No evidence of wear or damage was apparent on the VSIs or the heads. There were, however, several smooth indentations observed on the mating surface of the valve from both the composite composition V1 and standard KHD (ferrous alloy) VSI tests. These indentations were examined in-house and by Deutz and were likely the result of foreign material trapped between the VSI and the valve seat. Overall, Deutz found the valve to be in satisfactory condition.[9]

CONCLUSIONS

Conclusions that were drawn from the study are listed below:

1. The Lanxide™ directed oxidation process can be used to produce near-net-shape ceramic composite valve seat inserts. The process is capable of providing tight dimensional tolerances with minimal diamond machining required. Properties appear to be very reproducible.

2. The properties of composite compositions V1 and V2 compare favorably with metallic materials currently used for valve seat inserts. The Lanxide™ composites offer higher hardnesses at ambient and operat-

ing temperatures and comparable or higher thermal conductivities than metallic VSIs.

3. Lanxide™ composite VSIs can be successfully installed in an aluminum head by using standard installation procedures. Engine survival tests under steady state and cyclic testing resulted in no visible evidence of wear, and no spalling or cracking of the head or of the VSIs. No dropout of the VSIs was encountered.

4. Preliminary economic modelling of the process (not reported here) indicates that it can be cost effective.

ACKNOWLEDGMENTS

The authors thank Gary Smith, Tony Liszkiewicz, Toni Thomas, George Altland, and other Lanxide personnel for their assistance in the production, characterization, and engine testing. They also thank Cheryl Quarles and Lisa Martin for preparation of the manuscript, and Virgil Irick and Andrew Urquhart for helpful discussions and reviews of the manuscript.

REFERENCES

[1]E. P. Rothman, J. P. Clark, and H. K. Bowen, "Cost Modelling of Structural Ceramics," Adv. Ceram. Mat., **2** [1] 34-38 (1987).

[2]M. S. Newkirk, A. W. Urquhart, H. R. Zwicker, and E. Breval, "Formation of Lanxide™ Ceramic Composite Materials," J. Mat. Res., 1 [1] 81-89 (1986).

[3]M. S. Newkirk, H. D. Lesher, D. R. White, C. R. Kennedy, A. W. Urquhart, and T. D. Claar, "Preparation of Lanxide™ Ceramic Matrix Composites: Matrix Formation by the Directed Oxidation of Molten Metals," Ceram. Eng. Sci. Proc., **8** [7-8] 879-85 (1987).

[4]C. F. Taylor, "Engine Materials," The Internal Combustion Engine in Theory and Practice, Vol. II, MIT Press, 1985, 335.

[5]S. L. Narsimham, J. M. Larson, and E. P. Whelan, "Wear Characterization of New Nickel Base Alloys for Internal Combustion Engine Valve Seat Applications," SAE Paper, Int. Cong. & Expo., 1980, 210-218.

[6]T. M. Cadle, "Material and Process Development of Valve Seat Inserts," AE Symposium, Paper No. 11, 1978.

[7]R. Kamo and W. Bryzik, "Cummins/TACOM Advanced Adiabatic Engine," Ceram. Eng. Sci. Proc., **5** [5-6] 312-38 (1984).

[8]W. Weibull, "A Statistical Distribution Function of Wide Applicability," J. Appl. Mech., **18** [3] 293-97 (1951).

[9]J. E. Sauerteig, Manager, Product Regulation and Compliance, Deutz Corp., Atlanta, Georgia, private communication, 13 July 1987.

Table I: Dimensional Tolerance of Composite Composition V1

Batch #	OD (mm) Composite (Preform)	Wall Thickness (mm) Composite (Preform)	# of Samples
1	41.60 ± 0.053 (41.74 ± 0.044)	5.93 ± 0.027 (5.95 ± 0.035)	60
2	41.60 ± 0.048 (41.72 ± 0.031)	5.92 ± 0.023 (5.94 ± 0.026)	60
3	41.70 ± 0.033 (41.81 ± 0.047)	5.95 ± 0.019 (5.97 ± 0.034)	60
4	41.54 ± 0.048 (41.65 ± 0.027)	5.96 ± 0.050 (5.97 ± 0.037)	60
5	41.56 ± 0.062 (41.64 ± 0.017)	5.98 ± 0.065 (5.92 ± 0.027)	60
Average Total	41.60 ± 0.075 (41.71 ± 0.072)	5.95 ± 0.045 (5.95 ± 0.037)	300

Table II: Reproducibility of Composite Composition V1 Properties

Batch Number	1	2	3	4	5
Flexural Strength[a] (MPa)	314 ± 29	332 ± 31	315 ± 30	300 ± 50	311 ± 36
Rockwell A Hardness[a]	78 ± 2.5	76 ± 1.9	78 ± 1.7	78 ± 1.5	76 ± 2.8
Bulk Density[b]	3.546 ± 0.037	3.538 ± 0.033	3.531 ± 0.034	3.571 ± 0.046	3.565 ± 0.025

[a]10 samples/batch measured
[b]60 samples/batch measured

Table III: Properties of Composite Compositions V1 and V2 and Commercially Available VSI Materials

Property	Composition V1	Composition V2	Gray Cast Iron[a]	Stellite-6[b]	Eatonite[b]
Density (g/cm^3)	3.56	3.38	7.0-7.3	8.69	8.30
Young's Elastic Modulus (GPa)	301	324	100-140	214	205
Flexural Strength (MPa) (4-Point Bend)	312	334	-	-	-
Weibull Modulus	20[c]	22[d]	-	-	-
Fracture Toughness (K_{Ic}) ($MPa{\cdot}m^{1/2}$) (Chevron Notch)	5.9	6.9	37-45	50-100	50-100
Coeff. of Thermal Expansion ($x\ 10^6/K$)	9.0	7.5	10-12	14-16	12-15
Thermal Conductivity (W/mK)	36	82	36-60	15	11
Hot Hardness (GPa) (Vicker's 10 Kgf) RT	8.3[e]	6.4[e]	4.9	3.7	4.8
200°C	7.6	5.7	4.6	3.3	4.4
500°C	5.9	4.5	4.2	2.8	3.8
700°C	4.9	3.7	3.9	2.3	3.2

[a]Reference 4
[b]Reference 5
[c]Modulus determined from 33 samples, 11 different plates
[d]Modulus determined from 17 samples, 4 different plates
[e]Measurements by Bethlehem Steel Corporation, Bethlehem, PA

Table IV: Engine Operating Conditions During Steady State and Cyclic Engine Performance Tests

	Steady State Testing			Cycle Testing	
	KHD Seat	Composition V1 Seat	Composition V2 Seat	Full Throttle Full Load Condition Composition V1	Full Throttle No Load Condition Composition V2
Head Temp. Range (°C)	134-138	116-168	128-177	143-222	96-140
Exhaust Gas Temp. (°C)	432	424	445	585	313
rpm	1980	1880	1980	2950	3450
% of Rated Load	68	62	67	99.1	0
BMEP Bar	5.49	5.34	5.44	5.40	NA
Test Hours	100	100	100	7	1

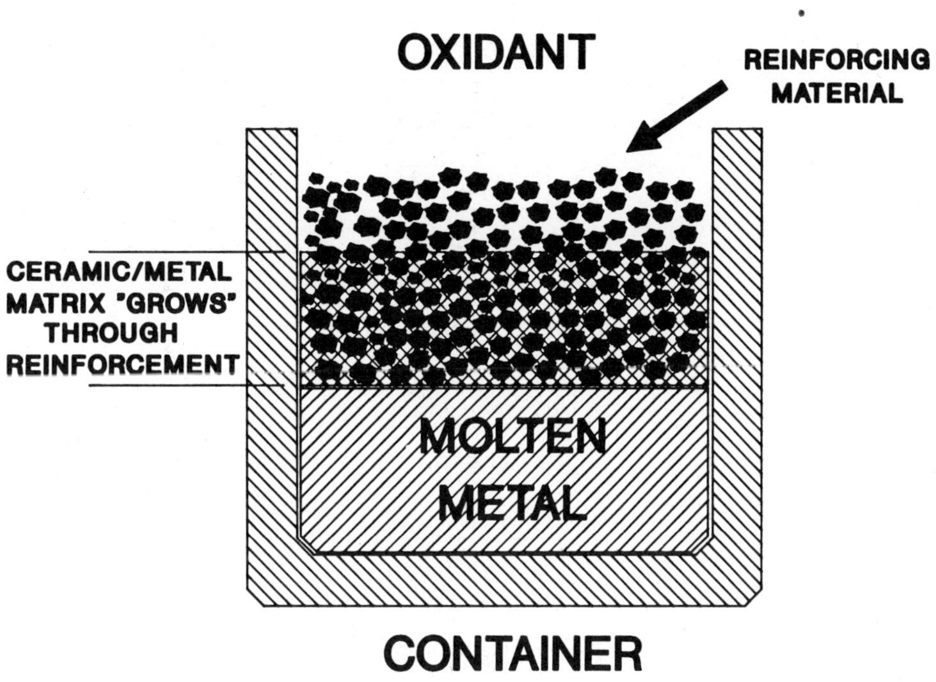

Figure 1: Formation of a ceramic composite by the outward growth of a ceramic/metal reaction product matrix through a mass of filler material.

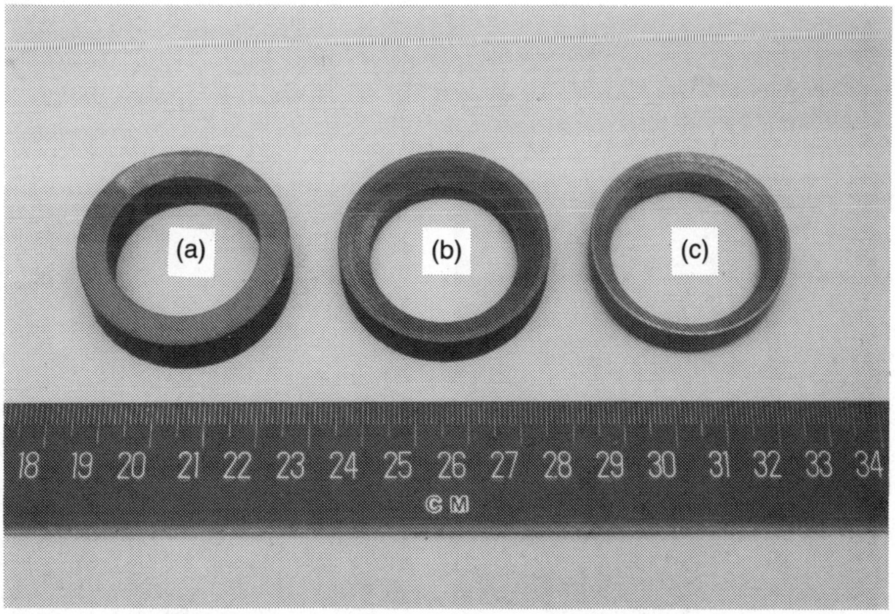

Figure 2: Composite composition V1 valve seat inserts: a) test blank; b) near-net shape prototype; c) machined valve seat insert.

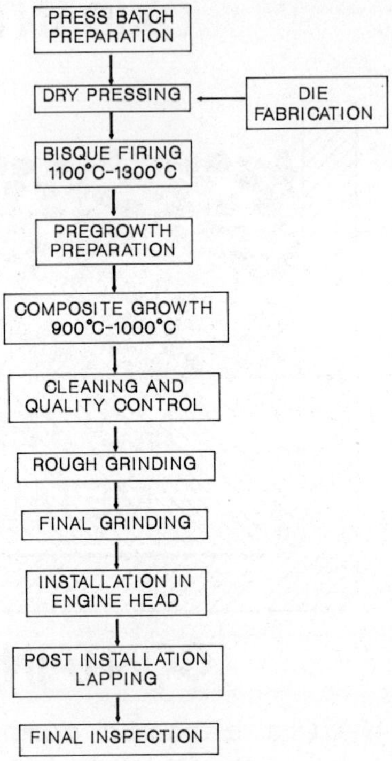

Figure 3: Schematic flow chart for the fabrication process used in the production of Lanxide™ composite valve seat inserts.

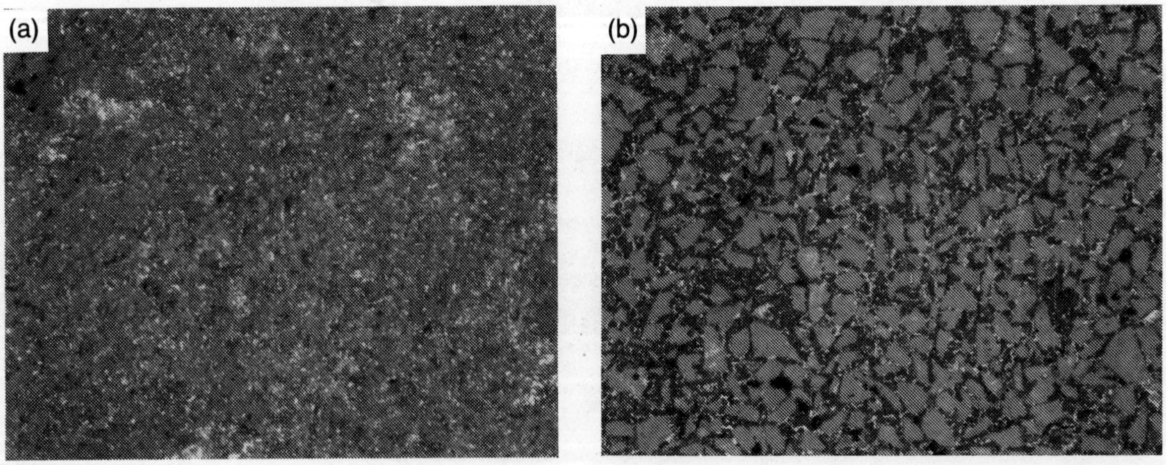

Figure 4: Microstructures of the compositions V1 and V2 composites: a) V1 (alumina filler); b) V2 (SiC filler).

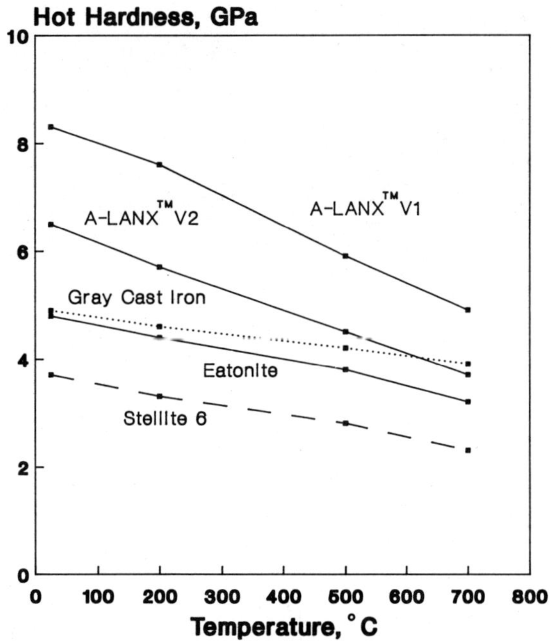

Figure 5: Hot hardness data for Lanxide™ composites and several metal alloys.

Index